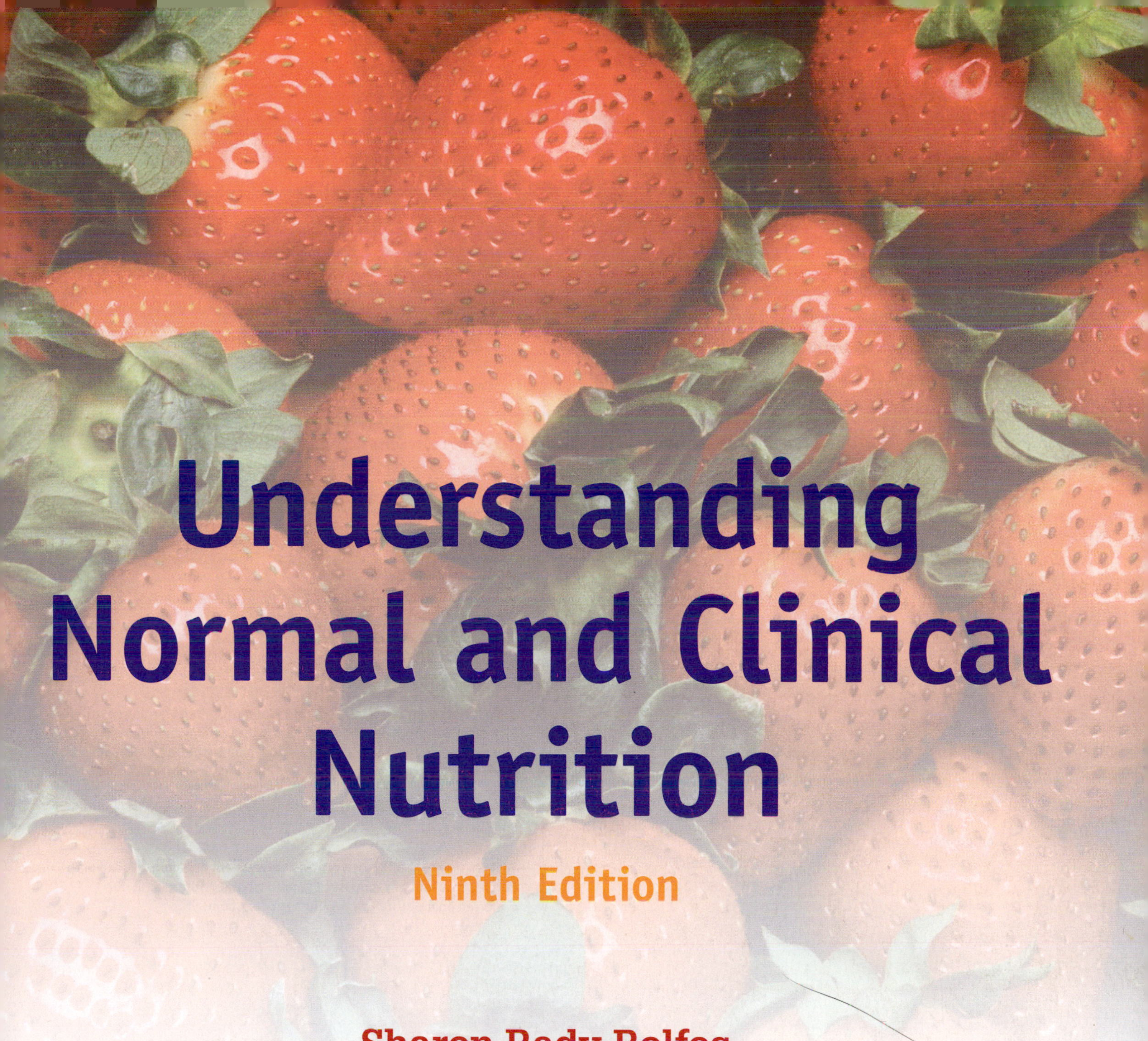

Understanding Normal and Clinical Nutrition

Ninth Edition

Sharon Rady Rolfes
Kathryn Pinna
Ellie Whitney

WADSWORTH
CENGAGE Learning™

Australia • Brazil • Japan • Korea • Mexico • Singapore
Spain • United Kingdom • United States

WADSWORTH
CENGAGE Learning

Understanding Normal and Clinical Nutrition, **Ninth Edition**
Sharon Rady Rolfes
Kathryn Pinna
Ellie Whitney

Publisher/Executive Editor: Yolanda Cossio

Acquisitions Editor: Peggy Williams

Developmental Editor: Elesha Feldman

Assistant Editor: Elesha Feldman

Editorial Assistant: Shana Baldassari

Media Editor: Miriam Myers

Marketing Manager: Laura McGinn

Marketing Communications Manager: Linda Yip

Content Project Manager: Carol Samet

Creative Director: Rob Hugel

Art Director: John Walker

Print Buyer: Karen Hunt

Rights Acquisitions Specialist: Dean Dauphinais

Production Service: Lachina Publishing Services

Text Designer: Gary Hespenheide

Photo Researcher: Scott Rosen, Bill Smith Group

Copy Editor: Lachina Publishing Services

Cover Designer: Riezebos Holzbaur/Brie Hattey

Cover Image: istock©mbbirdy

Compositor: Lachina Publishing Services

Library of Congress Control Number: 2011920459
ISBN-13: 978-0-8400-6845-3
ISBN-10: 0-8400-6845-X

Wadsworth
20 Davis Drive
Belmont, CA 94002-3098
USA

Cengage Learning is a leading provider of customized learning solutions with office locations around the globe, including Singapore, the United Kingdom, Australia, Mexico, Brazil, and Japan. Locate your local office at **www.cengage.com/global**.

Cengage Learning products are represented in Canada by Nelson Education, Ltd.

To learn more about Wadsworth visit **www.cengage.com/Wadsworth**
Purchase any of our products at your local college store or at our preferred online store **www.CengageBrain.com**.

Printed in Canada
1 2 3 4 5 6 7 15 14 13 12 11

To Ellie Whitney, my mentor, partner, and friend, with much appreciation for believing in me, sharing your wisdom, and giving me the opportunity to pursue a career more challenging and rewarding than any I could have imagined.

Sharon

To my mother, Tina Pinna, for her many years of love and support.

Kathryn

To the memory of Gary Woodruff, the editor who first encouraged me to write.

Ellie

About the Authors

Sharon Rady Rolfes received her M.S. in nutrition and food science from Florida State University. She is a founding member of Nutrition and Health Associates, an information resource center that maintains a research database on over 1000 nutrition-related topics. Her other publications include the college textbooks *Understanding Nutrition* and *Nutrition for Health and Health Care* and a multimedia CD-ROM called *Nutrition Interactive*. In addition to writing, she occasionally teaches at Florida State University and serves as a consultant for various educational projects. Her volunteer work includes serving on the board of Working Well, a community initiative dedicated to creating a healthy workforce. She maintains her registration as a dietitian and membership in the American Dietetic Association.

Kathryn Pinna received her M.S. and Ph.D. degrees in nutrition from the University of California at Berkeley. She has taught nutrition, food science, and human biology courses in the San Francisco Bay area for over 20 years and currently teaches nutrition classes at City College of San Francisco. She has also worked as an outpatient dietitian, Internet consultant, and freelance writer. Her other publications include the textbooks *Nutrition for Health and Health Care* and *Nutrition and Diet Therapy*. She is a registered dietitian and a member of the American Society for Nutrition and the American Dietetic Association.

Ellie Whitney, Ph.D., grew up in New York City and received her B.A. and Ph.D. degrees in English and biology at Radcliffe/Harvard University and Washington University, respectively. She has taught at both Florida State University and Florida A&M University, has written newspaper columns on environmental matters for the *Tallahassee Democrat*, and has authored almost a dozen college textbooks on nutrition, health, and related topics, many of which have been revised multiple times over the years. In addition to teaching and writing, she has spent the past three-plus decades exploring outdoor Florida and studying its ecology. Her latest book is *Priceless Florida: The Natural Ecosystems* (Pineapple Press, 2004).

Brief Contents

Contents

© imagebroker/Alamy

J. Helgason/Shutterstock.com

HGalina/Shutterstock.com

CHAPTER 9

Weight Management: Overweight, Obesity, and Underweight 271

CHAPTER 10

The Water-Soluble Vitamins: B Vitamins and Vitamin C 311

CHAPTER 11

The Fat-Soluble Vitamins: A, D, E, and K 355

Enur/Shutterstock.com

Natali Glado/shutterstock.com

Nayashkova Olga/shutterstock.com

© Daisy-Daisy/Alamy

Masterfile

© Siede Preis/Getty Images

CHAPTER 24
Lower Gastrointestinal Disorders 729

CHAPTER 25
Liver Disease and Gallstones 759

CHAPTER 26
Diabetes Mellitus 781

CHAPTER 27
Cardiovascular Diseases 811

© Image Source/JupiterImages

HOW TO

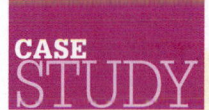

CASE STUDY

Preface

As we launch this ninth edition of *Understanding Normal and Clinical Nutrition*, nutrition research continues to uncover the many complex relationships between nutrition and health. Together with other lifestyle practices, sound nutrition remains a cornerstone of good health status and disease prevention and treatment. Our goals for this edition are therefore to incorporate current research findings into these pages while retaining the core information necessary for a beginning course in nutrition. As with previous editions, each chapter has been substantially revised and updated. New topics, such as functional foods, nutritional genomics, probiotics, and bariatric surgery, are introduced and some existing topics more fully explored. The chapters include practical information and valuable resources to help readers apply nutrition knowledge and skills to their daily lives and the clinical setting.

A main objective in writing this book has always been to share our enthusiasm about nutrition in a manner that motivates students to study and learn. Moreover, we seek to provide accurate information that is meaningful to the student or health professional. Students of nutrition often find the subject to be both fascinating and overwhelming; there are so many "details" to learn—new terms, new chemical structures, and new biological concepts. Taken one step at a time, however, the science of nutrition may seem less daunting and the "facts" more memorable. We hope that this book serves you well.

A Book Tour of This Edition

Understanding Normal and Clinical Nutrition presents updated, comprehensive coverage of the fundamentals of nutrition and nutrition therapy for an introductory nutrition course. The early chapters introduce the nutrients and their work in the body as well as recommendations about nutrition that are essential for maintaining health and preventing disease. The later chapters provide instruction in clinical nutrition—the pathophysiology and nutrition care for a wide range of medical conditions.

The Chapters Chapter 1 begins by exploring why we eat the foods we do and continues with a brief overview of the nutrients, the science of nutrition, recommended nutrient intakes, and important relationships between nutrition and health. Chapter 2 describes the menu-planning principles and food guides used to create diets that support good health and includes instructions on how to read a food label. In Chapter 3, readers follow the journey of digestion and absorption as the body breaks down foods into absorbable nutrients. Chapters 4 through 6 describe carbohydrates, fats, and proteins—their chemistry, roles in the body, and places in the diet. Chapter 7 shows how the body derives energy from these three nutrients. Chapters 8 and 9 continue the story with a look at energy balance, the factors associated with overweight and underweight, and the benefits and risks of weight loss and weight gain. Chapters 10 through 13 describe the vitamins, the minerals, and water—their roles in the body, deficiency and toxicity symptoms, and food sources. Chapters 14 through 16 complete the "normal" chapters by presenting the special nutrient needs of people through the life cycle—pregnancy and lactation, infancy, childhood, adolescence, and adulthood and the later years.

The remaining "clinical" chapters of the book focus on the nutrition care of individuals with health problems. Chapter 17 explains how illnesses and their treatments influence nutrient needs and describes the process of nutrition assessment. Chapter 18 discusses how nutrition care is implemented and introduces the different types of therapeutic diets used in patient care. Chapter 19 explores the potential interactions between nutrients and medications and examines the benefits and risks associated with herbal products. Chapters 20 and 21 describe special ways of feeding people who cannot eat conventional foods. Chapter 22 explains the inflammatory process and shows how metabolic and respiratory stress influence nutrient needs. Chapters 23 through 29 explore the pathology, medical treatment, and nutrition therapy for specific diseases, including gastrointestinal disorders, liver disease, diabetes mellitus, cardiovascular diseases, renal diseases, cancer, and HIV infection.

The Highlights Each chapter is followed by a highlight that provides readers with an in-depth look at a current, and often controversial, topic that may relate to its companion chapter. New to this edition is a highlight that examines the scientific evidence behind some of the current controversies surrounding carbohydrates and their role in weight gain and weight loss.

Special Features The art and layout in this edition have been carefully designed to be inviting while enhancing student learning. In addition, special features help readers identify key concepts and apply nutrition knowledge. For example, when a new term is introduced, it is printed in bold type, and a **definition** is provided. These definitions often include pronunciations and derivations to facilitate understanding. The glossary at the end of the text includes all defined terms.

definition (DEF-eh-NISH-en): the meaning of a word.
- **de** = from
- **finis** = boundary

Nutrition in Your Life/ Nutrition in the Clinical Setting

Chapters 1 through 16 begin with Nutrition in Your Life sections that introduce the essence of the chapter with a friendly and familiar scenario. Similarly, Chapters 17 through 29 begin with Nutrition in the Clinical Setting sections, which introduce real-life concerns associated with diseases or their treatments.

Nutrition Portfolio / Clinical Portfolio

At the end of Chapters 1 through 16, a Nutrition Portfolio section revisits the messages introduced in the chapter and prompts readers to consider whether their personal choices meet the dietary goals discussed. Chapters 17 through 29 finish with a Clinical Portfolio section, which enables readers to practice their clinical skills by addressing hypothetical clinical situations. New to this edition are instructions for using the Diet Analysis Plus computer program to complete this assignment.

IN SUMMARY **Each major section within a chapter concludes with a summary paragraph that reviews key concepts. Similarly, summary tables organize information in an easy-to-read format.**

Also featured in this edition are the *Dietary Guidelines for Americans*, which are introduced in Chapter 2 and presented throughout the text whenever their subjects are discussed. Look for the following design.

Dietary Guidelines for Americans 2010

These guidelines provide science-based advice to promote health and to reduce the risk of chronic disease through diet and physical activity.

HOW TO Many of the chapters include "How To" sections that guide readers through problem-solving tasks. For example, the "How To" in Chapter 1 takes students through the steps of calculating energy intake from the grams of carbohydrate, fat, and protein in a food; another "How To" in Chapter 18 describes how to estimate the energy requirements of a hospital patient.

TRY IT New to this edition are "Try It" activities that help readers practice the "How To" lessons. Additional activities can be found in CourseMate, the online student study tool that accompanies this text.

CASE STUDY The clinical chapters include case studies that present problems and pose questions that allow readers to apply chapter material to hypothetical situations. Readers who successfully master these exercises will be better prepared to face real-life challenges that arise in the clinical setting.

Nutrition Assessment Checklist

The clinical chapters close with Nutrition Assessment Checklists that help readers evaluate how various disorders impair nutrition status. These sections highlight the medical, dietary, anthropometric, biochemical, and physical findings most relevant to patients with specific diseases. Most of the clinical chapters also include a section on Diet-Drug Interactions that presents the nutrition-related concerns associated with the medications commonly used to treat the disorders described in the chapter.

Nutrition on the Net

Each chapter and many highlights conclude with Nutrition on the Net—a list of websites for further study of topics covered in the accompanying text. These lists do not imply an endorsement of the organizations or their programs. We have tried to provide reputable sources but cannot be responsible for the content of these sites. Read Highlight 1 to learn how to find reliable nutrition information on the Internet.

Study and Reference Cards New to this edition are study cards located at the back of the text—one for each chapter. Each study card presents a review list of the chapter's core concepts, and perhaps a table or figure to remind readers of key points. The backside of the study card provides essay and multiple-choice questions to help prepare students for exams. Following the study cards are a series of handy reference cards featuring the current nutrient recommendations; the Daily Values used on food labels and a glossary of nutrient measures; and aids to calculations and conversions.

The Appendixes The appendixes are valuable references for a number of purposes. Appendix A summarizes background information on the hormonal and nervous systems, complementing Appendixes B and C on basic chemistry, the chemical structures of nutrients, and major metabolic pathways. Appendix D describes measures of protein quality. Appendix E provides supplemental coverage of nutrition assessment, and Appendix F presents the estimated energy requirements for men and women at various levels of physical activity. Appendix G presents the 2008 *Choose Your Foods: Exchange List for Diabetes*. Appendix H is a 2000-item food composition table. Appendix I presents recommendations from the World Health Organization (WHO) and information for Canadians—the 2005 *Beyond the Basics* meal-planning system and guidelines to healthy eating and physical activities. Appendix J presents the Healthy People 2020 nutrition-related objectives. Appendix K provides examples of commercial enteral formulas commonly used in tube feedings or to supplement oral diets.

The Inside Covers The inside covers put commonly used information at your fingertips. The front covers (pp. A–C) present the current nutrient recommendations; and the inside back cover (p. Z on the right) shows the suggested weight ranges for various heights.

Notable Changes in This Edition

Much has changed in the world of nutrition and in our daily lives since the first edition. The connections between diet and disease have become more apparent—and consumer interest in making smart health choices has followed. More people are living longer and healthier lives. The science of nutrition has grown rapidly, with new "facts" emerging daily. In this edition, as with all previous editions, every chapter has been revised to enhance learning by presenting current information accurately and attractively.

For all chapters and highlights we have:

- Reviewed and updated content throughout the text, including the 2010 *Dietary Guidelines for Americans*, Healthy People 2020, and the 2011 Dietary Reference Intakes for calcium and vitamin D
- Created several new figures and tables and revised old ones to enhance learning
- Added activities to "How To" features
- Moved all Nutrition Calculation activities from the end of the chapters to the website
- Added *Diet Analysis Plus* activities to Nutrition Portfolios at the end of the chapters

Chapter 1

- Added a table describing the parts of a typical research article
- Updated the text and table for Healthy People 2020

Chapter 2

- Updated all text, tables, and figures to reflect the 2010 *Dietary Guidelines for Americans*
- Introduced the concept of nutrient profiling
- Created a new section that introduces the Healthy Eating Index and evaluates the current American diet against dietary guidelines
- Created a new figure comparing recommendations with actual intakes
- Created a new figure comparing low-fat milk and soy milk
- Created a new table listing good vegetarian sources of key nutrients

Chapter 3

- Created a new figure using gastrin as an example of a negative feedback loop (created similar figures for other GI hormones that are provided in the Instructor's Manual)
- Described how heart disease can damage the digestive tract (intestinal ischemia)

Chapter 4

- Added new photos to reinforce the concept that glucose is the primary fuel of the brain and to illustrate how consumers can read food labels to find whole-grain products
- Refocused organization on the monosaccharides, disaccharides, and polysaccharides (instead of "simple carbohydrates" and "complex carbohydrates")
- Added a little more on resistant starch; deleted a little from glycemic index
- Moved an introduction of alternative sweeteners from the highlight into the chapter; created a new table of alternative sweeteners; moved discussion on alternative sweeteners and weight control to Chapter 9
- Moved controversies surrounding sugars to the highlight and other chapters as appropriate
- Created a new highlight that explores the roles of carbohydrates in weight gain and in weight loss

Chapter 5

- Created a new figure diagramming how saturated fatty acids tend to "stack" and unsaturated fatty acids do not, helping explain why saturated fats tend to be solid in foods at room temperature and to clog arteries in the body
- Created a new figure summarizing lipid transport via lipoproteins
- Added discussion on endocrine role of adipose tissue and introduced adipokines

Chapter 6

- Introduced the terms *primary, secondary, tertiary,* and *quaternary* to the discussion on protein structure
- Simplified the discussion on fluid balance and added a photo of edema
- Moved the discussions on deamination, transamination, ammonia production, and urea excretion from Chapter 7 (metabolism) to here
- Created a new figure describing nutritional genomics
- Created a new figure illustrating nutrient regulation of gene expression

Chapter 7

- Moved the mitochondrion blowout from the figure of a typical cell to a new figure later in the chapter
- Revised the ATP figure illustrating the capture and release of energy
- Simplified the figure illustrating the glucose-to-energy pathway

- Moved the discussions on deamination, transamination, ammonia production, and urea excretion from here to Chapter 6 (proteins)
- Moved the discussions on low-carbohydrate diets from Highlight 9 to here
- Revised the table of alcoholic beverages to include grams of alcohol and additional drinks

Chapter 8

- Replaced the BMI figure with a BMI table of weights
- Created a new "How To" feature on using the BMI table to determine current BMI and a desired BMI
- Moved "How to Determine Body Weight Based on BMI" based on mathematical equations from the last edition to the Instructor's Manual

Chapter 9

- Introduced the term *nonexercise activity thermogenesis (NEAT)*
- Added discussion on adiponectin
- Created a table of proteins involved in the regulation of food intake and energy homeostasis
- Added discussion on phentermine and diethylpropion to section on drug treatments
- Added discussion on cognitive skills to support weight loss
- Revised Highlight 9 to examine fad diets in general (no longer tightly focused on low-carbohydrate diets)

Chapter 10

- Added discussion differentiating between "wet" and "dry" beriberi
- Rewrote section on folate and cancer
- Rewrote section on vitamin C and the common cold
- Created a figure illustrating dose levels and effects
- Moved the discussion on distinguishing symptoms and causes into the text

Chapter 11

- Created a new "How To" feature on converting IU to weight measurements
- Clarified vitamin A's roles in gene regulation, vision, and bone development
- Provided descriptions of vitamin D's many forms
- Added details on vitamin D's roles in gene regulation and disease development, including discussions on insufficiency, deficiency, and new recommendations
- Added details on vitamin K and osteocalcin

Chapter 12

- Clarified the details of the rennin-angiotensin-aldosterone pathway in the text and created a new figure

Chapter 13

- Simplified discussion of iron and zinc absorption and transport
- Revised the figure showing the fluoride map to reflect recent data
- Revised the table of phytochemicals

Chapter 14

- Elaborated on weight issues, including new weight-gain recommendations, weight gains compared with recommendations, pregnancy after gastric bypass, weight loss during pregnancy, and weight gains after pregnancy
- Added short discussion on maternal PKU and aspartame use

Chapter 15

- Created a new figure that includes MyPyramid for Preschoolers and MyPyramid for Kids
- Expanded discussion in the childhood obesity section
- Created a table of recommended eating and physical activity behaviors to prevent obesity
- Included the new *Physical Activity Guidelines for Americans, 2008* (specific to children)
- Created a table of foods and beverages that meet recommended school food standards
- Created a new figure of age-appropriate physical activities

Chapter 16

- Expanded discussion on immunity and inflammation in the aging process
- Added a figure illustrating a modified pyramid for older adults
- Added discussion on folate and zinc needs of older adults
- Created new figure illustrating the hunger-obesity paradox

Chapter 17

- Replaced the Mini Nutritional Assessment table with a figure
- Revised the table about historical information used in nutrition assessment
- Reorganized the discussions about anthropometric assessment in adults and physical examinations for nutrition assessment

Chapter 18

- Simplified the table of equations for estimating resting metabolic rate
- Incorporated the Mifflin–St. Jeor equation into the "How To" about estimating energy requirements in a hospital patient
- Reorganized the section about implementing nutrition care

Chapter 19

- Revised the introductory section about medications in disease treatment

- Simplified the table on terms prohibited in clinical documentation
- Added specific examples of drugs and drug names throughout the chapter
- Revised the tables listing grapefruit juice–drug interactions and foods high in tyramine
- Moved the section about herbal products to the end of the chapter; simplified the table on herb-drug interactions
- Moved the highlight about complementary and alternative medicine to here

Chapter 20

- Moved the discussion of formula selection to the section introducing enteral formulas
- Moved the discussion about meeting water needs to the section about tube feeding administration
- Revised and reorganized the table about the causes and management of tube feeding complications
- Added a new table to the highlight showing examples of nutrition-related inborn errors of metabolism

Chapter 21

- Revised the "How To" about expressing the osmolar concentration of a solution

Chapter 22

- Modified the discussions about the nutrient needs of patients undergoing acute stress and approaches to nutrition care
- Modified the "How To" about estimating energy needs in critical care patients by including the information in a table listing both the Ireton-Jones and Penn State equations
- Revised and reorganized the overall discussion about respiratory failure, including the description of acute respiratory distress syndrome and the nutrition therapy for acute respiratory failure

Chapter 23

- Revised the listings of foods to avoid in the various dysphagia diets
- Revised and expanded the section on bariatric surgery, including a revision of the "How To" about dietary habits after bariatric surgery

Chapter 24

- Revised or reorganized the tables describing laxatives and bulk-forming agents, foods that cause intestinal gas, foods that affect diarrhea, foods to include or restrict in a fat-controlled diet, and foods to include or restrict in a gluten-free diet
- Added sample menus for a fat-controlled diet and a gluten-free diet
- Revised or modified the sections on pancreatitis, cystic fibrosis, and short bowel syndrome

Chapter 25

- Expanded the descriptions of the different types of hepatitis viruses
- Revised the sections on ascites and the medical treatment and nutrition therapy for cirrhosis
- Revised several paragraphs about risk factors for gallstones
- Moved the highlight about anemia in illness to here

Chapter 26

- Added the glycated hemoglobin measure to the section about diagnosis of diabetes
- Revised the discussion about type 2 diabetes in children and adolescents
- Added a new figure outlining the acute effects of insulin insufficiency
- Revised some discussions related to the complications of diabetes
- Revised the section about physical activity and diabetes management

Chapter 27

- Reorganized and revised the sections on the causes of atherosclerosis
- Modified several sections related to therapeutic lifestyle changes for coronary heart disease (CHD)
- Reorganized the "How To" on the evaluation and treatment of high blood cholesterol levels; moved the discussion about drug therapies for CHD prevention from the text to the "How To"
- Expanded the section on stroke management and moved the sections about stroke to follow the discussion of CHD
- In the section on the treatment of hypertension, revised the paragraph about weight reduction and added a paragraph about physical activity
- Revised the discussion about the consequences of heart failure

Chapter 28

- Revised the discussions about the nephrotic syndrome and its consequences; revised the table describing a sodium-controlled diet
- Modified several paragraphs related to the consequences and treatment of acute kidney injury
- In the section on chronic kidney disease, revised discussions about the uremic syndrome and various aspects of nutrition therapy
- Added new tables showing the approximate potassium content of common fruits and vegetables, foods high in phosphorus, and an appropriate menu for a person with chronic kidney disease
- Revised the section on the nutrition therapy following a kidney transplant; eliminated the table on dietary guidelines following a transplant

- In the section on kidney stones, added a paragraph about calcium phosphate stones, revised the table on foods high in oxalate, and eliminated the table on foods high in purines
- In the section on the prevention and treatment of kidney stones, revised the discussion about calcium oxalate stones and added a paragraph about the medical treatment of kidney stones

Chapter 29

- Modified the discussion about nutrition and cancer risk and the table about recommendations for reducing risk
- Modified the section about hematopoietic stem cell transplantation; added a section about biological therapies for cancer
- Expanded the "How To" about increasing kcalories and protein in meals
- Added a section about the low-microbial diet for individuals with suppressed immunity
- Reorganized and revised the section on nutrition therapy for HIV infection, including a discussion about weight management and overweight/obesity
- Moved the highlight about food allergies to here

Student and Instructor Resources

- **CourseMate**: An intelligent, Web-based study system, CourseMate provides a completely integrated package of quizzes, personalized study, animations, videos, case studies, and more—along with the capability for instructors to track student progress. New to this edition is a set of 11 videos covering more difficult concepts. Videos are available as pop-up tutors in the e-book and can be downloaded to an iPod or other portable device.
- **Power Lecture DVD-ROM:** This one-stop course preparation and presentation resource makes it easy for you to assemble, edit, publish, and present custom lectures for your course, using PowerPoint®. The PowerLecture includes PowerPoint®, animations, BBC video clips, the instructor's manual, the test bank, "clicker" content, and ExamView computerized testing.
- **Test Bank:** The test bank features a large assortment of multiple-choice questions (categorized by knowledge or application), essay questions, and matching exercises, now organized by chapter section title for easier item selection.
- **Instructor's Manual:** New to this edition are assignable case studies, critical-thinking questions, and Internet exercises, all with grading rubrics. This comprehensive manual also includes chapter objectives, chapter outlines, answer keys for the new "How To" exercises from the text, assignment worksheets, handouts, and classroom activity suggestions. (For Canadian adopters, the manual contains a Canadian information section with equivalencies, nutrient recommendations, and more.)

Closing Comments

We have taken great care to provide accurate information and have included many references at the end of each chapter and highlight. However, to keep the number of references manageable, many statements that appeared in previous editions with references now appear without them. All statements reflect current nutrition knowledge and the authors will supply references upon request. In addition to supporting text statements, the end-of-chapter references provide readers with resources for finding a good overview or more details on the subject. Nutrition is a fascinating subject, and we hope our enthusiasm for it comes through on every page.

Sharon Rady Rolfes
Kathryn Pinna
Ellie Whitney
June 2011

Acknowledgments

To produce a book requires the coordinated effort of a team of people—and, no doubt, each team member has another team of support people as well. We salute, with a big round of applause, everyone who has worked so diligently to ensure the quality of this book.

We thank our partners and friends, Linda DeBruyne and Fran Webb, for their valuable consultations and contributions; working together over the past 25-plus years has been a most wonderful experience. We especially appreciate Linda's research assistance on several chapters. Special thanks to Elesha Feldman, whose critical eye, numerous suggestions, and unceasing support were especially helpful in revising the clinical chapters. Additional thank yous to our colleagues Gail Hammond for her Canadian perspective, and David Stone for his help in critiquing various sections in the clinical chapters. Thanks also to Alex Rodriguez for her work on manuscript preparation and to James Gegenheimer for his assistance in creating informative tables and descriptive figures.

We also thank the many professors who prepared the ancillaries that accompany this text: Harry Sitren and Ileana Trautwein for writing and enhancing the test bank; and Connie Goff, Jorunn Gran-Henriksen, Gail Hammond, Mandy Graves Hillstrom, Melissa Langone, Barbara Quinn, Kathleen Rourke, and Daryle Wane for contributing to the Instructor's Manual. Thanks also to Lauren Tarson, Shelly Ryan, Elesha Feldman, and the folks at Axxya Systems for their assistance in creating the food composition appendix and developing the computerized *Diet Analysis Plus* program that accompanies this book.

Our heartfelt thanks to our editorial team for their efforts in creating an outstanding nutrition textbook—Peggy Williams for her leadership and support; Nedah Rose for her thoughtful suggestions and efficient analysis of reviews; Carol Samet for her management of this project; Laura McGinn for her energetic efforts in marketing; Miriam Myers and Lauren Tarson for their dedication in developing online animations and study tools; Dean Dauphinais, Bill Jentzen, and Karyn Morrison for their assistance in obtaining permissions; and Elesha Feldman for her masterminding of ancillaries.

We also thank Gary Hespenheide for creatively designing these pages; Josh Garvin/Bill Smith Group for selecting photographs that deliver nutrition messages attractively; Susan Gall for copyediting more than 2000 manuscript pages; and Pat Lewis for proofreading close to 1000 final text pages. We would also like to extend our gratitude to the talented team at Lachina Publishing Services for their assistance with layout, production, proofreading, copyediting, and indexing. To the hundreds of others involved in production and sales, we tip our hats in appreciation.

We are especially grateful to our friends and families for their continued encouragement and support. We also thank our many reviewers for their comments and contributions.

Reviewers

Becky Alejandre
American River College

Janet B. Anderson
Utah State University

Paul E. Araujo
Baltimore City Community College

Sandra D. Baker
University of Delaware

Angelina Boyce
Hillsborough Community College

Lynn S. Brann
Syracuse University

Shalon Bull
Palm Beach Community College

Dorothy A. Byrne
University of Texas, San Antonio

John R. Capeheart
University of Houston, Downtown

Leah Carter
Bakersfield College

James F. Collins
University of Florida

Diane Curtis
Los Rios Community College District

Lisa K. Diewald
Montgomery County Community College

Kelly K. Eichmann
Fresno City College

Mary Flynn
Brown University

Sue Fredstrom
Minnesota State University, Mankato

Trish Froehlich
Palm Beach Community College

Stephen P. Gagnon
Hillsborough Community College

Jill Golden
Orange Coast College

Kathleen Gould
Towson University

Margaret Gunther
Palomar College

Charlene Hamilton
University of Delaware

D. J. Hennager
Kirkwood Community College

Catherine Hagen Howard
Texarkana College

Kathryn Hillstrom
California State University, Los Angeles

Ernest B. Izevbigie
Jackson State University

Craig Kasper
Hillsborough Community College

Younghee Kim
Bowling Green State University

Carrie King
University of Alaska, Anchorage

Rebecca A. Kleinschmidt
University of Alaska Southeast

Vicki Kloosterhouse
Oakland Community College

Donna M. Kopas
Pennsylvania State University

Susan M. Krueger
University of Wisconsin, Eau Claire

Melissa Langone
Pasco-Hernando Community College

Darlene M. Levinson
Oakland Community College, Orchard Ridge

Kimberly Lower
Collin County Community College

Nelda D. Malm
Seminole State College of Florida

Melissa B. McGuire
Maple Woods Community College

Diane L. McKay
Tufts University

Anne Miller
De Anza College

Anahita M. Mistry
Eastern Michigan University

Dawna Torres Mughal
Gannon University

Mithia Mukutmoni
Sierra College

Steven Nizielski
Grand Valley State University

Carmen L. Nochera
Grand Valley State University

Jane M. Osowski
University of Southern Mississippi

Sarah Panarello
Yakima Valley Community College

Ryan Paruch
Tulsa Community College

Jill Patterson
Pennsylvania State University

Julie Priday
Centralia College

Kathy L. Sedlet
Collin County Community College

Debra Barone Sheats
St. Catherine University

Melissa Shock
University of Central Arkansas

LuAnn Soliah
Baylor University

Kenneth Strothkamp
Lewis & Clark College

Alanna M. Tynes
Lone Star College, Tomball

Elizabeth Vargo
Community College of Allegheny County

Andrea Villarreal
Phoenix College

Terry Weideman
Oakland Community College, Highland Lake

H. Garrison Wilkes
University of Massachusetts, Boston

Lynne C. Zeman
Kirkwood Community College

Maureen Zimmerman
Mesa Community College

CHAPTER

1

© Radius Images/Jupiterimages Corporation

Nutrition in Your Life

Believe it or not, you have probably eaten at least 20,000 meals in your life. Without any conscious effort on your part, your body uses the nutrients from those foods to make all its components, fuel all its activities, and defend itself against diseases. How successfully your body handles these tasks depends, in part, on your food choices. Nutritious food choices support healthy bodies.

Throughout this chapter, the CourseMate icon indicates an opportunity for online self-study, linking you to activities to increase your understanding of chapter concepts. **www.cengagebrain.com** (search for ISBN 084006845X)

An Overview of Nutrition

CHAPTER OUTLINE

Welcome to the world of **nutrition**. Although you may not always have been aware of it, nutrition has played a significant role in your life. And it will continue to affect you in major ways, depending on the **foods** you select.

Every day, several times a day, you make food choices that influence your body's health for better or worse. Each day's choices may benefit or harm your health only a little, but when these choices are repeated over years and decades, the rewards or consequences become major. That being the case, paying close attention to good eating habits now supports health benefits later. Conversely, carelessness about food choices can contribute to many chronic diseases ♦ prevalent in later life, including heart disease, diabetes, and cancer. Of course, some people will become ill or die young no matter what choices they make, and others will live long lives despite making poor choices. For the majority of us, however, the food choices we make each and every day will benefit or impair our health in proportion to the wisdom of those choices.

Although most people realize that their food habits affect their health, they often choose foods for other reasons. After all, foods bring to the table a variety of pleasures, traditions, and associations as well as nourishment. The challenge, then, is to combine favorite foods and fun times with a nutritionally balanced **diet**.

♦ In general, a **chronic disease** progresses slowly or with little change and lasts a long time. By comparison, an **acute disease** develops quickly, produces sharp symptoms, and runs a short course.
- **chronos** = time
- **acute** = sharp

Food Choices

People decide what to eat, when to eat, and even whether to eat in highly personal ways, often based on behavioral or social motives rather than on an awareness of nutrition's importance to health. A variety of food choices can support good health, and an understanding of human nutrition helps you make sensible selections more often.

Personal Preference As you might expect, the number one reason people choose foods is taste—they like certain flavors. Two widely shared preferences are for the sweetness of sugar and the savoriness of salt. Liking high-fat foods also appears to be a universally common preference. Other preferences might be for the hot peppers common in Mexican cooking or the curry spices of Indian cuisine.

nutrition: the science of foods and the nutrients and other substances they contain, and of their actions within the body (including ingestion, digestion, absorption, transport, metabolism, and excretion). A broader definition includes the social, economic, cultural, and psychological implications of food and eating.

foods: products derived from plants or animals that can be taken into the body to yield energy and nutrients for the maintenance of life and the growth and repair of tissues.

diet: the foods and beverages a person eats and drinks.

Research suggests that genetics may influence taste perceptions and therefore food likes and dislikes.[1] Similarly, the hormones of pregnancy seem to influence food cravings and aversions (see Chapter 14).

Habit People sometimes select foods out of habit. They eat cereal every morning, for example, simply because they have always eaten cereal for breakfast. Eating a familiar food and not having to make any decisions can be comforting.

Ethnic Heritage or Tradition Among the strongest influences on food choices are ethnic heritage and tradition. People eat the foods they grew up eating. Every country, and in fact every region of a country, has its own typical foods and ways of combining them into meals. The "American diet" includes many ethnic foods from various countries, all adding variety to the diet. This is most evident when eating out: 60 percent of U.S. restaurants (excluding fast-food places) have an ethnic emphasis, most commonly Chinese, Italian, or Mexican.

© Corbis Premium RF/Alamy

An enjoyable way to learn about other cultures is to taste their ethnic foods.

Social Interactions Most people enjoy companionship while eating. It's fun to go out with friends for pizza or ice cream. Chapter 9 describes how people tend to eat more food when socializing with others. Meals are often social events, and sharing food is part of hospitality—regardless of hunger signals. Social customs invite people to accept food or drink offered by a host or shared by a group.

Availability, Convenience, and Economy People often eat foods that are accessible, quick and easy to prepare, and within their financial means. Consumers who value convenience frequently eat out, bring home ready-to-eat meals, or have food delivered. Even when they venture into the kitchen, they want to prepare a meal in 15 to 20 minutes, using less than a half dozen ingredients—and those "ingredients" are often semiprepared foods, such as canned soups. Alternatively, some consumers visit meal-preparation businesses where they can assemble several meals to feed their families from ingredients that have been purchased and portioned according to planned menus.[2] Those who frequently prepare their own meals eat fast-food less often and are more likely to meet dietary guidelines for fat, calcium, fruits, vegetables, and whole grains.[3]

Consumer emphasis on convenience limits food choices to the selections offered on menus and products designed for quick preparation. Whether decisions based on convenience meet a person's nutrition needs depends on the choices made. Eating a banana or a candy bar may be equally convenient, but the fruit provides more vitamins and minerals and less sugar and fat.

Rising food costs have shifted some consumers' priorities and changed their shopping habits. They are less likely to buy higher priced convenience foods and more likely to buy less-expensive store brand items and prepare home-cooked meals. In fact, more than 80 percent of U.S. consumers are eating home-cooked meals at least three times a week.[4]

Positive and Negative Associations People tend to like particular foods associated with happy occasions—such as hot dogs at ball games or cake and ice cream at birthday parties. By the same token, people can develop aversions and dislike foods that they ate when they felt sick or that they were forced to eat as a child. By using foods as rewards or punishments, parents may inadvertently teach their children to like and dislike certain foods.

Emotions Some people cannot eat when they are emotionally upset. Others may eat in response to a variety of emotional stimuli—for example, to relieve boredom or depression or to calm anxiety. A depressed person may choose to eat rather than to call a friend. A person who has returned home from an exciting evening out may unwind with a late-night snack. These people may find emotional comfort, in part, because foods can influence the brain's chemistry and the mind's response. Carbohydrate and alcohol, for example, tend to calm, whereas protein and caffeine are more likely to activate. Eating in response to emotions can easily lead to overeating and obesity, but it may be appropriate at times. For example, sharing food at times of bereavement serves both the giver's need to provide comfort and the receiver's need to be cared for and to interact with others, as well as to take nourishment.

Values Food choices may reflect people's religious beliefs, political views, or environmental concerns. For example, some Christians forgo meat on Fridays during Lent (the period prior to Easter), Jewish law includes an extensive set of dietary rules that govern the use of foods derived from animals, and Muslims fast between sunrise and sunset during Ramadan (the ninth month of the Islamic calendar). Some vegetarians select foods based on their concern for animal rights. A concerned consumer may boycott fruit picked by migrant workers who have been exploited. People may buy vegetables from local farmers to save the fuel and environmental costs of foods shipped from far away. They may also select foods packaged in containers that can be reused or recycled. Some consumers accept or reject foods that have been irradiated, grown organically, or genetically modified, depending on their approval of these processes.

© Wave Royalty Free/Alamy

To enhance your health, keep nutrition in mind when selecting foods. To protect the environment, shop at local markets and reuse cloth shopping bags.

Body Weight and Image Sometimes people select certain foods and supplements that they believe will improve their physical appearance and avoid those they believe might be detrimental. Such decisions can be beneficial when based on sound nutrition and fitness knowledge, but decisions based on fads or carried to extremes undermine good health, as pointed out in later discussions of eating disorders (Highlight 8) and dietary supplements commonly used by athletes.

Nutrition and Health Benefits Finally, of course, many consumers make food choices that will benefit health. Food manufacturers and restaurant chefs have responded to scientific findings linking health with nutrition by offering an abundant selection of health-promoting foods and beverages. Foods that provide health benefits beyond their nutrient contributions are called **functional foods.**[5] Whole foods—as natural and familiar as oatmeal or tomatoes—are the simplest functional foods. In other cases, foods have been modified to provide health benefits, perhaps by lowering the fat contents. In still other cases, manufacturers have fortified foods by adding nutrients or **phytochemicals** that provide health benefits (see Highlight 13). ♦ Examples of these functional foods include orange juice fortified with calcium to help build strong bones and margarine made with a plant sterol that lowers blood cholesterol.

♦ Functional foods may include whole foods, modified foods, or fortified foods.

Consumers typically welcome new foods into their diets, provided that these foods are reasonably priced, clearly labeled, easy to find in the grocery store, and convenient to prepare. These foods must also taste good—as good as the traditional choices. Of course, a person need not eat any "special" foods to enjoy a healthy diet; many "regular" foods provide numerous health benefits as well. In fact, "regular" foods such as whole grains; vegetables and legumes; fruits; seafood, meats, poultry, eggs, nuts, and seeds; and milk products are among the healthiest choices a person can make.

IN SUMMARY **A person selects foods for a variety of reasons. Whatever those reasons may be, food choices influence health. Individual food selections neither make nor break a diet's healthfulness, but the balance of foods selected over time can make an important difference to health.[6] For this reason, people are wise to think "nutrition" when making their food choices.**

The Nutrients

Biologically speaking, people eat to receive nourishment. Do you ever think of yourself as a biological being made of carefully arranged atoms, molecules, cells, tissues, and organs? Are you aware of the activity going on within your body

functional foods: foods that contain physiologically active compounds that provide health benefits beyond their nutrient contributions; sometimes called *designer foods* or *nutraceuticals*.

phytochemicals (FIE-toe-KEM-ih-cals): nonnutrient compounds found in plant-derived foods that have biological activity in the body.

- **phyto** = plant

© Masterfile

Foods bring pleasure—and nutrients.

even as you sit still? The atoms, molecules, and cells of your body continuously move and change, even though the structures of your tissues and organs and your external appearance remain relatively constant. Your skin, which has covered you since your birth, is replaced entirely by new cells every 7 years. The fat beneath your skin is not the same fat that was there a year ago. Your oldest red blood cell is only 120 days old, and the entire lining of your digestive tract is renewed every 3 to 5 days. To maintain your "self," you must continually replenish, from foods, the **energy** and the **nutrients** you deplete as your body maintains itself.

Nutrients in Foods and in the Body

Amazingly, our bodies can derive all the energy, structural materials, and regulating agents we need from the foods we eat. This section introduces the nutrients that foods deliver and shows how they participate in the dynamic processes that keep people alive and well.

Nutrient Composition of Foods Chemical analysis of a food such as a tomato shows that it is composed primarily of water (95 percent). Most of the solid materials are carbohydrates, lipids, ♦ and proteins. If you could remove these materials, you would find a tiny residue of vitamins, minerals, and other compounds. Water, carbohydrates, lipids, proteins, vitamins, and some of the minerals found in foods represent the six classes ♦ of nutrients—substances the body uses for the growth, maintenance, and repair of its tissues.

♦ As Chapter 5 explains, most lipids are fats.

♦ Six classes of nutrients:
- Carbohydrates
- Lipids (fats)
- Proteins
- Vitamins
- Minerals
- Water

This book focuses mostly on the nutrients, but foods contain other compounds as well—fibers, phytochemicals, pigments, additives, alcohols, and others. Some are beneficial, some are neutral, and a few are harmful. Later sections of the book touch on these compounds and their significance.

Nutrient Composition of the Body A chemical analysis of your body would show that it is made of materials similar to those found in foods (see Figure 1-1).

FIGURE 1-1 Body Composition of Healthy-Weight Men and Women

The human body is made of compounds similar to those found in foods—mostly water (60 percent) and some fat (13 to 21 percent for young men, 23 to 31 percent for young women), with carbohydrate, protein, vitamins, minerals, and other minor constituents making up the remainder. (Chapter 8 describes the health hazards of too little or too much body fat.)

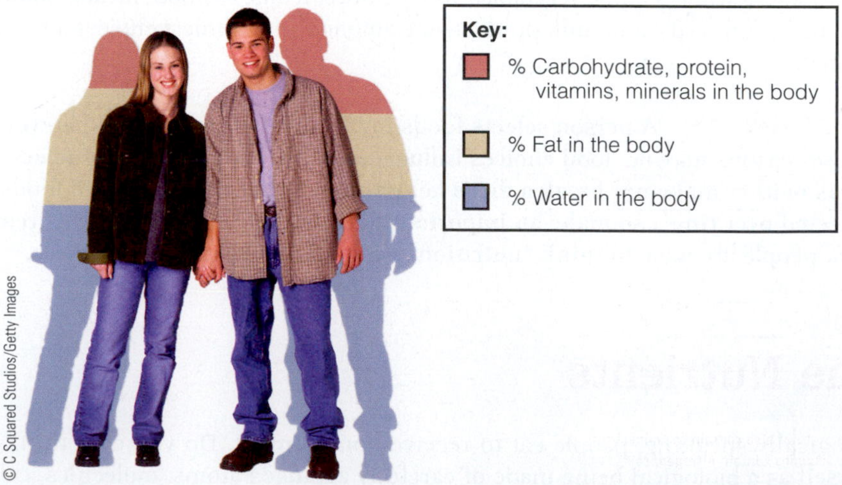

© C Squared Studios/Getty Images

energy: the capacity to do work. The energy in food is chemical energy. The body can convert this chemical energy to mechanical, electrical, or heat energy.

nutrients: chemical substances obtained from food and used in the body to provide energy, structural materials, and regulating agents to support growth, maintenance, and repair of the body's tissues. Nutrients may also reduce the risks of some diseases.

A healthy 150-pound body contains about 90 pounds of water and about 20 to 45 pounds of fat. The remaining pounds are mostly protein, carbohydrate, and the major minerals of the bones. Vitamins, other minerals, and incidental extras constitute a fraction of a pound.

Chemical Composition of Nutrients The simplest of the nutrients are the minerals. Each mineral is a chemical element; its atoms are all alike. As a result, its identity never changes. For example, iron may have different electrical charges, but the individual iron atoms remain the same when they are in a food, when a person eats the food, when the iron becomes part of a red blood cell, when the cell is broken down, and when the iron is lost from the body by excretion. The next simplest nutrient is water, a compound made of two elements—hydrogen and oxygen. Minerals and water are **inorganic** nutrients, which means they do not contain carbon.

The other four classes of nutrients (carbohydrates, lipids, proteins, and vitamins) are more complex. In addition to hydrogen and oxygen, they all contain carbon, an element found in all living things. They are therefore called **organic** compounds (meaning, literally, "alive"). This chemical definition of *organic* differs from the agricultural definition. *Organic farming* refers to growing crops and raising livestock according to standards set by the U.S. Department of Agriculture (USDA). Protein and some vitamins also contain nitrogen and may contain other elements such as sulfur as well (see Table 1-1).

Essential Nutrients The body can make some nutrients, but it cannot make all of them. Also, it makes some in insufficient quantities to meet its needs and, therefore, must obtain these nutrients from foods. The nutrients that foods must supply are **essential nutrients.** When used to refer to nutrients, the word *essential* means more than just "necessary"; it means "needed from outside the body"—normally, from foods.

The Energy-Yielding Nutrients: Carbohydrate, Fat, and Protein

In the body, three organic nutrients can be used to provide energy: carbohydrate, fat, and protein. ♦ In contrast to these **energy-yielding nutrients,** vitamins, minerals, and water do not yield energy in the human body.

Energy Measured in kCalories The energy released from carbohydrate, fat, and protein can be measured in **calories**—tiny units of energy so small that a single apple provides tens of thousands of them. To ease calculations, energy is expressed in 1000-calorie metric units known as kilocalories (shortened to kcalories, but commonly called "calories"). When you read in popular books or magazines

TABLE 1-1 Elements in the Six Classes of Nutrients

Notice that organic nutrients contain carbon.

	Carbon	Hydrogen	Oxygen	Nitrogen	Minerals
Inorganic nutrients					
Minerals					✓
Water		✓	✓		
Organic nutrients					
Carbohydrate	✓	✓	✓		
Lipid (fat)	✓	✓	✓		
Protein[a]	✓	✓	✓	✓	
Vitamins[b]	✓	✓	✓		

[a]Some proteins also contain the mineral sulfur.
[b]Some vitamins contain nitrogen; some contain minerals.

♦ Carbohydrate, fat, and protein are sometimes called **macronutrients** because the body requires them in relatively large amounts (many grams daily). In contrast, vitamins and minerals are **micronutrients,** required only in small amounts (milligrams or micrograms daily).

inorganic: not containing carbon or pertaining to living things.
- **in** = not

organic: in chemistry, a substance or molecule containing carbon-carbon bonds or carbon-hydrogen bonds. This definition excludes coal, diamonds, and a few carbon-containing compounds that contain only a single carbon and no hydrogen, such as carbon dioxide (CO_2), calcium carbonate ($CaCO_3$), magnesium carbonate ($MgCO_3$), and sodium cyanide (NaCN).

essential nutrients: nutrients a person must obtain from food because the body cannot make them for itself in sufficient quantity to meet physiological needs; also called **indispensable nutrients.** About 40 nutrients are currently known to be essential for human beings.

energy-yielding nutrients: the nutrients that break down to yield energy the body can use:
- Carbohydrate
- Fat
- Protein

calories: units by which energy is measured. Food energy is measured in *kilocalories* (1000 calories equal 1 kilocalorie), abbreviated **kcalories** or **kcal.** One kcalorie is the amount of heat necessary to raise the temperature of 1 kilogram (kg) of water 1°C. The scientific use of the term *kcalorie* is the same as the popular use of the term *calorie.*

◆ The international unit for measuring food energy is the **joule,** a measure of *work* energy. To convert kcalories to kilojoules, multiply by 4.2; to convert kilojoules to kcalories, multiply by 0.24.

that an apple provides "100 calories," it actually means 100 kcalories. This book uses the term **kcalorie** and its abbreviation **kcal** throughout, as do other scientific books and journals. ◆ The accompanying "How To" provides a few tips on "thinking metric."

HOW TO Think Metric

Like other scientists, nutrition scientists use metric units of measure. They measure food energy in kilocalories, people's height in centimeters, people's weight in kilograms, and the weights of foods and nutrients in grams, milligrams, or micrograms. For ease in using these measures, it helps to remember that the prefixes on the grams imply 1000. For example, a *kilo*gram is 1000 grams, a *milli*gram is 1/1000 of a gram, and a *micro*gram is 1/1000 of a milligram.

Most food labels and many recipe books provide "dual measures," listing both household measures, such as cups, quarts, and teaspoons, and metric measures, such as milliliters, liters, and grams. This practice gives people an opportunity to gradually learn to "think metric."

A person might begin to "think metric" by simply observing the measure—by noticing the amount of soda in a 2-liter bottle, for example. Through such experiences, a person can become familiar with a measure without having to do any conversions.

To facilitate communication, many members of the international scientific community have adopted a common system of measurement—the International System of Units (SI). In addition to using metric measures, the SI establishes common units of measurement. For example, the SI unit for measuring food energy is the joule (not the kcalorie). A joule is the amount of energy expended when 1 kilogram is moved 1 meter by a force of 1 newton. The joule is thus a measure of *work* energy, whereas the kcalorie is a measure of *heat* energy. While many scientists and journals report their findings in kilojoules (kJ), many others, particularly those in the United States, use kcalories (kcal). To convert energy measures from kcalories to kilojoules, multiply by 4.2. For example, a 50-kcalorie cookie provides 210 kilojoules:

$$50 \text{ kcal} \times 4.2 = 210 \text{ kJ}$$

Exact conversion factors for these and other units of measure are in the Aids to Calculation section on the last two pages of the book.

For additional practice, log on to **www.cengagebrain.com** and search for ISBN 084006845X.

Volume: Liters (L)

1 L = 1000 milliliters (mL)
0.95 L = 1 quart
1 mL = 0.03 fluid ounces
240 mL = 1 cup

© Felicia Martinez/Photo Edit

A liter of liquid is approximately one U.S. quart. (Four liters are only about 5 percent more than a gallon.)

© Felicia Martinez/Photo Edit

One cup is about 240 milliliters; a half-cup of liquid is about 120 milliliters.

Weight: Grams (g)

1 g = 1000 milligrams (mg)
1 g = 0.04 ounce (oz)
1 oz = 28.35 g (or 30 g)
100 g = 3½ oz
1 kilogram (kg) = 1000 g
1 kg = 2.2 pounds (lb)
454 g = 1 lb

A kilogram is slightly more than 2 lb; conversely, a pound is about ½ kg.

© Thomas Harm, Tom Peterson/Quest Photographic Inc.

A half-cup of vegetables weighs about 100 grams; one pea weighs about ½ gram.

© Tony Freeman/Photo Edit

A 5-pound bag of potatoes weighs about 2 kilograms, and a 176-pound person weighs 80 kilograms.

TRY IT Convert your body weight from pounds to kilograms and your height from inches to centimeters.

HOW TO Calculate the Energy Available from Foods

To calculate the energy available from a food, multiply the number of grams of carbohydrate, protein, and fat by 4, 4, and 9, respectively. Then add the results together. For example, 1 slice of bread with 1 tablespoon of peanut butter on it contains 16 grams carbohydrate, 7 grams protein, and 9 grams fat:

$$16 \text{ g carbohydrate} \times 4 \text{ kcal/g} = 64 \text{ kcal}$$
$$7 \text{ g protein} \times 4 \text{ kcal/g} = 28 \text{ kcal}$$
$$9 \text{ g fat} \times 9 \text{ kcal/g} = 81 \text{ kcal}$$
$$\text{Total} = 173 \text{ kcal}$$

From this information, you can calculate the percentage of kcalories each of the energy nutrients contributes to the total. To determine the percentage of kcalories from fat, for example, divide the 81 fat kcalories by the total 173 kcalories:

$$81 \text{ fat kcal} \div 173 \text{ total kcal} = 0.468 \text{ (rounded to 0.47)}$$

Then multiply by 100 to get the percentage:

$$0.47 \times 100 = 47\%$$

Dietary recommendations that urge people to limit fat intake to 20 to 35 percent of kcalories refer to the day's total energy intake, not to individual foods. Still, if the proportion of fat in each food choice throughout a day exceeds 35 percent of kcalories, then the day's total surely will, too. Knowing that this snack provides 47 percent of its kcalories from fat alerts a person to the need to make lower-fat selections at other times that day.

For additional practice, log on to **www.cengagebrain.com** and search for ISBN 084006845X.

TRY IT Calculate the energy available from a bean burrito with cheese (55 grams carbohydrate, 15 grams protein, and 12 grams fat). Determine the percentage of kcalories from each of the energy nutrients.

TABLE 1-2 kCalorie Values of Energy Nutrients[a]

Nutrients	Energy (kcal/g)
Carbohydrate	4
Fat	9
Protein	4

NOTE: Alcohol contributes 7 kcalories per gram that can be used for energy, but it is not considered a nutrient because it interferes with the body's growth, maintenance, and repair.
[a]For those using kilojoules: 1 g carbohydrate = 17 kJ; 1 g protein = 17 kJ; 1 g fat = 37 kJ; and 1 g alcohol = 29 kJ.

Energy from Foods The amount of energy a food provides depends on how much carbohydrate, fat, and protein it contains. ♦ When completely broken down in the body, a gram of carbohydrate yields about 4 kcalories of energy; a gram of protein also yields 4 kcalories; and a gram of fat yields 9 kcalories (see Table 1-2). The accompanying "How To" explains how to calculate the energy available from foods.

Because fat provides more energy per gram, it has a greater **energy density** than either carbohydrate or protein. Figure 1-2 (p. 10) compares the energy density of two breakfast options, and later chapters describe how considering a food's energy density can help with weight management. ♦

One other substance contributes energy—alcohol. Alcohol, however, is not considered a nutrient. Unlike the essential nutrients, alcohol does not sustain life. In fact, it interferes with the growth, maintenance, and repair of the body. Its only common characteristic with nutrients is that it yields energy (7 kcalories per gram) when metabolized in the body.

Most foods contain all three energy-yielding nutrients, as well as vitamins, minerals, water, and other substances. For example, meat contains water, fat, vitamins, and minerals as well as protein. Bread contains water, a trace of fat, a little protein, and some vitamins and minerals in addition to its carbohydrate. Only a few foods are exceptions to this rule—the common ones being sugar (pure carbohydrate) and oil (essentially pure fat).

Energy in the Body The body uses the energy-yielding nutrients to fuel all its activities. When the body uses carbohydrate, fat, or protein for energy, the bonds between the nutrient's atoms break. As the bonds break, they release energy. ♦ Some of this energy is released as heat, but some is used to send electrical impulses through the brain and nerves, to synthesize body compounds, and to move muscles. Thus the energy from food supports every activity from quiet thought to vigorous sports.

♦ The energy-yielding nutrients:
- Carbohydrate
- Fat
- Protein

♦ Foods with a high energy density help with weight gain, whereas those with a low energy density help with weight loss.

♦ The processes by which nutrients are broken down to yield energy or used to make body structures are known as **metabolism** (defined and described further in Chapter 7).

energy density: a measure of the energy a food provides relative to the amount of food (kcalories per gram).

FIGURE 1-2 Energy Density of Two Breakfast Options Compared

Gram for gram, ounce for ounce, and bite for bite, foods with a high energy density deliver more kcalories than foods with a low energy density. Both of these breakfast options provide 500 kcalories, but the cereal with milk, fruit salad, scrambled egg, turkey sausage, and toast with jam offers three times as much food as the doughnuts (based on weight); it has a lower energy density than the doughnuts. Selecting a variety of foods also helps to ensure nutrient adequacy.

LOWER ENERGY DENSITY
This 450-gram breakfast delivers 500 kcalories, for an energy density of 1.1 (500 kcal ÷ 450 g = 1.1 kcal/g).

HIGHER ENERGY DENSITY
This 144-gram breakfast delivers 500 kcalories, for an energy density of 3.5 (500 kcal ÷ 144 g = 3.5 kcal/g).

If the body does not use these nutrients to fuel its current activities, it converts them into storage compounds (such as body fat) to be used between meals and overnight when fresh energy supplies run low. If more energy is consumed than expended, the result is an increase in energy stores and weight gain. Similarly, if less energy is consumed than expended, the result is a decrease in energy stores and weight loss.

When consumed in excess of energy needs, alcohol, too, can be converted to body fat and stored. When alcohol contributes a substantial portion of the energy in a person's diet, the harm it does far exceeds the problems of excess body fat. (Highlight 7 describes the effects of alcohol on health and nutrition.)

Other Roles of Energy-Yielding Nutrients In addition to providing energy, carbohydrates, fats, and proteins provide the raw materials for building the body's tissues and regulating its many activities. In fact, protein's role as a fuel source is relatively minor compared with both the other two energy-yielding nutrients and its other roles. Proteins are found in structures such as the muscles and skin and help to regulate activities such as digestion and energy metabolism. (Chapter 6 presents a full discussion on proteins.)

The Vitamins

The **vitamins** are also organic, but they do not provide energy. Instead, they facilitate the release of energy from carbohydrate, fat, and protein and participate in numerous other activities throughout the body.

Each of the 13 vitamins has its own special roles to play.* One vitamin enables the eyes to see in dim light, another helps protect the lungs from air pollution, and still another helps make the sex hormones—among other things. When you cut yourself, one vitamin helps stop the bleeding and another helps repair the skin. Vitamins busily help replace old red blood cells and the lining of the digestive tract. Almost every action in the body requires the assistance of vitamins.

*The water-soluble vitamins are vitamin C and the eight B vitamins: thiamin, riboflavin, niacin, vitamins B_6 and B_{12}, folate, biotin, and pantothenic acid. The fat-soluble vitamins are vitamins A, D, E, and K. The water-soluble vitamins are the subject of Chapter 10 and the fat-soluble vitamins, of Chapter 11.

vitamins: organic, essential nutrients required in small amounts by the body for health.

Vitamins can function only if they are intact, but because they are complex organic molecules, they are vulnerable to destruction by heat, light, and chemical agents. This is why the body handles them carefully and why nutrition-wise cooks do, too. The strategies of cooking vegetables at moderate temperatures for short times and using small amounts of water help to preserve the vitamins.

The Minerals

In the body, some **minerals** are put together in orderly arrays in such structures as bones and teeth. Minerals are also found in the fluids of the body, which influences fluid balance and distribution. Whatever their roles, minerals do not yield energy.

Only 16 minerals are known to be essential in human nutrition.* Others are being studied to determine whether they play significant roles in the human body. Still other minerals, such as lead, are environmental contaminants that displace the nutrient minerals from their workplaces in the body, disrupting body functions. The problems caused by contaminant minerals are described in Chapter 13.

Because minerals are inorganic, they are indestructible and need not be handled with the special care that vitamins require. Minerals can, however, be bound by substances that interfere with the body's ability to absorb them. They can also be lost during food-refining processes or during cooking when they leach into water that is discarded.

Water

Water provides the environment in which nearly all the body's activities are conducted. It participates in many metabolic reactions and supplies the medium for transporting vital materials to cells and carrying waste products away from them. Water is discussed fully in Chapter 12, but it is mentioned in every chapter. If you watch for it, you cannot help but be impressed by water's participation in all life processes.

IN SUMMARY **Foods provide nutrients—substances that support the growth, maintenance, and repair of the body's tissues. The six classes of nutrients include:**

- **Carbohydrates**
- **Lipids (fats)**
- **Proteins**
- **Vitamins**
- **Minerals**
- **Water**

Foods rich in the energy-yielding nutrients (carbohydrate, fat, and protein) provide the major materials for building the body's tissues and yield energy for the body's use or storage. Energy is measured in kcalories. Vitamins, minerals, and water facilitate a variety of activities in the body.

Without exaggeration, nutrients provide the physical and metabolic basis for nearly all that we are and all that we do. The next section introduces the science of nutrition with emphasis on the research methods scientists have used in uncovering the wonders of nutrition.

© Ant Strack/Corbis

Water itself is an essential nutrient and naturally carries many minerals.

The Science of Nutrition

The science of nutrition is the study of the nutrients and other substances in foods and the body's handling of them. Its foundation depends on several other sciences, including biology, biochemistry, and physiology. As sciences go, nutrition is young, but as you can see from the size of this book, much has happened in nutrition's short life. And it is currently experiencing a tremendous growth spurt as scientists apply knowledge gained from sequencing the human **genome**. The

*The major minerals are calcium, phosphorus, potassium, sodium, chloride, magnesium, and sulfate. The trace minerals are iron, iodine, zinc, chromium, selenium, fluoride, molybdenum, copper, and manganese. Chapters 12 and 13 are devoted to the major and trace minerals, respectively.

minerals: inorganic elements. Some minerals are essential nutrients required in small amounts by the body for health.

genome (GEE-nome): the complete set of genetic material (DNA) in an organism or a cell. The study of genomes is called **genomics.**

integration of nutrition, genomics, and molecular biology has opened a whole new world of study called **nutritional genomics**—the science of how nutrients affect the activities of genes and how genes affect the interactions between diet and disease.[7] Highlight 6 describes how nutritional genomics is shaping the science of nutrition, and examples of nutrient–gene interactions appear throughout later sections of the book.

◆ A personal account of an experience or event is an **anecdote** and is not accepted as reliable scientific information.
- **anekdotos** = unpublished

Conducting Research

Consumers may depend on personal experience or reports from friends ◆ to gather information on nutrition, but researchers use the scientific method to guide their work (see Figure 1-3). As the figure shows, research always begins with a problem or a question. For example, "What foods or nutrients might protect against the common cold?" In search of an answer, scientists make an educated guess (**hypothesis**) such as "foods rich in vitamin C reduce the number of common colds." Then they systematically conduct research studies to collect data that will test the hypothesis (see the glossary for defini-

FIGURE 1-3 The Scientific Method

Research scientists follow the scientific method. Note that most research generates new questions, not final answers. Thus the sequence begins anew, and research continues in a somewhat cyclical way.

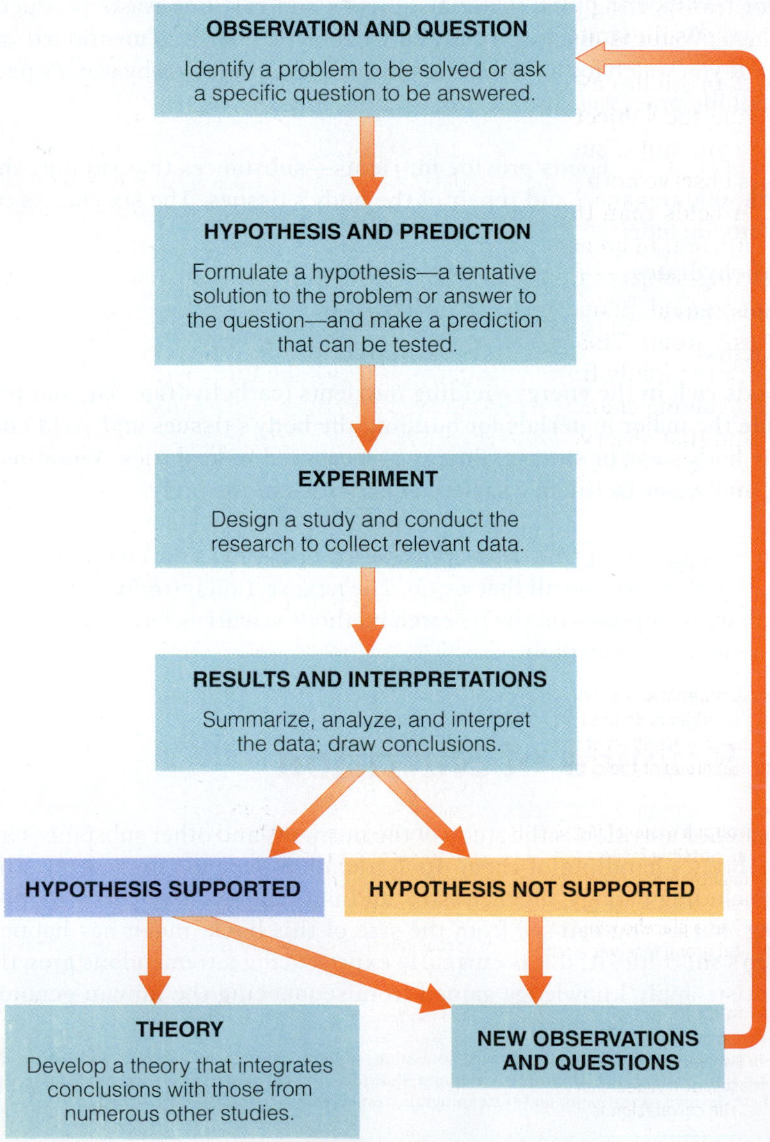

OBSERVATION AND QUESTION
Identify a problem to be solved or ask a specific question to be answered.
HYPOTHESIS AND PREDICTION
Formulate a hypothesis—a tentative solution to the problem or answer to the question—and make a prediction that can be tested.
EXPERIMENT
Design a study and conduct the research to collect relevant data.
RESULTS AND INTERPRETATIONS
Summarize, analyze, and interpret the data; draw conclusions.
HYPOTHESIS SUPPORTED
HYPOTHESIS NOT SUPPORTED
THEORY
Develop a theory that integrates conclusions with those from numerous other studies.
NEW OBSERVATIONS AND QUESTIONS

nutritional genomics: the science of how nutrients affect the activities of genes **(nutrigenomics)** and how genes affect the interactions between diet and disease **(nutrigenetics).**

TABLE 1-3 Strengths and Weaknesses of Research Designs

Type of Research	Strengths	Weaknesses
Epidemiological studies determine the incidence and distribution of diseases in a population. Epidemiological studies include cross-sectional, case-control, and cohort (see Figure 1-4).	• Can narrow down the list of possible causes • Can raise questions to pursue through other types of studies	• Cannot control variables that may influence the development or the prevention of a disease • Cannot prove cause and effect
Laboratory-based studies explore the effects of a specific variable on a tissue, cell, or molecule. Laboratory-based studies are often conducted in test tubes (in vitro) or on animals.	• Can control conditions • Can determine effects of a variable	• Cannot apply results from test tubes or animals to human beings
Human intervention or **clinical trials** involve human beings who follow a specified regimen.	• Can control conditions (for the most part) • Can apply findings to some groups of human beings	• Cannot generalize findings to all human beings • Cannot use certain treatments for clinical or ethical reasons

tions of research terms). Because each type of study has strengths and weaknesses, some provide stronger evidence than others (see Table 1-3). Some examples of various types of research designs are presented in Figure 1-4 (p. 14).

In attempting to discover whether a nutrient relieves symptoms or cures a disease, researchers deliberately manipulate one variable (for example, the amount of vitamin C in the diet) and measure any observed changes (perhaps the number of colds). As much as possible, all other conditions are held constant. The following paragraphs illustrate how this is accomplished.

Controls In studies examining the effectiveness of vitamin C, researchers typically divide the **subjects** into two groups. One group (the **experimental group**) receives a vitamin C supplement, and the other (the **control group**) does not. Researchers observe both groups to determine whether one group has fewer, milder, or shorter colds than the other. The following discussion describes some of the pitfalls inherent in an experiment of this kind and ways to avoid them.

In sorting subjects into two groups, researchers must ensure that each person has an equal chance of being assigned to either the experimental group or the control group. This is accomplished by **randomization**; that is, the subjects are chosen randomly from the same population by flipping a coin or some other method involving chance. Randomization helps to ensure that the two groups are "equal" and that observed differences reflect the treatment and not other factors.[8]

GLOSSARY OF RESEARCH TERMS

blind experiment: an experiment in which the subjects do not know whether they are members of the experimental group or the control group.

control group: a group of individuals similar in all possible respects to the experimental group except for the treatment. Ideally, the control group receives a placebo while the experimental group receives a real treatment.

correlation (CORE-ee-LAY-shun): the simultaneous increase, decrease, or change in two variables. If A increases as B increases, or if A decreases as B decreases, the correlation is *positive*. (This does not mean that A causes B or vice versa.) If A increases as B decreases, or if A decreases as B increases, the correlation is *negative*. (This does not mean that A prevents B or vice versa.) Some third factor may account for both A and B.

double-blind experiment: an experiment in which neither the subjects nor the researchers know which subjects are members of the experimental group and which are serving as control subjects, until after the experiment is over.

experimental group: a group of individuals similar in all possible respects to the control group except for the treatment. The experimental group receives the real treatment.

hypothesis (hi-POTH-eh-sis): an unproven statement that tentatively explains the relationships between two or more variables.

peer review: a process in which a panel of scientists rigorously evaluates a research study to assure that the scientific method was followed.

placebo (pla-SEE-bo): an inert, harmless medication given to provide comfort and hope; a sham treatment used in controlled research studies.

placebo effect: a change that occurs in response to expectations about the effectiveness of a treatment that actually has no pharmaceutical effects.

randomization (RAN-dom-ih-ZAY-shun): a process of choosing the members of the experimental and control groups without bias.

replication (REP-lih-KAY-shun): repeating an experiment and getting the same results.

subjects: the people or animals participating in a research project.

theory: a tentative explanation that integrates many and diverse findings to further the understanding of a defined topic.

validity (va-LID-ih-tee): having the quality of being founded on fact or evidence.

variables: factors that change. A variable may depend on another variable (for example, a child's height depends on his age), or it may be independent (for example, a child's height does not depend on the color of her eyes). Sometimes both variables correlate with a third variable (a child's height and eye color both depend on genetics).

FIGURE 1-4 **Examples of Research Designs**

EPIDEMIOLOGICAL STUDIES

CROSS-SECTIONAL STUDIES

Researchers observe how much and what kinds of foods a group of people eat and how healthy those people are. Their findings identify factors that might influence the incidence of a disease in various populations.

Example. Many people in the Mediterranean region drink more wine, eat more fat from olive oil, and yet have a lower incidence of heart disease than northern Europeans and North Americans.

CASE-CONTROL STUDIES

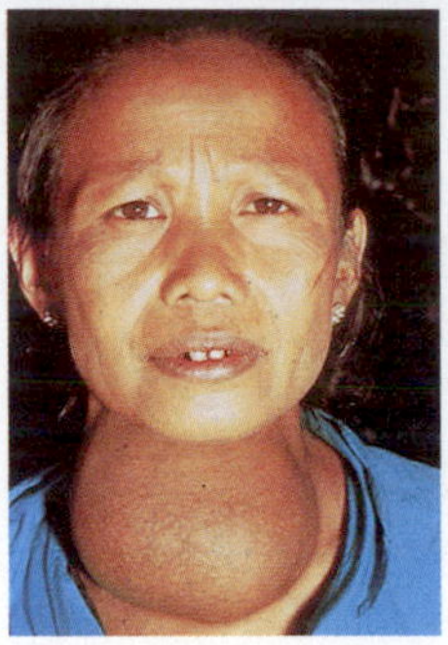

© Lester V. Bergman/Corbis

Researchers compare people who do and do not have a given condition such as a disease, closely matching them in age, gender, and other key variables so that differences in other factors will stand out. These differences may account for the condition in the group that has it.

Example. People with goiter lack iodine in their diets.

COHORT STUDIES

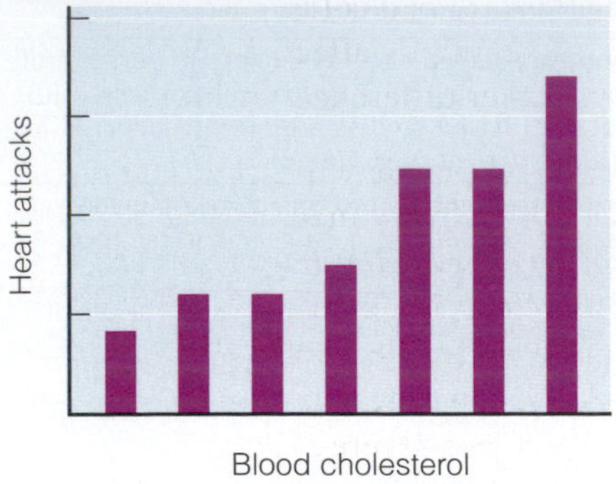

Researchers analyze data collected from a selected group of people (a cohort) at intervals over a certain period of time.

Example. Data collected periodically over the past several decades from over 5000 people randomly selected from the town of Framingham, Massachusetts, in 1948 have revealed that the risk of heart attack increases as blood cholesterol increases.

EXPERIMENTAL STUDIES

LABORATORY-BASED ANIMAL STUDIES

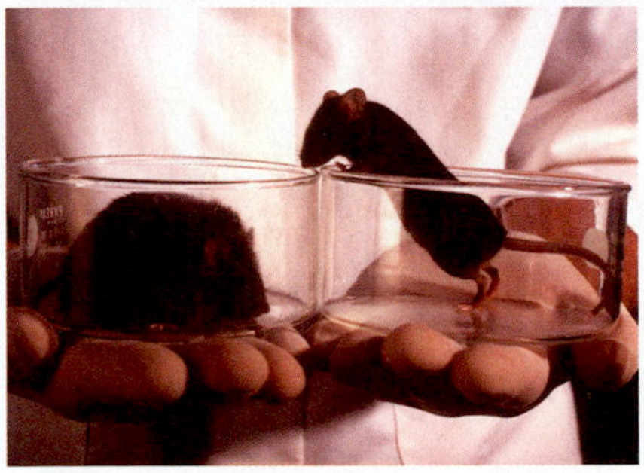

© R. Benali/Getty Images

Researchers feed animals special diets that provide or omit specific nutrients and then observe any changes in health. Such studies test possible disease causes and treatments in a laboratory where all conditions can be controlled.

Example. Mice fed a high-fat diet eat less food than mice given a lower-fat diet, so they receive the same number of kcalories—but the mice eating the fat-rich diet become severely obese.

LABORATORY-BASED IN VITRO STUDIES

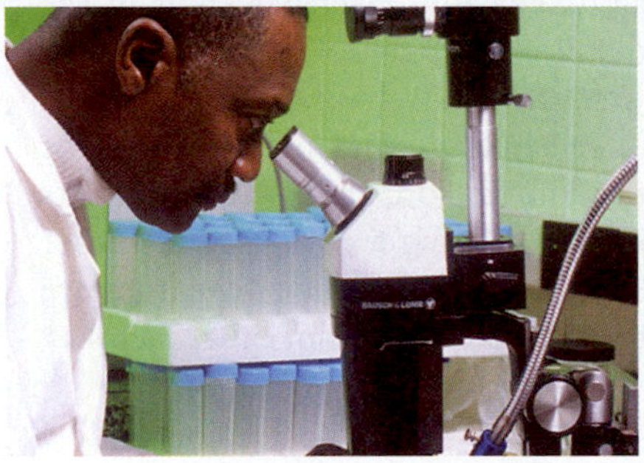

USDA Agricultural Research Service

Researchers examine the effects of a specific variable on a tissue, cell, or molecule isolated from a living organism.

Example. Laboratory studies find that fish oils inhibit the growth and activity of the bacteria implicated in ulcer formation.

HUMAN INTERVENTION (OR CLINICAL) TRIALS

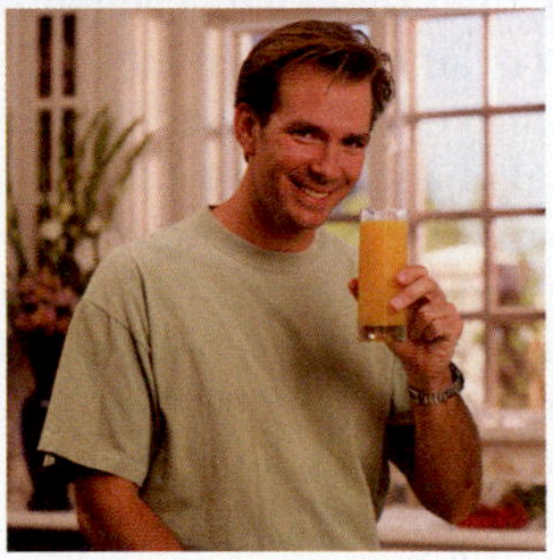

© PhotoDisc/Getty Images

Researchers ask people to adopt a new behavior (for example, eat a citrus fruit, take a vitamin C supplement, or exercise daily). These trials help determine the effectiveness of such interventions on the development or prevention of disease.

Example. Heart disease risk factors improve when men receive fresh-squeezed orange juice daily for two months compared with those on a diet low in vitamin C—even when both groups follow a diet high in saturated fat.

Importantly, the two groups of people must be similar and must have the same track record with respect to colds to rule out the possibility that observed differences in the rate, severity, or duration of colds might have occurred anyway. If, for example, the control group would normally catch twice as many colds as the experimental group, then the findings prove nothing.

In experiments involving a nutrient, the diets of both groups must also be similar, especially with respect to the nutrient being studied. If those in the experimental group were receiving less vitamin C from their usual diet, then any effects of the supplement may not be apparent.

Sample Size To ensure that chance variation between the two groups does not influence the results, the groups must be large. For example, if one member of a group of five people catches a bad cold by chance, he will pull the whole group's average toward bad colds; but if one member of a group of 500 catches a bad cold, she will not unduly affect the group average. Statistical methods are used to determine whether differences between groups of various sizes support a hypothesis.

Placebos If people who take vitamin C for colds *believe* it will cure them, their chances of recovery may improve. Taking anything believed to be beneficial may hasten recovery. This phenomenon, the result of expectations, is known as the **placebo effect.** In experiments designed to determine vitamin C's effect on colds, this mind-body effect must be rigorously controlled. Severity of symptoms is often a subjective measure, and people who believe they are receiving treatment may report less-severe symptoms.

One way experimenters control for the placebo effect is to give pills to all participants. Those in the experimental group, for example, receive pills containing vitamin C, and those in the control group receive a **placebo**—pills of similar appearance and taste containing an inactive ingredient. This way, the expectations of both groups will be equal. It is not necessary to convince all subjects that they are receiving vitamin C, but the extent of belief or unbelief must be the same in both groups. A study conducted under these conditions is called a **blind experiment**—that is, the subjects do not know (are blind to) whether they are members of the experimental group (receiving treatment) or the control group (receiving the placebo).

Double Blind When both the subjects and the researchers do not know which subjects are in which group, the study is called a **double-blind experiment.** Being fallible human beings and having an emotional and sometimes financial investment in a successful outcome, researchers might record and interpret results with a bias in the expected direction. To prevent such bias, the pills are coded by a third party, who does not reveal to the experimenters which subjects are in which group until all results have been recorded.

Analyzing Research Findings

Research findings must be analyzed and interpreted with an awareness of each study's limitations. Scientists must be cautious about drawing any conclusions until they have accumulated a body of evidence from multiple studies that have used various types of research designs. As evidence accumulates, scientists begin to develop a **theory** that integrates the various findings and explains the complex relationships.

Correlations and Causes Researchers often examine the relationships between two or more **variables**—for example, daily vitamin C intake and the number of colds or the duration and severity of cold symptoms. Importantly, researchers must be able to observe, measure, or verify the variables selected. Findings sometimes suggest no **correlation** between variables (regardless of the amount of vitamin C consumed, the number of colds remains the same). Other times, studies find either a **positive correlation** (the more vitamin C, the more colds) or a **negative correlation** (the more vitamin C, the fewer colds). Notice that in a positive correlation, both variables change in the same direction, regardless of whether the direction is "more" or "less"—"the more vitamin C, the more colds" is a positive correlation, just as is "the less vitamin C, the fewer colds." In a negative correlation, the two variables change in opposite directions: "the less vitamin C, the more colds" or "the more vitamin C, the fewer colds." Also notice that a positive correlation does not necessarily reflect a desired outcome, nor does a negative correlation always reflect an unwanted outcome.

Correlational evidence proves only that variables are associated, not that one is the cause of the other. People often jump to conclusions when they notice correlations, but their conclusions are often wrong. To actually prove that A causes B,

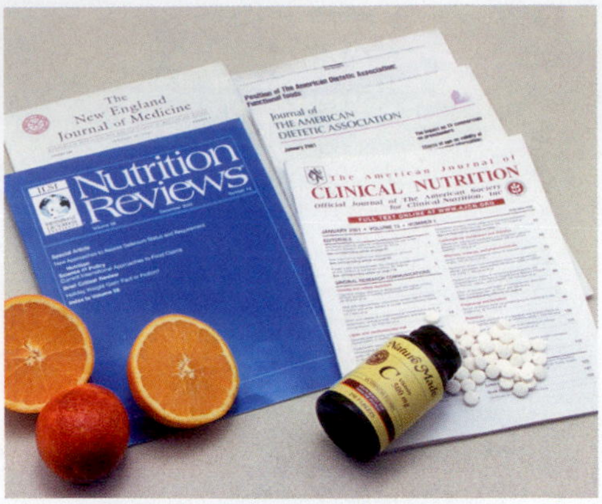

© Craig M. Moore

Knowledge about the nutrients and their effects on health comes from scientific studies.

scientists have to find evidence of the *mechanism*—that is, an explanation of how A might cause B.

Cautious Conclusions When researchers record and analyze the results of their experiments, they must exercise caution in their interpretation of the findings. For example, in an epidemiological study, scientists may use a specific segment of the population—say, men 18 to 30 years old. When the scientists draw conclusions, they are careful not to generalize the findings to all people. Similarly, scientists performing research studies using animals are cautious in applying their findings to human beings. Conclusions from any one research study are always tentative and take into account findings from studies conducted by other scientists as well. As evidence accumulates, scientists gain confidence about making recommendations that affect people's health and lives. Still, their statements are worded cautiously, such as "A diet high in fruits and vegetables *may* protect against *some* cancers."

Quite often, as scientists approach an answer to one research question, they raise several more questions, so future research projects are never lacking. Further scientific investigation then seeks to answer questions such as "What substance or substances within fruits and vegetables provide protection?" If those substances turn out to be the vitamins found so abundantly in fresh produce, then "How much is needed to offer protection?" "How do these vitamins protect against cancer?" "Is it their action as antioxidant nutrients?" "If not, might it be another action or even another substance that accounts for the protection fruits and vegetables provide against cancer?" (Highlight 11 explores the answers to these questions and reviews recent research on antioxidant nutrients and disease.)

Publishing Research

The findings from a research study are submitted to a board of reviewers composed of other scientists who rigorously evaluate the study to assure that the scientific method was followed—a process known as **peer review**. The reviewers critique the study's hypothesis, methodology, statistical significance, and conclusions. They also note the funding source, recognizing that financial support may bias scientific conclusions.[9] If the reviewers consider the conclusions to be well supported by the evidence—that is, if the research has **validity**—they endorse the work for publication in a scientific journal where others can read it. This raises an important point regarding information found on the Internet: much gets published without the rigorous scrutiny of peer review. Consequently, readers must assume greater responsibility for examining the data and conclusions presented—often without the benefit of journal citations. Highlight 1 offers guidance in determining whether website information is reliable. Table 1-4 describes the parts of a typical research article.

Even when a new finding is published or released to the media, it is still only preliminary and not very meaningful by itself. Other scientists will need to confirm or disprove the findings through **replication**. To be accepted into the body

TABLE 1-4 Parts of a Research Article

- *Abstract.* The abstract provides a brief overview of the article.
- *Introduction.* The introduction clearly states the purpose of the current study.
- *Review of literature.* A comprehensive review of the literature reveals all that science has uncovered on the subject to date.
- *Methodology.* The methodology section defines key terms and describes the instruments and procedures used in conducting the study.
- *Results.* The results report the findings and may include tables and figures that summarize the information.
- *Conclusions.* The conclusions drawn are those supported by the data and reflect the original purpose as stated in the introduction. Usually, they answer a few questions and raise several more.
- *References.* The references reflect the investigator's knowledge of the subject and should include an extensive list of relevant studies (including key studies several years old as well as current ones).

of nutrition knowledge, a finding must stand up to rigorous, repeated testing in experiments performed by several different researchers. What we "know" in nutrition results from years of replicating study findings. Communicating the latest finding in its proper context without distorting or oversimplifying the message is a challenge for scientists and journalists alike.

With each report from scientists, the field of nutrition changes a little—each finding contributes another piece to the whole body of knowledge. People who know how science works understand that single findings, like single frames in a movie, are just small parts of a larger story. Over years, the picture of what is "true" in nutrition gradually changes, and dietary recommendations change to reflect the current understanding of scientific research. Highlight 5 provides a detailed look at how dietary fat recommendations have evolved over the past several decades as researchers have uncovered the relationships between the various kinds of fat and their roles in supporting or harming health.

IN SUMMARY **Scientists learn about nutrition by conducting experiments that follow the protocol of scientific research. In designing their studies, researchers randomly assign control and experimental groups, seek large sample sizes, provide placebos, and remain blind to treatments. Their findings must be reviewed and replicated by other scientists before being accepted as valid.**

The characteristics of well-designed research have enabled scientists to study the actions of nutrients in the body. Such research has laid the foundation for quantifying how much of each nutrient the body needs.

Dietary Reference Intakes

Using the results of thousands of research studies, nutrition experts have produced a set of standards that define the amounts of energy, nutrients, other dietary components, and physical activity that best support health. These recommendations are called **Dietary Reference Intakes (DRI)**, and they reflect the collaborative efforts of researchers in both the United States and Canada.*[10] The inside front covers of this book provide a handy reference for DRI values.

© PhotoDisc/Getty Images

Don't let the DRI "alphabet soup" of nutrient intake standards confuse you. Their names make sense when you learn their purposes.

Establishing Nutrient Recommendations

The DRI Committee consists of highly qualified scientists who base their estimates of nutrient needs on careful examination and interpretation of scientific evidence. These recommendations apply to healthy people and may not be appropriate for people with diseases that increase or decrease nutrient needs. The next several paragraphs discuss specific aspects of how the committee goes about establishing the values that make up the DRI:

- Estimated Average Requirements (EAR)
- Recommended Dietary Allowances (RDA)
- Adequate Intakes (AI)
- Tolerable Upper Intake Levels (UL)

Estimated Average Requirements (EAR)

The committee reviews hundreds of research studies to determine the **requirement** for a nutrient—how much is needed in the diet. The committee selects a different criterion for each nutrient based on its roles in supporting various activities in the body and in reducing disease risks.[11]

An examination of all the available data reveals that each person's body is unique and has its own set of requirements. Men differ from women, and needs

*The DRI reports are produced by the Food and Nutrition Board, Institute of Medicine of the National Academies, with active involvement of scientists from Canada.

Dietary Reference Intakes (DRI): a set of nutrient intake values for healthy people in the United States and Canada. These values are used for planning and assessing diets and include:
- Estimated Average Requirements (EAR)
- Recommended Dietary Allowances (RDA)
- Adequate Intakes (AI)
- Tolerable Upper Intake Levels (UL)

requirement: the lowest continuing intake of a nutrient that will maintain a specified criterion of adequacy.

FIGURE 1-5 Estimated Average Requirements (EAR) and Recommended Dietary Allowances (RDA) Compared

Each square in the graphs below represents a person with unique nutritional requirements. (The text discusses three of these people—A, B, and C.) Some people require only a small amount of nutrient X and some require a lot. Most people, however, fall somewhere in the middle.

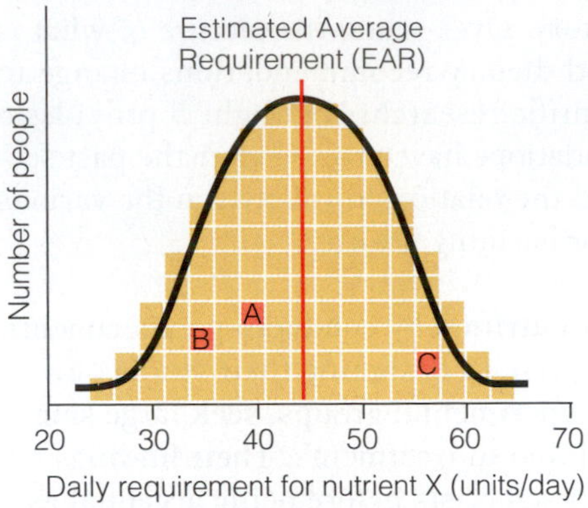

The Estimated Average Requirement (EAR) for a nutrient is the amount that covers half of the population (shown here by the red line).

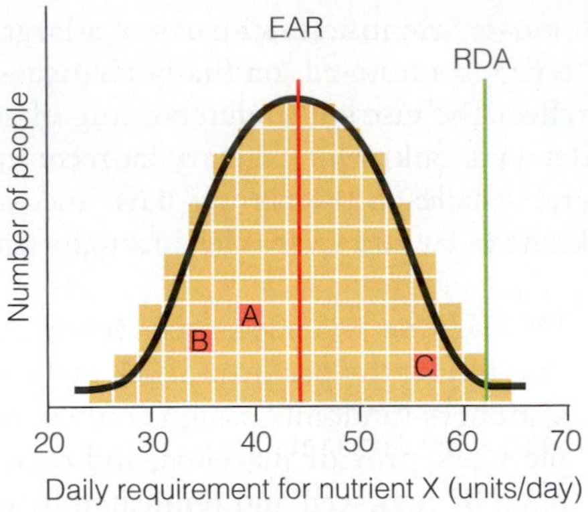

The Recommended Dietary Allowance (RDA) for a nutrient (shown here in green) is set well above the EAR, covering about 98% of the population.

change as people grow from infancy through old age. For this reason, the committee clusters its recommendations for people into groups based on age and gender. Even so, the exact requirements for people of the same age and gender are likely to be different. For example, person A might need 40 units of a particular nutrient each day; person B might need 35; and person C, 57. Looking at enough people might reveal that their individual requirements fall into a symmetrical distribution, with most near the midpoint and only a few at the extremes (see the left side of Figure 1-5). Using this information, the committee determines an **Estimated Average Requirement (EAR)** for each nutrient—the average amount that appears sufficient for half of the population. In Figure 1-5, the Estimated Average Requirement is shown as 45 units.

Recommended Dietary Allowances (RDA) Once a nutrient *requirement* is established, the committee must decide what intake to *recommend* for everybody—the **Recommended Dietary Allowance (RDA)**. As you can see by the distribution in Figure 1-5, the Estimated Average Requirement (shown in the figure as 45 units) is probably closest to everyone's need. However, if people consumed exactly the average requirement of a given nutrient each day, half of the population would develop deficiencies of that nutrient—in Figure 1-5, for example, person C would be among them. Recommendations are therefore set high enough above the Estimated Average Requirement to meet the needs of most healthy people.

Small amounts above the daily requirement do no harm, whereas amounts below the requirement may lead to health problems. When people's nutrient intakes are consistently **deficient** (less than the requirement), their nutrient stores decline, and over time this decline leads to poor health and deficiency symptoms. Therefore, to ensure that the nutrient RDA meet the needs of as many people as possible, the RDA are set near the top end of the range of the population's estimated requirements.

In this example, a reasonable RDA might be 63 units a day (see the right side of Figure 1-5). Such a point can be calculated mathematically so that it covers about 98 percent of a population. Almost everybody—including person C whose needs were higher than the average—would be covered if they met this dietary goal. Relatively few people's requirements would exceed this recommendation, and even then, they wouldn't exceed by much.

Estimated Average Requirement (EAR): the average daily amount of a nutrient that will maintain a specific biochemical or physiological function in half the healthy people of a given age and gender group.

Recommended Dietary Allowance (RDA): the average daily amount of a nutrient considered adequate to meet the known nutrient needs of practically all healthy people; a goal for dietary intake by individuals.

deficient: the amount of a nutrient below which almost all healthy people can be expected, over time, to experience deficiency symptoms.

Adequate Intakes (AI) For some nutrients there is insufficient scientific evidence to determine an Estimated Average Requirement (which is needed to set an RDA). In these cases, the committee establishes an **Adequate Intake (AI)** instead of an RDA. An AI reflects the average amount of a nutrient that a group of healthy people consumes. Like the RDA, the AI may be used as a nutrient goal for individuals.

Although both the RDA and the AI serve as nutrient intake goals for individuals, their differences are noteworthy. An RDA for a given nutrient is based on enough scientific evidence to expect that the needs of almost all healthy people will be met. An AI, on the other hand, must rely more heavily on scientific judgments because sufficient evidence is lacking. The percentage of people covered by an AI is unknown; an AI is expected to exceed average requirements, but it may cover more or fewer people than an RDA would cover (if an RDA could be determined). For these reasons, AI values are more tentative than RDA. The table on the inside front cover identifies which nutrients have an RDA and which have an AI. Later chapters present the RDA and AI values for the vitamins and minerals.

Tolerable Upper Intake Levels (UL) As mentioned earlier, the recommended intakes for nutrients are generous, and they do not necessarily cover every individual for every nutrient. Nevertheless, it is probably best not to exceed these recommendations by very much or very often. Individual tolerances for high doses of nutrients vary, and somewhere above the recommended intake is a point beyond which a nutrient is likely to become toxic.[12] This point is known as the **Tolerable Upper Intake Level (UL)**. It is naïve—and inaccurate—to think of recommendations as minimum amounts. A more accurate view is to see a person's nutrient needs as falling within a range, with marginal and danger zones both below and above it (see Figure 1-6).

Paying attention to upper levels is particularly useful in guarding against the overconsumption of nutrients, which may occur when people use large-dose dietary supplements and fortified foods regularly. Later chapters discuss the dangers associated with excessively high intakes of vitamins and minerals, and the inside front cover (p. C) presents tables of upper levels for selected nutrients.

Establishing Energy Recommendations

In contrast to the RDA and AI values for nutrients, the recommendation for energy is not generous. Excess energy cannot be readily excreted and is eventually stored as body fat. These reserves may be beneficial when food is scarce, but they can also lead to obesity and its associated health consequences.

Estimated Energy Requirement (EER) The energy recommendation—called the **Estimated Energy Requirement (EER)**—represents the average dietary energy intake (kcalories per day) that will maintain energy balance in a person who has a healthy body weight ♦ and level of physical activity. Balance is key to the energy recommendation. Enough energy is needed to sustain a healthy and active life, but too much energy can lead to weight gain and obesity. Because *any* amount in excess of energy needs will result in weight gain, no upper level for energy has been determined.

Acceptable Macronutrient Distribution Ranges (AMDR) People don't eat energy directly; they derive energy from foods containing carbohydrates, fats, and proteins. Each of these three energy-yielding nutrients contributes to the total energy intake, and those contributions vary in relation to one another. The DRI Committee has determined that the composition of a diet that provides adequate energy and nutrients and reduces the risk of chronic diseases is:

- 45 to 65 percent kcalories from carbohydrate
- 20 to 35 percent kcalories from fat
- 10 to 35 percent kcalories from protein

These values are known as **Acceptable Macronutrient Distribution Ranges (AMDR)**.

FIGURE 1-6 Inaccurate versus Accurate View of Nutrient Intakes

The RDA or AI for a given nutrient represents a point that lies within a range of appropriate and reasonable intakes between toxicity and deficiency. Both of these recommendations are high enough to provide reserves in times of short-term dietary inadequacies, but not so high as to approach toxicity. Nutrient intakes above or below this range may be equally harmful.

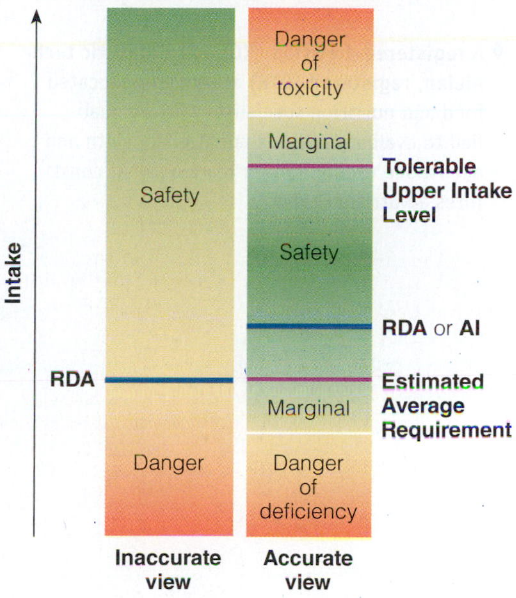

♦ Reference adults:
- Men: 19–30 yr, 5 ft 10 in, and 154 lb
- Women: 19–30 yr, 5 ft 4 in, and 126 lb

Adequate Intake (AI): the average daily amount of a nutrient that appears sufficient to maintain a specified criterion; a value used as a guide for nutrient intake when an RDA cannot be determined.

Tolerable Upper Intake Level (UL): the maximum daily amount of a nutrient that appears safe for most healthy people and beyond which there is an increased risk of adverse health effects.

Estimated Energy Requirement (EER): the average dietary energy intake that maintains energy balance and good health in a person of a given age, gender, weight, height, and level of physical activity.

Acceptable Macronutrient Distribution Ranges (AMDR): ranges of intakes for the energy nutrients that provide adequate energy and nutrients and reduce the risk of chronic diseases.

Using Nutrient Recommendations

Although the intent of nutrient recommendations seems simple, they are the subject of much misunderstanding and controversy. Perhaps the following facts will help put them in perspective:

1. Estimates of adequate energy and nutrient intakes apply to *healthy* people. They need to be adjusted for malnourished people or those with medical problems who may require supplemented or restricted dietary intakes.
2. *Recommendations* are not minimum requirements, nor are they necessarily optimal intakes for all individuals. Recommendations can target only "most" of the people and cannot account for individual variations in nutrient needs—yet. Given the recent explosion of knowledge about genetics, the day may be fast approaching when nutrition scientists will be able to determine an individual's optimal nutrient needs.[13] Until then, registered dietitians ♦ and other qualified health professionals can help determine if recommendations should be adjusted to meet individual needs.
3. Most nutrient goals are intended to be met through diets composed of a variety of *foods* whenever possible. Because foods contain mixtures of nutrients and nonnutrients, they deliver more than just those nutrients covered by the recommendations. Excess intakes of vitamins and minerals are unlikely when they come from foods rather than dietary supplements.
4. Recommendations apply to *average* daily intakes. Trying to meet the recommendations for every nutrient every day is difficult and unnecessary. The length of time over which a person's intake can deviate from the average without risk of deficiency or overdose varies for each nutrient, depending on how the body uses and stores the nutrient. For most nutrients (such as thiamin and vitamin C), deprivation would lead to rapid development of deficiency symptoms (within days or weeks); for others (such as vitamin A and vitamin B_{12}), deficiencies would develop more slowly (over months or years).
5. Each of the DRI categories serves a unique purpose. For example, the Estimated Average Requirements are most appropriately used to develop and evaluate nutrition programs for *groups* such as schoolchildren or military personnel. The RDA (or AI if an RDA is not available) can be used to set goals for *individuals*. Tolerable Upper Intake Levels serve as a reminder to keep nutrient intakes below amounts that increase the risk of toxicity—not a common problem when nutrients derive from foods, but a real possibility for some nutrients if supplements are used regularly.

♦ A **registered dietitian (RD)** and a **dietetic technician, registered (DTR)** are college-educated food and nutrition specialists who are qualified to evaluate people's nutritional health and needs. See Highlight 1 for more on what constitutes a nutrition expert.

With these understandings, professionals can use the DRI for a variety of purposes.

Comparing Nutrient Recommendations

At least 40 different nations and international organizations have published nutrient standards similar to those used in the United States and Canada. Slight differences may be apparent, reflecting differences both in the interpretation of the data from which the standards were derived and in the food habits and physical activities of the populations they serve.

Many countries use the recommendations developed by two international groups: FAO (Food and Agriculture Organization) and WHO (World Health Organization). ♦ The FAO/WHO recommendations are considered sufficient to maintain health in nearly all healthy people worldwide.

♦ Nutrient recommendations from FAO/WHO are provided in Appendix I.

IN SUMMARY **The Dietary Reference Intakes (DRI) are a set of nutrient intake values that can be used to plan and evaluate diets for healthy people. The Estimated Average Requirement (EAR) defines the amount of a nutrient that supports a specific function in the body for half of the population. The Recommended Dietary Allowance (RDA) is based on the Estimated Average Requirement and establishes a goal for dietary intake that will meet the needs of almost all healthy people. An Adequate Intake (AI) serves a similar pur-**

pose when an RDA cannot be determined. The Estimated Energy Requirement (EER) defines the average amount of energy intake needed to maintain energy balance, and the Acceptable Macronutrient Distribution Ranges (AMDR) define the proportions contributed by carbohydrate, fat, and protein to a healthy diet. The Tolerable Upper Intake Level (UL) establishes the highest amount that appears safe for regular consumption.

Nutrition Assessment

What happens when a person doesn't get enough or gets too much of a nutrient or energy? If the deficiency or excess is significant over time, the person experiences symptoms of **malnutrition.** With a deficiency of energy, the person may develop the symptoms of **undernutrition** by becoming extremely thin, losing muscle tissue, and becoming prone to infection and disease. With a deficiency of a nutrient, the person may experience skin rashes, depression, hair loss, bleeding gums, muscle spasms, night blindness, or other symptoms. With an excess of energy, the person may become obese and vulnerable to diseases associated with **overnutrition** such as heart disease and diabetes. With a sudden nutrient overdose, the person may experience hot flashes, yellowing skin, a rapid heart rate, low blood pressure, or other symptoms. Similarly, over time, regular intakes in excess of needs may also have adverse effects.

Malnutrition symptoms—such as diarrhea, skin rashes, and fatigue—are easy to miss because they resemble the symptoms of other diseases. But a person who has learned how to use assessment techniques to detect malnutrition can identify when these conditions are caused by poor nutrition and can recommend steps to correct it. This discussion presents the basics of nutrition assessment; many more details are offered in later chapters and in Appendix E.

Nutrition Assessment of Individuals

To prepare a **nutrition assessment**, a registered dietitian; dietetic technician, registered; or other trained health-care professional uses:

- Historical information
- Anthropometric measurements
- Physical examinations
- Laboratory tests

Each of these methods involves collecting data in various ways and interpreting each finding in relation to the others to create a total picture.

Historical Information One step in evaluating nutrition status is to obtain information about a person's history with respect to health status, socioeconomic status, drug use, and diet. The health history reflects a person's medical record and may reveal a disease that interferes with the person's ability to eat or the body's use of nutrients. The person's family history of major diseases is also noteworthy, especially for conditions such as heart disease that have a genetic tendency to run in families. Economic circumstances may show a financial inability to buy enough nutritious foods or inadequate kitchen facilities in which to prepare them. Social factors such as marital status, ethnic background, and educational level also influence food choices and nutrition status. A drug history, including all prescribed and over-the-counter medications, may highlight possible interactions that lead to nutrient deficiencies (as described in Chapter 19). A diet history that examines a person's intake of foods, beverages, and dietary supplements may reveal either a surplus or inadequacy of nutrients or energy.

To take a diet history, the assessor collects data about the foods a person eats. The data may be collected by recording the foods the person has eaten over a period of 24 hours, 3 days, or a week or more or by asking what foods the person typically eats and how much of each. The days in the record must be fairly typical of the person's diet, and portion sizes must be recorded accurately. To

malnutrition: any condition caused by excess or deficient food energy or nutrient intake or by an imbalance of nutrients.

- **mal** = bad

undernutrition: deficient energy or nutrients.

overnutrition: excess energy or nutrients.

nutrition assessment: a comprehensive analysis of a person's nutrition status that uses health, socioeconomic, drug, and diet histories; anthropometric measurements; physical examinations; and laboratory tests.

FIGURE 1-7 Using the DRI to Assess the Dietary Intake of a Healthy Individual

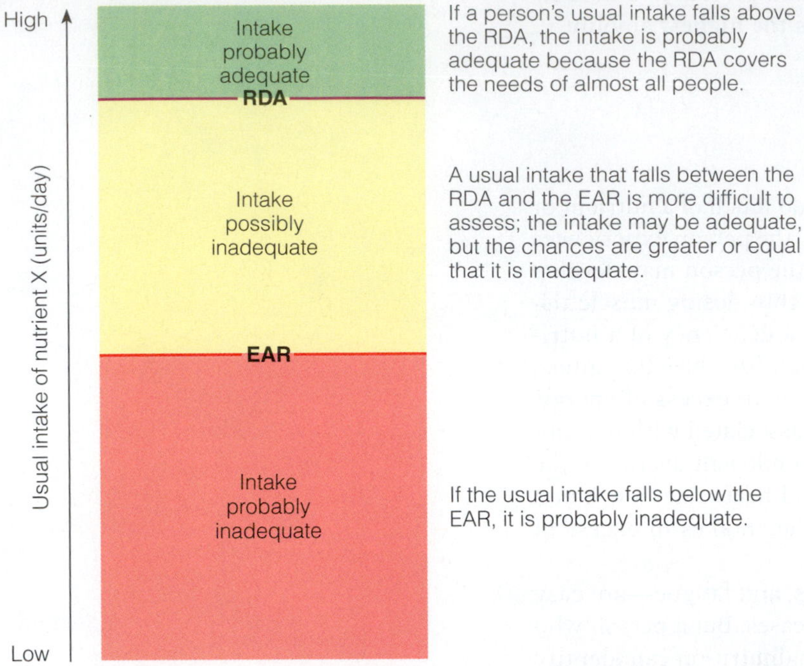

High
Low
Usual intake of nutrient X (units/day)
Intake probably adequate
RDA
Intake possibly inadequate
EAR
Intake probably inadequate
If a person's usual intake falls above the RDA, the intake is probably adequate because the RDA covers the needs of almost all people.
A usual intake that falls between the RDA and the EAR is more difficult to assess; the intake may be adequate, but the chances are greater or equal that it is inadequate.
If the usual intake falls below the EAR, it is probably inadequate.

determine the amounts of nutrients consumed, the assessor usually enters the foods and their portion sizes into a computer using a diet analysis program. This step can also be done manually by looking up each food in a table of food composition such as Appendix H in this book. The assessor then compares the calculated nutrient intakes with the DRI to determine the probability of adequacy (see Figure 1-7). Alternatively, the diet history might be compared against standards such as the USDA Food Guide or *Dietary Guidelines for Americans* (described in Chapter 2).

An estimate of energy and nutrient intakes from a diet history, when combined with other sources of information, can help confirm or rule out the *possibility* of suspected nutrition problems. A sufficient intake of a nutrient does not guarantee adequacy, and an insufficient intake does not always indicate a deficiency. Such findings, however, warn of possible problems.

Anthropometric Measurements A second technique that may help to reveal nutrition problems is taking **anthropometric** measurements such as those of height and weight. The assessor compares a person's measurements with standards specific for gender and age or with previous measures on the same individual. (Chapter 8 presents information on body weight and its standards, and Appendix E includes growth charts for children.)

Measurements taken periodically and compared with previous measurements reveal patterns and indicate trends in a person's overall nutrition status, but they provide little information about specific nutrients. Instead, measurements out of line with expectations may reveal such problems as growth failure in children, wasting or swelling of body tissues in adults, and obesity—conditions that may reflect energy or nutrient deficiencies or excesses.

Physical Examinations A third nutrition assessment technique is a physical examination looking for clues to poor nutrition status. Visual inspection of the hair, eyes, skin, posture, tongue, and fingernails can provide such clues. The examina-

© Blend Images/Alamy

A peek inside the mouth provides clues to a person's nutrition status. An inflamed tongue may indicate a deficiency of one of the B vitamins, and mottled teeth may reveal fluoride toxicity, for example.

anthropometric (AN-throw-poe-MET-rick): relating to measurement of the physical characteristics of the body, such as height and weight.

- **anthropos** = human
- **metric** = measuring

tion requires skill because many physical signs reflect more than one nutrient deficiency or toxicity—or even nonnutrition conditions. Like the other assessment techniques, a physical examination alone does not yield firm conclusions. Instead, physical examinations reveal possible imbalances that must be confirmed by other assessment techniques, or they confirm results from other assessment measures.

Laboratory Tests A fourth way to detect a developing deficiency, imbalance, or toxicity is to take samples of blood or urine, analyze them in the laboratory, and compare the results with normal values for a similar population. Laboratory tests are most useful in uncovering early signs of malnutrition before symptoms appear. In addition, they can confirm suspicions raised by other assessment methods.

Iron, for Example The mineral iron can be used to illustrate the stages in the development of a nutrient deficiency and the assessment techniques useful in detecting them. The **overt**, or outward, signs of an iron deficiency appear at the end of a long sequence of events. Figure 1-8 describes what happens in the body as a nutrient deficiency progresses and shows which assessment methods can reveal those changes.

FIGURE 1-8 Stages in the Development of a Nutrient Deficiency

Internal changes precede outward signs of deficiencies. However, outward signs of sickness need not appear before a person takes corrective measures. Laboratory tests can help determine nutrient status in the early stages.

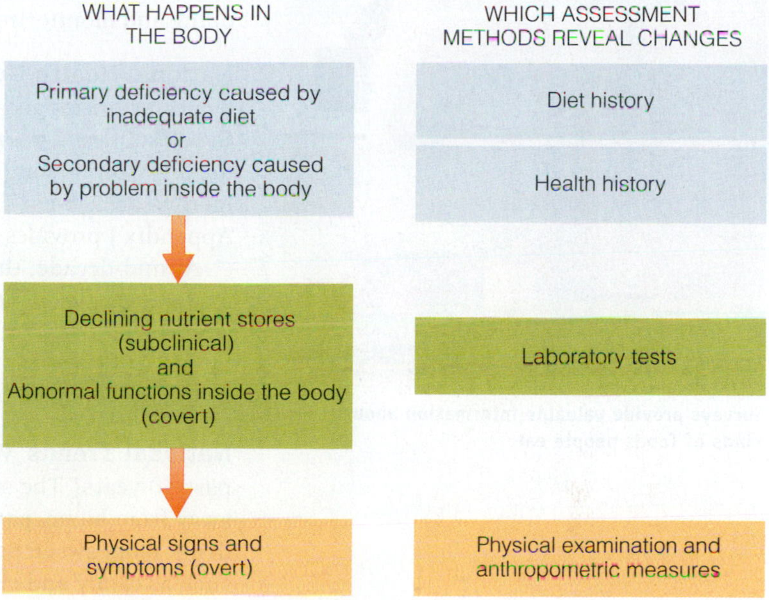

First, the body has too little iron—either because iron is lacking in the person's diet (a **primary deficiency**) or because the person's body doesn't absorb enough, excretes too much, or uses iron inefficiently (a **secondary deficiency**). A diet history provides clues to primary deficiencies; a health history provides clues to secondary deficiencies.

Next, the body begins to use up its stores of iron. At this stage, the deficiency might be described as a **subclinical deficiency**. It exists as a **covert** condition, and although it might be detected by laboratory tests, outward signs are not yet apparent.

Finally, the body's iron stores are exhausted. Now, it cannot make enough iron-containing red blood cells to replace those that are aging and dying. Iron is needed in red blood cells to carry oxygen to all the body's tissues. When iron is lacking, fewer red blood cells are made, the new ones are pale and small, and every part of the body feels the effects of oxygen shortage. Now the overt symptoms of deficiency appear—weakness, fatigue, pallor, and headaches, reflecting the iron-deficient state of the blood. A physical examination will reveal these symptoms.

Nutrition Assessment of Populations

To assess a population's nutrition status, researchers conduct surveys using techniques similar to those used on individuals. The data collected are then used by various agencies for numerous purposes, including the development of national health goals.

National Nutrition Surveys The National Nutrition Monitoring program coordinates the many nutrition-related surveys and research activities of various federal agencies. The integration of two major national surveys ♦ provides comprehensive data efficiently. One survey collects data on the kinds and amounts of foods people eat.* Then researchers calculate the energy and nutrients in the foods and compare the amounts consumed with a standard. The other survey examines the people themselves, using anthropometric measurements, physical examinations, and laboratory tests.** The data provide valuable information on several nutrition-related conditions such as growth retardation, heart disease, and nutrient deficiencies. National nutrition surveys often oversample high-risk groups (low-income families, pregnant women, adolescents, the elderly, African Americans, and Mexican Americans) to glean an accurate estimate of their health and nutrition status.

♦ The integrated survey is called *What We Eat in America*.

overt (oh-VERT): out in the open and easy to observe.
- **ouvrir** = to open

primary deficiency: a nutrient deficiency caused by inadequate dietary intake of a nutrient.

secondary deficiency: a nutrient deficiency caused by something other than an inadequate intake such as a disease condition or drug interaction that reduces absorption, accelerates use, hastens excretion, or destroys the nutrient.

subclinical deficiency: a deficiency in the early stages, before the outward signs have appeared.

covert (KOH-vert): hidden, as if under covers.
- **couvrir** = to cover

---

*This survey was formerly called the Continuing Survey of Food Intakes by Individuals (CSFII), conducted by the U.S. Department of Agriculture (USDA).

**This survey is known as the National Health and Nutrition Examination Survey (NHANES).

© 2010 Tetra Images/Jupiterimages Corporation

Surveys provide valuable information about the kinds of foods people eat.

The resulting wealth of information from the national nutrition surveys is used for a variety of purposes. For example, Congress uses this information to establish public policy on nutrition education, food assistance programs, and the regulation of the food supply. Scientists use the information to establish research priorities. The food industry uses these data to guide decisions in public relations and product development. The Dietary Reference Intakes and other major reports that examine the relationships between diet and health depend on information collected from these nutrition surveys. These data also provide the basis for developing and monitoring national health goals.

National Health Goals **Healthy People** is a program that identifies the nation's health priorities and guides policies that promote health and prevent disease. At the start of each decade, the program sets goals for improving the nation's health during the following ten years. Nutrition is one of many topic areas, each with numerous objectives. Table 1-5 lists the nutrition and weight status objectives, and Appendix J provides a table of nutrition-related objectives from other topic areas.

At mid-decade, the nation's progress toward meeting its nutrition and weight status Healthy People goals was somewhat bleak.[14] Trends in overweight and obesity worsened. Objectives to eat more fruits, vegetables, and whole grains and to increase physical activity showed little or no improvement. Clearly, "what we eat in America" must change if we hope to meet the Healthy People goals.

National Trends What do we eat in America and how has it changed over the past 40 years? The short answer to both questions is "a lot." We eat more meals away from home, particularly at fast-food restaurants. We eat larger portions. We drink more sweetened beverages and eat more energy-dense, nutrient-poor foods such as candy and chips. We snack frequently. As a result of these dietary habits,

TABLE 1-5 Healthy People 2020 Nutrition and Weight Status Objectives

- Increase the proportion of adults who are at a healthy weight.
- Reduce the proportion of adults who are obese.
- Reduce iron deficiency among young children and females of childbearing age.
- Reduce iron deficiency among pregnant females.
- Reduce the proportion of children and adolescents who are overweight or obese.
- Increase the contribution of fruits to the diets of the population aged 2 years and older.
- Increase the variety and contribution of vegetables to the diets of the population aged 2 years and older.
- Increase the contribution of whole grains to the diets of the population aged 2 years and older.
- Reduce consumption of saturated fat in the population aged 2 years and older.
- Reduce consumption of sodium in the population aged 2 years and older.
- Increase consumption of calcium in the population aged 2 years and older.
- Increase the proportion of worksites that offer nutrition or weight management classes or counseling.
- Increase the proportion of physician office visits that include counseling or education related to nutrition or weight.
- Eliminate very low food security among children in U.S. households.
- Prevent inappropriate weight gain in youth and adults.
- Increase the proportion of primary care physicians who regularly measure the body mass index of their patients.
- Reduce consumption of kcalories from solid fats and added sugars in the population aged 2 years and older.
- Increase the number of states that have state-level policies that incentivize food retail outlets to provide foods that are encouraged by the Dietary Guidelines.
- Increase the number of states with nutrition standards for foods and beverages provided to preschool-aged children in childcare.
- Increase the percentage of schools that offer nutritious foods and beverages outside of school meals.

NOTE: "Nutrition and Weight Status" is one of 38 topic areas, each with numerous objectives. Several of the other topic areas have nutrition-related objectives, and these are presented in Appendix J.
SOURCE: www.healthypeople.gov

Healthy People: a national public health initiative under the jurisdiction of the U.S. Department of Health and Human Services (DHHS) that identifies the most significant preventable threats to health and focuses efforts toward eliminating them.

our energy intake has risen and, consequently, so has the incidence of overweight and obesity. Overweight and obesity, in turn, profoundly influence our health—as the next section explains.

IN SUMMARY **People become malnourished when they get too little or too much energy or nutrients. Deficiencies, excesses, and imbalances of nutrients lead to malnutrition diseases. To detect malnutrition in individuals, health-care professionals use a combination of four nutrition assessment methods. Reviewing historical information on diet and health may suggest a possible nutrition problem. Laboratory tests may detect a possible nutrition problem in its earliest stages, whereas anthropometric measurements and physical examinations pick up on the problem only after it causes symptoms. National surveys use similar assessment methods to measure people's food consumption and to evaluate the nutrition status of populations.**

Diet and Health

Foods play a vital role in supporting health.[15] Early nutrition research focused on identifying the nutrients in foods that would prevent such common diseases as rickets and scurvy, the vitamin D– and vitamin C–deficiency diseases. With this knowledge, developed countries have successfully defended against nutrient deficiency diseases. World hunger and nutrient deficiency diseases still pose a major health threat in developing countries, however, but not because of a lack of nutrition knowledge. More recently, nutrition research has focused on **chronic diseases** associated with energy and nutrient excesses. Once thought to be "rich countries' problems," chronic diseases have now become epidemic in developing countries as well—contributing to three out of five deaths worldwide.[16]

Chronic Diseases

Table 1-6 lists the ten leading causes of death in the United States. These "causes" are stated as if a single condition such as heart disease caused death, but most chronic diseases arise from multiple factors over many years. A person who died of heart disease may have been overweight, had high blood pressure, been a cigarette smoker, and spent years eating a diet high in saturated fat and getting too little exercise.

Of course, not all people who die of heart disease fit this description, nor do all people with these characteristics die of heart disease. People who are overweight might die from the complications of diabetes instead, or those who smoke might die of cancer. They might even die from something totally unrelated to any of these factors such as an automobile accident. Still, statistical studies have shown that certain conditions and behaviors are linked to certain diseases.

Notice that Table 1-6 highlights four of the top six causes of death as having a link with diet. Since 1970, as knowledge about these diet and disease relationships grew, the death rates for three of these—heart disease, cancers, and strokes—decreased.[17] Death rates for diabetes—a chronic disease closely associated with obesity—increased.

TABLE 1-6 Leading Causes of Death in the United States

	Percentage of Total Deaths
1. **Heart disease**	26.5
2. **Cancers**	22.8
3. **Strokes**	5.9
4. Chronic lung diseases	5.3
5. Accidents	4.7
6. **Diabetes mellitus**	3.1
7. Alzheimer's disease	2.9
8. Pneumonia and influenza	2.6
9. Kidney diseases	1.8
10. Blood infections	1.4

NOTE: The diseases highlighted in bold have relationships with diet.
SOURCE: National Center for Health Statistics, www.cdc.gov/nchs

Risk Factors for Chronic Diseases

Factors that increase or reduce the *risk* of developing chronic diseases can be identified by analyzing statistical data. A strong association between a **risk factor** and a disease means that when the factor is present, the *likelihood* of developing the disease increases. It does not mean that all people with the risk factor will develop the disease. Similarly, a lack of risk factors does not guarantee freedom from a given disease. On the average, though, the more risk factors in a person's life, the greater that person's chances of developing the disease. Conversely, the fewer risk factors in a person's life, the better the chances for good health.

Risk Factors Persist

Risk factors tend to persist over time. Without intervention, a young adult with high blood pressure will most likely continue to have

chronic diseases: diseases characterized by a slow progression and long duration. Examples include heart disease, cancer, and diabetes.

risk factor: a condition or behavior associated with an elevated frequency of a disease but not proved to be causal. Leading risk factors for chronic diseases include obesity, cigarette smoking, high blood pressure, high blood cholesterol, physical inactivity, and a diet high in saturated fats and low in vegetables, fruits, and whole grains.

high blood pressure as an older adult, for example. Thus, to minimize the damage, early intervention is most effective.

Risk Factors Cluster Risk factors tend to cluster. For example, a person who is obese may be physically inactive, have high blood pressure, and have high blood cholesterol—all risk factors associated with heart disease. Intervention that focuses on one risk factor often benefits the others as well. For example, physical activity can help reduce weight. Physical activity and weight loss will, in turn, help to lower blood pressure and blood cholesterol.

Risk Factors in Perspective The most prominent factor contributing to death in the United States is tobacco use, ♦ followed closely by diet and activity patterns, and then alcohol use (see Table 1-7).[18] Risk factors such as smoking, poor dietary habits, physical inactivity, and alcohol consumption are personal behaviors that can be changed. Decisions to not smoke, to eat a well-balanced diet, to engage in regular physical activity, and to drink alcohol in moderation (if at all) improve the likelihood that a person will enjoy good health. Other risk factors, such as genetics, gender, and age, also play important roles in the development of chronic diseases, but they cannot be changed. Health recommendations acknowledge the influence of such factors on the development of disease, but they must focus on the factors that are changeable. For the two out of three Americans who do not smoke or drink alcohol excessively, the one choice that can influence long-term health prospects more than any other is diet.

♦ Cigarette smoking is responsible for one of every five deaths each year.

TABLE 1-7 Factors Contributing to Deaths in the United States

Factors	Percentage of Deaths
Tobacco	18
Poor diet/inactivity	15
Alcohol	4
Microbial agents	3
Toxic agents	2
Motor vehicles	2
Firearms	1
Sexual behavior	1
Illicit drugs	1

SOURCE: A. H. Mokdad and coauthors, Actual causes of death in the United States, 2000, *Journal of the American Medical Association* 291 (2004): 1238–1245, with corrections from *Journal of the American Medical Association* 293 (2005): 298.

IN SUMMARY **Within the range set by genetics, a person's choice of diet influences long-term health. Diet has no influence on some diseases but is linked closely to others. Personal life choices, such as engaging in physical activity and using tobacco or alcohol, also affect health for the better or worse.**

The next several chapters provide many more details about nutrients and how they support health. Whenever appropriate, the discussion shows how diet influences each of today's major diseases. Dietary recommendations appear again and again, as each nutrient's relationships with health are explored. Most people who follow the recommendations will benefit and can enjoy good health into their later years.

Nutrition Portfolio

Each chapter in this book ends with simple Nutrition Portfolio activities that invite you to review key messages and consider whether your personal choices are meeting the dietary goals introduced in the text. By using the information you are recording in Diet Analysis +, the dietary tracking software that accompanies this text, and keeping a journal of these Nutrition Portfolio assignments, you can examine how your knowledge and behaviors change as you progress in your study of nutrition.

Your food choices play a key role in keeping you healthy and reducing your risk of chronic diseases.

After you have recorded at least one day's foods in Diet Analysis +, please look at that day's choices and record your answers to the following in your journal:

- Identify the factors that most influence your food choices for meals and snacks.
- List the chronic disease risk factors and conditions (listed in the definition of *risk factors* p. 25) that you have.
- Describe lifestyle changes you can make to improve your chances of enjoying good health.

Diet Analysis PLUS To complete this exercise, go to your Diet Analysis Plus at www.cengage.com/sso.

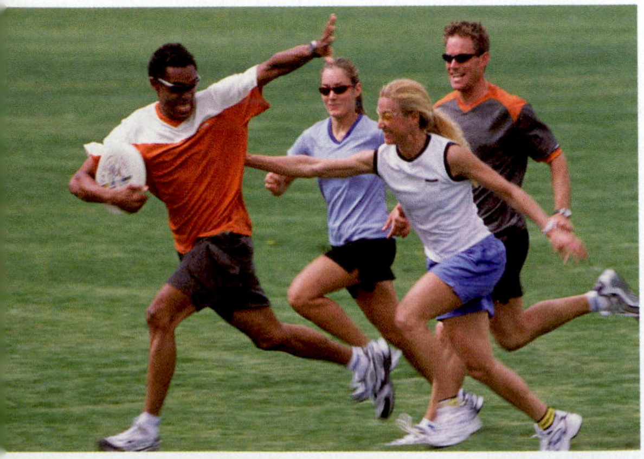

Physical activity can be both fun and beneficial.

Nutrition on the Net

For further study of topics covered in this chapter, log on to **www.cengagebrain.com** and search for ISBN 084006845X.

- Search for "nutrition" at the U.S. Government health and nutrition information sites: **www.healthfinder.gov** or **www.nutrition.gov**
- Learn more about basic science research from the National Science Foundation and Research!America: **www.nsf.gov** and **researchamerica.org**
- Review the Dietary Reference Intakes: **www.nap.edu**
- Review nutrition recommendations from the Food and Agriculture Organization and the World Health Organization: **www.fao.org** and **www.who.int**
- View progress on Healthy People 2010 and 2020: **www.healthypeople.gov**
- Visit the Food and Nutrition section of Health Canada: **www.hc-sc.gc.ca**
- Learn about the national nutrition survey: **www.cdc.gov/nchs/nhanes.htm**
- Create a chart of your family health history at the U.S. Surgeon General's site: **familyhistory.hhs.gov**
- Find credible health information from the Centers for Disease Control and Prevention: **www.cdc.gov/healthyliving**

References

1. B. J. Tepper, Nutritional implications of genetic taste variation: The role of PROP sensitivity and other taste phenotypes, *Annual Review of Nutrition* 28 (2008): 367–388; A. A. Bachmanov and G. K. Beauchamp, Taste receptor genes, *Annual Review of Nutrition* 27 (2007): 389–414; A. El-Sohemy and coauthors, Nutrigenomics of taste—Impact on food preferences and food production, *Forum of Nutrition* 60 (2007): 176–182; K. Keskitalo and coauthors, Same genetic components underlie different measures of sweet taste preference, *American Journal of Clinical Nutrition* 86 (2007): 1663–1669; M. R. Yeomans and coauthors, Human hedonic responses to sweetness: Role of taste genetics and anatomy, *Physiology and Behavior* 91 (2007): 264–273; A. Knaapila and coauthors, Food neophobia shows heritable variation in humans, *Physiology and Behavior* 91 (2007): 573–578; D. R. Reed, T. Tanaka, and A. H. McDaniel, Diverse tastes: Genetics of sweet and bitter perception, *Physiology and Behavior* 88 (2006): 215–226.
2. J. Mathieu, Hey everyone, dinner's ready . . . two weeks ago, *Journal of the American Dietetic Association* 107 (2007): 26–27.
3. N. Larson and coauthors, Food preparation by young adults is associated with better diet quality, *Journal of the American Dietetic Association* 106 (2006): 2001–2007.
4. Food Marketing Institute, www.fmi.org/research, accessed September 22, 2008.
5. Position of the American Dietetic Association: Functional foods, *Journal of the American Dietetic Association* 104 (2004): 814–826.
6. Position of the American Dietetic Association: Total diet approach to communicating food and nutrition information, *Journal of the American Dietetic Association* 107 (2007): 1224–1232.
7. L. Afman and M. Müller, Nutrigenomics: From molecular nutrition to prevention of disease, *Journal of the American Dietetic Association* 106 (2006): 569–576; J. Ordovas and V. Mooser, Nutrigenomics and nutrigenetics, *Current Opinion in Lipidology* 15 (2005): 101–108.
8. R. B. D'Agostino, Jr. and R. B. D'Agostino, Sr., Estimating treatment effects using observational data, *Journal of the American Medical Association* 297 (2007): 314–316.
9. L. I. Lesser and coauthors, Relationship between funding source and conclusion among nutrition-related scientific articles, *PLoS Medicine* 4 (2007): 0001–0006.
10. Committee on Dietary Reference Intakes, *Dietary Reference Intakes for Calcium and Vitamin D* (Washington, D.C.: National Academies Press, 2011); Committee on Dietary Reference Intakes, *Dietary Reference Intakes for Water, Potassium, Sodium, Chloride, and Sulfate* (Washington, D.C.: National Academies Press, 2005); Committee on Dietary Reference Intakes, *Dietary Reference Intakes for Energy, Carbohydrate, Fiber, Fat, Fatty Acids, Cholesterol, Protein, and Amino Acids* (Washington, D.C.: National Academies Press, 2005); Committee on Dietary Reference Intakes, *Dietary Reference Intakes for Vitamin A, Vitamin K, Arsenic, Boron, Chromium, Copper, Iodine, Iron, Manganese, Molybdenum, Nickel, Silicon, Vanadium, and Zinc* (Washington, D.C.: National Academies Press, 2001); Committee on Dietary Reference Intakes, *Dietary Reference Intakes for Vitamin C, Vitamin E, Selenium, and Carotenoids* (Washington, D.C.: National Academies Press, 2000); Committee on Dietary Reference Intakes, *Dietary Reference Intakes for Thiamin, Riboflavin, Niacin, Vitamin B_6, Folate, Vitamin B_{12}, Pantothenic Acid, Biotin, and Choline* (Washington, D.C.: National Academies Press, 1998); Committee on Dietary Reference Intakes, *Dietary Reference Intakes for Calcium, Phosphorus, Magnesium, Vitamin D, and Fluoride* (Washington, D.C.: National Academies Press, 1997).
11. R. M. Russell, Current framework for DRI development: What are the pros and cons? *Nutrition Reviews* 66 (2008): 455–458.
12. C. L. Taylor, Highlights of "a model for establishing upper levels of intake for nutrients and related substances: Report of a joint FAO/WHO technical workshop on nutrient risk assessment, May 26, 2005," *Nutrition Reviews* 65 (2007): 31–38.
13. P. J. Stover, Influence of human genetic variation on nutritional requirements, *American Journal of Clinical Nutrition* 83 (2006): 436S–442S.
14. U.S. Department of Health and Human Services, *Healthy People 2010 Midcourse Review* (Washington, D.C.: U.S. Government Printing Office, December 2006).
15. D. R. Jacobs and L. C. Tapsell, Food, not nutrients, is the fundamental unit in nutrition, *Nutrition Reviews* 65 (2007): 439–450.
16. B. M. Popkin, Global nutrition dynamics: The world is shifting rapidly toward a diet linked with noncommunicable disease, *American Journal of Clinical Nutrition* 84 (2006): 289–298.
17. A. Jemal and coauthors, Trends in the leading causes of death in the United States, 1970–2002, *Journal of the American Medical Association* 294 (2005): 1255–1259.
18. A. H. Mokdad and coauthors, Actual causes of death in the United States, 2000, *Journal of the American Medical Association* 291 (2004): 1238–1245.

HIGHLIGHT 1

Nutrition Information and Misinformation—On the Net and in the News

© Banana Stock/PhotoLibrary

How can people distinguish valid nutrition information from misinformation? One excellent approach is to notice *who* is providing the information. The "who" behind the information is not always evident, though, especially in the world of electronic media. Keep in mind that *people* develop CDs and DVDs and create websites on the Internet, just as people write books and report the news. In all cases, consumers need to determine whether the person is qualified to provide nutrition information.

This highlight begins by examining the unique potential as well as the problems of relying on the Internet and the media for nutrition information. It continues with a discussion of how to identify reliable nutrition information that applies to all resources, including the Internet and the news. (The accompanying glossary defines related terms.)

Nutrition on the Net

Got a question? The **Internet** has an answer. The Internet offers endless opportunities to obtain high-quality information, but it also delivers an abundance of incomplete, misleading, or inaccurate information.[1] Simply put: anyone can publish anything.

With hundreds of millions of **websites** on the **World Wide Web,** searching for nutrition information can be an overwhelming

GLOSSARY

accredited: approved; in the case of medical centers or universities, certified by an agency recognized by the U.S. Department of Education.

American Dietetic Association (ADA): the professional organization of dietitians in the United States. The Canadian equivalent is Dietitians of Canada, which operates similarly.

certified nutritionists or **certified nutritional consultants** or **certified nutrition therapists:** a person who has been granted a document declaring his or her authority as a nutrition professional. See also *nutritionist.*

dietetic technician: a person who has completed a minimum of an associate's degree from an accredited university or college and an approved dietetic technician program that includes a supervised practice experience. See also *dietetic technician, registered (DTR).*

dietetic technician, registered (DTR): a dietetic technician who has passed a national examination and maintains registration through continuing professional education.

dietitian: a person trained in nutrition, food science, and diet planning. See also *registered dietitian.*

DTR: see *dietetic technician, registered.*

fraudulent: the promotion, for financial gain, of devices, treatments, services, plans, or products (including diets and supplements) that alter or claim to alter a human condition without proof of safety or effectiveness.

Internet (the Net): a worldwide network of millions of computers linked together to share information.

license to practice: permission under state or federal law, granted on meeting specified criteria, to use a certain title (such as dietitian) and offer certain services. **Licensed dietitians** may use the initials **LD** after their names.

misinformation: false or misleading information.

nutritionist: a person who specializes in the study of nutrition. Note that this definition does not specify qualifications and may apply not only to registered dietitians but also to self-described experts whose training is questionable. Most states have licensing laws that define the scope of practice for those calling themselves nutritionists.

public health dietitians: dietitians who specialize in providing nutrition services through organized community efforts.

RD: see *registered dietitian.*

registered dietitian (RD): a person who has completed a minimum of a bachelor's degree from an accredited university or college, has completed approved course work and a supervised practice program, has passed a national examination, and maintains registration through continuing professional education.

registration: listing; with respect to health professionals, listing with a professional organization that requires specific course work, experience, and passing of an examination.

websites: Internet resources composed of text and graphic files, each with a unique URL (Uniform Resource Locator) that names the site (for example, www.usda.gov).

World Wide Web (the Web, commonly abbreviated **www**)**:** a graphical subset of the Internet.

experience—much like walking into an enormous bookstore with millions of books, magazines, newspapers, and videos. And like a bookstore, the Internet offers no guarantees of the accuracy of the information found there—much of which is pure fiction.

When using the Internet, keep in mind that the quality of health-related information available covers a broad range. You must evaluate websites for their accuracy, just like every other source. The accompanying "How To" provides tips for determining whether a website is reliable.

One of the most trustworthy sites used by scientists and others is the National Library of Medicine's PubMed, which provides free access to more than 10 million abstracts (short descriptions) of research papers published in scientific journals around the world. Many abstracts provide links to websites where full articles are available. Figure H1-1 (p. 30) introduces this valuable resource.

Did you receive the e-mail warning about Costa Rican bananas causing the disease "necrotizing fasciitis"? If so, you've been scammed by Internet misinformation. When nutrition information arrives in unsolicited e-mails, be suspicious if:

- The person sending it to you didn't write it and you cannot determine who did or if that person is a nutrition expert.
- The phrase "Forward this to everyone you know" appears.
- The phrase "This is not a hoax" appears because chances are good that it is.
- The news is sensational and you've never heard about it from legitimate sources.
- The language is emphatic and the text is sprinkled with capitalized words and exclamation marks.
- No references are given or, if present, are of questionable validity when examined.
- The message has been debunked on websites such as **www.quackwatch.org** or **urbanlegends.about.com**.

HOW TO Determine Whether a Website Is Reliable

To determine whether a website offers reliable nutrition information, ask the following questions:

- **Who?** Who is responsible for the site? Is it staffed by qualified professionals? Look for the authors' names and credentials. Have experts reviewed the content for accuracy?
- **When?** When was the site last updated? Because nutrition is an ever-changing science, sites need to be dated and updated frequently.
- **Where?** Where is the information coming from? The three letters following the dot in a Web address identify the site's affiliation. Addresses ending in "gov" (government), "edu" (educational institute), and "org" (organization) generally provide reliable information; "com" (commercial) sites represent businesses and, depending on their qualifications and integrity, may or may not offer dependable information.
- **Why?** Why is the site giving you this information? Is the site providing a public service or selling a product? Many commercial sites provide accurate information, but some do not. When money is the prime motivation, be aware that the information may be biased.

If you are satisfied with the answers to all of the previous questions, then ask this final question:

- **What?** What is the message, and is it in line with other reliable sources? Information that contradicts common knowledge should be questioned. Many reliable sites provide links to other sites to facilitate your quest for knowledge, but this provision alone does not guarantee a reputable intention. Be aware that any site can link to any other site without permission.

TRY IT Visit a nutrition website and answer the five "W" questions to determine whether it is a reliable resource.

Nutrition in the News

Consumers get much of their nutrition information from Internet websites, television news, and magazine articles, which have heightened awareness of how diet influences the development of diseases. Consumers benefit from news coverage of nutrition when they learn to make lifestyle changes that will improve their health. Sometimes, however, popular reports mislead consumers and create confusion. They often tell a lopsided story based on a few testimonials instead of presenting the results of research studies or a balance of expert opinions.

Tight deadlines and limited understanding sometimes make it difficult to provide a thorough report. Hungry for the latest news, the media often report scientific findings prematurely—without benefit of careful interpretation, replication, and peer review. Usually, the reports present findings from a single, recently released study, making the news current and controversial. Consequently, the public receives diet and health news quickly, but not always in perspective. Reporters may twist inconclusive findings into "meaningful discoveries" when pressured to write catchy headlines and sensational stories.

As a result, "surprising new findings" seem to contradict one another, and consumers feel frustrated and betrayed. Occasionally, the reports are downright false, but more often the apparent contradictions are simply the normal result of science at work. A single study contributes to the big picture, but when viewed alone, it can easily distort the image. To be meaningful, the conclusions of any study must be presented cautiously within the context of other research findings.

FIGURE H1-1 PUBMED (www.pubmed.gov): Internet Resource for Scientific Nutrition References

The U.S. National Library of Medicine's PubMed website offers tutorials to help teach beginners to use the search system effectively. Often, simply visiting the site, typing a query in the "Search for" box, and clicking "Go" will yield satisfactory results.

For example, to find research concerning calcium and bone health, typing "calcium bone" nets more than 30,000 results. Try setting limits on dates, types of articles, languages, and other criteria to obtain a more manageable number of abstracts to peruse.

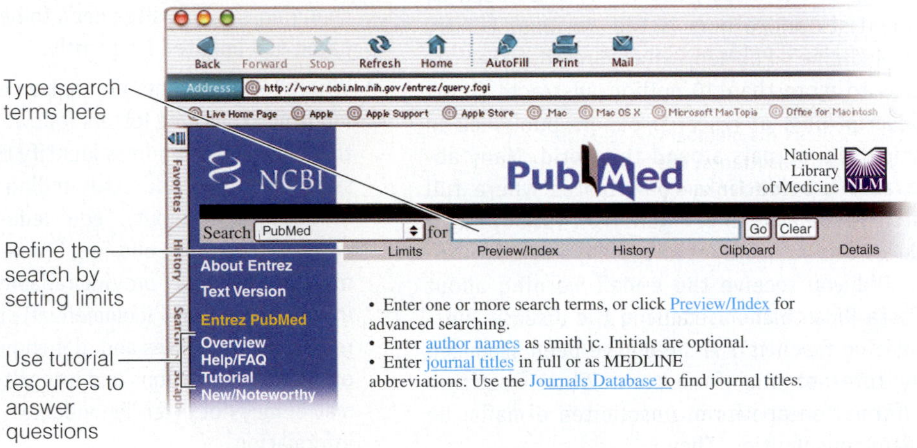

Identifying Nutrition Experts

Regardless of whether the medium is electronic, print, or video, consumers need to ask whether the person behind the information is qualified to speak on nutrition. If the creator of an Internet website recommends eating three pineapples a day to lose weight, a trainer at the gym praises a high-protein diet, or a health-store clerk suggests an herbal supplement, should you believe these people? Can you distinguish between accurate news reports and infomercials on television? Have you noticed that many televised nutrition messages are presented by celebrities, athletes, psychologists, food editors, and chefs—that is, almost anyone except a **dietitian**? When you are confused or need sound dietary advice, whom should you ask?

Physicians and Other Health-care Professionals

Many people turn to physicians or other health-care professionals for dietary advice, expecting them to know about all health-related matters. But are they the best sources of accurate and current information on nutrition? Only about 30 percent of all medical schools in the United States require students to take a separate nutrition course; less than half require the minimum 25 hours of nutrition instruction recommended by the National Academy of Sciences.[2] By comparison, most students reading this text are taking a nutrition class that provides an average of 45 hours of instruction.

The **American Dietetic Association (ADA)** asserts that standardized nutrition education should be included in the curricula for all health-care professionals: physicians, nurses, physician's assistants, dental hygienists, physical and occupational therapists, social workers, and all others who provide services directly to clients. When these professionals understand the relevance of nutrition in the treatment and prevention of diseases and have command of reliable nutrition information, then all the people they serve will also be better informed.

Most health-care professionals appreciate the connections between health and nutrition. Those who have specialized in clinical nutrition are especially well qualified to speak on the subject. Few, however, have the time or experience to develop diet plans and provide detailed diet instructions for clients. Often they wisely refer clients to a qualified nutrition expert—a **registered dietitian (RD).**

Registered Dietitian (RD)

A registered dietitian (RD) has the educational background necessary to deliver reliable nutrition advice and care.[3] To become an RD, a person must earn an undergraduate degree requiring about 60 credit hours in nutrition, food science, and other related subjects; complete a year's clinical internship or the equivalent; pass a national examination administered by the ADA; and maintain up-to-date knowledge and **registration** by participating in required continuing education activities such as attending seminars, taking courses, or conducting research.

Some states allow anyone to use the title dietitian or **nutritionist,** but others allow only an RD or people with specified qualifications to call themselves dietitians. Many states provide a further guarantee: a state registration, certification, or **license to practice.** In this way, states identify people who have met minimal standards of education and experience. Still, these state standards may fall short of those defining an RD. Similarly, some alternative educational programs qualify their graduates as **certified nutritionists, certified nutritional consultants,** or **certified nutrition therapists**—terms that sound authorita-

© Courtesy of eatright.org

Eddie displays his membership certificate to an association of nutritional consultants. His human companion, Connie Diekman, is a registered dietitian and past president of the American Dietetic Association.

tive but lack the credentials of an RD. In fact, even Eddie, an English cocker spaniel, was able to obtain a certificate of membership from the American Association of Nutritional Consultants.[4]

Dietitians perform a multitude of duties in many settings in most communities. They work in the food industry, pharmaceutical companies, home health agencies, long-term care institutions, private practice, public health departments, research centers, education settings, fitness centers, and hospitals. Depending on their work settings, dietitians can assume a number of different job responsibilities and positions. In hospitals, administrative dietitians manage the foodservice system; clinical dietitians provide client care; and nutrition support team dietitians coordinate nutrition care with other health-care professionals. In the food industry, dietitians conduct research, develop products, and market services.

Public health dietitians who work in government-funded agencies play a key role in delivering nutrition services to people in the community. Among their many roles, public health dietitians help plan, coordinate, and evaluate food assistance programs; act as consultants to other agencies; manage finances; and much more.

Dietetic Technician, Registered (DTR)

In some facilities, a **dietetic technician** assists registered dietitians in both administrative and clinical responsibilities. A dietetic technician has been educated and trained to work under the guidance of a registered dietitian; upon passing a national examination, the title changes to **dietetic technician, registered (DTR).**

Other Dietary Employees

In addition to the dietetic technician, other dietary employees may include clerks, aides, cooks, porters, and assistants. These dietary employees do not have extensive formal training in nutrition, and their ability to provide accurate information may be limited.

Identifying Fake Credentials

In contrast to registered dietitians, thousands of people obtain fake nutrition degrees and claim to be nutrition consultants or doctors of "nutrimedicine." These and other such titles may sound meaningful, but most of these people lack the established credentials and training of an ADA-sanctioned dietitian. If you look closely, you can see signs of their fake expertise.

Consider educational background, for example. The minimum standards of education for a dietitian specify a bachelor of science (BS) degree in food science and human nutrition or related fields from an **accredited** college or university.* Such a degree generally requires 4 to 5 years of study. Similarly, minimum standards of education for a dietetic technician specify an associate's degree that typically requires 2 years of study. In contrast, a fake nutrition expert may display a degree from a 6-month course. Such a degree simply falls short. In some cases, businesses posing as schools offer even less—they sell certificates to anyone who pays the fees. To obtain these "degrees," a candidate need not attend any classes, read any books, or pass any examinations.

To safeguard educational quality, an accrediting agency recognized by the U.S. Department of Education (DOE) certifies that certain schools meet criteria established to ensure that an institution provides complete and accurate schooling. Unfortunately, fake nutrition degrees are available from schools "accredited" by phony accrediting agencies. Acquiring false credentials is especially easy today, with **fraudulent** businesses operating via the Internet.

Knowing the qualifications of someone who provides nutrition information can help you determine whether that person's advice might be harmful or helpful. Don't be afraid to ask for credentials. Table H1-1 lists credible sources of nutrition information.

Red Flags of Nutrition Quackery

Figure H1-2 features eight red flags consumers can use to identify nutrition **misinformation.** Sales of unproven and dangerous products have always been a concern, but the Internet now provides merchants with an easy and inexpensive way to reach millions of customers around the world. Because of the difficulty in regulating the Internet, fraudulent and illegal sales of medical products have hit a bonanza. As is the case with the air, no one owns the Internet, and similarly, no one has control over the pollution. Countries

*To ensure the quality and continued improvement of nutrition and dietetics education programs, an ADA agency known as the Commission on Accreditation for Dietetics Education (CADE) establishes and enforces eligibility requirements and accreditation standards for programs preparing students for careers as registered dietitians or dietetic technicians. Programs meeting those standards are accredited by CADE.

HIGHLIGHT 1

TABLE H1-1 Credible Sources of Nutrition Information

Government agencies, volunteer associations, consumer groups, and professional organizations provide consumers with reliable health and nutrition information. Credible sources of nutrition information include:

- Nutrition and food science departments at a university or community college
- Local agencies such as the health department or County Cooperative Extension Service
- Government health agencies such as:
  - Department of Agriculture (USDA) — **www.usda.gov**
  - Department of Health and Human Services (DHHS) — **www.os.dhhs.gov**
  - Food and Drug Administration (FDA) — **www.fda.gov**
  - Health Canada — **www.hc-sc.gc.ca/nutrition**
- Volunteer health agencies such as:
  - American Cancer Society — **www.cancer.org**
  - American Diabetes Association — **www.diabetes.org**
  - American Heart Association — **www.americanheart.org**
- Reputable consumer groups such as:
  - American Council on Science and Health — **www.acsh.org**
  - Federal Citizen Information Center — **www.pueblo.gsa.gov**
  - International Food Information Council — **www.ific.org**
- Professional health organizations such as:
  - American Dietetic Association — **www.eatright.org**
  - American Medical Association — **www.ama-assn.org**
  - Dietitians of Canada — **www.dietitians.ca**
- Journals such as:
  - *American Journal of Clinical Nutrition* — **www.ajcn.org**
  - *Journal of the American Dietetic Association* — **www.adajournal.org**
  - *New England Journal of Medicine* — **www.nejm.org**
  - *Nutrition Reviews* — **www.ilsi.org**

FIGURE H1-2 Red Flags of Nutrition Quackery

Satisfaction guaranteed
Marketers may make generous promises, but consumers won't be able to collect on them.
One product does it all
No one product can possibly treat such a diverse array of conditions.
Time tested
Such findings would be widely publicized and accepted by health professionals.
Paranoid accusations
And this product's company doesn't want money? At least the drug company has scientific research proving the safety and effectiveness of its products.
Quick and easy fixes
Even proven treatments take time to be effective.
Natural
Natural is not necessarily better or safer; any product that is strong enough to be effective is strong enough to cause side effects.
Personal testimonials
Hearsay is the weakest form of evidence.
Meaningless medical jargon
Phony terms hide the lack of scientific proof.
Guaranteed! OR your money back!
"Cures gout, ulcers, diabetes and cancer"
Instant recovery, back to your everyday schedule
"Best pills around"
Wonder Pills
W
The natural way to becoming a better you
Revolutionary product, based on ancient medicine
Super Trim
S
Money grabbing drug companies further corporate means
"My friends feel good as new!"
Beats the hunger stimulation point (HSP)

have different laws regarding sales of drugs, dietary supplements, and other health products, but applying these laws to the Internet marketplace is almost impossible. Even if illegal activities could be defined and identified, finding the person responsible for a particular website is not always possible. Websites can open and close in a blink of a cursor. Now, more than ever, consumers must heed the caution "Buyer beware."

In summary, when you hear nutrition news, consider its source. Ask yourself these two questions: Is the person providing the information qualified to speak on nutrition? Is the information based on valid scientific research? If not, find a better source. After all, your health depends on it.

Nutrition on the Net

For further study of topics covered in this highlight, log on to **www.cengagebrain.com** and search for ISBN 084006845X.

- Visit the National Council Against Health Fraud: **www.ncahf.org**
- Find a registered dietitian in your area from the American Dietetic Association: **www.eatright.org**
- Find a nutrition professional in Canada from the Dietitians of Canada: **www.dietitians.ca**
- Find out whether a correspondence school is accredited from the Distance Education and Training Council: **www.detc.org**
- Find useful and reliable health information from the Health on the Net Foundation: **www.hon.ch**
- Find out whether a school is properly accredited for a dietetics degree from the American Dietetic Association: **www.eatright.org/cade**
- Obtain a listing of accredited institutions, professionally accredited programs, and candidates for accreditation from the American Council on Education: **www.acenet.edu**
- Learn more about quackery from Stephen Barrett's Quackwatch: **www.quackwatch.org**
- Check out health-related hoaxes and urban legends: **www.snopes.com**, **www.scambusters.org**, and **urbanlegends.about.com**
- Find reliable research articles: **www.pubmed.gov**

References

1. Position of the American Dietetic Association: Food and nutrition misinformation, *Journal of the American Dietetic Association* 106 (2006): 601–607.
2. K. M. Adams and coauthors, Status of nutrition education in medical schools, *American Journal of Clinical Nutrition* 83 (2006): 941S–944S.
3. Position of the American Dietetic Association: The roles of registered dietitians and dietetic technicians, registered in health promotion and disease prevention, *Journal of the American Dietetic Association* 106 (2006): 1875–1884.
4. Who's dishing out your nutrition advice? Consumers beware: Make sure your source is a registered dietitian, www.eatright.org, for release March 3, 2008.

CHAPTER

2

© Michael Seaman/Alamy

Nutrition in Your Life

You make food choices—deciding what to eat and how much to eat—more than 1000 times every year. We eat so frequently that it's easy to choose a meal without giving any thought to its nutrient contributions or health consequences. Even when we want to make healthy choices, we may not know which foods to select or how much to consume. With a few tools and tips, you can learn to plan a healthy diet.

Throughout this chapter, the CourseMate icon indicates an opportunity for online self-study, linking you to activities to increase your understanding of chapter concepts. **www.cengagebrain.com** (search for ISBN 084006845X)

Planning a Healthy Diet

CHAPTER OUTLINE

Chapter 1 explains that the body's many activities are supported by the nutrients delivered by the foods people eat. Food choices made over years influence the body's health, and consistently poor choices increase the risks of developing chronic diseases. This chapter shows how a person can select from the tens of thousands of available foods to create a diet that supports good health. Fortunately, most foods provide several nutrients, so one trick for wise diet planning is to select a combination of foods that deliver a full array of nutrients. This chapter begins by introducing the diet-planning principles and dietary guidelines that assist people in selecting foods that will deliver nutrients without excess energy (kcalories).

Principles and Guidelines

How well you nourish yourself does not depend on the selection of any one food. Instead, it depends on the selection of many different foods at numerous meals over days, months, and years. Diet-planning principles and dietary guidelines are key concepts to keep in mind whenever you are selecting foods—whether shopping at the grocery store, choosing from a restaurant menu, or preparing a home-cooked meal.

Diet-Planning Principles

Diet planners have developed several ways to select foods. Whatever plan or combination of plans they use, though, they keep in mind the six basic diet-planning principles ♦ listed in the margin.

Adequacy Adequacy means that the diet provides sufficient energy and enough of all the nutrients to meet the needs of healthy people. Take the essential nutrient iron, for example. Because the body loses some iron each day, people have to replace it by eating foods that contain iron. A person whose diet fails to provide enough iron-rich foods may develop the symptoms of iron-deficiency anemia: the person may feel weak, tired, and listless; have frequent headaches; and find that even the smallest amount of muscular work brings disabling fatigue. To prevent these deficiency symptoms, a person must include foods that supply adequate iron. The same is true for all the other essential nutrients introduced in Chapter 1.

♦ Diet-planning principles:
- Adequacy
- Balance
- kCalorie (energy) control
- Nutrient density
- Moderation
- Variety

adequacy (dietary): providing all the essential nutrients, fiber, and energy in amounts sufficient to maintain health.

© Polara Studios Inc.

To ensure an adequate and balanced diet, eat a variety of foods daily, choosing different foods from each group.

Balance The art of balancing the diet involves consuming enough—but not too much—of each type of food. The essential minerals calcium and iron, taken together, illustrate the importance of dietary **balance**. Meats, fish, and poultry are rich in iron but poor in calcium. Conversely, milk and milk products are rich in calcium but poor in iron. Use some meat (or meat alternatives) for iron; use some milk and milk products for calcium; and save some space for other foods, too, because a diet consisting of milk and meat alone would not be adequate. ♦ For the other nutrients, people need to consume whole grains, vegetables, and fruits.

♦ Balance in the diet helps to ensure adequacy.

kCalorie (Energy) Control Designing an adequate diet within a reasonable kcalorie allowance requires careful planning. Once again, balance plays a key role. The amount of energy coming into the body from foods should balance with the amount of energy being used by the body to sustain its metabolic and physical activities. Upsetting this balance leads to gains or losses in body weight. The discussion of energy balance and weight control in Chapters 8 and 9 examines this issue in more detail, but one key to **kcalorie control** is to select foods of high **nutrient density**.

Nutrient Density To eat well without overeating, select nutrient-dense foods—that is, foods that deliver the most nutrients for the least food energy.[1] Consider foods containing calcium, for example. You can get about 300 milligrams of calcium from either 1½ ounces of cheddar cheese or 1 cup of fat-free milk, but the cheese delivers about twice as much food energy (kcalories) as the milk. The fat-free milk, then, is twice as calcium dense as the cheddar cheese; it offers the same amount of calcium for half the kcalories. Both foods are excellent choices for adequacy's sake alone, but to achieve adequacy while controlling kcalories, ♦ the fat-free milk is the better choice. (Alternatively, a person could select a low-fat cheddar cheese with its kcalories similar to fat-free milk.) The many bar graphs that appear in Chapters 10 through 13 highlight the most nutrient-dense choices, and the accompanying "How To" describes how to compare foods based on nutrient density.

♦ Nutrient density promotes adequacy and kcalorie control.

Just as a financially responsible person pays for rent, food, clothes, and tuition on a limited budget, healthy people obtain iron, calcium, and all the other essential nutrients on a limited energy (kcalorie) allowance. Success depends on getting many nutrients for each kcalorie "dollar." For example, a can of cola and a handful of grapes may both provide about the same number of kcalories, but the grapes deliver many more nutrients. A person who makes nutrient-dense choices, such as fruit instead of cola, can meet daily nutrient needs on a lower energy budget. Such choices support good health.

Foods that are notably low in nutrient density—such as potato chips, candy, and colas—are sometimes called **empty-kcalorie foods**. The kcalories these foods provide are called "empty" because they deliver energy (from sugar, fat, or both) with little, or no, protein, vitamins, or minerals.

The concept of nutrient density is relatively simple when examining the contributions of one nutrient to a food or diet. With respect to calcium, milk ranks high and meats rank low. With respect to iron, meats rank high and milk ranks low. But it is a more complex task to answer the question, which food is more nutritious? To answer that question, we need to consider several nutrients—including nutrients that may harm health as well as those that may be beneficial. Ranking foods based on their overall nutrient composition is known as **nutrient profiling**.[2] Researchers have yet to agree on an ideal way to rate foods based on the nutrient profile, but when they do, nutrient profiling will be quite useful in helping consumers identify nutritious foods and plan healthy diets.[3]

♦ Moderation contributes to adequacy, balance, and kcalorie control.

Moderation Foods rich in fat and sugar provide enjoyment and energy but relatively few nutrients. In addition, they promote weight gain when eaten in excess. A person practicing **moderation** ♦ eats such foods only on occasion and regularly

balance (dietary): providing foods in proportion to one another and in proportion to the body's needs.

kcalorie (energy) control: management of food energy intake.

nutrient density: a measure of the nutrients a food provides relative to the energy it provides. The more nutrients and the fewer kcalories, the higher the nutrient density.

empty-kcalorie foods: a popular term used to denote foods that contribute energy but lack protein, vitamins, and minerals.

nutrient profiling: ranking foods based on their nutrient composition.

moderation (dietary): providing enough but not too much of a substance.

HOW TO Compare Foods Based on Nutrient Density

One way to evaluate foods is simply to notice their nutrient contribution *per serving*: 1 cup of milk provides about 300 milligrams of calcium, and ½ cup of fresh, cooked turnip greens provides about 100 milligrams. Thus a serving of milk offers three times as much calcium as a serving of turnip greens. To get 300 milligrams of calcium, a person could choose either 1 cup of milk or 1½ cups of turnip greens.

Another valuable way to evaluate foods is to consider their nutrient density—their nutrient contribution *per kcalorie*. Fat-free milk delivers about 85 kcalories with its 300 milligrams of calcium. To calculate the nutrient density, divide milligrams by kcalories:

$$\frac{300 \text{ mg calcium}}{85 \text{ kcal}} = 3.5 \text{ mg per kcal}$$

Do the same for the fresh turnip greens, which provide 15 kcalories with the 100 milligrams of calcium:

$$\frac{100 \text{ mg calcium}}{15 \text{ kcal}} = 6.7 \text{ mg per kcal}$$

The more milligrams per kcalorie, the greater the nutrient density. Turnip greens are more calcium dense than milk. They provide more calcium *per kcalorie* than milk, but milk offers more calcium *per serving*. Both approaches offer valuable information, especially when combined with a realistic appraisal. What matters most is which are you more likely to consume—1½ cups of turnip greens or 1 cup of milk? You can get 300 milligrams of calcium from either, but the greens will save you about 40 kcalories (the savings would be even greater if you usually use whole milk).

Keep in mind, too, that calcium is only one of the many nutrients that foods provide. Similar calculations for protein, for example, would show that fat-free milk provides more protein both *per kcalorie* and *per serving* than turnip greens—that is, milk is more protein dense. Combining variety with nutrient density helps to ensure the adequacy of all nutrients.

For additional practice, log on to **www.cengagebrain.com** and search for ISBN 084006845X.

TRY IT Compare the thiamin density of 3 ounces of lean t-bone steak (174 kcalories, 0.09 milligrams thiamin) with ½ cup of fresh cooked broccoli (27 kcalories, 0.05 milligrams thiamin).

selects foods low in solid fats and added sugars, a practice that automatically improves nutrient density. Returning to the example of cheddar cheese versus fat-free milk, the fat-free milk not only offers the same amount of calcium for less energy, but it also contains far less fat than the cheese.

Variety A diet may have all of the virtues just described and still lack **variety**, if a person eats the same foods day after day. People should select foods from each of the food groups daily and vary their choices within each food group from day to day for several reasons. First, different foods within the same group contain different arrays of nutrients. Among the fruits, for example, strawberries are especially rich in vitamin C while apricots are rich in vitamin A. Variety improves nutrient adequacy.[4] Second, no food is guaranteed to be entirely free of substances that, in excess, could be harmful. The strawberries might contain trace amounts of one contaminant, the apricots another. By alternating fruit choices, a person will ingest very little of either contaminant. Third, as the adage goes, variety is the spice of life. A person who eats beans frequently can enjoy pinto beans in Mexican burritos today, garbanzo beans in a Greek salad tomorrow, and baked beans with barbecued chicken on the weekend. Eating nutritious meals need never be boring.

variety (dietary): eating a wide selection of foods within and among the major food groups.

♦ A healthy diet:

- Emphasizes a variety of fruits, vegetables, whole grains, and fat-free and low-fat milk products.
- Includes lean meats, poultry, fish, legumes, eggs, and nuts.
- Is low in saturated and *trans* fats, cholesterol, salt (sodium), and added sugars.
- Stays within your daily energy needs for your recommended body weight.

♦ These key recommendations, along with additional recommendations for specific population groups, also appear throughout the text as their subjects are discussed.

Dietary Guidelines for Americans What should a person eat to stay healthy? ♦ The answers can be found in the *Dietary Guidelines for Americans*. These guidelines provide evidence-based advice to help people attain and maintain a healthy weight, reduce the risk of chronic diseases, and promote overall health through diet and physical activity.[5] Table 2-1 presents the key recommendations of the *Dietary Guidelines for Americans, 2010* clustered into four major topic areas. ♦ The first area focuses on balancing kcalories to manage a healthy body weight by improving eating habits and engaging in regular physical activ-

TABLE 2-1 Key Recommendations of the *Dietary Guidelines for Americans, 2010*

Balancing kCalories to Manage Weight

- Prevent and/or reduce overweight and obesity through improved eating and physical activity behaviors.
- Control total kcalorie intake to manage body weight. For people who are overweight or obese, this will mean consuming fewer kcalories from foods and beverages.
- Increase physical activity and reduce time spent in sedentary behaviors.
- Maintain appropriate kcalorie balance during each stage of life—childhood, adolescence, adulthood, pregnancy and breastfeeding, and older age.

Foods and Food Components to Reduce

- Reduce daily sodium intake to less that 2300 milligrams and further reduce intake to 1500 milligrams among persons who are 51 and older and those of any age who are African American or have hypertension, diabetes, or chronic kidney disease. The 1500 milligrams recommendation applies to about half of the U.S. population, including children and the majority of adults.
- Consume less than 10 percent of kcalories from saturated fatty acids by replacing them with monounsaturated and polyunsaturated fatty acids.
- Consume less than 300 milligrams per day of dietary cholesterol.
- Keep *trans* fatty acid consumption as low as possible by limiting foods that contain synthetic sources of *trans* fats, such as partially hydrogenated oils, and by limiting other solid fats.
- Reduce the intake of kcalories from solid fats and added sugars.
- Limit the consumption of foods that contain refined grains, especially refined grain foods that contain solid fats, added sugars, and sodium.
- If alcohol is consumed it should be consumed in moderation—up to one drink per day for women and two drinks per day for men—and only by adults of legal drinking age.

Foods and Nutrients to Increase

- Increase vegetable and fruit intake.
- Eat a variety of vegetables, especially dark-green and red and orange vegetables and beans and peas.
- Consume at least half of all grains as whole grains. Increase whole-grain intake by replacing refined grains with whole grains.
- Increase intake of fat-free or low-fat milk and milk products, such as milk, yogurt, cheese, or fortified soy beverages.
- Choose a variety of protein foods, which include seafood, lean meat and poultry, eggs, beans and peas, soy products, and unsalted nuts and seeds.
- Increase the amount and variety of seafood consumed by choosing seafood in place of some meat and poultry.
- Replace protein foods that are higher in solid fats with choices that are lower in solid fats and kcalories and/or are sources of oils.
- Use oils to replace solid fats where possible.
- Choose foods that provide more potassium, dietary fiber, calcium, and vitamin D, which are nutrients of concern in American diets. These foods include vegetables, fruits, whole grains, and milk and milk products.

Building Healthy Eating Patterns

- Select an eating pattern that meets nutrient needs over time at an appropriate kcalorie level.
- Account for all foods and beverages consumed and assess how they fit within a total healthy eating pattern.
- Follow food safety recommendations when preparing and eating foods to reduce the risk of foodborne illnesses.

NOTE: These guidelines are intended for adults and children ages 2 and older.

SOURCE: The *Dietary Guidelines for Americans*, available at www.dietaryguidelines.gov.

ity. The second area advises people to reduce their intakes of such foods and food components as sodium, saturated and *trans* fatty acids, cholesterol, solid fats, added sugars, refined grain products, and alcoholic beverages (for those who partake). The third area encourages consumers to select a variety of fruits and vegetables, whole grains, and low-fat milk products and protein foods (including seafood). The fourth area helps consumers build healthy eating patterns that meet energy and nutrient needs while reducing the risk of foodborne illnesses. Together, the *Dietary Guidelines for Americans 2010* point the way toward longer, healthier, and more active lives.

Some people might wonder why *dietary* guidelines include recommendations for physical activity. The simple answer is that most people who maintain a healthy body weight do more than eat right. They also exercise—the equivalent of 30 to 60 minutes or more of moderately intense physical activity on most days.[6] As you will see repeatedly throughout this text, food and physical activity choices are integral partners in supporting good health.

IN SUMMARY **A well-planned diet delivers adequate nutrients, a balanced array of nutrients, and an appropriate amount of energy. It is based on nutrient-dense foods, moderate in substances that can be detrimental to health, and varied in its selections. The *Dietary Guidelines* apply these principles, offering practical advice on how to eat for good health.**

Diet-Planning Guides

To plan a diet that achieves all of the dietary ideals just outlined, a person needs tools as well as knowledge. Among the most widely used tools for diet planning are **food group plans** that build a diet from clusters of foods that are similar in nutrient content. Thus each food group represents a set of nutrients that differs somewhat from the nutrients supplied by the other groups. Selecting foods from each of the groups eases the task of creating an adequate and balanced diet.

USDA Food Guide

The *Dietary Guidelines* encourage consumers to adopt a balanced eating plan, such as the USDA's Food Guide (see Figure 2-1 on pp. 40–41). The USDA Food Guide assigns foods to five major groups ♦ and recommends daily amounts of foods from each group to meet nutrient needs. In addition to presenting the food groups, the figure lists the most notable nutrients of each group, the serving equivalents, and the foods within each group sorted by nutrient density. Chapter 15 provides a food guide for young children, and Appendix I presents Canada's food group plan, *Eating Well with Canada's Food Guide*.

♦ Five food groups:
- Fruits
- Vegetables
- Grains
- Protein foods
- Milk and milk products

Dietary Guidelines for Americans 2010

Select an eating pattern (such as the USDA Food Guide or the DASH eating plan) that meets nutrient needs over time at an appropriate kcalorie level. (The DASH eating plan is presented in Chapters 12 and 27.)

Recommended Amounts

All food groups offer valuable nutrients, and people should make selections from each group daily. Table 2-2 (p. 42) specifies the amounts of foods from each group needed daily to create a healthful diet for several energy (kcalorie) levels. ♦ Estimated daily kcalorie needs for sedentary and active men and women are shown in Table 2-3 (p. 42). A sedentary young woman needing 2000 kcalories a day, for example, would select 2 cups of fruit; 2½ cups of vegetables (dispersed among the vegetable subgroups); 6 ounces of grain foods (with at least half coming from whole grains); 5½ ounces of protein foods (meat, poultry, or fish, or the equivalent of **legumes**, eggs, seeds, or nuts); and 3 cups of milk or yogurt, or the equivalent amount of cheese or fortified soy products. Additionally, a small amount of unsaturated oil, such as vegetable oil, or the oils of nuts, olives, or fatty fish, is required to supply needed nutrients.

♦ Chapter 8 explains how to determine energy needs. For an approximation, turn to the DRI Estimated Energy Requirement (EER) on the inside front cover.

food group plans: diet-planning tools that sort foods into groups based on nutrient content and then specify that people should eat certain amounts of foods from each group.

legumes (lay-GYOOMS, LEG-yooms): plants of the bean and pea family, with seeds that are rich in protein compared with other plant-derived foods.

FIGURE 2-1 USDA Food Guide

Key:
- ● Foods generally high in nutrient density (choose most often)
- ▲ Foods lower in nutrient density (limit selections)

FRUITS

© Polara Studios, Inc.

Consume a variety of fruits and no more than one-half of the recommended intake as fruit juice.

These foods contribute folate, vitamin A, vitamin C, potassium, and fiber.

1 c fruit is equivalent to 1 c fresh, frozen, or canned fruit; ½ c dried fruit; 1 c fruit juice.

● Apples, apricots, avocados, bananas, blueberries, cantaloupe, cherries, grapefruit, grapes, guava, kiwi, mango, nectarines, oranges, papaya, peaches, pears, pineapples, plums, raspberries, strawberries, tangerines, watermelon; dried fruit (dates, figs, raisins); unsweetened juices.

▲ Canned or frozen fruit in syrup; juices, punches, ades, and fruit drinks with added sugars; fried plantains.

VEGETABLES

© Polara Studios, Inc.

Choose a variety of vegetables each day, and choose from all five subgroups several times a week.

These foods contribute folate, vitamin A, vitamin C, vitamin K, vitamin E, magnesium, potassium, and fiber.

1 c vegetables is equivalent to 1 c cut-up raw or cooked vegetables; 1 c cooked legumes; 1 c vegetable juice; 2 c raw, leafy greens.

● Dark green vegetables: Broccoli and leafy greens such as arugula, beet greens, bok choy, collard greens, kale, mustard greens, romaine lettuce, spinach, and turnip greens.

● Red and orange vegetables: Carrots, carrot juice, pumpkin, red peppers, sweet potatoes, tomatoes, and winter squash (acorn, butternut).

● Legumes: Black beans, black-eyed peas, garbanzo beans (chickpeas), kidney beans, lentils, navy beans, pinto beans, soybeans and soy products such as tofu, and split peas.

● Starchy vegetables: Cassava, corn, green peas, hominy, lima beans, and potatoes.

● Other vegetables: Artichokes, asparagus, bamboo shoots, bean sprouts, beets, brussels sprouts, cabbages, cactus, cauliflower, celery, cucumbers, eggplant, green beans, iceberg lettuce, mushrooms, okra, onions, peppers, seaweed, snow peas, vegetable juices, zucchini.

▲ Baked beans, candied sweet potatoes, coleslaw, french fries, potato salad, refried beans, scalloped potatoes, tempura vegetables.

GRAINS

© Polara Studios, Inc.

Make at least half of the grain selections whole grains.

These foods contribute folate, niacin, riboflavin, thiamin, iron, magnesium, selenium, and fiber.

1 oz grains is equivalent to 1 slice bread; ½ c cooked rice, pasta, or cereal; 1 oz dry pasta or rice; 1 c ready-to-eat cereal; 3 c popped popcorn.

● Whole grains (amaranth, barley, brown rice, buckwheat, bulgur, millet, oats, quinoa, rye, wheat) and whole-grain, low-fat breads, cereals, crackers, and pastas; popcorn.

● Enriched bagels, breads, cereals, pastas (couscous, macaroni, spaghetti), pretzels, rice, rolls, tortillas.

▲ Biscuits, cakes, cookies, cornbread, crackers, croissants, doughnuts, french toast, fried rice, granola, muffins, pancakes, pastries, pies, presweetened cereals, taco shells, waffles.

FIGURE 2-1 USDA Food Guide, continued

PROTEIN FOODS

© Polara Studios, Inc.

Make lean or low-fat choices. Prepare them with little, or no, added fat.

Meat, poultry, fish, and eggs contribute protein, niacin, thiamin, vitamin B_6, vitamin B_{12}, iron, magnesium, potassium, and zinc; legumes and nuts are notable for their protein, folate, thiamin, vitamin E, iron, magnesium, potassium, zinc, and fiber.

1 oz meat is equivalent to 1 oz cooked lean meat, poultry, or seafood; 1 egg; ¼ c cooked legumes or tofu; 1 tbs peanut butter; ½ oz nuts or seeds.

- ● Seafood: Fish, shellfish
- ● Meat, poultry, eggs: Lean or low-fat meat (fat-trimmed beef, game, ham, lamb, pork), poultry (no skin), eggs
- ● Nuts, seeds, soy products: Unsalted nuts (almonds, filberts, pistachios, walnuts), seeds (flaxseeds, pumpkin seeds, sunflower seeds), legumes, low-fat tofu, tempeh, peanuts, peanut butter
- ▲ Bacon; baked beans; fried meat, fish, poultry, eggs, or tofu; refried beans; ground beef; hot dogs; luncheon meats; marbled steaks; poultry with skin; sausages; spare ribs.

MILK, YOGURT, AND CHEESE

© Polara Studios, Inc.

Make fat-free or low-fat choices. Choose lactose-free products or other calcium-rich foods if you don't consume milk.

These foods contribute protein, riboflavin, vitamin B_{12}, calcium, magnesium, potassium, and, when fortified, vitamin A and vitamin D.

1 c milk is equivalent to 1 c milk, fortified soy milk, or yogurt; 1½ oz natural cheese; 2 oz processed cheese.

- ● Fat-free milk and fat-free milk products such as buttermilk, cheeses, cottage cheese, yogurt; fat-free fortified soy milk.
- ▲ 1% low-fat milk, 2% reduced-fat milk, and whole milk; low-fat, reduced-fat, and whole-milk products such as cheeses, cottage cheese, and yogurt; milk products with added sugars such as chocolate milk, custard, ice cream, ice milk, milk shakes, pudding, sherbet; fortified soy milk.

OILS

Matthew Farruggio

Select the recommended amounts of oils from among these sources.

These foods contribute vitamin E and essential fatty acids (see Chapter 5), along with abundant kcalories.

1 tsp oil is equivalent to 1 tbs low-fat mayonnaise; 2 tbs light salad dressing; 1 tsp vegetable oil; 1 tsp soft margarine.

- ● Liquid vegetable oils such as canola, corn, flaxseed, nut, olive, peanut, safflower, sesame, soybean, and sunflower oils; mayonnaise, oil-based salad dressing, soft *trans*-free margarine.
- ● Unsaturated oils that occur naturally in foods such as avocados, fatty fish, nuts, olives, seeds (flaxseeds, sesame seeds), and shellfish.

SOLID FATS AND ADDED SUGARS

Matthew Farruggio

Limit intakes of food and beverages with solid fats and added sugars.

Solid fats deliver saturated fat and *trans* fat, and intake should be kept low. Solid fats and added sugars contribute abundant kcalories but few nutrients, and intakes should not exceed the discretionary kcalorie allowance—kcalories to meet energy needs after all nutrient needs have been met with nutrient-dense foods. Alcohol also contributes abundant kcalories but few nutrients, and its kcalories are counted among discretionary kcalories. See Table 2-2 for some discretionary kcalorie allowances.

- ▲ Solid fats that occur in foods naturally such as milk fat and meat fat (see ▲ in previous lists).
- ▲ Solid fats that are often added to foods such as butter, cream cheese, hard margarine, lard, sour cream, and shortening.
- ▲ Added sugars such as brown sugar, candy, honey, jelly, molasses, soft drinks, sugar, and syrup.
- ▲ Alcoholic beverages include beer, wine, and liquor.

TABLE 2-2 Recommended Daily Amounts from Each Food Group

	1600 kcal	1800 kcal	2000 kcal	2200 kcal	2400 kcal	2600 kcal	2800 kcal	3000 kcal
Fruits	1½ c	1½ c	2 c	2 c	2 c	2 c	2½ c	2½ c
Vegetables	2 c	2½ c	2½ c	3 c	3 c	3½ c	3½ c	4 c
Grains	5 oz	6 oz	6 oz	7 oz	8 oz	9 oz	10 oz	10 oz
Protein foods	5 oz	5 oz	5½ oz	6 oz	6½ oz	6½ oz	7 oz	7 oz
Milk	3 c	3 c	3 c	3 c	3 c	3 c	3 c	3 c
Oils	5 tsp	5 tsp	6 tsp	6 tsp	7 tsp	8 tsp	8 tsp	10 tsp
Discretionary kcalorie allowance	121 kcal	161 kcal	258 kcal	266 kcal	330 kcal	362 kcal	395 kcal	459 kcal

♦ **Phytochemicals** are the nonnutrient compounds found in plant-derived foods that have biological activity in the body.

TABLE 2-3 Estimated Daily kCalorie Needs for Adults

	Sedentary[a]	Active[b]
Women		
19–30 yr	1900	2400
31–50 yr	1800	2200
51+ yr	1600	2100
Men		
19–30 yr	2500	3000
31–50 yr	2300	2900
51+ yr	2100	2600

[a]Sedentary describes a lifestyle that includes only the activities typical of day-to-day life.
[b]Active describes a lifestyle that includes physical activity equivalent to walking more than 3 miles per day at a rate of 3 to 4 miles per hour, in addition to the activities typical of day-to-day life.
NOTE: kCalorie values for active people reflect the midpoint of the range appropriate for age and gender, but within each group, older adults may need fewer kcalories and younger adults may need more. In addition to gender, age, and activity level, energy needs vary with height and weight (see Chapter 8 and Appendix F).

All vegetables provide an array of nutrients, but some vegetables are especially good sources of certain vitamins, minerals, and beneficial phytochemicals. ♦ For this reason, the USDA Food Guide sorts the vegetable group into five subgroups. The dark green vegetables deliver the B vitamin folate; the red and orange vegetables provide vitamin A; legumes supply iron and protein; the starchy vegetables contribute carbohydrate energy; and the other vegetables fill in the gaps and add more of these same nutrients.

In a 2000-kcalorie diet, then, the recommended 2½ cups of daily vegetables should be varied among the subgroups over a week's time. In other words, consuming 2½ cups of potatoes or even nutrient-rich spinach every day for seven days does *not* meet the recommended vegetable intakes. Potatoes and spinach make excellent choices when consumed in balance with vegetables from other subgroups. One way to help ensure selections for all of the subgroups is to eat vegetables of various colors—for example, green broccoli, orange sweet potatoes, black beans, yellow corn, and red beets. Intakes of vegetables are appropriately averaged over a week's time—it is not necessary to include every subgroup every day.

For similar reasons, the USDA Food Guide sorts protein foods into three subgroups. Perhaps most notably, each of these subgroups contributes a different assortment of fats. Table 2-4 presents the recommended weekly amounts for each of the subgroups for vegetables and protein foods.

Notable Nutrients As Figure 2-1 notes, each food group contributes key nutrients. This feature provides flexibility in diet planning because a person can select any food from a food group and receive similar nutrients. For example, a person can choose milk, cheese, or yogurt and receive the same key nutrients. Importantly, foods provide not only these key nutrients, but small amounts of other nutrients and phytochemicals as well.

TABLE 2-4 Recommended Weekly Amounts from the Vegetable and Protein Foods Subgroups

Table 2-2 specifies the recommended amounts of total vegetables and protein foods per *day*. This table shows those amounts dispersed among five vegetable and three protein foods subgroups per *week*.

Vegetable Subgroups	1600 kcal	1800 kcal	2000 kcal	2200 kcal	2400 kcal	2600 kcal	2800 kcal	3000 kcal
Dark green	1½ c	1½ c	1½ c	2 c	2 c	2½ c	2½ c	2½ c
Red and orange	4 c	5½ c	5½ c	6 c	6 c	7 c	7 c	7½ c
Legumes	1 c	1½ c	1½ c	2 c	2 c	2½ c	2½ c	3 c
Starchy	4 c	5 c	5 c	6 c	6 c	7 c	7 c	8 c
Other	3½ c	4 c	4 c	5 c	5 c	5½ c	5½ c	7 c
Protein Foods Subgroups								
Seafood	8 oz	8 oz	8 oz	9 oz	10 oz	10 oz	11 oz	11 oz
Meat, poultry, eggs	24 oz	24 oz	26 oz	29 oz	31 oz	31 oz	34 oz	34 oz
Nuts, seeds, soy products	4 oz	4 oz	4 oz	4 oz	5 oz	5 oz	5 oz	5 oz

Because legumes contribute the same key nutrients—notably, protein, iron, and zinc—as meats, poultry, and fish, they are included in the same food group. For this reason, legumes are useful as meat alternatives, and they are also excellent sources of fiber and the B vitamin folate. To encourage frequent consumption, the USDA Food Guide also includes legumes as a subgroup of the vegetable group. Thus legumes count in either the vegetable group or the protein foods group.[7] In general, people who regularly eat meat, poultry, and fish count legumes as a vegetable, and vegetarians and others who seldom eat meat, poultry, or fish count legumes in the protein foods group.

The USDA Food Guide encourages greater consumption from certain food groups to provide the nutrients most often lacking in the diets of Americans. ♦ In general, most people need to eat:

- *More* dark green vegetables, red and orange vegetables, legumes, fruits, whole grains, seafood, and low-fat milk and milk products.
- *Fewer* refined grains, total fats (especially saturated fat, *trans* fat, and cholesterol), added sugars, and total kcalories.

♦ Nutrients of concern:
- Dietary fiber
- Vitamin D
- Calcium
- Potassium

Nutrient-Dense Choices The USDA Food Guide provides a foundation for a healthy diet by emphasizing nutrient-dense options within each food group. By consistently selecting nutrient-dense foods, a person can obtain all the nutrients needed and still keep kcalories under control. In contrast, eating foods that are low in nutrient density makes it difficult to get enough nutrients without exceeding energy needs and gaining weight. For this reason, consumers should select low-fat foods from each group and foods without added fats or sugars—for example, fat-free milk instead of whole milk, baked chicken without the skin instead of hot dogs, green beans instead of french fries, orange juice instead of fruit punch, and whole-wheat bread instead of biscuits. Notice that the key in Figure 2-1 indicates which foods *within each group* are high or low in nutrient density. Oil is a notable exception: even though oil is pure fat and therefore rich in kcalories, a small amount of oil from sources such as nuts, fish, or vegetable oils is necessary every day to provide nutrients lacking from other foods. Consequently, these high-fat foods are listed among the nutrient-dense foods (see Highlight 5 to learn why).

Dietary Guidelines for Americans 2010

Focus on eating the most nutrient-dense forms of foods from all food groups.

Discretionary kCalorie Allowance At each kcalorie level, people who consistently choose nutrient-dense foods may be able to meet most of their nutrient needs without consuming their full allowance of kcalories.[8] The difference between the kcalories needed to supply nutrients and those needed to maintain weight—known as the **discretionary kcalorie allowance**—is illustrated in Figure 2-2.

Table 2-2 includes the discretionary kcalorie allowance for several kcalorie levels. A person with discretionary kcalories available might choose to:

- Eat additional nutrient-dense foods, such as an extra serving of skinless chicken or a second ear of corn.
- Select a few foods with fats or added sugars, such as reduced-fat milk or sweetened cereal.
- Add a little fat or sugar to foods, such as butter or jelly on toast.
- Consume some alcohol. (Highlight 7 explains why this may not be a good choice for some individuals.)

Alternatively, a person wanting to lose weight might choose to:

- *Not* use the kcalories available from the discretionary kcalorie allowance.

Added fats and sugars are always counted as discretionary kcalories. The kcalories from the fat in higher-fat milks and meats are also counted among discretionary kcalories. It helps to think of fat-free milk as "milk" and whole milk or

FIGURE 2-2 Discretionary kCalorie Allowance for a 2000-kCalorie Diet Plan

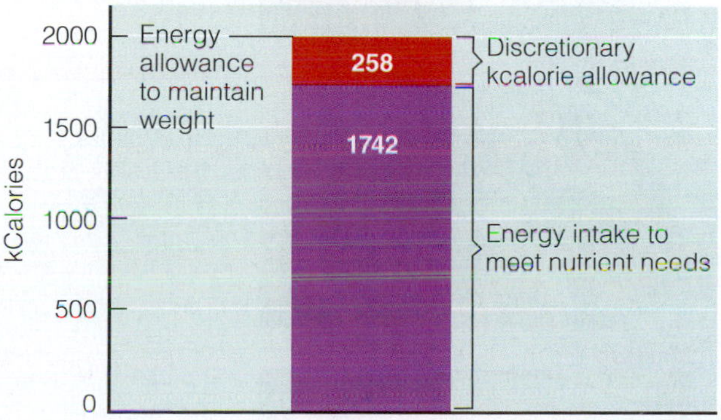

discretionary kcalorie allowance: the kcalories remaining in a person's energy allowance after consuming enough nutrient-dense foods to meet all nutrient needs for a day.

© Matthew Farruggio

Most bagels today weigh in at 4 ounces or more—meaning that a person eating one of these large bagels for breakfast is actually getting four or more grain servings, not one.

♦ For quick and easy estimates, visualize each portion as being about the size of a common object:

- 1 cup fruit or vegetables = a baseball
- ¼ cup dried fruit or nuts = a golf ball
- 3 ounces meat = a deck of cards
- 2 tablespoons peanut butter = a Ping-Pong ball
- 1 ounce cheese = 4 stacked dice
- ½ cup ice cream = a racquetball
- 4 small cookies = 4 poker chips

reduced-fat milk as "milk with added fat." Similarly, "meats" should be the leanest; other cuts are "meats with added fat." Puddings and other desserts made from whole milk provide discretionary kcalories from both the sugar added to sweeten them and the naturally occurring fat in the whole milk they contain. Even fruits, vegetables, and grains can carry discretionary kcalories into the diet in the form of peaches canned in syrup, scalloped potatoes, or high-fat crackers.

Discretionary kcalories are counted separately from the kcalories of the nutrient-dense foods of which they may be a part. A fried chicken leg, for example, provides discretionary kcalories from two sources: the naturally occurring fat of the chicken skin and the added fat absorbed during frying. The kcalories of the skinless chicken underneath are not discretionary kcalories—they are necessary to provide the nutrients of chicken.

Serving Equivalents Recommended serving amounts for fruits, vegetables, and milk are measured in cups, and those for grains and protein foods, in ounces. Figure 2-1 provides equivalent measures among the foods in each group specifying, for example, that 1 ounce of grains is equivalent to 1 slice of bread or ½ cup of cooked rice.

A person using the USDA Food Guide can become more familiar with measured portions by determining the answers to questions such as these: ♦ What portion of a cup is a small handful of raisins? Is a "helping" of mashed potatoes more or less than a half cup? How many ounces of cereal do you typically pour into the bowl? How many ounces is the steak at your favorite restaurant? How many cups of milk does your glass hold? Figure 2-1 includes the serving sizes and equivalent amounts for foods within each group.

Ethnic Food Choices People can use the USDA Food Guide and still enjoy a diverse array of culinary styles by sorting ethnic foods into their appropriate food groups. For example, a person eating Mexican foods would find tortillas in the grains group, jicama in the vegetable group, and guava in the fruit group. Table 2-5 features some ethnic food choices.

TABLE 2-5 Ethnic Food Choices

	Grains	Vegetables	Fruits	Protein Foods	Milk
Asian  © B.G. Smith/Shutterstock.com	Rice, noodles, millet	Amaranth, baby corn, bamboo shoots, chayote, bok choy, mung bean sprouts, sugar peas, straw mushrooms, water chestnuts, kelp	Carambola, guava, kumquat, lychee, persimmon, melons, mandarin orange	Soybeans and soy products such as soy milk and tofu, squid, duck eggs, pork, poultry, fish and other seafood, peanuts, cashews	Usually excluded
Mediterranean  © PhotoDisc/Getty Images	Pita pocket bread, pastas, rice, couscous, polenta, bulgur, focaccia, Italian bread	Eggplant, tomatoes, peppers, cucumbers, grape leaves	Olives, grapes, figs	Fish and other seafood, gyros, lamb, chicken, beef, pork, sausage, lentils, fava beans	Ricotta, provolone, parmesan, feta, mozzarella, and goat cheeses; yogurt
Mexican  © PhotoDisc/Getty Images	Tortillas (corn or flour), taco shells, rice	Chayote, corn, jicama, tomato salsa, cactus, cassava, tomatoes, yams, chilies	Guava, mango, papaya, avocado, plantain, bananas, oranges	Refried beans, fish, chicken, chorizo, beef, eggs	Cheese, custard

Vegetarian Food Guide Vegetarian diets rely mainly on plant foods: grains, vegetables, legumes, fruits, seeds, and nuts. Some vegetarian diets include eggs, milk products, or both. People who do not eat meats or milk products can still use the USDA Food Guide to create an adequate diet.[9] ♦ The food groups are similar, and the amounts for each serving remain the same. Highlight 2 defines vegetarian terms and provides details on planning healthy vegetarian diets.

♦ www.MyPyramid.gov offers information on vegetarian diets in its Tips & Resources section.

Mixtures of Foods Some foods—such as casseroles, soups, and sandwiches—fall into two or more food groups. With a little practice, users can learn to see these mixtures of foods as items from various food groups. For example, from the USDA Food Guide point of view, a taco represents four different food groups: the taco shell from the grains group; the onions, lettuce, and tomatoes from the "other vegetables" group; the ground beef from the protein foods group; and the cheese from the milk group.

MyPyramid—Steps to a Healthier You The USDA created an educational tool called MyPyramid to illustrate the concepts of the *Dietary Guidelines for Americans* and the USDA Food Guide. Figure 2-3 presents a graphic image of MyPyramid, which was designed to encourage consumers to make healthy food and physical activity choices every day. The recommendations in MyPyramid are supportive of, and consistent with, several other recommendations to control obesity and chronic diseases such as diabetes, heart disease, and cancer.[10]

The MyPyramid website (www.mypyramid.gov) helps consumers choose the kinds and amounts of foods to eat each day based on their height, weight, age, gender, and activity level.[11] Information is also available for pregnant and lactating women

FIGURE 2-3 MyPyramid: Steps to a Healthier You

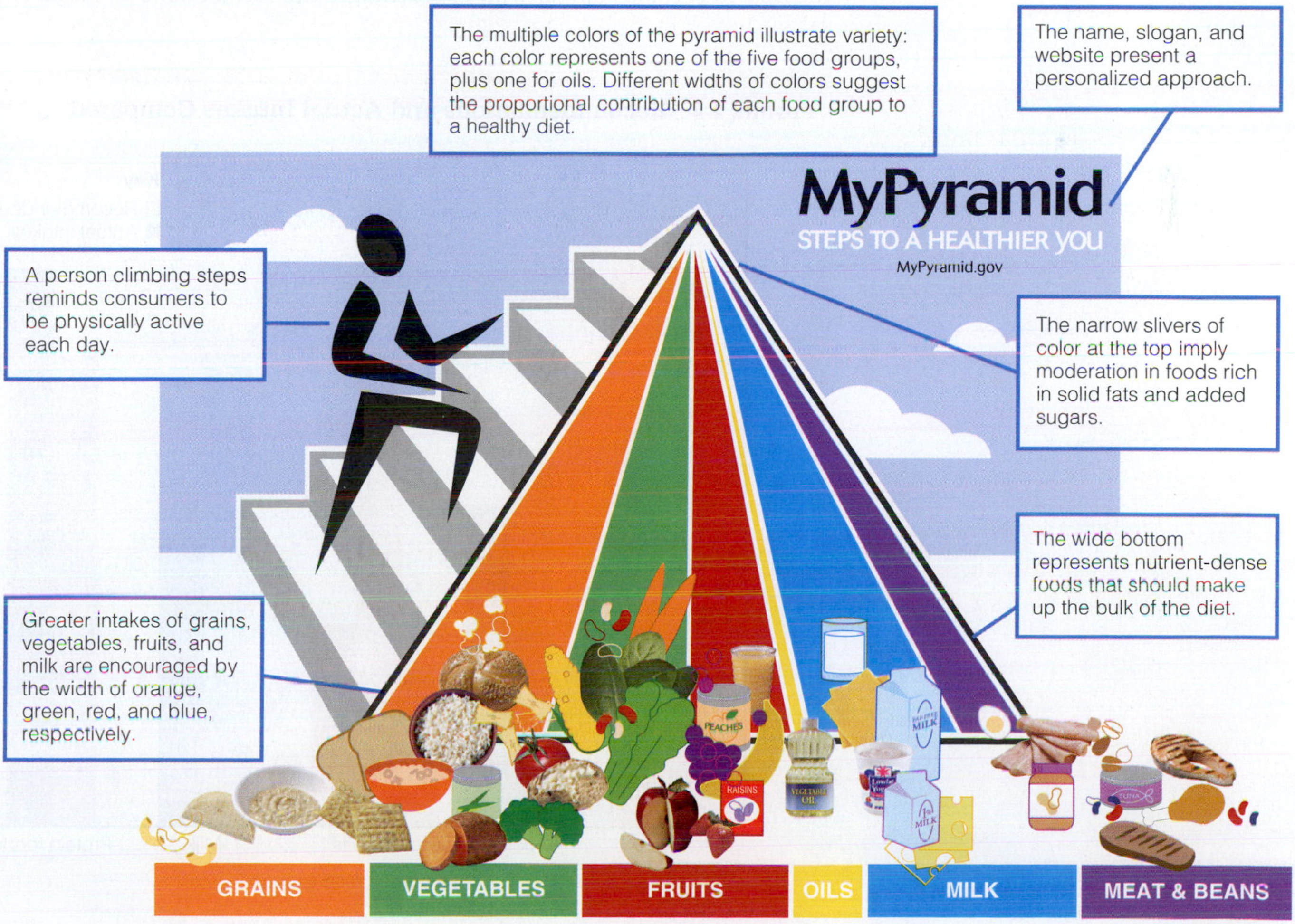
The multiple colors of the pyramid illustrate variety: each color represents one of the five food groups, plus one for oils. Different widths of colors suggest the proportional contribution of each food group to a healthy diet.
The name, slogan, and website present a personalized approach.
MyPyramid
STEPS TO A HEALTHIER YOU
MyPyramid.gov
A person climbing steps reminds consumers to be physically active each day.
The narrow slivers of color at the top imply moderation in foods rich in solid fats and added sugars.
The wide bottom represents nutrient-dense foods that should make up the bulk of the diet.
Greater intakes of grains, vegetables, fruits, and milk are encouraged by the width of orange, green, red, and blue, respectively.
PEACHES
RAISINS
VEGETABLE OIL
MILK
GRAINS
VEGETABLES
FRUITS
OILS
MILK
MEAT & BEANS

and for vegetarians. In addition to creating a personal plan, consumers can find tips to help them improve their diet and lifestyle by "taking small steps each day."

Recommendations versus Actual Intakes The USDA Daily Food Guide and MyPyramid were developed to help people choose a balanced and healthful diet. Are consumers actually eating according to these recommendations? The short answer is "not really." In general, consumers are not selecting the most nutrient-dense items from the food groups. Instead, they are consuming too many foods high in solid fats and added sugars—soft drinks, desserts, whole milk products, and fatty meats.[12] They are also not selecting the suggested quantities from each of the food groups, typically eating too few fruits, vegetables, whole grains, and milk products (see Figure 2-4).

An assessment tool, called the **Healthy Eating Index**, can be used to measure how well a diet meets the recommendations of the *Dietary Guidelines for Americans* and MyPyramid.[13] Various components of the diet are given scores that reflect the quantities consumed per 1000-kcalorie intake. For most components, higher intakes result in higher scores. For example, selecting at least 3 ounces of grains with at least half of them whole grains gives a score of 10 points, whereas selecting no grains gives a score of 0 points. For a few components, lower intakes provide higher scores. For example, less than 7 percent kcalories from saturated fat receives 10 points, but more than 15 percent gets 0 points. An assessment of recent nutrition surveys using the Healthy Eating Index reports that the American diet scores 58 out of a possible 100 points.[14]

Pyramid Shortcomings MyPyramid is not perfect and critics are quick to point out its flaws.[15] The first main criticism is that MyPyramid fails to convey enough information to help consumers make informed decisions about diet and health. MyPyramid contains no text and depends on its website to provide key information—which is wonderful for those who have Internet access and are willing to take the time to become familiar with its teachings. The second main criticism is that

FIGURE 2-4 Recommendations and Actual Intakes Compared

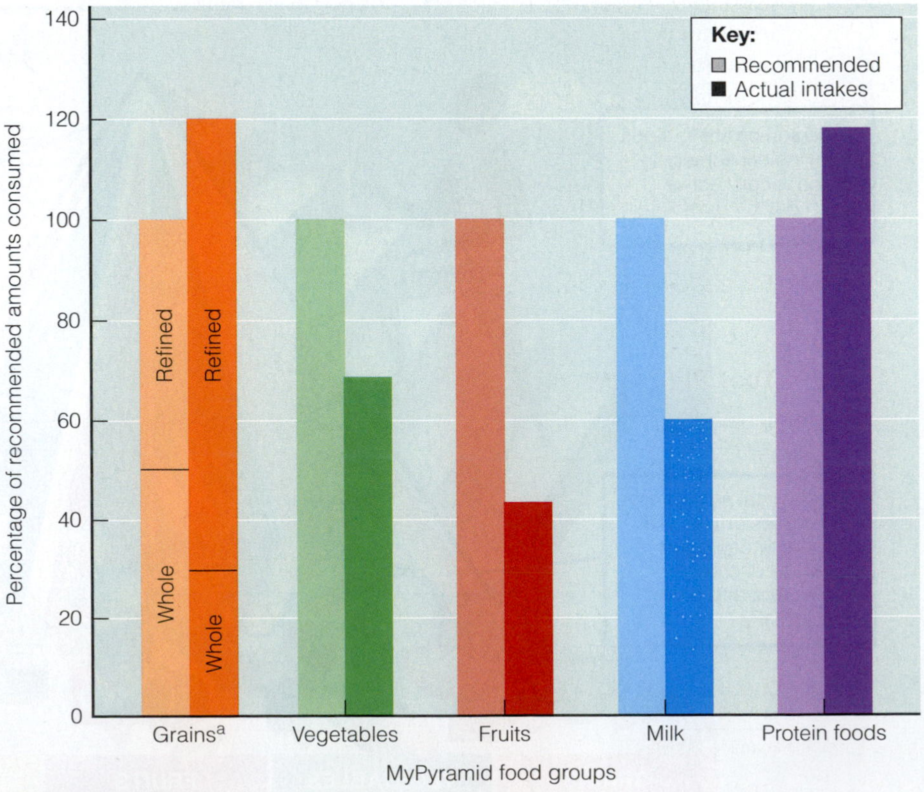

^a At least half of the grain selections should be whole grains.

Healthy Eating Index: a measure that assesses how well a diet meets the recommendations of the *Dietary Guidelines for Americans* and MyPyramid.

MyPyramid overemphasizes some foods and underemphasizes others, which may be detrimental to health. Critics assert that whole grains deserve more attention, that red meats differ from other protein sources and should be used sparingly, and that milk products offer no real benefits in preventing osteoporosis. Many of the upcoming chapters examine the links between diet and health.

Exchange Lists

Food group plans are particularly well suited to help a person achieve dietary adequacy, balance, and variety. **Exchange lists** provide additional help in achieving kcalorie control and moderation. Originally developed as a meal-planning guide for people with diabetes, exchange lists have proved useful for general diet planning as well.

Unlike the USDA Food Guide, which sorts foods primarily by their vitamin and mineral contents, the exchange system sorts foods according to their energy-nutrient contents. Consequently, foods do not always appear on the exchange list where you might first expect to find them. For example, cheeses are grouped with meats because, like meats, cheeses contribute energy from protein and fat but provide negligible carbohydrate. (In the USDA Food Guide presented earlier, cheeses are grouped with milk because they are milk products with similar calcium contents.)

For similar reasons, starchy vegetables such as corn, green peas, and potatoes are listed with grains on the starch list in the exchange system, rather than with the vegetables. Likewise, olives are not classed as a "fruit" as a botanist would claim; they are classified as a "fat" because their fat content makes them more similar to oil than to berries. Bacon and nuts are also on the fat list to remind users of their high fat content. These groupings highlight the characteristics of foods that are significant to energy intake. To learn more about this useful diet-planning tool, study Appendix G, which gives details of the exchange system used in the United States, and Appendix I, which provides details of *Beyond the Basics*, a similar diet-planning system used in Canada.

Putting the Plan into Action

Familiarizing yourself with each of the food groups is the first step in diet planning. Table 2-6 shows how to use the USDA Food Guide to plan a 2000-kcalorie diet. The amounts listed from each of the food groups (see the second column of the table) were taken from Table 2-2 (p. 42). The next step is to assign the food groups to meals (and snacks), as in the remaining columns of Table 2-6.

Now, a person can begin to fill in a plan with real foods to create a menu. For example, the breakfast calls for 1 ounce grain, ½ cup fruit, and 1 cup milk. A person might select a bowl of cereal with banana slices and milk:

1 cup cereal = 1 ounce grain

1 small banana = ½ cup fruit

1 cup fat-free milk = 1 cup milk

exchange lists: diet-planning tools that organize foods by their proportions of carbohydrate, fat, and protein. Foods on any single list can be used interchangeably.

TABLE 2-6 Diet Planning Using the USDA Food Guide

This diet plan is one of many possibilities. It follows the amounts of foods suggested for a 2000-kcalorie diet as shown in Table 2-2 (with a little less oil).

Food Group	Amounts	Breakfast	Lunch	Snack	Dinner	Snack
Fruits	2 c	½ c		½ c	1 c	
Vegetables	2½ c		1 c		1½ c	
Grains	6 oz	1 oz	2 oz	½ oz	2 oz	½ oz
Protein foods	5½ oz		2 oz		3½ oz	
Milk	3 c	1 c		1 c		1 c
Oils	6 tsp		1½ tsp		4 tsp	
Discretionary kcalorie allowance	258 kcal					

Or ½ bagel and a bowl of cantaloupe pieces topped with yogurt:

½ small bagel = 1 ounce grain

½ cup melon pieces = ½ cup fruit

1 cup fat-free plain yogurt = 1 cup milk

Then the person can continue to create a diet plan by creating menus for lunch, dinner, and snacks. The final plan might look like the one in Figure 2-5. With the addition of a small amount of oils, this sample diet plan provides about 1825 kcalories and adequate amounts of the essential nutrients.

As you can see, we all make countless food-related decisions daily—whether we have a plan or not. Following a plan, such as the USDA Food Guide, that incorporates health recommendations and diet-planning principles helps a person make wise decisions.

From Guidelines to Groceries

Dietary recommendations emphasize nutrient-rich foods such as whole grains, fruits, vegetables, lean meats, fish, poultry, and low-fat milk products. You can design such a diet for yourself, but how do you begin? Start with the foods you enjoy eating. Then try to make improvements, little by little. When shopping, think of the food groups and choose nutrient-dense foods within each group.

Be aware that many of the 50,000 food options available today are **processed foods** that have lost valuable nutrients and gained sugar, fat, and salt as they were transformed from farm-fresh foods to those found in the bags, boxes, and cans that line grocery-store shelves. Their value in the diet depends on the starting food and how it was prepared or processed. Sometimes these foods have been **fortified** to improve their nutrient contents.

Grains

When shopping for grain products, you will find them described as *refined, enriched,* or *whole grain.* These terms refer to the milling process and the making of grain products, and they have different nutrition implications (see Figure 2-6 on p. 50). **Refined** foods may have lost many nutrients during processing; **enriched** products may have had some nutrients added back; and **whole-grain**

When shopping for bread, look for the descriptive words *whole grain* or *whole wheat* and check the fiber contents on the Nutrition Facts panel of the label—the more fiber, the more likely the bread is a whole-grain product.

processed foods: foods that have been treated to change their physical, chemical, microbiological, or sensory properties.

fortified: the addition to a food of nutrients that were either not originally present or present in insignificant amounts. Fortification can be used to correct or prevent a widespread nutrient deficiency or to balance the total nutrient profile of a food.

refined: the process by which the coarse parts of a food are removed. When wheat is refined into flour, the bran, germ, and husk are removed, leaving only the endosperm.

enriched: the addition to a food of nutrients that were lost during processing so that the food will meet a specified standard.

whole grain: a grain that maintains the same relative proportions of starchy endosperm, germ, and bran as the original (all but the husk); not refined.

FIGURE 2-5 A Sample Diet Plan and Menu

This sample menu provides about 1825 kcalories and meets dietary recommendations to provide 45 to 65 percent of its kcalories from carbohydrate, 20 to 35 percent from fat, and 10 to 35 percent from protein. Some discretionary kcalories were spent on the fat in the low-fat cheese and in the sugar added to the graham crackers; about 175 discretionary kcalories remain available in this 2000-kcalorie diet plan.

© Polara Studios, Inc.

© Polara Studios, Inc.

© Polara Studios, Inc.

© Quest

© Quest

Amounts	SAMPLE MENU	Energy (kcal)
	Breakfast	
1 oz whole grains	1 c whole-grain cereal	108
1 c milk	1 c fat-free milk	83
½ c fruit	1 small banana (sliced)	105
	Lunch	
2 oz whole grains, 2 oz protein foods	1 turkey sandwich on whole-wheat roll	272
1½ tsp oils	1½ tbs low-fat mayonnaise	75
1 c vegetables	1 c vegetable juice	53
	Snack	
½ oz whole grains	4 whole-wheat, reduced-fat crackers	86
1 c milk	1½ oz low-fat cheddar cheese	74
½ c fruit	1 small apple	72
	Dinner	
½ c vegetables	1 c salad	8
1 oz protein foods	¼ c garbanzo beans	71
2 tsp oils	2 tbs oil-based salad dressing and olives	81
½ c vegetables, 2½ oz protein foods, 2 oz enriched grains	Spaghetti with meat sauce	425
½ c vegetables	½ c green beans	22
2 tsp oils	2 tsp soft margarine	67
1 c fruit	1 c strawberries	49
	Snack	
½ oz enriched grains	3 graham crackers	90
1 c milk	1 c fat-free milk	83

FIGURE 2-6 A Wheat Plant

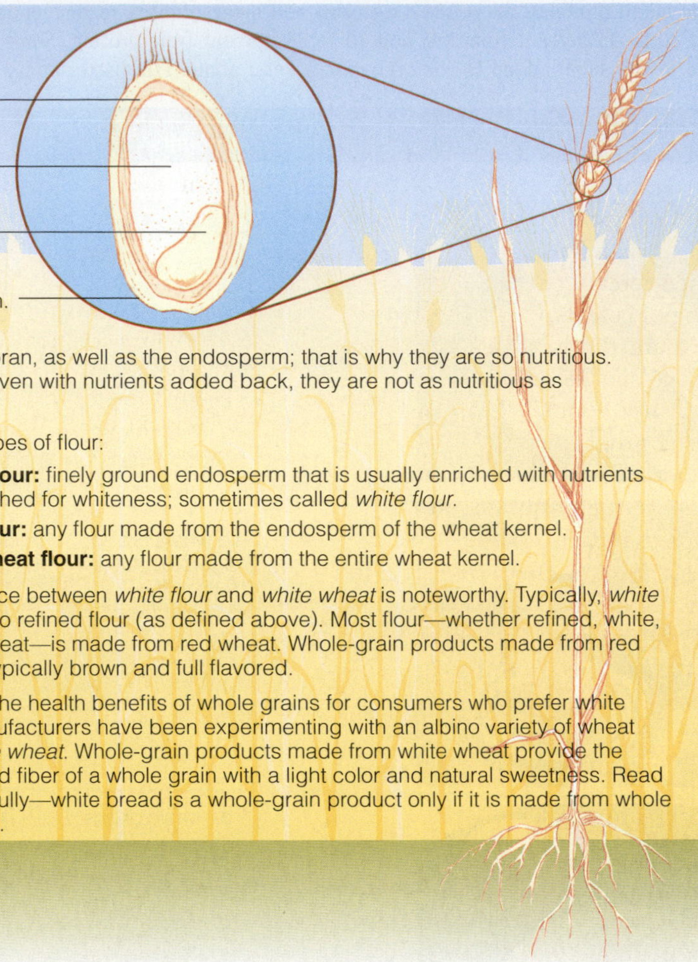

The protective coating of **bran** around the kernel of grain is rich in nutrients and fiber.
The **endosperm** contains starch and proteins.
The **germ** is the seed that grows into a wheat plant, so it is especially rich in vitamins and minerals to support new life.
The outer **husk** (or **chaff**) is the inedible part of a grain.

Whole-grain products contain much of the germ and bran, as well as the endosperm; that is why they are so nutritious. Refined grain products contain only the endosperm. Even with nutrients added back, they are not as nutritious as whole-grain products, as the next figure shows.

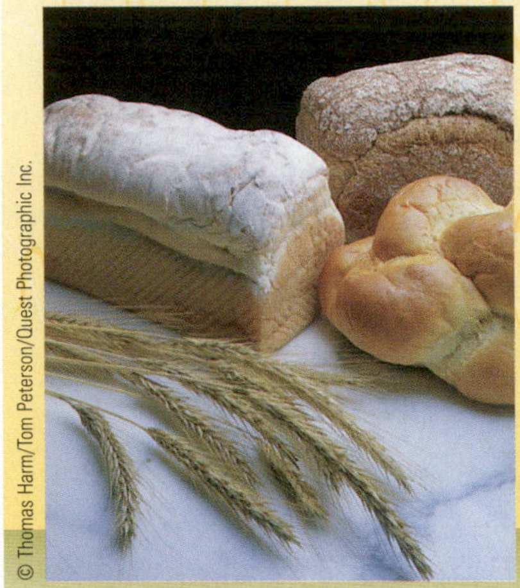

Common types of flour:

- **Refined flour:** finely ground endosperm that is usually enriched with nutrients and bleached for whiteness; sometimes called *white flour*.
- **Wheat flour:** any flour made from the endosperm of the wheat kernel.
- **Whole-wheat flour:** any flour made from the entire wheat kernel.

The difference between *white flour* and *white wheat* is noteworthy. Typically, *white flour* refers to refined flour (as defined above). Most flour—whether refined, white, or whole wheat—is made from red wheat. Whole-grain products made from red wheat are typically brown and full flavored.

To capture the health benefits of whole grains for consumers who prefer white bread, manufacturers have been experimenting with an albino variety of wheat called *white wheat*. Whole-grain products made from white wheat provide the nutrients and fiber of a whole grain with a light color and natural sweetness. Read labels carefully—white bread is a whole-grain product only if it is made from whole white wheat.

products may be rich in fiber and all the nutrients found in the original grain. As such, whole-grain products support good health and should account for at least half of the grains daily.[16] Adding more whole grains to the diet can be as easy as eating oatmeal for breakfast and popcorn for a snack or substituting brown rice for white rice and whole-wheat bread for white bread. To find whole-grain products, read food labels and select those that name a whole grain ♦ first in the ingredient list. Products described as "multi-grain," "stone-ground," or "100% wheat" are usually *not* whole-grain products. Brown color is also not a useful hint, but fiber content often is.

♦ Examples of whole grains:
- Amaranth
- Barley
- Buckwheat
- Bulgur
- Corn (and popcorn)
- Couscous
- Millet
- Oats (and oatmeal)
- Quinoa
- Rice (brown or wild)
- Whole rye
- Whole wheat

Dietary Guidelines for Americans 2010

Consume at least half of all grains as whole grains. Increase whole-grain intake by replacing refined grains with whole grains.

When it became a common practice to refine the wheat flour used for bread by milling it and throwing away the bran and the germ, consumers suffered a tragic loss of many nutrients. As a consequence, in the early 1940s Congress passed legislation requiring that all grain products that cross state lines be enriched with iron, thiamin, riboflavin, and niacin. In 1996, this legislation was amended to include folate, a vitamin considered essential in the prevention of some birth defects. Most grain products that have been refined, such as rice, wheat pastas like macaroni and spaghetti, and cereals (both cooked and ready-to-eat types), have subsequently been enriched. ♦ Food labels must specify that products have been enriched and include the enrichment nutrients in the ingredients list.

Enrichment doesn't make a slice of bread rich in these added nutrients, but people who eat several slices a day obtain significantly more of these nutrients than they would from unenriched bread. Even though the enrichment of flour helps to prevent deficiencies of these nutrients, it fails to compensate for losses of many other nutrients and fiber. As Figure 2-7 shows, whole-grain items still outshine the enriched ones. Only whole-grain flour contains all of the nutritive portions of the grain. Whole-grain products, such as brown rice and oatmeal, provide more nutrients and fiber and contain less salt and sugar than flavored, processed rice or sweetened cereals.

Speaking of cereals, ready-to-eat breakfast cereals are the most highly fortified foods on the market. Like an enriched food, a fortified food has had nutrients added during processing, but in a fortified food, the added nutrients may not have been present in the original product. (The terms *fortified* and *enriched* may be used interchangeably.[17]) Some breakfast cereals made from refined flour and fortified with high doses of vitamins and minerals are actually more like dietary supplements disguised as cereals than they are like whole grains. They may be nutritious—with respect to the nutrients added—but they still may fail to convey the full spectrum of nutrients that a whole-grain food or a mixture of such foods might provide. Still, fortified foods help people meet their vitamin and mineral needs.[18]

Vegetables Posters in the produce section of grocery stores encourage consumers to "eat five a day." Such efforts are part of a national educational campaign to increase fruit and vegetable consumption to five to nine servings every day (see Figure 2-8). To help consumers remember to eat a variety of fruits and vegetables, the campaign provides practical tips, such as selecting from each of five colors.

♦ Grain enrichment nutrients:
- Iron
- Thiamin
- Riboflavin
- Niacin
- Folate

FIGURE 2-7 Nutrients in Bread

Whole-grain bread is more nutritious than other breads, even enriched bread. For iron, thiamin, riboflavin, niacin, and folate, enriched bread provides about the same quantities as whole-grain bread and significantly more than unenriched bread. For fiber and the other nutrients (those shown here as well as those not shown), enriched bread provides less than whole-grain bread.

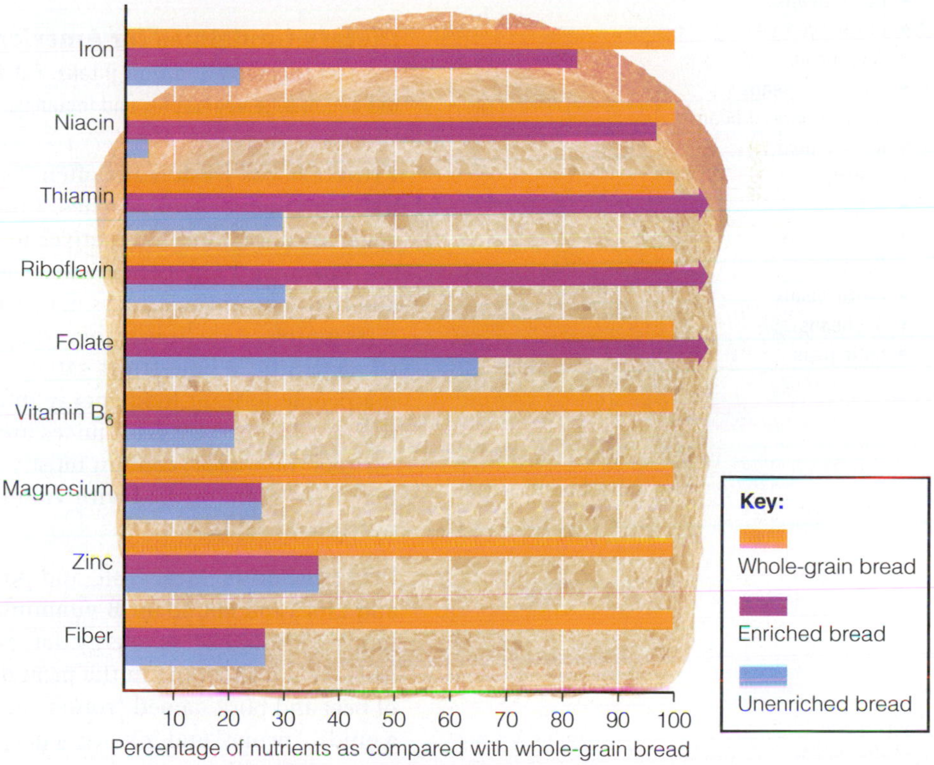

FIGURE 2-8 Fruits and Veggies—More Matters

Because "everyone benefits from eating more," the fruits and veggies matter campaign (www.fruitsandveggiesmatter.gov) encourages consumers to eat several servings of a variety of fruits and vegetables every day.

Combining legumes with foods from other food groups creates delicious meals.

Add rice to red beans for a hearty meal.

Enjoy a Greek salad topped with garbanzo beans for a little ethnic diversity.

A bit of meat and lots of spices turn kidney beans into chili con carne.

Choose fresh vegetables often, especially dark green leafy and yellow-orange vegetables like spinach, broccoli, and sweet potatoes. Cooked or raw, vegetables are good sources of vitamins, minerals, and fiber. Frozen and canned vegetables without added salt are acceptable alternatives to fresh. To control fat, energy, and sodium intakes, limit butter and salt on vegetables.

Choose often from the variety of legumes available. ♦ They are an economical, low-fat, as well as nutrient- and fiber-rich food choice.

♦ Legumes include a variety of beans and peas:
- Adzuki beans
- Black beans
- Black-eyed peas
- Fava beans
- Garbanzo beans
- Great northern beans
- Kidney beans
- Lentils
- Lima beans
- Navy beans
- Peanuts
- Pinto beans
- Soybeans
- Split peas

Dietary Guidelines for Americans 2010

Increase vegetable and fruit intake. Eat a variety of vegetables, especially dark-green and red and orange vegetables and legumes.

Fruit Choose fresh fruits often, especially citrus fruits and yellow-orange fruits like cantaloupes and peaches. Frozen, dried, and canned fruits without added sugar are acceptable alternatives to fresh. Fruits supply valuable vitamins, minerals, fibers, and phytochemicals. They add flavors, colors, and textures to meals, and their natural sweetness makes them enjoyable as snacks or desserts.

Fruit juices are healthy beverages but contain little dietary fiber compared with whole fruits. Whole fruits satisfy the appetite better than juices, thereby helping people to limit food energy intakes. For people who need extra food energy, though, 100 percent fruit juices are a good choice. Be aware that sweetened fruit "drinks" or "-ades" contain mostly water, sugar, and a little juice for flavor. Some may have been fortified with vitamin C or calcium but lack any other significant nutritional value.

Protein Foods Meat, fish, and poultry provide essential minerals, such as iron and zinc, and abundant B vitamins as well as protein. To buy and prepare these foods without excess energy, fat, and sodium takes a little knowledge and planning. When shopping in the meat department, choose fish, poultry, and lean cuts of beef and pork named "round" or "loin" (as in top round or pork tenderloin). As a guide, "prime" and "choice" cuts generally have more fat than "select" cuts. Restaurants usually serve prime cuts. Ground beef, even "lean" ground beef, derives most of its food energy from fat. Have the butcher trim and grind a lean round steak instead. Alternatively, **textured vegetable protein** can be used instead of ground beef in a casserole, spaghetti sauce, or chili, saving fat kcalories.

Weigh meat after it is cooked and the bones and fat are removed. In general, 4 ounces of raw meat is equal to about 3 ounces of cooked meat. Some examples of 3-ounce portions of meat include 1 medium pork chop, ½ chicken breast, or 1 steak or hamburger about the size of a deck of cards. To keep fat intake moderate, bake, roast, broil, grill, or braise meats (but do not fry them in fat); remove the skin from poultry after cooking; trim visible fat before cooking; and drain fat after cooking. Chapter 5 offers many additional strategies for moderating fat intake.

Dietary Guidelines for Americans 2010

Choose a variety of protein foods, which include seafood, lean meat and poultry, eggs, beans and peas, soy products, and unsalted nuts and seeds. Increase the amount and variety of seafood consumed by choosing seafood in place of some meat and poultry.

textured vegetable protein: processed soybean protein used in vegetarian products such as soy burgers.

Milk Shoppers find a variety of fortified foods in the dairy case. Examples are milk, to which vitamins A and D have been added, and soy milk, ♦ to which calcium, vitamin D, and vitamin B_{12} have been added. In addition, shoppers may find **imitation foods** (such as cheese products), **food substitutes** (such as egg substitutes), and functional foods ♦ (such as margarine with added plant sterols). As food technology advances, many such foods offer alternatives to traditional choices that may help people who want to reduce their fat and cholesterol intakes. Chapter 5 provides other examples.

When shopping, choose fat-free ♦ or low-fat milk, yogurt, and cheeses. Such selections help consumers meet their vitamin and mineral needs within their energy and fat allowances. Milk products are important sources of calcium, but can provide too much sodium and fat if not selected with care.

♦ Be aware that not all soy milks have been fortified. Read labels carefully.

♦ **Functional foods** contain physiologically active compounds that provide health benefits beyond basic nutrition.

♦ Milk descriptions:
- Fat-free milk = nonfat, skim, zero-fat, or no-fat
- Low-fat milk = 1% milk
- Reduced-fat milk = 2% milk or less-fat

Dietary Guidelines for Americans 2010

Increase intake of fat-free or low-fat milk and milk products, such as milk, yogurt, cheese, or fortified soy beverages.

IN SUMMARY **Food group plans such as the USDA Food Guide help consumers select the types and amounts of foods to provide adequacy, balance, and variety in the diet. They make it easier to plan a diet that includes a balance of grains, vegetables, fruits, protein foods, and milk products. In making any food choice, remember to view the food in the context of your total diet. The combination of many different foods provides the abundance of nutrients that is so essential to a healthy diet.**

Food Labels

Many consumers read food labels to help them make healthy choices.[19] Food labels appear on virtually all processed foods, and posters or brochures provide similar nutrition information for fresh meats, fruits, and vegetables (see Figure 2-9 on p. 54). A few foods need not carry nutrition labels: those contributing few nutrients, such as plain coffee, tea, and spices; those produced by small businesses; and those prepared and sold in the same establishment. Producers of some of these items, however, voluntarily use labels. Even markets selling nonpackaged items voluntarily present nutrient information, either in brochures or on signs posted at the point of purchase. Restaurants need not supply complete nutrition information for menu items unless claims such as "low fat" or "heart healthy" have been made. When ordering such items, keep in mind that restaurants tend to serve extra-large portions—two to three times standard serving sizes. A "low-fat" ice cream, for example, may have only 3 grams of fat per ½ cup, but you may be served 2 cups for a total of 12 grams of fat and all their accompanying kcalories.

The Ingredient List

All packaged foods must list *all* ingredients—including additives used to preserve or enhance foods, such as vitamins and minerals added to enrich or fortify products. The ingredients are listed on the label in descending order of predominance by weight. Knowing that the first ingredient predominates by weight, consumers can glean much information. Compare these products, for example:

- A beverage powder that contains "sugar, citric acid, natural flavors . . ." versus a juice that contains "water, tomato concentrate, concentrated juices of carrots, celery. . . ."
- A cereal that contains "puffed milled corn, sugar, corn syrup, molasses, salt . . ." versus one that contains "100 percent rolled oats."
- A canned fruit that contains "sugar, apples, water" versus one that contains simply "apples, water."

In each of these comparisons, consumers can see that the second product is more nutrient dense.

imitation foods: foods that substitute for and resemble another food, but are nutritionally inferior to it with respect to vitamin, mineral, or protein content. If the substitute is not inferior to the food it resembles and if its name provides an accurate description of the product, it need not be labeled "imitation."

food substitutes: foods that are designed to replace other foods.

FIGURE 2-9 Example of a Food Label

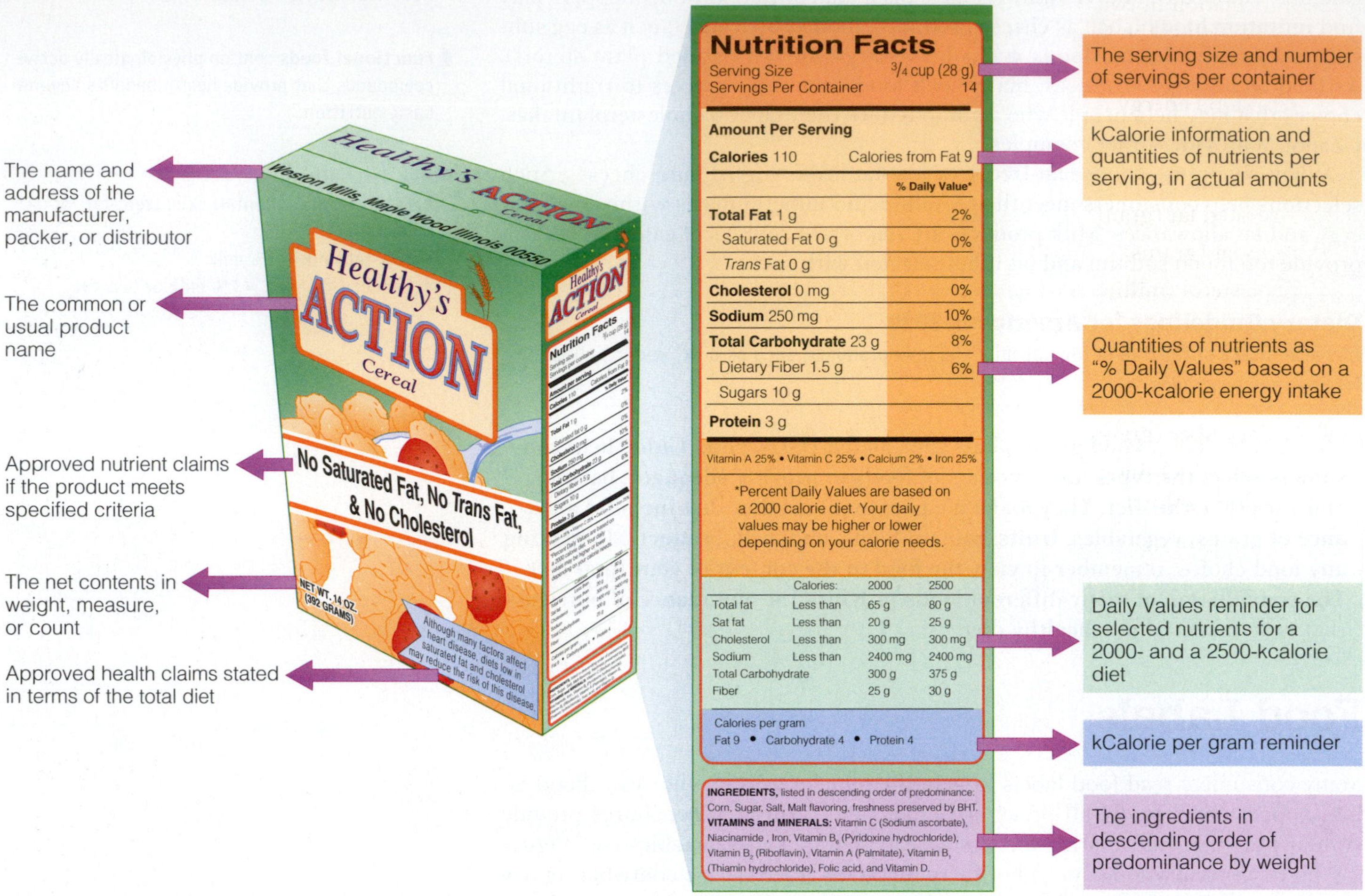

TABLE 2-7 Household and Metric Measures

- 1 teaspoon (tsp) = 5 milliliters (mL)
- 1 tablespoon (tbs) = 15 mL
- 1 cup (c) = 240 mL
- 1 fluid ounce (fl oz) = 30 mL
- 1 ounce (oz) = 28 grams (g)

NOTE: The Aids to Calculation section at the back of the book provides additional weights and measures.

Serving Sizes Because labels present nutrient information based on one serving, they must identify the size of the serving. The Food and Drug Administration (FDA) has established specific serving sizes for various foods and requires that all labels for a given product use the same serving size. For example, the standard serving size for all ice creams is ½ cup and for all beverages, 8 fluid ounces. This facilitates comparison shopping. Consumers can see at a glance which brand has more or fewer kcalories or grams of fat, for example. Standard serving sizes are expressed in both common household measures, such as cups, and metric measures, such as milliliters, to accommodate users of both types of measures (see Table 2-7).

When examining the nutrition facts on a food label, consumers need to compare the serving size on the label with how much they actually eat and adjust their calculations accordingly. For example, if the serving size is four cookies and you eat only two, then you need to cut the nutrient and kcalorie values in half; similarly, if you eat eight cookies, then you need to double the values. Notice, too, that small bags or individually wrapped items, such as chips or candy bars, may contain more than a single serving. The total number of servings per container is listed just below the serving size.

Be aware that serving sizes on food labels are not always the same as those of the USDA Food Guide. For example, a serving of rice on a food label is 1 cup, whereas in the USDA Food Guide it is ½ cup. Unfortunately, this discrepancy, coupled with each person's own perception (oftentimes misperception) of standard serving sizes, sometimes creates confusion for consumers trying to follow recommendations.

Nutrition Facts In addition to the serving size and the servings per container, the FDA requires that the Nutrition Facts panel on food labels present nutrient information in two ways—in quantities (such as grams) and as percentages of standards called the **Daily Values.** The Nutrition Facts panel must provide the nutrient amount, **percent Daily Value**, or both for the following:

- Total food energy (kcalories)
- Food energy from fat (kcalories)
- Total fat (grams and percent Daily Value)
- Saturated fat (grams and percent Daily Value)
- *Trans* fat (grams)
- Cholesterol (milligrams and percent Daily Value)
- Sodium (milligrams and percent Daily Value)
- Total carbohydrate, which includes starch, sugar, and fiber (grams and percent Daily Value)
- Dietary fiber (grams and percent Daily Value)
- Sugars, which includes both those naturally present in and those added to the food (grams)
- Protein (grams)

The labels must also present nutrient content information as a percent Daily Value for the following vitamins and minerals:

- Vitamin A
- Vitamin C
- Iron
- Calcium

Consumers read food labels to learn about the nutrient contents of a food or to compare similar foods.

The Daily Values Table 2-8 presents the Daily Value standards for nutrients that are required to provide this information. Food labels list the amount of a nutrient in a product as a percentage of its Daily Value. Comparing nutrient amounts against the Daily Values helps make the numbers more meaningful to consumers. A person reading a food label might wonder, for example, whether 1 milligram of iron or calcium is a little or a lot. As Table 2-8 shows, the Daily Value for iron is 18 milligrams, so 1 milligram of iron is enough to notice—it is more than 5 percent, and that is what the food label will say. But because the Daily Value for calcium on food labels is 1000 milligrams, 1 milligram of calcium is insignificant, and the food label will read "0%."

The Daily Values reflect dietary recommendations for nutrients and dietary components that have important relationships with health. The "% Daily Value" column on a label provides a ballpark estimate of how individual foods contribute to the total diet. It compares key nutrients in a serving of food with the goals of a person consuming 2000 kcalories per day. A 2000-kcalorie diet is considered about right for sedentary younger women, active older women, and sedentary older men. Young children and sedentary older women may need fewer kcalories. Most labels list, at the bottom, Daily Values for both a 2000-kcalorie and a 2500-kcalorie diet, but the "% Daily Value" column on all labels applies only to a 2000-kcalorie diet. A 2500-kcalorie diet is considered about right for many men, teenage boys, and active younger women. People who are exceptionally active may have still higher energy needs. Labels may also provide a reminder of the kcalories in a gram of carbohydrate, fat, and protein just below the Daily Value information (review Figure 2-9).

Daily Values (DV): reference values developed by the FDA specifically for use on food labels.

percent Daily Value (%DV): the percentage of a Daily Value recommendation found in a specified serving of food for key nutrients based on a 2000-kcalorie diet.

TABLE 2-8 Daily Values for Food Labels

Food labels must present the "% Daily Value" for these nutrients.

Food Component	Daily Value	Calculation Factors
Fat	65 g	30% of kcalories
Saturated fat	20 g	10% of kcalories
Cholesterol	300 mg	—
Carbohydrate (total)	300 g	60% of kcalories
Fiber	25 g	11.5 g per 1000 kcalories
Protein	50 g	10% of kcalories
Sodium	2400 mg	—
Potassium	3500 mg	—
Vitamin C	60 mg	—
Vitamin A	1500 µg	—
Calcium	1000 mg	—
Iron	18 mg	—

NOTE: Daily Values were established for adults and children more than 4 years old. The values for energy-yielding nutrients are based on 2000 kcalories a day. For fiber, the Daily Value was rounded up from 23.

HOW TO Calculate Personal Daily Values

The Daily Values on food labels are designed for a 2000-kcalorie intake, but you can calculate a personal set of Daily Values based on your energy allowance. Consider a 1500-kcalorie intake, for example. To calculate a daily goal for fat, multiply energy intake by 30 percent:

$$1500 \text{ kcal} \times 0.30 \text{ kcal from fat} = 450 \text{ kcal from fat}$$

The "kcalories from fat" are listed on food labels, so you can add all the "kcalories from fat" values for a day, using 450 as an upper limit. A person who prefers to count grams of fat can divide this 450 kcalories from fat by 9 kcalories per gram to determine the goal in grams:

$$450 \text{ kcal from fat} \div 9 \text{ kcal/g} = 50 \text{ g fat}$$

Alternatively, a person can calculate that 1500 kcalories is 75 percent of the 2000-kcalorie intake used for Daily Values:

$$1500 \text{ kcal} \div 2000 \text{ kcal} = 0.75$$
$$0.75 \times 100 = 75\%$$

Then, instead of trying to achieve 100 percent of the Daily Value, a person consuming 1500 kcalories will aim for 75 percent. Similarly, a person consuming 2800 kcalories would aim for 140 percent:

$$2800 \text{ kcal} \div 2000 \text{ kcal} = 1.40 \text{ or } 140\%$$

Table 2-8 includes a calculation column that can help you estimate your personal daily value for several nutrients.

For additional practice, log on to **www.cengagebrain.com** and search for ISBN 084006845X.

TRY IT Calculate the Daily Values for a 1800-kcalorie diet and revise the Daily Value percentages on the cereal label found on p. 54.

People who consume 2000 kcalories a day can simply add up all of the "% Daily Values" for a particular nutrient to see if their diet for the day fits recommendations. People who require more or less than 2000 kcalories daily must do some calculations to see how foods compare with their personal nutrition goals. They can use the calculation column in Table 2-8 or the suggestions presented in the accompanying "How To" feature.

Daily Values help consumers see easily whether a food contributes "a little" or "a lot" of a nutrient. ♦ For example, the "% Daily Value" column on a package of frozen macaroni and cheese may say 20 percent for fat. This tells the consumer that each serving of this food contains about 20 percent of the day's allotted 65 grams of fat. A person consuming 2000 kcalories a day could simply keep track of the percentages of Daily Values from foods eaten during a day and try not to exceed 100 percent. Be aware that for some nutrients (such as fat and sodium) you will want to select foods with a low "% Daily Value" and for others (such as calcium and fiber) you will want a high "% Daily Value." To determine whether a particular food is a wise choice, a consumer needs to consider its place in the diet among all the other foods eaten during the day.

♦ % Daily Values:
- ≥20% = high or excellent source
- 10–19% = good source
- 5% = low source

Daily Values also make it easy to compare foods. For example, a consumer might discover that frozen macaroni and cheese has a Daily Value for fat of 20 percent, whereas macaroni and cheese prepared from a boxed mix has a Daily Value of 15 percent. By comparing labels, consumers who are concerned about their fat intakes can make informed decisions.

The Daily Values used on labels are based in part on values from the 1968 Recommended Dietary Allowances. Since 1997, Dietary Reference Intakes that reflect recent scientific research on diet and health have been released. Efforts to update the Daily Values based on these current recommendations and to make labels more effective and easier to understand are under way.[20]

Nutrient Claims Have you noticed phrases such as "good source of fiber" on a box of cereal or "rich in calcium" on a package of cheese? These and other **nutrient claims** may be used on labels as long as they meet FDA definitions, which include the conditions under which each term can be used. For example, in addition to having less than 2 milligrams of cholesterol, a "cholesterol-free" product may not contain more than 2 grams of saturated fat and *trans* fat combined per serving. The accompanying glossary defines nutrient terms on food labels, including criteria for foods described as "low," "reduced," and "free." When nutrients have been added to enriched or fortified products, they must appear in the ingredients list.

Some descriptions *imply* that a food contains, or does not contain, a nutrient. Implied claims are prohibited unless they meet specified criteria. For example, a claim that a product "contains no oil" *implies* that the food contains no fat. If the product is truly fat-free, then it may make the no-oil claim, but if it contains another source of fat, such as butter, it may not.

Health Claims Until 2003, the FDA held manufacturers to the highest standards of scientific evidence before approving **health claims** on food labels.[21] Consumers reading "Diets low in sodium may reduce the risk of high blood pressure," for example, knew that the FDA had examined enough scientific evidence to establish a clear link between diet and health. Such reliable health claims make up the FDA's "A" list (see Table 2-9). The FDA refers to these health claims as

nutrient claims: statements that characterize the quantity of a nutrient in a food.

health claims: statements that characterize the relationship between a nutrient or other substance in a food and a disease or health-related condition.

GLOSSARY OF TERMS ON FOOD LABELS

GENERAL TERMS

free: "nutritionally trivial" and unlikely to have a physiological consequence; synonyms include *without, no,* and *zero.* A food that does not contain a nutrient naturally may make such a claim, but only as it applies to all similar foods (for example, "applesauce, a fat-free food").

good source of: the product provides between 10 and 19 percent of the Daily Value for a given nutrient per serving.

healthy: a food that is low in fat, saturated fat, cholesterol, and sodium and that contains at least 10 percent of the Daily Values for vitamin A, vitamin C, iron, calcium, protein, or fiber.

high: 20 percent or more of the Daily Value for a given nutrient per serving; synonyms include *rich in* or *excellent source.*

less: at least 25 percent less of a given nutrient or kcalories than the comparison food (see individual nutrients); synonyms include *fewer* and *reduced.*

light or **lite:** one-third fewer kcalories than the comparison food; 50 percent or less of the fat or sodium than the comparison food; any use of the term other than as defined must specify what it is referring to (for example, "light in color" or "light in texture").

low: an amount that would allow frequent consumption of a food without exceeding the Daily Value for the nutrient. A food that is naturally low in a nutrient may make such a claim, but only as it applies to all similar foods (for example, "fresh cauliflower, a low-sodium food"); synonyms include *little, few,* and *low source of.*

more: at least 10 percent more of the Daily Value for a given nutrient than the comparison food; synonyms include *added* and *extra.*

organic: on food labels, that at least 95 percent of the product's ingredients have been grown and processsed according to USDA regulations defining the use of fertilizers, herbicides, insecticides, fungicides, preservatives, and other chemical ingredients.

ENERGY

kcalorie-free: fewer than 5 kcalories per serving.

low kcalorie: 40 kcalories or less per serving.

reduced kcalorie: at least 25 percent fewer kcalories per serving than the comparison food.

FAT AND CHOLESTEROL[a]

percent fat-free: may be used only if the product meets the definition of *low fat* or *fat-free* and must reflect the amount of fat in 100 grams (for example, a food that contains 2.5 grams of fat per 50 grams can claim to be "95 percent fat free").

fat-free: less than 0.5 gram of fat per serving (and no added fat or oil); synonyms include *zero-fat, no-fat,* and *nonfat.*

low fat: 3 grams or less fat per serving.

less fat: 25 percent or less fat than the comparison food.

saturated fat-free: less than 0.5 gram of saturated fat and 0.5 gram of *trans* fat per serving.

low saturated fat: 1 gram or less saturated fat and less than 0.5 gram of *trans* fat per serving.

less saturated fat: 25 percent or less saturated fat and *trans* fat combined than the comparison food.

***trans* fat-free:** less than 0.5 gram of *trans* fat and less than 0.5 gram of saturated fat per serving.

cholesterol-free: less than 2 milligrams cholesterol per serving and 2 grams or less saturated fat and *trans* fat combined per serving.

low cholesterol: 20 milligrams or less cholesterol per serving and 2 grams or less saturated fat and *trans* fat combined per serving.

less cholesterol: 25 percent or less cholesterol than the comparison food (reflecting a reduction of at least 20 milligrams per serving), and 2 grams or less saturated fat and *trans* fat combined per serving.

extra lean: less than 5 grams of fat, 2 grams of saturated fat and *trans* fat combined, and 95 milligrams of cholesterol per serving and per 100 grams of meat, poultry, and seafood.

lean: less than 10 grams of fat, 4.5 grams of saturated fat and *trans* fat combined, and 95 milligrams of cholesterol per serving and per 100 grams of meat, poultry, and seafood. For mixed dishes such as burritos and sandwiches, less than 8 grams of fat, 3.5 grams of saturated fat, and 80 milligrams of cholesterol per reference amount customarily consumed.

CARBOHYDRATES: FIBER AND SUGAR

high fiber: 5 grams or more fiber per serving. A high-fiber claim made on a food that contains more than 3 grams fat per serving and per 100 grams of food must also declare total fat.

sugar-free: less than 0.5 gram of sugar per serving.

SODIUM

sodium-free and **salt-free:** less than 5 milligrams of sodium per serving.

low sodium: 140 milligrams or less per serving.

very low sodium: 35 milligrams or less per serving.

[a]Foods containing more than 13 grams total fat per serving or per 50 grams of food must indicate those contents immediately after a cholesterol claim. As you can see, all cholesterol claims are prohibited when the food contains more than 2 grams saturated fat and *trans* fat combined per serving.

TABLE 2-9 Reliable Health Claims on Food Labels—The "A" List

- Diets adequate in calcium may reduce the risk of osteoporosis.
- Diets low in sodium may reduce the risk of high blood pressure.
- Diets low in saturated fat and cholesterol, and as low as possible in *trans* fat, may reduce the risk of heart disease.
- Diets low in total fat may reduce the risk of some cancers.
- Low-fat diets rich in fiber-containing grain products, fruits, and vegetables may reduce the risk of some cancers.
- Diets low in saturated fat and cholesterol and rich in fruits, vegetables, and grain products that contain fiber, particularly soluble fiber, may reduce the risk of heart disease.
- Low-fat diets rich in fruits and vegetables may reduce the risk of some cancers.
- Diets adequate in folate may reduce a woman's risk of having a child with a neural tube defect.
- Sugar alcohols do not promote tooth decay.
- Diets low in saturated fat and cholesterol that include soluble fiber from foods may reduce the risk of heart disease.
- Diets low in saturated fat and cholesterol that include 25 grams of soy protein may reduce the risk of heart disease.
- Diets rich in whole grain foods and other plant foods and low in total fat, saturated fat, and cholesterol may reduce the risk of heart disease and some cancers.
- Diets low in saturated fat and cholesterol that include 3.4 grams of plant stanol esters may reduce the risk of heart disease.
- Diets containing foods that are rich in potassium and low in sodium may reduce the risk of high blood pressure and stroke.
- Drinking fluoridated water may reduce the risk of tooth decay.

TABLE 2-10 The FDA's Health Claims Report Card

Grade	Level of Confidence in Health Claim	Required Label Disclaimers
A	High: Significant scientific agreement	These health claims do not require disclaimers; see Table 2-9 for examples.
B	Moderate: Evidence is supportive but not conclusive	"[Health claim.] Although there is scientific evidence supporting this claim, the evidence is not conclusive."
C	Low: Evidence is limited and not conclusive	"Some scientific evidence suggests [health claim]. However, FDA has determined that this evidence is limited and not conclusive."
D	Very low: Little scientific evidence supporting this claim	"Very limited and preliminary scientific research suggests [health claim]. FDA concludes that there is little scientific evidence supporting this claim."

"unqualified"—not that they lack the necessary qualifications, but that they can stand alone without further explanation or qualification.

These reliable health claims still appear on some food labels, but finding them may be difficult now that the FDA has created three additional categories of claims based on scientific evidence that is less conclusive (see Table 2-10). These categories were added after a court ruled: "Holding only the highest scientific standard for claims interferes with commercial free speech." Food manufacturers had argued that they should be allowed to inform consumers about possible benefits based on less than clear and convincing evidence. The FDA must allow manufacturers to provide information about nutrients and foods that show preliminary promise in preventing disease. These health claims are "qualified"—not that they meet the necessary qualifications, but that they require a qualifying explanation. For example, "Very limited and preliminary research suggests that eating one-half to one cup of tomatoes and/or tomato sauce a week may reduce the risk of prostate cancer. FDA concludes that there is little scientific evidence supporting the claim." Consumer groups argue that such information is confusing. Even with required disclaimers for health claims graded "B," "C," or "D," distinguishing "A" claims from others is difficult, as the next section shows. (Health claims on supplement labels are presented in Highlight 10.)

Structure-Function Claims

Unlike health claims, which require food manufacturers to collect scientific evidence and petition the FDA, **structure-function claims** can be made without any FDA approval. Product labels can claim to "slow aging," "improve memory," and "build strong bones" without any proof. The only criterion for a structure-function claim is that it must not mention a disease or symptom. Unfortunately, structure-function claims can be deceptively similar to health claims. Consider these statements:

- "May reduce the risk of heart disease."
- "Promotes a healthy heart."

Most consumers do not distinguish between these two types of claims.[22] In the statements above, for example, the first is a health claim that requires FDA approval and the second is an unproven, but legal, structure-function claim. Table 2-11 lists examples of structure-function claims.

TABLE 2-11 Examples of Structure-Function Claims

- Builds strong bones
- Promotes relaxation
- Improves memory
- Boosts the immune system
- Supports heart health
- Defends health
- Slows aging
- Guards against colds
- Lifts spirits

NOTE: Structure-function claims cannot make statements about diseases. See Table 2-9 for examples of reliable health claims.

structure-function claims: statements that characterize the relationship between a nutrient or other substance in a food and its role in the body.

Consumer Education

Because labels are valuable only if people know how to use them, the FDA has designed several programs to educate consumers. Consumers who understand how to read labels are best able to apply the information to achieve and maintain healthful dietary practices.

Table 2-12 shows how the messages from the *Dietary Guidelines*, the USDA Food Guide, and food labels coordinate with one another. To promote healthy eating and physical activity, the "Healthier US Initiative" coordinates the efforts of national educational programs developed by government agencies.[23] The mission of this

TABLE 2-12 From Guidelines to Groceries

Dietary Guidelines	USDA Food Guide/MyPyramid	Food Labels
Balancing kcalories to manage weight	Enjoy your food, but eat less. Select the recommended amounts from each food group at the energy level appropriate for your energy needs; meet, but do not exceed, energy needs. Limit foods and beverages with solid fats and added sugars. Use appropriate portion sizes; avoid oversized portions. Increase physical activity and reduce time spent in sedentary behaviors.	Read the Nutrition Facts to see how many kcalories are in a serving and the number of servings that are in a package. Look for foods that describe their kcalorie contents as *free, low, reduced, light,* or *less*.
Foods and food components to reduce	Choose foods within each group that are low in salt or sodium. Choose foods within each group that are lean, low fat, or fat free and have little solid fat (sources of saturated and *trans* fats); use unsaturated oils instead of solid fats whenever possible. Choose foods and beverages within each group that have little added sugars; drink water instead of sugary beverages. If alcohol is consumed by adults, use in moderation (no more than one drink a day for women and two drinks a day for men).	Read the Nutrition Facts to see how much sodium, saturated fat, *trans* fat, and cholesterol is in a serving of food. Look for foods that describe their salt and sodium contents as *free, low,* or *reduced*; foods that describe their fat, saturated fat, *trans* fat, and cholesterol contents as *free, less, low, light, reduced, lean,* or *extra lean*; foods that describe their sugar contents as *free* or *reduced*. Look for foods that provide no more than 5 percent of the Daily Value for sodium, fat, saturated fat, and cholesterol. A food may be high in solid fats if its ingredients list begins with or contains several of the following: *beef fat (tallow, suet), butter, chicken fat, coconut oil, cream, hydrogenated oils, palm kernel oil, palm oil, partially hydrogenated oils, pork fat (lard), shortening,* or *stick margarine*. A food most likely contains *trans* fats if its ingredients list includes: *partially hydrogenated oils*. A food may be high in added sugars if its ingredients list begins with or contains several of the following: *brown sugar, confectioner's powdered sugar, corn syrup, dextrin, fructose, high-fructose corn syrup, honey, invert sugar, lactose, malt syrup, maltose, molasses, nectars, sucrose, sugar, syrup*. *Light* beverages contain fewer kcalories and less alcohol than regular versions.
Foods and nutrients to increase	Make half your plate fruits and vegetables. Choose a variety of vegetables from all five subgroups (dark green, red and orange, legumes, starchy vegetables, and other vegetables) several times a week. Choose a variety of fruits; consume whole or cut-up fruits more often than fruit juice. Choose potassium-rich foods such as fruits and vegetables often. Choose fiber-rich fruits, vegetables, and whole grains often. Choose whole grains; make at least half of the grain selections whole grains by replacing refined grains with whole grains whenever possible. Choose fat-free or low-fat milk and milk products. Choose a variety of protein foods; increase the amount and variety of seafood by choosing seafood in place of some meat and poultry.	Look for foods that describe their fiber, calcium, potassium, and vitamin D contents as *good, high,* or *excellent*. Look for foods that provide at least 10 percent of the Daily Value for fiber, calcium, potassium, and vitamin D from a variety of sources. A food may be a good source of whole grains if its ingredients list begins with or contains several of the following: *barley, brown rice, buckwheat, bulgur, corn, millet, oatmeal, popcorn, quinoa, rolled oats, rye, sorghum, triticale, whole wheat, wild rice*.
Building healthy eating patterns	Select nutrient-dense foods and beverages within and among the food groups. Be food safe.	Look for foods that describe their vitamin, mineral, or fiber contents as a *good source* or *high*. Follow the *safe handling instructions* on packages of meat and other safety instructions, such as *keep refrigerated*, on packages of perishable foods.

initiative is to deliver simple messages that will motivate consumers to make small changes in their eating and physical activity habits to yield big rewards.

IN SUMMARY **Food labels provide consumers with information they need to select foods that will help them meet their nutrition and health goals. When labels contain relevant information presented in a standardized, easy-to-read format, consumers are well prepared to plan and create healthful diets.**

Nutrition Portfolio

The secret to making healthy food choices is learning to incorporate the *Dietary Guidelines for Americans* and the USDA Food Guide into your decision-making process.

Go to Diet Analysis Plus and choose one of the days on which you have tracked your diet for the entire day. Choose the MyPyramid Report and, looking at it, record in your journal the answers to the following:

- How do the foods you consumed on the day you have chosen stack up with the daily goals (the percentages) in the MyPyramid breakdown? Which food groups are over- or under-represented?
- Think about your choices within each food group for the day you recorded. Are they typical of the foods you choose from day to day? Are there simple and realistic ways to enhance the variety in your diet?
- Write yourself a letter describing the dietary changes you can make to improve your chances of enjoying good health.

Diet Analysis PLUS

To complete this exercise, go to your Diet Analysis Plus at www.cengage.com/sso.

Nutrition on the Net

For further study of topics covered in this chapter, log on to **www.cengagebrain.com** and search for ISBN 084006845X.

- Learn more about the *Dietary Guidelines for Americans:* **www.dietaryguidelines.gov**
- Find Canadian information on nutrition guidelines and food labels at: **www.hc-sc.gc.ca**
- Learn more about the USDA Food Guide and MyPyramid: **www.mypyramid.gov**
- Visit the USDA Food Guide section (including its ethnic/cultural pyramids) of the U.S. Department of Agriculture: **www.nal.usda.gov/fnic**
- Visit the Traditional Diet Pyramids for various ethnic groups at Oldways Preservation and Exchange Trust: **www.oldwayspt.org**
- Learn more about food labeling from the Food and Drug Administration: **www.fda.gov/Food/default.htm**
- Search for "food labels" at the International Food Information Council: **www.ific.org**
- Learn more about the Healthy Eating Index: **www.cnpp.usda.gov**
- Get healthy eating tips from the Fruits and Veggies Matter campaign: **www.fruitsandveggiesmatter.gov**

References

1. Practice paper of the American Dietetic Association: Nutrient density: Meeting nutrient goals within calorie needs, *Journal of the American Dietetic Association* 107 (2007): 860–869.
2. A. Drewnowski and V. Fulgoni III, Nutrient profiling of foods: Creating a nutrient-rich food index, *Nutrition Reviews* 66 (2008): 23–39.
3. N. Darmon and coauthors, Nutrient profiles discriminate between foods according to their contribution to nutritionally adequate diets: A validation study using linear programming and the SAIN,LIM system, *American Journal of Clinical Nutrition* 89 (2009): 1227–1236; E. Kennedy, Food rating systems, diet quality, and health, *Nutrition Reviews* 66 (2008): 21–22.
4. S. P. Murphy and coauthors, Simple measures of dietary variety are associated with improved dietary quality, *Journal of the American Dietetic Association* 106 (2006): 425–429.
5. U.S. Department of Agriculture and U.S. Department of Health and Human Services, *Dietary Guidelines for Americans, 2010*, www.dietaryguidelines.gov.
6. U.S. Department of Health and Human Services, *2008 Physical Activity Guidelines for Americans*, available at www.health.gov/paguidelines; U.S. Department of Agriculture and U.S. Department of Health and Human Services, *Dietary Guidelines for Americans, 2010*, www.dietaryguidelines.gov.
7. www.pyramid.gov/pyramid/dry_beans_peas_table.html, accessed March 3, 2009.
8. X. Gao and coauthors, The 2005 USDA Food Guide Pyramid is associated with more adequate nutrient intakes within energy constraints than the 1992 pyramid, *The Journal of Nutrition* 136 (2006): 1341–1346.
9. Position of the American Dietetic Association: Vegetarian diets, *Journal of the American Dietetic Association* 109 (2009): 1266–1282.
10. S. M. Krebs-Smith and P. Kris-Etherton, How does MyPyramid compare to other population-based recommendations for controlling chronic disease? *Journal of the American Dietetic Association* 107 (2007): 830–837.
11. J. Haven and P. Britten, MyPyramid—The complete guide, *Nutrition Today* 41 (2006): 253–259.
12. J. L. Bachman and coauthors, Sources of food group intakes among the U.S. population, 2001–2002, *Journal of the American Dietetic Association* 108 (2008): 804–814.
13. P. M. Guenther, J. Reedy, and S. M. Krebs-Smith, Development of the Healthy Eating Index—2005, *Journal of the American Dietetic Associa-*

tion 108 (2008): 1896–1901; Center for Nutrition Policy and Promotion, Healthy Eating Index—2005, fact sheet revised June 2008, www.cnpp.usda.gov.

14. Center for Nutrition Policy and Promotion, Diet quality of Americans in 1994–1996 and 2001–02 as measured by the Healthy Eating Index—2005, nutrition insight 37 revised August 2008, www.cnpp.usda.gov.
15. Food pyramids, Harvard School of Public Health, www.hsph.harvard.edu/nutritionsource/pyramids.html, accessed April 17, 2008.
16. V. S. Malik and F. B. Hu, Dietary prevention of atherosclerosis: Go with whole grains, *American Journal of Clinical Nutrition* 85 (2007): 1444–1445.
17. As cited in 21 Code of Federal Regulations—Food and Drugs, Section 104.20, 45 *Federal Register* 6323, January 25, 1980, as amended in 58 *Federal Register* 2228, January 6, 1993.
18. Position of the American Dietetic Association: Fortification and nutritional supplements, *Journal of the American Dietetic Association* 105 (2005): 1300–1311.
19. C. L. Taylor and V. L. Wilkening, How the nutrition food label was developed, part 1: The nutrition facts panel, *Journal of the American Dietetic Association* 108 (2008): 437–442.
20. Dietary Reference Intakes (DRIs) for food labeling, *American Journal of Clinical Nutrition* 83 (2006): suppl; T. Philipson, Government perspective: Food labeling, *American Journal of Clinical Nutrition* 82 (2005): 262S–264S; The National Academy of Sciences, Dietary Reference Intakes: Guiding principles for nutrition labeling and fortification (2004), www.nap.edu/openbook/0309091438/html/R1.html.
21. C. L. Taylor and V. L. Wikening, How the nutrition food label was developed, part 2: The purpose and promise of nutrition claims, *Journal of the American Dietetic Association* 108 (2008): 618–623.
22. P. Williams, Consumer understanding and use of health claims for foods, *Nutrition Reviews* 63 (2005): 256–264.
23. K. A. Donato, National health education programs to promote healthy eating and physical activity, *Nutrition Reviews* 64 (2006): S65–S70.

HIGHLIGHT 2

© Polara Studios, Inc.

Vegetarian Diets

The waiter presents this evening's specials: a fresh spinach salad topped with mandarin oranges, raisins, and sunflower seeds, served with a bowl of pasta smothered in a mushroom and tomato sauce and topped with grated parmesan cheese. Then this one: a salad made of chopped parsley, scallions, celery, and tomatoes mixed with bulgur wheat and dressed with olive oil and lemon juice, served with a spinach and feta cheese pie. Do these meals sound good to you? Or is something missing . . . a pork chop or chicken breast, perhaps?

Would vegetarian fare be acceptable to you some of the time? Most of the time? Ever? Perhaps it is helpful to recognize that dietary choices fall along a continuum—from one end, where people eat no meat or foods of animal origin, to the other end, where they eat generous quantities daily. Meat's place in the diet has been the subject of much research and controversy, as this highlight will reveal. One of the missions of this highlight, in fact, is to identify the *range* of meat intakes most compatible with health. The health benefits of a primarily vegetarian diet seem to have encouraged many people to eat more vegetarian meals. The popular press refers to these "part-time vegetarians" who eat small amounts of meat, fish, or poultry from time to time as "flexitarians."

People who choose to exclude meat and other animal-derived foods from their diets today do so for many of the same reasons the Greek philosopher Pythagoras cited in the sixth century BC: physical health, ecological responsibility, and philosophical concerns. They might also cite world hunger issues, economic reasons, ethical concerns, or religious beliefs as motivating factors. Whatever their reasons—and even if they don't have a particular reason—people who exclude meat will be better prepared to plan well-balanced meals if they understand the nutrition and health implications of vegetarian diets.

Vegetarians generally are categorized, not by their motivations, but by the foods they choose to exclude (see the accompanying glossary). Some people exclude red meat only; some also exclude chicken or fish; others also exclude eggs; and still others exclude milk and milk products as well. In fact, finding agreement on the definition of the term *vegetarian* is a challenge.

As you will see, though, the foods a person *excludes* are not nearly as important as the foods a person *includes* in the diet. Vegetarian diets that include a variety of whole grains, vegetables, legumes, nuts, and fruits offer abundant complex carbohydrates and fibers, an assortment of vitamins and minerals, a mixture of phytochemicals, and little fat—characteristics that reflect current dietary recommendations aimed at promoting health and reducing obesity. Each of these foods—whole grains, vegetables, legumes, nuts, and fruits—independently reduces the risk for several chronic diseases. This highlight examines the health benefits and potential problems of vegetarian diets and shows how to plan a well-balanced vegetarian diet. A plant-based diet also offers many environmental benefits.[1]

Health Benefits of Vegetarian Diets

Research on the health implications of vegetarian diets would be relatively easy if vegetarians differed from other people only in not eating meat. Many vegetarians, however, have also adopted lifestyles that may differ from many **omnivores**: they often use no tobacco or illicit drugs, use little (if any) alcohol, and are physically active. Researchers must account for these lifestyle differ-

GLOSSARY

lacto-ovo-vegetarians: people who include milk, milk products, and eggs, but exclude meat, poultry, fish, and seafood from their diets.
- **ovo** = egg

lactovegetarians: people who include milk and milk products, but exclude meat, poultry, fish, seafood, and eggs from their diets.
- **lacto** = milk

macrobiotic diet: a philosophical approach of eating mostly plant-based foods such as whole grains, legumes, and vegetables, with small amounts of fish, fruits, nuts, and seeds.
- **macro** = large, great
- **biotic** = life

meat replacements: products formulated to look and taste like meat, fish, or poultry; usually made of textured vegetable protein.

omnivores: people who have no formal restriction on the eating of any foods.
- **omni** = all
- **vores** = to eat

tempeh (TEM-pay): a fermented soybean food, rich in protein and fiber.

tofu (TOE-foo): a curd made from soybeans, rich in protein and often fortified with calcium; used in many Asian and vegetarian dishes in place of meat.

vegans (VEE-gans): people who exclude all animal-derived foods (including meat, poultry, fish, eggs, and dairy products) from their diets; also called **pure vegetarians, strict vegetarians,** or **total vegetarians.**

vegetarians: a general term used to describe people who exclude meat, poultry, fish, or other animal-derived foods from their diets.

ences before they can determine which aspects of health correlate just with diet. Even then, *correlations* merely reveal what health factors *go with* the vegetarian diet, not what health effects may be *caused by* the diet. Despite these limitations, research findings suggest that well-planned vegetarian diets offer sound nutrition and health benefits to adults.[2] Dietary patterns that include very little, if any, meat may even increase life expectancy.

Weight Control

In general, weight gains are lowest for those eating the fewest animal-derived foods.[3] Vegetarians tend to maintain a lower and healthier body weight than nonvegetarians.[4] Vegetarians' lower body weights correlate with their high intakes of fiber and low intakes of fat. Because obesity impairs health in a number of ways, this gives vegetarians a health advantage.

Blood Pressure

Vegetarians tend to have lower blood pressure and lower rates of hypertension than nonvegetarians.[5] Appropriate body weight helps to maintain a healthy blood pressure, as does a diet low in total fat and saturated fat and high in fiber, fruits, vegetables, and soy protein.[6] Lifestyle factors also influence blood pressure: smoking and alcohol intake raise blood pressure, and physical activity lowers it.

Heart Disease

The incidence of heart disease and related deaths is slightly lower for vegetarians than for nonvegetarians, which could partly be explained by their avoidance of meat.[7] The dietary factor most directly related to heart disease is saturated animal fat, and in general, vegetarian diets are lower in total fat, saturated fat, and cholesterol than typical meat-based diets. The fats common in plant-based diets—the monounsaturated fats of olives, seeds, and nuts and the polyunsaturated fats of vegetable oils—are associated with a decreased risk of heart disease. Furthermore, vegetarian diets are generally higher in dietary fiber, antioxidant vitamins, and phytochemicals—all factors that help control blood lipids and protect against heart disease.

Many vegetarians include soy products such as **tofu** in their diets. Soy products may help to protect against heart disease because they contain polyunsaturated fats, fiber, vitamins, and minerals, and little saturated fat.[8] Even when intakes of energy, protein, carbohydrate, total fat, saturated fat, unsaturated fat, alcohol, and fiber are the same, people eating meals based on tofu have lower blood cholesterol and triglyceride levels than those eating meat. Some research suggests that soy protein and phytochemicals may be responsible for some of these health benefits (as Highlight 13 explains in greater detail).[9]

Cancer

Vegetarians have a significantly lower rate of cancer than the general population. Their low cancer rates may be due to their high intakes of fruits and vegetables (as Highlight 11 explains). In fact, the ratio of vegetables to meat may be the most relevant dietary factor responsible for cancer prevention.[10]

Some scientific findings indicate that vegetarian diets are associated not only with lower cancer mortality in general, but also with lower incidence of cancer at specific sites as well, most notably, colon cancer.[11] People with colon cancer seem to eat more meat, more saturated fat, and fewer vegetables than do people without colon cancer. High-protein, high-fat, low-fiber diets create an environment in the colon that promotes the development of cancer in some people. A high-meat diet has been associated with cancers of the esophagus, stomach, lungs, and liver as well as increased mortality.[12]

Other Diseases

In addition to obesity, hypertension, heart disease, and cancer, vegetarian diets may help prevent diabetes, osteoporosis, diverticular disease, gallstones, and rheumatoid arthritis.[13] These health benefits of a vegetarian diet depend on wise diet planning.

Vegetarian Diet Planning

The vegetarian has the same meal-planning task as any other person—using a variety of foods to deliver all the needed nutrients within an energy allowance that maintains a healthy body weight (as discussed in Chapter 2). Vegetarians who include milk products and eggs can meet recommendations for most nutrients about as easily as nonvegetarians. Such diets provide enough energy, protein, and other nutrients to support the health of adults and the growth of children and adolescents.

Vegetarians who exclude milk products and eggs can select legumes, nuts, and seeds and products made from them, such as peanut butter, **tempeh**, and tofu, from the protein foods group. Those who do not use milk can use soy "milk"—a product made from soybeans that provides similar nutrients if fortified with calcium, vitamin D, and vitamin B_{12} (see Figure H2-1 on p. 64). Similarly, "milks" made from rice, almonds, and oats are reasonable alternatives, if adequately fortified.

The MyPyramid resources include tips for planning vegetarian diets using the USDA Food Guide. In addition, several food guides have been developed specifically for vegetarian diets. They all address the particular nutrition concerns of vegetarians but differ slightly. Figure H2-2 (p. 64) presents one version. When selecting from the vegetable and fruit groups, vegetarians should emphasize particularly good sources of calcium and iron, respectively. Green leafy vegetables, for example, provide almost five times as much calcium per serving as other vegetables. Similarly, dried fruits deserve special notice in the fruit group because they deliver six times as much iron as other fruits. The milk group features fortified soy milks for those who do not use milk, cheese, or yogurt. The protein foods group includes legumes, soy products, nuts, and seeds. A group for oils encourages the use of vegetable oils, nuts, and seeds rich in unsaturated fats and omega-3 fatty acids. To ensure adequate intakes of vitamin B_{12}, vitamin D, and calcium, vegetarians need to select fortified foods or use supplements daily. The vegetarian food pyramid is flexible enough that a variety of people can use it: people who have adopted various vegetarian diets, those who want to make the transition to a vegetarian diet, and those who simply want to include more plant-based meals in their diets. Like MyPyramid, this vegetarian food pyramid also encourages physical activity.

Most vegetarians easily obtain large quantities of the nutrients that are abundant in plant foods: carbohydrate, fiber, thiamin, folate, vitamin B_6, vitamin C, vitamin A, and vitamin E. Vegetarian food guides help to ensure adequate intakes of the main nutrients vegetarian diets might otherwise lack: protein, iron, zinc, calcium, vitamin B_{12}, vitamin D, and omega-3 fatty acids. Table H2-1 presents good vegetarian sources of these key nutrients.

Protein

The protein RDA for vegetarians is the same as for others, although some have suggested that it should be higher because of the lower digestibility of plant proteins. **Lacto-ovo-vegetarians,** who use animal-derived foods such as milk and eggs, receive high-quality proteins and are likely to meet their protein needs. Even those who adopt only plant-based diets are likely to meet protein needs provided that their energy intakes are adequate and the protein sources varied.[14] The proteins of whole grains, legumes, seeds, nuts, and vegetables can provide adequate amounts of all the amino acids. An advantage of many vegetarian sources of protein is that they are generally lower in saturated fat than meats and are often higher in fiber and richer in some vitamins and minerals.

Vegetarians sometimes use **meat replacements** made of textured vegetable protein (soy protein). These foods are formulated to look and taste like meat, fish, or poultry. Many of these products are fortified to provide the vitamins and minerals found in animal sources of protein. A wise vegetarian learns to use a variety of whole, unrefined foods often and commercially prepared foods less frequently. Vegetarians may also use soy products such as tofu to bolster protein intake.

Iron

Getting enough iron can be a problem even for meat eaters, and those who eat no meat must pay special attention to their iron intake. The iron in plant foods such as legumes, dark green leafy vegetables, iron-fortified cereals, and whole-grain breads and cereals is poorly absorbed. Because iron absorption from a vegetarian diet is low, the iron RDA for vegetarians is higher than for others (see Chapter 13 for more details).

Fortunately, the body seems to adapt to a vegetarian diet by absorbing iron more efficiently. Furthermore, iron absorption is enhanced by vitamin C, and vegetarians typically eat many vita-

FIGURE H2-1 Low-Fat Milk and Soy Milk Compared

A comparison of low-fat milk and enriched soy milk shows that they provide similar amounts of key nutrients.

Low-Fat Milk

Nutrition Facts
Serving Size 1 cup (240mL)
Servings Per Container About 8

Amount Per Serving	
Calories 110	Calories from Fat 25
	% Daily Value*
Total Fat 2.5g	4%
Saturated Fat 1.5g	8%
Trans Fat 0g	
Polyunsaturated Fat 0.5g	
Monounsaturated Fat 0.5g	
Cholesterol 15mg	4%
Sodium 130mg	5%
Potassium 380mg	11%
Total Carbohydrate 13g	4%
Dietary Fiber 0g	0%
Sugars 12g	
Protein 8g	
Vitamin A 10%	Vitamin C 0%
Calcium 30%	Iron 0%
Vitamin D 25%	

Soy Milk

Nutrition Facts
Serving Size 1 cup (240mL)
Servings Per Container About 8

Amount Per Serving	
Calories 100	Calories from Fat 35
	% Daily Value*
Total Fat 4g	6%
Saturated Fat 0.5g	3%
Trans Fat 0g	
Polyunsaturated Fat 2.5g	
Monounsaturated Fat 1g	
Cholesterol 0mg	0%
Sodium 120mg	5%
Potassium 300mg	8%
Total Carbohydrate 8g	3%
Dietary Fiber 1g	4%
Sugars 6g	
Protein 7g	
Vitamin A 10%	Vitamin C 0%
Calcium 30%	Iron 6%
Vitamin D 30%	Riboflavin 30%
Folate 6%	Vitamin B12 50%

FIGURE H2-2 An Example of a Vegetarian Food Pyramid

Review Figure 2-1 and Table 2-2 to find recommended daily amounts from each food group, serving size equivalents, examples of common foods within each group, and the most notable nutrients for each group. Tips for planning a vegetarian diet can be found at www.mypyramid.gov.

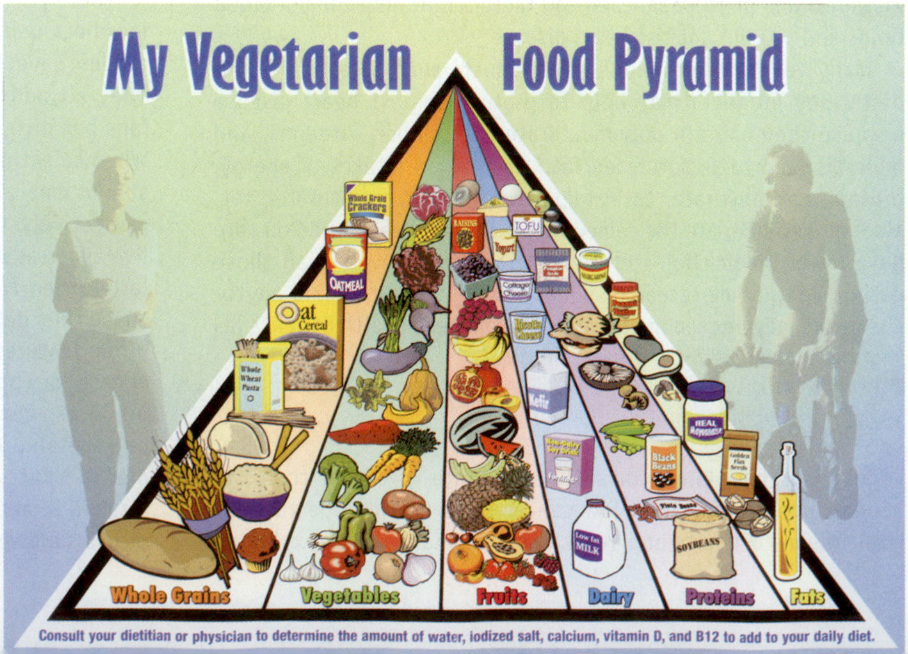

TABLE H2-1 Good Vegetarian Sources of Key Nutrients

	Food Groups					
Nutrients	*Grains*	*Vegetables*	*Fruits*	*Legumes and other protein-rich foods*	*Milk*	*Oils*
Protein	Whole grains[a]			Legumes, seeds, nuts, soy products (tempeh, tofu, veggie burgers)[a] Eggs (for ovo-vegetarians)	Milk, cheese, yogurt (for lacto-vegetarians)	
Iron	Fortified cereals, enriched and whole grains	Dark green leafy vegetables (spinach, turnip greens)	Dried fruits (apricots, prunes, raisins)	Legumes (black-eyed peas, kidney beans, lentils)		
Zinc	Fortified cereals, whole grains			Legumes (garbanzo beans, kidney beans, navy beans), nuts, seeds (pumpkin seeds)	Milk, cheese, yogurt (for lacto-vegetarians)	
Calcium	Fortified cereals	Dark green leafy vegetables (bok choy, broccoli, collard greens, kale, mustard greens, turnip greens, watercress)	Fortified juices, figs	Fortified soy products, nuts (almonds), seeds (sesame seeds)	Milk, cheese, yogurt (for lacto-vegetarians) Fortified soy milk	
Vitamin B_{12}	Fortified cereals			Eggs (for ovo-vegetarians) Fortified soy products	Milk, cheese, yogurt (for lacto-vegetarians Fortified soy milk	
Vitamin D					Milk, cheese, yogurt (for lacto-vegetarians Fortified soy milk	
Omega-3 Fatty acids				Flaxseed, walnuts, soybeans		Flaxseed oil, walnut oil, soybean oil

[a]As Chapter 6 explains, many plant proteins do not contain all the essential amino acids in the amounts and proportions needed by human beings. To improve protein quality, vegetarians can eat grains and legumes together, for example, although it is not necessary if protein intake is varied and energy intake is sufficient.

min C–rich fruits and vegetables. Consequently, vegetarians suffer no more iron deficiency than other people do.

Zinc

Zinc is similar to iron in that meat is its richest food source, and zinc from plant sources is not well absorbed. In addition, soy, which is commonly used as a meat alternative in vegetarian meals, interferes with zinc absorption. Nevertheless, most vegetarian adults are not zinc deficient. Perhaps the best advice to vegetarians regarding zinc is to eat a variety of nutrient-dense foods; include whole grains, nuts, and legumes such as black-eyed peas, pinto beans, and kidney beans; and maintain an adequate energy intake. For those who include seafood in their diets, oysters, crabmeat, and shrimp are rich in zinc.

Calcium

The calcium intakes of **lactovegetarians** are similar to those of the general population, but people who use no milk products risk deficiency. Careful planners select calcium-rich foods, such as calcium-fortified juices, soy milk, and breakfast cereals, in ample quantities regularly. This advice is especially important for children and adolescents. Soy formulas for infants are fortified with calcium and can be used in cooking, even for adults. Other good calcium sources include figs, some legumes, some green vegetables such as broccoli and turnip greens, some nuts such as almonds, certain seeds such as sesame seeds, and calcium-set tofu.* The choices should be varied because calcium absorption from some plant foods may be limited (as Chapter 12 explains).

Vitamin B_{12}

The requirement for vitamin B_{12} is small, but this vitamin is found only in animal-derived foods. Consequently, vegetarians, in general, and **vegans** who eat no foods of animal original, in particular, may not get enough vitamin B_{12} in their diets.[15] Fermented soy products such as tempeh may contain some vitamin B_{12} from the bacteria, but unfortunately, much of the vitamin B_{12} found in these products may be an inactive form. Seaweeds such as nori and chlorella supply some vitamin B_{12}, but not much, and excessive intakes of these foods can lead to iodine toxicity. To defend against vitamin B_{12} deficiency, vegans must rely on vitamin B_{12}–fortified sources (such as soy milk or breakfast cereals) or supplements. Without vitamin B_{12}, the nerves suffer damage, leading to such health consequences as loss of vision.

*Calcium salts are often added during processing to coagulate the tofu.

Vitamin D

The vitamin D status of vegetarians is similar to that of nonvegetarians.[16] People who do not use vitamin D–fortified foods and do not receive enough exposure to sunlight to synthesize adequate vitamin D may need supplements to defend against bone loss. This is particularly important for infants, children, and older adults. In northern climates during winter months, young children on vegan diets can readily develop rickets, the vitamin D–deficiency disease.

Omega-3 Fatty Acids

Both Chapter 5 and Highlight 5 describe the health benefits of unsaturated fats, most notably the omega-3 fatty acids commonly found in fatty fish. A diet that includes some meat and fish provides more omega-3 fatty acids than a vegetarian diet.[17] To obtain sufficient amounts of omega-3 fatty acids, vegetarians need to consume flaxseed, walnuts, soybeans, and their oils.

Healthy Food Choices

In general, adults who eat vegetarian diets have lowered their risks of mortality and several chronic diseases, including obesity, high blood pressure, heart disease, and cancer. But there is nothing mysterious or magical about the vegetarian diet. The quality of the diet depends not on whether it includes meat, but on whether the other food choices are nutritionally sound. A diet that includes ample fruits, vegetables, whole grains, legumes, nuts, and seeds is higher in fiber, antioxidant vitamins, and phytochemicals and lower in saturated fats than meat-based diets. Variety is key to nutritional adequacy in a vegetarian diet. Restrictive plans that limit selections to a few grains and vegetables cannot possibly deliver a full array of nutrients.

Vegetarianism is not a religion like Buddhism or Hinduism, but merely an eating plan that selects plant foods to deliver needed nutrients. That said, some vegetarians choose to follow a **macrobiotic diet**. Those following a macrobiotic diet select natural, organic foods and embrace a Zen-like spirituality. In other words, a macrobiotic diet represents a way of life, not just a meal plan. A macrobiotic diet emphasizes whole grains, legumes, and vegetables, with small amounts of fish, fruits, nuts, and seeds. Practices include selecting locally grown foods, eating foods in their most natural state, and balancing cold, sweet, and passive foods with hot, salty, and aggressive ones. Some items, such as processed foods, alcohol, hot spices, and potatoes, are excluded from the diet. Early versions of the macrobiotic diet followed a progression that ended with the "ultimate" diet of brown rice and water—a less-than-nutritiously balanced diet. Today's version reflects a modified vegetarian approach with an appreciation of how foods can enhance health. With careful planning, a macrobiotic diet can provide an array of nutrients that support good health.

If not properly balanced, any diet—vegetarian, macrobiotic, or otherwise—can lack nutrients. Poorly planned vegetarian diets typically lack iron, zinc, calcium, vitamin B_{12}, and vitamin D; without planning, the meat eater's diet may lack vitamin A, vitamin C, folate, and fiber, among others. Quite simply, the negative health aspects of any diet, including vegetarian diets, reflect poor diet planning. Careful attention to energy intake and specific problem nutrients can ensure adequacy.

Keep in mind, too, that diet is only one factor influencing health. Whatever a diet consists of, its context is also important: no smoking, alcohol consumption in moderation (if at all), regular physical activity, adequate rest, and medical attention when needed all contribute to a healthy life. Establishing these healthy habits early in life seems to be the most important step one can take to reduce the risks of later diseases (as Highlight 15 explains).

Nutrition on the Net

For further study of topics covered in this chapter, log on to **www.cengagebrain.com** and search for ISBN 084006845X.

- Search for "vegetarian" at the Food and Drug Administration's site: **www.fda.gov**
- Find tips for planning vegetarian diets at the USDA MyPyramid site: **www.mypyramid.gov**
- Visit the Vegetarian Resource Group: **www.vrg.org**
- Review another vegetarian diet pyramid developed by Oldways Preservation & Exchange Trust: **www.oldwayspt.org**

References

1. B. M. Popkin, Reducing meat consumption has multiple benefits for the world's health, *Archives of Internal Medicine* 169 (2009): 543–545.
2. G. E. Fraser, Vegetarian diets: What do we know of their effects on common chronic diseases? *American Journal of Clinical Nutrition* 89 (2009): 1607S–1612S; S. E. Berkow and N. Barnard, Vegetarian diets and weight status, *Nutrition Reviews* 64 (2006): 175–188; T. J. Key, P. N. Appleby, and M. S. Rosell, Health effects of vegetarian and vegan diets, *Proceedings of the Nutrition Society* 65 (2006): 35–41; Position of the American Dietetic Association: Vegetarian diets, *Journal of the American Dietetic Association* 109 (2009): 1266–1282.
3. M. Rosell and coauthors, Weight gain over 5 years in 21,966 meat-eating, fish-eating, vegetarian, and vegan men and women in EPIC-Oxford, *International Journal of Obesity* 30 (2006): 1389–1396.
4. Berkow and Barnard, 2006; P. K. Newby, K. L. Tucker, and A. Wolk, Risk of overweight and obesity among semivegetarian, lactovegetarian, and vegan women, *American Journal of Clinical Nutrition* 81 (2005): 1267–1274.

5. V. H. Myers and C. M. Champagne, Nutritional effects on blood pressure, *Current Opinion in Lipidology* 18 (2007): 20–24.
6. S. E. Berkow and N. D. Barnard, Blood pressure regulation and vegetarian diets, *Nutrition Reviews* 63 (2005): 1–8.
7. J. Chang-Claude and coauthors, Lifestyle determinants and mortality in German vegetarians and health-conscious persons: Results of a 21-year follow-up, *Cancer Epidemiology, Biomarkers, and Prevention* 14 (2005): 963–968.
8. F. M. Sacks and coauthors, Soy protein, isoflavones, and cardiovascular health: An American Heart Association Science Advisory for professionals from the Nutrition Committee, *Circulation* 113 (2006): 1034–1044.
9. D. Lukaczer and coauthors, Effect of a low glycemic index diet with soy protein and phytosterols on CVD risk factors in postmenopausal women, *Nutrition* 22 (2006): 104–113; B. L. McVeigh and coauthors, Effect of soy protein varying in isoflavone content on serum lipids in healthy young men, *American Journal of Clinical Nutrition* 83 (2006): 244–251.
10. M. Kapiszewska, A vegetable to meat consumption ratio as a relevant factor determining cancer preventive diet: The Mediterranean versus other European countries, *Forum of Nutrition* 59 (2006): 130–153.
11. M. H. Lewin and coauthors, Red meat enhances the colonic formation of the DNA adduct O6-carboxymethyl guanine: Implications for colorectal cancer risk, *Cancer Research* 66 (2006): 1859–1865.
12. R. Sinha and coauthors, Meat intake and mortality: A prospective study of over half a million people, *Archives of Internal Medicine* 169 (2009): 562–571; A. J. Cross and coauthors, A prospective study of red and processed meat intake in relation to cancer risk, *PLoS Medicine* 4 (2007): 1973–1984.
13. C. Leitzmann, Vegetarian diets: What are the advantages? *Forum of Nutrition* 57 (2005): 147–156.
14. Position of the American Dietetic Association, 2009.
15. I. Elmadfa and I. Singer, Vitamin B-12 and homocysteine status among vegetarians: A global perspective, *American Journal of Clinical Nutrition* 89 (2009): 1693S–1698S.
16. J. Chan, K. Jaceldo-Siegl, and G. E. Fraser, Serum 25-hydroxyvitamin D status of vegetarians, partial vegetarians, and nonvegetarians: The Adventist Health Study, *American Journal of Clinical Nutrition* 89 (2009): 1686S–1692S.
17. I. Mangat, Do vegetarians have to eat fish for optimal cardiovascular protection? *American Journal of Clinical Nutrition* 89 (2009): 1597S–1601S; N. Mann and coauthors, Fatty acid composition of habitual omnivore and vegetarian diets, *Lipids* 41 (2006): 637–646.

CHAPTER

3

© imagebroker/Alamy

Nutrition in Your Life

Have you ever wondered what happens to the food you eat after you swallow it? Or how your body extracts nutrients from food? Have you ever marveled at how it all just seems to happen? Follow foods as they travel through the digestive system. Learn how a healthy digestive system takes whatever food you give it—whether sirloin steak and potatoes or tofu and brussels sprouts—and extracts the nutrients that will nourish the cells of your body.

Throughout this chapter, the CourseMate icon indicates an opportunity for online self-study, linking you to activities to increase your understanding of chapter concepts. **www.cengagebrain.com** (search for ISBN 084006845X)

Digestion, Absorption, and Transport

CHAPTER OUTLINE

This chapter follows the journey that breaks down foods into the nutrients featured in the later chapters. Then it follows the nutrients as they travel through the intestinal cells and into the body to do their work. This introduction presents a general overview of the processes common to all nutrients; later chapters discuss the specifics of digesting and absorbing individual nutrients.

Digestion

Digestion is the body's ingenious way of breaking down foods into nutrients in preparation for **absorption.** In the process, it overcomes many challenges without any conscious effort. Consider these challenges:

1. Human beings breathe, eat, and drink through their mouths. Air taken in through the mouth must go to the lungs; food and liquid must go to the stomach. The throat must be arranged so that swallowing and breathing don't interfere with each other.
2. Below the lungs lies the diaphragm, a dome of muscle that separates the upper half of the major body cavity from the lower half. Food must pass through this wall to reach the stomach.
3. The materials within the digestive tract should be kept moving forward, slowly but steadily, at a pace that permits all reactions to reach completion.
4. To move through the system, food must be lubricated with fluids. Too much would form a liquid that would flow too rapidly; too little would form a paste too dry and compact to move at all. The amount of fluids must be regulated to keep the intestinal contents at the right consistency to move along smoothly.
5. Before the digestive enzymes can work, foods must be broken down into small particles and suspended in enough liquid so that every particle is accessible. Once digestion is complete and the needed nutrients have been absorbed out of the GI tract and into the body, the system must excrete the remaining waste. Excreting all the water along with the solid residue, however, would be both wasteful and messy. Some water must be withdrawn to leave a solid enough waste product to be smooth and easy to pass.

digestion: the process by which food is broken down into absorbable units.
- **digestion** = take apart

absorption: the uptake of nutrients by the cells of the small intestine for transport into either the blood or the lymph.
- **absorb** = suck in

Morgan Lane Photography/Shutterstock.com

The process of digestion breaks down all kinds of *foods* into *nutrients*.

6. The digestive enzymes are designed to digest carbohydrate, fat, and protein. The cells of the GI tract are also made of carbohydrate, fat, and protein. These cells need to be protected against the powerful digestive juices that they secrete.
7. Once waste matter has reached the end of the GI tract, it must be excreted, but it would be inconvenient and embarrassing if this function occurred continuously. Evacuation needs to occur periodically.

The following sections show how the body elegantly and efficiently handles these challenges. Each section follows the GI tract from one end to the other—first describing its anatomy, then its muscular actions, and finally its secretions.

Anatomy of the Digestive Tract

The **gastrointestinal (GI) tract** is a flexible muscular tube that extends from the mouth, through the esophagus, stomach, small intestine, large intestine, and rectum to the anus. Figure 3-1 traces the path followed by food from one end to the other. In a sense, the human body surrounds the GI tract. The inner space within the GI tract, called the **lumen**, is continuous from one end to the other. (GI anatomy terms appear in boldface type and are defined in the accompanying glossary.) Only when a nutrient or other substance finally penetrates the GI tract's wall does it enter the body proper; many materials pass through the GI tract without being digested or absorbed.

Mouth

The process of digestion begins in the **mouth**. During chewing, ♦ teeth crush large pieces of food into smaller ones, and fluids from foods, beverages, and salivary glands blend with these pieces to ease swallowing. Fluids also help dissolve the food so that the tongue can taste it; only particles in solution can react with taste buds. When stimulated, the taste buds detect one, or a combination, of the four basic taste sensations: sweet, sour, bitter, and salty. Some scientists also include the flavor associated with monosodium glutamate, sometimes called *savory* or its Asian name, *umami* (oo-MOM-ee). In addition to these chemical triggers, aroma, texture, and temperature also affect a food's flavor. In fact, the sense of smell is thousands of times more sensitive than the sense of taste.

♦ The process of chewing is called **mastication** (mass-tih-KAY-shun).

The tongue provides taste sensations and moves food around the mouth, facilitating chewing and swallowing. When a mouthful of food is swallowed, it passes through the **pharynx**, a short tube that is shared by both the **digestive system** and the respiratory system. To bypass the entrance to the lungs, the **epiglottis** closes off the airway so that choking doesn't occur when swallowing, thus resolving the

gastrointestinal (GI) tract: the digestive tract. The principal organs are the stomach and intestines.
- **gastro** = stomach
- **intestinalis** = intestine

digestive system: all the organs and glands associated with the ingestion and digestion of food.

GLOSSARY
OF GI ANATOMY TERMS

These terms are listed in order from start to end of the digestive system.

lumen (LOO-men): the space within a vessel, such as the intestine.

mouth: the oral cavity containing the tongue and teeth.

pharynx (FAIR-inks): the passageway leading from the nose and mouth to the larynx and esophagus, respectively.

epiglottis (epp-ih-GLOTT-iss): cartilage in the throat that guards the entrance to the trachea and prevents fluid or food from entering it when a person swallows.
- **epi** = upon (over)
- **glottis** = back of tongue

esophagus (ee-SOFF-ah-gus): the food pipe; the conduit from the mouth to the stomach.

sphincter (SFINK-ter): a circular muscle surrounding, and able to close, a body opening. Sphincters are found at specific points along the GI tract and regulate the flow of food particles.
- **sphincter** = band (binder)

esophageal (ee-SOF-ah-GEE-al) **sphincter:** a sphincter muscle at the upper or lower end of the esophagus. The *lower esophageal sphincter* is also called the *cardiac sphincter*.

stomach: a muscular, elastic, saclike portion of the digestive tract that grinds and churns swallowed food, mixing it with acid and enzymes to form chyme.

pyloric (pie-LORE-ic) **sphincter:** the circular muscle that separates the stomach from the small intestine and regulates the flow of partially digested food into the small intestine; also called *pylorus* or *pyloric valve*.
- **pylorus** = gatekeeper

small intestine: a 10-foot length of small-diameter intestine that is the major site of digestion of food and absorption of nutrients. Its segments are the duodenum, jejunum, and ileum.

gallbladder: the organ that stores and concentrates bile. When it receives the signal that fat is present in the duodenum, the gallbladder contracts and squirts bile through the bile duct into the duodenum.

pancreas: a gland that secretes digestive enzymes and juices into the duodenum. (The pancreas also secretes hormones into the blood that help to maintain glucose homeostasis.)

duodenum (doo-oh-DEEN-um, doo-ODD-num): the top portion of the small intestine (about "12 fingers' breadth" long in ancient terminology).
- **duodecim** = twelve

jejunum (je-JOON-um): the first two-fifths of the small intestine beyond the duodenum.

ileum (ILL-ee-um): the last segment of the small intestine.

ileocecal (ill-ee-oh-SEEK-ul) **valve:** the sphincter separating the small and large intestines.

large intestine or **colon** (COAL-un): the lower portion of intestine that completes the digestive process. Its segments are the ascending colon, the transverse colon, the descending colon, and the sigmoid colon.
- **sigmoid** = shaped like the letter S (sigma in Greek)

appendix: a narrow blind sac extending from the beginning of the colon that stores lymph cells.

rectum: the muscular terminal part of the intestine, extending from the sigmoid colon to the anus.

anus (AY-nus): the terminal outlet of the GI tract.

first challenge. (Choking is discussed on pp. 88–89.) After a mouthful of food has been chewed and swallowed, it is called a **bolus**.

Esophagus to the Stomach The **esophagus** has a **sphincter** muscle at each end. During a swallow, the upper **esophageal sphincter** opens. The bolus then slides down the esophagus, which passes through a hole in the diaphragm (challenge 2) to the **stomach**. The lower esophageal sphincter ♦ at the entrance to the stomach closes behind the bolus so that it proceeds forward and doesn't slip back into the

♦ The lower esophageal sphincter is also called the *cardiac sphincter* because of its proximity to the heart.

bolus (BOH-lus): a portion; with respect to food, the amount swallowed at one time.

- **bolos** = lump

FIGURE 3-1 **The Gastrointestinal Tract**

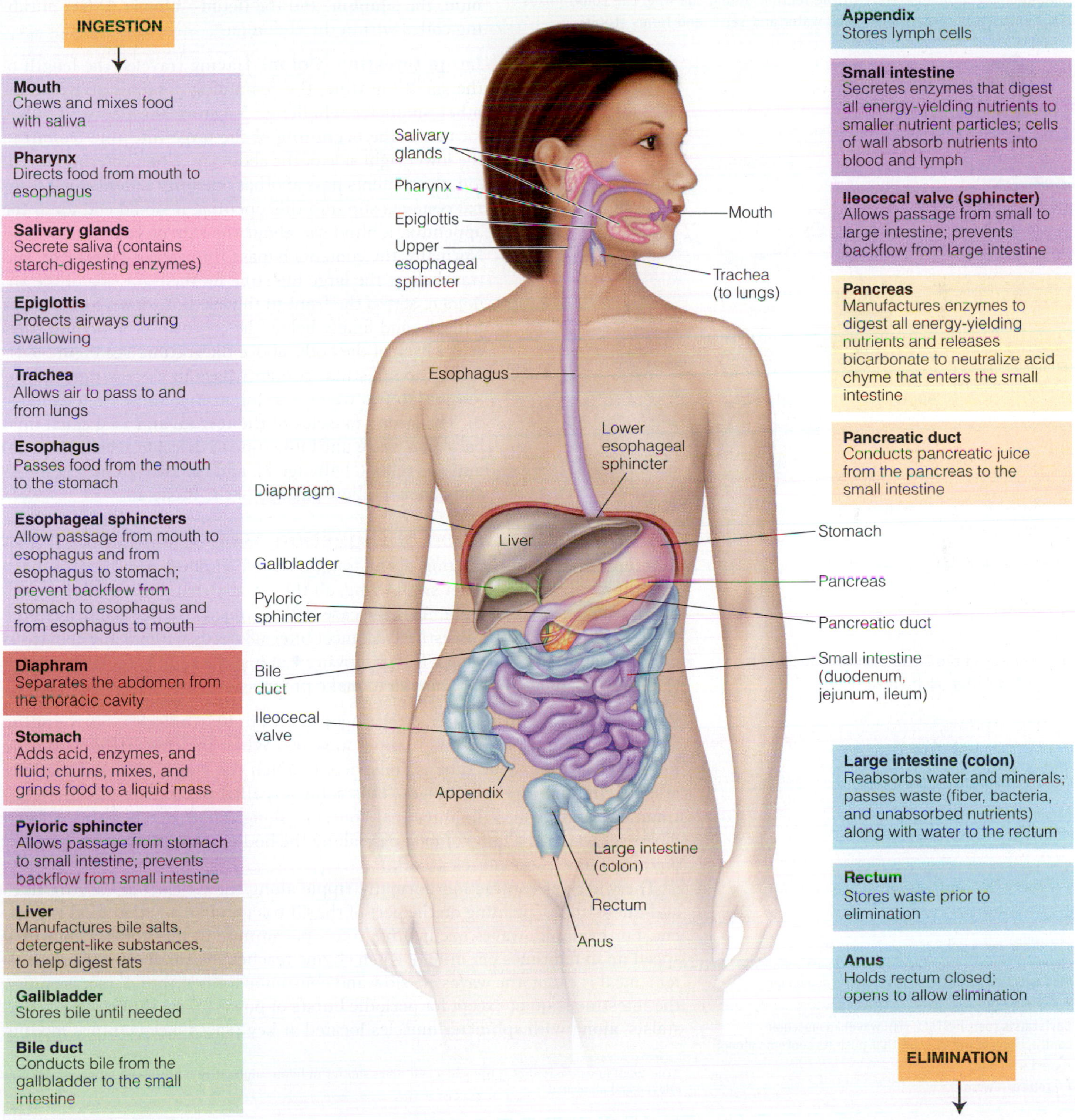

esophagus (challenge 3). The stomach retains the bolus for a while in its upper portion. Little by little, the stomach transfers the food to its lower portion, adds juices to it, and grinds it to a semiliquid mass called **chyme.** Then, bit by bit, the stomach releases the chyme through the **pyloric sphincter,** which opens into the **small intestine** and then closes behind the chyme.

FIGURE 3-2 The Colon

The colon begins with the ascending colon rising upward toward the liver. It becomes the transverse colon as it turns and crosses the body toward the spleen. The descending colon turns downward and becomes the sigmoid colon, which extends to the rectum. Along the way, the colon mixes the intestinal contents, absorbs water and salts, and forms stools.

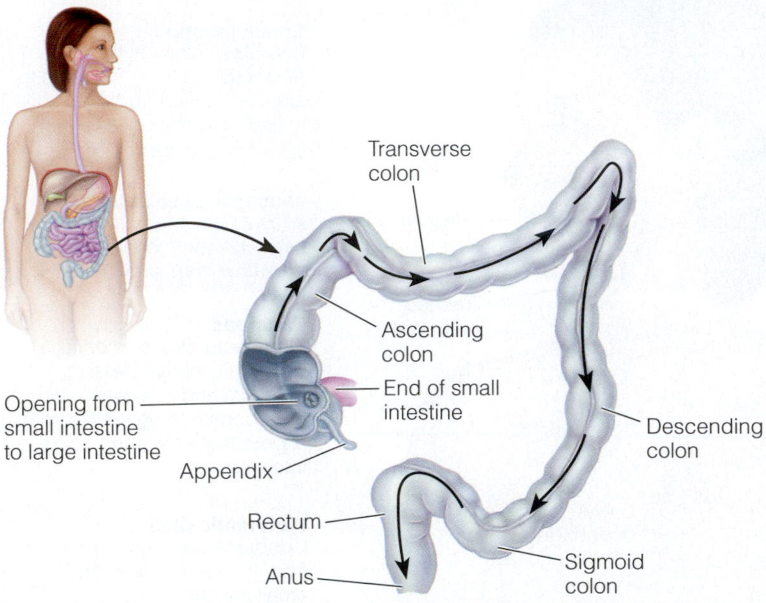

Small Intestine At the beginning of the small intestine, the chyme bypasses the opening from the common bile duct, which is dripping fluids (challenge 4) into the small intestine from two organs outside the GI tract—the **gallbladder** and the **pancreas.** The chyme travels on down the small intestine through its three segments—the **duodenum,** the **jejunum,** and the **ileum**—almost 10 feet of tubing coiled within the abdomen.*

Large Intestine (Colon) Having traveled the length of the small intestine, the remaining contents arrive at another sphincter (challenge 3 again): the **ileocecal valve,** located at the beginning of the **large intestine (colon)** in the lower right side of the abdomen. Upon entering the colon, the contents pass another opening. Should any intestinal contents slip into this opening, it would end up in the **appendix,** a blind sac about the size of your little finger. Normally, the contents bypass this opening, however, and travel along the large intestine up the right side of the abdomen, across the front to the left side, down to the lower left side, and finally below the other folds of the intestines to the back of the body, above the **rectum** (see Figure 3-2).

As the intestinal contents pass to the rectum, the colon withdraws water, leaving semisolid waste (challenge 5). The strong muscles of the rectum and anal canal hold back this waste until it is time to defecate. Then the rectal muscles relax (challenge 7), and the two sphincters of the **anus** open to allow passage of the waste.

The Muscular Action of Digestion

In the mouth, chewing, the addition of saliva, and the action of the tongue transform food into a coarse mash that can be swallowed. After swallowing, all the activity that follows occurs without much conscious thought. As is the case with so much else that happens in the body, the muscles of the digestive tract meet internal needs without any conscious effort on your part. They keep things moving ♦ at just the right pace, slow enough to get the job done and fast enough to make progress.

♦ The ability of the GI tract muscles to move is called **motility** (moh-TIL-ih-tee).

Peristalsis The entire GI tract is ringed with circular muscles. Surrounding these rings of muscle are longitudinal muscles. When the rings tighten and the long muscles relax, the tube is constricted. When the rings relax and the long muscles tighten, the tube bulges. This action—called **peristalsis**—occurs continuously and pushes the intestinal contents along (challenge 3 again). (If you have ever watched a lump of food pass along the body of a snake, you have a good picture of how these muscles work.)

The waves of contraction normally ripple along the GI tract at varying rates and intensities depending on the part of the GI tract and on whether food is present. For example, waves occur three times per minute in the stomach, but they speed up to ten times per minute when chyme reaches the small intestine. Just after a meal is eaten, the waves are slow and continuous; when the GI tract is empty, the intestine is quiet except for periodic bursts of powerful rhythmic waves. Peristalsis, along with sphincter muscles located at key places, keeps things moving

chyme (KIME): the semiliquid mass of partly digested food expelled by the stomach into the duodenum.
- **chymos** = juice

peristalsis (per-ih-STALL-sis): wavelike muscular contractions of the GI tract that push its contents along.
- **peri** = around
- **stellein** = wrap

*The small intestine is almost two and a half times shorter in living adults than it is at death, when muscles are relaxed and elongated.

along. Factors such as stress, medicines, and medical conditions may interfere with normal GI tract contractions.

Stomach Action The stomach has the thickest walls and strongest muscles of all the GI tract organs. In addition to the circular and longitudinal muscles, it has a third layer of diagonal muscles that also alternately contracts and relaxes (see Figure 3-3). These three sets of muscles work to force the chyme downward, but the pyloric sphincter usually remains tightly closed, preventing the chyme from passing into the duodenum of the small intestine. As a result, the chyme is churned and forced down, hits the pyloric sphincter, and remains in the stomach. Meanwhile, the stomach wall releases gastric juices. When the chyme is completely liquefied with gastric juices, the pyloric sphincter opens briefly, about three times a minute, to allow small portions of chyme to pass through. At this point, the chyme no longer resembles food in the least.

Segmentation The circular muscles of the intestines rhythmically contract and squeeze their contents (see Figure 3-4). These contractions, called **segmentation**, mix the chyme and promote close contact with the digestive

FIGURE 3-3 Stomach Muscles

The stomach has three layers of muscles.

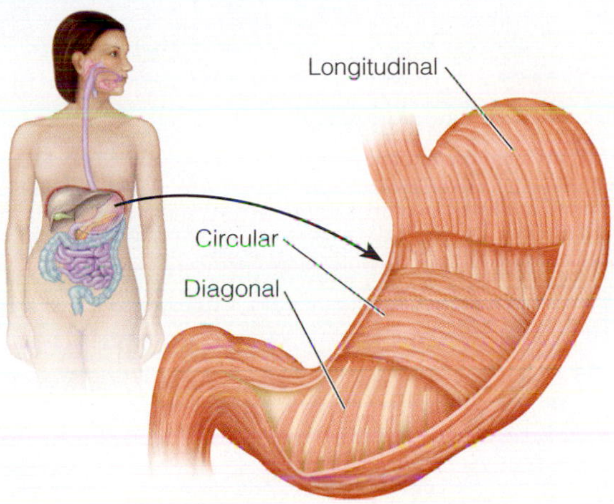

FIGURE 3-4 Peristalsis and Segmentation

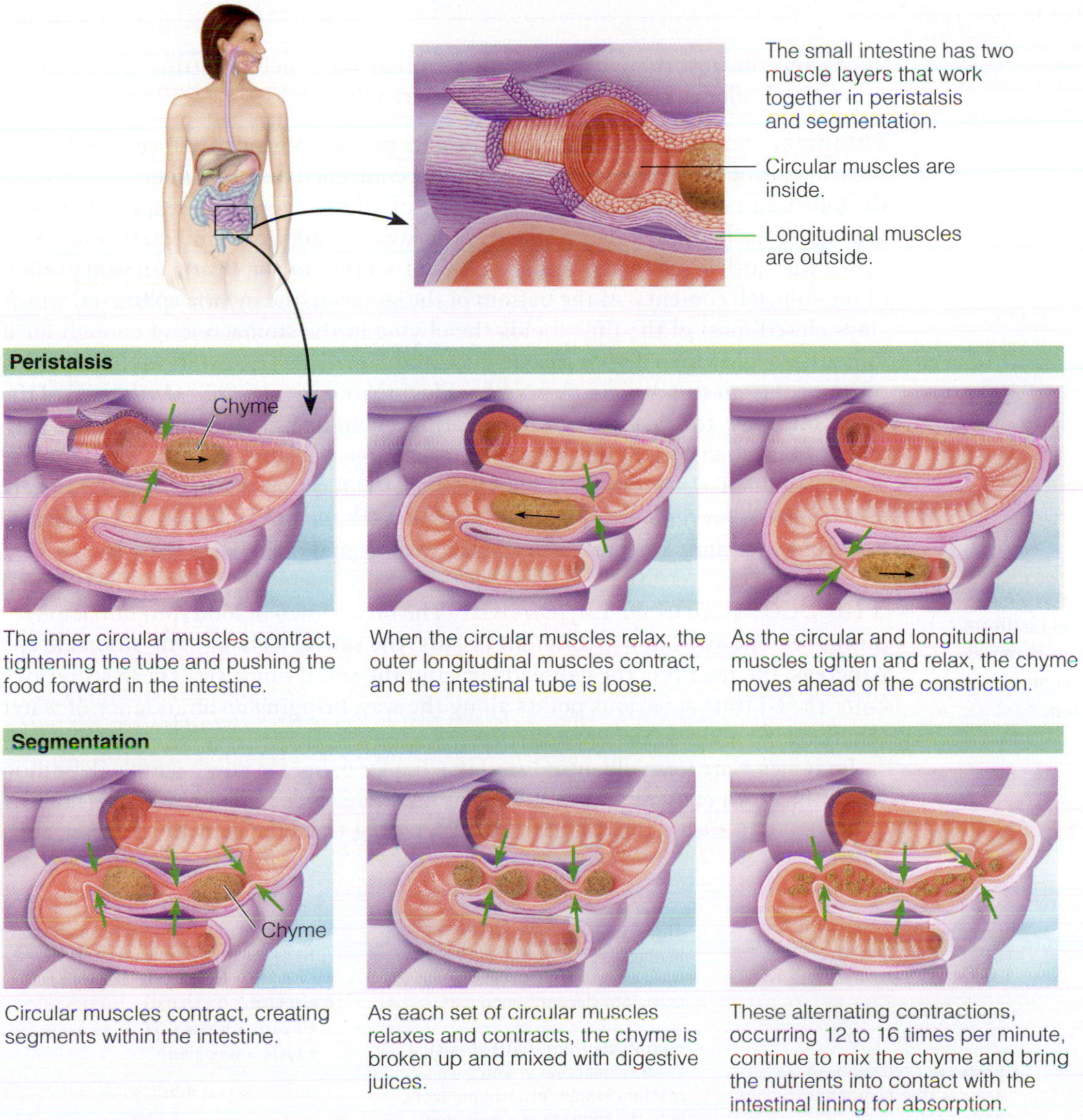

The small intestine has two muscle layers that work together in peristalsis and segmentation.
Circular muscles are inside.
Longitudinal muscles are outside.
Peristalsis
Chyme
The inner circular muscles contract, tightening the tube and pushing the food forward in the intestine.
When the circular muscles relax, the outer longitudinal muscles contract, and the intestinal tube is loose.
As the circular and longitudinal muscles tighten and relax, the chyme moves ahead of the constriction.
Segmentation
Chyme
Circular muscles contract, creating segments within the intestine.
As each set of circular muscles relaxes and contracts, the chyme is broken up and mixed with digestive juices.
These alternating contractions, occurring 12 to 16 times per minute, continue to mix the chyme and bring the nutrients into contact with the intestinal lining for absorption.

segmentation (SEG-men-TAY-shun): a periodic squeezing or partitioning of the intestine at intervals along its length by its circular muscles.

FIGURE 3-5 **An Example of a Sphincter Muscle**

When the circular muscles of a sphincter contract, the passage closes; when they relax, the passage opens.

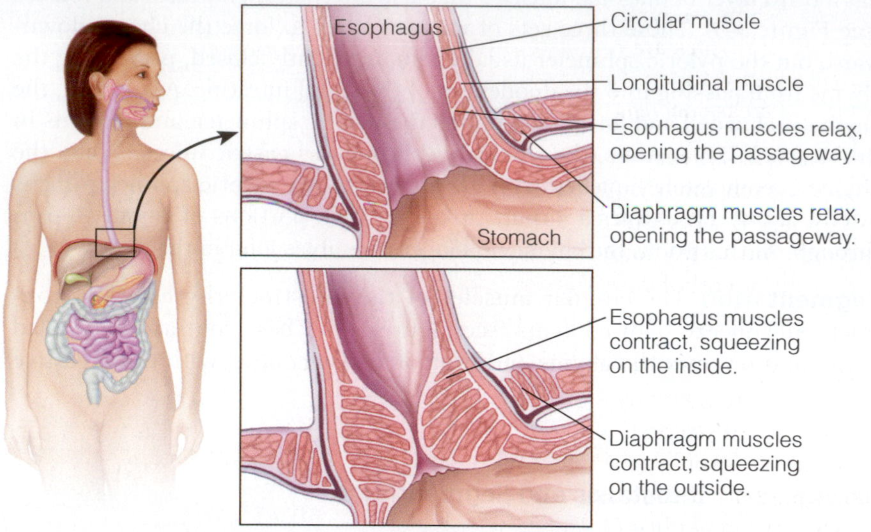

Esophagus
Circular muscle
Longitudinal muscle
Esophagus muscles relax, opening the passageway.
Diaphragm muscles relax, opening the passageway.
Stomach
Esophagus muscles contract, squeezing on the inside.
Diaphragm muscles contract, squeezing on the outside.

juices and the absorbing cells of the intestinal walls before letting the contents move slowly along. Figure 3-4 illustrates peristalsis and segmentation.

Sphincter Contractions Sphincter muscles periodically open and close, allowing the contents of the GI tract to move along at a controlled pace (challenge 3 again). At the top of the esophagus, the upper esophageal sphincter opens in response to swallowing. At the bottom of the esophagus, the lower esophageal sphincter (sometimes called the cardiac sphincter because of its proximity to the heart) prevents **reflux** of the stomach contents. At the bottom of the stomach, the pyloric sphincter, which stays closed most of the time, holds the chyme in the stomach long enough for it to be thoroughly mixed with gastric juice and liquefied. The pyloric sphincter also prevents the intestinal contents from backing up into the stomach. At the end of the small intestine, the ileocecal valve performs a similar function, allowing the contents of the small intestine to empty into the large intestine. Finally, the tightness of the rectal muscle acts as a kind of safety device; together with the two sphincters of the anus, it prevents continuous elimination (challenge 7). Figure 3-5 illustrates how sphincter muscles contract and relax to close and open passageways.

The Secretions of Digestion

The breakdown of food into nutrients requires secretions from five different organs: the salivary glands, the stomach, the pancreas, the liver (via the gallbladder), and the small intestine. These secretions enter the GI tract at various points along the way, bringing an abundance of water (challenge 4) and a variety of enzymes.

Enzymes ♦ are formally introduced in Chapter 6, but for now a simple definition will suffice. An enzyme is a protein that facilitates a chemical reaction—making a molecule, breaking a molecule apart, changing the arrangement of a molecule,

♦ All enzymes and some hormones are proteins, but enzymes are not hormones. Enzymes facilitate the making and breaking of bonds in chemical reactions; hormones act as chemical messengers, sometimes regulating enzyme action.

reflux: a backward flow.
- **re** = back
- **flux** = flow

GLOSSARY OF DIGESTIVE ENZYMES

-ase (ACE): a word ending denoting an enzyme. The word beginning often identifies the compounds the enzyme works on. Examples include:
- **carbohydrase** (KAR-boe-HIGH-drase), an enzyme that hydrolyzes carbohydrates.
- **lipase** (LYE-pase), an enzyme that hydrolyzes lipids (fats).
- **protease** (PRO-tee-ase), an enzyme that hydrolyzes proteins.

digestive enzymes: proteins found in digestive juices that act on food substances, causing them to break down into simpler compounds.

hydrolysis (high-DROL-ih-sis): a chemical reaction in which a major reactant is split into two products, with the addition of a hydrogen atom (H) to one and a hydroxyl group (OH) to the other (from water, H_2O). (The noun is **hydrolysis;** the verb is **hydrolyze.**)
- **hydro** = water
- **lysis** = breaking

or exchanging parts of molecules. As a **catalyst**, the enzyme itself remains unchanged. The enzymes involved in digestion facilitate a chemical reaction known as **hydrolysis**—the addition of water *(hydro)* to break *(lysis)* a molecule into smaller pieces. The glossary on p. 74 describes how to identify some of the common **digestive enzymes** and related terms; later chapters introduce specific enzymes. When learning about enzymes, it helps to know that the word ending *-ase* denotes an enzyme. Enzymes are often identified by the organ they come from and the compounds they work on. *Gastric lipase,* for example, is a stomach enzyme that acts on lipids, whereas *pancreatic lipase* comes from the pancreas (and also works on lipids).

Saliva The **salivary glands**, shown in Figure 3-6, squirt just enough **saliva** to moisten each mouthful of food so that it can pass easily down the esophagus (challenge 4). (Digestive **glands** and their secretions are defined in the glossary below.) The saliva contains water, salts, mucus, and enzymes that initiate the digestion of carbohydrates. Saliva also protects the teeth and the linings of the mouth, esophagus, and stomach from substances that might cause damage.

Gastric Juice In the stomach, **gastric glands** secrete **gastric juice**, a mixture of water, enzymes, and **hydrochloric acid**, which acts primarily in protein digestion. The acid is so strong that it causes the sensation of heartburn if it happens to reflux into the esophagus. Highlight 3, following this chapter, discusses heartburn, ulcers, and other common digestive problems.

The strong acidity of the stomach prevents bacterial growth and kills most bacteria that enter the body with food. It would destroy the cells of the stomach as well, but for their natural defenses. To protect themselves from gastric juice, the cells of the stomach wall secrete **mucus**, a thick, slippery, white substance that coats the cells, protecting them from the acid, enzymes, and disease-causing bacteria that might otherwise cause harm (challenge 6).

Figure 3-7 (p. 76) shows how the strength of acids is measured—in **pH** ♦ units. Note that the acidity of gastric juice registers below 2 on the pH scale—stronger than vinegar. The stomach enzymes work most efficiently in the stomach's strong acid, but the salivary enzymes, which are swallowed with food, do not work in acid this strong. Consequently, the salivary digestion of carbohydrates gradually ceases when the stomach acid penetrates each newly swallowed bolus of food. Once in the stomach, salivary enzymes just become other proteins to be digested.

Pancreatic Juice and Intestinal Enzymes By the time food leaves the stomach, digestion of all three energy nutrients (carbohydrates, fats, and proteins) has begun, and the action gains momentum in the small intestine. There the pancreas contributes digestive juices by way of ducts leading into the duodenum. The **pancreatic juice** contains enzymes that act on all three energy nutrients, and the cells of the intestinal wall also possess digestive enzymes on their surfaces.

FIGURE 3-6 The Salivary Glands

The salivary glands secrete saliva into the mouth and begin the digestive process. Given the short time food is in the mouth, salivary enzymes contribute little to digestion.

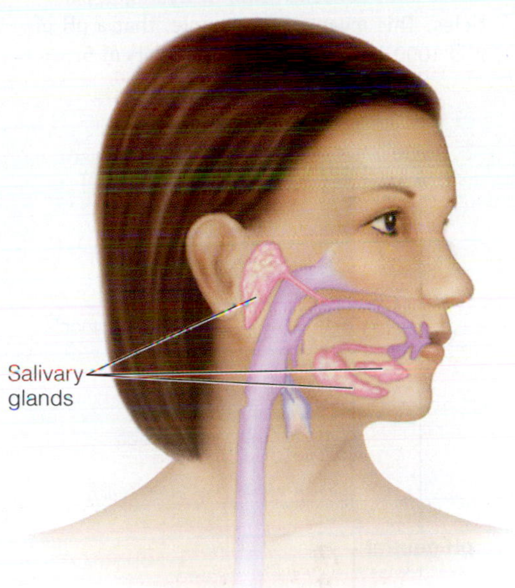

♦ The lower the pH, the higher the H^+ ion concentration and the stronger the acid. A pH above 7 is alkaline, or base (a solution in which OH^- ions predominate).

catalyst (CAT-uh-list): a compound that facilitates chemical reactions without itself being changed in the process.

pH: the unit of measure expressing a substance's acidity or alkalinity.

GLOSSARY OF DIGESTIVE GLANDS AND THEIR SECRETIONS

These terms are listed in order from start to end of the digestive tract.

glands: cells or groups of cells that secrete materials for special uses in the body. Glands may be **exocrine** (EKS-oh-crin) **glands,** secreting their materials "out" (into the digestive tract or onto the surface of the skin), or **endocrine** (EN-doe-crin) **glands,** secreting their materials "in" (into the blood).
- **exo** = outside
- **endo** = inside
- **krine** = to separate

salivary glands: exocrine glands that secrete saliva into the mouth.

saliva: the secretion of the salivary glands. Its principal enzyme begins carbohydrate digestion.

gastric glands: exocrine glands in the stomach wall that secrete gastric juice into the stomach.
- **gastro** = stomach

gastric juice: the digestive secretion of the gastric glands of the stomach.

hydrochloric acid: an acid composed of hydrogen and chloride atoms (HCl) that is normally produced by the gastric glands.

mucus (MYOO-kus): a slippery substance secreted by cells of the GI lining (and other body linings) that protects the cells from exposure to digestive juices (and other destructive agents). The lining of the GI tract with its coat of mucus is a **mucous membrane.** (The noun is **mucus;** the adjective is **mucous.**)

liver: the organ that manufactures bile. (The liver's many other functions are described in Chapter 7.)

bile: an emulsifier that prepares fats and oils for digestion; an exocrine secretion made by the liver, stored in the gallbladder, and released into the small intestine when needed.

emulsifier (ee-MUL-sih-fire): a substance with both water-soluble and fat-soluble portions that promotes the mixing of oils and fats in a watery solution.

pancreatic (pank-ree-AT-ic) **juice:** the exocrine secretion of the pancreas, containing enzymes for the digestion of carbohydrate, fat, and protein as well as bicarbonate, a neutralizing agent. The juice flows from the pancreas into the small intestine through the pancreatic duct. (The pancreas also has an endocrine function, the secretion of insulin and other hormones.)

bicarbonate: an alkaline compound with the formula HCO_3 that is secreted from the pancreas as part of the pancreatic juice. (Bicarbonate is also produced in all cell fluids from the dissociation of carbonic acid to help maintain the body's acid-base balance.)

FIGURE 3-7 The pH Scale

A substance's acidity or alkalinity is measured in pH units. The pH is the negative logarithm of the hydrogen ion concentration. Each increment represents a tenfold increase in concentration of hydrogen particles. This means, for example, that a pH of 2 is 1000 times stronger than a pH of 5.

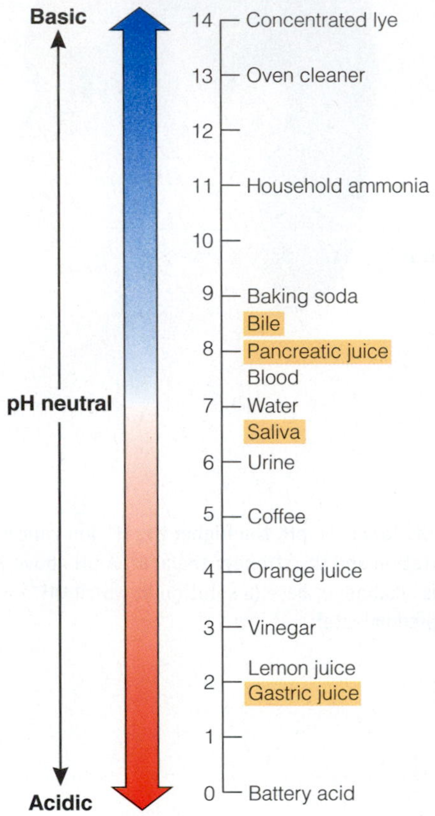

In addition to enzymes, the pancreatic juice contains sodium **bicarbonate**, which is basic or alkaline—the opposite of the stomach's acid (review Figure 3-7). The pancreatic juice thus neutralizes the acidic chyme arriving in the small intestine from the stomach. From this point on, the chyme remains at a neutral or slightly alkaline pH. The enzymes of both the intestine and the pancreas work best in this environment.

Bile **Bile** also flows into the duodenum. The **liver** continuously produces bile, which is then concentrated and stored in the gallbladder. The gallbladder squirts the bile into the duodenum of the small intestine when fat arrives there. Bile is not an enzyme; it is an **emulsifier** that brings fats into suspension in water so that enzymes can break them down into their component parts. A summary box of digestive secretions and their actions is presented below.

The Final Stage

At this point, the three energy-yielding nutrients—carbohydrate, fat, and protein—have been digested and are ready to be absorbed. Some vitamins and minerals are altered slightly during digestion, but most are absorbed as they are. Undigested residues, such as some fibers, are not absorbed. Instead, they continue through the digestive tract, carrying some minerals, bile acids, additives, and contaminants out of the body. This semisolid mass helps exercise the GI muscles and keep them strong enough to perform peristalsis efficiently. Fiber also retains water, accounting for the consistency of **stools**.

By the time the contents of the GI tract reach the end of the small intestine, little remains but water, a few dissolved salts and body secretions, and undigested materials such as fiber (with some fat, cholesterol, and a few minerals bound to it). All of this remaining matter enters the large intestine (colon).

In the colon, intestinal bacteria ferment some fibers, producing water, gas, and small fragments of fat that provide energy for the cells of the colon. The colon itself retrieves all materials that the body can recycle—water and dissolved salts. The waste that is finally excreted has little or nothing of value left in it. The body has extracted all that it can use from the food. Figure 3-8 summarizes digestion by following a sandwich through the GI tract and into the body.

IN SUMMARY As Figure 3-1 shows, food enters the mouth and travels down the esophagus and through the upper and lower esophageal sphincters to the stomach, then through the pyloric sphincter to the small intestine, on through the ileocecal valve to the large intestine, past the appendix to the rectum, ending at the anus. The wavelike contractions of peristalsis and the periodic squeezing of segmentation keep things moving at a reasonable pace. Along the way, secretions from the salivary glands, stomach, pancreas, liver (via the gallbladder), and small intestine deliver fluids and digestive enzymes.

Summary of Digestive Secretions and Their Major Actions

Organ or Gland	Target Organ	Secretion	Action
Salivary glands	Mouth	Saliva	Fluid eases swallowing; salivary enzyme breaks down some *carbohydrate*.*
Gastric glands	Stomach	Gastric juice	Fluid mixes with bolus; hydrochloric acid uncoils *proteins;* enzymes break down proteins; mucus protects stomach cells.*
Pancreas	Small intestine	Pancreatic juice	Bicarbonate neutralizes acidic gastric juices; pancreatic enzymes break down *carbohydrates*, *fats*, and *proteins*.
Liver	Gallbladder	Bile	Bile stored until needed.
Gallbladder	Small intestine	Bile	Bile emulsifies *fat* so that enzymes can have access to break it down.
Intestinal glands	Small intestine	Intestinal juice	Intestinal enzymes break down *carbohydrate*, *fat*, and *protein* fragments; mucus protects the intestinal wall.

*Saliva and gastric juice also contain lipases, but most fat breakdown occurs in the small intestine.

stools: waste matter discharged from the colon; also called **feces** (FEE-seez).

FIGURE 3-8 The Digestive Fate of a Sandwich

To review the digestive processes, follow a peanut butter and banana sandwich on whole-wheat, sesame seed bread through the GI tract. As the graph on the right illustrates, digestion of the energy nutrients begins in different parts of the GI tract, but all are ready for absorption by the time they reach the end of the small intestine.

Animated! figure
www.cengagebrain.com
(search for ISBN 084006845X)

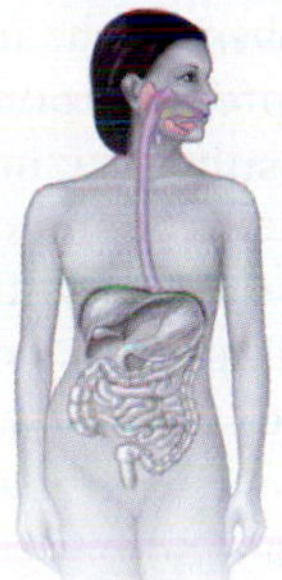

MOUTH: CHEWING AND SWALLOWING, WITH LITTLE DIGESTION

Carbohydrate digestion begins as the salivary enzyme starts to break down the starch from bread and peanut butter.
Fiber covering on the sesame seeds is crushed by the teeth, which exposes the nutrients inside the seeds to the upcoming digestive enzymes.

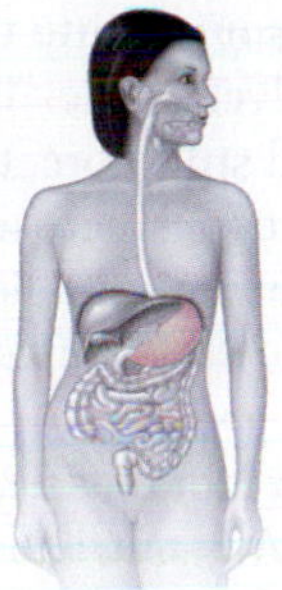

STOMACH: COLLECTING AND CHURNING, WITH SOME DIGESTION

Carbohydrate digestion continues until the mashed sandwich has been mixed with the gastric juices; the stomach acid of the gastric juices inactivates the salivary enzyme, and carbohydrate digestion ceases.
Proteins from the bread, seeds, and peanut butter begin to uncoil when they mix with the gastric acid, making them available to the gastric protease enzymes that begin to digest proteins.
Fat from the peanut butter forms a separate layer on top of the watery mixture.

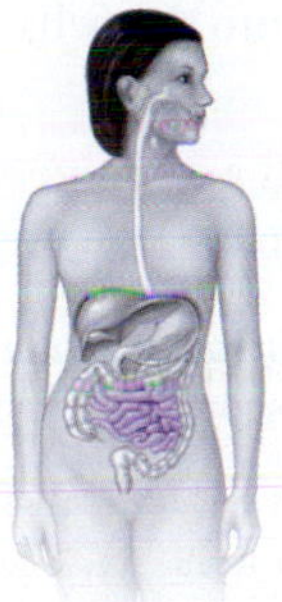

SMALL INTESTINE: DIGESTING AND ABSORBING

Sugars from the banana require so little digestion that they begin to traverse the intestinal cells immediately on contact.
Starch digestion picks up when the pancreas sends pancreatic enzymes to the small intestine via the pancreatic duct. Enzymes on the surfaces of the small intestinal cells complete the process of breaking down starch into small fragments that can be absorbed through the intestinal cell walls and into the hepatic portal vein.
Fat from the peanut butter and seeds is emulsified with the watery digestive fluids by bile. Now the pancreatic and intestinal lipases can begin to break down the fat to smaller fragments that can be absorbed through the cells of the small intestinal wall and into the lymph.
Protein digestion depends on the pancreatic and intestinal proteases. Small fragments of protein are liberated and absorbed through the cells of the small intestinal wall and into the hepatic portal vein.
Vitamins and minerals are absorbed.

Note: Sugars and starches are members of the carbohydrate family.

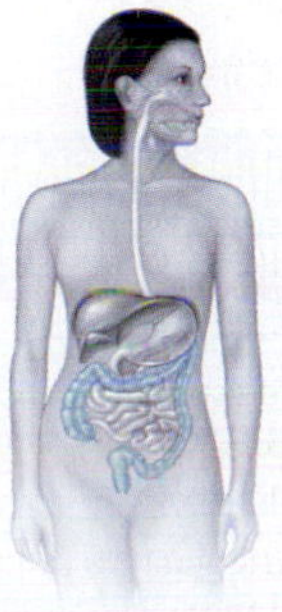

LARGE INTESTINE: REABSORBING AND ELIMINATING

Fluids and some minerals are absorbed.
Some fibers from the seeds, whole-wheat bread, peanut butter, and banana are partly digested by the bacteria living in the large intestine, and some of these products are absorbed.
Most fibers pass through the large intestine and are excreted as feces; some fat, cholesterol, and minerals bind to fiber and are also excreted.

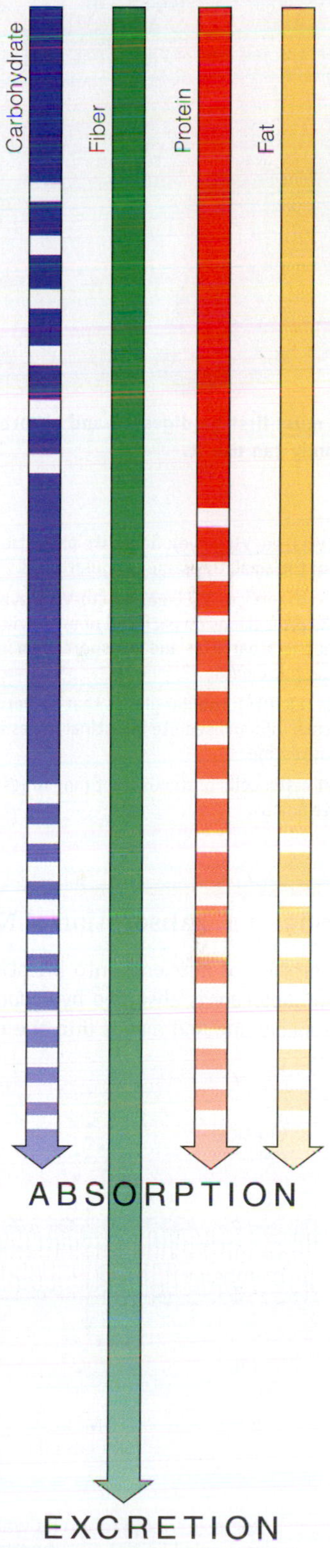

Food must first be digested and absorbed before the body can use it.

villi (VILL-ee, VILL-eye): fingerlike projections from the folds of the small intestine; singular *villus*.

microvilli (MY-cro-VILL-ee, MY-cro-VILL-eye): tiny, hairlike projections on each cell of every villus that can trap nutrient particles and transport them into the cells; singular *microvillus*.

crypts (KRIPTS): tubular glands that lie between the intestinal villi and secrete intestinal juices into the small intestine.

goblet cells: cells of the GI tract (and lungs) that secrete mucus.

Absorption

Within three or four hours after a person has eaten a dinner of beans and rice (or spinach lasagna, or steak and potatoes) with vegetable, salad, beverage, and dessert, the body must find a way to absorb the molecules derived from carbohydrate, protein, and fat digestion—and the vitamin and mineral molecules as well. Most absorption takes place in the small intestine, one of the most elegantly designed organ systems in the body. Within its 10-foot length, which provides a surface area equivalent to a tennis court, the small intestine traps and absorbs the nutrient molecules. To remove the absorbed molecules rapidly and provide room for more to be absorbed, a rush of circulating blood continuously washes the underside of this surface, carrying the absorbed nutrients away to the liver and other parts of the body. Figure 3-9 describes how most nutrients are absorbed by simple diffusion, facilitated diffusion, or active transport. Later chapters provide details on specific nutrients. Before following nutrients through the body, we must look more closely at the anatomy of the absorptive system.

Anatomy of the Absorptive System The inner surface of the small intestine looks smooth and slippery, but when viewed through a microscope, it turns out to be wrinkled into hundreds of folds. Each fold is contoured into thousands of fingerlike projections, as numerous as the hairs on velvet fabric. These small intestinal projections are the **villi.** A single villus, magnified still more, turns out to be composed of hundreds of cells, each covered with its own microscopic hairs, the **microvilli** (see Figure 3-10). In the crevices between the villi lie the **crypts**—tubular glands that secrete the intestinal juices into the small intestine. Nearby **goblet cells** secrete mucus.

The villi are in constant motion. Each villus is lined by a thin sheet of muscle, so it can wave, squirm, and wriggle like the tentacles of a sea anemone. Any nutrient molecule small enough to be absorbed is trapped among the microvilli that coat the cells, and then it is drawn into the cells. Some partially digested nutrients are caught in the microvilli, digested further by enzymes there, and then absorbed into the cells.

FIGURE 3-9 Absorption of Nutrients

Absorption of nutrients into intestinal cells typically occurs by simple diffusion, facilitated diffusion, or active transport. Occasionally, a large molecule is absorbed by *endocytosis*—a process in which the cell membrane engulfs the molecule, forming a sac that separates from the membrane and moves into the cell.

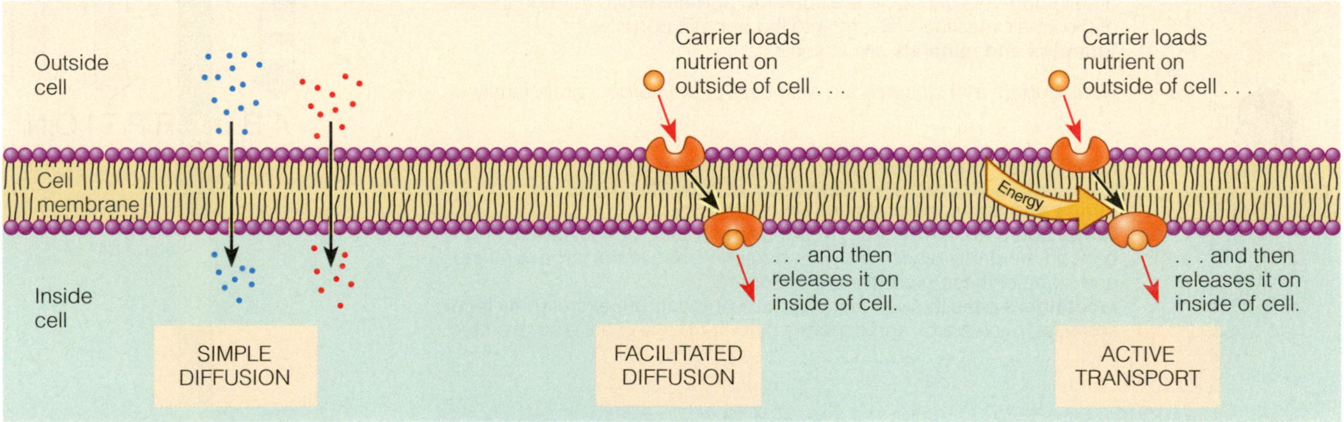

Some nutrients (such as water and small lipids) are absorbed by simple diffusion. They cross into intestinal cells freely.

Some nutrients (such as the water-soluble vitamins) are absorbed by facilitated diffusion. They need a specific carrier to transport them from one side of the cell membrane to the other. (Alternatively, facilitated diffusion may occur when the carrier changes the cell membrane in such a way that the nutrients can pass through.)

Some nutrients (such as glucose and amino acids) must be absorbed actively. These nutrients move against a concentration gradient, which requires energy.

A Closer Look at the Intestinal Cells The cells of the villi are among the most amazing in the body, for they recognize and select the nutrients the body needs and regulate their absorption. As already described, each cell of a villus is coated with thousands of microvilli, which project from the cell's membrane (review Figure 3-10). In these microvilli, and in the membrane, lie hundreds of different kinds of enzymes and "pumps," which recognize and act on different nutrients. Descriptions of specific enzymes and "pumps" for each nutrient are presented

FIGURE 3-10 The Small Intestinal Villi

Absorption of nutrients into intestinal cells typically occurs by simple diffusion or active transport.

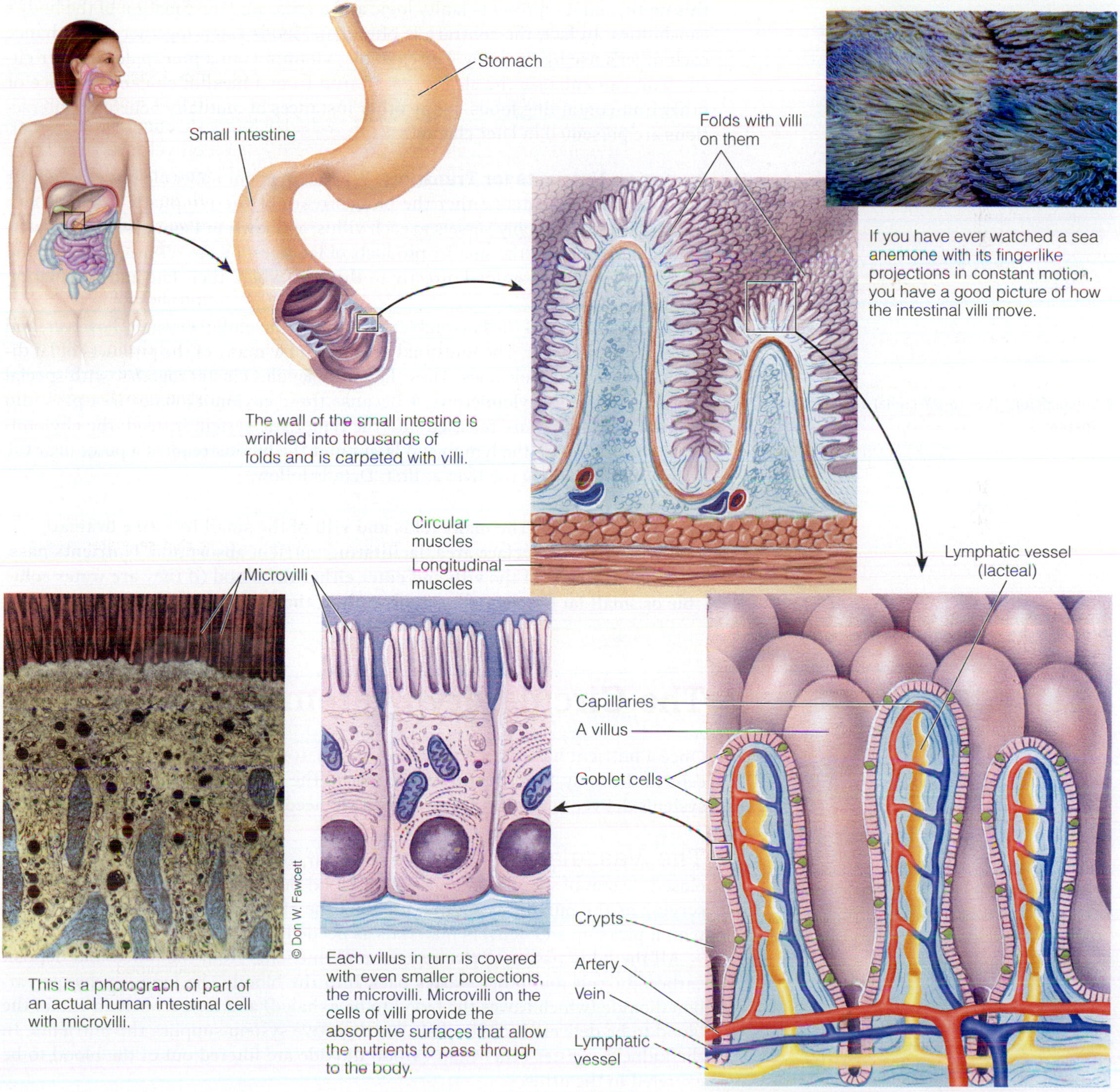

Stomach
Small intestine
The wall of the small intestine is wrinkled into thousands of folds and is carpeted with villi.
Folds with villi on them
Circular muscles
Longitudinal muscles
© littlesam/Shutterstock.com
If you have ever watched a sea anemone with its fingerlike projections in constant motion, you have a good picture of how the intestinal villi move.
Lymphatic vessel (lacteal)
Microvilli
© Don W. Fawcett
This is a photograph of part of an actual human intestinal cell with microvilli.
Each villus in turn is covered with even smaller projections, the microvilli. Microvilli on the cells of villi provide the absorptive surfaces that allow the nutrients to pass through to the body.
Capillaries
A villus
Goblet cells
Crypts
Artery
Vein
Lymphatic vessel

in the following chapters where appropriate; the point here is that the cells are equipped to handle all kinds and combinations of foods and their nutrients.

Specialization in the GI Tract A further refinement of the system is that the cells of successive portions of the intestinal tract are specialized to absorb different nutrients. The nutrients that are ready for absorption early are absorbed near the top of the GI tract; those that take longer to be digested are absorbed farther down. Registered dietitians and medical professionals who treat digestive disorders learn the specialized absorptive functions of different parts of the GI tract so that if one part becomes dysfunctional, the diet can be adjusted accordingly.

The Myth of "Food Combining" The idea that people should not eat certain food combinations (for example, fruit and meat) at the same meal, because the digestive system cannot handle more than one task at a time, is a myth. The art of "food combining" (which actually emphasizes "food separating") is based on this myth, and it represents faulty logic and a gross underestimation of the body's capabilities. In fact, the contrary is often true; foods eaten together can enhance each other's use by the body. For example, vitamin C in a pineapple or other citrus fruit can enhance the absorption of iron from a meal of chicken and rice or other iron-containing foods. Many other instances of mutually beneficial interactions are presented in later chapters.

Preparing Nutrients for Transport When a nutrient molecule has crossed the cell of a villus, it enters either the bloodstream or the lymphatic system. Both transport systems supply vessels to each villus, as shown in Figure 3-10. The water-soluble nutrients and the smaller products of fat digestion are released directly into the bloodstream and guided directly to the liver where their fate and destination will be determined.

The larger fats and the fat-soluble vitamins are insoluble in water, however, and blood is mostly water. The intestinal cells assemble many of the products of fat digestion into larger molecules. These larger molecules cluster together with special proteins, forming chylomicrons. ♦ Because these chylomicrons cannot pass into the **capillaries**, they are released into the lymphatic system instead; the chylomicrons move through the lymph and later enter the bloodstream at a point near the heart, thus bypassing the liver at first. Details follow.

♦ Chylomicrons (kye-lo-MY-cronz) are described in Chapter 5.

IN SUMMARY **The many folds and villi of the small intestine dramatically increase its surface area, facilitating nutrient absorption. Nutrients pass through the cells of the villi and enter either the blood (if they are water soluble or small fat fragments) or the lymph (if they are fat soluble).**

The Circulatory Systems

Once a nutrient has entered the bloodstream, it may be transported to any of the cells in the body, from the tips of the toes to the roots of the hair. The circulatory systems deliver nutrients wherever they are needed.

The Vascular System

The vascular, or blood circulatory, system is a closed system of vessels through which blood flows continuously, with the heart serving as the pump (see Figure 3-11). As the blood circulates through this system, it picks up and delivers materials as needed.

All the body tissues derive oxygen and nutrients from the blood and deposit carbon dioxide and other wastes back into the blood. The lungs exchange carbon dioxide (which leaves the blood to be exhaled) and oxygen (which enters the blood to be delivered to all cells). The digestive system supplies the nutrients. In the kidneys, wastes other than carbon dioxide are filtered out of the blood to be excreted in the urine.

capillaries (CAP-ill-aries): small vessels that branch from an artery. Capillaries connect arteries to veins. Exchange of oxygen, nutrients, and waste materials takes place across capillary walls.

FIGURE 3-11 **The Vascular System**

Animated! figure
www.cengagebrain.com
(search for ISBN 084006845X)

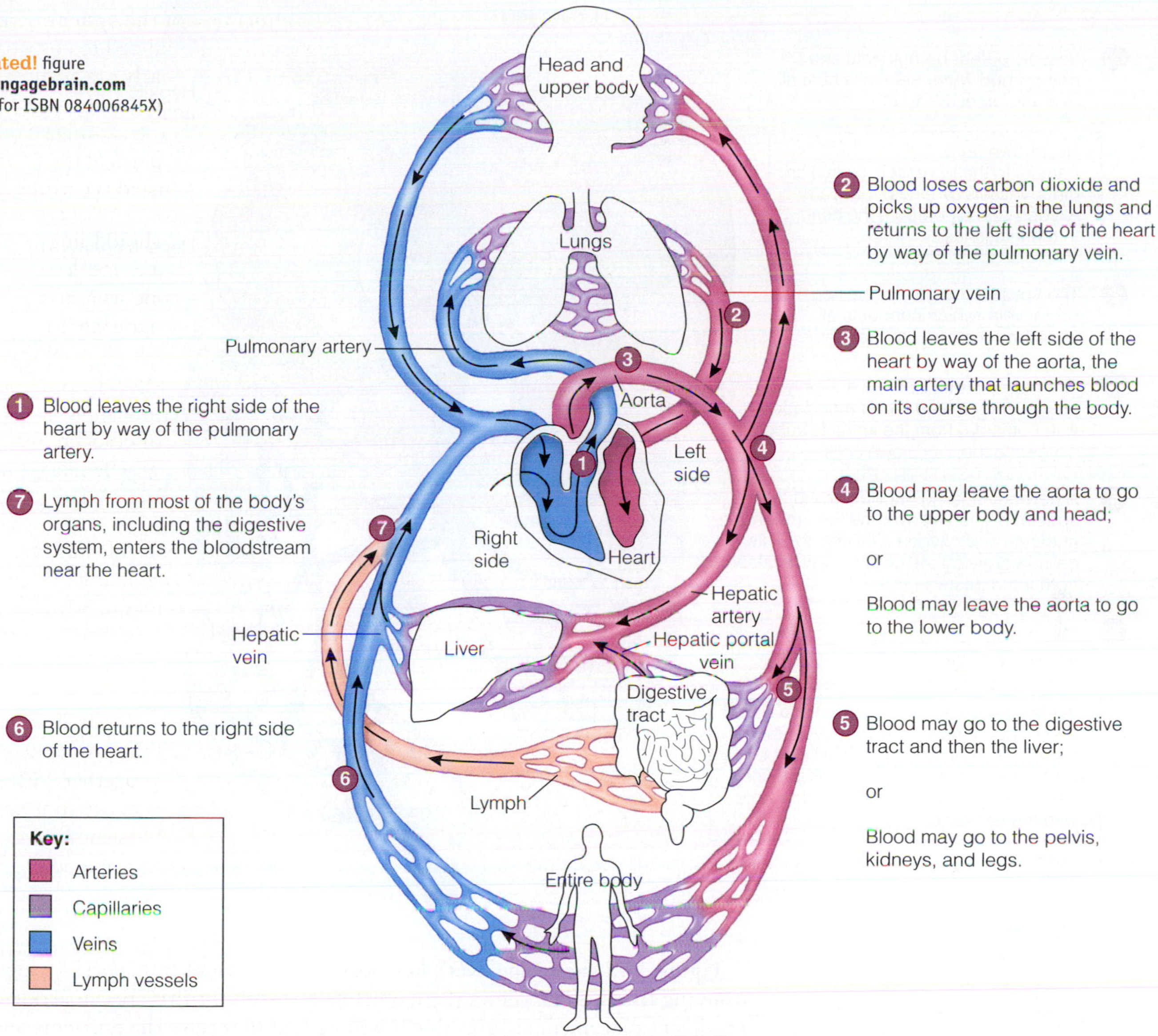
Head and upper body
Lungs
Pulmonary artery
Aorta
Left side
Right side
Heart
Pulmonary vein
Hepatic artery
Hepatic portal vein
Hepatic vein
Liver
Digestive tract
Lymph
Entire body
1 Blood leaves the right side of the heart by way of the pulmonary artery.
7 Lymph from most of the body's organs, including the digestive system, enters the bloodstream near the heart.
6 Blood returns to the right side of the heart.
2 Blood loses carbon dioxide and picks up oxygen in the lungs and returns to the left side of the heart by way of the pulmonary vein.
3 Blood leaves the left side of the heart by way of the aorta, the main artery that launches blood on its course through the body.
4 Blood may leave the aorta to go to the upper body and head;
or
Blood may leave the aorta to go to the lower body.
5 Blood may go to the digestive tract and then the liver;
or
Blood may go to the pelvis, kidneys, and legs.
Key:
Arteries
Capillaries
Veins
Lymph vessels

Blood leaving the right side of the heart circulates through the lungs and then back to the left side of the heart. The left side of the heart then pumps the blood out of the **aorta** through **arteries** to all systems of the body. The blood circulates in the capillaries, where it exchanges material with the cells and then collects into **veins**, which return it again to the right side of the heart. In short, blood travels this simple route:

- Heart to arteries to capillaries to veins to heart

The routing of the blood leaving the digestive system has a special feature. The blood is carried to the digestive system (as to all organs) by way of an artery, which (as in all organs) branches into capillaries to reach every cell. Blood leaving the digestive system, however, goes by way of a vein. The **hepatic portal vein** directs blood not back to the heart, but to another organ—the liver. This vein branches into a network of large capillaries and their spaces so that every cell of the liver has access to the blood. Blood leaving the liver then *again* collects into a vein, called the **hepatic vein**, which returns blood to the heart.

The route is:

- Heart to arteries to capillaries (in intestines) to hepatic portal vein to capillaries (in liver) to hepatic vein to heart

aorta (ay-OR-tuh): the large, primary artery that conducts blood from the heart to the body's smaller arteries.

arteries: vessels that carry blood from the heart to the tissues.

veins (VANES): vessels that carry blood to the heart.

hepatic portal vein: the vein that collects blood from the GI tract and conducts it to the liver.
- **portal** = gateway

hepatic vein: the vein that collects blood from the liver and returns it to the heart.
- **hepatic** = liver

FIGURE 3-12 The Liver

1. Vessels gather up nutrients and reabsorbed water and salts from all over the digestive tract.

Not shown here:
Parallel to these vessels (veins) are other vessels (arteries) that carry oxygen-rich blood from the heart to the intestines.

2. The vessels merge into the hepatic portal vein, which conducts all absorbed materials to the liver.

3. The hepatic artery brings a supply of freshly oxygenated blood (not loaded with nutrients) from the lungs to supply oxygen to the liver's own cells.

4. A network of large capillaries and their spaces branch all over the liver, making nutrients and oxygen available to all its cells and giving the cells access to blood from the digestive system.

5. The hepatic vein gathers up blood in the liver and returns it to the heart.

In contrast, nutrients absorbed into lymph do not go to the liver first. They go to the heart, which pumps them to all the body's cells. The cells remove the nutrients they need, and the liver then has to deal only with the remnants.

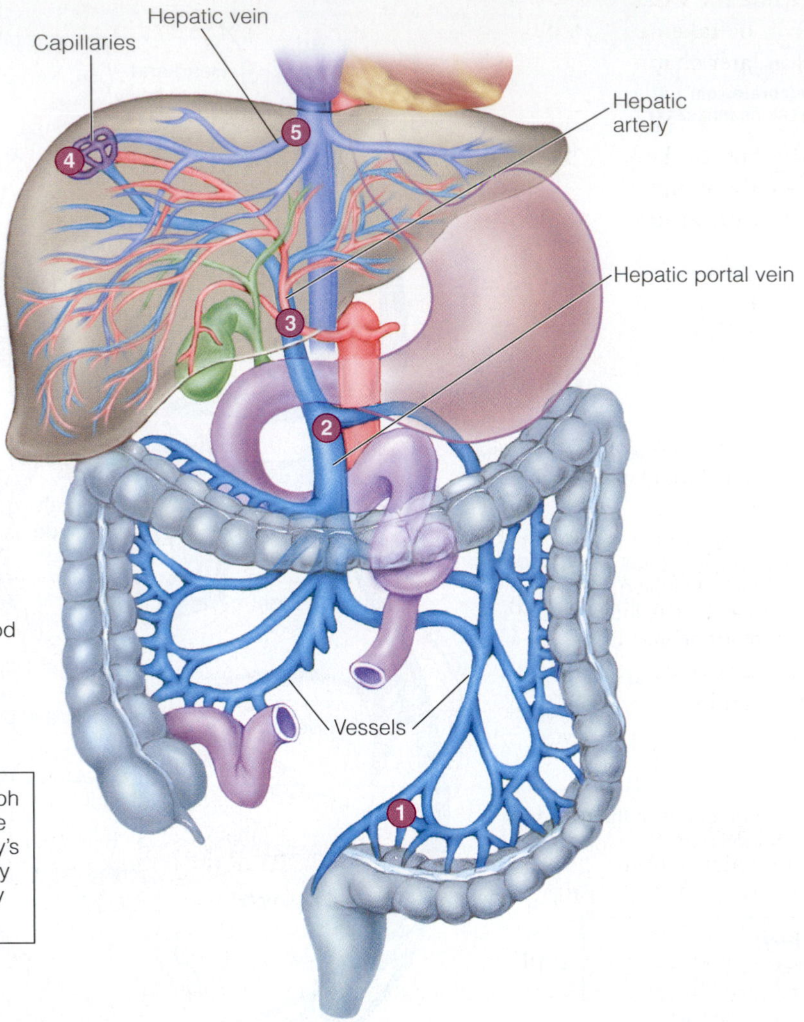

Figure 3-12 shows the liver's key position in nutrient transport. An anatomist studying this system knows there must be a reason for this special arrangement. The liver's placement ensures that it will be first to receive the nutrients absorbed from the GI tract. In fact, the liver has many jobs to do in preparing the absorbed nutrients for use by the body. It is the body's major metabolic organ.

In addition, the liver defends the body by detoxifying substances that might cause harm and preparing waste products for excretion. This is why, when people ingest poisons that succeed in passing the first barrier (the intestinal cells), the liver quite often suffers the damage—from viruses such as hepatitis, from drugs such as barbiturates or alcohol, from toxins such as pesticide residues, and from contaminants such as mercury. Perhaps, in fact, you have been undervaluing your liver, not knowing what heroic tasks it quietly performs for you.

The Lymphatic System

The **lymphatic system** provides a one-way route for fluid from the tissue spaces to enter the blood. Unlike the vascular system, the lymphatic system has no pump; instead, **lymph** circulates *between* the cells of the body and collects into tiny vessels. The fluid moves from one portion of the body to another as muscles contract and create pressure here and there. Ultimately, much of the lymph collects in the **thoracic duct** behind the heart. The thoracic duct opens into the **subclavian vein**, where the lymph enters the bloodstream. Thus nutrients from the GI tract that enter lymphatic vessels ♦ (large fats and fat-soluble vitamins) ultimately enter the bloodstream, circulating through arter-

♦ The lymphatic vessels of the intestine that take up nutrients and pass them to the lymph circulation are called **lacteals** (LACK-tee-als).

lymphatic (lim-FAT-ic) **system:** a loosely organized system of vessels and ducts that convey fluids toward the heart. The GI part of the lymphatic system carries the products of fat digestion into the bloodstream.

lymph (LIMF): a clear yellowish fluid that is similar to blood except that it contains no red blood cells or platelets. Lymph from the GI tract transports fat and fat-soluble vitamins to the bloodstream via lymphatic vessels.

thoracic (thor-ASS-ic) **duct:** the main lymphatic vessel that collects lymph and drains into the left subclavian vein.

subclavian (sub-KLAY-vee-an) **vein:** the vein that provides passage from the lymphatic system to the vascular system.

ies, capillaries, and veins like the other nutrients, with a notable exception—they bypass the liver at first.

Once inside the vascular system, the nutrients can travel freely to any destination and can be taken into cells, then used as needed. What becomes of them is described in later chapters.

IN SUMMARY **Nutrients leaving the digestive system via the blood are routed directly to the liver before being transported to the body's cells. Those leaving via the lymphatic system eventually enter the vascular system but bypass the liver at first.**

© Polara Sutdios Inc.

Eaten regularly, yogurt can alleviate common digestive problems.

The Health and Regulation of the GI Tract

This section describes the bacterial conditions and hormonal regulation of a healthy GI tract, but many factors ♦ can influence normal GI function. For example, peristalsis and sphincter action are poorly coordinated in newborns, so infants tend to "spit up" during the first several months of life. Older adults often experience constipation, in part because the intestinal wall loses strength and elasticity with age, which slows GI motility. Diseases can also interfere with digestion and absorption and often lead to malnutrition. Lack of nourishment, in general, and lack of certain dietary constituents such as fiber, in particular, alter the structure and function of GI cells. Quite simply, GI tract health depends on adequate nutrition.

♦ Factors influencing GI function:
- Physical immaturity
- Aging
- Illness
- Nutrition

Gastrointestinal Bacteria

An estimated 10 trillion bacteria ♦ representing some 400 or more different species and subspecies live in a healthy GI tract. The prevalence of different bacteria in various parts of the GI tract depends on such factors as pH, peristalsis, diet, and other microorganisms. Relatively few microorganisms can live in the low pH of the stomach with its relatively rapid peristalsis, whereas the neutral pH and slow peristalsis of the lower small intestine and the large intestine permit the growth of a diverse and abundant bacterial population.[1]

♦ Bacteria in the intestines are sometimes referred to as **flora.**

Most bacteria in the GI tract are not harmful; in fact, they may actually be beneficial. Provided that the normal intestinal flora are thriving, infectious bacteria have a hard time establishing themselves to launch an attack on the system.

Diet is one of several factors that influence the body's bacterial population and environment. Consider **yogurt**, for example. Yogurt contains *Lactobacillus* and other living bacteria. These microorganisms are considered **probiotics** because they change the conditions and native bacterial colonies in the GI tract in ways that seem to benefit health.[2] The potential GI health benefits of probiotics include helping to alleviate diarrhea, constipation, inflammatory bowel disease, ulcers, allergies, lactose intolerance, and infant colic; enhance immune function; and protect against colon cancer.[3] Some probiotics may have adverse effects under certain circumstances.[4] Research studies continue to explore how diet influences GI bacteria and which foods—with their probiotics—affect GI health. In addition, research studies are beginning to reveal several health benefits beyond the GI tract—such as improving blood pressure and immune responses.[5]

♦ Food components (such as fibers) that are not digested in the small intestine, but are used instead as food by bacteria to encourage their growth or activity are called **prebiotics.** A mixture of probiotics and prebiotics forms a **synbiotic.**

GI bacteria also digest fibers and complex proteins.[6] ♦ In doing so, the bacteria produce nutrients such as short fragments of fat that the cells of the colon use for energy. Bacteria in the GI tract also produce several vitamins, ♦ although the amount is insufficient to meet the body's total need for these vitamins.[7]

♦ Vitamins produced by bacteria include:
- Biotin
- Folate
- Pantothenic acid
- Riboflavin
- Thiamin
- Vitamin B_6
- Vitamin B_{12}
- Vitamin K

yogurt: milk product that results from the fermentation of lactic acid in milk by *Lactobacillus bulgaricus* and *Streptococcus thermophilus*.

probiotics: living microorganisms found in foods and dietary supplements that, when consumed in sufficient quantities, are beneficial to health.
- **pro** = for
- **bios** = life

Gastrointestinal Hormones and Nerve Pathways

The ability of the digestive tract to handle its ever-changing contents illustrates an important physiological principle that governs the way all living things function—the principle of **homeostasis**. Simply stated, survival depends on body conditions staying about the same; if they deviate too far from the norm, the body must "do something" to bring them back to normal. The body's regulation of digestion is one example of homeostatic regulation. The body also regulates its temperature, its blood pressure, and all other aspects of its blood chemistry in similar ways.

Two intricate and sensitive systems coordinate all the digestive and absorptive processes: the hormonal (or endocrine) system and the nervous system. Even before the first bite of food is taken, the mere thought, sight, or smell of food can trigger a response from these systems. Then, as food travels through the GI tract, it either stimulates or inhibits digestive secretions by way of messages that are carried from one section of the GI tract to another by both **hormones** ♦ and nerve pathways. (Appendix A presents a brief summary of the body's hormonal system and nervous system.)

♦ In general, any gastrointestinal hormone may be called an **enterogastrone** (EN-ter-oh-GAS-trone), but the term often refers specifically to the **gastric inhibitory peptide** that slows motility and inhibits gastric secretions.

Notice that the kinds of regulation described next are all examples of *feedback* mechanisms. A certain condition demands a response. The response changes that condition, and the change then cuts off the response. Thus the system is self-correcting. Examples follow:

- *The stomach normally maintains a pH between 1.5 and 1.7. How does it stay that way?* Food entering the stomach stimulates cells in the stomach wall to release the hormone **gastrin**. Gastrin, in turn, stimulates the stomach glands to secrete the components of hydrochloric acid. When pH 1.5 is reached, the acid itself turns off the gastrin-producing cells. They stop releasing gastrin, and the glands stop producing hydrochloric acid. Thus the system adjusts itself, as Figure 3-13 shows.

  Nerve receptors in the stomach wall also respond to the presence of food and stimulate the gastric glands to secrete juices and the muscles to contract. As the stomach empties, the receptors are no longer stimulated, the flow of juices slows, and the stomach quiets down.

- *The pyloric sphincter opens to let out a little chyme, then closes again. How does it know when to open and close?* When the pyloric sphincter relaxes, acidic chyme slips through. The cells of the pyloric muscle on the intestinal side sense the acid, causing the pyloric sphincter to close tightly. Only after the chyme has been neutralized by pancreatic bicarbonate and the juices surrounding the pyloric sphincter have become alkaline can the muscle relax again. This process ensures that the chyme will be released slowly enough to be neutralized as it flows through the small intestine. This is important because the small intestine has less of a mucous coating than the stomach does and so is not as well protected from acid.

- *As the chyme enters the small intestine, the pancreas adds bicarbonate to it so that the intestinal contents always remain at a slightly alkaline pH. How does the pancreas know how much to add?* The presence of chyme stimulates the cells of the duodenum wall to release the hormone **secretin** into the blood. When secretin reaches the pancreas, it stimulates the pancreas to release its bicarbonate-rich juices. Thus, whenever the duodenum signals that acidic chyme is present, the pancreas responds by sending bicarbonate to neutralize it. When the need has been met, the cells of the duodenum wall are no longer stimulated to release secretin, the hormone no longer flows through the blood, the pancreas no longer receives the message, and it stops sending pancreatic juice. Nerves also regulate pancreatic secretions.

- *Pancreatic secretions contain a mixture of enzymes to digest carbohydrate, fat, and protein. How does the pancreas know how much of each type of enzyme to provide?* This is one of the most interesting questions physiologists have asked. Clearly, the pancreas does know what its owner has been eating, and it secretes enzyme mixtures tailored to handle the food mixtures that have been arriving recently (over the last several days). Enzyme activity changes proportionately in response to the amounts of carbohydrate, fat, and protein

FIGURE 3-13 An Example of a Negative Feedback Loop

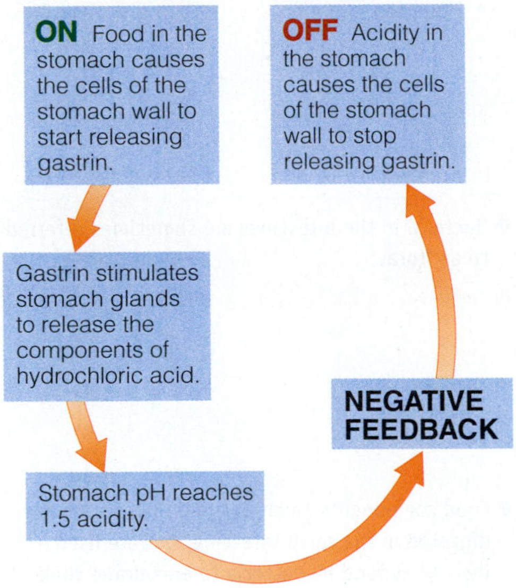
ON Food in the stomach causes the cells of the stomach wall to start releasing gastrin.
OFF Acidity in the stomach causes the cells of the stomach wall to stop releasing gastrin.
Gastrin stimulates stomach glands to release the components of hydrochloric acid.
NEGATIVE FEEDBACK
Stomach pH reaches 1.5 acidity.

homeostasis (HOME-ee-oh-STAY-sis): the maintenance of constant internal conditions (such as blood chemistry, temperature, and blood pressure) by the body's control systems. A homeostatic system is constantly reacting to external forces to maintain limits set by the body's needs.
- **homeo** = the same
- **stasis** = staying

hormones: chemical messengers. Hormones are secreted by a variety of glands in response to altered conditions in the body. Each hormone travels to one or more specific target tissues or organs, where it elicits a specific response to maintain homeostasis.

gastrin: a hormone secreted by cells in the stomach wall. Target organ: the glands of the stomach. Response: secretion of gastric acid.

secretin (see-CREET-in): a hormone produced by cells in the duodenum wall. Target organ: the pancreas. Response: secretion of bicarbonate-rich pancreatic juice.

in the diet. If a person has been eating mostly carbohydrates, the pancreas makes and secretes mostly carbohydrases; if the person's diet has been high in fat, the pancreas produces more lipases; and so forth. Presumably, hormones from the GI tract, secreted in response to meals, keep the pancreas informed as to its digestive tasks. The day or two lag between the time a person's diet changes dramatically and the time digestion of the new diet becomes efficient explains why dietary changes can "upset digestion" and should be made gradually.

- *Why don't the digestive enzymes damage the pancreas?* The pancreas protects itself from harm by producing an inactive form of the enzymes. ♦ It releases these proteins into the small intestine where they are activated to become enzymes. In pancreatitis, the digestive enzymes become active within the infected pancreas, causing inflammation and damaging the delicate pancreatic tissues.
- *When fat is present in the intestine, the gallbladder contracts to squirt bile into the intestine to emulsify the fat. How does the gallbladder get the message that fat is present?* Fat in the intestine stimulates cells of the intestinal wall to release the hormone **cholecystokinin (CCK)**. This hormone travels by way of the blood to the gallbladder and stimulates it to contract, which releases bile into the small intestine. Cholecystokinin also travels to the pancreas and stimulates it to secrete its juices, which releases bicarbonate and enzymes into the small intestine. Once the fat in the intestine is emulsified and enzymes have begun to work on it, the fat no longer provokes release of the hormone, and the message to contract is canceled. (By the way, fat emulsification can continue even after a diseased gallbladder has been surgically removed because the liver can deliver bile directly to the small intestine.)
- *Fat and protein take longer to digest than carbohydrate does. When fat or protein is present, intestinal motility slows to allow time for its digestion. How does the intestine know when to slow down?* Cholecystokinin is released in response to fat or protein in the small intestine. In addition to its role in fat emulsification and digestion, cholecystokinin slows GI tract motility. Slowing the digestive process helps to maintain a pace that allows all reactions to reach completion. Hormonal and nervous mechanisms like these account for much of the body's ability to adapt to changing conditions.

Table 3-1 summarizes the actions of these three GI hormones. Gastrin, secretin, and cholecystokinin are among the most studied GI hormones, but the GI tract releases more than 20 hormones.[8] In addition to assisting with digestion and absorption, many of these hormones regulate food intake and influence satiation. ♦ Current research is focusing on the roles these hormones may play in the development of obesity and its treatments (more details provided in Chapter 8).[9]

Once a person has started to learn the answers to questions like these, it may be hard to stop. Some people devote their whole lives to the study of physiology. For now, however, these few examples illustrate how all the processes throughout

♦ The inactive precursor of an enzyme is called a **zymogen** (ZYE-mo-jen).
- **zym** = concerning enzymes
- **gen** = to produce

♦ As Chapter 8 explains, **satiation** is the feeling of satisfaction and fullness that occurs during a meal and halts eating.

cholecystokinin (COAL-ee-SIS-toe-KINE-in), or **CCK:** a hormone produced by cells of the intestinal wall. Target organ: the gallbladder. Response: release of bile and slowing of GI motility.

TABLE 3-1 The Primary Actions of Selected GI Hormones

Hormone	Responds to	Secreted from	Stimulates	Response
Gastrin	Food in the stomach	Stomach wall	Stomach glands	Hydrochloric acid secreted into the stomach
Secretin	Acidic chyme in the small intestine	Duodenal wall	Pancreas	Bicarbonate-rich juices secreted into the small intestine
Cholecystokinin	Fat or protein in the small intestine	Intestinal wall	Gallbladder	Bile secreted into the duodenum
			Pancreas	Bicarbonate- and enzyme-rich juices secreted into the small intestine

the digestive system are precisely and automatically regulated without any conscious effort.

IN SUMMARY **A diverse and abundant bacteria population supports GI health. The regulation of GI processes depends on the coordinated efforts of the hormonal system and the nervous system. Together, digestion and absorption break down foods into nutrients for the body's use.**

The System at Its Best

This chapter describes the anatomy of the digestive tract on several levels: the sequence of digestive organs, the cells and structures of the villi, and the selective machinery of the cell membranes. The intricate architecture of the digestive system makes it sensitive and responsive to conditions in its environment. Several different kinds of GI tract cells confer specific immunity against intestinal diseases such as inflammatory bowel disease. In addition, secretions from the GI tract—saliva, mucus, gastric acid, and digestive enzymes—not only help with digestion, but also defend against foreign invaders. Together the GI's team of bacteria, cells, and secretions defend the body against numerous challenges.

One indispensable condition is good health of the digestive system itself. Like all the other organs of the body, the GI tract depends on a healthy supply of blood. The cells of the GI tract become weak and inflamed when blood flow is diminished, as may occur in heart disease when arteries become clogged or blood clots form. Just as a diminished blood flow to the heart or brain can cause a heart attack or stroke, respectively, too little blood to the intestines ♦ can also be damaging—or even fatal.

♦ A diminished blood flow to the intestines is called **intestinal ischemia** (is-KEY-me-ah) and is characterized by abdominal pain, forceful bowel movements, and blood in the stool.

The health of the digestive system is also affected by such lifestyle factors as sleep, physical activity, and state of mind. Adequate sleep allows for repair and maintenance of tissue and removal of wastes that might impair efficient functioning. Activity promotes healthy muscle tone. Mental state influences the activity of regulatory nerves and hormones; for healthy digestion, mealtimes should be relaxed and tranquil. Pleasant conversations and peaceful environments during meals ease the digestive process.

Another factor in GI health is the kind of foods eaten. Among the characteristics of meals that promote optimal absorption of nutrients are those mentioned in Chapter 2: balance, moderation, variety, and adequacy. Balance and moderation require having neither too much nor too little of anything. For example, too much fat can be harmful, but some fat is beneficial in slowing down intestinal motility and providing time for absorption of some of the nutrients that are slow to be absorbed.

Monkey Business Images/Shutterstock.com

Nourishing foods and pleasant conversations support a healthy digestive system.

Variety is important for many reasons, but one is that some food constituents interfere with nutrient absorption. For example, some compounds common in high-fiber foods such as whole-grain cereals, certain leafy green vegetables, and legumes bind with minerals. To some extent, then, the minerals in those foods may become unavailable for absorption. These high-fiber foods are still valuable, but they need to be balanced with a variety of other foods that can provide the minerals.

As for adequacy—in a sense, this entire book is about dietary adequacy. A diet must provide all the essential nutrients, fiber, and energy in amounts sufficient to maintain health. But here, at the end of this chapter, is a good place to emphasize the interdependence of the nutrients. It could almost be said that every nutrient depends on every other. All the nutrients work together, and all are present in the cells of a healthy digestive tract. To maintain health and promote the functions of the GI tract, make balance, moderation, variety, and adequacy features of every day's meals.

Nutrition Portfolio

A digestive system that is well cared for most of the time can adjust to handle almost any diet or combination of foods with ease on occasion.

Go to Diet Analysis Plus and choose one of the days on which you have tracked your diet for the entire day. Choose the day you thought you ate most poorly, and looking at it, record in your journal answers to the following:

- Describe the physical and emotional environment that typically surrounds your meals, including how it affects you and how it might be improved.
- Did you experience any GI discomforts on that day? Do you experience any GI discomforts regularly? If so, which of the foods that you ate might have contributed to your discomfort? What can you do to prevent or alleviate GI problems in the future? Use Table H3-1 (p. 94) as a guide.
- List any changes you can make in your eating habits to promote overall GI health.

Diet Analysis PLUS

To complete this exercise, go to your Diet Analysis Plus at www.cengage.com/sso.

Nutrition on the Net

For further study of topics covered in this chapter, log on to **www.cengagebrain.com** and search for ISBN 084006845X.

- Visit the patient information section of the American College of Gastroenterology: **www.acg.gi.org**

References

1. P. B. Eckburg and coauthors, Diversity of the human intestinal microbial flora, *Science* 308 (2005): 1635–1638.
2. C. C. Chen and W. A. Walker, Probiotics and prebiotics: Role in clinical disease states, *Advances in Pediatrics* 52 (2005): 77–113.
3. J. Rafter and coauthors, Dietary synbiotics reduce cancer risk factors in polypectomized and colon cancer patients, *American Journal of Clinical Nutrition* 85 (2007): 488–496; F. Savino and coauthors, *Lactobacillus reuteri* (American type culture collection strain 55730) versus simethicone in the treatment of infantile coli: A prospective randomized study, *Pediatrics* 119 (2007): e124; S. Santosa, E. Farnworth, and P. J. H. Jones, Probiotics and their potential health claims, *Nutrition Reviews* 64 (2006): 265–274; F. Guarner and coauthors, Should yoghurt cultures be considered probiotic? *British Journal of Nutrition* 93 (2005): 783–786.
4. J. Ezendam and H. van Loveren, Probiotics: Immunomodulation and evaluation of safety and efficacy, *Nutrition Reviews* 64 (2006): 1–14.
5. N. G. Hord, Eukaryotic-microbiota crosstalk: Potential mechanisms for health benefits of prebiotics and probiotics, *Annual Review of Nutrition* 28 (2008): 215–231; I. Lenoir-Wijnkoop and coauthors, Probiotic and prebiotic influence beyond the intestinal tract, *Nutrition Reviews* 65 (2007): 469–489; M. Liong, Probiotics: A critical review of their potential role as antihypertensives, immune modulators, hypocholesterolemics, and perimenopausal treatments, *Nutrition Reviews* 65 (2007): 316–328.
6. J. M. Wong and coauthors, Colonic health: Fermentation and short chain fatty acids, *Journal of Clinical Gastroenterology* 40 (2006): 235–243.
7. H. M. Said and Z. M. Mohammed, Intestinal absorption of water-soluble vitamins: An update, *Current Opinion Gastroenterology* 22 (2006): 140–146.
8. K. G. Murphy, W. S. Dhillo, and S. R. Bloom, Gut peptides in the regulation of food intake and energy homeostasis, *Endocrine Reviews* 27 (2006): 719–727.
9. D. E. Cummings and J. Overduin, Gastrointestinal regulation of food intake, *Journal of Clinical Investigation* 117 (2007): 13–23.

HIGHLIGHT 3

Common Digestive Problems

© Corbis Super RF/Alamy

The facts of anatomy and physiology presented in Chapter 3 permit easy understanding of some common problems that occasionally arise in the digestive tract. Food may slip into the airways instead of the esophagus, causing choking. Bowel movements may be loose and watery, as in diarrhea, or painful and hard, as in constipation. Some people complain about belching, while others are bothered by intestinal gas. Sometimes people develop medical problems such as ulcers. This highlight describes some of the symptoms of these common digestive problems and suggests strategies for preventing them (the accompanying glossary defines the relevant terms).

Choking

A person chokes when a piece of food slips into the **trachea** and becomes lodged so securely that it cuts off breathing (see Figure H3-1). Without oxygen, the person may suffer brain damage or die. For this reason, it is imperative that everyone learns to recognize a person grabbing his or her own throat as the international signal for choking (shown in Figure H3-2) and act promptly.

The choking scenario might read like this. A person is dining in a restaurant with friends. A chunk of food, usually meat, becomes lodged in his trachea so firmly that he cannot make a sound. No sound can be made because the **larynx** is in the trachea and makes sound only when air is pushed across it. Often he chooses to suffer

GLOSSARY

acid controllers: medications used to prevent or relieve indigestion by suppressing production of acid in the stomach; also called *H2 blockers*. Common brands include Pepcid AC, Tagamet HB, Zantac 75, and Axid AR.

antacids: medications used to relieve indigestion by neutralizing acid in the stomach. Common brands include Alka-Seltzer, Maalox, Rolaids, and Tums.

belching: the expulsion of gas from the stomach through the mouth.

colitis (ko-LYE-tis): inflammation of the colon.

colonic irrigation: the popular, but potentially harmful practice of "washing" the large intestine with a powerful enema machine.

constipation: the condition of having infrequent or difficult bowel movements.

defecate (DEF-uh-cate): to move the bowels and eliminate waste.
- **defaecare** = to remove dregs

diarrhea: the frequent passage of watery bowel movements.

diverticula (dye-ver-TIC-you-la): sacs or pouches that develop in the weakened areas of the intestinal wall (like bulges in an inner tube where the tire wall is weak).
- **divertir** = to turn aside

diverticulitis (DYE-ver-tic-you-LYE-tis): infected or inflamed diverticula.
- **itis** = infection or inflammation

diverticulosis (DYE-ver-tic-you-LOH-sis): the condition of having diverticula. About one in every six people in Western countries develops diverticulosis in middle or later life.
- **osis** = condition

enemas: solutions inserted into the rectum and colon to stimulate a bowel movement and empty the lower large intestine.

gastroesophageal reflux: the backflow of stomach acid into the esophagus, causing damage to the cells of the esophagus and the sensation of heartburn. **Gastroesophageal reflux disease (GERD)** is characterized by symptoms of reflux occurring two or more times a week.

heartburn: a burning sensation in the chest area caused by backflow of stomach acid into the esophagus.

Heimlich (HIME-lick) **maneuver (abdominal thrust maneuver):** a technique for dislodging an object from the trachea of a choking person (see Figure H3-2); named for the physician who developed it.

hemorrhoids (HEM-oh-royds): painful swelling of the veins surrounding the rectum.

hiccups (HICK-ups): repeated cough-like sounds and jerks that are produced when an involuntary spasm of the diaphragm muscle sucks air down the windpipe; also spelled *hiccoughs*.

indigestion: incomplete or uncomfortable digestion, usually accompanied by pain, nausea, vomiting, heartburn, intestinal gas, or belching.
- **in** = not

irritable bowel syndrome: an intestinal disorder of unknown cause. Symptoms include abdominal discomfort and cramping, diarrhea, constipation, or alternating diarrhea and constipation.

larynx (LAIR-inks): the entryway to the trachea that contains the vocal cords; also called the *voice box* (see Figure H3-1).

laxatives: substances that loosen the bowels and thereby prevent or treat constipation.

mineral oil: a purified liquid derived from petroleum and used to treat constipation.

peptic ulcer: a lesion in the mucous membrane of either the stomach (a gastric ulcer) or the duodenum (a duodenal ulcer).
- **peptic** = concerning digestion

trachea (TRAKE-ee-uh): the airway from the larynx to the lungs; also called the *windpipe*.

ulcer: a lesion of the skin or mucous membranes characterized by inflammation and damaged tissues. See also *peptic ulcer*.

vomiting: expulsion of the contents of the stomach up through the esophagus to the mouth.

FIGURE H3-1 Normal Swallowing and Choking

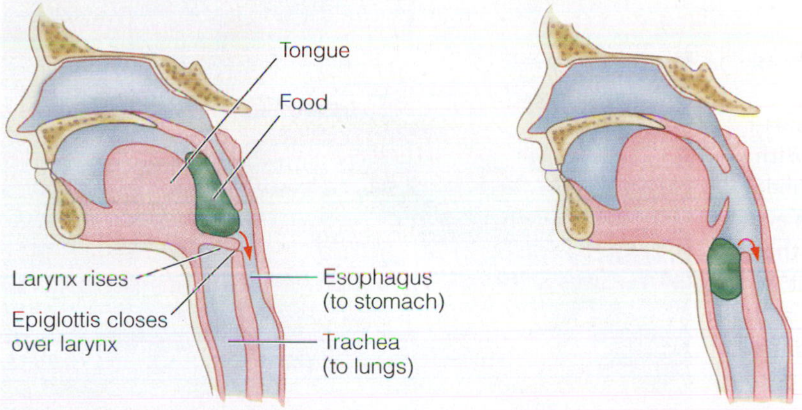

Swallowing. The epiglottis closes over the larynx, blocking entrance to the lungs via the trachea. The red arrow shows that food is heading down the esophagus normally.

Choking. A choking person cannot speak or gasp because food lodged in the trachea blocks the passage of air. The red arrow points to where the food should have gone to prevent choking.

alone rather than "make a scene in public." If he tries to communicate distress to his friends, he must depend on pantomime. The friends are bewildered by his antics and become terribly worried when he "faints" after a few minutes without air. They call for an ambulance, but by the time it arrives, he is dead from suffocation.

To help a person who is choking, first ask this critical question: "Can you make any sound at all?" If so, relax. You have time to decide what you can do to help. Whatever you do, *do not* hit him on the back—the particle may become lodged more firmly in his air passage. If the person cannot make a sound, shout for help and perform the **Heimlich maneuver** (described in Figure H3-2). You would do well to take a lifesaving course and practice these techniques because you will have no time for hesitation if you are called upon to perform this death-defying act.

Almost any food can cause choking, although some are cited more often than others: chunks of meat, hot dogs, nuts, whole grapes, raw carrots, marshmallows, hard or sticky candies, gum, popcorn, and peanut butter. These foods are particularly difficult for young children to safely chew and swallow. In 2000, more than 17,500 children (younger than 15 years old) in the United States choked; most of them choked on food, and 160 of them choked to death.[1] Always remain alert to the dangers of choking whenever young children are eating. To prevent choking, cut food into small pieces, chew thoroughly before swallowing, don't talk or laugh with food in your mouth, and don't eat when breathing hard.

Vomiting

Another common digestive mishap is **vomiting.** Vomiting can be a symptom of many different diseases or may arise in situations that upset the body's equilibrium, such as air or sea travel. For whatever reason, the contents of the stomach are propelled up through the esophagus to the mouth and expelled.

If vomiting continues long enough or is severe enough, the muscular contractions will extend beyond the stomach and carry the contents of the duodenum, with its green bile, into the stomach and then up the esophagus. Although certainly unpleasant and wearying for the nauseated person, vomiting such as this is no cause for alarm. Vomiting is one of the body's adaptive mechanisms to rid itself of something irritating.

FIGURE H3-2 First Aid for Choking

The first-aid strategy most likely to succeed is abdominal thrusts, sometimes called the Heimlich maneuver. Only if all else fails, open the person's mouth by grasping both his tongue and lower jaw and lifting. Then, and *only* if you can see the object, use your finger to sweep it out and begin rescue breathing.

The universal signal for choking is when a person grabs his throat. It alerts others to the need for assistance. If this happens, stand behind the person, and wrap your arms around him. Place the thumb side of one fist snugly against his body, slightly above the navel and below the rib cage. Grasp your fist with your other hand and give him a sudden strong hug inward and upward. Repeat thrusts as necessary.

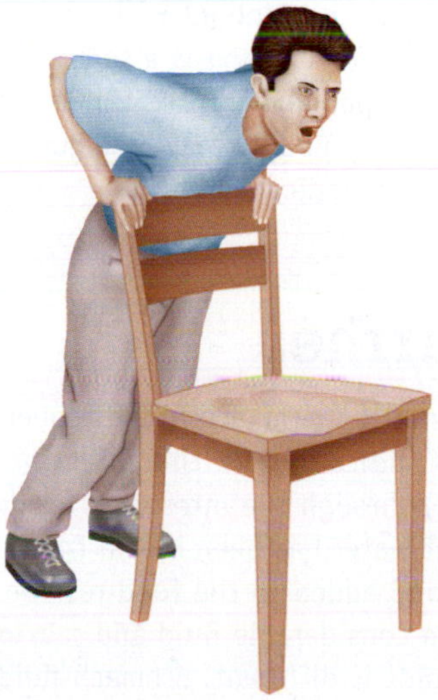

If you are choking and need to self-administer first aid, place the thumb side of one fist slightly above your navel and below your rib cage, grasp the fist with your other hand, and then press inward and upward with a quick motion. If this is unsuccessful, quickly press your upper abdomen over any firm surface such as the back of a chair, a countertop, or a railing.

The best advice is to rest and drink small amounts of liquids as tolerated until the nausea subsides.

A physician's care may be needed, however, when large quantities of fluid are lost from the GI tract, causing dehydration. With massive fluid loss from the GI tract, all of the body's other fluids redistribute themselves so that, eventually, fluid is taken from every cell of the body. Leaving the cells with the fluid are salts that are absolutely essential to the life of the cells, and they must be replaced. Replacement is difficult if the vomiting continues, and intravenous feedings of saline and glucose may be necessary while the physician diagnoses the cause of the vomiting and begins corrective therapy.

In an infant, vomiting is likely to become serious early in its course, and a physician should be contacted soon after onset. Infants have more fluid between their body cells than adults do, so more fluid can move readily into the digestive tract and be lost from the body. Consequently, the body water of infants becomes depleted and their body salt balance upset faster than in adults.

Self-induced vomiting, such as occurs in bulimia nervosa, also has serious consequences. In addition to fluid and salt imbalances, repeated vomiting can cause irritation and infection of the pharynx, esophagus, and salivary glands; erosion of the teeth and gums; and dental caries. The esophagus may rupture or tear, as may the stomach. Sometimes the eyes become red from pressure during vomiting. Bulimic behavior reflects underlying psychological problems that require intervention. (Bulimia nervosa is discussed fully in Highlight 8.)

Projectile vomiting is also serious. The contents of the stomach are expelled with such force that they leave the mouth in a wide arc like a bullet leaving a gun. This type of vomiting requires immediate medical attention.

Diarrhea

Diarrhea is characterized by frequent, loose, watery stools. Such stools indicate that the intestinal contents have moved too quickly through the intestines for fluid absorption to take place, or that water has been drawn from the cells lining the intestinal tract and added to the food residue. Like vomiting, diarrhea can lead to considerable fluid and salt losses, but the composition of the fluids is different. Stomach fluids lost in vomiting are highly acidic, whereas intestinal fluids lost in diarrhea are nearly neutral. When fluid losses require medical attention, correct replacement is crucial.

Diarrhea is a symptom of various medical conditions and treatments. It may occur abruptly in a healthy person as a result of infections (such as food poisoning) or as a side effect of medications. When used in large quantities, food ingredients such as the sugar alternative sorbitol and the fat alternative olestra may also cause diarrhea in some people. If a food is responsible, then that food must be omitted from the diet, at least temporarily. If medication is responsible, a different medicine, when possible, or a different form (injectable versus oral, for example) may alleviate the problem.

Mast3r/Shutterstock.com

Personal hygiene (such as regular hand washing with soap and water) and safe food preparation (as described in Highlight 18) are easy and effective steps to take in preventing diarrheal diseases.

Diarrhea may also occur as a result of disorders of the GI tract, such as irritable bowel syndrome or colitis. **Irritable bowel syndrome** is one of the most common GI disorders and is characterized by frequent or severe abdominal discomfort and a disturbance in the motility of the GI tract.[2] In most cases, GI contractions are stronger and last longer than normal, forcing intestinal contents through quickly and causing gas, bloating, and diarrhea. In some cases, however, GI contractions are weaker than normal, slowing the passage of intestinal contents and causing constipation. The exact cause of irritable bowel syndrome is not known, but researchers believe stress, genetics, and abnormal signals from the neurotransmitter serotonin are involved.[3] The condition seems to worsen for some people when they eat certain foods or during stressful events. These triggers seem to aggravate symptoms but not cause them. Dietary treatment hinges on identifying and avoiding individual foods that aggravate symptoms; small meals may also be beneficial. Other effective treatments include dietary fiber, antispasmodic drugs, and peppermint oil.[4]

People with **colitis,** an inflammation of the large intestine, may also suffer from severe diarrhea. They often benefit from complete bowel rest and medication. If treatment fails, surgery to remove the colon and rectum may be necessary.

Treatment for diarrhea depends on cause and severity, but it always begins with rehydration.[5] Mild diarrhea may subside with simple rest and extra liquids (such as clear juices and soups) to replace fluid losses. If diarrhea is bloody or if it worsens or persists—especially in an infant, young child, elderly person, or person with a compromised immune system—call a physician. Severe diarrhea can be life threatening.

Constipation

Like diarrhea, **constipation** describes a symptom, not a disease. Each person's GI tract has its own cycle of waste elimination, which depends on its owner's health, the type of food eaten, when it was eaten, and when the person takes time to **defecate.** What's normal for some people may not be normal for others. Some people have bowel movements three times a day; others may have them three times a week. The symptoms of constipation include straining during bowel movements, hard stools, and infrequent bowel movements (fewer than three per week). Abdominal discomfort, headaches, backaches, and the passing of gas sometimes accompany constipation.

Often a person's lifestyle may cause constipation. Being too busy to respond to the defecation signal is a common complaint. If a person receives the signal to defecate and ignores it, the signal may not return for several hours. In the meantime, fluids continue to be withdrawn from the fecal matter, so when the person does defecate, the stools are dry and hard. In such a case, a person's daily regimen may need to be revised to allow time to have a bowel movement when the body sends its signal. One possibility is to go to bed earlier in order to rise earlier, allowing ample time for a leisurely breakfast and a movement.

Although constipation usually reflects lifestyle habits, in some cases it may be a side effect of medication or may reflect a medical problem such as tumors that are obstructing the passage of waste. If discomfort is associated with passing fecal matter, seek medical advice to rule out disease. Once this has been done, simple treatments, such as increased fiber, fluids, and exercise are recommended before the use of medications.[6]

One dietary measure that may be appropriate is to increase dietary fiber to 20 to 25 grams per day gradually over the course of a week or two. Fibers found in fruits, vegetables, and whole grains help to prevent constipation by increasing fecal mass. In the GI tract, fiber attracts water, creating soft, bulky stools that stimulate bowel contractions to push the contents along. These contractions strengthen the intestinal muscles. The improved muscle tone, together with the water content of the stools, eases elimination, reducing the pressure in the rectal veins and helping to prevent **hemorrhoids.** Chapter 4 provides more information on fiber's role in maintaining a healthy colon and reducing the risks of colon cancer and diverticulosis. **Diverticulosis** is a condition in which the intestinal walls develop bulges in weakened areas, most commonly in the colon (see Figure H3-3). These bulging pockets, known as **diverticula,** can worsen constipation, entrap feces, and become painfully infected and inflamed **(diverticulitis).**[7] Treatment may require hospitalization, antibiotics, or surgery.

Drinking plenty of water in conjunction with eating high-fiber foods also helps to prevent constipation. The increased bulk physically stimulates the upper GI tract, promoting peristalsis throughout. Similarly, physical activity improves the muscle tone and motility of the digestive tract. As little as 30 minutes of physical activity a day can help prevent or alleviate constipation.

Eating prunes—or "dried plums" as some have renamed them—can also be helpful. Prunes are high in fiber and also contain a laxative substance.* If a morning defecation is desired, a person can drink prune juice at bedtime; if the evening is preferred, the person can drink prune juice with breakfast.

These suggested changes in lifestyle or diet should correct chronic constipation without the use of **laxatives, enemas,** or **mineral oil,** although television commercials often try to persuade people otherwise. One of the fallacies often perpetrated by advertisements is that one person's successful use of a product is a good recommendation for others to use that product. As a matter of fact, even dietary recommendations to relieve constipation may work for one person but may worsen the constipation of another. For instance, increasing fiber intake stimulates peristalsis and helps the person with a sluggish colon. Some people, though, have a spastic type of constipation, in which peristalsis promotes strong contractions that close off a segment of the colon and prevent passage; for these people, increasing fiber intake would be exactly the wrong thing to do.

A person who seems to need products such as laxatives frequently should seek a physician's advice. One potentially harmful but currently popular practice is **colonic irrigation**—the internal washing of the large intestine with a powerful enema machine. Such an extreme cleansing is not only unnecessary, but it can be hazardous, causing illness and death from equipment contamination, electrolyte depletion, and intestinal perforation. Less extreme practices can cause problems, too. Frequent use of laxatives and enemas can lead to dependency; upset the body's fluid, salt, and mineral balances; and, in the case of mineral oil, interfere with the absorption of fat-soluble vitamins. Mineral oil dissolves the vitamins but is not itself absorbed. Instead, it is excreted from the body, carrying the vitamins with it.

*This substance is dihydroxyphenyl isatin.

FIGURE H3-3 Diverticula in the Colon

Diverticula may develop anywhere along the GI tract, but they are most common in the colon.

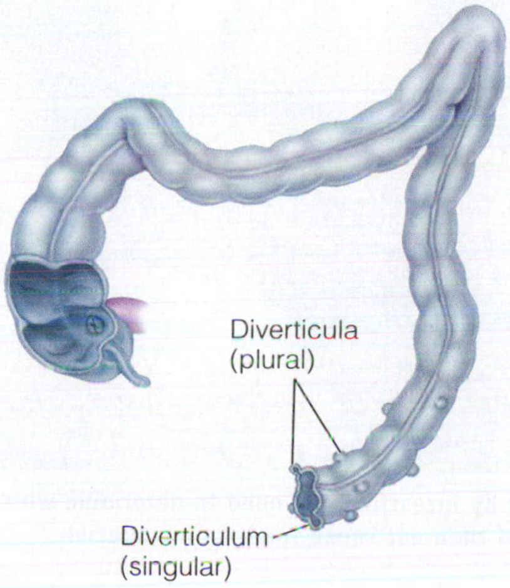

Belching and Gas

Many people complain of problems that they attribute to excessive gas. For some, **belching** is the complaint. Others blame intestinal gas for abdominal discomforts and embarrassment. Most people believe that the problems occur after they eat certain foods. This may be the case with intestinal gas, but belching results from swallowing air. The best advice for belching seems to be to eat slowly, chew thoroughly, and relax while eating.

Everyone swallows a little bit of air with each mouthful of food, but people who eat too fast may swallow too much air and then have to belch. Ill-fitting dentures, carbonated beverages, and chewing gum can also contribute to the swallowing of air with resultant belching. Occasionally, belching can be a sign of a more serious disorder, such as gallbladder disease or a peptic ulcer.

People who eat or drink too fast may also trigger **hiccups,** the repeated spasms that produce a cough-like sound and jerky movement. Normally, hiccups soon subside and are of no medical significance, but they can be bothersome. The most effective cure is to hold the breath for as long as possible, which helps to relieve the spasms of the diaphragm.

Although expelling intestinal gas can be a humiliating experience, it is quite normal. (People who experience painful bloating from malabsorption diseases, however, require medical treatment.) Healthy people expel several hundred milliliters of intestinal gas several times a day. Almost all (99 percent) of the gases expelled—nitrogen, oxygen, hydrogen, methane, and carbon dioxide—are odorless. The remaining "volatile" gases are the infamous ones.

Foods that produce gas usually must be determined individually. The most common offenders are foods rich in the carbohydrates—sugars, starches, and fibers. When partially digested carbohydrates reach the large intestine, bacteria digest them, giving off gas as a by-product. People can test foods suspected of forming gas by omitting them individually for a trial period to see if there is any improvement.

© Polara Studios Inc.

People troubled by intestinal gas need to determine which foods bother them and then eat those foods in moderation.

Heartburn and "Acid Indigestion"

Almost everyone has experienced **heartburn** at one time or another, usually soon after eating a meal. Medically known as **gastroesophageal reflux,** heartburn is the painful sensation a person feels behind the breastbone when the lower esophageal sphincter allows the stomach contents to reflux into the esophagus (see Figure H3-4).[8] This may happen if a person eats or drinks too much (or both). Tight clothing and even changes of position (lying down, bending over) can cause it, too, as can some medications and smoking. Weight gain and overweight increase the frequency, severity, and duration of heartburn symptoms.[9] A defect of the sphincter muscle itself is a possible, but less common, cause.

If heartburn is not caused by an anatomical defect, treatment is fairly simple. To avoid such misery in the future, the person needs to learn to eat less at a sitting, chew food more thoroughly, and eat more slowly. Additional strategies are presented in Table H3-1 at the end of this highlight.

As far as "acid indigestion" is concerned, recall from Chapter 3 that the strong acidity of the stomach is a desirable condition—television commercials for **antacids** and **acid controllers** notwithstanding. People who overeat or eat too quickly are likely to suffer from **indigestion.** The muscular reaction of the stomach to unchewed lumps or to being overfilled may be so violent that it upsets normal peristalsis. When this happens, overeaters may taste the stomach acid and feel pain. Responding to advertisements, they may reach for antacids or acid controllers. Both of these drugs were originally designed to treat GI illnesses such as ulcers. As is true of most over-the-counter medicines, antacids and acid controllers should be used only infrequently for occasional heartburn; they may mask or cause problems if used regularly. Acid-blocking drugs weaken the defensive mucous barrier of the GI tract, thereby increasing the risks of infections such as pneumonia, especially in vulnerable populations like the elderly. Instead of self-medicating, people who suffer from frequent and regular bouts of heartburn and indigestion should try the strategies presented in Table H3-1 later. If problems continue, they may need to see a physician, who can prescribe specific medication to control gastroesophageal reflux. Without treatment, the repeated splashes of acid can severely damage the cells of the

FIGURE H3-4 **Gastroesophageal Reflux**

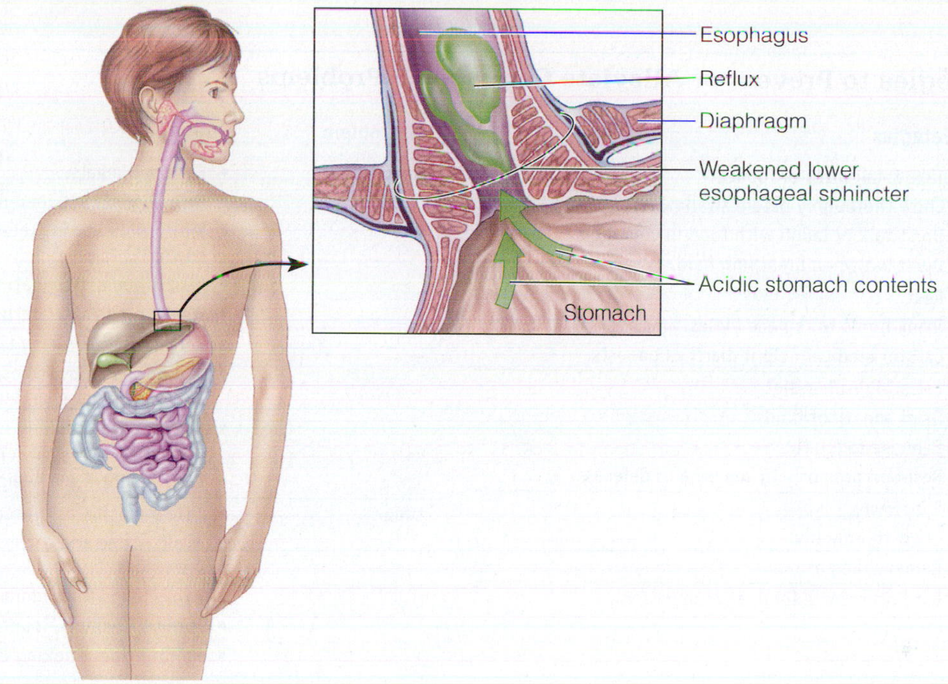

esophagus, creating a condition known as Barrett's esophagus. At that stage, the risk of cancer in the throat or esophagus increases dramatically.[10] To repeat, if symptoms persist, see a doctor—don't self-medicate.

Ulcers

Ulcers are another common digestive problem, affecting an estimated 1 out of every 12 adults in the United States.[11] An **ulcer** is a lesion (a sore), and a **peptic ulcer** is a lesion in the lining of the stomach (gastric ulcers) or the duodenum of the small intestine (duodenal ulcers). The compromised lining is left unprotected and exposed to gastric juices, which can be painful. In some cases, ulcers can cause internal bleeding. If GI bleeding is excessive, iron deficiency may develop. Ulcers that perforate the GI lining can pose life-threatening complications.

Many people naïvely believe that an ulcer is caused by stress or spicy foods, but this is not the case. The stomach lining in a healthy person is well protected by its mucous coat. What, then, causes ulcers to form?

Three major causes of ulcers have been identified: bacterial infection with *Helicobacter pylori* (commonly abbreviated *H. pylori*); the use of certain anti-inflammatory drugs such as aspirin, ibuprofen, and naproxen; and disorders that cause excessive gastric acid secretion. Most commonly, ulcers develop in response to *H. pylori* infection. The cause of the ulcer dictates the type of medication used in treatment. For example, people with ulcers caused by infection receive antibiotics, whereas those with ulcers caused by medicines discontinue their use.[12] In addition, all treatment plans aim to relieve pain, heal the ulcer, and prevent recurrence.

The regimen for ulcer treatment is to treat for infection, eliminate any food that routinely causes indigestion or pain, and avoid coffee and caffeine- and alcohol-containing beverages. Both regular and decaffeinated coffee stimulate acid secretion and so aggravate *existing* ulcers.

Ulcers and their treatments highlight the importance of not self-medicating when symptoms persist. People with *H. pylori* infection often take over-the-counter acid controllers to relieve the pain of their ulcers when, instead, they need physician-prescribed antibiotics. Suppressing gastric acidity not only fails to heal the ulcer, but it also actually worsens inflammation during an *H. pylori* infection. Furthermore, *H. pylori* infection has been linked with stomach cancer, making prompt diagnosis and appropriate treatment essential.[13]

Table H3-1 (p. 94) summarizes strategies to prevent or alleviate common GI problems. Many of these problems reflect hurried lifestyles. For this reason, many of their remedies require that people slow down and take the time to eat leisurely; chew food thoroughly to prevent choking, heartburn, and acid indigestion; rest until vomiting and diarrhea subside; and heed the urge to defecate. In addition, people must learn how to handle life's day-to-day problems and challenges without overreacting and becoming upset; learn how to relax, get enough sleep, and enjoy life. Remember, "what's eating you" may cause more GI distress than what you eat.

HIGHLIGHT 3

TABLE H3-1 Strategies to Prevent or Alleviate Common GI Problems

GI Problem	Strategies
Choking	• Take small bites of food. • Chew thoroughly before swallowing. • Don't talk or laugh with food in your mouth. • Don't eat when breathing hard.
Diarrhea	• Rest. • Drink fluids to replace losses. • Call for medical help if diarrhea persists.
Constipation	• Eat a high-fiber diet. • Drink plenty of fluids. • Exercise regularly. • Respond promptly to the urge to defecate.
Belching	• Eat slowly. • Chew thoroughly. • Relax while eating.
Intestinal gas	• Eat bothersome foods in moderation.
Heartburn	• Eat small meals. • Drink liquids between meals. • Sit up while eating; elevate your head when lying down. • Wait 3 hours after eating before lying down. • Wait 2 hours after eating before exercising. • Refrain from wearing tight-fitting clothing. • Avoid foods, beverages, and medications that aggravate your heartburn. • Refrain from smoking cigarettes or using tobacco products. • Lose weight if overweight.
Ulcer	• Take medicine as prescribed by your physician. • Avoid coffee and caffeine- and alcohol-containing beverages. • Avoid foods that aggravate your ulcer. • Minimize aspirin, ibuprofen, and naproxen use. • Refrain from smoking cigarettes.

Nutrition on the Net

For further study of topics covered in this highlight, log on to **www.cengagebrain.com** and search for ISBN 084006845X.

- Visit the Digestive Diseases section of the National Institute of Diabetes and Digestive and Kidney Diseases: **www.niddk.nih.gov/health/health.htm**
- Visit the patient information section of the American College of Gastroenterology: **www.acg.gi.org**
- Learn more about *H. pylori* from the Helicobacter Foundation: **www.helico.com**

References

1. K. Gotsch, J. L. Annest, and P. Holmgreen, Nonfatal choking-related episodes among children—United States, 2001, *Morbidity and Mortality Weekly Report* 51 (2002): 945–948.
2. E. A. Mayer, Irritable bowel syndrome, *New England Journal of Medicine* 358 (2008): 1692–1699.
3. A. Foxx-Orenstein, IBS—Review and what's new, *Medscape General Medicine* 8 (2006): 20.
4. A. C. Ford and coauthors, Effect of fibre, antispasmodics, and peppermint oil in the treatment of irritable bowel syndrome: Systematic review and meta-analysis, *British Journal of Medicine* 337 (2008): a2313.
5. L. R. Schiller, Management of diarrhea in clinical practice: Strategies for primary care physicians, *Reviews in Gastroenterological Disorders* 7 (2007): S27–S38.
6. J. F. Johanson, Review of the treatment options for chronic constipation, *Medscape General Medicine* 9 (2007): 25.
7. D. O. Jacobs, Diverticulitis, *New England Journal of Medicine* 357 (2007): 2057–2066; H. Salzman and D. Lillie, Diverticular disease: Diagnosis and treatment, *American Family Physician* 72 (2005): 1229–1234.
8. P. J. Kahrilas, Gastroesophageal reflux disease, *New England Journal of Medicine* 359 (2008): 1700–1707.
9. B. C. Jacobson and coauthors, Body-mass index and symptoms of gastroesophageal reflux in women, *New England Journal of Medicine* 354 (2006): 2340–2348.
10. M. J. Schuchert and J. D. Luketich, Management of Barrett's esophagus, *Oncology (Williston Park)* 21 (2007): 1382–1389.
11. Centers for Disease Control, National Center for Health Statistics, *Summary Health Statistics for U.S. Adults: National Health Interview Survey*, 2006, p. 6.
12. N. Vakil and D. Vaira, Sequential therapy for *Helicobacter pylori*—Time to consider making the switch? *Journal of the American Medical Association* 300 (2008): 1346–1347.
13. A. T. Axon, Relationship between *Helicobacter pylori* gastritis, gastric cancer and gastric acid secretion, *Advances in Medical Sciences* 52 (2007): 55–60.

CHAPTER

4

© Image Source Pink/Alamy

Nutrition in Your Life

Whether you are studying for an exam or daydreaming about your next vacation, your brain needs carbohydrate to power its activities. Your muscles need carbohydrate to fuel their work, too, whether you are racing up the stairs to class or moving on the dance floor to your favorite music. Where can you get carbohydrate? Are some foods healthier choices than others? As you will learn from this chapter, whole grains, vegetables, legumes, and fruits naturally deliver ample carbohydrate and fiber with valuable vitamins and minerals and little or no fat. Milk products typically lack fiber, but they also provide carbohydrate along with an assortment of vitamins and minerals.

Throughout this chapter, the CourseMate icon indicates an opportunity for online self-study, linking you to activities to increase your understanding of chapter concepts. **www.cengagebrain.com** (search for ISBN 084006845X)

The Carbohydrates: Sugars, Starches, and Fibers

CHAPTER OUTLINE

A student, quietly studying a textbook, is seldom aware that within his brain cells, billions of glucose molecules are splitting to provide the energy that permits him to learn. Yet glucose provides nearly all of the energy the human brain uses daily. Similarly, a marathon runner, bursting across the finish line in an explosion of sweat and triumph, seldom gives credit to the glycogen fuel her muscles have devoured to help her finish the race. Yet, together, these two **carbohydrates**—glucose and its storage form glycogen—provide about half of all the energy muscles and other body tissues use. The other half of the body's energy comes mostly from fat.

People don't eat glucose and glycogen directly. When they eat foods rich in carbohydrates, their bodies receive glucose for immediate energy and convert some glucose into glycogen for reserve energy. All plant foods—whole grains, vegetables, legumes, and fruits—provide ample carbohydrates. Milk also contains carbohydrates.

Many people mistakenly think of carbohydrates as "fattening" and avoid them when trying to lose weight. Such a strategy may be helpful if the carbohydrates are the concentrated sugars of soft drinks, candies, and cookies, but it is counterproductive if the carbohydrates are from whole grains, vegetables, and legumes. As the next section explains, not all carbohydrates are created equal.

The Chemist's View of Carbohydrates

The dietary carbohydrate family includes:[1] ♦

- Monosaccharides: single sugars
- Disaccharides: sugars composed of pairs of monosaccharides
- Polysaccharides: large molecules composed of chains of monosaccharides

To understand the structure of carbohydrates, look at the units of which they are made. The monosaccharides most important in nutrition ♦ each contain 6 carbon atoms, 12 hydrogens, and 6 oxygens (written in shorthand as $C_6H_{12}O_6$).

♦ Monosaccharides and disaccharides (the sugars) are sometimes called **simple carbohydrates,** and the polysaccharides (starches and fibers) are sometimes called **complex carbohydrates.**

♦ Most of the monosaccharides important in nutrition are **hexoses,** sugars with six atoms of carbon and the formula $C_6H_{12}O_6$.
- **hex** = six

carbohydrates: compounds composed of carbon, oxygen, and hydrogen arranged as monosaccharides or multiples of monosaccharides. Most, but not all, carbohydrates have a ratio of one carbon molecule to one water molecule: $(CH_2O)_n$.
- **carbo** = carbon (C)
- **hydrate** = with water (H_2O)

FIGURE 4-1 Atoms and Their Bonds

The four main types of atoms found in nutrients are hydrogen (H), oxygen (O), nitrogen (N), and carbon (C).

Each atom has a characteristic number of bonds it can form with other atoms.

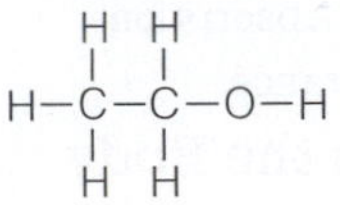

Notice that in this simple molecule of ethyl alcohol, each H has one bond, O has two, and each C has four.

FIGURE 4-2 Chemical Structure of Glucose

On paper, the structure of glucose has to be drawn flat, but in nature the five carbons and oxygen are roughly in a plane. The atoms attached to the ring carbons extend above and below the plane.

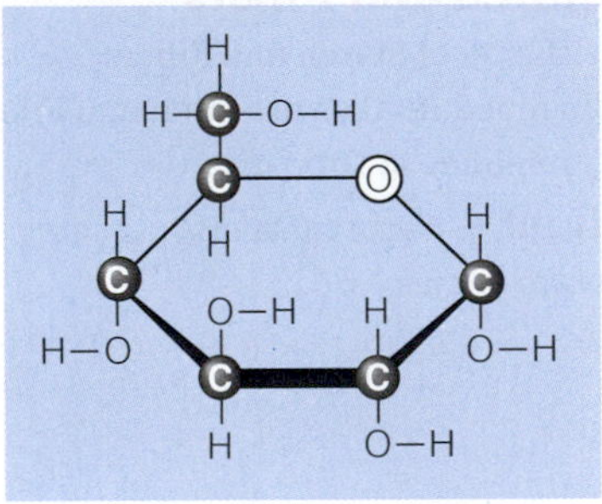

sugars: monosaccharides and disaccharides.

monosaccharides (mon-oh-SACK-uh-rides): carbohydrates of the general formula $C_nH_{2n}O_n$ that typically form a single ring. See Appendix C for the chemical structures of the monosaccharides.

- **mono** = one
- **saccharide** = sugar

glucose (GLOO-kose): a monosaccharide; sometimes known as *blood sugar* or *dextrose*.

- **ose** = carbohydrate
-  = glucose

fructose (FRUK-tose or FROOK-tose): a monosaccharide; sometimes known as *fruit sugar* or *levulose*. Fructose is found abundantly in fruits, honey, and saps.

- **fruct** = fruit
-  = fructose

Each atom can form a certain number of chemical bonds with other atoms:

- Carbon atoms, four
- Nitrogen atoms, three
- Oxygen atoms, two
- Hydrogen atoms, only one

Chemists represent the bonds as lines between the chemical symbols (such as C, N, O, and H) that stand for the atoms (see Figure 4-1).

Atoms form molecules in ways that satisfy the bonding requirements of each atom. Figure 4-1 includes the structure of ethyl alcohol, the active ingredient of alcoholic beverages, as an example. The two carbons each have four bonds represented by lines; the oxygen has two; and each hydrogen has one bond connecting it to other atoms. Chemical structures always bond according to these rules.

The following list of the most important **sugars** in nutrition symbolizes them as hexagons and pentagons of different colors.* Three are monosaccharides:

- Glucose 
- Fructose 
- Galactose 

Three are disaccharides:

- Maltose (glucose + glucose) 
- Sucrose (glucose + fructose) 
- Lactose (glucose + galactose) 

Monosaccharides

The three **monosaccharides** important in nutrition all have the same numbers and kinds of atoms, but in different arrangements. These chemical differences account for the differing sweetness of the monosaccharides. A pinch of purified glucose on the tongue gives only a mild sweet flavor, and galactose hardly tastes sweet at all. Fructose, however, is as intensely sweet as honey and, in fact, is the sugar primarily responsible for honey's sweetness.

Glucose Chemically, **glucose** is a larger and more complicated molecule than the ethyl alcohol shown in Figure 4-1, but it obeys the same rules of chemistry: each carbon atom has four bonds; each oxygen, two bonds; and each hydrogen, one bond. Figure 4-2 illustrates the chemical structure of a glucose molecule.

The diagram of a glucose molecule shows all the relationships between the atoms and proves simple on examination, but chemists have adopted even simpler ways to depict chemical structures. Figure 4-3 presents the chemical structure of glucose in a more simplified way by combining or omitting several symbols—yet it conveys the same information.

Commonly known as blood sugar, glucose serves as an essential energy source for all the body's activities. Its significance to nutrition is tremendous. Later sections explain that glucose is one of the two sugars in every disaccharide and the unit from which the polysaccharides are made almost exclusively. One of these polysaccharides, starch, is the chief food source of energy for all the world's people; another, glycogen, is an important storage form of energy in the body. Glucose reappears frequently throughout this chapter and all those that follow.

Fructose **Fructose** is the sweetest of the sugars. Curiously, fructose has exactly the same chemical *formula* as glucose—$C_6H_{12}O_6$—but its *structure* differs (see Figure 4-4). The arrangement of the atoms in fructose stimulates the taste buds on the tongue to produce the sweet sensation. Fructose occurs naturally in fruits

*Fructose is shown as a pentagon, but like the other monosaccharides, it has six carbons (as you will see in Figure 4-4).

FIGURE 4-3 Simplified Diagrams of Glucose

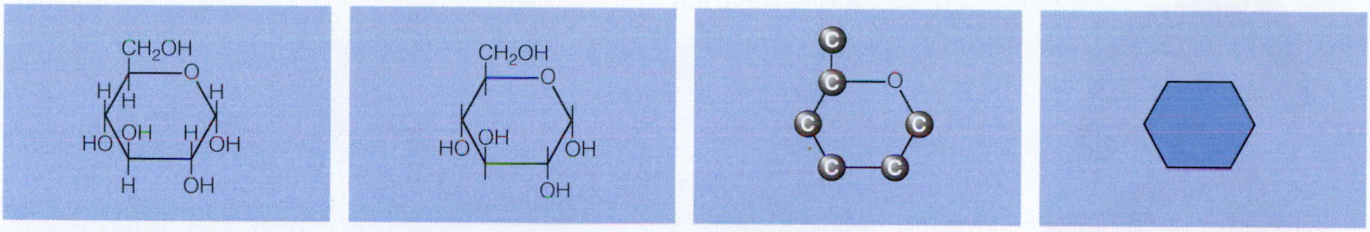

The lines representing some of the bonds and the carbons at the corners are not shown.

Now the single hydrogens are not shown, but lines still extend upward or downward from the ring to show where they belong.

Another way to look at glucose is to notice that its six carbon atoms are all connected.

In this and other illustrations throughout this book, glucose is represented as a blue hexagon.

and honey; other sources include products such as soft drinks, ready-to-eat cereals, and desserts that have been sweetened with high-fructose corn syrup (defined on p. 112).

Galactose The monosaccharide **galactose** occurs naturally as a single sugar in only a few foods. Galactose has the same numbers and kinds of atoms as glucose and fructose in yet another arrangement. Figure 4-5 shows galactose beside a molecule of glucose for comparison.

Disaccharides

The **disaccharides** are pairs of the three monosaccharides just described. Glucose occurs in all three; the second member of the pair is fructose, galactose, or another glucose. These carbohydrates—and all the other energy nutrients—are put together and taken apart by similar chemical reactions: condensation and hydrolysis.

Condensation To make a disaccharide, a chemical reaction known as **condensation** links two monosaccharides together (see Figure 4-6 on p. 100). A hydroxyl (OH) group from one monosaccharide and a hydrogen atom (H) from the other combine to create a molecule of water (H_2O). The two originally separate monosaccharides link together with a single oxygen (O).

Hydrolysis To break a disaccharide in two, a chemical reaction known as hydrolysis ♦ occurs (see Figure 4-7 on p. 100). A molecule of water splits to provide the H and OH needed to complete the resulting monosaccharides. Hydrolysis reactions commonly occur during digestion.

♦ A **hydrolysis** reaction splits a molecule into two, with H added to one and OH to the other (from water); Chapter 3 explained that hydrolysis reactions break down molecules during digestion

galactose (ga-LAK-tose): a monosaccharide; part of the disaccharide lactose.
-  = galactose

disaccharides (dye-SACK-uh-rides): pairs of monosaccharides linked together. See Appendix C for the chemical structures of the disaccharides.
- **di** = two

condensation: a chemical reaction in which water is released as two reactants combine to form one larger product.

FIGURE 4-4 Two Monosaccharides: Glucose and Fructose

Can you see the similarities? If you learned the rules in Figure 4-3, you will be able to "see" 6 carbons (numbered), 12 hydrogens (those shown plus one at the end of each single line), and 6 oxygens in both these compounds.

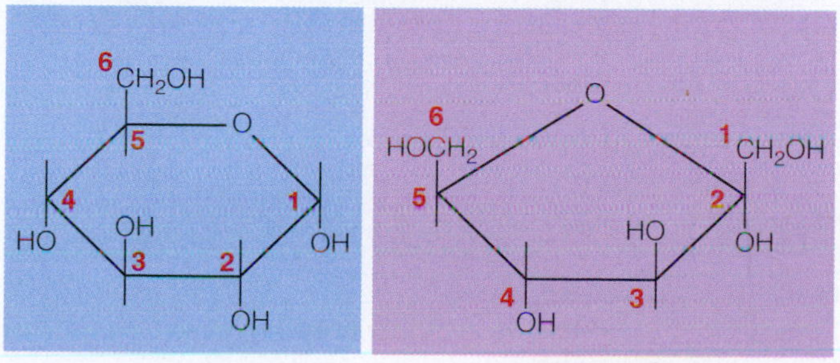

Glucose Fructose

FIGURE 4-5 Two Monosaccharides: Glucose and Galactose

Notice the similarities and the difference (highlighted in red) between glucose and galactose. Both have 6 carbons, 12 hydrogens, and 6 oxygens, but the position of one OH group differs slightly

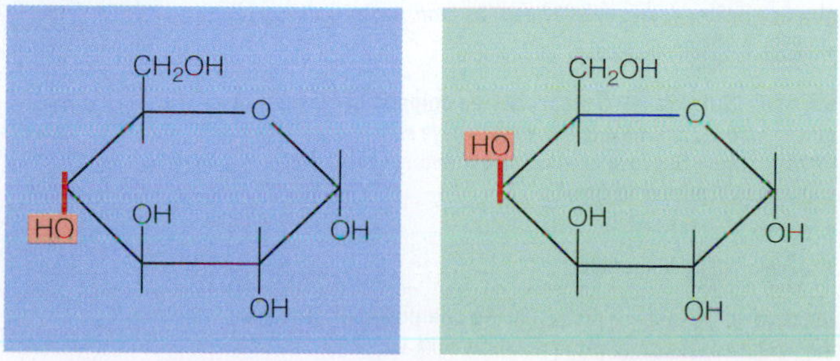

Glucose Galactose

iStockphoto.com/Digital Savant LLC

Fruits package their sugars with fibers, vitamins, and minerals, making them a sweet and healthy snack.

FIGURE 4-6 Condensation of Two Monosaccharides to Form a Disaccharide

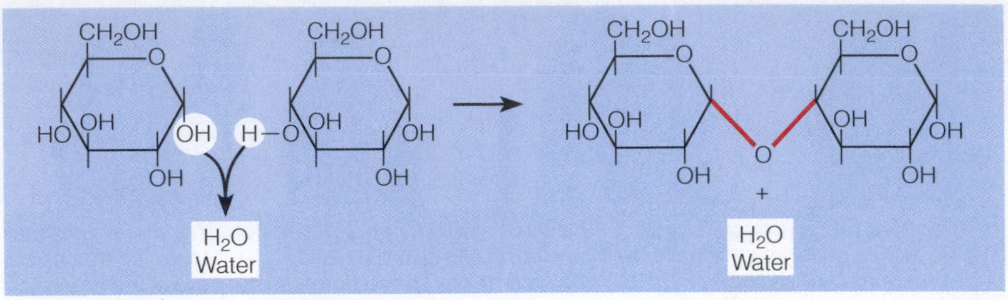

Glucose + glucose → Maltose

An OH group from one glucose and an H atom from another glucose combine to create a molecule of H_2O.

The two glucose molecules bond together with a single O atom to form the disaccharide maltose.

Maltose The disaccharide **maltose** consists of two glucose units. Maltose is produced whenever starch breaks down—as happens in human beings during carbohydrate digestion. It also occurs during the fermentation process that yields alcohol. Maltose is only a minor constituent of a few foods, most notably barley.

Sucrose Fructose and glucose together form **sucrose**. Because the fructose is accessible to the taste receptors, sucrose tastes sweet, accounting for some of the natural sweetness of fruits, vegetables, and grains. To make table sugar, sucrose is refined from the juices of sugarcane and sugar beets, then granulated. Depending on the extent to which it is refined, the product becomes the familiar brown, white, and powdered sugars available at grocery stores.

Lactose The combination of galactose and glucose makes the disaccharide **lactose**, the principal carbohydrate of milk. Known as milk sugar, lactose contributes half of the energy (kcalories) provided by fat-free milk.

IN SUMMARY **The carbohydrates are made of carbon (C), oxygen (O), and hydrogen (H). Each of these atoms can form a specified number of chemical bonds: carbon forms four, oxygen forms two, and hydrogen forms one. Six sugars are important in nutrition. The three monosaccharides (glucose, fructose, and galactose) all have the same chemical formula ($C_6H_{12}O_6$), but their structures differ. The three disaccharides (maltose, sucrose, and lactose) are pairs of monosaccharides, each containing a glucose paired with one of the three monosaccharides. The sugars derive primarily from plants, except for**

FIGURE 4-7 Hydrolysis of a Disaccharide

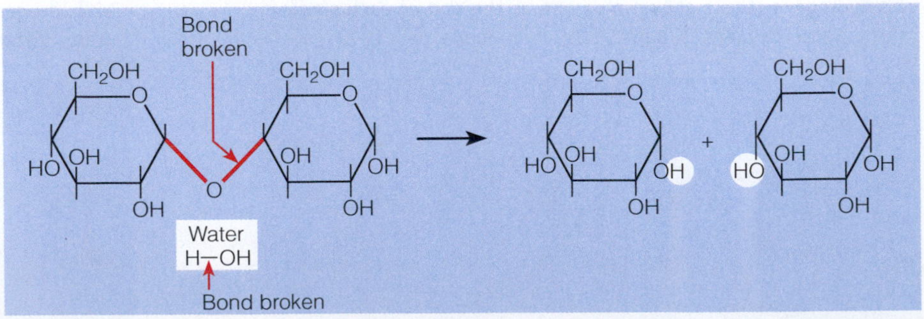

Maltose → Glucose + glucose

The disaccharide maltose splits into two glucose molecules with H added to one and OH to the other (from the water molecule).

maltose (MAWL-tose): a disaccharide composed of two glucose units; sometimes known as *malt sugar.*

-  = maltose

sucrose (SUE-krose): a disaccharide composed of glucose and fructose; commonly known as *table sugar, beet sugar,* or *cane sugar.* Sucrose also occurs in many fruits and some vegetables and grains.

- **sucro** = sugar
-  = sucrose

lactose (LAK-tose): a disaccharide composed of glucose and galactose; commonly known as *milk sugar.*

- **lact** = milk
-  = lactose

lactose and its component galactose, which come from milk and milk products. Two monosaccharides can be linked together by a condensation reaction to form a disaccharide and water. A disaccharide, in turn, can be broken into its two monosaccharides by a hydrolysis reaction using water.

Major sources of starch include grains (such as rice, wheat, millet, rye, barley, and oats), legumes (such as kidney beans, black-eyed peas, pinto beans, navy beans, and garbanzo beans), tubers (such as potatoes), and root crops (such as yams and cassava).

Polysaccharides

In contrast to the sugars just mentioned—the monosaccharides glucose, fructose, and galactose and the disaccharides maltose, sucrose, and lactose—the **polysaccharides** contain many glucose units and, in some cases, a few other monosaccharides strung together. Three types of polysaccharides are important in nutrition: glycogen, starches, and fibers.

Glycogen is a storage form of energy in the body; starch is the storage form of energy in plants; and fibers provide structure in stems, trunks, roots, leaves, and skins of plants. Both glycogen and starch are built of glucose units; fibers are composed of a variety of monosaccharides and other carbohydrate derivatives.

Glycogen

Glycogen is found to only a limited extent in meats and not at all in plants.* For this reason, food is not a significant source of this carbohydrate. However, glycogen performs an important role in the body: it stores glucose for future use. Glycogen is made of many glucose molecules linked together in highly branched chains (see the left side of Figure 4-8). When the hormonal message "release energy" arrives at the glycogen storage sites in a liver or muscle cell, enzymes respond by attacking the many branches of glycogen simultaneously, making a surge of glucose available.**

Starches

The human body stores glucose as glycogen, but plant cells store glucose as **starches**—long, branched or unbranched chains of hundreds or thousands of glucose molecules linked together (see the middle and right side of Figure 4-8). These giant starch molecules are packed side by side in grains such as wheat or rice, in root crops and tubers such as yams and potatoes, and in legumes such as peas and beans. When you eat the plant, your body hydrolyzes the starch to glucose and uses the glucose for its own energy purposes.

*Glycogen in animal muscles rapidly hydrolyzes after slaughter.
**Normally, liver cells produce glucose from glycogen to be sent *directly* to the blood; muscle cells can also produce glucose from glycogen, but must use it themselves. Muscle cells can restore the blood glucose level *indirectly*, however, as Chapter 7 explains.

polysaccharides: compounds composed of many monosaccharides linked together. An intermediate string of three to ten monosaccharides is an **oligosaccharide.**
- **poly** = many
- **oligo** = few

glycogen (GLY-ko-jen): an animal polysaccharide composed of glucose; manufactured and stored in the liver and muscles as a storage form of glucose. Glycogen is not a significant food source of carbohydrate and is not counted as a dietary carbohydrate in foods.
- **glyco** = glucose
- **gen** = gives rise to

starches: plant polysaccharides composed of glucose.

FIGURE 4-8 Glycogen and Starch Molecules Compared (Small Segments)

These units would have to be magnified millions of times to appear at the size shown in this figure. For details of the chemical structures, see Appendix C.

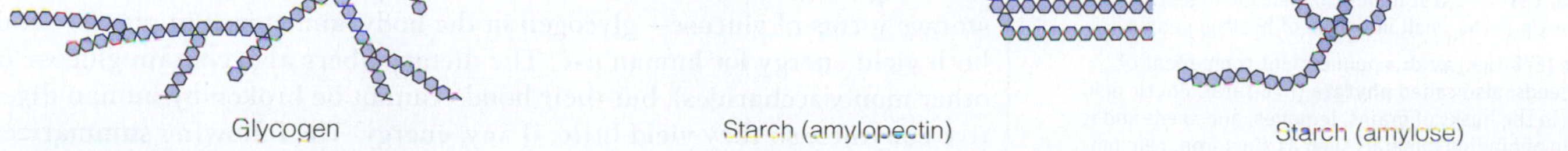

A glycogen molecule contains hundreds of glucose units in highly branched chains. Each new glycogen molecule needs a special protein for the attachment of the first glucose (shown here in red).

A starch molecule contains hundreds of glucose molecules in either occasionally branched chains (amylopectin) or unbranched chains (amylose).

FIGURE 4-9 Starch and Cellulose Molecules Compared (Small Segments)

The bonds that link the glucose molecules together in cellulose are different from the bonds in starch (and glycogen). Human enzymes cannot digest cellulose. See Appendix C for chemical structures and descriptions of linkages.

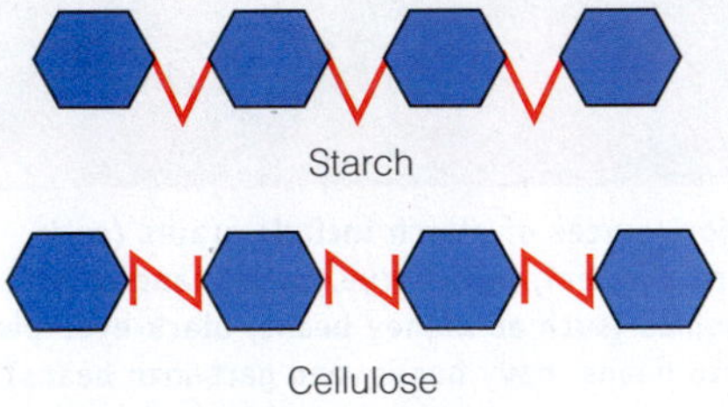

All starchy foods come from plants. Grains are the richest food source of starch, providing much of the food energy for people all over the world—rice in Asia; wheat in Canada, the United States, and Europe; corn in much of Central and South America; and millet, rye, barley, and oats elsewhere. Legumes and tubers are also important sources of starch.

Fibers **Dietary fibers** are the structural parts of plants and thus are found in all plant-derived foods—vegetables, fruits, whole grains, and legumes. Most dietary fibers are polysaccharides. As mentioned earlier, starches are also polysaccharides, but dietary fibers differ from starches in that the bonds between their monosaccharides cannot be broken down by digestive enzymes in the body. For this reason, dietary fibers are often described as *nonstarch polysaccharides*.* Figure 4-9 illustrates the difference in the bonds that link glucose molecules together in starch with those found in the fiber cellulose. Because dietary fibers pass through the body, they contribute no monosaccharides, and therefore little or no energy.

Even though most foods contain a variety of fibers, researchers often sort dietary fibers into two groups according to their solubility. Such distinctions help to explain their actions in the body.

Some dietary fibers dissolve in water **(soluble fibers)**, form gels **(viscous)**, and are easily digested by bacteria in the colon **(fermentable)**. Commonly found in oats, barley, legumes, and citrus fruits, soluble fibers are most often associated with protecting against heart disease and diabetes by lowering blood cholesterol and glucose levels, respectively.[2]

Other fibers do not dissolve in water **(insoluble fibers)**, do not form gels (nonviscous), and are less readily fermented. Found mostly in whole grains (bran) and vegetables, insoluble fibers promote bowel movements, alleviate constipation, and prevent diverticular disease.[3]

As mentioned, *dietary fibers* occur naturally in plants. When these fibers have been extracted from plants or are manufactured and then added to foods or used in supplements, they are called *functional fibers*—if they have beneficial health effects. Cellulose in cereals, for example, is a dietary fiber, but when consumed as a supplement to alleviate constipation, cellulose is considered a functional fiber. *Total fiber* refers to the sum of dietary fibers and functional fibers.

A few starches are classified as dietary fibers. Known as **resistant starches**, these starches escape digestion and absorption in the small intestine. Starch may resist digestion for several reasons, including the body's efficiency in digesting starches and the food's physical properties. Resistant starch is common in whole or partially milled grains, legumes, and just-ripened bananas. Cooked potatoes, pasta, and rice that have been chilled also contain resistant starch. Similar to insoluble fibers, resistant starch may support a healthy colon.[4]

Phytic acid is not a dietary fiber, but it is often found in the same foods. Because of this close association, researchers have been unable to determine whether it is the dietary fiber, the phytic acid, or both, that binds with minerals, preventing their absorption. This binding presents a risk of mineral deficiencies, but the risk is minimal when total fiber intake is reasonable (less than 40 grams a day) and mineral intake adequate. The nutrition consequences of mineral losses are described further in Chapters 12 and 13.

IN SUMMARY **The polysaccharides are chains of monosaccharides and include glycogen, starches, and dietary fibers. Both glycogen and starch are storage forms of glucose—glycogen in the body, and starch in plants—and both yield energy for human use. The dietary fibers also contain glucose (and other monosaccharides), but their bonds cannot be broken by human digestive enzymes, so they yield little, if any, energy. The following summarizes the carbohydrate family of compounds.**

dietary fibers: in plant foods, the *nonstarch polysaccharides* that are not digested by human digestive enzymes, although some are digested by GI tract bacteria. Dietary fibers include cellulose, hemicelluloses, pectins, gums, and mucilages as well as the nonpolysaccharides lignins, cutins, and tannins.

soluble fibers: nonstarch polysaccharides that dissolve in water to form a gel. An example is pectin from fruit, which is used to thicken jellies.

viscous: a gel-like consistency.

fermentable: the extent to which bacteria in the GI tract can break down fibers to fragments that the body can use.*

insoluble fibers: nonstarch polysaccharides that do not dissolve in water. Examples include the tough, fibrous structures found in the strings of celery and the skins of corn kernels.

resistant starches: starches that escape digestion and absorption in the small intestine of healthy people.

phytic (FYE-tick) **acid:** a nonnutrient component of plant seeds; also called **phytate** (FYE-tate). Phytic acid occurs in the husks of grains, legumes, and seeds and is capable of binding minerals such as zinc, iron, calcium, magnesium, and copper in insoluble complexes in the intestine, which the body excretes unused.

*Dietary fibers are fermented by bacteria in the colon to short-chain fatty acids, which are absorbed and metabolized by cells in the GI tract and liver (Chapter 5 describes fatty acids).

*The nonstarch polysaccharide fibers include cellulose, hemicelluloses, pectins, gums, and mucilages. Fibers also include some *nonpolysaccharides* such as lignins, cutins, and tannins.

The Carbohydrate Family

- Monosaccharides
  Glucose
  Fructose
  Galactose
- Disaccharides
  Maltose (glucose + glucose)
  Sucrose (glucose + fructose)
  Lactose (glucose + galactose)
- Polysaccharides:
  Glycogen[a]
  Starches (amylose and amylopectin)
  Fibers (soluble and insoluble)

[a]Glycogen is a polysaccharide, but not a dietary source of carbohydrate.

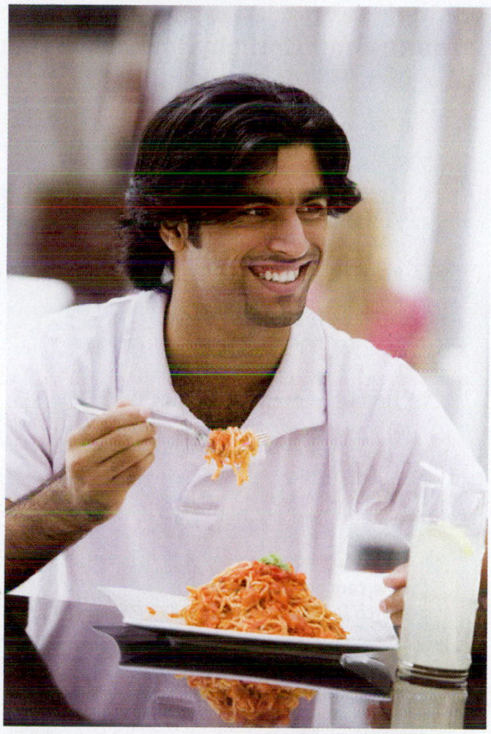

Monkey Business Images/Shutterstock.com

When a person eats carbohydrate-rich foods, the body receives a valuable commodity—glucose.

Digestion and Absorption of Carbohydrates

The ultimate goal of digestion and absorption of sugars and starches is to break them into small molecules—chiefly glucose—that the body can absorb and use. The large starch molecules require extensive breakdown; the disaccharides need only be broken once and the monosaccharides not at all. The details follow.

Carbohydrate Digestion

Figure 4-10 (p. 104) traces the digestion of carbohydrates through the GI tract. When a person eats foods containing starch, enzymes hydrolyze the long chains to shorter chains, ♦ the short chains to disaccharides, and, finally, the disaccharides to monosaccharides. This process begins in the mouth.

♦ The short chains of glucose units that result from the breakdown of starch are known as **dextrins.** The word sometimes appears on food labels because dextrins can be used as thickening agents in processed foods.

In the Mouth In the mouth, thoroughly chewing high-fiber foods slows eating and stimulates the flow of saliva. The salivary enzyme **amylase** starts to work, hydrolyzing starch to shorter polysaccharides and to the disaccharide maltose. In fact, you can taste the change if you chew a piece of starchy food like a cracker and hold it in your mouth for a few minutes without swallowing it—the cracker begins tasting sweeter as the enzyme acts on it. Because food is in the mouth for a relatively short time, very little carbohydrate digestion takes place there; it begins again in the small intestine.

In the Stomach The swallowed bolus ♦ mixes with the stomach's acid and protein-digesting enzymes, which inactivate salivary amylase. Thus the role of salivary amylase in starch digestion is relatively minor. To a small extent, the stomach's acid continues breaking down starch, but its juices contain no enzymes to digest carbohydrate. Fibers linger in the stomach and delay gastric emptying, thereby providing a feeling of fullness and **satiety**.

♦ A **bolus** is a portion of food swallowed at one time.

In the Small Intestine The small intestine performs most of the work of carbohydrate digestion. A major carbohydrate-digesting enzyme, pancreatic amylase, enters the intestine via the pancreatic duct and continues breaking down the polysaccharides to shorter glucose chains and maltose. The final step takes place on the outer membranes of the intestinal cells. There specific enzymes ♦ break down specific disaccharides:

- **Maltase** breaks maltose into two glucose molecules.
- **Sucrase** breaks sucrose into one glucose and one fructose molecule.
- **Lactase** breaks lactose into one glucose and one galactose molecule.

At this point, all polysaccharides and disaccharides have been broken down to monosaccharides—mostly glucose molecules, with some fructose and galactose molecules as well.

♦ In general, the word ending **-ase** identifies an enzyme, and the beginning of the word identifies the molecule that the enzyme works on.

amylase (AM-ih-lace): an enzyme that hydrolyzes amylose (a form of starch). Amylase is a **carbohydrase**, an enzyme that breaks down carbohydrates.

satiety (sah-TIE-eh-tee): the feeling of fullness and satisfaction that occurs after a meal and inhibits eating until the next meal. Satiety determines how much time passes between meals.
- **sate** = to fill

maltase: an enzyme that hydrolyzes maltose.

sucrase: an enzyme that hydrolyzes sucrose.

lactase: an enzyme that hydrolyzes lactose.

FIGURE 4-10 Carbohydrate Digestion in the GI Tract

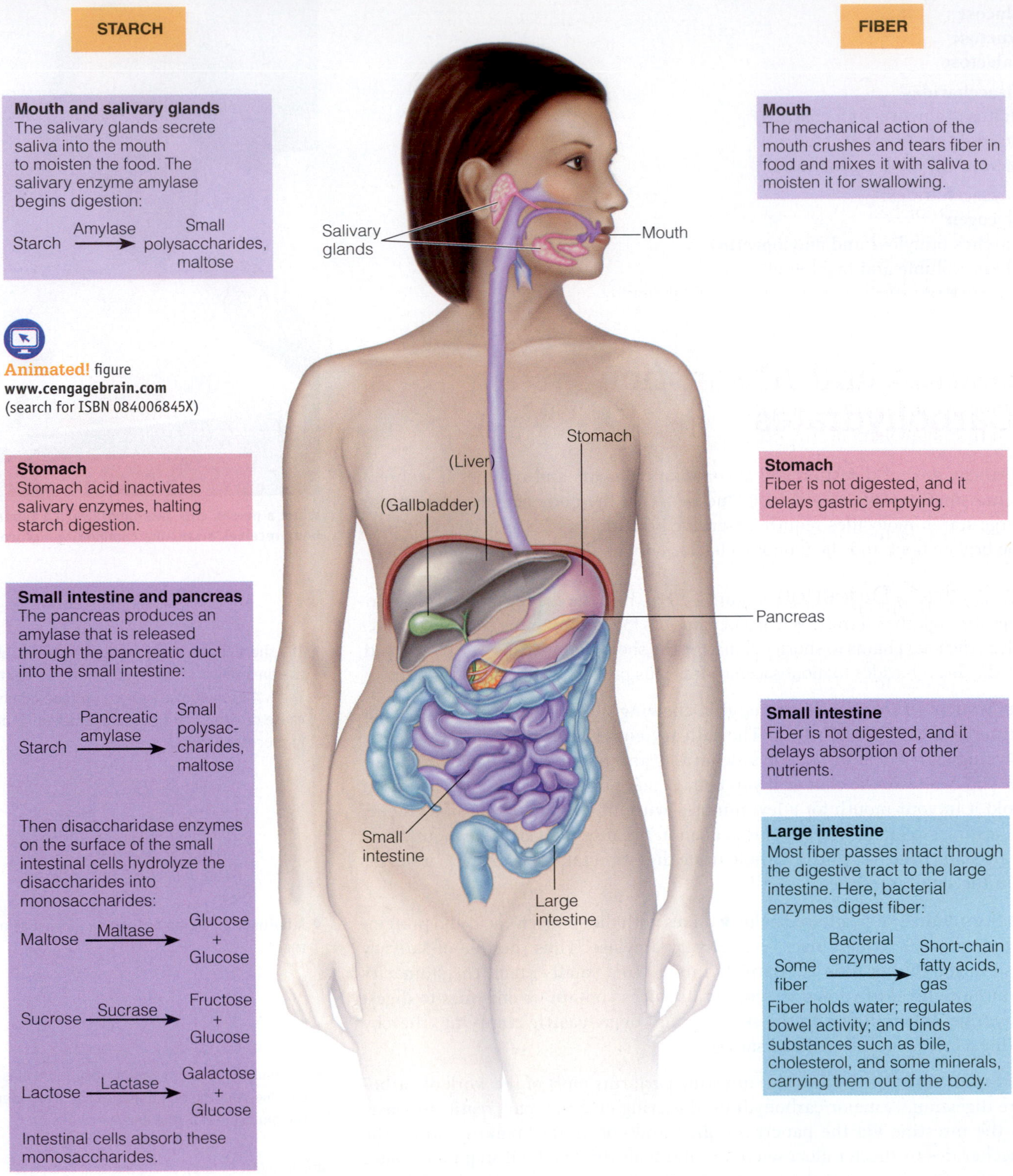

STARCH
FIBER
Mouth and salivary glands
The salivary glands secrete saliva into the mouth to moisten the food. The salivary enzyme amylase begins digestion:
Starch —Amylase→ Small polysaccharides, maltose
Stomach
Stomach acid inactivates salivary enzymes, halting starch digestion.
Small intestine and pancreas
The pancreas produces an amylase that is released through the pancreatic duct into the small intestine:
Starch —Pancreatic amylase→ Small polysaccharides, maltose
Then disaccharidase enzymes on the surface of the small intestinal cells hydrolyze the disaccharides into monosaccharides:
Maltose —Maltase→ Glucose + Glucose
Sucrose —Sucrase→ Fructose + Glucose
Lactose —Lactase→ Galactose + Glucose
Intestinal cells absorb these monosaccharides.
Salivary glands
Mouth
Stomach
(Liver)
(Gallbladder)
Pancreas
Small intestine
Large intestine
Mouth
The mechanical action of the mouth crushes and tears fiber in food and mixes it with saliva to moisten it for swallowing.
Stomach
Fiber is not digested, and it delays gastric emptying.
Small intestine
Fiber is not digested, and it delays absorption of other nutrients.
Large intestine
Most fiber passes intact through the digestive tract to the large intestine. Here, bacterial enzymes digest fiber:
Some fiber —Bacterial enzymes→ Short-chain fatty acids, gas
Fiber holds water; regulates bowel activity; and binds substances such as bile, cholesterol, and some minerals, carrying them out of the body.

Animated! figure
www.cengagebrain.com
(search for ISBN 084006845X)

◆ Starches and sugars are called **available carbohydrates** because human digestive enzymes break them down for the body's use. In contrast, fibers are called **unavailable carbohydrates** because human digestive enzymes cannot break their bonds.

In the Large Intestine Within one to four hours after a meal, all the sugars and most of the starches have been digested. ◆ Only the fibers remain in the digestive tract. Fibers in the large intestine attract water, which softens the stools for passage without straining. Also, bacteria in the GI tract ferment some fibers. This process generates water, gas, and short-chain fatty acids (described in Chapter 5).* The

*The short-chain fatty acids produced by GI bacteria are primarily acetic acid, propionic acid, and butyric acid.

cells of the colon use these small fat molecules for energy. Metabolism of short-chain fatty acids also occurs in the cells of the liver. Fibers, therefore, can contribute some energy (1.5 to 2.5 kcalories per gram), depending on the extent to which they are broken down by bacteria and the fatty acids are absorbed. How much energy fiber contributes to a person's daily intake remains unclear.[5]

Carbohydrate Absorption

Glucose is unique in that it can be absorbed to some extent through the lining of the mouth, but for the most part, nutrient absorption takes place in the small intestine. Glucose and galactose enter the cells lining the small intestine by active transport; fructose is absorbed by facilitated diffusion, which slows its entry and produces a smaller rise in blood glucose. Likewise, unbranched chains of starch are digested slowly and produce a smaller rise in blood glucose than branched chains, which have many more places for enzymes to attack and release glucose rapidly (review Figure 4-8 on p. 101).

As the blood from the small intestine circulates through the liver, cells there take up fructose and galactose and most often convert them to compounds within the same metabolic pathways as glucose. Figure 4-11 shows that fructose and galactose are mostly metabolized in the liver, whereas glucose is sent out to the body's cells for energy. In the end, disaccharides provide at least one glucose molecule directly, and they can provide the equivalent of another one indirectly—through the metabolism of fructose and galactose in the liver.

IN SUMMARY **In the digestion and absorption of carbohydrates, the body breaks down starches into the disaccharide maltose. Maltose and the other disaccharides (lactose and sucrose) from foods are broken down into monosaccharides, which are absorbed. The fibers help to regulate the passage of food through the GI system and slow the absorption of glucose, but they contribute little, if any, energy.**

Lactose Intolerance

Normally, the intestinal cells produce enough of the enzyme lactase to ensure that the disaccharide lactose found in milk is both digested and absorbed efficiently. Lactase activity is highest immediately after birth, as befits an infant whose first and only food for a while will be breast milk or infant formula. In the great majority of the world's populations, lactase activity

FIGURE 4-11 Absorption of Monosaccharides

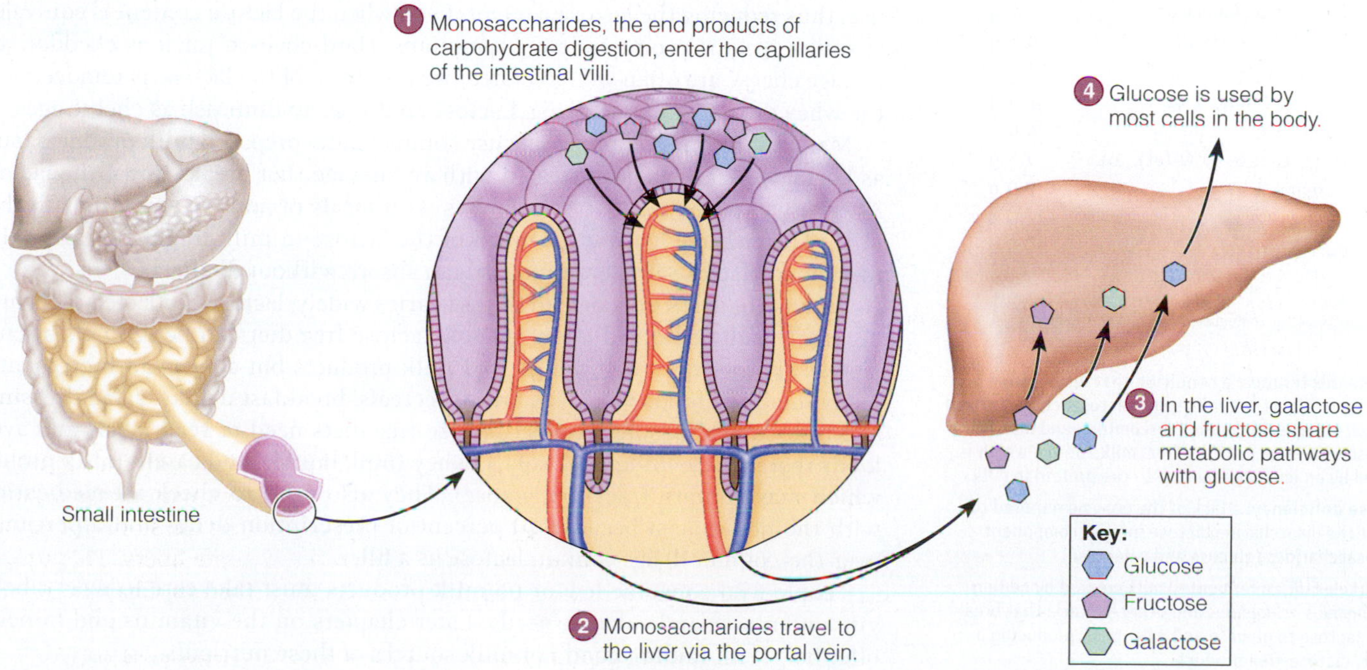

1 Monosaccharides, the end products of carbohydrate digestion, enter the capillaries of the intestinal villi.
Small intestine
2 Monosaccharides travel to the liver via the portal vein.
3 In the liver, galactose and fructose share metabolic pathways with glucose.
4 Glucose is used by most cells in the body.
Key:
Glucose
Fructose
Galactose

declines dramatically during childhood and adolescence to about 5 to 10 percent of the activity at birth. Only a relatively small percentage (about 30 percent) of the people in the world retain enough lactase to digest and absorb lactose efficiently throughout adult life.

Symptoms When more lactose is consumed than the available lactase can handle, lactose molecules remain in the intestine undigested, attracting water and causing bloating, abdominal discomfort, and diarrhea—the symptoms of **lactose intolerance.** The undigested lactose becomes food for intestinal bacteria, which multiply and produce irritating acid and gas, further contributing to the discomfort and diarrhea.

Causes As mentioned, lactase activity commonly declines with age. **Lactase deficiency** may also develop when the intestinal villi are damaged by disease, certain medicines, prolonged diarrhea, or malnutrition. Depending on the extent of the intestinal damage, lactose malabsorption may be temporary or permanent. In extremely rare cases, an infant is born with a lactase deficiency, making feeding a challenge.

Prevalence The prevalence ♦ of lactose intolerance varies widely among ethnic groups, indicating that the trait is genetically determined.[6] The prevalence of lactose intolerance is lowest among Scandinavians and other northern Europeans and highest among native North Americans and Southeast Asians. An estimated 30 to 50 million people in the United States are lactose intolerant.

♦ Estimated prevalence of lactose intolerance:
- 80% Southeast Asians
- 80% Native Americans
- 75% African Americans
- 70% Mediterranean peoples
- 60% Inuits
- 50% Hispanics
- 20% Caucasians
- 10% Northern Europeans

Dietary Changes Managing lactose intolerance requires some dietary changes, although total elimination of milk products usually is not necessary. Excluding all milk products from the diet can lead to nutrient deficiencies because these foods are a major source of several nutrients, notably the mineral calcium, vitamin D, and the B vitamin riboflavin. Fortunately, many people with lactose intolerance can consume foods containing up to 6 grams of lactose (½ cup milk) without symptoms. The most successful strategies are to increase intake of milk products gradually, take them with other foods in meals, and spread their intake throughout the day. In addition, yogurt containing live bacteria seems to improve lactose intolerance.[7] A change in the type, number, and activity of GI bacteria—not the reappearance of the missing enzyme—accounts for the ability to adapt to milk products.[8] Importantly, most lactose-intolerant individuals need to *manage* their dairy consumption rather than *restrict* it.

In many cases, lactose-intolerant people can tolerate fermented milk products such as yogurt and **kefir.** The bacteria in these products digest lactose for their own use, thus reducing the lactose content. Even when the lactose content is equivalent to milk's, yogurt produces fewer symptoms. Hard cheeses, such as cheddar, and cottage cheese are often well tolerated because most of the lactose is removed with the whey during manufacturing. Lactose continues to diminish as cheese ages.

Many lactose-intolerant people use commercially prepared milk products (such as Lactaid) that have been treated with an enzyme that breaks down the lactose. Alternatively, they take enzyme tablets with meals or add enzyme drops to their milk. The enzyme hydrolyzes much of the lactose in milk to glucose and galactose, which lactose-intolerant people can absorb without ill effects.

Because people's tolerance to lactose varies widely, lactose-restricted diets must be highly individualized. A completely lactose-free diet can be difficult because lactose appears not only in milk and milk products but also as an ingredient in many nondairy foods ♦ such as breads, cereals, breakfast drinks, salad dressings, and cake mixes. People on strict lactose-free diets need to read labels and avoid foods that include milk, milk solids, whey (milk liquid), and casein (milk protein, which may contain traces of lactose). They also need to check all medications with the pharmacist because 20 percent of prescription drugs and 5 percent of over-the-counter drugs contain lactose as a filler.

People who consume few or no milk products must take care to meet riboflavin, vitamin D, and calcium needs. Later chapters on the vitamins and minerals offer help with finding good nonmilk sources of these nutrients.

♦ Lactose in selected foods:

Whole-wheat bread, 1 slice	0.5 g
Dinner roll, 1	0.5 g
Cheese, 1 oz	
Cheddar or American	0.5 g
Parmesan or cream	0.8 g
Doughnut (cake type), 1	1.2 g
Chocolate candy, 1 oz	2.3 g
Sherbet, 1 c	4.0 g
Cottage cheese (low-fat), 1 c	7.5 g
Ice cream, 1 c	9.0 g
Milk, 1 c	12.0 g
Yogurt (low-fat), 1 c	15.0 g

NOTE: Yogurt is often enriched with nonfat milk solids, which increase its lactose content to a level higher than milk's.

lactose intolerance: a condition that results from inability to digest the milk sugar lactose; characterized by bloating, gas, abdominal discomfort, and diarrhea. Lactose intolerance differs from milk allergy, which is caused by an immune reaction to the protein in milk.

lactase deficiency: a lack of the enzyme required to digest the disaccharide lactose into its component monosaccharides (glucose and galactose).

kefir (keh-FUR): a fermented milk created by adding *Lactobacillus acidophilus* and other bacteria that break down lactose to glucose and galactose, producing a sweet, lactose-free product.

IN SUMMARY Lactose intolerance is a common condition that occurs when there is insufficient lactase to digest the disaccharide lactose found in milk and milk products. Symptoms include GI distress. Because treatment requires limiting milk intake, other sources of riboflavin, vitamin D, and calcium must be included in the diet.

Glucose in the Body

The primary role of the available carbohydrates in the body is to supply the cells with glucose for energy. Starch contributes most to the body's glucose supply, but as explained earlier, any of the monosaccharides can also provide glucose.

Scientists have long known that providing energy is glucose's primary role in the body, but they have recently uncovered additional roles that glucose and other sugars perform in the body. ♦ When sugar molecules adhere to the body's protein and fat molecules, the consequences can be dramatic. Sugars attached to a protein change the protein's shape and function; when they bind to lipids in a cell's membranes, sugars alter the way cells recognize one another. ♦

♦ The study of sugars is known as **glycobiology.**

♦ These combination molecules are known as **glycoproteins** and **glycolipids,** respectively.

A Preview of Carbohydrate Metabolism

Glucose plays the central role in carbohydrate metabolism. This brief discussion provides just enough information about carbohydrate metabolism to illustrate that the body needs and uses glucose as a chief energy nutrient. Chapter 7 provides a full description of energy metabolism, and Chapter 10 shows how the B vitamins participate.

Storing Glucose as Glycogen The liver stores about one-third of the body's total glycogen and releases glucose into the bloodstream as needed. After a meal, blood glucose rises, and liver cells link the excess glucose molecules by condensation reactions into long, branching chains of glycogen. When blood glucose falls, the liver cells break glycogen by hydrolysis reactions into single molecules of glucose and release them into the bloodstream. Thus glucose becomes available to supply energy to the brain and other tissues regardless of whether the person has eaten recently. Muscle cells can also store glucose as glycogen (the other two-thirds), but they hoard most of their supply, using it just for themselves during exercise. The brain maintains a small amount of glycogen, which is thought to provide an emergency energy reserve during times of severe glucose deprivation.

Glycogen holds water and, therefore, is rather bulky. The body can store only enough glycogen to provide energy for relatively short periods of time—less than a day during rest and a few hours at most during exercise. For its long-term energy reserves, for use over days or weeks of food deprivation, the body uses its abundant, water-free fuel, fat, as Chapter 5 describes.

Using Glucose for Energy Glucose fuels the work of most of the body's cells. Inside a cell, enzymes break glucose in half. These halves can be put back together to make glucose, or they can be further broken down into even smaller fragments (never again to be reassembled to form glucose). The small fragments can yield energy when broken down completely to carbon dioxide and water (see Chapter 7).

As mentioned, the liver's glycogen stores last only for hours, not for days. To keep providing glucose to meet the body's energy needs, a person has to eat dietary carbohydrate frequently. Yet people who do not always attend faithfully to their bodies' carbohydrate needs still survive. How do they manage without glucose from dietary carbohydrate? Do they simply draw energy from the other two energy-yielding nutrients, fat and protein? They do draw energy from them, but not simply.

Making Glucose from Protein Glucose is the preferred energy source for brain cells, other nerve cells, and developing red blood cells. The amino acids of protein can be converted to glucose to some extent, but amino acids and proteins have

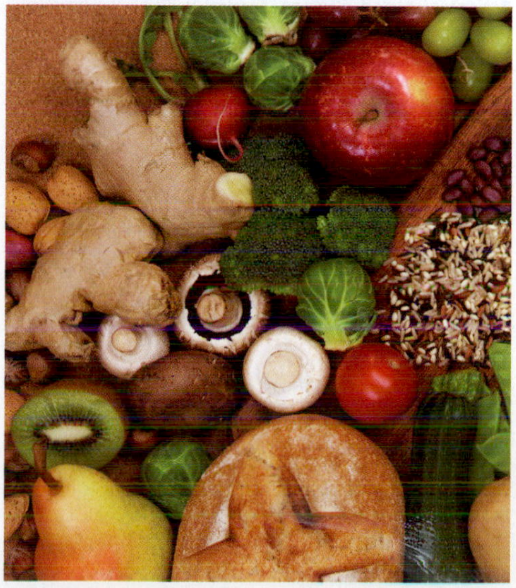
iStockphoto.com/Creativeye99

The carbohydrates of grains, vegetables, fruits, and legumes supply most of the energy in a healthful diet.

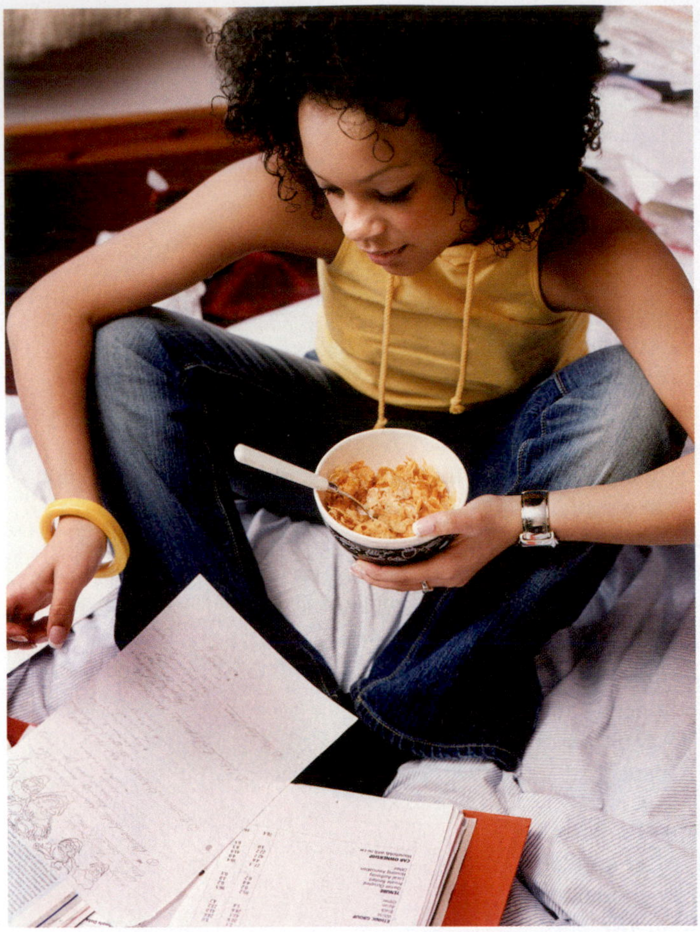

© JupiterImages/BananaStock/Alamy

The brain uses glucose as its primary fuel for energy.

jobs of their own that no other nutrient can perform. Fat cannot be converted to glucose to any significant extent. Thus, when a person does not replenish depleted glycogen stores by eating carbohydrate, body proteins are broken down to make glucose to fuel the brain and other special cells. These body proteins derive primarily from the liver and skeletal muscles.

The conversion of protein to glucose is called **gluconeogenesis**—literally, the making of new glucose. Only adequate dietary carbohydrate can prevent this use of protein for energy, and this role of carbohydrate is known as its **protein-sparing action.**

Making Ketone Bodies from Fat Fragments An inadequate supply of carbohydrate can shift the body's energy metabolism in a precarious direction. With less carbohydrate providing glucose to meet the brain's energy needs, fat takes an alternative metabolic pathway; instead of entering the main energy pathway, fat fragments combine with one another, forming **ketone bodies.** Ketone bodies provide an alternate fuel source during starvation, but when their production exceeds their use, they accumulate in the blood, causing **ketosis.** Because most ketone bodies are acidic, ketosis disturbs the body's normal **acid-base balance.** (Chapter 7 explores ketosis and the metabolic consequences of low-carbohydrate diets further.)

To spare body protein and prevent ketosis, the body needs at least 50 to 100 grams of carbohydrate a day. Dietary recommendations urge people to select abundantly from carbohydrate-rich foods to provide for considerably more.

Using Glucose to Make Fat After meeting its immediate energy needs and filling its glycogen stores to capacity, the body must find a way to handle any extra glucose. When glucose is abundant, energy metabolism shifts to use more glucose instead of fat. If that isn't enough to restore glucose balance, the liver breaks glucose into smaller molecules and puts them together into the more permanent energy-storage compound—fat. Thus, when carbohydrate is abundant, fat is either conserved (by using more carbohydrate in the fuel mix) or created (by using excess carbohydrate to make body fat). The fat then travels to the fatty tissues of the body for storage. Unlike the liver cells, which can store only enough glycogen to meet less than a day's energy needs, fat cells can store seemingly unlimited quantities of fat.

The Constancy of Blood Glucose

Every body cell depends on glucose for its fuel to some extent, and the cells of the brain and the rest of the nervous system depend almost exclusively on glucose for their energy. The activities of these cells never cease, and they have limited ability to store glucose. Day and night, they continually draw on the supply of glucose in the fluid surrounding them. To maintain the supply, a steady stream of blood moves past these cells bringing more glucose from either the small intestine (food) or the liver (via glycogen breakdown or gluconeogenesis).[9]

Maintaining Glucose Homeostasis To function optimally, the body must maintain blood glucose within limits that permit the cells to nourish themselves. If blood glucose falls below normal, ♦ a person may become dizzy and weak; if it rises above normal, a person may become fatigued. Left untreated, fluctuations to the extremes—either high or low—can be fatal.

The Regulating Hormones Blood glucose homeostasis ♦ is regulated primarily by two hormones: *insulin,* which moves glucose from the blood into the cells, and *glucagon,* which brings glucose out of storage when necessary. Figure 4-12 depicts these hormonal regulators at work.

♦ Normal blood glucose (fasting): 70 to 100 mg/dL (published values vary slightly).

♦ **Homeostasis** is the maintenance of constant internal conditions by the body's control systems.

gluconeogenesis (gloo-ko-nee-oh-JEN-ih-sis): the making of glucose from a noncarbohydrate source (described in more detail in Chapter 7).
- **gluco** = glucose
- **neo** = new
- **genesis** = making

protein-sparing action: the action of carbohydrate (and fat) in providing energy that allows protein to be used for other purposes.

ketone (KEE-tone) **bodies:** the metabolic products of the incomplete breakdown of fat when glucose is not available in the cells.

ketosis (kee-TOE-sis): an undesirably high concentration of ketone bodies in the blood and urine.

acid-base balance: the equilibrium in the body between acid and base concentrations (see Chapter 12).

FIGURE 4-12 Maintaining Blood Glucose Homeostasis

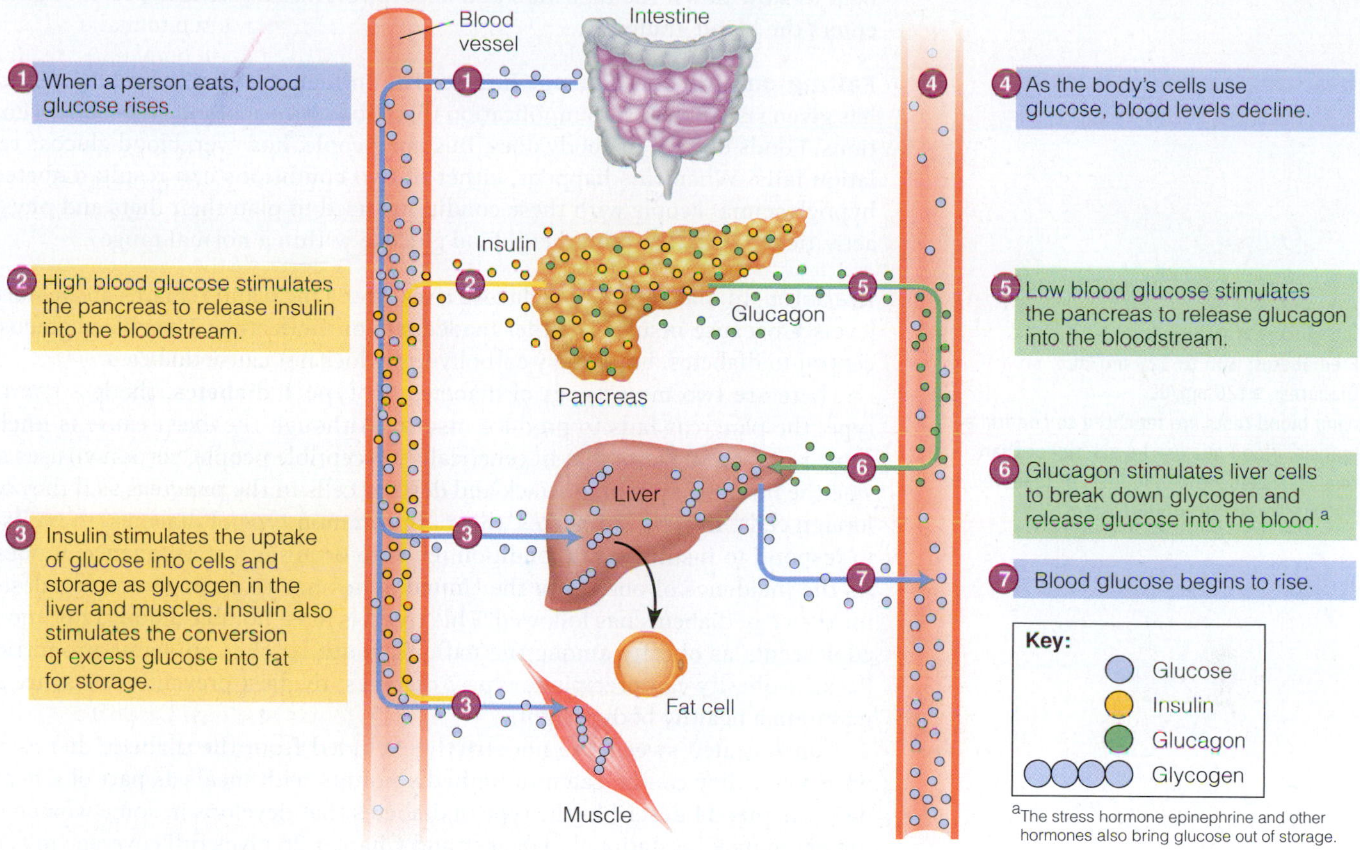

Blood vessel
Intestine
1 When a person eats, blood glucose rises.
2 High blood glucose stimulates the pancreas to release insulin into the bloodstream.
3 Insulin stimulates the uptake of glucose into cells and storage as glycogen in the liver and muscles. Insulin also stimulates the conversion of excess glucose into fat for storage.
Insulin
Pancreas
Glucagon
Liver
Fat cell
Muscle
4 As the body's cells use glucose, blood levels decline.
5 Low blood glucose stimulates the pancreas to release glucagon into the bloodstream.
6 Glucagon stimulates liver cells to break down glycogen and release glucose into the blood.[a]
7 Blood glucose begins to rise.
Key:
Glucose
Insulin
Glucagon
Glycogen

[a]The stress hormone epinephrine and other hormones also bring glucose out of storage.

After a meal, as blood glucose rises, special cells of the pancreas respond by secreting **insulin** into the blood.* In general, the amount of insulin secreted corresponds with the rise in glucose. As the circulating insulin contacts the receptors on the body's other cells, the receptors respond by ushering glucose from the blood into the cells. Most of the cells take only the glucose they can use for energy right away, but the liver and muscle cells can assemble the small glucose units into long, branching chains of glycogen for storage. The liver cells can also convert extra glucose to fat for export to other cells. Thus elevated blood glucose returns to normal levels as excess glucose is stored as glycogen and fat.

When blood glucose falls (as occurs between meals), other special cells of the pancreas respond by secreting **glucagon** into the blood.** Glucagon raises blood glucose by signaling the liver to break down its glycogen stores and release glucose into the blood for use by all the other body cells.

Another hormone that signals the liver cells to release glucose is the "fight-or-flight" hormone, **epinephrine**. When a person experiences stress, epinephrine acts quickly to ensure that all the body cells have energy fuel in emergencies. Among its many roles in the body, epinephrine works to release glucose from liver glycogen to the blood.

Balancing within the Normal Range The maintenance of normal blood glucose ordinarily depends on two processes. When blood glucose falls below normal, food can replenish it, or in the absence of food, glucagon can signal the liver to break down glycogen stores. When blood glucose rises above normal, insulin can signal the cells to take in glucose for energy. Eating balanced meals at regular

*The *beta* (BAY-tuh) *cells*, one of several types of cells in the pancreas, secrete insulin in response to elevated blood glucose concentration.

**The *alpha cells* of the pancreas secrete glucagon in response to low blood glucose.

insulin (IN-suh-lin): a hormone secreted by special cells in the pancreas in response to (among other things) increased blood glucose concentration. The primary role of insulin is to control the transport of glucose from the bloodstream into the muscle and fat cells.

glucagon (GLOO-ka-gon): a hormone that is secreted by special cells in the pancreas in response to low blood glucose concentration and elicits release of glucose from liver glycogen stores.

epinephrine (EP-ih-NEFF-rin): a hormone of the adrenal gland that modulates the stress response; formerly called *adrenaline*. When administered by injection, epinephrine counteracts anaphylactic shock by opening the airways and maintaining heartbeat and blood pressure.

intervals helps the body maintain a happy medium between the extremes. Balanced meals that provide abundant carbohydrates, including fibers, and a little fat help to slow down the digestion and absorption of carbohydrate so that glucose enters the blood gradually.

Falling outside the Normal Range The influence of foods on blood glucose has given rise to the oversimplification that foods *govern* blood glucose concentrations. Foods do not; the body does. In some people, however, blood glucose regulation fails. When this happens, either of two conditions can result: diabetes or hypoglycemia. People with these conditions need to plan their diets and physical activities to help maintain their blood glucose within a normal range.

Diabetes In **diabetes**, blood glucose rises after a meal and remains above normal levels ♦ because insulin is either inadequate or ineffective. Thus *blood* glucose is central to diabetes, but *dietary* carbohydrate does not cause diabetes.

♦ Blood glucose (fasting):
- Prediabetes: 100 to 125 mg/dL
- Diabetes: ≥126 mg/dL

Fasting blood tests are repeated to confirm a diagnosis. Blood glucose levels higher than normal, but below the diagnosis of diabetes, is sometimes called **prediabetes.**

There are two main types of diabetes. In **type 1 diabetes**, the less common type, the pancreas fails to produce insulin. Although the exact cause is unclear, some research suggests that in genetically susceptible people, certain viruses activate the immune system to attack and destroy cells in the pancreas as if they were foreign cells. In **type 2 diabetes**, the more common type of diabetes, the cells fail to respond to insulin. This condition tends to occur as a consequence of obesity. As the incidence of obesity in the United States has risen in recent decades, the incidence of diabetes has followed. This trend is most notable among children and adolescents as obesity among the nation's youth reaches epidemic proportions. Because obesity can precipitate type 2 diabetes, the best preventive measure is to maintain a healthy body weight.

Concentrated sweets are not strictly excluded from the diabetic diet as they once were; they can be eaten in limited amounts with meals as part of a healthy diet. Chapter 14 describes the type of diabetes that develops in some women during pregnancy (gestational diabetes), and Chapter 26 gives full coverage to type 1 and type 2 diabetes and their associated problems.

Hypoglycemia In healthy people, blood glucose rises after eating and then gradually falls back into the normal range. The transition occurs without notice. Should blood glucose drop below normal, a person would experience the symptoms of **hypoglycemia:** weakness, rapid heartbeat, sweating, anxiety, hunger, and trembling. Most commonly, hypoglycemia is a consequence of poorly managed diabetes: too much insulin, strenuous physical activity, inadequate food intake, or illness that causes blood glucose levels to plummet.

Hypoglycemia in healthy people is rare. Most people who experience hypoglycemia need only adjust their diets by replacing refined carbohydrates with fiber-rich carbohydrates and ensuring an adequate protein intake at each meal. In addition, smaller meals eaten more frequently may help. Hypoglycemia caused by certain medications, pancreatic tumors, overuse of insulin, alcohol abuse, uncontrolled diabetes, or other illnesses requires medical intervention.

The Glycemic Response The **glycemic response** refers to how quickly glucose is absorbed after a person eats, how high blood glucose rises, and how quickly it returns to normal. Slow absorption, a modest rise in blood glucose, and a smooth return to normal are desirable (a low glycemic response). Fast absorption, a surge in blood glucose, and an overreaction that plunges glucose below normal are less desirable (a high glycemic response). Different foods have different effects on blood glucose.

The rate of glucose absorption is particularly important to people with diabetes, who may benefit from limiting foods that produce too great a rise, or too sudden a fall, in blood glucose.[10] To aid their choices, they may be able to use the **glycemic index**, a method of classifying foods according to their potential to raise blood glucose. ♦ Figure 4-13 ranks selected foods by their glycemic index.[11] ♦ Some studies have shown that selecting foods with a low glycemic index is a practical way to improve glucose control.[12]

♦ A related term, **glycemic load,** reflects both the glycemic index and the amount of carbohydrate.

♦ Glycemic index generalizations:
- Low: Legumes, milk products
- Moderate: Whole grains, fruits
- High: Processed foods made from refined flour such as snack foods, breads, ready-to-eat cereals

diabetes (DYE-uh-BEET-eez): a chronic disorder of carbohydrate metabolism, usually resulting from insufficient or ineffective insulin.

type 1 diabetes: the less common type of diabetes in which the pancreas fails to produce insulin.

type 2 diabetes: the more common type of diabetes in which the cells fail to respond to insulin.

hypoglycemia (HIGH-po-gly-SEE-me-ah): an abnormally low blood glucose concentration.

glycemic (gly-SEEM-ic) **response:** the extent to which a food raises the blood glucose concentration and elicits an insulin response.

glycemic index: a method of classifying foods according to their potential for raising blood glucose.

FIGURE 4-13 **Glycemic Index of Selected Foods**

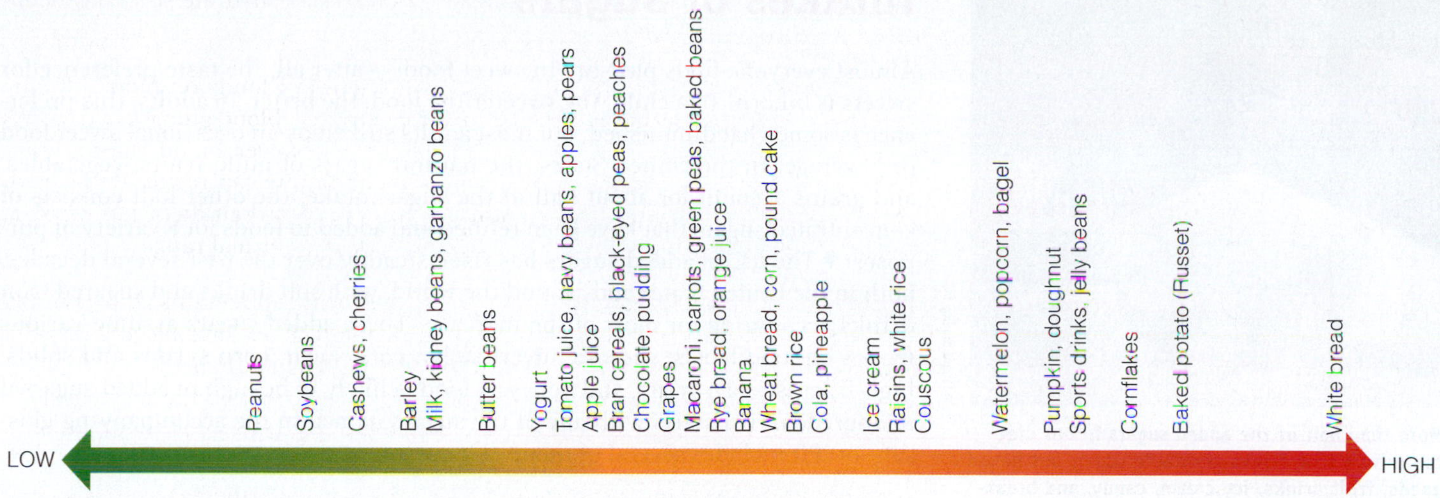

Lowering the glycemic index of the *diet* may improve blood lipids and reduce the risk of heart disease as well.[13] A low glycemic diet may also help with weight management, although research findings are mixed.[14]

Researchers debate whether selecting foods based on the glycemic index is practical or offers any real health benefits.[15] Those opposing the use of the glycemic index argue that it is not sufficiently supported by scientific research. The glycemic index has been determined for relatively few foods, and when the glycemic index has been established, it is based on an average of multiple tests with wide variations in their results. Values vary because of differences in the physical and chemical characteristics of foods, testing methods of laboratories, and digestive processes of individuals.[16]

Furthermore, the practical utility of the glycemic index is limited because this information is neither provided on food labels nor intuitively apparent. Indeed, a food's glycemic index is not always what one might expect. Ice cream, for example, is a high-sugar food but produces less of a glycemic response than baked potatoes, a high-starch food. Perhaps most relevant to real life, a food's glycemic effect differs depending on plant variety, food processing, cooking method, and whether it is eaten alone or with other foods. Most people eat a variety of foods, cooked and raw, that provide different amounts of carbohydrate, fat, and protein—all of which influence the glycemic index of a meal.

Paying attention to the glycemic index may not be necessary because current guidelines already suggest many low glycemic index choices: whole grains, legumes, vegetables, fruits, and milk products. In addition, eating frequent, small meals spreads glucose absorption across the day and thus offers similar metabolic advantages to eating foods with a low glycemic response. People wanting to follow a low glycemic diet should be careful not to adopt a low-carbohydrate diet as well. Highlight 4 explores the controversies surrounding low-carbohydrate and high glycemic diets.

IN SUMMARY **Dietary carbohydrates provide glucose that can be used by the cells for energy, stored by the liver and muscles as glycogen, or converted into fat if intakes exceed needs. All of the body's cells depend on glucose; those of the central nervous system are especially dependent on it. Without glucose, the body is forced to break down its protein tissues to make glucose and to alter energy metabolism to make ketone bodies from fats. Blood glucose regulation depends primarily on two pancreatic hormones: insulin to move glucose from the blood into the cells when levels are high and glucagon to free glucose from glycogen stores and release it into the blood when levels are low. The glycemic index measures how blood glucose responds to foods.**

© Polara Studios Inc.

More than half of the added sugars in our diet come from soft drinks and table sugar, but baked goods, fruit drinks, ice cream, candy, and breakfast cereals also make substantial contributions.

Health Effects and Recommended Intakes of Sugars

Almost everyone finds pleasure in sweet foods—after all, the taste preference for sweets is inborn. To a child, the sweeter the food, the better. In adults, this preference is somewhat diminished, but most adults still enjoy an occasional sweet food or beverage. In the United States, the natural sugars of milk, fruits, vegetables, and grains account for about half of the sugar intake; the other half consists of concentrated sugars that have been refined and added to foods for a variety of purposes. ♦ The use of **added sugars** has risen steadily over the past several decades, both in the United States and around the world, with soft drinks and sugared fruit drinks accounting for most of the increase. These added sugars assume various names on food labels: sucrose, invert sugar, corn sugar, corn syrups and solids, high-fructose corn syrup, and honey. A food is likely to be high in added sugars if its ingredient list starts with any of the sugars named in the accompanying glossary or if it includes several of them.

♦ As an additive, sugar:
- Enhances flavor
- Supplies texture and color to baked goods
- Provides fuel for fermentation, causing bread to rise or producing alcohol
- Acts as a bulking agent in ice cream and baked goods
- Acts as a preservative in jams
- Balances the acidity of tomato- and vinegar-based products

Health Effects of Sugars

In moderate amounts, sugars add pleasure to meals without harming health. In excess, however, they can be detrimental in two ways. One, sugars can contribute to nutrient deficiencies by supplying energy (kcalories) without providing nutrients. Two, sugars can contribute to tooth decay.

Nutrient Deficiencies

Empty-kcalorie foods that contain lots of added sugars such as cakes, candies, and sodas provide the body with glucose and energy, but few, if any, other nutrients. By comparison, foods such as whole grains, vegetables, legumes, and fruits that contain some natural sugars and lots of starches and fibers also provide protein, vitamins, and minerals.

A person spending 200 kcalories of a day's energy allowance on a 16-ounce soda gets little of value for those kcalories. In contrast, a person using 200 kcalories on three slices of whole-wheat bread gets 9 grams of protein, 6 grams of fiber, plus several of the B vitamins with those kcalories. For the person who wants something sweet, a reasonable compromise might be two slices of bread with a teaspoon of jam on each. The amount of sugar a person can afford to eat depends on how many discretionary kcalories are available beyond those needed to deliver indispensable vitamins and minerals.

By following MyPyramid and making careful food selections, a typical adult can obtain all the needed nutrients within an allowance of about 1500 kcalories. Some people have more generous energy allowances. For example, an active teen-

added sugars: sugars and syrups used as an ingredient in the processing and preparation of foods such as breads, cakes, beverages, jellies, and ice cream as well as sugars eaten separately or added to foods at the table.

GLOSSARY OF ADDED SUGARS

brown sugar: refined white sugar crystals to which manufacturers have added molasses syrup with natural flavor and color; 91 to 96 percent pure sucrose.

confectioners' sugar: finely powdered sucrose, 99.9 percent pure.

corn sweeteners: corn syrup and sugars derived from corn.

corn syrup: a syrup made from cornstarch that has been treated with acid, high temperatures, and enzymes that produce glucose, maltose, and dextrins. See also *high-fructose corn syrup (HFCS)*.

dextrose: an older name for glucose.

granulated sugar: crystalline sucrose; 99.9 percent pure.

high-fructose corn syrup (HFCS): a syrup made from cornstarch that has been treated with an enzyme that converts some of the glucose to the sweeter fructose; made especially for use in processed foods and beverages, where it is the predominant sweetener. With a chemical structure similar to sucrose, HFCS has a fructose content of 42, 55, or 90 percent, with glucose making up the remainder.

honey: sugar (mostly sucrose) formed from nectar gathered by bees. An enzyme splits the sucrose into glucose and fructose. Composition and flavor vary, but honey always contains a mixture of sucrose, fructose, and glucose.

invert sugar: a mixture of glucose and fructose formed by the hydrolysis of sucrose in a chemical process; sold only in liquid form and sweeter than sucrose. Invert sugar is used as a food additive to help preserve freshness and prevent shrinkage.

levulose: an older name for fructose.

maple sugar: a sugar (mostly sucrose) purified from the concentrated sap of the sugar maple tree.

molasses: the thick brown syrup produced during sugar refining. Molasses retains residual sugar and other by-products and a few minerals; blackstrap molasses contains significant amounts of calcium and iron.

raw sugar: the first crop of crystals harvested during sugar processing. Raw sugar cannot be sold in the United States because it contains too much filth (dirt, insect fragments, and the like). Sugar sold as "raw sugar" domestically has actually gone through more than half of the refining steps.

turbinado (ter-bih-NOD-oh) **sugar:** sugar produced using the same refining process as white sugar, but without the bleaching and anticaking treatment. Traces of molasses give turbinado its sandy color.

white sugar: pure sucrose or "table sugar," produced by dissolving, concentrating, and recrystallizing raw sugar.

TABLE 4-1 Sample Nutrients in Sugar and Other Foods

The indicated portion of any of these foods provides approximately 100 kcalories. Notice that for a similar number of kcalories and grams of carbohydrate, milk, legumes, fruits, grains, and vegetables offer more of the other nutrients than do the sugars.

	Size of 100 kcal Portion	Carbohydrate (g)	Protein (g)	Calcium (mg)	Iron (mg)	Vitamin A (µg)	Vitamin C (mg)
Foods							
Milk, 1% low-fat	1 c	12	8	300	0.1	144	2
Kidney beans	½ c	20	7	30	1.6	0	2
Apricots	6	24	2	30	1.1	554	22
Bread, whole-wheat	1½ slices	20	4	30	1.9	0	0
Broccoli, cooked	2 c	20	12	188	2.2	696	148
Sugars							
Sugar, white	2 tbs	24	0	trace	trace	0	0
Molasses, blackstrap	2½ tbs	28	0	343	12.6	0	0.1
Cola beverage	1 c	26	0	6	trace	0	0
Honey	1½ tbs	26	trace	2	0.2	0	trace

age boy may need as many as 3000 kcalories a day. If he eats mostly nutritious foods, then he may have discretionary kcalories available for cola beverages and other "extras." In contrast, an inactive older woman who is limited to fewer than 1500 kcalories a day can afford to eat only the most nutrient-dense foods—with few, or no, discretionary kcalories available.

Some people believe that because honey is a natural food, it is nutritious—or, at least, more nutritious than sugar.* A look at their chemical structures reveals the truth. Honey, like table sugar, contains glucose and fructose. The primary difference is that in table sugar the two monosaccharides are bonded together as the disaccharide sucrose, whereas in honey some of them are free. Whether a person eats monosaccharides individually, as in honey, or linked together, as in table sugar, they end up the same way in the body: as glucose and fructose.

Honey does contain a few vitamins and minerals, but not many. Honey is denser than crystalline sugar, too, so it provides more energy per spoonful. Table 4-1 shows that honey and white sugar are similar nutritionally—and both fall short of milk, legumes, fruits, grains, and vegetables. Honey may offer some health benefits, however: it seems to relieve nighttime coughing in children and reduce the severity of mouth ulcers in cancer patients undergoing chemotherapy or radiation.[17]

While the body cannot distinguish whether fructose and glucose derive from honey or table sugar, this is not to say that all sugar sources are alike. Some sugar sources are more nutritious than others. Consider a fruit, say, an orange. The fruit may give you the same amounts of fructose and glucose and the same number of kcalories as a spoonful of sugar or honey, but the packaging is more valuable nutritionally. The fruit's sugars arrive in the body diluted in a large volume of water, packaged in fiber, and mixed with essential vitamins, minerals, and phytochemicals.

As these comparisons illustrate, the significant difference between sugar sources is not between "natural" honey and "purified" sugar but between concentrated sugars and the dilute, naturally occurring sugars that sweeten foods. You can suspect an exaggerated nutrition claim when someone asserts that one product is more nutritious than another because it contains honey.

© Matthew Farrugio

You receive about the same amount and kinds of sugars from an orange as from a tablespoon of honey, but the packaging makes a big nutrition difference.

*Honey should never be fed to infants because of the risk of botulism. Chapter 15 and Highlight 18 provide more details.

FIGURE 4-14 **Dental Caries**

Dental caries begins when acid dissolves the enamel that covers the tooth. If not repaired, the decay may penetrate the dentin and spread into the pulp of the tooth, causing inflammation, abscess, and possible loss of the tooth.

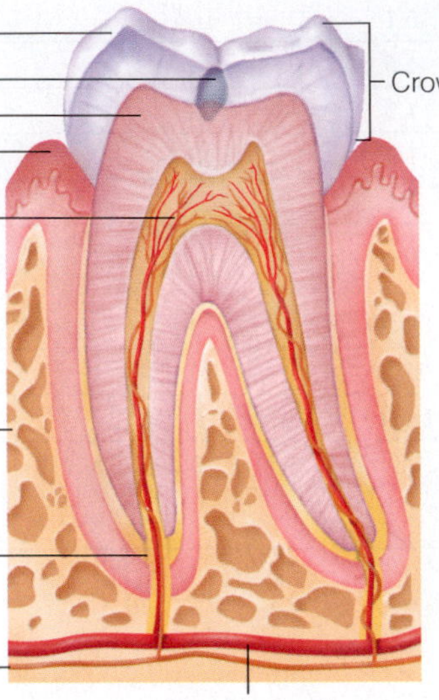

Added sugars contribute to nutrient deficiencies by displacing nutrients.[18] For nutrition's sake, the appropriate attitude to take is not that sugar is "bad" and must be avoided, but that nutritious foods must come first. If nutritious foods crowd sugar out of the diet, that is fine—but not the other way around. As always, balance, variety, and moderation guide healthy food choices.

Dental Caries Sugars from foods and from the breakdown of starches in the mouth can contribute to tooth decay. Bacteria in the mouth ferment the sugars and, in the process, produce an acid that erodes tooth enamel (see Figure 4-14), causing **dental caries**, or tooth decay. People can eat sugar without this happening, though, for much depends on how long foods stay in the mouth. Sticky foods stay on the teeth longer and continue to yield acid longer than foods that are readily cleared from the mouth. For that reason, sugar in a juice consumed quickly, for example, is less likely to cause dental caries than sugar in a pastry. By the same token, the sugar in sticky foods such as dried fruits can be more detrimental than its quantity alone would suggest.

Another concern is how often people eat sugar. Bacteria produce acid for 20 to 30 minutes after each exposure. If a person eats three pieces of candy at one time, the teeth will be exposed to approximately 30 minutes of acid destruction. But, if the person eats three pieces at half-hour intervals, the time of exposure increases to 90 minutes. Likewise, slowly sipping a sugary sports beverage may be more harmful than drinking quickly and clearing the mouth of sugar. Nonsugary foods can help remove sugar from tooth surfaces; hence, it is better to eat sugar with meals than between meals. Foods such as milk and cheese may be particularly helpful in protecting against dental caries by neutralizing acids, stimulating salivary flow, inhibiting bacterial activity, and promoting remineralization of damaged enamel.[19]

Beverages such as soft drinks, orange juice, and sports drinks not only contain sugar but also have a low pH. These acidic drinks can erode tooth enamel and may explain why the prevalence of dental erosion is growing steadily.[20]

The development of caries depends on several factors: the bacteria that reside in **dental plaque**, the saliva that cleanses the mouth, the minerals that form the teeth, and the foods that remain after swallowing. For most people, good oral hygiene will prevent ♦ dental caries. In fact, regular brushing (twice a day, with a fluoride toothpaste) and flossing may be more effective in preventing dental caries than restricting sugary foods. Still nutrition is a key component of dental health.[21] *The Dietary Guidelines for Americans* recommend a combined approach to prevent dental caries—practicing good oral hygiene, drinking fluoridated water, and consuming sugar- and starch-containing foods and beverages less frequently.

♦ To prevent dental caries:

- Limit between-meal juices and snacks containing sugars and starches.
- Brush with a fluoride toothpaste and floss teeth regularly.
- If brushing and flossing are not possible, at least rinse with water.
- Get a dental checkup regularly.
- Drink fluoridated water.

Recommended Intakes of Sugars

Estimates indicate that, on average, each person in the United States consumes about 105 pounds (almost 50 kilograms) of added sugars per year, or about 30 teaspoons (about 120 grams) of added sugars a day.[22] Most of the sugars in the average American diet are added to foods and beverages by manufacturers during processing; major sources of added sugars include sugar-sweetened beverages (sodas, energy drinks, sports drinks, fruit drinks), desserts, candy, and ready-to-eat cereals. Some sugars are also added by consumers during food preparation and at the table. Because added sugars deliver kcalories, but few or no nutrients or fiber, the *Dietary Guidelines for Americans* urge consumers to "reduce the intake of kcalories from added sugars." These added sugar kcalories (and those from solid fats and alcohol) are considered discretionary kcalories—and most people need to limit their intake. By reducing the intake of foods and beverages with added sugars, consumers can lower the kcalorie content of the diet without compromising the nutrient content.

Dietary Guidelines for Americans 2010

Reduce the intake of kcalories from added sugars.

Estimating the *added* sugars in a diet is not always easy for consumers. Food labels list the *total* grams of sugar a food provides, but this total reflects both added

dental caries: decay of teeth.
- **caries** = rottenness

dental plaque: a gummy mass of bacteria that grows on teeth and can lead to dental caries and gum disease.

sugars and those occurring naturally in foods. To help estimate sugar and energy intakes accurately, the list in the margin ♦ shows the amounts of concentrated sweets that are equivalent to 1 teaspoon of white sugar. These sugars all provide *about* 5 grams of carbohydrate and *about* 20 kcalories per teaspoon. Some are lower (16 kcalories for table sugar), and others are higher (22 kcalories for honey), but a 20-kcalorie average is an acceptable approximation. For a person who uses ketchup liberally, it may help to remember that 1 tablespoon of ketchup supplies about 1 teaspoon of sugar.

The DRI Committee did not publish an Upper Level for sugar, but as mentioned, excessive intakes can interfere with sound nutrition and dental health. Few people can eat lots of sugary treats and still meet all of their nutrient needs without exceeding their kcalorie allowance. Specifically, the DRI suggests that added sugars should account for no more than 25 percent of the day's total energy intake.[23] When added sugars occupy this much of a diet, however, intakes from the five food groups usually fall below recommendations. For a person consuming 2000 kcalories a day, 25 percent represents 500 kcalories (that is, 125 grams, or 31 teaspoons) from concentrated sugars—and that's a lot of sugar. ♦ Perhaps an athlete in training whose energy needs are high can afford the added sugars from sports drinks without compromising nutrient intake, but most people do better by limiting their use of added sugars. The World Health Organization (WHO) and the Food and Agriculture Organization (FAO) suggest restricting consumption of added sugars to less than 10 percent of total energy.

♦ 1 tsp white sugar =
- 1 tsp brown sugar
- 1 tsp candy
- 1 tsp corn sweetener or corn syrup
- 1 tsp honey
- 1 tsp jam or jelly
- 1 tsp maple sugar or maple syrup
- 1 tsp molasses
- 1½ oz carbonated soda
- 1 tbs ketchup

♦ For perspective, each of these concentrated sugars provides about 500 kcal:
- 40 oz cola
- ½ c honey
- 125 jelly beans
- 23 marshmallows
- 30 tsp sugar

How many kcalories from sugar does your favorite beverage or snack provide?

IN SUMMARY **Sugars pose no major health threat except for an increased risk of dental caries. Excessive intakes, however, may displace needed nutrients and fiber and may contribute to obesity when energy intake exceeds needs. A person deciding to limit daily sugar intake should recognize that not all sugars need to be restricted, just concentrated sweets, which are relatively empty of other nutrients and high in kcalories. Sugars that occur naturally in fruits, vegetables, and milk are acceptable.**

© Matthew Farruggio

Consumers use artificial sweeteners to help them limit kcalories and minimize sugar intake.

Alternative Sweeteners

To control weight gain, blood glucose, and dental caries, many consumers turn to alternative sweeteners to help them limit kcalories and minimize sugar intake. In doing so, they encounter three sets of alternative sweeteners: artificial sweeteners, herbal products, and sugar alcohols.

Artificial Sweeteners

Artificial sweeteners are sometimes called **nonnutritive sweeteners** because they provide virtually no energy. Table 4-2 (pp. 116–117) provides general details about each of the sweeteners. Chapter 9 includes a discussion of their use in weight control. ♦ Considering that all substances are toxic at some dose, it is little surprise that large doses of artificial sweeteners (or their components or metabolic by-products) may have adverse effects. The question to ask is whether their ingestion is safe for human beings in quantities people normally use (and potentially abuse).

♦ The estimated amount of a sweetener that individuals can safely consume each day over the course of a lifetime without adverse effect is known as the **Acceptable Daily Intake (ADI).**

Stevia—An Herbal Product

The herb stevia derives from a plant whose leaves have long been used by the people of South America to sweeten their beverages. Until recently, stevia was sold in the United States only as a dietary supplement. Having recently been granted the status of "generally recognized as safe," stevia can now be used as an additive in a variety of foods and beverages.

Sugar Alcohols

Some "sugar-free" or reduced-kcalorie products contain sugar alcohols. The **sugar alcohols** (or polyols) provide bulk and sweetness in cookies, hard candies, sugarless gums, jams, and jellies. These products claim to

artificial sweeteners: sugar substitutes that provide negligible, if any, energy; sometimes called *nonnutritive sweeteners*.

nonnutritive sweeteners: sweeteners that yield no energy (or insignificant energy in the case of aspartame).

sugar alcohols: sugarlike compounds that can be derived from fruits or commercially produced from dextrose; also called **polyols.** Sugar alcohols are absorbed more slowly than other sugars and metabolized differently in the human body; they are not readily utilized by ordinary mouth bacteria. Examples are *maltitol, mannitol, sorbitol, xylitol, isomalt,* and *lactitol.*

TABLE 4-2 Alternative Sweeteners

Sweetener	Chemical Composition	Body's Response	Relative Sweetness[a]	Energy (kcal/g)	Acceptable Daily Intake (ADI) and (Estimated Equivalent[b])	Approval Status
			Artificial Sweeteners			
Acesulfame potassium or Acesulfame K[c] (AY-sul-fame)	Potassium salt	Not digested or absorbed	200	0	15 mg/kg body weight[d] (30 cans diet soda)	Approved for use in the United States and Canada
Aspartame[e] (ah-SPAR-tame or ASS-par-tame)	Amino acids (phenylalanine and aspartic acid) and a methyl group	Digested and absorbed	200	4[f]	50 mg/kg body weight[g] (18 cans diet soda)	Approved for use in the United States and Canada; warning for PKU
Cyclamate (SIGH-kla-mate)	Sodium or calcium salt of cyclamic acid	Incompletely absorbed; absorbed cyclamate is excreted unchanged; unabsorbed cyclamate may be metabolized by bacteria in the GI tract	30	0	11 mg/kg body weight (8 cans of diet soda)	Approval pending in the United States; approved for use in Canada
Neotame (NEE-oh-tame)	Aspartame with an additional side group attached	Not digested or absorbed	8000	0	18 mg/day	Approved for use in the United States; no warning for PKU
Saccharin[h] (SAK-ah-ren)	Benzoic sulfimide	Rapidly absorbed and excreted	450	0	5 mg/kg body weight (10 packets of sweetener)	Approved for use in the United States; restricted use as a tabletop sweetener in Canada
Sucralose[i] (SUE-kra-lose)	Sucrose with Cl atoms instead of OH groups	Not digested or absorbed	600	0	5 mg/kg body weight (6 cans diet soda)	Approved for use in the United States and Canada
Tagatose[j] (TAG-ah-tose)	Monosaccharide similar in structure to fructose; naturally occurring or derived from lactose	Mostly not absorbed; some short-chain fatty acids absorbed	0.8	1.5	7.5 g/day	GRAS[k] approved; does not promote dental caries and may carry a health claim
			Herbal Sweeteners			
Stevia[l] (STEE-vee-ah)	Glycosides found in the leaves of the *Stevia rebaudiana* herb	Digested and absorbed	300	0	4 mg/kg body weight	GRAS approved
			Sugar Alcohols			
Erythritol	Sugar alcohol	Partially absorbed in small intestine; unabsorbed sugar alcohols may be metabolized by bacteria in the GI tract	0.7	0.2	—[m]	GRAS approved

(continued)

be "sugar-free" on their labels, but in this case, "sugar-free" does not mean free of kcalories. Sugar alcohols do provide kcalories, but fewer than their carbohydrate cousins, the sugars. Because sugar alcohols yield energy, they are sometimes referred to as **nutritive sweeteners.** Table 4-2 includes their energy values. Sugar alcohols occur naturally in fruits and vegetables; manufacturers also use sugar alcohols in many processed foods to add bulk and texture, to provide a cooling effect or taste, to inhibit browning from heat, and to retain moisture.

Sugar alcohols evoke a low glycemic response. The body absorbs sugar alcohols slowly; consequently, they are slower to enter the bloodstream than other sugars. Common side effects include intestinal gas, abdominal discomfort, and diarrhea. For this reason, regulations require food labels to state "Excess consumption may have a laxative effect" if reasonable consumption of that food could result in the daily ingestion of 50 grams of a sugar alcohol.

nutritive sweeteners: sweeteners that yield energy, including both sugars and sugar alcohols.

TABLE 4-2 Alternative Sweeteners (*continued*)

Sweetener	Chemical Composition	Body's Response	Relative Sweetness[a]	Energy (kcal/g)	Acceptable Daily Intake (ADI) and (Estimated Equivalent[b])	Approval Status
			Sugar Alcohols			
Isomalt	Sugar alcohol	Partially absorbed in small intestine; unabsorbed sugar alcohols may be metabolized by bacteria in the GI tract	0.5	2.0	—[m]	GRAS approved
Lactitol	Sugar alcohol	Partially absorbed in small intestine; unabsorbed sugar alcohols may be metabolized by bacteria in the GI tract	0.4	2.0	—[m]	GRAS approved
Maltitol	Sugar alcohol	Partially absorbed in small intestine; unabsorbed sugar alcohols may be metabolized by bacteria in the GI tract	0.9	2.1	—[m]	GRAS approved
Mannitol	Sugar alcohol	Partially absorbed in small intestine; unabsorbed sugar alcohols may be metabolized by bacteria in the GI tract	0.7	1.6	—[m]	Approved for use in the United States
Sorbitol	Sugar alcohol	Partially absorbed in small intestine; unabsorbed sugar alcohols may be metabolized by bacteria in the GI tract	0.5	2.6	—[m]	GRAS approved
Xylitol	Sugar alcohol	Partially absorbed in small intestine; unabsorbed sugar alcohols may be metabolized by bacteria in the GI tract	1.0	2.4	—[m]	Approved for use in the United States

[a]Relative sweetness is determined by comparing the approximate sweetness of a sugar substitute with the sweetness of pure sucrose, which has been defined as 1.0. Chemical structure, temperature, acidity, and other flavors of the foods in which the substance occurs all influence relative sweetness.

[b]Based on a person weighing 70 kg (154 lb).

[c]Marketed under the trade names Sunett, Sweet One.

[d]Recommendations from the World Health Organization limit acesulfame-K intake to 9 mg per kilogram of body weight per day.

[e]Marketed under the trade names NutraSweet, Equal, NatraTaste, Canderel.

[f]Aspartame provides 4 kcal per gram, as does protein, but because so little is used, its energy contribution is negligible. In powdered form, it is sometimes mixed with lactose, however, so a 1 g packet may provide 4 kcal.

[g]Recommendations from the World Health Organization and in Europe and Canada limit aspartame intake to 40 mg per kilogram of body weight per day.

[h]Marketed under the trade names Sweet'N Low, Necta Sweet.

[i]Marketed under the trade names Splenda, SucraPlus.

[j]Marketed under the trade names Nutralose, Nutrilatose, Tagatesse.

[k]GRAS = food additives that are generally recognized as safe. First established by the FDA in 1958, the GRAS list is subject to revision as new facts become known.

[l]Marketed under the trade names Sweetleaf, Purevia, Truvia, Honey Leaf.

[m]An ADI is "not specified" for sugar alcohols, indicating the highest safety category. They require a warning label, however, that states "Excess consumption may have a laxative effect" if reasonable consumption could result in the daily ingestion of 50 g of a sugar alcohol.

The real benefit of using sugar alcohols is that they do not contribute to dental caries. Bacteria in the mouth cannot metabolize sugar alcohols as rapidly as sugar. Sugar alcohols are therefore valuable in chewing gums, breath mints, and other products that people keep in their mouths for a while. Figure 4-15 (p. 118) presents labeling information for products using sugar alternatives.

For consumers choosing to use alternative sweeteners, the American Dietetic Association wisely advises that they be used in moderation and only as part of a well-balanced nutritious diet.[24] When used in moderation, these sweeteners will do no harm. In fact, they may even help, by providing an alternative to sugar for people with diabetes, by inhibiting caries-causing bacteria, and by limiting energy intake. People may find it appropriate to use any of the sweeteners at times: artificial sweeteners, herbal products, sugar alcohols, and sugar itself.

FIGURE 4-15 Sugar Alternatives on Food Labels

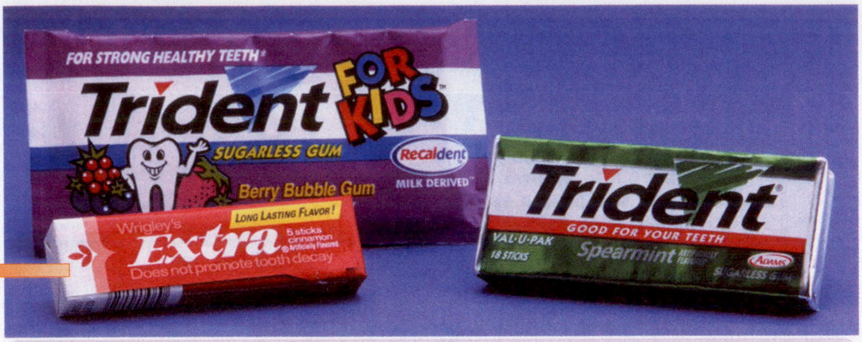

Products containing sugar replacers may claim to "not promote tooth decay" if they meet FDA criteria for dental plaque activity.

INGREDIENTS: SORBITOL, MALTITOL, GUM BASE, MANNITOL, ARTIFICIAL AND NATURAL FLAVORING, ACACIA, SOFTENERS, TITANIUM DIOXIDE (COLOR), ASPARTAME, ACESULFAME POTASSIUM AND CANDELILLA WAX.
PHENYLKETONURICS: CONTAINS PHENYLALANINE.

This ingredient list includes both sugar alcohols and artificial sweeteners.

Products containing aspartame must carry a warning for people with phenylketonuria.

35% FEWER CALORIES THAN SUGARED GUM.

Nutrition Facts

Serving Size 2 pieces (3g)
Servings 6
Calories 5

Amount per serving	**% DV***
Total Fat 0g	0%
Sodium 0mg	0%
Total Carb. 2g	1%
Sugars 0g	
Sugar Alcohol 2g	
Protein 0g	

*Percent Daily Values (DV) are based on a 2,000 calorie diet.

Not a significant source of other nutrients.

Products containing less than 0.5 g of sugar per serving can claim to be "sugarless" or "sugar-free."

Products that claim to be "reduced kcalories" must provide at least 25% fewer kcalories per serving than the comparison item.

Health Effects and Recommended Intakes of Starch and Fibers

Carbohydrates and fats are the two major sources of energy in the diet. When one is high, the other is usually low—and vice versa. A diet that provides abundant carbohydrates (45 to 65 percent of energy intake) and some fat (20 to 35 percent of energy intake) within a reasonable energy allowance best supports good health. To increase carbohydrates in the diet, focus on whole grains, vegetables, legumes, and fruits—foods noted for their starch, fibers, and naturally occurring sugars.

Health Effects of Starch and Fibers

In addition to starch, fibers, and natural sugars, whole grains, vegetables, legumes, and fruits supply valuable vitamins and minerals and little or no fat. The following paragraphs describe some of the health benefits of diets that include a variety of these foods daily.

Heart Disease Unlike high-carbohydrate diets rich in sugars that can alter blood lipids to favor heart disease, those rich in whole grains and soluble fibers may protect against heart disease and stroke, by lowering blood pressure, improving blood lipids, and reducing inflammation.[25] Such diets are low in animal fat and cholesterol and high in dietary fibers, vegetable proteins, and phytochemicals—all factors associated with a lower risk of heart disease. (The role of animal fat and cholesterol in heart disease is discussed in Chapter 5. The role of vegetable proteins in heart disease is presented in Chapter 6. The benefits of phytochemicals in disease prevention are featured in Highlight 13.)

Oatmeal was the first food recognized for its ability to reduce cholesterol and the risk of heart disease.[26] Foods rich in soluble fibers (such as oat bran, barley, and legumes) lower blood cholesterol ♦ by binding with bile acids in the GI tract and thereby increasing their excretion. Consequently, the liver must use its cholesterol to make new bile acids. In addition, the bacterial by-products of fiber fermentation in the colon also inhibit cholesterol synthesis in the liver. The net result is lower blood cholesterol.

♦ Consuming 5 to 10 g of soluble fiber daily reduces blood cholesterol by 3 to 5%. For perspective, ½ c dry oat bran provides 8 g of fiber, and 1 c cooked barley or ½ c cooked legumes provides about 6 g of fiber.

Several researchers have speculated that fiber may also exert its effect by displacing fats in the diet. Whereas this is certainly helpful, even when dietary fat is low, fibers exert a separate and significant cholesterol-lowering effect. In other words, a high-fiber diet helps to decrease the risk of heart disease independent of fat intake.

Diabetes High-fiber foods—especially whole grains—play a key role in reducing the risk of type 2 diabetes.[27] When soluble fibers trap nutrients and delay their transit through the GI tract, glucose absorption is slowed, which helps to prevent the glucose surge and rebound that seem to be associated with diabetes onset.

GI Health Dietary fibers may enhance the health of the large intestine. The healthier the intestinal walls, the better they can block absorption of unwanted constituents. Insoluble fibers such as cellulose (as in cereal brans, fruits, and vegetables) increase stool weight, easing passage, and reduce transit time. In this way, the fibers help to alleviate or prevent constipation.

Taken with ample fluids, fibers help to prevent several GI disorders. Large, soft stools ease elimination for the rectal muscles and reduce the pressure in the lower bowel, making it less likely that rectal veins will swell (hemorrhoids). Fiber prevents compaction of the intestinal contents, which could obstruct the appendix and permit bacteria to invade and infect it (appendicitis). In addition, fiber stimulates the GI tract muscles so that they retain their strength and resist bulging out into pouches known as diverticula (illustrated in Figure H3-3 on p. 91).[28]

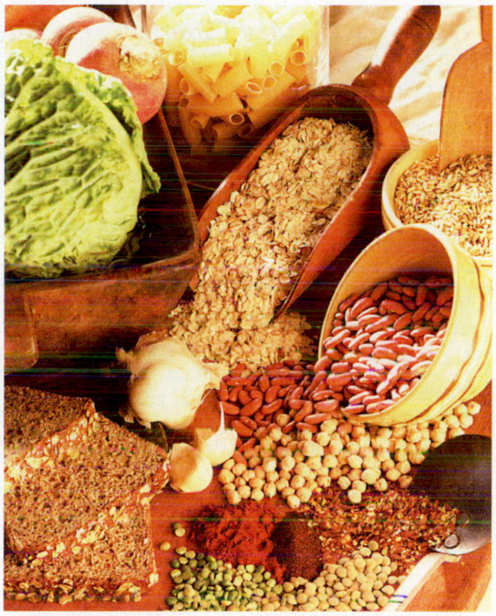

Foods rich in starch and fiber offer many health benefits.

Cancer Many, but not all, research studies suggest that increasing dietary fiber protects against colon cancer.[29] When the largest study of diet and cancer to date examined the diets of more than a half million people in ten countries for four and a half years, the researchers found an inverse association between dietary fiber and colon cancer. People who ate the most dietary fiber (35 grams per day) reduced their risk of colon cancer by 40 percent compared with those who ate the least fiber (15 grams per day). Importantly, the study focused on dietary fiber, not fiber supplements or additives, which lack valuable nutrients and phytochemicals that also help protect against cancer. Plant foods—vegetables, fruits, and whole-grain products—reduce the risks of colon and rectal cancers.

Fibers may help prevent colon cancer by diluting, binding, and rapidly removing potential cancer-causing agents from the colon. In addition, soluble fibers stimulate bacterial fermentation of resistant starch and fiber in the colon, a process that produces short-chain fatty acids that lower the pH. These small fat molecules activate cancer-killing enzymes and inhibit inflammation in the colon.[30]

Weight Management High-fiber and whole-grain foods may help a person to maintain a healthy body weight. Foods rich in fibers tend to be low in fat and added sugars and can therefore prevent weight gains and promote weight loss by delivering less energy ♦ per bite.[31] In addition, as fibers absorb water from the digestive juices, they swell, creating feelings of fullness, lowering food intake, and delaying hunger.[32]

♦ Remember:
- Carbohydrate: 4 kcal/g
- Fat: 9 kcal/g

Many weight-loss products on the market today contain bulk-inducing fibers such as methylcellulose, but buying pure fiber compounds like this is neither necessary nor advisable. Instead of fiber supplements, consumers should select whole grains, legumes, fruits, and vegetables. High-fiber foods not only add bulk to the diet but are economical and nutritious as well.

Dietary fiber provides numerous health benefits.[33] Table 4-3 (p. 120) summarizes fiber characteristics, food sources, actions in the body, and health benefits.

Harmful Effects of Excessive Fiber Intake Despite fiber's benefits to health, a diet excessively high in fiber also has a few drawbacks. A person who has a small capacity and eats mostly high-fiber foods may not be able to take in enough food to meet energy or nutrient needs. The malnourished, the elderly, and young children adhering to all-plant (vegan) diets are especially vulnerable to this problem.

TABLE 4-3 Dietary Fibers: Their Characteristics, Food Sources, and Health Effects in the Body

Fiber Characteristics	Major Food Sources	Actions in the Body	Health Benefits
Soluble, viscous, more fermentable • Gums and mucilages • Pectins • Psyllium[a] • Some hemicelluloses	Whole-grain products (barley, oats, oat bran, rye), fruits (apples, citrus), legumes, seeds and husks, vegetables; also extracted and used as food additives	• Lower blood cholesterol by binding bile • Slow glucose absorption • Slow transit of food through upper GI tract • Hold moisture in stools, softening them • Yield small fat molecules after fermentation that the colon can use for energy	• Lower risk of heart disease • Lower risk of diabetes
Insoluble, nonviscous, less fermentable • Cellulose • Lignins • Psyllium[a] • Resistant starch • Many hemicelluloses	Brown rice, fruits, legumes, seeds, vegetables (cabbage, carrots, brussels sprouts), wheat bran, whole grains; also extracted and used as food additives	• Increase fecal weight and speed fecal passage through colon • Provide bulk and feelings of fullness	• Alleviate constipation • Lower risks of diverticulosis, hemorrhoids, and appendicitis • May help with weight management

[a]Psyllium, a fiber laxative and cereal additive, has both soluble and insoluble properties.

Launching suddenly into a high-fiber diet can cause temporary bouts of abdominal discomfort, gas, and diarrhea and, more seriously, can obstruct the GI tract. To prevent such complications, a person adopting a high-fiber diet can take the following precautions:

- Increase fiber intake gradually over several weeks to give the GI tract time to adapt.
- Drink plenty of liquids to soften the fiber as it moves through the GI tract.
- Select fiber-rich foods from a variety of sources—fruits, vegetables, legumes, and whole-grain breads and cereals.

Some fibers can limit the absorption of nutrients by speeding the transit of foods through the GI tract and by binding to minerals. When mineral intake is adequate, however, a *reasonable* intake of high-fiber foods (less than 40 grams a day) does not compromise mineral balance.

Clearly, fiber is like all nutrients in that "more" is "better" only up to a point. Again, the key dietary goals are balance, moderation, and variety.

IN SUMMARY **Adequate intake of fiber:**

- **Fosters weight management**
- **Lowers blood cholesterol**
- **May help prevent colon cancer**
- **Helps prevent and control diabetes**
- **Helps prevent and alleviate hemorrhoids**
- **Helps prevent appendicitis**
- **Helps prevent diverticulosis**

Excessive intake of fiber:

- **Displaces energy- and nutrient-dense foods**
- **Causes intestinal discomfort and distention**
- **May interfere with mineral absorption**

♦ Acceptable Macronutrient Distribution Ranges (AMDR):
- Carbohydrate: 45–65%
- Fat: 20–35%
- Protein: 10–35%

♦ The Aids to Calculation section at the end of this book explains how to solve such problems.

♦ RDA for carbohydrate:
- 130 g/day
- 45 to 65% of energy intake

♦ Daily Value:
- 300 g carbohydrate (based on 60% of 2000 kcal diet)

Recommended Intakes of Starch and Fibers

The DRI suggests that carbohydrates provide about half (45 to 65 percent) of the energy requirement. ♦ A person consuming 2000 kcalories a day should therefore have 900 to 1300 kcalories of carbohydrate, or about 225 to 325 grams. ♦ This amount is more than adequate to meet the RDA ♦ for carbohydrate, which is set at 130 grams per day, based on the average minimum amount of glucose used by the brain.[34]

On food labels, the Food and Drug Administration (FDA) uses a 60 percent of kcalories guideline in setting the Daily Value ♦ for carbohydrate at 300 grams per day. For most people, this means increasing total carbohydrate intake. To this end, the *Dietary Guidelines* encourage people to choose a variety of whole grains, vegetables, fruits, and legumes daily.

Dietary Guidelines for Americans 2010

Increase vegetable and fruit intake. Eat a variety of vegetables, especially dark-green and red and orange vegetables and beans and peas. Increase whole-grain intake by replacing refined grains with whole grains.

Recommendations for fiber ♦ suggest the same foods just mentioned: whole grains, vegetables, fruits, and legumes, which also provide minerals and vitamins. The FDA sets the Daily Value ♦ for fiber at 25 grams, rounding up from the recommended 11.5 grams per 1000 kcalories for a 2000-kcalorie intake. The DRI recommendation is slightly higher, at 14 grams per 1000-kcalorie intake—roughly 25 to 35 grams of dietary fiber daily. These recommendations are about two times higher than the usual intake in the United States.[35] An effective way to add fiber while lowering fat is to substitute plant sources of proteins (legumes) for animal sources (meats). Table 4-4 presents a list of fiber sources.

Because high-fiber foods are so filling, they are not likely to be eaten in excess. Too much fiber can cause GI problems for some people, but it generally does not have adverse effects in most healthy people. For these reasons, an upper level ♦ has not been set for fiber.

♦ To increase your fiber intake:
- Eat raw vegetables.
- Eat fresh and dried fruit for snacks.
- Add legumes to soups, salads, and casseroles.
- Eat whole-grain breads that contain ≥3 g fiber per serving.
- Eat whole-grain cereals that contain ≥5 g fiber per serving.
- Eat fruits (such as pears) and vegetables (such as potatoes) with their skins.

♦ Daily Value:
- 25 g fiber (based on 11.5 g/1000 kcal)

♦ National Cancer Institute advises ≤35 g/day. World Health Organization advises ≤40 g/day.

Dietary Guidelines for Americans 2010

Choose foods that provide more dietary fiber, which is a nutrient of concern in American diets. These foods include vegetables, fruits, and whole grains.

From Guidelines to Groceries

A diet following the USDA Food Guide, which includes several servings of fruits, vegetables, and whole grains daily, can

TABLE 4-4 Fiber in Selected Foods

Grains

Whole-grain products provide about 1 to 2 g (or more) of fiber per serving:
- 1 slice whole-wheat, pumpernickel, rye bread
- 1 oz ready-to-eat cereal (100% bran cereals contain 10 g or more)
- ½ c cooked barley, bulgur, grits, oatmeal

© Polara Sutdios Inc.

Vegetables

Most vegetables contain about 2 to 3 g of fiber per serving:
- 1 c raw bean sprouts
- ½ c cooked broccoli, brussels sprouts, cabbage, carrots, cauliflower, collards, corn, eggplant, green beans, green peas, kale, mushrooms, okra, parsnips, potatoes, pumpkin, spinach, sweet potatoes, swiss chard, winter squash
- ½ c chopped raw carrots, peppers

© Polara Studios, Inc.

Fruit

Fresh, frozen, and dried fruits have about 2 g of fiber per serving:
- 1 medium apple, banana, kiwi, nectarine, orange, pear
- ½ c applesauce, blackberries, blueberries, raspberries, strawberries
- Fruit juices contain very little fiber

© Polara Sutdios Inc.

Legumes

Many legumes provide about 6 to 8 g of fiber per serving:
- ½ c cooked baked beans, black beans, black-eyed peas, kidney beans, navy beans, pinto beans

Some legumes provide about 5 g of fiber per serving:
- ½ c cooked garbanzo beans, great northern beans, lentils, lima beans, split peas

© PhotoDisc Blue/Getty Images

NOTE: Appendix H provides fiber grams for more than 2000 foods.

Courtesy Oldways and the Whole Grain Council, © wholegrainscouncil.org

Some food labels use a "whole-grain stamp" to help consumers identify whole-grain foods.

♦ Some food labels include information about the whole-grain contents (often in grams—and 16 grams of whole-grain ingredients are equivalent to one whole-grain serving).

easily supply the recommended amount of carbohydrates and fiber. In selecting high-fiber foods, keep in mind the principle of variety. The fibers in oats lower cholesterol, whereas those in bran help promote GI tract health. (Review Table 4-3 to see the diverse health effects of various fibers.)

Grains An ounce-equivalent of most foods in the grain group (for example, one slice of bread) provides about 15 grams of carbohydrate, mostly as starch. Be aware that some foods in this group, especially snack crackers and baked goods such as biscuits, croissants, and muffins, contain added sugars, added fat, or both. When selecting from the grain group, be sure to include at least half as whole-grain products (see Figure 4-16). The "three are key" message may help consumers to remember to choose a whole-grain cereal for breakfast, a whole-grain bread for lunch, and a whole-grain pasta or rice for dinner. ♦

Vegetables The amount of carbohydrate a serving of vegetables provides depends primarily on its starch content. Starchy vegetables—a half-cup of cooked corn, peas, or potatoes—provide about 15 grams of carbohydrate per serving. A serving of most other *nonstarchy* vegetables—such as a half-cup of broccoli, green beans, or tomatoes—provides about 5 grams.

FIGURE 4-16 Bread Labels Compared

Although breads may appear similar, their ingredients vary widely. Breads made mostly from whole-grain flours provide more benefits to the body than breads made of enriched, refined, wheat flour.

Some "high-fiber" breads may contain purified cellulose or more nutritious whole grains. "Low-carbohydrate" breads may be regular white bread, thinly sliced to reduce carbohydrates per serving, or may contain soy flour, barley flour, or flaxseed to reduce starch content.

A trick for estimating a bread's content of a nutritious ingredient, such as whole-grain flour, is to read the ingredients list (ingredients are listed in order of predominance). Bread recipes generally include one teaspoon of salt per loaf. Therefore, when a bulky nutritious ingredient, such as whole grain, is listed after the salt, you'll know that less than a teaspoonful of the nutritious ingredient was added to the loaf—not enough to significantly improve the nutrient value of one slice of bread.

Whole Grain
WHOLE WHEAT

Nutrition Facts	
Serving size 1 slice (30g) Servings Per Container 18	
Amount per serving	
Calories 90	Calories from Fat 14
	% Daily Value*
Total Fat 1.5g	2%
Trans Fat 0g	
Sodium 135mg	6%
Total Carbohydrate 15g	5%
Dietary fiber 2g	8%
Sugars 2g	
Protein 4g	

MADE FROM: UNBROMATED STONE GROUND 100% WHOLE WHEAT FLOUR, WATER, CRUSHED WHEAT, HIGH FRUCTOSE CORN SYRUP, PARTIALLY HYDROGENATED VEGETABLE SHORTENING (SOYBEAN AND COTTONSEED OILS), RAISIN JUICE CONCENTRATE, WHEAT GLUTEN, YEAST, WHOLE WHEAT FLAKES, UNSULPHURED MOLASSES, SALT, HONEY, VINEGAR, ENZYME MODIFIED SOY LECITHIN, CULTURED WHEY, UNBLEACHED WHEAT FLOUR AND SOY LECITHIN.

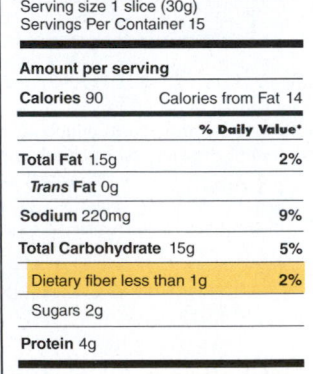

Natural
Wheat Bread

Nutrition Facts	
Serving size 1 slice (30g) Servings Per Container 15	
Amount per serving	
Calories 90	Calories from Fat 14
	% Daily Value*
Total Fat 1.5g	2%
Trans Fat 0g	
Sodium 220mg	9%
Total Carbohydrate 15g	5%
Dietary fiber less than 1g	2%
Sugars 2g	
Protein 4g	

INGREDIENTS: UNBLEACHED ENRICHED WHEAT FLOUR [MALTED BARLEY FLOUR, NIACIN, REDUCED IRON, THIAMIN MONONITRATE (VITAMIN B1), RIBOFLAVIN (VITAMIN B2), FOLIC ACID], WATER, HIGH FRUCTOSE CORN SYRUP, MOLASSES, PARTIALLY HYDROGENATED SOYBEAN OIL, YEAST, CORN FLOUR, SALT, GROUND CARAWAY, WHEAT GLUTEN, CALCIUM PROPIONATE (PRESERVATIVE), MONOGLYCERIDES, SOY LECITHIN.

Multi-fiber
Low carb

Nutrition Facts	
Serving size 1 slice (30g) Servings Per Container 21	
Amount per serving	
Calories 60	Calories from Fat 15
	% Daily Value*
Total Fat 1.5g	2%
Trans Fat 0g	
Sodium 135mg	6%
Total Carbohydrate 9g	3%
Dietary fiber 3g	12%
Sugars 0g	
Protein 5g	

INGREDIENTS: UNBLEACHED ENRICHED WHEAT FLOUR, WATER, WHEAT GLUTEN, CELLULOSE, YEAST, SOYBEAN OIL, CRACKED WHEAT, SALT, BARLEY, NATURAL FLAVOR PRESERVATIVES, MONOCALCIUM PHOSPHATE, MILLET, CORN, OATS, SOYBEANS, BROWN RICE, FLAXSEED, SUCRALOSE.

Fruits A typical fruit serving—a small banana, apple, or orange or a half-cup of most canned or fresh fruit—contains an average of about 15 grams of carbohydrate, mostly as sugars, including the fruit sugar fructose. Fruits vary greatly in their water and fiber contents and, therefore, in their sugar concentrations.

Milks and Milk Products A serving (a cup) of milk or yogurt provides about 12 grams of carbohydrate. Cottage cheese provides about 6 grams of carbohydrate per cup, but most other cheeses contain little, if any, carbohydrate.

Protein Foods With two exceptions, protein foods deliver almost no carbohydrate to the diet. The exceptions are nuts, which provide a little starch and fiber along with their abundant fat, and legumes, which provide an abundance of both starch and fiber. Just a half-cup serving of legumes provides about 20 grams of carbohydrate, a third from fiber.

Read Food Labels Food labels list the amount, in grams, of *total* carbohydrate—including starch, fibers, and sugars—per serving (review Figure 4-16). Fiber grams are also listed separately, as are the grams of sugars. (With this information, you can calculate starch grams ♦ by subtracting the grams of fibers and sugars from the total carbohydrate.) Sugars reflect both added sugars and those that occur naturally in foods. Total carbohydrate and dietary fiber are also expressed as "% Daily Values" for a person consuming 2000 kcalories; there is no Daily Value for sugars.

♦ To calculate starch grams using the first label in Figure 4-16:
15 g total − 4 g (dietary fiber + sugars) = 11 g starch

IN SUMMARY **Clearly, a diet rich in starches and fibers supports efforts to control body weight and prevent heart disease, cancer, diabetes, and GI disorders. For these reasons, recommendations urge people to eat plenty of whole grains, vegetables, legumes, and fruits—enough to provide 45 to 65 percent of the daily energy intake from carbohydrate.**

In today's world, there is one other reason why plant foods rich in complex carbohydrates and natural sugars are a better choice than animal foods or foods high in concentrated sugars: in general, less energy and fewer resources are required to grow and process plant foods than to produce sugar or foods derived from animals.

Nutrition Portfolio

Foods that derive from plants—whole grains, vegetables, legumes, and fruits—naturally provide ample carbohydrates and fiber with little or no fat. Refined foods often contain added sugars and fat.

Go to Diet Analysis Plus and choose one of the days on which you have tracked your diet for the entire day. Go to the Intake Spreadsheet report. Scroll down until you see: carb (g).

- Which of your foods for this day were highest in carbohydrate? Which of these foods also contain added sugars and fats? List better alternatives.
- List the types and amounts of grain products you ate on that day, making note of which are whole-grain or refined foods and how your choices could include more whole-grain options.
- List the types and amounts of fruits and vegetables you ate on that day, making note of how many are dark green, orange, or deep yellow, how many are starchy or legumes, and how your choices could include more of these options.
- Describe choices you can make in selecting and preparing foods and beverages to lower your intake of added sugars.

Diet Analysis PLUS

To complete this exercise, go to your Diet Analysis Plus at www.cengage.com/sso.

Nutrition on the Net

For further study of topics covered in this chapter, log on to **www.cengagebrain.com** and search for ISBN 084006845X.

- Learn more about lactose intolerance from the National Institute of Diabetes and Digestive and Kidney Diseases: **digestive.niddk.nih.gov/ddiseases/pubs/lactoseintolerance**
- Search for "sugars" and "fiber" at the International Food Information Council site: **www.ific.org**
- Learn more about dental caries from the American Dental Association and the National Institute of Dental and Craniofacial Research: **www.ada.org** and **www.nidcr.nih.gov**
- Learn more about diabetes from the American Diabetes Association, the Canadian Diabetes Association, and the National Institute of Diabetes and Digestive and Kidney Diseases: **www.diabetes.org**, **www.diabetes.ca**, and **www2.niddk.nih.gov**

References

1. J. H. Cummings and A. M. Stephen, Carbohydrate terminology and classification, *European Journal of Clinical Nutrition* 61 (2007): S5–S18.
2. L. Van Horn and coauthors, The evidence for dietary prevention and treatment of cardiovascular disease, *Journal of the American Dietetic Association* 108 (2008): 287–331; M. O. Weickert and A. F. Pfeiffer, Metabolic effects of dietary fiber consumption and prevention of diabetes, *Journal of Nutrition* 138 (2008): 439–442; N. R. Sahyoan and coauthors, Whole-grain intake is inversely associated with metabolic syndrome and mortality in older adults, *American Journal of Clinical Nutrition* 83 (2006): 124–131.
3. J. R. Korzenik, Case closed? Diverticulitis: Epidemiology and fiber, *Journal of Clinical Gastroenterology* 40 (2006): S112–S116.
4. M.Nofrarías and coauthors, Long-term intake of resistant starch improves colonic mucosal integrity and reduces gut apoptosis and blood immune cells, *Nutrition* 23 (2007): 861–870.
5. Committee on Dietary Reference Intakes, *Dietary Reference Intakes: Energy, Carbohydrate, Fiber, Fat, Fatty Acids, Cholesterol, Protein, and Amino Acids* (Washington, D.C.: National Academies Press, 2005).
6. C. C. Robayo-Torres and B. L. Nichols, Molecular differentiation of congenital lactase deficiency from adult-type hypolactasia, *Nutrition Reviews* 65 (2007): 95–98; A. K. Campbell, J. P. Waud, and S. B. Matthews, The molecular basis of lactose intolerance, *Science Progress* 88 (2005): 157–202.
7. F. Guarner and coauthors, Should yoghurt cultures be considered probiotic? *British Journal of Nutrition* 93 (2005): 783–786.
8. T. He and coauthors, Effects of yogurt and bifidobacteria supplementation on the colonic microbiotia in lactose-intolerant subjects, *Journal of Applied Microbiology* 104 (2008): 595–604.
9. J. Wahren and K. Ekberg, Splanchnic regulation of glucose production, *Annual Review of Nutrition* 27 (2007): 329–345.
10. G. Riccardi, A. A. Rivellese, and R. Giacco, Role of glycemic index and glycemic load in the healthy state, in prediabetes, and in diabetes, *American Journal of Clinical Nutrition* 87 (2008): 269S–274S.
11. K. Foster-Powell, S.H.A. Holt, and J. C. Brand-Miller, International table of glycemic index and glycemic load values: 2002, *American Journal of Clinical Nutrition* 76 (2002): 5–56.
12. G. Livesey and coauthors, Glycemic response and health—A systematic review and meta-analysis: Relations between dietary glycemic properties and health outcomes, *American Journal of Clinical Nutrition* 87 (2008): 258S–268S.
13. A. W. Barclay and coauthors, Glycemic index, glycemic load, and chronic disease risk: A meta-analysis of observational studies, *American Journal of Clinical Nutrition* 87 (2008): 627–637; J. Howlett and M. Ashwell, Glycemic response and health: Summary of a workshop, *American Journal of Clinical Nutrition* 87 (2008): 212S–216S; A. Mosdøl and coauthors, Dietary glycemic index and glycemic load are associated with high-density-lipoprotein cholesterol at baseline but not with increased risk of diabetes in the Whitehall II study, *American Journal of Clinical Nutrition* 86 (2007): 988–994; T. L. Halton and coauthors, Low-carbohydrate-diet score and the risks of coronary heart disease in women, *New England Journal of Medicine* 355 (2006): 1991–2002; C. B. Ebbeling and coauthors, Effects of an ad libitum low-glycemic load diet on cardiovascular disease risk factors in obese young adults, *American Journal of Clinical Nutrition* 81 (2005): 976–982; S. Dickinson and J. Brand-Miller, Glycemic index, postprandial glycemia and cardiovascular disease, *Current Opinion in Lipidology* 16 (2005): 69–75.
14. C. B. Ebbeling and coauthors, Effects of a low-glycemic load vs low-fat diet in obese young adults: A randomized trial, *Journal of the American Medical Association* 297 (2007): 2092–2102; K. C. Maki and coauthors, Effects of a reduced-glycemic-load diet on body weight, body composition, and cardiovascular disease risk markers in overweight and obese adults, *American Journal of Clinical Nutrition* 85 (2007): 724–734; R. Sichieri and coauthors, An 18-mo randomized trial of a low-glycemic-index diet and weight change in Brazilian women, *American Journal of Clinical Nutrition* 86 (2007): 707–713; A. Flint and coauthors, Glycemic and insulinemic responses as determinants of appetite in humans, *American Journal of Clinical Nutrition* 84 (2006): 1365–1373; H. Hare-Bruun, A. Flint, and B. L. Heitmann, Glycemic index and glycemic load in relation to changes in body weight, body fat distribution, and body composition in adult Danes, *American Journal of Clinical Nutrition* 84 (2006): 871–879; M. A. Pereira, Weighing in on glycemic index and body weight, *American Journal of Clinical Nutrition* 84 (2006): 677–679; G. Livesey, Low-glycaemic diets and health: Implications for obesity, *Proceedings of the Nutrition Society* 64 (2005): 105–113.
15. H. Hare-Bruun and coauthors, Should glycemic index and glycemic load be considered in dietary recommendations? *Nutrition Reviews* 66 (2008): 569–590.
16. T.M.S. Wolever and coauthors, Measuring the glycemic index of foods: Interlaboratory study, *American Journal of Clinical Nutrition* 87 (2008): 247S–257S.
17. H. V. Worthington, J. E. Clarkson, and O. B. Eden, Interventions for preventing oral mucositis for patients with cancer receiving treatment, *Cochrane Database of Systematic Reviews* 4 (2007): CD000978; I. M. Paul and coauthors, Effect of honey, dextromethorphan, and no treatment on nocturnal cough and sleep quality for coughing children and their parents, *Archives of Pediatric and Adolescent Medicine* 161 (2007): 1140–1146.
18. A. Bhargava and A. Amialchuk, Added sugars displaced the use of vital nutrients in the National Food Stamp Program Survey, *Journal of Nutrition* 137 (2007): 453–460.
19. K. J. Cross, N. L. Huq, and E. C. Reynolds, Casein phosphopeptides in oral health: Chemistry and clinical applications, *Current Pharmaceutical Design* 13 (2007): 793–800; B. T. Amaechi and S. M. Higham, Dental erosion, Possible approaches to prevention and control, *Journal of Dentistry* 33 (2005): 243–252.
20. T. Jaeggi and A. Lussi, Prevalence, incidence and distribution of erosion, *Monographs in Oral Science* 20 (2006): 44–65; S. Wongkhantee and coauthors, Effect of acidic food and drinks on surface hardness of enamel, dentine, and tooth-coloured filling materials, *Journal of Den-

tistry 34 (2006): 214–220; W. K. Seow and K. M. Thong, Erosive effects of common beverages on extracted premolar teeth, *Australian Dental Journal* 50 (2005): 173–178.

21. Position of the American Dietetic Association: Oral health and nutrition, *Journal of the American Dietetic Association* 107 (2007): 1418–1428.
22. What We Eat in America, NHANES, 2005–2006, www.ars.usda.gov/ba/bhnrc/fsrg, published 2008; S. Haley and coauthors, Sweetener consumption in the United States, *Economic Research Service, USDA*, August 2005.
23. Committee on Dietary Reference Intakes, 2005.
24. Position of the American Dietetic Association: Use of nutritive and nonnutritive sweeteners, *Journal of the American Dietetic Association* 104 (2004): 255–275.
25. M. T. Streppel and coauthors, Dietary fiber intake in relation to coronary heart disease and all-cause mortality over 40 y: The Zutphen Study, *American Journal of Clinical Nutrition* 88 (2008): 1119–1125; M. F. Chong, B. A. Fielding, and K. N. Frayn, Metabolic interaction of dietary sugars and plasma lipids with a focus on mechanisms and de novo lipogenesis, *Proceedings of the Nutrition Society* 66 (2007): 52–59; P. B. Mellen, T. F. Walsh, and D. M. Herrington, Whole grain intake and cardiovascular disease: A meta-analysis, *Nutrition, Metabolism and Cardiovascular Diseases* (2007): 283–290; R. Solà and coauthors, Effects of soluble fiber (*Plantago ovata* husk) on plasma lipids, lipoproteins, and apolipoproteins in men with ischemic heart disease, *American Journal of Clinical Nutrition* 85 (2007): 1157–1163; M. K. Jenson and coauthors, Whole grains, bran and germ in relation to homocysteine and markers of glycemic control, lipids, and inflammation, *American Journal of Clinical Nutrition* 83 (2006): 275–283.
26. M. B. Andon and J. W. Anderson, State of the art reviews: The oatmeal-cholesterol connection: 10 Years later, *American Journal of Lifestyle Medicine* 2 (2008): 51–57.
27. Jenson and coauthors, 2006.
28. D. O. Jacobs, Diverticulitis, *New England Journal of Medicine* 357 (2007): 2057–2066; H. Salzman and D. Lillie, Diverticular disease: Diagnosis and treatment, *American Family Physician* 72 (2005): 1229–1234.
29. A. Schatzkin and coauthors, Dietary fiber and whole-grain consumption in relation to colorectal cancer in the NIH-AARP Diet and Health Study, *American Journal of Clinical Nutrition* 85 (2007): 1353–1360; S. Bingham, Symposium on "Plant foods and public health": The fibre-folate debate in colo-rectal cancer, *Proceedings of the Nutrition Society* 65 (2006): 19–23; K. B. Michels and coauthors, Fiber intake and incidence of colorectal cancer among 76,947 women and 47,279 men, *Cancer Epidemiology Biomarkers and Prevention* 14 (2005): 842–849; Y. Park and coauthors, Dietary fiber intake and risk of colorectal cancer, *Journal of the American Medical Association* 294 (2005): 2849–2857.
30. D. J. Rose and coauthors, Influence of dietary fiber on inflammatory bowel disease and colon cancer: Importance of fermentation pattern, *Nutrition Reviews* 65 (2007): 51–62.
31. L. A. Tucker and K. S. Thomas, Increasing total fiber intake reduces risk of weight and fat gains in women, *Journal of Nutrition* 139 (2009): 567–581.
32. R. A. Samra and G. H. Anderson, Insoluble cereal fiber reduces appetite and short-term food intake and glycemic response to food consumed 75 min later by healthy men, *American Journal of Clinical Nutrition* 86 (2007): 972–979.
33. J. W. Anderson and coauthors, Health benefits of dietary fiber, *Nutrition Reviews* 67 (2009): 188–205.
34. Committee on Dietary Reference Intakes, 2005.
35. What We Eat in America, 2008; Position of the American Dietetic Association: Health implications of dietary fiber, *Journal of the American Dietetic Association* 108 (2008): 1716–1731.

HIGHLIGHT 4

© Susan Van Etten/PhotoEdit

Carbs, kCalories, and Controversies

Carbohydrate-rich foods are easy to like. Mashed potatoes, warm muffins, blueberry pancakes, freshly baked bread, and tasty rice or pasta dishes tempt most people's palates. In recent years, such homey foods have been blamed for causing weight gain and harming health. Popular writers have persuaded consumers that carbohydrates are "bad."[1] In contrast, the *Dietary Guidelines for Americans* urge people to consume plenty of fruits, vegetables, legumes, and whole grains—all carbohydrate-rich foods.

Do carbohydrate-rich foods cause obesity and related health problems?[2] Should people "cut carbs" to lose weight and protect their health? Many popular diet books espouse a carbohydrate-restricted or carbohydrate-modified diet. Some claim that all or some types of carbohydrates are bad. Some go so far as to equate carbohydrates with toxic poisons or addictive drugs. "Bad" carbohydrates—such as sugar, white flour, and potatoes—are considered evil because they are absorbed easily and raise blood glucose. The pancreas then responds by secreting insulin—and insulin is touted as the real villain responsible for our nation's epidemic of obesity. Whether restricting overall carbohydrate intake or replacing certain "bad" carbohydrates with "good" carbohydrates, many of these popular diets tend to distort the facts. This highlight examines the scientific evidence behind some of the current controversies surrounding carbohydrates and their kcalories.

Carbohydrates' kCalorie Contributions

The incidence of obesity in the United States has risen dramatically over the past several decades.[3] Popular diet books often blame carbohydrates for this increase in obesity. One way researchers can explore whether the amount of carbohydrate in the diet contributes to increases in body weight over time is by reviewing national food intake survey records, such as NHANES (introduced in Chapter 1). Figure H4-1 presents a summary of energy nutrient data over the past three decades. Since the 1970s, kcalories from carbohydrates increased from 42 percent to 49 percent today.[4] At the same time, kcalories from fat dropped from 41 percent to 34 percent. The percentage of protein intake stayed about the same.

A closer look at the data reveals that, as the percentage of kcalories from the three energy nutrients shifted slightly, total daily energy intake increased significantly. In general, as food became more readily available in this nation, consumers began to eat more than they had in the past.[5] Since the 1970s, total energy intakes have increased by about 300 kcalories a day (see Figure H4-2).[6] Almost all of the increase in kcalories came from an increase in carbohydrate kcalories. At the same time, most people were not active enough to use up those extra kcalories; in fact, activity levels declined.[7] Consequently, the average body weight for adults increased over these decades by about 20 pounds (see Figure H4-3).

Might too many carbohydrates in the diet be to blame for weight gains? Interestingly, epidemiological studies find an *inverse* relationship between carbohydrate intake and body weight.[8] Those with the highest carbohydrate intake have the lowest body weight and vice versa. Dietary fiber, which favors a healthy body weight, explains some but not all of this relationship.

Might a low-carbohydrate diet support weight losses? Studies report that people following low-carbohydrate diets do lose

FIGURE H4-1 Energy Nutrients over Time

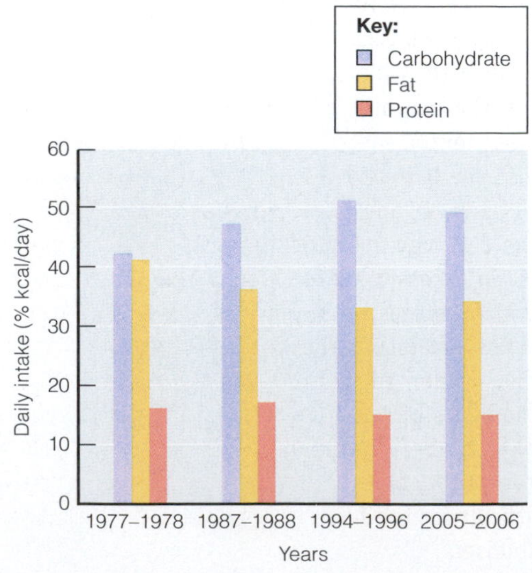

FIGURE H4-2 Daily Energy Intake over Time

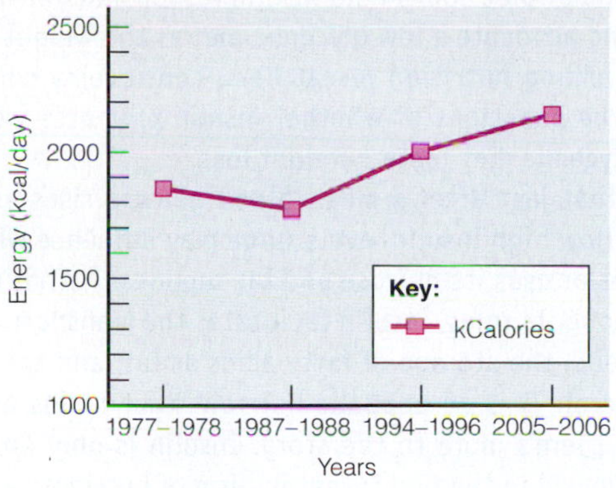

FIGURE H4-3 Increases in Adult Body Weight over Time

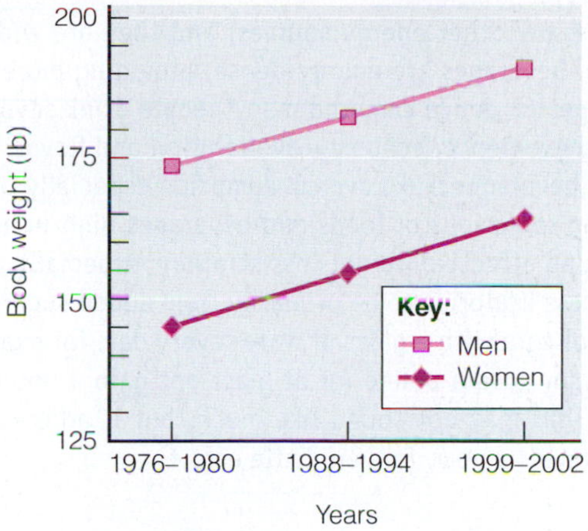

weight.[9] In fact, they lose more than people following conventional high-carbohydrate, low-fat diets—but only for the first six months. Their later gains make up the difference, so total weight loss is no different after one year.[10] For the most part, weight loss is similar for people following either a low-carbohydrate diet or a high-carbohydrate diet.[11] This is an important point. Weight losses reflect restricted kcalories—not the proportion of energy nutrients in the diet.[12] Any diet can produce weight loss, at least temporarily, if energy intake is restricted.

Sugars' Share in the Problem

Over the past several decades, as obesity rates increased sharply, consumption of added sugars reached an all-time high—much of it because high-fructose corn syrup use, especially in beverages, surged.[13] High-fructose corn syrup is composed of fructose and glucose in a ratio of roughly 50:50. Compared with sucrose, high-fructose corn syrup is less expensive, easier to use, and more stable. In addition to being used in beverages, high-fructose corn syrup sweetens candies, baked goods, and hundreds of other foods. Fructose contributes about half of the added sugars in the U. S. food supply and accounts for about 10 percent of the average energy intake in the United States.[14]

Although the use of high-fructose corn syrup sweetener parallels unprecedented increases in the incidence of obesity, does it mean that the increasing sugar intakes are responsible for the increase in body fat and its associated health problems?[15] Excess sugar in the diet may be associated with more fat on the body.[16] When eaten in excess of need, energy from added sugars contributes to body fat stores, just as excess energy from other sources does.[17] Added sugars provide excess energy, raising the risk of weight gain.[18] When total energy intake is controlled, however, *moderate* amounts of sugar do not *cause* obesity. Yet moderating sugar intake can be a challenge. Some claim sugar is addictive. Others assert sugary beverages are particularly easy to swallow and make it difficult for the body to regulate appetite control and energy metabolism.

Cravings and Addictions

Do sugars cause cravings and addictions? Foods in general, and carbohydrates and sugars more specifically, are not physically addictive in the ways that drugs are. Yet some people describe themselves as having "carbohydrate cravings" or being "sugar addicts." One frequently noted theory is that people seek carbohydrates as a way to increase their levels of the brain neurotransmitter serotonin, which elevates mood. Interestingly, when those with self-described carbohydrate cravings indulge, they tend to eat more of everything; the percentage of energy from carbohydrates remains unchanged.

One reasonable explanation for the carbohydrate cravings that some people experience involves the self-imposed labeling of a food as both "good" and "bad"—that is, one that is desirable but should be eaten with restraint. Restricting intake heightens the desire further (a "craving"). Then "addiction" is used to explain why resisting the food is so difficult and, sometimes, even impossible. But the "addiction" is not physiological or pharmacological.

Simple to Swallow

One added sugar in particular—the liquid high-fructose corn syrup—is used to sweeten beverages. In general, the energy intake of people who drink soft drinks, fruit punches, and other sugary beverages is greater than those who choose differently. Adolescents, for example, who drink as much as 26 ounces or more (about two cans) of sugar-sweetened soft drinks daily, consume 400 more kcalories a day than teens who don't. Not too surprisingly, they also tend to weigh more.[19] Overweight children and adolescents consume more sweet desserts and soft drinks than their normal-weight peers.[20] Review of the research confirms that consuming sugary beverages correlates with both increased food energy and being overweight.[21]

The liquid form of sugar in soft drinks makes it especially easy to overconsume kcalories. Swallowing liquid kcalories requires little effort. The sugar kcalories of sweet beverages also cost less than many other energy sources, and they are widely available. Also, beverages are energy-dense, providing more than 150 kcalories per 12-ounce can, and many people drink several cans a day. The convenience, economy, availability, and flavors of sugary foods and beverages make overconsumption especially likely.

Limiting selections of foods and beverages high in added sugars can be an effective weight-loss strategy, especially for people whose excess kcalories come primarily from added sugars. Replacing a can of cola with a glass of water every day, for example, can help a person lose a pound (or at least not gain a pound) in one month.[22] That may not sound like much, but it adds up to more than 10 pounds a year, for very little effort.

Appetite Control

Recall from Chapter 4 that glucose stimulates the release of insulin from the pancreas. Insulin, in turn, sets off a sequence of hormonal actions that suppress the appetite.[23] (Appetite regulation is discussed fully in Chapter 8.) Fructose, in contrast, does not stimulate the release of insulin, and therefore does not suppress appetite. Theoretically, then, eating lots of fructose would never satisfy a person's appetite. Although this idea sounds plausible, a major flaw exists: people don't typically eat pure fructose. They eat sucrose or high-fructose corn syrup, and both of these sugars contain sufficient glucose to stimulate the release of insulin and suppress appetite accordingly.

Whether the meal or snack is liquid or solid may also affect appetite. Even when kcaloric intake is the same, a fresh apple suppresses appetite more than apple juice.[24] Consequently, beverages can influence weight gains both by providing energy and by not satisfying hunger.

Energy Regulation

One explanation of why it is so easy to overconsume sugary beverages is that perhaps the body's energy regulation system cannot detect the kcalories of sugar in liquid form. Consequently, a person would not compensate for energy excesses by reducing food intake at other times. A research study tested this hypothesis by giving students 450 kcalories' worth of either solid sugars (roughly 40 jelly beans) or liquid sugars (about three 12-ounce cans of soft drinks) to consume daily whenever they chose.[25] Sure enough, the energy intake from other foods during "jelly-bean weeks" was lower—the students ate less food to compensate for the kcalories received from the jelly beans. By comparison, energy intake from other foods during the "soft-drink weeks" did *not* decrease—the students ate their meals without compensating for the kcalories in the beverages. Consequently, body weight increased during the beverage weeks, but not during the candy weeks. Other studies, however, have found no differences between liquid and solid sugars when examining appetite, energy intake, or body weight.[26]

Insulin's Response

Several popular diet books hold insulin responsible for the obesity problem and advocate a low glycemic diet as the weight-loss solution. Yet, among nutrition researchers, controversy continues to surround the questions of whether insulin promotes weight gain or a low glycemic diet fosters weight loss.[27]

Recall that just after a meal, blood glucose rises and insulin responds. How high insulin levels surge may influence whether the body stores or uses its glucose and fat supplies.[28] What does insulin do? Among its roles, insulin facilitates the transport of glucose into the cells, the storage of fatty acids as fat, and the synthesis of cholesterol. It is an anabolic hormone that builds and stores. True—but there's more to the story. Insulin is only one of many factors involved in the body's metabolism of nutrients and regulation of body weight.

Furthermore, as Chapter 4's discussion of the glycemic index pointed out, the glycemic effect of a particular food varies (see Figure 4-13 on p. 111)—diet books often mislead people by claiming that each food has a set glycemic effect. The glycemic effect of a food depends on how the food is ripened, processed, and cooked; the time of day the food is eaten; the other foods eaten with it; and the presence or absence of certain diseases such as type 2 diabetes in the person eating the food.[29]

Most importantly, insulin is critical to maintaining health, as any person with type 1 diabetes can attest. Insulin causes problems only when a person develops insulin resistance—that is, when the body's cells do not respond to the large quantities of insulin that the pancreas continues to pump out in an effort to get a response. Insulin resistance is a major health problem—but it is not caused by carbohydrate, or by protein, or by fat. It results from being overweight. Importantly, when a person loses weight, insulin response improves, regardless of the diet.

The Glycemic Index and Body Weight

As Chapter 4 mentions, the glycemic index identifies foods that raise blood glucose and stimulate insulin secretion. What is the relationship between a diet's glycemic index and fat storage? Studies find that diets with a high glycemic index are positively associated with body weight.[30] Because fructose does not stimulate insulin secretion, it has a low glycemic index.[31] Yet some research suggests that fructose favors the fat-making pathways and impairs the fat-clearing pathways in the body.[32] As the liver busily makes lipids, its handling of glucose becomes unbalanced and insulin resistance develops.[33] Research is beginning to find links between high fructose intake and prediabetes and the metabolic syndrome.[34]

Might a low glycemic diet foster weight loss?[35] When obese people followed one of three low-kcalorie diets—high glycemic diet, low glycemic diet, or high-fat diet—for 9 months, they all lost about 20 pounds.[36] Furthermore, insulin sensitivity improved for all of them. Other studies confirm that overweight people experience similar weight losses on a low-kcalorie diet regardless of whether it has a high or a low glycemic index.[37] In other words, all low-kcalorie diets support weight loss; defining the type or amount of carbohydrate does not enhance losses. A low glycemic meal may, however, prompt less energy intake at the next meal.[38]

Clearly, if kcalories are low, obese people on either a low glycemic diet or a traditional low-fat diet can lose weight. Overweight people can lose as much or more weight by emphasizing low glycemic foods as they can by following a typical low-fat, portion-controlled weight-loss diet.[39]

The Individual's Response to Foods

The body's insulin response depends not only on a food, but also on a person's metabolism. Some people react to dietary carbohydrate with a low insulin response. Others have a high insulin response. One study reports that increases in body weight over 6 years were similar in people following either a high-carbohydrate diet or a low-carbohydrate diet. But those with a high insulin response gained more weight, especially when they were on a high-carbohydrate diet.[40] By the same token, for those with a higher insulin response, weight loss may be greater on a low glycemic diet.[41]

How energy is stored after a meal depends in part on how the body responds to insulin. After eating a high-carbohydrate meal, normal-weight people who are insulin resistant tend to synthesize about half as much glycogen in muscles and make about twice as much fat in the liver as people who are insulin sensitive.[42] Some research suggests that restricting carbohydrate intake may improve glucose control, insulin response, and blood lipids.[43]

In Summary

As might be expected given the similarity in their chemical composition, high-fructose corn syrup and sucrose produce similar effects in appetite control and energy metabolism.[44] In fact, high-fructose corn syrup is more like sucrose than it is like fructose. Furthermore, people don't eat pure fructose; they eat foods and drink beverages that contain added sugars—either high-fructose corn syrup or sucrose. Limiting these sugars is a helpful strategy when trying to control body weight, but restricting all carbohydrates would be unwise.

The quality of the diet suffers when carbohydrates are restricted.[45] Without fruits, vegetables, and whole grains, low-carbohydrate diets lack not only carbohydrate, but fiber, vitamins, minerals, and phytochemicals as well—all dietary factors protective against disease.[46] The DRI recommends that carbohydrates contribute between 45 and 65 percent of daily energy intake. Intakes within this range can support healthy body weight and do not contribute to obesity—when total energy intake is appropriate. Similarly, added sugars increase energy intake, but need not contribute to obesity—when total energy intake is appropriate.

Research results on the glycemic index of diets are mixed, but when results are clear, a low glycemic diet has the greatest advantages. A low glycemic meal seems to curb appetite and limit energy intake of the next meal. Low glycemic diets are also more likely to be rich in nutrients and fiber than high glycemic diets.[47] A healthy diet includes a variety of carbohydrate-rich sources: whole-grain cereals, vegetables, legumes, and fruits.[48]

References

1. G. A. Bray, Viewpoint: *Good Calories, Bad Calories* by Gary Taubes, *Obesity Reviews* 9 (2008): 251–263.
2. D. B. Allison and R. D. Mattes, Nutritively sweetened beverage consumption and obesity: The need for solid evidence on a fluid issue, *Journal of the American Medical Association* 301 (2009): 318–320.
3. C. L. Ogden and coauthors, Prevalence of overweight and obesity in the United States, 1999–2004, *Journal of the American Medical Association* 295 (2006): 1549–1555.
4. U.S. Department of Agriculture, Agricultural Research Service, 2008, Nutrient intakes from food, www.ars.usda.gov/ba/bhnrc/fsrg, accessed October 2008.
5. K. Sinventoinen and coauthors, Trends in obesity an energy supply in the WHO MONICA project, *International Journal of Obesity and Related Metabolic Disorders* 28 (2004): 710–718.
6. Centers for Disease Control and Prevention, Trend in intake of kcalories and macronutrients: United States, 1971–2000, *Morbidity and Mortality Weekly Report* 53 (2004): 80–82.
7. P. M. Barnes and C. A. Schoenborn, *Physical Activity among Adults: United States, 2000,* www.cdc.gov/nchs/about/major/nhis/released200306.htm#7, 2003.
8. G. A. Gaesser, Carbohydrate quantity and quality in relation to body mass index, *Journal of the American Dietetic Association* 107 (2007): 1768–1780.
9. A. Astrup, T. M. Larsen, and A. Harper, Atkins and other low-carbohydrate diets: Hoax or an effective tool for weight loss? *The Lancet* 364 (2004): 897–899; E. C. Westman and coauthors, Effect of 6-month adherence to a very low carbohydrate diet program, *American Journal of Medicine* 113 (2002): 30–36.
10. C. Erlanson-Albertsson and J. Mei, The effect of low carbohydrate on energy metabolism, *International Journal of Obesity* 29 (2005): S26–S30; L. Stern and coauthors, The effects of low-carbohydrate versus conventional weight loss diets in severely obese adults: One-year follow-up of a randomized trial, *Annals of Internal Medicine* 140 (2004): 778–785; G. D. Foster and coauthors, A randomized trial of a low-carbohydrate diet for obesity, *New England Journal of Medicine* 348 (2003): 2082–2090.
11. I. Shai and coauthors, Weight loss with a low-carbohydrate, Mediterranean, or low-fat diet, *New England Journal of Medicine* 359 (2008): 229–241; R. F. Kushner and B. Doerfier, Low-carbohydrate, high-protein diets revisited, *Current Opinion in Gastroenterology* 24 (2008): 198–203; A. K. Halyburton and coauthors, Low- and high-carbohydrate weight-loss diets have similar effects on mood but not cognitive performance, *American Journal of Clinical Nutrition* 86 (2007): 580–587; T. McLaughlin and coauthors, Effects of moderate variations in macronutrient composition on weight loss and reduction in cardiovascular disease risk in obese, insulin-resistant adults, *American Journal of Clinical Nutrition* 84 (2006): 813–821.
12. F. M. Sacks and coauthors, Comparison of weight-loss diets with different compositions of fat, protein, and carbohydrates, *New England Journal of Medicine* 360 (2009): 859-873; R. M. van Dam and J. C. Seidell, Carbohydrate intake and obesity, *European Journal of Clinical Nutrition* 61 (2007): S75–S99.

13. V. S. Malik, M. B. Schulze, and F. B. Hu, Intake of sugar-sweetened beverages and weight gain: A systematic review, *American Journal of Clinical Nutrition* 84 (2006): 274–288.
14. J. P. Bantle, Is fructose the optimal low glycemic index sweetener? *Nestlé Nutrition Workshop Series: Clinical & Performance Program* 11 (2006): 83–91.
15. G. A. Bray, How bad is fructose? *American Journal of Clinical Nutrition* 86 (2007): 895–896; R. Dhingra and coauthors, Soft drink consumption and risk of developing cardiometabolic risk factors and the metabolic syndrome in middle-aged adults in the community, *Circulation* 116 (2007): 480–488; R. J. Johnson and coauthors, Potential role of sugar (fructose) in the epidemic of hypertension, obesity and the metabolic syndrome, diabetes, kidney disease, and cardiovascular disease, *American Journal of Clinical Nutrition* 86 (2007): 899–906; S. C. Larsson, L. Bergkvist, and A. Wolk, Consumption of sugar and sugar-sweetened foods and the risk of pancreatic cancer in a prospective study, *American Journal of Clinical Nutrition* 84 (2006): 1171–1176.
16. J. N. Davis and coauthors, Associations of dietary sugar and glycemic index with adiposity and insulin dynamics in overweight Latino youth, *American Journal of Clinical Nutrition* 86 (2007): 1331–1338.
17. R. A. Forshee and coauthors, A critical examination of the evidence relating high fructose corn syrup and weight gain, *Critical Reviews in Food Science and Nutrition* 47 (2007): 561–582.
18. L. R. Vartanian, M. B. Schwartz, and K. D. Brownell, Effects of soft drink consumption on nutrition and health: A systematic review and meta-analysis, *American Journal of Public Health* 97 (2007): 667–675.
19. R. Dhingra and coauthors, Soft drink consumption and risk of developing cardiometabolic risk factors and the metabolic syndrome in middle-aged adults in the community, *Circulation* 116 (2007): 480–488; L. R. Vartanian, M. B. Schwartz, and K. D. Brownell, Effects of soft drink consumption on nutrition and health: A systematic review and meta-analysis, *American Journal of Public Health* 97 (2007): 667–675.
20. I. Aeberli and coauthors, Fructose intake is a predictor of LDL particle size in overweight schoolchildren, *American Journal of Clinical Nutrition* 86 (2007): 1174–1178.
21. A. Drewnowski and F. Bellisle, Liquid kcalories, sugar, and body weight, *American Journal of Clinical Nutrition* 85 (2007): 651–661; V. S. Malik, M. B. Schulze, and F. B. Hu, Intake of sugar-sweetened beverages and weight gain: A systematic review, *American Journal of Clinical Nutrition* 84 (2006): 274–288.
22. L. Chen and coauthors, Reduction in consumption of sugar-sweetened beverages is associated with weight loss: The PREMIER trial, *American Journal of Clinical Nutrition* 89 (2009): 1299–1306.
23. K. J. Melanson and coauthors, High-fructose corn syrup, energy intake, and appetite regulation, *American Journal of Clinical Nutrition* 88 (2008): 1738S–1744S.
24. R. D. Mattes and W. W. Campbell, Effects of food form and timing of ingestion on appetite and energy intake in lean young adults and in young adults with obesity, *Journal of the American Dietetic Association* 109 (2009): 430–437.
25. D. P. DiMeglio and R. D. Mattes, Liquid versus solid carbohydrate: Effects on food intake and body weight, *International Journal of Obesity and Related Metabolic Disorders* 24 (2000): 794–800.
26. T. Akhavan and G. H. Anderson, Effects of glucose-to-fructose ratios in solutions on subjective satiety, food intake, and satiety hormones in young men, *American Journal of Clinical Nutrition* 86 (2007): 1354–1363; K. J. Melanson and coauthors, Effects of high-fructose corn syrup and sucrose consumption on circulating glucose, insulin, leptin, and ghrelin on appetite in normal weight-women, *Nutrition* 23 (2007): 103–112S; S. Soenen and M. S. Weterterp-Plantenga, No differences in satiety or energy intake after high-fructose corn syrup, sucrose, or milk preloads, *American Journal of Clinical Nutrition* 86 (2007): 1586–1594.
27. R. Clemens and P. Pressman, Clinical value of glycemic index unclear, *Food Technology* 58 (2004): 18; M. A. Pereira and coauthors, Effects of a low-glycemic load diet on resting energy expenditure and heart disease risk factors during weight loss, *Journal of the American Medical Association* 292 (2004): 2482–2490; A. Raben, Should obese patients be counselled to follow a low-glycaemic index diet? No, *Obesity Reviews* 3 (2002): 245–256; D. B. Pawlak, C. B. Ebbeling, and D. S. Ludwig, Should obese patients be counselled to follow a low-glycaemic index diet? Yes, *Obesity Reviews* 3 (2002): 235–243.
28. M. A. Pereira, Weighing in on glycemic index and body weight, *American Journal of Clinical Nutrition* 84 (2006): 677–679.
29. F. X. Pi-Sunyer, Glycemic index and disease, *American Journal of Clinical Nutrition* 76 (2002): 290S–298S.
30. H. Hare-Bruun, A. Flint, and B. L. Heitmann, Glycemic index and glycemic load in relation to changes in body weight, body fat distribution, and body composition in adult Danes, *American Journal of Clinical Nutrition* 84 (2006): 871–879.
31. M. S. Segal, E. Gollub, and R. J. Johnson, Is the fructose index more relevant with regards to cardiovascular disease than the glycemic index? *European Journal of Nutrition* 46 (2007): 406–417.
32. E. J. Parks and coauthors, Dietary sugars stimulate fatty acid synthesis in adults, *Journal of Nutrition* 138 (2008): 1039–1046; M. F. Chong, B. A. Fielding, and K. N. Frayn, Mechanisms for the acute effect of fructose on postprandial lipemia, *American Journal of Clinical Nutrition* 85 (2007): 1511–1520; Bray, 2007; Bantle, 2006; P. J. Havel, Dietary fructose: Implications for dysregulation of energy homeostasis and lipid/carbohydrate metabolism, *Nutrition Reviews* 63 (2005): 133–157.
33. K. A. Lê and L. Tappy, Metabolic effects of fructose, *Current Opinion in Clinical and Metabolic Care* 9 (2006): 469–475.
34. A. Miller and K. Adeli, Dietary fructose and the metabolic syndrome, *Current Opinion in Gastroenterology* 24 (2008): 204–209.
35. J. Brand-Miller and coauthors, Carbohydrates: The good, the bad and the whole grain, *Asia Pacific Journal of Clinical Nutrition* 17 (2008): 16–19.
36. S. K. Raatz and coauthors, Reduced glycemic index and glycemic load diets do not increase the effects of energy restriction on weight loss and insulin sensitivity in obese men and women, *Journal of Nutrition* 135 (2005): 2387–2391.
37. R. Sichieri and coauthors, An 18-mo randomized trial of a low-glycemic-index diet and weight change in Brazilian women, *American Journal of Clinical Nutrition* 86 (2007): 707–713; S. K. Das and coauthors, Long-term effects of 2 energy-restricted diets differing in glycemic load on dietary adherence, body composition, and metabolism in CALERIE: A 1-y randomized controlled trial, *American Journal of Clinical Nutrition* 85 (2007): 1023–1030.
38. A. Flint and coauthors, Glycemic and insulinemic responses as determinants of appetite in humans, *American Journal of Clinical Nutrition* 84 (2006): 1365–1373.
39. K. C. Maki and coauthors, Effects of a reduced-glycemic-load diet on body weight, body composition, and cardiovascular disease risk markers in overweight and obese adults, *American Journal of Clinical Nutrition* 85 (2007): 724–734.
40. J. P. Chaput and coauthors, A novel interaction between dietary composition and insulin secretion: Effects on weight gain in the Quebec Family Study, *American Journal of Clinical Nutrition* 87 (2008): 303–309.
41. C. B. Ebberling and coauthors, Effects of a low-glycemic load vs low-fat diet in obese young adults: A randomized trial, *Journal of the American Medical Association* 297 (2007): 2092–2102.
42. K. F. Petersen and coauthors, The role of skeletal muscle insulin resistance in the pathogenesis of the metabolic syndrome, *Proceedings of the National Academy of Sciences* 104 (2007): 12587-12594.
43. R. J. Wood and M. L. Fernandez, Carbohydrate-restricted versus low-glycemic-index diets for the treatment of insulin resistance and metabolic syndrome, *Nutrition Reviews* 67 (2009): 179–183; J. S. Volek

and R. D. Feinman, Carbohydrate restriction improves the features of metabolic syndrome. Metabolic syndrome may be defined by the response to carbohydrate restriction, *Nutrition and Metabolism* 2 (2005): 31–47.

44. K. J. Melanson and coauthors, High-fructose corn syrup, energy intake, and appetite regulation, *American Journal of Clinical Nutrition* 88 (2008): 1738S–1744S.
45. L. S. Greene-Finestone and coauthors, Adolescents' low-carbohydrate-density diets are related to poorer dietary intakes, *Journal of the American Dietetic Association* 105 (2005): 1783–1788; E. T. Kennedy and coauthors, Popular diets: Correlation to health, nutrition, and obesity, *Journal of the American Dietetic Association* 101 (2001): 411–420.
46. W. Cunningham and D. Hyson, The skinny on high-protein, low-carbohydrate diets, *Preventive Cardiology* 9 (2006): 166–171.
47. Pereira, 2006.
48. Van Dam and Seidell, 2007.

CHAPTER

5

J. Helgason/Shutterstock.com

Nutrition in Your Life

Most likely, you know what you don't like about body fat, but do you appreciate how it insulates you against the cold or powers your hike around a lake? And what about food fat? You're right to credit fat for providing the delicious flavors and aromas of buttered popcorn and fried chicken—and to criticize it for contributing to the weight gain and heart disease so common today. The challenge is to strike a healthy balance of enjoying some fat, but not too much. Learning which kinds of fats are beneficial and which are most harmful will help you make wise decisions.

Throughout this chapter, the CourseMate icon indicates an opportunity for online self-study, linking you to activities to increase your understanding of chapter concepts. **www.cengagebrain.com** (search for ISBN 084006845X)

The Lipids: Triglycerides, Phospholipids, and Sterols

CHAPTER OUTLINE

Most people are surprised to learn that fat has some virtues. Only when people consume either too much or too little fat, or too much of some kinds of fat, does poor health develop. It is true, though, that in our society of abundance, people are likely to consume too much fat.

Fat refers to the class of nutrients known as **lipids**. The lipid family includes triglycerides (**fats** and **oils**), phospholipids, and sterols. The triglycerides ♦ are most abundant, both in foods and in the body.

♦ Of the lipids in foods, 95% are fats and oils (triglycerides); of the lipids stored in the body, 99% are triglycerides.

The Chemist's View of Fatty Acids and Triglycerides

Like carbohydrates, lipids are composed of carbon (C), hydrogen (H), and oxygen (O). Because lipids have many more carbons and hydrogens in proportion to their oxygens, they can supply more energy per gram than carbohydrates can (Chapter 7 provides details).

The many names and relationships in the lipid family can seem overwhelming—like meeting a friend's extended family for the first time. To ease the introductions, this chapter first presents each of the lipids from a chemist's point of view using both words and diagrams. Then the chapter follows the lipids through digestion and absorption and into the body to examine their roles in health and disease. For people who think more easily in words than in chemical symbols, this *preview* of the upcoming chemistry may be helpful:

1. Every triglyceride contains one molecule of glycerol and three fatty acids (basically, chains of carbon atoms).
2. Fatty acids may be 4 to 24 (even numbers of) carbons long, the 18-carbon ones being the most common in foods and especially noteworthy in nutrition.
3. Fatty acids may be saturated or unsaturated. Unsaturated fatty acids may have one or more points of unsaturation. (That is, they may be *mono*unsaturated or *poly*unsaturated.)

lipids: a family of compounds that includes triglycerides, phospholipids, and sterols. Lipids are characterized by their insolubility in water. (Lipids also include the fat-soluble vitamins, described in Chapter 11.)

fats: lipids that are solid at room temperature (77°F or 25°C).

oils: lipids that are liquid at room temperature (77°F or 25°C).

4. Of special importance in nutrition are the polyunsaturated fatty acids whose *first* point of unsaturation is next to the third carbon (known as omega-3 fatty acids) or next to the sixth carbon (omega-6 fatty acids).
5. The 18-carbon fatty acids that fit this description are linolenic acid (omega-3) and linoleic acid (omega-6). Each is the primary member of a family of longer-chain fatty acids that help to regulate blood pressure, blood clotting, and other body functions important to health.

The paragraphs, definitions, and diagrams that follow present this information again in much more detail.

FIGURE 5-1 Acetic Acid

Acetic acid is a two-carbon organic acid.

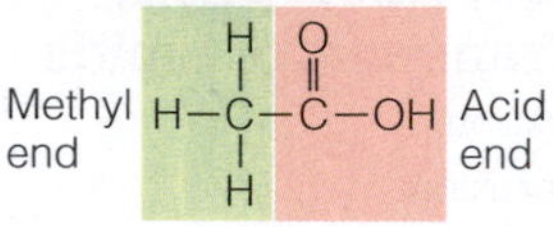

Fatty Acids

A **fatty acid** is an organic acid—a chain of carbon atoms with hydrogens attached—that has an acid group (COOH) at one end and a methyl group (CH_3) at the other end. The organic acid shown in Figure 5-1 is acetic acid, the compound that gives vinegar its sour taste. Acetic acid is the shortest such acid, with a "chain" only two carbon atoms long. (Fatty acid and related terms are defined in the accompanying glossary.)

The Length of the Carbon Chain

Most naturally occurring fatty acids contain even numbers of carbons in their chains—up to 24 carbons in length. This discussion begins with the 18-carbon fatty acids, which are abundant in our food supply. Stearic acid is the simplest of the 18-carbon fatty acids; the bonds between its carbons are all alike:

Stearic acid, an 18-carbon saturated fatty acid

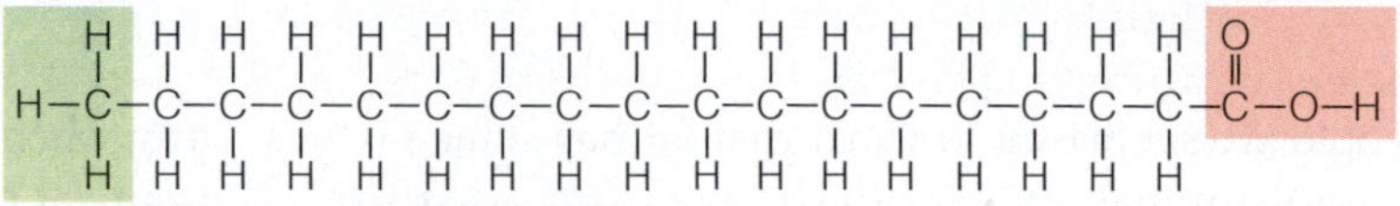

As you can see, stearic acid is 18 carbons long, and each atom meets the rules of chemical bonding described in Figure 4-1 on p. 98. The following structure also depicts stearic acid, but in a simpler way, with each "corner" on the zigzag line representing a carbon atom with two attached hydrogens:

Stearic acid (simplified structure)

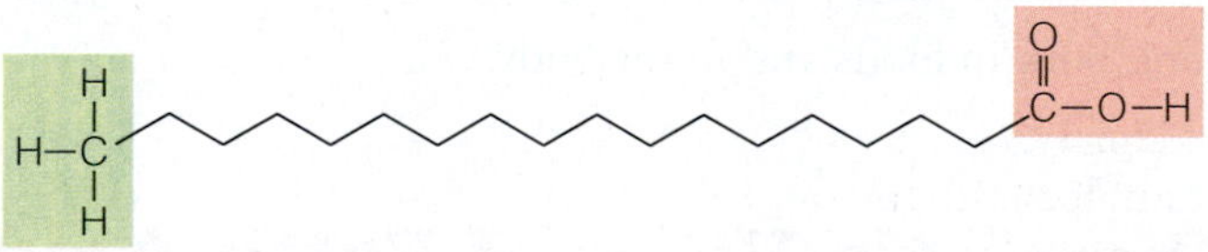

As mentioned, the carbon chains of fatty acids vary in length. The long-chain (12 to 24 carbons) fatty acids of meats, fish, and vegetable oils are most common in the diet. Smaller amounts of medium-chain (6 to 10 carbons) and short-chain (fewer than 6 carbons) fatty acids also occur, primarily in dairy products. (Tables C-1 and C-2 in Appendix C provide the names, chain lengths, and sources of fatty acids commonly found in foods.)

The Degree of Unsaturation

Stearic acid (described and shown previously) is a **saturated fatty acid**. A saturated fatty acid is fully loaded with hydrogen atoms

GLOSSARY OF FATTY ACID TERMS

fatty acid: an organic compound composed of a carbon chain with hydrogens attached and an acid group (COOH) at one end and a methyl group (CH_3) at the other end.

monounsaturated fatty acid (MUFA): a fatty acid that lacks two hydrogen atoms and has one double bond between carbons—for example, oleic acid. A *monounsaturated fat* is composed of triglycerides in which most of the fatty acids are monounsaturated.
- **mono** = one

point of unsaturation: the double bond of a fatty acid, where hydrogen atoms can easily be added to the structure.

polyunsaturated fatty acid (PUFA): a fatty acid that lacks four or more hydrogen atoms and has two or more double bonds between carbons—for example, linoleic acid (two double bonds) and linolenic acid (three double bonds). A *polyunsaturated fat* is composed of triglycerides in which most of the fatty acids are polyunsaturated.
- **poly** = many

saturated fatty acid: a fatty acid carrying the maximum possible number of hydrogen atoms—for example, stearic acid. A *saturated fat* is composed of triglycerides in which most of the fatty acids are saturated.

unsaturated fatty acid: a fatty acid that lacks hydrogen atoms and has at least one double bond between carbons (includes monounsaturated and polyunsaturated fatty acids). An *unsaturated fat* is composed of triglycerides in which most of the fatty acids are unsaturated.

and contains only single bonds between its carbon atoms. If two hydrogens were missing from the middle of the carbon chain, the remaining structure might be:

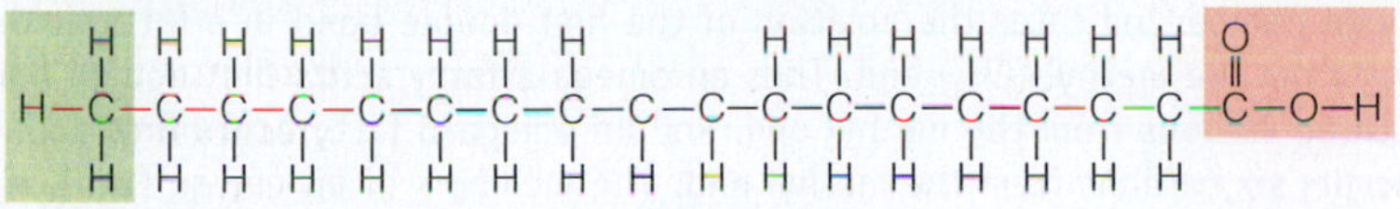

An impossible chemical structure

Such a compound cannot exist, however, because two of the carbons have only three bonds each, and every carbon must have four bonds. The two carbons therefore form a double bond:

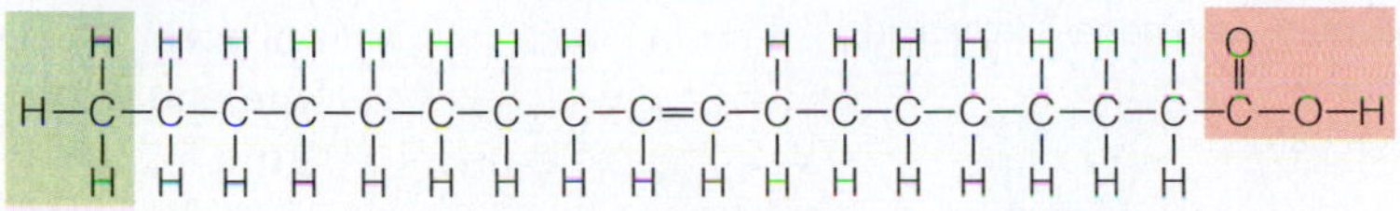

Oleic acid, an 18-carbon monounsaturated fatty acid

The same structure drawn more simply looks like this:*

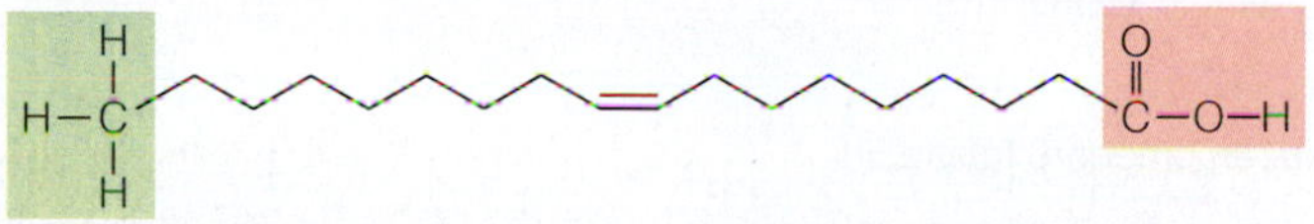

Oleic acid (simplified structure)

Although drawn straight here, the actual shape bends at the double bond (as shown in the left side of Figure 5-8 on p. 139). The double bond is a **point of unsaturation**. A fatty acid like this—with two hydrogens missing and a double bond—is an **unsaturated fatty acid**. This one is the 18-carbon **monounsaturated fatty acid** oleic acid, which is abundant in olive oil and canola oil.

A **polyunsaturated fatty acid** has two or more carbon-to-carbon double bonds. **Linoleic acid**, the 18-carbon fatty acid common in vegetable oils, lacks four hydrogens and has two double bonds:

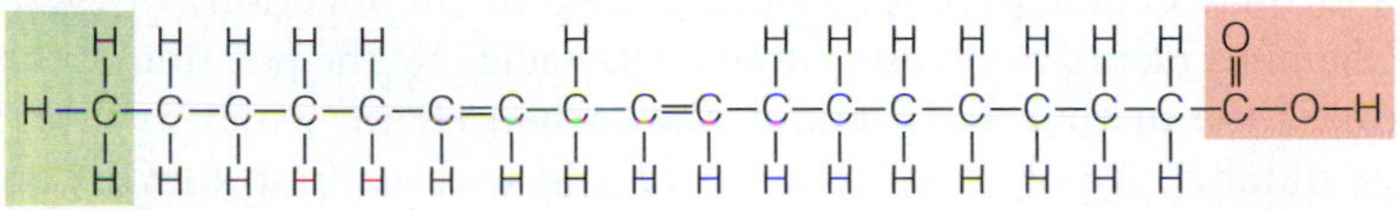

Linoleic acid, an 18-carbon polyunsaturated fatty acid

Drawn more simply, linoleic acid looks like this (though the actual shape would bend at the double bonds, as shown in the left side of Figure 5-8 on p. 139):

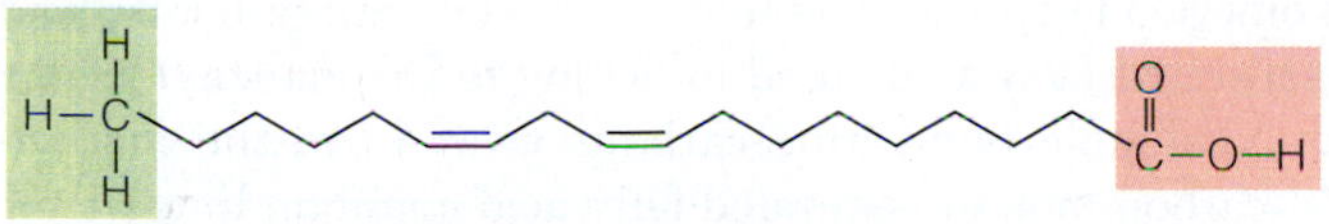

Linoleic acid (simplified structure)

A fourth 18-carbon fatty acid is **linolenic acid**, which has three double bonds. Table 5-1 presents the 18-carbon fatty acids. ♦

♦ Chemists use a shorthand notation to describe fatty acids. The first number indicates the number of carbon atoms; the second, the number of the double bonds. For example, the notation for stearic acid is 18:0.

*Remember that each "corner" on the zigzag line represents a carbon atom with two attached hydrogens.

TABLE 5-1 18-Carbon Fatty Acids

Name	Number of Carbon Atoms	Number of Double Bonds	Saturation	Common Food Sources
Stearic acid	18	0	Saturated	Most animal fats
Oleic acid	18	1	Monounsaturated	Olive, canola oils
Linoleic acid	18	2	Polyunsaturated	Sunflower, safflower, corn, and soybean oils
Linolenic acid	18	3	Polyunsaturated	Soybean and canola oils, flaxseed, walnuts

linoleic (lin-oh-LAY-ick) **acid:** an essential fatty acid with 18 carbons and two double bonds.

linolenic (lin-oh-LEN-ick) **acid:** an essential fatty acid with 18 carbons and three double bonds.

FIGURE 5-2 **Omega-3 and Omega-6 Fatty Acids Compared**

The omega number indicates the position of the first double bond in a fatty acid, counting from the methyl (CH_3) end. Thus an omega-3 fatty acid's first double bond occurs three carbons from the methyl end, and an omega-6 fatty acid's first double bond occurs six carbons from the methyl end. The members of an omega family may have different lengths and different numbers of double bonds, but the first double bond occurs at the same point in all of them. These structures are drawn linearly here to ease counting carbons and locating double bonds, but their shapes actually bend at the double bonds, as shown in Figure 5-8 (p. 139).

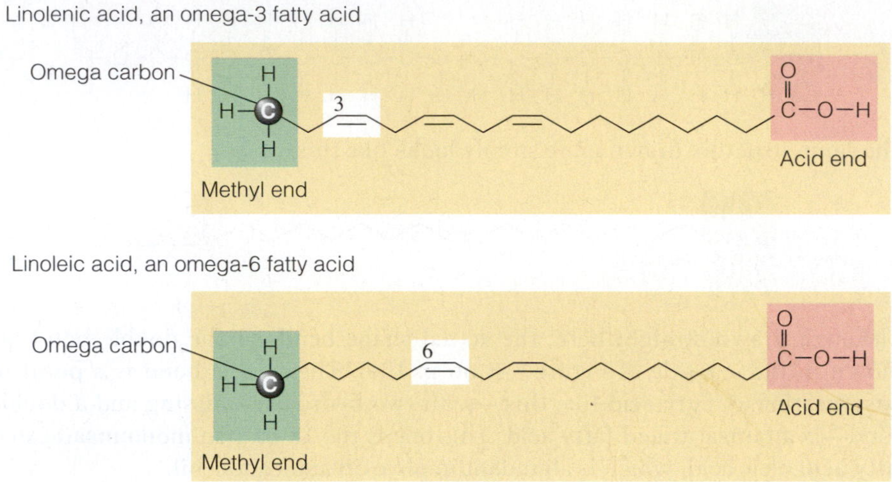

FIGURE 5-3 **Glycerol**

When glycerol is free, an OH group is attached to each carbon. When glycerol is part of a triglyceride, each carbon is attached to a fatty acid by a carbon-oxygen bond.

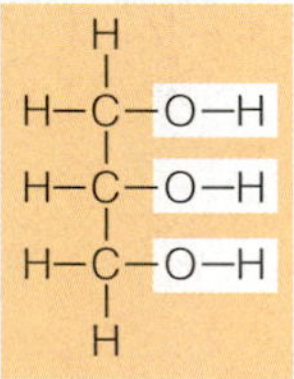

The Location of Double Bonds Fatty acids differ not only in the length of their chains and their degree of saturation, but also in the locations of their double bonds. Chemists identify polyunsaturated fatty acids by the position of the double bond nearest the methyl (CH_3) end of the carbon chain, which is described by an **omega** number. A polyunsaturated fatty acid with its first double bond three carbons away from the methyl end is an **omega-3 fatty acid**. Similarly, an **omega-6 fatty acid** is a polyunsaturated fatty acid with its first double bond six carbons away from the methyl end. Figure 5-2 compares two 18-carbon fatty acids—linolenic acid (an omega-3 fatty acid) and linoleic acid (an omega-6 fatty acid).

Monounsaturated fatty acids tend to belong to the omega-9 group, with their first (and only) double bond nine carbons away from the methyl end. Oleic acid—the 18-carbon monounsaturated fatty acid common in olive oil mentioned earlier—is an omega-9 fatty acid. It is also the most predominant monounsaturated fatty acid in the diet.

Triglycerides

Few fatty acids occur free in foods or in the body. Most often, they are incorporated into **triglycerides**—lipids composed of three fatty acids attached to a **glycerol**. (Figure 5-3 presents a glycerol molecule.) To make a triglyceride, a series of condensation reactions combine a hydrogen atom (H) from the glycerol and a hydroxyl (OH) group from a fatty acid, forming a molecule of water (H_2O) and leaving a bond between the two molecules (see Figure 5-4). Most triglycerides contain a mixture of more than one type of fatty acid (as shown on the right side of Figure 5-4).

Degree of Unsaturation Revisited

The chemistry of a fatty acid—whether it is short or long, saturated or unsaturated, with its first double bond at carbon 3 or carbon 6—influences the characteristics of foods and the health of the body. A section later in this chapter explains how these features affect health; this section describes how the chemistry influences the fats and oils in foods.

Firmness The degree of unsaturation influences the firmness of fats at room temperature (see Figure 5-5). Generally speaking, most polyunsaturated vegetable

omega: the last letter of the Greek alphabet (ω), used by chemists to refer to the position of the first double bond from the methyl (CH_3) end of a fatty acid.

omega-3 fatty acid: a polyunsaturated fatty acid in which the first double bond is three carbons away from the methyl (CH_3) end of the carbon chain.

omega-6 fatty acid: a polyunsaturated fatty acid in which the first double bond is six carbons from the methyl (CH_3) end of the carbon chain.

triglycerides (try-GLISS-er-rides): the chief form of fat in the diet and the major storage form of fat in the body; composed of a molecule of glycerol with three fatty acids attached; also called **triacylglycerols** (try-ay-seel-GLISS-er-ols).*

- **tri** = three
- **glyceride** = of glycerol
- **acyl** = a carbon chain

glycerol (GLISS-er-ol): an alcohol composed of a three-carbon chain, which can serve as the backbone for a triglyceride.

- **ol** = alcohol

*Research scientists commonly use the term *triacylglycerols;* this book continues to use the more familiar term *triglycerides,* as do many other health and nutrition books and journals.

FIGURE 5-4 **Condensation of Glycerol and Fatty Acids to Form a Triglyceride**

To make a triglyceride, three fatty acids attach to glycerol in condensation reactions.

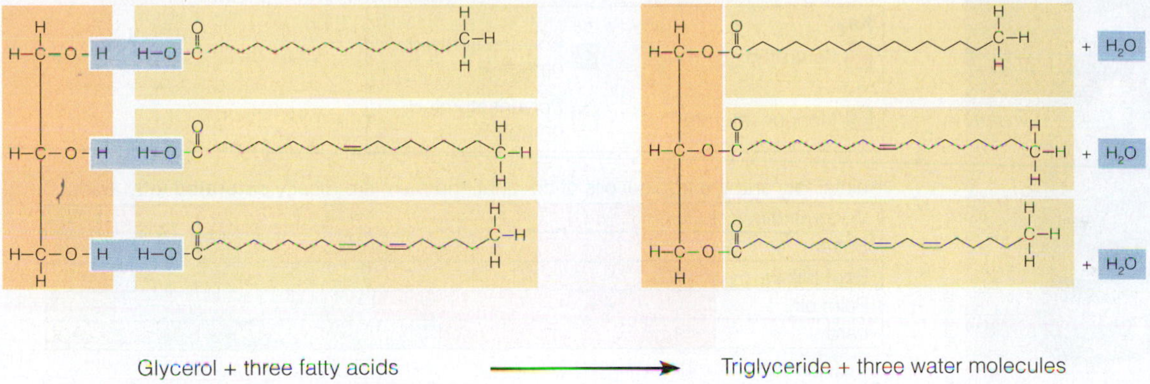

An H atom from glycerol and an OH group from a fatty acid combine to create water, leaving the O on the glycerol and the C at the acid end of each fatty acid to form a bond.

Three fatty acids attached to a glycerol form a triglyceride and yield water. In this example, the triglyceride includes a saturated fatty acid, a monounsaturated fatty acid, and a polyunsaturated fatty acid, respectively.

FIGURE 5-5 **Diagram of Saturated and Unsaturated Fatty Acids Compared**

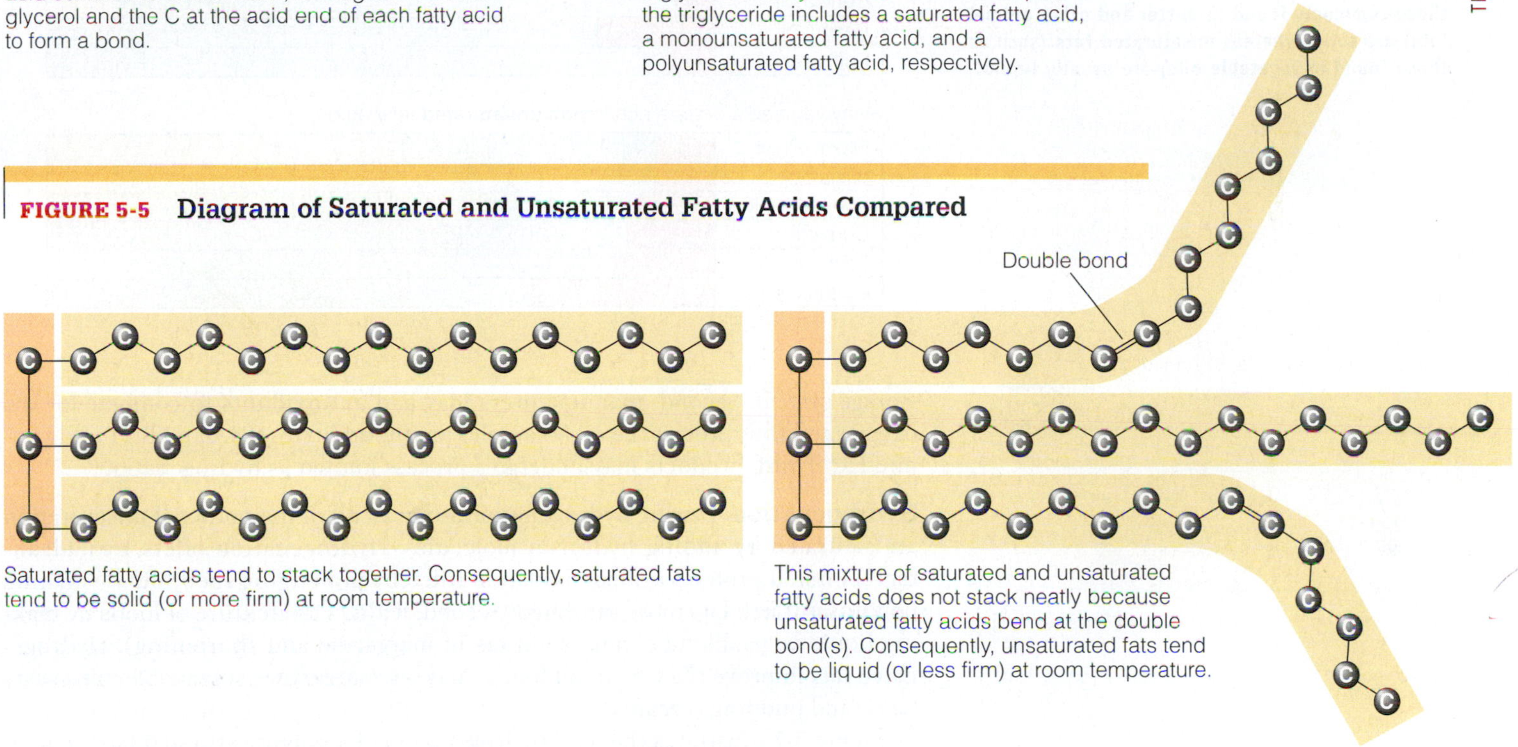

Saturated fatty acids tend to stack together. Consequently, saturated fats tend to be solid (or more firm) at room temperature.

This mixture of saturated and unsaturated fatty acids does not stack neatly because unsaturated fatty acids bend at the double bond(s). Consequently, unsaturated fats tend to be liquid (or less firm) at room temperature.

oils are liquid at room temperature, and the more saturated animal fats are solid. Some vegetable oils—notably cocoa butter, palm oil, palm kernel oil, and coconut oil—are saturated ♦; they are firmer than most vegetable oils because of their saturation, but softer than most animal fats because of their shorter carbon chains (8 to 14 carbons long). Generally, the shorter the carbon chain, the softer the fat is at room temperature. Fatty acid compositions of selected fats and oils are shown in Figure 5-6 (p. 138), and Appendix H provides the fat and fatty acid contents of many other foods.

♦ The food industry often refers to these saturated vegetable oils as the "tropical oils."

Stability Saturation also influences stability. All fats become spoiled when exposed to oxygen. The **oxidation** of fats produces a variety of compounds that smell and taste rancid. (Other types of spoilage can occur due to microbial growth.) Polyunsaturated fats spoil most readily because their double bonds are unstable; monounsaturated fats are slightly less susceptible. Saturated fats are most resistant to oxidation and thus least likely to become rancid.

Manufacturers can protect fat-containing products against rancidity in three ways—none of them perfect. First, products may be sealed in air-tight, nonmetallic containers, protected from light, and refrigerated—an expensive and inconvenient

oxidation (OKS-ee-day-shun): the process of a substance combining with oxygen; oxidation reactions involve the loss of electrons.

© Polara Studios Inc.

At room temperature, saturated fats (such as those commonly found in butter and other animal fats) are solid, whereas unsaturated fats (such as those found in vegetable oils) are usually liquid.

FIGURE 5-6 Comparison of Dietary Fats

Most fats are a mixture of saturated, monounsaturated, and polyunsaturated fatty acids.

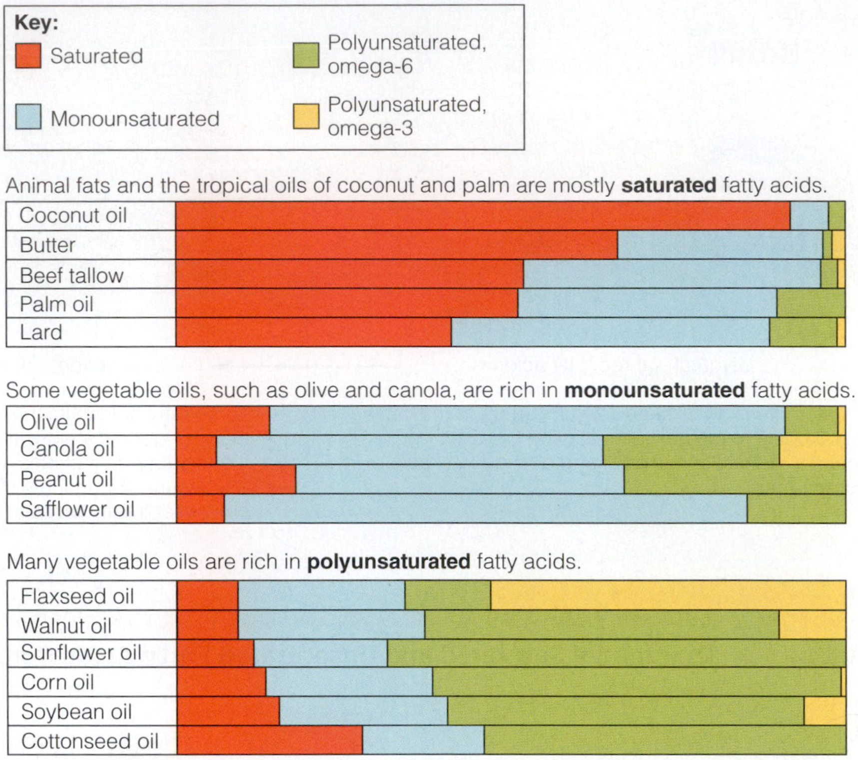

storage system. Second, manufacturers may add **antioxidants** to compete for the oxygen and thus protect the oil (examples are the additives BHA and BHT and vitamin E).* Third, products may undergo a process known as hydrogenation.

Hydrogenation During **hydrogenation**, some or all of the points of unsaturation are saturated by adding hydrogen molecules. Hydrogenation offers two advantages. First, it protects against oxidation (thereby prolonging shelf life) by making polyunsaturated fats more saturated. Second, it alters the texture of foods by making liquid vegetable oils more solid (as in margarine and shortening). Hydrogenated fats improve the texture of foods, making margarines spreadable, pie crusts flaky, and puddings creamy.

Figure 5-7 illustrates the *total* hydrogenation of a polyunsaturated fatty acid to a saturated fatty acid. Total hydrogenation rarely occurs during food processing.

*BHA is butylated hydroxyanisole; BHT is butylated hydroxytoluene.

FIGURE 5-7 Hydrogenation

Double bonds carry a slightly negative charge and readily accept positively charged hydrogen atoms, creating a saturated fatty acid. Most often, fat is *partially* hydrogenated, creating a *trans*-fatty acid (shown in Figure 5-8).

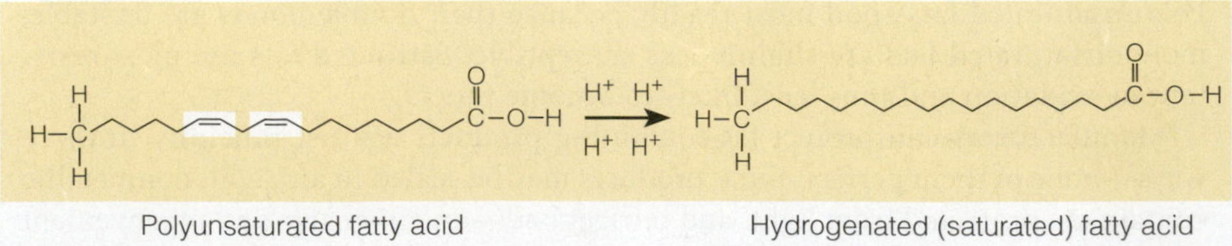

antioxidants: as a food additive, preservatives that delay or prevent rancidity of fats in foods and other damage to food caused by oxygen.

hydrogenation (HIGH-dro-jen-AY-shun or high-DROJ-eh-NAY-shun): a chemical process by which hydrogens are added to monounsaturated or polyunsaturated fatty acids to reduce the number of double bonds, making the fats more saturated (solid) and more resistant to oxidation (protecting against rancidity). Hydrogenation produces *trans*-fatty acids.

FIGURE 5-8 ***Cis*- and *Trans*-Fatty Acids Compared**

This example shows the *cis* configuration for an 18-carbon monounsaturated fatty acid (oleic acid) and its corresponding *trans* configuration (elaidic acid).

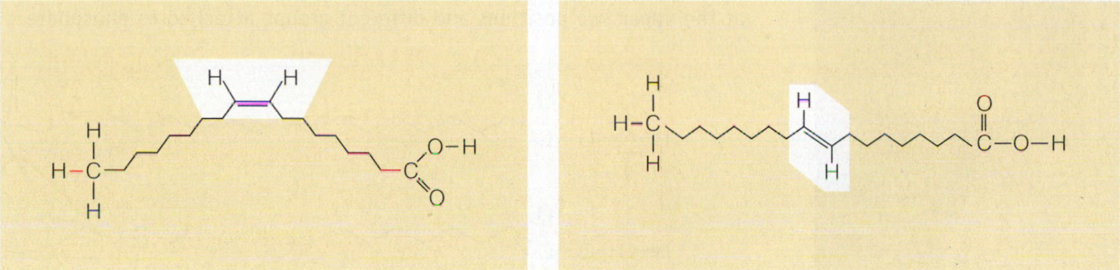

cis-fatty acid

A *cis*-fatty acid has its hydrogens on the same side of the double bond; *cis* molecules bend into a U-like formation. Most naturally occuring unsaturated fatty acids in foods are *cis*.

trans-fatty acid

A *trans*-fatty acid has its hydrogens on the opposite sides of the double bond; *trans* molecules are more linear. The *trans* form typically occurs in partially hydrogenated foods when hydrogen atoms shift around some double bonds and change the configuration from *cis* to *trans*.

Most often, a fat is *partially* hydrogenated, and some of the double bonds that remain after processing change their configuration from ***cis*** to ***trans***.

***Trans*-Fatty Acids** In nature, most double bonds are *cis*—meaning that the hydrogens next to the double bonds are on the same side of the carbon chain. Only a few fatty acids (notably a small percentage of those found in milk and meat products) naturally occur as ***trans*-fatty acids**—meaning that the hydrogens next to the double bonds are on opposite sides of the carbon chain (see Figure 5-8).* In the body, *trans*-fatty acids that derive from hydrogenation behave more like saturated fats than like unsaturated fats. The relationship between *trans*-fatty acids and heart disease has been the subject of much recent research, as a later section describes. In contrast, naturally occurring fatty acids that have a *trans* configuration, such as **conjugated linoleic acids**, may have health benefits.[1] Conjugated linoleic acids are not counted as *trans* fats on food labels.

IN SUMMARY **The predominant lipids both in foods and in the body are triglycerides: glycerol backbones with three fatty acids attached. Fatty acids vary in the length of their carbon chains, their degrees of unsaturation, and the location of their double bond(s). Those that are fully loaded with hydrogens are saturated; those that are missing hydrogens and therefore have double bonds are unsaturated (monounsaturated or polyunsaturated). The vast majority of triglycerides contain more than one type of fatty acid. Fatty acid saturation affects fats' physical characteristics and storage properties. Hydrogenation, which makes polyunsaturated fats more saturated, creates *trans*-fatty acids, altered fatty acids that may damage health in ways similar to those of saturated fatty acids.**

The Chemist's View of Phospholipids and Sterols

The preceding pages have been devoted to one of the classes of lipids, the triglycerides, and their component parts, glycerol and the fatty acids. The other lipids, the phospholipids and sterols, make up only 5 percent of the lipids in the diet.

*For example, most dairy products contain less than 0.5 grams *trans* fat per serving.

cis: on the near side of; refers to a chemical configuration in which the hydrogen atoms are located on the same side of a double bond.

trans: on the other side of; refers to a chemical configuration in which the hydrogen atoms are located on opposite sides of a double bond.

***trans*-fatty acids:** fatty acids with hydrogens on opposite sides of the double bond.

conjugated linoleic acids: several fatty acids that have the same chemical formula as linoleic acid (18 carbons, two double bonds) but with different configurations (the double bonds occur on adjacent carbons).

Without help from emulsifiers, fats and water don't mix.

FIGURE 5-9 Lecithin

Lecithin is one of the phospholipids. Notice that a molecule of lecithin is similar to a triglyceride but contains only two fatty acids. The third position is occupied by a phosphate group and a molecule of choline. Other phospholipids have different fatty acids at the upper two positions and different groups attached to phosphate.

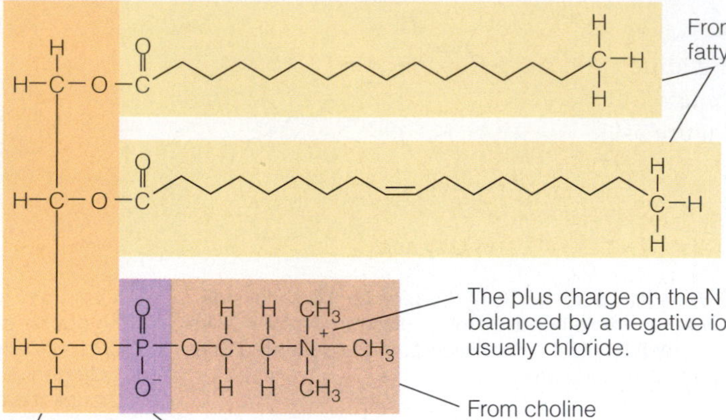

Phospholipids

The best-known **phospholipid** is **lecithin** (see Figure 5-9). Notice that lecithin has one glycerol with two of its three attachment sites occupied by fatty acids like those in triglycerides. The third site is occupied by a phosphate group and a molecule of **choline**. The fatty acids make phospholipids soluble in fat; the phosphate group allows them to dissolve in water. Such versatility enables the food industry to use phospholipids as emulsifiers ♦ to mix fats with water in such products as mayonnaise and candy bars.

♦ **Emulsifiers** are substances with both water-soluble and fat-soluble portions that promote the mixing of oils and fats in watery solutions.

Phospholipids in Foods

In addition to the phospholipids used by the food industry as emulsifiers, phospholipids are also found naturally in foods. The richest food sources of lecithin are eggs, liver, soybeans, wheat germ, and peanuts.

Roles of Phospholipids

Lecithin and other phospholipids are constituents of cell membranes (see Figure 5-10). Because phospholipids are soluble in both water and fat, they can help fat-soluble substances, including vitamins and hormones, to pass easily in and out of cells. The phospholipids also act as emulsifiers in the body, helping to keep fats suspended in the blood and body fluids.

Lecithin periodically receives attention in the popular press. Its advocates claim that it is a major constituent of cell membranes (true), that cell membranes are essential to the integrity of cells (true), and that consumers must therefore take lecithin supplements (false). The liver makes from scratch all the lecithin a person needs. As for lecithin taken as a supplement, the digestive enzyme lecithinase ♦ in the intestine hydrolyzes most of it before it passes into the body, so little lecithin reaches the tissues intact. In other words, lecithin is *not an essential nutrient;* it is just another lipid. Like other lipids, lecithin contributes 9 kcalories per gram—an unexpected "bonus" many people taking lecithin supplements fail to realize. Furthermore, large doses of lecithin may cause GI distress, sweating, and loss of appetite. Perhaps

FIGURE 5-10 Phospholipids of a Cell Membrane

A cell membrane is made of phospholipids assembled into an orderly formation called a bilayer. The fatty acid "tails" orient themselves away from the watery fluid inside and outside of the cell. The glycerol and phosphate "heads" are attracted to the watery fluid.

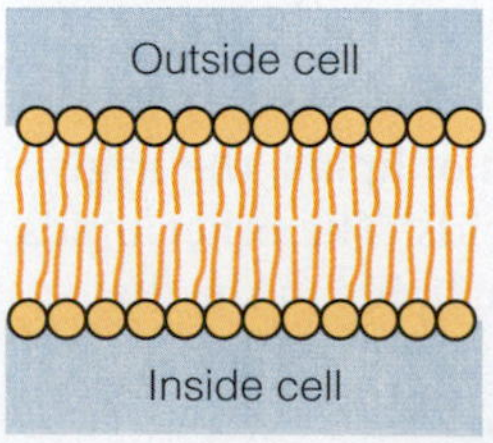

♦ The word ending **-ase** denotes an enzyme. Hence, lecithinase is an enzyme that works on lecithin.

phospholipid (FOS-foe-LIP-id): a compound similar to a triglyceride but having a phosphate group (a phosphorus-containing salt) and choline (or another nitrogen-containing compound) in place of one of the fatty acids.

lecithin (LESS-uh-thin): one of the phospholipids. Both nature and the food industry use lecithin as an emulsifier to combine water-soluble and fat-soluble ingredients that do not ordinarily mix, such as water and oil.

choline (KOH-leen): a nitrogen-containing compound found in foods and made in the body from the amino acid methionine. Choline is part of the phospholipid lecithin and the neurotransmitter acetylcholine.

these symptoms can be considered beneficial—if they serve to warn people to stop taking lecithin supplements.

IN SUMMARY **Phospholipids, including lecithin, have a unique chemical structure that allows them to be soluble in both water and fat. In the body, phospholipids are part of cell membranes; the food industry uses phospholipids as emulsifiers to mix fats with water.**

Sterols

In addition to triglycerides and phospholipids, the lipids include the **sterols**, compounds with a multiple-ring structure.* The most famous sterol is **cholesterol**; Figure 5-11 shows its chemical structure.

Sterols in Foods Foods derived from both plants and animals contain sterols, but only those from animals contain significant amounts of cholesterol—meats, eggs, fish, poultry, and dairy products. Some people, confused about the distinction between dietary cholesterol and blood cholesterol, have asked which foods contain the "good" cholesterol. "Good" cholesterol is not a type of cholesterol found in foods, but it refers to the way the body transports cholesterol in the blood, as explained in a later section of this chapter.

Sterols other than cholesterol are naturally found in plants. Being structurally similar to cholesterol, plant sterols interfere with cholesterol absorption. By inhibiting cholesterol absorption, a diet rich in plant sterols lowers blood cholesterol levels.[2] Food manufacturers have fortified foods such as margarine with plant sterols, creating a functional food that helps to reduce blood cholesterol.[3]

Roles of Sterols Many vitally important body compounds are sterols. Among them are bile acids, the sex hormones (such as testosterone), the adrenal hormones (such as cortisol), and vitamin D, as well as cholesterol itself. Cholesterol in the body can serve as the starting material for the synthesis of these compounds ♦ or as a structural component of cell membranes; more than 90 percent of all the body's cholesterol resides in the cells. Despite popular impressions to the contrary, cholesterol is not a villain lurking in some evil foods—it is a compound the body makes and uses. ♦ Right now, as you read, your liver is manufacturing cholesterol from fragments of carbohydrate, protein, and fat. In fact, the liver makes about 800 to 1500 milligrams of cholesterol per day, ♦ thus contributing much more to the body's total than does the diet.

Cholesterol's harmful effects in the body occur when it accumulates in the artery walls and contributes to the formation of **plaque**. These plaque deposits lead to **atherosclerosis**, a disease that causes heart attacks and strokes. (Chapter 27 provides many more details.)

IN SUMMARY **Sterols have a multiple-ring structure that differs from the structure of other lipids. In the body, sterols include cholesterol, bile, vitamin D, and some hormones. Animal-derived foods are rich sources of cholesterol.**

To summarize, the members of the lipid family include:

- ***Triglycerides*** **(fats and oils), which are made of:**
  - ***Glycerol*** **(1 per triglyceride) and**
  - ***Fatty acids*** **(3 per triglyceride); depending on the number of double bonds, fatty acids may be:**
    - ***Saturated*** **(no double bonds)**
    - ***Monounsaturated*** **(one double bond)**
    - ***Polyunsaturated*** **(more than one double bond); depending on the location of the double bonds, polyunsaturated fatty acids may be:**
      - ***Omega-3*** **(first double bond 3 carbons away from methyl end)**
      - ***Omega-6*** **(first double bond 6 carbons away from methyl end)**
- ***Phospholipids*** **(such as lecithin)**
- ***Sterols*** **(such as cholesterol)**

*The four-ring core structure identifies a steroid; sterols are alcohol derivatives with a steroid ring structure.

FIGURE 5-11 Cholesterol

The fat-soluble vitamin D is synthesized from cholesterol; notice the many structural similarities. The only difference is that cholesterol has a closed ring (highlighted in red), whereas vitamin D's is open, accounting for its vitamin activity. Notice, too, how different cholesterol is from the triglycerides and phospholipids.

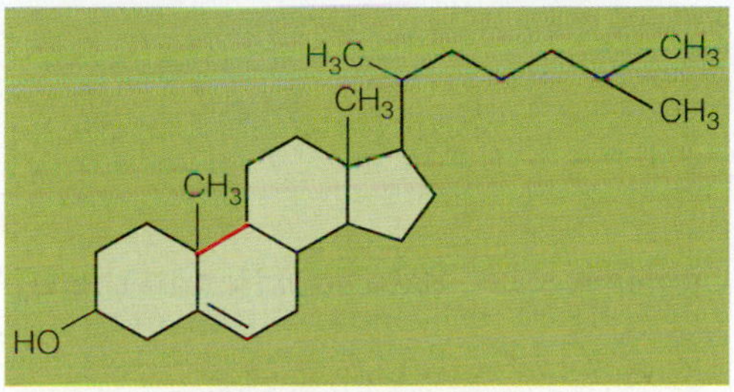

Cholesterol

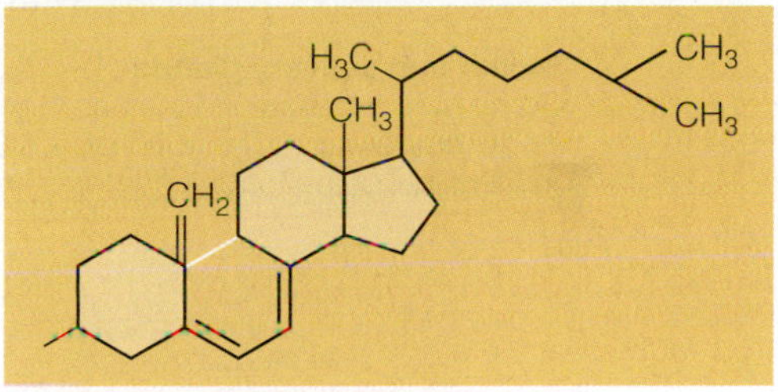

Vitamin D_3

♦ Compounds made from cholesterol:
- Bile acids
- Steroid hormones (testosterone, androgens, estrogens, progesterones, cortisol, cortisone, and aldosterone)
- Vitamin D

♦ The chemical structure is the same, but cholesterol that is made in the body is called **endogenous** (en-DOGDE-eh-nus), whereas cholesterol from outside the body (from foods) is called **exogenous** (eks-ODGE-eh-nus).
- **endo** = within
- **gen** = arising
- **exo** = outside (the body)

♦ For perspective, the Daily Value for cholesterol is 300 mg/day.

sterols (STARE-ols or STEER-ols): compounds containing a four-ring carbon structure with any of a variety of side chains attached.

cholesterol (koh-LESS-ter-ol): one of the sterols containing a four-ring carbon structure with a carbon side chain.

plaque (PLACK): an accumulation of fatty deposits, smooth muscle cells, and fibrous connective tissue that develops in the artery walls in atherosclerosis. Plaque associated with atherosclerosis is known as *atheromatous* (ATH-er-OH-ma-tus) *plaque*.

atherosclerosis (ATH-er-oh-scler-OH-sis): a type of artery disease characterized by plaques (accumulations of lipid-containing material) on the inner walls of the arteries (see Chapter 27).

hydrophobic (high-dro-FOE-bick): a term referring to water-fearing, or non-water-soluble, substances; also known as *lipophilic* (fat loving).

- **hydro** = water
- **phobia** = fear
- **lipo** = lipid
- **phile** = love

hydrophilic (high-dro-FIL-ick): a term referring to water-loving, or water-soluble, substances.

Digestion, Absorption, and Transport of Lipids

Each day, the GI tract receives, on average from the food we eat, 50 to 100 grams of triglycerides, 4 to 8 grams of phospholipids, and 200 to 350 milligrams of cholesterol. The body faces a challenge in digesting and absorbing these lipids. Fats are **hydrophobic**—that is, they tend to separate from the watery fluids of the GI tract—whereas the enzymes for digesting fats are **hydrophilic**. The challenge is keeping the fats mixed in the watery fluids of the GI tract.

Lipid Digestion

Figure 5-12 traces the digestion of fat through the GI tract. The goal of fat digestion is to dismantle triglycerides into small molecules that the

FIGURE 5-12 Fat Digestion in the GI Tract

FAT

Mouth and salivary glands
Some hard fats begin to melt as they reach body temperature. The sublingual salivary gland in the base of the tongue secretes lingual lipase.

Stomach
The acid-stable lingual lipase initiates lipid digestion by hydrolyzing one bond of triglycerides to produce diglycerides and fatty acids. The degree of hydrolysis by lingual lipase is slight for most fats but may be appreciable for milk fats. The stomach's churning action mixes fat with water and acid. A gastric lipase accesses and hydrolyzes (only a very small amount of) fat.

Small intestine
Bile flows in from the gallbladder (via the common bile duct):

Fat —(Bile)→ Emulsified fat

Pancreatic lipase flows in from the pancreas (via the pancreatic duct):

Emulsified fat (triglycerides) —(Pancreatic (and intestinal) lipase)→ Monoglycerides, glycerol, fatty acids (absorbed)

Large intestine
Some fat and cholesterol, trapped in fiber, exit in feces.

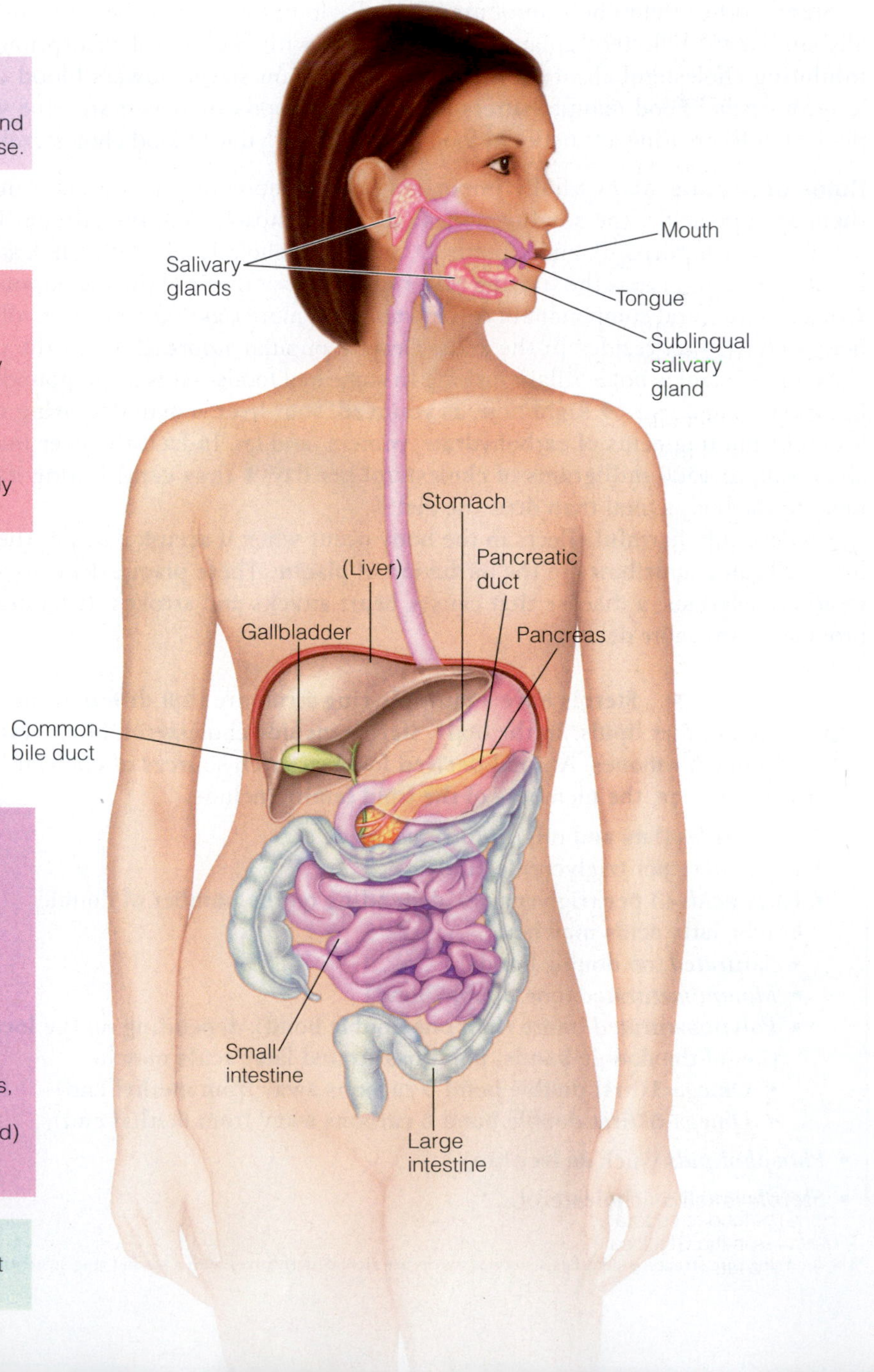

body can absorb and use—namely, **monoglycerides**, fatty acids, and glycerol. The following paragraphs provide the details.

In the Mouth Fat digestion starts off slowly in the mouth. Some hard fats begin to melt as they reach body temperature. A salivary gland at the base of the tongue releases an enzyme (lingual lipase) ♦ that plays a minor role in fat digestion in adults and an active role in infants. In infants, this enzyme efficiently digests the short- and medium-chain fatty acids found in milk.

In the Stomach In a quiet stomach, fat would float as a layer above the watery components of swallowed food. But the strong muscle contractions of the stomach propel the stomach contents toward the pyloric sphincter. Some chyme passes through the pyloric sphincter periodically, but the remaining partially digested food is propelled back into the body of the stomach. This churning grinds the solid pieces to finer particles, mixes the chyme, and disperses the fat into small droplets. These actions help to expose the fat for attack by the gastric lipase enzyme—an enzyme that performs best in the acidic environment ♦ of the stomach. Still, little fat digestion takes place in the stomach; most of the action occurs in the small intestine.

♦ An enzyme that hydrolyzes lipids is called a **lipase; lingual** refers to the tongue.

♦ The pH of the stomach is just below 2.

♦ In addition to bile acids and bile salts, bile contains cholesterol, phospholipids (especially lecithin), antibodies, water, electrolytes, and bilirubin and biliverdin (pigments resulting from the breakdown of heme).

monoglycerides: molecules of glycerol with one fatty acid attached. A molecule of glycerol with two fatty acids attached is a **diglyceride.**

- **mono** = one
- **di** = two

In the Small Intestine When fat enters the small intestine, it triggers the release of the hormone cholecystokinin (CCK), which signals the gallbladder to release its stores of bile. (Remember that the liver makes bile, and the gallbladder stores it until it is needed.) Among bile's many ingredients ♦ are bile acids, which are made in the liver from cholesterol and have a similar structure. In addition, bile acids often pair up with an amino acid (a building block of protein). The amino acid end is attracted to water, and the sterol end is attracted to fat (see Figure 5-13). This structure improves bile's ability to act as an emulsifier, drawing fat molecules into the surrounding watery fluids. There, the fats are fully digested as they encounter lipase enzymes from the pancreas and small intestine. The process of emulsification is diagrammed in Figure 5-14.

Most of the hydrolysis of triglycerides occurs in the small intestine. The major fat-digesting enzymes are pancreatic lipases; some intestinal lipases are also active. These enzymes remove one, then the other, of each triglyceride's outer fatty

FIGURE 5-13 A Bile Acid

This is one of several bile acids the liver makes from cholesterol. It is then bound to an amino acid to improve its ability to form spherical complexes of emulsified fat (micelles). Most bile acids occur as bile salts, usually in association with sodium, but sometimes with potassium or calcium.

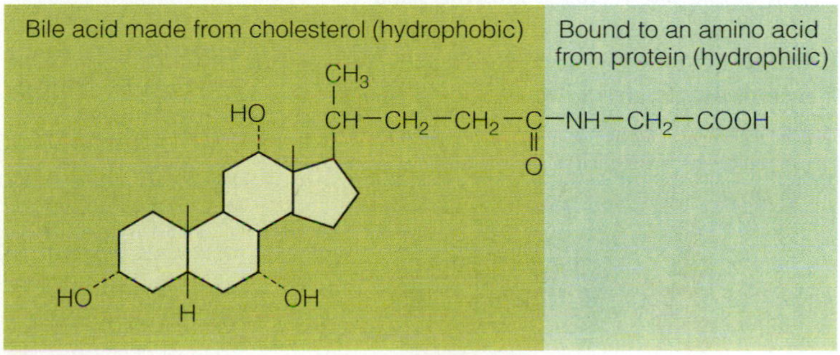

FIGURE 5-14 Emulsification of Fat by Bile

Like bile, detergents are emulsifiers and work the same way, which is why they are effective in removing grease spots from clothes. Molecule by molecule, the grease is dissolved out of the spot and suspended in the water, where it can be rinsed away.

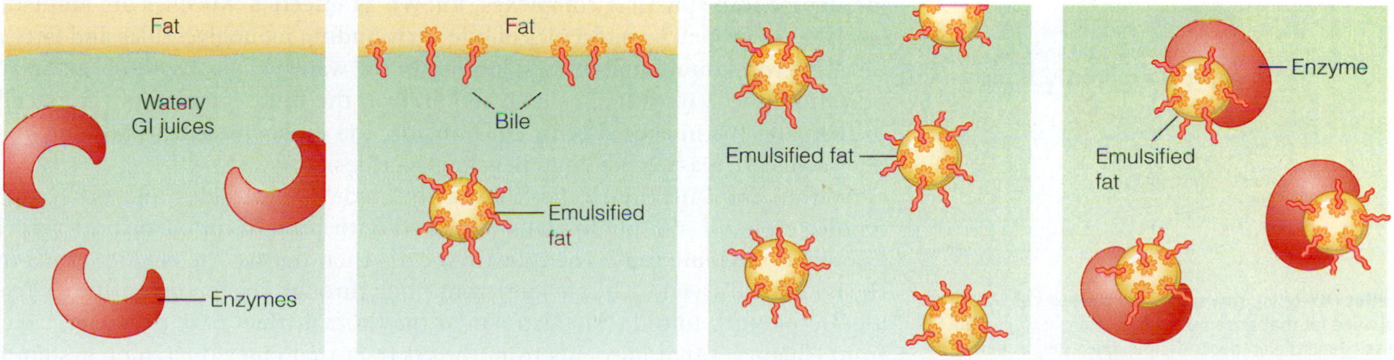

In the stomach, the fat and watery GI juices tend to separate. The enzymes in the GI juices can't get at the fat.

When fat enters the small intestine, the gallbladder secretes bile. Bile has an affinity for both fat and water, so it can bring the fat into the water.

Bile's emulsifying action converts large fat globules into small droplets that repel one another.

After emulsification, more fat is exposed to the enzymes, making fat digestion more efficient.

FIGURE 5-15 Digestion (Hydrolysis) of a Triglyceride

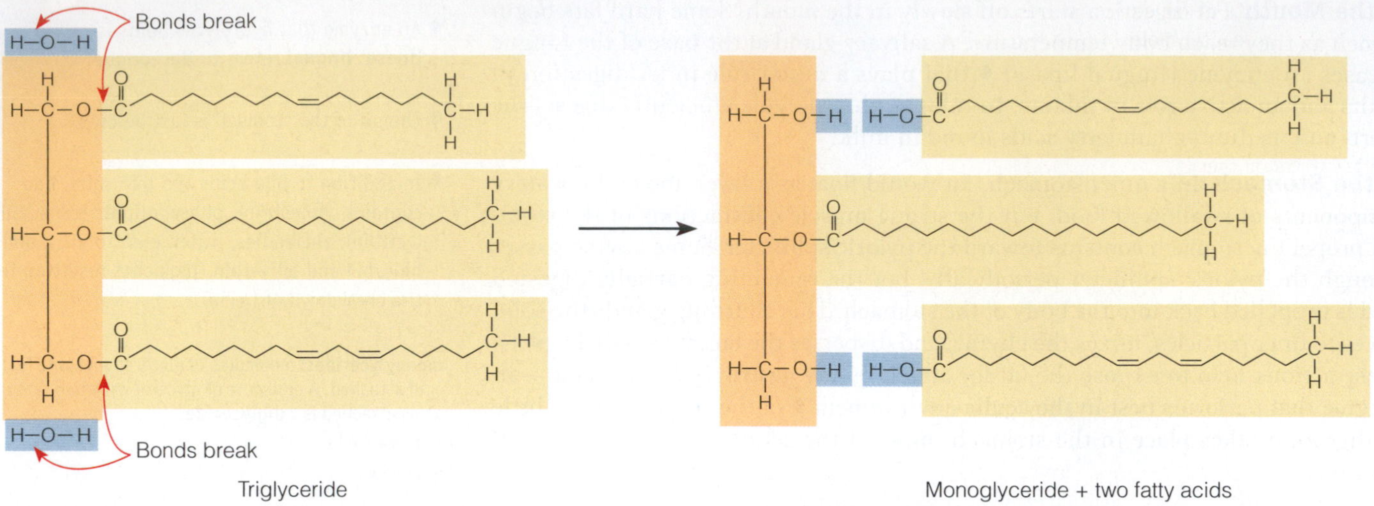

The triglyceride and two molecules of water are split. The H and OH from water complete the structures of two fatty acids and leave a monoglyceride.

These products may pass into the intestinal cells, but sometimes the monoglyceride is split with another molecule of water to give a third fatty acid and glycerol. Fatty acids, monoglycerides, and glycerol are absorbed into intestinal cells.

acids, leaving a monoglyceride. Occasionally, enzymes remove all three fatty acids, leaving a free molecule of glycerol. Hydrolysis of a triglyceride is shown in Figure 5-15.

Phospholipids are digested similarly—that is, their fatty acids are removed by hydrolysis. The two fatty acids and the remaining phospholipid fragment are then absorbed. Most sterols can be absorbed as is; if any fatty acids are attached, they are first hydrolyzed off.

Bile's Routes After bile enters the small intestine and emulsifies fat, it has two possible destinations, illustrated in Figure 5-16. Most of the bile is reabsorbed from the small intestine and recycled. The other possibility is that some of the bile can be trapped by dietary fibers in the large intestine and excreted. Because cholesterol is needed to make bile, the excretion of bile effectively reduces blood cholesterol. As Chapter 4 explains, the dietary fibers most effective at lowering blood cholesterol this way are the soluble fibers commonly found in fruits, whole grains, and legumes.

FIGURE 5-16 Enterohepatic Circulation

Most of the bile released into the small intestine is reabsorbed and sent back to the liver to be reused. This cycle is called the **enterohepatic circulation** of bile. Some bile is excreted.

- **enteron** = intestine
- **hepat** = liver

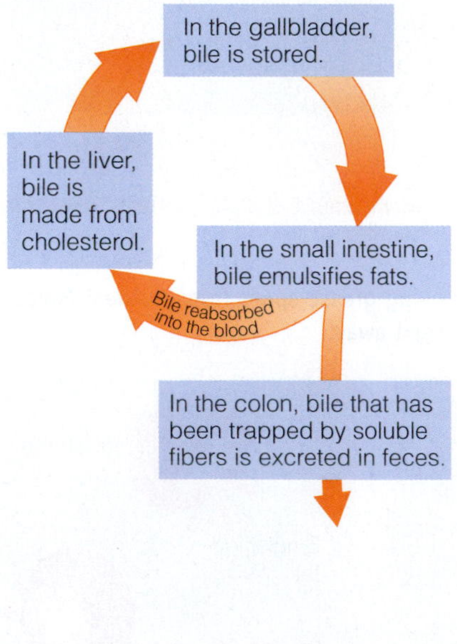

In the gallbladder, bile is stored.
In the liver, bile is made from cholesterol.
In the small intestine, bile emulsifies fats.
Bile reabsorbed into the blood
In the colon, bile that has been trapped by soluble fibers is excreted in feces.

Lipid Absorption

Figure 5-17 illustrates the absorption of lipids. Small molecules of digested triglycerides (glycerol and short- and medium-chain fatty acids) can diffuse easily into the intestinal cells; they are absorbed directly into the bloodstream. Larger molecules (the monoglycerides and long-chain fatty acids) merge into spherical complexes, known as **micelles.** Micelles are emulsified fat droplets formed by molecules of bile surrounding monoglycerides and fatty acids. This configuration permits solubility in the watery digestive fluids and transportation to the intestinal cells. Upon arrival, the lipid contents of the micelles diffuse into the intestinal cells. Once inside, the monoglycerides and long-chain fatty acids are reassembled into new triglycerides.

Within the intestinal cells, the newly made triglycerides and other lipids (cholesterol and phospholipids) are packed with protein into transport vehicles known as **chylomicrons.** The intestinal cells then release the chylomicrons into the lymphatic system. The chylomicrons glide through the lymph until they reach a point of entry into the bloodstream at the thoracic duct near the heart. (Recall from Chapter 3 that nutrients from the GI tract that enter the lymph system bypass the liver at first.) The blood carries these lipids to the rest of the body for immediate use or storage. A look at these lipids in the body reveals the kinds of fat the diet has been delivering.[4] The blood, fat stores, and muscle cells of people who

micelles (MY-cells): tiny spherical complexes of emulsified fat that arise during digestion; most contain bile salts and the products of lipid digestion, including fatty acids, monoglycerides, and cholesterol.

chylomicrons (kye-lo-MY-cronz): the class of lipoproteins that transport lipids from the intestinal cells to the rest of the body.

FIGURE 5-17 Absorption of Fat

The end products of fat digestion are mostly monoglycerides, some fatty acids, and very little glycerol. Their absorption differs depending on their size. (In reality, molecules of fatty acid are too small to see without a powerful microscope, whereas villi are visible to the naked eye.)

Animated! figure
www.cengagebrain.com
(search for ISBN 084006845X)

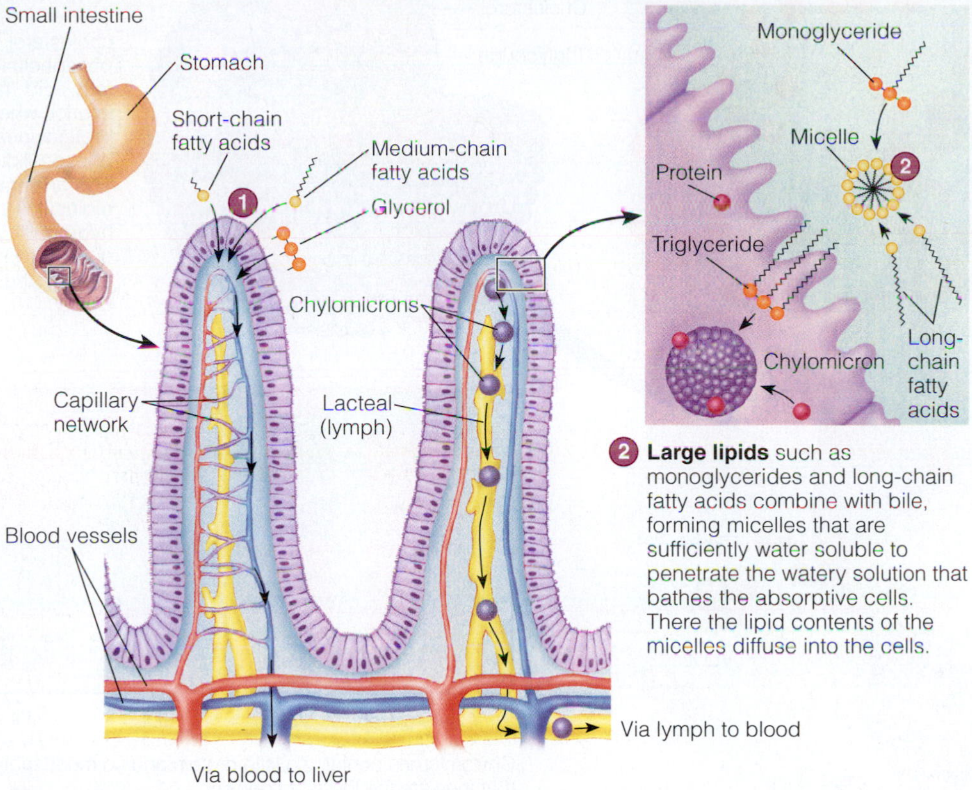

② **Large lipids** such as monoglycerides and long-chain fatty acids combine with bile, forming micelles that are sufficiently water soluble to penetrate the watery solution that bathes the absorptive cells. There the lipid contents of the micelles diffuse into the cells.

① **Glycerol and small lipids** such as short- and medium-chain fatty acids can move directly into the bloodstream.

eat a diet rich in unsaturated fats, for example, contain more unsaturated fats than those of people who select a diet high in saturated fats.

IN SUMMARY **The body makes special arrangements to digest and absorb lipids. It provides the emulsifier bile to make them accessible to the fat-digesting lipases that dismantle triglycerides, mostly to monoglycerides and fatty acids, for absorption by the intestinal cells. The intestinal cells assemble freshly absorbed lipids into chylomicrons, lipid packages with protein escorts, for transport so that cells all over the body may select needed lipids from them.**

Lipid Transport

The chylomicrons are only one of several clusters of lipids and proteins that are used as transport vehicles for fats. As a group, these vehicles are known as **lipoproteins**, and they solve the body's challenge of transporting fat through the watery bloodstream. The body makes four main types of lipoproteins, distinguished by their size and density.* Each type contains different kinds and amounts of lipids and proteins. ♦ Figure 5-18 (p. 146) shows the relative compositions and sizes of the lipoproteins.

♦ The more lipids, the lower the density; the more proteins, the higher the density.

Chylomicrons The chylomicrons are the largest and least dense of the lipoproteins. They transport *diet*-derived lipids (mostly triglycerides) from the small intestine (via the lymph system) to the rest of the body. Cells all over the body

*Chemists can identify the various lipoproteins by their density. They place a blood sample below a thick fluid in a test tube and spin the tube in a centrifuge. The most buoyant particles (highest in lipids) rise to the top and have the lowest density; the densest particles (highest in proteins) remain at the bottom and have the highest density. Others distribute themselves in between.

lipoproteins (LIP-oh-PRO-teenz): clusters of lipids associated with proteins that serve as transport vehicles for lipids in the lymph and blood.

FIGURE 5-18 Sizes and Compositions of the Lipoproteins

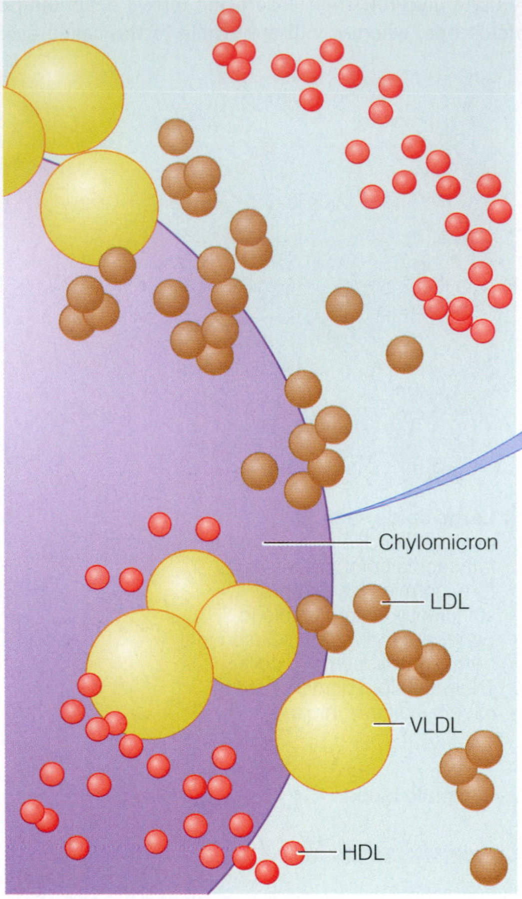

This solar system of lipoproteins shows their relative sizes. Notice how large the fat-filled chylomicron is compared with the others and how the others get progressively smaller as their proportion of fat declines and protein increases.

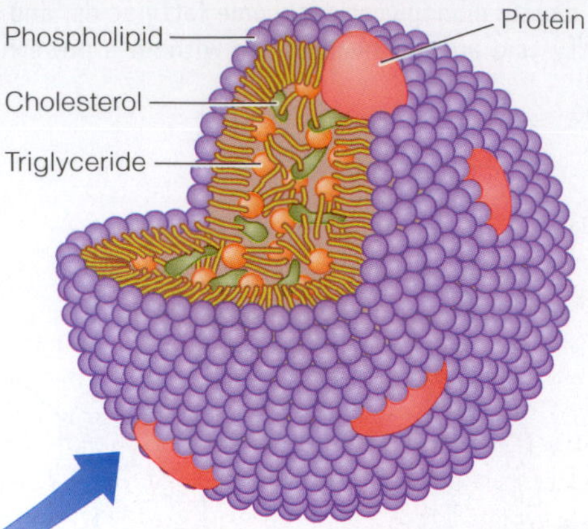

A typical lipoprotein contains an interior of triglycerides and cholesterol surrounded by phospholipids. The phospholipids' fatty acid "tails" point toward the interior, where the lipids are. Proteins near the outer ends of the phospholipids cover the structure. This arrangement of hydrophobic molecules on the inside and hydrophilic molecules on the outside allows lipids to travel through the watery fluids of the blood.

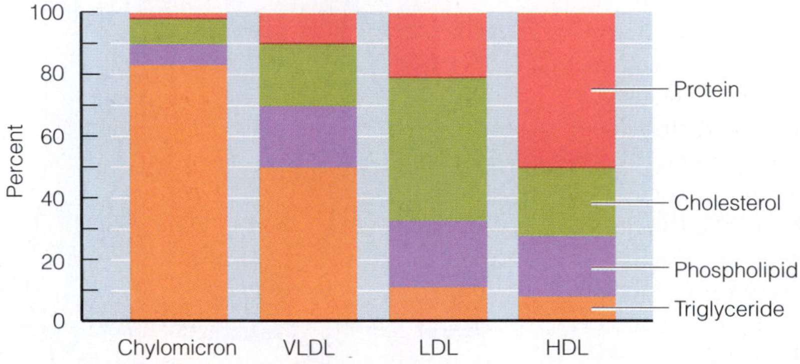

Chylomicrons contain so little protein and so much triglyceride that they are the lowest in density.

Very-low-density lipoproteins (VLDL) are half triglycerides, accounting for their very low density.

Low-density lipoproteins (LDL) are half cholesterol, accounting for their implication in heart disease.

High-density lipoproteins (HDL) are half protein, accounting for their high density.

remove triglycerides from the chylomicrons as they pass by, so the chylomicrons get smaller and smaller. Within 14 hours after absorption, most of the triglycerides have been depleted, and only a few remnants of protein, cholesterol, and phospholipid remain. Special protein receptors on the membranes of the liver cells recognize and remove these chylomicron remnants from the blood. After collecting the remnants, the liver cells first dismantle them and then either use or recycle the pieces.

VLDL (Very-Low-Density Lipoproteins) Meanwhile, in the liver—the most active site of lipid synthesis—cells are making cholesterol, fatty acids, and other lipid compounds. Ultimately, the lipids made in the liver and those collected from chylomicron remnants are packaged with proteins as **VLDL (very-low-density lipoproteins)** and shipped to other parts of the body.

As the VLDL travel through the body, cells remove triglycerides, causing the VLDL to shrink. As VLDL lose triglycerides, the proportion of lipids shifts. Cholesterol becomes the predominant lipid, and the lipoprotein density increases. The VLDL becomes an **LDL (low-density lipoprotein)**.* This transformation explains why LDL contain few triglycerides but are loaded with cholesterol.

VLDL (very-low-density lipoprotein): the type of lipoprotein made primarily by liver cells to transport lipids to various tissues in the body; composed primarily of triglycerides.

LDL (low-density lipoprotein): the type of lipoprotein derived from very-low-density lipoproteins (VLDL) as VLDL triglycerides are removed and broken down; composed primarily of cholesterol.

*Before becoming LDL, the VLDL are first transformed into intermediate-density lipoproteins (IDL), sometimes called VLDL remnants. Some IDL may be picked up by the liver and rapidly broken down; those IDL that remain in circulation continue to deliver triglycerides to the cells and eventually become LDL. Researchers debate whether IDL are simply transitional particles or a separate class of lipoproteins; normally, IDL do not accumulate in the blood. Measures of blood lipids include IDL with LDL.

LDL (Low-Density Lipoproteins) The LDL circulate throughout the body, making their contents available to the cells of all tissues—muscles (including the heart muscle), fat stores, the mammary glands, and others. The cells take triglycerides, cholesterol, and phospholipids to build new membranes, make hormones or other compounds, or store for later use. Special LDL receptors on the liver cells play a crucial role in the control of blood cholesterol concentrations by removing LDL from circulation.

HDL (High-Density Lipoproteins) The liver makes **HDL (high-density lipoprotein)** to remove cholesterol from the cells and carry it back to the liver for recycling or disposal. In addition, HDL have anti-inflammatory properties that seem to keep atherosclerotic plaque from breaking apart and causing heart attacks.[5] Figure 5-19 summarizes lipid transport via the lipoproteins.

Health Implications The distinction between LDL and HDL has implications for the health of the heart and blood vessels. The blood cholesterol linked to heart disease is LDL cholesterol. As mentioned, HDL also carry cholesterol, but elevated HDL represent cholesterol returning ♦ from the rest of the body to the liver for breakdown and excretion. High LDL cholesterol is associated with a high risk of heart attack, whereas high HDL cholesterol seems to have a protective effect. This is why some people refer to LDL as "bad," and HDL as "good," cholesterol. ♦ Keep in mind that the cholesterol itself is the same, and that the differences between LDL and HDL reflect the *proportions* and *types* of lipids and proteins within them—not the type of cholesterol. The margin ♦ lists factors that influence LDL and HDL, and Chapter 27 provides many more details.

Not too surprisingly, numerous genes influence how the body handles the synthesis, transport, and degradation of lipids and lipoproteins. Much current research is focused on how nutrient-gene interactions may direct the progression of heart disease.

♦ The transport of cholesterol from the tissues to the liver is sometimes called *reverse cholesterol transport* or the *scavenger pathway*.

♦ Think of **H**DL as **H**ealthy and **L**DL as **L**ess healthy.

♦ Factors that lower LDL and/or raise HDL:
- Weight control
- Monounsaturated or polyunsaturated, instead of saturated, fat in the diet
- Soluble dietary fibers (see Chapter 4)
- Phytochemicals (see Highlight 13)
- *Moderate* alcohol consumption
- Physical activity

HDL (high-density lipoprotein): the type of lipoprotein that transports cholesterol back to the liver from the cells; composed primarily of protein.

FIGURE 5-19 Lipid Transport via Lipoproteins

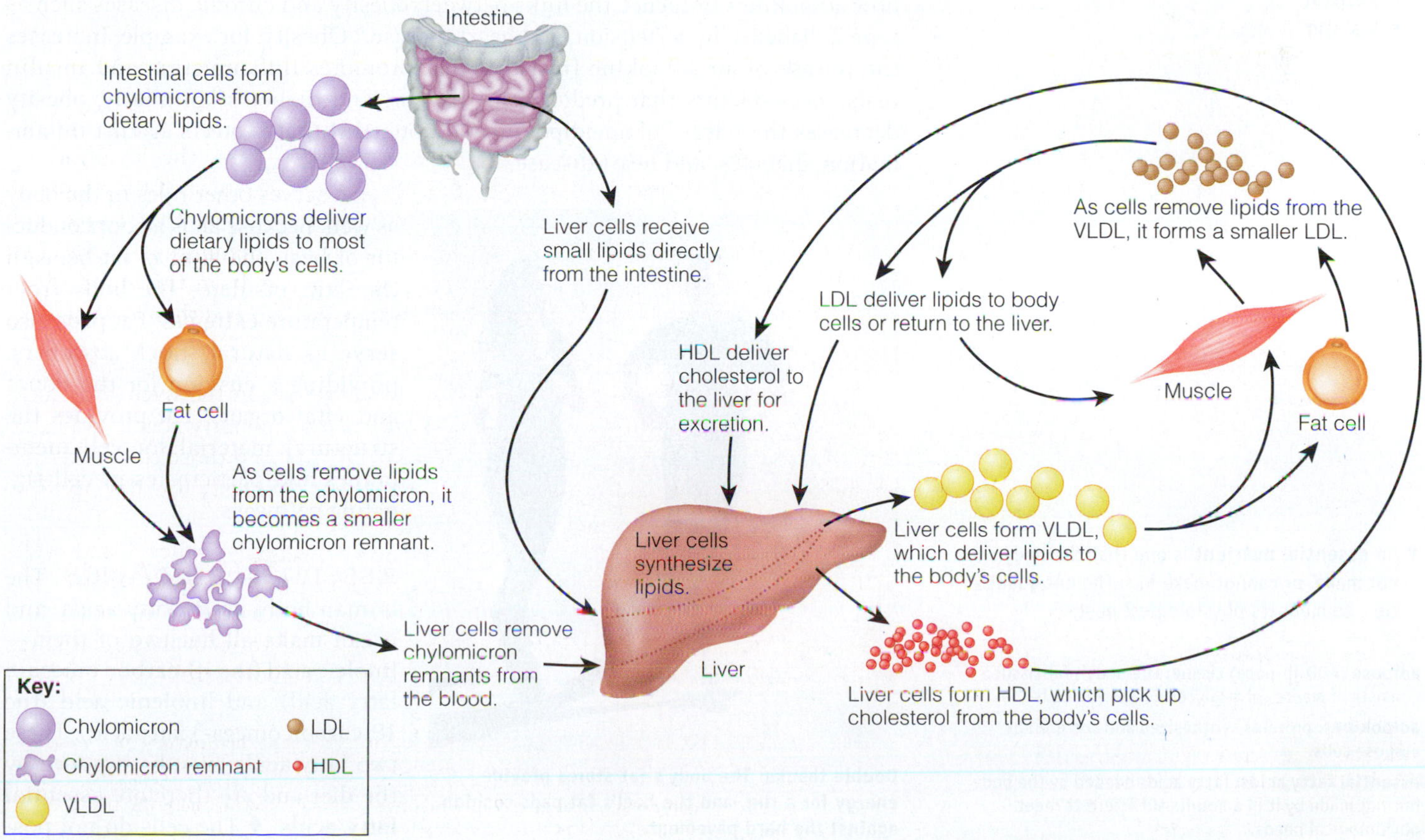
Intestine
Intestinal cells form chylomicrons from dietary lipids.
Chylomicrons deliver dietary lipids to most of the body's cells.
Liver cells receive small lipids directly from the intestine.
Muscle
Fat cell
As cells remove lipids from the chylomicron, it becomes a smaller chylomicron remnant.
Liver cells remove chylomicron remnants from the blood.
Liver cells synthesize lipids.
Liver
HDL deliver cholesterol to the liver for excretion.
LDL deliver lipids to body cells or return to the liver.
As cells remove lipids from the VLDL, it forms a smaller LDL.
Muscle
Fat cell
Liver cells form VLDL, which deliver lipids to the body's cells.
Liver cells form HDL, which pick up cholesterol from the body's cells.
Key:
Chylomicron
Chylomicron remnant
VLDL
LDL
HDL

FIGURE 5-20 An Adipose Cell

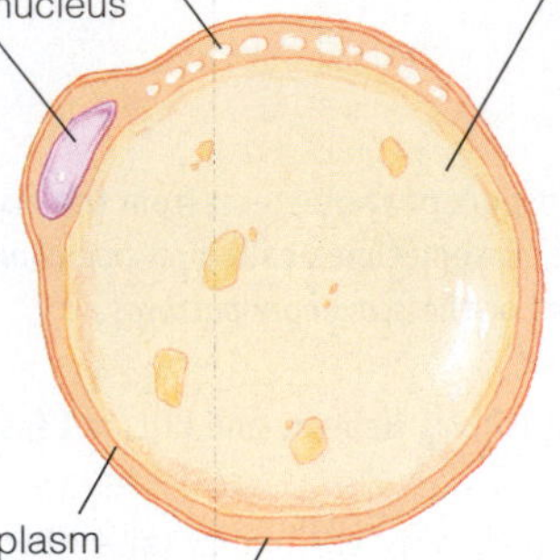

Newly imported triglycerides first form small droplets at the periphery of the cell, then merge with the large, central globule.
Large central globule of (pure) fat
Cell nucleus
Cytoplasm
As the central globule enlarges, the fat cell membrane expands to accommodate its swollen contents.

IN SUMMARY Lipoproteins transport lipids around the body. All four types of lipoproteins carry all classes of lipids (triglycerides, phospholipids, and cholesterol), but the chylomicrons are the largest and contain mostly triglycerides; VLDL are smaller and are about half triglycerides; LDL are smaller still and contain mostly cholesterol; and HDL are the densest and are rich in protein.

Lipids in the Body

In the body, lipids provide energy, insulate against temperature extremes, protect against shock, and maintain cell membranes. This section provides an overview of the roles of triglycerides and fatty acids and then of the metabolic pathways they can follow within the body's cells.

Roles of Triglycerides

First and foremost, the triglycerides—either from food or from the body's fat stores—provide the cells with energy. When a person dances all night, her dinner's triglycerides provide some of the fuel that keeps her moving. When a person loses his appetite, his stored triglycerides fuel much of his body's work until he can eat again.

Fat provides more than twice the energy of carbohydrate and protein, ♦ making it an extremely efficient storage form of energy. Unlike the liver's glycogen stores, the body's fat stores have virtually unlimited capacity, thanks to the special cells of the **adipose tissue**. Unlike most body cells, which can store only limited amounts of fat, the fat cells of the adipose tissue readily take up and store triglycerides. An adipose cell is depicted in Figure 5-20.

♦ Gram for gram, fat provides more than twice as much energy (9 kcal) as carbohydrate or protein (4 kcal).

Adipose tissue is more than just a storage depot for fat. Adipose tissue actively secretes several hormones known as **adipokines**—proteins that help regulate energy balance and influence several body functions.[6] ♦ When body fat is markedly reduced or excessive, the type and quantity of adipokine secretions change, with consequences for the body's health.[7] Researchers are currently exploring how adipokines influence the links between obesity and chronic diseases such as type 2 diabetes, hypertension, and heart disease.[8] Obesity, for example, increases the release of an adipokine (resistin) that promotes inflammation and insulin resistance—factors that predict heart disease and diabetes.[9] Similarly, obesity decreases the release of an adipokine (adiponectin) that protects against inflammation, diabetes, and heart disease.[10]

♦ Examples of adipokines:
- Leptin
- Adiponectin
- Resistin
- Visfatin

Fat serves other roles in the body as well. Because fat is a poor conductor of heat, the layer of fat beneath the skin insulates the body from temperature extremes. Fat pads also serve as natural shock absorbers, providing a cushion for the bones and vital organs. Fat provides the structural material for cell membranes and participates in cell signaling pathways.

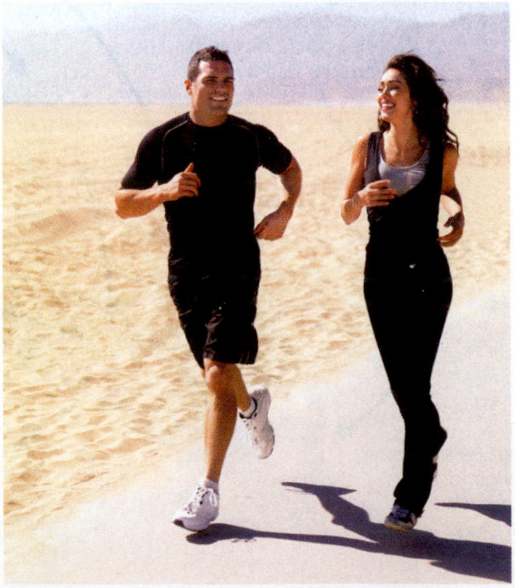
Yuri Arcurs/Shutterstock.com

Double thanks: The body's fat stores provide energy for a run, and the heel's fat pads cushion against the hard pavement.

Essential Fatty Acids

The human body needs fatty acids, and it can make all but two of them—linoleic acid (the 18-carbon omega-6 fatty acid) and linolenic acid (the 18-carbon omega-3 fatty acid). These two fatty acids must be supplied by the diet and are therefore **essential fatty acids**. ♦ The cells do not pos-

♦ An **essential nutrient** is one that the body cannot make, or cannot make in sufficient quantities, to meet its physiological needs.

adipose (ADD-ih-poce) **tissue:** the body's fat tissue; consists of masses of triglyceride-storing cells.

adipokines: proteins synthesized and secreted by adipose cells.

essential fatty acids: fatty acids needed by the body but not made by it in amounts sufficient to meet physiological needs.

sess the enzymes to make any of the omega-6 or omega-3 fatty acids from scratch, nor can they convert an omega-6 fatty acid to an omega-3 fatty acid or vice versa. Cells *can*, however, start with the 18-carbon member of an omega family and make the longer fatty acids of that family by forming double bonds (desaturation) and lengthening the chain two carbons at a time (elongation), as shown in Figure 5-21. This is a slow process because the omega-3 and omega-6 families compete for the same enzymes. Too much of a fatty acid from one family can create a deficiency of the other family's longer fatty acids, which becomes critical only when the diet fails to deliver adequate supplies. Therefore, the most effective way to maintain body supplies of all the omega-6 and omega-3 fatty acids is to obtain them directly from foods—most notably, from vegetable oils, seeds, nuts, fish, and other marine foods.

Linoleic Acid and the Omega-6 Family Linoleic acid is the primary member of the omega-6 fatty acid family. When the body receives linoleic acid from the diet, it can make other members of the omega-6 family—such as the 20-carbon polyunsaturated fatty acid **arachidonic acid.** Should a linoleic acid deficiency develop, arachidonic acid, and all other fatty acids that derive from linoleic acid, would also become essential and have to be obtained from the diet. ♦ Normally, vegetable oils and meats supply enough omega-6 fatty acids to meet the body's needs.

Linolenic Acid and the Omega-3 Family Linolenic acid is the primary member of the omega-3 fatty acid family.* Like linoleic acid, linolenic acid cannot be made in the body and must be supplied by foods. Given the 18-carbon linolenic acid, the body can make small amounts of the 20- and 22-carbon members of the omega-3 series, **eicosapentaenoic acid** (EPA) and **docosahexaenoic acid** (DHA), respectively. These omega-3 fatty acids are found in the eyes and brain and are essential for normal growth and cognitive development.[11] They may also play an important role in the prevention and treatment of heart disease, as later sections explain.[12]

Eicosanoids The body uses arachidonic acid and EPA to make substances known as **eicosanoids.** Eicosanoids are a diverse group of compounds that are sometimes described as "hormonelike," but they differ from hormones in important ways. For one, hormones are secreted in one location and travel to affect cells all over the body, whereas eicosanoids appear to affect only the cells in which they are made or nearby cells in the same localized environment. For another, hormones elicit the same response from all their target cells, whereas eicosanoids often have different effects on different cells.

The actions of various eicosanoids sometimes oppose one another. For example, one causes muscles to relax and blood vessels to dilate, whereas another causes muscles to contract and blood vessels to constrict. Certain eicosanoids participate in the immune response to injury and infection, producing fever, inflammation, and pain. One of the ways aspirin relieves these symptoms is by slowing the synthesis of these eicosanoids.

Eicosanoids that derive from EPA differ from those that derive from arachidonic acid, with those from EPA providing greater health benefits. The EPA eicosanoids help lower blood pressure, prevent blood clot formation, protect against irregular heartbeats, and reduce inflammation.

Fatty Acid Deficiencies Most diets in the United States and Canada meet the minimum essential fatty acid requirement adequately. Historically, deficiencies have developed only in infants and young children who have been fed fat-free milk and low-fat diets or in hospital clients who have been mistakenly fed formulas that provided no polyunsaturated fatty acids for long periods of time. Classic deficiency symptoms include growth retardation, reproductive failure, skin lesions, kidney and liver disorders, and subtle neurological and visual problems.

*This omega-3 linolenic acid is known as alpha-linolenic acid and is the fatty acid referred to in this chapter. Another fatty acid, also with 18 carbons and three double bonds, belongs to the omega-6 family and is known as gamma-linolenic acid.

FIGURE 5-21 The Pathway from One Omega-6 Fatty Acid to Another

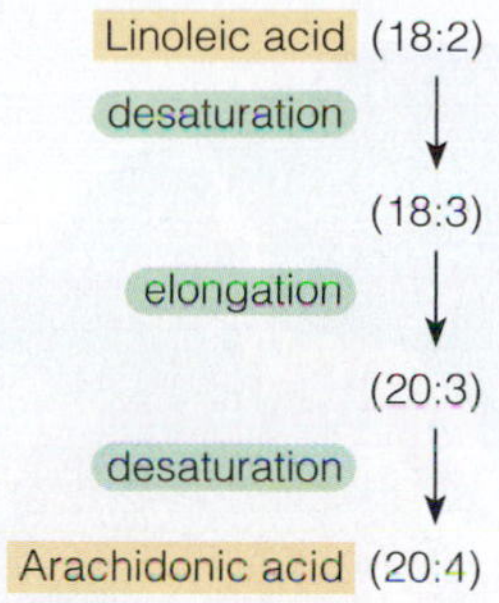

The first number indicates the number of carbons and the second, the number of double bonds. Similar reactions occur when the body makes the omega-3 fatty acids EPA and DHA from linolenic acid.

♦ A nonessential nutrient (such as arachidonic acid) that must be supplied by the diet in special circumstances (as in a linoleic acid deficiency) is considered **conditionally essential.**

arachidonic (a-RACK-ih-DON-ic) **acid:** an omega-6 polyunsaturated fatty acid with 20 carbons and four double bonds; present in small amounts in meat and other animal products and synthesized in the body from linoleic acid.

eicosapentaenoic (EYE-cossa-PENTA-ee-NO-ick) **acid (EPA):** an omega-3 polyunsaturated fatty acid with 20 carbons and five double bonds; present in fatty fish and synthesized in limited amounts in the body from linolenic acid.

docosahexaenoic (DOE-cossa-HEXA-ee-NO-ick) **acid (DHA):** an omega-3 polyunsaturated fatty acid with 22 carbons and six double bonds; present in fatty fish and synthesized in limited amounts in the body from linolenic acid.

eicosanoids (eye-COSS-uh-noyds): derivatives of 20-carbon fatty acids; biologically active compounds that help to regulate blood pressure, blood clotting, and other body functions. They include *prostaglandins* (PROS-tah-GLAND-ins), *thromboxanes* (throm-BOX-ains), and *leukotrienes* (LOO-ko-TRY-eens).

IN SUMMARY In the body, triglycerides:

- Provide energy
- Insulate against temperature extremes
- Protect against shock
- Help the body use carbohydrate and protein efficiently

Linoleic acid (18 carbons, omega-6) and linolenic acid (18 carbons, omega-3) are essential nutrients. They serve as structural parts of cell membranes and as precursors to the longer fatty acids that can make eicosanoids—powerful compounds that participate in blood pressure regulation, blood clot formation, and the immune response to injury and infection, among other functions. Because essential fatty acids are common in the diet and stored in the body, deficiencies are unlikely.

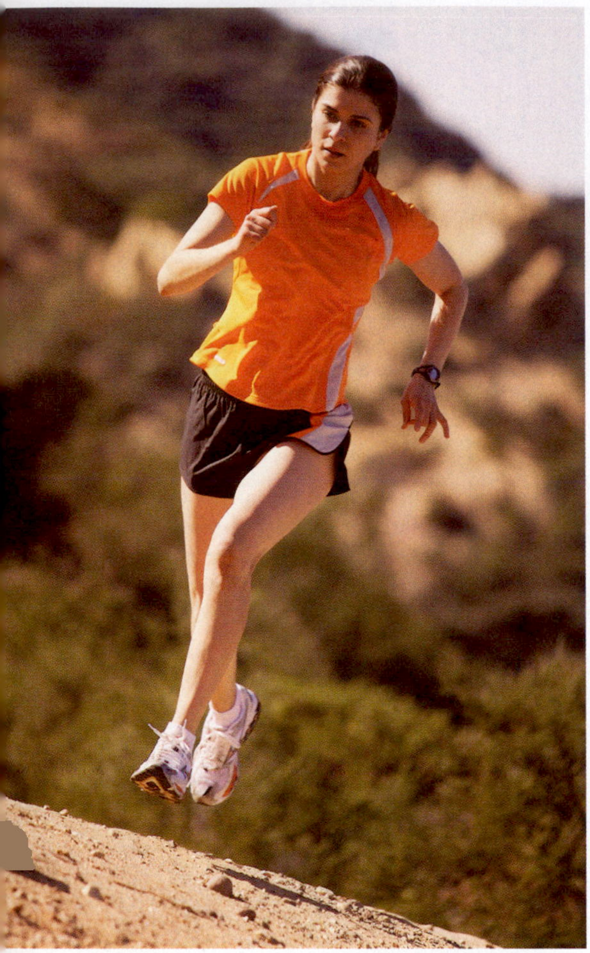
© UpperCut Images/Alamy

Fat supplies most of the energy during a long-distance run.

A Preview of Lipid Metabolism

This preview of fat metabolism describes how the cells store and release energy from fat. Chapter 7 provides details.

Storing Fat as Fat Adipose cells store fat after meals when a heavy traffic of chylomicrons and VLDL loaded with triglycerides passes by. An enzyme—**lipoprotein lipase (LPL)**—hydrolyzes triglycerides from lipoproteins, releasing fatty acids, diglycerides, and monoglycerides that enter the adipose cells. Inside the cells, other enzymes reassemble these lipids into triglycerides again for storage. Earlier, Figure 5-4 showed how the body can make a triglyceride from glycerol and fatty acids. Triglycerides fill the adipose cells, storing a lot of energy in a relatively small space.

Using Fat for Energy Efficient energy metabolism depends on the energy nutrients—carbohydrate, fat, and protein—supporting one another. Glucose fragments combine with fat fragments during energy metabolism, and fat and carbohydrate help spare protein, providing energy so that protein can be used for other important tasks.

Fat supplies 60 percent of the body's ongoing energy needs during rest. During prolonged light to moderately intense exercise or extended periods of food deprivation, fat stores may make a slightly greater contribution to energy needs.

During energy deprivation, several lipase enzymes (most notably **hormone-sensitive lipase**) inside the adipose cells respond by dismantling stored triglycerides and releasing the glycerol and fatty acids directly into the blood.[13] Energy-hungry cells anywhere in the body can then capture these compounds and take them through a series of chemical reactions to yield energy, carbon dioxide, and water.

A person who fasts (drinking only water) will rapidly metabolize body fat. A pound of body fat provides 3500 kcalories, ♦ so you might think a fasting person who expends 2000 kcalories a day could lose more than half a pound of body fat each day.* Actually, the person has to obtain some energy from lean tissue because the brain, nerves, and red blood cells need glucose. Also, the complete breakdown of fat requires carbohydrate or protein. Even on a total fast, a person cannot lose more than half a pound of pure fat per day. Still, in conditions of forced starvation—say, during a siege or a famine—a fatter person can survive longer than a thinner person thanks to this energy reserve.

♦ 1 lb body fat = 3500 kcal

Although fat provides energy during a fast, it can provide very little glucose to give energy to the brain and nerves. Only the small glycerol molecule can be converted to glucose; fatty acids cannot be. (Figure 7-12 on p. 217 illustrates how only 3 of the 50 or so carbon atoms in a molecule of fat can yield glucose.) After prolonged glucose deprivation, brain and nerve cells develop the ability to derive about two-thirds of their minimum energy needs from the ketone bodies that the body makes from fat fragments. Ketone bodies cannot sustain life by themselves,

lipoprotein lipase (LPL): an enzyme that hydrolyzes triglycerides passing by in the bloodstream and directs their parts into the cells, where they can be metabolized for energy or reassembled for storage.

hormone-sensitive lipase: an enzyme inside adipose cells that responds to the body's need for fuel by hydrolyzing triglycerides so that their parts (glycerol and fatty acids) escape into the general circulation and thus become available to other cells for fuel. The signals to which this enzyme responds include epinephrine and glucagon, which oppose insulin (see Chapter 4).

*The reader who knows that 1 pound = 454 grams and that 1 gram of fat = 9 kcalories may wonder why a pound of body fat does not equal 4086 (9 × 454) kcalories. The reason is that body fat contains some cell water and other materials; it is not quite pure fat.

however. As Chapter 7 explains, fasting for too long will cause death, even if the person still has ample body fat.

IN SUMMARY **The body can easily store unlimited amounts of fat if given excesses, and this body fat is used for energy when needed. (Remember that the liver can also convert excess carbohydrate and protein into fat.) Fat breakdown requires simultaneous carbohydrate breakdown for maximum efficiency; without carbohydrate, fats break down to ketone bodies.**

Health Effects and Recommended Intakes of Lipids

Of all the nutrients, fat is most often linked with heart disease, some types of cancer, and obesity. Fortunately, the same recommendation can help with all of these health problems: choose a diet that is low in saturated fats, *trans* fats, and cholesterol and moderate in total fat.

Health Effects of Lipids

Hearing a physician say "Your blood lipid profile looks fine" is reassuring. The **blood lipid profile** ♦ reveals the concentrations of various lipids in the blood, notably triglycerides and cholesterol, and their lipoprotein carriers (VLDL, LDL, and HDL). This information alerts people to possible disease risks and perhaps to a need for changing their exercise and eating habits. Both the amounts and types of fat in the diet influence people's risk for disease.[14]

♦ Desirable blood lipid profile:
- Total cholesterol: <200 mg/dL
- LDL cholesterol: <100 mg/dL
- HDL cholesterol: ≥60 mg/dL
- Triglycerides: <150 mg/dL

Heart Disease

Most people realize that elevated blood cholesterol is a major risk factor for **cardiovascular disease (CVD)**. Cholesterol accumulates in the arteries, restricting blood flow and raising blood pressure. The consequences are deadly; in fact, heart disease is the nation's number one killer of adults. Blood cholesterol level is often used to predict the likelihood of a person's suffering a heart attack or stroke; the higher the cholesterol, the earlier and more likely the tragedy. Much of the effort to prevent heart disease focuses on lowering blood cholesterol.

Risks from Saturated Fats

As mentioned earlier, LDL cholesterol raises the risk of heart disease. Saturated fats are most often implicated in raising LDL cholesterol. In general, the more saturated fat in the diet, the more LDL cholesterol in the body. Not all saturated fats have the same cholesterol-raising effect, however. Most notable among the saturated fatty acids that raise blood cholesterol are lauric, myristic, and palmitic acids (12, 14, and 16 carbons, respectively). In contrast, stearic acid (18 carbons) does not seem to raise blood cholesterol.[15] However, making such distinctions may be impractical in diet planning because these saturated fatty acids typically appear together in the same foods. In addition to raising blood cholesterol, saturated fatty acids contribute to heart disease by promoting blood clotting.[16]

Fats from animal sources (meats and milk products) are the main sources of saturated fats ♦ in most people's diets.[17] Some vegetable fats (coconut and palm) and hydrogenated fats provide smaller amounts of saturated fats. Selecting skinless poultry or fish and fat-free milk products helps to lower saturated fat intake and the risk of heart disease. Using nonhydrogenated margarine and unsaturated cooking oil is another simple change that can dramatically lower saturated fat intake.

♦ Major sources of saturated fats:
- Whole milk, cream, butter, cheese, ice cream
- Fatty cuts of beef and pork
- Coconut, palm, and palm kernel oils (the tropical oils and products containing them such as candies, pastries, pies, doughnuts, and cookies)

Risks from *Trans* Fats

Research also suggests an association between dietary *trans*-fatty acids and heart disease.[18] In the body, *trans*-fatty acids alter blood cholesterol the same way some saturated fats do: they raise LDL cholesterol and, at high intakes, lower HDL cholesterol.[19] Limiting the intake of *trans*-fatty acids can improve blood cholesterol and lower the risk of heart disease. To that end, many restaurants and manufacturers have taken steps to eliminate or greatly reduce *trans*

blood lipid profile: results of blood tests that reveal a person's total cholesterol, triglycerides, and various lipoproteins.

cardiovascular disease (CVD): a general term for all diseases of the heart and blood vessels. Atherosclerosis is the main cause of CVD. When the arteries that carry blood to the heart muscle become blocked, the heart suffers damage known as **coronary heart disease (CHD)**.
- **cardio** = heart
- **vascular** = blood vessels

♦ Major sources of *trans* fats:
- Cakes, cookies, doughnuts, pastry, crackers
- Meat and dairy products
- Margarine
- Deep-fried foods (vegetable shortening)
- Snack chips

♦ When selecting margarine, look for:
- Soft (liquid or tub) instead of hard (stick)
- ≤2 g saturated fat
- Liquid vegetable oil (not hydrogenated or partially hydrogenated) as first ingredient
- "*Trans* fat free"

♦ Major sources of cholesterol:
- Eggs
- Milk products
- Meat, poultry, shellfish

♦ Major sources of monounsaturated fats:
- Olive oil, canola oil, peanut oil
- Avocados

♦ Major sources of polyunsaturated fats:
- Vegetable oils (safflower, sesame, soy, corn, sunflower)
- Nuts and seeds

♦ Major sources of omega-3 fats:
- Vegetable oils (canola, soybean, flaxseed)
- Walnuts, flaxseeds
- Fatty fish (mackerel, salmon, sardines)

fats in foods.[20] The average daily intake of *trans*-fatty acids in the United States is about 6 grams per day—mostly from products that have been hydrogenated. ♦

Reports on *trans*-fatty acids raise the question whether margarine or butter is a better choice for heart health. The American Heart Association has stated that because butter is rich in both saturated fat and cholesterol whereas margarine is made from vegetable fat with no dietary cholesterol, margarine is still preferable to butter. Be aware that soft margarines (liquid or tub) ♦ are less hydrogenated and relatively lower in *trans*-fatty acids; consequently, they do not raise blood cholesterol as much as the saturated fats of butter or the *trans* fats of hard (stick) margarines do. Many manufacturers are now offering nonhydrogenated margarines that are "*trans* fat free." The last section of this chapter describes how to read food labels and compares butter and margarines. Whichever you decide to use, remember to use them sparingly.

Risks from Cholesterol Although its effect is not as strong as that of saturated fat or *trans* fat, dietary cholesterol also raises blood cholesterol and increases the risk of heart disease. To maximize the effect on blood cholesterol, limit dietary cholesterol as well.

Recall that cholesterol is found in all foods derived from animals. Consequently, eating less fat from meats, eggs, and milk products helps lower dietary cholesterol intake ♦ (as well as total and saturated fat intakes).

Most foods that are high in cholesterol are also high in saturated fat, but eggs are an exception. An egg contains only 1 gram of saturated fat but just over 200 milligrams of cholesterol—roughly two-thirds of the recommended daily limit. For people with a healthy lipid profile, eating one egg a day is not detrimental. People with high blood cholesterol, however, may benefit from limiting daily cholesterol intake to less than 200 milligrams.[21] When eggs are included in the diet, other sources of cholesterol may need to be limited on that day. Eggs are a valuable part of the diet because they are inexpensive, useful in cooking, and a source of high-quality protein and other nutrients. Low saturated fat, high omega-3 fat eggs are now available, and food manufacturers have produced several fat-free, cholesterol-free egg substitutes.

Benefits from Monounsaturated Fats and Polyunsaturated Fats Replacing both saturated and *trans* fats with monounsaturated ♦ and polyunsaturated ♦ fats may be the most effective dietary strategy in preventing heart disease.[22] The lower rate of heart disease among people in the Mediterranean region of the world is often attributed to their liberal use of olive oil, a rich source of monounsaturated fatty acids. Olive oil, especially virgin olive oil, also delivers valuable phytochemicals that help to protect against heart disease.[23] Replacing saturated fats with the polyunsaturated fatty acids of other vegetable oils also lowers blood cholesterol. Highlight 5 examines various types of fats and their roles in supporting or harming heart health.

Benefits from Omega-3 Fats Research on the different types of fats has spotlighted the beneficial effects of the omega-3 ♦ polyunsaturated fatty acids in reducing the risks of heart disease and stroke.[24] Regular consumption of omega-3 fatty acids helps to prevent blood clots, protect against irregular heartbeats, and lower blood pressure, especially in people with hypertension or atherosclerosis.[25] In addition, omega-3 fatty acids support a healthy immune system and defend against inflammatory disorders.[26]

Table 5-2 provides sources of omega-6 and omega-3 fatty acids. Fatty fish are among the best sources of omega-3 fatty acids, and Highlight 5 features their role in supporting heart health.[27] The American Heart Association recommends two servings of fish a week, with an emphasis on fatty fish (salmon, herring, and mackerel, for example).[28] Eating fish supports heart health, especially when combined with physical activity. When preparing fish, grill, bake, or broil, but do not fry. Fried fish does not benefit heart disease.[29] Fried fish from fast-food restaurants and frozen fried fish products are often low in omega-3 fatty acids and high

TABLE 5-2 Sources of Omega-3 and Omega-6 Fatty Acids

Omega-6	
Linoleic acid	Vegetable oils (corn, sunflower, safflower, soybean, cottonseed), poultry fat, nuts, seeds
Arachidonic acid	Meats, poultry, eggs (or can be made from linoleic acid)
Omega-3	
Linolenic acid	Oils (flaxseed, canola, walnut, wheat germ, soybean) Nuts and seeds (butternuts, flaxseeds, walnuts, soybean kernels) Vegetables (soybeans)
EPA and DHA	Human milk Pacific oysters and fish[a] (or can be made from linolenic acid)

[a]All fish contain some EPA and DHA; the amounts vary among species and within a species depending on such factors as diet, season, and environment (see Table H5-1 on p. 166).

in *trans*- and saturated fatty acids. Fish provides many minerals (except iron) and vitamins. Because fish is leaner than most other animal-protein sources, it can help with weight-loss efforts. The combination of losing weight and eating fish improves blood lipids even more effectively than can be explained by either the weight loss or the omega-3 fats of the fish.

Consumers may be concerned about the adverse consequences of mercury ♦ and other environmental contaminants common in some fish. To maximize the benefits and minimize the risks, most healthy people should eat two servings of fish a week.[30]

In addition to fish, other functional foods ♦ are being developed to help consumers improve their omega-3 fatty acid intake.[31] For example, hens fed flaxseed produce eggs rich in omega-3 fatty acids. Including even one enriched egg in the diet daily can significantly increase a person's intake of omega-3 fatty acids. Another option may be to select wild game or pasture-fed cattle or bison, which provide more omega-3 fatty acids and less saturated fat than grain-fed cattle.

Omega-3 fatty acids are also available in capsules of fish oil supplements. Routine supplementation, however, is not recommended.[32] High intakes of omega-3 polyunsaturated fatty acids may increase bleeding time, interfere with wound healing, raise LDL cholesterol, and suppress immune function.* Such findings reinforce the concept that too much of a good thing can sometimes be harmful. People with heart disease, however, may benefit from doses greater than can be achieved through diet alone. They should always consult a physician first because including supplements as part of a treatment plan may be contraindicated for some patients.[33] Because high intakes of omega-3 fatty acids can cause excessive bleeding, intakes should not exceed 3 grams a day without close medical supervision.[34]

Omega-6 to Omega-3 Ratio Because omega-6 and omega-3 fatty acids compete for the same enzymes and their actions often oppose each other, researchers have studied whether there is an ideal ratio that best supports cardiovascular health. Suggested ratios range from 5:1 to 10:1, but little scientific evidence supports such recommendations.[35] In fact, the emerging consensus is that the omega-6 to omega-3 ratio is of little value in improving health or predicting risk.[36] Simply increasing the amount of omega-3 fatty acids in the diet is most beneficial.[37] Reducing the amount of omega-6 fatty acids in the diet to "improve" the ratio is not beneficial, and could even be harmful.[38]

Cancer The evidence for links between dietary fats and cancer ♦ is less convincing than for heart disease. Dietary fat does not seem to *initiate* cancer development but, instead, may *promote* cancer once it has arisen.

♦ Fish relatively high in mercury:
- Tilefish (also called golden snapper or golden bass), swordfish, king mackerel, shark

Fish relatively low in mercury:
- Cod, haddock, pollock, salmon, sole, tilapia
- Most shellfish

♦ **Functional foods** contain physiologically active compounds that provide health benefits beyond basic nutrition. (See Highlight 13 for a full discussion.)

♦ Other risk factors for cancer include smoking, alcohol, and environmental contaminants. Chapter 29 provides many more details about these risk factors and the development of cancer.

*Suppressed immune function is seen with daily intake of 0.9 to 9.4 grams EPA and 0.6 to 6.0 grams DHA for 3 to 24 weeks.

© Polara Studios Inc.

Well-balanced, healthy meals provide some fat with an emphasis on monounsaturated and polyunsaturated fats.

The relationship between dietary fat and the risk of cancer differs for various types of cancers. In the case of breast cancer, evidence has been weak and inconclusive. Some studies indicate an association between dietary fat and breast cancer; more convincing evidence indicates that body fat contributes to the risk.[39] In the case of colon cancer, limited evidence suggests a harmful association with foods containing animal fats.

The relationship between dietary fat and the risk of cancer differs for various types and combinations of fats as well.[40] The increased risk in cancer from fat appears to be due primarily to saturated fats or dietary fat from meats (which is mostly saturated). Fat from milk or fish has not been implicated in cancer risk.[41] In fact, the omega-3 fatty acids of fatty fish may protect against some cancers, perhaps by suppressing inflammation.[42] Thus dietary advice to reduce cancer risks parallels that given to reduce heart disease risks: reduce saturated fats and increase omega-3 fatty acids. Evidence does not support omega-3 supplementation.

Obesity Fat contributes more than twice as many kcalories ♦ per gram as either carbohydrate or protein. Consequently, people who eat high-fat diets regularly may exceed their energy needs and gain weight, especially if they are inactive.[43] Because fat boosts energy intake, cutting fat from the diet can be an effective strategy in cutting kcalories. In some cases, though, choosing a fat-free food offers no kcalorie savings. Fat-free frozen desserts, for example, often have so much sugar added that the kcalorie count can be as high as in the regular-fat product. In this case, cutting fat and adding carbohydrate offers no kcalorie savings or weight-loss advantage. In fact, it may even raise energy intake and exacerbate weight problems. Later chapters revisit the role of dietary fat in the development of obesity.

♦ Fat is a more concentrated energy source than the other energy nutrients: 1 g carbohydrate or protein = 4 kcal, but 1 g fat = 9 kcal.

IN SUMMARY **High blood LDL cholesterol poses a risk of heart disease, and high intakes of saturated and *trans* fats, specifically, contribute most to high LDL. Omega-3 fatty acids appear to be protective.**

Recommended Intakes of Fat

Some fat in the diet is essential for good health, but too much fat, especially saturated and *trans* fat, increases the risks for chronic diseases. Defining the exact amount of fat, saturated fat, *trans* fat, or cholesterol that benefits health or begins to harm health, however, is not possible. For this reason, no RDA or Upper Level has been set. Instead, the DRI and *Dietary Guidelines for Americans* suggest a diet that is low in saturated fat, *trans* fat, and cholesterol and provides 20 to 35 percent of the daily energy intake from fat. ♦ These recommendations recognize that diets with up to 35 percent of kcalories from fat can be compatible with good health if energy intake is reasonable and saturated and *trans* fat intakes are low. When total fat exceeds 35 percent, saturated fat increases to unhealthy levels.[44] For a 2000-kcalorie diet, 20 to 35 percent represents 400 to 700 kcalories from fat (roughly 45 to 75 grams). Part of this fat allowance should provide for the essential fatty acids—linoleic acid and linolenic acid. For this reason, an Adequate Intake (AI) has been established for these two fatty acids. Recommendations suggest that linoleic acid ♦ provide 5 to 10 percent of the daily energy intake and linolenic acid ♦ 0.6 to 1.2 percent.[45]

To help consumers meet the dietary fat goals, the Food and Drug Administration (FDA) established Daily Values ♦ on food labels using 30 percent of energy intake as the guideline for fat and 10 percent for saturated fat. The Daily Value for cholesterol is 300 milligrams regardless of energy intake. There is no Daily Value for *trans* fat, but consumers should try to keep intakes as low as possible and within the 10 percent allotted for saturated fat. According to surveys, diets in the United States provide about 34 percent of their total energy from fat, with saturated fat contributing about 12 percent of the total.[46] Cholesterol intakes in the United States average 237 milligrams a day for women and 358 for men.

♦ DRI and *Dietary Guidelines* for fat:
20 to 35% of energy intake (from mostly polyunsaturated and monounsaturated fat sources such as fish, nuts, and vegetable oils)

♦ Linoleic acid (omega-6) AI:
Men:
- 19–50 yr: 17 g/day
- 51+ yr: 14 g/day

Women:
- 19–50 yr: 12 g/day
- 51+ yr: 11 g/day

♦ Linolenic acid (omega-3) AI:
- Men: 1.6 g/day
- Women: 1.1 g/day

♦ Daily Values:
- 65 g fat (based on 30% of 2000 kcal diet)
- 20 g saturated fat (based on 10% of 2000 kcal diet)
- 300 mg cholesterol

Dietary Guidelines for Americans 2010

Consume less than 10 percent of kcalories from saturated fatty acids by replacing them with monounsaturated and polyunsaturated fatty acids. Consume less than 300 milligrams per day of dietary cholesterol. Keep *trans*-fatty acid consumption as low as possible by limiting foods that contain synthetic sources of *trans* fats, such as partially hydrogenated oils, and by limiting other solid fats.

Because solid fats ♦ deliver an abundance of saturated fatty acids, MyPyramid counts them as discretionary kcalories. The fats of fish, nuts, and vegetable oils are *not* counted as discretionary kcalories because they provide valuable omega-3 fatty acids, essential fatty acids, and vitamin E. When discretionary kcalories are available, they may be used to add fats in cooking or at the table or to select higher fat items from the food groups.

♦ Solid fats include meat and poultry fats (as in poultry skin, luncheon meats, sausage); milk fat (as in whole milk, cheese, butter); shortening (as in fried foods and baked goods); and hard margarines.

Although it is very difficult to do, some people actually manage to eat too little fat—to their detriment. Among them are people with eating disorders, described in Highlight 8, and athletes. Athletes following a diet too low in fat (less than 20 percent of total kcalories) fall short on energy, vitamins, minerals, and essential fatty acids as well as on performance.[47] As a practical guideline, it is wise to include the equivalent of at least a teaspoon of fat in every meal—a little peanut butter on toast or mayonnaise in tuna salad, for example. Dietary recommendations that limit fat are designed for healthy people over age 2; Chapter 15 discusses the fat needs of infants and young children.

As the photos in Figure 5-22 show, fat accounts for much of the energy in foods, and removing the fat from foods cuts energy and saturated fat intakes dramatically. To reduce dietary fat, eliminate fat as a seasoning and in cooking; remove the fat from high-fat foods; replace high-fat foods with low-fat alternatives; and

FIGURE 5-22 Cutting Fat Cuts kCalories—and Saturated Fat

Pork chop with fat (340 kcal, 19 g fat, 7 g saturated fat)

Potato with 1 tbs butter and 1 tbs sour cream (350 kcal, 14 g fat, 10 g saturated fat)

Whole milk, 1 c (150 kcal, 8 g fat, 5 g saturated fat)

Pork chop with fat trimmed off (230 kcal, 9 g fat, 3 g saturated fat)

Plain potato (200 kcal, <1 g fat, 0 g saturated fat)

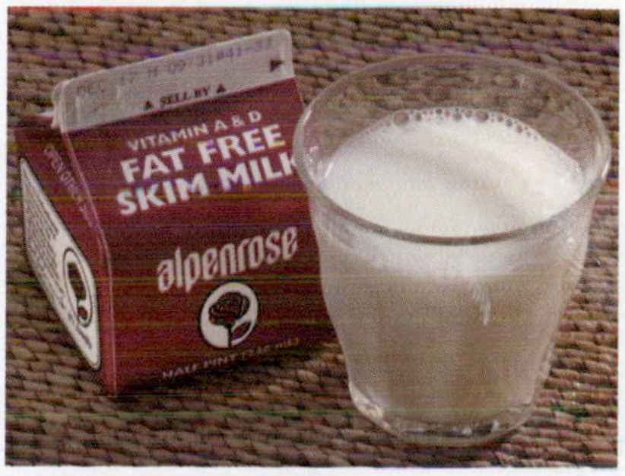

Fat-free milk, 1 c (90 kcal, <1 g fat, <1 g saturated fat)

HOW TO Make Heart-Healthy Choices—by Food Group

Breads and Cereals

- Select breads, cereals, and crackers that are low in saturated and *trans* fat (for example, bagels instead of croissants).
- Prepare pasta with a tomato sauce instead of a cheese or cream sauce.

Vegetables and Fruits

- Enjoy the natural flavor of steamed vegetables (without butter) for dinner and fruits for dessert.
- Eat at least two vegetables (in addition to a salad) with dinner.
- Snack on raw vegetables or fruits instead of high-fat items like potato chips.
- Buy frozen vegetables without sauce.

Milk and Milk Products

- Switch from whole milk to reduced-fat, from reduced-fat to low-fat, and from low-fat to fat-free (nonfat).
- Use fat-free and low-fat cheeses (such as part-skim ricotta and low-fat mozzarella) instead of regular cheeses.
- Use fat-free or low-fat yogurt or sour cream instead of regular sour cream.
- Use evaporated fat-free milk instead of cream.
- Enjoy fat-free frozen yogurt, sherbet, or ice milk instead of ice cream.

Protein Foods

- Fat adds up quickly, even with lean meat; limit intake to about 6 ounces (cooked weight) daily.
- Eat at least two servings of fish per week (particularly fish such as mackerel, lake trout, herring, sardines, and salmon).
- Choose fish, poultry, or lean cuts of pork or beef; look for unmarbled cuts named *round* or *loin* (eye of round, top round, bottom round, round tip, tenderloin, sirloin, center loin, and top loin).
- Choose processed meats such as lunch meats and hot dogs that are low in saturated fat and cholesterol.
- Trim the fat from pork and beef; remove the skin from poultry.
- Grill, roast, broil, bake, stir-fry, stew, or braise meats; don't fry. When possible, place food on a rack so that fat can drain.
- Use lean ground turkey or lean ground beef in recipes; brown ground meats without added fat, then drain off fat.
- Select tuna, sardines, and other canned meats packed in water; rinse oil-packed items with hot water to remove much of the fat.
- Fill kabob skewers with lots of vegetables and slivers of meat; create main dishes and casseroles by combining a little meat, fish, or poultry with a lot of pasta, rice, or vegetables.
- Use legumes often.
- Eat a meatless meal or two daily.
- Use egg substitutes in recipes instead of whole eggs or use two egg whites in place of each whole egg.

Fats and Oils

- Use butter or stick margarine sparingly; select soft margarines instead of hard margarines.
- Use fruit butters, reduced-kcalorie margarines, or butter replacers instead of butter.
- Use low-fat or fat-free mayonnaise and salad dressing instead of regular.
- Limit use of lard and meat fat.
- Limit use of products made with coconut oil, palm kernel oil, and palm oil (read labels on bakery goods, processed foods, popcorn oils, and nondairy creamers).
- Reduce use of hydrogenated shortenings and stick margarines and products that contain them (read labels on crackers, cookies, and other commercially prepared baked goods); use vegetable oils instead.

Miscellaneous

- Use a nonstick pan or coat the pan lightly with vegetable oil.
- Refrigerate soups and stews; when the fat solidifies, remove it before re-heating.
- Use wine; lemon, orange, or tomato juice; herbs; spices; fruits; or broth instead of butter or margarine when cooking.
- Stir-fry in a small amount of oil; add moisture and flavor with broth, tomato juice, or wine.
- Use variety to enhance enjoyment of the meal: vary colors, textures, and temperatures—hot cooked versus cool raw foods—and use garnishes to complement food.
- Omit high-fat meat gravies and cheese sauces.
- Order pizzas with lots of vegetables, a little lean meat, and half the cheese.

SOURCE: Adapted from *Third Report of the National Cholesterol Education Program (NCEP) Expert Panel on Detection, Evaluation, and Treatment of High Blood Cholesterol in Adults (Adult Treatment Panel III)*, NIH publication no. 02-5215 (Bethesda, Md.: National Heart, Lung, and Blood Institute, 2002), pp. V-25–V-27.

For additional practice, log on to **www.cengagebrain.com** and search for ISBN 084006845X.

TRY IT Compare the total kcalories, grams of fat, and percent kcalories from fat for 1 cup of whole milk, reduced-fat milk, low-fat milk, and nonfat milk.

emphasize whole grains, fruits, and vegetables. The accompanying "How To" suggests additional heart-healthy choices by food group.

From Guidelines to Groceries Fats accompany protein in foods derived from animals, such as meat, fish, poultry, and eggs, and fats accompany carbohydrate in foods derived from plants, such as avocados and coconuts. Fats carry with them the four fat-soluble vitamins—A, D, E, and K—together with many of the compounds that give foods their flavor, texture, and palatability. Fat is responsible for the delicious aromas associated with sizzling bacon, hamburgers on the grill, onions being sautéed, and vegetables in a stir-fry. The essential oils of many spices are fat soluble. Of course, these wonderful characteristics lure people into eating too much from time to time. With careful selections, a diet following MyPyramid can support good health and still meet fat recommendations.

Dietary Guidelines for Americans 2010

Replace protein foods that are higher in solid fats with choices that are lower in solid fats and kcalories and/or are sources of oils.

Protein Foods Many protein foods ♦ contain fat, saturated fat, and cholesterol but also provide high-quality protein and valuable vitamins and minerals. They can be included in a healthy diet if a person makes lean choices, prepares them using the suggestions outlined in the "How To" feature, and eats small portions. Selecting "free-range" meats from grass-fed instead of grain-fed livestock offers the nutrient advantages of being lower in fat, and the fat has more polyunsaturated fatty acids, including the omega-3 type. Another strategy to lower blood cholesterol is to prepare meals using soy protein instead of animal protein.[48]

♦ Very lean options:
- Chicken (white meat, no skin); cod, flounder, trout; tuna (canned in water); legumes

Lean options:
- Beef or pork "round" or "loin" cuts; chicken (dark meat, no skin); herring or salmon; tuna (canned in oil)

Medium-fat options:
- Ground beef, eggs, tofu

High-fat options:
- Sausage, bacon, luncheon meats, hot dogs, peanut butter, nuts

Milks and Milk Products Like meats, milks and milk products ♦ should also be selected with an awareness of their fat, saturated fat, and cholesterol contents. Fat-free and low-fat milk products provide as much or more protein, calcium, and other nutrients as their whole-milk versions—but with little or no saturated fat. Selecting fermented milk products, such as yogurt, may also help to lower blood cholesterol.[49] These foods increase the population and activity of bacteria in the colon that use cholesterol.[50]

♦ Fat-free and low-fat options:
- Fat-free or 1% milk or yogurt (plain); fat-free and low-fat cheeses

Reduced-fat options:
- 2% milk, low-fat yogurt (plain)

High-fat options:
- Whole milk, regular cheeses

Vegetables, Fruits, and Grains Choosing vegetables, fruits, whole grains, and legumes also helps lower the saturated fat, cholesterol, and total fat content of the diet. Most vegetables and fruits naturally contain little or no fat. Although avocados and olives are exceptions, most of their fat is unsaturated, which is not harmful to heart health. Most grains contain only small amounts of fat. Consumers need to read food labels, though, because some grain *products* such as fried taco shells, croissants, and biscuits are high in saturated fat, and pastries, crackers, and cookies may be high in *trans* fats. Similarly, many people add butter, margarine, or cheese sauce to grains and vegetables, which raises their saturated- and *trans*-fat contents. Because fruits are often eaten without added fat, a diet that includes several servings of fruit daily can help a person meet the dietary recommendations for fat.

A diet rich in vegetables, fruits, whole grains, and legumes also offers abundant vitamin C, folate, vitamin A, vitamin E, and dietary fiber—all important in supporting health. Consequently, such a diet protects against disease by reducing saturated fat, cholesterol, and total fat as well as by increasing nutrients. It also provides valuable phytochemicals, which help defend against heart disease.

© Matthew Farruggio

Beware of fast-food meals delivering too much fat, especially saturated fat. This double bacon cheeseburger, fries, and milk shake provide more than 1600 kcalories, with almost 90 grams of fat and more than 30 grams of saturated fat—far exceeding dietary fat guidelines for the entire day.

Invisible Fat *Visible* fat, such as butter and the fat trimmed from meat, is easy to see. *Invisible* fat is less apparent and can be present in foods in surprisingly high amounts. Invisible fat "marbles" a steak or is hidden in foods such as cheese. Any *fried* food contains abundant fat—potato chips, French fries, fried wontons, and fried fish. Many *baked* goods, too, are high in fat—pie crusts, pastries, crackers, biscuits, cornbread, doughnuts, sweet rolls, cookies, and cakes. Most chocolate bars deliver more kcalories from fat than from sugar. Even cream of mushroom soup prepared with water derives two-thirds of its energy from fat. Keep invisible fats in mind when making food selections.

Choose Wisely Consumers can find an abundant array of fresh, unprocessed foods that are naturally low in saturated fat, *trans* fat, cholesterol, and total fat. In addition, many familiar foods have been processed to provide less fat. For example, fat can be removed by skimming milk or trimming meats. Manufacturers can dilute fat by adding water or whipping in air. They can use fat-free milk in creamy desserts and lean meats in frozen entrées. Sometimes manufacturers simply prepare the products differently. For example, fat-free potato chips may be baked instead of fried. Beyond lowering the fat content, manufacturers have developed margarines fortified with plant sterols that lower blood cholesterol.*[51] (Highlight 13 explores these and other functional foods designed to support health.) Such choices make heart-healthy eating easy.

Dietary Guidelines for Americans 2010

Reduce the intake of kcalories from solid fats.

To replace saturated fats with unsaturated fats, sauté foods in olive oil instead of butter, garnish salads with sunflower seeds instead of bacon, snack on mixed nuts instead of potato chips, use avocado instead of cheese on a sandwich, and eat salmon instead of steak. Table 5-3 shows how these simple substitutions can lower the saturated fat and raise the unsaturated fat in a meal. Highlight 5 provides more details about the benefits of healthy fats in the diet.

Fat Replacers Some foods are made with **fat replacers**—ingredients that provide some of the taste and texture of fats, but with fewer kcalories. Because the body may digest and absorb some of these fat replacers, they may contribute energy, although significantly less energy than fat's 9 kcalories per gram.

Some fat replacers are derived from carbohydrate, protein, or fat. Carbohydrate-based fat replacers are used primarily as thickeners or stabilizers in foods such as soups and salad dressings. Protein-based fat replacers provide a creamy feeling in the mouth and are often used in foods such as ice creams and yogurts. Fat-based replacers act as emulsifiers and are heat stable, making them most versatile in shortenings used in cake mixes and cookies.

Fat replacers offering the sensory and cooking qualities of fats but none of the kcalories are called **artificial fats**. A familiar example of an artificial fat that has been approved for use in snack foods such as potato chips, crackers, and tortilla chips is **olestra**. Olestra's chemical structure is similar to that of a regular fat (a triglyceride) but with important differences. A triglyceride is composed of a glycerol molecule with three fatty acids attached, whereas olestra is made of a sucrose molecule with six to eight fatty acids attached. Enzymes in the digestive tract cannot break the bonds of olestra, so unlike sucrose or fatty acids, olestra passes through the system unabsorbed.

*Margarines that lower blood cholesterol contain plant sterols and are marketed under the brand names Benecol and Take Control.

TABLE 5-3 Choosing Unsaturated Fat instead of Saturated Fat

Portion sizes have been adjusted so that each of these foods provides approximately 100 kcalories. Notice that for a similar number of kcalories and grams of fat, the first choices offer less saturated fat and more unsaturated fat.

Foods (100 kcal portions)	Saturated Fat (g)	Unsaturated Fat (g)	Total Fat (g)
Olive oil (1 tbs) vs butter (1 tbs)	2 vs 7	9 vs 4	11 vs 11
Sunflower seeds (2 tbs) vs bacon (2 slices)	1 vs 3	7 vs 6	8 vs 9
Mixed nuts (2 tbs) vs potato chips (10 chips)	1 vs 2	8 vs 5	9 vs 7
Avocado (6 slices) vs cheese (1 slice)	2 vs 4	8 vs 4	10 vs 8
Salmon (2 oz) vs steak (1½ oz)	1 vs 2	3 vs 3	4 vs 5
Totals	**7 vs 18**	**35 vs 22**	**42 vs 40**

fat replacers: ingredients that replace some or all of the functions of fat and may or may not provide energy.

artificial fats: zero-energy fat replacers that are chemically synthesized to mimic the sensory and cooking qualities of naturally occurring fats but are totally or partially resistant to digestion.

olestra: a synthetic fat made from sucrose and fatty acids that provides 0 kcalories per gram; also known as *sucrose polyester.*

HOW TO Calculate a Personal Daily Value for Fat

The % Daily Value for fat on food labels is based on 65 grams. To know how your intake compares with this recommendation, you can either count grams until you reach 65, or add the "% Daily Values" until you reach 100 percent—if your energy intake is 2000 kcalories a day. If your energy intake is more or less, you can calculate your personal daily fat allowance in grams. Suppose your energy intake is 1800 kcalories per day and your goal is 30 percent kcalories from fat. Multiply your total energy intake by 30 percent, then divide by 9:

$$1800 \text{ total kcal} \times 0.30 \text{ from fat} = 540 \text{ fat kcal}$$
$$540 \text{ fat kcal} \div 9 \text{ kcal/g} = 60 \text{ g fat}$$

(In familiar measures, 60 grams of fat is about the same as ⅔ stick of butter or ¼ cup of oil.)

The accompanying table shows the numbers of grams of fat allowed per day for various energy intakes. With one of these numbers in mind, you can quickly evaluate the number of fat grams in foods you are considering eating.

For additional practice, log on to **www.cengagebrain.com** and search for ISBN 084006845X.

Energy (kcal/day)	20% kCal from Fat	35% kCal from Fat	Fat (g/day)
1200	240	420	27–47
1400	280	490	31–54
1600	320	560	36–62
1800	360	630	40–70
2000	400	700	44–78
2200	440	770	49–86
2400	480	840	53–93
2600	520	910	58–101
2800	560	980	62–109
3000	600	1050	67–117

TRY IT Calculate a personal daily fat allowance for a person with an energy intake of 2100 kcalories and a goal of 25 percent kcalories from fat.

The FDA's evaluation of olestra's safety addressed two questions. First, is olestra toxic? Research on both animals and human beings supports the safety of olestra as a partial replacement for dietary fats and oils, with no reports of cancer or birth defects. Second, does olestra affect either nutrient absorption or the health of the digestive tract? When olestra passes through the digestive tract unabsorbed, it binds with some of the fat-soluble vitamins A, D, E, and K and carries them out of the body, robbing the person of these valuable nutrients. To compensate for these losses, the FDA requires the manufacturer to fortify olestra with vitamins A, D, E, and K. Saturating olestra with these vitamins does not make the product a good source of vitamins, but it does block olestra's ability to bind with the vitamins from other foods. An asterisk in the ingredients list informs consumers that these added vitamins are "dietarily insignificant."

Some consumers experience digestive distress after eating olestra, such as cramps, gas, bloating, and diarrhea. The FDA initially required a warning label stating that "olestra may cause abdominal cramping and loose stools" and that it "inhibits the absorption of some vitamins and other nutrients" but has since concluded that such a statement is no longer warranted.

Consumers need to keep in mind that low-fat and fat-free foods still deliver kcalories. Alternatives to fat can help to lower energy intake and support weight loss only when they actually *replace* fat and energy in the diet.[52]

Read Food Labels Labels list total fat, saturated fat, *trans* fat, and cholesterol contents of foods in addition to fat kcalories per serving (see Figure 5-23 on p. 160). Because each package provides information for a single serving and because serving sizes are standardized, consumers can easily compare similar products.

Total fat, saturated fat, and cholesterol are also expressed as "% Daily Values" for a person consuming 2000 kcalories. People who consume more or less than 2000 kcalories daily can calculate their personal Daily Value for fat as described in the accompanying "How To" feature. *Trans* fats do not have a Daily Value.

FIGURE 5-23 Butter and Margarine Labels Compared

Food labels list the kcalories from fat; the quantities and Daily Values for fat, saturated fat, and cholesterol; and the quantities for *trans* fat. Information on polyunsaturated and monounsaturated fats is optional. Products that contain 0.5 g or less of *trans* fat and 0.5 g or less of saturated fat may claim "no *trans* fat." Similarly, products that contain 2 mg or less of cholesterol and 2 g or less of saturated fat may claim to be "cholesterol-free."

If the list of ingredients includes hydrogenated oils, you know the food contains *trans* fat. Chapter 2 explained that foods list their ingredients in descending order of predominance by weight. As you can see from this example, the closer "partially hydrogenated oils" is to the beginning of the ingredients list, the more *trans* fats the product contains. Notice that most of the fat in butter is saturated, whereas most of the fat in margarine is unsaturated; partially hydrogenated margarines tend to have more *trans* fat than hydrogenated liquid margarines.

Butter

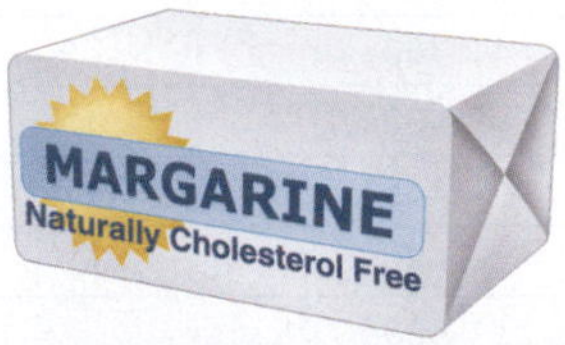

Margarine (stick)

Margarine (tub)

Margarine (liquid)

Butter

Nutrition Facts	
Serving Size 1 Tbsp (14g)	
Servings per container about 32	
Amount per serving	
Calories 100 Calories from Fat 100	
	%Daily Value*
Total Fat 11g	17%
Saturated Fat 7g	37%
Trans Fat 0g	
Cholesterol 30mg	10%
Sodium 95mg	4%
Total Carbohydrate 0g	0%
Protein 0g	
Vitamin A 8%	
Not a significant source of dietary fiber, sugars, vitamin C, calcium, and iron.	
*Percent Daily Values are based on a 2,000 calorie diet.	

INGREDIENTS: Cream, salt.

Margarine (stick)

Nutrition Facts	
Serving Size 1 Tbsp (14g)	
Servings per container about 32	
Amount per serving	
Calories 100 Calories from Fat 100	
	%Daily Value*
Total Fat 11g	17%
Saturated Fat 2g	11%
Trans Fat 2.5g	
Polyunsaturated Fat 3.5g	
Monounsaturated Fat 2.5g	
Cholesterol 0mg	0%
Sodium 105mg	4%
Total Carbohydrate 0g	0%
Protein 0g	
Vitamin A 10%	
Not a significant source of dietary fiber, sugars, vitamin C, calcium, and iron.	
*Percent Daily Values are based on a 2,000 calorie diet.	

INGREDIENTS: Liquid soybean oil, partially hydrogenated soybean oil, water, buttermilk, salt, soy lecithin, sodium benzoate (as a preservative), vegetable mono and diglycerides, artificial flavor, vitamin A palmitate, colored with beta carotene (provitamin A).

Margarine (tub)

Nutrition Facts	
Serving size 1 Tbsp (14g)	
Servings per container about 32	
Amount per serving	
Calories 100 Calories from Fat 100	
	%Daily Value*
Total Fat 11g	17%
Saturated Fat 2.5g	13%
Trans Fat 2g	
Polyunsaturated Fat 4g	
Monounsaturated Fat 2.5g	
Cholesterol 0mg	0%
Sodium 80mg	3%
Total Carbohydrate 0g	0%
Protein 0g	
Vitamin A 10%	
Not a significant source of dietary fiber, sugars, vitamin C, calcium, and iron.	
*Percent Daily Values are based on a 2,000 calorie diet.	

INGREDIENTS: Liquid soybean oil, partially hydrogenated soybean oil, buttermilk, water, butter (cream, salt), salt, soy lecithin, vegetable mono and diglycerides, sodium benzoate added as a preservative, artificial flavor, vitamin A palmitate, colored with beta carotene.

Margarine (liquid)

Nutrition Facts	
Serving size 1 Tbsp (14g)	
Servings per container about 24	
Amount per serving	
Calories 70 Calories from Fat 70	
	%Daily Value*
Total Fat 8g	13%
Saturated Fat 1.5g	7%
Trans Fat 0g	
Polyunsaturated Fat 4.5g	
Monounsaturated Fat 2g	
Cholesterol 0mg	0%
Sodium 110mg	8%
Total Carbohydrate 0g	0%
Protein 0g	
Vitamin A 10%	
Not a significant source of dietary fiber, sugars, vitamin C, calcium, and iron.	
*Percent Daily Values are based on a 2,000 calorie diet.	

INGREDIENTS: Liquid soybean oil, water, salt, hydrogenated cottonseed oil, vegetable monoglycerides and soy lecithin (emulsifiers), potassium sorbate and sodium benzoate (to preserve freshness), artificial flavor, phosphoric acid (acidulant), colored with beta carotene (source of vitamin A), vitamin A palmitate.

Be aware that the "% Daily Value" for fat is not the same as "% kcalories from fat." This important distinction is explained in the accompanying "How To" feature. Because recommendations apply to average daily intakes rather than individual food items, food labels do not provide "% kcalories from fat." Still, you can get an idea of whether a particular food is high or low in fat.

IN SUMMARY In foods, triglycerides:

- Deliver fat-soluble vitamins, energy, and essential fatty acids
- Contribute to the sensory appeal of foods and stimulate appetite

HOW TO Understanding "% Daily Value" and "% kCalories from Fat"

The "% Daily Value" that is used on food labels to describe the amount of fat in a food is not the same as the "% kcalories from fat" that is used in dietary recommendations to describe the amount of fat in the diet. They may appear similar, but their difference is worth understanding. Consider, for example, a piece of lemon meringue pie that provides 140 kcalories and 12 grams of fat. Because the Daily Value for fat is 65 grams for a 2000-kcalorie intake, 12 grams represent about 18 percent:

$$12 \text{ g} \div 65 \text{ g} = 0.18$$
$$0.18 \times 100 = 18\%$$

The pie's "% Daily Value" is 18 percent, or almost one-fifth, of the day's fat allowance.

Uninformed consumers may mistakenly believe that this food meets recommendations to limit fat to "20 to 35 percent kcalories," but it doesn't—for two reasons. First, the pie's 12 grams of fat contribute 108 of the 140 kcalories, for a total of 77 percent kcalories from fat:

$$12 \text{ g fat} \times 9 \text{ kcal/g} = 108 \text{ kcal}$$
$$108 \text{ kcal} \div 140 \text{ kcal} = 77\%$$

Second, the "percent kcalories from fat" guideline applies to a day's total intake, not to an individual food. Of course, if every selection throughout the day exceeds 35 percent kcalories from fat, you can be certain that the day's total intake will, too.

For additional practice, log on to **www.cengagebrain.com** and search for ISBN 084006845X.

Michael C. Gray/Shutterstock.com

Whether a person's energy and fat allowance can afford a piece of lemon meringue pie depends on the other food and activity choices made that day.

TRY IT Calculate the percent Daily Value and the percent kcalories from fat for ½ cup frozen yogurt that provides 115 kcalories and 4 grams of fat.

Although some fat in the diet is necessary, health authorities recommend a diet moderate in total fat and low in saturated fat, *trans* fat, and cholesterol. They also recommend replacing saturated fats with monounsaturated and polyunsaturated fats, particularly omega-3 fatty acids from foods such as fatty fish, not from supplements. Many selection and preparation strategies can help bring these goals within reach, and food labels help to identify foods consistent with these guidelines.

If people were to make only one change in their diets, they would be wise to limit their intakes of saturated fat. Sometimes these choices can be difficult, though, because fats make foods taste delicious. To maintain good health, must a person give up all high-fat foods forever—never again to eat marbled steak, hollandaise sauce, or gooey chocolate cake? Not at all. These foods bring pleasure to a meal and can be enjoyed as part of a healthy diet when eaten occasionally in small quantities; but they should not be everyday foods. The key dietary principle for fat is *moderation,* not *deprivation.* Appreciate the energy and enjoyment that fat provides, but take care not to exceed your needs.

Nutrition Portfolio

To maintain good health, eat enough, but not too much, fat and select the right kinds.

Go to Diet Analysis Plus and choose one of the days on which you have tracked your diet for the entire day. Go to the Intake Spreadsheet report. Scroll down until you see: fat (g), sat fat (g), mono fat (g), poly fat (g) and chol (g), which stand for grams of total fat, saturated fat, monounsaturated fat, polyunsaturated fat, and cholesterol, respectively. Use these columns to answer the following questions:

- List the types and amounts of fats and oils you ate on that day, making note of which are saturated, monounsaturated, or polyunsaturated and how your choices could include fewer saturated options.

- List the types and amounts of milk products, meats, fish, and poultry you eat daily, noting how your choices could include more low-fat options.
- Describe choices you can make in selecting and preparing foods to lower your intake of solid fats.

Diet Analysis PLUS To complete this exercise, go to your Diet Analysis Plus at www.cengage.com/sso.

Nutrition on the Net

For further study of topics covered in this chapter, log on to **www.cengagebrain.com** and search for ISBN 084006845X.

- Search for "fat" at the International Food Information Council site: **www.ific.org**
- Find dietary strategies to prevent heart disease at the National Heart, Lung, and Blood Institute: **www.nhlbi.nih.gov**
- Visit the nutrition section, especially Face the Fats, of the American Heart Association: **www.americanheart.org**

References

1. J. M. Chardigny and coauthors, Do *trans* fatty acids from industrially produced sources and from natural sources have the same effect on cardiovascular disease risk factors in healthy subjects? Results of the *trans* Fatty Acids Collaboration (TRANSFACT) study, *American Journal of Clinical Nutrition* 87 (2008): 558–566; M. A. Zulet and coauthors, Inflammation and conjugated linoleic acid: Mechanisms of action and implications for human health, *Journal of Physiology and Biochemistry* 61 (2005): 483–494.
2. S. Klingberg and coauthors, Inverse relation between dietary intake of naturally occurring plant sterols and serum cholesterol in northern Sweden, *American Journal of Clinical Nutrition* 87 (2008): 993–1001.
3. J. Plat and R. P. Mensink, Plant stanol and sterol esters in the control of blood cholesterol levels: Mechanism and safety aspects, *American Journal of Cardiology* 96 (2005): 15D–22D.
4. Q. Sun and coauthors, Comparison between plasma and erythrocyte fatty acid content as biomarkers of fatty acid intake in U.S. women, *American Journal of Clinical Nutrition* 86 (2007): 74–81; W. S. Harris and coauthors, Comparison of the effects of fish and fish-oil capsules on the n-3 fatty acid content of blood cells and plasma phospholipids, *American Journal of Clinical Nutrition* 86 (2007): 1621–1625.
5. T. Hampton, New clues to HDL's benefits revealed, *Journal of the American Medical Association* 297 (2007): 1537.
6. T. Yamada and H. Katagiri, Avenues of communication between the brain and tissues/organs involved in energy homeostasis, *Endocrine Journal* 54 (2007): 497–505.
7. A. Garg, Adipose tissue dysfunction in obesity and lipodystrophy, *Clinical Cornerstone* 8 (2006): S7–S13.
8. G. Govindarajan, M. A. Alpert, and L. Tejwani, Endocrine and metabolic effects of fat: Cardiovascular implications, *American Journal of Medicine* 121 (2008): 366–370; G. Fantuzzi and T. Mazzone, Adipose tissue and atherosclerosis: Exploring the connection, *Arteriosclerosis, Thrombosis, and Vascular Biology* 27 (2007): 996–1003; P. Trayhurn, C. Bing, and I. S. Wood, Adipose tissue and adipokines: Energy regulation from the human perspective, *Journal of Nutrition* 136 (2006): 1935S–1939S; C. Bulcão and coauthors, The new adipose tissue and adipocytokines, *Current Diabetes Reviews* 2 (2006): 19–28; T. J. Guzik, D. Mangalat, and R. Korbut, Adipocytokines: Novel link between inflammation and vascular function? *Journal of Physiology and Pharmacology* 57 (2006): 505–528.
9. R. N. Redinger, The physiology of adiposity, *Journal of the Kentucky Medical Association* 106 (2008): 53–62; P. G. McTernan, C. M. Kusminski, and S. Kumar, Resistin, *Current Opinion in Lipidology* 17 (2006): 170–175.
10. Y. Okamoto and coauthors, Adiponectin: A key adipocytokine in metabolic syndrome, *Clinical Science* 110 (2006): 267–278.
11. M. A. Beydoun and coauthors, Plasma n-3 fatty acids and the risk of cognitive decline in older adults: The Atherosclerosis Risk in Communities Study, *American Journal of Clinical Nutrition* 85 (2007): 1103–1111; C. Dullemeijer and coauthors, n-3 Fatty acid proportions in plasma and cognitive performance in older adults, *American Journal of Clinical Nutrition* 86 (2007): 1479–1485; S. M. Innis, Dietary (n-3) fatty acids and brain development, *Journal of Nutrition* 137 (2007): 855–859; B. M. van Gelder and coauthors, Fish consumption, n-3 fatty acids, and subsequent 5-y cognitive decline in elderly men: The Zutphen Elderly Study, *American Journal of Clinical Nutrition* 85 (2007): 1142–1147; E. Nurk and coauthors, Cognitive performance among the elderly and dietary fish intake: The Hordaland Health Study, *American Journal of Clinical Nutrition* 86 (2007): 1470–1478; R. Uauy and A. D. Dangour, Nutrition in brain development and aging: Role of essential fatty acids, *Nutrition Reviews* 64 (2006): S24–S33; W. C. Heird and A. Lapillonne, The role of essential fatty acids in development, *Annual Review of Nutrition* 25 (2005): 549–571.
12. A. H. Stark, M. A. Crawford, and R. Reifen, Update on alpha-linolenic acid, *Nutrition Reviews* 66 (2008): 326–332; J. L. Breslow, n-3 Fatty acids and cardiovascular disease, *American Journal of Clinical Nutrition* 83 (2006): 1477S–1482S.
13. P. Arner and D. Langin, The role of neutral lipases in human adipose tissue lipolysis, *Current Opinion in Lipidology* 18 (2007): 246–250; R. E. Duncan and coauthors, Regulation of lipolysis in adipocytes, *Annual Review of Nutrition* 27 (2007): 79–101; M. Rydén and coauthors, Comparative studies of the role of hormone-sensitive lipase and adipose triglyceride lipase in human fat cell lipolysis, *American Journal of Physiology—Endocrinology and Metabolism* 292 (2007): E1847–E1855.
14. P. J. Nestel and coauthors, Relation of diet to cardiovascular disease risk factors in subjects with cardiovascular disease in Australia and New Zealand: Analysis of the Long-Term Intervention with Pravastatin in Ischaemic Disease trial, *American Journal of Clinical Nutrition* 81 (2005): 1322–1329.
15. S.E.E. Berry, G. J. Miller, and T.A.B. Sanders, The solid fat content of stearic acid–rich fats determines their postprandial effects, *American Journal of Clinical Nutrition* 85 (2007): 1486–1494.
16. J. Delgado-Lista and coauthors, Chronic dietary fat intake modifies the postprandial response of hemostatic markers to a single fatty test meal, *American Journal of Clinical Nutrition* 87 (2008): 317–322.

17. Position of the American Dietetic Association and Dietitians of Canada: Dietary fatty acids, *Journal of the American Dietetic Association* 107 (2007): 1599–1611.
18. D. Mozaffarian and coauthors, *Trans* fatty acids and cardiovascular disease, *New England Journal of Medicine* 354 (2006): 1601–1613.
19. J. E. Hunter, Dietary trans fatty acids: Review of recent human studies and food industry responses, *Lipids* 41 (2006): 967–992.
20. M. J. Albers and coauthors, 2006 Marketplace survey of *trans*–fatty acid content of margarines and butters, cookies and snack cakes, and savory snacks, *Journal of the American Dietetic Association* 108 (2008): 367–370; S. Borra and coauthors, An update of *trans*-fat reduction in the American diet, *Journal of the American Dietetic Association* 107 (2007): 2048–2050; S. Okie, New York to trans fats: You're out! *New England Journal of Medicine* 356 (2007): 2017–2021.
21. Expert Panel on Detection, Evaluation, and Treatment of High Blood Cholesterol in Adults (Adult Treatment Panel III), *Third Report of the National Cholesterol Education Program* (NCEP), NIH publication no. 02–5215 (Bethesda, Md.: National Heart, Lung, and Blood Institute, 2002), p. v-10.
22. L. Berglund and coauthors, Comparison of monounsaturated fat with carbohydrates as a replacement for saturated fat in subjects with a high metabolic risk profile: Studies in the fasting and postprandial states, *American Journal of Clinical Nutrition* 86 (2007): 1611–1620; M. P. St-Onge and coauthors, Snack chips fried in corn oil alleviate cardiovascular disease risk factors when substituted for low-fat or high-fat snacks, *American Journal of Clinical Nutrition* 85 (2007): 1503–1510.
23. J. Ruano and coauthors, Intake of phenol-rich virgin olive oil improves the postprandial prothrombotic profile in hypercholesterolemic patients, *American Journal of Clinical Nutrition* 86 (2007): 341–346.
24. J. L. Breslow, n-3 Fatty acids and cardiovascular disease, *American Journal of Clinical Nutrition* 83 (2006): 1477S–1482S.
25. Breslow, 2006; P.J.H. Jones and V.W.Y. Lau, Effect of n-3 polyunsaturated fatty acids on risk reduction of sudden death, *Nutrition Reviews* 60 (2002): 407–413.
26. K. Fritsche, Fatty acids as modulators of the immune response, *Annual Review of Nutrition* 26 (2006): 45–73; S. M. Innis and K. Jacobson, Dietary lipids in early development and intestinal inflammatory disease, *Nutrition Reviews* 65 (2007): S188–S193; S. R. Shaikh and M. Edidin, Polyunsaturated fatty acids, membrane organization, T cells, and antigen presentation, *American Journal of Clinical Nutrition* 84 (2006): 1277–1289.
27. A. Philibert and coauthors, Fish intake and serum fatty acid profiles from freshwater fish, *American Journal of Clinical Nutrition* 84 (2006): 1299–1307.
28. AHA Scientific statement: Diet and lifestyle recommendations revision 2006, *Circulation* 114 (2006): 82–96.
29. K. He and coauthors, Intakes of long-chain n-3 polyunsaturated fatty acids and fish in relation to measurement of subclinical atherosclerosis, *American Journal of Clinical Nutrition* 88 (2008): 1111–1118.
30. M. C. Nesheim and A. L. Yaktine, eds., Seafood, *Seafood Choices: Balancing Benefits and Risks* (Washington, D.C.: National Academies Press, 2007), p. 12; C. W. Levenson and D. M. Axelrad, Too much of a good thing? Update on fish consumption and mercury exposure, *Nutrition Reviews* 64 (2006): 139–145.
31. J. Whelan and C. Rust, Innovative dietary sources of n-3 fatty acids, *Annual Review of Nutrition* 26 (2006): 75–103.
32. P. M. Kris-Etherton and A. M. Hill, n-3 Fatty acids: Foods or supplements? *American Dietetic Association* 108 (2008): 1125–1130.
33. M. H. Raitt and coauthors, Fish oil supplementation and risk of ventricular tachycardia and ventricular fibrillation in patients with implantable defibrillators: A randomized control study, *Journal of the American Medical Association* 293 (2005): 2884–2891.
34. Fish and omega-3 fatty acids, www.americanheart.org, accessed 2009.
35. Committee on Dietary Reference Intakes, *Dietary Reference Intakes for Energy, Carbohydrate, Fiber, Fat, Fatty Acids, Cholesterol, Protein, and Amino Acids* (Washington, D.C.: National Academies Press, 2005).
36. J. C. Stanley and coauthors, UK Food Standards Agency Workshop report: The effects of the dietary n-6:n-3 fatty acid ratio on cardiovascular health, *British Journal of Nutrition* 98 (2007): 1305–1310; W. S. Harris, The omega-6/omega-3 ratio and cardiovascular disease risk: Uses and abuses, *Current Atherosclerosis Reports* 8 (2006): 453–459.
37. B. A. Griffin, How relevant is the ratio of dietary n-6 to n-3 polyunsaturated fatty acids to cardiovascular disease risk? Evidence from the OPTILIP study, *Current Opinion in Lipidology* 19 (2008): 57–62.
38. W. S. Harris and coauthors, Omega-6 fatty acids and risk for cardiovascular disease: A Science Advisory from the American Heart Association Nutrition Subcommittee of the Council on Nutrition, Physical Activity, and Metabolism; Council on Cardiovascular Nursing, and Council on Epidemiology and Prevention, *Circulation* 108 (2009): 902–907; W. C. Willett, The role of dietary n-6 fatty acids in the prevention of cardiovascular disease, *The Journal of Cardiovascular Medicine* 8 (2007): S42–S45.
39. A.C.M. Thiébaut and coauthors, Dietary fat and postmenopausal invasive breast cancer in the National Institutes of Health: AARP Diet and Health Study Cohort, *Journal of the National Cancer Institute* 99 (2007): 451–462; World Cancer Research Fund and American Institute for Cancer Research, *Food, Nutrition, Physical Activity, and the Prevention of Cancer: A Global Perspective*, www.dietandcancerreport.org, accessed, 2007.
40. E. Theodoratou and coauthors, Dietary fatty acids and colorectal cancer: A case-control study, *American Journal of Epidemiology* 166 (2007): 181–195; P. Bougnoux, B. Giraudeau, and C. Couet, Diet, cancer, and the lipidome, *Cancer Epidemiology Biomarkers and Prevention* 15 (2006): 416–421.
41. P. W. Parodi, Dairy product consumption and the risk of breast cancer, *Journal of the American College of Nutrition* 24 (2005): 556S–568S; J. Zhang and H. Kesteloot, Milk consumption in relation to incidence of prostate, breast, colon, and rectal cancers: Is there an independent effect? *Nutrition and Cancer* 53 (2005): 65–72.
42. R. S. Chapkin, D. N. McMurray, and J. R. Lupton, Colon cancer, fatty acids and anti-inflammatory compounds, *Current Opinion in Gastroenterology* 23 (2007): 48–54; A. Geelen and coauthors, Fish consumption, n-3 fatty acids, and colorectal cancer: A meta-analysis of prospective cohort studies, *American Journal of Epidemiology* 166 (2007): 1116–1125; J. Shannon and coauthors, Erythrocyte fatty acids and breast cancer risk: A case-control study in Shanghai, China, *American Journal of Clinical Nutrition* 85 (2007): 1090–1097; C. H. MacLean and coauthors, Effects of omega-3 fatty acids on cancer risk: A systematic review, *Journal of the American Medical Association* 295 (2006): 403–415.
43. Committee on Dietary Reference Intakes, 2005.
44. Committee on Dietary Reference Intakes, 2005.
45. Committee on Dietary Reference Intakes, 2005.
46. U.S. Department of Agricultural Research Service, Nutrient intakes from food: Mean amounts consumed per individual, one day, 2005–2006, www.ars.usda.gov/ba/bhnrc/fsrg, accessed 2008.
47. Position of the American Dietetic Association, Dietitians of Canada, and the American College of Sports Medicine: Nutrition and athletic performance, *Journal of the American Dietetic Association* 100 (2000): 1543–1556.
48. C. W. Xiao, J. Mei, and C. M. Wood, Effect of soy proteins and isoflavones on lipid metabolism and involved gene expression, *Frontiers in Bioscience* 13 (2008): 2660–2673; K. Reynolds and coauthors, A meta-analysis of the effect of soy protein supplementation on serum lipids, *American Journal of Cardiology* 98 (2006): 633–640.
49. E. Fabian and I. Elmadfa, Influence of daily consumption of probiotic and conventional yoghurt on the plasma lipid profile in young healthy women, *Annals of Nutrition and Metabolism* 50 (2006): 387–393.
50. A. Dilmi-Bouras, Assimilation *(in vitro)* of cholesterol by yogurt bacteria, *Annals of Agricultural and Environmental Medicine* 13 (2006): 49–53.
51. C. S. Patch, L. C. Tapsell, and P. G. Williams, Plant sterol/stanol prescription is an effective treatment strategy for managing hypercholesterolemia in outpatient clinical practice, *Journal of the American Dietetic Association* 105 (2005): 46–52.
52. Position of the American Dietetic Association: Fat replacers, *Journal of the American Dietetic Association* 105 (2005): 266–275.

HIGHLIGHT 5

High-Fat Foods—Friend or Foe?

ElenaGaak/Shutterstock.com

Eat less fat. Eat more fatty fish. Give up butter. Use margarine. Give up margarine. Use olive oil. Steer clear of saturated. Seek out omega-3. Stay away from *trans*. Stick with mono- and polyunsaturated. Keep fat intake moderate. Today's fat messages seem to be forever multiplying and changing. No wonder some people feel confused about dietary fat. The confusion stems in part from the complexities of fat and in part from the nature of recommendations. As Chapter 5 explained, *dietary fat* refers to several kinds of fats. Some fats support health whereas others impair it, and foods typically provide a mixture of fats in varying proportions. Researchers have spent decades sorting through the relationships among the various kinds of fat and their roles in supporting or harming health. Translating these research findings into dietary recommendations is challenging. Too little information can mislead consumers, but too much detail can overwhelm them. As research findings accumulate, recommendations slowly evolve and become more refined. Fortunately, that's where we are with fat recommendations today—refining them from the general to the specific. Though they may seem to be "forever multiplying and changing," in fact, they are becoming more meaningful.

This highlight begins with a look at the dietary guidelines for fat intake. It continues by identifying which foods provide which fats and presenting the Mediterranean diet, an example of a food plan that embraces the heart-healthy fats. It closes with strategies to help consumers choose the right amounts of the right kinds of fats for a healthy diet.

Guidelines for Fat Intake

Dietary recommendations for fat have shifted emphasis from lowering total fat, in general, to limiting saturated and *trans* fat, specifically. Instead of urging people to cut back on all fats, recommendations suggest carefully replacing the "bad" saturated fats with the "good" unsaturated fats and enjoying them in moderation.[1] The goal is to create a diet moderate in kcalories that provides enough of the fats that support good health, but not too much of those that harm health. (Turn to pp. 151–153 for a review of the health consequences of each type of fat.)

With these findings and goals in mind, the DRI Committee suggests a healthy range of 20 to 35 percent of energy intake from fat. This range appears to be compatible with low rates of heart disease, diabetes, obesity, and cancer.[2] Heart-healthy recommendations suggest that within this range, consumers should try to minimize their intakes of saturated fat, *trans* fat, and cholesterol and use monounsaturated and polyunsaturated fats instead.[3]

Asking consumers to limit their total fat intake is less than perfect advice, but it is straightforward—find the fat and cut back. Asking consumers to keep their intakes of saturated fats, *trans* fats, and cholesterol low and to use monounsaturated and polyunsaturated fats instead is more on target with heart health, but it also makes diet planning a bit more complicated. To make appropriate selections, consumers must first learn which foods contain which fats.

High-Fat Foods and Heart Health

Avocados, bacon, walnuts, potato chips, and mackerel are all high-fat foods, yet some of these foods have detrimental effects on heart health when consumed in excess, whereas others seem neutral or even beneficial. This section presents some of the accumulating evidence that helped to distinguish which high-fat foods belong in a healthy diet and which ones need to be kept to a minimum. As you will see, fat in the diet can be compatible with heart health, but only if the great majority of it is the unsaturated kind.

Cook with Olive Oil

As it turns out, the traditional diets of Greece and other countries in the Mediterranean region offer an excellent example of eating patterns that use "good" fats liberally. The primary fat in these diets is olive oil, which seems to play a key role in providing health benefits.[4] A classic study of the world's people, the Seven Countries Study, found that death rates from heart disease were strongly associated with diets high in saturated fats but only weakly linked with total fat.[5] In fact, the two countries with the highest fat intakes, Finland and the Greek island of Crete, had the highest (Finland) and lowest (Crete) rates of heart disease deaths. In both countries, the people consumed 40 percent or more of their kcalories from fat. Clearly, a high-fat diet was not the primary problem, so researchers refocused their attention on the *type* of fat. They began to notice the benefits of olive oil.

A diet that uses olive oil instead of other cooking fats, especially butter, stick margarine, and meat fats may offer numerous health benefits. Olive oil and other oils rich in monounsaturated fatty acids help to protect against heart disease by:

- Lowering total and LDL cholesterol and not lowering HDL cholesterol or raising triglyceride[6]
- Lowering LDL cholesterol susceptibility to oxidation[7]
- Lowering blood-clotting factors[8]
- Providing phytochemicals that act as antioxidants (see Highlight 11)[9]
- Lowering blood pressure[10]
- Interfering with the inflammatory response[11]

When compared with other fats, olive oil seems to be a wise choice, but controlled clinical trials are too scarce to support population-wide recommendations to switch to a high-fat diet rich in olive oil. Importantly, olive oil is not a magic potion; drizzling it on foods does not make them healthier. Like other fats, olive oil delivers 9 kcalories per gram, which can contribute to weight gain in people who fail to balance their energy intake with physical activity. Its role in a healthy diet is to *replace* the saturated fats. Other vegetable oils, such as canola or safflower oil, are also generally low in saturated fats and high in unsaturated fats. For this reason, heart-healthy diets use these unsaturated vegetable oils as substitutes for the more saturated fats of butter, hydrogenated stick margarine, lard, or shortening. (Remember that the tropical oils—coconut, palm, and palm kernel—are too saturated to be included with the heart-healthy vegetable oils.)

Matthew Farruggio

Olives and their oil may benefit heart health.

Matthew Farruggio

For heart health, snack on a few nuts instead of potato chips. Because nuts are energy dense (high in kcalories per ounce), it is especially important to keep portion size in mind when eating them.

Nibble on Nuts

Tree nuts and peanuts are traditionally excluded from low-fat diets, and for good reasons. Nuts provide up to 80 percent of their kcalories from fat, and a quarter cup (about an ounce) of mixed nuts provides more than 200 kcalories. In a recent review of the literature, however, researchers found that people who ate a 1-ounce serving of nuts on five or more days a week had lower LDL cholesterol and a reduced risk of heart disease compared with people who consumed no nuts.[12] A smaller positive association was noted for any amount greater than one serving of nuts a week. The nuts in this study were those commonly eaten in the United States: almonds, Brazil nuts, cashews, hazelnuts, macadamia nuts, pecans, pistachios, walnuts, and even peanuts. On average, these nuts contain mostly monounsaturated fat (59 percent), some polyunsaturated fat (27 percent), and little saturated fat (14 percent).

Including nuts may be a wise diet strategy against heart disease. Nuts may protect against heart disease because they provide:

- Monounsaturated and polyunsaturated fats in abundance, but few saturated fats
- Fiber, vegetable protein, and other valuable nutrients, including the antioxidant vitamin E (see Highlight 11)
- Phytochemicals that act as antioxidants (see Highlight 13)[13]
- Plant sterols

Before advising consumers to include nuts in their diets, however, a caution is in order. As mentioned, most of the energy nuts provide comes from fats. Consequently, they deliver many kcalories per bite. In studies examining the effects of nuts on heart disease, researchers carefully adjust diets to make room for the nuts without increasing the total kcalories—that is, they use nuts *instead of, not in addition to,* other foods (such as meats, potato chips, oils, margarine, and butter). Consumers who do not make similar replacements could end up gaining weight if they simply add nuts on top of their regular diets. Weight gain, in turn, elevates blood lipids and raises the risks of heart disease.

Feast on Fish

Research into the health benefits of the long-chain omega-3 polyunsaturated fatty acids began with a simple observation: the native peoples of Alaska, northern Canada, and Greenland, who eat a traditional diet rich in omega-3 fatty acids, notably EPA and DHA, have a remarkably low rate of heart disease even though their diets are relatively high in fat.[14] These omega-3 fatty acids help to protect against heart disease by:[15]

- Reducing blood triglycerides
- Preventing blood clots
- Protecting against irregular heartbeats[16]
- Lowering blood pressure
- Defending against inflammation
- Serving as precursors to eicosanoids

Stephen VanHorn/Shutterstock.com

Fish is a good source of the omega-3 fatty acids.

For people with hypertension or atherosclerosis, these actions can be life saving.

Research studies have provided strong evidence that increasing omega-3 fatty acids in the diet supports heart health and lowers the rate of deaths from heart disease. For this reason, the American Heart Association recommends including fish in a heart-healthy diet. People who eat some fish each week can lower their risks of heart attack and stroke. Table H5-1 ranks commonly eaten fish by their omega-3 fatty acid content.

Fish is the best source of EPA and DHA in the diet, but it is also a major source of mercury, an environmental contaminant. Most fish contain at least trace amounts of mercury, but some have especially high levels. For this reason, the FDA advises pregnant and lactating women, women of childbearing age who may become pregnant, and young children to include fish in their diets, but to avoid:

- Tilefish (also called golden snapper or golden bass), swordfish, king mackerel, marlin, and shark

And to limit average weekly consumption of:

- A variety of ocean fish and shellfish to 12 ounces (cooked or canned)
- White (albacore) tuna to 6 ounces (cooked or canned)

Commonly eaten seafood relatively low in mercury include shrimp, catfish, pollock, salmon, and canned light tuna.

In addition to the direct toxic effects of mercury, some (but not all) research suggests that mercury may diminish the health benefits of omega-3 fatty acids. Such findings serve as a reminder that our health depends on the health of our planet. The protective effect of fish in the diet is available, provided that the fish and their surrounding waters are not heavily contaminated.

TABLE H5-1 Omega-3 Fatty Acid Content of Commonly Eaten Fish

>500 mg per 3.5 oz Serving	150–500 mg per 3.5 oz Serving	<150 mg per 3.5 oz Serving
Bronzini	Black bass	Cod (Pacific)
Herring (Atlantic and Pacific)	Catfish (wild and farmed)	Corvina
Mackerel	Clam	Grouper
Oyster (Pacific)	Cod (Atlantic)	Lobster
Salmon (chinook, coho, Copper River, farmed, pink, sockeye, wild Atlantic)	Crab (Alaskan king)	Mahi-mahi
Sardines	Croakers	Monkfish
Toothfish	Escolar	Red snapper
Trout (wild and farmed)	Flounder	Skate
	Haddock	Triggerfish
	Hake	Tuna
	Halibut	Wahoo
	Oyster (eastern and farmed)	
	Perch	
	Scallop	
	Shrimp (mixed varieties)	
	Sole	
	Swordfish	
	Tilapia (farmed)	

SOURCE: K. L. Weaver and coauthors, The content of favorable and unfavorable polyunsaturated fatty acids found in commonly eaten fish, *Journal of the American Dietetic Association* 108 (2008): 1178–1185; P. M. Kris-Etherton, W. S. Harris, and L. J. Appel, Fish consumption, fish oil, omega-3 fatty acids, and cardiovascular disease, *Circulation* 106 (2002): 2747–2757.

In an effort to limit exposure to pollutants, some consumers choose farm-raised fish. Compared with fish caught in the wild, farm-raised fish tend to be lower in mercury, but they are also lower in omega-3 fatty acids. When selecting fish, keep the diet strategies of variety and moderation in mind. Varying choices and eating moderate amounts helps to limit the intake of contaminants such as mercury.

High-Fat Foods and Heart Disease

The number one dietary determinant of LDL cholesterol is saturated fat. Figure H5-1 shows that each 1 percent increase in energy from saturated fatty acids in the diet may produce a 2 percent jump in heart disease risk by elevating blood LDL cholesterol. Conversely, reducing saturated fat intake by 1 percent can be expected to produce a 2 percent drop in heart disease risk by the same mechanism. Even a 2 percent drop in LDL represents a significant improvement for the health of the heart.[17] Like saturated fats, *trans* fats also raise heart disease risk by elevating LDL cholesterol. A heart-healthy diet limits foods rich in these two types of fat.

Limit Fatty Meats, Whole-Milk Products, and Tropical Oils

The major sources of saturated fats in the U.S. diet are fatty meats, whole milk products, tropical oils, and products made from any of these foods. To limit saturated fat intake, consumers must choose carefully among these high-fat foods. More than a third of the fat in most meats is saturated. Similarly, more than half of the fat is saturated in whole milk and other high-fat dairy products, such as cheese, butter, cream, half-and-half, cream cheese, sour cream, and ice cream. The tropical oils of palm, palm kernel, and coconut, which are rarely used by consumers in the kitchen, are used heavily by food manufacturers, and are commonly found in many commercially prepared foods.

When choosing meats, milk products, and commercially prepared foods, look for those lowest in saturated fat. Labels provide a useful guide for comparing products in this regard, and Appendix H lists the saturated fat in several thousand foods.

Even with careful selections, a nutritionally adequate diet will provide some saturated fat. Zero saturated fat is not possible even when experts design menus with the mission to keep saturated fat as low as possible.[18] Because most saturated fats come from animal foods, vegetarian diets can, and usually do, deliver fewer saturated fats than mixed diets.

Limit Hydrogenated Foods

Chapter 5 explains that solid shortening and margarine are made from vegetable oil that has been hardened through hydrogenation. This process both saturates some of the unsaturated fatty acids and introduces *trans*-fatty acids. Many convenience foods contain *trans* fats, including:

- Fried foods such as french fries, chicken, and other commercially fried foods
- Commercial baked goods such as cookies, doughnuts, pastries, breads, and crackers
- Snack foods such as chips
- Imitation cheeses

To keep *trans*-fat intake low, use these foods sparingly as an occasional taste treat.

Table H5-2 summarizes which foods provide which fats. Substituting unsaturated fats for saturated fats at each meal and snack can help protect against heart disease. Figure H5-2 compares two meals and shows how such substitutions can lower saturated fat and raise unsaturated fat—even when total fat and kcalories remain unchanged.

FIGURE H5-1 **Potential Relationships among Dietary Saturated Fatty Acids, LDL Cholesterol, and Heart Disease Risk**

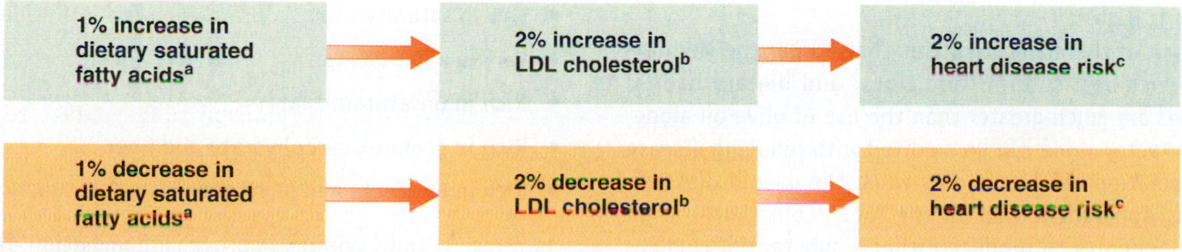

[a]Percentage of change in total dietary energy from saturated fatty acids.
[b]Percentage of change in blood LDL cholesterol.
[c]Percentage of change in an individual's risk of heart disease; the percentage of change in risk may increase when blood lipid changes are sustained over time.
SOURCE: *Third Report of the National Cholesterol Education Program (NCEP) Expert Panel on Detection, Evaluation, and Treatment of High Blood Cholesterol in Adults (Adult Treatment Panel III)*, NIH publication no. 02-5215 (Bethesda, Md.: National Heart, Lung, and Blood Institute, 2002), pp. V-8 and II-4.

TABLE H5-2 Major Sources of Various Fatty Acids

Healthful Fatty Acids		
Monounsaturated	*Omega-6 polyunsaturated*	*Omega-3 polyunsaturated*
Avocado	Margarine (nonhydrogenated)	Fatty fish (herring, mackerel, salmon, tuna)
Oils (canola, olive, peanut, sesame)	Oils (corn, cottonseed, safflower, soybean)	Flaxseed
Nuts (almonds, cashews, filberts, hazelnuts, macadamia nuts, peanuts, pecans, pistachios)	Nuts (pine nuts, walnuts)	Nuts (walnuts)
Olives	Mayonnaise	
Peanut butter	Salad dressing	
Seeds (sesame)	Seeds (pumpkin, sunflower)	

Harmful Fatty Acids	
Saturated	*Trans*
Bacon	Fried foods (hydrogenated shortening)
Butter	Margarine (hydrogenated or partially hydrogenated)
Chocolate	Nondairy creamers
Coconut	Many fast foods
Cream cheese	Shortening
Cream, half-and-half	Commercial baked goods (including doughnuts, cakes, cookies)
Lard	Many snackfoods (including microwave popcorn, chips, crackers)
Meat	
Milk and milk products (whole)	
Oils (coconut, palm, palm kernel)	
Shortening	
Sour cream	

NOTE: Keep in mind that foods contain a mixture of fatty acids.

The Mediterranean Diet

The links between good health and traditional Mediterranean diets of the mid-1900s were introduced earlier with regard to olive oil. For people who eat these diets, the incidence of heart disease, some cancers, and other chronic inflammatory diseases is low, and life expectancy is high.[19]

Although each of the many countries that border the Mediterranean Sea has its own culture, traditions, and dietary habits, their similarities are much greater than the use of olive oil alone. In fact, no one factor alone can be credited with reducing disease risks—the association holds true only when the overall diet pattern is present. Apparently, each of the foods contributes small benefits that harmonize to produce either a substantial cumulative or a synergistic effect.

The Mediterranean diet features fresh, whole foods.[20] The people select crusty breads, whole grains, potatoes, and pastas; a variety of vegetables (including wild greens) and legumes; feta and mozzarella cheeses and yogurt; nuts; and fruits (especially grapes and figs). They eat some fish, other seafood, poultry, a few eggs, and little meat. Along with olives and olive oil, their principal sources of fat are nuts and fish; they rarely use butter or encounter hydrogenated fats. Consequently, traditional Mediterranean diets are:

- Low in saturated fat
- Very low in *trans* fat
- Rich in unsaturated fat
- Rich in complex carbohydrate and fiber
- Rich in nutrients and phytochemicals that support good health

As a result, lipid profiles improve, inflammation diminishes, and the risk of heart disease declines.[21]

People following the traditional Mediterranean diet can receive as much as 40 percent of a day's kcalories from fat, but their limited consumption of dairy products and meats provides less than

FIGURE H5-2 Two Meals Compared: Replacing Saturated Fat with Unsaturated Fat

Examples of ways to replace saturated fats with unsaturated fats include sautéing vegetables in olive oil instead of butter, garnishing salads with avocado and sunflower seeds instead of bacon and blue cheese, and eating salmon instead of steak. Each of these meals provides roughly the same number of kcalories and grams of fat, but the one on the left has almost four times as much saturated fat and only half as many omega-3 fatty acids.

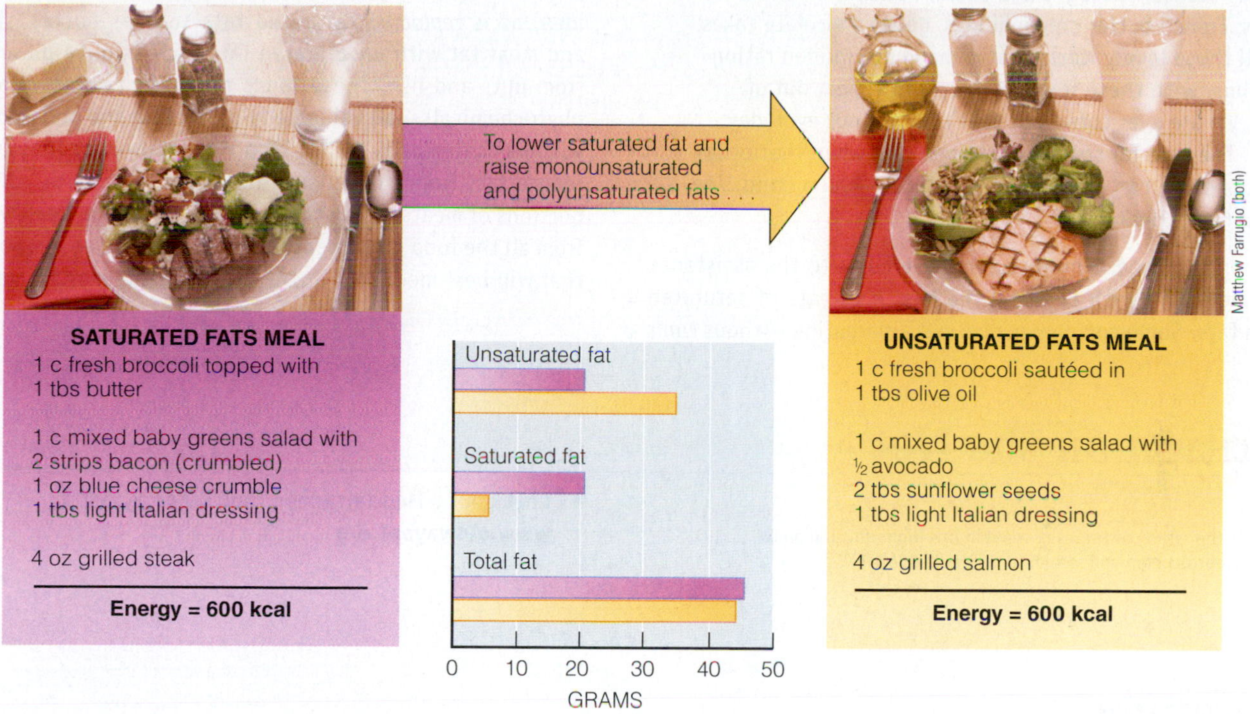

10 percent from saturated fats. In addition, because the animals in the Mediterranean region graze, the meat, dairy products, and eggs are richer in omega-3 fatty acids than those from animals fed grain. Other foods typical of the Mediterranean, such as wild plants and snails, provide omega-3 fatty acids as well. All in all, the traditional Mediterranean diet has gained a reputation for its health benefits as well as its delicious flavors, but beware of the typical Mediterranean-style cuisine available in U.S. restaurants. It has been adjusted to popular tastes, meaning that it is often much higher in saturated fats and meats—and much lower in the potentially beneficial constituents—than the traditional fare. Unfortunately, it appears that people in the Mediterranean region who are replacing some of their traditional dietary habits with those of the United States are losing the health benefits previously enjoyed.[22]

Conclusion

Are some fats "good," and others "bad" from the body's point of view? The saturated and *trans* fats indeed seem mostly bad for the health of the heart. Aside from providing energy, which unsaturated fats can do equally well, saturated and *trans* fats bring no indispensable benefits to the body. Furthermore, no harm can come from consuming diets low in them. Still, foods rich in these fats are often delicious, giving them a special place in the diet.

In contrast, the unsaturated fats are mostly good for the health of the heart when consumed in moderation. To date, their one proven fault seems to be that they, like all fats, provide abundant energy to the body and so may promote obesity if they drive kcalorie intakes higher than energy needs.[23] Obesity, in turn, often begets many body ills, as Chapter 8 makes clear.

When judging foods by their fatty acids, keep in mind that the fat in foods is a mixture of "good" and "bad," providing both saturated and unsaturated fatty acids. Even predominantly monounsaturated olive oil delivers some saturated fat. Consequently, even when a person chooses foods with mostly unsaturated fats, saturated fat can still add up if total fat is high. For this reason, fat must be kept below 35 percent of total kcalories if the diet is to be moderate in saturated fat. Even experts run into difficulty when attempting to create nutritious diets from a variety of foods that are low in saturated fats when kcalories from fat exceed 35 percent of the total.[24]

Does this mean that you must forever go without favorite cheeses, ice cream cones, or grilled steak? The famous chef Julia Child made this point about moderation:

An imaginary shelf labeled INDULGENCES is a good idea. It contains the best butter, jumbo-size eggs, heavy cream, marbled steaks, sausages and pâtés, hollandaise and butter sauces, French butter-cream fillings, gooey chocolate cakes, and all those lovely items that demand disciplined rationing. Thus, with these items high up and almost out of reach, we are ever conscious that they are not everyday foods. They are for special occasions, and when that occasion comes we can enjoy every mouthful.—Julia Child, *The Way to Cook,* 1989

Additionally, food manufacturers have come to the assistance of consumers who wish to avoid the health threats of saturated and *trans* fats. Some companies now make margarine without *trans* fats, and many snack manufacturers have reduced the saturated and *trans* fats in their products and offer snack foods in 100-kcalorie packages. Other companies are following as consumers respond favorably.

Adopting some of the Mediterranean eating habits may serve those who enjoy a little more fat in the diet. Including vegetables, fruits, and legumes as part of a balanced daily diet is a good idea, as is *replacing* saturated fats such as butter, shortening, and meat fat with unsaturated fats such as olive oil and the oils from nuts and fish. These foods provide vitamins, minerals, and phytochemicals—all valuable in protecting the body's health. The authors of this book do not stop there, however. They urge you to reduce fats from convenience foods and fast foods; choose small portions of meats, fish, and poultry; and include fresh whole foods from all the food groups each day. Take care to select portion sizes that will best meet your energy needs. Also, exercise daily.

Nutrition on the Net

For further study of topics covered in this highlight, log on to **www.cengagebrain.com** and search for ISBN 084006845X.

- Check out a Mediterranean food guide pyramid: **www.oldwayspt.org**

References

1. *Third Report of the National Cholesterol Education Program (NCEP) Expert Panel on Detection, Evaluation, and Treatment of High Blood Cholesterol in Adults (Adult Treatment Panel III),* publication NIH no. 02-5215 (Bethesda, Md.: National Heart, Lung, and Blood Institute, 2002); Committee on Dietary Reference Intakes, *Dietary Reference Intakes for Energy, Carbohydrate, Fiber, Fat, Fatty Acids, Cholesterol, Protein, and Amino Acids* (Washington, D.C.: National Academies Press, 2005).
2. Committee on Dietary Reference Intakes, 2005, p. 769.
3. American Heart Association Scientific statement: Diet and lifestyle recommendations revision 2006, *Circulation* 114 (2006): 82–96; *Third Report of the National Cholesterol Education Program (NCEP) Expert Panel on Detection, Evaluation, and Treatment of High Blood Cholesterol in Adults (Adult Treatment Panel III),* 2002; Committee on Dietary Reference Intakes, *Dietary Reference Intakes for Energy, Carbohydrate, Fiber, Fat, Fatty Acids, Cholesterol, Protein, and Amino Acids* (Washington, D.C.: National Academies Press, 2002, 2005).
4. M. A. Carluccio and coauthors, Vasculoprotective potential of olive oil components, *Molecular Nutrition and Food Research* 51 (2007): 1225–1234.
5. A. Keys, *Seven Countries: A Multivariate Analysis of Death and Coronary Heart Disease* (Cambridge: Harvard University Press, 1980).
6. M. I. Covas and coauthors, The effect of polyphenols in olive oil on the heart disease risk factors, *Annals of Internal Medicine* 145 (2006): 333–341.
7. M. Fitó, R. de la Torre, and M. I. Covas, Olive oil and oxidative stress, *Molecular Nutrition and Food Research* 51 (2007): 1215–1224; F. Visioli and coauthors, Virgin olive oil study (VOLOS): Vasoprotective potential of extra virgin olive oil in mildly dyslipidemic patients, *European Journal of Nutrition* 44 (2005): 121–127.
8. J. López-Miranda, Monounsaturated fat and cardiovascular risk, *Nutrition Reviews* 64 (2006): S2-S12.
9. J. Ruano and coauthors, Intake of phenol-rich virgin olive oil improves the postprandial prothrombotic profile in hypercholesterolemic patients, *American Journal of Clinical Nutrition* 86 (2007): 341–346; M. Covas and coauthors, The effect of polyphenols in olive oil on heart disease risk factors, *Annals of Internal Medicine* 145 (2006): 333–341.
10. B. M. Rasmussen and coauthors, Effects of dietary saturated, monounsaturated, and n-3 fatty acids on blood pressure in healthy subjects, *American Journal of Clinical Nutrition* 83 (2006): 221–226.
11. F. Pérez-Jiménez and coauthors, The influence of olive oil on human health: Not a question of fat alone, *Molecular Nutrition and Food Research* 51 (2007): 1199–1208.
12. A. E. Griel and P. M. Kris-Etherton, Tree nuts and the lipid profile: A review of clinical studies, *British Journal of Nutrition* 96 (2006): S68–S78; J. H. Kelly and J. Sabate, Nuts and coronary heart disease: An epidemiological perspective, *British Journal of Nutrition* 96 (2006): S61–S67.
13. P. E. Milbury and coauthors, Determination of flavonoids and phenolics and their distribution in almonds, *Journal of Agricultural and Food Chemistry* 54 (2006): 5027–5033.
14. A. Bersamin and coauthors, Westernizing diets influence fat intake, red blood cell fatty acid composition, and health in remote Alaskan native communities in the Center for Alaska Native Health Study, *Journal of the American Dietetic Association* 108 (2008): 266–273.

15. J. L. Breslow, n-3 Fatty acids and cardiovascular disease, *American Journal of Clinical Nutrition* 83 (2006): 1477S–1482S.
16. C. Chrysohoou and coauthors, Long-term fish consumption is associated with protection against arrhythmia in healthy persons in a Mediterranean region—The ATTICA study, *American Journal of Clinical Nutrition* 85 (2007): 1385–1391.
17. *Third Report of the National Cholesterol Education Program (NCEP) Expert Panel on Detection, Evaluation, and Treatment of High Blood Cholesterol in Adults (Adult Treatment Panel III)*, 2002, pp. v-8.
18. Committee on Dietary Reference Intakes, 2005, p. 835.
19. M. De Lorgeril, Essential polyunsaturated fatty acids, inflammation, atherosclerosis and cardiovascular diseases, *Subcellular Biochemistry* 42 (2007): 283–297; D. Lairon, Intervention studies on Mediterranean diet and cardiovascular risk, *Molecular Nutrition and Food Research* 51 (2007): 1209–1214; L. Serra-Majem, B. Roman, and R. Estruch, Scientific evidence of interventions using the Mediterranean diet: A systematic review, *Nutrition Reviews* 64 (2006): S27–S47; C. Pitsavos and coauthors, Adherence to the Mediterranean diet is associated with total antioxidant capacity in healthy adults: The ATTICA study, *American Journal of Clinical Nutrition* 82 (2005): 694–699; M. Meydani, A Mediterranean-style diet and metabolic syndrome, *Nutrition Reviews* 63 (2005): 312–314.
20. J. M. Ordovas, J. Kaput, and D. Corella, Nutrition in the genomics era: Cardiovascular disease risk and the Mediterranean diet, *Molecular Nutrition and Food Research* 51 (2007): 1293–1299.
21. M. Fitó and coauthors, Effect of a traditional Mediterranean diet on lipoprotein oxidation, *Archives of Internal Medicine* 167 (2007): 1195–1203; K. Esposito, M. Ciotola, and D. Giugliano, Mediterranean diet, endothelial function and vascular inflammatory markers, *Public Health Nutrition* 9 (2006): 1073–1076.
22. P. A. Gilbert and S. Khokhar, Changing dietary habits of ethnic groups in Europe and implications for health, *Nutrition Reviews* 66 (2008): 203–215; F. Sofi and coauthors, Dietary habits, lifestyle, and cardiovascular risk factors in a clinically healthy Italian population: The "Florence" diet is not Mediterranean, *European Journal of Clinical Nutrition* 59 (2005): 584–591.
23. Committee on Dietary Reference Intakes, 2005, pp. 796–797.
24. Committee on Dietary Reference Intakes, 2005, pp. 799–802.

CHAPTER

6

Gelinshu/Shutterstock.com

Nutrition in Your Life

The versatility of proteins in the body is impressive. They help your muscles to contract, your blood to clot, and your eyes to see. They keep you alive and well by facilitating chemical reactions and defending against infections. Without them, your bones, skin, and hair would have no structure. No wonder they were named *proteins,* meaning "of prime importance." Does that mean proteins deserve top billing in your diet as well? Are the best sources of protein beef, beans, or broccoli? Learn which foods will supply you with enough, but not too much, high-quality protein.

Throughout this chapter, the CourseMate icon indicates an opportunity for online self-study, linking you to activities to increase your understanding of chapter concepts. **www.cengagebrain.com** (search for ISBN 084006845X)

Protein: Amino Acids

CHAPTER OUTLINE

A few misconceptions surround the roles of protein in the body and the importance of protein in the diet. For example, people who associate meat with protein and protein with strength may eat steak to build muscles. Their thinking is only partly correct, however. Protein is a vital structural and working substance in all cells—not just muscle cells. To build strength, muscles cells need physical activity and all the nutrients—not just protein. Furthermore, protein is found in milk, eggs, legumes, and many grains and vegetables—not just meat. By overvaluing protein and overemphasizing meat in the diet, a person may mistakenly crowd out other, equally important nutrients and foods. As this chapter describes the various roles of protein in the body and food sources in the diet, keep in mind that protein is one of many nutrients needed to maintain good health.

The Chemist's View of Proteins

Chemically, **proteins** contain the same atoms as carbohydrates and lipids—carbon (C), hydrogen (H), and oxygen (O)—but proteins also contain nitrogen (N) atoms. These nitrogen atoms give the name *amino* (nitrogen containing) to the amino acids—the links in the chains of proteins.

Amino Acids All **amino acids** have the same basic structure—a central carbon (C) atom with a hydrogen atom (H), an amino group (NH_2), and an acid group (COOH) attached to it. However, carbon atoms need to form four bonds, ♦ so a fourth attachment is necessary. This fourth site distinguishes each amino acid from the others. Attached to the central carbon at the fourth bond is a distinct atom, or group of atoms, known as the *side group* or *side chain* (see Figure 6-1 on p. 174).

Unique Side Groups The side groups on the central carbon vary from one amino acid to the next, making proteins more complex than either carbohydrates or lipids. A polysaccharide (starch, for example) may be several thousand units long, but each unit is a glucose molecule just like all the others. A protein, on the

♦ Reminder:
- H forms one bond.
- O forms two bonds.
- N forms three bonds.
- C forms four bonds.

proteins: compounds composed of carbon, hydrogen, oxygen, and nitrogen atoms, arranged into amino acids linked in a chain. Some amino acids also contain sulfur atoms.

amino (a-MEEN-oh) **acids:** building blocks of proteins. Each contains an amino group, an acid group, a hydrogen atom, and a distinctive side group, all attached to a central carbon atom.

- **amino** = containing nitrogen

FIGURE 6-1 Amino Acid Structure

All amino acids have a central carbon with an amino group (NH_2), an acid group (COOH), a hydrogen (H), and a side group attached. The side group is a unique chemical structure that differentiates one amino acid from another.

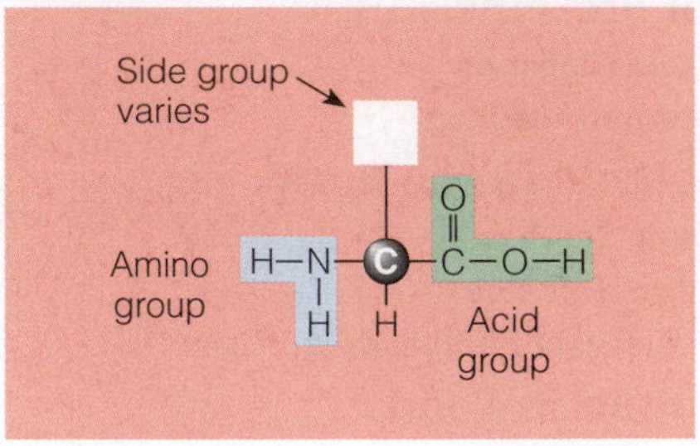

TABLE 6-1 Amino Acids

Proteins are made up of about 20 common amino acids. The first column lists the essential amino acids for human beings (those the body cannot make—that must be provided in the diet). The second column lists the nonessential amino acids. In special cases, some nonessential amino acids may become conditionally essential (see the text). In a newborn, for example, only five amino acids are truly nonessential; the other nonessential amino acids are conditionally essential until the metabolic pathways are developed enough to make those amino acids in adequate amounts.

Essential Amino Acids		Nonessential Amino Acids	
Histidine	(HISS-tuh-deen)	Alanine	(AL-ah-neen)
Isoleucine	(eye-so-LOO-seen)	Arginine	(ARJ-ih-neen)
Leucine	(LOO-seen)	Asparagine	(ah-SPAR-ah-geen)
Lysine	(LYE-seen)	Aspartic acid	(ah-SPAR-tic acid)
Methionine	(meh-THIGH-oh-neen)	Cysteine	(SIS-teh-een)
Phenylalanine	(fen-il-AL-ah-neen)	Glutamic acid	(GLU-tam-ic acid)
Threonine	(THREE-oh-neen)	Glutamine	(GLU-tah-meen)
Tryptophan	(TRIP-toe-fan, TRIP-toe-fane)	Glycine	(GLY-seen)
Valine	(VAY-leen)	Proline	(PRO-leen)
		Serine	(SEER-een)
		Tyrosine	(TIE-roe-seen)

other hand, is made up of about 20 different amino acids, each with a different side group. Table 6-1 lists the amino acids most common in proteins.*

The simplest amino acid, glycine, has a hydrogen atom as its side group. A slightly more complex amino acid, alanine, has an extra carbon with three hydrogen atoms. Other amino acids have more complex side groups (see Figure 6-2 for examples). Thus, although all amino acids share a common structure, they differ in size, shape, electrical charge, and other characteristics because of differences in these side groups.

Nonessential Amino Acids More than half of the amino acids are *nonessential*, meaning that the body can synthesize them for itself. Proteins in foods usually deliver these amino acids, but it is not essential that they do so. The body can make all **nonessential amino acids**, given nitrogen to form the amino group and fragments from carbohydrate or fat to form the rest of the structure.

*Besides the 20 common amino acids, which can all be components of proteins, others do not occur in proteins but can be found individually (for example, taurine and ornithine). Some amino acids occur in related forms (for example, proline can acquire an OH group to become hydroxyproline).

FIGURE 6-2 Examples of Amino Acids

Note that all amino acids have a common chemical structure but that each has a different side group. Appendix C presents the chemical structures of the 20 amino acids most common in proteins.

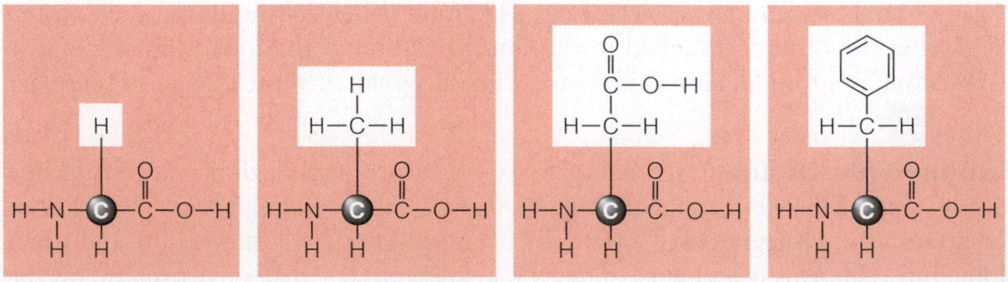

Glycine — Alanine — Aspartic acid — Phenylalanine

nonessential amino acids: amino acids that the body can synthesize (see Table 6-1).

FIGURE 6-3 **Condensation of Two Amino Acids to Form a Dipeptide**

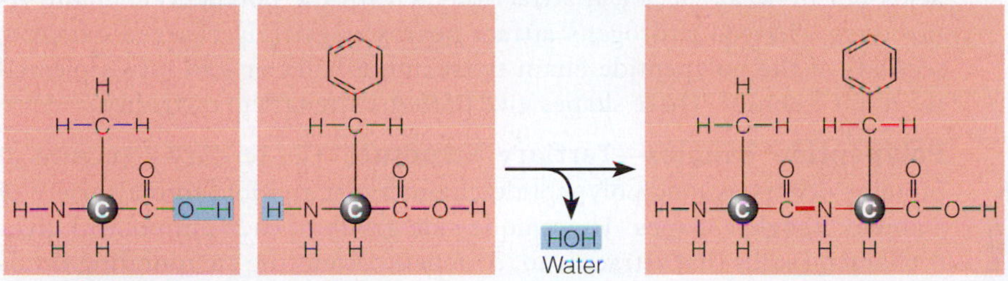

Amino acid
Amino acid
HOH
Water
Dipeptide
An OH group from the acid end of one amino acid and an H atom from the amino group of another join to form a molecule of water.
A peptide bond (highlighted in red) forms between the two amino acids, creating a dipeptide.

Essential Amino Acids There are nine amino acids that the human body either cannot make at all or cannot make in sufficient quantity to meet its needs. These nine amino acids must be supplied by the diet; they are *essential.* ♦ The first column in Table 6-1 presents the **essential amino acids.**

♦ Some researchers refer to essential amino acids as **indispensable** and to nonessential amino acids as **dispensable.**

Conditionally Essential Amino Acids Sometimes a nonessential amino acid becomes essential under special circumstances. For example, the body normally uses the essential amino acid phenylalanine to make tyrosine (a nonessential amino acid). But if the diet fails to supply enough phenylalanine, or if the body cannot make the conversion for some reason (as happens in the inherited disease phenylketonuria), then tyrosine becomes a **conditionally essential amino acid.**

Proteins

Cells link amino acids end-to-end in a variety of sequences to form thousands of different proteins. A **peptide bond** unites each amino acid to the next.

Amino Acid Chains Condensation reactions connect amino acids, just as they combine two monosaccharides to form a disaccharide and three fatty acids with a glycerol to form a triglyceride. Two amino acids bonded together form a **dipeptide** (see Figure 6-3). By another such reaction, a third amino acid can be added to the chain to form a **tripeptide.** As additional amino acids join the chain, a **polypeptide** is formed. Most proteins are a few dozen to several hundred amino acids long. Figure 6-4 illustrates the protein insulin.

FIGURE 6-4 **Amino Acid Sequence of Human Insulin**

Human insulin is a relatively small protein that consists of 51 amino acids in two short polypeptide chains. (For amino acid abbreviations, see Appendix C.) Two bridges link the two chains. A third bridge spans a section within the short chain. Known as *disulfide bridges,* these links always involve the amino acid cysteine (Cys), whose side group contains sulfur (S). Cysteines connect to each other when bonds form between these side groups.

Amino Acid Sequence—Primary Structure The primary structure of a protein is determined by the sequence of amino acids. If a person could walk along a carbohydrate molecule like starch, the first stepping stone would be a glucose. The next stepping stone would also be a glucose, and it would be followed by a glucose, and yet another glucose. But if a person were to walk along a polypeptide chain, each stepping stone would be one of 20 different amino acids. The first stepping stone might be the amino acid methionine. The second might be an alanine. The third might be a glycine, the fourth a tryptophan, and so on. Walking along another polypeptide path, a person might step on a phenylalanine, then a valine, then a glutamine. In other words, amino acid sequences within proteins vary.

The amino acids can act somewhat like the letters in an alphabet. If you had only the letter G, all you could write would be a string of Gs: G–G–G–G–G–G–G. But with 20 different letters available, you can create poems, songs, and novels. Similarly, the 20 amino acids can be linked together in a variety of sequences—even more than are possible for letters in a word or words in a sentence. Thus the variety of possible sequences for polypeptide chains is tremendous.

essential amino acids: amino acids that the body cannot synthesize in amounts sufficient to meet physiological needs (see Table 6-1).

conditionally essential amino acid: an amino acid that is normally nonessential, but must be supplied by the diet in special circumstances when the need for it exceeds the body's ability to produce it.

peptide bond: a bond that connects the acid end of one amino acid with the amino end of another, forming a link in a protein chain.

dipeptide (dye-PEP-tide): two amino acids bonded together.
- **di** = two
- **peptide** = amino acid

tripeptide: three amino acids bonded together.
- **tri** = three

polypeptide: many (ten or more) amino acids bonded together.
- **poly** = many

FIGURE 6-5 The Structure of Hemoglobin

Four highly folded polypeptide chains form the globular hemoglobin protein.

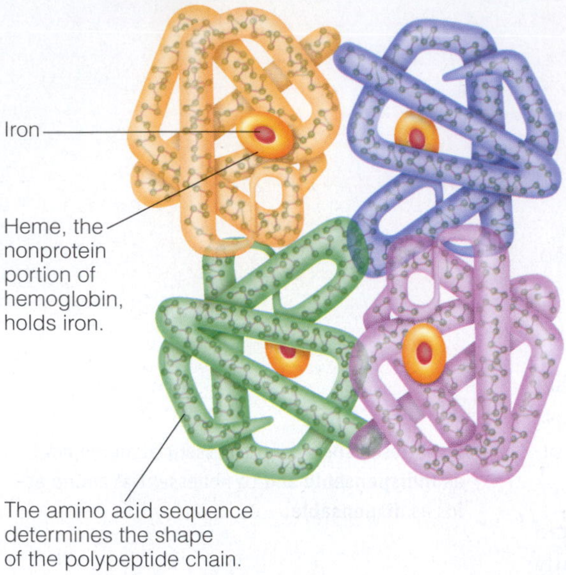
Iron
Heme, the nonprotein portion of hemoglobin, holds iron.
The amino acid sequence determines the shape of the polypeptide chain.

Polypeptide Shapes—Secondary Structure The secondary structure of proteins is determined not by chemical bonds as between the amino acids but by weak electrical attractions within the polypeptide chain. As positively charged hydrogens attract nearby negatively charged oxygens, sections of the polypeptide chain twist into a helix or fold into a pleated sheet, for example. These shapes give proteins strength and rigidity.

Polypeptide Tangles—Tertiary Structure The tertiary structure of proteins occurs as long polypeptide chains twist and fold into a variety of complex, tangled shapes. The unique side group of each amino acid gives it characteristics that attract it to, or repel it from, the surrounding fluids and other amino acids. Some amino acid side groups are attracted to water molecules; they are *hydrophilic*. Other side groups are repelled by water; they are *hydrophobic*. As amino acids are strung together to make a polypeptide, the chain folds so that its hydrophilic side groups are on the outer surface near water; the hydrophobic groups tuck themselves inside, away from water. Similarly, the disulfide bridges in insulin (see Figure 6-4) determine its tertiary structure. The extraordinary and unique shapes of proteins enable them to perform their various tasks in the body. Some form globular or spherical structures that can carry and store materials within them, and some, such as those of tendons, form linear structures that are more than ten times as long as they are wide. The intricate shape a protein finally assumes gives it maximum stability.

Multiple Polypeptide Interactions—Quaternary Structures Some polypeptides are functioning proteins just as they are; others need to associate with other polypeptides to form larger working complexes. The quaternary structure of proteins involves the interactions between two or more polypeptides. One molecule of **hemoglobin**—the large, globular protein molecule that, by the billions, packs the red blood cells and carries oxygen—is made of four associated polypeptide chains, each holding the mineral iron (see Figure 6-5).

© Matthew Farruggio

Cooking an egg denatures its proteins.

Protein Denaturation When proteins are subjected to heat, acid, or other conditions that disturb their stability, they undergo **denaturation**—that is, they uncoil and lose their shapes and, consequently, also lose their ability to function. Past a certain point, denaturation is irreversible. Familiar examples of denaturation include the hardening of an egg when it is cooked, the curdling of milk when acid is added, and the stiffening of egg whites when they are whipped. In the body, proteins are denatured when they are exposed to stomach acid.

IN SUMMARY **Chemically speaking, proteins are more complex than carbohydrates or lipids; they are made of some 20 different amino acids, 9 of which the body cannot make (the essential amino acids). Each amino acid contains an amino group, an acid group, a hydrogen atom, and a distinctive side group, all attached to a central carbon atom. Cells link amino acids together in a series of condensation reactions to create proteins. The distinctive sequence of amino acids in each protein determines its unique shape and function.**

Digestion and Absorption of Proteins

Proteins in foods do not become body proteins directly. Instead, they supply the amino acids from which the body makes its own proteins. When a person eats foods containing protein, enzymes break the long polypeptide strands into shorter strands, the short strands into tripeptides and dipeptides, and, finally, the tripeptides and dipeptides into amino acids.

Protein Digestion

Figure 6-6 illustrates the digestion of protein through the GI tract. Proteins are crushed and moistened in the mouth, but the real action begins in the stomach.

hemoglobin (HE-moh-GLO-bin): the globular protein of the red blood cells that carries oxygen from the lungs to the cells throughout the body.

- **hemo** = blood
- **globin** = globular protein

denaturation (dee-NAY-chur-AY-shun): the change in a protein's shape and consequent loss of its function brought about by heat, agitation, acid, base, alcohol, heavy metals, or other agents.

FIGURE 6-6 Protein Digestion in the GI Tract

Animated! figure
www.cengagebrain.com
(search for ISBN 084006845X)

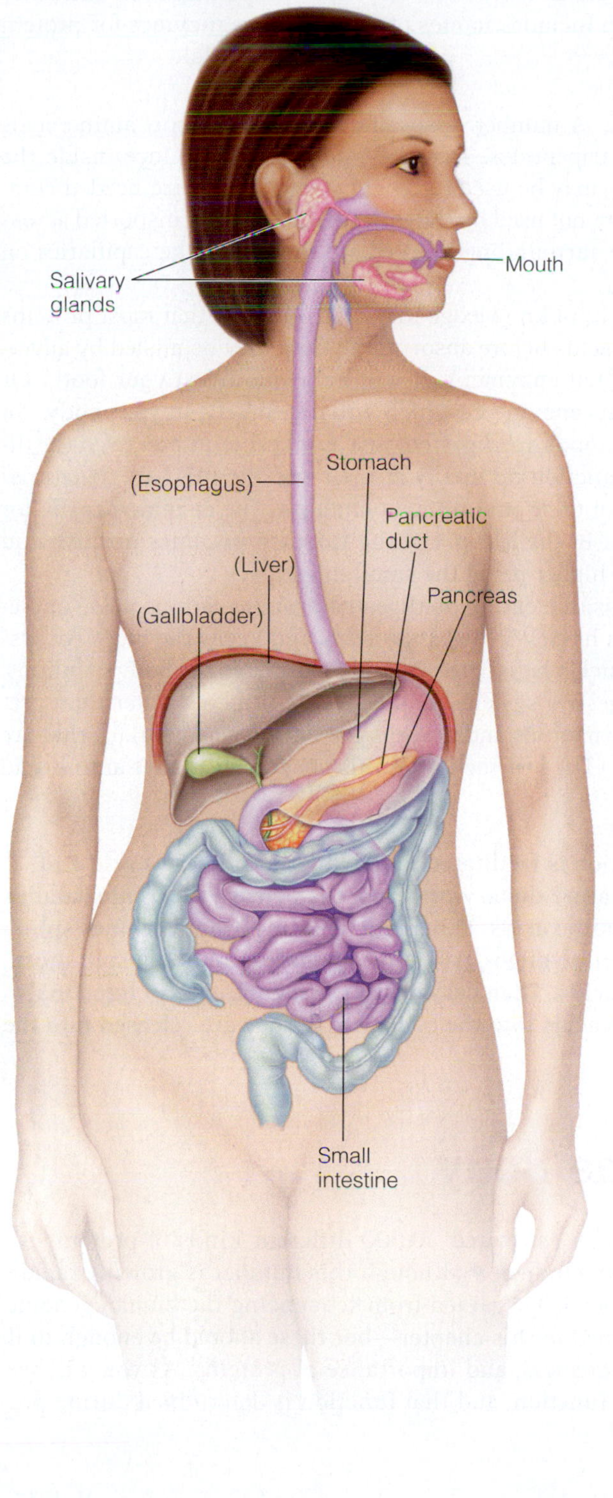

PROTEIN	HYDROCHLORIC ACID AND THE DIGESTIVE ENZYMES
Mouth and salivary glands Chewing and crushing moisten protein-rich foods and mix them with saliva to be swallowed	
Stomach Hydrochloric acid (HCl) uncoils protein strands and activates stomach enzymes: Protein —(Pepsin, HCl)→ Smaller polypeptides	**In the stomach:** **Hydrochloric acid (HCl)** • Denatures protein structure • Activates pepsinogen to pepsin **Pepsin** • Cleaves proteins to smaller polypeptides and some free amino acids • Inhibits pepsinogen synthesis
Small intestine and pancreas Pancreatic and small intestinal enzymes split polypeptides further: Polypeptides —(Pancreatic and intestinal proteases)→ Tripeptides, dipeptides, amino acids Then enzymes on the surface of the small intestinal cells hydrolyze these peptides and the cells absorb them: Peptides —(Intestinal tripeptidases and dipeptidases)→ Amino acids (absorbed)	**In the small intestine:** **Enteropeptidase**[a] • Converts pancreatic trypsinogen to trypsin **Trypsin** • Inhibits trypsinogen synthesis • Cleaves peptide bonds next to the amino acids lysine and arginine • Converts pancreatic procarboxypeptidases to carboxypeptidases • Converts pancreatic chymotrypsinogen to chymotrypsin **Chymotrypsin** • Cleaves peptide bonds next to the amino acids phenylalanine, tyrosine, tryptophan, methionine, asparagine, and histidine **Carboxypeptidases** • Cleave amino acids from the acid (carboxyl) ends of polypeptides **Elastase and collagenase** • Cleave polypeptides into smaller polypeptides and tripeptides **Intestinal tripeptidases** • Cleave tripeptides to dipeptides and amino acids **Intestinal dipeptidases** • Cleave dipeptides to amino acids **Intestinal aminopeptidases** • Cleave amino acids from the amino ends of small polypeptides (oligopeptides) [a]Enteropeptidase was formerly known as *enterokinase*.

In the Stomach The major event in the stomach is the partial breakdown (hydrolysis) of proteins. Hydrochloric acid uncoils (denatures) each protein's tangled strands so that digestive enzymes can attack the peptide bonds. The hydrochloric acid also converts the inactive form ♦ of the enzyme pepsinogen to its active form, **pepsin.** Pepsin cleaves proteins—large polypeptides—into smaller polypeptides and some amino acids.

In the Small Intestine When polypeptides enter the small intestine, several pancreatic and intestinal proteases hydrolyze them further into short peptide

♦ The inactive form of an enzyme is called a **proenzyme** or a **zymogen** (ZYE-moh-jen).

pepsin: a gastric enzyme that hydrolyzes protein. Pepsin is secreted in an inactive form, **pepsinogen,** which is activated by hydrochloric acid in the stomach.

♦ A string of four to nine amino acids is an **oligopeptide** (OL-ee-go-PEP-tide).
- **oligo** = few

chains, ♦ tripeptides, dipeptides, and amino acids. Then **peptidase** enzymes on the membrane surfaces of the intestinal cells split most of the dipeptides and tripeptides into single amino acids. Only a few peptides escape digestion and enter the blood intact. Figure 6-6 includes names of the digestive enzymes for protein and describes their actions.

Protein Absorption

A number of specific carriers transport amino acids (and some dipeptides and tripeptides) into the intestinal cells. Once inside the intestinal cells, amino acids may be used for energy or to synthesize needed compounds. Amino acids that are not used by the intestinal cells are transported across the cell membrane into the surrounding fluid where they enter the capillaries on their way to the liver.

Consumers lacking nutrition knowledge may fail to realize that most proteins are broken down to amino acids before absorption. They may be misled by advertisements urging them to "Eat enzyme A. It will help you digest your food." Or "Don't eat food B. It contains enzyme C, which will digest cells in your body." In reality, though, enzymes in foods are digested, just as all proteins are. Even the digestive enzymes—which function optimally at their specific pH—are denatured and digested when the pH of their environment changes. The enzyme pepsin, for example, which works best in the low pH of the stomach becomes inactive and digested when it enters the higher pH of the small intestine.

Another misconception is that eating predigested proteins (amino acid supplements) saves the body from having to digest proteins and keeps the digestive system from "overworking." Such a belief grossly underestimates the body's abilities. As a matter of fact, the digestive system handles whole proteins *better* than predigested ones because it dismantles and absorbs the amino acids at rates that are optimal for the body's use. (The last section of this chapter discusses amino acid supplements further.)

IN SUMMARY **Digestion is facilitated mostly by the stomach's acid and enzymes, which first denature dietary proteins, then cleave them into smaller polypeptides and some amino acids. Pancreatic and intestinal enzymes split these polypeptides further, to oligo-, tri-, and dipeptides, and then split most of these to single amino acids. Then carriers in the membranes of intestinal cells transport the amino acids into the cells, where they are released into the bloodstream.**

Proteins in the Body

♦ The study of the body's proteins is called **proteomics.**

♦ The **human genome** is the full set of chromosomes, including all of the genes and associated DNA.

The human body contains an estimated 30,000 different kinds of proteins. Of these, about 3000 have been studied, ♦ although this number is growing rapidly with the recent surge in knowledge gained from sequencing the human genome. ♦ Only about 10 are described in this chapter—but these should be enough to illustrate the versatility, uniqueness, and importance of proteins. As you will see, each protein has a specific function, and that function is determined during protein synthesis.

Protein Synthesis

Each human being is unique because of small differences in the body's proteins. These differences are determined by the amino acid sequences of proteins, which, in turn, are determined by genes. The following paragraphs describe in words the ways cells synthesize proteins; Figure 6-7 provides a pictorial description. Protein synthesis depends on a diet that provides adequate protein and essential amino acids.

The instructions for making every protein in a person's body are transmitted by way of the genetic information received at conception. This body of knowledge, which is filed in the DNA (deoxyribonucleic acid) within the nucleus of every cell, never leaves the nucleus.

peptidase: a digestive enzyme that hydrolyzes peptide bonds. *Tripeptidases* cleave tripeptides; *dipeptidases* cleave dipeptides. *Endopeptidases* cleave peptide bonds within the chain to create smaller fragments, whereas *exopeptidases* cleave bonds at the ends to release free amino acids.
- **tri** = three
- **di** = two
- **endo** = within
- **exo** = outside

FIGURE 6-7 Protein Synthesis

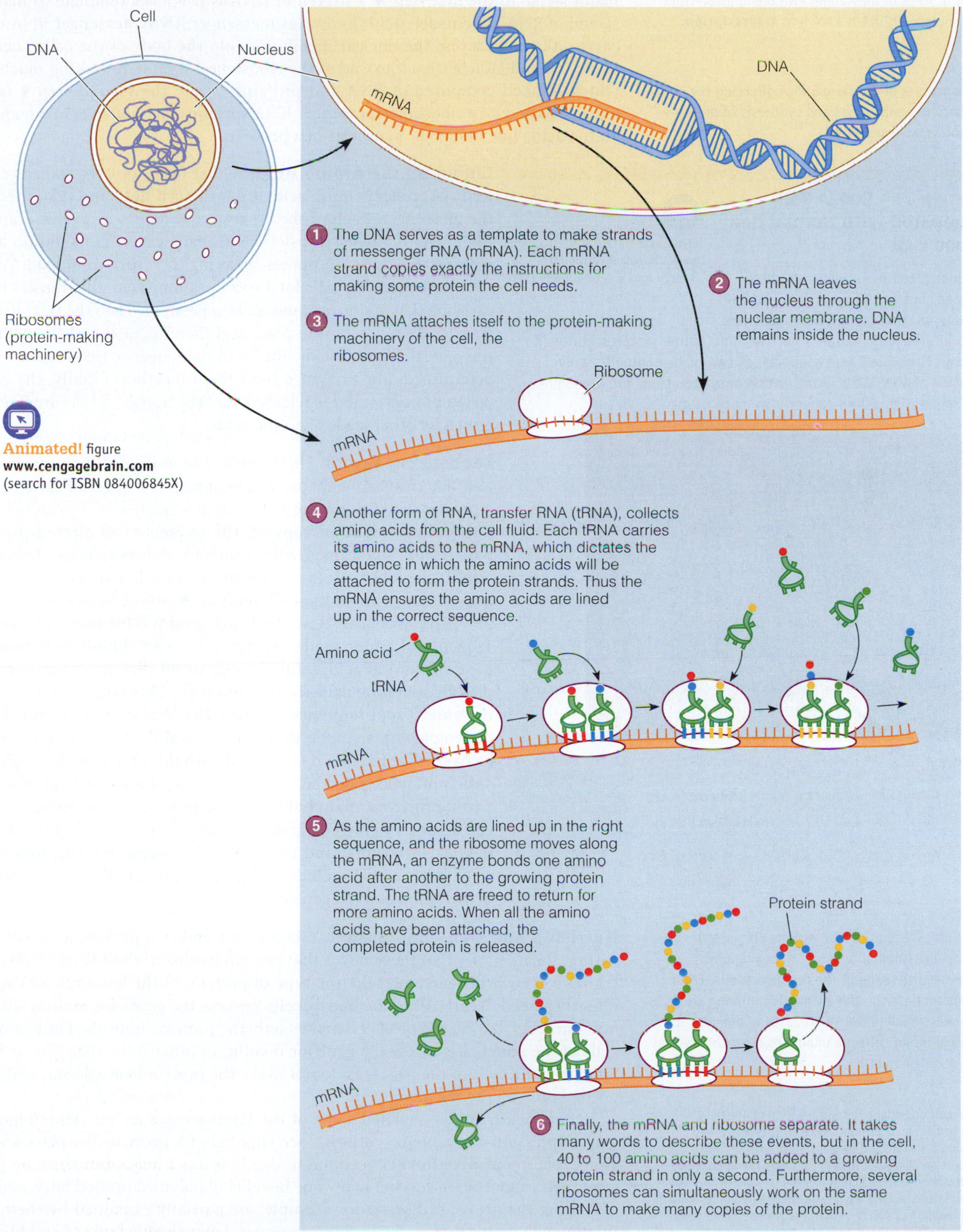

Cell
DNA
Nucleus
Ribosomes
(protein-making
machinery)
mRNA
DNA
1 The DNA serves as a template to make strands of messenger RNA (mRNA). Each mRNA strand copies exactly the instructions for making some protein the cell needs.
2 The mRNA leaves the nucleus through the nuclear membrane. DNA remains inside the nucleus.
3 The mRNA attaches itself to the protein-making machinery of the cell, the ribosomes.
Ribosome
mRNA
4 Another form of RNA, transfer RNA (tRNA), collects amino acids from the cell fluid. Each tRNA carries its amino acids to the mRNA, which dictates the sequence in which the amino acids will be attached to form the protein strands. Thus the mRNA ensures the amino acids are lined up in the correct sequence.
Amino acid
tRNA
mRNA
5 As the amino acids are lined up in the right sequence, and the ribosome moves along the mRNA, an enzyme bonds one amino acid after another to the growing protein strand. The tRNA are freed to return for more amino acids. When all the amino acids have been attached, the completed protein is released.
Protein strand
mRNA
6 Finally, the mRNA and ribosome separate. It takes many words to describe these events, but in the cell, 40 to 100 amino acids can be added to a growing protein strand in only a second. Furthermore, several ribosomes can simultaneously work on the same mRNA to make many copies of the protein.

Animated! figure
www.cengagebrain.com
(search for ISBN 084006845X)

♦ This process of messenger RNA being made from a template of DNA is known as **transcription.**

♦ This process of messenger RNA directing the sequence of amino acids and synthesis of proteins is known as **translation.**

Delivering the Instructions Transforming the information in DNA into the appropriate sequence of amino acids needed to make a specific protein requires two major steps. In the first step, ♦ a stretch of DNA is used as a template to make a strand of RNA (ribonucleic acid) known as messenger RNA. Messenger RNA then carries the code across the nuclear membrane into the body of the cell. There it seeks out and attaches itself to one of the ribosomes (a protein-making machine, which is itself composed of RNA and protein), where the second step ♦ takes place. Situated on a ribosome, messenger RNA specifies the sequence in which the amino acids line up for the synthesis of a protein.

FIGURE 6-8 Sickle Cell Compared with Normal Red Blood Cell

Animated! figure
www.cengagebrain.com
(search for ISBN 084006845X)

Normally, red blood cells are disc-shaped, but in the inherited disorder sickle-cell anemia, red blood cells are sickle- or crescent-shaped. This alteration in shape occurs because valine replaces glutamic acid in the amino acid sequence of two of hemoglobin's polypeptide chains. As a result of this one alteration, the hemoglobin has a diminished capacity to carry oxygen.

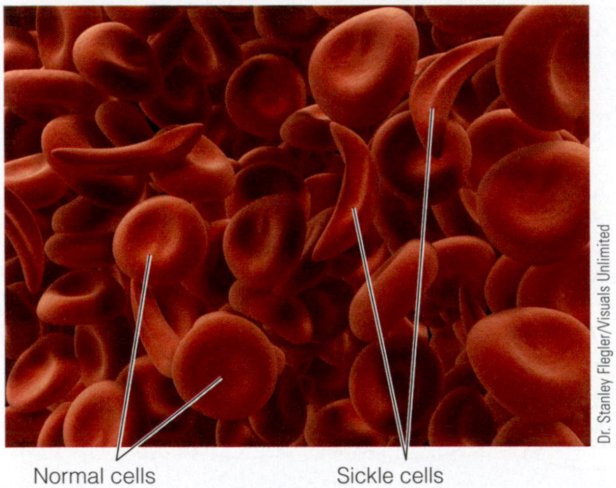

Amino acid sequence of normal hemoglobin:
Val – His – Leu – Thr – Pro – **Glu** – Glu

Amino acid sequence of sickle-cell hemoglobin:
Val – His – Leu – Thr – Pro – **Val** – Glu

Lining Up the Amino Acids Other forms of RNA, called transfer RNA, collect amino acids from the cell fluid and take them to the messenger. Each of the 20 amino acids has a specific transfer RNA. Thousands of transfer RNAs, each carrying its amino acid, cluster around the ribosomes, awaiting their turn to unload. When the messenger's list calls for a specific amino acid, the transfer RNA carrying that amino acid moves into position. Then the next loaded transfer RNA moves into place and then the next and the next. In this way, the amino acids line up in the sequence that is genetically determined, and enzymes bind them together. Finally, the completed protein strand is released, and the transfer RNAs are freed to return for other loads of amino acids.

Sequencing Errors The sequence of amino acids in each protein determines its shape, which supports a specific function. If a genetic error alters the amino acid sequence of a protein, or if a mistake is made in copying the sequence, an altered protein will result, sometimes with dramatic consequences. The protein hemoglobin offers one example of such a genetic variation. In a person with **sickle-cell anemia,** ♦ two of hemoglobin's four polypeptide chains (described earlier on p. 176) have the normal sequence of amino acids, but the other two chains do not—they have the amino acid valine in a position that is normally occupied by glutamic acid (see Figure 6-8). This single alteration in the amino acid sequence changes the characteristics and shape of hemoglobin so much that it loses its ability to carry oxygen effectively. The red blood cells filled with this abnormal hemoglobin stiffen into elongated sickle, or crescent, shapes instead of maintaining their normal pliable disc shape—hence the name, sickle-cell anemia. Sickle-cell anemia raises energy needs, causes many medical problems, and can be fatal.[1] Caring for children with sickle-cell anemia includes diligent attention to their water needs; dehydration can trigger a crisis.

♦ Anemia is a symptom of various diseases. In sickle-cell anemia, a defect in hemoglobin changes the shape of the red blood cells. Later chapters describe the anemias of vitamin and mineral deficiencies. In all cases of anemia, abnormal blood cells are unable to meet the body's oxygen demands.

Nutrients and Gene Expression When a cell makes a protein as described earlier, scientists say that the gene for that protein has been "expressed." Cells can regulate **gene expression** to make the type of protein, in the amounts and at the rate, they need. Nearly all of the body's cells possess the genes for making all human proteins, but each type of cell makes only the proteins it needs. For example, cells of the pancreas express the gene for insulin; in other cells, that gene is idle. Similarly, the cells of the pancreas do not make the protein hemoglobin, which is needed only by the red blood cells.

Recent research has unveiled some of the fascinating ways nutrients regulate gene expression and protein synthesis (see Highlight 6). Because diet plays an ongoing role in our lives from conception to death, it has a major influence on gene expression and disease development. The benefits of polyunsaturated fatty acids in defending against heart disease, for example, are partially explained by their role in influencing gene expression for lipid enzymes. Later chapters provide additional examples of relationships among nutrients, genes, and disease development.

sickle-cell anemia: a hereditary form of anemia characterized by abnormal sickle- or crescent-shaped red blood cells. Sickled cells interfere with oxygen transport and blood flow. Symptoms are precipitated by dehydration and insufficient oxygen (as may occur at high altitudes) and include hemolytic anemia (red blood cells burst), fever, and severe pain in the joints and abdomen.

gene expression: the process by which a cell converts the genetic code into RNA and protein.

IN SUMMARY Cells synthesize proteins according to the genetic information provided by the DNA in the nucleus of each cell. This information dictates the sequence in which amino acids are linked together to form a given protein. Sequencing errors occasionally occur, sometimes with significant consequences.

Roles of Proteins

Whenever the body is growing, repairing, or replacing tissue, proteins are involved. Sometimes their role is to facilitate or to regulate; other times it is to become part of a structure. Versatility is a key feature of proteins.

As Building Materials for Growth and Maintenance From the moment of conception, proteins form the building blocks of muscles, blood, and skin—in fact, of most body structures. For example, to build a bone or a tooth, cells first lay down a **matrix** of the protein **collagen** and then fill it with crystals of calcium, phosphorus, magnesium, fluoride, and other minerals.

Collagen also provides the material of ligaments and tendons and the strengthening "glue" between the cells of the artery walls that enables the arteries to withstand the pressure of the blood surging through them with each heartbeat. Also made of collagen are scars that knit the separated parts of torn tissues together.

Proteins are also needed for replacing dead or damaged cells. The life span of a skin cell is only about 30 days. As old skin cells are shed, new cells made largely of protein grow from underneath to replace them. Cells in the deeper skin layers synthesize new proteins to form hair and fingernails. Muscle cells make new proteins to grow larger and stronger in response to exercise.[2] Cells of the GI tract are replaced every few days. Both inside and outside, then, the body continuously deposits protein into the new cells that replace those that have been lost.

As Enzymes Some proteins act as **enzymes.** Digestive enzymes have appeared in every chapter since Chapter 3, but digestion is only one of the many processes facilitated by enzymes. Enzymes not only break down substances, but they also build substances (such as bone) ♦ and transform one substance into another (amino acids into glucose, for example). Figure 6-9 diagrams a synthesis reaction.

♦ Breaking down reactions are **catabolic,** whereas building up reactions are **anabolic.** (Chapter 7 provides more details.)

FIGURE 6-9 Enzyme Action

Each enzyme facilitates a specific chemical reaction. In this diagram, an enzyme enables two compounds to make a more complex structure, but the enzyme itself remains unchanged.

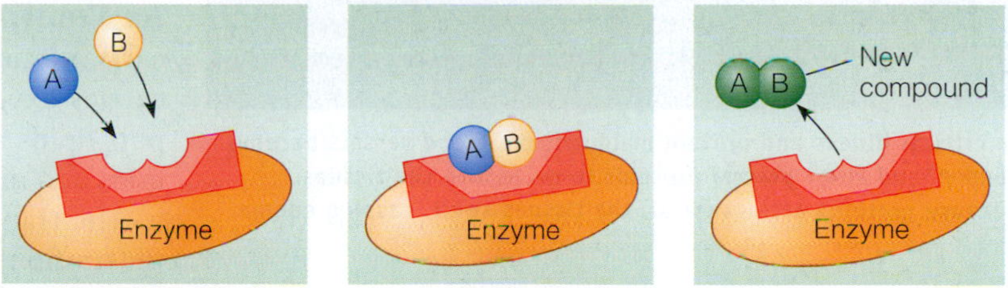

The separate compounds, A and B, are attracted to the enzyme's active site, making a reaction likely. The enzyme forms a complex with A and B. The enzyme is unchanged, but A and B have formed a new compound, AB.

An analogy may help to clarify the role of enzymes. Enzymes are comparable to the clergy and judges who make and dissolve marriages. When a minister marries two people, they become a couple, with a new bond between them. They are joined together—but the minister remains unchanged. The minister represents enzymes that synthesize large compounds from smaller ones. One minister can perform thousands of marriage ceremonies, just as one enzyme can perform billions of synthetic reactions.

Similarly, a judge who lets married couples separate may decree many divorces before retiring. The judge represents enzymes that hydrolyze larger compounds to smaller ones; for example, the digestive enzymes. The point is that, like the minister and the judge, enzymes themselves are not altered by the reactions they facilitate. They are catalysts, permitting reactions to occur more quickly and efficiently than if substances depended on chance encounters alone.

As Hormones The body's many hormones are messenger molecules, and *some* hormones are proteins. ♦ Various endocrine glands in the body release hormones in response to changes that challenge the body. The blood carries the hormones from these glands to their target tissues, where they elicit the appropriate responses to restore and maintain normal conditions.

♦ Some hormones, such as estrogen and testosterone, derive from the lipid cholesterol.

The hormone insulin provides a familiar example. After a meal, when blood glucose rises, the pancreas releases its insulin. Insulin stimulates the transport

matrix (MAY-tricks): the basic substance that gives form to a developing structure; in the body, the formative cells from which teeth and bones grow.

collagen (KOL-ah-jen): the protein from which connective tissues such as scars, tendons, ligaments, and the foundations of bones and teeth are made.

enzymes: proteins that facilitate chemical reactions without being changed in the process; protein catalysts.

TABLE 6-2 Examples of Hormones and Their Actions

Hormones	Actions
Growth hormone	Promotes growth
Insulin and glucagon	Regulate blood glucose (see Chapter 4)
Thyroxin	Regulates the body's metabolic rate (see Chapter 8)
Calcitonin and parathyroid hormone	Regulate blood calcium (see Chapter 12)
Antidiuretic hormone	Regulates fluid and electrolyte balance (see Chapter 12)

NOTE: Hormones are chemical messengers that are secreted by endocrine glands in response to altered conditions in the body. Each travels to one or more specific target tissues or organs, where it elicits a specific response. For descriptions of many hormones important in nutrition, see Appendix A.

proteins of the muscles and adipose tissue to pump glucose into the cells faster than it can leak out. After acting on the message, the cells destroy the insulin. As blood glucose falls, the pancreas slows its release of insulin. Many other proteins act as hormones, regulating a variety of actions in the body (see Table 6-2 for examples).

As Regulators of Fluid Balance Proteins help to maintain the body's **fluid balance**. Normally, proteins are found primarily within the cells and in the plasma (essentially blood without its red blood cells). Being large, proteins do not normally cross the walls of the blood vessels. During times of critical illness or protein malnutrition, however, plasma proteins leak out of the blood vessels into the tissues (between the cells). Because proteins attract water, fluid accumulates and causes swelling. Swelling due to an excess of fluid in the tissues is known as **edema**. The protein-related causes of edema include:

- Excessive protein losses caused by inflammation and critical illnesses
- Inadequate protein synthesis caused by liver disease
- Inadequate dietary intake of protein

Whatever the cause of edema, the result is the same: a diminished capacity to deliver nutrients and oxygen to the cells and to remove wastes from them. As a consequence, cells fail to function adequately.

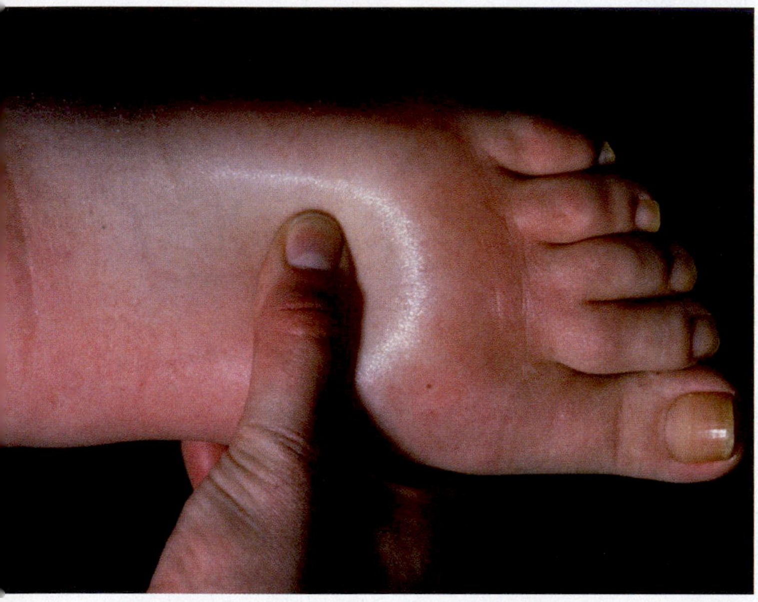

SPL/Photo Researchers, Inc.

In critical illness and protein malnutrition, blood vessels become "leaky" and allow plasma proteins to move into the tissues. Because proteins attract water, the tissues swell, causing edema.

As Acid-Base Regulators Proteins also help to maintain the balance between **acids** and **bases** within the body fluids. Normal body processes continually produce acids and bases, which the blood carries to the kidneys and lungs for excretion. The challenge is to do this without upsetting the blood's acid-base balance.

In an acid solution, hydrogen ions (H^+) abound; the more hydrogen ions, the more concentrated the acid. Proteins, which have negative charges on their surfaces, attract hydrogen ions, which have positive charges. By accepting and releasing hydrogen ions, ♦ proteins maintain the acid-base balance of the blood and body fluids.

The blood's acid-base balance is tightly controlled. The extremes of **acidosis** and **alkalosis** lead to coma and death, largely because they denature working proteins. Disturbing a protein's shape renders it useless. To give just one example, denatured hemoglobin loses its capacity to carry oxygen.

As Transporters Some proteins move about in the body fluids, carrying nutrients and other molecules. The protein hemoglobin carries oxygen from the lungs to the cells. The lipoproteins transport lipids around the body. Special transport proteins carry vitamins and minerals.

The transport of the mineral iron provides an especially good illustration of these proteins' specificity and precision. When iron is absorbed, it is captured in an intestinal cell by a protein. Before leaving the intestinal cell, iron is attached to another protein that carries it through the bloodstream to the cells. Once iron enters a cell, it is attached to a storage protein that will hold the iron until it is needed. When it is needed, iron is incorporated into proteins in the red blood cells and muscles that assist in oxygen transport and use. (Chapter 13 provides more details on how these protein carriers transport and store iron.)

Some transport proteins reside in cell membranes and act as "pumps," picking up compounds on one side of the membrane and releasing them on the other as needed. Each transport protein is specific for a certain compound or group of related compounds. Figure 6-10 illustrates how a membrane-bound transport protein helps to maintain the sodium and potassium concentrations in the fluids inside and outside cells. The balance of these two minerals is critical to nerve transmissions and muscle contractions; imbalances can cause irregular heartbeats, muscular weakness, kidney failure, and even death.

♦ Compounds that keep a solution's pH constant when acids or bases are added are called **buffers.**

fluid balance: maintenance of the proper types and amounts of fluid in each compartment of the body fluids (see also Chapter 12).

edema (eh-DEEM-uh): the swelling of body tissue caused by excessive amounts of fluid in the interstitial spaces; seen in protein deficiency (among other conditions).

acids: compounds that release hydrogen ions in a solution.

bases: compounds that accept hydrogen ions in a solution.

acidosis (assi-DOE-sis): higher-than-normal acidity in the blood and body fluids.

alkalosis (alka-LOE-sis): higher-than-normal alkalinity (base) in the blood and body fluids.

FIGURE 6-10 **An Example of a Transport Protein**

This transport protein resides within a cell membrane and acts as a two-door passageway. Molecules enter on one side of the membrane and exit on the other, but the protein doesn't leave the membrane. This example shows how the transport protein moves sodium and potassium in opposite directions across the membrane to maintain a high concentration of potassium and a low concentration of sodium within the cell. This active transport system requires energy.

Animated! figure
www.cengagebrain.com
(search for ISBN 084006845X)

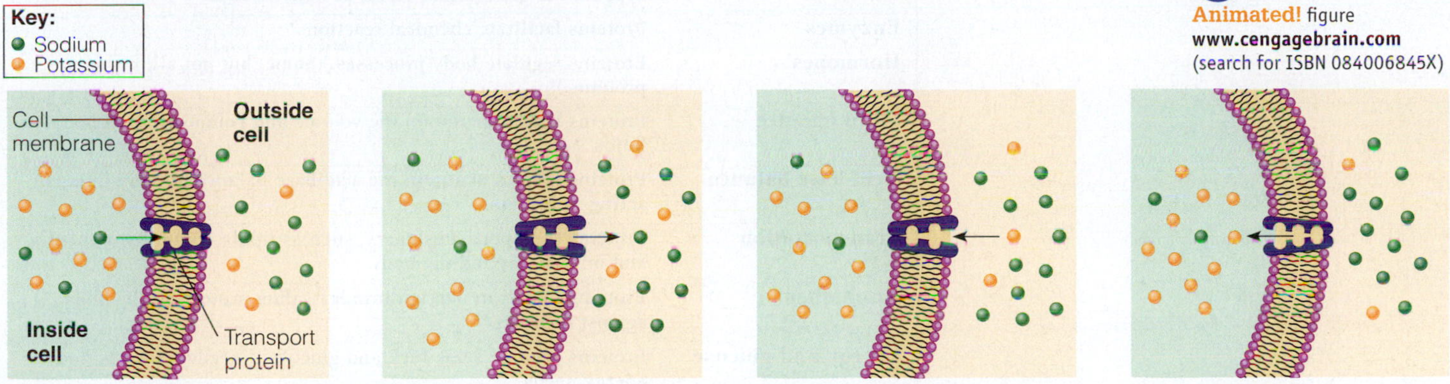

The transport protein picks up sodium from inside the cell. The protein changes shape and releases sodium outside the cell. The transport protein picks up potassium from outside the cell. The protein changes shape and releases potassium inside the cell.

As Antibodies Proteins also defend the body against disease. A virus—whether it is one that causes flu, smallpox, measles, or the common cold—enters the cells and multiplies there. One virus may produce 100 replicas of itself within an hour or so. Each replica can then burst out and invade 100 different cells, soon yielding 10,000 viruses, which invade 10,000 cells. Left free to do their worst, they will soon overwhelm the body with disease.

Fortunately, when the body detects these invading **antigens**, it manufactures **antibodies**, giant protein molecules designed specifically to combat them. The antibodies work so swiftly and efficiently that in a normal, healthy individual, most diseases never have a chance to get started. Without sufficient protein, though, the body cannot maintain its army of antibodies to resist infectious diseases.

Each antibody is designed to destroy a specific antigen. Once the body has manufactured antibodies against a particular antigen (such as the measles virus), it "remembers" how to make them. Consequently, the next time the body encounters that same antigen, it produces antibodies even more quickly. In other words, the body develops a molecular memory, known as **immunity**. (Chapter 15 describes food allergies—the immune system's response to food antigens.)

As a Source of Energy and Glucose Without energy, cells die; without glucose, the brain and nervous system falter. Even though proteins are needed to do the work that only they can perform, they will be sacrificed to provide energy ♦ and glucose ♦ during times of starvation or insufficient carbohydrate intake. The body will break down its tissue proteins to make amino acids available for energy or glucose production. In this way, protein can maintain blood glucose levels, but at the expense of losing lean body tissue. Chapter 7 provides many more details on energy metabolism.

Other Roles As mentioned earlier, proteins form integral parts of most body structures such as skin, muscles, and bones. They also participate in some of the body's most amazing activities such as blood clotting and vision. When a tissue is injured, a rapid chain of events leads to the production of fibrin, a stringy, insoluble mass of protein fibers that forms a solid clot from liquid blood. Later, more slowly, the protein collagen forms a scar to replace the clot and permanently heal the wound. The light-sensitive pigments in the cells of the eye's retina are molecules of the protein opsin. Opsin responds to light by changing its shape, thus initiating the nerve impulses that convey the sense of sight to the brain.

♦ Protein provides 4 kcal/g. Return to p. 9 for a refresher on how to calculate the protein kcalories from foods.

♦ The making of glucose from noncarbohydrate sources such as amino acids is **gluconeogenesis.**

antigens: substances that elicit the formation of antibodies or an inflammation reaction from the immune system. A bacterium, a virus, a toxin, and a protein in food that causes allergy are all examples of antigens.

antibodies: large proteins of the blood and body fluids, produced by the immune system in response to the invasion of the body by foreign molecules (usually proteins called *antigens*). Antibodies combine with and inactivate the foreign invaders, thus protecting the body.

immunity: the body's ability to defend itself against diseases (see also Highlight 17).

IN SUMMARY The protein functions discussed here are summarized in the accompanying table. They are only a few of the many roles proteins play, but they convey some sense of the immense variety of proteins and their importance in the body.

Growth and maintenance	Proteins form integral parts of most body structures such as skin, tendons, membranes, muscles, organs, and bones. As such, they support the growth and repair of body tissues.
Enzymes	Proteins facilitate chemical reactions.
Hormones	Proteins regulate body processes. (Some, but not all, hormones are proteins.)
Fluid balance	Proteins help to maintain the volume and composition of body fluids.
Acid-base balance	Proteins help to maintain the acid-base balance of body fluids by acting as buffers.
Transportation	Proteins transport substances, such as lipids, vitamins, minerals, and oxygen, around the body.
Antibodies	Proteins inactivate foreign invaders, thus protecting the body against diseases.
Energy and glucose	Proteins provide some fuel, and glucose if needed, for the body's energy needs.

A Preview of Protein Metabolism

This section previews protein metabolism; Chapter 7 provides a full description. Cells have several metabolic options, depending on their protein and energy needs.

Protein Turnover and the Amino Acid Pool Within each cell, proteins are continually being made and broken down, a process known as **protein turnover**. When proteins break down, they free amino acids. ♦ These amino acids mix with amino acids from dietary protein to form an "**amino acid pool**" within the cells and circulating blood. The rate of protein degradation and the amount of protein intake may vary, but the pattern of amino acids within the pool remains fairly constant. Regardless of their source, any of these amino acids can be used to make body proteins or other nitrogen-containing compounds, or they can be stripped of their nitrogen and used for energy (either immediately or stored as fat for later use).

♦ Amino acids (or proteins) that derive from within the body are **endogenous** (en-DODGE-eh-nus). In contrast, those that derive from foods are **exogenous** (eks-ODGE-eh-nus).
- **endo** = within
- **gen** = arising
- **exo** = outside (the body)

Nitrogen Balance Protein turnover and **nitrogen balance** go hand in hand. ♦ In healthy adults, protein synthesis balances with degradation, and protein intake from food balances with nitrogen excretion in the urine, feces, and sweat. When nitrogen intake equals nitrogen output, the person is in nitrogen equilibrium, or zero nitrogen balance. Researchers use nitrogen balance studies to estimate protein requirements.

♦ Nitrogen balance:
- Nitrogen equilibrium (zero nitrogen balance): N in = N out
- Positive nitrogen: N in > N out
- Negative nitrogen: N in < N out

If the body synthesizes more than it degrades and adds protein, nitrogen status becomes positive. Nitrogen status is positive in growing infants, children, adolescents, pregnant women, and people recovering from protein deficiency or illness; their nitrogen intake exceeds their nitrogen output. They are retaining protein in new tissues as they add blood, bone, skin, and muscle cells to their bodies.

If the body degrades more than it synthesizes and loses protein, nitrogen status becomes negative. Nitrogen status is negative in people who are starving or suffering other severe stresses such as burns, injuries, infections, and fever; their nitrogen output exceeds their nitrogen intake. During these times, the body loses nitrogen as it breaks down muscle and other body proteins for energy.

Using Amino Acids to Make Other Compounds Cells can also use amino acids to make other compounds. For example, the amino acid tyrosine is used to make the **neurotransmitters** norepinephrine and epinephrine, which relay nervous system messages throughout the body. Tyrosine can also be made into the pigment melanin, which is responsible for brown hair, eye, and skin color, or into the hormone thyroxin, which helps to regulate the metabolic rate. For another example, the amino acid tryptophan serves as a precursor for the vitamin niacin and

protein turnover: the degradation and synthesis of protein.

amino acid pool: the supply of amino acids derived from either food proteins or body proteins that collect in the cells and circulating blood and stand ready to be incorporated in proteins and other compounds or used for energy.

nitrogen balance: the amount of nitrogen consumed (N in) as compared with the amount of nitrogen excreted (N out) in a given period of time.*

neurotransmitters: chemicals that are released at the end of a nerve cell when a nerve impulse arrives there. They diffuse across the gap to the next cell and alter the membrane of that second cell to either inhibit or excite it.

*The genetic materials DNA and RNA contain nitrogen, but the quantity is insignificant compared with the amount in protein. Protein is 16 percent nitrogen. Said another way, the average protein weighs about 6.25 times as much as the nitrogen it contains, so scientists can estimate the amount of protein in a sample of food, body tissue, or other material by multiplying the weight of the nitrogen in it by 6.25.

© 2010 Ross Anania/Jupiterimages Corporation

Growing children end each day with more bone, blood, muscle, and skin cells than they had at the beginning of the day.

for serotonin, a neurotransmitter important in sleep regulation, appetite control, and sensory perception.

Using Amino Acids for Energy and Glucose As mentioned earlier, when glucose or fatty acids are limited, cells are forced to use amino acids for energy and glucose. The body does not make a specialized storage form of protein as it does for carbohydrate and fat. Glucose is stored as glycogen in the liver and fat as triglycerides in adipose tissue, but protein in the body is available only from the working and structural components of the tissues. When the need arises, the body breaks down its tissue proteins and uses their amino acids for energy or glucose. Thus, over time, energy deprivation (starvation) always causes wasting of lean body tissue as well as fat loss. An adequate supply of carbohydrates and fats spares amino acids from being used for energy and allows them to perform their unique roles.

Using Amino Acids to Make Fat Amino acids may be used to make fat when energy and protein intakes exceed needs and carbohydrate intake is adequate. When protein is abundant, energy metabolism shifts to use more protein instead of fat. Excess amino acids can also be converted to fat and stored for later use. Consequently, protein-rich foods can contribute to weight gain.

Deaminating Amino Acids When amino acids are broken down (as occurs when they are used for energy or to make glucose or fat), they are first deaminated—stripped of their nitrogen-containing amino groups. Two products result from **deamination**: one is **ammonia** (NH_3); the other product is the carbon structure without its amino group—often a **keto acid**. Keto acids may enter a number of metabolic pathways—for example, they may be used for energy or for the production of glucose, ketones, cholesterol, or fat.* They may also be used to make nonessential amino acids.

*Chemists sometimes classify amino acids according to the destinations of their carbon fragments after deamination. If the fragment leads to the production of glucose, the amino acid is called *glucogenic;* if it leads to the formation of ketone bodies, fats, and sterols, the amino acid is called *ketogenic*. There is no sharp distinction between glucogenic and ketogenic amino acids, however. A few are both, most are considered glucogenic, only one (leucine) is clearly ketogenic.

deamination (dee-AM-ih-NAY-shun): removal of the amino (NH_2) group from a compound such as an amino acid.

ammonia: a compound with the chemical formula NH_3; produced during the deamination of amino acids.

keto (KEY-toe) **acid:** an organic acid that contains a carbonyl group (C=O).

FIGURE 6-11 Deamination and Synthesis of a Nonessential Amino Acid

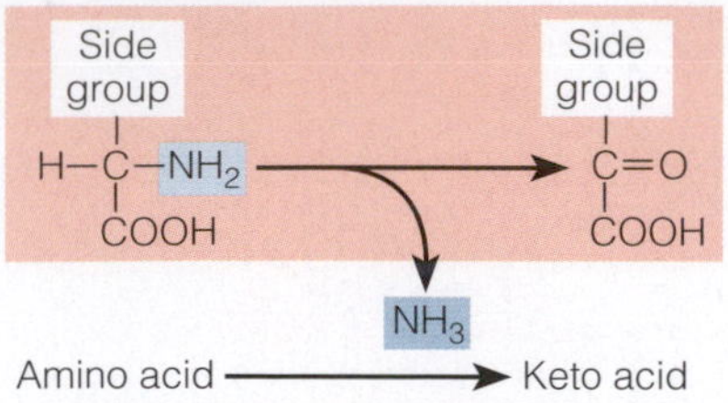

The deamination of an amino acid produces ammonia (NH_3) and a keto acid.

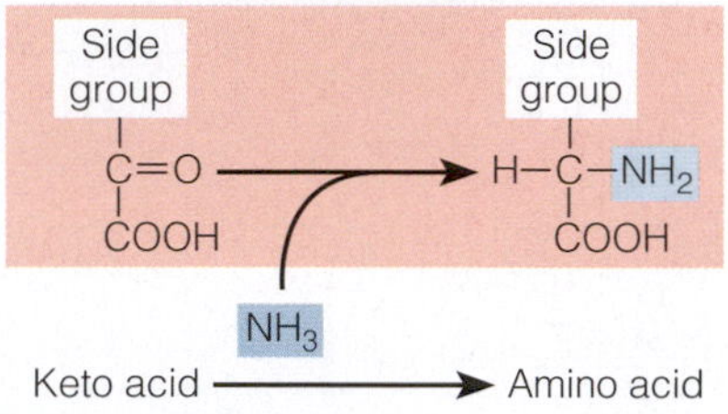

Given a source of NH_3, the body can make nonessential amino acids from keto acids.

FIGURE 6-12 Transamination and Synthesis of a Nonessential Amino Acid

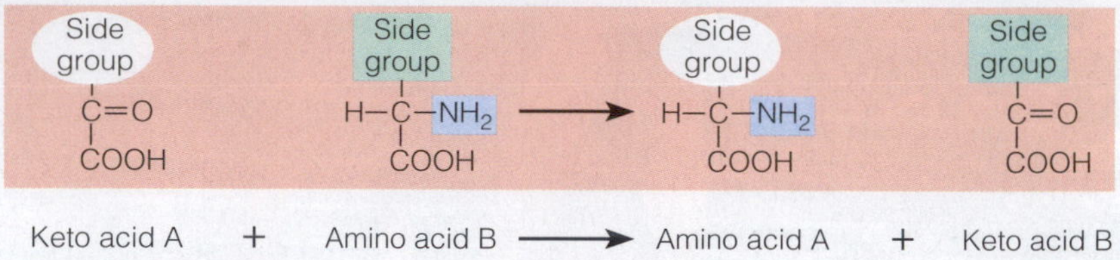

The body can transfer amino groups (NH_2) from an amino acid to a keto acid, forming a new *nonessential* amino acid and a new keto acid. Transamination reactions require the vitamin B_6 coenzyme.

Using Amino Acids to Make Proteins and Nonessential Amino Acids As mentioned, cells can assemble amino acids into the proteins they need to do their work. If an essential amino acid is missing, the body may break down some of its own proteins to obtain it. If a particular nonessential amino acid is not readily available, cells can make it from a keto acid—if a nitrogen source is available. Ammonia provides some of the nitrogen needed for the synthesis of nonessential amino acids from keto acids (see Figure 6-11). Cells can also make a nonessential amino acid by transferring an amino group from one amino acid to its corresponding keto acid, as shown in Figure 6-12. Through many such **transamination** reactions, involving many different keto acids, the liver cells can synthesize the nonessential amino acids.

Converting Ammonia to Urea Ammonia is a toxic compound chemically identical to the strong-smelling ammonia in bottled cleaning solutions. Because ammonia is a base, the blood's critical acid-base balance becomes upset if the cells produce larger quantities than the liver can handle.

To prevent such a crisis, the liver combines ammonia with carbon dioxide to make **urea**, a much less toxic compound. Figure 6-13 provides a greatly oversimplified diagram of urea synthesis; details are shown in Appendix C. The production of urea increases as dietary protein increases, until production hits its maximum rate at intakes approaching 250 grams per day.

Excreting Urea Liver cells release urea into the blood, where it circulates until it passes through the kidneys (see Figure 6-14). The kidneys then filter urea out of the blood for excretion in the urine. Normally, the liver efficiently captures all the ammonia, makes urea from it, and releases the urea into the blood; then the kidneys clear all the urea from the blood. This division of labor allows easy diagnosis of diseases of both organs. In liver disease, blood ammonia will be high; in kidney disease, blood urea will be high.

Urea is the body's principal vehicle for excreting unused nitrogen, and the amount of urea produced increases with protein intake. To keep urea in solution, the body needs water. For this reason, a person who regularly consumes a high-protein diet (say, 100 grams a day or more) must drink plenty of water to dilute and excrete urea from the body. Without extra water, a person on a high-protein diet risks dehydration because the body uses its water to rid itself of urea. This explains some of the water loss that accompanies high-protein diets. Such losses may make high-protein diets *appear* to be effective, but water loss, of course, is of no value to the person who wants to lose body fat (as Highlight 8 explains).

FIGURE 6-13 Urea Synthesis

When amino acids are deaminated, ammonia is produced. The liver detoxifies ammonia before releasing it into the bloodstream by combining it with another waste product, carbon dioxide, to produce urea. See Appendix C for details.

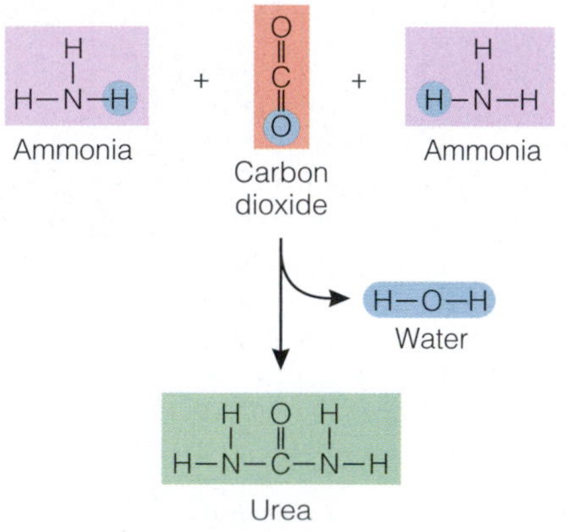

transamination (TRANS-am-ih-NAY-shun): the transfer of an amino group from one amino acid to a keto acid, producing a new nonessential amino acid and a new keto acid.

urea (you-REE-uh): the principal nitrogen-excretion product of protein metabolism. Two ammonia fragments are combined with carbon dioxide to form urea.

IN SUMMARY **Proteins are constantly being synthesized and broken down as needed. The body's assimilation of amino acids into proteins and its release of amino acids via protein degradation and excretion can be tracked by measuring nitrogen balance, which should be positive during growth and**

steady in adulthood. An energy deficit or an inadequate protein intake may force the body to use amino acids as fuel, creating a negative nitrogen balance. Protein eaten in excess of need is degraded and stored as body fat.

Protein in Foods

In the United States and Canada, where nutritious foods are abundant, most people eat protein in such large quantities that they receive all the amino acids they need. In countries where food is scarce and the people eat only marginal amounts of protein-rich foods, however, the *quality* of the protein becomes crucial.

Protein Quality

The protein quality of the diet determines, in large part, how well children grow and how well adults maintain their health. Put simply, **high-quality proteins** provide enough of all the essential amino acids needed to support the body's work, and low-quality proteins don't. Two factors influence protein quality—the protein's digestibility and its amino acid composition.

Digestibility As explained earlier, proteins must be digested before they can provide amino acids. **Protein digestibility** depends on such factors as the protein's source and the other foods eaten with it. The digestibility of most animal proteins is high (90 to 99 percent); plant proteins are less digestible (70 to 90 percent for most, but more than 90 percent for soy and legumes).

Amino Acid Composition To make proteins, a cell must have all the needed amino acids available simultaneously. The liver can make any nonessential amino acid that may be in short supply so that the cells can continue linking amino acids into protein strands. If an essential amino acid is missing, though, a cell must dismantle its own proteins to obtain it. Therefore, to prevent protein breakdown in the body, dietary protein must supply at least the nine essential amino acids plus enough nitrogen-containing amino groups and energy for the synthesis of the nonessential ones. If the diet supplies too little of any essential amino acid, protein synthesis will be limited. The body makes whole proteins only; if one amino acid is missing, the others cannot form a "partial" protein. An essential amino acid supplied in less than the amount needed to support protein synthesis is called a **limiting amino acid.**

Reference Protein The quality of a food protein is determined by comparing its amino acid composition with the essential amino acid requirements of preschool-age children. Such a standard is called a **reference protein.** ♦ The rationale behind using the requirements of this age group is that if a protein will effectively support a young child's growth and development, then it will meet or exceed the requirements of older children and adults.

High-Quality Proteins As mentioned earlier, a high-quality protein contains all the essential amino acids in relatively the same amounts and proportions that human beings require; it may or may not contain all the nonessential amino acids. Proteins that are low in an essential amino acid cannot, by themselves, support protein synthesis. Generally, foods derived from animals (meat, fish, poultry, cheese, eggs, yogurt, and milk) provide high-quality proteins, although gelatin is an exception. (It lacks tryptophan and cannot support growth and health as a diet's sole protein.) Proteins from plants (vegetables, nuts, seeds, grains, and legumes) have more diverse amino acid patterns and tend to be limiting in one or more essential amino acids. Some plant proteins are notoriously low quality (for example, corn protein). A few others are high quality (for example, soy protein).

Researchers have developed several methods for evaluating the quality of food proteins and identifying high-quality proteins. Appendix D provides details.

FIGURE 6-14 Urea Excretion

The liver and kidneys both play a role in disposing of excess nitrogen. Can you see why the person with liver disease has high blood ammonia, whereas the person with kidney disease has high blood urea? (Figure 12-2 provides details of how the kidneys work.)

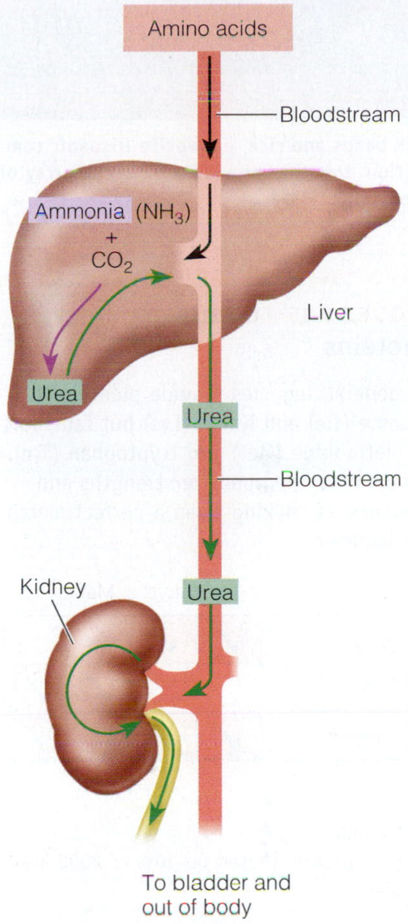

♦ In the past, egg protein was commonly used as the reference protein. Table D-1 in Appendix D presents the amino acid profile of egg. As the reference protein, egg was assigned the value of 100; Table D-3 includes scores of other food proteins for comparison.

high-quality proteins: dietary proteins containing all the essential amino acids in relatively the same amounts that human beings require. They may also contain nonessential amino acids.

protein digestibility: a measure of the amount of amino acids absorbed from a given protein intake.

limiting amino acid: the essential amino acid found in the shortest supply relative to the amounts needed for protein synthesis in the body. Four amino acids are most likely to be limiting:

- Lysine
- Methionine
- Threonine
- Tryptophan

reference protein: a standard against which to measure the quality of other proteins.

© Polara Studios, Inc.

Black beans and rice, a favorite Hispanic combination, together provide a balanced array of amino acids.

Complementary Proteins In general, plant proteins are lower quality than animal proteins, and plants also offer less protein (per weight or measure of food). For this reason, many vegetarians improve the quality of proteins in their diets by combining plant-protein foods that have different but complementary amino acid patterns. This strategy yields **complementary proteins** that together contain all the essential amino acids in quantities sufficient to support health. The protein quality of the combination is greater than for either food alone (see Figure 6-15).

Many people have long believed that combining plant proteins at every meal is critical to protein nutrition. For most healthy vegetarians, though, it is *not* necessary to balance amino acids at each meal if protein intake is varied and energy intake is sufficient.[3] Vegetarians can receive all the amino acids they need over the course of a day by eating a variety of whole grains, legumes, seeds, nuts, and vegetables. Protein deficiency will develop, however, when fruits and certain vegetables make up the core of the diet, severely limiting both the *quantity* and *quality* of protein. Highlight 2 describes how to plan a nutritious vegetarian diet.

FIGURE 6-15 Complementary Proteins

In general, legumes provide plenty of isoleucine (Ile) and lysine (Lys) but fall short in methionine (Met) and tryptophan (Trp). Grains have the opposite strengths and weaknesses, making them a perfect match for legumes.

	Ile	Lys	Met	Trp
Legumes	✓	✓		
Grains			✓	✓
Together	✓	✓	✓	✓

IN SUMMARY **A diet that supplies all of the essential amino acids in adequate amounts ensures protein synthesis. The best guarantee of amino acid adequacy is to eat foods containing high-quality proteins or mixtures of foods containing complementary proteins that can each supply the amino acids missing in the other. In addition to its amino acid content, the quality of protein is measured by its digestibility and its ability to support growth. Such measures are of great importance in dealing with malnutrition worldwide, but in the United States and Canada, where protein deficiency is not common, protein quality of individual foods deserves little emphasis.**

Protein Regulations for Food Labels

All food labels must state the *quantity* of protein in grams. The "% Daily Value" ♦ for protein is not mandatory on all labels but is required whenever a food makes a protein claim or is intended for consumption by children younger than four years old.* Whenever the Daily Value percentage is declared, researchers must determine the *quality* of the protein. Thus, when a % Daily Value is stated for protein, it reflects both quantity and quality.

♦ Daily Value:
- 50 g protein (based on 10% of 2000 kcal diet)

Health Effects and Recommended Intakes of Protein

As you know by now, protein is indispensable to life. It should come as no surprise that protein deficiency can have devastating effects on people's health. But, like the other nutrients, protein in excess can also be harmful. This section examines the health effects and recommended intakes of protein.

Protein-Energy Malnutrition

When people are deprived of protein, energy, or both, the result is **protein-energy malnutrition (PEM)**. Although PEM touches many adult lives, it most often strikes early in childhood. It is one of the most prevalent and devastating forms of malnutrition in the world, afflicting one of every four children worldwide. Most of the 33,000 children who die each day are malnourished.[4]

Inadequate food intake leads to poor growth in children and to weight loss and wasting in adults. Children who are underweight for their height may be suffering from **acute PEM** (recent severe food deprivation), whereas children who are short for their age have experienced **chronic PEM** (long-term food deprivation).

complementary proteins: two or more dietary proteins whose amino acid assortments complement each other in such a way that the essential amino acids missing from one are supplied by the other.

protein-energy malnutrition (PEM): a deficiency of protein, energy, or both, including kwashiorkor, marasmus, and instances in which they overlap; also called **protein-kcalorie malnutrition (PCM).**

acute PEM: protein-energy malnutrition caused by recent severe food restriction; characterized in children by underweight for height (wasting).

chronic PEM: protein-energy malnutrition caused by long-term food deprivation; characterized in children by short height for age (stunting).

*For labeling purposes, the Daily Values for protein are as follows: for infants, 14 grams; for children younger than age four, 16 grams; for older children and adults, 50 grams; for pregnant women, 60 grams; and for lactating women, 65 grams.

Poor growth due to PEM is easy to overlook because a small child may look quite normal, but it is the most common sign of malnutrition.

PEM is most prevalent in Africa, Central America, South America, and East and Southeast Asia. In the United States, homeless people and those living in substandard housing in inner cities and rural areas have been diagnosed with PEM. In addition to those living in poverty, elderly people who live alone and adults who are addicted to drugs and alcohol are frequently victims of PEM. PEM can develop in young children when parents mistakenly provide "health-food beverages" ♦ that lack adequate energy or protein instead of milk, most commonly because of nutritional ignorance, perceived milk intolerance, or food faddism. Adult PEM is also seen in people hospitalized with infections such as AIDS or tuberculosis; these infections deplete body proteins, demand extra energy, induce nutrient losses, and alter metabolic pathways. Furthermore, poor nutrient intake during hospitalization worsens malnutrition and impairs recovery, whereas nutrition intervention often improves the body's response to other treatments and the chances of survival. PEM is also common in those suffering from the eating disorder anorexia nervosa (discussed in Highlight 8). Prevention emphasizes frequent, nutrient-dense, energy-dense meals and, equally important, resolution of the underlying causes of PEM—poverty, infections, and illness.

♦ Rice drinks are often sold as milk alternatives, but they fail to provide adequate protein, vitamins, and minerals.

Classifying PEM PEM occurs in two forms: marasmus and kwashiorkor, which differ in their clinical features (see Table 6-3). The following paragraphs present three clinical syndromes—marasmus, kwashiorkor, and the combination of the two.

Marasmus Appropriately named from the Greek word meaning "dying away," marasmus reflects a severe deprivation of food over a long time (chronic PEM). Put simply, the person is starving and suffering from an inadequate energy *and* protein intake (and inadequate essential fatty acids, vitamins, and minerals as well). Marasmus occurs most commonly in children from 6 to 18 months of age in all the overpopulated and impoverished areas of the world. Children living in poverty simply do not have enough to eat. They subsist on diluted cereal drinks that supply scant energy and protein of low quality; such food can barely sustain life, much less support growth. Consequently, marasmic children look like little old people—just "skin and bones."

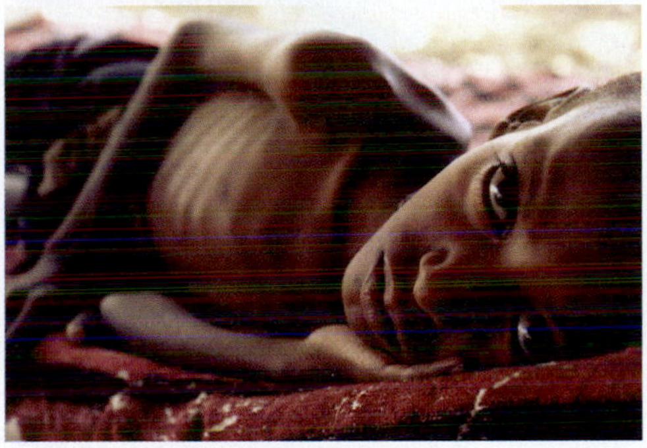

The severe wasting characteristic of marasmus is apparent in this child's "matchstick" arms.

marasmus (ma-RAZ-mus): a form of PEM that results from a severe deprivation, or impaired absorption, of energy, protein, vitamins, and minerals.

TABLE 6-3 Features of Marasmus and Kwashiorkor in Children

Separating PEM into two classifications oversimplifies the condition, but at the extremes, marasmus and kwashiorkor exhibit marked differences. Marasmus-kwashiorkor mix presents symptoms common to both marasmus and kwashiorkor. In all cases, children are likely to develop diarrhea, infections, and multiple nutrient deficiencies.

Marasmus	Kwashiorkor
Infancy (less than 2 yr)	Older infants and young children (1 to 3 yr)
Severe deprivation, or impaired absorption, of protein, energy, vitamins, and minerals	Inadequate protein intake or, more commonly, infections
Develops slowly; chronic PEM	Rapid onset; acute PEM
Severe weight loss	Some weight loss
Severe muscle wasting, with no body fat	Some muscle wasting, with retention of some body fat
Growth: <60% weight-for-age	Growth: 60 to 80% weight-for-age
No detectable edema	Edema
No fatty liver	Enlarged fatty liver
Anxiety, apathy	Apathy, misery, irritability, sadness
Good appetite possible	Loss of appetite
Hair is sparse, thin, and dry; easily pulled out	Hair is dry and brittle; easily pulled out; changes color; becomes straight
Skin is dry, thin, and easily wrinkles	Skin develops lesions

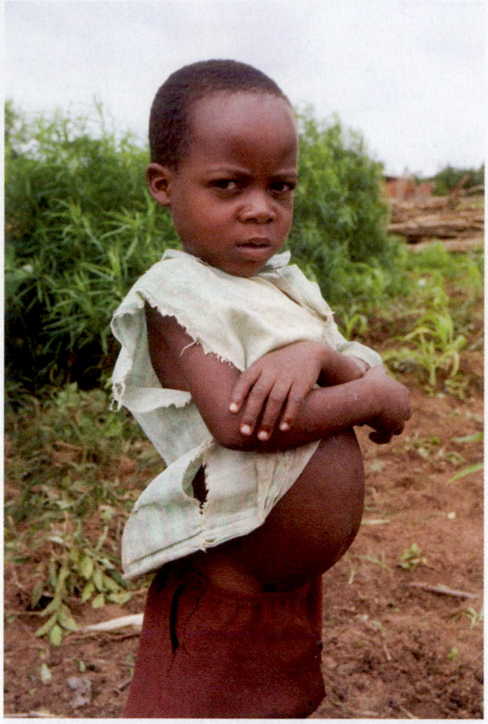

The edema characteristic of kwashiorkor is apparent in this child's swollen belly. Malnourished children commonly have an enlarged abdomen from parasites as well.

Without adequate nutrition, muscles, including the heart, waste and weaken. Because the brain normally grows to almost its full adult size within the first two years of life, marasmus impairs brain development and learning ability. Reduced synthesis of key hormones slows metabolism and lowers body temperature. There is little or no fat under the skin to insulate against cold. Hospital workers find that children with marasmus need to be clothed, covered, and kept warm. Because these children often suffer delays in their mental and behavioral development, they also need loving care, a stimulating environment, and parental attention.

The starving child faces this threat to life by engaging in as little activity as possible—not even crying for food. The body musters all its forces to meet the crisis, so it cuts down on any expenditure of energy not needed for the functioning of the heart, lungs, and brain. Growth ceases; the child is no larger at age four than at age two. Enzymes are in short supply and the GI tract lining deteriorates. Consequently, what little food is eaten can't be digested and absorbed.

Kwashiorkor **Kwashiorkor** typically reflects a sudden and recent deprivation of food (acute PEM). *Kwashiorkor* is a Ghanaian word that refers to the birth position of a child and is used to describe the illness a child develops when the next child is born. When a mother who has been nursing her first child bears a second child, she weans the first child and puts the second one on the breast. The first child, suddenly switched from nutrient-dense, protein-rich breast milk to a starchy, protein-poor cereal, soon begins to sicken and die. Kwashiorkor typically sets in between 18 months and 2 years.

Kwashiorkor usually develops rapidly as a result of protein deficiency or, more commonly, is precipitated by an illness such as measles or other infection.[5] Other factors, such as aflatoxins (a contaminant sometimes found in moldy grains), may also contribute to the development of, or symptoms that accompany, kwashiorkor.

The loss of weight and body fat is usually not as severe in kwashiorkor as in marasmus, but some muscle wasting may occur. Without proteins to maintain fluid balance, the child's limbs and abdomen become swollen with edema—a distinguishing feature of kwashiorkor. ♦ A fatty liver develops due to a lack of the protein carriers to transport lipids out of the liver. The fatty liver lacks enzymes to clear metabolic toxins from the body, so their harmful effects are prolonged. Inflammation in response to these toxins and to infections further contributes to the edema that accompanies kwashiorkor. Without sufficient tyrosine to make melanin, hair loses its color, and inadequate protein synthesis leaves the skin patchy and scaly, often with sores that fail to heal. The lack of proteins to carry or store iron leaves iron free. Unbound iron is common in kwashiorkor and may contribute to illness and death by promoting bacterial growth and free-radical damage. (Free-radical damage is discussed fully in Highlight 11.)

♦ For this reason, kwashiorkor is sometimes referred to as "wet" PEM and marasmus as "dry" PEM.

Marasmus-Kwashiorkor Mix The combination of marasmus and kwashiorkor is characterized by the edema of kwashiorkor with the wasting of marasmus. Most often, the child suffers the effects of both malnutrition and infections.

Infections In PEM, antibodies to fight off invading bacteria are degraded to provide amino acids for other uses, leaving the malnourished child vulnerable to infections. Blood proteins, including hemoglobin, are no longer synthesized, so the child becomes anemic and weak. **Dysentery**, an infection of the digestive tract, causes diarrhea, further depleting the body of nutrients and fluids. In the marasmic child, once infection sets in, kwashiorkor often follows, and the immune response weakens further.

The combination of infections, fever, fluid imbalances, and anemia often leads to heart failure and occasionally sudden death. Infections combined with malnutrition are responsible for two-thirds of the deaths of young children in developing countries. Measles, which might make a healthy child sick for a week or two, kills a child with PEM within two or three days.

Rehabilitation If caught in time, the life of a starving child may be saved with rehydration and nutrition intervention.[6] In severe cases, diarrhea will have incurred dramatic fluid and mineral losses that need to be replaced during the first

kwashiorkor (kwash-ee-OR-core, kwash-ee-or-CORE): a form of PEM that results from inadequate protein intake and infections.

dysentery (DISS-en-terry): an infection of the digestive tract that causes diarrhea.

24 to 48 hours to help raise the blood pressure and strengthen the heartbeat. After that, protein and food energy may be given in *small* quantities several times a day, with intakes *gradually* increased as tolerated.[7] Severely malnourished people, especially those with edema, recover better with an initial diet that is relatively low in protein (10 percent of energy intake).

Experts assure us that we possess the knowledge, technology, and resources to end hunger. Programs that tailor interventions to the local people and involve them in the process of identifying problems and devising solutions have the most success. To win the war on hunger, those who have the food, technology, and resources must make fighting hunger a priority (see Highlight 16 for more on hunger).

© dbimages/Alamy

Donated food saves some people from starvation, but it is usually insufficient to meet nutrient needs or even to defend against hunger.

Health Effects of Protein

While many of the world's people struggle to obtain enough food energy and protein, in developed countries both are so abundant that problems of excess are seen. Overconsumption of protein offers no benefits and may pose health risks. High-protein diets have been implicated in several chronic diseases, including heart disease, cancer, osteoporosis, obesity, and kidney stones, but evidence is insufficient to establish an Upper Level (UL).[8]

Researchers attempting to clarify the relationships between excess protein and chronic diseases face several obstacles. Population studies have difficulty determining whether diseases correlate with animal proteins or with their accompanying saturated fats, for example. Studies that rely on data from vegetarians must sort out the many lifestyle factors, in addition to a "no-meat diet," that might explain relationships between protein and health.

Heart Disease A high-protein diet may contribute to the progression of heart disease. As Chapter 5 mentions, foods rich in animal protein also tend to be rich in saturated fats. Consequently, it is not surprising to find a correlation between animal-protein intake (red meats and dairy products) and heart disease.[9] On the other hand, substituting vegetable protein for animal protein may improve blood lipids and decrease heart disease mortality.[10]

Research suggests that elevated levels of the amino acid homocysteine may be an independent risk factor for heart disease, heart attacks, and sudden death in patients with heart disease.[11] Researchers do not yet fully understand the many factors—including a diet high in saturated fatty acids—that can raise homocysteine in the blood or whether elevated levels are a cause or an effect of heart disease.[12] Elevated homocysteine is associated with increased oxidative stress and inflammation.[13] Until researchers can determine the exact role homocysteine plays in heart disease, they are following several leads in pursuit of the answers. Coffee's role in heart disease has been controversial, but research suggests it is among the most influential factors in raising homocysteine, which may explain some of the adverse health effects of heavy consumption.[14] Elevated homocysteine levels are among the many adverse health consequences of smoking cigarettes and drinking alcohol as well.[15] Homocysteine is also elevated with inadequate intakes of B vitamins and can usually be lowered with fortified foods or supplements of vitamin B_{12}, vitamin B_6, and folate.[16] Lowering homocysteine, however, may not help in preventing heart attacks.[17] Supplements of the B vitamins do not always benefit those with heart disease and, in fact, may actually increase the risks.[18]

In contrast to homocysteine, the amino acid arginine may help protect against heart disease by lowering blood pressure and homocysteine levels.[19] Additional research is needed to confirm the benefits of arginine. In the meantime, it is unwise for consumers to use supplements of arginine, or any other amino acid for that matter (as p. 195 explains). Physicians, however, may find it beneficial to add arginine supplements to their heart patients' treatment plan.[20]

Cancer Protein does not seem to increase the risk of cancer, but some protein-rich foods do. For example, evidence suggests a strong correlation between high intakes of red meat and processed meats ♦ with cancer of the colon. Chapter 29 discusses dietary links with cancer, and Highlight 18 presents food-safety issues of processed meats.

♦ Processed meats include ham, bacon, pastrami, salami, sausage, bratwurst, and hot dogs; they have been preserved by smoking, curing, salting, or adding preservatives.

Adult Bone Loss (Osteoporosis) Chapter 12 presents calcium metabolism, and Highlight 12 elaborates on the main factors that influence osteoporosis. This section briefly describes the relationships between protein intake and bone loss. When protein intake is high, calcium excretion increases. Whether excess protein depletes the bones of their chief mineral may depend upon the ratio of calcium intake to protein intake. After all, bones need both protein and calcium. An ideal ratio has not been determined, but a young woman whose intake meets recommendations for both nutrients has a calcium-to-protein ratio of more than 20 to 1 (milligrams to grams), which probably provides adequate protection for the bones. For most women in the United States, however, average calcium intakes are lower and protein intakes are higher, yielding a 9-to-1 ratio, which may produce calcium losses significant enough to compromise bone health. In other words, the problem may reflect too little calcium, not too much protein. In establishing recommendations, the DRI Committee considered protein's effect on calcium metabolism and bone health, but it did not find sufficient evidence to warrant an adjustment for calcium or a UL for protein.[21]

Some (but not all) research suggests that animal protein may be more detrimental to calcium metabolism and bone health than vegetable protein.[22] Importantly, *inadequate* intakes of protein may compromise bone health. Osteoporosis is particularly common in elderly women and in adolescents with anorexia nervosa—groups who typically receive less protein than they need. For these people, increasing protein intake may be just what they need to protect their bones.[23]

Weight Control Fad weight-loss diets that encourage a high-protein, low-carbohydrate diet may be effective, but only because they are low-kcalorie diets. Diets that provide adequate protein, moderate fat, and sufficient energy from carbohydrates can better support weight loss and good health. Including protein at each meal may help with weight loss by providing satiety.[24] Selecting too many protein-rich foods, such as meat and milk, may crowd out fruits, vegetables, and whole grains, making the diet inadequate in other nutrients.

Kidney Disease Excretion of the end products of protein metabolism depends, in part, on an adequate fluid intake and healthy kidneys. A high protein intake does not cause kidney disease, but it does increase the work of the kidneys and accelerate kidney deterioration in people with chronic kidney disease.[25] Restricting dietary protein may help to slow the progression of kidney disease in people who have this condition.[26]

IN SUMMARY **Protein deficiencies arise from both energy-poor and protein-poor diets and lead to the devastating diseases of marasmus and kwashiorkor. Together, these diseases are known as PEM (protein-energy malnutrition), a major form of malnutrition causing death in children worldwide. Excesses of protein offer no advantage; in fact, overconsumption of protein-rich foods may incur health problems as well.**

Recommended Intakes of Protein

As mentioned earlier, the body continuously breaks down and loses some protein and cannot store amino acids. To replace protein, the body needs dietary protein for two reasons. First, dietary protein is the only source of the *essential* amino acids, and second, it is the only practical source of *nitrogen* with which to build the nonessential amino acids and other nitrogen-containing compounds the body needs.

Given recommendations that people's fat intakes should contribute 20 to 35 percent of total food energy and carbohydrate intakes should contribute 45 to 65 percent, that leaves 10 to 35 percent for protein. In a 2000-kcalorie diet, that represents 200 to 700 kcalories from protein, or 50 to 175 grams. Average intakes in the United States and Canada fall within this range.

Protein RDA The protein RDA ♦ for adults is 0.8 grams per kilogram of healthy body weight per day. For infants and children, the RDA is slightly higher. The table on the inside front cover lists the RDA for males and females at various ages

♦ RDA for protein:
- 0.8 g/kg/day
- 10 to 35% of energy intake

HOW TO Calculate Recommended Protein Intakes

To figure your protein RDA:

- Look up the healthy weight for a person of your height (inside back cover). If your present weight falls within that range, use it for the following calculations. If your present weight falls outside the range, use the midpoint of the healthy weight range as your reference weight.
- Convert pounds to kilograms, if necessary (pounds divided by 2.2 equals kilograms).
- Multiply kilograms by 0.8 to get your RDA in grams per day. (Teens 14 to 18 years old, multiply by 0.85.) Example:

$$\text{Weight} = 150 \text{ lb}$$

$$150 \text{ lb} \div 2.2 \text{ lb/kg} = 68 \text{ kg (rounded off)}$$

$$68 \text{ kg} \times 0.8 \text{ g/kg} = 54 \text{ g protein (rounded off)}$$

For additional practice, log on to **www.cengagebrain.com** and search for ISBN 084006845X.

TRY IT Calculate your protein RDA.

in two ways—grams per day based on reference body weights and grams per kilogram body weight per day. Some evidence suggests that intakes greater than the protein RDA may be beneficial.[27]

The RDA covers the needs for replacing worn-out tissue, so it increases for larger people; it also covers the needs for building new tissue during growth, so it increases for infants, children, adolescents, and pregnant and lactating women. The protein RDA is the same for athletes as for others, even though athletes may need more protein and many fitness authorities recommend a higher range of protein intakes for athletes pursuing different activities.[28] ♦ Even so, "higher" intakes still fall within the 10 to 35 percent Acceptable Macronutrient Distribution Range (AMDR).[29] The accompanying "How To" explains how to calculate your RDA for protein.

♦ Protein recommendations for athletes: 1.2–1.7 g/kg/day

In setting the RDA, the DRI Committee assumes that people are healthy and do not have unusual metabolic needs for protein, that the protein eaten will be of mixed quality (from both high- and low-quality sources), and that the body will use the protein efficiently. In addition, the committee assumes that the protein is consumed along with sufficient carbohydrate and fat to provide adequate energy and that other nutrients in the diet are also adequate.

Adequate Energy Note the qualification "adequate energy" in the preceding statement, and consider what happens if energy intake falls short of needs. An intake of 50 grams of protein provides 200 kcalories, which represents 10 percent of the total energy from protein, if the person receives 2000 kcalories a day. But if the person cuts energy intake drastically—to, say, 800 kcalories a day—then an intake of 200 kcalories from protein is suddenly 25 percent of the total; yet it's still the same amount of protein (number of grams). The protein intake is reasonable, but the energy intake is not. The low energy intake forces the body to use the protein to meet energy needs rather than to replace lost body protein. Similarly, if the person's energy intake is high—say, 4000 kcalories—the 50 gram protein intake represents only 5 percent of the total; yet it *still* is a reasonable protein intake. Again, the energy intake is unreasonable for most people, but in this case, it permits the protein to be used to meet the body's needs.

Be careful when judging protein (or carbohydrate or fat) intake as a percentage of energy. Always consider the number of grams as well, and compare it with the RDA or another standard stated in grams. A recommendation stated as a percentage of energy intake is useful only if the energy intake is within reason.

For many people, this 5-ounce steak provides almost all of the meat and much of the protein recommended for a day's intake.

Vegetarians obtain their protein from whole grains, legumes, nuts, vegetables, and, in some cases, eggs and milk products.

Protein in Abundance Most people in the United States and Canada receive more protein than they need. Even athletes in training typically don't need to increase their protein intakes because the additional foods they eat to meet their high energy needs deliver protein as well. That protein intake is high is not surprising considering the abundance of food eaten and the central role meats hold in the North American diet. A single ounce of meat (or ½ cup legumes) delivers about 7 grams of protein, so 8 ounces of meat alone supply more than the RDA for an average-size person. Besides meat, well-fed people eat many other nutritious foods, many of which also provide protein. A cup of milk provides 8 grams of protein. Grains and vegetables provide small amounts of protein, but they can add up to significant quantities; fruits and fats provide no protein.

To illustrate how easy it is to overconsume protein, consider the amounts recommended by MyPyramid for a 2000-kcalorie diet. Six ounces of grains provide about 18 grams of protein; 2½ cups of vegetables deliver about 10 grams; 3 cups of milk offer 24 grams; and 5½ ounces of meat supply 38 grams. This totals 90 grams of protein—higher than the protein RDA for most people and yet still lower than the average intake of people in the United States.

People in the United States and Canada typically get more protein than they need. If they have an adequate *food* intake, they have a more-than-adequate protein intake. The key diet-planning principle to emphasize for protein is moderation. Even though most people receive plenty of protein, some feel compelled to take supplements as well, as the next section describes.

IN SUMMARY **The optimal diet is adequate in energy from carbohydrate and fat and delivers 0.8 grams of protein per kilogram of healthy body weight each day. U.S. and Canadian diets are typically more than adequate in this respect.**

Protein and Amino Acid Supplements

Websites, health-food stores, and popular magazine articles advertise a wide variety of protein supplements, and people take these supplements for many different reasons. Athletes take protein powders to build muscle. Dieters take them to spare their bodies' protein while losing weight. Women take them to strengthen their fingernails. People take individual amino acids, too—to cure herpes, to make themselves sleep better, to lose weight, and to relieve pain and depression.* Like many other magic solutions to health problems, protein and amino acid supplements ♦ don't work these miracles. Furthermore, they may be harmful.

♦ Use of amino acids as dietary supplements is *inappropriate,* especially for:
- All women of childbearing age
- Pregnant or lactating women
- Infants, children, and adolescents
- Elderly people
- People with inborn errors of metabolism that affect their bodies' handling of amino acids
- Smokers
- People on low-protein diets
- People with chronic or acute mental or physical illnesses who take amino acids without medical supervision

*Canada allows only single amino acid supplements to be sold as drugs or used as food additives.

Protein Powders Because the body builds muscle protein from amino acids, many athletes take protein powders with the false hope of stimulating muscle growth. Muscle work builds muscle; protein supplements do not, and athletes do not need them. Taking protein supplements does not improve athletic performance.[30] Protein powders can supply amino acids to the body, but nature's protein sources—lean meat, milk, eggs, and legumes—supply all these amino acids and more.

Whey protein appears to be particularly popular among athletes hoping to achieve greater muscle gains. A waste product of cheese manufacturing, whey protein is a common ingredient in many low-cost protein powders. When combined with strength training, whey supplements may increase protein synthesis slightly, but they do not seem to enhance athletic performance. To build stronger muscles, athletes need to eat food with adequate energy and protein to support the weight-training work that does increase muscle mass. Those who still think they need more whey can drink a glass of milk; one cup provides 1.5 grams of whey.

Purified protein preparations contain none of the other nutrients needed to support the building of muscle, and the protein they supply is not needed by athletes who eat food. It is excess protein, and the body dismantles it and uses it for energy or stores it as body fat. The deamination of excess amino acids places an extra burden on the kidneys to excrete unused nitrogen.

Amino Acid Supplements Single amino acids do not occur naturally in foods and offer no benefit to the body; in fact, they may be harmful. The body was not designed to handle the high concentrations and unusual combinations of amino acids found in supplements. Large doses of amino acids cause diarrhea.[31] An excess of one amino acid can create such a demand for a carrier that it limits the absorption of another amino acid, presenting the possibility of a deficiency. Those amino acids winning the competition enter in excess, creating the possibility of toxicity. Toxicity of single amino acids in animal studies raises concerns about their use in human beings. Anyone considering taking amino acid supplements should be cautious not to exceed levels normally found in foods.[32]

Most healthy athletes eating well-balanced diets do not need amino acid supplements. Advertisers point to research that identifies the **branched-chain amino acids** ♦ as the main ones used as fuel by exercising muscles. What the ads leave out is that compared to glucose and fatty acids, branched-chain amino acids provide very little fuel and that ordinary foods provide them in abundance anyway. Large doses of branched-chain amino acids can raise plasma ammonia concentrations, which can be toxic to the brain. Branched-chain amino acid supplements may be beneficial in conditions such as liver disease, but otherwise, they are not routinely recommended.[33]

♦ The branched-chain amino acids are leucine, isoleucine, and valine.

In two cases, recommendations for single amino acid supplements have led to widespread public use—lysine to prevent or relieve the infections that cause herpes cold sores on the mouth or genital organs, and tryptophan to relieve pain, depression, and insomnia. In both cases, enthusiastic popular reports preceded careful scientific experiments and health recommendations. Research is insufficient to determine whether lysine suppresses herpes infections, but it appears safe (up to 3 grams per day) when taken in divided doses with meals.[34]

Tryptophan may be effective with respect to pain and sleep, but its use for these purposes is experimental. About 20 years ago, more than 1500 people who elected to take tryptophan supplements developed a rare blood disorder known as eosinophilia-myalgia syndrome (EMS). EMS is characterized by severe muscle and joint pain, extremely high fever, and, in more than three dozen cases, death. Treatment for EMS usually involves physical therapy and low doses of corticosteroids to relieve symptoms temporarily. The Food and Drug Administration implicated impurities in the supplements, issued a recall of all products containing manufactured tryptophan, and warned that high-dose supplements of tryptophan might provoke symptoms of EMS even in the absence of impurities.[35]

whey protein: a by-product of cheese production; falsely promoted as increasing muscle mass. Whey is the watery part of milk that separates from the curds.

branched-chain amino acids: the essential amino acids leucine, isoleucine, and valine, which are present in large amounts in skeletal muscle tissue; falsely promoted as fuel for exercising muscles.

IN SUMMARY Normal, healthy people never need protein or amino acid supplements. It is safest to obtain lysine, tryptophan, and all other amino acids from protein-rich foods, eaten with abundant carbohydrate and some fat to facilitate their use in the body. With all that we know about science, it is hard to improve on nature.

Nutrition Portfolio

Foods that derive from animals—meats, fish, poultry, eggs, and milk products—provide plenty of protein but are often accompanied by fat. Those that derive from plants—whole grains, vegetables, and legumes—may provide less protein but also less fat.

Go to Diet Analysis Plus and choose one of the days on which you have tracked your diet for the entire day. Go to the Intake Spreadsheet report. Scroll down until you see: protein (g).

- Which of your food choices provided you with the most protein on that day? Does that food also have a lot of fat? Refer to the fat (g) column for this information.
- Describe your dietary sources of proteins and whether you use mostly plant-based or animal-based protein foods in your diet.

Now take a look at the Intake vs. Goals report.

- How do your protein needs compare with your protein intake? Consider whether you receive enough, but not too much, protein daily. Remember, 100 percent means your intake is meeting your needs based on your intake and profile information.
- If your protein intake exceeds 100 percent, consider the possible negative consequences of a high protein intake over many years.
- Debate the risks and benefits of taking protein or amino acid supplements.

Diet Analysis PLUS **To complete this exercise, go to your Diet Analysis Plus at www.cengage.com/sso.**

Nutrition on the Net

For further study of topics covered in this chapter, log on to **www.cengagebrain.com** and search for ISBN 084006845X.

- Learn more about sickle-cell anemia from the National Heart, Lung, and Blood Institute or the Sickle Cell Disease Association of America: **www.nhlbi.nih.gov** or **www.sicklecelldisease.org**
- Learn more about protein-energy malnutrition and world hunger from the World Health Organization Nutrition Programme or the National Institute of Child Health and Human Development: **www.who.int/nutrition/en** or **www.nichd.nih.gov**
- Highlight 16 offers many more websites on malnutrition and world hunger.

References

1. M. T. Gladwin and E. Vichinsky, Pulmonary complications of sickle cell disease, *New England Journal of Medicine* 359 (2008): 2254–2265; F. J. Kirkham, Therapy insight: Stroke risk and its management in patients with sickle cell disease, *Nature Clinical Practice. Neurology* 3 (2007): 264–278.
2. M. K. C. Hesselink, R. Minnaard, and P. Schrauwen, Eat the meat or feed the meat: Protein turnover in remodeling muscle, *Current Opinion in Clinical Nutrition and Metabolic Care* 9 (2006): 672–676.
3. Position of the American Dietetic Association and Dietitians of Canada: Vegetarian diets, *Journal of the American Dietetic Association* 103 (2003): 748–765.
4. Data from www.unicef.org, posted April 2005 and May 2006.
5. N. S. Scrimshaw, Fifty-five-year personal experience with human nutrition worldwide, *Annual Review of Nutrition* 27 (2007): 1–18.
6. J. F. Desjeux, Recent issues in energy-protein malnutrition in children, *Nestlé Nutrition Workshop Series: Pediatric Program* 58 (2006): 177–184.
7. D. R. Brewster, Critical appraisal of the management of severe malnutrition: 2. Dietary management, *Journal of Paediatrics and Child Health* 42 (2006): 575–582; A. A. Jackson, A. Ashworth, and S. Khanum, Improving child survival: Malnutrition Task Force and the paediatrician's responsibility, *Archives of Diseases in Childhood* 91 (2006): 706–710.

8. Committee on Dietary Reference Intakes, *Dietary Reference Intakes for Energy, Carbohydrate, Fiber, Fat, Fatty Acids, Cholesterol, Protein, and Amino Acids* (Washington, D.C.: National Academies Press, 2005), p. 694.
9. L. E. Kelemen and coauthors, Associations of dietary protein with disease and mortality in a prospective study of postmenopausal women, *American Journal of Epidemiology* 161 (2005): 239–249.
10. N. R. Matthan and coauthors, Effect of soy protein from differently processed products on cardiovascular disease risk factors and vascular endothelial function in hypercholesterolemic subjects, *American Journal of Clinical Nutrition* 85 (2007): 960–966; B. L. McVeigh and coauthors, Effect of soy protein varying in isoflavone content on serum lipids in healthy young men, *American Journal of Clinical Nutrition* 83 (2006): 244–251; Kelemen and coauthors, 2005.
11. M. Haim and coauthors, Serum homocysteine and long-term risk of myocardial infarction and sudden death in patients with coronary heart disease, *Cardiology* 107 (2006): 52–56; M. B. Kazemi and coauthors, Homocysteine level and coronary artery disease, *Angiology* 57 (2006): 9–14.
12. P. Berstad and coauthors, Dietary fat and plasma total homocysteine concentrations in 2 adult age groups: The Hordaland Homocysteine Study, *American Journal of Clinical Nutrition* 85 (2007): 1598–1605; J. Selhub, The many facets of hyperhomocysteinemia: Studies from the Framingham cohorts, *Journal of Nutrition* 136 (2006): 1726S–1730S.
13. C. Antoniades and coauthors, Asymmetrical dimethylarginine regulates endothelial function in methionine-induced but not in chronic homocystinemia in humans: Effect of oxidative stress and proinflammatory cytokines, *American Journal of Clinical Nutrition* 84 (2006): 781–788.
14. S. E. Chiuve and coauthors, Alcohol intake and methylenetetrahydrofolate reductase polymorphism modify the relation of folate intake to plasma homocysteine, *American Journal of Clinical Nutrition* 82 (2005): 155–162.
15. J. A. Troughton and coauthors, Homocysteine and coronary heart disease risk in the PRIME study, *Atherosclerosis* 191(2006): 90–97; Chiuve and coauthors, 2005.
16. T. J. Green and coauthors, Lowering homocysteine with B vitamins has no effect on biomarkers of bone turnover in older persons: A 2-y randomized controlled trial, *American Journal of Clinical Nutrition* 85 (2007): 460–464; D. Genser and coauthors, Homocysteine, folate and vitamin B_{12} in patients with coronary heart disease, *Annals of Nutrition & Metabolism* 50 (2006): 413–419; D. S. Wald and coauthors, Folic acid, homocysteine, and cardiovascular disease: Judging causality in the face of inconclusive trial evidence, *British Medical Journal* 333 (2006): 1114–1117.
17. B-Vitamin Treatment Trialists' Collaboration, Homocysteine-lowering trials for prevention of cardiovascular events: A review of the design and power of the large randomized trials, *American Heart Journal* 151 (2006): 282–287.
18. C. Baigent and R. Clarke, B Vitamins for the prevention of vascular diseases: Insufficient evidence to justify treatment, *Journal of the American Medical Association* 298 (2007): 1212–1214; R. L. Jamison and coauthors, Effect of homocysteine lowering on mortality and vascular disease in advanced chronic kidney disease and end-stage renal disease: A randomized controlled trial, *Journal of the American Medical Association* 298 (2007): 1163–1170; L. A. Bazzano and coauthors, Effect of folic acid supplementation on risk of cardiovascular diseases: A meta-analysis of randomized controlled trials, *Journal of the American Medical Association* 296 (2006): 2720–2726; K. H. Bonaa and coauthors, Homocysteine lowering and cardiovascular events after acute myocardial infarction, *New England Journal of Medicine* 354 (2006): 1578–1588; C. M. Carlsson, Homocysteine lowering with folic acid and vitamin B supplements: Effects on cardiovascular disease in older adults, *Drugs and Aging* 23 (2006): 491–502; E. Lonn and coauthors, Homocysteine lowering with folic acid and B vitamins in vascular disease, *New England Journal of Medicine* 354 (2006): 1567–1577.
19. S. G. West and coauthors, Oral L-arginine improves hemodynamic responses to stress and reduces plasma homocysteine in hypercholesterolemic men, *Journal of Nutrition* 135 (2005): 212–217.
20. B. S. Kendler, Supplemental conditionally essential nutrients in cardiovascular disease therapy, *Journal of Cardiovascular Nursing* 21 (2006): 9–16.
21. Committee on Dietary Reference Intakes, 2005, p. 841; Committee on Dietary Reference Intakes, *Dietary Reference Intakes for Calcium and Vitamin D* (Washington, D.C.: National Academies Press, 2011).
22. J. P. Bonjour, Dietary protein: An essential nutrient for bone health, *Journal of the American College of Nutrition* 24 (2005): 526S–536S; C. Weikert and coauthors, The relation between dietary protein, calcium and bone health in women: Results from the EPIC-Potsdam cohort, *Annals of Nutrition & Metabolism* 49 (2005): 312–318.
23. A. D. Conigrave, E. M. Brown, and R. Rizzoli, Dietary protein and bone health: Roles of amino acid-sensing receptors in the control of calcium metabolism and bone homeostasis, *Annual Review of Nutrition* 28 (2008): 131–155; A. Devine and coauthors, Protein consumption is an important predictor of lower limb bone mass in elderly women, *American Journal of Clinical Nutrition* 81 (2005): 1423–1428.
24. R. F. Kushner and B. Doerfler, Low-carbohydrate, high-protein diets revisited, *Current Opinion in Gastroenterology* 24 (2008): 198–203; A. Astrup, The satiating power of protein—A key to obesity prevention? *American Journal of Clinical Nutrition* 82 (2005): 1–2; D. S. Weigle and coauthors, A high-protein diet induces sustained reductions in appetite, ad libitum caloric intake, and body weight despite compensatory changes in diurnal plasma leptin and ghrelin concentrations, *American Journal of Clinical Nutrition* 82 (2005): 41–48.
25. A. M. Bernstein, L. Treyzon, and Z. Li, Are high-protein, vegetable-based diets safe for kidney function? A review of the literature, *Journal of the American Dietetic Association* 107 (2007): 644–650.
26. R. Pecoits-Filho, Dietary protein intake and kidney disease in Western diet, *Contributions to Nephrology* 155 (2007): 102–112; S. Mandayam and W. E. Mitch, Dietary protein restriction benefits patients with chronic kidney disease, *Nephrology* 11 (2006): 53–57.
27. R. R. Wolfe and S. L. Miller, The recommended dietary allowance of protein: A misunderstood concept, *Journal of the American Medical Association* 299 (2008): 2891–2893.
28. Position of the American Dietetic Association, Dietitians of Canada, and the American College of Sports Medicine, Nutrition and athletic performance, *Journal of the American Dietetic Association* 109 (2009): 509–527; S. M. Phillips, Dietary protein for athletes: From requirements to metabolic advantage, *Applied Physiology, Nutrition, and Metabolism* 31 (2006): 647–654.
29. S. M. Phillips, Dietary protein for athletes: From requirements to metabolic advantage, *Applied Physiology, Nutrition, and Metabolism* 31 (2006): 647–654.
30. L. L. Andersen and coauthors, The effect of resistance training combined with timed ingestion of protein on muscle fiber size and muscle strength, *Metabolism: Clinical and Experimental* 54 (2005): 151–156.
31. G. K. Grimble, Adverse gastrointestinal effects of arginine and related amino acids, *Journal of Nutrition* 137 (2007): 1693S–1701S.
32. *Dietary Reference Intakes—The Essential Guide to Nutrient Requirements*, (Washington, D.C.: National Academies Press, 2006), p. 152.
33. S. Takeshita and coauthors, A snack enriched with oral branched-chain amino acids prevents a fall in albumin in patients with liver cirrhosis undergoing chemoembolization for hepatocellular carcinoma, *Nutrition Research* 29 (2009): 89–93; T. Kawaguchi and coauthors, Branched-chain amino acid–enriched supplementation improves insulin resistance in patients with chronic liver disease, *International Journal of Molecular Medicine* (2008): 105–112; S. Khanna and S. Gopalan, Role of branched-chain amino acids in liver disease: The evidence for and against, *Current Opinion in Clinical Nutrition and Metabolic Care* (2007): 297–303; M. Charlton, Branched-chain amino acid enriched supplements as therapy for liver disease, *Journal of Nutrition* 136 (2006): 295S–298S; Y. Shimomura and coauthors, Branched-chain amino acid catabolism in exercise and liver disease, *Journal of Nutrition* 136 (2006): 250S–253S.
34. M. M. Perfect and coauthors, Use of complementary and alternative medicine for the treatment of genital herpes, *Herpes* 12 (2005): 38–41.
35. M. J. Smith and R. H. Garrett, A heretofore undisclosed crux of eosinophilia-myalgia syndrome: Compromised histamine degradation, *Inflammation Research* 54 (2005): 435–450.

HIGHLIGHT 6

Nutritional Genomics

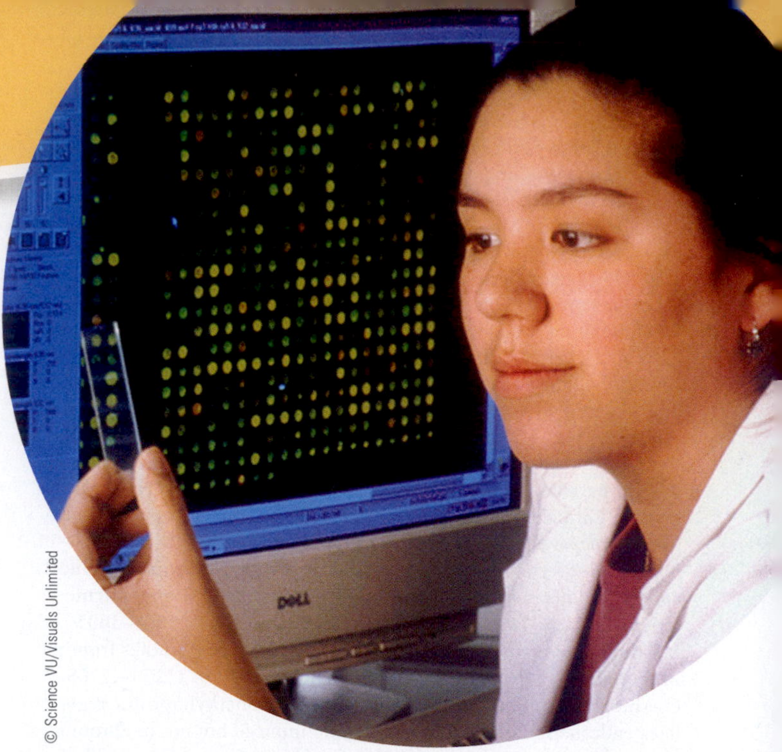

© Science VU/Visuals Unlimited

Imagine this scenario: A physician scrapes a sample of cells from inside your cheek and submits it to a **genomics** lab. The lab returns a report based on your genetic profile that reveals which diseases you are most likely to develop and makes recommendations for specific diet and lifestyle changes that can help you maintain good health. You may also be given a prescription for a dietary supplement that will best meet your personal nutrient requirements. Such a scenario may one day become reality as scientists uncover the relationships between **genetics,** diet, and disease. Until then, however, consumers need to know that current genetic test kits commonly available on the Internet are unproven and may create more problems than they resolve.[1]

Figure H6-1 introduces **nutritional genomics,** a new field of study that examines how nutrients influence gene activity (*nutrigenomics*) and how **genes** influence the activities of nutrients (*nutrigenetics*).[2] The accompanying glossary defines related terms.

Unlike sciences in the 20th century, nutritional genomics takes a comprehensive approach in analyzing information from several fields of study, providing an integrated understanding of the findings.[3] Consider how multiple disciplines contributed to our understanding of vitamin A over the past several decades, for example. Biochemistry revealed vitamin A's three chemical structures. Immunology identified the anti-infective properties of one of these structures while physiology focused on another structure and its role in vision. Epidemiology has reported improvements in the death rates and vision of malnourished children given vitamin A supplements, and biology has explored how such effects might be possible. A comprehensive understanding was slow to develop as researchers collected information on one gene, one action, and one nutrient at a time. Today's research in nutritional genomics involves all of the sciences, coordinating their multiple findings and explaining their interactions among several genes, actions, and nutrients in relatively little time. As a result, nutrition knowledge is growing at an incredibly fast pace.

The recent surge in genomics research grew from the Human Genome Project, an international effort by industry and government scientists to identify and describe all of the genes in the **human genome**—that is, all the genetic information contained within a person's cells. Completed in 2003, this project developed many of the research technologies needed to study genes and genetic variation. Scientists are now working to identify the individual proteins made by the genes, the genes associated with aging and diseases, and the dietary and lifestyle choices that most influence the expression of those genes.[4] Such information will have major implications for society in general, and for health care in particular.[5]

GLOSSARY

chromosomes: structures within the nucleus of a cell made of DNA and associated proteins. Human beings have 46 chromosomes in 23 pairs. Each chromosome has many genes.

DNA (deoxyribonucleic acid): the double helix molecules of which genes are made.

epigenetics: the study of heritable changes in gene function that occur without a change in the DNA sequence.

gene expression: the process by which a cell converts the genetic code into RNA and protein.

genes: sections of chromosomes that contain the instructions needed to make one or more proteins.

genetics: the study of genes and inheritance.

genomics: the study of all the genes in an organism and their interactions with environmental factors.

human genome (GEE-nome): the complete set of genetic material (DNA) in a human being.

methylation: the addition of a methyl group (CH_3).

microarray technology: research tools that analyze the expression of thousands of genes simultaneously and search for particular gene changes associated with a disease. DNA microarrays are also called *DNA chips.*

mutations: permanent changes in the DNA that can be inherited.

nucleotide bases: the nitrogen-containing building blocks of DNA and RNA—cytosine (C), thymine (T), uracil (U), guanine (G), and adenine (A). In DNA, the base pairs are A–T and C–G and in RNA, the base pairs are A–U and C–G.

nucleotides: the subunits of DNA and RNA molecules, composed of a phosphate group, a 5-carbon sugar (deoxyribose for DNA and ribose for RNA), and a nitrogen-containing base.

nutritional genomics: the science of how food (and its components) interacts with the genome. The study of how nutrients affect the activities of genes is called *nutrigenomics.* The study of how genes affect the activities of nutrients is called *nutrigenetics.*

phenylketonuria (FEN-il-KEY-toe-NEW-ree-ah) or **PKU:** an inherited disorder characterized by failure to metabolize the amino acid phenylalanine to tyrosine.

RNA (ribonucleic acid): a compound similar to DNA, but RNA is a single strand with a ribose sugar instead of a deoxyribose sugar and uracil instead of thymine as one of its bases.

FIGURE H6-1 Nutritional Genomics

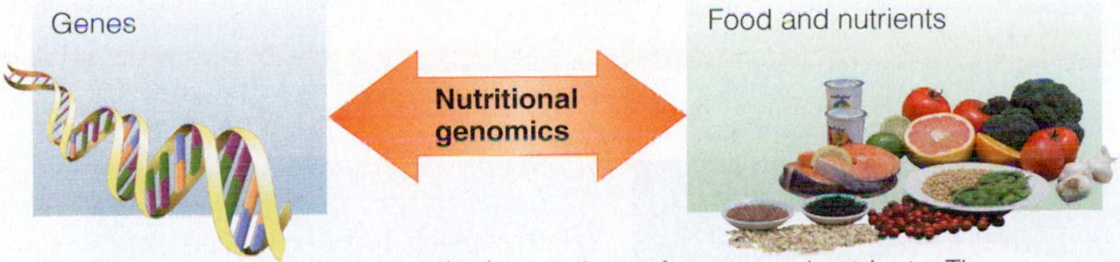

Nutritional genomics examines the interactions of genes and nutrients. These interactions include both nutrigenetics and nutrigenomics.

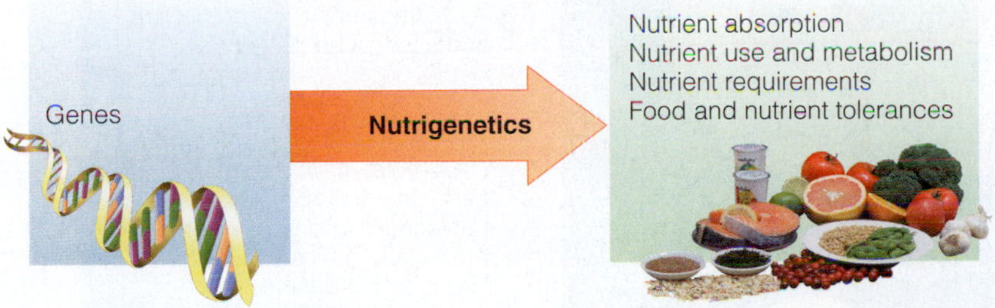

Nutrigenetics (or nutritional genetics) examines how genes influence the activities of nutrients.

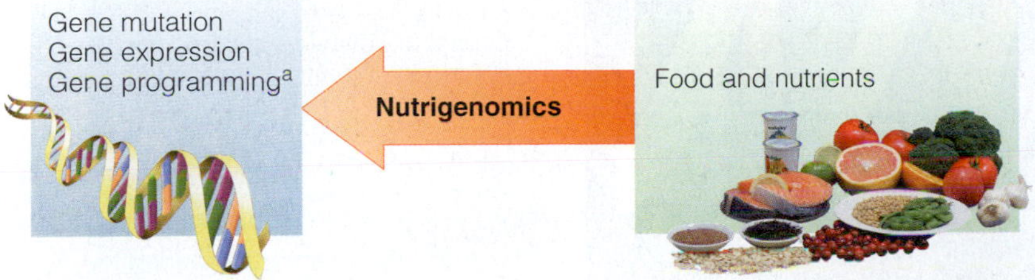

Nutrigenomics, which includes epigenetics, examines how nutrients influence the activities of genes.

[a]Chapter 14 introduces programming and describes how a mother's nutrition can permanently change gene expression in the fetus with consequences for future generations.

A Genomics Primer

Figure H6-2 (p. 200) shows the relationships among the materials that comprise the genome. As Chapter 6's discussion of protein synthesis points out, genetic information is encoded in DNA molecules within the nucleus of cells. The **DNA (deoxyribonucleic acid)** molecules and associated proteins are packed within 46 **chromosomes.** The genes are segments of a DNA strand that can eventually be translated into one or more proteins. The sequence of **nucleotide bases** within each gene determines the amino acid sequence of a particular protein. Scientists currently estimate that there are between 20,000 and 25,000 genes in the human genome.

As Figure 6-7 (p. 179) explains, when cells make proteins, a DNA sequence is used to make messenger **RNA (ribonucleic acid).** The **nucleotide** sequence in messenger RNA then determines the amino acid sequence to make a protein. This process—from genetic information to protein synthesis—is known as **gene expression.** Gene expression can be determined by measuring the amounts of messenger RNA in a tissue sample. **Microarray technology** (see photo on p. 198) allows researchers to detect messenger RNA and analyze the expression of thousands of genes simultaneously.

Simply having a certain gene does not determine that its associated trait will be expressed; the gene has to be activated. (Similarly, owning lamps does not ensure you will have light in your home unless you turn them on.) Nutrients are among many environmental factors that play key roles in either activating or silencing genes. Switching genes on and off does not change the DNA itself, but it can have dramatic consequences for a person's health.

The area of study that examines how environmental factors influence gene expression without changing the DNA is known as **epigenetics.**[6] To turn genes on, enzymes attach proteins near the beginning of a gene. If enzymes attach a methyl group (CH_3) instead, the protein is blocked from binding to the gene, and the gene remains switched off. Other factors influence gene expression as well, but methyl groups are currently the most well understood.[7] They also are known to have dietary connections.

The accompanying photo of two mice illustrates epigenetics and how diet can influence genetic traits such as hair color and body weight. Both mice have a gene that tends to produce fat, yellow pups, but their mothers were given different diets during pregnancy. The mother of the mouse on the right was given a dietary supplement containing the B vitamins folate and vitamin B_{12}. These nutrients silenced the gene for "yellow and fat," resulting in brown pups with normal appetites. As Chapter 10 explains, one of the main roles of these B vitamins is to transfer methyl groups. In the

Both of these mice have the gene that tends to produce fat, yellow pups, but their mothers had different diets. The mother of the mouse on the right received a dietary supplement, which silenced the gene, resulting in brown pups with normal appetites.

FIGURE H6-2 The Human Genome

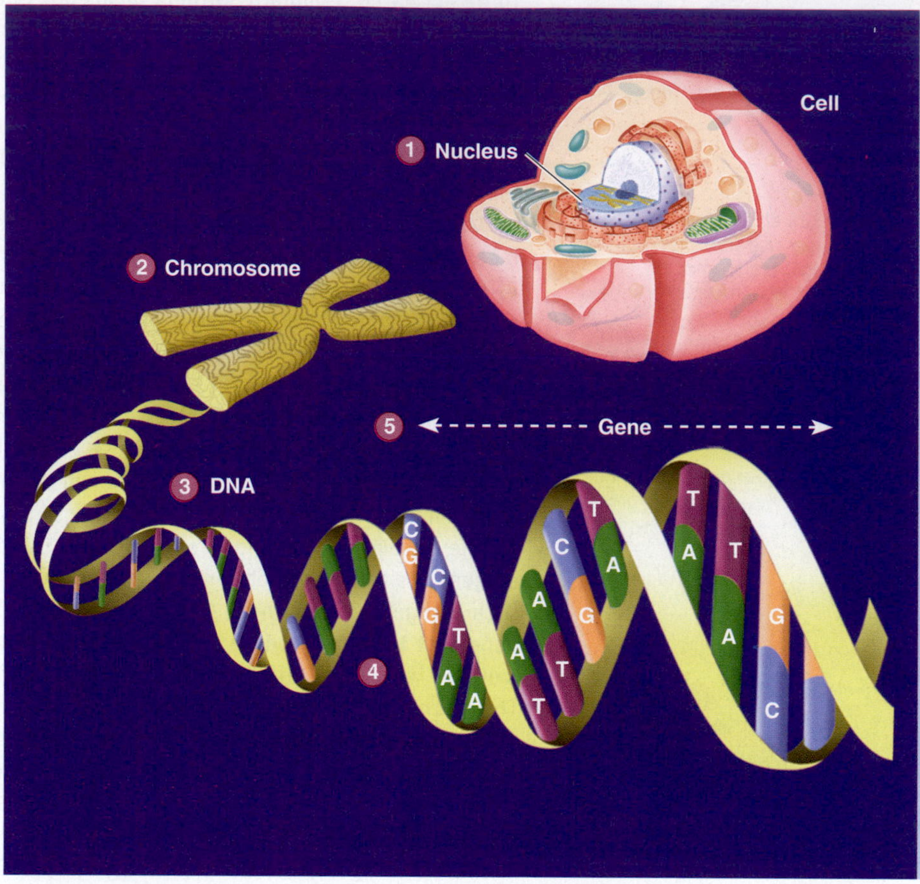

1 The human genome is a complete set of genetic material organized into 46 chromosomes, located within the nucleus of a cell.

2 A chromosome is made of DNA and associated proteins.

3 The double helical structure of a DNA molecule is made up of two long chains of nucleotides. Each nucleotide is composed of a phosphate group, a 5-carbon sugar, and a base.

4 The sequence of nucleotide bases (C, G, A, T) determines the amino acid sequence of proteins. These bases are connected by hydrogen bonding to form base pairs—adenine (A) with thymine (T) and guanine (G) with cytosine (C).

5 A gene is a segment of DNA that includes the information needed to synthesize one or more proteins.

SOURCE: Adapted from "A Primer: From DNA to Life," Human Genome Project, U.S. Department of Energy Office of Science; www.orn.gov/sci/techresources/Human_Genome/primer-pic.shtml.

case of the supplemented mice, methyl groups migrated onto DNA and silenced several genes, thus producing brown coats and protecting against the development of obesity and some related diseases. Keep in mind that these changes occurred epigenetically. In other words, the DNA sequence within the genes of the mice remained the same. Nutrition and other environmental factors can influence genes in a way that creates inheritable changes in the body's metabolism and susceptibility to disease.[8] In this way, the dietary habits of parents, and even grandparents, can influence future generations.

Many nutrients and phytochemicals regulate gene expression and influence health through their involvement in DNA **methylation.**[9] Some, such as folate, silence genes and protect against some cancers by providing methylation.[10] Others, such as a phytochemical found in green tea, activate genes and protect against some cancers by inhibiting methylation activity.[11] Whether silencing or activating a gene is beneficial or harmful depends on what the gene does. Silencing a gene that stimulates cancer growth, for example, would be beneficial, but silencing a gene that suppresses cancer growth would be harmful. Similarly, activating a gene that defends against obesity would be beneficial, but activating a gene that promotes obesity would be harmful. Figure H6-3 illustrates how nutrient regulation of gene expression can influence a person's health. Much research is under way to determine which nutrients activate or silence which genes.

Genetic Variation and Disease

Except for identical twins, no two persons are genetically identical. Even then, over the years a particular gene may become active in one twin and silenced in the other because of epigenetic changes.

The variation in the genomes of any two persons is only about 0.1 percent, a difference of only one nucleotide base in every 1000. Yet it is this incredibly small difference that makes each of us unique and explains why, given the same environmental influences, some of us develop certain diseases and others do not.

FIGURE H6-3 Nutrient Regulation of Gene Expression

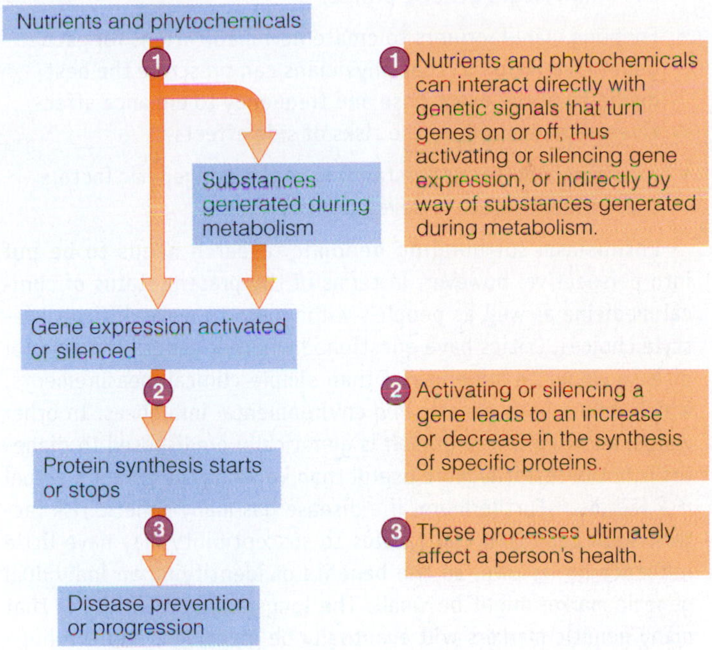

Nutrients and phytochemicals
1
Substances generated during metabolism
Gene expression activated or silenced
2
Protein synthesis starts or stops
3
Disease prevention or progression
1 Nutrients and phytochemicals can interact directly with genetic signals that turn genes on or off, thus activating or silencing gene expression, or indirectly by way of substances generated during metabolism.
2 Activating or silencing a gene leads to an increase or decrease in the synthesis of specific proteins.
3 These processes ultimately affect a person's health.

Similarly, genetic variation explains why some of us respond to interventions such as diet and others do not. For example, following a diet low in saturated fats will significantly lower LDL cholesterol for most people, but the degree of change varies dramatically among individuals, with some people having only a small decrease or even a slight increase.[12] In other words, dietary factors may be more helpful or more harmful depending on a person's particular genetic variations.[13] Such findings help to explain some of the conflicting results from research studies. The goal of nutritional genomics is to custom design *specific* recommendations that fit the needs of *each* individual. Such personalized recommendations are expected to provide more effective disease prevention and treatment solutions.

Diseases characterized by a single-gene disorder are genetically predetermined, usually exert their effects early in life, and greatly affect those touched by them; such diseases are relatively rare. The cause and effect of single-gene disorders is clear—those with the genetic defect get the disease and those without it don't. In contrast, the more common diseases, such as heart disease and cancer, are influenced by many genes and typically develop over several decades. These chronic diseases have multiple genetic components that *predispose* the prevention or development of a disease, depending on a variety of environmental factors (such as smoking, diet, and physical activity).[14] Both types are of interest to researchers studying nutritional genomics.

Single-Gene Disorders

Some disorders are caused by **mutations** in single genes that are inherited at birth. The consequences of a missing or malfunctioning protein can seriously disrupt metabolism and may require significant dietary or medical intervention. A classic example of a diet-related, single-gene disorder is **phenylketonuria,** or **PKU.**

Approximately one in every 15,000 infants in the United States is born with PKU. PKU arises from mutations in the gene that codes for the enzyme that converts the essential amino acid phenylalanine to the amino acid tyrosine. Without this enzyme, phenylalanine and its metabolites accumulate and damage the nervous system, resulting in mental retardation, seizures, and behavior abnormalities. At the same time, the body cannot make tyrosine or compounds made from it (such as the neurotransmitter epinephrine). Consequently, tyrosine becomes an essential amino acid: because the body cannot make it, the diet must supply it.

Although the most debilitating effect is on brain development, other symptoms of PKU become evident if the condition is left untreated. Infants with PKU may have poor appetites and grow slowly. They may be irritable or have tremors or seizures. Their bodies and urine may have a musty odor. Their skin coloring may be unusually pale, and they may develop skin rashes.

The effect of nutrition intervention in PKU is remarkable. In fact, the only current treatment for PKU is a diet that restricts phenylalanine and supplies tyrosine to maintain blood levels of these amino acids within safe ranges. Because all foods containing protein provide phenylalanine, the diet must depend on a special formula to supply a phenylalanine-free source of energy, protein, vitamins, and minerals. If the restricted diet is conscientiously followed, the symptoms can be prevented. Because phenylalanine is an essential amino acid, the diet cannot exclude it completely. Children with PKU need phenylalanine to grow, but they cannot handle excesses without detrimental effects. Therefore, their diets must provide enough phenylalanine to support normal growth and health but not enough to cause harm. The diet must also provide tyrosine. To ensure that blood concentrations of phenylalanine and tyrosine are close to normal, children and adults who have PKU must have blood tests periodically and adjust their diets as necessary.

Multigene Disorders

In multigene disorders, several genes can influence the progression of a disease, but no single gene causes the disease on its own. For this reason, genomics researchers must study the expression and interactions of *multiple* genes. Because multigene disorders are often sensitive to interactions with environmental influences, they are not as straightforward as single-gene disorders.

Heart disease provides an example of a chronic disease with multiple gene and environmental influences. Consider that major risk factors for heart disease include elevated blood cholesterol levels, obesity, diabetes, and hypertension, yet the underlying genetic and environmental causes of any of these individual risk factors is not completely understood. Genomic research can reveal details about each of these risk factors.[15] For example, tests could determine whether blood cholesterol levels are high due to increased cholesterol absorption or production or because of decreased cholesterol degradation. This information could then guide physicians and dietitians to prescribe the most appropriate medical and dietary interventions from among many possible solutions.[16] Today's dietary recommendations advise a low-fat diet, which helps people with a small type of LDL but not those with the large type. In fact, a low-fat diet is actually more harmful for people with the large type. Finding the best option for each person will be a challenge given the many possible interactions between genes and environmental factors and the millions of possible gene variations in the human genome that make each individual unique.[17]

The results of genomic research are helping to explain findings from previous nutrition research. Consider dietary fat and heart disease, for example. As Highlight 5 explains, epidemiological and clinical studies have found that a diet high in unsaturated fatty acids often helps to maintain a healthy blood lipid profile. Now genetic studies offer an underlying explanation of this relationship: diets rich in polyunsaturated fatty acids activate genes responsible for making enzymes that break down fats and silence genes responsible for making enzymes that synthesize fats.[18] Both actions change fat metabolism in the direction of lowering blood lipids.

To learn more about how individuals respond to diet, researchers examine the genetic differences between people. The most common genetic differences involve a change in a single nucleotide base located in a particular region of a DNA strand—thymine replacing cytosine, for example. Such variations are called single nucleotide polymorphisms (SNPs), and they commonly occur throughout the genome. Many SNPs (commonly pronounced "snips") have no effect on cell activity. In fact, SNPs are significant only if they affect the amino acid sequence of a protein in a way that alters its function *and* if that function is critical to the body's well-being. Research on a gene that plays a key role in lipid metabolism reveals differences in a person's response to diet depending on whether the gene has a common SNP. People with the SNP have lower LDL when eating a diet rich in polyunsaturated fatty acids—and higher LDL with a low intake—than those without the SNP.[19] These findings clearly show how diet (in this case, polyunsaturated fat) interacts with a gene (in this case, a fat metabolism gene with a SNP) to influence the development of a disease (changing blood lipids implicated in heart disease).

Clinical Concerns

Because multigene, chronic diseases are common, an understanding of the human genome will have widespread ramifications for health care.[20] This new understanding of the human genome is expected to change health care by:

- Providing knowledge of an individual's genetic predisposition to specific diseases.
- Allowing physicians to develop "designer" therapies—prescribing the most effective schedule of screening, behavior changes (including diet), and medical interventions based on each individual's genetic profile.
- Enabling manufacturers to create new medications for each genetic variation so that physicians can prescribe the best medicine in the exact dose and frequency to enhance effectiveness and minimize the risks of side effects.[21]
- Providing a better understanding of the nongenetic factors that influence disease development.

Enthusiasm surrounding genomic research needs to be put into perspective, however, in terms of the present status of clinical medicine as well as people's willingness to make difficult lifestyle choices. Critics have questioned whether genetic markers for disease would be more useful than simple clinical measurements, which reflect both genetic *and* environmental influences. In other words, knowing that a person is genetically predisposed to diabetes is not necessarily more useful than knowing the person's actual risk factors.[22] Furthermore, if a disease has many genetic risk factors, each gene that contributes to susceptibility may have little influence on its own, so the benefits of identifying an individual genetic marker might be small. The long-range possibility is that many genetic markers will eventually be identified, and the hope is that the combined information will be a useful and accurate predictor of disease.

Having the knowledge to prevent disease and actually taking action do not always coincide. Despite the abundance of current dietary recommendations, many people are unwilling to make behavior changes known to improve their health. For example, it has been estimated that heart disease and type 2 diabetes are 90 percent preventable when people adopt an appropriate diet, maintain a healthy body weight, and exercise regularly. Yet these two diseases remain among the leading causes of death. Given the difficulty that many people have with current recommendations, it may be unrealistic to expect that they will enthusiastically adopt an even more detailed list of lifestyle modifications. Then again, compliance may be better when it is supported by information based on a person's own genetic profile and the knowledge that the epigenetic profile can be changed.

The debate over nature versus nurture—whether genes or the environment are more influential—has quieted. The focus has shifted. Scientists acknowledge the important roles of each and understand the real answers lie within the myriad interactions. Current research is sorting through how nutrients and other dietary factors interact with genes to confer health benefits or risks. Answers from genomic research may not become apparent for years to come, but the opportunities and rewards may prove well worth the efforts.[23]

Nutrition on the Net

For further study of topics covered in this highlight, log on to **www.cengagebrain.com** and search for ISBN 084006845X.

- Get information about human genomic discoveries and how they can be used to improve health from the Office of Public Health Genomics site of the Centers for Disease Control: **www.cdc.gov/genomics**

References

1. A. L. McGuire and W. Burke, An unwelcome side effect of direct-to-customer personal genome testing: Raiding the medical commons, *Journal of the American Medical Association* 300 (2008): 2669–2671.
2. P. J. Stover and M. A. Caudill, Genetic and epigenetic contributions to human nutrition and health: Managing genome-diet interactions, *Journal of the American Dietetic Association* 108 (2008): 1480–1487; J. Kaput, Nutrigenomics–2006 update, *Clinical Chemistry and Laboratory Medicine* 45 (2007): 279–287.
3. G. T. Keusch, What do *-omics* mean for the science and policy of the nutritional sciences? *American Journal of Clinical Nutrition* 83 (2006): 520S–522S.
4. G. W. Duff, Influence of genetics on disease susceptibility and progression, *Nutrition Reviews* 65 (2007): S177–S181; P. J. Gillies, Preemptive nutrition of pro-inflammatory states: A nutrigenomic model, *Nutrition Reviews* 65 (2007): S217–S220; T. A. Manolio, Study designs to enhance identification of genetic factors in healthy aging, *Nutrition Reviews* 65 (2007): S228–S233.
5. A. P. Feinberg, Epigenetics at the epicenter of modern medicine, *Journal of the American Medical Association* 299 (2008): 1345–1350.
6. G. P. Kauwell, Epigenetics: What it is and how it can affect dietetics practice, *Journal of the American Dietetic Association* 108 (2008): 1056–1059.
7. M. Esteller, Epigenetics in cancer, *New England Journal of Medicine* 358 (2008): 1148–1159; L. Cobiac, Epigenomics and nutrition, *Forum of Nutrition* 60 (2007): 31–41; T. M. Edwards and J. P. Myers, Environmental exposures and gene regulation in disease etiology, *Environmental Health Perspectives* 115 (2007): 1264–1270.
8. S. A. Ross and coauthors, Introduction: Diet, epigenetic events and cancer prevention, *Nutrition Reviews* 66 (2008): S1–S6; R. L. Jirtle and M. K. Skinner, Environmental epigenomics and disease susceptibility, *Nature Reviews: Genetics* 8 (2007): 253–262; R. A. Waterland and K. B. Michels, Epigenetic epidemiology of the developmental origins hypothesis, *Annual Review of Nutrition* 27 (2007): 363–388.
9. M. P. Lee and B. K. Dunn, Influence of genetic inheritance on global epigenetic states and cancer risk prediction with DNA methylation signature: Challenges in technology and data analysis, *Nutrition Reviews* 66 (2008): S69–S72.
10. C. M. Ulrich, M. C. Reed, and H. F. Nijhout, Modeling folate, one-carbon metabolism, and DNA methylation, *Nutrition Reviews* 66 (2008): S27–S30.
11. C. S. Yang and coauthors, Reverse of hypermethylation and reactivation of genes by dietary polyphenolic compounds, *Nutrition Reviews* 66 (2008): S18–S20.
12. D. Corella and J. M. Ordovas, Single nucleotide polymorphisms that influence lipid metabolism: Interaction with dietary factors, *Annual Review of Nutrition* 25 (2005): 341–390.
13. B. Fontaine-Bisson and coauthors, Genetic polymorphisms of tumor necrosis factor-α modify the association between dietary polyunsaturated fatty acids and fasting HDL-cholesterol and apo A-I concentrations, *American Journal of Clinical Nutrition* 86 (2007): 768–774; E. A. Ruiz-Narváez, P. Kraft, and H. Campos, Ala12 variant of the peroxisome proliferator-activated receptor-γ gene (*PPARG*) is associated with higher polyunsaturated fat in adipose tissue and attenuates the protective effect of polyunsaturated fat intake on the risk of myocardial infarction, *American Journal of Clinical Nutrition* 86 (2007): 1238–1242; E. Trujillo, C. Davis, and J. Milner, Nutrigenomics, proteomics, metabolomics, and the practice of dietetics, *Journal of the American Dietetic Association* 106 (2006): 403–413.
14. L. R. Ferguson and M. Philpott, Nutrition and mutagenesis, *Annual Review of Nutrition* 28 (2008): 313–329; J. Kaput and coauthors, The case for strategic international alliances to harness nutritional genomics for public and personal health, *British Journal of Nutrition* 94 (2005): 623–632.
15. J. H. Hardy and A. Singleton, Genomewide association studies and human disease, *New England Journal of Medicine* 360 (2009): 1759–1768.
16. R. M. DeBusk and coauthors, Nutritional genomics in practice: Where do we begin? *Journal of the American Dietetic Association* 105 (2005): 589–597.
17. J. M. Ordovas, Nutrigenetics, plasma lipids, and cardiovascular risk, *Journal of the American Dietetic Association* 106 (2006): 1074–1081.
18. H. Sampath and J. M. Ntambi, Polyunsaturated fatty acid regulation of genes of lipid metabolism, *Annual Review of Nutrition* 25 (2005): 317–340.
19. E. S. Tai and coauthors, Polyunsaturated fatty acids interact with PPARA–L162V polymorphism to affect plasma triglyceride apolipoprotein C-III concentrations in the Framingham Heart Study, *Journal of Nutrition* 135 (2005): 397–403.
20. M. M. Bergmann, U. Görman, and J. C. Mathers, Bioethical considerations for human nutrigenomics, *Annual Review of Nutrition* 28 (2008): 447–467; J. P. Evans, Health care in the age of genetic medicine, *Journal of the American Medical Association* 298 (2007): 2670–2672; S. Vakili and M. A. Caudill, Personalized nutrition: Nutritional genomics as a potential tool for targeted medical nutrition therapy, *Nutrition Reviews* 65 (2007): 301–315.
21. S. B. Shurin and E. G. Nabel, Pharmacogenomics—Ready for prime time? *New England Journal of Medicine* 358 (2008): 1061–1063.
22. J. B. Meigs and coauthors, Genotype score in addition to common risk factors for prediction of type 2 diabetes, *New England Journal of Medicine* 359 (2008): 2208–2219.
23. A. E. Guttmacher and F. S. Collins, Realizing the promise of genomics in biomedical research, *Journal of the American Medical Association* 294 (2005): 1399–1402.

CHAPTER

7

Elnur/Shutterstock.com

Nutrition in Your Life

You eat breakfast and hustle off to class. After lunch, you study for tomorrow's exam. Dinner is followed by an evening of dancing. Do you ever think about how the food you eat powers the activities of your life? What happens when you don't eat—or when you eat too much? Learn how the cells of your body transform carbohydrates, fats, and proteins into energy—and what happens when you give your cells too much or too little of any of these nutrients. Discover the metabolic pathways that lead to body fat and those that support physical activity. It's really quite fascinating.

Throughout this chapter, the CourseMate icon indicates an opportunity for online self-study, linking you to activities to increase your understanding of chapter concepts. **www.cengagebrain.com** (search for ISBN 084006845X)

Metabolism: Transformations and Interactions

CHAPTER OUTLINE

Energy makes it possible for people to breathe, ride bicycles, compose music, and do everything else they do. As Chapter 1 explains, *energy* is the capacity to do work. Although every aspect of our lives depends on energy, the concept of energy can be difficult to grasp because it cannot be seen or touched, and it manifests in various forms, including heat, mechanical, electrical, and chemical energy. In the body, heat energy maintains a constant body temperature, mechanical energy moves muscles, and electrical energy sends nerve impulses. Energy is stored in foods and in the body as chemical energy. This chemical energy powers the myriad activities of all cells.

All the energy that sustains human life initially comes from the sun—the ultimate source of energy. During **photosynthesis**, plants make simple sugars from carbon dioxide and capture the sun's light energy in the chemical bonds of those sugars. Then human beings eat either the plants or animals that have eaten the plants. These foods provide energy, but how does the body obtain that energy from foods? This chapter answers that question by following the nutrients that provide the body with **fuel** through a series of reactions that release energy from their chemical bonds. As the bonds break, they release energy in a controlled version of the same process by which wood burns in a fire. Both wood and food have the potential to provide energy. When wood burns in the presence of oxygen, it generates heat and light (energy), steam (water), and some carbon dioxide and ash (waste). Similarly, during **metabolism**, the body releases energy, water, and carbon dioxide (and other waste products).

By studying metabolism, you will understand how the body uses foods to meet its needs and why some foods meet those needs better than others. Readers who are interested in weight control will discover which foods contribute most to body fat and which to select when trying to gain or lose weight safely. Readers who are physically active will discover which foods best support endurance activities and which to select when trying to build lean body mass.

photosynthesis: the process by which green plants use the sun's energy to make carbohydrates from carbon dioxide and water.
- **photo** = light
- **synthesis** = put together (making)

fuel: compounds that cells can use for energy. The major fuels include glucose, fatty acids, and amino acids; other fuels include ketone bodies, lactate, glycerol, and alcohol.

metabolism: the sum total of all the chemical reactions that go on in living cells. Energy metabolism includes all the reactions by which the body obtains and expends the energy from food.
- **metaballein** = change

Chemical Reactions in the Body

Earlier chapters introduce some of the body's chemical reactions: the making and breaking of the bonds in carbohydrates, lipids, and proteins. Metabolism is the sum of these and all the other chemical reactions that go on in living cells; *energy metabolism* includes all the ways the body obtains and uses energy from food.

The Site of Metabolic Reactions—Cells The human body is made up of trillions of cells, and each cell busily conducts its metabolic work all the time. (Appendix A presents a brief summary of the structure and function of the cell.) Figure 7-1 depicts a typical cell and shows where the major reactions of energy metabolism take place. The type and extent of metabolic activities vary depending on the type of cell, but of all the body's cells, the liver cells are the most versatile and metabolically active. Table 7-1 offers insights into the liver's work.

The Building Reactions—Anabolism Earlier chapters describe how condensation reactions combine the basic units of energy-yielding nutrients to build body compounds. Glucose molecules may be joined together to make glycogen chains. Glycerol and fatty acids may be assembled into triglycerides. Amino acids may be linked together to make proteins. Each of these reactions starts with small, simple compounds and uses them as building blocks to form larger, more complex structures. Because such reactions involve doing work, they require energy. The building up of body compounds is known as **anabolism.** Anabolic reactions are represented in this book, wherever possible, with "up" arrows in chemical diagrams (such as those shown at the top of Figure 7-2).

anabolism (an-AB-o-lism): reactions in which small molecules are put together to build larger ones. Anabolic reactions require energy.
- **ana** = (build) up

catabolism (ca-TAB-o-lism): reactions in which large molecules are broken down to smaller ones. Catabolic reactions release energy.
- **kata** = (break) down

The Breakdown Reactions—Catabolism The breaking down of body compounds is known as **catabolism**; catabolic reactions release energy and are represented, wherever possible, by "down" arrows in chemical diagrams (as in the bottom of Figure 7-2). Earlier chapters describe how hydrolysis reactions break

FIGURE 7-1 **A Typical Cell (Simplified Diagram)**

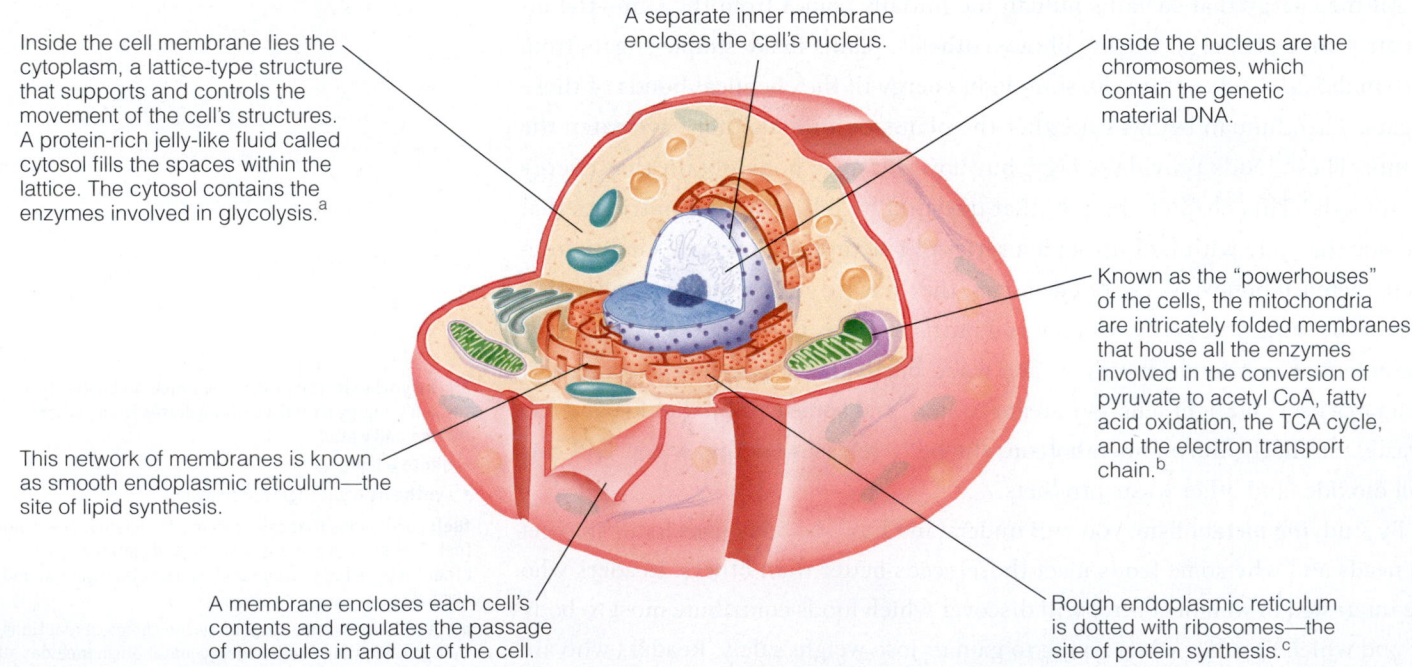

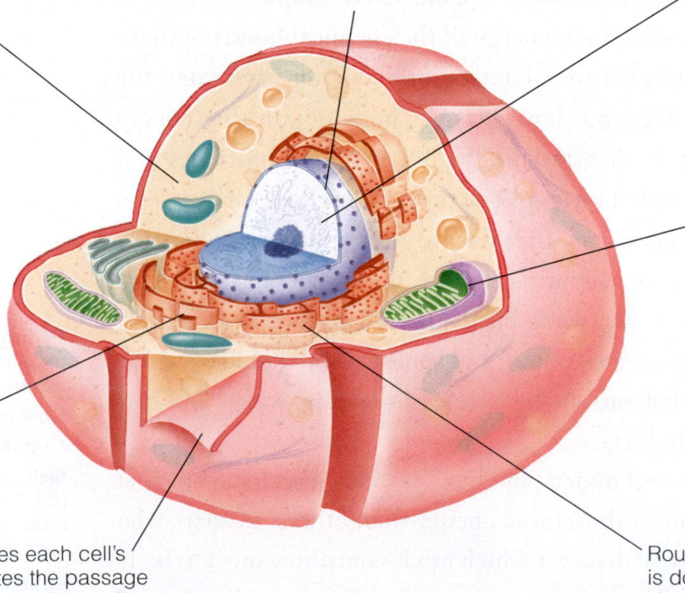

Inside the cell membrane lies the cytoplasm, a lattice-type structure that supports and controls the movement of the cell's structures. A protein-rich jelly-like fluid called cytosol fills the spaces within the lattice. The cytosol contains the enzymes involved in glycolysis.[a]
A separate inner membrane encloses the cell's nucleus.

A membrane encloses each cell's contents and regulates the passage of molecules in and out of the cell.

[a]Glycolysis is introduced on p. 211.
[b]The conversion of pyruvate to acetyl CoA, fatty acid oxidation, the TCA cycle, and the electron transport chain are described later in the chapter.
[c]Figure 6-7 on p. 179 describes protein synthesis.

TABLE 7-1 Metabolic Work of the Liver

The liver is the most active processing center in the body. When nutrients enter the body from the digestive tract, the liver receives them first; then it metabolizes, packages, stores, or ships them out for use by other organs. When alcohol, drugs, or poisons enter the body, they are also sent directly to the liver; here they are detoxified and their by-products shipped out for excretion. An enthusiastic anatomy and physiology professor once remarked that given the many vital activities of the liver, we should express our feelings for others by saying, "I love you with all my liver" instead of "with all my heart." Granted, this declaration lacks romance, but it makes a valid point. Here are just some of the many jobs performed by the liver. To renew your appreciation for this remarkable organ, review Figure 3-12 (p. 82).

Carbohydrates

- Metabolizes fructose, galactose, and glucose
- Makes and stores glycogen
- Breaks down glycogen and releases glucose
- Breaks down glucose for energy when needed
- Makes glucose from some amino acids and glycerol when needed
- Converts excess glucose to fatty acids

Lipids

- Builds and breaks down triglycerides, phospholipids, and cholesterol as needed
- Breaks down fatty acids for energy when needed
- Packages extra lipids in lipoproteins for transport to other body organs
- Manufactures bile to send to the gallbladder for use in fat digestion
- Makes ketone bodies when necessary

Proteins

- Manufactures nonessential amino acids that are in short supply
- Removes from circulation amino acids that are present in excess of need and converts them to other amino acids or deaminates them and converts them to glucose or fatty acids
- Removes ammonia from the blood and converts it to urea to be sent to the kidneys for excretion
- Makes other nitrogen-containing compounds the body needs (such as bases used in DNA and RNA)
- Makes many proteins

Other

- Detoxifies alcohol, other drugs, and poisons; prepares waste products for excretion
- Helps dismantle old red blood cells and captures the iron for recycling
- Stores most vitamins and many minerals
- Activates vitamin D

FIGURE 7-2 Anabolic and Catabolic Reactions Compared

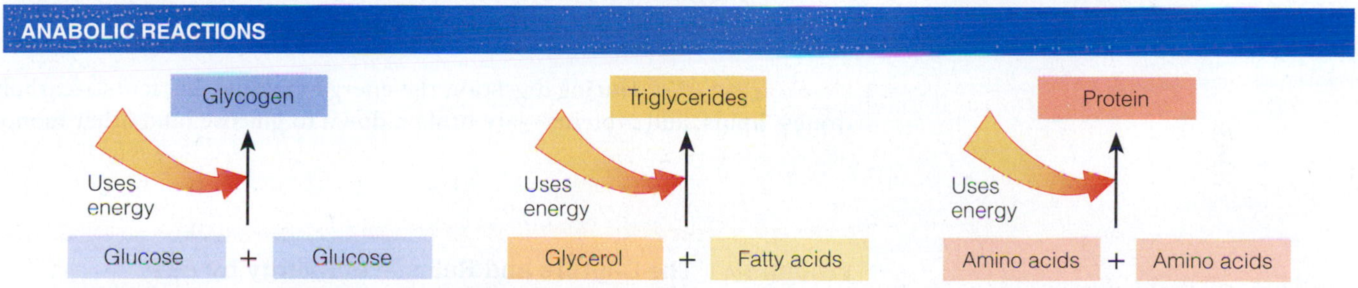

Anabolic reactions include the making of glycogen, triglycerides, and protein; these reactions require differing amounts of energy.

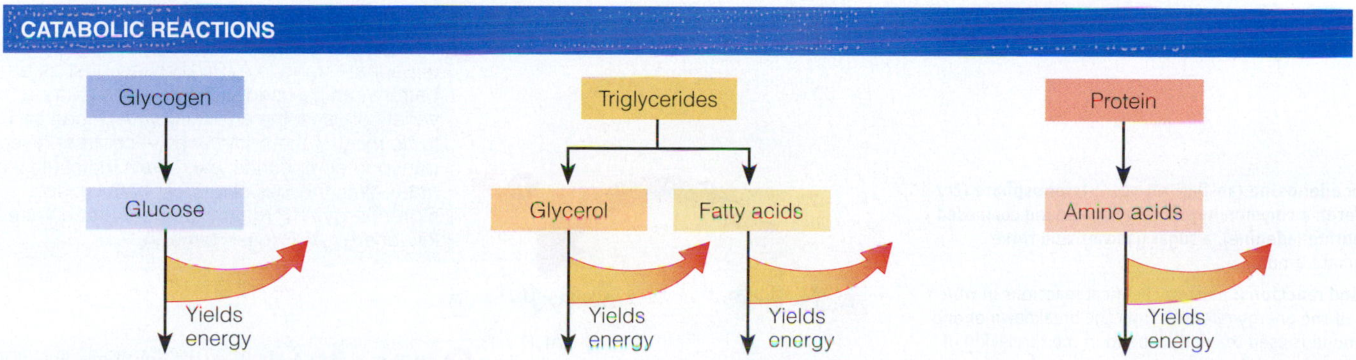

Catabolic reactions include the breakdown of glycogen, triglycerides, and protein; the further catabolism of glucose, glycerol, fatty acids, and amino acids releases differing amounts of energy. Much of the energy released is captured in the bonds of adenosine triphosphate (ATP).

NOTE: You need not memorize a color code to understand the figures in this chapter, but you may find it helpful to know that blue is used for carbohydrates, yellow for fats, and red for proteins.

FIGURE 7-3 **ATP (Adenosine Triphosphate)**

ATP is one of the body's high-energy molecules. Notice that the bonds connecting the three phosphate groups have been drawn as wavy lines, indicating a high-energy bond. When these bonds are broken, energy is released.

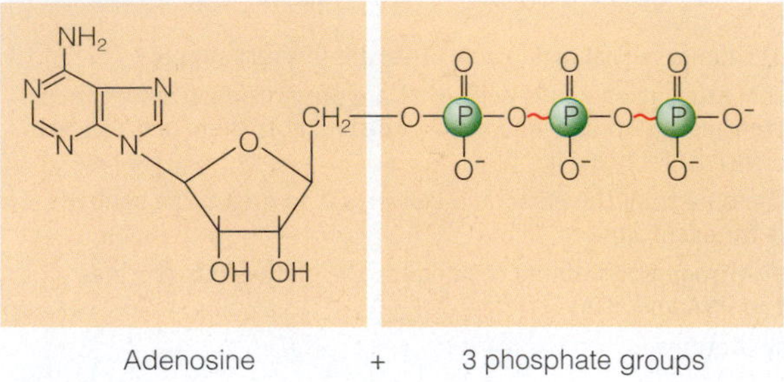

♦ ATP = A—P~P~P.
Each ~ denotes a "high-energy" bond.

down glycogen to glucose, triglycerides to fatty acids and glycerol, and proteins to amino acids. When the body needs energy, it breaks down any or all of these four basic units into even smaller units, as described later.

The Transfer of Energy in Reactions—ATP Some of the energy released during the breakdown of glucose, glycerol, fatty acids, and amino acids from foods is captured in the high-energy storage compound **ATP (adenosine triphosphate)**. ATP, as its name indicates, contains three phosphate groups (see Figure 7-3). ♦ The negative charges on the phosphate groups make ATP vulnerable to hydrolysis. When the bonds between the phosphate groups are hydrolyzed, they readily break, splitting off one or two phosphate groups and releasing energy. In this way, ATP provides the energy that powers all the activities of living cells. Figure 7-4 describes how the body captures and releases energy in the bonds of ATP.

Quite often, the hydrolysis of ATP occurs simultaneously with reactions that will use that energy—a metabolic duet known as **coupled reactions.** In essence, the body uses ATP to transfer the energy released during catabolic reactions to power its anabolic reactions. The body converts the chemical energy of food to the chemical energy of ATP with about 50 percent efficiency, radiating the rest as heat.[1] Some energy is lost as heat again when the body uses the chemical energy of ATP to do its work—moving muscles, synthesizing compounds, or transporting nutrients, for example.

The Helpers in Metabolic Reactions—Enzymes and Coenzymes Metabolic reactions almost always require enzymes ♦ to facilitate their action. In many cases, the enzymes need assistants to help them. Enzyme helpers are called **coenzymes.** ♦

♦ **Enzymes** are proteins that act as catalysts. **Catalysts** facilitate chemical reactions without being changed in the process.

♦ The general term for substances that facilitate enzyme action is **cofactors;** they include both organic coenzymes made from vitamins and inorganic substances such as minerals.

Coenzymes are complex organic molecules that associate closely with most enzymes but are not proteins themselves. The relationships between various coenzymes and their respective enzymes may differ in detail, but one thing is true of all: without its coenzyme, an enzyme cannot function. Some of the B vitamins serve as coenzymes that participate in the energy metabolism of glucose, glycerol, fatty acids, and amino acids. (Chapter 10 provides more details.)

IN SUMMARY **During digestion the energy-yielding nutrients—carbohydrates, lipids, and proteins—are broken down to glucose (and other mono-**

FIGURE 7-4 **The Capture and Release of Energy by ATP**

It may help to think of ATP as a rechargeable battery—capturing and releasing energy as it does the body's work.

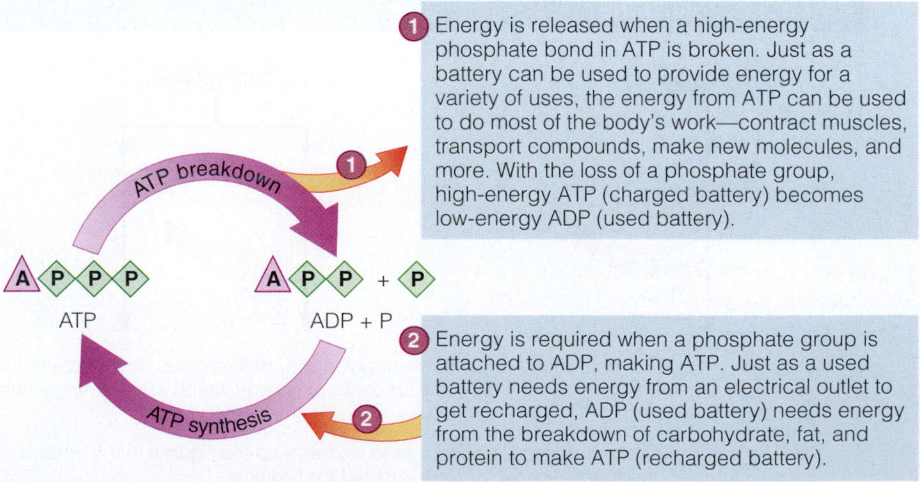

1 Energy is released when a high-energy phosphate bond in ATP is broken. Just as a battery can be used to provide energy for a variety of uses, the energy from ATP can be used to do most of the body's work—contract muscles, transport compounds, make new molecules, and more. With the loss of a phosphate group, high-energy ATP (charged battery) becomes low-energy ADP (used battery).
ATP breakdown
1
A P P P
ATP
A P P + P
ADP + P
ATP synthesis
2
2 Energy is required when a phosphate group is attached to ADP, making ATP. Just as a used battery needs energy from an electrical outlet to get recharged, ADP (used battery) needs energy from the breakdown of carbohydrate, fat, and protein to make ATP (recharged battery).

ATP or **adenosine** (ah-DEN-oh-seen) **triphosphate** (try-FOS-fate): a common high-energy compound composed of a purine (adenine), a sugar (ribose), and three phosphate groups.

coupled reactions: pairs of chemical reactions in which some of the energy released from the breakdown of one compound is used to create a bond in the formation of another compound.

coenzymes: complex organic molecules that work with enzymes to facilitate the enzymes' activity. Many coenzymes have B vitamins as part of their structures. (Figure 10-2 on p. 315 illustrates coenzyme action.)

- **co** = with

saccharides), glycerol, fatty acids, and amino acids. With the help of enzymes and coenzymes, the cells use these products of digestion to build more complex compounds (anabolism) or break them down further to release energy (catabolism). High-energy compounds such as ATP may capture the energy released during catabolism.

Breaking Down Nutrients for Energy

Chapters 4, 5, and 6 lay the groundwork for the study of metabolism; a brief review may be helpful. During digestion, the body breaks down the three energy-yielding nutrients—carbohydrates, lipids, and proteins—into four basic units that can be absorbed into the blood:

- From carbohydrates—glucose (and other monosaccharides)
- From fats (triglycerides)—glycerol and fatty acids
- From proteins—amino acids

The body uses carbohydrates and fats for most of its energy needs. Amino acids are used primarily as building blocks for proteins, but they also enter energy pathways, contributing about 10 to 15 percent of the day's energy use. Look for these four basic units—glucose, glycerol, fatty acids, and amino acids—to appear again and again in the metabolic reactions described in this chapter. Alcohol also enters many of the metabolic pathways; Highlight 7 focuses on how alcohol disrupts metabolism and how the body handles it.

Glucose, glycerol, fatty acids, and amino acids are the basic units derived from food, but a molecule of each of these compounds is made of still smaller units, the atoms—carbons, nitrogens, oxygens, and hydrogens. During catabolism, the body separates these atoms from one another. To follow this action, recall how many carbons are in the "backbones" of these compounds:

- Glucose has 6 carbons:

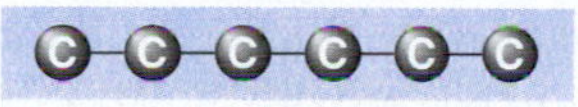

- Glycerol has 3 carbons:

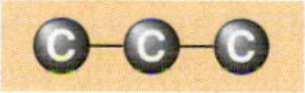

- A fatty acid usually has an even number of carbons, commonly 16 or 18 carbons:*

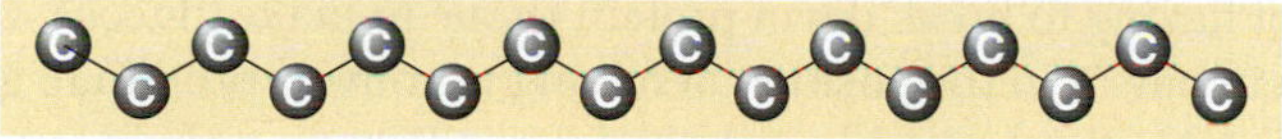

- An amino acid has 2, 3, or more carbons with a nitrogen attached:**

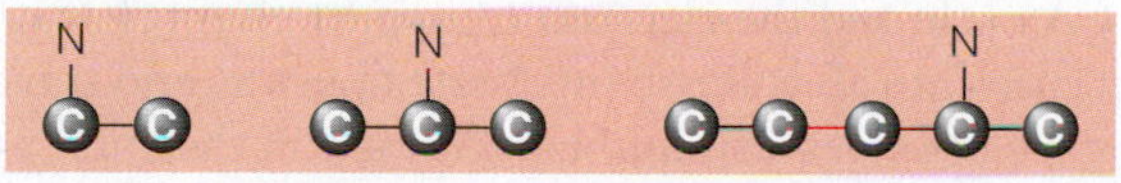

Full chemical structures and reactions appear both in the earlier chapters and in Appendix C; this chapter diagrams the reactions using just the compounds' carbon and nitrogen backbones.

As you will see, each of the compounds—glucose, glycerol, fatty acids, and amino acids—starts down a different path. Along the way, two new names appear—**pyruvate** (a 3-carbon structure) and **acetyl CoA** (a 2-carbon structure with a

*The figures in this chapter show 16- or 18-carbon fatty acids. Fatty acids may have 4 to 20 or more carbons, with chain lengths of 16 and 18 carbons most prevalent.

**The figures in this chapter usually show amino acids as compounds of 2, 3, or 5 carbons arranged in a straight line, but in reality amino acids may contain other numbers of carbons and assume other structural shapes (see Appendix C).

pyruvate (PIE-roo-vate): a 3-carbon compound that plays a key role in energy metabolism.

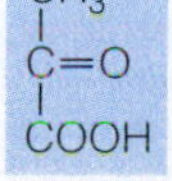

acetyl CoA (ASS-eh-teel, or ah-SEET-il, coh-AY): a 2-carbon compound (*acetate,* or *acetic acid,* shown in Figure 5-1 on p. 134) to which a molecule of CoA is attached.

All the energy used to keep the heart beating, the brain thinking, and the legs running comes from the carbohydrates, fats, and proteins in foods.

coenzyme, **CoA**, attached)—and the rest of the story falls into place around them.* Two major points to notice in the following discussion:

- Pyruvate can be used to make glucose.
- Acetyl CoA cannot be used to make glucose.

A key to understanding these metabolic pathways is learning which fuels can be converted to glucose and which cannot. The parts of protein and fat that can be converted to pyruvate *can* provide glucose for the body, whereas the parts that are converted to acetyl CoA *cannot* provide glucose but can readily provide fat. The body must have glucose to fuel the activities of the central nervous system and red blood cells. Without glucose from food, the body will devour its own lean (protein-containing) tissue to get the amino acids needed to make glucose. Therefore, to keep this from happening, the body needs foods that can provide glucose—primarily carbohydrate. Giving the body only fat, which delivers mostly acetyl CoA, puts it in the position of having to break down protein tissue to make glucose. Giving the body only protein puts it in the position of having to convert protein to glucose. Clearly, the best diet ♦ provides ample carbohydrate, adequate protein, and some fat.

♦ A healthy diet provides:
- 45–65% kcalories from carbohydrate
- 10–35% kcalories from protein
- 20–35% kcalories from fat

Eventually, all of the energy-yielding nutrients can enter the final energy pathways of the **TCA cycle** and the **electron transport chain.** ♦ (Similarly, people from three different cities can all enter an interstate highway and travel to the same destination.) Figure 7-5 provides a simplified overview of the energy-yielding pathways. The next sections of the text describe how each of the energy-yielding nutrients is broken down to acetyl CoA and other compounds in preparation for their entrance into the TCA cycle and electron transport chain. These final energy pathways have central roles in energy metabolism and receive full attention later in the chapter.

♦ The TCA cycle is also called the **citric acid cycle** or the **Kreb's cycle.** The electron transport chain is also called the **respiratory chain.**

CoA (coh-AY): coenzyme A; the coenzyme derived from the B vitamin pantothenic acid and central to energy metabolism.

TCA cycle or **tricarboxylic** (try-car-box-ILL-ick) **acid cycle:** a series of metabolic reactions that break down molecules of acetyl CoA to carbon dioxide and hydrogen atoms; also called the *citric acid cycle* or the *Kreb's cycle* after the biochemist who elucidated its reactions.

electron transport chain: the final pathway in energy metabolism that transports electrons from hydrogen to oxygen and captures the energy released in the bonds of ATP.

Glucose

What happens to glucose, glycerol, fatty acids, and amino acids during energy metabolism can best be understood by starting with glucose. This discussion features glucose because of its central role in carbohydrate metabolism and because liver cells can convert the monosaccharides fructose and galactose to compounds that enter the same energy pathways.

*The term *pyruvate* means a salt of *pyruvic acid*. (Throughout this book, the ending *–ate* is used interchangeably with *–ic acid;* for our purposes they mean the same thing.)

FIGURE 7-5 Simplified Overview of Energy-Yielding Pathways

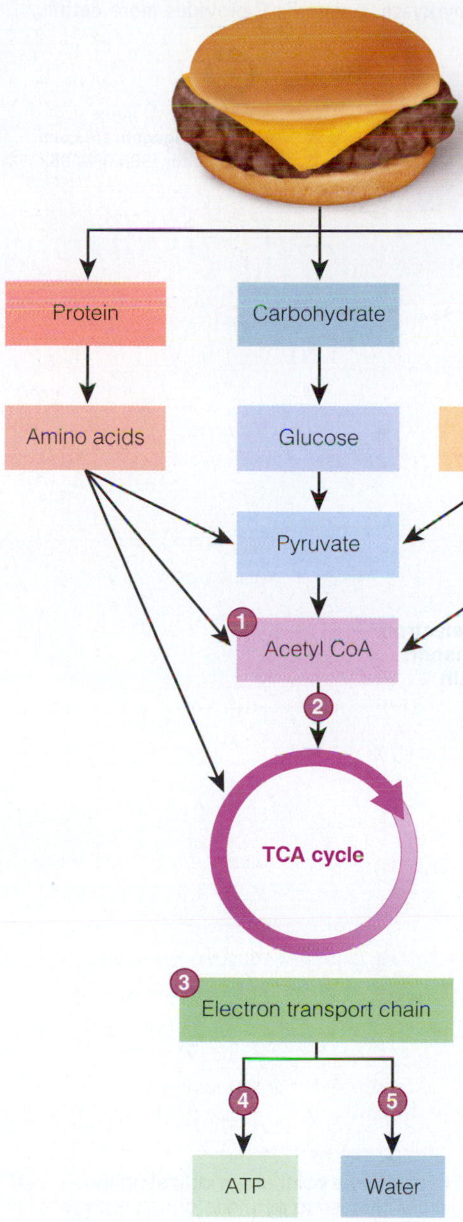

1 All of the energy-yielding nutrients—protein, carbohydrate, and fat—can be broken down to acetyl CoA.

2 Acetyl CoA can enter the TCA cycle.

3 Most of the reactions above release hydrogen atoms with their electrons, which are carried by coenzymes to the electron transport chain.

4 ATP is synthesized.

5 Hydrogen atoms react with oxygen to produce water.

Glucose-to-Pyruvate The first pathway glucose takes on its way to yield energy is called **glycolysis** (glucose splitting).* Figure 7-6 (p. 212) shows a simplified drawing of glycolysis. (This pathway actually involves several steps and several enzymes, which are detailed in Appendix C.) In a series of reactions, the 6-carbon glucose is converted to similar 6-carbon compounds before being split in half, forming two 3-carbon compounds. These 3-carbon compounds continue along the pathway until they are converted to pyruvate. Thus the net yield of one glucose molecule is two pyruvate molecules. The net yield of energy at this point is small; to start glycolysis, the cell uses a little energy and then produces only a little more than it had to invest initially.** In addition, as glucose breaks down to pyruvate, hydrogen atoms with their electrons are released and carried to the electron transport chain by coenzymes made from the B vitamin niacin. A later section of the chapter explains how oxygen accepts the electrons and combines

*Glycolysis takes place in the cytosol of the cell (see Figure 7-1, p. 206).

**The cell uses 2 ATP to begin the breakdown of glucose to pyruvate, but it then gains 4 ATP for a net gain of 2 ATP.

glycolysis (gly-COLL-ih-sis): the metabolic breakdown of glucose to pyruvate. Glycolysis does not require oxygen (anaerobic).

- **glyco** = glucose
- **lysis** = breakdown

FIGURE 7-6 Glycolysis: Glucose-to-Pyruvate

This simplified overview of glycolysis illustrates the steps in the process of converting glucose to pyruvate. Appendix C provides more details.

Animated! figure
www.cengagebrain.com
(search for ISBN 084006845X)

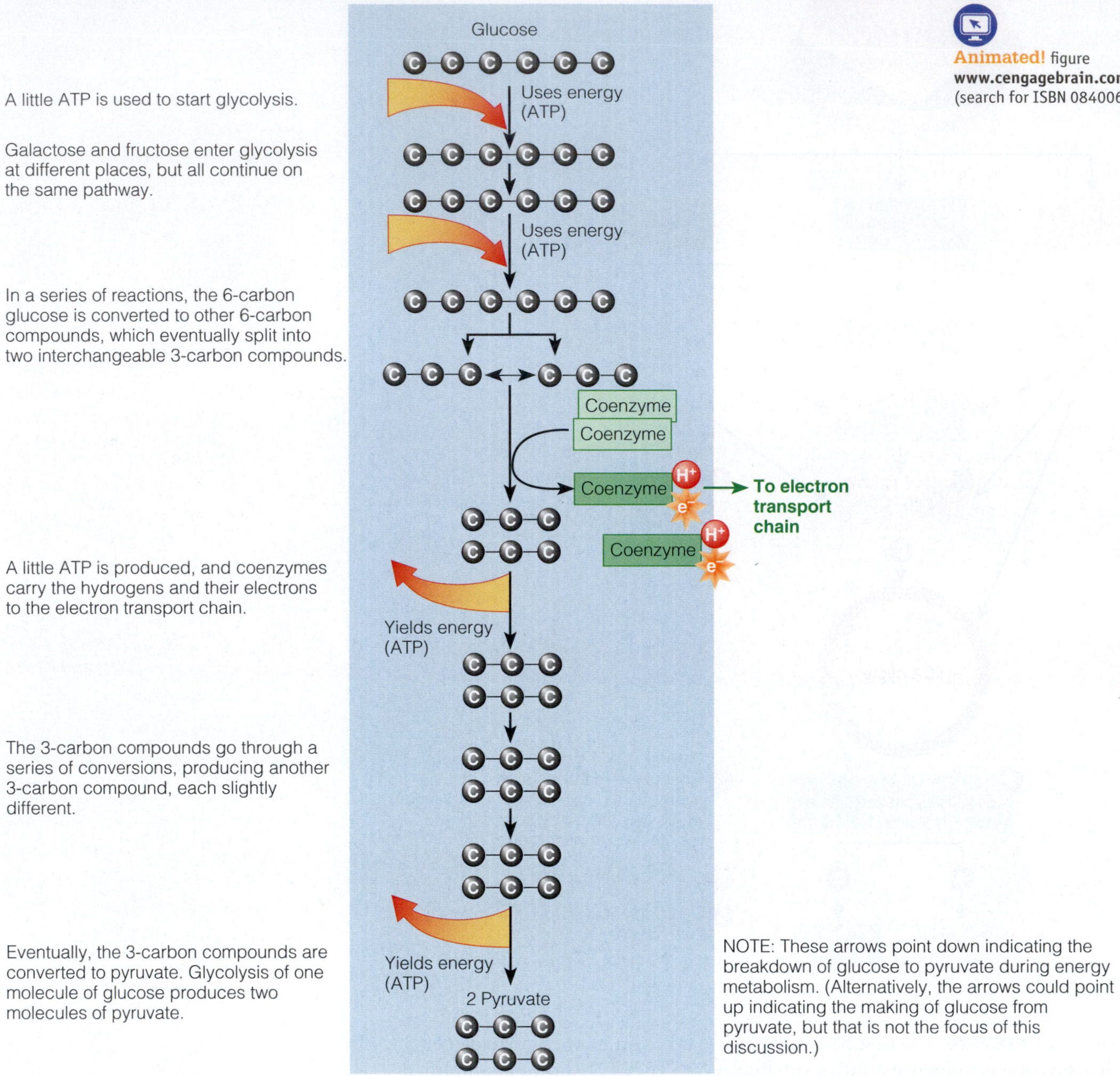

with the hydrogens to form water and how the process captures energy in the bonds of ATP.

A little ATP is used to start glycolysis.
Galactose and fructose enter glycolysis at different places, but all continue on the same pathway.
In a series of reactions, the 6-carbon glucose is converted to other 6-carbon compounds, which eventually split into two interchangeable 3-carbon compounds.
A little ATP is produced, and coenzymes carry the hydrogens and their electrons to the electron transport chain.
The 3-carbon compounds go through a series of conversions, producing another 3-carbon compound, each slightly different.
Eventually, the 3-carbon compounds are converted to pyruvate. Glycolysis of one molecule of glucose produces two molecules of pyruvate.
Glucose
Uses energy (ATP)
Uses energy (ATP)
Coenzyme
Coenzyme
Coenzyme
Coenzyme
To electron transport chain
Yields energy (ATP)
Yields energy (ATP)
2 Pyruvate
NOTE: These arrows point down indicating the breakdown of glucose to pyruvate during energy metabolism. (Alternatively, the arrows could point up indicating the making of glucose from pyruvate, but that is not the focus of this discussion.)

♦ Glucose may go "down" to make pyruvate, or pyruvate may go "up" to make glucose, depending on the cell's needs.

This discussion focuses primarily on the breakdown of glucose for energy, but if needed, cells in the liver (and to some extent, the kidneys) can make glucose again from pyruvate in a process similar to the reversal of glycolysis. Making glucose requires energy, however, and a few different enzymes. Still, glucose can be made from pyruvate, so the arrows between glucose and pyruvate could point up as well as down. ♦

Pyruvate's Options—Anaerobic or Aerobic Whenever carbohydrates, fats, or proteins are broken down to provide energy, oxygen is always ultimately involved in the process. The role of oxygen in metabolism is worth noticing, for it helps our understanding of physiology and metabolic reactions.

When the body needs energy quickly—as occurs when you run a quarter mile as fast as you can—pyruvate is converted to lactate. The breakdown of glucose-to-pyruvate-to-lactate proceeds without oxygen—it is **anaerobic**. This anaerobic pathway yields energy quickly, but it cannot be sustained for long—a couple of minutes at most.

When energy expenditure proceeds at a slower pace—as occurs when you jog around the track for an hour—pyruvate breaks down to acetyl CoA in an **aerobic** pathway. Aerobic pathways produce energy more slowly, but because they can be sustained for a long time, their total energy yield is greater. The following paragraphs explain these pathways.

Pyruvate-to-Lactate (Anaerobic) As mentioned earlier, coenzymes carry the hydrogens from glucose breakdown to the electron transport chain. If the electron transport chain is unable to accept these hydrogens, as may occur when cells lack sufficient **mitochondria** (review Figure 7-1) or in the absence of sufficient oxygen, pyruvate can accept the hydrogens. As Figure 7-7 shows, by accepting the hydrogens, pyruvate becomes **lactate**, and the coenzymes are freed to return to glycolysis to pick up more hydrogens. In this way, glucose can continue providing energy anaerobically for a while (see the left side of Figure 7-7).

The production of lactate occurs to a limited extent even at rest. During high-intensity exercise, however, the muscles rely heavily on anaerobic glycolysis to produce ATP quickly, and the concentration of lactate increases dramatically. The rapid rate of glycolysis produces abundant pyruvate and releases

anaerobic (AN-air-ROE-bic): not requiring oxygen.
- **an** = not

aerobic (air-ROE-bic): requiring oxygen.

mitochondria (my-toh-KON-dree-uh): the cellular organelles responsible for producing ATP aerobically; made of membranes (lipid and protein) with enzymes mounted on them. (The singular is *mitochondrion*.)
- **mitos** = thread (referring to their slender shape)
- **chondros** = cartilage (referring to their external appearance)

lactate: a 3-carbon compound produced from pyruvate during anaerobic metabolism.

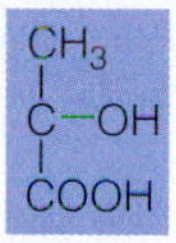

FIGURE 7-7 Pyruvate-to-Lactate

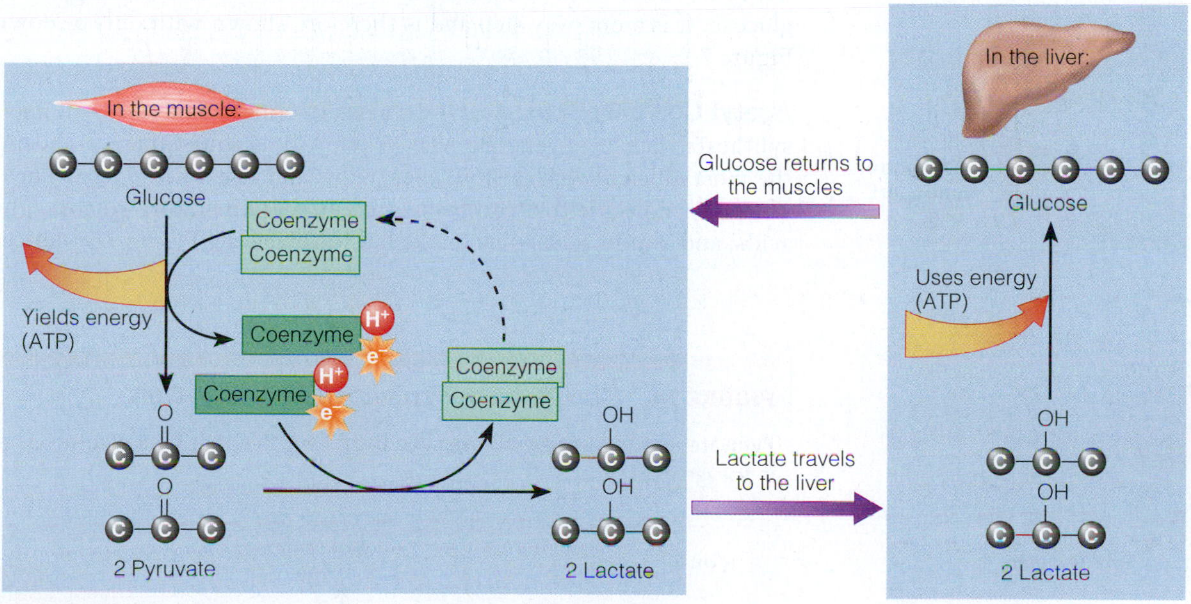

Working muscles break down most of their glucose molecules anaerobically to pyruvate. If the cells lack sufficient mitochondria or in the absence of sufficient oxygen, pyruvate can accept the hydrogens from glucose breakdown and become lactate. This conversion frees the coenzymes so that glycolysis can continue.

NOTE: Other figures in this chapter focus narrowly on the carbons of pyruvate. Its oxygen group is included in this figure to more clearly illustrate this reaction. See definitions for the chemical structures of pyruvate and lactate.

Liver enzymes can convert lactate to glucose, but this reaction requires energy. The process of converting lactate from the muscles to glucose in the liver that can be returned to the muscles is known as the Cori cycle.

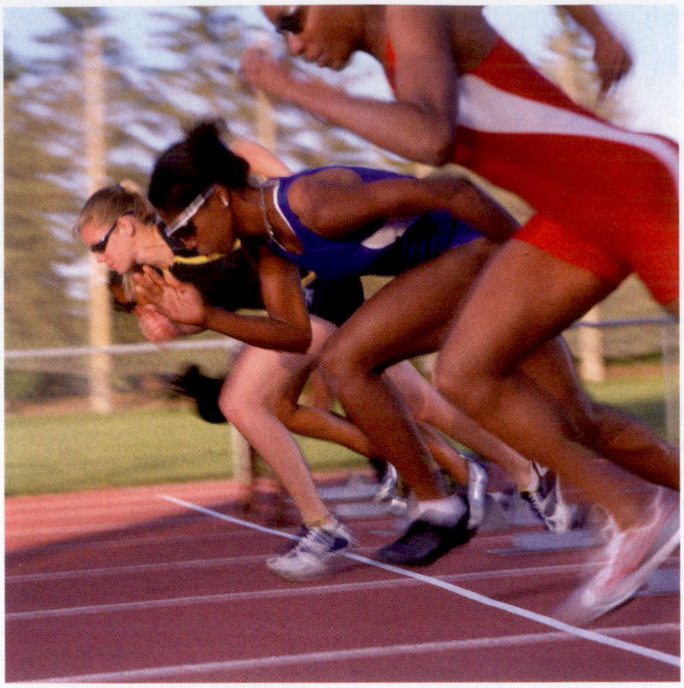

The anaerobic breakdown of glucose-to-pyruvate-to-lactate is the major source of energy for short, intense exercise.

hydrogen-carrying coenzymes more rapidly than the mitochondria can handle them. To enable exercise to continue at this intensity, pyruvate is converted to lactate and coenzymes are released, which allows glycolysis to continue. The accumulation of lactate in the muscles coincides with—but does not seem to be the cause of—the subsequent drop in blood pH, burning pain, and fatigue that are commonly associated with intense exercise.[2] In fact, making lactate from pyruvate consumes two hydrogen ions, which actually diminishes acidity and improves the performance of tired muscles. A person performing the same exercise following endurance training actually experiences less discomfort—in part because the number of mitochondria in the muscle cells has increased. This adaptation improves the mitochondrias' ability to keep pace with the muscles' demand for energy.

One possible fate of lactate is to be transported from the muscles to the liver. There the liver can convert the lactate produced in muscles to glucose, which can then be returned to the muscles. This recycling process is called the **Cori cycle** (see Figure 7-7). (Muscle cells cannot recycle lactate to glucose because they lack a necessary enzyme.)

Pyruvate-to-Acetyl CoA (Aerobic) If the cell needs energy and oxygen is available, pyruvate molecules enter the mitochondria of the cell (review Figure 7-1, p. 206). There a carbon group (COOH) from the 3-carbon pyruvate is removed to produce a 2-carbon compound that bonds with a molecule of CoA, becoming acetyl CoA. The carbon group from pyruvate becomes carbon dioxide, which is released into the blood, circulated to the lungs, and breathed out. Figure 7-8 diagrams the pyruvate-to-acetyl CoA reaction.

The step from pyruvate to acetyl CoA is metabolically irreversible: a cell cannot retrieve the shed carbons from carbon dioxide to remake pyruvate and then glucose. It is a one-way step and is therefore shown with only a "down" arrow in Figure 7-9.

Acetyl CoA's Options Acetyl CoA has two main functions—it may be used to synthesize fats or to generate ATP. When ATP is abundant, acetyl CoA makes fat, the most efficient way to store energy for later use when energy may be needed. Thus any molecule that can make acetyl CoA—including glucose, glycerol, fatty acids, and amino acids—can make fat. In reviewing Figure 7-9, notice that acetyl

FIGURE 7-8 Pyruvate-to-Acetyl CoA

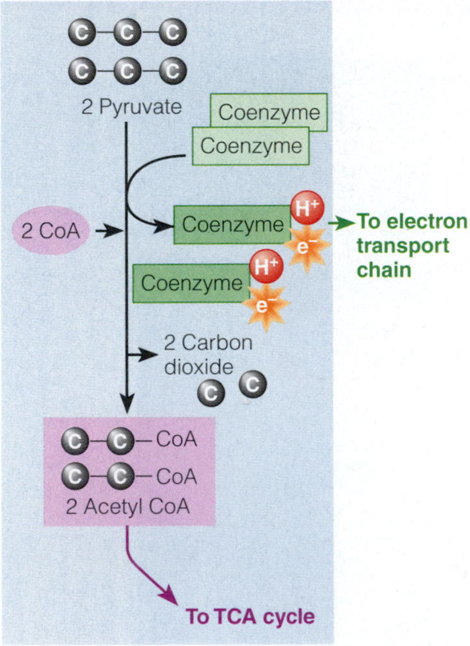

Each pyruvate loses a carbon as carbon dioxide and picks up a molecule of CoA, becoming acetyl CoA. The arrow goes only one way (down) because the step is not reversible.

Cori cycle: the path from muscle glycogen to glucose to pyruvate to lactate (which travels to the liver) to glucose (which can travel back to the muscle) to glycogen; named after the scientist who elucidated this pathway.

FIGURE 7-9 The Paths of Pyruvate and Acetyl CoA

Pyruvate may follow several reversible paths, but the path from pyruvate to acetyl CoA is irreversible.

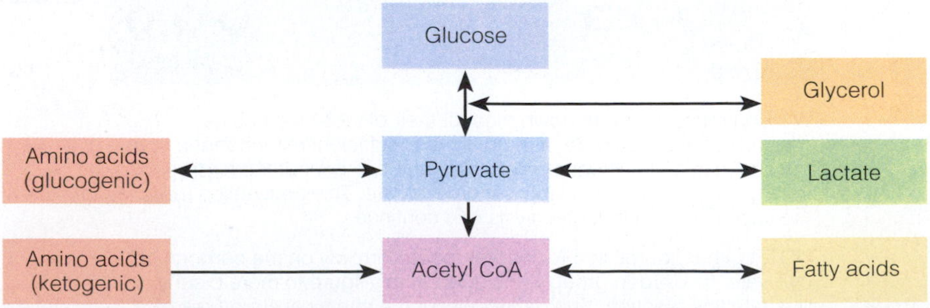

NOTE: Amino acids that can be used to make glucose are called *glucogenic;* amino acids that are converted to acetyl CoA are called *ketogenic.*

CoA can be used as a building block for fatty acids, but it cannot be used to make glucose or amino acids. When ATP is low and the cell needs energy, acetyl CoA may proceed through the TCA cycle, releasing hydrogens, with their electrons, to the electron transport chain.

The story of acetyl CoA continues on p. 218 after a discussion of how fat and protein arrive at the same crossroads. For now, know that when acetyl CoA from the breakdown of glucose enters the aerobic pathways of the TCA cycle and electron transport chain, much more ATP is produced than during glycolysis. The role of glycolysis is to provide energy for short bursts of activity and to prepare glucose for the later energy pathways.

IN SUMMARY **The breakdown of glucose to energy begins with glycolysis, a pathway that produces pyruvate. Keep in mind that glucose can be synthesized only from pyruvate or compounds earlier in the pathway. Pyruvate may be converted to lactate anaerobically or to acetyl CoA aerobically. Once the commitment to acetyl CoA is made, glucose is not retrievable; acetyl CoA cannot go back to glucose. Figure 7-10 summarizes the breakdown of glucose.**

FIGURE 7-10 Glucose Enters the Energy Pathway

This figure summarizes the breakdown of glucose-to-pyruvate-to-acetyl CoA. Details of the TCA cycle and the electron transport chain are given later and in Appendix C.

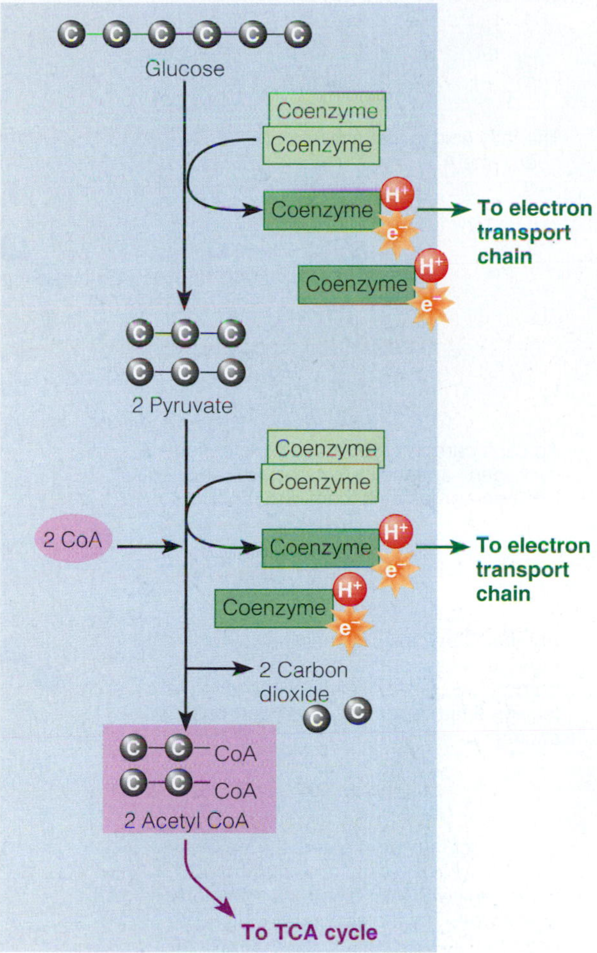

IN SUMMARY
1 glucose yields 2 pyruvate, which yield 2 acetyl CoA.

Glycerol and Fatty Acids

Once glucose breakdown is understood, fat and protein breakdown are easily learned, for all three eventually enter the same energy pathways. Recall that triglycerides can break down to glycerol and fatty acids.

Glycerol-to-Pyruvate Glycerol is a 3-carbon compound like pyruvate but with a different arrangement of H and OH on the C. As such, glycerol can easily be converted to another 3-carbon compound that can go either "up" the pathway to form glucose or "down" to form pyruvate and then acetyl CoA (review Figure 7-9 on p. 214).

Fatty Acids-to-Acetyl CoA Fatty acids are taken apart 2 carbons at a time in a series of reactions known as **fatty acid oxidation.*** Figure 7-11 (p. 216) illustrates fatty acid oxidation and shows that in the process, each 2-carbon fragment splits off and combines with a molecule of CoA to make acetyl CoA. As each 2-carbon fragment breaks off from a fatty acid during oxidation, hydrogens and their electrons are released and carried to the electron transport chain by coenzymes made from the B vitamins riboflavin and niacin. Figure 7-12 (p. 217) summarizes the breakdown of fats.

Fatty Acids Cannot Be Used to Synthesize Glucose Red blood cells and the brain and nervous system depend primarily on glucose as fuel. When carbohydrate is unavailable, the liver cells can make glucose from pyruvate and other 3-carbon compounds, such as glycerol. Importantly, cells cannot make glucose from the 2-carbon fragments of fatty acids. In chemical diagrams, the arrow between pyruvate and acetyl CoA always points only one way—down—and fatty acids enter the metabolic path below this arrow (review Figure 7-9). The down arrow indicates that fatty acids cannot be used to make glucose.

Remember that almost all dietary fats are triglycerides and that triglycerides contain only one small molecule of glycerol with three fatty acids. The glycerol can yield glucose, ♦ but that represents only 3 of the 50 or so carbon atoms in a triglyceride—about 5 percent of its weight (see Figure 7-13 on p. 217). The other 95 percent cannot be converted to glucose.

♦ Making glucose from noncarbohydrate sources is called **gluconeogenesis.** The glycerol portion of a triglyceride and most amino acids can be used to make glucose (review Figure 7-9). The liver is the major site of gluconeogenesis, but the kidneys become increasingly involved under certain circumstances, such as starvation.

fatty acid oxidation: the metabolic breakdown of fatty acids to acetyl CoA; also called *beta oxidation.*

*Oxidation of fatty acids occurs in the mitochondria of the cells (review Figure 7-1, p. 206).

FIGURE 7-11 **Fatty Acid-to-Acetyl CoA**

Fatty acids are broken apart into 2-carbon fragments that combine with CoA to make acetyl CoA.

Animated! figure
www.cengagebrain.com
(search for ISBN 084006845X)

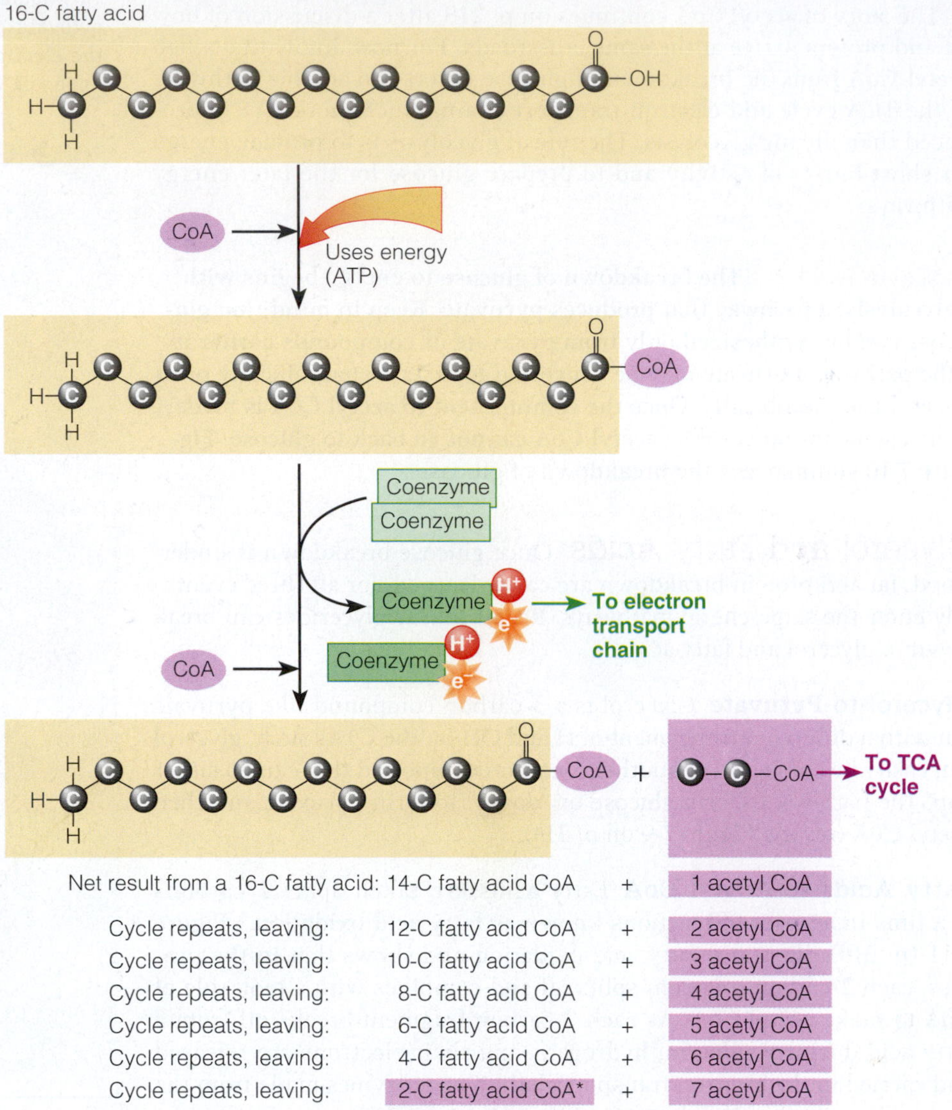

The fatty acid is first activated by coenzyme A.

As each carbon-carbon bond is cleaved, hydrogens and their electrons are released, and coenzymes pick them up.

Another CoA joins the chain, and the bond at the second carbon (the beta-carbon) weakens. Acetyl CoA splits off, leaving a fatty acid that is two carbons shorter.

The shorter fatty acid enters the pathway and the cycle repeats, releasing more hydrogens with their electrons and more acetyl CoA. The molecules of acetyl CoA enter the TCA cycle, and the coenzymes carry the hydrogens and their electrons to the electron transport chain.

*Notice that 2-C fatty acid CoA = acetyl CoA, so that the final yield from a 16-C fatty acid is 8 acetyl CoA.

IN SUMMARY The body can convert the small glycerol portion of a triglyceride to either pyruvate (and then glucose) or acetyl CoA. The fatty acids of a triglyceride, on the other hand, cannot make glucose, but they can provide abundant acetyl CoA. Acetyl CoA may then enter the TCA cycle to release energy or combine with other molecules of acetyl CoA to make body fat.

Amino Acids

The preceding two sections have described how the breakdown of carbohydrate and fat produces acetyl CoA, which can enter the pathways that provide energy for the body's use. One energy-yielding nutrient remains: protein or, rather, the amino acids of protein. Before entering the metabolic pathways, amino acids are deaminated (that is, they lose their nitrogen-containing amino group). Chapter 6 describes how deamination produces ammonia (NH_3), which provides the nitrogen needed to make nonessential amino acids and other nitrogen-containing compounds. Any remaining ammonia is cleared from the body via urea synthesis in the liver and excretion in the kidneys.

FIGURE 7-12 **Fats Enter the Energy Pathway**

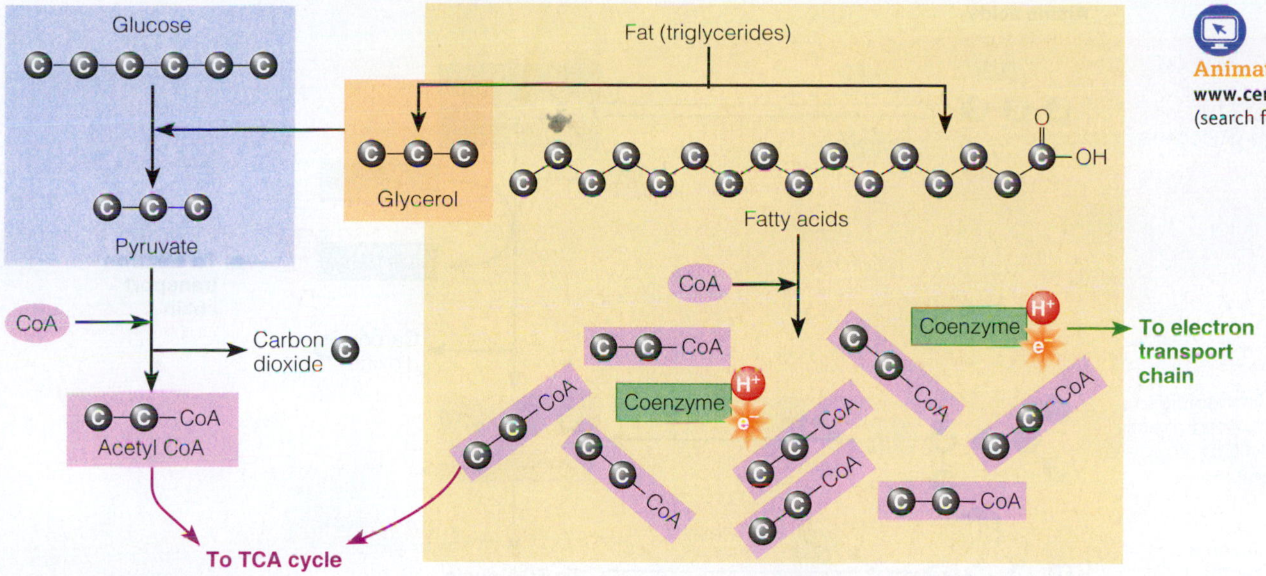

Glycerol enters the glycolysis pathway about midway between glucose and pyruvate. Fatty acids are broken down into 2-carbon fragments that combine with CoA to form acetyl CoA (shown in Figure 7-11).

IN SUMMARY A 16-carbon fatty acid yields 8 acetyl CoA.

Animated! figure
www.cengagebrain.com
(search for ISBN 084006845X)

Amino Acids-to-Energy Amino acids can enter the energy pathways in several ways. As Figure 7-14 (p. 218) illustrates, some amino acids can be converted to pyruvate, others are converted to acetyl CoA, and still others enter the TCA cycle directly as compounds other than acetyl CoA.

Amino Acids-to-Glucose As you might expect, amino acids that are used to make pyruvate can provide glucose, whereas those used to make acetyl CoA can provide additional energy or make body fat but cannot make glucose. ♦ Amino acids entering the TCA cycle directly can continue in the cycle and generate energy; alternatively, they can generate glucose.[3] Thus protein, unlike fat, is a fairly good source of glucose when carbohydrate is not available.

♦ Amino acids that can make glucose via either pyruvate or TCA cycle intermediates are **glucogenic;** amino acids that are degraded to acetyl CoA are **ketogenic.**

IN SUMMARY **The body can use some amino acids to produce glucose, whereas others can be used either to provide energy or to make fat. Before an amino acid enters any of these metabolic pathways, its nitrogen-containing amino group must be removed through deamination.**

FIGURE 7-13 **The Carbons of a Typical Triglyceride**

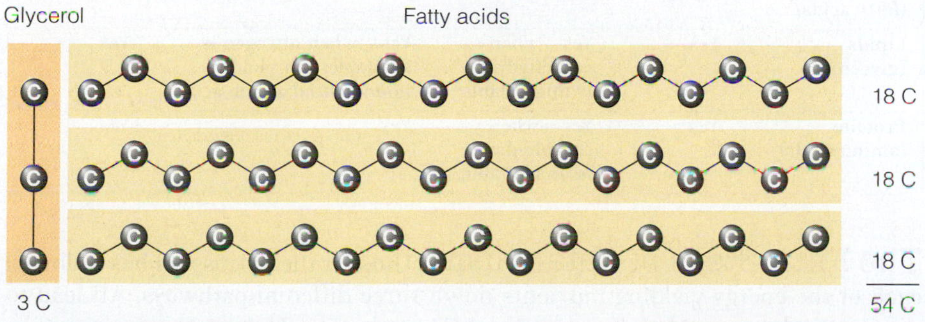

A typical triglyceride contains only one small molecule of glycerol (3 C) but has three fatty acids (each commonly 16 C or 18 C, or about 48 C to 54 C in total). Only the glycerol portion of a triglyceride can yield glucose.

FIGURE 7-14 Amino Acids Enter the Energy Pathway

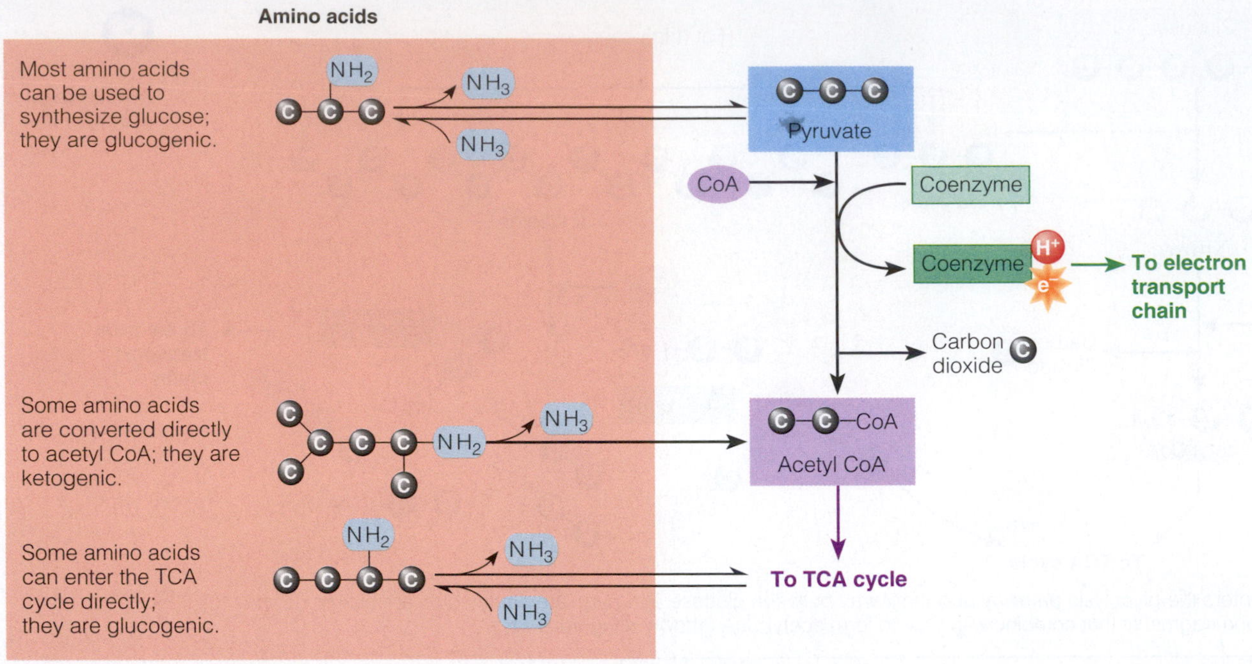

NOTE: Deamination and the synthesis of urea are discussed and illustrated in Chapter 6, Figure 6-13 (p. 186). The arrows from pyruvate and the TCA cycle to amino acids are possible only for *nonessential* amino acids; remember, the body cannot make essential amino acids.

Breaking Down Nutrients for Energy—In Summary

To review the ways the body can use the energy-yielding nutrients, see the following summary table. To obtain energy, the body uses glucose and fatty acids as its primary fuels and amino acids to a lesser extent. To make glucose, the body can use all carbohydrates and most amino acids, but it can convert only 5 percent of fat (the glycerol portion) to glucose. Fatty acids cannot make glucose. To make proteins, the body needs amino acids. It can use glucose and glycerol to make some nonessential amino acids when nitrogen is available; it cannot use fatty acids to make body proteins. Finally, when energy intake exceeds the body's needs, all three energy-yielding nutrients can contribute to body fat stores.

IN SUMMARY

Nutrient	Yields Energy?	Yields Glucose?	Yields Amino Acids and Body Proteins?	Yields Fat Stores?
Carbohydrates (glucose)	Yes	Yes	Yes—when nitrogen is available, can yield *nonessential* amino acids	Yes
Lipids (fatty acids)	Yes	No	No	Yes
Lipids (glycerol)	Yes	Yes—when carbohydrate is unavailable	Yes—when nitrogen is available, can yield *nonessential* amino acids	Yes
Proteins (amino acids)	Yes	Yes—when carbohydrate is unavailable	Yes	Yes

The Final Steps of Catabolism

Thus far the discussion has followed each of the energy-yielding nutrients down three different pathways. All lead to the point where acetyl CoA enters the TCA cycle. The TCA cycle reactions take place in the inner compartment of the mitochondria. Examine the structure of the

FIGURE 7-15 A Mitochondrion

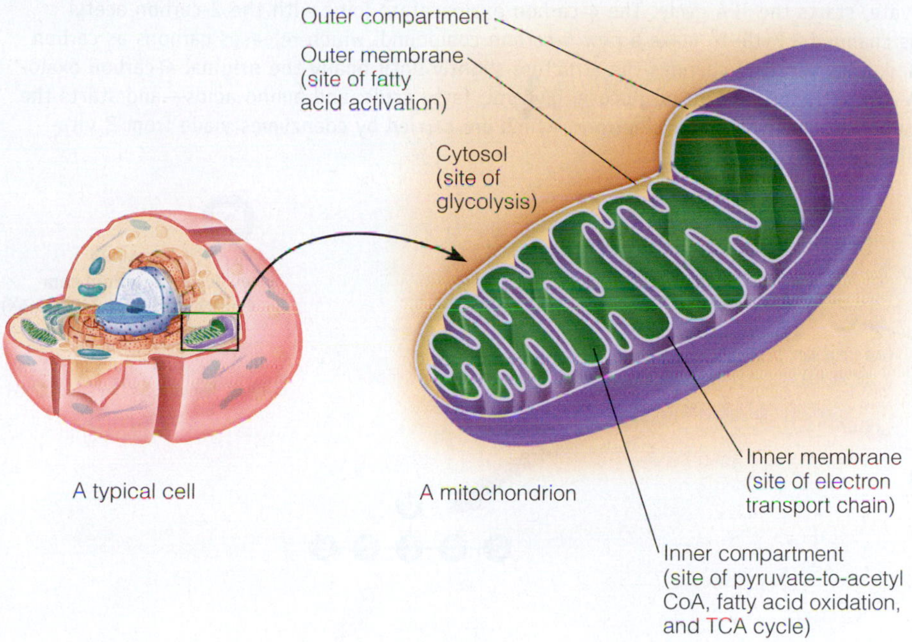

mitochondria shown in Figure 7-15. The significance of its structure will become evident as details unfold.

The TCA Cycle Acetyl CoA enters the TCA cycle, a busy metabolic traffic center. The TCA cycle is a circular path, but that doesn't mean it regenerates acetyl CoA. Acetyl CoA goes one way only—down to two carbon dioxide molecules and a coenzyme (CoA). The TCA cycle is a circular path because a 4-carbon compound known as **oxaloacetate** is needed in the first step and it is synthesized in the last step.

Oxaloacetate's role in replenishing the TCA cycle is critical. When oxaloacetate is insufficient, the TCA cycle slows down, and the cells face an energy crisis. Oxaloacetate is made primarily from pyruvate, although it can also be made from certain amino acids. Importantly, oxaloacetate cannot be made from fat. That oxaloacetate must be available for acetyl CoA to enter the TCA cycle underscores the importance of carbohydrates in the diet. A diet that provides ample carbohydrate ensures an adequate supply of oxaloacetate (because glucose produces pyruvate during glycolysis). (The chapter closes with more information on the consequences of low-carbohydrate diets.)

As Figure 7-16 (p. 220) shows, oxaloacetate is the first 4-carbon compound to enter the TCA cycle. Oxaloacetate picks up acetyl CoA (a 2-carbon compound), drops off one carbon (as carbon dioxide), then another carbon (as carbon dioxide), and returns to pick up another acetyl CoA. As for the acetyl CoA, its carbons go only one way—to carbon dioxide (see Appendix C for additional details).*

As acetyl CoA molecules break down to carbon dioxide, hydrogen atoms with their electrons are removed from the compounds in the cycle. Each turn of the TCA cycle releases a total of eight electrons. Coenzymes made from the B vitamins niacin and riboflavin receive the hydrogens and their electrons from the TCA cycle and transfer them to the electron transport chain—much like a taxi cab that picks up passengers in one location and drops them off in another.

*Actually, the carbons that enter the cycle in acetyl CoA may not be the exact ones that are given off as carbon dioxide. In one of the steps of the cycle, a 6-carbon compound of the cycle becomes symmetrical, both ends being identical. Thereafter it loses carbons to carbon dioxide at one end or the other. Thus only half of the carbons from acetyl CoA are given off as carbon dioxide in any one turn of the cycle; the other half become part of the compound that returns to pick up another acetyl CoA. It is true to say, though, that for each acetyl CoA that enters the TCA cycle, two carbons are given off as carbon dioxide. It is also true that with each turn of the cycle, the energy equivalent of one acetyl CoA is released.

oxaloacetate (OKS-ah-low-AS-eh-tate): a carbohydrate intermediate of the TCA cycle.

FIGURE 7-16 The TCA Cycle

Oxaloacetate, a compound made primarily from pyruvate, starts the TCA cycle. The 4-carbon oxaloacetate joins with the 2-carbon acetyl CoA to make a 6-carbon compound. This compound is changed a little to make a new 6-carbon compound, which releases carbons as carbon dioxide, becoming a 5- and then a 4-carbon compound. Each reaction changes the structure slightly until finally the original 4-carbon oxaloacetate forms again and picks up another acetyl CoA—from the breakdown of glucose, glycerol, fatty acids, and amino acids—and starts the cycle over again. The breakdown of acetyl CoA releases hydrogens with their electrons, which are carried by coenzymes made from B vitamins to the electron transport chain. (For more details, see Appendix C.)

Animated! figure
www.cengagebrain.com
(search for ISBN 084006845X)

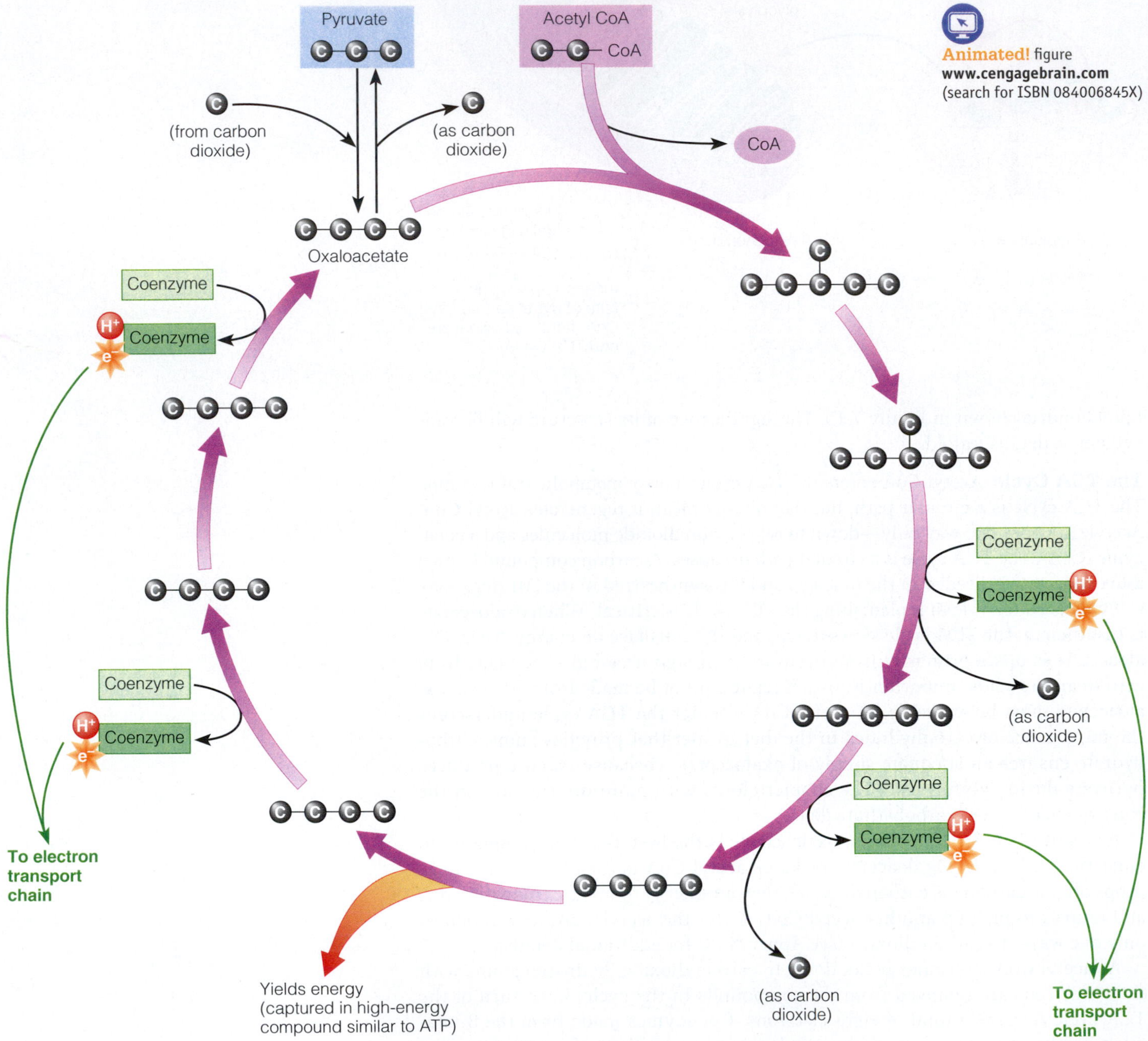

NOTE: Knowing that glucose produces pyruvate during glycolysis and that oxaloacetate must be available to start the TCA cycle, you can understand why the complete oxidation of fat requires carbohydrate.

The Electron Transport Chain In the final pathway, the electron transport chain, energy is captured in the high-energy bonds of ATP. The electron transport chain consists of a series of proteins that serve as electron "carriers." These carriers are mounted in sequence on the inner membrane of the mitochondria (review Figure 7-15). As the coenzymes deliver their electrons from the TCA cycle, glycolysis, and fatty acid oxidation to the electron transport chain, each carrier receives

FIGURE 7-17 **Electron Transport Chain and ATP Synthesis**

Animated! figure
www.cengagebrain.com (search for ISBN 084006845X)

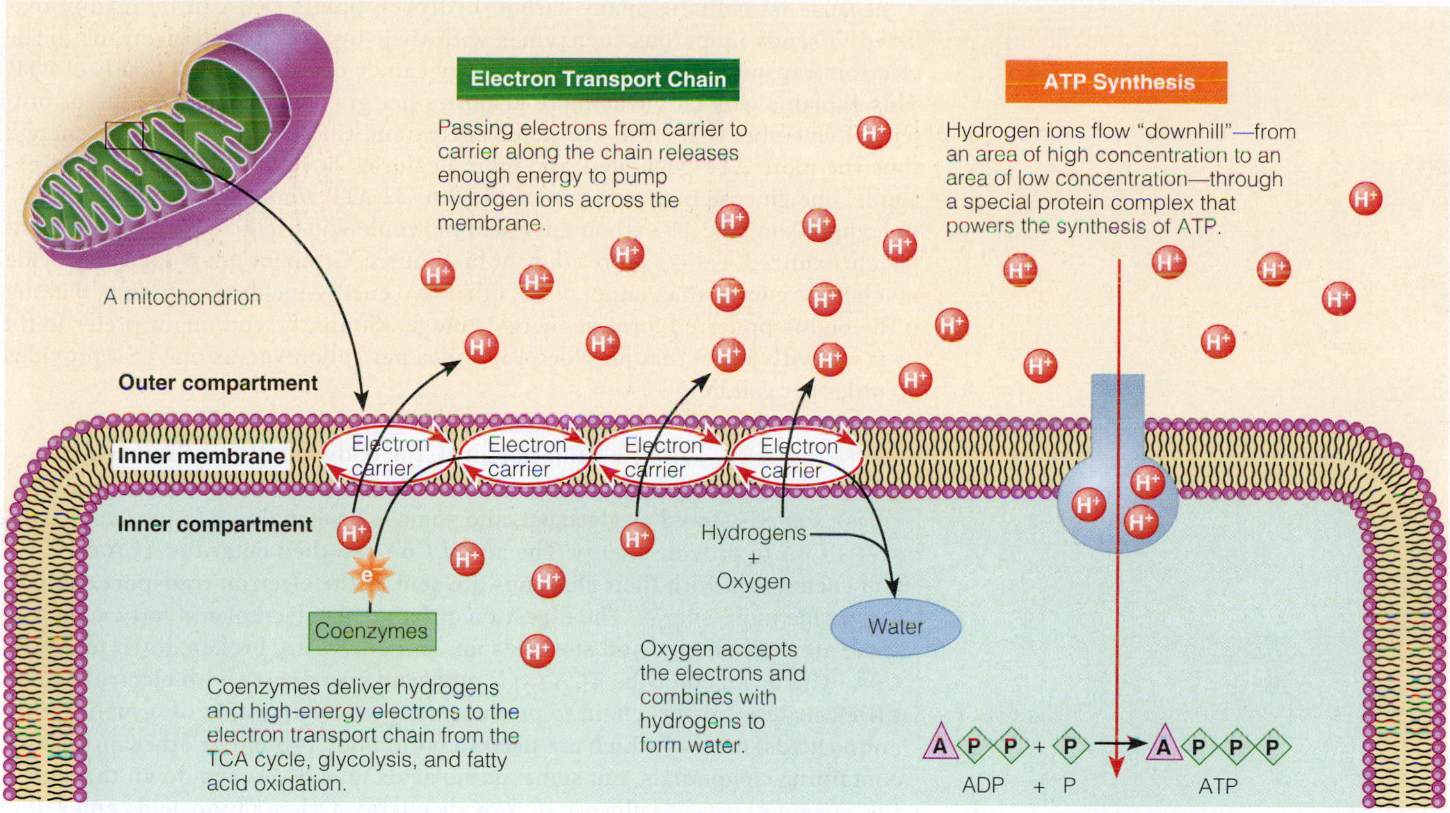
Electron Transport Chain
Passing electrons from carrier to carrier along the chain releases enough energy to pump hydrogen ions across the membrane.
ATP Synthesis
Hydrogen ions flow "downhill"—from an area of high concentration to an area of low concentration—through a special protein complex that powers the synthesis of ATP.
A mitochondrion
Outer compartment
Inner membrane
Inner compartment
Electron carrier
Electron carrier
Electron carrier
Electron carrier
Hydrogens + Oxygen
Water
Coenzymes
Coenzymes deliver hydrogens and high-energy electrons to the electron transport chain from the TCA cycle, glycolysis, and fatty acid oxidation.
Oxygen accepts the electrons and combines with hydrogens to form water.
ADP + P
ATP

the electrons and passes them on to the next carrier. These electron carriers continue passing the electrons down until they reach oxygen at the end of the chain. Oxygen (O) accepts the electrons and combines with hydrogen atoms (H) to form water (H_2O). That oxygen must be available for energy metabolism explains why it is essential to life.

As electrons are passed from carrier to carrier, hydrogen ions are pumped across the membrane to the outer compartment of the mitochondria. The rush of hydrogen ions back into the inner compartment powers the synthesis of ATP. In this way, energy is captured in the bonds of ATP. The ATP leaves the mitochondria and enters the cytoplasm, where it can be used for energy. Figure 7-17 provides a simple diagram of the electron transport chain (see Appendix C for details).

The kCalories-per-Gram Secret Revealed Of the three energy-yielding nutrients, fat provides the most energy per gram. ♦ The reason may be apparent in Figure 7-18, which compares a fatty acid with a glucose molecule. Notice that nearly all the bonds in the fatty acid are between carbons and hydrogens. Oxygen can be added to all of them—forming carbon dioxide with the carbons and water with the hydrogens. As this happens, hydrogens are released to coenzymes heading for

♦ Fat = 9 kcal/g
Carbohydrate = 4 kcal/g
Protein = 4 kcal/g

FIGURE 7-18 **Chemical Structures of a Fatty Acid and Glucose Compared**

To ease comparison, the structure shown here for glucose is not the ring structure shown in Chapter 4, but an alternative way of drawing its chemical structure.

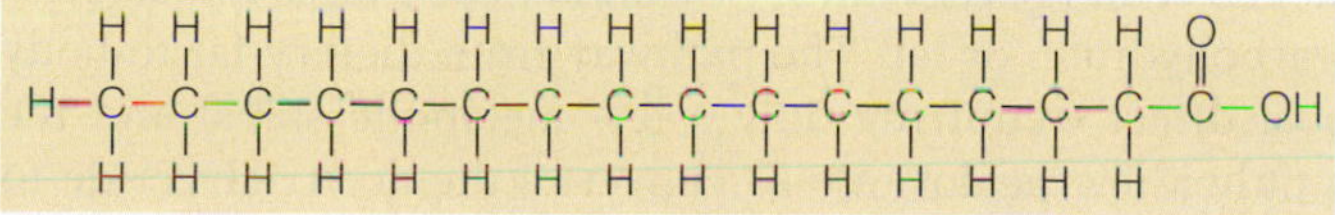

Fatty acid

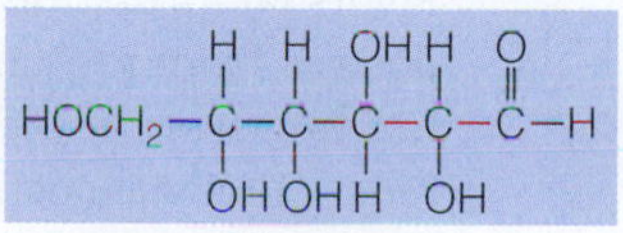

Glucose

the electron transport chain. In glucose, on the other hand, an oxygen is already bonded to each carbon. Thus there is less potential for oxidation, and fewer hydrogens are released when the remaining bonds are broken.

Because fat contains many carbon-hydrogen bonds that can be readily oxidized, it sends numerous coenzymes with their hydrogens and electrons to the electron transport chain where that energy can be captured in the bonds of ATP. This explains why fat yields more kcalories per gram than carbohydrate or protein. (Remember that each ATP holds energy and that kcalories measure energy; thus the more ATP generated, the more kcalories have been collected.) For example, one glucose molecule will yield 30 to 32 ATP when completely oxidized.[4] In comparison, one 16-carbon fatty acid molecule will yield 129 ATP when completely oxidized. Fat is a more efficient fuel source. Gram for gram, fat can provide much more energy than either of the other two energy-yielding nutrients, making it the body's preferred form of energy storage. (Similarly, you might prefer to fill your car with a fuel that provides 130 miles per gallon versus one that provides 30 miles per gallon.)

IN SUMMARY **After a balanced meal, the body handles the nutrients as follows. The digestion of carbohydrate yields glucose (and other monosaccharides); some is stored as glycogen, and some is broken down to pyruvate and acetyl CoA to provide energy. The acetyl CoA can then enter the TCA cycle and coenzymes with their electrons are sent to the electron transport chain to provide more energy. The digestion of fat yields glycerol and fatty acids; some are reassembled and stored as fat, and others are broken down to acetyl CoA, which can enter the TCA cycle and send coenzymes with electrons to the electron transport chain to provide energy. The digestion of protein yields amino acids, most of which are used to build body protein or other nitrogen-containing compounds, but some amino acids may be broken down through the same pathways as glucose to provide energy. Other amino acids enter directly into the TCA cycle, and these, too, can be broken down to yield energy.**

In summary, although carbohydrate, fat, and protein enter the TCA cycle by different routes, the final pathways are common to all energy-yielding nutrients. These pathways, which are shown as a simplified overview in Figure 7-5 (p. 211), are shown again in more detail in Figure 7-19. Instead of dismissing this figure as "too busy," take a few moments to appreciate the busyness of it all. Consider that this figure is merely an overview of energy metabolism, and then imagine how busy a cell really is during the metabolism of hundreds of compounds, each of which may be involved in several reactions, each requiring an enzyme.

Energy Balance

Every day, a healthy diet delivers more than a thousand kcalories from foods, and the active body uses most of them to do its work. As a result, body weight changes little, if at all. Maintaining body weight reflects that the body's energy budget is balanced. Some people, however, eat too much or exercise too little and get fat; others eat too little or exercise too much and get thin. The metabolic details have already been described; the next sections review them from the perspective of the body fat gained or lost. The possible reasons why people gain or lose weight are explored in Chapter 8.

Feasting—Excess Energy

When a person eats too much, metabolism favors fat formation. Fat cells enlarge regardless of whether the excess in kcalories derives from protein, carbohydrate, or fat. The pathway from dietary fat to body fat, however, is the most direct (requiring only a few metabolic steps) and the most efficient (costing only a few kcalories). To convert a dietary triglyceride to

FIGURE 7-19 The Central Pathways of Energy Metabolism

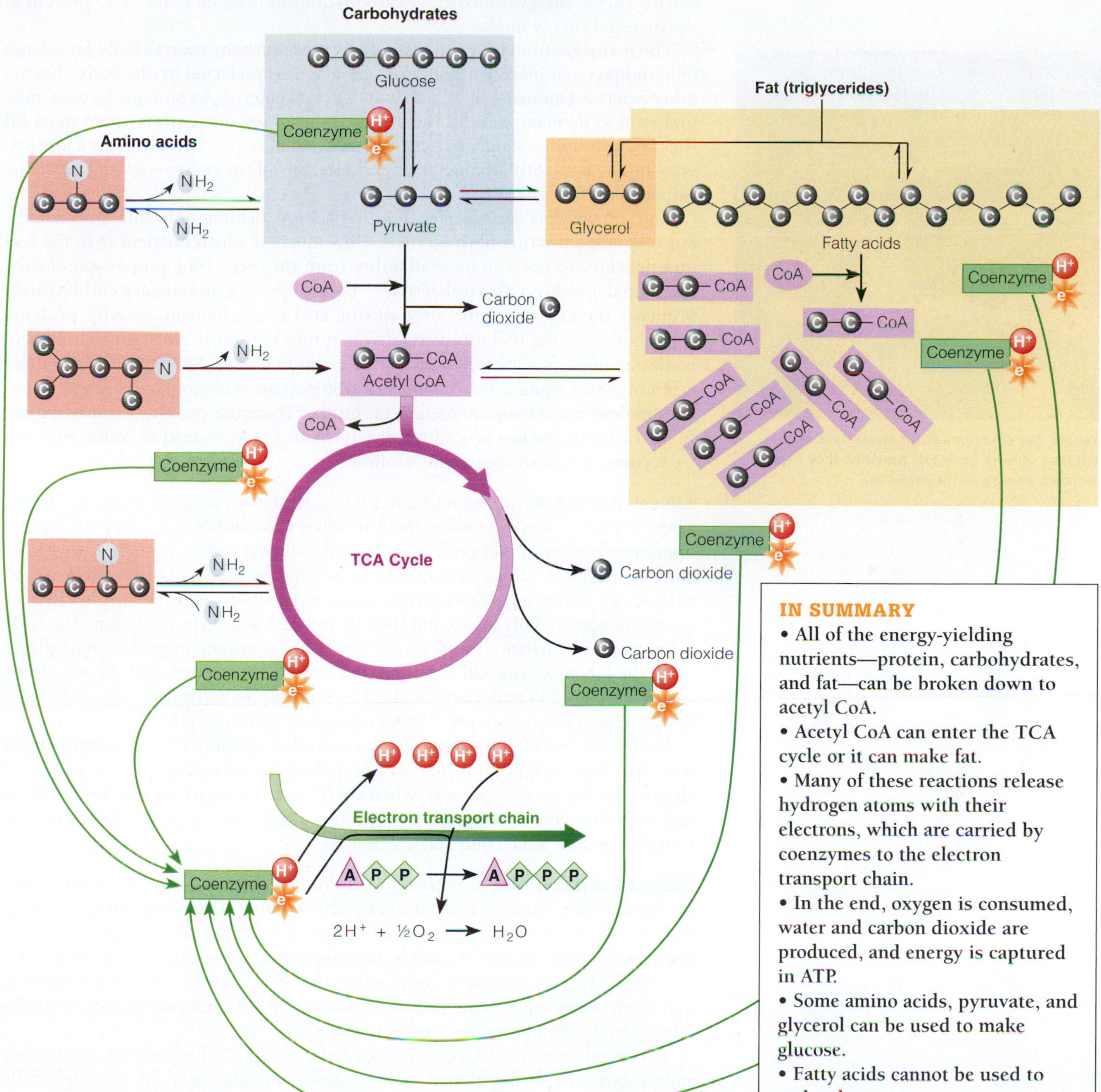

IN SUMMARY

- All of the energy-yielding nutrients—protein, carbohydrates, and fat—can be broken down to acetyl CoA.
- Acetyl CoA can enter the TCA cycle or it can make fat.
- Many of these reactions release hydrogen atoms with their electrons, which are carried by coenzymes to the electron transport chain.
- In the end, oxygen is consumed, water and carbon dioxide are produced, and energy is captured in ATP.
- Some amino acids, pyruvate, and glycerol can be used to make glucose.
- Fatty acids cannot be used to make glucose.

a triglyceride in adipose tissue, the body removes two of the fatty acids from the glycerol backbone, absorbs the parts, and puts them (and others) together again. By comparison, to convert a molecule of sucrose, the body has to split glucose from fructose, absorb them, dismantle them to pyruvate and acetyl CoA, assemble many acetyl CoA molecules into fatty acid chains, and finally attach fatty acids to a glycerol backbone to make a triglyceride for storage in adipose tissue. Quite simply, the body uses much less energy to convert dietary fat to body fat than it does to convert dietary carbohydrate to body fat. On average, storing excess energy from dietary fat as body fat uses only 5 percent of the ingested energy intake, but storing excess energy from dietary carbohydrate as body fat requires 25 percent of the ingested energy intake.

The pathways from excess protein and excess carbohydrate to body fat are not only indirect and inefficient, but they are also less preferred by the body (having other priorities for using these nutrients). Before entering fat storage, protein must first tend to its many roles in the body's lean tissues, and carbohydrate must fill the glycogen stores. Simply put, using these two nutrients to make fat is a low priority for the body. Still, if eaten in abundance, any of the energy-yielding nutrients can be made into fat.

iStockphoto.com/sack

People can enjoy bountiful meals such as this without storing body fat, provided they expend as much energy as they take in.

This chapter has described each of the energy-yielding nutrients individually, but cells use a mixture of these fuels. How much of which nutrient is in the fuel mix depends, in part, on its availability from the diet.[5] (The proportion of each fuel also depends on physical activity.) Dietary protein and dietary carbohydrate influence the mixture of fuel used during energy metabolism. Usually, protein's contribution to the fuel mix is relatively minor and fairly constant, but protein oxidation does increase when protein is eaten in excess. Similarly, carbohydrate eaten in excess significantly enhances carbohydrate oxidation. In contrast, fat oxidation does *not* respond to dietary fat intake. The more protein or carbohydrate in the fuel mix, the less fat contributes to the fuel mix. Instead of being oxidized, fat accumulates in storage. Details follow.

Excess Protein Recall from Chapter 6 that the body cannot store excess amino acids as such; it has to convert them to other compounds. Contrary to popular opinion, a person cannot grow muscle simply by overeating protein. Lean tissue such as muscle develops in response to a stimulus such as hormones or physical activity. When a person overeats protein, the body uses the surplus first by replacing normal daily losses and then by increasing protein oxidation. The body achieves protein balance this way, but any increase in protein oxidation displaces fat in the fuel mix. Any additional protein is then deaminated, and the remaining carbons are used to make fatty acids, which are stored as triglycerides in adipose tissue. Thus a person can grow fat by eating too much protein.

People who eat huge portions of meat and other protein-rich foods may wonder why they have weight problems. Not only does the fat in those foods lead to fat storage, but the protein can, too, when energy intake exceeds energy needs. Many fad weight-loss diets encourage high protein intakes based on the false assumption that protein builds only muscle, not fat.

Excess Carbohydrate Compared with protein, the proportion of carbohydrate in the fuel mix changes more dramatically when a person overeats. The body handles abundant carbohydrate by first storing it as glycogen, but glycogen storage areas are limited and fill quickly. Because maintaining glucose balance is critical, the body uses glucose frugally when the diet provides only small amounts and freely when supplies are abundant. In other words, glucose oxidation rapidly adjusts to the dietary intake of carbohydrate.

Excess glucose can also be converted to fat directly.[6] This pathway is relatively minor, however. As mentioned earlier, converting glucose to fat is energetically expensive and does not occur until after glycogen stores have been filled. Even then, only a little, if any, new fat is made from carbohydrate.

Nevertheless, excess dietary carbohydrate can displace fat in the fuel mix.[7] When this occurs, carbohydrate spares both dietary fat and body fat from oxidation—an effect that may be more pronounced in overweight people than in lean people. The net result: excess carbohydrate contributes to obesity or at least to the maintenance of an overweight body.

Excess Fat Unlike excess protein and carbohydrate, which both enhance their own oxidation, eating too much fat does not promote fat oxidation. Instead, excess dietary fat moves efficiently into the body's fat stores; almost all of the excess is stored.

IN SUMMARY If energy intake exceeds the body's energy needs, the result will be weight gain—regardless of whether the excess intake is from protein, carbohydrate, or fat. The difference is that the body is much more efficient at storing energy when the excess derives from dietary fat.

The Transition from Feasting to Fasting

Figure 7-20 shows the metabolic pathways operating in the body as it shifts from feasting (part A) to

FIGURE 7-20 Feasting and Fasting

A When a person overeats (feasting): When a person eats in excess of energy needs, the body stores a small amount of glycogen and much larger quantities of fat.

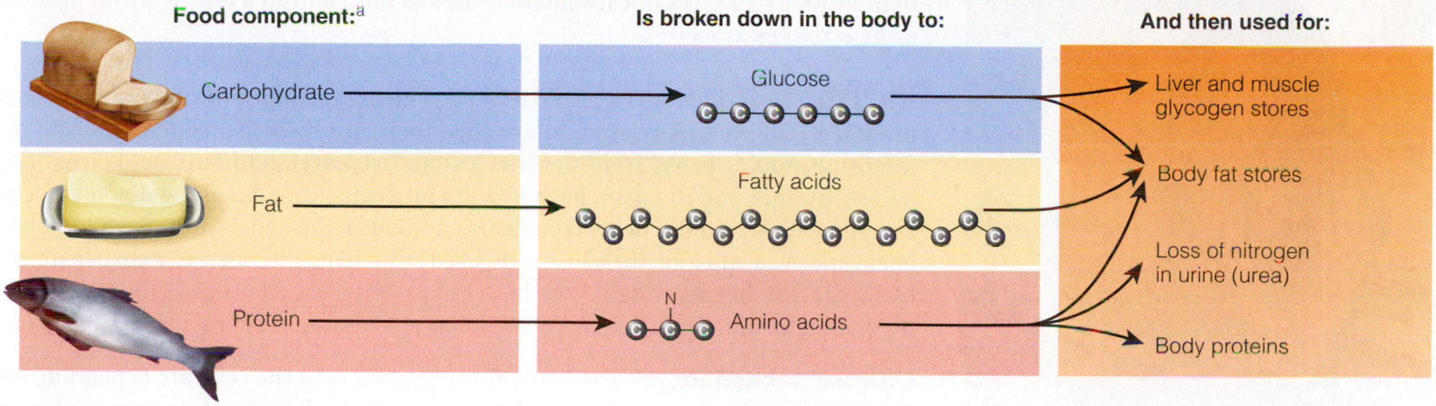

B When a person draws on stores (fasting): When nutrients from a meal are no longer available to provide energy (about 2 to 3 hours after a meal), the body draws on its glycogen and fat stores for energy.

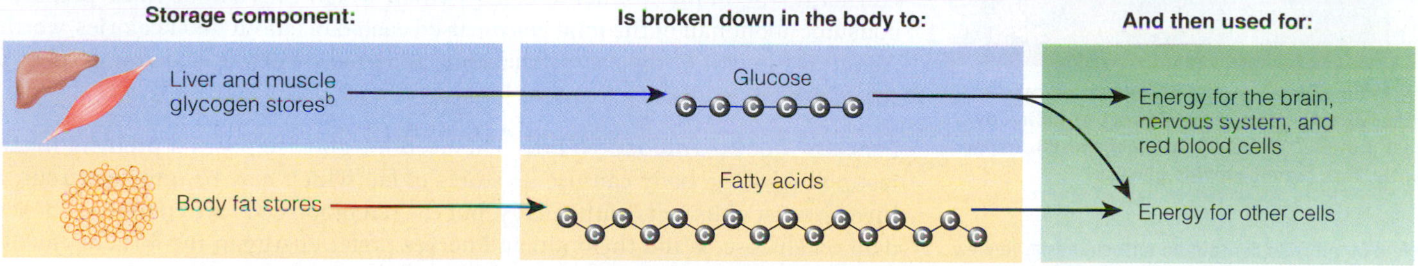

C If the fast continues beyond glycogen depletion: As glycogen stores dwindle (after about 24 hours of starvation), the body begins to break down its protein (muscle and lean tissue) to amino acids to synthesize glucose needed for brain and nervous system energy. In addition, the liver converts fats to ketone bodies, which serve as an alternative energy source for the brain, thus slowing the breakdown of body protein.

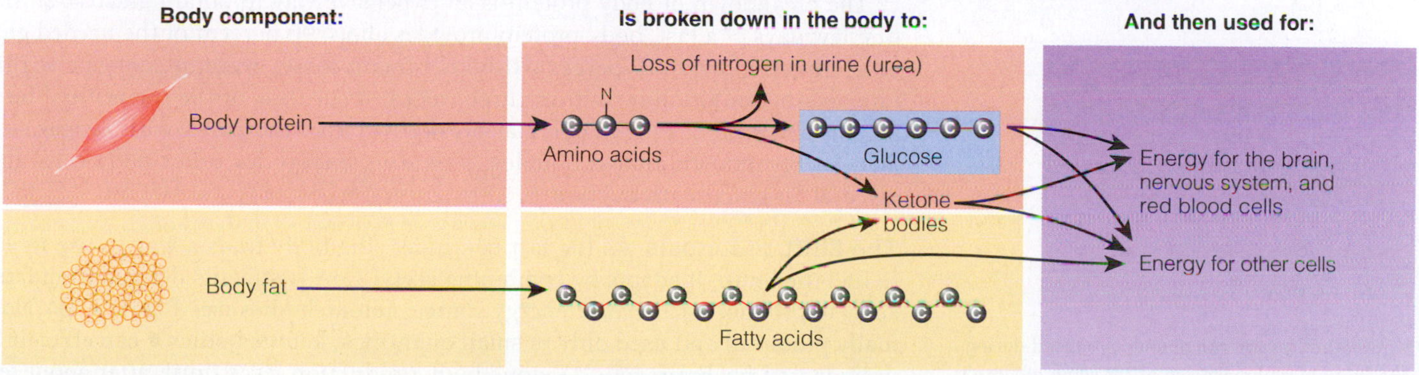

[a]Alcohol is not included because it is a toxin and not a nutrient, but it does contribute energy to the body. After detoxifying the alcohol, the body uses the remaining two carbon fragments to build fatty acids and stores them as fat.

[b]The muscles' stored glycogen provides glucose only for the muscle in which the glycogen is stored.

fasting (parts B and C). After a meal, glucose, glycerol, and fatty acids from foods are used as needed and then stored. Later, as the body shifts from a fed state to a fasting one, it begins drawing on these stores. Glycogen and fat are released from storage to provide more glucose, glycerol, and fatty acids for energy.

Energy is needed all the time. Even when a person is asleep and totally relaxed, the cells of many organs are hard at work. In fact, this work—the cells' work that maintains all life processes ♦ without any conscious effort—represents about two-thirds of the total energy a person expends in a day. The small remainder is the work that a person's muscles perform voluntarily during waking hours.

♦ The cells' work that maintains all life processes refers to the body's **basal metabolism**, which is described in Chapter 8.

The body's top priority is to meet the cells' needs for energy, and it normally does this by periodic refueling—that is, by eating several times a day. When food is not available, the body turns to its own tissues for other fuel sources. If people choose not to eat, we say they are fasting; if they have no choice, we say they are starving. The body makes no such distinction. In either case, the body is forced to draw on its reserves of carbohydrate and fat and, within a day or so, on its vital protein tissues as well.

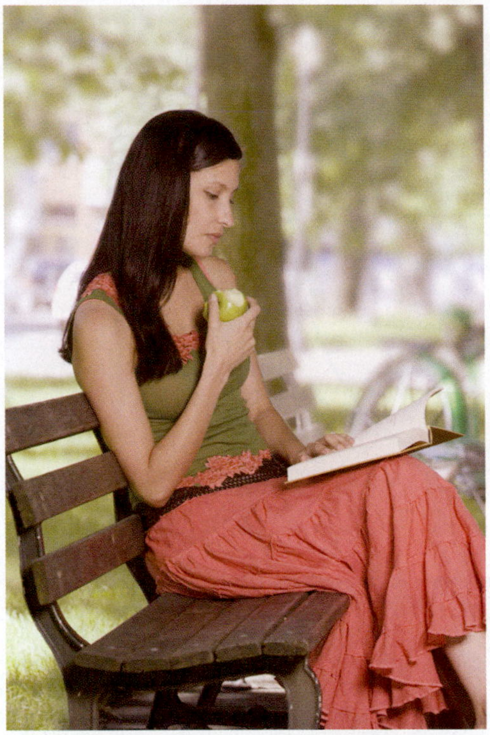

© David Buffington/PhotoLibrary

The brain and nerve cells depend on glucose—either directly from carbohydrates or indirectly from proteins (through gluconeogenesis). Importantly, fats cannot provide glucose.

Fasting—Inadequate Energy

During fasting, carbohydrate, fat, and protein are all eventually used for energy—fuel must be delivered to every cell. As the fast begins, glucose from the liver's stored glycogen and fatty acids from the adipose tissue's stored fat are both flowing into cells, then breaking down to yield acetyl CoA, and finally delivering energy to power the cells' work. Several hours later, however, most of the glucose is used up—liver glycogen is exhausted and blood glucose begins to fall. Low blood glucose serves as a signal that promotes further fat breakdown and release of amino acids from muscles.

Glucose Needed for the Brain

At this point, most of the cells are depending on fatty acids to continue providing their fuel. But, as mentioned earlier, red blood cells and the cells of the nervous system need glucose. Glucose is their primary energy fuel, and even when other energy fuels are available, glucose must be present to permit the energy-metabolizing machinery of the nervous system to work. Normally, the brain and nerve cells—which weigh only about three pounds—consume about half of the total *glucose* used each day (about 500 kcalories' worth). About one-fourth of the *energy* the adult body uses when it is at rest is spent by the brain.[8]

Protein Meets Glucose Needs

The need for glucose poses a problem for the fasting body. The body can use its stores of fat, which may be quite generous, to furnish most of its cells with energy, but the red blood cells are completely dependent on glucose, ♦ and the brain and nerves prefer energy in the form of glucose. Amino acids that yield pyruvate can be used to make glucose. ♦ To obtain the amino acids, body proteins must be broken down. For this reason, body protein tissues such as muscle and liver always break down to some extent during fasting. The amino acids that can't be used to make glucose are used as an energy source for other body cells.

♦ Red blood cells contain no mitochondria. Review Figure 7-1, p. 206, to fully appreciate why red blood cells must depend on glucose for energy.

♦ 1 g protein can make ½ g glucose

The breakdown of body protein is an expensive way to obtain glucose. In the first few days of a fast, body protein provides about 90 percent of the needed glucose; glycerol, about 10 percent. If body protein losses were to continue at this rate, death would follow within three weeks, regardless of the quantity of fat a person had stored. Fortunately, fat breakdown also increases with fasting—in fact, fat breakdown almost doubles, providing energy for other body cells and glycerol for glucose production.

The Shift to Ketosis

As the fast continues, the body finds a way to use its fat to fuel the brain. It adapts by combining acetyl CoA fragments derived from fatty acids to produce an alternate energy source, ketone bodies (see Figure 7-21). Normally produced and used only in small quantities, ketone bodies ♦ can efficiently provide fuel for brain cells.[9] Ketone body production rises until, after about ten days of fasting, it is meeting much of the nervous system's energy needs. Still,

♦ **Ketone bodies** are compounds produced during the incomplete breakdown of fat when glucose is not available.

many areas of the brain rely exclusively on glucose, and to produce it, the body continues to sacrifice protein—albeit at a slower rate than in the early days of fasting.

When ketone bodies contain an acid group (COOH), they are called **keto acids.** Small amounts of keto acids are a normal part of the blood chemistry, but when their concentration rises, the pH of the blood drops. This is ketosis, a sign that the body's chemistry is going awry. Acidic blood denatures proteins, leaving them unable to function. Elevated blood ketones (ketonemia) are excreted in the urine (ketonuria). A fruity odor on the breath (known as acetone breath) develops, reflecting the presence of the ketone acetone.

FIGURE 7-21 Ketone Body Formation

1. The first step in the formation of ketone bodies is the condensation of two molecules of acetyl CoA and the removal of the CoA to form a compound that is converted to the first ketone body.
2. This ketone body may lose a molecule of carbon dioxide to become another ketone.
3. Or, the acetoacetate may add two hydrogens, becoming another ketone body (beta-hydroxybutyrate). See Appendix C for more details.

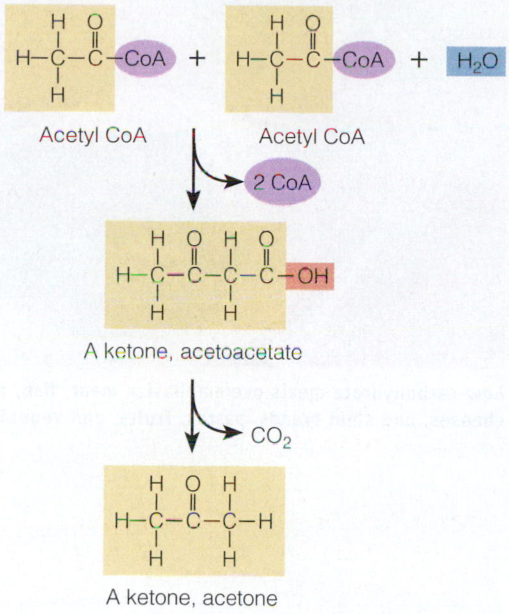

Suppression of Appetite Ketosis also induces a loss of appetite. As starvation continues, this loss of appetite becomes an advantage to a person without access to food because the search for food would be a waste of energy. When the person finds food and eats again, the body shifts out of ketosis, the hunger center gets the message that food is again available, and the appetite returns.

Slowing of Metabolism In an effort to conserve body tissues for as long as possible, the hormones of fasting slow metabolism. As the body shifts to the use of ketone bodies, it simultaneously reduces its energy output and conserves both its fat and its lean tissue. Still the lean (protein-containing) tissues shrink and perform less metabolic work, reducing energy expenditures. As the muscles waste, they can do less work and so demand less energy, reducing expenditures further. Although fasting may promote dramatic *weight* loss, a low-kcalorie diet and physical activity better support *fat* loss while retaining lean tissue.

Symptoms of Starvation The adaptations just described—slowing of energy output and reduction in fat loss—occur in the starving child, the hungry homeless adult, the fasting religious person, the adolescent with anorexia nervosa, and the malnourished hospital patient. Such adaptations help to prolong their lives and explain the physical symptoms of starvation: wasting; slowed heart rate, respiration, and metabolism; lowered body temperature; impaired vision; organ failure; and reduced resistance to disease. Psychological effects of food deprivation include depression, anxiety, and food-related dreams.

The body's adaptations to fasting are sufficient to maintain life for a long time—up to two months. Mental alertness need not be diminished, and even some physical energy may remain unimpaired for a surprisingly long time. These remarkable adaptations, however, should not prevent anyone from recognizing the very real hazards that fasting presents.

IN SUMMARY **When fasting, the body makes a number of adaptations: increasing the breakdown of fat to provide energy for most of the cells, using glycerol and amino acids to make glucose for the red blood cells and central nervous system, producing ketones to fuel the brain, suppressing the appetite, and slowing metabolism. All of these measures conserve energy and minimize losses.**

Low-Carbohydrate Diets

When a person consumes a low-carbohydrate diet, a metabolism similar to that of fasting prevails. With little dietary carbohydrate coming in, the body uses its glycogen stores to provide glucose for the cells of the brain, nerves, and blood. Once the body depletes its glycogen reserves, it

keto (KEY-toe) **acids:** organic acids that contain a carbonyl group (C=O).

Low-carbohydrate meals overemphasize meat, fish, poultry, eggs, and cheeses, and shun breads, pastas, fruits, and vegetables.

begins making glucose from the amino acids of protein (gluconeogenesis). A low-carbohydrate diet may provide abundant protein from food, but the body still uses some protein from body tissues.

Dieters can know glycogen depletion has occurred and gluconeogenesis has begun by monitoring their urine. Whenever glycogen or protein is broken down, water is released and urine production increases. Low-carbohydrate diets also induce ketosis, and ketones can be detected in the urine. Ketones form whenever glucose is lacking and fat breakdown is incomplete.

Many fad diets regard ketosis as the key to losing weight, but studies comparing weight-loss diets find no relation between ketosis and weight loss.[10] People in ketosis may experience a loss of appetite and a dramatic weight loss within the first few days.[11] They should know that much of this weight loss reflects the loss of glycogen and protein together with large quantities of body fluids and important minerals. They need to appreciate the difference between loss of *fat* and loss of *weight*. Fat losses on ketogenic diets are no greater than on other diets providing the same number of kcalories. Once the dieter returns to well-balanced meals that provide adequate energy, carbohydrate, fat, protein, vitamins, and minerals, the body avidly retains these needed nutrients. The weight will return, quite often to a level higher than the starting point. Table 7-2 lists other consequences of a ketogenic diet.

TABLE 7-2 Adverse Side Effects of Low-Carbohydrate, Ketogenic Diets

- Nausea
- Fatigue (especially if physically active)
- Constipation
- Low blood pressure
- Elevated uric acid (which may exacerbate kidney disease and cause inflammation of the joints in those predisposed to gout)
- Stale, foul taste in the mouth (bad breath)
- In pregnant women, fetal harm and stillbirth

This chapter has probed the intricate details of metabolism at the level of the cells, exploring the transformations of nutrients to energy and to storage compounds. Several chapters and highlights build on this information. The highlight that follows this chapter shows how alcohol disrupts normal metabolism. Chapter 8 describes how a person's intake and expenditure of energy are reflected in body weight and body composition. Chapter 9 examines the consequences of unbalanced energy budgets—overweight and underweight. Chapter 10 shows the vital roles the B vitamins play as coenzymes assisting all the metabolic pathways described here.

Nutrition Portfolio

All day, every day, your cells dismantle carbohydrates, fats, and proteins, with the help of vitamins, minerals, and water, releasing energy to meet your body's immediate needs or storing it as fat for later use.

Go to Diet Analysis Plus and choose one of the days on which you have tracked your diet for the entire day. Go to the Intake vs. Goals report and answer the following questions. Keep in mind that in this report 100 percent means you are meeting your needs perfectly.

- How close were you to 100 percent for: carbohydrates, fats, proteins, vitamins, minerals, and water? In general, which category was lowest? Which category was highest?
- Describe what types of foods best support aerobic and anaerobic activities.
- Consider whether you eat more protein, carbohydrate, or fat than your body needs.
- Explain how a low-carbohydrate diet forces your body into ketosis.

Diet Analysis PLUS **To complete this exercise, go to your Diet Analysis Plus at www.cengage.com/sso.**

References

1. R. H. Garrett and C. M. Grisham, *Biochemistry* (Belmont, Calif.: Thomson Brooks/Cole, 2005), p. 73.
2. S. P. Cairns, Lactic acid and exercise performance: Culprit or friend? *Sports Medicine* 36 (2006): 279–291; J. P. Weir and coauthors, Is fatigue all in your head? A critical review of the central governor model, *British Journal of Sports Medicine* 40 (2006): 573–586; A. Philp, A. L. Macdonald, and P. W. Watt, Lactate: A signal coordinating cell and systemic function, *Journal of Experimental Biology* 208 (2005): 4561–4575.
3. S. S. Gropper, J. L. Smith, and J. L. Groff, *Advanced Nutrition and Human Metabolism* (Belmont, Calif.: Thomson Wadsworth, 2005), p. 198.
4. Garrett and Grisham, 2005, p. 669.
5. A. Wise, Transcriptional switches in the control of macronutrient metabolism, *Nutrition Reviews* 66 (2008): 321–325.
6. M. F. Chong and coauthors, Parallel activation of de novo lipogenesis and stearoyl-CoA desaturase activity after 3 d of high-carbohydrate feeding, *American Journal of Clinical Nutrition* 87 (2008): 817–823.
7. R. Roberts and coauthors, Reduced oxidation of dietary fat after a short term high-carbohydrate diet, *American Journal of Clinical Nutrition* 87 (2008): 824–831.
8. W. R. Leonard, J. J. Snodgrass, and M. L. Robertson, Effects of brain evolution on human nutrition and metabolism, *Annual Review of Nutrition* 27 (2007): 311–327.
9. G. F. Cahill, Fuel metabolism in starvation, *Annual Review of Nutrition* 26 (2006): 1–22.
10. M. D. Coleman and S. M. Nickols-Richardson, Urinary ketones reflect serum ketone concentration but do not relate weight loss in overweight premenopausal women following a low-carbohydrate/high-protein diet, *Journal of the American Dietetic Association* 105 (2005): 608–611.
11. A. M. Johnstone and coauthors, Effects of a high-protein ketogenic diet on hunger, appetite, and weight loss in obese men feeding ad libitum, *American Journal of Clinical Nutrition* 87 (2008): 44–55.

HIGHLIGHT 7

Alcohol and Nutrition

Yellowj/Shutterstock.com

With the understanding of metabolism gained from Chapter 7, you are in a position to understand how the body handles alcohol, how alcohol interferes with metabolism, and how alcohol impairs health and nutrition. Before examining alcohol's damaging effects, it may be appropriate to mention that drinking alcohol in *moderation* may have some health benefits, including reduced risks of heart attacks, strokes, dementia, diabetes, and osteoporosis.[1] Moderate alcohol consumption may lower mortality from all causes, but only in adults aged 35 and older.[2] No health benefits are evident before middle age.[3] Similarly, health benefits begin to disappear in older age, as metabolism changes and organs become more sensitive to toxic substances.[4] Importantly, any benefits of moderate alcohol use must be weighed against the many harmful effects of excessive alcohol use described in this highlight, as well as the possibility of alcohol abuse.

Alcohol in Beverages

To the chemist, **alcohol** refers to a class of organic compounds containing hydroxyl (OH) groups (the accompanying glossary defines alcohol and related terms). The glycerol to which fatty acids are attached in triglycerides is an example of an alcohol to a chemist. To most people, though, *alcohol* refers to the intoxicat-

GLOSSARY

acetaldehyde (ass-et-AL-duh-hide): an intermediate in alcohol metabolism.

alcohol: a class of organic compounds containing hydroxyl (OH) groups.

alcohol abuse: a pattern of drinking that includes failure to fulfill work, school, or home responsibilities; drinking in situations that are physically dangerous (as in driving while intoxicated); recurring alcohol-related legal problems (as in aggravated assault charges); or continued drinking despite ongoing social problems that are caused by or worsened by alcohol.

alcohol dehydrogenase (dee-high-DROJ-eh-nayz): an enzyme active in the stomach and the liver that converts ethanol to acetaldehyde.

alcoholism: a pattern of drinking that includes a strong craving for alcohol, a loss of control and an inability to stop drinking once begun, withdrawal symptoms (nausea, sweating, shakiness, and anxiety) after heavy drinking, and the need for increasing amounts of alcohol to feel "high."

antidiuretic hormone (ADH): a hormone produced by the pituitary gland in response to dehydration (or a high sodium concentration in the blood). It stimulates the kidneys to reabsorb more water and therefore prevents water loss in urine (also called *vasopressin*). (This ADH should not be confused with the enzyme alcohol dehydrogenase, which is also sometimes abbreviated ADH.)

beer: an alcoholic beverage traditionally brewed by fermenting malted barley and adding hops for flavor.

binge drinking: four or more drinks for women and five or more drinks for men of alcohol in a row (within a couple of hours).

cirrhosis (seer-OH-sis): advanced liver disease in which liver cells turn orange, die, and harden, permanently losing their function; often associated with alcoholism.

- **cirrhos** = an orange

distilled liquor or **hard liquor:** an alcoholic beverage traditionally made by fermenting and distilling a carbohydrate source such as molasses, potatoes, rye, beets, barley, or corn; sometimes called *distilled spirits.*

drink: a dose of any alcoholic beverage that delivers ½ ounce of pure ethanol:

- 5 ounces of wine
- 10 ounces of wine cooler
- 12 ounces of beer
- 1½ ounces of hard liquor (80 proof whiskey, scotch, rum, or vodka)

drug: a substance that can modify one or more of the body's functions.

ethanol: a particular type of alcohol found in beer, wine, and distilled liquor; also called *ethyl alcohol* (see Figure H7-1). Ethanol is the most widely used—and abused—drug in our society. It is also the only legal, nonprescription drug that produces euphoria.

excessive drinking: heavy drinking, binge drinking, or both.

fatty liver: an early stage of liver deterioration seen in several diseases, including kwashiorkor and alcoholic liver disease. Fatty liver is characterized by an accumulation of fat in the liver cells.

fibrosis (fye-BROH-sis): an intermediate stage of liver deterioration seen in several diseases, including viral hepatitis and alcoholic liver disease. In fibrosis, the liver cells lose their function and assume the characteristics of connective tissue cells (fibers).

heavy drinking: more than one drink per day on average for women and more than two drinks per day on average for men.

MEOS or **microsomal** (my-krow-SO-mal) **ethanol-oxidizing system:** a system of enzymes in the liver that oxidize not only alcohol but also several classes of drugs.

moderation: in relation to alcohol consumption, not more than two drinks a day for the average-size man and not more than one drink a day for the average-size woman.

NAD (nicotinamide adenine dinucleotide): the main coenzyme form of the vitamin niacin. Its reduced form is NADH.

narcotic (nar-KOT-ic): a drug that dulls the senses, induces sleep, and becomes addictive with prolonged use.

proof: a way of stating the percentage of alcohol in distilled liquor. Liquor that is 100 proof is 50 percent alcohol; 90 proof is 45 percent, and so forth.

Wernicke-Korsakoff (VER-nee-key KORE-sah-kof) **syndrome:** a neurological disorder typically associated with chronic alcoholism and caused by a deficiency of the B vitamin thiamin; also called *alcohol-related dementia.*

wine: an alcoholic beverage traditionally made by fermenting a sugar source such as grape juice.

FIGURE H7-1 Two Alcohols: Glycerol and Ethanol

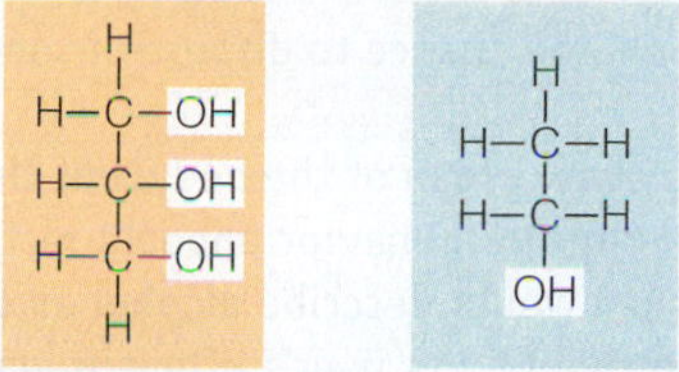

Glycerol is the alcohol used to make triglycerides.

Ethanol is the alcohol in beer, wine, and distilled liquor.

ing ingredient in **beer, wine,** and **distilled liquor (hard liquor).** The chemist's name for this particular alcohol is *ethyl alcohol,* or **ethanol.** Glycerol has three carbons with three hydroxyl groups attached; ethanol has only two carbons and one hydroxyl group (see Figure H7-1). The remainder of this highlight talks about the particular alcohol ethanol but refers to it simply as *alcohol.*

Alcohols affect living things profoundly, partly because they act as lipid solvents. Their ability to dissolve lipids out of cell membranes allows alcohols to penetrate rapidly into cells, destroying cell structures and thereby killing the cells. For this reason, most alcohols are toxic in relatively small amounts; by the same token, because they kill microbial cells, they are useful as skin disinfectants.

Ethanol is less toxic than the other alcohols. Sufficiently diluted and taken in small enough doses, its action in the brain produces an effect that people seek—not with zero risk, but with a low enough risk (if the doses are low enough) to be tolerable. Used in this way, alcohol is a **drug**—that is, a substance that modifies body functions. Like all drugs, alcohol both offers benefits and poses hazards. The *Dietary Guidelines for Americans* advise "if alcohol is consumed, it should be consumed in moderation."

Dietary Guidelines for Americans 2010

- If alcohol is consumed, it should be consumed in moderation: up to one drink per day for women and two drinks per day for men.
- Alcoholic beverages should not be consumed by some individuals, including those who cannot restrict their alcohol intake, women of childbearing age who may become pregnant, pregnant and lactating women, children and adolescents, individuals taking medications that can interact with alcohol, and those with specific medical conditions.
- Alcoholic beverages should be avoided by individuals engaging in activities that require attention, skill, or coordination, such as driving or operating machinery.

The term **moderation** is important when describing alcohol use. How many drinks constitute moderate use, and how much is "a drink"? First, a **drink** is any alcoholic beverage that delivers ½ ounce of *pure ethanol:*

© Polara Studios, Inc.

Each of these servings equals one drink.

- 5 ounces of wine
- 10 ounces of wine cooler
- 12 ounces of beer
- 1½ ounces of distilled liquor (80 proof whiskey, scotch, rum, or vodka)

As a practical tip, prevent overpouring by measuring liquids and using tall, narrow glasses.[5]

Beer, wine, and liquor deliver different amounts of alcohol. The amount of alcohol in distilled liquor is stated as **proof:** 100 proof liquor is 50 percent alcohol, 80 proof is 40 percent alcohol, and so forth. Wine and beer have less alcohol than distilled liquor, although some fortified wines and beers have more alcohol than the regular varieties (see photo caption below).

Matthew Farruggio

Wines contain 7 to 24 percent alcohol by volume; those containing 14 percent or more must state their alcohol content on the label, whereas those with less than 14 percent may simply state "table wine" or "light wine." Beers typically contain less than 5 percent alcohol by volume and malt liquors, 5 to 8 percent; regulations vary, with some states requiring beer labels to show the alcohol content and others prohibiting such statements.

Second, because people have different tolerances for alcohol, it is impossible to name an exact daily amount of alcohol that is appropriate for everyone. Authorities have attempted to identify amounts that are acceptable for most healthy people. An accepted definition of moderation is up to two drinks per day for men and up to one drink per day for women. (Pregnant women are advised to abstain from alcohol, as Highlight 14 explains.) Notice that this advice is stated as a maximum, not as an average; seven drinks one night a week would not be considered moderate, even though one a day would be. Doubtless, some people could consume slightly more; others could not handle nearly so much without risk. The amount a person can drink safely is highly individual, depending on genetics, health, gender, body composition, age, and family history.

Alcohol in the Body

From the moment an alcoholic beverage enters the body, alcohol is treated as if it has special privileges. Unlike foods, which require time for digestion, alcohol needs no digestion and is quickly absorbed across the walls of an empty stomach, reaching the brain within a few minutes. Consequently, a person can immediately feel euphoric when drinking, especially on an empty stomach.

When the stomach is full of food, alcohol has less chance of touching the walls and diffusing through, so its influence on the brain is slightly delayed. This information leads to another practical tip: eat snacks when drinking alcoholic beverages. Carbohydrate snacks slow alcohol absorption and high-fat snacks slow peristalsis, keeping the alcohol in the stomach longer. Salty snacks make a person thirsty; to quench thirst, drink water instead of more alcohol.

The stomach begins to break down alcohol with its **alcohol dehydrogenase** enzyme. Women produce less of this stomach enzyme than men; consequently, more alcohol reaches the intestine for absorption into the bloodstream. As a result, women absorb more alcohol than men of the same size who drink the same amount of alcohol. Consequently, they are more likely to become more intoxicated on less alcohol than men. Such differences between men and women help explain why women have a lower alcohol tolerance and a lower recommendation for moderate intake.

In the small intestine, alcohol is rapidly absorbed. From this point on, alcohol receives priority treatment: it gets absorbed and metabolized before most nutrients. Alcohol's priority status helps to ensure a speedy disposal and reflects two facts: alcohol cannot be stored in the body, and it is potentially toxic.

Alcohol Arrives in the Liver

As Chapter 3 explains, the capillaries of the digestive tract merge into veins that carry blood first to the liver. These veins branch and rebranch into a capillary network that touches every liver cell. Consequently, liver cells are the first to receive alcohol-laden blood. Liver cells are also the only other cells in the body that can make enough of the alcohol dehydrogenase enzyme to oxidize alcohol at an appreciable rate. The routing of blood through the liver cells gives them the chance to dispose of some alcohol before it moves on.

Alcohol affects every organ of the body, but the most dramatic evidence of its disruptive behavior appears in the liver. If liver cells could talk, they would describe alcohol as demanding, egocentric, and disruptive of the liver's efficient way of running its business. For example, liver cells normally prefer fatty acids as their fuel, and they like to package excess fatty acids into triglycerides and ship them out to other tissues. When alcohol is present, however, the liver cells are forced to metabolize alcohol and let the fatty acids accumulate, sometimes in huge stockpiles. Alcohol metabolism can also permanently change liver cell structure, impairing the liver's ability to metabolize fats. As a result, heavy drinkers develop fatty livers.

The liver is the primary site of alcohol metabolism.[6] It can process about ½ ounce of *ethanol* per hour (the amount in a typical drink), depending on the person's body size, previous drinking experience, food intake, and general health. This maximum rate of alcohol breakdown is set by the amount of alcohol dehydrogenase available. If more alcohol arrives at the liver than the enzymes can handle, the extra alcohol travels to all parts of the body, circulating again and again until liver enzymes are finally available to process it. Another practical tip derives from this information: drink slowly enough to allow the liver to keep up—no more than one drink per hour.

The amount of alcohol dehydrogenase enzyme present in the liver varies with individuals, depending on the genes they have inherited and on how recently they have eaten. Fasting for as little as a day forces the body to degrade its proteins, including the alcohol-processing enzymes, and this can slow the rate of alcohol metabolism by half. Drinking after not eating all day thus causes the drinker to feel the effects more promptly for two reasons: rapid absorption and slowed breakdown. By maintaining higher blood alcohol concentrations for longer times, alcohol can anesthetize the brain more completely (as described later in this highlight).

The alcohol dehydrogenase enzyme breaks down alcohol by removing hydrogens in two steps. (Figure H7-2 provides a simplified diagram of alcohol metabolism; Appendix C provides the chemical details.) In the first step, alcohol dehydrogenase oxidizes alcohol to **acetaldehyde**—a highly reactive and toxic compound. High concentrations of acetaldehyde in the brain and other tissues are responsible for many of the damaging effects of **alcohol abuse.**

In the second step, a related enzyme, acetaldehyde dehydrogenase, converts acetaldehyde to acetate, which is then converted to either carbon dioxide (CO_2) or acetyl CoA—the "crossroads" compound introduced in Chapter 7. The reactions from alcohol to acetaldehyde to acetate produce hydrogens (H^+) and electrons. The B vitamin niacin, in its role as the coenzyme **NAD (nicotinamide adenine dinucleotide),** helpfully picks up these hydrogens and electrons (becoming NADH) and escorts them through the electron transport chain. Thus, whenever the body breaks down alcohol, NAD diminishes and NADH accumulates. (Chapter 10 presents information on NAD and the other coenzyme roles of the B vitamins.)

FIGURE H7-2 Alcohol Metabolism

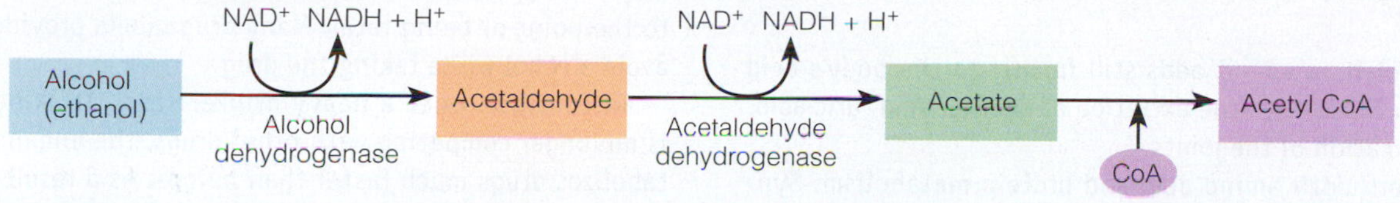

The conversion of alcohol to acetyl CoA requires the B vitamin niacin in its role as the coenzyme NAD. When the enzymes oxidize alcohol, they remove H atoms and attach them to NAD. Thus NAD is used up and NADH accumulates. NOTE: More accurately, NAD^+ is converted to NADH + H^+.

Alcohol Disrupts the Liver

During alcohol metabolism, the multitude of other metabolic processes for which NAD is required, including glycolysis, the TCA cycle, and the electron transport chain, falter. Its presence is sorely missed in these energy pathways because it is the chief carrier of the hydrogens that travel with their electrons along the electron transport chain. Without adequate NAD, these energy pathways cannot function. Traffic either backs up or an alternate route is taken. Such changes in the normal flow of energy pathways have striking physical consequences.

For one, the accumulation of hydrogen ions during alcohol metabolism shifts the body's acid-base balance toward acid. For another, the accumulation of NADH slows the TCA cycle, so pyruvate and acetyl CoA build up. Excess acetyl CoA then takes the route to fatty acid synthesis (as Figure H7-3 illustrates), and fat clogs the liver.

As you might expect, a liver overburdened with fat cannot function properly. Liver cells become less efficient at performing a number of tasks. Much of this inefficiency impairs a person's nutritional health in ways that cannot be corrected by diet alone. For example, the liver has difficulty activating vitamin D, as well as producing and releasing bile. To overcome such problems, a person needs to stop drinking alcohol.

The synthesis of fatty acids accelerates with exposure to alcohol. Fat accumulation can be seen in the liver after a single night of heavy drinking. **Fatty liver,** the first stage of liver deterioration seen in heavy drinkers, interferes with the distribution of nutrients and oxygen to the liver cells. Fatty liver is reversible with abstinence from alcohol. If fatty liver lasts long enough, however, the liver cells will die and form fibrous scar tissue. This second stage of liver deterioration is called **fibrosis.** Some liver cells can regenerate with good nutrition and abstinence from alcohol, but in the most advanced stage, **cirrhosis,** damage is the least reversible.

The fatty liver has difficulty making glucose from protein. Without gluconeogenesis, blood glucose can plummet, leading to irreversible damage to the central nervous system.

The lack of glucose together with the overabundance of acetyl CoA sets the stage for ketosis. The body uses excess acetyl CoA to make ketone bodies; their acidity pushes the acid-base balance further toward acid and suppresses nervous system activity.

Excess NADH also promotes the making of lactate from pyruvate. The conversion of pyruvate to lactate uses the hydrogens from NADH and restores some NAD, but a lactate build-up has serious

FIGURE H7-3 Alternate Route for Acetyl CoA: To Fat

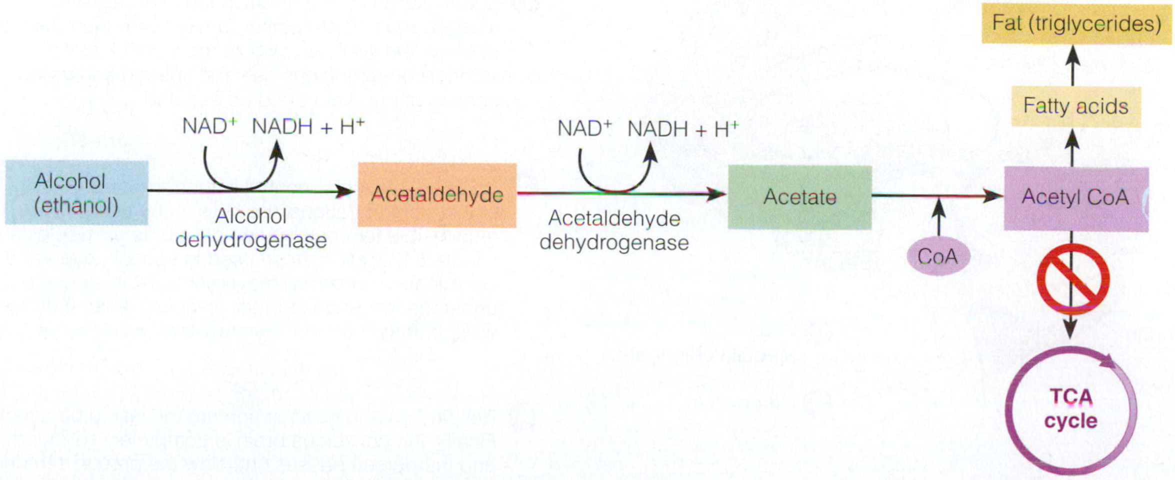

Acetyl CoA molecules are blocked from getting into the TCA cycle by the low level of NAD. Instead of being used for energy, the acetyl CoA molecules become building blocks for fatty acids.

consequences of its own—it adds still further to the body's acid burden and interferes with the excretion of another acid, uric acid, causing inflammation of the joints.

Alcohol alters both amino acid and protein metabolism. Synthesis of proteins important in the immune system slows down, weakening the body's defenses against infections. Evidence of protein deficiency becomes apparent, both from a diminished synthesis of proteins and from a poor diet. Normally, the cells would at least use the amino acids from the protein foods a person eats, but the drinker's liver deaminates the amino acids and uses the carbon fragments primarily to make fat or ketones. Eating well does not protect the drinker from protein depletion; a person has to stop drinking alcohol.

The liver's priority treatment of alcohol affects its handling of drugs as well as nutrients. In addition to the dehydrogenase enzyme already described, the liver possesses an enzyme system that metabolizes *both* alcohol and several other types of drugs. Called the **MEOS (microsomal ethanol-oxidizing system)**, this system handles about one-fifth of the total alcohol a person consumes. At high blood concentrations or with repeated exposures, alcohol stimulates the synthesis of enzymes in the MEOS. The result is a more efficient metabolism of alcohol and tolerance to its effects.

As a person's blood alcohol rises, alcohol competes with—and wins out over—other drugs whose metabolism also relies on the MEOS. If a person drinks and uses another drug at the same time, the MEOS will dispose of alcohol first and metabolize the drug more slowly. While the drug waits to be handled later, the dose may build up so that its effects are greatly amplified—sometimes to the point of being fatal. Many drug labels provide warnings to avoid alcohol while taking the drug.

In contrast, once a heavy drinker stops drinking and alcohol is no longer competing with other drugs, the enhanced MEOS metabolizes drugs much faster than before. As a result, determining the correct dosages of medications can be challenging.

This discussion has emphasized the major way that the blood is cleared of alcohol—metabolism by the liver—but there is another way. About 10 percent of the alcohol leaves the body through the breath and in the urine. This is the basis for the breath and urine tests for drunkenness. The amounts of alcohol in the breath and in the urine are in proportion to the amount still in the bloodstream and brain. In nearly all states, legal drunkenness is set at 0.10 percent or less, reflecting the relationship between alcohol use and traffic and other accidents.

Alcohol Arrives in the Brain

Figure H7-4 describes alcohol's effects on the brain. Alcohol is a **narcotic.** People used it for centuries as an anesthetic because it can deaden pain. But alcohol was a poor anesthetic because one could never be sure how much a person would need and how much would be a fatal dose. Consequently, new, more predictable anesthetics have replaced alcohol. Nonetheless, alcohol continues to be used today as a kind of social anesthetic to help people relax

FIGURE H7-4 Alcohol's Effects on the Brain

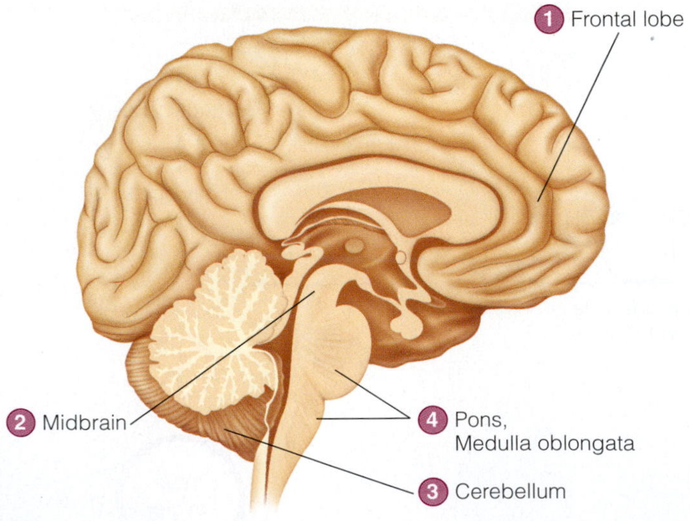

1. Judgment and reasoning centers are most sensitive to alcohol. When alcohol flows to the brain, it first sedates the frontal lobe, the center of all conscious activity. As the alcohol molecules diffuse into the cells of these lobes, they interfere with reasoning and judgment.

2. Speech and vision centers in the midbrain are affected next. If the drinker drinks faster than the rate at which the liver can oxidize the alcohol, blood alcohol concentrations rise: the speech and vision centers of the brain become sedated.

3. Voluntary muscular control is then affected. At still higher concentrations, the cells in the cerebellum responsible for coordination of voluntary muscles are affected, including those used in speech, eye-hand coordination, and limb movements. At this point people under the influence stagger or weave when they try to walk, or they may slur their speech.

4. Respiration and heart action are the last to be affected. Finally, the conscious brain is completely subdued, and the person passes out. Now the person can drink no more; this is fortunate because higher doses would anesthetize the deepest brain centers that control breathing and heartbeat, causing death.

TABLE H7-1 Alcohol Doses and Approximate Blood Level Percentages for Men and Women

Drinks[a]	Body Weight in Pounds—Men								
	100	120	140	160	180	200	220	240	
	00	00	00	00	00	00	00	00	ONLY SAFE DRIVING LIMIT
1	.04	.03	.03	.02	.02	.02	.02	.02	IMPAIRMENT BEGINS
2	.08	.06	.05	.05	.04	.04	.03	.03	
3	.11	.09	.08	.07	.06	.06	.05	.05	DRIVING SKILLS SIGNIFICANTLY AFFECTED
4	.15	.12	.11	.09	.08	.08	.07	.06	
5	.19	.16	.13	.12	.11	.09	.09	.08	
6	.23	.19	.16	.14	.13	.11	.10	.09	
7	.26	.22	.19	.16	.15	.13	.12	.11	LEGALLY INTOXICATED
8	.30	.25	.21	.19	.17	.15	.14	.13	
9	.34	.28	.24	.21	.19	.17	.15	.14	
10	.38	.31	.27	.23	.21	.19	.17	.16	

Drinks[a]	Body Weight in Pounds—Women									
	90	100	120	140	160	180	200	220	240	
	00	00	00	00	00	00	00	00	00	ONLY SAFE DRIVING LIMIT
1	.05	.05	.04	.03	.03	.03	.02	.02	.02	IMPAIRMENT BEGINS
2	.10	.09	.08	.07	.06	.05	.05	.04	.04	DRIVING SKILLS SIGNIFICANTLY AFFECTED
3	.15	.14	.11	.10	.09	.08	.07	.06	.06	
4	.20	.18	.15	.13	.11	.10	.09	.08	.08	
5	.25	.23	.19	.16	.14	.13	.11	.10	.09	
6	.30	.27	.23	.19	.17	.15	.14	.12	.11	LEGALLY INTOXICATED
7	.35	.32	.27	.23	.20	.18	.16	.14	.13	
8	.40	.36	.30	.26	.23	.20	.18	.17	.15	
9	.45	.41	.34	.29	.26	.23	.20	.19	.17	
10	.51	.45	.38	.32	.28	.25	.23	.21	.19	

NOTE: In some states, driving under the influence is proved when an adult's blood contains 0.08 percent alcohol, and in others, 0.10. Many states have adopted a "zero-tolerance" policy for drivers under age 21, using 0.02 percent as the limit.

[a]Taken within an hour or so; each drink equivalent to ½ ounce pure ethanol.

SOURCE: National Clearinghouse for Alcohol and Drug Information.

or to relieve anxiety. People think that alcohol is a stimulant because it seems to relieve inhibitions. Actually, though, it accomplishes this by sedating *inhibitory* nerves, which are more numerous than excitatory nerves. Ultimately, alcohol acts as a depressant and affects all the nerve cells.

It is lucky that the brain centers respond to a rising blood alcohol concentration in the order described in Figure H7-4 because a person usually passes out before managing to drink a lethal dose. It is possible, though, to drink so fast that the effects of alcohol continue to accelerate after the person has passed out. Occasionally, a person drinks so much as to stop breathing and die. Table H7-1 shows the blood alcohol levels that correspond to progressively greater intoxication, and Table H7-2 shows the brain responses that occur at these blood levels.

TABLE H7-2 Alcohol Blood Levels and Brain Responses

Blood Alcohol Concentration	Effect on Brain
0.05	Impaired judgment, relaxed inhibitions, altered mood, increased heart rate
0.10	Impaired coordination, delayed reaction time, exaggerated emotions, impaired peripheral vision, impaired ability to operate a vehicle
0.15	Slurred speech, blurred vision, staggered walk, seriously impaired coordination and judgment
0.20	Double vision, inability to walk
0.30	Uninhibited behavior, stupor, confusion, inability to comprehend
0.40 to 0.60	Unconsciousness, shock, coma, death (cardiac or respiratory failure)

NOTE: Blood alcohol concentration depends on a number of factors, including alcohol in the beverage, the rate of consumption, the person's gender, and body weight. For example, a 100-pound female can become legally drunk ($\geq$0.10 concentration) by drinking three beers in an hour, whereas a 220-pound male consuming that amount at the same rate would have a 0.05 blood alcohol concentration.

Like liver cells, brain cells die with excessive exposure to alcohol. Liver cells may be replaced, but not all brain cells can regenerate. Thus some heavy drinkers suffer permanent brain damage. Whether alcohol impairs cognition in moderate drinkers is unclear.

People who drink alcoholic beverages may notice that they urinate more, but they may be unaware of the vicious cycle that results. Alcohol depresses production of **antidiuretic hormone (ADH)**, a hormone produced by the pituitary gland that retains water—consequently, with less ADH, more water is lost. Loss of body water leads to thirst, and thirst leads to more drinking. Water will relieve dehydration, but the thirsty drinker may drink alcohol instead, which only worsens the problem. Such information provides another practical tip: drink water when thirsty and before each alcoholic drink. Drink an extra glass or two before going to bed. This strategy will help lessen the effects of a hangover.

Water loss is accompanied by the loss of important minerals. As Chapters 12 and 13 explain, these minerals are vital to the body's fluid balance and to many chemical reactions in the cells, including muscle action. Detoxification treatment includes restoration of mineral balance as quickly as possible.

Alcohol and Malnutrition

For many moderate drinkers, alcohol does not suppress food intake and may actually stimulate appetite. Moderate drinkers usually consume alcohol as *added* energy—on top of their normal food intake. In addition, alcohol in moderate doses is efficiently metabolized. Consequently, alcohol can contribute to body fat and weight gain—either by inhibiting oxidation or by being converted to fat.[7] Metabolically, alcohol is almost as efficient as fat in promoting obesity; each ounce of alcohol represents about a half-ounce of fat. Alcohol's contribution to body fat is most evident in the central obesity that commonly accompanies alcohol consumption, popularly—and appropriately—known as the "beer belly."[8] Alcohol in heavy doses, though, is not efficiently metabolized, generating more heat than fat. Heavy drinkers usually consume alcohol as *substituted* energy—instead of their normal food intake. They tend to eat poorly and suffer malnutrition.

Alcohol is rich in energy (7 kcalories per gram), but as with pure sugar or fat, the kcalories are empty of nutrients. The more alcohol people drink, the less likely that they will eat enough food to obtain adequate nutrients. The more kcalories used on alcohol, the fewer kcalories available to use on nutritious foods. Table H7-3 shows the kcalorie amounts of typical alcoholic beverages.

Chronic alcohol abuse not only displaces nutrients from the diet, but it also interferes with the body's metabolism of nutrients. Most dramatic is alcohol's effect on the B vitamin folate. The liver loses its ability to retain folate, and the kidneys increase their excretion of it. Alcohol abuse creates a folate deficiency that devastates digestive system function. The intestine normally releases and retrieves folate continuously, but it becomes damaged by folate deficiency and alcohol toxicity, so it fails to retrieve its own folate and misses any that may trickle in from food as well. Alcohol also interferes with the action of folate in converting the amino acid homocysteine to methionine. The result is an excess of homocysteine, which has been linked to heart disease, and an inadequate supply of methionine, which slows the production of new cells, especially the rapidly dividing cells of the intestine and the blood. The combination of poor folate status and alcohol consumption has also been implicated in promoting colorectal cancer.[9]

The inadequate food intake and impaired nutrient absorption that accompany chronic alcohol abuse frequently lead to a deficiency of another B vitamin—thiamin. In fact, the cluster of thiamin-deficiency symptoms commonly seen in chronic **alcoholism** has its own name—the **Wernicke-Korsakoff syndrome.**[10] This syndrome is characterized by paralysis of the eye muscles, poor muscle coordination, impaired memory, and damaged nerves; it and other alcohol-related memory problems may respond to thiamin supplements.

Acetaldehyde, an intermediate in alcohol metabolism (review Figure H7-2, p. 233), interferes with nutrient use, too. For example, acetaldehyde dislodges vitamin B_6 from its protective binding protein so that it is destroyed, causing a vitamin B_6 deficiency and, thereby, lowered production of red blood cells.

Malnutrition occurs not only because of lack of intake and altered metabolism but because of direct toxic effects as well. Alcohol causes stomach cells to oversecrete both gastric acid and histamine, an immune system agent that produces inflammation. Beer in particular stimulates gastric acid secretion, irritating the linings of the stomach and esophagus and making them vulnerable to ulcer formation.

Overall, nutrient deficiencies are virtually inevitable in alcohol abuse, not only because alcohol displaces food but also because alcohol directly interferes with the body's use of nutrients, making them ineffective even if they are present. Intestinal cells fail to absorb B vitamins, notably, thiamin, folate, and vitamin B_{12}. Liver cells lose efficiency in activating vitamin D. Cells in the retina of the eye, which normally process the alcohol form of vitamin A (retinol) to the aldehyde form needed in vision (retinal), find themselves processing ethanol to acetaldehyde instead. Likewise, the liver cannot convert the aldehyde form of vitamin A to its acid form (retinoic acid), which is needed to support the growth of its (and all) cells.

Regardless of dietary intake, excessive drinking over a lifetime creates deficits of all the nutrients mentioned in this discussion and more. No diet can compensate for the damage caused by heavy alcohol consumption.

TABLE H7-3 kCalories in Alcoholic Beverages and Mixers

Beverage	Amount (oz)	Energy (kcal)	Alcohol (g)
Beer			
Regular	12	153	14
Light	12	103	11
Nonalcoholic	12	32	0
Cocktails			
Daiquiri, canned	6.8	259	20
Daiquiri, from recipe	4.5	223	28
Piña colada, canned	6.8	526	20
Piña colada, from recipe	4.5	245	14
Tequila sunrise, canned	6.8	232	20
Whiskey sour, canned	6.8	249	20
Distilled liquor (gin, rum, vodka, whiskey)			
80 proof	1.5	97	14
86 proof	1.5	105	15
90 proof	1.5	110	16
94 proof	1.5	116	17
100 proof	1.5	124	18
Sake	1.5	58	7
Liqueurs			
Coffee and cream liqueur, 34 proof	1.5	154	7
Coffee liqueur, 53 proof	1.5	170	11
Coffee liqueur, 63 proof	1.5	160	14
Crème de menthe, 72 proof	1.5	186	15
Mixers			
Club soda	12	0	0
Cola	12	136	0
Cranberry juice cocktail	4	72	0
Ginger ale or tonic water	12	124	0
Grapefruit juice	4	48	0
Orange juice	4	56	0
Tomato or vegetable juice	4	21	0
Wine			
Champagne	5	105	13
Cooking	5	72	5
Dessert, dry	5	224	23
Dessert, sweet	5	236	23
Red or rosé	5	125	16
White	5	121	15
Wine cooler	10	150	11

Alcohol's Short-Term Effects

The effects of abusing alcohol may be apparent immediately, or they may not become evident for years to come. Among the immediate consequences, all of the following involve alcohol use:[11]

- 20 percent of all boating fatalities
- 23 percent of all suicides
- 39 percent of all traffic fatalities
- 40 percent of all residential fire fatalities
- 47 percent of all homicides
- 65 percent of all domestic violence incidents

These statistics are sobering. The consequences of heavy drinking touch all races and all segments of society—men and women, young and old, rich and poor. One group particularly hard hit by heavy drinking is college students—not because they are prone to alcoholism, but because they live in an environment and are in a developmental stage of life in which risk-taking behaviors are common and heavy drinking is acceptable.

Excessive drinking—including both **heavy drinking** and **binge drinking**—is widespread on college campuses and poses serious health and social consequences to drinkers and nondrinkers alike.[12] In fact, binge drinking can kill: the respiratory center of the brain becomes anesthetized, and breathing stops. Acute alcohol intoxication can cause coronary artery spasms, leading to a heart attack.

Binge drinking is especially common among college students, especially males.[13] Compared with nondrinkers or moderate drinkers, people who frequently binge drink (at least three times within two weeks) are more likely to engage in unprotected sex, have multiple sex partners, damage property, and assault others. On average, *every day* alcohol is involved in the:[14]

- Death of 5 college students
- Sexual assault of 266 college students
- Injury of 1641 college students
- Assault of 1907 college students

Binge drinkers skew the statistics on college students' alcohol use. The median number of drinks consumed by college students is 1.5 per week, but for binge drinkers, it is 14.5. Nationally, only 20 percent of all students are frequent binge drinkers; yet they account for two-thirds of all the alcohol students report consuming and most of the alcohol-related problems.

Binge drinking is not limited to college campuses, of course, but it is most common among 18 to 24 year olds.[15] That age group and campus environment seem most accepting of such behavior despite its problems. Social acceptance may make it difficult for binge drinkers to recognize themselves as problem drinkers. For this reason, interventions must focus both on educating individuals and on changing the campus social environment. The damage alcohol causes only becomes worse if the pattern is not broken. Alcohol abuse sets in much more quickly in young people than in adults. Those who start drinking at an early age more often suffer from alcoholism than people who start later on. Table H7-4 lists the key signs of alcoholism.

TABLE H7-4 Signs of Alcoholism

- Tolerance: the person needs higher and higher intakes of alcohol to achieve intoxication.
- Withdrawal: the person who stops drinking experiences anxiety, agitation, increased blood pressure, or seizures, or seeks alcohol to relieve these symptoms.
- Impaired control: the person intends to have 1 or 2 drinks, but has 9 or 10 instead, or the person tries to control or quit drinking, but fails.
- Disinterest: the person neglects important social, family, job, or school activities because of drinking.
- Time: the person spends a great deal of time obtaining and drinking alcohol or recovering from excessive drinking.
- Impaired ability: the person's intoxication or withdrawal symptoms interfere with work, school, or home.
- Problems: the person continues drinking despite physical hazards or medical, legal, psychological, family, employment, or school problems.

The presence of three or more of these conditions is required to make a diagnosis.

SOURCE: Adapted from *Diagnostic and Statistical Manual of Mental Disorders,* 4th ed. (Washington, D.C.: American Psychiatric Association, 1994).

Alcohol's Long-Term Effects

The most devastating long-term effect of alcohol is the damage done to a child whose mother abused alcohol during pregnancy. The effects of alcohol on the unborn and the message that pregnant women should not drink alcohol are presented in Highlight 14.

For nonpregnant adults, a drink or two sets in motion many destructive processes in the body, but the next day's abstinence reverses them. As long as the doses are moderate, the time between them is ample, and nutrition is adequate, recovery is probably complete.

If the doses of alcohol are heavy and the time between them short, complete recovery cannot take place. Repeated onslaughts of alcohol gradually take a toll on all parts of the body (see Table H7-5 on p. 238). Compared with nondrinkers and moderate drinkers, heavy drinkers have significantly greater risks of dying from all causes. Excessive alcohol consumption is the third leading preventable cause of death in the United States.

Personal Strategies

One obvious option available to people attending social gatherings is to enjoy the conversation, eat the food, and drink nonalcoholic beverages. Several nonalcoholic beverages are available that mimic the look and taste of their alcoholic counterparts. For those who enjoy champagne or beer, sparkling ciders and beers without alcohol are available. Instead of drinking a cocktail, a person can sip tomato juice with a slice of lime and a stalk of celery or just a plain cola beverage. Any of these drinks can ease conversation.

The person who chooses to drink alcohol should sip each drink slowly with food. The alcohol should arrive at the liver cells slowly enough that the enzymes can handle the load. It is best to space drinks, too, allowing about an hour or so to metabolize each drink.

TABLE H7-5 Health Effects of Heavy Alcohol Consumption

Health Problem	Effects of Alcohol
Arthritis	Increases the risk of inflamed joints.
Cancer	Increases the risk of cancer of the liver, rectum, breast, mouth, pharynx, larynx, and esophagus.
Fetal alcohol syndrome	Causes physical and behavioral abnormalities in the fetus (see Highlight 14).
Heart disease	In heavy drinkers, raises blood pressure, blood lipids, and the risk of stroke and heart disease; when compared with those who abstain, heart disease risk is generally lower in light-to-moderate drinkers.
Hyperglycemia	Raises blood glucose.
Hypoglycemia	Lowers blood glucose, especially in people with diabetes.
Infertility	Increases the risks of menstrual disorders and spontaneous abortions (in women); suppresses luteinizing hormone (in women) and testosterone (in men).
Kidney disease	Enlarges the kidneys, alters hormone functions, and increases the risk of kidney failure.
Liver disease	Causes fatty liver, alcoholic hepatitis, and cirrhosis.
Malnutrition	Increases the risk of protein-energy malnutrition; low intakes of protein, calcium, iron, vitamin A, vitamin C, thiamin, vitamin B_6, and riboflavin; and impaired absorption of calcium, phosphorus, vitamin D, and zinc.
Nervous disorders	Causes neuropathy and dementia; impairs balance and memory.
Obesity	Increases energy intake, but is not a primary cause of obesity.
Psychological disturbances	Causes depression, anxiety, and insomnia.

NOTE: This list is by no means all-inclusive. Alcohol has direct toxic effects on all body systems.

If you want to help sober up a friend who has had too much to drink, don't bother walking arm in arm around the block. Walking muscles have to work harder, but muscle cells can't metabolize alcohol; only liver cells can. Remember that each person has a limited amount of the alcohol dehydrogenase enzyme that clears the blood at a steady rate. Time alone will do the job. Nor will it help to give your friend a cup of coffee. Caffeine is a stimulant, but it won't speed up alcohol metabolism. Table H7-6 presents other alcohol myths.

People who have passed out from drinking need 24 hours to sober up completely. Let them sleep, but watch over them. Encourage them to lie on their sides, instead of their backs. That way, if they vomit, they won't choke.

TABLE H7-6 Myths and Truths Concerning Alcohol

Myth:	Hard liquors such as rum, vodka, and tequila are more harmful than wine and beer.
Truth:	The damage caused by alcohol depends largely on the *amount* consumed. Compared with hard liquor, beer and wine have relatively low percentages of alcohol, but they are often consumed in larger quantities.
Myth:	Consuming alcohol with raw seafood diminishes the likelihood of getting hepatitis.
Truth:	People have eaten contaminated oysters while drinking alcoholic beverages and not gotten as sick as those who were not drinking. But do not be misled: hepatitis is too serious an illness for anyone to depend on alcohol for protection.
Myth:	Alcohol stimulates the appetite.
Truth:	For some people, alcohol may stimulate appetite, but it seems to have the opposite effect in heavy drinkers. Heavy drinkers tend to eat poorly and suffer malnutrition.
Myth:	Drinking alcohol is healthy.
Truth:	Moderate alcohol consumption is associated with a lower risk for heart disease. Higher intakes, however, raise the risks for high blood pressure, stroke, heart disease, some cancers, accidents, violence, suicide, birth defects, and deaths in general. Furthermore, excessive alcohol consumption damages the liver, pancreas, brain, and heart. No authority recommends that nondrinkers begin drinking alcoholic beverages to obtain health benefits.
Myth:	Wine increases the body's absorption of minerals.
Truth:	Wine may increase the body's absorption of potassium, calcium, phosphorus, magnesium, and zinc, but the alcohol in wine also promotes the body's excretion of these minerals, so no benefit is gained.
Myth:	Alcohol is legal and, therefore, not a drug.
Truth:	Alcohol is legal for adults 21 years old and older, but it is also a drug—a substance that alters one or more of the body's functions.
Myth:	A shot of alcohol warms you up.
Truth:	Alcohol diverts blood flow to the skin making you *feel* warmer, but it actually cools the body.
Myth:	Wine and beer are mild; they do not lead to alcoholism.
Truth:	Alcoholism is not related to the kind of beverage, but rather to the quantity and frequency of consumption.
Myth:	Mixing different types of drinks gives you a hangover.
Truth:	Too much alcohol in any form produces a hangover.
Myth:	Alcohol is a stimulant.
Truth:	People think alcohol is a stimulant because it seems to relieve inhibitions, but it does so by depressing the activity of the brain. Alcohol is medically defined as a depressant drug.
Myth:	Beer is a great source of carbohydrate, vitamins, minerals, and fluids.
Truth:	Beer does provide some carbohydrate, but most of its kcalories come from alcohol. The few vitamins and minerals in beer cannot compete with rich food sources. And the diuretic effect of alcohol causes the body to lose more fluid in urine than is provided by the beer.

Don't drive too soon after drinking. The lack of glucose for the brain to function and the length of time to clear the blood of alcohol make alcohol's adverse effects linger long after its blood concentration has fallen. Driving coordination is still impaired the morning *after* a night of drinking, even if the drinking was moderate. Responsible aircraft pilots know that they must allow 24 hours for their bodies to clear alcohol completely, and they do not fly any sooner. The Federal Aviation Administration and major airlines enforce this rule.

Look again at the drawing of the brain in Figure H7-4 (p. 234), and note that when someone drinks, judgment fails first. Judgment might tell a person to limit alcohol consumption to two drinks at a party, but if the first drink takes judgment away, many more drinks may follow. The failure to stop drinking as planned, on repeated occasions, is a warning sign that the person should not drink at all. The accompanying Nutrition on the Net provides websites for organizations that offer information about alcohol and alcohol abuse.

Ethanol interferes with a multitude of chemical and hormonal reactions in the body—many more than have been enumerated here. With heavy alcohol consumption, the potential for harm is great. If you drink alcoholic beverages, do so with care, and in moderation.

Nutrition on the Net

For further study of topics covered in this highlight, log on to **www.cengagebrain.com** and search for ISBN 084006845X.

- Gather information on alcohol and drug abuse from the National Clearinghouse for Alcohol and Drug Information (NCADI): **ncadi.samhsa.gov**
- Learn more about alcoholism and drug dependence from the National Council on Alcoholism and Drug Dependence (NCADD): **www.ncadd.org**
- Visit the National Institute on Alcohol Abuse and Alcoholism: **www.collegedrinkingprevention.gov**
- Find help for a family alcohol problem from Alateen and Al-Anon Family support groups: **www.al-anon.alateen.org**
- Find help for an alcohol or drug problem from Alcoholics Anonymous (AA) or Narcotics Anonymous: **www.aa.org** or **www.na.org**

References

1. J. H. O'Keefe, K. A. Bybee, and C. J. Lavie, Alcohol and cardiovascular health: The razor-sharp double-edged sword, *Journal of the American College of Cardiology* 50 (2007): 1009–1014; D. J. Meyerhoff and coauthors, Health risks of chronic moderate and heavy alcohol consumption: How much is too much? *Alcoholism, Clinical and Experimental Research* 29 (2005): 1334–1340.
2. M. P. Ferreira and D. Willoughby, Alcohol consumption: The good, the bad, and the indifferent, *Applied Physiology, Nutrition, and Metabolism* 33 (2008): 12–20; V. Arndt and coauthors, Age, alcohol consumption, and all-cause mortality, *Annals of Epidemiology* 14 (2004): 750–753.
3. J. Connor and coauthors, The burden of death, disease, and disability due to alcohol in New Zealand, *New Zealand Medical Journal* 118 (2005): U1412.
4. P. Meier and H. K. Seitz, Age, alcohol metabolism and liver disease, *Current Opinion in Clinical Nutrition and Metabolic Care* 11 (2008): 21–26; H. K. Seitz and F. Stickel, Alcoholic liver disease in the elderly, *Clinics in Geriatric Medicine* 23 (2007): 905–921.
5. B. Wansink and K. van Ittersum, Shape of glass and amount of alcohol poured: Comparative study of effect of practice and concentration, *British Medicine Journal* 331 (2005): 1512–1514.
6. S. Zakhari, Overview: How is alcohol metabolized by the body? *Alcohol Research & Health* 29 (2006): 245–254.
7. R. A. Breslow and B. A. Smothers, Drinking patterns and body mass index in never smokers: National Health Interview Survey, 1997–2001, *American Journal of Epidemiology* 161 (2005): 368–376.
8. S. G. Wannamethee, A. G. Shaper, and P. H. Whincup, Alcohol and adiposity: Effects of quantity and type of drink and time relation with meals, *International Journal of Obesity and Related Metabolic Disorders* 29 (2005): 1436–1444.
9. M. Ryan-Harshman and W. Aldoori, Diet and colorectal cancer: Review of the evidence, *Canadian Family Physician* 53 (2007): 1913–1920.
10. A. D. Thomson and coauthors, Wernicke's encephalopathy: "Plus ça change, plus c'est la même chose," *Alcohol and Alcoholism* 43 (2008): 180–186.
11. Centers for Disease Control and Prevention, National Center for Injury Prevention and Control (NCIPC), www.cdc.gov, accessed June 23, 2008.
12. R. D. Brewer and M. H. Swahn, Binge drinking and violence, *Journal of the American Medical Association* 294 (2005): 616–618.
13. A. M. White, C. L. Kraus, and H. Swartzwelder, Many college freshmen drink at levels far beyond the binge threshold, *Alcoholism, Clinical and Experimental Research* 30 (2006): 1006–1010.
14. R. W. Hingson and coauthors, Magnitude of alcohol-related mortality and morbidity among U.S. college students ages 18–24: Changes from 1998 to 2001, *Annual Review of Public Health* 26 (2005): 259–279.
15. National Center for Health Statistics, *Chartbook on Trends in the Health of Americans,* Alcohol consumption by adults 18 years of age and over, according to selected characteristics: United States, selected years 1997–2003, (2005): 264–266.

CHAPTER

8

© Image Source/JupiterImages

Nutrition in Your Life

It's a simple equation: energy in + energy out = energy balance. The reality, of course, is much more complex. One day you may devour a dozen doughnuts at midnight and sleep through your morning workout—tipping the scales toward weight gain. Another day you may snack on veggies and train for a 10K race—shifting the balance toward weight loss. Your body weight—especially as it relates to your body fat—and your level of fitness have consequences for your health. So, how are you doing? Are you ready to see how your "energy in" and "energy out" balance and whether your body weight and fat measures are consistent with good health?

Throughout this chapter, the CourseMate icon indicates an opportunity for online self-study, linking you to activities to increase your understanding of chapter concepts. **www.cengagebrain.com** (search for ISBN 084006845X)

Energy Balance and Body Composition

The body's remarkable machinery can cope with many extremes of diet. As Chapter 7 explains, excess carbohydrate (glucose), excess protein (amino acids), and excess fat all can contribute to body fat. To some extent, amino acids can be used to make glucose. To a very limited extent, even fat (the glycerol portion) can be used to make glucose. But a grossly unbalanced diet imposes hardships on the body. If energy intake is too low or if too little carbohydrate or protein is supplied, the body must degrade its own lean tissue to meet its glucose and protein needs. If energy intake is too high, the body stores fat.

Both excessive and deficient body fat result from an energy imbalance. The simple picture is as follows. People who have consumed more food energy than they have expended bank the surplus as body fat. To reduce body fat, they need to expend more energy than they take in from food. In contrast, people who have consumed too little food energy to support their bodies' activities have relied on their bodies' fat stores and possibly some of their lean tissues as well. To gain weight, these people need to take in more food energy than they expend. As you will see, though, the details of the body's weight regulation are quite complex.[1] This chapter describes energy balance and body composition and examines the health problems associated with having too much or too little body fat. The next chapter presents strategies toward resolving these problems.

Energy Balance

People expend energy continuously and eat periodically to refuel. Ideally, their energy intakes cover their energy expenditures without too much excess. Excess energy is stored as fat, and stored fat is used for energy between meals. The amount of body fat a person deposits in, or withdraws from, storage on any given day depends on the energy balance for that day—the amount consumed (energy in) versus the amount expended (energy out). When a person is maintaining weight, energy in equals energy out. When the balance shifts, weight changes. For each 3500 kcalories eaten in excess, a pound of body fat is stored; similarly, a pound of fat is lost for each 3500 kcalories expended beyond those consumed. ♦ The fat

♦ 1 lb body fat = 3500 kcal
Body fat, or adipose tissue, is composed of a mixture of mostly fat, some protein, and water. A pound of body fat (454 g) is approximately 87% fat, or (454 × 0.87) 395 g, and 395 g × 9 kcal/g = 3555 kcal.

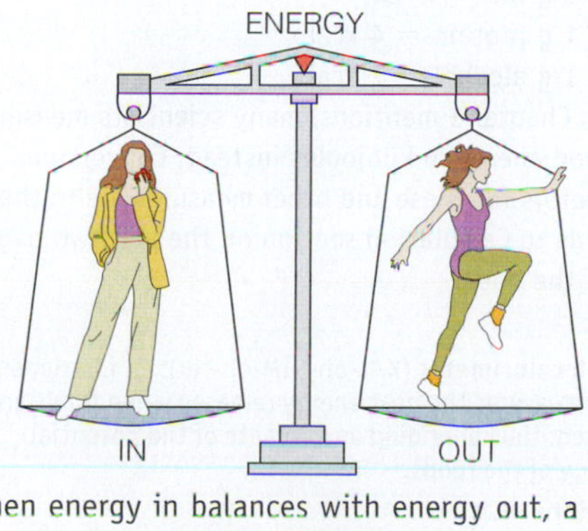

When energy in balances with energy out, a person's body weight is stable.

stores of even a healthy-weight adult represent an ample reserve of energy—50,000 to 200,000 kcalories.

Dietary Guidelines for Americans 2010
Prevent and/or reduce overweight and obesity through improved eating and physical activity behaviors.

Quick changes in body weight are not simple changes in fat stores. Weight gained or lost rapidly includes some fat, large amounts of fluid, and some lean tissues such as muscle proteins and bone minerals. (Because water constitutes about 60 percent of an adult's body weight, retention or loss of water can greatly influence body weight.) Even over the long term, the composition of weight gained or lost is normally about 75 percent fat and 25 percent lean. During starvation, losses of fat and lean are about equal. (Recall from Chapter 7 that without adequate carbohydrate, protein-rich lean tissues break down to provide glucose.) Invariably, though, *fat* gains and losses are gradual. The next two sections examine the two sides of the energy-balance equation: energy in and energy out.

IN SUMMARY **When the energy consumed equals the energy expended, a person is in energy balance and body weight is stable. If more energy is taken in than is expended, a person gains weight. If more energy is expended than is taken in, a person loses weight.**

Energy In: The kCalories Foods Provide

Foods and beverages provide the "energy in" part of the energy-balance equation. How much energy a person receives depends on the composition of the foods and beverages and on the amount the person eats and drinks.

Food Composition

To find out how many kcalories a food provides, a scientist can burn the food in a **bomb calorimeter** (see Figure 8-1). When the food burns, energy is released in the form of heat. The amount of heat given off provides a *direct* measure of the food's energy value (remember that kcalories are units of heat energy). In addition to releasing heat, these reactions generate carbon dioxide and water—just as the body's cells do when they metabolize the energy-yielding nutrients from foods. Details of the chemical reactions in a calorimeter and in the body differ, but the overall process is similar: when the food burns and the chemical bonds break, the carbons (C) and hydrogens (H) combine with oxygens (O) to form carbon dioxide (CO_2) and water (H_2O). The amount of oxygen consumed gives an *indirect* measure ♦ of the amount of energy released.

A bomb calorimeter measures the available energy in foods but overstates the amount of energy that the human body ♦ derives from foods. The body is less efficient than a calorimeter and cannot metabolize all of the energy-yielding nutrients in a food completely. Researchers can correct for this discrepancy mathematically to create useful tables of the energy values of foods (such as Appendix H). These values provide reasonable estimates, but they do not reflect the *precise* amount of energy a person will derive from the foods consumed.

The energy values of foods can also be computed from the amounts of carbohydrate, fat, and protein (and alcohol, if present) in the foods.* For example, a food ♦ containing 12 grams of carbohydrate, 5 grams of fat, and 8 grams of protein will provide 48 carbohydrate kcalories, 45 fat kcalories, and 32 protein kcalories, for a total of 125 kcalories. (To review how to calculate the energy foods provide, turn to p. 9.)

*Some of the food energy values in the table of food composition in Appendix H were derived by bomb calorimetry, and many were calculated from their energy-yielding nutrient contents.

♦ Food energy values can be determined by:
- **Direct calorimetry,** which measures the amount of heat released.
- **Indirect calorimetry,** which measures the amount of oxygen consumed.

♦ The number of kcalories that the body derives from a food, in contrast to the number of kcalories determined by calorimetry, is the **physiological fuel value.**

♦
- 1 g carbohydrate = 4 kcal
- 1 g fat = 9 kcal
- 1 g protein = 4 kcal
- 1 g alcohol = 7 kcal

As Chapter 1 mentions, many scientists measure food energy in kilojoules instead. Conversion factors for these and other measures are in the Aids to Calculation section on the last two pages of the book.

bomb calorimeter (KAL-oh-RIM-eh-ter): an instrument that measures the heat energy released when foods are burned, thus providing an estimate of the potential energy of the foods.
- **calor** = heat
- **metron** = measure

Food Intake To achieve energy balance, the body must meet its needs without taking in too much or too little energy. **Appetite** prompts a person to eat—or not to eat.[2] Somehow the body decides how much and how often to eat—when to start eating and when to stop. As you will see, many signals—from both the environment and genetics—initiate or delay eating.[3]

Hunger People eat for a variety of reasons, most obviously (although not necessarily most commonly) because they are hungry. Most people recognize **hunger** as an irritating feeling that prompts thoughts of food and motivates them to start eating. In the body, hunger is the physiological response to a need for food triggered by nerve signals and chemical messengers originating and acting in the brain, primarily in the **hypothalamus.**[4] Hunger can be influenced by the presence or absence of nutrients in the bloodstream, the size and composition of the preceding meal, customary eating patterns, climate (heat reduces food intake; cold increases it), exercise, hormones, and physical and mental illnesses. Hunger determines what to eat, when to eat, and how much to eat.

The stomach is ideally designed to handle periodic batches of food, and people typically eat meals at roughly four-hour intervals. Four hours after a meal, most, if not all, of the food has left the stomach. Most people do not feel like eating again until the stomach is either empty or almost so. Even then, a person may not feel hungry for quite a while.

Satiation During the course of a meal, as food enters the GI tract and hunger diminishes, **satiation** develops. As receptors in the stomach stretch and hormones such as cholecystokinin become active, the person begins to feel full. The response: satiation occurs and the person stops eating.

Satiety After a meal, the feeling of **satiety** continues to suppress hunger and allows a person to not eat again for a while. Whereas *satiation* tells us to "stop eating," *satiety* reminds us to "not start eating again." Figure 8-2 (p. 244) summarizes the relationships among hunger, satiation, and satiety. Of course, people can override these signals, especially when presented with stressful situations or favorite foods.

Overriding Hunger and Satiety Not surprisingly, eating can be triggered by signals other than hunger, even when the body does not need food. Some people experience food cravings when they are bored or anxious. In fact, they may eat in response to any kind of stress, ♦ negative or positive. ("What do I do when I'm grieving? Eat. What do I do when I'm celebrating? Eat!") Not too surprisingly, repeatedly eating to relieve chronic stress can lead to overeating and weight gain.[5]

Many people respond to external cues such as the time of day ("It's time to eat") or the availability, sight, and taste of food ("I'd love a piece of chocolate even though I'm stuffed"). Environmental influences such as large portion sizes, favorite

FIGURE 8-1 Bomb Calorimeter

When food is burned, energy is released in the form of heat. Heat energy is measured in kcalories.

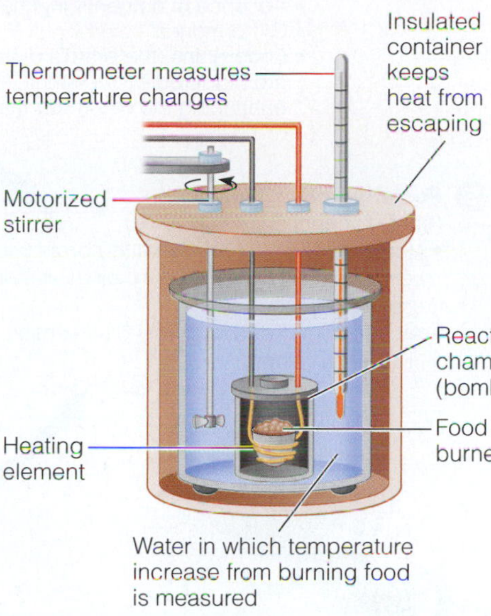

♦ Eating in response to arousal is called **stress eating.**

iStockphoto.com/art-4-art

Regardless of hunger, people typically overeat when offered the abundance and variety of a buffet. To limit unhealthy weight gains, listen to hunger and satiety signals.

appetite: the integrated response to the sight, smell, thought, or taste of food that initiates or delays eating.

hunger: the painful sensation caused by a lack of food that initiates food-seeking behavior.

hypothalamus (high-po-THAL-ah-mus): a brain center that controls activities such as maintenance of water balance, regulation of body temperature, and control of appetite.

satiation (say-she-AY-shun): the feeling of satisfaction and fullness that occurs during a meal and halts eating. Satiation determines how much food is consumed during a meal.

satiety: the feeling of fullness and satisfaction that occurs after a meal and inhibits eating until the next meal. Satiety determines how much time passes between meals.

FIGURE 8-2 Hunger, Satiation, and Satiety

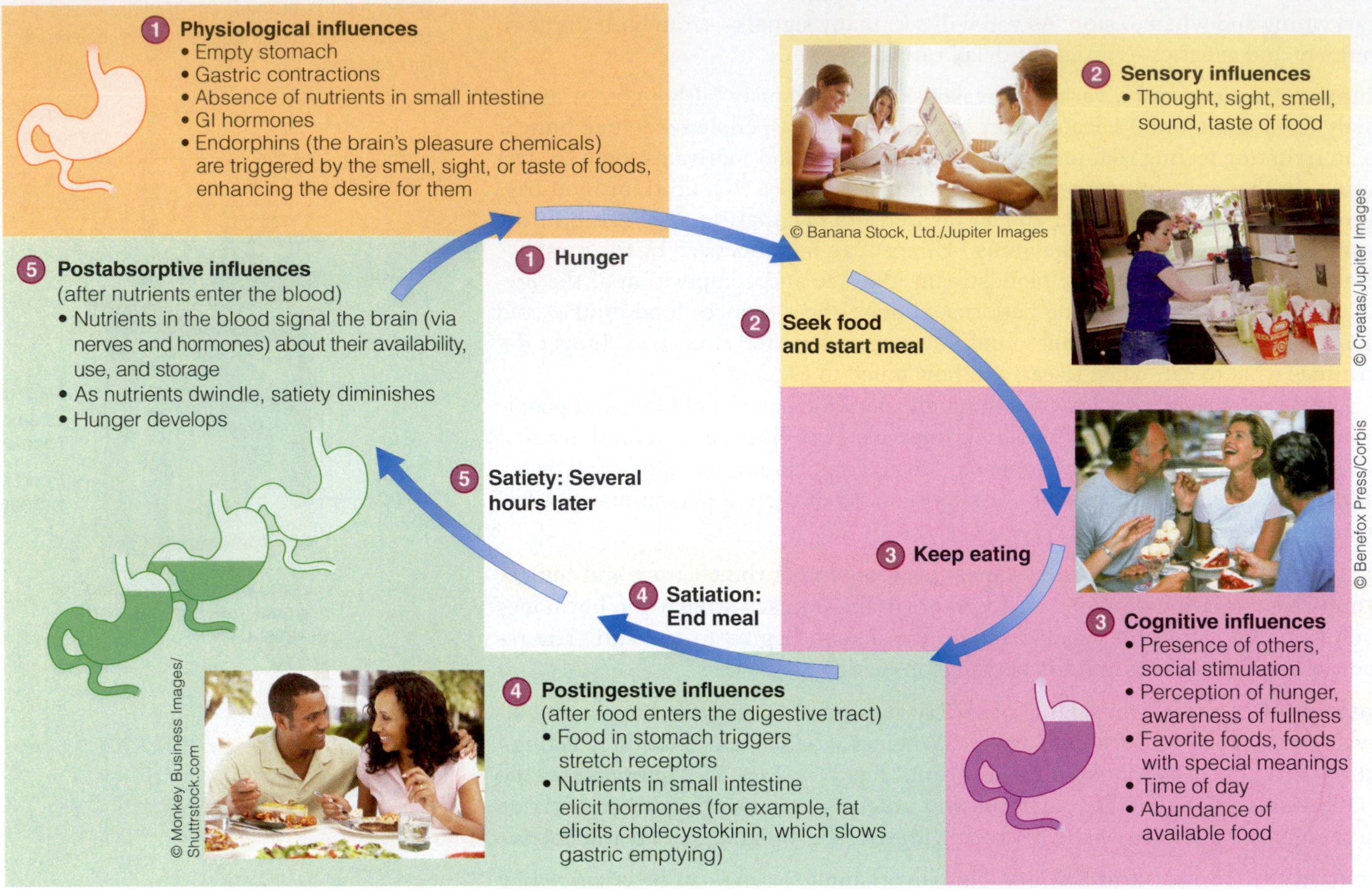

foods, or an abundance or variety of foods stimulate eating and increase energy intake. These cognitive influences ♦ can easily lead to weight gain.

1 Physiological influences
• Empty stomach
• Gastric contractions
• Absence of nutrients in small intestine
• GI hormones
• Endorphins (the brain's pleasure chemicals) are triggered by the smell, sight, or taste of foods, enhancing the desire for them
2 Sensory influences
• Thought, sight, smell, sound, taste of food
© Banana Stock, Ltd./Jupiter Images
© Creatas/Jupiter Images
1 Hunger
2 Seek food and start meal
3 Keep eating
4 Satiation: End meal
5 Satiety: Several hours later
5 Postabsorptive influences
(after nutrients enter the blood)
• Nutrients in the blood signal the brain (via nerves and hormones) about their availability, use, and storage
• As nutrients dwindle, satiety diminishes
• Hunger develops
© Benefox Press/Corbis
3 Cognitive influences
• Presence of others, social stimulation
• Perception of hunger, awareness of fullness
• Favorite foods, foods with special meanings
• Time of day
• Abundance of available food
4 Postingestive influences
(after food enters the digestive tract)
• Food in stomach triggers stretch receptors
• Nutrients in small intestine elicit hormones (for example, fat elicits cholecystokinin, which slows gastric emptying)
© Monkey Business Images/Shutterstock.com

♦ Cognitive influences include perceptions, memories, intellect, and social interactions.

Eating can also be suppressed by signals other than satiety, even when a person is hungry. People with the eating disorder anorexia nervosa, for example, use tremendous discipline to ignore the pangs of hunger. Some people simply cannot eat during times of stress, negative or positive. ("I'm too sad to eat." "I'm too excited to eat!") Why some people overeat in response to stress and others cannot eat at all remains a bit of a mystery, although researchers are beginning to understand the connections between stress hormones, brain activity, and "comfort foods." Factors that appear to be involved include how the person perceives the stress and whether usual eating behaviors are restrained. (Highlight 8 features anorexia nervosa and other eating disorders.)

Sustaining Satiation and Satiety The extent to which foods produce satiation and sustain satiety depends in part on the nutrient composition of a meal. Of the three energy-yielding nutrients, protein is considered the most **satiating.** In fact, too little protein in the diet can leave a person feeling hungry.[6]

Foods low in energy density ♦ are also more satiating. High-fiber foods effectively provide satiation by filling the stomach and delaying the absorption of nutrients. For this reason, eating a large salad as a first course helps a person eat less during the meal. In contrast, fat has a weak effect on satiation; consequently, eating high-fat foods may lead to passive overconsumption. High-fat foods are flavorful, which stimulates the appetite and entices people to eat more. High-fat foods are also energy dense; consequently, they deliver more kcalories per bite. (Chapter 1 introduces the concept of energy density, and Chapter 9 describes how considering a food's energy density can help with weight management.) Although

♦ Energy density is a measure of the energy a food provides relative to the amount of food (kcalories per gram). Foods with a low energy density provide fewer kcalories, and those with high energy density provide more kcalories, for the same amount of food.

satiating: having the power to suppress hunger and inhibit eating.

FIGURE 8-3 How Fat Influences Portion Sizes

For the same size portion, peanuts deliver more than 15 times the kcalories and 20 times the fat of popcorn.

For the same number of kcalories, a person can have a few high-fat peanuts or almost 2 cups of high-fiber popcorn. (This comparison used oil-based popcorn; using air-popped popcorn would double the amount of popcorn in this example.)

fat provides little satiation during a meal, it produces strong satiety signals once it enters the intestine. Fat in the intestine triggers the release of cholecystokinin—a hormone that signals satiety and inhibits food intake.

Eating high-fat foods while trying to limit energy intake requires small portion sizes, which can leave a person feeling unsatisfied. Portion size correlates directly with a food's satiety. Instead of eating small portions of high-fat foods and feeling deprived, a person can feel satisfied by eating large portions of low-fat, high-fiber, and low energy density foods. Figure 8-3 illustrates how fat influences portion size.

Message Central—The Hypothalamus As you can see, eating is a complex behavior controlled by a variety of psychological, social, metabolic, and physiological factors. The hypothalamus appears to be the control center, integrating messages about energy intake, expenditure, and storage from other parts of the brain and from the mouth, GI tract, and liver. Some of these messages influence satiation, which helps control the size of a meal; others influence satiety, which helps determine the frequency of meals.[7]

Dozens of gastrointestinal hormones ♦ influence appetite control and energy balance.[8] By understanding the action of these hormones, researchers may one day be able to develop anti-obesity treatments.[9] The greatest challenge now is to sort out the many actions of these brain chemicals. For example, one of these chemicals, **neuropeptide Y**, causes carbohydrate cravings, initiates eating, decreases energy expenditure, and increases fat storage—all factors favoring a positive energy balance and weight gain.

♦ Gastrointestinal hormones that regulate food intake:
- Amylin
- Cholecystokinin (CCK)
- Enterostatin
- Ghrelin
- Glucagon-like peptide-1 (GLP-1)
- Oxyntomodulin
- Pancreatic polypeptide (PP)
- Peptide YY (PYY)

IN SUMMARY A mixture of signals governs a person's eating behaviors. Hunger and appetite initiate eating, whereas satiation and satiety stop and delay eating, respectively. Each responds to messages from the nervous and hormonal systems. Superimposed on these signals are complex factors involving emotions, habits, and other aspects of human behavior.

Energy Out: The kCalories the Body Expends

Chapter 7 explains that heat is released whenever the body breaks down carbohydrate, fat, or protein for energy and again when that energy is used to do work. The generation of heat, known as **thermogenesis**, can be measured to determine

neuropeptide Y: a chemical produced in the brain that stimulates appetite, diminishes energy expenditure, and increases fat storage.

thermogenesis: the generation of heat; used in physiology and nutrition studies as an index of how much energy the body is expending.

♦ Energy expenditure, like food energy, can be determined by:

- **Direct calorimetry,** which measures the amount of heat released.
- **Indirect calorimetry,** which measures the amount of oxygen consumed and carbon dioxide expelled.

the amount of energy expended. ♦ The total energy a body expends reflects three main categories of thermogenesis:

- Energy expended for basal metabolism
- Energy expended for physical activity
- Energy expended for food consumption

A fourth category is sometimes involved:

- Energy expended for adaptation

Components of Energy Expenditure

People expend energy when they are physically active, of course, but they also expend energy when they are resting quietly. In fact, quiet metabolic activities account for the lion's share of most people's energy expenditures, as Figure 8-4 shows.

Basal Metabolism

About two-thirds of the energy the average person expends in a day supports the body's **basal metabolism.** Metabolic activities maintain the body temperature, keep the lungs inhaling and exhaling air, the bone marrow making new red blood cells, the heart beating 100,000 times a day, and the kidneys filtering wastes—in short, they support all the basic processes of life.

The **basal metabolic rate (BMR)** is the rate at which the body expends energy for these life-sustaining activities. ♦ The rate may vary from person to person and may vary for the same individual with a change in circumstance or physical condition. The rate is slowest when a person is sleeping undisturbed, but it is usually measured in a room with a comfortable temperature when the person is awake, but lying still, after a restful sleep and an overnight (12 to 14 hour) fast. A similar measure of energy output—called the **resting metabolic rate (RMR)**—is slightly higher than the BMR because its criteria for recent food intake and physical activity are not as strict. When energy needs cannot be measured, equations ♦ can provide reasonably accurate estimates.

♦ Quick and easy estimates for basal energy needs:

- Men: Slightly >1 kcal/min (1.1 to 1.3 kcal/min) or 24 kcal/kg/day
- Women: Slightly <1 kcal/min (0.8 to 1.0 kcal/min) or 23 kcal/kg/day

For perspective, a burning candle or a 75-watt light bulb releases about 1 kcal/min.

♦ BMR equations use actual weight in kilograms, height in centimeters, and age in years:

- Men: $(10 \times \text{wt}) + (6.25 \times \text{ht}) - (5 \times \text{age}) + 5$
- Women: $(10 \times \text{wt}) + (6.25 \times \text{ht}) - (5 \times \text{age}) - 161$

In general, the more a person weighs, the more *total* energy is expended on basal metabolism, but the amount of energy *per pound* of body weight may be lower. For example, an adult's BMR might be 1500 kcalories per day and an infant's only 500, but compared to body weight, the infant's BMR is more than twice as fast. Similarly, a normal-weight adult may have a metabolic rate one and a half times that of an obese adult when compared to body weight because lean tissue is metabolically more active than body fat.

Table 8-1 summarizes the factors that raise and lower the BMR. For the most part, the BMR is highest in people who are growing (children, adolescents, and pregnant women) and in those with considerable **lean body mass** (physically fit people and males). One way to increase the BMR then is to participate in endurance and strength-training activities regularly to maximize lean body mass. The BMR is also high in people with fever or under stress and in people with highly active thyroid glands. The BMR slows down with a loss of lean body mass and during fasting and malnutrition.

Physical Activity

The second component of a person's energy output is physical activity: voluntary movement of the skeletal muscles and support systems. Physical activity is the most variable—and the most changeable—component of energy expenditure. Consequently, its influence on both weight gain and weight loss can be significant.

During physical activity, the muscles need extra energy to move, and the heart and lungs need extra energy to deliver nutrients and oxygen and dispose of wastes. The amount of energy needed for any activity, whether playing tennis or studying for an exam, depends on three factors: muscle mass, body weight, and activity. The larger the muscle mass and the heavier the weight of the body part being moved, the more energy is expended. Table 8-2 gives average energy expenditures for various activities. The activity's duration, frequency, and intensity also influence energy expenditure: the longer, the more frequent, and the more intense the activity, the more

basal metabolism: the energy needed to maintain life when a body is at complete digestive, physical, and emotional rest.

basal metabolic rate (BMR): the rate of energy use for metabolism under specified conditions: after a 12-hour fast and restful sleep, without any physical activity or emotional excitement, and in a comfortable setting. It is usually expressed as kcalories per kilogram body weight per hour.

resting metabolic rate (RMR): similar to the basal metabolic rate (BMR), a measure of the energy use of a person at rest in a comfortable setting, but with less stringent criteria for recent food intake and physical activity. Consequently, the RMR is slightly higher than the BMR.

lean body mass: the body minus its fat.

TABLE 8-1 Factors that Affect the BMR

Factor	Effect on BMR
Age	Lean body mass diminishes with age, slowing the BMR.[a]
Height	In tall, thin people, the BMR is higher.[b]
Growth	In children, adolescents, and pregnant women, the BMR is higher.
Body composition (gender)	The more lean tissue, the higher the BMR (which is why males usually have a higher BMR than females). The more fat tissue, the lower the BMR.
Fever	Fever raises the BMR.[c]
Stresses	Stresses (including many diseases and certain drugs) raise the BMR.
Environmental temperature	Both heat and cold raise the BMR.
Fasting/starvation	Fasting/starvation lowers the BMR.[d]
Malnutrition	Malnutrition lowers the BMR.
Hormones (gender)	The thyroid hormone thyroxin, for example, can speed up or slow down the BMR.[e] Premenstrual hormones slightly raise the BMR.
Smoking	Nicotine increases energy expenditure.
Caffeine	Caffeine increases energy expenditure.
Sleep	BMR is lowest when sleeping.

[a] The BMR begins to decrease in early adulthood (after growth and development cease) at a rate of about 2 percent/decade. A reduction in voluntary activity as well brings the total decline in energy expenditure to 5 percent/decade.

[b] If two people weigh the same, the taller, thinner person will have the faster metabolic rate, reflecting the greater skin surface, through which heat is lost by radiation, in proportion to the body's volume (see the margin drawing on p. 249).

[c] Fever raises the BMR by 7 percent for each degree Fahrenheit.

[d] Prolonged starvation reduces the total amount of metabolically active lean tissue in the body, although the decline occurs sooner and to a greater extent than body losses alone can explain. More likely, the neural and hormonal changes that accompany fasting are responsible for changes in the BMR.

[e] The thyroid gland releases hormones that travel to the cells and influence cellular metabolism. Thyroid hormone activity can speed up or slow down the rate of metabolism by as much as 50 percent.

FIGURE 8-4 Components of Energy Expenditure

The amount of energy spent in a day differs for each individual, but in general, basal metabolism is the largest component of energy expenditure and the thermic effect of food is the smallest. The amount spent in voluntary physical activities has the greatest variability, depending on a person's activity patterns. For a sedentary person, physical activities may account for less than half as much energy as basal metabolism, whereas an extremely active person may expend as much on activity as for basal metabolism.

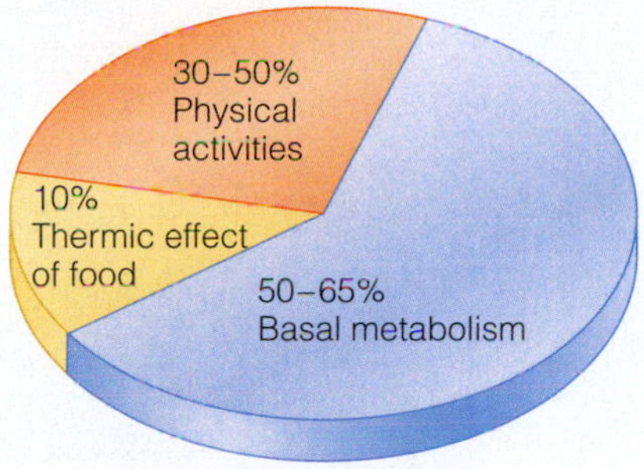

TABLE 8-2 Energy Expended on Various Activities

The values listed in this table reflect both the energy expended in physical activity *and* the amount used for BMR. To calculate kcalories spent per minute of activity for your own body weight, multiply kcal/lb/min (or kcal/kg/min) by your exact weight and then multiply that number by the number of minutes spent in the activity. For example, if you weigh 142 pounds, and you want to know how many kcalories you spent doing 30 minutes of vigorous aerobic dance: 0.062 × 142 = 8.8 kcalories per minute; 8.8 × 30 minutes = 264 total kcalories spent.

Activity	kCal/lb min	kCal/kg min
Aerobic dance (vigorous)	.062	.136
Basketball (vigorous, full court)	.097	.213
Bicycling		
13 mph	.045	.099
15 mph	.049	.108
17 mph	.057	.125
19 mph	.076	.167
21 mph	.090	.198
23 mph	.109	.240
25 mph	.139	.306
Canoeing, flat water, moderate pace	.045	.099
Cross-country skiing		
8 mph	.104	.229
Gardening	.045	.099
Golf (carrying clubs)	.045	.099
Handball	.078	.172
Horseback riding (trot)	.052	.114
Rowing (vigorous)	.097	.213
Running		
5 mph	.061	.134
6 mph	.074	.163
7.5 mph	.094	.207
9 mph	.103	.227
10 mph	.114	.251
11 mph	.131	.288
Soccer (vigorous)	.097	.213
Studying	.011	.024
Swimming		
20 yd/min	.032	.070
45 yd/min	.058	.128
50 yd/min	.070	.154
Table tennis (skilled)	.045	.099
Tennis (beginner)	.032	.070
Vacuuming and other household tasks	.030	.066
Walking (brisk pace)		
3.5 mph	.035	.077
4.5 mph	.048	.106
Weight lifting		
light-to-moderate effort	.024	.053
vigorous effort	.048	.106
Wheelchair basketball	.084	.185
Wheeling self in wheelchair	.030	.066
Wii games		
bowling	.021	.046
boxing	.021	.047
tennis	.022	.048

It feels like work and it may make you tired, but studying requires only one or two kcalories per minute.

kcalories expended. (An activity's duration, frequency, and intensity also influence the body's use of the energy-yielding nutrients.)

Thermic Effect of Food When a person eats, the GI tract muscles speed up their rhythmic contractions, the cells that manufacture and secrete digestive juices begin their tasks, and some nutrients are absorbed by active transport. This acceleration of activity requires energy and produces heat; it is known as the **thermic effect of food (TEF).**

The thermic effect of food is proportional to the food energy taken in and is usually estimated at 10 percent of energy intake. Thus a person who ingests 2000 kcalories probably expends about 200 kcalories on the thermic effect of food. The proportions vary for different foods, however, and are also influenced by factors such as meal size and frequency. In general, the thermic effect of food is greater for high-protein foods than for high-fat foods ♦ and for a meal eaten all at once rather than spread out over a couple of hours. For most purposes, however, the thermic effect of food can be ignored when estimating energy expenditure because its contribution to total energy output is smaller than the probable errors involved in estimating overall energy intake and output.

♦ Thermic effect of foods:
- Carbohydrate: 5–10%
- Fat: 0–5%
- Protein: 20–30%
- Alcohol: 15–20%

The percentages are calculated by dividing the energy expended during digestion and absorption (above basal) by the energy content of the food.

Adaptive Thermogenesis Some additional energy is spent when a person must adapt to dramatically changed circumstances **(adaptive thermogenesis).** When the body has to adapt to physical conditioning, extreme cold, overfeeding, starvation, trauma, or other types of stress, it has extra work to do, building the tissues and producing the enzymes and hormones necessary to cope with the demand. In some circumstances, this energy makes a considerable difference in the total energy expended. Because this component of energy expenditure is so variable and specific to individuals, it is not included when calculating energy requirements.

Estimating Energy Requirements

In estimating energy requirements, the DRI Committee developed equations based on research measuring total daily energy expenditure. These equations consider how the following factors influence energy expenditure: ♦

- *Gender.* In general, women have a lower BMR than men, in large part because men typically have more lean body mass. Two sets of energy equations—one

♦ Note that Table 8-1 (p. 247) lists these factors among those that influence BMR and consequently energy expenditure.

thermic effect of food (TEF): an estimation of the energy required to process food (digest, absorb, transport, metabolize, and store ingested nutrients); also called the *specific dynamic effect (SDE)* of food or the *specific dynamic activity (SDA)* of food. The sum of the TEF and any increase in the metabolic rate due to overeating is known as *diet-induced thermogenesis (DIT).*

adaptive thermogenesis: adjustments in energy expenditure related to changes in environment such as extreme cold and to physiological events such as overfeeding, trauma, and changes in hormone status.

for men and one for women—were developed to accommodate the influence of gender on energy expenditure.

- *Growth.* The BMR is high in people who are growing. For this reason, pregnant and lactating women, infants, children, and adolescents have their own sets of energy equations (provided in Appendix F).
- *Age.* The BMR declines during adulthood as lean body mass diminishes. This change in body composition occurs, in part, because some hormones that influence appetite, body weight, and metabolism become more, or less, active with age.[10] Physical activities tend to decline as well, bringing the average reduction in energy expenditure to about 5 percent per decade. The decline in BMR that occurs when a person becomes less active reflects the loss of lean body mass and may be minimized with ongoing physical activity. Because age influences energy expenditure, it is also factored into the energy equations.
- *Physical activity.* Using individual values for various physical activities (as in Table 8-2 on p. 247) is time-consuming and impractical for estimating the energy needs of a population. Instead, various activities are clustered according to the typical intensity of a day's efforts. Energy equations include a physical activity factor for various levels of intensity for each gender.
- *Body composition and body size.* The BMR is high in people who are tall and so have a large surface area. ♦ Similarly, the more a person weighs, the more energy is expended on basal metabolism. For these reasons, the energy equations include a factor for both height and weight.

As just explained, energy needs vary between individuals depending on such factors as gender, growth, age, physical activity, and body size and composition. Even when two people are similarly matched, however, their energy needs still differ because of genetic differences. Perhaps one day genetic research will reveal how to estimate requirements for each individual. For now, the "How To" on p. 250 provides instructions on calculating your estimated energy requirements using the DRI equations and physical activity factors. ♦

♦ Each of these structures is made of eight blocks. They weigh the same, but they are arranged differently. The short, wide structure has 24 sides exposed and the tall, thin one has 34. Because the tall, thin structure has a greater surface area, it will lose more heat (expend more energy) than the short, wide one. Similarly, two people of different heights might weigh the same, but the taller, thin one will have a higher BMR (expending more energy) because of the greater skin surface.

♦ Appendix F presents DRI tables that provide a shortcut to estimating total energy expenditure and instructions to help you determine the appropriate physical activity factor to use in the equation.

IN SUMMARY **A person in energy balance takes in energy from food and expends much of it on basal metabolic activities, some of it on physical activities, and a little on the thermic effect of food. Because energy requirements vary from person to person, such factors as gender, age, weight, and height as well as the intensity and duration of physical activity must be considered when estimating energy requirements.**

Body Weight, Body Composition, and Health

A person 5 feet 10 inches tall who weighs 150 pounds ♦ may carry only about 30 of those pounds as fat. The rest is mostly water and lean tissues—muscles, organs such as the heart and liver, and the bones of the skeleton. Direct measures of **body composition** are impossible in living human beings; instead, researchers assess body composition indirectly based on the following assumption:

$$\text{Body weight} = \text{fat} + \text{lean tissue (including water)}$$

♦ In metric terms, a person 1.78 meters tall who weighs 68 kilograms may carry only about 14 of those kilograms as fat.

Weight gains and losses tell us nothing about how the body's composition may have changed, yet weight is the measure most people use to judge their "fatness." For many people, overweight means overfat, but this is not always the case. Athletes with dense bones and well-developed muscles may be overweight by some standards but have little body fat. Conversely, inactive people may seem to have acceptable weights, when, in fact, they may have too much body fat.

body composition: the proportions of muscle, bone, fat, and other tissue that make up a person's total body weight.

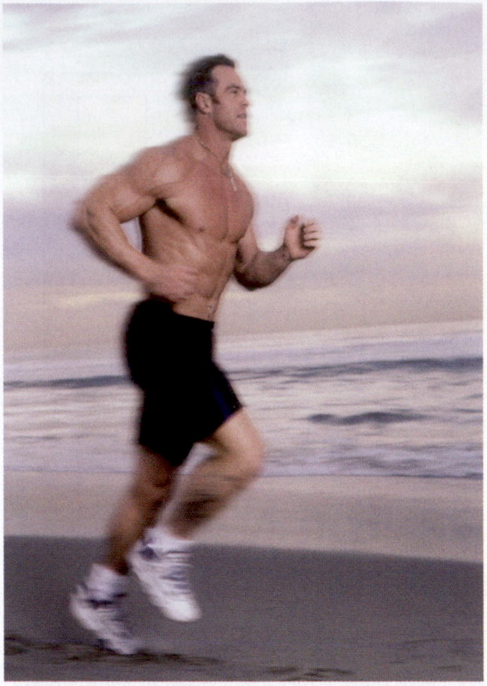

At 6 feet 4 inches tall and 250 pounds (1.93 meters and 113 kilograms), this runner would be considered over*weight* by most standards. Yet he is clearly not over*fat*.

Defining Healthy Body Weight

How much should a person weigh? How can a person know if her weight is appropriate for her height? How can a person know if his weight is jeopardizing his health? Such questions seem so simple, yet the answers can be complex—and quite different depending on whom you ask.

The Criterion of Fashion

In asking what is ideal, people often mistakenly turn to fashion for the answer and judge body weight by appearances. No doubt our society sets unrealistic ideals for body weight, especially for women. Miss America, our nation's icon of beauty, has never been overweight, and until recently, she has grown progressively thinner over the years (see Figure 8-5). Magazines, movies,

HOW TO Estimate Energy Requirements

To determine your estimated energy requirement (EER), use the appropriate equation, inserting your age in years, weight (wt) in kilograms, height (ht) in meters, and physical activity (PA) factor from the accompanying table. (To convert pounds to kilograms, divide by 2.2; to convert inches to meters, divide by 39.37.)

- For men 19 years and older:
  EER = [662 − (9.53 × age)] + PA × [(15.91 × wt) + (539.6 × ht)]
- For women 19 years and older:
  EER = [354 − (6.91 × age)] + PA × [(9.36 × wt) + (726 × ht)]

For example, consider an active 30-year-old male who is 5 feet 11 inches tall and weighs 178 pounds. First, he converts his weight from pounds to kilograms and his height from inches to meters, if necessary:

178 lb ÷ 2.2 = 80.9 kg
71 in ÷ 39.37 = 1.8 m

Next, he considers his level of daily physical activity and selects the appropriate PA factor from the accompanying table. (In this example, 1.25 for an active male.) Then, he inserts his age, PA factor, weight, and height into the appropriate equation:

EER = [662 − (9.53 × 30)] + 1.25 × [(15.91 × 80.9) + (539.6 × 1.8)]

(A reminder: Do calculations within the parentheses first.) He calculates:

EER = [662 − 286] + 1.25 × [1287 + 971]

(Another reminder: Do calculations within the brackets next.)

EER = 376 + 1.25 × 2258

(One more reminder: Do multiplication before addition.)

EER = 376 +2823
EER = 3199

The estimated energy requirement for an active 30-year-old male who is 5 feet 11 inches tall and weighs 178 pounds is about 3200 kcalories/day. His actual requirement probably falls within a range ♦ of 200 kcalories above and below this estimate.

NOTE: Appendix F provides tables of energy expenditure for adults at various levels of activity and various heights and weights. It also includes EER equations for infants, children, adolescents, and pregnant women.

For additional practice, log on to **www.cengagebrain.com** and search for ISBN 084006845X.

Physical Activity (PA) Factors for EER Equations

	Men	Women	Physical Activity
Sedentary	1.0	1.0	Typical daily living activities
Low active	1.11	1.12	Plus 30–60 min moderate activity
Active	1.25	1.27	Plus ≥ 60 min moderate activity
Very active	1.48	1.45	Plus ≥ 60 min moderate activity and 60 min vigorous or 120 min moderate activity

NOTE: Moderate activity is equivalent to walking at 3 to 4½ mph.

TRY IT Estimate your energy requirement based on your current age, weight, height, and activity level.

♦ For *most* people, the actual energy requirement falls within these ranges:
- For men, EER ± 200 kcal
- For women, EER ± 160 kcal

For *almost all* people, the actual energy requirement falls within these ranges:
- For men, EER ± 400 kcal
- For women, EER ± 320 kcal

FIGURE 8-5 The Declining Weight of Miss America

As explained on p. 252, the body mass index (BMI) describes relative weight for height. Until the past decade, the BMI of Miss America declined steadily. Notice how many fell below 18.5, the cutoff point indicating underweight with its associated health problems.

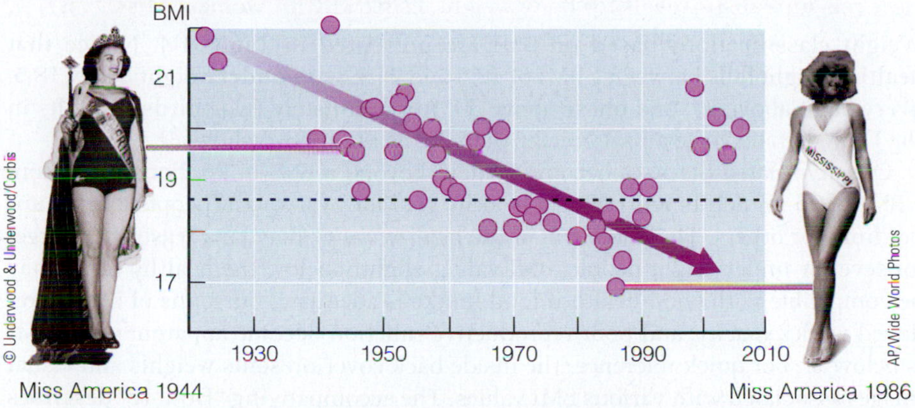

SOURCE: S. Rubenstein and B. Caballero, Is Miss America an undernourished role model? *Journal of the American Medical Association* 283 (2000): 1569.

and television all convey the message that to be thin is to be beautiful and happy. As a result, the media have a great influence on the weight concerns and dieting patterns of people of all ages, but most tragically on young, impressionable children and adolescents.[11]

Importantly, perceived body image often has little to do with actual body weight or size. People of all shapes, sizes, and ages—including extremely thin fashion models with anorexia nervosa and fitness instructors with ideal body composition—have learned to be unhappy with their "overweight" bodies. Such dissatisfaction can lead to damaging behaviors, such as starvation diets, diet pill abuse, and health-care avoidance. The first step toward making healthy changes may be self-acceptance. Keep in mind that fashion is fickle; the body shapes valued by our society change with time. Furthermore, body shapes valued by our society differ from those of other societies. The standards defining "ideal" are subjective and frequently have little in common with health. Table 8-3 offers some tips for adopting health as an ideal, rather than society's misconceived image of beauty.

The Criterion of Health Even if our society were to accept fat as beautiful, obesity would still be a major risk factor for several life-threatening diseases. For this reason, the most important criterion for determining how much a person should weigh and how much body fat a person needs is not appearance but good health and longevity. Ideally, a person has enough fat to meet basic needs ♦ but not so

♦ Fat in the body:
- Provides energy
- Insulates against temperature extremes
- Protects against physical shock
- Forms cell membranes
- Makes compounds such as hormones, vitamin D, and bile

TABLE 8-3 Tips for Accepting a Healthy Body Weight

- Value yourself and others for human attributes other than body weight. Realize that prejudging people by weight is as harmful as prejudging them by race, religion, or gender.
- Use positive, nonjudgmental descriptions of your body.
- Accept positive comments from others.
- Focus on your whole self including your intelligence, social grace, and professional and scholastic achievements.
- Accept that no magic diet exists.
- Stop dieting to lose weight. Adopt a lifestyle of healthy eating and physical activity permanently.
- Follow the USDA Food Guide. Never restrict food intake below the minimum levels that meet nutrient needs.
- Become physically active, not because it will help you get thin but because it will make you feel good and enhance your health.
- Seek support from loved ones. Tell them of your plan for a healthy life in the body you have been given.
- Seek professional counseling, *not* from a weight-loss counselor, but from someone who can help you make gains in self-esteem without weight as a factor.
- Appreciate body weight for its influence on health, not appearance.

much as to incur health risks. This range of healthy body weights has been identified using a common measure of weight and height—the body mass index.

Body Mass Index The **body mass index (BMI)** describes relative weight for height: ♦

$$\text{BMI} = \frac{\text{weight (kg)}}{\text{height (m)}^2} \quad \text{or} \quad \frac{\text{weight (lb)}}{\text{height (in)}^2} \times 703.$$

♦ To convert pounds to kilograms:
lb ÷ 2.2 lb/kg = kg
To convert inches to meters:
in ÷ 39.37 in/m = m

Weight classifications based on BMI are presented in Table 8-4. Notice that healthy weight falls between a BMI of 18.5 and 24.9, with **underweight** below 18.5, **overweight** above 25, and **obese** above 30. Approximately two-thirds of adults in the United States have a BMI greater than 25, as Figure 8-6 shows.[12]

Obesity-related diseases become evident beyond a BMI of 25. For this reason, a BMI of 25 for adults represents a healthy goal for overweight people and an upper limit for others. The lower end of the healthy range may be a reasonable target for severely underweight people. BMI values slightly below the healthy range may be compatible with good health if food intake is adequate, but signs of illness, reduced work capacity, and poor reproductive function become apparent when BMI is below 17. For quick reference, the inside back cover presents weights and visual images associated with various BMI values. The accompanying "How To" describes how to determine your BMI and how to find a goal weight based on a desired BMI.

Keep in mind that BMI reflects height and weight measures and not body composition. Consequently, muscular athletes may be classified as over*weight* by BMI standards and not be over*fat*.[13] At the peak of his bodybuilding career, Arnold

body mass index (BMI): an index of a person's weight in relation to height; determined by dividing the weight (in kilograms) by the square of the height (in meters).

underweight: body weight below some standard of acceptable weight that is usually defined in relation to height (such as BMI); BMI below 18.5.

overweight: body weight above some standard of acceptable weight that is usually defined in relation to height (such as BMI); BMI 25 to 29.9.

obese: overweight with adverse health effects; BMI 30 or higher.

TABLE 8-4 Body Mass Index (BMI)

	Underweight (<18.5)	Healthy Weight (18.5–24.9)						Overweight (25–29.9)					Obese (≥30)										
	18	19	20	21	22	23	24	25	26	27	28	29	30	31	32	33	34	35	36	37	38	39	40
Height	*Body weight (pounds)*																						
4′10″	86	91	96	100	105	110	115	119	124	129	134	138	143	148	153	158	162	167	172	177	181	186	191
4′11″	89	94	99	104	109	114	119	124	128	133	138	143	148	153	158	163	168	173	178	183	188	193	198
5′0″	92	97	102	107	112	118	123	128	133	138	143	148	153	158	163	168	174	179	184	189	194	199	204
5′1″	95	100	106	111	116	122	127	132	137	143	148	153	158	164	169	174	180	185	190	195	201	206	211
5′2″	98	104	109	115	120	126	131	136	142	147	153	158	164	169	175	180	186	191	196	202	207	213	218
5′3″	102	107	113	118	124	130	135	141	146	152	158	163	169	175	180	186	191	197	203	208	214	220	225
5′4″	105	110	116	122	128	134	140	145	151	157	163	169	174	180	186	192	197	204	209	215	221	227	232
5′5″	108	114	120	126	132	138	144	150	156	162	168	174	180	186	192	198	204	210	216	222	228	234	240
5′6″	112	118	124	130	136	142	148	155	161	167	173	179	186	192	198	204	210	216	223	229	235	241	247
5′7″	115	121	127	134	140	146	153	159	166	172	178	185	191	198	204	211	217	223	230	236	242	249	255
5′8″	118	125	131	138	144	151	158	164	171	177	184	190	197	203	210	216	223	230	236	243	249	256	262
5′9″	122	128	135	142	149	155	162	169	176	182	189	196	203	209	216	223	230	236	243	250	257	263	270
5′10″	126	132	139	146	153	160	167	174	181	188	195	202	209	216	222	229	236	243	250	257	264	271	278
5′11″	129	136	143	150	157	165	172	179	186	193	200	208	215	222	229	236	243	250	257	265	272	279	286
6′0″	132	140	147	154	162	169	177	184	191	199	206	213	221	228	235	242	250	258	265	272	279	287	294
6′1″	136	144	151	159	166	174	182	189	197	204	212	219	227	235	242	250	257	265	272	280	288	295	302
6′2″	141	148	155	163	171	179	186	194	202	210	218	225	233	241	249	256	264	272	280	287	295	303	311
6′3″	144	152	160	168	176	184	192	200	208	216	224	232	240	248	256	264	272	279	287	295	303	311	319
6′4″	148	156	164	172	180	189	197	205	213	221	230	238	246	254	263	271	279	287	295	304	312	320	328
6′5″	151	160	168	176	185	193	202	210	218	227	235	244	252	261	269	277	286	294	303	311	319	328	336
6′6″	155	164	172	181	190	198	207	216	224	233	241	250	259	267	276	284	293	302	310	319	328	336	345

A healthy body contains enough lean tissue to support health and the right amount of fat to meet body needs.

FIGURE 8-6 Distribution of Body Weights in U.S. Adults

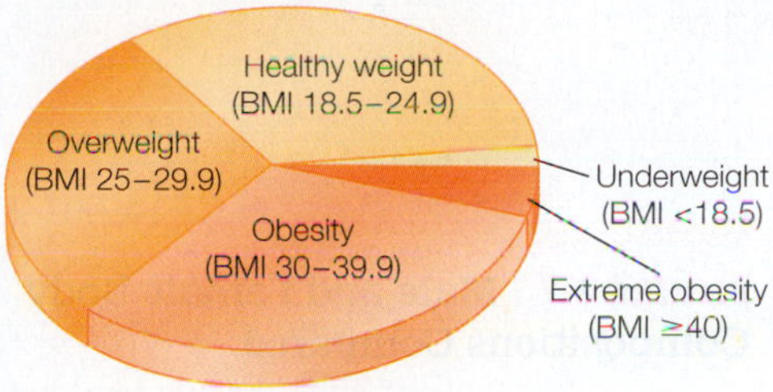

Schwarzenegger won the Mr. Olympia competition with a BMI of 31; the runner on p. 250 also has a BMI greater than 30. Yet neither would be considered obese. Striking differences in body composition are also apparent among people of different ages and various ethnic and racial groups, making standard BMI guidelines inappropriate for some populations.[14] For example, blacks tend to have a greater bone

HOW TO Determine BMI

To calculate your body mass index (BMI), use one of the following equations:

$$BMI = \frac{\text{weight (lb)}}{\text{height (in)}^2} \times 703$$

or

$$BMI = \frac{\text{weight (kg)}}{\text{height (m)}^2}$$

Consider, for example, a person who is 5′5″ (1.65 m) tall and weighs 174 lb (79 kg):

$$BMI = \frac{174 \text{ lb}}{65 \text{ in}^2} \times 703 = 29$$

or

$$BMI = \frac{79 \text{ kg}}{1.65 \text{ m}^2} = 29$$

This person has a BMI of 29 and is considered overweight.

You could also use Table 8-4 to determine your BMI. Locate your height in the first column (in this example, 5′5″). Then look across the row until you find the number that is closest to your weight (in this example, 174). The number at the top of that column identifies your BMI (in this example, 29).

A reasonable initial target for most overweight people is a BMI 2 units below their current one. To determine a goal weight based on a desired BMI, locate your height in the first column and then look across the row until you reach the column with the desired BMI at the top. In this example, to reach a BMI of 27, this person's goal weight is 162 pounds, which represents a 12-pound weight loss. Such a determination can help a person set realistic weight goals using health risk as a guide.

For additional practice, log on to **www.cengagebrain.com** and search for ISBN 084006845X.

TRY IT Calculate your BMI and determine whether you are underweight, healthy weight, overweight, or obese. If your BMI is less than 18.5 or greater than 25, identify a weight that takes your BMI 2 units closer to the healthy weight range.

density and protein content than whites; consequently, using BMI as the standard may overestimate the prevalence of overweight and obesity among blacks.

IN SUMMARY Current standards for body weight are based on a person's weight in relation to height, called the body mass index (BMI), and reflect disease risks. To its disadvantage, BMI does not reflect body fat, and it may misclassify very muscular people as overweight.

Body Fat and Its Distribution

Although weight measures are inexpensive, easy to take, and highly accurate, they fail to reveal two valuable pieces of information in assessing disease risk: how much of the weight is fat and where the fat is located. The ideal amount of body fat depends partly on the person. A normal-weight man may have from 13 to 21 percent body fat; a woman, because of her greater quantity of essential fat, 23 to 31 percent. In general, health problems typically develop when body fat exceeds 22 percent in young men, 25 percent in men older than age 40, 32 percent in young women, and 35 percent in women older than age 40. Body fat may contribute as much as 70 percent in excessively obese adults. Figure 8-7 compares the body composition of healthy-weight men and women.

FIGURE 8-7 Male and Female Body Compositions Compared

The differences between male and female body compositions become apparent during adolescence. Lean body mass (primarily muscle) increases more in males than in females. Fat assumes a larger percentage of female body composition as essential body fat is deposited in the mammary glands and pelvic region in preparation for childbearing. Both men and women have essential fat associated with the bone marrow, the central nervous system, and the internal organs.

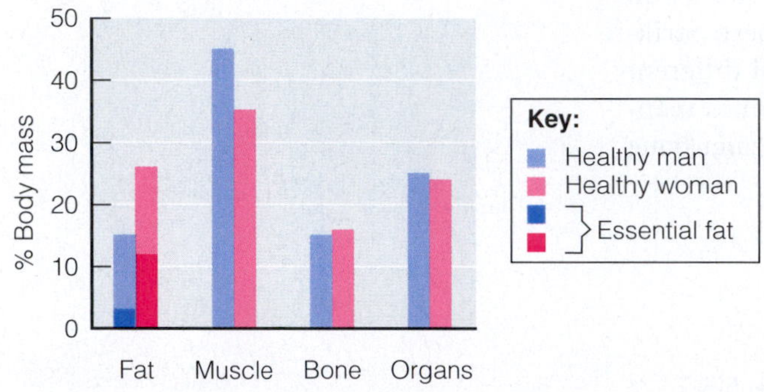

SOURCE: R. E. C. Wildman and D. M. Medeiros, *Advanced Human Nutrition* (Boca Raton, Fla.: CRC Press, 2000), pp. 321–323. Copyright © 2000 Taylor and Francis Books LLC. Used with permission.

Some People Need Less Body Fat

For many athletes, a lower percentage of body fat may be ideal—just enough fat to provide fuel, insulate and protect the body, assist in nerve impulse transmissions, and support normal hormone activity, but not so much as to burden the body with excess bulk. For some athletes, then, ideal body fat might be 5 to 10 percent for men and 15 to 20 percent for women. (Review the photo on p. 250 to appreciate what 8 percent body fat looks like.)

Some People Need More Body Fat

For an Alaska fisherman, a higher percentage of body fat is probably beneficial because fat provides an insulating blanket to prevent excessive loss of body heat in cold climates. A woman starting a pregnancy needs sufficient body fat to support conception and fetal growth. Below a certain threshold for body fat, hormone synthesis falters, and individuals may become infertile, develop depression, experience abnormal hunger regulation, or become unable to keep warm. These thresholds differ for each function and for each individual; much remains to be learned about them.

Fat Distribution

The distribution of fat on the body may be more critical than the total amount of fat alone. **Visceral fat** that is stored around the organs of the abdomen is referred to as **central obesity** or upper-body fat (see Figure 8-8). Independently of BMI or total body fat, central obesity is associated with increased risks of heart disease, stroke, diabetes, insulin resistance, hypertension, gallstones, and some types of cancer.[15]

Visceral fat is most common in men and to a lesser extent in women past menopause. Even when total body fat is similar, men have more visceral fat than women. Regardless of gender, the risks of cardiovascular disease, diabetes, and mortality are increased for those with excessive visceral fat. Interestingly, smokers tend to have more visceral fat than nonsmokers even though they typically have lower BMI.[16]

Subcutaneous fat around the hips and thighs, sometimes referred to as lower-body fat, is most common in women during their reproductive years and seems relatively harmless. In fact, overweight people who have little visceral fat are less susceptible to health problems than overweight people with visceral fat. Figure 8-9 compares the body shapes of people with upper-body fat and lower-body fat.

Waist Circumference

A person's **waist circumference** is a good indicator of fat distribution and central obesity.[17] In general, women with a waist circumference

visceral fat: fat stored within the abdominal cavity in association with the internal abdominal organs; also called *intra-abdominal fat.*

central obesity: excess fat around the trunk of the body; also called *abdominal fat* or *upper-body fat.*

subcutaneous fat: fat stored directly under the skin.
- **sub** = beneath
- **cutaneous** = skin

waist circumference: an anthropometric measurement used to assess a person's abdominal fat.

FIGURE 8-8 Abdominal Fat

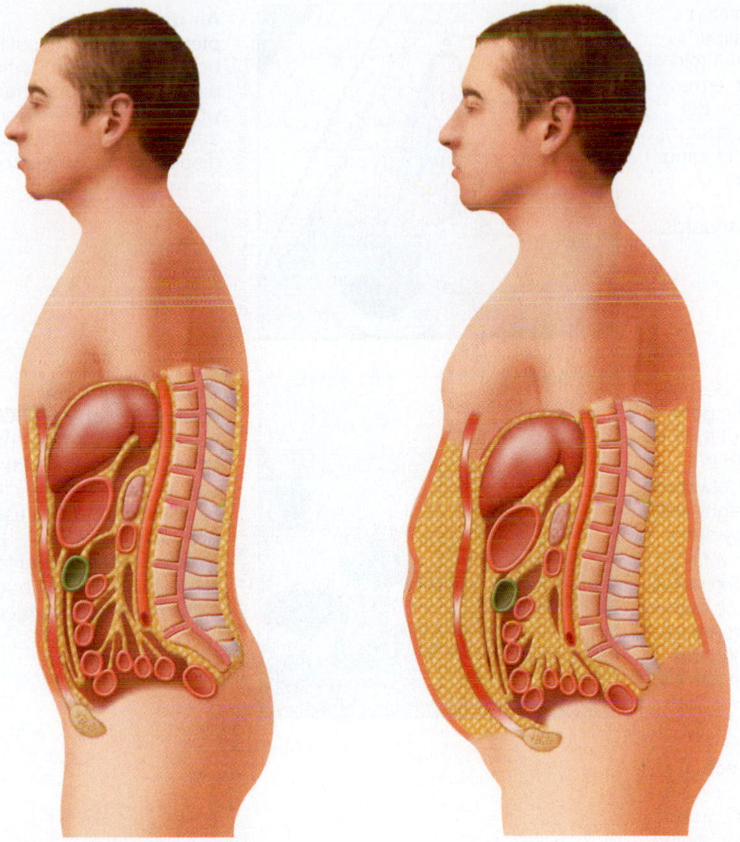

In healthy-weight people, some fat is stored around the organs of the abdomen.
In overweight people, excess abdominal fat increases the risks of diseases.

FIGURE 8-9 "Apple" and "Pear" Body Shapes Compared

Popular articles sometimes call bodies with upper-body fat "apples" and those with lower-body fat, "pears." Researchers sometimes refer to upper-body fat as "android" (manlike) obesity and to lower-body fat as "gynoid" (womanlike) obesity.

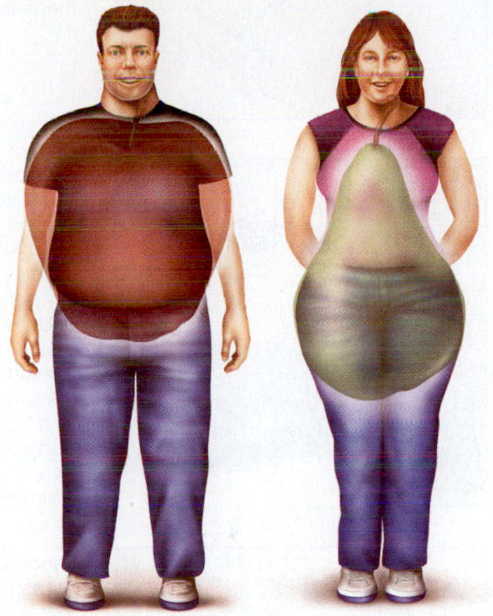

of greater than 35 inches (88 centimeters) and men with a waist circumference of greater than 40 inches (102 centimeters) have a high risk of central obesity–related health problems, such as diabetes and cardiovascular disease. As waist circumference increases, disease risks increase. Appendix E includes instructions for measuring waist circumference and assessing abdominal fat.

Upper-body fat is more common in men than in women and is closely associated with heart disease, stroke, diabetes, hypertension, and some types of cancer.
Lower-body fat is more common in women than in men and is not usually associated with chronic diseases.

Waist circumference is a better indicator of abdominal fat, but some researchers use the waist-to-hip ratio when studying disease risks. The ratio requires another step or two (measuring the hips and comparing that measure to the waist measure), but it does not provide any additional information. Therefore, waist circumference alone is the preferred method for assessing abdominal fat in a clinical setting.*

Other Measures of Body Composition Health-care professionals commonly use BMI and waist circumference measures because they are relatively easy and inexpensive. Together, these two measures prove most valuable in assessing a person's health risks and monitoring changes over time.[18] Researchers needing more precise measures of body composition may choose any of several other techniques to estimate body fat and its distribution (see Figure 8-10 on p. 256). Mastering these techniques requires proper instruction and practice to ensure reliability. In addition to the methods shown in Figure 8-10, researchers sometimes estimate body composition using these methods: total body water, radioactive potassium count, near-infrared spectrophotometry, ultrasound, computed tomography, and

*The National Heart, Lung, and Blood Institute recommends using the waist circumference instead of the waist-to-hip ratio to assess obesity health risks.

FIGURE 8-10 Common Methods Used to Assess Body Fat

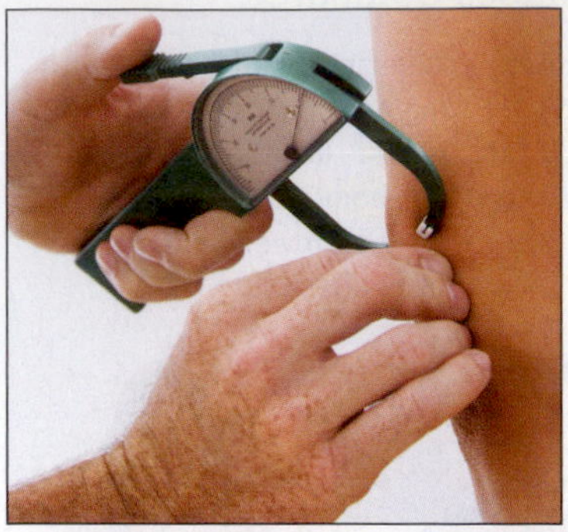

iofoto/Shutterstock.com

Skinfold measures estimate body fat by using a caliper to gauge the thickness of a fold of skin on the back of the arm (over the triceps), below the shoulder blade (subscapular), and in other places (including lower-body sites), and then comparing these measurements with standards.

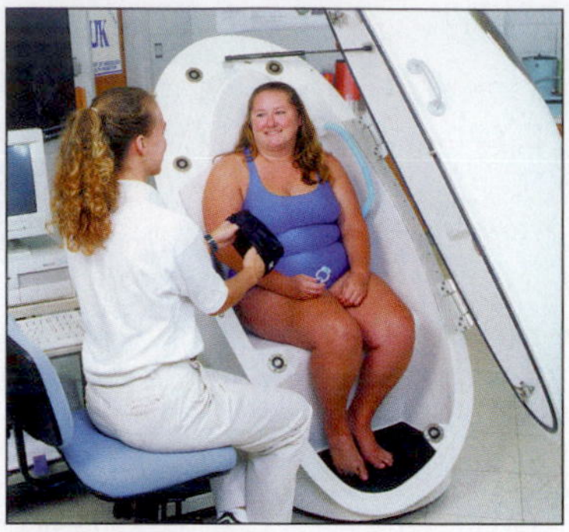

© Photo Courtesy of Live Measurement, Inc.

Air displacement plethysmography estimates body composition by having a person sit inside a chamber while computerized sensors determine the amount of air displaced by the person's body.

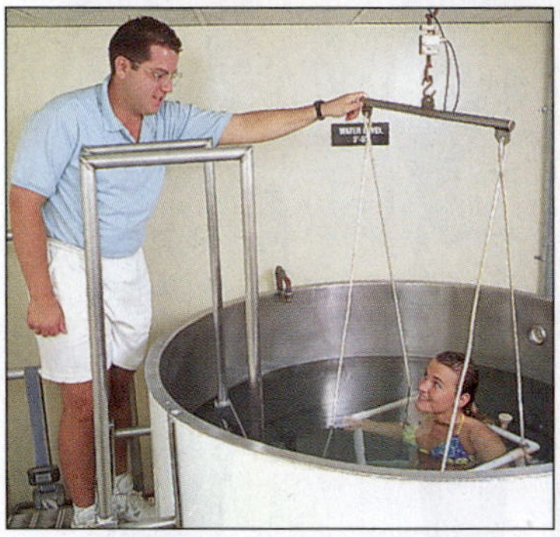

© David Young-Wolff/PhotoEdit

Hydrodensitometry measures body density by weighing the person first on land and then again while submerged in water. The difference between the person's actual weight and underwater weight provides a measure of the body's volume. A mathematical equation using the two measurements (volume and actual weight) determines body density, from which the percentage of body fat can be estimated.

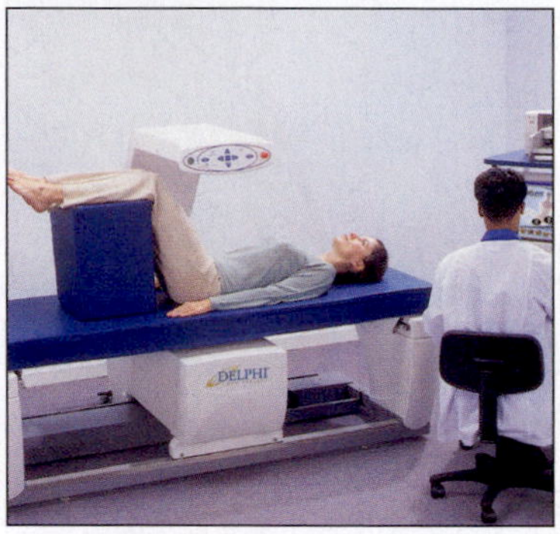

© Photo Courtesy of Hologic, Inc.

Dual energy X-ray absorptiometry (DEXA) uses two low-dose X-rays that differentiate among fat-free soft tissue (lean body mass), fat tissue, and bone tissue, providing a precise measurement of total fat and its distribution in all but extremely obese subjects.

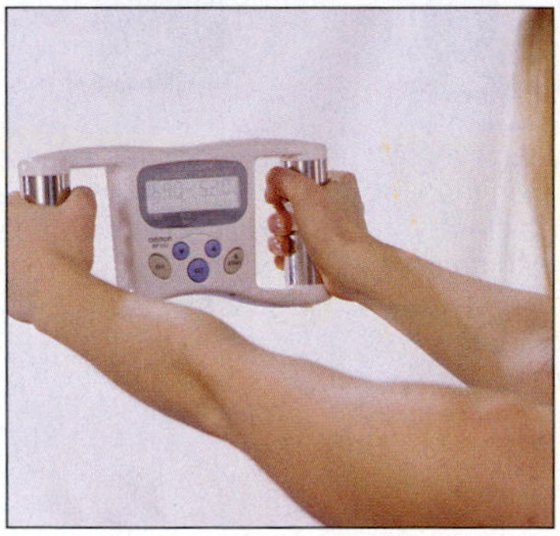

May/Photo Researchers, Inc.

Bioelectrical impedance measures body fat by using a low-intensity electrical current. Because electrolyte-containing fluids, which readily conduct an electrical current, are found primarily in lean body tissues, the leaner the person, the less resistance to the current. The measurement of electrical resistance is then used in a mathematical equation to estimate the percentage of body fat.

magnetic resonance imaging. Each method has advantages and disadvantages with respect to cost, technical difficulty, and precision of estimating body fat (see Appendix E for a comparison). Appendix E provides additional details and includes many of the tables and charts routinely used in assessment procedures.

IN SUMMARY **The ideal amount of body fat varies from person to person, but researchers have found that body fat in excess of 22 percent for young men and 32 percent for young women (the levels rise slightly with age) poses health risks. Central obesity, in which excess abdominal fat is distributed around the trunk of the body, presents greater health risks than excess fat distributed on the lower body.**

Health Risks Associated with Body Weight and Body Fat

Insurance data indicate that body weight and fat distribution correlate with disease risks and life expectancy. The correlation suggests a greater *likelihood* of developing a chronic disease and shortening life expectancy. Not all overweight and underweight people will get sick and die before their time nor will all normal-weight people live long healthy lives. *Correlations* are not *causes*. For the most part, people with a BMI between 18.5 and 24.9 have relatively few weight-related health risks; risks increase as BMI falls below or rises above this range, indicating that both too little and too much body fat impair health.[19] Epidemiological data show a J- or U-shaped relationship between body weights and mortality (see Figure 8-11).[20] People who are extremely underweight or extremely obese carry higher risks of early deaths ♦ than those whose weights fall within the healthy, or even the slightly overweight, range.[21] These mortality risks decline with age.[22]

Independently of BMI, factors such as smoking habits raise health risks, and physical fitness lowers them. A man with a BMI of 22 who smokes two packs of cigarettes a day is jeopardizing his health, whereas a woman with a BMI of 32 who walks briskly for an hour a day is improving her health.

♦ BMI and mortality:
- BMI 22.5–25 = optimal survival
- BMI 30–35 = 3 years' loss of life
- BMI >40 = 10 years' loss of life (equivalent to lifetime of smoking)

Health Risks of Underweight Some underweight people enjoy an active, healthy life, but others are underweight because of malnutrition, smoking habits, substance abuse, or illnesses. Weight and fat measures alone would not reveal these underlying causes, but a complete assessment that includes a diet and medical history, physical examination, and biochemical analysis would.

An underweight person, especially an older adult, may be unable to preserve lean tissue during the fight against a wasting disease such as cancer or a digestive disorder, especially when the disease is accompanied by malnutrition. Without adequate nutrient and energy reserves, an underweight person will have a particularly tough battle against such medical stresses. Underweight women develop menstrual irregularities and become infertile. Those who do conceive may give birth to unhealthy infants. An underweight woman can improve her chances of having a healthy infant by gaining weight prior to conception, during pregnancy, or both. Underweight and significant weight loss are also associated with osteoporosis and bone fractures. For all these reasons, underweight people may benefit from enough of a weight gain to provide an energy reserve and protective amounts of all the nutrients that can be stored.

FIGURE 8-11 BMI and Mortality

This J-shaped curve describes the relationship between body mass index (BMI) and mortality and shows that both underweight and overweight present risks of a premature death.

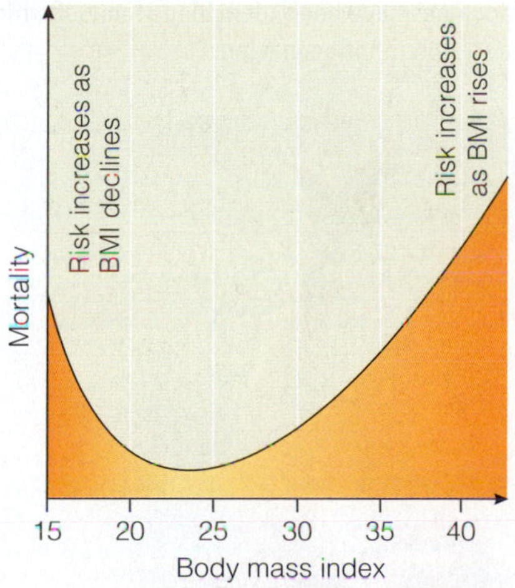

Health Risks of Overweight As for excessive body fat, the health risks are so many that it has been designated a disease—obesity. Among the health risks associated with obesity are diabetes, hypertension, cardiovascular disease, sleep apnea (abnormal ceasing of breathing during sleep), osteoarthritis, some cancers, gallbladder disease, kidney stones, respiratory problems (including Pickwickian syndrome, a breathing blockage linked with sudden death), infertility, and complications in pregnancy and surgery. Obese people are more likely to be disabled in their later years.[23] Each year, these obesity-related illnesses cost our nation billions of dollars—in fact, as much as, or more than, the medical costs of smoking.[24]

The cost in terms of lives is also great: an estimated 300,000 people die each year from obesity-related diseases. In fact, obesity is second only to tobacco in causing preventable illnesses and premature deaths. Mortality increases as excess weight increases.[25] People with a BMI greater than 35 are about twice as likely to die of heart disease as others.[26] The risks associated with a high BMI appear to be greater for whites than for blacks; in fact, the health risks associated with obesity do not become apparent in black women until a BMI of 37. In contrast, health risks appear to be greater for Asians than for Caucasians at the same BMI.[27]

Equally important, both central obesity and weight gains of more than 20 pounds (9 kilograms) between early and middle adulthood correlate with increased disease risks.[28] Fluctuations in body weight, as typically occur with "yo-yo" dieting, may also increase the risks of chronic diseases and premature death. In contrast, sustained weight loss improves physical well-being, reduces disease risks, and increases life expectancy.

Smoking is the leading cause of preventable illnesses and early deaths. Obesity is a close second.

Cardiovascular Disease The relationship between obesity and cardiovascular disease risk is strong, with links to both elevated blood cholesterol and hypertension. Central obesity may raise the risk of heart attack and stroke as much as

♦ Cardiovascular disease risk factors associated with obesity:
- High LDL cholesterol
- Low HDL cholesterol
- High blood pressure (hypertension)
- Diabetes

Chapter 27 provides many more details.

♦ Metabolic syndrome is a cluster of at least three of the following risk factors:
- High blood pressure
- High blood glucose
- High blood triglycerides
- Low HDL cholesterol
- High waist circumference

♦ Chapter 5 introduces **adipokines**—proteins released from adipose tissue that signal changes in the body's fat and energy status. More than 50 adipokines have been identified, some of which play a role in inflammation.

© Joe Sampson, Courtesy of Jennifer Portnick

Being active—even if overweight—is healthier than being sedentary. With a BMI of 36, aerobics instructor Jennifer Portnick is considered obese, but her daily workout routine helps to keep her in good health.

insulin resistance: the condition in which a normal amount of insulin produces a subnormal effect in muscle, adipose, and liver cells, resulting in an elevated fasting glucose; a metabolic consequence of obesity that precedes type 2 diabetes.

inflammation: an immunological response to cellular injury characterized by an increase in white blood cells.

the three leading risk factors (high LDL cholesterol, hypertension, and smoking) do. ♦ In addition to body fat and its distribution, weight gain also increases the risk of cardiovascular disease. Weight loss, on the other hand, can effectively lower both blood cholesterol and blood pressure in overweight and obese people. Of course, lean and normal-weight people may also have high blood cholesterol and blood pressure, and these factors are just as dangerous in lean people as in obese people.

Diabetes Most adults with type 2 diabetes are overweight or obese. Diabetes (type 2) is three times more likely to develop in an obese person than in a nonobese person. Furthermore, the person with type 2 diabetes often has central obesity. Central-body fat cells appear to be larger and more insulin-resistant than lower-body fat cells.[29] The association between **insulin resistance** and obesity is strong. Both are major risk factors for the development of type 2 diabetes.

Diabetes appears to be influenced by weight gains as well as by body weight. A weight gain of more than 10 pounds (4.5 kilograms) after the age of 18 doubles the risk of developing diabetes, even in women of average weight. In contrast, weight loss is effective in improving glucose tolerance and insulin resistance.[30]

Inflammation and the Metabolic Syndrome Chronic **inflammation** accompanies obesity, and inflammation contributes to chronic diseases.[31] As a person grows fatter, lipids first fill the adipose tissue and then migrate into other tissues such as the muscles and liver.[32] This accumulation of fat, especially in the abdominal region, changes the body's metabolism, resulting in insulin resistance, low HDL, high triglycerides, and high blood pressure.[33] This cluster of symptoms—collectively known as the metabolic syndrome—increases the risks for diabetes, hypertension, and atherosclerosis. ♦ Fat accumulation, especially in the abdominal region, activates genes that code for proteins ♦ involved in inflammation.[34] Furthermore, although relatively few immune cells are commonly found in adipose tissue, weight gain significantly increases their number and their role in inflammation.[35] Elevated blood lipids—whether due to obesity or to a high-fat diet—also promote inflammation.[36] Together, these factors help to explain why chronic inflammation accompanies obesity and how obesity contributes to the metabolic syndrome and the progression of chronic diseases.[37] Even in healthy youngsters, body fat correlates positively with chronic inflammation.[38] As might be expected, weight loss reduces the number of immune cells in adipose tissue and changes gene expression to reduce inflammation.[39]

Cancer The risk of some cancers increases with both body weight and weight gain, but researchers do not fully understand the relationships. One possible explanation may be that obese people have elevated levels of hormones that could influence cancer development. For example, adipose tissue is the major site of estrogen synthesis in women, obese women have elevated levels of estrogen, and estrogen has been implicated in the development of cancers of the female reproductive system—cancers that account for half of all cancers in women.

Fit and Fat versus Sedentary and Slim Importantly, BMI and weight gains and losses do not tell the whole story. Cardiorespiratory and muscular fitness play major roles in health and longevity, independently of body weight.[40] Normal-weight people who are fit have a lower risk of mortality than normal-weight people who are unfit. Furthermore, overweight but fit people have lower risks than normal-weight, unfit ones.[41] Fit people are also likely to gain less weight over the years.[42] Clearly, a healthy body weight is good, but it may not be good enough. Fitness, in and of itself, offers many health benefits. The next chapter explores weight management and the benefits of choosing nutritious foods and exercising regularly.

IN SUMMARY **The weight appropriate for an individual depends largely on factors specific to that individual, including body fat distribution, family health history, and current health status. At the extremes, both overweight and underweight carry clear risks to health.**

Nutrition Portfolio

When combined with fitness, a healthy body weight will help you to defend against chronic diseases.

Go to Diet Analysis Plus and choose one of the days on which you have tracked your diet for the entire day. Go to the Energy Balance report; use this report to help you answer the following questions:

- Describe how your daily food intake and physical activity balance with each other.
- What did the diet analysis program estimate as your daily energy requirement? What information was this based on?
- Describe any health risks that may be of concern for a person who continuously has very high "net kcalories" or very low "net kcalories" for many years?

Diet Analysis PLUS To complete this exercise, go to your Diet Analysis Plus at www.cengage.com/sso.

Nutrition on the Net

For further study of topics covered in this chapter, log on to **www.cengagebrain.com** and search for ISBN 084006845X.

- Obtain food composition data from the USDA Nutrient Data Laboratory: **www.ars.usda.gov/main/site_main.htm?modecode=12-35-45-00**
- Learn about the 10,000 Steps Program at Shape Up America: **www.shapeup.org**
- Visit the special web pages and interactive applications that Aim for a Healthy Weight: **www.nhlbi.nih.gov/health/public/heart/obesity/lose_wt**

References

1. G. A. Bray and C. M. Champagne, Beyond energy balance: There is more to obesity than kilocalories, *Journal of the American Dietetic Association* 105 (2005): S17–S23.
2. R. D. Mattes and coauthors, Appetite: Measurement and manipulation misgivings, *Journal of the American Dietetic Association* 105 (2005): S87–S97.
3. T. Rankinen and C. Bouchard, Genetics of food intake and eating behavior phenotypes in humans, *Annual Review of Nutrition* 26 (2006): 413–434.
4. C. D. Morrison and H. Berthoud, Neurobiology of nutrition and obesity, *Nutrition Reviews* 65 (2007): 517–534; M. J. Wolfgang and M. D. Lane, Control of energy homeostasis: Role of enzymes and intermediates of fatty acid metabolism in the central nervous system, *Annual Review of Nutrition* 26 (2006): 23–44.
5. T. C. Adam and E. S. Epel, Stress, eating and the reward system, *Physiology and Behavior* 91 (2007): 449–458; S. J. Torres and C. A. Nowson, Relationship between stress, eating behavior, and obesity, *Nutrition* 23 (2007): 887–894.
6. J. W. Apolzan and coauthors, Inadequate dietary protein increases hunger and desire to eat in young and older men, *Journal of Nutrition* 137 (2007): 1478–1482.
7. D. E. Cummings and J. Overduin, Gastrointestinal regulation of food intake, *Journal of Clinical Investigation* 117 (2007): 13–23.
8. E. Valassi, M. Scacchi, and F. Cavagnini, Neuroendocrine control of food intake, *Nutrition, Metabolism, and Cardiovascular Disease* 18 (2008): 158–168; Cummings and Overduin, 2007; K. G. Murphy, W. S. Dhillo, and S. R. Bloom, Gut peptides in the regulation of food intake and energy homeostasis, *Endocrine Reviews* 27 (2006): 719–727.
9. J. A. Harrold and J. C. Halford, The hypothalamus and obesity, *Recent Patents on CNS Drug Discovery* 1 (2006): 305–314.
10. N. Meunier and coauthors, Basal metabolic rate and thyroid hormones of late-middle-aged and older human subjects: The ZENITH study, *European Journal of Clinical Nutrition* 59 (2005): S53–S57.
11. A. J. Hill, Motivation for eating behaviour in adolescent girls: The body beautiful, *Proceedings of the Nutrition Society* 65 (2006): 376–384.
12. C. L. Ogden and coauthors, Prevalence of overweight and obesity in the United States, 1999–2004, *Journal of the American Medical Association* 295 (2006): 1549–1555.
13. K. A. Witt and E. A. Bush, College athletes with an elevated body mass index often have a high upper arm muscle area, but not elevated triceps and subscapular skinfolds, *Journal of the American Dietetic Association* 105 (2005): 599–602.
14. R. Huxley and coauthors, Ethnic comparisons of the cross-sectional relationships between measures of body size with diabetes and hypertension, *Obesity Reviews* 9 (2008): 53–61.
15. K. F. Adams and coauthors, Body mass and colorectal cancer risk in the NIH-AARP cohort, *American Journal of Epidemiology* 166 (2007): 36–45; G. Hu and coauthors, Body mass index, waist circumference, and waist-hip ratio on the risk of total and type-specific stroke, *Archives of Internal Medicine* 167 (2007): 1420–1427; C. D. Lee and coauthors, Abdominal obesity and coronary artery calcification in young adults: The Coronary Artery Risk Development in Young Adults (CARDIA) Study, *American Journal of Clinical Nutrition* (2007): 48–54; S. B. Votruba and M. D. Jensen, Regional fat deposition as a factor in FFA metabolism, *Annual Review of Nutrition* 27 (2007): 149–163; G. R. Dagenais and coauthors, Prognostic impact of body weight and abdominal obesity in women and men with cardiovascular disease, *American Heart Journal* 149 (2005): 54–60; Y. Wang and coauthors, Comparison of abdominal adiposity and overall obesity in predicting risk of type 2 diabetes among men, *American Journal of Clinical Nutrition* 81 (2005): 555–563.

16. D. Canoy and coauthors, Cigarette smoking and fat distribution in 21,828 British men and women: A population-based study, *Obesity Research* 13 (2005): 1466–1475.
17. J. P. Després and coauthors, Abdominal obesity and the metabolic syndrome: Contribution to global cardiometabolic risk, *Arteriosclerosis, Thrombosis, and Vascular Biology* 28 (2008): 1039–1049; S. Klein and coauthors, Waist circumference and cardiometabolic risk: A consensus statement from Shaping America's Health: Association for Weight Management and Obesity Prevention; NAASO, The Obesity Society; The American Society for Nutrition; and the American Diabetes Association, *American Journal of Clinical Nutrition* 85 (2007): 1197–1202; Wang and coauthors, 2005.
18. Position of the American Dietetic Association: Weight management, *Journal of the American Dietetic Association* 109 (2009): 330–346.
19. K. M. Flegal and B. I. Graubard, Estimates of excess deaths associated with body mass index and other anthropometric variables, *American Journal of Clinical Nutrition* 89 (2009): 1213–1219; D. M. Freedman and coauthors, Body mass index and all-cause mortality in a nationwide U.S. cohort, *International Journal of Obesity* 30 (2006): 822–829.
20. G. Whitlock and coauthors, Body-mass index and cause-specific mortality in 900000 adults: Collaborative analyses of 57 prospective studies, *Lancet* 373 (2009): 1083–1096; T. Pischon and coauthors, General and abdominal adiposity and risk of death in Europe, *New England Journal of Medicine* 359 (2008): 2105–2120.
21. K. M. Flegal and coauthors, Cause-specific excess deaths associated with underweight, overweight, and obesity, *Journal of the American Medical Association* 298 (2007): 2028–2037; C. L. Ogden and coauthors, The epidemiology of obesity, *Gastroenterology* 132 (2007): 2087–2102; G. M. Price and coauthors, Weight, shape, and mortality risk in older persons: Elevated waist-hip ratio, not high body mass index, is associated with a greater risk of death, *American Journal of Clinical Nutrition* 84 (2006): 449–460; K. M. Flegal and coauthors, Excess deaths associated with underweight, overweight, and obesity, *Journal of the American Medical Association* 293 (2005): 1861–1867; D. L. McGee, Body mass index and mortality: A meta-analysis based on person-level data from twenty-six observational studies, *Annals of Epidemiology* 15 (2005): 87–97.
22. Price and coauthors, 2006.
23. D. E. Alley and V. W. Chang, The changing relationship of obesity and disability, 1988–2004, *Journal of the American Medical Association* 298 (2007): 2020–2027.
24. K. D. Bertakis and R. Azari, The influence of obesity, alcohol abuse, and smoking on utilization of health care services, *Family Medicine* 38 (2006): 427–434.
25. R. P. Gelber and coauthors, Body mass index and mortality in men: Evaluating the shape of the association, *International Journal of Obesity* 31 (2007): 1240–1247.
26. A. Romero-Corral and coauthors, Association of bodyweight with total mortality and with cardiovascular events in coronary artery disease: A systematic review of cohort studies, *The Lancet* 368 (2006): 666–678; Flegal and coauthors, 2005.
27. Huxley and coauthors, 2008.
28. A. Scheinkiewitz and coauthors, Body mass index history and risk of type 2 diabetes: Results from the European Prospective Investigation into Cancer and Nutrition (EPIC)—Potsdam Study, *American Journal of Clinical Nutrition* 84 (2006): 427–433.
29. E. H. Livingston, Lower body subcutaneous fat accumulation and diabetes mellitus risk, *Surgery for Obesity and Related Diseases* 2 (2006): 362–368.
30. G. M. Reaven, The insulin resistance syndrome: Definition and dietary approaches to treatment, *Annual Review of Nutrition* 25 (2005): 391–406.
31. A. W. Fogarty and coauthors, A prospective study of weight change and systemic inflammation over 9 y, *American Journal of Clinical Nutrition* 87 (2008): 30–35; R. DeCaterina and coauthors, Nutritional mechanisms that influence cardiovascular disease, *American Journal of Clinical Nutrition* 83 (2006): 421S–426S.
32. E. N. Hansen, A. Torquati, and N. N. Abumrad, Results of bariatric surgery, *Annual Review of Nutrition* 26 (2006): 481–511.
33. S. L. Gray and A. J. Vidal-Puig, Adipose tissue expandability in the maintenance of metabolic homeostasis, *Nutrition Reviews* 65 (2007): S7–S12; C. S. Fox and coauthors, Abdominal visceral and subcutaneous adipose tissue compartments: Association with metabolic risk factors in the Framingham Heart Study, *Circulation* 116 (2007): 39–48; A. Pradhan, Obesity, metabolic syndrome, and type 2 diabetes: Inflammatory basis of glucose metabolic disorders, *Nutrition Reviews* 65 (2007): S152–S156; A. Tchernof, Visceral adipocytes and the metabolic syndrome, *Nutrition Reviews* 65 (2007): S24–S29; J. P. Despres, Is visceral obesity the cause of the metabolic syndrome, *Annals of Medicine* 38 (2006): 52–63.
34. P. Trayhurn, C. Bing, and I. S. Wood, Adipose tissue and adipokines—Energy regulation from the human perspective, *Journal of Nutrition* 136 (2006): 1935S–1939S.
35. A. H. Berg and P. E. Scherer, Adipose tissue, inflammation, and cardiovascular disease, *Circulation Research* 96 (2005): 939–968.
36. G. Boden, Fatty acid-induced inflammation and insulin resistance in skeletal muscle and liver, *Current Diabetes Reports* 6 (2006): 177–181.
37. P. Calabro and E. T. Yeh, Intra-abdominal adiposity, inflammation, and cardiovascular risk: New insight into global cardiometabolic risk, *Current Hypertension Reports* 10 (2008): 32–38; V. Z. Rocha and P. Libby, The multiple facets of the fat tissue, *Thyroid* 18 (2008): 175–183; D. C. W. Lau and coauthors, Adipokines: Molecular links between obesity and atherosclerosis, *American Journal of Physiology—Heart and Circulatory Physiology* 288 (2005): H2031–H2041.
38. A. Sbarbati and coauthors, Obesity and inflammation: Evidence for an elementary lesion, *Pediatrics* 117 (2006): 220–223; J. Warnberg and coauthors, Inflammatory proteins are related to total and abdominal adiposity in a healthy adolescent population: The AVENA Study, *American Journal of Clinical Nutrition* 84 (2006): 505–512.
39. J. P. Bastard and coauthors, Recent advances in the relationship between obesity, inflammation, and insulin resistance, *European Cytokine Network* 17 (2006): 4–12.
40. T. A. Lakka and D. E. Laaksonen, Physical activity in prevention and treatment of the metabolic syndrome, *Applied Physiology, Nutrition, and Metabolism* 32 (2007): 76–88; X. Sui and coauthors, Cardiorespiratory fitness and adiposity as mortality predictors in older adults, *Journal of the American Medical Association* 298 (2007): 2507–2516; R. D. Telford, Low physical activity and obesity: Causes of chronic disease or simply predictors? *Medicine & Science in Sports & Exercise* 39 (2007): 1233–1240.
41. R. P. Wildman and coauthors, The obese without cardiometabolic risk factor clustering and the normal weight with cardiometabolic risk factor clustering, *Archives of Internal Medicine* 168 (2008): 1617–1624.
42. C. Mason and coauthors, Musculoskeletal fitness and weight gain in Canada, *Medicine & Science in Sports & Exercise* 39 (2007): 38–43; P. T. Williams, Maintaining vigorous activity attenuates 7-yr weight gain in 8340 runners, *Medicine & Science in Sports & Exercise* 39 (2007): 801–809.

HIGHLIGHT 8

Eating Disorders

© BananaStock/SuperStock

For some people, the struggle with body weight manifests itself as an **eating disorder.** (The accompanying glossary defines this and related terms.) Three eating disorders—anorexia nervosa, bulimia nervosa, and binge eating disorder—are relatively uncommon, but present real concerns because of their health consequences. Findings from a large national survey suggest that 0.9 percent of women and 0.3 percent of men suffer from anorexia nervosa at some time in their lives.[1] Prevalence of bulimia nervosa is slightly higher, with 1.5 percent of women and 0.5 percent of men. Binge eating disorder is higher still, with 3.5 percent of women and 2 percent of men. Many more suffer from other unspecified conditions that, even though they do not meet the strict criteria for an eating disorder, imperil a person's well-being.

Why do so many people in our society suffer from eating disorders? Most experts agree that the causes include multiple factors: sociocultural, psychological, and perhaps neurochemical. Excessive pressure to be thin is at least partly to blame. Comments and criticisms about body weight and shape from family and friends can have lasting effects.[2] Young people may have learned to identify discomforts such as anger, jealousy, or disappointment with "feeling fat." They often have other psychological issues such as depression, anxiety, or substance abuse. As weight issues become more of a focus, psychological problems worsen, and the likelihood of developing eating disorders intensifies. Unfortunately, few seek health care for eating disorders.[3] Athletes are among those most likely to develop eating disorders.

The Female Athlete Triad

At age 14, Suzanne was a top contender for a spot on the state gymnastics team. Each day her coach reminded team members that they must weigh no more than their assigned weights to qualify for competition. The coach chastised gymnasts who gained weight, and Suzanne was terrified of being singled out. Convinced that the less she weighed the better she would perform, Suzanne weighed herself several times a day to confirm that she had not exceeded her 80-pound limit. Driven to excel in her sport, Suzanne kept her weight down by eating very little and training very hard. Unlike many of her friends, Suzanne never began to menstruate. A few months before her fifteenth birthday, Suzanne's coach dropped her back to the second-level team. Suzanne blamed her poor performance on a slow-healing stress fracture. Mentally stressed and physically exhausted, she quit gymnastics and began overeating between periods of self-starvation. Suzanne had developed the dangerous combination of problems that characterize the **female athlete triad**—disordered eating, amenorrhea, and osteoporosis (see Figure H8-1 on p. 262).[4]

Disordered Eating

Part of the reason many athletes engage in **disordered eating** behaviors may be that they and their coaches have embraced unsuitable weight standards. An athlete's body must be heavier for a given height than a nonathlete's body because the athlete's body is dense, containing more healthy bone and muscle and less fat. When athletes rely only on the scales, they may mistakenly believe

GLOSSARY

amenorrhea (ay-MEN-oh-REE-ah): the absence of or cessation of menstruation. *Primary amenorrhea* is menarche delayed beyond 16 years of age. *Secondary amenorrhea* is the absence of three to six consecutive menstrual cycles.

anorexia (an-oh-RECK-see-ah) **nervosa:** an eating disorder characterized by a refusal to maintain a minimally normal body weight and a distortion in perception of body shape and weight.

- **an** = without
- **orex** = mouth
- **nervos** = of nervous origin

binge-eating disorder: an eating disorder with criteria similar to those of bulimia nervosa, excluding purging or other compensatory behaviors.

bulimia (byoo-LEEM-ee-ah) **nervosa:** an eating disorder characterized by repeated episodes of binge eating usually followed by self-induced vomiting, misuse of laxatives or diuretics, fasting, or excessive exercise.

- **buli** = ox

cathartic (ka-THAR-tik): a strong laxative.

disordered eating: eating behaviors that are neither normal nor healthy, including restrained eating, fasting, binge eating, and purging.

eating disorders: disturbances in eating behavior that jeopardize a person's physical or psychological health.

emetic (em-ETT-ic): an agent that causes vomiting.

female athlete triad: a potentially fatal combination of three medical problems—disordered eating, amenorrhea, and osteoporosis.

muscle dysmorphia (dis-MORE-fee-ah): a psychiatric disorder characterized by a preoccupation with building body mass.

stress fractures: bone damage or breaks caused by stress on bone surfaces during exercise.

unspecified eating disorders: eating disorders that do not meet the defined criteria for specific eating disorders.

FIGURE H8-1 The Female Athlete Triad

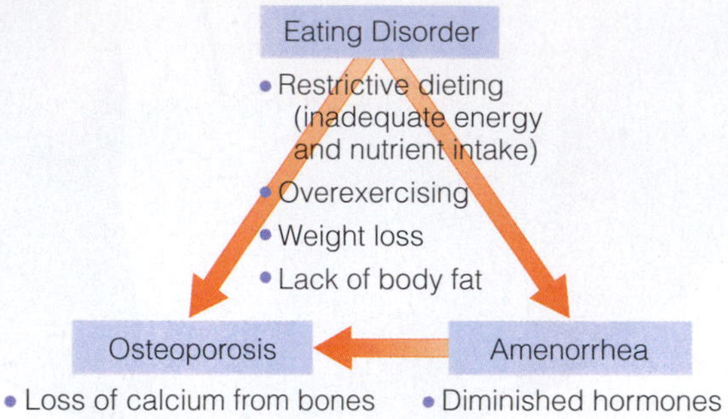

they are too fat because weight standards, such as the BMI, do not provide adequate information about body composition.

Many young athletes severely restrict energy intakes to improve performance, enhance the aesthetic appeal of their performance, or meet the weight guidelines of their specific sports. They fail to realize that the loss of lean tissue that accompanies energy restriction actually impairs their physical performance. The increasing incidence of abnormal eating habits among athletes is cause for concern. Male athletes, especially wrestlers and gymnasts, are affected by these disorders as well, but females are most vulnerable. Risk factors for eating disorders among athletes include:

- Young age (adolescence)
- Pressure to excel at a chosen sport
- Focus on achieving or maintaining an "ideal" body weight or body fat percentage
- Participation in sports or competitions that emphasize a lean appearance or judge performance on aesthetic appeal such as gymnastics, wrestling, figure skating, or dance[5]
- Weight-loss dieting at an early age
- Unsupervised dieting

© Reuters NewMedia Inc./Corbis

A few years ago, this Olympic gold medalist was weak and malnourished from anorexia nervosa. However, she recovered and set a world record in the cycling road race.

Amenorrhea

The prevalence of **amenorrhea** among premenopausal women in the United States is about 2 to 5 percent overall, but among female athletes, it may be as high as 66 percent. Contrary to previous notions, amenorrhea is *not* a normal adaptation to strenuous physical training: it is a symptom of something going wrong. Amenorrhea is characterized by low blood estrogen, infertility, and often bone mineral losses. Excessive training, depleted body fat, low body weight, and inadequate nutrition all contribute to amenorrhea. However amenorrhea develops, it threatens the integrity of the bones. Bone losses remain significant even after recovery. (Women with bulimia frequently have menstrual irregularities, but because they rarely cease menstruating, they may be spared this loss of bone integrity.)

Osteoporosis

For most people, weight-bearing physical activity, dietary calcium, and (for women) the hormone estrogen protect against the bone loss of osteoporosis. For young women with disordered eating and amenorrhea, strenuous activity can impair bone health.[6] Vigorous training combined with inadequate food intake disrupts metabolic and hormonal balances. These disturbances compromise bone health, greatly increasing the risks of **stress fractures.**[7] Stress fractures, a serious form of bone injury, commonly occur among dancers and other competitive athletes with amenorrhea, low calcium intakes, and disordered eating. Many underweight young athletes have bones like those of postmenopausal women, and they may never recover their lost bone even after diagnosis and treatment—which makes prevention critical. Young athletes should be encouraged to consume 1300 milligrams of calcium each day, to eat nutrient-dense foods, and to obtain enough energy to support both weight gain and the energy expended in physical activity. Nutrition is critical to bone recovery.[8]

Other Dangerous Practices of Athletes

Only females face the threats of the female athlete triad, of course, but many male athletes face pressure to achieve a certain body weight and may develop eating disorders.[9] Each week throughout the season, David drastically restricts his food and fluid intake before a wrestling match in an effort to "make weight." Wrestlers and their coaches believe that competing in a lower weight class will give them a competitive advantage over smaller opponents. To that end, David practices in rubber suits, sits in saunas, and takes diuretics and laxatives to lose 4 to 6 pounds. He hopes to replenish the lost fluids, glycogen, and lean tissue during the hours between his weigh-in and competition, but the body needs days to correct this metabolic mayhem. Reestablishing fluid and electrolyte balances may take a day or two, replenishing glycogen stores may take two to three days, and replacing lean tissue may take even longer.

Ironically, the combination of food deprivation and dehydration impairs physical performance by reducing muscle strength, decreasing anaerobic power, and reducing endurance capacity. For optimal performance, wrestlers need to first achieve their competitive weight during the off-season and then eat well-balanced meals and drink plenty of fluids during the competitive season.

Some athletes, usually males, go to extreme measures to bulk up and *gain* weight. People afflicted with **muscle dysmorphia** eat high-protein diets, take dietary supplements, weight train for hours at a time, and often abuse steroids in an attempt to increase muscle mass. Their bodies are large and muscular, yet they see themselves as puny 90-pound weaklings. They are preoccupied with the idea that their bodies are too small or inadequately muscular. Like others with distorted body images, people with muscle dysmorphia weigh themselves frequently and center their lives on diet and exercise. Paying attention to diet and pumping iron for fitness is admirable, but obsessing over it can cause serious social, occupational, and physical problems.

Preventing Eating Disorders in Athletes

To prevent eating disorders in athletes and dancers, the performers, their coaches, and their parents must learn about inappropriate body weight ideals, improper weight-loss techniques, eating disorder development, proper nutrition, and safe weight-control methods. Young people naturally search for identity and will often follow the advice of a person in authority without question. Therefore, coaches and dance instructors should never encourage unhealthy weight loss to qualify for competition or to conform to distorted artistic ideals. Athletes who truly need to lose weight should try to do so during the off-season and under the supervision of a health-care professional. Frequent weigh-ins can push young people who are striving to lose weight into a cycle of starving to confront the scale, then bingeing uncontrollably afterward. The erosion of self-esteem that accompanies these events can interfere with normal psychological development and set the stage for serious problems later on.

Table H8-1 includes suggestions to help athletes and dancers protect themselves against developing eating disorders. The remaining sections describe eating disorders that anyone, athlete or nonathlete, may experience.

Anorexia Nervosa

Julie, 18 years old, is a superachiever in school. She watches her diet with great care, and she exercises daily, maintaining a rigorous schedule of self-discipline. She is thin, but she is determined to lose more weight. She is 5 feet 6 inches tall and weighs 85 pounds (roughly 1.68 meters and 39 kilograms). She has **anorexia nervosa.**

Characteristics of Anorexia Nervosa

Julie is unaware that she is undernourished, and she sees no need to obtain treatment. She developed amenorrhea several months ago and has become moody and chronically depressed. She views normal healthy body weight as too fat and insists that she needs to lose weight, although her eyes are sunk in deep hollows in her face. Julie denies that she is ever tired, although she is close to physical exhaustion and no longer sleeps easily. Her family is concerned, and though reluctant to push her, they have finally insisted that she see a psychiatrist. Julie's psychiatrist has diagnosed anorexia nervosa (see Table H8-2 on p. 264) and prescribed group therapy as a start. If she does not begin to gain weight soon, she may need to be hospitalized.

Central to the diagnosis of anorexia nervosa is a distorted body image that overestimates personal body fatness. When Julie looks at herself in the mirror, she sees a "fat" 85-pound body. The more Julie overestimates her body size, the more resistant she is to treatment, and the more unwilling to examine her faulty values and misconceptions. In fact, she finds value in her condition.[10] Malnutrition is known to affect brain functioning and judgment in this way, causing lethargy, confusion, and delirium.

Anorexia nervosa cannot be self-diagnosed. Many people in our society are engaged in the pursuit of thinness, and denial runs high among people with anorexia nervosa. Some women have all the attitudes and behaviors associated with the condition, but without the dramatic weight loss.

How can a person as thin as Julie continue to starve herself? Julie uses tremendous discipline against her hunger to strictly limit her portions of low-fat, high-fiber, low-kcalorie foods.[11] She

TABLE H8-1 Tips for Combating Eating Disorders

General Guidelines

- Never restrict food amounts to below those suggested for adequacy by the USDA Food Guide (see Table 2-2).
- Eat frequently. Include healthy snacks between meals. The person who eats frequently never gets so hungry as to allow hunger to dictate food choices.
- If not at a healthy weight, establish a reasonable weight goal based on a healthy body composition.
- Allow a reasonable time to achieve the goal. A reasonable loss of excess fat can be achieved at the rate of about 10 percent of body weight in six months.
- Establish a weight-maintenance support group with people who share interests.

Specific Guidelines for Athletes and Dancers

- Replace weight-based goals with performance-based goals.
- Restrict weight-loss activities to the off-season.
- Remember that eating disorders impair physical performance. Seek confidential help in obtaining treatment if needed.
- Focus on proper nutrition as an important facet of your training, as important as proper technique.

TABLE H8-2 Criteria for Diagnosis of Anorexia Nervosa

A person with anorexia nervosa demonstrates the following:

A. Refusal to maintain body weight at or above a minimal normal weight for age and height (e.g., weight loss leading to maintenance of body weight less than 85 percent of that expected; or failure to make expected weight gain during period of growth, leading to body weight less than 85 percent of that expected).

B. Intense fear of gaining weight or becoming fat, even though underweight.

C. Disturbance in the way in which one's body weight or shape is experienced, undue influence of body weight or shape on self-evaluation, or denial of the seriousness of the current low body weight.

D. In females past puberty, amenorrhea—the absence of at least three consecutive menstrual cycles. (A woman is considered to have amenorrhea if her periods occur only following hormone, e.g., estrogen, administration.)

Two types:

- *Restricting type:* During the episode of anorexia nervosa, the person does not regularly engage in binge eating or purging behavior (i.e., self-induced vomiting or the misuse of laxatives, diuretics, or enemas).
- *Binge eating/purging type:* During the episode of anorexia nervosa, the person regularly engages in binge eating or purging behavior (i.e., self-induced vomiting or the misuse of laxatives, diuretics, or enemas).

SOURCE: Reprinted with permission from American Psychiatric Association, *Diagnostic and Statistical Manual of Mental Disorders*, 4th ed. Text Revision (Washington, D.C.: American Psychiatric Association, 2000).

will deny her hunger, and having adapted to so little food, she feels full after eating only a half-dozen carrot sticks. She knows the kcalorie intake of various foods and the kcalorie expenditure of different exercises. If she feels that she has gained an ounce of weight, she runs or jumps rope until she is sure she has exercised it off. If she fears that the food she has eaten outweighs the exercise, she may take laxatives to hasten the passage of food from her system. She drinks water incessantly to fill her stomach, risking dangerous mineral imbalances. She is desperately hungry. In fact, she is starving, but she doesn't eat because her need for self-control dominates.

Many people, on learning of this disorder, say they wish they had "a touch" of it to get thin. They mistakenly think that people with anorexia nervosa feel no hunger. They also fail to recognize the pain of the associated psychological and physical trauma.

The starvation of anorexia nervosa damages the body just as the starvation of war and poverty does. In fact, after a few months, most people with anorexia nervosa have protein-energy malnutrition (PEM) that is similar to marasmus (described in Chapter 6). Their bodies have been depleted of both body fat and protein. Victims are dying to be thin—quite literally. In young people, growth ceases and normal development falters. They lose so much lean tissue that basal metabolic rate slows. In addition, the heart pumps inefficiently and irregularly, the heart muscle becomes weak and thin, the chambers diminish in size, and the blood pressure falls. Minerals that help to regulate heartbeat become unbalanced. Many deaths occur due to multiple organ system failure when the heart, kidneys, and liver cease to function.

Starvation brings other physical consequences as well, such as loss of brain tissue, impaired immune response, anemia, and a loss of digestive functions that worsens malnutrition. Peristalsis becomes sluggish, the stomach empties slowly, and the lining of the intestinal tract atrophies. The pancreas slows its production of digestive enzymes. The deteriorated GI tract fails to provide sufficient digestive enzymes and absorptive surfaces for handling any food that is eaten. The person may suffer from diarrhea, further worsening malnutrition.

Other effects of starvation include altered blood lipids, high blood vitamin A and vitamin E, low blood proteins, dry thin skin, abnormal nerve functioning, reduced bone density, low body temperature, low blood pressure, and the development of fine body hair (the body's attempt to keep warm). The electrical activity of the brain becomes abnormal, and insomnia is common. Both women and men lose their sex drives.

Women with anorexia nervosa develop amenorrhea. (It is one of the diagnostic criteria.) In young girls, the onset of menstruation is delayed. Menstrual periods typically resume with recovery, although some women never restart even after they have gained weight. Should an underweight woman with anorexia nervosa become pregnant, she is likely to give birth to an underweight baby—and low-birthweight babies face many health problems (as Chapter 14 explains). Mothers with anorexia nervosa may underfeed their children, who then fail to grow and may also suffer the other consequences of starvation.

Treatment of Anorexia Nervosa

Treatment of anorexia nervosa requires a multidisciplinary approach.[12] Teams of physicians, nurses, psychiatrists, family therapists, and dietitians work together to resolve two sets of issues and behaviors: those relating to food and weight and those involving relationships with oneself and others. The first dietary objective is to stop weight loss while establishing regular eating patterns. Appropriate diet is crucial to recovery and must be tailored to each individual's needs. Because body weight is low and fear of weight gain is high, initial food intake may be small—perhaps only 1200 kcalories per day.[13] A variety of foods and foods with a higher energy density help to ensure greater success.[14] As eating becomes more comfortable, clients should gradually increase energy intake. Initially, clients may be unwilling to eat for themselves. Those who do eat will have a good chance of recovering without additional interventions. Even after recovery, however, energy intakes and eating behaviors may not fully return to normal.[15] Furthermore, weight gains may be slow because energy needs may be slightly elevated due to anxiety, abdominal pain, and cigarette smoking.

Because anorexia nervosa is like starvation physically, health-care professionals classify clients based on indicators of PEM.* Low-risk clients need nutrition counseling. Intermediate-risk cli-

*Indicators of protein-energy malnutrition: a low percentage of body fat, low serum albumin, low serum transferrin, and impaired immune reactions.

© David Young-Wolff/Alamy

People with anorexia nervosa see themselves as fat, even when they are dangerously underweight.

ents may need supplements such as high-kcalorie, high-protein formulas in addition to regular meals. High-risk clients may require hospitalization and may need to be fed by tube at first to prevent death.[16] This step may cause psychological trauma. Although drugs are commonly prescribed, they play a limited role in treatment.

Denial runs high among those with anorexia nervosa. Few seek treatment on their own. About half of the women who are treated can maintain their body weight at 85 percent or more of a healthy weight, and at that weight, many of them may begin menstruating again. The other half have poor to fair treatment outcomes, relapse into abnormal eating behaviors, or die. Anorexia nervosa has one of the highest mortality rates among psychiatric disorders—most commonly from cardiac complications or by suicide.[17]

Before drawing conclusions about someone who is extremely thin or who eats very little, remember that diagnosis requires professional assessment. Several national organizations offer information for people who are seeking help with anorexia nervosa, either for themselves or for others.*

*Internet sites are listed at the end of this highlight.

Bulimia Nervosa

Kelly is a charming, intelligent, 30-year-old flight attendant of normal weight who thinks constantly about food. She alternates between starving herself and secretly bingeing, and when she has eaten too much, she makes herself vomit. Most readers recognize these symptoms as those of **bulimia nervosa.**

Characteristics of Bulimia Nervosa

Bulimia nervosa is distinct from anorexia nervosa and is more prevalent, although the true incidence is difficult to establish because bulimia nervosa is not as physically apparent. More men suffer from bulimia nervosa than from anorexia nervosa, but bulimia nervosa is still more common in women than in men. The secretive nature of bulimic behaviors makes recognition of the problem difficult, but once it is recognized, diagnosis is based on the criteria listed in Table H8-3.

Like the typical person with bulimia nervosa, Kelly is single, female, and white. She is well educated and close to her ideal body weight, although her weight fluctuates over a range of 10 pounds or so every few weeks. She prefers to weigh less than the weight that her body maintains naturally.

Kelly seldom lets her eating disorder interfere with work or other activities, although a third of all bulimics do. From early childhood, she has been a high achiever and emotionally dependent on her parents. As a young teen, Kelly frequently followed severely

TABLE H8-3 Criteria for Diagnosis of Bulimia Nervosa

A person with bulimia nervosa demonstrates the following:

A. Recurrent episodes of binge eating. An episode of binge eating is characterized by both of the following:
   1. Eating, in a discrete period of time (e.g., within any two-hour period), an amount of food that is definitely larger than most people would eat during a similar period of time and under similar circumstances.
   2. A sense of lack of control over eating during the episode (e.g., a feeling that one cannot stop eating or control what or how much one is eating).

B. Recurrent inappropriate compensatory behavior to prevent weight gain, such as self-induced vomiting; misuse of laxatives, diuretics, enemas, or other medications; fasting; or excessive exercise.

C. Binge eating and inappropriate compensatory behaviors both occur, on average, at least twice a week for three months.

D. Self-evaluation unduly influenced by body shape and weight.

E. The disturbance does not occur exclusively during episodes of anorexia nervosa.

Two types:

- *Purging type:* The person regularly engages in self-induced vomiting or the misuse of laxatives, diuretics, or enemas.
- *Nonpurging type:* The person uses other inappropriate compensatory behaviors, such as fasting or excessive exercise, but does not regularly engage in self-induced vomiting or the misuse of laxatives, diuretics, or enemas.

SOURCE: Reprinted with permission from American Psychiatric Association, *Diagnostic and Statistical Manual of Mental Disorders,* 4th ed. Text Revision (Washington, D.C.: American Psychiatric Association, 2000).

restricted diets but could never maintain the weight loss. Kelly feels anxious at social events and cannot easily establish close personal relationships. She is usually depressed, is often impulsive, and has low self-esteem. When crisis hits, Kelly responds by replaying events, worrying excessively, and blaming herself but never asking for help—behaviors that interfere with effective coping.

Like the person with anorexia nervosa, the person with bulimia nervosa spends much time thinking about body weight and food. The preoccupation with food manifests itself in secret binge-eating episodes, which usually progress through several emotional stages: anticipation and planning, anxiety, urgency to begin, rapid and uncontrollable consumption of food, relief and relaxation, disappointment, and finally shame or disgust.

A bulimic binge is characterized by a sense of lacking control over eating. During a binge, the person consumes food for its emotional comfort and cannot stop eating or control what or how much is eaten. A typical binge occurs periodically, in secret, usually at night, and lasts an hour or more. Because a binge frequently follows a period of rigid dieting, eating is accelerated by intense hunger. Energy restriction followed by bingeing can set in motion a pattern of weight cycling, which may make weight loss and maintenance more difficult over time.

During a binge, Kelly consumes thousands of kcalories of easy-to-eat, low-fiber, high-fat, and, especially, high-carbohydrate foods. Typically, she chooses cookies, cakes, and ice cream—and she eats the entire bag of cookies, the whole cake, and every last spoonful in a carton of ice cream. After the binge, Kelly pays the price with swollen hands and feet, bloating, fatigue, headache, nausea, and pain.

To purge the food from her body, Kelly may use a **cathartic**—a strong laxative that can injure the lower intestinal tract. Or she may induce vomiting, with or without the use of an **emetic**—a drug intended as first aid for poisoning. These purging behaviors are often accompanied by feelings of shame or guilt. Hence a vicious cycle develops: negative self-perceptions followed by dieting, bingeing, and purging, which in turn lead to negative self-perceptions (see Figure H8-2).

On first glance, purging seems to offer a quick and easy solution to the problems of unwanted kcalories and body weight. Many people perceive such behavior as neutral or even positive, when, in fact, binge eating and purging have serious physical consequences. Signs of subclinical malnutrition are evident in a compromised immune system. Fluid and mineral imbalances caused by vomiting or diarrhea can lead to abnormal heart rhythms and injury to the kidneys. Urinary tract infections can lead to kidney failure. Vomiting causes irritation and infection of the pharynx, esophagus, and salivary glands; erosion of the teeth; and dental caries. The esophagus may rupture or tear, as may the stomach. Sometimes the eyes become red from pressure during vomiting. The hands may be calloused or cut by the teeth while inducing vomiting. Overuse of emetics depletes potassium concentrations and can lead to death by heart failure.

FIGURE H8-2 The Vicious Cycle of Restrictive Dieting and Binge Eating

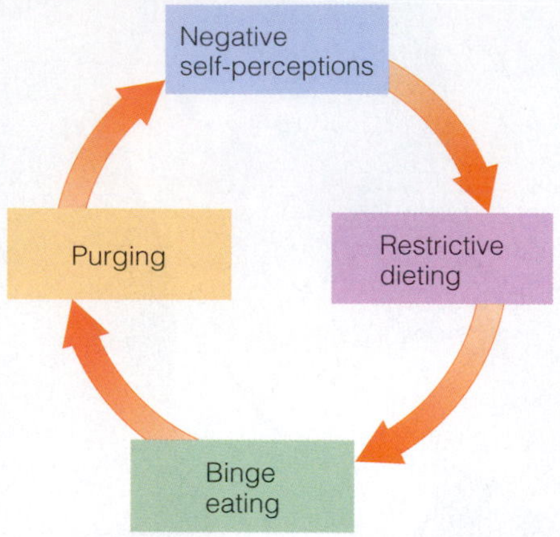

Unlike Julie, Kelly is aware that her behavior is abnormal, and she is deeply ashamed of it. She wants to recover, and this makes recovery more likely for her than for Julie, who clings to denial. Feeling inadequate ("I can't even control my eating"), Kelly tends to be passive and to look to others for confirmation of her sense of worth. When she experiences rejection, either in reality or in her imagination, her bulimia nervosa becomes worse. If Kelly's depression deepens, she may seek solace in drug or alcohol abuse or in other addictive behaviors. Clinical depression is common in people with bulimia nervosa, and the rates of substance abuse are high.

Treatment of Bulimia Nervosa

Kelly needs to establish regular eating patterns. She may also benefit from a regular exercise program. Weight maintenance, rather than cyclic weight gains and losses, is the treatment goal. Major steps toward recovery include discontinuing purging and restrictive dieting habits and learning to eat three meals a day plus snacks. Initially, energy intake should provide enough food to satisfy hunger and maintain body weight. Table H8-4 offers diet strategies to correct the eating problems of bulimia nervosa. Most women diagnosed with bulimia nervosa recover within five to ten years, with or without treatment, but treatment probably speeds the recovery process.

A mental health professional should be on the treatment team to help clients with their depression and addictive behaviors. Some physicians prescribe the antidepressant drug fluoxetine in the treatment of bulimia nervosa.* Another drug that may be useful in the management of bulimia nervosa is naloxone, an opiate antagonist that suppresses the consumption of sweet and high-fat foods in binge eaters.

Anorexia nervosa and bulimia nervosa are distinct eating disorders, yet they sometimes overlap in important ways. Anorexia victims may purge, and victims of both disorders may be overly concerned with body weight and have a tendency to drastically

*Fluoxetine is marketed under the trade name Prozac.

TABLE H8-4 Diet Strategies for Combating Bulimia Nervosa

Planning Principles

- Plan meals and snacks; record plans in a food diary prior to eating.
- Plan meals and snacks that require eating at the table and using utensils.
- Refrain from finger foods.
- Refrain from "dieting" or skipping meals.

Nutrition Principles

- Eat a well-balanced diet and regularly timed meals consisting of a variety of foods.
- Include raw vegetables, salad, or raw fruit at meals to prolong eating times.
- Choose whole-grain, high-fiber breads, pasta, rice, and cereals to increase bulk.
- Consume adequate fluid, particularly water.

Other Tips

- Choose foods that provide protein and fat for satiety and bulky, fiber-rich carbohydrates for immediate feelings of fullness.
- Try including soups and other water-rich foods for satiety.
- Choose portions that meet the definition of "a serving" according to the Daily Food Guide (pp. 40–41).
- For convenience (and to reduce temptation) select foods that naturally divide into portions. Select one potato, rather than rice or pasta that can be overloaded onto the plate; purchase yogurt and cottage cheese in individual containers; look for small packages of precut steak or chicken; choose frozen dinners with measured portions.
- Include 30 minutes of physical activity every day—exercise may be an important tool in defeating bulimia.

undereat. Many perceive foods as "forbidden" and "give in" to an eating binge. The two disorders can also appear in the same person, or one can lead to the other. Treatment is challenging and relapses are not unusual. Other people have **unspecified eating disorders** that fall short of the criteria for anorexia nervosa or bulimia nervosa but share some of their features. One such condition is binge-eating disorder.

Binge-Eating Disorder

Charlie is a 40-year-old schoolteacher who has been overweight all his life. His friends and family are forever encouraging him to lose weight, and he has come to believe that if he only had more willpower, dieting would work. He periodically gives dieting his best shot—restricting energy intake for a day or two only to succumb to uncontrollable cravings, especially for high-fat foods. Like Charlie, up to half of the obese people who try to lose weight periodically binge; unlike people with bulimia nervosa, however, they typically do not purge. Such an eating disorder does not meet the criteria for either anorexia nervosa or bulimia nervosa—yet such compulsive overeating is a problem and occurs in people of normal weight as well as those who are severely overweight. Table H8-5 lists criteria for unspecified eating disorders, including binge eating. Obesity alone is not an eating disorder.

TABLE H8-5 Unspecified Eating Disorders, Including Binge-Eating Disorder

Criteria for Diagnosis of Unspecified Eating Disorders, in General

Many people have eating disorders but do not meet all the criteria to be classified as having anorexia nervosa or bulimia nervosa. Some examples include those who:

A. Meet all of the criteria for anorexia nervosa, except irregular menses.
B. Meet all of the criteria for anorexia nervosa, except that their current weights fall within the normal ranges.
C. Meet all of the criteria for bulimia nervosa, except that binges occur less frequently than stated in the criteria.
D. Are of normal body weight and who compensate inappropriately for eating small amounts of food (example: self-induced vomiting after eating two cookies).
E. Repeatedly chew food but spit it out without swallowing.
F. Have recurrent episodes of binge eating but do not compensate as do those with bulimia nervosa.

Criteria for Diagnosis of Binge-Eating Disorder, Specifically

A person with a binge-eating disorder demonstrates the following:

A. Recurrent episodes of binge eating. An episode of binge eating is characterized by both of the following:
   1. Eating, in a discrete period of time (e.g., within any two-hour period) an amount of food that is definitely larger than most people would eat in a similar period of time under similar circumstances.
   2. A sense of lack of control over eating during the episode (e.g., a feeling that one cannot stop eating or control what or how much one is eating).
B. Binge-eating episodes are associated with at least three of the following:
   1. Eating much more rapidly than normal.
   2. Eating until feeling uncomfortably full.
   3. Eating large amounts of food when not feeling physically hungry.
   4. Eating alone because of being embarrassed by how much one is eating.
   5. Feeling disgusted with oneself, depressed, or very guilty after overeating.
C. The binge eating causes marked distress.
D. The binge eating occurs, on average, at least twice a week for six months.
E. The binge eating is not associated with the regular use of inappropriate compensatory behaviors (e.g., purging, fasting, excessive exercise) and does not occur exclusively during the course of anorexia nervosa or bulimia nervosa.

SOURCE: Reprinted with permission from American Psychiatric Association, *Diagnostic and Statistical Manual of Mental Disorders,* 4th ed. Text Revision (Washington, D.C.: American Psychiatric Association, 2000).

Clinicians note differences between people with bulimia nervosa and those with binge-eating disorder. People with **binge-eating disorder** consume less during a binge, rarely purge, and exert less restraint during times of dieting. Similarities also exist, including feeling out of control, disgusted, depressed, embarrassed, guilty, or distressed because of their self-perceived gluttony.

There are also differences between obese binge eaters and obese people who do not binge. Those with the binge-eating disorder report higher rates of self-loathing, disgust about body size, depression, and anxiety. Their eating habits differ as well. Obese binge eaters tend to consume more kcalories and more dessert and snack-type foods during regular meals and binges than obese people who do not binge.

Binge eating is a behavioral disorder that can be resolved with treatment. Even a simple Internet-based treatment program can help.[18] Reducing binge eating makes participation in weight-control programs easier. It also improves physical health, mental health, and the chances of success in breaking the cycle of rapid weight losses and gains.

Eating Disorders in Society

Society plays a central role in eating disorders. Adolescent girls who read magazine articles on dieting and weight loss are likely to engage in unhealthy eating habits.[19] Further proof of society's influence is found in the demographic distribution of eating disorders—they are known only in developed nations, and they become more prevalent as wealth increases and food becomes plentiful. Some people point to the vomitoriums of ancient times and claim that bulimia nervosa is not new, but the two are actually distinct. Ancient people were eating for pleasure, without guilt, and in the company of others; they vomited so that they could rejoin the feast. Bulimia nervosa is a disorder of isolation and is often accompanied by low self-esteem.

Chapter 8 describes how our society sets unrealistic ideals for body weight, especially in women, and devalues those who do not conform to them. Anorexia nervosa and bulimia nervosa are not a form of rebellion against these unreasonable expectations, but rather an exaggerated acceptance of them. In fact, body dissatisfaction is a primary factor in the development of eating disorders. Not everyone who is dissatisfied will develop an eating disorder, but everyone with an eating disorder is dissatisfied.

Characteristics of disordered eating such as restrained eating, fasting, binge eating, purging, fear of fatness, and distortion of body image are extraordinarily common among young girls. Most are "on diets," and many are poorly nourished. Some eat too little food to support normal growth; thus they miss out on their adolescent growth spurts and may never catch up. Many eat so little that hunger propels them into binge-purge cycles.

Perhaps a person's best defense against these disorders is to learn to appreciate his or her own uniqueness. When people discover and honor their body's real physical needs, they become unwilling to sacrifice health for conformity. To respect and value oneself may be lifesaving.

Nutrition on the Net

For further study of topics covered in this highlight, log on to **www.cengagebrain.com** and search for ISBN 084006845X.

- Find resources and referrals from the National Association of Anorexia Nervosa and Associated Disorders: **www.anad.org**
- Learn more about eating disorders from the National Eating Disorders Association: **www.nationaleatingdisorders.org**
- Get facts about eating disorders from the National Institute of Mental Health: **www.nimh.nih.gov/health/topics/eating-disorders/index.shtml**
- Take the Eating Attitudes Test to determine if you might need to seek medical advice regarding an eating disorder: **psychcentral.com/quizzes/eat.htm**

Be aware that some websites encourage disordered eating behaviors and may have a negative impact.

References

1. J. I. Hudson and coauthors, The prevalence and correlates of eating disorders in the National Comorbidity Survey Replication, *Biological Psychiatry* 61 (2007): 348–358.
2. C. B. Taylor and coauthors, The adverse effect of negative comments about weight and shape from family and siblings on women at high risk for eating disorders, *Pediatrics* 118 (2006): 731–738.
3. H. W. Hoek, Incidence, prevalence and mortality of anorexia nervosa and other eating disorders, *Current Opinion in Psychiatry* 19 (2006): 389–394.
4. American College of Sports Medicine, Position stand: The female athlete triad, *Medicine & Science in Sports & Exercise* 39 (2007): 1867–1882; C. M. Lebrun, The female athlete triad: What's a doctor to do? *Current Sports Medicine Reports* 6 (2007): 397–404.
5. M. F. Reinking and L. E. Alexander, Prevalence of disordered-eating behaviors in undergraduate female collegiate athletes and nonathletes, *Journal of Athletic Training* 40 (2005): 47–51; M. K. Torstveit and J. Sundgot-Borgen, The female athlete triad: Are elite athletes at increased risk? *Medicine & Science in Sports & Exercise* 37 (2005): 184–193.
6. M. T. Barrack and coauthors, Dietary restraint and low bone mass in female adolescent endurance runners, *American Journal of Clinical Nutrition* 87 (2008): 36–43.
7. J. L. Kelsey and coauthors, Risk factors for stress fracture among young female cross-country runners, *Medicine and Science in Sports and Exercise* 39 (2007): 1457–1463.
8. J. Dominguez and coauthors, Treatment of anorexia nervosa is associated with increases in bone mineral density, and recovery is a biphasic process involving both nutrition and return of menses, *American Journal of Clinical Nutrition* 86 (2007): 92–99.
9. M. Vertalino and coauthors, Participation in weight-related sports is associated with higher use of unhealthful weight-control behaviors and steroid use, *Journal of the American Dietetic Association* 107 (2007): 434–440.

10. R. H. S. Nordbø and coauthors, The meaning of self-starvation: Qualitative study of patients' perception of anorexia nervosa, *International Journal of Eating Disorders* 39 (2006): 556–564.
11. M. Misra and coauthors, Nutrient intake in community-dwelling adolescent girls with anorexia nervosa and in healthy adolescents, *American Journal of Clinical Nutrition* 84 (2006): 698–706.
12. Position of the American Dietetic Association: Nutrition intervention in the treatment of anorexia nervosa, bulimia nervosa, and other eating disorders, *Journal of the American Dietetic Association* 106 (2006): 2073–2082.
13. J. Yager and A. E. Andersen, Anorexia nervosa, *New England Journal of Medicine* 353 (2005): 1481–1488.
14. J. E. Schebendach and coauthors, Dietary energy density and diet variety as predictors of outcome in anorexia nervosa, *American Journal of Clinical Nutrition* 87 (2008): 810–816.
15. R. Sysko and coauthors, Eating behavior among women with anorexia nervosa, *American Journal of Clinical Nutrition* 82 (2005): 296–301.
16. E. Attia and B. T. Walsh, Behavioral management for anorexia nervosa, *New England Journal of Medicine* 360 (2009): 500–506.
17. J. M. Holm-Denoma and coauthors, Deaths by suicide among individuals with anorexia as arbiters between competing explanations of the anorexia-suicide link, *Journal of Affective Disorders* 107 (2008): 231–236; G. Murialdo and coauthors, Alterations in the autonomic control of heart rate variability in patients with anorexia or bulimia nervosa: Correlations between sympathovagal activity, clinical features, and leptin levels, *Journal of Endocrinological Investigation* 30 (2007): 356–362; D. Casiero and W. H. Frishman, Cardiovascular complications of eating disorders, *Cardiology in Review* 14 (2006): 227–231; M. Pompili and coauthors, Suicide and attempted suicide in eating disorders, obesity and weight-image concern, *Eating Behaviors* 7 (2006): 384–394.
18. M. Jones and coauthors, Randomized, controlled trial of an Internet-facilitated intervention for reducing binge eating and overweight in adolescents, *Pediatrics* 121 (2008): 453–462.
19. P. Van den Berg and coauthors, Is dieting advice from magazines helpful or harmful? Five-year associations with weight-control behaviors and psychological outcomes in adolescents, *Pediatrics* 119 (2007): e30–e37.

CHAPTER

9

© Masterfile

Nutrition in Your Life

Are you pleased with your body weight? If so, you are a rare individual. Most people in our society think they should weigh more or less (mostly less) than they do. Usually, their primary concern is appearance, but they often understand that physical health is also somehow related to body weight. One does not necessarily cause the other—that is, an ideal body weight does not ensure good health. Instead, both depend on diet and physical activity. A well-balanced diet and active lifestyle support good health—and help maintain body weight within a reasonable range.

Throughout this chapter, the CourseMate icon indicates an opportunity for online self-study, linking you to activities to increase your understanding of chapter concepts. **www.cengagebrain.com** (search for ISBN 084006845X)

Weight Management: Overweight, Obesity, and Underweight

CHAPTER OUTLINE

The previous chapter described how body weight is stable when energy in equals energy out. Weight gains occur when energy intake exceeds energy expended, and conversely, weight losses occur when energy expended exceeds energy intake. At the extremes, both overweight and underweight present health risks. **Weight management** is a key component of good health.

This chapter emphasizes overweight and obesity, partly because they have been more intensively studied and partly because they represent a major health problem in the United States and a growing concern worldwide. Information on underweight is presented at the end of the chapter. The highlight that follows this chapter examines fad diet plans.

Overweight and Obesity

Despite our preoccupation with body image and weight loss, the prevalence of overweight and obesity in the United States continues to rise dramatically.[1] In the past two decades, obesity increased in every state, in both genders, and across all ages, races, and educational levels (see Figure 9-1 on p. 272). An estimated 66 percent of the adults in the United States are now considered overweight or obese, as defined by a BMI of 25 or greater.[2] ♦ The prevalence of overweight is especially high among women, the poor, blacks, and Hispanics.

The prevalence of overweight among children in the United States has also risen at an alarming rate. An estimated 33 percent of children and adolescents aged 2 to 19 years are either overweight or obese.[3] Chapter and Highlight 15 present information on overweight during childhood and adolescence.

♦ BMI:
- Underweight: <18.5
- Healthy weight: 18.5–24.9
- Overweight: 25.0–29.9
- Obese: ≥30

weight management: maintaining body weight in a healthy range by preventing gradual weight gain over time and losing weight if overweight.

FIGURE 9-1 Increasing Prevalence of Obesity (BMI ≥ 30) among U.S. Adults

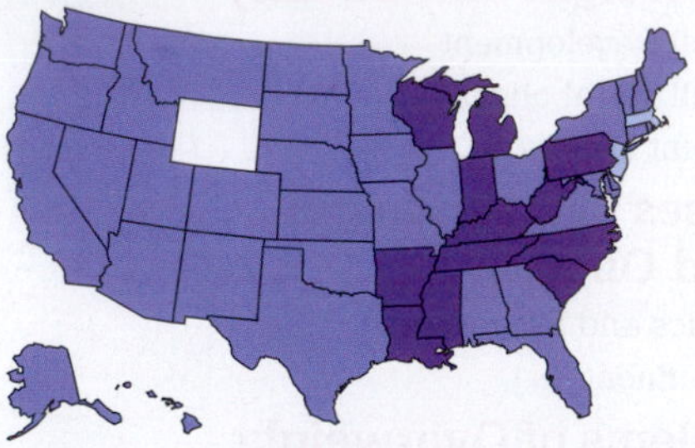

1993: Most states had prevalence rates less than 15 percent, with a couple reporting rates less than 10 percent; no state had prevalence rates greater than or equal to 20 percent.

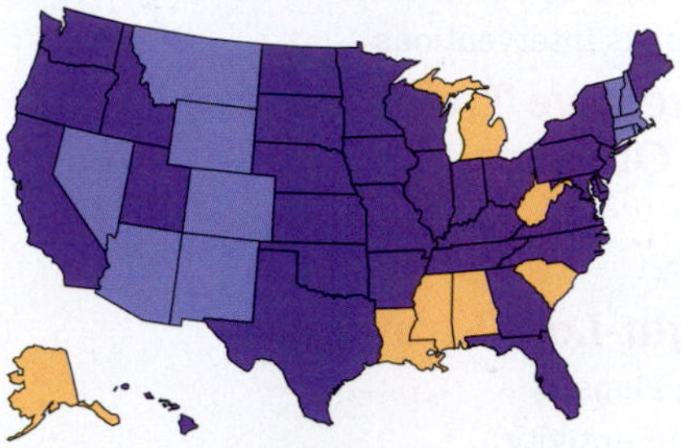

1998: Most states had prevalence rates less than 20 percent, with none reporting rates less than 10 percent; seven states had prevalence rates greater than or equal to 20 percent.

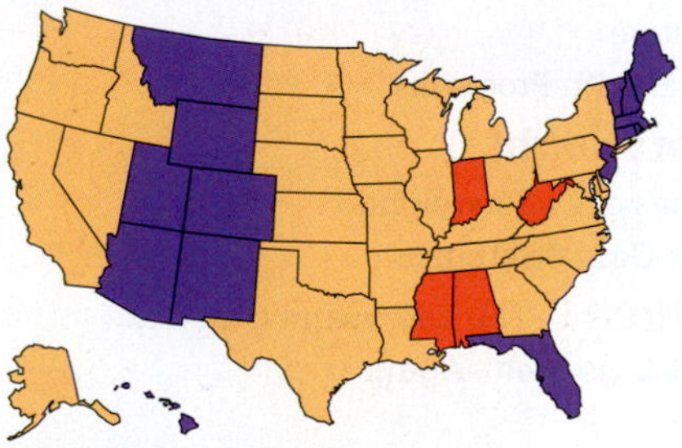

2003: More than half the states had prevalence rates greater than 20 percent, with four states reporting prevalence rates greater than or equal to 25 percent.

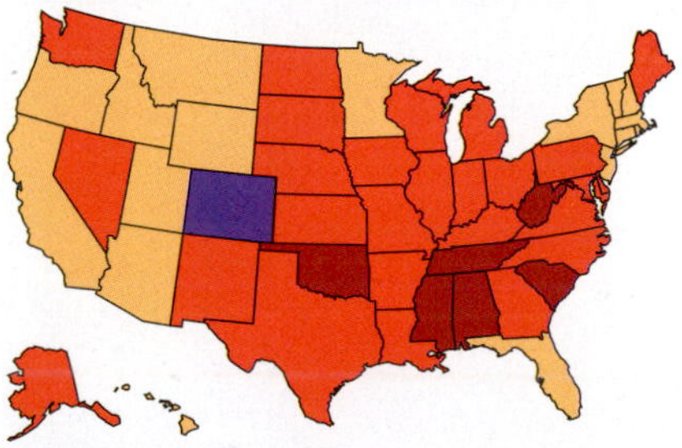

2008: Only one state had prevalence rates less than 20 percent; more than half the states had prevalence rates greater than 25 percent, with six states reporting prevalence rates greater than or equal to 30 percent.

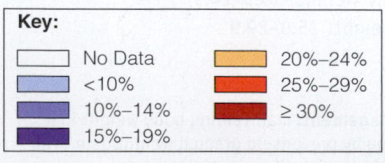

Key:
- No Data
- <10%
- 10%–14%
- 15%–19%
- 20%–24%
- 25%–29%
- ≥ 30%

SOURCE: www.cdc.nccdphp/dnpa/obesity/trend/maps/index.htm

Obesity in the United States is widespread. Prevalence increased rapidly over the past four decades, but seems to have leveled out in the past year or so.[4] This **epidemic** of obesity has spread worldwide, affecting more than 300 million adults and 155 million children.[5] Contrary to popular opinion, obesity is not limited to industrialized nations; more than 115 million people in developing countries suffer from obesity-related problems. Before examining the suspected causes of obesity and the various strategies used to treat it, it is helpful to understand the development and metabolism of body fat.

Fat Cell Development

When "energy in" exceeds "energy out," much of the excess energy is stored in the fat cells of adipose tissue. The amount of fat in a person's body reflects both the *number* and the *size* of the fat cells. The number of fat cells increases most rapidly during the growing years of late childhood and early puberty. After growth ceases, fat cell number may continue to increase whenever energy balance is positive. Obese people have more fat cells than healthy-weight people; their fat cells are also larger.

As fat cells accumulate triglycerides, they expand in size (review Figure 5-20 on p. 148). When the cells enlarge, they stimulate cell proliferation so that their numbers increase again. Thus obesity develops ♦ when a person's fat cells increase in number, in size, or quite often both. Figure 9-2 illustrates fat cell development.

When "energy out" exceeds "energy in," the size of fat cells dwindles, but not their number. People with extra fat cells tend to regain lost weight rapidly; with weight gain, their many fat cells readily fill. In contrast, people with an average number of enlarged fat cells may be more successful in maintaining weight losses; when their cells shrink, both cell size and number are normal. Prevention of obesity is most critical, then, during the growing years of childhood and adolescence when fat cells increase in number.[6] Researchers are exploring ways to induce fat cell death—which would decrease the number.[7] ♦

As mentioned, excess fat first fills the body's natural storage site—adipose tissue. Then it becomes necessary to deposit excess fat in organs such as the heart or liver, which plays a key role in the development of diseases such as heart failure or fatty liver.[8] ♦ Metabolic changes such as insulin resistance become apparent, and chronic inflammation develops as adipose tissue produces its adipokines.[9]

Fat Cell Metabolism

The enzyme lipoprotein lipase (LPL) ♦ removes triglycerides from the blood for storage in both adipose tissue and muscle cells. Obese people generally have much more LPL activity in their adipose cells than lean people do (their muscle cell LPL activity is similar, though). This high LPL activity makes fat storage especially efficient. Consequently, even modest excesses in energy intake have a more dramatic impact on obese people than on lean people.

The activity of LPL in different regions of the body is partially influenced by gender.[10] In women, fat cells in the breasts, hips, and thighs produce abundant LPL, putting fat away in those body sites; in men, fat cells in the abdomen produce abundant LPL. This enzyme activity explains why men tend to develop central obesity around the abdomen (apple-shaped) whereas women more readily develop lower-body fat around the hips and thighs (pear-shaped).

Gender differences are also apparent in the activity of the enzymes ♦ controlling the release and breakdown of fat in various parts of the body. The release of lower-body fat is less active in women than in men, whereas the release of upper-body fat is similar. Furthermore, the rate of fat breakdown is lower in women than in men. Consequently, women may have a more difficult time losing fat in general, and from the hips and thighs in particular.

Enzyme activity may also explain why some people who lose weight regain it so easily. After weight loss, adipose LPL activity increases.[11] Apparently, weight loss serves as a signal to the gene that produces the LPL enzyme, saying "Make more of the enzyme that stores fat." People easily regain weight after having lost it because they are battling against enzymes that want to store fat. Not only is fat storage efficient, but fat oxidation is not. Dietary fat oxidation correlates negatively with

FIGURE 9-2 Fat Cell Development

Fat cells are capable of increasing their size by 20-fold and their number by several thousandfold.

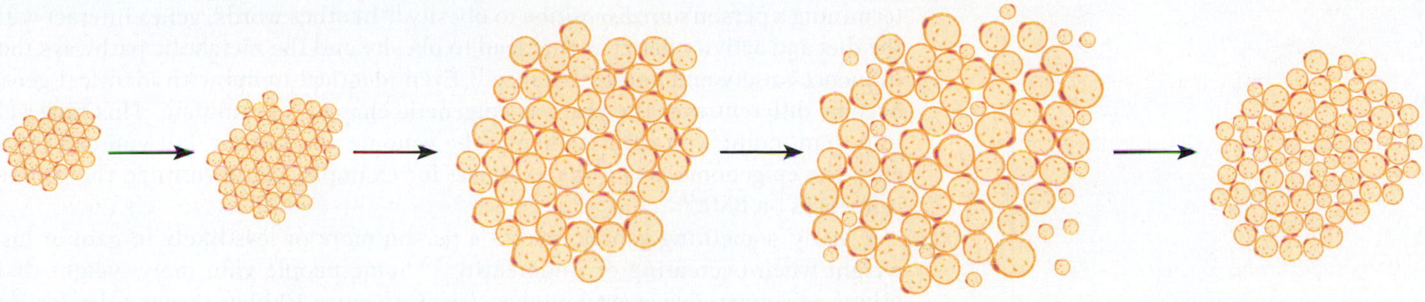

body fatness: obese people have the least activity.[12] The activities of these and other proteins provide an explanation for the observation that some inner mechanism seems to set a person's weight or body composition at a fixed point; the body will adjust to restore that **set point** if the person tries to change it.

During growth, fat cells increase in number.
When energy intake exceeds expenditure, fat cells increase in size.
When fat cells have enlarged and energy intake continues to exceed energy expenditure, fat cells increase in number again.
With fat loss, the size of the fat cells shrinks but not the number.

Set-Point Theory

Many internal physiological variables, such as blood glucose, blood pH, and body temperature, remain fairly stable under a variety of conditions. The hypothalamus and other regulatory centers constantly monitor and delicately adjust conditions to maintain homeostasis. The stability of such complex systems may depend on set-point regulators that maintain variables within specified limits.

Researchers have confirmed that after weight gains or losses, the body adjusts its metabolism to restore the original weight. Energy expenditure increases after weight gain and decreases after weight loss. These changes in energy expenditure differ from those that would be expected based on body composition alone, and they help to explain why it is so difficult for an underweight person to maintain weight gains and an overweight person to maintain weight losses.[13]

IN SUMMARY Fat cells develop by increasing in number and size. Prevention of excess weight gain depends on maintaining a reasonable number of fat cells. With weight gains or losses, the body adjusts in an attempt to return to its previous status.

◆ Obesity due to an increase in the *number* of fat cells is **hyperplastic obesity.** Obesity due to an increase in the *size* of fat cells is **hypertrophic obesity.**

◆ Cell death is known as **apoptosis.**

◆ The adverse effects of fat in nonadipose tissues are known as **lipotoxicity.**

◆ **Lipoprotein lipase (LPL)** is an enzyme that hydrolyzes triglycerides passing by in the bloodstream and directs their parts into the cells, where they can be metabolized or reassembled for storage.

◆ Enzymes involved in the breakdown of fat include **hormone-sensitive lipase** and **adipose tissue lipase.**

Causes of Overweight and Obesity

Why do people accumulate excess body fat? The obvious answer is that they take in more food energy than they expend. But that answer falls short of explaining why they do this. Is it genetic? Environmental? Cultural? Behavioral? Socioeconomic? Psychological? Metabolic? All of these? Most likely, obesity has many interrelated causes. Why an imbalance between energy intake and energy expenditure occurs remains a bit of a mystery; the next sections summarize possible explanations.

Genetics and Epigenetics

Genetics plays a true causative role in relatively few cases of obesity, for example, in Prader-Willi syndrome—a genetic disorder characterized by excessive appetite, massive obesity, short stature, and often mental retardation. Most cases of obesity, however, do not stem from a single gene, yet genetic influences do seem to be involved. Highlight 6 describes epigenetics—the influence of environmental factors ◆ on gene expression. Obesity provides a classic example.[14]

◆ Environmental factors include diet and physical activity.

epidemic (ep-ih-DEM-ick): the appearance of a disease (usually infectious) or condition that attacks many people at the same time in the same region.
- **epi** = upon
- **demos** = people

set point: the point at which controls are set (for example, on a thermostat). The set-point theory that relates to body weight proposes that the body tends to maintain a certain weight by means of its own internal controls.

Researchers have found that adopted children tend to be more similar in weight to their biological parents than to their adoptive parents. Studies of twins yield similar findings: compared with fraternal twins, identical twins are twice as likely to weigh the same.[15] These findings suggest an important role for genetics in determining a person's *predisposition* to obesity.[16] In other words, genes interact with the diet and activity patterns that lead to obesity and the metabolic pathways that influence satiety and energy balance.[17] Even identical twins with identical genes become different over the years as epigenetic changes accumulate. This raises an important point: you cannot change the genome you inherit, but you can influence the epigenome. Vigorous exercise, for example, can minimize the genetic influences on BMI.[18]

Clearly, something genetic makes a person more or less likely to gain or lose weight when overeating or undereating.[19] Some people gain more weight than others on comparable energy intakes. Given an extra 1000 kcalories a day for 100 days, some pairs of identical twins gain less than 10 pounds while others gain up to 30 pounds. Within each pair, the amounts of weight gained, percentages of body fat, and locations of fat deposits are similar. Similarly, some people lose more weight than others following comparable exercise routines.

Researchers have been examining the human genome in search of genetic and epigenetic answers to obesity questions.[20] As the section on protein synthesis in Chapter 6 describes, each cell expresses only the genes for the proteins it needs, and each protein performs a unique function. The following paragraphs describe some recent research involving proteins that might help explain appetite control, energy regulation, and obesity development.

Leptin Researchers have identified an obesity gene, called *ob,* that is expressed primarily in the adipose tissue and codes for the protein **leptin**. Leptin acts as a hormone, primarily in the hypothalamus. Research suggests that leptin from adipose tissue signals sufficient energy stores and promotes a negative energy balance by suppressing appetite and increasing energy expenditure.[21] Changes in energy expenditure primarily reflect changes in basal metabolism but may also include changes in physical activity patterns. Leptin is also released from stomach cells in response to the presence of food, suggesting a role for both short-term and long-term regulation of food intake and energy storage.[22]

Mice with a defective *ob* gene do not produce leptin and can weigh up to three times as much as normal mice and have five times as much body fat (see Figure 9-3). When injected with a synthetic form of leptin, the mice rapidly lose body fat. (Because leptin is a protein, it would be destroyed during digestion if given

FIGURE 9-3 Mice with and without Leptin Compared

Both of these mice have a defective *ob* gene. Consequently, they do not produce leptin. They both became obese, but the one on the right received daily injections of leptin, which suppressed food intake and increased energy expenditure, resulting in weight loss.

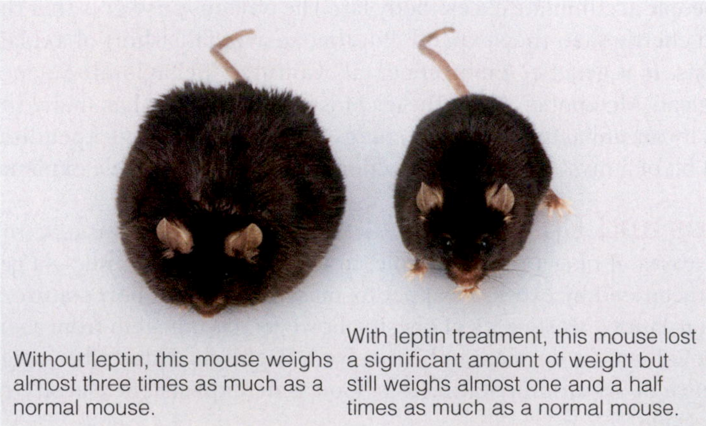
Without leptin, this mouse weighs almost three times as much as a normal mouse.
With leptin treatment, this mouse lost a significant amount of weight but still weighs almost one and a half times as much as a normal mouse.

leptin: a protein produced by fat cells under direction of the *ob* gene that decreases appetite and increases energy expenditure.
- **leptos** = thin

orally; consequently, it must be given by injection.) The fat cells not only lose fat, but they self-destruct (reducing cell number), which may explain why weight gains are delayed when the mice are fed again.

Although extremely rare, a genetic deficiency of leptin or genetic mutation of its receptor has been identified in human beings as well.[23] Extremely obese children with barely detectable blood levels of leptin have little appetite control; they are constantly hungry and eat considerably more than their siblings or peers. Given daily injections of leptin, these children lose a substantial amount of weight, confirming leptin's role in regulating appetite and body weight.

Not too surprisingly, leptin injections are effective in suppressing appetite and supporting weight loss only when overeating and obesity are the result of a leptin deficiency. Very few obese people have a leptin deficiency, however. In fact, leptin levels increase as BMI increases.[24] Leptin rises but fails to suppress appetite or enhance energy expenditure—a condition researchers describe as leptin resistance.[25] Interestingly, excessive fructose consumption seems to induce leptin resistance and accelerate fat storage.[26]

Some researchers have reexamined the evidence on leptin from another point of view—one of undernutrition. Instead of focusing on leptin's role as a satiety signal that might help prevent obesity by diminishing appetite, they view leptin as a starvation hormone that signals energy deficits. When energy intake is low, leptin levels decline, and metabolism slows in an effort to reduce energy demands. Clearly, leptin plays a major role in energy regulation, but additional research is needed to clarify its actions when intake is either excessive or deficient.

In addition to its involvement in energy regulation, leptin plays several other roles in the body. For example, leptin may inform the female reproductive system about body fat reserves; stimulate growth of new blood vessels, especially in the cornea of the eye; enhance the maturation of bone marrow cells; promote formation of red blood cells; and help support a normal immune response. Elevated leptin levels may be partially responsible for the early maturation that commonly occurs in obese children.

Adiponectin In addition to leptin, adipose tissue secretes another protein known as **adiponectin**. Unlike leptin, however, adiponectin correlates inversely with body fat: lean people have higher amounts than obese people—which helps to explain some of the relationships between obesity and diseases.[27] Adiponectin seems to have the beneficial effects of inhibiting inflammation and protecting against insulin resistance, type 2 diabetes, and cardiovascular disease.[28] Researchers are hopeful that if they can find ways to raise adiponectin levels or enhance its activity, they will be able to reduce the disease risks associated with obesity.[29]

Ghrelin Another protein, known as **ghrelin**, also acts as a hormone primarily in the hypothalamus.[30] In contrast to leptin or adiponectin, ghrelin is secreted primarily by the stomach cells and promotes a positive energy balance by stimulating appetite and promoting efficient energy storage.[31] The role ghrelin plays in regulating food intake and body weight is currently the subject of much intense research.[32]

Ghrelin triggers the desire to eat. Blood levels of ghrelin typically rise before and fall after a meal in proportion to the kcalories ingested—reflecting the hunger and satiety that precede and follow eating. In general, fasting blood levels correlate inversely with body weight: lean people have high ghrelin levels and obese people have low levels.[33]

Ghrelin fights to maintain a stable body weight. In fact, some researchers speculate that its role is to maximize fat stores during times of famine.[34] On average, ghrelin levels are high whenever the body is in negative energy balance, as occurs during low-kcalorie diets, for example. This response may help explain why weight loss is so difficult to maintain. Weight loss is more successful with exercise and after gastric bypass surgery, in part because ghrelin levels are relatively low.[35] Ghrelin levels decline again whenever the body is in positive energy balance, as occurs with weight gains.

Like leptin, ghrelin plays roles in the body beyond energy regulation. In fact, it was first recognized for its participation in growth hormone activity. Some research

adiponectin: a protein produced by the fat cells that inhibits inflammation and protects against insulin resistance, type 2 diabetes, and cardiovascular disease.

ghrelin (GRELL-in): a protein produced by the stomach cells that enhances appetite and decreases energy expenditure.

- **ghre** = growth

TABLE 9-1 Proteins Involved in Regulation of Food Intake and Energy Homeostasis

Protein	Concentration	Secreted from	Action
Adiponectin	Lower in obesity	Adipose tissue	Increases insulin sensitivity
Ghrelin	Increases with fasting Decreases after a meal	Stomach	Stimulates appetite
Leptin	Higher in obesity	Adipose tissue	Suppresses appetite Increases energy expenditure
Oxyntomodulin	Increases after a meal	Central nervous system GI tract	Suppresses appetite
Pancreatic peptide (PP)	Increases after a meal	Pancreas	Suppresses appetite
PYY	Lower in obesity Increases after a meal	Small intestine	Suppresses appetite
Resistin	Higher in obesity	Adipose tissue, bone marrow, and immune system cells	Provides short-term satiety Opposes insulin
Visfatin	Higher in obesity	Adipose tissue (specifically visceral)	Mimics glucose-lowering effects of insulin

SOURCES: Adapted from S. S. Gropper, J. L. Smith, and J. L. Groff, *Advanced Nutrition and Human Metabolism,* 5th ed. (Belmont, Calif.: Thomson Cengage, 2009) p. 299; M. H. Rokling-Andersen and coauthors, Effects of long-term exercise and diet intervention on plasma adipokine concentrations, *American Journal of Clinical Nutrition* 86 (2007): 1293–1301; H. Xie and coauthors, Insulin-like effects of visfatin on human osteoblasts, *Calcified Tissue International* 80 (2007): 201–210; M. E. Shills and coauthors, *Modern Nutrition in Health and Disease,* 10th ed. (Philadelphia: Lippincott Williams and Wilkins, 2006); S. Tovar and coauthors, Central administration of resistin promots short-term satiety in rats: A review, *European Journal of Endocrinology* 153 (2005): R1–R5; J. Berndt and coauthors, Plasma visfatin concentrations and fat depot-specific mRNA expression in humans, *Diabetes* 54 (2005): 2911–2916.

also indicates that ghrelin promotes sleep. Interestingly, a lack of sleep increases the hunger hormone ghrelin and decreases the satiety hormone leptin—which may help to explain epidemiological evidence finding an association between short sleep duration and high BMI.[36] Researchers are trying to understand the relationships among genes, sleep disorders, eating habits, and other related factors that may influence body weight and weight gain.[37]

PYY Ghrelin levels also decline in response to high levels of PYY, a peptide that the GI cells secrete after a meal in proportion to the kcalories ingested. In one study, people who were given PYY and then offered buffet meals consumed 30 percent fewer kcalories in the day than the control group. Like the hormone leptin, PYY signals satiety and decreases food intake, but unlike leptin, PYY may be an effective treatment for obesity. An ideal diet would maintain the satiating hormones (leptin, PYY, and cholecystokinin) and minimize the appetite stimulating hormone (ghrelin); fortunately, the diet that seems to do that best is one that is low in fat and rich in fiber.[38]

Table 9-1 summarizes the actions of the proteins just described. It also includes a few more to illustrate the many complex factors involved in the regulation of food intake and energy homeostasis.

Uncoupling Proteins Genes also code for proteins involved in energy metabolism. These proteins may influence the storing or expending of energy with different efficiencies or in different types of fat. The body has two types of fat: white and **brown adipose tissue.**[39] White adipose tissue stores fat for other cells to use for energy; brown adipose tissue releases stored energy as heat. Recall from Chapter 7 that when fat is oxidized, some of the energy is released in heat and some is captured in ATP. In brown adipose tissue, oxidation is uncoupled ♦ from ATP formation, producing heat only.[40] By radiating energy away as heat, the body expends, rather than stores, energy. In contrast, efficient coupling leads to fat storage.[41] In other words, weight gains or losses may depend on whether the body dissipates the energy from an ice cream sundae as heat or stores it in body fat.

♦ In **coupled reactions,** the energy released from the breakdown of one compound is used to create a bond in the formation of another compound. In **uncoupled reactions,** the energy is released as heat.

Brown fat and heat production is particularly important in newborns and in animals exposed to cold weather, especially those that hibernate.[42] They have plenty of brown adipose tissue. In contrast, most human adults have little brown fat—less than 1 percent of all fat cells and interspersed among the white fat cells.[43] Brown fat activity is most apparent during exposure to cold.[44] Importantly, brown fat quantity is inversely related with BMI; overweight and obese individuals have

brown adipose tissue: masses of specialized fat cells packed with pigmented mitochondria that produce heat instead of ATP.

less brown fat activity than others.[45] The role of brown fat in body weight regulation is not yet understood, but such an understanding may prove most useful in developing obesity treatments.[46]

Uncoupling proteins are active not only in brown fat, but also in white fat and many other tissues. Their actions seem to influence the basal metabolic rate (BMR) and oppose the development of obesity. Animals with abundant amounts of these uncoupling proteins resist weight gain, whereas those with minimal amounts gain weight easily. Similarly, people with a genetic variant of an uncoupling protein have lower metabolic rates and are more overweight than others.

Environment

With obesity rates rising and the **gene pool** remaining relatively unchanged, environment must also play a role in obesity. Obesity reflects the interactions between genes and the environment.[47] The *environment* includes all of the circumstances that we encounter daily that push us toward fatness or thinness. Over the past four decades, the demand for physical activity has decreased as the abundance of food has increased.[48]

Keep in mind that genetic and environmental factors are not mutually exclusive; in fact, their *interactions* create the epigenetics that provide a greater understanding of obesity and related diseases.[49] Genes can influence eating behaviors, for example, and food and activity behaviors influence the genes that regulate body weight. Interestingly, even social relationships can influence the development of obesity.[50] The likelihood that a person will become obese increases when a friend, sibling, or spouse becomes obese.

Overeating

One explanation for obesity is that overweight people overeat, although diet histories may not always reflect high intakes. Diet histories are not always accurate records of actual intakes; both normal-weight and obese people commonly misreport their dietary intakes.[51] Most importantly, current dietary intakes may not reflect the eating habits that led to obesity. Obese people who had a positive energy balance for years and accumulated excess body fat may not currently have a positive energy balance. This reality highlights an important point: the energy-balance equation must consider time. Both present *and* past eating and activity patterns influence current body weight.

We live in an environment that exposes us to an abundance of high-kcalorie, high-fat foods that are readily available, relatively inexpensive, heavily advertised, ♦ and reasonably tasty. Food is available everywhere, all the time—thanks largely to fast food. Our highways are lined with fast-food restaurants, and convenience stores and service stations offer fast food as well. Fast food is available in our schools, malls, and airports. It's convenient and it's available morning, noon, and night—and all times in between.

♦ The food industry spends billions of dollars a year on advertising. The message? "Eat more."

Most alarming are the extraordinarily large serving sizes and ready-to-go meals that offer supersize ♦ combinations. People buy the large sizes and combinations, perceiving them to be a good value, but then they eat more than they need—a bad deal. In fact, one research study calculated that for the 67 cents extra to upsize a meal, consumers receive an extra 400 kcalories, an extra 36 grams of body fat, and an extra $1 to $7 in health-care costs.[52]

♦ "Want fries with that?" A supersize portion delivers more than 600 kcalories.

Large package or portion sizes can increase consumption—even when the food is not particularly appealing. Moviegoers given stale popcorn ate more when eating from a huge container than from a large container (both sizes were greater than anyone could finish).[53] Simply put, large portion sizes deliver more kcalories. And portion sizes of virtually all foods and beverages have increased markedly in the past several decades, most notably at fast-food restaurants. Not only have portion sizes increased over time, but they are now two to eight times larger than standard serving sizes. The trend toward large portion sizes parallels the increasing prevalence of overweight and obesity in the United States, beginning in the 1970s, increasing sharply in the 1980s, and continuing today.

Restaurant food, especially fast food, contributes significantly to the development of obesity. Fast food is often energy-dense food, which increases energy intake, BMI, and body fatness.[54] The combination of large portions and energy-dense

gene pool: all the genetic information of a population at a given time.

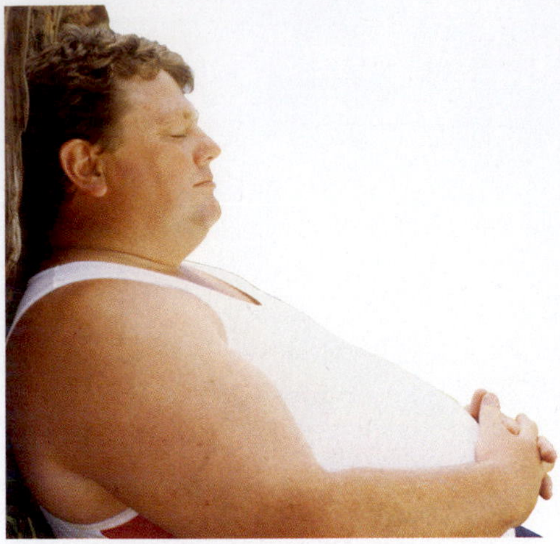

Lack of physical activity fosters obesity.

foods is a double whammy.[55] Reducing portion sizes is somewhat helpful, but the real kcalorie savings come from lowering the energy density. After all, large portions of foods with low energy density such as lean meats, fruits, and vegetables can help with weight loss. Unfortunately, low-energy-dense foods tend to be more expensive and less convenient than energy-dense foods.[56] The financial interests of the food industry do not always align with consumer health goals.[57] Consumers' health would benefit from restaurants providing appropriate portion sizes and offering more fruits, vegetables, legumes, and whole grains. Restaurant portion size decisions, however, are based on the chefs' plate presentations, food costs, and customer expectations—not customer health needs.[58]

Physical Inactivity Our environment fosters physical inactivity as well.[59] Life requires little exertion—escalators carry us up stairs, automobiles take us across town, buttons roll down windows, and remote controls change television channels from a distance. Modern technology has replaced physical activity at home, at work, and in transportation. Inactivity contributes to weight gain and poor health. In turn, watching television, playing video games, and using the computer may contribute most to physical inactivity. The more time people spend in these sedentary activities, the more likely they are to be overweight.

Sedentary activities contribute to weight gain in several ways. First, they require little energy beyond the resting metabolic rate. Second, they replace time spent in more vigorous activities. Third, watching television influences food purchases and correlates with between-meal snacking on the high-kcalorie, high-fat foods most heavily advertised.

Some obese people are so extraordinarily inactive that even when they eat less than lean people, they still have an energy surplus. Reducing their food intake further would incur nutrient deficiencies and jeopardize health. Physical activity is a necessary component of nutritional health. People must be physically active if they are to eat enough food to deliver all the nutrients they need without unhealthy weight gain. In fact, *to prevent weight gain,* the DRI ♦ suggests an accumulation of 60 minutes of moderately intense physical activities every day in addition to the less intense activities of daily living. Recommendations *to lose weight* encourage even greater duration, intensity, or frequency of physical activity (as a later section of the chapter discusses).

♦ DRI for physical activity: 60 min/day (moderate intensity)

People may be obese, therefore, not because they eat too much, but because they move too little—both in purposeful exercise and in the activities of daily life. Studies report that the differences in the time obese and lean people spent lying, sitting, standing, and moving accounts for about 350 kcalories a day.[60] In general, lean people tend to be more spontaneously active in their occupations and their leisure time.[61] The energy expended ♦ in these everyday spontaneous activities plays a pivotal role in energy balance and weight management.[62]

♦ The energy expenditure associated with everyday spontaneous activities is called **nonexercise activity thermogenesis (NEAT).**

IN SUMMARY **Obesity has many causes and different combinations of causes in different people. Some causes, such as overeating and physical inactivity, may be within a person's control, and some, such as genetics, may be beyond it.**

Problems of Overweight and Obesity

An estimated 59 percent of U.S. adults are trying to lose weight at any given time.[63] Some of these people do not even need to lose weight. Others may benefit from weight loss, but they are not successful. Relatively few people succeed in losing weight, and even fewer succeed permanently. Whether a person needs to lose weight is a question of health.

Health Risks

Chapter 8 describes some of the health problems that commonly accompany obesity. In evaluating the risks to health from obesity, health-care professionals use three indicators:[64]

- Body mass index ♦ (BMI, as Chapter 8 describes)
- Waist circumference ♦ (as Chapter 8 describes)
- Disease risk profile

Importantly, the disease risk profile takes into account life-threatening diseases, family history, and common risk factors for chronic diseases (such as blood lipid profile).[65] The higher the BMI, the greater the waist circumference, and the more risk factors—the greater the urgency to treat obesity.

People can best decide whether weight loss might be beneficial by considering their health status and motivation. People who are overweight by BMI standards, but otherwise in good health, might not benefit from losing weight; they might focus on preventing further weight gains instead. In contrast, those who are obese and suffering from a life-threatening disease such as diabetes might improve their health substantially by adopting a diet and exercise plan that supports weight loss. Motivation is a key component; to lose weight, a person needs to be ready and willing to make lifestyle changes for a lifetime.

Overweight in Good Health Often a person's motivations for weight loss have nothing to do with health. A healthy young woman with a BMI of 26 ♦ might want to lose a few pounds for spring break, but doing so might not improve her health. In fact, if she opts for a starvation diet or diet pills, she would be healthier *not* trying to lose weight.

Obese or Overweight with Risk Factors Weight loss is recommended for people who are obese and those who are overweight (or who have a high waist circumference) with two or more risk factors for chronic diseases. ♦ A 50-year-old man with a BMI of 28 ♦ who has high blood pressure and a family history of heart disease can improve his health by adopting a diet low in saturated fat and a regular exercise plan.

Obese or Overweight with Life-Threatening Condition Weight loss is also recommended for a person who is either obese or overweight and suffering from a life-threatening condition such as heart disease, diabetes, or sleep apnea. ♦ The health benefits of weight loss are clear. For example, a 30-year-old man with a BMI of 40 ♦ might be able to prevent or control diabetes by losing 75 pounds. Although the effort required to do so may be great, it may be no greater than the effort and consequences of living with diabetes.

Perceptions and Prejudices

Many people assume that every obese person can achieve slenderness and should pursue that goal. First consider that most obese people do not—for whatever reason—successfully lose weight and maintain their losses. Then consider the prejudice involved in that assumption. People come with varying weight tendencies, just as they come with varying potentials for height and degrees of health, yet we do not expect tall people to shrink or healthy people to get sick in an effort to become "normal."

Social Consequences Large segments of our society place such enormous value on thinness that obese people face prejudice and discrimination on the job, at school, and in social situations: they are judged on their appearance more than on their character. Socially, obese people are stereotyped as lazy and lacking in self-control. Such a critical view of overweight is not prevalent in many other cultures, including segments of our own society. Instead, overweight is simply accepted or even embraced as a sign of robust health and beauty. Many overweight people today are tired of the focus on weight control and simply want to be accepted as they are. To free society of its obsession with body weight and prejudice against obesity, people must first learn to judge others for who they are and not for what they weigh.

Psychological Problems Psychologically, obese people may suffer embarrassment when others treat them with hostility and contempt, and some have even

♦ BMI 25.0–29.9 = overweight
BMI ≥30 = obese

♦ Men: >40 in (>102 cm)
Women: >35 in (>88 cm)

♦ For reference, a woman with a BMI of 26 might be:
- 5 ft 3 in, 146 lb (1.60 m, 66.2 kg)
- 5 ft 5 in, 156 lb (1.65 m, 70.8 kg)
- 5 ft 7 in, 166 lb (1.70 m, 75.3 kg)

♦ Obese people and overweight people with two or more of these risk factors require aggressive treatment:
- Hypertension
- Cigarette smoking
- High LDL
- Low HDL
- Impaired glucose tolerance
- Family history of heart disease
- Men ≥45 yr; women ≥55 yr

♦ For reference, a man with a BMI of 28 might be:
- 5 ft 8 in, 184 lb (1.73 m, 83.5 kg)
- 5 ft 10 in, 195 lb (1.78 m, 88.5 kg)
- 6 ft, 206 lb (1.83 m, 93.4 kg)

♦ Obese people and overweight people with any of these diseases require aggressive treatment:
- Heart disease
- Diabetes (type 2)
- Sleep apnea (a disturbance of breathing during sleep, including temporarily stopping)

♦ For reference, a man with a BMI of 40 might be:
- 5 ft 8 in, 265 lb (1.73 m, 120.2 kg)
- 5 ft 10 in, 280 lb (1.78 m, 127 kg)
- 6 ft, 295 lb (1.83 m, 133.8 kg)

FIGURE 9-4 The Psychology of Weight Cycling

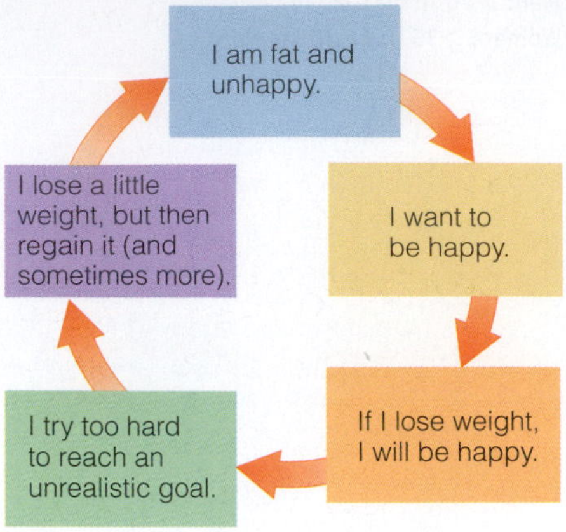

come to view their own bodies as grotesque and loathsome. Feelings of rejection, shame, or depression are common among obese people.

Most weight-loss programs assume that the problem can be solved simply by applying willpower and hard work. If determination were the only factor involved, though, the success rate would be far greater than it is. Overweight people may readily assume blame for failure to lose weight and maintain the losses when, in fact, it is the programs that have failed. Ineffective treatment and its associated sense of failure add to a person's psychological burden. Figure 9-4 illustrates how the devastating psychological effects of obesity and dieting perpetuate themselves.

Dangerous Interventions

Some people attach so many dreams of happiness to weight loss that they willingly risk huge sums of money for the slightest chance of success. As a result, weight-loss schemes flourish. Of the tens of thousands of claims, treatments, and theories for losing weight, few are effective—and many are downright dangerous. The negative effects must be carefully considered before embarking on any weight-loss program. Some interventions ♦ entail greater dangers than the risk of being overweight. Physical problems may arise from fad diets, "yo-yo" dieting, and drug use, and psychological problems may emerge from repeated "failures."

♦ Scrutinize fad diets, magic potions, and wonder gizmos with a healthy dose of skepticism.

Some of the nation's most popular diet books and weight-loss programs have misled consumers with unsubstantiated claims and deceptive testimonials. Furthermore, they fail to provide an assessment of the short- and long-term results of their treatment plans, even though such evaluations are possible and would permit consumers to make informed decisions. Of course, some weight-loss programs are better than others in terms of cost, approach, and customer satisfaction, but few are particularly successful in helping people keep off lost weight. Clients can expect reputable programs to abide by a consumer bill of rights that explains the risks associated with weight-loss programs and provides honest predictions of success (see Table 9-2).

© Richard Levine/Alamy

So many promises, so little success.

Fad Diets

Fad diets often sound good, but they typically fall short of delivering on their promises. They espouse exaggerated or false theories of weight loss and advise consumers to follow inadequate diets. Some fad diets are hazardous to health as Highlight 9 explains. Adverse reactions can be as minor as headaches, nausea, and dizziness or as serious as death. Table H9-3 (p. 308) offers guidelines for identifying unsound weight-loss schemes and fad diets.

Weight-Loss Products

Millions of people in the United States use nonprescription weight-loss products. Most of them are women, especially young overweight women, but almost 10 percent are of normal weight.

fad diets: popular eating plans that promise quick weight loss. Most fad diets severely limit certain foods or overemphasize others (for example, never eat potatoes or pasta or eat cabbage soup daily).

TABLE 9-2 Weight-Loss Consumer Bill of Rights (An Example)

1. *Warning:* Rapid weight loss may cause serious health problems. Rapid weight loss is weight loss of more than 1½ to 2 pounds per week or weight loss of more than 1 percent of body weight per week after the second week of participation in a weight-loss program.
2. Consult your personal physician before starting any weight-loss program.
3. Only permanent lifestyle changes, such as making healthful food choices and increasing physical activity, promote long-term weight loss and successful maintenance.
4. Qualifications of this provider are available upon request.
5. *You have a right to:*
   - Ask questions about the potential health risks of this program and its nutritional content, psychological support, and educational components.
   - Receive an itemized statement of the actual or estimated price of the weight-loss program, including extra products, services, supplements, examinations, and laboratory tests.
   - Know the actual or estimated duration of the program.
   - Know the name, address, and qualifications of the dietitian or nutritionist who has reviewed and approved the weight-loss program.

TABLE 9-3 Selected Herbal and Other Dietary Supplements Marketed for Weight Loss

Product	Claims	Research Findings	Risks
Bitter orange[a] (*Citrus aurantium,* a natural flavoring that contains synephrine, a compound structurally similiar to epinephrine)	Stimulates weight loss; provides an alternative to ephedra	Little evidence available	May increase blood pressure; may interact with drugs
Chitosan[b] (pronounced KITE-oh-san; derived from chitin, the substance that forms the hard shells of lobsters, crabs, and other crustaceans)	Binds to dietary fat, preventing digestion and absorption	Ineffective	Impaired absorption of fat-soluble vitamins
Chromium (trace mineral)	Eliminates body fat	Ineffective; weight gain reported when not accompanied by exercise	Headaches, sleep disturbances, and mood swings; hexavalent form is toxic and carcinogenic
Conjugated linoleic acid (CLA; a group of fatty acids related to linoleic acid, but with different *cis-* and *trans-*configurations)	Reduces body fat and suppresses appetite	Some evidence in animal studies, modest fat loss in human studies	None known
Ephedrine[c] (amphetamine-like substance derived from the Chinese ephedra herb *ma huang*)	Speeds body's metabolism	Short-term weight loss and dangerous side effects	Insomnia, tremors, heart attacks, strokes, and death; FDA has banned the sale of these products
Fucoxanthin[d] (derived from seaweed)	Speeds metabolism; burns fat	No evidence available	None known
Hoodia (derived from cactus)	Suppresses appetite	Little evidence available	None known
Hydroxycitric acid[e] (active ingredient derived from the rind of the tropical fruit *garcinia cambogia*)	Inhibits the enzyme that converts citric acid to fat; suppresses appetite	Ineffective	Toxicity symptoms reported in animal studies; headaches, respiratory and gastrointestinal distress in humans
Pyruvate[f] (3-carbon compound produced during glycolysis)	Speeds body's metabolism	Modest weight loss with high doses	GI distress
Yohimbine (derived from the bark of a West African tree)	Promotes weight loss	Ineffective	Nervousness, insomnia, anxiety, dizziness, tremors, headaches, nausea, vomiting, hypertension

[a] Marketed under the trade names Xenadrine EFX, Metabolife Ultra, NOW Diet Support.
[b] Marketed under the trade names Chitorich, Exofat, Fat Breaker, Fat Blocker, Fat Magnet, Fat Trapper, and Fatsorb.
[c] Marketed under the trade names Diet Fuel, Metabolife, and Nature's Nutrition Formula One.
[d] Marketed under the trade name FucoThin.
[e] Marketed under the trade names Ultra Burn, Citralean, CitriMax, Citrin, Slim Life, Brindleslim, Medislim, and Beer Belly Busters.
[f] Marketed under the trade names Exercise in a Bottle, Pyruvate Punch, Pyruvate-c, and Provate.
NOTE: The FDA has not approved the use of any of these products; most products are used in conjunction with a 1000- to 1800-kcalorie diet.

In their search for weight-loss magic, some consumers turn to "natural" herbal products and dietary supplements, even though few have proved to be effective.[66] St. John's wort, for example, contains substances that inhibit the uptake of **serotonin** and thus suppress appetite. In addition to the many cautions that accompany the use of all herbal remedies, consumers should be aware that St. John's wort is often prepared in combination with the herbal stimulant ephedrine. ♦ Ephedrine-containing supplements promote modest short-term weight loss (about 2 pounds a month), but the associated risks are high. These supplements have been implicated in several cases of heart attacks and seizures and have been linked to about 100 deaths. For this reason, the FDA has banned the sale of dietary supplements containing ephedrine, but they are still readily available on the Internet.* Table 9-3 presents the claims and the dangers behind ephedrine and several other dietary supplements commonly used for weight loss.[67]

♦ Ephedrine is an amphetamine-like substance extracted from the Chinese ephedra herb *ma huang*.

Herbal laxatives containing senna, aloe, rhubarb root, cascara, castor oil, and buckthorn (or various combinations) are commonly sold as "dieter's tea." Such concoctions commonly cause nausea, vomiting, diarrhea, cramping, and fainting and may have contributed to the deaths of four women who had drastically reduced their food intakes. Consumers mistakenly believe that laxatives will diminish nutrient absorption and reduce kcalorie intake, but remember that absorption occurs primarily in the small intestine and laxatives act on the large intestine. Chapter and Highlight 19 explore the possible benefits and potential dangers of herbal products and other alternative therapies. Current laws do not

serotonin (ser-oh-TONE-in): a neurotransmitter important in sleep regulation, appetite control, and sensory perception, among other roles. Serotonin is synthesized in the body from the amino acid tryptophan with the help of vitamin B_6.

**Ma huang* (ephedrine) is illegal in Canada.

require manufacturers of dietary supplements to test the safety or effectiveness of any product. Consumers cannot assume that an herb or dietary supplement of any kind is safe or effective just because it is available on the market. The FDA has identified more than 70 tainted dietary supplements that contain undeclared active pharmaceutical ingredients that can have serious consequences such as seizures and heart attacks. In addition, many weight-loss supplements do not contain the amounts of active ingredients listed on the labels. Anyone using dietary supplements for weight loss should first consult with a physician.

Other Gimmicks Other gimmicks don't help with weight loss either. Hot baths do not speed up metabolism so that pounds can be lost in hours. Steam and sauna baths do not melt the fat off the body, although they may dehydrate people so that they lose water weight. Brushes, sponges, wraps, creams, and massages intended to move, burn, or break up "**cellulite**" do nothing of the kind because there is no such thing as cellulite.

IN SUMMARY **The question of whether a person should lose weight depends on many factors: among them are the extent of overweight, age, health, and genetic makeup. Not all obesity will cause disease or shorten life expectancy. Just as there are unhealthy, normal-weight people, there are healthy, obese people. Some people may risk more in the process of losing weight than in remaining overweight. Fad diets and weight-loss supplements can be as physically and psychologically damaging as excess body weight.**

Aggressive Treatments for Obesity

The appropriate strategies for weight reduction depend on the degree of obesity and the risk of disease. An overweight person in good health may need only to improve eating habits and increase physical activity, but someone with **clinically severe obesity** may need more aggressive treatment ♦ options—drugs or surgery. Drugs appear to be modestly effective and safe, at least in the short term; surgery appears to be dramatically effective but can have severe complications, at least for some people.

♦ The field of medicine that specializes in treating obesity is called **bariatrics.**
- **bar** = weight

Drugs

Based on new understandings of obesity's genetic basis and its classification as a chronic disease, much research effort has focused on drug treatments for obesity. Experts reason that if obesity is a chronic disease, it should be treated as such—and the treatment of most chronic diseases includes drugs. The challenge, then, is to develop an effective drug—or more likely, a combination of drugs—that can be used over time without adverse side effects or the potential for abuse.[68]

Several drugs for weight loss have been tried over the years. When used as part of a long-term, comprehensive weight-loss program, drugs ♦ can help with modest weight loss.[69] Because weight regain commonly occurs with the discontinuation of drug therapy, treatment must be long term. Yet the long-term use of drugs poses risks. We don't yet know whether a person would be harmed more from maintaining a 100-pound excess or from taking a drug for a decade to keep the 100 pounds off. Physicians must prescribe drugs appropriately, inform consumers of the potential risks, and monitor side effects carefully. The FDA has approved four drugs to treat obesity.[70]

♦ Drugs may be an option for people with all of the following conditions:
- Unable to achieve adequate weight loss with diet and exercise
- BMI ≥30 or BMI ≥27 with weight-related health problems
- No medical contraindications

Sibutramine **Sibutramine** suppresses appetite.* The drug is most effective when used in combination with a reduced-kcalorie diet and increased physical activity. Side effects include dry mouth, headache, constipation, rapid heart rate, and high blood pressure. The FDA warns those with high blood pressure not to use sibutramine and advises others to monitor their blood pressure.

*Sibutramine is marketed under the trade name Meridia.

cellulite (SELL-you-light or SELL-you-leet): supposedly, a lumpy form of fat; actually, a fraud. Fatty areas of the body may appear lumpy when the strands of connective tissue that attach the skin to underlying muscles pull tight where the fat is thick. The fat itself is the same as fat anywhere else in the body. If the fat in these areas is lost, the lumpy appearance disappears.

clinically severe obesity: a BMI of 40 or greater or a BMI of 35 or greater with additional medical problems. A less preferred term used to describe the same condition is *morbid obesity*.

sibutramine (sigh-BYOO-tra-mean): a drug used in the treatment of obesity that slows the reabsorption of serotonin in the brain, thus suppressing appetite and creating a feeling of fullness.

Orlistat Orlistat takes a different approach to weight control.* It inhibits pancreatic lipase activity in the GI tract, thus blocking dietary fat digestion and absorption by about 30 percent. The drug is taken with meals and is most effective when accompanied by a reduced-kcalorie, low-fat diet. Side effects include gas, frequent bowel movements, and reduced absorption of fat-soluble vitamins. An over-the-counter, low-dose version of orlistat is also available.**

Phentermine and Diethylpropion Phentermine and diethylproprion enhance the release of the neurotransmitter norepinephrine, which tends to reduce food intake.† Weight reduction is modest. Side effects include increased blood pressure and insomnia.

Other Drugs Some physicians prescribe drugs that have not been approved for weight loss, a practice known as "off-label" use. These drugs have been approved for other conditions (such as seizures) and incidentally cause modest weight loss.[71] Physicians using off-label drugs must be well-informed of the drugs' use and effects and monitor their patients' responses closely.

Surgery

Surgery ♦ as an approach to weight loss is justified in some specific cases of clinically severe obesity.[72] An estimated 200,000 such surgeries are performed annually.[73] As Figure 9-5 shows, surgical procedures effectively limit food intake by reducing the capacity of the stomach. In addition, they suppress hunger by reducing production of the hormone ghrelin.[74] The results are significant: depending on the type of surgery, initial weight loss is 20 to 32 percent of body

*Orlistat is marketed under the trade name Xenical.
**The low-dose, over-the-counter version of orlistat is marketed under the trade name Alli (AL-eye).
†Phentermine is marketed under the trade names Adipex-P, Fastin, and Ionamin. Diethylproprion is marketed under the trade name Tenuate.

♦ Surgery may be an option for people with all of the following conditions:

- Unable to achieve adequate weight loss with diet and exercise
- BMI ≥40 or BMI ≥35 with weight-related health problems (such as diabetes or hypertension)
- No medical or psychological contraindications
- Understanding of risks and strong motivation to comply with post-surgery treatment plan

FIGURE 9-5 Gastric Surgery Used in the Treatment of Severe Obesity

Both of these surgical procedures limit the amount of food that can be comfortably eaten.

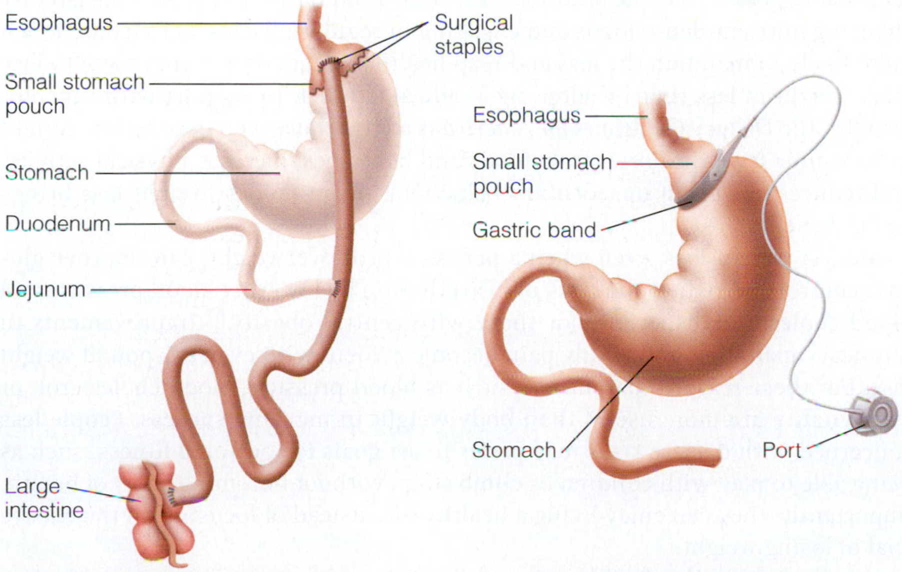

In gastric bypass, the surgeon constructs a small stomach pouch and creates an outlet directly to the small intestine, bypassing most of the stomach, the entire duodenum, and some of the jejunum. (Dark areas highlight the flow of food through the GI tract; pale areas indicate bypassed sections.)

In gastric banding, the surgeon uses a gastric band to reduce the opening from the esophagus to the stomach. The size of the opening can be adjusted by inflating or deflating the band by way of a port placed in the abdomen just beneath the skin.

orlistat (OR-leh-stat): a drug used in the treatment of obesity that inhibits the absorption of fat in the GI tract, thus limiting kcaloric intake.

weight and 14 to 25 percent after ten years.[75] Importantly, most people experience dramatic improvements in their diabetes, blood lipids, and blood pressure.[76] Whether surgery is a reasonable option for obese teens is the subject of much debate among pediatricians and bariatric surgeons (see Chapter 15).

The long-term safety and effectiveness of gastric surgery depend, in large part, on compliance with dietary instructions. Common immediate post-surgical complications include infections, nausea, vomiting, and dehydration. In the long term, vitamin and mineral deficiencies are common.[77] Weight regain and psychological problems may also occur. Lifelong medical supervision is necessary for those who choose the surgical route, but in suitable candidates, the possible health benefits of weight loss—improved blood lipid profile, blood pressure, and insulin sensitivity—may balance the risks.[78] Overall mortality is lower for obese people after surgery than for other obese people.[79]

Another surgical procedure is used, not to treat obesity, but to remove some of the evidence. Plastic surgeons can extract some fat deposits by suction lipectomy, or "liposuction." This cosmetic procedure has little effect on body weight (less than 10 pounds), but can alter body shape slightly in specific areas. Liposuction is a popular procedure in part because of its perceived safety, but, in fact, serious complications can occasionally result in death. Furthermore, removing adipose tissue by way of liposuction does not provide the health benefits that typically accompany weight loss.[80] In other words, liposuction does not improve blood pressure, inflammation, blood lipid profile, or insulin sensitivity. Furthermore, as with other weight-loss attempts, fat deposits will return when dietary intake exceeds needs.

IN SUMMARY **Obese people with high risks of medical problems may need aggressive treatment, including drugs or surgery. Others may benefit most from improving eating and exercise habits.**

Weight-Loss Strategies

Successful weight-loss strategies embrace small changes, moderate losses, and reasonable goals.[81] People who lose 10 to 20 pounds in a year by consistently choosing nutrient-dense foods and engaging in regular physical activity are much more likely to maintain the loss and reap health benefits than if they were to lose more weight in less time by adopting a radical fad diet. In keeping with this philosophy, the *Dietary Guidelines for Americans* advise those who need to lose weight to "consume fewer kcalories from foods and beverages, increase physical activity, and reduce time spent in sedentary behaviors." Even modest weight loss brings health benefits.

Modest weight loss, even when a person is still overweight, can improve glucose control and reduce the risks of heart disease by lowering blood pressure and blood cholesterol, especially for those with central obesity.[82] Improvements in physical capabilities and bodily pain become evident with even a 5-pound weight loss. For these reasons, parameters such as blood pressure, blood cholesterol, or even vitality are more useful than body weight in marking success. People less concerned with disease risks may prefer to set goals for personal fitness, such as being able to play with children or climb stairs without becoming short of breath. Importantly, they can enjoy living a healthy life instead of focusing on the elusive goal of losing weight.

Whether the goal is health or fitness, expectations need to be reasonable. Unreachable targets ensure frustration and failure. When realistic, yet moderately challenging, goals are achieved or exceeded, people enjoy rewards instead of finding disappointment.

Research findings highlight the great disparity between lofty expectations and reasonable success.[83] Before beginning a weight-loss program, obese women identified the weights they would describe as "dream," "happy," "acceptable," and "disappointing" (see Figure 9-6). All of these weights were below their starting

FIGURE 9-6 Reasonable Weight Goals and Expectations Compared

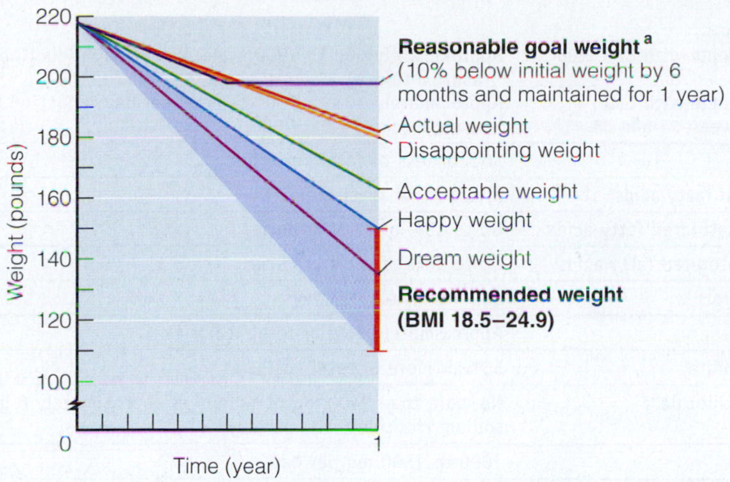

[a]Reasonable goal weights reflect pounds lost over time. Given more time, reasonable goals may eventually fall within the recommended weight range.

SOURCE: Adapted from G. D. Foster and coauthors, What is a reasonable weight loss? Patients' expectations and evaluations of obesity treatment outcomes, *Journal of Consulting and Clinical Psychology* 65 (1997): 79–85

weight. Their goal weights far exceeded the 5 to 10 percent recommended by experts, or even the 15 percent reported by the most successful weight-loss studies. Even their "disappointing" weights exceeded recommended goals. Close to a year later, and after an average loss of 35 pounds, almost half of the women did not achieve even their "disappointing" weights. They did, however, experience more physical, social, and psychological benefits than they had predicted for that weight. Still, in a culture that overvalues thinness, these women were not satisfied with a 16 percent reduction in weight—not because their efforts were unsuccessful, but because their expectations were unrealistic.

Depending on initial body weight, a reasonable rate of weight loss for overweight people is ½ to 2 pounds a week, ♦ or 10 percent of body weight over six months.[84] For a person weighing 250 pounds, a 10 percent loss is 25 pounds, or about 1 pound a week for six months. Such gradual weight losses are more likely to be maintained than rapid losses. Keep in mind that pursuing good health is a lifelong journey. Most adults are keenly aware of their body weights and shapes and realize that what they eat and what they do can make a difference to some extent. Those who are most successful at weight management seem to have fully incorporated healthful eating and physical activity into their daily lives. Such advice—to reduce energy intake and increase physical activity—would hardly surprise anyone, yet relatively few people trying to control their weight follow these recommendations.

♦ Safe rate for weight loss:
- ½ to 2 lb/week (0.2 to 0.9 kg)
- 10% body weight/6 mo

For a person weighing 110 kg, a 10% loss is 11 kg, or about 0.5 kg a week for six months.

Eating Plans

Contrary to the claims of fad diets, no single food plan is magical, and no specific food must be included or avoided in a weight-management program. In designing a plan, people need only consider foods that they like or can learn to like, that are available, and that are within their means.

Be Realistic about Energy Intake The main characteristic of a weight-loss diet is that it provides less energy than the person needs to maintain present body weight. If food energy is restricted too severely, dieters may not receive sufficient nutrients and may lose lean tissue. Rapid weight loss usually means excessive loss of lean tissue, a lower BMR, and a rapid weight gain to follow. In addition, restrictive eating may set in motion the unhealthy behaviors of eating disorders as described in Highlight 8.

Table 9-4 (p. 286) outlines the recommendations of a weight-loss diet. Energy intake should provide nutritional adequacy without excess—that is, somewhere between deprivation and complete freedom to eat whatever, whenever. A reasonable

TABLE 9-4 Recommendations for a Weight-Loss Diet

Nutrient	Recommended Intake
kCalories	
For people with BMI ≥35	Approximately 500 to 1000 kcalories per day reduction from usual intake
For people with BMI between 27 and 35	Approximately 300 to 500 kcalories per day reduction from usual intake
Total fat	30% or less of total kcalories
Saturated fatty acids[a]	8 to 10% of total kcalories
Monounsaturated fatty acids	Up to 15% of total kcalories
Polyunsaturated fatty acids	Up to 10% of total kcalories
Cholesterol[a]	300 mg or less per day
Protein[b]	Approximately 15% of total kcalories
Carbohydrate[c]	55% or more of total kcalories
Sodium chloride	No more than 2400 mg of sodium or approximately 6 g of sodium chloride (salt) per day
Calcium	1000 to 1500 mg per day
Fiber[c]	20 to 30 g per day

[a]People with high blood cholesterol should aim for less than 7 percent kcalories from saturated fat and 200 milligrams of cholesterol per day.

[b]Protein should be derived from plant sources and lean sources of animal protein.

[c]Carbohydrates and fiber should be derived from vegetables, fruits, and whole grains.

SOURCE: National Institutes of Health Obesity Education Initiative, *The Practical Guide: Identification, Evaluation, and Treatment of Overweight and Obesity in Adults* (Washington, D.C.: U.S. Department of Health and Human Services, 2000), p. 27.

suggestion is that an adult needs to increase activity and reduce food intake enough to create a deficit of 500 to 1000 kcalories per day. Such a deficit produces a weight loss of 1 to 2 pounds per week—a rate that supports the loss of fat efficiently while retaining lean tissue.[85] In general, weight-loss diets provide about 1200 kcalories per day for women and 1600 kcalories a day for men.[86]

Some people skip meals, typically breakfast, in an effort to reduce energy intake, but research suggests such a strategy may be counterproductive. Breakfast frequency is inversely associated with obesity—that is, people who frequently eat breakfast have lower BMI than those who tend to skip breakfast.[87] Furthermore, when people eat breakfast, overall diet quality is better and daily energy density is lower—two factors that support healthy body weight.[88]

Emphasize Nutritional Adequacy Nutritional adequacy is difficult to achieve on fewer than 1200 kcalories a day, and most healthy adults need never consume any less. A plan that provides an adequate intake supports a healthier and more successful weight loss than a restrictive plan that creates feelings of starvation and deprivation, which can lead to an irresistible urge to binge.

Table 9-5 specifies the amounts of foods from each food group for diets providing 1200 to 1600 kcalories. Such an intake would allow most people to lose weight and still meet their nutrient needs with careful, nutrient-dense food selections. (Women might need iron or calcium supplements.) Keep in mind, too, that well-balanced diets that emphasize fruits, vegetables, whole grains, lean meats or meat alternates, and low-fat milk products offer many health rewards even when they don't result in weight loss.[89] A dietary supplement providing vitamins and minerals at or below 100 percent of the Daily Values can help people following low-kcalorie diets to achieve nutrient adequacy.[90]

Eat Small Portions As mentioned earlier, portion sizes at markets, at restaurants, and even at home have increased dramatically over the years. We have come to expect large portions, and we have learned to clean our plates. Many of us pay more attention to these external cues defining how much to eat than to our internal cues of hunger and satiety. For health's sake, we may need to learn to eat less food at each meal—one piece of chicken for dinner instead of two, a teaspoon of butter on vegetables instead of a tablespoon, and one cookie for dessert instead of

TABLE 9-5 Daily Amounts from Each Food Group for 1200- to 1600-kCalorie Diets

Food Group	1200 kCalories	1400 kCalories	1600 kCalories
Fruit	1 c	1½ c	1½ c
Vegetables	1½ c	1½ c	2 c
Grains	4 oz	5 oz	5 oz
Protein foods	3 oz	4 oz	5 oz
Milk	3 c	3 c	3 c
Oils	3 tsp	3 tsp	4 tsp

NOTE: The USDA Food Guide patterns for 1200 and 1400 kcalories were designed for children and provided 2 cups milk. They were modified here to include an additional cup of milk, as 3 cups per day is recommended for all adults. The discretionary kcalorie allowance for these patterns is about 100 kcalories.

six. The goal is to eat enough food for adequate energy, abundant vitamins and minerals, and some pleasure, but not more. This amount should leave a person feeling satisfied—not stuffed.

Keep in mind that even fat-free and low-fat foods can deliver a lot of kcalories when a person eats large quantities. A low-fat cookie or two can be a sweet treat even on a weight-loss diet, but larger portions defeat the savings.

People who have difficulty making low-kcalorie selections or controlling portion sizes may find it easier to use structured meal replacement plans. Meal replacements that provide low-kcalorie, nutritious meals or snacks can support weight loss while easing the task of diet planning.[91] Ideally, those using a meal replacement plan will seek advice from a registered dietitian to learn how to select appropriately from conventional food choices as well.

Lower Energy Density Most people take their cues about how much to eat based on portion sizes, and the larger the portion size, the more they eat—even when the food is not particularly tasty.[92] To lower energy intake, a person can either reduce the portion size or reduce the energy density.[93] Reducing energy density while maintaining food quantity, especially by including fruits and vegetables, seems to be a successful strategy to control hunger and lose weight.[94] Figure 9-7 illustrates how water, fiber, and fat influence energy density, and the "How To" feature on p. 288 compares foods based on their energy density. Foods containing water, those rich in fiber, and those low in fat help to lower energy density, providing more satiety for fewer kcalories.[95] Because a low-energy-density diet is a low-fat, high-fiber diet rich in many vitamins and minerals, it supports good

FIGURE 9-7 Energy Density

Decreasing the energy density (kcal/g) of foods allows a person to eat satisfying portions while still reducing energy intake. To lower energy density, select foods high in water or fiber and low in fat.

Selecting grapes with their high water content instead of raisins increases the volume and cuts the energy intake in half.

Even at the same weight and similar serving sizes, the fiber-rich broccoli delivers twice the fiber of the potatoes for about one-fourth the energy.

By selecting the water-packed tuna (on the right) instead of the oil-packed tuna (on the left), a person can enjoy the same amount for fewer kcalories.

HOW TO Compare Foods Based on Energy Density

Chapter 2 describes how to evaluate foods based on their nutrient density—their nutrient contribution per kcalorie. Another way to evaluate foods is to consider their energy density—their energy contribution per gram. This example compares carrot sticks with french fries. The conclusion is no surprise, but understanding the mathematics may offer valuable insight into the concept of energy density. A carrot weighing 72 grams delivers 31 kcalories. To calculate the energy density, divide kcalories by grams:

$$\frac{31 \text{ kcal}}{72 \text{ g}} = 0.43 \text{ kcal/g}$$

Do the same for french fries weighing 50 grams and contributing 167 kcalories:

$$\frac{167 \text{ kcal}}{50 \text{ g}} = 3.34 \text{ kcal/g}$$

The more kcalories per gram, the greater the energy density. French fries are more energy dense than carrots. They provide more energy per gram—and per bite. Considering a food's energy density is especially useful in planning diets for weight management. Foods with a high energy density help with weight gain, whereas foods with a low energy density help with weight loss.

For additional practice, log on to **www.cengagebrain.com** and search for ISBN 084006845X.

© Matthew Farruggio

TRY IT Compare the energy density of a hard-boiled egg (50 grams and 78 kcalories) with light tuna canned in water (57 grams and 66 kcalories).

health in addition to weight loss.[96] Unfortunately, low-energy-density foods tend to be relatively expensive.[97]

Remember Water Water helps with weight management in several ways. For one, foods with high water content (such as broth-based soups) increase fullness, reduce hunger, and consequently reduce energy intake.[98] For another, drinking a large glass of water before a meal may ease hunger, fill the stomach, and reduce energy intake.[99] Importantly, water adds no kcalories. The average U.S diet delivers an estimated 75 to 150 kcalories a day from sweetened beverages.[100] Consuming large portions of kcaloric beverages increases energy intake.[101] Simply replacing nutrient-poor, energy-dense beverages with water could save a person up to 15 pounds a year. Water also helps the GI tract adapt to a high-fiber diet.

Focus on Fiber Healthy meals and snacks center on high-fiber foods. Fresh fruits, vegetables, legumes, and whole grains offer abundant vitamins, minerals, and fiber but little fat. Consequently, diets rich in fiber tend to be relatively low in energy and high in nutrients.

High-fiber foods also require effort to eat—an added bonus. Eating fiber-rich fruits and vegetables reduces energy density, lowers kcalorie intake, and promotes satiety. A person who slows down and savors each bite eats less before the satiety signal reaches the brain. Consequently, energy intake is lower when meals are eaten slowly.[102] Savoring each bite also activates the pleasure centers of the brain. Some research suggests that people may overeat when the brain doesn't sense enough gratification from food.[103] Faster eating correlates with higher weights.[104]

Choose Fats Sensibly Ideally, a weight-loss diet is both high in fiber and low in fat. Lowering the fat content of a food lowers its energy density—for example, selecting fat-free milk instead of whole milk. That way, a person can consume the usual amount (say, a cup of milk) at a lower energy intake (85 instead of 150 kcalories).

Fat has a weak satiating effect, and satiation plays a key role in determining food intake during a meal. Consequently, a person eating a high-fat meal raises energy intake by adding more food and more fat kcalories. For these reasons, measure fat with extra caution. (Review p. 156 for strategies to lower fat in the diet.) Be careful not to take this advice to extremes, however; too little fat in the diet or in the body carries health risks as well, as Chapter 5 explains.

Whether a low-fat diet is the best option for weight loss is the subject of some controversy and much debate. An important point to notice in any discussion on weight-loss diets is total energy intake. *Low fat* simply means the energy derived from fat is relatively low compared with the total energy intake; it does not mean total energy intake is low. And reducing energy intake to less than expended is essential for weight loss. One way to lower energy intake is to lower fat intake. In these cases, adopting a low-fat diet can help with weight loss.

Select Carbohydrates Carefully Another currently popular way to lower energy intake is to lower carbohydrate intake. Highlight 4's discussion of carbohydrate-restricted and carbohydrate-modified diets reached the same conclusions as the previous paragraph on low-fat diets: they only work when kcalorie intake is less than kcalorie output.

People trying to control weight often use foods and beverages sweetened with artificial sweeteners. Using artificial sweeteners instead of sugar lowers the energy density of foods and beverages. Most studies find no difference in hunger and no compensation in food intakes when diet soft drinks replace regular ones.[105] In this way, people who eat or drink artificially sweetened products can lower their energy intakes and expect modest weight losses.[106]

To what extent artificial sweeteners can help someone lose weight depends in part on the person's motivations and actions. For example, one person might drink an artificially sweetened beverage now so as to be able to eat a high-kcalorie food later. This person's energy intake might stay the same or increase. A person trying to control food energy intake might drink an artificially sweetened beverage now and choose a low-kcalorie food later. This plan would help reduce the person's total energy intake. Using artificial sweeteners will not automatically lower energy intake. To control energy intake successfully, a person needs to make informed diet and activity decisions throughout the day.

Watch for Other Empty kCalories A person trying to achieve or maintain a healthy weight needs to pay attention not only to fat, but to sugar and alcohol, too. Using sugar or alcohol for pleasure on occasion is compatible with health as long as most daily choices are of nutrient-dense foods. Not only does alcohol add kcalories, but accompanying mixers can also add both kcalories and fat, especially in creamy drinks such as piña coladas (review Table H7-3 on p. 236). Furthermore, drinking alcohol reduces a person's inhibitions, which can sabotage weight-control efforts—at least temporarily.

IN SUMMARY **A person who adopts a lifelong "eating plan for good health" rather than a "diet for weight loss" will be more likely to keep the lost weight off. Table 9-6 (p. 290) provides several tips for successful weight management.**

Physical Activity

The best approach to weight management includes physical activity.[107] The greater the energy used in exercise, the greater the body fat loss.[108] Yet among people trying to lose weight, fewer than half are physically active and only half of the active group meet minimal recommendations.[109] To prevent weight gains and support weight losses, current recommendations advise 60 minutes of moderately intense physical activity a day in addition to activities of daily life.[110] People who combine diet and exercise typically lose more fat, retain more muscle, and regain less weight than those who only follow a weight-loss diet. Even when people who include physical activity in their weight-management program

If you want to lose weight, steer clear of the empty kcalories in fancy coffee drinks. A 16-ounce café mocha delivers 400 kcalories—half of them from fat.

TABLE 9-6 Weight-Management Strategies

In General

- Focus on healthy eating and activity habits, not on weight losses or gains.
- Adopt reasonable expectations about health and fitness goals and about how long it will take to achieve them.
- Make nutritional adequacy a high priority.
- Learn, practice, and follow a healthful eating plan for the rest of your life.
- Participate in some form of physical activity regularly.
- Adopt permanent lifestyle changes to achieve and maintain a healthy weight.

For Weight Loss

- Energy out should exceed energy in by about 500 kcalories/day. Increase your physical activity enough to spend more energy than you consume from foods.
- Emphasize foods with a low energy density and a high nutrient density.
- Eat small portions. Share a restaurant meal with a friend or take home half for lunch tomorrow.
- Eat slowly.
- Limit high-fat foods. Make legumes, whole grains, vegetables, and fruits central to your diet plan.
- Limit low-fat treats to the serving size on the label.
- Limit concentrated sweets and alcoholic beverages.
- Drink a glass of water before you begin to eat and another while you eat. Drink plenty of water throughout the day.
- Keep a record of diet and exercise habits; it reveals problem areas, the first step toward improving behaviors.
- Learn alternative ways to deal with emotions and stresses.
- Attend support groups regularly or develop supportive relationships with others.

For Weight Gain

- Energy in should exceed energy out by at least 500 kcalories/day. Increase your food intake enough to store more energy than you expend in exercise. Exercise and eat to build muscles.
- Expect weight gain to take time (1 pound per month would be reasonable).
- Emphasize energy-dense foods.
- Eat at least three meals a day.
- Eat large portions of foods and expect to feel full.
- Eat snacks between meals.
- Drink plenty of juice and milk.

do not lose more weight, they seem to follow their diet plans more closely and maintain their losses better than those who do not exercise. Consequently, they benefit from taking in a little less energy as well as from expending a little more energy in physical activity. Importantly, those who exercise reduce abdominal obesity and improve their blood pressure, insulin resistance, and cardiorespiratory fitness, regardless of weight loss.[111] Of course there are many health benefits of physical activity; the focus here is on its role in weight management.

iStockphoto.com/Erel Photography

The key to good health is to combine sensible eating with regular exercise.

Dietary Guidelines for Americans 2010

To achieve and maintain a healthy body weight, adults should do the equivalent of 150 [to 300] minutes of moderate-intensity aerobic activity each week.

Activity and Energy Expenditure Table 8-2 (p. 247) shows how much energy each of several activities uses. The number of kcalories spent in an activity depends on body weight, intensity, and duration. For example, a person who weighs 150 pounds and walks 3½ miles in 60 minutes expends about 315 kcalories. That same person running 3 miles in 30 minutes uses a similar amount. By comparison, a 200-pound person running 3 miles in 30 minutes expends an additional 100 kcalories or so. The goal is to expend as much energy as your time allows. The greater the energy deficit created by exercise, the greater the fat loss. And be care-

FIGURE 9-8 **Influence of Physical Activity on Discretionary kCalorie Allowance**

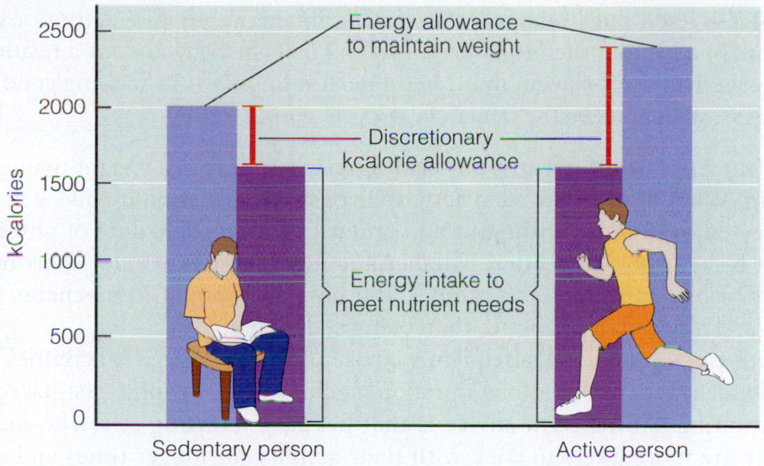

Animated! figure
www.cengagebrain.com
(search for ISBN 084006845X)

ful not to compensate for the energy expended in exercise by eating more food. Otherwise, energy balance won't shift and fat loss will be less significant.

Activity and Discretionary kCalorie Allowance Chapter 2 introduces the discretionary kcalorie allowance as the difference between the kcalories needed to supply nutrients and those needed to maintain energy balance. Because exercise expends energy, the energy allowance to maintain weight increases with increased physical activity—yet the energy needed to deliver needed nutrients remains about the same. In this way, physical activity increases the discretionary kcalorie allowance (see Figure 9-8). Having a larger discretionary kcalorie allowance puts a little more wiggle room in a weight-loss diet for such options as second helpings, sweet treats, or alcoholic beverages on occasion. Of course, selecting nutrient-dense foods and *not* using discretionary kcalories will maximize weight loss.

Activity and Metabolism Activity also contributes to energy expenditure in an indirect way—by speeding up metabolism. It does this both immediately and over the long term. On any given day, metabolism remains slightly elevated for several hours after intense and prolonged exercise.[112] ♦ Over the long term, a person who engages in daily vigorous activity gradually develops more lean tissue. Metabolic rate rises accordingly, and this supports continued weight loss or maintenance.

♦ This postexercise effect may raise the energy expenditure of exercise up to 15 percent.

Activity and Body Composition Physically active people have less body fat than sedentary people do—even if they have the same BMI. Physical activity, even without weight loss, changes body composition: body fat decreases and lean body mass increases. Furthermore, strength training exercises specifically prevent increases in body fat and abdominal fat.[113]

Activity and Appetite Control Many people think that exercising will make them eat more, but this is not entirely true. Active people do have healthy appetites, but *immediately* after an intense workout, most people do not feel like eating. They may be thirsty and want to shower, but they are not hungry.[114] The body has released fuels from storage to support the exercise, so glucose and fatty acids are abundant in the blood. At the same time, the body has suppressed its digestive functions. Hard physical work and eating are not compatible. A person must calm down, put energy fuels back in storage, and relax before eating. At that time, a physically active person may eat more than a sedentary person, but not so much as to fully compensate for the kcalories expended in exercise.

Exercise may help curb the inappropriate appetite that accompanies boredom, anxiety, or depression. Weight-management programs encourage people who feel the urge to eat when not hungry to go out and exercise instead. The activity passes time, relieves anxiety, and prevents inappropriate eating.

Activity and Psychological Benefits Activity also helps reduce stress. Because stress itself cues inappropriate eating for many people, activity can help here, too. In addition, the fit person looks and feels healthy and, as a result, gains self-esteem. High self-esteem motivates a person to persist in seeking good health and fitness, which keeps the beneficial ♦ cycle going.

♦ Benefits of physical activity in a weight-management program:

- Short-term increase in energy expenditure (from exercise and from a slight rise in metabolism)
- Long-term increase in BMR (from an increase in lean tissue)
- Improved body composition
- Appetite control
- Stress reduction and control of stress eating
- Physical, and therefore psychological, well-being
- Improved self-esteem

Choosing Activities Clearly, physical activity is a plus in a weight-management program. What kind of physical activity is best? People should choose activities that they enjoy and are willing to do regularly. ♦ What schedule of physical activity is best? It doesn't matter; whether a person chooses several short bouts of exercise or one continuous workout, the fitness and weight-loss benefits are the same—and any activity is better than being sedentary.

♦ For an active life, limit sedentary activities, engage in strength and flexibility activities, enjoy leisure activities often, engage in vigorous activities regularly, and be as active as possible every day.

Health-care professionals frequently advise people to engage in activities of low-to-moderate intensity for a long duration, such as an hour-long, fast-paced walk. The reasoning behind such advice is that people exercising at low-to-moderate intensity are more likely to stick with their activity for longer times and are less likely to injure themselves. A person who stays with an activity routine long enough to enjoy the rewards will be less inclined to give it up and will, over the long term, reap many health benefits. Activity of low-to-moderate intensity ♦ that expends at least 2000 kcalories per week is especially helpful for weight management. Higher levels produce even greater losses.

♦ Estimated energy expended when walking at a moderate pace = 1 kcal/mi/kg body weight.

In addition to exercise, a person can incorporate hundreds of energy-expending activities into daily routines: take the stairs instead of the elevator, walk to the neighbor's apartment instead of making a phone call, and rake the leaves instead of using a blower. Remember that sitting uses more kcalories than lying down, standing uses more kcalories than sitting, and moving uses more kcalories than standing. A 175-pound person who replaces a 30-minute television program with a 2-mile walk a day can expend enough energy to lose (or at least not gain) 18 pounds in a year. Meeting an activity goal of 10,000 steps a day helps to support a healthy BMI.[115] By wearing a pedometer, a person can easily increase physical activity, lose weight, and lower blood pressure without measuring miles or watching the clock.[116] The point is to be active. Walk. Run. Swim. Dance. Cycle. Climb. Skip. Do whatever you enjoy doing—and do it often.

Spot Reducing People sometimes ask about "spot reducing." Unfortunately, muscles do not "own" the fat that surrounds them. Fat cells all over the body release fat in response to the demand of physical activity for use by whatever muscles are active. Specific exercises—whether moderate or intense—do not influence the site of adipose tissue loss.[117]

Exercise can help with trouble spots in another way, though. The "trouble spot" for most men is the abdomen, their primary site of fat storage. During aerobic exercise, abdominal fat readily releases its stores, providing fuel to the physically active body. With regular exercise and weight loss, men will deplete these abdominal fat stores before those in the lower body. Women may also deplete abdominal fat with exercise, but their "trouble spots" are more likely to be their hips and thighs.

In addition to aerobic activity, strength training can help to improve the tone of muscles in a trouble area, and stretching to gain flexibility can help with associated posture problems. A combination of aerobic, strength, and flexibility workouts best improves fitness and physical appearance.

IN SUMMARY **Physical activity should be an integral part of a weight-control program. Physical activity can increase energy expenditure, improve body composition, help control appetite, reduce stress and stress eating, and enhance physical and psychological well-being.**

Environmental Influences Chapter 8 describes how hormones regulate hunger, satiety, and satiation, but people don't always pay close attention to such internal signals. Instead, their eating behaviors are often dictated by environmental factors. Environmental factors include those surrounding the eating experience as well as those pertaining to the food itself. Changing any of these factors can influence how much a person eats.

Atmosphere The environment surrounding a meal or snack influences its duration. When the lighting, décor, aromas, and sounds of an environment are pleasant and comfortable, people tend to spend more time eating and thus eat more. A person needn't eat under neon lights with offensive music to eat less, of course. Instead, after completing a meal, remove food from the table and enjoy the ambience—without the presence of visual cues to stimulate additional eating.

Accessibility Among the strongest influences on how much we eat are the accessibility, ease, and convenience of obtaining food. In general, the less effort needed to obtain food, the more likely food will be eaten. Are you more likely to eat if half a leftover pizza is in your refrigerator or if you have to drive to the grocery store, buy a frozen pizza, and bake it for 45 minutes? Having food nearby and visible encourages eating. In one study, secretaries ate more chocolates when the candy was on their desks than when they had to walk six feet.[118] Interestingly, the secretaries underestimated the amount of chocolates they had eaten when the candy was on their desk and overestimated when it was a short distance away. The message is clear for people wanting to eat less candy (or any other tempting item)—keep it out of sight and in an inconvenient place (or don't even buy it).

Socializing People tend to eat more when socializing with others. Pleasant conversations extend the duration of a meal, allowing a person more time to eat more, and research confirms that the longer the meal, the greater the consumption.[119] In addition, by taking a visual cue from companions, a person might eat more when others at the table clean their plates or go to the buffet line for seconds. One way to eat less is to pace yourself with the person who seems to be eating the least and slowest.

Social interactions also distract a person from paying attention to how much has been eaten. In some cases, socializing with friends during a meal may provide comfort and lower a person's motivation to limit consumption. In other cases, socializing with unfamiliar people during a meal—during a job interview or blind date, for example—may create stress and reduce food consumption. To eat less while socializing, pay attention to portion size.

Distractions Distractions influence food intake by initiating eating, interfering with internal controls to stop eating, and extending the duration of eating. Some people start eating dinner when a favorite television program comes on, regardless of hunger. Other people continue eating breakfast until they finish reading the newspaper. Such mindless eating can easily become overeating. Distractions also interfere with a person's ability to monitor and regulate how much is consumed.[120] If distractions are a part of the eating experience, extra care is needed to control portion sizes.

Presence The mere sight (or smell, or even thought) of a food can prompt a person to start eating—regardless of hunger. The chocolates in the clear candy dishes on the secretaries' desks were eaten much faster than those in opaque containers.[121]

Multiple Choices When offered a large assortment of foods, or several flavors of the same food, people tend to eat more. Interestingly, they tend to eat more even when the number of choices is only *perceived*. Given six flavors of jelly beans, people will eat more when offered an assorted mixture than when presented with the exact same flavors and quantities sorted in a sectioned container.

Ivonne Wierink/Shutterstock.com

Eating foods with added fats from a large bowl while distracted by television is a weight-gaining combination.

Having multiple choices is both pleasing and distracting—two factors that slow the eating experience and delay satiation.[122] To limit intake, then, focus on a limited number of foods per meal. Be careful not to misunderstand and abandon variety in diet planning. Eating a variety of nutrient-dense foods from each of the food groups is still a healthy plan.

Package and Portion Sizes As noted earlier, the sizes of packages in grocery stores and portion sizes at restaurants and at home have increased dramatically in recent decades, contributing to the increase in obesity in the United States. Put simply, we tend to clean our plates and finish the package. The larger the bag of potato chips, the greater the intake. To keep from overeating, repackage snacks into smaller containers or eat them from a plate, not directly from the package.

Serving Containers We often use plates, utensils, and glasses as visual cues to guide our decisions on how much to eat and drink.[123] If you plan to eat a bowl of ice cream, it matters whether the bowl you select holds 8 ounces or 24 ounces. Even the size of the serving container matters. Students took more—and ate more—snacks when serving from two large bowls instead of from four medium bowls.[124]

Large dinner plates and wide glasses create illusions and misperceptions about quantities consumed. A scoop of mashed potatoes on a small plate looks larger than the same-size scoop on a large plate, leading a person to underestimate the amount of food eaten.[125] To control portion sizes, use small bowls and plates, small serving spoons, and tall, narrow glasses.[126] Of course, using a small plate will not result in less food eaten if multiple servings are taken.[127]

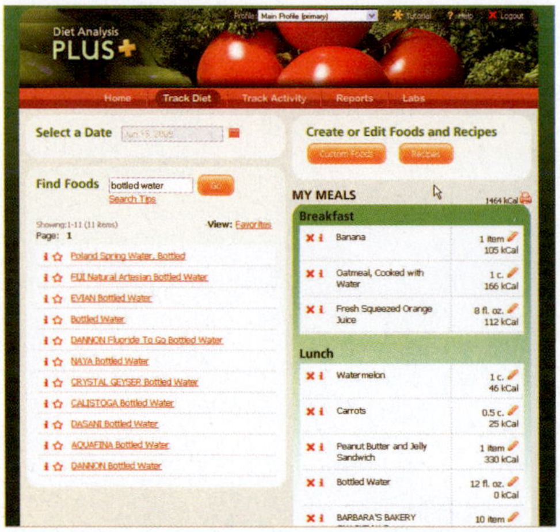

Diet analysis programs help people identify high-kcalorie foods and monitor their eating habits.

Behavior and Attitude

Changes in behavior and attitude can be very effective in supporting efforts to achieve and maintain appropriate body weight and composition.[128] **Behavior modification** focuses on how to change behaviors to increase energy expenditure and decrease energy intake.[129] A person must commit to taking action.

Adopting a positive, matter-of-fact attitude helps to ensure success. Healthy eating and activity choices are an essential part of healthy living and should simply be incorporated into the day—much like brushing one's teeth or wearing a safety belt.

Become Aware of Behaviors To solve a problem, a person must first identify all the behaviors that created the problem. Keeping a record ♦ will help to identify eating and exercise behaviors that may need changing (see Figure 9-9). It will also establish a baseline against which to measure future progress.

♦ Visit mypyramid.gov to find menu planning and diet assessment resources.

Change Behaviors Behavior modification strategies ♦ focus on learning desired eating and exercise behaviors and eliminating unwanted behaviors. With so many possible behavior changes, a person can feel overwhelmed. Start with small time-specific goals for each behavior—for example, "I'm going to take a 30 minute walk after dinner every evening" instead of "I'm going to run in a marathon someday." Practice desired behaviors until they become routine. Addressing multiple behaviors that focus on a common goal simultaneously may better support changes than taking on one at a time.[130] Using a reward system also seems to effectively support weight-loss efforts.[131]

♦ Examples of behavioral strategies to support weight change:

- Do not grocery shop when hungry.
- Eat slowly (pause during meals, chew thoroughly, put down utensils between bites).
- Exercise when watching television.

Cognitive Skills Successful behavior changes depend in part on two cognitive skills—problem solving and cognitive restructuring.[132] Problem solving requires a person to identify the problem, generate potential solutions, list the pros and cons and the risks and benefits of each, implement the most feasible solution, and evaluate to determine whether behaviors should be continued or abandoned. Cognitive restructuring requires a person to replace negative thoughts that derail success with positive thoughts that support behavior change.

Personal Attitude For many people, overeating and being overweight have become an integral part of their identity. Those who fully understand their personal relationships with food are best prepared to make healthful changes in eating and exercise behaviors.

behavior modification: the changing of behavior by the manipulation of antecedents (cues or environmental factors that trigger behavior), the behavior itself, and consequences (the penalties or rewards attached to behavior).

FIGURE 9-9 Food Record

The entries in a food record should include the times and places of meals and snacks, the types and amounts of foods eaten, and a description of the individual's feelings when eating. The diary should also record physical activities: the kind, the intensity level, the duration, and the person's feelings about them.

Time	Place	Activity or food eaten	People present	Mood
10:30–10:40	School vending machine	6 peanut butter crackers and 12 oz. cola	by myself	starved
12:15–12:30	Restaurant	Sub sandwich and 12 oz. cola	friends	relaxed & friendly
3:00–3:45	Gym	Weight training	work out partner	tired
4:00–4:10	Snack bar	Small frozen yogurt	by myself	OK

Sometimes habitual behaviors that are hazardous to health, such as smoking or drinking alcohol, contribute positively by helping people adapt to stressful situations. Similarly, many people overeat to cope with the stresses of life. To break out of that pattern, they must first identify the particular stressors that trigger the urge to overeat. Then, when faced with these situations, they must learn and practice problem-solving skills that will help them to respond appropriately.

All this is not to imply that psychological therapy holds the magic answer to a weight problem. Still, efforts to improve one's general well-being may result in healthy eating and activity habits even when weight loss is not the primary goal. When the problems that trigger the urge to overeat are resolved in alternative ways, people may find they eat less. They may begin to respond appropriately to internal cues of hunger rather than inappropriately to external cues of stress. Sound emotional health supports a person's ability to take care of physical health in all ways—including nutrition, weight management, and fitness.

Support Groups Group support can prove helpful when making life changes. Some people find it useful to join a group such as Take Off Pounds Sensibly (TOPS), Weight Watchers (WW), Overeaters Anonymous (OA), or others. Some dieters prefer to form their own self-help groups or find support online. The Internet offers numerous opportunities for weight-loss education and counseling that may be effective alternatives to face-to-face programs.[133] As always, consumers need to choose wisely and avoid rip-offs.

IN SUMMARY **A surefire remedy for obesity has yet to be found, although many people find a combination of the approaches just described to be most effective. Diet and exercise shift energy balance so that more energy is being expended than is taken in. Physical activity increases energy expenditure, builds lean tissue, and improves health. Energy intake should be reduced by 500 to 1000 kcalories per day, depending on starting body weight and usual food intake. Behavior modification and cognitive restructuring retrain habits to support a healthy eating and exercise plan. This treatment package requires time, individualization, and sometimes the assistance of a registered dietitian or support group.**

Maintaining a healthy body weight requires maintaining the vigorous physical activities and careful eating habits that supported weight loss.

Weight Maintenance

People who are successful often experience much of their weight loss within half a year and then reach a plateau. This slowdown can be disappointing, but it should be recognized as an opportunity for the body to adjust to its new weight. Reaching a plateau provides a little relief from the distraction of weight-loss dieting. An appropriate goal at this point is to continue the eating and activity behaviors that will maintain weight. Attempting to lose additional weight at this point would require major effort and would almost certainly meet with failure.

The prevalence of **successful weight-loss maintenance** is difficult to determine, in part because researchers have used different criteria. Some look at success after one year and others after five years; some quantify success as 10 or more pounds lost and others as 5 or 10 percent of initial body weight lost. Furthermore, most research studies examine the success of one episode of weight loss in a structured program, but this scenario does not necessarily reflect the experiences of the general population. In reality, most people have lost weight several times in their lifetimes and did so on their own, not in a formal program. Almost 50 percent of people who intentionally lost weight have successfully maintained the loss for at least a year.[134]

Those who are successful in maintaining their weight loss have established vigorous exercise regimens and careful eating patterns, taking in less energy and a lower percentage of kcalories from fat than the national average. Because these people are more efficient at storing fat, they do not have the same flexibility in their food and activity habits as their friends who have never been overweight. With weight loss, metabolism shifts downward so that formerly overweight people require less energy than might be expected given their current body weight and body composition. This decrease in energy expenditure persists over time.[135] Consequently, to keep weight off, they must either eat less or exercise more than people the same size who have never been obese. Put simply, it takes more to prevent weight *regain* than to prevent weight gain.[136]

Physical activity plays a key role in maintaining weight loss. Those who exercise vigorously are far more successful than those who are inactive. Weight maintenance may require a person to expend at least 2500 kcalories in physical activity per week.[137] To accomplish this, a person might exercise either moderately (such as brisk walking at 4 miles per hour) for 60 minutes a day or vigorously (such as fast bicycling at 18 miles per hour) for 30 minutes a day, for example. Being active during both work hours and leisure time also helps a person to maintain weight loss.[138]

In addition to limiting energy intake and exercising regularly, one other strategy helps with weight maintenance: frequent self-monitoring.[139] People who weigh themselves periodically and monitor their eating and exercise habits regularly can detect weight gains in the early stages and promptly initiate changes to prevent relapse.[140]

Losing weight and maintaining the loss may not be easy, but it is possible. Strategies of those who have been successful may differ in the details, but in general, most do the following:

- Eat a low-kcalorie diet (usually small portions four to five times a day).
- Follow a diet that is high in nutrient density and low in energy density.
- Eat a consistent diet day to day (simplicity helps to keep the focus).
- Eat breakfast (curbs hunger).
- Be very physically active (at least 60 minutes of moderate activity daily).
- Weigh frequently (take prompt action with small gains).
- Watch only a limited amount of television (less than 10 hours a week).
- Don't allow small weight gains to become large ones.

Importantly, people who are successful losing weight find that it gets easier with time—the changes in diet and activity patterns become permanent.[141]

successful weight-loss maintenance: achieving a weight loss of at least 10 percent of initial body weight and maintaining the loss for at least one year.

Prevention

Given the information presented up to this point in the chapter, the adage "An ounce of prevention is worth a pound of cure" seems particularly apropos. Preventing weight gain would benefit almost everybody.[142] Obesity is a major risk factor for numerous diseases, and losing weight is challenging and often temporary. Strategies for preventing weight gain ♦ are very similar to those for losing weight, with one exception: they begin early. Over the years, they become an integral part of a person's life. It is much easier for a person to resist doughnuts for breakfast if he rarely eats them. Similarly, a person will have little trouble walking each morning if she has always been active.

♦ To prevent weight gain:

- Eat regular meals and limit snacking.
- Drink water instead of high-kcalorie beverages.
- Select sensible portion sizes and limit daily energy intake to no more than energy expended.
- Become physically active and limit sedentary activities.

Dietary Guidelines for Americans 2010

Maintain appropriate kcalorie balance during each stage of life—childhood, adolescence, adulthood, pregnancy and breastfeeding, and older age.

Public Health Programs

Has anyone in the United States *not* heard the message that obesity raises the risks of chronic diseases and that overweight people should aim for a healthy weight by eating sensibly and becoming physically active? Not likely. Yet implementing such advice is difficult in an environment of abundant food and physical inactivity. To successfully treat obesity, we may have to change the environment in which we live through public health law.[143] Table 9-7 provides examples of proposed legislation and public health strategies that have been suggested to improve our nation's nutrition environment.[144] Some of these strategies may seem radical, but dramatic measures may be needed if we are to curb the obesity epidemic that is sweeping across the nation. Dozens of bills and resolutions are pending in Congress.[145] Whether changes in public policy—such as a tax on sugared beverages and snack foods—will influence diet habits or simply generate revenues remains to be seen.[146] Clearly, effective strategies will need to reach beyond individuals to address social networks, community institutions, and government policies.[147]

IN SUMMARY Preventing weight gains and maintaining weight losses require vigilant attention to diet and physical activity. Taking care of oneself is a lifelong responsibility.

TABLE 9-7 Suggested Public Health Strategies

Strategies	Examples of Suggested Nutritional Strategies	Examples of Successful Nonnutritional Strategies
Impose safety standards to reduce the potential for harm.	• Regulate the energy or fat density of foods. • Regulate the size of packages of high-fat foods.	• Mandate safety glass in automobiles. • Regulate the lead content of paint.
Control commercial advertising to limit the influence of harmful products.	• Improve nutrition labeling and product packaging. • Restrict the promotion of high-fat foods (especially when directed at children).	• Restrict cigarette advertising (especially when directed at children). • Add health warnings to alcoholic beverages.
Control the conditions under which products are sold to limit exposure to hazardous substances.	• Remove high-fat, low-nutrient density foods from school vending machines. • Restrict the number of vendors licensed to sell high-fat foods.	• Mandate minimum-age laws for the use of tobacco, alcohol, and automobiles. • Restrict the number of vendors licensed to sell alcohol.
Control prices to reduce consumption.	• Tax soft drinks and other foods high in kcalories, fat, or sugar.	• Tax alcohol and tobacco.

SOURCES: Adapted from L. O. Gostin, Law as a tool to facilitate healthier lifestyles and prevent obesity, *Journal of the American Medical Association* 297 (2007): 87–90; M. Nestle and M. F. Jacobson, Halting the obesity epidemic: A public health policy approach, *Public Health Reports* 115 (2000): 12–24.

♦ **Underweight** is a body weight so low as to have adverse health effects; it is generally defined as BMI <18.5.

Underweight

Underweight ♦ is a far less prevalent problem than overweight, affecting no more than 5 percent of U.S. adults (review Figure 8-6 on p. 253). Whether the underweight person needs to gain weight is a question of health and, like weight loss, a highly individual matter. People who are healthy at their present weight may stay there; there are no compelling reasons to try to gain weight. Those who are thin because of malnourishment or illness, however, might benefit from a diet that supports weight gain. Medical advice can help make the distinction.

Thin people may find gaining weight difficult. Those who wish to gain weight for appearance's sake or to improve their athletic performance need to be aware that healthful weight gains can be achieved only by physical conditioning combined with high energy intakes. On a high-kcalorie diet alone, a person may gain weight, but it will be mostly fat. Even if the gain improves appearance, it can be detrimental to health and might impair athletic performance. Therefore, in weight gain, as in weight loss, physical activity and energy intake are essential components of a sound plan.

Problems of Underweight

The causes of underweight may be as diverse as those of overweight—genetic tendencies, hunger, appetite, and satiety irregularities; psychological traits; and metabolic factors. Habits learned early in childhood, especially food aversions, may perpetuate themselves.

The demand for energy to support physical activity and growth often contributes to underweight. An active, growing boy may need more than 4000 kcalories a day to maintain his weight and may be too busy to take time to eat adequately. Underweight people find it hard to gain weight due, in part, to their expenditure of energy in adaptive thermogenesis. So much energy may be expended adapting to a higher food intake that at first as many as 750 to 800 extra kcalories a day may be needed to gain a pound a week. Like those who want to lose weight, people who want to gain must learn new habits and learn to like new foods. They are also similarly vulnerable to potentially harmful schemes and would be wise to review the consumer bill of rights on p. 280, using "weight gain" instead of "weight loss" where appropriate.

As described in Highlight 8, the underweight condition anorexia nervosa sometimes develops in people who employ self-denial to control their weight. They go to such extremes that they become severely undernourished, achieving final body weights of 70 pounds or even less. One difference between a person with anorexia nervosa and other underweight people is that starvation is intentional. Another difference is the levels of hormones such as leptin and ghrelin.[148] (See Highlight 8 for a review of anorexia nervosa and other eating disorders.)

Weight-Gain Strategies

Adequacy and balance are the key diet-planning strategies for weight gain. Meals focus on energy-dense foods to provide many kcalories in a small volume and exercise to build muscle. By using the USDA Food Guide recommendations for the higher kcalorie levels (see Table 2-3), a person can gain weight while meeting nutrient needs.

Energy-Dense Foods

Energy-dense foods (the very ones eliminated from a successful weight-loss diet) hold the key to weight gain. Pick the highest-kcalorie items from each food group—that is, milk shakes instead of fat-free milk, salmon instead of snapper, avocados instead of cucumbers, a cup of grape juice instead of a small apple, and whole-wheat muffins instead of whole-wheat bread. Because fat provides more than twice as many kcalories per teaspoon as sugar does, fat adds kcalories without adding much bulk.

Although eating high-kcalorie, high-fat foods is not healthy for most people, it may be essential for an underweight individual who needs to gain weight. An underweight person who is physically active and eating a nutritionally adequate diet can afford a few extra kcalories from fat. For health's sake, it is wise to select foods with monounsaturated and polyunsaturated fats instead of those with saturated or *trans* fats: for example, sautéing vegetables in olive oil instead of butter or hydrogenated margarine.

Regular Meals Daily People who are underweight need to make meals a priority and take the time to plan, prepare, and eat each meal. They should eat at least three healthy meals every day. Another suggestion is to eat meaty appetizers or the main course first and leave the soup or salad until later.

Large Portions Underweight people need to learn to eat more food at each meal. For example, they can add extra slices of ham and cheese on a sandwich for lunch, drink milk from a larger glass, and eat cereal from a larger bowl.

The person should expect to feel full. Most underweight individuals are accustomed to small quantities of food. When they begin eating significantly more, they feel uncomfortable. This is normal and passes over time.

Extra Snacks Because a substantially higher energy intake is needed each day, in addition to eating more food at each meal, it is necessary to eat more frequently. Between-meal snacks do not interfere with later meals; they can readily lead to weight gains. For example, a student might make three sandwiches in the morning and eat them between classes in addition to the day's three regular meals. Snacking on dried fruit, nuts, and seeds is also an easy way to add kcalories.

Juice and Milk Beverages provide an easy way to increase energy intake. Consider that 6 cups of cranberry juice add almost 1000 kcalories to the day's intake. kCalories can be added to milk by mixing in powdered milk or packets of instant breakfast.

For people who are underweight due to illness, concentrated liquid formulas are often recommended because a weak person can swallow them easily. A physician or registered dietitian can recommend high-protein, high-kcalorie formulas to help an underweight person maintain or gain weight. Used in addition to regular meals, these supplements can help considerably.

Exercising to Build Muscles To gain weight, use strength training primarily, and increase energy intake to support that exercise. Eating extra food will then support a gain of both muscle and fat. An additional 500 to 1000 kcalories a day above normal energy needs is enough to support the exercise as well as the building of muscle.[149]

IN SUMMARY **Both the incidence of underweight and the health problems associated with it are less prevalent than overweight and its associated problems. To gain weight, a person must train physically and increase energy intake by selecting energy-dense foods, eating regular meals, taking larger portions, and consuming extra snacks and beverages. Table 9-6 (p. 290) includes a summary of weight-gain strategies.**

Nutrition Portfolio

To enjoy good health and maintain a reasonable body weight, combine sensible eating habits and regular physical activity.

Go to Diet Analysis Plus and choose one of the days on which you have tracked your diet for the entire day. Go to the Energy Balance and Intake vs. Goals reports.

- Calculate your BMI and consider whether you need to lose or gain weight for the sake of good health. If you do need to gain or lose weight, do the Diet Analysis reports give you insight into why you may be overweight or underweight?
- Reflect on your weight over the past year or so and explain any weight gains or losses. Using the Intake vs. Goals report, can you identify areas in which you need to adjust your food intake, perhaps eating more or less?

- Describe the potential risks and possible benefits of fad diets and over-the-counter weight-loss drugs or herbal supplements.

Diet Analysis PLUS To complete this exercise, go to your Diet Analysis Plus at www.cengage.com/sso.

Nutrition on the Net

For further study of topics covered in this chapter, log on to **www.cengagebrain.com** and search for ISBN 084006845X.

- Review the Clinical Guidelines on the Identification, Evaluation, and Treatment of Overweight and Obesity in Adults: **www.nhlbi.nih.gov/guidelines/obesity/ob_home.htm**
- Learn about weight control and the WIN program from the Weight-control Information Network: **www.win.niddk.nih.gov**
- Visit weight-loss support groups, such as Take Off Pounds Sensibly (TOPS), Overeaters Anonymous (OA), and Weight Watchers: **www.tops.org, www.oa.org,** and **www.weightwatchers.com**
- See what the obesity professionals think at the Obesity Society and the American Society for Metabolic and Bariatric Surgery: **www.obesity.org** and **www.asbs.org**
- Learn about the 10,000 Steps Program from Shape Up America!: **www.shapeup.org**
- Find helpful information on achieving and maintaining a healthy weight from the Calorie Control Council: **www.caloriecontrol.org**
- Learn how to end size discrimination from the National Association to Advance Fat Acceptance: **www.naafa.org**
- Explore the Weight Management section of the U.S. government site: **www.nutrition.gov**
- Consider ways to live a healthy life at any weight: **www.bodypositive.com**

References

1. C. L. Ogden and coauthors, Prevalence of overweight and obesity in the United States, 1999–2004, *Journal of the American Medical Association* 295 (2006): 1549–1555; State-specific prevalence of obesity among adults: United States, 2005, *Morbidity and Mortality Weekly Report* 55 (2006): 985–988.
2. Ogden and coauthors, 2006; National Center for Health Statistics, *Chartbook on Trends in the Health of Americans,* 2005, www.cdc.gov/nchs, accessed January 18, 2006.
3. Ogden and coauthors, 2006.
4. C. L. Ogden and coauthors, Obesity among adults in the United States: No statistically significant change since 2003–2004, HCHS Data Brief November 2007, pp. 1–6.
5. P. Hossain, B. Kawar, and M. El Nahas, Obesity and diabetes in the developing world—A growing challenge, *New England Journal of Medicine* 356 (2007): 213–215.
6. A. McCarthy and coauthors, Birth weight; postnatal, infant, and childhood growth; and obesity in young adulthood: Evidence from the Barry Caerphilly Growth Study, *American Journal of Clinical Nutrition* 86 (2007): 907–913.
7. C. Nelson-Dooley and coauthors, Novel treatments for obesity and osteoporosis: Targeting apoptotic pathways in adipocytes, *Current Medicinal Chemistry* 12 (2005): 2215–2225.
8. P. Angulo, Obesity and nonalcoholic fatty liver disease, *Nutrition Reviews* 65 (2007): S57–S63; E. Yan and coauthors, Nonalcoholic fatty liver disease: Pathogenesis, identification, progression, and management, *Nutrition Reviews* 65 (2007): 376–384.
9. W. L. Holland and coauthors, Lipid mediators of insulin resistance, *Nutrition Reviews* 65 (2007): S39–S46; R. Weiss, Fat distribution and storage: How much, where, and how? *European Journal of Endocrinology* 157 (2007): S39–S45.
10. S. B. Votruba and M. D. Jensen, Sex differences in abdominal, gluteal, and thigh LPL activity, *American Journal of Physiology: Endocrinology and Metabolism* 292 (2007): E1823–E1828.
11. M. Patalay and coauthors, The lowering of plasma lipids following a weight reduction program is related to increased expression of the LDL receptor and lipoprotein lipase, *Journal of Nutrition* 135 (2007): 735–739.
12. K. R. Westerterp and coauthors, Dietary fat oxidation as a function of body fat, *American Journal of Clinical Nutrition* 87 (2008): 132–135.
13. G. C. Major and coauthors, Clinical significance of adaptive thermogenesis, *International Journal of Obesity* 31 (2007): 204–212.
14. R. A. Waterland, Epigenetic epidemiology of obesity: Application of epigenomic technology, *Nutrition Reviews* 66 (2008): S21–S23.
15. J. Wardle and coauthors, Evidence for a strong genetic influence on childhood adiposity despite the force of the obesogenic environment, *American Journal of Clinical Nutrition* 87 (2008): 398–404.
16. C. M. Lindgren and M. I. McCarthy, Mechanisms of disease: Genetic insights into the etiology of type 2 diabetes and obesity, *Nature Clinical Practice. Endocrinology & Metabolism* 4 (2008): 156–163.
17. A. Newell and coauthors, Addressing the obesity epidemic: A genomics perspective, *Preventing Chronic Disease* 4 (2007): 1–6; R. J. F. Loos and T. Rankinen, Gene-diet interactions on body weight changes, *Journal of the American Dietetic Association* 105 (2005): S29–S34; S. Tholin and coauthors, Genetic and environmental influences on eating behavior: The Swedish Young Male Twins Study, *American Journal of Clinical Nutrition* 81 (2005): 564–569.
18. J. M. McCaffery and coauthors, Gene x environment interaction of vigorous exercise and body mass index among male Vietnam-era twins, *American Journal of Clinical Nutrition* 89 (2009): 1011–1018.
19. T. Rankinen and C. Bouchard, Genetics of food intake and eating behavior phenotypes in humans, *Annual Review of Nutrition* 26 (2006): 413–434; H. N. Lyon and J. N. Hirschhorn, Genetics of common forms of obesity: A brief overview, *American Journal of Clinical Nutrition* 82 (2005): 215S–217S.
20. R. J. Loos and C. Bouchard, FTO: The first gene contributing to common forms of human obesity, *Obesity Reviews* 9 (2008): 246–250; N. J. Timpson and coauthors, The fat mass- and obesity-associated locus

and dietary intake in children, *American Journal of Clinical Nutrition* 88 (2008): 971–978; A. Körner and coauthors, Polygenic contribution to obesity: Genome-wide strategies reveal new targets, *Frontiers of Hormone Research* 36 (2008): 12–36; R. L. Leibel, Energy in, energy out, and the effects of obesity-related genes, *New England Journal of Medicine* (2008): 2603–2604; C. L. Saunders and coauthors, Meta-analysis of genome-wide linkage studies in BMI and obesity, *Obesity* (2007): 2263–2275.

21. M. Rosenbaum and coauthors, Leptin reverses weight loss—Induced changes in regional neural activity responses to visual food stimuli, *Journal of Clinical Investigation* 118 (2008): 2583–2591; I. S. Farooqi and coauthors, Leptin regulates striatal regions and human eating behavior, *Science* 317 (2007): 1355.
22. P. G. Cammisotto and M. Bendayan, Leptin secretion by white adipose tissue and gastric mucosa, *Histology and Histopathology* 22 (2007): 199–210.
23. I. S. Farooqi and coauthors, Clinical and molecular genetic spectrum of congenital deficiency of the leptin receptor, *New England Journal of Medicine* 356 (2007): 237–247.
24. C. E. Ruhl and coauthors, Body mass index and serum leptin concentration independently estimate percentage body fat in older adults, *American Journal of Clinical Nutrition* 85 (2007): 1121–1126; V. Monti and coauthors, Relationship of ghrelin and leptin hormones with body mass index and waist circumference in a random sample of adults, *Journal of the American Dietetic Association* 106 (2006): 822–828.
25. M. G. Myers, M. A. Cowley, and H. Münzberg, Mechanisms of leptin action and leptin resistance, *Annual Review of Physiology* 70 (2008): 537–556; P. J. Enriori and coauthors, Leptin resistance and obesity, *Obesity* (2006): 254S–258S.
26. A. Shapiro and coauthors, Fructose-induced leptin resistance exacerbates weight gain in response to subsequent high fat feeding, *American Journal of Physiology. Regulatory, Integrative and Comparative Physiology* 295 (2008): R1370–R1375.
27. J. Beltowski, A. Jamroz-Wisniewska, and S. Widomska, Adiponectin and its role in cardiovascular disease, *Cardiovascular & Hematological Disorders Drug Targets* 8 (2008): 7–46.
28. G. Wolf, New insights into thiol-mediated regulation of adiponectin secretion, *Nutrition Reviews* 66 (2008): 642–645; M. Garaulet and coauthors, Adiponectin, the controversial hormone, *Public Health Nutrition* 10 (2007): 1145–1150; Y. Takemura, K. Walsh, and N. Ouchi, Adiponectin and cardiovascular inflammatory responses, *Current Atherosclerosis Report* 9 (2007): 238–243.
29. M. Guerre-Millo, Adiponectin: An update, *Diabetes & Metabolism* 34 (2008): 12–18.
30. V. Popovic and L. H. Duntas, Brain somatic cross-talk: Ghrelin, leptin, and ultimate challengers of obesity, *Nutritional Neuroscience* 8 (2005): 1–5.
31. D. E. Cummings, K. E. Foster-Schubert, and J. Overduin, Gherlin and energy balance: Focus on current controversies, *Current Drug Targets* 6 (2005): 153–169.
32. Cummings, Foster-Schubert, and Overduin, 2005.
33. Monti and coauthors, 2006.
34. Cummings, Foster-Schubert, and Overduin, 2005.
35. D. R. Broom and coauthors, Exercise induced suppression of acylated ghrelin in humans, *Journal of Applied Physiology* 102 (2007): 2165–2171; Cummings, Foster-Schubert, and Overduin, 2005.
36. N. D. Kohatsu and coauthors, Sleep duration and body mass index in rural population, *Archives of Internal Medicine* 166 (2006): 1701–1705; S. R. Patel and coauthors, Association between reduced sleep and weight gain in women, *American Journal of Epidemiology* 164 (2006): 947–954; R. D. Verona and coauthors, Overweight and obese patients in a primary care population report less sleep than patients with a normal body mass index, *Archives of Internal Medicine* 165 (2005): 25–34.
37. P. Hamet and J. Tremblay, Genetics of sleep-wake cycles and its disorders, *Metabolism* 55 (2006): S7–S12.
38. J. Orr and B. Davy, Dietary influences on peripheral hormones regulating energy intake: Potential applications for weight management, *Journal of the American Dietetic Association* 105 (2005): 1115–1124.
39. S. Enerbäck, The origins of brown adipose tissue, *New England Journal of Medicine* 360 (2009): 2021–2023; A. S. Avram, M. M. Avram, and W. D. James, Subcutaneous fat in normal and diseased states: 2. Anatomy and physiology of white and brown adipose tissue, *Journal of the American Academy of Dermatology* 53 (2005): 671–673
40. J. S. Kim-Han and L. L. Dugan, Mitochondrial uncoupling proteins in the central nervous system, *Antioxidants and Redox Signaling* 7 (2005): 1173–1181; R. J. F. Roos and T. Rankinen, Gene-diet interactions on body weight changes, *Journal of the American Dietetic Association* 105 (2005): S29–S34; P. Trayhurn, The biology of obesity, *Proceedings of the Nutrition Society* 64 (2005): 31–38.
41. M. Harper, K. Green, and M. D. Brand, The efficiency of cellular energy transduction and its implications for obesity, *Annual Review of Nutrition* 28 (2008): 13–33.
42. P. Laurberg, S. Andersen, and J. Karmisholt, Cold adaptation and thyroid hormone metabolism, *Hormone and Metabolic Research* 37 (2005): 545–549.
43. Avram, Avram, and James, 2005.
44. W. D. van Marken Lichtenbelt and coauthors, Cold-activated brown adipose tissue in healthy men, *New England Journal of Medicine* 360 (2009): 1500–1508.
45. A. M. Cypess and coauthors, Identification and importance of brown adipose tissue in adult humans, *New England Journal of Medicine* 360 (2009): 1509–1517.
46. K. A. Virtanen and coauthors, Functional brown adipose tissue in healthy adults, *New England Journal of Medicine* 360 (2009): 1518–1525.
47. L. Qi and Y. A. Cho, Gene-environment interaction and obesity, *Nutrition Reviews* 66 (2008): 684–694.
48. W. P. James, The fundamental drivers of the obesity epidemic, *Obesity Reviews* 9 (2008): S6–S13.
49. I. Romao and J. Roth, Genetic and environmental interactions in obesity and type 2 diabetes, *Journal of the American Dietetic Association* 108 (2008): S24–S28.
50. N. A. Christakis and J. H. Fowler, The spread of obesity in a large social network over 32 years, *New England Journal of Medicine* 357 (2007): 370–379.
51. J. M. Abbot and coauthors, Psychosocial and behavioral profile and predictors of self-reported energy underreporting in obese middle-aged women, *Journal of the American Dietetic Association* 108 (2008): 114–119; A. Amend and coauthors, Validation of dietary intake data in black women with type 2 diabetes, *Journal of the American Dietetic Association* 107 (2007): 112–117; R. L. Bailey and coauthors, Assessing the effect of underreporting energy intake on dietary patterns and weight status, *Journal of the American Dietetic Association* 107 (2007): 64–71; S. Hendrickson and R. Mattes, Financial incentive for diet recall accuracy does not affect reported energy intake or number of underreporters in a sample of overweight females, *Journal of the American Dietetic Association* 107 (2007): 118–121; J. Maurer and coauthors, The psychological and behavioral characteristics related to energy misreporting, *Nutrition Reviews* 64 (2006): 53–66.
52. R. N. Close and D. A. Schoeller, The financial reality of overeating, *Journal of the American College of Nutrition* 25 (2006): 203–209.
53. B. Wansink and J. Kim, Bad popcorn in big buckets: Portion size can influence intake as much as taste, *Journal of Nutrition Education and Behavior* 37 (2005): 242–245.
54. L. Johnson and coauthors, Energy-dense, low-fiber, high-fat dietary pattern is associated with increased fatness in childhood, *American Journal of Clinical Nutrition* 87 (2008): 846–854; N. C. Howarth and coauthors, Dietary energy density is associated with overweight status among 5 ethnic groups in the Multiethnic Cohort Study, *Journal of Nutrition* 136 (2006): 2243–2248.
55. B. J. Rolls, L. S. Roe, and J. S. Meengs, Reductions in portion size and energy density of foods are additive and lead to sustained decreases in energy intake, *American Journal of Clinical Nutrition* 83 (2006): 11–17.
56. L. H. Epstein and coauthors, Price and maternal obesity influence purchasing of low- and high-energy-dense foods, *American Journal of Clinical Nutrition* 86 (2007): 914–922; P. Monsivais and A. Drewnowski, The rising cost of low-energy-density foods, *Journal of the American*

Dietetic Association 107 (2007): 2071–2076; A. Drewnowski and N. Darmon, The economics of obesity: Dietary energy density and energy cost, *American Journal of Clinical Nutrition* 82 (2005): 265S–273S.

57. D. S. Ludwig and M. Nestle, Can the food industry play a constructive role in the obesity epidemic? *Journal of the American Medical Association* 300 (2008): 1808–1811.
58. M. Condrasky and coauthors, Chefs' opinions of restaurant portion sizes, *Obesity* (2007): 2086–2094.
59. J. L. Black and J. Macinko, Neighborhoods and obesity, *Nutrition Reviews* 66 (2008): 2–20; M. C. Nelson and coauthors, Built and social environments: Associations with adolescent overweight and activity, *American Journal of Prevention Medicine* 31 (2006): 109–117; K. M. Booth, M. M. Pinkston, and W.S.C. Poston, Obesity and the built environment, *Journal of the American Dietetic Association* 105 (2005): S110–S117.
60. J. A. Levine and coauthors, Non-exercise activity thermogenesis: The Crouching Tiger Hidden Dragon of society weight gain, *Arteriosclerosis, Thrombosis, and Vascular Biology* 26 (2006): 729–736; J. A. Levine and coauthors, Interindividual variation in posture allocation: Possible role in human obesity, *Science* 307 (2005): 584–586.
61. J. A. Teske, C. J. Billington, and C. M. Kotz, Neuropeptidergic mediators of spontaneous physical activity and non-exercise activity thermogenesis, *Neuroendocrinology* 87 (2008): 71–90.
62. J. A. Levine, Nonexercise activity thermogenesis—Liberating the life-force, *Journal of Internal Medicine* 262 (2007): 273–287; M. Marra and coauthors, BMR variability in women of different weight, *Clinical Nutrition* 26 (2007): 567–572.
63. American on the move: Steps to a healthier way of life, press release, September 10, 2007.
64. R. F. Kushner and D. J. Blatner, Risk assessment of the overweight and obese patient, *Journal of the American Dietetic Association* 105 (2005): S53–S62; National Institutes of Health Obesity Education Initiative, *The Practical Guide: Identification, Evaluation, and Treatment of Overweight and Obesity in Adults,* NIH publication no. 00-4084 (Washington, D.C.: U.S. Department of Health and Human Services, 2000).
65. National Institutes of Health Obesity Education Initiative, 2000.
66. Ethics opinion: Weight loss products and medications, *Journal of the American Dietetic Association* 108 (2008): 2109–2113; H. M. Blanck and coauthors, Use of nonprescription dietary supplements for weight loss is common among Americans, *Journal of the American Dietetic Association* 107 (2007): 441–447.
67. J. T. Dwyer, D. B. Allison, and P. M. Coates, Dietary supplements in weight reduction, *Journal of the American Dietetic Association* 105 (2005): S80–S86.
68. R. F. Kushner, Anti-obesity drugs, *Expert Opinion on Pharmacotherapy* 9 (2008): 1339–1350.
69. D. Rucker and coauthors, Long-term pharmacotherapy for obesity and overweight: Updated meta-analysis, *British Medical Journal* 335 (2007): 1194–1199.
70. R. H. Eckel, Nonsurgical management of obesity in adults, *New England Journal of Medicine* 358 (2008): 1941–1950.
71. S. B. Moyers, Medications as adjunct therapy for weight loss: Approved and off-label agents in use, *Journal of the American Dietetic Association* 105 (2005): 948–959.
72. E. J. DeMaria, Bariatric surgery for morbid obesity, *New England Journal of Medicine* 356 (2007): 2176–2183.
73. American Society for Metabolic & Bariatric Surgery, Fact Sheet, www.asbs.org/Newsite07/media/fact-sheet1_bariatric-surgery.pdf, accessed July 14, 2008.
74. Cummings, Foster-Schubert, and Overduin, 2005.
75. L. Sjöström and coauthors, Effects of bariatric surgery on mortality in Swedish obese subjects, *New England Journal of Medicine* 357 (2007): 741–752; G. L. Blackburn, Solutions in weight control: Lessons from gastric surgery, *American Journal of Clinical Nutrition* 82 (2005): 248S–252S.
76. J. B. Dixon and coauthors, Adjustable gastric banding and conventional therapy for type 2 diabetes: A randomized controlled trial, *Journal of the American Medical Association* 299 (2008): 316–323; E. N. Hansen, A. Torquati, and N. N. Abumrad, Results of bariatric surgery, *Annual Review of Nutrition* 26 (2006): 481–511.
77. S. Singh and A. Kumar, Wernicke encephalopathy after obesity surgery, *Neurology* 68 (2007): 807–811; M. Shah, V. Simha, and A. Garg, Long-term impact of bariatric surgery on body weight, co-morbidities, and nutritional status: A review, *Journal of Clinical Endocrinology and Metabolism* 91 (2006): 4223–4231.
78. J. A. Vogel and coauthors, Reduction in predicted coronary heart disease risk after substantial weight reduction after bariatric surgery, *American Journal of Cardiology* 99 (2007): 222–226; Hansen, Torquati, and Abumrad, 2006.
79. T. D. Adams and coauthors, Long-term mortality after gastric bypass surgery, *New England Journal of Medicine* 357 (2007): 753–761; Sjöström and coauthors, 2007.
80. Hansen, Torquati, and Abumrad, 2006.
81. C. A. Nonas and G. D. Foster, Setting achievable goals for weight loss, *Journal of the American Dietetic Association* 105 (2005): S118–S123.
82. D. R. Jacobs and coauthors, Association of 1-y changes in diet pattern with cardiovascular disease risk factors and adipokines: Results from the 1-y randomized Oslo Diet and Exercise Study, *American Journal of Clinical Nutrition* 89 (2009): 509–517; M. L. Fernandez, The metabolic syndrome, *Nutrition Reviews* 65 (2007): S30–S34; C. Galani and H. Schneider, Prevention and treatment of obesity with lifestyle interventions: Review and meta-analysis, *International Journal of Public Health* 52 (2007): 348–359.
83. G. D. Foster and coauthors, Obese patients' perceptions of treatment outcomes and the factors that influence them, *Archives of Internal Medicine* 161 (2001): 2133–2139.
84. National Institutes of Health Obesity Education Initiative, 2000, p. 2.
85. Position of the American Dietetic Association: Weight management, *Journal of the American Dietetic Association* 109 (2009): 330–346.
86. National Institutes of Health Obesity Education Initiative, 2000, pp. 26–27.
87. M. T. Timlin and coauthors, Breakfast eating and weight change in a 5-year prospective analysis of adolescents: Project EAT (Eating Among Teens), *Pediatrics* 121 (2008): e638; M. T. Timlin and M. A. Pereira, Breakfast frequency and quality in the etiology of adult obesity and chronic diseases, *Nutrition Reviews* 65 (2007): 268–281.
88. A. K. Kant and coauthors, Association of breakfast energy density with diet quality and body mass index in American adults: National Health and Nutrition Examination Surveys, 1999–2004, *American Journal of Clinical Nutrition* 88 (2008): 1396–1404.
89. M. Bulló and coauthors, Inflammation, obesity and comorbidities: The role of diet, *Public Health Nutrition* 10 (2007): 1164–1172.
90. J. T. Dwyer, D. B. Allison, and P. M. Coates, Dietary supplements in weight reduction, *Journal of the American Dietetic Association* 105 (2005): S80–S86.
91. Position of the American Dietetic Association, 2009.
92. B. J. Rolls, L. S. Roe, and J. S. Meengs, The effect of large portion sizes on energy intake is sustained for 11 days, *Obesity* 15 (2007): 1535–1543; Wansink and Kim, 2005.
93. M. P. Mattson, Energy intake, meal frequency, and health: A neurobiological perspective, *Annual Review of Nutrition* 25 (2005): 237–260.
94. K. E. Leahy, L. L. Birch, and B. J. Rolls, Reducing the energy density of multiple meals decreases the energy intake of preschool-age children, *American Journal of Clinical Nutrition* 88 (2008): 1459–1468; J. A. Ello-Martin and coauthors, Dietary energy density in the treatment of obesity: A year-long trial comparing 2 weight-loss diets, *American Journal of Clinical Nutrition* 85 (2007): 1465–1477; M. J. Franz and coauthors, Weight-loss outcomes: A systematic review and meta-analysis of weight-loss clinical trials with a minimum 1-year follow-up, *Journal of the American Dietetic Association* 107 (2007): 1755–1767; J. H. Ledikwe, J. A. Ello-Martin, and B. J. Rolls, Portion sizes and the obesity epidemic, *Journal of Nutrition* 135 (2005): 905–909; J. A. Ello-Martin, J. H. Ledikwe, and B. J. Rolls, The influence of food portion size and energy density on energy intake: Implications for weight management, *American Journal of Clinical Nutrition* 82 (2005): 236S–241S.
95. B. J. Rolls, A. Drewnowski, and J. H. Ledikwe, Changing the energy density of the diet as a strategy for weight management, *Journal of the American Dietetic Association* 105 (2005): S98–S103.

96. J. H. Ledikwe and coauthors, Reductions in dietary energy density are associated with weight loss in overweight and obese participants in the PREMIER trial, *American Journal of Clinical Nutrition* 85 (2007): 1212–1221; J. H. Ledikwe and coauthors, Low-energy-density diets are associated with high diet quality in adults in the United States, *Journal of the American Dietetic Association* 106 (2006): 1172–1180.
97. P. Monsivais and A. Drewnowski, The rising cost of low-energy-density foods, *Journal of the American Dietetic Association* 107 (2007): 2071–2076.
98 J. E. Flood and B. J. Rolls, Soup preloads in a variety of forms reduce meal energy intake, *Appetite* 49 (2007): 626–634.
99. B. M. Davy and coauthors, Water consumption reduces energy intake at a breakfast meal in obese older adults, *Journal of the American Dietetic Association* 108 (2008): 1236–1239; E. L. Van Walleghen and coauthors, Pre-meal water consumption reduces meal energy intake in older but not younger subjects, *Obesity* 15 (2007): 93–99.
100. B. M. Popkin and coauthors, A new proposed guidance system for beverage consumption in the United States, *American Journal of Clinical Nutrition* 83 (2006): 529–542.
101. J. E. Flood and coauthors, The effect of increased beverage portion size on energy intake at a meal, *Journal of the American Dietetic Association* 106 (2006): 1984–1990.
102. A. M. Andrade and coauthors, Eating slowly led to decreases in energy intake within meals in healthy women, *Journal of the American Dietetic Association* 108 (2008): 1186–1191.
103. E. Stice and coauthors, Relation between obesity and blunted striatal response to food is moderated by *Taq*IA A1 allele, *Science* 322 (2008): 449–452.
104. C. H. Llewellyn and coauthors, Eating rate is a heritable phenotype related to weight in children, *American Journal of Clinical Nutrition* 88 (2008): 1560–1566.
105. F. Bellisle and A. Drewnowski, Intense sweeteners, energy intake and the control of body weight, *European Journal of Clinical Nutrition* 61 (2007): 691–700; P. Monsivais, M. M. Perrigue, and A. Drewnowski, Sugars and satiety: Does the type of sweetener make a difference? *American Journal of Clinical Nutrition* 86 (2007): 116–123.
106. Bellisle and Drewnowski, 2007.
107. J. M. Jakicic and A. D. Otto, Treatment and prevention of obesity: What is the role of exercise? *Nutrition Reviews* 64 (2006): S57–S61.
108. S. J. Elder and S. B. Roberts, The effects of exercise on food intake and body fatness: A summary of published studies, *Nutrition Reviews* 65 (2007): 1–19.
109. J. Kruger, M. M. Yore, and H. W. Kohl, III, Leisure-time physical activity patterns by weight control status: 1999-2002 NHANES, *Medicine & Science in Sports & Exercise* 39 (2007): 788–795.
110. Committee on Dietary Reference Intakes, *Dietary Reference Intakes for Energy, Carbohydrate, Fiber, Fat, Fatty Acids, Cholesterol, Protein, and Amino Acids* (Washington, D.C.: National Academies Press, 2005).
111. L. L. Frank and coauthors, Effects of exercise on metabolic risk variables in overweight postmenopausal women: A randomized clinical trial, *Obesity Research* 13 (2005): 615–625.
112. K. Ohkawara and coauthors, Twenty-four-hour analysis of elevated energy expenditure after physical activity in a metabolic chamber: Models of daily total energy expenditure, *American Journal of Clinical Nutrition* 87 (2008): 1268–1276.
113. K. H. Schmitz and coauthors, Strength training and adiposity in premenopausal women: Strong, Healthy, and Empowered study, *American Journal of Clinical Nutrition* 86 (2007): 566–572.
114. S. J. Elder and S. B. Roberts, The effects of exercise on food intake and body fatness: A summary of published studies, *Nutrition Reviews* 65 (2007): 1–19.
115. H. R. Wyatt and coauthors, A Colorado statewide survey of walking and its relation to excessive weight, *Medicine and Science in Sports and Exercise* 37 (2005): 724–730.
116. D. M. Bravata and coauthors, Using pedometers to increase physical activity and improve health: A systematic review, *Journal of the American Medical Association* 298 (2007): 2296–2304.
117. B. J. Nicklas and coauthors, Effect of exercise intensity on abdominal fat loss during calorie restriction in overweight and obese postmenopausal women: A randomized, controlled trial, *American Journal of Clinical Nutrition* 89 (2009): 1043–1052.
118. B. Wansink, J. E. Painter, and Y. K. Lee, The office candy dish: Proximity's influence on estimated and actual consumption, *International Journal of Obesity* 30 (2006): 871–875.
119. P. Pliner and coauthors, Meal duration mediates the effect of "social facilitation" on eating in humans, *Appetite* 46 (2006): 189–198.
120. J. M. Poothullil, Recognition of oral sensory satisfaction and regulation of the volume of intake in humans, *Nutritional Neuroscience* 8 (2005): 245–250.
121. Wansink, Painter, and Lee, 2006.
122. M. M. Hetherington and coauthors, Understanding variety: Tasting different foods delays satiation, *Physiology and Behavior* 87 (2006): 263–271.
123. B. Wansink, J. E. Painter, and J. North, Bottomless bowls: Why visual cues of portion size may influence intake, *Obesity Research* 13 (2005): 93–100.
124. B. Wansink and M. M. Cheney, Super bowls: Serving bowl size and food consumption, *Journal of the American Medical Association* 293 (2005): 1727–1728.
125. B. Wansink, K. van Ittersum, and J. E. Painter, Ice cream illusions bowls, spoons, and self-served portion sizes, *American Journal of Preventive Medicine* 31 (2006): 240–243.
126. B. Wansink and K. van Ittersum, Shape of glass and amount of alcohol poured: Comparative study of effect of practice and concentration, *British Medical Journal* 331 (2005): 1512–1514.
127. B. J. Rolls and coauthors, Using a smaller plate did not reduce energy intake at meals, *Appetite* 49 (2007): 652–660.
128. J. W. Anderson, S. B. Conley, and A. S. Nicholas, One hundred-pound weight losses with an intensive behavioral program: Changes in risk factors in 118 patients with long-term follow-up, *American Journal of Clinical Nutrition* 86 (2007): 301–307.
129. L. A. Berkel and coauthors, Behavioral interventions for obesity, *Journal of the American Dietetic Association* 105 (2005): S35–S43; G. D. Foster, A. P. Makris, and B. A. Bailer, Behavioral treatment of obesity, *American Journal of Clinical Nutrition* 82 (2005): 230S–235S.
130. D. J. Hyman and coauthors, Simultaneous vs sequential counseling for multiple behavior change, *Archives of Internal Medicine* 167 (2007): 1152–1158.
131. K. G. Volpp and coauthors, Financial incentive-based approaches for weight loss: A randomized trial, *Journal of the American Medical Association* 300 (2008): 2631–2637.
132. A. N. Fabricatore, Behavior therapy and cognitive-behavioral therapy of obesity: Is there a difference? *Journal of the American Dietetic Association* 107 (2007): 92–99.
133. L. P. Svetkey and coauthors, Comparison of strategies for sustaining weight loss: The weight loss maintenance randomized controlled trial, *Journal of the American Medical Association* 299 (2008): 1139–1148.
134. G. L. Blackburn, and B. A. Waltman, Expanding the limits of treatment: New strategic initiatives, *Journal of the American Dietetic Association* 105 (2005): S131–S135.
135. M. Rosenbaum and coauthors, Long-term persistence of adaptive thermogenesis in subjects who have maintained a reduced body weight, *American Journal of Clinical Nutrition* 88 (2008): 906–912.
136. S. Phelan and coauthors, Empirical evaluation of physical activity recommendations for weight control in women, *Medicine & Science in Sports & Exercise* 39 (2007): 1832–1836.
137. V. A. Catenacci and coauthors, Physical activity patterns in the National Weight Control Registry, *Obesity* 16 (2008): 153–161; D. F. Tate and coauthors, Long-term weight losses associated with prescription of higher physical activity goals. Are higher levels of physical activity protective against weight regain? *American Journal of Clinical Nutrition* 85 (2007): 954–959.
138. E. C. Weiss and coauthors, Weight regain in U.S. adults who experienced substantial weight loss, 1999–2002, *American Journal of Preventive Medicine* 33 (2007): 34–40; D. A. Raynor and coauthors,

Television viewing and long-term weight maintenance: Results from the National Weight Control Registry, *Obesity* 14 (2006): 1816–1824.

139. D. L. Helsel, J. M. Jakicic, and A. D. Otto, Comparison of techniques for self-monitoring eating and exercise behaviors on weight loss in a correspondence-based intervention, *Journal of the American Diet Association* 107 (2007): 1807–1810; L. F. Tinker and coauthors, Predictors of dietary change and maintenance in the Women's Health Initiative Dietary Modification Trial, *Journal of the American Dietetic Association* 107 (2007): 1155–1165.
140. R. R. Wing and coauthors, A self-regulation program for maintenance of weight loss, *New England Journal of Medicine* 355 (2006): 1563–1571.
141. R. R. Wing and S. Phelan, Long-term weight loss maintenance, *American Journal of Clinical Nutrition* 82 (2005): 222S–225S.
142. J. O. Hill, H. Thompson, and H. Wyatt, Weight maintenance: What's missing? *Journal of the American Dietetic Association* 105 (2005): S63–S66.
143. W. P. James, The epidemiology of obesity: The size of the problem, *Journal of Internal Medicine* 263 (2008): 336–352; L. O. Gostin, Law as a tool to facilitate healthier lifestyles and prevent obesity, *Journal of the American Medical Association* 297 (2007): 87–90; M. M. Mello, D. M. Studdert, and T. A. Brennan, Obesity: The new frontier of public health law, *New England Journal of Medicine* 354 (2006): 2601–2610.
144. Gostin, 2007.
145. R. Smith, Passing an effective obesity bill, *Journal of the American Dietetic Association* 106 (2006): 1349–1350.
146. K. D. Brownell and T. R. Frieden, Ounces of prevention—The public policy case for taxes on sugared beverages, *New England Journal of Medicine* 360 (2009): 1805–1808; F. Kuchler, A. Tegene, and J. M. Harris, Taxing snack foods: What to expect for diet and tax revenues, *Current Issues in Economics of Food Markets,* Agriculture Information Bulletin No. 747–08, August 2004.
147. T. T. Huang and T. A. Glass, Transforming research strategies for understanding and preventing obesity, *Journal of the American Medical Association* 300 (2008): 1811–1813.
148. N. Germain and coauthors, Constitutional thinness and lean anorexia nervosa display opposite concentrations of peptide YY, glucagon-like peptide 1, ghrelin, and leptin, *American Journal of Clinical Nutrition* 85 (2007): 967–971.
149. Position Paper: Nutrition and athletic performance: Position of the American Dietetic Association, Dietitians of Canada, and the American College of Sports Medicine, *Journal of the American Dietetic Association* 100 (2000): 1543–1556.

HIGHLIGHT 9

The Latest and Greatest Weight-Loss Diet—Again

© Geri Engberg

To paraphrase William Shakespeare, "a fad diet by any other name would still be a fad diet." And the names are legion: the Atkins Diet, the Calories Don't Count Diet, the Cheater's Diet, the South Beach Diet, the Zone Diet.* Year after year, "new and improved" diets appear on bookstore shelves and circulate among friends. People of all sizes eagerly try the best diet on the market ever, hoping that this one will really work. Sometimes these diets seem to work for a while, but more often than not, their success is short-lived. Then another diet takes the spotlight. Here's how Dr. K. Brownell, an obesity researcher at Yale University, describes this phenomenon: "When I get calls about the latest diet fad, I imagine a trick birthday cake candle that keeps lighting up and we have to keep blowing it out."

Realizing that fad diets do not offer a safe and effective long-term plan for weight loss, health professionals speak out, but they never get the candle blown out permanently. New fad diets can keep making outrageous claims because no one requires their advocates to prove what they say. Fad diet gurus do not have to conduct credible research on the benefits or dangers of their diets. They can simply make recommendations and then later, if questioned, search for bits and pieces of research that support the conclusions they have already reached. That's backward. Diet and health recommendations should *follow* years of sound scientific research *before* being offered to the public.

Because anyone can publish anything—in books or on the Internet—peddlers of fad diets can make unsubstantiated statements that fall far short of the truth but sound impressive to the uninformed. They often offer distorted bits of legitimate research. They may start with one or more actual facts but then leap from one erroneous conclusion to the next. Anyone who wants to believe these claims has to wonder how the thousands of scientists working on obesity research over the past century could possibly have missed such obvious connections. Table H9-1 (p. 306) presents some of the claims and truths of fad diets.

Fad diets come in almost as many shapes and sizes as the people who search them out. Some restrict fats or carbohydrates, some limit portion sizes, some focus on food combinations, and some claim that a person's genetic type or blood type determines the foods best suited to manage weight and prevent disease. Table H9-2 (p. 307) compares some of today's more popular diets. Table H9-3 (p. 308) offers guidelines for identifying fad diets and other weight-loss scams; it includes the hallmarks of a reasonable weight-loss program as well.

Fad Diets' Appeal

With more than half of our nation's adults overweight and many more concerned about their weight, the market for a weight-loss book, product, or program is huge (no pun intended). Americans spend an estimated $33 billion a year on weight-loss books and products. Even a plan that offers only minimal weight-loss success easily attracts a following.

Perhaps the greatest appeal of fad diets is that they tend to ignore dietary recommendations. Foods such as meats and milk products that need to be selected carefully to limit saturated fat can be eaten with abandon. Whole grains, legumes, vegetables, and fruits that should be eaten in abundance can now be bypassed. For some people, this is a dream come true: steaks without the potatoes, ribs without the coleslaw, and meatballs without the pasta. Who can resist the promise of weight loss while eating freely from a list of favorite foods?

Dieters are also lured into fad diets by sophisticated—yet often erroneous—explanations of the metabolic consequences of eating certain foods. Terms such as *eicosanoids* and *de novo lipogenesis* are scattered about, often intimidating readers into believing that the authors must be right given their brilliance in understanding the body.

*The following sources offer comparisons and evaluations of various fad diets for your review: New diet winners, *Consumer Reports*, June 2007, pp. 12–17; Battle of the diet books II, *Nutrition Action Healthletter*, July/August 2006, pp. 10–11; B. Liebman, Weighing the diet books, *Nutrition Action Healthletter*, January/February 2004, pp. 1–8; S. T. St. Jeor and coauthors, Dietary protein and weight reduction. A statement for healthcare professionals from the nutrition committee of the Council on Nutrition, Physical Activity, and Metabolism of the American Heart Association, *Circulation* 104 (2001): 1869–1874.

If fad diets were as successful as some people claim, then consumers who tried them would lose weight, and their obesity problems would be solved. But this is not the case. Similarly, if fad diets were as worthless as others claim, then consumers would eventually stop pursuing them. Clearly, this is not happening either. Most fad diets have enough going for them that they work for some people at least for a short time, but they fail to produce long-lasting results for most people.

Don't Count kCalories

Who wants to count kcalories? Even experienced dieters find counting kcalories burdensome, not to mention timeworn. They want a new, easy way to lose weight, and fad diet plans seem to offer this boon. But, though fad diets often claim to disregard kcalories, their design typically ensures a low energy intake. Most of the sample menu plans, especially in the early stages, are designed to deliver an average of 1200 kcalories a day.

Even when counting kcalories is truly not necessary, total kcalories tend to be low simply because food intake is so limited. Diets that omit hundreds of foods and several food groups limit a person's options and lack variety. Chapter 2 praises variety as a valuable way to ensure an adequate intake of nutrients, but variety also entices people to eat more food and gain more weight. Without variety, some people lose interest in eating, which further reduces energy intake. Even if the allowed foods are favorites, eating the same foods week after week can become monotonous.

Without its refried beans, tortilla wrapping, and chopped vegetables, a burrito is reduced to a pile of ground beef. Without the baked potato, there's no need for butter and sour cream. Weight loss occurs because of the low energy intake. This is an important point. Any diet can produce weight loss, at least temporarily, if intake is restricted. The real value of a diet is determined by its ability to maintain weight loss and support good health over the long term. The goal is not simply weight loss, but health gains—and most fad diets cannot support optimal health over time.

When food choices are limited, nutrient intakes may be inadequate. To help shore up some of these inadequacies, fad diets often recommend a dietary supplement. Conveniently, many of the

TABLE H9-1 The Claims and Truths of Fad Diets

The Claim:	You can lose weight "easily."
The Truth:	Most fad diet plans have complicated rules that require you to calculate protein requirements, count carbohydrate grams, combine certain foods, time meal intervals, purchase special products, plan daily menus, and measure serving sizes.
The Claim:	You can lose weight by eating a specific ratio of carbohydrate, protein, and fat.
The Truth:	Weight loss depends on expending more energy than you take in, not on the proportion of energy nutrients.
The Claim:	This "revolutionary diet" can "reset your genetic code."
The Truth:	You inherited your genes and cannot alter your genetic code.
The Claim:	High-protein diets are popular, selling more than 20 million books, because they work.
The Truth:	Weight-loss books are popular because people grasp for quick fixes and simple solutions to their weight problems. If book sales were an indication of weight-loss success, we would be a lean nation—but they're not, and neither are we.
The Claim:	People gain weight on low-fat diets.
The Truth:	People can gain weight on low-fat diets if they overindulge in carbohydrates and proteins while cutting fat; low-fat diets are not necessarily low-kcalorie diets. But people can also lose weight on low-fat diets if they cut kcalories as well as fat.
The Claim:	High-protein diets energize the brain.
The Truth:	The brain depends on glucose for its energy; the primary dietary source of glucose is carbohydrate, not protein.
The Claim:	Thousands of people have been successful with this plan.
The Truth:	Authors of fad diets have not published their research findings in scientific journals. Success stories are anecdotal and failures are not reported.
The Claim:	Carbohydrates raise blood glucose levels, triggering insulin production and fat storage.
The Truth:	Insulin promotes fat storage when energy intake exceeds energy needs. Furthermore, insulin is only one hormone involved in the complex processes of maintaining the body's energy balance and health.
The Claim:	Eat protein and lose weight.
The Truth:	For every complicated problem, there is a simple—and wrong—solution.

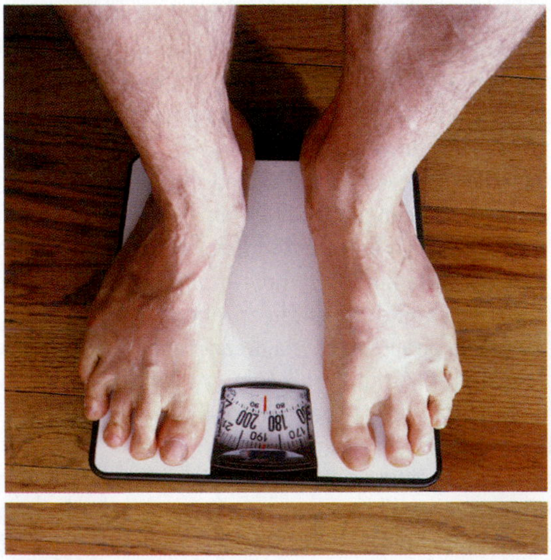

The wise consumer seeks a diet that supports not only weight loss, but health gains.

TABLE H9-2 Popular Diets Compared

Diet	Major Premise Promoted	Strong Point(s)	Weak Point(s)
Atkins Diet	• People are overweight or obese because they have metabolic imbalances caused by eating too many carbohydrates; by restricting carbohydrates, these imbalances can be corrected. • You can lose weight without lowering kcalorie intake.	• Quick, short-term weight loss is achieved.	• Restricts carbohydrates to a level that induces ketosis. • Ketosis can cause nausea, lightheadedness, and fatigue. • Ketosis can worsen existing medical problems such as kidney disease. • A diet high in fat such as Atkins can increase the risk of heart disease and some cancers.
Cheater's Diet	• Successful weight loss depends on eliminating boredom and allowing indulgences. • Cheating on weekends "stokes your metabolism."	• Meals are proportioned one-half fruit or vegetables, one-fourth lean protein, and one-fourth whole grains. • Encourages as much exercise as possible.	• No scientific data on cheating boosting metabolism or supporting weight loss.
Eat Right 4 Your Type	• Your blood type determines which foods you should eat or not eat.	None	• Food groups or individual foods are excluded, depending on blood type. • No scientific data on the relationship between blood type and food choices.
Glucose Revolution	• Low glycemic index foods satisfy hunger, control blood glucose, and promote weight loss.	• Emphasizes fiber-rich vegetables, legumes, fruits, and whole grains. • Minimizes saturated fat intake.	• Difficult to know the glycemic index of some foods.
Ornish Diet	• By strictly limiting fat (both animal and vegetable), you eat fewer kcalories without eating less food.	• High-fiber, low-fat foods in this plan can lower blood cholesterol and blood pressure.	• So little fat that essential fatty acids may be lacking. • Limits fish, nuts, and olive oil, which may protect against heart disease.
Pritikin Program	• By eating low-fat, mainly plant-based foods, you can eat more food and still feel satisfied.	• No food group is completely eliminated in this high-fiber, low-fat diet program. • Some use of foods rich in omega-3 fatty acids is encouraged.	• For some people, very low-fat diets may be unsatisfying and therefore difficult to adhere to.
Sonoma Diet	• Enjoying portion-controlled Mediterranean-style foods supports weight loss and promotes good health.	• Emphasizes nutrient-dense foods.	• Initial phase restricts fruits and limits milk products.
South Beach Diet	• Eating "good carbohdrates" such as vegetables, whole-wheat pastas, and brown rice will maintain satiety and resist cravings for "bad carbohydrates" such as white rice and potatoes.	• Encourages consumption of vegetables, lean meats, and fish, and the use of unsaturated oils when cooking. • Restricts fatty meats and cheeses as well as sweets.	• Starchy carbohydrates and all fruits are completely excluded during the first two weeks.
Ultimate Weight Solution Diet	• Foods that require great effort to prepare and eat are nutrient-dense; eating these kinds of foods (raw vegetables, vegetable soups, whole grains, beans, meats, poultry, and fish) will lead to weight loss. • Foods that take little effort to prepare and eat provide excess kcalories relative to nutrients; eating these kinds of foods (fast foods, puddings, high-kcalorie convenience foods, processed foods) leads to uncontrolled eating and weight gain.	• Encourages consumption of lean meats and fish; whole grains; vegetables; fruit; and low-fat milk, yogurt, and cheese. • Restricts fatty meats and cheeses as well as sweets. • Encourages exercise.	• Confusing as to exactly what to eat or how much.
Zone Diet	• Eating the correct proportions of carbohydrates, fat, and protein leads to hormonal balance, weight loss, disease prevention, and increased vitality.	• Promotes weight loss because it is a low-kcalorie diet.	• The diet is rigid, restrictive, and complicated, making it difficult for most people to follow accurately. • The overblown health claims of the diet's proponents are based on misinterpreted science and remain unsubstantiated.

TABLE H9-3 Guidelines for Identifying Fad Diets and Other Weight-Loss Scams

Fad Diets and Weight-Loss Scams	Healthy Diet Guidelines
1. They promise dramatic, rapid weight loss.	1. Weight loss should be gradual and not exceed 2 pounds per week.
2. They promote diets that are nutritionally unbalanced or extremely low in kcalories.	2. Diets should provide: • A reasonable number of kcalories (not fewer than 1000 kcalories per day for women and 1200 kcalories per day for men) • Enough, but not too much, protein (between the RDA and twice the RDA) • Enough, but not too much, fat (between 20 and 35% of daily energy intake from fat) • Enough carbohydrates to spare protein and prevent ketosis (at least 100 grams per day) and 20 to 30 grams of fiber from food sources • A balanced assortment of vitamins and minerals from a variety of foods from each of the food groups • At least 1 liter (about 1 quart) of water daily or 1 milliliter per kcalorie daily—whichever is more
3. They use liquid formulas rather than foods.	3. Foods should accommodate a person's ethnic background, taste preferences, and financial means.
4. They attempt to make clients dependent upon special foods or devices.	4. Programs should teach clients how to make good choices from the conventional food supply.
5. They fail to encourage permanent, realistic lifestyle changes.	5. Programs should teach physical activity plans that involve expending at least 300 kcalories a day and behavior-modification strategies that help to correct poor eating habits.
6. They misrepresent salespeople as "counselors" supposedly qualified to give guidance in nutrition and/or general health.	6. Even if adequately trained, such "counselors" would still be objectionable because of the obvious conflict of interest that exists when providers profit directly from products they recommend and sell.
7. They collect large sums of money at the start or require that clients sign contracts for expensive, long-term programs.	7. Programs should be reasonably priced and run on a pay-as-you-go basis.
8. They fail to inform clients of the risks associated with weight loss in general or the specific program being promoted.	8. They should provide information about dropout rates, the long-term success of their clients, and possible diet side effects.
9. They promote unproven or spurious weight-loss aids such as human chorionic gonadotropin hormone (HCG), starch blockers, diuretics, sauna belts, body wraps, passive exercise, ear stapling, acupuncture, electric muscle-stimulating (EMS) devices, spirulina, amino acid supplements (e.g., arginine, ornithine), glucomannan, methylcellulose (a "bulking agent"), "unique" ingredients, and so forth.	9. They should focus on nutrient-rich foods and regular excercise.
10. They fail to provide for weight maintenance after the program ends.	10. They should provide a plan for weight maintenance after successful weight loss.

SOURCES: Adapted from American College of Sports Medicine, *ACSM's Guidelines for Exercise Testing and Prescription* (Baltimore: Williams & Wilkins, 1995), pp. 218–219; J. T. Dwyer, Treatment of obesity: Conventional programs and fad diets, in *Obesity*, ed. P. Björntorp and B. N. Brodoff (Philadelphia: J.B. Lippincott, 1992), p. 668; *National Council Against Health Fraud Newsletter*, March/April 1987, National Council Against Health Fraud, Inc.

companies selling fad diets also peddle these supplements. But as Highlights 10 and 11 explain, foods offer many more health benefits than any supplement can provide. Quite simply, if the diet is inadequate, it needs to be improved, not supplemented.

Follow a Plan

Most people need specific instructions and examples to make dietary changes. Popular diets offer dieters a plan. The user doesn't have to decide what foods to eat, how to prepare them, or how much to eat. Unfortunately, these instructions serve only short-term weight-loss needs. They do not provide for long-term changes in lifestyle that will support weight maintenance or health goals.

The success of any weight-loss diet depends on the person adopting the plan and sticking with it.[1] People who prefer a high-protein, low-carbohydrate diet over a high-carbohydrate, low-fat diet, for example, may have more success at sticking with it. Keep in mind, though, that weight loss occurs because of the duration of a low-kcalorie plan—not the proportion of energy nutrients.

The Real Deal

Fad diets attribute magical powers to their weight-loss plans, but in reality, the magic is in tipping the energy balance so that activities expend more kcalories than foods bring in. Because new diets emerge in the market regularly, it can be challenging to sort the fad diets from the healthy options. Furthermore, it can be difficult determining how a diet's overall quality rates and how it compares with others. One study used the Healthy Eating Index (introduced in Chapter 2) to compare popular weight-loss plans—based on a healthful approach, not on weight loss.[2] Of the eight popular plans examined, Ornish ranked highest and Atkins ranked lowest.

Keep in mind that healthy weight loss requires long-term lifestyle changes in eating and activity habits—not quick, short-term fixes. A healthy plan may not be quick but it allows for flexibility and a variety of foods, including some favorite treats on occasion.

Fad diets may not harm healthy people if used for only a little while, but they cannot support optimal health for long. Chapter 9 includes reasonable approaches to weight management and concludes that the ideal diet is one you can live with for the rest of your life. Keep that criterion in mind when you evaluate the next "latest and greatest weight-loss diet" that comes along.

References

1. M. L. Dansinger and coauthors, Comparison of the Atkins, Ornish, Weight Watchers, and Zone Diets for weight loss and heart disease risk reduction: A randomized trial, *Journal of the American Medical Association* 293 (2005): 43–53.
2. Y. Ma and coauthors, A dietary quality comparison of popular weight-loss plans, *Journal of the American Dietetic Association* 107 (2007): 1786–1791.

CHAPTER

10

HGalina/Shutterstock.com

Nutrition in Your Life

If you were playing a word game and your partner said "vitamins," how would you respond? If "pills" and "supplements" immediately come to mind, you may be missing the main message of the vitamin story—that hundreds of foods deliver more than a dozen vitamins that participate in thousands of activities throughout your body. Quite simply, foods supply vitamins to support all that you are and all that you do—and supplements of any one of them, or even a combination of them, can't compete with foods in keeping you healthy.

Throughout this chapter, the CourseMate icon indicates an opportunity for online self-study, linking you to activities to increase your understanding of chapter concepts. **www.cengagebrain.com** (search for ISBN 084006845X)

The Water-Soluble Vitamins: B Vitamins and Vitamin C

CHAPTER OUTLINE

Earlier chapters focused on the energy-yielding nutrients—carbohydrates, fats, and proteins. This chapter begins with an overview of the vitamins and then examines each of the water-soluble vitamins and a nonvitamin compound named choline; the next chapter features the fat-soluble vitamins.

The Vitamins—An Overview

Researchers first recognized that foods contain substances that are "vital to life" in the early 1900s. Since then, the world of vitamins has opened up dramatically. The vitamins ♦ are powerful substances, as their *absence* attests. Vitamin A deficiency can cause blindness; a lack of the B vitamin niacin can cause dementia; and a lack of vitamin D can retard bone growth. The consequences of deficiencies are so dire, and the effects of restoring the needed vitamins so dramatic, that people spend billions of dollars every year in the belief that vitamin pills will cure a host of ailments (see Highlight 10). Vitamins certainly support sound nutritional health, but they do not cure all ills. Furthermore, vitamin supplements do not offer the many benefits that come from vitamin-rich foods.

The *presence* of the vitamins also attests to their power. The B vitamin folate helps to prevent birth defects and vitamin K allows blood to clot. As you will see, the vitamins' roles in supporting optimal health extend far beyond preventing deficiency diseases. In fact, some of the credit given to low-fat diets in preventing disease actually belongs to the vitamins found in vegetables, fruits, and whole grains (see Highlight 11 for more on vitamins in disease prevention).

The vitamins differ from carbohydrates, fats, and proteins in the following ways:

- *Structure.* Vitamins are individual units; they are not linked together (as are molecules of glucose or amino acids). Appendix C presents the chemical structure for each of the vitamins.
- *Function.* Vitamins do not yield usable energy when broken down; they assist the enzymes that release energy from carbohydrates, fats, and proteins.

♦ The **vitamins** are organic, essential nutrients required in tiny amounts to perform specific functions that promote growth, reproduction, or the maintenance of health and life.
- **vita** = life
- **amine** = containing nitrogen (the first vitamins discovered contained nitrogen)

♦ For perspective, a dollar bill weighs about 1 g.
1 g = 1000 mg
1 mg = 1000 μg
The Aids to Calculation section at the back of this book explains how to convert a measurement from one unit of measure to another.

- *Food contents*. The amounts of vitamins people ingest daily from foods and the amounts they require are measured in *micrograms* (μg) or *milligrams* (mg), rather than grams (g). ♦

The vitamins are similar to the energy-yielding nutrients, though, in that they are vital to life, organic, and available from foods.

Bioavailability The amount of vitamins available from foods depends not only on the quantity provided by a food but also on the amount absorbed and used by the body—referred to as the vitamins' **bioavailability**. The quantity of vitamins in a food can be determined relatively easily. Researchers analyze foods to determine their vitamin contents and publish the results in tables of food composition such as Appendix H. Determining the bioavailability of a vitamin is a more complex task because it depends on many factors, including:

- Efficiency of digestion and time of transit through the GI tract
- Previous nutrient intake and nutrition status
- Method of food preparation (raw, cooked, or processed)
- Source of the nutrient (synthetic, fortified, or naturally occurring)
- Other foods consumed at the same time

Chapters 10–13 describe factors that inhibit or enhance the absorption of individual vitamins and minerals. Experts consider these factors when estimating recommended intakes.

Precursors Some of the vitamins are available from foods in inactive forms known as **precursors**, or provitamins. Once inside the body, the precursor is converted to an active form of the vitamin. Thus, in measuring a person's vitamin intake, it is important to count both the amount of the active vitamin and the potential amount available from its precursors. The discussions and summary tables throughout this chapter and the next indicate which vitamins have precursors.

♦ Organic nutrients contain carbon.

Organic Nature Being organic, ♦ vitamins can be destroyed and left unable to function. Therefore, they must be handled with care during storage and in cooking. Prolonged heating may destroy much of the thiamin in food. Because riboflavin can be destroyed by the ultraviolet rays of the sun or by fluorescent light, foods stored in transparent glass containers are most likely to lose riboflavin. Oxygen destroys vitamin C, so losses occur when foods are cut, processed, and stored; these losses may be enough to reduce its action in the body.[1] Table 10-1 summarizes ways to minimize nutrient losses in the kitchen.

♦ **Water-soluble vitamins:**
- B vitamins:
  Thiamin
  Riboflavin
  Niacin
  Biotin
  Pantothenic acid
  Vitamin B_6
  Folate
  Vitamin B_{12}
- Vitamin C

Fat-soluble vitamins:
- Vitamin A
- Vitamin D
- Vitamin E
- Vitamin K

Solubility As you may recall, carbohydrates and proteins are hydrophilic and lipids are hydrophobic. The vitamins divide along the same lines—the hydrophilic, water-soluble ones ♦ are the eight B vitamins and vitamin C; the hydrophobic, fat-soluble ones are vitamins A, D, E, and K. As each vitamin was discovered, it was given a name and sometimes a letter and number as well. Many of the water-soluble vitamins have multiple names, which has led to some confusion. The margin lists the standard names, and summary tables throughout this chapter provide the common alternative names.

TABLE 10-1 Minimizing Nutrient Losses

- To slow the degradation of vitamins, refrigerate (most) fruits and vegetables.
- To minimize the oxidation of vitamins, store fruits and vegetables that have been cut in airtight wrappers, and store juices that have been opened in closed containers (and refrigerate them).
- To prevent losses during washing, rinse fruits and vegetables before cutting (not after).
- To minimize losses during cooking, use a microwave oven or steam vegetables in a small amount of water. Add vegetables after water has come to a boil. Use the cooking water in mixed dishes such as casseroles and soups. Avoid high temperatures and long cooking times.

bioavailability: the rate at and the extent to which a nutrient is absorbed and used.

precursors: substances that precede others; with regard to vitamins, compounds that can be converted into active vitamins; also known as *provitamins*.

Solubility is apparent in the food sources of the different vitamins, and it affects their absorption, transport, storage, and excretion by the body. The water-soluble vitamins are found in the watery compartments of foods; the fat-soluble vitamins usually occur together in the fats and oils of foods. On being absorbed, the water-soluble vitamins move directly into the blood. Like fats, however, the fat-soluble vitamins must first enter the lymph, then the blood. Once in the blood, many of the water-soluble vitamins travel freely, whereas many of the fat-soluble vitamins require transport proteins. Upon reaching the cells, water-soluble vitamins freely circulate in the water-filled compartments of the body, but fat-soluble vitamins are held in fatty tissues and the liver until needed. The kidneys, monitoring the blood that flows through them, detect and remove small excesses of water-soluble vitamins; large excesses, however, may overwhelm the system, creating adverse effects. Fat-soluble vitamins tend to remain in fat-storage sites in the body rather than being excreted, and so are more likely to reach toxic levels when consumed in excess.

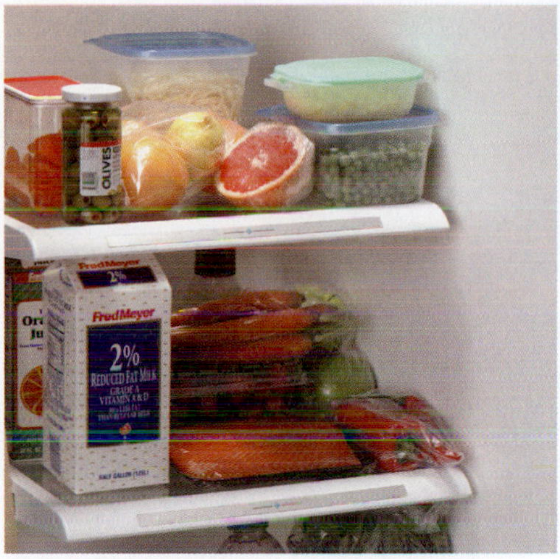

To minimize vitamin losses, wrap cut fruits and vegetables or store them in airtight containers.

Because the body stores fat-soluble vitamins, they can be eaten in large amounts once in a while and still meet the body's needs over time. Water-soluble vitamins are retained for varying lengths of time in the body. Although a single day's omission from the diet does not bring on a deficiency, the water-soluble vitamins must still be eaten more regularly than the fat-soluble vitamins.

Toxicity Knowledge about some of the amazing roles of vitamins has prompted many people to assume that "more is better" and take vitamin supplements. Just as an inadequate intake can cause harm, so can an excessive intake. Even some of the water-soluble vitamins have adverse effects when taken in large doses.

That a vitamin can be both essential and harmful may seem surprising, but the same is true of most nutrients. The effects of every substance depend on its dose, and this is one reason consumers should not self-prescribe supplements for their ailments. Figure 10-1 shows three possible relationships between dose levels and effects. The third diagram represents the situation with nutrients—more is better up to a point, but beyond that point, still more can be harmful.

The Committee on Dietary Reference Intakes (DRI) addresses the possibility of adverse effects from high doses of nutrients by establishing Tolerable Upper Intake Levels (UL). The UL defines the highest amount of a nutrient that is likely not to cause harm for most healthy people when consumed daily. The risk of harm increases as intakes rise above the UL. Of the nutrients discussed in this chapter, niacin, vitamin B_6, folate, choline, and vitamin C have UL, and these values are presented in their respective summary tables. Data are lacking to establish UL for

FIGURE 10-1 Dose Levels and Effects

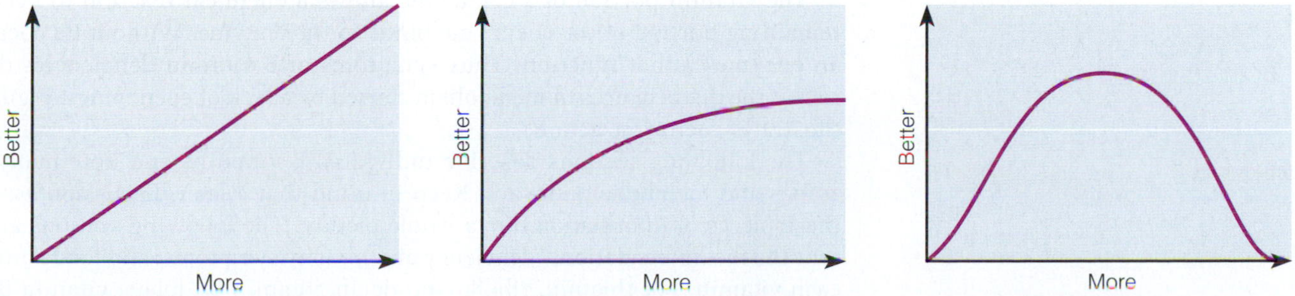

As you progress in the direction of more, the effect gets better and better, with no end in sight (real life is seldom, if ever, like this).

As you progress in the direction of more, the effect reaches a maximum and then a plateau, becoming no better with higher doses.

As you progress in the direction of more, the effect reaches an optimum at some intermediate dose and then declines, showing that more is better up to a point and then harmful. That too much can be as harmful as too little represents the situation with most nutrients.

the remaining B vitamins, but this does not mean that excessively high intakes would be without risk. (The inside front cover pages present UL for the vitamins and minerals.)

IN SUMMARY **The vitamins are essential nutrients needed in tiny amounts in the diet both to prevent deficiency diseases and to support optimal health. The water-soluble vitamins are the B vitamins and vitamin C; the fat-soluble vitamins are vitamins A, D, E, and K. The accompanying table provides a summary of the differences between the water-soluble and fat-soluble vitamins.**

	Water-Soluble Vitamins: B Vitamins and Vitamin C	Fat-Soluble Vitamins: Vitamins A, D, E, and K
Absorption	Directly into the blood	First into the lymph, then the blood
Transport	Travel freely	Many require transport proteins
Storage	Circulate freely in water-filled parts of the body	Stored in the cells associated with fat
Excretion	Kidneys detect and remove excess in urine	Less readily excreted; tend to remain in fat-storage sites
Toxicity	Possible to reach toxic levels when consumed from supplements	Likely to reach toxic levels when consumed from supplements
Requirements	Needed in frequent doses (perhaps 1 to 3 days)	Needed in periodic doses (perhaps weeks or even months)

NOTE: Exceptions occur, but these differences between the water-soluble and fat-soluble vitamins are valid generalizations.

The discussion of B vitamins that follows begins with a brief description of each of them, then offers a look at the ways they work together. Thus, a preview of the individual vitamins is followed by a survey of how they work together, in concert.

The B Vitamins—As Individuals

Despite supplement advertisements that claim otherwise, the vitamins do not provide the body with fuel for energy. It is true, though, that without B vitamins the body would lack energy. The energy-yielding nutrients—carbohydrate, fat, and protein—are used for fuel; the B vitamins help the body to use that fuel. Several of the B vitamins—thiamin, riboflavin, niacin, pantothenic acid, and biotin—form part of the coenzymes ♦ that assist enzymes in the release of energy from carbohydrate, fat, and protein. Other B vitamins play other indispensable roles in metabolism. Vitamin B_6 assists enzymes that metabolize amino acids. Folate and vitamin B_{12} help cells to multiply. Among these cells are the red blood cells and the cells lining the GI tract—cells that deliver energy to all the others.

♦ A **coenzyme** is a small organic molecule that associates closely with certain enzymes; many B vitamins form an integral part of coenzymes.

The vitamin portion of a coenzyme allows a chemical reaction to occur; the remaining portion of the coenzyme binds to the enzyme. Without its coenzyme, an enzyme cannot function. Thus symptoms of B vitamin deficiencies directly reflect the disturbances of metabolism caused by a lack of coenzymes. Figure 10-2 illustrates coenzyme action.

The following sections describe individual B vitamins and note many coenzymes and metabolic pathways. Keep in mind that a later discussion assembles these pieces of information into a whole picture. The following sections also present the recommendations, deficiency and toxicity symptoms, and food sources for each vitamin. For thiamin, riboflavin, niacin, vitamin B_6, folate, vitamin B_{12}, and vitamin C, sufficient data were available to establish an RDA; for biotin, pantothenic acid, and choline, an Adequate Intake (AI) was set; only niacin, vitamin B_6, folate, choline, and vitamin C have Tolerable Upper Intake Levels (UL).[2] These values appear in the summary tables and figures that follow and on the pages of the inside front cover.

FIGURE 10-2 Coenzyme Action

Some vitamins form part of the coenzymes that enable enzymes either to synthesize compounds (as illustrated by the lower enzymes in this figure) or to dismantle compounds (as illustrated by the upper enzymes).

Animated! figure
www.cengagebrain.com
(search for ISBN 084006845X)

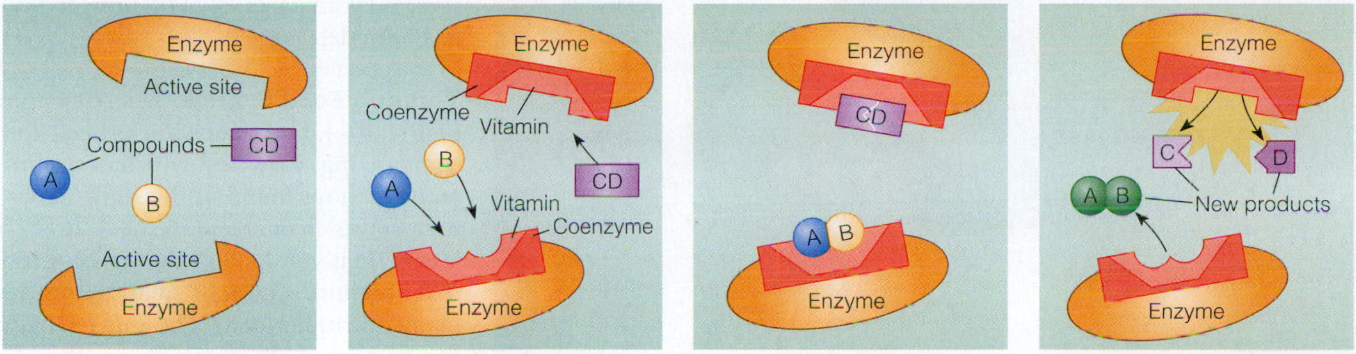

Without coenzymes, compounds A, B, and CD don't respond to their enzymes.

With the coenzymes in place, compounds are attracted to their sites on the enzymes . . .

. . . and the reactions proceed instantaneously. The coenzymes often donate or accept electrons, atoms, or groups of atoms.

The reactions are completed with either the formation of a new product, AB, or the breaking apart of a compound into two new products, C and D, and the release of energy.

Thiamin

Thiamin is the vitamin part of the coenzyme TPP (thiamin pyrophosphate), which assists in energy metabolism. The TPP coenzyme participates in the conversion of pyruvate to acetyl CoA (described in Chapter 7). The reaction removes one carbon from the 3-carbon pyruvate to make the 2-carbon acetyl CoA and carbon dioxide (CO_2). In a similar step in the TCA cycle, TPP helps convert a 5-carbon compound to a 4-carbon compound. Besides playing these pivotal roles in energy metabolism, thiamin occupies a special site on the membranes of nerve cells. Consequently, nerve activity and muscle activity in response to nerves depend heavily on thiamin.

Thiamin Recommendations Dietary recommendations are based primarily on thiamin's role in enzyme activity. Generally, thiamin needs will be met if a person eats enough food to meet energy needs—if that energy comes from nutritious foods. The average thiamin intake in the United States and Canada meets or exceeds recommendations.

Thiamin Deficiency and Toxicity People who fail to eat enough food to meet energy needs risk nutrient deficiencies, including thiamin deficiency. Inadequate thiamin intakes have been reported among the nation's malnourished and homeless people. Similarly, people who derive most of their energy from empty-kcalorie foods and beverages risk thiamin deficiency. Alcohol is a good example. ♦ It contributes energy but provides few, if any, nutrients and often displaces food. In addition, alcohol impairs thiamin absorption and enhances thiamin excretion in the urine, doubling the risk of deficiency. An estimated four out of five alcoholics are thiamin deficient.

Prolonged thiamin deficiency can result in the disease **beriberi**, which was first observed in Indonesia when the custom of polishing rice became widespread.[3] Rice provided 80 percent of the energy intake of the people of that area, and the germ and bran of the rice grain had been their principal source of thiamin. When the germ and bran were removed in the preparation of white rice, beriberi became rampant.

Beriberi is often described as "dry" or "wet." Dry beriberi reflects damage to the nervous system and is characterized by muscle weakness in the arms and legs. Wet beriberi reflects damage to the cardiovascular system and is characterized by dilated blood vessels, which cause the heart to work harder and the

♦ Severe thiamin deficiency in alcohol abusers is called the **Wernicke-Korsakoff** (VER-nee-key KORE-sah-kof) **syndrome.** Symptoms include disorientation, loss of short-term memory, jerky eye movements, and staggering gait.

thiamin (THIGH-ah-min): a B vitamin. The coenzyme form is **TPP (thiamin pyrophosphate).**

beriberi: the thiamin-deficiency disease.

- **beri** = weakness
- **beriberi** = "I can't, I can't"

FIGURE 10-3 Thiamin-Deficiency Symptom—The Edema of Beriberi

Physical examination confirms that this person has wet beriberi. Notice how the impression of the physician's thumb remains on the leg.

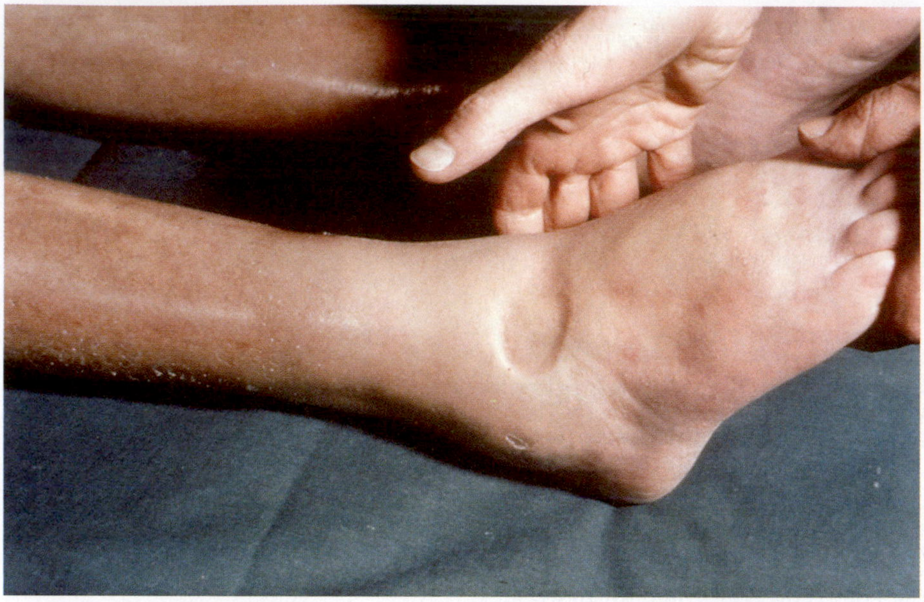

kidneys to retain salt and water, resulting in edema. Typically, both types of beriberi appear together, with one set of symptoms predominating. Figure 10-3 presents the edema of beriberi. No adverse effects have been associated with excesses of thiamin and no UL has been determined.

Thiamin Food Sources Before examining Figure 10-4, you may want to read the accompanying "How To," which describes the many features found in this and similar figures in this chapter and the next three chapters. When you look at Figure 10-4, notice that thiamin occurs in small quantities in many nutritious foods. The long red bar near the bottom of the graph shows that meats in the pork family are exceptionally rich in thiamin. Yellow bars confirm that enriched grains are a reliable source of thiamin.

As mentioned earlier, prolonged cooking can destroy thiamin. Also, like other water-soluble vitamins, thiamin leaches into water when foods are boiled or blanched. Cooking

HOW TO Evaluate Foods for Their Nutrient Contributions

Figure 10-4 is the first of a series of figures in this and the next three chapters that present the vitamins and minerals in foods. Each figure presents the same 24 foods, which were selected to ensure a variety of choices representative of each of the food groups as suggested by MyPyramid. For example, a bread, a cereal, and a pasta were chosen from the grain group. The suggestion to include a variety of vegetables was also considered: dark green, leafy vegetables (broccoli); deep orange and yellow vegetables (carrots); starchy vegetables (potatoes); legumes (pinto beans); and other vegetables (tomato juice). The selection of fruits followed suggestions to use whole fruits (bananas); citrus fruits (oranges); melons (watermelon); and berries (strawberries). Items were selected from the milk group and protein foods in a similar way. In addition to the 24 foods that appear in all of the figures, three different foods were selected for each of the nutrients to add variety and often reflect excellent, and sometimes unusual, sources.

Notice that the figures list the food, the serving size, and the food energy (kcalories) on the left. The amount of the nutrient per serving is presented in the graph on the right along with the RDA (or AI) for adults, so you can see how many servings would be needed to meet recommendations.

The colored bars show at a glance which food groups best provide a nutrient: yellow for breads and cereals; green for vegetables; purple for fruits; white for milk and milk products; brown for legumes; and red for meat, fish, and poultry. Because MyPyramid includes legumes with both the protein foods and the vegetable group and because legumes are especially rich in many vitamins and minerals, they have been given their own color to highlight their nutrient contributions.

Notice how the bar graphs shift in the various figures. Careful study of all of the figures taken together will confirm that variety is the key to nutrient adequacy.

Another way to evaluate foods for their nutrient contributions is to consider their nutrient density (their thiamin *per 100 kcalories,* for example). Quite often, vegetables rank higher on a nutrient-per-kcalorie list than they do on a nutrient-per-serving list (see p. 37 to review how to evaluate foods based on nutrient density). The left column in the figure highlights about five foods that offer the best nutrient density. Notice how many of them are vegetables.

Realistically, people cannot eat for single nutrients. Fortunately, most foods deliver more than one nutrient, allowing people to combine foods into nourishing meals.

For additional practice, log on to **www.cengagebrain.com** and search for ISBN 084006845X.

TRY IT Calculate which food provides more riboflavin per 1-ounce serving—a pork chop (3 oz, 291 kcal, 0.25 mg riboflavin) or cheddar cheese (1½ oz, 165 kcal, 0.11 mg riboflavin). Which food is more nutrient dense with respect to riboflavin?

FIGURE 10-4 Thiamin in Selected Foods

See the "How To" section on p. 316 for more information on using this figure.

Food	Serving size (kcalories)
Bread, whole wheat	1 oz slice (70 kcal)
Cornflakes, fortified	1 oz (110 kcal)
Spaghetti pasta	½ c cooked (99 kcal)
Tortilla, flour	1 10"-round (234 kcal)
Broccoli	½ c cooked (22 kcal)
Carrots	½ c shredded raw (24 kcal)
Potato	1 medium baked w/skin (133 kcal)
Tomato juice	¾ c (31 kcal)
Banana	1 medium raw (109 kcal)
Orange	1 medium raw (62 kcal)
Strawberries	½ c fresh (22 kcal)
Watermelon	1 slice (92 kcal)
Milk	1 c reduced-fat 2% (121 kcal)
Yogurt, plain	1 c low-fat (155 kcal)
Cheddar cheese	1½ oz (171 kcal)
Cottage cheese	½ c low-fat 2% (101 kcal)
Pinto beans	½ c cooked (117 kcal)
Peanut butter	2 tbs (188 kcal)
Sunflower seeds	1 oz dry (165 kcal)
Tofu (soybean curd)	½ c (76 kcal)
Ground beef, lean	3 oz broiled (244 kcal)
Chicken breast	3 oz roasted (140 kcal)
Tuna, canned in water	3 oz (99 kcal)
Egg	1 hard cooked (78 kcal)
Excellent, and sometimes unusual, sources:	
Pork chop, lean	3 oz broiled (169 kcal)
Soy milk	1 c (81 kcal)
Squash, acorn	½ c baked (69 kcal)

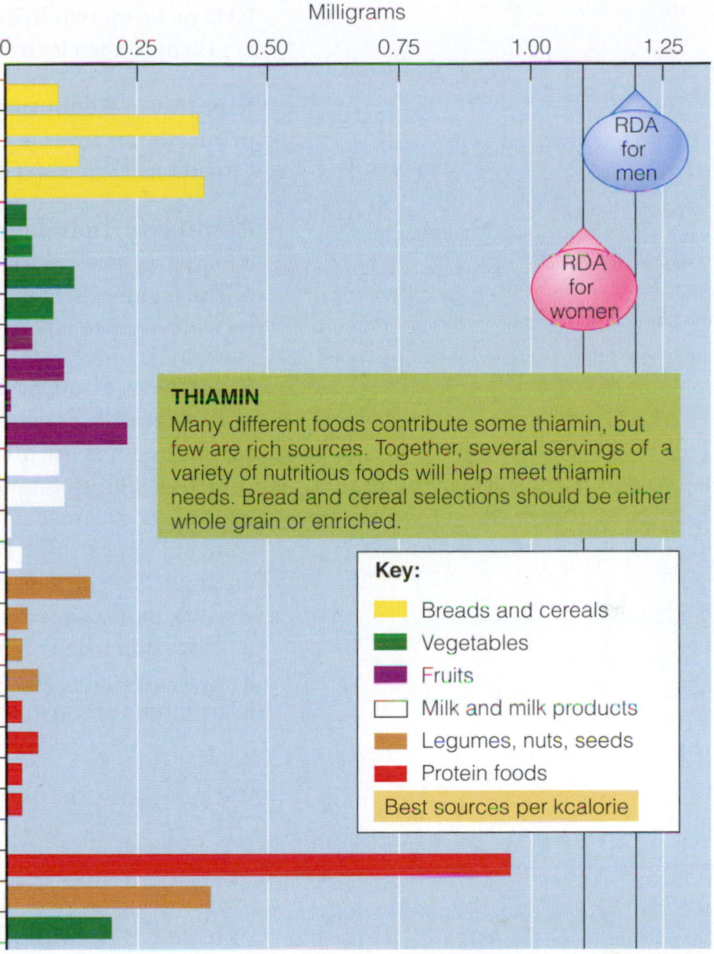

methods that require little or no water such as steaming and microwave heating conserve thiamin and other water-soluble vitamins. The accompanying table provides a summary of thiamin.

Pork is the richest source of thiamin, but enriched or whole-grain products typically make the greatest contribution to a day's intake because of the quantities eaten. Legumes such as split peas are also valuable sources of thiamin.

IN SUMMARY Thiamin

Other Names

Vitamin B_1

RDA

Men: 1.2 mg/day

Women: 1.1 mg/day

Chief Functions in the Body

Part of coenzyme TPP (thiamin pyrophosphate) used in energy metabolism

Significant Sources

Whole-grain, fortified, or enriched grain products; moderate amounts in all nutritious food; pork

Easily destroyed by heat

Deficiency Disease

Beriberi (wet, with edema; dry, with muscle wasting)

Deficiency Symptoms[a]

Enlarged heart, cardiac failure; muscular weakness; apathy, poor short-term memory, confusion, irritability; anorexia, weight loss

Toxicity Symptoms

None reported

[a]Severe thiamin deficiency is often related to heavy alcohol consumption with limited food consumption (Wernicke-Korsakoff syndrome).

Riboflavin Like thiamin, **riboflavin** serves as a coenzyme in many reactions, most notably in energy metabolism. The coenzyme forms of riboflavin are FMN (flavin mononucleotide) and FAD (flavin adenine dinucleotide); both can accept and then donate two hydrogens (see Figure 10-5). During energy metabolism, FAD picks up two hydrogens (with their electrons) from the TCA cycle and delivers them to the electron transport chain (described in Chapter 7).

Riboflavin Recommendations Like thiamin's RDA, riboflavin's RDA is based primarily on its role in enzyme activity. Most people in the United States and Canada meet or exceed riboflavin recommendations.

Riboflavin Deficiency and Toxicity Riboflavin deficiency ♦ most often accompanies other nutrient deficiencies. Lack of the vitamin causes inflammation of the membranes of the mouth, skin, eyes, and GI tract. Excesses of riboflavin appear to cause no harm and no UL has been established.

♦ Riboflavin deficiency is called **ariboflavinosis** (ay-RYE-boh-FLAY-vin-oh-sis).
- **a** = not
- **osis** = condition

Riboflavin Food Sources The greatest contributions of riboflavin come from milk and milk products (see Figure 10-6). Whole-grain or enriched bread and cereal products are also valuable sources because of the quantities typically consumed. When riboflavin sources are ranked by nutrient density (per kcalorie), ♦ many dark green, leafy vegetables (such as broccoli, turnip greens, asparagus, and spinach) appear high on the list. Vegans and others who don't use milk must rely on ample servings of dark greens and enriched grains for riboflavin. Nutritional yeast is another good source.

♦ Turn to p. 37 for a review of how to evaluate foods based on nutrient density (per kcalorie).

Ultraviolet light and irradiation destroy riboflavin. For these reasons, milk is sold in cardboard or opaque plastic containers, instead of clear glass bottles. Precautions are also taken when vitamin D is added to milk by irradiation.* In

*Vitamin D can be added to milk by feeding cows irradiated yeast or by irradiating the milk itself.

FIGURE 10-5 Riboflavin Coenzyme, Accepting and Donating Hydrogens

This figure shows the chemical structure of the riboflavin portion of the coenzyme only; the remainder of the coenzyme structure is represented by dotted lines (see Appendix C for the complete chemical structures of FAD and FMN). The reactive sites that accept and donate hydrogens are highlighted in white.

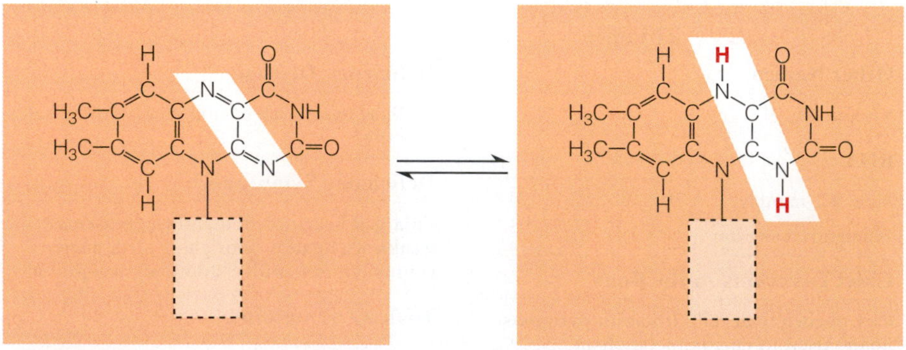

FAD ⇌ $FADH_2$

During the TCA cycle, compounds release hydrogens, and the riboflavin coenzyme FAD picks up two of them. As it accepts two hydrogens, FAD becomes $FADH_2$.

$FADH_2$ carries the hydrogens to the electron transport chain. At the end of the electron transport chain, the hydrogens are accepted by oxygen, creating water, and $FADH_2$ becomes FAD again. For every $FADH_2$ that passes through the electron transport chain, two ATP are generated.

riboflavin (RYE-boh-flay-vin): a B vitamin. The coenzyme forms are **FMN (flavin mononucleotide)** and **FAD (flavin adenine dinucleotide)**.

FIGURE 10-6 Riboflavin in Selected Foods

See the "How To" section on p. 316 for more information on using this figure.

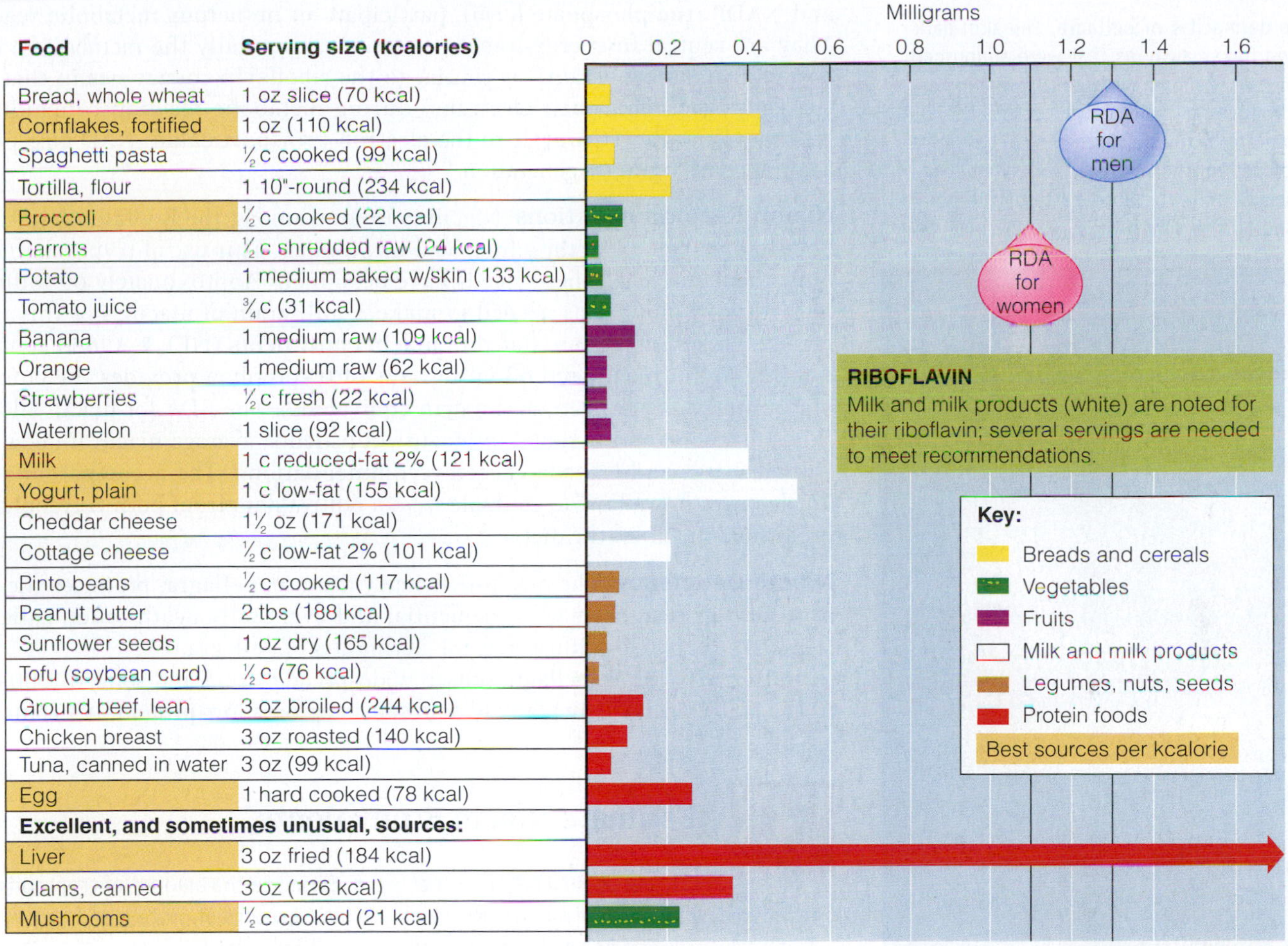

contrast, riboflavin is stable to heat, so cooking does not destroy it. The accompanying table provides a summary of riboflavin.

Milligrams
Food
Serving size (kcalories)
0
0.2
0.4
0.6
0.8
1.0
1.2
1.4
1.6
Bread, whole wheat — 1 oz slice (70 kcal)
Cornflakes, fortified — 1 oz (110 kcal)
Spaghetti pasta — ½ c cooked (99 kcal)
Tortilla, flour — 1 10"-round (234 kcal)
Broccoli — ½ c cooked (22 kcal)
Carrots — ½ c shredded raw (24 kcal)
Potato — 1 medium baked w/skin (133 kcal)
Tomato juice — ¾ c (31 kcal)
Banana — 1 medium raw (109 kcal)
Orange — 1 medium raw (62 kcal)
Strawberries — ½ c fresh (22 kcal)
Watermelon — 1 slice (92 kcal)
Milk — 1 c reduced-fat 2% (121 kcal)
Yogurt, plain — 1 c low-fat (155 kcal)
Cheddar cheese — 1½ oz (171 kcal)
Cottage cheese — ½ c low-fat 2% (101 kcal)
Pinto beans — ½ c cooked (117 kcal)
Peanut butter — 2 tbs (188 kcal)
Sunflower seeds — 1 oz dry (165 kcal)
Tofu (soybean curd) — ½ c (76 kcal)
Ground beef, lean — 3 oz broiled (244 kcal)
Chicken breast — 3 oz roasted (140 kcal)
Tuna, canned in water — 3 oz (99 kcal)
Egg — 1 hard cooked (78 kcal)
Excellent, and sometimes unusual, sources:
Liver — 3 oz fried (184 kcal)
Clams, canned — 3 oz (126 kcal)
Mushrooms — ½ c cooked (21 kcal)
RDA for men
RDA for women
RIBOFLAVIN
Milk and milk products (white) are noted for their riboflavin; several servings are needed to meet recommendations.
Key:
Breads and cereals
Vegetables
Fruits
Milk and milk products
Legumes, nuts, seeds
Protein foods
Best sources per kcalorie

IN SUMMARY Riboflavin

Other Names

Vitamin B_2

RDA

Men: 1.3 mg/day

Women: 1.1 mg/day

Chief Functions in the Body

Part of coenzymes FMN (flavin mononucleotide) and FAD (flavin adenine dinucleotide) used in energy metabolism

Significant Sources

Milk products (yogurt, cheese); whole-grain, fortified, or enriched grain products; liver

Easily destroyed by ultraviolet light and irradiation

Deficiency Disease

Ariboflavinosis (ay-RYE-boh-FLAY-vin-oh-sis)

Deficiency Symptoms

Sore throat; cracks and redness at corners of mouth;[a] painful, smooth, purplish red tongue;[b] inflammation characterized by skin lesions covered with greasy scales

Toxicity Symptoms

None reported

[a]Cracks at the corners of the mouth are called *angular stomatitis* or *cheilosis* (kye-LOH-sis or kee-LOH-sis).
[b]Smoothness of the tongue is caused by loss of its surface structures and is termed *glossitis* (gloss-EYE-tis).

All of these foods are rich in riboflavin, but milk and milk products provide much of the riboflavin in the diets of most people.

FIGURE 10-7 Niacin-Deficiency Symptom—The Dermatitis of Pellagra

In the dermatitis of pellagra, the skin darkens and flakes away as if it were sunburned. The protein-deficiency disease kwashiorkor also produces a "flaky paint" dermatitis, but the two are easily distinguished. The dermatitis of pellagra is bilateral and symmetrical and occurs only on those parts of the body exposed to the sun.

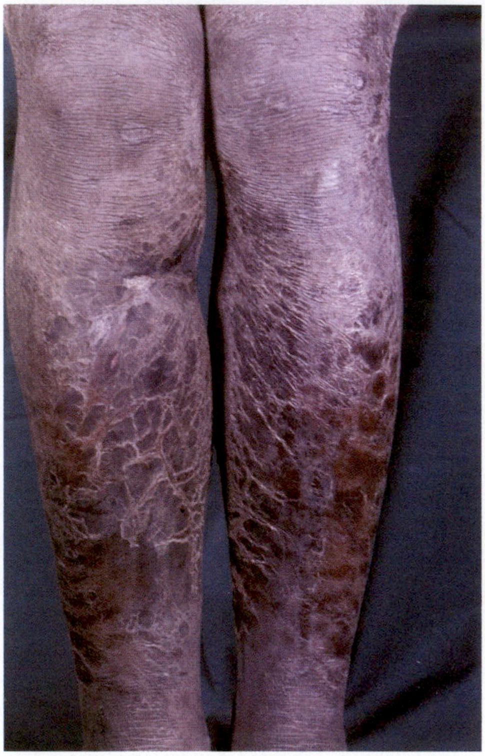

Niacin

The name **niacin** describes two chemical structures: nicotinic acid and nicotinamide (also known as niacinamide). The body can easily convert nicotinic acid to nicotinamide, which is the major form of niacin in the blood.

The two coenzyme forms of niacin, NAD (nicotinamide adenine dinucleotide) and NADP (the phosphate form), participate in numerous metabolic reactions. They are central in energy-transfer reactions, especially the metabolism of glucose, fat, and alcohol. NAD is similar to the riboflavin coenzymes in that it carries hydrogens (and their electrons) during metabolic reactions, including the pathway from the TCA cycle to the electron transport chain. NAD also protects against neurological degeneration.[4]

Niacin Recommendations Niacin is unique among the B vitamins in that the body can make it from the amino acid tryptophan. This use of tryptophan occurs only after protein synthesis needs have been met.[5] Approximately 60 milligrams of dietary tryptophan is needed to make 1 milligram of niacin. For this reason, recommended intakes are stated in **niacin equivalents (NE)**. ♦ A food containing 1 milligram of niacin and 60 milligrams of tryptophan provides the equivalent of 2 milligrams of niacin, or 2 niacin equivalents. The RDA for niacin allows for this conversion and is stated in niacin equivalents; average niacin intakes in the United States and Canada exceed recommendations. The accompanying "How To" feature shows how to estimate niacin equivalents from both tryptophan and preformed niacin in the diet.

♦ 1 NE = 1 mg niacin or 60 mg tryptophan

Niacin Deficiency The niacin-deficiency disease, **pellagra**, produces the symptoms of diarrhea, dermatitis, dementia, and eventually death (often called "the four Ds"). Figure 10-7 illustrates the dermatitis of pellagra.

In the early 1900s, pellagra caused widespread misery and some 87,000 deaths in the U.S. South, where many people subsisted on a low-protein diet centered on

HOW TO Estimate Niacin Equivalents

Niacin recommendations are expressed as niacin equivalents (NE), but diet analysis programs and food composition tables report only preformed niacin. To estimate niacin equivalents from the tryptophan in dietary protein:

- Assume that most dietary proteins contain about 1 percent tryptophan. To determine the amount of tryptophan in protein, divide grams of protein by 100.
- Multiply by 1000 to convert grams of tryptophan to milligrams.
- Because it takes 60 milligrams of tryptophan to make 1 milligram of niacin, divide milligrams of tryptophan by 60 to get niacin equivalents.
- Add the amount of preformed niacin obtained in the diet.

Consider, for example, a person who consumes 80 grams of protein and 5 milligrams of preformed niacin.

- Estimate the amount of tryptophan in 80 grams of protein and convert to milligrams:

80 g protein ÷ 100 = 0.8 g tryptophan
0.8 g tryptophan × 1000 = 800 mg tryptophan

- Convert milligrams of tryptophan to niacin equivalents:

800 mg tryptophan ÷ 60 = 13 mg NE

To determine the total amount of niacin available from the diet, add the amount available from tryptophan to the amount preformed in the diet.

13 mg NE + 5 mg preformed niacin = 18 mg NE

For additional practice, log on to **www.cengagebrain.com** and search for ISBN 084006845X.

TRY IT Calculate how many niacin equivalents a person receives from a diet that delivers 60 grams protein and 6 milligrams niacin.

niacin (NIGH-a-sin): a B vitamin. The coenzyme forms are **NAD (nicotinamide adenine dinucleotide)** and **NADP (the phosphate form of NAD).** Niacin can be eaten preformed or made in the body from its precursor, tryptophan, an essential amino acid.

niacin equivalents (NE): the amount of niacin present in food, including the niacin that can theoretically be made from its precursor, tryptophan, present in the diet.

pellagra (pell-AY-gra): the niacin-deficiency disease.

- **pellis** = skin
- **agra** = rough

corn. This diet supplied neither enough niacin nor enough tryptophan. At least 70 percent of the niacin in corn is bound to complex carbohydrates and small peptides, making it unavailable for absorption. Furthermore, corn is high in the amino acid leucine, which interferes with the tryptophan-to-niacin conversion, thus further contributing to the development of pellagra.

Pellagra was originally believed to be caused by an infection. Medical researchers spent many years and much effort searching for infectious microbes until they realized that the problem was not what was *present* in the food but what was *absent* from it. That a disease such as pellagra could be caused by diet—and not by pathogens—was a groundbreaking discovery. It contradicted commonly held medical opinions that diseases were caused only by infectious agents. By carefully following the scientific method (as described in Chapter 1), researchers advanced the science of nutrition dramatically.

Niacin Toxicity Naturally occurring niacin from foods ♦ causes no harm, but large doses from supplements or drugs produce a variety of adverse effects, most notably "**niacin flush.**" Niacin flush occurs when nicotinic acid is taken in doses only three to four times the RDA. It dilates the capillaries and causes a tingling sensation that can be painful. The nicotinamide form does not produce this effect.

Large doses of nicotinic acid have been used to lower LDL cholesterol, raise HDL cholesterol, and increase adiponectin levels—all factors that help to protect against heart disease.[6] Such therapy must be closely monitored. People with the following conditions may be particularly susceptible to the toxic effects of niacin: liver disease, diabetes, peptic ulcers, gout, irregular heartbeats, inflammatory bowel disease, migraine headaches, and alcoholism. The nicotinamide form does not improve blood cholesterol levels.

♦ When a normal dose of a nutrient (levels commonly found in foods) provides a normal blood concentration, the nutrient is having a **physiological** effect. When a large dose (levels commonly available only from supplements) overwhelms some body system and acts like a drug, the nutrient is having a **pharmacological** effect.

- **physio** = natural
- **pharma** = drug

Niacin Food Sources Tables of food composition typically list preformed niacin only, but as mentioned, niacin can also be made in the body from the amino acid tryptophan. Dietary tryptophan could meet about half the daily niacin need for most people, but the average diet easily supplies enough preformed niacin.

Figure 10-8 (p. 322) presents niacin in selected foods. Meat, poultry, legumes, and enriched and whole grains contribute about half the niacin people consume. Mushrooms, potatoes, and tomatoes are among the richest vegetable sources, and they can provide abundant niacin when eaten in generous amounts.

Niacin is less vulnerable to losses during food preparation and storage than other water-soluble vitamins. Being fairly heat resistant, niacin can withstand reasonable cooking times, but like other water-soluble vitamins, it will leach into cooking water. The accompanying table provides a summary of niacin.

niacin flush: a temporary burning, tingling, and itching sensation that occurs when a person takes a large dose of nicotinic acid; often accompanied by a headache and reddened face, arms, and chest.

IN SUMMARY Niacin

Other Names

Nicotinic acid, nicotinamide, niacinamide, vitamin B_3; precursor is dietary tryptophan (an amino acid)

RDA

Men: 16 mg NE/day

Women: 14 mg NE/day

Upper Level

Adults: 35 mg/day

Chief Functions in the Body

Part of coenzymes NAD (nicotinamide adenine dinucleotide) and NADP (its phosphate form) used in energy metabolism

Significant Sources

Milk, eggs, meat, poultry, fish; whole-grain, fortified, and enriched grain products; nuts and all protein-containing foods

Deficiency Disease

Pellagra

Deficiency Symptoms

Diarrhea, abdominal pain, vomiting; inflamed, swollen, smooth, bright red tongue;[a] depression, apathy, fatigue, loss of memory, headache; bilateral symmetrical rash on areas exposed to sunlight

Toxicity Symptoms

Painful flush, hives, and rash ("niacin flush"); nausea and vomiting; liver damage, impaired glucose tolerance

[a]Smoothness of the tongue is caused by loss of its surface structures and is termed *glossitis* (gloss-EYE-tis).

Protein-rich foods such as meat, fish, poultry, and peanut butter contribute much of the niacin in people's diets. Enriched breads and cereals and a few vegetables are also rich in niacin.

FIGURE 10-8 Niacin in Selected Foods

See the "How To" section on p. 316 for more information on using this figure.

Food	Serving size (kcalories)
Bread, whole wheat	1 oz slice (70 kcal)
Cornflakes, fortified	1 oz (110 kcal)
Spaghetti pasta	½ c cooked (99 kcal)
Tortilla, flour	1 10"-round (234 kcal)
Broccoli	½ c cooked (22 kcal)
Carrots	½ c shredded raw (24 kcal)
Potato	1 medium baked w/skin (133 kcal)
Tomato juice	¾ c (31 kcal)
Banana	1 medium raw (109 kcal)
Orange	1 medium raw (62 kcal)
Strawberries	½ c fresh (22 kcal)
Watermelon	1 slice (92 kcal)
Milk	1 c reduced-fat 2% (121 kcal)
Yogurt, plain	1 c low-fat (155 kcal)
Cheddar cheese	1½ oz (171 kcal)
Cottage cheese	½ c low-fat 2% (101 kcal)
Pinto beans	½ c cooked (117 kcal)
Peanut butter	2 tbs (188 kcal)
Sunflower seeds	1 oz dry (165 kcal)
Tofu (soybean curd)	½ c (76 kcal)
Ground beef, lean	3 oz broiled (244 kcal)
Chicken breast	3 oz roasted (140 kcal)
Tuna, canned in water	3 oz (99 kcal)
Egg	1 hard cooked (78 kcal)
Excellent, and sometimes unusual, sources:	
Liver	3 oz fried (184 kcal)
Peanuts	1 oz roasted (165 kcal)
Mushrooms	½ c cooked (21 kcal)

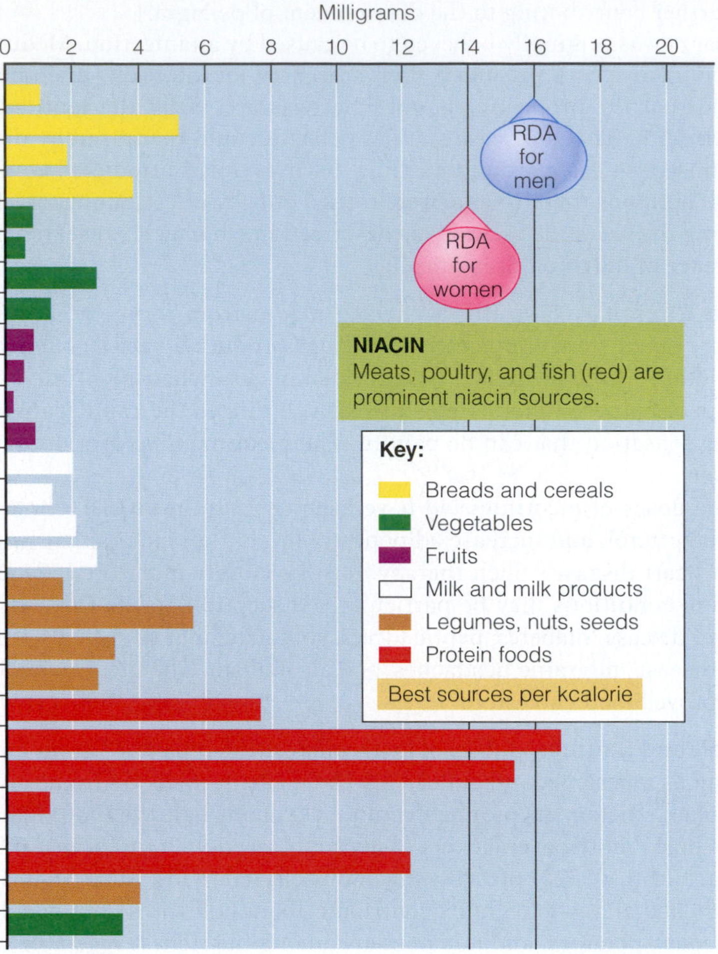

Biotin

Biotin plays an important role in metabolism as a coenzyme that carries activated carbon dioxide. This role is critical in the TCA cycle: biotin delivers a carbon to 3-carbon pyruvate, thus replenishing oxaloacetate, the 4-carbon compound needed to combine with acetyl CoA to keep the TCA cycle turning (review Figure 7-16 on p. 220). The biotin coenzyme also participates in gluconeogenesis, ♦ fatty acid synthesis, and the breakdown of certain fatty acids and amino acids.

♦ **Gluconeogenesis** is the synthesis of glucose from noncarbohydrate sources such as amino acids or glycerol.

Biotin Recommendations Biotin is needed in very small amounts. Because there is insufficient research on biotin requirements, an Adequate Intake (AI) has been determined, instead of an RDA.

Biotin Deficiency and Toxicity Biotin deficiencies rarely occur. Researchers can induce a biotin deficiency in animals or human beings by feeding them raw egg whites, which contain a protein ♦ that binds biotin and thus prevents its absorption. Biotin-deficiency symptoms include skin rash, hair loss, and neurological impairment. More than two dozen raw egg whites must be consumed daily for several months to produce these effects; cooking eggs denatures the binding protein. No adverse effects from high biotin intakes have been reported. Biotin does not have a UL.

♦ The protein **avidin** (AV-eh-din) in egg whites binds biotin.
- **avid** = greedy

Biotin Food Sources Biotin is widespread in foods (including egg yolks), so eating a variety of foods protects against deficiencies. Some biotin is also synthesized by GI tract bacteria, but this amount may not contribute much to the biotin absorbed. The accompanying table provides a summary of biotin.

biotin (BY-oh-tin): a B vitamin that functions as a coenzyme in metabolism.

IN SUMMARY Biotin

Adequate Intake (AI)	Deficiency Symptoms
Adults: 30 μg/day	Depression, lethargy, hallucinations, numb or tingling sensation in the arms and legs; red, scaly rash around the eyes, nose, and mouth; hair loss
Chief Functions in the Body	**Toxicity Symptoms**
Part of a coenzyme used in energy metabolism, fat synthesis, amino acid metabolism, and glycogen synthesis	None reported
Significant Sources	
Widespread in foods; liver, egg yolks, soybeans, fish, whole grains; also produced by GI bacteria	

Pantothenic Acid

Pantothenic acid is part of the chemical structure of coenzyme A—the same CoA that forms acetyl CoA, the "crossroads" compound in several metabolic pathways, including the TCA cycle. (Appendix C presents the chemical structures of these two molecules and shows that coenzyme A is made up in part of pantothenic acid.) As such, it is involved in more than 100 different steps in the synthesis of lipids, neurotransmitters, steroid hormones, and hemoglobin.

Pantothenic Acid Recommendations An Adequate Intake (AI) for pantothenic acid has been set. It reflects the amount needed to replace daily losses.

Pantothenic Acid Deficiency and Toxicity Pantothenic acid deficiency is rare. Its symptoms involve a general failure of all the body's systems and include fatigue, GI distress, and neurological disturbances. The "burning feet" syndrome that affected prisoners of war in Asia during World War II is thought to have been caused by pantothenic acid deficiency. No toxic effects have been reported, and no UL has been established.

Pantothenic Acid Food Sources Pantothenic acid is widespread in foods, and typical diets seem to provide adequate intakes. Beef, poultry, whole grains, potatoes, tomatoes, and broccoli are particularly good sources. Losses of pantothenic acid during food production can be substantial because it is readily destroyed by the freezing, canning, and refining processes. The accompanying table provides a summary of pantothenic acid.

IN SUMMARY Pantothenic Acid

Adequate Intake (AI)	Deficiency Symptoms
Adults: 5 mg/day	Vomiting, nausea, stomach cramps; insomnia, fatigue, depression, irritability, restlessness, apathy; hypoglycemia, increased sensitivity to insulin; numbness, muscle cramps, inability to walk
Chief Functions in the Body	**Toxicity Symptoms**
Part of coenzyme A, used in energy metabolism	None reported
Significant Sources	
Widespread in foods; chicken, beef, potatoes, oats, tomatoes, liver, egg yolk, broccoli, whole grains	
Easily destroyed by food processing	

Vitamin B_6

Vitamin B_6 occurs in three forms—pyridoxal, pyridoxine, and pyridoxamine. All three can be converted to the coenzyme PLP (pyridoxal phosphate), which is active in amino acid metabolism. Because PLP can transfer amino groups (NH_2) from an amino acid to a keto acid, the body can make nonessential amino acids (review Figure 6-12 on p. 186). The ability to add and remove amino groups makes PLP valuable in protein and urea metabolism as well. The conversions of the amino acid tryptophan to niacin or to the neurotransmitter serotonin ♦ also depend on PLP. In addition, PLP participates in the synthesis of heme (the nonprotein portion of hemoglobin), nucleic acids (such as DNA and RNA), and lecithin (a phospholipid).

♦ **Serotonin** is a neurotransmitter important in appetite control, sleep regulation, and sensory perception, among other roles; it is synthesized in the body from the amino acid tryptophan with the help of vitamin B_6.

pantothenic (PAN-toe-THEN-ick) **acid:** a B vitamin. The principal active form is part of coenzyme A, called "CoA" throughout Chapter 7.
- **pantos** = everywhere

vitamin B_6: a family of compounds—pyridoxal, pyridoxine, and pyridoxamine. The primary active coenzyme form is **PLP (pyridoxal phosphate).**

© Polara Studios Inc.

Most protein-rich foods such as meat, fish, and poultry provide ample vitamin B_6; some vegetables and fruits are good sources, too.

A surge of research in the last decade has revealed that vitamin B_6 influences cognitive performance, immune function, and steroid hormone activity. Unlike other water-soluble vitamins, vitamin B_6 is stored extensively in muscle tissue.

Vitamin B_6 Recommendations Because vitamin B_6 coenzymes participate in amino acid metabolism, previous RDA were expressed in terms of protein intakes; the current RDA for vitamin B_6, however, is not. Research does not support claims that large doses of vitamin B_6 enhance muscle strength or physical endurance. Dietary supplements cannot compete with a nutritious diet and physical training.

Vitamin B_6 Deficiency Without adequate vitamin B_6, synthesis of key neurotransmitters diminishes, and abnormal compounds produced during tryptophan metabolism accumulate in the brain. Early symptoms of vitamin B_6 deficiency include depression and confusion; advanced symptoms include abnormal brain wave patterns and convulsions.

Alcohol contributes to the destruction and loss of vitamin B_6 from the body. As Highlight 7 describes, when the body breaks down alcohol, it produces acetaldehyde. If allowed to accumulate, acetaldehyde dislodges the PLP coenzyme from its enzymes; once loose, PLP breaks down and is excreted.

Another drug that acts as a vitamin B_6 **antagonist** is isoniazid, a medication that inhibits the growth of the tuberculosis bacterium.* This drug has saved countless lives, but because isoniazid binds and inactivates vitamin B_6, it can induce a deficiency. Whenever isoniazid is used to treat tuberculosis, vitamin B_6 supplements must be given to protect against deficiency.

Vitamin B_6 Toxicity The first major report of vitamin B_6 toxicity appeared in the early 1980s. Until that time, most researchers and dietitians believed that, like the other water-soluble vitamins, vitamin B_6 could not reach toxic concentrations in the body. The report described neurological damage in people who had been taking more than 2 *grams* of vitamin B_6 daily (20 times the current UL of 100 *milligrams* per day) for two months or more.

Some people have taken vitamin B_6 supplements in an attempt to cure **carpal tunnel syndrome** even though such treatment appears to be ineffective.[7] Self-prescribing is ill advised because large doses of vitamin B_6 may cause irreversible nerve degeneration.

Vitamin B_6 Food Sources As you can see from the colors in Figure 10-9, meats, fish, and poultry (red bars), potatoes and a few other vegetables (green bars), and fruits (purple bars) offer vitamin B_6. As is true of most of the other vitamins, fruits and vegetables rank considerably higher when foods are judged by nutrient density (vitamin B_6 per kcalorie). Several servings of vitamin B_6–rich foods are needed to meet recommended intakes.

Foods lose vitamin B_6 when heated. Information is limited, but vitamin B_6 bioavailability from plant-derived foods seems to be lower than from animal-derived foods. Fiber does not appear to interfere with vitamin B_6 absorption. The accompanying table provides a summary of vitamin B_6.

IN SUMMARY Vitamin B_6

Other Names	
Pyridoxine, pyridoxal, pyridoxamine	
RDA	
Adults (19–50 yr): 1.3 mg/day	
Upper Level	
Adults: 100 mg/day	

antagonist: a competing factor that counteracts the action of another factor. When a drug displaces a vitamin from its site of action, the drug renders the vitamin ineffective and thus acts as a vitamin antagonist.

carpal tunnel syndrome: a pinched nerve at the wrist, causing pain or numbness in the hand. It is often caused by repetitive motion of the wrist.

*Isoniazid (eye-so-NYE-uh-zid) is also known as INH (isonicotinic acid hydrazide).

Chief Functions in the Body

Part of coenzymes PLP (pyridoxal phosphate) and PMP (pyridoxamine phosphate) used in amino acid and fatty acid metabolism; helps to convert tryptophan to niacin and to serotonin; helps to make red blood cells

Significant Sources

Meats, fish, poultry, potatoes and other starchy vegetables, legumes, noncitrus fruits, fortified cereals, liver, soy products

Easily destroyed by heat

Deficiency Symptoms

Scaly dermatitis; anemia (small-cell type);[a] depression, confusion, convulsions

Toxicity Symptoms

Depression, fatigue, irritability, headaches, nerve damage causing numbness and muscle weakness leading to an inability to walk and convulsions; skin lesions

[a]Small-cell-type anemia is called *microcytic anemia*.

Folate

Folate, also known as folacin or folic acid, has a chemical name that would fit a flying dinosaur: pteroylglutamic acid (PGA for short). Its primary coenzyme form, THF (tetrahydrofolate), serves as part of an enzyme complex that transfers 1-carbon compounds that arise during metabolism. This action converts vitamin B_{12} to one of its coenzyme forms, synthesizes the DNA required for all rapidly growing cells, and regenerates the amino acid methionine from homocysteine.

folate (FOLE-ate): a B vitamin; also known as folic acid, folacin, or pteroylglutamic (tare-o-EEL-glue-TAM-ick) acid (PGA). The coenzyme forms are **DHF (dihydrofolate)** and **THF (tetrahydrofolate).**

FIGURE 10-9 Vitamin B_6 in Selected Foods

See the "How To" section on p. 316 for more information on using this figure.

Food	Serving size (kcalories)
Bread, whole wheat	1 oz slice (70 kcal)
Cornflakes, fortified	1 oz (110 kcal)
Spaghetti pasta	½ c cooked (99 kcal)
Tortilla, flour	1 10"-round (234 kcal)
Broccoli	½ c cooked (22 kcal)
Carrots	½ c shredded raw (24 kcal)
Potato	1 medium baked w/skin (133 kcal)
Tomato juice	¾ c (31 kcal)
Banana	1 medium raw (109 kcal)
Orange	1 medium raw (62 kcal)
Strawberries	½ c fresh (22 kcal)
Watermelon	1 slice (92 kcal)
Milk	1 c reduced-fat 2% (121 kcal)
Yogurt, plain	1 c low-fat (155 kcal)
Cheddar cheese	1½ oz (171 kcal)
Cottage cheese	½ c low-fat 2% (101 kcal)
Pinto beans	½ c cooked (117 kcal)
Peanut butter	2 tbs (188 kcal)
Sunflower seeds	1 oz dry (165 kcal)
Tofu (soybean curd)	½ c (76 kcal)
Ground beef, lean	3 oz broiled (244 kcal)
Chicken breast	3 oz roasted (140 kcal)
Tuna, canned in water	3 oz (99 kcal)
Egg	1 hard cooked (78 kcal)
Excellent, and sometimes unusual, sources:	
Prune juice	¾ c (137 kcal)
Bluefish	3 oz baked (135 kcal)
Squash, acorn	½ c baked (69 kcal)

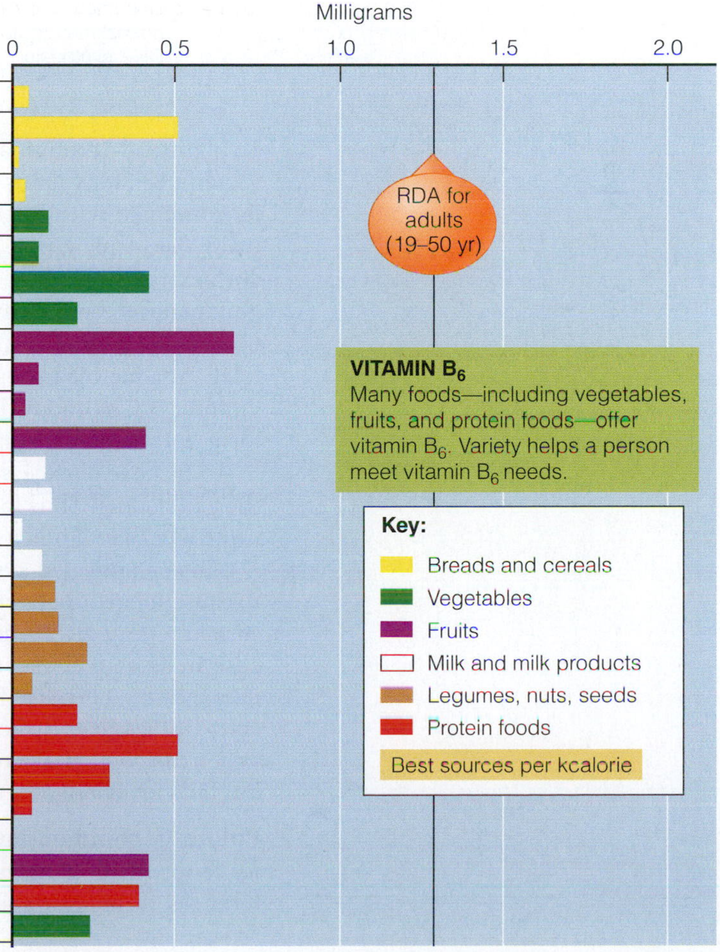

FIGURE 10-10 **Folate's Absorption and Activation**

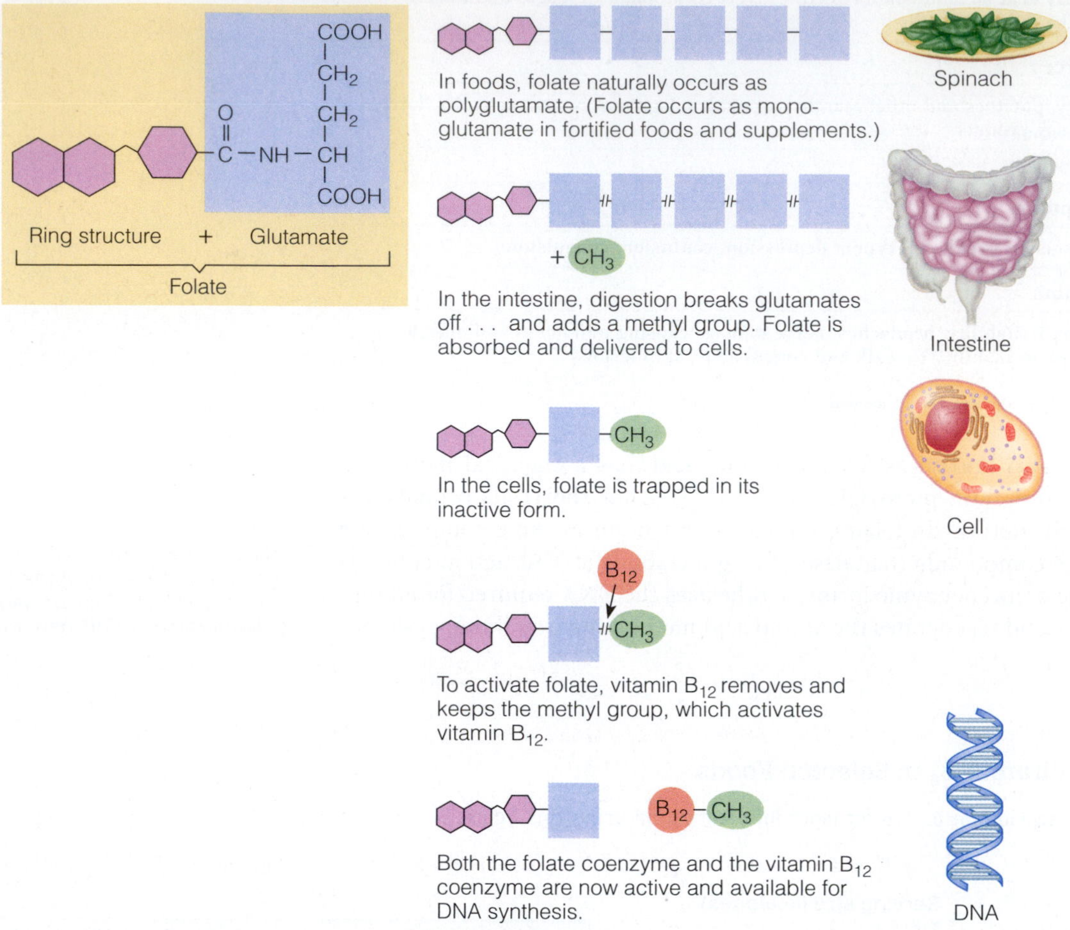

COOH
CH_2
O
CH_2
C—NH—CH
COOH
Ring structure + Glutamate
Folate
In foods, folate naturally occurs as polyglutamate. (Folate occurs as monoglutamate in fortified foods and supplements.)
Spinach
+ CH_3
In the intestine, digestion breaks glutamates off . . . and adds a methyl group. Folate is absorbed and delivered to cells.
Intestine
CH_3
In the cells, folate is trapped in its inactive form.
Cell
B_{12}
CH_3
To activate folate, vitamin B_{12} removes and keeps the methyl group, which activates vitamin B_{12}.
B_{12} CH_3
Both the folate coenzyme and the vitamin B_{12} coenzyme are now active and available for DNA synthesis.
DNA

Figure 10-10 summarizes folate's absorption, activation, and relationship with vitamin B_{12}. It explains that foods deliver folate mostly in the "bound" form—that is, combined with a string of amino acids (all glutamate), known as polyglutamate. (See Appendix C for the chemical structure.) Enzymes on the intestinal cell surfaces hydrolyze the polyglutamate to monoglutamate—folate with only one glutamate attached—and several glutamates. The monoglutamate is then attached to a methyl group (CH_3) and delivered to the liver and other body cells. To activate folate, the methyl group must be removed by an enzyme that requires the help of vitamin B_{12}. Without that help, folate becomes trapped inside cells in its methyl form, unavailable to support DNA synthesis and cell growth.

To dispose of excess folate, the liver secretes most of it into bile that is sent to the gallbladder. Thus folate returns to the intestine in an enterohepatic circulation route like that of bile itself (review Figure 5-16 on p. 144).

This complicated system for handling folate is vulnerable to GI tract injuries. Because folate is actively secreted back into the GI tract with bile, it can be reabsorbed repeatedly. If the GI tract cells are damaged, then folate is lost. Such is the case in alcohol abuse; folate deficiency rapidly develops and, ironically, further damages the GI tract. Remember, folate is active in cell multiplication—and the cells lining the GI tract are among the most rapidly replaced cells in the body. When unable to make new cells, the GI tract deteriorates and not only loses folate, but fails to absorb other nutrients as well.

Folate Recommendations The bioavailability of folate ranges from 50 percent for foods to 100 percent for supplements taken on an empty stomach. These differences in bioavailability were considered when establishing the folate RDA.

HOW TO Estimate Dietary Folate Equivalents

Folate is expressed in terms of DFE (dietary folate equivalents) because synthetic folate from supplements and fortified foods is absorbed at almost twice (1.7 times) the rate of naturally occurring folate from other foods. Use the following equation to calculate:

$$\text{DFE} = \mu g \text{ food folate} + (1.7 \times \mu g \text{ synthetic folate})$$

Consider, for example, a pregnant woman who takes a supplement and eats a bowl of fortified cornflakes, 2 slices of fortified bread, and a cup of fortified pasta. From the supplement and fortified foods, she obtains synthetic folate:

Supplement	100 μg folate
Fortified cornflakes	100 μg folate
Fortified bread	40 μg folate
Fortified pasta	60 μg folate
	300 μg folate

To calculate the DFE, multiply the amount of synthetic folate by 1.7:

$$300\ \mu g \times 1.7 = 510\ \mu g \text{ DFE}$$

Now add the naturally occurring folate from the other foods in her diet—in this example, another 90 μg of folate.

$$510\ \mu g \text{ DFE} + 90\ \mu g = 600\ \mu g \text{ DFE}$$

Notice that if we had not converted synthetic folate from supplements and fortified foods to DFE, then this woman's intake would appear to fall short of the 600 μg recommendation for pregnancy (300 μg + 90 μg = 390 μg). But as our example shows, her intake does meet the recommendation. At this time, supplement and fortified food labels list folate in μg only, not μg DFE, making such calculations necessary.

For additional practice, log on to **www.cengagebrain.com** and search for ISBN 084006845X.

TRY IT Calculate how many dietary folate equivalents a person receives from 200 μg folate from a supplement, 75 μg folate from fortified cereal, and 120 μg folate from other foods.

Naturally occurring folate from foods is given full credit. Synthetic folate from fortified foods and supplements is given extra credit because, on average, it is 1.7 times more available than naturally occurring food folate. Thus a person consuming 100 micrograms of folate from foods and 100 micrograms from a supplement receives 270 **dietary folate equivalents (DFE)**. ♦ (The "How To" describes how to estimate dietary folate equivalents.) The need for folate rises considerably during pregnancy and whenever cells are multiplying, so the recommendations for pregnant women are considerably higher than for other adults.

Folate and Neural Tube Defects The brain and spinal cord develop from the **neural tube**, and defects in its orderly formation during the early weeks of pregnancy may result in various central nervous system disorders and death. (Chapter 14 includes photos of neural tube development and an illustration of a neural tube defect.)

Folate supplements taken one month before conception and continued throughout the first trimester of pregnancy can help prevent **neural tube defects**. ♦ For this reason, all women of childbearing age ♦ who are capable of becoming pregnant should consume 0.4 milligram (400 micrograms) of folate daily. ♦ This recommendation can be met through a diet that includes at least five servings of fruits and vegetables daily, but most women receive about half this amount from foods.[8] Furthermore, the bioavailability of food folate is less than that of synthetic folate.[9] Consequently, supplementation or fortification improves folate status significantly. Women who have given birth to infants with neural tube defects previously should consume 4 milligrams of folate daily before conception and throughout the first trimester of pregnancy.

Because half of the pregnancies each year are unplanned and because neural tube defects occur early in development before most women realize they are pregnant, the Food and Drug Administration (FDA) has mandated that grain products be fortified to deliver folate to the U.S. population.* Labels on fortified products

*Bread products, flour, corn grits, cornmeal, farina, rice, macaroni, and noodles must be fortified with 140 micrograms of folate per 100 grams of grain. For perspective, 100 grams is roughly 3 slices of bread; 1 cup of flour; ½ cup of corn grits, cornmeal, farina, or rice; or ¾ cup of macaroni or noodles.

♦ To calculate DFE:
DFE = μg food folate + (1.7 × μg synthetic folate)
Using the example in the text:

	100 μg food
+	170 μg supplement (1.7 × 100 μg)
	270 μg DFE

♦ The two main types of neural tube defects are **spina bifida** (literally, "split spine") and **anencephaly** ("no brain").

♦ Women of childbearing age (15 to 45 yr) should:
- Eat folate-rich foods
- Eat folate-fortified foods
- Take a multivitamin daily (most provide 400 μg folate)

♦ A milligram (mg) is one-thousandth of a gram. A microgram (μg) is one-thousandth of a milligram (or one-millionth of a gram).
- 0.4 mg = 400 μg

dietary folate equivalents (DFE): the amount of folate available to the body from naturally occurring sources, fortified foods, and supplements, accounting for differences in the bioavailability from each source.

neural tube: the embryonic tissue that forms the brain and spinal cord.

neural tube defects: malformations of the brain, spinal cord, or both during embryonic development that often result in lifelong disability or death.

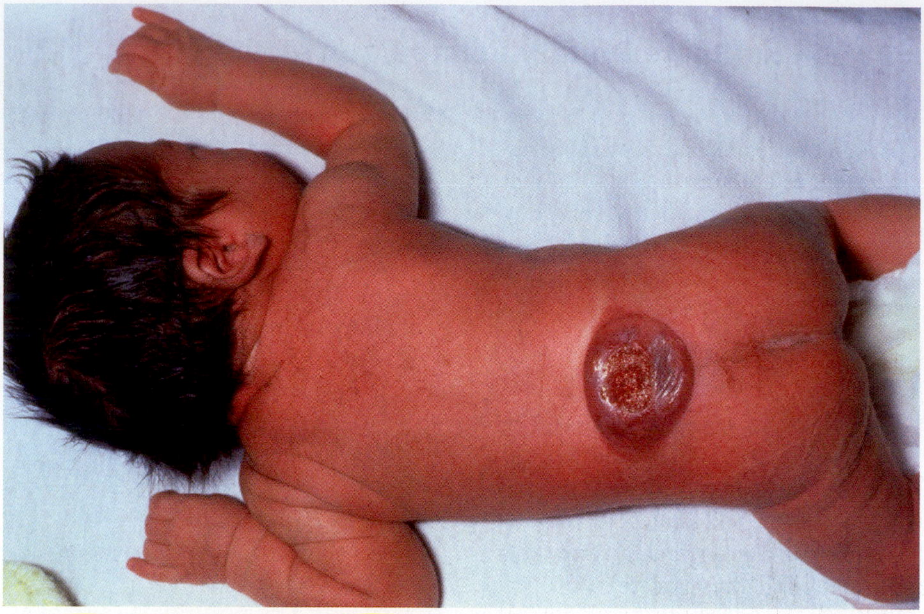

Folate helps to protect against spina bifida, a neural tube defect characterized by the incomplete closure of the spinal cord and its bony encasement.

may claim that "adequate intake of folate has been shown to reduce the risk of neural tube defects." Fortification has improved folate status in women of childbearing age and lowered the number of neural tube defects that occur each year, as Figure 10-11 shows.[10]

Some research suggests that folate may also prevent other congenital birth defects, such as cleft lip and palate.[11] Such findings strengthen recommendations for pregnant women to pay attention to their folate needs.

Folate fortification raises safety concerns as well. Because high intakes of folate can mask a vitamin B_{12} deficiency, folate consumption should not exceed 1 milligram daily without close medical supervision.[12] The risks and benefits of folate fortification continue to be a topic of current debate.[13]

Folate and Heart Disease The FDA's decision to fortify grain products with folate was strengthened by research suggesting a role for folate in protecting against heart disease.[14] One of folate's key roles in the body is to break down the amino acid homocysteine. Without folate, homocysteine accumulates, which seems to enhance formation of blood clots and atherosclerotic lesions. Fortified foods and folate supplements raise blood folate and reduce blood homocysteine, but do not seem to reduce the risk of heart attacks, strokes, or death from cardiovascular causes.[15]

FIGURE 10-11 Decreasing Spina Bifida Rates since Folate Fortification

Neural tube defects have declined since folate fortification began in 1996.

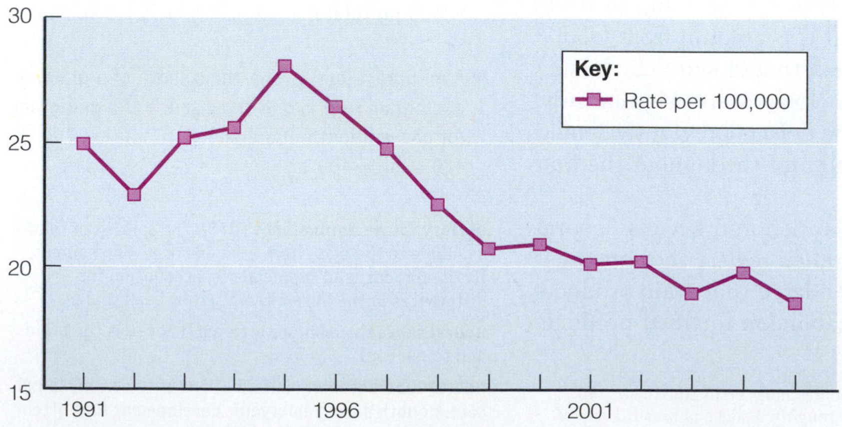

SOURCE: National Vital Statistics System, National Center for Health Statistics, Centers for Disease Control, December 2007.

Folate and Cancer Because the synthesis of DNA and the transfer of methyl groups depend on folate, its relationships with cancer are complex, depending on the type of cancer and the timing of folate supplementation. Sufficient folate may protect against the initiation of cancer, but it may enhance progression once cancer has begun.[16] In general, foods containing folate probably reduce the risk of pancreatic cancer.[17] Limited evidence suggests folate may also reduce the risk of esophageal and colorectal cancer.[18]

Folate Deficiency Folate deficiency impairs cell division and protein synthesis—processes critical to growing tissues. In a folate deficiency, the replacement of red blood cells and GI tract cells falters. Not surprisingly, then, two of the first symptoms of a folate deficiency are **anemia** and GI tract deterioration.

The anemia of folate deficiency is characterized by large, ♦ immature red blood cells. Without folate, DNA damage destroys many of the red blood cells as they attempt to divide and mature. The result is fewer, but larger, red blood cells that cannot carry oxygen or travel through the capillaries as efficiently as normal red blood cells.

♦ Large-cell anemia is known as **macrocytic** or **megaloblastic** anemia.
- **macro** = large
- **cyte** = cell
- **mega** = large

Primary folate deficiencies may develop from inadequate intake and have been reported in infants who were fed goat's milk, which is notoriously low in folate. Secondary folate deficiencies may result from impaired absorption or an unusual metabolic need for the vitamin. Metabolic needs increase in situations where cell multiplication must speed up, such as pregnancies involving twins and triplets; cancer; skin-destroying diseases such as chicken pox and measles; and burns, blood loss, GI tract damage, and the like.

Of all the vitamins, folate appears to be most vulnerable to interactions with drugs, which can also lead to a secondary deficiency. Some medications, notably anticancer drugs, have a chemical structure similar to folate's structure and can displace the vitamin from enzymes and interfere with normal metabolism. Like all cells, cancer cells need the real vitamin to multiply—without it, they die. Unfortunately, anticancer drugs affect both cancerous cells and healthy cells, creating a folate deficiency for all cells. (Chapter 19 discusses nutrient-drug interactions and includes a figure illustrating the similarities between the vitamin folate and the anticancer drug methotrexate.)

Aspirin and antacids also interfere with the body's folate status: aspirin inhibits the action of folate-requiring enzymes, and antacids limit the absorption of folate. Healthy adults who use these drugs to relieve an occasional headache or upset stomach need not be concerned, but people who rely heavily on aspirin or antacids should be aware of the nutrition consequences. Oral contraceptives may also impair folate status, as may smoking.[19]

Leafy dark green vegetables (such as spinach and broccoli), legumes (such as black beans, kidney beans, and black-eyed peas), liver, and some fruits (notably citrus fruits and juices) are naturally rich in folate.

Folate Toxicity Naturally occurring folate from foods alone appears to cause no harm. Excess folate from fortified foods or supplements, however, can reach levels that are high enough to obscure a vitamin B_{12} deficiency and delay diagnosis of neurological damage. For this reason, a UL has been established for folate from fortified foods or supplements (see the inside front cover).

Folate Food Sources Figure 10-12 (p. 330) shows that folate is especially abundant in legumes, fruits, and vegetables. The vitamin's name suggests the word *foliage,* and indeed, leafy green vegetables are outstanding sources. With fortification, grain products also contribute folate. The small red and white bars in Figure 10-12 indicate that meats and milk products are poor folate sources. Heat and oxidation during cooking and storage can destroy as much as half of the folate in foods. The accompanying table provides a summary of folate.

IN SUMMARY Folate

Other Names
Folic acid, folacin, pteroylglutamic acid (PGA)

RDA
Adults: 400 µg/day

Upper Level
Adults: 1000 µg/day

Chief Functions in the Body
Part of coenzymes THF (tetrahydrofolate) and DHF (dihydrofolate) used in DNA synthesis and therefore important in new cell formation

Significant Sources
Fortified grains, leafy green vegetables, legumes, seeds, liver

Easily destroyed by heat and oxygen

Deficiency Symptoms
Anemia (large-cell type);[a] smooth, red tongue;[b] mental confusion, weakness, fatigue, irritability, headache; shortness of breath; elevated homocysteine

Toxicity Symptoms
Masks vitamin B_{12}–deficiency symptoms

[a]Large-cell-type anemia is known as either *macrocytic* or *megaloblastic anemia*.

[b]Smoothness of the tongue is caused by loss of its surface structures and is termed *glossitis* (gloss-EYE-tis).

anemia (ah-NEE-me-ah): literally, "too little blood." Anemia is any condition in which too few red blood cells are present, or the red blood cells are immature (and therefore large) or too small or contain too little hemoglobin to carry the normal amount of oxygen to the tissues. It is not a disease itself but can be a consequence of many different disease conditions, including many nutrient deficiencies, bleeding, excessive red blood cell destruction, and defective red blood cell formation.
- **an** = without
- **emia** = blood

FIGURE 10-12 **Folate in Selected Foods**

See the "How To" section on p. 316 for more information on using this figure.

Food	Serving size (kcalories)
Bread, whole wheat	1 oz slice (70 kcal)
Cornflakes, fortified	1 oz (110 kcal)
Spaghetti pasta	½ c cooked (99 kcal)
Tortilla, flour	1 10"-round (234 kcal)
Broccoli	½ c cooked (22 kcal)
Carrots	½ c shredded raw (24 kcal)
Potato	1 medium baked w/skin (133 kcal)
Tomato juice	½ c (31 kcal)
Banana	1 medium raw (109 kcal)
Orange	1 medium raw (62 kcal)
Strawberries	½ c fresh (22 kcal)
Watermelon	1 slice (92 kcal)
Milk	1 c reduced-fat 2% (121 kcal)
Yogurt, plain	1 c low-fat (155 kcal)
Cheddar cheese	1½ oz (171 kcal)
Cottage cheese	½ c low-fat 2% (101 kcal)
Pinto beans	½ c cooked (117 kcal)
Peanut butter	2 tbs (188 kcal)
Sunflower seeds	1 oz dry (165 kcal)
Tofu (soybean curd)	½ c (76 kcal)
Ground beef, lean	3 oz broiled (244 kcal)
Chicken breast	3 oz roasted (140 kcal)
Tuna, canned in water	3 oz (99 kcal)
Egg	1 hard cooked (78 kcal)
Excellent, and sometimes unusual, sources:	
Lentils	½ c cooked (115 kcal)
Asparagus	½ c cooked (22 kcal)
Orange juice	¾ c fresh (84 kcal)

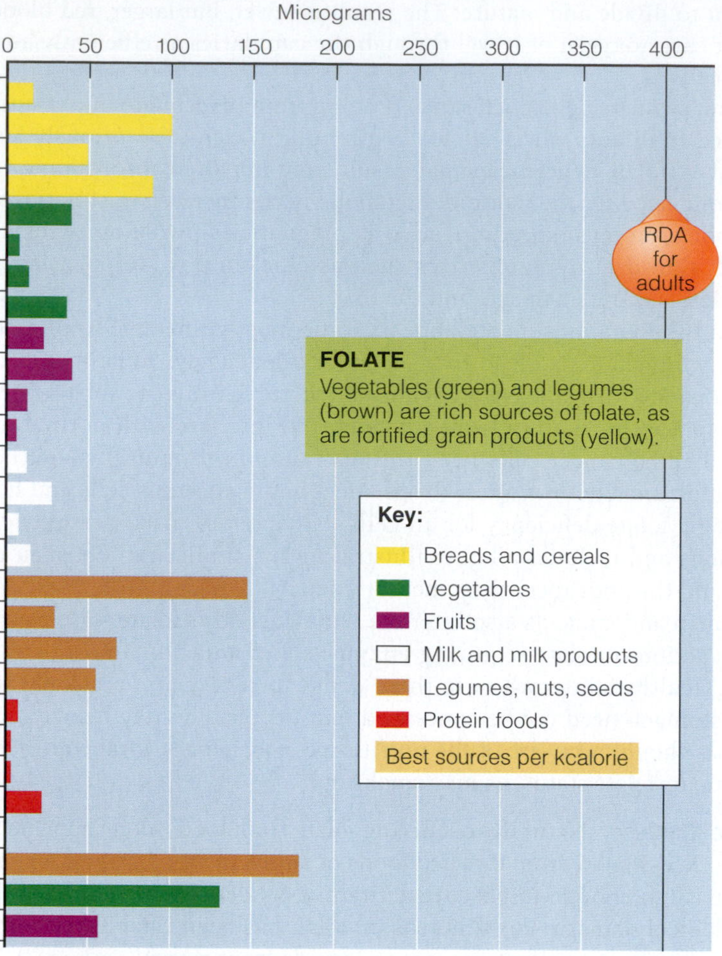

Vitamin B_{12}

Vitamin B_{12} and folate are closely related: each depends on the other for activation. Recall that vitamin B_{12} removes a methyl group to activate the folate coenzyme. When folate gives up its methyl group, the vitamin B_{12} coenzyme becomes activated (review Figure 10-10 on p. 326).

The regeneration of the amino acid methionine and the synthesis of DNA and RNA depend on both folate and vitamin B_{12}.* In addition, without any help from folate, vitamin B_{12} maintains the sheath that surrounds and protects nerve fibers and promotes their normal growth. Bone cell activity and metabolism also depend on vitamin B_{12}.

The digestion and absorption of vitamin B_{12} depend on several steps. In the stomach, hydrochloric acid and the digestive enzyme pepsin release vitamin B_{12} from the proteins to which it is attached in foods. Then as vitamin B_{12} passes from the stomach to the small intestine, it binds with a stomach secretion called **intrinsic factor**. Bound together, intrinsic factor and vitamin B_{12} travel to the end of the small intestine, where receptors recognize the complex. Importantly, the receptors do not recognize vitamin B_{12} without intrinsic factor. The vitamin is gradually absorbed into the bloodstream as the intrinsic factor is degraded. Transport of vitamin B_{12} in the blood depends on specific binding proteins.

Like folate, vitamin B_{12} enters the enterohepatic circulation—continually being secreted into bile and delivered to the intestine where it is reabsorbed. Because

*In the body, methionine serves as a methyl (CH_3) donor. In doing so, methionine can be converted to other amino acids. Some of these amino acids can regenerate methionine, but methionine is still considered an essential amino acid that is needed in the diet.

vitamin B_{12}: a B vitamin characterized by the presence of cobalt (see Figure 13-13 on p. 444). The active forms of coenzyme B_{12} are methylcobalamin and deoxyadenosylcobalamin.

intrinsic factor: a glycoprotein (a protein with short polysaccharide chains attached) secreted by the stomach cells that binds with vitamin B_{12} in the small intestine to aid in the absorption of vitamin B_{12}.

- **intrinsic** = on the inside

most vitamin B_{12} is reabsorbed, healthy people rarely develop a deficiency even when their intake is minimal.

Vitamin B_{12} Recommendations The RDA for adults is only 2.4 micrograms of vitamin B_{12} a day—just over two-millionths of a gram. The ink in the period at the end of this sentence may weigh about 2.4 micrograms. As tiny as this amount appears to the human eye, it contains billions of molecules of vitamin B_{12}, enough to provide coenzymes for all the enzymes that need its help.

Vitamin B_{12} Deficiency and Toxicity Most vitamin B_{12} deficiencies reflect inadequate absorption, not poor intake. Inadequate absorption typically occurs for one of two reasons: a lack of hydrochloric acid or a lack of intrinsic factor. Without hydrochloric acid, the vitamin is not released from the dietary proteins and so is not available for binding with the intrinsic factor. Without the intrinsic factor, the vitamin cannot be absorbed.

Vitamin B_{12} deficiency is common among the elderly.[20] Many people, especially those older than 50, develop **atrophic gastritis**, a condition that damages the cells of the stomach. Atrophic gastritis may also develop in response to iron deficiency or infection with *Helicobacter pylori,* the bacterium implicated in ulcer formation. Without healthy stomach cells, production of hydrochloric acid and intrinsic factor diminishes. Even with an adequate intake from foods, vitamin B_{12} status suffers. The vitamin B_{12} deficiency caused by atrophic gastritis and a lack of intrinsic factor is known as **pernicious anemia.**

Some people inherit a defective gene for the intrinsic factor. In such cases, or when the stomach has been injured and cannot produce enough of the intrinsic factor, vitamin B_{12} must be injected to bypass the need for intestinal absorption. Alternatively, the vitamin may be delivered by nasal spray; absorption is rapid, high, and well tolerated.

A prolonged inadequate intake, as can occur with a vegan diet, ♦ may also create a vitamin B_{12} deficiency. People who stop eating animal-derived foods containing vitamin B_{12} may take several years to develop deficiency symptoms because the body recycles much of its vitamin B_{12}, reabsorbing it over and over again. Even when the body fails to absorb vitamin B_{12}, deficiency may take up to three years to develop because the body conserves its supply. Neurological degeneration, a sign of vitamin B_{12} deficiency, appears more rapidly in infants born to mothers with unsupplemented vegan diets or untreated pernicious anemia.[21]

♦ Vitamin B_{12} is found primarily in foods derived from animals.

Because vitamin B_{12} is required to convert folate to its active form, one of the most obvious vitamin B_{12}–deficiency symptoms is the anemia of folate deficiency. This anemia is characterized by large, immature red blood cells, which indicate slow DNA synthesis and an inability to divide (see Figure 10-13 on p. 332). When folate is trapped in its inactive (methyl folate) form due to vitamin B_{12} deficiency or is unavailable due to folate deficiency itself, DNA synthesis slows.

First to be affected in a vitamin B_{12} or folate deficiency are the rapidly growing blood cells. Either vitamin B_{12} or folate will clear up the anemia, but if folate is given when vitamin B_{12} is needed, the result is disastrous: devastating neurological symptoms. Remember that vitamin B_{12}, but not folate, maintains the sheath that surrounds and protects nerve fibers and promotes their normal growth. Folate "cures" the *blood* symptoms of a vitamin B_{12} deficiency, but cannot stop the *nerve* symptoms from progressing.[22] By doing so, folate "masks" a vitamin B_{12} deficiency.

Marginal vitamin B_{12} deficiency impairs cognition.[23] Advanced neurological symptoms include a creeping paralysis that begins at the extremities and works inward and up the spine. Early detection and correction are necessary to prevent permanent nerve damage and paralysis. With sufficient folate in the diet, the neurological symptoms of vitamin B_{12} deficiency can develop without evidence of anemia. Such interactions between folate and vitamin B_{12} highlight some of the safety issues surrounding the use of supplements and the fortification of foods.[24] No adverse effects have been reported for excess vitamin B_{12}, and no UL has been set.

atrophic (a-TRO-fik) **gastritis** (gas-TRY-tis): chronic inflammation of the stomach accompanied by a diminished size and functioning of the mucous membrane and glands.
- **atrophy** = wasting
- **gastro** = stomach
- **itis** = inflammation

pernicious (per-NISH-us) **anemia:** a blood disorder that reflects a vitamin B_{12} deficiency caused by lack of intrinsic factor and characterized by abnormally large and immature red blood cells. Other symptoms include muscle weakness and irreversible neurological damage.
- **pernicious** = destructive

Vitamin B_{12} Food Sources Vitamin B_{12} is unique among the vitamins in being found almost exclusively in foods derived from animals. Its bioavailability is

FIGURE 10-13 **Normal and Anemic Blood Cells**

The anemia of folate deficiency is indistinguishable from that of vitamin B_{12} deficiency. Appendix E describes the biochemical tests used to differentiate the two conditions.

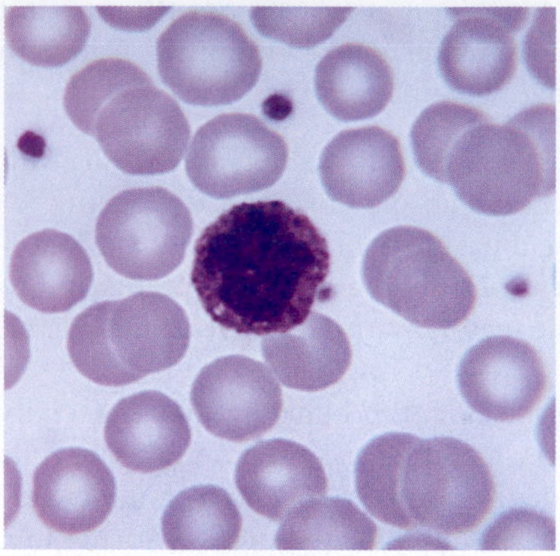

Normal blood cells. The size, shape, and color of these red blood cells show that they are normal.

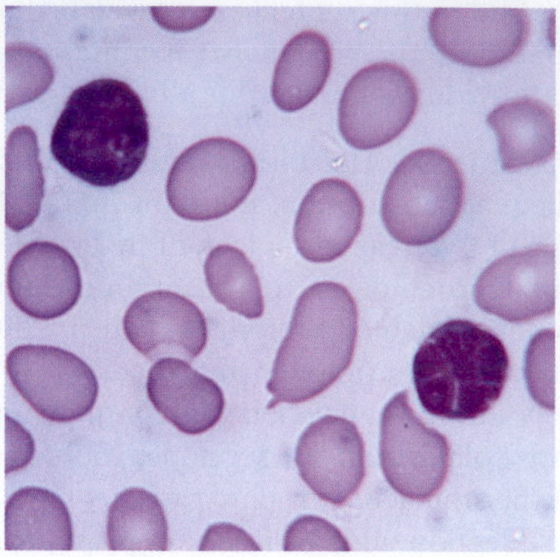

Blood cells in pernicious anemia (megaloblastic). These megaloblastic blood cells are slightly larger (macrocytic) than normal red blood cells, and their shapes are irregular.

greatest from milk and fish.[25] Anyone who eats reasonable amounts of animal-derived foods is most likely to have an adequate intake, including vegetarians who use milk products or eggs. Vegans, who restrict all foods derived from animals, need a reliable source, such as vitamin B_{12}–fortified soy milk or vitamin B_{12} supplements. Yeast grown on a vitamin B_{12}–enriched medium and mixed with that medium provides some vitamin B_{12}, but yeast itself does not contain active vitamin B_{12}. Fermented soy products such as miso (a soybean paste) and sea algae such as spirulina also do *not* provide active vitamin B_{12}. Extensive research shows that the amounts listed on the labels of these plant products are inaccurate and misleading because the vitamin B_{12} is in an inactive, unavailable form.

As mentioned earlier, the water-soluble vitamins are particularly vulnerable to losses in cooking. For most of these nutrients, microwave heating minimizes losses as well as, or better than, traditional cooking methods. Such is not the case for vitamin B_{12}, however. Microwave heating inactivates vitamin B_{12}. To preserve this vitamin, use the oven or stovetop instead of a microwave to cook meats and milk products (major sources of vitamin B_{12}). The accompanying table provides a summary of vitamin B_{12}.

IN SUMMARY Vitamin B_{12}

Other Names

Cobalamin (and related forms)

RDA

Adults: 2.4 µg/day

Chief Functions in the Body

Part of coenzymes methylcobalamin and deoxyadenosylcobalamin used in new cell synthesis; helps to maintain nerve cells; reforms folate coenzyme; helps to break down some fatty acids and amino acids

Significant Sources

Foods of animal origin (meat, fish, poultry, shellfish, milk, cheese, eggs), fortified cereals

Easily destroyed by microwave cooking

Deficiency Disease

Pernicious anemia[a]

Deficiency Symptoms

Anemia (large-cell type);[b] fatigue, degeneration of peripheral nerves progressing to paralysis; sore tongue, loss of appetite, constipation

Toxicity Symptoms

None reported

[a]The name *pernicious anemia* refers to the vitamin B_{12} deficiency caused by atrophic gastritis and a lack of intrinsic factor, but not to that caused by inadequate dietary intake.

[b]Large-cell-type anemia is known as either *macrocytic* or *megaloblastic anemia.*

Vitamin-Like Compounds Nutrition scientists debate whether other dietary compounds might also be considered vitamins. These compounds may have functions in the body, but to be a vitamin, a compound must be dietarily essential. In some cases, the compounds may be conditionally essential—that is, needed by the body from foods when synthesis becomes insufficient to support normal growth and metabolism. In other cases, the compounds may simply not be needed from the diet under any circumstances.

Choline Determining whether the nitrogen-containing compound choline is an essential nutrient has been blurry for decades, in part because the body can make choline from the amino acid methionine. Furthermore, choline is commonly found in foods such as milk, eggs, and peanuts and as part of lecithin, a food additive commonly used as an emulsifying agent (review Figure 5-9 on p. 140). Consequently, choline deficiencies are rare. Without any dietary choline, however, synthesis alone appears to be insufficient to meet the body's needs, making choline a conditionally essential nutrient. For this reason, the DRI Committee established an Adequate Intake (AI) for choline. The body uses choline to make the neurotransmitter acetylcholine and the phospholipid lecithin. During fetal development, choline supports the structure and function of the brain and spinal cord, by supporting neural tube closure and enhancing learning performance.[26] The UL for choline is based on its critical effect in lowering blood pressure. The accompanying table provides a summary of choline.

IN SUMMARY Choline

Adequate Intake (AI)	Deficiency Symptoms
Men: 550 mg/day	Liver damage
Women: 425 mg/day	
Toxicity Symptoms	
	Body odor, sweating, salivation, reduced growth rate, low blood pressure, liver damage
Upper Level	
Adults: 3500 mg/day	
Chief Functions in the Body	**Significant Sources**
Needed for the synthesis of the neurotransmitter acetylcholine and the phospholipid lecithin	Milk, liver, eggs, peanuts

Inositol and Carnitine **Inositol** is a part of cell membrane structures, and **carnitine** transports long-chain fatty acids from the cytosol to the mitochondria for oxidation. Like choline, these two substances can be made by the body, but unlike choline, no recommendations have been established. Researchers continue to explore the possibility that these substances may be essential. Even if they are essential, though, supplements are unnecessary because these compounds are widespread in foods.

Some vitamin companies include choline, inositol, and carnitine in their formulations to make their vitamin pills look more "complete" than others, but this strategy offers no real advantage. For a rational way to compare vitamin-mineral supplements, read Highlight 10.

Non-Vitamins Other substances have been mistaken for essential nutrients for human beings because they are needed for growth by bacteria or other forms of life. Among them are PABA (para-aminobenzoic acid, a component of folate's chemical structure), the bioflavonoids (vitamin P or hesperidin), pyrroloquinoline quinone (methoxatin), orotic acid, lipoic acid, and ubiquinone (coenzyme Q_{10}). Other names erroneously associated with vitamins are "vitamin O" (oxygenated saltwater), "vitamin B_5" (another name for pantothenic acid), "vitamin B_{15}" (also called "pangamic acid," a hoax), and "vitamin B_{17}" (laetrile, an alleged "cancer cure" and not a vitamin or a cure by any stretch of the imagination—in fact, laetrile is a potentially dangerous substance).

IN SUMMARY The B vitamins serve as coenzymes that facilitate the work of every cell. They are active in carbohydrate, fat, and protein metabolism and

inositol (in-OSS-ih-tall): a nonessential nutrient that can be made in the body from glucose. Inositol is a part of cell membrane structures.

carnitine (CAR-neh-teen): a nonessential, nonprotein amino acid made in the body from lysine that helps transport fatty acids across the mitochondrial membrane.

in the making of DNA and thus new cells. Historically famous B vitamin–deficiency diseases are beriberi (thiamin), pellagra (niacin), and pernicious anemia (vitamin B_{12}). Pellagra can be prevented by adequate protein because the amino acid tryptophan can be converted to niacin in the body. A high intake of folate can mask the blood symptoms of a vitamin B_{12} deficiency, but it will not prevent the associated nerve damage. Vitamin B_6 participates in amino acid metabolism and can be harmful in excess. Biotin and pantothenic acid serve important roles in energy metabolism and are common in a variety of foods. Many substances that people claim as B vitamins are not.

The B Vitamins—In Concert

This chapter has described some of the impressive ways that vitamins work individually, as if their many actions in the body could easily be disentangled. In fact, it is often difficult to tell which vitamin is truly responsible for a given effect because the nutrients are interdependent; the presence or absence of one affects another's absorption, metabolism, and excretion. You have already seen this interdependence with folate and vitamin B_{12}.

Riboflavin and vitamin B_6 provide another example. One of the riboflavin coenzymes, FMN, assists the enzyme that converts vitamin B_6 to its coenzyme form PLP. Consequently, a severe riboflavin deficiency can impair vitamin B_6 activity. Thus a deficiency of one nutrient may alter the action of another. Furthermore, a deficiency of one nutrient may create a deficiency of another. For example, both riboflavin and vitamin B_6 (as well as iron) are required for the conversion of tryptophan to niacin. Consequently, an inadequate intake of either riboflavin or vitamin B_6 can diminish the body's niacin supply. These interdependent relationships are evident in many of the roles B vitamins play in the body.

B Vitamin Roles Figure 10-14 summarizes the metabolic pathways introduced in Chapter 7 and conveys an *impression* of the many ways B vitamins assist in metabolic pathways. Metabolism is the body's work, and the B vitamin coenzymes are indispensable to every step. In scanning the pathways of metabolism depicted in the figure, note the many abbreviations for the coenzymes that keep the processes going.

Look at the now-familiar pathway of glucose breakdown. To break down glucose to pyruvate, the cells must have certain enzymes. For the enzymes to work, they must have the niacin coenzyme NAD. Cells can make NAD, but only if they have enough niacin (or enough of the amino acid tryptophan to make niacin).

The next step is the breakdown of pyruvate to acetyl CoA. The enzymes involved in this step require both NAD and the thiamin and riboflavin coenzymes TPP and FAD, respectively. The cells can manufacture the enzymes they need from the vitamins, if the vitamins are in the diet.

Another coenzyme needed for this step is CoA. Predictably, the cells can make CoA except for an essential part that must be obtained in the diet—pantothenic acid. Another coenzyme requiring biotin serves the enzyme complex involved in converting pyruvate to oxaloacetate, the compound that combines with acetyl CoA to start the TCA cycle.

These and other coenzymes participate throughout all the metabolic pathways. Vitamin B_6 is an indispensable part of PLP—a coenzyme required for many amino acid conversions, for a crucial step in the making of the iron-containing portion of hemoglobin for red blood cells, and for many other reactions. Folate becomes THF—the coenzyme required for the synthesis of new genetic material and therefore new cells. The vitamin B_{12} coenzyme, in turn, regenerates THF to its active form; thus vitamin B_{12} is also necessary for the formation of new cells.

Thus each of the B vitamin coenzymes is involved, directly or indirectly, in energy metabolism. Some facilitate the energy-releasing reactions themselves; others help build new cells to deliver the oxygen and nutrients that allow the energy reactions to occur.

FIGURE 10-14 Metabolic Pathways Involving B Vitamins

These metabolic pathways are introduced in Chapter 7 and are presented here to highlight the many coenzymes that facilitate the reactions. These coenzymes depend on the following vitamins:

- NAD and NADP: niacin
- TPP: thiamin
- CoA: pantothenic acid
- B_{12}: vitamin B_{12}
- FMN and FAD: riboflavin
- THF: folate
- PLP: vitamin B_6
- Biotin

Pathways leading toward acetyl CoA and the TCA cycle are catabolic, and those leading toward amino acids, glycogen, and fat are anabolic. For further details, see Appendix C.

Animated! figure
www.cengagebrain.com
(search for ISBN 084006845X)

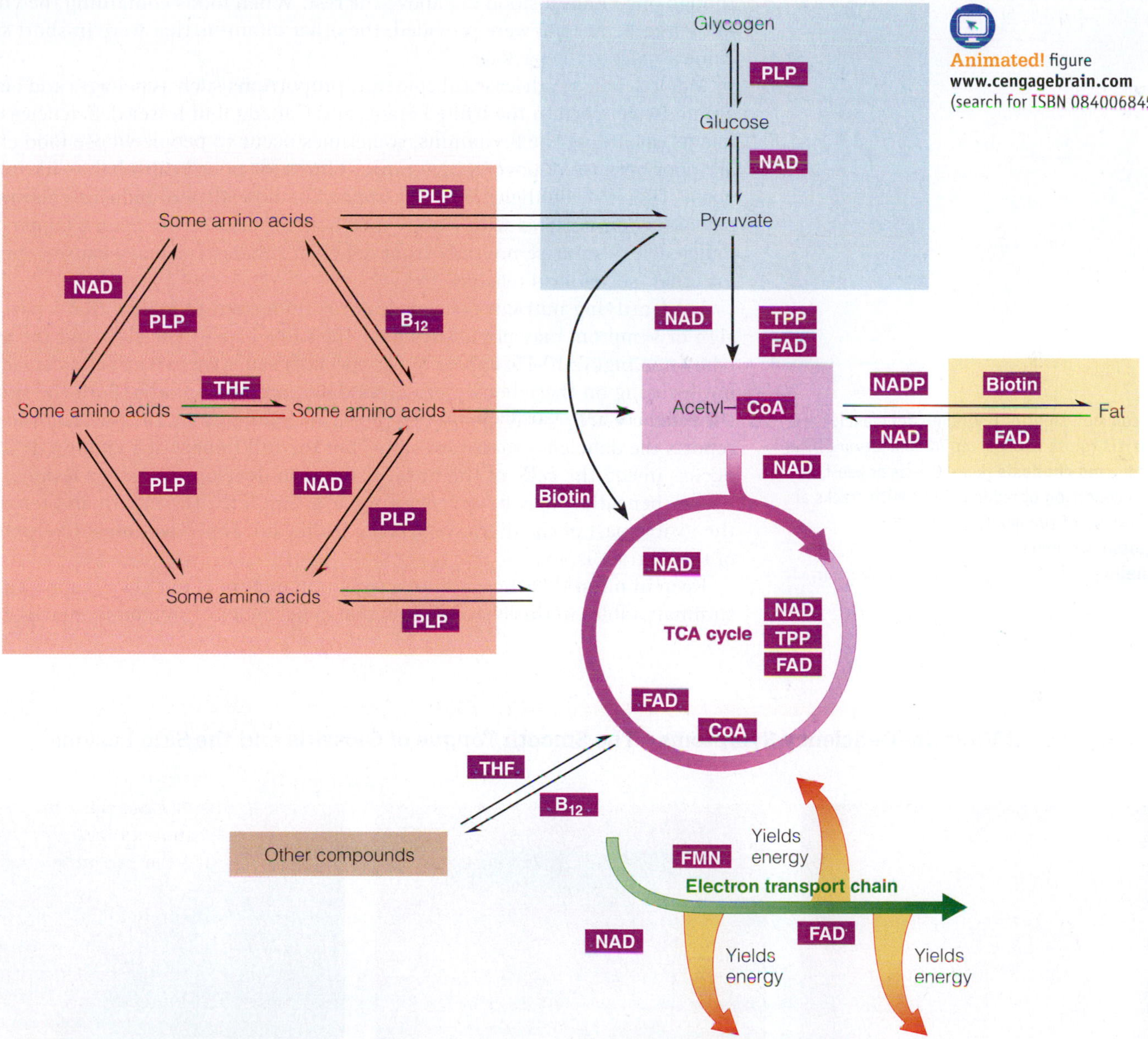

B Vitamin Deficiencies

Now suppose the body's cells lack one of these B vitamins—niacin, for example. Without niacin, the cells cannot make NAD. Without NAD, the enzymes involved in every step of the glucose-to-energy pathway cannot function. Then, because all the body's activities require energy, literally everything begins to grind to a halt. This is no exaggeration. The deadly disease pellagra, caused by niacin deficiency, produces the "devastating four *D*s": dermatitis, which reflects a failure of the skin; dementia, a failure of the nervous system; diarrhea, a failure of digestion and absorption; and eventually, as would be the case for any severe nutrient deficiency, death. These symptoms are the obvious ones, but a niacin deficiency affects all other organs, too, because all are dependent on the energy pathways.

All the vitamins are as essential as niacin. With any B vitamin deficiency, many body systems become deranged, and similar symptoms may appear. A lack of any of them can have disastrous and far-reaching effects.

Deficiencies of single B vitamins seldom show up in isolation, however. After all, people do not eat nutrients singly; they eat foods, which contain mixtures of nutrients. Only in two cases described earlier—beriberi and pellagra—have dietary deficiencies associated with single B vitamins been observed on a large scale in human populations. Even in these cases, several vitamins were lacking even though one vitamin stood out above the rest. When foods containing the vitamin known to be needed were provided, the other vitamins that were in short supply came as part of the package.

Major deficiency diseases of epidemic proportions such as pellagra and beriberi are no longer seen in the United States and Canada, but lesser deficiencies of nutrients, including the B vitamins, sometimes occur in people whose food choices are poor because of poverty, ignorance, illness, or poor health habits like alcohol abuse. (Review Highlight 7 to fully appreciate how alcohol induces vitamin deficiencies and interferes with energy metabolism.) Remember from Chapter 1 that deficiencies can arise not only from deficient intakes (primary causes), but also for other (secondary) reasons.

In identifying nutrient deficiencies, it is important to realize that a particular sign or symptom may not always have the same cause. The skin and the tongue (shown in Figure 10-15) appear to be especially sensitive to B vitamin deficiencies, but focusing on these body parts gives them undue emphasis. Both the skin and the tongue ♦ are readily visible in a physical examination. The physician sees and reports the deficiency's outward signs, but the full impact of a vitamin deficiency occurs inside the cells of the body. If the skin develops a rash or lesions, other tissues beneath it may be degenerating, too. Similarly, the mouth and tongue are the visible part of the digestive system; if they are abnormal, most likely the rest of the GI tract is, too.

♦ Two common signs of B vitamin deficiencies are **glossitis** (gloss-EYE-tis), an inflammation of the tongue, and **cheilosis** (kye-LOH-sis or kee-LOH-sis), a condition of reddened lips with cracks at the corners of the mouth.

- **glossa** = tongue
- **cheilos** = lip

Keep in mind that the cause of a sign or symptom is not always apparent. The summary tables in this chapter show that deficiencies of riboflavin, niacin, biotin,

FIGURE 10-15 B Vitamin–Deficiency Symptoms—The Smooth Tongue of Glossitis and the Skin Lesions of Cheilosis

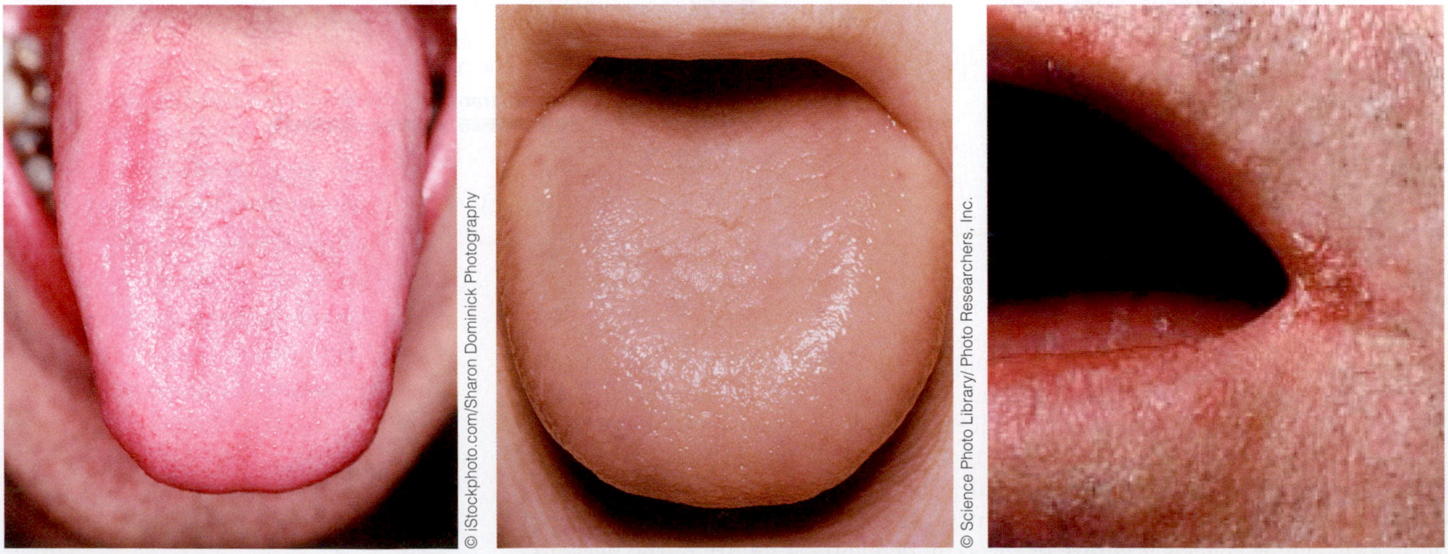

A healthy tongue has a rough and somewhat bumpy surface.

In a B vitamin deficiency, the tongue becomes smooth and swollen due to atrophy of the tissue (glossitis).

In a B vitamin deficiency, the corners of the mouth become irritated and inflamed (cheilosis).

and vitamin B_6 can all cause skin rashes. So can a deficiency of protein, linoleic acid, or vitamin A. Because skin is on the outside and easy to see, it is a useful indicator of "things going wrong inside cells." By itself, a skin condition says nothing about its possible cause.

The same is true of anemia. Anemia is often caused by iron deficiency, but it can also be caused by a folate or vitamin B_{12} deficiency; by digestive tract failure to absorb any of these nutrients; or by such nonnutritional causes as infections, parasites, cancer, or loss of blood. No single nutrient will always cure a given symptom.

A person who feels chronically tired may be tempted to self-diagnose iron-deficiency anemia and self-prescribe an iron supplement. But this will relieve tiredness only if the cause is indeed iron-deficiency anemia. If the cause is a folate deficiency, taking iron will only prolong the fatigue. A person who is better informed may decide to take a vitamin supplement with iron, covering the possibility of a vitamin deficiency. But the symptom may have a nonnutritional cause. If the cause of the tiredness is actually hidden blood loss due to cancer, the postponement of a diagnosis may be fatal. When fatigue is caused by a lack of sleep, of course, no nutrient or combination of nutrients can replace a good night's rest. A person who is chronically tired should see a physician rather than self-prescribe. If the condition is nutrition related, a registered dietitian should be consulted as well.

B Vitamin Toxicities Toxicities of the B vitamins from foods alone are unknown, but they can occur when people overuse dietary supplements. With supplements, the quantities can quickly overwhelm the cells. Consider that one small capsule can easily deliver 2 milligrams of vitamin B_6, but it would take more than 3000 bananas, 6600 cups of rice, or 3600 chicken breasts to supply an equivalent amount. When the cells become oversaturated with a vitamin, they must work to eliminate the excess. The cells dispatch water-soluble vitamins to the urine for excretion, but sometimes they cannot keep pace with the onslaught. Homeostasis becomes disturbed and symptoms of toxicity develop.

B Vitamin Food Sources Significantly, deficiency diseases, such as beriberi and pellagra, were eliminated by supplying foods—not pills. Vitamin pill advertisements make much of the fact that vitamins are indispensable to life, but human beings obtained their nourishment from foods for centuries before vitamin pills existed. If the diet lacks a vitamin, the first solution is to adjust food intake to obtain that vitamin.

The bar graphs of selected foods in this chapter, taken together, sing the praises of a balanced diet. The grains deliver thiamin, riboflavin, niacin, and folate. The fruit and vegetable groups excel in folate. Protein foods serve thiamin, niacin, vitamin B_6, and vitamin B_{12} well. The milk group stands out for riboflavin and vitamin B_{12}. A diet that offers a variety of foods from each group, prepared with reasonable care, serves up ample B vitamins.

IN SUMMARY **The B vitamin coenzymes work together in energy metabolism. Some facilitate the energy-releasing reactions themselves; others help build cells to deliver the oxygen and nutrients that permit the energy pathways to run. These vitamins depend on one another to function optimally; a deficiency of any of them creates multiple problems. Fortunately, a variety of foods from each of the food groups provides an adequate supply of all of the B vitamins.**

Vitamin C

Two hundred and sixty years ago, any man who joined the crew of a seagoing ship knew he had at best a 50–50 chance of returning alive—not because he might be slain by pirates or die in a storm, but because he might contract **scurvy**. As many as two-thirds of a ship's crew could die of scurvy during a long voyage. Only men

scurvy: the vitamin C–deficiency disease.

on short voyages, especially around the Mediterranean Sea, were free of scurvy. No one knew the reason: that on long ocean voyages, the ship's cook used up the fresh fruits and vegetables early and then served only cereals and meats until the return to port.

The first nutrition experiment ever performed on human beings was devised in the mid-1700s to find a cure for scurvy. James Lind, a British physician, divided 12 sailors with scurvy into 6 pairs. Each pair received a different supplemental ration: cider, vinegar, sulfuric acid, seawater, oranges and lemons, or a strong laxative mixed with spices. Those receiving the citrus fruits quickly recovered, but sadly, it was 50 years before the British navy required all vessels to provide every sailor ♦ with lime juice daily.

The antiscurvy "something" in limes and other foods was dubbed the **antiscorbutic factor.** Nearly 200 years later, the factor was isolated and found to be a 6-carbon compound similar to glucose; it was named **ascorbic acid.**

♦ The tradition of providing British sailors with citrus juice daily to prevent scurvy gave them the nickname "limeys."

Vitamin C Roles

Vitamin C parts company with the B vitamins in its mode of action. In some settings, vitamin C serves as a cofactor ♦ helping a specific enzyme perform its job, but in others, it acts as an antioxidant participating in more general ways.

♦ A **cofactor** is a small, inorganic or organic substance that facilitates the action of an enzyme.

As an Antioxidant

Vitamin C loses electrons easily, a characteristic that allows it to perform as an **antioxidant.** ♦ In the body, antioxidants defend against **free radicals.** Free radicals are discussed fully in Highlight 11, but for now, a simple definition will suffice. A free radical is a molecule with one or more unpaired electrons, which makes it unstable and highly reactive. By donating an electron or two, antioxidants neutralize free radicals and protect other substances from their damage. Figure 10-16 illustrates how vitamin C can give up electrons to stop free-radical damage and then accept them again to become reactivated. This recycling of vitamin C is key to limiting losses and maintaining a reserve of antioxidants in the body. Transporting and concentrating vitamin C in the cells enhances its role as an antioxidant.[27]

♦ Key antioxidant nutrients:
- Vitamin C, vitamin E, beta-carotene
- Selenium

Vitamin C is like a bodyguard for water-soluble substances; it stands ready to sacrifice its own life to save theirs. In the cells and body fluids, vitamin C protects tissues from **oxidative stress** and thus may play an important role in preventing diseases. In the intestines, vitamin C enhances iron absorption by protecting iron from oxidation. (Chapter 13 provides more details about the relationship between vitamin C and iron.)

As a Cofactor in Collagen Formation

Vitamin C helps to form the fibrous structural protein of connective tissues known as collagen. ♦ Collagen serves as

♦ **Collagen** is the structural protein from which connective tissues such as scars, tendons, ligaments, and the foundations of bones and teeth are made.

FIGURE 10-16 Active Forms of Vitamin C

The two hydrogens highlighted in yellow give vitamin C its acidity and its ability to act as an antioxidant.

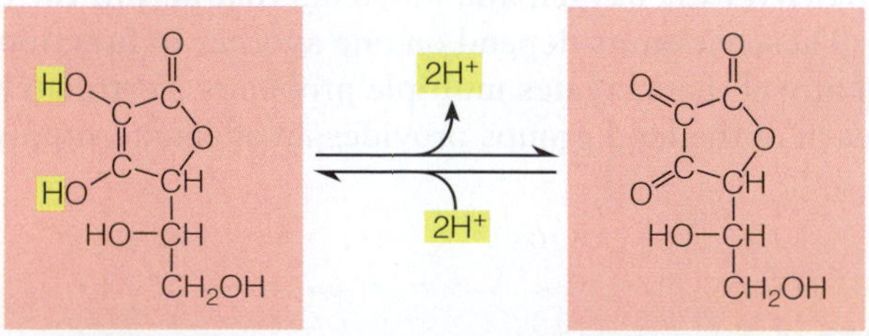

Ascorbic acid protects against oxidative damage by donating its two hydrogens with their electrons to free radicals (molecules with unpaired electrons). In doing so, ascorbic acid becomes dehydroascorbic acid.

Dehydroascorbic acid can readily accept hydrogens to become ascorbic acid. The reversibility of this reaction is key to vitamin C's role as an antioxidant.

antiscorbutic (AN-tee-skor-BUE-tik) **factor:** the original name for vitamin C.
- **anti** = against
- **scorbutic** = causing scurvy

ascorbic acid: one of the two active forms of vitamin C (see Figure 10-16). Many people refer to vitamin C by this name.
- **a** = without
- **scorbic** = having scurvy

antioxidant: a substance in foods that significantly decreases the adverse effects of free radicals on normal physiological functions in the human body.

free radicals: unstable molecules with one or more unpaired electrons.

oxidative stress: a condition in which the production of oxidants and free radicals exceeds the body's ability to handle them and prevent damage.

the matrix on which bones and teeth are formed. When a person is wounded, collagen glues the separated tissues together, forming scars. Cells are held together largely by collagen; this is especially important in the walls of the blood vessels, which must withstand the pressure of blood surging with each beat of the heart.

Chapter 6 describes how the body makes proteins by stringing together chains of amino acids. During the synthesis of collagen, each time a proline or lysine is added to the growing protein chain, an enzyme hydroxylates it (adds an OH group to it), making the amino acid hydroxyproline or hydroxylysine, respectively. These two special amino acids facilitate the binding together of collagen fibers to make strong, ropelike structures. The conversion of proline to hydroxyproline requires both vitamin C and iron. Iron works as a cofactor in the reaction, and vitamin C protects iron from oxidation, thereby allowing iron to perform its duty. Without vitamin C and iron, the hydroxylation step does not occur.

As a Cofactor in Other Reactions Vitamin C also serves as a cofactor in the synthesis of several other compounds. As in collagen formation, vitamin C helps in the hydroxylation of carnitine, a compound that transports fatty acids, especially long-chain fatty acids, across the inner membrane of mitochondria in cells. It participates in the conversions of the amino acids tryptophan and tyrosine to the neurotransmitters serotonin and norepinephrine, respectively. Vitamin C also assists in the making of hormones, including thyroxin, which regulates the metabolic rate; when metabolism speeds up in times of extreme physical stress, the body's use of vitamin C increases.

In Stress The adrenal glands contain more vitamin C than any other organ in the body. ♦ During stress, the adrenal glands release vitamin C, together with hormones, into the blood.[28] The vitamin's exact role in the stress reaction remains unclear, but physical stresses raise vitamin C needs. Among the stresses known to increase vitamin C needs are infections; burns; extremely high or low temperatures; intakes of toxic heavy metals such as lead, mercury, and cadmium; the chronic use of certain medications, including aspirin, barbiturates, and oral contraceptives; and cigarette smoking.

When immune system cells are called into action, they use a great deal of oxygen and produce free radicals. In this case, free radicals are helpful. They act as ammunition in an "oxidative burst" that demolishes the offending viruses and bacteria and destroys the damaged cells. Vitamin C steps in as an antioxidant to control this oxidative activity.

♦ Vitamin C is found in:

- Adrenal glands, pituitary glands
- Liver, spleen, heart, kidneys, lungs, pancreas, white blood cells
- Muscles, red blood cells

In the Prevention and Treatment of the Common Cold Vitamin C has been a popular option for the prevention and treatment of the common cold for decades, but research supporting such claims has been conflicting and controversial. Some studies find no relationship between vitamin C and the occurrence of the common cold, whereas others report modest benefits—fewer colds, fewer days, and shorter duration of severe symptoms, especially for those exposed to physical and environmental stresses.[29] A review of the research on vitamin C in the treatment and prevention of the common cold reveals a slight, but consistent reduction (of 8 percent) in the duration of the common cold in favor of those taking a daily dose of at least 200 milligrams of vitamin C.[30] The question for consumers to consider is, "Is this enough to warrant routine daily supplementation?" Findings from one study show that consumers want their colds to be at least 25 percent less severe to justify the costs of taking vitamin C supplements regularly.[31]

Discoveries about how vitamin C works in the body provide possible links between the vitamin and the common cold. Anyone who has ever had a cold knows the discomfort of a runny or stuffed-up nose. Nasal congestion develops in response to elevated blood **histamine**, and people commonly take antihistamines for relief. Like an antihistamine, vitamin C comes to the rescue and deactivates histamine.

In Disease Prevention Whether vitamin C may help in preventing or treating cancer, heart disease, cataract, and other diseases is still being studied, and findings are presented in Highlight 11's discussion on antioxidants. Conducting

histamine (HISS-tah-mean or HISS-tah-men): a substance produced by cells of the immune system as part of a local immune reaction to an antigen; participates in causing inflammation.

research in the United States and Canada can be difficult, however, because diets typically contribute enough vitamin C to provide optimal health benefits.

Vitamin C Recommendations For decades, vitamin C ranked at the top of dietary supplement sales. How much vitamin C does a person need? As is true of all the vitamins, recommendations are set generously above the minimum requirement to prevent deficiency disease and well below the toxicity level (see Figure 10-17).[32]

The requirement—the amount needed to prevent the overt symptoms of scurvy—is only 10 milligrams daily. However, 10 milligrams a day does not saturate all the body tissues; higher intakes will increase the body's total vitamin C. At about 100 milligrams ♦ per day, 95 percent of the population probably reaches tissue saturation. Recommendations are slightly lower, ♦ based on the amounts needed to provide antioxidant protection. At about 200 milligrams, absorption reaches a maximum, and there is little, if any, increase in blood concentrations at higher doses. Excess vitamin C is readily excreted.

♦ For perspective, 1 c orange juice provides >100 mg vitamin C.

♦ Vitamin C RDA
- Men: 90 mg/day
- Women: 75 mg/day

As mentioned earlier, cigarette smoking increases the need for vitamin C. Cigarette smoke contains oxidants, which greedily deplete this potent antioxidant. Exposure to cigarette smoke, especially when accompanied by low dietary intakes of vitamin C, depletes the body's vitamin C in both active and passive smokers. People who chew tobacco also have low levels of vitamin C. Because people who smoke cigarettes regularly suffer significant oxidative stress, their requirement for vitamin C is increased an additional 35 milligrams; nonsmokers regularly exposed to cigarette smoke should also be sure to meet their RDA for vitamin C.

FIGURE 10-17 Vitamin C Intake (mg/day)

Recommendations for vitamin C are set generously above the minimum requirement and well below the toxicity level.

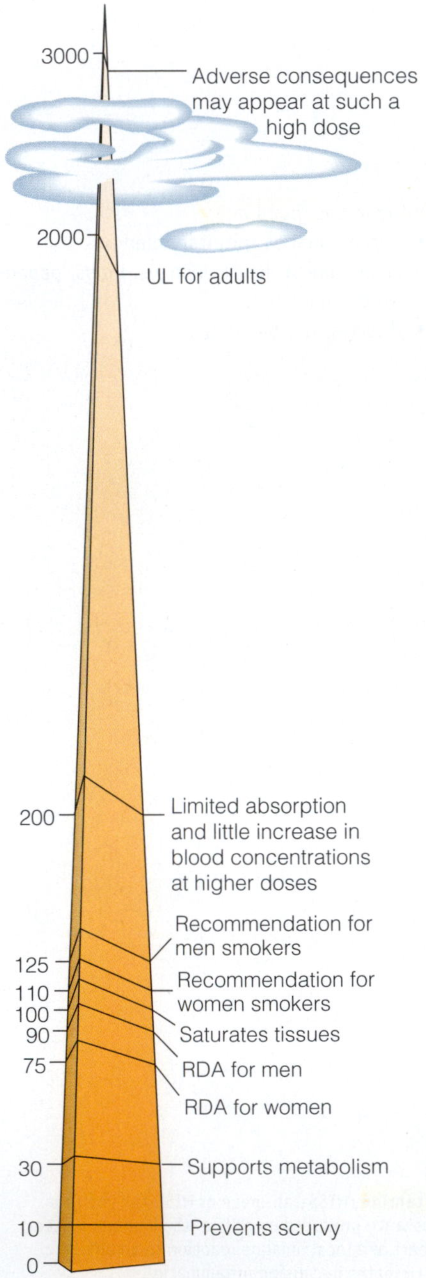

Vitamin C Deficiency Two of the most notable signs of a vitamin C deficiency reflect its role in maintaining the integrity of blood vessels. The gums bleed easily around the teeth, and capillaries under the skin break spontaneously, producing pinpoint hemorrhages (see Figure 10-18).

When vitamin C concentrations fall to about a fifth of optimal levels (this may take more than a month on a diet lacking vitamin C), scurvy symptoms begin to appear. Inadequate collagen synthesis causes further hemorrhaging. Muscles, including the heart muscle, degenerate. The skin becomes rough, brown, scaly, and dry. Wounds fail to heal because scar tissue will not form. Bone rebuilding falters; the ends of the long bones become softened, malformed, and painful, and fractures develop. The teeth become loose as the cartilage around them weakens. Anemia and infections are common. There are also characteristic psychological signs, including hysteria and depression. Sudden death is likely, caused by massive internal bleeding.

Once diagnosed, scurvy is readily resolved by vitamin C. Moderate doses in the neighborhood of 100 milligrams per day are sufficient, curing the scurvy within

FIGURE 10-18 Vitamin C–Deficiency Symptoms—Scorbutic Gums and Pinpoint Hemorrhages

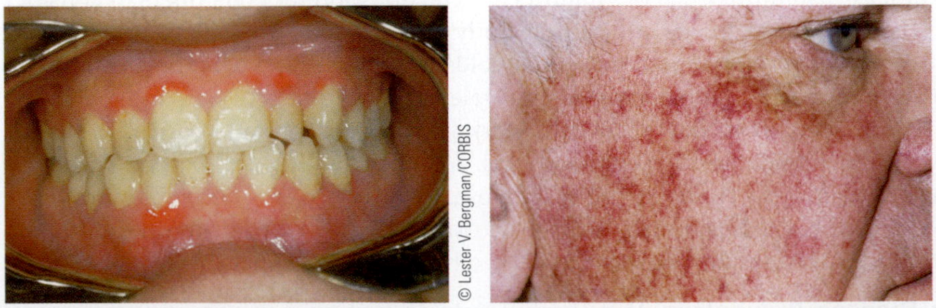

© Lester V. Bergman/CORBIS

© Dr. P. Marazzi/Photo Researchers Inc.

Scorbutic gums. Unlike other lesions of the mouth, scurvy presents a symmetrical appearance without infection.

Pinpoint hemorrhages. Small red spots appear in the skin, indicating spontaneous bleeding internally.

about five days. Such an intake is easily achieved by including vitamin C–rich foods in the diet.

Vitamin C Toxicity

The availability of vitamin C supplements and the publication of books recommending vitamin C to prevent colds and cancer have led thousands of people to take large doses of vitamin C. Not surprisingly, side effects of vitamin C supplementation such as gastrointestinal distress and diarrhea have been reported. The UL for vitamin C was established based on these symptoms.

Several instances of interference with medical regimens are also known. Large amounts of vitamin C excreted in the urine obscure the results of tests used to detect glucose or ketones in the diagnosis of diabetes. In some instances, excess vitamin C gives a **false positive** result; in others, a **false negative.** People taking anticlotting medications may unwittingly counteract the effect if they also take massive doses of vitamin C. Those with kidney disease, a tendency toward gout, ♦ or a genetic abnormality that alters vitamin C's breakdown to its excretion products are prone to forming kidney stones if they take large doses of vitamin C.* Vitamin C supplements may adversely affect people with iron overload. As Chapter 13 explains, vitamin C enhances iron absorption and releases iron from body stores; too much free iron causes the kind of cellular damage typical of free radicals. These adverse consequences of vitamin C's effects on iron have not been seen in clinical studies, but they illustrate how vitamin C can act as a *pro*oxidant when quantities exceed the body's needs.[33]

♦ **Gout** is a metabolic disease in which uric acid crystals precipitate in the joints.

Vitamin C Food Sources

Fruits and vegetables can easily provide a generous amount of vitamin C. A cup of orange juice at breakfast, a salad for lunch, and a stalk of broccoli and a potato for dinner alone provide more than 300 milligrams. ♦ Clearly, a person making such food choices does not need vitamin C supplements.

♦ For perspective, review Figure 10-17.

Figure 10-19 (p. 342) shows the amounts of vitamin C in various common foods. The overwhelming abundance of purple and green bars reveals not only that the citrus fruits are justly famous for being rich in vitamin C, but that other fruits and vegetables are in the same league. A half cup of broccoli, bell pepper, or strawberries provides more than 50 milligrams of the vitamin (and an array of other nutrients). Because vitamin C is vulnerable to heat, raw fruits and vegetables usually

false positive: a test result indicating that a condition is present (positive) when in fact it is not present (therefore false).

false negative: a test result indicating that a condition is not present (negative) when in fact it is present (therefore false).

*Vitamin C is inactivated and degraded by several routes, and sometimes oxalate, which can form kidney stones, is produced along the way. People may also develop oxalate crystals in their kidneys regardless of vitamin C status.

Edyta Pawlowska/Shutterstock.com

When dietitians say "vitamin C," people think "citrus fruits" . . .

© Polara Studios Inc.

. . . but these foods are also rich in vitamin C.

FIGURE 10-19 **Vitamin C in Selected Foods**

See the "How To" section on p. 316 for more information on using this figure.

Food	Serving size (kcalories)
Bread, whole wheat	1 oz slice (70 kcal)
Cornflakes, fortified	1 oz (110 kcal)
Spaghetti pasta	½ c cooked (99 kcal)
Tortilla, flour	1 10"-round (234 kcal)
Broccoli	½ c cooked (22 kcal)
Carrots	½ c shredded raw (24 kcal)
Potato	1 medium baked w/skin (133 kcal)
Tomato juice	¾ c (31 kcal)
Banana	1 medium raw (109 kcal)
Orange	1 medium raw (62 kcal)
Strawberries	½ c fresh (22 kcal)
Watermelon	1 slice (92 kcal)
Milk	1 c reduced-fat 2% (121 kcal)
Yogurt, plain	1 c low-fat (155 kcal)
Cheddar cheese	1½ oz (171 kcal)
Cottage cheese	½ c low-fat 2% (101 kcal)
Pinto beans	½ c cooked (117 kcal)
Peanut butter	2 tbs (188 kcal)
Sunflower seeds	1 oz dry (165 kcal)
Tofu (soybean curd)	½ c (76 kcal)
Ground beef, lean	3 oz broiled (244 kcal)
Chicken breast	3 oz roasted (140 kcal)
Tuna, canned in water	3 oz (99 kcal)
Egg	1 hard cooked (78 kcal)
Excellent, and sometimes unusual, sources:	
Red bell pepper	½ c raw chopped (20 kcal)
Kiwi	1 (46 kcal)
Brussels sprouts	½ c cooked (30 kcal)

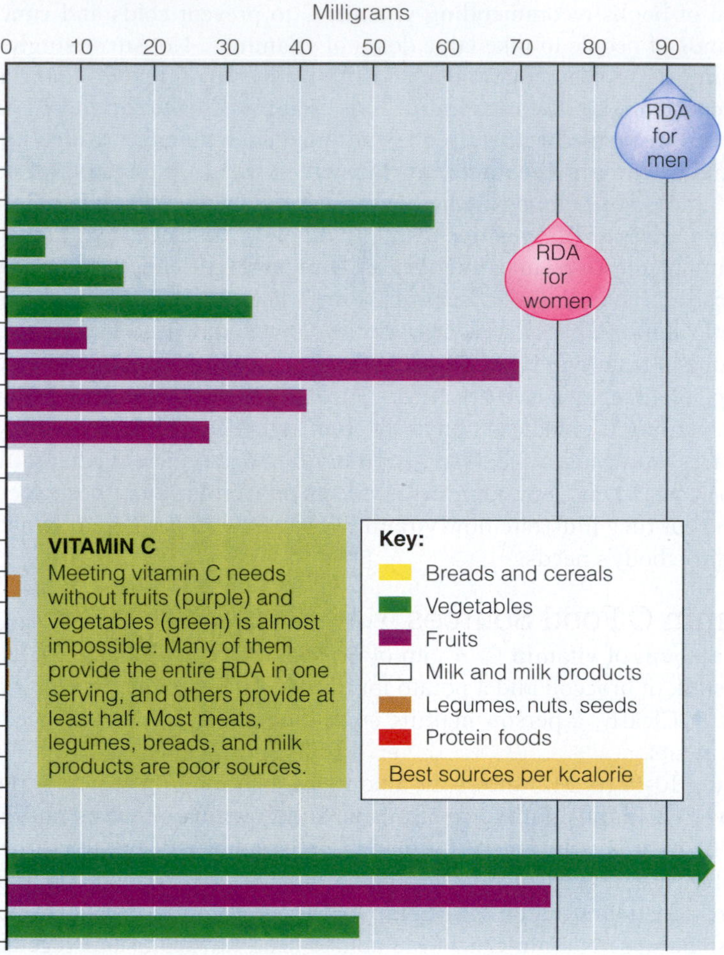

have a higher nutrient density than their cooked counterparts. Similarly, because vitamin C is readily destroyed by oxygen, foods and juices should be stored properly and consumed within a week of opening.

The potato is an important source of vitamin C, not because one potato by itself meets the daily need, but because potatoes are such a common staple that they make significant contributions. In fact, scurvy was unknown in Ireland until the potato blight of the mid-1840s when some two million people died of malnutrition and infection.

The lack of yellow, white, brown, and red bars in Figure 10-19 confirms that grains, milk (except breast milk), legumes, and meats are notoriously poor sources of vitamin C. Organ meats (liver, kidneys, and others) and raw meats contain some vitamin C, but most people don't eat large quantities of these foods. Raw meats and fish contribute enough vitamin C to be significant sources in parts of Alaska, Canada, and Japan, but elsewhere fruits and vegetables are necessary to supply sufficient vitamin C.

Because of vitamin C's antioxidant property, food manufacturers sometimes add a variation of vitamin C to some beverages and most cured meats, such as luncheon meats, to prevent oxidation and spoilage. This compound safely preserves these foods, but it does not have vitamin C activity in the body. Simply put, "ham and bacon cannot replace fruits and vegetables." The accompanying table provides a summary of vitamin C.

IN SUMMARY Vitamin C

Other Names

Ascorbic acid

RDA

Men: 90 mg/day

Women: 75 mg/day

Smokers: +35 mg/day

Upper Level

Adults: 2000 mg/day

Chief Functions in the Body

Collagen synthesis (strengthens blood vessel walls, forms scar tissue, provides matrix for bone growth), antioxidant, thyroxin synthesis, amino acid metabolism, strengthens resistance to infection, helps in absorption of iron

Significant Sources

Citrus fruits, cabbage-type vegetables (such as brussels sprouts and cauliflower), dark green vegetables (such as bell peppers and broccoli), cantaloupe, strawberries, lettuce, tomatoes, potatoes, papayas, mangoes

Easily destroyed by heat and oxygen

Deficiency Disease

Scurvy

Deficiency Symptoms

Anemia (small-cell type),[a] atherosclerotic plaques, pinpoint hemorrhages; bone fragility, joint pain; poor wound healing, frequent infections; bleeding gums, loosened teeth; muscle degeneration, pain, hysteria, depression; rough skin, blotchy bruises

Toxicity Symptoms

Nausea, abdominal cramps, diarrhea; headache, fatigue, insomnia; hot flashes; rashes; interference with medical tests, aggravation of gout symptoms, urinary tract problems, kidney stones[b]

[a]Small-cell-type anemia is *microcytic anemia*.

[b]People with kidney disease, a tendency toward gout, or a genetic abnormality that alters the breakdown of vitamin C are prone to forming kidney stones. Vitamin C is inactivated and degraded by several routes, sometimes producing oxalate, which can form stones in the kidneys.

Vita means life. After this discourse on the vitamins, who could dispute that they deserve their name? Their regulation of metabolic processes makes them vital to the normal growth, development, and maintenance of the body. The accompanying summary table condenses the information provided in this chapter for a quick review. The remarkable roles of the vitamins continue in the next chapter.

IN SUMMARY The Water-Soluble Vitamins

Vitamin and Chief Functions	Deficiency Symptoms	Toxicity Symptoms	Food Sources
Thiamin Part of coenzyme TPP in energy metabolism	Beriberi (edema or muscle wasting), anorexia and weight loss, neurological disturbances, muscular weakness, heart enlargement and failure	None reported	Enriched, fortified, or whole-grain products; pork
Riboflavin Part of coenzymes FAD and FMN in energy metabolism	Inflammation of the mouth, skin, and eyelids	None reported	Milk products; enriched, fortified, or whole-grain products; liver
Niacin Part of coenzymes NAD and NADP in energy metabolism	Pellagra (diarrhea, dermatitis, and dementia)	Niacin flush, liver damage, impaired glucose tolerance	Protein-rich foods
Biotin Part of coenzyme in energy metabolism	Skin rash, hair loss, neurological disturbances	None reported	Widespread in foods; GI bacteria synthesis
Pantothenic acid Part of coenzyme A in energy metabolism	Digestive and neurological disturbances	None reported	Widespread in foods
Vitamin B_6 Part of coenzymes used in amino acid and fatty acid metabolism	Scaly dermatitis, depression, confusion, convulsions, anemia	Nerve degeneration, skin lesions	Protein-rich foods
Folate Activates vitamin B_{12}; helps synthesize DNA for new cell growth	Anemia, glossitis, neurological disturbances, elevated homocysteine	Masks vitamin B_{12} deficiency	Legumes, vegetables, fortified grain products
Vitamin B_{12} Activates folate; helps synthesize DNA for new cell growth; protects nerve cells	Anemia; nerve damage and paralysis	None reported	Foods derived from animals
Vitamin C Synthesis of collagen, carnitine, hormones, neurotransmitters; antioxidant	Scurvy (bleeding gums, pinpoint hemorrhages, abnormal bone growth, and joint pain)	Diarrhea, GI distress	Fruits and vegetables

Nutrition Portfolio

To obtain all the vitamins you need each day, be sure to select from a variety of foods from all the food groups.

Go to Diet Analysis Plus and choose one of the days on which you have tracked your diet for the entire day. Go to the Intake vs. Goals report. Near the bottom of this report, you will see all of the vitamins grouped together; using this section of the report for reference, answer the following questions:

- How was your vitamin intake overall? Did you consume too much or too little of any vitamin? Which vitamins concerned you most?

Next go to the Intake Spreadsheet report, and looking at each of the vitamins, answer the following questions:

- Which of your foods provided high intakes of vitamins?
- Which of your foods provided few or no vitamins?
- Examine your daily choices of whole or enriched grains, dark green leafy vegetables, citrus fruits, and legumes, then evaluate their contributions to your vitamin intakes.
- If you are a woman of childbearing age, calculate the dietary folate equivalents you receive from folate-rich foods, fortified foods, and supplements, then compare that to your RDA.
- Compare your vitamin intakes from supplements with their UL.

Diet Analysis PLUS

To complete this exercise, go to your Diet Analysis Plus at www.cengage.com/sso.

Nutrition on the Net

For further study of topics covered in this chapter, log on to **www.cengagebrain.com** and search for ISBN 084006845X.

- Search for "vitamins" at the American Dietetic Association: **www.eatright.org**
- Visit the World Health Organization to learn about "vitamin deficiencies" around the world: **www.who.int**
- Learn more about neural tube defects from the Spina Bifida Association of America: **www.spinabifidaassociation.org**
- Read about Dr. Joseph Goldberger and his groundbreaking discovery linking pellagra to diet by searching for his name at: **www.nih.gov** or **www.pbs.org**
- Learn how fruits and vegetables support a healthy diet rich in vitamins from the Fruits and Veggies Matter program: **www.fruitsandveggiesmatter.gov**

References

1. C. S. Johnston and J. C. Hale, Oxidation of ascorbic acid in stored orange juice is associated with reduced plasma vitamin C concentrations and elevated lipid peroxides, *Journal of the American Dietetic Association* 105 (2005): 106–109.
2. Committee on Dietary Reference Intakes, *Dietary Reference Intakes for Vitamin C, Vitamin E, Selenium, and Carotenoids* (Washington, D.C.: National Academies Press, 2000); Committee on Dietary Reference Intakes, *Dietary Reference Intakes for Thiamin, Riboflavin, Niacin, Vitamin B_6, Folate, Vitamin B_{12}, Pantothenic Acid, Biotin, and Choline* (Washington, D.C.: National Academies Press, 1998).
3. K. J. Carpenter, *Beriberi, White Rice, and Vitamin B: A Disease, a Cause, and a Cure* (Berkeley: University of California Press, 2000).
4. K. L. Bogan and C. Brenner, Nicotinic acid, nicotinamide, and nicotinamide riboside: A molecular evaluation of NAD+ precursor vitamins in human nutrition, *Annual Review of Nutrition* 28 (2008): 115–130.
5. Committee on Dietary Reference Intakes, 1998, pp. 128–129.
6. L. Zhang and coauthors, Niacin inhibits surface expression of ATP synthase β chain in HepG2 cells: Implications for raising HDL, *Journal of Lipid Research* 49 (2008): 1195–1201; P. L. Canner, C. D. Furberg, and M. E. McGovern, Benefits of niacin in patients with versus without the metabolic syndrome and healed myocardial infarction (from the Coronary Drug Project), *American Journal of Cardiology* 97 (2006): 477–479; S. Westphal and coauthors, Adipokines and treatment with niacin, *Metabolism* 55 (2006): 1283–1285.
7. D. B. Piazzini and coauthors, A systematic review of conservative treatment of carpal tunnel syndrome, *Clinical Rehabilitation* 21 (2007): 299–314.
8. Q. Yang and coauthors, Race-ethnicity differences in folic acid intake in women of childbearing age in the United States after folic acid fortification: Findings from the National Health and Nutrition Examination Survey, 2001–2002, *American Journal of Clinical Nutrition* 85 (2007): 1409–1416; Use of dietary supplements containing folic acid among women of childbearing age: United States, 2005, *Morbidity and Mortality Weekly Report* 54 (2005): 955–957.
9. R. M. Winkels and coauthors, Bioavailability of food folates is 80% of that of folic acid, *American Journal of Clinical Nutrition* 85 (2007): 465–473.
10. P. D. Wals and coauthors, Reduction in neural-tube defects after folic acid fortification in Canada, *New England Journal of Medicine* 357 (2007): 135–142; T.G.K. Bentley and coauthors, Population-level

changes in folate intake by age, gender, and race/ethnicity after folic acid fortification, *American Journal of Public Health* 96 (2006): 2040–2047; T. Tamura and M. F. Picciano, Folate and human reproduction, *American Journal of Clinical Nutrition* 83 (2006): 993–1016.

11. A. J. Wilcox and coauthors, Folic acid supplements and risk of facial clefts: National population-based case-control study, *British Medical Journal* 334 (2007): 464–469; Y. I. Goh and coauthors, Prenatal multivitamin supplementation and rates of congenital anomalies: A meta-analysis, *Journal of Obstetrics and Gynaecology Canada* 28 (2006): 680–689.
12. Committee on Dietary Reference Intakes, 1998.
13. O. Dary, Nutritional interpretation of folic acid interventions, *Nutrition Reviews* 67 (2009): 235–244; A. D. Smith, Y. I. Kim, and H. Refsum, Is folic acid good for everyone? *American Journal of Clinical Nutrition* 87 (2008): 517–533; Y. Kim, Folic acid fortification and supplementation: Good for some but not so good for others, *Nutrition Reviews* 65 (2007): 504–511; I. H. Rosenberg, Folic acid fortification, *Nutrition Reviews* 65 (2007): 503; N. W. Solomons, Food fortification with folic acid: Has the other shoe dropped? *Nutrition Reviews* 65 (2007): 512–515; R. L. Brent and G. P. Oakley, The folate debate, *Pediatrics* 117 (2006): 1418–1419; J. I. Rader and B. O. Schneeman, Prevalence of neural tube defects, folate status, and folate fortification of enriched cereal-grain products in the United States, *Pediatrics* 117 (2006): 1394–1399.
14. A. de Bree, L. A. van Mierlo, and R. Draijer, Folic acid improves vascular reactivity in humans: A meta-analysis of randomized controlled trials, *American Journal of Clinical Nutrition* 86 (2007): 610–617; D. S. Wald and coauthors, Folic acid, homocysteine, and cardiovascular disease: Judging causality in the face of inconclusive trial evidence, *British Journal of Medicine* 333 (2006): 1114–1117.
15. C. M. Albert and coauthors, Effect of folic acid and B vitamins on risk of cardiovascular events and total mortality among women at high risk for cardiovascular disease: A randomized trial, *Journal of the American Medical Association* 299 (2008): 2027–2036; M. Ebbing and coauthors, Mortality and cardiovascular events in patients treated with homocysteine-lowering B vitamins after coronary angiography, *Journal of the American Medical Association* 300 (2008): 795–801; E. Lonn, Homocysteine-lowering B vitamin therapy in cardiovascular prevention—Wrong again? *Journal of the American Medical Association* 299 (2008): 2086–2087; C. Baigent and R. Clarke, B Vitamins for the prevention of vascular disease: Insufficient evidence to justify treatment, *Journal of the American Medical Association* 298 (2007): 1212–1214; R. L. Jamison and coauthors, Effect of homocysteine lowering on mortality and vascular disease in advanced chronic kidney disease and end-stage renal disease: A randomized trial, *Journal of the American Medical Association* 298 (2007): 1163–1170; L. A. Bazzano and coauthors, Effect of folic acid supplementation on risk of cardiovascular diseases—A meta-analysis of randomized controlled trials, *Journal of the American Medical Association* 296 (2006): 2720–2726; C. M. Carlsson, Homocysteine lowering with folic acid and vitamin B supplements: Effects on cardiovascular disease in older adults, *Drugs and Aging* 23 (2006): 491–502; The Heart Outcomes Prevention Evaluation (HOPE) 2 Investigators, Homocysteine lowering with folic acid and B vitamins in vascular disease, *New England Journal of Medicine* 354 (2006): 1567–1577.
16. J. B. Mason, Folate, cancer risk, and the Greek god, Proteus: A tale of two chameleons, *Nutrition Reviews* 67 (2009): 206–212; Smith, Kim, and Refsum, 2008; C. M. Ulrich, Folate and cancer prevention: A closer look at a complex picture, *American Journal of Clinical Nutrition* 86 (2007): 271–273.
17. A. R. Hart, H. Kennedy, and I. Harvey, Pancreatic cancer: A review of the evidence on causation, *Clinical Gastroenterology Hepatology* 6 (2008): 275–282; World Cancer Research Fund and American Institute for Cancer Research, *Food, Nutrition, Physical Activity, and the Prevention of Cancer: A Global Perspective* (Washington, D.C.: AICR, 2007), pp. 106–107.
18. World Cancer Research Fund and American Institute for Cancer Research, 2007.
19. K. D. Stark and coauthors, Status of plasma folate after folic acid fortification of the food supply in pregnant African American women and the influences of diet, smoking, and alcohol consumption, *American Journal of Clinical Nutrition* 81 (2005): 669–671.
20. L. H. Allen, How common is vitamin B-12 deficiency? *American Journal of Clinical Nutrition* 89 (2009): 693S–696S; S. P. Stabler and coauthors, Elevated serum S-adenosylhomocysteine in cobalamin-deficient elderly and response to treatment, *American Journal of Clinical Nutrition* 84 (2006): 1422–1429.
21. D. K. Dror and L. H. Allen, Effect of vitamin B_{12} deficiency on neurodevelopment in infants: Current knowledge and possible mechanisms, *Nutrition Reviews* 66 (2008): 250–255.
22. K. F. Wykoff and V. Ganji, Proportion of individuals with low serum vitamin B-12-concentrations without macrocytosis is higher in the post-folic acid fortification period than in the pre-folic acid fortification period, *American Journal of Clinical Nutrition* 86 (2007): 1187–1192.
23. R. Clarke and coauthors, Low vitamin B-12; status and risk of cognitive decline in older adults, *American Journal of Clinical Nutrition* 86 (2007): 1384–1391; C. McCracken and coauthors, Methylmalonic acid and cognitive function in the Medical Research Council Cognitive Function and Ageing Study, *American Journal of Clinical Nutrition* 84 (2006): 1406–1411.
24. M. A. Johnson, If high folic acid aggravates vitamin B_{12} deficiency what should be done about it? *Nutrition Reviews* 65 (2007): 451–458.
25. A. Vogiatzoglou and coauthors, Dietary sources of vitamin B-12 and their association with plasma vitamin B-12 concentrations in the general population: The Hordaland Homocysteine Study, *American Journal of Clinical Nutrition* 89 (2009): 1078–1087.
26. L. M. Sanders and S. H. Zeisel, Choline: Dietary requirements and role in brain development, *Nutrition Today* 42 (2007): 181–186; S. H. Zeisel, Choline: Critical role during fetal development and dietary requirements in adults, *Annual Review of Nutrition* 26 (2006): 229–250.
27. J. X. Wilson, Regulation of vitamin C transport, *Annual Review of Nutrition* 25 (2005): 105–125.
28. S. J. Padayatty and coauthors, Human adrenal glands secrete vitamin C in response to adrenocorticotrophic hormone, *American Journal of Clinical Nutrition* 86 (2007): 145–149.
29. M. Simasek and D. A. Blandino, Treatment of the common cold, *American Family Physician* 75 (2007): 515–520; S. Sasazuki and coauthors, Effect of vitamin C on common cold: Randomized controlled trial, *European Journal of Clinical Nutrition* 60 (2006): 9–17; B. Arroll, Non-antibiotic treatments for upper-respiratory tract infections (common cold), *Respiratory Medicine* 99 (2005): 1477–1484.
30. R. M. Douglas and coauthors, Vitamin C for preventing and treating the common cold, *Cochrane Database of Systematic Reviews* 3 (2007): CD000980; E. S. Wintergerst, S. Maggini, and D. H. Hornig, Immune-enhancing role of vitamin C and zinc and effect on clinical conditions, *Annals of Nutrition and Metabolism* 50 (2006): 85–94.
31. B. Barrett and coauthors, Sufficiently important difference for common cold: Severity reduction, *Annals of Family Medicine* 5 (2007): 216–223.
32. Committee on Dietary Reference Intakes, 2000.
33. J. N. Hathcock and coauthors, Vitamins E and C are safe across a broad range of intakes, *American Journal of Clinical Nutrition* 81 (2005): 736–745.

HIGHLIGHT 10

Vitamin and Mineral Supplements

© Alex Segre/Alamy

An estimated 75,000 supplements are currently on the market. More than half of the adults in the United States take a **dietary supplement** regularly, spending almost $24 billion each year.[1] Many people take supplements as dietary insurance—in case they are not meeting their nutrient needs from foods alone. Others take supplements as health insurance—to protect against certain diseases.[2]

One out of every three people takes multinutrient pills daily. Others take large doses of single nutrients, most commonly, vitamin C, vitamin E, beta-carotene, iron, and calcium. In many cases, taking supplements is a costly but harmless practice; sometimes, it is both costly and harmful to health.

For the most part, people self-prescribe supplements, taking them on the advice of friends, advertisements, websites, or books that may or may not be reliable. Sometimes, they take supplements on the recommendation of a physician. When such advice follows a valid nutrition assessment, supplementation may be warranted, but even then the preferred course of action is to improve food choices and eating habits.[3] Without an assessment, the advice to take supplements may be inappropriate. A registered dietitian can help with the decision.[4]

When people think of dietary supplements, they often think of vitamins, but a diet that lacks vitamins probably lacks several minerals as well. This highlight asks several questions related to vitamin-mineral supplements. (The accompanying glossary defines dietary supplements and related terms.) What are the arguments *for* taking supplements? What are the arguments *against* taking them? Finally, if people do take supplements, how can they choose the appropriate ones? (Amino acid supplements and herbal supplements are discussed in Chapter 6 and Chapter 19, respectively.)

Arguments for Supplements

Vitamin-mineral supplements may be appropriate in some circumstances. In some cases, they can prevent or correct deficiencies; in others, they can reduce the risk of diseases. Consumers should discuss supplement use with their health-care professionals who can help monitor for adverse effects or nutrient-drug interactions.[5]

Correct Overt Deficiencies

In the United States and Canada, adults rarely suffer nutrient deficiency diseases such as scurvy, pellagra, and beriberi, but nutrient deficiencies do still occur. To correct an overt deficiency disease, a physician may prescribe therapeutic doses two to ten times the RDA (or AI) of a nutrient. At such high doses, the supplement is acting as a drug.

Support Increased Nutrient Needs

As Chapters 14 through 16 explain, nutrient needs increase during certain stages of life, making it difficult to meet some of those needs without supplementation. For example, women who lose a lot of blood and therefore a lot of iron during menstruation each month may need an iron supplement. Women of childbearing age need folate supplements to reduce the risks of neural tube defects. Similarly, pregnant women and women who are breastfeeding their infants have exceptionally high nutrient needs and so usually need special supplements. Newborns routinely receive a single dose of vitamin K at birth to prevent abnormal bleeding. Infants may need other supplements as well, depending on whether they are breastfed or receiving formula, and on whether their water contains fluoride.

GLOSSARY

dietary supplement: any pill, capsule, tablet, liquid, or powder that contains vitamins, minerals, herbs, or amino acids; intended to increase dietary intake of these substances.

FDA (Food and Drug Administration): a part of the Department of Health and Human Services' Public Health Service that is responsible for ensuring the safety and wholesomeness of all dietary supplements and food processed and sold in interstate commerce except meat, poultry, and eggs (which are under the jurisdiction of the USDA); inspecting food plants and imported foods; and setting standards for food composition and product labeling.

high potency: 100% or more of the Daily Value for the nutrient in a single supplement and for at least two-thirds of the nutrients in a multinutrient supplement.

Improve Nutrition Status

In contrast to the classical deficiencies, which present a multitude of symptoms and are relatively easy to recognize, subclinical deficiencies are subtle and easy to overlook—and they are also more likely to occur. People who do not eat enough food to deliver the needed amounts of nutrients, such as habitual dieters and the elderly, risk developing subclinical deficiencies.[6] Similarly, vegetarians who restrict their use of entire food groups without appropriate substitutions may fail to fully meet their nutrient needs. If there is no way for these people to eat enough nutritious foods to meet their needs, then vitamin-mineral supplements may be appropriate to help prevent nutrient deficiencies.

Improve the Body's Defenses

Health-care professionals may provide special supplementation to people being treated for addictions to alcohol or other drugs and to people with prolonged illnesses, extensive injuries, or other severe stresses such as surgery. Illnesses that interfere with appetite, eating, or nutrient absorption impair nutrition status. For example, the stomach condition atrophic gastritis often creates a vitamin B_{12} deficiency. In addition, nutrient needs are often heightened by diseases or medications. In all these cases, supplements are appropriate.

Reduce Disease Risks

Few people consume the optimal amounts of all the vitamins and minerals by diet alone. Inadequate intakes have been linked to chronic diseases such as heart disease, some cancers, and osteoporosis. For this reason, some physicians recommend that all adults take vitamin-mineral supplements. Such regular supplementation would provide an optimum intake to enhance metabolic harmony and prevent disease at relatively little cost. Others recognize the lack of conclusive evidence and the potential harm of supplementation and advise against such a recommendation.[7] The most recent statement from the National Institutes of Health acknowledges that evidence is insufficient to recommend either for or against the use of supplements to prevent chronic diseases.[8]

Highlight 11 reviews the relationships between supplement use and disease prevention. It describes some of the accumulating evidence suggesting that intakes of certain nutrients at levels much higher than can be attained from foods alone may be beneficial in reducing disease risks. It also presents research confirming the associated risks.[9] Clearly, consumers must be cautious in taking supplements to prevent disease.

Who Needs Supplements?

In summary, the following list acknowledges that in these specific conditions, these people may need to take supplements:

- People with specific nutrient deficiencies need specific nutrient supplements.
- People whose energy intakes are particularly low (fewer than 1600 kcalories per day) need multivitamin-mineral supplements.
- Vegetarians who eat all-plant diets (vegans) and older adults with atrophic gastritis need vitamin B_{12}.
- People who have lactose intolerance or milk allergies or who otherwise do not consume enough milk products to forestall extensive bone loss need calcium.
- People in certain stages of the life cycle who have increased nutrient requirements need specific nutrient supplements. For example, infants need iron and fluoride, women of childbearing age and pregnant women need folate and iron, and the elderly need vitamin B_{12} and vitamin D.
- People who have inadequate milk intakes, limited sun exposure, or heavily pigmented skin need vitamin D.
- People who have diseases, infections, or injuries or who have undergone surgery that interferes with the intake, absorption, metabolism, or excretion of nutrients may need specific nutrient supplements.
- People taking medications that interfere with the body's use of specific nutrients may need specific nutrient supplements.

Except for people in these circumstances, most adults can normally get all the nutrients they need by eating a varied diet of nutrient-dense foods. Even athletes can meet their nutrient needs without the help of supplements.

Arguments against Supplements

Foods rarely cause nutrient imbalances or toxicities, but supplements can. The higher the dose, the greater the risk of harm. People's tolerances for high doses of nutrients vary, just as their risks of deficiencies do. Amounts that some can tolerate may be harmful for others, and no one knows who falls where along the spectrum. It is difficult to determine just how much of a nutrient is enough—or too much. The Tolerable Upper Intake Levels of the DRI answer the question "How much is too much?" by defining the highest amount that appears safe for most healthy people. Table H10-1 (p. 348) presents the Upper Levels and Daily Values for selected vitamins and minerals and the quantities typically found in supplements.

Toxicity

Supplement users are more likely to have excessive intakes of certain nutrients—notably iron, zinc, vitamin A, and niacin.[10] The extent and severity of supplement toxicity remain unclear. Only a few alert health-care professionals can recognize toxicity, even when it is acute. When it is chronic, with the effects developing subtly and progressing slowly, it often goes unrecognized. In view of the potential hazards, some authorities believe supplements should bear warning labels, advising consumers that large doses may be toxic.

HIGHLIGHT 10

Toxic overdoses of vitamins and minerals in children are more readily recognized and, unfortunately, fairly common. Fruit-flavored, chewable vitamins shaped like cartoon characters entice young children to eat them like candy in amounts that can cause poisoning. Iron supplements (30 milligrams of iron or more per tablet) are especially toxic and are the leading cause of accidental ingestion fatalities among children. Even mild overdoses cause GI distress, nausea, and black diarrhea, which reflects gastric bleeding. Severe overdoses result in bloody diarrhea, shock, liver damage, coma, and death.

Life-Threatening Misinformation

Another problem arises when people who are ill come to believe that high doses of vitamins or minerals can be therapeutic. Not only can high doses be toxic, but the person may take them instead of seeking medical help. Furthermore, there are no guarantees that the supplements will be effective. Taking B vitamin

TABLE H10-1 Vitamin and Mineral Intakes for Adults

Nutrient	Tolerable Upper Intake Levels[a]	Daily Values	Typical Multivitamin-Mineral Supplement	Average Single-Nutrient Supplement
Vitamins				
Vitamin A	3000 μg (10,000 IU)	5000 IU	5000 IU	8000 to 10,000 IU
Vitamin D	100 μg (4000 IU)	400 IU	400 IU	400 IU
Vitamin E	1000 mg (1500 to 2200 IU)[b]	30 IU	30 IU	100 to 1000 IU
Vitamin K	—[c]	80 μg	40 μg	—[e]
Thiamin	—[c]	1.5 mg	1.5 mg	50 mg
Riboflavin	—[c]	1.7 mg	1.7 mg	25 mg
Niacin (as niacinamide)	35 mg[b]	20 mg	20 mg	100 to 500 mg
Vitamin B_6	100 mg	2 mg	2 mg	100 to 200 mg
Folate	1000 μg[b]	400 μg	400 μg	400 μg
Vitamin B_{12}	—[c]	6 μg	6 μg	100 to 1000 μg
Pantothenic acid	—[c]	10 mg	10 mg	100 to 500 mg
Biotin	—[c]	300 μg	30 μg	300 to 600 μg
Vitamin C	2000 mg	60 mg	10 mg	500 to 2000 mg
Choline	3500 mg	—	10 mg	250 mg
Minerals				
Calcium	2500 mg	1000 mg	160 mg	250 to 600 mg
Phosphorus	4000 mg	1000 mg	110 mg	—[e]
Magnesium	350 mg[d]	400 mg	100 mg	250 mg
Iron	45 mg	18 mg	18 mg	18 to 30 mg
Zinc	40 mg	15 mg	15 mg	10 to 100 mg
Iodine	1100 μg	150 μg	150 μg	—[e]
Selenium	400 μg	70 μg	10 μg	50 to 200 μg
Fluoride	10 mg	—	—	—[e]
Copper	10 mg	2 mg	0.5 mg	—[e]
Manganese	11 mg	2 mg	5 mg	—[e]
Chromium	—[c]	120 μg	25 μg	200 to 400 μg
Molybdenum	2000 μg	75 μg	25 μg	—[e]

[a]Unless otherwise noted, Upper Levels represent total intakes from food, water, and supplements.
[b]Upper Levels represent intakes from supplements, fortified foods, or both.
[c]These nutrients have been evaluated by the DRI Committee for Tolerable Upper Intake Levels, but none were established because of insufficient data. No adverse effects have been reported with intakes of these nutrients at levels typical of supplements, but caution is still advised, given the potential for harm that accompanies excessive intakes.
[d]Upper Levels represent intakes from supplements only.
[e]Available as a single supplement by prescription.

supplements instead of medication may sound appealing, but they do not protect against the progression of atherosclerosis.[11] Marketing materials for supplements often make health statements that are required to be "truthful and not misleading," but they often fall far short of both. Chapter and Highlight 19 revisit this topic and include a discussion of herbal preparations and other alternative therapies.

Unknown Needs

Another argument against the use of supplements is that no one knows exactly how to formulate the "ideal" supplement. What nutrients should be included? Which, if any, of the phytochemicals should be included? How much of each? On whose needs should the choices be based? Surveys have repeatedly shown little relationship between the supplements people take and the nutrients they actually need.

False Sense of Security

Another argument against supplement use is that it may lull people into a false sense of security. A person might eat irresponsibly, thinking, "My supplement will cover my needs." Or, experiencing a warning symptom of a disease, a person might postpone seeking a diagnosis, thinking, "I probably just need a supplement to make this go away." Such self-diagnosis is potentially dangerous.

Other Invalid Reasons

Other invalid reasons people might use for taking supplements include:

- The belief that the food supply or soil contains inadequate nutrients
- The belief that supplements can provide energy
- The belief that supplements can enhance athletic performance or build lean body tissues without physical work or faster than work alone
- The belief that supplements will help a person cope with stress
- The belief that supplements can prevent, treat, or cure conditions ranging from the common cold to cancer

Ironically, people with health problems are more likely to take supplements than other people, yet today's health problems are more likely to be due to overnutrition and poor lifestyle choices than to nutrient deficiencies. The truth—that most people would benefit from improving their eating and exercise habits—is harder to swallow than a supplement pill.

Bioavailability and Antagonistic Actions

In general, the body absorbs nutrients best from foods in which the nutrients are diluted and dispersed among other substances that may facilitate their absorption. Taken in pure, concentrated form, nutrients are likely to interfere with one another's absorption or with the absorption of nutrients in foods eaten at the same time. Documentation of these effects is particularly extensive for minerals: zinc hinders copper and calcium absorption, iron hinders zinc absorption, calcium hinders magnesium and iron absorption, and magnesium hinders the absorption of calcium and iron. Similarly, binding agents in supplements limit mineral absorption.

Although minerals provide the most-familiar and best-documented examples, interference among vitamins is now being seen as supplement use increases. The vitamin A precursor beta-carotene, long thought to be nontoxic, interferes with vitamin E metabolism when taken over the long term as a dietary supplement. Vitamin E, on the other hand, antagonizes vitamin K activity and so should not be used by people being treated for blood-clotting disorders. Consumers who want the benefits of optimal absorption of nutrients should eat ordinary foods, selected for nutrient density and variety.

Whenever the diet is inadequate, the person should first attempt to improve it so as to obtain the needed nutrients from foods. If that is truly impossible, then the person needs a multivitamin-mineral supplement that supplies between 50 and 150 percent of the Daily Value for each of the nutrients. These amounts reflect the ranges commonly found in foods and therefore are compatible with the body's normal handling of nutrients (its physiologic tolerance). The next section provides some pointers to assist in the selection of an appropriate supplement.

Selection of Supplements

Whenever a physician or registered dietitian recommends a supplement, follow the directions carefully. When selecting a supplement yourself, look for a single, balanced vitamin-mineral supplement. Supplements with a USP verification logo have been tested by the U.S. Pharmacopeia (USP) to assure that the supplement:

- Contains the declared ingredients and amounts listed on the label
- Does not contain harmful levels of contaminants
- Will disintegrate and release ingredients in the body
- Was made under safe and sanitary conditions

If you decide to take a vitamin-mineral supplement, ignore the eye-catching art and meaningless claims. Pay attention to the form the supplements are in, the list of ingredients, and the price. Here's where the truth lies, and from it you can make a rational decision based on facts. You have two basic questions to answer.

Form

The first question: What form do you want—chewable, liquid, or pills? If you'd rather drink your supplements than chew them, fine. If you choose a chewable form, though, be aware that chewable vitamin C can dissolve tooth enamel. If you choose pills, look for statements about the disintegration time. The USP suggests that supplements should completely disintegrate within 30 to 45 minutes. Obviously, supplements that don't dissolve have little chance of entering the bloodstream, so look for a brand that claims to meet USP disintegration standards.

Contents

The second question: What vitamins and minerals do *you* need? Generally, an appropriate supplement provides vitamins and minerals in amounts that do not exceed recommended intakes. Avoid supplements that, in a daily dose, provide more than the Tolerable Upper Intake Level for *any* nutrient. Avoid preparations with more than 10 milligrams of iron per dose, except as prescribed by a physician. Iron is hard to get rid of once it's in the body, and an excess of iron can cause problems, just as a deficiency can (see Chapter 13).

Misleading Claims

Manufacturers of *organic* or natural vitamins boast that their pills are purified from real foods rather than synthesized in a laboratory. These supplements are no more effective than others and often cost more. The word *synthetic* may sound like "fake," but to synthesize just means to put together. Think back on the course of human evolution; it is not natural to take any kind of pill. In reality, the finest, most natural vitamin "supplements" available are whole grains, vegetables, fruits, meat, fish, poultry, eggs, legumes, nuts, and milk and milk products.

Avoid products that make **"high potency"** claims. More is not better (review Figure 10-1 on p. 313). Remember that foods are also providing these nutrients. Nutrients can build up and cause unexpected problems. For example, a man who takes vitamins and begins to lose his hair may think his hair loss means he needs *more* vitamins, when in fact it may be the early sign of a vitamin A overdose. (Of course, it may be completely unrelated to nutrition as well.)

Be aware that fake vitamins and preparations that contain items not needed in human nutrition, such as carnitine and inositol, reflect a marketing strategy aimed at your pocket, not at your health. The manufacturer wants you to believe that its pills contain the latest "new" nutrient that other brands omit, but in reality, these substances are not known to be needed by human beings.

Realize that the claim that supplements "relieve stress" is another marketing ploy. If you give even passing thought to what people mean by "stress," you'll realize manufacturers could never design a supplement to meet everyone's needs. Is it stressful to take an exam? Well, yes. Is it stressful to survive a major car wreck with third-degree burns and multiple bone fractures? Definitely, yes. The body's responses to these stresses are different. The body does use vitamins and minerals in mounting a stress response, but a body fed a well-balanced diet can meet the needs of most minor stresses. For the major ones, medical intervention is needed. In any case, taking a dietary supplement won't make life any less stressful.

Other marketing tricks to sidestep are "green" pills that contain dehydrated, crushed parsley, alfalfa, and other fruit and vegetable extracts. The nutrients and phytochemicals advertised can be obtained from a serving of vegetables more easily and for less money. Such pills may also provide enzymes, but enzymes are inactivated in the stomach during protein digestion.

Recognize the latest nutrition buzzwords. Manufacturers were marketing "antioxidant" supplements before the print had time to dry on the first scientific reports of antioxidant vitamins' action in preventing cancer and cardiovascular disease. Remember, too, that high doses can alter a nutrient's action in the body. An antioxidant in physiological quantities may be beneficial, but in pharmacological quantities, it may act as a prooxidant and produce harmful by-products. Highlight 11 explores antioxidants and supplement use in more detail.

Finally, be aware that advertising on the Internet is cheap and not closely regulated. Promotional e-mails can be sent to millions of people in an instant. Internet messages can easily cite references and provide links to other sites, implying an endorsement when in fact none has been given. Be cautious when examining unsolicited information and search for a balanced perspective.

Cost

When shopping for supplements, remember that local or store brands may be just as good as nationally advertised brands. If they are less expensive, it may be because the price does not have to cover the cost of national advertising.

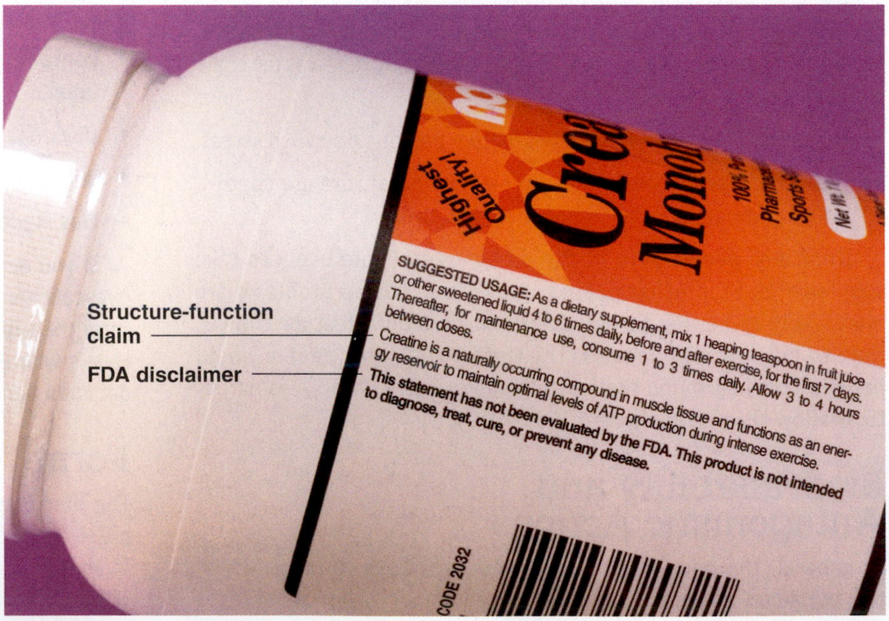

Structure-function claims do not need FDA authorization, but they must be accompanied by a disclaimer.

Regulation of Supplements

Dietary supplements are regulated by the **FDA (Food and Drug Administration)** as foods. Details of supplement regulation are defined in the Dietary Supplement Health and Education Act of 1994, which was intended to enable consumers to make informed choices about nutrient supplements. The act subjects supplements to the same general labeling requirements that apply to foods. Specifically:

- Nutrition labeling for dietary supplements is required.
- Labels may make nutrient claims (as "high" or "low") according to specific criteria (for example, "an excellent source of vitamin C").
- Labels may claim that the lack of a nutrient can cause a deficiency disease, but if they do, they must also include the prevalence of that deficiency disease in the United States.
- Labels may make health claims that are supported by significant scientific agreement and are not brand specific (for example, "folate protects against neural tube defects").
- Labels may claim to diagnose, treat, cure, or relieve common complaints such as menstrual cramps or memory loss, but may *not* make claims about specific diseases (except as noted previously).
- Labels may make structure-function claims about the role a nutrient plays in the body, how the nutrient performs its function, and how consuming the nutrient is associated with general well-being. These claims must be accompanied by an FDA disclaimer statement: "This statement has not been evaluated by the Food and Drug Administration. This product is not intended to diagnose, treat, cure, or prevent any disease." Figure H10-1 provides an example of a supplement label that complies with the requirements.

The multibillion-dollar-a-year supplement industry spends much money and effort influencing these regulations. The net effect of the Dietary Supplement Health and Education Act was a deregulation of the supplement industry. Unlike food additives or drugs, supplements do not need to be proved safe and effective, nor do they need the FDA's approval before being marketed. Furthermore, there are no standards for potency or dosage and no requirements for providing warnings of potential side effects. The FDA can only require good manufacturing practices: that dietary supplements be produced and packaged in a quality manner, do not contain contaminants or impurities, and are accurately labeled to reflect the actual contents.[12]

Should a problem arise, the burden falls to the FDA to prove that the supplement poses a "significant or unreasonable risk of illness or injury." Only then would it be removed from the market. When asked, most Americans express support for greater regulation of dietary supplements. Health professionals agree.

If all the nutrients we need can come from food, why not just eat food? Foods have so much more to offer than supplements do. Nutrients in foods come in an infinite variety of combinations with a multitude of different carriers and absorption enhancers. They come with water, fiber, and an array of beneficial phytochemicals. Foods stimulate the GI tract to keep it healthy. They provide energy, and as long as you need energy each day, why not have nutritious foods deliver it? Foods offer pleasure, satiety, and opportunities for socializing while eating. Quite simply, foods meet human health needs far better than dietary supplements. For further proof, read Highlight 11.

FIGURE H10-1 An Example of a Supplement Label

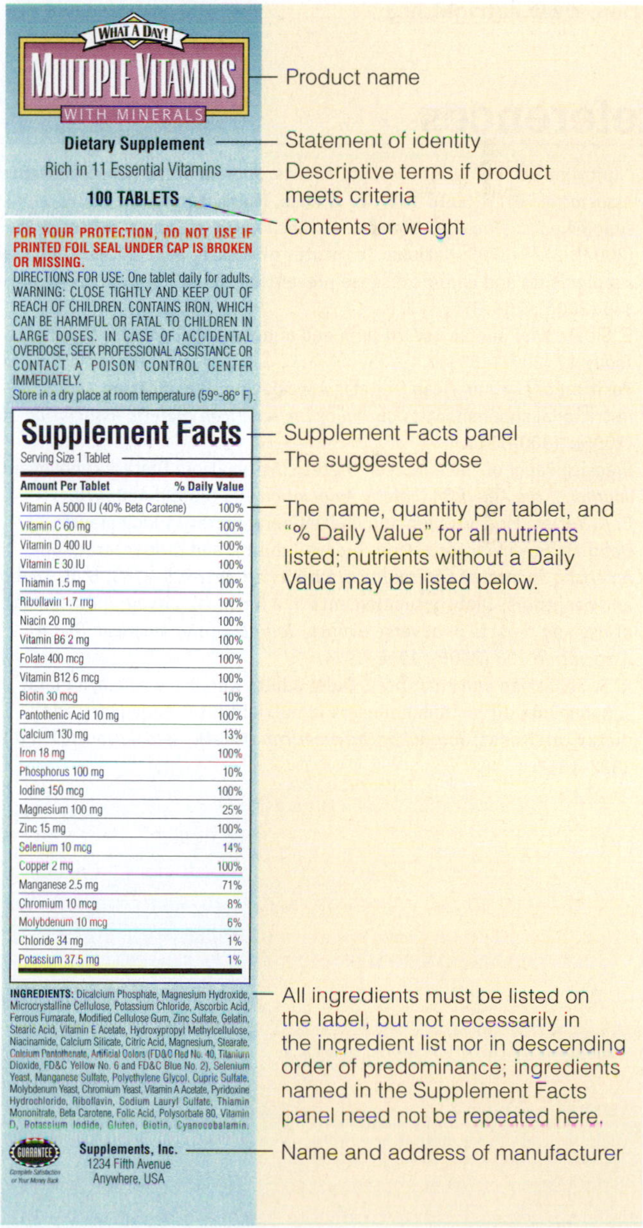

Nutrition on the Net

For further study of topics covered in this highlight, log on to **www.cengagebrain.com** and search for ISBN 084006845X.

- Gather information from the Office of Dietary Supplements or Health Canada: **http://ods.od.nih.gov** or **www.hc-sc.gc.ca**
- Report adverse reactions associated with dietary supplements to the FDA's MedWatch program: **www.fda.gov/Safety/MedWatch**
- Search for "supplements" at the American Dietetic Association: **www.eatright.org**
- Learn more about supplements from the FDA Center for Food Safety and Applied Nutrition: **www.fda.gov/Food/DietarySupplements**
- Obtain consumer information on dietary supplements from the U.S. Pharmacopeia: **www.usp.org**
- Review the Federal Trade Commission policies for dietary supplement advertising: **www.ftc.gov/bcp**

References

1. Capitol health call: Dietary supplements, *Journal of the American Medical Association* 301 (2009): 1427; C. L. Rock, Multivitamin-multimineral supplements: Who uses them? *American Journal of Clinical Nutrition* 85 (2007): 277S–279S; National Institutes of Health, Multivitamin/mineral supplements and chronic disease prevention, *Annals of Internal Medicine* 145 (2006): 364–371.
2. E. Sloan, Why people use vitamin and mineral supplements, *Nutrition Today* 42 (2007): 55–61.
3. Position of the American Dietetic Association: Fortification and nutritional supplements, *Journal of the American Dietetic Association* 105 (2005): 1300–1311.
4. Practice Paper of the American Dietetic Association: Dietary supplements, *Journal of the American Dietetic Association* 105 (2005): 460–470.
5. E. A. Yetley, Multivitamin and multimineral dietary supplements: Definitions, characterization, bioavailability, and drug interactions, *American Journal of Clinical Nutrition* 85 (2007): 269S–276S; B. B. Timbo and coauthors, Dietary supplements in a National Survey: Prevalence of use and reports of adverse events, *Journal of the American Dietetic Association* 106 (2006): 1966–1974.
6. R. S. Sebastian and coauthors, Older adults who use vitamin/mineral supplements differ from nonusers in nutrient intake adequacy and dietary attitudes, *Journal of the American Dietetic Association* 107 (2007): 1322–1332.
7. H. Huang and coauthors, The efficacy and safety of multivitamin and mineral supplement use to prevent cancer and chronic disease in adults: A systematic review for a National Institutes of Health State-of-the-Science Conference, *Annals of Internal Medicine* 145 (2006): 372–385.
8. National Institutes of Health, 2006.
9. G. Bjelakovic and coauthors, Mortality in randomized trials of antioxidant supplements for primary and secondary prevention, *Journal of the American Medical Association* 297 (2007): 842–857; K. A. Lawson and coauthors, Multivitamin use and risk of prostate cancer in the National Institutes of Health-AARP Diet and Health Study, *Journal of the National Cancer Institute* 99 (2007): 754–764.
10. S. P. Murphy and coauthors, Multivitamin-multimineral supplements' effect on total nutrient intake, *American Journal of Clinical Nutrition* 85 (2007): 280S–284S.
11. J. Bleys and coauthors, Vitamin-mineral supplementation and the progression of atherosclerosis: A meta-analysis of randomized controlled trials, *American Journal of Clinical Nutrition* 84 (2006): 880–887; D. B. McCormick, The dubious use of vitamin-mineral supplements in relation to cardiovascular disease, *American Journal of Clinical and Nutrition* 84 (2006): 680–681.
12. Final rule promotes safe use of dietary supplements, www.fda.gov/consumer/updates/dietarysupps062207.html.

CHAPTER

11

Natali Glado/shutterstock.com

Nutrition in Your Life

Realizing that vitamin A from vegetables participates in vision, a mom encourages her children to "eat your carrots" because "they're good for your eyes." A dad takes his children outside to "enjoy the fresh air and sunshine" because they need the vitamin D that is made with the help of the sun. A physician recommends that a patient use vitamin E to slow the progression of heart disease. Another physician gives a newborn a dose of vitamin K to protect against life-threatening blood loss. These common daily occurrences highlight some of the heroic work of the fat-soluble vitamins.

Throughout this chapter, the CourseMate icon indicates an opportunity for online self-study, linking you to activities to increase your understanding of chapter concepts. **www.cengagebrain.com** (search for ISBN 084006845X)

The Fat-Soluble Vitamins: A, D, E, and K

CHAPTER OUTLINE

The fat-soluble vitamins A, D, E, and K differ from the water-soluble vitamins in several significant ways (review the table on p. 314). Being insoluble in the watery GI juices, the fat-soluble vitamins require bile for their digestion and absorption. Upon absorption, fat-soluble vitamins travel through the lymphatic system within chylomicrons before entering the bloodstream, where many of them require protein carriers for transport. The fat-soluble vitamins participate in numerous activities throughout the body, but excesses are stored primarily in the liver and adipose tissue. The body maintains blood concentrations by retrieving these vitamins from storage as needed; thus people can eat less than their daily need for days, weeks, or even months or years without ill effects. They need only ensure that, over time, *average* daily intakes approximate recommendations. By the same token, because fat-soluble vitamins are not readily excreted, the risk of toxicity is greater than it is for the water-soluble vitamins.

Vitamin A and Beta-Carotene

Vitamin A was the first fat-soluble vitamin to be recognized. A century later, vitamin A and its precursor, **beta-carotene**, ♦ continue to intrigue researchers with their diverse roles and profound effects on health.

Three different forms of vitamin A are active in the body: retinol, retinal, and retinoic acid. Collectively, these compounds are known as **retinoids**. Foods derived from animals provide compounds (retinyl esters) that are readily digested and absorbed as retinol in the intestine.[1] Foods derived from plants provide **carotenoids**, ♦ some of which have **vitamin A activity**.* The most studied of the carotenoids is beta-carotene, which can be split to form retinol in the intestine and liver. Beta-carotene's absorption and conversion are significantly less efficient than those of the retinoids. Figure 11-1 (p. 356) illustrates the structural similarities and differences of these vitamin A compounds and the cleavage of beta-carotene.

The cells can convert retinol and retinal to the other active forms of vitamin A as needed. The conversion of retinol to retinal is reversible, but the further

*Carotenoids with vitamin A activity include alpha-carotene, beta-carotene, and beta-cryptoxanthin; carotenoids with no vitamin A activity include lycopene, lutein, and zeaxanthin.

♦ A compound that can be converted into an active vitamin is called a **precursor.**

♦ Carotenoids are among the best-known phytochemicals.

vitamin A: all naturally occurring compounds with the biological activity of **retinol** (RET-ih-nol), the alcohol form of vitamin A.

beta-carotene (BAY-tah KARE-oh-teen): one of the carotenoids; an orange pigment and vitamin A precursor found in plants.

retinoids (RET-ih-noyds): chemically related compounds with biological activity similar to that of retinol; metabolites of retinol.

carotenoids (kah-ROT-eh-noyds): pigments commonly found in plants and animals, some of which have vitamin A activity. The carotenoid with the greatest vitamin A activity is beta-carotene.

vitamin A activity: a term referring to both the active forms of vitamin A and the precursor forms in foods without distinguishing between them.

FIGURE 11-1 Forms of Vitamin A

In this diagram, corners represent carbon atoms, as in all previous diagrams in this book. A further simplification here is that methyl groups (CH_3) are understood to be at the ends of the lines extending from corners. (See Appendix C for complete structures.)

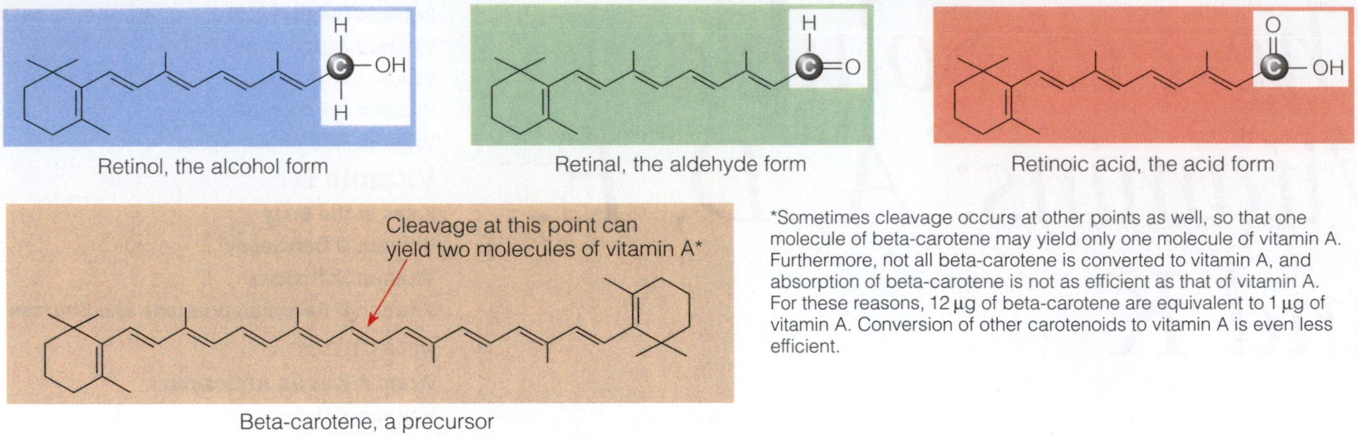

*Sometimes cleavage occurs at other points as well, so that one molecule of beta-carotene may yield only one molecule of vitamin A. Furthermore, not all beta-carotene is converted to vitamin A, and absorption of beta-carotene is not as efficient as that of vitamin A. For these reasons, 12 µg of beta-carotene are equivalent to 1 µg of vitamin A. Conversion of other carotenoids to vitamin A is even less efficient.

conversion of retinal to retinoic acid is irreversible (see Figure 11-2). This irreversibility is significant because each form of vitamin A performs a function that the others cannot.

Several proteins participate in the digestion and absorption of vitamin A.[2] After absorption via the lymph system, vitamin A eventually arrives at the liver, where it is stored. There, a special transport protein, **retinol-binding protein (RBP)**, picks up vitamin A from the liver and carries it in the blood. Cells that use vitamin A have special protein receptors for it, and its action within each cell may differ depending on the receptor.[3] For example, retinoic acid can stimulate cell growth in the skin and inhibit cell growth in tumors.[4]

Roles in the Body

Vitamin A is a versatile vitamin, known to regulate the expression of several hundred genes.[5] Its major roles include:

- Promoting vision
- Participating in protein synthesis and cell differentiation, thereby maintaining the health of epithelial tissues and skin
- Supporting reproduction and growth

As mentioned, each form of vitamin A performs specific tasks. Retinol supports reproduction and is the major transport and storage form of the vitamin. Retinal is active in vision and is also an intermediate in the conversion of retinol to retinoic acid (review Figure 11-2). Retinoic acid acts like a hormone, regulating cell differentiation, growth, and embryonic development. Animals raised on

retinol-binding protein (RBP): the specific protein responsible for transporting retinol.

FIGURE 11-2 Conversion of Vitamin A Compounds

Notice that the conversion from retinol to retinal is reversible, whereas the pathway from retinal to retinoic acid is not.

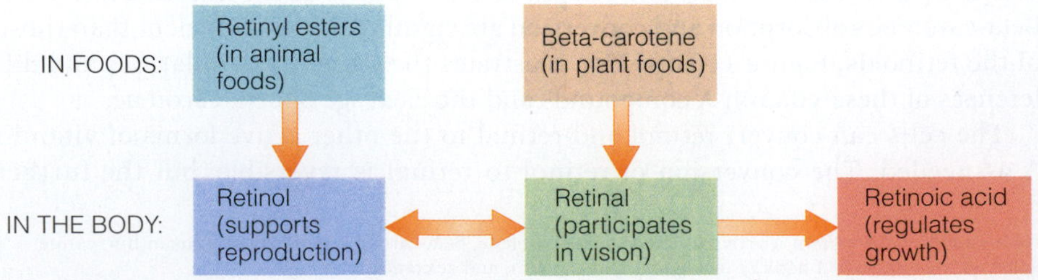

retinoic acid as their sole source of vitamin A can grow normally, but they become blind because retinoic acid cannot be converted to retinal (review Figure 11-2).

Vitamin A in Vision Vitamin A plays two indispensable roles in the eye: it helps maintain a crystal-clear outer window, the **cornea**, and it participates in the conversion of light energy into nerve impulses at the **retina** (see Figure 11-3 for details). Some of the photosensitive cells ♦ of the retina contain **pigment** molecules called **rhodopsin**; each rhodopsin molecule is composed of a protein called **opsin** bonded to a molecule of retinal. ♦ When light passes through the cornea of the eye and strikes the retina, rhodopsin responds by changing shape and becoming bleached. As it does, the retinal shifts from a *cis* to a *trans* configuration, just as fatty acids do during hydrogenation (see pp. 138–139). The bleached *trans*-retinal cannot remain bonded to opsin. When retinal is released, opsin changes shape, thereby disturbing the membrane of the cell and generating an electrical impulse that travels along the cell's length. At the other end of the cell, the impulse is transmitted to a nerve cell, which conveys the message to the brain. Much of the retinal is then converted back to its active *cis* form and combined with the opsin protein to regenerate the pigment rhodopsin. Some retinal, however, may be oxidized to retinoic acid, a biochemical dead end for the visual process. Visual activity leads to repeated small losses of retinal, necessitating its constant replenishment either directly from foods or indirectly from retinol stores.

Vitamin A in Protein Synthesis and Cell Differentiation Despite its important role in vision, only one-thousandth of the body's vitamin A is in the retina. Much more is in the cells lining the body's surfaces. There, the vitamin participates in protein synthesis and **cell differentiation**, a process by which each type of cell develops to perform a specific function.

All body surfaces, both inside and out, are covered by layers of cells known as **epithelial cells**. The **epithelial tissue** on the outside of the body is, of course, the skin—and vitamin A helps to protect against skin damage from sunlight. The epithelial tissues that line the inside of the body are the **mucous membranes**: the linings of the mouth, stomach, and intestines; the linings of the lungs and the passages leading to them; the linings of the urinary bladder and urethra; the linings of the uterus and vagina; and the linings of the eyelids and sinus passageways. Within the body, the mucous membranes of the GI tract alone line an area larger than a quarter of a football field, and vitamin A helps to maintain their integrity (see Figure 11-4 on p. 358).

Vitamin A promotes differentiation of epithelial cells and goblet cells, one-celled glands that synthesize and secrete mucus. Mucus coats and protects the

♦ Photosensitive cells of the retina:
- Rods contain the rhodopsin pigment and respond to faint light.
- Cones contain the iodopsin pigment and function in color vision.

♦ More than 100 million cells reside in the retina, and each contains about 30 million molecules of vitamin A–containing visual pigments.

cornea (KOR-nee-uh): the transparent membrane covering the outside of the eye.

retina (RET-in-uh): the innermost membrane of the eye, composed of several layers including one that contains the rods and cones.

pigment: a molecule capable of absorbing certain wavelengths of light so that it reflects only those that we perceive as a certain color.

rhodopsin (ro-DOP-sin): a light-sensitive pigment of the retina; contains the retinal form of vitamin A and the protein opsin.
- **rhod** = red (pigment)
- **opsin** = visual protein

opsin (OP-sin): the protein portion of the visual pigment molecule.

cell differentiation (DIF-er-EN-she-AY-shun): the process by which immature cells develop specific functions different from those of the original that are characteristic of their mature cell type.

epithelial (ep-i-THEE-lee-ul) **cells:** cells on the surface of the skin and mucous membranes.

epithelial tissue: the layer of the body that serves as a selective barrier between the body's interior and the environment. (Examples are the cornea of the eyes, the skin, the respiratory lining of the lungs, and the lining of the digestive tract.)

mucous (MYOO-kus) **membranes:** the membranes, composed of mucus-secreting cells, that line the surfaces of body tissues.

FIGURE 11-3 Vitamin A's Role in Vision

The retina receives the image formed by the lens and converts light energy into chemical energy and nerve signals that reach the brain via the optic nerve.

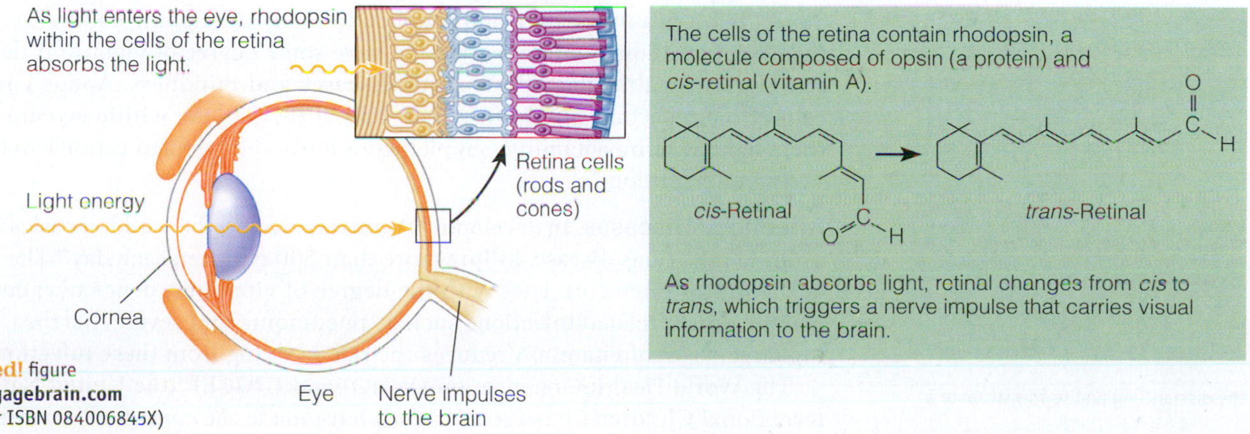
As light enters the eye, rhodopsin within the cells of the retina absorbs the light.
Retina cells (rods and cones)
Light energy
Cornea
Eye
Nerve impulses to the brain
The cells of the retina contain rhodopsin, a molecule composed of opsin (a protein) and cis-retinal (vitamin A).
cis-Retinal
trans-Retinal
As rhodopsin absorbs light, retinal changes from cis to trans, which triggers a nerve impulse that carries visual information to the brain.

Animated! figure
www.cengagebrain.com
(search for ISBN 084006845X)

FIGURE 11-4 Mucous Membrane Integrity

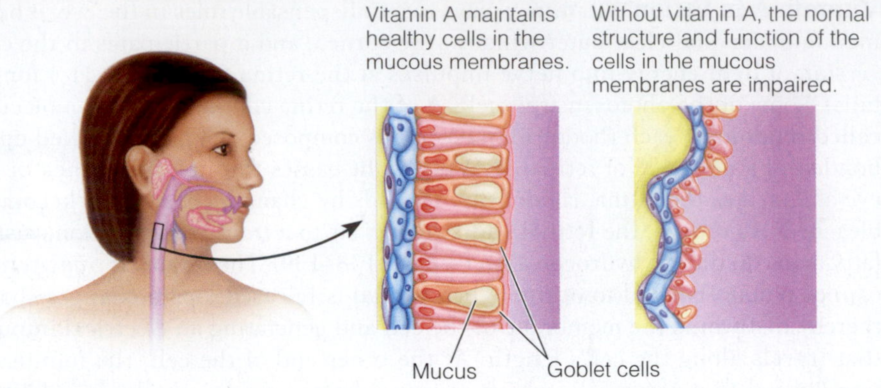

Vitamin A maintains healthy cells in the mucous membranes.
Without vitamin A, the normal structure and function of the cells in the mucous membranes are impaired.
Mucus
Goblet cells

epithelial cells from invasive microorganisms and other potentially damaging substances, such as gastric juices.

Vitamin A in Reproduction and Growth As mentioned, vitamin A also supports reproduction and growth. In men, retinol participates in sperm development, and in women, vitamin A supports normal fetal development during pregnancy. Children lacking vitamin A fail to grow. When given vitamin A supplements, these children gain weight and grow taller.

The growth of bones illustrates that growth is a complex phenomenon of **remodeling.** To convert a small bone into a large bone, the bone-remodeling cells must "undo" some parts of the bone as they go, ♦ and vitamin A participates in the dismantling. The cells that break down bone contain sacs of degradative enzymes. ♦ With the help of vitamin A, these enzymes eat away at selected sites in the bone, removing the parts that are not needed.

♦ The cells that destroy bone during growth are **osteoclasts;** those that build bone are **osteoblasts.**
- **osteo** = bone
- **clast** = break
- **blast** = build

♦ The sacs of degradative enzymes are **lysosomes** (LYE-so-zomes).

Beta-Carotene as an Antioxidant In the body, beta-carotene serves primarily as a vitamin A precursor.[6] Not all dietary beta-carotene is converted to active vitamin A, however. Some beta-carotene may act as an antioxidant ♦ capable of protecting the body against disease. (See Highlight 11 for details.) In fact, a diet rich in fruits and vegetables containing beta-carotene and other carotenoids helps to defend against some cancers.

♦ Key antioxidant nutrients:
- Vitamin C, vitamin E, beta-carotene
- Selenium

Vitamin A Deficiency

Vitamin A status depends mostly on the adequacy of vitamin A stores, 90 percent of which are in the liver. Vitamin A status also depends on a person's protein status because retinol-binding protein serves as the vitamin's transport carrier inside the body.

If a person were to stop eating vitamin A–containing foods, deficiency symptoms would not begin to appear until after stores were depleted—one to two years for a healthy adult but much sooner for a growing child. Then the consequences would be profound and severe.[7] Vitamin A deficiency is uncommon in the United States, but it is a major nutrition problem in many developing countries. An estimated 250 million children worldwide have some degree of vitamin A deficiency and thus are vulnerable to infectious diseases and blindness. About 1 to 2 percent of them become blind every year, half of them dying within a year of losing their sight. Routine vitamin A supplementation and food fortification can be a lifesaving intervention.[8]

Infectious Diseases In developing countries around the world, measles is a devastating infectious disease, killing more than 500 children each day.[9] The severity of the illness often correlates with the degree of vitamin A deficiency; deaths are usually due to related infections such as pneumonia and severe diarrhea. Providing large doses of vitamin A reduces the risk of dying from these infections.

The World Health Organization (WHO) and UNICEF (the United Nations International Children's Emergency Fund) have made the control of vitamin A defi-

remodeling: the dismantling and re-formation of a structure.

FIGURE 11-5 Vitamin A–Deficiency Symptom—Night Blindness

These photographs illustrate the eyes' slow recovery in response to a flash of bright light at night. In animal research studies, the response rate is measured with electrodes.

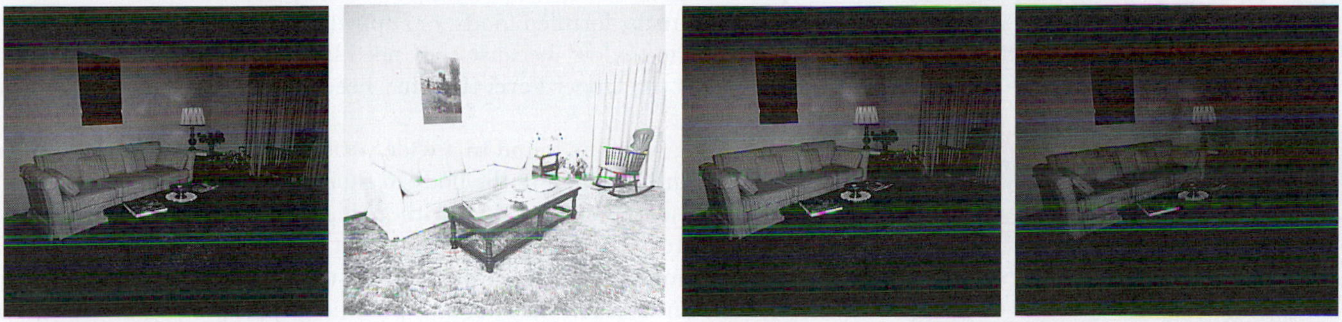

In dim light, you can make out the details in this room. You are using your rods for vision.

A flash of bright light momentarily blinds you as the pigment in the rods is bleached.

You quickly recover and can see the details again in a few seconds.

With inadequate vitamin A, you do not recover but remain blinded for many seconds.

© David Farr/Image Smythe (all)

ciency a major goal in their quest to improve child health and survival throughout the developing world. They recommend routine vitamin A supplementation for all children with measles in areas where vitamin A deficiency is a problem or where the measles death rate is high. In the United States, the American Academy of Pediatrics recommends vitamin A supplementation for certain groups of measles-infected infants and children. Vitamin A supplementation also protects against the complications of other life-threatening infections, including malaria, lung diseases, and HIV (human immunodeficiency virus, the virus that causes AIDS).

Night Blindness **Night blindness** is one of the first detectable signs of vitamin A deficiency and permits early diagnosis. In night blindness, the retina does not receive enough retinal to regenerate the visual pigments bleached by light. The person loses the ability to recover promptly from the temporary blinding that follows a flash of bright light at night or to see after the lights go out. In many parts of the world, after the sun goes down, vitamin A–deficient people become night-blind. They often cling to others or sit still, afraid that they may trip and fall or lose their way if they try to walk alone. Figure 11-5 shows the eyes' slow recovery in response to a flash of bright light in night blindness.

Blindness (Xerophthalmia) Beyond night blindness is total blindness—failure to see at all. Night blindness is caused by a lack of vitamin A at the back of the eye, the retina; total blindness is caused by a lack at the front of the eye, the cornea. Severe vitamin A deficiency is the major cause of childhood blindness in the world, causing more than half a million preschool children to lose their sight each year. Blindness due to vitamin A deficiency, known as **xerophthalmia**, develops in stages. At first, the cornea becomes dry and hard because of inadequate mucus production—a condition known as **xerosis.** Then, corneal xerosis can quickly progress to **keratomalacia**, the softening of the cornea that leads to irreversible blindness.

Keratinization Elsewhere in the body, vitamin A deficiency affects other surfaces. On the body's outer surface, the epithelial cells change shape and begin to secrete the protein **keratin**—the hard, inflexible protein of hair and nails. As Figure 11-6 shows, the skin becomes dry, rough, and scaly as lumps of keratin accumulate (**keratinization**). Without vitamin A, the goblet cells in the GI tract diminish in number and activity, limiting the secretion of mucus. With less mucus, normal digestion and absorption of nutrients falter, and this, in turn, worsens malnutrition by limiting the absorption of whatever nutrients the diet may deliver. Similar changes in the cells of other epithelial tissues weaken defenses, making infections of the respiratory tract, the GI tract, the urinary tract, the vagina, and inner ear likely.

FIGURE 11-6 Vitamin A–Deficiency Symptom—The Rough Skin of Keratinization

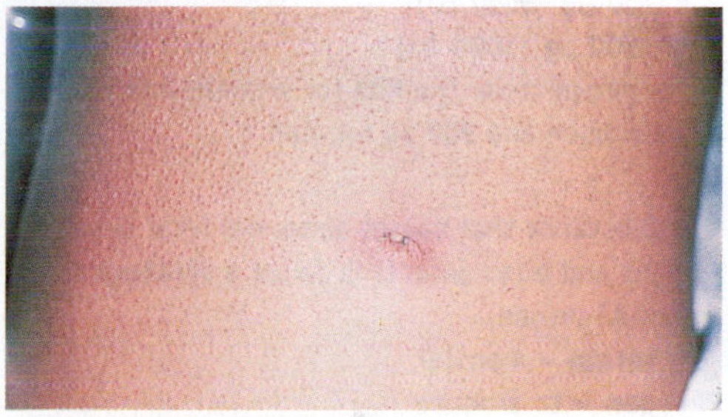

Ken Greer/Visuals Unlimited

In vitamin A deficiency, the epithelial cells secrete the protein keratin in a process known as *keratinization.* (Keratinization doesn't occur in the GI tract, but mucus-producing cells dwindle and mucus production declines.) The extreme of this condition is *hyperkeratinization* or *hyperkeratosis.* When keratin accumulates around hair follicles, the condition is known as *follicular hyperkeratosis.*

night blindness: slow recovery of vision after flashes of bright light at night or an inability to see in dim light; an early symptom of vitamin A deficiency.

xerophthalmia (zer-off-THAL-mee-uh): progressive blindness caused by inadequate tear production due to severe vitamin A deficiency.

- **xero** = dry
- **ophthalm** = eye

xerosis (zee-ROW-sis): abnormal drying of the skin and mucous membranes; a sign of vitamin A deficiency.

keratomalacia (KARE-ah-toe-ma-LAY-shuh): softening of the cornea that leads to irreversible blindness; seen in severe vitamin A deficiency.

keratin (KARE-uh-tin): a water-insoluble protein; the normal protein of hair and nails.

keratinization: accumulation of keratin in a tissue; a sign of vitamin A deficiency.

FIGURE 11-7 Symptom of Beta-Carotene Excess—Discoloration of the Skin

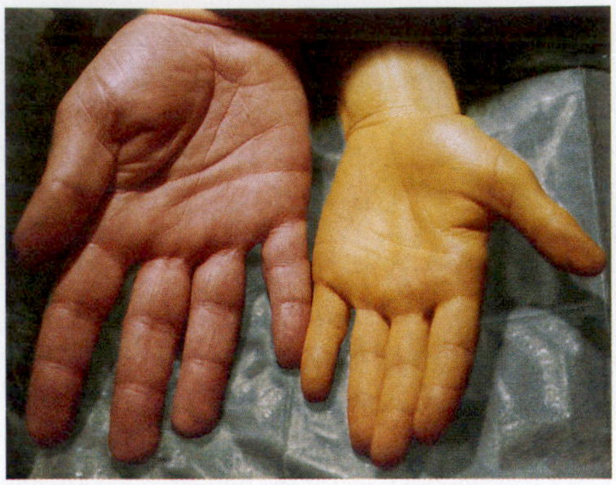

The hand on the right shows the skin discoloration that occurs when blood levels of beta-carotene rise in response to a low-kcalorie diet that features carrots, pumpkins, and orange juice. (The hand on the left belongs to someone else and is shown here for comparison.)

Vitamin A Toxicity

Just as a deficiency of vitamin A affects all body systems, so does a toxicity. Symptoms of toxicity begin to develop when all the binding proteins are swamped, and free vitamin A damages the cells. Such effects are unlikely when a person depends on a balanced diet for nutrients, but toxicity is a real possibility when concentrated amounts of **preformed vitamin A** in foods derived from animals, fortified foods, or supplements is consumed.[10] Children are most vulnerable to toxicity because they need less vitamin A and are more sensitive to overdoses. An Upper Level (UL) has been set for preformed vitamin A (see inside front cover).

Beta-carotene, which is found in a wide variety of fruits and vegetables, is not converted efficiently enough in the body to cause vitamin A toxicity; instead, it is stored in the fat just under the skin. Although overconsumption of beta-carotene from foods may turn the skin yellow, this is not harmful (see Figure 11-7). In contrast, overconsumption of beta-carotene from supplements may be quite harmful. In excess, this antioxidant may act as a prooxidant, promoting cell division and destroying vitamin A. Furthermore, the adverse effects of beta-carotene supplements are most evident in people who drink alcohol and smoke cigarettes.

Bone Defects Excessive intake of vitamin A over the years may weaken the bones and contribute to fractures and osteoporosis.[11] Vitamin A suppresses bone-building activity, stimulates bone-dismantling activity, and interferes with vitamin D's ability to maintain normal blood calcium.[12] Research findings suggest that most people should not take vitamin A supplements.[13] Even multivitamin supplements ♦ provide more vitamin A than most people need.

Birth Defects Excessive vitamin A during pregnancy leads to abnormal cell death in the spinal cord, which increases the risk of birth defects.[14] ♦ High intakes (10,000 IU ♦ of supplemental vitamin A daily) before the seventh week of pregnancy appear to be the most damaging. For this reason, vitamin A is not given as a supplement in the first trimester of pregnancy without specific evidence of deficiency, which is rare.

Not for Acne Adolescents need to know that massive doses of vitamin A have no beneficial effect on **acne.** The prescription medicine Accutane is made from vitamin A but is chemically different.* Taken orally, Accutane is effective against the deep lesions of cystic acne. It is highly toxic, however, especially during growth, and has caused birth defects in infants when women have taken it during their pregnancies. For this reason, women taking Accutane must begin using two effective forms of contraception at least one month before taking the drug and continue using contraception at least one month after discontinuing its use. They should also refrain from taking any supplements containing vitamin A to avoid additive toxic effects.

Another vitamin A relative, Retin-A, fights acne, the wrinkles of aging, and other skin disorders.** Applied topically, this ointment smooths and softens skin; it also lightens skin that has become darkly pigmented after inflammation. During treatment, the skin becomes red and tender and peels.

Vitamin A Recommendations

Because the body can derive vitamin A from various retinoids and carotenoids, its content in foods and its recommendations are expressed as **retinol activity equivalents (RAE).** One microgram of retinol counts as 1 RAE, ♦ as does 12 micrograms of dietary beta-carotene. Most food and supplement labels report their vitamin A contents using International Units (IU), ♦ a measure of vitamin activity used before direct chemical analysis was possible. The accompanying "How To" feature explains how to convert IU to a weight measurement.

Vitamin A in Foods

The richest sources of the retinoids are foods derived from animals—liver, fish liver oils, milk and milk products, butter, and eggs.

*The generic name for Accutane is isotretinoin.
**The generic name for Retin-A is tretinoin topical.

♦ Multivitamin supplements typically provide:
- 750 µg (2500 IU)
- 1500 µg (5000 IU)

For perspective, the RDA for vitamin A is 700 µg for women and 900 µg for men.

♦ A substance that causes abnormal fetal development and birth defects is called a **teratogen** (ter-AT-oh-jen).
- **terato** = monster
- **gen** = to produce

♦ For perspective, 10,000 IU ≈ 3000 µg vitamin A, roughly four times the RDA for women.

♦ 1 µg RAE = 1 µg retinol
= 2 µg beta-carotene (supplement)
= 12 µg beta-carotene (dietary)
= 24 µg of other vitamin A precursor carotenoids

♦ 1 IU retinol = 0.3 µg retinol or 0.3 µg RAE
1 IU beta-carotene (supplement) = 0.5 IU retinol or 0.15 µg RAE
1 IU beta-carotene (dietary) = 0.165 IU retinol or 0.05 µg RAE
1 IU other vitamin A precursor carotenoids = 0.025 µg RAE

preformed vitamin A: dietary vitamin A in its active form.

acne: a chronic inflammation of the skin's follicles and oil-producing glands, which leads to an accumulation of oils inside the ducts that surround hairs; usually associated with the maturation of young adults.

retinol activity equivalents (RAE): a measure of vitamin A activity; the amount of retinol that the body will derive from a food containing preformed retinol or its precursor beta-carotene.

HOW TO Convert International Units (IU) to Weight Measurements

Supplement labels often list the amount of fat-soluble vitamins in International Units (IU), a universally accepted measure of a vitamin's biological effect. Such a measure allows scientists to compare the potency of substances and was most useful decades ago before chemicals could be purified and weighed accurately.

Because IU measures biological activity and not weight, the conversion factors differ for each fat-soluble vitamin:

For vitamin A (retinol):

- 1 IU = 0.3 μg
- 1 μg = 3.33 IU

For vitamin D (cholecalciferol):

- 1 IU = 0.025 μg
- 1 μg = 40 IU

For vitamin E (natural α-tocopherol):

- 1 IU = 0.67 mg
- 1 mg = 1.49 IU

To convert from IU to a weight measurement, multiply IU by the appropriate equivalent. For example, for a supplement listing 5000 IU vitamin A, 400 IU vitamin D, and 30 IU vitamin E:

5000 IU × 0.3 μg/IU = 1500 μg retinol

400 IU × 0.025 μg/IU = 10 μg cholecalciferol

30 IU × 0.67 mg/IU = 20 mg α-tocopherol

For additional practice, log on to **www.cengagebrain.com** and search for ISBN 084006845X.

TRY IT Convert these values on a supplement label from IU to weight measurements: 4000 IU vitamin A, 600 IU vitamin D, and 12 IU vitamin E.

Because vitamin A is fat soluble, it is lost when milk is skimmed. To compensate, reduced-fat, low-fat, and fat-free milks are often fortified so as to supply 6 to 10 percent of the Daily Value per cup.* Margarine is usually fortified to provide the same amount of vitamin A as butter.

Plants contain no retinoids, but many vegetables and some fruits contain vitamin A precursors—the carotenoids, red and yellow pigments of plants. Only a few carotenoids have vitamin A activity; the carotenoid with the greatest vitamin A activity is beta-carotene. The bioavailability of carotenoids depends in part on fat accompanying the meal.[15] More carotenoids are absorbed when salads have regular dressing than when reduced-fat dressing is used, and essentially no carotenoid absorption occurs when fat-free dressing is used.

The Colors of Vitamin A Foods Dark leafy greens (like spinach—not celery or cabbage) and rich yellow or deep orange vegetables and fruits (such as winter squash, cantaloupe, carrots, and sweet potatoes—not corn or bananas) help people meet their vitamin A needs (see Figure 11-8 on p. 362). A diet including several servings of such carotene-rich sources helps to ensure a sufficient intake.

An attractive meal that includes foods of different colors most likely supplies vitamin A as well. Most foods with vitamin A activity are brightly colored—green, yellow, orange, and red. Any plant-derived food with significant vitamin A activity must have some color because beta-carotene is a rich, deep yellow, almost orange compound. The beta-carotene in dark green, leafy vegetables is abundant but masked by large amounts of the green pigment **chlorophyll**.

Bright color is not always a sign of vitamin A activity, however. Beets and corn, for example, derive their colors from the red and yellow **xanthophylls**, which have

The carotenoids in foods bring colors to meals; the retinoids in our eyes allow us to see them.

chlorophyll (KLO-row-fil): the green pigment of plants, which absorbs light and transfers the energy to other molecules, thereby initiating photosynthesis.

xanthophylls (ZAN-tho-fills): pigments found in plants; responsible for the color changes seen in autumn leaves.

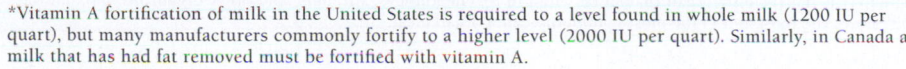

*Vitamin A fortification of milk in the United States is required to a level found in whole milk (1200 IU per quart), but many manufacturers commonly fortify to a higher level (2000 IU per quart). Similarly, in Canada all milk that has had fat removed must be fortified with vitamin A.

FIGURE 11-8 Vitamin A in Selected Foods

See the "How To" section on p. 316 for more information on using this figure.

Food	Serving size (kcalories)
Bread, whole wheat	1 oz slice (70 kcal)
Cornflakes, fortified	1 oz (110 kcal)
Spaghetti pasta	½c cooked (99 kcal)
Tortilla, flour	1 10"-round (234 kcal)
Broccoli	½c cooked (22 kcal)
Carrots	½c shredded raw (24 kcal)
Potato	1 medium baked w/skin (133 kcal)
Tomato juice	¾c (31 kcal)
Banana	1 medium raw (109 kcal)
Orange	1 medium raw (62 kcal)
Strawberries	½c fresh (22 kcal)
Watermelon	1 slice (92 kcal)
Milk, fortified	1 c reduced-fat 2% (121 kcal)
Yogurt, plain	1 c low-fat (155 kcal)
Cheddar cheese	1½ oz (171 kcal)
Cottage cheese	½c low-fat 2% (101 kcal)
Pinto beans	½c cooked (117 kcal)
Peanut butter	2 tbs (188 kcal)
Sunflower seeds	1 oz dry (165 kcal)
Tofu (soybean curd)	½c (76 kcal)
Ground beef, lean	3 oz broiled (244 kcal)
Chicken breast	3 oz roasted (140 kcal)
Tuna, canned in water	3 oz (99 kcal)
Egg	1 hard cooked (78 kcal)
Excellent, and sometimes unusual, sources:	
Beef liver	3 oz fried (184 kcal)
Sweet potatoes	½c cooked (116 kcal)
Mango	1 (135 kcal)

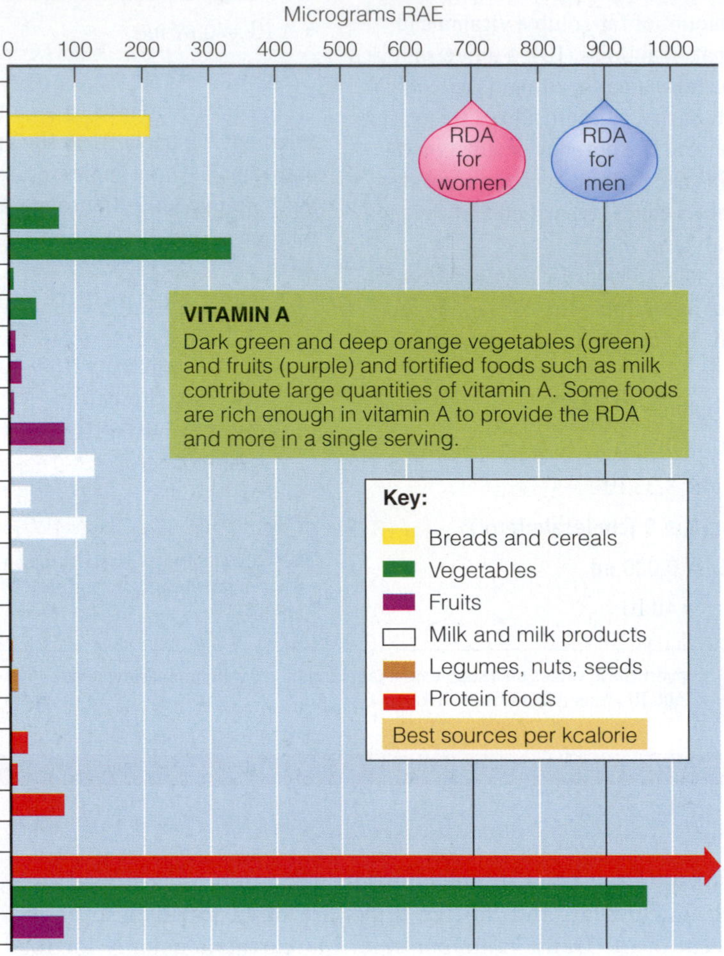

no vitamin A activity. As for white plant foods such as potatoes, cauliflower, pasta, and rice, they also offer little or no vitamin A. Similarly, fast foods often lack vitamin A. Anyone who dines frequently on hamburgers, french fries, and colas is wise to emphasize colorful vegetables and fruits at other meals.

Vitamin A–Rich Liver People sometimes wonder if eating liver too frequently can cause vitamin A toxicity. Liver is a rich source because vitamin A is stored in the livers of animals, just as in humans.* Arctic explorers who have eaten large quantities of polar bear liver have become ill with symptoms suggesting vitamin A toxicity, as have young children who regularly ate a chicken liver spread that provided three times their daily recommended intake. Liver offers many nutrients, and eating it periodically may improve a person's nutrition status. But caution is warranted not to eat too much too often, especially for pregnant women. With one ounce of beef liver providing more than three times the RDA for vitamin A, intakes can rise quickly.

IN SUMMARY **Vitamin A is found in the body in three forms: retinol, retinal, and retinoic acid. Together, they are essential to vision, healthy epithelial tissues, and growth. Vitamin A deficiency is a major health problem**

*The liver is not the only organ that stores vitamin A. The kidneys, adrenals, and other organs do, too, but the liver stores the most and is the most commonly eaten organ meat.

world-wide, leading to infections, blindness, and keratinization. Toxicity can also cause problems and is most often associated with supplement abuse. Animal-derived foods such as liver and whole or fortified milk provide retinoids, whereas brightly colored plant-derived foods such as spinach, carrots, and pumpkins provide beta-carotene and other carotenoids. In addition to serving as a precursor for vitamin A, beta-carotene may act as an antioxidant in the body. The accompanying table provides a summary of vitamin A.

Vitamin A

Other Names	**Deficiency Disease**
Retinol, retinal, retinoic acid; precursors are carotenoids such as beta-carotene	Hypovitaminosis A
2001 RDA	**Deficiency Symptoms**
Men: 900 μg RAE/day Women: 700 μg RAE/day	Night blindness, corneal drying (xerosis), triangular gray spots on eye (Bitot's spots), softening of the cornea (keratomalacia), and corneal degeneration and blindness (xerophthalmia); impaired immunity (infectious diseases); plugging of hair follicles with keratin, forming white lumps (hyperkeratosis)
Upper Level	
Adults: 3000 μg/day	
Chief Functions in the Body	**Toxicity Disease**
Vision; maintenance of cornea, epithelial cells, mucous membranes, skin; bone and tooth growth; reproduction; immunity	Hypervitaminosis A[a]
Significant Sources	**Chronic Toxicity Symptoms**
Retinol: fortified milk, cheese, cream, butter, fortified margarine, eggs, liver	Increased activity of osteoclasts[b] causing reduced bone density; liver abnormalities; birth defects
Beta-carotene: spinach and other dark leafy greens; broccoli, deep orange fruits (apricots, cantaloupe) and vegetables (squash, carrots, sweet potatoes, pumpkin)	**Acute Toxicity Symptoms**
	Blurred vision, nausea, vomiting, vertigo; increase of pressure inside skull, mimicking brain tumor; headaches; muscle incoordination

[a]A related condition, *hypercarotenemia*, is caused by the accumulation of too much of the vitamin A precursor beta-carotene in the blood, which turns the skin noticeably yellow. Hypercarotenemia is not, strictly speaking, a toxicity symptom.

[b]*Osteoclasts* are the cells that destroy bone during its growth. Those that build bone are *osteoblasts*.

FIGURE 11-9 Vitamin D Synthesis and Activation

The precursor of vitamin D is made in the liver from cholesterol (see Figure 5-11 on p. 141 and Appendix C). The activation of vitamin D is closely regulated by parathyroid hormone. The final product, active vitamin D, is also known as 1,25-dihydroxycholecalciferol (or calcitriol).

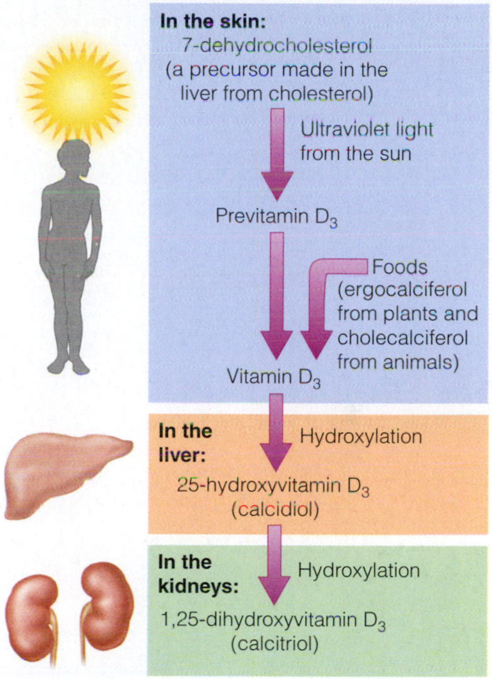

Animated! figure
www.cengagebrain.com
(search for ISBN 084006845X)

Vitamin D

Vitamin D (calciferol) ♦ is different from all the other nutrients in that the body can synthesize it, with the help of sunlight, from a precursor that the body makes from cholesterol. Therefore, vitamin D is not an essential nutrient; given enough time in the sun, people need no vitamin D from foods.

♦ Vitamin D comes in many forms, but the two most important in the diet are a plant version called **vitamin D_2** or **ergocalciferol** (ER-go-kal-SIF-er-ol) and an animal version called **vitamin D_3** or **cholecalciferol** (KO-lee-kal-SIF-er-ol).

Figure 11-9 diagrams the pathway for making and activating vitamin D in the body. Ultraviolet rays from the sun hit the precursor in the skin and convert it to previtamin D_3. This compound diffuses from the skin into the blood and is converted to its active form with the help of the body's heat. The biological activity of the active vitamin is 500- to 1000-fold greater than that of its precursor.

Regardless of whether the body manufactures vitamin D or obtains it directly from foods, two hydroxylation reactions must occur before the vitamin becomes fully active.[16] First, the liver adds an OH group, and then the kidneys add another OH group to produce the active vitamin. ♦ A review of Figure 11-9 reveals how diseases affecting either the liver or the kidneys can interfere with the activation of vitamin D and produce symptoms of deficiency.

♦ Before hydroxylation, vitamin D in the blood is known as **calciol;** after hydroxylation in the liver, vitamin D is known as **calcidiol** (or **25-hydroxyvitamin D**); and after hydroxylation in the kidneys, **active vitamin D** is known as **calcitriol** (or **1,25-dihydroxyvitamin D**).

Roles in the Body

Though called a vitamin, the active form of vitamin D is actually a hormone—a compound manufactured by one part of the body that travels through the blood and causes another body part to respond.[17] Like vitamin A, vitamin D has a binding protein that carries it to the target organs—most notably, the intestines, the kidneys, and the bones. All respond to vitamin D by making the minerals needed for bone growth and maintenance available.

Vitamin D in Bone Growth Vitamin D is a member of a large and cooperative bone-making and maintenance team ♦ composed of nutrients and other compounds, including vitamins A and K; the hormones parathyroid hormone and calcitonin; the protein collagen; and the minerals calcium, phosphorus, magnesium, and fluoride. Vitamin D's special role in bone health is to assist in the absorption of calcium and phosphorus, thus helping to maintain blood concentrations of these minerals.[18] The bones grow denser and stronger as they absorb and deposit these minerals. Details of calcium balance and mineral deposition appear in Chapter 12.

♦ Key bone nutrients:
- Vitamin D, vitamin K, vitamin A
- Calcium, phosphorus, magnesium, fluoride

Vitamin D raises blood concentrations of bone minerals in three ways. When the diet is sufficient, vitamin D enhances their absorption from the GI tract. When the diet is insufficient, vitamin D provides the needed minerals from other sources: reabsorption by the kidneys and mobilization from the bones into the blood.[19] The vitamin may work alone, as it does in the GI tract, or in combination with parathyroid hormone, as it does in the bones and kidneys.[20]

Vitamin D in Other Roles Scientists have discovered many other tissues that respond to vitamin D, including cells of the immune system, brain and nervous system, pancreas, skin, muscles and cartilage, and reproductive organs.[21] In many cases, vitamin D enhances or suppresses the activity of genes that regulate cell growth.[22] As such, it may be valuable in treating a number of diseases.[23] Recent evidence suggests that vitamin D may protect against tuberculosis, inflammation, multiple sclerosis, hypertension, and some cancers.[24]

Vitamin D Deficiency

Overt signs of vitamin D deficiency are relatively rare, but vitamin D insufficiency is remarkably common.[25] Factors that contribute to vitamin D deficiency include dark skin, breastfeeding without supplementation, lack of sunlight, and not using fortified milk. In vitamin D deficiency, production of a protein that binds calcium ♦ in the intestinal cells slows. Thus, even when calcium in the diet is adequate, it passes through the GI tract unabsorbed, leaving the bones undersupplied. Consequently, a vitamin D deficiency creates a calcium deficiency and increases the risks of several chronic diseases.[26] Vitamin D–deficient adolescents may not reach their peak bone mass.[27] Low blood calcium due to a vitamin D deficiency can also trigger seizures.[28]

♦ Synthesis of **calbindin,** a calcium-binding transport protein, requires vitamin D.

Rickets Worldwide, the prevalence of the vitamin D–deficiency disease **rickets** is extremely high, affecting more than half of the children in countries such as Mongolia, Tibet, and the Netherlands.[29] In the United States, rickets is not common, but when it occurs black children and adolescents—especially females and overweight teens—are the ones most likely to be affected.[30] To prevent rickets, the American Academy of Pediatrics recommends a supplement for all infants, children, and adolescents who do not receive enough vitamin D.[31] ♦ In rickets, the bones fail to calcify normally, causing growth retardation and skeletal abnormalities. The bones become so weak that they bend when they have to support the body's weight (see Figure 11-10). A child with rickets who is old enough to walk characteristically develops bowed legs, often the most obvious sign of the disease. Another sign is the beaded ribs ♦ that result from the poorly formed attachments of the bones to the cartilage.

♦ For perspective, 1 cup of fortified milk provides about 100 IU vitamin D.

♦ Because the poorly formed rib attachments resemble rosary beads, this symptom is commonly known as **rachitic** (ra-KIT-ik) **rosary** ("the rosary of rickets").

Osteomalacia In adults, the poor mineralization of bone results in the painful bone disease **osteomalacia.**[32] The bones become increasingly soft, flexible, brittle, and deformed.

rickets: the vitamin D–deficiency disease in children characterized by inadequate mineralization of bone (manifested in bowed legs or knock-knees, outward-bowed chest, and knobs on ribs). A rare type of rickets, not caused by vitamin D deficiency, is known as *vitamin D–refractory rickets.*

osteomalacia (OS-tee-oh-ma-LAY-shuh): a bone disease characterized by softening of the bones. Symptoms include bending of the spine and bowing of the legs. The disease occurs most often in adult women.
- **osteo** = bone
- **malacia** = softening

FIGURE 11-10 **Vitamin D–Deficiency Symptoms—Bowed Legs and Beaded Ribs of Rickets**

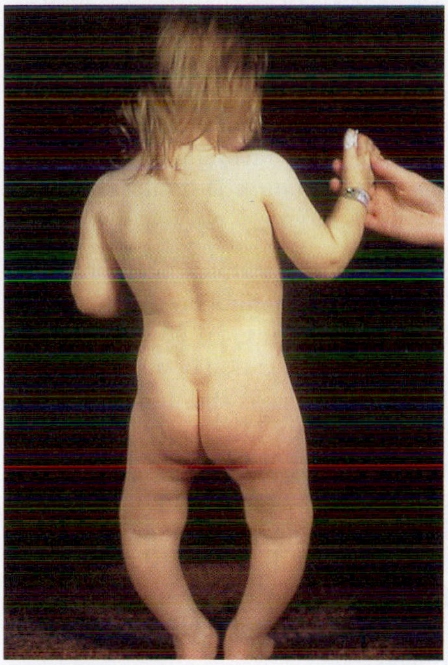

Biophoto Associates/Photo Researchers, Inc.

Bowed legs. In rickets, the poorly formed long bones of the legs bend outward as weight-bearing activities such as walking begin.

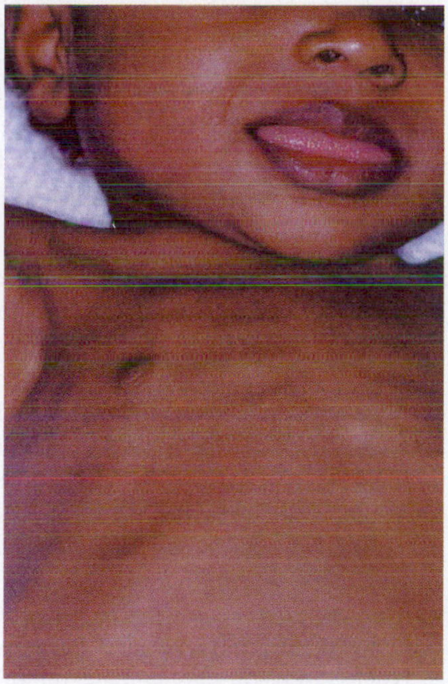

© Photo Courtesy of Dr. Normal Carvalho at Children's Healthcare of Atlanta

Beaded ribs. In rickets, a series of "beads" develop where the cartilages and bones attach.

Osteoporosis Any failure to synthesize adequate vitamin D or obtain enough from foods sets the stage for a loss of calcium from the bones, which can result in fractures. Highlight 12 describes the many factors that lead to osteoporosis, a condition of reduced bone density.

The Elderly Vitamin D deficiency is especially likely in older adults for several reasons. For one, the skin, liver, and kidneys lose their capacity to make and activate vitamin D with advancing age. For another, older adults typically drink little or no milk—the main dietary source of vitamin D. And finally, older adults typically spend much of the day indoors, and when they do venture outside, many of them cautiously wear protective clothing or apply sunscreen to all sun-exposed areas of their skin. Dark-skinned people living in northern regions are particularly vulnerable. All of these factors increase the likelihood of vitamin D deficiency and its consequences: bone losses and fractures. Vitamin D supplementation helps to raise blood levels, reduce bone loss, improve muscle performance, and lower the risks of falls and fractures in elderly persons.[33]

Vitamin D Toxicity

Vitamin D clearly illustrates how nutrients in optimal amounts support health, but both inadequacies and excesses cause trouble. Vitamin D is the most likely of the vitamins to have toxic effects when consumed in excessive amounts. The amounts of vitamin D made by the skin and found in foods are well within the safe limits set by the UL, but supplements containing the vitamin in concentrated form should be kept out of the reach of children and used cautiously, if at all, by adults.

Excess vitamin D raises the concentration of blood calcium.[34] ♦ Excess blood calcium tends to precipitate in the soft tissue, forming stones, especially in the kidneys where calcium is concentrated in an effort to excrete it. Calcification may

♦ High blood calcium is known as **hypercalcemia** and may develop from a variety of disorders, including vitamin D toxicity. It does *not* develop from a high calcium intake.

also harden the blood vessels and is especially dangerous in the major arteries of the brain, heart, and lungs, where it can cause death.

Vitamin D Recommendations and Sources

Only a few foods—notably oily fish and egg yolks—contain vitamin D naturally. Fortunately, the body can make vitamin D with the help of a little sunshine. In setting dietary recommendations, however, the DRI Committee assumed that no vitamin D was available from skin synthesis. In order to reach sufficient levels of vitamin D in the blood without contributions from the sun, DRI recommendations were recently increased.[35] ♦ Some research suggests that vitamin D intake should be higher still.[36]

♦ Vitamin D RDA for adults <71: 15 μg/day

Vitamin D in Foods Most adults, especially in sunny regions, need not make special efforts to obtain vitamin D from food. People who are not outdoors much or who live in northern or predominantly cloudy or smoggy areas are advised to drink at least 2 cups of vitamin D–fortified milk a day. The fortification of milk with vitamin D is the best guarantee that people will meet their needs and underscores the importance of milk in a well-balanced diet.* Despite vitamin D fortification, the average intake in the United States falls short of recommendations. Oily fish such as salmon, mackerel, and sardines are the best natural sources of vitamin D.

Meeting vitamin D needs is difficult without adequate sunshine, fortification, or supplementation.[37] Vegetarians who do not include milk in their diets may use vitamin D–fortified soy milk and cereals. Importantly, feeding infants and young children nonfortified "health beverages" instead of milk or infant formula can create severe nutrient deficiencies, including rickets.

© Westend61 GmbH/Alamy

A cold glass of milk refreshes as it replenishes vitamin D and other bone-building nutrients.

Vitamin D from the Sun Most of the world's population relies on natural exposure to sunlight to maintain adequate vitamin D nutrition. The sun imposes no risk of vitamin D toxicity; prolonged exposure to sunlight degrades the vitamin D precursor in the skin, preventing its conversion to the active vitamin. Even lifeguards on southern beaches are safe from vitamin D toxicity from the sun.

Prolonged exposure to sunlight can, however, prematurely wrinkle the skin and cause skin cancer. Sunscreens help reduce these risks, but sunscreens with a sun protection factor (SPF) of 8 and higher can also reduce vitamin D synthesis. Still, even with an SPF 15–30 sunscreen, sufficient vitamin D synthesis can be obtained in 10 to 20 minutes of sun exposure.[38] Alternatively, a person could apply sunscreen after enough time has elapsed to provide sufficient vitamin D synthesis. For most people, exposing hands, face, and arms on a clear summer day for 5 to 10 minutes two or three times a week should be sufficient to maintain vitamin D nutrition.

The pigments of dark skin provide some protection from the sun's damage, but they also reduce vitamin D synthesis. Dark-skinned people require longer sunlight exposure than light-skinned people: heavily pigmented skin achieves the same amount of vitamin D synthesis in three hours as fair skin in a half hour. Latitude, season, and time of day ♦ also have dramatic effects on vitamin D synthesis and status (see Figure 11-11).[39] Heavy clouds, smoke, or smog block the ultraviolet (UV) rays of the sun that promote vitamin D synthesis. Differences in skin pigmentation, latitude, and smog may account for the finding that African American people, especially those in northern, smoggy cities, are most likely to be vitamin D deficient and develop rickets. Vitamin D deficiency is especially prevalent in the winter.[40] To ensure an adequate vitamin D status, supplements may be needed.[41] The body's vitamin D supplies from summer synthesis alone are insufficient to meet winter needs.

♦ Factors that may limit sun exposure and, therefore, vitamin D synthesis:
- Geographic location
- Season of the year
- Time of day
- Air pollution
- Clothing
- Tall buildings
- Indoor living
- Sunscreens

Dietary Guidelines for Americans 2010

Choose foods that provide more vitamin D, which is a nutrient of concern in American diets. These foods include fortified milk and milk products.

*Vitamin D fortification of milk in the United States is 10 micrograms cholecalciferol (400 IU) per quart; in Canada, 9 to 12 micrograms (350 to 470 IU) per liter, with a current proposal to raise it slightly.

FIGURE 11-11 Vitamin D Synthesis and Latitude

Above 40° north latitude (and below 40° south latitude in the southern hemisphere), vitamin D synthesis essentially ceases for the four months of winter. Synthesis increases as spring approaches, peaks in summer, and declines again in the fall. People living in regions of extreme northern (or extreme southern) latitudes may miss as much as six months of vitamin D production.

Benis Arapovic/Shutterstock.com

The sunshine vitamin—vitamin D.

Depending on the radiation used, the UV rays from tanning lamps and tanning beds may also stimulate vitamin D synthesis and increase bone density. The potential hazards of skin damage, however, may outweigh any possible benefits.* The Food and Drug Administration (FDA) warns that if the lamps are not properly filtered, people using tanning booths risk burns, damage to the eyes and blood vessels, and skin cancer.

IN SUMMARY **Vitamin D can be synthesized in the body with the help of sunlight or obtained from fortified milk. It sends signals to three primary target sites: the GI tract to absorb more calcium and phosphorus, the bones to release more, and the kidneys to retain more. These actions maintain blood calcium concentrations and support bone formation. A deficiency causes rickets in childhood and osteomalacia in later life. The accompanying table provides a summary of vitamin D.**

Vitamin D

Other Names

- **ergocalciferol (vitamin D_2):** vitamin D derived from plants in the diet and made from the yeast and plant sterol ergosterol.
- **cholecalciferol (vitamin D_3 or calciol):** vitamin D derived from animals in the diet or made in the skin from 7-dehydrocholesterol, a precursor of cholesterol, with the help of sunlight.
- **calcidiol (25-hydroxyvitamin D):** vitamin D found in the blood that is made from the hydroxylation of cholecalciferol in the liver.
- **calcitriol (1,25-dihydroxyvitamin D):** vitamin D that is made from the hydroxylation of calcidiol in the kidneys; the biologically active hormone, sometimes called **active vitamin D.**

2011 RDA

Adults:	15 μg/day (19–70 yr)
	20 μg/day (>70 yr)

Upper Level

Adults:	50 μg/day

Chief Functions in the Body

Mineralization of bones (raises blood calcium and phosphorus by increasing absorption from digestive tract, withdrawing calcium from bones, stimulating retention by kidneys)

*The best wavelengths for vitamin D synthesis are UV-B rays between 290 and 310 nanometers. Some tanning parlors advertise "UV-A rays only, for a tan without the burn," but UV-A rays can damage the skin.

Significant Sources

Synthesized in the body with the help of sunlight; fortified milk, margarine, butter, juices, cereals, and chocolate mixes; veal, beef, egg yolks, liver, fatty fish (herring, salmon, sardines) and their oils

Deficiency Diseases

Rickets, osteomalacia

Deficiency Symptoms

Rickets in Children

Inadequate calcification, resulting in misshapen bones (bowing of legs); enlargement of ends of long bones (knees, wrists); deformities of ribs (bowed, with beads or knobs);[a] delayed closing of fontanel, resulting in rapid enlargement of head (see figure below); lax muscles resulting in protrusion of abdomen; muscle spasms

Osteomalacia or Osteoporosis in Adults

Loss of calcium, resulting in soft, flexible, brittle, and deformed bones; progressive weakness; pain in pelvis, lower back, and legs

Toxicity Disease

Hypervitaminosis D

Toxicity Symptoms

Elevated blood calcium; calcification of soft tissues (blood vessels, kidneys, heart, lungs, tissues around joints)

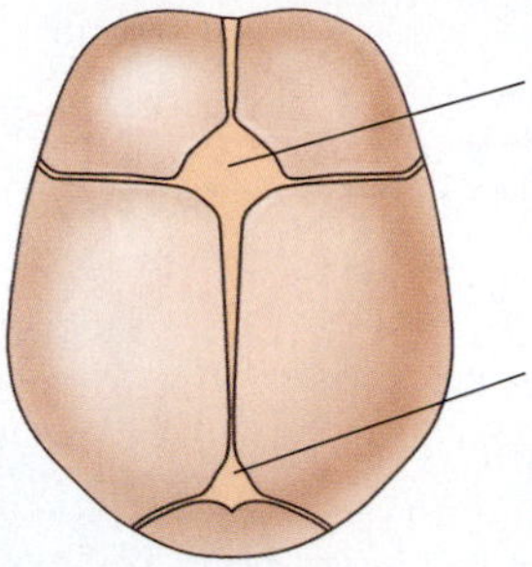

Fontanel
A fontanel is an open space in the top of a baby's skull before the bones have grown together. In rickets, closing of the fontanel is delayed.
Anterior fontanel normally closes by the end of the second year.
Posterior fontanel normally closes by the end of the first year.

[a]Bowing of the ribs causes the symptoms known as *pigeon breast*. The beads that form on the ribs resemble rosary beads; thus this symptom is known as *rachitic* (ra-KIT-ik) *rosary* ("the rosary of rickets").

Vitamin E

Researchers discovered a component of vegetable oils necessary for reproduction in rats and named this antisterility factor **tocopherol**, which means "to bring forth offspring." When chemists isolated four different tocopherol compounds, they designated them by the first four letters of the Greek alphabet: alpha, beta, gamma, and delta. The tocopherols consist of a complex ring structure and a long saturated side chain. (Appendix C provides the chemical structures.) The positions of methyl groups (CH_3) on the side chain and their chemical rotations distinguish one tocopherol from another. **Alpha-tocopherol** is the only one with vitamin E activity in the human body.[42] The other tocopherols are not readily converted to alpha-tocopherol in the body, nor do they perform the same roles. Whether these other tocopherols might be beneficial in other ways is the subject of current research.[43]

Vitamin E as an Antioxidant

Vitamin E is a fat-soluble antioxidant ♦ and one of the body's primary defenders against the adverse effects of free radicals. Its main action is to stop the chain reaction of free radicals producing more free radicals (see Highlight 11). In doing so, vitamin E protects the vulnerable components of the cells and their membranes from destruction. Most notably, vitamin E prevents the oxidation of the polyunsaturated fatty acids, but it protects other lipids and related compounds (for example, vitamin A) as well.

Accumulating evidence suggests that vitamin E may reduce the risk of heart disease by protecting low-density lipoproteins (LDL) against oxidation and reducing inflammation.[44] The oxidation of LDL and inflammation have been implicated as key factors in the development of heart disease. Highlight 11 explains how vitamin E and other antioxidants might protect against chronic diseases, such as heart disease and cancer, and explores whether foods or supplements might be most helpful—or harmful.

♦ Key antioxidant nutrients:
- Vitamin C, vitamin E, beta-carotene
- Selenium

tocopherol (tuh-KOFF-er-ol): a general term for several chemically related compounds, one of which has vitamin E activity. (See Appendix C for chemical structures.)

alpha-tocopherol: the active vitamin E compound.

Vitamin E Deficiency

A primary deficiency of vitamin E (from poor dietary intake) is rare; deficiency is usually associated with diseases of fat malabsorption such as cystic fibrosis. Without vitamin E, the red blood cells break open and spill their contents, probably due to oxidation of the polyunsaturated fatty acids in their membranes. This classic sign of vitamin E deficiency, known as **erythrocyte hemolysis**, is seen in premature infants, born before the transfer of vitamin E from the mother to the infant that takes place in the last weeks of pregnancy. Vitamin E treatment corrects **hemolytic anemia.**

Prolonged vitamin E deficiency also causes neuromuscular dysfunction involving the spinal cord and retina of the eye. Common symptoms include loss of muscle coordination and reflexes and impaired vision and speech. Vitamin E treatment corrects these neurological symptoms of vitamin E deficiency.

Two other conditions seem to respond to vitamin E treatment, although results are inconsistent. One is **fibrocystic breast disease**, a nonmalignant breast disease. The other is **intermittent claudication**, an abnormality of blood flow that causes cramping in the legs.

Vitamin E Toxicity

Vitamin E supplement use has risen in recent years as its protective actions against chronic diseases have been recognized. Fortunately, the liver carefully regulates vitamin E concentrations.[45] Toxicity is rare, and vitamin E appears safe across a broad range of intakes.[46] The UL for vitamin E (1000 milligrams) is more than 65 times greater than the recommended intake for adults (15 milligrams). Extremely high doses of vitamin E may interfere with the blood-clotting action of vitamin K and enhance the effects of drugs used to oppose blood clotting, causing hemorrhage. Additional research is needed to determine whether vitamin E supplements increase the risk of hemorrhagic stroke.[47]

Fat-soluble vitamin E is found predominantly in vegetable oils, seeds, and nuts.

Vitamin E Recommendations

The current RDA for vitamin E is based on the alpha-tocopherol form only. As mentioned earlier, the other tocopherols cannot be converted to alpha-tocopherol, nor can they perform the same metabolic roles in the body. A person who consumes large quantities of polyunsaturated fatty acids needs more vitamin E. Fortunately, vitamin E and polyunsaturated fatty acids tend to occur together in the same foods. Current research suggests that most adults in the United States fall short of recommended intakes for vitamin E and that smokers may have a higher requirement.[48]

Vitamin E in Foods

Vitamin E is widespread in foods. Much of the vitamin E in the diet comes from vegetable oils and products made from them, such as margarine and salad dressings. Wheat germ oil is especially rich in vitamin E.

Because vitamin E is readily destroyed by heat processing (such as deep-fat frying) and oxidation, fresh or lightly processed foods are preferable sources. Most processed and convenience foods do not contribute enough vitamin E to ensure an adequate intake.

Prior to 2000, values of the vitamin E in foods reflected all of the tocopherols and were expressed in "milligrams of tocopherol equivalents." ♦ These measures overestimated the amount of alpha-tocopherol. To estimate the alpha-tocopherol content of foods stated in tocopherol equivalents, multiply by 0.8.[49]

IN SUMMARY **Vitamin E acts as an antioxidant, defending lipids and other components of the cells against oxidative damage. Deficiencies are rare, but they do occur in premature infants, the primary symptom being erythrocyte hemolysis. Vitamin E is found predominantly in vegetable oils and appears to be one of the least toxic of the fat-soluble vitamins. The accompanying table (p. 370) provides a summary of vitamin E.**

♦ Appendix H accurately presents vitamin E data in milligrams of alpha-tocopherol.

erythrocyte (eh-RITH-ro-cite) **hemolysis** (he-MOLL-uh-sis): the breaking open of red blood cells (erythrocytes); a symptom of vitamin E–deficiency disease in human beings.

- **erythro** = red
- **cyte** = cell
- **hemo** = blood
- **lysis** = breaking

hemolytic (HE-moh-LIT-ick) **anemia:** the condition of having too few red blood cells as a result of erythrocyte hemolysis.

fibrocystic (FYE-bro-SIS-tik) **breast disease:** a harmless condition in which the breasts develop lumps, sometimes associated with caffeine consumption. In some, it responds to abstinence from caffeine; in others, it can be treated with vitamin E.

- **fibro** = fibrous tissue
- **cyst** = closed sac

intermittent claudication (klaw-dih-KAY-shun): severe calf pain caused by inadequate blood supply. It occurs when walking and subsides during rest.

- **intermittent** = at intervals
- **claudicare** = to limp

Vitamin E

Other Names

Alpha-tocopherol

2000 RDA

Adults: 15 mg/day

Upper Level

Adults: 1000 mg/day

Chief Functions in the Body

Antioxidant (stabilization of cell membranes, regulation of oxidation reactions, protection of polyunsaturated fatty acids [PUFA] and vitamin A)

Significant Sources

Polyunsaturated plant oils (margarine, salad dressings, shortenings), leafy green vegetables (spinach, turnip greens, collard greens, broccoli), wheat germ, whole grains, liver, egg yolks, nuts, seeds, fatty meats

Easily destroyed by heat and oxygen

Deficiency Symptoms

Red blood cell breakage,[a] nerve damage

Toxicity Symptoms

Augments the effects of anticlotting medication

[a]The breaking of red blood cells is called *erythrocyte hemolysis*.

Vitamin K

Like vitamin D, vitamin K can be obtained from a nonfood source. Bacteria in the GI tract synthesize vitamin K that the body can absorb. Vitamin K ♦ acts primarily in blood clotting, where its presence can make the difference between life and death. Blood has a remarkable ability to remain liquid, but it can clot within seconds when the integrity of that system is disturbed.

♦ *K* stands for the Danish word *koagulation* ("coagulation" or "clotting").

Roles in the Body

More than a dozen different proteins and the mineral calcium are involved in making a blood clot. Vitamin K is essential for the activation of several of these proteins, among them prothrombin, made by the liver as a precursor of the protein thrombin (see Figure 11-12). When any of the blood-clotting factors is lacking, **hemorrhagic disease** results. If an artery or vein is cut or broken, bleeding goes unchecked. Of course, this is not to say that hemorrhaging is always caused by vitamin K deficiency. Another cause is the hereditary disorder **hemophilia**, ♦ which is not curable with vitamin K.

♦ Hemophilia is caused by a genetic defect and has no relation to vitamin K.

Vitamin K also participates in the metabolism of bone proteins, most notably **osteocalcin**. Without vitamin K, osteocalcin cannot bind to the minerals that nor-

FIGURE 11-12 Blood-Clotting Process

When blood is exposed to air, foreign substances, or secretions from injured tissues, platelets (small, cell-like structures in the blood) release a phospholipid known as thromboplastin. Thromboplastin catalyzes the conversion of the inactive protein prothrombin to the active enzyme thrombin. Thrombin then catalyzes the conversion of the precursor protein fibrinogen to the active protein fibrin that forms the clot.

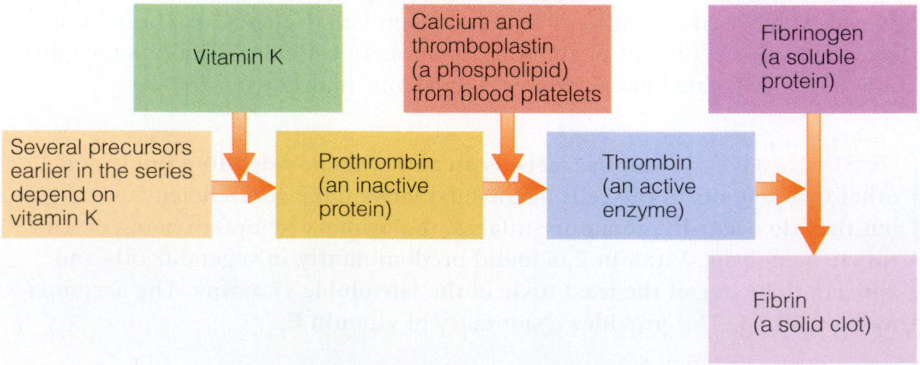

hemorrhagic (hem-oh-RAJ-ik) **disease:** a disease characterized by excessive bleeding.

hemophilia (HE-moh-FEEL-ee-ah): a hereditary disease in which the blood is unable to clot because it lacks the ability to synthesize certain clotting factors.

osteocalcin (os-teo-KAL-sen): a calcium-binding protein in bones, essential for normal mineralization.

mally form bones, resulting in low bone density.*[50] An adequate intake of vitamin K helps to decrease bone turnover and protect against fractures.[51] The effectiveness of vitamin K supplements on bone health is inconclusive.[52]

Vitamin K is historically known for its role in blood clotting, and more recently for its participation in bone building, but researchers continue to discover proteins needing vitamin K's assistance.[53] These proteins have been identified in the plaques of atherosclerosis, the kidneys, and the nervous system.

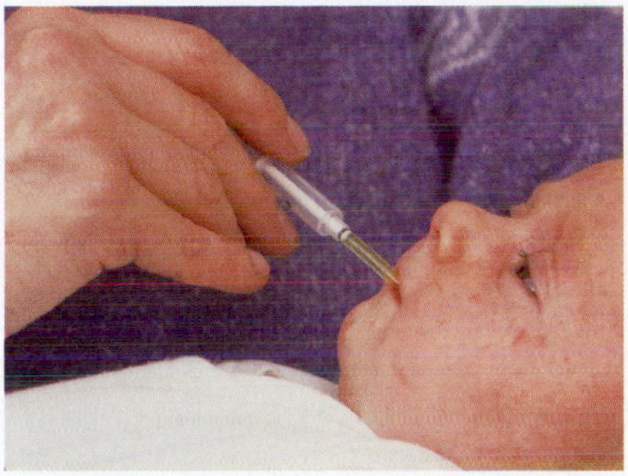

© Simon Fraser/Photo Researchers, Inc.

Soon after birth, newborn infants receive a dose of vitamin K to prevent hemorrhagic disease.

Vitamin K Deficiency

A primary deficiency ♦ of vitamin K is rare, but a secondary deficiency may occur in two circumstances. First, whenever fat absorption falters, as occurs when bile production fails, vitamin K absorption diminishes. Second, some drugs disrupt vitamin K's synthesis and action in the body: antibiotics kill the vitamin K–producing bacteria in the intestine, and anticoagulant drugs interfere with vitamin K metabolism and activity. Excessive bleeding due to a vitamin K deficiency can be fatal.

♦ A **primary deficiency** develops in response to an inadequate dietary intake whereas a **secondary deficiency** occurs for other reasons.

Newborn infants present a unique case of vitamin K nutrition because they are born with a **sterile** intestinal tract, and the vitamin K–producing bacteria take weeks to establish themselves. At the same time, plasma prothrombin concentrations are low. This reduces the likelihood of fatal blood clotting during the stress of birth. To prevent hemorrhagic disease in the newborn, a single dose of vitamin K ♦ is given at birth either orally or by intramuscular injection. Concerns that vitamin K given at birth raises the risks of childhood cancer are unproved and unlikely.

♦ Vitamin K is usually given in the naturally occurring form known as **phylloquinone** (FILL-oh-KWIN-own); the synthetic form of vitamin K is **menadione** (men-uh-DYE-own). See Appendix C for the chemistry of these structures.

Vitamin K Toxicity

Toxicity is not common, and no adverse effects have been reported with high intakes of vitamin K. Therefore, a UL has not been established. High doses of vitamin K can reduce the effectiveness of anticoagulant drugs used to prevent blood clotting.[54] People taking these drugs should eat vitamin K–rich foods in moderation and keep their intakes consistent from day to day.

Vitamin K Recommendations and Sources

As mentioned earlier, vitamin K is made in the GI tract by the billions of bacteria that normally reside there. Once synthesized, vitamin K is absorbed and stored in the liver. This source provides only about half of a person's needs. Vitamin K–rich foods such as green vegetables and vegetable oils can easily supply the rest.

© Matthew Farruggio

Notable food sources of vitamin K include green vegetables such as collards, spinach, bib lettuce, brussels sprouts, and cabbage and vegetable oils such as soybean oil and canola oil.

IN SUMMARY **Vitamin K helps with blood clotting, and its deficiency causes hemorrhagic disease (uncontrolled bleeding). Bacteria in the GI tract can make the vitamin; people typically receive about half of their requirements from bacterial synthesis and half from foods such as green vegetables and vegetable oils. Because people depend on bacterial synthesis for vitamin K, deficiency is most likely in newborn infants and in people taking antibiotics. The accompanying table provides a summary of vitamin K.**

Vitamin K

Other Names	Phylloquinone, menaquinone, menadione, naphthoquinone
2001 Adequate Intake (AI)	Men: 120 µg/day Women: 90 µg/day
Chief Functions in the Body	Synthesis of blood-clotting proteins and bone proteins
Significant Sources	Bacterial synthesis in the digestive tract;[a] liver; leafy green vegetables, cabbage-type vegetables; milk
Deficiency Symptoms	Hemorrhaging
Toxicity Symptoms	None known

[a]Vitamin K needs cannot be met from bacterial synthesis alone; however, it is a potentially important source in the small intestine, where absorption efficiency ranges from 40 to 70 percent.

*Vitamin K is a cofactor for a carboxylase enzyme. When vitamin K is inadequate, osteocalcin is undercarboxylated and therefore less effective in binding calcium.

sterile: free of microorganisms, such as bacteria.

The Fat-Soluble Vitamins—In Summary

The four fat-soluble vitamins play many specific roles in the growth and maintenance of the body. Their presence affects the health and function of the eyes, skin, GI tract, lungs, bones, teeth, nervous system, and blood; their deficiencies become apparent in these same areas. Toxicities of the fat-soluble vitamins are possible, especially when people use supplements, because the body stores excesses.

As with the water-soluble vitamins, the function of one fat-soluble vitamin often depends on the presence of another. Recall that vitamin E protects vitamin A from oxidation. In vitamin E deficiency, vitamin A absorption and storage are impaired. Three of the four fat-soluble vitamins—A, D, and K—play important roles in bone growth and remodeling. As mentioned, vitamin K helps synthesize a specific bone protein, and vitamin D regulates that synthesis. Vitamin A, in turn, may control which bone-building genes respond to vitamin D. Vitamin E interferes with vitamin K activity, especially blood clotting, but mechanisms for the interaction remain a mystery.[55]

Fat-soluble vitamins also interact with minerals. Vitamin D and calcium cooperate in bone formation, and zinc is required for the synthesis of vitamin A's transport protein, retinol-binding protein. Zinc also assists the enzyme that regenerates retinal from retinol in the eye. Vitamin A deficiency and iron deficiency often occur together, and each seems to worsen the other's metabolism.[56]

The roles of the fat-soluble vitamins differ from those of the water-soluble vitamins, and they appear in different foods—yet they are just as essential to life. The need for them underlines the importance of eating a wide variety of nourishing foods daily. The accompanying table provides a summary of the fat-soluble vitamins.

IN SUMMARY The Fat-Soluble Vitamins

Vitamin and Chief Functions	Deficiency Symptoms	Toxicity Symptoms	Significant Sources
Vitamin A Vision; maintenance of cornea, epithelial cells, mucous membranes, skin; bone and tooth growth; reproduction; immunity	Infectious diseases, night blindness, blindness (xerophthalmia), keratinization	Reduced bone mineral density, liver abnormalities, birth defects	Retinol: milk and milk products Beta-carotene: dark green leafy and deep yellow/orange vegetables
Vitamin D Mineralization of bones (raises blood calcium and phosphorus by increasing absorption from digestive tract, withdrawing calcium from bones, stimulating retention by kidneys)	Rickets, osteomalacia	Calcium imbalance (calcification of soft tissues and formation of stones)	Synthesized in the body with the help of sunshine; fortified milk
Vitamin E Antioxidant (stabilization of cell membranes, regulation of oxidation reactions, protection of polyunsaturated fatty acids [PUFA] and vitamin A)	Erythrocyte hemolysis, nerve damage	Hemorrhagic effects	Vegetable oils
Vitamin K Synthesis of blood-clotting proteins and bone proteins	Hemorrhage	None known	Synthesized in the body by GI bacteria; green leafy vegetables

Nutrition Portfolio

For the fat-soluble vitamins, select colorful fruits and vegetables, fortified milk or soy products, and vegetable oils; use supplements with caution, if at all. Go to Diet Analysis Plus and choose one of the days on which you tracked your diet for an entire day. Select the MyPyramid report and then consider the following questions:

- How was your overall intake in the vegetable group? Do you need improvement in this area? If so, what are some changes you could make?

Now look at the report titled Intake Spreadsheet to answer the following questions:

- Examine your weekly choices of vegetables and evaluate whether you meet the recommendations for dark green or orange and deep yellow vegetables.
- Consider whether you drink enough vitamin D–fortified milk or go outside in the sunshine regularly.
- Describe the vegetable oils you use when you cook and their vitamin contributions.

Diet Analysis PLUS

To complete this exercise, go to your Diet Analysis Plus at www.cengage.com/sso.

Nutrition on the Net

For further study of topics covered in this chapter, log on to **www.cengagebrain.com** and search for ISBN 084006845X.

- Search for "vitamins" at the American Dietetic Association: **www.eatright.org**
- Review the Dietary Reference Intakes for vitamins A, D, E, and K and the carotenoids by searching for "DRI": **www.nap.edu**
- Visit the World Health Organization to learn about "vitamin deficiencies" around the world: **www.who.int**
- Search for "vitamins" at the U.S. Government health information site: **www.healthfinder.gov**

References

1. E. H. Harrison, Mechanisms of digestion and absorption of dietary vitamin A, *Annual Review of Nutrition* 25 (2005): 87–103.
2. Harrison, 2005.
3. G. Wolf, Identification of a membrane receptor for retinol-binding protein functioning in the cellular uptake of retinol, *Nutrition Reviews* 65 (2007): 385–388.
4. G. Wolf, Retinoic acid as cause of cell proliferation or cell growth inhibition depending on activation of one of two different nuclear receptors, *Nutrition Reviews* 66 (2008): 55–59.
5. R. Blomhoff and H. K. Blomhoff, Overview of retinoid metabolism and function, *Journal of Neurobiology* 66 (2006): 606–630.
6. Committee on Dietary Reference Intakes, *Dietary Reference Intakes for Vitamin C, Vitamin E, Selenium, and Carotenoids* (Washington, D.C.: National Academies Press, 2000).
7. A. Sommer, Vitamin A deficiency and clinical disease: An historical overview, *Journal of Nutrition* 138 (2008): 1835–1839.
8. S. A. Abrams and D. C. Hilmers, Postnatal vitamin A supplementation in developing countries: An intervention whose time has come? *Pediatrics* 122 (2008): 180–181; K. Kraemer and coauthors, Are low tolerable upper intake levels for vitamin A undermining effective food fortification efforts? *Nutrition Reviews* 66 (2008): 517–525.
9. www.who.int/mediacentre/factsheets, f286, revised December 2008.
10. A. Sheth, R. Khurana, and V. Khurana, Potential liver damage associated with over-the-counter vitamin supplements, *Journal of the American Dietetic Association* 108 (2008): 1536–1537; K. L. Penniston and S. A. Tanumihardjo, The acute and chronic toxic effects of vitamin A, *American Journal of Clinical Nutrition* 83 (2006): 191–201.
11. J. D. Ribaya-Mercado and J. B. Blumberg, Vitamin A: Is it a risk factor for osteoporosis and bone fracture? *Nutrition Reviews* 65 (2007): 425–438; M. Kneissel and coauthors, Retinoid-induced bone thinning is caused by subperiosteal osteoclast activity in adult rodents, *Bone* 36 (2005): 202–214.
12. P. S. Genaro and L. A. Martini, Vitamin A supplementation and risk of skeletal fracture, *Nutrition Reviews* 62 (2004): 65–72.
13. H. A. Jackson and A. H. Sheehan, Effect of vitamin A on fracture risk, *The Annals of Pharmacotherapy* 39 (2005): 2086–2090.
14. J. Zhao and coauthors, Retinoic acid downregulates microRNAs to induce abnormal development of spinal cord in spina bifida rat model, *Child's Nervous System* 24 (2008): 485–492; S. Reijntjes and coauthors, The control of morphogen signaling: Regulation of the synthesis and catabolism of retinoic acid in the developing embryo, *Developmental Biology* 285 (2005): 224–237.
15. J. D. Ribaya-Mercado and coauthors, Carotene-rich plant foods ingested with minimal dietary fat enhance the total-body vitamin A pool size in Filipino schoolchildren as assessed by stable-isotope-dilution methodology, *American Journal of Clinical Nutrition* 85 (2007): 1041–1049.
16. P. Lips, Vitamin D physiology, *Progress in Biophysics and Molecular Biology* 92 (2006): 4–8.
17. H. F. Deluca, Evolution of our understanding of vitamin D, *Nutrition Reviews* 66 (2008): S73–S87; A. W. Norman, From vitamin D to hormone D: Fundamentals of the vitamin D endocrine system essential for good health, *American Journal of Clinical Nutrition* 88 (2008): 491S–499S.
18. R. P. Heaney, Vitamin D and calcium interactions: Functional outcomes, *American Journal of Clinical Nutrition* 88 (2008): 541S–544S.
19. Deluca, 2008; M. R. Haussler and coauthors, Vitamin D receptor: Molecular signaling and actions of nutritional ligands in disease prevention, *Nutrition Reviews* 66 (2008): S98–S112.
20. R. C. Khanal and I. Nemere, Regulation of intestinal calcium transport, *Annual Review of Nutrition* 28 (2008): 179–196.
21. A. W. Norman, Minireview: Vitamin D receptor—new assignments for an already busy receptor, *Endocrinology* 147 (2006): 5542–5548.
22. S. Samual and M. D. Sitrin, Vitamin D's role in cell proliferation and differentiation, *Nutrition Reviews* 66 (2008): S116–S124.
23. E. van Etten and coauthors, Regulation of vitamin D homeostasis: Implications for the immune system, *Nutrition Reviews* 66 (2008): S125–S134.
24. M. T. Cantorna, Vitamin D and multiple sclerosis: An update, *Nutrition Reviews* 66 (2008): S135–S138; M. F. Holick, Vitamin D: A d-lightful health perspective, *Nutrition Reviews* 66 (2008): S182–S194; S. E. Judd and coauthors, Optimal vitamin D status attenuates the age-associated increase in systolic blood pressure in white Americans: Results from the third National Health and Nutrition Examination Survey, *American*

Journal of Clinical Nutrition 87 (2008): 136–141; G. E. Mullin and A. Dobs, Vitamin D and its role in cancer and immunity: A prescription for sunlight, *Nutrition in Clinical Practice* 22 (2007): 305–322; Y. Cui and T. E. Rohan, Vitamin D, calcium, and breast cancer risk: A review, *Cancer Epidemiological, Biomarkers and Prevention* 15 (2006): 1427–1437; A. F. Gombart, Q. T. Luong, and H. P. Koeffler, Vitamin D compounds: Activity against microbes and cancer, *Anticancer Research* 26 (2006): 2531–2542; P. T. Liu and coauthors, Toll-like receptor triggering of a vitamin D–mediated human antimicrobial response, *Science* 311 (2006): 1770–1773; L. A. Martini and R. J. Wood, Vitamin D status and the metabolic syndrome, *Nutrition Reviews* 64 (2006): 479–486; K. L. Munger and coauthors, Serum 25-hydroxyvitamin D levels and risk of multiple sclerosis, *Journal of the American Medical Association* 296 (2006): 2832–2838; S. S. Schleithoff and coauthors, Vitamin D supplementation improves cytokine profiles in patients with congestive heart failure: A double-blind, randomized, placebo-controlled trial, *American Journal of Clinical Nutrition* 83 (2006): 754–759; T. Dietrich and coauthors, Association between serum concentrations of 25-hydroxyvitamin D and gingival inflammation, *American Journal of Clinical Nutrition* 82 (2005): 575–580.

25. A. C. Looker and coauthors, Serum 25-hydroxyvitamin D status of the US population: 1988–1994 compared with 2000–2004, *American Journal of Clinical Nutrition* 88 (2008): 1519–1527; S. A. Bowden and coauthors, Prevalence of vitamin D deficiency and insufficiency in children with osteopenia or osteoporosis referred to a pediatric metabolic bone clinic, *Pediatrics* 121 (2008): e1585–e1590; M. L. Neuhouser and coauthors, Vitamin D insufficiency in a multiethnic cohort of breast cancer survivors, *American Journal of Clinical Nutrition* 88 (2008): 133–139.
26. L. A. Martini and R. J. Wood, Vitamin D and blood pressure connection: Update on epidemiologic, clinical, and mechanistic evidence, *Nutrition Reviews* 66 (2008): 291–297; T. J. Wang and coauthors, Vitamin D deficiency and risk of cardiovascular disease, *Circulation* 117 (2008): 503–511; M. F. Holick, Vitamin D deficiency, *New England Journal of Medicine* 357 (2007): 266–281.
27. K. D. Cashman and coauthors, Low vitamin D status adversely affects bone health parameters in adolescents, *American Journal of Clinical Nutrition* 87 (2008): 1039–1044.
28. D. Schnadower and coauthors, Hypocalcemic seizures and secondary bilateral femoral fractures in an adolescent with primary vitamin D deficiency, *Pediatrics* 118 (2006): 2226–2230.
29. A. Prentice, Vitamin D deficiency: A global perspective, *Nutrition Reviews* 66 (2008): S153–S164.
30. S. Saintonge, H. Bang, and L. M. Gerber, Implications of a new definition of vitamin D deficiency in a multiracial US adolescent population: The National Health and Nutrition Examination Survey III, *Pediatrics* 123 (2009): 797–803; F. R. Greer, 25-Hydroxyvitamin D: Functional outcome in infants and young children, *American Journal of Clinical Nutrition* (2008): 529S–533S; S. Y. Huh and C. M. Gordon, Vitamin D deficiency in children and adolescents: Epidemiology, impact and treatment, *Reviews in Endocrine and Metabolic Disorders* 9 (2008): 161–170; M. Misra and coauthors, Vitamin D deficiency in children and its management: Review of current knowledge and recommendations, *Pediatrics* 122 (2008): 398–417.
31. C. L. Wagner, F. R. Greer, and the Section on Breastfeeding and Committee on Nutrition, Prevention of rickets and vitamin D deficiency in infants, children, and adolescents, *Pediatrics* 122 (2008): 1142–1152.
32. M. F. Holick, High prevalence of vitamin D inadequacy and implications for health, *Mayo Clinic Proceedings* 81 (2006): 353–373.
33. B. Dawson-Hughes, Serum 25-hydroxyvitamin D and functional outcomes in the elderly, *American Journal of Clinical Nutrition* 88 (2008): 537S–540S; S. A. Talwar and coauthors, Dose response to vitamin D supplementation among postmenopausal African American women, *American Journal of Clinical Nutrition* 86 (2007): 1657–1662; H. A. Bischoff-Ferrari and coauthors, Fracture prevention with vitamin D supplementation: A meta-analysis of randomized controlled trials, *Journal of the American Medical Association* 293 (2005): 2257–2264.
34. G. Jones, Pharmacokinetics of vitamin D toxicity, *American Journal of Clinical Nutrition* 88 (2008): 582S–586S.
35. Dietary Reference Intake Committee, *Dietary Reference Intakes for Calcium and Vitamin D* (Washington, D.C.: National Academies Press, 2011).
36. K. D. Cashman and coauthors, Estimation of the dietary requirement for vitamin D in healthy adults, *American Journal of Clinical Nutrition* 88 (2008): 1535–1542; J. F. Aloia and coauthors, Vitamin D intake to attain a desired serum 25-hydroxyvitamin D concentration, *American Journal of Clinical Nutrition* 87 (2008): 1952–1958.
37. A. Burgaz and coauthors, Associations of diet, supplement use, and ultraviolet B radiation exposure with vitamin D status in Swedish women during winter, *American Journal of Clinical Nutrition* 86 (2007): 1399–1404; R. M. van Dam and coauthors, Potentially modifiable determinants of vitamin D status in an older population in the Netherlands: The Hoorn Study, *American Journal of Clinical Nutrition* 85 (2007): 755–761.
38. B. A. Gilchrest, Sun exposure and vitamin D sufficiency, *American Journal of Clinical Nutrition* 88 (2008): 570S–577S.
39. M. J. Bolland and coauthors, The effects of seasonal variation of 25-hydroxyvitamin D and fat mass on a diagnosis of vitamin D sufficiency, *American Journal of Clinical Nutrition* 86 (2007): 959–964.
40. E. Hyppönen and C. Power, Hypovitaminosis D in British adults at age 45 y: Nationwide cohort study of dietary and lifestyle predictors, *American Journal of Clinical Nutrition* 85 (2007): 860–868; F. L. Weng and coauthors, Risk factors for low serum 25-hydroxyvitamin D concentrations in otherwise healthy children and adolescents, *American Journal of Clinical Nutrition* 86 (2007): 150–158.
41. L. Steingrimsdottir and coauthors, Relationship between serum parathyroid hormone levels, vitamin D sufficiency, and calcium intake, *Journal of the American Medical Association* 294 (2005): 2336–2341.
42. Committee on Dietary Reference Intakes, 2000.
43. S. Devaraj and I. Jialal, Failure of vitamin E in clinical trials: Is gamma-tocopherol the answer? *Nutrition Reviews* 63 (2005): 290–293; M. C. Morris and coauthors, Relation of the tocopherol forms to incident Alzheimer disease and to cognitive change, *American Journal of Clinical Nutrition* 81 (2005): 508–514.
44. D. L. Rainwater and coauthors, Vitamin E dietary supplementation significantly affects multiple risk factors for cardiovascular disease in baboons, *American Journal of Clinical Nutrition* 86 (2007): 597–603; U. Singh, S. Devaraj, and I. Jialal, Vitamin E, oxidative stress, and inflammation, *Annual Review of Nutrition* 25 (2005): 151–174.
45. M. G. Traber, Vitamin E regulatory mechanisms, *Annual Review of Nutrition* 27 (2007): 347–362.
46. J. N. Hathcock and coauthors, Vitamins E and C are safe across a broad range of intakes, *American Journal of Clinical Nutrition* 81 (2005): 736–745.
47. Committee on Dietary Reference Intakes, 2000, p. 252.
48. R. S. Bruno and coauthors, α-Tocopherol disappearance is faster in cigarette smokers and is inversely related to their ascorbic acid status, *American Journal of Clinical Nutrition* 81 (2005): 95–103.
49. Committee on Dietary Reference Intakes, 2000.
50. H. M. Macdonald and coauthors, Vitamin K_1 intake is associated with higher bone mineral density and reduced bone resorption in early postmenopausal Scottish women: No evidence of gene-nutrient interaction with apolipoprotein E polymorphisms, *American Journal of Clinical Nutrition* 87 (2008): 1513–1520; K. L. Berkner, The vitamin K-dependent carboxylase, *Annual Review of Nutrition* 25 (2005): 127–149.
51. S. Cockayne and coauthors, Vitamin K and the prevention of fractures: Systematic review and meta-analysis of randomized controlled trials, *Archives of Internal Medicine* 166 (2006): 1256–1261; J. Iwamoto, T. Takeda, and Y. Sato, Menatetrenone (vitamin K_2) and bone quality

in the treatment of postmenopausal osteoporosis, *Nutrition Reviews* 64 (2006): 509–517; K. D. Cashman, Vitamin K status may be an important determinant of childhood bone health, *Nutrition Reviews* 63 (2005): 284–293.
52. K. D. Cashman and E. O'Connor, Does high vitamin K_1 intake protect against bone loss in later life? *Nutrition Reviews* 66 (2008): 532–538; M. K. Shea and S. L. Booth, Update on the role of vitamin K in skeletal health, *Nutrition Reviews* 66 (2008): 549–557.
53. Berkner, 2005.
54. M. A. Johnson, Influence of vitamin K on anticoagulant therapy depends on vitamin K status and the source and chemical forms of vitamin K, *Nutrition Reviews* 63 (2005): 91–100.
55. M. G. Traber, Vitamin E and K interactions: A 50-year-old problem, *Nutrition Reviews* 66 (2008): 624–629.
56. J. M. Oliveira and coauthors, Influence of iron on vitamin A nutritional status, *Nutrition Reviews* 66 (2008): 141–147.

HIGHLIGHT 11

Antioxidant Nutrients in Disease Prevention

Count on supplement manufacturers to exploit the day's hot topics in nutrition. The moment bits of research news surface, new supplements appear—and terms like *antioxidants* and *lycopene* become household words. Friendly faces in TV commercials try to persuade us that these supplements hold the magic in the fight against aging and disease. New supplements hit the market and cash registers ring. Vitamin C, for years the leading single nutrient supplement, gains new popularity, and sales of lutein, beta-carotene, and vitamin E supplements soar as well.

In the meantime, scientists and medical experts around the world continue their work to clarify and confirm the roles of antioxidants in preventing chronic diseases. This highlight summarizes some of the accumulating evidence. It also revisits the advantages of foods over supplements. But first it is important to introduce the troublemakers—the **free radicals.** (The accompanying glossary defines free radicals and related terms.)

Free Radicals and Disease

Chapter 7 describes how the body's cells use oxygen in metabolic reactions. In the process, oxygen sometimes reacts with body compounds and produces highly unstable molecules known as free radicals. In addition to normal body processes, environmental factors such as ultraviolet radiation, air pollution, and tobacco smoke generate free radicals.

A free radical is a molecule with one or more unpaired electrons.* An electron without a partner is unstable and highly reactive. To regain its stability, the free radical quickly finds a stable but vulnerable compound from which to steal an electron.

With the loss of an electron, the formerly stable molecule becomes a free radical itself and steals an electron from another nearby molecule. Thus, an electron-snatching chain reaction is under way with free radicals producing more free radicals.

Antioxidants neutralize free radicals by donating one of their own electrons, thus ending the chain reaction. When they lose electrons, antioxidants do not become free radicals because they are stable in either form. (Review Figure 10-16 on p. 338 to see how ascorbic acid can give up two hydrogens with their electrons and become dehydroascorbic acid.)

Once formed, free radicals attack. Occasionally, these free-radical attacks are helpful. For example, cells of the immune system use free radicals as ammunition in an "oxidative burst" that demolishes disease-causing viruses and bacteria. Most often, however, free-radical attacks cause widespread damage. They commonly damage the polyunsaturated fatty acids in lipoproteins and in cell membranes, disrupting the transport of substances into and out of cells. Free radicals also alter DNA, RNA, and proteins, creating excesses and deficiencies of specific proteins, impairing cell functions, and eliciting an inflammatory response. All of these actions contribute to cell damage, disease progression, and aging (see Figure H11-1).

The body's natural defenses and repair systems try to control the destruction caused by free radicals, but these systems are not 100 percent effective. In fact, they become less effective with age, and the unrepaired damage accumulates. To some extent, dietary antioxidants defend the body against **oxidative stress,** but if antioxidants are unavailable or if free-radical production becomes excessive, health problems may develop.[1] Oxygen-derived free radicals may cause diseases, not only by indiscriminately destroying the valuable components of cells, but also by serving as signals for specific activities within the cells. Scientists have identified oxidative stress as a causative factor and antioxidants as a protective factor in cognitive performance and the aging process as well as in the development of diseases such as cancer, arthritis, cataracts, diabetes, and heart disease.

*Many free radicals exist, but oxygen-derived free radicals are most common in the human body. Examples of oxygen-derived free radicals include superoxide radical ($O_2^{\cdot -}$), hydroxyl radical ($OH^{\cdot}$), and nitric oxide ($NO^{\cdot}$). (The dots in the symbols represent the unpaired electrons.) Technically, hydrogen peroxide (H_2O_2) and singlet oxygen are not free radicals because they contain paired electrons, but the unstable conformation of their electrons makes radical-producing reactions likely. Scientists sometimes use the term *reactive oxygen species (ROS)* to describe all of these compounds.

GLOSSARY

antioxidants: substances typically found in foods that significantly decrease the adverse effects of free radicals on normal functions in the body.

free radicals: unstable and highly reactive atoms or molecules that have one or more unpaired electrons in the outer orbital. (See Appendix B for a review of basic chemistry concepts.)

oxidants (OKS-ih-dants): compounds (such as oxygen itself) that oxidize other compounds. Compounds that prevent oxidation are called *antioxidants,* whereas those that promote it are called *prooxidants.*

- **anti** = against
- **pro** = for

oxidative stress: a condition in which the production of oxidants and free radicals exceeds the body's ability to handle them and prevent damage.

phytochemicals: nonnutrient compounds found in plant-derived foods that have biological activity in the body.

prooxidants: substances that significantly induce oxidative stress.

Defending against Free Radicals

The body maintains a couple lines of defense against free-radical damage. A system of enzymes disarms the most harmful **oxidants.*** The action of these enzymes depends on the minerals selenium, copper, manganese, and zinc. If the diet fails to provide adequate supplies of these minerals, this line of defense weakens. The body also uses the antioxidant vitamins—vitamin E, beta-carotene, and vitamin C. Vitamin E defends the body's lipids (cell membranes and lipoproteins, for example) by efficiently stopping the free-radical chain reaction. Beta-carotene also acts as an antioxidant in lipid membranes. Vitamin C protects other tissues, such as the skin and fluid of the blood, against free-radical attacks.[2] Vitamin C seems especially adept at neutralizing free radicals from polluted air and cigarette smoke; it may also restore oxidized vitamin E to its active state.

Dietary antioxidants may also include some of the **phytochemicals** (featured in Highlight 13). Together, nutrients and phytochemicals with antioxidant activity minimize damage in the following ways:

- Limiting free-radical formation
- Destroying free radicals or their precursors
- Stimulating antioxidant enzyme activity
- Repairing oxidative damage
- Stimulating repair enzyme activity
- Supporting a healthy immune system[3]

These actions play key roles in defending the body against chronic diseases such as cancer and heart disease.

*These enzymes include glutathione peroxidase, thioredoxin reductase, superoxide dismutase, and catalase.

FIGURE H11-1 Free-Radical Damage

Free radicals are highly reactive. They might attack the polyunsaturated fatty acids in a cell membrane, which generates lipid radicals that damage cells and accelerate disease progression. Free radicals might also attack and damage DNA, RNA, and proteins, which interferes with the body's ability to maintain normal cell function, causing disease and premature aging.

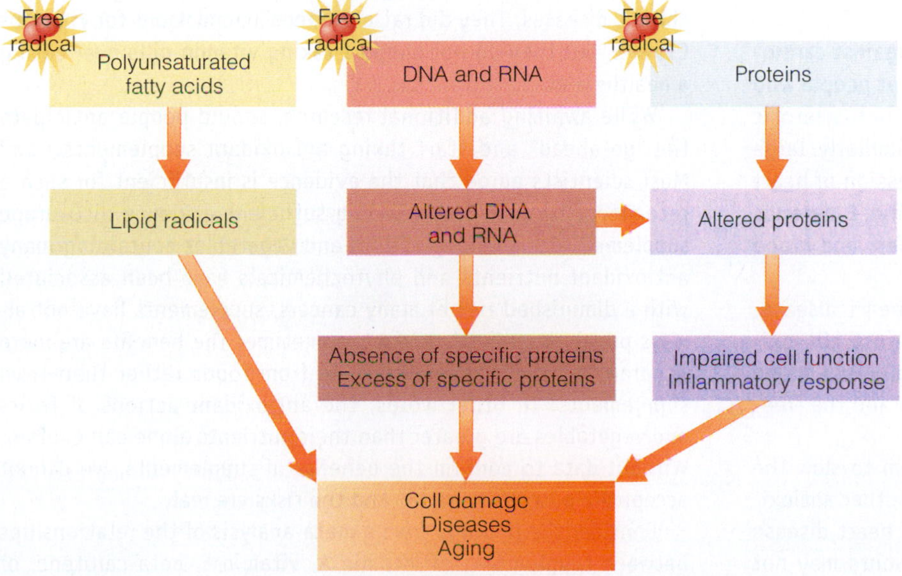

Defending against Cancer

Cancers arise when cellular DNA is damaged—sometimes by free-radical attacks. Antioxidants may reduce cancer risks by protecting DNA from this damage. Many researchers have reported low rates of cancer in people whose diets include abundant vegetables and fruits, rich in antioxidants.[4] Preliminary reports suggest an inverse relationship between DNA damage and vegetable intake and a positive relationship with beef and pork intake. Laboratory studies with animals and with cells in tissue culture also seem to support such findings.

Foods rich in vitamin C seem to protect against certain types of cancers, especially those of the esophagus. Such a correlation may reflect the benefits of a diet rich in fruits and vegetables and low in fat; it does not necessarily support taking vitamin C supplements to treat or prevent cancer. At high doses vitamin C acts as a **prooxidant,** generating free radicals. Limited research suggests this action may be useful in destroying cancer cells.[5]

Researchers hypothesize that vitamin E might inhibit cancer formation by attacking free radicals that damage DNA. Evidence that vitamin E helps guard against cancer, however, is contradictory and inconclusive.[6]

Several studies report a cancer-preventing benefit of vegetables and fruits rich in beta-carotene and the other carotenoids as well. Carotenoids seem to protect against oxidative damage to DNA.[7] High concentrations of beta-carotene are associated with a lower mortality from all causes and lower rates of some cancers.[8] These benefits may simply reflect a healthy diet abundant in fruits and vegetables.[9]

Defending against Heart Disease

High blood cholesterol carried in LDL is a major risk factor for cardiovascular disease, but how do LDL exert their damage? One scenario is that free radicals within the arterial walls oxidize LDL, changing their structure and function. The oxidized LDL then accelerate the formation of artery-clogging plaques. These free radicals also oxidize the polyunsaturated fatty acids of the cell membranes, sparking additional changes in the arterial walls, which impede the flow of blood. Susceptibility to such oxidative damage within the arterial walls is heightened by a diet high in saturated fat or cigarette smoke. In contrast, diets that include plenty of fruits and vegetables, especially when combined with little saturated fat, strengthen antioxidant defenses against LDL oxidation.

Antioxidants, especially vitamin E, may protect against cardiovascular disease. Epidemiological studies suggest that people who eat foods rich in vitamin E have relatively few atherosclerotic plaques and low rates of death from heart disease. Similarly, large doses of vitamin E supplements may slow the progression of heart disease.[10] Among its many protective roles, vitamin E defends against LDL oxidation, inflammation, arterial injuries, and blood clotting.[11]

Vitamin C supplements may reduce the risk of heart disease. Some studies suggest that vitamin C protects against LDL oxidation, raises HDL, lowers total cholesterol, and improves blood pressure. Vitamin C may also minimize inflammation and the free-radical action within the arterial wall.[12]

Antioxidant nutrients taken as supplements seem to slow the early progression of atherosclerosis. Less clear is whether antioxidant supplements benefit people who already have heart disease or multiple risk factors for it. Antioxidant supplements may not be beneficial and, in fact, may even be harmful for these people.[13]

Foods, Supplements, or Both?

In the process of scavenging and quenching free radicals, antioxidants themselves become oxidized. To some extent, they can be regenerated, but losses still occur and free radicals attack continuously. To maintain defenses, a person must replenish dietary antioxidants regularly. But should antioxidants be replenished from foods or from supplements?

Foods—especially fruits and vegetables—offer not only antioxidants, but an array of other valuable vitamins and minerals as well. Importantly, deficiencies of these nutrients can damage DNA as readily as free radicals can. Eating fruits and vegetables in abundance protects against both deficiencies and diseases—and is associated with reduced mortality.[14] A major review of the evidence gathered from metabolic studies, epidemiologic studies, and dietary intervention trials identified three dietary strategies most effective in preventing heart disease:

- Use unsaturated fats (that have not been hydrogenated) instead of saturated or *trans* fats (see Highlight 5).
- Select foods rich in omega-3 fatty acids (see Chapter 5).
- Consume a diet high in fruits, vegetables, nuts, and whole grains and low in refined grain products.

Such a diet combined with exercise, weight control, and not smoking serves as the best prescription for health. Notably, taking supplements is not among these disease-prevention recommendations.

Some research suggests a protective effect from as little as a daily glass of orange juice or carrot juice (rich sources of vitamin C and beta-carotene, respectively). Other intervention studies, however, have used levels of nutrients that far exceed current recommendations and can be achieved only by taking supplements. In making their recommendations for the antioxidant nutrients, members of the DRI Committee considered whether these studies support substantially higher intakes to help protect against chronic diseases. They did raise the recommendations for vitamins C and E, but they do not support taking vitamin pills over eating a healthy diet.

While awaiting additional research, should people anticipate the "go-ahead" and start taking antioxidant supplements now? Most scientists agree that the evidence is insufficient for such a recommendation.[15] Those finding sufficient evidence discourage supplement use.[16] Though fruits and vegetables containing many antioxidant nutrients and phytochemicals have been associated with a diminished risk of many cancers, supplements have not always proved beneficial. In fact, sometimes the benefits are more apparent when the vitamins come from foods rather than from supplements. In other words, the antioxidant actions of fruits and vegetables are greater than their nutrients alone can explain. Without data to confirm the benefits of supplements, we cannot accept the potential risks.[17] And the risks are real.

Consider the findings from a meta-analysis of the relationships between supplements of vitamin A, vitamin E, beta-carotene, or combinations and total mortality. Researchers concluded that supplements either had *no benefit* or *increased* mortality and should be avoided.[18] In fact, beta-carotene *enhances* the risk of lung cancer in smokers by increasing the formation of free radicals.[19]

Even if research clearly proves that a particular nutrient is the ultimate protective ingredient in foods, supplements would not be the answer because their contents are limited. Vitamin E supplements, for example, usually contain alpha-tocopherol, but foods provide an assortment of tocopherols among other nutrients, many of which provide valuable protection against free-radical damage. In addition to a full array of nutrients, foods provide phytochemicals that also fight against many diseases. Supplements shortchange users. Furthermore, supplements should be used only as an adjunct to other measures such as smoking cessation, weight control, physical activity, and medication as needed.

Clearly, much more research is needed to define optimal and dangerous levels of intake. This much we know: antioxidants behave differently under various conditions. At physiological levels typical of a healthy diet, they act as antioxidants, but at pharmacological doses typical of supplements, they may act as pro-oxidants, stimulating the production of free radicals and altering metabolism in a way that may promote disease. A high intake of vitamin C from supplements, for example, may *increase* the risk of heart disease in women with diabetes. Until the optimum intake of antioxidant nutrients can be determined, the risks of supplement use remain unclear. Table H11-1 presents a summary of the relationships between antioxidants and chronic diseases—sorted by foods or supplements.[20] As you can see, many studies report either no effect or inconsistent results. Any decrease in risk is attributed to foods 9 out of 10 times. Any increase in risk is always from supplements, and often in smokers. Clearly, the best way to add antioxidants to the diet is to eat generous servings of fruits and vegetables daily.

TABLE H11-1 Antioxidants and Chronic Disease Risk

Antioxidant	Disease	Risk from Foods	Risk from Supplements
Vitamin C	Coronary heart disease	Inconsistent results	Inconsistent results
	Breast cancer	Inconsistent results	—
	Colorectal cancer	Inconsistent results	—
	Gastrointestinal cancer	—	Not known
	Lung cancer	No effect	Not known
Vitamin E	Coronary heart disease	Inconsistent results	No effect or possible increased risk
	Breast cancer	—	No effect
	Colorectal cancer	Inconsistent results	—
	Gastrointestinal cancer	—	No effect
	Lung cancer	No effect	No effect
	Prostate cancer	Decreased risk	Decreased risk in smokers
Beta-carotene	Coronary heart disease	Decreased risk	No effect in nonsmokers, increased risk in smokers
	Lung cancer	Inconsistent results	No effect in nonsmokers, increased risk in smokers
	Colorectal cancer	Decreased risk	—
	Gastrointestinal cancer	—	No effect
	Prostate cancer	No effect	—
Other carotenoids	Lung cancer	Decreased risk for beta-cryptoxanthin	—
	Colorectal cancer	Decreased risk	—
	Prostate cancer	Decreased risk for lycopene	—
Fruits and vegetables	Coronary heart disease	Decreased risk	
	Breast cancer	No effect	
	Colorectal cancer	Inconsistent results	
	Gastric and esophageal cancer	Decreased risk	
	Lung cancer	Decreased risk for fruits, no effect for vegetables	
	Prostate cancer	No effect	
Supplement containing a combination of antioxidants	Coronary heart disease		Possibly increased risk
	Gastroinestinal cancer		Possibly increased risk
	Lung cancer		No effect in nonsmokers, increased risk in smokers

SOURCE: Adapted from H. Verhagen and coauthors, The state of antioxidant affairs, *Nutrition Today* 41 (2006): 244–249.

It should be clear by now that we cannot know the identity and action of every chemical in every food. Even if we did, why create a supplement to replicate a food? Why not eat foods and enjoy the pleasure, nourishment, and health benefits they provide? The beneficial constituents in foods are widespread among plants. Among the fruits, pomegranates, berries, and citrus rank high in antioxidants; top antioxidant vegetables include kale, spinach, and brussels sprouts; millet and oats contain the most antioxidants among the grains; pinto beans and soybeans are the outstanding legumes; and walnuts outshine the other nuts. But don't try to single out one particular food for its "magical" nutrient, antioxidant, or phytochemical. Instead, eat a wide variety of fruits, vegetables, grains, legumes, and nuts every day—and get *all* the benefits these foods have to offer.

Oledjio/Shutterstock.com

Many cancer-fighting products are available now at your local produce counter.

References

1. H. Verhagen and coauthors, The state of antioxidant affairs, *Nutrition Today* 41 (2006): 244–250; A. J. McEligot, S. Yang, and F. L. Meyskens, Redox regulation by intrinsic species and extrinsic nutrients in normal and cancer cells, *Annual Review of Nutrition* 25 (2005): 261–295.
2. M. V. Catani and coauthors, Biological role of vitamin C in keratinocytes, *Nutrition Reviews* 63 (2005): 81–90.
3. A. L. Webb and E. Villamor, Update: Effects of antioxidant and non-antioxidant vitamin supplementation on immune function, *Nutrition Reviews* 65 (2007): 181–217.
4. D. P. Hayes, The protective role of fruits and vegetables against radiation-induced cancer, *Nutrition Reviews* 63 (2005): 303–311.
5. Q. Chen and coauthors, Pharmacologic doses of ascorbate act as a prooxidant and decrease growth of aggressive tumor xenografts in mice, *Proceedings of the National Academy of Sciences* 105 (2008): 11105–11109.
6. D. Q. Pham and R. Plakogiannis, Vitamin E supplementation in cardiovascular disease and cancer prevention: Part 1, *Annals of Pharmacotherapy* 39 (2005): 1870–1878.
7. X. Zhao and coauthors, Modification of lymphocyte DNA damage by carotenoid supplementation in postmenopausal women, *American Journal of Clinical Nutrition* 83 (2006): 163–169.
8. S. C. Larsson and coauthors, Vitamin A, retinol, and carotenoids and the risk of gastric cancer: A prospective cohort study, *American Journal of Clinical Nutrition* 85 (2007): 497–503; B. Buijsse and coauthors, Plasma carotene and α-tocopherol in relation to 10-y all-cause and cause-specific mortality in European elderly: The Survey in Europe on Nutrition and the Elderly, a Concerted Action (SENECA), *American Journal of Clinical Nutrition* 82 (2005): 879–886.
9. L. Gallicchio and coauthors, Carotenoids and the risk of developing lung cancer: A systematic review, *American Journal of Clinical Nutrition* 88 (2008): 372–383.
10. J. Zingg, A. Azzi, and M. Meydani, Genetic polymorphisms as determinants for disease-preventive effects of vitamin E, *Nutrition Reviews* 66 (2008): 406–414; S. Devaraj and coauthors, Effect of high-dose α-tocopherol supplementation on biomarkers of oxidative stress and inflammation and carotid atherosclerosis in patients with coronary artery disease, *American Journal of Clinical Nutrition* 86 (2007): 1392–1398.
11. U. Singh, S. Devaraj, and I. Jialal, Vitamin E, oxidative stress, and inflammation, *Annual Review of Nutrition* 25 (2005): 151–174.
12. S. G. Wannamethee and coauthors, Associations of vitamin C status, fruit and vegetable intakes, and markers of inflammation and hemostasis, *American Journal of Clinical Nutrition* 83 (2006): 567–574.
13. N. R. Cook and coauthors, A randomized factorial trial of vitamins C and E and beta carotene in the secondary prevention of cardiovascular events in women: Results from the Women's Antioxidant Cardiovascular Study, *Archives of Internal Medicine* 167 (2007): 1610–1618; The HOPE and HOPE-TOO Investigators, Effects of long-term vitamin E supplementation on cardiovascular events and cancer: A randomized controlled trial, *Journal of the American Medical Association* 293 (2005): 1338–1347.
14. A. Agudo and coauthors, Fruit and vegetable intakes, dietary antioxidant nutrients, and total mortality in Spanish adults: Findings from the Spanish cohort of the European Prospective Investigation into Cancer and Nutrition (EPIC-Spain), *American Journal of Clinical Nutrition* 85 (2007): 1634–1642.
15. H. Y. Huang and coauthors, The efficacy and safety of multivitamin and mineral supplement use to prevent cancer and chronic disease in adults: A systematic review for a National Institutes of Health state-of-the-science conference, *Annals of Internal Medicine* 145 (2006): 372–385.
16. D. Q. Pham and R. Plakogiannis, Vitamin E supplementation in cardiovascular disease and cancer prevention: Part 1, *Annals of Pharmacotherapy* 39 (2005): 1870–1878.
17. S. Hercberg, The history of β-Carotene and cancers: From observational to intervention studies. What lessons can be drawn for future research on polyphenols? *American Journal of Clinical Nutrition* 81 (2005): 218S–222S.
18. G. Bjelakovic and coauthors, Mortality in randomized trials of antioxidant supplements for primary and secondary prevention: Systematic review and meta-analysis, *Journal of the American Medical Association* 297 (2007): 842–857; I. Lee and coauthors, Vitamin E in the primary prevention of cardiovascular disease and cancer: The Women's Health Study: A randomized controlled trial, *Journal of the American Medical Association* 294 (2005): 56–65; E. R. Miller and coauthors, Meta-analysis: High-dosage vitamin E supplementation may increase all-cause mortality, *Annals of Internal Medicine* 142 (2005): 37–46.
19. Y. G. J. van Helden and coauthors, β-Carotene metabolites enhance inflammation-induced oxidative DNA damage in lung epithelial cells, *Free Radical Biology and Medicine* 46 (2009): 299–304.
20. Verhagen and coauthors, 2006.

CHAPTER

12

Nayashkova Olga/shutterstock.com

Nutrition in Your Life

What's your beverage of choice? If you said water, then congratulate yourself for recognizing its importance in maintaining your body's fluid balance. If you answered milk, then pat yourself on the back for taking good care of your bones. Without water, you would realize within days how vital it is to your survival. The consequences of a lack of milk (or other calcium-rich foods) are also dramatic, but may not become apparent for decades. Water, calcium, and all the other major minerals support fluid balance and bone health. Before getting too comfortable reading this chapter, pour yourself a glass of water or milk. Your body will thank you.

Throughout this chapter, the CourseMate icon indicates an opportunity for online self-study, linking you to activities to increase your understanding of chapter concepts. **www.cengagebrain.com** (search for ISBN 084006845X)

Water and the Major Minerals

CHAPTER OUTLINE

Water is an essential nutrient, more important to life than any of the others. The body needs more water each day than any other nutrient. Furthermore, you can survive only a few days without water, whereas a deficiency of the other nutrients may take weeks, months, or even years to develop.

This chapter begins with a look at water and the body's fluids. The body maintains an appropriate balance and distribution of fluids with the help of another class of nutrients—the minerals. In addition to introducing the minerals that help regulate body fluids, this chapter describes many of the other important functions minerals perform in the body.

Water and the Body Fluids

Water constitutes about 60 percent of an adult's body weight and a higher percentage of a child's (see Figure 1-1, p. 6). Because water makes up about three-fourths of the weight of lean tissue and less than one-fourth of the weight of fat, a person's body composition influences how much of the body's weight is water. The proportion of water is generally smaller in females, obese people, and the elderly because of their smaller proportion of lean tissue.

In the body, water is the fluid in which all life processes occur. The water in the body fluids:

- Carries nutrients and waste products throughout the body
- Maintains the structure of large molecules such as proteins and glycogen
- Participates in metabolic reactions
- Serves as the solvent for minerals, vitamins, amino acids, glucose, and many other small molecules so that they can participate in metabolic activities
- Acts as a lubricant and cushion around joints and inside the eyes, the spinal cord, and, in pregnancy, the amniotic sac surrounding the fetus in the womb
- Aids in the regulation of normal body temperature, as the evaporation of sweat from the skin removes excess heat from the body
- Maintains blood volume

Water is the most indispensable nutrient.

♦ Water balance: intake = output

To support these and other vital functions, the body actively maintains an appropriate **water balance.** ♦

♦ Fluids in the body:
- Intracellular (inside cells)
- Extracellular (outside cells)
  - Interstitial (between cells)
  - Intravascular (inside blood vessels)

Water Balance and Recommended Intakes

Every cell contains fluid of the exact composition that is best for that cell. Fluid inside cells is called **intracellular fluid**, whereas fluid outside cells is called **extracellular fluid.** ♦ The extracellular fluid that surrounds each cell is called **interstitial fluid.** Figure 12-1 illustrates a cell and its associated fluids. The composition of intercellular and extracellular fluids differ from one another. They continually lose and replace their components, yet the composition in each compartment remains remarkably constant under normal conditions. Because imbalances can be devastating, the body quickly responds by adjusting both water intake and excretion as needed. Consequently, the entire system of cells and fluids remains in a delicate, but controlled, state of homeostasis.

♦ The **hypothalamus** is a brain center that controls activities such as maintenance of water balance, regulation of body temperature, and control of appetite.

Water Intake

Thirst and satiety influence water intake, apparently in response to changes sensed by the mouth, hypothalamus, ♦ and nerves.[1] When water intake is inadequate, the blood becomes concentrated (having lost water but not the dissolved substances within it), the mouth becomes dry, and the hypothalamus initiates drinking behavior. When water intake is excessive, the stomach expands and stretch receptors send signals to stop drinking. Similar signals are sent from receptors in the heart as blood volume increases.

Thirst drives a person to seek water, but it lags behind the body's need. When too much water is lost from the body and not replaced, **dehydration** develops. A first sign of dehydration is thirst, the signal that the body has already lost some of its fluid. If a person is unable to obtain fluid or, as in many elderly people, fails to perceive the thirst message, the symptoms of dehydration may progress rapidly from thirst to weakness, exhaustion, and delirium—and end in death if not corrected (see Table 12-1). Dehydration may easily develop with either water deprivation or excessive water losses.

Water intoxication, on the other hand, is rare but can occur with excessive water ingestion and kidney disorders that reduce urine production. The symptoms may include confusion, convulsions, and even death in extreme cases. Excessive water ingestion (10 to 20 liters) within a few hours dilutes the sodium concentration of the blood and contributes to a dangerous condition known as **hyponatremia.** For this reason, guidelines suggest limiting fluid intake during times of heavy sweating to between 1 and 1.5 liters per hour.

♦ Water generated during metabolism is called **metabolic water.**

Water Sources

The obvious dietary sources of water are water itself and other beverages, but nearly all foods also contain water. Most fruits and vegetables contain up to 90 percent water, and many meats and cheeses contain at least 50 percent. See Table 12-2 for selected foods and Appendix H for many more. Also, water is produced as an end-product ♦ of condensation reactions and during the oxidation of energy-yielding nutrients. Recall from Chapter 7 that when the energy-yielding nutrients break down, their carbons and hydrogens combine with oxygen

water balance: the balance between water intake and output (losses).

intracellular fluid: fluid within the cells, usually high in potassium and phosphate. Intracellular fluid accounts for approximately two-thirds of the body's water.
- **intra** = within

extracellular fluid: fluid outside the cells. Extracellular fluid includes two main components—the interstitial fluid between cells and the intravascular fluid of plasma. Extracellular fluid accounts for approximately one-third of the body's water.
- **extra** = outside

interstitial (IN-ter-STISH-al) **fluid:** fluid between the cells (intercellular), usually high in sodium and chloride. Interstitial fluid is a large component of extracellular fluid.
- **inter** = in the midst, between

thirst: a conscious desire to drink.

dehydration: the condition in which body water output exceeds water input. Symptoms include thirst, dry skin and mucous membranes, rapid heartbeat, low blood pressure, and weakness.

water intoxication: the rare condition in which body water contents are too high in all body fluid compartments.

hyponatremia (HIGH-po-na-TREE-me-ah): a decreased concentration of sodium in the blood.

TABLE 12-1 Signs of Dehydration

Body Weight Lost (%)	Symptoms
1–2	Thirst, fatigue, weakness, vague discomfort, loss of appetite
3–4	Impaired physical performance, dry mouth, reduction in urine, flushed skin, impatience, apathy
5–6	Difficulty concentrating, headache, irritability, sleepiness, impaired temperature regulation, increased respiratory rate
7–10	Dizziness, spastic muscles, loss of balance, delirium, exhaustion, collapse

NOTE: The onset and severity of symptoms at various percentages of body weight lost depend on the activity, fitness level, degree of acclimation, temperature, and humidity. If not corrected, dehydration can lead to death.

TABLE 12-2 Percentage of Water in Selected Foods

100%	Water
90–99%	Fat-free milk, strawberries, watermelon, lettuce, cabbage, celery, spinach, broccoli
80–89%	Fruit juice, yogurt, apples, grapes, oranges, carrots
70–79%	Shrimp, bananas, corn, potatoes, avocados, cottage cheese, ricotta cheese
60–69%	Pasta, legumes, salmon, ice cream, chicken breast
50–59%	Ground beef, hot dogs, feta cheese
40–49%	Pizza
30–39%	Cheddar cheese, bagels, bread
20–29%	Pepperoni sausage, cake, biscuits
10–19%	Butter, margarine, raisins
1–9%	Crackers, cereals, pretzels, taco shells, peanut butter, nuts
0%	Oils, sugars

to yield carbon dioxide (CO_2) and water (H_2O). As Table 12-3 shows, the water derived daily from these three sources averages about 2½ liters (roughly 2½ quarts or 10½ cups).

Water Losses The body must excrete a minimum of about 500 milliliters (about 2 cups) of water each day ♦ as urine—enough to carry away the waste products generated by a day's metabolic activities. Above this amount, excretion adjusts to balance intake. If a person drinks more water, the kidneys excrete more urine, and the urine becomes more dilute. In addition to urine, water is lost from the lungs as vapor and from the skin as sweat; some is also lost in feces.* The amount of fluid lost from each source varies, depending on the environment (such as heat or humidity) and physical conditions (such as exercise or fever). On average, daily losses total about 2.5 liters. Table 12-3 shows how daily water losses and intakes balance; maintaining this balance requires healthy kidneys and an adequate intake of fluids.

Water Recommendations Because water needs vary depending on diet, activity, environmental temperature, and humidity, a general water requirement is difficult to establish. Recommendations ♦ are sometimes expressed in proportion to the amount of energy expended under average environmental conditions.[2] The recommended water intake for a person who expends 2000 kcalories a day, for example, is 2 to 3 liters of water (about 8 to 12 cups). This recommendation is in line with the Adequate Intake (AI) for *total* water set by the DRI Committee. ♦ Total water includes not only drinking water, but water in other beverages and in foods as well.

*Water lost from the lungs and skin accounts for almost one-half of the daily losses even when a person is not visibly perspiring; these losses are commonly referred to as *insensible water losses.*

TABLE 12-3 Water Balance

Water Sources	Amount (mL)	Water Losses	Amount (mL)
Liquids	550 to 1500	Kidneys (urine)	500 to 1400
Foods	700 to 1000	Skin (sweat)	450 to 900
Metabolic water	200 to 300	Lungs (breath)	350
		GI tract (feces)	150
Total	1450 to 2800	Total	1450 to 2800

NOTE: For perspective, 100 milliliters is a little less than ½ cup and 1000 milliliters is a little more than 1 quart (1 mL = 0.03 oz).

FIGURE 12-1 One Cell and Its Associated Fluids

Fluids are found within the cells (intracellular) or outside the cells (extracellular). Extracellular fluids include plasma (the fluid portion of blood in the intravascular spaces of blood vessels) and interstitial fluids (the tissue fluid that fills the intercellular spaces between the cells).

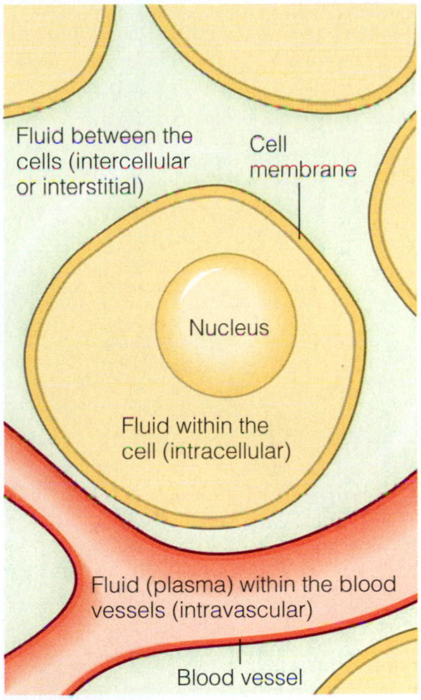

♦ The amount of water the body has to excrete each day to dispose of its wastes is the **obligatory** (ah-BLIG-ah-TORE-ee) **water excretion**—about 500 mL (about 2 c, or a pint).

♦ Water recommendation:
- 1.0 to 1.5 mL/kcal expended (adults)*
- 1.5 mL/kcal expended (infants and athletes)

Conversion factors:
- 1 mL = 0.03 fluid oz
- 125 mL ≈ ½ c

Easy estimation: ½ c per 100 kcal expended

♦ AI for *total* water:
- Men: 3.7 L/day
- Women: 2.7 L/day

Conversion factors:
- 1 L ≈ 1 qt ≈ 32 oz ≈ 4 c

*For those using kilojoules: 4.2 to 6.3 mL/kJ expended.

Because a wide range of water intakes will prevent dehydration and its harmful consequences, the AI is based on average intakes. People who are physically active or who live in hot environments may need more.[3]

Which beverages are best? Any beverage can readily meet the body's fluid needs, but those with few or no kcalories do so without contributing to weight gain. Given that obesity is a major health problem and that beverages currently represent more than 20 percent of the total energy intake in the United States, water is the best choice for most people. Other choices include tea, coffee, nonfat and low-fat milk and soy milk, artificially sweetened beverages, fruit and vegetable juices, sports drinks, and lastly, sweetened nutrient-poor beverages.[4]

Some research indicates that people who drink caffeinated beverages lose a little more fluid than when drinking water because caffeine acts as a diuretic. The DRI Committee considered such findings in their recommendations for water intake and concluded: "Caffeinated beverages contribute to the daily total water intake similar to that contributed by non-caffeinated beverages."[5] In other words, it doesn't seem to matter whether people rely on caffeine-containing beverages or other beverages to meet their fluid needs.

As Highlight 7 explains, alcohol acts as a diuretic and can impair a person's health. Alcohol should not be used to meet fluid needs.

Health Effects of Water In addition to meeting the body's fluid needs, drinking plenty of water may protect against urinary stones and constipation.[6] Even mild dehydration seems to interfere with daily tasks involving concentration, alertness, and short-term memory.[7]

The kind of water a person drinks may also make a difference to health. Water is usually either hard or soft. **Hard water** has high concentrations of calcium and magnesium; sodium or potassium is the principal mineral of **soft water.** (See the accompanying glossary for these and other common terms used to describe water.) In practical terms, soft water makes more bubbles with less soap; hard water leaves a ring on the tub, a crust of rocklike crystals in the teakettle, and a gray residue in the laundry.

Soft water may seem more desirable around the house, and some homeowners purchase water softeners that replace magnesium and calcium with sodium. In the body, however, soft water with sodium may aggravate hypertension and heart disease. In contrast, the minerals in hard water may benefit these conditions.

Soft water also more easily dissolves certain contaminant minerals, such as cadmium and lead, from old plumbing pipes. As Chapter 13 explains, these contaminant minerals harm the body by displacing the nutrient minerals from their normal sites of action. People who live in buildings with old plumbing should run the cold water tap a minute or two to flush out harmful minerals whenever the water faucet has been off for more than six hours.[8]

GLOSSARY OF TYPES OF WATER

artesian water: water drawn from a well that taps a confined aquifer in which the water is under pressure.

carbonated water: water that contains carbon dioxide gas, either naturally occurring or added, that causes bubbles to form in it; also called *bubbling* or *sparkling water.* Seltzer, soda, and tonic waters are legally soft drinks and are not regulated as water.

distilled water: water that has been vaporized and recondensed, leaving it free of dissolved minerals.

filtered water: water treated by filtration, usually through *activated carbon filters* that reduce the lead in tap water, or by *reverse osmosis* units that force pressurized water across a membrane removing lead, arsenic, and some microorganisms from tap water.

hard water: water with a high calcium and magnesium content.

mineral water: water from a spring or well that naturally contains at least 250 parts per million (ppm) of minerals. Minerals give water a distinctive flavor. Many mineral waters are high in sodium.

natural water: water obtained from a spring or well that is certified to be safe and sanitary. The mineral content may not be changed, but the water may be treated in other ways such as with ozone or by filtration.

public water: water from a municipal or county water system that has been treated and disinfected.

purified water: water that has been treated by distillation or other physical or chemical processes that remove dissolved solids. Because purified water contains no minerals or contaminants, it is useful for medical and research purposes.

soft water: water with a high sodium or potassium content.

spring water: water originating from an underground spring or well. It may be bubbly (carbonated), or "flat" or "still," meaning not carbonated. Brand names such as "Spring Pure" do not necessarily mean that the water comes from a spring.

well water: water drawn from groundwater by tapping into an aquifer.

IN SUMMARY Water makes up about 60 percent of the adult body's weight. It assists with the transport of nutrients and waste products throughout the body, participates in chemical reactions, acts as a solvent, serves as a shock absorber, and regulates body temperature. To maintain water balance, intake from liquids, foods, and metabolism must equal losses from the kidneys, skin, lungs, and GI tract. The amount and type of water a person drinks may have positive or negative health effects.

Blood Volume and Blood Pressure

Fluids maintain the blood volume, which in turn influences blood pressure. The kidneys are central to the regulation of blood volume and blood pressure.[9] All day, every day, the kidneys reabsorb needed substances and water and excrete wastes with some water in the urine (see Figure 12-2). The kidneys meticulously adjust the volume and the concentration of the urine to accommodate changes in the body, including variations in the day's food and beverage intakes. Instructions on whether to retain or release substances or water come from ADH, renin, angiotensin, and aldosterone.

ADH Whenever blood volume or blood pressure falls too low, or whenever the extracellular fluid becomes too concentrated, the hypothalamus signals the pituitary gland to release antidiuretic hormone (ADH). ♦ ADH is a water-conserving hormone ♦ that stimulates the kidneys to reabsorb water. Consequently, the more water you need, the less your kidneys excrete. These events also trigger thirst. Drinking water and retaining fluids raise the blood volume and dilute the concentrated fluids, thus helping to restore homeostasis.

Renin Cells in the kidneys respond to low blood pressure by releasing an enzyme called **renin**. Through a complex series of events, renin causes the kidneys

♦ **Antidiuretic hormone (ADH)** is a hormone produced by the pituitary gland in response to dehydration (or a high sodium concentration in the blood). It stimulates the kidneys to reabsorb more water and therefore to excrete less.

♦ Recall from Highlight 7 that alcohol depresses ADH activity, thus promoting fluid losses and dehydration. In addition to its antidiuretic effect, ADH elevates blood pressure and so is also called **vasopressin** (VAS-oh-PRES-in).
- **vaso** = vessel
- **press** = pressure

FIGURE 12-2 A Nephron, One of the Kidney's Many Functioning Units

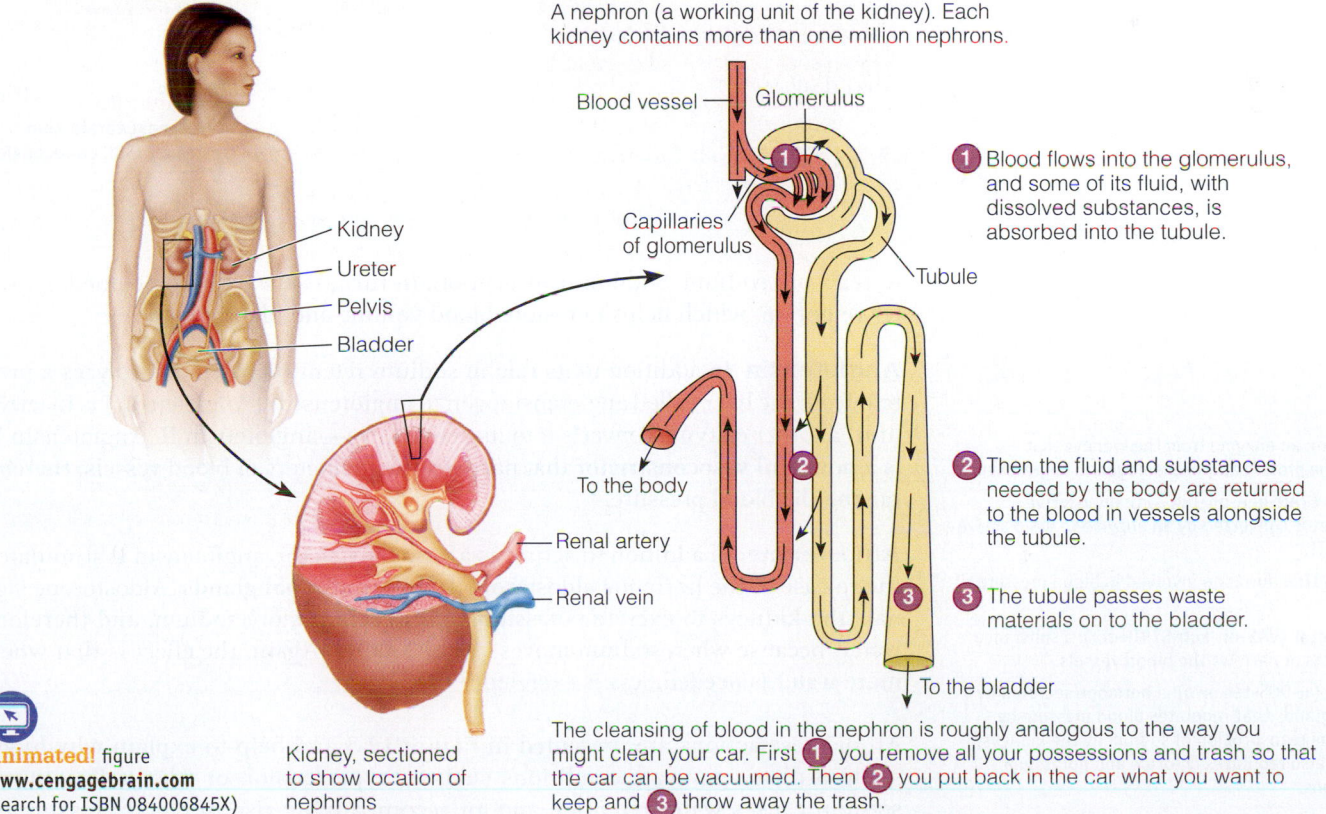

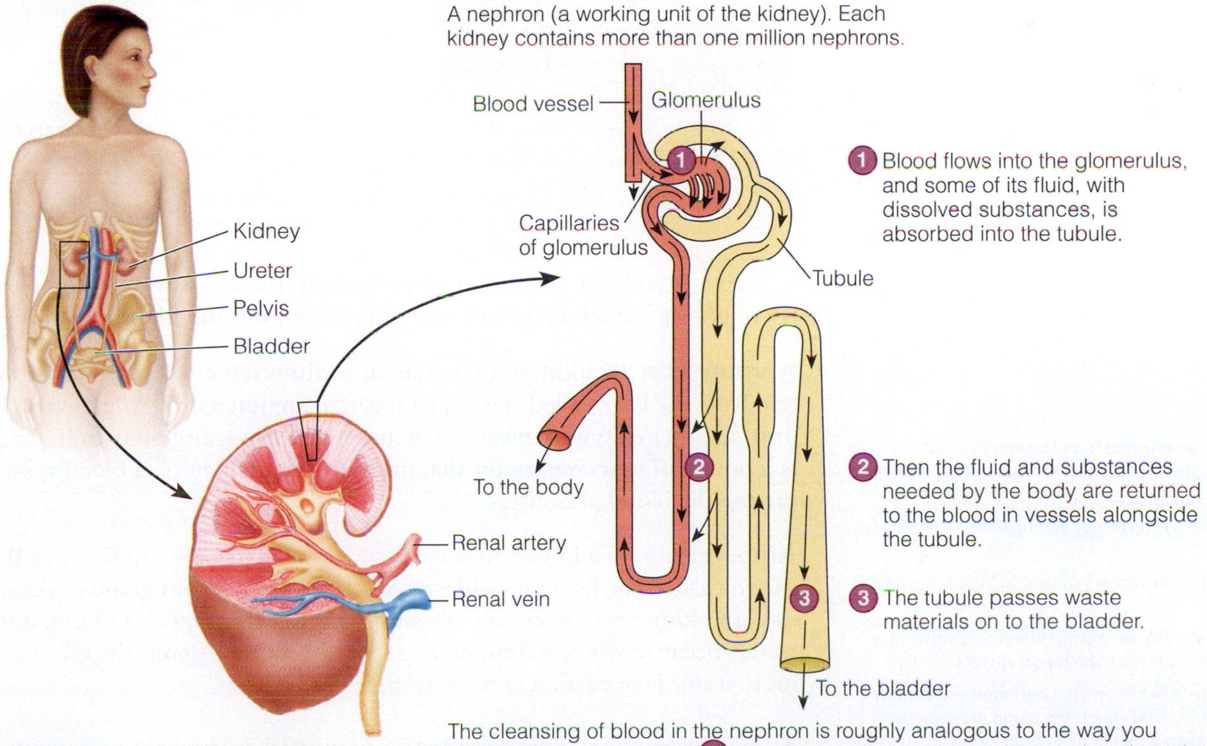
A nephron (a working unit of the kidney). Each kidney contains more than one million nephrons.
Blood vessel
Glomerulus
Capillaries of glomerulus
Tubule
Kidney
Ureter
Pelvis
Bladder
To the body
Renal artery
Renal vein
To the bladder
1 Blood flows into the glomerulus, and some of its fluid, with dissolved substances, is absorbed into the tubule.
2 Then the fluid and substances needed by the body are returned to the blood in vessels alongside the tubule.
3 The tubule passes waste materials on to the bladder.
Kidney, sectioned to show location of nephrons

The cleansing of blood in the nephron is roughly analogous to the way you might clean your car. First (1) you remove all your possessions and trash so that the car can be vacuumed. Then (2) you put back in the car what you want to keep and (3) throw away the trash.

Animated! figure
www.cengagebrain.com
(search for ISBN 084006845X)

FIGURE 12-3 How the Body Regulates Blood Volume

The renin-angiotensin-aldosterone system helps regulate blood volume and therefore blood pressure.

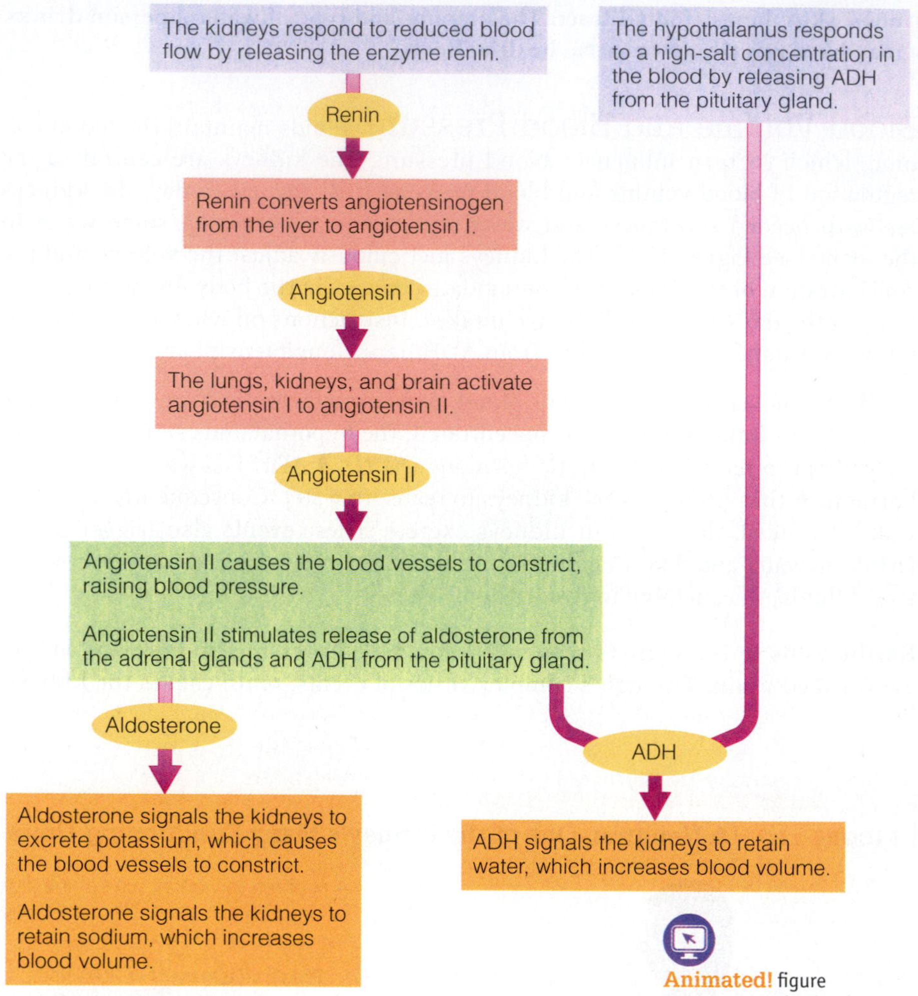

The kidneys respond to reduced blood flow by releasing the enzyme renin.
The hypothalamus responds to a high-salt concentration in the blood by releasing ADH from the pituitary gland.
Renin
Renin converts angiotensinogen from the liver to angiotensin I.
Angiotensin I
The lungs, kidneys, and brain activate angiotensin I to angiotensin II.
Angiotensin II
Angiotensin II causes the blood vessels to constrict, raising blood pressure.
Angiotensin II stimulates release of aldosterone from the adrenal glands and ADH from the pituitary gland.
Aldosterone
ADH
Aldosterone signals the kidneys to excrete potassium, which causes the blood vessels to constrict.
Aldosterone signals the kidneys to retain sodium, which increases blood volume.
ADH signals the kidneys to retain water, which increases blood volume.

Animated! figure
www.cengagebrain.com
(search for ISBN 084006845X)

to reabsorb sodium. Sodium reabsorption, in turn, is always accompanied by water retention, which helps to restore blood volume and blood pressure.

Angiotensin In addition to its role in sodium retention, renin hydrolyzes a protein from the liver called angiotensinogen to **angiotensin I**. Angiotensin I is inactive until another enzyme converts it to its active form—**angiotensin II**. Angiotensin II is a powerful **vasoconstrictor** that narrows the diameters of blood vessels, thereby raising the blood pressure.

Aldosterone In addition to acting as a vasoconstrictor, angiotensin II stimulates the release of the hormone **aldosterone** from the **adrenal glands**. Aldosterone signals the kidneys to excrete potassium and to retain more sodium, and therefore water, because when sodium moves, water follows. Again, the effect is that when more water is needed, less is excreted.

All of these actions are presented in Figure 12-3 and help to explain why high-sodium diets aggravate conditions such as hypertension or edema. Too much sodium causes water retention and an accompanying rise in blood pressure or swelling in the interstitial spaces. Chapter 27 discusses hypertension in detail.

renin (REN-in): an enzyme from the kidneys that hydrolyzes the protein angiotensinogen to angiotensin I.

angiotensin I (AN-gee-oh-TEN-sin): an inactive precursor that is converted by an enzyme to yield active angiotensin II.

angiotensin II: a hormone involved in blood pressure regulation.

vasoconstrictor (VAS-oh-kon-STRIK-tor): a substance that constricts or narrows the blood vessels.

aldosterone (al-DOS-ter-own): a hormone secreted by the adrenal glands that regulates blood pressure by increasing the reabsorption of sodium by the kidneys. Aldosterone also regulates chloride and potassium concentrations.

adrenal glands: glands adjacent to, and just above, each kidney.

IN SUMMARY The body responds to low blood volume, low blood pressure, or highly concentrated body fluids by producing:

- ADH, which stimulates the kidneys to reabsorb water.
- Renin, which initiates the pathway that leads to the production of angiotensin II.
- Angiotensin II, which constricts blood vessels and stimulates the release of aldosterone and ADH.
- Aldosterone, which regulates potassium and sodium levels.

All these actions combine to effectively restore homeostasis. Water balance can be maintained only if a person drinks enough water.

Fluid and Electrolyte Balance

Maintaining a balance of about two-thirds of the body fluids inside the cells and one-third outside is vital to the life of the cells. If too much water were to enter the cells, they might rupture; if too much water were to leave, they would collapse. To control the movement of water, the cells direct the movement of the major minerals. ♦

Dissociation of Salt in Water

When a mineral **salt** such as sodium chloride (NaCl) dissolves in water, it separates (**dissociates**) into **ions**—positively and negatively charged particles (Na^+ and Cl^-). The positive ions are **cations**; the negative ones are **anions**. ♦ Unlike pure water, which conducts electricity poorly, ions dissolved in water carry electrical current. For this reason, salts that dissociate into ions are called **electrolytes**, and fluids that contain them are **electrolyte solutions**.

In all electrolyte solutions, anion and cation concentrations are balanced (the number of negative and positive charges are equal). If a fluid contains 1000 negative charges, it must contain 1000 positive charges, too. If an anion enters the fluid, a cation must accompany it or another anion must leave so that electrical neutrality will be maintained. Thus, whenever sodium (Na^+) ions leave a cell, potassium (K^+) ions enter, for example. In fact, it's a good bet that whenever Na^+ and K^+ ions are moving, they are going in opposite directions.

Table 12-4 shows that, indeed, the positive and negative charges inside and outside cells are perfectly balanced even though the numbers of each kind of ion differ

TABLE 12-4 Important Body Electrolytes

Electrolytes	Intracellular (inside cells) Concentration (mEq/L)	Extracellular (outside cells) Concentration (mEq/L)
Cations (positively charged ions)		
Sodium (Na^+)	10	142
Potassium (K^+)	150	5
Calcium (Ca^{++})	2	5
Magnesium (Mg^{++})	40	3
	202	155
Anions (negatively charged ions)		
Chloride (Cl^-)	2	103
Bicarbonate (HCO_3^-)	10	27
Phosphate ($HPO_4^=$)	103	2
Sulfate ($SO_4^=$)	20	1
Organic acids (lactate, pyruvate)	10	6
Proteins	57	16
	202	155

NOTE: The numbers of positive and negative charges in a given fluid are the same. For example, in extracellular fluid, the cations and anions both equal 155 milliequivalents per liter (mEq/L). Of the cations, sodium ions make up 142 mEq/L; and potassium, calcium, and magnesium ions make up the remainder. Of the anions, chloride ions number 103 mEq/L; bicarbonate ions number 27; and the rest are provided by phosphate ions, sulfate ions, organic acids, and protein.

♦ The major minerals:
- Sodium
- Chloride
- Potassium
- Calcium
- Phosphorus
- Magnesium
- Sulfur

♦ To remember the difference between cations and anions, think of the "t" in cations as a "plus" (+) sign and the "n" in anions as a "negative."

salt: a compound composed of a positive ion other than H^+ and a negative ion other than OH^-. An example is sodium chloride ($Na^+ Cl^-$).
- **Na** = sodium
- **Cl** = chloride

dissociates (dis-SO-see-aites): physically separates.

ions (EYE-uns): atoms or molecules that have gained or lost electrons and therefore have electrical charges. Examples include the positively charged sodium ion (Na^+) and the negatively charged chloride ion (Cl^-). For a closer look at ions, see Appendix B.

cations (CAT-eye-uns): positively charged ions.

anions (AN-eye-uns): negatively charged ions.

electrolytes: salts that dissolve in water and dissociate into charged particles called ions.

electrolyte solutions: solutions that can conduct electricity.

FIGURE 12-4 Water Dissolves Salts and Follows Electrolytes

The structural arrangement of the two hydrogen atoms and one oxygen atom enables water to dissolve salts. Water's role as a solvent is one of its most valuable characteristics.

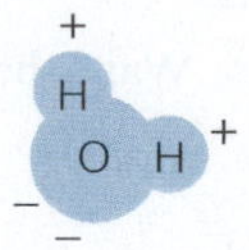

The negatively charged electrons that bond the hydrogens to the oxygen spend most of their time near the oxygen atom. As a result, the oxygen is slightly negative, and the hydrogens are slightly positive (see Appendix B).

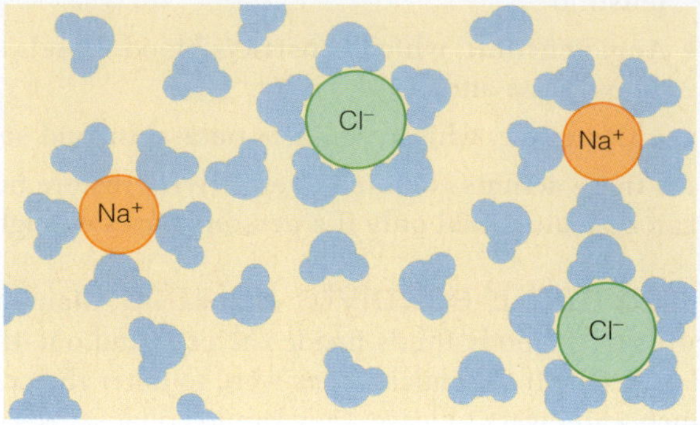

In an electrolyte solution, water molecules are attracted to both anions and cations. Notice that the negative oxygen atoms of the water molecules are drawn to the sodium cation (Na^+), whereas the positive hydrogen atoms of the water molecules are drawn to the chloride ions (Cl^-).

♦ The concentration of electrolytes in a volume of solution is expressed as **milliequivalents per liter (mEq/L).** Milliequivalents are a useful measure when considering ions because the number of charges reveals characteristics about the solution that are not evident when the concentration is expressed in terms of weight.

♦ A neutral molecule, such as water, that has opposite charges spatially separated within the molecule is **polar.** See Appendix B for more details.

solutes (SOLL-yutes): the substances that are dissolved in a solution. The number of molecules in a given volume of fluid is the *solute concentration.*

osmosis: the movement of water across a membrane *toward* the side where the solutes are more concentrated.

osmotic pressure: the amount of pressure needed to prevent the movement of water across a membrane.

over a wide range. Inside the cells, the positive charges total 202 and the negative charges balance these perfectly. Outside the cells, the amounts and proportions of the ions differ from those inside, but again the positive and negative charges balance. Scientists count these charges in milliequivalents, mEq. ♦

Electrolytes Attract Water Electrolytes attract water. Each water molecule has a net charge of zero, ♦ but the oxygen side of the molecule has a slight negative charge, and the hydrogens have a slight positive charge. Figure 12-4 shows the result in an electrolyte solution: both positive and negative ions attract clusters of water molecules around them. This attraction dissolves salts in water and enables the body to move fluids into appropriate compartments.

Water Follows Electrolytes As Figure 12-5 shows, some electrolytes reside primarily outside the cells (notably, sodium and chloride), whereas others reside predominantly inside the cells (notably, potassium, magnesium, phosphate, and sulfate). Cell membranes are *selectively permeable,* meaning that they allow the passage of some molecules, but not others. Whenever electrolytes move across the membrane, water follows.

The movement of water across a membrane toward the more concentrated **solutes** is called **osmosis.** The amount of pressure needed to prevent the movement of water across a membrane is called the **osmotic pressure.** Figure 12-6 presents osmosis, and the photos of salted eggplant and rehydrated raisins provide familiar examples.

Proteins Regulate Flow of Fluids and Ions Chapter 6 describes how proteins attract water and help to regulate fluid movement. In addition, transport proteins in the cell membranes regulate the passage of positive ions and other substances from one side of the membrane to the other. Negative ions follow positive ions, and water flows toward the more concentrated solution.

A protein that regulates the flow of fluids and ions in and out of cells is the sodium-potassium pump. The pump actively exchanges sodium for potassium across the cell membrane, using ATP as an energy source. Figure 6-10 on p. 183 illustrates this action.

Regulation of Fluid and Electrolyte Balance The amounts of various minerals in the body must remain nearly constant. Regulation occurs chiefly at two sites: the GI tract and the kidneys.

FIGURE 12-5 A Cell and Its Electrolytes

All of these electrolytes are found both inside and outside the cells, but each can be found mostly on one side or the other of the cell membrane.

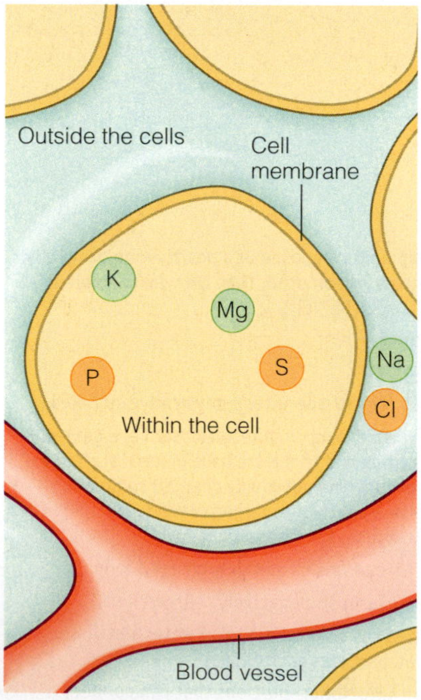

Chemical symbols:
- K = potassium
- P = phosphorus
- Mg = magnesium
- S = sulfate
- Na = sodium
- Cl = chloride

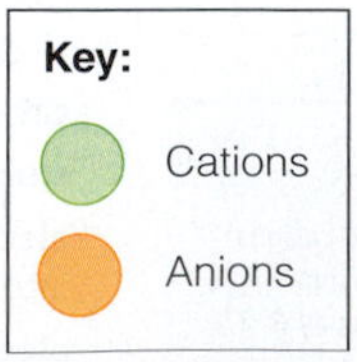

FIGURE 12-6 Osmosis

Water flows in the direction of the more highly concentrated solution.

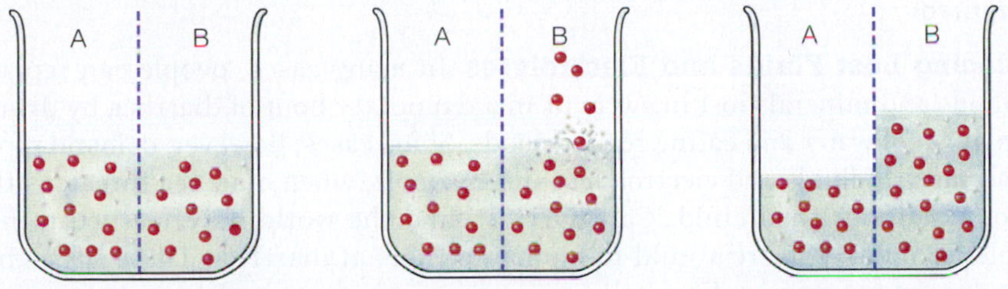

1. With equal numbers of solute particles on both sides of the semi-permeable membrane, the concentrations are equal, and the tendency of water to move in either direction is about the same.
2. Now additional solute is added to side B. Solute cannot flow across the divider (in the case of a cell, its membrane).
3. Water can flow both ways across the divider, but has a greater tendency to move from side A to side B, where there is a greater concentration of solute. The volume of water becomes greater on side B, and the concentrations on side A and B become equal.

The digestive juices of the GI tract contain minerals. These minerals and those from foods are reabsorbed in the large intestine as needed. Each day, 8 liters of fluids and associated minerals are recycled this way, providing ample opportunity for the regulation of electrolyte balance.

The kidneys' control of the body's *water* content by way of the hormone ADH has already been described (see p. 387). The kidneys regulate the *electrolyte* contents by responding to the hormone aldosterone (explained on p. 388). If the body's sodium is low, aldosterone stimulates sodium reabsorption from the kidneys. As sodium is reabsorbed, potassium (another positive ion) is excreted in accordance with the rule that total positive charges must remain in balance with total negative charges.

Fluid and Electrolyte Imbalance

Normally, the body defends itself successfully against fluid and electrolyte imbalances. Certain situations and some medications, however, may overwhelm the body's ability to compensate. Severe, prolonged vomiting and diarrhea as well as heavy sweating, burns, and traumatic wounds may incur such great fluid and electrolyte losses as to precipitate a medical emergency.

Different Solutes Lost by Different Routes

Different solutes are lost depending on why fluid is lost. If fluid is lost by vomiting or diarrhea, sodium is lost indiscriminately. If the adrenal glands oversecrete aldosterone, as may occur when

When immersed in water, raisins become plump because water moves toward the higher concentration of sugar inside the raisins.

When sprinkled with salt, vegetables "sweat" because water moves toward the higher concentration of salt outside the eggplant.

Physically active people must remember to replace their body fluids.

♦ Health-care workers use **oral rehydration therapy (ORT)**—a simple solution of sugar, salt, and water, taken by mouth—to treat dehydration caused by diarrhea. A simple ORT recipe (cool before giving):

- ½ L boiling water
- A small handful of sugar (4 tsp)
- 3 pinches of salt (½ tsp)

♦ **pH** is the unit of measure expressing a substance's acidity or alkalinity.

they develop a tumor, the kidneys may excrete too much potassium. Also, the person with uncontrolled diabetes may lose glucose, a solute not normally excreted, and large amounts of fluid with it. Each situation results in dehydration, but drinking water alone cannot restore electrolyte balance. Medical intervention is required.

Replacing Lost Fluids and Electrolytes In many cases, people can replace the fluids and minerals lost in sweat or in a temporary bout of diarrhea by drinking plain cool water and eating regular foods. Some cases, however, demand rapid replacement of fluids and electrolytes—for example, when diarrhea threatens the life of a malnourished child. Caregivers around the world have learned to use simple formulas ♦ to treat mild-to-moderate cases of diarrhea. These lifesaving formulas do not require hospitalization and can be prepared from ingredients available locally. Caregivers need only learn to measure ingredients carefully and use sanitary water. Once rehydrated, a person can begin eating foods.

Acid-Base Balance

The body uses its ions not only to help maintain fluid and electrolyte balance, but also to regulate the acidity (pH) ♦ of its fluids. The pH scale introduced in Chapter 3 is repeated here, in Figure 12-7, with the normal and abnormal pH ranges of the blood added. As you can see, the body must maintain the pH within a narrow range to avoid life-threatening consequences. Slight deviations in either direction can denature proteins, rendering them useless. Enzymes couldn't catalyze reactions and hemoglobin couldn't carry oxygen—to name just two examples.

FIGURE 12-7 The pH Scale

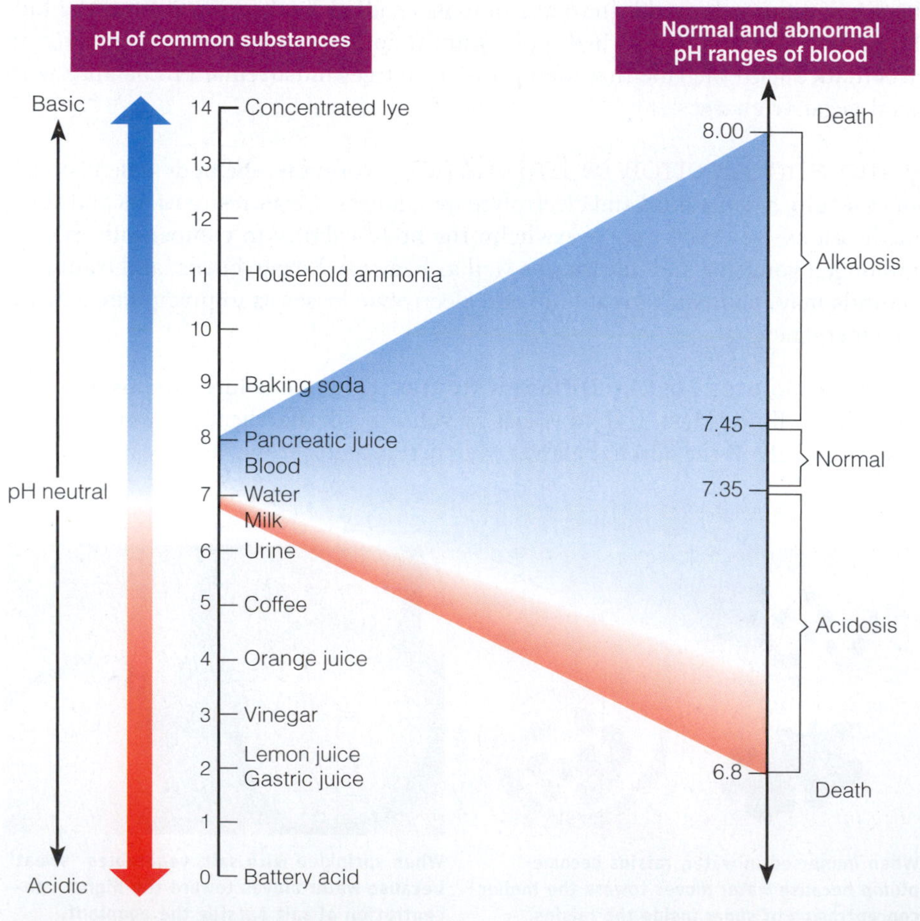

NOTE: Each step is ten times as concentrated in base (1/10 as much acid, or H^+) as the one below it.

The acidity of the body's fluids is determined by the concentration of hydrogen ions (H^+). ♦ A high concentration of hydrogen ions is very acidic. Normal energy metabolism generates hydrogen ions, as well as many other acids, that must be neutralized. Three systems defend the body against fluctuations in pH—buffers in the blood, respiration in the lungs, and excretion in the kidneys.

♦ The lower the pH, the higher the H^+ ion concentration and the stronger the acid. A pH above 7 is alkaline, or base—a solution in which OH^- ions predominate.

Regulation by the Buffers Bicarbonate ♦ (a base) and **carbonic acid** (an acid) in the body fluids, as well as some proteins, protect the body against changes in acidity by acting as buffers—substances that can neutralize acids or bases. Carbon dioxide, which is formed all the time during energy metabolism, dissolves in water to form carbonic acid in the blood. Carbonic acid, in turn, dissociates to form hydrogen ions and bicarbonate ions. The appropriate balance between carbonic acid and bicarbonate is essential to maintaining optimal blood pH. Figure 12-8 presents the chemical reactions of this buffer system, which is primarily under the control of the lungs and kidneys.

♦ **Bicarbonate** is an alkaline compound with the formula HCO_3. It is produced in all cell fluids from the dissociation of carbonic acid to help maintain the body's acid-base balance. (Bicarbonate is also secreted from the pancreas during digestion as part of the pancreatic juice.)

FIGURE 12-8 Bicarbonate–Carbonic Acid Buffer System

The reversible reactions of the bicarbonate-carbonic acid buffer system help to regulate the body's pH. Recall from Chapter 7 that carbon dioxide and water are formed during energy metabolism.

Carbon dioxide (CO_2) is a volatile gas that quickly dissolves in water (H_2O), forming carbonic acid (H_2CO_3):

$$CO_2 + H_2O \longleftrightarrow H_2CO_3$$

carbon dioxide + water ⟷ carbonic acid

Carbonic acid readily dissociates to a hydrogen ion (H^+) and a bicarbonate ion (HCO_3^-):

$$H_2CO_3 \longleftrightarrow H^+ + HCO_3^-$$

carbonic acid ⟷ hydrogen ion + bicarbonate ion

Regulation in the Lungs The lungs control the concentration of carbonic acid by raising or slowing the respiration rate, depending on whether the pH needs to be increased or decreased. If too much carbonic acid builds up, the respiration rate speeds up; this hyperventilation increases the amount of carbon dioxide exhaled, thereby lowering the carbonic acid concentration and restoring homeostasis. Conversely, if bicarbonate builds up, the respiration rate slows; carbon dioxide is retained and forms more carbonic acid. Again, homeostasis is restored.

Regulation in the Kidneys The kidneys control the concentration of bicarbonate by either reabsorbing or excreting it, depending on whether the pH needs to be increased or decreased, respectively. Their work is complex, but the net effect is easy to sum up. The *body's* total acid burden remains nearly constant; the acidity of the *urine* fluctuates to accommodate that balance.

IN SUMMARY **Electrolytes (charged minerals) in the fluids help distribute the fluids inside and outside the cells, thus ensuring the appropriate water balance and acid-base balance to support all life processes. Excessive losses of fluids and electrolytes upset these balances, and the kidneys play a key role in restoring homeostasis.**

The Minerals—An Overview

Figure 12-9 (p. 394) shows the amounts of the **major minerals** found in the body and, for comparison, some of the trace minerals. The distinction between the major and trace minerals does not mean that one group is more important than the other—all minerals are vital. The major minerals are so named because they are present, and needed, in larger amounts in the body. They are shown at the top of the figure and are discussed in this chapter. The trace minerals, shown at the bottom, are discussed in Chapter 13. A few generalizations pertain to all of the minerals and distinguish them from the vitamins. Especially notable is their chemical nature.

Inorganic Elements Unlike the organic vitamins, which are easily destroyed, minerals are inorganic elements ♦ that always retain their chemical identity. Once minerals enter the body proper, they remain there until excreted; they cannot be changed into anything else. Iron, for example, may temporarily combine with other charged elements in salts, but it is always iron. Neither can minerals be destroyed by heat, air, acid, or mixing. Consequently, little care is needed to preserve minerals

♦ An **inorganic** substance does not contain carbon.

carbonic acid: a compound with the formula H_2CO_3 that results from the combination of carbon dioxide (CO_2) and water (H_2O); of particular importance in maintaining the body's acid-base balance.

major minerals: essential mineral nutrients the human body requires in relatively large amounts (greater than 100 milligrams per day); sometimes called *macrominerals*.

FIGURE 12-9 Minerals in a 60-kilogram (132-pound) Human Body

Not only are the major minerals needed by the body in larger amounts, but they are also present in the body in larger amounts than the trace minerals.

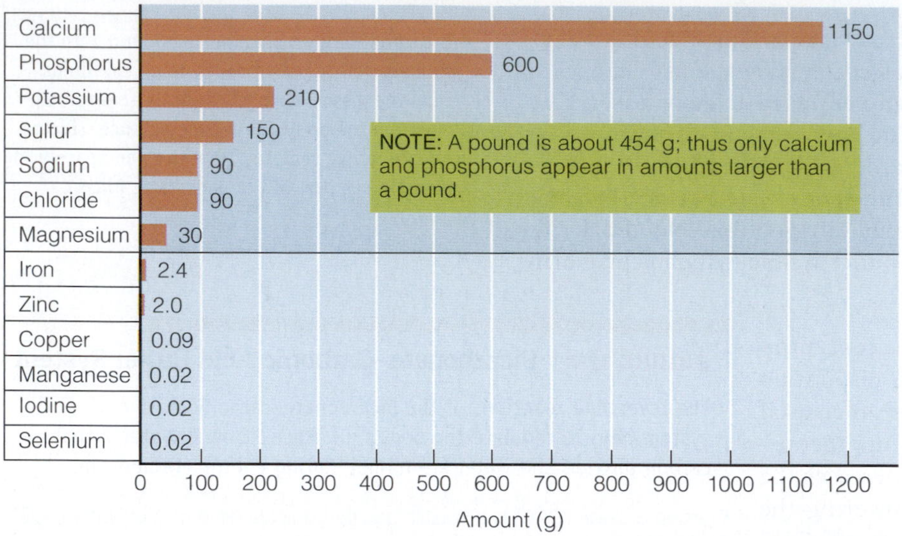

during food preparation. In fact, the ash that remains when a food is burned contains all the minerals that were in the food originally. Minerals can be lost from food only when they leach into cooking water that is then poured down the drain.

The Body's Handling of Minerals The minerals also differ from the vitamins in the amounts the body can absorb and in the extent to which they must be specially handled. Some minerals, such as potassium, are easily absorbed into the blood, transported freely, and readily excreted by the kidneys, much like the water-soluble vitamins. Other minerals, such as calcium, are more like fat-soluble vitamins in that they must have carriers to be absorbed and transported. And, like some of the fat-soluble vitamins, minerals taken in excess can be toxic.

Variable Bioavailability The bioavailability ♦ of minerals varies. Some foods contain **binders** that combine chemically with minerals, preventing their absorption and carrying them out of the body with other wastes. Examples of binders include phytates, which are found primarily in legumes and grains, and oxalates, which are present in rhubarb and spinach, among other foods. These foods contain more minerals than the body actually receives for use.

♦ **Bioavailability** refers to the rate at and the extent to which a nutrient is absorbed and used.

Nutrient Interactions Chapter 10 describes how the presence or absence of one vitamin can affect another's absorption, metabolism, and excretion. The same is true of the minerals. The interactions between sodium and calcium, for example, cause both to be excreted when sodium intakes are high. Phosphorus binds with magnesium in the GI tract, so magnesium absorption is limited when phosphorus intakes are high. These are just two examples of the interactions involving minerals featured in this chapter. Discussions in both this chapter and the next point out additional problems that arise from such interactions. Notice how often they reflect an excess of one mineral creating an inadequacy of another and how supplements—not foods—are most often to blame.

Varied Roles Although all the major minerals help to maintain the body's fluid balance as described earlier, sodium, chloride, and potassium are most noted for that role. ♦ For this reason, these three minerals are discussed first here. Later sections describe the minerals most noted for their roles in bone growth and health—calcium, phosphorus, and magnesium.

♦ Key fluid balance nutrients:
- Sodium, potassium, chloride

IN SUMMARY **The major minerals are found, and needed, in larger quantities in the body, whereas the trace minerals occur in smaller amounts. Minerals are inorganic elements that retain their chemical identities. They usually receive special handling and regulation in the body, and they may bind with other substances or interact with other minerals, thus limiting their absorption.**

Sodium

People have held salt (sodium chloride) in high regard throughout recorded history. We describe someone we admire as "the salt of the earth" and people we consider worthless as "not worth their salt." Even the word *salary* comes from the Latin word for salt.

binders: chemical compounds in foods that combine with nutrients (especially minerals) to form complexes the body cannot absorb. Examples include *phytates* (FYE-tates) and *oxalates* (OCK-sa-lates).

Cultures vary in their use of salt, but most people find its taste innately appealing. Salt brings its own tangy taste and enhances other flavors, most likely by suppressing the bitter flavors. You can taste this effect for yourself: tonic water with its bitter quinine tastes sweeter with a little salt added.

barbaraaaa/shutterstock.com

Fresh herbs add flavor to a recipe without adding salt.

Sodium Roles in the Body **Sodium** is the principal cation of the extracellular fluid and the primary regulator of its volume. Sodium also helps maintain acid-base balance and is essential to nerve impulse transmission and muscle contraction.*

Sodium is readily absorbed by the intestinal tract and travels freely in the blood until it reaches the kidneys, which filter all the sodium out of the blood. Then, with great precision, the kidneys return to the bloodstream the exact amount of sodium the body needs. Normally, the amount excreted is approximately equal to the amount ingested on a given day. When blood sodium rises, as when a person eats salted foods, thirst signals the person to drink until the appropriate sodium-to-water concentration is restored. Then the kidneys excrete both the excess water and the excess sodium together.

Sodium Recommendations Diets rarely lack sodium, and even when intakes are low, the body adapts by reducing sodium losses in urine and sweat, thus making deficiencies unlikely. Sodium recommendations ♦ are set low enough to protect against high blood pressure, but high enough to allow an adequate intake of other nutrients with a typical diet. Because high sodium intakes correlate with high blood pressure, the Upper Level (UL) for adults is set at 2300 milligrams per day, slightly lower than the Daily Value used on food labels (2400 milligrams). The average sodium intake for adults in the United States exceeds the UL—and most adults will develop hypertension at some point in their lives.

♦ AI for sodium:
- 1500 mg/day (19–50 yr)
- 1300 mg/day (51–70 yr)
- 1200 mg/day (>70 yr)

Sodium and Hypertension For years, a high *sodium* intake was considered the primary factor responsible for high blood pressure. Then research pointed to *salt* (sodium chloride) ♦ as the dietary culprit. Salt has a greater effect on blood pressure than either sodium or chloride alone or in combination with other ions.

♦ Salt (sodium chloride) is about 40% sodium.
1 g salt contributes about 400 mg sodium
6 g salt = 1 tsp
1 tsp salt contributes about 2300 mg sodium

For some individuals—most notably, those with hypertension, African Americans, and people older than 40 years of age—blood pressure increases in response to excesses in salt intake.[10] **Salt sensitivity** is apparent in about 25 percent of those with normal blood pressure and in about 50 percent of those with high blood pressure.[11] For them, a high salt intake correlates strongly with heart disease, and salt restriction (to no more than 1500 milligrams of sodium per day) helps to lower blood pressure.

In fact, a salt-restricted diet lowers blood pressure in people without hypertension as well. Because reducing salt intake causes no harm and diminishes the risk of hypertension and heart disease, the *Dietary Guidelines for Americans* advise limiting daily *salt* intake to about 1 teaspoon (the equivalent of 2.3 grams or 2300 milligrams of *sodium*).[12] The "How To" on p. 396 offers strategies for cutting salt (and therefore sodium) intake.

Dietary Guidelines for Americans 2010

Reduce daily sodium intake to less than 2300 milligrams and further reduce intake to 1500 milligrams among persons who are 51 and older and those of any age who are African American or have hypertension, diabetes, or chronic kidney disease.

One diet plan, known as the DASH (Dietary Approaches to Stop Hypertension) diet, may also lower blood pressure. The DASH approach emphasizes fruits, vegetables, and low-fat milk products; includes whole grains, nuts, poultry, and fish; and calls for reduced intakes of red meat, butter, and other high-fat foods. The DASH diet in combination with a reduced sodium intake is even more effective in lowering blood pressure than either strategy alone. Chapter 27 offers a complete discussion of hypertension and the dietary recommendations for its prevention and treatment.

*One of the ways the kidneys regulate acid-base balance is by excreting hydrogen ions (H^+) in exchange for sodium ions (Na^+).

sodium: the principal cation in the extracellular fluids of the body; critical to the maintenance of fluid balance, nerve impulse transmissions, and muscle contractions.

salt sensitivity: a characteristic of individuals who respond to a high salt intake with an increase in blood pressure or to a low salt intake with a decrease in blood pressure.

HOW TO Cut Salt (and Sodium) Intake

Most people eat more salt (and therefore sodium) than they need. Some people can lower their blood pressure by avoiding highly salted foods and removing the salt shaker from the table. Foods eaten without salt may seem less tasty at first, but with repetition, people can learn to enjoy the natural flavors of many unsalted foods. Strategies to cut salt intake include:

- Select fresh, unprocessed foods.
- Cook with little or no added salt.
- Prepare foods with sodium-free spices such as basil, bay leaves, curry, garlic, ginger, mint, oregano, pepper, rosemary, and thyme; lemon juice; vinegar; or wine.
- Add little or no salt at the table; taste foods before adding salt.
- Read labels with an eye open for sodium. (See the glossary on p. 57 for terms used to describe the sodium contents of foods on labels.)
- Select low-salt or salt-free products when available.

Use these foods sparingly:

- Foods prepared in brine, such as pickles, olives, and sauerkraut
- Salty or smoked meats, such as bologna, corned or chipped beef, bacon, frankfurters, ham, lunchmeats, salt pork, sausage, and smoked tongue
- Salty or smoked fish, such as anchovies, caviar, salted and dried cod, herring, sardines, and smoked salmon
- Snack items such as potato chips, pretzels, salted popcorn, salted nuts, and crackers
- Condiments such as bouillon cubes; seasoned salts; MSG; soy, teriyaki, Worcestershire, and barbeque sauces; prepared horseradish, ketchup, and mustard
- Cheeses, especially processed types
- Canned and instant soups

For additional practice, log on to **www.cengagebrain.com** and search for ISBN 084006845X.

TRY IT Compare the sodium contents of 1 ounce of the following foods: a plain bagel, potato chips, and animal crackers.

Sodium and Bone Loss (Osteoporosis) A high salt intake is also associated with increased calcium excretion, but its influence on bone loss is less clear. In addition, potassium may prevent the calcium excretion caused by a high-salt diet. For these reasons, dietary advice to prevent bone loss parallels that suggested for hypertension—a DASH diet that is low in sodium and abundant in potassium-rich fruits and vegetables and calcium-rich low-fat milk.

Sodium in Foods In general, processed foods have the most sodium, whereas unprocessed foods such as fresh fruits, vegetables, milk, and meats have the least. In fact, as much as 75 percent of the sodium in people's diets comes from salt added to foods by manufacturers; about 15 percent comes from salt added during cooking and at the table; and only 10 percent comes from the natural content in foods. To help consumers limit their intake, public health organizations and policymakers are calling for regulations to reduce sodium in the nation's food supply.[13]

Because processed foods may contain sodium without chloride, as in additives such as sodium bicarbonate or sodium saccharin, they do not always taste salty. Most people are surprised to learn that 1 ounce of some cereals contains more sodium than 1 ounce of salted peanuts—and that ½ cup of instant chocolate pudding contains still more. The peanuts taste saltier because the salt is all on the surface, where the tongue's taste receptors immediately pick it up.

Figure 12-10 shows that processed foods not only contain more sodium than their less-processed counterparts but also have less potassium. Low potassium may be as significant as high sodium when it comes to blood pressure regulation, so processed foods have two strikes against them.[14]

Dietary Guidelines for Americans 2010

Limit the consumption of foods that contain refined grains, especially refined grain foods that contain sodium.

Sodium Deficiency If blood sodium drops, as may occur with vomiting, diarrhea, or heavy sweating, both sodium and water must be replenished. Under

FIGURE 12-10 What Processing Does to the Sodium and Potassium Contents of Foods

People who eat foods high in salt often happen to be eating fewer potassium-containing foods at the same time. Notice how potassium is lost and sodium is gained as foods become more processed, causing the potassium-to-sodium ratio to fall dramatically. Even when potassium isn't lost, the addition of sodium still lowers the potassium-to-sodium ratio. Limiting sodium intake may help in two ways, then—by lowering blood pressure in salt-sensitive individuals and by indirectly raising potassium intakes in all individuals.

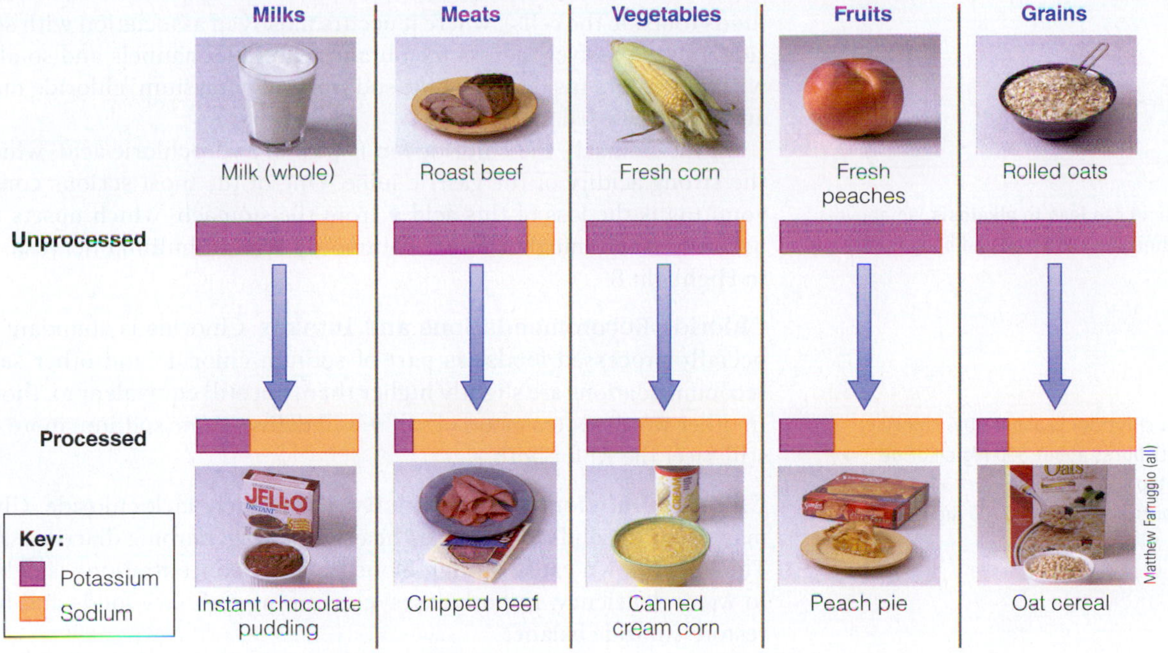

normal conditions of sweating due to physical activity, salt losses can easily be replaced later in the day with ordinary foods. Salt tablets are not recommended because too much salt, especially if taken with too little water, can induce dehydration. During intense activities, such as ultra-endurance events, athletes can lose so much sodium and drink so much water that they develop hyponatremia—the dangerous condition ♦ of having too little sodium in the blood.

♦ Symptoms of hyponatremia:
- Headache, confusion, stupor
- Seizures, coma

Sodium Toxicity and Excessive Intakes The immediate symptoms of acute sodium toxicity are edema and high blood pressure. Prolonged excessive sodium intake ♦ may contribute to hypertension in some people, as explained earlier.

♦ UL for sodium: 2300 mg/day

IN SUMMARY Sodium is the main cation outside cells and one of the primary electrolytes responsible for maintaining fluid balance. Dietary deficiency is rare, and excesses may aggravate hypertension in some people. For this reason, health professionals advise a diet moderate in salt and sodium. The accompanying table provides a summary of sodium.

Sodium

Adequate Intake (AI)

Adults: 1500 mg/day (19–50 yr)
1300 mg/day (51–70 yr)
1200 mg/day (>70 yr)

Upper Level

Adults: 2300 mg/day

Chief Functions in the Body

Maintains normal fluid and electrolyte balance; assists in nerve impulse transmission and muscle contraction

Deficiency Symptoms

Muscle cramps, mental apathy, loss of appetite

Toxicity Symptoms

Edema, acute hypertension

Significant Sources

Table salt, soy sauce; moderate amounts in meats, milks, breads, and vegetables; large amounts in processed foods

Chloride

The element *chlorine* (Cl_2) is a poisonous gas. When chlorine reacts with sodium or hydrogen, however, it forms the negative chloride ion (Cl^-). *Chloride,* an essential nutrient, is required in the diet.

Chloride Roles in the Body **Chloride** is the major anion of the extracellular fluids (outside the cells), where it occurs mostly in association with sodium. Chloride moves passively across membranes through channels and so also associates with potassium inside cells. Like sodium and potassium, chloride maintains fluid and electrolyte balance.

In the stomach, the chloride ion is part of hydrochloric acid, which maintains the strong acidity of the gastric juice. One of the most serious consequences of vomiting is the loss of this acid ♦ from the stomach, which upsets the acid-base balance.* Such imbalances are commonly seen in bulimia nervosa, as described in Highlight 8.

♦ The loss of acid can lead to **alkalosis,** an above-normal alkalinity in the blood and body fluids.

Chloride Recommendations and Intakes Chloride is abundant in foods (especially processed foods) as part of sodium chloride and other salts. Chloride recommendations are slightly higher than, but still equivalent to, those of sodium. In other words, ¾ teaspoon of salt ♦ will deliver some sodium, more chloride, and still meet the AI for both.

♦ Salt (sodium chloride) is about 60% chloride.
1 g salt contributes about 600 mg chloride
6 g salt = 1 tsp
1 tsp salt contributes about 3700 mg chloride

Chloride Deficiency and Toxicity Diets rarely lack chloride. Chloride losses may occur in conditions such as heavy sweating, chronic diarrhea, and vomiting. The only known cause of high blood chloride concentrations is dehydration due to water deficiency. In both cases, consuming ordinary foods and beverages can restore chloride balance.

IN SUMMARY **Chloride is the major anion outside cells, and it associates closely with sodium. In addition to its role in fluid balance, chloride is part of the stomach's hydrochloric acid. The accompanying table provides a summary of chloride.**

Chloride

Adequate Intake (AI)	Deficiency Symptoms
Adults: 2300 mg/day (19–50 yr) 2000 mg/day (51–70 yr) 1800 mg/day (>70 yr)	Do not occur under normal circumstances
Upper Level	**Toxicity Symptoms**
Adults: 3600 mg/day	Vomiting
Chief Functions in the Body	**Significant Sources**
Maintains normal fluid and electrolyte balance; part of hydrochloric acid found in the stomach, necessary for proper digestion	Table salt, soy sauce; moderate amounts in meats, milks, eggs; large amounts in processed foods

Potassium

Like sodium, **potassium** is a positively charged ion. In contrast to sodium, potassium is the body's principal intracellular cation, *inside* the body cells.

Potassium Roles in the Body Potassium plays a major role in maintaining fluid and electrolyte balance and cell integrity. During nerve impulse transmis-

chloride (KLO-ride): the major anion in the extracellular fluids of the body. Chloride is the ionic form of chlorine, Cl^-. See Appendix B for a description of the chlorine-to-chloride conversion.

potassium: the principal cation within the body's cells; critical to the maintenance of fluid balance, nerve impulse transmissions, and muscle contractions.

*Hydrochloric acid secretion into the stomach involves the addition of bicarbonate ions (base) to the plasma. These bicarbonate ions (HCO_3^-) are neutralized by hydrogen ions (H^+) from the gastric secretions that are reabsorbed into the plasma. When hydrochloric acid is lost during vomiting, these hydrogen ions are no longer available for reabsorption, and so, in effect, the concentrations of bicarbonate ions in the plasma are increased. In this way, excessive vomiting of acidic gastric juices leads to *metabolic alkalosis.*

sion and muscle contraction, potassium and sodium briefly trade places across the cell membrane. The cell then quickly pumps them back into place. Controlling potassium distribution is a high priority for the body because it affects many aspects of homeostasis, including a steady heartbeat.

Potassium Recommendations and Intakes Potassium is abundant in all living cells. Because cells remain intact unless foods are processed, the richest sources of potassium are *fresh* foods—as Figure 12-11 shows. In contrast, most processed foods such as canned vegetables, ready-to-eat cereals, and luncheon meats contain less potassium—and more sodium (recall Figure 12-10 on p. 397). To meet the AI for potassium, most people need to increase their intake of fruits and vegetables to five to nine servings daily.

Potassium and Hypertension Diets low in potassium seem to play an important role in the development of high blood pressure. Low potassium intakes, especially when combined with high sodium intakes, raise blood pressure and increase the risk of death from heart disease.[15] In contrast, high potassium intakes, especially when combined with low sodium intakes, appear to both prevent and correct hypertension. ♦ Potassium-rich fruits and vegetables also appear to reduce the risk of stroke—more so than can be explained by the reduction in blood pressure alone.

Potassium Deficiency Potassium deficiency is characterized by an increase in blood pressure, salt sensitivity, kidney stones, and bone turnover. As deficiency progresses, symptoms include irregular heartbeats, muscle weakness, and glucose intolerance.

Fresh foods, especially fruits and vegetables, provide potassium in abundance.

♦ The DASH diet, used to lower blood pressure, emphasizes potassium-rich foods such as fruits and vegetables.

FIGURE 12-11 Potassium in Selected Foods

See the "How To" section on p. 316 for more information on using this figure.

Food	Serving size (kcalories)
Bread, whole wheat	1 oz slice (70 kcal)
Cornflakes, fortified	1 oz (110 kcal)
Spaghetti pasta	½ c cooked (99 kcal)
Tortilla, flour	1 10"-round (234 kcal)
Broccoli	½ c cooked (22 kcal)
Carrots	½ c shredded raw (24 kcal)
Potato	1 medium baked w/skin (133 kcal)
Tomato juice	¾ c (31 kcal)
Banana	1 medium raw (109 kcal)
Orange	1 medium raw (62 kcal)
Strawberries	½ c fresh (22 kcal)
Watermelon	1 slice (92 kcal)
Milk	1 c reduced-fat 2% (121 kcal)
Yogurt, plain	1 c low-fat (155 kcal)
Cheddar cheese	1½ oz (171 kcal)
Cottage cheese	½ c low-fat 2% (101 kcal)
Pinto beans	½ c cooked (117 kcal)
Peanut butter	2 tbs (188 kcal)
Sunflower seeds	1 oz dry (165 kcal)
Tofu (soybean curd)	½ c (76 kcal)
Ground beef, lean	3 oz broiled (244 kcal)
Chicken breast	3 oz roasted (140 kcal)
Tuna, canned in water	3 oz (99 kcal)
Egg	1 hard cooked (78 kcal)
Excellent, and sometimes unusual, sources:	
Squash, acorn	½ c baked (69 kcal)
Soybeans	½ c cooked (149 kcal)
Artichoke	1 (60 kcal)

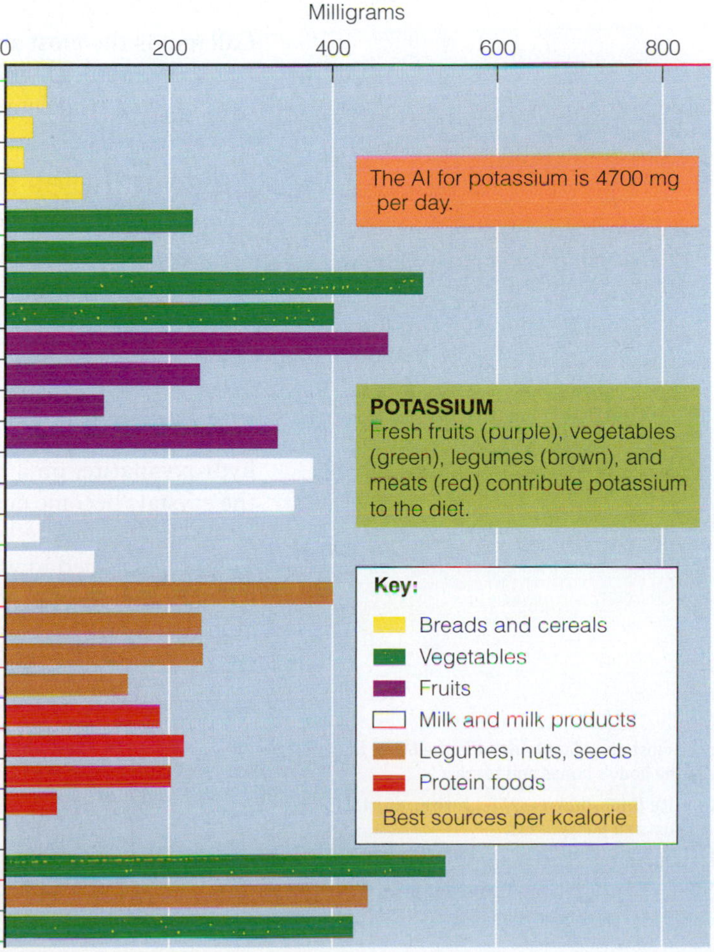

Potassium Toxicity Potassium toxicity does not result from overeating foods high in potassium; therefore a UL has not been set. It can result from overconsumption of potassium salts or supplements (including some "energy fitness shakes") and from certain diseases or treatments. Given more potassium than the body needs, the kidneys accelerate their excretion. If the GI tract is bypassed, however, and potassium is injected directly into a vein, it can stop the heart.

IN SUMMARY **Potassium, like sodium and chloride, is an electrolyte that plays an important role in maintaining fluid balance. Potassium is the primary cation inside cells; fresh foods, notably fruits and vegetables, are its best sources. The accompanying table provides a summary of potassium.**

Potassium

Adequate Intake (AI)	Toxicity Symptoms
Adults: 4700 mg/day	Muscular weakness; vomiting; if given into a vein, can stop the heart
Chief Functions in the Body	**Significant Sources**
Maintains normal fluid and electrolyte balance; facilitates many reactions; supports cell integrity; assists in nerve impulse transmission and muscle contractions	All whole foods: meats, milks, fruits, vegetables, grains, legumes
Deficiency Symptoms[a]	
Irregular heatbeat, muscular weakness, glucose intolerance	

[a]Deficiency accompanies dehydration.

Calcium

Calcium is the most abundant mineral in the body. It receives much emphasis in this chapter and in the highlight that follows because an adequate intake helps grow a healthy skeleton in early life and minimize bone loss in later life.

Calcium Roles in the Body

Ninety-nine percent of the body's calcium is in the bones (and teeth), where it plays two roles. First, it is an integral part of bone structure, providing a rigid frame that holds the body upright and serves as attachment points for muscles, making motion possible. Second, it serves as a calcium bank, offering a readily available source of the mineral to the body fluids should a drop in blood calcium occur. The remaining 1 percent of the body's calcium is in the body fluids.

Calcium in Bones As bones begin to form, calcium salts form crystals, called **hydroxyapatite**, on a matrix of the protein collagen. During **mineralization**, as the crystals become denser, they give strength and rigidity to the maturing bones. As a result, the long leg bones of children can support their weight by the time they have learned to walk.

Many people have the idea that once a bone is built, it is inert like a rock. Actually, the bones are gaining and losing minerals continuously in an ongoing process of remodeling. Growing children gain more bone than they lose, and healthy adults maintain a reasonable balance. When withdrawals substantially exceed deposits, problems such as osteoporosis develop (as described in Highlight 12).

The formation of teeth follows a pattern similar to that of bones. The turnover of minerals in teeth is not as rapid as in bone, however; fluoride hardens and stabilizes the crystals of teeth, opposing the withdrawal of minerals from them.

Calcium in Body Fluids Although only 1 percent of the body's calcium circulates in the extracellular and intracellular fluids, its presence there is vital to

calcium: the most abundant mineral in the body; found primarily in the body's bones and teeth.

hydroxyapatite (high-drox-ee-APP-ah-tite): crystals made of calcium and phosphorus.

mineralization: the process in which calcium, phosphorus, and other minerals crystallize on the collagen matrix of a growing bone, hardening the bone.

life. Cells throughout the body can detect calcium in the extracellular fluids and respond accordingly. Many of calcium's actions help to maintain normal blood pressure, perhaps by stabilizing the smooth muscle cells of the blood vessels or by releasing relaxing factors from the blood vessel cell walls.[16] Extracellular calcium also participates in blood clotting.

The calcium in intracellular fluids binds to proteins within the cells and activates them. ♦ These proteins participate in the regulation of muscle contractions, the transmission of nerve impulses, the secretion of hormones, and the activation of some enzyme reactions.

♦ An example of a protein that calcium binds with and activates is **calmodulin** (cal-MOD-you-lin). One of calmodulin's roles is to activate the enzymes involved in breaking down glycogen, which releases energy for muscle contractions.

Calcium and Disease Prevention Calcium may protect against hypertension, although research results are inconsistent and inconclusive.[17] Considering the success of the DASH diet in lowering blood pressure, restricting sodium to treat hypertension may be narrow advice. The DASH diet is not particularly low in sodium, but it is rich in calcium, as well as in magnesium and potassium. As mentioned earlier, the DASH diet, together with a reduced sodium intake, is more effective in lowering blood pressure than either strategy alone. Some research also suggests protective relationships between dietary calcium and blood cholesterol, diabetes, and colon cancer.[18] Highlight 12 explores calcium's role in preventing osteoporosis.

Calcium and Obesity Calcium may also play a role in maintaining a healthy body weight.[19] Epidemiological studies suggest an inverse relationship between calcium intake and body weight: the higher the calcium intake, the lower the prevalence of overweight. In particular, calcium from dairy foods, but *not* from supplements, seems to influence body weight.[20] An adequate dietary calcium intake may help prevent excessive fat accumulation by stimulating hormonal action that targets the breakdown of stored fat. Not all research suggests that consumption of calcium or dairy foods alters fat metabolism or energy expenditure or improves body weight or composition.[21] Large, well-designed clinical studies are needed to clarify the effects of dietary calcium intake on body weight.

Calcium Balance Calcium homeostasis involves a system of hormones and vitamin D.[22] Whenever blood calcium falls too low or rises too high, three organ systems respond: the intestines, bones, and kidneys. Figure 12-12 illustrates how vitamin D and two hormones—**parathyroid hormone** and **calcitonin**—return blood calcium to normal.

FIGURE 12-12 Calcium Balance

Blood calcium is regulated in part by vitamin D and two hormones—calcitonin and parathyroid hormone. Bone serves as a reservoir when blood calcium is high and as a source of calcium when blood calcium is low. Osteoclasts break down bone and release calcium into the blood; osteoblasts build new bone using calcium from the blood.

Animated! figure **www.cengagebrain.com** (search for ISBN 084006845X)

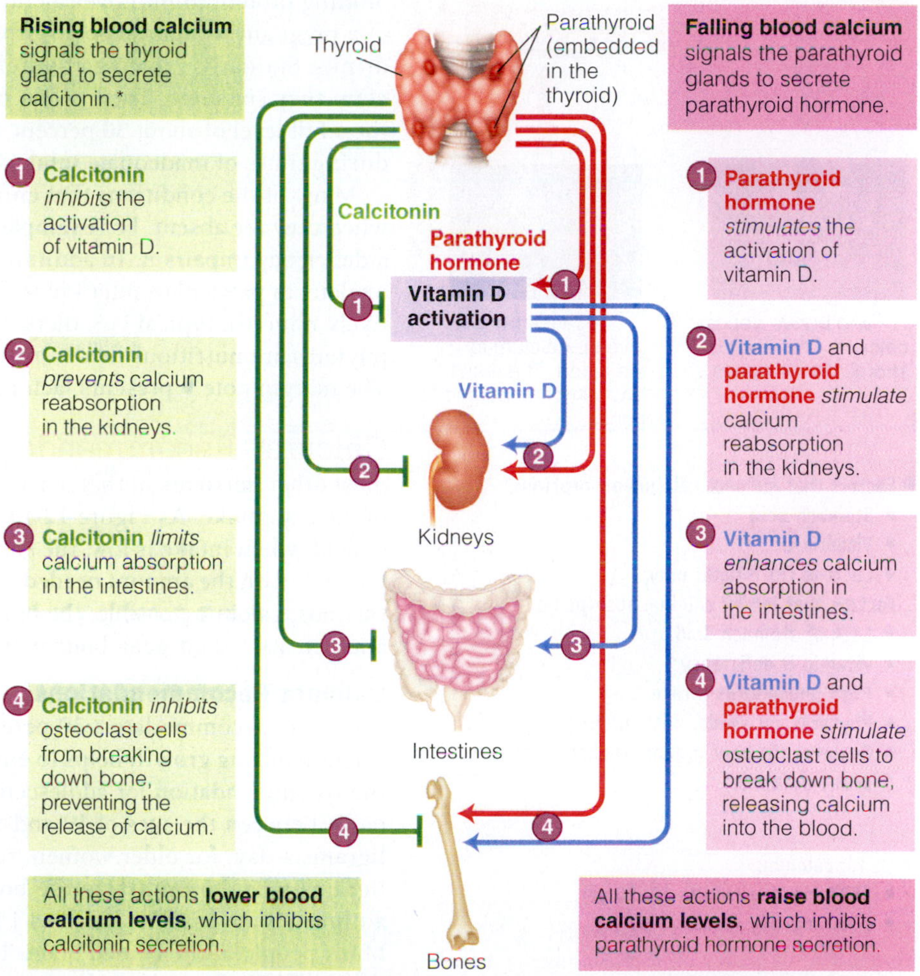
Rising blood calcium signals the thyroid gland to secrete calcitonin.*
Thyroid
Parathyroid (embedded in the thyroid)
Falling blood calcium signals the parathyroid glands to secrete parathyroid hormone.
1 **Calcitonin** *inhibits* the activation of vitamin D.
Calcitonin
Parathyroid hormone
Vitamin D activation
1 **Parathyroid hormone** *stimulates* the activation of vitamin D.
2 **Calcitonin** *prevents* calcium reabsorption in the kidneys.
Vitamin D
Kidneys
2 **Vitamin D** and **parathyroid hormone** *stimulate* calcium reabsorption in the kidneys.
3 **Calcitonin** *limits* calcium absorption in the intestines.
3 **Vitamin D** *enhances* calcium absorption in the intestines.
Intestines
4 **Calcitonin** *inhibits* osteoclast cells from breaking down bone, preventing the release of calcium.
4 **Vitamin D** and **parathyroid hormone** *stimulate* osteoclast cells to break down bone, releasing calcium into the blood.
All these actions **lower blood calcium levels**, which inhibits calcitonin secretion.
Bones
All these actions **raise blood calcium levels**, which inhibits parathyroid hormone secretion.

*Calcitonin plays a major role in defending infants and young children against the dangers of rising blood calcium that can occur when regular feedings of milk deliver large quantities of calcium to a small body. In contrast, calcitonin plays a relatively minor role in adults because their absorption of calcium is less efficient and their bodies are larger, making elevated blood calcium unlikely.

The calcium in bone provides a nearly inexhaustible bank of calcium for the blood. The blood borrows and returns calcium as needed so that even with a dietary deficiency, *blood* calcium remains normal—even as *bone* calcium diminishes (see Figure 12-13 on p. 402). Blood calcium changes only in response to abnormal

parathyroid hormone: a hormone from the parathyroid glands that regulates blood calcium by raising it when levels fall too low; also known as *parathormone* (PAIR-ah-THOR-moan).

calcitonin (KAL-seh-TOE-nin): a hormone secreted by the thyroid gland that regulates blood calcium by lowering it when levels rise too high.

FIGURE 12-13 **Maintaining Blood Calcium from the Diet and from the Bones**

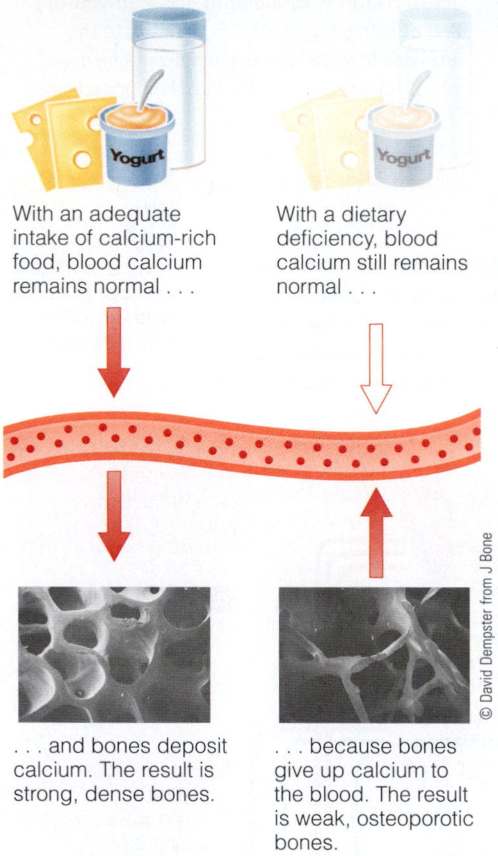

Yogurt
Yogurt
With an adequate intake of calcium-rich food, blood calcium remains normal . . .
With a dietary deficiency, blood calcium still remains normal . . .
. . . and bones deposit calcium. The result is strong, dense bones.
. . . because bones give up calcium to the blood. The result is weak, osteoporotic bones.

© David Dempster from J Bone Miner Res, 1986 (both)

♦ Factors that *enhance* calcium absorption:

- Stomach acid
- Vitamin D
- Lactose (in infants only)

Factors that *inhibit* calcium absorption:

- Lack of stomach acid
- Vitamin D deficiency
- High phosphorus intake
- Phytates (in seeds, nuts, grains)
- Oxalates (in beet greens, rhubarb, spinach, sweet potatoes)

♦ UL for calcium:

- 2500 mg/day (adults 19–50 yr)
- 2000 mg/day (adults >50 yr)

calcium rigor: hardness or stiffness of the muscles caused by high blood calcium concentrations.

calcium tetany (TET-ah-nee): intermittent spasm of the extremities due to nervous and muscular excitability caused by low blood calcium concentrations.

calcium-binding protein: a protein in the intestinal cells, made with the help of vitamin D, that facilitates calcium absorption.

peak bone mass: the highest attainable bone density for an individual, developed during the first three decades of life.

regulatory control, not to diet. A person can have an inadequate calcium intake for years and suffer no noticeable symptoms. Only later in life does it become apparent that bone integrity has been compromised.

Blood calcium above normal results in **calcium rigor**: the muscles contract and cannot relax. Similarly, blood calcium below normal causes **calcium tetany**—also characterized by uncontrolled muscle contraction. These conditions do *not* reflect a *dietary* excess or lack of calcium; they are caused by a lack of vitamin D or by abnormal secretion of the regulatory hormones. A chronic *dietary* deficiency of calcium, or a chronic deficiency due to poor absorption over the years, depletes the bones. Again: the *bones,* not the blood, are robbed by a calcium deficiency.

Calcium Absorption Because many factors affect calcium absorption, the most effective way to ensure adequacy is to increase calcium intake.[23] On average, adults absorb about 30 percent of the calcium they ingest. The stomach's acidity helps to keep calcium soluble, and vitamin D helps to make the **calcium-binding protein** needed for absorption. This explains why calcium-rich milk is the best food for vitamin D fortification.

Whenever calcium is needed, the body increases its production of the calcium-binding protein to improve calcium absorption. The result is obvious in the case of a pregnant woman, who absorbs 50 percent of the calcium from the milk she drinks. Similarly, growing children and teens absorb 50 to 60 percent of the calcium they consume. Then, when bone growth slows or stops, absorption falls to the adult level of about 30 percent. In addition, absorption becomes more efficient during times of inadequate intakes.

Many of the conditions that enhance calcium absorption inhibit its absorption when they are absent. For example, sufficient vitamin D supports absorption, and a deficiency impairs it. In addition, fiber, in general, and the binders phytate and oxalate, in particular, interfere with calcium absorption, but their effects are relatively minor in typical U.S. diets. Vegetables with oxalates and whole grains with phytates are nutritious foods, of course, but they are not useful calcium sources. The margin note ♦ presents factors that influence calcium balance.

Calcium Recommendations and Sources

Calcium is unlike most other nutrients in that hormones maintain its *blood* concentration regardless of dietary intake. As Figure 12-13 shows, when calcium intake is high, the *bones* benefit; when intake is low, the *bones* suffer. Calcium recommendations are therefore based on the amount needed to retain the most calcium in bones. By retaining the most calcium possible, the bones can develop to their fullest potential in size and density—their **peak bone mass**—within genetic limits.

Calcium Recommendations Calcium recommendations have been set high enough to accommodate a 30 percent absorption rate. Because obtaining enough calcium during growth helps to ensure that the skeleton will be strong and dense, the recommendation for adolescents up to the age of 18 years is 1300 milligrams daily. Between the ages of 19 and 50, recommendations are lowered to 1000 milligrams a day; for older women, recommendations are raised again to 1200 milligrams a day to minimize the bone loss that tends to occur later in life. Some authorities advocate as much as 1500 milligrams a day for women older than 50. Many people in the United States have calcium intakes below current recommendations.[24] High intakes of calcium from supplements may have adverse effects such as kidney stone formation.[25] For this reason, a UL has been established. ♦

High intakes of both dietary protein and sodium increase calcium losses, but whether these losses impair bone development remains unclear. In the case of protein, high intakes of either animal or plant proteins may be problematic, but the effects are minimized by the beneficial effects of other nutrients in the food and diet—for example, by the potassium in legumes and the calcium in milk. In establishing an RDA for calcium, the DRI Committee considered these nutrient interactions and did not adjust dietary recommendations based on this information.

Calcium in Milk Products Figure 12-14 shows that calcium is found most abundantly in a single class of foods—milk. ♦ The person who doesn't like to drink milk may prefer to eat cheese or yogurt. Alternatively, milk and milk products can be concealed in foods. Powdered fat-free milk can be added to casseroles, soups, and other mixed dishes during preparation; 5 heaping tablespoons offer the equivalent of 1 cup of milk. This simple step is an excellent way for older women not only to obtain extra calcium, but more protein, vitamins, and minerals as well.

It is especially difficult for children who don't drink milk to meet their calcium needs.[26] Children who don't drink milk have lower calcium intakes and poorer bone health than those who drink milk regularly. The consequences of drinking too little milk during childhood and adolescence persist into adulthood. Women who seldom drank milk as children or teenagers have lower bone density and greater risk of fractures than those who drank milk regularly.[27] It is possible for people who do not drink milk to obtain adequate calcium, but only if they carefully select other calcium-rich foods.

♦ Suggested daily amounts:
- Young children (2 to 8 yr): 2 c
- Older children, teenagers, and all adults: 3 c

Calcium in Other Foods Many people, for a variety of reasons, cannot or do not drink milk. Some cultures do not use milk in their cuisines; some vegetarians exclude milk as well as meat; and some people are allergic to milk protein or are lactose intolerant. ♦ Others simply do not enjoy the taste of milk. These people need to find nonmilk sources of calcium to help meet their calcium needs. Some brands of tofu, corn tortillas, some nuts (such as almonds), and some seeds (such as sesame seeds) can supply calcium for the person who doesn't use milk products.

♦ People with lactose intolerance may be able to consume small quantities of milk, as Chapter 4 explains.

FIGURE 12-14 **Calcium in Selected Foods**

See the "How To" section on p. 316 for more information on using this figure.

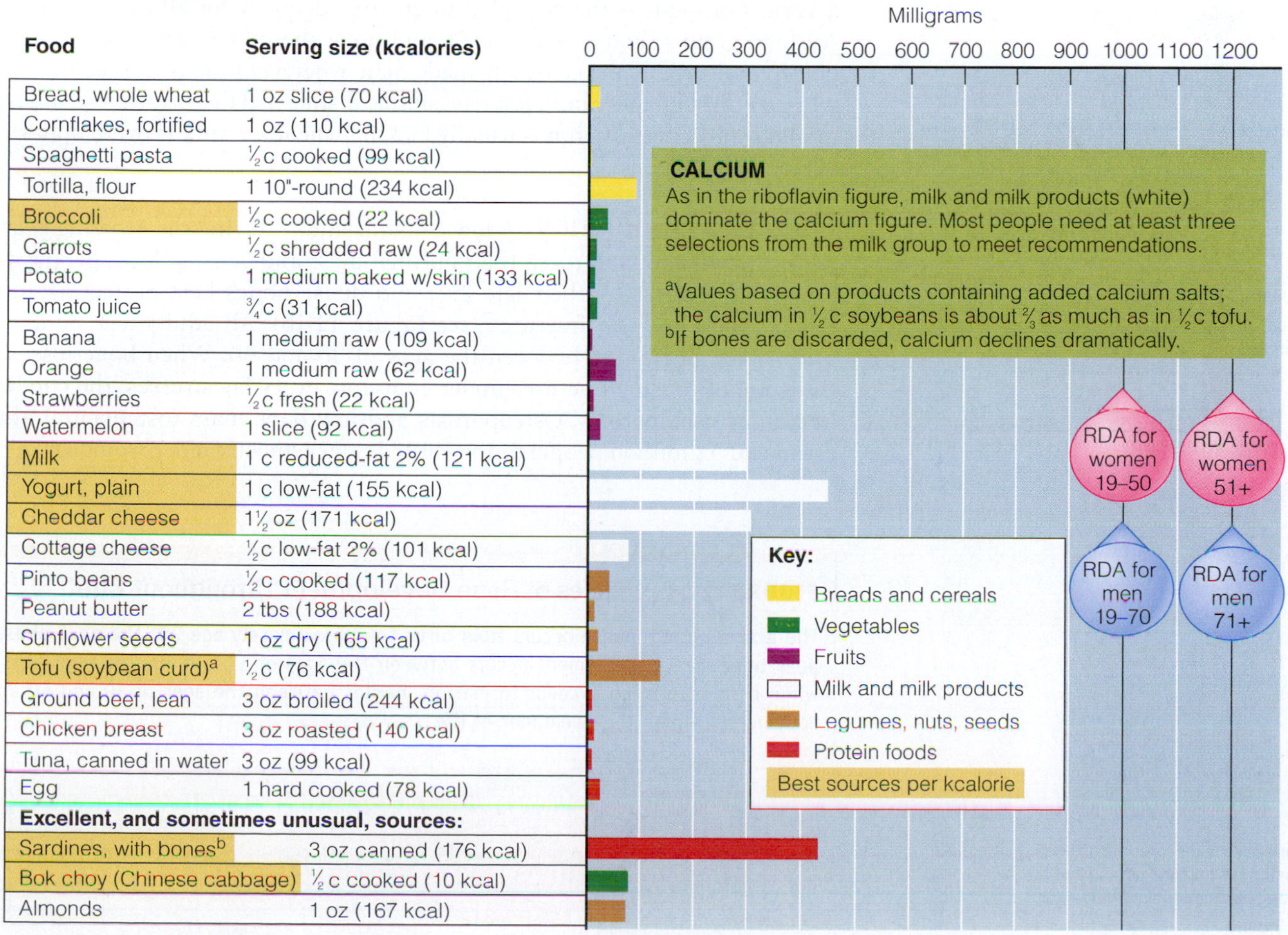

Food
Serving size (kcalories)
Milligrams
0 100 200 300 400 500 600 700 800 900 1000 1100 1200
Bread, whole wheat | 1 oz slice (70 kcal)
Cornflakes, fortified | 1 oz (110 kcal)
Spaghetti pasta | ½ c cooked (99 kcal)
Tortilla, flour | 1 10"-round (234 kcal)
Broccoli | ½ c cooked (22 kcal)
Carrots | ½ c shredded raw (24 kcal)
Potato | 1 medium baked w/skin (133 kcal)
Tomato juice | ¾ c (31 kcal)
Banana | 1 medium raw (109 kcal)
Orange | 1 medium raw (62 kcal)
Strawberries | ½ c fresh (22 kcal)
Watermelon | 1 slice (92 kcal)
Milk | 1 c reduced-fat 2% (121 kcal)
Yogurt, plain | 1 c low-fat (155 kcal)
Cheddar cheese | 1½ oz (171 kcal)
Cottage cheese | ½ c low-fat 2% (101 kcal)
Pinto beans | ½ c cooked (117 kcal)
Peanut butter | 2 tbs (188 kcal)
Sunflower seeds | 1 oz dry (165 kcal)
Tofu (soybean curd)a | ½ c (76 kcal)
Ground beef, lean | 3 oz broiled (244 kcal)
Chicken breast | 3 oz roasted (140 kcal)
Tuna, canned in water | 3 oz (99 kcal)
Egg | 1 hard cooked (78 kcal)
Excellent, and sometimes unusual, sources:
Sardines, with bonesb | 3 oz canned (176 kcal)
Bok choy (Chinese cabbage) | ½ c cooked (10 kcal)
Almonds | 1 oz (167 kcal)
CALCIUM
As in the riboflavin figure, milk and milk products (white) dominate the calcium figure. Most people need at least three selections from the milk group to meet recommendations.
aValues based on products containing added calcium salts; the calcium in ½ c soybeans is about ⅔ as much as in ½ c tofu.
bIf bones are discarded, calcium declines dramatically.
RDA for women 19–50
RDA for women 51+
RDA for men 19–70
RDA for men 71+
Key:
Breads and cereals
Vegetables
Fruits
Milk and milk products
Legumes, nuts, seeds
Protein foods
Best sources per kcalorie

Matthew Farruggio

Milk and milk products are well known for their calcium, but calcium-set tofu, bok choy, kale, calcium-fortified orange juice, and broccoli are also rich in calcium.

A slice of most breads contains only about 5 to 10 percent of the calcium found in milk, but it can be a major source for people who eat many slices because the calcium is well absorbed. Oysters are also a rich source of calcium, as are small fish eaten with their bones, such as canned sardines.

Among the vegetables, mustard and turnip greens, bok choy, kale, parsley, watercress, and broccoli are good sources of available calcium. So are some seaweeds such as the nori popular in Japanese cooking. Some dark green, leafy vegetables—notably spinach and Swiss chard—appear to be calcium-rich but actually provide little, if any, calcium to the body because of the binders they contain. It would take 8 cups of spinach—containing six times as much calcium as 1 cup of milk—to deliver the equivalent in *absorbable* calcium.

With the exception of foods such as spinach that contain calcium binders, however, the calcium content of foods is usually more important than bioavailability. Consequently, recognizing that people eat a variety of foods containing calcium, the DRI Committee did not adjust for calcium bioavailability when setting recommendations. Figure 12-15 ranks selected foods according to their calcium bioavailability.

Some mineral waters provide as much as 500 milligrams of calcium per liter, offering a convenient way to meet both calcium and water needs.[28] Similarly, calcium-fortified orange juice and other fruit and vegetable juices allow a person to obtain both calcium and vitamins easily. Other examples of calcium-fortified foods include high-calcium milk (milk with extra calcium added) and calcium-fortified cereals. Fortified juices and foods help consumers increase calcium intakes, but depending on the calcium sources, the bioavailability may be significantly less than quantities listed on food labels.[29] The accompanying "How To" describes a shortcut method for estimating your calcium intake. Highlight 12 discusses calcium supplements.

A generalization that has been gaining strength throughout this book is supported by the information given here about calcium. A balanced diet that supplies a variety of foods is the best plan to ensure adequacy for all essential nutrients. All food groups should be included, and none should be overemphasized. In our culture, calcium intake is usually inadequate wherever milk is lacking in the diet—whether through ignorance, poverty, simple dislike, fad dieting, lactose intolerance, or allergy. By contrast, iron is usually lacking whenever milk is overemphasized, as Chapter 13 explains.

FIGURE 12-15 **Bioavailability of Calcium from Selected Foods**

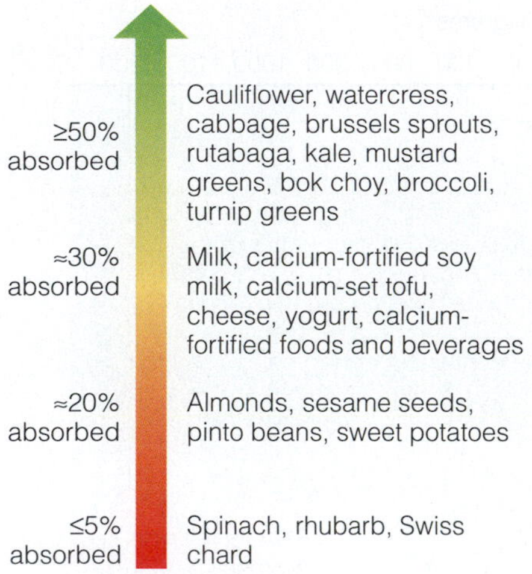

Calcium Deficiency

A low calcium intake during the growing years limits the bones' ability to reach their optimal mass and density. Most people achieve a peak bone mass by their late 20s, and dense bones best protect against age-related bone loss and fractures (see Figure 12-16). All adults lose bone as they grow older, beginning between the ages of 30 and 40. When bone losses reach the point of causing fractures under common, everyday stresses, the condition is known as **osteoporosis**. Osteoporosis and low bone mass (osteopenia) affect an estimated 52 million people in the United States, mostly older women.[30]

FIGURE 12-16 **Phases of Bone Development throughout Life**

The active growth phase occurs from birth to approximately age 20. The next phase of peak bone mass development occurs between the ages of 12 and 30. The final phase, when bone resorption exceeds formation, begins between the ages of 30 and 40 and continues through the remainder of life.

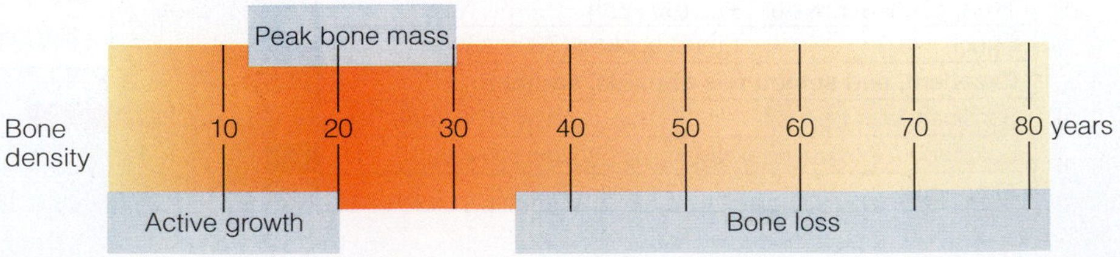

osteoporosis (OS-tee-oh-pore-OH-sis): a disease in which the bones become porous and fragile due to a loss of minerals; also called *adult bone loss.*

- **osteo** = bone
- **porosis** = porous

HOW TO Estimate Your Calcium Intake

Most dietitians have developed useful shortcuts to help them estimate nutrient intakes and "see" inadequacies in the diet. They can tell at a glance whether a day's meals fall short of calcium recommendations, for example.

To estimate calcium intakes, keep two bits of information in mind:

- A cup of milk provides about 300 milligrams of calcium.
- Adults need between 1000 and 1200 milligrams of calcium per day, which represents 3 to 4 cups of milk—or the equivalent:

$$1000 \text{ mg} \div 300 \text{ mg/c} = 3\tfrac{1}{3} \text{ c}$$
$$1200 \text{ mg} \div 300 \text{ mg/c} = 4 \text{ c}$$

If a person drinks 3 to 4 cups of milk a day, it's easy to see that calcium needs are being met. If not, it takes some detective work to identify the other sources and estimate total calcium intake.

To estimate a person's daily calcium intake, use this shortcut, which compares the calcium in calcium-rich foods to the calcium content of milk. The calcium in a cup of milk is assigned 1 point, and the goal is to attain 3 to 4 points per day. Foods are given points as follows:

- 1 c milk, yogurt, or fortified soy milk or 1½ oz cheese = 1 point
- 4 oz canned fish with bones (sardines) = 1 point
- 1 c ice cream, cottage cheese, or calcium-rich vegetable (see the text) = ½ point

Then, because other foods also contribute small amounts of calcium, together they are given a point.

- Well-balanced diet containing a variety of foods = 1 point

Now consider a day's meals with calcium in mind. Cereal with 1 cup of milk for breakfast (1 point for milk), a ham and cheese sub sandwich for lunch (1 point for cheese), and a cup of broccoli and lasagna for dinner (½ point for calcium-rich vegetable and 1 point for cheese in lasagna)—plus 1 point for all other foods eaten that day—adds up to 4½ points. This shortcut estimate indicates that calcium recommendations have been met, and a diet analysis of these few foods reveals a calcium intake of more than 1000 milligrams. By knowing the best sources of each nutrient, you can learn to scan the day's meals and quickly see if you are meeting your daily goals.

For additional practice, log on to **www.cengagebrain.com** and search for ISBN 084006845X.

TRY IT Compare the calcium contents of ½ cup of the following foods: almonds, broccoli, and yogurt.

Unlike many diseases that make themselves known through symptoms such as pain, shortness of breath, skin lesions, tiredness, and the like, osteoporosis is silent. The body sends no signals saying bones are losing their calcium and, as a result, their integrity. Blood samples offer no clues because blood calcium remains normal regardless of bone content, and measures of bone density may not be routinely taken until later in life. Highlight 12 suggests strategies to protect against bone loss, of which eating calcium-rich foods is only one.

IN SUMMARY Most of the body's calcium is in the bones where it provides a rigid structure and a reservoir of calcium for the blood. Blood calcium participates in muscle contraction, blood clotting, and nerve impulses, and it is closely regulated by a system of hormones and vitamin D. Calcium is found predominantly in milk and milk products, but some other foods including certain vegetables and tofu also provide calcium. Even when calcium intake is inadequate, blood calcium remains normal, but at the expense of bone loss, which can lead to osteoporosis. The following table provides a summary of calcium.

Calcium

2011 RDA	Deficiency Symptoms
1000 mg/day (adults 19–50 yr) 1000 mg/day (men 51–70 yr) 1200 mg/day (women 51–70 yr) 1200 mg/day (adults ≥71 yr)	Stunted growth in children; bone loss (osteoporosis) in adults
Upper Level	**Toxicity Symptoms**
Adults: 2500 mg/day	Constipation; increased risk of urinary stone formation and kidney dysfunction; interference with absorption of other minerals
Chief Functions in the Body	**Significant Sources**
Mineralization of bones and teeth; also involved in muscle contraction and relaxation, nerve functioning, blood clotting, blood pressure	Milk and milk products, small fish (with bones), calcium-set tofu (bean curd), greens (bok choy, broccoli, chard, kale), legumes

Phosphorus

Phosphorus is the second most abundant mineral in the body. About 85 percent of it is found combined with calcium in the hydroxyapatite crystals of bones and teeth.

Phosphorus Roles in the Body Phosphorus salts (phosphates) are found not only in bones and teeth, but in all body cells as part of a major buffer system (phosphoric acid and its salts). Phosphorus is also part of DNA and RNA and is therefore necessary for all growth.

Phosphorus assists in energy metabolism. Many enzymes and the B vitamins become active only when a phosphate group is attached. The high-energy compound ATP uses three phosphate groups to do its work.

Lipids containing phosphorus as part of their structures (phospholipids) help to transport other lipids in the blood. Phospholipids are also the major structural components of cell membranes, where they control the transport of nutrients into and out of the cells. Some proteins, such as the casein in milk, contain phosphorus as part of their structures (phosphoproteins).

Phosphorus Recommendations and Intakes Because phosphorus is commonly found in almost all foods, dietary deficiencies are unlikely. As Figure 12-17 shows, foods rich in proteins are the best sources of phosphorus. Milk and cheese contribute about one-fourth of the phosphorus in the U.S. diet.

In the past, researchers emphasized the importance of an ideal calcium-to-phosphorus ratio from the diet to support calcium metabolism, but there is little or no evidence to support this concept. The quantities of calcium and phosphorus in the diet are far more important than their ratio to each other. A high phosphorus intake has been blamed for bone loss when, in fact, a low calcium intake—not a phosphorus toxicity or an improper ratio—is responsible. Research shows that the displacement of milk in the diet by cola drinks, not the phosphoric acid content of the beverages, limits bone density. No adverse effects of high dietary phosphorus intakes have been reported; still, a UL has been established (see inside front cover).

IN SUMMARY **Phosphorus accompanies calcium both in the crystals of bone and in many foods such as milk. Phosphorus is also important in energy metabolism, as part of phospholipids, and as part of the genetic materials DNA and RNA. The accompanying table provides a summary of phosphorus.**

Phosphorus

RDA
Adults: 700 mg/day
Upper Level
Adults (19–70 yr): 4000 mg/day

phosphorus: a major mineral found mostly in the body's bones and teeth.

FIGURE 12-17 **Phosphorus in Selected Foods**

See the "How To" section on p. 316 for more information on using this figure.

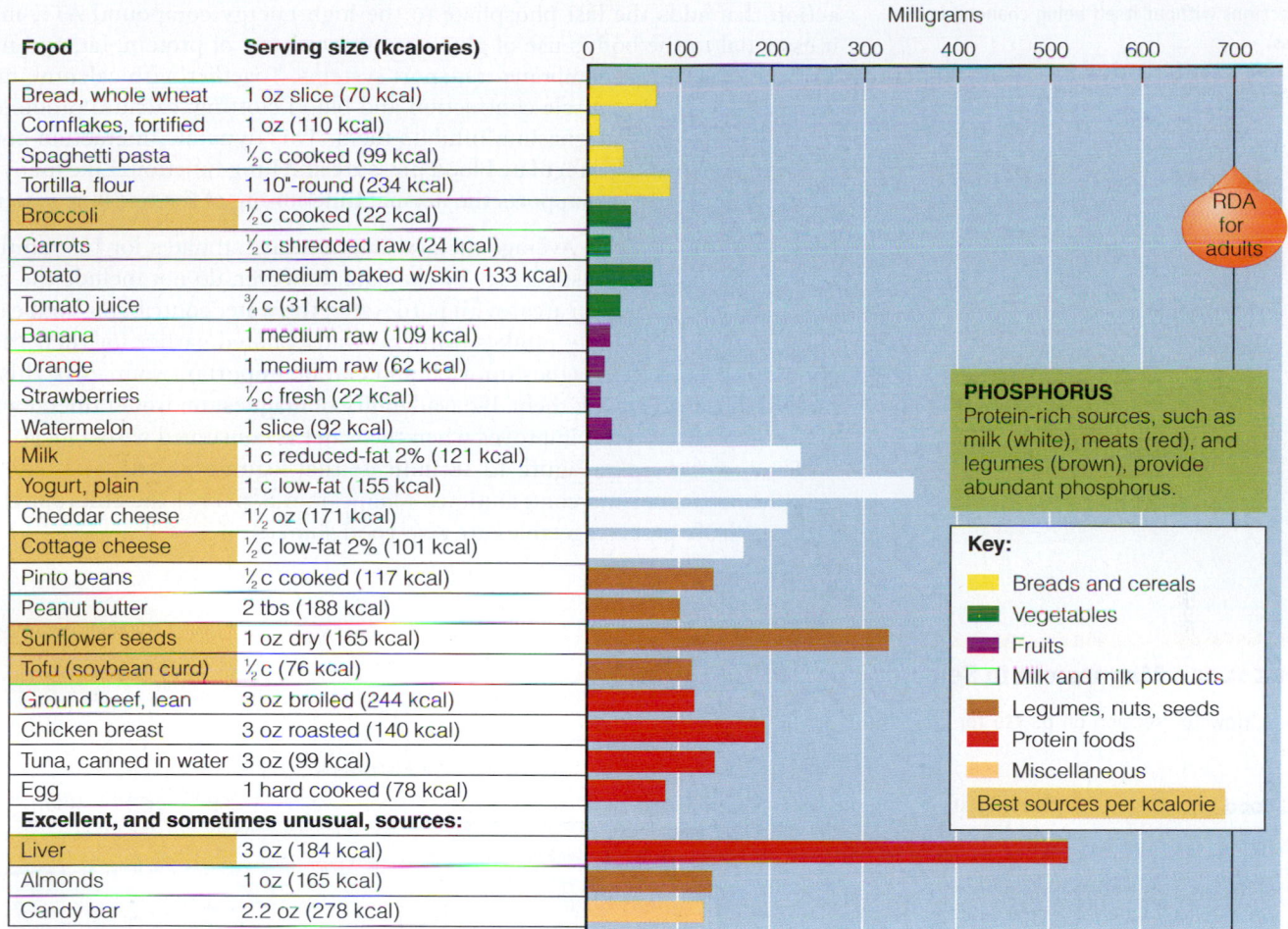

Food	Serving size (kcalories)
Bread, whole wheat	1 oz slice (70 kcal)
Cornflakes, fortified	1 oz (110 kcal)
Spaghetti pasta	½ c cooked (99 kcal)
Tortilla, flour	1 10"-round (234 kcal)
Broccoli	½ c cooked (22 kcal)
Carrots	½ c shredded raw (24 kcal)
Potato	1 medium baked w/skin (133 kcal)
Tomato juice	¾ c (31 kcal)
Banana	1 medium raw (109 kcal)
Orange	1 medium raw (62 kcal)
Strawberries	½ c fresh (22 kcal)
Watermelon	1 slice (92 kcal)
Milk	1 c reduced-fat 2% (121 kcal)
Yogurt, plain	1 c low-fat (155 kcal)
Cheddar cheese	1½ oz (171 kcal)
Cottage cheese	½ c low-fat 2% (101 kcal)
Pinto beans	½ c cooked (117 kcal)
Peanut butter	2 tbs (188 kcal)
Sunflower seeds	1 oz dry (165 kcal)
Tofu (soybean curd)	½ c (76 kcal)
Ground beef, lean	3 oz broiled (244 kcal)
Chicken breast	3 oz roasted (140 kcal)
Tuna, canned in water	3 oz (99 kcal)
Egg	1 hard cooked (78 kcal)
Excellent, and sometimes unusual, sources:	
Liver	3 oz (184 kcal)
Almonds	1 oz (165 kcal)
Candy bar	2.2 oz (278 kcal)

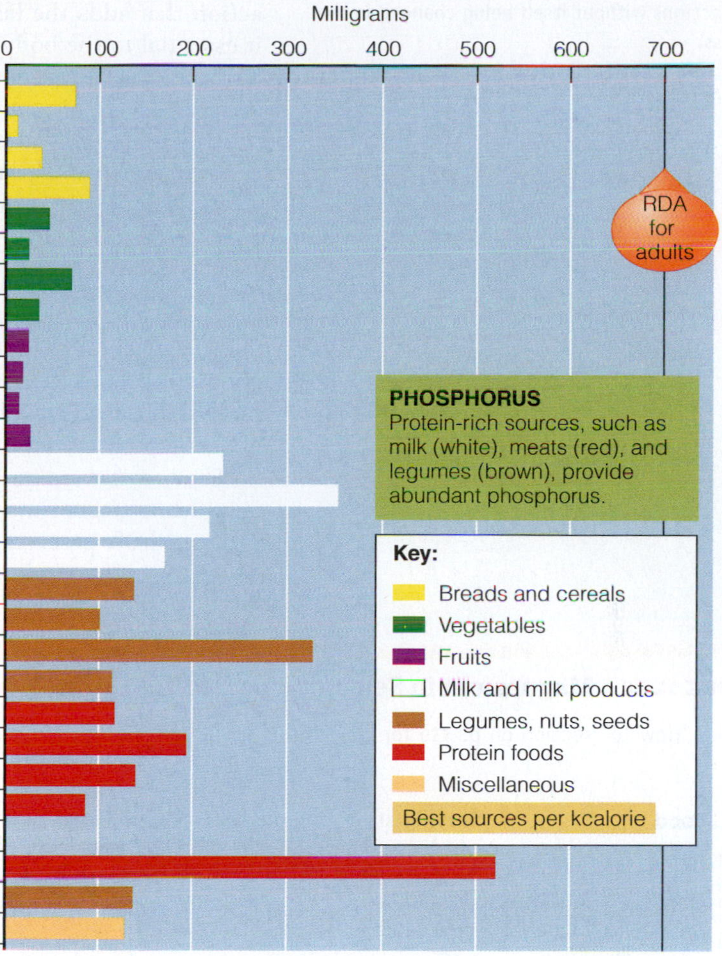

Chief Functions in the Body

Mineralization of bones and teeth; part of every cell; important in genetic material, part of phospholipids, used in energy transfer and in buffer systems that maintain acid-base balance

Deficiency Symptoms

Muscular weakness, bone pain[a]

Toxicity Symptoms

Calcification of nonskeletal tissues, particularly the kidneys

Significant Sources

All animal tissues (meat, fish, poultry, eggs, milk)

[a]Dietary deficiency rarely occurs, but some drugs can bind with phosphorus making it unavailable and resulting in bone loss that is characterized by weakness and pain.

Magnesium

Only about 1 ounce of **magnesium** is present in the body of a 130-pound person. More than half of the body's magnesium is in the bones. Much of the rest is in the muscles and soft tissues, with only 1 percent in the extracellular fluid. As with calcium, bone magnesium may serve as a reservoir to ensure normal blood concentrations.

magnesium: a cation within the body's cells, active in many enzyme systems.

♦ A **catalyst** is a compound that facilitates chemical reactions without itself being changed in the process.

Magnesium Roles in the Body In addition to maintaining bone health, magnesium acts in all the cells of the soft tissues, where it forms part of the protein-making machinery and is necessary for energy metabolism. It participates in hundreds of enzyme systems. A major role of magnesium is as a catalyst ♦ in the reaction that adds the last phosphate to the high-energy compound ATP, making it essential to the body's use of glucose; the synthesis of protein, fat, and nucleic acids; and the cells' membrane transport systems. Together with calcium, magnesium is involved in muscle contraction and blood clotting: calcium promotes the processes, whereas magnesium inhibits them. This dynamic interaction between the two minerals helps regulate blood pressure and lung function. Like many other nutrients, magnesium supports the normal functioning of the immune system.

Magnesium Intakes Average dietary magnesium estimates for U.S. adults fall below recommendations. Dietary intake data, however, do not include the contribution made by water. In areas with hard water, the water contributes both calcium and magnesium to daily intakes. Mineral waters noted earlier for their calcium content may also be magnesium-rich and can be important sources of this mineral for those who drink them. Bioavailability of magnesium from mineral water is about 50 percent, but it improves when the water is consumed with a meal.

The brown bars in Figure 12-18 indicate that legumes, seeds, and nuts make significant magnesium contributions. Magnesium is part of the chlorophyll molecule, so leafy green vegetables are also good sources.

FIGURE 12-18 **Magnesium in Selected Foods**

See the "How To" section on p. 316 for more information on using this figure.

Food	Serving size (kcalories)
Bread, whole wheat	1 oz slice (70 kcal)
Cornflakes, fortified	1 oz (110 kcal)
Spaghetti pasta	½ c cooked (99 kcal)
Tortilla, flour	1 10"-round (234 kcal)
Broccoli	½ c cooked (22 kcal)
Carrots	½ c shredded raw (24 kcal)
Potato	1 medium baked w/skin (133 kcal)
Tomato juice	¾ c (31 kcal)
Banana	1 medium raw (109 kcal)
Orange	1 medium raw (62 kcal)
Strawberries	½ c fresh (22 kcal)
Watermelon	1 slice (92 kcal)
Milk	1 c reduced-fat 2% (121 kcal)
Yogurt, plain	1 c low-fat (155 kcal)
Cheddar cheese	1½ oz (171 kcal)
Cottage cheese	½ c low-fat 2% (101 kcal)
Pinto beans	½ c cooked (117 kcal)
Peanut butter	2 tbs (188 kcal)
Sunflower seeds	1 oz dry (165 kcal)
Tofu (soybean curd)	½ c (76 kcal)
Ground beef, lean	3 oz broiled (244 kcal)
Chicken breast	3 oz roasted (140 kcal)
Tuna, canned in water	3 oz (99 kcal)
Egg	1 hard cooked (78 kcal)
Excellent, and sometimes unusual, sources:	
Halibut	3 oz baked (199 kcal)
Cashews	1 oz (161 kcal)
Artichoke	1 (60 kcal)

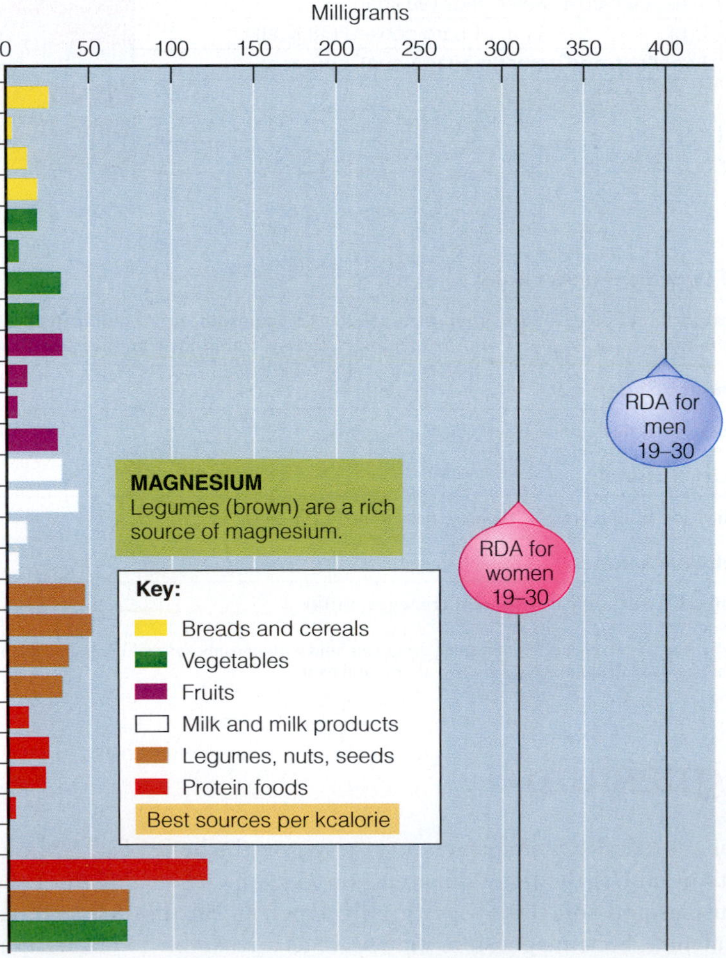

Magnesium Deficiency Even with average magnesium intakes below recommendations, deficiency symptoms rarely appear except with diseases. Magnesium deficiency may develop in cases of alcohol abuse, protein malnutrition, kidney disorders, and prolonged vomiting or diarrhea. People using diuretics may also show symptoms. A severe magnesium deficiency causes a tetany similar to the calcium tetany described earlier. Magnesium deficiencies also impair central nervous system activity and may be responsible for the hallucinations experienced during alcohol withdrawal.

Magnesium and Hypertension Magnesium is critical to heart function and seems to protect against hypertension and heart disease. Interestingly, people living in areas of the country with hard water, which contains high concentrations of calcium and magnesium, tend to have low rates of heart disease. With magnesium deficiency, the walls of the arteries and capillaries tend to constrict—a possible explanation for the hypertensive effect.

Magnesium Toxicity Magnesium toxicity is rare, but it can be fatal. The UL for magnesium applies only to nonfood sources such as supplements or magnesium salts.

IN SUMMARY **Like calcium and phosphorus, magnesium supports bone mineralization. Magnesium is also involved in numerous enzyme systems and in heart function. It is found abundantly in legumes and leafy green vegetables and, in some areas, in water. The accompanying table provides a summary of magnesium.**

Magnesium

RDA	**Deficiency Symptoms**
Men (19–30 yr): 400 mg/day Women (19–30 yr): 310 mg/day	Weakness; confusion; if extreme, convulsions, bizarre muscle movements (especially of eye and face muscles), hallucinations, and difficulty in swallowing; in children, growth failure[a]
Upper Level	**Toxicity Symptoms**
Adults: 350 mg nonfood magnesium/day	From nonfood sources only; diarrhea, alkalosis, dehydration
Chief Functions in the Body	**Significant Sources**
Bone mineralization, building of protein, enzyme action, normal muscle contraction, nerve impulse transmission, maintenance of teeth, and functioning of immune system	Nuts, legumes, whole grains, dark green vegetables, seafood, chocolate, cocoa

[a]A still more severe deficiency causes tetany, an extreme, prolonged contraction of the muscles similar to that caused by low blood calcium.

Sulfate

Sulfate is the oxidized form of the mineral **sulfur**, as it exists in food and water. The body's need for sulfate is easily met by a variety of foods and beverages. In addition, the body receives sulfate from the amino acids methionine and cysteine, which are found in dietary proteins. These sulfur-containing amino acids help determine the contour of protein molecules. The sulfur-containing side chains in cysteine molecules can link to each other via disulfide bridges, which stabilize the protein structure. (See the drawing of insulin with its disulfide bridges on p. 175.) Skin, hair, and nails contain some of the body's more rigid proteins, which have a high sulfur content.

Because the body's sulfate needs are easily met with normal protein intakes, there is no recommended intake for sulfate. Deficiencies do not occur when diets contain protein. Only when people lack protein to the point of severe deficiency will they lack the sulfur-containing amino acids.

sulfate: a salt produced from the oxidation of sulfur.

sulfur: a mineral present in the body as part of some proteins.

IN SUMMARY Like the other nutrients, minerals' actions are coordinated to get the body's work done. The major minerals, especially sodium, chloride, and potassium, influence the body's fluid balance; whenever an anion moves, a cation moves—always maintaining homeostasis. Sodium, chloride, potassium, calcium, and magnesium are key members of the team of nutrients that direct nerve impulse transmission and muscle contraction. They are also the primary nutrients involved in regulating blood pressure. Phosphorus and magnesium participate in many reactions involving glucose, fatty acids, amino acids, and the vitamins. Calcium, phosphorus, and magnesium combine to form the structure of the bones and teeth. Each major mineral also plays other specific roles in the body. The accompanying table provides a summary of major minerals.

The Major Minerals

Mineral and Chief Functions	Deficiency Symptoms	Toxicity Symptoms	Significant Sources
Sodium Maintains normal fluid and electrolyte balance; assists in nerve impulse transmission and muscle contraction	Muscle cramps, mental apathy, loss of appetite	Edema, acute hypertension	Table salt, soy sauce; moderate amounts in meats, milks, breads, and vegetables; large amounts in processed foods
Chloride Maintains normal fluid and electrolyte balance; part of hydrochloric acid found in the stomach, necessary for proper digestion	Do not occur under normal circumstances	Vomiting	Table salt, soy sauce; moderate amounts in meats, milks, eggs; large amounts in processed foods
Potassium Maintains normal fluid and electrolyte balance; facilitates many reactions; supports cell integrity; assists in nerve impulse transmission and muscle contractions	Irregular heartbeat, muscular weakness, glucose intolerance	Muscular weakness; vomiting; if given into a vein, can stop the heart	All whole foods; meats, milks, fruits, vegetables, grains, legumes
Calcium Mineralization of bones and teeth; also involved in muscle contraction and relaxation, nerve functioning, blood clotting, and blood pressure	Stunted growth in children; bone loss (osteoporosis) in adults	Constipation; increased risk of urinary stone formation and kidney dysfunction; interference with absorption of other minerals	Milk and milk products, small fish (with bones), tofu, greens (bok choy, broccoli, chard), legumes
Phosphorus Mineralization of bones and teeth; part of every cell; important in genetic material, part of phospholipids, used in energy transfer and in buffer systems that maintain acid-base balance	Muscular weakness, bone pain[a]	Calcification of nonskeletal tissues, particularly the kidneys	All animal tissues (meat, fish, poultry, eggs, milk)
Magnesium Bone mineralization, building of protein, enzyme action, normal muscle contraction, nerve impulse transmission, maintenance of teeth, and functioning of immune system	Weakness; confusion; if extreme, convulsions, bizarre muscle movements (especially of eye and face muscles), hallucinations, and difficulty in swallowing; in children, growth failure[b]	From nonfood sources only; diarrhea, alkalosis, dehydration	Nuts, legumes, whole grains, dark green vegetables, seafood, chocolate, cocoa
Sulfate As part of proteins, stabilizes their shape by forming disulfide bridges; part of the vitamins biotin and thiamin and the hormone insulin	None known; protein deficiency would occur first	Toxicity would occur only if sulfur-containing amino acids were eaten in excess; this (in animals) suppresses growth	All protein-containing foods (meats, fish, poultry, eggs, milk, legumes, nuts)

[a]Dietary deficiency rarely occurs, but some drugs can bind with phosphorus making it unavailable and resulting in bone loss that is characterized by weakness and pain.

[b]A still more severe deficiency causes tetany, an extreme, prolonged contraction of the muscles similar to that caused by low blood calcium.

With all of the tasks these minerals perform, they are of great importance to life. Consuming enough of each of them every day is easy, given a variety of foods from each of the food groups. Whole-grain breads supply magnesium; fruits, vegetables, and legumes provide magnesium and potassium, too; milks offer calcium and phosphorus; meats offer phosphorus and sulfate as well; all foods provide sodium and chloride, with excesses being more problematic than inadequacies. The message is quite simple and has been repeated throughout this text: for an adequate intake of all the nutrients, including the major minerals, choose different foods from each of the five food groups. And drink plenty of water.

Nutrition Portfolio

Many people may miss the mark when it comes to drinking enough water to keep their bodies well hydrated or obtaining enough calcium to promote strong bones; in contrast, sodium intakes often exceed those recommended for health.

Go to Diet Analysis Plus and choose one of the days on which you tracked your diet for an entire day. Select the Intake vs. Goals report and then consider the following questions. Remember that scoring 100 percent on this report means you met your goal.

- Were you above, below, or at your goal for water intake? Was that a typical day for you? Describe your strategy for ensuring that you drink plenty of water—about eight glasses—every day.
- Take a look at your sodium intake in this report. Most people in the United States exceed the UL. Did you? Explain the importance of selecting and preparing foods with less salt.
- How was your intake of calcium for that day? If you are not getting enough calcium, consult Chapter 12 for ideas to help you get more, then list at least three foods or beverages you would be willing to eat or drink that would improve your intake.

Diet Analysis PLUS **To complete this exercise, go to your Diet Analysis Plus at www.cengage.com/sso.**

Nutrition on the Net

For further study of topics covered in this chapter, log on to **www.cengagebrain.com** and search for ISBN 084006845X.

- Search for "minerals" at the American Dietetic Association site: **www.eatright.org**
- Learn about sodium from the American Heart Association: **www.americanheart.org**
- Find tips and recipes for including more milk in the diet: **www.whymilk.com**
- Learn about the benefits of calcium from the National Dairy Council: **www.nationaldairycouncil.org**
- Learn more about the DASH diet: **www.nhlbi.nih.gov/health/public/heart/hbp/dash/new_dash.pdf**

References

1. A. K. Johnson, The sensory psychobiology of thirst and salt appetite, *Medicine & Science in Sports & Exercise* 39 (2007): 1388–1400.
2. F. Manz and A. Wentz, Hydration status in the United States and Germany, *Nutrition Reviews* 63 (2005): S55–S62.
3. M. N. Sawka, S. N. Cheuvront, and R. Carter III, Human water needs, *Nutrition Reviews* 63 (2005): S30–S39.
4. B. M. Popkin and coauthors, A new proposed guidance system for beverage consumption in the United States, *American Journal of Clinical Nutrition* 83 (2006): 529–542.
5. Committee on Dietary Reference Intakes, *Dietary Reference Intakes for Water, Potassium, Sodium, Chloride, and Sulfate* (Washington, D.C.: National Academies Press, 2004), p. 67.
6. F. Manz and A. Wentz, The importance of good hydration for the prevention of chronic diseases, *Nutrition Reviews* 63 (2005): S2–S5.
7. P. Ritz and G. Berrut, The importance of good hydration for day-to-day health, *Nutrition Reviews* 63 (2005): S6–S13.
8. Actions You Can Take to Reduce Lead in Drinking Water, www.epa.gov/ogwdw/lead/lead1.html, updated April 2008.
9. K. M. O'Shaughnessy and F. E. Karet, Salt handling and hypertension, *Annual Review of Nutrition* 26 (2006): 343–365.
10. Centers for Disease Control and Prevention, Application of lower sodium intake recommendations to adults: United States, 1999–2006, *Morbidity and Mortality Weekly Report* 58 (2009): 281–283.
11. B. Rodriguez-Iturbe and N. D. Vaziri, Salt-sensitive hyptertension—Update on novel findings, *Nephrology, Dialysis, Transplantation: Official Publication of the European Dialysis and Transplant Association* 22 (2007): 992–995.
12. N. R. Cook and coauthors, Long-term effects of dietary sodium reduction on cardiovascular disease outcomes: Observational follow-up of the trials of hypertension prevention (TOHP), *British Medical Journal* 334 (2007): 885–888.
13. R. A. Forshee, Innovative regulatory approaches to reduce sodium consumption: Could a cap-and-trade system work? *Nutrition Reviews* 66 (2008): 280–285; S. Havas, B. D. Dickinson, and M. Wilson, The urgent need to reduce sodium consumption, *Journal of the American Medical Association* 298 (2007): 1439–1441; O'Shaughnessy and Karet, 2006.
14. H. J. Adrogué and N. E. Madias, Sodium and potassium in the pathogenesis of hypertension, *New England Journal of Medicine* 356 (2007): 1966–1978.
15. M. Umesawa and coauthors, Relations between dietary sodium and potassium intakes and mortality from cardiovascular disease: The Japan Collaborative Cohort Study for Evaluation of Cancer Risks, *American Journal of Clinical Nutrition* 88 (2008): 195–202; Adrogué and Madias, 2007.
16. P. R. Trumbo and K. C. Ellwood, Supplemental calcium and risk reduction of hypertension, pregnancy-induced hypertension, and

preeclampsia: An evidence-based review by the US Food and Drug Administration, *Nutrition Reviews* 65 (2007): 78–87.

17. Trumbo and Ellwood, 2007.
18. J. Ishihara and coauthors, Dietary calcium, vitamin D, and the risk of colorectal cancer, *American Journal of Clinical Nutrition* 88 (2008): 1576–1583; C. S. Guerreiro and coauthors, The *D1822V APC* polymorphism interacts with fat, calcium, and fiber intakes in modulating the risk of colorectal cancer in Portuguese persons, *American Journal of Clinical Nutrition* 85 (2007): 1592–1597; M. E. Martínez and E. T. Jacobs, Calcium supplementation and prevention of colorectal neoplasia: Lessons from clinical trials, *Journal of the National Cancer Institute* 99 (2007): 99–100; S. C. Larsson and coauthors, Calcium and dairy food intakes are inversely associated with colorectal cancer risk in the Cohort of Swedish Men, *American Journal of Clinical Nutrition* 83 (2006): 667–673; A. Flood and coauthors, Calcium from diet and supplements is associated with reduced risk of colorectal cancer in a prospective cohort of women, *Cancer Epidemiology, Biomarkers, and Prevention* 14 (2005): 126–132.
19. R. P. Heaney and K. Rafferty, Preponderance of the evidence: An example from the issue of calcium intake and body composition, *Nutrition Reviews* 67 (2009): 32–39; G. C. Major and coauthors, Recent developments in calcium-related obesity research, *Obesity Reviews* 9 (2008): 428–445; G. Barba and P. Russo, Dairy foods, dietary calcium and obesity: A short review of the evidence, *Nutrition, Metabolism, and Cardiovascular Diseases* 16 (2006): 445–451; S. Schrager, Dietary calcium intake and obesity, *Journal of the American Board of Family Practice* 18 (2005): 205–210.
20. J. K. Lorenzen and coauthors, Calcium supplementation for 1 y does not reduce body weight or fat mass in young girls, *American Journal of Clinical Nutrition* 83 (2006): 18–23.
21. M. Bortolotti and coauthors, Dairy calcium supplementation in overweight or obese persons: Its effect on markers of fat metabolism, *American Journal of Clinical Nutrition* 88 (2008): 877–885; S. N. Rajpathak and coauthors, Calcium and dairy intakes in relation to long-term weight gain in US men, *American Journal of Clinical Nutrition* 83 (2006): 559–566; C. W. Gunther and coauthors, Dairy products do not lead to alterations in body weight or fat mass in young women in a 1-y intervention, *American Journal of Clinical Nutrition* 81 (2005): 751–756.
22. R. C. Khanal and I. Nemere, Regulation of intestinal calcium transport, *Annual Review of Nutrition* 28 (2008): 179–196.
23. F. Bronner, Recent developments in intestinal calcium absorption, *Nutrition Reviews* 67 (2009): 109–113.
24. J. Ma, R. A. Johns, and R. S. Stafford, Americans are not meeting current calcium recommendations, *American Journal of Clinical Nutrition* 85 (2007): 1361–1366.
25. R. D. Jackson and coauthors, Calcium plus vitamin D supplementation and the risk of fractures, *New England Journal of Medicine* 354 (2006): 669–683.
26. X. Gao and coauthors, Meeting adequate intake for dietary calcium without dairy foods in adolescents aged 9 to 18 years (National Health and Nutrition Examination Survey 2001–2002), *Journal of the American Dietetic Association* 106 (2006): 1759–1765.
27. F. R. Greer, N. F. Krebs, and the Committee on Nutrition, Optimizing bone health and calcium intakes of infants, children, and adolescents, *Pediatrics* 117 (2006): 578–585.
28. R. P. Heaney, Absorbability and utility of calcium in mineral waters, *American Journal of Clinical Nutrition* 84 (2006): 371–374.
29. R. P. Heaney and coauthors, Calcium fortification systems differ in bioavailability, *Journal of the American Dietetic Association* 105 (2005): 807–809.
30. National Osteoporosis Foundation, America's bone health: The state of osteoporosis and low bone mass, 2008, www.nof.org, accessed January 2009; S. Khosla and L. J. Melton III, Osteopenia, *New England Journal of Medicine* 356 (2007): 2293–2300.

HIGHLIGHT 12

© ONOKY-Photononstop/Alamy

Osteoporosis and Calcium

Osteoporosis becomes apparent during the later years, but it develops much earlier—and without warning. Few people are aware that their bones are being robbed of their strength. The problem often first becomes evident when someone's hip suddenly gives way. People say, "She fell and broke her hip," but in fact the hip may have been so fragile that it broke *before* she fell. Even bumping into a table may be enough to shatter a porous bone into fragments so numerous and scattered that they cannot be reassembled. Removing them and replacing them with an artificial joint requires major surgery. An estimated 300,000 people in the United States are hospitalized each year because of hip fractures related to osteoporosis. About a fourth die of complications within a year. A fourth of those who survive will never walk or live independently again. Their quality of life slips downward.

This highlight examines osteoporosis, one of the most prevalent diseases of aging, affecting an estimated 52 million people in the United States—most of them women older than 50.[1] It reviews the many factors that contribute to the 1.5 million breaks in the bones of the hips, vertebrae, wrists, arms, and ankles each year. And it presents strategies to reduce the risks, paying special attention to the role of dietary calcium.

Bone Development and Disintegration

Bone has two compartments: the outer, hard shell of **cortical bone** and the inner, lacy matrix of **trabecular bone.** (The glossary defines these and other bone-related terms.) Both can lose minerals, but in different ways and at different rates. The photograph on p. 414 shows a human leg bone sliced lengthwise, exposing the lacy, calcium-containing crystals of trabecular bone. These crystals give up calcium to the blood when the diet runs short, and they take up calcium again when the supply is plentiful (review Figure 12-13 on p. 402). For people who have eaten calcium-rich foods throughout the bone-forming years of their youth, these deposits make bones dense and provide a rich reservoir of calcium.

Surrounding and protecting the trabecular bone is a dense, ivorylike exterior shell—the cortical bone. Cortical bone composes the shafts of the long bones, and a thin cortical shell caps the end of the bone, too. Both compartments confer strength on bone: cortical bone provides the sturdy outer wall, and trabecular bone provides support along the lines of stress.

The two types of bone play different roles in calcium balance and osteoporosis. Supplied with blood vessels and metabolically active, trabecular bone is sensitive to hormones that govern day-to-day deposits and withdrawals of calcium. It readily gives up minerals whenever blood calcium needs replenishing. Losses of trabecular bone start becoming significant for men and women in their 30s, although losses can occur whenever calcium withdrawals exceed deposits.

GLOSSARY

antacids: medications used to relieve indigestion by neutralizing acid in the stomach. Calcium-containing preparations (such as Tums) contain available calcium. Antacids with aluminum or magnesium hydroxides (such as Rolaids) can accelerate calcium losses.

bone meal or **powdered bone:** crushed or ground bone preparations intended to supply calcium to the diet. Calcium from bone is not well absorbed and is often contaminated with toxic minerals such as arsenic, mercury, lead, and cadmium.

bone density: a measure of bone strength. When minerals fill the bone matrix (making it dense), they give it strength.

cortical bone: the very dense bone tissue that forms the outer shell surrounding trabecular bone and comprises the shaft of a long bone.

dolomite: a compound of minerals (calcium magnesium carbonate) found in limestone and marble. Dolomite is powdered and is sold as a calcium-magnesium supplement. However, it may be contaminated with toxic minerals, is not well absorbed, and interacts adversely with absorption of other esssential minerals.

osteoporosis: a disease characterized by porous and fragile bones.

oyster shell: a product made from the powdered shells of oysters that is sold as a calcium supplement, but it is not well absorbed by the digestive system.

trabecular (tra-BECK-you-lar) **bone:** the lacy inner structure of calcium crystals that supports the bone's structure and provides a calcium storage bank.

type I osteoporosis: osteoporosis characterized by rapid bone losses, primarily of trabecular bone.

type II osteoporosis: osteoporosis characterized by gradual losses of both trabecular and cortical bone.

Cortical bone also gives up calcium, but slowly and at a steady pace. Cortical bone losses typically begin at about age 40 and continue slowly but surely thereafter.

Losses of trabecular and cortical bone reflect two types of osteoporosis, which cause two types of bone breaks. **Type I osteoporosis** involves losses of trabecular bone (see Figure H12-1). These losses sometimes exceed three times the expected rate, and bone breaks may occur suddenly. Trabecular bone becomes so fragile that even the body's own weight can overburden the spine—vertebrae may suddenly disintegrate and crush down, painfully pinching major nerves.[2] Wrists may break as bone ends weaken, and teeth may loosen or fall out as the trabecular bone of the jaw recedes. Women are most often the victims of this type of osteoporosis, outnumbering men six to one.

In **type II osteoporosis,** the calcium of both cortical and trabecular bone is drawn out of storage, but slowly over the years. As old age approaches, the vertebrae may compress into wedge shapes, forming what is often called a "dowager's hump," the posture many older people assume as they "grow shorter." Figure H12-2 shows the effect of compressed spinal bone on a woman's height and posture. Because both the cortical shell and the trabecular interior weaken, breaks most often occur in the hip, as mentioned in the introductory paragraph. A woman is twice as likely as a man to suffer type II osteoporosis.

Table H12-1 summarizes the differences between the two types of osteoporosis. Physicians can diagnose osteoporosis and assess the risk of bone fractures by measuring **bone density** using dual-energy X-ray absorptiometry (DEXA scan) or ultrasound. They also consider risk factors that predict bone fractures, including age, personal and family history of fracture, BMI, and physical inactiv-

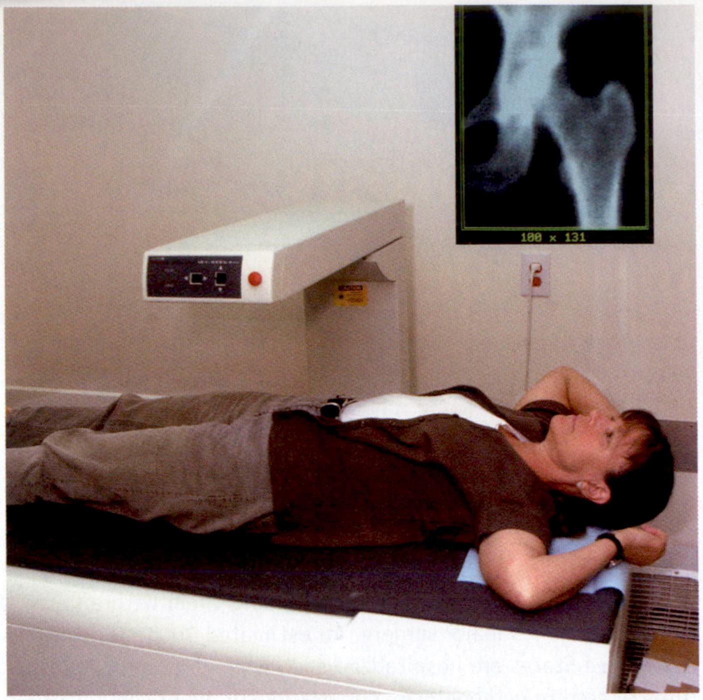

© David Young Wolff/PhotoEdit

Using a DEXA (dual-energy X-ray absorpiometry) test to measure bone mineral density identifies osteoporosis, determines risks for fractures, and tracks responses to treatment.

FIGURE H12-1 Healthy and Osteoporotic Trabecular Bones

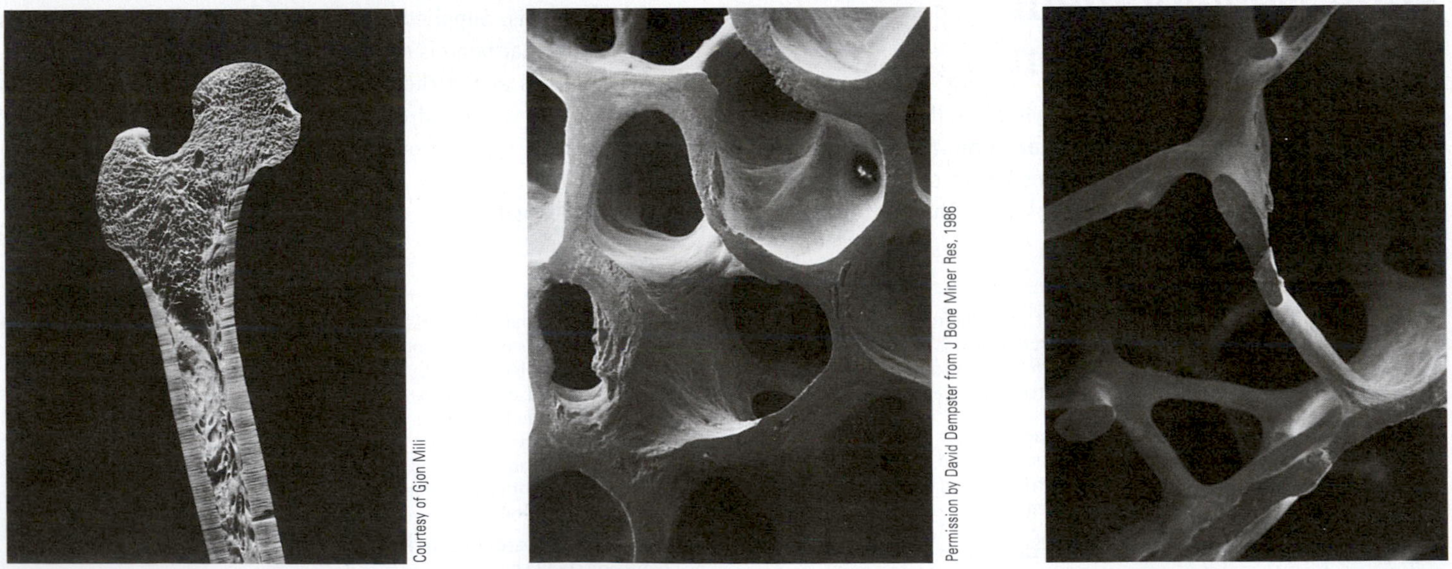

Courtesy of Gjon Mili

Trabecular bone is the lacy network of calcium-containing crystals that fills the interior. Cortical bone is the dense, ivorylike bone that forms the exterior shell.

Permission by David Dempster from J Bone Miner Res, 1986

Electron micrograph of healthy trabecular bone.

Permission by David Dempster from J Bone Miner Res, 1986 (both)

Electron micrograph of trabecular bone affected by osteoporosis.

FIGURE H12-2 Loss of Height in a Woman Caused by Osteoporosis

The woman on the left is about 50 years old. On the right, she is 80 years old. Her legs have not grown shorter. Instead, her back has lost length due to collapse of her spinal bones (vertebrae). Collapsed vertebrae cannot protect the spinal nerves from pressure that causes excruciating pain.

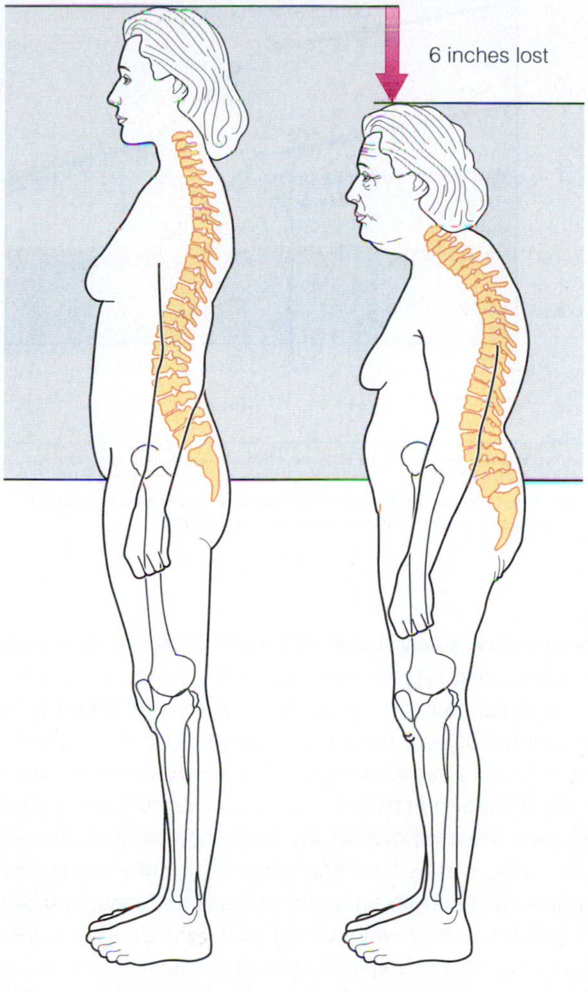

ity.[3] Table H12-2 summarizes the major risk factors and protective factors for osteoporosis. The more risk factors that apply to a person, the greater the chances of bone loss. Notice that several risk factors that are influential in the development of osteoporosis—such as age, gender, and genetics—cannot be changed. Other risk factors—such as diet, physical activity, body weight, smoking, and alcohol use—are personal behaviors that can be changed. By eating a calcium-rich, well-balanced diet, being physically active, abstaining from smoking, and drinking alcohol in moderation (if at all), people can defend themselves against osteoporosis. These decisions are particularly important for those with other risk factors that cannot be changed.

TABLE H12-1 Types of Osteoporosis Compared

	Type I	Type II
Other name	Postmenopausal osteoporosis	Senile osteoporosis
Age of onset	50 to 70 years old	70 years and older
Bone loss	Trabecular bone	Both trabecular and cortical bone
Fracture sites	Wrist and spine	Hip
Gender incidence	6 women to 1 man	2 women to 1 man
Primary causes	Rapid loss of estrogen in women following menopause; loss of testosterone in men with advancing age	Reduced calcium absorption, increased bone mineral loss, increased propensity to fall

TABLE H12-2 Risk Factors and Protective Factors for Osteoporosis

Risk Factors	Protective Factors
• Older age	• Younger age
• Low BMI	• High BMI
• Caucasian, Asian, or Hispanic heritage	• African American heritage
• Cigarette smoking	• No smoking
• Alcohol consumption in excess	• Alcohol consumption in moderation
• Sedentary lifestyle	• Regular weight-bearing exercise
• Use of glucocorticoids or anticonvulsants	• Use of diuretics
• Female gender	• Male gender
• Maternal history of osteoporosis fracture or personal history of fracture	• Bone density assessment and treatment (if necessary)
• Estrogen deficiency in women (amenorrhea or menopause, especially early or surgically induced); testosterone deficiency in men	• Use of estrogen therapy
• Lifetime diet inadequate in calcium and vitamin D	• Lifetime diet rich in calcium and vitamin D

Whether a person develops osteoporosis seems to depend on the interactions of several factors, including nutrition. Age is the strongest predictor of bone density: osteoporosis is responsible for 90 percent of the hip fractures in women and 80 percent in men older than the age of 65.

Age and Bone Calcium

Two major stages of life are critical in the development of osteoporosis. The first is the bone-acquiring stage of childhood and adolescence. The second is the bone-losing decades of late adulthood, especially in women after menopause. The bones gain strength and density all through the growing years and into young adulthood. As people age, the cells that build bone gradually become less active, but those that dismantle bone continue working. The result is that bone loss exceeds bone formation. Some bone loss is inevitable, but losses can be curtailed by maximizing bone mass.

Maximizing Bone Mass

To maximize bone mass, the diet must deliver an adequate supply of calcium during the first three decades of life. Children and teens who get enough calcium and vitamin D have denser bones than those with inadequate intakes.[4] With little or no calcium from the diet, the body must depend on bone to supply calcium to the blood—bone mass diminishes, and bones lose their density and strength. When people reach the bone-losing years of middle age, those who formed dense bones during their youth have the advantage. They simply have more bone starting out and can lose more before suffering ill effects. Figure H12-3 demonstrates this effect.

FIGURE H12-3 Bone Losses over Time Compared

Peak bone mass is achieved by age 30. Women gradually lose bone mass until menopause, when losses accelerate dramatically and then gradually taper off.

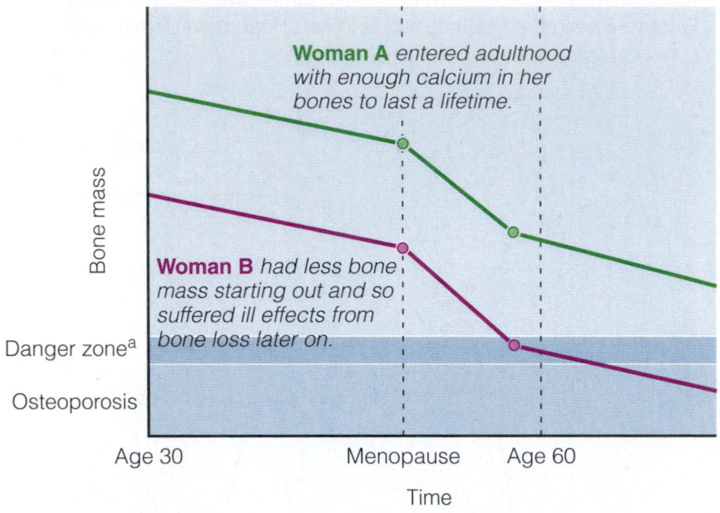

Woman A entered adulthood with enough calcium in her bones to last a lifetime.
Woman B had less bone mass starting out and so suffered ill effects from bone loss later on.
Bone mass
Danger zone[a]
Osteoporosis
Age 30
Menopause
Age 60
Time

[a]People with a moderate degree of bone mass reduction are said to have *osteopenia* and are at increased risk of fractures.

SOURCE: Data from Committee on Dietary Reference Intakes, *Dietary Reference Intakes for Calcium and Vitamin D* (Washington, D.C.: National Academies Press, 2011).

Minimizing Bone Loss

Not only does dietary calcium build strong bones in youth, but it remains important in protecting against losses in the later years. Unfortunately, calcium intakes of older adults are typically low, and calcium absorption declines after menopause. The kidneys do not activate vitamin D as well as they did earlier (recall that active vitamin D enhances calcium absorption). Also, sunlight is needed to form vitamin D, and many older people spend little or no time outdoors in the sunshine. For these reasons, and because intakes of vitamin D are typically low anyway, blood vitamin D declines.

Some of the hormones that regulate bone and calcium metabolism—parathyroid hormone, calcitonin, and estrogen—also change with age and accelerate bone mineral withdrawal. Together, these age-related factors contribute to bone loss: inefficient bone remodeling, reduced calcium intakes, impaired calcium absorption, poor vitamin D status, and hormonal changes that favor bone mineral withdrawal.

Gender and Hormones

After age, gender is the next strongest predictor of osteoporosis. Men have greater bone density than women at maturity, and women have greater losses than men in later life. Consequently, men develop bone problems about 10 years later than women, and women account for four out of five cases of osteoporosis. Menopause imperils women's bones. Bone dwindles rapidly when the hormone estrogen diminishes and menstruation ceases. The lack of estrogen contributes to the release of cytokines that produce inflammation and accelerate bone loss.[5] Women may lose up to 20 percent of their bone mass during the six to eight years following menopause. Eventually, losses taper off so that women again lose bone at the same rate as men their age. Losses of bone minerals continue throughout the remainder of a woman's lifetime, but not at the free-fall pace of the menopause years (review Figure H12-3).

Rapid bone losses also occur when *young* women's ovaries fail to produce enough estrogen, causing menstruation to cease. In some cases, diseased ovaries are to blame and must be removed; in others, the ovaries fail to produce sufficient estrogen because the women suffer from anorexia nervosa and have unreasonably restricted their body weight (see Highlight 8). The amenorrhea and low body weights explain much of the bone loss seen in these young women, even years after diagnosis and treatment. Estrogen therapy can help nonmenstruating women prevent further bone loss and reduce the incidence of fractures. Because estrogen therapy may increase the risks for breast cancer, women must carefully weigh any potential benefits against the possible dangers.

The two main classes of drugs used to prevent or treat osteoporosis are antiresorptive agents that block bone resorption by inhibiting osteoclast activity (examples include raloxifene, alendronate, risedronate, and calcitonin) and anabolic agents that stimulate bone formation by acting on osteoblasts (an example is parathyroid hormone).*[6] A combination of these drugs or of hormone replacement and a drug may be most beneficial.[7]

*Raloxifene (rah-LOX-ih-feen) is a selective estrogen-receptor modulator (SERM), marketed as Evista; alendronate (a-LEN-droe-nate) is a bisphosphonate, marketed as Fosamax; risedronate (rih-SEH-droe-nate) is a bisphosphonate, marketed as Actonel; and calcitonin is a hormone, marketed as Calcimar and Miacalcin.

Some women who choose not to use estrogen therapy turn to soy as an alternative treatment. Interestingly, the phytochemicals commonly found in soybeans mimic the actions of estrogen in the body. When natural estrogen is lacking, as after menopause, these phytochemicals may step in to stimulate estrogen-sensitive tissues. By way of this action, soy and its phytochemicals may help to prevent the rapid bone losses of the menopause years.[8] Unfortunately, soy photochemicals may also stimulate breast cancer cell growth in some women.[9] Because the risks and benefits vary depending on life-stage and prior history of breast cancer, women should discuss soy options with their physicians.[10]

If estrogen deficiency is a major cause of osteoporosis in women, what is the cause of bone loss in men? The male sex hormone testosterone appears to play a role. Men with low levels of testosterone, as occurs after removal of diseased testes or when testes lose function with aging, suffer more fractures. Other common causes of osteoporosis in men include corticosteroid use and alcohol abuse.[11] Treatment for men with osteoporosis includes testosterone replacement therapy. Thus both male and female sex hormones participate in the development and treatment of osteoporosis.

Genetics and Ethnicity

Osteoporosis may, in part, be hereditary, and family history of osteoporosis or fracture is a risk factor.[12] The exact role of genetics is unclear, but it most likely influences both the peak bone mass achieved during growth and the bone loss incurred during the later years. The extent to which a given genetic potential is realized, however, depends on many outside factors. Diet and physical activity, for example, can maximize peak bone density during growth, whereas alcohol and tobacco abuse can accelerate bone losses later in life.

Risks of osteoporosis appear to run along racial lines and reflect genetic differences in bone development. African Americans, for example, seem to use and retain calcium more efficiently than Caucasians.[13] Consequently, even though their calcium intakes are typically lower, black people have denser bones than white people do. Greater bone density expresses itself in less bone loss, fewer fractures, and a lower rate of osteoporosis among blacks.[14] Fractures, for example, are about twice as likely in white women age 65 or older as in black women.

Other ethnic groups have a high risk of osteoporosis. Asians from China and Japan, Mexican Americans, Hispanic people from Central and South America, and Inuit people from St. Lawrence Island typically have lower bone density than Caucasians. One might expect that these groups would suffer more bone fractures, but this is not always the case. Again, genetic differences may explain why. Asians, for example, generally have small, compact hips, which makes them less susceptible to fractures.

Findings from around the world demonstrate that although a person's genes may lay the groundwork for bone health, environmental factors influence the genes' ultimate expression. Diet in general, and calcium in particular, are among those environmental factors. Others include physical activity, body weight, smoking, and alcohol. Importantly, all of these factors are within a person's control.

Physical Activity and Body Weight

Physical activity may be the single most important factor supporting bone growth during adolescence.[15] Furthermore, physical activity during the adolescent years of bone growth appears to have lasting benefits for older women.[16] Muscle strength and bone strength go together. When muscles work, they pull on the bones, stimulating them to develop more trabeculae and grow denser. The hormones that promote new muscle growth also favor the building of bone. As a result, active bones are denser and stronger than sedentary bones.[17]

To keep bones healthy, a person should engage in weight training or weight-bearing endurance activities (such as tennis and jogging or vigorous walking) regularly.[18] Regular physical activity combined with an adequate calcium intake helps to maximize bone density in adolescence.[19] Adults can also maximize and maintain bone density with a regular program of weight training. Even past menopause, when most women are losing bone, weight training improves bone density.

Heavier body weights and weight gains place a similar stress on the bones and promote their density. In fact, overweight may protect bones against the negative effects of a low-calcium diet.[20] Interestingly, leptin may play a key role in the relationship between body weight and bone mass.[21] Obese mice that are deficient

© Tetra Images/Alamy

Strength training helps to build strong bones.

in leptin show increases in bone formation. In contrast, weight losses reduce bone density and increase the risk of fractures—in part because energy restriction diminishes calcium absorption and compromises calcium balance. When calcium intake meets recommendations, however, calcium absorption is sufficient to maintain bone density during weight loss.[22] As mentioned in Highlight 8, the combination of underweight, severely restricted energy intake, extreme daily exercise, and amenorrhea reliably predicts bone loss.

Smoking and Alcohol

Add bone damage to the list of ill consequences associated with smoking. The bones of smokers are less dense than those of nonsmokers—even after controlling for differences in age, body weight, and physical activity habits.[23] Fortunately, the damaging effects can be reversed with smoking cessation. Blood indicators of beneficial bone activity are apparent six weeks after a person stops smoking. In time, bone density is similar for former smokers and nonsmokers.

People who abuse alcohol often suffer from osteoporosis and experience more bone breaks than others. Several factors appear to be involved. Alcohol enhances fluid excretion, leading to excessive calcium losses in the urine; upsets the hormonal balance required for healthy bones; slows bone formation, leading to lower bone density; stimulates bone breakdown; and increases the risk of falling. Limited research suggests that *moderate* alcohol consumption increases bone mineral density.[24]

Dietary Calcium

For older adults, an adequate calcium intake alone cannot protect against bone fractures.[25] Bone strength later in life depends most on how well the bones were built during childhood and adolescence. Adequate calcium nutrition during the growing years is essential to achieving optimal peak bone mass. Simply put, growing children who do not get enough calcium do not have strong bones. Neither do adults who did not get enough calcium during their childhood and adolescence. To that end, the DRI Committee recommends 1300 milligrams of calcium per day for everyone 9 through 18 years of age. Unfortunately, few girls meet the recommendations for calcium during these bone-forming years. (Boys generally obtain intakes close to those recommended because they eat more food.) Consequently, most girls start their adult years with less-than-optimal bone density. As adults, women rarely meet their recommended intakes of 1000 to 1200 milligrams from food. Some authorities suggest 1500 milligrams of calcium for postmenopausal women who are not receiving estrogen, but they warn that intakes exceeding the UL could cause health problems.

Other Nutrients

Much research has focused on calcium, but other nutrients support bone health, too.[26] Adequate protein protects bones and reduces the likelihood of hip fractures.[27] As mentioned earlier, vitamin D is needed to maintain calcium metabolism and optimal bone health.[28] Supplementation with vitamin D reduces bone loss and the risk of fractures.[29] Vitamin K decreases bone turnover and protects against hip fractures. Vitamin C may slow bone losses.[30] The minerals magnesium and potassium also help to maintain bone mineral density. Vitamin A is needed in the bone-remodeling process, but too much vitamin A may be associated with osteoporosis.[31] Omega-3 fatty acids may help preserve bone integrity.[32] Additional research points to the bone benefits not of a specific nutrient, but of a diet rich in fruits and vegetables such as the DASH diet.[33] Phytochemicals such as lycopene reduce oxidative stress, which may help to defend against osteoporosis.[34] In contrast, diets containing too much salt are associated with bone losses. Similarly, diets containing too many colas or commercially baked snack and fried foods are associated with low bone mineral density.[35] Clearly, a well-balanced diet that depends on all the food groups to supply a full array of nutrients is central to bone health.

A Perspective on Supplements

Bone health depends, in part, on calcium. People who do not consume milk products or other calcium-rich foods in amounts that provide even half the recommended calcium should consider consulting a registered dietitian who can assess the diet and suggest food choices to correct any inadequacies. Calcium from foods may support bone health better than calcium from supplements.[36] For those who are unable to consume enough calcium-rich foods, however, taking calcium supplements—especially in combination with vitamin D—may help to enhance bone density and protect against bone loss and fractures.[37] Because calcium supplements may increase the risk of heart attacks, women should consult their physicians when making this decision.[38]

Selecting a calcium supplement requires a little investigative work to sort through the many options. Before examining calcium supplements, recognize that multivitamin-mineral pills contain little or no calcium. The label may list a few milligrams of calcium, but remember that the recommended intake is a gram or more for adults.

Calcium supplements are typically sold as compounds of calcium carbonate (common in **antacids** and fortified chocolate candies), citrate, gluconate, lactate, malate, or phosphate. These supplements often include magnesium, vitamin D, or both. In addition, some calcium supplements are made from **bone meal, oyster shell,** or **dolomite** (limestone). Many calcium supplements, especially those derived from these natural products, contain lead—which impairs health in numerous ways, as Chapter 13 points out. Fortunately, calcium interferes with the absorption and action of lead in the body.

The first question to ask is how much calcium the supplement provides. Most calcium supplements provide between 250 and 1000 milligrams of calcium. To be safe, total calcium intake from both foods and supplements should not exceed 2500 milligrams a day. Read the label to find out how much a dose supplies. Unless the label states otherwise, supplements of calcium carbonate are 40 percent calcium; those of calcium citrate are 21 percent; lactate, 13 percent; and gluconate, 9 percent. Select a low-dose supplement and take it several times a day rather than taking a large-dose supplement all at once. Taking supplements in doses of 500 milligrams or less improves absorption. Small doses also help ease the GI distress (constipation, intestinal bloating, and excessive gas) that sometimes accompanies calcium supplement use.

The next question to ask is how well the body absorbs and uses the calcium from various supplements. Most healthy people absorb calcium equally well (and as well as from milk) from any of these supplements: calcium carbonate, citrate, or phosphate. More important than supplement solubility is tablet disintegration. When manufacturers compress large quantities of calcium into small pills, the stomach acid has difficulty penetrating the pill. To test a supplement's ability to dissolve, drop it into a 6-ounce cup of vinegar, and stir occasionally. A high-quality formulation will dissolve within half an hour.

Finally, people who choose supplements must take them regularly. Furthermore, consideration should be given to the best time to take the supplements. To circumvent adverse nutrient interactions, take calcium supplements between, not with, meals. (Importantly, do not take calcium supplements with iron supplements or iron-rich meals; calcium inhibits iron absorption.) To enhance calcium absorption, take supplements with meals. If such contradictory advice drives you crazy, reconsider the benefits of food sources of calcium. Most experts agree that foods are the best source of most nutrients.

Some Closing Thoughts

Unfortunately, many of the strongest risk factors for osteoporosis are beyond people's control: age, gender, and genetics. But several strategies are still effective for prevention. First, ensure an optimal peak bone mass during childhood and adolescence by eating a balanced diet rich in calcium and engaging in regular physical activity. Then, maintain that bone mass by continuing those healthy diet and activity habits, abstaining from cigarette smoking, and using alcohol moderately, if at all. Finally, minimize bone loss by maintaining an adequate nutrition and exercise regimen, and, for women, consult a physician about calcium supplements or other drug therapies that may be effective both in preventing bone loss and in restoring lost bone. The reward is the best possible chance of preserving bone health throughout life.

Nutrition on the Net

For further study of topics covered in this highlight, log on to **www.cengagebrain.com** and search for ISBN 084006845X.

- Search for "falls and fractures" at the National Institute on Aging: **www.nia.nih.gov**
- Visit the National Institutes of Health Osteoporosis and Related Bone Diseases' National Resource Center: **www.niams.nih.gov/Health_Info/Bone**
- Obtain additional information from the National Osteoporosis Foundation: **www.nof.org**
- Review the Surgeon General's Report on Bone Health and Osteoporosis: **http://surgeongeneral.gov/library/bonehealth/content.html**

References

1. National Osteoporosis Foundation, America's bone health: The state of osteoporosis and low bone mass, 2008, www.nof.org, accessed January 2009; U.S. Department of Health and Human Services, *Bone Health and Osteoporosis: A Report of the Surgeon General* (Rockville, Md.: U.S. Department of Health and Human Services, Office of the Surgeon General, 2004).
2. A. M. Cheung and A. S. Detsky, Osteoporosis and fractures: Missing the bridge? *Journal of the American Medical Association* 299 (2008): 1468–1470.
3. L. G. Raisz, Screening for osteoporosis, *New England Journal of Medicine* 353 (2005): 164–171.
4. F. R. Greer, N. F. Krebs, and the Committee on Nutrition, Optimizing bone health and calcium intakes of infants, children, and adolescents, *Pediatrics* 117 (2006): 578–585.
5. G. R. Mundy, Osteoporosis and inflammation, *Nutrition Reviews* 65 (2007): S147–S151.
6. C. J. Rosen, Postmenopausal osteoporosis, *New England Journal of Medicine* 353 (2005): 595–603.
7. R. P. Heaney and R. R. Recker, Combination and sequential therapy of osteoporosis, *New England Journal of Medicine* 353 (2005): 624–625.
8. A. Atmaca and coauthors, Soy isoflavones in the management of postmenopausal osteoporosis, *Menopause* 15 (2008): 748–757; R. C. Poulsen and M. C. Kruger, Soy phytoestrogens: Impact on postmenopausal bone loss and mechanisms of action, *Nutrition Reviews* 66 (2008): 359–374.
9. W. G. Helferich, J. E. Andrade, and M. S. Hoagland, Phytoestrogens and breast cancer: A complex story, *Inflammopharmacology* 16 (2008): 219–226; Y. Zhang and coauthors, Soy isoflavones and their bone protective effect, *Inflammopharmacology* 16 (2008): 213–215.
10. S. Reinwald and C. M. Weaver, Soy isoflavones and bone health: A double-edged sword? *Journal of Natural Products* 69 (2006): 450–459.
11. P. R. Ebeling, Osteoporosis in men, *New England Journal of Medicine* 358 (2008): 1474–1482.

12. S. H. Ralston and B. de Crombrugghe, Genetic regulation of bone mass and susceptibility to osteoporosis, *Genes Development* 20 (2006): 2492–2506.
13. M. Braun and coauthors, Racial differences in skeletal calcium retention in adolescent girls with varied controlled calcium intakes, *American Journal of Clinical Nutrition* 85 (2007): 1657–1663; K. Wigertz and coauthors, Racial differences in calcium retention in response to dietary salt in adolescent girls, *American Journal of Clinical Nutrition* 81 (2005): 845–850.
14. J. A. Cauley and coauthors, Longitudinal study of changes in hip bone mineral density in Caucasian and African-American women, *Journal of the American Geriatrics Society* 53 (2005): 183–189; J. A. Cauley and coauthors, Bone mineral density and the risk of incident nonspinal fractures in black and white women, *Journal of the American Medical Association* 293 (2005): 2102–2108.
15. A. J. Lanou, S. E. Berkow, and N. D. Barnard, Calcium, dairy products, and bone health in children and young adults: A reevaluation of the evidence, *Pediatrics* 115 (2005): 736–743.
16. C. Rideout, H. McKay, and S. Barr, Self-reported lifetime physical activity and areal bone mineral density in healthy postmenopausal women: The importance of teenage activity, *Calcified Tissue International* 79 (2006): 214–222.
17. K. T. Borer, Physical activity in the prevention and amelioration of osteoporosis in women: Interaction of mechanical, hormonal and dietary factors, *Sports Medicine* 35 (2005): 779–830; F. R. Greer, Bone health: It's more than calcium intake, *Pediatrics* 115 (2005): 792–794.
18. American College of Sports Medicine Position Stand, Physical activity and bone health, *Medicine & Science in Sports & Exercise* 36 (2004): 1985–1996.
19. J. M. Welch and C. M. Weaver, Calcium and exercise affect the growing skeleton, *Nutrition Reviews* 63 (2005): 361–373.
20. M. Varenna and coauthors, Effects of dietary calcium intake on body weight and prevalence of osteoporosis in early postmenopausal women, *American Journal of Clinical Nutrition* 86 (2007): 639–644.
21. G. Wolf, Energy regulation by the skeleton, *Nutrition Reviews* 66 (2008): 229–233.
22. C. S. Riedt and coauthors, Premenopausal overweight women do not lose bone during moderate weight loss with adequate or higher calcium intake, *American Journal of Clinical Nutrition* 85 (2007): 972–980.
23. M. Lorentzon and coauthors, Smoking is associated with lower bone mineral density and reduced cortical thickness in young men, *Journal of Clinical Endocrinology and Metabolism* 92 (2007): 497–503.
24. R. Jugdaohsingh and coauthors, Moderate alcohol consumption and increased bone mineral density: Potential ethanol and non-ethanol mechanisms, *Proceedings of the Nutrition Society* 65 (2006): 291–310.
25. H. A. Bischoff-Ferrari and coauthors, Calcium intake and hip fracture risk in men and women: A meta-analysis of prospective cohort studies and randomized controlled trials, *American Journal of Clinical Nutrition* 86 (2007): 1780–1790.
26. C. Palacios, The role of nutrients in bone health, from A to Z, *Critical Reviews of Food Science and Nutrition* 46 (2006): 621–628; J. W. Nieves, Osteoporosis: The role of micronutrients, *American Journal of Clinical Nutrition* 81 (2005): 1232S–1239S.
27. A. D. Conigrave, E. M. Brown, and R. Rizzoli, Dietary protein and bone health: Roles of amino acid–sensing receptors in the control of calcium metabolism and bone homeostasis *Annual Review of Nutrition* 28 (2008): 131–155.
28. K. D. Cashman and coauthors, Low vitamin D status adversely affects bone health parameters in adolescents, *American Journal of Clinical Nutrition* 87 (2008): 1039–1044; L. Steingrimsdottir and coauthors, Relationship between serum parathyroid hormone levels, vitamin D sufficiency, and calcium intake, *Journal of the American Medical Association* 294 (2005): 2336–2341.
29. H. A. Bischoff-Ferrari and coauthors, Fracture prevention with vitamin D supplementation: A meta-analysis of randomized controlled trials, *Journal of the American Medical Association* 293 (2005): 2257–2264.
30. S. Sahni and coauthors, High vitamin C intake is associated with lower 4-year bone loss in elderly men, *Journal of Nutrition* 138 (2008): 1931–1938.
31. J. D. Ribaya-Mercado and J. B. Blumberg, Vitamin A: Is it a risk factor for osteoporosis and bone fracture? *Nutrition Reviews* 65 (2007): 425–438; K. L. Penniston and coauthors, Serum retinyl esters are not elevated in postmenopausal women with and without osteoporosis whose preformed vitamin A intakes are high, *American Journal of Clinical Nutrition* 84 (2006): 1350–1356.
32. A. E. Griel and coauthors, An increase in dietary n-3 fatty acids decreases a marker of bone resorption in humans, *Nutrition Journal* 6 (2007): 2; M. Högström, P. Nordström, and A. Nordström, n-3 Fatty acids are positively associated with peak bone mineral density and bone accrual in healthy men: The NO_2 Study, *American Journal of Clinical Nutrition* 85 (2007): 803–807; L. A. Weiss, E. Barrett-Connor, and D. von Mühlen, Ratio of n-6 to n-3 fatty acids and bone mineral density in older adults: The Rancho Bernardo Study, *American Journal of Clinical Nutrition* 81 (2005): 934–938.
33. H. Vatanparast and coauthors, Positive effects of vegetable and fruit consumption and calcium intake on bone mineral accrual in boys during growth from childhood to adolescence: The University of Saskatchewan Pediatric Bone Mineral Accrual Study, *American Journal of Clinical Nutrition* 82 (2005): 700–706.
34. L. G. Rao and coauthors, Lycopene consumption decreases oxidative stress and bone resorption markers in postmenopausal women, *Osteoporosis International* 18 (2007): 109–115.
35. L. M. Troy and coauthors, Dihydrophylloquinone intake is associated with low bone mineral density in men and women, *American Journal of Clinical Nutrition* 86 (2007): 504–508; K. L. Tucker and coauthors, Colas, but not other carbonated beverages, are associated with low bone mineral density in older women: The Framingham Osteoporosis Study, *American Journal of Clinical Nutrition* 84 (2006): 936–942.
36. Y. Manios and coauthors, Changes in biochemical indexes of bone metabolism and bone mineral density after a 12-mo dietary intervention program: The Postmenopausal Health Study, *American Journal of Clinical Nutrition* 86 (2007): 781–789; N. Napoli and coauthors, Effects of dietary calcium compared with calcium supplements on estrogen metabolism and bone mineral density, *American Journal of Clinical Nutrition* 85 (2007): 1428–1433.
37. H. A. Bischoff-Ferrari and coauthors, Effect of calcium supplementation on fracture risk: A double-blind randomized controlled trial, *American Journal of Clinical Nutrition* 87 (2008): 1945–1951; R. M. Daly and coauthors, The skeletal benefits of calcium- and vitamin D_3-fortified milk are sustained in older men after withdrawal of supplementation: An 18-mo follow-up study, *American Journal of Clinical Nutrition* 87 (2008): 771–777; M. F. Hitz, J. B. Jensen, and P. C. Eskildsen, Bone mineral density and bone markers in patients with a recent low-energy fracture: Effect of 1 y of treatment with calcium and vitamin D, *American Journal of Clinical Nutrition* 86 (2007): 251–259; B. M. P. Tang and coauthors, Use of calcium or calcium in combination with vitamin D supplementation to prevent fractures and bone loss in people aged 50 years and older: A meta-analysis, *Lancet* 370 (2007): 657–666; V. Matkovic and coauthors, Calcium supplementation and bone mineral density in females from childhood to young adulthood: A randomized controlled trial, *American Journal of Clinical Nutrition* 81 (2005): 175–188; R. P. Dodiuk-Gad and coauthors, Sustained effect of short-term calcium supplementation on bone mass in adolescent girls with low calcium intake, *American Journal of Clinical Nutrition* 81 (2005): 168–174.
38. M. J. Bolland and coauthors, Vascular events in healthy older women receiving calcium supplementation: Randomised controlled trial, *British Medical Journal* 336 (2008): 262–266.

CHAPTER

13

© B&Y Photography/Alamy

Nutrition in Your Life

Trace—barely a perceptible amount. But the trace minerals tackle big jobs. Your blood can't carry oxygen without iron, and insulin can't deliver glucose without chromium. Teeth become decayed without fluoride, and thyroid glands develop goiter without iodine. Together, the trace minerals keep you healthy and strong. Where can you get these amazing minerals? A variety of foods, especially those from the protein foods group, sprinkled with a little iodized salt and complemented by a glass of fluoridated water will do the trick. It's remarkable what your body can do with only a few milligrams—or even micrograms—of the trace minerals.

Throughout this chapter, the CourseMate icon indicates an opportunity for online self-study, linking you to activities to increase your understanding of chapter concepts. **www.cengagebrain.com** (search for ISBN 084006845X)

The Trace Minerals

CHAPTER OUTLINE

Figure 12-9 in Chapter 12 (p. 394) showed the tiny quantities of **trace minerals** in the human body. The trace minerals are so named because they are present, and needed, in relatively small amounts in the body. All together, they would produce only a bit of dust, hardly enough to fill a teaspoon. Yet they are no less important than the major minerals or any of the other nutrients. Each of the trace minerals performs a vital role. A deficiency of any of them may be fatal, and an excess of many is equally deadly. Remarkably, people's diets normally supply just enough of these minerals to maintain health.

The Trace Minerals—An Overview

The body requires the trace minerals in minuscule quantities. They participate in diverse tasks all over the body, each having special duties that only it can perform.

Food Sources The trace mineral contents of foods depend on soil and water composition and on how foods are processed. Furthermore, many factors in the diet and within the body affect the minerals' bioavailability. ♦ Still, outstanding food sources for each of the trace minerals, just like those for the other nutrients, include a wide variety of foods.

♦ **Bioavailability** refers to the rate at and the extent to which a nutrient is absorbed and used.

Deficiencies Severe deficiencies of the better-known minerals are easy to recognize. Deficiencies of the others may be harder to diagnose, and for all minerals, mild deficiencies are easy to overlook. Because the minerals are active in many body systems—digestive, cardiovascular, circulatory, muscular, skeletal, and nervous—deficiencies can have wide-reaching effects and can affect people of all ages. The most common result of a deficiency in children is failure to grow and thrive.

Toxicities Most of the trace minerals are toxic at intakes only two and a half to seven times above the estimated requirements (see Figure 13-1, p. 424). Thus it is important not to habitually exceed the Upper Level (UL) of recommended intakes (see inside front cover). Many dietary supplements contain trace minerals, making it easy for users to exceed their needs. Highlight 10 discusses supplement use and some of the regulations included in the Dietary Supplement Health and Education Act. As that discussion notes, consumers have demanded the freedom to choose their own doses of nutrients. By law, the Food and Drug Administration

trace minerals: essential mineral nutrients the human body requires in relatively small amounts (less than 100 milligrams per day); sometimes called *microminerals*.

FIGURE 13-1 Estimated Requirements and UL Compared for Selected Trace Minerals

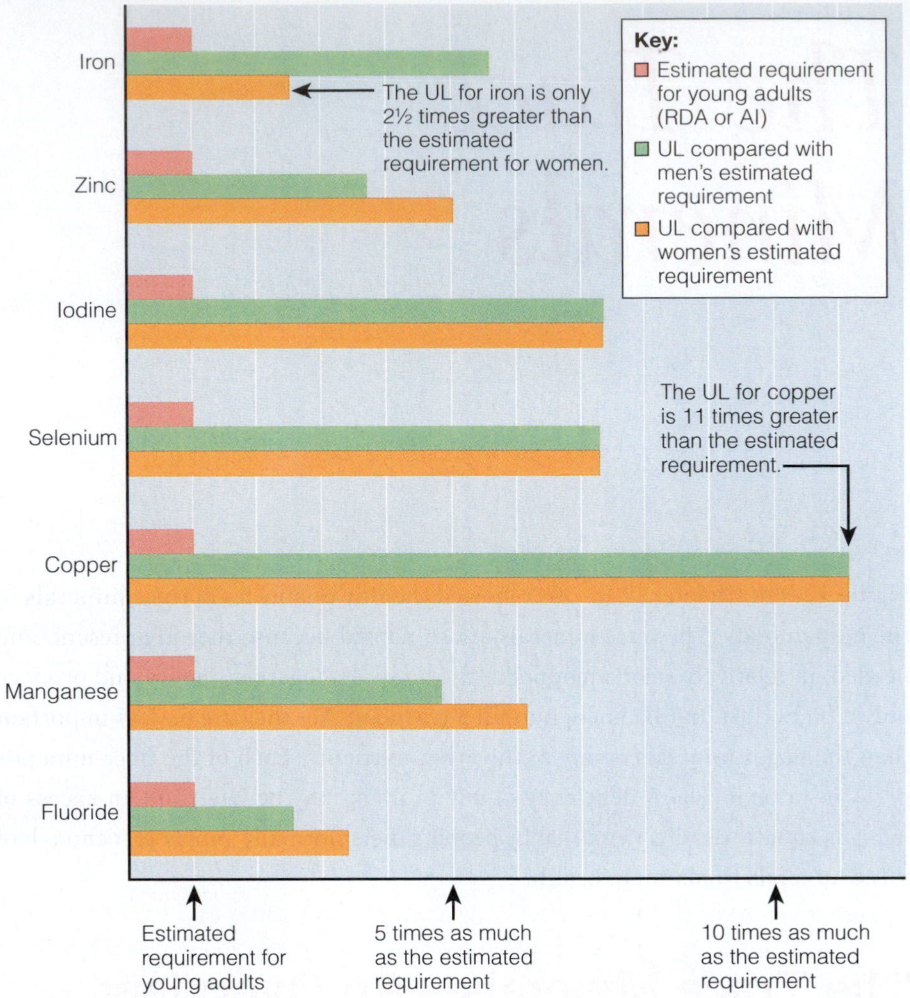

(FDA) has no authority to limit the amounts of trace minerals in supplements.* Individuals who take supplements must therefore be aware of the possible dangers and select supplements that contain no more than 100 percent of the Daily Value. It would be easier and safer to meet nutrient needs by selecting a variety of foods than by combining an assortment of supplements.

Interactions Interactions among the trace minerals are common and often well coordinated to meet the body's needs. For example, several of the trace minerals support insulin's work, influencing its synthesis, storage, release, and action.

At other times, interactions lead to nutrient imbalances. An excess of one may cause a deficiency of another. (A slight manganese overload, for example, may aggravate an iron deficiency.) A deficiency of one may interfere with the work of another. (A selenium deficiency halts the activation of the iodine-containing thyroid hormones.) A deficiency of a trace mineral may even open the way for a contaminant mineral to cause a toxic reaction. (Iron deficiency, for example, makes the body more vulnerable to lead poisoning.) These examples reinforce the need to balance intakes and to use supplements wisely, if at all.

A good food source of one nutrient may be a poor food source of another, and factors that enhance the action of some trace minerals may interfere with others. (Meats are a good source of iron but a poor source of calcium; vitamin C enhances

*Canada regulates the amounts of trace minerals in supplements.

the absorption of iron but hinders that of copper.) Research on the trace minerals is active, suggesting that we have much more to learn about them.

IN SUMMARY **Although the body uses only tiny amounts of the trace minerals, they are vital to health. Because so little is required, the trace minerals can be toxic at levels not far above estimated requirements—a consideration for supplement users. Like the other nutrients, the trace minerals are best obtained by eating a variety of foods.**

Iron

Iron is an essential nutrient, vital to many of the cells' activities, but it poses a problem for millions of people. Some people simply don't eat enough iron-containing foods to support their health optimally, whereas others absorb so much iron that it threatens their health. Iron exemplifies the principle that both too little and too much of a nutrient in the body can be harmful. In its wisdom, the body has several ways to achieve iron homeostasis, protecting against both deficiency and overload.

Iron Roles in the Body

Iron has the knack of switching back and forth between two ionic states. ♦ In the reduced state, iron has lost two electrons and therefore has a net positive charge of two; it is known as *ferrous iron*. In the oxidized state, iron has lost a third electron, has a net positive charge of three, and is known as *ferric iron*. Ferrous iron can be oxidized to ferric iron, and ferric iron can be reduced to ferrous iron. Thus iron can serve as a cofactor ♦ to enzymes involved in oxidation-reduction reactions—reactions so widespread in metabolism that they occur in all cells. Enzymes involved in making amino acids, collagen, hormones, and neurotransmitters all require iron. (For details about ions, oxidation, and reduction, see Appendix B.)

Iron forms a part of the electron carriers that participate in the electron transport chain (discussed in Chapter 7).* In this pathway, these carriers transfer hydrogens and electrons to oxygen, forming water, and in the process, make ATP for the cells' energy use.

Most of the body's iron is found in two proteins: hemoglobin ♦ in the red blood cells and **myoglobin** in the muscle cells. In both, iron helps accept, carry, and then release oxygen.

♦ Iron's two ionic states:
- Ferrous iron (reduced): Fe^{++}
- Ferric iron (oxidized): Fe^{+++}

♦ A **cofactor** is a substance that works with an enzyme to facilitate a chemical reaction.

♦ **Hemoglobin** is the oxygen-carrying protein of the red blood cells that transports oxygen from the lungs to tissues throughout the body; hemoglobin accounts for 80% of the body's iron.

Iron Absorption and Metabolism

The body conserves iron. Because it is difficult to excrete iron once it is in the body, balance is maintained primarily through absorption. More iron is absorbed when stores are empty and less is absorbed when stores are full.[1]

Iron Absorption

Special proteins help the body absorb iron from food (see Figure 13-2 on p. 426). The iron-storage protein **ferritin** captures iron from food and stores it in the cells of the small intestine. When the body needs iron, ferritin releases some iron to an iron transport protein called **transferrin.** If the body does not need iron, it is carried out when the intestinal cells are shed and excreted in the feces; intestinal cells are replaced about every three to five days. By holding iron temporarily, these cells control iron absorption by either delivering iron when the day's intake falls short or disposing of it when intakes exceed needs.

Heme and Nonheme Iron

Iron absorption depends in part on its dietary source. Iron occurs in two forms in foods: as **heme iron**, which is found only in foods derived from the flesh of animals, such as meats, poultry, and fish, and as **nonheme iron**,

myoglobin: the oxygen-holding protein of the muscle cells.
- **myo** = muscle

ferritin (FAIR-ih-tin): the iron-storage protein.

transferrin (trans-FAIR-in): the iron transport protein.

heme (HEEM) **iron:** the iron in foods that is bound to the hemoglobin and myoglobin proteins; found only in meat, fish, and poultry.

nonheme iron: the iron in foods that is not bound to proteins; found in both plant-derived and animal-derived foods.

*The iron-containing electron carriers of the electron transport chain are known as *cytochromes*. See Appendix C for details of this pathway.

FIGURE 13-2 Iron Absorption

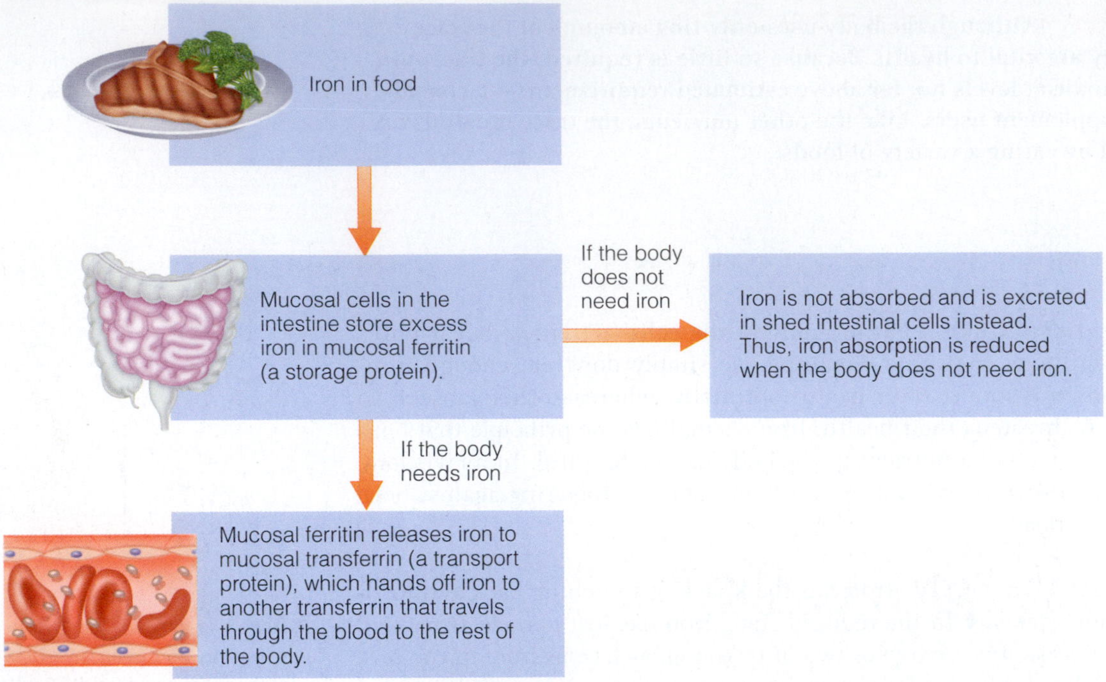

♦ About 40% of the iron in meat, fish, and poultry is bound into heme; the other 60% is nonheme iron. All of the iron in plant foods is nonheme iron.

♦ Factors that *enhance* nonheme iron absorption:
- MFP factor
- Vitamin C (ascorbic acid)

MFP factor: a peptide released during the digestion of **m**eat, **f**ish, and **p**oultry that enhances nonheme iron absorption.

which is found in both plant-derived and animal-derived foods (see Figure 13-3). ♦ On average, heme iron represents about 10 percent of the iron a person consumes in a day. Even though heme iron accounts for only a small proportion of the intake, it is so well absorbed that it contributes significant iron. About 25 percent of heme iron and 17 percent of nonheme iron is absorbed, depending on dietary factors and the body's iron stores.[2] In iron deficiency, absorption increases.[3] In iron overload, absorption declines. Researchers disagree as to whether heme iron absorption responds to iron stores as sensitively as nonheme iron absorption does.

Absorption-Enhancing Factors Meat, fish, and poultry contain not only the well-absorbed heme iron, but also a peptide (called the **MFP factor**) that promotes the absorption of nonheme iron ♦ from other foods eaten at the same meal.[4] Vitamin C also enhances nonheme iron absorption from foods eaten in the same meal by capturing the iron and keeping it in the reduced ferrous form, ready for absorption. Some acids and sugars also enhance nonheme iron absorption.

FIGURE 13-3 Heme and Nonheme Iron in Foods

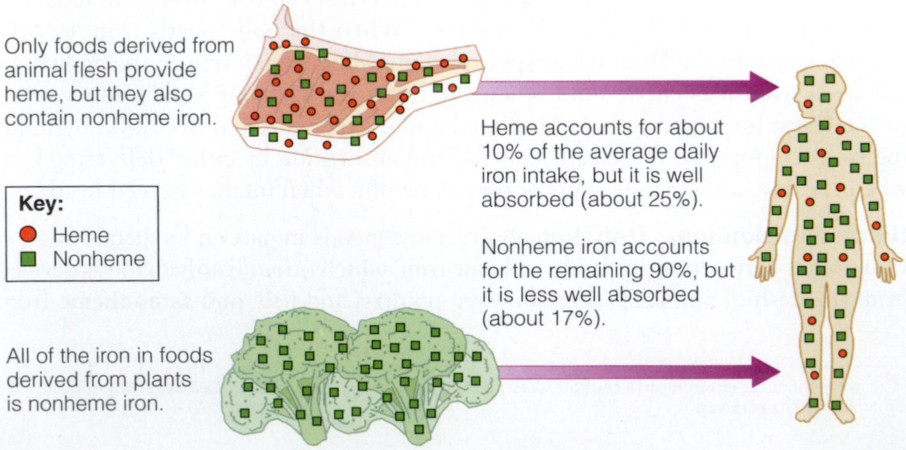

Only foods derived from animal flesh provide heme, but they also contain nonheme iron.
Key:
Heme
Nonheme
Heme accounts for about 10% of the average daily iron intake, but it is well absorbed (about 25%).
Nonheme iron accounts for the remaining 90%, but it is less well absorbed (about 17%).
All of the iron in foods derived from plants is nonheme iron.

Absorption-Inhibiting Factors Some dietary factors bind with nonheme iron, inhibiting absorption. ♦ These factors include the phytates in legumes, whole grains, and rice; the vegetable proteins in soybeans, other legumes, and nuts; the calcium in milk; and the polyphenols (such as tannic acid) in tea, coffee, grain products, oregano, and red wine.

♦ Factors that *inhibit* nonheme iron absorption:
- Phytates (legumes, grains, and rice)
- Vegetable proteins (soybeans, legumes, nuts)
- Calcium (milk)
- Tannic acid (and other polyphenols in tea and coffee)

Dietary Factors Combined The many dietary enhancers, inhibitors, and their combined effects make it difficult to estimate iron absorption.[5] Most of these factors exert a strong influence individually, but not when combined with the others in a meal. Furthermore, the impact of the combined effects diminishes when a diet is evaluated over several days. When multiple meals are analyzed together, three factors appear to be most relevant: MFP and vitamin C as enhancers and phytates as inhibitors.

Individual Variation Overall, about 18 percent of dietary iron is absorbed from mixed diets and only about 10 percent from vegetarian diets.[6] As you might expect, vegetarian diets do not have the benefit of easy-to-absorb heme iron or the help of MFP in enhancing absorption. In addition to dietary influences, iron absorption also depends on an individual's health, stage in the life cycle, and iron status. Absorption can be as low as 2 percent in a person with GI disease or as high as 35 percent in a rapidly growing, healthy child. The body adapts to absorb more iron when a person's iron stores fall short or when the need increases for any reason (such as pregnancy). The body makes more ferritin to absorb more iron from the small intestine and more transferrin to carry more iron around the body. Similarly, when iron stores are sufficient, the body adapts to absorb less iron.

Mark Stout Photography/Shutterstock.com

This chili dinner provides several factors that may enhance iron absorption: heme and nonheme iron and MFP from meat, nonheme iron from legumes, and vitamin C from tomatoes.

Iron Transport and Storage The blood transport protein transferrin delivers iron to the bone marrow and other tissues. The bone marrow uses large quantities to make new red blood cells, whereas other tissues use less. Surplus iron is stored in the protein ferritin, primarily in the liver, but also in the bone marrow and spleen. When dietary iron has been plentiful, ferritin is constantly and rapidly made and broken down, providing an ever-ready supply of iron. When iron concentrations become abnormally high, the liver converts some ferritin into another storage protein called **hemosiderin.** Hemosiderin releases iron more slowly than ferritin does. Storing excess iron in hemosiderin protects the body against the damage that free iron can cause. Free iron acts as a free radical, attacking cell lipids, DNA, and protein. (See Highlight 11 for more information on free radicals and the damage they can cause.)

Iron Recycling The average red blood cell lives about four months; then the spleen and liver cells remove it from the blood, take it apart, and prepare the degradation products for excretion or recycling. The iron is salvaged: the liver attaches it to transferrin, which transports it back to the bone marrow to be reused in making new red blood cells. Thus, although red blood cells live for only about four months, the iron recycles through each new generation of cells (see Figure 13-4 on p. 428). The body loses some iron daily via the GI tract and, if bleeding occurs, in blood. Only tiny amounts of iron are lost in urine, sweat, and shed skin.*

Iron Balance Maintaining iron balance depends on the careful regulation of iron absorption, transport, storage, recycling, and losses. The hormone **hepcidin** is central to the regulation of iron balance.[7] Produced by the liver, hepcidin helps to maintain blood iron within the normal range by limiting absorption from the small intestine and controlling release from the liver, spleen, and bone marrow.

Iron Deficiency

Worldwide, **iron deficiency** is the most common nutrient deficiency, with **iron-deficiency anemia** affecting more than 1.6 billion people—almost half of preschool children and pregnant women.[8] In the United States, iron deficiency is less prevalent, but it still affects 10 percent of toddlers, adolescent girls, and women of childbearing age. Iron deficiency is also relatively common

hemosiderin (heem-oh-SID-er-in): an iron-storage protein primarily made in times of iron overload.

hepcidin: a hormone produced by the liver that regulates iron balance.

iron deficiency: the state of having depleted iron stores.

iron-deficiency anemia: severe depletion of iron stores that results in low hemoglobin and small, pale red blood cells. Anemias that impair hemoglobin synthesis are *microcytic* (small cell).
- **micro** = small
- **cytic** = cell

*Adults lose about 1.0 milligram of iron per day. Women lose additional iron in menses. Menstrual losses vary considerably, but over a month, they average about 0.5 milligram per day.

FIGURE 13-4 **Iron Recycled in the Body**

Once iron enters the body, most of it is recycled. Some is lost with body tissues and must be replaced by eating iron-containing food.

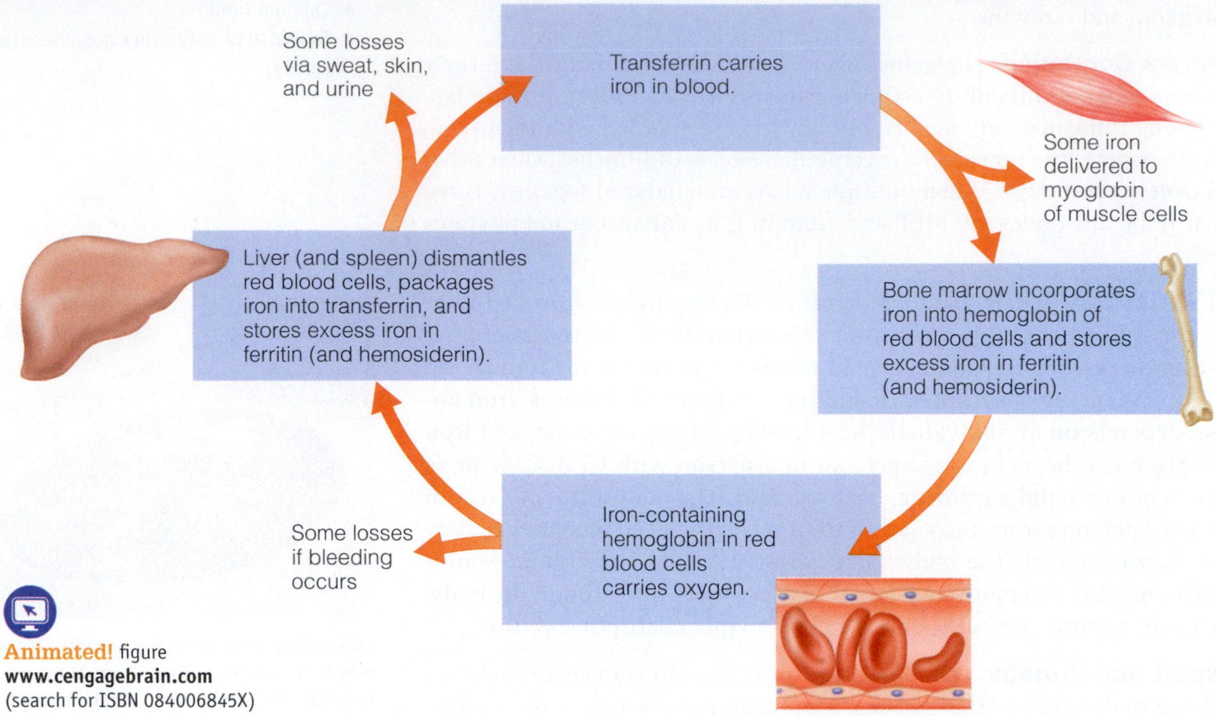
Transferrin carries iron in blood.
Some iron delivered to myoglobin of muscle cells
Bone marrow incorporates iron into hemoglobin of red blood cells and stores excess iron in ferritin (and hemosiderin).
Iron-containing hemoglobin in red blood cells carries oxygen.
Some losses if bleeding occurs
Liver (and spleen) dismantles red blood cells, packages iron into transferrin, and stores excess iron in ferritin (and hemosiderin).
Some losses via sweat, skin, and urine

Animated! figure
www.cengagebrain.com
(search for ISBN 084006845X)

among overweight children and adolescents compared with those who are normal weight.[9] The association between iron deficiency and obesity has yet to be explained, but researchers are currently examining the relationships between the inflammation that develops with excess body fat and reduced iron absorption.[10] Preventing and correcting iron deficiency are high priorities.

♦ High risk for iron deficiency:
- Women in their reproductive years
- Pregnant women
- Infants and young children
- Teenagers

♦ The condition of developing iron-deficiency anemia because iron-poor milk displaces iron-rich foods is called **milk anemia.**

Vulnerable Stages of Life Some stages of life ♦ demand more iron but provide less, making deficiency likely. Women in their reproductive years are especially prone to iron deficiency because of repeated blood losses during menstruation. Pregnancy demands additional iron to support the added blood volume, growth of the fetus, and blood loss during childbirth. Infants and young children receive little iron from their high-milk diets, ♦ yet need extra iron to support their rapid growth and brain development.[11] Iron deficiency among toddlers in the United States is common.[12] The rapid growth of adolescence, especially for males, and the menstrual losses of females also demand extra iron that a typical teen diet may not provide. An adequate iron intake is especially important during these stages of life.

♦ The iron content of blood is about 0.5 mg/100 mL blood. A person donating a pint of blood (approximately 500 mL) loses about 2.5 mg of iron.

Blood Losses Bleeding ♦ from any site incurs iron losses. In some cases, such as an active ulcer, the bleeding may not be obvious, but even small chronic blood losses significantly deplete iron reserves. In developing countries, blood loss is often brought on by malaria and parasitic infections of the GI tract. People who donate blood regularly also incur losses and may benefit from iron supplements. As mentioned, menstrual losses can be considerable as they tap women's iron stores regularly.

♦ Stages of iron deficiency:
- Iron stores diminish
- Transport iron decreases
- Hemoglobin production declines

Assessment of Iron Deficiency Iron deficiency develops in stages. ♦ This section provides a brief overview of how to detect these stages, and Appendix E provides more details. In the first stage of iron deficiency, iron stores diminish. Measures of serum ferritin (in the blood) reflect iron stores and are most valuable in assessing iron status at this earliest stage.[13]

The second stage of iron deficiency is characterized by a decrease in transport iron: serum iron falls, and the iron-carrying protein transferrin *increases* (an adaptation that enhances iron absorption). Together, measurements of serum iron and transferrin can determine the severity of the deficiency—the more transferrin and the less iron in the blood, the more advanced the deficiency is. Transferrin saturation—the percentage of transferrin that is saturated with iron—decreases as iron stores decline.

The third stage of iron deficiency occurs when the lack of iron limits hemoglobin production. Now the hemoglobin precursor, **erythrocyte protoporphyrin**, begins to accumulate as hemoglobin and **hematocrit** values decline.

Hemoglobin and hematocrit tests are easy, quick, and inexpensive, so they are the tests most commonly used in evaluating iron status. Their usefulness in detecting iron deficiency is limited, however, because they are late indicators. Furthermore, other nutrient deficiencies and medical conditions can influence their values.

Iron Deficiency and Anemia Notice that iron deficiency and iron-deficiency anemia are not the same: people may be iron deficient without being anemic. The term *iron deficiency* refers to depleted body iron stores without regard to the degree of depletion or to the presence of anemia. The term *iron-deficiency anemia* refers to the severe depletion of iron stores that results in a low hemoglobin concentration. In iron-deficiency anemia, hemoglobin synthesis decreases, resulting in red blood cells that are pale (hypochromic) and small (microcytic), ♦ as shown in Figure 13-5. These cells can't carry enough oxygen from the lungs to the tissues. Without adequate iron, energy metabolism in the cells falters. The result is fatigue, weakness, headaches, apathy, pallor, and poor resistance to cold temperatures. Because hemoglobin is the bright red pigment of the blood, the skin of a fair person who is anemic may become noticeably pale. In a dark-skinned person, the tongue and eye lining, normally pink, is very pale.

The fatigue that accompanies iron-deficiency anemia differs from the tiredness a person experiences from a simple lack of sleep. People with anemia feel fatigue only when they exert themselves. Iron supplementation can relieve the fatigue and improve the body's response to physical activity.

Iron Deficiency and Behavior Long before the red blood cells are affected and anemia is diagnosed, a developing iron deficiency affects behavior.[14] Even at slightly lowered iron levels, energy metabolism is impaired and neurotransmitter synthesis is altered, reducing physical work capacity and mental productivity. Without the physical energy and mental alertness to work, plan, think, play, sing,

♦ Iron-deficiency anemia is a **microcytic** (my-cro-SIT-ic) **hypochromic** (high-po-KROME-ic) **anemia.**

- **micro** = small
- **cytic** = cell
- **hypo** = too little
- **chrom** = color

erythrocyte protoporphyrin (PRO-toe-PORE-fe-rin): a precursor to hemoglobin.

hematocrit (hee-MAT-oh-krit): measurement of the volume of the red blood cells packed by centrifuge in a given volume of blood.

FIGURE 13-5 Normal and Anemic Blood Cells

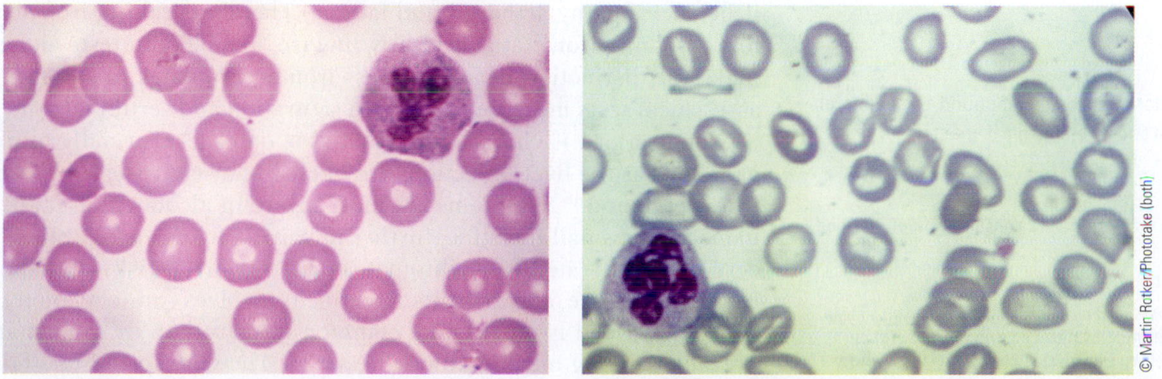

Both size and color are normal in these blood cells.

Blood cells in iron-deficiency anemia are small (microcytic) and pale (hypochromic) because they contain less hemoglobin.

or learn, people simply do these things less. They have no obvious deficiency symptoms; they just appear unmotivated, apathetic, and less physically fit. Work productivity and voluntary activities decline.

Many of the symptoms associated with iron deficiency are easily mistaken for behavioral or motivational problems. A restless child who fails to pay attention in class might be thought contrary. An apathetic homemaker who has let housework pile up might be thought lazy. No responsible dietitian would ever claim that all behavioral problems are caused by nutrient deficiencies, but poor nutrition is always a possible contributor to problems like these. When investigating a behavioral problem, check the adequacy of the diet and seek a routine physical examination before undertaking more expensive, and possibly more harmful, treatment options. Treatment with long-term, low-dose iron supplements may improve cognitive skills and physical development.[15] The effects of iron deficiency on children's behavior are discussed further in Chapter 15.

Iron Deficiency and Pica A curious behavior seen in some iron-deficient people, especially in women and children of low-income groups, is **pica**—the craving and consumption of ice, chalk, starch, and other nonfood substances. ♦ These substances contain no iron and cannot remedy a deficiency; in fact, clay actually inhibits iron absorption, which may explain the iron deficiency that accompanies such behavior. Pica is poorly understood. Its cause is unknown, but researchers hypothesize that it may be motivated by hunger, nutrient deficiencies, or an attempt to protect against toxins or microbes.[16] The consequence of pica is anemia.

♦ Pica is known as *geophagia* (gee-oh-FAY-gee-uh) when referring to eating clay, baby powder, chalk, ash, ceramics, paper, paint chips, or charcoal; *pagophagia* (pag-oh-FAY-gee-uh) when referring to eating large quantities of ice; and *amylophagia* (AM-ee-low-FAY-gee-ah) when referring to eating uncooked starch (flour, laundry starch, or raw rice).

Iron Toxicity

In general, even a diet that includes fortified foods poses no special risk for iron toxicity. The body normally absorbs less iron when its stores are full, but some individuals are poorly defended against excess iron. Once considered rare, **iron overload** has emerged as an important disorder of iron metabolism and regulation.

Iron Overload The iron overload disorder known as **hemochromatosis** is usually caused by a genetic failure to prevent unneeded iron in the diet from being absorbed.[17] Recent research suggests that just as insulin supports normal glucose homeostasis and its absence or ineffectiveness causes diabetes, the hormone hepcidin supports iron homeostasis and its absence or ineffectiveness causes hemochromatosis.

Hereditary hemochromatosis is the most common genetic disorder in the United States, affecting some 1.5 million people. Other causes of iron overload include repeated blood transfusions (which bypass the intestinal defense), massive doses of supplementary iron (which overwhelm the intestinal defense), and other rare metabolic disorders. Excess iron may cause **hemosiderosis**, a condition characterized by deposits of the iron-storage protein hemosiderin in the liver, heart, joints, and other tissues.

Some of the signs and symptoms of iron overload are similar to those of iron deficiency: apathy, lethargy, and fatigue. Therefore, taking iron supplements before assessing iron status is clearly unwise; hemoglobin tests alone would fail to make the distinction because excess iron accumulates in storage. Iron overload assessment tests measure transferrin saturation and serum ferritin.

Iron overload is characterized by tissue damage, especially in iron-storing organs such as the liver. Infections are likely because viruses and bacteria thrive on iron-rich blood.[18] Symptoms are most severe in alcohol abusers because alcohol damages the small intestine, further impairing its defenses against absorbing excess iron. Untreated hemochromatosis increases the risks of diabetes, liver cancer, heart disease, and arthritis. Treatment involves iron-chelation therapy.[19] ♦

♦ Chelation therapy uses a compound to sequester a toxic substance, rendering it inactive or less harmful.

Iron overload is much more common in men than in women and is twice as prevalent among men as iron deficiency.[20] The widespread fortification of foods with iron makes it difficult for people with hemochromatosis to follow a low-iron diet, and greater dangers lie in the indiscriminate use of iron and vitamin C supplements. Vitamin C not only enhances iron absorption, but also releases iron

pica (PIE-ka): a craving for and consumption of nonfood substances.

iron overload: toxicity from excess iron.

hemochromatosis (HE-moh-KRO-ma-toe-sis): a genetically determined failure to prevent absorption of unneeded dietary iron that is characterized by iron overload and tissue damage.

hemosiderosis (HE-moh-sid-er-OH-sis): a condition characterized by the deposition of hemosiderin in the liver and other tissues.

from ferritin, allowing free iron to wreak the damage typical of free radicals. Thus vitamin C acts as a *pro*oxidant when taken in high doses. (See Highlight 11 for a discussion of free radicals and their effects on disease development.)

Iron and Heart Disease Some research suggests a link between heart disease and excess iron, especially when accompanied by alcohol consumption.[21] As mentioned, free radicals can attack ferritin, causing it to release iron from storage. Free iron, in turn, acts as an oxidant that can generate more free radicals. Reducing iron stores, however, does not appear to reduce the risks of heart attack, stroke, or mortality.[22]

Iron and Cancer There may be an association between iron and some cancers.[23] Explanations for how iron might be involved in causing cancer focus on its free-radical activity, which can damage DNA (see Highlight 11). One of the benefits of a high-fiber diet may be that the accompanying phytates bind iron, making it less available for such reactions.

Iron Poisoning Large doses of iron supplements cause GI distress, including constipation, nausea, vomiting, and diarrhea. These effects may not be as serious as other consequences of iron toxicity, but they are consistent enough to establish a UL of 45 milligrams per day for adults.

Ingestion of iron-containing supplements remains a leading cause of accidental poisoning in young children.[24] Symptoms of toxicity include nausea, vomiting, diarrhea, a rapid heartbeat, a weak pulse, dizziness, shock, and confusion. As few as five iron tablets containing as little as 200 milligrams of iron have caused the deaths of dozens of young children. The exact cause of these deaths is uncertain, but excessive free-radical damage is thought to play a role in heart failure and respiratory distress. Autopsy reports reveal iron deposits and cell death in the stomach, small intestine, liver, and blood vessels (which can cause internal bleeding). As with medicines and other potentially toxic substances, keep iron-containing tablets out of the reach of children. If you suspect iron poisoning, call the nearest poison control center or a physician immediately.

Iron Recommendations and Sources

To obtain enough iron, people must first select iron-rich foods—both naturally occurring and enriched or fortified—and then take advantage of factors that maximize iron absorption. This discussion begins by identifying iron-rich foods and then reviews the factors affecting absorption.

Recommended Iron Intakes The usual diet in the United States provides about 6 to 7 milligrams of iron for every 1000 kcalories. The recommended daily intake for men is 8 milligrams, and because most men eat more than 2000 kcalories a day, they can meet their iron needs with little effort. Women in their reproductive years, however, need 18 milligrams a day. The "How To" on p. 432 explains how to calculate the recommended intake.

Vegetarians need 1.8 times as much iron ♦ to make up for the low bioavailability typical of their diets.[25] To maximize iron absorption, vegetarians should incorporate iron-rich foods into a diet that is low in inhibitors (foods such as leavened breads and fermented soy products such as miso and tempeh) and high in enhancers (foods rich in vitamin C and the organic acids found in fruits and vegetables). Good vegetarian sources of iron include soy foods (such as soybeans and tofu), legumes (such as lentils and kidney beans), nuts (such as cashews and almonds), seeds (such as pumpkin seeds and sunflower seeds), cereals (such as cream of wheat and oatmeal), dried fruit (such as apricots and raisins), vegetables (such as mushrooms and potatoes), and blackstrap molasses.

♦ To calculate the RDA for vegetarians, multiply by 1.8:
- 8 mg × 1.8 = 14 mg/day (vegetarian men)
- 18 mg × 1.8 = 32 mg/day (vegetarian women, 19 to 50 yr)

Because women have higher iron needs and lower energy needs, they sometimes have trouble obtaining enough iron. On average, women receive only 12 to 13 milligrams of iron per day, which is not enough iron for women until after menopause. To meet their iron needs from foods, premenopausal women need to select iron-rich foods at every meal.

HOW TO Estimate the Recommended Daily Intake for Iron

To calculate the recommended daily iron intake, the DRI Committee considers a number of factors. For example, for a woman of childbearing age (19 to 50):

- Losses from feces, urine, sweat, and shed skin: 1.0 milligram
- Losses through menstruation: 0.5 milligram (about 14 milligrams total averaged over 28 days)

These losses reflect an average daily need (total) of 1.5 milligrams of *absorbed* iron.

An estimated average requirement is determined based on the daily need and the assumption that an average of 18 percent of ingested iron is absorbed:

1.5 mg iron (needed)
÷ 0.18 (percent iron absorbed)
= 8 mg iron (estimated average requirement)

Then, a margin of safety is added to cover the needs of essentially all women of childbearing age, and the RDA is set at 18 milligrams.

For additional practice, log on to **www.cengagebrain.com** and search for ISBN 084006845X.

TRY IT Calculate how many slices of whole-wheat bread, cups of broccoli, ounces of hamburger meat, and cups of milk it takes to provide 18 milligrams of iron.

Dietary Guidelines for Americans 2010

Women capable of becoming pregnant should choose foods that supply heme iron, which is more readily absorbed by the body, additional iron sources, and enhancers of iron absorption such as vitamin C. If pregnant, they should take an iron supplement, as recommended by a health care provider.

Iron in Foods Figure 13-6 shows the amounts of iron in selected foods. Meats, fish, and poultry contribute the most iron per serving; other protein-rich foods such as legumes and eggs are also good sources. Although an indispensable part of the diet, foods in the milk group are notoriously poor in iron. Grain products vary, with whole-grain, enriched, and fortified breads and cereals contributing significantly to iron intakes. Finally, dark greens (such as broccoli) and dried fruits (such as raisins) contribute some iron.

Iron-Enriched Foods The FDA does not mandate iron enrichment, but most states require manufacturers to enrich flour and grain products with iron.* One serving of enriched bread or cereal provides only a little iron, but because people eat many servings of these foods, the contribution can be significant. Iron added to foods is nonheme iron, which is not absorbed as well as heme iron, but when eaten with absorption-enhancing foods, enrichment iron can increase iron stores and reduce iron deficiency.[26] In cases of iron overload, enrichment may exacerbate the problem.

Maximizing Iron Absorption In general, the bioavailability of iron is high in meats, fish, and poultry, intermediate in grains and legumes, and low in most vegetables, especially those containing oxalates such as spinach. As mentioned earlier, the amount of iron ultimately absorbed from a meal depends on the combined effects of several enhancing and inhibiting factors. For maximum absorption of nonheme iron, eat meat for MFP and fruits or vegetables for vitamin C. The iron of baked beans, for example, will be enhanced by the MFP in a piece of ham served with them. The iron of bread will be enhanced by the vitamin C in a slice of tomato on a sandwich.

When the label on a grain product says "enriched," it means iron and several B vitamins have been added to meet FDA standards.

*FDA standards require that each pound of enriched flour contain 20 milligrams iron.

FIGURE 13-6 **Iron in Selected Foods**

See the "How To" section on p. 316 for more information on using this figure.

Food	Serving size (kcalories)
Bread, whole wheat	1 oz slice (70 kcal)
Cornflakes, fortified	1 oz (110 kcal)
Spaghetti pasta	½ c cooked (99 kcal)
Tortilla, flour	1 10"-round (234 kcal)
Broccoli	½ c cooked (22 kcal)
Carrots	½ c shredded raw (24 kcal)
Potato	1 medium baked w/skin (133 kcal)
Tomato juice	½ c (31 kcal)
Banana	1 medium raw (109 kcal)
Orange	1 medium raw (62 kcal)
Strawberries	½ c fresh (22 kcal)
Watermelon	1 slice (92 kcal)
Milk	1 c reduced-fat 2% (121 kcal)
Yogurt, plain	1 c low-fat (155 kcal)
Cheddar cheese	1½ oz (171 kcal)
Cottage cheese	½ c low-fat 2% (101 kcal)
Pinto beans	½ c cooked (117 kcal)
Peanut butter	2 tbs (188 kcal)
Sunflower seeds	1 oz dry (165 kcal)
Tofu (soybean curd)	½ c (76 kcal)
Ground beef, lean	3 oz broiled (244 kcal)
Chicken breast	3 oz roasted (140 kcal)
Tuna, canned in water	3 oz (99 kcal)
Egg	1 hard cooked (78 kcal)
Excellent, and sometimes unusual, sources:	
Clams, canned	3 oz (126 kcal)
Beef liver	3 oz fried (184 kcal)
Parsley	1 c raw (22 kcal)

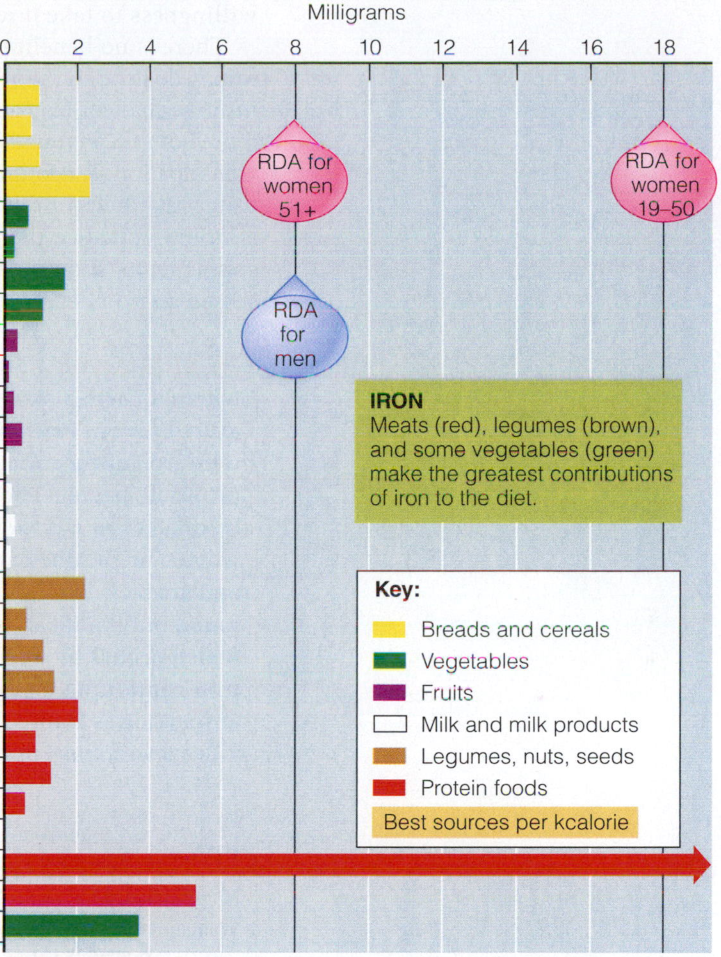

Iron Contamination and Supplementation

In addition to the iron from foods, **contamination iron** from nonfood sources of inorganic iron salts can contribute to the day's intakes. People can also get iron from supplements.

Contamination Iron Foods cooked in iron cookware take up iron salts. The more acidic the food and the longer it is cooked in iron cookware, the higher the iron content. ♦ The iron content of eggs can triple in the time it takes to scramble them in an iron pan. Admittedly, the absorption of this iron may be poor (perhaps only 1 to 2 percent), but every little bit helps a person who is trying to increase iron intake.

Iron Supplements People who are iron deficient may need supplements as well as an iron-rich, absorption-enhancing diet. Many physicians routinely recommend iron supplements to pregnant women, infants, and young children. Iron from supplements is less well absorbed than that from food, so the doses must be high. The absorption of iron taken as ferrous sulfate or as an iron **chelate** is better than that from other iron

An old-fashioned iron skillet adds iron to foods.

♦ Increase in iron content (mg) for selected foods (3 oz) after cooking in iron skillet:

Beef stew	0.66 → 3.40
Chili	0.96 → 6.27
Cornbread	0.67 → 0.86
Hamburger	1.49 → 2.29
Pancake	0.63 → 1.31
Rice	0.67 → 1.97
Scrambled egg	1.49 → 4.76
Spaghetti sauce	0.61 → 5.77

contamination iron: iron found in foods as the result of contamination by inorganic iron salts from iron cookware, iron-containing soils, and the like.

chelate (KEY-late): a substance that can grasp the positive ions of a mineral.

- **chele** = claw

supplements. Absorption also improves when supplements are taken between meals, at bedtime on an empty stomach, and with liquids (other than milk, tea, or coffee, which inhibit absorption). Taking iron supplements in a single dose instead of several doses per day is equally effective and may improve a person's willingness to take it regularly.

There is no benefit to taking iron supplements with orange juice because vitamin C does not enhance absorption from supplements as it does from foods. Vitamin C enhances iron absorption by converting insoluble ferric iron in foods to the more soluble ferrous iron, and supplemental iron is already in the ferrous form. Constipation is a common side effect of iron supplementation; drinking plenty of water may help to relieve this problem. The best strategy to ensure compliance is to individualize the dose, formulation, and schedule.[27] Most importantly, iron supplements should be taken only when prescribed by a physician who has assessed an iron deficiency.

IN SUMMARY Most of the body's iron is in hemoglobin and myoglobin where it carries oxygen for use in energy metabolism; some iron is also required for enzymes involved in a variety of reactions. Special proteins assist with iron absorption, transport, and storage—all helping to maintain an appropriate balance, because both too little and too much iron can be damaging. Iron deficiency is most common among infants and young children, teenagers, women of childbearing age, and pregnant women. Symptoms include fatigue and anemia. Iron overload is most common in men. Heme iron, which is found only in meat, fish, and poultry, is better absorbed than nonheme iron, which occurs in most foods. Nonheme iron absorption is improved by eating iron-containing foods with foods containing the MFP factor and vitamin C; absorption is limited by phytates and oxalates. The accompanying table provides a summary of iron.

Iron

RDA
Men: 8 mg/day
Women: 18 mg/day (19–50 yr) 8 mg/day (51+)
Upper Level
Adults: 45 mg/day
Chief Functions in the Body
Part of the protein hemoglobin, which carries oxygen in the blood; part of the protein myoglobin in muscles, which makes oxygen available for muscle contraction; necessary for the utilization of energy as part of the cells' metabolic machinery

Significant Sources
Red meats, fish, poultry, shellfish, eggs, legumes, dried fruits
Deficiency Symptoms
Anemia: weakness, fatigue, headaches; impaired work performance and cognitive function; impaired immunity; pale skin, nail beds, mucous membranes, and palm creases; concave nails; inability to regulate body temperature; pica
Toxicity Symptoms
GI distress
Iron overload: infections, fatigue, joint pain, skin pigmentation, organ damage

Zinc

Zinc is a versatile trace element required as a cofactor by more than 100 enzymes. Virtually all cells contain zinc, but the highest concentrations are found in muscle and bone.[28]

Zinc Roles in the Body

Zinc supports the work of numerous proteins in the body, such as the **metalloenzymes**, ♦ which are involved in a variety of metabolic processes, including the regulation of gene expression.* In addition, zinc

♦ Metalloenzymes that require zinc:
- Help make parts of the genetic materials DNA and RNA
- Manufacture heme for hemoglobin
- Participate in essential fatty acid metabolism
- Release vitamin A from liver stores
- Metabolize carbohydrates
- Synthesize proteins
- Metabolize alcohol in the liver
- Dispose of damaging free radicals

metalloenzymes (meh-tal-oh-EN-zimes): enzymes that contain one or more minerals as part of their structures.

*Among the metalloenzymes requiring zinc are carbonic anhydrase, deoxythymidine kinase, DNA and RNA polymerase, and alkaline phosphatase.

stabilizes cell membranes, helping to strengthen their defense against free-radical attacks. Zinc also assists in immune function and in growth and development. Zinc participates in the synthesis, storage, and release of the hormone insulin in the pancreas, although it does not appear to play a direct role in insulin's action. Zinc interacts with platelets in blood clotting, affects thyroid hormone function, and influences behavior and learning performance. It is needed to produce the active form of vitamin A (retinal) in visual pigments and the retinol-binding protein that transports vitamin A. It is essential to normal taste perception, wound healing, sperm production, and fetal development. A zinc deficiency impairs all these and other functions, underlining the vast importance of zinc in supporting the body's proteins.

Zinc Absorption and Metabolism

The body's handling of zinc resembles that of iron in some ways and differs in others. A key difference is the circular passage of zinc from the small intestine to the body and back again.

Zinc Absorption The rate of zinc absorption varies from about 15 to 40 percent, depending on a person's zinc status—if more is needed, more is absorbed. Also, dietary factors influence zinc absorption. For example, phytates bind zinc, thus limiting its bioavailability.[29]

Upon absorption into an intestinal cell, zinc has two options. Zinc may participate in the metabolic functions of the intestinal cell itself, or it may be retained within the intestinal cells by **metallothionein** until the body needs zinc.

Zinc Recycling Some zinc eventually reaches the pancreas, where it is incorporated into many of the digestive enzymes that the pancreas releases into the small intestine at mealtimes. The small intestine thus receives two doses of zinc with each meal—one from foods and the other from the zinc-rich pancreatic secretions. The recycling of zinc in the body from the pancreas to the small intestine and back to the pancreas is referred to as the **enteropancreatic circulation** of zinc. Each time zinc circulates through the small intestine, it may be excreted in shed intestinal cells or reabsorbed into the body (see Figure 13-7). The body loses zinc

metallothionein (meh-TAL-oh-THIGH-oh-neen): a sulfur-rich protein that avidly binds with and transports metals such as zinc.

- **metallo** = containing a metal
- **thio** = containing sulfur
- **ein** = a protein

enteropancreatic (EN-ter-oh-PAN-kree-AT-ik) **circulation:** the circulatory route from the pancreas to the small intestine and back to the pancreas.

FIGURE 13-7 Enteropancreatic Circulation of Zinc

Some zinc from food is absorbed by the small intestine and sent to the pancreas to be incorporated into digestive enzymes that return to the small intestine. This cycle is called the enteropancreatic circulation of zinc.

Animated! figure
www.cengagebrain.com
(search for ISBN 084006845X)

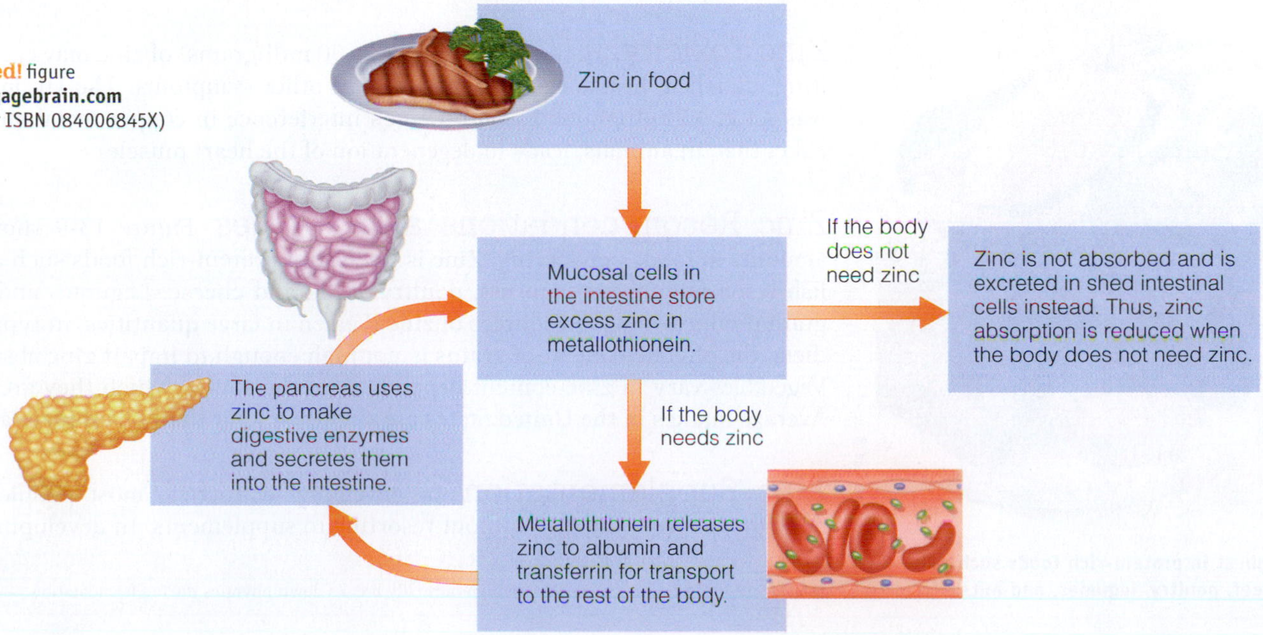

Zinc in food
If the body does not need zinc
Zinc is not absorbed and is excreted in shed intestinal cells instead. Thus, zinc absorption is reduced when the body does not need zinc.
Mucosal cells in the intestine store excess zinc in metallothionein.
The pancreas uses zinc to make digestive enzymes and secretes them into the intestine.
If the body needs zinc
Metallothionein releases zinc to albumin and transferrin for transport to the rest of the body.

FIGURE 13-8 Zinc-Deficiency Symptom—The Stunted Growth of Dwarfism

The growth retardation, known as dwarfism, is rightly ascribed to zinc deficiency because it is partially reversible when zinc is restored to the diet.

© H. Sanstead, University of Texas at Galveston

The Egyptian man on the right is an adult of average height. The Egyptian boy on the left is 17 years old but is only 4 feet tall, like a 7-year-old in the United States. His genitalia are like those of a 6-year-old.

© Polara Sutdios Inc.

Zinc is highest in protein-rich foods such as oysters, beef, poultry, legumes, and nuts.

primarily in feces. Smaller losses occur in urine, shed skin, hair, sweat, menstrual fluids, and semen.

Zinc Transport Zinc's main transport vehicle in the blood is the protein albumin. Some zinc also binds to transferrin—the same transferrin that carries iron in the blood. In healthy individuals, transferrin is usually less than 50 percent saturated with iron, but in iron overload, it is more saturated. Diets that deliver more than twice as much iron as zinc leave too few transferrin sites available for zinc. The result is poor zinc absorption. The converse is also true: large doses of zinc inhibit iron absorption.

Large doses of zinc create a similar problem with another essential mineral, copper. These nutrient interactions highlight one of the many reasons why people should use supplements conservatively, if at all: supplementation can easily create imbalances.

Zinc Deficiency

Severe zinc deficiencies are not widespread in developed countries, but they do occur in vulnerable groups—pregnant women, young children, the elderly, and the poor.[30] Human zinc deficiency was first reported in the 1960s in children and adolescent boys in Egypt, Iran, and Turkey. Children have especially high zinc needs because they are growing rapidly and synthesizing many zinc-containing proteins, and the native diets among those populations were not meeting these needs. Middle Eastern diets are typically low in the richest zinc source, meats. Furthermore, the staple foods in these diets are legumes, unleavened breads, and other whole-grain foods—all high in fiber and phytates, which inhibit zinc absorption.*

Figure 13-8 shows the severe growth retardation and mentions the immature sexual development characteristic of zinc deficiency. In addition, zinc deficiency hinders digestion and absorption, causing diarrhea, which worsens malnutrition not only for zinc, but for all nutrients. It also impairs the immune response, making infections likely—among them, GI tract infections, which worsen malnutrition, including zinc malnutrition (a classic downward spiral of events). Chronic zinc deficiency damages the central nervous system and brain and may lead to poor motor development and cognitive performance. Because zinc deficiency directly impairs vitamin A metabolism, vitamin A–deficiency symptoms often appear. Zinc deficiency also disturbs thyroid function and the metabolic rate. It alters taste, causes loss of appetite, and slows wound healing—in fact, its symptoms are so pervasive that generalized malnutrition and sickness are more likely to be the diagnosis than simple zinc deficiency.

Zinc Toxicity

High doses (more than 50 milligrams) of zinc may cause vomiting, diarrhea, headaches, exhaustion, and other symptoms. The UL for adults was set at 40 milligrams based on zinc's interference in copper metabolism—an effect that, in animals, leads to degeneration of the heart muscle.

Zinc Recommendations and Sources

Figure 13-9 shows zinc amounts in foods per serving. Zinc is highest in protein-rich foods such as shellfish (especially oysters), meats, poultry, milk, and cheese. Legumes and whole-grain products are good sources of zinc if eaten in large quantities; in typical U.S. diets, the phytate content of grains is not high enough to impair zinc absorption. Vegetables vary in zinc content depending on the soil in which they are grown. Average intakes in the United States are slightly higher than recommendations.

Zinc Supplementation

In developed countries, most people obtain enough zinc from the diet without resorting to supplements. In developing coun-

*Unleavened bread contains no yeast, which normally breaks down phytates during fermentation.

FIGURE 13-9 Zinc in Selected Foods

See the "How To" section on p. 316 for more information on using this figure.

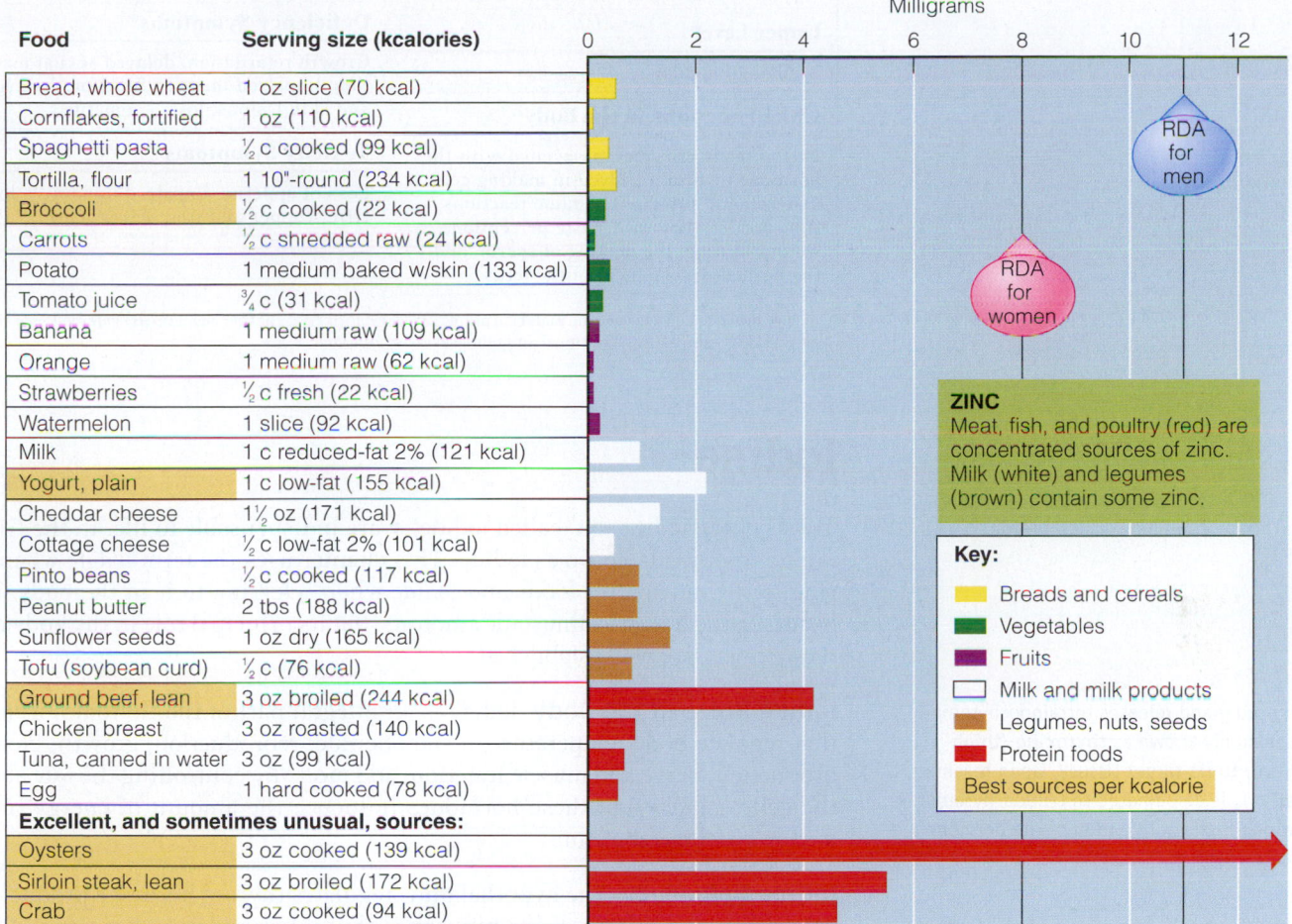

Food	Serving size (kcalories)
Bread, whole wheat	1 oz slice (70 kcal)
Cornflakes, fortified	1 oz (110 kcal)
Spaghetti pasta	½ c cooked (99 kcal)
Tortilla, flour	1 10"-round (234 kcal)
Broccoli	½ c cooked (22 kcal)
Carrots	½ c shredded raw (24 kcal)
Potato	1 medium baked w/skin (133 kcal)
Tomato juice	¾ c (31 kcal)
Banana	1 medium raw (109 kcal)
Orange	1 medium raw (62 kcal)
Strawberries	½ c fresh (22 kcal)
Watermelon	1 slice (92 kcal)
Milk	1 c reduced-fat 2% (121 kcal)
Yogurt, plain	1 c low-fat (155 kcal)
Cheddar cheese	1½ oz (171 kcal)
Cottage cheese	½ c low-fat 2% (101 kcal)
Pinto beans	½ c cooked (117 kcal)
Peanut butter	2 tbs (188 kcal)
Sunflower seeds	1 oz dry (165 kcal)
Tofu (soybean curd)	½ c (76 kcal)
Ground beef, lean	3 oz broiled (244 kcal)
Chicken breast	3 oz roasted (140 kcal)
Tuna, canned in water	3 oz (99 kcal)
Egg	1 hard cooked (78 kcal)
Excellent, and sometimes unusual, sources:	
Oysters	3 oz cooked (139 kcal)
Sirloin steak, lean	3 oz broiled (172 kcal)
Crab	3 oz cooked (94 kcal)

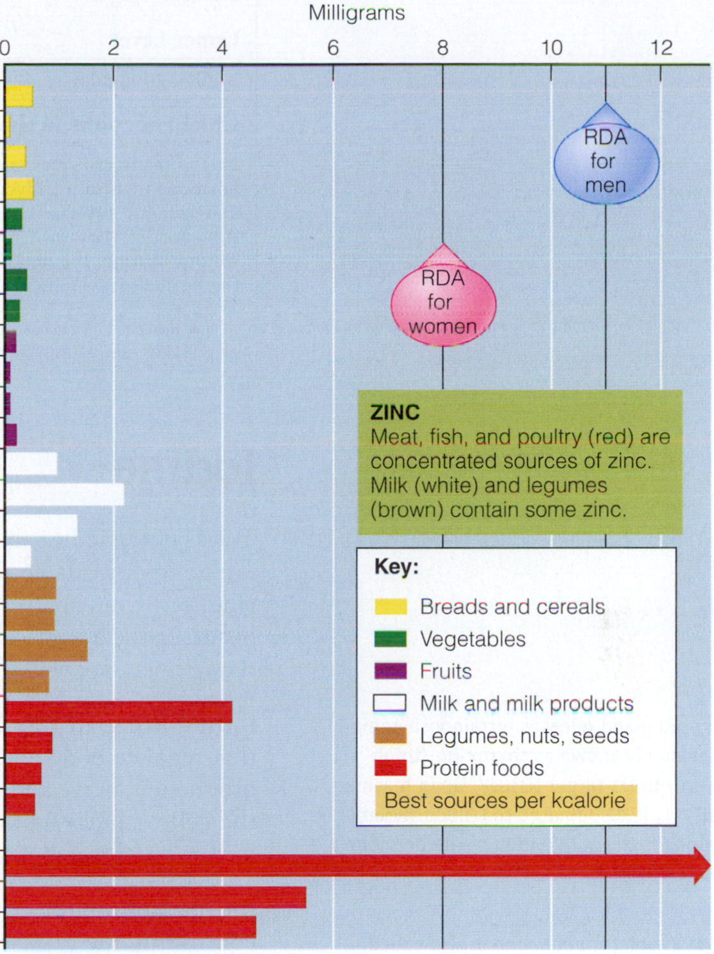

tries, zinc supplementation plays a major role in the treatment of childhood infectious diseases.[31] Zinc supplements effectively reduce the incidence of disease and death associated with diarrhea in children.[32] Similarly, zinc supplements reduce the incidence of pneumonia and associated deaths in older adults.[33]

The use of zinc lozenges to treat the common cold has been controversial and inconclusive, with some studies finding them effective and others not.[34] The different study results may reflect the effectiveness of various zinc compounds. Some studies using zinc gluconate report shorter duration of cold symptoms, whereas most studies using other combinations of zinc report no effect. Common side effects of zinc lozenges include nausea and bad taste reactions.

IN SUMMARY **Zinc-requiring enzymes participate in a multitude of reactions affecting growth, vitamin A activity, and pancreatic digestive enzyme synthesis, among others. Both dietary zinc and zinc-rich pancreatic secretions (via enteropancreatic circulation) are available for absorption. Absorption is monitored by a special binding protein (metallothionein) in the small intestine. Protein-rich foods derived from animals are the best sources of bioavailable zinc. Fiber and phytates in cereals bind zinc, limiting absorption. Growth retardation and sexual immaturity are hallmark symptoms of zinc deficiency. The following table provides a summary of zinc.**

Zinc

RDA	Significant Sources
Men: 11 mg/day Women: 8 mg/day	Protein-containing foods: red meats, shellfish, whole grains; some fortified cereals
Upper Level	**Deficiency Symptoms[a]**
Adults: 40 mg/day	Growth retardation, delayed sexual maturation, impaired immune function, hair loss, eye and skin lesions, loss of appetite
Chief Functions in the Body	**Toxicity Symptoms**
Part of many enzymes; associated with the hormone insulin; involved in making genetic material and proteins, immune reactions, transport of vitamin A, taste perception, wound healing, the making of sperm, and the normal development of the fetus	Loss of appetite, impaired immunity, low HDL, copper and iron deficiencies

[a]A rare inherited disease of zinc malabsorption, *acrodermatitis* (AK-roh-der-ma-TIE-tis) *enteropathica* (EN-ter-oh-PATH-ick-ah), causes additional and more severe symptoms.

Iodine

Traces of the iodine ion (called iodide) ♦ are indispensable to life. In the GI tract, iodine from foods becomes iodide. This chapter uses the term *iodine* when referring to the nutrient in foods and *iodide* when referring to it in the body. Iodide occurs in the body in minuscule amounts, but its principal role in the body and its requirement are well established.

♦ The ion form of iodine is called **iodide.**

Iodide Roles in the Body Iodide is an integral part of the thyroid hormones ♦ that regulate body temperature, metabolic rate, reproduction, growth, blood cell production, nerve and muscle function, and more. By controlling the rate at which the cells use oxygen, these hormones influence the amount of energy released during basal metabolism.

♦ The thyroid gland releases tetraiodothyronine (T_4), commonly known as **thyroxine** (thigh-ROCKS-in), to its target tissues. Upon reaching the cells, T_4 is deiodinated to triiodothyronine (T_3), which is the active form of the hormone.

Iodine Deficiency The hypothalamus regulates thyroid hormone production by controlling the release of the pituitary's thyroid-stimulating hormone (TSH). ♦ With iodine deficiency, thyroid hormone production declines, and the body responds by secreting more TSH in a futile attempt to accelerate iodide uptake by the thyroid gland. If a deficiency persists, the cells of the thyroid gland enlarge to trap as much iodide as possible. Sometimes the gland enlarges until it makes a visible lump in the neck, a **goiter** (shown in Figure 13-10).

♦ Thyroid-stimulating hormone is also called **thyrotropin.**

Goiter afflicts about 200 million people the world over, many of them in South America, Asia, and Africa. In all but 4 percent of these cases, the cause is iodine deficiency. As for the 4 percent (8 million), most have goiter because they regularly eat excessive amounts of foods ♦ that contain an antithyroid substance (**goitrogen**) whose effect is not counteracted by dietary iodine. The goitrogens present in plants remind us that even natural components of foods can cause harm when eaten in excess.

♦ Examples of goitrogen-containing foods:
- Cabbage, spinach, radishes, rutabagas
- Soybeans, peanuts
- Peaches, strawberries

Goiter may be the earliest and most obvious sign of iodine deficiency, but the most tragic and prevalent damage occurs in the brain. Iodine deficiency is the most common cause of *preventable* mental retardation and brain damage in the world. Children with even a mild iodine deficiency typically have goiters and perform poorly in school. With sustained treatment, however, mental performance in the classroom as well as thyroid function improves.

A severe iodine deficiency during pregnancy causes the extreme and irreversible mental and physical retardation known as **cretinism.** ♦ Cretinism affects approximately six million people worldwide and can be averted by the early diagnosis and treatment of maternal iodine deficiency. A worldwide effort to provide iodized salt to people living in iodine-deficient areas has been dramatically successful. Because iron deficiency is common among people with iodine deficiency

♦ The underactivity of the thyroid gland is known as **hypothyroidism** and may be caused by iodine deficiency or any number of other causes. Without treatment, an infant with **congenital hypothyroidism** will develop the physical and mental retardation of **cretinism.**

goiter (GOY-ter): an enlargement of the thyroid gland due to an iodine deficiency, malfunction of the gland, or overconsumption of a goitrogen. Goiter caused by iodine deficiency is sometimes called *simple goiter.*

goitrogen (GOY-troh-jen): a substance that enlarges the thyroid gland and causes *toxic goiter.* Goitrogens occur naturally in such foods as cabbage, kale, brussels sprouts, cauliflower, broccoli, and kohlrabi.

cretinism (CREE-tin-ism): a congenital disease characterized by mental and physical retardation and commonly caused by maternal iodine deficiency during pregnancy.

and because iron deficiency reduces the effectiveness of iodized salt, dual fortification with both iron and iodine may be most beneficial.[35]

Iodine Toxicity Excessive intakes of iodine can interfere with thyroid function and enlarge the gland, just as deficiency can.[36] During pregnancy, exposure to excessive iodine from foods, prenatal supplements, or medications is especially damaging to the developing infant. An infant exposed to toxic amounts of iodine during gestation may develop a goiter so severe as to block the airways and cause suffocation. The UL is 1100 micrograms ♦ per day for an adult—several times higher than average intakes.

Iodine Recommendations and Sources The ocean is the world's major source of iodine. In coastal areas, kelp, seafood, water, and even iodine-containing sea mist are dependable iodine sources. Further inland, the amount of iodine in foods is variable and generally reflects the amount present in the soil in which plants are grown or on which animals graze. Landmasses that were once under the ocean have soils rich in iodine; those in flood-prone areas where water leaches iodine from the soil are poor in iodine. In the United States and Canada, the iodization of salt ♦ has eliminated the widespread misery caused by iodine deficiency during the 1930s, but iodized salt is not available in many parts of the world. Some countries add iodine to bread, fish paste, or drinking water instead. Families that do not use iodized foods have a higher prevalence of child malnutrition and mortality.[37]

Although average consumption of iodine in the United States exceeds recommendations, it falls below toxic levels. Some of the excess iodine in the U.S. diet stems from fast foods, which use iodized salt liberally. Some iodine comes from bakery products and from milk. The baking industry uses iodates (iodine salts) as dough conditioners, and most dairies feed cows iodine-containing medications and use iodine to disinfect milking equipment. Now that these sources have been identified, food industries have reduced their use of these compounds, but the sudden emergence of this problem points to a need for continued surveillance of the food supply. Processed foods in the United States use regular salt, not iodized salt.

The recommended intake of iodine for adults is a minuscule amount. The need for iodine is easily met by consuming seafood, vegetables grown in iodine-rich soil, and iodized salt. ♦ In the United States, labels indicate whether salt is iodized; in Canada, all table salt is iodized.

IN SUMMARY **Iodide, the ion of the mineral iodine, is an essential component of the thyroid hormone. An iodine deficiency can lead to simple goiter (enlargement of the thyroid gland) and can impair fetal development, causing cretinism. Iodization of salt has largely eliminated iodine deficiency in the United States and Canada. The accompanying table provides a summary of iodine.**

Iodine

RDA

Adults: 150 µg/day

Upper Level

1100 µg/day

Chief Functions in the Body

A component of two thyroid hormones that help to regulate growth, development, and metabolic rate

Significant Sources

Iodized salt, seafood, bread, dairy products, plants grown in iodine-rich soil and animals fed those plants

Deficiency Disease

Simple goiter, cretinism

Deficiency Symptoms

Underactive thyroid gland, goiter, mental and physical retardation in infants (cretinism)

Toxicity Symptoms

Underactive thyroid gland, elevated TSH, goiter

FIGURE 13-10 **Iodine-Deficiency Symptom—The Enlarged Thyroid of Goiter**

Scott Camazine/Photo Researchers, Inc.

In iodine deficiency, the thyroid gland enlarges—a condition known as simple goiter. Iodine toxicity also enlarges the thyroid gland, creating a similar-looking goiter.

♦ For perspective, most foods provide 3 to 75 µg iodine per serving.

♦ Iodized salt contains about 60 µg iodine per gram salt.

♦ On average, ½ tsp iodized salt provides the RDA for iodine.

© Craig M. Moore

Only "iodized salt" has had iodine added.

Selenium

The essential mineral **selenium** shares some of the chemical characteristics of the mineral sulfur. This similarity allows selenium to substitute for sulfur in the amino acids methionine, cysteine, and cystine.

Selenium Roles in the Body Selenium is one of the body's antioxidant nutrients, ♦ working primarily as a part of proteins—most notably, the enzyme glutathione peroxidase.[38] Glutathione peroxidase and vitamin E work in tandem. Glutathione peroxidase prevents free-radical formation, thus blocking the chain reaction before it begins; if free radicals do form and a chain reaction starts, vitamin E stops it. (Highlight 11 describes free-radical formation, chain reactions, and antioxidant action in detail.) Another enzyme that converts the thyroid hormone to its active form also contains selenium.

♦ Key antioxidant nutrients:
- Vitamin C, vitamin E, beta-carotene
- Selenium

Selenium Deficiency Selenium deficiency is associated with a heart disease ♦ that is prevalent in regions of China where the soil and foods lack selenium. Although the primary cause of this heart disease is probably a virus, selenium deficiency appears to predispose people to it, and adequate selenium seems to prevent it.

♦ The heart disease associated with selenium deficiency is named **Keshan** (KESH-an or ka-SHAWN) **disease** for one of the provinces of China where it was first studied. Keshan disease is characterized by heart enlargement and insufficiency; fibrous tissue replaces the muscle tissue that normally composes the middle layer of the walls of the heart.

Selenium and Cancer Some research suggests that selenium may protect against some types of cancers. Given the potential for harm and the lack of conclusive evidence, however, recommendations to take selenium supplements would be premature—and perhaps ineffective as well. Selenium from foods appears to be more effective in inhibiting cancer growth than selenium from supplements. Such a finding reinforces a theme that has been repeated throughout this text—foods offer many more health benefits than supplements.

Selenium Recommendations and Sources Selenium is found in the soil, and therefore in the crops grown for consumption.[39] People living in regions with selenium-poor soil may still get enough selenium, partly because they eat vegetables and grains transported from other regions and partly because they eat meats, milk, and eggs, which are reliable sources of selenium. Eating as few as two Brazil nuts a day effectively improves selenium status.[40] Average intakes in the United States and Canada are above the RDA, which is based on the amount needed to maximize glutathione peroxidase activity.

Selenium Toxicity Because high doses of selenium are toxic, a UL has been set. Selenium toxicity causes loss and brittleness of hair and nails, garlic breath odor, and nervous system abnormalities.

IN SUMMARY **Selenium is an antioxidant nutrient that works closely with the glutathione peroxidase enzyme and vitamin E. Selenium is found in association with protein in foods. Deficiencies are associated with a predisposition to a type of heart abnormality known as Keshan disease. The accompanying table provides a summary of selenium.**

Selenium

RDA	**Deficiency Symptoms**
Adults: 55 µg/day	Predisposition to heart disease characterized by cardiac tissue becoming fibrous (Keshan disease)
Upper Level	**Toxicity Symptoms**
Adults: 400 µg/day	Loss and brittleness of hair and nails; skin rash, fatigue, irritability, and nervous system disorders; garlic breath odor
Chief Functions in the Body	
Defends against oxidation; regulates thyroid hormone	
Significant Sources	
Seafood, meat, whole grains, fruits, and vegetables (depending on soil content)	

selenium (se-LEEN-ee-um): a trace element.

Copper

The body contains about 100 milligrams of copper in a variety of cells and tissues. Copper balance and transport depend on a system of proteins.[41]

Copper Roles in the Body Copper serves as a constituent of several enzymes. The copper-containing enzymes have diverse metabolic roles with one common characteristic: all involve reactions that consume oxygen or oxygen radicals. For example, copper-containing enzymes catalyze the oxidation of ferrous iron to ferric iron, which allows iron to bind to transferrin. Copper's role in iron metabolism makes it a key factor in hemoglobin synthesis. Copper- and zinc-containing enzymes participate in the body's natural defense against the oxidative damage of free radicals. Still other copper enzymes help to manufacture collagen, inactivate histamine, and degrade serotonin. Copper, like iron, is needed in many of the metabolic reactions related to the release of energy.

Copper Deficiency and Toxicity Typical U.S. diets provide adequate amounts of copper and deficiency is rare. In animals, copper deficiency raises blood cholesterol and damages blood vessels, raising questions about whether low dietary copper might contribute to cardiovascular disease in humans.

Some genetic disorders create a copper toxicity, but excessive intakes from foods are unlikely. Excessive intakes from supplements may cause liver damage, and therefore a UL has been set.

Two rare genetic disorders affect copper status in opposite directions. In Menkes disease, the intestinal cells absorb copper, but cannot release it into circulation, causing a life-threatening deficiency. Treatment involves giving copper intravenously. In Wilson's disease, copper accumulates in the liver and brain, creating a life-threatening toxicity. Wilson's disease can be controlled by reducing copper intake, using chelating agents such as penicillamine, and taking zinc supplements, which interfere with copper absorption.

Copper Recommendations and Sources The richest food sources of copper are legumes, whole grains, nuts, shellfish, and seeds. More than half of the copper from foods is absorbed, and the major route of elimination appears to be bile. Water may also provide copper, depending on the type of plumbing pipe and the hardness of the water.

IN SUMMARY **Copper is a component of several enzymes, all of which are involved in some way with oxygen or oxidation. Some act as antioxidants; others are essential to iron metabolism. Legumes, whole grains, and shellfish are good sources of copper. The accompanying table provides a summary of copper.**

Copper

RDA	Significant Sources
Adults: 900 µg/day	Seafood, nuts, whole grains, seeds, legumes
Upper Level	**Deficiency Symptoms**
Adults: 10,000 µg/day (10 mg/day)	Anemia, bone abnormalities
Chief Functions in the Body	**Toxicity Symptoms**
Necessary for the absorption and use of iron in the formation of hemoglobin; part of several enzymes	Liver damage

Manganese

The human body contains a tiny 20 milligrams of manganese. Most of it can be found in the bones and metabolically active organs such as the liver, kidneys, and pancreas.

Manganese Roles in the Body Manganese acts as a cofactor for many enzymes that facilitate the metabolism of carbohydrate, lipids, and amino acids. In addition, manganese-containing metalloenzymes assist in bone formation and the conversion of pyruvate to a TCA cycle compound.

Manganese Deficiency and Toxicity Manganese requirements are low, and many plant foods contain significant amounts of this trace mineral, so deficiencies are rare. As is true of other trace minerals, however, dietary factors such as phytates inhibit its absorption. In addition, high intakes of iron and calcium limit manganese absorption, so people who use supplements of those minerals regularly may impair their manganese status.

Toxicity is more likely to occur from an environment contaminated with manganese than from dietary intake. Miners who inhale large quantities of manganese dust on the job over prolonged periods show symptoms of a brain disease, along with abnormalities in appearance and behavior. Still, a UL has been established based on intakes from food, water, and supplements.

Manganese Recommendations and Sources Grain products make the greatest contribution of manganese to the diet. With insufficient information to establish an RDA, an AI was set based on average intakes.

IN SUMMARY **Manganese-dependent enzymes are involved in bone formation and various metabolic processes. Because manganese is widespread in plant foods, deficiencies are rare, although regular use of calcium and iron supplements may limit manganese absorption. The accompanying table provides a summary of manganese.**

Manganese

AI	Significant Sources
Men: 2.3 mg/day	Nuts, whole grains, leafy vegetables, tea
Women: 1.8 mg/day	**Deficiency Symptoms**
Upper Level	Rare
Adults: 11 mg/day	**Toxicity Symptoms**
Chief Functions in the Body	Nervous system disorders
Cofactor for several enzymes; bone formation	

FIGURE 13-11 **U.S. Population with Access to Fluoridated Water through Public Water Systems**

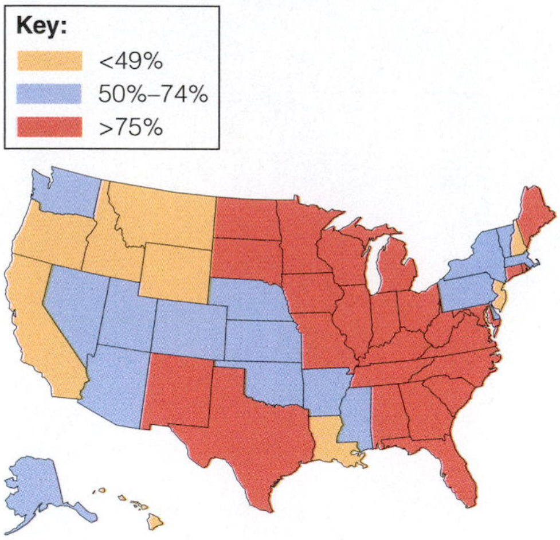

♦ Key bone nutrients:
- Vitamin D, vitamin K, vitamin A
- Calcium, phosphorus, magnesium, fluoride

fluorapatite (floor-APP-uh-tite): the stabilized form of bone and tooth crystal, in which fluoride has replaced the hydroxyl groups of hydroxyapatite.

Fluoride

Fluoride is present in virtually all soils, water supplies, plants, and animals. Only a trace of fluoride occurs in the human body, but with this amount, the crystalline deposits in bones and teeth are larger and more perfectly formed.

Fluoride Roles in the Body As Chapter 12 explains, during the mineralization of bones ♦ and teeth, calcium and phosphorus form crystals called hydroxyapatite. Then fluoride replaces the hydroxyl (OH) portions of the hydroxyapatite crystal, forming **fluorapatite**, which makes the bones stronger and the teeth more resistant to decay.

Dental caries ranks as the nation's most widespread health problem: an estimated 95 percent of the population have decayed, missing, or filled teeth. By interfering with a person's ability to chew and eat a wide variety of foods, these dental problems can quickly lead to a multitude of nutrition problems. Where fluoride is lacking, dental decay is common.

Drinking water is usually the best source of fluoride, and almost 70 percent of the U.S. population served by public water systems receives optimal levels of fluoride (see Figure 13-11).[42] (Most bottled waters lack fluoride.) Fluoridation of drinking water (to raise the concentration to 1 part fluoride per 1 million ♦ parts water) offers the greatest protection against dental caries at virtually no risk of toxicity.[43] By fluoridating the drinking water, a community offers its residents, particularly the children, a safe, economical, practical, and effective way to defend against dental caries.

♦ For perspective, 1 part per million (1 ppm) is approximately 1 mg per liter.

Fluoride Toxicity Too much fluoride can damage the teeth, causing **fluorosis.**[44] For this reason, a UL has been established. In mild cases, the teeth develop small white specks; in severe cases, the enamel becomes pitted and permanently stained (as shown in Figure 13-12). Fluorosis occurs only during tooth development and cannot be reversed, making its prevention ♦ a high priority. To limit fluoride ingestion, take care not to swallow fluoride-containing dental products such as toothpaste and mouthwash.

♦ To prevent fluorosis:
- Monitor the fluoride content of the local water supply.
- Supervise toddlers when they brush their teeth—using only a little toothpaste (pea-size amount).
- Use fluoride supplements only as prescribed by a physician.

Fluoride Recommendations and Sources As mentioned earlier, much of the U.S. population has access to water with an optimal fluoride concentration, which typically delivers about 1 milligram per person per day.[45] Fish and most teas contain appreciable amounts of natural fluoride.

IN SUMMARY **Fluoride makes bones stronger and teeth more resistant to decay. Fluoridation of public water supplies can significantly reduce the incidence of dental caries, but excess fluoride during tooth development can cause fluorosis—discolored and pitted tooth enamel. The accompanying table provides a summary of fluoride.**

Fluoride

AI	Significant Sources
Men: 4 mg/day Women: 3 mg/day	Drinking water (if fluoride containing or fluoridated), tea, seafood
Upper Level	**Deficiency Symptoms**
Adults: 10 mg/day	Susceptibility to tooth decay
Chief Functions in the Body	**Toxicity Symptoms**
Maintains health of bones and teeth; helps to make teeth resistant to decay	Fluorosis (pitting and discoloration of teeth)

FIGURE 13-12 **Fluoride-Toxicity Symptom—The Mottled Teeth of Fluorosis**

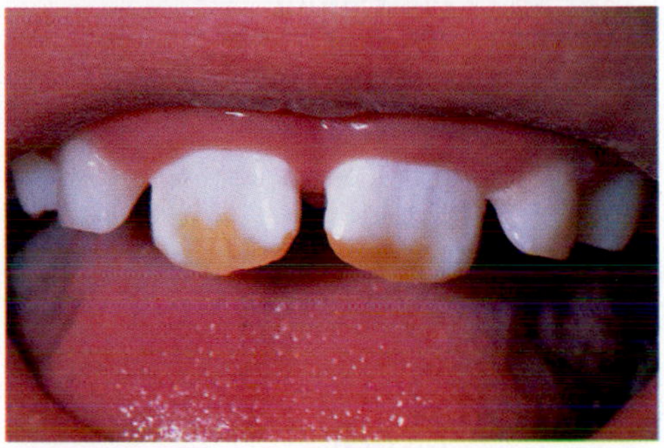

Chromium

Chromium is an essential mineral that participates in carbohydrate and lipid metabolism. Like iron, chromium assumes different charges. In chromium, the Cr^{+++} ion is the most stable and most commonly found in foods.

Chromium Roles in the Body Chromium helps maintain glucose homeostasis by enhancing the activity of the hormone insulin. ♦ When chromium is lacking, a diabetes-like condition may develop with elevated blood glucose and impaired glucose tolerance, insulin response, and glucagon response. Some research findings suggest that chromium supplements improve glucose or insulin responses in diabetes, but the FDA finds these relationships "highly uncertain."[46]

♦ Small organic compounds that enhance insulin's action are called **glucose tolerance factors (GTF).** Some glucose tolerance factors contain chromium.

Chromium Recommendations and Sources Chromium is present in a variety of foods. The best sources are unrefined foods, particularly liver, brewer's yeast, and whole grains. The more refined foods people eat, the less chromium they ingest.

Chromium Supplements Supplement advertisements have succeeded in convincing consumers that they can lose fat and build muscle by taking chromium picolinate. Whether chromium supplements (either picolinate or plain) reduce body fat or improve muscle strength remains controversial.

fluorosis (floor-OH-sis): discoloration and pitting of tooth enamel caused by excess fluoride during tooth development.

IN SUMMARY Chromium enhances insulin's action. A deficiency can result in a diabetes-like condition. Chromium is widely available in unrefined foods including brewer's yeast, whole grains, and liver. The accompanying table provides a summary of chromium.

Chromium

AI	**Significant Sources**
Men: 35 µg/day Women: 25 µg/day	Meats (especially liver), whole grains, brewer's yeast
Chief Functions in the Body	**Deficiency Symptoms**
Enhances insulin action and may improve glucose tolerance	Diabetes-like condition
	Toxicity Symptoms
	None reported

Molybdenum

Molybdenum acts as a working part of several metalloenzymes. Dietary deficiencies of molybdenum are unknown because the amounts needed are minuscule—as little as 0.1 part per million parts of body tissue. Legumes, breads and other grain products, leafy green vegetables, milk, and liver are molybdenum-rich foods. Average daily intakes fall within the suggested range of intakes.

Molybdenum toxicity in people is rare. It has been reported in animal studies, and a UL has been established. Characteristics of molybdenum toxicity include kidney damage and reproductive abnormalities. The accompanying table provides a summary of molybdenum.

IN SUMMARY Molybdenum

RDA	**Significant Sources**
Adults: 45 µg/day	Legumes, cereals, nuts
Upper Level	**Deficiency Symptoms**
Adults: 2 mg/day	Unknown
Chief Functions in the Body	**Toxicity Symptoms**
Cofactor for several enzymes	None reported; reproductive effects in animals

FIGURE 13-13 Cobalt in Vitamin B_{12}

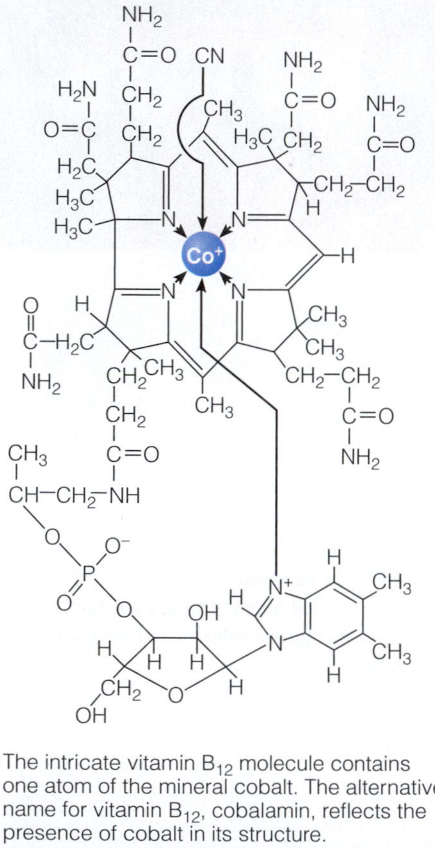

The intricate vitamin B_{12} molecule contains one atom of the mineral cobalt. The alternative name for vitamin B_{12}, cobalamin, reflects the presence of cobalt in its structure.

molybdenum (mo-LIB-duh-num): a trace element.

Other Trace Minerals

Research to determine whether other trace minerals are essential is difficult because their quantities in the body are so small and also because human deficiencies are unknown. Guessing their functions in the body can be particularly problematic. Much of the available knowledge comes from research using animals.

Nickel may serve as a cofactor for certain enzymes. Silicon is involved in the formation of bones and collagen. Vanadium, too, is necessary for growth and bone development and for normal reproduction. Cobalt is a key mineral in the large vitamin B_{12} molecule (see Figure 13-13), but it is not an essential nutrient and no recommendation has been established. Boron may play a key role in bone health, brain activities, and immune response.[47]

In the future, we may discover that many other trace minerals play key nutritional roles. Even arsenic—famous as a poison used by murderers and known to be a carcinogen—may turn out to be essential for human beings in tiny quantities. It has already proved useful in the treatment of some types of leukemia.[48]

Contaminant Minerals

Chapter 12 and this chapter explain the many ways minerals serve the body—maintaining fluid and electrolyte balance, providing structural support to the bones, transporting oxygen, and assisting enzymes. In contrast to the minerals that the body requires, contaminant minerals impair the body's growth, work capacity, and general health. Contaminant minerals include the **heavy metals** lead, mercury, and cadmium that enter the food supply by way of soil, water, and air pollution. This section focuses on lead poisoning because it is a serious environmental threat to young children and because reducing blood lead levels in children is a goal of the Healthy People initiative.[49] Much of the information on lead applies to the other contaminant minerals as well—they all disrupt body processes and impair nutrition status similarly.

Like other minerals, lead is indestructible and the body cannot change its chemistry. Chemically similar to nutrient minerals like iron, calcium, and zinc (cations with two positive charges), lead displaces them from some of the metabolic sites they normally occupy but is then unable to perform their roles. For example, lead competes with iron in heme, but it cannot carry oxygen. Similarly, lead competes with calcium in the brain, but it cannot signal messages from nerve cells. Excess lead in the blood also deranges the structure of red blood cell membranes, making them leaky and fragile. Lead interacts with white blood cells, too, impairing their ability to fight infection, and it binds to antibodies, thwarting their effort to resist disease. Chapter 15 examines the damaging effects of lead toxicity on a child's growth and development.

Lead typifies the ways all heavy metals behave in the body: they interfere with nutrients that are trying to do their jobs. The "good guy" nutrients are shoved aside by the "bad guy" contaminants. Then, when the contaminants cannot perform the roles of the nutrients, health diminishes. To safeguard our health, we must defend ourselves against contamination by eating nutrient-rich foods and preserving a clean environment.

Closing Thoughts on the Nutrients

This chapter completes the introductory lessons on the nutrients. Each nutrient from the amino acids to zinc has been described rather thoroughly—its chemistry, roles in the body, sources in the diet, symptoms of deficiency and toxicity, and influences on health and disease. Such a detailed examination is informative, but it can also be misleading. It is important to step back from the detailed study of the individual nutrients to look at them as a whole. After all, people eat foods, not nutrients, and most foods deliver dozens of nutrients. Furthermore, nutrients work cooperatively with one another in the body; their actions are most often *inter*actions. This chapter alone mentioned how iron depends on vitamin C to keep it in its active form and copper to incorporate it into hemoglobin, how zinc is needed to activate and transport vitamin A, and how both iodine and selenium are needed for the synthesis of thyroid hormone. The following table provides a summary of the trace minerals for your review. Highlight 13 explores the benefits of phytochemicals.

heavy metals: mineral ions such as mercury and lead, so called because they are of relatively high atomic weight. Many heavy metals are poisonous.

IN SUMMARY The Trace Minerals

Mineral and Chief Functions	Deficiency Symptoms	Toxicity Symptoms[a]	Significant Sources
Iron Part of the protein hemoglobin, which carries oxygen in the blood; part of the protein myoglobin in muscles, which makes oxygen available for muscle contraction; necessary for energy metabolism	Anemia: weakness, fatigue, headaches; impaired work performance; impaired immunity; pale skin, nail beds, mucous membranes, and palm creases; concave nails; inability to regulate body temperature; pica	GI distress; iron overload: infections, fatigue, joint pain, skin pigmentation, organ damage	Red meats, fish, poultry, shellfish, eggs, legumes, dried fruits
Zinc Part of insulin and many enzymes; involved in making genetic material and proteins, immune reactions, transport of vitamin A, taste perception, wound healing, the making of sperm, and normal fetal development	Growth retardation, delayed sexual maturation, impaired immune function, hair loss, eye and skin lesions, loss of appetite	Loss of appetite, impaired immunity, low HDL, copper and iron deficiencies	Protein-containing foods: red meats, fish, shellfish, poultry, whole grains; fortified cereals
Iodine A component of the thyroid hormones that help to regulate growth, development, and metabolic rate	Underactive thyroid gland, goiter, mental and physical retardation (cretinism)	Underactive thyroid gland, elevated TSH, goiter	Iodized salt; seafood; plants grown in iodine-rich soil and animals fed those plants
Selenium Part of an enzyme that defends against oxidation; regulates thyroid hormone	Associated with Keshan disease	Nail and hair brittleness and loss; fatigue, irritability, and nervous system disorders, skin rash, garlic breath odor	Seafoods, organ meats; other meats, whole grains, fruits, and vegetables (depending on soil content)
Copper Helps form hemoglobin; part of several enzymes	Anemia, bone abnormalities	Liver damage	Seafood, nuts, legumes, whole grains, seeds
Manganese Cofactor for several enzymes; bone formation	Rare	Nervous symptom disorders	Nuts, whole grains, leafy vegetables, tea
Fluoride Maintains health of bones and teeth; confers decay resistance on teeth	Susceptibility to tooth decay	Fluorosis (pitting and discoloration) of teeth	Drinking water (if fluoridated), tea, seafood
Chromium Enhances insulin action, may improve glucose intolerance	Diabetes-like condition	None reported	Meats (liver), whole grains, brewer's yeast
Molybdenum Cofactor for several enzymes	Unknown	None reported	Legumes, cereals, nuts

[a]Acute toxicities of many minerals cause abdominal pain, nausea, vomiting, and diarrhea.

How much of each particular nutrient does the body need? Estimates fall somewhere between intakes that are inadequate and cause illness and intakes that are excessive and cause illness. A wide range of intakes that support health, to varying degrees, lies between deficiency and toxicity. In the past, nutrient needs were determined by how much was needed to prevent deficiency symptoms. If lack of a nutrient caused illness, it was defined as essential. Today, nutrient needs are based on how much is needed to support optimal health. The amount of vitamin C needed to prevent scurvy is much less than the amount correlated with reducing the risk of cancer, for example. Furthermore, nutrients are being examined within the context of the whole diet. Health benefits are not credited to vitamin C alone, but also to the vitamin C–rich fruits and vegetables that provide many other nutrients—and nonnutrients (phytochemicals)—important to health.

People can also improve their health with physical activity. Energy expenditure is unlike money expenditure: it is desirable to *spend* energy, not to save it (within reason, of course). The more energy people spend, the more food they can afford to eat—food that delivers both nutrients and pleasure. The next chapter presents details on nutrition and physical activity.

Nutrition Portfolio

Trace minerals from a variety of foods, especially those in the protein foods group, support many of your body's activities.

Go to Diet Analysis Plus and choose one of the days on which you tracked your diet for an entire day. Select the Intake vs. Goals report and then consider the following questions. Remember that scoring 100 percent on this report means you met your goal.

- Your Intake vs. Goals report may only display your intake for two of the trace minerals: iron and zinc. How was your intake for these two trace minerals?

Now click on the Intake Spreadsheet report and consider the following questions:

- Examine the variety in your food intake, taking particular notice of how often you include meats, seafood, poultry, legumes, and enriched or fortified grain products weekly. These foods often contain trace minerals.
- Describe the advantages of using iodized salt.
- Determine whether your community provides fluoridated water.

Diet Analysis PLUS To complete this exercise, go to your Diet Analysis Plus at www.cengage.com/sso.

Nutrition on the Net

For further study of topics covered in this chapter, log on to **www.cengagebrain.com** and search for ISBN 084006845X.

- Search for "minerals" at the American Dietetic Association: **www.eatright.org**
- Search for the individual minerals by name at the U.S. Department of Health and Human Services information site: **www.healthfinder.gov**
- Learn more about iron overload from the Iron Overload Diseases Association: **www.ironoverload.org**
- Learn more about iodine and thyroid disease from the American Thyroid Association: **www.thyroid.org**
- Learn more about lead in paint, dust, and soil from the Centers for Disease Control or Environmental Protection Agency: **www.cdc.gov/lead** or **www.epa.gov/lead**

References

1. R. E. Fleming and B. R. Bacon, Orchestration of iron homeostasis, *New England Journal of Medicine* 352 (2005): 1741–1744.
2. Committee on Dietary Reference Intakes, *Dietary Reference Intakes for Vitamin A, Vitamin K, Arsenic, Boron, Chromium, Copper, Iodine, Iron, Manganese, Molybdenum, Nickel, Silicon, Vanadium, and Zinc* (Washington, D.C.: National Academies Press, 2001), p. 315.
3. P. Thankachan and coauthors, Iron absorption in young Indian women: The interaction of iron status with the influence of tea and ascorbic acid, *American Journal of Clinical Nutrition* 87 (2008): 881–886.
4. R. F. Hurrell and coauthors, Meat protein fractions enhance nonheme iron absorption in humans, *Journal of Nutrition* 136 (2006): 2808–2812.
5. R. E. Conway, J. J. Powell, and C. A. Geissler, A food-group based algorithm to predict non-heme iron absorption, *International Journal of Food Sciences and Nutrition* 58 (2007): 29–41.
6. Committee on Dietary Reference Intakes, 2001, p. 351.
7. M. D. Knutson, Into the matrix: Regulation of the iron regulatory hormone hepcidin by matriptase-2, *Nutrition Reviews* 67 (2009): 284–288; M. F. Young and coauthors, Serum hepcidin is significantly associated with iron absorption from food and supplemental sources in healthy young women, *American Journal of Clinical Nutrition* 89 (2009): 533–538; M. U. Muckenthaler, B. Galy, and M. W. Hentze, Systemic iron homeostasis and the iron-responsive element/iron-regulatory protein (IRE/IRP) regulatory network, *Annual Review of Nutrition* 28 (2008): 197–213; E. Nemeth and T. Ganz, Regulation of iron metabolism by hepcidin, *Annual Review of Nutrition* 26 (2006): 323–342.
8. Worldwide prevalence of anaemia 1993–2005: WHO Global Database on Anaemia, published 2008, available at www.who.org.
9. L. M. Tussing-Humphreys and coauthors, Excess adiposity, inflammation, and iron-deficiency in female adolescents, *Journal of the American Dietetic Association* 109 (2009): 297–302.
10. J. P. McClung and J. P. Karl, Iron deficiency and obesity: The contribution of inflammation and diminished iron absorption, *Nutrition Reviews* 67 (2008): 100–104.
11. J. L. Beard, Why iron deficiency is important in infant development, *Journal of Nutrition* 138 (2008): 2534–2536.
12. K. C. White, Anemia is a poor predictor of iron deficiency among toddlers in the United States: For heme the bell tolls, *Pediatrics* 115 (2005): 315–320.
13. Z. Yang and coauthors, Comparison of plasma ferritin concentration with the ratio of plasma transferrin receptor to ferritin in estimating body iron stores: Results of 4 intervention trials, *American Journal of Clinical Nutrition* 87 (2008): 1892–1898.
14. J. C. McCann and B. N. Ames, An overview of evidence for a causal relation between iron deficiency during development and deficits in cognitive or behavioral function, *American Journal of Clinical Nutrition* 85 (2007): 931–945.

15. L. L. Iannotti and coauthors, Iron supplementation in early childhood: Health benefits and risks, *American Journal of Clinical Nutrition* 84 (2006): 1261–1276.
16. S. L. Young and coauthors, Toward a comprehensive approach to the collection and analysis of pica substances, with emphasis on geophagic materials, *PLoS ONE* 3 (2008): e3147.
17. A. Pietrangelo, Hereditary hemochromatosis, *Annual Review of Nutrition* 26 (2006): 251–270.
18. H. Drakesmith and A. Prentice, Viral infection and iron metabolism, *Nature Reviews. Microbiology* 6 (2008): 541–552; A. M. Prentice, Iron metabolism, malaria, and other infections: What is all the fuss about? *Journal of Nutrition* 138 (2008): 2537–2541; C. Ratledge, Iron metabolism and infection, *Food and Nutrition Bulletin* 28 (2007): S515–S523.
19. F. Dreyfus, The deleterious effects of iron overload in patients with myelodysplastic syndromes, *Blood Reviews* 22 (2008): S29–S34.
20. K. J. Allen and coauthors, Iron-overload: Related disease in *HFE* hereditary hemochromatosis, *New England Journal of Medicine* 358 (2008): 221–230.
21. F. B. Hu, The iron-heart hypothesis: Search for the ironclad evidence, *Journal of the American Medical Association* 297 (2007): 639–641; D. Lee, A. R. Folsom, and D. R. Jacobs, Iron, zinc, and alcohol consumption and mortality from cardiovascular diseases: The Iowa Women's Health Study, *American Journal of Clinical Nutrition* 81 (2005): 787–791.
22. L. R. Zacharski and coauthors, Reduction of iron stores and cardiovascular outcomes in patients with peripheral arterial disease: A randomized controlled trial, *Journal of the American Medical Association* 297 (2007): 603–610.
23. A. G. Mainous and coauthors, Iron, lipids, and risk of cancer in the Framingham Offspring Cohort, *American Journal of Epidemiology* 160 (2005): 1115–1122.
24. A. S. Manoguerra and coauthors, Iron ingestion: An evidence-based consensus guideline for out-of-hospital management, *Clinical Toxicology* 43 (2005): 553–570.
25. Committee on Dietary Reference Intakes, 2001, p. 351.
26. D. Moretti and coauthors, Extruded rice fortified with micronized ground ferric pyrophosphate reduces iron deficiency in Indian schoolchildren: A double-blind randomized controlled trial, *American Journal of Clinical Nutrition* 84 (2006): 822–829; F. Pizarro and coauthors, Ascorbyl palmitate enhances iron bioavailability in iron-fortified bread, *American Journal of Clinical Nutrition* 84 (2006): 830–834.
27. M. Alleyne, M. K. Horne, and J. L. Miller, Individualized treatment for iron-deficiency anemia in adults, *American Journal of Medicine* 121 (2008): 943–948.
28. H. Tapiero and K. D. Tew, Trace elements in human physiology and pathology: Zinc and metallothioneins, *Biomedicine and Pharmacotherapy* 57 (2003): 399–411.
29. J. R. Hunt, J. M. Beiseigel, and L. K. Johnson, Adaptation in human zinc absorption as influenced by dietary zinc and bioavailability, *American Journal of Clinical Nutrition* 87 (2008): 1336–1345.
30. S. N. Meydani and coauthors, Serum zinc and pneumonia in nursing home elderly, *American Journal of Clinical Nutrition* 86 (2007): 1167–1173; J. M. Schneider and coauthors, The prevalence of low serum zinc and copper levels and dietary habits associated with serum zinc and copper in 12- to 36-month-old children from low-income families at risk for iron deficiency, *Journal of the American Dietetic Association* 107 (2007): 1924–1929.
31. D. E. Roth and coauthors, Acute lower respiratory infections in childhood: Opportunities for reducing the global burden through nutritional interventions, *Bulletin of the World Health Organization* 86 (2008): 321–416.
32. M. Lukacik, R. L. Thomas, and J. V. Aranda, A meta-analysis of the effects of oral zinc in the treatment of acute and persistent diarrhea, *Pediatrics* 121 (2008): 326–336; S. E. Wuehler, F. Sempértegui, and K. H. Brown, Dose-response trial of prophylactic zinc supplements, with or without copper, in young Ecuadorian children at risk of zinc deficiency, *American Journal of Clinical Nutrition* 87 (2008): 723–733; R. Aggarwal, J. Sentz, and M. A. Miller, Role of zinc administration in prevention of childhood diarrhea and respiratory illnesses: A meta-analysis, *Pediatrics* 119 (2007): 1120–1130; J.M.M. Gardner and coauthors, Zinc supplementation and psychosocial stimulation: Effects on the development of undernourished Jamaican children, *American Journal of Clinical Nutrition* 82 (2005): 399–405.
33. Meydani and coauthors, 2007.
34. G. A. Eby and W. W. Halcomb, Ineffectiveness of zinc gluconate nasal spray and zinc orotate lozenges in common-cold treatment: A double-blind placebo-controlled clinical trial, *Alternative Therapies in Health and Medicine* 12 (2006): 34–48; B. Arroll, Non-antibiotic treatments for upper-respiratory tract infections (common cold), *Respiratory Medicine* 99 (2005): 1477–1484.
35. M. B. Zimmerman, The influence of iron status on iodine utilization and thyroid function, *Annual Review of Nutrition* 26 (2006): 367–389.
36. W. Teng and coauthors, Effect of iodine intake on thyroid diseases in China, *New England Journal of Medicine* 354 (2006): 2783–2793.
37. R. D. Semba and coauthors, Child malnutrition and mortality among families not utilizing adequately iodized salt in Indonesia, *American Journal of Clinical Nutrition* 87 (2008): 438–444.
38. X. G. Lei, W. Cheng, and J. P. McClung, Metabolic regulation and function of glutathione peroxidase-1, *Annual Review of Nutrition* 27 (2007): 41–61; R. F. Burk and K. E. Hill, Selenoprotein P: An extracellular protein with unique physical characteristics and a role in selenium homeostasis, *Annual Review of Nutrition* 25 (2005): 215–235.
39. J. W. Finley, Selenium accumulation in plant foods, *Nutrition Reviews* 63 (2005): 196–202.
40. C. D. Thomson and coauthors, Brazil nuts: An effective way to improve selenium status, *American Journal of Clinical Nutrition* 87 (2008): 379–384.
41. J. R. Prohaska, Role of copper transporters in copper homeostasis, *American Journal of Clinical Nutrition* 88 (2008): 826S–829S.
42. Populations receiving optimally fluoridated public drinking water: United States, 1992–2006, *Morbidity and Mortality Weekly Report* 57 (2008): 737–741.
43. Position of the American Dietetic Association: The impact of fluoride on health, *Journal of the American Dietetic Association* 105 (2005): 1620–1628.
44. Surveillance for dental caries, dental sealants, tooth retention, edentulism, and enamel fluorosis: United States, 1988–1994 and 1999–2002, *Morbidity and Mortality Weekly Report* 54 (2005): 1–44.
45. Populations receiving optimally fluoridated public drinking water: United States, 1992–2006, 2008.
46. E. M. Balk and coauthors, Effect of chromium supplementation on glucose metabolism and lipids: A systematic review of randomized controlled trials, *Diabetes Care* 30 (2007): 2154–2163; C. L. Broadhurst and P. Domenico, Clinical studies on chromium picolinate supplementation in diabetes mellitus—A review, *Diabetes Technology and Therapeutics* 8 (2006): 677–687; P. R. Trumbo and K. C. Ellwood, Chromium picolinate intake and risk of type 2 diabetes: An evidence-based review by the United States Food and Drug Administration, *Nutrition Reviews* 64 (2006): 357–363.
47. F. H. Nielsen, Is boron nutritionally relevant? *Nutrition Reviews* 66 (2008): 183–191.
48. M. S. Tallman, What is the role of arsenic in newly diagnosed APL? *Best Practice and Research Clinical Haematology* 21 (2008): 659–666.
49. Committee on Environmental Health, Lead exposure in children: Prevention, detection, and management, *Pediatrics* 116 (2005): 1036–1046; Blood lead levels: United States, 1999–2002, *Morbidity and Mortality Weekly Report* 54 (2005): 513–527.

HIGHLIGHT 13

Phytochemicals and Functional Foods

© Ross Durant/Jupiterimages

Chapter 13 completes the introductory discussions on the six classes of nutrients—carbohydrates, lipids, proteins, vitamins, minerals, and water. In addition to these nutrients, foods contain thousands of nonnutrient compounds, including the **phytochemicals.** Chapter 1 introduces the phytochemicals as compounds found in plant-derived foods (*phyto* means plant) that have biological activity in the body. Research on phytochemicals is unfolding daily, adding to our knowledge of their roles in human health, but there are still many questions and only tentative answers. Just a few of the tens of thousands of phytochemicals have been researched at all, and only a sampling are mentioned in this highlight—enough to illustrate their wide variety of food sources and roles in supporting health.

The concept that foods provide health benefits beyond those of the nutrients emerged from numerous epidemiological studies showing the protective effects of plant-based diets on cancer and heart disease. People have been using foods to maintain health and prevent disease for years, but now these foods have been given a name—they are called **functional foods.**[1] (The accompanying glossary defines this and other terms.) As Chapter 1 explains, functional foods include all foods (whole, fortified, or modified foods) that have a potentially beneficial effect on health.[2] Much of this text touts the benefits of nature's functional foods—whole grains rich in dietary fibers, oily fish rich in omega-3 fatty acids, and fresh fruits rich in phytochemicals, for example. This highlight begins with a look at some of these familiar functional foods, the phytochemicals they contain, and their roles in disease prevention. Then the discussion turns to examine the most controversial of functional foods—novel foods to which phytochemicals have been added to promote health. How these foods fit into a healthy diet is still unclear.

The Phytochemicals

In foods, phytochemicals impart tastes, aromas, colors, and other characteristics. They give hot peppers their burning sensation, garlic its pungent flavor, and tomatoes their dark red color. In the body, phytochemicals can have profound physiological effects—acting as antioxidants, mimicking hormones, stimulating enzymes, interfering with DNA replication, destroying bacteria, and binding physically to cell walls. Any of these actions may suppress the development of diseases. They might also have adverse effects when consumed in excess.[3] Table H13-1 (p. 450) presents the names, possible effects, and food sources of some of the better-known phytochemicals.

Defending against Cancer

A variety of phytochemicals from a variety of foods appear to protect against DNA damage and defend the body against cancer. A few examples follow.

Soybeans and products made from them correlate with low rates of breast and prostate cancers.[4] Soybeans—as well as other legumes, **flaxseeds,** whole grains, fruits, and vegetables—are a

GLOSSARY

flavonoids (FLAY-von-oyds): yellow pigments in foods; phytochemicals that may exert physiological effects on the body.

flaxseeds: the small brown seeds of the flax plant; valued as a source of linseed oil, fiber, and omega-3 fatty acids.

functional foods: foods that contain physiologically active compounds that provide health benefits beyond basic nutrition.

lignans: phytochemicals present in flaxseed, but not in flax oil, that are converted to phytosterols by intestinal bacteria and are under study as possible anticancer agents.

lutein (LOO-teen): a plant pigment of yellow hue; a phytochemical believed to play roles in eye functioning and health.

lycopene (LYE-koh-peen): a pigment responsible for the red color of tomatoes and other red-hued vegetables; a phytochemical that may act as an antioxidant in the body.

phytochemicals: nonnutrient compounds found in plant-derived foods that have biological activity in the body.

phytoestrogens: plant-derived compounds that have structural and functional similarities to human estrogen. Phytoestrogens include the isoflavones genistein, daidzein, and glycitein.

phytosterols: plant-derived compounds that have structural similarities to cholesterol and lower blood cholesterol by competing with cholesterol for absorption. Phytosterols include sterol esters and stanol esters.

TABLE H13-1 Phytochemicals—Their Food Sources and Actions

Name	Possible Effects	Food Sources
Alkylresorcinols (phenolic lipids)	May contribute to the protective effect of grains in reducing the risks of diabetes, heart disease, and some cancers.	Whole-grain wheat and rye
Allicin (organosulfur compound)	Antimicrobial that may reduce ulcers; may lower blood cholesterol.	Chives, garlic, leeks, onions
Capsaicin	Modulates blood clotting, possibly reducing the risk of fatal clots in heart and artery disease.	Hot peppers
Carotenoids (include beta-carotene, lycopene, lutein, and hundreds of related compounds)	Act as antioxidants, possibly reducing risks of cancer and other diseases.	Deeply pigmented fruits and vegetables (apricots, broccoli, cantaloupe, carrots, pumpkin, spinach, sweet potatoes, tomatoes)
Curcumin	Acts as an antioxidant and anti-inflammatory agent; may reduce blood clot formation; may inhibit enzymes that activate carcinogens.	Turmeric, a yellow-colored spice
Flavonoids (include flavones, flavonols, isoflavones, catechins, and others)	Act as antioxidants; scavenge carcinogens; bind to nitrates in the stomach, preventing conversion to nitrosamines; inhibit cell proliferation.	Berries, black tea, celery, citrus fruits, green tea, olives, onions, oregano, purple grapes, purple grape juice, soybeans and soy products, vegetables, whole wheat, wine
Genistein and daidzein (isoflavones)	Phytoestrogens that inhibit cell replication in GI tract; may reduce risk of breast, colon, ovarian, prostate, and other estrogen-sensitive cancers; may reduce cancer cell survival; may reduce risk of osteoporosis.	Soybeans, soy flour, soy milk, tofu, textured vegetable protein, other legume products
Indoles (organosulfur compound)	May trigger production of enzymes that block DNA damage from carcinogens; may inhibit estrogen action.	Cruciferous vegetables such as broccoli, brussels sprouts, cabbage, cauliflower; horseradish, mustard greens, kale
Isothiocyanates (organosulfur compounds that include sulforaphane)	Act as antioxidants; inhibit enzymes that activate carcinogens; activate enzymes that detoxify carcinogens; may reduce risk of breast cancer, prostate cancer.	Cruciferous vegetables such as broccoli, brussels sprouts, cabbage, cauliflower; horseradish, mustard greens, kale
Lignans	Phytoestrogens that block estrogen activity in cells possibly reducing the risk of cancer of the breast, colon, ovaries, and prostate.	Flaxseed and its oil, whole grains
Monoterpenes (including limonene)	May trigger enzyme production to detoxify carcinogens; inhibit cancer promotion and cell proliferation.	Citrus fruit peels and oils
Phenolic acids	May trigger enzyme production to make carcinogens water soluble, facilitating excretion.	Coffee beans, fruits (apples, blueberries, cherries, grapes, oranges, pears, prunes), oats, potatoes, soybeans
Phytic acid	Binds to minerals, preventing free-radical formation, possibly reducing cancer risk.	Whole grains
Resveratrol	Acts as antioxidant; may inhibit cancer growth; reduce inflammation, LDL oxidation, and blood clot formation.	Red wine, peanuts, grapes, raspberries
Saponins (glucosides)	May interfere with DNA replication, preventing cancer cells from multiplying; stimulate immune response.	Alfalfa sprouts, other sprouts, green vegetables, potatoes, tomatoes
Tannins	Act as antioxidants; may inhibit carcinogen activation and cancer promotion.	Black-eyed peas, grapes, lentils, red and white wine, tea

rich source of an array of phytochemicals, among them the **phytoestrogens.** Because the chemical structure of these phytochemicals is similar to the steroid hormone estrogen, they can weakly mimic or modulate the effects of estrogen in the body. They also have antioxidant activity that appears to slow the growth of some cancers.[5] However, the use of phytoestrogen supplements is ill-advised as they may stimulate the growth of estrogen-dependent cancers (such as breast cancer).[6] Soy foods may be most effective when consumed in moderation throughout life. The role of soy foods for breast cancer survivors is less certain. The American Cancer Society recommends: "Breast cancer survivors should consume only moderate amounts of soy foods as part of a healthy plant-based diet and should not intentionally ingest very high levels of soy products."[7]

Limited evidence suggests that tomatoes may offer protection against some cancers.[8] Among the phytochemicals thought to be responsible for this effect is **lycopene,** one of beta-carotene's many carotenoid relatives.[9] Lycopene is the pigment that gives apricots, guava, papaya, pink grapefruits, and watermelon their red color—and it is especially abundant in tomatoes and cooked tomato products. Lycopene is a powerful antioxidant that seems to inhibit the growth of cancer cells.[10] Importantly, the benefits of lycopene are not consistently evident.[11] When benefits have been seen, it has been from people having eaten *foods* containing lycopene.[12]

Soybeans and tomatoes are only two of the many fruits and vegetables credited with providing anticancer activity. Strong and convincing evidence shows that the risk of many cancers, and perhaps of cancer in general, decreases when diets include an abundance of fruits and vegetables.[13] To that end, current recommendations urge consumers to eat five to nine servings of fruits and vegetables a day.

Defending against Heart Disease

Diets based primarily on unprocessed foods appear to support heart health better than those founded on highly refined foods—perhaps because of the abundance of nutrients, fibers, or phytochemicals such as the **flavonoids.** Flavonoids, a large group of phytochemicals known for their health-promoting qualities, are found in whole grains, legumes, soy, vegetables, fruits, herbs, spices, teas, chocolate, nuts, olive oil, and red wines. Flavonoids are powerful antioxidants that may help to protect LDL cholesterol against oxidation, minimize inflammation, and reduce blood platelet stickiness, thereby slowing the progression of atherosclerosis and making blood clots less likely.[14] Whereas an abundance of flavonoid-containing *foods* in the diet may lower the risks of chronic diseases, no claims can be made for flavonoids themselves as the protective factor, particularly when they are extracted from foods and sold as supplements. In fact, some research suggests that the antioxidant activity of flavonoid-rich foods is *not* because of the flavonoids themselves.[15]

In addition to flavonoids, fruits and vegetables are rich in carotenoids such as beta-carotene and **lutein.** Studies suggest that a diet rich in carotenoids is also associated with a lower risk of heart disease.[16]

Nature offers a variety of functional foods that provide us with many health benefits.

The **phytosterols** of soybeans and the **lignans** of flaxseed may also protect against heart disease.[17] These cholesterol-like molecules are naturally found in all plants and inhibit cholesterol absorption in the body.[18] As a result, blood cholesterol levels decline.[19] These phytochemicals also seem to protect against heart disease by reducing inflammation and lowering blood pressure.[20]

The Phytochemicals in Perspective

Because foods deliver thousands of phytochemicals in addition to dozens of nutrients, researchers must be careful in giving credit for particular health benefits to any one compound. Diets rich in whole grains, legumes, vegetables, fruits, and nuts seem to protect against heart disease and cancer, but identifying *the* specific foods or components of foods that are responsible is difficult. Each food possesses a unique array of phytochemicals—citrus fruits provide monoterpenes; grapes, resveratrol; and flaxseed, lignans. (Review Table H13-1 for the possible effects and other food sources of these phytochemicals.) Broccoli may contain as many as 10,000 different phytochemicals—each with the potential to influence some action in the body. Beverages such as wine, spices such as oregano, and oils such as olive oil (especially virgin olive oil) contain many phytochemicals that may explain, in part, why people who live in the Mediterranean region have reduced risks of heart disease and cancer.[21] Phytochemicals might also explain why the DASH diet is so effective in lowering blood pressure and blood lipids. Even identifying all of the phytochemicals and their effects doesn't answer all the questions because the actions of phytochemicals may be complementary or overlapping—which reinforces the principle of variety in diet planning. For an appreciation of the array of phytochemicals offered by a variety of foods, see Figure H13-1 (p. 452).

Functional Foods

Because foods naturally contain thousands of phytochemicals that are biologically active in the body, virtually all of them have some special value in supporting health. In other words, even simple, whole foods, in reality, are functional foods. Cranberries may help prevent urinary tract infections; garlic may lower blood cholesterol; and green tea may inhibit ulcer infections, just to name a few examples.[22] But that hasn't stopped food manufacturers from trying to create functional foods as well. The creation of more functional foods has become the fastest-growing trend and the greatest influence transforming the global food supply.[23]

Many processed foods become functional foods when they are fortified with nutrients or enhanced with phytochemicals or herbs (calcium-fortified orange juice, for example). Less frequently, an entirely new food is created, as in the case of a meat substitute made of mycoprotein—a protein derived from a fungus.* This functional food not only provides dietary fiber, polyunsaturated fats, and high-quality protein, but it lowers LDL cholesterol, raises

*This mycoprotein product is marketed under the trade name Quorn (pronounced KWORN).

HDL cholesterol, improves glucose response, and prolongs satiety after a meal. Such a novel functional food raises the question—is it a food or a drug?

Foods as Pharmacy

Not too long ago, most of us could agree on what was a food and what was a drug. Today, functional foods blur the distinctions.[24] They have characteristics similar to both foods and drugs, but do not fit neatly into either category. Consider margarine, for example.

Eating nonhydrogenated margarine sparingly instead of butter generously may lower blood cholesterol slightly over several months and clearly falls into the food category. Taking a statin drug, on the other hand, lowers blood cholesterol significantly within weeks and clearly falls into the drug category. But margarine enhanced with a phytosterol that lowers blood cholesterol is in a gray area between the two. The margarine looks and tastes like a food, but it acts like a drug.

The use of functional foods as drugs creates a whole new set of diet-planning challenges. Not only must foods provide an ad-

FIGURE H13-1 An Array of Phytochemicals in a Variety of Fruits and Vegetables

Broccoli and broccoli sprouts contain an abundance of the cancer-fighting phytochemical sulforaphane.

The phytochemical resveratrol found in grapes (and nuts) protects against cancer by inhibiting cell growth and against heart disease by limiting clot formation and inflammation.

The flavonoids in cocoa and chocolate defend against oxidation and reduce the tendency of blood to clot.

An apple a day—rich in flavonoids—may protect against lung cancer.

The ellagic acid of strawberries may inhibit certain types of cancer.

Tomatoes, with their abundant lycopene, may defend against cancer by protecting DNA from oxidative damage.

Spinach and other colorful vegetables contain the carotenoids lutein and zeaxanthin, which help protect the eyes against macular degeneration.

The phytoestrogens of soybeans seem to starve cancer cells and inhibit tumor growth; the phytosterols may lower blood cholesterol and protect cardiac arteries.

The monoterpenes of citrus fruits (and cherries) may inhibit cancer growth.

Flaxseed, the richest source of lignans, may prevent the spread of cancer.

Garlic, with its abundant organosulfur compounds, may lower blood cholesterol and protect against stomach cancer.

The flavonoids in black tea may protect against heart disease, whereas those in green tea may defend against cancer.

Blueberries, a rich source of flavonoids, improve memory in animals.

equate intake of all the nutrients to support good health, but they must also deliver drug-like ingredients to protect against disease. Like drugs used to treat chronic diseases, functional foods may need to be eaten several times a day for several months or years to have a beneficial effect. Sporadic users may be disappointed in the results. Margarine enriched with 2 to 3 grams of phytosterols may reduce cholesterol by up to 15 percent, much more than regular margarine does, but not nearly as much as the more than 30 percent reduction seen with cholesterol-lowering drugs.[25] For this reason, functional foods may be more useful for prevention and mild cases of disease than for intervention and more severe cases.

Foods and drugs differ dramatically in cost as well. Functional foods such as fruits and vegetables incur no added costs, of course, but foods that have been manufactured with added phytochemicals can be expensive, costing up to six times as much as their conventional counterparts. The price of functional foods typically falls between that of traditional foods and medicines.

© Craig M. Moore

Functional foods currently on the market promise to "enhance mood," "promote relaxation and good karma," "increase alertness," and "improve memory," among other claims.

Unanswered Questions

To achieve a desired health effect, which is the better choice: to eat a food designed to affect some body function or simply to adjust the diet? Does it make more sense to use a margarine enhanced with a phytosterol that lowers blood cholesterol or simply to limit the amount of butter eaten?* Is it smarter to eat eggs enriched with omega-3 fatty acids or to restrict egg consumption? Might functional foods offer a sensible solution for improving our nation's health—if done correctly? Perhaps so, but the problem is that the food industry is moving too fast for either scientists or the Food and Drug Administration to keep up. Consumers were able to buy soup with St. John's wort that claimed to enhance mood and fruit juice with echinacea that was supposed to fight colds while scientists were still conducting their studies on these ingredients. Research to determine the safety and effectiveness of these substances is still in progress. Until this work is complete, consumers are on their own in finding the answers to the following questions:

- *Does it work?* Research is generally lacking and findings are often inconclusive.
- *How much does it contain?* Food labels are not required to list the quantities of added phytochemicals. Even if they were, consumers have no standard for comparison and cannot deduce whether the amounts listed are a little or a lot. Most importantly, until research is complete, food manufacturers do not know what amounts (if any) are most effective—or most toxic.
- *Is it safe?* Functional foods can act like drugs. They contain ingredients that can alter body functions and cause allergies, drug interactions, drowsiness, and other side effects. Yet, unlike drug labels, food labels do not provide instructions for the dosage, frequency, or duration of treatment.
- *Is it healthy?* Adding phytochemicals to a food does not magically make it a healthy choice. A candy bar may be fortified with phytochemicals, but it is still made mostly of sugar and fat.

Critics suggest that the designation "functional foods" may be nothing more than a marketing tool. After all, even the most experienced researchers cannot yet identify the perfect combination of nutrients and phytochemicals to support optimal health. Yet manufacturers are freely experimenting with various concoctions as if they possessed that knowledge. Is it okay for them to sprinkle phytochemicals on fried snack foods or caramel candies and label them "functional," thus implying health benefits?

Future Foods

Nature has elegantly designed foods to provide us with a complex array of dozens of nutrients and thousands of additional compounds that may benefit health—most of which we have yet to identify or understand. Over the years, we have taken those foods, deconstructed them, and then reconstructed them in an effort to "improve" them. With new scientific understandings of how nutrients—and the myriad other compounds in foods—interact with genes, we may someday be able to design foods to meet the *exact* health needs of *each* individual. Indeed, our knowledge of the human genome and of human nutrition may well merge to allow specific recommendations for individuals based on their predisposition to diet-related diseases.

If the present trend continues, someday physicians may be able to prescribe the perfect foods to enhance your health, and farmers will be able to grow them. Scientists have already developed gene technology to alter the composition of food crops. They can grow rice enriched with vitamin A and tomatoes containing a hepatitis vaccine, for example. It seems quite likely that foods can be created to meet every possible human need. But then, in a sense, that was largely true 100 years ago when we relied on the bounty of nature.

*Margarine products that lower blood cholesterol contain either sterol esters from vegetable oils, soybeans, and corn or stanol esters from wood pulp.

Nutrition on the Net

For further study of topics covered in this highlight, log on to **www.cengagebrain.com** and search for ISBN 084006845X.

- Search for "functional foods" at the International Food Information Council: **www.ific.org**
- Search for "functional foods" at the Center for Science in the Public Interest: **www.cspinet.org**
- Find out if warnings have been issued for any food ingredients at the FDA website: **www.fda.gov**

References

1. W. A. Walker and coauthors, Functional foods for health promotion: Microbes and health extended abstracts from the 11th Annual Conference on Functional Foods for Health Promotion, April 2008, *Nutrition Reviews* 67 (2009): 40–48.
2. Position of the American Dietetic Association: Functional foods, *Journal of the American Dietetic Association* 104 (2004): 814–826.
3. J. D. Lambert, S. Sang, and C. S. Yang, Possible controversy over dietary polyphenols: Benefits vs risks, *Chemical Research in Toxicology* 20 (2007): 583–585.
4. E. Cheung and coauthors, Diet and prostate cancer risk reduction, *Expert Review of Anticancer Therapy* 8 (2008): 43–50; E. Linos, and W. C. Willett, Diet and breast cancer risk reduction, *Journal of the National Comprehensive Cancer Network* 5 (2007): 711–718; M. B. Schabath and coauthors, Dietary phytoestrogens and lung cancer risk, *Journal of the American Medical Association* 294 (2005): 1493–1504.
5. T. A. Ryan-Borchers and coauthors, Soy isoflavones modulate immune function in healthy postmenopausal women, *American Journal of Clinical Nutrition* 83 (2006): 1118–1125.
6. M. Messina, W. McCaskill-Stevens, and J. W. Lampe, Addressing the soy and breast cancer relationship: Review, commentary, and workshop proceedings, *Journal of the National Cancer Institute* 98 (2006): 1275–1284.
7. G. Maskarinec, Soy foods for breast cancer survivors and women at high risk for breast cancer? *Journal of the American Dietetic Association* 105 (2005): 1524–1528.
8. C. J. Kavanaugh, P. R. Trumbo, and K. C. Ellwood, The U.S. Food and Drug Administration's evidence-based review for qualified health claims: Tomatoes, lycopene, and cancer, *Journal of the National Cancer Institute* 99 (2007): 1074–1085.
9. J. R. Mein, F. Lian, and X. Wang, Biological activity of lycopene metabolites: Implications for cancer prevention, *Nutrition Reviews* 66 (2008): 667–683; A. Vrieling and coauthors, Lycopene supplementation elevates circulating insulin-like growth factor-binding protein-1 and -2 concentrations in persons at greater risk of colorectal cancer, *American Journal of Clinical Nutrition* 86 (2007): 1456–1462.
10. N. Khan, F. Afaq, and H. Mukhtar, Cancer chemoprevention through dietary antioxidants: Progress and promise, *Antioxidants and Redox Signaling* 10 (2008): 475–510.
11. U. Peters and coauthors, Serum lycopene, other carotenoids, and prostate cancer risk: A nested case-control study in the Prostate, Lung, Colorectal, and Ovarian Cancer Screening Trial, *Cancer Epidemiology, Biomarkers, and Prevention* 16 (2007): 962–968.
12. S. Ellinger, J. Ellinger, and P. Stehle, Tomatoes, tomato products and lycopene in the prevention and treatment of prostate cancer: Do we have the evidence from intervention studies? *Current Opinion in Clinical Nutrition and Metabolic Care* 9 (2006): 722–727.
13. C. A. Gonzalez, Nutrition and cancer: The current epidemiological evidence, *British Journal of Nutrition* 96 (2006): S42–S45; H. Vainio and E. Weiderpass, Fruit and vegetables in cancer prevention, *Nutrition and Cancer* 54 (2006): 111–142.
14. R. di Giuseppe and coauthors, Regular consumption of dark chocolate is associated with low serum concentrations of C-reactive protein in a healthy Italian population, *Journal of Nutrition* 138 (2008): 1939–1945; I. Erlund and coauthors, Favorable effects of berry consumption on platelet function, blood pressure, and HDL cholesterol, *American Journal of Clinical Nutrition* 87 (2008): 323–331; L. Hooper and coauthors, Flavonoids, flavonoid-rich foods, and cardiovascular risk: A meta-analysis of randomized controlled trials, *American Journal of Clinical Nutrition* 88 (2008): 38–50; W. M. Loke and coauthors, Pure dietary flavonoids quercetin and (–)-epicatechin augment nitric oxide products and reduce endothelin-1 acutely in healthy men, *American Journal of Clinical Nutrition* 88 (2008): 1018–1025; S. Baba and coauthors, Continuous intake of polyphenolic compounds containing cocoa powder reduces LDL oxidative susceptibility and has beneficial effects on plasma HDL-cholesterol concentrations in humans, *American Journal of Clinical Nutrition* 85 (2007): 709–717; A. Basu and E. A. Lucas, Mechanisms and effects of green tea on cardiovascular health, *Nutrition Reviews* 65 (2007): 361–375; D. R. Jacobs, L. F. Andersen, and R. Blomhoff, Whole-grain consumption is associated with a reduced risk of noncardiovascular, noncancer death attributed to inflammatory diseases in the Iowa Women's Health Study, *American Journal of Clinical Nutrition* 85 (2007): 1606–1614; D. L. McKay and J. B. Blumberg, Cranberries *(Vaccinium macrocarpon)* and cardiovascular disease risk factors, *Nutrition Reviews* 65 (2007): 490–502; K. Taku and coauthors, Soy isoflavones lower serum total and LDL cholesterol in humans: A meta-analysis of 11 randomized controlled trials, *American Journal and Clinical Nutrition* 85 (2007): 1148–1156; S. S. Wijeratne, M. M. Abou-Zaid, and F. Shahidi, Antioxidant polyphenols in almond and its coproducts, *Journal of Agricultural and Food Chemistry* 54 (2006): 312–318.
15. S. B. Lotito and B. Frei, Consumption of flavonoid-rich foods and increased plasma antioxidant capacity in humans: Cause, consequence, or epiphenomenon? *Free Radical Biology and Medicine* 41 (2006): 1727–1746.
16. S. Voutilainen and coauthors, Carotenoids and cardiovascular health, *American Journal of Clinical Nutrition* 83 (2006): 1265–1271.
17. N. Lee, Phytoestrogens as bioactive ingredients in functional foods: Canadian regulatory update, *Journal of the AOAC International* 89 (2006): 1135–1137.
18. R. E. Ostlund, Jr., Phytosterols, cholesterol absorption and healthy diets, *Lipids* 42 (2007): 41–45.
19. B. Hansel and coauthors, Effect of low-fat, fermented milk enriched with plant sterols on serum lipid profile and oxidative stress in moderate

hypercholesterolemia, *American Journal of Clinical Nutrition* 86 (2007): 790–796; L. H. Ellegård and coauthors, Dietary plant sterols and cholesterol metabolism, *Nutrition Reviews* 65 (2007): 39–45; V. W. Y. Lau, M. Journoud, and P. J. H. Jones, Plant sterols are efficacious in lowering plasma LDL and non-HDL cholesterol in hypercholesterolemic type 2 diabetic and nondiabetic persons, *American Journal of Clinical Nutrition* 81 (2005): 1351–1358; S. Zhan and S. C. Ho, Meta-analysis of the effects of soy protein containing isoflavones on the lipid profile, *American Journal of Clinical Nutrition* 81 (2005): 397–408.

20. D. Fuchs and coauthors, Proteomic biomarkers of peripheral blood mononuclear cells obtained from postmenopausal women undergoing an intervention with soy isoflavones, *American Journal of Clinical Nutrition* 86 (2007): 1369–1375; S. Devaraj, B. C. Autret, and I. Jialal, Reduced-calorie orange juice beverage with plant sterols lowers C-reactive protein concentrations and improves the lipid profile in human volunteers, *American Journal of Clinical Nutrition* 84 (2006): 756–761.
21. D. M. Minich and J. S. Bland, Dietary management of the metabolic syndrome beyond macronutrients, *Nutrition Reviews* 66 (2008): 429–444; M. I. Covas and coauthors, The effect of polyphenols in olive oil on heart disease risk factors: A randomized trial, *Annals of Internal Medicine* 145 (2006): 333–341.
22. S. Y. Lee, Y. W. Shin, and K. B. Hahm, Phytoceuticals: Mighty but ignored weapons against *Helicobacter pylori* infection, *Journal of Digestive Diseases* 9 (2008): 129–139; Y. Liu and coauthors, Cranberry changes the physicochemical surface properties of *E. coli* and adhesion with uroepithelial cells, *Colloids and Surfaces. B, Biointerfaces* 65 (2008): 35–42; S. Gorinstein and coauthors, The atherosclerotic heart disease and protecting properties of garlic: Contemporary data, *Molecular Nutrition and Food Research* 51 (2007): 1365–1381; A. B. Howell, Bioactive compounds in cranberries and their role in prevention of urinary tract infections, *Molecular Nutrition and Food Research* 51 (2007): 732–737.
23. I. Siró and coauthors, Functional food. Product development, marketing and consumer acceptance—A review, *Appetite* 51 (2008): 456–467; Position of the American Dietetic Association, 2004.
24. P. J. Jones and K. A. Varady, Are functional foods redefining nutritional requirements? *Applied Physiology, Nutrition, and Metabolism* 33 (2008): 118–123.
25. C. S. Patch, L. C. Tapsell, and P. G. Williams, Plant sterol/stanol prescription is an effective treatment strategy for managing hypercholesterolemia in outpatient clinical practice, *Journal of the American Dietetic Association* 105 (2005): 46–52.

CHAPTER

14

Nutrition in Your Life

Food choices have consequences. Sometimes they are immediate, as when you get heartburn after eating a pepperoni pizza. Other times they sneak up on you, as when you gain weight after repeatedly overindulging in double hot fudge sundaes. Quite often, they are temporary and easily resolved, as when hunger pangs strike after you skip lunch. During pregnancy, however, the consequences of a woman's food choices are dramatic. They affect not only her health, but also the growth and development of another human being—and not just for today, but for years to come. Making smart food choices is a huge responsibility, but fortunately, it's fairly simple.

Throughout this chapter, the CourseMate icon indicates an opportunity for online self-study, linking you to activities to increase your understanding of chapter concepts. **www.cengagebrain.com** (search for ISBN 084006845X)

Life Cycle Nutrition: Pregnancy and Lactation

CHAPTER OUTLINE

All people—pregnant and lactating women, infants, children, adolescents, and adults—need the same nutrients, but the amounts they need vary depending on their stage of life. This chapter focuses on nutrition in preparation for, and support of, pregnancy and lactation. The next two chapters address the needs of infants, children, adolescents, and older adults.

Nutrition prior to Pregnancy

A section on nutrition prior to pregnancy must, by its nature, focus mainly on women. Both a man's and a woman's nutrition may affect **fertility** and possibly the genetic contributions they make to their children, but it is the woman's nutrition that has the most direct influence on the developing fetus. Her body provides the environment for the growth and development of a new human being. Prior to pregnancy, a woman has a unique opportunity to prepare herself physically, mentally, and emotionally for the many changes to come. In preparation for a healthy pregnancy, a woman can establish the following habits:[1]

- *Achieve and maintain a healthy body weight.* Both underweight and overweight are associated with infertility.[2] Overweight and obese men have low sperm counts and hormonal changes that reduce fertility.[3] Excess body fat in women disrupts menstrual regularity and ovarian hormone production.[4] Should a pregnancy occur, mothers, both underweight and overweight, and their newborns, face increased risks of complications.
- *Choose an adequate and balanced diet.* Malnutrition reduces fertility and impairs the early development of an infant should a woman become pregnant. In contrast, a healthy diet that emphasizes monounsaturated fats instead of *trans* fats, vegetable proteins instead of animal proteins, and low glycemic carbohydrates instead of simple sugars can favorably influence fertility.[5] Men with diets rich in antioxidant nutrients have higher sperm numbers and motility.[6]
- *Be physically active.* A woman who wants to be physically active *when* she is pregnant needs to become physically active *beforehand.*

fertility: the capacity of a woman to produce a normal ovum periodically and of a man to produce normal sperm; the ability to reproduce.

iStockphoto.com/Isaac L. Koval

Young adults can prepare for a healthy pregnancy by taking care of themselves today.

- *Receive regular medical care.* Regular health-care visits can help ensure a healthy start to pregnancy.
- *Manage chronic conditions.* Conditions such as diabetes, HIV/AIDS, phenylketonuria (PKU), and sexually transmitted diseases can adversely affect a pregnancy and need close medical attention to help ensure a healthy outcome.
- *Avoid harmful influences.* Both maternal and paternal ingestion of or exposure to harmful substances (such as cigarettes, alcohol, drugs, or environmental contaminants) can cause miscarriage or abnormalities, alter genes or their expression, and interfere with fertility.[7]

Young adults who nourish and protect their bodies do so not only for their own sakes, but also for future generations.

Dietary Guidelines for Americans 2010

Women capable of becoming pregnant should:

- Choose foods that supply heme iron, which is more readily absorbed by the body, additional iron sources, and enhancers of iron absorption such as vitamin C-rich foods.
- Consume 400 micrograms per day of synthetic folate (from fortified foods and/or supplements) in addition to food forms of folate from a varied diet.

Growth and Development during Pregnancy

A whole new life begins at **conception**. Organ systems develop rapidly, and nutrition plays many supportive roles. This section describes placental development and fetal growth, paying close attention to times of intense developmental activity.

Placental Development

In the early days of pregnancy, a spongy structure known as the **placenta** develops in the **uterus**. Two associated structures also form (see Figure 14-1). One is the **amniotic sac**, a fluid-filled balloonlike structure that houses the developing fetus. The other is the **umbilical cord**, a ropelike structure containing fetal blood vessels that extends through the fetus's "belly button" (the umbilicus) to the placenta. These three structures play crucial roles during pregnancy, and then are expelled from the uterus during childbirth.

The placenta develops as an interweaving of fetal and maternal blood vessels embedded in the uterine wall. The maternal blood transfers oxygen and nutrients to the fetus's blood and picks up fetal waste products. By exchanging oxygen, nutrients, and waste products, the placenta performs the respiratory, absorptive, and excretory functions that the fetus's lungs, digestive system, and kidneys will provide after birth.

The placenta is a versatile, metabolically active organ. Like all body tissues, the placenta uses energy and nutrients to support its work. It produces an array of hormones that maintain pregnancy and prepare the mother's breasts for lactation (making milk). A healthy placenta is essential for the developing fetus to attain its full potential.[8]

Fetal Growth and Development

Fetal development begins with the fertilization of an **ovum** by a **sperm**. Three stages follow: the zygote, the embryo, and the fetus (see Figure 14-2).

The Zygote

The newly fertilized ovum is called a **zygote**. It begins as a single cell and rapidly divides to become a **blastocyst**. During that first week, the blastocyst floats down into the uterus where it will embed itself in the inner uterine

conception: the union of the male sperm and the female ovum; fertilization.

placenta (plah-SEN-tuh): the organ that develops inside the uterus early in pregnancy, through which the fetus receives nutrients and oxygen and returns carbon dioxide and other waste products to be excreted.

uterus (YOU-ter-us): the muscular organ within which the infant develops before birth.

amniotic (am-nee-OTT-ic) **sac:** the "bag of waters" in the uterus, in which the fetus floats.

umbilical (um-BILL-ih-cul) **cord:** the ropelike structure through which the fetus's veins and arteries reach the placenta; the route of nourishment and oxygen to the fetus and the route of waste disposal from the fetus. The scar in the middle of the abdomen that marks the former attachment of the umbilical cord is the *umbilicus* (um-BILL-ih-cus), commonly known as the "belly button."

ovum (OH-vum): the female reproductive cell, capable of developing into a new organism upon fertilization; commonly referred to as an egg.

sperm: the male reproductive cell, capable of fertilizing an ovum.

zygote (ZY-goat): the initial product of the union of ovum and sperm; a fertilized ovum.

blastocyst (BLASS-toe-sist): the developmental stage of the zygote when it is about five days old and ready for implantation.

FIGURE 14-1 The Placenta and Associated Structures

To understand how placental villi absorb nutrients without maternal and fetal blood interacting directly, think of how the intestinal villi work. The GI side of the intestinal villi is bathed in a nutrient-rich fluid (chyme). The intestinal villi absorb the nutrient molecules and release them into the body via capillaries. Similarly, the maternal side of the placental villi is bathed in nutrient-rich maternal blood. The placental villi absorb the nutrient molecules and release them to the fetus via fetal capillaries.

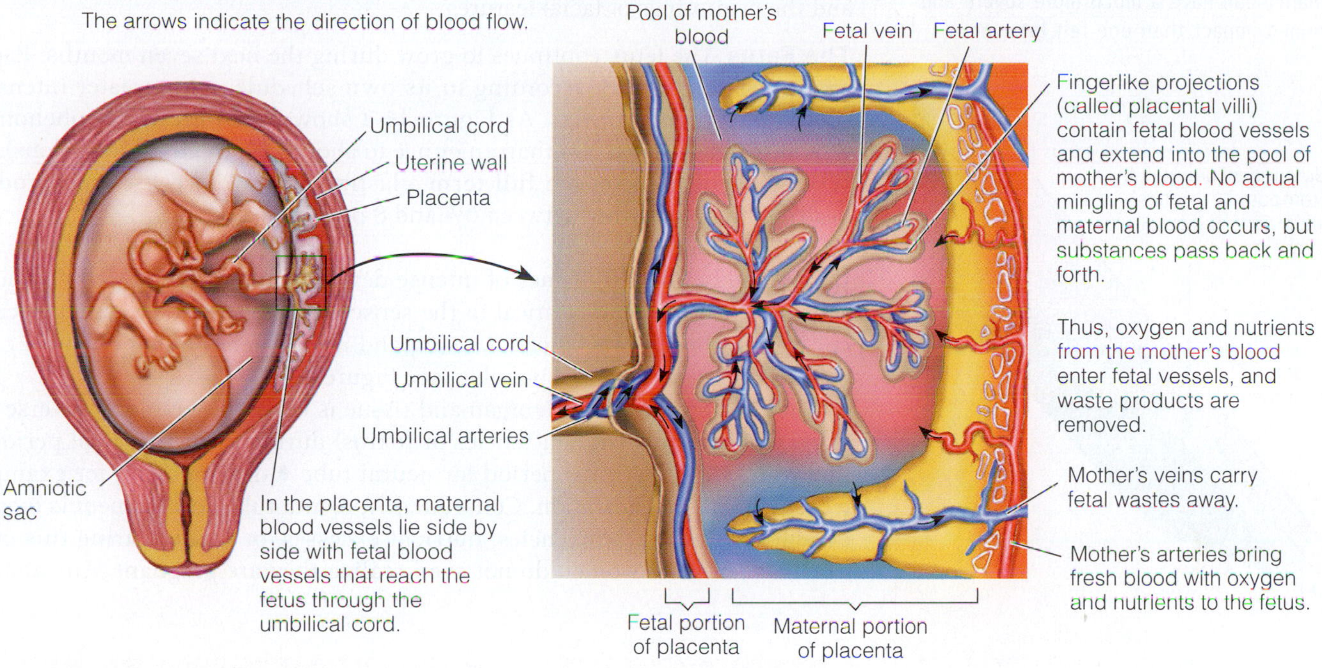

The arrows indicate the direction of blood flow.
Umbilical cord
Uterine wall
Placenta
Amniotic sac
In the placenta, maternal blood vessels lie side by side with fetal blood vessels that reach the fetus through the umbilical cord.
Pool of mother's blood
Fetal vein
Fetal artery
Umbilical cord
Umbilical vein
Umbilical arteries
Fingerlike projections (called placental villi) contain fetal blood vessels and extend into the pool of mother's blood. No actual mingling of fetal and maternal blood occurs, but substances pass back and forth.
Thus, oxygen and nutrients from the mother's blood enter fetal vessels, and waste products are removed.
Mother's veins carry fetal wastes away.
Mother's arteries bring fresh blood with oxygen and nutrients to the fetus.
Fetal portion of placenta
Maternal portion of placenta

FIGURE 14-2 Stages of Embryonic and Fetal Development

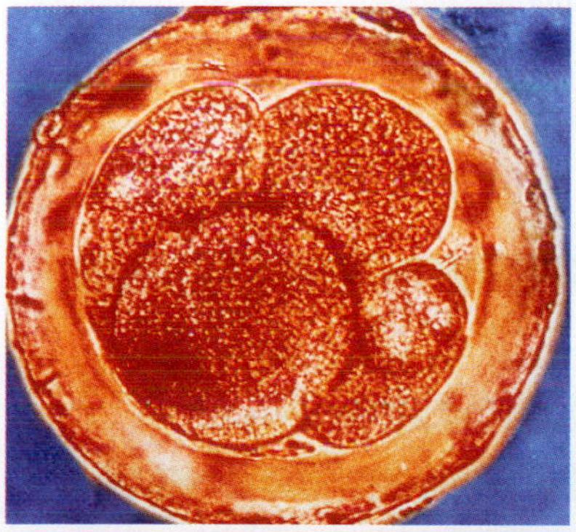

1. A newly fertilized ovum is called a **zygote** and is about the size of a period at the end of this sentence. Less than one week after fertilization, these cells have rapidly divided multiple times and are ready for implantation.

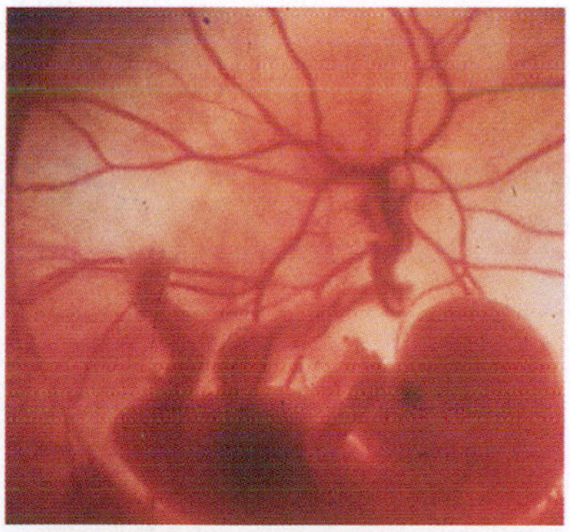

3. A **fetus** after 11 weeks of development is just over an inch long. Notice the umbilical cord and blood vessels connecting the fetus with the placenta.

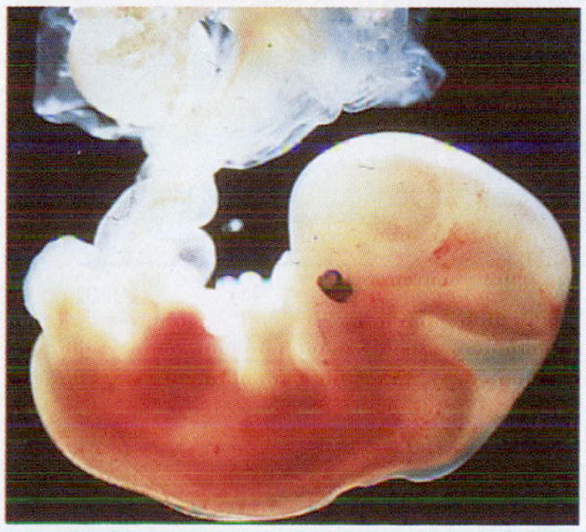

2. After implantation, the placenta develops and begins to provide nourishment to the developing embryo. An **embryo** 5 weeks after fertilization is about 1/2 inch long.

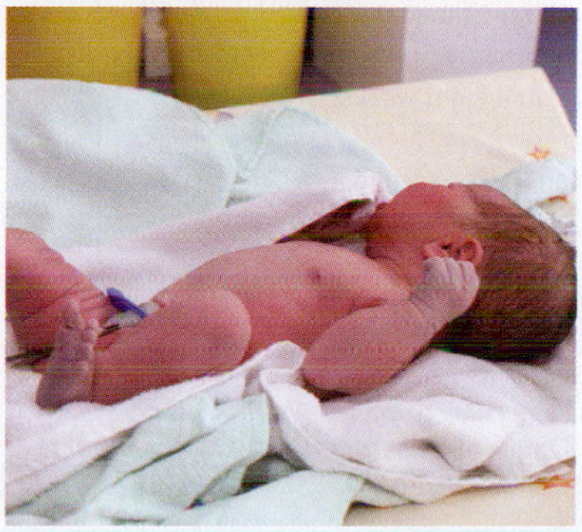

4. A **newborn infant** after nine months of development measures close to 20 inches in length. From 8 weeks to term, this infant grew 20 times longer and 50 times heavier.

FIGURE 14-3 The Concept of Critical Periods in Fetal Development

Critical periods occur early in fetal development. An adverse influence felt early in pregnancy can have a much more severe and prolonged impact than one felt later on.

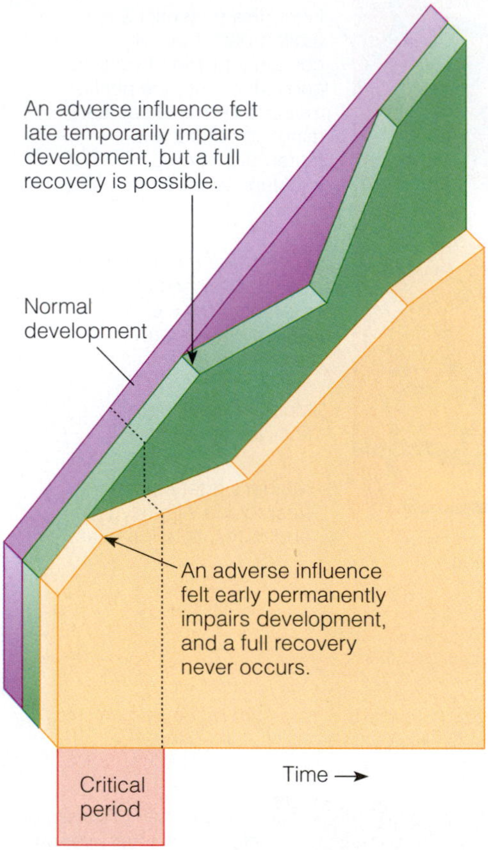

♦ The **neural tube** is the structure that eventually becomes the brain and spinal cord.

wall—a process known as **implantation**. Cell division continues at an amazing rate as each set of cells divides into many other cells.

The Embryo At first, the number of cells in the **embryo** doubles approximately every 24 hours; later the rate slows, and only one doubling occurs during the final 10 weeks of pregnancy. At 8 weeks, the 1¼-inch embryo has a complete central nervous system, a beating heart, a digestive system, well-defined fingers and toes, and the beginnings of facial features.

The Fetus The **fetus** continues to grow during the next seven months. Each organ grows to maturity according to its own schedule, with greater intensity at some times than at others. As Figure 14-2 shows, fetal growth is phenomenal: weight increases from less than an ounce to about 7½ pounds (3500 grams). Most successful pregnancies are **full term**—lasting 38 to 42 weeks—and produce a healthy infant weighing between 6½ and 8 pounds.

Critical Periods

Times of intense development and rapid cell division are called **critical periods**—critical in the sense that those cellular activities can occur only at those times. If cell division and number are limited during a critical period, full recovery is not possible (see Figure 14-3).

The development of each organ and tissue is most vulnerable to adverse influences (such as nutrient deficiencies or toxins) during its own critical period (see Figure 14-4). The critical period for neural tube ♦ development, for example, is from 17 to 30 days **gestation**. Consequently, neural tube development is most vulnerable to nutrient deficiencies, nutrient excesses, or toxins during this critical time—when most women do not even realize they are pregnant. Any abnormal

FIGURE 14-4 Critical Periods of Development

During embryonic development (from two to eight weeks), many of the tissues are in their critical periods (purple area of the bars); events occur that will have irreversible effects on the development of those tissues. In the later stages of development (green area of the bars), the tissues continue to grow and change, but the events are less critical in that they are relatively minor or reversible.

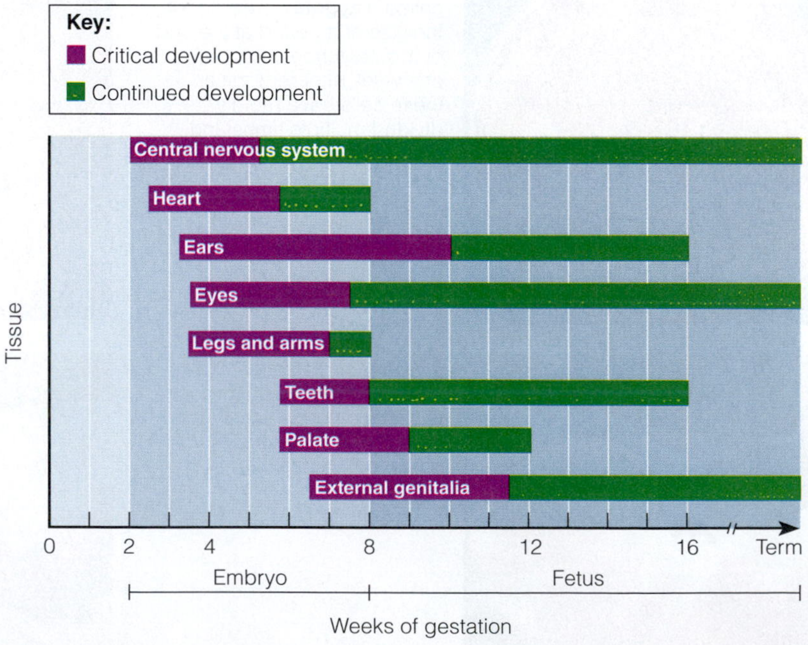

SOURCE: Adapted from *Before We Are Born: Essentials of Embryology and Birth Defects* by K. L. Moore and T. V. N. Persaud (W. B. Saunders, 2003).

implantation (IM-plan-TAY-shun): the embedding of the blastocyst in the inner lining of the uterus.

embryo (EM-bree-oh): the developing infant from two to eight weeks after conception.

fetus (FEET-us): the developing infant from eight weeks after conception until term.

full term: between the thirty-eighth and forty-second week of pregnancy.

critical periods: finite periods during development in which certain events occur that will have irreversible effects on later developmental stages; usually a period of rapid cell division.

gestation (jes-TAY-shun): the period from conception to birth. For human beings, the average length of a healthy gestation is 40 weeks. Pregnancy is often divided into three-month periods, called *trimesters*.

FIGURE 14-5 Neural Tube Development

The neural tube is the beginning structure of the brain and spinal cord. Any failure of the neural tube to close or to develop normally results in central nervous system disorders such as spina bifida and anencephaly. Successful development of the neural tube depends, in part, on the vitamin folate.

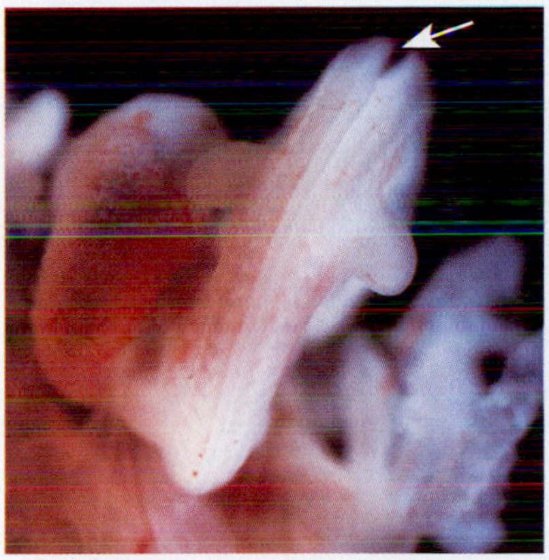

At 4 weeks, the neural tube has yet to close (notice the gap at the top).

At 6 weeks, the neural tube (outlined by the delicate red vertebral arteries) has successfully closed.

development of the neural tube or its failure to close completely can cause a major defect in the central nervous system. Figure 14-5 shows photos of neural tube development in the early weeks of gestation.

Neural Tube Defects In the United States, approximately 30 of every 100,000 newborns are born with a neural tube defect; ♦ some 1000 or so infants are affected each year.* Many other pregnancies with neural tube defects end in abortions or stillbirths.

The two most common types of neural tube defects are anencephaly and spina bifida. In **anencephaly**, the upper end of the neural tube fails to close. Consequently, the brain is either missing or fails to develop. Pregnancies affected by anencephaly often end in miscarriage; infants born with anencephaly die shortly after birth.

Spina bifida is characterized by incomplete closure of the spinal cord and its bony encasement (see Figure 14-6). The meninges membranes covering the spinal cord often protrude as a sac, which may rupture and lead to meningitis, a life-threatening infection. Spina bifida is accompanied by varying degrees of paralysis, depending on the extent of the spinal cord damage. Mild cases may not even be noticed, but severe cases lead to death. Common problems include clubfoot, dislocated hip, kidney disorders, curvature of the spine, muscle weakness, mental handicaps, and motor and sensory losses.

The cause of neural tube defects is unknown, but researchers are examining several gene-gene, gene-nutrient, and gene-environment interactions.[9] A pregnancy affected by a neural tube defect can occur in any woman, but these factors make it more likely:[10]

- A personal or family history of a pregnancy affected by a neural tube defect
- Maternal diabetes
- Maternal use of certain antiseizure medications
- Mutations in folate-related enzymes
- Maternal obesity

Folate supplementation reduces the risk.

*Worldwide, some 300,000 to 400,000 infants are born with neural tube defects each year.

♦ A **neural tube defect** is a malformation of the brain, spinal cord, or both during embryonic development. The two main types of neural tube defects are **spina bifida** (literally, "split spine") and **anencephaly** ("no brain").

anencephaly (AN-en-SEF-a-lee): an uncommon and always fatal type of neural tube defect; characterized by the absence of a brain.

- **an** = not (without)
- **encephalus** = brain

spina (SPY-nah) **bifida** (BIFF-ih-dah): one of the most common types of neural tube defects; characterized by the incomplete closure of the spinal cord and its bony encasement.

- **spina** = spine
- **bifida** = split

FIGURE 14-6 **Spina Bifida**

Spina bifida, a common neural tube defect, occurs when the vertebrae of the spine fail to close around the spinal cord, leaving it unprotected. The B vitamin folate—consumed prior to and during pregnancy—helps prevent spina bifida and other neural tube defects.

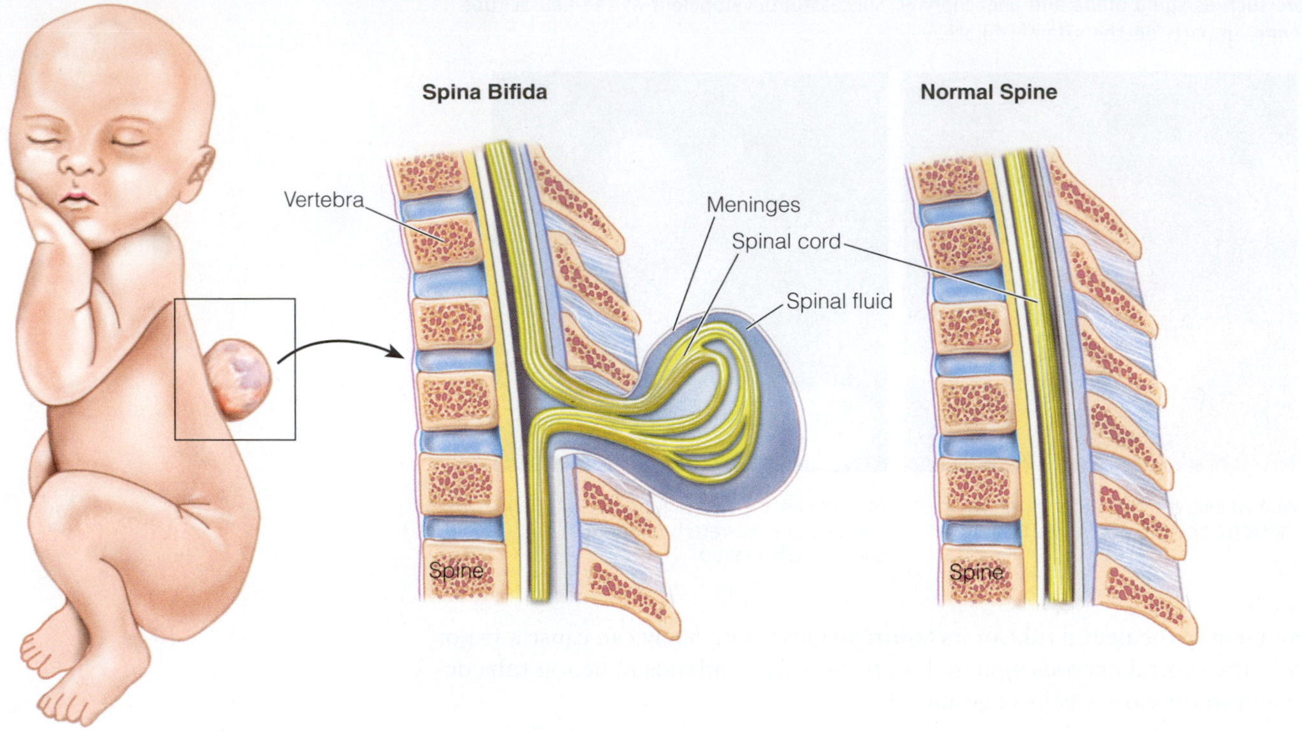

♦ Folate RDA:
- For women: 400 µg (0.4 mg)/day
- During pregnancy: 600 µg (0.6 mg)/day

Folate Supplementation Chapter 10 describes how folate supplements taken one month before conception and continued throughout the first trimester can help support a healthy pregnancy, prevent neural tube defects, and reduce the severity of those that do occur.[11] For this reason, all women of childbearing age ♦ who are capable of becoming pregnant should consume 400 micrograms (0.4 milligrams) of folate daily. A woman who has previously had an infant with a neural tube defect may be advised by her physician to take folate supplements in doses ten times larger—4 milligrams daily. Because high doses of folate can mask the symptoms of the pernicious anemia of a vitamin B_{12} deficiency, quantities of 1 milligram or more require a prescription. Most over-the-counter multivitamin-mineral supplements contain 400 micrograms of folate; prenatal supplements usually contain 800 micrograms.

Dietary Guidelines for Americans 2010

Women who are pregnant are advised to consume 600 micrograms of dietary folate equivalents daily from all sources.

Because half of the pregnancies each year are unplanned and because neural tube defects occur early in development before most women realize they are pregnant, grain products in the United States are fortified with folate to help ensure an adequate intake. Labels on fortified products may claim that an "adequate intake of folate has been shown to reduce the risk of neural tube defects." Fortification has improved folate status in women of childbearing age and lowered the number of neural tube defects that occur each year, as shown in Figure 10-11 on p. 328.[12] Whether folate fortification should be increased further is still the subject of much debate.[13]

Chronic Diseases Much research suggests that adverse influences at critical times during fetal development set the stage for the infant to develop chronic diseases in adult life.[14] Poor maternal diet or health during pregnancy may alter the infant's bodily functions such as blood pressure, cholesterol metabolism, and immune functions that influence disease development.[15] For example, maternal malnutrition may alter blood vessel growth and program lipid metabolism and lean body mass development in such a way that the infant will develop risk factors for cardiovascular disease as an adult.[16]

Malnutrition during the critical period of pancreatic cell growth provides an example of how type 2 diabetes may develop in adulthood. The pancreatic cells responsible for producing insulin (the beta cells) normally increase more than 130-fold between 12 weeks gestation and 5 months after birth. Nutrition is a primary determinant of beta cell growth, and infants who have suffered prenatal malnutrition have significantly fewer beta cells than well-nourished infants. They are also more likely to be low-birthweight infants—and low birthweight and premature birth correlate with insulin resistance later in life.[17] One hypothesis suggests that diabetes may develop from the interaction of inadequate nutrition early in life with abundant nutrition later in life: the small mass of beta cells developed in times of undernutrition during fetal development may be insufficient in times of overnutrition during adulthood when the body needs more insulin.

Hypertension may develop from a similar scenario of inadequate growth during placental and gestational development followed by accelerated growth during early childhood: the small mass of kidney cells developed during malnutrition may be insufficient to handle the excessive demands of later life.[18] Low-birthweight infants who gain weight rapidly as young children are likely to develop hypertension and heart disease as adults.

Fetal Programming Recent genetic research may help to explain the phenomenon of substances such as nutrients influencing the development of obesity and diseases later on in adulthood—a process known as **fetal programming**. In the case of pregnancy, the mother's nutrition can change gene expression in the fetus.[19] Such epigenetic changes during pregnancy can affect the infant's development of obesity and related adult diseases.[20] Some research suggests that fetal programming may influence succeeding generations.[21]

IN SUMMARY **Maternal nutrition before and during pregnancy affects both the mother's health and the infant's growth. As the infant develops through its three stages—the zygote, embryo, and fetus—its organs and tissues grow, each on its own schedule. Times of intense development are critical periods that depend on nutrients to proceed smoothly. Without folate, for example, the neural tube fails to develop completely during the first month of pregnancy, prompting recommendations that all women of childbearing age take folate daily.**

Because critical periods occur throughout pregnancy, a woman should continuously take good care of her health. That care should include achieving and maintaining a healthy body weight prior to pregnancy and gaining sufficient weight during pregnancy to support a healthy infant.

Maternal Weight

Birthweight is the most reliable indicator of an infant's health. As a later section of this chapter explains, compared with a normal-weight infant, an underweight infant is more likely to have physical and mental defects, become ill, and die. In

fetal programming: the influence of substances during fetal growth on the development of diseases in later life.

general, higher birthweights present fewer risks for infants. Two characteristics of the mother's weight influence an infant's birthweight: her weight *prior* to conception and her weight gain *during* pregnancy.

Weight prior to Conception

A woman's weight ♦ prior to conception influences fetal growth. Even with the same weight gain during pregnancy, underweight women tend to have smaller babies than heavier women. Ideally, before a woman becomes pregnant, she will have established diet and activity habits to support an adequate, and not excessive, weight gain during pregnancy.[22]

♦ BMI is introduced in Chapter 8.
- Underweight = BMI <18.5
- Normal weight = BMI 18.5 to 24.9
- Overweight = BMI 25 to 29.9
- Obesity = BMI ≥30

Underweight An underweight woman has a high risk of having a low-birthweight infant, especially if she is malnourished or unable to gain sufficient weight during pregnancy. In addition, the rates of **preterm** births and infant deaths are higher for underweight women. An underweight woman improves her chances of having a healthy infant by gaining sufficient weight prior to conception or by gaining extra pounds during pregnancy. To gain weight and ensure nutrient adequacy, an underweight woman can follow the dietary recommendations for pregnant women (described later in the chapter).

Overweight and Obesity An estimated one-third of all pregnant women in the United States are obese, which can create problems related to pregnancy and childbirth.[23] Obese women have an especially high risk of medical complications such as hypertension, gestational diabetes, and postpartum infections. Compared with other women, obese women are also more likely to have other complications of labor and delivery.[24] Complications in women after gastric bypass surgery and weight loss are lower than in obese women.[25]

Overweight women have the lowest rate of low-birthweight infants. In fact, infants of overweight women are more likely to be born **post term** and to weigh more than 9 pounds. ♦ Large newborns increase the likelihood of a difficult labor and delivery, birth trauma, and **cesarean section**.[26] Consequently, these infants have a greater risk of poor health and death than infants of normal weight.

♦ The term *macrosomia* (mak-roh-SO-me-ah) describes high-birthweight infants (roughly 9 lb, or 4000 g, or more); macrosomia results from prepregnancy obesity, excessive weight gain during pregnancy, or uncontrolled diabetes.
- **macro** = large
- **oma** = body

Of greater concern than infant birthweight is the poor development of infants born to obese mothers.[27] Obesity may double the risk for neural tube defects. Folate's role has been examined, but a more likely explanation seems to be poor glycemic control.[28] Undiagnosed diabetes might also explain why obese women have a greater risk of giving birth to infants with heart defects and other abnormalities.[29]

Health-care providers have traditionally advised against weight-loss dieting during pregnancy. Limited research, however, suggests that obese women who follow a well-balanced, kcalorie-restricted diet and regular exercise program can gain little or no weight without adverse consequences.[30] Ideally, overweight women will achieve a healthy body weight before becoming pregnant, avoid excessive weight gain during pregnancy, and postpone weight loss until after childbirth.[31]

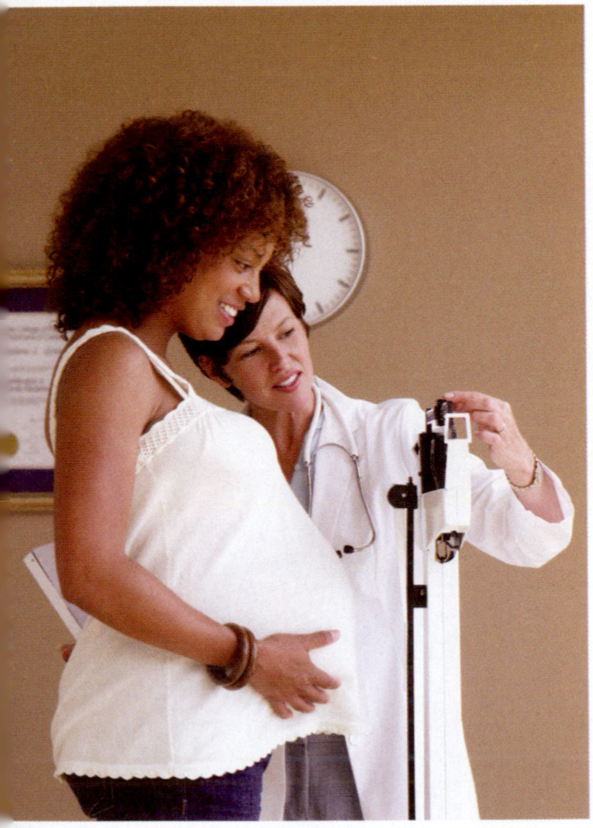

© Larry Williams/Corbis

Fetal growth and maternal health depend on a sufficient weight gain during pregnancy.

Weight Gain during Pregnancy

Fetal growth and maternal health depend on a sufficient weight gain during pregnancy. Maternal weight gain during pregnancy correlates closely with infant birthweight, which is a strong predictor of the health and subsequent development of the infant.

Dietary Guidelines for Americans 2010

Pregnant women are encouraged to gain weight within the gestational weight gain guidelines (see Table 14-1, p. 465).

Recommended Weight Gains Table 14-1 presents recommended weight gains for various prepregnancy weights. The recommended gain for a woman who begins pregnancy at a healthy weight and is carrying a single fetus is 25 to 35 pounds.[32] An underweight woman needs to gain between 28 and 40 pounds; and an overweight woman, between 15 and 25 pounds. About one-third of U.S. women gain

preterm (premature): prior to the thirty-eighth week of pregnancy.

post term: after the forty-second week of pregnancy.

cesarean (si-ZAIR-ee-un) **section:** a surgically assisted birth involving removal of the fetus by an incision into the uterus, usually by way of the abdominal wall.

TABLE 14-1 Recommended Weight Gains Based on Prepregnancy Weight

	Recommended Weight Gain	
Prepregnancy Weight	*For single birth*	*For twin birth*
Underweight (BMI <18.5)	28 to 40 lb (12.5 to 18.0 kg)	Insufficient data to make recommendation
Healthy weight (BMI 18.5 to 24.9)	25 to 35 lb (11.5 to 16.0 kg)	37 to 54 lb (17.0 to 25.0 kg)
Overweight (BMI 25.0 to 29.9)	15 to 25 lb (7.0 to 11.5 kg)	31 to 50 lb (14.0 to 23.0 kg)
Obese (BMI ≥30)	11 to 20 lb (5.0 to 9.0 kg)	25 to 42 lb (11.0 to 19.0 kg)

SOURCE: Institute of Medicine, *Weight Gain during Pregnancy: Reexamining the Guidelines* (Washington, D.C.: National Academies Press, 2009).

weight within these recommended ranges; most gain more than recommended.[33] Appropriate weight gains help women limit weight retention after pregnancy and help their infants prevent obesity during childhood.[34] To limit excessive weight gains, pregnant women can select foods with a high nutrient density but a low energy density. ♦[35]

♦ Nutrient density = nutrient/kcal
Energy density = kcal/g

Weight-Gain Patterns For the normal-weight woman, weight gain ideally follows a pattern of 3½ pounds during the first trimester and 1 pound per week thereafter. Health-care professionals monitor weight gain using a prenatal weight-gain grid (see Figure 14-7).

If a woman gains more than is recommended early in pregnancy, she should not restrict her energy intake later in order to lose weight. A large weight gain over a short time, however, indicates excessive fluid retention and may be the first sign of the serious medical complication preeclampsia, which is discussed later.

Components of Weight Gain Women often express concern about the weight gain that accompanies a healthy pregnancy. They may find comfort by remembering that most of the gain supports the growth and development of the placenta, uterus, blood, and breasts, the increase in blood supply and fluid volume, as well as a healthy 7½-pound infant. A small amount goes into maternal fat stores, and

FIGURE 14-7 Recommended Prenatal Weight Gain Based on Prepregnancy Weight

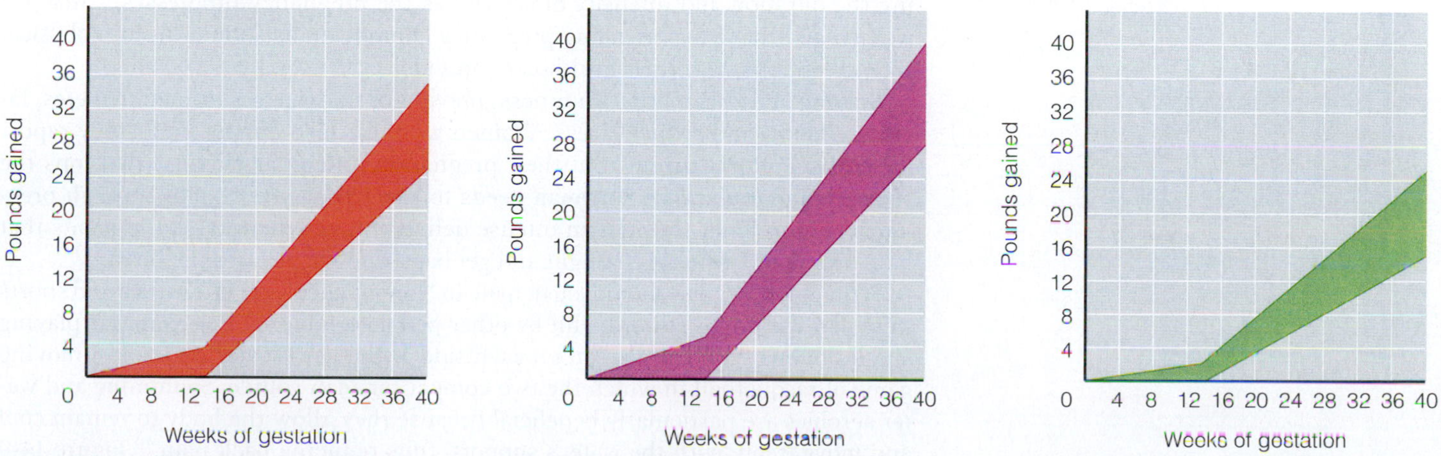

Normal-weight women should gain about 3½ pounds in the first trimester and just under 1 pound/week thereafter, achieving a total gain of 25 to 35 pounds by term.

Underweight women should gain about 5 pounds in the first trimester and just over 1 pound/week thereafter, achieving a total gain of 28 to 40 pounds by term.

Overweight women should gain about 2 pounds in the first trimester and ⅔ pound/week thereafter, achieving a total gain of 15 to 25 pounds.

FIGURE 14-8 Components of Weight Gain during Pregnancy

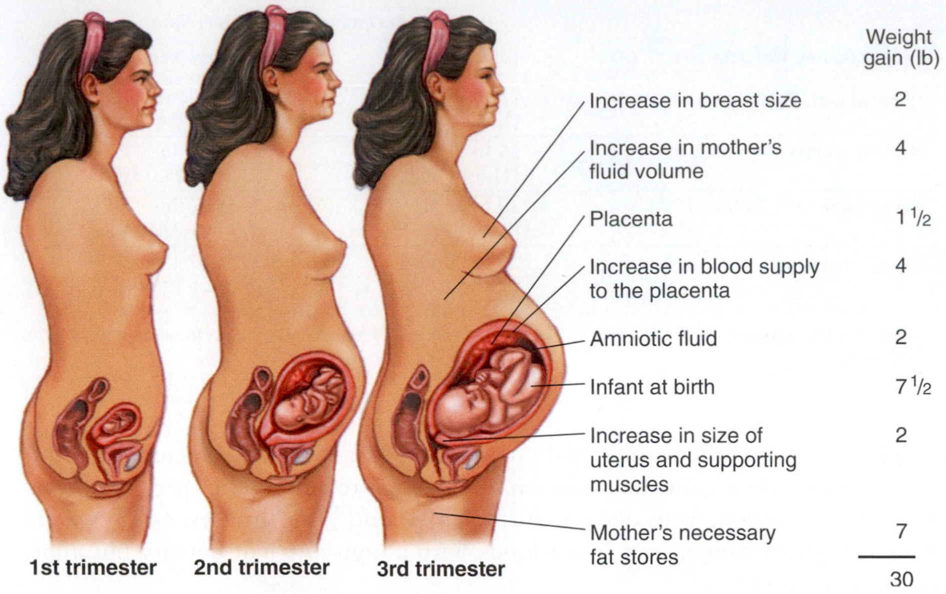

even that fat is there for a special purpose—to provide energy for labor and lactation. Figure 14-8 shows the components of a healthy 30-pound weight gain.

Weight Loss after Pregnancy The pregnant woman loses some weight at delivery. In the following weeks, she loses more as her blood volume returns to normal and she sheds accumulated fluids. The typical woman does not, however, return to her prepregnancy weight. In general, the more weight a woman gains beyond the needs of pregnancy, the more she retains—mostly as body fat.[36] Even with an average weight gain during pregnancy, most women tend to retain a couple of pounds with each pregnancy. When those couple of pounds become seven or more and BMI increases by a unit or more, complications such as diabetes and hypertension in future pregnancies as well as chronic diseases in later life can increase—even for women who are not overweight.[37] Those who are successful in losing their pregnancy weight are more likely to limit weight gains through middle adulthood.[38]

Exercise during Pregnancy

An active, physically fit woman experiencing a normal pregnancy can continue to exercise throughout pregnancy, adjusting the duration and intensity of activity as the pregnancy progresses.[39] Inactive women and those experiencing pregnancy complications should discuss physical activity options with their health-care provider.

Staying active can improve fitness, prevent or manage gestational diabetes, facilitate labor, and reduce stress. Women who exercise during pregnancy report fewer discomforts throughout their pregnancies. Regular exercise develops the strength and endurance a woman needs to carry the extra weight through pregnancy and to labor through an intense delivery. It also maintains the habits that help a woman lose excess weight and get back into shape after the birth.

A pregnant woman should participate in "low-impact" activities and avoid sports in which she might fall or be hit by other people or objects. For example, playing singles tennis with one person on each side of the net is safer than a fast-moving game of racquetball in which the two competitors can collide. Swimming and water aerobics are particularly beneficial because they allow the body to remain cool and move freely with the water's support, thus reducing back pain.[40] Figure 14-9 provides some guidelines for exercise during pregnancy. Several of the guidelines are aimed at preventing excessively high internal body temperature and dehydration, both of which can harm fetal development. To this end, pregnant women should also stay out of saunas, steam rooms, and hot tubs or hot whirlpool baths.

FIGURE 14-9 **Exercise Guidelines during Pregnancy**

DO

Do begin to exercise gradually.

Do exercise regularly (most, if not all, days of the week).

Do warm up with 5 to 10 minutes of light activity.

Do 30 minutes or more of moderate physical activity.

Do cool down with 5 to 10 minutes of slow activity and gentle stretching.

Do drink water before, after, and during exercise.

Do eat enough to support the needs of pregnancy plus exercise.

Do rest adequately.

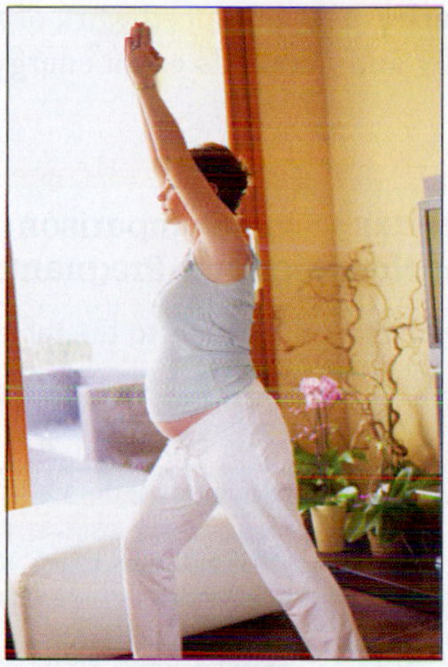

Diego Cervo/Shutterstock.com

Pregnant women can enjoy the benefits of exercise.

DON'T

Don't exercise vigorously after long periods of inactivity.

Don't exercise in hot, humid weather.

Don't exercise when sick with fever.

Don't exercise while lying on your back after the 1st trimester of pregnancy or stand motionless for prolonged periods.

Don't exercise if you experience any pain, discomfort, or fatigue.

Don't participate in activities that may harm the abdomen or involve jerky, bouncy movements.

Don't scuba dive.

Physical Activity Guidelines for Americans, 2008

- Healthy women who are not already highly active or doing vigorous-intensity activity should get at least 150 minutes (2 hours and 30 minutes) of moderate-intensity aerobic activity per week during pregnancy and the postpartum period. Preferably, this activity should be spread throughout the week.
- Pregnant women who habitually engage in vigorous-intensity aerobic activity or are highly active can continue physical activity during pregnancy and the postpartum period, provided that they remain healthy and discuss with their health-care provider how and when activity should be adjusted over time.

IN SUMMARY A healthy pregnancy depends on a sufficient weight gain. Women who begin their pregnancies at a healthy weight need to gain about 30 pounds, which covers the growth and development of the placenta, uterus, blood, breasts, and infant. By remaining active throughout pregnancy, a woman can develop the strength she needs to carry the extra weight and maintain habits that will help her lose it after the birth.

Nutrition during Pregnancy

A woman's body changes dramatically during pregnancy. Her uterus and its supporting muscles increase in size and strength; her blood volume increases by half to carry the additional nutrients and other materials; her joints become more flexible in preparation for childbirth; her feet swell in response to high concentrations of the hormone estrogen, which promotes water retention and helps to ready the uterus for delivery; and her breasts enlarge in preparation for lactation. The hormones that mediate all these changes may influence her mood. She can best prepare to handle these changes given a nutritious diet, regular physical activity, plenty of rest, and caring companions. This section highlights the role of nutrition.

Energy and Nutrient Needs during Pregnancy

From conception to birth, all parts of the infant—bones, muscles, blood cells, skin, and all other tissues—are made from nutrients in the foods the mother eats. For most women, nutrient needs during pregnancy and lactation ♦ are higher than at any other time (see Figure 14-10). To meet the high nutrient demands of pregnancy,

♦ The Dietary Reference Intakes (DRI) table on the inside front cover provides separate listings for women during pregnancy and lactation, reflecting their heightened nutrient needs.

A pregnant woman's food choices support both her health and her infant's growth and development.

a woman will need to make careful food choices, but her body will also help by maximizing absorption and minimizing losses.

Energy The enhanced work of pregnancy raises the basal metabolic rate dramatically and demands extra energy.[41] Energy needs of pregnant women are greater

FIGURE 14-10 Comparison of Nutrient Recommendations for Nonpregnant, Pregnant, and Lactating Women

For actual values, turn to the table on the inside front cover.

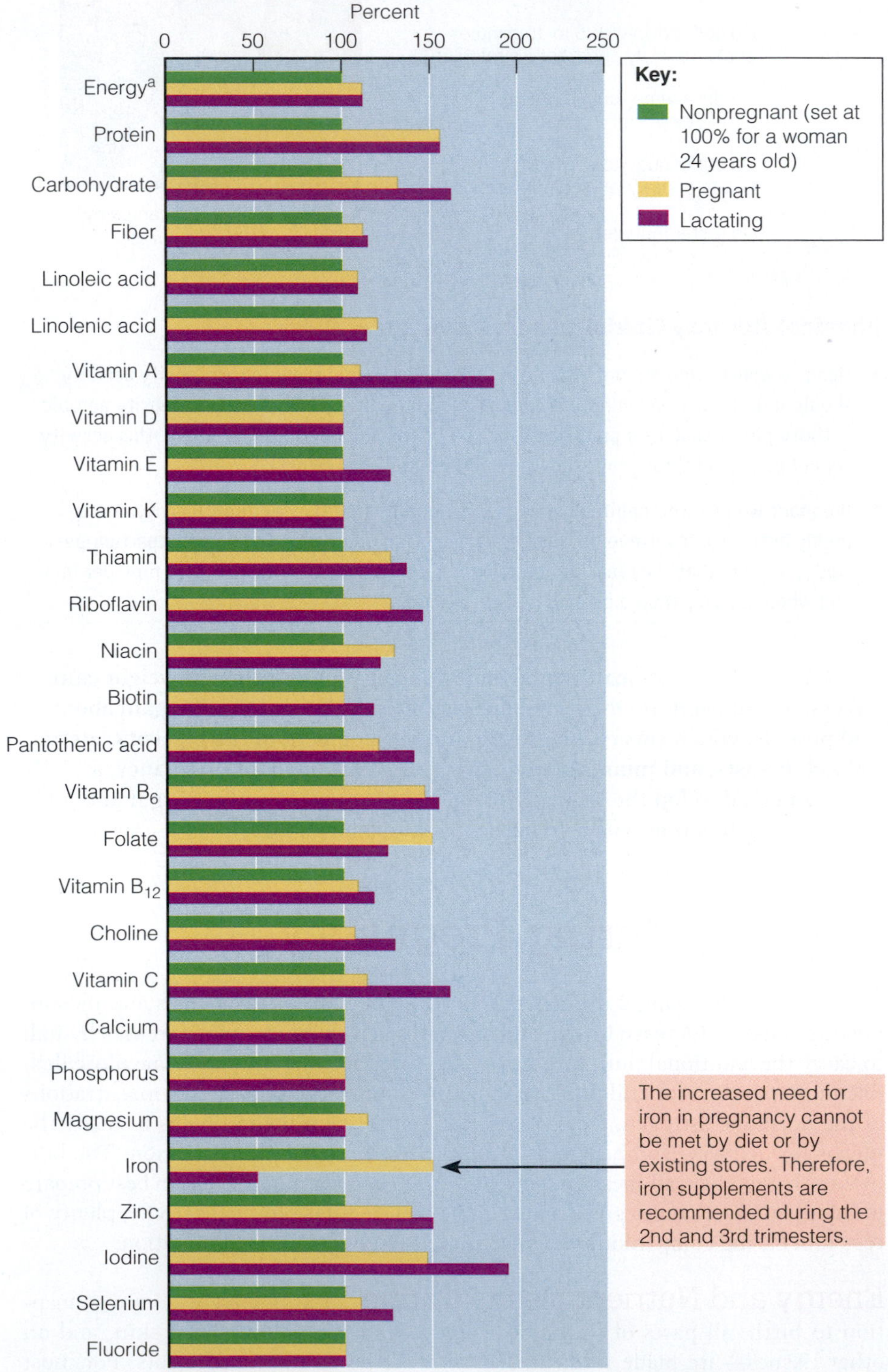

[a]Energy allowance during pregnancy is for second trimester; energy allowance during the third trimester is slightly higher; no additional allowance is provided during the first trimester. Energy allowance during lactation is for the first six months; energy allowance during the second six months is slightly higher.

than those of nonpregnant women—an additional 340 kcalories per day during the second trimester and an extra 450 kcalories per day during the third. ♦ A woman can easily get these added kcalories with nutrient-dense selections from the five food groups. See Table 2-2 (p. 42) for suggested dietary patterns for several kcalorie levels and Figure 14-11 for a sample menu for pregnant and lactating women.

♦ Energy requirement during pregnancy:
- 1st trimester: +0 kcal/day
- 2nd trimester: +340 kcal/day
- 3rd trimester: +450 kcal/day

For a 2000-kcalorie daily intake, these added kcalories represent about 15 to 20 percent more food energy than before pregnancy. The increase in nutrient needs is often greater than this, so nutrient-dense foods should be chosen to supply the extra kcalories: foods such as whole-grain breads and cereals, legumes, dark green vegetables, citrus fruits, low-fat milk and milk products, and lean meats, fish, poultry, and eggs.

Carbohydrate Ample carbohydrate (ideally, 175 grams or more per day and certainly no less than 135 grams) is necessary to fuel the fetal brain. Sufficient carbohydrate also ensures that the protein needed for growth will not be broken down and used to make glucose.

Protein The protein RDA ♦ for pregnancy is an additional 25 grams per day higher than for nonpregnant women. Pregnant women can easily meet their protein needs by selecting meats, milk products, and protein-containing plant foods such as legumes, whole grains, nuts, and seeds. Because use of high-protein supplements during pregnancy may be harmful to the infant's development, it is discouraged.

♦ Protein RDA during pregnancy:
- +25 g/day

Essential Fatty Acids The high nutrient requirements of pregnancy leave little room in the diet for excess fat, but the essential long-chain polyunsaturated fatty acids are particularly important to the growth and development of the fetus.[42] The brain is largely made of lipid material, and it depends heavily on the long-chain omega-3 and omega-6 fatty acids for its growth, function, and structure.[43] (See Table 5-2 on p. 153 for a list of good food sources of the omega fatty acids.)

Nutrients for Blood Production and Cell Growth New cells are laid down at a tremendous pace as the fetus grows and develops. At the same time, the mother's red blood cell mass expands. All nutrients are important in these processes, but for folate, vitamin B_{12}, iron, and zinc, the needs are especially great due to their key roles in the synthesis of DNA and new cells.

The requirement for folate increases dramatically during pregnancy. ♦ It is best to obtain sufficient folate from a combination of supplements, fortified foods, and a diet that includes fruits, juices, green vegetables, and whole grains.[44] The

♦ Folate RDA during pregnancy:
- 600 µg/day

FIGURE 14-11 Daily Food Choices for Pregnancy (2nd and 3rd trimesters) and Lactation

Food Group	Amount
Fruits	2 c
Vegetables	3 c
Grains	8 oz
Protein foods	6½ oz
Milk	3 c

SAMPLE MENU

Breakfast
1 whole-wheat English muffin
2 tbs peanut butter
1 c low-fat vanilla yogurt
½ c fresh strawberries
1 c orange juice

Midmorning snack
½ c cranberry juice
1 oz pretzels

Lunch
Sandwich (tuna salad on whole-wheat bread)
½ carrot (sticks)
1 c low-fat milk

Dinner
Chicken cacciatore
3 oz chicken
½ c stewed tomatoes
1 c rice
½ c summer squash
1½ c salad (spinach, mushrooms, carrots)
1 tbs salad dressing
1 slice Italian bread
2 tsp soft margarine
1 c low-fat milk

NOTE: This sample meal plan provides about 2500 kcalories (55% from carbohydrate, 20% from protein, and 25% from fat) and meets most of the vitamin and mineral needs of pregnant and lactating women.

"How To" feature in Chapter 10 on p. 327 describes how folate from each of these sources contributes to a day's intake.

♦ Vitamin B_{12} RDA during pregnancy:
- 2.6 μg/day

The pregnant woman also has a slightly greater need for the B vitamin that activates the folate enzyme—vitamin B_{12}. ♦ Generally, even modest amounts of meat, fish, eggs, or milk products together with body stores easily meet the need for vitamin B_{12}. Vegans who exclude all foods of animal origin, however, need daily supplements of vitamin B_{12} or vitamin B_{12}–fortified foods to prevent the neurological complications of a deficiency.

♦ Iron RDA during pregnancy:
- 27 mg/day

Pregnant women need iron ♦ to support their enlarged blood volume and to provide for placental and fetal needs.[45] The developing fetus draws on maternal iron stores to create sufficient stores of its own to last through the first four to six months after birth. Even women with inadequate iron stores transfer significant amounts of iron to the fetus, suggesting that the iron needs of the fetus have priority over those of the mother. In addition, blood losses are inevitable at birth, especially during a cesarean section, and can further drain the mother's supply.*

During pregnancy, the body makes several adaptations to help meet the exceptionally high need for iron. Menstruation, the major route of iron loss in women, ceases, and iron absorption improves thanks to an increase in transferrin, the body's iron-absorbing and iron-carrying protein. Without sufficient intake, though, iron stores would quickly dwindle.

Few women enter pregnancy with adequate iron stores, so a daily iron supplement is recommended during the second and third trimesters for all pregnant women. For this reason, most prenatal supplements provide 30 to 60 milligrams of iron a day. To enhance iron absorption, the supplement should be taken between meals or at bedtime and with liquids other than milk, coffee, or tea, which inhibit iron absorption. Drinking orange juice does not enhance iron absorption from supplements as it does from foods; vitamin C enhances iron absorption by converting iron from ferric to ferrous, but supplemental iron is already in the ferrous form. Vitamin C is helpful, however, in preventing the premature rupture of amniotic membranes.[46]

♦ Zinc RDA during pregnancy:
- 12 mg/day (≤18 yr)
- 11 mg/day (19–50 yr)

Zinc ♦ is required for DNA and RNA synthesis and thus for protein synthesis and cell development. Typical zinc intakes for pregnant women are lower than recommendations, but fortunately, zinc absorption increases when zinc intakes are low.[47] Routine supplementation is not advised.[48] Women taking iron supplements (more than 30 milligrams per day), however, may need zinc supplementation because large doses of iron can interfere with the body's absorption and use of zinc.

Nutrients for Bone Development Vitamin D and the bone-building minerals calcium, phosphorus, magnesium, and fluoride are in great demand during pregnancy. Insufficient intakes may produce abnormal fetal bones and teeth.

♦ The RDA for vitamin D does not increase during pregnancy.

Vitamin D ♦ plays a vital role in calcium absorption and utilization. Consequently, severe maternal vitamin D deficiency interferes with normal calcium metabolism, resulting in rickets in the infant and osteomalacia in the mother. Regular exposure to sunlight and consumption of vitamin D–fortified milk are usually sufficient to provide the recommended amount of vitamin D during pregnancy, although some researchers question whether current recommendations, even with prenatal supplements, are adequate.[49]

♦ The RDA for calcium does not increase during pregnancy.

Calcium absorption and retention increase dramatically in pregnancy, helping the mother to meet the calcium needs of pregnancy. ♦ During the last trimester, as the fetal bones begin to calcify, more than 300 milligrams a day are transferred to the fetus. Recommendations to ensure an adequate calcium intake during pregnancy help to conserve maternal bones while supplying fetal needs.[50]

♦ The USDA Food Guide suggests consuming 3 cups per day of fat-free or low-fat milk or the equivalent in milk products.

Calcium intakes for pregnant women ♦ typically fall below recommendations. Because bones are still actively depositing minerals until about age 30, adequate calcium is especially important for young women. Pregnant women younger than age 25 who receive less than 600 milligrams of dietary calcium daily need to in-

*On average, almost twice as much blood is lost during a cesarean delivery as during the average vaginal delivery of a single fetus.

crease their intake of milk, cheese, yogurt, and other calcium-rich foods. Alternatively, and less preferably, they may need a daily supplement of 600 milligrams of calcium.

Other Nutrients The nutrients mentioned here are those most intensely involved in blood production, cell growth, and bone growth. Of course, other nutrients are also needed during pregnancy to support the growth and health of both fetus and mother. Even with adequate nutrition, repeated pregnancies within a short time span can deplete nutrient reserves. When this happens, fetal growth may be compromised, and maternal health may decline. The optimal interval between pregnancies is 18 to 23 months.

Nutrient Supplements Pregnant women who make wise food choices can meet most of their nutrient needs, with the possible exception of iron. Even so, physicians routinely recommend daily multivitamin-mineral supplements for pregnant women. Prenatal supplements typically contain greater amounts of folate, iron, and calcium than regular multivitamin-mineral supplements. These supplements are particularly beneficial for women who do not eat adequately and for those in high-risk groups: women carrying multiple fetuses, cigarette smokers, and alcohol and drug abusers. The use of prenatal supplements may help reduce the risks of preterm delivery, low infant birthweights, and birth defects. Supplement use *prior* to conception also seems to reduce the risk of preterm births. Figure 14-12 presents a label from a standard prenatal supplement.

FIGURE 14-12 Example of a Prenatal Supplement

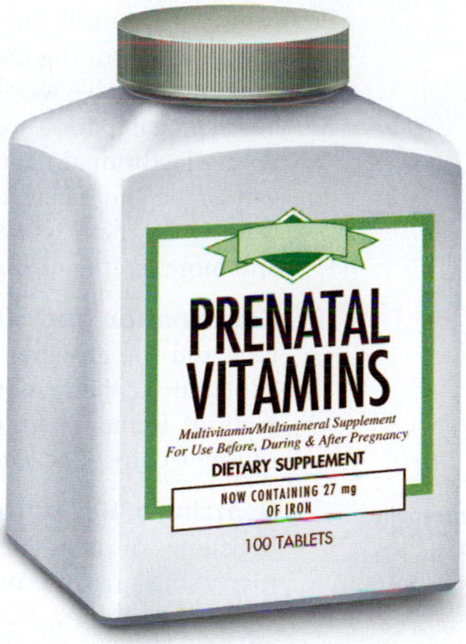

Supplement Facts
Serving Size 1 Tablet

Amount per Tablet	% Daily Value for Pregnant/Lactating Women
Vitamin A 4000 IU	50%
Vitamin C 100 mg	167%
Vitamin D 400 IU	100%
Vitamin E 11 IU	37%
Thiamin 1.84 mg	108%
Riboflavin 1.7 mg	85%
Niacin 18 mg	90%
Vitamin B6 2.6 mg	104%
Folate 800 mcg	100%
Vitamin B12 4 mcg	50%
Calcium 200 mg	15%
Iron 27 mg	150%
Zinc 25 mg	167%

INGREDIENTS: calcium carbonate, microcrystalline cellulose, dicalcium phosphate, ascorbic acid, ferrous fumarate, zinc oxide, acacia, sucrose ester, niacinamide, modified cellulose gum, di-alpha tocopheryl acetate, hydroxypropyl methylcellulose, hydroxypropyl cellulose, artificial colors (FD&C blue no. 1 lake, FD&C red no. 40 lake, FD&C yellow no. 6 lake, titanium dioxide), polyethylene glycol, starch, pyridoxine hydrochloride, vitamin A acetate, riboflavin, thiamin mononitrate, folic acid, beta carotene, cholecalciferol, maltodextrin, gluten, cyanocobalamin, sodium bisulfite.

Vegetarian Diets during Pregnancy and Lactation

In general, a well-planned vegetarian diet can support a healthy pregnancy and successful lactation if it provides adequate energy and contains a wide variety of legumes, whole grains, nuts, seeds, fruits, and vegetables.[51] Many vegetarian women are well nourished, with nutrient intakes from diet alone exceeding the RDA for all vitamins and minerals except iron, which is low for most women. In contrast, vegan women who restrict themselves to an exclusively plant-based diet generally have low food energy intakes and are thin. For pregnant women, this can be a problem. Women with low prepregnancy weights and small weight gains during pregnancy jeopardize a healthy pregnancy.

Vegan diets may require supplementation with vitamin B_{12}, calcium, and vitamin D, or the addition of foods fortified with these nutrients. Infants of vegan parents may suffer spinal cord damage and develop severe psychomotor retardation due to a lack of vitamin B_{12} in the mother's diet during pregnancy. Breastfed infants of vegan mothers have been reported to develop vitamin B_{12} deficiency and severe movement disorders. Giving the infants vitamin B_{12} supplements corrects the blood and neurological symptoms of deficiency, as well as the structural abnormalities, but cognitive and language development delays may persist. A vegan mother needs a regular source of vitamin B_{12}–fortified foods or a supplement that provides 2.6 micrograms daily.

A pregnant woman who cannot meet her calcium needs through diet alone may need 600 milligrams of supplemental calcium daily, taken with meals. Pregnant women who do not receive sufficient dietary vitamin D or enough exposure to sunlight may need a supplement.

Common Nutrition-Related Concerns of Pregnancy

Nausea, constipation, heartburn, and food sensitivities are common nutrition-related concerns during pregnancy. A few simple strategies can help alleviate maternal discomforts (see Table 14-2 on p. 472).

TABLE 14-2 Strategies to Alleviate Maternal Discomforts

To Alleviate the Nausea of Pregnancy	To Prevent or Alleviate Constipation	To Prevent or Relieve Heartburn
• On waking, arise slowly. • Eat dry toast or crackers. • Chew gum or suck hard candies. • Eat small, frequent meals. • Avoid foods with offensive odors. • When nauseated, drink carbonated beverages instead of citrus juice, water, milk, coffee, or tea.	• Eat foods high in fiber (fruits, vegetables, and whole grains). • Exercise regularly. • Drink at least eight glasses of liquids a day. • Respond promptly to the urge to defecate. • Use laxatives only as prescribed by a physician; do not use mineral oil, because it interferes with absorption of fat-soluble vitamins.	• Relax and eat slowly. • Chew food thoroughly. • Eat small, frequent meals. • Drink liquids between meals. • Avoid spicy or greasy foods. • Sit up while eating; elevate the head while sleeping. • Wait 3 hours after eating before lying down. • Wait 2 hours after eating before exercising.

Nausea Not all women have queasy stomachs in the early months of pregnancy, but many do. The nausea of "morning sickness" may actually occur anytime and ranges from mild queasiness to debilitating nausea and vomiting. Severe and continued vomiting may require hospitalization if it results in acidosis, dehydration, or excessive weight loss. The hormonal changes of early pregnancy seem to be responsible for a woman's sensitivities to the appearance, texture, or smell of foods. Traditional strategies for quelling nausea are listed in Table 14-2, but many women benefit most from simply resting when nauseous and eating the foods they want when they feel like eating. They may also find comfort in a cleaner, quieter, and more temperate environment.

Constipation and Hemorrhoids As the hormones of pregnancy alter muscle tone and the growing fetus crowds intestinal organs, an expectant mother may experience constipation. She may also develop hemorrhoids (swollen veins of the rectum). Hemorrhoids can be painful, and straining during bowel movements may cause bleeding. She can gain relief by following the strategies listed in Table 14-2.

Heartburn Heartburn ♦ is another common complaint during pregnancy. The hormones of pregnancy relax the digestive muscles, and the growing fetus puts increasing pressure on the mother's stomach. This combination causes gastroesophageal reflux, the painful sensation a person feels behind the breastbone when stomach acid splashes back up into the lower esophagus. Tips to help relieve heartburn are included in Table 14-2.

♦ Heartburn, medically known as **gastroesophageal reflux,** is discussed in Highlight 3.

Food Cravings and Aversions Some women develop cravings for, or aversions to, particular foods and beverages during pregnancy. **Food cravings** and **food aversions** are fairly common, but they do not seem to reflect real physiological needs. In other words, a woman who craves pickles does not necessarily need salt. Similarly, cravings for ice cream are common in pregnancy but do not signify a calcium deficiency. Cravings and aversions that arise during pregnancy are most likely due to hormone-induced changes in sensitivity to taste and smell.

Nonfood Cravings Some pregnant women develop cravings for nonfood items ♦ such as freezer frost, laundry starch, clay, soil, or ice—a practice known as pica. Pica is a cultural phenomenon that reflects a society's folklore; it is especially common among African American women. Pica is often associated with iron-deficiency anemia, but whether iron deficiency leads to pica or pica leads to iron deficiency is unclear. Eating clay or soil may interfere with iron absorption and displace iron-rich foods from the diet.

♦ **Pica** is the general term for eating nonfood items. The specific craving for nonfood items that come from the earth, such as clay or dirt, is known as **geophagia.**

IN SUMMARY **Energy and nutrient needs are high during pregnancy. A balanced diet that includes an extra serving from each of the five food groups can usually meet these needs, with the possible exception of iron and folate (supplements are recommended). The nausea, constipation, and heartburn that sometimes accompany pregnancy can usually be alleviated with a few simple strategies. Food cravings do not typically reflect physiological needs.**

food cravings: strong desires to eat particular foods.

food aversions: strong desires to avoid particular foods.

High-Risk Pregnancies

Some pregnancies jeopardize the life and health of the mother and infant. Table 14-3 identifies several characteristics of a **high-risk pregnancy**. A woman with none of these risk factors is said to have a **low-risk pregnancy**. The more factors that apply, the higher the risk. All pregnant women, especially those in high-risk categories, need prenatal medical care, including dietary ♦ advice.

♦ Nutrition advice in prenatal care:
- Eat well-balanced meals.
- Gain enough weight to support fetal growth.
- Take prenatal supplements as prescribed.
- Stop drinking alcohol.

The Infant's Birthweight

A high-risk pregnancy is likely to produce an infant with **low birthweight (LBW)**. Low-birthweight infants, defined as infants who weigh 5½ pounds or less, are classified according to their gestational age. Preterm infants are born before they are fully developed; they are often underweight and have trouble breathing because their lungs are immature. Preterm infants may be small, but if their size and weight are appropriate for their age, ♦ they can catch up in growth given adequate nutritional support. In contrast, small-for-gestational-age infants have suffered growth failure in the uterus and do not catch up as well. For the most part, survival improves with increased gestational age and birthweight.

♦ The weight of some preterm infants is **appropriate for gestational age (AGA);** others are **small for gestational age (SGA),** often reflecting malnutrition.

Low-birthweight infants are more likely to experience complications during delivery than normal-weight babies. They also have a statistically greater chance of having physical and mental birth defects, contracting diseases, and dying early in life. Of infants who die before their first birthdays, about two-thirds were low-birthweight newborns. Very-low-birthweight infants (3½ pounds or less) struggle not only for their immediate physical health and survival, but for their future cognitive development and abilities as well.

A strong relationship is evident between socioeconomic disadvantage and low birthweight. Low socioeconomic status impairs fetal development by causing stress and by limiting access to medical care and to nutritious foods. Low socioeconomic status often accompanies teen pregnancies, smoking, and alcohol and drug abuse—all predictors of low birthweight.

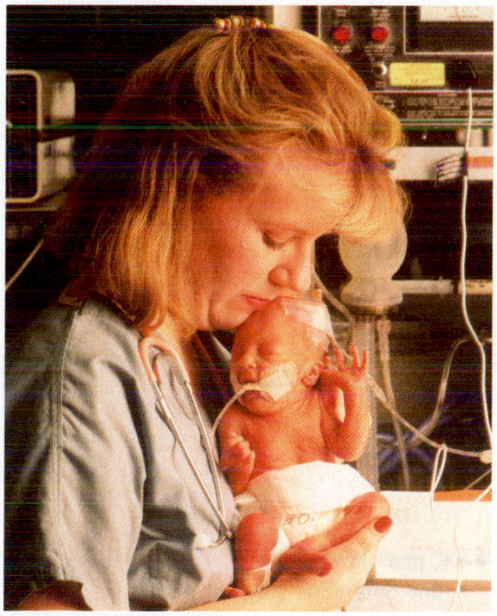

© Terry Vine/Getty Images

Low-birthweight babies need special care and nourishment.

Malnutrition and Pregnancy

Good nutrition clearly supports a healthy pregnancy. In contrast, malnutrition interferes with the ability to conceive,

TABLE 14-3 High-Risk Pregnancy Factors

Factor	Condition That Raises Risk
Maternal weight	
• Prior to pregnancy	Prepregnancy BMI either <18.5 or ≥ 25
• During pregnancy	Insufficient or excessive pregnancy weight gain
Maternal nutrition	Nutrient deficiencies or toxicities; eating disorders
Socioeconomic status	Poverty, lack of family support, low level of education, limited food available
Lifestyle habits	Smoking, alcohol or other drug use
Age	Teens, especially 15 years or younger; women 35 years or older
Previous pregnancies	
• Number	Many previous pregnancies (3 or more to mothers under age 20; 4 or more to mothers age 20 or older)
• Interval	Short or long intervals between pregnancies (<18 months or >59 months)
• Outcomes	Previous history of problems
• Multiple births	Twins or triplets
• Birthweight	Low- or high-birthweight infants
Maternal health	
• High blood pressure	Development of gestational hypertension
• Diabetes	Development of gestational diabetes
• Chronic diseases	Diabetes; heart, respiratory, and kidney disease; certain genetic disorders; special diets and medications

high-risk pregnancy: a pregnancy characterized by risk factors that make it likely the birth will be surrounded by problems such as premature delivery, difficult birth, retarded growth, birth defects, and early infant death.

low-risk pregnancy: a pregnancy characterized by factors that make it likely the birth will be normal and the infant healthy.

low birthweight (LBW): a birthweight of 5½ pounds (2500 grams) or less; indicates probable poor health in the newborn and poor nutrition status in the mother during pregnancy, before pregnancy, or both. Optimal birthweight for a full-term baby is 6.8 to 7.9 pounds (about 3100 to 3600 grams).

the likelihood of implantation, and the subsequent development of a fetus should conception and implantation occur.[52]

Malnutrition and Fertility The nutrition habits and lifestyle choices people make can influence the course of a pregnancy they are not even planning at the time. Severe malnutrition and food deprivation can reduce fertility because women may develop amenorrhea, ♦ and men may be unable to produce viable sperm. Furthermore, both men and women lose sexual interest during times of starvation. Starvation arises predictably during famines, wars, and droughts, but it can also occur amid peace and plenty. Many young women who diet excessively are starving and suffering from malnutrition (see Highlight 8).

♦ **Amenorrhea** is the temporary or permanent absence of menstrual periods. Amenorrhea is normal before puberty, after menopause, during pregnancy, and during lactation; otherwise it is abnormal.

Malnutrition and Early Pregnancy If a malnourished woman does become pregnant, she faces the challenge of supporting both the growth of a baby and her own health with inadequate nutrient stores. Malnutrition prior to and around conception prevents the placenta from developing fully. A poorly developed placenta cannot deliver optimum nourishment to the fetus, and the infant will be born small and possibly with physical and cognitive abnormalities. If this small infant is a female, she may develop poorly and have an elevated risk of developing a chronic condition that could impair her ability to give birth to a healthy infant. Thus a woman's malnutrition can adversely affect not only her children but her *grandchildren*.

Malnutrition and Fetal Development Without adequate nutrition during pregnancy, fetal growth and infant health are compromised. In general, consequences of malnutrition during pregnancy include fetal growth retardation, congenital malformations (birth defects), spontaneous abortion and stillbirth, preterm birth, and low infant birthweight. Malnutrition, coupled with low birthweight, is a factor in more than half of all deaths of children younger than four years of age worldwide.

Food Assistance Programs

Women in high-risk pregnancies can find assistance from the WIC program—a high-quality, cost-effective health-care and nutrition services program for women, infants, and children in the United States. Formally known as the Special Supplemental Nutrition Program for Women, Infants, and Children, WIC provides nutrition education and nutritious foods to infants, children up to age five, and pregnant and breastfeeding women who qualify financially and have a high risk of medical or nutritional problems. ♦ The program is both remedial and preventive: services include health-care referrals, nutrition education, and food packages or vouchers for specific foods. These foods supply nutrients known to be lacking in the diets of the target population—most notably, protein, calcium, iron, vitamin A, and vitamin C. WIC-sponsored foods include tuna fish, tofu, fruits, vegetables, eggs, milk, iron-fortified cereal, whole-grain breads, vitamin C–rich juices, cheeses, legumes, peanut butter, and iron-fortified infant formula and cereal.

♦ WIC participants:
- One-third of all pregnant women
- One-half of all infants
- One-quarter of all children ages 1–4 yr

More than eight million people—most of them young children—receive WIC benefits each month. Prenatal WIC participation can effectively reduce iron deficiency, infant mortality, low birthweight, and maternal and newborn medical costs.[53] In 2006, Congress appropriated more than $5.2 billion for WIC. For every dollar spent on WIC, an estimated $3 in medical costs are saved in the first two months after birth.

Maternal Health

Medical disorders can threaten the life and health of both mother and fetus. If diagnosed and treated early, many diseases can be managed to ensure a healthy outcome—another strong argument for early prenatal care. Furthermore, the changes in pregnancy can reveal disease risks, making screening important and early intervention possible.[54]

Preexisting Diabetes The risks of diabetes depend on how well it is managed before and during pregnancy. Without proper management of maternal diabetes, women face high infertility rates, and those who do conceive may experience episodes of severe hypoglycemia or hyperglycemia, preterm labor, and pregnancy-related hypertension. Infants may be large, suffer physical and mental abnormalities, and experience other complications such as severe hypoglycemia or respiratory distress, both of which can be fatal. Signs of fetal health problems are apparent even when maternal glucose is above normal but still below the diagnosis of diabetes.[55] To minimize complications, a woman needs to achieve glucose control before conception and continued glucose control throughout pregnancy.[56]

Gestational Diabetes For every 25 women entering pregnancy, one will develop a condition known as **gestational diabetes** during pregnancy. Gestational diabetes usually develops during the second half of pregnancy, with subsequent return to normal after childbirth. Some women with gestational diabetes, however, develop diabetes (usually type 2) after pregnancy, especially if they are overweight.[57] For this reason, health-care professionals strongly advise against excessive weight gain during pregnancy.

The most common consequences of gestational diabetes are complications during labor and delivery and a high infant birthweight. Birth defects associated with gestational diabetes include heart damage, limb deformities, and neural tube defects. To ensure that the problems of gestational diabetes are dealt with promptly, physicians screen for the risk factors ♦ listed in the margin and test high-risk women for glucose intolerance immediately and average-risk women between 24 and 28 weeks gestation.[58]

♦ Risk factors for gestational diabetes:
- Age 25 or older
- BMI ≥25 or excessive weight gain
- Complications in previous pregnancies, including gestational diabetes or high-birthweight infant
- Prediabetes or symptoms of diabetes
- Family history of diabetes
- Hispanic American, African American, Native American, Asian American, Pacific Islander

Dietary recommendations should meet the needs of pregnancy and maternal blood glucose goals.[59] Diet and moderate exercise may control gestational diabetes, but if blood glucose fails to normalize, insulin or other drugs may be required. Importantly, treatment reduces birth complications, infant deaths, and maybe even postpartum depression.[60]

Chronic Hypertension Hypertension complicates pregnancy and affects its outcome in different ways, depending on when the hypertension first develops and on how severe it becomes.[61] In addition to the threats hypertension always carries (such as heart attack and stroke), high blood pressure increases the risks of a low-birthweight infant or the separation of the placenta from the wall of the uterus before the birth, resulting in stillbirth. Ideally, before a woman with hypertension becomes pregnant, her blood pressure is under control.

Gestational Hypertension Some women develop **gestational hypertension**—high blood pressure during the second half of pregnancy.* For 50 percent of the women with gestational hypertension, the rise in blood pressure is mild and does not affect the pregnancy adversely. Blood pressure usually returns to normal during the first few weeks after childbirth. For the other 50 percent, gestational hypertension is an early sign of the most serious maternal complication of pregnancy—preeclampsia.

Preeclampsia **Preeclampsia** is a condition characterized not only by gestational hypertension but also by protein in the urine. The cause of preeclampsia remains unclear, but it usually occurs with first pregnancies ♦ and most often after 20 weeks gestation. Symptoms typically regress within two days of delivery. Both men and women who were born of pregnancies complicated by preeclampsia are more likely to have a child born of a pregnancy complicated by preeclampsia, suggesting a genetic predisposition. Black women have a much greater risk of preeclampsia than white women.

♦ Signs and symptoms of preeclampsia:
- Hypertension
- Protein in the urine
- Upper abdominal pain
- Severe headaches
- Swelling of hands, feet, and face
- Vomiting
- Blurred vision
- Sudden weight gain (1 lb/day)
- Fetal growth retardation

Preeclampsia affects almost all of the mother's organs—the circulatory system, liver, kidneys, and brain.[62] Blood flow through the vessels that supply oxygen and nutrients to the placenta diminishes. For this reason, preeclampsia often retards fetal growth. It also seems to increase the risk of epilepsy for the infant.[63] In some cases, the placenta separates from the uterus, resulting in preterm birth or stillbirth.

gestational diabetes: glucose intolerance with onset or first recognition during pregnancy.

gestational hypertension: high blood pressure that develops in the second half of pregnancy and resolves after childbirth, usually without affecting the outcome of the pregnancy.

preeclampsia (PRE-ee-KLAMP-see-ah): a condition characterized by hypertension and protein in the urine.

*Blood pressure of 140/90 millimeters mercury or greater during the second half of pregnancy in a woman who has not previously exhibited hypertension indicates high blood pressure.

Preeclampsia can progress rapidly to **eclampsia**—a condition characterized by seizures and coma. Maternal death during pregnancy and childbirth is extremely rare in developed countries, but when it does occur, eclampsia is a common cause. The rate of death for black women with eclampsia is more than four times the rate for white women.

Preeclampsia demands prompt medical attention. Treatment focuses on controlling blood pressure and preventing seizures. If preeclampsia develops early and is severe, induced labor or cesarean section may be necessary, regardless of gestational age. The infant will be preterm, with all of the associated problems, including poor lung development and special care needs. Several dietary factors have been studied, but none have proved beneficial in preventing preeclampsia.[64] Limited research suggests that exercise may protect against preeclampsia by stimulating placenta growth and vascularity and reducing oxidative stress.[65]

The Mother's Age

Maternal age also influences the course of a pregnancy. Compared with women of the physically ideal childbearing age of 20 to 25, both younger and older women face more complications of pregnancy.

Pregnancy in Adolescents

Many adolescents become sexually active before age 19, and approximately 757,000 adolescent girls face pregnancies each year in the United States; slightly more than half of them give birth.[66] Nourishing a growing fetus adds to a teenage girl's nutrition burden, especially if her growth is still incomplete. Simply being young and physically immature increases the risks of pregnancy complications. Pregnant teens are less likely to receive early prenatal care and are more likely to smoke during pregnancy—two factors that predict low birthweight and infant death.[67]

Common complications among adolescent mothers include iron-deficiency anemia (which may reflect poor diet and inadequate prenatal care) and prolonged labor (which reflects the mother's physical immaturity). On a positive note, maternal death is lowest for mothers under age 20.

The rates of stillbirths, preterm births, and low-birthweight infants are high for teenagers—both for teen moms and for teen dads.[68] Many of these infants suffer physical problems, require intensive care, and die within the first year. The care of infants born to teenagers costs our society an estimated $1 billion annually. Because teenagers have few financial resources, they cannot pay these costs. Furthermore, their low economic status contributes significantly to the complications surrounding their pregnancies. At a time when prenatal care is most important, it is less accessible. And the pattern of teenage pregnancies continues from generation to generation, with almost 40 percent of the daughters born to teenage mothers becoming teenage mothers themselves. Clearly, teenage pregnancy is a major public health problem.

To support the needs of both mother and fetus, young teenagers (13 to 16 years old) are encouraged to strive for the highest weight gains recommended for pregnancy. For a teen who enters pregnancy at a healthy body weight, a weight gain of approximately 35 pounds is recommended; this amount minimizes the risk of delivering a low-birthweight infant. Pregnant and lactating teenagers can use the USDA Food Guide presented in Table 2-2, making sure to select a high enough kcalorie level to support adequate weight gain.

Without the appropriate economic, social, and physical support, a young mother will not be able to care for herself during her pregnancy and for her child after the birth. To improve her chances for a successful pregnancy and a healthy infant, she must seek prenatal care. WIC provides health-care referrals and helps pregnant teenagers obtain adequate food for themselves and their infants. (WIC is introduced earlier in the chapter.)

Pregnancy in Older Women

In the last several decades, many women have delayed childbearing while they pursue education and careers. As a result, the number of first births to women 35 and older has increased dramatically. Most of these women, even those older than age 50, have healthy pregnancies.

eclampsia (eh-KLAMP-see-ah): a severe stage of preeclampsia characterized by seizures.

The few complications associated with later childbearing often reflect chronic conditions such as hypertension and diabetes, which can complicate an otherwise healthy pregnancy. These complications may result in a cesarean section, which is twice as common in women older than 35 as among younger women. For all these reasons, maternal death rates are higher in women older than 35 than in younger women.

The babies of older mothers face problems of their own including higher rates of preterm births and low birthweight. Their rates of birth defects are also high. Because 1 out of 50 pregnancies in older women produces an infant with genetic abnormalities, obstetricians routinely screen women older than 35. For a 40-year-old mother, the risk of having a child with **Down syndrome**, for example, is about 1 in 100 compared with 1 in 300 for a 35 year old and 1 in 10,000 for a 20 year old. In addition, fetal death is twice as high for women 35 years and older than for younger women. Why this is so remains a bit of a mystery. One possibility is that the uterine blood vessels of older women may not fully adapt to the increased demands of pregnancy.

Practices Incompatible with Pregnancy

Besides malnutrition, a variety of lifestyle factors can have adverse effects on pregnancy, and some may be teratogenic. ♦ People who are planning to have children can make the choice to practice healthy behaviors.

♦ The word *teratogenic* describes a factor that causes abnormal fetal development and birth defects.

Alcohol One out of 12 pregnant women drinks alcohol at some time during her pregnancy; 1 out of 30 drinks frequently.[69] Alcohol consumption during pregnancy can cause irreversible mental and physical retardation of the fetus—fetal alcohol syndrome (FAS). Of the leading causes of mental retardation, FAS is the only one that is totally *preventable*. To that end, the surgeon general urges all pregnant women to refrain from drinking alcohol. Fetal alcohol syndrome is the topic of Highlight 14, which includes mention of how alcohol consumption by men may also affect fertility and fetal development.

Medicinal Drugs Drugs other than alcohol can also cause complications during pregnancy, problems in labor, and serious birth defects. For these reasons, pregnant women should not take any medicines without consulting their physicians, who must weigh the benefits against the risks.

Herbal Supplements Similarly, pregnant women should seek a physician's advice before using herbal supplements. Women sometimes seek herbal preparations during their pregnancies to quell nausea, induce labor, aid digestion, promote water loss, support restful sleep, and fight depression. As Chapter 19 explains, some herbs may be safe, but many others are definitely harmful.

Illicit Drugs The recommendation to avoid drugs during pregnancy also includes illicit drugs, of course. Unfortunately, use of illicit drugs, such as cocaine and marijuana, is common among some pregnant women.

Drugs of abuse, such as cocaine, easily cross the placenta and impair fetal growth and development. Furthermore, they are responsible for preterm births, low-birthweight infants, perinatal deaths, ♦ and sudden infant deaths. If these newborns survive, central nervous system damage is evident: their cries, sleep, and behaviors early in life are abnormal, and their cognitive development later in life is impaired.[70] They may be hypersensitive or underaroused; those who test positive for drugs suffer the greatest effects of toxicity and withdrawal. Their growth throughout childhood continues at a slow rate.[71]

♦ The word *perinatal* refers to the time between the twenty-eighth week of gestation and one month after birth.

Smoking and Chewing Tobacco Unfortunately, an estimated 10 percent of pregnant women in the United States smoke.[72] Smoking cigarettes and chewing tobacco at any time exert harmful effects, and pregnancy dramatically magnifies the hazards of these practices. Smoking restricts the blood supply to the growing fetus and thus limits oxygen and nutrient delivery and waste removal. A mother who smokes is more likely to have a complicated birth and a low-birthweight infant. Indeed, of all preventable causes of low birthweight in the United States, smoking

Down syndrome: a genetic abnormality that causes mental retardation, short stature, and flattened facial features.

is at the top of the list. Although, most infants born to cigarette smokers are low birthweight, some are not, suggesting that the effect of smoking on birthweight also depends, in part, on genes involved in the metabolism of smoking toxins.

In addition to contributing to low birthweight, smoking interferes with lung growth and increases the risks of poor lung function, respiratory infections, and childhood asthma.[73] It can also cause death in an otherwise healthy fetus or newborn. A positive relationship exists between **sudden infant death syndrome (SIDS)** and both cigarette smoking during pregnancy and postnatal exposure to passive smoke. Smoking during pregnancy may reduce brain size and impair the intellectual and behavioral development of the child later in life.[74] The margin ♦ lists complications of smoking during pregnancy.

♦ Complications associated with smoking during pregnancy:

- Fetal growth retardation
- Low birthweight
- Complications at birth (prolonged final stage of labor)
- Mislocation of the placenta
- Premature separation of the placenta
- Vaginal bleeding
- Spontaneous abortion
- Fetal death
- Sudden infant death syndrome (SIDS)
- Middle ear diseases
- Cardiac and respiratory diseases

Infants of mothers who chew tobacco also have low birthweights and high rates of fetal deaths. Any woman who smokes cigarettes or chews tobacco and is considering pregnancy or who is already pregnant needs to quit.

Environmental Contaminants Proving that environmental contaminants cause reproductive damage is difficult, but evidence is established for wildlife and seems likely for human beings. Infants and young children of pregnant women exposed to environmental contaminants such as lead show signs of delayed mental and psychomotor development. During pregnancy, lead readily moves across the placenta, inflicting severe damage on the developing fetal nervous system. In addition, infants exposed to even low levels of lead during gestation weigh less at birth and consequently struggle to survive. For these reasons, it is particularly important that pregnant women receive foods and beverages grown and prepared in environments free of contamination. A diet high in calcium will help to defend against lead contamination, and breastfeeding may help to counterbalance developmental damage incurred from contamination during pregnancy.[75]

Mercury is among the contaminants of concern. As Chapter 5 mentions, fatty fish are a good source of omega-3 fatty acids, but some fish contain large amounts of the pollutant mercury, which can impair fetal growth and harm the developing brain and nervous system.[76] Because the benefits of seafood consumption seem to outweigh the risks, pregnant (and lactating) women should do the following:[77]

- Avoid shark, swordfish, king mackerel, and tilefish (also called golden snapper or golden bass).
- Limit average weekly consumption to 12 ounces (cooked or canned) of seafood *or* to 6 ounces (cooked or canned) of white (albacore) tuna.

Supplements of fish oil are not recommended because they may contain concentrated toxins and because their effects on pregnancy remain unknown.

Foodborne Illness Foodborne illnesses arise when people eat foods that contain infectious microbes or microbes that produce toxins. At best, the vomiting and diarrhea associated with these illnesses can leave a pregnant woman exhausted and dehydrated; at worst, foodborne illnesses can cause meningitis, pneumonia, or even fetal death. Pregnant women are about 20 times more likely than other healthy adults to get the foodborne illness **listeriosis**. The margin ♦ presents tips to prevent listeriosis, and Highlight 18 includes precautions to minimize the risks of other common foodborne illnesses.

♦ Listeriosis can be prevented in the following ways:

- Use only pasteurized juices and dairy products; avoid Mexican soft cheeses, feta cheese, brie, Camembert, and blue-veined cheeses such as Roquefort.
- Thoroughly cook meat, poultry, eggs, and seafood.
- Thoroughly reheat hot dogs, luncheon meats, and deli meats, including cured meats such as salami.
- Wash all fruits and vegetables.
- Avoid refrigerated pâté, meat spreads, smoked seafood such as salmon or trout, and any fish labeled "nova," "lox," or "kippered," unless prepared in a cooked dish.

Dietary Guidelines for Americans 2010

Women who are pregnant should:

- Only eat foods with seafood, meat, poultry, or eggs that have been cooked to recommended safe minimum internal temperatures.
- Take special precautions not to consume unpasteurized juice or milk products.
- Reheat deli and luncheon meats and hot dogs to steaming hot and not eat raw sprouts.

Vitamin-Mineral Megadoses The pregnant woman who is trying to eat well may mistakenly assume that more is better when it comes to multivitamin-mineral

sudden infant death syndrome (SIDS): the unexpected and unexplained death of an apparently well infant; the most common cause of death of infants between the second week and the end of the first year of life; also called *crib death.*

listeriosis (lis-TEAR-ee-OH-sis): an infection caused by eating food contaminated with the bacterium *Listeria monocytogenes,* which can be killed by pasteurization and cooking but can survive at refrigerated temperatures; certain ready-to-eat foods, such as hot dogs and deli meats, may become contaminated after cooking or processing, but before packaging.

supplements. This is simply not true; many vitamins and minerals are toxic when taken in excess. Excessive vitamin A is particularly infamous for its role in fetal malformations of the cranial nervous system. Intakes before the seventh week appear to be the most damaging. (Review Figure 14-4 on p. 460 to see how many tissues are in their critical periods prior to the seventh week.) For this reason, vitamin A supplements are not given during pregnancy unless there is specific evidence of deficiency, which is rare. A pregnant woman can obtain all the vitamin A and most of the other vitamins and minerals she needs by making wise food choices. She should take supplements only on the advice of a registered dietitian or physician.

Caffeine Caffeine crosses the placenta, and the developing fetus has a limited ability to metabolize it. Research studies have not proved that caffeine (even in high doses) causes birth defects in human infants (as it does in animals), but limited evidence suggests that heavy use increases the risk of miscarriage and fetal death.[78] (In these studies, heavy caffeine use is defined as the equivalent of three or more cups of coffee a day.) Depending on the quantities consumed and the mother's metabolism, caffeine may also interfere with fetal growth.[79] All things considered, it is most sensible to limit caffeine consumption to the equivalent of a cup of coffee or two 12-ounce cola beverages a day. (The caffeine contents of selected beverages, foods, and drugs are listed at the beginning of Appendix H.)

Weight-Loss Dieting Weight-loss dieting, even for short periods, can be hazardous during pregnancy. Low-carbohydrate diets or fasts that cause ketosis deprive the fetal brain of needed glucose and may impair cognitive development. Such diets are also likely to lack other nutrients vital to fetal growth. Regardless of prepregnancy weight, pregnant women need an adequate diet and sufficient weight gain to support healthy fetal development.

Sugar Substitutes Artificial sweeteners have been extensively investigated and found to be acceptable during pregnancy if used within the FDA's guidelines.[80] Still, it is prudent for pregnant women to use sweeteners in moderation and within an otherwise nutritious and well-balanced diet. Women with the inherited disease phenylketonuria (PKU) should not use the artificial sweetener aspartame. Aspartame contains the amino acid phenylalanine, and people with PKU are unable to dispose of any excess phenylalanine. The accumulation of phenylalanine and its by-products is toxic to the developing nervous system, causing irreversible brain damage.

IN SUMMARY **High-risk pregnancies, especially for teenagers, threaten the life and health of both mother and infant. Proper nutrition and abstinence from smoking, alcohol, and other drugs improve the outcome. In addition, prenatal care includes monitoring pregnant women for gestational diabetes and preeclampsia.**

In general, the following guidelines will allow most women to enjoy a healthy pregnancy:[81]

- Strive for good nutrition and health prior to pregnancy and get prenatal care during pregnancy.
- Gain a healthy amount of weight.
- Eat a balanced diet, safely prepared, and engage in physical activity regularly.
- Take prenatal vitamin and mineral supplements as prescribed.
- Refrain from cigarettes, alcohol, and drugs (including herbal remedies, unless prescribed by a physician).

Childbirth marks the end of pregnancy and the beginning of a new set of parental responsibilities—including feeding the newborn.

Nutrition during Lactation

◆ To learn about breastfeeding, a pregnant woman can read at least one of the many books available. At the end of this chapter, Nutrition on the Net provides a list of resources, including La Leche League International.

Before the end of her pregnancy, a woman needs to consider whether to feed her infant breast milk, ◆ infant formula, or both. These options are the only recommended foods for an infant during the first four to six months of life. The rate of breastfeeding is close to the Healthy People goal of 75 percent at birth, but it falls far short of goals at three months, six months, and a year.[82] This section focuses on how the mother's nutrition supports the making of breast milk, and the next chapter describes how the infant benefits from drinking breast milk.

In many countries around the world, a woman breastfeeds her newborn without considering the alternatives or making a conscious decision. In other parts of the world, a woman feeds her newborn formula simply because she knows so little about breastfeeding. She may have misconceptions or feel uncomfortable about a process she has never seen or experienced. Breastfeeding offers many health benefits to both mother and infant, and every pregnant woman should seriously consider it (see Table 14-4).[83] Even so, women's choices are often influenced by factors other than health and science—factors such as culture, politics, religion, and marketing.[84] Mothers may have valid reasons for not breastfeeding, and formula-fed infants grow and develop into healthy children.

Lactation: A Physiological Process

Lactation naturally follows pregnancy, as the mother's body continues to nourish the infant. The **mammary glands** secrete milk for this purpose. The mammary glands develop during puberty but remain fairly inactive until pregnancy. During pregnancy, hormones promote the growth and branching of a duct system in the breasts and the development of the milk-producing cells.

The hormones **prolactin** and **oxytocin** finely coordinate lactation. The infant's demand for milk stimulates the release of these hormones, which signal the mammary glands to supply milk. Prolactin is responsible for milk production. As long as the infant is nursing, prolactin concentrations remain high, and milk production continues.

TABLE 14-4 Benefits of Breastfeeding

For Infants

- Provides the appropriate composition and balance of nutrients with high bioavailability
- Provides hormones that promote physiological development
- Improves cognitive development
- Protects against a variety of infections
- May protect against some chronic diseases—such as diabetes (both types), obesity, atherosclerosis, asthma, and hypertension—later in life
- Protects against food allergies

For Mothers

- Contracts the uterus
- Delays the return of regular ovulation, thus lengthening birth intervals (is not, however, a dependable method of contraception)
- Conserves iron stores (by prolonging amenorrhea)
- May protect against breast and ovarian cancer and reduce the risk of diabetes (type 2)

Other

- Cost savings from not needing medical treatment for childhood illnesses or time off work to care for them
- Cost savings from not needing to purchase formula (even after adjusting for added foods in the diet of a lactating mother)[a]
- Environmental savings to society from not needing to manufacture, package, and ship formula and dispose of the packaging
- Convenience of not having to shop for and prepare formula

[a]A nursing mother produces more than 35 gallons of milk during the first six months, saving roughly $450 in formula costs.

lactation: production and secretion of breast milk for the purpose of nourishing an infant.

mammary glands: glands of the female breast that secrete milk.

prolactin (pro-LAK-tin): a hormone secreted from the anterior pituitary gland that acts on the mammary glands to promote the production of milk. The release of prolactin is mediated by *prolactin-inhibiting hormone (PIH)*.
- **pro** = promote
- **lacto** = milk

oxytocin (OCK-see-TOH-sin): a hormone that stimulates the mammary glands to eject milk during lactation and the uterus to contract during childbirth.

TABLE 14-5 Ten Steps to Successful Breastfeeding

To promote breastfeeding, every maternity facility should:

- Develop a written breastfeeding policy that is routinely communicated to all health-care staff
- Train all health-care staff in the skills necessary to implement the breastfeeding policy
- Inform all pregnant women about the benefits and management of breastfeeding
- Help mothers initiate breastfeeding within ½ hour of birth
- Show mothers how to breastfeed and how to maintain lactation, even if they need to be separated from their infants
- Give newborn infants no food or drink other than breast milk, unless medically indicated
- Practice rooming-in, allowing mothers and infants to remain together 24 hours a day
- Encourage breastfeeding on demand
- Give no artificial nipples or pacifiers to breastfeeding infants[a]
- Foster the establishment of breastfeeding support groups and refer mothers to them at discharge from the facility

[a]Compared with nonusers, infants who use pacifiers breastfeed less frequently and stop breastfeeding at a younger age.
SOURCE: United Nations Children's Fund and World Health Organization, *Protecting, Promoting and Supporting Breastfeeding: The Special Role of Maternity Services.*

iStockphoto.com/Goldmund Photography

A woman who decides to breastfeed offers her infant a full array of nutrients and protective factors to support optimal health and development.

The hormone oxytocin causes the mammary glands to eject milk into the ducts, a response known as the **let-down reflex**. The mother feels this reflex as a contraction of the breast, followed by the flow of milk and the release of pressure. By relaxing and eating well, the nursing mother promotes easy let-down of milk and greatly enhances her chances of successful lactation.

Breastfeeding: A Learned Behavior

Lactation is an automatic physiological process that virtually all mothers are capable of doing. Breastfeeding, on the other hand, is a learned behavior that not all mothers decide to do. Of women who do breastfeed, those who receive early and repeated information and support breastfeed their infants longer than others. Health-care professionals ♦ play an important role in providing encouragement and accurate information on breastfeeding.[85] Women who have been successful breastfeeding can offer advice and dispel misperceptions about lifestyle issues. Table 14-5 lists ten steps maternity facilities and health-care professionals can take to promote successful breastfeeding among new mothers.[86]

The mother's partner also plays an important role in encouraging breastfeeding.[87] When partners support the decision, mothers are more likely to start and continue breastfeeding. Clearly, educating those closest to the mother could change attitudes and promote breastfeeding.

Most healthy women who want to breastfeed can do so with a little preparation. Physical obstacles to breastfeeding are rare, although most nursing mothers quit before the recommended six months because of perceived difficulties. Obese mothers seem to have a particularly difficult time because of both biological and sociocultural factors.[88] Successful breastfeeding requires adequate nutrition and rest. This, plus the support of all who care, will help to enhance the well-being of mother and infant.

♦ Some hospitals employ **certified lactation consultants** who specialize in helping new mothers establish a healthy breastfeeding relationship with their newborn. These consultants are often registered nurses with specialized training in breast and infant anatomy and physiology.

Maternal Energy and Nutrient Needs during Lactation

Ideally, the mother who chooses to breastfeed her infant will continue to eat nutrient-dense foods throughout lactation. An adequate diet is needed to support the stamina, patience, and self-confidence that nursing an infant demands.

Energy Intake and Exercise

A nursing mother produces about 25 ounces of milk per day, with considerable variation from woman to woman and in the same woman from time to time, depending primarily on the infant's demand for milk. To produce an adequate supply of milk, a woman needs extra energy—almost 500 kcalories a day above her regular need during the first six months of lactation. To meet this energy need, ♦ she can eat an extra 330 kcalories of food each day and let the fat reserves she accumulated during pregnancy provide the rest. Most

♦ Energy requirement during lactation:
- 1st 6 mo: +330 kcal/day
- 2nd 6 mo: +400 kcal/day

let-down reflex: the reflex that forces milk to the front of the breast when the infant begins to nurse.

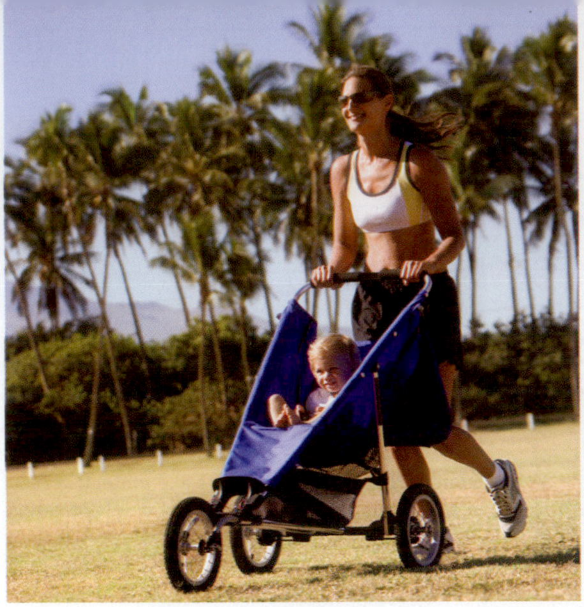

A jog through the park provides an opportunity for physical activity and fresh air.

women need at least 1800 kcalories a day to receive all the nutrients required for successful lactation. Severe energy restriction may hinder milk production.

After the birth of the infant, many women actively try to lose the extra weight and body fat they accumulated during pregnancy. How much weight a woman retains after pregnancy depends on her gestational weight gain and the duration and intensity of breastfeeding. Many women who follow recommendations for gestational weight gain and breastfeeding can readily return to prepregnancy weight by six months.[89] Neither the quality nor the quantity of breast milk is adversely affected by moderate weight loss, and infants grow normally.

Women often exercise to lose weight and improve fitness, and this is compatible with breastfeeding and infant growth. Because intense physical activity can raise the lactate concentration of breast milk and influence the milk's taste, some infants may prefer milk produced prior to exercise. In these cases, mothers can either breastfeed before exercise or express their milk before exercise for use afterward.

Dietary Guidelines for Americans 2010

- Maintain appropriate kcalorie balance during breastfeeding.

Energy Nutrients Recommendations for protein and fatty acids intakes remain about the same during lactation as during pregnancy, but they increase for carbohydrates and fibers. Nursing mothers need additional carbohydrate to replace the glucose used to make the lactose in breast milk. The fiber recommendation is 1 gram higher simply because it is based on kcalorie intake, which increases during lactation.

Vitamins and Minerals A question often raised is whether a mother's milk may lack a nutrient if she fails to get enough in her diet. The answer differs from one nutrient to the next, but in general, nutritional inadequacies reduce the *quantity,* not the *quality,* of breast milk. Women can produce milk with adequate protein, carbohydrate, fat, and most minerals, even when their own supplies are limited. For these nutrients and for the vitamin folate as well, milk quality is maintained at the expense of maternal stores. This is most evident in the case of calcium: dietary calcium has no effect on the calcium concentration of breast milk, but maternal bones lose some density during lactation if calcium intakes are inadequate.[90] Bone density increases again when lactation ends; breastfeeding has no long-term harmful effects on bones. The nutrients in breast milk that are most likely to decline in response to prolonged inadequate intakes are the vitamins—especially vitamins B_6, B_{12}, A, and D. Review Figure 14-10 (p. 468) to compare a lactating woman's nutrient needs with those of pregnant and nonpregnant women.

Water Despite misconceptions, a mother who drinks more fluid does not produce more breast milk. To protect herself from dehydration, however, a lactating woman needs to drink plenty of fluids. ♦ A sensible guideline is to drink a glass of milk, juice, or water at each meal and each time the infant nurses.

♦ AI for *total* water (including drinking water, other beverages, and foods) during lactation: 3.8 L/day. Because foods provide about 20 percent of total water intake, beverages—including drinking water—should provide 3.1 L/day (≈13 cups).

Nutrient Supplements Most lactating women can obtain all the nutrients they need from a well-balanced diet without taking multivitamin-mineral supplements. Nevertheless, some may need iron supplements, not to enhance the iron in their breast milk, but to refill their depleted iron stores. The mother's iron stores dwindle during pregnancy as she supplies the developing fetus with enough iron to last through the first four to six months of the infant's life. In addition, childbirth may have incurred blood losses. Thus a woman may need iron supplements during lactation even though, until menstruation resumes, her iron requirement is about half that of other nonpregnant women her age.

Food Assistance Programs In general, women most likely to participate in the food assistance program WIC—those who are poor and have little education—are less likely to breastfeed. Furthermore, WIC provides infant formula at no cost. Because WIC recognizes the many benefits of breastfeeding, efforts are made to overcome this dilemma.[91] In addition to nutrition education and encouragement, breastfeeding mothers receive the following WIC incentives:

- Higher priority in certification into WIC
- Longer eligibility to participate in WIC
- More foods and larger quantities
- Breast pumps and other support materials

Together, these efforts help to provide nutrition support and encourage WIC mothers to breastfeed.

BlueOrange Studio/Shutterstock.com

Nutritious foods support successful lactation.

Particular Foods Foods with strong or spicy flavors (such as garlic) may alter the flavor of breast milk. A sudden change in the taste of the milk may annoy some infants. Familiar flavors may enhance enjoyment.

Current evidence does not support a major role for maternal dietary restrictions during lactation to prevent or delay the onset of food allergy in infants.[92] Infants who develop symptoms of food allergy, however, may be more comfortable if the mother's diet excludes the most common offenders—cow's milk, eggs, fish, peanuts, and tree nuts. Generally, infants with a strong family history of food allergies benefit from breastfeeding.[93]

A nursing mother can usually eat whatever nutritious foods she chooses. If she suspects a particular food is causing the infant discomfort, her physician may recommend a dietary challenge: eliminate the food from the diet to see if the infant's reactions subside; then return the food to the diet and again monitor the infant's reactions. If a food must be eliminated for an extended time, appropriate substitutions must be made to ensure nutrient adequacy.

Maternal Health

If a woman has an ordinary cold, she can continue nursing without worry. If susceptible, the infant will catch it from her anyway. (Thanks to the immunological protection of breast milk, the baby may be less susceptible than a formula-fed baby would be.) With appropriate treatment, a woman who has an infectious disease such as tuberculosis or hepatitis can breastfeed; transmission is rare. Women with HIV (human immunodeficiency virus) infections, however, should consider other options.

HIV Infection and AIDS Mothers with HIV infections can transmit the virus (which causes AIDS) to their infants through breast milk, especially during the early months of breastfeeding. In developed countries such as the United States, where safe alternatives are available, HIV-positive women should *not* breastfeed their infants.[94] In developing countries, where the feeding of inappropriate or contaminated formulas causes 1.5 million infant deaths each year, breastfeeding can be critical to infant survival. Thus the decision of whether HIV-infected women in developing countries should breastfeed must consider the potential risks and benefits. The World Health Organization (WHO) recommends exclusive breastfeeding for infants of HIV-infected women for the first six months of life unless formula feeding is acceptable, feasible, affordable, sustainable, and safe before that time.[95] Alternatively, HIV-exposed infants may be protected by receiving antiretroviral treatment while being breastfed.[96]

Diabetes Women with diabetes (type 1) may need careful monitoring and counseling to ensure successful lactation. These women need to adjust their energy intakes and insulin doses to meet the heightened needs of lactation. Maintaining good glucose control helps to initiate lactation and support milk production.

Postpartum Amenorrhea Women who breastfeed experience prolonged **postpartum amenorrhea.** Absent menstrual periods, however, do not protect a woman from pregnancy. To prevent pregnancy, a couple must use some form of contraception. Breastfeeding women who use oral contraceptives should use progestin-only agents for at least the first six months.[97] Estrogen-containing oral contraceptives reduce the volume and the protein content of breast milk.

Breast Health Some women fear that breastfeeding will cause their breasts to sag. The breasts do swell and become heavy and large immediately after the birth, but even when they produce enough milk to nourish a thriving infant, they eventually shrink back to their prepregnant size. Given proper support, diet, and exercise, breasts often return to their former shape and size when lactation ends. Breasts change their shape as the body ages, but breastfeeding does not accelerate this process.

Whether the physical and hormonal events of pregnancy and lactation protect women from later breast cancer is an area of active research.[98] Some research suggests no association between breastfeeding and breast cancer, whereas other research suggests a protective effect. Protection against breast cancer is most apparent for premenopausal women who were young when they breastfed and who breastfed for a long time.

Practices Incompatible with Lactation

Some substances impair milk production or enter breast milk and interfere with infant development. This section discusses practices that a breastfeeding mother should avoid.

Alcohol Alcohol easily enters breast milk, and its concentration peaks within an hour of ingestion. Infants drink less breast milk when their mothers have consumed even small amounts of alcohol (equivalent to a can of beer). Three possible reasons, acting separately or together, may explain why. For one, the alcohol may have altered the flavor of the breast milk and thereby the infants' acceptance of it. For another, because infants metabolize alcohol inefficiently, even low doses may be potent enough to suppress their feeding and cause sleepiness. Third, the alcohol may have interfered with lactation by inhibiting the hormone oxytocin.

In the past, alcohol has been recommended to mothers to facilitate lactation despite a lack of scientific evidence that it does so. The research summarized here suggests that alcohol actually hinders breastfeeding. An occasional alcoholic beverage may be within safe limits, but breastfeeding should be delayed for at least two hours afterward.

Medicinal Drugs Most medicines are compatible with breastfeeding, but some are contraindicated, either because they suppress lactation or because they are secreted into breast milk and can harm the infant. As a precaution, a nursing mother should consult with her physician prior to taking any drug, including herbal supplements.

Illicit Drugs Illicit drugs, of course, are harmful to the physical and emotional health of both the mother and the nursing infant. Breast milk can deliver such high doses of illicit drugs as to cause irritability, tremors, hallucinations, and even death in infants. Women whose infants have overdosed on illicit drugs contained in breast milk have been convicted of murder. Women on methadone maintenance can safely breastfeed their infants.[99]

Smoking Because cigarette smoking reduces milk volume, smokers may produce too little milk to meet their infants' energy needs. The milk they do produce contains nicotine, which alters its smell and flavor. Furthermore, infants of breastfeeding mothers who smoke sleep less than infants of those who do not smoke.[100] Infant exposure to passive smoke negates the protective effect breastfeeding offers against SIDS and increases the risks dramatically.

Environmental Contaminants Some environmental contaminants, such as DDT, PCBs, and dioxin, can find their way into breast milk. Inuit mothers living in Arctic Québec who eat seal and beluga whale blubber have high concentrations

postpartum amenorrhea (ay-MEN-oh-REE-ah): the normal temporary absence of menstrual periods immediately following childbirth.

of DDT and PCBs in their breast milk, but the impact on infant development is unclear. Preliminary studies indicate that the children of these Inuit mothers are developing normally. Researchers speculate that the abundant omega-3 fatty acids of the Inuit diet may protect against damage to the central nervous system. Breast milk tainted with dioxin interferes with tooth development during early infancy, producing soft, mottled teeth that are vulnerable to dental caries. To limit mercury intake, lactating women should heed the fish restrictions mentioned earlier for pregnant women.

Caffeine Caffeine enters breast milk and may make an infant irritable and wakeful. As during pregnancy, caffeine consumption should be moderate—the equivalent of one to two cups of coffee a day. Larger doses of caffeine may interfere with the bioavailability of iron from breast milk and impair the infant's iron status.

IN SUMMARY **The lactating woman needs extra fluid and enough energy and nutrients to produce about 25 ounces of milk a day. Breastfeeding is contraindicated for those with HIV/AIDS. Alcohol, other drugs, smoking, and contaminants may reduce milk production or enter breast milk and impair infant development.**

This chapter has focused on the nutrition needs of the mother during pregnancy and lactation. The next chapter explores the dietary needs of infants, children, and adolescents.

Nutrition Portfolio

The choices a woman makes in preparation for, and in support of, pregnancy and lactation can influence both her health and her infant's development—today and for decades to come.

Go to Diet Analysis Plus and choose one of the days on which you tracked your diet and activity for an entire day. Select the Intake vs. Goals report to help you answer the following questions:

- For women of childbearing age, determine whether you consume at least 400 micrograms of dietary folate equivalents daily.
- For women who are pregnant, evaluate whether you are meeting your nutrition needs and gaining the amount of weight recommended.
- For women who are about to give birth, carefully consider all the advantages of breastfeeding your infant and obtain the needed advice to support you.

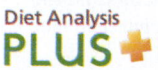

Diet Analysis PLUS **To complete this exercise, go to your Diet Analysis Plus at www.cengage.com/sso.**

Nutrition on the Net

For further study of topics covered in this chapter, log on to **www.cengagebrain.com** and search for ISBN 084006845X.

- Learn more about having a healthy baby and about birth defects from the March of Dimes and the National Center on Birth Defects and Developmental Disabilities: **www.marchofdimes.com** and **www.cdc.gov/ncbddd**
- Learn more about neural tube defects from the Spina Bifida Association of America: **www.spinabifidaassociation.org**
- Search for "birth defect," "pregnancy," "teenage pregnancy," "maternal health," and "breastfeeding" at the U.S. Government health information site: **www.healthfinder.gov**
- Search for "pregnancy" at the American Dietetic Association and the Mayo Clinic: **www.eatright.org** and **www.mayoclinic.com**
- Learn more about the WIC program: **www.fns.usda.gov/fns**
- Visit the American College of Obstetricians and Gynecologists: **www.acog.org**

- Learn more about gestational diabetes from the American Diabetes Association: **www.diabetes.org**
- Learn more about breastfeeding from La Leche League International: **www.lalecheleague.org**
- Obtain prenatal nutrition guidelines from Health Canada: **www.hc-sc.gc.ca**
- Visit the Women's Health Information Center: **www .womenshealth.gov**

References

1. A Report of the CDC/ATSDR Preconception Care Work Group and the Select Panel on Preconception Care prepared by K. Johnson and coauthors, Recommendations to improve preconception health and health care: United States, *Morbidity and Mortality Weekly Report* 55 (2006): 1–23.
2. M. Jokela, M. Elovainio, and M. Kivimäki, Lower fertility associated with obesity and underweight: The U.S. National Longitudinal Survey of Youth, *American Journal of Clinical Nutrition* 88 (2008): 886–893; M. J. Davies, Evidence for effects of weight on reproduction in women, *Reproductive Biomedicine Online* 12 (2006): 552–561.
3. M. Sallmén and coauthors, Reduced fertility among overweight and obese men, *Epidemiology* 17 (2006): 520–523; H. I. Kort and coauthors, Impact of body mass index values on sperm quantity and quality, *Journal of Andrology* 27 (2006): 450–452.
4. R. Pasquali and A. Gambineri, Metabolic effects of obesity on reproduction, *Reproductive Biomedicine Online* 12 (2006): 542–551.
5. B. Eskenazi and coauthors, Antioxidant intake is associated with semen quality in healthy men, *Human Reproduction* 20 (2005): 1006–1012.
6. J. Mendiola and coauthors, A low intake of antioxidant nutrients is associated with poor semen quality in patients attending fertility clinics, *Fertility and Sterility,* http://doi:10.1016/j.fertnstert.2008.10.075; J. E. Chavarro and coauthors, Diet and lifestyle in the prevention of ovulatory disorder infertility, *Obstetrics and Gynecology* 110 (2007): 1050–1058.
7. S. Cordier, Evidence for a role of paternal exposures in developmental toxicity, *Basic and Clinical Pharmacology and Toxicology* 102 (2008): 176–181; S. Sépaniak, T. Forges, and P. Monnier-Barbarino, Cigarette smoking and fertility in women and men, *Gynécologie, Obstétrique and Fertilité* 34 (2006): 945–949.
8. J. C. Cross and L. Mickelson, Nutritional influences on implantation and placental development, *Nutrition Reviews* 64 (2006): S12–S18.
9. R. Padmanabhan, Etiology, pathogenesis and prevention of neural tube defects, *Congenital Anomalies* 46 (2006): 55–67.
10. U.S. Preventive Services Task Force, Folic acid for the prevention of neural tube defects: U.S. Preventive Services Task Force recommendation statement, *Annals of Internal Medicine* 150 (2009): 626–631.
11. K. A. Bol, J. S. Collins, and R. S. Kirby, Survival of infants with neural tube defects in the presence of folic acid fortification, *Pediatrics* 117 (2006): 803–813; T. Tamura and M. F. Picciano, Folate and human reproduction, *American Journal of Clinical Nutrition* 83 (2006): 993–1016; L. B. Bailey and R. J. Berry, Folic acid supplementation and the occurrence of congenital heart defects, orofacial clefts, multiple births, and miscarriage, *American Journal of Clinical Nutrition* 81 (2005): 1213S–1217S.
12. P. Mersereau and coauthors, Spina bifida and anencephaly before and after folic acid mandate: United States, 1995–1996 and 1999–2000, *Morbidity and Mortality Weekly Report* 53 (2004): 362–365; J. Erickson, Folic acid and prevention of spina bifida and anencephaly, *Morbidity and Mortality Weekly Report* 51 (2002): 1–3.
13. R. L. Brent and G. P. Oakley, The folate debate, *Pediatrics* 117 (2006): 1418–1419; J. I. Rader and B. O. Schneeman, Prevalence of neural tube defects, folate status, and folate fortification of enriched cereal-grain products in the United States, *Pediatrics* 117 (2006): 1394–1399.
14. P. D. Gluckman and coauthors, Effect of in utero and early-life conditions on adult health and disease, *New England Journal of Medicine* 359 (2008): 61–73.
15. W. Palinski and coauthors, Developmental programming: Maternal hypercholesterolemia and immunity influence susceptibility to atherosclerosis, *Nutrition Reviews* 65 (2007): S182–S187.
16. F. Lussana and coauthors, Prenatal exposure to the Dutch famine is associated with a preference for fatty foods and a more atherogenic lipid profile, *American Journal of Clinical Nutrition* 88 (2008): 1648–1652; R. C. Painter and coauthors, Early onset of coronary artery disease after prenatal exposure to the Dutch famine, *American Journal of Clinical Nutrition* 84 (2006): 322–327; O. A. Kensara and coauthors, Fetal programming of body composition: Relation between birth weight and body composition measured with dual-energy X-ray absorptiometry and anthropometric methods in older Englishmen, *American Journal of Clinical Nutrition* 82 (2005): 980–987.
17. J. Rotteveel and coauthors, Infant and childhood growth patterns, insulin sensitivity, and blood pressure in prematurely born young adults, *Pediatrics* 122 (2008): 313–321.
18. L. Adair and D. Dahly, Developmental determinants of blood pressure in adults, *Annual Review of Nutrition* 25 (2005): 407–434.
19. J. C. Mathers, Early nutrition: Impact on epigenetics, *Forum of Nutrition* 60 (2007): 42–48.
20. B. Delage and R. H. Dashwood, Dietary manipulation of histone structure and function, *Annual Review of Nutrition* 28 (2008): 347–366; W. Kiess and coauthors, Adipocytes and adipose tissues, *Best Practice and Research. Clinical Endocrinology and Metabolism* 22 (2008): 135–153; W. S. Cutfield and coauthors, Could epigenetics play a role in the developmental origins of health and disease? *Pediatric Research* 61 (2007): 68R–75R; C. Junien and P. Nathanielsz, Report on the IASO Stock Conference 2006: Early and lifelong environmental epigenomic programming of metabolic syndrome, obesity and type II diabetes, *Obesity Review* 8 (2007): 487–502; D. A. Lawlor and coauthors, Epidemiologic evidence for the fetal overnutrition hypothesis: Findings from the Mater-University Study of Pregnancy and its Outcomes, *American Journal of Epidemiology* 165 (2007): 418–424; R. A. Waterland and K. B. Michels, Epigenetic epidemiology of the developmental origins hypothesis, *Annual Review of Nutrition* 27 (2007): 363–388.
21. G. K. Swamy, T. Østbye, and R. Skjærven, Association of preterm birth with long-term survival, reproduction, and next-generation preterm birth, *Journal of the American Medical Association* 299 (2008): 1429–1436.
22. Position of the American Dietetic Association and American Society for Nutrition: Obesity, reproduction, and pregnancy outcomes, *Journal of the American Dietetic Association* 109 (2009): 918–927; P. Brawarsky and coauthors, Pre-pregnancy and pregnancy-related factors and the risk of excessive or inadequate gestational weight gain, *International Journal of Gynaecology and Obestetrics* 91 (2005): 125–131.
23. T. Henriksen, Nutrition and pregnancy outcome, *Nutrition Reviews* 64 (2006): S19–S23; J. C. King, Maternal obesity, metabolism, and pregnancy outcomes, *Annual Review of Nutrition* 26 (2006): 271–291; D. B. Sarwer and coauthors, Pregnancy and obesity: A review and agenda for future research, *Journal of Women's Health* 15 (2006): 720–733.
24. S. Y. Chu and coauthors, Association between obesity during pregnancy and increased use of health care, *New England Journal of Medicine* 358 (2008): 1444–1453.
25. M. A. Maggard and coauthors, Pregnancy and fertility following bariatric surgery: A systematic review, *Journal of the American Medical Association* 300 (2008): 2286–2296.
26. E. A. Nohr and coauthors, Combined associations of prepregnancy body mass index and gestational weight gain with the outcome of pregnancy, *American Journal of Clinical Nutrition* 87 (2008): 1750–1759.
27. K. J. Stothard and coauthors, Maternal overweight and obesity and the risk of congenital anomalies: A systematic review and meta-analysis, *Journal of the American Medical Association* 301 (2009): 636–650.
28. King, 2006.

29. D. K. Waller and coauthors, Prepregnancy obesity as a risk factor for structural birth defects, *Archives of Pediatric and Adolescent Medicine* 161 (2007): 745–750.
30. R. Artal and coauthors, A lifestyle intervention of weight-gain restriction: Diet and exercise in obese women with gestational diabetes mellitus, *Applied Physiology, Nutrition, and Metabolism* 32 (2007): 596–601.
31. J. H. Cohen and H. Kim, Sociodemographic and health characteristics associated with attempting weight loss during pregnancy, *Preventing Chronic Disease* 6 (2009): A07.
32. Institute of Medicine, *Weight Gain during Pregnancy: Reexamining the Guidelines* (Washington, D.C.: National Academies Press, 2009).
33. C. M. Olson, Achieving a healthy weight gain during pregnancy, *Annual Review of Nutrition* 28 (2008): 411–423.
34. B. H. Wrotniak and coauthors, Gestational weight gain and risk of overweight in the offspring at age 7 y in a multicenter, multiethnic cohort study, *American Journal of Clinical Nutrition* 87 (2008): 1818–1824.
35. A. L. Deierlein, A. M. Siega-Riz, and A. Herring, Dietary energy density but not glycemic load is associated with gestational weight gain, *American Journal of Clinical Nutrition* 88 (2008): 693–699.
36. J. L. Baker and coauthors, Breastfeeding reduces postpartum weight retention, *American Journal of Clinical Nutrition* 88 (2008): 1543–1551; N. F. Butte and coauthors, Composition of gestational weight gain impacts maternal fat retention and infant birth weight, *American Journal of Obstetrics and Gynecology* 189 (2003): 1423–1432.
37. D. A. Krummel, Postpartum weight control: A vicious cycle, *Journal of the American Dietetic Association* 107 (2007): 37–40; E. Villamor and S. Cnattingius, Interpregnancy weight change and risk of adverse pregnancy outcomes: A population-based study, *Lancet* 368 (2006): 1164–1170.
38. A. R. Amorium and coauthors, Does excess pregnancy weight gain constitute a major risk for increasing long-term BMI? *Obesity* 15 (2007): 1278–1286.
39. Position of the American Dietetic Association: Nutrition and lifestyle for a healthy pregnancy outcome, *Journal of the American Dietetic Association* 108 (2008): 553–561; U.S. Department of Health and Human Services, *2008 Physical Activity Guidelines for Americans*, www.health.gov/paguidelines/guidelines/chapter7.aspx, accessed June 26, 2009.
40. A. B. Granath, M. S. Hellgren, and R. K. Gunnarsson, Water aerobics reduces sick leave due to low back pain during pregnancy, *Journal of Obstetrics, Gynecology, and Neonatal Nursing* 35 (2006): 465–471; S. A. Smith and Y. Michel, A pilot study on the effects of aquatic exercises on discomforts of pregnancy, *Journal of Obstetrics, Gynecology, and Neonatal Nursing* 35 (2006): 315–323.
41. E. Forsum and M. Löf, Energy metabolism during human pregnancy, *Annual Review of Nutrition* 27 (2007): 277–292; M. Löf and coauthors, Changes in basal metabolic rate during pregnancy in relation to changes in body weight and composition, cardiac output, insulin-like growth factor I, and thyroid hormones and in relation to fetal growth, *American Journal of Clinical Nutrition* 81 (2005): 678–685.
42. S. M. Innis and R. W. Freisen, Essential n-3 fatty acids in pregnant women and early visual acuity maturation in term infants, *American Journal of Clinical Nutrition* 87 (2008): 548–557; M. von Eijsden and coauthors, Maternal n-3, n-6, and *trans* fatty acid profile early in pregnancy and term birth weight: A prospective cohort study, *American Journal of Clinical Nutrition* 87 (2008): 887–895.
43. R. Uauy and A. D. Dangour, Nutrition in brain development and aging: Role of essential fatty acids, *Nutrition Reviews* 64 (2006): S24–S33.
44. Committee on Dietary Reference Intakes, *Dietary Reference Intakes for Thiamin, Riboflavin, Niacin, Vitamin B_6, Folate, Vitamin B_{12}, Pantothenic Acid, Biotin, and Choline* (Washington, D.C.: National Academies Press, 1998), pp. 196–305.
45. T. O. Scholl, Iron status during pregnancy: Setting the stage for mother and infant, *American Journal of Clinical Nutrition* 81 (2005): 1218S–1222S.
46. E. Casanueva and coauthors, Vitamin C supplementation to prevent premature rupture of the chorioamniotic membranes: A randomized trial, *American Journal of Clinical Nutrition* 81 (2005): 859–863.
47. C. M. Donangelo and coauthors, Zinc absorption and kinetics during pregnancy and lactation in Brazilian women, *American Journal of Clinical Nutrition* 82 (2005): 118–124.
48. D. Shah and H. P. S. Sachdev, Zinc deficiency in pregnancy and fetal outcome, *Nutrition Reviews* 64 (2006): 15–30.
49. C. S. Kovacs, Vitamin D in pregnancy and lactation: Maternal, fetal, and neonatal outcomes from human and animal studies, *American Journal of Clinical Nutrition* 88 (2008): 520S–528S; L. M. Bodnar and coauthors, High prevalence of vitamin D insufficiency in black and white pregnant women residing in the northern United States and their neonates, *Journal of Nutrition* 137 (2007): 447–452.
50. K. O. O'Brien and coauthors, Bone calcium turnover during pregnancy and lactation in women with low calcium diets is associated with calcium intake and circulating insulin-like growth factor 1 concentrations, *American Journal of Clinical Nutrition* 83 (2006): 317–323.
51. Position of the American Dietetic Association: Vegetarian diets, *Journal of the American Dietetic Association* 109 (2009): 1266–1282.
52. L. H. Allen, Multiple micronutrients in pregnancy and lactation: An overview, *American Journal of Clinical Nutrition* 81 (2005): 1206S–1212S.
53. J. M. Schneider and coauthors, The use of multiple logistic regression to identify risk factors associated with anemia and iron deficiency in a convenience sample of 12–36-mo-old children from low-income families, *American Journal of Clinical Nutrition* 87 (2008): 614–620.
54. R. J. Kaaja and I. A. Greer, Manifestations of chronic disease during pregnancy, *Journal of the American Medical Association* 294 (2005): 2751–2757.
55. The HAPO Study Cooperative Research Group, Hyperglycemia and adverse pregnancy outcomes, *New England Journal of Medicine* 358 (2008): 1991–2002.
56. J. L. Kitzmiller and coauthors, Managing preexisting diabetes for pregnancy: Summary of evidence and consensus recommendations for care, *Diabetes Care* 31 (2008): 1060–1079; C. Mulholland and coauthors, Comparison of guidelines available in the United States for diagnosis and management of diabetes before, during, and after pregnancy, *Journal of Women's Health* 16 (2007): 790–801.
57. Y. Yogev and G. H. Visser, Obesity, gestational diabetes and pregnancy outcome, *Seminars in Fetal and Neonatal Medicine,* available online October 2008.
58. American Diabetes Association, Diagnosis and classification of diabetes mellitus, *Diabetes Care* 31 (2008): S55–S60.
59. Position of the American Diabetes Association: Nutrition recommendations and interventions for diabetes, *Diabetes Care* 31 (2008): S61–S78.
60. C. A. Crowther and coauthors, Effect of treatment of gestational diabetes mellitus on pregnancy outcomes, *New England Journal of Medicine* 352 (2005): 2477–2486; O. Langer and coauthors, Overweight and obese in gestational diabetes: The impact on pregnancy outcomes, *American Journal of Obstetrics and Gynecology* 192 (2005): 1768–1776.
61. P. E. Marik, Hypertensive disorders of pregnancy, *Postgraduate Medicine* 121 (2009): 69–76.
62. B. E. Vikse and coauthors, Preeclampsia and the risk of end-stage renal disease, *New England Journal of Medicine* 359 (2008): 800–809.
63. C. S. Wu and coauthors, Preeclampsia and risk for epilepsy in offspring, *Pediatrics* 122 (2008): 1072–1078.
64. P. R. Trumbo and K. C. Ellwood, Supplemental calcium and risk reduction of hypertension, pregnancy-induced hypertension, and preeclampsia: An evidence-based review by the U.S. Food and Drug Administration, *Nutrition Reviews* 65 (2007): 78–87; L. Poston and coauthors, Vitamin C and vitamin E in pregnant women at risk for preeclampsia (VIP trial): Randomized placebo-controlled trial, *Lancet* 367 (2006): 1145–1154.
65. C. B. Rudra and coauthors, A prospective analysis of recreational physical activity and preeclampsia risk, *Medicine & Science in Sports & Exercise* 40 (2008): 1581–1588.
66. J. A. Martin and coauthors, Births: Final data for 2006, *National Vital Statistics Reports* 57 (2009): 1–102; E. B. Hamilton and coauthors, Annual summary of vital statistics: 2005, *Pediatrics* 119 (2007): 345–360; S. J. Ventura and coauthors, Recent trends in teenage pregnancy in the United States, 1990–2002, *Health E-stats* (Hyattsville, Md.: National Center for Health Statistics), released December 13, 2006.
67. Quickstats—Birthrates among females aged 15–19 years, by state: United States, 2004, *Morbidity and Mortality Weekly Report* 55 (2007): 1383.

68. X. Chen and coauthors, Paternal age and adverse birth outcomes: Teenager or 40+, who is at risk? *Human Reproduction* 23 (2008): 1290–1296.
69. Alcohol use and pregnancy, *Department of Health and Human Services,* www.cdc.gov/ncbddd, created September 15, 2005.
70. M. J. Rivkin and coauthors, Volumetric MRI study of brain in children with intrauterine exposure to cocaine, alcohol, tobacco, and marijuana, *Pediatrics* 121 (2008): 741–750; H. S. Bada and coauthors, Impact of prenatal cocaine exposure on child behavior problems through school age, *Pediatrics* 119 (2007): e348; B. A. Lewis and coauthors, Prenatal cocaine and tobacco effects on children's language trajectories, *Pediatrics* 120 (2007): e78.
71. G. A. Richardson, L. Goldschmidt, and C. Larkby, Effects of prenatal cocaine exposure on growth: A longitudinal analysis, *Pediatrics* 120 (2007): e1017.
72. J. A. Martin and coauthors, Annual summary of vital statistics: 2008, *Pediatrics* 121 (2008): 788–801.
73. H. Moshhammer and coauthors, Parental smoking and lung function in children: An international study, *American Journal of Respiratory and Critical Care Medicine* 173 (2006): 1255–1263.
74. Rivkin and coauthors, 2008.
75. N. Ribas-Fitó and coauthors, Breastfeeding, exposure to organochlorine compounds, and neurodevelopment in infants, *Pediatrics* 111 (2003): e580–e585.
76. T. I. Halldorsson and coauthors, Is high consumption of fatty fish during pregnancy a risk factor for fetal growth retardation? A study of 44,824 Danish pregnant women, *American Journal of Epidemiology* 166 (2007): 687–696.
77. E. Oken and coauthors, Maternal fish intake during pregnancy, blood mercury levels, and child cognition at age 3 years in a U.S. cohort, *American Journal of Epidemiology* 167 (2008): 1171–1181; J. R. Hibbeln and coauthors, Maternal seafood consumption in pregnancy and neurodevelopmental outcomes in childhood (ALSPAC study): An observational cohort study, *Lancet* 369 (2007): 578–585; D. Mozaffarian and E. B. Rimm, Fish intake, contaminants, and human health: Evaluating the risks and the benefits, *Journal of the American Medical Association* 296 (2006): 1885–1899; Institute of Medicine report brief, *Seafood Choices: Balancing Benefits and Risks,* October 2006.
78. X. Weng, R. Odouli, and D. Li, Maternal caffeine consumption during pregnancy and the risk of miscarriage: A prospective cohort study, *American Journal of Obstetrics and Gynecology* 198 (2008): 279.e1–279.e8; M. L. Browne, Maternal exposure to caffeine and risk of congenital anomalies: A systematic review, *Epidemiology* 17 (2006): 324–331; A. Matijasevich and coauthors, Maternal caffeine consumption and fetal death: A case-control study in Uruguay, *Paediatric and Perinatal Epidemiology* 20 (2006): 100–109; B. H. Bech and coauthors, Coffee and fetal death: A cohort study with prospective data, *American Journal of Epidemiology* 162 (2005): 983–990.
79. CARE Study Group, Maternal caffeine intake during pregnancy and risk of fetal growth restriction: A large prospective observational study, *British Medical Journal* 337 (2008): a2332.
80. Position of the American Dietetic Association: Use of nutritive and nonnutritive sweeteners, *Journal of the American Dietetic Association* 104 (2004): 255–275.
81. Position of the American Dietetic Association, 2008.
82. K. S. Scanlon and coauthors, Breastfeeding trends and updated national health objectives for exclusive breastfeeding: United States, birth years 2000–2004, *Morbidity and Mortality Weekly Report* 56 (2007): 760–763.
83. American Academy of Pediatrics, Breastfeeding and the use of human milk, *Pediatrics* 115 (2005): 496–506; Position of the American Dietetic Association: Promoting and supporting breastfeeding, *Journal of the American Dietetic Association* 105 (2005): 810–818.
84. D. Thulier, Breastfeeding in America: A history of influencing factors, *Journal of Human Lactation* 25 (2009): 85–94.
85. K. A. Bonuck and coauthors, Randomized, controlled trial of a prenatal and postnatal lactation consultant intervention on duration and intensity of breastfeeding up to 12 months, *Pediatrics* 116 (2005): 1413–1426; J. Labarere and coauthors, Efficacy of breastfeeding support provided by trained clinicians during an early, routine, preventive visit: A prospective, randomized, open trial of 226 mother-infant pairs, *Pediatrics* 115 (2005): e139.
86. S. Merten, J. Dratva, and U. Ackermann-Liebrich, Do baby-friendly hospitals influence breastfeeding duration on a national level? *Pediatrics* 116 (2005): e702; A. Merewood and coauthors, Breastfeeding rates in U.S. baby-friendly hospitals: Results of a national survey, *Pediatrics* 116 (2005): 628–634.
87. A. Pisacane and coauthors, A controlled trial of the father's role in breastfeeding promotion, *Pediatrics* 116 (2005): e494.
88. K. M. Rasmussen, Association of maternal obesity before conception with poor lactation performance, *Annual Review of Nutrition* 27 (2007): 103–121; C. A. Lovelady, Is maternal obesity a cause of poor lactation performance? *Nutrition Reviews* 63 (2005): 352–355.
89. Baker and coauthors, 2008.
90. O'Brien and coauthors, 2006.
91. A. Jacknowitz, D. Novillo, and L. Tiehen, Special supplemental nutrition program for women, infants, and children and infant feeding practices, *Pediatrics* 119 (2007): 281–289.
92. F. R. Greer, S. H. Sicherer, A. Wesley Burks, and the Committee on Nutrition and Section on Allergy and Immunology, Effects of early nutritional interventions on the development of atopic disease in infants and children: The role of maternal dietary restriction, breastfeeding, timing of introduction of complementary foods, and hydrolyzed formulas, *Pediatrics* 121 (2008): 183-191.
93. Greer, Sicherer, Wesley Burks, and the Committee on Nutrition and Section on Allergy and Immunology, 2008.
94. P. L. Havens, L. M. Mofenson, and the Committee on Pediatric AIDS, Evaluation and management of the infant exposed to HIV-1 in the United States, *Pediatrics* 123 (2009): 175-187.
95. World Health Organization, *HIV and Infant Feeding,* www.who.int/child_adolescent_health/topics/prevention_care/child/nutrition/hivif/en/, accessed June 1, 2009; M. W. Kline, Early exclusive breastfeeding: Still the cornerstone of child survival, *American Journal of Clinical Nutrition* 89 (2009): 1281–1282.
96. G. E. Gray and H. Saloojee, Breast-feeding, antiretroviral prophylaxis, and HIV, *New England Journal of Medicine* 359 (2008): 189–191; N. I. Kumwenda and coauthors, Extended antiretroviral prophylaxis to reduce breast-milk HIV-1 transmission, *New England Journal of Medicine* 359 (2008): 119–129; W. T. Shearer, Breastfeeding and HIV infection, *Pediatrics* 121 (2008): 1046–1047.
97. R. Lesnewski and L. Prine, Initiating hormonal contraception, *American Family Physician* 74 (2006): 105–112.
98. S. Cnattingius and coauthors, Pregnancy characteristics and maternal risk of breast cancer, *Journal of the American Medical Association* 294 (2005): 2474–2480.
99. L. M. Jansson and coauthors, Methadone maintenance and breastfeeding in the neonatal period, *Pediatrics* 121 (2008): 106–114.
100. J. A. Mennella, L. M. Yourshaw, and L. K. Morgan, Breastfeeding and smoking: Short-term effects on infant feeding and sleep, *Pediatrics* 120 (2007): 497–502.

HIGHLIGHT 14

Fetal Alcohol Syndrome

Copyright © David Young-Wolff/Photo Edit

As Chapter 14 mentions, drinking alcohol during pregnancy endangers the fetus. Alcohol crosses the placenta freely and deprives the developing fetus of both nutrients and oxygen. The damaging effects of alcohol on the developing fetus cover a range of abnormalities referred to as **fetal alcohol spectrum disorder** (see the accompanying glossary).[1] Those at the most severe end of the spectrum are described as having **fetal alcohol syndrome (FAS),** a cluster of physical, mental, and neurobehavioral symptoms that includes:

- Prenatal and postnatal growth retardation
- Impairment of the brain and central nervous system, with consequent mental retardation, poor motor skills and coordination, and hyperactivity
- Abnormalities of the face and skull (see Figure H14-1)
- Increased frequency of major birth defects: cleft palate, heart defects, and defects in ears, eyes, genitals, and urinary system

Those with more severe physical abnormalities have more cognitive limitations.[2] Tragically, the damage evident at birth persists: children with FAS never fully recover.

Each year, as many as 6000 infants are born with FAS because their mothers drank too much alcohol during pregnancy.[3] In addition, some four million infants are born with **prenatal alcohol exposure.** The cluster of mental problems associated with prenatal alcohol exposure is known as **alcohol-related neurodevelopmental disorder (ARND),** and the physical malformations are referred to as **alcohol-related birth defects (ARBD).** Some children with ARBD and ARND have no outward signs; others may be short or have only minor facial abnormalities. They often go undiagnosed even when they develop learning difficulties in the early school years. Mood disorders and problem behaviors, such as aggression, are common. They typically need support and guidance to function and participate in daily activities.[4]

The surgeon general states that pregnant women should abstain from alcohol. Abstinence from alcohol is the best policy for pregnant women both because alcohol consumption during pregnancy has such severe consequences and because FAS can only be prevented—it cannot be treated. Further, because the most severe damage occurs around the time of conception—*before a woman may even realize that she is pregnant*—the warning to abstain includes women who may become pregnant.

Drinking during Pregnancy

As mentioned in Chapter 14, 1 out of 12 pregnant women drinks alcohol at some time during her pregnancy; 1 out of 30 uses alcohol frequently and admits to binge drinking.[5] When a woman drinks during pregnancy, she causes damage in two ways: directly, by intoxication, and indirectly, by malnutrition. Prior to the complete formation of the placenta (approximately 12 weeks), alcohol diffuses directly into the tissues of the developing embryo, causing incredible damage. (Review Figure 14-4 on p. 460 and note that the critical periods for most tissues occur during embryonic development.) Alcohol interferes with the orderly development of tissues during their critical periods, reducing the number of cells and damaging those that are produced. The damage of alcohol toxicity during brain development is apparent in its reduced size and impaired function.

When alcohol crosses the placenta, fetal blood alcohol rises until it reaches equilibrium with maternal blood alcohol. The mother may not even appear drunk, but the fetus may be poisoned. The

GLOSSARY

alcohol-related birth defects (ARBD): malformations in the skeletal and organ systems (heart, kidneys, eyes, ears) associated with prenatal alcohol exposure.

alcohol-related neurodevelopmental disorder (ARND): abnormalities in the central nervous system and cognitive development associated with prenatal alcohol exposure.

fetal alcohol spectrum disorder: a range of physical, behavioral, and cognitive abnormalities caused by prenatal alcohol exposure.

fetal alcohol syndrome (FAS): a cluster of physical, behavioral, and cognitive abnormalities associated with prenatal alcohol exposure, including facial malformations, growth retardation, and central nervous disorders.

prenatal alcohol exposure: subjecting a fetus to a pattern of excessive alcohol intake characterized by substantial regular use or heavy episodic drinking.

Note: See Highlight 7 for other alcohol-related terms and information.

fetus's body is small, its detoxification system is immature, and alcohol remains in fetal blood long after it has disappeared from maternal blood.

A pregnant woman harms her unborn child not only by consuming alcohol but also by not consuming food. This combination enhances the likelihood of malnutrition and a poorly developed infant. It is important to realize, however, that malnutrition is not the cause of FAS. It is true that mothers of FAS children often have unbalanced diets and nutrient deficiencies. It is also true that nutrient deficiencies may exacerbate the clinical signs seen in these children, but it is the *alcohol* that causes the damage.[6] An adequate diet alone will not prevent FAS if alcohol abuse continues.

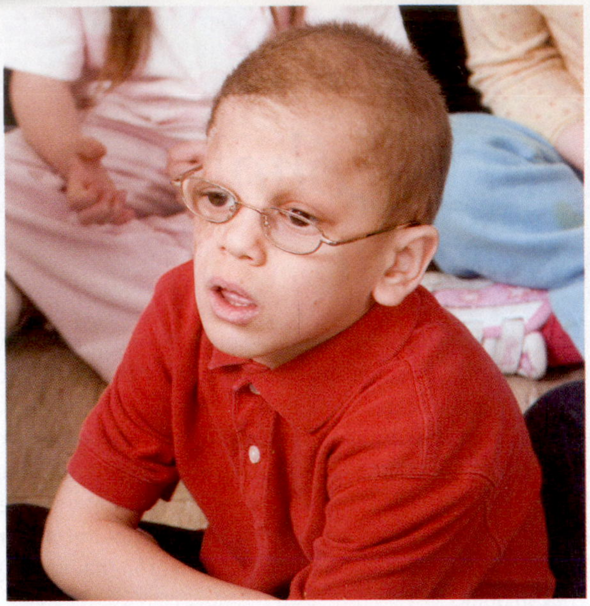

© Ellen B. Senisi/The Image Works

Characteristic facial features may diminish with time, but children with FAS typically continue to be short and underweight for their age.

How Much Is Too Much?

A pregnant woman need not have an alcohol-abuse problem to give birth to a baby with FAS. She need only drink in excess of her liver's capacity to detoxify alcohol. Even one drink a day threatens neurological development and behaviors. Four drinks a day dramatically increase the risk of having an infant with physical malformations.

FIGURE H14-1 Typical Facial Characteristics of FAS

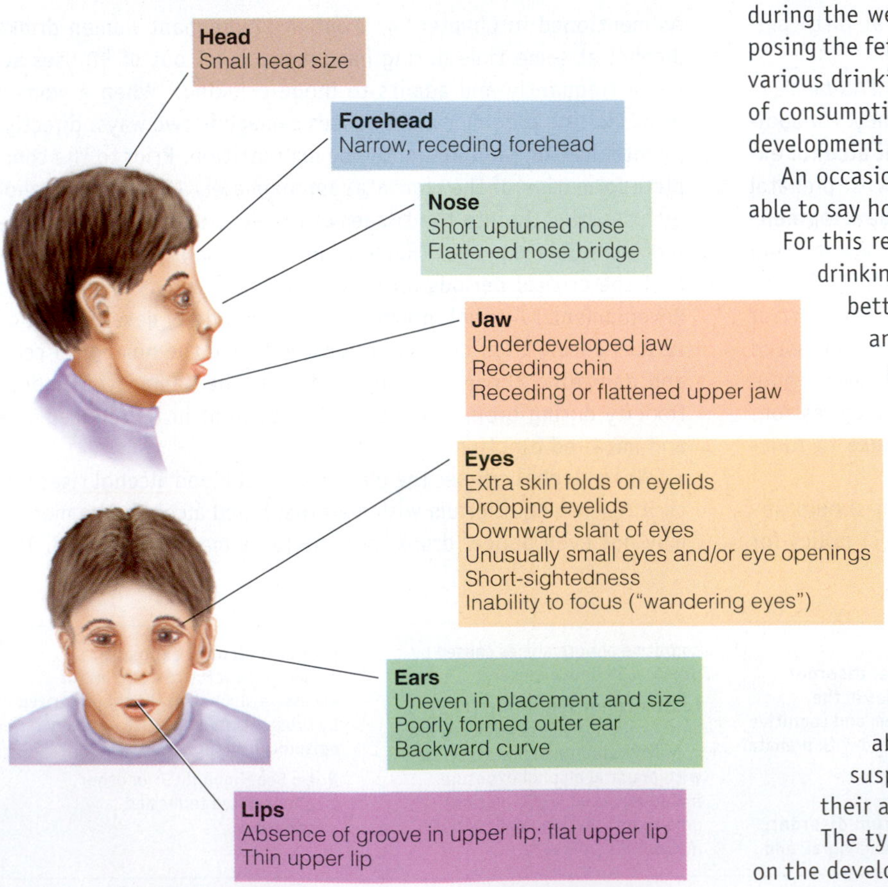

In addition to total alcohol intake, drinking patterns play an important role. Most FAS studies report their findings in terms of average intake per day, but people usually drink more heavily on some days than on others. For example, a woman who drinks an *average* of 1 ounce of alcohol (2 drinks) a day may not drink at all during the week, but then have 10 drinks on Saturday night, exposing the fetus to extremely toxic quantities of alcohol. Whether various drinking patterns incur damage depends on the frequency of consumption, the quantity consumed, and the stage of fetal development at the time of each drinking episode.

An occasional drink may be innocuous, but researchers are unable to say how much alcohol is safe to consume during pregnancy. For this reason, health-care professionals urge women to stop drinking alcohol as soon as they realize they are pregnant, or better, as soon as they *plan* to become pregnant. Why take any risk? The only sure way to protect an infant from alcohol damage is for the mother to abstain.[7]

When Is the Damage Done?

The first month or two of pregnancy is a critical period of fetal development. Because pregnancy usually cannot be confirmed before five to six weeks, a woman may not even realize she is pregnant during that critical time. Therefore, it is advisable for women who are trying to conceive, or who suspect they might be pregnant, to abstain or curtail their alcohol intakes to ensure a healthy start.

The type of abnormality observed in an FAS infant depends on the developmental events occurring at the times of alcohol ex-

Betty Udesen/Newscom

Children born with FAS must live with the long-term consequences of prenatal brain damage.

Matthew Farruggio

All containers of beer, wine, and liquor warn women not to drink alcoholic beverages during pregnancy because of the risk of birth defects.

posure. During the first trimester, developing organs such as the brain, heart, and kidneys may be malformed. During the second trimester, the risk of spontaneous abortion increases. During the third trimester, body and brain growth may be retarded.

Male alcohol ingestion may also affect fertility and fetal development. Animal studies have found smaller litter sizes, lower birthweights, reduced survival rates, and impaired learning ability in the offspring of males consuming alcohol prior to conception. An association between paternal alcohol intake one month prior to conception and low infant birthweight is also apparent in human beings. (Paternal alcohol intake was defined as an average of two or more drinks daily or at least five drinks on one occasion.) This relationship was independent of either parent's smoking and of the mother's use of alcohol, caffeine, or other drugs.

In view of the damage caused by FAS, prevention efforts focus on educating women not to drink during pregnancy. Everyone should know of the potential dangers. Women who drink alcohol and who are sexually active may benefit from counseling and effective contraception to prevent pregnancy. Almost half of all pregnancies are unintended, with many conceived during a binge-drinking episode.

Public service announcements and alcohol beverage warning labels help to raise awareness. Everyone should hear the message loud and clear: don't drink alcohol prior to conception or during pregnancy.

Nutrition on the Net

For further study of topics covered in this highlight, log on to **www.cengagebrain.com** and search for ISBN 084006845X.

- Visit the National Organization on Fetal Alcohol Syndrome: **www.nofas.org**
- Search for "fetal alcohol syndrome" at the U.S. Government health information site: **www.healthfinder.gov**
- Request information on fetal alcohol syndrome from the National Clearinghouse for Alcohol and Drug Information: **ncadi.samhsa.gov**
- Request information on drinking during pregnancy from the National Institute on Alcohol Abuse and Alcoholism: **www.niaaa.nih.gov**
- Gather facts on fetal alcohol syndrome from the March of Dimes: **www.marchofdimes.com**

References

1. H. E. Hoyme and coauthors, A practical clinical approach to diagnosis of fetal alcohol spectrum disorders: Clarification of the 1996 Institute of Medicine Criteria, *Pediatrics* 115 (2005): 39–47.
2. N. Ervalahti and coauthors, Relationship between dysmorphic features and general cognitive function in children with fetal alcohol spectrum disorders, *American Journal of Medical Genetics* 143A (2007): 2916–2923.
3. Fetal Alcohol Spectrum Disorders, *Department of Health and Human Services*, www.cdc.gov/ncbddd, created September 9, 2005; Guidelines for identifying and referring persons with fetal alcohol syndrome, *Morbidity and Mortality Weekly Report* 54 (2005): 1–10.
4. T. Jirikowic, D. Kartin, and H. C. Olsen, Children with fetal alcohol spectrum disorders: A descriptive profile of adaptive function, *Canadian Journal of Occupational Therapy* 75 (2008): 238–248.
5. Alcohol use and pregnancy, *Department of Health and Human Services*, www.cdc.gov/ncbddd, created September 15, 2005.
6. R. C. Carter and coauthors, Fetal alcohol exposure, iron-deficiency anemia, and infant growth, *Pediatrics* 120 (2007): 559–567.
7. R. A. S. Mukherjee and coauthors, Low level alcohol consumption and the fetus, *British Medical Journal* 330 (2005): 375–376.

CHAPTER

15

Siede Preis/Getty Images

Nutrition in Your Life

Much of this book has focused on you—your food choices and how they might affect your health. This chapter shifts the focus from you the recipient to you the caregiver. One day (if not already), children will depend on you to feed them well and teach them wisely. The responsibility of nourishing children can seem overwhelming at times, but the job is fairly simple. Offer children a variety of nutritious foods to support their growth, and teach them how to make healthy food and activity choices. Presenting foods in a relaxed and supportive environment nourishes both physical and emotional well-being.

Throughout this chapter, the CourseMate icon indicates an opportunity for online self-study, linking you to activities to increase your understanding of chapter concepts. **www.cengagebrain.com** (search for ISBN 084006845X)

Life Cycle Nutrition: Infancy, Childhood, and Adolescence

CHAPTER OUTLINE

The first year of life (infancy) is a time of phenomenal growth and development. After the first year, a child continues to grow and change, but more slowly. Still, the cumulative effects over the next decade are remarkable. Then, as the child enters the teen years, the pace toward adulthood accelerates dramatically. This chapter examines the special nutrient needs of infants, children, and adolescents.

Nutrition during Infancy

Initially, the infant drinks only breast milk or formula but later begins to eat some foods, as appropriate. Common sense in the selection of infant foods along with a nurturing, relaxed environment support an infant's health and well-being.

Energy and Nutrient Needs

An infant grows fast during the first year, as Figure 15-1 shows. Growth directly reflects nutrient intake and is an important parameter in assessing the nutrition status of infants and children. Health-care professionals measure the heights and weights of infants and children at intervals and compare the measurements with standard growth curves for gender and age and with previous measures of each child (see the "How To" on the next page).

Energy Intake and Activity A healthy infant's birthweight doubles by about 5 months of age and triples by 1 year, typically reaching 20 to 25 pounds. The infant's length changes more slowly than weight, increasing about 10 inches from birth to 1 year. By the end of the first year, infant growth slows considerably; during the second year, an infant typically gains less than 10 pounds and grows about 5 inches in length.

Not only do infants grow rapidly, but their energy requirement is remarkably high—about twice that of an adult, based on body weight. A newborn baby requires about 450 kcalories per day, whereas most adults require about 2000 kcalories per day. In terms of body weight, the difference is remarkable. Infants require

FIGURE 15-1 Weight Gain of Infants in Their First Five Years of Life

In the first year, an infant's birthweight may triple, but over the following several years, the rate of weight gain gradually diminishes.

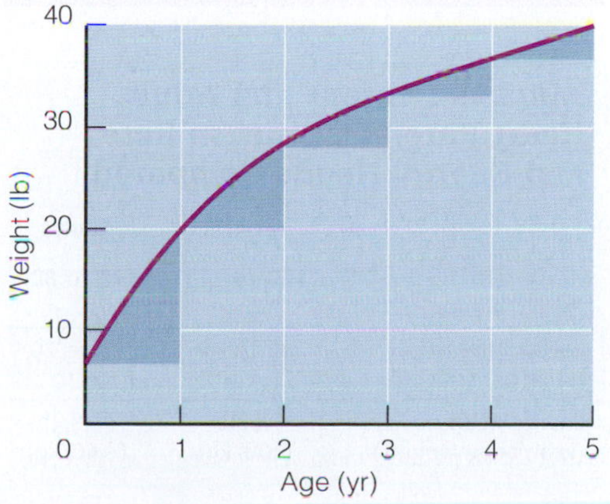

HOW TO Plot Measures on a Growth Chart

You can assess the growth of infants and children by plotting their measurements on a percentile graph. Percentile graphs divide the measures of a population into 100 equal divisions so that half of the population falls at or above the 50th percentile and half falls below. Using percentiles allows for comparisons among people of the same age and gender.

To plot measures on a growth chart, follow these steps:

- Select the appropriate chart based on age and gender. For this example, use the accompanying chart, which gives percentiles for weight for girls from birth to 36 months. (Appendix E provides other growth charts for both boys and girls of various ages.)
- Locate the infant's age along the horizontal axis at the bottom of the chart (in this example, 6 months).
- Locate the infant's weight in pounds or kilograms along the vertical axis of the chart (in this example, 17 pounds or 7.7 kilograms).
- Mark the chart where the age and weight lines intersect (shown here with a red dot), and follow the curved line to find the percentile.

This 6-month-old infant is at the 75th percentile. Her pediatrician will weigh her again over the next few months and expect the growth curve to follow the same percentile throughout the first year. In general, dramatic changes or measures much above the 80th percentile or much below the 10th percentile may be cause for concern.

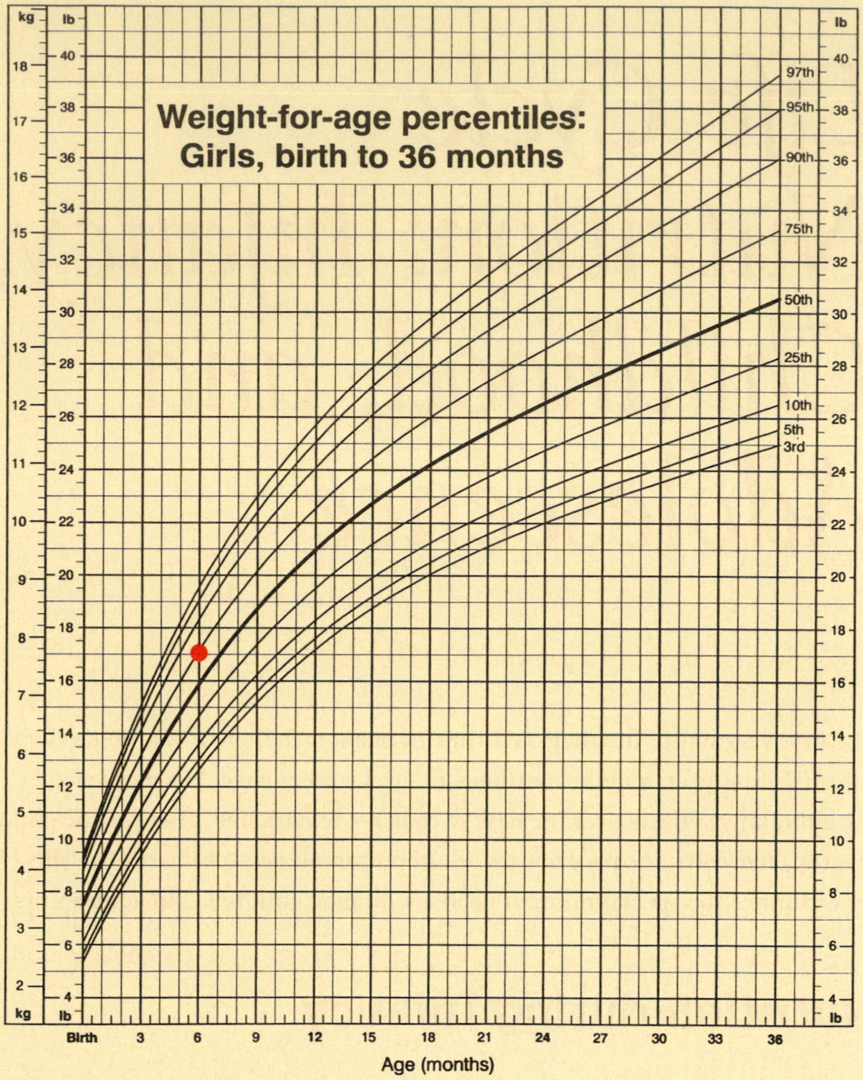

SOURCE: Developed by the National Center for Health Statistics in collaboration with the National Center for Chronic Disease Prevention and Health Promotion (2000).

For additional practice, log on to www.cengagebrain.com and search for ISBN 084006845X.

TRY IT Determine the percentile for a 12-month-old girl who weighs 21 pounds.

TABLE 15-1 Infant and Adult Heart Rate, Respiration Rate, and Energy Needs Compared

	Infants	Adults
Heart rate (beats/minute)	120 to 140	70 to 80
Respiration rate (breaths/minute)	20 to 40	15 to 20
Energy needs (kcal/body weight)	45/lb (100/kg)	<18/lb (<40/kg)

about 100 kcalories per kilogram of body weight per day, whereas most adults need fewer than 40 (see Table 15-1). If an infant's energy needs were applied to an adult, a 170-pound adult would require more than 7000 kcalories a day. After 6 months, the infant's energy needs decline as the growth rate slows, but some of the energy saved by slower growth is spent in increased activity.

Energy Nutrients Recommendations for the energy nutrients—carbohydrate, fat, and protein—during the first six months of life are based on the average intakes of healthy, full-term infants fed breast milk.[1] During the second six months of life, recommendations reflect typical intakes from solid foods as well as breast milk.

As Chapter 4 discusses, carbohydrates provide energy to all the cells of the body, especially those in the brain, which depend primarily on glucose to fuel activities. Relative to the size of the body, ♦ the size of an infant's brain is greater than that of an adult's. Thus, an infant's brain uses *relatively* more glucose—about 60 percent of the day's total energy intake.[2]

♦ An infant's brain weight is about 12% of its body weight, whereas an adult's brain weight is about 2%.

Fat provides most of the energy in breast milk and standard infant formula. Its high energy density supports the rapid growth of early infancy.

No single nutrient is more essential to growth than protein. All of the body's cells and most of its fluids contain protein; it is the basic building material of the body's tissues. Chapter 6 details the problems inadequate protein can cause. Excess dietary protein can cause problems, too, especially in a small infant. Too much protein stresses the liver and kidneys, which have to metabolize and excrete the excess nitrogen. Signs of protein overload include acidosis, dehydration, diarrhea, elevated blood ammonia, elevated blood urea, and fever. Such problems are not common, but they have been observed in infants fed inappropriate foods, such as fat-free milk or concentrated formula.

Vitamins and Minerals An infant's needs for most nutrients, in proportion to body weight, are more than double those of an adult. Figure 15-2 illustrates this

FIGURE 15-2 Recommended Intakes of an Infant and an Adult Compared on the Basis of Body Weight

Because infants are small, they need smaller total amounts of the nutrients than adults do, but when comparisons are based on body weight, infants need more than twice as much of many nutrients. Infants use large amounts of energy and nutrients, in proportion to their body size, to keep all their metabolic processes going.

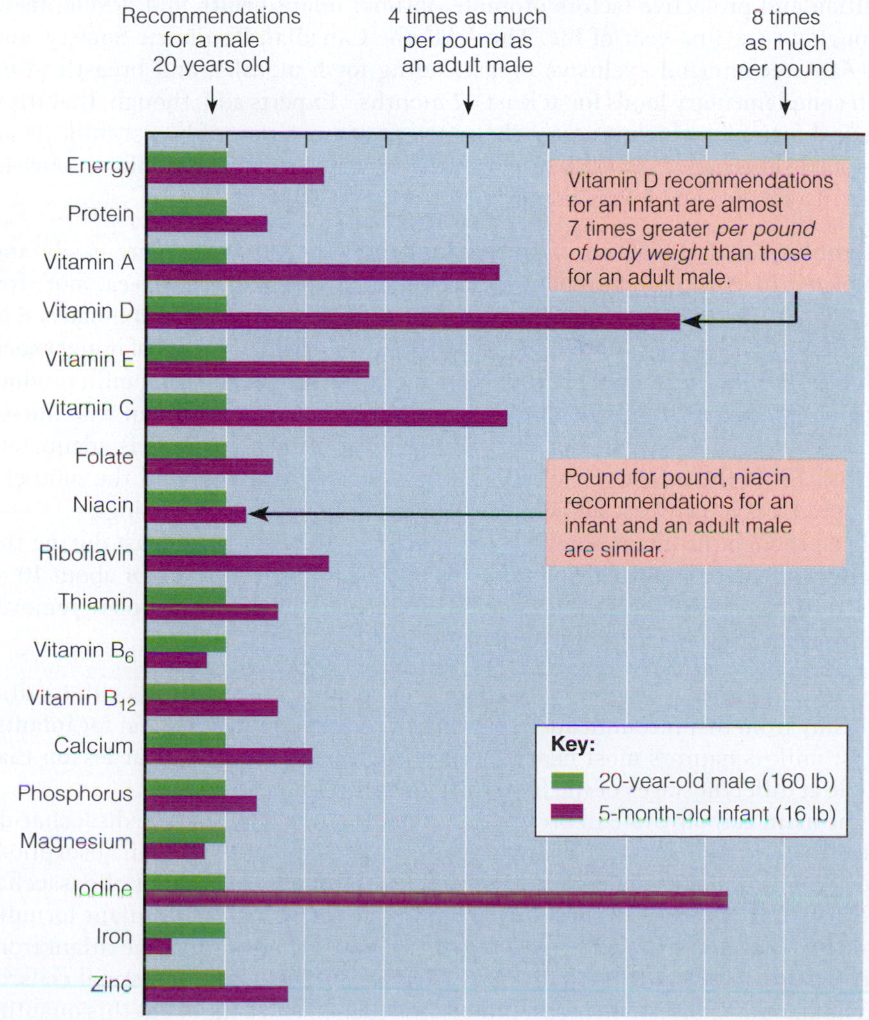

David Lees, Getty Images

After 6 months, energy saved by slower growth is spent in increased activity.

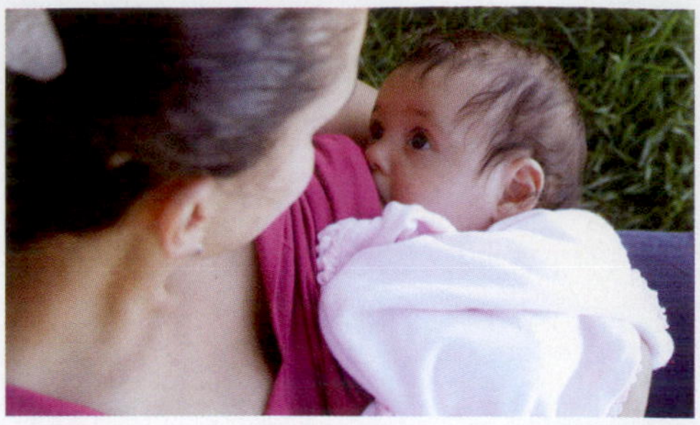

Women are encouraged to breastfeed whenever possible because breast milk offers infants many nutrient and health advantages.

by comparing a 5-month-old infant's needs per unit of body weight with those of an adult man. Some of the differences are extraordinary. Infant recommendations are based on the average amount of nutrients consumed by thriving infants breastfed by well-nourished mothers.

Water One of the most essential nutrients for infants, as for everyone, is water. The younger the infant, the greater the percentage of body weight is water. During early infancy, breast milk or infant formula normally provides enough water to replace fluid losses in a healthy infant. If the environmental temperature is extremely high, however, infants need supplemental water.[3] Because much of the fluid in an infant's body is located *outside* the cells—between the cells and in the blood vessels—rapid fluid losses and the resulting dehydration can be life-threatening. Conditions that cause rapid fluid loss, such as diarrhea or vomiting, require treatment with an electrolyte solution designed for infants.

Breast Milk

In the United States and Canada, the two dietary practices that have the most significant effect on an infant's nutrition are the milk the infant receives and the age at which solid foods are introduced. A later section discusses the introduction of solid foods, but as to the milk, both the American Academy of Pediatrics (AAP) and the Canadian Paediatric Society strongly recommend breastfeeding for healthy full-term infants, except where specific contraindications exist. The American Dietetic Association (ADA) also advocates breastfeeding for the nutritional health it confers on the infant as well as for the many other benefits it provides both infant and mother (review Table 14-4 on p. 480).[4]

Breast milk excels as a source of nutrients for infants. Its unique nutrient composition and protective factors promote optimal infant health and development throughout the first year of life. The AAP, the Canadian Paediatric Society, and the ADA recommend exclusive breastfeeding for 6 months, and breastfeeding with complementary foods for at least 12 months.[5] Experts add, though, that iron-fortified formula, which imitates the nutrient composition of breast milk, is an acceptable alternative. After all, the primary goal is to provide the infant nourishment in a relaxed and loving environment. ♦

♦ Chapter 14 discusses breastfeeding, breastfeeding support, reasons why some women choose not to breastfeed, and contraindications to breastfeeding.

Frequency and Duration of Breastfeeding Breast milk is more easily and completely digested than formula, so breastfed infants usually need to eat more frequently than formula-fed infants do. During the first few weeks, approximately 8 to 12 feedings a day, on demand, as soon as the infant shows early signs of hunger such as increased alertness, activity, or suckling motions, promote optimal milk production and infant growth.[6] Crying is a late indicator of hunger. An infant who nurses every two to three hours and sleeps contentedly between feedings is adequately nourished. As the infant gets older, stomach capacity enlarges and the mother's milk production increases, allowing for longer intervals between feedings.

Even though the infant obtains about half the milk from the breast during the first two or three minutes of sucking, breastfeeding is encouraged for about 10 to 15 minutes on each breast. The infant's sucking, as well as the complete removal of milk from the breast, stimulates lactation.

FIGURE 15-3 Percentages of Energy-Yielding Nutrients in Breast Milk and in Recommended Adult Diets

The proportions of energy-yielding nutrients in human breast milk differ from those recommended for adults.[a]

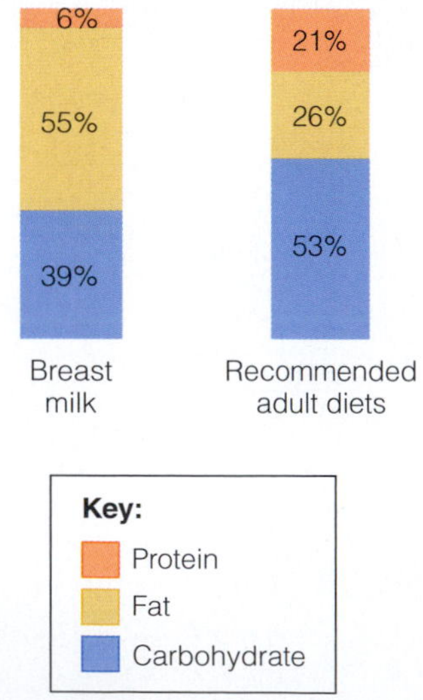

[a]The values listed for adults represent approximate midpoints of the acceptable ranges for protein (10 to 35 percent), fat (20 to 35 percent), and carbohydrate (45 to 65 percent).

Energy Nutrients The energy-nutrient composition of breast milk differs dramatically from that recommended for adult diets (see Figure 15-3). Yet for infants, breast milk is nature's most nearly perfect food, providing the clear lesson that people at different stages of life have different nutrient needs.

The main carbohydrate in breast milk (and infant formula) is the disaccharide lactose. In addition to being easily digested, lactose enhances calcium absorption. The carbohydrate component of breast milk also contains abundant oligosaccharides, which are present only in trace amounts in cow's milk and infant formula made from cow's milk.[7] Human milk oligosaccharides help protect the infant from infection by preventing the binding of pathogens to the infant's intestinal cells.[8]

The amount of protein in breast milk is less than in cow's milk, but this quantity is actually beneficial because it places less stress on the infant's immature kidneys to excrete the major end product of protein metabolism, urea. Much of the protein in breast milk is **alpha-lactalbumin**, which is efficiently digested and absorbed.

As for the lipids, breast milk contains a generous proportion of the essential fatty acids linoleic acid and linolenic acid, as well as their longer-chain derivatives arachidonic acid and DHA (docosahexaenoic acid). In the past, infant formula provided only linoleic acid and linolenic acid, but now arachidonic acid and DHA are also added. Infants can make arachidonic acid and DHA from linoleic acid and linolenic acid, respectively, but some infants may need more than they can make.

As Chapter 5 mentions, DHA is the most abundant fatty acid in the brain and is also present in the retina of the eye. DHA accumulation in the brain is greatest during fetal development and early infancy.[9] Research has focused on the mental and visual development of breastfed infants and infants fed standard formula with and without DHA added.[10] One group of researchers found that infants fed formula fortified with DHA had sharper vision at 1 year of age than those who were fed standard formula.[11] Most studies, however, show no beneficial effect of DHA supplementation of formula for term infants.[12] Adding DHA to standard infant formulas has no adverse effects, however, and most standard formulas are currently fortified with both DHA and arachidonic acid.

Vitamins With the exception of vitamin D, the vitamins in breast milk are ample to support infant growth. The vitamin D in breast milk is low, and vitamin D deficiency impairs bone mineralization. Vitamin D deficiency is most likely in infants who are not exposed to sunlight daily, have darkly pigmented skin, and receive breast milk without vitamin D supplementation.[13] Reports of infants in the United States developing the vitamin D–deficiency disease rickets and recommendations by the AAP to keep infants under 6 months of age out of direct sunlight prompted revisions in vitamin D guidelines. The AAP currently recommends a vitamin D supplement for all infants who are breastfed exclusively, and for any infants who do not receive at least one liter (1000 milliliters, roughly one quart, or 32 ounces) of vitamin D–fortified formula daily.[14]

Minerals The calcium content of breast milk is ideal for infant bone growth, and the calcium is well absorbed. Breast milk contains relatively small amounts of iron, but the iron has a high bioavailability. Zinc also has a high bioavailability, thanks to the presence of a zinc-binding protein. Breast milk is low in sodium, another benefit for immature kidneys. Fluoride promotes the development of strong teeth, but breast milk is not a good source.

Supplements Pediatricians may routinely prescribe liquid supplements containing vitamin D, iron, and fluoride. Table 15-2 offers a schedule of supplements that are recommended during infancy. In addition, the AAP recommends giving a single dose of vitamin K to infants at birth to protect them from bleeding to death. (See Chapter 11 for a description of vitamin K's role in blood clotting.)

TABLE 15-2 Supplements for Full-Term Infants

	Vitamin D[a]	Iron[b]	Fluoride[c]
Breastfed infants			
Birth to 6 months of age	✓		
6 months to 1 year	✓	✓	✓
Formula-fed infants			
Birth to 6 months of age			
6 months to 1 year		✓	✓

[a]Vitamin D supplements are recommended for all infants who are exclusively breastfed and for any infants who do not receive at least 1 liter (1000 milliliters) or 1 quart (32 ounces) of vitamin D–fortified formula per day.
[b]All infants 6 months of age need additional iron, preferably in the form of iron-fortified infant cereal and/or infant meats. Formula-fed infants need iron-fortified infant formula.
[c]At 6 months of age, breastfed infants and formula-fed infants who receive ready-to-use formulas (these are prepared with water low in fluoride) or formula mixed with water that contains little or no fluoride (less than 0.3 ppm) need supplements.
SOURCE: Adapted from Committee on Nutrition, American Academy of Pediatrics, *Pediatric Nutrition Handbook*, 6th ed., ed. R. E. Kleinman (Elk Grove Village, Ill.: American Academy of Pediatrics, 2009).

Immunological Protection In addition to its nutritional benefits, breast milk offers immunological protection. Not only is breast milk sterile, but it actively fights disease and protects infants from illnesses.[15] Such protection is most valuable during the first year, when the infant's immune system is not fully prepared to mount a response against infections.

During the first two or three days after delivery, the breasts produce **colostrum**, a premilk substance containing mostly serum with antibodies and white blood cells. Colostrum (like breast milk) helps protect the newborn from infections against which the mother has developed immunity. The maternal antibodies in the breast milk inactivate disease-causing bacteria within the infant's digestive

alpha-lactalbumin (lact-AL-byoo-min): a major protein in human breast milk, as opposed to **casein** (CAY-seen), a major protein in cow's milk.

colostrum (ko-LAHS-trum): a milklike secretion from the breast, present during the first few days after delivery before milk appears; rich in protective factors.

tract before they can start infections.[16] This explains, in part, why breastfed infants have fewer intestinal infections than formula-fed infants.

In addition to antibodies, colostrum and breast milk provide other powerful agents ♦ that help to fight against bacterial infection. Among them are the oligosaccharides, described earlier, that prevent pathogens from binding to intestinal cells. Also present are **bifidus factors**, which favor the growth of the "friendly" bacterium *Lactobacillus bifidus* in the infant's digestive tract, so that other, harmful bacteria cannot become established. An iron-binding protein in breast milk, **lactoferrin**, keeps bacteria from getting the iron they need to grow, helps absorb iron into the infant's intestinal cells, and kills some bacteria directly.[17] The protein **lactadherin** in breast milk binds to, and inhibits replication of, the virus that causes most infant diarrhea.[18] Breastfeeding also protects against other common illnesses of infancy such as middle ear infection and respiratory illness.[19] In addition, a growth factor that is present in breast milk stimulates the development and maintenance of the infant's digestive tract and its protective factors. Several breast milk enzymes such as lipase also help protect the infant against infection. Clearly, breast milk is a very special substance.

♦ Protective factors in breast milk:

- Antibodies
- Oligosaccharides
- Bifidus factors
- Lactoferrin
- Lactadherin
- Growth factor
- Lipase enzyme

Allergy and Disease Protection In addition to protection against infection, breast milk may offer protection against the development of allergies.[20] Compared with formula-fed infants, breastfed infants have a lower incidence of allergic reactions, such as recurrent wheezing and skin rashes.[21] This protection is especially noticeable among infants with a family history of allergies.[22] Similarly, breast milk may offer protection against the development of cardiovascular disease. Compared with formula-fed infants, breastfed infants have lower blood pressure and lower blood cholesterol as adults.[23]

Other Potential Benefits Breastfeeding may offer some protection against excessive weight gain later, although findings are inconsistent.[24] One extensive review suggests that initial breastfeeding protects against obesity in later life.[25] Another study confirms this finding and adds that the longer the duration of breastfeeding, the lower the risk of overweight in childhood.[26] Still another review reports a protective effect, a protective effect only in certain groups, or no effect.[27] Researchers note that many other factors—socioeconomic status, other infant and child feeding practices, and especially the mother's weight—strongly predict a child's body weight.

Many studies suggest a beneficial effect of breastfeeding on intelligence, but when subjected to strict standards of methodology (for example, large sample size and appropriate intelligence testing), the evidence is less convincing.[28] Nevertheless, the possibility that breastfeeding may positively affect later intelligence is intriguing. It may be that some specific component of breast milk, such as DHA, stimulates brain development or that certain factors associated with the feeding process itself promote intellect.[29] Most likely, a combination of factors is involved. More large, well-controlled studies are needed to confirm the effects, if any, of breastfeeding on later intelligence.

Breast Milk Banks Similar to blood banks that collect blood from individuals to give to others in need, **breast milk banks** receive milk from lactating women who have an abundant supply to give to infants whose own mothers' milk is unavailable or insufficient. The women who donate breast milk are carefully screened to exclude those who smoke cigarettes, use illicit drugs, take medications (including high doses of dietary supplements), drink more than two alcoholic beverages a day, or have communicable diseases. The breast milk from several donors is pooled to ensure an even distribution of all components, pasteurized to destroy bacteria, checked for contamination, and frozen before being shipped overnight to hospitals, where it is dispensed by physician prescription. In the absence of a mother's own breast milk, donor milk may be the life-saving solution for fragile infants, most notably those with very low birthweight or unusual medical conditions.[30]

bifidus (BIFF-id-us, by-FEED-us) **factors:** factors in colostrum and breast milk that favor the growth of the "friendly" bacterium *Lactobacillus* (lack-toh-ba-SILL-us) *bifidus* in the infant's intestinal tract, so that other, less desirable intestinal inhabitants will not flourish.

lactoferrin (lack-toh-FERR-in): a protein in breast milk that binds iron and keeps it from supporting the growth of the infant's intestinal bacteria.

lactadherin (lack-tad-HAIR-in): a protein in breast milk that attacks diarrhea-causing viruses.

breast milk bank: a service that collects, screens, processes, and distributes donated human milk.

Infant Formula

A woman who breastfeeds for a year can **wean** her infant to cow's milk, bypassing the need for infant formula. However, a woman who decides to feed her infant formula from birth, to wean to formula after less than a year of breastfeeding, or to substitute formula for breastfeeding on occasion must select an appropriate infant formula and learn to prepare it. Cow's milk is inappropriate.

Infant Formula Composition Formula manufacturers attempt to copy the nutrient composition of breast milk as closely as possible. Figure 15-4 illustrates the energy-nutrient balance of both. The AAP recommends that all formula-fed infants receive iron-fortified infant formulas.[31] The increasing use of iron-fortified formulas during the past few decades is a major reason for the decline in iron-deficiency anemia among infants in the United States.

FIGURE 15-4 Percentages of Energy-Yielding Nutrients in Breast Milk, Infant Formula, and Cow's Milk

The average proportions of energy-yielding nutrients in human breast milk and formula differ slightly. In contrast, cow's milk provides too much protein and too little carbohydrate.

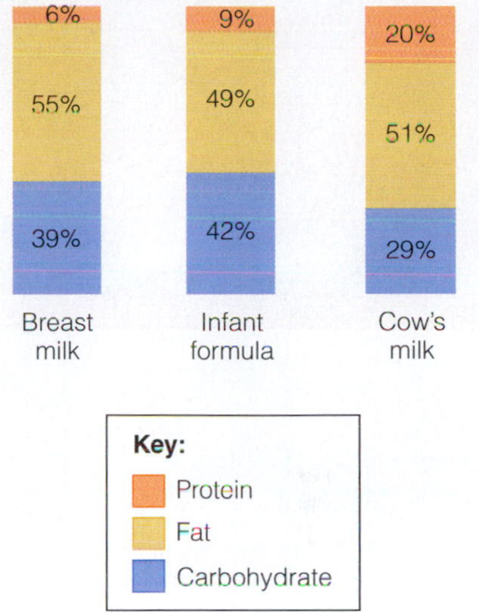

Risks of Formula Feeding Infant formulas contain no protective antibodies for infants, but in general, vaccinations, purified water, and clean environments in developed countries help protect infants from infections. Formulas can be prepared safely by following the rules of proper food handling and by using water that is free of contamination. Of particular concern is lead-contaminated water, a major source of lead poisoning in infants. Because the first water drawn from the tap each day is highest in lead, a person living in a house with old, lead-soldered plumbing should let the water run a few minutes before drinking or using it to prepare formula or food.

In developing countries and in poor areas of the United States, formula may be unavailable, prepared with contaminated water, or overdiluted in an attempt to save money. Contaminated formulas often cause infections, leading to diarrhea, dehydration, and malabsorption. Without sterilization and refrigeration, formula is an ideal breeding ground for bacteria. Whenever such risks are present, breastfeeding can be a life-saving option: breast milk is sterile, and its antibodies enhance an infant's resistance to infections.

Infant Formula Standards National and international standards have been set for the nutrient contents of infant formulas. In the United States, the standard developed by the AAP reflects "human milk taken from well-nourished mothers during the first or second month of lactation, when the infant's growth rate is high." The Food and Drug Administration (FDA) mandates the safety and nutritional quality of infant formulas. Formulas meeting these standards have similar nutrient compositions. Small differences among formulas are sometimes confusing, but they are usually unimportant.

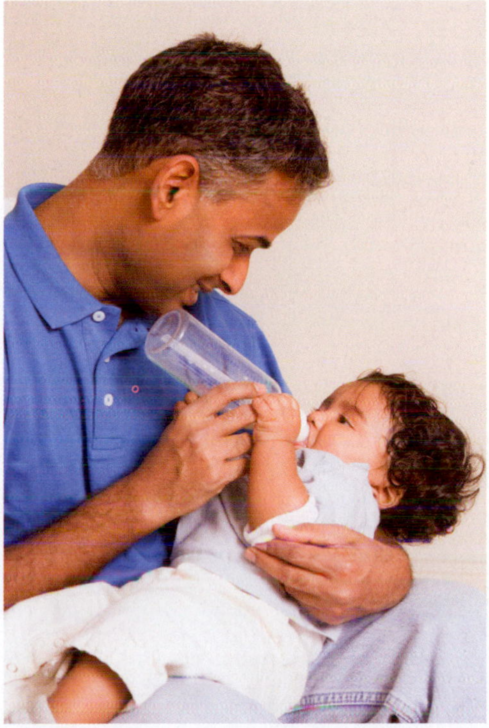

The infant thrives on infant formula offered with affection.

Special Formulas Standard formulas are inappropriate for some infants. Special formulas have been designed to meet the dietary needs of infants with specific conditions such as prematurity or inherited diseases. Most infants allergic to milk protein can drink formulas based on soy protein.[32] Soy formulas also use cornstarch and sucrose instead of lactose and so are recommended for infants with lactose intolerance as well. They are also useful as an alternative to milk-based formulas for vegan families. Despite these limited uses, soy formulas account for one-fourth of the infant formulas sold today. Although soy formulas support the normal growth and development of infants, for infants who don't need them, they offer no advantage over milk formulas.

Some infants who are allergic to cow's milk protein may also be allergic to soy protein.[33] For these infants, special formulas based on hydrolyzed protein are available.

Inappropriate Formulas Caregivers must use only products designed for infants; soy *beverages,* for example, are nutritionally incomplete and inappropriate for infants. Goat's milk is also inappropriate for infants in part because of its low folate content. An infant receiving goat's milk is likely to develop "goat's milk anemia," an anemia characteristic of folate deficiency.

Nursing Bottle Tooth Decay An infant cannot be allowed to sleep with a bottle because of the potential damage to developing teeth. Salivary flow, which normally

wean: to gradually replace breast milk with infant formula or other foods appropriate to an infant's diet.

FIGURE 15-5 Nursing Bottle Tooth Decay

This child was frequently put to bed sucking on a bottle filled with apple juice, so the teeth were bathed in carbohydrate for long periods of time—a perfect medium for bacterial growth. The upper teeth show signs of decay.

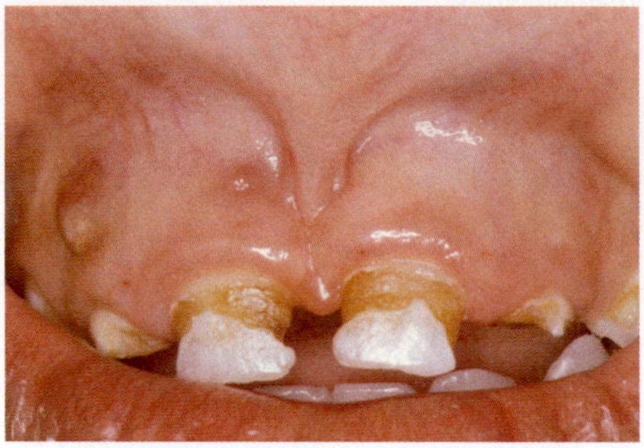

© E. H. Gill/Custom Medical Stock Photo

cleanses the mouth, diminishes as the infant falls asleep. Prolonged sucking on a bottle of formula, milk, or juice bathes the upper teeth in a carbohydrate-rich fluid that nourishes decay-producing bacteria. (The tongue covers and protects most of the lower teeth, but they, too, may be affected.) The result is extensive and rapid tooth decay (see Figure 15-5). To prevent **nursing bottle tooth decay**, no infant should be put to bed with a bottle of nourishing fluid.

Special Needs of Preterm Infants

An estimated one out of eight pregnancies in the United States results in a preterm birth.[34] The terms *preterm* and *premature* imply incomplete fetal development, or immaturity, of many body systems. As might be expected, preterm birth is a leading cause of infant deaths. Preterm infants face physical independence from their mothers before some of their organs and body tissues are ready. The rate of weight gain in the fetus is greater during the last trimester of gestation than at any other time. Therefore, a preterm infant is most often a low-birthweight infant as well. A premature birth deprives the infant of the nutritional support of the placenta during a time of maximal growth.

The last trimester of gestation is also a time of building nutrient stores. Being born with limited nutrient stores intensifies the already precarious situation for the infant. The physical and metabolic immaturity of preterm infants further compromises their nutrition status. Nutrient absorption, especially of fat and calcium, from an immature GI tract is limited. Consequently, preterm, low-birthweight infants are candidates for nutrient imbalances. Deficiencies of the fat-soluble vitamins, calcium, iron, and zinc are common.

Preterm breast milk is well suited to meet a preterm infant's needs. During early lactation, preterm breast milk contains higher concentrations of protein and is lower in volume than term breast milk. The low milk volume is advantageous because preterm infants consume small quantities of milk per feeding, and the higher protein concentration allows for better growth. In many instances, supplements of nutrients specifically designed for preterm infants are added to the mother's expressed breast milk and fed to the infant from a bottle. When fortified with a preterm supplement, preterm breast milk supports growth at a rate that approximates the growth rate that would have occurred within the uterus.[35]

Introducing Cow's Milk

The age at which whole cow's milk should be introduced to the infant's diet has long been a source of controversy. The AAP advises that whole cow's milk is not appropriate during the first year.[36] Children 1 to 2 years of age should not be given reduced-fat, low-fat, or fat-free milk routinely; they need whole milk. Between the ages of 2 and 5 years, a gradual transition from whole milk to the lower-fat milks can take place, but care should be taken to avoid excessive restriction of dietary fat.

Dietary Guidelines for Americans 2010

Children 2 to 3 years should consume 2 cups and children 4 to 8 years should consume $2^1/_2$ cups per day of fat-free or low-fat milk or equivalent milk products.

In some infants, particularly those younger than 6 months of age, whole cow's milk may cause intestinal bleeding, which can lead to iron deficiency. Cow's milk is also a poor source of iron. Consequently, it both causes iron loss and fails to replace iron. Furthermore, the bioavailability of iron from infant cereal and other foods is reduced when cow's milk replaces breast milk or iron-fortified formula during the first year. Compared with breast milk or iron-fortified formula, cow's milk is higher in calcium and lower in vitamin C, characteristics that reduce iron absorption. Furthermore, the higher protein concentration of cow's milk can stress the infant's kidneys. In short, cow's milk is a poor choice during the first year of life; infants need breast milk or iron-fortified infant formula.

Introducing Solid Foods

The high nutrient needs of infancy are met first by breast milk or formula only and then by the limited addition of selected foods over time. Infants gradually develop the ability to chew, swallow, and digest

nursing bottle tooth decay: extensive tooth decay due to prolonged tooth contact with formula, milk, fruit juice, or other carbohydrate-rich liquid offered to an infant in a bottle.

the wide variety of foods available to adults. The caregiver's selection of appropriate foods at the appropriate stages of development is prerequisite to the infant's optimal growth and health.

When to Begin In addition to breast milk or formula, an infant can begin eating solid foods between 4 and 6 months.[37] The AAP supports exclusive breastfeeding for 6 months but recognizes that infants are often developmentally ready to accept complementary foods between 4 and 6 months of age. The main purpose of introducing solid foods is to provide needed nutrients that are no longer supplied adequately by breast milk or formula alone. The foods ♦ chosen must be those that the infant is developmentally capable of handling both physically and metabolically. The exact timing depends on the individual infant's needs and developmental readiness (see Table 15-3), ♦ which vary from infant to infant because of differences in growth rates, activities, and environmental conditions. In addition to the infant's nutrient needs and physical readiness to handle different forms of foods, the introduction of solid foods should consider the need to detect and control allergic reactions.

♦ The German word *beikost* (BYE-cost) describes any nonmilk foods given to an infant.

♦ Digestive secretions gradually increase throughout the first year of life, making the digestion of solid foods more efficient.

Food Allergies To prevent allergy and to facilitate its prompt identification should it occur, experts recommend introducing single-ingredient foods, one at a time, in small portions, and waiting three to five days before introducing the next new food.[38] For example, rice cereal is usually the first cereal introduced because it is the least allergenic. When it is clear that rice cereal is not causing an allergy, another grain, perhaps barley or oat, is introduced. Wheat cereal is offered last because it is the most common offender. If a cereal causes an allergic reaction such as a skin rash, digestive upset, or respiratory discomfort, it should be discontinued before introducing the next food. A later section in this chapter offers more information about food allergies.

Choice of Infant Foods Infant foods should be selected to provide variety, balance, and moderation. Commercial baby foods offer a wide variety of palatable, nutritious foods in a safe and convenient form. Homemade infant foods can be as

TABLE 15-3 Infant Development and Recommended Foods

Because each stage of development builds on the previous stage, the foods from an earlier stage continue to be included in all later stages.

Age (mo)	Feeding Skill	Appropriate Foods Added to the Diet
0–4	Turns head toward any object that brushes cheek. Initially swallows using back of tongue; gradually begins to swallow using front of tongue as well. Strong reflex (extrusion) to push food out during first 2 to 3 months.	Feed breast milk or infant formula.
4–6	Extrusion reflex diminishes, and the ability to swallow nonliquid foods develops. Indicates desire for food by opening mouth and leaning forward. Indicates satiety or disinterest by turning away and leaning back. Sits erect with support at 6 months. Begins chewing action. Brings hand to mouth. Grasps objects with palm of hand.	Begin iron-fortified cereal mixed with breast milk, formula, or water. Begin pureed meats, legumes, vegetables, and fruits.
6–8	Able to self-feed finger foods. Develops pincer (finger to thumb) grasp. Begins to drink from cup.	Begin textured vegetables and fruits. Begin unsweetened, diluted fruit juices from cup.
8–10	Begins to hold own bottle. Reaches for and grabs food and spoon. Sits unsupported.	Begin breads and cereals from table. Begin yogurt. Begin pieces of soft, cooked vegetables and fruit from table. Gradually begin finely cut meats, fish, casseroles, cheese, eggs, and mashed legumes.
10–12	Begins to master spoon, but still spills some.	Add variety. Gradually increase portion sizes.[a]

[a]Portion sizes for infants and young children are smaller than those for an adult. For example, a grain serving might be ½ slice of bread instead of 1 slice, or ¼ cup rice instead of ½ cup.

SOURCE: Adapted in part from Committee on Nutrition, American Academy of Pediatrics, *Pediatric Nutrition Handbook*, 6th ed., ed. R. E. Kleinman (Elk Grove Village, Ill.: American Academy of Pediatrics, 2009), pp. 113–142.

nutritious as commercially prepared ones, as long as the cook minimizes nutrient losses during preparation. Ingredients for homemade foods should be fresh, whole foods without added salt, sugar, or seasonings. Pureed food can be frozen in ice cube trays, providing convenient-size blocks of food that can be thawed, warmed, and fed to the infant. To guard against foodborne illnesses, hands and equipment must be kept clean.

Because recommendations to restrict fat do not apply to children younger than age 2, labels on foods for children younger than 2 (such as infant meats and cereals) cannot carry information about fat. Fat information is omitted from infant food labels to prevent parents from restricting fat in infants' diets. Fearing that their infant will become overweight, parents may unintentionally malnourish the infant by limiting fat. In fact, infants and young children, because of their rapid growth, need more fat than older children and adults.

Foods to Provide Iron Rapid growth demands iron. At about 4 to 6 months of age, the infant begins to need more iron than body stores plus breast milk or iron-fortified formula can provide. In addition to breast milk or iron-fortified formula, infants can receive iron from iron-fortified cereals and, once they readily accept solid foods, from meat or meat alternates such as legumes. Iron-fortified cereals contribute a significant amount of iron to an infant's diet, but the iron's bioavailability is poor.[39] Caregivers can enhance iron absorption from iron-fortified cereals by serving vitamin C–rich foods with meals.

© Polara Studios Inc.

Foods such as iron-fortified cereals and formulas, mashed legumes, and strained meats provide iron.

Foods to Provide Vitamin C The best sources of vitamin C are fruits and vegetables (see pp. 341–342 in Chapter 10). It has been suggested that infants who are introduced to fruits before vegetables may develop a preference for sweets and find the vegetables less palatable, but there is no evidence to support offering these foods in a particular order.[40]

Fruit juice is a good source of vitamin C, but drinking too much juice can lead to diarrhea in infants and young children.[41] AAP recommendations limit juice consumption for infants and young children (1 to 6 years of age) to between 4 and 6 ounces per day.[42] Too much fruit juice contributes excessive kcalories and displaces other nutrient-rich foods. Fruit juices should be diluted and served in a cup, not a bottle, once the infant is 6 months of age or older.

Foods to Omit Concentrated sweets, including baby food "desserts," have no place in an infant's diet. They convey no nutrients to support growth, and the extra food energy can promote obesity. Products containing sugar alcohols such as sorbitol should also be limited, as they may cause diarrhea. Canned vegetables are also inappropriate for infants, as they often contain too much sodium. Honey and corn syrup should never be fed to infants because of the risk of **botulism**.* Infants and young children are vulnerable to foodborne illnesses, and the *Dietary Guidelines for Americans* address this risk.

Dietary Guidelines for Americans 2010

Infants and young children should not eat or drink unpasteurized milk, milk products, or juices; raw or undercooked eggs, meat, poultry, fish, or shellfish; or raw sprouts.

Infants and even young children cannot safely chew and swallow any of the foods listed in the margin; ♦ they can easily choke on these foods, a risk not worth taking. Nonfood items may present even greater choking hazards to infants and young children. Parents and caregivers must pay careful attention to eliminate choking hazards in children's environments.

♦ To prevent choking, do not give infants or young children:
- Cherries
- Gum
- Hard or gel-type candies
- Hot dog slices
- Marshmallows
- Nuts
- Peanut butter
- Popcorn
- Raw carrots
- Raw celery
- Whole beans
- Whole grapes

Keep these nonfood items out of their reach:
- Balloons
- Coins
- Pen tops
- Small balls and marbles

Vegetarian Diets during Infancy The newborn infant is a lacto-vegetarian. As long as the infant has access to sufficient quantities of either iron-fortified infant

botulism (BOT-chew-lism): an often fatal foodborne illness caused by the ingestion of foods containing a toxin produced by bacteria that grow without oxygen.

*In infants, but not in older individuals, ingestion of *Clostridium botulinum* spores can cause illness when the spores germinate in the intestine and produce a toxin, which is absorbed. Symptoms include poor feeding, constipation, loss of tension in the arteries and muscles, weakness, and respiratory compromise. Infant botulism has been implicated in 5 percent of cases of sudden infant death syndrome (SIDS).

formula or breast milk (plus a vitamin D supplement) from a mother who eats an adequate diet, the infant will thrive during the early months. "Health-food beverages," such as rice milk, are inappropriate choices because they lack the protein, vitamins, and minerals infants and toddlers need; in fact, their use can lead to severe nutritional deficiencies.

Infants beyond about 6 months of age present a greater challenge in terms of meeting nutrient needs by way of vegetarian and, especially, vegan diets. Continued breastfeeding or formula feeding is recommended, but supplementary feedings are necessary to ensure adequate energy and iron intakes. Infants and young children in vegetarian families should be given iron-fortified infant cereals well into the second year. Mashed or pureed legumes, tofu, and cooked eggs can be added to their diets in place of meat.

© Stefen Kiefer/imagebroker.net/Photolibrary

Ideally, a 1 year old eats many of the same foods as the rest of the family.

The risks of malnutrition in infants increase with weaning and reliance on table foods. Infants who receive a well-balanced vegetarian diet that includes milk products and a variety of other foods can easily meet their nutritional requirements for growth. This is not always true for vegan infants; the growth of vegan infants slows significantly around the time of transition from breast milk to solid foods. Protein-energy malnutrition and deficiencies of vitamin D, vitamin B_{12}, iron, and calcium have been reported in infants fed vegan diets. Vegan diets that are high in fiber, other complex carbohydrates, and water will fill infants' stomachs before meeting their energy needs. This problem can be partially alleviated by providing more energy-dense foods, such as nut butters, legumes, dried fruit spreads, and mashed avocado. Using soy formulas (or milk) fortified with calcium, vitamin B_{12}, and vitamin D and including vitamin C–containing foods at meals to enhance iron absorption will help prevent some nutrient deficiencies in vegan diets. Parents or caregivers who choose to feed their infants vegan diets should consult with their pediatrician and a registered dietitian frequently to ensure a nutritionally adequate diet that will support growth.

Foods at 1 Year At 1 year of age, whole cow's milk can become a primary source of most of the nutrients an infant needs; 2 to 3 cups a day meets those needs sufficiently. Ingesting more milk than this can displace iron-rich foods, which can lead to **milk anemia.** If powdered milk is used, it should contain fat.

Other foods—meats, iron-fortified cereals, enriched or whole-grain breads, fruits, and vegetables—should be supplied in variety and in amounts sufficient to round out total energy needs. Ideally, a 1 year old will sit at the table, eat many of the same foods everyone else eats, and drink liquids from a cup, not a bottle. Figure 15-6 shows a meal plan that meets a 1 year old's requirements.

FIGURE 15-6 Sample Meal Plan for a 1 Year Old

SAMPLE MENU

Breakfast	1 scrambled egg 1 slice whole-wheat toast ½ c whole milk
Morning snack	½ c yogurt ¼ c fruit[a]
Lunch	½ grilled cheese sandwich: 1 slice whole-wheat bread with 1 slice cheese ½ c vegetables[b] (steamed carrots) ¼ c 100% fruit juice
Afternoon snack	½ c fruit[a] ½ c toasted oat cereal
Dinner	1 oz chopped meat or ¼ c well-cooked mashed legumes ½ c rice or pasta ½ c vegetables[b] (chopped broccoli) ½ c whole milk

NOTE: This sample menu provides about 1000 kcalories.
[a]Include citrus fruits, melons, and berries.
[b]Include dark green, leafy and deep yellow vegetables.

Mealtimes with Toddlers

The nurturing of a young child involves more than nutrition. Those who care for young children are responsible for providing not only nutritious foods, milk, and water, but also a safe, loving, secure environment in which the children may grow and develop. In light of toddlers' developmental and nutrient needs and their often contrary and willful behavior, a few feeding guidelines may be helpful:

- *Discourage unacceptable behavior, such as standing at the table or throwing food.* Be consistent and firm, not punitive. For example, instead of saying "You make me mad when you don't sit down," say "The fruit salad tastes good, please sit down and eat some with me." The child will soon learn to sit and eat.
- *Let toddlers explore and enjoy food, even if this means eating with fingers for a while.* Learning to use a spoon will come in time. Children who are allowed to touch, mash, and smell their food while exploring it are more likely to accept it.

milk anemia: iron-deficiency anemia that develops when an excessive milk intake displaces iron-rich foods from the diet.

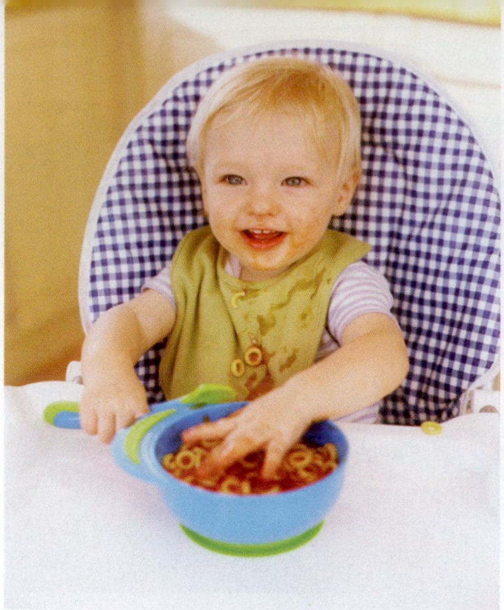

Let toddlers explore and enjoy their food.

- *Don't force food on children.* Rejecting new foods is normal and acceptance is more likely as children become familiar with new foods through repeated opportunities to taste them. Instead of saying "You cannot go outside to play until you taste your carrots," say "You can try the carrots again another time."
- *Provide nutritious foods, and let children choose which ones, and how much, they will eat.* Gradually, they will acquire a taste for different foods.
- *Limit sweets.* Infants and young children have little room for empty-kcalorie foods in their daily energy allowance. Do not use sweets as a reward for eating meals.
- *Don't turn the dining table into a battleground.* Make mealtimes enjoyable. Teach healthy food choices and eating habits in a pleasant environment. Mealtimes are not the time to fight, argue, or scold.

IN SUMMARY **The primary food for infants during the first 12 months is either breast milk or iron-fortified formula. In addition to nutrients, breast milk also offers immunological protection. At about 4 to 6 months of age, infants should gradually begin eating solid foods. By 1 year, they are drinking from a cup and eating many of the same foods as the rest of the family.**

Nutrition during Childhood

Each year from age 1 to adolescence, a child typically grows taller by 2 to 3 inches and heavier by 5 to 6 pounds. Growth charts provide valuable clues to a child's health. Weight gains out of proportion to height gains may reflect overeating and inactivity, whereas measures significantly below the standard suggest malnutrition.

Increases in height and weight are only two of the many changes growing children experience (see Figure 15-7). At age 1, children can stand alone and are beginning to toddle; by 2, they can walk and are learning to run; and by 3, they can jump and climb with confidence. Bones and muscles increase in mass and density to make these accomplishments possible. Thereafter, lengthening of the long bones and increases in musculature proceed unevenly and more slowly until adolescence.

FIGURE 15-7 **Body Shape of 1 Year Old and 2 Year Old Compared**

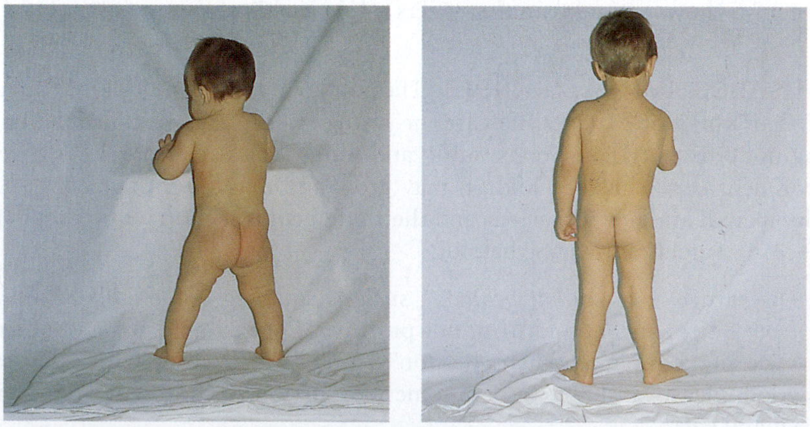

The body shape of a 1 year old (left) changes dramatically by age 2 (right). The 2 year old has lost much of the baby fat; the muscles (especially in the back, buttocks, and legs) have firmed and strengthened; and the leg bones have lengthened.

Energy and Nutrient Needs Children's appetites begin to diminish around 1 year, consistent with the slowing growth. Thereafter, children spontaneously vary their food intakes to coincide with their growth patterns; they demand more food during periods of rapid growth than during slow growth. Sometimes they seem insatiable, and other times they seem to live on air and water.

Children's energy intakes also vary widely from meal to meal. Even so, their total daily intakes remain remarkably constant.[43] If children eat less at one meal, they typically eat more at the next, and vice versa. Overweight children do not always adjust their energy intakes appropriately, however, and may eat in response to external cues, disregarding hunger and satiety signals.[44]

Energy Intake and Activity Individual children's energy needs vary widely, depending on their growth and physical activity. A 1-year-old child needs about 800 kcalories a day; an active 6-year-old child needs twice as many kcalories a day. By age 10, an active child needs about 2000 kcalories a day. Total energy needs increase slightly with age, but energy needs per kilogram of body weight actually decline gradually.

Physically active children of any age need more energy because they expend more, and inactive children can become obese even when they eat less food than the average. Unfortunately, our nation's children are becoming less and less active; schools would serve our children well by offering activities to promote physical fitness.[45] Children who learn to enjoy physical play and exercise, both at home and at school, are best prepared to maintain active lifestyles as adults.

Dietary Guidelines for Americans 2010

Children ages 6 years and older should engage in 60 minutes or more of physical activity per day.

Some children, notably those adhering to a vegan diet, may have difficulty meeting their energy needs. Grains, vegetables, and fruits provide plenty of fiber, adding bulk, but may provide too little energy to support growth. Soy products, other legumes, and nut or seed butters offer more concentrated sources of energy to support optimal growth and development.[46]

Carbohydrate and Fiber Carbohydrate recommendations are based on glucose use by the brain. After 1 year of age, brain glucose use remains fairly constant and is within the adult range. Carbohydrate recommendations for children after 1 year are therefore the same as for adults (see inside front cover).[47]

Fiber recommendations ◆ derive from adult intakes shown to reduce the risk of coronary heart disease and are based on energy intakes. Consequently, fiber recommendations for younger children with low energy intakes are less than those for older ones with high energy intakes.[48]

◆ Fiber recommendations for children:

Age (yr)	AI (g)
1–3	19
4–8	25
9–13	
Boys	31
Girls	26
14–18	
Boys	38
Girls	26

Fat and Fatty Acids No RDA for total fat has been established, but the DRI Committee recommends a fat intake of 30 to 40 percent of energy for children 1 to 3 years of age and 25 to 35 percent for children 4 to 18 years of age.[49] As long as children's energy intakes are adequate, however, fat intakes below 30 percent of total energy do not impair growth.[50] Children who eat low-fat diets, however, tend to have low intakes of some vitamins and minerals. Recommended intakes of the essential fatty acids are based on average intakes (see inside front cover).

Dietary Guidelines for Americans 2010

Keep total fat intake between 30 to 40 percent of kcalories for children 1 to 3 years of age and between 25 and 35 percent of kcalories for children and adolescents 4 to 18 years of age, with most fats coming from sources of polyunsaturated and monounsaturated fatty acids, such as fish, nuts, and vegetable oils.

TABLE 15-4 Recommended Daily Amounts from Each Food Group (1000 to 1800 kCalories)

Food Group	1000 kcal	1200 kcal	1400 kcal	1600 kcal	1800 kcal
Fruits	1 c	1 c	1½ c	1½ c	1½ c
Vegetables	1 c	1½ c	1½ c	2 c	2½ c
Grains	3 oz	4 oz	5 oz	5 oz	6 oz
Protein foods	2 oz	3 oz	4 oz	5 oz	5 oz
Milk	2 c	2 c	2 c	3 c	3 c
Oils	3 tsp	3 tsp	3 tsp	4 tsp	5 tsp

NOTE: The discretionary kcalorie allowance for these patterns is about 100 kcalories.

Protein Like energy needs, total protein needs increase slightly with age, but when the child's body weight is considered, the protein requirement actually declines slightly (see inside front cover). Protein recommendations must consider the requirements for maintaining nitrogen balance, the quality of protein consumed, and the added needs of growth.

Vitamins and Minerals The vitamin and mineral needs of children increase with age (see inside front cover). A balanced diet of nutritious foods can meet children's needs for these nutrients, with the notable exception of iron, and possibly vitamin D. Iron-deficiency anemia is a major problem worldwide, and is prevalent among both U.S. and Canadian children, especially toddlers 1 to 3 years of age.[51] During the second year of life, toddlers progress from a diet of iron-rich infant foods such as breast milk, iron-fortified formula, and iron-fortified infant cereal to a diet of adult foods and iron-poor cow's milk. In addition, their appetites often fluctuate—some become finicky about the foods they eat, and others prefer milk and juice to solid foods. These situations can interfere with children eating iron-rich foods at a critical time for brain growth and development.

To prevent iron deficiency, children's foods must deliver 7 to 10 milligrams of iron per day. To achieve this goal, snacks and meals should include iron-rich foods, and milk intake should be reasonable so that it will not displace lean meats, fish, poultry, eggs, legumes, and whole-grain or enriched products. (Chapter 13 describes iron-rich foods and ways to maximize iron absorption.)

As for vitamin D, the American Academy of Pediatrics recently updated guidelines for children. Children who do not obtain enough vitamin D by drinking vitamin–D fortified milk (2.5 micrograms per 1 cup serving) and eating fortified foods such as dry cereals (1 microgram per ½ cup serving) should receive a vitamin D supplement.[52]

Supplements With the exception of specific recommendations for fluoride, iron, and vitamin D during infancy and childhood, the AAP and other professional groups agree that well-nourished children do not need vitamin and mineral supplements. Despite this, many children and adolescents take supplements.[53] Ironically, children with poor nutrient intakes typically do not receive supplements, and those who do take supplements typically receive extra nutrients they do not need.[54] Furthermore, researchers are still studying the safety of supplement use by children. The Federal Trade Commission has warned parents about giving supplements advertised to prevent or cure childhood illnesses such as colds, ear infections, and asthma. Dietary supplements on the market today include many herbal products that have not been tested for safety and effectiveness in children.

Planning Children's Meals Table 15-4 lists recommended amounts from each food group for several kcalorie levels. Estimated daily kcalorie needs for active and sedentary children of various ages are shown in Table 15-5. To provide all the needed nutrients, children's meals should include a variety of foods from each food group—in amounts suited to their appetites and needs. Figure 15-8 presents MyPyramid for Preschoolers, designed for children 2 to 5 years of age, and MyPyramid

TABLE 15-5 Estimated Daily kCalorie Needs for Children

Children	Sedentary[a]	Active[b]
2 to 3 yr	1000	1400
Females		
4 to 8 yr	1200	1800
9 to 13 yr	1600	2200
Males		
4 to 8 yr	1400	2000
9 to 13 yr	1800	2600

[a]*Sedentary* describes a lifestyle that includes only the activities typical of day-to-day life.
[b]*Active* describes a lifestyle that includes at least 60 minutes per day of moderate physical activity (equivalent to walking more than 3 miles per day at 3 to 4 miles per hour) in addition to the activities of day-to-day life.

for Kids, ♦ designed for children 6 to 11 years of age. The figure includes the recommended amounts of food for a 1200-kcalorie intake (appropriate for many preschoolers) and for an 1800-kcalorie intake (appropriate for many older children). ♦

Children whose diets follow the pattern presented in Figure 15-8 meet their nutrient needs fully, but few children eat according to these recommendations.

♦ www.mypyramid.gov/kids

♦ For kcalorie levels more than 1800, see Table 2-2 on p. 42.

FIGURE 15-8 Food Guide Pyramid for Young Children

Grains Make half your grains whole	Vegetables Vary your veggies	Fruits Focus on fruits	Milk Get your calcium-rich foods	Meat & Beans Go lean with protein
Start smart with breakfast. Look for whole-grain cereals. Just because bread is brown doesn't mean it's whole grain. Search the ingredients list to make sure the first word is "whole" (like "whole wheat").	Color your plate with all kinds of great-tasting veggies. What's green and orange and tastes good? Veggies! Go dark green with broccoli and spinach, or try orange ones like carrots and sweet potatoes.	Fruits are nature's treats—sweet and delicious. Go easy on juice and make sure it's 100%.	Move to the milk group to get your calcium. Calcium builds strong bones. Look at the carton or container to make sure your milk, yogurt, or cheese is low fat or fat-free.	Eat lean or low-fat meat, chicken, turkey, and fish. Ask for it baked, broiled, or grilled—not fried. It's nutty, but true. Nuts, seeds, peas, and beans are all great sources of protein, too.
For a 1200-kcalorie diet (suitable for many preschoolers ages 2 to 5), include the amounts below from each food group.				
Eat 4 oz. every day; at least half should be whole	Eat 1 1/2 cups every day	Eat 1 cup every day	Get 2 cups every day	Eat 3 oz. every day
For a 1800-kcalorie diet (suitable for many children ages 6 to 11), include the amounts below from each food group.				
Eat 6 oz. every day; at least half should be whole	Eat 2 1/2 cups every day	Eat 1 1/2 cups every day	Get 3 cups every day; for kids ages 6 to 8, it's 2 cups	Eat 5 oz. every day

Oils Oils are not a food group, but you need some for good health. Get your oils from fish, nuts, and liquid oils such as corn oil, soybean oil, and canola oil.

Find your balance between food and fun

- Move more. Aim for at least 60 minutes every day, or most days.
- Walk, dance, rollerblade—it all counts. How great is that!

Fats and sugars—know your limits

- Get your fat facts and sugar smarts from the Nutrition Facts label.
- Limit solid fats as well as foods that contain them.
- Choose food and beverages low in added sugars and other kcaloric sweeteners.

Healthy, well-nourished children are alert in the classroom and energetic at play.

One analysis of the quality of children's diets found that most (up to 88 percent) children between 2 and 9 years of age have diets that need substantial improvement.[55] A comprehensive survey, called the Feeding Infants and Toddlers Study (FITS), assessed the food and nutrient intakes of more than 3000 infants and toddlers.[56] The survey found that fruit and vegetable intakes of infants and toddlers are limited, and in fact, about 25 percent of infants and toddlers older than 9 months did not eat a single serving of fruits or vegetables in a day.[57] By 15 to 18 months of age, the most commonly consumed vegetable was french fries and the most commonly consumed fruit was bananas—neither particularly rich sources of vitamins or minerals. Parents and caregivers of infants and toddlers thus need to offer a much greater variety of nutrient-dense vegetables and fruits at meals and snacks to help ensure adequate nutrition. Among other nutrition concerns for U.S. children are inadequate intakes of vitamin E, calcium, magnesium, potassium, and fiber, and excessive intakes of sodium.[58]

Hunger and Malnutrition in Children

Most children in the United States and Canada have access to regular meals, but hunger and malnutrition do appear in certain circumstances. Children in very low-income families, for example, are more likely to be hungry and malnourished. More than 12 million U.S. children are hungry at least some of the time and are living in poverty.[59] Highlight 16 examines the causes and consequences of hunger in the United States.

Hunger and Behavior

Even when hunger is temporary, as when a child misses one meal, behavior and academic performance are affected. Children who eat nutritious breakfasts improve their school performance and are tardy or absent significantly less often than their peers who do not.[60] A nutritious breakfast is a central feature of a diet that meets the needs of children and supports their healthy growth and development.[61] Children who skip breakfast typically do not make up the deficits at later meals—they simply have lower intakes of energy, vitamins, and minerals than those who eat breakfast. Without breakfast, children perform poorly in tasks requiring concentration, their attention spans are shorter, and they even score lower on intelligence tests than their well-fed peers. Malnourished children are particularly vulnerable. Common sense dictates that it is unreasonable to expect anyone to learn and perform without fuel. For the child who hasn't had breakfast, the morning's lessons may be lost altogether. Even if a child has eaten breakfast, discomfort from hunger may become distracting by late morning. Teachers aware of the late-morning slump in their classrooms wisely request that midmorning snacks be provided; snacks improve classroom performance all the way to lunchtime.

Iron Deficiency and Behavior

Iron deficiency has well-known and widespread effects on children's behavior and intellectual performance.[62] In addition to carrying oxygen in the blood, iron transports oxygen within cells, which use it for energy metabolism. Iron is also used to make neurotransmitters—most notably, those that regulate the ability to pay attention, which is crucial to learning. Consequently, iron deficiency not only causes an energy crisis, but also directly impairs attention span and learning ability.

Iron deficiency is often diagnosed by a quick, easy, inexpensive hemoglobin or hematocrit test that detects a deficit of iron in the *blood*. A child's *brain*, however, is sensitive to low iron concentrations long before the blood effects appear. Iron deficiency lowers the "motivation to persist in intellectually challenging tasks" and impairs overall intellectual performance. Anemic children perform poorly on tests and are disruptive in the classroom; iron supplementation improves learning and memory. When combined with other nutrient deficiencies, iron-deficiency anemia has synergistic effects that are especially detrimental to learning. Furthermore, children who had iron-deficiency anemia *as infants* continue to perform poorly as they grow older, even if their iron status improves.[63] The long-term damaging effects on mental development make prevention and treatment of iron deficiency during infancy and early childhood a high priority.

Other Nutrient Deficiencies and Behavior A child with any of several nutrient deficiencies may be irritable, aggressive, and disagreeable, or sad and withdrawn. Such a child may be labeled "hyperactive," "depressed," or "unlikable," when in fact these traits may be due to simple, even marginal, malnutrition. Parents and medical practitioners often overlook the possibility that malnutrition may account for abnormalities of appearance and behavior. Any departure from normal healthy appearance and behavior is a sign of possible poor nutrition (see Table 15-6). In any such case, inspection of the child's diet by a registered dietitian or other qualified health-care professional is in order. Any suspicion of dietary inadequacies, no matter what other causes may be implicated, should prompt steps to correct those inadequacies immediately.

The Malnutrition-Lead Connection

Children who are malnourished are vulnerable to lead poisoning. They absorb more lead if their stomachs are empty; if they have low intakes of calcium, zinc, vitamin C, or vitamin D; and, of greatest concern because it is so common, if they have an iron deficiency. Iron deficiency weakens the body's defenses against lead absorption, and lead poisoning can cause iron deficiency. Common to both iron deficiency and lead poisoning are a low socioeconomic background and a lack of immunizations against infectious diseases. Another common factor is pica—a craving for nonfood items. Many children with lead poisoning eat dirt or chips of old paint, two common sources of lead.

The anemia brought on by lead poisoning may be mistaken for a simple iron deficiency and therefore may be incorrectly treated. Like iron deficiency, mild lead toxicity has nonspecific symptoms, including diarrhea, irritability, and fatigue. Adding iron to the diet does not reverse the symptoms; exposure to lead must stop and treatment for lead poisoning must begin. With further exposure, the symptoms become more pronounced, and children develop learning disabilities and behavioral problems. Still more severe lead toxicity can cause irreversible nerve damage, paralysis, mental retardation, and death.

TABLE 15-6 Physical Signs of Malnutrition in Children

	Well-Nourished	Malnourished	Possible Nutrient Deficiencies
Hair	Shiny, firm in the scalp	Dull, brittle, dry, loose; falls out	PEM
Eyes	Bright, clear pink membranes; adjust easily to light	Pale membranes; spots; redness; adjust slowly to darkness	Vitamin A, the B vitamins, zinc, and iron
Teeth and gums	No pain or caries, gums firm, teeth bright	Missing, discolored, decayed teeth; gums bleed easily and are swollen and spongy	Minerals and vitamin C
Face	Clear complexion without dryness or scaliness	Off-color, scaly, flaky, cracked skin	PEM, vitamin A, and iron
Glands	No lumps	Swollen at front of neck, cheeks	PEM and iodine
Tongue	Red, bumpy, rough	Sore, smooth, purplish, swollen	B vitamins
Skin	Smooth, firm, good color	Dry, rough, spotty; "sandpaper" feel or sores; lack of fat under skin	PEM, essential fatty acids, vitamin A, B vitamins, and vitamin C
Nails	Firm, pink	Spoon-shaped, brittle, ridged	Iron
Internal systems	Regular heart rhythm, heart rate, and blood pressure; no impairment of digestive function, reflexes, or mental status	Abnormal heart rate, heart rhythm, or blood pressure; enlarged liver, spleen; abnormal digestion; burning, tingling of hands, feet; loss of balance, coordination; mental confusion, irritability, fatigue	PEM and minerals
Muscles and bones	Muscle tone; posture, long bone development appropriate for age	"Wasted" appearance of muscles; swollen bumps on skull or ends of bones; small bumps on ribs; bowed legs or knock-knees	PEM, minerals, and vitamin D

Recent research suggests that childhood lead exposure disrupts normal brain development—a finding that may partially explain the impaired cognitive and behavioral abilities of lead-exposed children. For six years, researchers measured blood lead levels at intervals in young children who lived in lead-contaminated houses.[64] Years later, brain images revealed that the higher the blood lead concentrations during childhood, the smaller the brain size as a young adult. The brains of boys were more affected than the brains of girls.

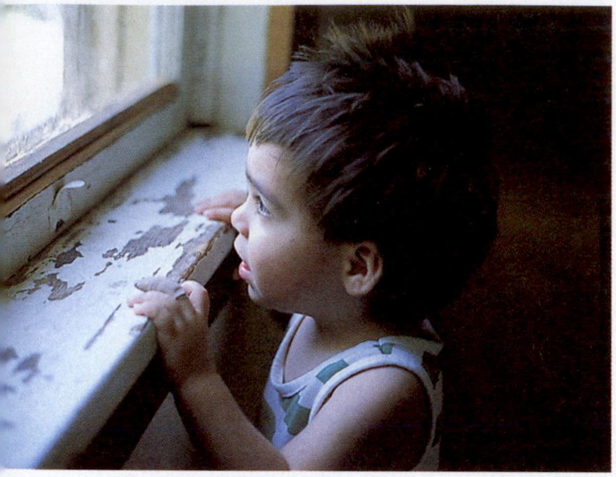
© Tony Freeman/PhotoEdit

Old, lead-based paint threatens the health of an exploring child.

More than 300,000 children in the United States—most of them under age 6—have blood lead concentrations high enough to cause mental, behavioral, and other health problems.[65] Lead toxicity in young children comes from their own behaviors and activities—putting their hands in their mouths, playing in dirt and dust, and chewing on nonfood items.[66] Unfortunately, the body readily absorbs lead during times of rapid growth and hoards it possessively thereafter. Lead is not easily excreted and accumulates mainly in the bones, but also in the brain, teeth, and kidneys. Tragically, a child's neuromuscular system is also maturing during these first few years of life. No wonder children with elevated lead levels experience impairment of balance, motor development, and the relaying of nerve messages to and from the brain. Deficits in intellectual development are only partially reversed when lead levels decline.[67]

Federal laws mandating reductions in leaded gasoline, lead-based solder, and other products over the past four decades have helped to reduce the amounts of lead in food and in the environment in the United States. As a consequence, the prevalence of lead toxicity in children has declined dramatically for most of the United States, but lead exposure is still a threat in certain communities. The accompanying "How To" presents strategies for defending children against lead toxicity.

Hyperactivity and "Hyper" Behavior

All children are naturally active, and many of them become overly active on occasion—for example, in anticipation of a birthday party. Such behavior is markedly different from true **hyperactivity**.

Hyperactivity Hyperactive children have trouble sleeping, cannot sit still for more than a few minutes at a time, act impulsively, and have difficulty paying attention. These behaviors interfere with social development and academic progress. The cause of hyperactivity remains unknown, but it affects about 5 to 10 percent of young school-age children.[68] To resolve the problems surrounding hyperactivity, physicians often recommend specific behavioral strategies, special educational programs, and psychological counseling. In many cases, they prescribe medication.[69]

Researchers have debated whether hyperactivity is due to a brain deficit or to a delay in brain development. Recent research supports the view that some areas in the brains of children with hyperactivity develop more slowly than those of other children.[70] Such findings remain controversial, but further support comes from the fact that hyperactivity symptoms tend to improve as children get older.

Parents of hyperactive children often blame sugar as the cause. They mistakenly believe that simply eliminating candy and other sweet treats will solve the problem. This dietary change will not solve the problem, however, and studies have consistently found no convincing evidence that sugar causes hyperactivity or worsens behavior. Such speculation has been based on personal stories. No scientific evidence supports a relationship between sugar and hyperactivity or other misbehaviors.

Food additives have also been blamed for hyperactivity and other behavior problems in children, but scientific evidence to substantiate the connection has been elusive—until recently. A well-controlled study of close to 300 children suggests that food additives such as artificial colors or sodium benzoate preservative (or both) exacerbate hyperactive symptoms such as inattention and impulsivity.[71] Additional studies are needed to confirm the findings and to determine which additives studied might be responsible for negative behaviors.

hyperactivity: inattentive and impulsive behavior that is more frequent and severe than is typical of others a similar age; professionally called **attention-deficit/hyperactivity disorder (ADHD).**

Misbehaving Even a child who is not truly hyperactive can be difficult to manage at times. Michael may act unruly out of a desire for attention, Jessica may be cranky because of a lack of sleep, Christopher may react violently after watching

HOW TO Protect against Lead Toxicity

Researchers simultaneously made three major discoveries about lead toxicity: lead poisoning has *subtle* effects, the effects are *permanent*, and they occur at *low levels of exposure*. The amount of lead recognized to cause harm is only 10 micrograms per 100 milliliters of blood. Some research shows that blood lead concentrations *below* this amount may adversely affect children's physical and mental development.[a] Consequently, consumers should take ultraconservative measures to protect themselves, and especially their infants and young children, from lead poisoning. The American Academy of Pediatrics and the Centers for Disease Control recommend screening in communities with a substantial number of houses built before 1950 and in those with a substantial number of children with elevated lead levels. In addition to screening children most likely to be exposed, pediatricians should alert all parents to the possible dangers of lead exposure and explain prevention strategies.

Preventive strategies include:

- In contaminated environments, keep small children from putting dirty or old painted objects in their mouths, and make sure children wash their hands before eating. Similarly, keep small children from eating any nonfood items. Lead poisoning has been reported in young children who have eaten crayons or pool cue chalk.
- Wet-mop floors and damp-sponge walls regularly. Children's blood lead levels decline when the homes they live in are cleaned regularly.
- Be aware that other countries do not have the same regulations protecting consumers against lead. Children have been poisoned by eating crayons made in China and drinking fruit juice canned in Mexico.
- Do not use lead-contaminated water to make infant formula.
- Once you have opened canned food, store it in a lead-free container to prevent lead migration into the food.
- Do not store acidic foods or beverages (such as vinegar or orange juice) in ceramic dishware or alcoholic beverages in pewter or crystal decanters.
- Many manufacturers are now making lead-safe products. Old, handmade, or imported ceramic cups and bowls may contain lead and should not be used to heat coffee or tea or acidic foods such as tomato soup.
- Feed children nutritious meals regularly.
- Before using your newspaper to wrap food, mulch garden plants, or add to your compost, confirm with the publisher that the paper uses no lead in its ink.

The Environmental Protection Agency (EPA) also publishes a booklet, *Lead and Your Drinking Water,* in which the following cautions appear:

- Have the water in your home tested by a competent laboratory.
- Use only cold water for drinking, cooking, and making formula (cold water absorbs less lead).
- When water has been standing in pipes for more than two hours, flush the cold-water pipes by running water through them for 30 seconds before using it for drinking, cooking, or mixing formulas.
- If lead contamination of your water supply seems probable, obtain additional information and advice from the EPA and your local public health agency.

By taking these steps, parents can protect themselves and their children from this preventable danger.[b]

For additional practice, log on to **www.cengagebrain.com** and search for ISBN 084006845X.

[a]Centers for Disease Control and Prevention, Interpreting and managing blood lead levels <10μg/dL in children and reducing childhood exposures to lead: Recommendations of CDC's Advisory Committee on Childhood Lead Poisoning Prevention, *Morbidity and Mortality Weekly Report* 56/RR-8 (2007): 1–16; Policy of Committee on Environmental Health, American Academy of Pediatrics: Lead exposure in children: Prevention, detection, and management, *Pediatrics* 116 (2005): 1036–1046.

[b]Call the National Lead Information Center hotline at (800) 424-LEAD (424-5323) for general information.

TRY IT Visit the website for the Environmental Protection Agency (www.epa.gov/lead) and identify the most common sources of lead poisoning.

too much television, and Sheila may be unable to sit still in class due to a lack of exercise. All of these children may benefit from more consistent care—regular hours of sleep, regular mealtimes, and regular outdoor activity.

Food Allergy and Intolerance

Food allergy is frequently blamed for physical and behavioral abnormalities in children, but just 6 to 8 percent of children younger than 4 years of age are diagnosed with true food allergies.[72] Food allergies diminish with age, until in adulthood they affect less than 4 percent of the population.[73] The prevalence of food allergy, especially peanut allergy, is on the rise, however.[74] Reasons for an increase in peanut allergy are not yet clear, but possible contributing factors include genetics, food preparation methods (roasting peanuts at very high temperatures makes them more allergenic), and exposure to medicinal skin creams containing peanut oil.[75]

A true food allergy occurs when fractions of a food protein or other large molecule are absorbed into the blood and elicit an immunologic response. (Recall that

food allergy: an adverse reaction to food that involves an immune response; also called **food-hypersensitivity reaction.**

proteins are normally dismantled in the digestive tract to amino acids that are absorbed without such a reaction.) The body's immune system reacts to these large food molecules as it does to other antigens—by producing antibodies, histamines, or other defensive agents.

Detecting Food Allergy Allergies may have one or two components. They always involve antibodies, but they may or may not involve symptoms. ♦ This means that allergies can be diagnosed only by testing for antibodies. Even symptoms exactly like those of an allergy may not be caused by an allergy. Once a food allergy has been diagnosed, the required treatment is strict elimination of the offending food. Children with allergies, like all children, need all their nutrients, so it is important to include other foods that offer the same nutrients as the omitted foods.[76]

♦ A person who produces antibodies *without* having any symptoms has an **asymptomatic allergy;** a person who produces antibodies *and* has symptoms has a **symptomatic allergy.**

Allergic reactions to food may be immediate or delayed. In either case, the antigen interacts immediately with the immune system, but the timing of symptoms varies from minutes to 24 hours after consumption of the antigen. Identifying the food that causes an immediate allergic reaction is fairly easy because the symptoms appear shortly after the food is eaten. Identifying the food that causes a delayed reaction is more difficult because the symptoms may not appear until much later. By this time, many other foods may have been eaten, complicating the picture.

© Polara Studios Inc.

These normally wholesome foods may cause life-threatening symptoms in people with allergies.

Anaphylactic Shock The life-threatening food allergy reaction of **anaphylactic shock** is most often caused by peanuts, tree nuts, milk, eggs, wheat, soybeans, fish, or shellfish. Among these foods, eggs, milk, soy, and peanuts most often cause problems in children.[77] Children are more likely to outgrow allergies to eggs, milk, and soy than allergies to peanuts. Peanuts cause more life-threatening reactions than do all other food allergies combined. Research is currently under way to help people with peanut allergies tolerate small doses, thus saving lives and minimizing reactions.[78] One possible solution depends on finding a natural, hypoallergenic peanut among the 14,000 varieties of peanuts. Families of children with a life-threatening food allergy and the school personnel who supervise those children must guard them against any exposure to the allergen. The child must learn to identify which foods pose a problem and then learn and use refusal skills for all foods that may contain the allergen.

Parents of children with allergies can pack safe foods for lunches and snacks and ask school officials to strictly enforce a "no swapping" policy in the lunchroom. The child must be able to recognize the symptoms of impending anaphylactic shock, ♦ such as a tingling of the tongue, throat, or skin, or difficulty breathing. Any person with food allergies severe enough to cause anaphylactic shock should wear a medical alert bracelet or necklace. Finally, the responsible child and the school staff should be prepared with injections of epinephrine, ♦ which prevents anaphylaxis after exposure to the allergen. Many preventable deaths occur each year when people with food allergies accidentally ingest the allergen but have no epinephrine available.

♦ Symptoms of impending anaphylactic shock:

- Tingling sensation in mouth
- Swelling of the tongue and throat
- Irritated, reddened eyes
- Difficulty breathing, asthma
- Hives, swelling, rashes
- Vomiting, abdominal cramps, diarrhea
- Drop in blood pressure
- Loss of consciousness
- Death

♦ **Epinephrine** is a hormone of the adrenal gland that modulates the stress response; formerly called *adrenaline*. When administered by injection, epinephrine counteracts anaphylactic shock by opening the airways and maintaining heartbeat and blood pressure.

anaphylactic (ana-fill-LAC-tic) **shock:** a life-threatening, whole-body allergic reaction to an offending substance.

Food Labeling Food labels must list the presence of common allergens in plain language, using the names of the eight most common allergy-causing foods.[79] For example, a food containing "textured vegetable protein" must say "soy" on its label. Similarly, "casein" must be identified as "milk," and so forth. Food producers must also prevent cross-contamination during production and clearly label foods in which it is likely to occur. For example, equipment used for making peanut butter must be scrupulously clean before being used to pulverize cashew nuts for cashew butter to protect unsuspecting cashew butter consumers from peanut allergens.

Technology may soon offer new solutions. New drugs are being developed that may interfere with the immune response that causes allergic reactions.[80] Also, through genetic engineering, scientists may one day create allergen-free peanuts, soybeans, and other foods to make them safer.

Food Intolerances Not all **adverse reactions** to foods are food allergies, although even physicians may describe them as such. Signs of adverse reactions to foods include stomachaches, headaches, rapid pulse rate, nausea, wheezing, hives, bronchial irritation, coughs, and other such discomforts. Among the causes may be reactions to chemicals in foods, such as the flavor enhancer monosodium glutamate (MSG), the natural laxative in prunes, or the mineral sulfur; digestive diseases, obstructions, or injuries; enzyme deficiencies, such as lactose intolerance; and even psychological aversions. These reactions involve symptoms but no antibody production. Therefore, they are **food intolerances**, not allergies.

Pesticides on produce may also cause adverse reactions. Pesticides that were applied in the fields may linger on the foods. Health risks from pesticide exposure may be low for healthy adults, but children are vulnerable. Therefore, government agencies have set a **tolerance level** for each pesticide by first identifying foods that children commonly eat in large amounts and then considering the effects of pesticide exposure during each developmental stage.

Hunger, lead poisoning, hyperactivity, and allergic reactions can all adversely affect a child's nutrition status and health. Fortunately, each of these problems has solutions. They may not be easy solutions, but at least we have a reasonably good understanding of the problems and ways to correct them. Such is not the case with the most pervasive health problem for children in the United States—obesity.

Childhood Obesity

The number of overweight children has increased dramatically over the past three decades (see Figure 15-9). Like their parents, children in the United States are becoming fatter. An estimated 32 percent of U.S. children and adolescents 2 to 19 years of age are overweight and 16 percent are obese.[81] Based on data from the BMI-for-age growth charts, children and adolescents are categorized as *overweight* above the 85th percentile and as *obese* at the 95th percentile and above.[82] There are exceptions to the use of the 85th and 95th percentile cutoff points. For older adolescents, a BMI at the 95th percentile is higher than a BMI of 30, the adult obesity cutoff point. Therefore, obesity is defined as a BMI at the 95th percentile or a BMI of 30 or greater, whichever is lower. For children younger than 2 years of age, BMI values are not available. For this age group, weight-for-height values above the 95th percentile are classified as overweight. Figure 15-10 (p. 514) presents the BMI for children and adolescents, indicating cutoff points for obesity and overweight.

The Expert Committee of the American Medical Association recommends a third cutoff point (99th percentile) to define severe obesity in childhood.[83] Unfortunately, severe obesity in children is becoming more prevalent. Many of these children have multiple risk factors for cardiovascular disease and a high risk of severe obesity in adulthood.[84] The special risks and treatment needs of severely obese children need to be recognized.

The problem of obesity in children is especially troubling because overweight children have the potential of becoming obese adults with all the social, economic, and medical ramifications that often accompany obesity. They have additional problems, too, arising from differences in their growth, physical health, and psychological development. In trying to explain the rise in childhood obesity, researchers point to both genetic and environmental factors.

Genetic and Environmental Factors Parental obesity predicts an early increase in a young child's BMI, and it more than doubles the chances that a young child will become an obese adult. Children with neither parent obese have a less than 10 percent chance of becoming obese in adulthood, whereas overweight teens with at least one obese parent have a greater than 80 percent chance of be-

FIGURE 15-9 Trends in Childhood Obesity

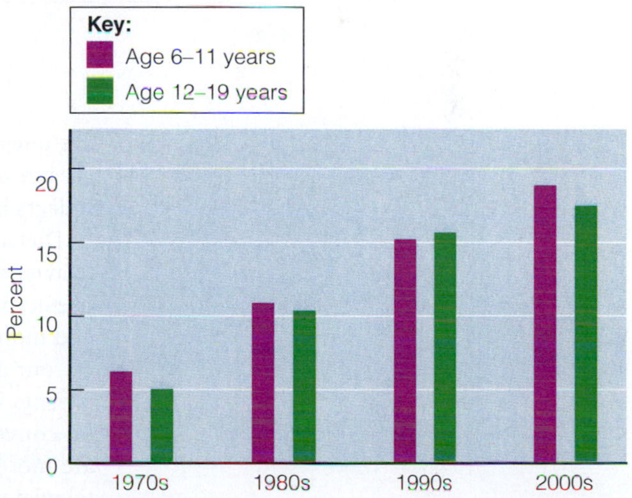

adverse reactions: unusual responses to food (including intolerances and allergies).

food intolerances: adverse reactions to foods that do not involve the immune system.

tolerance level: the maximum amount of residue permitted in a food when a pesticide is used according to the label directions.

FIGURE 15-10 Body Mass Index-for-Age Percentiles: Boys and Girls, Age 2 to 20

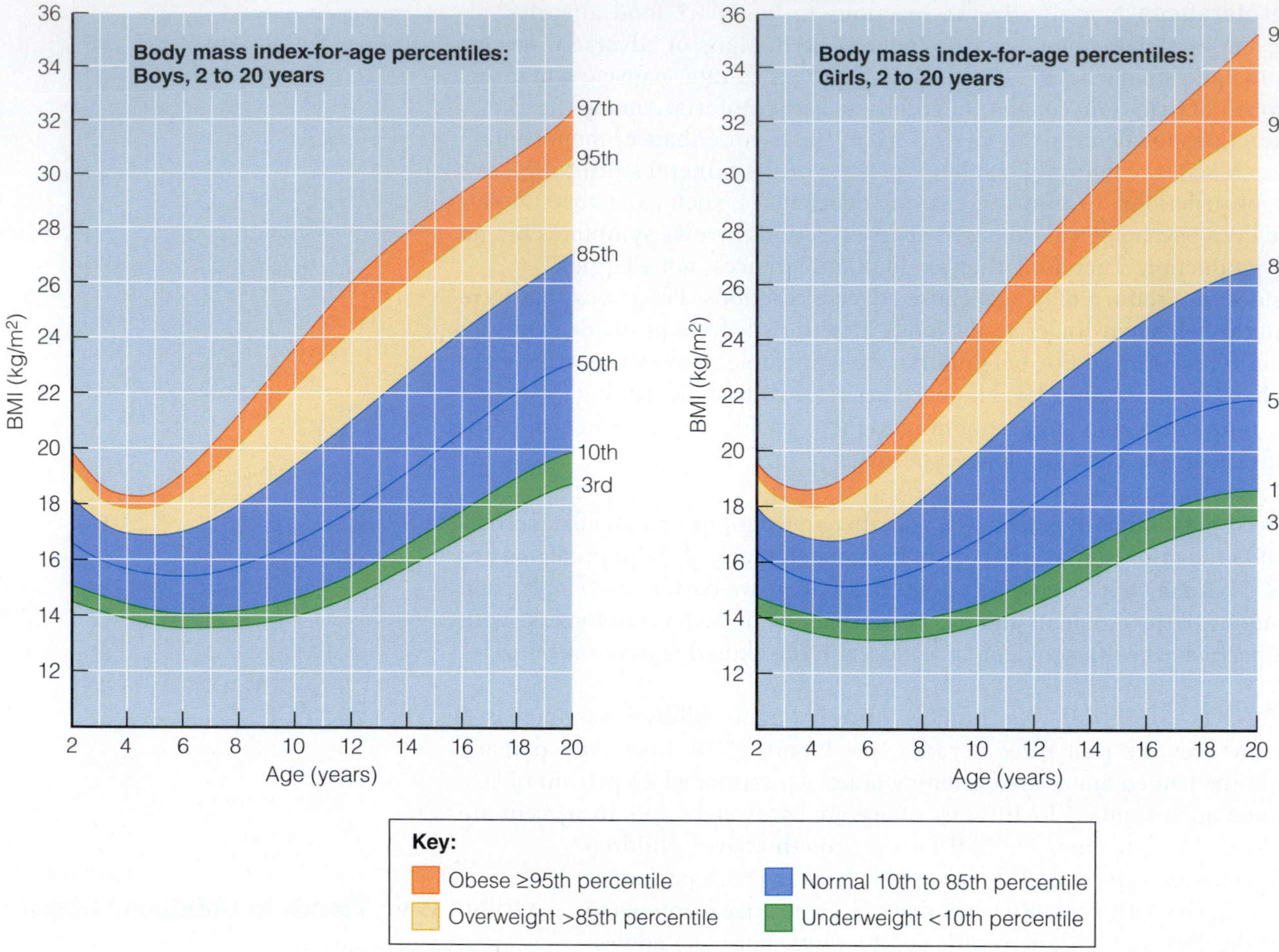

ing obese adults. The chances of an obese child becoming an obese adult grow greater as the child grows older.[85] The link between parental and child obesity reflects both genetic and environmental factors (as described in Chapter 9).

Diet and physical inactivity must also play a role in explaining why children are heavier today than they were 40 or so years ago. As the prevalence of childhood obesity throughout the United States has more than doubled for young children and more than tripled for children 6 to 11 years of age and adolescents, the society our children live in has changed considerably.[86] In many families today, both parents work outside the home and work longer hours; more emphasis is placed on convenience foods and foods eaten away from home; meal choices at school are more diverse and often less nutritious; sedentary activities such as watching television and playing video or computer games occupy much of children's free time; and opportunities for physical activity and outdoor play both during and after school have declined.[87] All of these factors—and many others—influence children's eating and activity patterns.

Children learn food behaviors from their families, and research confirms the significant roles parents play in teaching their children about healthy food choices, providing nutrient-dense foods, and serving as role models.[88] When parents eat fruits and vegetables frequently, their children do, too.[89] The more fruits and vegetables children eat, the more vitamins, minerals, and fibers, and the less saturated fat in their diets.[90]

Research shows that one in four toddlers (19 to 24 months of age) exceeds estimated energy requirements as a result of eating such foods as candy, pizza, chicken nuggets, soda, sweet tea, and salty snacks such as cheese puffs and chips.[91] Not surprisingly, when researchers ask, "Are today's children eating more kcalories than those of 40 years ago?" the answer is, "Yes."

As Highlight 4 discusses, as the prevalence of obesity among both children and adults has surged over the past four decades, so has the consumption of added sugars and, especially, high-fructose corn syrup—the easily consumed, energy-dense liquid sugar added to soft drinks.[92] Each 12-ounce can of soft drink provides the equivalent of about 10 teaspoons of sugar and 150 kcalories. More than half of school-age children consume at least one soft drink each day at school; adolescent males consume the most—four or more cans daily.[93] Research shows that soft drink consumption is associated with increased energy intake and body weight.[94] According to one estimate, the risk of obesity increases by 60 percent with each sugared soft drink consumed daily.

No doubt, the tremendous increase in soft drink consumption plays a role, but much of the obesity epidemic can be explained by lack of physical activity. Children have become more sedentary, and sedentary children are more often overweight.[95] Television watching ♦ may contribute most to physical inactivity. Longer television time is linked with overweight in children.[96] A child who spends more than an hour or two each day in front of a television, computer monitor, or other media can become overweight even while eating fewer kcalories than a more active child. Too much screen time and not enough activity time also contributes to a child's psychological distress.[97]

♦ TV fosters obesity because it:
- Requires no energy beyond basal metabolism
- Replaces vigorous activities
- Encourages snacking
- Promotes a sedentary lifestyle

Playing video games influences children's activity patterns similarly.

Children who have television sets in their bedrooms spend more time watching TV, are less physically active, and are more likely to be overweight than children who do not have televisions in their rooms.[98] Watching television influences food intake as well as physical activity.[99] Children who watch a great deal of television are most likely to be overweight and least likely to eat family meals or fruits and vegetables.[100] They often snack on the nutrient-poor, energy-dense foods that are advertised.[101] The average child sees an estimated 40,000 TV commercials a year—many peddling foods high in sugar, saturated fat, and salt such as sugar-coated breakfast cereals, candy bars, chips, fast foods, and carbonated beverages.[102] More than half of all food advertisements are aimed specifically at children and market their products as fun and exciting.[103] Not surprisingly, the more time children spend watching television, the more they request these advertised foods and beverages—and they get their requests about half of the time.[104] The most popular foods and beverages are marketed to children and adolescents on the Internet as well, using "advergaming" (advertised product as part of a game), cartoon characters or "spokes-characters," and designated children's areas.[105]

The physically inactive time spent watching television is second only to time spent sleeping. Children also spend more time playing computer and video games. These activities use no more energy than resting, displace participation in more vigorous activities, and foster snacking on high-fat foods.[106] Simply reducing the amount of time spent watching television (and playing video games) can improve a child's BMI. The American Academy of Pediatrics (AAP) now recommends no television viewing before 2 years of age and thereafter limiting television and video time to two hours per day as a strategy to help prevent childhood obesity.[107]

Growth Overweight children develop a characteristic set of physical traits. They typically begin puberty earlier and so grow taller than their peers at first, but then they stop growing at a shorter height. They develop greater bone and muscle mass in response to the demand of having to carry more weight—both fat and lean weight. Consequently, they appear "stocky" even when they lose their excess fat.

Physical Health Like overweight adults, overweight children display a blood lipid profile indicating that atherosclerosis is beginning to develop—high levels of total cholesterol, triglycerides, and LDL cholesterol. Overweight children also tend to have high blood pressure; in fact, obesity is a leading cause of pediatric hypertension.[108] Their risks for developing type 2 diabetes and respiratory diseases (such as asthma) are also exceptionally high.[109] These relationships between childhood obesity and chronic diseases are discussed fully in Highlight 15.

Excessive television watching promotes physical inactivity and poor snacking habits.

Psychological Development In addition to the physical consequences, childhood obesity brings a host of emotional and social problems.[110] Because people frequently judge others on appearance more than on character, overweight children are often victims of prejudice. Many suffer discrimination by adults and rejection by their peers. They may have poor self-images, a sense of failure, and a passive approach to life. Television shows, which are a major influence in children's lives, often portray the fat person as the bumbling misfit.[111] Overweight children may come to accept this negative stereotype in themselves and in others, which can lead to additional emotional and social problems. Researchers investigating children's reactions to various body types find that both normal-weight and underweight children respond unfavorably to overweight bodies.

Prevention and Treatment of Obesity Medical science has worked wonders in preventing or curing many of even the most serious childhood diseases, but obesity remains a challenge.[112] Once excess fat has been stored, it is difficult to lose. In light of all this, parents are encouraged to make major efforts to prevent childhood obesity, starting at birth, or to begin treatment early—before adolescence.[113] The Expert Committee of the American Medical Association recommends specific eating and physical activity behaviors to prevent obesity, for all children (see Table 15-7).

Treatment of obesity must consider the many aspects of the problem and possible solutions. The main goal of obesity treatment is to improve long-term physical health through permanent healthy lifestyle habits.[114] The most successful approach integrates diet, physical activity, psychological support, and behavioral changes.[115] As a first step, the Expert Committee recommends that overweight and obese children and their families adopt the same healthy eating and activity behaviors presented in Table 15-7 for obesity prevention. The goal for overweight and obese children is to improve BMI. If the child's BMI does not improve after several months, the Expert Committee recommends increasing the intensity of the treatment. The level of intensity depends on treatment response, age, degree of obesity, health risks, and the family's readiness to change. Advanced treatment involves close follow-up monitoring by a health-care provider and greater support and structure for the child.[116]

Diet The initial goal for overweight children is to reduce the rate of weight gain; that is, to maintain weight as the child grows taller. Continued growth will then accomplish the desired change in BMI. Weight loss is usually not recommended

TABLE 15-7 Recommended Eating and Physical Activity Behaviors to Prevent Obesity

The Expert Committee of the American Medical Association recommends the following healthy habits for children 2 to 18 years of age to help prevent childhood obesity:

- Limit consumption of sugar-sweetened beverages, such as soft drinks and fruit flavored punches.
- Eat the recommended amounts of fruits and vegetables every day (2 to 4.5 cups per day based on age).
- Learn to eat age-appropriate portions of foods.
- Eat foods low in energy density such as those high in fiber and/or water and modest in fat.
- Eat a nutritious breakfast every day.
- Eat a diet rich in calcium.
- Eat a diet balanced in recommended proportions for carbohydrate, fat, and protein.
- Eat a diet high in fiber.
- Eat together as a family as often as possible.
- Limit the frequency of restaurant meals.
- Limit television watching or other screen time to no more than 2 hours per day and do not have televisions or computers in bedrooms.
- Engage in at least 60 minutes of moderate to vigorous physical activity every day.

SOURCE: S. E. Barlow, Expert Committee recommendations regarding the prevention, assessment, and treatment of child and adolescent overweight and obesity: Summary report, *Pediatrics* 120 (2007): S164–S192. Used by permission.

because diet restriction can interfere with growth and development. Intervention for some overweight children with accompanying medical conditions may warrant weight loss, but this treatment requires an individualized approach based on the degree of overweight and severity of the medical conditions.[117] Dietary strategies begin with those listed in Table 15-7 and progress to more structured family meal plans when necessary. For example, the child or the parent may be instructed to keep detailed records of dietary intake and physical activity.

Dietary Guidelines for Americans 2010

Children are encouraged to maintain kcalorie balance to support normal growth and development without promoting excess weight gain.

Physical Activity The many benefits of physical activity are well known but often are not enough to motivate overweight people, especially children. Yet regular vigorous activity can improve a child's weight, body composition, and physical fitness.[118] Ideally, parents will limit sedentary activities and encourage at least one hour of daily physical activity to promote strong skeletal, muscular, and cardiovascular development and instill in their children the desire to be physically active throughout life. Opportunities to be physically active can include team, individual, and recreational activities (see Figure 15-11). Most importantly, parents need to set a good example. Physical activity is a natural and lifelong behavior of healthy living. It can be as simple as riding a bike, playing tag, jumping rope, or doing chores. The AAP supports the efforts of schools to include more physical activity in the curriculum and encourages parents to support their children's participation.[119]

Physical Activity Guidelines for Americans 2008

- Children and adolescents should do 60 minutes (1 hour) or more of physical activity daily.
- *Aerobic.* Most of the 60 or more minutes a day should be either moderate- or vigorous-intensity aerobic physical activity and should include vigorous-intensity physical activity at least three days a week.
- *Muscle-strengthening.* As part of their 60 or more minutes of daily physical activity, children and adolescents should include muscle-strengthening physical activity on at least three days of the week.
- *Bone-strengthening.* As part of their 60 or more minutes of daily physical activity, children and adolescents should include bone-strengthening physical activity on at least three days of the week.

FIGURE 15-11 **Physical Activity Pyramid for Kids**

Preschoolers (2 to 5 years)
Games in the yard or park
Family walks after dinner
Playing with the dog
Dancing freestyle
Tumbling and gymnastics
T-ball
Playing catch
Family bike rides
Building a snowman
Family swimming at the pool or beach
Playing hide and seek
Older children (6 to 12 years)
Throwing a Frisbee
Jumping rope
Bicycling
Playing games and sports such as soccer, softball, baseball, and basketball
Rollerblading
Running
Weight training with light weights
Dancing
Competitive swimming
Snowboarding or skiing
Family kayaking, canoeing, or surfing

Psychological Support Weight-loss programs that involve parents and other caregivers in treatment report greater success than those without parental involvement. Because obesity in parents and their children tends to be positively correlated, both benefit when parents participate in a weight-loss program. Parental attitudes about food greatly influence children's eating behavior, so it is important that the influence be positive. Otherwise, eating problems may become exacerbated.

Behavioral Changes In contrast to traditional weight-loss programs that focus on *what* to eat, behavioral programs focus on *how* to eat. These techniques involve changing learned habits that lead a child to eat excessively.

Drugs The use of weight-loss drugs to treat obesity in children merits special concern because the long-term effects of these drugs on growth and development have not been studied.[120] The drugs may be used in addition to structured lifestyle changes for carefully selected children or adolescents who are at high risk for severe obesity in adulthood. Only two obesity drugs, orlistat and sibutramine (see Chapter 9), have been approved for limited use in children and adolescents.

Surgery The use of surgery to treat severe obesity in adults (see Chapter 9) has created interest in its use for adolescents. Limited research shows that after surgery extremely obese adolescents lose significant weight and experience improvements in type 2 diabetes and cardiovascular risk factors.[121] The selection criteria for surgery to treat obesity in adolescents ♦ are based on recommendations of a panel of pediatricians and surgeons.[122]

♦ Surgery may be an option for adolescents who meet the following criteria:
- Have reached physical maturity
- BMI ≥50 or BMI ≥40 with significant weight-related health problems
- Have experienced failure in a formal, six-month weight-loss program
- Are capable of adhering to the long-term lifestyle changes required after surgery

Obesity is prevalent in our society. Because treatment of obesity is frequently unsuccessful, it is most important to prevent its onset. Above all, be sensible in teaching children how to maintain appropriate body weight. Children can easily get the impression that their worth is tied to their body weight. Parents and the media are most influential in shaping self-concept, weight concerns, and dieting practices.[123] Some parents fail to realize that society's ideal of slimness can be perilously close to starvation and that a child encouraged to "diet" cannot obtain the energy and nutrients required for normal growth and development. Even healthy children without diagnosable eating disorders have been observed to limit their growth through "dieting." Weight loss in truly overweight children can be managed without compromising growth, but it should be overseen by a health-care professional.

Mealtimes at Home

Traditionally, parents served as **gatekeepers**, determining what foods and activities were available in their children's lives. Then the children made their own selections. Gatekeepers who wanted to promote nutritious choices and healthful habits provided access to nutrient-dense, delicious foods and opportunities for active play at home.

In today's consumer-oriented society, children have greater influence over family decisions concerning food—the fast-food restaurant the family chooses when eating out, the type of food the family eats at home, and the specific brands the family purchases at the grocery store. Parental guidance in food choices is still necessary, but teaching children consumer skills to help them make informed choices is equally important.

© Masterfile

Eating is more fun for children when friends are there.

gatekeepers: with respect to nutrition, key people who control other people's access to foods and thereby exert profound impacts on their nutrition. Examples are the spouse who buys and cooks the food, the parent who feeds the children, and the caregiver in a day-care center.

Honoring Children's Preferences Researchers attempting to explain children's food preferences encounter contradictions. Children say they like colorful foods, yet they most often reject green and yellow vegetables in favor of brown peanut butter and white potatoes, apple wedges, and bread. They seem to like raw vegetables better than cooked ones, so it is wise to offer vegetables that are raw or slightly undercooked, served separately, and easy to eat. Foods should be warm, not hot, because a child's mouth is much more sensitive than an adult's. The flavor should be mild because a child has more taste buds, and smooth foods such as mashed potatoes or split-pea soup should contain no lumps (a child wonders, with some disgust, what the lumps might be).

Make mealtimes fun for children. Young children like to eat at little tables and to be served small portions of food. They like sandwiches cut in different geometric shapes and common foods called silly names. They also like to eat with other children, and they tend to eat more when in the company of their friends. Children are also more likely to give up their prejudices against foods when they see their peers eating them.

Learning through Participation Allowing children to help plan and prepare the family's meals provides enjoyable learning experiences and encourages children to eat the foods they have prepared. Vegetables are pretty, especially when fresh, and provide opportunities for children to learn about color, seeds, growing vegetables, and shapes and textures—all of which are fascinating to young children. Measuring, stirring, washing, and arranging foods are skills that even a young child can practice with enjoyment and pride (see Table 15-8).

Avoiding Power Struggles Problems over food often arise during the second or third year, when children begin asserting their independence. Many of these problems stem from the conflict between children's developmental stages and capabilities and parents who, in attempting to do what they think is best for their children, try to control every aspect of eating. Such conflicts can disrupt children's abilities to regulate their own food intakes or to determine their own likes and dislikes. For example, many people share the misconception that children must be persuaded or coerced to try new foods. In fact, the opposite is true. When children are forced to try new foods, even by way of rewards, they are less likely to try those foods again than are children who are left to decide for themselves. Similarly, when children are restricted from eating their favorite foods, they are more likely to want those foods.[124] Wise parents provide healthful foods and allow their child to determine *how much* and even *whether* to eat.

When introducing new foods, offer them one at a time and only in small amounts such as one bite at first. The more often a food is presented to a young child, the more likely the child will accept that food.[125] Offer the new food at the beginning of the meal, when the child is hungry, and allow the child to make the decision to accept or reject it. Never make an issue of food acceptance.

Choking Prevention Parents must always be alert to the dangers of choking. A choking child is silent, so an adult should be present whenever a child is eating. Make sure the child sits when eating; choking is more likely when a child is running or falling. (See the margin list on p. 502 for foods and nonfood items most likely to cause choking.)

Playing First Children may be more relaxed and attentive during meals if outdoor play or other fun activities are scheduled before, rather than immediately after, mealtimes. Otherwise children "hurry up and eat" so that they can go play.

Snacking Parents may find that when their children snack, they aren't hungry at mealtimes. Instead of teaching children *not* to snack, parents are wise to teach them *how* to snack. Provide snacks that are as nutritious as the foods served at mealtime. Snacks can even be mealtime foods served individually over time, instead of all at once on one plate. When providing snacks to children, think of the five food groups and offer such snacks as pieces of cheese, tangerine slices, and egg salad on whole-wheat crackers (see Table 15-9, p. 520). Snacks that are easy to prepare should be readily available to children, especially if they arrive home from school before their parents.

To ensure that children have healthy appetites and plenty of room for nutritious foods when they are hungry, parents and teachers must limit access to candy, soft drinks, and other concentrated sweets. Limiting access includes limiting the amount of pocket money children have to buy such foods themselves. If these foods are permitted in large quantities, the only possible outcomes are nutrient deficiencies, obesity, or both. The preference for sweets is innate; most children do not naturally select nutritious foods on the basis of taste. When children are allowed to create meals freely from a variety of foods, they typically select foods that provide a lot of sugar. When their parents are watching, or even when they only think their parents are watching, children improve their selections.

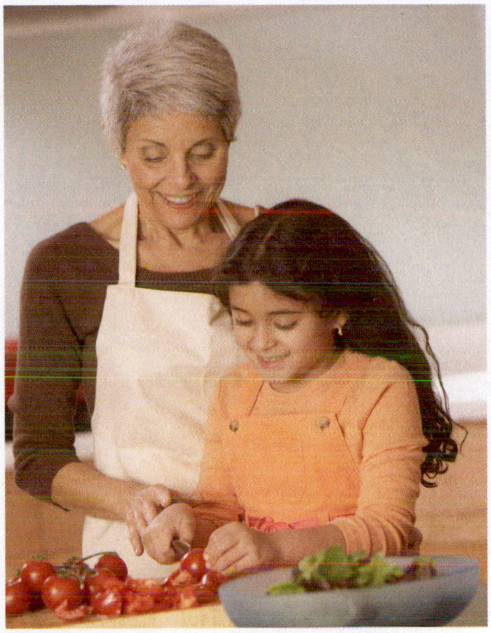

© Masterfile

Children enjoy eating the foods they help to prepare.

TABLE 15-8 Food Skills of Preschool Children[a]

Age 2 years, when large muscles develop:

- Uses a spoon
- Helps feed self
- Lifts and drinks from a cup
- Helps scrub fruits and vegetables, tear lettuce or greens, snap green beans, or dip foods
- Wipes table
- Places items in recycle bin or trash

Age 3 years, when medium hand muscles develop:

- Spears food with a fork
- Feeds self independently
- Adds ingredients to pancake batters, cookie recipes, salads or other mixed dishes
- Helps wrap, pour, mix, shake, stir, or spread foods
- Helps crack nuts with supervision

Age 4 years, when small finger muscles develop:

- Uses all utensils and napkin
- Helps roll, juice, or mash foods
- Helps measure dry ingredients
- Cracks egg shells
- Helps make sandwiches and toss salads
- Peels foods such as hard-boiled eggs and bananas

Age 5 years, when fine coordination of fingers and hands develops:

- Measures liquids
- Helps grind, grate, and cut (soft foods with dull knife)
- Uses hand mixer with supervision

[a]These ages are approximate. Healthy, normal children develop at their own pace.

TABLE 15-9 Healthful Snack Ideas—Think Food Groups, Alone and in Combination

Selecting two or more foods from different food groups adds variety and nutrient balance to snacks. The combinations are endless, so be creative. Whenever possible, choose whole grains, low-fat or reduced-fat milk products, and lean meats.

Grains

Grain products are filling snacks, especially when combined with other foods:

- Cereal with fruit and milk
- Crackers and cheese
- Whole-grain toast with peanut butter
- Popcorn with grated cheese
- Oatmeal raisin cookies with milk

Vegetables

Cut-up, fresh, raw vegetables make great snacks alone or in combination with foods from other food groups:

- Celery with peanut butter
- Broccoli, cauliflower, and carrot sticks with a flavored cottage cheese dip

Fruits

Fruits are delicious snacks and can be eaten alone—fresh, dried, or juiced—or combined with other foods:

- Apples and cheese
- Bananas and peanut butter
- Peaches with yogurt
- Raisins mixed with sunflower seeds or nuts

Protein foods

Seafood, meat, poultry, eggs, legumes, nuts, seeds, and soy products add protein to snacks:

- Refried beans with nachos and cheese
- Tuna on crackers
- Luncheon meat on whole-grain bread

Milk and Milk Products

Milk can be used as a beverage with any snack, and many other milk products, such as yogurt and cheese, can be eaten alone or with other foods as listed above.

Sweets need not be banned altogether. Children who are exceptionally active can enjoy high-kcalorie foods such as ice cream or pudding from the milk group or pancakes from the bread group. Sedentary children need to become more active so they can also enjoy some of these foods without unhealthy weight gain.

Preventing Dental Caries Children frequently snack on sticky, sugary foods that stay on the teeth and provide an ideal environment for the growth of bacteria that cause dental caries. Teach children to brush and floss after meals, to brush or rinse after eating snacks, to avoid sticky foods, and to select crisp or fibrous foods frequently.

Serving as Role Models In an effort to practice these many tips, parents may overlook perhaps the single most important influence on their children's food habits—themselves.[126] Parents who don't eat carrots shouldn't be surprised when their children refuse to eat carrots. Likewise, parents who comment negatively on the smell of brussels sprouts may not be able to persuade children to try them. Children learn much through imitation. It is not surprising that children prefer the foods other family members enjoy and dislike foods that are never offered to them.[127] Parents, older siblings, and other caregivers set an irresistible example by sitting with younger children, eating the same foods, and having pleasant conversations during mealtimes.

While serving and enjoying food, caregivers can promote both physical and emotional growth at every stage of a child's life. They can help their children develop both a positive self-concept and a positive attitude toward food. With good beginnings, children will grow without the conflicts and confusions about food that can lead to nutrition and health problems.

Nutrition at School

While parents are doing what they can to establish good eating habits in their children at home, others are preparing and serving foods to their children at day-care centers and schools. In addition, children begin to learn about food and nutrition in the classroom. Meeting the nutrition ♦ and education needs of children is critical to supporting their healthy growth and development.[128]

♦ The American Dietetic Association has set nutrition standards for child-care programs. Among them, meal plans should include the following:
- Be nutritionally adequate and consistent with the *Dietary Guidelines for Americans*
- Involve parents in planning
- Follow recommended meal patterns that balance energy and nutrients with children's ages, appetites, activity levels, and special needs while respecting cultural and ethnic differences
- Minimize added fat, sugar, and sodium
- Emphasize fresh fruit, fresh and frozen vegetables, and whole grains
- Provide furniture and eating utensils that are age appropriate and developmentally suitable to encourage children to accept and enjoy mealtime

Meals at School The U.S. government assists schools financially so that every student can receive nutritious meals at school. Both the School Breakfast Program and the National School Lunch Program provide meals at a reasonable cost to children from families without the financial means to pay. Meals are available free or at reduced cost to children from low-income families. In addition, schools can obtain food commodities. Nationally, the U.S. Department of Agriculture (USDA) administers the programs; on the state level, state departments of education operate them.* The programs usually cost local school districts little, but the educational rewards are great. Several studies have reported that children who participate in school food programs perform better in the classroom.[129]

More than 30 million children receive lunches through the National School Lunch Program—more than half of them free or at a reduced price.[130] School lunches offer a variety of food choices and help children meet at least one-third of their recommended intakes for energy, protein, vitamin A, vitamin C, iron, and calcium. Table 15-10 shows school lunch patterns for children of different ages and specifies the numbers of servings of milk, protein-rich foods (meat, poultry, fish, cheese, eggs, legumes, or peanut butter), vegetables, fruits, and breads or

*School lunches in Canada are administered locally and therefore vary from area to area.

TABLE 15-10 School Lunch Patterns for Different Ages[a]

Food Group	Preschool (Age)		Grade School through High School (Grade)		
	1 to 2	*3 to 4*	*K to 3*	*4 to 6*	*7 to 12*
Protein foods					
1 serving:					
Lean meat, poultry, or fish	1 oz	1½ oz	1½ oz	2 oz	3 oz
Cheese	1 oz	1½ oz	1½ oz	2 oz	3 oz
Large egg(s)	½	¾	¾	1	1½
Cooked dry beans or peas	¼ c	⅜ c	⅜ c	½ c	¾ c
Peanut butter	2 tbs	3 tbs	3 tbs	4 tbs	6 tbs
Yogurt	½ c	¾ c	¾ c	1 c	1½ c
Peanuts, soynuts, tree nuts, or seeds[b].	½ oz	¾ oz	¾ oz	1 oz	1½ oz
Vegetable and/or fruit					
2 or more servings, both to total	½ c	½ c	½ c	¾ c	¾ c
Bread or bread alternate[c]					
Servings	5/week	8/week	8/week	8/week	10/week
Milk					
1 serving of fluid milk	¾ c	¾ c	1 c	1 c	1 c

[a]The quantities listed represent per-lunch minimums for each age and grade except those for the oldest group, which are recommendations. Schools unable to serve the recommended quantities for grades 7 to 12 must provide at least the amount shown for grades 4 to 6.

[b]These meat alternates may be used to meet no more than half of the meat or meat alternate requirement; therefore, they must be used in a meal with another meat or meat alternate.

[c]Schools must serve daily at least ½ serving of bread or bread alternate to the youngest age group and at least 1 serving to older children.

SOURCE: U.S. Department of Agriculture, National School Lunch Program Regulations, revised January 1, 1998.

other grain foods. In an effort to help reduce disease risk, all government-funded meals served at schools must follow the *Dietary Guidelines for Americans*.

Parents often rely on school lunches to meet a significant part of their children's nutrient needs on school days. Indeed, students who regularly eat school lunches have higher intakes of many nutrients and fiber than students who do not.[131]

The School Breakfast Program ♦ is available in more than 80 percent of the nation's schools that offer school lunch, and close to 9 million children participate in it.[132] Nevertheless, for many children who need it, the School Breakfast Program is either unavailable, or the children do not participate in it.[133] The majority of children who eat school breakfasts are from low-income families. As research results continue to emphasize the positive impact breakfast has on school performance and health, vigorous campaigns to expand and improve school breakfast programs are under way.[134]

♦ The school breakfast must contain at a minimum:
- One serving of fluid milk
- One serving of fruit or vegetable or full-strength juice
- Two servings of bread or bread alternates; or two servings of meat or meat alternates; or one of each

Another federal program, the Child and Adult Care Food Program (CACFP), operates similarly and provides funds to organized child-care programs. All eligible children, centers, and family day-care homes may participate. Sponsors are reimbursed for most meal costs and may also receive USDA commodity foods.

iStockphoto.com/Monkey Business Images

School lunches provide children with nourishment at little or no charge.

Competing Influences at School Serving healthful lunches is only half the battle; students need to eat them, too. Short lunch periods and long waiting lines prevent some students from eating a school lunch and leave others with too little time to complete their meals.[135] Nutrition efforts at schools are also undermined when students can buy what the USDA labels "competitive" or "nonreimbursable" foods—meals from fast-food restaurants or a la carte foods such as pizza or snack foods and carbonated beverages from snack bars, school stores, and vending machines.[136] In one study, students who selected competitive foods in addition to, or instead of, school meals consumed more energy and fat and less calcium and vitamin A than those who selected only the school lunch.[137]

Increasingly, school-based nutrition issues are being addressed by legislation. Some states restrict the sale of competitive foods and have higher rates of participation in school meal programs than the national average. Federal legislation mandates that all school districts that participate in the USDA's National School Lunch Program develop and put in place a local wellness policy.[138] By law, wellness policies must:[139]

- Set goals for nutrition education, physical activity, and other school-based activities.
- Establish nutrition guidelines for all foods available on school campuses during the school day.
- Develop a plan to measure policy implementation.

School districts across the nation have made progress toward meeting these goals, but implementation is inconsistent, and because wellness policies are established locally, a great deal of variety exists among them. Some are well defined and detailed, while others are vague and less detailed.[140] To enhance local wellness policies, standards for competitive foods and beverages served in schools spell out appropriate fat, saturated fat, kcalorie, sugar, and sodium contents (see Table 15-11).[141] Establishing and implementing these nutrition standards for competitive foods and beverages helps to ensure that all foods served in schools are consistent and comply with the *Dietary Guidelines for Americans*.

IN SUMMARY **Children's appetites and nutrient needs reflect their stage of growth. Those who are chronically hungry and malnourished suffer growth retardation; when hunger is temporary and nutrient deficiencies are mild, the problems are usually more subtle—such as poor academic performance. Iron deficiency is widespread and has many physical and behavioral consequences. "Hyper" behavior is not caused by poor nutrition; misbehavior may be due to lack of sleep, too little physical activity, or too much television,**

TABLE 15-11 Foods and Beverages That Meet Recommended School Food Standards

Preferred Foods for All Students

Foods	Beverages

These foods are fruits, vegetables, whole grains and related combination products,[a] and nonfat and low-fat milk products that are limited to 200 kcalories or less per serving and:
• No more than 35 percent of total kcalories from fat.
• Less than 10 percent of total kcalories from saturated fats.
• *Trans* fat-free (≤0.5 g per serving).
• 35 percent or less of kcalories from total sugars, except for yogurt with no more than 30 g of total sugars, per 8 oz portion as packaged.
• Sodium content of 200 mg or less per portion as packaged.[b]
	These beverages are:
	• Water without flavoring, additives, or carbonation.
	• Low-fat[c] and nonfat milk:
	  • Lactose-free and soy beverages are included.
	  • Flavored milk with no more than 22 g of total sugars per 8 oz serving.
	• 100 percent fruit juice in 4 oz portion for elementary/middle school and 8 oz for high school.
	• Caffeine-free, with the exception of trace amounts of naturally occurring caffeine substances.

Snacks for High School Students after School

Foods	Beverages
Snack foods are those that do not exceed 200 kcalories per portion as packaged and: • No more than 35 percent of total kcalories from fat. • Less than 10 percent of total kcalories from saturated fats. • *Trans* fat-free (≤0.5 g per serving). • 35 percent or less of kcalories from total sugars. • Sodium content of 200 mg or less per portion as packaged.	These beverages are: • Nonnutritive-sweetened, noncaffeinated, nonfortified beverages with less than 5 kcalories per portion as packaged.

[a]Combination products must contain a total of one or more servings as packaged of fruit, vegetables, or whole-grain products per portion.
[b]Á la carte entrée items meet fat and sugar limits and have a sodium content of 480 mg or less.
[c]1 percent milk fat.
SOURCE: V. A. Stallings and A. L. Yaktine, eds., *Nutrition Standards for Foods in Schools: Leading the Way Toward Healthier Youth* (Washington, D.C.: National Academies Press, 2007), p. 5.

among other factors. Childhood obesity has become a major health problem. Adults at home and at school need to provide children with nutrient-dense foods and teach them how to make healthful diet and activity choices.

Nutrition during Adolescence

Teenagers make many more choices for themselves than they did as children. They are not fed, they eat; they are not sent out to play, they choose to go. At the same time, social pressures thrust choices at them, such as whether to drink alcoholic beverages and whether to develop their bodies to meet extreme ideals of slimness or athletic prowess. Their interest in nutrition—both valid information and misinformation—derives from personal, immediate experiences. They are concerned with how diet can improve their lives now—they engage in fad dieting in order to fit into a new bathing suit, avoid greasy foods in an effort to clear acne, or eat a pile of spaghetti to prepare for a big sporting event. In presenting information on the nutrition and health of adolescents, this section includes many topics of interest to teens.

Growth and Development

With the onset of **adolescence**, the steady growth of childhood speeds up abruptly and dramatically, and the growth patterns of females and males become distinct. Hormones direct the intensity of the adolescent growth spurt, profoundly affecting every organ of the body, including the brain. After two to three years of intense growth and a few more at a slower pace, physically mature adults emerge.

In general, the adolescent growth spurt begins at age 10 or 11 for females and at 12 or 13 for males. It lasts about two and a half years. Before **puberty**, male and female body compositions differ only slightly, but during the adolescent spurt, differences between the genders become apparent in the skeletal system, lean body mass, and fat stores. In females, fat assumes a larger percentage of total

adolescence: the period from the beginning of puberty until maturity.

puberty: the period in life in which a person becomes physically capable of reproduction.

body weight, and in males, the lean body mass—principally muscle and bone—increases much more than in females (review Figure 8-7 on p. 254). On average, males grow 8 inches taller, and females, 6 inches taller. Males gain approximately 45 pounds, and females, about 35 pounds.

© Image Source/Getty Images

Nutritious snacks contribute valuable nutrients and energy to an active teen's diet.

Energy and Nutrient Needs

Energy and nutrient needs are greater during adolescence than at any other time of life, except pregnancy and lactation. In general, nutrient needs rise throughout childhood, peak in adolescence, and then level off or even diminish as the teen becomes an adult.

Energy Intake and Activity The energy needs of adolescents vary greatly, depending on their current rate of growth, gender, body composition, and physical activity.[142] Boys' energy needs may be especially high; they typically grow faster than girls and, as mentioned, develop a greater proportion of lean body mass. An exceptionally active boy of 15 may need 3500 kcalories or more a day just to maintain his weight. Girls start growing earlier than boys and attain shorter heights and lower weights, so their energy needs peak sooner and decline earlier than those of their male peers. A sedentary girl of 15 whose growth is nearly at a standstill may need fewer than 1800 kcalories a day if she is to avoid excessive weight gain. Thus adolescent girls need to pay special attention to being physically active and selecting foods of high nutrient density so as to meet their nutrient needs without exceeding their energy needs.

Dietary Guidelines for Americans 2010

Adolescents should engage in 60 minutes or more of physical activity per day.

The insidious problem of obesity becomes ever more apparent in adolescence and often continues into adulthood. The problem is most evident in females of African American descent and in Hispanic children of both genders. Without intervention, overweight adolescents face numerous physical and socioeconomic consequences for years to come. The consequences of obesity are so dramatic and our society's attitude toward obese people is so negative that even teens of normal or below-normal weight may perceive a need to lose weight. When taken to extremes, restrictive diets bring dramatic physical consequences of their own, as Highlight 8 explains.

Vitamins The RDA (or AI) for most vitamins increases during the adolescent years (see the table on the inside front cover). Several of the vitamin recommendations for adolescents are similar to those for adults, including the recommendations for vitamin D. Vitamin D is essential for bone growth and development. Recent studies of vitamin D status in adolescents show that as many as half of adolescents are vitamin D deficient; blacks, females, and overweight adolescents are most at risk.[143] Such information led to revised recommendations.[144] The previous recommendation of 5 micrograms per day was increased to the current RDA of 15 micrograms per day. Adolescents who do not receive enough vitamin D from vitamin D–fortified milk (2.5 micrograms per cup) and vitamin D–fortified foods such as cereals each day should take a vitamin D supplement.[145] Although drinking one quart of vitamin D–fortified milk will provide the recommended amount, the majority of adolescents in the United States drink much less than this.

Iron The need for iron increases during adolescence for both females and males, but for different reasons. Iron needs increase for females as they start to lose blood through menstruation and for males as their lean body mass develops. Hence, the RDA increases at age 14 for both males and females. For females, the RDA remains high into late adulthood. For males, the RDA returns to preadolescent values in early adulthood.

In addition, iron needs increase when the adolescent growth spurt begins, whether that occurs before or after age 14. Therefore, boys in a growth spurt need an additional 2.9 milligrams of iron per day above the RDA for their age; girls need an additional 1.1 milligrams per day.[146]

Furthermore, iron recommendations for girls before age 14 do not reflect the iron losses of menstruation. The average age of menarche (first menstruation) in the United States is 12.5 years. Therefore, for girls younger than the age of 14 who have started to menstruate, an additional 2.5 milligrams of iron per day is recommended.[147] Thus the RDA for iron depends not only on age and gender but also on whether the individual is in a growth spurt or has begun to menstruate, as listed in the margin. ♦

Iron intakes often fail to keep pace with increasing needs, especially for females, who typically consume fewer iron-rich foods such as meat and fewer total kcalories than males. Not surprisingly, iron deficiency is most prevalent among adolescent girls. Iron-deficient children and teens score lower on standardized tests than those who are not iron deficient.

♦ Iron RDA for males:
- 9–13 yr: 8 mg/day
- 9–13 yr in growth spurt: 10.9 mg/day
- 14–18 yr: 11 mg/day
- 14–18 yr in growth spurt: 13.9 mg/day

Iron RDA for females:
- 9–13 yr: 8 mg/day
- 9–13 yr in menarche: 10.5 mg/day
- 9–13 yr in menarche and growth spurt: 11.6 mg/day
- 14–18 yr: 15 mg/day
- 14–18 yr in growth spurt: 16.1 mg/day

Calcium Adolescence is a crucial time for bone development, and the requirement for calcium reaches its peak during these years.[148] Unfortunately, low calcium intakes among adolescents have reached crisis proportions: 90 percent of females and 70 percent of males aged 12 to 19 years have calcium intakes below recommendations.[149] ♦ Low calcium intakes during times of active growth, especially if paired with physical inactivity, can compromise the development of peak bone mass, which is considered the best protection against adolescent fractures and adult osteoporosis. Increasing milk products in the diet to meet calcium recommendations greatly increases bone density.[150] Once again, however, teenage girls are most vulnerable, for their milk—and therefore their calcium—intakes begin to decline at the time when their calcium needs are greatest. Furthermore, women have much greater bone losses than men in later life. In addition to dietary calcium, bones grow stronger with physical activity. However, because most high schools do not require students to attend physical education classes, many adolescents are not as physically active as healthy bones demand.

♦ Calcium RDA for males and females:
- 9–13 yr: 1300 mg/day

Dietary Guidelines for Americans 2010

Children 9 years of age and older should consume 3 cups per day of fat-free or low-fat milk or equivalent milk products.

Food Choices and Health Habits

Teenagers like the freedom to come and go as they choose. They eat what they want if it is convenient and if they have the time. With a multitude of after school, social, and job activities, they almost inevitably fall into irregular eating habits. At any given time on any given day, a teenager may be skipping a meal, eating a snack, preparing a meal, or consuming food prepared by a parent or restaurant. Adolescents who frequently eat meals with their families, however, eat more fruits, vegetables, grains, and calcium-rich foods, and drink fewer soft drinks, than those who seldom eat with their families.[151] Some research shows that the more often teenagers eat dinner with their families, the less likely they are to smoke, drink, or use drugs; other research supports these findings only in teenage girls.[152] Many adolescents also begin to skip breakfast on a regular basis, missing out on important nutrients that are not made up at later meals during the day. Compared with those who skip breakfast, teenagers who do eat breakfast have higher intakes of vitamin A, vitamin C, and riboflavin, as well as calcium, iron, and zinc.[153] Teenagers who eat breakfast are therefore more likely to meet their nutrient recommendations.

Breakfast skipping may also lead to weight gain in adolescents. Research shows a dose-response, inverse relationship between breakfast eating and BMI.[154] As adolescents make the transition to adulthood, not only do they skip breakfast more often, they also eat fast food more often. Both skipping breakfast and eating fast foods lead to weight gain.[155]

Ideally, in light of adolescents' busy schedules and desire for freedom, parents continue to play the role of gatekeepers, controlling the type and availability of food in the teenager's environment. Teenagers should find plenty of nutritious, easy-to-grab foods in the refrigerator (meats for sandwiches; low-fat cheeses;

fresh, raw vegetables and fruits; fruit juices; and milk) and more in the cabinets (whole-grain breads and crackers, peanut butter, nuts, popcorn, and cereal). In many households today, with adults working outside the home, teenagers perform some of the gatekeepers' roles, such as shopping for groceries or choosing fast or prepared foods.

Snacks Snacks typically provide at least one-fourth of the average teenager's daily food energy intake. Often, favorite snacks are too high in added sugars, saturated fat, and sodium and too low in fiber.[156] A survey of more than 4000 adolescents, however, found that those who ate snacks more often had higher intakes of fruit compared with those who ate snacks less often.[157] Table 15-9 on p. 520 shows how to combine foods from different food groups to create healthy snacks.

Beverages Most frequently, adolescents drink soft drinks instead of fruit juice or milk with lunch, supper, and snacks. About the only time they select fruit juices is at breakfast. When teens drink milk, they are more likely to consume it with a meal (especially breakfast) than as a snack. Because of their greater food intakes, boys are more likely than girls to drink enough milk to meet their calcium needs.

Soft drinks, when chosen as the primary beverage, may affect bone density, partly because they displace milk from the diet.[158] Over the past three decades, teens (especially girls) have been drinking more soft drinks and less milk. Adolescents who drink soft drinks regularly have a higher energy intake and a lower calcium intake than those who do not; they are also more likely to be overweight.[159]

Soft drinks containing caffeine present a different problem if caffeine ♦ intake becomes excessive. Caffeine seems to be relatively harmless when used in moderate doses (the equivalent of fewer than three 12-ounce cola beverages a day). In greater amounts, however, it can cause the symptoms associated with anxiety, such as sweating, tenseness, and inability to concentrate.

♦ For perspective, caffeine-containing soft drinks typically deliver between 30 and 55 mg of caffeine per 12-ounce can. A pharmacologically active dose of caffeine is defined as 200 mg. Appendix H starts with a table listing the caffeine contents of selected foods, beverages, and drugs.

Eating Away from Home Adolescents eat about one-third of their meals away from home, and their nutritional welfare is enhanced or hindered by the choices they make. A lunch consisting of a hamburger, a chocolate shake, and french fries supplies substantial quantities of many nutrients at a kcalorie cost of about 800, an energy intake some adolescents can afford. When they eat this sort of lunch, teens can adjust their breakfast and dinner choices to include fruits and vegetables for vitamin A, vitamin C, folate, and fiber and lean meats and legumes for iron and zinc. (See Appendix H for the nutrient contents of fast foods.) Fortunately, many fast-food restaurants are offering more nutritious choices than the standard hamburger meal.

Peer Influence Physical maturity and growing independence present adolescents with new choices. The consequences of those choices will influence their health and nutrition status both today and throughout life. Many of the food and health choices adolescents make reflect the opinions and actions of their peers. When others perceive milk as "babyish," a teen may choose soft drinks instead; when others skip lunch and hang out in the parking lot, a teen may join in for the camaraderie, regardless of hunger. Some teenagers begin using drugs, alcohol, and tobacco; others wisely refrain. Adults can set up the environment so that nutritious foods are available and can stand by with reliable information and advice about health and nutrition, but the rest is up to the adolescents. Ultimately, they make the choices. (Highlight 8 examines the influence of social pressures on the development of eating disorders.)

Drug Abuse The nutrition problems associated with drugs vary in degree, but drug abusers in general face multiple nutrition problems. ♦ During withdrawal from drugs, an important part of treatment is to identify and correct nutrient deficiencies.

♦ Nutrition problems of drug abusers:
- They buy drugs with money that could be spent on food.
- They lose interest in food during "highs."
- They use drugs that suppress appetite.
- Their lifestyle fails to promote good eating habits.
- If they use intravenous (IV) drugs, they may contract AIDS, hepatitis, or other infectious diseases, which increase their nutrient needs. Hepatitis also causes taste changes and loss of appetite.
- Medicines used to treat drug abuse may alter nutrition status.

Alcohol Abuse Sooner or later all teenagers face the decision of whether to drink alcohol. The law forbids the sale of alcohol to people younger than 21, but most adolescents who want it can get it. By the end of high school, 77 percent of students have tried alcohol, and about half have been drunk at least once.[160] Highlight 7 describes how alcohol affects nutrition status. To sum it up, alcohol provides energy but no nutrients, and it can displace nutritious foods from the diet. Alcohol alters nutrient absorption and metabolism, so imbalances develop. People who cannot keep their alcohol use moderate must abstain to maintain their health. Highlight 7 lists resources for people with alcohol-related problems.

iStockphoto.com/kali9

Because their lunches rarely include fruits, vegetables, or milk, many teens fail to get all the vitamins and minerals they need each day.

Smoking Slightly less than 30 percent of U.S. high school students report smoking a cigarette in the previous month.[161] This is the lowest rate of smoking among high school students since 1991. Cigarette smoking is a pervasive health problem causing thousands of people to suffer from cancer and diseases of the cardiovascular, digestive, and respiratory systems. These effects are beyond the scope of nutrition, but smoking cigarettes does influence hunger, body weight, and nutrient status.

Smoking a cigarette eases feelings of hunger. When smokers receive a hunger signal, they can quiet it with cigarettes instead of food. Such behavior ignores body signals and postpones energy and nutrient intake. Indeed, smokers tend to weigh less than nonsmokers and to gain weight when they stop smoking. People contemplating giving up cigarettes should know that the average weight gain is about 10 pounds in the first year. Smokers wanting to quit should prepare for the possibility of weight gain and adjust their diet and activity habits so as to maintain weight during and after quitting. Smoking cessation programs need to include strategies for weight management.

Nutrient intakes of smokers and nonsmokers differ. Smokers tend to have lower intakes of dietary fiber, vitamin A, beta-carotene, folate, and vitamin C. The association between smoking and low intakes of fruits and vegetables rich in these nutrients may be noteworthy, considering their protective effect against lung cancer (see Highlight 11).

Smokeless Tobacco Like cigarettes, smokeless tobacco use is linked to many health problems, from minor mouth sores to tumors in the nasal cavities, cheeks, gums, and throat. The risk of mouth and throat cancers is even greater than for smoking tobacco. Other drawbacks to tobacco chewing and snuff dipping include bad breath, stained teeth, and blunted senses of smell and taste. Tobacco chewing also damages the gums, tooth surfaces, and jawbones, making tooth loss later in life likely.

The nutrition and lifestyle choices people make as children and adolescents have long-term, as well as immediate, effects on their health. Highlight 15 describes how sound choices and good habits during childhood and adolescence can help prevent chronic diseases later in life.

Nutrition Portfolio

Encouraging children to eat nutritious foods today helps them learn how to make healthy food choices tomorrow.

- If there are children in your life, think about the food they eat and consider whether they receive enough food for healthy growth, but not so much as to lead to obesity.
- Describe the advantages of physical activity to children's health and well-being.
- Plan a day's menu for a child 4 to 8 years of age, making sure to include foods that provide enough calcium and iron.
- Now, go to Diet Analysis Plus and create a profile for a child 4 to 8 years of age. Enter the day's menu you suggested in the previous exercise and see if you met the basic requirements for that child.

Diet Analysis PLUS To complete this exercise, go to your Diet Analysis Plus at www.cengage.com/sso.

Nutrition on the Net

For further study of topics covered in this chapter, log on to **www.cengagebrain.com** and search for ISBN 084006845X.

- Learn more about breast milk banks from the HumanMilk Banking Association of North America: **www.hmbana.org**
- Search for "infant health," "baby bottle tooth decay," "premature birth," "hyperactivity," "food allergies," and "teenage health," at the U.S. Government health information site: **www.healthfinder.gov**
- Learn how to care for infants, children, and adolescents from the American Academy of Pediatrics and the Canadian Paediatric Society: **www.aap.org** and **www.cps.ca**
- Download growth charts and learn more about them: **www.cdc.gov/growthcharts**
- Get information on the Food Guide Pyramid for young children from the USDA: **www.mypyramid.gov/kids**
- Get tips for feeding children from the American Dietetic Association: **www.eatright.org**
- Get tips for keeping children healthy from the Nemours Foundation: **www.kidshealth.org**
- Visit the National Center for Education in Maternal and Child Health and the National Institute of Child Health and Human Development: **www.ncemch.org** and **www.nichd.nih.gov**
- Learn about child nutrition programs: **www.fns.usda.gov/fns**
- Learn how UNICEF works to protect children: **www.unicef.org**
- Learn how to reduce lead exposure in your home from the U.S. Department of Housing and Urban Development Office of Lead Hazard Control: **www.hud.gov/lead**
- Learn more about food allergies from the American Academy of Allergy, Asthma, and Immunology; the Food Allergy and Anaphylaxis Network; and the International Food Information Council: **www.aaaai.org**, **www.foodallergy.org**, and **www.ific.org**
- Learn more about hyperactivity from Children and Adults with Attention Deficit/Hyperactivity Disorder: **www.chadd.org**
- Visit the Milk Matters section of the National Institute of Child Health and Human Development: **www.nichd.nih.gov/milk**
- Learn more about caffeine from the International Food Information Council: **www.ific.org**
- To learn about healthy foods and to find recipes and ideas for physical activities, visit: **www.kidnetic.com**
- Get weight-loss tips for children and adolescents: **www.shapedown.com**
- Get help quitting smoking at QuitNet: **www.quitnet.com**
- Visit the Tobacco Information and Prevention Source (TIPS) of the Centers for Disease Control and Prevention: **www.cdc.gov/tobacco**

References

1. Committee on Dietary Reference Intakes, *Dietary Reference Intakes for Energy, Carbohydrate, Fiber, Fat, Fatty Acids, Cholesterol, Protein, and Amino Acids* (Washington, D.C.: National Academies Press, 2005).
2. Committee on Dietary Reference Intakes, 2005, pp. 280–281.
3. Formula feeding of term infants, in *Pediatric Nutrition Handbook*, 6th ed., ed. R. E. Kleinman (Elk Grove Village, Ill.: American Academy of Pediatrics, 2009), pp. 61–78.
4. Position of the American Dietetic Association: Promoting and supporting breastfeeding, *Journal of the American Dietetic Association* 105 (2005): 810–818.
5. Breastfeeding, in *Pediatric Nutrition Handbook*, 6th ed., ed. R. E. Kleinman (Elk Grove Village, Ill.: American Academy of Pediatrics, 2009), pp. 29–59; Position of the American Academy of Pediatrics: Breastfeeding and the use of human milk, *Pediatrics* 115 (2005): 496–506; M. Boland, Exclusive breastfeeding should continue to six months, *Paediatrics and Child Health* 10 (2005): 148–149; Position of the American Dietetic Association: Promoting and supporting breastfeeding, 2005.
6. American Academy of Pediatrics, 2005.
7. L. Bode, Recent advances on structure, metabolism, and function of human milk oligosaccharides, *Journal of Nutrition* 136 (2006): 2127–2130.
8. S. M. Donovan, Human milk oligosaccharides: The plot thickens, *British Journal of Nutrition* 101 (2009): 1267–1269.
9. S. M. Innis, Dietary (n-3) fatty acids and brain development, *Journal of Nutrition* 137 (2007): 855–859.
10. K. Simmer, S. K. Patole, and S. C. Rao, Longchain polyunsaturated fatty acid supplementation in infants born at term, *Cochrane Database of Systematic Reviews*, January 23, 2008, CD000376; Innis, 2007; M. S. Fewtrell, Long-chain polyunsaturated fatty acids in early life: Effects on multiple health outcomes, in *Primary Prevention by Nutrition Intervention in Infancy and Childhood*, eds. A. Lucas and H. A. Sampson, *Nestle Nutrition Workshop Series Pediatric Program* 57 (2006): 203–221; W. C. Heird and A. Lapillonne, The role of essential fatty acids in development, *Annual Review of Nutrition* 25 (2005): 549–571; J. C. McCann and B. N. Ames, Is docosahexaenoic acid, an n-3 long-chain polyunsaturated fatty acid, required for development of normal brain function? An overview of evidence from cognitive and behavioral tests in humans and animals, *American Journal of Clinical Nutrition* 82 (2005): 281–295.
11. E. E. Birch and coauthors, Visual maturation of term infants fed long-chain polyunsaturated fatty acid-supplemented or control formula for 12 mo, *American Journal of Clinical Nutrition* 81 (2005): 871–879.
12. Simmer, Patole, and Rao, 2008.
13. Fat-soluble vitamins, in *Pediatric Nutrition Handbook*, 6th ed., ed. R. E. Kleinman (Elk Grove Village, Ill.: American Academy of Pediatrics, 2009), pp. 461–474.
14. C. L. Wagner, F. R. Greer, and the Section on Breastfeeding and Committee on Nutrition, Prevention of rickets and vitamin D deficiency in infants, children, and adolescents, *Pediatrics* 122 (2008): 1142–1152.
15. Breastfeeding, in *Pediatric Nutrition Handbook*, 2009; American Academy of Pediatrics, 2005; Position of the American Dietetic Association: Promoting and supporting breastfeeding, 2005.
16. K. Sadeharju and coauthors, Maternal antibodies in breast milk protect the child from enterovirus infections, *Pediatrics* 119 (2007): 941–946.
17. L. A. Hanson, Session 1: Feeding and infant development breast-feeding and immune function, *Proceedings of the Nutrition Society* 66 (2007): 384–396.

18. D. S. Newburg, G. M. Ruiz-Palacios, and A. L. Morrow, Human milk glycans protect infants against enteric pathogens, *Annual Review of Nutrition* 25 (2005): 37–58.
19. Breastfeeding, in *Pediatric Nutrition Handbook,* 2009; Hanson, 2007; C. J. Chantry, C. R. Howard, and P. Auinger, Full breastfeeding duration and associated decrease in respiratory tract infection in U.S. children, *Pediatrics* 117 (2006): 425–432; American Academy of Pediatrics, 2005; Position of the American Dietetic Association: Promoting and supporting breastfeeding, 2005.
20. F. R. Greer, S. H. Sicherer, A. W. Burks, and the Committee on Nutrition and Section on Allergy and Immunology, Effects of early nutritional interventions on the development of atopic disease in infants and children: The role of maternal dietary restriction, breastfeeding, timing of introduction of complementary foods, and hydrolyzed formulas, *Pediatrics* 121 (2008): 183–191.
21. Greer, Sicherer, Burks, and the Committee on Nutrition and Section on Allergy and Immunology, 2008; R. S. Zeiger and N. J. Friedman, The relationship of breastfeeding to the development of atopic disorders, *Nestle Nutrition Workshop Series: Pediatric Program* 57 (2006): 93–108.
22. Greer, Sicherer, Burks, and the Committee on Nutrition and Section on Allergy and Immunology, 2008; A. C. Krakowski and coauthors, Management of atopic dermatitis in the pediatric population, *Pediatrics* 122 (2008): 812–824.
23. C. G. Owen and coauthors, Does initial breastfeeding lead to lower blood cholesterol in adult life? A quantitative review of the evidence, *American Journal of Clinical Nutrition* 88 (2008): 305–314; L. Schack-Nielsen and K. F. Michaelsen, Advances in our understanding of the biology of human milk and its effects on the offspring, *Journal of Nutrition* 137 (2007): 503S–510S; R. A. Singhal, Early nutrition and long-term cardiovascular health, *Nutrition Reviews* 64 (2006): S44–S49; M. Martin, D. Gunnell, and G. D. Smith, Breastfeeding in infancy and blood pressure in later life: Systematic review and meta-analysis, *American Journal of Epidemiology* 161 (2005): 15–26.
24. A. S. Ryan, Breastfeeding and the risk of childhood obesity, *Collegium Antropologicum* 31 (2007): 19–28; S. Scholtens and coauthors, Breastfeeding, weight gain in infancy, and overweight at seven years of age: The prevention and incidence of asthma and mite allergy birth cohort study, *American Journal of Epidemiology* 165 (2007): 919–926; A. M. Toschke and coauthors, Infant feeding method and obesity: Body mass index and dual-energy X-ray absorptiometry measurements at 9–10 y of age from the Avon Longitudinal Study of Parents and Children (ALSPAC), *American Journal of Clinical Nutrition* 85 (2007): 1578–1585; R. Novotny and coauthors, Breastfeeding is associated with lower body mass index among children of the Commonwealth of the Northern Mariana Islands, *Journal of the American Dietetic Association* 107 (2007): 1743–1746; K. B. Michels and coauthors, A longitudinal study of infant feeding and obesity throughout life course, *International Journal of Obesity* advance online publication, April 24, 2007.
25. C. G. Owen and coauthors, Effect of infant feeding on the risk of obesity across the life course: A quantitative review of published evidence, *Pediatrics* 115 (2005): 1367–1377.
26. T. Harder and coauthors, Duration of breastfeeding and risk of overweight: A meta-analysis, *Journal of Epidemiology* 162 (2005): 397–403.
27. Ryan, 2007.
28. Schack-Nielsen and Michaelsen, 2007; G. Der, G. D. Batty, and I. J. Deary, Effect of breast feeding on intelligence in children; prospective study, sibling pairs analysis, and meta-analysis, *British Medical Journal* 333 (2006): 929–930; M. C. Daniels and L. S. Adair, Breastfeeding influences cognitive development in Filipino children, *Journal of Nutrition* 135 (2005): 2589–2595.
29. Schack-Nielsen and Michaelsen, 2007.
30. K. Woo and D. Spatz, Human milk donation: What do you know about it? *American Journal of Maternal and Child Nursing* 32 (2007): 150–155.
31. Formula feeding of term infants, in *Pediatric Nutrition Handbook,* 6th ed., ed. R. E. Kleinman (Elk Grove Village, Ill.: American Academy of Pediatrics, 2009), pp. 61–78.
32. Formula feeding of term infants, in *Pediatric Nutrition Handbook,* 2009; L. Seppo and coauthors, A follow-up study of nutrient intake, nutritional status, and growth in infants with cow milk allergy fed either a soy formula or an extensively hydrolyzed whey formula, *American Journal of Clinical Nutrition* 82 (2005): 140–145.
33. Formula feeding of term infants, in *Pediatric Nutrition Handbook,* 2009; Seppo and coauthors, 2005.
34. J. A. Maretin and coauthors, Annual summary of vital statistics: 2006, *Pediatrics* 121 (2008): 788–801.
35. D. L. O'Connor and coauthors, Growth and nutrient intakes of human milk-fed preterm infants provided with extra energy and nutrients after hospital discharge, *Pediatrics* 121 (2008): 766–776.
36. Formula feeding of term infants, in *Pediatric Nutrition Handbook,* 2009.
37. Complementary feeding, in *Pediatric Nutrition Handbook,* 6th ed., ed. R. E. Kleinman (Elk Grove Village, Ill.: American Academy of Pediatrics, 2009), pp. 113–142.
38. Complementary feeding, in *Pediatric Nutrition Handbook,* 2009; A. Fiocchi, A. Assa'ad, and S. Bahna, Food allergy and the introduction of solid foods to infants: A consensus document, *Annals of Allergy, Asthma and Immunology* 97 (2006): 10–21.
39. Iron, in *Pediatric Nutrition Handbook,* 6th ed., ed. R. E. Kleinman (Elk Grove Village, Ill.: American Academy of Pediatrics, 2009), pp. 403–422.
40. Complementary feeding, in *Pediatric Nutrition Handbook,* 2009.
41. Feeding the child, in *Pediatric Nutrition Handbook,* 6th ed., ed. R. E. Kleinman (Elk Grove Village, Ill.: American Academy of Pediatrics, 2009), pp. 145–174.
42. Feeding the child, in *Pediatric Nutrition Handbook,* 6th ed., ed. R. E. Kleinman (Elk Grove Village, Ill.: American Academy of Pediatrics, 2009), pp. 145–174.
43. M. K. Fox and coauthors, Relationship between portion size and energy intake among infants and toddlers: Evidence of self-regulation, *Journal of the American Dietetic Association* 106 (2006): S77–S83.
44. Position of the American Dietetic Association: Nutrition guidance for healthy children ages 2 to 11 years, *Journal of the American Dietetic Association* 108 (2008): 1038–1047; Position of the American Dietetic Association: Individual-, family-, school-, and community-based interventions for pediatric overweight, *Journal of the American Dietetic Association* 106 (2006): 925–945.
45. American Academy of Pediatrics, Council on Sports Medicine and Fitness and Council on School Health, Active healthy living: Prevention of childhood obesity through increased physical activity, *Pediatrics* 117 (2006): 1834–1842.
46. Nutritional aspects of vegetarian diets, in *Pediatric Nutrition Handbook,* 6th ed., ed. R. E. Kleinman (Elk Grove Village, Ill.: American Academy of Pediatrics, 2009), pp. 201–224.
47. Committee on Dietary Reference Intakes, 2005, Chapter 6.
48. Committee on Dietary Reference Intakes, 2005, Chapter 7.
49. Committee on Dietary Reference Intakes, 2005, Chapter 11.
50. Committee on Dietary Reference Intakes, 2005, Chapter 8.
51. J. M. Brotanek and coauthors, Iron deficiency in early childhood in the United States: Risk factors and racial/ethnic disparities, *Pediatrics* 120 (2007): 568–575; K. C. White, Anemia is a poor predictor of iron deficiency among toddlers in the United States: For heme the bell tolls, *Pediatrics* 115 (2005): 315–320.
52. Wagner and Greer, and the Section on Breastfeeding and Committee on Nutrition, 2008.
53. Feeding the child, in *Pediatric Nutrition Handbook,* 2009.
54. U. Shaikh, R. S. Byrd, and P. Auinger, Vitamin and mineral supplement use by children and adolescents in the 1999–2004 National Health and Nutrition Examination survey: Relationship with nutrition, food security, physical activity, and health care access, *Archives of Pediatrics and Adolescent Medicine* 163 (2009): 150–157; R. Briefel and coauthors, Feeding Infants and Toddlers Study: Do vitamin and mineral supplements contribute to nutrient adequacy or excess among U.S. infants and toddlers? *Journal of the American Dietetic Association* 106 (2006): S52–S65.
55. Position of the American Dietetic Association, 2008.
56. P. Ziegler and coauthors, Feeding infants and toddlers study (FITS): Development of the FITS Survey in comparison to other dietary survey methods, *Journal of the American Dietetic Association* 106 (2006): S12–S27.

57. J. Stang, Improving the eating patterns of infants and toddlers, *Journal of the American Dietetic Association* 106 (2006): S7–S9; M. K. Fox and coauthors, Feeding infants and toddlers study: What foods are infants and toddlers eating? *Journal of the American Dietetic Association* 104 (2004): S22–S30.
58. Position of the American Dietetic Association, 2008.
59. Bread for the World, Hunger facts: Domestic, available at www.bread.org/learn/hunger-basics/hunger-facts-domestic.html, updated February, 2008; accessed on January 23, 2009.
60. K. Widenhorn-Muller and coauthors, Influence of having breakfast on cognitive performance and mood in 13- to 20-year-old high school students: Results of a crossover trial, *Pediatrics* 122 (2008): 279–284.
61. W. O. Song and coauthors, Ready-to-eat breakfast cereal consumption enhances milk and calcium intake in the U.S. population, *Journal of the American Dietetic Association* 106 (2006): 1783–1789; S. G. Affenito and coauthors, Breakfast consumption by African-American and white adolescent girls correlates positively with calcium and fiber intake and negatively with body mass index, *Journal of the American Dietetic Association* 105 (2005): 938–945; G. C. Rampersaud and coauthors, Breakfast habits, nutritional status, body weight, and academic performance in children and adolescents, *Journal of the American Dietetic Association* 105 (2005): 743–760.
62. J. C. McCann and B. N. Ames, An overview of evidence for a causal relation between iron deficiency during development and deficits in cognitive or behavioral function, *American Journal of Clinical Nutrition* 85 (2007): 931–945.
63. B. Lozhoff and coauthors, Long-lasting neural and behavioral effects of iron deficiency in infancy, *Nutrition Reviews* 64 (2006): S34–S43.
64. K. M. Cecil and coauthors, Decreased brain volume in adults with childhood lead exposure, *PLoS Medicine* 27 (2008): e112.
65. Centers for Disease Control and Prevention, General lead information: Questions and answers, available at www.cdc.gov/nceh/lead/faq/about.htm, accessed January 23, 2009.
66. Centers for Disease control and Prevention, Interpreting and managing blood lead levels 10μg/dL in children and reducing childhood exposures to lead: Recommendations of DCD's Advisory Committee on Childhood Lead Poisoning Prevention, *Morbidity and Mortality Weekly Report* 56/RR-8 (2007): 1–16; Position of the Committee on Environmental Health, American Academy of Pediatrics: Lead exposure in children: Prevention, detection, and management, *Pediatrics* 116 (2005): 1036–1046.
67. Committee on Environmental Health, American Academy of Pediatrics, 2005.
68. P. N. Pastor and C. A. Reuben, Diagnosed attention deficit hyperactivity disorder and learning disability: United States, 2004–2006, *Vital and Health Statistics, Series 10, Data from the National Health Survey* 237 (2008): 1–14; E. Romano and coauthors, Development and prediction of hyperactive symptoms from 2 to 7 years in a population-based sample, *Pediatrics* 117 (2006): 2101–2109.
69. M. L. Wolraich and coauthors, Attention-deficit/hyperactivity disorder among adolescents: A review of the diagnosis, treatment, and clinical implications, *Pediatrics* 115 (2005): 1734–1746.
70. K. Rubia, Neuro-anatomic evidence for the maturational delay hypothesis of ADHD, *Proceedings of the National Academy of Sciences* 104 (2007): 19663–19664.
71. D. McCann and coauthors, Food additives and hyperactive behaviour in 3-year-old and 8/9-year-old children in the community: A randomized, double-blinded, placebo-controlled trial, *Lancet* 370 (2007): 1560–1567.
72. National Institutes of Health, National Institute of Allergy and Infectious Diseases, *Food Allergy: Report of the NIH Expert Panel on Food Allergy Research,* March 13–14, 2006, available at www.naid.nih.gov.
73. National Institutes of Health, National Institute of Allergy and Infectious Diseases, 2006.
74. U.S. Food and Drug Administration, Food allergies; Reducing the risks, January 23, 2009, available at www.fda.gov/consumer/updates/foodallergies012209.html.
75. A. Boulay and coauthors, A EuroPrevall review of factors affecting incidence of peanut allergy: Priorities for research and policy, *Allergy* 63 (2008): 797–809; L. A. Lee and A. W. Burks, Food allergies: Prevalence, molecular characterization, and treatment/prevention strategies, *Annual Review of Nutrition* 26 (2006): 539–565.
76. M. Boguniewicz, N. Moore, and K. Paranto, Allergic diseases, quality of life, and the role of the dietitian, *Nutrition Today* 43 (2008): 6–10.
77. S. Ramesh, Food allergy overview in children, *Clinical Reviews in Allergy & Immunology* 34 (2008): 217–230.
78. Lee and Burks, 2006.
79. Lee and Burks, 2006.
80. National Institutes of Health, National Institute of Allergy and Infectious Diseases, 2006.
81. C. L. Ogden, M. D. Carroll, and K. M. Flegal, High body mass index for age among U.S. children and adolescents, 2003–2006, *Journal of the American Medical Association* 299 (2008): 2401–2405.
82. S. E. Barlow and the Expert Committee, Expert Committee recommendations regarding the prevention, assessment, and treatment of child and adolescent overweight and obesity: Summary report, *Pediatrics* 120 (2007): S164–S192.
83. Barlow and the Expert Committee, 2007.
84. D. S. Freedman and coauthors, Cardiovascular risk factors and excess adiposity among overweight children and adolescents: The Bogalusa Heart Study, *Journal of Pediatrics* 150 (2007): 12–17.
85. N. F. Krebs and coauthors, Assessment of child and adolescent overweight and obesity, *Pediatrics* 120 (2007): S193–S228.
86. Barlow and the Expert Committee, 2007.
87. B. A. Spear and coauthors, Recommendations for treatment of child and adolescent overweight and obesity, *Pediatrics* 120 (2007): S254–S287; S. L. Martin, S. M. Lee, and R. Lowry, National prevalence and correlates of walking and bicycling to school, *American Journal of Preventive Medicine* 33 (2007): 98–105; American Academy of Pediatrics, Council on Sports Medicine and Fitness and Council on School Health, Active healthy living: Prevention of childhood obesity through increased physical activity, *Pediatrics* 117 (2006): 1834–1842; J. P. Koplan, C. T. Liverman, and V. I. Kraak, eds., *Preventing Childhood Obesity: Health in the Balance* (Washington, D.C.: National Academies Press, 2005), pp. 79–123.
88. K. J. Campbell and coauthors, Associations between the home food environment and obesity-promoting eating behaviors in adolescence, *Obesity* 15 (2007): 719–730; Barlow and the Expert Committee, 2007.
89. K. S. Geller and D. A. Dzewaltowski, Longitudinal and cross-sectional influences on youth fruit and vegetable consumption, *Nutrition Reviews* 67 (2009): 65–76; J. Brug and coauthors, Taste preferences, liking and other factors related to fruit and vegetable intakes among schoolchildren: Results from observational studies, *British Journal of Nutrition* 99 (2008): S7–S14; C. A. Forestell and J. A. Mennella, Early determinants of fruit and vegetable acceptance, *Pediatrics* 120 (2007): 1247–1254; C. Arcan and coauthors, Parental eating behaviours, home food environment and adolescent intakes of fruits, vegetables and dairy foods: Longitudinal finding form Project EAT, *Public Health Nutrition* 11 (2007): 1257–1265; M. Wind and coauthors, Correlates of fruit and vegetable consumption among 11-year-old Belgina-Flemish and Dutch schoolchildren, *Journal of Nutrition Education and Behavior* 38 (2006): 211–221.
90. K. E. Leahy, L. L. Birch, and B. Rolls, Reducing the energy density of multiple meals decreases the energy intake of preschool-age children, *American Journal of Clinical Nutrition* 88 (2008): 1459–1468.
91. S. A. Lederman and coauthors, Summary of the presentations at the Conference on Preventing Childhood Obesity, December 8, 2003, *Pediatrics* 114 (2004): 1146–1173.
92. S. N. Bleich and coauthors, Increasing consumption of sugar-sweetened beverages among U.S. adults: 1988–1994 to 1999–2004, *American Journal of Clinical Nutrition* 89 (2009): 372–381; Spear and coauthors, 2007; L. Dubois and coauthors, Regular sugar-sweetened beverage consumption between meals increases risk of overweight among preschool-aged children, *Journal of the American Dietetic Association* 107 (2007): 924–934; V. S. Malik, M. B. Schultz, and F. B. Hu, Intake of sugar-sweetened beverages and weight gain: A systematic review, *American Journal of Clinical Nutrition* 84 (2006): 274–288.

93. S. Harrington, The role of sugar-sweetened beverage consumption in adolescent obesity: A review of the literature, *Journal of School Nursing* 24 (2008): 3–12; Committee on School Health, American Academy of Pediatrics, Soft drinks in schools, *Pediatrics* 113 (2004): 152–154.
94. L. R. Vartanian, M. B. Schwartz, and K. D. Brownell, Effects of soft drink consumption on nutrition and health: A systematic review and meta-analysis, *American Journal of Public Health* 97 (2007): 667–675.
95. American Academy of Pediatrics, Council on Sports Medicine and Fitness and Council on School Health, Active healthy living: Prevention of childhood obesity through increased physical activity, *Pediatrics* 117 (2006): 1834–1842.
96. S. Gable, Y. Chang, and J. L. Krull, Television watching and frequency of family meals are predictive of overweight onset and persistence in a national sample of school-aged children, *Journal of the American Dietetic Association* 107 (2007): 53–61; Spear and coauthors, 2007.
97. M. Hamer, E. Stamsatakis, and G. Mishra, Psychological distress, television viewing, and physical activity in children aged 4 to 12 years, *Pediatrics* 123 (2009): 1263–1268.
98. D. J. Barr-Anderson and coauthors, Characteristics associated with older adolescents who have a television in their bedrooms, *Pediatrics* 121 (2008): 718–724; A. M. Adachi-Mejia and coauthors, Children with a TV in their bedroom at higher risk for being overweight, *International Journal of Obesity* 31 (2007): 644–651.
99. D. M. Jackson and coauthors, Increased television viewing is associated with elevated body fatness but not with lower total energy expenditure in children, *American Journal of Clinical Nutrition* 89 (2009): 1031–1036.
100. L. Dubois and coauthors, Social factors and television use during meals and snacks is associated with higher BMI among pre-school children, *Public Health Nutrition* 11 (2008): 1267–1279; Gable, Chang, and Krull, 2007.
101. J. L. Wiecha and coauthors, When children eat what they watch: Impact of television viewing on dietary intake in youth, *Archives of Pediatrics & Adolescent Medicine* 160 (2006): 436–442; S. C. Folta and coauthors, Food advertising targeted at school-age children: A content analysis, *Journal of Nutrition Education and Behavior* 38 (2006): 244–248.
102. A. Batada and coauthors, Nine out of 10 food advertisements shown during Saturday morning children's television programming are for foods high in fat, sodium, or added sugars, or low in nutrients, *Journal of the American Dietetic Association* 108 (2008): 673–678; L. M. Powell and coauthors, Nutritional content of television food advertisements seen by children and adolescents in the United States, *Pediatrics* 120 (2007): 576–583; Wiecha and coauthors, 2006.
103. S. M. Connor, Food-related advertising on preschool television: Building brand recognition in young viewers, *Pediatrics* 118 (2006): 1478–1485; Folta and coauthors, 2006.
104. L. J. Chamberlain, Y. Wang, and T. N. Robinson, Does children's screen time predict requests for advertised products? Cross-sectional and prospective analyses, *Archives of Pediatrics & Adolescent Medicine* 160 (2006): 363–368; Y. Aktas-Arnas, the effects of television food advertisement on children's food purchasing requests, *Pediatrics International* 48 (2006): 138–145; M. O'Dougherty, M. Story, and J. Stang, Observations of parent-child co-shoppers in supermarkets: Children's involvement in food selections, parental yielding, and refusal strategies, *Journal of Nutrition Education and Behavior* 38 (2006): 183–188.
105. K. Weber, M. Story, and L. Harnack, Internet food marketing strategies aimed at children and adolescents: A content analysis of food and beverage brand web sites, *Journal of the American Dietetic Association* 106 (2006): 1463–1466.
106. Spear and coauthors, 2007.
107. Barlow and the Expert Committee, 2007.
108. M. Salvadori and coauthors, Elevated blood pressure in relation to overweight and obesity among children in a rural Canadian community, *Pediatrics* 122 (2008): e821–e827; R. Jago and coauthors, Prevalence of abnormal lipid and blood pressure values among an ethnically diverse population of eighth-grade adolescents and screening implications, *Pediatrics* 117 (2006): 2065–2073.
109. S. Cook and coauthors, Metabolic Syndrome rates in United States adolescents, from the National Health and Nutrition Examination Survey, 1999–2002, *Journal of Pediatrics* 152 (2008): 165–170; K. L. Jones, Role of obesity in complicating and confusing the diagnosis and treatment of diabetes in children, *Pediatrics* 121 (2008): 361–368; C. L. Carroll and coauthors, Childhood overweight increases hospital admission rates for asthma, *Pediatrics* 120 (2007): 734–740; M. L. Cruz and coauthors, Pediatric obesity and insulin resistance: Chronic disease risk and implications for treatment and prevention beyond body weight modification, *Annual Review of Nutrition* 25 (2005):v 435–468.
110. D. S. Ludwig, Childhood obesity: The shape of things to come, *New England Journal of Medicine* 357 (2007): 2325–2326; A. J. Daley and coauthors, Exercise therapy as a treatment for psychopathologic conditions in obese and morbidly obese adolescents: A randomized, controlled trial, *Pediatrics* 118 (2006): 2126–2134; J. Franklin and coauthors, Obesity and risk of low self-esteem: A statewide survey of Australian children, *Pediatrics* 118 (2006): 2481–2487.
111. S. M. Himes and J. K. Thompson, Fat stigmatization in television shows and movies: A content analysis, *Obesity* 15 (2007): 712–718.
112. Barlow and the Expert Committee, 2007.
113. M. M. Davis and coauthors, Recommendations for prevention of childhood obesity, *Pediatrics* 120 (2007): S229–S253; Position of the American Dietetic Association: Individual-, family-, school-, and community-based interventions for pediatric overweight, 2006.
114. Barlow and the Expert Committee, 2007.
115. Spear and coauthors, 2007.
116. Barlow and the Expert Committee, 2007.
117. Barlow and the Expert Committee, 2007.
118. American Academy of Pediatrics, 2006.
119. American Academy of Pediatrics, 2006.
120. Spear and coauthors, 2007.
121. T. H. Inge and coauthors, Reversal of type 2 diabetes mellitus and improvements in cardiovascular risk factors after surgical weight loss in adolescents, *Pediatrics* 123 (2009): 214–222.
122. Spear and coauthors, 2007.
123. Position of the American Dietetic Association, 2008; Position of the American Dietetic Association: Individual-, family-, school-, and community-based interventions for pediatric overweight, 2006.
124. E. Jansen and coauthors, From the Garden of Eden to the land of plenty: Restriction of fruit and sweets intake leads to increased fruit and sweets consumption in children, *Appetite* 51 (2008): 570–575; E. Jansen, S. Mulkens, and A. Jansen, Do not eat the red food: Prohibition of snacks leads to their relatively higher consumption in children, *Appetite* 49 (2007): 572–577.
125. Position of the American Dietetic Association, 2008.
126. Position of the American Dietetic Association, 2008; J. Wardle, S. Carnell, and L. Cooke, Parental control over feeding and children's fruit and vegetable intake: How are they related? *Journal of the American Dietetic Association* 105 (2005): 227–232; A. T. Galloway and coauthors, Parental pressure, dietary patterns, and weight status among girls who are "picky eaters," *Journal of the American Dietetic Association* 105 (2005): 541–548.
127. L. Cooke, The importance of exposure for healthy eating in childhood: A review, *Journal of Human Nutrition and Dietetics* 20 (2007): 294–301.
128. Position of the American Dietetic Association: Benchmarks for nutrition programs in child care settings, *Journal of the American Dietetic Association* 105 (2005): 979–986; Position of the American Dietetic Association, Society of Nutrition Education, and American School Food Service Association—Nutrition services: An essential component of comprehensive school health programs, *Journal of the American Dietetic Association* 103 (2003): 505–514.
129. Position of the American Dietetic Association, Society of Nutrition Education, and American School Food Service Association, 2003.
130. Position of the American Dietetic Association, 2008; K. Ralston and coauthors, The National School Lunch Program: Background, issues, and trends, USDA Economic Research Service, July 2008, available at www.ers.usda.gov/publications/err61; Position of the American

Dietetic Association: Local support for nutrition integrity in schools, *Journal of the American Dietetic Association* 106 (2006): 122–133.
131. Position of the American Dietetic Association, 2008.
132. Position of the American Dietetic Association: Local support for nutrition integrity in schools, 2006.
133. Position of the American Dietetic Association: Local support for nutrition integrity in schools, 2006.
134. M. K. Crepinsek and coauthors, Meals offered and served in U.S. public schools: do they meet nutrient standards? *Journal of the American Dietetic Association* 109 (2009): S31–S43; Widenhorn-Muller and coauthors, 2008.
135. Position of the American Dietetic Association: Local support for nutrition integrity in schools, 2006.
136. Centers for Disease Control and Prevention, Competitive foods and beverages available for purchase in secondary schools—selected sites, United States, 2006, *Morbidity and Mortality Weekly Report* 57 (2008): 935–938; Position of the American Dietetic Association: Local support for nutrition integrity in schools, 2006; C. Probart and coauthors, Competitive foods available in Pennsylvania public high schools, *Journal of the American Dietetic Association* 105 (2005): 1243–1249.
137. S. B. Templeton and coauthors, Competitive foods increase the intake of energy and decrease the intake of certain nutrients by adolescents consuming school lunch, *Journal of the American Dietetic Association* 105 (2005): 215–220.
138. Position of the American Dietetic Association: Local support for nutrition integrity in schools, 2006.
139. Institute of Medicine, Food and Nutrition Board, Committee on Nutrition Standards for Foods in Schools, eds. V. A. Stallings and A. L. Yaktine, *Nutrition Standards for Foods in Schools: Leading the Way toward Healthier Youth* (Washington, D.C.: National Academies Press, 2007).
140. Institute of Medicine, Food and Nutrition Board, Committee on Nutrition Standards for Foods in Schools, 2007.
141. Institute of Medicine, Food and Nutrition Board, Committee on Nutrition Standards for Foods in Schools, 2007.
142. Committee on Dietary Reference Intakes, 2005, Chapter 5.
143. S. Saintonge, H. Bang, and L. M. Gerber, Implications of a new definition of vitamin D deficiency in a multiracial US adolescent population: The National Health and Nutrition Examination Survey III, *Pediatrics* 123 (2009): 797-803; Wagner, Greer, and the Section on Breastfeeding and Committee on Nutrition, 2008.
144. Committee on Dietary Reference Intakes, *Dietary Reference Intakes for Calcium and Vitamin D* (Washington, D.C.: National Academies Press, 2011).
145. Wagner, Greer, and the Section on Breastfeeding and Committee on Nutrition, 2008.
146. Committee on Dietary Reference Intakes, *Dietary Reference Intakes for Vitamin A, Vitamin K, Arsenic, Boron, Chromium, Copper, Iodine, Iron, Manganese, Molybdenum, Nickel, Silicon, Vanadium, and Zinc* (Washington, D.C.: National Academies Press, 2001), pp. 290–393.
147. Committee on Dietary Reference Intakes, 2001.
148. F. R. Greer, N. F. Krebs, and the Committee on Nutrition, American Academy of Pediatrics, Optimizing bone health and calcium intakes of infants, children, and adolescents, *Pediatrics* 117 (2006): 578–585.
149. Greer, Krebs, and the Committee on Nutrition, 2006.
150. L. Esterie and coauthors, Milk, rather than other foods, is associated with vertebral bone mass and circulating IGF-1 in female adolescents, *Osteoporosis International* 2008 20 (2009): 567–575; M. M. Murphy and coauthors, Drinking flavored or plain milk is positively associated with nutrient intake and is not associated with adverse effects on weight status in U.S. children and adolescents, *Journal of the American Dietetic Association* 108 (2008): 631–639; M. L. Savaiano and coauthors, Perceived milk intolerance is related to bone mineral content in 10- to 13-year-old female adolescents, *Pediatrics* 120 (2007): e669–e677.
151. N. I. Larson and coauthors, Family meals during adolescence are associated with higher diet quality and healthful meal patterns during young adulthood, *Journal of the American Dietetic Association* 107 (2007): 1502–1510.
152. M. E. Eisenberg and coauthors, Family meals and substance use: Is there a long-term association? *Journal of Adolescent Health* 43 (2008): 151–156; National Center on Addiction and Substance Abuse (CASA) at Columbia University, *The Importance of Family Dinners,* September, 2003.
153. Rampersaud and coauthors, 2005.
154. M. T. Timlin and coauthors, Breakfast eating and weight change in a 5-year prospective analysis of adolescents: Project EAT (Eating Among Teens), *Pediatrics* 121 (2008): e638–e645.
155. C. M. McDonald and coauthors, Overweight is more prevalent than stunting and is associated with socioeconomic status, maternal obesity, and a snacking dietary pattern in school children from Bogota, Colombia, *Journal of Nutrition* 139 (2009): 370–376; H. M. Niemeier and coauthors, Fast food consumption and breakfast skipping: Predictors of weight gain from adolescence to adulthood in a nationally representative sample, *Journal of Adolescent Health* 39 (2006): 842–849.
156. American Heart Association, S. S. Gidding, and coauthors, Dietary recommendations for children and adolescents: A guide for practitioners, *Pediatrics* 117 (2006): 544–559
157. R. S. Sebastian, L. E. Cleveland, and J. D. Goldman, Effect of snacking frequency on adolescents' dietary intakes and meeting national recommendations, *Journal of Adolescent Health* 42 (2008): 503–511.
158. L. Libuda and coauthors, Association between long-term consumption of soft drinks and variables of bone modeling and remodeling in a sample of healthy German children and adolescents, *American Journal of Clinical Nutrition* 88 (2008): 1670–1677; Greer, Krebs, and the Committee on Nutrition, 2006; H. Vatanparast and coauthors, Positive effects of vegetable and fruit consumption and calcium intake on bone mineral accrual in boys during growth from childhood to adolescence: The University of Saskatchewan Pediatric Bone Mineral Accrual Study, *American Journal of Clinical Nutrition* 82 (2005): 700–706.
159. Vartanian, Schwartz, and Brownell, 2007; Malik, Schulze, and Hu, 2006.
160. J. W. Kulig and the Committee on Substance Abuse, Tobacco, alcohol, and other drugs: The role of the pediatrician in prevention, identification, and management of substance abuse, *Pediatrics* 115 (2005): 816–821.
161. Centers for Disease Control and Prevention, Youth tobacco surveillance: United States, 2001–2002, *Morbidity and Mortality Weekly Report* 55 (2006): entire supplement.

HIGHLIGHT 15

Childhood Obesity and the Early Development of Chronic Diseases

When people think about the health problems of children and adolescents, they typically think of ear infections, colds, and acne—not heart disease, diabetes, or hypertension. Today, however, unprecedented numbers of U.S. children are being diagnosed with obesity and the serious "adult diseases," such as type 2 diabetes, that accompany overweight.[1] When type 2 diabetes develops before the age of 20, the incidence of diabetic kidney disease and death in middle age increases dramatically, largely because of the long duration of the disease.[2] For children born in the United States in the year 2000, the risk of developing type 2 diabetes sometime in their lives is estimated to be 30 percent for boys and 40 percent for girls.[3] U.S. children are not alone—rapidly rising rates of obesity threaten the health of an alarming number of children around the globe.[4] Without immediate intervention, millions of children are destined to develop type 2 diabetes and hypertension in childhood followed by **cardiovascular disease (CVD)** in early adulthood.[5] (See the accompanying glossary for this and related terms.)

This highlight focuses on efforts to prevent childhood obesity and the development of heart disease and type 2 diabetes, but the benefits extend to other obesity-related diseases as well. The years of childhood (ages 2 to 18) are emphasized here, because the earlier in life health-promoting habits become established, the better they will stick.

Invariably, questions arise as to what extent genetics is involved in disease development. For heart disease and type 2 diabetes, genetics does not appear to play a *determining* role; that is, a person is not simply destined at birth to develop these diseases. Instead, genetics appears to play a *permissive* role—the potential is inherited and will develop if given a push by poor health choices such as excessive weight gain, poor diet, sedentary lifestyle, and cigarette smoking.

Many experts agree that preventing or treating obesity in childhood will reduce the rate of chronic diseases in adulthood. Without intervention, most overweight children become overweight adolescents who become overweight adults, and being overweight exacerbates every chronic disease that adults face.[6] Fatty liver, a condition that correlates directly with BMI, was not even recognized in pediatric research until recently. Today, fatty liver disease affects about one in three obese children.[7]

Early Development of Type 2 Diabetes

In recent years, type 2 diabetes, a chronic disease closely linked with obesity, has been on the rise among children and adolescents as the prevalence of obesity in U.S. youth has increased.[8] Obesity is the most important risk factor for type 2 diabetes—most of the children diagnosed with it are obese.[9] Most are diagnosed during puberty, but as younger children become more obese and less active, the trend is shifting to younger ages. Type 2 diabetes is most likely to occur in those who are obese and sedentary and have a family history of diabetes.

GLOSSARY

atherosclerosis (ATH-er-oh-scler-OH-sis): a type of artery disease characterized by plaques (accumulations of lipid-containing material) on the inner walls of the arteries (see Chapter 27).
- **athero** = porridge or soft
- **scleros** = hard
- **osis** = condition

cardiovascular disease (CVD): a general term for all diseases of the heart and blood vessels. Atherosclerosis is the main cause of CVD. When the arteries that carry blood to the heart muscle become blocked, the heart suffers damage known as **coronary heart disease (CHD).**
- **cardio** = heart
- **vascular** = blood vessels

fatty streaks: accumulations of cholesterol and other lipids along the walls of the arteries.

plaque (PLACK): an accumulation of fatty deposits, smooth muscle cells, and fibrous connective tissue that develops in the artery walls in atherosclerosis. Plaque associated with atherosclerosis is known as **atheromatous** (ATH-er-OH-ma-tus) **plaque.**

HIGHLIGHT 15

In type 2 diabetes, the cells become insulin-resistant—that is, the cells become less sensitive to insulin, reducing the amount of glucose entering the cells from the blood. The combination of obesity and insulin resistance produces a cluster of symptoms, including high blood cholesterol and high blood pressure, which, in turn, promotes the development of atherosclerosis and the early development of CVD.[10] Other common problems evident by early adulthood include kidney disease, blindness, and miscarriages. The complications of diabetes, especially when encountered at a young age, can shorten life expectancy.

Prevention and treatment of type 2 diabetes depend on weight management, which can be particularly difficult in a youngster's world of food advertising, video games, and pocket money for candy bars. The activity and dietary suggestions to help defend against heart disease later in this highlight apply to type 2 diabetes as well.

Early Development of Heart Disease

Most people consider heart disease to be an adult disease because its incidence rises with advancing age, and symptoms rarely appear before age 30. The disease process actually begins much earlier.

Atherosclerosis

Most cardiovascular disease involves **atherosclerosis.** Atherosclerosis develops when regions of an artery's walls become progressively thickened with **plaque**—an accumulation of fatty deposits, smooth muscle cells, and fibrous connective tissue. If it progresses, atherosclerosis may eventually block the flow of blood to the heart and cause a heart attack or cut off blood flow to the brain and cause a stroke. Infants are born with healthy, smooth, clear arteries, but within the first decade of life, **fatty streaks** may begin to appear (see Figure H15-1). During adolescence, these fatty streaks may begin to accumulate fibrous connective tissue. By early adulthood, the fibrous plaques may begin to calcify and become raised lesions, especially in boys and young men. As the lesions grow more numerous and enlarge, the heart disease rate begins to rise, most dramatically at about age 45 in men and 55 in women. From this point on, arterial damage and blockage progress rapidly, and heart attacks and strokes threaten life. In short, the consequences of atherosclerosis, which become apparent only in adulthood, have their beginnings in the first decades of life.[11]

Atherosclerosis is not inevitable; people can grow old with relatively clear arteries. Early lesions may either progress or regress, depending on several factors, many of which reflect lifestyle behaviors. Smoking, for example, is strongly associated with the prevalence of fatty streaks and raised lesions, even in young adults.

Blood Cholesterol

As blood cholesterol rises, atherosclerosis worsens. Cholesterol values at birth are similar in all populations; differences emerge in early childhood. Standard values for cholesterol in children and adolescents (ages 2 to 18 years) are listed in Table H15-1. Cholesterol concentrations change with age in children and adolescents, however, and are especially variable during puberty.[12] Thus, using a single cut point for all pediatric age groups has limitations.

In general, blood cholesterol tends to rise as dietary saturated fat intakes increase. Blood cholesterol also correlates with childhood obesity, especially abdominal obesity.[13] LDL cholesterol rises with obesity, and HDL declines. These relationships are apparent throughout childhood, and their magnitude increases with age.

Children who are both overweight and have high blood cholesterol are likely to have parents who develop heart disease early. For this reason, selective screening is recommended for children and adolescents who are overweight or obese; those whose parents (or grandparents) have premature (≤55 years of age for men

FIGURE H15-1 The Formation of Plaques in Atherosclerosis

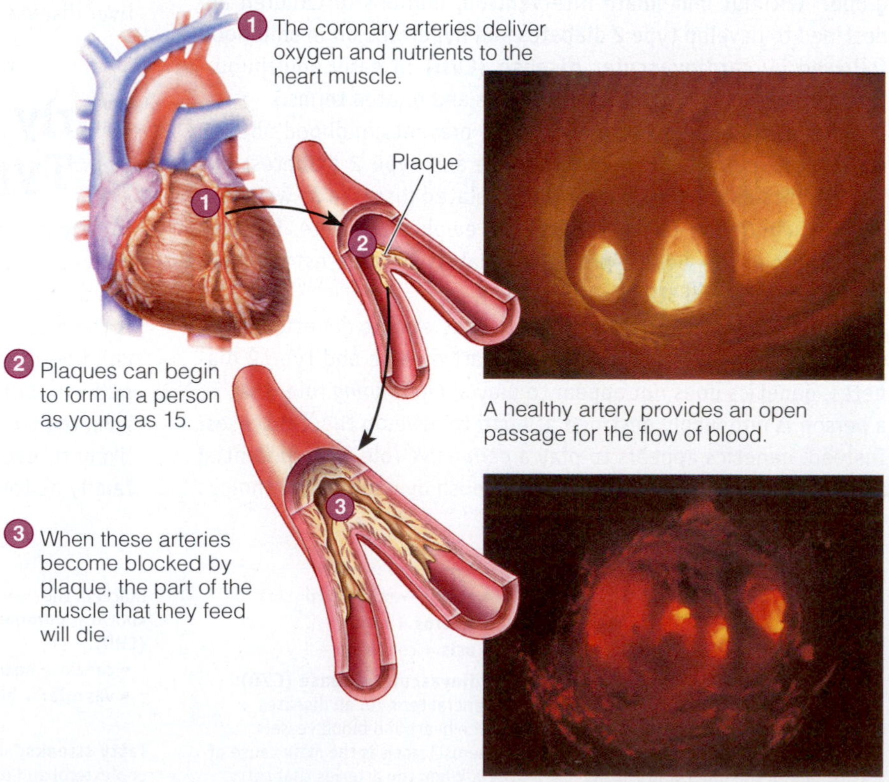

1 The coronary arteries deliver oxygen and nutrients to the heart muscle.
Plaque
1
2
2 Plaques can begin to form in a person as young as 15.
3
3 When these arteries become blocked by plaque, the part of the muscle that they feed will die.
A healthy artery provides an open passage for the flow of blood.

Plaques form along the artery's inner wall, reducing blood flow. Clots can form, aggravating the problem.

TABLE H15-1 Cholesterol Values for Children and Adolescents

Disease Risk	Total Cholesterol (mg/dL)	LDL Cholesterol (mg/dL)
Acceptable	<170	<110
Borderline	170–199	110–129
High	≥200	≥130

NOTE: Adult values appear in Chapter 27.

and ≤65 years of age for women) heart disease; those whose parents have elevated blood cholesterol; those who have other risk factors for heart disease such as hypertension, cigarette smoking, or diabetes; and those whose family history is unavailable.[14] Because blood cholesterol in children is a good predictor of adult values, some experts recommend universal screening for all children, and particularly for those who are overweight, smoke, are sedentary, or consume diets high in saturated fat.

Early—but not advanced—atherosclerotic lesions are reversible, making screening and education a high priority. Both those with family histories of heart disease and those with multiple risk factors need intervention. Children with the highest risks of developing heart disease are sedentary and obese, with high blood pressure and high blood cholesterol. In contrast, children with the lowest risks of heart disease are physically active and of normal weight, with low blood pressure and favorable lipid profiles. Routine pediatric care should identify these known risk factors and provide intervention when needed.

Blood Pressure

Pediatricians routinely monitor blood pressure in children and adolescents. High blood pressure may signal an underlying disease or the early onset of hypertension. Hypertension accelerates the development of atheroscerlosis.[15] Diagnosing hypertension in children and adolescents requires consideration of age, gender, and height and cannot be assessed using simple tables applied to adults.[16]

Like atherosclerosis and high blood cholesterol, hypertension may develop in the first decades of life, especially among obese children, and worsen with time. Children can control their hypertension by participating in regular aerobic activity and by losing weight or maintaining their weight as they grow taller. Restricting dietary sodium also causes an immediate drop in most children's and adolescents' blood pressure.[17]

Physical Activity

Research has also confirmed an association between blood lipids and physical activity in children, similar to that seen in adults. Physically active children have a better lipid profile and lower blood pressure than physically inactive children, and these positive findings often persist into adulthood. The *Physical Activity Guidelines for Americans, 2008* recommendations for children and adolescents are listed in Chapter 15 on p. 517.

Just as blood cholesterol and obesity track over the years, so does a child's level of physical activity. Those who are inactive now are likely to still be inactive years later. Similarly, those who are physically active now tend to remain so. Compared with inactive teens, those who are physically active weigh less, smoke less, eat a diet lower in saturated fats, and have better blood lipid profiles. Both obesity and blood cholesterol correlate with the inactive pastime of watching television. The message is clear: physical activity offers numerous health benefits, and children who are active today are most likely to be active for years to come.

Dietary Recommendations for Children

Regardless of family history, experts agree that all children older than age 2 should eat a variety of foods and maintain desirable weight (see Table H15-2). Children (4 to 18 years of age) should receive at least 25 percent and no more than 35 percent of total energy from fat, less than 10 percent from saturated fat, and less than 300 milligrams of cholesterol per day.[18] Recommendations limiting fat and cholesterol are not intended for infants or children younger than 2 years old. Infants and toddlers need a higher percentage of

TABLE H15-2 American Heart Association Dietary Guidelines and Strategies for Children[a]

- Balance dietary kcalories with physical activity to maintain normal growth.
- Every day, engage in 60 minutes of moderate to vigorous play or physical activity.
- Eat vegetables and fruits daily. Use fresh, frozen, and canned vegetables and fruits and serve at every meal; limit those with added fats, salt, and sugar.
- Limit juice intake (4 to 6 ounces per day for children 1 to 6 years of age, 8 to 12 ounces for children 7 to 18 years of age).
- Use vegetable oils (canola, soybean, olive, safflower, or other unsaturated oils) and soft margarines low in saturated fat and *trans*-fatty acids instead of butter or most other animal fats in the diet.
- Choose whole-grain breads and cereals rather than refined products; read labels and make sure that "whole grain" is the first ingredient.
- Reduce the intake of sugar-sweetened beverages and foods.
- Consume low-fat and non-fat milk and milk products daily.
- Include two servings of fish per week, especially fatty fish such as broiled or baked salmon.
- Choose legumes and tofu in place of meat for some meals.
- Choose only lean cuts of meat and reduced-fat meat products; remove the skin from poultry.
- Use less salt, including salt from processed foods. Breads, breakfast cereals, and soups may be high in salt and/or sugar so read food labels and choose high-fiber, low-salt, low-sugar alternatives.
- Limit the intake of high-kcalorie add-ons such as gravy, Alfredo sauce, cream sauce, cheese sauce, and hollandaise sauce.
- Serve age-appropriate portion sizes on appropriately sized plates and bowls.

[a] These guidelines are for children 3 years of age and older.
SOURCE: Adapted from American Heart Association, Samuel S. Gidding, and coauthors, Dietary recommendations for children and adolescents: A guide for practitioners, *Pediatrics* 117 (2006): 544–559.

fat to support their rapid growth. For children between 1 year of age and 2 years of age who are overweight or obese, or have a family history of heart disease, obesity, or abnormal blood lipids, however, the use of reduced-fat milk is recommended.[19]

Moderation, Not Deprivation

Healthy children older than age 2 can begin the transition to eating according to recommendations by eating fewer foods high in saturated fat and selecting more fruits and vegetables. Healthy meals can occasionally include moderate amounts of a child's favorite foods, even if they are high in saturated fat such as french fries and ice cream. A steady diet from some "children's menus" in restaurants such as chicken nuggets, hot dogs, and french fries easily exceeds a prudent intake of saturated fat, *trans* fat, and kcalories, however, and invites both nutrient shortages and weight gains.[20] Fortunately, most restaurants chains are changing children's menus to include steamed vegetables, fruit cups, and broiled or grilled chicken—additions welcomed by busy parents who often dine out or purchase take-out foods.

Other fatty foods, such as nuts, vegetable oils, and some varieties of fish such as tuna or salmon, contribute essential fatty acids. Low-fat milk and milk products also deserve special attention in a child's diet for the needed calcium and other nutrients they supply.[21]

Parents and caregivers play a key role in helping children establish healthy eating habits. Balanced meals need to provide lean meat, poultry, fish, and legumes; fruits and vegetables; whole grains; and low-fat milk products. Such meals can provide enough energy and nutrients to support growth and maintain blood cholesterol within a healthy range.

Pediatricians warn parents to avoid extremes. Although intentions may be good, excessive food restriction may create nutrient deficiencies and impair growth. Furthermore, parental control over eating may instigate battles and foster attitudes about foods that can lead to inappropriate eating behaviors.

Diet First, Drugs Later

Experts agree that children with high blood cholesterol should first be treated with diet. If high blood cholesterol persists despite dietary intervention in children 8 years of age and older, then drugs may be necessary to lower blood cholesterol. Drugs can effectively lower blood cholesterol without interfering with adolescent growth or development.[22]

Smoking

Even though the focus of this text is nutrition, another risk factor for heart disease that starts in childhood and carries over into adulthood must also be addressed—cigarette smoking. Each day 3000 children light up for the first time—typically in grade school. Among high school students, almost two out of three have tried smoking, and one in five smokes regularly.[23] Approximately 80 percent of all adult smokers began smoking before the age of 18.

Of those teenagers who continue smoking, half will eventually die of smoking-related causes. Efforts to teach children about the dangers of smoking need to be aggressive. Children are not likely to consider the long-term health consequences of tobacco use. They are more likely to be struck by the immediate health consequences, such as shortness of breath when playing sports, or social consequences, such as having bad breath. Whatever the context, the message to all children and teens should be clear: don't start smoking. If you've already started, quit.

In conclusion, *adult* heart disease is a major *pediatric* problem. Without intervention, some 60 million children are destined to suffer its consequences within the next 30 years. Optimal prevention efforts focus on children, especially on those who are overweight.[24] Just as young children receive vaccinations against infectious diseases, they need screening for, and education about, chronic diseases. Many health education programs have been implemented in schools around the country. These programs are most effective when they include education in the classroom, heart-healthy meals in the lunch room, fitness activities on the playground, and parental involvement at home.

Cigarette smoking is the number-one preventable cause of deaths.

Nutrition on the Net

For further study of topics covered in this highlight, log on to **www.cengagebrain.com** and search for ISBN 084006845X.

- Get weight-loss tips for children and adolescents: **www.shapedown.com**
- Visit the Nemours Foundation: **www.kidshealth.org**
- Find information on diabetes in children at the American Diabetes Association and Juvenile Diabetes Research Foundation: **www.diabetes.org** and **www.jdrf.org**

References

1. D. S. Freedman and coauthors, Risk factors and adult body mass index among overweight children: The Bogalusa Heart Study, *Pediatrics* 123 (2009): 750–757; S. Cook and coauthors, Metabolic syndrome rates in United States adolescents, from the National Health and Nutrition Examination survey, 1999–2002, *Journal of Pediatrics* 152 (2008): 165–170; M. Gardner, D. W. Gardner, and J. R. Sowers, The cardiometabolic syndrome in the adolescent, *Pediatric Endocrinology Reviews* Suppl. 4 (2008): 964–968; K. L. Jones, Role of obesity in complicating and confusing the diagnosis and treatment of diabetes in children, *Pediatrics* 121 (2008): 361–368; C. L. Ogden, M. D. Carroll, and K. M. Flegal, High body mass index for age among U.S. children and adolescents, 2003–2006, *Journal of the American Medical Association* 299 (2008): 2401–2405; D. S. Ludwig, Childhood obesity: The shape of things to come, *New England Journal of Medicine* 357 (2007): 2325–2327; J. P. Kaplan, C. T. Liverman, and V. I. Kraak, eds., *Preventing Childhood Obesity: Health in the Balance* (Washington, D.C.: National Academies Press, 2005), pp. 1–20; M. L. Cruz and coauthors, Pediatric obesity and insulin resistance: Chronic disease risk and implications for treatment and prevention beyond body weight modification, *Annual Review of Nutrition* 25 (2005): 435–468.
2. M. E. Pavkov and coauthors, Effect of youth-onset type 2 diabetes mellitus on incidence of end-stage renal disease and mortality in young and middle-aged Pima Indians, *Journal of the American Medical Association* 296 (2006): 421–426.
3. Kaplan, Liverman, and Kraak, 2005.
4. W. Maziak, K. D. Ward, and M. B. Stockton, Childhood obesity: Are we missing the big picture? *Obesity Reviews* 9 (2008): 35–42; Y. Wang and T. Lobstein, Worldwide trends in childhood overweight and obesity, *International Journal of Pediatric Obesity* 1 (2006): 11–25; Cruz and coauthors, 2005.
5. Freedman and coauthors, 2009; H. Zhu and coauthors, Relationships of cardiovascular phenotypes with healthy weight, at risk of overweight, and overweight in U.S. youths, *Pediatrics* 121 (2008): 115–122; D. S. Freedman and coauthors, Cardiovascular risk factors and excess adiposity among overweight children and adolescents: The Bogalusa Heart Study, *Journal of Pediatrics* 150 (2007): 12–17.
6. S. E. Barlow, and the Expert Committee, Expert Committee recommendations regarding the prevention, assessment, and treatment of child and adolescent overweight and obesity: Summary report, *Pediatrics* 120 (2007): S164–S192; Ludwig, 2007.
7. Ludwig, 2007.
8. Jones, 2008; T. S. Hannon, G. Rao, and S. A. Arslanian, Childhood obesity and type 2 diabetes mellitus, *Pediatrics* 116 (2005): 473–480; Cruz and coauthors, 2005.
9. Jones, 2008; Hannon, Rao, and Arslanian, 2005; Cruz and coauthors, 2005.
10. Cook and coauthors, 2008; G. S. Boyd and coauthors, Effect of obesity and high blood pressure on plasma lipid levels in children and adolescents, *Pediatrics* 116 (2005): 473–480.
11. Zhu and coauthors, 2008; Freedman and coauthors, 2007; D. R. Thompson and coauthors, Childhood overweight and cardiovascular disease risk factors: The National Heart, Lung, and Blood Institute Growth and Health Study, *Journal of Pediatrics* 150 (2007): 18–25; J. Botton and coauthors, Cardiovascular risk factor levels and their relationships with overweight and fat distribution in children: The Fleurbaix Laventie Ville Sante II Study, *Metabolism* 56 (2007): 614–622; J. C. Eisenmann and coauthors, Fatness, fitness, and cardiovascular disease risk factors in children and adolescents, *Medicine & Science in Sports & Exercise* 39 (2007): 1251–1256.
12. S. R. Daniels, F. R. Greer, and the Committee on Nutrition, Lipid screening and cardiovascular health in childhood, *Pediatrics* 122 (2008): 198–208.
13. Freedman and coauthors, 2009; Cook and coauthors, 2008; Botton and coauthors, 2007; Thompson and coauthors, 2007.
14. Daniels, Greer, and the Committee on Nutrition, 2008.
15. National High Blood Pressure Education Program Working Group on High Blood Pressure in Children and Adolescents, The fourth report on the diagnosis, evaluation, and treatment of high blood pressure in children and adolescents, *Pediatrics* 114 (2004): 555S–576S.
16. National High Blood Pressure Education Program Working Group on High Blood Pressure in Children and Adolescents, 2004.
17. F. J. He and G. A. MacGregor, Importance of salt in determining blood pressure in children: Meta-analysis of controlled trials, *Hypertension* 48 (2006): 861–869.
18. Committee on Dietary Reference Intakes, *Dietary Reference Intakes for Energy, Carbohydrate, Fiber, Fat, Fatty Acids, Cholesterol, Protein, and Amino Acids* (Washington, D.C.: National Academies Press, 2005), pp. 769–879.
19. Daniels, Greer, and the Committee on Nutrition, 2008.
20. L. Johnson and coauthors, Energy-dense, low-fiber, high-fat dietary pattern is associated with increased fatness in childhood, *American Journal of Clinical Nutrition* 87 (2008): 846–854; J. Hurley and B. Liebman, Kids' cuisine: "What would you like with your fries?" *Nutrition Action Healthletter* 31 (2004): 12–15.
21. F. R. Greer, N. F. Krebs, and the Committee on Nutrition, American Academy of Pediatrics, Optimizing bone health and calcium intakes of infants, children, and adolescents, *Pediatrics* 117 (2006): 578–585.
22. Daniels, Greer, and the Committee on Nutrition, 2008.
23. Centers for Disease Control and Prevention, Youth tobacco surveillance: United States, 2001–2002, *Morbidity and Mortality Weekly Report* 55 (2006): entire supplement.
24. Daniels, Greer, and the Committee on Nutrition, 2008; Barlow and the Expert Committee, 2007.

CHAPTER

16

Seth C Fisher/Shutterstock.com

Nutrition in Your Life

Take a moment to envision yourself at age 60, 75, or even 90. Are you physically fit and healthy? Can you see yourself walking on the beach with friends or tossing a ball with children? Are you able to climb stairs and carry your own groceries? Importantly, are you enjoying life? If you're lucky, you will enjoy old age in good health. Making nutritious foods and physical activities a priority in your life can help bring rewards of continued health and enjoyment throughout life.

Throughout this chapter, the CourseMate icon indicates an opportunity for online self-study, linking you to activities to increase your understanding of chapter concepts. **www.cengagebrain.com** (search for ISBN 084006845X)

Life Cycle Nutrition: Adulthood and the Later Years

CHAPTER OUTLINE

Wise food choices, made throughout adulthood, can support a person's ability to meet physical, emotional, and mental challenges and to enjoy freedom from disease.[1] Two goals motivate adults to pay attention to their diets: promoting health and slowing aging. Much of this text has focused on nutrition to support health. This chapter focuses on aging and the nutrition needs of older adults. As you will see, the same diet and behaviors that reduce disease risks also slow aging.

The U.S. population is growing older. The majority is now middle-aged, and the ratio of old people to young is increasing, as Figure 16-1 (p. 540) shows. In 1900, only 1 out of 25 people was 65 or older. In 2000, 1 out of 8 had reached age 65. Projections for 2030 are 1 out of 5.

Our society uses the arbitrary age of 65 years ♦ to define the transition point between middle age and old age, but growing "old" happens day by day, with changes occurring gradually over time. Since 1950 the population of those older than 65 has almost tripled. Remarkably, the fastest-growing age group has been people older than 85 years; since 1950 their numbers have increased sevenfold. The number of people in the United States age 100 or older doubled in the last decade. Similar trends are occurring in populations worldwide.

♦ Commonly used age groups:
- young old (65–74 years)
- old old (75–84 years)
- oldest old (≥85 years)

Life expectancy in the United States is 78 years: 81 years for white women and 77 years for black women, 76 years for white men and 70 years for black men. All of these record highs are much higher than the average life expectancy of 47 years in 1900.[2] Women who live to 70 can expect to survive an additional 16 years, on average; men, an additional 14 years. Advances in medical science—antibiotics and other treatments—are largely responsible for almost doubling the life expectancy in the 20th century. Improved nutrition and an abundant food supply have also

life expectancy: the average number of years lived by people in a given society.

FIGURE 16-1 **The Aging of the U.S. Population**

In general, the percentage of older people in the population has increased over the decades whereas the percentage of younger people has decreased.

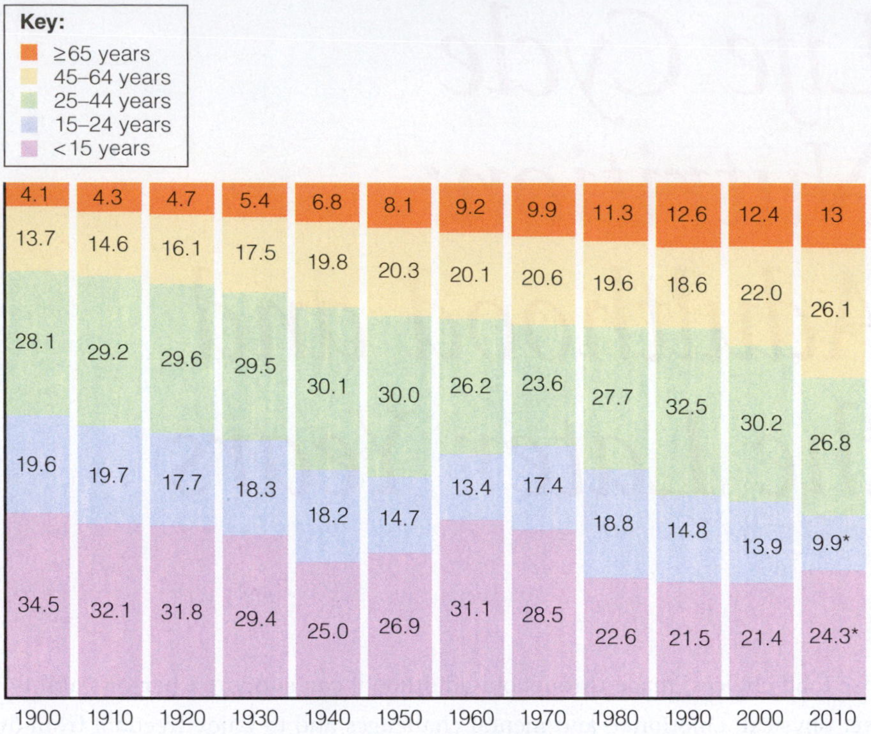

* Projections for 2010 split age groups slightly differently. Blue represents 18–24 years and purple represents <18 years.
SOURCE: U.S. Census Bureau, Decennial census of population, 1900 to 2000.

contributed to lengthening life expectancy. Ironically, an abundant food supply has also jeopardized the chances of lengthening life expectancy as obesity rates increase.[3]

The **life span** has not lengthened as dramatically; human **longevity** appears to have an upper limit. The maximum potential human life span is currently about 130 years. The verifiably oldest person died in 1997 at age 122. With recent advances in medical technology and genetic knowledge, researchers may one day be able to extend the life span even further by slowing, or perhaps preventing, aging and its accompanying diseases.[4]

Nutrition and Longevity

Research in the field of aging is active—and difficult. Researchers are challenged by the diversity of older adults. When older adults experience health problems, it is hard to know whether to attribute these problems to genetics, aging, or environmental factors such as nutrition. The idea that nutrition can influence the aging process is particularly appealing because people can control and change their eating habits. The questions being asked include:

- To what extent is aging inevitable, and can it be slowed through changes in lifestyle and environment?
- What role does nutrition play in the aging process, and what role can it play in slowing aging?

With respect to the first question, it seems that aging is an inevitable, natural process, programmed into the genes at conception. People can, however, slow the

life span: the maximum number of years of life attainable by a member of a species.

longevity: long duration of life.

process within genetic limits by adopting healthy lifestyle habits such as eating nutritious foods and engaging in physical activities. In fact, an estimated 70 to 80 percent of the average person's life expectancy may depend on individual health-related behaviors; genes determine the remaining 20 to 30 percent.

With respect to the second question, good nutrition helps to maintain a healthy body and can therefore ease the aging process in many significant ways. Clearly, nutrition can improve the **quality of life** in the later years.

© Corbis Super RF/Alamy

Growing old can be enjoyable for people who take care of their health and live each day fully.

Observation of Older Adults

The strategies adults use to meet the two goals mentioned at the start of this chapter—promoting health and slowing aging—are actually very much the same. What to eat, how physically active to be, and other lifestyle choices greatly influence both physical health and the aging process.

Healthy Habits A person's **physiological age** reflects his or her health status and may or may not reflect the person's **chronological age**. Quite simply, some people seem younger, ♦ and others older, than their years. Six lifestyle behaviors seem to have the greatest influence on people's health and therefore on their physiological age:[5]

- Eating well-balanced meals rich in fruits and vegetables regularly
- Engaging in physical activity regularly
- Not smoking
- Not using alcohol, or using it in moderation
- Maintaining a healthy body weight
- Sleeping regularly and adequately

♦ Older adults who lead active lives contrary to stereotypes of diminished abilities are sometimes referred to as the *young old*.

Over the years, the effects of these lifestyle choices accumulate—that is, people who follow most of these practices live longer and have fewer disabilities as they age. They are in better health, even when older in chronological age, than people who do not adopt these behaviors.[6] Even though people cannot change their birth dates, they may be able to add years to, and enhance the quality of, their lives.[7] Physical activity seems to be most influential in preventing or slowing the many changes that define a stereotypical "old" person. After all, many of the physical limitations that accompany aging occur because people become inactive, not because they become older.

Physical Activity The many remarkable benefits of regular physical activity are not limited to the young. Compared with those who are inactive, older adults who are active weigh less; have greater flexibility, more endurance, better balance, and better health; and live longer.[8] They reap additional benefits from various activities as well: aerobic activities improve cardiorespiratory endurance, blood pressure, and blood lipid concentrations; moderate endurance activities improve the quality of sleep; and strength training improves posture and mobility. In fact, regular physical activity is the most powerful predictor of a person's mobility in the later years. Physical activity also increases blood flow to the brain, thereby preserving mental ability, alleviating depression, supporting independence, and improving quality of life.[9]

Muscle mass and muscle strength tend to decline with aging, making older people vulnerable to falls and immobility. Falls are a major cause of fear, injury, disability, and even death among older adults.[10] Many lose their independence as a result of falls. Regular physical activity tones, firms, and strengthens muscles, helping to improve balance, restore confidence, reduce the risk of falling, and lessen the risk of injury should a fall occur.

Even without a fall, older adults may become so weak that they can no longer perform life's daily tasks, such as climbing stairs, carrying packages, and opening jars. Resistance training helps older adults to maintain independence by

quality of life: a person's perceived physical and mental well-being.

physiological age: a person's age as estimated from her or his body's health and probable life expectancy.

chronological age: a person's age in years from his or her date of birth.

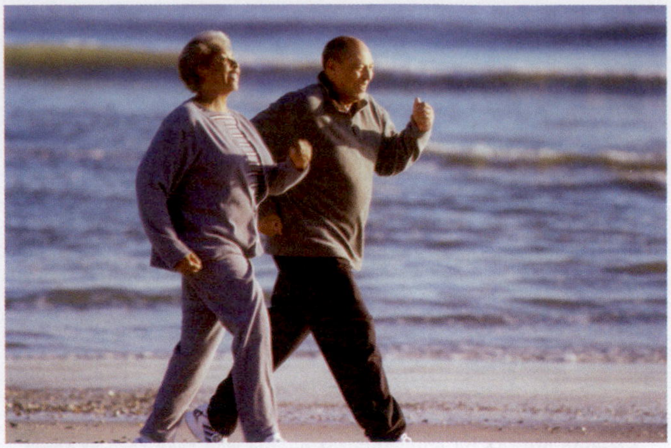
Purestock/Getty Images

Regular physical activity promotes a healthy, independent lifestyle.

improving muscle strength to perform these tasks. Even in frail, elderly people older than 85 years of age, strength training not only improves balance, muscle strength, and mobility, but it also increases energy expenditure and energy intake, thereby enhancing nutrient intakes. This finding highlights another reason to be physically active: a person expending energy can afford to eat more food and thus receives more nutrients. People who are committed to an ongoing fitness program can benefit from higher energy and nutrient intakes and still maintain their body weights.

Ideally, physical activity should be part of each day's schedule and should be intense enough to prevent muscle atrophy and to speed the heartbeat and respiration rate. Although aging reduces both speed and endurance to some degree, older adults can still train and achieve exceptional performances. Healthy older adults who have not been active can ease into a suitable routine, becoming as physically active as their abilities allow. They can start by walking short distances until they are walking at least 10 minutes continuously and then gradually increase their distance to a 30- to 45-minute workout at least 5 days a week. Table 16-1 provides exercise goals and guidelines for seniors.[11] Only 1 out of 12 older adults meets the goals for both aerobic and strength training activities.[12] People with medical conditions should check with a physician before beginning an exercise routine, as should sedentary men older than 40 and sedentary women older than 50 who want to participate in a vigorous program.

Dietary Guidelines for Americans 2010

Older adults should be as physically active as their abilities and conditions allow.

Manipulation of Diet

In their efforts to understand longevity, researchers have not only observed people, but they have also manipulated influencing factors, such as diet, in animals. This research has given rise to some interesting and suggestive findings.

Energy Restriction in Animals

Animals live longer and have fewer age-related diseases when their energy intakes are restricted. These life-prolonging benefits become evident when the diet provides enough food to prevent malnutrition and an energy intake of about 70 percent of normal; benefits decline as the age of starting the energy restriction is delayed.[13] Exactly how energy restriction prolongs life remains unexplained, although gene activity appears to play a key role. The genetic activity of old mice differs from that of young mice, with some genes becoming more active with age and others less active. With an energy-restricted diet, many of the genetic activities of older mice revert to those of younger mice. These "slow-aging" genetic changes are apparent in as little as one month on an energy-restricted, but still nutritionally adequate, diet.

The consequences of energy restriction in animals include a delay in the onset, or prevention, of chronic diseases such as cancer and atherosclerosis and age-related conditions such as neuron degeneration; prolonged growth and development; and improved blood glucose, insulin sensitivity, and blood lipids.[14] In addition, energy metabolism slows and body temperature drops—indications of a reduced rate of oxygen consumption. Oxygen consumption is lower in mice prone to obesity, and they benefit more from energy-restricted diets than other mice.[15] As Highlight 11 explains, the use of oxygen during energy metabolism produces free radicals, which have been implicated in the aging process. Restricting energy intake in animals not only produces fewer free radicals but also increases antioxidant activity and enhances DNA repair. Reducing oxidative stress may at least partially explain how restricting energy intake lengthens life expectancy.

Interestingly, longevity appears to depend on restricting energy intake and not on the amount of body fat. Genetically obese rats live longer when given a re-

TABLE 16-1 Exercise Guidelines for Older Adults

	Aerobic	Strength	Balance	Flexibility
Examples	 © Geoff Manasse/ PhotoDisc/Jupiter Images	 © IT Stock Free/Jupiter Images	 © IT Stock Free/Jupiter Images	 © Fon Chapple/ Thinkstock/Jupiter Images
Start easy and progress gradually	Be active 5 minutes on most or all days	Using 0- to 2-pound weights, do 1 set of 8–12 repetitions twice a week	Hold onto table or chair with one hand, then with one finger	Hold stretch for 10 seconds; do each stretch 3 times
Frequency	At least 5 days per week of moderate activity or at least 3 days per week of vigorous activity	At least 2 (nonconsecutive) days per week	3 days each week	At least 2 days per week; preferably on all days that aerobic or strength activities are performed
Intensity[a]	Moderate, vigorous, or combination	Moderate to high; 10 to 15 repetitions per exercise		At least 10 minutes per day
Duration	At least 30 minutes of moderate activity in bouts of at least 10 minutes each or at least 20 minutes of continuous vigorous activity	8 to 10 exercises involving the major muscle groups		Stretch major muscle groups for 10–30 seconds, repeating each stretch 3 to 4 times
Cautions and comments	Stop if you are breathing so hard you can't talk or if you feel dizziness or chest pain	Breathe out as you contract and in as you relax (do not hold breath); use smooth, steady movements	Incorporate balance techniques with strength exercises as you progress	Stretch after strength and endurance exercises for 20 minutes, 3 times a week; use slow, steady movements; bend joints slightly

[a]On a 10-point scale, where sitting = 0 and maximum effort = 10, moderate intensity = 5 to 6 and vigorous intensity = 7 to 8.

NOTE: Activity recommendations are in addition to routine activities of daily living (such as getting dressed, cooking, grocery shopping) and moderate activities lasting less than 10 minutes.

SOURCE: M. E. Nelson and coauthors, Physical activity and public health in older adults: Recommendation from the American College of Sports Medicine and the American Heart Association, *Medicine & Science in Sports & Exercise* 39 (2007): 1435–1445.

stricted diet even though their body fat is similar to that of other rats allowed to eat freely.

Energy Restriction in Human Beings Research on a variety of animals ♦ confirms the relationship between energy restriction and longevity. Applying the results of animal studies to human beings is problematic, however, and conducting studies on human beings raises numerous questions—beginning with how to define energy restriction. Does it mean eating less or just weighing less? Is it less than you want or less than the average? Does eating less have to result in weight loss? Does it matter whether weight loss results from more exercise or from less food? Or whether weight loss is intentional or unintentional? Answers await research.

♦ kCalorie-restricted research has been conducted on various species, including mice, rats, rhesus monkeys, cynomolgus monkeys, spiders, and fish.

Extreme starvation to extend life, like any extreme, is rarely, if ever, worth the price. Hunger is persistent when energy is restricted by 30 percent. Furthermore, using animal data to extrapolate to humans, researchers estimate that it would take 30 years of such energy-restricted dieting to increase life expectancy by less than 3 years.[16]

Moderation, on the other hand, may be valuable. Many of the physiological responses to energy restriction seen in animals also occur in people whose intakes are *moderately* restricted. When people cut back on their usual energy intake by 10 to 20 percent, ♦ body weight, body fat, and blood pressure drop, and blood lipids and insulin response improve—favorable changes for preventing chronic diseases.[17] Some research suggests that fasting on alternative days may provide similar benefits.[18]

♦ For perspective, a person with a usual energy intake of 2000 kcalories might cut back to 1600 to 1800 kcalories.

The reduction in oxidative damage that occurs with energy restriction in animals also occurs in people whose diets include antioxidant nutrients and phytochemicals. Diets, such as the Mediterranean diet, which include an abundance of fruits, vegetables, olive oil, and red wine—with their array of phytochemicals that have antioxidant activity—support good health and long life. Clearly, nutritional adequacy is essential to living a long and healthy life.

IN SUMMARY **Life expectancy in the United States increased dramatically in the 20th century. Factors that enhance longevity include limited or no alcohol use, regular balanced meals, weight control, abstinence from smoking, regular physical activity, and adequate sleep. Energy restriction in animals seems to lengthen their lives. Whether such dietary intervention in human beings is beneficial remains unknown. At the very least, nutrition—especially when combined with regular physical activity—can influence aging and longevity in human beings by supporting good health and preventing disease.**

The Aging Process

As people get older, each person becomes less and less like anyone else. The older people are, the more time has elapsed for such factors as nutrition, genetics, physical activity, and everyday **stress** to influence physical and psychological aging.

Stress contributes to a variety of age-related diseases.[19] Both physical **stressors** (such as alcohol abuse, other drug abuse, smoking, pain, and illness) and psychological stressors (such as exams, divorce, moving, and the death of a loved one) elicit the body's **stress response**. The body responds to such stressors with an elaborate series of physiological steps, as the nervous and hormonal systems bring about defensive readiness in every body part. These effects favor physical action—the classic fight-or-flight response. Prolonged or severe stress can drain the body of its reserves and leave it weakened, aged, and vulnerable to illness, especially if physical action is not taken. As people age, they lose their ability to adapt to both external and internal disturbances. When disease strikes, the reduced ability to adapt makes the aging individual more vulnerable to death than a younger person. Measures to preserve health forestall disease, disability, and death.[20]

Because the stress response is mediated by hormones, it differs between men and women. The fight-or-flight response may be more typical of men than of women. Women's reactions to stress more typically follow a pattern of "tend-and-befriend." Women *tend* by nurturing and protecting themselves and their children. These actions promote safety and reduce stress. Women *befriend* by creating and maintaining a social group that can help in the process.

Highlight 11 describes the oxidative stresses and cellular damage that occur when free radicals exceed the body's ability to defend itself. Increased free-radical activity and decreased antioxidant protection are common features of aging—and foods rich in antioxidants seem to help slow the aging process and improve cognition.[21] Such findings seem to suggest that the fountain of youth may actually be a cornucopia of fruits and vegetables rich in antioxidants. (Return to Highlight 11 for more details on the antioxidant action of fruits and vegetables in defending against oxidative stress.)

Physiological Changes

As aging progresses, inevitable changes in each of the body's organs contribute to the body's declining function. These physiological changes influence nutrition status, just as growth and development do in the earlier stages of the life cycle.

Body Weight

Two-thirds of older adults in the United States are now considered overweight or obese. Chapter 8 presents the many health problems that accompany obesity and the BMI guidelines for a healthy body weight (18.5 to 24.9). These guidelines apply to all adults, regardless of age, but they may be too restrictive for older adults. The importance of body weight in defending against chronic diseases differs for older adults. Being moderately *overweight* may not be harmful. For adults older than 65, health risks do not become apparent until BMI reaches

stress: any threat to a person's well-being; a demand placed on the body to adapt.

stressors: environmental elements, physical or psychological, that cause stress.

stress response: the body's response to stress, mediated by both nerves and hormones.

at least 27—and the relationship tends to diminish with age until it disappears by age 75. Older adults who are *obese,* however, face serious medical complications and can significantly improve their quality of life with weight loss.[22]

For some older adults, a low body weight may be more detrimental than a high one. Low body weight often reflects malnutrition and the trauma associated with a fall. Many older adults experience unintentional weight loss, in large part because of an inadequate food intake.[23] Without adequate nutrient reserves, an underweight person may be unprepared to fight against diseases. For underweight people, even a slight weight loss (5 percent) increases the likelihood of disease and premature death, making every meal a life-saving event. Snacking between meals can help older adults obtain needed nutrients and energy.[24]

Body Composition In general, older people tend to lose bone and muscle and gain body fat. Many of these changes occur because some hormones that regulate appetite and metabolism become less active with age, whereas others become more active.*

Loss of muscle, known as **sarcopenia**, can be significant in the later years, and its consequences can be quite dramatic (see Figure 16-2). As muscles diminish and weaken, people lose the ability to move and maintain balance—making falls likely. The limitations that accompany the loss of muscle mass and strength play a key role in the diminishing health that often accompanies aging.[25] Optimal nutrition with sufficient protein and regular physical activity can help maintain muscle mass and strength and minimize the changes in body composition associated with aging.[26]

Risk factors for sarcopenia include weight loss, little physical activity, and cigarette smoking.[27] Obesity and the inflammation that accompanies it may also contribute to sarcopenia.[28]

FIGURE 16-2 Sarcopenia

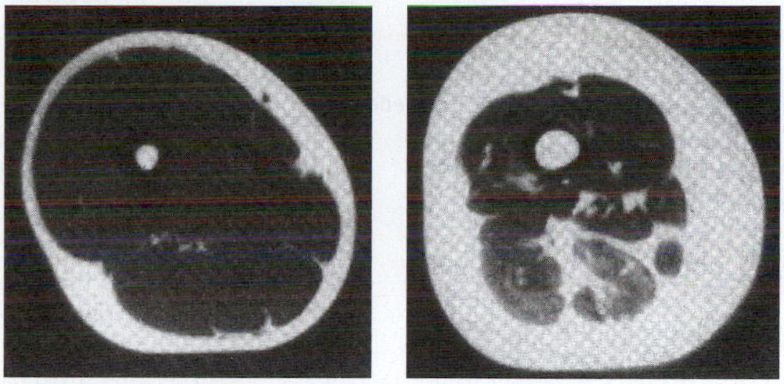

Courtesy of Dr. William Evans (both)

These cross sections of two women's thighs may appear to be about the same size from the outside, but the 20-year-old woman's thigh (left) is dense with muscle tissue. The 64-year-old woman's thigh (right) has lost muscle and gained fat, changes that may be largely preventable with strength-building physical activities.

Immunity and Inflammation As people age, the immune system loses function. As they become ill, the immune system becomes overstimulated. The combination of an inefficient and overactive response in aging—known as "inflammaging"—results in a chronic inflammation that accompanies frailty, illness, and death.[29]

Most diseases common in older adults—such as atherosclerosis, Alzheimer's disease, obesity, and rheumatoid arthritis—are different in obvious ways, but they all reflect an underlying inflammatory process.[30] Because of this association with diseases, inflammation is often perceived as a harmful process, yet it is critical in supporting health as the immune system destroys invading organisms and repairs damaged tissues.[31] Thus inflammation presents a challenge to identify factors that will both protect the beneficial effects and limit the harmful consequences.

In addition to aging and diseases, the immune system is compromised by nutrient deficiencies. Thus the combination of age, illness, and malnutrition makes older people particularly vulnerable to infectious diseases. Adding insult to injury, antibiotics often are not effective against infections in people with compromised immune systems. Consequently, infectious diseases are a major cause of death in older adults. Older adults may improve their immune system responses with regular physical activity.[32]

GI Tract In the GI tract, the intestinal wall loses strength and elasticity with age, and GI hormone secretions change. All of these actions slow motility. Constipation is much more common in the elderly than in the young. Changes in GI hormone

*Causes of diminished appetite in older adults include increased cholecystokinin, leptin, and cytokines and decreased ghrelin and testosterone. Additional examples of hormones that change with age include growth hormone and androgens, which decline with advancing age, thus contributing to the decrease in lean body mass, and prolactin, which increases with age, helping to maintain body fat. Insulin sensitivity also diminishes as people grow older, most likely because of increases in body fat and decreases in physical activity.

sarcopenia (SAR-koh-PEE-nee-ah): loss of skeletal muscle mass, strength, and quality.

- **sarco** = flesh
- **penia** = loss or lack

secretions also diminish appetite, leading to decreased energy intake and unintentional weight loss.

Atrophic gastritis, a condition that affects almost one-third of those older than 60, is characterized ♦ by an inflamed stomach, bacterial overgrowth, and a lack of hydrochloric acid and intrinsic factor. All of these can impair the digestion and absorption of nutrients, most notably, vitamin B_{12}, but also biotin, folate, calcium, iron, and zinc.

♦ Consequences of atrophic gastritis:
- Inflamed stomach
- Increased bacterial growth
- Reduced hydrochloric acid
- Reduced intrinsic factor
- Increased risk of nutrient deficiencies, notably of vitamin B_{12}

Difficulty swallowing, medically known as **dysphagia**, occurs in all age groups, but especially in the elderly. Being unable to swallow a mouthful of food can be scary, painful, and dangerous. Even swallowing liquids can be a problem for some people. Consequently, the person may eat less food and drink fewer beverages, resulting in weight loss, malnutrition, and dehydration. Dietary intervention for dysphagia is highly individualized based on the person's abilities and tolerances. The diet typically provides moist, soft-textured, tender-cooked, or pureed foods and thickened liquids.

Tooth Loss Regular dental care over a lifetime protects against tooth loss and gum disease, which are common in old age. These conditions make chewing difficult or painful. Dentures, even when they fit properly, are less effective than natural teeth, and inefficient chewing can cause choking. Inefficient chewing can also interfere with protein digestibility.[33]

People with tooth loss, ♦ gum disease, and ill-fitting dentures tend to limit their food selections to soft foods. If foods such as corn on the cob, apples, and hard rolls are replaced by creamed corn, applesauce, and rice, then nutrition status may not be greatly affected. However, when food groups are eliminated and variety is limited, poor nutrition follows. People without teeth typically eat fewer fruits and vegetables and have less variety in their diets. Consequently, they have low intakes of fiber and vitamins, which exacerbates their dental and overall health problems. To determine whether a visit to the dentist is needed, an older adult can check the conditions listed in the margin. ♦

♦ The medical term for lack of teeth is **edentulous** (ee-DENT-you-lus).
- **e** = without
- **dens** = teeth

♦ Conditions requiring dental care:
Dry mouth
Eating difficulty
No dental care within two years
Tooth or mouth pain
Altered food selections
Lesions, sores, or lumps in mouth

Sensory Losses and Other Physical Problems Sensory losses and other physical problems can also interfere with an older person's ability to obtain adequate nourishment. Failing eyesight, for example, can make driving to the grocery store impossible and shopping for food a frustrating experience. It may become so difficult to read food labels and count money that the person doesn't buy needed foods. Carrying bags of groceries may be an unmanageable task. Similarly, a person with limited mobility may find cooking and cleaning up too hard to do. Not too surprisingly, the prevalence of undernutrition is high among those who are homebound.

Sensory losses can also interfere with a person's ability or willingness to eat. Taste and smell sensitivities tend to diminish with age and may make eating less enjoyable. If a person eats less, then weight loss and nutrient deficiencies may follow. Loss of vision and hearing may contribute to social isolation, and eating alone may lead to poor intake.

Other Changes

In addition to the physiological changes that accompany aging, adults change in many other ways that influence their nutrition status.[34] Psychological, economic, and social factors play big roles in a person's ability and willingness to eat.

Psychological Changes Although not an inevitable component of aging, depression is common among older adults. Depressed people, even those without disabilities, lose their ability to perform simple physical tasks. They frequently lose their appetite and the motivation to cook or even to eat. An overwhelming sense of grief and sadness at the death of a spouse, friend, or family member may leave a person, especially an elderly person, feeling powerless to overcome depression. When a person is suffering the heartache and loneliness of bereavement, cooking meals may not seem worthwhile. The support and companionship

dysphagia (dis-FAY-jah): difficulty swallowing.

© MBI/Alamy

Shared meals can brighten the day and enhance the appetite.

of family and friends, especially at mealtimes, can help overcome depression and enhance appetite.

Economic Changes Overall, older adults today have higher incomes than their cohorts of previous generations. Still, 10 percent of the people older than age 65 live in poverty. Factors such as living arrangements and income make significant differences in the food choices, eating habits, and nutrition status of older adults, especially those older than age 80. People of low socioeconomic means are likely to have inadequate food and nutrient intakes. Only about one-third of eligible seniors participate in the Supplemental Nutrition Assistance Program (SNAP). ♦

♦ SNAP is the new name for the federal Food Stamp Program.

Social Changes Malnutrition among older adults is most common in hospitals and nursing homes.[35] In the community, malnutrition is most likely to occur among those living alone, especially men; those with the least education; those living in federally funded housing (an indicator of low income); and those who have recently experienced a change in lifestyle. Adults who live alone do not necessarily make poor food choices, but they often consume too little food. Loneliness is directly related to nutritional inadequacies, especially of energy intake.

IN SUMMARY Many changes that accompany aging can impair nutrition status. Among physiological changes, hormone activity alters body composition, immune system changes raise the risk of infections, atrophic gastritis interferes with digestion and absorption, and tooth loss limits food choices. Psychological changes such as depression, economic changes such as loss of income, and social changes such as loneliness contribute to poor food intake.

Energy and Nutrient Needs of Older Adults

Knowledge about the nutrient needs and nutrition status of older adults has grown considerably in recent years. The Dietary Reference Intakes (DRI) cluster people older than 50 into two age categories—one group of 51 to 70 years and one of 71 and older.

To ensure adequate hydration, keep a glass of water next to you at home, drink from water fountains whenever you walk by, and put a bottle of water in your car.

Setting standards for older people is difficult because individual differences become more pronounced as people grow older.[36] People start out with different genetic predispositions and ways of handling nutrients, and the effects of these differences become magnified with years of unique dietary habits. For example, one person may tend to omit fruits and vegetables from his diet, and by the time he is old, he may have a set of nutrition problems associated with a lack of fiber and antioxidants. Another person may have omitted milk and milk products all her life—her nutrition problems may be related to a lack of calcium. Also, as people age, they suffer different chronic diseases and take various medicines—both of which will affect nutrient needs. For all of these reasons, researchers have difficulty even defining "healthy aging," a prerequisite to developing recommendations to meet the "needs of practically all healthy persons." The following discussion gives special attention to the nutrients of greatest concern.

Water

Despite real fluid needs, many older people do not seem to feel thirsty or notice mouth dryness. Many nursing home employees say it is hard to persuade their elderly clients to drink enough water and fruit juices. Older adults may find it difficult and bothersome to get a drink or to get to a bathroom. Those who have lost bladder control may be afraid to drink too much water.

Dehydration is a risk for older adults.[37] Total body water decreases as people age, so even mild stresses such as fever or hot weather can precipitate rapid dehydration in older adults. Dehydrated older adults seem to be more susceptible to urinary tract infections, pneumonia, **pressure ulcers**, and confusion and disorientation. To prevent dehydration, older adults need to drink *at least* six glasses of water or other beverages every day. ♦ Emphasizing foods with high-water content, such as melons and soups, can also be helpful.

♦ Beverage recommendation for adults 51+ yr:
- Men: 13 c/day
- Women: 9 c/day

Energy and Energy Nutrients

On average, energy needs decline an estimated 5 percent per decade. One reason is that people usually reduce their physical activity as they age, although they need not do so. Another reason is that basal metabolic rate declines 1 to 2 percent per decade in part because lean body mass and thyroid hormones diminish.[38]

The lower energy expenditure of older adults means that they need to eat less food to maintain their weights. Accordingly, the estimated energy requirements for adults decrease steadily after age 19. The accompanying "How To" explains how to estimate energy requirements for older adults.

Older adults need fewer kcalories as they age, but their nutrient needs remain high. For this reason, it is most important that they select mostly nutrient-dense foods. There is little leeway for added sugars, solid fats, or alcohol. Such nutrient-poor selections can easily lead to weight gain and malnutrition. The USDA Food Guide (introduced in Chapter 2) offers a dietary framework for adults of all ages. A modified pyramid for older adults has been developed to supplement the USDA MyPyramid (see Figure 16-3).[39]

Protein

Because energy needs decrease, protein must be obtained from low-kcalorie sources of high-quality protein, such as lean meats, poultry, fish, and eggs; fat-free and low-fat milk products; and legumes. Protein is especially important for the elderly to support a healthy immune system, prevent muscle wasting, and optimize bone mass.[40] Maintaining muscles helps to support protein metabolism and immune function.

Underweight or malnourished older adults need protein- and energy-dense snacks such as hard-boiled eggs, tuna salad, peanut butter on wheat toast, and hearty soups. Drinking liquid nutritional formulas between meals can also boost energy and nutrient intakes. Importantly, the diet should provide enjoyment as well as nutrients.[41]

pressure ulcers: damage to the skin and underlying tissues as a result of compression and poor circulation; commonly seen in people who are bedridden or chairbound.

HOW TO Estimate Energy Requirements for Older Adults

The "How To" on p. 250 describes how to estimate the energy requirements for adults using an equation that accounts for age, physical activity, weight, and height. Alternatively, energy requirements for older adults can be "guesstimated" by using the values listed in the tables in Appendix F for adults 30 years of age and subtracting 7 kcalories for women and 10 kcalories for men per day for each year older than 30.

For example, Table F-4 lists 2556 kcalories per day for a 30-year-old woman who is 5 feet 5 inches tall, weighs 150 pounds, and has a low activity level. To estimate the energy requirements of a similar 50-year-old woman, subtract 7 kcalories per day for each year over 30:

50 − 30 = 20 yr
20 yr × 7 kcal/day = 140 kcal/day
2556 kcal/day (at age 30) − 140 kcal/day = 2416 kcal/day (at age 50)

Similarly, using Table F-5 to estimate the energy requirements of a sedentary 65-year-old man who is 5 feet 11 inches tall and weighs 250 pounds, subtract 10 kcalories per day for each year over 30:

65 − 30 = 35 yr
35 yr × 10 kcal/day = 350 kcal/day
3088 kcal/day (at age 30) − 350 kcal/day = 2738 kcal/day (at age 65)

Adults between the ages of 19 and 30 can also use the values listed in the tables in Appendix F by adding 7 kcalories for women and 10 kcalories for men per day for each year below 30.

For additional practice, log on to **www.cengagebrain.com** and search for ISBN 084006845X.

TRY IT Use Appendix F to calculate your energy requirements.

FIGURE 16-3 Modified MyPyramid for Older Adults

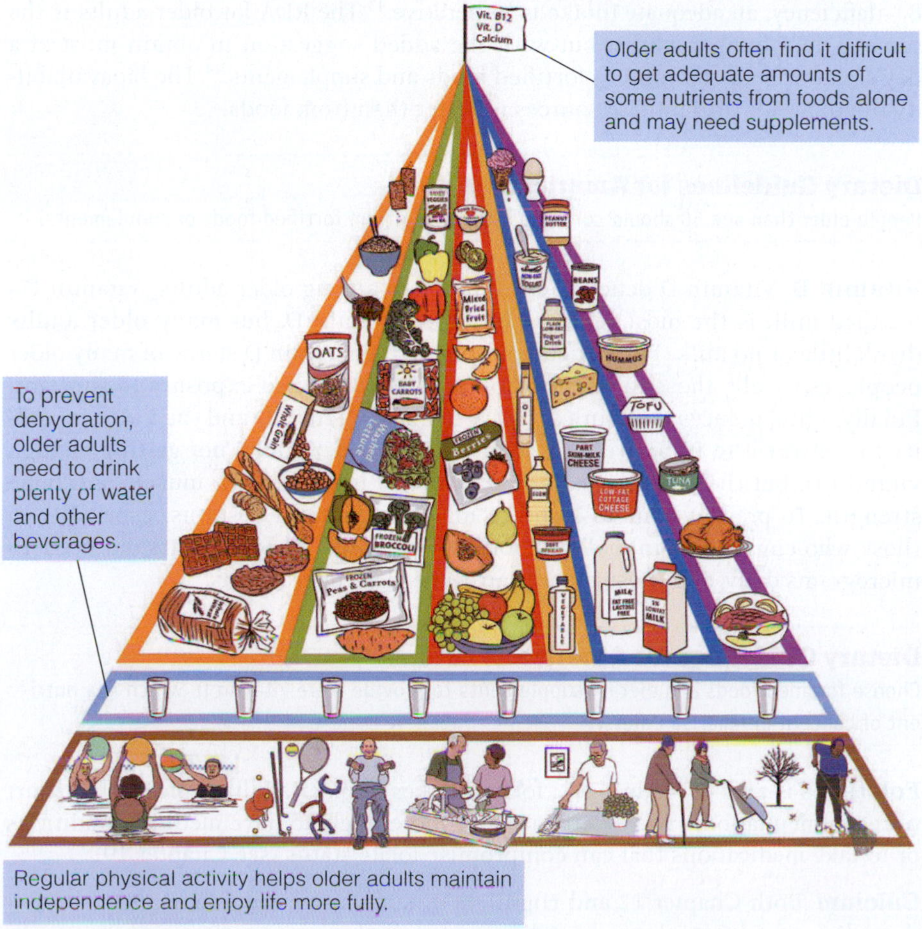

Vit. B12
Vit. D
Calcium
Older adults often find it difficult to get adequate amounts of some nutrients from foods alone and may need supplements.
To prevent dehydration, older adults need to drink plenty of water and other beverages.
Regular physical activity helps older adults maintain independence and enjoy life more fully.

SOURCE: Copyright 2007 Tufts University. For full article, see Lichtenstein AH, Rasmussen H, Yu WW, Epstein SR, Russell RM. Modified MyPyramid for Older Adults. *J Nutr.* 2008; 138:78-82.

Carbohydrate and Fiber As always, abundant carbohydrate is needed to protect protein from being used as an energy source. Carbohydrate-rich foods such as legumes, vegetables, whole grains, and fruits are also rich in fiber and essential vitamins and minerals. Average fiber intakes among older adults are lower than current recommendations (14 grams per 1000 kcalories).[42] Eating high-fiber foods and drinking water can alleviate constipation—a condition common among older adults, especially nursing home residents. Physical inactivity and medications also contribute to the high incidence of constipation.

Fat As is true for people of all ages, fat intake needs to be moderate in the diets of most older adults—enough to enhance flavors and provide valuable nutrients, but not so much as to raise the risks of atherosclerosis and other degenerative diseases. This recommendation should not be taken too far; limiting fat too severely may lead to nutrient deficiencies and weight loss—two problems that carry greater health risks in the elderly than overweight.

Vitamins and Minerals

Most people can achieve adequate vitamin and mineral intakes simply by including foods from all food groups in their diets, but older adults often omit fruits and vegetables. Similarly, few older adults consume the recommended amounts of milk or milk products.

♦ **Atrophic gastritis** is a chronic inflammation of the stomach characterized by inadequate hydrochloric acid and intrinsic factor—two key players in vitamin B_{12} absorption.

Vitamin B_{12} An estimated 10 to 30 percent of adults older than 50 have atrophic gastritis. ♦ As Chapter 10 explains, people with atrophic gastritis are particularly vulnerable to vitamin B_{12} deficiency. The bacterial overgrowth that accompanies this condition uses up the vitamin, and without hydrochloric acid and intrinsic factor, digestion and absorption of vitamin B_{12} are inefficient. Given the poor cognition, anemia, and devastating neurological effects associated with a vitamin B_{12} deficiency, an adequate intake is imperative.[43] The RDA for older adults is the same as for younger adults, but with the added suggestion to obtain most of a day's intake from vitamin B_{12}–fortified foods and supplements.[44] The bioavailability of vitamin B_{12} from these sources is better than from foods.

Dietary Guidelines for Americans 2010

People older than age 50 should consume vitamin B_{12} from fortified foods or supplements.

Vitamin D Vitamin D deficiency is a problem among older adults. Vitamin D–fortified milk is the most reliable source of vitamin D, but many older adults drink little or no milk. Further compromising the vitamin D status of many older people, especially those in nursing homes, is their limited exposure to sunlight. Finally, aging reduces the skin's capacity to make vitamin D and the kidneys' ability to convert it to its active form. Not only are older adults not getting enough vitamin D, but they may actually need more to improve both muscle and bone strength. To prevent bone loss and to maintain vitamin D status, especially in those who engage in minimal outdoor activity, adults 51 to 70 years old need 15 micrograms daily, and those older than 70 need 20 micrograms.[45]

Dietary Guidelines for Americans 2010

Choose fortified foods and dietary supplements to provide more vitamin D, which is a nutrient of concern in American diets.

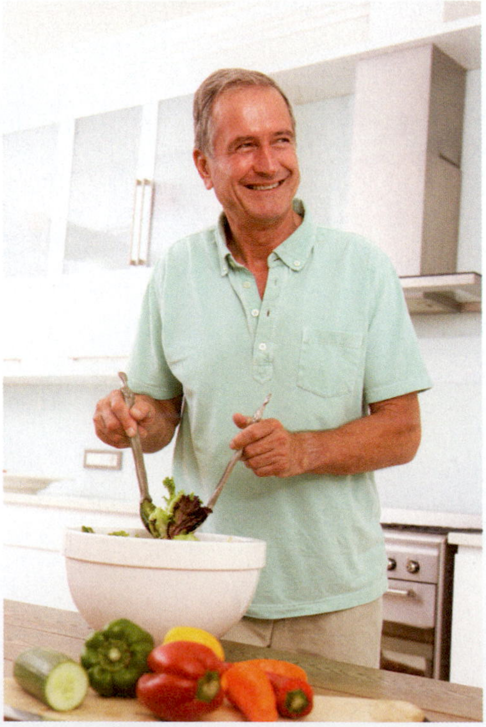

iStockphoto.com/Monkey Business Images

Taking time to nourish your body well is a gift you give yourself.

Folate As is true of vitamin B_{12}, folate intakes of older adults typically fall short of recommendations. The elderly are also more likely to have medical conditions or to take medications that can compromise folate status (see Chapter 19).

Calcium Both Chapter 12 and Highlight 12 emphasize the importance of abundant dietary calcium throughout life, especially for women after menopause, to protect against osteoporosis. The DRI Committee recommends 1200 milligrams of calcium daily for women over age 50 and men over age 70, but the calcium intakes of older people in the United States are well below recommendations.[46]

Some older adults avoid milk and milk products because they dislike these foods or associate them with stomach discomfort. Simple solutions include using calcium-fortified juices, adding powdered milk to recipes, and taking supplements. Chapter 12 offers many other strategies for including nonmilk sources of calcium for those who do not drink milk.

Iron The iron needs of men remain unchanged throughout adulthood. For women, iron needs decrease substantially when blood loss through menstruation ceases. Consequently, iron-deficiency anemia is less common in older adults than in younger people. In fact, elevated iron stores are more likely than deficiency in older people, especially those who take iron supplements, eat red meat regularly, and include vitamin C–rich fruits in their daily diet.

Nevertheless, iron deficiency may develop in older adults, especially when their food energy intakes are low. Aside from diet, two other factors may lead to iron deficiency in older people: chronic blood loss from diseases and medicines and poor iron absorption due to reduced stomach acid secretion and antacid use. Iron deficiency impairs immunity and leaves older adults vulnerable to infectious diseases. Anyone concerned with older people's nutrition should keep these possibilities in mind.

Zinc Zinc intake is commonly low in older people. Zinc deficiency can depress the appetite and blunt the sense of taste, thereby reducing food intake and worsening zinc status. Many medications that older adults commonly use can impair zinc absorption or enhance its excretion and thus lead to deficiency.

Nutrient Supplements

People judge for themselves how to manage their nutrition, and more than half of older adults turn to dietary supplements. When recommended by a physician or registered dietitian, vitamin D and calcium supplements for osteoporosis or vitamin B_{12} for pernicious anemia may be beneficial. Many health-care professionals recommend a daily multivitamin-mineral supplement that provides 100 percent or less of the Daily Value for the listed nutrients. They reason that such a supplement is more likely to be beneficial than to cause harm. Supplement use does not seem to help older adults meet their calcium, vitamin C, and magnesium needs.[47]

People with small energy allowances would do well to become more active so they can afford to eat more food. Food is the best source of nutrients for everybody. Supplements are just that—supplements to foods, not substitutes for them. For anyone who is motivated to obtain the best possible health, it is never too late to learn to eat well, drink water, exercise regularly, and adopt other lifestyle habits such as quitting smoking and moderating alcohol use.

IN SUMMARY **The accompanying table provides a summary of the nutrient concerns of aging. Although some nutrients need special attention in the diet, supplements are not routinely recommended. The ever-growing number of older people creates an urgent need to learn more about how their nutrient requirements differ from those of others and how such knowledge can enhance their health.**

Nutrient	Effect of Aging	Comments
Water	Lack of thirst and decreased total body water make dehydration likely.	Mild dehydration is a common cause of confusion. Difficulty obtaining water or getting to the bathroom may compound the problem.
Energy	Need decreases as muscle mass decreases (sarcopenia).	Physical activity moderates the decline.
Fiber	Likelihood of constipation increases with low intakes and changes in the GI tract.	Inadequate water intakes and lack of physical activity, along with some medications, compound the problem.
Protein	Needs may stay the same or increase slightly.	Low-fat, high-fiber legumes and grains meet both protein and other nutrient needs.
Vitamin B_{12}	Atrophic gastritis is common.	Deficiency causes neurological damage; supplements may be needed.
Vitamin D	Increased likelihood of inadequate intake; skin synthesis declines.	Daily sunlight exposure in moderation or supplements may be beneficial.
Calcium	Intakes may be low; osteoporosis is common.	Stomach discomfort commonly limits milk intake; calcium substitutes or supplements may be needed.
Iron	In women, status improves after menopause; deficiencies are linked to chronic blood losses and low stomach acid output.	Adequate stomach acid is required for absorption; antacid or other medicine use may aggravate iron deficiency; vitamin C and meat increase absorption.

Nutrition-Related Concerns of Older Adults

Nutrition may play a greater role than has been realized in preventing many changes once thought to be inevitable consequences of growing older. The following discussions of vision, arthritis, and the aging brain show that nutrition may provide at least some protection against some of the conditions associated with aging.

Vision

One key aspect of healthy aging is maintaining good vision.[48] Age-related eye diseases that impair vision, such as cataract and macular degeneration, correlate with poor survival that cannot be explained by other risk factors.[49] Following a healthy diet as described by the *Dietary Guidelines for Americans* is one way to protect against these age-related vision problems. Foods containing phytochemicals that act as antioxidants or anti-inflammatory agents may be especially beneficial.[50]

Cataracts

Cataracts are age-related clouding of the lenses of the eyes that impairs vision. If not surgically removed, they ultimately lead to blindness. Cataracts may develop as a result of ultraviolet light exposure, oxidative stress, injury, viral infections, toxic substances, and genetic disorders. Most cataracts, however, are vaguely called senile cataracts—meaning "caused by aging." In the United States, more than half of all adults 65 and older have a cataract.

Oxidative stress appears to play a significant role in the development of cataracts, and the antioxidant nutrients may help minimize the damage. Studies have reported an inverse relationship between cataracts and dietary intakes of vitamin C, vitamin E, and carotenoids; taking supplements or eating fruits and vegetables rich in these antioxidant nutrients seems to slow the progression or reduce the risk of developing cataracts.[51]

One other diet-related factor may play a role in the development of cataracts—obesity. Obesity appears to be associated with cataracts, but its role has not been identified. Risk factors that typically accompany overweight, such as inactivity, diabetes, or hypertension, do not explain the association.

Macular Degeneration

The leading cause of visual loss among older people is age-related **macular degeneration**, a deterioration of the macular region of the retina.[52] As with cataracts, risk factors for age-related macular degeneration include oxidative stress from sunlight, and preventive factors may include supplements of the omega-3 fatty acid DHA and the carotenoids lutein and zeaxanthin.[53]

Arthritis

More than 46 million people in the United States have some form of **arthritis.**[54] As the population ages, it is expected that the prevalence will increase to 60 million by 2020.

Osteoarthritis

The most common type of arthritis that disables older people is **osteoarthritis**, a painful deterioration of the cartilage in the joints. During movement, the ends of bones are normally protected from wear by cartilage and by small sacs of fluid that act as a lubricant. With age, the cartilage sometimes disintegrates, and the joints become malformed and painful to move.

One known connection between osteoarthritis ♦ and nutrition is overweight. Weight loss may relieve some of the pain for overweight persons with osteoarthritis, partly because the joints affected are often weight-bearing joints that are stressed and irritated by having to carry excess pounds. Interestingly, though, weight loss often relieves much of the pain of arthritis in the hands as well, even though they are not weight-bearing joints. Jogging and other weight-bearing exercises do not worsen arthritis. In fact, both aerobic activity and strength training offer improvements in physical performance and pain relief, especially when accompanied by even modest weight loss.[55]

♦ Risk factors for osteoarthritis:
- Age
- Smoking
- High BMI at age 40
- Lack of hormone therapy (in women)

cataracts (KAT-ah-rakts): clouding of the eye lenses that impairs vision and can lead to blindness.

macular (MACK-you-lar) **degeneration:** deterioration of the macular area of the eye that can lead to loss of central vision and eventual blindness. The **macula** is a small, oval, yellowish region in the center of the retina that provides the sharp, straight-ahead vision so critical to reading and driving.

arthritis: inflammation of a joint, usually accompanied by pain, swelling, and structural changes.

osteoarthritis: a painful, degenerative disease of the joints that occurs when the cartilage in a joint deteriorates; joint structure is damaged, with loss of function; also called *degenerative arthritis*.

Rheumatoid Arthritis Another type of arthritis known as **rheumatoid arthritis** has possible links to diet through the immune system. In rheumatoid arthritis, the immune system mistakenly attacks the bone coverings as if they were made of foreign tissue. In some individuals, certain foods, notably a Mediterranean-type diet of fish, vegetables, and olive oil, may moderate the inflammatory response and provide some relief.[56]

The omega-3 fatty acids commonly found in fatty fish reduce joint tenderness and improve mobility in some people with rheumatoid arthritis. The same diet recommended for heart health—one low in saturated fat from meats and milk products and high in omega-3 fats from fish—helps prevent or reduce the inflammation in the joints that makes arthritis so painful.

Another possible link between nutrition and rheumatoid arthritis involves the oxidative damage to the membranes within joints that causes inflammation and swelling. The antioxidant vitamins C and E and the carotenoids defend against oxidation, and increased intakes of these nutrients may help prevent or relieve the pain of rheumatoid arthritis.[57]

Gout Another form of arthritis, which most commonly affects men, is **gout**, a condition characterized by deposits of uric acid crystals in the joints. Uric acid derives from the breakdown of **purines**, primarily from those made by the body but also from those found in foods.[58] Foods such as meat and seafood that are rich in purines increase uric acid levels and the risk of gout, whereas milk products seem to lower uric acid levels and the risk of gout.[59]

Treatment Treatment for arthritis—dietary or otherwise—may help relieve discomfort and improve mobility, but it does not cure the condition. Traditional medical intervention for arthritis includes medication and surgery. Alternative therapies to treat arthritis abound, but none have proved safe and effective in scientific studies. Popular supplements—glucosamine, chondroitin, or a combination—may relieve pain and improve mobility as well as over-the-counter pain relievers, but mixed reports from studies emphasize the need for additional research.[60] Drugs and supplements used to relieve arthritis can impose nutrition risks; many affect appetite and alter the body's use of nutrients, as Chapter 19 explains.

The Aging Brain

The brain, like all of the body's organs, responds to both genetic and environmental factors ♦ that can enhance or diminish its amazing capacities. One of the challenges researchers face when studying the human brain is to distinguish among normal age-related physiological changes, changes caused by diseases, and changes that result from cumulative, environmental factors such as diet.

♦ Factors that protect brain function:
- Physical activities
- Intellectual challenges
- Social interactions
- Balanced diet rich in antioxidants

The brain normally changes in some characteristic ways as it ages. For one thing, its blood supply decreases. For another, the number of **neurons**, the brain cells that specialize in transmitting information, diminishes as people age. When the number of nerve cells in one part of the cerebral cortex diminishes, hearing and speech are affected. Losses of neurons in other parts of the cortex can impair memory and cognitive function. When the number of neurons in the cerebellum diminishes, balance and posture are affected. Losses of neurons in other parts of the brain affect still other functions. Some of the cognitive loss and forgetfulness generally attributed to aging may be due in part to environmental, and therefore controllable, factors—including nutrient deficiencies.

Nutrient Deficiencies and Brain Function Nutrients influence the development and activities of the brain. The ability of neurons to synthesize specific neurotransmitters depends in part on the availability of precursor nutrients that are obtained from the diet. The neurotransmitter serotonin, for example, derives from the amino acid tryptophan. To function properly, the enzymes involved in neurotransmitter synthesis require vitamins and minerals. Thus nutrient deficiencies may contribute to the loss of memory and cognition that some older adults experience. Such losses may be preventable or at least diminished or delayed through

rheumatoid (ROO-ma-toyd) **arthritis:** a disease of the immune system involving painful inflammation of the joints and related structures.

gout (GOWT): a common form of arthritis characterized by deposits of uric acid crystals in the joints.

purines: compounds of nitrogen-containing bases such as adenine, guanine, and caffeine. Purines that originate from the body are *endogenous* and those that derive from foods are *exogenous*.

neurons: nerve cells; the structural and functional units of the nervous system. Neurons initiate and conduct nerve impulse transmissions.

TABLE 16-2 Summary of Nutrient-Brain Relationships

Brain Function	Adequate Intake of
Short-term memory	Vitamin B_{12}, vitamin C, vitamin E
Performance in problem-solving tests	Riboflavin, folate, vitamin B_{12}, vitamin C
Mental health	Thiamin, niacin, zinc, folate
Cognition	Folate, vitamin B_6, vitamin B_{12}, iron, vitamin E
Vision	Essential fatty acids, vitamin A
Neurotransmitter synthesis	Tyrosine, tryptophan, choline

TABLE 16-3 Common Signs of Dementia

- Agitated behavior
- Becoming lost in familiar surroundings or circumstances
- Confusion
- Delusions
- Loss of interest in daily activities
- Loss of memory
- Loss of problem-solving skills
- Unclear thinking

diet and exercise. Table 16-2 summarizes some of the better-known connections between brain function and nutrients.

In some instances, the degree of cognitive loss is extensive. Such **senile dementia** may be attributable to a specific disorder such as a brain tumor or Alzheimer's disease. Table 16-3 lists common signs of dementia.

Alzheimer's Disease Much attention has focused on the *abnormal* deterioration of the brain called **Alzheimer's disease**, which affects about 10 percent of U.S. adults older than age 70. Diagnosis of Alzheimer's disease depends on its characteristic symptoms: the victim gradually loses memory and reasoning, the ability to communicate, physical capabilities, and eventually life itself. Nerve cells in the brain die, and communication between the cells breaks down.

Researchers are closing in on the exact cause of Alzheimer's disease.* Clearly, genetic factors are involved.[61] Free radicals and oxidative stress also seem to be involved.[62] Nerve cells in the brains of people with Alzheimer's disease show evidence of free-radical attack—damage to DNA, cell membranes, and proteins.[63] They also show evidence of the minerals that trigger free-radical attacks—iron, copper, zinc, and aluminum. Increasing evidence also suggests that overweight and obesity in middle age are associated with dementia in general, and with Alzheimer's disease in particular.[64]

In Alzheimer's disease, the brain develops **senile plaques** and **neurofibrillary tangles.** Senile plaques are clumps of a protein fragment called beta-amyloid, whereas neurofibrillary tangles are snarls of the fibers that extend from the nerve cells. Both seem to occur in response to oxidative stress.[65] Researchers question whether these characteristics are the cause or the result of Alzheimer's disease.[66] In fact, scientists are unsure whether these plaques and tangles are causing the damage, serving as markers, or even protecting by sequestering the proteins that begin the dementia process.[67] In any case, treatment research focuses on lowering beta-amyloid levels.[68]

Late in the course of the disease there is a decline in the activity of the enzyme that assists in the production of the neurotransmitter acetylcholine from choline and acetyl CoA. Acetylcholine is essential to memory, but supplements of choline (or of lecithin, which contains choline) have no effect on memory or on the progression of the disease. Drugs, such as donepezil, that inhibit the breakdown of acetylcholine, on the other hand, have proved beneficial.[69]

Research suggests that cardiovascular disease risk factors such as high blood pressure, diabetes, and elevated levels of homocysteine may be related to the development of Alzheimer's disease.[70] Diets designed to support a healthy heart, including the omega-3 fatty acids of oily fish, may benefit brain health as well.[71] Similarly, physical activity supports heart health and slows the cognitive decline of Alzheimer's disease.[72]

Treatment for Alzheimer's disease involves providing care to clients and support to their families. Drugs are used to improve or at least to slow the loss of short-term memory and cognition, but they do not cure the disease. Other drugs may be used to control depression, anxiety, and behavior problems.

Maintaining appropriate body weight may be the most important nutrition concern for the person with Alzheimer's disease. Depression and forgetfulness can lead to changes in eating behaviors and poor food intake. Furthermore, changes in the body's weight-regulation system may contribute to weight loss. Perhaps the best that a caregiver can do nutritionally for a person with Alzheimer's disease is to supervise food planning and mealtimes. Providing well-liked and well-balanced meals and snacks in a cheerful atmosphere encourages food consumption. To minimize confusion, offer a few ready-to-eat foods, in bite-size pieces, with seasonings and sauces. To avoid mealtime disruptions, control distractions such as music, television, children, and the telephone.

senile dementia: the loss of brain function beyond the normal loss of physical adeptness and memory that occurs with aging.

Alzheimer's (AHLZ-high-merz) **disease:** a degenerative disease of the brain involving memory loss and major structural changes in neuron networks; also known as *senile dementia of the Alzheimer's type (SDAT), primary degenerative dementia of senile onset,* or *chronic brain syndrome.*

senile plaques: clumps of the protein fragment beta-amyloid on the nerve cells, commonly found in the brains of people with Alzheimer's dementia.

neurofibrillary tangles: snarls of the threadlike strands that extend from the nerve cells, commonly found in the brains of people with Alzheimer's dementia.

*A report on the genetic and other aspects of Alzheimer's is available from Alzheimer's Disease Education and Referral Center, P.O. Box 8250, Silver Spring, MD 20907-8250.

IN SUMMARY Senile dementia and other losses of brain function afflict millions of older adults, and others face loss of vision due to cataracts or macular degeneration or cope with the pain of arthritis. As the number of people older than age 65 continues to grow, the need for solutions to these problems becomes urgent. Some problems may be inevitable, but others are preventable and good nutrition may play a key role.

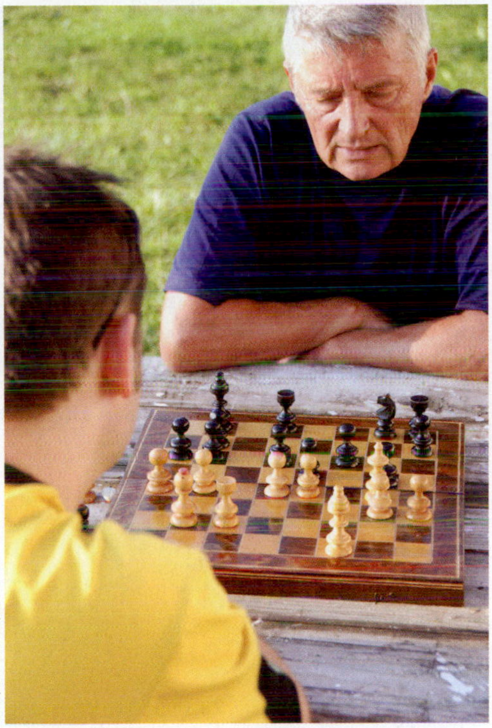

iStockphoto.com/by_nicholas

Both foods and mental challenges nourish the brain.

Food Choices and Eating Habits of Older Adults

Older people are an incredibly diverse group, and for the most part, they are independent, socially sophisticated, mentally lucid, fully participating members of society who report themselves to be happy and healthy. In fact, the quality of life among the elderly has improved, and their chronic disabilities have declined dramatically in recent years. By practicing stress-management skills, maintaining physical fitness, participating in activities of interest, and cultivating spiritual health, as well as obtaining adequate nourishment, people can support a high quality of life into old age (see Table 16-4 for some strategies).

Older people spend more money per person on foods to eat at home than other age groups and less money on foods away from home. Manufacturers would be wise to cater to the preferences of older adults by providing good-tasting, nutritious foods in easy-to-open, single-serving packages with labels that are easy to read. Such services enable older adults to maintain their independence and to feel a sense of control and involvement in their own lives. Another way older adults can take care of themselves is by remaining or becoming physically active. As mentioned earlier, physical activity helps preserve one's ability to perform daily tasks and so promotes independence.

Familiarity, taste, and health beliefs are most influential on older people's food choices. Eating foods that are familiar, especially ethnic foods that recall family meals and pleasant times, can be comforting. People 65 and older are less likely to diet to lose weight than younger people are, but they are more likely to diet in pursuit of medical goals such as controlling blood glucose and cholesterol.

TABLE 16-4 Strategies for Growing Old Healthfully

- Choose nutrient-dense foods.
- Be physically active. Walk, run, dance, swim, bike, or row for aerobic activity. Lift weights, do calisthenics, or pursue some other activity to tone, firm, and strengthen muscles. Practice balancing on one foot or doing simple movements with your eyes closed. Modify activities to suit changing abilities and preferences.
- Maintain appropriate body weight.
- Reduce stress—cultivate self-esteem, maintain a positive attitude, manage time wisely, know your limits, practice assertiveness, release tension, and take action.
- For women, discuss with a physician the risks and benefits of estrogen replacement therapy.
- For people who smoke, discuss with a physician strategies and programs to help you quit.
- Expect to enjoy sex, and learn new ways of enhancing it.
- Use alcohol only moderately, if at all; use drugs only as prescribed.
- Take care to prevent accidents.
- Expect good vision and hearing throughout life; obtain glasses and hearing aids if necessary.
- Take care of your teeth; obtain dentures if necessary.
- Be alert to confusion as a disease symptom, and seek diagnosis.
- Take medications as prescribed; see a physician before self-prescribing medicines or herbal remedies and a registered dietitian before self-prescribing supplements.
- Control depression through activities and friendships; seek professional help if necessary.
- Drink six to eight glasses of water every day.
- Practice mental skills. Keep on solving math problems and crossword puzzles, playing cards or other games, reading, writing, imagining, and creating.
- Make financial plans early to ensure security.
- Accept change. Work at recovering from losses; make new friends.
- Cultivate spiritual health. Cherish personal values. Make life meaningful.
- Go outside for sunshine and fresh air as often as possible.
- Be socially active—play bridge, join an exercise or dance group, take a class, teach a class, eat with friends, volunteer time to help others.
- Stay interested in life—pursue a hobby, spend time with grandchildren, take a trip, read, grow a garden, or go to the movies.
- Enjoy life.

Social interactions at a congregate meal site can be as nourishing as the foods served.

Food Assistance Programs

The Nutrition Screening Initiative is part of a national effort to identify and treat nutrition problems in older persons; it uses a screening checklist. To *determine* the risk of malnutrition in older clients, health-care professionals can keep in mind the characteristics and questions listed in Table 16-5.

An integral component of the Older Americans Act (OAA) is the OAA Nutrition Program, formerly known as the Elderly Nutrition Program. Its services are designed to improve older people's nutrition status and enable them to avoid medical problems, continue living in communities of their own choice, and stay out of institutions. Its specific goals are to provide low-cost, nutritious meals; opportunities for social interaction; homemaker education and shopping assistance; counseling and referral to social services; and transportation. The program's mission has always been to provide "more than a meal."

The OAA Nutrition Program provides for **congregate meals** at group settings such as community centers. Administrators try to select sites for congregate meals where as many eligible people as possible can participate. Volunteers may also deliver meals to those who are homebound either permanently or temporarily; these home-delivered meals are known as **Meals on Wheels.** Although the home-delivery program ensures nutrition, its recipients miss out on the social benefits of the congregate meals. Therefore, every effort is made to persuade older people to come to the shared meals, if they can. All persons aged 60 years and older and their spouses are eligible to receive meals from these programs, regardless of their income. Priority is given to those who are economically and socially needy. An estimated three million of our nation's older adults benefit from these meals.

These programs provide at least one meal a day that meets one-third of the RDA for this age group, and they must operate five or more days a week. Many programs voluntarily offer additional services designed to appeal to older adults: provisions for special diets (to meet medical needs or religious preferences), food pantries, ethnic meals, and delivery of meals to the homeless. Adding breakfast to the service increases energy and nutrient intakes, which helps to relieve hunger and depression.

Older adults can also take advantage of the Senior Farmers Market Nutrition Program, which provides low-income older adults with coupons that can be ex-

TABLE 16-5 Risk Factors for Malnutrition in Older Adults

	These questions help *determine* the risk of malnutrition in older adults:
Disease	• Do you have an illness or condition that changes the types or amounts of foods you eat?
Eating poorly	• Do you eat fewer than two meals a day? Do you eat fruits, vegetables, and milk products daily?
Tooth loss or mouth pain	• Is it difficult or painful to eat?
Economic hardship	• Do you have enough money to buy the food you need?
Reduced social contact	• Do you eat alone most of the time?
Multiple medications	• Do you take three or more different prescribed or over-the-counter medications daily?
Involuntary weight loss or gain	• Have you lost or gained 10 pounds or more in the last 6 months?
Needs assistance	• Are you physically able to shop, cook, and feed yourself?
Elderly person	• Are you older than 80?

NOTE: A complete description of DETERMINE and its scoring system are available online from the American Academy of Family Physicians: www.aafp.org/afp/980301ap/edits.html

congregate meals: nutrition programs that provide food for the elderly in conveniently located settings such as community centers.

Meals on Wheels: a nutrition program that delivers food for the elderly to their homes.

changed for fresh fruits, vegetables, and herbs at community-supported farmers' markets and roadside stands. This program increases fresh fruit and vegetable consumption, provides nutrition information, and even reaches the homebound elderly, a group of people who normally do not have access to farmers' markets.

Older adults can learn about the available programs in their communities by looking in the Yellow Pages of the telephone book under "Social Services" or "Senior Citizens' Organizations."* In addition, the local senior center and hospital can usually direct people to programs that provide nutrition and other health-related services.

In addition to programs designed specifically for older adults, the Supplemental Nutrition Assistance Program (formerly called the Food Stamp Program) offers services to eligible people of all ages. As mentioned earlier, though, the participation rate for eligible seniors is only about 30 percent.

Meals for Singles

Many older adults live alone, and singles of all ages face challenges in purchasing, storing, and preparing food. Large packages of meat and vegetables are often intended for families of four or more, and even a head of lettuce can spoil before one person can use it all. Many singles live in small dwellings and have little storage space for foods. A limited income presents additional obstacles. This section offers suggestions that can help to solve some of the problems singles face, beginning with a special note about the dangers of foodborne illness.

Foodborne Illness The risk of older adults getting a foodborne illness is greater than for other adults. The consequences of an upset stomach, diarrhea, fever, vomiting, abdominal cramps, and dehydration are oftentimes more severe, sometimes leading to paralysis, meningitis, or even death. For these reasons, older adults need to carefully follow the food-safety suggestions presented in Highlight 18.

Dietary Guidelines for Americans 2010

- Older adults should not eat or drink unpasteurized milk, milk products, or juices; raw or undercooked eggs, meat, poultry, fish, or shellfish; or raw sprouts.
- Older adults should only eat deli meats and frankfurters that have been reheated to steaming hot.

Spend Wisely People who have the means to shop and cook for themselves can cut their food bills simply by being wise shoppers. Large supermarkets are usually less expensive than convenience stores. A grocery list helps reduce impulse buying, and specials and coupons can save money when the items featured are those that the shopper needs and uses.

Buying the right amount so as not to waste any food is a challenge for people eating alone. They can buy fresh milk in the size best suited for personal needs. Pint-size and even cup-size boxes ♦ of milk are available and can be stored unopened on a shelf for as long as three months without refrigeration.

♦ Boxes of milk that can be stored at room temperature have been exposed to temperatures above those of pasteurization just long enough to sterilize the milk—a process called **ultrahigh temperature (UHT).**

Many foods that offer a variety of nutrients for practically pennies have a long shelf life; staples such as rice, pastas, dry powdered milk, and dried legumes can be purchased in bulk and stored for months at room temperature. Other foods that are usually a good buy include whole pieces of cheese rather than sliced or shredded cheese, fresh produce in season, variety meats such as chicken livers, and cereals that require cooking instead of ready-to-serve cereals.

A person who has ample freezer space can buy large packages of meat, such as pork chops, ground beef, or chicken, when they are on sale. Then the meat can be immediately wrapped into individual servings for the freezer. All the individual servings can be put in a bag marked appropriately with the contents and the date.

*To find a local provider, call Eldercare Locator at (800) 677-1116.

Noel Hendrickson/Getty Images

Buy only what you will use.

Frozen vegetables are more economical in large bags than in small boxes. After the amount needed is taken out, the bag can be closed tightly with a twist tie or rubber band. If the package is returned quickly to the freezer each time, the vegetables will stay fresh for a long time.

Finally, breads and cereals usually must be purchased in larger quantities. Again the amount needed for a few days can be taken out and the rest stored in the freezer.

Grocers will break open a package of wrapped meat and rewrap the portion needed. Similarly, eggs can be purchased by the half-dozen. Eggs do keep for long periods, though, if stored properly in the refrigerator.

Fresh fruits and vegetables can be purchased individually. A person can buy fresh fruit at various stages of ripeness: a ripe one to eat right away, a semiripe one to eat soon after, and a green one to ripen on the windowsill. If vegetables are packaged in large quantities, the grocer can break open the package so that a smaller amount can be purchased. Small cans of fruits and vegetables, even though they are more expensive per unit, are a reasonable alternative, considering that it is expensive to buy a regular-size can and let the unused portion spoil.

Be Creative Creative chefs think of various ways to use foods when only large amounts are available. For example, a head of cauliflower can be divided into thirds. Then one-third is cooked and eaten hot. Another third is put into a vinegar and oil marinade for use in a salad. And the last third can be used in a casserole or stew.

A variety of vegetables and meats can be enjoyed stir-fried; inexpensive vegetables such as cabbage, celery, and onion are delicious when crisp cooked in a little oil with herbs or lemon added. Interesting frozen vegetable mixtures are available in larger grocery stores. Cooked, leftover vegetables can be dropped in at the last minute. A bonus of a stir-fried meal is that there is only one pan to wash. Similarly, a microwave oven allows a chef to use fewer pots and pans. Meals and leftovers can also be frozen or refrigerated in microwavable containers to reheat as needed.

Many frozen dinners offer nutritious options. Adding a fresh salad, a whole-wheat roll, and a glass of milk can make a nutritionally balanced meal.

Also, single people shouldn't hesitate to invite someone to share meals with them whenever there is enough food. It's likely that the person will return the invitation, and both parties will get to enjoy companionship and a meal prepared by others.

IN SUMMARY Older people can benefit from both the nutrients provided and the social interaction available at congregate meals. Other government programs deliver meals to those who are homebound. With creativity and careful shopping, those living alone can prepare nutritious, inexpensive meals. Physical activity, mental challenges, stress management, and social activities can also help people grow old comfortably.

Kuznetsov Alexey/Shutterstock.com

Invite guests to share a meal.

Nutrition Portfolio

By eating a balanced diet, maintaining a healthy body weight, and engaging in a variety of physical, social, and mental activities, you can enjoy good health in later life.

Visit older adults in your community and . . .

- Consider whether they have the financial means, physical ability, and social support they need to eat adequately.
- Note whether they have experienced an unintentional loss of weight recently.
- Discuss how they occupy their time physically, socially, and mentally.
- Offer to analyze the diet of an older adult who you care about using Diet Analysis Plus. Have the person write down one, two, or even three days of their food and beverage intake, and then enter it into the program and print out a three-day average report of the results. It will be fun and educational to go over it together. Remind the person that you are not a doctor and that this is a learning tool for an introductory nutrition course.

Diet Analysis PLUS To complete this exercise, go to your Diet Analysis Plus at www.cengage.com/sso.

Nutrition on the Net

For further study of topics covered in this chapter, log on to **www.cengagebrain.com** and search for ISBN 084006845X.

- Search for "aging," "arthritis," and "Alzheimer's" on the U.S. Government health information site: **www.healthfinder.gov**
- Visit the National Aging Information Center of the Administration on Aging: **www.aoa.gov**
- Visit the American Geriatrics Society Foundation for Health in Aging: **www.healthinaging.org**
- Visit the National Institute on Aging: **www.nia.nih.gov**
- Visit the American Association of Retired Persons: **www.aarp.org**
- Get nutrition tips for growing older in good health from the American Dietetic Association: **www.eatright.org**
- Learn more about cataracts and macular degeneration from the National Eye Institute, the Macular Degeneration Partnership, and the American Society of Cataract and Refractive Surgery: **www.nei.nih.gov**, **www.amd.org**, and **www.ascrs.org**
- Learn more about arthritis from the Arthritis Society, the Arthritis Foundation, and the National Institute of Arthritis and Musculoskeletal and Skin Diseases: **www.arthritis.ca**, **www.arthritis.org**, and **www.niams.nih.gov**
- Learn more about Alzheimer's disease from the NIA Alzheimer's Disease Education and Referral Center and the Alzheimer's Association: **www.nia.nih.gov/alzheimers** and **www.alz.org**
- Find out about federal government programs designed to help senior citizens maintain good health: **www.seniors.gov**
- Learn more about cognitive impairment from the National Institute of Neurological Disorders and Stroke: **www.ninds.nih.gov**
- Visit the National Council on Aging: **www.ncoa.org**

References

1. D. E. King, A. G. Mainous, and M. E. Geesey, Turning back the clock: Adopting a healthy lifestyle in middle age, *American Journal of Medicine* 120 (2007): 598–603.
2. H. Kung and coauthors, Deaths: Final data for 2005, *National Vital Statistics Reports* 56 (2008): 1–120; S. Harper and coauthors, Trends in the black-white life expectancy gap in the United States, 1983–2003, *Journal of the American Medical Association* 297 (2007): 1224–1232.
3. R. S. Blacklow, Actuarially speaking: An overview of life expectancy. What can we expect? *American Journal of Clinical Nutrition* 86 (2007): 1560S–1562S.

4. Living well to 100: Nutrition, genetics, inflammation, supplement to *American Journal of Clinical Nutrition* 83 (2006): 401S–490S.
5. N. M. Peel, R. J. McClure, and H. P. Bartlett, Behavioral determinants of healthy aging, *American Journal of Preventive Medicine* 28 (2005): 298–304.
6. K. Khaw and coauthors, Combined impact of health behaviours and mortality in men and women: The EPIC-Norfolk Prospective Population Study, *PLoS Medicine* 5 (2008): e12.
7. L. B. Yates and coauthors, Exceptional longevity in men—Modifiable factors associated with survival and function to age 90 years, *Archives of Internal Medicine* 168 (2008): 284–290.
8. P. Kokkinos and coauthors, Exercise capacity and mortality in black and white men, *Circulation* 117 (2008): 614–622.
9. R. C. Cassilhas and coauthors, The impact of resistance exercise on the cognitive function of the elderly, *Medicine & Science in Sports & Exercise* 39 (2007): 1401–1407.
10. J. A. Stevens, G. Ryan, and M. Kresnow, Fatalities and injuries from falls among older adults: United States, 1993–2003 and 2001–2005, *Morbidity and Mortality Weekly Report* 55 (2006): 1221–1224.
11. M. E. Nelson and coauthors, Physical activity and public health in older adults: Recommendation from the American College of Sports Medicine and the American Heart Association, *Medicine & Science in Sports & Exercise* 39 (2007): 1435–1445.
12. J. Kruger, S. A. Carlson, and D. Buchner, How active are older Americans? *Preventing Chronic Disease* 4 (2007): 1–12.
13. J. R. Speakman and C. Hambly, Starving for life: What animal studies can and cannot tell us about the use of caloric restriction to prolong human lifespan, *Journal of Nutrition* 137 (2007): 1078–1086.
14. C. W. Levenson and N. J. Rich, Eat less, live longer? New insights into the role of caloric restriction in the brain, *Nutrition Reviews* 65 (2007): 412–415; S. R. Spindler and J. M. Dhahbi, Conserved and tissue-specific genic and physiologic responses to caloric restriction and altered IGFI signaling in mitotic and postmitotic tissues, *Annual Review of Nutrition* 27 (2007): 193–217.
15. R. S. Sohal and coauthors, Life span extension in mice by food restriction depends on an energy imbalance, *Journal of Nutrition* 139 (2009): 533–539.
16. Speakman and Hambly, 2007.
17. L. Fontana and S. Klein, Aging, adiposity, and calorie restriction, *Journal of the American Medical Association* 297 (2007): 986–994; G. Wolf, Calorie restriction increases life span: A molecular mechanism, *Nutrition Reviews* 64 (2006): 89–92.
18. K. A. Varaday and M. K. Hellerstein, Do calorie restriction or alternate-day fasting regimens modulate adipose tissue physiology in a way that reduces chronic disease risk? *Nutrition Reviews* 66 (2008): 333–342; L. K. Heilbronn and coauthors, Alternate-day fasting in nonobese subjects: Effects on body weight, body composition, and energy metabolism, *American Journal of Clinical Nutrition* 81 (2005): 69–73; M. P. Mattson, Energy intake, meal frequency, and health: A neurobiological perspective, *Annual Review of Nutrition* 25 (2005): 237–260.
19. S. Cohen, D. Janicki-Deverts, and G. E. Miller, Psychological stress and disease, *Journal of the American Medical Association* 298 (2007): 1685–1687.
20. R. S. Rivlin, Keeping the young-elderly healthy: Is it too late to improve our health through nutrition? *American Journal of Clinical Nutrition* 86 (2007): 1572S–1576S.
21. L. M. Willis, B. Shukitt-Hale, and J. A. Joseph, Recent advances in berry supplementation and age-related cognitive decline, *Current Opinion in Clinical Nutrition and Metabolic Care* 12 (2009): 91–94; E. Head, Combining an antioxidant-fortified diet with behavioral enrichment leads to cognitive improvement and reduced brain pathology in aging canines: Strategies for healthy aging, *Annals of the New York Academy of Sciences* 1114 (2007): 398–406; D. P. Jones, Extracellular redox state: Refining the definition of oxidative stress in aging, *Rejuvenation Research* 9 (2006): 169–181; F. Sierra, Is (your cellular response to) stress killing you? *Journals of Gerontology. Series A, Biological Sciences and Medical Sciences* 61 (2006): 557–561; B. P. Yu and H. Y. Chung, Adaptive mechanisms to oxidative stress during aging, *Mechanisms of Ageing and Development* 127 (2006): 436–443.
22. D. T. Villareal and coauthors, Obesity in older adults: Technical review and position statement of the American Society for Nutrition and NAASO, The Obesity Society, *American Journal of Clinical Nutrition* 82 (2005): 923–934.
23. S. M. H. Alibhai, C. Greenwood, and H. Payette, An approach to the management of unintentional weight loss in elderly people, *Canadian Medical Association Journal* 172 (2005): 773–780.
24. C. A. Zizza, F. A. Tayie, and M. Lino, Benefits of snacking in older Americans, *Journal of the American Dietetic Association* 107 (2007): 800–806.
25. M. Cesari and coauthors, Frailty syndrome and skeletal muscle: Results from the Invecchiare in Chianti Study, *American Journal of Clinical Nutrition* 83 (2006): 1142–1148.
26. D. K. Houston and coauthors, Dietary protein intake is associated with lean mass change in older, community-dwelling adults: The Health, Aging, and Body Composition (Health ABC) Study, *American Journal of Clinical Nutrition* 87 (2008): 150–155; D. Paddon-Jones and coauthors, Role of dietary protein in the sarcopenia of aging, *American Journal of Clinical Nutrition* 87 (2008): 1562S–1566S; H. B. Iglay and coauthors, Resistance training and dietary protein: Effects on glucose tolerance and contents of skeletal muscle insulin signaling proteins in older persons, *American Journal of Clinical Nutrition* 85 (2007): 1005–1013; K. S. Nair, Aging muscle, *American Journal of Clinical Nutrition* 81 (2005): 953–963.
27. A. B. Newman and coauthors, Weight change and the conservation of lean mass in old age: The Health, Aging and Body Composition Study, *American Journal of Clinical Nutrition* 82 (2005): 872–878; P. Szulc and coauthors, Hormonal and lifestyle determinants of appendicular skeletal muscle mass in men: The MINOS Study, *American Journal of Clinical Nutrition* 80 (2004): 496–503.
28. M. Cesari and coauthors, Sarcopenia, obesity, and inflammation—Results from the Trial of Angiotensin Converting Enzyme Inhibition and Novel Cardiovascular Risk Factors Study, *American Journal of Clinical Nutrition* 82 (2005): 428–434.
29. C. Franceschi, Inflammaging as a major characteristic of old people: Can it be prevented or cured? *Nutrition Reviews* 65 (2007): S173–S176.
30. P. Libby, Inflammatory mechanisms: The molecular basis of inflammation and disease, *Nutrition Reviews* 65 (2007): S140–S146.
31. J. Gauldie, Inflammation and the aging process: Devil or angel, *Nutrition Reviews* 65 (2007): S167–S169.
32. R. Roubenoff, Physical activity, inflammation, and muscle loss, *Nutrition Reviews* 65 (2007): S208–S212.
33. D. Rémond and coauthors, Postprandial whole-body protein metabolism after a meat meal is influenced by chewing efficiency in elderly subjects, *American Journal of Clinical Nutrition* 85 (2007): 1286–1292.
34. Position paper of the American Dietetic Association: Nutrition across the spectrum of aging, *Journal of the American Dietetic Association* 105 (2005): 616–633.
35. N. Kagansky and coauthors, Poor nutritional habits are predictors of poor outcome in very old hospitalized patients, *American Journal of Clinical Nutrition* 82 (2005): 784–791.
36. R. Chernoff, Micronutrient requirements in older women, *American Journal of Clinical Nutrition* 81 (2005): 1240S–1245S.
37. M. Ferry, Strategies for ensuring good hydration in the elderly, *Nutrition Reviews* 63 (2005): S22–S29.
38. N. Meunier and coauthors, Basal metabolic rate and thyroid hormones of late-middle-aged and older human subjects: The ZENITH Study, *European Journal of Clinical Nutrition* 59 (2005): S53–S57.
39. A. H. Lichtenstein and coauthors, Modified MyPyramid for older adults, *Journal of Nutrition* 138 (2008): 5–11.
40. R. P. Heaney and D. K. Layman, Amount and type of protein influences bone health, *American Journal of Clinical Nutrition* 87 (2008): 1567S–1570S; A. E. Thalacker-Mercer and coauthors, Inadequate protein intake affects skeletal muscle transcript profiles in older humans, *American Journal of Clinical Nutrition* 85 (2007): 1344–1352; R. R. Wolfe, The underappreciated role of muscle in health and disease, *American Journal of Clinical Nutrition* 84 (2006): 475–482.
41. Position of the American Dietetic Association: Liberalization of the diet prescription improves quality of life for older adults in long-term care, *Journal of the American Dietetic Association* 105 (2005): 1955–1965.

42. Committee on Dietary Reference Intakes, *Dietary Reference Intakes for Energy, Carbohydrate, Fiber, Fat, Fatty Acids, Cholesterol, Protein, and Amino Acids* (Washington, DC: National Academies Press, 2002).
43. R. Clarke and coauthors, Low vitamin B-12 status and risk of cognitive decline in older adults, *American Journal of Clinical Nutrition* 86 (2007): 1384–1391.
44. Committee on Dietary Reference Intakes, *Dietary Reference Intakes for Thiamin, Riboflavin, Niacin, Vitamin B_6, Folate, Vitamin B_{12}, Pantothenic Acid, Biotin, and Choline* (Washington, DC: National Academies Press, 2000), p. 338.
45. Committee on Dietary Reference Intakes, *Dietary Reference Intakes for Calcium and Vitamin D* (Washington, DC: National Academies Press, 2011).
46. Committee on Dietary Reference Intakes, 2011.
47. A. N. Burnett-Hartman and coauthors, Supplement use contributes to meeting recommended dietary intakes for calcium, magnesium, and vitamin C in four ethnicities of middle-aged and older Americans: The Multi-Ethnic Study of Atherosclerosis, *Journal of the American Dietetic Association* 109 (2009): 422–429.
48. T. Ostbye and coauthors, Ten dimensions of health and their relationships with overall self-reported health and survival in a predominately religiously active elderly population: The Cache County Memory Study, *Journal of the American Geriatrics Society* 54 (2006): 199–209.
49. M. D. Knudtson, B. E. Klein, and R. Klein, Age-related disease, visual impairment, and survival: The Beaver Dam Eye Study, *Archives of Ophthalmology* 124 (2006): 243–249.
50. M. Rhone and A. Basu, Phytochemicals and age-related eye diseases, *Nutrition Reviews* 66 (2008): 465–472.
51. W. G. Christen and coauthors, Dietary carotenoids, vitamins C and E, and risk of cataract in women. A Prospective Study, *Archives of Ophthalmology* 126 (2008): 102–109; A. G. Tan and coauthors, Antioxidant nutrient intake and the long-term incidence of age-related cataract: The Blue Mountains Eye Study, *American Journal of Clinical Nutrition* 87 (2008): 1899–1905.
52. R. D. Jager, W. F. Mieler, and J. W. Miller, Age-related macular degeneration, *New England Journal of Medicine* 358 (2008): 2606–2617.
53. C. Augood and coauthors, Oily fish consumption, dietary docosahexaenoic acid and eicosapentaenoic acid intakes, and associations with neovascular age-related macular degeneration, *American Journal of Clinical Nutrition* 88 (2008): 398–406; E. J. Johnson and coauthors, The influence of supplemental lutein and docosahexaenoic acid on serum, lipoproteins, and macular pigmentation, *American Journal of Clinical Nutrition* 87 (2008): 1521–1529; E. D. O'Connell and coauthors, Diet and risk factors for age-related maculopathy, *American Journal of Clinical Nutrition* 87 (2008): 712–722.
54. J. Hootman and coauthors, Prevalence of doctor-diagnosed arthritis and arthritis-attributable activity limitation—United States, 2003–2005, *Morbidity and Mortality Weekly* 55 (2006): 1089–1092.
55. L. Devos-Comby, T. Cronan, and S. C. Roesch, Do exercise and self-management interventions benefit patients with osteoarthritis of the knee? A metaanalytic review, *Journal of Rheumatology* 33 (2006): 744–756.
56. G. McKellar and coauthors, A pilot study of a Mediterranean-type diet intervention in female patients with rheumatoid arthritis living in areas of social deprivation in Glasgow, *Annals of the Rheumatic Disease* 66 (2007): 1239–1243.
57. D. J. Pattison and coauthors, Dietary β-cryptoxanthin and inflammatory polyarthritis: Results from a population-based prospective study, *American Journal of Clinical Nutrition* 82 (2005): 451–455.
58. N. Schlesinger, Dietary factors and hyperuricaemia, *Current Pharmaceutical Design* 11 (2005): 4133–4138.
59. H. K. Choi, S. Liu, and G. Curhan, Intake of purine-rich foods, protein, and dairy products and relationship to serum levels of uric acid: The Third National Health and Nutrition Examination Survey, *Arthritis and Rheumatism* 52 (2005): 283–289.
60. D. O. Clegg and coauthors, Glucosamine, chondroitin sulfate, and the two in combination for painful knee osteoarthritis, *New England Journal of Medicine* 354 (2006): 795–808.
61. T. D. Bird, Genetic factors in Alzheimer's disease, *New England Journal of Medicine* 352 (2005): 862–864; P. M. Kidd, Neurodegeneration from mitochondrial insufficiency: Nutrients, stem cells, growth factors, and prospects for brain rebuilding using integrative management, *Alternative Medicine Review* 10 (2005): 268–293.
62. P. I. Moreira and coauthors, Oxidative stress: The old enemy in Alzheimer's disease pathophysiology, *Current Alzheimer Research* 2 (2005): 403–408.
63. P. I. Moreira and coauthors, Alzheimer disease and the role of free radicals in the pathogenesis of the disease, *CNS and Neurological Disorders Drug Targets* 7 (2008): 3–10; A. Nunomura and coauthors, Involvement of oxidative stress in Alzheimer disease, *Journal of Neuropathology and Experimental Neurology* 65 (2006): 631–641.
64. J. A. Luchsinger and D. R. Gustafon, Adiposity and Alzheimer's disease, *Current Opinion in Clinical Nutrition and Metabolic Care* 12 (2009): 15–21; D. B. Miller and J. P. O'Callaghan, Do early-life insults contribute to the late-life development of Parkinson's and Alzeimer disease? *Metabolism* 57 (2008): S44–S49; R. A. Whitmer, The epidemiology of adiposity and dementia, *Current Alzheimer Research* 4 (2007): 117–122.
65. R. J. Castellani and coauthors, Antioxidant protection and neurodegenerative disease: The role of amyloid-beta and tau, *American Journal of Alzheimer's Disease and Other Dementias* 21 (2006): 126–130; P. Zafrilla and coauthors, Oxidative stress in Alzheimer patients in different stages of the disease, *Current Medicinal Chemistry* 13 (2006): 1075–1083.
66. R. A. Armstrong, Plaques and tangles and the pathogenesis of Alzheimer's disease, *Folia Neuropathologica* 44 (2006): 1–11; G. L. Wenk, Neuropathologic changes in Alzheimer's disease: Potential targets for treatment, *Journal of Clinical Psychiatry* 67 (2006): 3–7.
67. A. Nunomura and coauthors, Neuropathology in Alzheimer's disease: Awaking from a hundred-year-old dream, *Science of Aging Knowledge Environment* (2006): pe10; R. E. Tanzi, Tangles and neurodegenerative disease: A surprising twist, *New England Journal of Medicine* 353 (2005): 1853–1855.
68. D. J. Selkoe, Developing preventive therapies for chronic diseases: Lessons learned from Alzheimer's disease, *Nutrition Reviews* 65 (2007): S239–S243.
69. B. Benjamin and A. Burns, Donepezil for Alzheimer's disease, *Expert Review of Neurotherapeutics* 7 (2007): 1243–1249.
70. G. Ravaglia and coauthors, Homocysteine and folate as risk factors for dementia and Alzheimer disease, *American Journal of Clinical Nutrition* 82 (2005): 636–643; K. L. Tucker and coauthors, High homocysteine and low B vitamins predict cognitive decline in aging men: The Veterans Affairs Normative Aging Study, *American Journal of Clinical Nutrition* 82 (2005): 627–635.
71. L. J. Whalley and coauthors, n-3 Fatty acid erythrocyte membrane content, APOE ε4, and cognitive variation: An observational follow-up study in late adulthood, *American Journal of Clinical Nutrition* 87 (2008): 449–454; M. A. Beydoun and coauthors, Plasma n-3 fatty acids and the risk of cognitive decline in older adults: The Atherosclerosis Risk in Communities Study, *American Journal of Clinical Nutrition* 85 (2007): 1103–1111; W. E. Connor and S. L. Connor, The importance of fish and docosahexaenoic acid in Alzheimer disease, *American Journal of Clinical Nutrition* 85 (2007): 929–930; C. Dullemeijer and coauthors, n-3 Fatty acid proportions in plasma and cognitive performance in older adults, *American Journal of Clinical Nutrition* 86 (2007): 1479–1485; B.M.V. Gelder and coauthors, Fish consumption, n-3 fatty acids, and subsequent 5-y cognitive decline in elderly men: The Zutphen Elderly Study, *American Journal of Clinical Nutrition* 85 (2007): 1142–1147; E. Nurk and coauthors, Cognitive performance among the elderly and dietary fish intake: The Hordaland Health Study, *American Journal of Clinical Nutrition* 86 (2007): 1470–1478.
72. E. B. Larson, Physical activity for older adults at risk for Alzheimer disease, *Journal of the American Medical Association* 300 (2008): 1077–1079; N. T. Lautenschlager and coauthors, Effect of physical activity on cognitive function in older adults at risk for Alzheimer disease, *Journal of the American Medical Association* 300 (2008): 1027–1037.

HIGHLIGHT 16

Hunger and Community Nutrition

© Jim West/Alamy

Worldwide, one person in every seven experiences persistent hunger—not the healthy appetite triggered by anticipation of a hearty meal, but the painful sensation caused by a lack of food.[1] In this chapter, **hunger** takes on the greater meaning—hunger that develops from prolonged, recurrent, and involuntary lack of food and results in discomfort, illness, weakness, or pain that exceeds the usual uneasy sensation (the accompanying glossary defines hunger and related terms). Such hunger deprives a person of the physical and mental energy needed to enjoy a full life and often leads to severe malnutrition and death. Tens of thousands of people die of hunger-related causes each day—one child every five seconds.

Resolving the hunger problem may seem at first beyond the influence of the ordinary person. Can one person's choice to limit family size or to recycle a bottle or to volunteer at a food recovery program make a difference? In truth, such choices produce several benefits. For one, a person's action may influence many other people over time. For another, a repeated action becomes a habit, with compounded benefits. For still another, making choices with an awareness of the consequences gives a person a sense of personal control, hope, and effectiveness. The daily actions of many concerned people can help solve the problems of hunger in their own neighborhoods or on the other side of the world.

Hunger in the United States

Ideally, all people at all times would have access to enough food to support an active, healthy life. In other words, they would experience **food security**. Unfortunately, more than 35 million people in the United States, including almost 13 million children, live in poverty and cannot afford to buy enough food to maintain good health.[2] Said another way, one out of nine households experiences hunger or the threat of hunger. Given the agricultural bounty and enormous wealth in this country, do these numbers surprise you? The limited or uncertain availability of nutritionally adequate and safe foods is known as **food insecurity** and is a major social problem in our nation today. Inadequate diets lead to poor health in adults and impaired physical, psychological, and cognitive development in children.

Table H16-1 presents the questions used in national surveys to identify food insecurity in the United States, and Figure H16-1

GLOSSARY

emergency shelters: facilities that are used to provide temporary housing.

food bank: a facility that collects and distributes food donations to authorized organizations feeding the hungry.

food insecurity: limited or uncertain access to foods of sufficient quality or quantity to sustain a healthy and active life. Food insecurity categories include **low food security**, which involves reduced quality of life with little or no indication of reduced food intake (formerly known as *food insecurity without hunger*) and **very low food security**, which involves multiple indications of disrupted eating patterns and reduced food intake (formerly known as *food insecurity with hunger*).

food insufficiency: an inadequate amount of food due to a lack of resources.

food pantries: programs that provide groceries to be prepared and eaten at home.

food poverty: hunger resulting from inadequate access to available food for various reasons, including inadequate resources, political obstacles, social disruptions, poor weather conditions, and lack of transportation.

food recovery: collecting wholesome food for distribution to low-income people who are hungry. Four common methods of food recovery include **field gleaning**, which involves collecting crops from fields that either have already been harvested or are not profitable to harvest; **perishable food rescue or salvage**, which involves collecting perishable produce from wholesalers and markets; **prepared food rescue**, which involves collecting prepared foods from commercial kitchens; and **nonperishable food collection**, which involves collecting processed foods from wholesalers and markets.

food security: access to enough food to sustain a healthy and active life. Food security categories include **high food security**, which reflects no indications of food-access problems or limitations, and **marginal food security**, which reflects one or two indications of food-access problems but with little or no change in food intake.

hunger: consequence of food insecurity that, because of prolonged, involuntary lack of food, results in discomfort, illness, weakness, or pain that goes beyond the usual uneasy sensation.

soup kitchens: programs that provide prepared meals to be eaten on site.

FIGURE H16-1 Prevalence of Food Insecurity and Hunger in U.S. Households, 2009

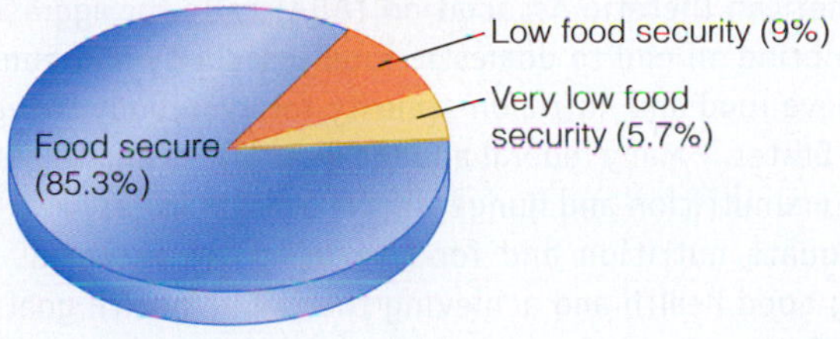

SOURCE: Economic Research Service, U.S. Department of Agriculture, www.ers.usda.gov.

shows recent findings. Responses to these questions provide crude, but necessary, data to estimate the degree of hunger in this country. Specific questions and measures focus on food insecurity in children.[3]

Defining Hunger in the United States

At its most extreme, people experience hunger because they have absolutely no food. More often, they have too little food (**food insufficiency**) and try to stretch their limited resources by eating small meals or skipping meals—often for days at a time. Sometimes hungry people obtain enough food to satisfy their hunger, perhaps by seeking food assistance or finding food through socially unacceptable ways—begging from strangers, stealing from markets, or scavenging through garbage cans, for example. Sometimes obtaining food raises concerns for food safety—for example, when rot, slime, mold, or insects have damaged foods or when people eat others' leftovers or meat from roadkill.

Hunger has many causes, but in developed countries, the primary cause is **food poverty**. People are hungry not because there is no food nearby to purchase, but because they lack money. The rate and severity of U.S. poverty has increased sharply over the past decade.[4] An estimated one out of eight people in the United States lives in poverty. Even those above the poverty line may not have food security. Physical and mental illnesses and disabilities, unemployment, low-paying jobs, unexpected or ongoing medical expenses, and high living expenses threaten financial stability. When money is tight, people are forced to choose between food and life's other necessities—utilities, housing, and medical care. Food costs are more variable and flexible; people can purchase fewer groceries to lower the monthly food bill, but they usually can't pay only a portion of the bills for electricity, rent, or medication. Other problems further contribute to food poverty, such as abuse of alcohol and other drugs; lack of awareness of available food assistance programs; and the reluctance of people, particularly the elderly, to accept what they perceive as "welfare" or "charity." Lack of resources remains the major cause of food poverty in developed countries, and solving this problem would do a lot to relieve hunger.

In the United States, poverty and hunger reach across various segments of society, touching some more than others—notably,

TABLE H16-1 Questions to Identify Food Insecurity in a U.S. Household

To determine the extent of food insecurity in a household, surveys ask questions about behaviors and conditions known to characterize households having difficulty meeting basic food needs during the past 12 months. Most often, adults tend to protect their children from hunger. In the most severe cases, children also suffer from hunger and eat less.

1. Did you worry whether food would run out before you got money to buy more?
2. Did you find that the food you bought just didn't last and you didn't have money to buy more?
3. Were you unable to afford to eat balanced meals?
4. Did you or other adults in your household ever cut the size of your meals or skip meals because there wasn't enough food?
5. Did this happen in three or more months during the previous year?
6. Did you ever eat less than you felt you should because there wasn't enough money for food?
7. Were you ever hungry but didn't eat because you couldn't afford enough food?
8. Did you ever lose weight because you didn't have enough money to buy food?
9. Did you or other adults in your household ever not eat for a whole day because you were running out of money to buy food?
10. Did this happen in three or more months during the previous year?
11. Did you rely on only a few kinds of low-cost food to feed your children because you were running out of money to buy food?
12. Were you unable to feed your children a balanced meal because you couldn't afford it?
13. Were your children not eating enough because you just couldn't afford enough food?
14. Did you ever cut the size of your children's meals because there wasn't enough money for food?
15. Were your children ever hungry but you just couldn't afford enough food?
16. Did your children ever skip a meal because there wasn't enough money for food?
17. Did this happen in three or more months during the previous year?
18. Did your children ever not eat for a whole day because there wasn't enough money for food?

The more positive responses, the greater the food insecurity. Households with children answer all of the questions and are categorized as follows:

<2 positive responses = food secure

3–7 positive responses = low food security

>8 positive responses = very low food security

Households without children answer the first 10 questions and are categorized as follows:

<2 positive responses = food secure

3–5 positive responses = low food security

>6 positive responses = very low food security

Figure H16-1 shows the results of the 2009 surveys.

SOURCE: United States Department of Agriculture, *Household Food Security in the United States, 2008*, available at www.ers.usda.gov

Relieving Hunger in the United States

single parents living in households with their children, Hispanics and African Americans, and those living in the inner cities. People living in poverty are simply unable to buy sufficient amounts of nourishing foods, even if they are wise shoppers. Consequently, their diets tend to be inadequate.[5] For many of the children in these families, school lunch (and breakfast, where available) may be the only nourishment for the day. Otherwise they go hungry, waiting for an adult to find money for food. Not surprisingly, these children are more likely to have health problems than those who eat regularly.[6] They also tend to perform poorly in school and in social situations.[7]

Ironically, hunger and obesity exist side by side—sometimes within the same household or even the same person.[8] That hunger reflects an inadequate food intake and obesity implies an excessive intake seems paradoxical, but research studies have confirmed the relationship.[9] The highest rates of obesity occur among those living in the greatest poverty—the same people who live with food insecurity.[10] Unfortunately, many healthful food choices, such as fruits and vegetables, are not readily available in low-income or rural neighborhoods.[11] Furthermore, fruits and vegetables tend to cost more than the energy-dense foods that foster weight gain but offer few, if any, nutrients. Foods such as doughnuts, pizzas, and hamburgers provide the most energy and satiety for the least cost. Quite simply, poor-quality diets deliver more kcalories, but fewer nutrients, for less money; high-quality diets deliver fewer kcalories, but more nutrients, for more money.[12] People who are unsure about their next meal may overeat when food or money are available. Interestingly, food insecure people who do not participate in food assistance programs have a greater risk of obesity than those who do participate—illustrating that providing food actually helps to prevent obesity.[13] Figure H16-2 shows how poverty and food insecurity can lead to both malnutrition and obesity.[14]

FIGURE H16-2 The Poverty-Obesity Paradox

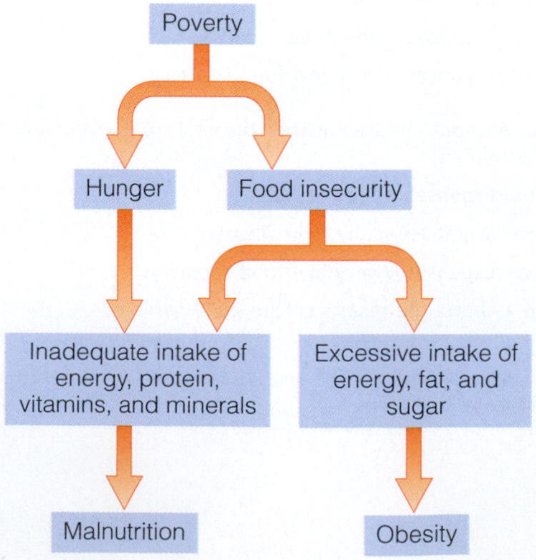

The American Dietetic Association (ADA) calls for aggressive action to bring an end to domestic food insecurity and hunger and to achieve food and nutrition security for everybody living in the United States.[15] Many federal and local programs aim to prevent or relieve malnutrition and hunger in the United States.

Adequate nutrition and food security are essential in supporting good health and achieving the public health goals of the United States. To that end, an extensive network of federal assistance programs provides life-giving food to millions of U.S. citizens daily. One out of every five Americans receives food assistance of some kind, at a total cost of $60 billion per year. Even so, the programs are not fully successful in preventing hunger, but they do seem to improve the nutrient intakes of those who participate. Programs described in earlier chapters include the WIC program for low-income pregnant women, breastfeeding mothers, and their young children (Chapter 14); the school lunch, breakfast, and child-care food programs for children (Chapter 15); and the food assistance programs for older adults such as congregate meals and Meals on Wheels (Chapter 16).

The Supplemental Nutrition Assistance Program (SNAP), administered by the U.S. Department of Agriculture (USDA), is the largest of the federal food assistance programs, both in amount of money spent and in number of people served. Formerly known as the Food Stamp Program, SNAP provides assistance to more than 28 million people at a cost of more than $34 billion per year; about half of the recipients are children.[16] The USDA issues debit cards through state agencies to households—people who buy and prepare food together. The amount a household receives depends on its size, resources, and income. The average monthly benefit is about $100 per person. Recipients may use the cards to purchase food and food-bearing plants and seeds, but not to buy tobacco, cleaning items, alcohol, or other nonfood items. The accompanying "How To" offers shopping tips for those on a limited budget.

Food assistance programs improve nutrient intakes significantly, but hunger continues to plague the United States. Of the estimated two million homeless people in the United States who are eligible for food assistance, only 15 percent of single adults and 50 percent of families receive food stamps. For some, reading, understanding, and completing the application can be difficult. For others, having to show identification and proof of homelessness can be frustrating. For many, accepting hunger is simply easier than meeting these challenges.

Efforts to resolve the problem of hunger in the United States do not depend solely on federal assistance programs. National **food recovery** programs have made a dramatic difference. The largest program, Feeding America, coordinates the efforts of more than 70,000 **food pantries**, **emergency shelters**, and **soup kitchens** that feed an estimated 37 million people a year.

Each year, an estimated one-fifth of our food supply is wasted in fields, commercial kitchens, grocery stores, and restaurants—that's enough food to feed 49 million people. Food recovery programs collect and distribute good food that would otherwise go to waste. Volunteers might pick corn left in an already harvested field, a grocer might deliver ripe bananas to a local **food bank**, and a caterer might take leftover chicken salad to a community shelter, for example. All of these efforts help to feed the hungry in the United States.

Food recovery programs depend on volunteers. Concerned citizens work through local agencies and churches to feed the hungry. Community-based food pantries provide groceries, and soup kitchens serve prepared meals. Meals often deliver adequate nourishment, but most homeless people receive fewer than one and a half meals a day, so many are still inadequately nourished. A combination of various strategies helps to build food security in a community.[17]

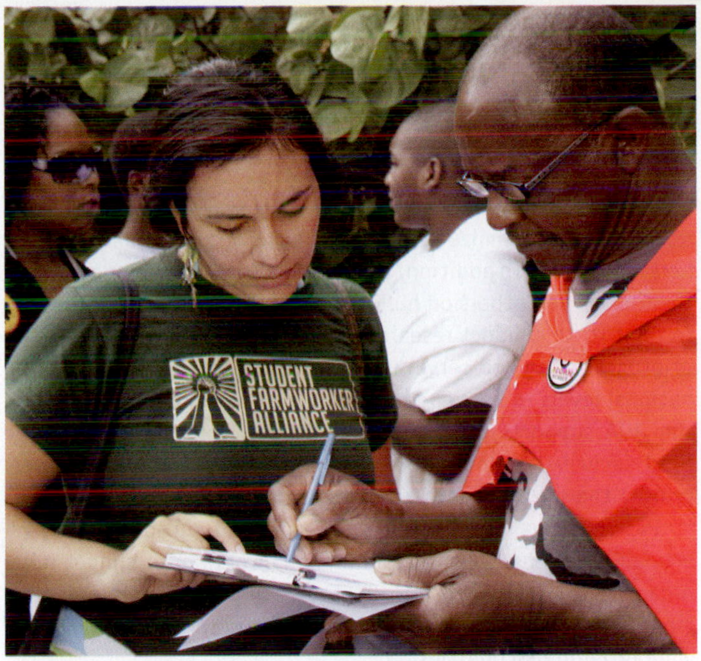

© Jeff Greenberg/Alamy

Each person's choice to get involved and be heard can help lead to needed change.

Sustainable Actions

Every segment of our society can join in the fight against hunger and poverty. The federal government, the states, local communities, big business and small companies, educators, and all individuals have many opportunities to resolve these problems.

HOW TO Plan Healthy, Thrifty Meals

Chapter 2 introduces the USDA MyPyramid Food Guide and principles for planning a healthy diet. Meeting that goal on a limited budget adds to the challenge. To save money and spend wisely, plan and shop for healthy meals with the following tips in mind:

Planning

- Make a grocery list before going to the store to avoid expensive "impulse" items.
- Do not shop when hungry.
- Use leftovers.
- Center meals on rice, noodles, and other grains.
- Use small quantities of meat, poultry, fish, or eggs.
- Use legumes instead of meat, poultry, fish, or eggs several times a week.
- Use cooked cereals such as oatmeal instead of ready-to-eat breakfast cereals.
- Cook large quantities when time and money allow.
- Check for sales and clip coupons for products you need; plan meals to take advantage of sale items.

Shopping

- Buy day-old bread and other products from the bakery outlet.
- Select whole foods instead of convenience foods (potatoes instead of instant mashed potatoes, for example).
- Try store brands.
- Buy fresh produce that is in season; buy canned or frozen items at other times.
- Buy only the amount of fresh foods that you will eat before it spoils. Buy large bags of frozen items or dry goods; when cooking, take out the amount needed and store the remainder.
- Buy fat-free dry milk; mix and refrigerate quantities needed for a day or two. Buy fresh milk by the gallon or half-gallon.
- Buy less expensive cuts of meat. Chuck and bottom round roast are usually inexpensive; cover during cooking and cook long enough to make meat tender. Buy whole chickens instead of pieces.
- Compare the unit price (cost per ounce, for example) of similar foods so that you can select the least expensive brand or size.
- Buy nonfood items such as toilet paper and laundry detergent at discount stores instead of grocery stores.

For daily menus and recipes for healthy, thrifty meals, visit the USDA Center for Nutrition Policy and Promotion: **www.cnpp.usda.gov**

TRY IT Search for "thrifty meals" at the USDA website (www.cnpp.usda.gov) and select a day's meals to analyze using your personal profile in a diet analysis program.

Dietitians and foodservice managers have a special role to play, and their efforts can make an impressive difference. Their professional organization, the ADA, urges members to conserve resources and minimize waste in both their professional and their personal lives.[18] In addition, the ADA urges its members to educate themselves and others on hunger, its consequences, and programs to fight it; to conduct research on the effectiveness and benefits of programs; and to serve as advocates on the local, state, and national levels to help end hunger in the United States.[19] Globally, the ADA supports programs that combat malnutrition, provide food security, promote self-sufficiency, respect local cultures, protect the environment, and sustain the economy.[20]

Individuals can assist the global community in solving its poverty and hunger problems by joining and working for hunger-relief organizations (see Table H16-2). They can also support organizations that lobby for the needed changes in economic policies toward developing countries.

Most importantly, be part of the solution, not part of the problem. In other words, don't waste time or energy moaning and groaning about how bad things are: do something to improve them. They are our problems: human beings created them, and human beings must solve them.

TABLE H16-2 Hunger-Relief Organizations

Action without Borders www.idealist.org
Bread for the World www.bread.org
Catholic Relief Services www.crs.org
Community Food Security Coalition www.foodsecurity.org
Congressional Hunger Center www.hungercenter.org
Feeding America www.feedingamerica.org
Food and Agriculture Organization (FAO) of the United Nations www.fao.org
Oxfam America www.oxfamamerica.org
Pan American Health Organization www.paho.org
Society of St. Andrew www.endhunger.org
The Hunger Project www.thp.org
United Nations Children's Fund (UNICEF) www.unicef.org
World Food Program www.wfp.org
World Health Organization (WHO) www.who.int
World Hunger Year (WHY) www.whyhunger.org

Nutrition on the Net

For further study of topics covered in this chapter, log on to **www.cengagebrain.com** and search for ISBN 084006845X.

- Explore the problems of hunger, malnutrition, and food insecurity at the Feeding Minds, Fighting Hunger site: **www.feedingminds.org**
- Learn about constructive, community-based solutions to the problems of poverty and hunger within and between the public and private sectors from the National Hunger Clearinghouse: **www.whyhunger.org**
- Visit the USDA Supplemental Nutrition Assistance Program: **www.fns.usda.gov/snap**
- Download recipes, sample menus, and numerous tips for planning, shopping for, and cooking healthy meals on a tight budget from the USDA cookbook entitled "Recipes and Tips for Healthy, Thrifty Meals": **www.cnpp.usda.gov**
- Review the Best Practices Manual for Food Recovery and Gleaning at the USDA Food and Nutrition Service site: **www.fns.usda.gov/fdd/gleaning/gleanintro.htm**
- Find information on feeding the hungry from the Emergency Food and Shelter National Board Program: **www.efsp.unitedway.org**
- Donate free food at The Hunger Site: **www.thehungersite.com**

References

1. *State of Food Insecurity in the World, 2008,* Food Security Statistics, available from Food and Agriculture Organization, www.fao.org.
2. U.S. Department of Agriculture, *Household Food Security in the United States, 2005,* ERS Research Briefs, November 2006, available at www.ers.usda.gov/publications.
3. M. Nord and H. Hopwood, Recent advances provide improved tools for measuring children's food security, *Journal of Nutrition* 137 (2007): 533–536.

4. S. H. Woolf, R. E. Johnson, and H. J. Geiger, The rising prevalence of severe poverty in America: A growing threat to public health, *American Journal of Preventive Medicine* 31 (2006): 332–341.
5. C. M. Champagne and coauthors, Poverty and food intake in rural America: Diet quality is lower in food insecure adults in the Mississippi Delta, *Journal of the American Dietetic Association* 107 (2007): 1886–1894; M. L. Kropf and coauthors, Food security status and produce intake and behaviors of Special Supplemental Nutrition Program for Women, Infants, and Children and Farmer's Market Nutrition Program participants, *Journal of the American Dietetic Association* 107 (2007): 1903–1908.
6. R. Rose-Jacobs and coauthors, Household food insecurity: Associations with at-risk infant and toddler development, *Pediatrics* 121 (2008): 65–72.
7. D. F. Jyoti, E. A. Frongillo, and S. J. Jones, Food insecurity affects school children's academic performance, weight gain, and social skills, *Journal of Nutrition* 135 (2005): 2831–2839.
8. E. T. Kennedy, The global face of nutrition: What can governments and industry do? *Journal of Nutrition* 135 (2005): 913–915.
9. S. A. Tanumihardjo and coauthors, Poverty, obesity, and malnutrition: An international perspective recognizing the paradox, *Journal of the American Dietetic Association* 107 (2007): 1966–1972; P. H. Casey and coauthors, The association of child and household food insecurity with childhood overweight status, *Pediatrics* 118 (2006): e1406; L. M. Scheier, What is the hunger-obesity paradox? *Journal of the American Dietetic Association* 105 (2005): 883–886.
10. L. M. Dinour, D. Bergen, and M. Yeh, The food insecurity-obesity paradox: A review of the literature and the role food stamps may play, *Journal of the American Dietetic Association* 107 (2007): 1952–1961; P. E. Wilde and J. N. Peterman, Individual weight change is associated with household food security status, *Journal of Nutrition* 136 (2006): 1395–1400.
11. A. D. Liese and coauthors, Food store types, availability, and cost of foods in a rural environment, *Journal of the American Dietetic Association* 107 (2007): 1916–1923; J. P. Stimpson and coauthors, Neighborhood deprivation is associated with lower levels of serum carotenoids among adults participating in the Third National Health and Nutrition Examination Survey, *Journal of the American Dietetic Association* 107 (2007): 1895–1902; S. N. Zenk and coauthors, Fruit and vegetable intake in African Americans—Income and store characteristics, *American Journal of Preventive Medicine* 29 (2005): 1–9.
12. M. S. Townsend and coauthors, Less-energy-dense diets of low-income women in California are associated with higher energy-adjusted diet costs, *American Journal of Clinical Nutrition* 89 (2009): 1220–1226.
13. S. J. Jones and E. A. Frongillo, The modifying effects of Food Stamp program participation on the relation between food insecurity and weight change in women, *Journal of Nutrition* 136 (2006): 1091–1094; S. J. Jones and coauthors, Lower risk of overweight in school-aged food insecure girls who participate in food assistance, *Archives of Pediatrics and Adolescent Medicine* 157 (2003): 780–784.
14. Tanumihardjo and coauthors, 2007.
15. Position of the American Dietetic Association: Food insecurity and hunger in the United States, *Journal of the American Dietetic Association* 106 (2006): 446–458.
16. *Leading the Fight Against Hunger—Federal Nutrition Assistance*, June 2008, www.fns.usda.gov/fns.hunger.pdf; P. S. Landers, The Food Stamp Program: History, nutrition education, and impact, *Journal of the American Dietetic Association* 107 (2007): 1945–1951.
17. C. McCullum and coauthors, Evidence-based strategies to build community food security, *Journal of the American Dietetic Association* 105 (2005): 278–283.
18. Position of the American Dietetic Association: Food and nutrition professionals can implement practices to conserve natural resources and support ecological sustainability, *Journal of the American Dietetic Association* 107 (2007): 1033–1043.
19. Position of the American Dietetic Association, 2006.
20. Position of the American Dietetic Association: Addressing world hunger, malnutrition, and food insecurity, *Journal of the American Dietetic Association* 103 (2003): 1046–1057.

CHAPTER

17

© 2010 Image Source/Jupiterimages Corporation

Nutrition in the Clinical Setting

For a busy health practitioner, it can be easy to put a patient's nutritional needs on the back burner. After all, the benefits of nutrition therapy are not always as obvious or immediate as those of other medical treatments. Health practitioners who want to provide the best care for their patients, however, soon learn that nutrition status can affect both short-term and long-term outcomes of many disease treatments. Moreover, patients are often concerned about the diet they need to improve their health.

Throughout this chapter, the CourseMate icon indicates an opportunity for online self-study, linking you to activities to increase your understanding of chapter concepts. **www.cengagebrain.com** (search for ISBN 084006845X)

Nutrition Care and Assessment

CHAPTER OUTLINE

Previous chapters introduced the nutrients and described how appropriate dietary choices can support good health. Turning now to clinical nutrition, the remaining chapters explain how various illnesses influence nutrition status and how nutrition therapy contributes to medical care. This chapter introduces the process used for providing nutrition care and describes the strategies used for assessing nutrition status.

Nutrition in Health Care

Malnutrition is frequently reported in patients hospitalized with acute illness, and acutely ill individuals without nutrition problems on admission often exhibit a subsequent decline in nutrition status. In the past few decades, estimates of malnutrition in hospital patients have ranged from 38 to 62 percent.[1] Poor nutrition status weakens immune function and compromises a person's healing ability, influencing both the course of disease and the body's response to treatment. Thus, preventing and correcting nutrition problems can improve the outcome of disease treatments and can also help to prevent complications.

Effects of Illness on Nutrition Status An illness, its symptoms, and its treatments can lead to malnutrition by reducing food intake, interfering with digestion and absorption, or altering nutrient metabolism and excretion (see Figure 17-1). For example, the nausea associated with some illnesses and disease

FIGURE 17-1 **Ways in Which Illness Can Affect Nutrition Status**

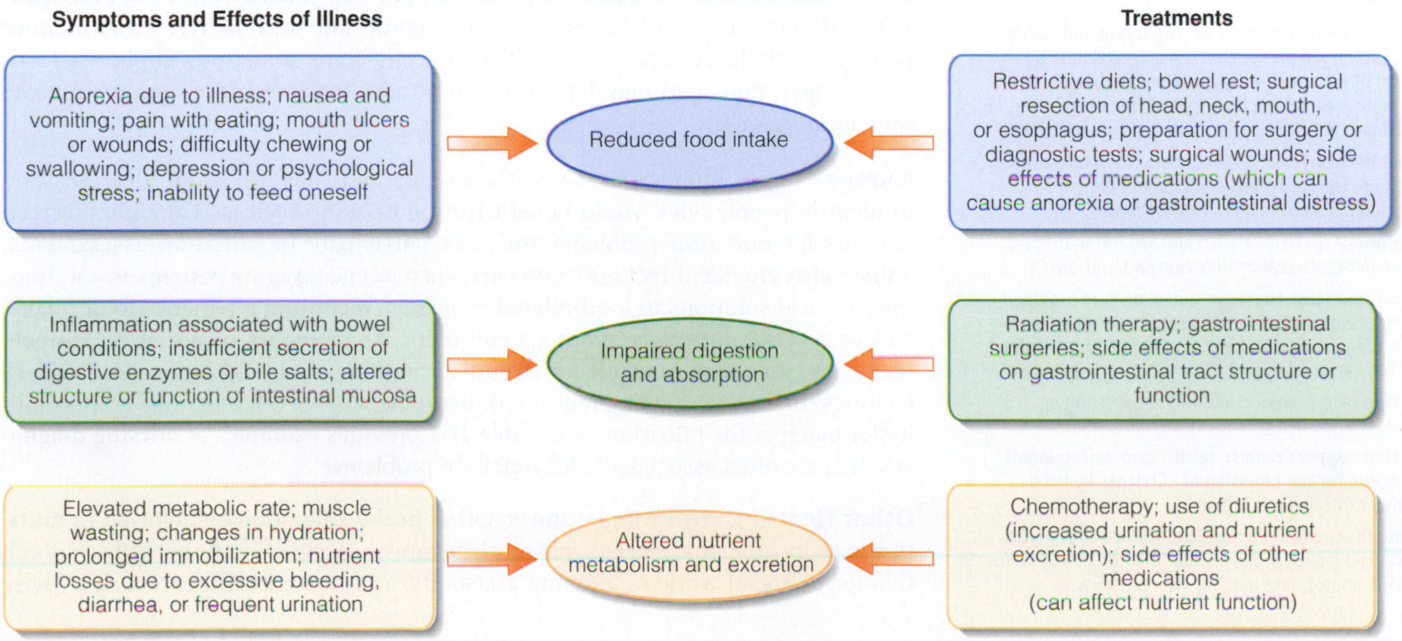

◆ Chapter 22 discusses the nutrition needs of patients undergoing acute metabolic stress.

treatments can diminish appetite and reduce food intake; similarly, an inflamed mouth or esophagus can make the physical act of eating uncomfortable. Certain medications can cause anorexia or gastrointestinal discomfort or can interfere with nutrient function and metabolism. Prolonged bed rest often results in **pressure sores**, which increase metabolic stress and raise protein and energy needs. ◆

The dietary changes required during an acute illness are usually temporary and can be tailored to accommodate an individual's preferences and lifestyle. Conversely, chronic illnesses may necessitate long-term dietary modifications. For example, diabetes treatment requires lifelong changes in diet and lifestyle that some people may find difficult to maintain. The challenge for health professionals is to help their patients appreciate the potential benefits of treatment and accept dietary changes that can improve their health.

Responsibility for Nutrition Care

The members of a health care team work together to ensure that the nutritional needs of patients are met during illness. The roles of health professionals may vary in different institutions and their responsibilities can sometimes overlap. Sometimes a patient's nutrition care is incorporated into the medical care plan developed by the entire health care team. Such plans, called **critical pathways**, outline coordinated plans of care for specific medical diagnoses, procedures, or treatments.

Physicians

Physicians are responsible for meeting all of a patient's medical needs, including nutrition. They prescribe **diet orders** (also called *nutrition prescriptions)* and other orders related to nutrition care, including referrals for nutrition therapy and dietary counseling. Physicians rely on nurses, registered dietitians, and other health professionals to alert them to nutrition problems, suggest strategies for handling these problems, and provide nutrition services.

◆ Reminder: A *registered dietitian (RD)* has completed the education and training specified by the the American Dietetic Association (or Dietitians of Canada), including an undergraduate degree in nutrition or dietetics, a supervised internship, and a national registration examination.

Registered Dietitians

Registered dietitians ◆ are food and nutrition experts who are qualified to provide **medical nutrition therapy**. They conduct nutrition and dietary assessments; diagnose nutrition problems; develop, implement, and evaluate **nutrition care plans** (described later); plan and approve menus; and provide dietary counseling and nutrition education services. Registered dietitians may also manage food and cafeteria services in health care institutions.

Registered Dietetic Technicians

Registered dietetic technicians often work in partnership with registered dietitians and assist in the implementation and monitoring of nutrition services. Depending on their background and experience, they may screen patients for nutrition problems, provide patient education and counseling, develop menus and recipes, ensure appropriate meal delivery, and monitor patients' food choices and intakes. Dietetic technicians sometimes supervise foodservice operations and may have roles in purchasing, inventory, quality control, sanitation, or safety.

Nurses

Nurses interact closely with patients and thus are in an ideal position to identify people who would benefit from nutrition services. They often screen patients for nutrition problems and may participate in nutrition assessments. Nurses also provide direct nutrition care, such as encouraging patients to eat, finding practical solutions to food-related problems, recording a patient's food intake, and answering questions about special diets. As members of **nutrition support teams**, nurses are responsible for administering tube and intravenous feedings. In facilities that do not employ registered dietitians, nurses often assume responsibility for much of the nutrition care. Table 17-1 provides examples of **nursing diagnoses** that are often associated with nutrition problems.

Other Health Care Professionals

Other health practitioners involved in nutrition care include pharmacists, physical therapists, occupational therapists, speech therapists, social workers, nursing assistants, and home health care aides. These

pressure sores: regions of damaged skin and tissue due to prolonged pressure on the affected area by an external object, such as a bed, wheelchair, or cast; vulnerable areas of the body include buttocks, hips, and heels. Also called *decubitus* (deh-KYU-bih-tus) *ulcers*.

critical pathways: coordinated programs of treatment that merge the care plans of different health practitioners; also called *clinical pathways*.

diet orders: specific instructions regarding dietary management; also called *nutrition prescriptions*.

medical nutrition therapy: nutrition care provided by a registered dietitian; includes assessing nutrition status, diagnosing nutrition problems, and providing nutrition care.

nutrition care plans: strategies for meeting an individual's nutritional needs.

nutrition support teams: health care professionals responsible for the provision of nutrients by tube feeding or intravenous infusion.

nursing diagnoses: clinical judgments about actual or potential health problems that provide the basis for selecting appropriate nursing interventions.

TABLE 17-1 Nursing Diagnoses with Nutritional Implications

- Chronic confusion
- Chronic pain
- Constipation
- Diarrhea
- Disturbed body image
- Feeding self-care deficit
- Imbalanced nutrition: less than body requirements
- Imbalanced nutrition: more than body requirements
- Impaired dentition
- Impaired oral mucous membrane
- Impaired physical mobility
- Impaired swallowing
- Ineffective breastfeeding
- Nausea
- Readiness for enhanced nutrition
- Risk for aspiration
- Risk for deficient fluid volume
- Risk for unstable blood glucose

SOURCE: NANDA International, *Nursing Diagnoses: Definitions and Classification 2009–2011* (Oxford: Wiley-Blackwell, 2009).

individuals can be instrumental in alerting dietitians or nurses to nutrition problems or may share relevant information about a patient's health status or personal needs.

Nutrition Screening

To identify patients who are malnourished or at risk for malnutrition, a **nutrition screening** is conducted within 24 hours of a patient's admission to a hospital or other extended-care facility. It may also be included in outpatient services and community health programs. ♦ A nutrition screening involves collecting a limited amount of health-related information that can indicate the presence of protein-energy malnutrition (PEM) ♦ or other nutrition problems. Although the screening should be sensitive enough to identify the patients who require nutrition care, it must be simple enough to be completed within 10 to 15 minutes. Usually a nurse, nursing assistant, registered dietitian, or dietetic technician performs and documents the screening. In some instances, a screening may be repeated (or followed up with a more comprehensive screening) during a patient's stay.

♦ Reminder: The Nutrition Screening Initiative, which addresses malnutrition risk in older adults, is described in Chapter 16 (p. 556).

♦ Reminder: *Protein-energy malnutrition* (*PEM*) is a deficiency of protein and food energy and is characterized by weight loss and loss of muscle tissue.

The information gathered during a nutrition screening varies according to the patient population, the type of care offered by the health care facility, and the patient's medical problem. Usually included are the admitting diagnosis, physical measurements, laboratory test results, relevant symptoms, and information about diet and health status provided by the patient or caregiver (see Table 17-2 for examples). Screening tools that use different combinations of these variables have become popular in recent years; these tools include the *Mini Nutritional Assessment* and the *Subjective Global Assessment*, outlined in Figure 17-2 and Table 17-3, respectively. The Mini Nutritional Assessment was developed to detect risk of malnutrition in adults over 65 years of age, whereas the Subjective Global Assessment has been found to be applicable to a variety of patient populations. Briefer screening methods use just two or three variables; for example, several tools screen for malnutrition risk by evaluating unintentional weight changes and reduced appetite or food intake.[2] Note that health care institutions often develop specific techniques that meet their particular needs.

A nutrition or health screening may lead to a referral for nutrition care. The following section describes the next stage of the process: the method used by dietitians to address nutritional concerns.

nutrition screening: a brief assessment of health-related variables to identify patients who are malnourished or at risk for malnutrition.

TABLE 17-2 Information Included in a Nutrition Screening

- Age, medical diagnosis, severity of illness
- Height and weight, BMI, unintentional weight changes
- Tissue wasting, loss of subcutaneous fat
- Changes in appetite or food intake
- Problems that interfere with food intake (such as chewing or swallowing difficulty, or nausea and vomiting)
- Food allergies or intolerances, extensive dietary restrictions
- Laboratory test results that indicate poor health status
- History of diabetes, renal disease, or other chronic illness
- Presence of anemia or pressure sores
- Use of medications that can impair nutrition status
- Depression, social isolation, dementia

TABLE 17-3 Subjective Global Assessment

The Subjective Global Assessment evaluates a person's risk of malnutrition by ranking key variables of the medical history and physical examination. These variables are each given an A, B, or C rating: A for well nourished, B for potential or mild malnutrition, and C for severe malnutrition. Patients are classified according to the final numbers of A, B, and C ratings.

Medical History

- Body weight changes: percentage change in past six months; weight change in past two weeks
- Dietary changes: suboptimal, low kcalorie, liquid diet, or starvation
- GI symptoms: nausea, diarrhea, vomiting, or anorexia for more than two weeks
- Functional ability: full capacity versus suboptimal, walking versus bedridden
- Degree of disease-related metabolic stress: low, medium, or high

Physical Examination

- Subcutaneous fat loss (triceps or chest)
- Muscle loss (quadriceps or deltoids)
- Ankle edema
- Sacral (lower spine) edema
- Ascites (abdominal edema)

Classification

A: Well nourished: if no significant loss of weight, fat, or muscle tissue and no dietary difficulties, functional impairments, or GI symptoms; also applies to patients with recent weight gain and improved appetite, functioning, or medical prognosis

B: Moderate malnutrition: if 5 to 10 percent weight loss, mild loss of muscle or fat tissue, decreased food intake, and digestive or functional difficulties that impair food intake; the B classification usually applies to patients with an even mix of A, B, and C ratings

C: Severe malnutrition: if more than 10 percent weight loss, severe loss of muscle or fat tissue, edema, multiple GI symptoms, and functional impairments

SOURCES: R. S. Gibson, *Principles of Nutritional Assessment* (New York: Oxford University Press, 2005), pp. 809–826; A. S. Detsky and coauthors, What is subjective global assessment of nutritional status? *Journal of Parenteral and Enteral Nutrition* 11 (1987): 8–13.

The Nutrition Care Process

Registered dietitians use a systematic approach to medical nutrition therapy called the **nutrition care process.** ♦ Figure 17-3 presents the four distinct, yet interrelated, steps of the nutrition care process:[3]

1. Nutrition assessment
2. Nutrition diagnosis
3. Nutrition intervention
4. Nutrition monitoring and evaluation

Although the nutrition care process is easiest to visualize as a series of steps, the steps are frequently revisited in order to reassess and revise diagnoses and intervention strategies. Note that each step of the nutrition care process must be docu-

♦ As a comparison, the *nursing process* consists of five steps:
1. Assessment
2. Nursing diagnosis
3. Planning
4. Implementation
5. Evaluation

nutrition care process: a problem-solving method that dietetics professionals use to evaluate and treat nutrition-related problems.

FIGURE 17-2 Mini Nutritional Assessment

Nestlé Nutrition INSTITUTE

Mini Nutritional Assessment MNA®

Last name: First name:

Sex: Age: Weight, kg: Height, cm: Date:

Complete the screen by filling in the boxes with the appropriate numbers. Total the numbers for the final screening score.

Screening

A Has food intake declined over the past 3 months due to loss of appetite, digestive problems, chewing or swallowing difficulties?
0 = severe decrease in food intake
1 = moderate decrease in food intake
2 = no decrease in food intake ☐

B Weight loss during the last 3 months
0 = weight loss greater than 3 kg (6.6 lbs)
1 = does not know
2 = weight loss between 1 and 3 kg (2.2 and 6.6 lbs)
3 = no weight loss ☐

C Mobility
0 = bed or chair bound
1 = able to get out of bed / chair but does not go out
2 = goes out ☐

D Has suffered psychological stress or acute disease in the past 3 months?
0 = yes 2 = no ☐

E Neuropsychological problems
0 = severe dementia or depression
1 = mild dementia
2 = no psychological problems ☐

F1 Body Mass Index (BMI) (weight in kg) / (height in m^2)
0 = BMI less than 19
1 = BMI 19 to less than 21
2 = BMI 21 to less than 23
3 = BMI 23 or greater ☐

IF BMI IS NOT AVAILABLE, REPLACE QUESTION F1 WITH QUESTION F2.
DO NOT ANSWER QUESTION F2 IF QUESTION F1 IS ALREADY COMPLETED.

F2 Calf circumference (CC) in cm
0 = CC less than 31
3 = CC 31 or greater ☐

Screening score
(max. 14 points) ☐☐

12-14 points: Normal nutritional status
8-11 points: At risk of malnutrition
0-7 points: Malnourished

For a more in-depth assessment, complete the full MNA® which is available at **www.mna-elderly.com**

Ref. Vellas B, Villars H, Abellan G, et al. *Overview of the MNA® - Its History and Challenges.* J Nutr Health Aging 2006;10:456-465.
Rubenstein LZ, Harker JO, Salva A, Guigoz Y, Vellas B. *Screening for Undernutrition in Geriatric Practice: Developing the Short-Form Mini Nutritional Assessment (MNA-SF).* J. Geront 2001;56A: M366-377.
Guigoz Y. *The Mini-Nutritional Assessment (MNA®) Review of the Literature - What does it tell us?* J Nutr Health Aging 2006; 10:466-487.

For more information: www.mna-elderly.com

FIGURE 17-3 The Nutrition Care Process

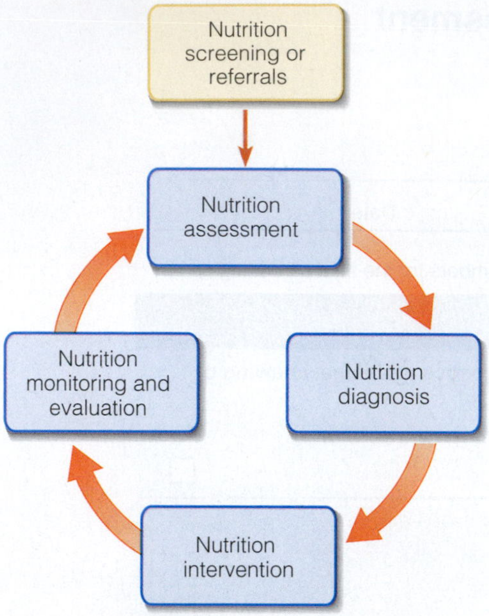

♦ This format is called a *PES statement* because it includes the *P*roblem, the *E*tiology, and the *S*igns and symptoms.

♦ The American Dietetic Association maintains an Evidence Analysis Library to keep members updated about recent developments in nutrition and dietetics research.

mented in the medical record, providing a record for future reference and facilitating communication among members of the health care team. Chapter 18 provides additional information about documentation.

Nutrition Assessment A nutrition assessment involves the collection and analysis of health-related information for the purpose of identifying specific nutrition problems and their underlying causes. A well-conducted assessment allows the dietitian to devise a plan of action to prevent or correct nutrient imbalances or to evaluate whether a particular care plan is working. Information may be obtained from the medical record, physical examination, laboratory analyses, medical procedures, interview with the patient or caregiver, and consultation with other health professionals. If applicable, the data are compared with reliable standards to help with their interpretation. The second half of this chapter describes the components of nutrition assessment in detail.

Nutrition Diagnosis After completing a nutrition assessment, the dietitian identifies existing and potential nutrition problems, a step that requires a careful and objective analysis of the patterns and relationships among the assessment data. Each nutrition problem receives a separate diagnosis, which includes the specific problem, etiology or cause, and signs and symptoms that provide evidence of the problem.[4] ♦ For example, a potential nutrition diagnosis might be "Involuntary weight gain (*the problem*) related to chronic use of a medication (corticosteroids) that causes weight gain (*the etiology or cause*) as evidenced by an unintentional weight gain of 10 percent of body weight over the past six months (*the sign or symptom*)." Note that a nutrition diagnosis is likely to change over the course of illness due to either a successful nutrition intervention or resolution of the medical problem.

Nutrition diagnoses fall into three main categories: *intake*, *clinical*, and *behavioral-environmental*. Intake-related diagnoses involve either the inadequate or excessive ingestion of nutrients, energy, fluid, alcohol, dietary supplements, and food ingredients. Clinical diagnoses involve medical or physical conditions that disrupt nutrition status, such as disruptions in physiological or mechanical functioning, altered nutrient metabolism, and body weight problems. Behavioral-environmental diagnoses include problems related to the patient's knowledge, attitudes, or beliefs; the physical environment; access to food; and food safety. Table 17-4 lists examples of nutrition diagnoses in each of these categories.

Nutrition Intervention After nutrition problems are identified, the appropriate nutrition care can be planned and implemented. Nutrition interventions attempt to modify dietary and lifestyle practices or environmental conditions that interfere with nutrition status or health. When possible, the intervention targets the etiology or cause of the problem as identified in the nutrition diagnosis. Nutrition interventions may include counseling or education about appropriate dietary and lifestyle practices, changes in a medication or other treatment, or modifications in the meals offered to a hospital patient. To be successful, the intervention must consider the patient's food habits, lifestyle, and other personal factors. Note that nutrition interventions used by dietitians are *evidence based*; that is, they are based on a scientific rationale and supported by the results of high-quality research.[5] ♦

The goals of nutrition interventions are stated in terms of measurable outcomes, such as the results of laboratory or anthropometric tests. For example, goals for an overweight person with diabetes might include target ranges for blood glucose levels and body weight. Other desirable outcomes may include positive changes in dietary behaviors and lifestyle; for example, the diabetes patient may need to learn how to control carbohydrate intake or portion sizes and may benefit from regular exercise. These outcomes can be assessed during an interview with the patient.

Although many aspects of nutrition care fall within the scope of dietetics practice, others require the assistance of other health professionals. For example, a physician's help would be required if a medication interfered with food intake; the

TABLE 17-4 Examples of Nutrition Diagnoses

Intake diagnoses

- Excessive alcohol intake
- Inadequate energy intake
- Inadequate fluid intake
- Inappropriate infusion of parenteral nutrition
- Increased calcium needs
- Inconsistent carbohydrate intake

Clinical diagnoses

- Altered blood potassium levels
- Altered GI function (constipation)
- Breastfeeding difficulty
- Food-medication interaction
- Involuntary weight gain
- Swallowing difficulty

Behavioral-environmental diagnoses

- Disordered eating pattern
- Impaired ability to prepare meals
- Limited access to food
- Physical inactivity
- Self-feeding difficulty
- Undesirable food choices

SOURCE: American Dietetic Association, *Nutrition Diagnosis and Intervention: Standardized Language for the Nutrition Care Process* (Chicago: American Dietetic Association, 2007).

nursing or foodservice staff might be involved if the feeding environment or meal delivery required adjustment. Chapter 18 provides additional information about nutrition intervention.

Nutrition Monitoring and Evaluation The effectiveness of the nutrition care plan must be evaluated periodically: the patient's progress should be monitored closely, and updated assessment data or diagnoses may require adjustments in goals or outcome measures. Sometimes a new situation alters nutritional needs; for example, a change in the medical treatment or a new medication may alter a person's tolerance to certain foods. The nutrition care plan must be flexible enough to adapt to the new situation.

If progress is slow or a patient is unable or unwilling to make the suggested changes, the care plan should be redesigned and take into account the reasons why the earlier plan was not successful. The new plan may need to include motivational techniques or additional patient education. If the patient remains unwilling to modify behaviors despite the expected benefits, the health care provider can try again at a later time when the patient may be more receptive.

IN SUMMARY **Illnesses and their treatments can affect food intake and nutrient needs, leading to malnutrition. In turn, poor nutrition status can influence the course of illness and reduce the effectiveness of medical treatments. The combined efforts of each member of the health care team ensure that patients receive optimal nutrition care. Nutrition screening identifies individuals who can benefit from nutrition assessment and follow-up nutrition care. The nutrition care process includes four interrelated steps: nutrition assessment, nutrition diagnosis, nutrition intervention, and nutrition monitoring and evaluation.**

Nutrition Assessment

As described earlier, a nutrition assessment provides the information needed for diagnosing nutrition problems, designing a nutrition care plan, or determining whether a care plan has been effective. Ideally, the assessment should be sensitive enough to detect subtle nutrition problems and specific enough to identify problem nutrients. For most nutrient imbalances, a variety of tests are necessary to identify nutrition problems.

Historical Information

Historical information provides valuable clues about nutrition status and nutrient requirements; it also reveals personal preferences that need consideration when developing a nutrition care plan. Table 17-5 summarizes the various types of historical information that contribute to a nutrition assessment.[6] This information can be obtained from the medical record or by interviewing the patient or caregiver. ♦

♦ The terms used in this section are based on the standardized language developed by the American Dietetic Association; other resources may use different terminology.

Medical History A substantial number of medical problems and their treatments may either interfere with food intake or require dietary changes; Table 17-6 lists examples. The medical history generally includes the family medical history as well; this information reveals a person's genetic susceptibilities for diseases that can potentially be prevented with dietary and lifestyle changes.

Medication and Supplement History A number of medications may have detrimental effects on nutrition status, and various dietary components can alter the absorption or metabolism of drugs. Ingredients in dietary and herbal supplements can also interact with medications. Chapter 19 provides examples of notable diet-drug interactions that may need consideration when planning nutrition care.

Personal and Social History Personal and social factors influence food choices as well as a person's ability to manage health and nutrition problems. For example, financial concerns may restrict access to health care and nutritious foods. Cultural background or religious beliefs can affect food preferences. Some individuals may depend on others to prepare or procure food. An individual who is depressed or lives alone may eat poorly or be uninterested in following complex dietary instructions. Use of alcohol, tobacco, or illegal drugs may alter food intake and have disruptive effects on health and nutrition status.

Food and Nutrition History A food and nutrition history (often called a *diet history*) is a detailed account of a person's dietary practices. It includes food intake data, lifestyle habits, and information about the various factors that may influ-

TABLE 17-5 Historical Information Used in Nutrition Assessment[a]

Medical History	Medication and Supplement History	Personal and Social History	Food and Nutrition History
Age	Prescription drugs	Employment status	Food intake
Current complaint(s)	Over-the-counter drugs	Educational level	Alcohol consumption
Past medical conditions	Dietary and herbal supplements	Socioeconomic status	Dietary restrictions
Surgical history		Cultural/ethnic identity	Food allergies and intolerances
Family medical history		Religious beliefs	Nutrition and health knowledge
Chronic disease risk		Home/family situation	Food availability
Allergies		Cognitive abilities	Physical activity and exercise habits
Mental/emotional health status		Use of tobacco or illegal drugs	

[a]Historical information is classified in different ways among medical institutions.
SOURCE: American Dietetic Association, *Nutrition Diagnosis and Intervention: Standardized Language for the Nutrition Care Process* (Chicago: American Dietetic Association, 2007).

ence dietary choices, such as the person's knowledge or beliefs about nutrition and health. The procedure often includes an interview about recent food intake (for example, a *24-hour recall*) and a survey of usual food choices (such as a *food frequency questionnaire*). The food and nutrition history helps the dietitian detect current or potential nutrition problems, as well as patterns of behavior that contribute to health problems. The following section describes the most common methods of gathering food intake information.

Food Intake Data Obtaining accurate food intake data is challenging, and results may vary depending on the individual's memory and honesty and the assessor's skill and training. Each method has its own strengths and weaknesses, so best results are obtained by using a combination of methods. Table 17-7 summarizes the methods commonly used and each method's advantages and disadvantages.

After food intake data are collected, the nutrient intake can be estimated using dietary analysis software or a table of food composition (such as that in Appendix H), and then the nutrient intake levels can be compared with RDA and AI values. Another option is to compare the food list with a diet-planning guide such as the USDA Food Guide (see Chapter 1). A food list also reveals the person's food preferences, which are helpful for developing an appropriate nutrition care plan, planning menus, or providing dietary counseling.

The 24-Hour Recall The **24-hour recall** is a guided interview in which an individual recounts all of the foods and beverages consumed in the past 24 hours or during the previous day. The interviewer includes questions about the times when meals or snacks were eaten, amounts consumed, and ways in which foods were pre-

TABLE 17-6 Medical Problems Often Associated with Malnutrition

- Acquired immune deficiency syndrome (AIDS)
- Alcoholism
- Anorexia nervosa
- Bulimia
- Burns (extensive or severe)
- Cancer and cancer treatments
- Cardiovascular diseases
- Celiac disease
- Chewing or swallowing difficulties
- Chronic kidney disease
- Dementia
- Diabetes mellitus
- Feeding disabilities
- Infections
- Inflammatory bowel diseases
- Liver disease
- Malabsorption
- Mental illness
- Pressure sores
- Surgery (major)
- Vomiting (prolonged or severe)

24-hour recall: a record of foods consumed in the previous 24 hours; sometimes modified to include foods consumed in a typical day.

TABLE 17-7 Methods for Obtaining Food Intake Data

Method	Description	Advantages	Disadvantages
24-hour recall	Guided interview in which the foods and beverages consumed in a 24-hour period are described in detail.	• Results are not dependent on literacy or educational level of respondent. • Interview occurs after food is consumed, so method does not influence dietary choices. • Results are obtained quickly; method is relatively easy to conduct.	• Process is reliant on memory. • Underestimation and overestimation of food intakes are common. • Food items that cause embarrassment (alcohol, desserts) may be omitted. • Data from a single day cannot accurately represent the respondent's usual intake. • Seasonal variations may not be addressed. • Skill of interviewer affects outcome.
Food frequency questionnaire	Written survey of food consumption during a specific period of time, often a one-year period.	• Process examines long-term food intake, so day-to-day and seasonal variability should not affect results. • Questionnaire is completed after food is consumed, so method does not influence food choices. • Method is inexpensive to administer.	• Process is reliant on memory. • Food lists often include common foods only. • Serving sizes are often difficult for respondents to evaluate without assistance. • Calculated nutrient intakes may not be accurate. • Food lists for the general population are of limited value in special populations. • Method is not effective for monitoring short-term changes in food intake.
Food record	Written account of food consumed during a specified period, usually several consecutive days. Accuracy is improved by including weights or measures of foods.	• Process does not rely on memory. • Recording foods as they are consumed may improve accuracy of food intake data. • Process is useful for controlling intake because keeping records increases awareness of food choices.	• Recording process itself influences food intake. • Underreporting and portion size errors are common. • Process is time-consuming and burdensome for respondent; requires high degree of motivation. • Method requires literacy and the physical ability to write. • Seasonal changes in diet are not taken into account.
Direct observation	Observation of meal trays or shelf inventories before and after eating; possible only in residential facilities.	• Process does not rely on memory. • Method does not influence food intake. • Method can be used to evaluate the acceptability of a prescribed diet.	• Process is possible only in residential situations. • Method is labor intensive.

pared. Accuracy can be improved by prompting the respondent to recall food items that are often forgotten, such as condiments, snacks, and beverages.[7]

In a typical interview, the assessor may begin by asking, "What is the first thing you ate or drank yesterday morning?" After the first food items are described, the follow-up questions might be, "What time was that?" and "How much did you eat?" Questioning continues until the intake record for the day is complete. Food models or measuring cups and spoons can be used to help the individual visualize and describe the amounts consumed. After the day's intake is recounted, the interviewer asks whether the intake that day is fairly typical and, if not, how it varies from the person's usual intake. A recall interview may be conducted on several nonconsecutive days to obtain a better representation of a person's usual diet.

© Nathan Benn/Corbis

Food models and measuring utensils can help an individual visualize portion sizes.

A recall interview provides useful data for developing an acceptable nutrition care plan and identifying food items that may need to be restricted due to illness. It is a poor technique, however, for determining the adequacy of a diet, because it does not take into account fluctuations in food intake or seasonal variations. Moreover, food intakes are often underestimated because the process relies on an individual's memory and reporting accuracy. People often forget to mention alcohol, soft drinks, snack foods, and desserts unless specifically prompted to do so, and some individuals find it embarrassing to report consumption of foods such as chocolate, butter, and red meat.[8]

Food Frequency Questionnaire A **food frequency questionnaire** surveys the foods and beverages regularly consumed during a specific time period. Some questionnaires are qualitative only: food lists contain common foods, organized by food group, with check boxes to indicate frequency of consumption. Other types of questionnaires provide semiquantitative information by including portion sizes as well. Figure 17-4 shows a sample section of a semiquantitative questionnaire that surveys fruit intake over the previous year. Because the respondent is often asked to estimate food intakes over a one-year period, the results should not be affected by seasonal changes in diet. Conversely, a disadvantage of this method is its inability

food frequency questionnaire: a survey of foods routinely consumed. Some questionnaires ask about the types of food eaten and yield only qualitative information; others include questions about portions consumed and yield semiquantitative data as well.

FIGURE 17-4 Sample Section of a Food Frequency Questionnaire

FRUIT	HOW OFTEN								HOW MUCH			
	Never or less than once per month	1 per mon.	2–3 per mon.	1 per week	2 per week	3–4 per week	5–6 per week	Every day	MEDIUM SERVING	YOUR SERVING SIZE S	M	L
EXAMPLE: Bananas	○	○	○	●	○	○	○	○	1 medium	○ 1/2	● 1	○ 2
Bananas	○	○	○	○	○	○	○	○	1 medium	○ 1/2	○ 1	○ 2
Apples, applesauce	○	○	○	○	○	○	○	○	1 medium or 1/2 cup	○ 1/2	○ 1	○ 2
Oranges (not including juice)	○	○	○	○	○	○	○	○	1 medium	○ 1/2	○ 1	○ 2
Grapefruit (not including juice)	○	○	○	○	○	○	○	○	1/2 medium	○ 1/4	○ 1/2	○ 1
Cantaloupe	○	○	○	○	○	○	○	○	1/4 medium	○ 1/8	○ 1/4	○ 1/2
Peaches, apricots (fresh, in season)	○	○	○	○	○	○	○	○	1 medium	○ 1/2	○ 1	○ 2
Peaches, apricots (canned or dried)	○	○	○	○	○	○	○	○	1 medium or 1/2 cup	○ 1/2	○ 1	○ 2
Prunes, or prune juice	○	○	○	○	○	○	○	○	1/2 cup	○ 1/4	○ 1/2	○ 1
Watermelon (in season)	○	○	○	○	○	○	○	○	1 slice	○ 1/2	○ 1	○ 2
Strawberries, other berries (in season)	○	○	○	○	○	○	○	○	1/2 cup	○ 1/4	○ 1/2	○ 1
Any other fruit, including kiwi, fruit cocktail, grapes, raisins, mangoes	○	○	○	○	○	○	○	○	1/2 cup	○ 1/4	○ 1/2	○ 1

to determine recent changes in food intake. Another limitation is that the questionnaires typically list only common food items, so the accuracy of food intake data is reduced if an individual consumes atypical foods.

Simple versions of food frequency questionnaires focus on food categories relevant to a person's medical condition. For example, a questionnaire designed to evaluate calcium intake may include only milk products, fortified foods, certain fruits and vegetables, and dietary supplements that contain calcium. A computer analysis can then quickly estimate the individual's calcium intake and compare it with recommendations.

Food Record A **food record** is a written account of foods and beverages consumed during a specified time period, usually several consecutive days. Foods are recorded as they are consumed in order to obtain the most complete and accurate record possible; thus the process does not rely on memory. A detailed food record includes the types and amounts of foods and beverages consumed, times of consumption, and methods of preparation. For weight-management purposes, it may also include information about a person's mood, the occasion, activities engaged in while eating, and daily physical activity. For establishing blood glucose control, the record may include information about medication use, physical activity, and the results of blood glucose monitoring.

The food record provides valuable information about food intake, as well as a person's response to and compliance with nutrition therapy. Unfortunately, food records require a great deal of time to complete, and people need to be highly motivated to keep accurate records. Another drawback is that the recording process itself may influence food intake. Furthermore, it is difficult to obtain accurate estimates of nutrient intakes in just a few days or even a week due to day-to-day and seasonal variations in food intake.

Direct Observation In facilities that serve meals, food intakes can be directly observed and analyzed. This method can also reveal a person's food preferences, changes in appetite, and any problems with a prescribed diet. Health practitioners use direct observation to conduct patients' **kcalorie counts**, which are estimates of the food energy (and often protein) consumed by patients during a single day or several consecutive days. To perform a kcalorie count, the clinician records the dietary items that a patient is given at meals and subtracts the amounts remaining after meals are completed; this procedure allows an estimate of the kcaloric content of foods and beverages actually consumed. Although a useful means of discerning patients' intakes, direct observation requires regular and careful documentation and can be labor intensive and costly.

Anthropometric Data

Anthropometric data ♦ can reveal problems related to both overnutrition and PEM. Height (or length) and weight are the most widely used anthropometric measurements and are used to evaluate growth in children and nutrition status in adults. Other helpful values include body composition tests (described in Chapter 8 and Appendix E) and circumferences of the head, waist, and limbs.

♦ Reminder: *Anthropometric* refers to physical measurements of the body.

Height (or Length) Poor growth in children can be a sign of malnutrition. In adults, height measurements alone do not reflect current nutrition status but can be used for estimating a person's energy needs or appropriate body weight. Length is measured in infants and children younger than 24 months of age, and height is usually measured in older children and adults. ♦ Length can also be measured in adults and children who cannot stand unassisted due to physical or medical reasons. The "How To" describes some standard techniques for measuring length and height.

♦ *Length* is measured while a person is recumbent (lying down), whereas *height* is measured while a person is standing upright.

In adults who are bedridden or unable to stand, height can be estimated from equations that include either the knee height or the full arm span, both of which correlate well with height.[9] Knee height, which extends from the heel to the top of the knee when the leg is bent at a 90-degree angle, can be measured in either a sitting or supine position with a knee-height caliper; specific formulas are available for various age, gender, and ethnic groups. The full arm span is the distance from the

food record: a detailed log of food eaten during a specified time period, usually several days; also called a *food diary*. A food record may also include information regarding medications, disease symptoms, and physical activity.

kcalorie counts: estimates of food energy (and often, protein) consumed by patients for one or more days.

HOW TO Measure Length and Height

To improve the accuracy of length and height measurements, keep the following in mind:

- Always measure—never ask! Self-reported heights are less accurate than measured heights. If height is not measured, document that the height is self-reported.
- Measure the length of infants and young children by using a measuring board with a fixed headboard and a movable footboard. It generally takes two people to measure length: one person gently holds the infant's head against the headboard; the other straightens the infant's legs and moves the footboard to the bottom of the infant's feet.
- Measure height next to a wall on which a nonstretchable measuring tape or board has been fixed. Ask the person to stand erect without shoes and with heels together. The person's eyes and head should be facing forward, with heels, buttocks, and shoulder blades touching the wall. Place a ruler or other flat, stiff object on the top of the head at a right angle to the wall and carefully note the height measurement. Immediately record length and height measurements to the nearest 1/8 inch or 0.1 centimeter.
- For evaluating growth rate in young children, use the appropriate growth chart (Appendix E) when plotting results. If length is measured, use the growth chart for children between 0 and 36 months; if height is measured, use the chart for individuals between 2 and 20 years.
- Higher values are obtained from supine measurements than from vertical height measurements due to gravity.

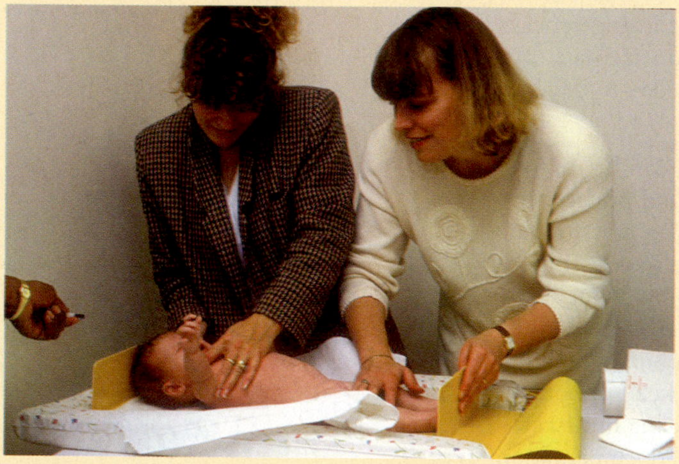

Janine Wiedel Photolibrary/Alamy

It generally takes two people to measure the length of an infant.

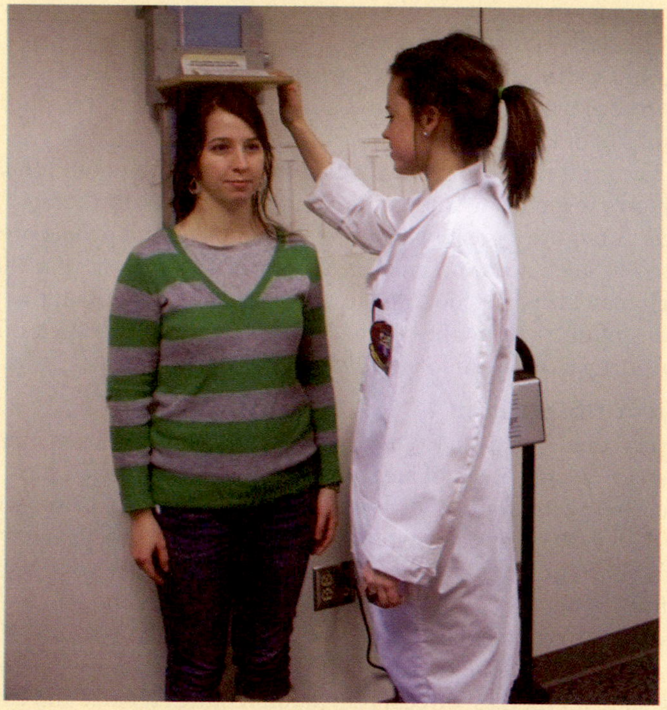

Courtesy of Marcia Nahikian Nelms

Standing erect allows for an accurate height measurement.

TRY IT Affix a measuring tape to a wall and measure a friend's height. Then, compare the vertical height measurement with one taken while the friend is lying in a supine position.

tip of one middle finger to the other while the arms are extended horizontally. In children with disabilities that affect stature, alternative measures of linear growth include the full arm span, lower-leg lengths (knee to heel, similar to the knee height measure), and upper-arm lengths (shoulder to elbow), all of which can be compared with reference percentiles.

Body Weight During clinical care, health care providers monitor body weights closely: weight changes may reflect changes in hydration status, and an involuntary weight loss can be a sign of PEM. Body weights can be compared with healthy ranges on height-weight tables and growth charts or used to calculate the body mass index (BMI). ♦ A healthy body weight typically falls within a BMI range of 18.5 to 25; thus, an appropriate body weight can usually be estimated by using a BMI table or graph (see the inside back cover of this book).[10] The "How To" includes suggestions for improving the accuracy of weight measurements.

♦ Reminder: $\text{BMI} = \dfrac{\text{weight (kg)}}{\text{height (m)}^2}$

HOW TO Measure Weight

Tips for measuring weight include:

- Always measure—never ask! Self-reported weights are often inaccurate. If weight is not measured, document that the weight is self-reported.
- Valid weight measurements require scales that have been carefully maintained, calibrated, and checked for accuracy at regular intervals. Beam balance and electronic scales are the most accurate. Bathroom scales are inaccurate and inappropriate for clinical use.
- Measure an infant's weight with a scale that allows the infant to sit or lie down. The tray should be large enough to support an infant or young child up to 40 pounds, and weight graduations should be in $^1/_2$-ounce or 10-gram increments. For accurate results, weigh infants without clothes or diapers. Excessive movement by the infant can reduce accuracy.
- Children who can stand are weighed in the same way as adults, using beam balance or electronic scales with platforms large enough for standing comfortably. If repeated weight measurements are needed, each weighing should take place at the same time of day (preferably before breakfast), in the same amount of clothing, after the person has voided, and using the same scale. Record weights to the nearest $^1/_4$ pound or 0.1 kilogram.
- Special scales and hospital beds with built-in scales are available for weighing people who are bedridden.

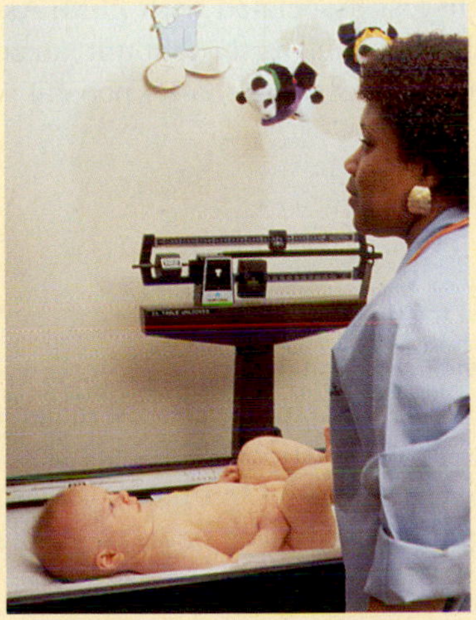

G. Degrazia/Custom Medical Stock Photo

Infants are weighed on scales that allow them to sit or lie down.

© 2010 Image Source/Jupiterimages Corporation

Beam balance scales allow accurate weight measurements for older children and adults.

TRY IT Measure your weight using a beam balance or electronic scale.

Head Circumference A measurement of head circumference can help to assess brain growth and malnutrition in children up to three years of age, although this measure is not necessarily reduced in a malnourished child. Head circumference values can also track brain development in premature and small-for-gestational-age infants. To measure head circumference, the assessor encircles the largest circumference measure of a child's head with a nonstretchable measuring tape: the tape is placed just above the eyebrows and ears and around the occipital prominence at the back of the head (see the photo). The measurement is read to the nearest 1/8th inch or 0.1 centimeter.

Head circumference measurements can help to assess brain growth.

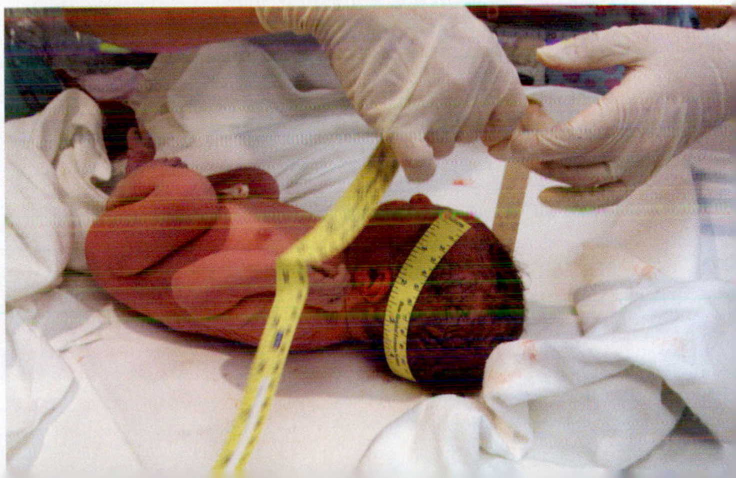

© Eric Fowke/Alamy

Circumferences of Waist and Limbs The waist circumference correlates with intra-abdominal fat and can help in assessing overnutrition. Circumferences of the mid-upper arm, mid-thigh, and mid-calf regions can help in evaluating the effects of illness, aging, and PEM on skeletal muscle tissue. For improved accuracy, circumference measurements are often used together with skinfold measurements to correct for the subcutaneous fat in limbs.

Anthropometric Assessment in Infants and Children To evaluate growth patterns, anthropometric data can be plotted on growth charts, such as those provided in Appendix E. ♦ The most commonly used growth charts compare height (or length) to age, weight to age, head circumference to age, weight to length, and BMI to age. Although individual growth patterns vary, a child's growth will generally stay at about the same percentile throughout childhood; a sharp drop in a previously steady growth pattern suggests malnutrition. Growth patterns below the 5th percentile may also be cause for concern, although genetic influences must be considered when interpreting low values. Growth charts with BMI-for-age percentiles can be used to assess risk of underweight and overweight in children over two years of age: the 5th and 85th percentiles are used as cutoffs to identify children who may be malnourished or overweight, respectively.[11] Chapter 15 provides additional information about growth during infancy and childhood.

♦ The Centers for Disease Control and Prevention provides complete sets of growth charts at its website: **http://cdc.gov/growthcharts/**

Anthropometric Assessment in Adults To evaluate the nutritional risks associated with illness, clinicians monitor both the total reduction in weight and the rate of weight loss over time.[12] Weight changes must be evaluated carefully: although unintentional weight *loss* can indicate malnutrition, weight *gain* may result from fluid retention rather than an increase in muscle mass or overnutrition. Furthermore, fluid retention can mask the weight loss associated with PEM. As mentioned, the rate of weight loss must be considered as well as the amount: an adult is at risk of PEM if there is an involuntary weight loss of more than 5 percent within

HOW TO Estimate and Evaluate %UBW and %IBW

%UBW: To estimate %UBW, compare an individual's current weight with the weight that the person generally maintains:

$$\%UBW = \frac{\text{current weight}}{\text{usual weight}} \times 100$$

For example, if a man loses 32 pounds during illness and his usual weight is 180 pounds, his current weight would be 148 pounds. These values can be incorporated into the previous equation:

$$\%UBW = \frac{148}{180} \times 100 = 82.2\%$$

The man in this example weighs 82.2 percent of his usual weight. A look at Table 17-8 shows that a person who is at 82 percent of UBW may be moderately malnourished.

%IBW: To estimate %IBW, compare an individual's current weight with a reasonable (ideal) weight from a BMI table or other appropriate reference:

$$\%IBW = \frac{\text{current weight}}{\text{ideal weight}} \times 100$$

For example, suppose you wish to calculate the %IBW for a woman who is 5 feet 8 inches tall and weighs 116 pounds. The midpoint of the healthy BMI range is approximately 22, so using a BMI table (as shown on the inside back cover of this book), you estimate that a reasonable weight for this woman would be about 144 pounds:

$$\%IBW = \frac{116}{144} \times 100 = 80.6\%$$

The woman in this example weighs about 80.6 percent of her ideal body weight. A look at Table 17-8 suggests that, at 80.6 percent of IBW, she may be mildly malnourished. Keep in mind that the calculation of "ideal body weight" is somewhat arbitrary, because the BMI table and various other references provide a range of weights for individuals of a given height.

TRY IT Calculate your current %IBW.

TABLE 17-8 Body Weight and Nutritional Risk

%UBW	%IBW	Nutritional Risk
85–95	80–90	Risk of mild malnutrition
75–84	70–79	Risk of moderate malnutrition
<75	<70	Risk of severe malnutrition

one month or more than 10 percent within six months.[13] ♦ Note that certain types of medications can also contribute to weight changes (see Chapter 19).

♦ Percent weight loss = $\frac{\text{usual weight} - \text{current weight}}{\text{usual weight}} \times 100$

Weight data are often expressed as a percentage of usual body weight (%UBW) or ideal body weight (%IBW). The %UBW is more effective than %IBW for interpreting weight changes that occur in underweight, overweight, or obese individuals. In overweight persons, the %IBW may fail to identify significant weight loss. Conversely, in underweight individuals, the %IBW can overstate the degree of weight loss due to illness. The "How To" explains how to estimate %UBW and %IBW, and Table 17-8 shows how to interpret these values (Reminder: Chapter 8 provides guidelines for determining a healthy body weight; see pp. 252–253).

Some illnesses discussed in later chapters are associated with losses in muscle tissue that resist nutrition intervention. In older adults, losses in both lean tissue and height are common even though body weights may remain stable. Including measures such as skinfold measurements and limb circumferences can help the clinician evaluate changes in body composition that need to be addressed in the treatment plan.

Biochemical Data

Biochemical data provide information about protein-energy nutrition, vitamin and mineral status, fluid and electrolyte balance, and organ function. Most tests are based on analyses of blood or urine samples, which contain proteins, nutrients, and metabolites that reflect nutrition and health status. Table 17-9 lists and describes common blood tests ♦ that have nutritional implications. Laboratory tests relevant to specific diseases will be discussed in the chapters that follow.

♦ Blood test results are reported in terms of either *plasma* or *serum* levels. *Plasma* is the yellow fluid that remains after cells are removed and still contains clotting factors. *Serum* is the fluid remaining after both cells and clotting factors are removed.

Interpreting laboratory values can be challenging because a number of factors influence the test results. For example, serum protein values can be affected by fluid imbalances, ♦ pregnancy, infections, and some medications. Similarly, serum levels of vitamins and minerals are often poor indicators of nutrient deficiency because the values are affected by multiple variables; therefore, a variety of tests are generally needed to diagnose a nutrition problem. Taken together with other assessment data, however, laboratory test results help to present a clearer picture than is possible to obtain otherwise.

♦ Fluid retention can cause lab results that are deceptively low. Dehydration may cause lab results to be deceptively high.

Plasma Proteins Plasma protein levels can aid in the assessment of protein-energy status, but the levels may fluctuate for other reasons as well.[14] For example, plasma proteins are synthesized in the liver, so plasma levels of these proteins can reflect liver function. Metabolic stress can alter plasma protein levels because the liver responds by increasing its synthesis of some proteins and reducing the synthesis of others. Values may also be influenced by pregnancy, kidney function, zinc status, and some medications. Because plasma proteins are affected by so many factors, their values must be considered along with other data to evaluate nutrition status.

Albumin Albumin is the most abundant plasma protein and its levels are routinely monitored during illness. Although many medical conditions influence albumin, it is slow to reflect changes in nutrition status because of its large body pool and slow rate of degradation. ♦ In people with chronic PEM, albumin levels remain normal for long periods of time despite depletion of body proteins, and levels fall only after prolonged malnutrition. Similarly, albumin concentrations rise slowly when malnutrition is treated, so albumin is not a sensitive indicator of effective treatment.

♦ In blood tests, the term *half-life* defines the length of time that a substance remains in plasma. The albumin in plasma has a half-life of 14 to 20 days, meaning that half of the amount circulating in plasma is degraded in this time period.

TABLE 17-9 Routine Laboratory Tests with Nutritional Implications

This table presents a partial listing of some uses of commonly performed lab tests that have implications for nutritional problems.

Laboratory Test	Acceptable Range	Description
Hematology		
Red blood cell (RBC) count	Male: 4.3–5.7 million/μL Female: 3.8–5.1 million/μL	Number of RBC; aids anemia diagnosis.
Hemoglobin (Hb)	Male: 13.5–17.5 g/dL Female: 12.0–16.0 g/dL	Hemoglobin content of RBC; aids anemia diagnosis.
Hematocrit (Hct)	Male: 39–49% Female: 35–45%	Percentage RBC in total blood volume; aids anemia diagnosis.
Mean corpuscular volume (MCV)	80–100 fL	RBC size; helps to distinguish between microcytic and macrocytic anemia.
Mean corpuscular hemoglobin concentration (MCHC)	31–37% Hb/cell	Hb concentration within RBC; helps to distinguish iron-deficiency anemia.
White blood cell (WBC) count	4500–11,000 cells/μL	Number of WBC; general assessment of immunity.
Serum Proteins		
Total protein	6.4–8.3 g/dL	Protein levels are not specific to disease or highly sensitive; they can reflect body protein, illness or infections, changes in hydration or metabolism, pregnancy, or use of certain medications.
Albumin	3.4–4.8 g/dL	May reflect illness or PEM; slow to respond to improvement or worsening of disease.
Transferrin	200–400 mg/dL >60 yr: 180–380 mg/dL	May reflect illness, PEM, or iron deficiency; slightly more sensitive to health status changes than albumin.
Prealbumin (transthyretin)	10–40 mg/dL	May reflect illness or PEM; more responsive to health status changes than albumin or transferrin.
C-reactive protein	68–8200 ng/mL	Indicator of inflammation or disease.
Serum Enzymes		
Creatine kinase (CK)	Male: 38–174 U/L Female: 26–140 U/L	Different forms of CK are found in muscle, brain, and heart. High levels in blood may indicate heart attack, brain tissue damage, or skeletal muscle injury.
Lactate dehydrogenase (LDH)	208–378 U/L	LDH is found in many tissues. Specific types may be elevated after heart attack, lung damage, or liver disease.
Alkaline phosphatase	25–100 U/L	Found in many tissues; often measured to evaluate liver function.
Aspartate aminotransferase (AST, formerly SGOT)	10–30 U/L	Usually monitored to assess liver damage; elevated in most liver diseases. Levels are somewhat increased after muscle injury.
Alanine aminotransferase (ALT, formerly SGPT)	Male: 10–40 U/L Female: 7–35 U/L	Usually monitored to assess liver damage; elevated in most liver diseases. Levels are somewhat increased after muscle injury.
Serum Electrolytes		
Sodium	136–146 mEq/L	Helps to evaluate hydration status or neuromuscular, kidney, and adrenal functions.
Potassium	3.5–5.1 mEq/L	Helps to evaluate acid-base balance and kidney function; can detect potassium imbalances.
Chloride	98–106 mEq/L	Helps to evaluate hydration status and detect acid-base and electrolyte imbalances.
Other		
Glucose (fasting)[a]	74–106 mg/dL >60 yr: 80–115 mg/dL	Detects risk of glucose intolerance, diabetes mellitus, and hypoglycemia; helps to monitor diabetes treatment.
Glycated hemoglobin (HbA_{1c})	5.0–7.5% of total Hb	Used to monitor long-term blood glucose control (approximately 1 to 3 months prior).
Blood urea nitrogen (BUN)	6–20 mg/dL	Primarily used to monitor kidney function; value is altered by liver failure, dehydration, or shock.
Uric acid	Male: 3.5–7.2 mg/dL Female: 2.6–6.0 mg/dL	Used for detecting gout or changes in kidney function; levels affected by age and diet; varies among different ethnic groups.
Creatinine (serum or plasma)	Male: 0.7–1.3 mg/dL Female: 0.6–1.1 mg/dL	Used to monitor renal function.

[a]Fasting glucose levels that repeatedly exceed 100 mg/dL suggest prediabetes.

NOTE: μL = microliter; dL = deciliter; fL = femtoliter; ng = nanogram; U/L = units per liter; mEq = milliequivalents.

SOURCE: L. Goldman and D. Ausiello, coeditors, *Cecil Medicine* (Philadelphia: Saunders, 2008).

Transferrin Transferrin is an iron-transport protein, and its concentrations respond to both iron status and PEM. Transferrin levels rise as iron status worsens and fall as iron status improves, so using transferrin values to evaluate protein-energy status is difficult if an iron deficiency is also present. Transferrin degrades more rapidly than albumin, ♦ but its levels change relatively slowly in response to nutrition therapy.

♦ Transferrin's half-life in plasma is approximately 8 to 10 days.

Prealbumin and Retinol-Binding Protein Levels of prealbumin (also called transthyretin) and retinol-binding protein decrease rapidly during PEM and respond quickly to improved protein intakes. ♦ Thus these proteins are more sensitive than albumin to changes in protein status. Like other plasma proteins, their usefulness in nutrition assessment is limited because they are affected by a number of different factors, including metabolic stress, zinc deficiency, and various medical conditions. Prealbumin and retinol-binding protein are more expensive to measure than albumin, so they are not routinely included during nutrition assessment.

♦ Half-lives of prealbumin and retinol-binding protein are 2 days and 12 hours, respectively.

Physical Examinations

As with other assessment methods, interpreting physical signs of malnutrition requires skill and clinical judgment. Most physical signs are nonspecific; they can reflect any of several nutrient deficiencies, as well as conditions unrelated to nutrition. For example, cracked lips may be caused by several B vitamin deficiencies but may also be caused by sunburn, windburn, or dehydration. Dietary and laboratory data are usually needed as additional evidence to confirm suspected nutrient deficiencies.

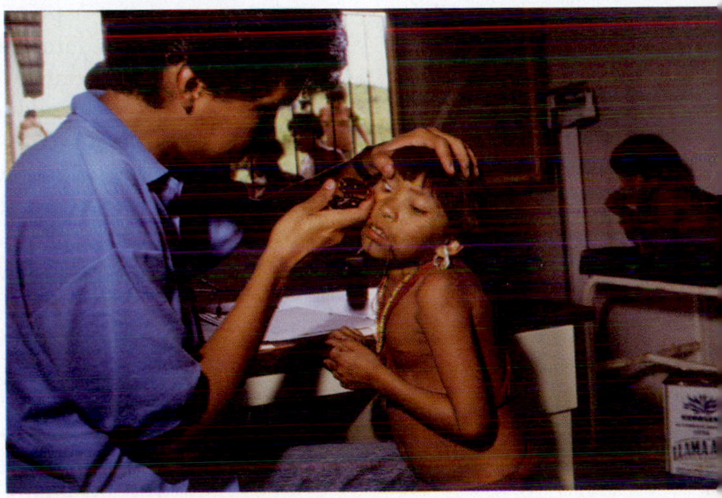

Mark Edwards/Peter Arnold, Inc./PhotoLibrary

Physical signs of malnutrition are often evident in parts of the body where cells are replaced at a rapid rate.

Clinical Signs of Malnutrition Signs of malnutrition tend to appear most often in parts of the body where cell replacement occurs at a rapid rate, such as the hair, skin, and digestive tract (including the mouth and tongue). Table 17-10 lists some clinical signs of nutrient deficiencies. Many of the symptoms listed occur only in advanced stages of deficiency. The summary tables in Chapters 10 through 13 provide additional examples of the physical signs of nutrient imbalances.

Hydration State As mentioned, fluid imbalances may accompany some illnesses and can also result from the use of certain medications.

TABLE 17-10 Clinical Signs of Nutrient Deficiencies

Body System	Acceptable Appearance	Signs of Malnutrition	Other Possible Causes
Hair	Shiny, firm in scalp	Dull, brittle, dry, loose; falls out (PEM); corkscrew hair (vitamin C)	Excessive hair bleaching, hair loss from aging, chemotherapy, or radiation therapy
Eyes	Bright; clear; shiny; pink, moist membranes; adjust easily to light	Pale membranes (iron); spots, dryness, night blindness (vitamin A); redness at corners of eyes (B vitamins)	Anemia that is unrelated to nutrition; eye disorders; allergies
Lips	Smooth	Dry, cracked, or with sores in the corners of the lips (B vitamins)	Sunburn, windburn, excessive salivation from ill-fitting dentures or various disorders
Mouth and gums	Oral tissues without lesions, swelling, or bleeding; red tongue; normal sense of taste; teeth without caries; ability to chew and swallow	Bleeding gums (vitamin C); smooth or magenta tongue (B vitamins), poor taste sensation (zinc)	Medications, periodontal disease (poor oral hygiene)
Skin	Smooth, firm, good color	Poor wound healing (PEM, vitamin C, zinc); dry, rough, lack of fat under skin (essential fatty acids, PEM, vitamin A, B vitamins); bruising or bleeding under skin (vitamins C and K); pale (iron)	Poor skin care, diabetes mellitus, aging, medications
Nails	Smooth, firm, pink	Ridged (PEM); spoon shaped, pale (iron)	—
Other	—	Dementia, peripheral neuropathy (B vitamins); swollen glands at front of neck (PEM, iodine); bowed legs (vitamin D)	Disorders of aging (dementia), diabetes mellitus (peripheral neuropathy)

Therefore, recognizing the patient's hydration state is necessary for the correct interpretation of blood tests and the body weight measurement.

Fluid retention (also called *edema*) can accompany malnutrition, infection, injury, or the use of certain medications. It can be caused by disorders of the heart and blood vessels, kidneys, liver, and lungs. Physical signs of fluid retention include weight gain, facial puffiness, swelling of limbs, abdominal distention, and tight-fitting shoes.

Dehydration can result from vomiting, diarrhea, sweating, fever, excessive urination, and skin injuries or burns (due to fluid loss through skin lesions). Risk of dehydration is especially high in older adults, who have a reduced thirst response and various other impairments in fluid regulation.[15] Symptoms include thirst, dry skin or mouth, reduced skin tension, dark yellow or amber urine, and low urine volume.

Functional Assessment Nutrient deficiencies sometimes impair normal physiological functions, so functional tests or procedures can be used to evaluate some aspects of malnutrition. For example, both PEM and zinc deficiency can depress immunity, which can be evaluated by testing the skin's response to antigens that cause redness and swelling when immune function is adequate. Muscle weakness due to **wasting**, or loss of muscle tissue, may be assessed by testing hand-grip strength. Exercise tolerance, which is reduced in heart and lung disorders, is sometimes evaluated using a treadmill or cycle ergometer. The Case Study can help you review the different components of a nutrition assessment.

wasting: the gradual atrophy (loss) of body tissues; associated with protein-energy malnutrition or chronic illness.

IN SUMMARY **Nutrition assessments include historical information, anthropometric and biochemical data, and physical examinations. Historical information includes the medical history, medication and supplement history, personal and social history, and food and nutrition history. Health care providers assess food intake using 24-hour recall interviews, food frequency questionnaires, food records, and direct observation. Anthropometric measurements help to evaluate growth patterns, overnutrition and undernutrition, and body composition. Biochemical analyses help in the assessment of nutrient imbalances but are influenced by various other medical problems. Physical examinations can help the assessor detect signs of nutrient deficiency, fluid imbalances, and functional deficits.**

CASE STUDY

Nutrition Screening and Assessment

Elise Walden is an 80-year-old retired businesswoman who has been a widow for 10 years. She uses a walker and has poorly fitting dentures. She was recently admitted to the hospital with pneumonia and also has congestive heart failure and diabetes. She routinely takes several medications to control blood glucose, hypertension, and heart function. In addition to these medications, the physician recently ordered antibiotics to treat the pneumonia. During an initial nutrition screening, Mrs. Walden stated that she had been eating very poorly over the past two weeks. She said that she usually weighs about 125 pounds; a fact that was documented in her medical chart from a previous visit. Although she felt she was losing weight, she didn't know how much weight she may have lost or when she started losing weight. Upon admission to the hospital, Mrs. Walden weighed 110 pounds and was 5 feet 3 inches tall. Her serum albumin level was 3.0 grams per deciliter. A physical exam revealed edema, and several other laboratory tests confirmed that she was retaining fluid. As a result of the nutrition screening, Mrs. Walden was referred to a registered dietitian for a nutrition assessment.

1. From the brief description provided, which items in Mrs. Walden's medical history, personal and social history, and food and nutrition history might alert the dietitian that this patient is at risk of malnutrition?
2. Identify a healthy body weight for Mrs. Walden, and calculate her %UBW and %IBW. What do the results reveal? What effect does fluid retention have on Mrs. Walden's weight?
3. How can fluid retention alter Mrs. Walden's serum protein levels? What physical symptoms may have suggested that she was retaining excess fluid?
4. What tools can be used to estimate Mrs. Walden's food intake? What medical, physical, and personal factors are likely to influence her diet?
5. Describe other types of assessment information the dietitian may need before developing a nutrition care plan.

Clinical Portfolio

1. Describe the potential nutritional implications of these findings from a patient's medical, personal, and social histories: age 78, lives alone, recently lost spouse, uses a walker, has no natural teeth or dentures, has a history of hypertension and diabetes, uses medications that cause frequent urination.
2. Calculate the %UBW and %IBW for a man who is 5 feet 11 inches tall with a current weight of 150 pounds and a usual body weight of 180 pounds. What additional information do you need to interpret the implications of his weight loss?
3. Nurses often shoulder much of the responsibility for collecting food intake data for kcalorie counts because they typically deliver food trays and snacks and later retrieve them. Why is it important to verify and record both what the patient receives (foods and amounts) and the foods that remain uneaten? When might patients be enlisted in the coallection of food intake data, and when might such a course be unwise?

Nutrition on the Net

For further study of topics covered in this chapter, log on to **www.cengagebrain.com** and search for ISBN 084006845X.

- Learn about careers in nutrition and dietetics at the website of the American Dietetic Association: **www.eatright.org**
- Find additional information about the Mini Nutritional Assessment: **www.mna-elderly.com**
- Obtain food composition data from the USDA Nutrient Data Laboratory: **www.nal.usda.gov/fnic/foodcomp/search/**
- Analyze your diet following the process described at this website: **www.mypyramid.gov**

References

1. D. C. Heimburger, Adulthood, in M. E. Shils and coeditors, *Modern Nutrition in Health and Disease* (Baltimore: Lippincott Williams & Wilkins, 2006), pp. 830–842.
2. P. Charney and M. Marian, Nutrition screening and nutrition assessment, in P. Charney and A. M. Malone, eds., *ADA Pocket Guide to Nutrition Assessment* (Chicago: American Dietetic Association, 2009), pp. 1–19.
3. Writing Group of the Nutrition Care Process/Standardized Language Committee, Nutrition care process and model part I: The 2008 update, *Journal of the American Dietetic Association* 108 (2008): 1113–1117; K. Lacey and E. Pritchett, Nutrition care process and model: ADA adopts road map to quality care and outcomes management, *Journal of the American Dietetic Association* 103 (2003): 1061–1072.
4. American Dietetic Association, *Nutrition Diagnosis and Intervention: Standardized Language for the Nutrition Care Process* (Chicago: American Dietetic Association, 2007).
5. American Dietetic Association, 2007.
6. American Dietetic Association, 2007.
7. F. E. Thompson and A. F. Subar, Dietary assessment methodology, in A. M. Coulston and C. J. Boushey, eds., *Nutrition in the Prevention and Treatment of Disease* (Burlington, MA: Elsevier Academic Press, 2008), pp. 3–39.
8. R. S. Gibson, *Principles of Nutritional Assessment* (New York: Oxford University Press, 2005), pp. 809–826; L. C. Tapsell, V. Brenninger, and J. Barnard, Applying conversation analysis to foster accurate reporting in the diet history interview, *Journal of the American Dietetic Association* 100 (2000): 818–824.
9. E. Saltzman and M. A. McCrory, Physical assessment of nutritional status, in A. M. Coulston and C. J. Boushey, eds., *Nutrition in the Prevention and Treatment of Disease* (Burlington, MA: Elsevier Academic Press, 2008), pp. 57–73.
10. B. Shah, K. Sucher, and C. B. Hollenbeck, Comparison of ideal body weight equations and published height-weight tables with body mass index tables for healthy adults in the United States, *Nutrition in Clinical Practice* 21 (2006): 312–319.
11. S. E. Barlow and the Expert Committee, Expert Committee recommendations regarding the prevention, assessment, and treatment of child and adolescent overweight and obesity: Summary Report, *Pediatrics* 120 (2007): S164–S192.
12. S. B. Heymsfield and R. N. Baumgartner, Body composition and anthropometry, in M. E. Shils and coeditors, *Modern Nutrition in Health and Disease* (Baltimore: Lippincott Williams & Wilkins, 2006), pp. 751–770.
13. American Dietetic Association, *Nutrition Care Manual* (Chicago: American Dietetic Association, 2010).
14. C. W. Thompson, Laboratory assessment, in P. Charney and A. M. Malone, eds., *ADA Pocket Guide to Nutrition Assessment* (Chicago: American Dietetic Association, 2009), pp. 62–153.
15. Standing Committee on the Scientific Evaluation of Dietary Reference Intakes, Food and Nutrition Board, Institute of Medicine, *Dietary Reference Intakes for Water, Potassium, Sodium, Chloride, and Sulfate* (Washington, DC: National Academies Press, 2005), pp. 147–150.

HIGHLIGHT 17

Nutrition and Immunity

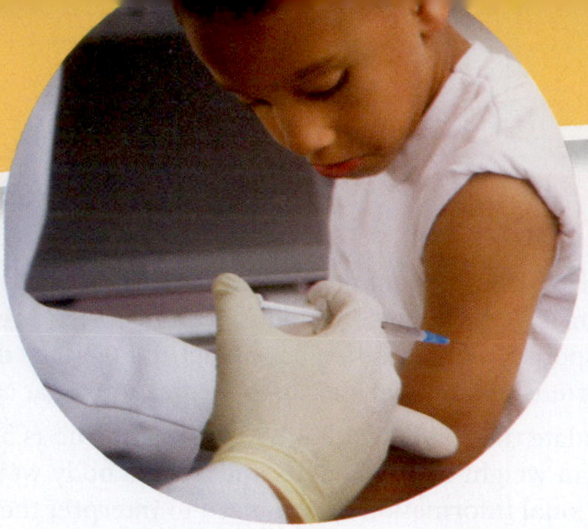

Matt Meadows/Peter Arnold, Inc./PhotoLibrary

The **immune system** protects the body by fighting infectious agents and eliminating abnormal or "worn-out" cells. Its elaborate network of interacting cells and molecules works to block invading organisms from entering the body and destroys those that do gain entry. Substances that elicit an immune response are called antigens; common examples include foreign proteins produced by bacteria, viruses, parasites, or fungi. Because the immune system can usually distinguish between the body's cells and proteins and those of invading organisms, the body's own tissues are protected.

This highlight introduces the immune system and its relationships to malnutrition and illness; the glossary defines relevant terms. Later chapters examine some of the relationships between specific illnesses and immune processes. Some diseases result from inadequate immune responses, as when infections spread, causing sepsis (Chapter 22), or when malignant cells develop into tumors (Chapter 29). Other conditions, such as inflammatory bowel diseases (Chapter 24) and atherosclerosis (Chapter 27), result from **inflammation** (Chapter 22). Most of the time, however, the immune system's carefully orchestrated actions are quietly working to preserve health.

Tissues of the Immune System

The immune system resides in no single organ, but depends on the physical and chemical interactions of a loosely organized network of cells and tissues scattered throughout the body.[1] The tissues and organs involved in immunity are collectively known as the lymphatic system (see Figure H17-1). **Lymphoid tissues** include the thymus gland and bone marrow (where **lymphocytes** are made), and the spleen, tonsils, adenoids, and lymph nodes (where foreign materials and debris are filtered out and discarded). Additional lymphoid tissue is dispersed in various loca-

GLOSSARY

acute-phase proteins: plasma proteins released from the liver at the onset of acute infection. An example is **C-reactive protein**, which is considered one of the main indicators of severe infection and has antimicrobial effects.

adaptive immunity: immunity that is specific for particular antigens; it adapts to antigens in an individual's environment and is characterized by "memory" for particular antigens. Also called **acquired immunity**.

allergen: any substance that triggers an inappropriate immune response.

allergy: an excessive and inappropriate immune reaction to a harmless substance.

autoimmune diseases: diseases characterized by an attack of immune defenses on the body's own cells.

B cell: a lymphocyte that produces antibodies.

cell-mediated immunity: immunity conferred by T cells and macrophages.

complement: a group of plasma proteins that assist the activities of antibodies.

cytokines (SIGH-toe-kines): signaling proteins produced by the body's cells; those produced by white blood cells regulate immune cell development and immune responses.

humoral immunity: immunity conferred by B cells, which produce and release antibodies into body fluids.

- **humor** = fluid

hypersensitivity: immune responses that are excessive or inappropriate. One type of hypersensitivity is *allergy*.

immune system: the body's defense system against foreign substances.

immunoglobulins (IM-you-no-GLOB-you-linz): large globular proteins produced by B cells that function as antibodies.

inflammation: a nonspecific response to injury or infection; a type of innate immune response.

innate immunity: immunity that is present at birth, unchanging throughout life, and nonspecific for particular antigens; also called **natural immunity**.

leukocytes: blood cells that function in immunity; also called **white blood cells**.

lymph (limf): the body fluid carried in lymphatic vessels; lymph is collected from the extracellular fluids of body tissues and ultimately transported to the bloodstream.

lymphatic vessels: vessels through which lymph travels.

lymphocytes (LIM-foe-sites): white blood cells that recognize specific antigens and therefore function in adaptive immunity; include *T cells* and *B cells*.

lymphoid tissues: tissues that contain lymphocytes.

lysozyme (LYE-so-zyme): an enzyme with antibacterial properties; found in immune cells and body secretions such as tears, saliva, and sweat.

macrophages (MAK-roe-fay-jez): monocytes that have left circulation and settled in a tissue, where they serve as scavengers and activate the immune response.

monocytes (MON-oh-sites): cells released from the bone marrow that move into tissues and mature into macrophages.

natural killer cells: lymphocytes that confer nonspecific immunity by destroying a wide array of viruses and tumor cells.

neutrophils (NEW-tro-fills): the most common type of white blood cell. Neutrophils destroy antigens by phagocytosis.

phagocytes (FAG-oh-sites): white blood cells (primarily neutrophils and macrophages) that have the ability to engulf and destroy antigens.

- **phagein** = to eat

phagocytosis (FAG-oh-sigh-TOE-sis): the process by which phagocytes engulf and destroy antigens.

T cell: a lymphocyte that attacks antigens; functions in cell-mediated immunity.

tions throughout the body, especially within the body's mucosal linings, where antigens are most likely to enter the body—in the gastrointestinal tract, the respiratory tract, and the genitourinary tract.

FIGURE H17-1 The Lymphatic System

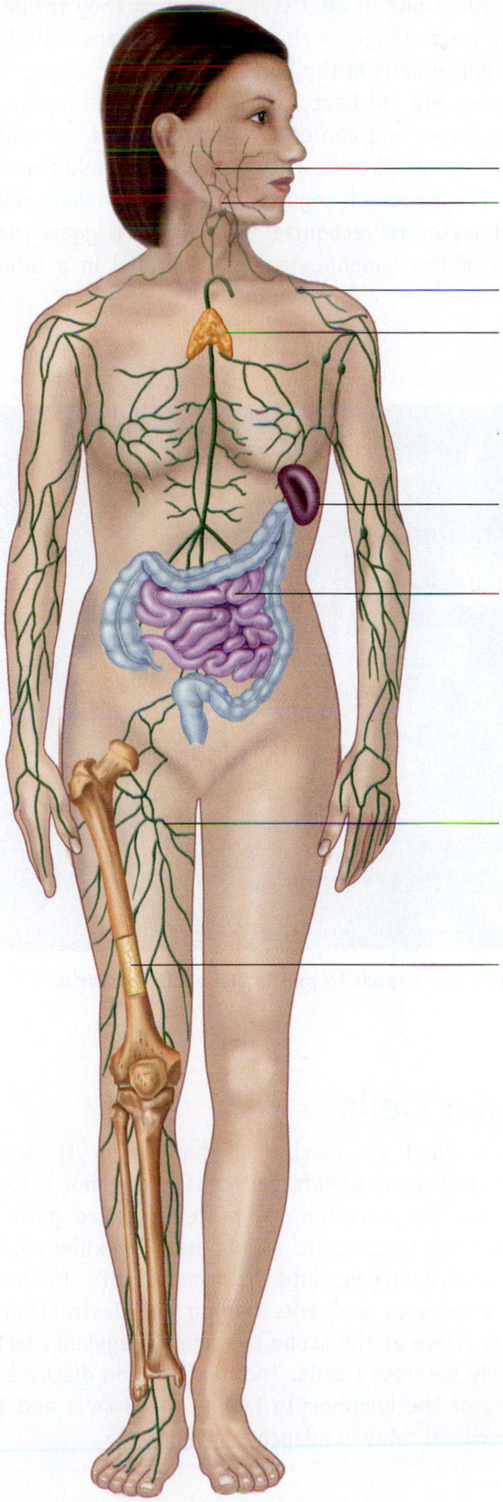

TABLE H17-1 Cells of the Immune System

White Blood Cells	Cell Type	Function
Lymphocytes	T cells	Activate macrophages Assist B cells Destroy virally infected cells
	B cells	Produce and secrete antibodies
	Natural killer cells	Destroy virally infected cells
Phagocytes	Monocytes/ macrophages[a]	Present antigen fragments to T cells Engulf pathogens and cellular debris
	Neutrophils	Engulf pathogens and cellular debris
	Eosinophils	Release proteins that damage parasites Suppress inflammatory reactions
Accessory Cells	**Cell Type**	**Function**
Inflammatory mediators	Basophils	Release mediators that regulate inflammation
	Mast cells	Release mediators that regulate inflammation
	Platelets	Have primary role in blood clotting Release mediators that regulate inflammation

[a]Monocytes circulate in blood and become macrophages after they enter tissues.

The cells active in immunity are the **leukocytes** (commonly known as **white blood cells**) and several types of accessory cells, as described in Table H17-1 and discussed in the following pages. These cells act by releasing chemicals such as enzymes, prostaglandins, and histamine, as well as proteins called **cytokines** that bind to receptors on target cells. White blood cells travel between the tissues and blood in **lymph**, a body fluid carried by the **lymphatic vessels**. Lymph is collected from the extracellular fluids that bathe tissues and is eventually transported to the bloodstream.

Examples of Innate Immunity

The immune protection present at birth is called **innate**, or **natural, immunity**. Innate immunity is nonspecific—it deters and destroys a wide range of pathogens. Nonspecific defenses include physical barriers to invading organisms, actions of defensive proteins, and activities of phagocytes and natural killer cells.

Physical Barriers to Infection

The body's first line of defense—the skin and mucous membranes—prevents the entry of infectious agents, which might otherwise gain easy access to tissues and blood. Skin not only provides an impenetrable physical barrier but also contains its own lymphoid tissue and a variety of immune cells interspersed in its outer layers. Mucous membranes lining the gastrointestinal, respiratory, and genitourinary tracts also act as barriers to infection: the mucous layers of these tissues trap microorganisms and prevent them from attaching to tissue surfaces.[2]

Microbes that arrive in the stomach face possible destruction from acidic gastric juices and enzymes. Those that survive enter the small intestine, where digestive secretions and specialized cells, antibodies, and lymphoid tissue protect against infection. The large intestine also contains defensive cells and antibodies, as well as stable bacterial populations that help to maintain mucosal tissue and create a hostile environment for invasive bacteria.[3]

Defensive Proteins

Proteins contribute to nonspecific immune defenses by serving as enzymes or signaling molecules. The liver releases **acute-phase proteins** in response to trauma, infection, or inflammation. Some acute-phase proteins, such as **C-reactive protein**, have antimicrobial activities that destroy some types of bacteria. C-reactive protein is considered a "marker" of acute inflammation and becomes elevated only when the body is fighting disease. Other acute-phase proteins include **complement**, a group of about 25 plasma proteins, so named because the proteins "complement" the activities of antibodies. When an antibody interacts with an antigen, a complex is formed that starts a series of reactions between the complement proteins. These actions may render microbes more susceptible to phagocytosis (described later), puncture a target cell's membrane, or help rid the body of antigen-antibody complexes. Another protein called **lysozyme** attacks bacteria by breaking down carbohydrates on bacterial cell walls, causing the bacteria to burst.

Phagocytes

Upon entering the body, pathogens may encounter **phagocytes**, the scavenger cells of the immune system. Phagocytes engulf and digest bacteria, cellular debris (from damaged cells), and foreign particles in a process called **phagocytosis**. Phagocytes are attracted to their targets by the presence of common microbial products, complement fragments, or chemical signals produced by cells. They pull in their prey by extending pseudopods ("false feet") and then douse it with a mix of potent chemicals that include hydrolytic enzymes, lysozyme, and free radicals.

The two main types of phagocytes are neutrophils and macrophages. **Neutrophils** are the predominant leukocytes in the blood, making up about 50 to 65 percent of the total. They also have the shortest life spans, surviving only a day or two after they are released from bone marrow. Neutrophils migrate into tissues in response to injury or infection and accumulate in large numbers during the inflammatory process (discussed in Chapter 22). **Macrophages** are initially released from bone marrow as **monocytes**; after about a day in circulation, a monocyte migrates into one particular tissue, where it develops into a macrophage and survives for several months or longer. Each tissue has its own resident macrophages, and although their names may vary, they have similar functions in all tissues in which they reside. Examples of tissue macrophages include the Langerhans cells in the skin and the Kupffer cells in the liver.

Macrophages move and kill bacteria more slowly than neutrophils, but they are larger and can engulf larger targets, such as the body's dead and damaged cells. They also have the additional ability to display fragments of engulfed antigens on their cell surfaces for lymphocytes to recognize. This action triggers the immune responses of the lymphocytes, as described in a later section.

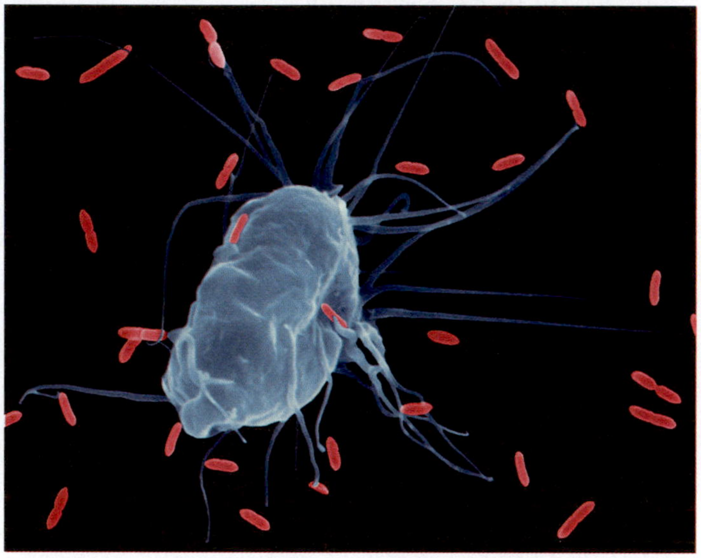

A macrophage extends pseudopods to pull in and engulf bacteria.

Natural Killer Cells

Natural killer cells, which are members of the lymphocyte family, recognize and destroy virus-infected cells and tumor cells. These killer cells produce pore-forming proteins (called perforins) that puncture their target cells' membranes. The killer cells then transfer destructive enzymes into the damaged cells, further destroying their structures and encouraging self-destruction. Finally, phagocytes arrive at the scene to remove fragments left behind by the newly destroyed cells. The next section discusses the other members of the lymphocyte family, the B cells and T cells, which have critical roles in adaptive immunity.

Examples of Adaptive Immunity

In **adaptive**, or **acquired, immunity**, immune cells and proteins recognize *specific* pathogens as being foreign. Each **B cell** or **T cell** generates antibodies or receptors that can recognize only one type of antigen. Once activated, a lymphocyte produces other cells just like itself so that the newly formed army can attack the invading antigens and combat the infection. The lymphocytes are able to recognize a huge and diverse number of foreign molecules. Some of the lymphocytes serve as "memory cells," which survive for many years, enabling the immune system to respond rapidly if the same infection recurs.

B Cells

The B cells confer **humoral immunity**, so named because the cells' secretions, not the cells themselves, mount the defense within bodily fluids. B cells respond to antigens by producing antibodies that travel in the blood or tissue fluids to the site of infection. Antibodies, also known as **immunoglobulins**, are literally large globular proteins that provide immune protection. Each B cell expresses thousands of identical antibodies on its surface. Once an antigen binds, the B cell multiplies. Its daughter cells produce large numbers of the same antibody and secrete them into the surrounding fluids. The free antibodies then attach to the surfaces of antigens to neutralize them or make them an easy target for attack by phagocytes. The antibodies can also bind to viral proteins to prevent viruses from entering cells.

T Cells

T cells participate in **cell-mediated immunity**, so named because the cells themselves direct an immune response. A T cell has thousands of identical receptors on its cell surface (called T-cell receptors) that can recognize only one type of antigen. The antigens are displayed on the surfaces of antigen-presenting cells, specialized cells designed for this task (such as macrophages and B cells). After a *helper T cell* binds to an antigen fragment on an antigen-presenting cell, it recruits a *cytotoxic T cell* to the region to attack and destroy the local antigens. The actions of cytotoxic T cells are similar to those of natural killer cells: they perforate cell membranes and deliver powerful chemicals that eventually lead to a cell's destruction. Helper T cells can also activate B cells to produce antibodies and can activate macrophages to destroy the pathogens they have engulfed.

Undesirable Effects of Immunity

The body's immune function can sometimes create problems. Exaggerated or inappropriate immune reactions, referred to as **hypersensitivity**, can lead to discomfort or illness. **Allergy** is an example of an exaggerated response to an **allergen**, a harmless protein that may be eaten or inhaled. (Food allergy, introduced in Chapter 16, is discussed further in Highlight 29.) As another example, the immune complexes formed from antigens and antibodies can cause damage to tissues if not readily cleared by phagocytes. **Autoimmune diseases**, including such familiar diseases as type 1 diabetes mellitus and pernicious anemia, develop when immune responses are mounted against the body's own cells. Although the effects of the immune system are lifesaving when directed at harmful pathogens, they can be life threatening when turned against the body.

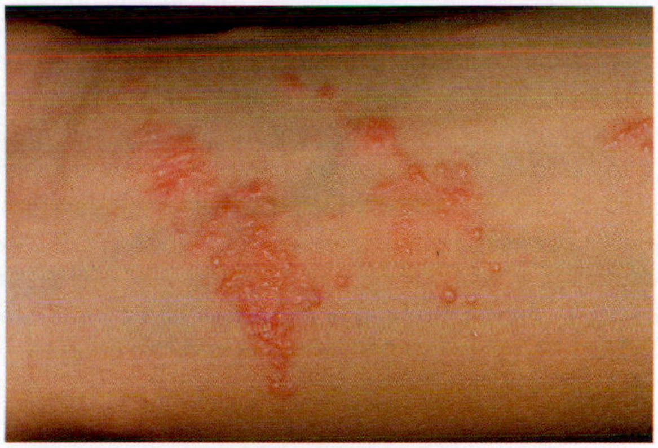

The rash that appears after contact with poison oak is an example of skin hypersensitivity.

Malnutrition and Immunity

Malnutrition affects all aspects of immunity, including both innate and adaptive immune defenses. For example, both PEM and vitamin A deficiency can result in damage to skin and mucous membranes, allowing microorganisms easier entry into the body. Deficiencies of protein and various micronutrients can affect the synthesis of hydrolytic enzymes, complement, antibodies, and other proteins important for immune function. Cell-mediated immunity is impaired in numerous ways by both PEM and zinc deficiencies.[4]

Because PEM is usually associated with multiple micronutrient deficiencies, it has been difficult for researchers to separate out the influences of individual nutrients. Nevertheless, zinc, iron, and vitamin A deficiencies are among the predominant micronutrient deficiencies worldwide, and a large body of research has demonstrated that each has a strong, independent influence on immunity. Notably, supplementation of various micronutrients (especially zinc and vitamin A) in malnourished populations has been found to reduce the incidence and severity of illness.[5]

Malnutrition and Infection

Because malnutrition impairs immune function in numerous ways, it is frequently associated with an increased risk of infection.

In addition, an infection itself can worsen malnutrition (as described in the following section). The result is a downward spiral in immunity and overall health. Malnourished populations have higher-than-normal incidences of infectious diseases such as measles, malaria, acute respiratory infections, diarrheal diseases, and tuberculosis.[6] These diseases are major causes of morbidity and mortality in developing countries.

Infection and Nutrition Status

As mentioned, the effects of infection can be detrimental to nutrition status.[7] Anorexia often develops and is worse when an infection is severe, resulting in weight loss, negative nitrogen balance, and delayed growth and healing. Intestinal infections can cause nutrient malabsorption, atrophy of intestinal tissue, substantial blood loss, and diarrhea. Furthermore, infections generally stimulate metabolic processes, raising metabolic rate and nutrient needs as well. Chapter 22 delves further into the consequences of severe infection and discusses the nutrient needs of individuals who suffer from these conditions.

References

1. P. C. Calder, Immunological parameters: What do they mean? *Journal of Nutrition* 137 (2007): 773S–780S.
2. P. Winkler and coauthors, Molecular and cellular basis of microflora-host interactions, *Journal of Nutrition* 137 (2007): 756S–772S.
3. Winkler and coauthors, 2007.
4. G. Fernandes, C. A. Jolly, and R. A. Lawrence, Nutrition and the immune system, in M. E. Shils and coeditors, *Modern Nutrition in Health and Disease* (Baltimore: Lippincott Williams & Wilkins, 2006), pp. 670–684; R. D. Semba, Nutrition and infection, in M. E. Shils and coeditors, *Modern Nutrition in Health and Disease* (Baltimore: Lippincott Williams & Wilkins, 2006), pp. 1401–1413.
5. N. S. Scrimshaw, Historical concepts of interactions, synergism and antagonism between nutrition and infection, *Journal of Nutrition* 133 (2003): 316S–321S; K. H. Brown, Diarrhea and malnutrition, *Journal of Nutrition* 133 (2003): 328S–332S.
6. Semba, 2006.
7. U. E. Schaible and S. H. E. Kaufmann, Malnutrition and infection: Complex mechanisms and global impacts, *PloS Medicine* 4 (2007) e115, available at doi: 10.1371/journal.pmed.0040115; G. T. Keusch, The history of nutrition: Malnutrition, infection and immunity, *Journal of Nutrition* 133 (2003): 336S–340S; Scrimshaw, 2003.

8

CHAPTER

18

Nutrition in the Clinical Setting

When working with patients, remember to establish a caring environment. Use familiar language, maintain eye contact, and be a good listener. Showing your interest can go a long way toward winning a patient's trust. A patient may greet even an ideal dietary plan with resentment and bitterness, for it may restrict favorite foods and make it more difficult to forget about an illness. When their interactions with health practitioners are positive and encouraging, individuals are more likely to make the dietary changes that benefit their health.

Throughout this chapter, the CourseMate icon indicates an opportunity for online self-study, linking you to activities to increase your understanding of chapter concepts. **www.cengagebrain.com** (search for ISBN 084006845X)

CHAPTER OUTLINE

Nutrition Intervention

Chapter 17 discusses the interactions between illness and nutrition status and describes the process of nutrition assessment. As that chapter explains, the results of the nutrition assessment allow dietitians to diagnose both potential and actual nutrition problems. This chapter describes how dietitians and other health care professionals address nutrition problems and provide nutrition care. Ensuring that dietary needs are met is a key part of this process, so the chapter includes methods for estimating energy intakes, descriptions of common dietary modifications, and a review of the procedures and challenges involved in foodservice delivery.

Implementing Nutrition Care

After formulating nutrition diagnoses, the dietitian determines the appropriate nutrition interventions. Table 18-1 shows how nutrition interventions are categorized.[1] Most nutrition interventions include a **nutrition prescription**, which provides specific dietary recommendations regarding food, nutrient, or energy intake or feeding method. Many interventions include nutrition education and counseling,

nutrition prescription: specific dietary recommendations related to food, nutrient, or energy intake or feeding method; also called the *diet order.*

TABLE 18-1 Examples of Nutrition Interventions

Intervention	Examples
Food and/or nutrient delivery	Providing appropriate meals, snacks, and dietary supplements Providing specialized nutrition support (tube and intravenous feedings) Determining the need for feeding assistance or adjustment in feeding environment Managing nutrition-related medication problems
Nutrition education	Providing basic nutrition-related instruction Providing in-depth training to increase dietary knowledge or skills
Nutrition counseling	Helping the individual set priorities and establish goals Motivating the individual to change behaviors Solving problems that interfere with the nutrition care plan
Coordination of nutrition care	Providing referrals or consulting other health professionals or agencies that can assist with treatment Organizing treatments that involve other health professionals or health care facilities Arranging transfer of nutrition care to another professional or location

SOURCE: American Dietetic Association, *Nutrition Diagnosis and Intervention: Standardized Language for the Nutrition Care Process* (Chicago: American Dietetic Association, 2007).

TABLE 18-2 Elements of Nutrition Interventions

Planning Nutrition Care
• Prioritizing nutrition diagnoses • Consulting dietetics practice guidelines • Reviewing the policies of the health care facility • Determining specific dietary recommendations • Conferring with the patient or caregivers • Establishing goals and expected outcomes
Implementing the Nutrition Care Plan
• Documenting the nutrition care plan in the medical record • Discussing the nutrition care plan with the patient or caregivers • Individualizing treatment as warranted • Continuing data collection and documentation • Revising the nutrition care plan as warranted

SOURCE: American Dietetic Association, *Nutrition Diagnosis and Intervention: Standardized Language for the Nutrition Care Process* (Chicago: American Dietetic Association, 2007).

which provide the knowledge, skills, and motivation that enable the patient to make necessary dietary and lifestyle changes. Some nutrition interventions require coordination with a number of other health professionals or facilities.

A nutrition intervention always includes two interrelated components: the planning process and the plan's implementation.[2] As Table 18-2 shows, the planning phase includes prioritizing the nutrition problems that were identified, determining their proper treatments, and setting goals. Implementing the plan involves communication with the patient, caregiver, and colleagues; carrying out the necessary treatments; and adjusting the plan when necessary.

Approaches to Nutrition Care

A nutrition care plan often involves significant dietary modifications. To ensure better compliance, the plan must be compatible with the desires and abilities of the person it is designed to help. The challenge is greater if dietary adjustments are required for extended periods.

Long-Term Dietary Intervention

When long-term changes are necessary, a care plan must take into account a person's current food habits, lifestyle, and degree of motivation. Behavior change is a process that occurs in stages; therefore, more than one consultation is usually necessary. The following approaches may be helpful in implementing long-term dietary changes:[3]

- *Determine the individual's readiness for change.* Some people have little desire to change their dietary behaviors, and even those who are willing may not be fully prepared to take the necessary steps. The health practitioner needs to consider a patient's readiness to adopt new dietary behaviors before attempting to implement an ambitious care plan.
  - *Emphasize what to eat, rather than what not to eat.* Emphasizing foods to include in the diet, rather than those to restrict, can make dietary changes more appealing. For example, encouraging additional fruits and vegetables is a more attractive message than telling the patient to restrict butter, cream sauces, and ice cream.
  - *Suggest only one or two changes at a time.* People are more likely to adopt a dietary plan that does not deviate too much from their usual diet. If they succeed in adopting one or two changes, they are more likely to stick to the plan and be open to additional suggestions. Stricter plans may yield quicker results but are useful only for highly motivated people.

Dietary counseling requires sensitivity to cultural orientation, educational background, and motivation for change.

© ACE STOCK LIMITED / Alamy

Nutrition Education

Nutrition education allows patients to learn about the dietary factors that affect their particular medical condition. Ideally, this knowledge can motivate them to change their diet and lifestyle in order to improve their health status.

A nutrition education program should be tailored to a person's age, level of literacy, and cultural background. Learning style must also be considered: some people learn best by discussion supplemented with written materials, whereas others prefer visual examples, such as food models and measuring devices.[4] Information can be provided in one-on-one sessions or group discussions. The meeting should include an assessment of the person's understanding of the material and commitment to making changes. Follow-up sessions can reveal whether the person has successfully adopted a dietary plan. For example, a dietitian who counsels a woman who is lactose intolerant and hesitant to use milk products might proceed as follows:

- The dietitian provides sample menus of a nutritionally adequate diet that limits milk and milk products. Together, the dietitian and the woman design menus that consider her food preferences.
- The dietitian describes the types and amounts of milk products that would likely be tolerated without causing symptoms and explains how to gradually incorporate these foods into the diet.
- Using diet analysis software, the dietitian demonstrates how altering intakes of calcium-containing foods changes a meal's calcium content.
- The dietitian explains how to use the Daily Values on food labels to estimate the calcium content of packaged foods.
- The dietitian provides information about the advantages and disadvantages of different calcium supplements.
- The dietitian assesses the woman's understanding by having her identify non-milk products that are high in calcium.

Ideally, the dietitian would be able to monitor the woman's progress in a subsequent counseling session.

Follow-up Care For optimal success, dietitians should monitor the patient's progress and periodically evaluate the effectiveness of the nutrition care plan. Doing so usually involves comparing relevant outcome measures (such as the results of blood tests) with initial values and meeting with the patient to learn whether the plan has been satisfactory from the patient's point of view. Such follow-up efforts can reveal whether the care plan needs to be revised or updated, as is often the case when a person's situation changes. For example, after a pregnant woman delivers her baby, she may need instructions on how to feed her infant or how to modify her diet to support lactation (if she is breastfeeding) so that she can return to a healthy body weight. If a follow-up meeting with a dietitian is not possible, a dietetic technician or other qualified health practitioner should provide additional guidance and education.

Documenting Nutrition Care

Each step of the nutrition care process must be documented in the patient's medical record. The entries should be as succinct as possible so that they can be easily read and quickly understood by other members of the health care team. In addition, electronic (computerized) data systems, which have been widely adopted in the past decade, have standardized templates that require concise language. Before making entries in patients' medical records, health care professionals need to learn the particular charting methods preferred by their medical facility. Although a variety of charting styles are in use, the content is more relevant than the particular format used. The following sections describe some popular formats used for documenting nutrition care.[5]

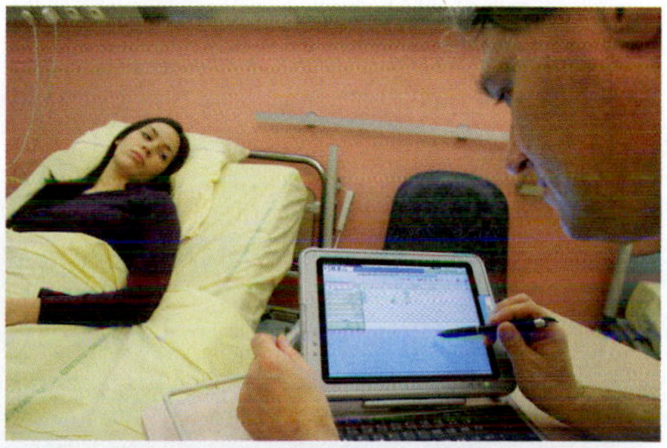

Many health care facilities maintain computerized medical records, which have standardized templates that require concise language.

ADIME Format The ADIME format closely reflects the steps of the nutrition care process. The letters represent the different steps: *A*ssessment, *D*iagnosis, *I*ntervention, and *M*onitoring and *E*valuation. Using this format, the nutrition care plan is recorded as follows:

- *Assessment*. The assessment section summarizes relevant assessment results, such as the medical problem, historical information, height, weight, BMI, laboratory test results, and relevant symptoms.
- *Diagnosis*. The diagnosis section lists and prioritizes the nutrition diagnoses.
- *Intervention*. The intervention section describes treatment goals and expected outcomes, specific interventions, and the patient's responses to nutrition care.
- *Monitoring and evaluation*. The monitoring and evaluation section records the patient's progress, changes in the patient's condition, and adjustments in the care plan.

SOAP Format The SOAP format is the oldest method used for documenting nutrition care and is still in popular use. The letters represent the types of information included in each section: *S*ubjective, *O*bjective, *A*ssessment, and the *P*lan for care.

- *Subjective* information is obtained in an interview with the patient or caregiver and includes the chief medical problem and relevant symptoms.
- *Objective* information includes nutrition screening or assessment data, such as the results of anthropometric and laboratory tests and the physical examination.
- The *Assessment* section contains a brief evaluation of the subjective and objective data and provides concise diagnoses of the nutrition problems.
- The *Plan* includes recommendations that can help solve the problem, including the nutrition prescription, plan for nutrition education and counseling, and referrals to other professionals or agencies.

Figure 18-1 shows an example of a SOAP note, although there are many possible variations.

PES Statement The PES statement, introduced in Chapter 17 (see p. 574), is the general structure used for formatting nutrition diagnoses and can be used in any formatting style. The PES statement is so named because it includes the *P*roblem, the *E*tiology or cause of the problem, and the *S*igns and symptoms that provide evidence for the problem. The SOAP note in Figure 18-1 includes two PES statements (review the nutrition diagnoses in the "Assessment" section).

IN SUMMARY **Nutrition interventions are designed to correct the nutrition problems associated with illness. An intervention should take into account a person's food practices, lifestyle, cultural orientation, educational background, and degree of motivation. Nutrition education must be individualized to accommodate a patient's needs and learning style. The care plan can be evaluated by reviewing relevant outcome measures of health status and determining the patient's understanding and acceptance of the intervention. Each step of nutrition care should be clearly documented in the medical record; the ADIME and SOAP formats are popular styles of documentation.**

Energy Intakes in Hospital Patients

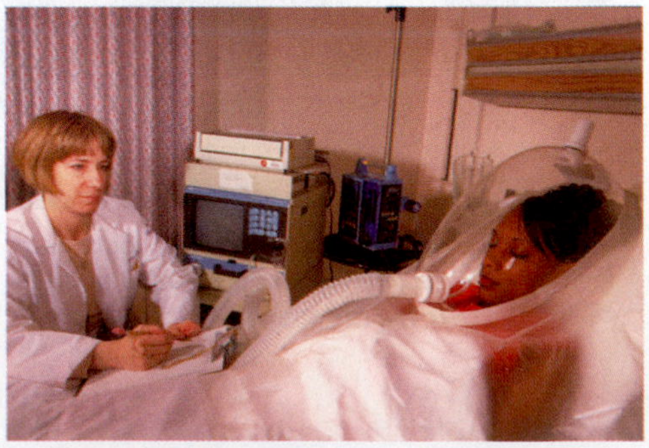
Richard T. Nowitz/Photo Researchers, Inc.

Indirect calorimetry is performed using equipment that analyzes the oxygen and carbon dioxide of inhaled and exhaled air.

To estimate energy intakes that are appropriate for hospital patients, clinicians typically measure or calculate the resting metabolic rate (RMR) and then adjust the RMR value with "stress factors" that account for medical problems and, in some cases, medical treatments. In ambulatory patients, a factor for activity level may also be applied. The standard clinical procedure for determining RMR is indirect calorimetry, which measures oxygen consumption and carbon dioxide production (thereby determining kcalories burned) during a period of rest.[6] The procedure is labor intensive, so clinicians more often use predictive equations that yield similar results. Table 18-3 lists examples of RMR equations in common use; the "How To" presents an example of this method.

FIGURE 18-1 Example of a SOAP Note

SOAP NOTE

Patient Name: James Steiner **Date:** Sept. 15, 2011

Age: 58 **Gender:** Male **Medical diagnosis:** Hypercholesterolemia

Subjective:

Mr. Steiner recently learned of his hypercholesterolemia; wants to try dietary/lifestyle changes to reduce need for the medication. Reports frequent snacking and little time for exercise. Willing to attempt weight loss.

Objective:

Total cholesterol: 288 mg/dL Height: 6'1"; Weight: 268 lb.
LDL-C: 214 mg/dL; HDL-C: 48 mg/dL BMI: 35.4
Triglycerides: 132 mg/dL Waist circumference: 45"

Assessment:

Abdominal obesity; analysis reveals intake of approximately 4200 kcal per day, about 1500 kcal above estimated needs; snack food choices are high in kcal and saturated fat.
Nutrition Diagnoses:
1. Obesity related to excess energy intake of 1500 kcal/day and physical inactivity as evidenced by BMI of 35.4
2. Undesirable food choices related to inadequate access to appropriate foods at work as evidenced by elevated body weight and LDL cholesterol

Plan:

Goal: 15 lb. weight loss over next 6 months.
Mr. Steiner to start 45-minute walking program, evenings.
Nutrition prescription: reduction of food intake to about 2400 kcal per day with about 30% kcal from fat, and 7% of kcal from saturated fat.
Initial education: appropriate food portions, low-kcal foods and snacks, food sources of saturated fat, pre-planning lunches at work.
Referral: Heart-healthy workshop on Sept. 22 (one week); Mr. Steiner to attend with wife.
Follow-up visit: Oct 15 (one month); Mr. Steiner to keep 3-day food record before visit; will identify appropriate food portions and between-meal snacks.

Form completed by: Genevieve Johnson, MPH, RD **Position:** Dietitian, Nutrition Services

In overweight and obese individuals who are not critically ill, the Mifflin–St. Jeor equation has been found to yield the most accurate results.[7] In other equations, adjusted body weights are sometimes used in place of actual body weights in an attempt to improve accuracy. For example, some research studies have suggested

TABLE 18-3 Selected Equations for Estimating Resting Metabolic Rate (RMR)

Harris–Benedict[a]

Women: RMR = 655.1 + [9.563 × weight (kg)] + [1.85 × height (cm)] − [4.676 × age (years)].
Men: RMR = 66.5 + [13.75 × weight (kg)] + [5.003 × height (cm)] − [6.755 × age (years)].

Mifflin–St. Jeor

Women: RMR = [9.99 × weight (kg)] + [6.25 × height (cm)] − [4.92 × age (years)] − 161
Men: RMR = [9.99 × weight (kg)] + [6.25 × height (cm)] − [4.92 × age (years)] + 5

[a]Although these equations are sometimes used for estimating basal metabolic rate (BMR), they were derived from data measured during resting conditions in most cases.

HOW TO Estimate Appropriate Energy Intakes for Hospital Patients

To estimate the appropriate energy intake for a hospital patient, the health practitioner measures or calculates the patient's resting metabolic rate (RMR) and then applies a "stress factor" to accommodate the additional energy needs imposed by illness. The stress factor 1.25 has been shown to be reasonably accurate for many hospitalized patients; other examples are listed in Table 22-2 on p. 687.

The following example uses the Mifflin–St. Jeor equation (shown in Table 18-3) and the stress factor 1.25 to determine the energy needs of a 57-year-old female patient who is 5 feet 3 inches tall, weighs 115 pounds, and is confined to bed.

Step 1: The patient's weight and height are converted to the units used in the equation:

$$\text{Weight in kilograms} = 115 \text{ lb} \div 2.2 \text{ lb/kg} = 52.3 \text{ kg}$$

$$\text{Height in centimeters} = 63 \text{ in} \times 2.54 \text{ cm/in} = 160 \text{ cm}$$

Step 2: Using the Mifflin–St. Jeor equation for estimating RMR in women:

$$\begin{aligned} \text{RMR} &= \\ &[9.99 \times \text{weight (kg)}] + [6.25 \times \text{height (cm)}] - [4.92 \times \text{age (years)}] - 161 \\ &= (9.99 \times 52.3) + (6.25 \times 160) - (4.92 \times 57) - 161 \\ &= 522 + 1000 - 280 - 161 = 1081 \text{ kcal} \end{aligned}$$

Step 3: The RMR value is multiplied by the appropriate stress factor:

$$\text{RMR} \times \text{stress factor} = 1081 \times 1.25 = 1351 \text{ kcal}$$

Thus, an appropriate energy intake for this patient would be approximately 1351 kcal. Her weight should be monitored to determine if her actual needs are higher or lower.

For a patient who is not confined to bed, an additional activity factor can be applied to accommodate the extra energy needs. For example, if the patient in the example begins limited activity while in the hospital, an activity factor of 1.2 can be multiplied by the results obtained in Step 3:

$$1351 \times \text{activity factor} = 1351 \times 1.2 = 1621 \text{ kcal}$$

The activity factor for a hospitalized patient often falls between 1.1 and 1.4, and it is likely to change as the patient's condition improves.

TRY IT Use the Mifflin–St. Jeor equation to calculate an appropriate energy intake for a 45-year-old female hospital patient who is 5 feet 2 inches tall and weighs 135 pounds.

that the Harris–Benedict equation may be more appropriate for obese patients if the body weight used in the equation falls between an estimated ideal weight and the patient's actual weight. ♦ Other studies, however, have been unable to confirm the usefulness of body weight adjustments in predictive equations.[8]

♦ A method sometimes used for adjusting body weight: Adjusted body weight = ideal weight + 0.25 (actual weight − ideal weight)

Critical care patients may have higher-than-normal energy needs due to fever, mechanical ventilation, restlessness, or the presence of open wounds. Patients who are critically ill are usually bedridden and inactive, however, so the energy needed for physical activity is minimal. Energy requirements for critical care patients are discussed further in Chapter 22.

IN SUMMARY **Energy intakes appropriate for hospital patients can be estimated by multiplying a person's resting metabolic rate (RMR) by factors that account for the medical condition, medical treatments, and activity level. The RMR value can be obtained from indirect calorimetry or a predictive equation. Energy needs of critical care patients may be higher than normal due to fever, mechanical ventilation, restlessness, and open wounds.**

Dietary Modifications

During illness, many patients can meet energy and nutrient needs by following a **regular diet**. Other patients may require a **modified diet**, which is altered by changing food consistency or texture, nutrient content, or the foods included in the diet. If a patient's medical condition makes it difficult to meet nutrient needs orally, two options remain: *tube feedings* and *parenteral nutrition*. This section introduces the

regular diet: a diet that includes all foods and meets the nutrient needs of healthy people; also called a *standard diet* or *house diet*.

modified diet: a diet that contains foods altered in texture, consistency, or nutrient content or that includes or omits specific foods; sometimes called a *therapeutic diet*.

use of modified diets and alternative feeding routes in clinical care. Later chapters describe other types of modified diets and additional dietary strategies for treating nutrition problems.

Modified Diets

Table 18-4 lists examples of modified diets that are often prescribed during illness.[9] Diets that contain foods with altered texture and consistency may be prescribed for individuals with chewing and swallowing difficulties. Diets with modified nutrient or food content are frequently used to correct malnutrition, relieve disease symptoms, or reduce the risk of developing complications. Some patients may have several medical problems and need a number of dietary changes. Keep in mind that modified diets should be adjusted to satisfy individual preferences and tolerances and may also need to be altered as a patient's condition changes.

Mechanically Altered Diets

Mechanically altered diets are helpful for individuals who have difficulty chewing or swallowing. Chewing difficulties usually result from dental problems. Impaired swallowing, or **dysphagia**, may result from neurological disorders, surgical procedures involving the head and neck, and various physiological or anatomical abnormalities that restrict the movement of food within the throat or esophagus. Dysphagia diets are highly individualized because

TABLE 18-4 Examples of Modified Diets

Type of Diet	Description of Diet	Appropriate Uses
Modified Texture and Consistency		
Mechanically altered diets	Contain foods that are modified in texture. Pureed diets include only pureed foods; mechanical soft diets may include solid foods that are mashed, minced, ground, or soft.	Pureed diets are used for people with swallowing difficulty, poor lip and tongue control, or oral hypersensitivity. Mechanical soft diets are appropriate for people with limited chewing ability or certain swallowing impairments.
Blenderized liquid diet	Contains fluids and foods that are blenderized to liquid form.	For people who cannot chew, swallow easily, or tolerate solid foods.
Clear liquid diet	Contains clear fluids or foods that are liquid at room temperature and leave minimal residue in the colon.	For preparation for bowel surgery or colonoscopy, for acute GI disturbances (such as after GI surgeries), or as a transition diet after intravenous feeding. For short-term use only.
Modified Nutrient or Food Content		
Fat-controlled diet	Limits dietary fat to low (<50 g/day) or very low (<25 g/day) intakes.	For people who have certain malabsorptive disorders or symptoms of diarrhea, flatulence, or steatorrhea (fecal fat) resulting from dietary fat intolerance.
Fiber-restricted diet	Limits dietary fiber; degree of restriction depends on the patient's condition and reason for restriction.	For acute phases of intestinal disorders or to reduce fecal output before surgery. Not recommended for long-term use.
Sodium-controlled diet	Limits dietary sodium; degree of restriction depends on symptoms and disease severity.	To help lower blood pressure or prevent fluid retention; used in hypertension, congestive heart failure, renal disease, and liver disease.
High-kcalorie, high-protein diet	Contains foods that are kcalorie and protein dense.	Used for patients with high kcalorie and protein requirements (due to cancer, AIDS, burns, trauma, and other conditions); also used to reverse malnutrition, improve nutritional status, or promote weight gain.

American Dietetic Association, *Nutrition Care Manual* (Chicago: American Dietetic Association, 2010).

dysphagia: difficulty swallowing.

TABLE 18-5 Foods Included in Mechanically Altered Diets

Depending on the feeding problem, a mechanically altered diet may include foods that are pureed, mashed, ground, minced, or soft textured. Foods vary according to tolerance.

Pureed Diets	Mechanical Soft Diets
Milk products: Milk, smooth yogurt, pudding, custard	**Milk products:** Milk, yogurt with soft fruit, pudding, cottage cheese
Fruits: Pureed fruits and juices without pulp, seeds, skins, or chunks; well-mashed fresh bananas; applesauce	**Fruits:** Canned or cooked fruits without seeds or skin, fruit juices with small amounts of pulp, ripe bananas
Vegetables: Pureed cooked vegetables without seeds, skins, or chunks; mashed potatoes, pureed potatoes with gravy	**Vegetables:** Soft, well-cooked vegetables that are not rubbery or fibrous; well-cooked, moist potatoes
Meats and meat substitutes: Pureed meats; smooth, homogenous soufflés; hummus or other pureed legume spreads	**Meats and meat substitutes:** Ground, minced, or tender meat, poultry, or fish with gravy or sauce; tofu; well-cooked, moist legumes; scrambled or soft-cooked eggs
Breads and cereals: Smooth cooked cereals such as Cream of Wheat, slurried breads and pancakes,[a] pureed rice and pasta	**Breads and cereals:** Cooked cereals or moistened dry cereals with minimal texture, soft pancakes or breads, well-cooked noodles or dumplings in sauce or gravy

[a]Slurried foods are mixed with liquid until the consistency is appropriate; they may be gelled and shaped to improve their appearance.

FIGURE 18-2 Menu—Clear Liquid Diet

SAMPLE MENU

Breakfast	Strained orange juice
	Flavored gelatin
	Ginger ale
	Coffee or tea, sugar
Lunch	Bouillon or consommé
	Flavored gelatin
	Frozen juice bars
	Apple or grape juice
	Coffee or tea, sugar
Supper	Bouillon or consommé
	Flavored gelatin
	Fruit ice
	Cranberry juice
	Coffee or tea, sugar
Snacks	Soft drinks
	Fruit ices
	Hard candy

the nature and severity of swallowing problems vary greatly. Furthermore, patients must be monitored regularly because swallowing ability can fluctuate over time. Chapter 23 provides details about the specific diets used for treating dysphagia.

Table 18-5 lists examples of foods included in mechanically altered diets. Although the names for these diets vary, a more restrictive diet may contain mostly pureed foods (*pureed diet*), whereas a less restrictive diet may include moist, soft-textured foods that easily form a bolus (*mechanical soft diet*, or simply *soft diet*). Diets for people with chewing problems typically include foods that are ground or minced (*ground/minced diet*). Note that the foods used in these diets can overlap, and individual tolerances should ultimately determine whether foods are included or excluded.

Blenderized Liquid Diet Blenderized diets may be prescribed following oral or facial surgeries (for example, jaw wiring) or be recommended to individuals with chewing problems. Soft or tender foods that can be blenderized (often with added liquid) are available from all food groups, and include cereals and breads; cooked vegetables; fresh or cooked fruits without skins and seeds; cooked, tender meats and fish; and potatoes, rice, and pasta. Foods that do not blend well should be excluded; these include hard or rubbery foods such as nuts and seeds, dried fruits, coconut, hard cheeses, sausages and frankfurters, and some raw vegetables.

Clear Liquid Diet Clear liquids, which require minimal digestion and are easily tolerated by the gastrointestinal (GI) tract, are often the foods recommended before some GI procedures (such as GI examinations, X-rays, or surgeries), after GI surgery, or after fasting or intravenous feeding. The **clear liquid diet** consists of clear fluids and foods that are liquid at body temperature and leave little undigested material (called **residue**) in the colon. Permitted foods include clear or pulp-free fruit juices, carbonated beverages, clear meat and vegetable broths (such as consommé and bouillon), fruit-flavored gelatin, fruit ices made from clear juices, frozen juice bars, and plain hard candy. Although the clear liquid diet provides fluid and electrolytes, its nutrient and energy contents are extremely limited. If used for longer than a day or two, this diet should be supplemented with commercially prepared low-residue formulas that provide required nutrients. Figure 18-2 gives an example of a one-day clear liquid menu.

Sometimes a **full liquid diet**, a liquid diet that is not limited to clear liquids, is used as a transitional diet between liquids and solid foods. In addition to clear

clear liquid diet: a diet that consists of foods that are liquid at room temperature, require minimal digestion, and leave little residue (undigested material) in the colon.

residue: material left in the intestine after digestion; includes mostly dietary fiber and undigested starches and proteins.

full liquid diet: a liquid diet that includes clear liquids, milk, yogurt, ice cream, and liquid nutritional supplements (such as Ensure).

liquids, a full liquid diet may include milk, eggnog, cream soups, and thin cereal gruels. Because the diet contains milk products, it may be inappropriate for patients with significant lactose intolerance. Moreover, a gradual progression from clear liquids to solid foods is generally unnecessary, ♦ so the usefulness of this diet is in question.

♦ A change in diet as a patient's food tolerance improves is called *diet progression*.

Fat-Controlled Diet A fat-controlled diet may be necessary for reducing the symptoms of fat malabsorption, which often accompanies diseases of the liver, gallbladder, pancreas, and intestines. Controlling fat intake may also alleviate the symptoms of heartburn. Although the fat intake is occasionally limited to as little as 25 grams daily, it should not be restricted more than necessary because fat is an important source of kcalories.

Most foods included in a fat-controlled diet provide less than 1 gram of fat per serving. The diet includes fat-free milk products, most breads and cooked grains, fat-free broths and soups, vegetables prepared without fats, most fruits, and fat-free candies and sweets. Restricted foods include low-fat and whole-milk products, baked products with added fat (like muffins), and most prepared desserts. Lean meat and meat substitutes are permitted but may be restricted to 4 to 6 ounces per day, depending on the degree of restriction. Some patients with malabsorptive disorders cannot tolerate large amounts of lactose or dietary fiber, so foods that include these substances may also need to be excluded from the diet. Chapter 24 provides additional information about fat-controlled diets.

Fiber-Restricted Diet Fiber restriction is recommended during acute phases of intestinal disorders, when the presence of fiber may exacerbate intestinal discomfort or cause diarrhea or blockages. Fiber-restricted diets are sometimes used before surgery to minimize fecal volume and after surgery during transition to a regular diet. Long-term fiber restriction is discouraged, however, because it is associated with constipation, diverticulosis, and other problems.

Fiber-restricted diets often eliminate whole-grain breads and cereals, nuts and nut butters, most fresh fruits (except peeled apples, ripe bananas, and melons), dried fruits, dried beans and peas, and many vegetables (including broccoli, cabbage, corn, onions, peppers, spinach, and winter squash). ♦ If required, even greater reductions in colonic residue can be achieved by following a **low-residue diet**, which excludes most fruits and vegetables, foods high in resistant starch (see p. 102), milk products that contain significant lactose, and foods that contain fructose or sugar alcohols (such as sorbitol). These foods contribute to colonic residue because some of their nutrients may be poorly digested (such as the lactose in milk) or poorly absorbed (such as sorbitol and fructose). Note that the terms "low-fiber diet" and "low-residue diet" are often used interchangeably.

♦ Specific information about the fiber content of foods can be found in Chapter 4 (p. 121) and Appendix H.

Sodium-Controlled Diet A sodium-controlled diet can help to prevent or correct fluid retention and may be prescribed for treatment of hypertension, congestive heart failure, kidney disease, and liver disease. The sodium intake recommended depends on the illness, the severity of symptoms, and the specific drug treatment prescribed. In most cases, sodium is restricted to 2000 or 3000 milligrams daily, although more severe restrictions may be used in the hospital setting. Many patients find it difficult to significantly reduce their sodium intake, so while the sodium-controlled diet is prescribed in an attempt to improve the patient's medical problem, the recommended sodium intake may sometimes exceed the Tolerable Upper Intake Level (UL) for sodium of 2300 milligrams. ♦

♦ The average sodium intake in the United States is approximately 3400 mg per day. The sodium UL was set at 2300 mg to help prevent hypertension.

The sodium-controlled diet limits the use of salt when cooking and at the table, eliminates most prepared foods and condiments, and limits consumption of milk and milk products (if excessive). Because so many processed foods are high in sodium, people following a sodium-controlled diet should check food labels and consume only low-sodium products. Sodium restrictions are difficult to implement on a long-term basis because many people find reduced-sodium diets unpalatable and fail to adhere to them. Additional information about controlling dietary sodium is provided in Chapters 27 and 28.

low-residue diet: a diet low in fiber and other food constituents that contribute to colonic residue.

♦ Reminder: Current guidelines recommend a total fat intake between 20 and 35 percent of kcalories, with most fats coming from polyunsaturated and monounsaturated fats.

High-kCalorie, High-Protein Diet The high-kcalorie, high-protein diet is used to increase kcalorie and protein intakes in patients who have unusually high requirements or in those who are eating poorly. High-fat foods are added to increase energy intakes; consequently, the diet may exceed 35 percent kcalories from fat. ♦ Consuming small, frequent meals and commercial liquid supplements (such as Ensure or Boost) can also help a patient meet increased energy, protein, and other nutrient needs.

Examples of foods included in high-kcalorie, high-protein diets are listed in Table 18-6. Some of these foods are high in saturated fat, which is limited in heart-healthy diets. These foods are used liberally in diets for malnourished patients to help correct their immediate nutrition problems—weight loss and muscle wasting. Chapter 29 offers additional suggestions for increasing the kcalorie and protein contents of meals.

Alternative Feeding Routes

In most cases, patients can meet their nutrient needs by consuming regular foods. If their nutrient needs are high or their appetites poor, liquid supplements can be added to their diets to improve their intakes. Sometimes, however, a person's medical condition makes it difficult to meet nutrient needs orally. Two options remain: **tube feedings** and **parenteral nutrition**, described in detail in Chapters 20 and 21.

- *Tube feedings*. Nutritionally complete formulas can be delivered through a tube placed directly into the stomach or intestine. Tube feedings are preferred to parenteral nutrition if the GI tract is functioning normally.
- *Parenteral nutrition*. A person's medical condition sometimes prohibits the use of the GI tract to deliver nutrients. If the person is malnourished and the GI tract cannot be used for a significant period of time, parenteral nutrition, in which nutrients are supplied intravenously, can meet nutritional needs.

TABLE 18-6 Foods Included in High-kCalorie, High-Protein Diets

Milk products	Whole milk, half-and-half, cream
	Milk shakes, eggnog
	Cheese
	Ice cream, whipped cream
Fruits	Dried fruit
	Canned fruit in heavy syrup
	Avocado
Vegetables	High-kcalorie vegetables such as potatoes, corn, and peas
	Vegetables prepared with butter, margarine, sour cream, cheese sauces, mayonnaise, or salad dressing
	Cream of vegetable soups
Meats and high-protein foods	All meats, fish, and poultry, including bacon, frankfurters, and luncheon meats; eggs; beans; tofu
	All meats, prepared fried or covered in cream sauces or gravies
	Protein bars
	Nuts and seeds, peanut and other nut butters, coconut
Breads and cereals	Granola and dry cereals prepared with whole milk or cream and dried fruit
	Hot cereals with whole milk or cream, or added fat
	Pasta, rice, and biscuits with added fat
	Pancakes, waffles, French toast
Beverages	Fruit juices, sweetened beverages
	Meal replacement drinks
	Beverages with added protein powders

tube feedings: liquid formulas delivered through a tube placed in the stomach or intestine.

parenteral nutrition: the provision of nutrients by vein, bypassing the intestine.

Nothing by Mouth (NPO)

An order to not give a patient anything at all—food, beverages, or medications—is indicated by NPO, an abbreviation for *non per os*, meaning "nothing by mouth." For example, an order may read "NPO for 24 hours" or "NPO until after X-ray." The NPO order is commonly used during certain acute illnesses or diagnostic tests involving the GI tract.

IN SUMMARY **Dietary modifications prescribed during illness include changes in food texture or consistency, nutrient content, or food content. Mechanically altered diets may be prescribed for people with swallowing and chewing difficulties. Clear liquid diets may be used briefly after acute gastrointestinal disturbances or parenteral nutrition or before various diagnostic tests. Some medical conditions may require control or restriction of specific nutrients, such as fat, fiber, or sodium. A high-kcalorie, high-protein diet may help to prevent or reverse malnutrition, improve nutrition status, or promote weight gain. In some cases, nutrients need to be delivered via tube feedings or intravenously.**

Foodservice

The work of a foodservice department may appear deceptively simple: appropriate foods are delivered to patients who need specific types of diets. Behind the scenes, however, a complex system is at work. The foodservice department faces a daily challenge in planning, producing, and delivering hundreds of nutritious meals and accommodating a large variety of diets and food preferences.

BaranaStock/Jupiter Images

Foodservice departments strive to prepare appetizing and nutritious meals and may accommodate dozens of special diets.

Menu Planning

When designing menus for modified diets, the dietary and foodservice personnel refer to a **diet manual**, which details the exact foods or preparation methods to include or exclude in a modified diet. The diet manual may also outline the rationale and indications for use of the diets and include sample menus. The manual may be compiled by the dietetics staff or adopted from another health care facility or a dietetics organization.

Food Selection

Most hospitals provide **selective menus** from which patients can select their meals (see Figure 18-3). A patient following a modified diet receives menus that include only the foods specified in the hospital's diet manual for that particular diet. The use of selective menus allows patients to choose the foods they prefer and are most likely to eat. An added advantage is that patients who must follow a modified diet can become familiar with the foods permitted on their particular diet.

In hospitals that provide selective menus, patients may need to make menu selections a day or two in advance so that the foodservice department can estimate the amounts and types of food they need to purchase and prepare. If a menu is not marked correctly or is misplaced, the patient may receive a meal selected by the foodservice department. Other potential problems that may arise when selective menus are used include the following:

- Patients may have difficulty seeing, reading, understanding, or physically marking menus.
- Patients may not understand that their selections will be for the next (or another) day.
- Patients may be out of their rooms (for tests, procedures, or physical activity) or asleep when the menus arrive and may miss the menu pickup time.
- Patients may be too ill or too uninterested in food to make menu selections.

Problems with menu procedures can often be corrected by explaining the system or by taking the time to help patients mark menus.

diet manual: a resource that specifies the foods to include or exclude in modified diets and provides sample menus.

selective menus: menus that provide choices in some or all menu categories.

FIGURE 18-3 Sample Selective Menu

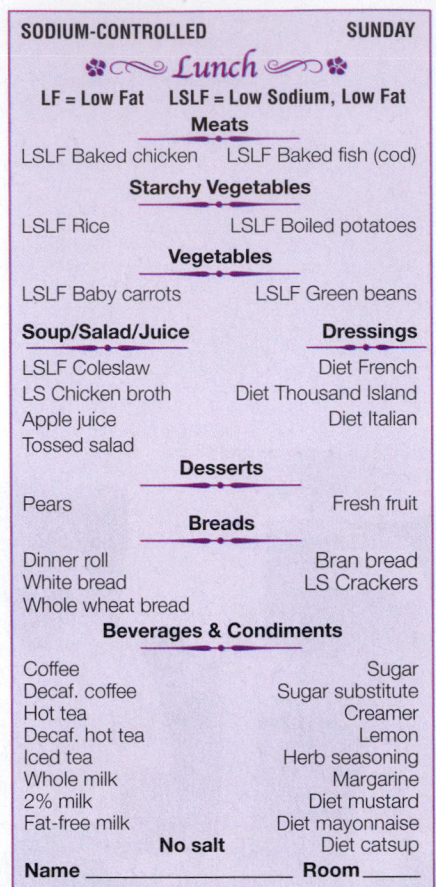

SODIUM-CONTROLLED	SUNDAY
Lunch	
LF = Low Fat	LSLF = Low Sodium, Low Fat
Meats	
LSLF Baked chicken	LSLF Baked fish (cod)
Starchy Vegetables	
LSLF Rice	LSLF Boiled potatoes
Vegetables	
LSLF Baby carrots	LSLF Green beans
Soup/Salad/Juice	**Dressings**
LSLF Coleslaw	Diet French
LS Chicken broth	Diet Thousand Island
Apple juice	Diet Italian
Tossed salad	
Desserts	
Pears	Fresh fruit
Breads	
Dinner roll	Bran bread
White bread	LS Crackers
Whole wheat bread	
Beverages & Condiments	
Coffee	Sugar
Decaf. coffee	Sugar substitute
Hot tea	Creamer
Decaf. hot tea	Lemon
Iced tea	Herb seasoning
Whole milk	Margarine
2% milk	Diet mustard
Fat-free milk	Diet mayonnaise
No salt	Diet catsup
Name ______________	Room _____

Some hospitals do not offer selective menus. Instead, they may provide **nonselective menus** (menus with preselected food items) or menus that include some elements of both systems (**semiselective menus**). Nonselective menus have been gaining popularity in hospital foodservice because they simplify operations and may help to cut costs.

Food Preparation and Delivery

The logistics of preparing foods tailored to each modified diet can be overwhelming. For this reason, foodservice departments use systems designed to limit costs and minimize errors. Meals may be produced in a central kitchen and delivered directly to patients' rooms, using serving equipment that keeps hot foods hot and cold foods cold. Another popular practice is to produce meals beforehand, deliver trays to the nursing unit, and then reheat hot food items in areas close to the patients' rooms. Generally, foodservice personnel deliver food carts directly to the nursing unit, and then either nursing or foodservice personnel take trays to patients.

If a patient receives the wrong type of meal or receives foods that differ substantially from those requested, the dietetics or nursing staff should contact the foodservice department. A person's experience with foodservice strongly influences perception of the overall hospital stay;[10] thus, good communication between patients and the hospital staff regarding meals and food quality can substantially improve a patient's satisfaction.

Food Safety

Each institution has protocols for handling food products based on the identification of potential hazards and critical control points in food preparation, usually referred to as **Hazard Analysis and Critical Control Points (HACCP)**.[11] Generally, a HACCP program addresses food handling, cooking, and storage procedures; cleaning and disinfecting of utensils, surfaces, and equipment; and staff sanitation issues. Personnel involved with preparing or delivering meals need to be aware of the specific HACCP systems at their facility.

Improving Food Intake

People in hospitals and other medical facilities often lose their appetites as a result of their medical condition, treatment, or emotional distress. Furthermore, some medications and other treatments can dramatically alter taste perceptions. Patients usually receive meals at specified times whether they are hungry or not and often must eat in bed without companionship; under these conditions, eating can be more of a chore than a pleasurable experience. Meals may also be unwelcome if the person is in pain or has been sedated.

Nurses, dietitians, and dietetic technicians often have central roles in helping patients to eat. Whenever possible, the patient's room should remain calm and quiet during mealtime. Excessive activity, like room maintenance or ward rounds, can distract patients and reduce appetite. If the patient's appetite or sense of taste is affected by illness, the health practitioner should work with the patient to identify foods that are the most enjoyable. When meals are served, the nurse can help the patient wash up before eating and check to see that foods and utensils are arranged attractively. Placing an occasional "surprise" on the tray—a decoration or funny card, for example—may help patients look forward to meals or perk up sagging spirits. The "How To" lists additional suggestions that may help to improve food intake at mealtimes.

IN SUMMARY **Hospital foodservice departments may accommodate the special needs of hundreds of patients daily. Diet manuals specify the foods to include in modified diets. Many hospitals provide selective menus from which patients can choose meals that are appropriate for their medical condition. Hospital patients may need assistance at mealtime and encouragement to consume adequate amounts of food. The Case Study provides an opportunity for you to review the implementation of nutrition care.**

nonselective menus: menus that do not allow choices and list only preselected food items.

semiselective menus: menus that combine aspects of both selective and nonselective menus.

Hazard Analysis and Critical Control Points (HACCP): systems of food or formula preparation that identify food safety hazards and critical control points during foodservice procedures.

HOW TO Help Hospital Patients Improve Their Food Intakes

1. Empathize with the patient. Show that you understand how difficult eating may be when a person feels too sick to move or too tired to sit up. Help to motivate the patient by explaining how important good nutrition is to recovery.
2. Help patients select the foods they like and mark menus appropriately. When appropriate and permissible, let friends or family members bring favorite foods from outside the hospital.
3. For patients who are weak, suggest foods that require little effort to eat. Eating a roast beef sandwich, for example, requires less effort than cutting and eating a steak. Drinking soup from a cup may be easier than eating it with a spoon.
4. During mealtimes, make sure the patient's room is quiet and has sufficient lighting for viewing the food. See that the room is free of odors that may interfere with the appetite.
5. Help patients prepare for meals. Help them get comfortable, either in bed or in a chair. Adjust the extension table to a comfortable distance and height and make sure it is clean. Take these steps before the tray arrives so that the meal can be served promptly and at the right temperature.
6. When the food cart arrives, check the patient's tray. Confirm that the patient is receiving the right diet, the foods on the tray are those selected from the menu, and the foods look appealing. Order a new tray if the foods are not appropriate.
7. Help with eating, if necessary. Help patients open containers or cut foods, and assist with feeding if patients cannot feed themselves. Encourage patients with little appetite to eat the most nutritious foods first and to drink liquids between meals.
8. Take a positive attitude toward the hospital's food. Never say something like "I couldn't eat this either." Instead, say, "The foodservice department really tries to make foods appetizing. I'm sure we can find a solution."

TRY IT For the next day or two, briefly note what you eat and how long it takes you to complete each meal or snack. Which foods take the least time and effort to eat?

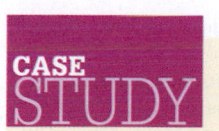

CASE STUDY Implementing Nutrition Care

Max is an 11-year-old boy who was admitted to the hospital after he passed out while playing with friends. Tests confirm a diagnosis of type 1 diabetes mellitus. Max remains in the hospital for several days until his blood glucose and ketone levels are under control. During this time, he and his family learn about diabetes, the diet Max needs to follow, the use of insulin, the monitoring of blood glucose levels, and the required coordination of diet, insulin, and physical activity. The details of diabetes mellitus are reserved for Chapter 26, but for now you can consider the steps that are necessary for implementing Max's nutrition care.

1. Given the chronic nature of Max's illness and his age, what approaches should the health care provider use when discussing the required dietary and medical treatments with Max and his family? What factors need consideration when designing a nutrition education program for Max and his parents?
2. After a first visit with Max and his family, what information should the dietitian add to Max's medical record? Is enough information given above for designing a SOAP note? If not, what elements are missing?
3. Max will need additional care to learn more about diabetes and to make the adjustments that will allow him to cope with his condition. Why is it important to address follow-up care before Max leaves the hospital?

Clinical Portfolio

1. David is a 29-year-old male who is 6 feet 2 inches tall and has a usual body weight of 180 pounds. He was admitted to the hospital following an automobile accident and was treated for minor injuries. Using the method described in the "How To" on p. 600, estimate an appropriate energy intake for David using both the Harris–Benedict and Mifflin–St. Jeor equations. Use the stress factor 1.25, with no additional activity factor.
2. A healthy 70-year-old woman, admitted to the hospital for a hip replacement surgery, develops an infection after the surgery and recovery is slower than expected. You notice that she seems uninterested in meals and has eaten only small amounts of food for several days. What steps can be taken to uncover and address problems that the woman might be having with food?

Nutrition on the Net

For further study of topics covered in this chapter, log on to **www.cengagebrain.com** and search for ISBN 084006845X.

- Visit the website of the American Dietetic Association to learn more about nutrition services in clinical care: **www.eatright.org**
- Design patient education materials with information from the MedlinePlus website: **www.nlm.nih.gov/medlineplus/**
- This Food and Drug Administration website discusses issues of clinical interest and is targeted to health professionals: **www.fda.gov/ForHealthProfessionals/**

References

1. American Dietetic Association, *Nutrition Diagnosis and Intervention: Standardized Language for the Nutrition Care Process* (Chicago: American Dietetic Association, 2007).
2. American Dietetic Association, *Nutrition Diagnosis and Intervention*, 2007.
3. K. Glanz, Current theoretical bases for nutrition intervention and their uses, in A. M. Coulston and C. J. Boushey, eds., *Nutrition in the Prevention and Treatment of Disease* (Burlington, MA: Elsevier Academic Press, 2008), pp. 127–138.
4. L. M. Delahanty and J. M. Heins, Tools and techniques to facilitate nutrition intervention, in A. M. Coulston and C. J. Boushey, eds., *Nutrition in the Prevention and Treatment of Disease* (Burlington, MA: Elsevier Academic Press, 2008), pp. 149–167.
5. American Dietetic Association, *Nutrition Diagnosis and Intervention*, 2007.
6. J. Boullata and coauthors, Accurate determination of energy needs in hospitalized patients, *Journal of the American Dietetic Association* 107 (2007): 393–401.
7. Boullata and coauthors, 2007.
8. Boullata and coauthors, 2007.
9. American Dietetic Association, *Nutrition Care Manual* (Chicago: American Dietetic Association, 2010).
10. C. A. Watters and coauthors, Exploring patient satisfaction with foodservice through focus groups and meal rounds, *Journal of the American Dietetic Association* 103 (2003): 1347–1349.
11. K. W. McClusky, Implementing hazard analysis critical control points, *Journal of the American Dietetic Association* 104 (2004): 1699–1700.

HIGHLIGHT 18

Foodborne Illnesses

Kzenon /Shutterstock.com

Preparing meals to meet the special dietary needs of patients during times of sickness requires careful attention to food safety. **Foodborne illness** is the leading food safety concern because **outbreaks** of food poisoning far outnumber episodes of any other kind of food contamination. An estimated 48 million people experience foodborne illness each year in the United States.[1] For some 3000 people each year, the symptoms can be so severe as to cause death. Most vulnerable are pregnant women; very young, very old, sick, or malnourished people; and those with a weakened immune system (as in AIDS).

The potential **hazard** of foods differs from the **toxicity** of a substance—a distinction worth understanding. Anything can be toxic. Toxicity simply means that a substance *can* cause harm *if* enough is consumed. We consume many substances that are toxic, without **risk**, because the amounts are so small. The term *hazard*, on the other hand, is more relevant to our daily lives because it refers to the harm that is *likely* under real-life conditions. Consumers rely on government agencies to set **safety** standards and can minimize their chances of contracting foodborne illnesses by taking the proper precautions. The accompanying glossary defines related terms.

Foodborne Infections and Food Intoxications

Foodborne illness can be caused by either an infection or intoxication. Table H18-1 summarizes the most common and most severe foodborne illnesses, along with their food sources, general symptoms, and prevention methods.

Foodborne Infections

Foodborne infections are caused by eating foods contaminated by infectious microbes. Among foodborne infections, Norovirus is the major cause of illness and *Salmonella* is the major cause of death.[2] These **pathogens** enter the GI tract in contaminated

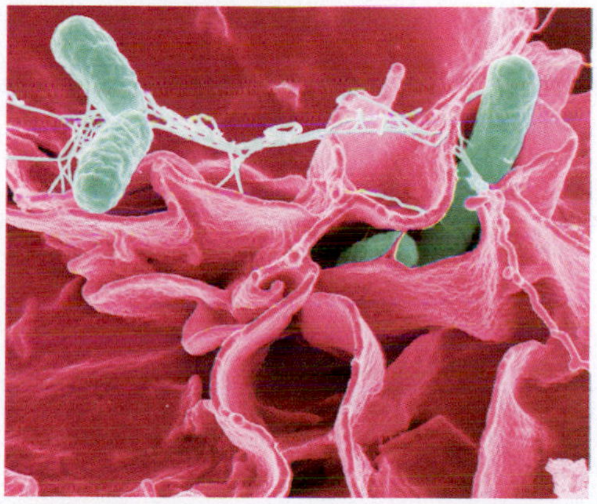

Science Source/Photo Researchers, Inc.

An infection with *Salmonella* bacteria typically causes diarrhea, fever, and abdominal cramps for 12 to 72 hours.

GLOSSARY

cross-contamination: the contamination of food by bacteria that occurs when the food comes into contact with surfaces previously touched by raw meat, poultry, or seafood.

foodborne illness: illness transmitted to human beings through food and water, caused by either an infectious agent (foodborne infection) or a poisonous substance (food intoxication); commonly known as *food poisoning*.

hazard: a source of danger; used to refer to circumstances in which harm is possible under normal conditions of use.

Hazard Analysis Critical Control Points (HACCP): a systematic plan to identify and correct potential microbial hazards in the manufacturing, distribution, and commercial use of food products; commonly referred to as "HASS-ip."

outbreaks: two or more cases of a similar illness resulting from the ingestion of a common food.

pasteurization: heat processing of food that inactivates some, but not all, microorganisms in the food; not a sterilization process. Bacteria that cause spoilage are still present.

pathogens (PATH-oh-jenz): a microorganism capable of producing disease.

risk: a measure of the probability and severity of harm.

safety: the condition of being free from harm or danger.

sushi: vinegar-flavored rice and seafood, typically wrapped in seaweed and stuffed with colorful vegetables. Some sushi is stuffed with raw fish; other varieties contain cooked seafood.

toxicity: the ability of a substance to harm living organisms. All substances are toxic if high enough concentrations are used.

TABLE H18-1 Foodborne Illnesses

Disease and Organism That Causes It	Most Frequent Food Sources	Onset and General Symptoms	Prevention Methods[a]
Foodborne Infections			
Campylobacteriosis (KAM-pee-loh-BAK-ter-ee-OH-sis) *Campylobacter* bacterium	Raw and undercooked poultry, unpasturized milk, contaminated water	Onset: 2 to 5 days. Diarrhea, vomiting, abdominal cramps, fever; sometimes bloody stools; lasts 2 to 10 days.	Cook foods thoroughly; use pasteurized milk; use sanitary food-handling methods.
Cryptosporidiosis (KRIP-toe-spo-rid-ee-OH-sis) *Crytosporidium parvum* parasite	Contaminated swimming or drinking water, even from treated sources; highly chlorine-resistant; contaminated raw produce and unpasteurized juices and ciders	Onset: 2 to 10 days. Diarrhea, stomach cramps, upset stomach, slight fever; symptoms may come and go for weeks or months.	Wash all raw vegetables and fruits before peeling; use pasteurized milk and juice; do not swallow drops of water while using pools, hot tubs, ponds, lakes, rivers, or streams for recreation.
Cyclosporiasis (sigh-clo-spore-EYE-uh-sis) *Cyclospora cayetanensis* parasite	Contaminated water, contaminated fresh produce	Onset: 1 to 14 days. Diarrhea, loss of appetite, weight loss, stomach cramps, nausea, vomiting, fatigue; symptoms may come and go for weeks or months.	Use treated, boiled, or bottled water; cook foods thoroughly; peel fruits.
***E. coli* infection** *Escherichia coli*[b] bacterium	Undercooked ground beef, unpasteurized milk and juices, raw fruits and vegetables, contaminated water, and person-to-person contact	Onset: 1 to 8 days. Severe bloody diarrhea, abdominal cramps, vomiting; lasts 5 to 10 days.	Cook ground beef thoroughly; use pasteurized milk; use sanitary food-handling methods; use treated, boiled, or bottled water.
Gastroenteritis[c] Norovirus	Person-to-person contact; raw foods, salads, sandwiches	Onset: 1 to 2 days. Vomiting, diarrhea, stomach pains; lasts 1 to 2 days.	Use sanitary food-handling methods.
Giardiasis (JYE-are-DYE-ah-sis) *Giardia intestinalis* parasite	Contaminated water; uncooked foods	Onset: 7 to 14 days. Diarrhea (but occasionally constipation), abdominal pain, gas.	Use sanitary food-handling methods; avoid raw fruits and vegetables where parasites are endemic; dispose of sewage properly.
Hepatitis (HEP-ah-TIE-tis) Hepatitis A virus	Undercooked or raw shellfish	Onset: 15 to 50 days (28 days average). Diarrhea, dark urine, fever, headache, nausea, abdominal pain, jaundice (yellowed skin and eyes from build-up of wastes); lasts 2 to 12 weeks.	Cook foods thoroughly.
Listeriosis (lis-TER-ee-OH-sis) *Listeria monocytogenes* bacterium	Unpasteurized milk; fresh soft cheeses; luncheon meats, hot dogs	Onset: 1 to 21 days. Fever, muscle aches; nausea, vomiting, blood poisoning, complications in pregnancy, and meningitis (stiff neck, severe headache, and fever).	Use sanitary food-handling methods; cook foods thoroughly; use pasteurized milk.
Perfringens (per-FRINGE-enz) **food poisoning** *Clostridium perfringens* bacterium	Meats and meat products stored at temperatures between 120°F and 130°F	Onset: 8 to 16 hours. Abdominal pain, diarrhea, nausea; lasts 1 to 2 days	Use sanitary food-handling methods; use pasteurized milk; cook foods thoroughly; refrigerate foods promptly and properly.
Salmonellosis (sal-moh-neh-LOH-sis) *Salmonella* bacteria (>2300 types)	Raw or undercooked eggs, meats, poultry, raw milk and other dairy products, shrimp, frog legs, yeast, coconut, pasta, and chocolate	Onset: 1 to 3 days. Fever, vomiting, abdominal cramps, diarrhea; lasts 4 to 7 days; can be fatal.	Use sanitary food-handling methods; use pasteurized milk; cook foods thoroughly; refrigerate foods promptly and properly.
Shigellosis (shi-gel-LOH-sis) *Shigella* bacteria (>30 types)	Person-to-person contact, raw foods, salads, sandwiches, and contaminated water	Onset: 1 to 2 days. Bloody diarrhea, cramps, fever; lasts 4 to 7 days.	Use sanitary food-handling methods; cook foods thoroughly; use proper refrigeration.
Vibrio (VIB-ree-oh) **infection** *Vibrio vulnificus*[d] bacterium	Raw or undercooked seafood, contaminated water	Onset: 1 to 7 days. Diarrhea, abdominal cramps, nausea, vomiting; lasts 2 to 5 days; can be fatal.	Use sanitary food-handling methods; cook foods thoroughly.
Yersiniosis (yer-SIN-ee-OH-sis) *Yersinia enterocolitica* bacterium	Raw and undercooked pork, unpasteurized milk	Onset: 1 to 2 days. Diarrhea, vomiting, fever, abdominal pain; lasts 1 to 3 weeks.	Cook foods thoroughly; use pasteurized milk; use treated, boiled, or bottled water.

TABLE H18-1 Foodborne Illnesses (*continued*)

Disease and Organism That Causes It	Most Frequent Food Sources	Onset and General Symptoms	Prevention Methods[a]
Food Intoxications			
Botulism (BOT-chew-lizm) Botulinum toxin produced by *Clostridium botulinum* bacterium, which grows without oxygen, in low-acid foods, and at temperatures between 40°F and 120°F; the **botulinum** (BOT-chew-line-um) **toxin** responsible for botulism is called **botulin** (BOT-chew-lin).	Anaerobic environment of low acidity (canned corn, peppers, green beans, soups, beets, asparagus, mushrooms, ripe olives, spinach, tuna, chicken, chicken liver, liver pâté, luncheon meats, ham, sausage, stuffed eggplant, lobster, and smoked and salted fish)	Onset: 4 to 36 hours. Nervous system symptoms, including double vision, inability to swallow, speech difficulty, and progressive paralysis of the respiratory system; often fatal; leaves prolonged symptoms in survivors.	Use proper canning methods for low-acid foods; refrigerate homemade garlic and herb oils; avoid commercially prepared foods with leaky seals or with bent, bulging, or broken cans. Do not give infants honey because it may contain spores of *Clostridium botulinum*, which is a common source of infection for infants.
Staphylococcal (STAF-il-oh-KOK-al) **food poisoning** Staphylococcal toxin (produced by *Staphylococcus aureus* bacterium)	Toxin produced in improperly refrigerated meats; egg, tuna, potato, and macaroni salads; cream-filled pastries	Onset: 1 to 6 hours. Diarrhea, nausea, vomiting, abdominal cramps, fever; lasts 1 to 2 days.	Use sanitary food-handling methods; cook food thoroughly; refrigerate foods promptly and properly; use proper home-canning methods.

NOTE: Travelers' diarrhea is most commonly caused by *E. coli, Campylobacter jejuni, Shigella,* and *Salmonella.*

[a]The "How To" on pp. 614–615 provides more details on the proper handling, cooking, and refrigeration of foods.

[b]The most serious strain is *E. coli* STEC 0157.

[c]Gastroenteritis refers to an inflammation of the stomach and intestines but is the most common name used for illnesses caused by Noroviruses.

[d]Most cases of *Vibrio vulnificus* infection occur in persons with underlying illness, particularly those with liver disorders, diabetes, cancer, and AIDS, and those who require long-term steroid use. The fatality rate is 50 percent for this population.

foods such as undercooked poultry and unpasteurized milk. Symptoms generally include abdominal cramps, fever, vomiting, and diarrhea.

Food Intoxications

Food intoxications are caused by eating foods containing natural toxins or, more likely, microbes that produce toxins. The most common food toxin is produced by *Staphylococcus aureus;* it affects more than one million people each year. Less common, but more infamous, is *Clostridium botulinum*, an organism that produces a deadly toxin in anaerobic conditions such as improperly canned (especially home-canned) foods and homemade garlic or herb-flavored oils stored at room temperature. Because the toxin paralyzes muscles, a person with botulism has difficulty seeing, speaking, swallowing, and breathing. Because death can occur within 24 hours of onset, botulism demands immediate medical attention. Even then, survivors may suffer the effects for months or years.

Food Safety in the Marketplace

Transmission of foodborne illness has changed as our food supply and lifestyles have changed.[3] In the past, foodborne illness was caused by one person's error in a small setting, such as improperly refrigerated egg salad at a family picnic, and affected only a few victims. Today, we eat more foods that have been prepared and packaged by others. Consequently, when a food manufacturer or restaurant chef makes an error, foodborne illness can become epidemic. An estimated 80 percent of reported foodborne illnesses are caused by errors in a commercial setting, such as the improper **pasteurization** of milk at a large dairy.

In the early 2000s, a national food company had to recall more than 4 million pounds of poultry products after *Listeria* poisoning killed 7 people and made more than 50 others sick. In the 2006 *Escherichia coli* outbreak due to contaminated fresh spinach, nearly 200 people became sick, and 2 elderly women and a 2-year-old boy died before consumers got the FDA message to not eat fresh spinach. In 2009, *Salmonella* was found in peanut butter that had been used in more than 2100 products made by more than 200 companies. These incidents and others focus the national spotlight on two important safety issues: disease-causing organisms are commonly found in foods, and safe food-handling practices can minimize harm from most of these foodborne pathogens.

Industry Controls

The USDA, the FDA, and the food-processing industries have developed and implemented programs to control foodborne illness.* The **Hazard Analysis Critical Control Points (HACCP)** system requires food manufacturers to identify points of contamination and implement controls to prevent foodborne disease. For example, after tracing two large outbreaks of *Salmonella* to imported cantaloupe, producers began using chlorinated water

* In addition to HACCP, these programs include the Emerging Infections Program (EIP), the Foodborne Diseases Active Surveillance Network (FoodNet), and the Food Safety Inspection Service (FSIS).

to wash the melons and to make ice for packing and shipping. Safety procedures such as this prevent hundreds of thousands of foodborne illnesses each year and are responsible for the decline in infections over the past decade.[4]

This example raises another issue regarding the safety of imported foods. FDA inspectors cannot keep pace with the increasing numbers of imported foods.[5] The FDA is working with other countries to help them adopt the safe food-handling practices used in the United States.

Consumer Awareness

Canned and packaged foods sold in grocery stores are easily controlled, but rare accidents do happen. Batch numbering makes it possible to recall contaminated foods through public announcements via Internet, newspapers, television, and radio. In the grocery store, consumers can buy items before the "sell by" date and inspect the safety seals and wrappers of packages. A broken seal, bulging can lid, or mangled package fails to protect the consumer against microbes, insects, spoilage, or even vandalism.

State and local health regulations provide guidelines on the cleanliness of facilities and the safe preparation of foods for restaurants, cafeterias, and fast-food establishments. Even so, consumers can also take these actions to help prevent foodborne illnesses when dining out:

- Wash hands with hot, soapy water before meals.
- Expect clean tabletops, dinnerware, utensils, and food preparation areas.

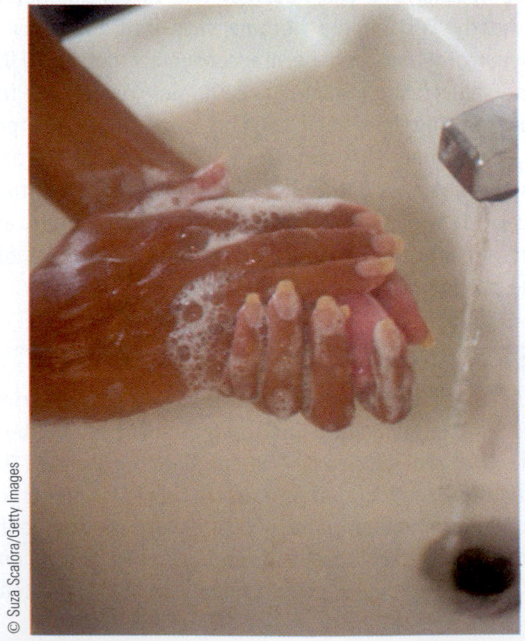

Wash your hands with warm water and soap for at least 20 seconds before preparing or eating food to reduce the chance of microbial contamination.

FIGURE H18-1 Fight BAC!

Four ways to keep food safe. The Fight BAC! website is at www.fightbac.org.

- Expect cooked foods to be served piping hot and salads to be fresh and cold.
- Refrigerate doggy bags within two hours.

Improper handling of foods can occur anywhere along the line from commercial manufacturers to large supermarkets to small restaurants to private homes. Maintaining a safe food supply requires everyone's efforts.

Food Safety in the Kitchen

Whether microbes multiply and cause illness depends, in part, on a few key food-handling behaviors in the kitchen—whether the kitchen is in your home, a school cafeteria, a gourmet restaurant, or a canning facility. Figure H18-1 summarizes the four simple things that can help most to prevent foodborne illness:

- *Keep a clean, safe kitchen.* Wash countertops, cutting boards, hands, sponges, and utensils in hot, soapy water before and after each step of food preparation. Hand sanitizers are as effective as hand washing in reducing bacterial contamination on hands.[6]
- *Avoid cross-contamination.* Keep raw eggs, meat, poultry, and seafood separate from other foods. Wash all utensils and surfaces (such as cutting boards or platters) that have been in contact with these foods with hot, soapy water before using them again. Bacteria inevitably left on the surfaces from the raw meat can recontaminate the cooked meat or other foods—a problem known as **cross-contamination**. Washing raw eggs, meat, and poultry is not recommended because the extra handling increases the risk of cross-contamination.
- *Keep hot foods hot.* Cook foods long enough to reach internal temperatures that will kill microbes, and maintain adequate temperatures to prevent bacterial growth until the foods are served.

- *Keep cold foods cold.* Go directly home upon leaving the grocery store and immediately place foods in the refrigerator or freezer. After a meal, refrigerate any leftovers immediately.

Unfortunately, consumers commonly fail to follow these simple food-handling recommendations. See the "How To" on pp. 614–615 for additional food safety tips.

Safe Handling of Meats and Poultry

Figure H18-2 presents label instructions for the safe handling of meat and poultry and two types of USDA seals. Meats and poultry contain bacteria and provide a moist, nutrient-rich environment that favors microbial growth. Ground meat is especially susceptible because it receives more handling than other kinds of meat and has more surface exposed to bacterial contamination. Consumers cannot detect the harmful bacteria in or on meat. For safety's sake, cook meat thoroughly, using a thermometer to test the internal temperature (see Figure H18-3).

FIGURE H18-2 Meat and Poultry Safety, Grading, and Inspection Seals

Inspection is mandatory; grading is voluntary. Neither guarantees that the product will not cause foodborne illnesses, but consumers can help to prevent foodborne illnesses by following the safe handling instructions.

U.S.D.A
SELECT
The voluntary "Graded by USDA" seal indicates that the product has been graded for tenderness, juiciness, and flavor. Beef is graded Prime (abundant marbling of the meat muscle), Choice (less marbling), and Select (lean). Similarly, poultry is graded A, B, and C.
38
U. S.
INSP'D & P'S'D
The mandatory "Inspected and Passed by the USDA" seal ensures that meat and poultry products are safe, wholesome, and correctly labeled. Inspection does not guarantee that the meat is free of potentially harmful bacteria.
Safe Handling Instructions
THIS PRODUCT WAS PREPARED FROM INSPECTED AND PASSED MEAT AND/OR POULTRY. SOME FOOD PRODUCTS MAY CONTAIN BACTERIA THAT CAN CAUSE ILLNESS IF THE PRODUCT IS MISHANDLED OR COOKED IMPROPERLY. FOR YOUR PROTECTION, FOLLOW THESE SAFE HANDLING INSTRUCTIONS.
KEEP REFRIGERATED OR FROZEN. THAW IN REFRIGERATOR OR MICROWAVE.
KEEP RAW MEAT AND POULTRY SEPARATE FROM OTHER FOODS. WASH WORKING SURFACES (INCLUDING CUTTING BOARDS), UTENSILS, AND HANDS AFTER TOUCHING RAW MEAT OR POULTRY.
COOK THOROUGHLY.
KEEP HOT FOODS HOT. REFRIGERATE LEFTOVERS IMMEDIATELY OR DISCARD.
The USDA requires that safe handling instructions appear on all packages of meat and poultry.

Mad Cow Disease

Reports on mad cow disease from dozens of countries, including Canada and the United States, have sparked consumer concerns.[7] Mad cow disease is a slowly progressive, fatal condition that affects the central nervous system of cattle.* A similar disease develops in people who have eaten contaminated beef from infected cows (milk products appear to be safe).** Approximately 150 cases have been reported worldwide, primarily in the United Kingdom. The USDA has taken numerous steps to prevent the transmission of mad cow disease in cattle, and if these measures are followed, then risks from U.S. cattle are low. Because the infectious agents occur in the intestines, central nervous system, and other organs, but not in muscle meat, concerned consumers may want to select whole cuts of meat instead of ground beef or sausage. A few reports of hunters developing fatal neurological disorders have raised concerns about a similar disease in wild game. Hunters and consumers who regularly eat elk, deer, or antelope should check the advisories of their state department of agriculture.

FIGURE H18-3 Recommended Safe Temperatures (Fahrenheit)

Bacteria multiply rapidly at temperatures between 40°F and 140°F. Cook foods to the temperatures shown on this thermometer and hold them at 140°F or higher.

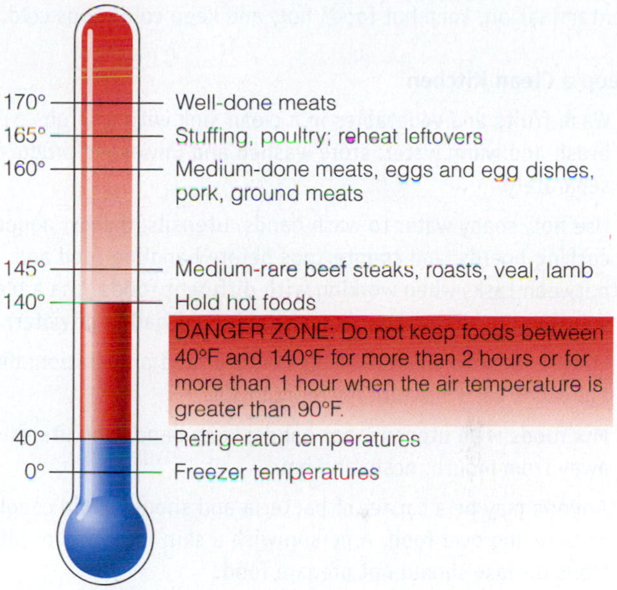
170° Well-done meats
165° Stuffing, poultry; reheat leftovers
160° Medium-done meats, eggs and egg dishes, pork, ground meats
145° Medium-rare beef steaks, roasts, veal, lamb
140° Hold hot foods
DANGER ZONE: Do not keep foods between 40°F and 140°F for more than 2 hours or for more than 1 hour when the air temperature is greater than 90°F.
40° Refrigerator temperatures
0° Freezer temperatures

© Eric O'Connell/Getty Images

Cook hamburgers to 160°F; color alone cannot determine doneness. Some burgers will turn brown before reaching 160°F, whereas others may retain some pink color, even when cooked to 175°F.

* Mad cow disease is technically known as bovine spongiform encephalopathy (BSE).

** The human form of BSE is called variant Creutzfeldt-Jakob Disease (vCJD).

HOW TO Prevent Foodborne Illnesses

Most foodborne illnesses can be prevented by following four simple rules: keep a clean kitchen, avoid cross-contamination, keep hot foods hot, and keep cold foods cold.

Keep a Clean Kitchen

- Wash fruits and vegetables in a clean sink with a scrub brush and warm water; store washed and unwashed produce separately.
- Use hot, soapy water to wash hands, utensils, dishes, nonporous cutting boards, and countertops before handling food and between tasks when working with different foods. Use a bleach solution on cutting boards (one capful per gallon of water).
- Cover cuts with clean bandages before food preparation; dirty bandages carry harmful microorganisms.
- Mix foods with utensils, not hands; keep hands and utensils away from mouth, nose, and hair.
- Anyone may be a carrier of bacteria and should avoid coughing or sneezing over food. A person with a skin infection or infectious disease should not prepare food.
- Wash or replace sponges and towels regularly.
- Clean up food spills and crumb-filled crevices.

Avoid Cross-Contamination

- Wash all surfaces that have been in contact with raw meats, poultry, eggs, fish, and shellfish before reusing.
- Serve cooked foods on a clean plate with a clean utensil. Separate raw foods from those that have been cooked.
- Don't use marinade that was in contact with raw meat for basting or sauces.

Keep Hot Foods Hot

- When cooking meats or poultry, use a thermometer to test the internal temperature. Insert the thermometer between the thigh and the body of a turkey or into the thickest part of other meats, making sure the tip of the thermometer is not in contact with bone or the pan. Cook to the temperature indicated for that particular meat (see Figure H18-3 on p. 613); cook hamburgers to at least medium well done. If you have safety questions, call the USDA Meat and Poultry Hotline: (800) 535-4555.
- Cook stuffing separately, or stuff poultry just prior to cooking.
- Do not cook large cuts of meat or turkey in a microwave oven; it leaves some parts undercooked while overcooking others.
- Cook eggs before eating them (soft-boiled for at least 3½ minutes; scrambled until set, not runny; fried for at least 3 minutes on one side and 1 minute on the other).
- Cook seafood thoroughly. If you have safety questions about seafood, call the FDA hotline: (800) FDA-4010.
- When serving foods, maintain temperatures at 140°F or higher.
- Heat leftovers thoroughly to at least 165°F.

Keep Cold Foods Cold

- When running errands, stop at the grocery store last. When you get home, refrigerate the perishable groceries (such as meats and dairy products) immediately. Do not leave perishables in the car any longer than it takes for ice cream to melt.
- Put packages of raw meat, fish, or poultry on a plate before refrigerating to prevent juices from dripping on food stored below.
- Buy only foods that are solidly frozen in store freezers.
- Keep cold foods at 40°F or less; keep frozen foods at 0°F or less (keep a thermometer in the refrigerator).
- Marinate meats in the refrigerator, not on the counter.
- Look for "Keep Refrigerated" or "Refrigerate After Opening" on food labels.
- Refrigerate leftovers promptly; use shallow containers to cool foods faster; use leftovers within three to four days.

(*continued*)

Safe Handling of Seafood

Most seafood available in the United States and Canada is safe, but eating it undercooked or raw can cause severe illnesses—hepatitis, worms, parasites, viral intestinal disorders, and other diseases.[†] Rumor has it that freezing fish will make it safe to eat raw, but this is only partly true. Commercial freezing kills mature parasitic worms, but only cooking can kill all worm eggs and other microorganisms that can cause illness. For safety's sake, all seafood should be cooked until it is opaque. Even **sushi** can be safe to eat when chefs combine cooked seafood and other ingredients into these delicacies.

Eating raw oysters can be dangerous for anyone, but people with liver disease and weakened immune systems are most vulnerable.[8] At least 10 species of bacteria found in raw oysters can cause serious illness and even death.[††] Raw oysters may also carry

[†] Diseases caused by toxins from the sea include ciguatera poisoning, scombroid poisoning, and paralytic and neurotoxic shellfish poisoning.

[††] Raw oysters can carry the bacterium *Vibrio vulnificus*; see Table H18-1 for details.

HOW TO Prevent Foodborne Illnesses (*continued*)

- Thaw meats or poultry in the refrigerator, not at room temperature. If you must hasten thawing, use cool water (changed every 30 minutes) or a microwave oven.
- Freeze meat, fish, or poultry immediately if not planning to use within a few days.

In General

- Do not reuse disposable containers; use nondisposable containers or recycle instead.
- Do not taste food that is suspect. "If in doubt, throw it out."
- Throw out foods with danger-signaling odors. Be aware, though, that most food-poisoning bacteria are odorless, colorless, and tasteless.
- Do not buy or use items that have broken seals or mangled packaging; such containers cannot protect against microbes, insects, spoilage, or even vandalism. Check safety seals, buttons, and expiration dates.
- Follow label instructions for storing and preparing packaged and frozen foods; throw out foods that have been thawed or refrozen.
- Discard foods that are discolored, moldy, or decayed or that have been contaminated by insects or rodents.

For Specific Food Items

- *Canned goods.* Carefully discard food from cans that leak or bulge so that other people and animals will not accidentally ingest it; before canning, seek professional advice from the USDA Extension Service (check your phone book under U.S. government listings, or ask directory assistance).
- *Milk and cheeses.* Use only pasteurized milk and milk products. Aged cheeses, such as cheddar and Swiss, do well for an hour or two without refrigeration, but they should be refrigerated or stored in an ice chest for longer periods.
- *Eggs.* Use clean eggs with intact shells. Do not eat eggs, even pasteurized eggs, raw; raw eggs are commonly found in Caesar salad dressing, eggnog, cookie dough, hollandaise sauce, and key lime pie. Cook eggs until whites are firmly set and yolks begin to thicken.
- *Honey.* Honey may contain dormant bacterial spores, which can awaken in the human body to produce botulism. In adults, this poses little hazard, but infants younger than 1 year of age should never be fed honey. Honey can accumulate enough toxin to kill an infant; it has been implicated in several cases of sudden infant death. (Honey can also be contaminated with environmental pollutants picked up by the bees.)
- *Mayonnaise.* Commercial mayonnaise may actually help a food to resist spoilage because of the acid content. Still, keep it refrigerated after opening.
- *Mixed salads.* Mixed salads of chopped ingredients spoil easily because they have extensive surface area for bacteria to invade, and they have been in contact with cutting boards, hands, and kitchen utensils that easily transmit bacteria to food (regardless of their mayonnaise content). Chill them well before, during, and after serving.
- *Picnic foods.* Choose foods that last without refrigeration, such as fresh fruits and vegetables, breads and crackers, and canned spreads and cheeses that can be opened and used immediately. Pack foods cold, layer ice between foods, and keep foods out of water.
- *Seafood.* Buy only fresh seafood that has been properly refrigerated or iced. Cooked seafood should be stored separately from raw seafood to avoid cross-contamination.

TRY IT After cutting the fat from a pork loin, you rinse the wooden cutting board under warm water before using it to chop vegetables. Discuss whether this precaution is adequate to protect against cross-contamination.

the hepatitis A virus, which can cause liver disease. Some hot sauces can kill many of these bacteria, but not the virus; alcohol inactivates some bacteria, but not enough to guarantee protection (or to recommend drinking alcohol).[9] Pasteurization of raw oysters—holding them at a specified temperature for a specified time—holds promise for killing bacteria without cooking the oyster or altering its texture or flavor.

As population density increases along the shores of seafood-harvesting waters, pollution inevitably invades the sea life there. Preventing seafood-borne illness is in large part a task of controlling water pollution. To help ensure a safe seafood market, the FDA requires processors to adopt food safety practices based on the HACCP system mentioned earlier.

Chemical pollution and microbial contamination lurk not only in the water, but also in the boats and warehouses where seafood is cleaned, prepared, and refrigerated. Because seafood is one of the most perishable foods, time and temperature are critical to its freshness, flavor, and safety. To keep seafood as fresh as possible, people in the industry must "keep it cold, keep it clean, and keep it moving." Wise consumers eat it cooked.

Other Precautions and Procedures

Fresh food generally smells fresh. Not all types of food poisoning are detectable by odor, but some bacterial wastes produce "off" odors. If an abnormal odor exists, the food is spoiled. Throw it out or, if it was recently purchased, return it to the grocery store. Do not taste it. Table H18-2 lists safe refrigerator storage times for selected foods.

TABLE H18-2 Safe Refrigerator Storage Times (≤40°F)

One to Two Days
Raw ground meats, breakfast or other raw sausages, raw fish or poultry; gravies
Three to Five Days
Raw steaks, roasts, or chops; cooked meats, poultry, vegetables, and mixed dishes; lunchmeats (packages opened); mayonnaise salads (chicken, egg, pasta, tuna)
One Week
Hard-cooked eggs, bacon or hot dogs (opened packages); smoked sausages or seafood
Two to Four Weeks
Raw eggs (in shells); lunchmeats, bacon, or hot dogs (packages unopened); dry sausages (pepperoni, hard salami); most aged and processed cheeses (Swiss, brick)
Two Months
Mayonnaise (opened jar); most dry cheeses (Parmesan, Romano)

Local health departments and the USDA Extension Service can provide additional information about food safety. If precautions fail and a mild foodborne illness develops, drink clear liquids to replace fluids lost through vomiting and diarrhea. If serious foodborne illness is suspected, first call a physician. Then wrap the remainder of the suspected food and label the container so that the food cannot be mistakenly eaten, place it in the refrigerator, and hold it for possible inspection by health authorities.

Dietary Guidelines for Americans

To avoid microbial foodborne illness:

- Clean hands, food contact surfaces, and fruit and vegetables.
- Separate raw, cooked, and ready-to-eat foods while shopping, preparing, or storing foods.
- Cook foods to a safe temperature to kill microorganisms.
- Chill (refrigerate) perishable food promptly and defrost foods properly.
- Meat and poultry should *not* be washed or rinsed.
- Avoid raw (unpasteurized) milk or any products made from unpasteurized milk, raw or partially cooked eggs or foods containing raw eggs, raw or undercooked meat and poultry, unpasteurized juices, and raw sprouts.

Millions of people suffer mild to life-threatening symptoms caused by foodborne illnesses (review Table H18-1). As the "How To" on pp. 614–615 describes, most of these illnesses can be prevented by storing and cooking foods at their proper temperatures and by preparing them in sanitary conditions.

Nutrition on the Net

For further study of topics covered in this chapter, log on to **www.cengagebrain.com** and search for ISBN 084006845X.

- Get food-safety tips from the Gateway to Government Food Safety Information site or from the Fight BAC! campaign of the Partnership for Food Safety Education: **www.foodsafety.gov** or **www.fightbac.org**
- Learn about the various types of food thermometers and how and when to use them from the USDA Thermy campaign: **www.fsis.usda.gov/thermy**
- Report adverse reactions to the FDA MedWatch program at (800) 332-1088 or: **www.fda.gov/Safety/MedWatch**
- Get fish advisories from the Environmental Protection Agency: **www.epa.gov/waterscience/fish**
- Visit the Canadian Food Inspection Agency CFIA): **www.inspection.gc.ca**
- Learn more about food safety in the marketplace from the Food Safety and Inspection Service: **www.fsis.usda.gov/**
- Learn more about organic foods and national organic food standards from the National Organic Program: **www.ams.usda.gov/nop**
- Find information on foodborne illnesses and safe food handling from the American Dietetic Association: **www.homefoodsafety.org**
- Learn more about safe drinking water from the Environmental Protection Agency: **www.epa.gov/safewater**
- Enjoy the humor and music of food toxicologist Carl Winter at: **http://foodsafe.ucdavis.edu/music.html**

References

1. Centers for Disease Control and Prevention, 2011 Estimates of Foodborne Illness, www.cdc.gov, updated December 2010.
2. Centers for Disease Control and Prevention, 2010.
3. Position of the American Dietetic Association: Food and water safety, *Journal of the American Dietetic Association* 103 (2003): 1203–1218.
4. Centers for Disease Control and Prevention, June 2006.
5. U.S. Food and Drug Administration, Food protection plan: An integrated strategy for protecting the nation's food supply, November 2007, www.fda.gov.
6. D. W. Schaffner and K. M. Schaffner, Management of risk of microbial cross-contamination from uncooked frozen hamburgers by alcohol-based hand sanitizer, *Journal of Food Protection* 70 (2007): 109–113.
7. U.S. Food and Drug Administration, Consumer asked questions about BSE in products regulated by FDA's Center for Food Safety and Applied Nutrition (CFSAN), www.cfsan.fda.gov/~comm/bsefaq.html, updated September 14, 2005, accessed December 6, 2006; U.S. Department of Agriculture, Bovine spongiform encephalopathy (BSE) Q & A's, www.aphis.usda.gov/lpa/issues/bse/bse_q&a.html, updated January 21, 2004, accessed December 6, 2006.
8. S. M. Haq and H. H. Dayal, Chronic liver disease and consumption of raw oysters: A potentially lethal combination: A review of *Vibrio vulnificus* septicemia, *American Journal of Gastroenterology* 100 (2005): 1195–1199.
9. C. Liu, R. Chen, and Y. C. Su, Bactericidal effects of wine on *Vibrio parahaemolyticus* in oysters, *Journal of Food Protection* 69 (2006): 1823–1828.

CHAPTER

19

© Masterfile

Nutrition in the Clinical Setting

Most individuals require medications at some point in their lives, although some may not realize that these drugs can have dangerous effects when taken incorrectly. Hence, health practitioners should confirm that their patients are taking prescription drugs as directed and they fully understand the prescription directions. They should also monitor their patients' use of nonprescription drugs and herbal supplements and inform patients about the potentially risky interactions among these substances.

Throughout this chapter, the CourseMate icon indicates an opportunity for online self-study, linking you to activities to increase your understanding of chapter concepts. **www.cengagebrain.com** (search for ISBN 084006845X)

Medications, Diet-Drug Interactions, and Herbal Products

CHAPTER OUTLINE

People often rely on medications to prevent and treat their health problems. Because any ingested chemical can affect metabolism and potentially disrupt body processes, medications can occasionally produce serious side effects. This chapter introduces the use of medications in clinical care, describes potential diet-drug interactions, and discusses herbal products, which some individuals use in hope of treating their medical problems.

© Bill Aron/PhotoEdit

Although over-the-counter drugs are considered safe enough for self-medication, they can cause adverse effects when used inappropriately.

Medications in Disease Treatment

Drugs must be proved to be safe and effective before they can be marketed in the United States. The Food and Drug Administration (FDA) is responsible for approving sales of new drugs and inspecting facilities where drugs are manufactured. By law, drugs are divided into two categories:[1]

- *Prescription drugs* are usually given to treat serious conditions and may cause severe side effects. For these reasons, they are sold by prescription only, which ensures that a physician has evaluated the patient's medical condition and determined that the benefits of using the medication outweigh the risks of incurring side effects.
- *Over-the-counter (OTC) drugs* are those that individuals can use safely and effectively without medical supervision. People use them to treat less serious illnesses that are easily self-diagnosed. Examples include aspirin to treat headaches or pain and antacids to combat indigestion. The FDA regulates labels on OTC drugs to make sure they provide accurate information about the drugs' appropriate uses and dosages and potential adverse effects. Prescription drugs considered safe enough for self-medication are often given OTC status, sometimes in smaller doses than are available by prescription.

Brand-name drugs are usually given patent protection for 20 years after the patent is submitted. After the patent expires, less-expensive generic versions of the drugs can be sold. ♦ To gain FDA approval, a generic drug must have similar biological effects as compared to the original drug: it must contain the same active ingredients; be identical in strength, dosage form, and route of administration; and meet the same requirements for purity and quality. In some cases, the bioavailability (amount absorbed) of a brand-name drug and generic drug may differ due to

♦ Generic vs. brand-name drugs:
- Generic name: diazepam; brand name: Valium (an antianxiety drug)
- Generic name: furosemide; brand name: Lasix (a diuretic)

differences in the drugs' solubility or the types of inactive ingredients present; thus, greater benefit may be obtained by using the brand-name drug.[2] Most often, however, consumers can be confident that generic drugs are as safe and effective as the brand-name products they replace.

Medication Administration

Drugs are introduced into the body by a number of different routes. Although most drugs are taken orally due to convenience, a substantial fraction of the drug is then lost due to incomplete absorption or metabolism by intestinal or liver enzymes (these losses are termed *first-pass elimination*). Drugs may also be provided by injection into a vein (intravenous route) or muscle (intramuscular route) or beneath the skin (subcutaneous route). Other common methods include administration under the tongue (sublingual route), into the rectum, across the skin (transdermal route), or via inhalation. Each method has specific advantages and disadvantages.

Risks from Medications

The risk of an adverse reaction always accompanies the use of a medicine. Thus, a medication should be used only when the benefits of using it outweigh the potential risks. The risks become greater when a drug is incorrectly prescribed or administered. This section discusses the types of risks associated with medications and suggests some steps for managing risk.

Side Effects

By the time a drug reaches the marketplace, large-scale clinical trials have revealed the majority of side effects associated with its use. However, rare side effects are sometimes detected only after a drug has been more widely used. In some instances, these effects occur because drugs are used for longer periods or in different circumstances than originally anticipated. The FDA monitors adverse events after drugs are marketed. Manufacturers are required to submit periodic reports, and individuals using the drugs are encouraged to report unexpected effects directly to the FDA. ♦ The FDA may decide to change labeling information or even withdraw drugs from the marketplace if they are believed to cause unacceptable risks to health.

♦ The FDA's MedWatch program encourages health professionals and consumers to report medication problems by mail, fax, telephone, or the Internet (**www.fda.gov/medwatch**).

Because OTC drugs are available without a prescription, patients may not realize that adverse effects can occur if the drugs are used inappropriately. Under certain circumstances, the active ingredients in these drugs may worsen a medical condition, produce complications, or interact with other medications. Furthermore, people who use products with several active ingredients may inadvertently take toxic amounts of a substance when using several drugs simultaneously. For example, a person with a cold may take one medication to treat a cough and another medication for a headache without realizing that both contain an analgesic (pain medication).

Drug-Drug Interactions

When a person uses multiple drugs, one drug may alter the effects of another drug, and the risk of side effects increases. These problems are common in older adults, who often use several medications daily over long periods. Although primary care physicians typically supervise medication use, some individuals use drugs prescribed by a number of different physicians. Others may use OTC medications and dietary supplements in addition to prescription drugs without being aware of the risks associated with certain combinations.

Diet-Drug Interactions

Substances in the diet may alter the effectiveness of drugs, and drugs may affect food intake or the digestion, absorption, metabolism, or excretion of nutrients. Later sections of this chapter describe these interactions in detail.

Medication Errors

A medication error is any preventable action that causes inappropriate drug use or patient harm due to mistakes made by a health professional or patient. Many errors leading to patient harm involve the use of incorrect drugs or improper dosages.[3] The wrong drug is sometimes administered when two different drugs have names that look or sound alike or have similar packaging. ♦ In other cases, the physician's prescription is misread or misinterpreted; for example,

♦ Drugs that have similar names:
- Bupropion (for depression) and buspirone (for anxiety)
- Hydralazine (for hypertension) and hydroxyzine (for anxiety)

TABLE 19-1 Terms Prohibited on Clinical Documentation

Prohibited Terms	Intended Meaning	Potential Problem	Correct Term for Documentation
U	Unit	Can be misread as the number 0 or 4; may cause 10-fold overdose or higher.	Write out "unit."
IU	International unit	Can be misread as IV (intravenous) or 10.	Write out "international unit."
Trailing zero (1.0 mg) or lack of leading zero (.1 mg)	1 mg; 0.1 mg	Decimal point can be missed, leading to 10-fold error in dosages.	Never use zero by itself after a decimal. Always use zero before a decimal point.
μg	Microgram	Can be misread as mg (milligram).	Write out "microgram."
Q.D. (q.d.), Q.O.D. (q.o.d.)	*Q.D.* means "every day"; *Q.O.D.* means "every other day."	Can be mistaken for one another or misread as "q.i.d." (four times daily).	Write out "daily" or "every other day."

one patient died after receiving 10 milliliters of morphine solution instead of 10 milligrams—a 20-fold overdose.

Several policy changes and programs are helping to reduce medication errors. The bar codes currently used on medications and patient identification bracelets allow health practitioners to verify that the correct medication and dosage are administered: error messages alert personnel if the drug, dose, or timing of administration is inappropriate. In addition, a national education campaign is attempting to eliminate one of the most common but preventable sources of medication errors—the use of ambiguous medical abbreviations (see examples in Table 19-1). Because terms such as these are easily misread or misinterpreted, they can no longer be used in clinical documentation related to patient care.

Patients at High Risk of Adverse Effects

Health care professionals should be aware that some patients are more vulnerable than others to adverse effects from drugs. This category includes the populations that rarely participate in clinical trials that determine product safety: pregnant and lactating women, children, and people with medical conditions that are not the main focus of the study. In these groups, side effects may be discovered only after a drug has been marketed. Children may react in different ways to drugs than adults do, and the appropriate dosage for their age and size is often unknown. Also, limited data are available on drug safety in older adults. Elderly people with chronic diseases that require multiple medications are especially susceptible to adverse effects. They are also more likely to have impaired function of the liver or kidneys—the two organs critical to metabolizing and eliminating drugs from the body. The "How To" on p. 622 provides suggestions that may help to reduce the risks of adverse effects from medications.

IN SUMMARY Both prescription and OTC drugs must be shown to be safe and effective before they are sold. Drugs are introduced into the body by a number of different routes, each with specific advantages and disadvantages. The benefits of using a medication should be greater than the risks associated with its use. Potential risks include side effects, drug-drug and diet-drug interactions, and medication errors. The most common types of medication errors involve incorrect dosages or use of the wrong drug. Patients at highest risk of experiencing adverse effects from medications include pregnant and nursing women, children, and the elderly. Health professionals should discuss the risks

Elderly people using multiple medications are especially susceptible to adverse effects from drugs.

HOW TO Reduce the Risks of Adverse Effects from Medications

To reduce the likelihood of adverse effects from medications, health care professionals can take the following steps:

- Advise the patient that drugs should not be taken unless absolutely necessary. Discuss dietary or lifestyle practices that have benefits similar to those of drugs. For example, laxatives may not be necessary if an individual increases consumption of foods high in fiber and begins exercising regularly.
- Request a complete list of prescription medications, OTC drugs, and dietary supplements that the patient is taking. Ensure that at least one physician is coordinating the patient's drug use. Encourage the patient to purchase all medications at the same pharmacy so that the pharmacist can alert physicians and patients to potential problems.
- Verify that the patient understands how to take medications properly. Alert the patient to potential drug-drug and diet-drug interactions.
- Encourage the patient to keep track of side effects. Inform the patient that new or unusual symptoms may be due to a new medication rather than the medical condition. In some cases, other medications that treat the condition may have fewer side effects.

TRY IT Using the Medline website **www.nlm.nih.gov/medlineplus/druginformation.html**, look up one OTC and one prescription medication that either you or someone you know has taken. What advice would you give to help a patient using these drugs to do so safely?

and benefits of medications with patients and alert them to potential dangers and possible solutions.

Diet-Drug Interactions

When working with patients, medical personnel should be alert to possible interactions between drugs and dietary substances. These interactions can raise health care costs and result in serious, and sometimes fatal, complications. Accordingly, health professionals must learn to take steps to prevent or lessen their adverse consequences. Diet-drug interactions generally fall into the following categories:

- Drugs may alter food intake by reducing the appetite or by causing complications that make food consumption difficult or unpleasant. Other drugs may increase the appetite and cause weight gain.
- Drugs may alter the absorption, metabolism, and excretion of nutrients. Conversely, nutrients and other food components may alter the absorption, metabolism, and excretion of drugs.
- Some interactions between dietary components and drugs can be toxic.

Examples of these types of diet-drug interactions are shown in Table 19-2.[4]

Drug Effects on Food Intake Some drugs can make food intake difficult or unpleasant: they may suppress the appetite, induce nausea or vomiting, cause mouth dryness, alter the sense of taste, or lead to inflammation or lesions in the mouth or GI tract. Certain side effects of drugs, including abdominal discomfort, constipation, and diarrhea, may be worsened by food consumption. Medications that cause drowsiness, such as sedatives and some painkillers, can make a person too tired to eat.

TABLE 19-2 Examples of Diet-Drug Interactions

Drugs may alter food intake by:
• Altering the appetite (amphetamines suppress appetite; corticosteroids increase appetite). • Interfering with taste or smell (amphetamines change taste perceptions). • Inducing nausea or vomiting (digitalis may do both). • Interfering with oral function (some antidepressants may cause dry mouth). • Causing sores or inflammation in the mouth (methotrexate may cause painful mouth ulcers).
Drugs may alter nutrient absorption by:
• Changing the acidity of the digestive tract (antacids may interfere with iron and folate absorption). • Damaging mucosal cells (cancer chemotherapy may damage mucosal cells). • Binding to nutrients (bile acid binders bind to fat-soluble vitamins).
Foods and nutrients may alter drug absorption by:
• Stimulating the secretion of gastric acid (the antifungal agent ketoconazole is absorbed better with meals due to increased acid secretion). • Altering the rate of gastric emptying (intestinal absorption of drugs may be delayed when they are taken with food). • Binding to drugs (calcium binds to tetracycline, reducing the absorption of both substances). • Competing for absorption sites in the intestine (dietary amino acids interfere with levodopa absorption).
Drugs and nutrients may interact and alter metabolism by:
• Acting as structural analogs (as do warfarin and vitamin K). • Using similar enzyme systems (phenobarbital induces liver enzymes that increase the metabolism of folate, vitamin D, and vitamin K). • Competing for transport on plasma proteins (fatty acids and drugs may compete for the same sites on the plasma protein albumin).
Drugs may alter nutrient excretion by:
• Altering nutrient reabsorption in the kidneys (some diuretics increase the excretion of sodium and potassium). • Causing diarrhea or vomiting (diarrhea and vomiting may cause electrolyte losses).
Food substances may alter drug excretion by:
• Inducing the activities of liver enzymes that metabolize drugs, increasing drug excretion (components of charcoal-broiled meats increase the metabolism of warfarin, theophylline, and acetaminophen).
Food substances and drugs may interact and cause toxicity by:
• Increasing side effects of the drug (the caffeine in beverages can increase the adverse effects of stimulants). • Increasing drug action to excessive levels (grapefruit components inhibit the enzymes that degrade certain drugs, increasing drug concentrations in the body).

Drug complications that reduce food intake are significant only when they continue for a long period. Although many drugs can cause nausea in some individuals, the nausea often subsides after the first few doses of the medication and thus has little effect on nutrition status. If side effects persist, other medications may be used to treat them; for example, antinauseants and antiemetics may help to reduce nausea and vomiting and thereby improve food intake.

Some medications stimulate the appetite and encourage weight gain. Unintentional weight gain may result from the use of some antidepressants, antipsychotics, antidiabetic drugs, and corticosteroids (such as prednisone).[5] For some conditions, however, weight gain is desirable. Patients with diseases that cause wasting, such as cancer or AIDS, are sometimes prescribed appetite enhancers such as megestrol acetate (Megace), a progesterone analog, or dronabinol (Marinol), which is derived from the active ingredient in marijuana.

Drug Effects on Nutrient Absorption

The medications that most often cause widespread nutrient malabsorption are those that upset gastrointestinal function or damage the intestinal mucosa. Antineoplastic and antiretroviral drugs ♦ are especially detrimental, although nonsteroidal anti-inflammatory drugs (NSAIDs) and some antibiotics can have similar, though milder, effects. This section describes additional ways in which medications may alter nutrient absorption.

♦ *Antineoplastic drugs* combat tumor growth. *Antiretroviral drugs* treat HIV infection.

Drug-Nutrient Binding Some medications bind to nutrients in the GI tract, preventing their absorption. For example, bile acid binders, which are used to reduce

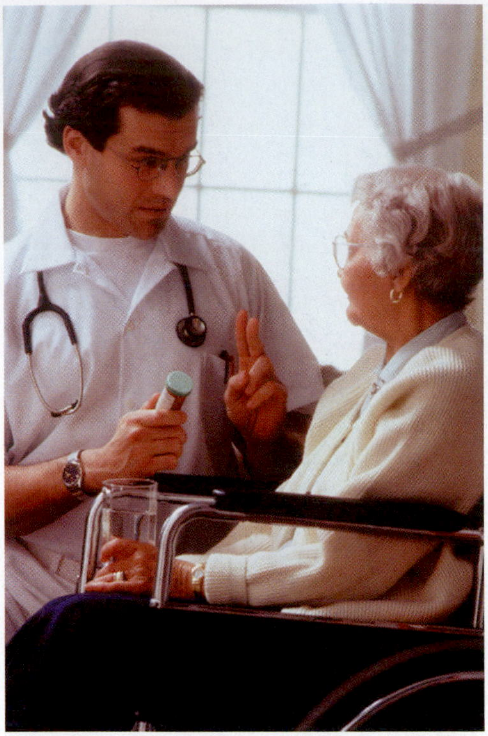

© Jose L. Pelaez/Corbis

To help prevent diet-drug interactions, ask about *all* of the drugs and supplements the patient takes, including prescription and over-the-counter medications, herbal products, and other dietary supplements.

cholesterol levels, may bind to fat-soluble vitamins. Some antibiotics, notably tetracycline and ciprofloxacin (Cipro), bind to the calcium in foods and supplements, reducing the absorption of both the calcium and the antibiotic. Other antibiotics can bind to minerals such as iron, magnesium, and zinc. Consumers are advised to use dairy products and all mineral supplements at least two hours apart from these medications.

Altered Stomach Acidity Medications that reduce stomach acidity can impair the absorption of vitamin B_{12}, folate, and iron. Examples include antacids, which neutralize stomach acid by acting as weak bases, and antiulcer drugs (such as proton pump inhibitors and H2 blockers), which interfere with acid secretion.

Direct Inhibition Several drugs impede nutrient absorption by interfering with their intestinal metabolism or transport into mucosal cells. For example, the antibiotics trimethoprim (Proloprim) and pyrimethamine (Daraprim) compete with folate for absorption into intestinal cells.

Dietary Effects on Drug Absorption

Major influences on drug absorption include the stomach-emptying rate, the level of acidity in the stomach, and direct interactions with dietary components. The drug's formulation may also influence its absorption. The instructions included with medications typically advise whether food should be included or avoided with use.

Stomach-Emptying Rate Drugs reach the small intestine more quickly when the stomach is empty. Therefore, taking a medication with meals may delay its absorption, even though the total amount absorbed may not be lower. As an example, aspirin works faster when taken on an empty stomach, although taking it with food is often encouraged to reduce stomach irritation.

Slow stomach emptying can sometimes enhance drug absorption because the drug's absorption sites in the small intestine are less likely to become saturated. However, a slow drug absorption rate (due to slow stomach emptying) can be a problem if high drug concentrations are needed for effectiveness, as when a hypnotic is taken to induce sleep.

Stomach Acidity Some drugs are better absorbed in an acidic environment, whereas others are better absorbed under alkaline conditions. For example, reduced stomach acidity (due to secretory disorders or antacid medications) may reduce the absorption of ketoconazole (an antifungal medication) and atazanavir (an antiretroviral medication), but increase the absorption of digoxin (Lanoxin, which treats heart failure) and alendronate (Fosamax, which treats osteoporosis).[6] Some drugs can be damaged by acid and are available in coated forms that resist the stomach's acidity.

Interactions with Dietary Components Some dietary substances can bind to drugs and inhibit their absorption. For example, the phytates ♦ in foods can bind to digoxin. High-fiber diets can decrease the absorption of some tricyclic antidepressants due to binding between the fiber and the drugs. As mentioned earlier, minerals can bind to some antibiotics, reducing absorption of both the minerals and the drugs.

♦ Reminder: *Phytates* are compounds found in many plant foods, including whole grains and legumes. Phytates can bind to minerals and reduce their absorption.

Drug Effects on Nutrient Metabolism

Drugs and nutrients share similar enzyme systems in the small intestine and liver. Consequently, some drugs may enhance or inhibit the activities of enzymes needed for nutrient metabolism. For example, the anticonvulsants phenobarbital and phenytoin increase levels of the liver enzymes that metabolize folate, vitamin D, and vitamin K; therefore, persons using these drugs may require supplements of these vitamins.

The drug methotrexate, which treats cancer (and some inflammatory conditions), acts by interfering with folate metabolism and thus depriving rapidly dividing cancer cells of the folate they need to multiply. Methotrexate resembles folate in structure (see Figure 19-1) and competes with folate for the enzyme that converts folate to its active form. The adverse effects of using methotrexate therefore include

FIGURE 19-1 Folate and Methotrexate

By competing for the enzyme that activates folate, methotrexate prevents cancer cells from obtaining the folate they need to multiply. In the process, normal cells are also deprived of the folate they need.

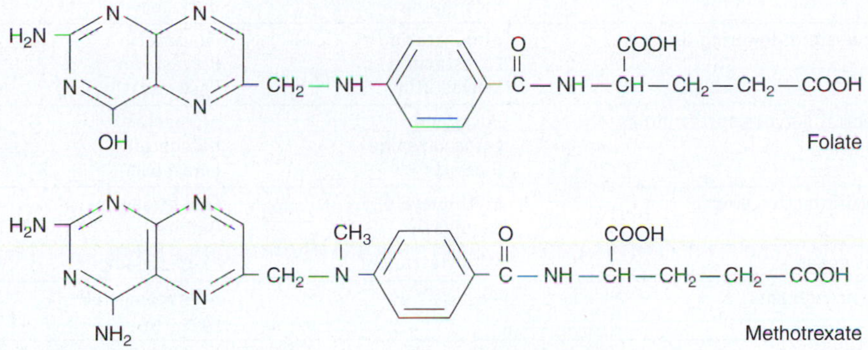

symptoms of folate deficiency. These adverse effects can be reduced by using a preactivated form of folate (called leucovorin), which is often prescribed along with methotrexate to ensure that the body's rapidly dividing cells (cells of the digestive tract, skin cells, and red blood cells) receive adequate folate.

Corticosteroids, used as anti-inflammatory agents and immunosuppressants, have actions that mimic those of the hormone cortisol. ♦ Long-term corticosteroid use can have broad effects on nutritional health and may cause weight gain, muscle wasting, bone loss, and hyperglycemia, with eventual development of osteoporosis and diabetes.

♦ Cortisol is a steroid hormone secreted by the adrenal cortex as part of the body's stress response.

Dietary Effects on Drug Metabolism

Some food components alter the activities of enzymes that metabolize drugs or may counteract drug effects in other ways. Compounds in grapefruit juice (or whole grapefruit) have been found to inhibit or inactivate enzymes that metabolize a number of different drugs. As a result of the reduced enzyme action, blood concentrations of the drugs increase, leading to stronger physiological effects. The effect of the grapefruit juice lasts for a substantial period after the juice is consumed; for example, in experiments with a drug prescribed for heart disease, the juice's effect had an estimated half-life of 12 hours.[7] ♦ Table 19-3 provides examples of drugs that interact with grapefruit juice, as well as some common drugs that are unaffected.

♦ Reminder: The term *half-life* defines the time period of a chemical effect. If the grapefruit effect has a 12-hour half-life, this means that after 12 hours, its biological effect is half of the maximum effect measured.

A number of dietary substances can alter the activity of the anticoagulant drug warfarin (Coumadin). One important interaction is with vitamin K, which is structurally similar to warfarin. Warfarin acts by blocking the enzyme that activates vitamin K, thereby preventing the synthesis of blood-clotting factors. ♦ The amount of warfarin prescribed is dependent, in part, on how much vitamin K is in the diet. If vitamin K consumption from foods or supplements changes substantially, it can alter the effect of the drug. Individuals using warfarin are advised to consume similar amounts of vitamin K daily to keep warfarin activity stable. The dietary sources highest in vitamin K are green leafy vegetables.

♦ Reminder: Vitamin K is required for the synthesis of prothrombin and various other blood-clotting proteins.

Several popular herbs contain natural compounds that enhance the activity of warfarin and therefore should be avoided during warfarin treatment. These herbs include St. John's wort, garlic, ginseng, dong quai, danshen, and others.[8]

Drug Effects on Nutrient Excretion

Drugs that increase urine production may reduce nutrient reabsorption in the kidneys, ♦ resulting in greater urinary losses of the nutrients. For example, some diuretics can increase losses of calcium, potassium, magnesium, and thiamin; thus, dietary supplements may be necessary to avoid deficiency. Risk of nutrient depletion is higher if multiple drugs with the same effect are used, if kidney function is impaired, or if the medications

♦ When the kidneys reabsorb a substance, they retain it in the blood. Substances that are not reabsorbed are excreted in urine.

TABLE 19-3 Examples of Grapefruit Juice–Drug Interactions

Drug Category	Drugs Affected by Grapefruit Juice	Drugs Unaffected by Grapefruit Juice
Cardiovascular drugs	Amiodarone Felodipine Nicardipine	Amlodipine Digoxin Diltiazem
Cholesterol-lowering drugs	Atorvastatin Lovastatin Simvastatin	Fluvastatin Pravastatin Rosuvastatin
Central nervous system drugs	Buspirone Carbamazepine Diazepam	Alprazolam Haloperidol Lorazepam
Anti-infective drugs	Erythromycin Saquinavir	Clarithromycin Quinine
Estrogens	Ethinylestradiol	17-β-estradiol
Anticoagulants	—	Acenocoumarol Warfarin
Immunosuppressants	Cyclosporine Tacrolimus	Prednisone

are used for a long time. Note that some diuretics can cause certain minerals to be retained, rather than excreted.[9]

A number of drugs can increase the excretion of vitamin B_6. One example is isoniazid (INH), an antituberculosis drug similar in structure to vitamin B_6. This drug induces excretion of vitamin B_6 and therefore may lead to a vitamin B_6 deficiency.[10] Because the drug must be taken for at least six months to treat infection, vitamin B_6 supplements are often given simultaneously to prevent deficiency.

Dietary Effects on Drug Excretion

Inadequate excretion of medications can cause toxicity, whereas excessive losses may reduce the amount available for therapeutic effect. Some food components influence drug excretion by altering the amount reabsorbed in the kidneys. For example, the amount of lithium (a mood stabilizer) reabsorbed in the kidneys is similar to the amount of sodium that is reabsorbed. Consequently, both dehydration and sodium depletion, which promote sodium reabsorption, can result in lithium retention. Similarly, a person with a high sodium intake will excrete more sodium in the urine and, therefore, more lithium. Individuals using lithium are advised to maintain a consistent sodium intake from day to day to maintain stable blood concentrations of lithium.

Urine acidity can affect drug excretion due to the effects of pH on a compound's ionic (chemical) form. The medication quinidine, used to treat arrhythmias, is excreted more readily in acidic urine. Foods or drugs that cause urine to become more alkaline may reduce quinidine excretion and raise blood levels of the medication.

Diet-Drug Interactions and Toxicity

Interactions between food components and drugs can cause toxicity or exacerbate a drug's side effects. The combination of tyramine, a food component, and monoamine oxidase (MAO) inhibitors, which treat depression and Parkinson's disease, can be fatal. MAO inhibitors block an enzyme that normally inactivates tyramine, as well as the hormones epinephrine and norepinephrine. When people who take MAO inhibitors consume excessive tyramine, the increased tyramine in the blood can induce a sudden release of stored norepinephrine. This surge in norepinephrine results in severe headaches, rapid heartbeat, and a dangerous rise in blood pressure. For this reason, people taking MAO inhibitors are advised to restrict their intakes of foods rich in tyramine.

TABLE 19-4 Examples of Foods with a High Tyramine Content[a]

- Aged cheeses (cheddar, Gruyère)
- Aged or cured meats (sausage, salami)
- Beer
- Fermented vegetables (sauerkraut, kim chee)
- Fish, smoked or pickled
- Mushrooms
- Prepared soy foods (miso, tempeh, tofu)
- Soy sauce
- Wine (red)
- Yeast extract (Marmite, Vegemite)

[a] The tyramine content of foods depends on storage conditions and processing; thus the amounts in similar products can vary substantially.

Tyramine occurs naturally in foods and is also formed when bacteria degrade the protein in foods. Thus, the tyramine content of a food usually increases when a food ages or spoils. Individuals at risk of tyramine toxicity are advised to buy mainly fresh foods and consume them promptly.[11] Foods that often contain substantial amounts of tyramine are listed in Table 19-4.

Considering the many ways in which drugs and dietary substances can interact, health professionals should attempt to understand the mechanisms underlying diet-drug interactions, identify them when they occur, and prevent them whenever possible. The accompanying "How To" offers some practical advice about preventing diet-drug interactions.

HOW TO Prevent Diet-Drug Interactions

The Joint Commission, an accreditation agency for health care organizations, has recommended that all patients be educated about potential diet-drug interactions. Health professionals can help by informing patients of precautions related to medications and watching for signs of problems that may arise.

To prevent diet-drug interactions, first list the types and amounts of over-the-counter drugs, prescription drugs, and dietary supplements that the patient uses on a regular basis. Look up each drug in a drug reference and make a note of:

- The appropriate method of administration (twice daily or at bedtime, for example).
- How the drug should be administered with respect to foods, beverages, and specific nutrients (for example, take on an empty stomach, take with food, do not take with milk, or do not drink alcoholic beverages while using the medication).
- How the drug should be used with respect to other medications.
- The side effects that may influence food intake (nausea and vomiting, diarrhea, constipation, or sedation, for example) or nutrient needs (interference with nutrient absorption or metabolism, for example).

A similar process can be used to review the dietary supplements that a person is taking. A reliable reference may list their appropriate uses, possible side effects, and potential interactions with food and medications.

Patients who take multiple medications may need to time their intakes carefully to avoid drug-drug or diet-drug interactions. The health professional can use information from a patient's food and nutrition history (see Chapter 17) to help the patient coordinate meals and drugs so as to avoid interactions.

Some medications have well-known effects on nutritional status. The health professional should remain alert for signs of problems, especially when:

- Nutritional problems are a frequent result of using the medication.
- A patient requires multiple medications.
- The patient is in a high-risk group; for example, a child, a pregnant or lactating woman, an older adult, or a person who is malnourished, abuses alcohol, or has impaired liver or kidney function.
- The patient needs to use the medications for an extended period.

Check with the pharmacist for additional information about drugs and their potential adverse effects.

TRY IT The drug *levodopa*, used for Parkinson's disease, interacts with several different nutrients. Using a drug reference, identify one or more clinically relevant diet-drug interactions that the patient using this drug should be made aware of.

IN SUMMARY Medications can alter food intake and affect the absorption, metabolism, and excretion of nutrients; components of foods can similarly affect drug activity. The accompanying table summarizes the various types of diet-drug interactions.

Affected Body Function	Effects of Drugs	Effects of Food Components
Food intake	May increase or decrease appetite, alter taste sensation, cause GI discomfort	—
Absorption	May bind to nutrients, alter stomach acidity, interfere with nutrient transport into intestinal cells	May alter stomach emptying rate, alter stomach acidity, bind to drugs
Metabolism	May alter activity of enzymes that metabolize nutrients	May alter activity of enzymes that metabolize drugs
Excretion	May increase or decrease nutrient losses in the urine	May increase or decrease drug losses in the urine
Varies	May interact with nutrients and cause toxic side effects	May interact with drugs and cause toxic side effects

Herbal Products

The use of herbal products has grown rapidly in the past decade. Currently, about 17 percent of adults in the United States report using herbal supplements regularly.[12] Consumers use these products in the hope of improving their general health and preventing or treating specific diseases. ♦ Top-selling herbal supplements include echinacea, garlic, ginkgo biloba, ginseng, and St. John's wort.[13] Table 19-5 lists these and other popular herbal products along with their common uses and potential risks associated with their use.

♦ Reminder: Highlight 10 discusses vitamin and mineral supplements.

Effectiveness and Safety of Herbal Products

Despite the popularity of herbal products in the United States, the benefits of their use are uncertain. Although many medicinal herbs contain naturally occurring compounds that exert physiological effects, few herbal products have been rigorously tested, many make unfounded claims, and some may contain contaminants or produce toxic effects.[14]

Efficacy Herbs have been used for centuries to treat medical conditions, and many have acquired reputations for being beneficial for people with specific diseases. Unfortunately, only a limited number of clinical studies support the traditional uses, and the results of studies that suggest little or no benefit are rarely publicized by the supplement industry. The National Center for Complementary and Alternative Medicine (a division of the National Institutes of Health) is currently funding large, controlled trials of several popular herbal treatments in an effort to obtain reliable efficacy and safety data.

Although labels on herbal products cannot make claims about preventing or treating specific diseases, suggestive statements are common. For example, a label may claim that a product "promotes restful sleep" but cannot state that it cures insomnia. Stores often shelve herbal products by health condition; for example, posted signs may indicate the supplements suggested for "liver health" or "men's health." The reading materials positioned close to those shelves often suggest that the products can improve one's health.

Consistency of Herbal Ingredients Herbs contain numerous compounds, and it is often unclear which of these ingredients, if any, might produce the implied beneficial effects. Because the compounds in herbs vary among species and are affected by a plant's growing conditions, different samples of an herb can have different chemical compositions. The preparation method may also cause variations in the composition of an herbal product. Some manufacturers attempt to standardize

TABLE 19-5 Popular Herbal Products, Their Common Uses, and Adverse Effects[a]

Herb	Scientific Name	Common Uses	Adverse Effects
Black cohosh	*Cimicifuga racemosa*	Relief of menopausal symptoms	Rare; occasional stomach upset, headache, weight gain
Chaparral	*Larrea tridentata*	General tonic; treatment of infection, cancer, and arthritis	Hepatitis, liver failure
Comfrey	*Symphytum officinale*	Wound healing (topical use), treatment of lung and GI disorders	Liver damage
Echinacea	*Echinacea augustifolia, E. pallida, E. purpurea*	Prevention and treatment of upper respiratory infections	Rare; stomach upset, headache, occasional allergic reactions
Feverfew	*Tanacetum parthenium*	Prevention of migraine headache	Mouth and tongue sores, swelling of lips, GI upset
Garlic	*Allium sativum*	Reduction of blood clotting, atherosclerosis, blood pressure, and blood cholesterol	Bad breath, body odor, occasional stomach upset or flatulence, excessive bleeding
Ginger	*Zingiber officinale*	Prevention and treatment of nausea and motion sickness	Rare; occasional heartburn
Ginkgo	*Ginkgo biloba*	Treatment of dementia, memory defects, and circulatory impairment	Rare; occasional stomach upset, headache, skin hypersensitivity, excessive bleeding
Ginseng	*Panax ginseng, P. quinquefolius*	General tonic, reduction of blood glucose levels	Rare
Kava kava	*Piper methysticum*	Treatment of anxiety, stress, and insomnia	Dyspepsia, restlessness, drowsiness, tremor, headache, dermatitis (with heavy use), occasional hepatitis and liver failure
St. John's wort	*Hypericum perforatum*	Treatment of mild to moderate depression	Rare; occasional stomach upset, fatigue, dizziness, headache, dry mouth, dermatitis, skin photosensitivity
Saw palmetto	*Serenoa repens*	Reduction of symptoms associated with enlarged prostate	Rare
Valerian	*Valeriana officinalis*	Sedation, treatment of insomnia	Rare
Yohimbe	*Pausinystalia yohimbe*	Treatment of erectile dysfunction	Anxiety, headache, dizziness, nausea, rapid heartbeat, hypertension, increased urinary frequency; isolated reports of renal failure, blood disorders, and airway constriction

[a] An *herb* is a nonwoody, seed-producing plant; *herbal products* include other types of plant products, such as garlic and ginkgo.

the herbal extracts they sell so that the compound believed to be beneficial is more likely to be obtained from each dose.

Even when the active ingredients in an herbal supplement have been shown to be effective, the dosage suggested on the label might not provide the quantity of active ingredients required for benefit. For example, a consumer group (ConsumerLab.com) tested seven ginkgo biloba products and found that four of the products, when consumed at the recommended dosage, lacked the amounts of the compounds indicated on the labels.[15] In a university study of echinacea preparations, 10 percent of the 59 products tested contained no measurable echinacea, and only 52 percent of the samples contained the variety of echinacea listed on the label.[16]

Safety Issues Consumers often assume that because plants are "natural," herbal products must be harmless. Many herbal remedies have toxic effects, however. The most common adverse effects of herbs include diarrhea, nausea, and vomiting. The popular herbs chaparral, comfrey, and kava have caused liver damage. The use of yohimbe (promoted for bodybuilding) has been linked to alterations in blood pressure and heart arrhythmias.[17] Note that the adverse effects of herbs are seldom listed on supplement labels.

Contamination of herbal products is another safety concern.[18] Some products have been found to contain lead and other toxic metals in excessive amounts. Other contaminants frequently found in herbal supplements include molds, bacteria, and pesticides that have been banned for use on food crops.[19] Adulteration of imported products is a serious concern: several studies found that some herbal products imported from China and India contained synthetic drugs that were not declared

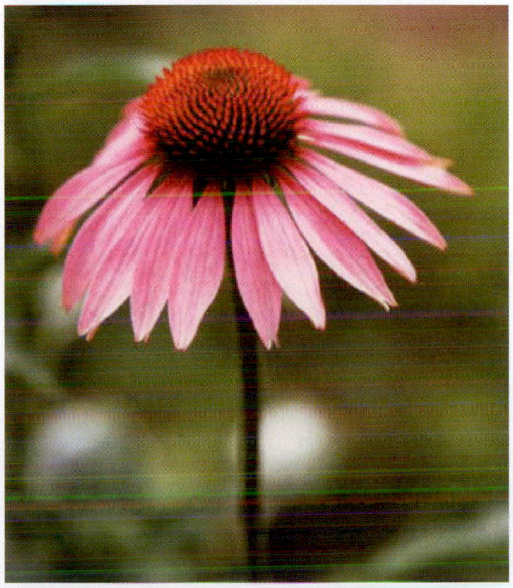

Despite the popularity of echinacea, its benefits for treating the common cold have not been supported by some well-designed clinical studies.

TABLE 19-6 Examples of Herb-Drug Interactions

Herb	Drugs	Interaction
Echinacea	Immunosuppressant drugs	Suppresses drug effects
Feverfew	Anticoagulants, aspirin	Enhances drug effects
Garlic, ginkgo, ginseng	Anticoagulants, antiplatelet drugs	Increases risk of hemorrhaging
Kava kava	Acetaminophen, antifungal drugs, methotrexate, steroids	Increases risk of liver damage
St. John's wort	Various	Suppresses drug effects
Valerian	Antidepressants, antiepileptics, anesthetics, analgesics, barbiturates	May enhance drug effects, increase sedation

on the label.[20] There have also been reports of illnesses and fatalities occurring from the intentional or accidental substitutions of one plant species for another.

Unlike drugs, herbal products do not need FDA approval before they are marketed. According to the Dietary Supplement Health and Education Act (DSHEA) of 1994, the companies that produce or distribute dietary (including herbal) supplements are responsible for determining their safety, yet these companies are not required to provide any evidence or conduct safety studies. If a company receives reports of illness or injury related to the use of its products, it is not required to submit this information to the FDA. In addition, the FDA must show that a supplement is unsafe before it can take action to remove the product from the marketplace.

Herb-Drug Interactions Like drugs, herbs may either intensify or interfere with the effects of other herbs and drugs or they may raise the risk of toxicity.[21] For example, garlic, ginkgo, and ginseng may increase the risk of bleeding when used with anticoagulant drugs. St. John's wort has been found to inhibit the actions of oral contraceptives, anticoagulants, and other drugs. Ginseng contains compounds that raise blood pressure and may increase the toxicity of drugs that have a similar side effect.[22] Unfortunately, information about herb-drug interactions is limited, and much of what is known has been obtained from case studies rather than controlled clinical trials. Table 19-6 provides some examples of herb-drug interactions.

Use of Herbal Products in Illness

When people self-medicate or ask the advice of store clerks instead of seeking effective medical treatment, the consequences are sometimes serious and irreversible. Purchasing an herbal remedy may be less stressful than a visit to the doctor, but it may delay getting an appropriate treatment and allow an illness to progress. Although retailers are not legally permitted to provide medical advice, a 2010 investigation found that sellers of herbal products routinely made improper claims that their herbal products were able to treat, prevent, or cure specific illnesses.[23]

Patients are often unaware that herbal products may be unsafe or can interact with medications. Elderly individuals (65 years old and older) are at highest risk of herb-drug interactions because most individuals in this age group take three or more prescription drugs over the course of a year.[24] Some pharmacology textbooks and handbooks now contain information about herbal supplements and potential herb-drug interactions, and various consumer websites and periodicals provide information about the safety of brand-name herbal products. Health professionals should turn to these resources to help patients who plan to use herbal supplements.

IN SUMMARY **Herbal products are not reliable treatments for medical conditions; there is little evidence demonstrating their effvectiveness and safety, and the concentrations of active ingredients may vary greatly. Safety concerns include adverse effects, contamination, and herb-drug interactions. Manufacturers and distributors of herbal supplements are responsible for determining**

product safety but are not required to conduct safety studies. The FDA must prove that a supplement is unsafe before removing it from the market. Consumers using herbs may delay getting an appropriate treatment for their condition and may receive questionable advice from supplement retailers.

Clinical Portfolio

1. An elderly woman in a residential home has been losing weight since her arrival there. She has been taking several medications to treat both a heart problem and a mild case of bronchitis. You notice that she eats only a few bites at mealtimes and seems uninterested in food. Describe several steps you can take to learn whether the medications are interfering with her food intake in some way.
2. A patient mentions that he regularly takes six to seven dietary and herbal supplements and that he has not told the physician that he uses them. His prescription medications include an antihypertensive agent (to reduce blood pressure) and warfarin. What approach might you take to learn the details about his supplement use and his reasons for taking them? If you discover that some of the supplements may pose a risk for diet-drug or herb-drug interactions with his prescription medications, what steps should you take?

Nutrition on the Net

For further study of topics covered in this chapter, log on to **www.cengagebrain.com** and search for ISBN 084006845X.

- Visit the home page of the U.S. Food and Drug Administration (FDA), the agency that regulates all drugs in the United States: **www.fda.gov**
- The FDA provides safety information about drugs and other medical products on the MedWatch website: **www.fda.gov/medwatch**
- The MedlinePlus website provides information about drugs, dietary supplements, and herbal products: **www.nlm.nih.gov/medlineplus/druginformation.html**
- Find information about dietary supplements at the Office of Dietary Supplements, a division of the National Institutes of Health: **http://ods.od.nih.gov/**

References

1. R. L. Corelli, Therapeutic and toxic potential of over-the-counter agents, in B. G. Katzung, ed., *Basic and Clinical Pharmacology* (New York: Lange Medical Books/McGraw-Hill, 2007), pp. 1041–1049.
2. P. W. Lofholm and B. G. Katzung, Rational prescribing and prescription writing, in B. G. Katzung, ed., *Basic and Clinical Pharmacology* (New York: Lange Medical Books/McGraw-Hill, 2007), pp. 1063–1072.
3. Institute of Medicine, Committee on Identifying and Preventing Medication Errors, *Preventing Medication Errors* (Washington, DC: National Academy Press, 2007).
4. L.-N. Chan, Drug–nutrient interactions, in M. E. Shils and coeditors, *Modern Nutrition in Health and Disease* (Baltimore: Lippincott Williams & Wilkins, 2006), pp. 1539–1553.
5. W. S. Leslie, C. R. Hankey, and M. E. J. Lean, Weight gain as an adverse effect of some commonly prescribed drugs: A systematic review, *QJM: An International Journal of Medicine* 100 (2007): 395–404.
6. E. Lahner and coauthors, Systematic review: Impaired drug absorption related to the co-administration of antisecretory therapy, *Alimentary Pharmacology and Therapeutics* 29 (2009): 1219–1229.
7. D. G. Bailey, Grapefruit juice–drug interaction issues, in J. I. Boullata and V. T. Armenti, eds., *Handbook of Drug-Nutrient Interactions* (Totowa, NJ: Humana Press, 2004), pp. 175–194.
8. M. L. Chavez, M. A. Jordan, and P. I. Chavez, Evidence-based drug–herbal interactions, *Life Sciences* 78 (2006): 2146–2157.
9. S. A. Shapses, Y. R. Schlussel, and M. Cifuentes, Drug-nutrient interactions that impact mineral status, in J. I. Boullata and V. T. Armenti, eds., *Handbook of Drug-Nutrient Interactions* (Totowa, NJ: Humana Press, 2004), pp. 301–328; B. J. McCabe, E. H. Frankel, and J. J. Wolfe, Monitoring nutritional status in drug regimens, in B. J. McCabe, E. H. Frankel, and J. J. Wolfe, eds., *Handbook of Food-Drug Interactions* (Boca Raton, FL: CRC Press, 2003), pp. 73–108.
10. H. F. Chambers, Antimycobacterial drugs, in B. G. Katzung, ed., *Basic and Clinical Pharmacology* (New York: Lange Medical Books/McGraw-Hill, 2007), pp. 771–780.
11. B. J. McCabe, Dietary counseling to prevent food-drug interactions, in B. J. McCabe, E. H. Frankel, and J. J. Wolfe, eds., *Handbook of Food-Drug Interactions* (Boca Raton, FL: CRC Press, 2003), pp. 295–324.
12. M. E. Gershwin and coauthors, Public safety and dietary supplementation, *Annals of the New York Academy of Sciences* 1190 (2010): 104–117.
13. U.S. Government Accountability Office (GAO), Herbal dietary supplements: Examples of deceptive or questionable marketing practices and potentially dangerous advice, GAO-10-662T (Washington, DC: May 26, 2010); J. Kennedy, Herb and supplement use in the U.S. adult population, *Clinical Therapeutics* 27 (2005): 1847–1858.
14. Gershwin and coauthors, 2010; U.S. Government Accountability Office (GAO), 2010.

15. ConsumerLab.com, Product review: Supplements for memory and cognition enhancement (ginkgo, huperzine-A, and acetyl-L-carnitine), www.consumerlab.com, posted December 30, 2009, accessed June 8, 2010.
16. C. M. Gilroy and coauthors, Echinacea and truth in labeling, *Archives of Internal Medicine* 163 (2003): 699–704.
17. J. A. Rindfleisch and B. Barrett, Herbs and other dietary supplements, in R. E. Rakel, ed., *Textbook of Family Medicine* (Philadelphia: Saunders, 2007), pp. 243–266.
18. Gershwin and coauthors, 2010; ConsumerLab.com, 2010.
19. V. H. Tournas, E. Katsoudas, and E. J. Miracco, Moulds, yeasts, and aerobic plate counts in ginseng supplements, *International Journal of Food Microbiology* 108 (2006): 178–181; K. S. Leung and coauthors, Systematic evaluation of organochlorine pesticide residues in Chinese materia medica, *Phytotherapy Research* 19 (2005): 514–518.
20. R. J. Ko, A U.S. perspective on the adverse reactions from traditional Chinese medicines, *Journal of the Chinese Medical Association* 67 (2004): 109–116; E. Ernst, Toxic heavy metals and undeclared drugs in Asian herbal medicines, *Trends in Pharmacological Sciences* 23 (2002): 136–139.
21. C. E. Dennehy and C. Tsourounis, Botanicals ("herbal medications") and nutritional supplements, in B. G. Katzung, ed., *Basic and Clinical Pharmacology* (New York: McGraw-Hill, 2007), pp. 1050–1062; Rindfleisch and Barrett, 2007; J. J. Mucksavage and L.-N. Chan, Dietary supplement interactions with medication, in J. I. Boullata and V. T. Armenti, eds., *Handbook of Drug-Nutrient Interactions* (Totowa, NJ: Humana Press, 2004), pp. 217–233.
22. J. Barnes, L. A. Anderson, and J. D. Phillipson, *Herbal Medicines: A Guide for Healthcare Professionals* (Chicago: Pharmaceutical Press, 2002).
23. U.S. Government Accountability Office (GAO), 2010.
24. U.S. Government Accountability Office (GAO), 2010.

HIGHLIGHT 19

© Art Montes De Oca/Taxi/Getty Images

Complementary and Alternative Medicine

The medical treatments described in the clinical chapters are based upon current scientific understanding of human physiology and biochemistry and are generally supported by well-conducted clinical research. This highlight examines therapies that have *not* been scientifically validated and are therefore not currently utilized by conventional medical professionals; these therapies fall into a category called *complementary and alternative medicine* (*CAM*). When the therapies are used together with conventional medicine, they are called *complementary*; when used in place of conventional medicine, they are called *alternative*.[1] The term "alternative" may be misleading in that it inappropriately implies that unproven methods of treatment are valid alternatives to conventional treatments.

Use of CAM in the United States

An estimated 38 percent of adults in the United States use some form of CAM (excluding the use of prayer).[2] CAM is most prevalent among people with chronic, debilitating diseases; for example, 84 percent of AIDS patients reportedly use CAM.[3] Many patients use CAM as an adjunct to conventional medicine—often for symptoms or illnesses that are not sufficiently helped by conventional treatments. CAM therapies remain popular despite the dearth of evidence demonstrating their effectiveness. Reasons for their popularity include consumers' growing interest in self-help measures, the noninvasive nature of many CAM therapies, and the positive interactions consumers have with CAM practitioners.[4]

In response to the enormous popularity of CAM in the United States, in 1998 Congress established the National Center for Complementary and Alternative Medicine (NCCAM), which is now one of the 27 institutes that make up the National Institutes of Health (NIH). NCCAM's missions are to investigate complementary and alternative therapies by funding well-designed scientific studies and to provide authoritative information for consumers and health professionals. If enough evidence is found to support the use of a complementary or alternative therapy, it will likely be incorporated into mainstream medical practice.[5]

Due to substantial consumer interest, many health professionals are interested in learning about CAM therapies so that they can better communicate with patients regarding their medical care and advise them when an alternative approach conflicts with standard therapy or presents a danger to health. To provide medical students with objective information about CAM, many U.S. medical schools now offer elective courses about alternative forms of treatment. Physicians who practice *integrative medicine* refer patients for complementary therapies while continuing to provide standard treatments.

Overview of CAM Therapies

CAM encompasses any and all therapies that are not normally part of conventional medicine. Consequently, the list of CAM approaches includes hundreds of advertised therapies purchased and used by consumers. Unfortunately, CAM has become a marketing buzzword and is used by unscrupulous sellers of worthless treatments. NCCAM categorizes CAM therapies as shown in Table H19-1 and defined in the Glossary of Alternative Therapies on p. 634. Several popular examples are described in this Highlight; other examples are discussed on the NCCAM website (**http://nccam.nih.gov/health**).

Alternative Medical Systems

Alternative medical systems are based on beliefs that lack the scientific basis of the theories underlying conventional medicine. Virtually all of these alternative systems were developed well over 100 years ago, before our bodies' biochemical and physiological processes were well understood. The alternative treatments may appeal to consumers because the interventions are nontechnical and seem nonthreatening. In general, however, the alternative theories and practices remain rooted in the past and have not been updated to include current knowledge.

Naturopathic Medicine

Naturopathic medicine proposes that a person's natural "life force" can foster self-healing. This life force is allegedly stimulated by certain health-promoting factors and suppressed by excesses and deficiencies. Naturopaths believe that ill health results from an internal disruption rather than from external disease-causing agents. Naturopathic therapies aim to enhance the natural healing powers of the body and may include special diets or fasting, herbal remedies and other dietary

TABLE H19-1 Examples of Complementary and Alternative Medicine

Alternative Medical Systems	Biologically Based Therapies	Energy Therapies
• Naturopathic medicine • Homeopathic medicine • Traditional Chinese medicine • Ayurveda	• Dietary supplements • Foods and special diets • Herbal products • Hormones • Aromatherapy	• Biofield therapies (including acupuncture, qi gong, and therapeutic touch) • Bioelectrical therapies (including electrical and magnetic fields)
Mind-Body Interventions	**Manipulative and Body-Based Methods**	
• Biofeedback • Meditation • Mental healing (including hypnotherapy) • Music, art, and dance therapy • Faith healing (prayer)	• Chiropractic • Massage therapy • Osteopathic manipulation • Reflexology	

supplements, acupuncture, homeopathy, massage, and various other interventions.

Homeopathic Medicine

Homeopathic medicine is based on the dubious theory that "like cures like." Homeopaths believe that a substance that causes a particular set of symptoms can be used to cure a disease that has similar symptoms. Homeopathic medicines are usually natural substances that are substantially diluted in the belief that dilution increases potency, and most remedies are so extremely diluted that the original substance is no longer present. Homeopaths theorize that even though their remedies no longer con-

GLOSSARY OF ALTERNATIVE THERAPIES

acupuncture (AK-you-PUNK-chur): a therapy that involves inserting thin needles into the skin at specific anatomical points, allegedly to correct disruptions in the flow of energy within the body.

aromatherapy: inhalation of oil extracts from plants to cure illness or enhance health.

ayurveda: a traditional medical system from India that promotes the use of diet, herbs, meditation, massage, and yoga for preventing and treating illness.

bioelectrical or **bioelectromagnetic therapies:** therapies that involve the unconventional use of electric or magnetic fields to cure illness.

biofeedback: a technique in which individuals are trained to gain voluntary control of certain physiological processes, such as skin temperature or brain wave activity, to help reduce stress and anxiety.

biofield therapies: healing methods based on the belief that illnesses can be healed by manipulating energy fields that purportedly surround and penetrate the body. Examples include *acupuncture*, *qi gong*, and *therapeutic touch*.

chiropractic (KYE-roh-PRAK-tic): a method of treatment based on the unproven theory that spinal manipulation can restore health.
- A *subluxation* is a misaligned vertebra or other spinal alteration that may cause illness.
- *Adjustment* is the manipulative therapy practiced by chiropractors.

faith healing: the use of prayer or belief in divine intervention to promote healing.

homeopathic (HO-mee-oh-PATH-ic) **medicine:** a practice based on the theory that "like cures like"; that is, substances believed to cause certain symptoms are prescribed for curing the same symptoms, but are given in extremely diluted amounts.
- **homeo** = like
- **pathos** = suffering

hypnotherapy: a technique that uses hypnosis and the power of suggestion to improve health behaviors, relieve pain, and promote healing.

imagery: the use of mental images of things or events to aid relaxation or promote self-healing.

massage therapy: manual manipulation of muscles to reduce tension, increase blood circulation, improve joint mobility, and promote healing of injuries.

meditation: a self-directed technique of calming the mind and relaxing the body.

naturopathic (NAY-chur-oh-PATH-ic) **medicine:** an approach to health care using practices alleged to enhance the body's natural healing abilities. Treatments may include a variety of alternative therapies including dietary supplements, herbal remedies, exercise, and homeopathy.

osteopathic (OS-tee-oh-PATH-ic) **manipulation:** a CAM technique performed by a doctor of osteopathy (D.O., or osteopath) that includes deep tissue massage and manipulation of the joints, spine, and soft tissues. A D.O. is a fully trained and licensed medical physician, although osteopathic manipulation has not been proved to be an effective treatment.

qi gong (chee-GUNG): a traditional Chinese system that combines movement, meditation, and breathing techniques and allegedly cures illness by enhancing the flow of qi (energy) within the body.

reflexology: a technique that applies pressure or massage on areas of the hands or feet to allegedly cure disease or relieve pain in other areas of the body; sometimes called *zone therapy*.

therapeutic touch: a technique of passing hands over a patient to purportedly identify energy imbalances and transfer healing power from therapist to patient; also called *laying on of hands*.

traditional Chinese medicine (TCM): an approach to health care based on the concept that illness can be cured by enhancing the flow of qi (energy) within a person's body. Treatments may include herbal therapies, physical exercises, meditation, acupuncture, and remedial massage.

tain a diluted substance, they still have powerful healing effects because the water structure is somehow altered during the dilution process used to prepare homeopathic medicines. This theory, however, conflicts with scientific understanding of water structure and properties.

Traditional Chinese Medicine

Traditional Chinese medicine (TCM) includes a large number of folk practices that originated in China. TCM is based on the theory that the body has pathways (called *meridians*) that conduct energy (called *qi*; pronounced "chee"). The interrupted flow of qi is believed to cause illness. TCM practices allegedly improve the flow of qi and include acupuncture, qi gong, herbal remedies, dietary practices, and massage. (Acupuncture and qi gong are described in a later section on energy therapies.) Ironically, TCM is used by relatively few in the Chinese population, as Chinese physicians have largely adopted the Western approach to managing illness.[6]

Mind-Body Interventions

Mind-body interventions attempt to improve a person's sense of psychological or spiritual well-being despite the presence of illness. The treatments are also used in the hope of reducing stress, dealing with pain, or lowering blood pressure. Some of these therapies have been incorporated into mainstream medicine for stress reduction or relaxation. For example, **biofeedback** training, in which individuals learn to monitor skin temperature, muscle tension, or brain wave activity while practicing relaxation techniques, is frequently taught by behavioral medicine specialists to help patients reduce stress or anxiety. Other techniques to reduce stress and promote relaxation include **meditation**, art and music therapy, and prayer.

The clinical applications of other mind-body therapies are far more questionable. One example is guided **imagery**, in which a person tries to reverse the disease process (for example, shrink a tumor) by using mental pictures. Another example is the use of **faith healing** in place of proven conventional treatments to cure disease.

Biofeedback training is a stress reduction and relaxation technique.

Biologically Based Therapies

Biological therapies include the use of natural products, such as vitamin supplements, herbal and plant extracts, and special foods (Chapter 19 provides information about herbal products). Because the FDA does not regulate these products, there is no way of knowing whether they are safe or effective. Moreover, the amount of active ingredient in a dose (as listed on the label) may not be accurate, and the potential hazards of using many of these products are unknown.

Hormones

Some hormones or hormone-like products derived from foods are considered dietary supplements and can be sold over the counter. One example is melatonin, a hormone made by the pineal gland and alleged to correct sleep disorders and prevent jet lag. Another example is the adrenal hormone dehydroepiandrosterone (DHEA), which is promoted to enhance immunity, increase muscle mass, improve memory, and defend against aging.

Glucosamine-Chondroitin Supplements

The use of glucosamine and chondroitin supplements is an example of a CAM therapy that is being considered for adoption by mainstream medicine, depending on the outcome of studies of their safety and effectiveness. Glucosamine and chondroitin are produced in the body and help to maintain joint cartilage. Early studies suggested that glucosamine and chondroitin supplements reduced moderate to severe symptoms of osteoarthritis better than a placebo, prompting some physicians to recommend using these supplements for pain relief.[7] Recent studies have cast doubt on the earlier findings, however, and several trials are still in progress.[8]

Aromatherapy

Aromatherapy is the practice of inhaling aromatic substances derived from plants, called *essential oils*. Aromatherapy allegedly improves health and enhances natural healing processes. Popular examples of essential oils include those from eucalyptus, lavender, peppermint, rosemary, and lemon.

Manipulative and Body-Based Methods

Manipulative interventions include physical touch, forceful movement of different parts of the body, and the application of pressure. Some practitioners maintain that special energy fields are manipulated during the physical treatment and that proper energy flow induces healing.

Chiropractic

Chiropractic theory proposes that keeping the nervous system free from obstruction allows the body to heal itself, because the healing process stems from the brain and is conducted via the spinal cord and nerves to all parts of the body. Chiropractors claim to diagnose illnesses by detecting subluxations in the spine, which are variously described as misaligned vertebrae or pinched nerves that allegedly cause subtle interferences within the nervous system. The main treatment is the adjustment, a manual manipulation that is said to correct a subluxation and restore the body's natural healing ability. Although spinal manipulation has mainly been found to be helpful for improving back pain, many chiropractors still assert that chiropractic can cure disease rather than simply relieve symptoms.[9] For example, many chiropractors promote spinal manipulation to treat infectious diseases and prevent cancer, even though the nervous system and spinal alignment do not play roles in the pathology of these conditions.

Massage Therapy

Massage therapy is the manipulation of muscle and connective tissue to improve muscle function, reduce pain, or promote relaxation. Massage therapists may also apply heat or cold and give advice about exercises that may improve muscle tone and range of motion. Massage is often integrated into conventional physical therapy, although some massage therapists may incorrectly suggest that massage is a valid treatment for a wide range of medical conditions.

Energy Therapies

Two categories of therapies involve the alleged curative power of "energy." **Biofield therapies** are said to influence the energy that surrounds or pervades the human body, and their proponents claim that an energy therapy can strengthen or restore a person's "energy flow" and induce healing. Acupuncture, qi gong, and therapeutic touch are among the therapies that subscribe to these theories. Note that CAM adherents often use the term "energy" unscientifically and that there is no objective evidence of this sort of energy flow. **Bioelectrical** or **bioelectromagnetic therapies** use electric or magnetic fields to allegedly promote healing; for example, magnets have been marketed with claims that they can improve circulation, reduce inflammation, and speed recovery from injuries.

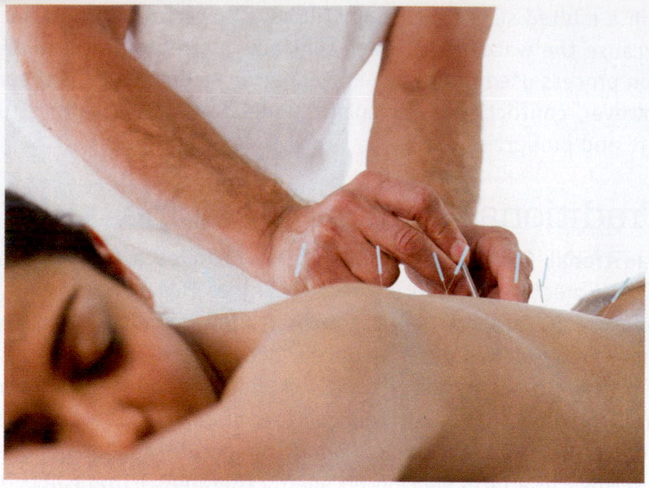

Acupuncture involves the shallow insertion of stainless steel needles into the skin, sometimes accompanied by a low-frequency current.

Acupuncture

Acupuncture, a component of traditional Chinese medicine, is based on the theory that disease is caused by the disrupted flow of qi through the body. Acupuncture allegedly corrects such disruptions and restores health. The practice involves the shallow insertion of stainless steel needles into the skin at designated points on the body, sometimes accompanied by a low-frequency current to produce greater stimulation.

Qi Gong

Qi gong is another therapy originating in China that is said to improve the flow of qi within the body. Qi gong masters allegedly cure disease by releasing energy from their body and passing it to the person being treated. Self-help practices include deep breathing, certain types of physical exercise, and concentration and relaxation techniques.

Therapeutic Touch

Therapeutic touch is based on the premise that the "healing force" of a practitioner can be used to cure disease. Practitioners claim to identify and correct energy imbalances by passing their hands above a patient's body and transferring "excess energy" to the patient.

Is CAM Safe and Effective?

As mentioned earlier, CAM treatments are generally excluded from mainstream medical practice because there is no evidence proving that they are effective for treating the diseases and medical conditions for which they are used. Many consumers think otherwise and seem satisfied that these treatments "work." How is this dichotomy to be explained?

Does CAM Work?

Surveys suggest that consumers perceive their visits to CAM therapists as far more pleasant than visits to conventional health practitioners. CAM therapists often spend more time with patients, are more attentive, and use less invasive interventions.[10] Self-help measures are encouraged, so the consumer has more control over the treatment. The therapies appear to be more "natural" and to have fewer side effects. Possible explanations for "cures" include the following:

- A person may seem cured because of misdiagnosis; that is, the condition diagnosed by the CAM practitioner may not have actually existed.
- The condition may have been self-limiting, or it may have gone into temporary remission after the treatment.
- Undue credit may be inappropriately assigned to the CAM therapy when the improvement was actually due to a previous or concurrent conventional treatment.
- The placebo effect may have had an influence on the course of disease.

The central question remains: Do the CAM therapies merely make people *feel* better, or do they really *get* better? This question can be answered only by well-controlled research studies.

Potential Hazards of CAM

One of the attractions of alternative therapies is the assumption that they are safe. Recall, however, the concerns associated with the use of herbal products discussed in Chapter 19, which include the potential toxicity of herbal ingredients, product contamination or adulteration, and interactions with conventional medications. Another concern is that use of CAM therapies may delay the use of reliable treatments that have demonstrable benefits.[11] Various reports have described people with treatable medical conditions who suffered permanent disability or death when they were misdiagnosed or improperly treated by CAM practitioners. For example, a rare but well-known risk of spinal cord injury or stroke is associated with a type of cervical manipulation performed by chiropractors.[12] Unfortunately, because most CAM therapies are not regulated or monitored, there are no accurate estimates of their adverse effects.

Working with Patients Who Use CAM

CAM therapies may have consequences that influence the course of a disease and its treatment. Accordingly, it is important that health practitioners routinely inquire about the use of CAM therapies and educate patients about the hazards of postponing or stopping conventional treatment. Patients should also be told about potential interactions between conventional treatments and CAM therapies. Some patients may want to learn about differences between evidence-based medical practices and untested CAM theories and may be interested in the integrative medicine options available.

All alternative therapies have one characteristic in common: their effectiveness is, for the most part, unproven.[13] Because patients often choose CAM therapies because of positive interactions with CAM practitioners, health care practitioners should realize that empathizing with patients may go a long way toward winning their trust and improving their compliance with therapy. Furthermore, health practitioners should obtain reliable, objective resources to update their knowledge about unconventional practices so that they can knowledgeably discuss these options with patients.

References

1. S. E. Straus, Complementary and alternative medicine, in L. Goldman and D. Ausiello, eds., *Cecil Medicine* (Philadelphia: Saunders, 2008), pp. 206–209.
2. P. M. Barnes, B. Bloom, and R. Nahin, Complementary and alternative medicine use among adults and children: United States, 2007, *CDC National Health Statistics Report* 12 (December 10, 2008): 1–24.
3. J. D. Berman and S. E. Straus, Implementing a research agenda for complementary and alternative medicine, *Annual Review of Medicine* 55 (2004): 239–254.
4. A. M. McCaffrey, G. F. Pugh, and B. B. O'Connor, Understanding patient preference for integrative medical care: Results from patient focus groups, *Journal of General Internal Medicine* 22 (2007): 1500–1505; E. Ernst, The role of complementary and alternative medicine, *British Medical Journal* 321 (2000): 1133–1135.
5. Straus, 2008.
6. D. Normile, The new face of Chinese medicine, *Science* 299 (2003): 188–190.
7. A. D. Sawitzke and coauthors, The effect of glucosamine and/or chondroitin sulfate on the progression of knee osteoarthritis: A report from the glucosamine/chondroitin arthritis intervention trial, *Arthritis and Rheumatism* 58 (2008): 3183–3191; O. Bruyere and J. Y. Reginster, Glucosamine and chondroitin sulfate as therapeutic agents for knee and hip osteoarthritis, *Drugs and Aging* 24 (2007): 573–580.
8. R. M. Rozendaal and coauthors, Effect of glucosamine sulfate on hip osteoarthritis, *Annals of Internal Medicine* 148 (2008): 268–277; S. Reichenbach and coauthors, Meta-analysis: Chondroitin for osteoarthritis of the knee or hip, *Annals of Internal Medicine* 146 (2007): 580–590.
9. J. C. Keating and coauthors, Subluxation: Dogma or science? *Chiropractic and Osteopathy* (August 2005) epub 10, available at doi: 10.1186/1746-1340-13-17.
10. McCaffrey, Pugh, and O'Connor, 2007; B. Barrett and coauthors, What complementary and alternative medicine practitioners say about health and health care, *Annals of Family Medicine* 2 (2004): 253–259.
11. Straus, 2008.
12. W.-L. Chen and coauthors, Vertebral artery dissection and cerebellar infarction following chiropractic manipulation, *Emergency Medical Journal* 23 (2006): e1, available at doi: 10.1136/emj.2004.015636; W. S. Smith and coauthors, Spinal manipulative therapy is an independent risk factor for vertebral artery dissection, *Neurology* 13 (2003): 1424–1428; R. Dziewas and coauthors, Cervical artery dissection—Clinical features, risk factors, therapy and outcome in 126 patients, *Journal of Neurology* 250 (2003): 1179–1184.
13. Barnes, Bloom, and Nahin, 2008.

CHAPTER

20

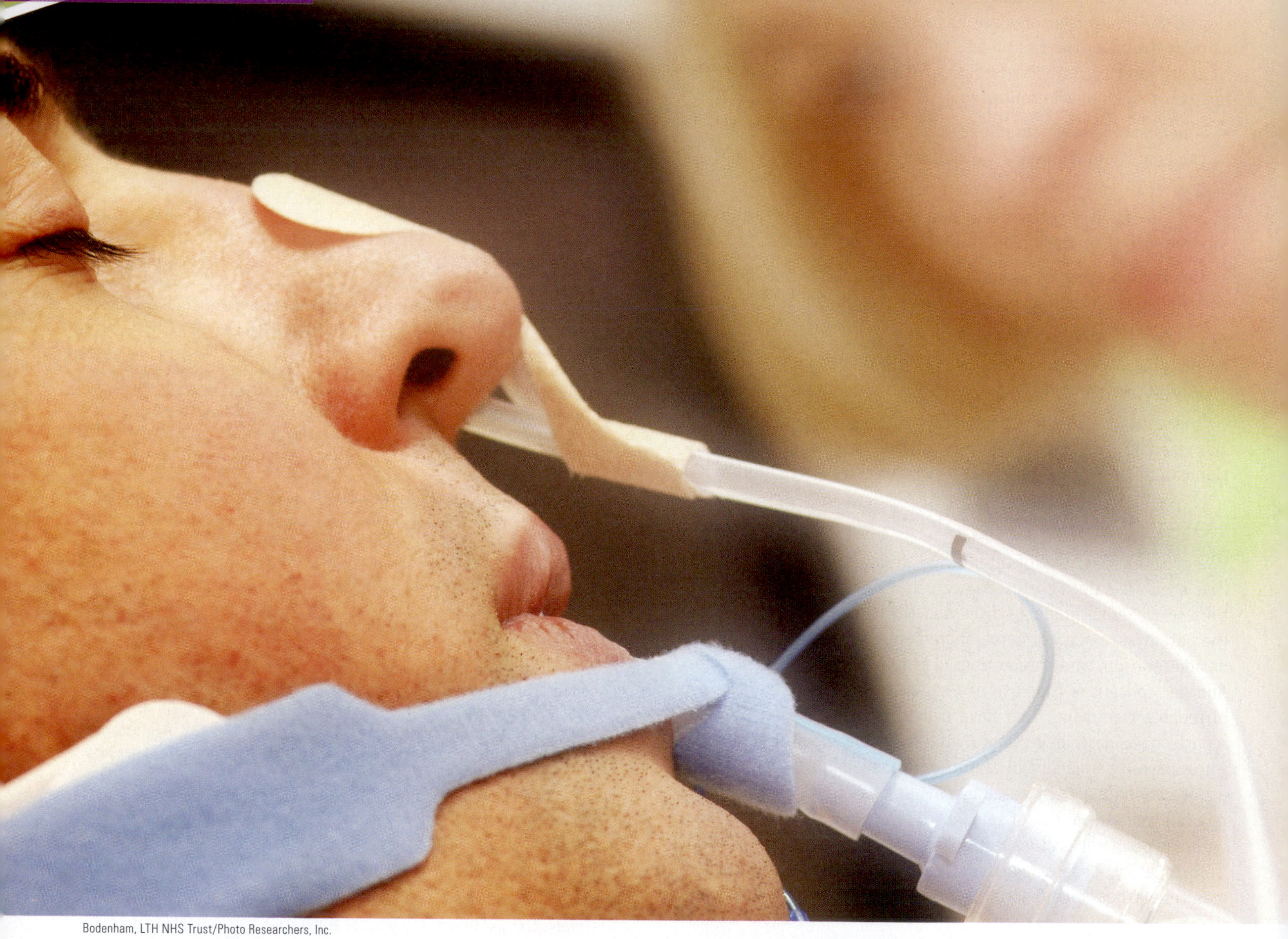

Bodenham, LTH NHS Trust/Photo Researchers, Inc.

Nutrition in the Clinical Setting

Patients are often too sick to obtain the energy and nutrients they need by consuming foods. In such cases, enteral nutrition support can help many patients regain health. Because some enteral formulas are grocery items, patients usually feel comfortable using them as oral supplements or meal substitutes. Tube feedings, however, are unfamiliar to most people, so patients and caregivers may be anxious about them at first. Showing understanding and carefully explaining the procedure can help to alleviate patients' concerns.

Throughout this chapter, the CourseMate icon indicates an opportunity for online self-study, linking you to activities to increase your understanding of chapter concepts. **www.cengagebrain.com** (search for ISBN 084006845X)

Enteral Nutrition Support

CHAPTER OUTLINE

Some illnesses may interfere with eating, digestion, or absorption to such a degree that conventional foods cannot supply the necessary nutrients. In such cases, **nutrition support**—the delivery of nutrients using a feeding tube or intravenous infusions—can meet a patient's nutritional needs. **Enteral nutrition** provides nutrients using the gastrointestinal (GI) tract. Enteral nutrition includes oral diets or supplements, but the term more often refers to the use of tube feedings, which supply nutrients directly to the stomach or intestine via a thin, flexible tube. **Parenteral nutrition**, discussed in Chapter 21, provides nutrients intravenously to patients who do not have adequate gastrointestinal function to handle enteral feedings. If the GI tract remains functional, enteral nutrition support is preferred, partly to avoid the expense and complications associated with intravenous infusions and partly to preserve healthy GI function.

If gastrointestinal function is normal and a poor appetite is the primary nutrition problem, enteral formulas can be provided as oral supplements to the usual diet. If patients cannot consume enough food or drink enough formula to meet nutrient needs, tube feedings may be used to deliver the required nutrients.

Enteral Formulas

More than 100 enteral formulas are currently marketed.[1] Most formulas can supply all of an individual's nutrient requirements when consumed in sufficient volume, a necessity for the patient who is using a tube feeding or oral liquid diet for more than a few days. Thus, an enteral formula can be considered a liquid form of a standard or modified diet.

Types of Enteral Formulas

Enteral formulas are categorized according to their macronutrient sources. ♦ **Standard formulas** usually contain intact proteins and polysaccharides, whereas **elemental formulas** contain macronutrients that have been broken down to some extent and require less digestion. **Specialized formulas** include nutrient combinations that can assist in the treatment of certain illnesses. When an ideal formula is unavailable, a **modular formula** can be prepared in the hospital pharmacy by combining individual macronutrient preparations (called *modules*). Examples of enteral formulas are provided in Appendix K.

Standard Formulas Standard formulas, also called *polymeric formulas*, are provided to individuals who can digest and absorb nutrients without difficulty. They contain intact proteins extracted from milk or soybeans (called **protein isolates**) or a combination of such proteins. The carbohydrate sources include modified starches, glucose polymers (such as maltodextrin), and sugars. A few formulas, called **blenderized formulas**, are made from whole foods and derive their protein primarily from pureed meat or poultry.

♦ Reminder: The *macronutrients* are carbohydrates, fats, and proteins.

nutrition support: the delivery of nutrients using a feeding tube or intravenous infusions.

enteral (EN-ter-al) **nutrition:** the provision of nutrients using the GI tract, including the use of tube feedings and oral diets.

parenteral (par-EN-ter-al) **nutrition:** the intravenous provision of nutrients that bypasses the GI tract.

- **par** = beside
- **entero** = intestine

standard formulas: enteral formulas that contain mostly intact proteins and polysaccharides; also called *polymeric formulas*.

elemental formulas: enteral formulas that contain carbohydrates and proteins that are partially or fully hydrolyzed; also called *hydrolyzed, chemically defined*, or *monomeric formulas*.

specialized formulas: enteral formulas designed to meet the nutrient needs of patients with specific illnesses; also called *disease-specific formulas*.

modular formulas: enteral formulas prepared in the hospital from *modules* that contain single macronutrients; used for people with unique nutrient needs.

protein isolates: proteins that have been isolated from foods.

blenderized formulas: enteral formulas that are prepared by using a food blender to mix and puree whole foods.

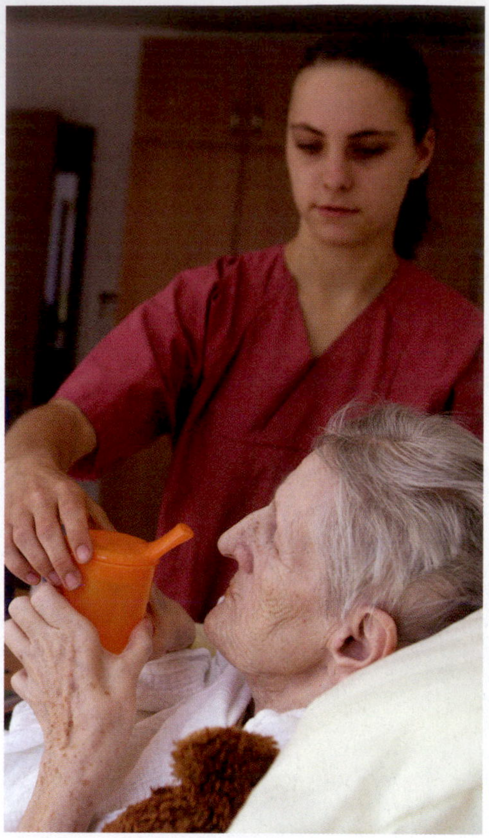
© Caro/Alamy

Patients can drink enteral formulas when they are unable to consume enough food from a conventional diet.

TABLE 20-1 Macronutrient Sources in Standard and Elemental Formulas

Type of Formula	Protein Sources	Carbohydrate Sources	Fat Sources
Standard formulas	Intact proteins, such as casein, whey, lactalbumin, and soy protein isolates Milk protein concentrate Egg white	Corn syrup solids Hydrolyzed cornstarch Sucrose Fructose	Vegetable oils (such as corn oil, soybean oil, and canola oil) MCT Palm kernel oil
Elemental formulas	Hydrolyzed casein, whey, lactalbumin, or soy protein Crystalline amino acids	Hydrolyzed cornstarch Maltodextrin Fructose	Vegetable oils (such as corn oil, soybean oil, and canola oil) MCT

NOTE: MCT = medium-chain triglycerides.

Elemental Formulas Elemental formulas, also called *hydrolyzed*, *chemically defined*, or *monomeric formulas*, are prescribed for patients who have compromised digestive or absorptive functions. Elemental formulas contain proteins and carbohydrates that have been partially or fully broken down to fragments that require little (if any) digestion. The formulas are often low in fat and may contain **medium-chain triglycerides (MCT)** to ease digestion and absorption. Table 20-1 compares the sources of macronutrients in standard and elemental formulas.

Specialized Formulas Specialized formulas, also called *disease-specific formulas*, are designed to meet the nutrient needs of patients with particular illnesses. Products have been developed for individuals with liver, kidney, and lung diseases; glucose intolerance; and metabolic stress (later chapters provide details). Specialized formulas are generally expensive, and their effectiveness is controversial.[2]

Modular Formulas Modular formulas are sometimes prepared for patients who require specific nutrient combinations to treat their illnesses. Vitamin and mineral preparations are also included in these formulas so that they can meet all of a person's nutrient needs. In some cases, one or more modules are added to other enteral formulas to adjust their nutrient composition.

Formula Characteristics

Formulas vary in their nutrient and energy densities so that they can supply the required nutrients in different volumes of fluid. The fiber content influences intestinal function and blood glucose control. These properties affect the administration of tube feedings, as well as the side effects that patients may experience.

Macronutrient Composition The amounts of protein, carbohydrate, and fat in enteral formulas vary substantially (see Appendix K for details). ♦ The protein content of most formulas ranges from 12 to 20 percent of total kcalories;[3] note that protein needs are high in patients with severe metabolic stress, whereas protein restrictions are necessary for patients with chronic kidney disease. Carbohydrate and fat provide most of the energy in enteral formulas; standard formulas generally provide 40 to 60 percent of kcalories from carbohydrate and 30 to 40 percent of kcalories from fat.[4]

♦ Reminder: Chapter 1 includes calculations for determining the macronutrient composition and energy content of foods and diets (see p. 9).

Energy Density The energy density of enteral formulas ranges from 0.5 to 2.0 kcalories per milliliter of fluid. Standard formulas provide 1.0 to 1.2 kcalories per milliliter and are appropriate for patients with average fluid requirements. Formulas that have higher energy densities can meet energy and nutrient needs in a smaller volume of fluid and therefore benefit patients who have high nutrient needs or fluid restrictions. Individuals with high fluid needs can be given a formula with low energy density or be supplied with additional water via the feeding tube or intravenously.

medium-chain triglycerides (MCT): triglycerides that contain fatty acids that are 8 to 10 carbons in length. MCT do not require digestion and can be absorbed in the absence of lipase or bile.

Fiber Content Fiber-containing formulas can be helpful for improving fecal bulk and colonic function, treating diarrhea or constipation, and maintaining blood glucose control. Conversely, fiber-containing formulas are avoided in patients with acute intestinal conditions or pancreatitis, and before or after some intestinal examinations and surgeries.

Osmolality **Osmolality** refers to the moles of osmotically active solutes (or *osmoles*) per kilogram of solvent. ♦ An enteral formula with an osmolality similar to that of blood serum (about 300 milliosmoles per kilogram) is an **isotonic formula**, whereas a **hypertonic formula** has an osmolality greater than that of blood serum.

Most enteral formulas have osmolalities between 300 and 700 milliosmoles per kilogram; generally, elemental formulas and nutrient-dense formulas have higher osmolalities than standard formulas. Most people are able to tolerate both isotonic and hypertonic feedings without difficulty.[5] When medications are infused along with enteral feedings, however, the osmotic load increases substantially and may contribute to the diarrhea experienced by many tube-fed patients.

♦ Reminder: Osmotically active solutes affect *osmosis*, the movement of water across biological membranes (see pp. 390–391).

osmolality (OZ-moe-LAL-ih-tee): the concentration of osmotically active solutes in a solution, expressed as milliosmoles (mOsm) per kilogram of solvent.

isotonic formula: a formula with an osmolality similar to that of blood serum (about 300 milliosmoles per kilogram).

- **iso** = equal
- **tono** = pressure

hypertonic formula: a formula with an osmolality greater than that of blood serum.

Formula Selection

The formula is selected after careful assessment of the patient's medical problems, fluid and nutrition status, and ability to digest and absorb nutrients; some of the factors considered are shown in Figure 20-1.

FIGURE 20-1 Selecting a Formula

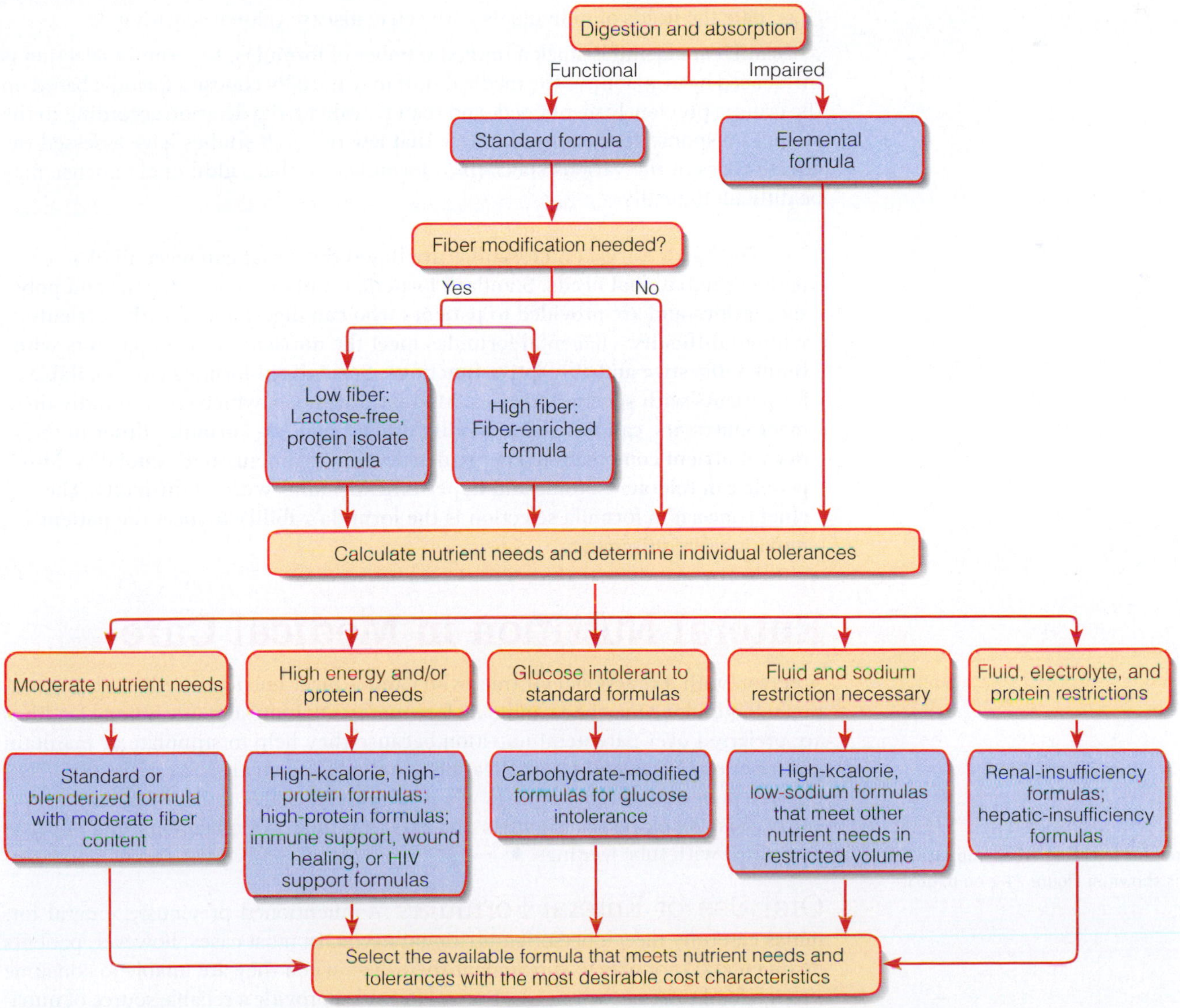

Generally, the best formula is one that meets the patient's medical and nutrient needs with the lowest risk of complications and the lowest cost. The vast majority of patients can use standard formulas. A person with a functional, but impaired, GI tract may require an elemental formula. Factors that influence formula selection include:

- *Nutrient and energy needs.* As with patients consuming regular diets, an adjustment in macronutrient and energy intakes may be necessary for tube-fed patients. For example, patients with diabetes may need to control carbohydrate intake, critical-care patients may have high protein and energy requirements, and patients with chronic kidney disease may need to limit their intakes of protein and several minerals.
- *Fluid requirements.* High nutrient needs must be met using the volume of formula a patient can tolerate. If fluids are restricted, the formula should have adequate nutrient content and energy density to deliver the required nutrients in the volume prescribed.
- *The need for fiber modifications.* The choice of formulas is narrower if fiber intake needs to be high or low. Formulas that provide fiber may be helpful for managing diarrhea, constipation, or hyperglycemia in some patients; other patients may need to avoid fiber due to an increased risk of GI obstructions.[6]
- *Individual tolerances (food allergies and sensitivities).* Most formulas are lactose-free, because many patients who need enteral formulas have some degree of lactose intolerance. Many formulas are also gluten-free and can accommodate the needs of individuals with celiac disease (gluten sensitivity).

Health care facilities stock a limited number of formulas, so formula selection is influenced by availability. The medical staff may initially choose a formula based on the criteria previously mentioned, and then reevaluate the decision according to the patient's response to the formula. Note that few research studies have assessed the effectiveness of the various specialized formulas, so their additional expense may be difficult to justify.

IN SUMMARY **Enteral formulas are liquid diets that can meet all of a patient's nutritional needs. Standard formulas contain intact proteins and polysaccharides and are provided to patients who can digest and absorb nutrients without difficulty; elemental formulas meet the nutrient needs of patients with limited digestive and absorptive functions. Specialized formulas are available for patients with specific diseases. Modular formulas, which contain individual macronutrients, can be used to modify other formulas. Formulas differ in their macronutrient composition, energy density, fiber content, and osmolality. Most people can tolerate isotonic and hypertonic formulas without difficulty. The chief concern in formula selection is the formula's ability to meet the patient's nutritional requirements.**

Enteral Nutrition in Medical Care

A person with a functional GI tract who cannot meet nutrient needs with conventional foods alone may be a candidate for enteral nutrition support. Enteral feedings are preferred over parenteral nutrition because they help to stimulate or maintain gut function, cause fewer complications, and are less costly.[7] Similarly, oral feedings are preferred to tube feedings when the person is able to drink enteral formulas, because drinking the formulas prevents the stress, complications, and expense associated with tube feedings. ♦

♦ A decision tree for selecting an appropriate feeding method is shown in Figure 21-1 on p. 663.

Oral Use of Enteral Formulas

As mentioned previously, enteral formulas can fully meet a person's nutritional needs. In most cases, however, patients drink enteral formulas to supplement their diets when they are unable to consume enough food to meet their needs. Enteral formulas provide a reliable source of nutri-

HOW TO Help Patients Accept Oral Formulas

People using enteral formulas are often quite ill and have poor appetites. Even when a person enjoys a formula, the taste can become monotonous in time. Elemental formulas are usually less palatable than standard formulas, and patients may find them difficult to drink. Health practitioners can motivate patients to drink formulas by trying these suggestions:

- Let the patient sample different formulas that are appropriate for his or her needs, and use only those that the patient enjoys.
- Serve formulas attractively, and remind patients to drink them. Formulas offered in a glass on an attractive plate may be more appealing than those served from a can with an unfamiliar name.
- If a patient finds the smell of a formula unappealing, it may help to cover the top of the glass with plastic wrap or a lid, leaving just enough room for a straw.
- Provide easy access. Keep the formula close to the patient's bed where it can be reached with little effort and within sight so that the patient is reminded to drink it. Patients who are very ill may lack the motivation to reach for the formula, let alone drink it.
- Try keeping the formula in an ice bath so that it will be cool and refreshing when the patient drinks it. Check with the patient to make sure the colder temperature is suitable.
- For patients with little appetite, offer the formula in small amounts that are easy to tolerate, and serve it more frequently during the day.
- If the patient stops enjoying the formula, recommend different flavors or try other formulas. Check with the pharmacy to see if alternative enteral products are available, such as milkshakes, puddings, or snack bars.

TRY IT Take a trip to your local grocery store to learn about the availability of different enteral formulas. If available, compare the nutrient content, flavors, and prices of two or three different products. For one of the formulas, calculate the volume required to meet your estimated daily energy needs.

ents and add energy and protein to the diets of malnourished patients. Those who are weak or debilitated may also find it easier to manage formulas than meals.

When a patient drinks a formula, taste becomes an important consideration. Allowing patients to sample different products and flavors and to select the ones they prefer helps to promote acceptance. The "How To" above offers additional suggestions for helping patients to accept and enjoy oral formulas.

Several enteral products are sold in pharmacies and grocery stores for home use; examples include Ensure, Boost, and Carnation Instant Breakfast. These products are sometimes used as nutrition supplements or convenient meal replacements by healthy individuals. The products are available in ready-to-drink liquid form or in powdered forms that must be reconstituted with water or milk.

Indications for Tube Feedings

Tube feedings are typically recommended for patients at risk of protein-energy malnutrition who are unable to consume adequate food or formula for at least seven days.[8] The following medical conditions may indicate the need for tube feedings:

- Severe swallowing disorders
- Impaired motility in the upper GI tract
- Gastrointestinal obstructions and **fistulas** that can be bypassed with a feeding tube
- Certain types of intestinal surgeries
- Mechanical ventilation
- Extremely high nutrient requirements
- Little or no appetite for extended periods, especially if the patient is malnourished
- Mental incapacitation due to confusion, neurological disorders, or coma

Contraindications for tube feedings include severe GI bleeding, high-output fistulas, **intractable** vomiting or diarrhea, complete intestinal obstruction, and severe malabsorption.[9] In addition, various clinical studies have suggested that tube

fistulas (FIST-you-luz): abnormal passages between organs or tissues (or between an internal organ and the body's surface) that permit the passage of fluids or secretions.

intractable: not easily managed or controlled.

FIGURE 20-2 Tube Feeding Routes

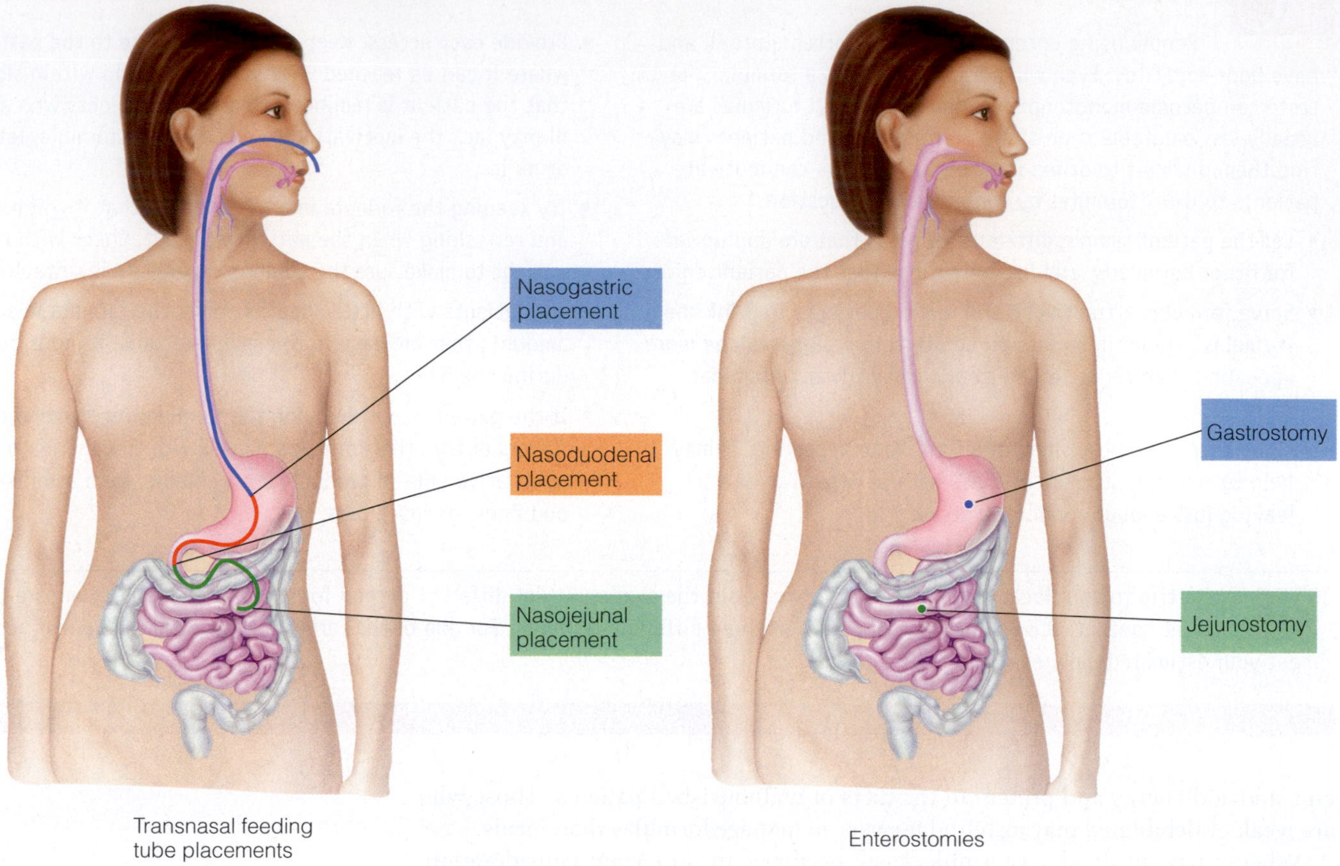

feedings are not always effective in some of the patient populations in which they are routinely used; thus, the decision to use tube feedings should be considered in light of the most recent research evidence.[10]

Feeding Routes The feeding route chosen depends on the patient's medical condition, the expected duration of tube feeding, and the potential complications of a particular route. Figure 20-2 illustrates the main feeding routes, and the Glossary of Tube Feeding Routes describes each route.

Gastrointestinal Access When a patient is expected to be tube fed for less than four weeks, a **nasogastric** or **nasoenteric** route is generally chosen; for these routes, the feeding tube is passed into the GI tract via the nose. The patient is frequently

GLOSSARY OF TUBE FEEDING ROUTES

For each type of tube placement, the terms are listed in order from the upper to lower organs of the digestive system.

transnasal: a *transnasal feeding tube* is one that is inserted through the nose.

- **nasogastric (NG):** tube is placed into the stomach via the nose.
- **nasoenteric:** tube is placed into the GI tract via the nose. (*Nasoenteric feedings* usually refer to *nasoduodenal* and *nasojejunal* feedings.)
- **nasoduodenal (ND):** tube is placed into the duodenum via the nose.
- **nasojejunal (NJ):** tube is placed into the jejunum via the nose.

orogastric: tube is inserted into the stomach through the mouth. This method is often used to feed infants because a nasogastric tube may hinder the infant's breathing.

enterostomy (EN-ter-AH-stoe-mee): an opening into the GI tract through the abdominal wall.

- **gastrostomy** (gah-STRAH-stoe-mee): an opening into the stomach through which a feeding tube can be passed. A nonsurgical technique for creating a gastrostomy under local anesthesia is called *percutaneous endoscopic gastrostomy* (*PEG*).
- **jejunostomy** (JEH-ju-NAH-stoe-mee): an opening into the jejunum through which a feeding tube can be passed. A nonsurgical technique for creating a jejunostomy is called *percutaneous endoscopic jejunostomy* (*PEJ*). The tube can either be guided into the jejunum via a gastrostomy or passed directly into the jejunum (*direct PEJ*).

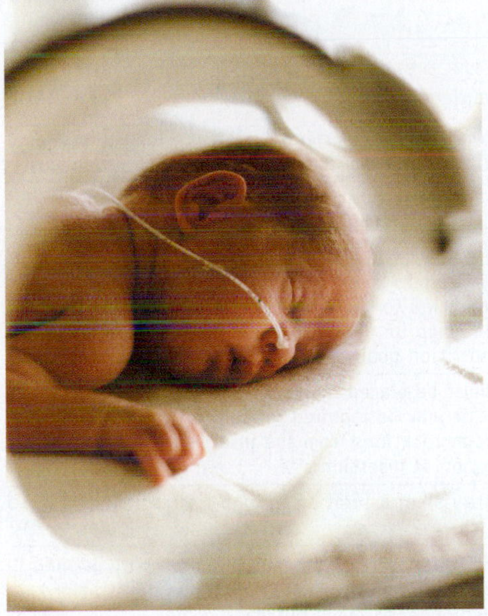
Ursula Markus/Photo Researchers, Inc.

A transnasal feeding tube accesses the GI tract via the nose.

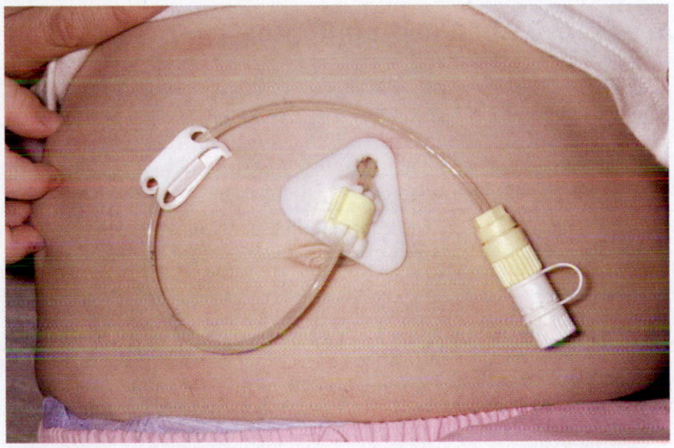
Dr. P. Marazzi/Photo Researchers, Inc.

In a gastrostomy, the feeding tube accesses the GI tract through the abdominal wall.

awake during **transnasal** (through-the-nose) placement of a feeding tube. While the patient is in a slightly upright position with head tilted, the tube is inserted into a nostril and passed into the stomach (nasogastric placement), duodenum (**naso-duodenal** placement), or jejunum (**nasojejunal** placement). If the patient is awake and alert, he or she can swallow water to ease the tube's passage. The final position of the feeding tube tip is verified by abdominal X-ray or other means. In infants, **orogastric** placement, in which the feeding tube is passed into the stomach via the mouth, is sometimes preferred over transnasal routes; this placement allows the infant to breathe more normally during feedings.

When a patient will be tube fed for longer than four weeks or if the nasoenteric route is inaccessible due to an obstruction or other medical reasons, a direct route to the stomach or intestine may be created by passing the tube through an **enterostomy**, an opening in the abdominal wall that leads to the stomach (**gastrostomy**) or jejunum (**jejunostomy**). An enterostomy can be made by either surgical incision or needle puncture.

Selecting a Feeding Route As mentioned, transnasal access is usually preferred when the tube feeding duration is expected to be less than four weeks, and enterostomies are often appropriate when tube feedings are planned for longer periods. Gastric feedings (nasogastric and gastrostomy routes) are preferred whenever possible. These feedings are more easily tolerated and less complicated to deliver than intestinal feedings because the stomach controls the rate at which nutrients enter the intestine. Gastric feedings are not possible, however, if patients have gastric obstructions or motility disorders that interfere with the stomach's ability to empty.

Gastric feedings are often avoided in patients at high risk of **aspiration**, ♦ a common complication in which formula or GI secretions enter the lungs, possibly from the backflow of stomach contents. **Aspiration pneumonia**, a lung disease that is sometimes fatal, may result. Although health practitioners frequently administer nasoenteric feedings to minimize the possibility of aspiration, studies have not consistently shown that gastric feedings are associated with increased aspiration risk.[11] Table 20-2 summarizes the advantages and disadvantages of the various tube feeding routes.

Feeding Tubes Feeding tubes are made from soft, flexible materials (usually silicone, polyurethane, or polyvinyl) and come in a variety of lengths and diameters. The tube selected largely depends on the patient's age and size, the feeding

♦ Aspiration risk is high in patients with esophageal disorders, neuromuscular diseases, and conditions that reduce consciousness or cause dementia.

aspiration: drawing in by suction or breathing; a common complication of enteral feedings in which foreign material enters the lungs, often from GI secretions or the reflux of stomach contents.

aspiration pneumonia: a lung disease resulting from the abnormal entry of foreign material; caused by either bacterial infection or irritation of the lower airways.

TABLE 20-2 Comparison of Tube Feeding Routes[a]

Insertion Method or Feeding Site	Advantages	Disadvantages
Transnasal	Does not require surgery or incisions for placement; tubes can be placed by a nurse or trained dietitian.	Easy to remove by disoriented patients; long-term use may irritate the nasal passages, throat, and esophagus.
Nasogastric	Easiest to insert and confirm placement; least expensive method; feedings can often be given intermittently and without an infusion pump.	Highest risk of aspiration in compromised patients;[b] risk of tube migration to small intestine.
Nasoduodenal and nasojejunal	Lower risk of aspiration in compromised patients;[b] allows for earlier tube feedings than gastric feedings during severe stress; may allow enteral feedings even when obstructions, fistulas, or other medical conditions prevent gastric feedings.	More difficult to insert and confirm placement; risk of tube migration to stomach; tube feedings require an infusion pump for administration; may take longer to reach nutrition goals.
Tube enterostomies	Allow the lower esophageal sphincter to remain closed, reducing the risk of aspiration;[b] more comfortable than transnasal insertion for long-term use; site is not visible under clothing.	Tubes must be placed by physician or surgeon; general anesthesia may be required for surgically placed tubes; risk of complications from the insertion procedure; risk of infection at insertion site.
Gastrostomy	Feedings can often be given intermittently and without a pump; easier insertion procedure than a jejunostomy.	Moderate risk of aspiration in high-risk patients.[b]
Jejunostomy	Lowest risk of aspiration;[b] allows for earlier tube feedings than gastrostomy during severe stress; may allow enteral feedings even when obstructions, fistulas, or medical conditions prevent gastric feedings.	Most difficult insertion procedure; most costly method; feedings require an infusion pump for administration; may take longer to reach nutrition goals.

[a]Relative to other tube feeding routes. The actual advantages and disadvantages of different insertion procedures depend on the person's medical condition.
[b]The risk of aspiration associated with the different feeding routes is controversial and still under investigation.

route, and the formula's viscosity. In many cases, the tube selected is the smallest-diameter tube through which the formula will flow without clogging.

The outer diameter of a feeding tube is measured in **French units**, in which each unit equals $1/3$ millimeter; thus, a "12 French" feeding tube has a 4-millimeter diameter. ♦ The inner diameter depends on the thickness of the tubing material. Double-lumen tubes are also available; these allow a single tube to be used for both intestinal feedings and **gastric decompression**, a procedure in which the stomach contents of patients with motility disorders are removed by suction.

♦ 1 French = $1/3$ mm
12 French = 12 × $1/3$ mm = 4 mm

IN SUMMARY **Enteral formulas are provided to patients with functional GI tracts who cannot meet nutritional needs with conventional foods alone. A nasoenteric feeding route is preferred for short-term tube feedings, whereas enterostomies are usually used for longer-term feedings. Because the stomach delivers nutrients into the intestine at a controlled rate, gastric feedings are typically preferred, although they are frequently avoided in patients at risk of aspiration. The selection of feeding tubes is based on patient age and size, the feeding route, and formula viscosity.**

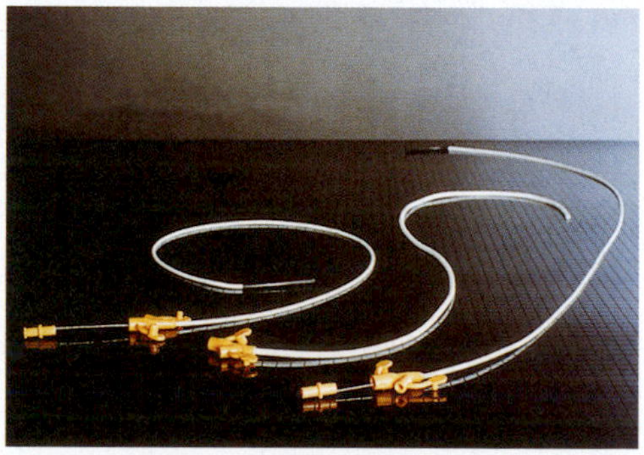

The thin wires protruding from the ends of these feeding tubes are stylets, which stiffen the tubes to ease insertion and are discarded thereafter. The orange Y-connectors provide ports for administering water or medications without disrupting the feeding.

Administration of Tube Feedings

After the feeding route and formula have been selected, the formula must be safely delivered. The methods of tube feeding administration vary somewhat from one health care facility to the next. The procedures presented in the following sections are suggested guidelines.

Safe Handling

Individuals who are ill or malnourished often have suppressed immune systems, making them vulnerable to infection from foodborne illness. Thus, the personnel involved with preparing or delivering formula—usually individuals in the foodservice department or pharmacy—should be aware of the specific protocols at their facility that prevent formula contamination.

French units: units of measure used to indicate the size of a feeding tube's outer diameter; 1 French unit equals $1/3$ millimeter.

gastric decompression: the removal of the stomach contents (including swallowed saliva, stomach secretions, and gas) of patients with motility disorders or obstructions that prevent stomach emptying.

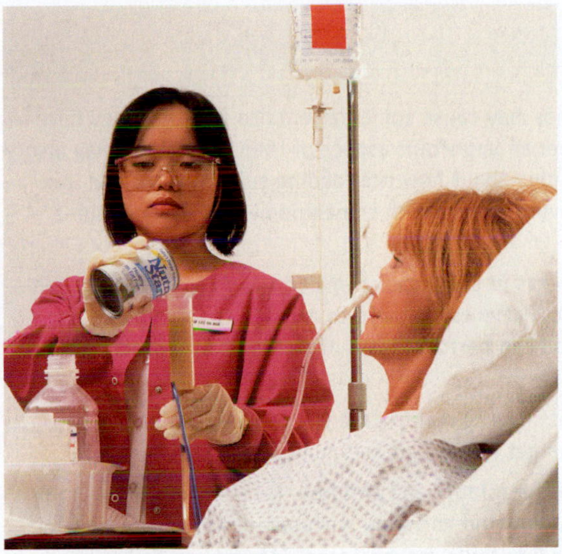

© Ed Eckstein/Photozake

In an open feeding system, the formula is transferred from its original packaging to a feeding container.

© Courtesy of Abbott Nutrition

In a closed feeding system, the formula is prepackaged in a container that can be attached directly to a feeding tube, such as the bottle shown on the left. The formula in the can at right can be used in an open feeding system.

Safety Guidelines As mentioned in earlier chapters, health care facilities have specific protocols for handling food products and formulas based on the potential hazards and critical control points in food preparation, referred to as *Hazard Analysis and Critical Control Points (HACCP)* systems. Personnel involved with preparing or delivering formula should be aware of the specific HACCP systems at their facility related to formula preparation and administration.

Feeding Systems Formulas are available in open feeding systems and closed feeding systems. With an **open feeding system**, the formula needs to be transferred from its original packaging to a feeding container. Examples include formulas that are packaged in cans or bottles, concentrates that need to be diluted, and powders that require reconstitution. In a **closed feeding system**, the formula is prepackaged in a container that can be connected directly to a feeding tube. Closed systems are less likely to become contaminated, require less nursing time, and can hang for longer periods of time than open systems. Although closed systems cost more initially, they may be less expensive in the long run because they prevent bacterial contamination and thus avoid the costs of treating infections.

At the Nursing Station After the formula reaches the nursing station, the nursing staff assumes responsibility for its safe handling. Hands should be carefully washed before handling formulas and feeding containers. Some facilities require that nonsterile gloves be worn whenever formulas are handled. The following steps can reduce the risk of formula contamination when using open feeding systems:

- Before opening a can of formula, clean the lid with a disposable alcohol wipe and wash the can opener with detergent and hot water. (Check HACCP protocols for details.) If you do not use the entire can at one feeding, label the can with the date and time it was opened.
- Store opened cans or mixed formulas in clean, closed containers. Refrigerate the unused portion of formula promptly.
- Discard unlabeled or improperly labeled containers and all opened containers of formula that are not used within 24 hours.

At the Bedside To reduce the risk of bacterial infections in tube-fed patients, the nurse should hang no more than an 8-hour supply of formula when using an

open feeding system: a delivery system that requires the formula to be transferred from its original packaging to a feeding container before being administered through the feeding tube.

closed feeding system: a delivery system in which the formula comes prepackaged in a container that can be attached directly to the feeding tube for administration.

HOW TO Help Patients Cope with Tube Feedings

Although many patients are initially apprehensive about receiving tube feedings, they may be less resistant once they understand the insertion method, the expected duration of the tube feeding, and the strategic role that nutrition plays in recovery from disease. The pointers that follow can help health practitioners prepare patients for transnasal tube feedings:

- Allow the patient to see and touch the feeding tube. Understanding that the tube is soft and narrow (only about half the diameter of a pencil) often alleviates anxiety.
- Show the patient how the feeding equipment is attached to the feeding tube, and explain how the feeding will work. For young children, use dolls or stuffed toys to demonstrate tube insertion and feeding procedures.
- Explain that the patient remains fully alert during the procedure and helps pass the tube by swallowing. A numbing solution sprayed on the back of the throat minimizes discomfort and prevents gagging during the procedure.
- Inform the patient that after the tube has been inserted, most people become accustomed to its presence within a few hours. In most cases, the patient can continue to swallow foods and beverages with the tube in place.

Many patients may be relieved to know that they can receive sound nutrition without any effort. As they feel better and begin to eat again, the volume of the feeding can be reduced and then discontinued when oral intake is adequate.

Tube feedings may cause some patients to feel that they have lost control over an important aspect of their lives. They may also feel self-conscious about how the feeding tube looks or feel awkward when moving around with the equipment. A few measures can help:

- Involve patients in the decision-making and care process whenever possible. Patients can help to arrange their daily feeding schedules and can perform some of the feeding procedures themselves.
- Show patients how to manipulate the feeding equipment so that they can get out of bed and move around.

When caring for infants and children, keep the developmental age of the child in mind and work with parents to ensure that appropriate feeding skills are mastered. Infants can be provided with a pacifier during feedings to help maintain the associations between sucking, swallowing, and fullness. When possible, the formula can be provided by bottle to an infant, or by spoon to a child, to further develop skills.

The more complex the procedure, the easier it becomes for health care professionals to focus on the procedure and disregard a patient's emotional response. No matter how many technicalities you have to keep in mind, remember to stay focused on the person receiving your care.

TRY IT Ask a friend who is unfamiliar with tube feedings what his or her reaction would be if told that his/her medical condition required this type of nutrition care. Make a list of your friend's main concerns, and see which concerns are eliminated after you fully describe the procedure.

open feeding system. The nurse should discard any formula that remains, rinse out the feeding bag and tubing, and add fresh formula to the feeding bag. A new feeding container and tubing (except for the feeding tube itself) is necessary every 24 hours.[12]

For closed feeding systems, the hang time should be no longer than 24 to 48 hours. Contamination is more likely with the longer time periods.

Initiating and Progressing a Tube Feeding

Before starting a tube feeding, health practitioners can ease fears by fully discussing the procedure with the patient and family members, who may feel anxious about the use of a feeding tube. The discussion should address the reasons why tube feeding is appropriate and the benefits and risks of the procedure. The "How To" offers suggestions that may help to ease the concerns of patients who may benefit from tube feeding.

Tube Placement Serious complications can develop if a transnasal tube is accidentally inserted into the respiratory tract or if formula or GI secretions are aspirated into the lungs. To minimize the risk of incorrect tube placement, clinicians usually use X-rays to verify the position of the feeding tube before a feeding is initiated. After the tube's placement has been confirmed, the nurse secures the tube to the patient's nose and cheek with tape and monitors the position of the tubing throughout the day. Tube placement can also be monitored by testing the pH of a sample of bodily fluid drawn into the feeding tube; recall that the pH of stomach fluid is much lower than the pH of fluid obtained from the intestine or respiratory tract. ♦

To reduce the risk of aspiration, the patient's upper body is elevated to a 30- to 45-degree angle during the feeding and for 30 minutes after the feeding whenever possible. The addition of blue food coloring to formula was formerly suggested as a means of identifying aspirated formula in lung secretions; however, this practice is now discouraged because several deaths have been attributed to its use.[13]

♦ A fasting gastric sample usually has a pH of 5 or lower. A sample from the intestine or respiratory tract has a pH of about 7 or higher.

Formula Delivery Methods A day's nutrient needs can be met by delivering relatively large amounts of formula several times per day (**intermittent feedings**) or smaller amounts continuously during the day (**continuous feedings**). A patient may also start with continuous feedings and gradually transition to intermittent feedings. Each method has specific uses, advantages, and disadvantages.

Intermittent feedings are best tolerated when they are delivered into the stomach (not the intestine). Generally, a total of about 250 to 400 milliliters of formula is delivered over 30 to 45 minutes using a gravity drip method or an infusion pump. The exact amount is determined by dividing the required volume of formula into several daily feedings, as shown in the "How To" below. Due to the relatively high volume of formula delivered at one time, intermittent feedings may be difficult for some patients to tolerate, and the risk of aspiration may be higher than with continuous feedings. An advantage of intermittent feedings is that they are similar to the usual pattern of eating and allow the patient freedom of movement between meals.

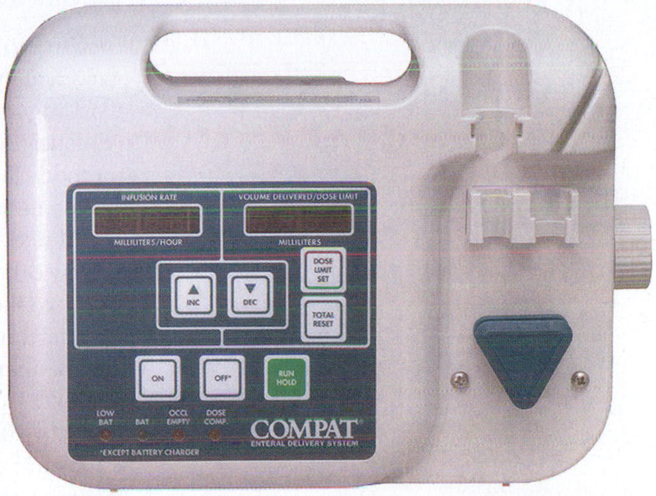

The delivery of intermittent and continuous feedings can be controlled with an infusion pump.

Rapid delivery of a large volume of formula into the stomach (250 to 500 milliliters in less than 20 minutes) is called a **bolus feeding**. This type of feeding may be given every 3 to 4 hours using a syringe. Bolus feedings can cause abdominal discomfort, nausea, and cramping in some patients, and the risk of aspiration is

HOW TO Plan a Tube Feeding Schedule

After selecting a suitable formula, the clinician must determine the volume of formula that meets the patient's nutritional needs. Consider a patient who needs 2000 kcalories daily and is receiving a standard formula that provides 1.0 kcalorie per milliliter. The total volume of formula required would be 2000 milliliters per day:

$$x \text{ mL} \times 1.0 \text{ kcal/mL} = 2000 \text{ kcal}$$

$$x \text{ mL} = \frac{2000 \text{ kcal}}{1.0 \text{ kcal/mL}} = 2000 \text{ mL}$$

If the patient is to receive intermittent feedings six times a day, he will need about 333 milliliters of formula at each feeding:

$$2000 \text{ mL} \div 6 \text{ feedings} = 333 \text{ mL/feeding}$$

Alternatively, if he is to receive intermittent feedings eight times a day, he will need 250 milliliters (or about one can of ready-to-feed formula) at each feeding:

$$2000 \text{ mL} \div 8 \text{ feedings} = 250 \text{ mL/feeding}$$

If the patient is to receive the formula continuously over 24 hours, he will need about 83 milliliters of formula each hour:

$$2000 \text{ mL} \div 24 \text{ hours} = 83 \text{ mL/hr}$$

TRY IT A patient who requires 1920 kcalories per day is receiving a standard formula that provides 1.2 kcalories per milliliter in six intermittent feedings daily. Calculate the volume of formula required at each feeding.

intermittent feedings: delivery of about 250 to 400 milliliters of formula over 20 to 40 minutes.

continuous feedings: slow delivery of formula at a constant rate over an 8- to 24-hour period.

bolus (BOH-lus) **feeding:** delivery of about 250 to 500 milliliters of formula in less than 20 minutes.

greater than with other methods of feeding. For these reasons, bolus feedings are used only in patients who are not critically ill.

Continuous feedings are delivered slowly and at a constant rate over a period of 8 to 24 hours. Continuous feedings are used to deliver intestinal feedings, and are generally recommended for critically ill patients because the slower delivery rate may be easier to tolerate. Continuous feedings may also be recommended for patients who cannot tolerate intermittent feedings. An infusion pump is required to ensure accurate and steady flow rates; consequently, the feedings can limit the patient's freedom of movement and are also more costly.

Initiating and Advancing Tube Feedings Formula administration techniques vary widely among institutions, so protocols should be reviewed carefully before working with patients. In addition, patient tolerance must be monitored when adjusting formula delivery rates. Note that few studies have evaluated the various methods for initiating and advancing enteral feedings.

Formulas are typically provided full-strength, although they may occasionally be diluted if the patient's fluid requirements are high and water needs cannot be met by other means.[14] In addition, formula dilution may sometimes be necessary to improve the flow of a viscous formula.

Intermittent feedings may start with 60 to 120 milliliters at the initial feeding and be increased by 60 to 120 milliliters at each feeding until the goal volume is reached. Continuous feedings may start at 10 to 40 milliliters per hour and be raised by 10 to 20 milliliters per hour every 8 to 12 hours as tolerated.[15] Concentrated formulas are often started at the slower rates. For both intermittent and continuous feedings, the delivery rate and amount of increase depend on the patient's tolerance to the formula. If the patient cannot tolerate an increased rate of delivery, the feeding rate is slowed until the person adapts. In some patients, formula delivery can be started at the goal rate immediately, which significantly improves the patient's overall caloric intake.[16]

Checking the Gastric Residual Volume When a patient receives a gastric feeding, the nurse regularly measures the **gastric residual volume** (the volume of formula remaining in the stomach after feeding) to ensure that the stomach is emptying properly. The gastric residual volume is measured by gently withdrawing the gastric contents through the feeding tube using a syringe, usually before each intermittent feeding and every 4 to 6 hours during continuous feedings.[17] Although opinions vary, some experts recommend that an evaluation be conducted if the gastric residual volume exceeds 200 milliliters and that feedings be withheld if it exceeds 500 milliliters.[18] If the tendency to accumulate fluids persists, the physician may recommend intestinal feedings or begin drug therapy to stimulate gastric emptying.

Meeting Water Needs

Although water needs vary ♦, many adults require about 2000 milliliters (about 2 quarts) of water daily. Fluids may be restricted in persons with kidney, liver, or heart disease. Additional water is required in patients with fever, high urine output, diarrhea, excessive sweating, severe vomiting, fistula drainage, high-output ostomies, blood loss, or open wounds.

♦ To estimate fluid requirements in adults and children:
- Adults: allow 30 to 40 mL/kg; 25 to 30 mL/kg in adults ≥65 years old
- Children: allow 50 to 60 mL/kg
- Infants: allow 100 to 150 mL/kg

In alert adults, thirst is often a good indicator of water needs. People who complain of thirst may be given more water unless medical orders restrict fluid intake. In the elderly, however, thirst may be slow to develop in response to dehydration. Health professionals routinely monitor patients' weight changes, record fluid intake and output, and measure urine specific gravity to evaluate hydration status. Chapter 17 provides additional information about evaluating hydration.

Formula Water Content The water in formulas meets a substantial portion of water needs. Standard formulas contain about 85 percent water, or about 850 milliliters of water per liter of formula. Nutrient-dense formulas contain about 70 to 75 percent water; exact amounts can be obtained from the product label or manufacturer's information sheet.

gastric residual volume: the volume of formula remaining in the stomach from a previous feeding.

HOW TO Administer Medications to Patients Receiving Tube Feedings

The pharmacist is your best resource for learning how and when medications can be administered via feeding tubes, especially when you are dealing with an unfamiliar drug. Check with the pharmacist to learn the following:

- Whether a particular medication is known to be incompatible with formulas.
- The proper timing of medication administration to avoid drug-nutrient interactions.
- For patients using intestinal feedings, whether a medication can be absorbed without exposure to stomach acid.
- Whether a liquid form of a medication is available and, if so, the appropriate dosage of the liquid form.
- If only tablets are available, whether the tablets can be crushed and mixed with water. Enteric-coated and sustained-release medications should not be crushed due to the potential for adverse effects.

In general, it is best to give medications by mouth instead of by tube whenever possible. In some cases, the injectable form of a medication may be the best option. For medications that must be given by feeding tube:

- Do not mix medications with enteral formulas. Do not mix medications together.
- Before administering medications, ensure that the feeding tube is placed correctly, that it is not clogged, and that the gastric residual volume is not excessive.
- Position the patient in a semi-upright position (30 degrees or higher) to prevent aspiration.
- Flush the feeding tube with 30 milliliters of warm water before and after administering a medication. When more than one medication is administered, flush the feeding tube with water between medications.
- Use liquid forms of medications whenever possible. Dilute viscous or hypertonic liquid medications with at least 30 milliliters of water before administering them through the feeding tube.
- If tablets are used, crush tablets to a fine powder and mix with about 30 milliliters of warm water before administering.

TRY IT Imagine the physician has just prescribed a drug and a tube feeding for a malnourished patient with impaired gastric motility. What questions should you ask the pharmacist when planning administration of the formula and the medication?

Water Flushes In addition to the water in formulas, water can be provided by flushing water separately through the feeding tube. Water flushes are also conducted to prevent feeding tubes from clogging; the tubes are flushed with about 30 milliliters of warm water about every 4 hours during continuous feedings and before and after each intermittent feeding. The water used for routine flushes should be included when estimating fluid intakes.

Medication Delivery through Feeding Tubes

Patients receiving tube feedings sometimes require one or more medications that need to be delivered through feeding tubes. Because medications can interact with substances in enteral formulas in the same ways that they interact with substances in foods, potential diet-drug interactions must be considered. In addition, some medications may need to be exposed to the acidic stomach environment and thus cannot be administered via an intestinal feeding tube. Medications can also cause feeding tubes to clog. The "How To" provides some guidelines that may help to prevent complications.

Medications and Continuous Feedings Continuous feedings are ordinarily stopped during medication administration so that the components of enteral formulas do not interfere with the medication's absorption. The feeding is typically halted for 15 minutes before and 15 minutes after medication delivery. Some medications may require a longer formula-free interval; for example, feedings need to be stopped for at least one hour before and after administering phenytoin, a medication that controls seizures.[19] In such cases, the formula's delivery rate needs to be increased so that the correct amount of formula can be delivered.

Diarrhea Medications are a major cause of the diarrhea that frequently accompanies tube feedings. Diarrhea is especially associated with the administration of sorbitol-containing medications, laxatives, and some types of antibiotics.[20] The high osmolality of many liquid medications can also cause diarrhea, so dilution of hypertonic medications may be helpful.

TABLE 20-3 Causes and Management of Tube Feeding Complications

Complications	Possible Causes	Preventive/Corrective Measures
Aspiration of formula	Inappropriate tube placement	Ensure correct placement of feeding tube.
	Delayed gastric emptying	Elevate head of bed during and after feeding; decrease formula delivery rate if gastric residual volume is excessive; consider using intestinal feedings in high-risk patients.
	Excessive sedation	Minimize use of medications that cause sedation.
Clogged feeding tube	Excessive formula viscosity	Ensure that tube size is appropriate; flush tubing with water before and after giving formula. Remedies to unclog feeding tubes include flushes with warm water or solutions that contain pancreatic enzymes and sodium bicarbonate; consult pharmacist for more options.
	Improper administration of medications	Use oral, liquid, or injectable medications whenever possible; flush tubing with water before and after a medication is given; avoid mixing medications with formula; dilute thick or sticky liquid medications before administering; crush tablets to a fine powder and mix with water (except enteric-coated or sustained-release medications).
Constipation	Inadequate dietary fiber	Use a formula with appropriate fiber content.
	Dehydration	Provide additional fluids.
	Lack of exercise	Encourage walking and other activities, if appropriate.
	Medication side effect	Consult physician about minimizing or replacing medications that cause constipation.
Diarrhea	Medication intolerance	Dilute hypertonic medications before administering; avoid using poorly tolerated medications.
	Infection in GI tract	Consult physician about specific diagnosis and appropriate treatment.
	Formula contamination	Review safety guidelines for formula preparation and delivery.
	Excessively rapid formula administration	Decrease formula delivery rate or use continuous feedings.
	Lactose or gluten intolerance	Use lactose-free or gluten-free formula in patients with intolerances.
	Unknown cause	Review medical record for conditions that predispose to diarrhea; try an alternative formula that contains adequate fiber; consult physician about using antidiarrheal or antimotility medications.
Fluid and electrolyte imbalances	Diarrhea	See items under *Diarrhea*.
	Inappropriate fluid intake or excessive losses	Monitor daily weights, intake and output records, serum electrolyte levels, and clinical signs that indicate dehydration or overhydration; ensure that water intake and formula delivery rates are appropriate.
	Inappropriate insulin, diuretic, or other therapy	Ensure that medication doses are appropriate.
	Inappropriate nutrient intake	Use a formula with appropriate nutrient content; ensure that malnourished patients do not receive excessive nutrients.[a]
Irritation or inflammation of skin or mucous membranes	Inappropriate feeding tube	Use small-bore tube made from soft materials.
	Friction from feeding tube movement	Tape feeding tube securely to prevent excessive movement.
	Infection at insertion site	Keep site clean; inspect area for redness, tenderness, and drainage; use protective dressing or ointment; treat with antibiotics, if necessary.
Nausea and vomiting, cramps	Delayed stomach emptying	Decrease formula delivery rate or use continuous feedings; halt feeding if gastric residual volume is excessive (>500 mL); evaluate for obstruction; consider use of medications to improve emptying rate.
	Formula intolerance	Ensure that formula is at room temperature, delivery rate is appropriate, and formula odor is not objectionable; consider using formula that is low in fat, low in fiber, or elemental.
	Medication intolerance	Consult physician about replacing medications that are poorly tolerated.
	Response to disease or disease treatment	Consider use of medications that control nausea and vomiting.

[a]An excessive nutrient intake in malnourished patients may cause *refeeding syndrome*, a disorder that can lead to fluid and electrolyte imbalances (see Chapter 21).

Tube Feeding Complications Complications are a frequent occurrence during tube feedings. Possible complications include gastrointestinal problems, such as aspiration and diarrhea; mechanical problems related to the tube feeding process; and metabolic problems, such as biochemical alterations and nutrient deficiencies. Examples of the most common complications, along with some preventive and corrective measures, are summarized in Table 20-3.

Gastrointestinal Complications Diarrhea may be caused by malabsorption problems, medications, bacterial overgrowth, malnutrition, or more rarely, hypertonic formulas. Constipation sometimes occurs due to dehydration, motility impairments, obstructions, and low fiber intakes. Impaired gastric motility or inadequate functioning of the lower esophageal sphincter may result in aspiration of GI secretions or formula. Other GI complications include abdominal discomfort, nausea, and vomiting.

Mechanical Complications Mechanical problems include clogged feeding tubes, malfunctioning feeding pumps, and feeding tubes that become dislodged after placement. The feeding tube itself may be a physical irritant and may warrant a change to a different type of tubing or a different feeding route. Nasoenteric tube placement may cause a number of side effects, such as dry mouth from increased mouth breathing and reduced salivary secretions, blocked eustachian tubes and resultant middle ear infections, and sinus infections due to blocking of the sinus tract. Sometimes ostomies are associated with leakage of gastrointestinal secretions at the site of tube insertion.

Metabolic Complications Common metabolic complications include fluid imbalances (either dehydration or overhydration), electrolyte imbalances, and glucose intolerance. Routine blood tests are used to monitor levels of potassium, phosphorus, sodium, and glucose until a patient has stabilized. Some patients may need insulin or medications to reverse hyperglycemia. Vitamin K and essential fatty acid deficiencies may result if formulas lacking these nutrients are used for a prolonged period.

Monitoring Tube Feedings Many complications of tube feeding can be prevented by choosing the most appropriate feeding route, formula, and delivery method. Attention to a patient's primary medical condition and medication use is important as well. The health practitioners responsible for the patient's day-to-day care monitor body weight, hydration status, and results of laboratory tests to detect problems before complications develop. Table 20-4 provides a monitoring schedule that may help with the early detection of common tube feeding problems.

TABLE 20-4 Monitoring Patients on Tube Feedings[a]

Before starting a new feeding:	Conduct a complete nutrition assessment. Check tube placement.
Before each intermittent feeding:	Check patient's position. Check tube placement. Check gastric residual volume. Flush feeding tube with water.
After each intermittent feeding:	Flush feeding tube with water.
Every hour:	Check infusion pump rate, when applicable.
Every 4 hours:	Check vital signs, including blood pressure, temperature, pulse, and respiration.
Every 4 to 6 hours of continuous feeding:	Check patient's position. Check gastric residual volume. Flush feeding tube with water.
Every 24 hours:	Check intake and output and hydration status. Check blood glucose; once stable, check blood glucose weekly (individuals without diabetes). Change feeding container and attached tubing. Clean feeding equipment.
Twice weekly:	Check body weight (check daily if patient is nutritionally unstable).
As necessary:	Observe patient for undesirable responses to tube feeding, such as delayed gastric emptying, nausea, vomiting, or diarrhea. Check results of laboratory tests. Check nitrogen balance.

[a]Guidelines vary among institutions. Monitoring frequency depends on the patient's medical condition. Patients beginning tube feedings and patients who are medically or nutritionally unstable need more intense monitoring.

CASE STUDY: Injured Hiker Requiring Enteral Nutrition Support

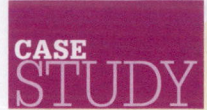

Sharyn Eschler is a 24-year-old student who suffered multiple fractures when she fell from a cliff while hiking. She has been in the hospital for two weeks and has no appetite. Due to her injuries, she is in traction and is immobile, although the head of her bed can be elevated 45 degrees. Sharyn weighed 140 pounds upon her arrival in the hospital, but she has lost 8 pounds over the course of her hospitalization. The health care team agrees that nasoduodenal tube feeding should be instituted before her nutrition status deteriorates further. The standard formula selected for the feeding is lactose-free, and Sharyn's nutrient requirements can be met with 2200 milliliters of the formula per day.

1. What steps can be taken to prepare Sharyn for tube feeding? What are some general reasons why nasoduodenal placement of the feeding tube might be preferred over nasogastric placement?
2. The physician's orders specify that the feeding should be given continuously over 18 hours. Using the method shown in the "How To" on p. 649, determine an appropriate tube feeding schedule.
3. Estimate Sharyn's fluid needs using her current weight and the fluid intake range suggested in the margin on p. 650. Will Sharyn receive enough water from the formula she is given daily? If not, estimate the additional fluid she would need and explain how it could be provided.
4. What steps can the health care team take to prevent aspiration? Describe precautions that should be taken if Sharyn is to receive medications through the feeding tube.
5. After three days of feeding, Sharyn develops diarrhea. Check Table 20-3 to determine the possible causes. What measures can be taken to correct the diarrhea?

Transition to Table Foods After the patient's condition improves, the volume of formula can be tapered off as the patient gradually shifts to an oral diet. The steps in the transition depend on the patient's medical condition and the type of feeding the patient is receiving. Individuals using continuous feedings are often switched to intermittent feedings initially. In some patients, swallowing function may need to be evaluated before oral feedings begin. Patients receiving elemental formulas may begin the transition by using a standard formula, either orally or via tube feeding. If the patient has not consumed lactose for several weeks, a low-lactose diet may be better tolerated. Oral intake should supply about two-thirds of estimated nutrient needs before the tube feeding is discontinued completely.[21] The Case Study allows you to consider the many factors involved in tube feedings.

IN SUMMARY **Enteral formulas should be prepared and administered using food safety protocols that reduce the risk of contamination. Tube placement must be verified and monitored to reduce the risks of aspiration and inadvertent placement into the respiratory tract. Depending on the patient's medical condition and the feeding route, the formula can be delivered in bolus feedings, intermittently, or continuously. Although enteral formulas meet a substantial portion of the water requirements, additional water can be provided by flushing water through the feeding tube. Medications should be given separately and accompanied by water flushes to prevent tube clogging. Complications of tube feedings can be gastrointestinal, mechanical, or metabolic in nature. Tube feedings are tapered off when the patient begins consuming an oral diet.**

Clinical Portfolio

1. Appendix K provides examples of enteral formulas on the market and lists their energy and macronutrient contents. Select one standard formula and one elemental formula from Tables K-1 and K-2, respectively. For the two formulas you selected, calculate the volume of formula that would meet the energy needs of a patient who requires about 1750 kcalories daily. Use these results in answering the following questions:
   a. What is the amount of protein, carbohydrate, and fat that the patient would obtain in a typical day? Determine the percentages of kcalories that come from carbohydrate and fat (see Chapter 1, p. 9). Do these percentages fall within the Acceptable Macronutrient Distribution Ranges described in Chapter 1 (p. 19)?
   b. Tables K-1 and K-2 show the formula volumes that would meet the Reference Daily Intakes (RDI). Would the volumes you obtained meet typical vitamin and mineral needs?
2. The administration of tube feedings requires attention to many technical details, which makes it easy to focus on the procedure rather than the patient. Imagine that your brother, sister, or a parent requires a transnasal tube feeding. How might this person react to the need for a tube feeding? How would you explain the benefits and possible problems associated with the procedure? Think about the ways you would want the health practitioner to help your relative.

Nutrition Assessment Checklist for People Receiving Enteral Nutrition Support

Medical History

Check the medical record for medical conditions that:

- Alter nutrient needs and influence the formula selection
- Influence the selection of tube placement sites and feeding routes
- Suggest the length of time that the tube feeding will be needed

Monitor the medical record for complications or risks that may influence the formula selection or delivery technique, including:

- Aspiration
- Constipation
- Fluid and electrolyte imbalances
- Diarrhea
- Hyperglycemia
- Nausea and vomiting
- Skin irritation

Medications

Check medications for those that can cause side effects similar to the adverse effects associated with the tube feeding, such as:

- Nausea and vomiting
- Diarrhea
- Constipation
- GI discomfort

For medications delivered through the feeding tube, check:

- Form of medication and possible alternatives
- Viscosity of liquid medications
- Potential for diet-drug interactions

Dietary Intake

To assess nutritional adequacy, check to see whether:

- The formula is appropriate for patient's needs
- Supplemental water is provided to meet needs
- The formula is administered as prescribed

Anthropometric Data

Measure baseline height and weight, and monitor body weight regularly. If weight is not appropriate:

- Determine whether energy needs have been correctly assessed.
- Check to see if the formula is being delivered as prescribed.
- Check for signs of dehydration or overhydration.

Laboratory Tests

Check serum and urine tests for signs of:

- Fluid and electrolyte imbalances
- Glucose intolerance
- Inadequate protein status (serum protein levels)
- Improvement or deterioration of the medical condition

Physical Signs

Look for physical signs of:

- Dehydration or overhydration
- Delayed gastric emptying (gastric residual volume)
- Malnutrition

Nutrition on the Net

For further study of topics covered in this chapter, log on to **www.cengagebrain.com** and search for ISBN 084006845X.

- To learn more about the appropriate uses of enteral and parenteral nutrition, visit the websites of these organizations:

  American Society for Parenteral and Enteral Nutrition: **www.nutritioncare.org**

  British Association for Parenteral and Enteral Nutrition: **www.bapen.org.uk**

- Information about enteral formulas is available from these manufacturers' websites:

  Abbott Nutrition: **www.abbottnutrition.com**

  Nestlé Nutrition: **www.nestle-nutrition.com**

- To learn about home nutrition support, visit the website of the Oley Foundation, a national nonprofit organization that provides information, outreach services, and emotional support for consumers of home enteral and parenteral services: **www.oley.org**

References

1. A. M. Malone, Enteral formula selection, in P. Charney and A. Malone, eds., *ADA Pocket Guide to Enteral Nutrition* (Chicago: American Dietetic Association, 2006), pp. 63–122.
2. J. L. Rombeau, Enteral nutrition, in L. Goldman and D. Ausiello, eds., *Cecil Medicine* (Philadelphia: Saunders, 2008), pp. 1617–1621.
3. M. Shike, Enteral feeding, in M. E. Shils and coeditors, *Modern Nutrition in Health and Disease* (Baltimore: Lippincott Williams & Wilkins, 2006), pp. 1554–1566.
4. Shike, 2006.
5. Malone, 2006; C. R. Parrish and S. McCray, Enteral feeding: Dispelling myths, *Practical Gastroenterology* 27 (September 2003): 33–50.
6. A. M. Malone, Enteral formulations, in G. Cresci, ed., *Nutrition Support for the Critically Ill Patient: A Guide to Practice* (Boca Raton, FL: Taylor & Francis Group, 2005), pp. 253–277.
7. N. Gupta and R. G. Martindale, Parenteral vs. enteral nutrition, in G. Cresci, ed., *Nutrition Support for the Critically Ill Patient: A Guide to Practice* (Boca Raton, FL: Taylor & Francis Group, 2005), pp. 193–208.
8. Rombeau, 2008.
9. M. Marian and P. Charney, Patient selection and indications for enteral feedings, in P. Charney and A. Malone, eds., *ADA Pocket Guide to Enteral Nutrition* (Chicago: American Dietetic Association, 2006), pp. 1–25; Shike, 2006.
10. R. L. Koretz, Do data support nutrition support? Part II. Enteral artificial nutrition, *Journal of the American Dietetic Association* 107 (2007): 1374–1380.
11. K. K. Kattelmann and coauthors, Preliminary evidence for a medical nutrition therapy protocol: Enteral feedings for critically ill patients, *Journal of the American Dietetic Association* 106 (2006): 1226–1241; B. Taylor and J. E. Mazuski, Enteral feeding access in the critically ill, in G. Cresci, ed., *Nutrition Support for the Critically Ill Patient: A Guide to Practice* (Boca Raton, FL: Taylor & Francis Group, 2005), pp. 235–252.
12. P. D. Wohlt and coauthors, Recommendations for the use of medications with continuous enteral nutrition, *American Journal of Health-Systems Pharmacy* 66 (2009): 1458–1467.
13. Kattelmann and coauthors, 2006.
14. C. Thompson, Initiation, advancement, and transition of enteral feedings, in P. Charney and A. Malone, eds., *ADA Pocket Guide to Enteral Nutrition* (Chicago: American Dietetic Association, 2006), pp. 123–154.
15. Wohlt and coauthors, 2009.
16. A. Desachy and coauthors, Initial efficacy and tolerability of early enteral nutrition with immediate or gradual introduction in intubated patients, *Intensive Care Medicine* 34 (2008): 1054–1059.
17. M. K. Russell, Monitoring complications of enteral feedings, in P. Charney and A. Malone, eds., *ADA Pocket Guide to Enteral Nutrition* (Chicago: American Dietetic Association, 2006), pp. 155–192.
18. S. A. McClave and coauthors, Guidelines for the provision and assessment of nutrition support therapy in the adult critically ill patient: Society of Critical Care Medicine (SCCM) and American Society for Parenteral and Enteral Nutrition (A.S.P.E.N.), *Journal of Parenteral and Enteral Nutrition* 33 (2009): 277–316; Russell, 2006.
19. Wohlt and coauthors, 2009.
20. Russell, 2006.
21. Thompson, 2006.

HIGHLIGHT 20

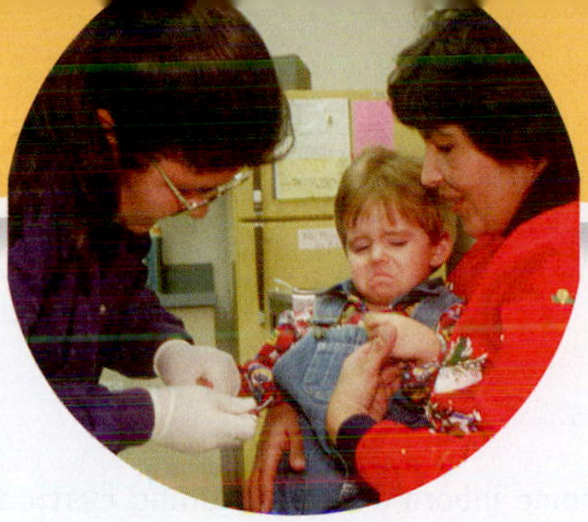

© Miguel Gandert/Corbis

Inborn Errors of Metabolism

Chapter 20 describes the use of enteral formulas for patients who are unable to meet their nutrient needs with conventional foods. Such is the case for individuals with some inborn errors of metabolism; for them, enteral formulas play a vital role in disease management. This highlight describes some inborn errors of metabolism and discusses the role of diet in two of these disorders: phenylketonuria and galactosemia. The accompanying glossary defines terms related to inborn errors of metabolism.

Inborn Errors of Metabolism

An **inborn error of metabolism** is an inherited trait, caused by a genetic **mutation**, that results in the absence, deficiency, or malfunction of a protein that has a critical metabolic role.[1] The protein may function as an enzyme, receptor, transport protein, or structural protein. When the body fails to make a protein, the functions that depend on that protein are impaired. For example, when an enzyme is missing or malfunctioning in a metabolic pathway that typically converts compound A to compound B, compound A will accumulate and compound B will not be made. The excess of compound A and the lack of compound B may have harmful effects. Furthermore, the imbalances in one pathway may affect other pathways and ultimately cause a number of metabolic and physiologic disturbances. The severity of the inborn error's effects are ultimately related to the degree of impairment caused by the altered or missing protein. Table H20-1 lists some examples of inborn errors related to defects in nutrient metabolism.

Treatment for Inborn Errors of Metabolism

Successful treatment for an inborn error of metabolism depends on the ability to screen newborns and diagnose metabolic diseases before irreversible damage can occur. After a genetic defect is identified, family members undergo **genetic counseling** to evaluate the likelihood that they may pass on the disorder to future offspring. During counseling, couples may learn about reproductive options such as artificial insemination, *in vitro* fertilization, or prenatal monitoring after conception.

Medical nutrition therapy is the primary treatment for many inborn errors that involve nutrient metabolism. Once the biochemical pathway affected by a mutation is identified, a health practitioner may be able to manipulate elements of the diet to compensate for deficiencies and excesses. Dietary intervention generally involves restricting substances that cannot be properly metabolized and supplying substances that cannot be produced. Thus, dietary changes may be able to improve outcomes of some inborn errors by:

- Preventing the accumulation of toxic metabolites
- Replacing nutrients that are deficient as a result of a defective metabolic pathway
- Providing a diet that supports normal growth and development and maintains health

Nondietary therapies can treat some inborn errors of metabolism, although the options are somewhat limited. In some cases, the missing protein is infused; this is the primary means of treating **hemophilia**, caused by deficiency of one of the plasma proteins needed for clotting blood. Drug therapy is the main

GLOSSARY

cystic fibrosis: an inherited disorder that affects the transport of chloride across epithelial cell membranes; primarily affects the gastrointestinal and respiratory systems.

galactosemia (ga-LACK-toe-SEE-me-ah): an inherited disorder that affects galactose metabolism. Accumulated galactose causes damage to the liver, kidneys, and brain in untreated patients.

gene therapy: treatment for inherited disorders, in which DNA sequences are introduced into the chromosomes of affected cells, prompting the cells to express the protein needed to correct the disease.

genetic counseling: support for families at risk of genetic disorders; involves diagnosis of disease, identification of inheritance patterns within the family, and review of reproductive options.

hemophilia (HE-moh-FEEL-ee-ah): inherited bleeding disorders characterized by deficiency or malfunction of plasma proteins needed for clotting blood.

inborn error of metabolism: an inherited trait (one that is present at birth) that causes the absence, deficiency, or malfunction of a protein that has a critical metabolic role.

metabolites: products of metabolism; compounds produced by a biochemical pathway.

mutation: an inheritable alteration in the DNA sequence of a gene.

phenylketonuria (FEN-il-KEY-toe-NU-ree-ah) or **PKU:** an inherited disorder that affects the conversion of the essential amino acid phenylalanine to the amino acid tyrosine.

treatment for some inborn errors, including **cystic fibrosis** (discussed in Chapter 24), which is characterized by a defect that prevents normal chloride transport across cell membranes. Future approaches may include **gene therapy**, a treatment that introduces DNA sequences into the chromosomes of affected cells, prompting the cells to express the protein needed to correct the abnormality. The following sections of this highlight describe two examples of inborn errors that benefit primarily from medical nutrition therapy.

Phenylketonuria

One of many inborn errors affecting amino acid metabolism, **phenylketonuria (PKU)** affects approximately 1 out of every 10,000 births in the United States each year.[2] The screening of newborns for PKU is one of the most common genetic tests in the United States and many other countries. The early detection and treatment of PKU have successfully prevented most of the damaging consequences of this disorder.

The Error in PKU

In PKU, the missing or defective protein is a liver enzyme that converts the essential amino acid phenylalanine to the amino acid tyrosine (see Figure H20-1). Without this enzyme, phenylalanine and its **metabolites** (metabolic products) accumulate and damage the developing nervous system. The impairment in the metabolic pathway also prevents liver synthesis of tyrosine and tyrosine-derived compounds (such as the neurotransmitter epinephrine). Under these conditions, tyrosine becomes essential: the body cannot produce tyrosine, and therefore the diet must supply it.

Although PKU's most debilitating effect is on brain development, other symptoms may manifest if the condition is untreated. Infants with PKU may have poor appetites and grow slowly. They may be irritable or have tremors or seizures. Their bodies and urine may have a musty odor. Their skin may be unusually pale, and they may develop skin rashes. In older children and adults who discontinue treatment, neurological and psychological problems are common.

Detecting PKU

PKU must be diagnosed soon after birth so that early treatment can prevent its devastating effects. For this reason, newborns

TABLE H20-1 Nutrition-Related Inborn Errors of Metabolism

Disorder	Affected Nutrient(s) or Substance	Metabolic Defect	Nutritional Treatment
Amino acid metabolism			
Maple syrup urine disease	Branched-chain amino acids (isoleucine, leucine, and valine)	Impaired metabolism of branched-chain amino acids	Restriction of branched-chain amino acids; thiamin supplementation
Phenylketonuria	Phenylalanine	Impaired conversion of phenylalanine to tyrosine	Phenylalanine-restricted diet; tyrosine supplementation
Carbohydrate metabolism			
Galactosemia	Galactose	Impaired conversion of galactose to glucose	Galactose-restricted diet
Glycogen storage disease	Glycogen	Impaired metabolism or transport of glycogen causing glycogen accumulation in tissues	Varies; may require frequent feedings, cornstarch supplementation, high-protein diet
Lipid metabolism			
Carnitine transporter deficiency	Fatty acids	Impaired transport of fatty acids into mitrochondria for oxidation	Carnitine supplementation; avoidance of fasting and strenuous exercise
X-adrenoleukodystrophy[a]	Very long-chain fatty acids	Impaired breakdown of very long-chain fatty acids in peroxisomes	Under investigation; limited benefit from restriction of very long-chain fatty acids and supplementation with various fatty acid mixtures[a]
Mineral metabolism			
Hemochromatosis	Iron	Excessive iron absorption (causes iron accumulation)	Avoidance of iron and vitamin C supplements and alcoholic beverages (routine blood draws remove excess iron from the body)
Wilson's disease	Copper	Impaired copper excretion (causes copper accumulation)	Avoidance of copper-rich foods; zinc therapy (reduces copper absorption)

[a]The disease X-adrenoleukodystrophy was featured in the 1992 film *Lorenzo's Oil*.

FIGURE H20-1 Biochemical Alterations in PKU

Normal:

Normally, the amino acid phenylalanine follows two pathways, one in the liver and the other in the kidneys. In the liver, the enzyme phenylalanine hydroxylase adds a hydroxyl group (OH) to produce the amino acid tyrosine. Tyrosine, in turn, produces melanin, the pigmented compound found in skin and brain cells; the neurotransmitters epinephrine and norepinephrine; and the hormone thyroxine. In the kidneys, enzymes convert phenylalanine to by-products that are excreted.

In the liver:

Phenylalanine —(Phenylalanine hydroxylase)→ Tyrosine → Melanin, Epinephrine, Norepinephrine, Thyroxine

In the kidneys:

Phenylalanine → Phenylpyruvic acid (a ketone body) → Other phenyl acids (excreted)

In PKU:

Individuals with PKU lack the liver enzyme phenylalanine hydroxylase, impairing conversion of phenylalanine to tyrosine. Phenylalanine accumulates in the liver and blood, reaching the kidneys in abnormally high concentrations. In the kidneys, an aminotransferase enzyme converts phenylalanine to the ketone body phenylpyruvic acid, which spills into the urine—thus the name phenylketonuria.

In the liver:

Phenylalanine (accumulates) —(Phenylalanine hydroxylase (deficient))→ Tyrosine (deficient)

In the kidneys:

Phenylalanine (accumulates) → Phenylpyruvic acid (accumulates) → Other phenyl acids (accumulate)

are screened for PKU in all 50 states.[3] A standard blood test for phenylalanine is typically conducted by heel puncture after the infant has consumed several meals containing protein. Abnormal results require further testing. Before widespread newborn screening, infants with PKU demonstrated developmental delays (for example, inability to crawl) by six to nine months of age. By the time parents recognized the problem, the damage was irreversible.

Medical Nutrition Therapy for PKU

The only current treatment for PKU is a diet that restricts phenylalanine and supplies tyrosine so that the blood levels of these amino acids are maintained within safe ranges. Because phenylalanine is an essential amino acid, the diet cannot exclude it completely. Children with PKU need phenylalanine to grow, but they cannot handle excesses without detrimental effects. Therefore, their diets must provide enough phenylalanine to support growth and health but not so much as to cause harm. The diets must also provide tyrosine, which is an essential nutrient for individuals with PKU. To ensure that blood concentrations of phenylalanine and tyrosine are close to normal, blood tests are performed periodically, and diets are adjusted when necessary. If the dietary treatment is conscientiously followed, it can prevent the symptoms described earlier. Adults must continue to follow the PKU diet as well to prevent deterioration in brain function.

The PKU Diet

Central to the PKU diet (for all ages) is the use of an enteral formula that is phenylalanine-free yet supplies energy, amino acids, vitamins, and minerals. In infants, the phenylalanine-free formula is supplemented with measured amounts of breast milk or regular infant formula to provide the phenylalanine that an infant needs for growth. Low-phenylalanine formulas are available for infants who meet most or all of their nutrient needs by consuming formula. Formula requirements need to be recalculated periodically to accommodate the growing infant's shifting needs for protein, phenylalanine, tyrosine, and energy.

Once food consumption begins, a phenylalanine-free formula supplies the needed amino acids, and foods that contain phenylalanine are carefully monitored. All proteins contain some phenylalanine; therefore, high-protein foods such as meat, fish, poultry, milk, cheese, legumes, and nuts (including peanut butter) are omitted. Fruits, vegetables, and cereals also contain phenylalanine, so only limited amounts are allowed. Low-protein flours and mixes are available for making low-phenylalanine breads, pasta, cakes, and cookies. Foods that do not contain phenylalanine, such as jams, jellies, and most sweeteners, can be used freely. Growth rates and nutrition status are monitored to ensure that the diet is adequate. Older children, teens, and adults with PKU should continue to use the phenylalanine-free formulas to help meet their protein and energy needs.

Individuals with PKU should be encouraged to develop creative ways to make their diets enjoyable. The formula can be flavored or combined with fruits or juices to make smoothies or frozen juice bars. Sandwiches can include low-phenylalanine breads and fillings such as mashed bananas or avocados, shredded carrots and olives, or tomato slices with mayonnaise. Children often enjoy creating special recipes with permitted foods to make their choices more varied and to share meals with friends.

Continuing Dietary Restrictions

Lifelong adherence to a phenylalanine-restricted diet is currently recommended for all individuals with PKU. Elevated phenylalanine levels can adversely affect cognitive function at any age. Case studies have suggested that individuals with PKU who discontinue dietary management may have problems with attention span, concentration, and memory. It is especially important that women with PKU maintain safe phenylalanine concentrations during pregnancy. Elevated phenylalanine levels, especially during

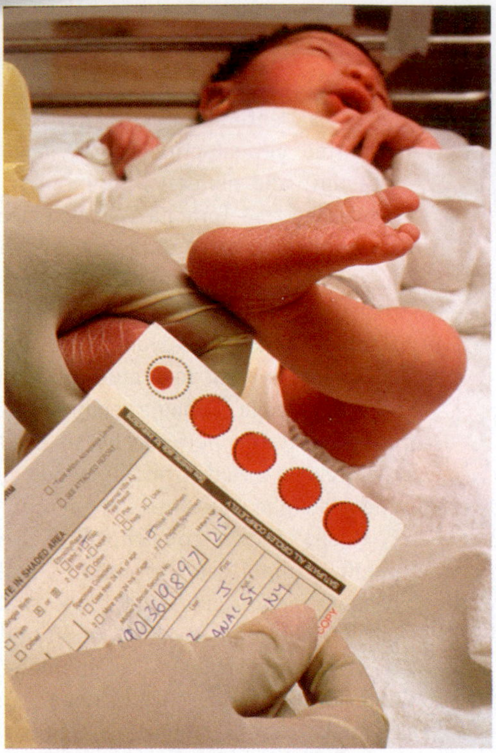

A simple blood test screens newborns for PKU—a common inborn error of metabolism.

the first trimester, have been associated with mental retardation and organ malformations in the offspring of PKU mothers who have discontinued dietary treatment.[4]

Galactosemia

Galactosemia is an example of an inborn error of carbohydrate metabolism. Individuals with galactosemia are deficient in one of the enzymes needed to metabolize galactose, a sugar that is primarily found in milk products (recall that each lactose molecule contains a molecule of galactose). An accumulation of galactose can cause damage in multiple tissues. Infants with galactosemia who are given milk react with severe vomiting and liver jaundice within days of the initial feeding. Serious liver damage can develop and progress to symptomatic cirrhosis. Other complications may include kidney failure, cataracts, and brain damage. Treatment in the first weeks of life can prevent the most detrimental effects of galactose accumulation, but if treatment is delayed, the damage to the brain is irreversible.[5]

The Galactosemia Diet

The diet for galactosemia is much simpler than the diet for PKU. For one thing, galactose is not an essential nutrient. The galactosemia diet essentially eliminates galactose from the diet and does not need to provide a carefully determined amount of any nutrient, as the PKU diet does. In addition, dietary galactose is primarily obtained from lactose (the milk sugar), so the main focus of dietary treatment is the exclusion of milk and milk products. A number of other foods that contain galactose in substantial amounts, such as organ meats and some legumes, fruits, and vegetables, must also be avoided or restricted. Patients receive food lists that identify the galactose content of common foods.

Infants diagnosed with galactosemia are given lactose-free formulas to meet their nutrient needs. Once a child can consume adequate amounts of regular foods, special formulas are unnecessary. However, care must be taken to ensure that the diet supplies adequate calcium.

Long-Term Complications

Although the early introduction of a galactose-restricted diet can eliminate the acute toxic effects of galactosemia, complications of the disease may develop despite an individual's compliance with diet therapy. For example, most patients experience delays in speech and language development. Ovarian failure occurs in up to 85 percent of women who have galactosemia.[6] In addition, some evidence suggests that IQ declines as a person with galactosemia ages. The reasons for these long-term complications are not fully understood.

As our scientific understanding of human genetics and biochemistry increases, more inborn errors of metabolism are being recognized. Mainstays of management for these diseases include effective diagnosis, early treatment, and control of environmental factors that cause toxicity. In some cases, dietary changes are central to treatment and can prevent serious complications. Not all inborn errors are easily treated, however. Future developments in biotechnology may someday allow medical practitioners to correct genetic errors using gene therapy.

References

1. L. J. Elsas II, Approach to inborn errors of metabolism, in L. Goldman and D. Ausiello, eds., *Cecil Medicine* (Philadelphia: Saunders, 2008), pp. 1539–1546.
2. S. D. Cederbaum, Disorders of phenylalanine and tyrosine metabolism, in L. Goldman and D. Ausiello, eds., *Cecil Medicine* (Philadelphia: Saunders, 2008), pp. 1573–1576.
3. L. J. Elsas II and P. B. Acosta, Inherited metabolic disease: Amino acids, organic acids, and galactose, in M. E. Shils and coeditors, *Modern Nutrition in Health and Disease* (Baltimore: Lippincott Williams & Wilkins, 2006), pp. 909–959.
4. Cederbaum, 2008.
5. L. J. Elsas II, Galactosemia, in L. Goldman and D. Ausiello, eds., *Cecil Medicine* (Philadelphia: Saunders, 2008), pp. 1555–1558.
6. Elsas, Galactosemia, 2008.

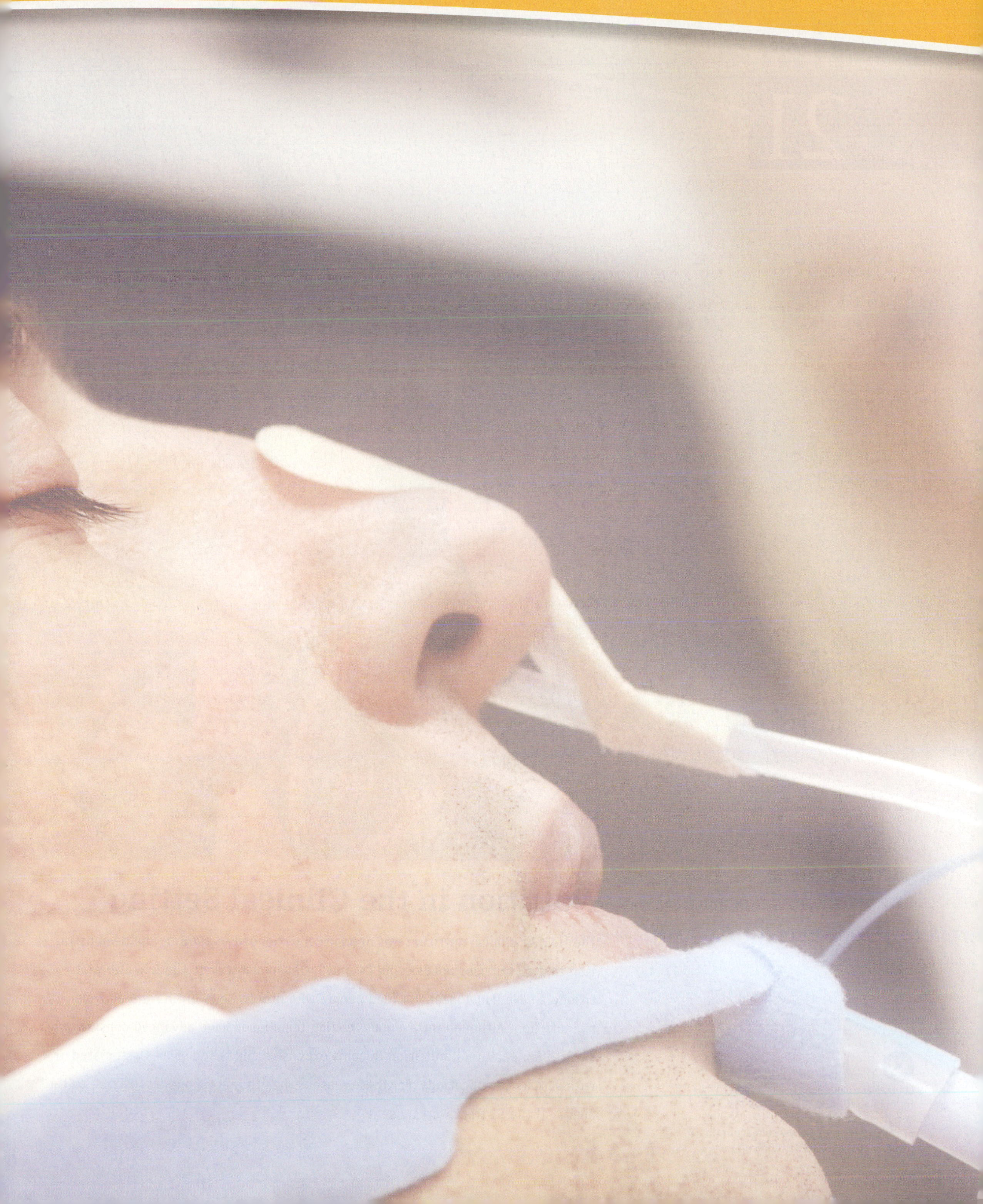

CHAPTER

21

Custom Medical Stock Photo/Newscom

Nutrition in the Clinical Setting

The field of clinical nutrition was dramatically changed in 1968 by the demonstration that all nutrient needs could be met intravenously. Since then, health practitioners have had a means to feed people who otherwise might have died from malnutrition. Although intravenous feeding techniques have advanced considerably since 1968, parenteral nutrition remains expensive and is sometimes associated with serious complications. For these reasons, health practitioners subscribe to the adage "If the GI tract works, use it."

Throughout this chapter, the CourseMate icon indicates an opportunity for online self-study, linking you to activities to increase your understanding of chapter concepts. **www.cengagebrain.com** (search for ISBN 084006845X)

Parenteral Nutrition Support

CHAPTER OUTLINE

Chapter 20 described how enteral formulas can supplement or replace conventional foods to meet nutritional needs. Because enteral formulas cannot be used if intestinal function is inadequate, the ability to meet nutrient needs intravenously is a life-saving option for critically ill persons. The procedure is costly, however, and associated with a number of potentially dangerous complications. Therefore, enteral nutrition support is preferred over parenteral nutrition if the gastrointestinal (GI) tract is functional, partly to avoid the expense and complications associated with intravenous therapy and partly to preserve healthy GI function. Figure 21-1 summarizes the decision-making process for selecting the most appropriate feeding method.

FIGURE 21-1 Selecting a Feeding Route

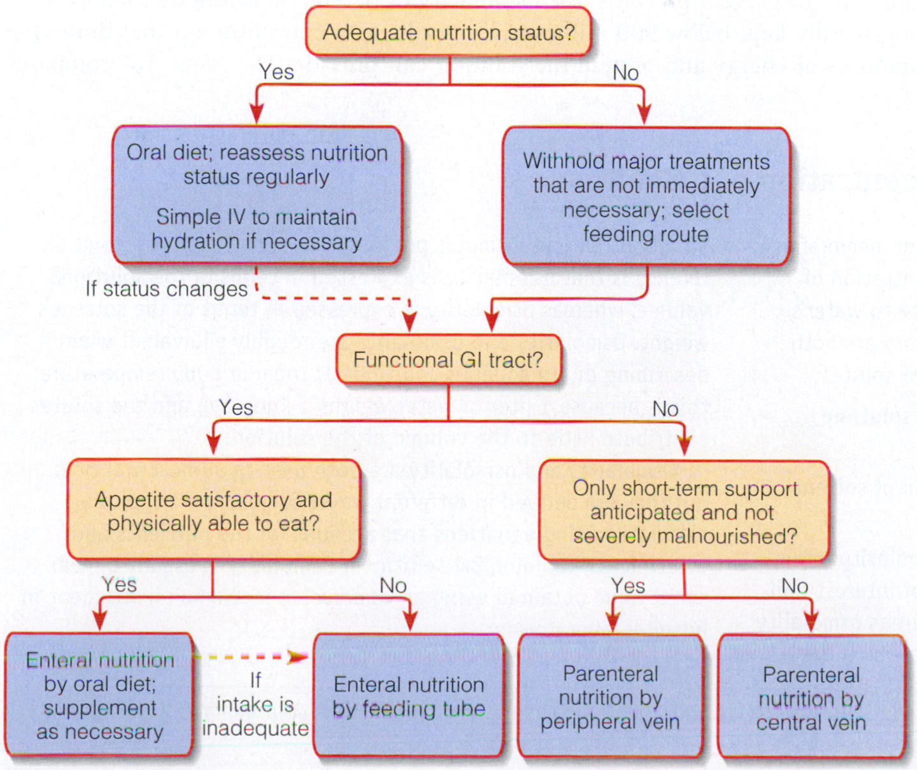

Indications for Parenteral Nutrition

As with other nutrition therapies, the decision to use parenteral nutrition is based on a thorough assessment of the patient's medical condition and nutrient needs. Generally, parenteral nutrition is indicated for patients who are unable to use the GI tract and who are either malnourished or likely to become so. Parenteral support may also be of benefit if using the GI tract would cause harm to the patient, as when severe tissue damage in the small intestine requires bowel rest for an extended period. Thus, patients with the following conditions are often considered candidates for parenteral nutrition:

- Intestinal obstructions or fistulas
- Paralytic ileus (intestinal paralysis)
- Short bowel syndrome (a substantial portion of the small intestine has been removed)
- Intractable vomiting or diarrhea
- Bone marrow transplants
- Severe malnutrition and intolerance to enteral nutrition

Some clinical studies suggest that parenteral nutrition is not always effective in the patient populations in which it is routinely used; thus, the decision to use it should be considered in light of the most recent research evidence.[1] In addition, parenteral nutrition is unlikely to be beneficial when used for periods of less than 7 to 14 days.[2]

Once the decision to use parenteral nutrition has been made, the access site must be selected. The access sites for intravenous feedings fall into two main categories: the **peripheral veins** located in the arms and legs, and the large-diameter **central veins** located near the heart.

peripheral veins: the small-diameter veins that carry blood from the arms and legs.

central veins: the large-diameter veins located close to the heart.

peripheral parenteral nutrition (PPN): the infusion of nutrient solutions into peripheral veins, usually a vein in the arm or back of the hand.

osmolarity: the concentration of osmotically active solutes in a solution, expressed as milliosmoles per liter of solution (mOsm/L). *Osmolality* (mOsm/kg) is an alternative term used to describe a solution's osmotic properties.

Peripheral Parenteral Nutrition

In **peripheral parenteral nutrition (PPN)**, nutrients are delivered using only the peripheral veins. Peripheral veins can be damaged by overly concentrated solutions, however: phlebitis (inflammation of the vein) may result, characterized by redness, swelling, and tenderness at the infusion site. To prevent phlebitis, the **osmolarity** of parenteral solutions used for PPN is generally kept below 900 milliosmoles per liter,[3] a concentration that limits the amounts of energy and protein the solution can provide. The "How To" compares

HOW TO Express the Osmolar Concentration of a Solution

The movement of water across biological membranes (called *osmosis*) is influenced by a solution's concentration of *milliosmoles*, the ions and molecules that contribute to water's osmotic pressure. The terms *osmolarity* and *osmolality* are both used to express the concentration of these types of solutes:

- *Osmolarity* refers to the milliosmoles per liter of solution (mOsm/L).
- *Osmolality* refers to the milliosmoles per kilogram of solvent (mOsm/kg).

The main difference between the terms is that osmolarity refers to a volume of solution that includes the solutes of interest (milliosmoles contained in a liter of the solution), whereas osmolality refers to the solutes separately from the solvent in which they are dissolved (milliosmoles per kilogram of solvent). A second difference is that osmolarity is expressed in terms of the solution's volume, whereas osmolality is expressed in terms of the solvent's weight. Osmolarity and osmolality are roughly equivalent when describing dilute aqueous solutions at room or body temperature; this is because 1 liter of water weighs 1 kilogram, and the solutes contribute little to the volume of the solution.

Osmolarity and osmolality are both used in clinical practice, but they are derived in different ways. Osmolarity is typically calculated using equations that account for the nutrients and electrolytes in biological solutions. Osmolality is usually a measured value obtained using an *osmometer*, a common instrument in hospital laboratories.

TRY IT Draw a simple diagram to illustrate the concepts of osmolarity and osmolality, using normal saline as your example.

FIGURE 21-2 Accessing Central Veins for Total Parenteral Nutrition

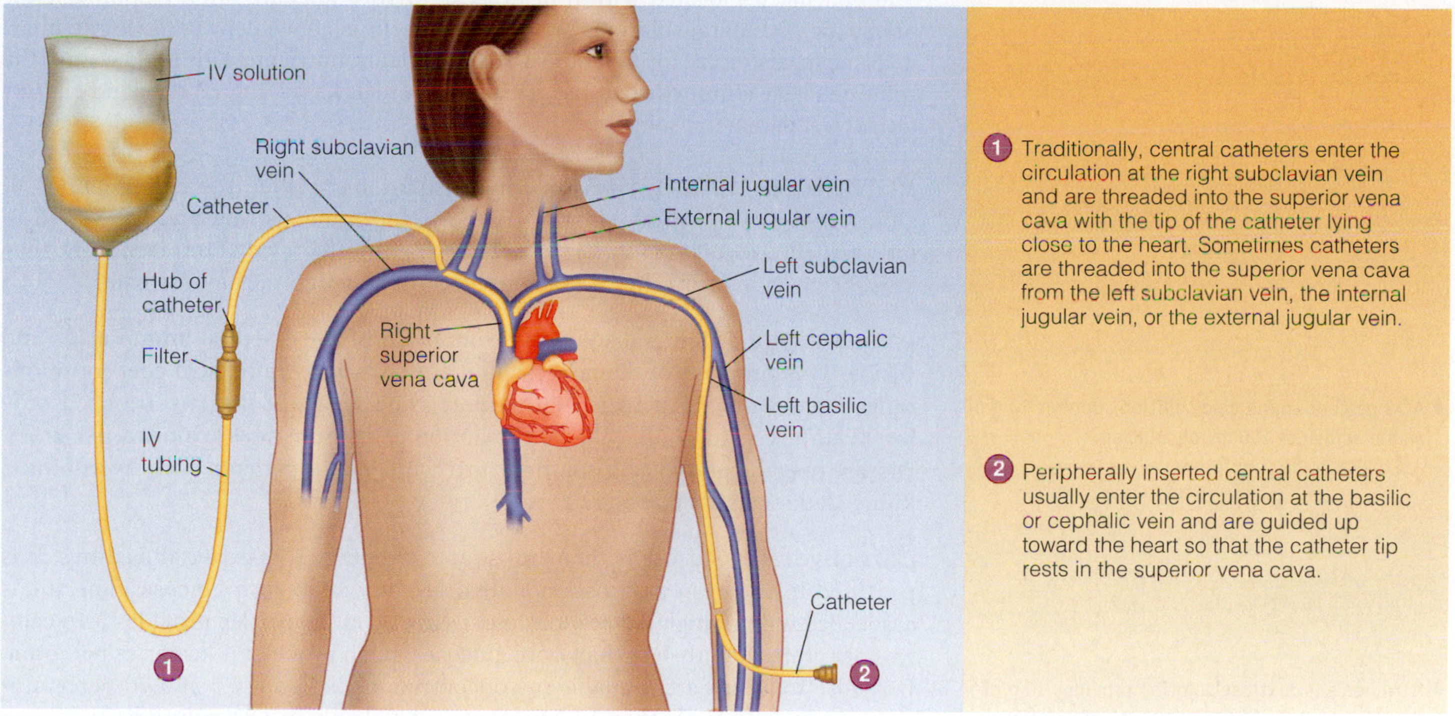

IV solution
Right subclavian vein
Catheter
Hub of catheter
Filter
IV tubing
Right superior vena cava
Internal jugular vein
External jugular vein
Left subclavian vein
Left cephalic vein
Left basilic vein
Catheter
1
2
1 Traditionally, central catheters enter the circulation at the right subclavian vein and are threaded into the superior vena cava with the tip of the catheter lying close to the heart. Sometimes catheters are threaded into the superior vena cava from the left subclavian vein, the internal jugular vein, or the external jugular vein.
2 Peripherally inserted central catheters usually enter the circulation at the basilic or cephalic vein and are guided up toward the heart so that the catheter tip rests in the superior vena cava.

the expressions *osmolarity* and *osmolality*, both of which can be used to express the osmolar concentration of a solution.

PPN is used most often in patients who require short-term nutrition support (about 7 to 10 days) and who do not have high nutrient needs or fluid restrictions. The use of PPN is not possible if the peripheral veins are too weak to tolerate the procedure. In many cases, clinicians must rotate venous access sites to avoid damaging veins.

Total Parenteral Nutrition

Most patients meet their nutrient needs using the larger, central veins, where blood volume is greater and nutrient concentrations do not need to be limited. Because this method can reliably meet a person's complete nutrient requirements, it is called **total parenteral nutrition (TPN)**. Central veins lie close to the heart, where the large volume of blood rapidly dilutes parenteral solutions. Therefore, patients with very high nutrient needs or fluid restrictions are able to receive the nutrient-dense solutions they require. TPN is also preferred for patients who require long-term parenteral nutrition.

There are several ways to access central veins. The tip of a central venous **catheter** can be placed directly into a large-diameter central vein or threaded into a central vein through a peripheral vein (see Figure 21-2). Peripheral insertion of central catheters is less invasive and lower in cost than the direct insertion of catheters into central veins; this method is usually preferred for short-term venous access (about two months or less in duration).[4]

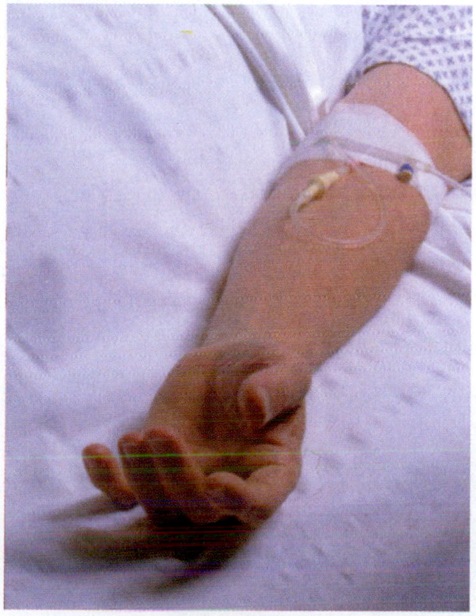

The peripheral veins can provide access to the blood for the delivery of parenteral solutions.

IN SUMMARY Parenteral nutrition support delivers nutrients intravenously; it is used in patients whose GI tract is not functioning and who may readily become malnourished. Patients receiving parenteral nutrition typically have intestinal disorders or are critically ill. If nutrients are infused directly into peripheral veins (peripheral parenteral nutrition), nutrient concentrations must be limited to avoid inflammation of the veins. The infusion of nutrients into central veins (total parenteral nutrition) can supply nutrient-dense solutions and is used for long-term intravenous feedings.

total parenteral nutrition (TPN): the infusion of nutrient solutions into a central vein.

catheter: a thin tube placed within a narrow lumen (such as a blood vessel) or body cavity; can be used to infuse or withdraw fluids or keep a passage open.

Parenteral Solutions

The pharmacies located within health care institutions are often responsible for preparing parenteral solutions. This arrangement is convenient because the pharmacist can customize formulations to meet patients' nutrient needs and because the solutions have a limited shelf life. This section reviews the ingredients and characteristics of parenteral solutions.

Parenteral Nutrients

Parenteral solutions provide the combinations of amino acids, carbohydrate, lipids, vitamins, and minerals that are best suited to meet patients' requirements. Because the nutrients are provided intravenously, they must be given in forms that are safe to inject directly into the bloodstream.

Amino Acids

Parenteral solutions contain all of the essential amino acids and various combinations of the nonessential amino acids. Amino acid concentrations range from 3.5 to 15 percent; ♦ the more concentrated solutions are used only for TPN. Just as in regular foods, the amino acids provide 4 kcalories per gram. Disease-specific amino acid solutions are available for patients with liver failure, kidney failure, and metabolic stress.

♦ A 10 percent amino acid solution supplies 10 g of amino acids per 100 mL of solution.

Carbohydrate

Glucose is the main source of energy in parenteral feedings. It is provided in the form dextrose monohydrate, in which each glucose molecule is associated with a single water molecule. Dextrose monohydrate provides 3.4 kcalories per gram, slightly less than pure glucose, which provides 4 kcalories per gram. Dextrose solutions are available in concentrations between 2.5 and 70 percent. ♦ Concentrations greater than 12.5 percent are used only in TPN solutions.[5]

♦ A 10 percent dextrose solution provides 10 g of dextrose monohydrate per 100 mL of solution.

In parenteral solutions, the dextrose concentration is indicated by a "D" followed by its concentration in water (W) or normal saline (NS). For example, D5 or D5W indicates that a solution contains 5 percent dextrose in water. Similarly, D5/NS means that a solution contains 5 percent dextrose in normal saline.

Lipids

Lipid emulsions supply essential fatty acids and are a significant source of energy. The emulsions usually contain triglycerides from soybean oil and safflower oil, phospholipids to serve as emulsifying agents, and glycerol to make the solutions isotonic. Lipid emulsions are available in 10, 20, and 30 percent solutions, containing 1.1, 2.0, and 3.0 kcalories per milliliter, respectively. Therefore, a 500-milliliter container of 10 percent lipid emulsion would provide 550 kcalories; the same volume of a 20 percent lipid emulsion would provide 1000 kcalories. ♦ In the United States, the 30 percent lipid emulsion can be used for preparing mixed parenteral solutions but cannot be directly infused into patients.[6]

♦ 500 mL of a 10% lipid emulsion:
$500\ mL \times 1.1\ kcal/mL = 550\ kcal$
500 mL of a 20% lipid emulsion:
$500\ mL \times 2\ kcal/mL = 1000\ kcal$

A lipid emulsion gives a parenteral solution a milky white color.

© Steve Gerard/Science Photo Library/Photo Researchers, Inc.

Lipid emulsions are often provided daily and may supply 20 to 30 percent of total kcalories. Including lipids as an energy source reduces the need for energy from dextrose and lowers the risk of hyperglycemia in glucose-intolerant patients. Lipid infusions must be restricted in patients with hypertriglyceridemia, however. There is also some concern that lipid emulsions that contain excessive linoleic acid can suppress some aspects of the immune response.

Fluids and Electrolytes

Daily fluid needs usually range from 30 to 40 milliliters per kilogram of body weight in young adults and 25 to 30 milliliters per kilogram of body weight in adults who are 65 years and older, averaging between 1500 and 2500 milliliters for most people. The amounts are adjusted according to daily fluid losses and the results of hydration assessment.

The electrolytes added to parenteral solutions include sodium, potassium, chloride, calcium, magnesium, and phosphate. The amounts in parenteral solutions differ from DRI values because the nutrients are infused directly into the blood and are not influenced by absorption, as they are when consumed orally. Because electrolyte imbalances can be lethal, electrolyte management by experienced professionals is necessary whenever intravenous therapies are used. Blood tests are administered daily to monitor electrolyte levels until patients have stabilized.

The electrolyte content of parenteral solutions is expressed in *milliequivalents* (*mEq*), which are units indicating the number of ionic charges provided by electrolytes. ♦ The body's fluids and parenteral solutions are neutral solutions that contain equal numbers of positive and negative charges.

♦ Milliequivalents are determined by dividing an ion's molecular weight (MW) by its number of charges. For example:

- For calcium, MW = 40, and the ion has 2 positive charges: 40 ÷ 2 = 20. Thus, 1 mEq of Ca^{++} is equivalent to 20 mg of calcium.
- For sodium, MW = 23, and the ion has 1 positive charge: 23 ÷ 1 = 23. Thus, 1 mEq of Na^{+} is equivalent to 23 mg of sodium.
- *1 mEq of* Ca^{++} has the same number of charges as *1 mEq of* Na^{+}.

Vitamins and Trace Minerals Commercial multivitamin and trace mineral preparations are added to parenteral solutions to meet micronutrient needs. All of the vitamins are usually included, although a preparation without vitamin K is available for patients using warfarin therapy.[7] ♦ The trace minerals typically added to parenteral solutions include zinc, copper, chromium, selenium, and manganese. Iron is excluded because it alters the stability of other ingredients in parenteral mixtures; therefore, special forms of iron need to be injected separately.

♦ Reminder: The anticoagulant warfarin works by interfering with vitamin K's blood-clotting function (see Chapter 19).

Osmolarity Recall that the osmolarity of PPN solutions is limited to 900 milliosmoles per liter because peripheral veins are sensitive to high nutrient concentrations, whereas TPN solutions may be as nutrient dense as necessary. The components of a solution that contribute most to its osmolarity are amino acids, dextrose, and electrolytes: as concentrations of these nutrients increase, the osmolarity of a solution increases. Because lipids contribute little to osmolarity, lipid emulsions are used to increase the energy provided in PPN solutions. The "How To" shows a method for estimating the osmolarity of a parenteral solution.

Medications To avoid the need for a separate infusion site, medications are occasionally added directly to parenteral solutions or infused through a separate port in the catheter (attached via a Y-connector). The administration of a second solution using a separate port is called a **piggyback**. Insulin, for example, is sometimes added by piggyback to improve glucose tolerance. Heparin (an anticoagulant) may be added to prevent clotting at the catheter tip. In practice, few medications are added to parenteral solutions so that potential drug-nutrient interactions can be avoided.

Solution Preparation

The prescription for a parenteral solution must take into account the patient's medical condition and nutrition status and the method of venous access. Parenteral solutions are highly individualized and may need to

piggyback: the administration of a second solution using a separate port in an intravenous catheter.

HOW TO Estimate the Osmolarity of a Parenteral Solution

For a quick estimate of the osmolarity (mOsm/L) of a 1-liter parenteral solution, follow these steps:

- Multiply the grams of amino acids in the solution by 10.
- Multiply the grams of dextrose in the solution by 5.
- Multiply the milliequivalents (mEq) of electrolytes in the solution by 2.
- Add the three values to determine the approximate osmolarity.

Example:
A liter of a TPN solution has the composition shown. Determine the approximate osmolarity of the solution.

Amino acids: 40 g	Sodium: 40 mEq	Calcium: 5 mEq
Dextrose: 250 g	Potassium: 35 mEq	Magnesium: 8 mEq
Lipids:[a] 40 g	Chloride: 77 mEq	Phosphate: 21 mEq

Answer:

Amino acids: 40 g × 10 = 400 mOsm/L

Dextrose: 250 g × 5 = 1250 mOsm/L

Electrolytes: (40 + 35 + 77 + 5 + 8 + 21) × 2 = 372 mOsm/L

Approximate osmolarity: 400 + 1250 + 372 = 2022 mOsm/L

[a]Because lipid emulsions have little or no effect on osmolarity, they are usually excluded from the calculation.

TRY IT Estimate the osmolarity of a 1-liter solution that contains 35 grams of amino acids, 225 grams of dextrose, and 170 mEq of electrolytes.

FIGURE 21-3 Sample Parenteral Nutrition Order Form

Physician Orders

PARENTERAL NUTRITION (PN) – ADULT

Primary Diagnosis: ______________________ **Ht:** ________ cm **Dosing Wt:** ________ kg

PN Indication: ______________________ **Allergies** ______________

Instructions: This form must be completed for a new order or continuation of PN and faxed to the Pharmacy by [Insert Time] to receive same day preparation. PN administration begins at [Insert Time]. Contact the Nutrition Support Service at (XXX) XXX-XXXX for additional information.

Administration Route: CVC or PICC *Note: Proper tip placement of the CVC or PICC must be confirmed prior to PN infusion*

Peripheral IV (PIV) (*Final PN Osmolarity ≤ ________ mOsm/L*)

Monitoring: Daily weights, Strict input & output, Bedside glucose monitoring every ______ hours

Na, K, Cl, CO_2, Glucose, BUN, Scr, Mg, PO_4 every ____________

T, Bili, Alk Phos, AST, ALT, Albumin, Triglycerides, Calcium every ____________

Base Solution: *Select one*

Parenteral nutrition MUST be administered through a dedicated infusion port and filtered with a 1.2-micron in-line filter at all times. Discard any unused volume after 24 hours.

PERIPHERAL 2-in-1	CENTRAL 2-in-1	CENTRAL 3-in-1
Dextrose ______ g	Dextrose ______ g	Dextrose ______ g
Amino Acids (*Brand* ________) ______ g	Amino Acids (*Brand* ________) ______ g	Amino Acids (*Brand* ________) ______ g
		Fat Emulsion (*Brand* ________) ______ g
For patients with PIV and established glucose tolerance; Provides ______ kcal; Maximum Rate not to exceed ______ mL/hour	*For patients with CVC or PICC and established glucose tolerance; Provides ______ kcal; Maximum Rate not to exceed ______ mL/hour*	*For patients with CVC or PICC and established glucose/fat emulsion tolerance; Provides ______ kcal; Maximum Rate not to exceed ______ mL/hour*
		Use of additional fat emulsion not required with 3-in-1 base solution

RATE & VOLUME: ______ mL/hour for ______ hours = ______ mL/day
Must specify

or **CYCLIC INFUSION:** ______ mL/hour for ______ hours, then ______ mL/hour for ______ hours = ______ mL/day

Fat Emulsion ***(Brand*** ____________***) – via PIV or CVC with 2-in-1 base solutions*** (*Select caloric density & volume*)

10% 250 mL Infuse at ______ mL/hour over ______ hours Frequency ______________

20% 500 mL (*Note: infusions < 4 or > 12 hours not recommended*) *Discard any unused volume after 12 hours.*

Additives: (*per day*)			***Normal Dosages***
Sodium Chloride	______	mEq	*1-2 mEq Sodium/kg/day*
as Acetate	______	mEq	*pH or CO_2 dependent*
as Phosphate	______	mmol of PO_4	*Consider if hyperkalemic*
Potassium Chloride	______	mEq	*1-2 mEq Potassium/kg/day*
as Acetate	______	mEq	*pH or CO_2 dependent*
as Phosphate	______	mmol of PO_4	*20-40 mmol/day (1 mmol Phos = 1.5 mEq K)*
Calcium **Gluconate**	______	mEq	*5-15 mEq/day*
Magnesium **Sulfate**	______	mEq	*8-24 mEq/day*
Adult **Multivitamins**	______	mL/day	*Contains Vitamin K 150 mcg*
Adult **Trace Elements**	______	mL/day	*Zn ___ mg, Cu ___ mg, Mn ___ mg, Cr ___ mcg, Se ___ mcg (with normal hepatic function)*
H_2 Antagonist ________	______	mg	*____ mg/day with normal renal function*
Other:			

Additives: (*per day*)
Regular Insulin ______ units
Recommend if hyperglycemic, start with 1 unit for every 10 g of dextrose

Pharmacy Use Only: **Ca/PO4 Limit Checked** ____
(*Note: Some brands of amino acids contain phosphate*)

Physician's Signature: ______________________ **Pager Number:** __________ **Date/time:** __________

Orders transcribed by: __________ **Date/time:** __________ **Orders verified by:** __________ **Date/time:** __________

SEND COMPLETED ORDERS TO PHARMACY

be recalculated daily until the patient's condition is stable. Figure 21-3 provides an example of a parenteral nutrition order form.

total nutrient admixture (TNA): a parenteral solution that contains dextrose, amino acids, and lipids; also called a **3-in-1 solution** or an **all-in-one solution.**

2-in-1 solution: a parenteral solution that contains dextrose and amino acids, but excludes lipids.

Parenteral Formulations When a parenteral solution contains dextrose, amino acids, and lipids, it is called a **total nutrient admixture** (TNA), a **3-in-1 solution**, or an **all-in-one solution**. A **2-in-1 solution** excludes lipids, and the lipid emulsion is administered separately, often by piggyback administration. Although the adminis-

HOW TO Calculate the Macronutrient and Energy Content of a Parenteral Solution

Suppose a patient is receiving 1.25 liters (1250 milliliters) of a parenteral solution that contains 5 percent amino acids and 30 percent dextrose, supplemented with 250 milliliters of a 20 percent lipid emulsion daily. How many grams of protein and carbohydrate is the person receiving, and what is the total energy intake for the day?

Amino acids:

$$5\% \text{ amino acids} = \frac{5 \text{ g amino acids}}{100 \text{ mL}}$$

$$\frac{5 \text{ g amino acids}}{100 \text{ mL}} \times 1250 \text{ mL} = 62.5 \text{ g of amino acids}$$

$$62.5 \text{ g amino acids} \times 4.0 \text{ kcal/g} = 250 \text{ kcal}$$

Carbohydrate:

$$30\% \text{ dextrose} = \frac{30 \text{ g dextrose}}{100 \text{ mL}}$$

$$\frac{30 \text{ g dextrose}}{100 \text{ mL}} \times 1250 \text{ mL} = 375 \text{ g of dextrose}$$

$$375 \text{ g dextrose} \times 3.4 \text{ kcal/g} = 1275 \text{ kcal}$$

Lipids:

Recall that a 20 percent lipid emulsion provides 2.0 kcalories per milliliter. If the patient is given 250 milliliters of the emulsion:

$$250 \text{ mL} \times 2.0 \text{ kcal/mL} = 500 \text{ kcal}$$

Total energy intake:

$$250 \text{ kcal} + 1275 \text{ kcal} + 500 \text{ kcal} = 2025 \text{ kcal}$$

TRY IT Calculate the energy content of 1 liter of a parenteral solution that contains 10 percent amino acids and 20 percent dextrose supplemented with 250 milliliters of a 10 percent lipid emulsion.

tration of TNA solutions is simpler because only one infusion pump is required, the addition of lipid emulsions may reduce solution stability, potentially resulting in the formation of enlarged lipid droplets and particulates that can obstruct capillaries or have other damaging effects.[8] Thus, lipids are often administered separately when they are not a major energy source and are used only to provide essential fatty acids. The "How To" above describes a method for calculating the macronutrient and energy content of a parenteral solution.

Nonprotein kCalorie-to-Nitrogen Ratio Some practitioners calculate the **nonprotein kcalorie-to-nitrogen ratio** to assess whether the nitrogen provided by the solution is sufficient for maintaining muscle tissue. A ratio of 150:1 to 200:1 is

nonprotein kcalorie-to-nitrogen ratio: a ratio between the nonprotein kcalories and nitrogen content of the diet; used to assess whether the nitrogen intake is sufficient for maintaining muscle tissue.

HOW TO Calculate the Nonprotein kCalorie-to-Nitrogen Ratio

The nonprotein kcalorie-to-nitrogen ratio of the diet is sometimes used to determine whether a patient is receiving adequate nitrogen to maintain muscle tissue. A ratio between 150:1 and 200:1 is often adequate for stable patients, whereas ratios of 100:1 and below may be necessary for patients who are critically ill.

To calculate the nonprotein kcalorie-to-nitrogen ratio, determine the energy intake from carbohydrates and lipids and compare this amount to the nitrogen intake. To determine the nitrogen intake, multiply the amino acid or protein intake by 16% (protein contains about 16% nitrogen by weight).

The "How To" above describes a parenteral solution that provides 1275 kcalories from dextrose, 500 kcalories from lipids, and 62.5 grams of amino acids. Using these values to calculate the nonprotein kcalorie-to-nitrogen ratio:

Nonprotein kcalories:

$$1275 \text{ kcal (dextrose)} + 500 \text{ kcal (lipids)} = 1775 \text{ kcal}$$

Nitrogen content:

$$62.5 \text{ g amino acids} \times 16\% \text{ nitrogen} = 62.5 \text{ g} \times 0.16 = 10 \text{ g}$$

Nonprotein kcalorie-to-nitrogen ratio:

$$1775 \div 10 = 178{:}1$$

Thus, the parenteral solution in the example has a nonprotein kcalorie-to-nitrogen ratio of 178:1, which is likely to be adequate for a stable hospital patient.

TRY IT Using the nutrient and energy values you obtained in the Try It problem in the "How To" box above, calculate the nonprotein kcalorie-to-nitrogen ratio of the parenteral solution described in the problem.

considered adequate for stable patients, whereas a ratio of 100:1 or less is preferred for critically ill patients who have difficulty maintaining muscle mass. The ratio is sometimes used as a guideline for preparing parenteral solutions in conjunction with information related to the patient's clinical condition and nutrition status.[9] The "How To" on p. 669 shows how to calculate the nonprotein kcalorie-to-nitrogen ratio.

Safety Concerns Intravenous infusions are similar to tube feedings in that careful attention to solution preparation and handling can minimize complications. To prevent bacterial contamination and maintain stability, parenteral solutions are compounded in the pharmacy under aseptic conditions, shielded from light, and refrigerated. Prior to infusion, the solutions are removed from the refrigerator and allowed to reach room temperature. During feedings, the solution and catheter need to be checked frequently for signs of contamination.

IN SUMMARY **Prescriptions for parenteral solutions are individualized to meet each patient's needs. The solutions are compounded in hospital pharmacies using commercial nutrient preparations and include amino acids, dextrose, electrolytes, vitamins, and trace minerals. Few medications are added to parenteral solutions due to the potential for drug-nutrient interactions. Parenteral solutions that include lipids are called total nutrient admixtures, 3-in-1 solutions, or all-in-one solutions; solutions that exclude lipids are called 2-in-1 solutions. Parenteral solutions are prepared and handled using aseptic techniques to prevent contamination.**

Administering Parenteral Nutrition

Parenteral nutrition is a complex treatment that requires skills from a variety of disciplines. Many hospitals organize nutrition support teams, ♦ consisting of physicians, nurses, dietitians, and pharmacists, that specialize in the provision of both intravenous and tube feedings. Members of the team may serve as advisers to other clinicians or may manage nutrition support directly. They may also have administrative responsibilities, such as receiving patients, purchasing supplies, developing guidelines, and keeping records. Figure 21-4 describes the typical roles of each member of the nutrition support team.

♦ Reminder: A *nutrition support team* is a multidisciplinary team of health care professionals who are responsible for the provision of nutrients by tube feeding or intravenous infusion.

Insertion and Care of Intravenous Catheters

Although skilled nurses can place catheters into peripheral veins, only qualified physicians can insert catheters directly into central veins. Patients may be awake for the procedure and given local anesthesia. Unnecessary apprehension can be avoided by explaining the procedure to the patient beforehand.

Catheter-related problems frequently cause complications (see Table 21-1). Catheters may be improperly positioned or may dislodge after placement. Air can leak into catheters and escape into the bloodstream, obstructing blood flow. Catheters in peripheral veins may cause phlebitis, necessitating reinsertion at an alternate site. A catheter may become clogged from blood clotting or from a buildup of scar tissue around the catheter tip. Catheters are also a leading cause of infection: contamination may be introduced during insertion or may develop at the placement site.

To reduce the risk of complications, nurses use aseptic techniques when inserting catheters, changing tubing, or changing a dressing that covers the catheter site. Unusual bleeding or a wet dressing suggests a problem with catheter placement. A change in infusion rate may indicate a clogged catheter. Infection may be indicated by redness or swelling around the catheter site or by an unexplained fever. Routine inspections of equipment and frequent monitoring of patients' symptoms help to minimize the problems associated with catheter use.

Administration of Parenteral Solutions

The method used to initiate and advance parenteral nutrition depends on the patient's condition and the potential for complications. In addition, infusion protocols vary among institutions.

FIGURE 21-4 The Nutrition Support Team

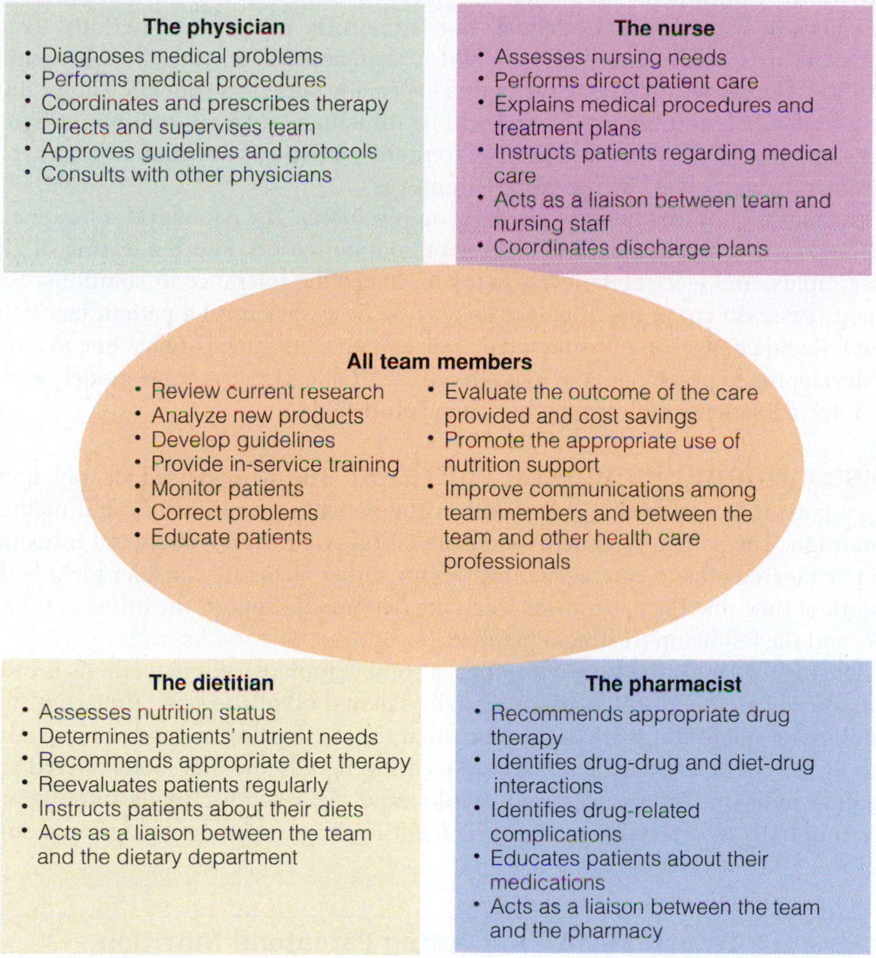

The physician
• Diagnoses medical problems
• Performs medical procedures
• Coordinates and prescribes therapy
• Directs and supervises team
• Approves guidelines and protocols
• Consults with other physicians
The nurse
• Assesses nursing needs
• Performs direct patient care
• Explains medical procedures and treatment plans
• Instructs patients regarding medical care
• Acts as a liaison between team and nursing staff
• Coordinates discharge plans
All team members
• Review current research
• Analyze new products
• Develop guidelines
• Provide in-service training
• Monitor patients
• Correct problems
• Educate patients
• Evaluate the outcome of the care provided and cost savings
• Promote the appropriate use of nutrition support
• Improve communications among team members and between the team and other health care professionals
The dietitian
• Assesses nutrition status
• Determines patients' nutrient needs
• Recommends appropriate diet therapy
• Reevaluates patients regularly
• Instructs patients about their diets
• Acts as a liaison between the team and the dietary department
The pharmacist
• Recommends appropriate drug therapy
• Identifies drug-drug and diet-drug interactions
• Identifies drug-related complications
• Educates patients about their medications
• Acts as a liaison between the team and the pharmacy

One approach is to start the infusion at a slow rate (with a solution that is either full strength or nutrient dilute) and increase the rate gradually over a 2- to 3-day period. For example, 40 milliliters per hour can be infused during the first 24 hours of administration (supplying 960 milliliters), and the volume increased to the goal rate on the second day. Another method is to give the full volume of a nutrient-dilute solution on the first day and advance nutrient concentrations as tolerated. Solutions can often be started at full volume and full strength unless there is a risk of hyperglycemia or other complications.[10]

TABLE 21-1 Potential Complications of Parenteral Nutrition

Catheter-Related	Metabolic
Air embolism	Abnormal liver function
Blood clotting at catheter tip	Electrolyte imbalances
Clogging of catheter	Gallbladder disease
Dislodgment of catheter	Hyperglycemia, hypoglycemia
Improper placement	Hypertriglyceridemia
Infection, sepsis	Metabolic bone disease
Phlebitis	Nutrient deficiencies
Tissue injury	Refeeding syndrome

Parenteral solutions can be infused continuously over 24 hours (**continuous parenteral nutrition**) or during 8- to 16-hour periods only (**cyclic parenteral nutrition**). Continuous infusions are given to critically ill and malnourished patients who cannot receive adequate nutrients in the shorter time periods. Cyclic infusions are sometimes provided at night so that patients can participate in routine activities during the day. This method is especially suited to patients who require long-term parenteral support or who will be infusing parenteral solutions at home. Patients may begin with continuous parenteral nutrition and transition to cyclic parenteral nutrition as their condition improves.

Regular monitoring helps to prevent complications. The parenteral solution and tubing are checked frequently for signs of contamination. Routine testing of glucose, lipids, and electrolyte levels helps to determine tolerance to solutions. Frequent reassessment of nutrition status may be necessary until a patient has stabilized. Rapid changes in infusion rate are discouraged in some patients due to a risk of developing hyperglycemia or hypoglycemia.[11] Table 21-2 lists some guidelines for monitoring patients undergoing intravenous infusions.

Discontinuing Parenteral Nutrition

The method used for transitioning a patient to oral feedings depends on the patient's overall health and medical condition. The patient must have adequate GI function before parenteral infusions can be tapered off and enteral feedings begun. Other factors to consider include the length of time that the patient was receiving parenteral support, the infusion schedule, and the follow-up treatment planned.

During the transition to oral feedings, a combination of feeding methods is often necessary. Parenteral infusions are usually tapered off at the same time that tube feedings or oral feedings are begun, such that the two feeding methods can together supply the needed nutrients. Clear liquids are generally the first foods offered and include pulp-free fruit juices, soft drinks, and clear broths; small amounts are given initially to determine tolerance. ♦ Later feedings include beverages and solid

♦ Chapter 18 provides more information about the clear liquid diet.

TABLE 21-2 Patient Monitoring during Parenteral Nutrition

Before starting:
• Perform a nutrition assessment • Record body height, weight, and body mass index • Confirm catheter placement by X-ray • Check laboratory values, including a complete blood count, blood glucose levels, blood triglycerides, serum bilirubin and liver enzyme levels, serum proteins, blood urea nitrogen, serum creatinine, and serum electrolytes (sodium, potassium, calcium, magnesium, chloride, phosphate, bicarbonate)
Every 4 to 8 hrs:
• Check vital signs, including body temperature • Inspect catheter site for signs of inflammation or infection (frequency depends on patient condition) • Check pump infusion rate and appearance of parenteral solution and tubing • Check blood glucose levels (once stabilized, check daily)
Daily:
• Replace parenteral solution and tubing • Monitor weight changes • Record fluid intake and output • Check blood glucose levels, blood urea nitrogen, serum creatinine, and serum electrolytes until stabilized
Several times weekly (or as needed):
• Reassess nutrition status • Check laboratory values to monitor blood chemistry

continuous parenteral nutrition: continuous administration of parenteral solutions over a 24-hour period.

cyclic parenteral nutrition: administration of parenteral solutions over an 8- to 16-hour period each day.

foods that are unlikely to cause discomfort; a low-fat lactose-free diet may be recommended. If gastrointestinal symptoms (such as nausea, vomiting, bloating, or diarrhea) develop, oral feedings are limited in size or frequency until the intestines adapt. Once about two-thirds to three-fourths of nutrient needs can be provided enterally, the parenteral infusions may be discontinued.

Transitioning to an oral diet is sometimes difficult because a person's appetite remains suppressed for several weeks after parenteral nutrition is terminated. Patients receiving continuous parenteral nutrition may have better appetites during the day if they are switched to nocturnal cyclic feedings before beginning oral intakes.

Managing Metabolic Complications

As discussed previously, the catheters used for intravenous infusions may cause a number of serious complications. This section describes some metabolic complications that may result from parenteral nutrition (review Table 21-1) and some suggestions for managing them.[12]

Hyperglycemia Hyperglycemia ♦ most often occurs in patients who are glucose intolerant, receiving excessive energy or dextrose, or undergoing severe metabolic stress.[13] It may be prevented by providing insulin along with parenteral solutions, avoiding overfeeding or overly rapid infusion rates, or restricting the amount of dextrose in parenteral solutions. Dextrose infusions are generally limited to less than 5 milligrams per kilogram of body weight per minute in critically ill adult patients so that the carbohydrate intake does not exceed the maximum glucose oxidation rate.

♦ For most patients receiving parenteral nutrition, blood glucose levels should not exceed 200 mg/dL.

Hypoglycemia Although uncommon, hypoglycemia sometimes occurs when feedings are interrupted or discontinued or if excessive insulin is given. In patients at risk, such as young infants, feedings may be tapered off over several hours before discontinuation. Another option is to infuse a 10 percent dextrose solution at the same time that the parenteral feedings are interrupted or stopped.[14]

Hypertriglyceridemia Hypertriglyceridemia may develop in critically ill patients who cannot tolerate the amount of lipid emulsion supplied. Patients at risk include those with severe infection, liver disease, kidney failure, or hyperglycemia, and those using immunosuppressant or corticosteroid medications. If blood triglyceride levels exceed 500 milligrams per deciliter, lipid infusions should be reduced or stopped.[15]

Refeeding Syndrome Severely malnourished patients who are fed aggressively (parenterally or otherwise) may develop **refeeding syndrome**, characterized by electrolyte and fluid imbalances and hyperglycemia. These effects occur because dextrose infusions raise circulating insulin levels, which promote anabolic processes that quickly remove potassium, phosphate, and magnesium from the blood. The altered electrolyte levels can lead to fluid retention and life-threatening changes in organ systems. Heart failure and respiratory failure are possible consequences.

Refeeding syndrome generally develops within two weeks of beginning parenteral infusions.[16] The patients at highest risk are those who have experienced chronic malnutrition or substantial weight loss. Symptoms include edema, cardiac arrhythmias, muscle weakness, and confusion. To prevent refeeding syndrome, health practitioners start parenteral feedings slowly and carefully monitor electrolyte and glucose levels when malnourished patients begin receiving nutrition support.

Abnormal Liver Function Fatty liver often results from parenteral support, but it is usually corrected after the parenteral infusions are discontinued. Long-term parenteral nutrition, however, may result in progressive liver disease. The cause of the liver abnormalities is unclear.

Liver enzyme levels are monitored weekly during parenteral support, and abnormal values are often seen within weeks of beginning the infusions. The patients at highest risk are those with preexisting GI or liver disorders, malnutrition, or severe

refeeding syndrome: a condition that sometimes develops when a severely malnourished person is aggressively fed; characterized by electrolyte and fluid imbalances and hyperglycemia.

infection.[17] To minimize the risk, practitioners may avoid giving the patient excess energy, dextrose, or lipids. Cyclic infusions may be less problematic than continuous infusions. If appropriate, some oral feedings may be encouraged to reduce the amount of parenteral support necessary. Note that various critical illnesses and disease treatments can also cause liver complications, so parenteral nutrition cannot be assumed to be the underlying cause.[18]

Gallbladder Disease Gallbladder problems frequently develop when the GI tract remains unused for long periods. When parenteral nutrition continues for more than four weeks, sludge (thickened bile) may build up in the gallbladder and eventually lead to gallstone formation. Prevention is sometimes possible by initiating enteral feedings before problems develop. Patients requiring long-term parenteral nutrition may be given medications to stimulate gallbladder contraction or improve bile flow or may have their gallbladders removed surgically.

Metabolic Bone Disease Long-term parenteral nutrition is associated with reduced bone mineralization and lower bone density, which may be related to altered calcium, phosphorus, magnesium, and sodium metabolism. Inappropriate intakes of vitamin D, vitamin K, and phosphorus may also contribute to the disorder. The ideal intervention varies among patients; it may include dietary adjustments, nutrient supplements, medications, and physical activity.[19]

IN SUMMARY **A nutrition support team, made up of physicians, nurses, dietitians, and pharmacists, may administer parenteral nutrition support or serve as advisers to other clinicians. Parenteral solutions may be initiated gradually or provided at full volume and full strength in selected patients. Critically ill patients may require continuous infusions, whereas healthier patients and long-term users may prefer cyclic infusions. Catheters are frequently the cause of complications, which include improper placement or dislodgment, infection, clotting, embolism, and phlebitis. Metabolic complications include hyperglycemia and hypoglycemia; hypertriglyceridemia; fluid and electrolyte imbalances; and diseases affecting the liver, gallbladder, and bone. When the need for parenteral nutrition resolves, patients are transitioned to an enteral diet as the volume of parenteral nutrition is gradually reduced. The Case Study can be used to check your understanding of the concepts introduced in this chapter.**

Nutrition Support at Home

Some individuals may require nutrition support—either tube feedings or parenteral nutrition—after a medical condition has stabilized and they no longer require hospital services. For such a person, home nutrition support may be a suitable option. Current medical technology allows for the safe administration of nutrition support in home settings, and insurance coverage often pays a substantial portion of the costs. Medical equipment providers and home infusion pharmacies can provide the supplies, enteral formulas or parenteral solutions, and services necessary for home nutrition care. Most important, patients using these services can continue to receive specialized nutrition care while leading normal lives.

Candidates for Home Nutrition Support

Individuals referred for home nutrition support usually need long-term nutrition care for chronic medical conditions. Users of home nutrition services (or their families and other caregivers) must be capable of learning the required procedures and managing any complications that arise. The home should be clean and have adequate storage for formulas or solutions and equipment. The costs should be clearly explained to families who cannot get insurance reimbursement. Candidates for home nutrition support include the following:[20]

- For home enteral nutrition, individuals who have disorders that prevent food from reaching the intestines or interfere with nutrient absorption.

CASE STUDY

Patient with Intestinal Disease Requiring Parenteral Nutrition

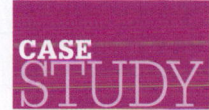

Jerry Huang, a 27-year-old man with an inflammatory intestinal disease, underwent a surgical procedure in which a substantial portion of his small intestine was removed. He had received TPN prior to surgery and continued to receive it afterwards. After 10 days, tube feeding was begun and initially delivered very small feedings.

1. List some reasons why the nutrition support team initially chose TPN as a means of nutrition support for this patient. How would you explain the need for parenteral nutrition to Jerry?
2. Describe the components of a typical TPN solution. Calculate the energy content of 1 liter of a solution that provides 140 grams of dextrose monohydrate, 45 grams of amino acids, and 90 milliliters of 20 percent lipid emulsion. Then calculate the nonprotein kcalorie-to-nitrogen ratio in the solution. If Jerry's energy requirement is 2100 kcalories per day, how many liters of solution will he need each day?
3. Why is it important that Jerry begin enteral feedings as soon as possible? Assuming that Jerry eventually tolerates a tube feeding, in what ways can the health care team help Jerry make the transition from parenteral nutrition to tube feedings? Consider some of the physiological problems that Jerry might face when he begins eating an oral diet.
4. If Jerry is unable to meet his nutrient needs orally, he may need to continue tube feeding or TPN at home. As you read through the section on nutrition support at home, consider the factors that would make Jerry a good candidate for a home nutrition support program. Consider both the benefits of a proposed program and the problems he could encounter.

Examples include people with head and neck cancers, severe dysphagia, gastric outlet obstructions, and pancreatic or intestinal conditions that cause malabsorption.

- For home parenteral nutrition, individuals who have disorders that severely impede nutrient absorption or interfere with intestinal motility. Examples include people with short bowel syndrome, inflammatory bowel diseases, and intestinal obstructions.

Planning Home Nutrition Care

As with the nutrition support provided in health care facilities, planning for home nutrition care involves decisions about access sites, formulas, and nutrient delivery methods. Users of home services should be involved in the decision making to ensure long-term compliance and satisfaction.

Home Enteral Nutrition Access to the GI tract is possible using either nasal tubes or enterostomies. Although people can learn to place nasogastric tubes themselves, active children and adults often prefer low-profile gastrostomy or jejunostomy tubes, which allow them to lead a more normal lifestyle. Jejunostomy tubes are generally less convenient because the frequent feedings required can interfere with daytime activities.

The advantages and disadvantages associated with the different administration methods should be fully discussed with patients. For gastric feedings, bolus infusions are simplest and can be quickly delivered. If intermittent feedings require slow or reliable delivery rates, infusion pumps may be necessary. Portable pumps can free individuals from the need to infuse formula at home and can also be used when traveling.

Portable pumps and convenient carrying cases allow people who require home nutrition support to move about freely.

Courtesy of Kendall Healthcare

The formula chosen for home use is influenced by its cost and availability. Insurance reimbursements do not always include the cost of enteral formulas, which are considered to be "food" products. For this reason, some people choose to prepare simple formulas at home. Blenderizing home-cooked foods is possible, but the foods need to be strained to remove particles and clumps that may obstruct the tube. Closed (ready-to-hang) feeding systems are useful for avoiding contamination risk but are not appropriate for intermittent feedings that require smaller amounts of formula.

Home Parenteral Nutrition Although both peripheral parenteral nutrition and total parenteral nutrition (TPN) can be provided at home, long-term therapy requires access to the larger, central veins that are appropriate for TPN. The catheter's exit site is generally placed in a region accessible to the patient. Most people prefer cyclic infusions over continuous infusions and transition to cyclic infusions before discharge from the hospital. Because infusion pumps are required for home TPN, sufficient battery backup in necessary in case electrical service is interrupted. Portable pumps are useful for individuals who lead active lifestyles or prefer to infuse during the day.

Parenteral solutions need to be sterile and aseptically prepared, and individuals who mix their own solutions must be carefully trained. Ready-made parenteral solutions require refrigeration and are stable for limited periods; for example, 3-in-1 solutions may be stable for only one week when refrigerated.

Quality-of-Life Issues

Although home nutrition programs can help to improve health and extend life, consumers of these services and their families may struggle with the lifestyle adjustments required. In addition to the economic impact of nutrition support, home infusions are often time consuming and inconvenient. Activities and work schedules must accommodate the feeding schedule. Extra planning is necessary and precautions must be taken when a person wants to travel or participate in sports activities. Explaining one's medical needs to friends and acquaintances may be embarrassing.

Among physical difficulties, people receiving nocturnal feedings often cite disturbed sleep as a major problem. Disruptions may be due to multiple nighttime bathroom visits, noisy infusion pumps, or difficulty finding a comfortable sleeping position when "hooked up." People using parenteral support sometimes prefer infusing solutions during the day to improve their sleeping patterns.

Among social issues, the inability to share meals with family and friends is often a great concern.[21] Many individuals miss the enjoyment, comfort, and socialization they previously experienced from food and mealtimes. Joining friends at restaurants and attending certain types of social events may become a source of stress for individuals who cannot consume food.

People who depend on nutrition support face many challenges that can affect quality of life. Support groups or counseling resources can help patients cope with the demands of treatment. The Oley Foundation (www.oley.org) is an excellent source of current information and emotional support for individuals who require home nutrition support.

IN SUMMARY **Candidates for home enteral nutrition services have disorders that interfere with swallowing ability, GI motility, or nutrient absorption. Candidates for home parenteral nutrition have disorders that severely impair nutrient absorption or cause intestinal motility problems. Patients and caregivers should participate in decisions about access sites, formulas, and nutrient delivery methods. Enteral formulas and parenteral solutions can be purchased or prepared in the home. The use of portable pumps may help individuals lead a normal lifestyle. Nevertheless, lifestyle adjustments to nutrition support may be difficult and stressful.**

Clinical Portfolio

1. A liter of a TPN solution contains 500 milliliters of 50 percent dextrose solution and 500 milliliters of 5 percent amino acid solution. Determine the daily energy and protein intakes of a person who receives 2 liters per day of such a solution. Calculate the average daily energy intake if the person also receives 500 milliliters of a 20 percent fat emulsion three times a week.
2. Consider the social, psychological, clinical, and financial ramifications of using home parenteral nutrition, with no foods allowed by mouth, in answering the following questions:
   a. What would be the advantages of living at home instead of in a hospital or other residential facility? Can you think of some disadvantages?
   b. Think about how you, as the patient, might manage daily infusions: consider the time, cost, and commitment required to maintain the therapy.
   c. If not allowed to consume foods, what possible difficulties might you encounter? How would you handle holidays and special occasions that center around food?

Nutrition on the Net

For further study of topics covered in this chapter, log on to **www.cengagebrain.com** and search for ISBN 084006845X.

- To learn more about the appropriate uses of enteral and parenteral nutrition, visit the websites of these organizations:
  American Society for Parenteral and Enteral Nutrition: **www.nutritioncare.org**
  British Association for Parenteral and Enteral Nutrition: **www.bapen.org.uk**
- To learn about home parenteral nutrition, visit the website of the Oley Foundation, a national nonprofit organization that provides information, outreach services, and emotional support for consumers of home enteral and parenteral services: **www.oley.org**

Nutrition Assessment Checklist for People Receiving Parenteral Nutrition Support

Medical History

Check the medical record for medical conditions that:

- Prevent the use of enteral nutrition
- Indicate the appropriate infusion route (peripheral versus central)
- Suggest the length of time that parenteral nutrition will be required

Monitor the medical record for complications or risks that may influence the parenteral solution formulation or delivery technique, including:

- Acid-base imbalances
- Fluid and electrolyte imbalances
- Hyperglycemia or hypoglycemia
- Hypertriglyceridemia
- Preexisting liver disease
- Refeeding syndrome

Medications

For medications added to the parenteral solution, determine the:

- Medication's compatibility with the parenteral solution
- Length of time that the medication can remain stable in solution

For medications infused separately, determine:

- Length of time that the feeding may need to be stopped
- Necessary adjustments in parenteral infusions to compensate for medication delivery

Dietary Intake

To assess nutritional adequacy, check to see whether:

- Patient's nutrient needs were correctly determined
- Solution is administered as prescribed
- Infusion pump is operating correctly

Anthropometric Data

Measure baseline height and weight, and monitor daily weights. If weight is not appropriate:

- Determine whether energy needs have been correctly assessed
- Check to see if the parenteral solution is being delivered as prescribed
- Check for signs of dehydration or overhydration

Laboratory Tests

Check serum and urine tests for signs of:

- Fluid, electrolyte, and acid-base imbalances
- Hyperglycemia or hypoglycemia
- Hypertriglyceridemia
- Abnormal liver function
- Adequacy of protein intake (serum protein levels)
- Improvement or deterioration of medical condition

Physical Signs

Routinely monitor the following:

- Catheter insertion site for signs of infection or inflammation
- Blood pressure, temperature, pulse, and respiration for signs of fluid, electrolyte, and acid-base imbalances

Look for physical signs of:

- Dehydration or overhydration
- Protein-energy malnutrition
- Malnutrition

References

1. R. L. Koretz, Do data support nutrition support? Part I. Intravenous nutrition, *Journal of the American Dietetic Association* 107 (2007): 988–996.
2. M. Russell and M. Marian, Patient selection and indications for parenteral nutrition, in P. Charney and A. Malone, eds., *ADA Pocket Guide to Parenteral Nutrition* (Chicago: American Dietetic Association, 2007), pp. 1–32.
3. C. Hamilton, Vascular access, in P. Charney and A. Malone, eds., *ADA Pocket Guide to Parenteral Nutrition* (Chicago: American Dietetic Association, 2007), pp. 33–51.
4. Hamilton, 2007.
5. M. L. Christensen, Parenteral formulations, in G. Cresci, ed., *Nutrition Support for the Critically Ill Patient: A Guide to Practice* (Boca Raton, FL: Taylor & Francis Group, 2005), pp. 279–302.
6. R. O. Brown and G. Minard, Parenteral nutrition, in M. E. Shils and coeditors, *Modern Nutrition in Health and Disease* (Baltimore: Lippincott Williams & Wilkins, 2006), pp. 1567–1597; Christensen, 2005.
7. A. M. Malone, Parenteral nutrients and formulations, in P. Charney and A. Malone, eds., *ADA Pocket Guide to Parenteral Nutrition* (Chicago: American Dietetic Association, 2007), pp. 52–63.
8. G. Hardy and M. Puzovic, Formulation, stability, and administration of parenteral nutrition with new lipid emulsions, *Nutrition in Clinical Practice* 24 (2009): 616–625.
9. E. E. Szeszycki and S. Benjamin, Complications of parenteral feeding, in G. Cresci, ed., *Nutrition Support for the Critically Ill Patient: A Guide to Practice* (Boca Raton, FL: Taylor & Francis Group, 2005), pp. 303–319.
10. S. Roberts, Initiation, advancement, and acute complications, in P. Charney and A. Malone, eds., *ADA Pocket Guide to Parenteral Nutrition* (Chicago: American Dietetic Association, 2007), pp. 76–102.
11. Szeszycki and Benjamin, 2005.
12. M. P. Fuhrman, Complications of long-term parenteral nutrition, in P. Charney and A. Malone, eds., *ADA Pocket Guide to Parenteral Nutrition* (Chicago: American Dietetic Association, 2007), pp. 103–117; Roberts, 2007; Szeszycki and Benjamin, 2005.
13. Roberts, 2007.
14. A. Wilmer and G. Van den Berghe, Parenteral nutrition, in L. Goldman and D. Ausiello, eds., *Cecil Medicine* (Philadelphia: Saunders, 2008), pp. 1621–1626; Szeszycki and Benjamin, 2005.
15. Roberts, 2007.
16. Wilmer and Van den Berghe, 2008.
17. Roberts, 2007.
18. Szeszycki and Benjamin, 2005.
19. Fuhrman, 2007; Brown and Minard, 2006.
20. C. Hamilton and T. Austin, Home parenteral nutrition, in P. Charney and A. Malone, eds., *ADA Pocket Guide to Parenteral Nutrition* (Chicago: American Dietetic Association, 2007), pp. 118–146; A. Pattinson and J. Buchholtz, Home enteral nutrition, in P. Charney and A. Malone, eds., *ADA Pocket Guide to Enteral Nutrition* (Chicago: American Dietetic Association, 2006), pp. 193–227.
21. M. F. Winkler, Living with enteral and parenteral nutrition: How food and eating contribute to quality of life, *Journal of the American Dietetic Association* 110 (2010): 169–177.

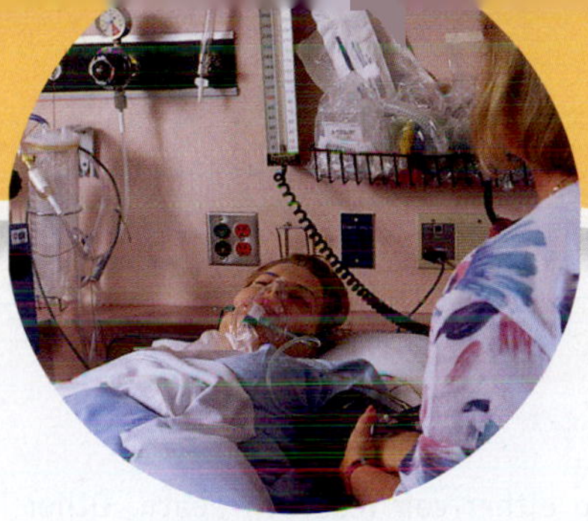

Ethical Issues in Nutrition Care

As with other medical technologies, the availability of specialized nutrition support forces health care professionals and members of our society to face difficult **ethical** issues. When medical treatments prolong life by merely delaying death, the lifetime that remains may be of extremely low quality. This highlight examines the ethical dilemmas that clinicians must face when dealing with patients in critical care. The glossary defines the relevant terms.

Ethical Considerations

If providing nutrition care can do little to promote recovery, is it morally and legally appropriate to withhold or to withdraw nutrition support? Do patients and family members have the rights to make these types of decisions themselves? How important is the input of the health professional? In attempting to answer questions such as these, health professionals must consider the following ethical principles:[1]

- A patient has the right to make decisions concerning his or her own well-being (**patient autonomy**), even if refusing treatment could result in death. It is generally accepted that a patient's preferences should take precedence over the desires of others.[2]
- A patient should be fully informed of a treatment's benefits and risks in a fair and honest manner (**disclosure**). A patient's acceptance of a treatment that has been adequately disclosed is considered **informed consent**.
- A patient must have the mental capacity to make appropriate health care decisions (**decision-making capacity**). If a patient is mentally incapable of doing so, a person designated by the patient should serve as a **surrogate** decision maker.
- The potential benefits (**beneficence**) of any treatment should outweigh its potential harm (**maleficence**).
- Health care providers must determine whether the provision of health care to one patient would unfairly limit the care of other patients (**distributive justice**).

Although these principles may seem simple and obvious, it is often difficult to determine the appropriate action to take during intensive care.[3] When clinicians and families disagree, the courts may be asked to decide.

When a patient's preferences are unknown, the medical staff is obligated to provide any and all available care that is likely to sustain the patient's life. Nutrition support and hydration are both considered life-sustaining treatments because withholding

GLOSSARY

advance directive: written or oral instruction regarding one's preferences for medical treatment to be used in the event of becoming incapacitated.

beneficence (be-NEF-eh-sense): the act of performing beneficial services rather than harmful ones.

cardiopulmonary resuscitation (CPR): life-sustaining treatment that supplies oxygen and restores a person's ability to breathe and pump blood.

decision-making capacity: the ability to understand pertinent information and make appropriate decisions; known as *decision-making competency* within the legal system.

defibrillation: life-sustaining treatment in which an electronic device is used to shock the heart and reestablish a pattern of normal contractions. Defibrillation is used when the heart has arrhythmias or has experienced cardiac arrest.

dialysis: life-sustaining treatment in which a patient's blood is filtered using selective diffusion through a semipermeable membrane; substitutes for kidney function.

disclosure: the act of revealing pertinent information. For example, clinicians should accurately describe proposed tests and procedures, their benefits and risks, and alternative approaches.

distributive justice: the equitable distribution of resources.

do-not-resuscitate (DNR) order: a request by a patient or surrogate to withhold cardiopulmonary resuscitation.

durable power of attorney: a legal document (sometimes called a *health care proxy*) that gives legal authority to another (a *health care agent*) to make medical decisions in the event of incapacitation.

ethical: in accordance with accepted principles of right and wrong.

futile: medical care that will not improve the medical circumstances of a patient.

health care agent: a person given legal authority to make medical decisions for another in the event of incapacitation.

informed consent: a patient's or caregiver's agreement to undergo a treatment that has been adequately disclosed. Persons must be mentally competent in order to make the decision.

living will: a written statement that specifies the medical procedures desired or not desired in the event that a person is unable to communicate or is incapacitated; also called a *medical directive*.

maleficence (mah-LEF-eh-sense): the act of doing evil or harm.

mechanical ventilation: life-sustaining treatment in which a mechanical ventilator is used to substitute for a patient's failing lungs.

patient autonomy: a principle of self-determination, such that patients (or surrogate decision makers) are free to choose the medical interventions that are acceptable to them, even if they choose to refuse interventions that may extend their lives.

persistent vegetative state: a vegetative mental state resulting from brain injury that persists for at least one month. Individuals lose awareness and the ability to think but retain noncognitive brain functions, such as motor reflexes and normal sleep patterns.

surrogate: a substitute; a person who takes the place of another.

or withdrawing either can result in death. Other life-sustaining treatments include **cardiopulmonary resuscitation (CPR)**, which supplies oxygen and restores a person's ability to breathe and pump blood; **defibrillation**, in which an electronic device shocks the heart and reestablishes normal contractions; **mechanical ventilation**, which substitutes for lung function; and **dialysis**, which substitutes for kidney function.

Ethical Dilemmas

Although life-sustaining treatments are readily provided to patients who have a reasonable chance of recovering from illness, it is sometimes difficult to determine the best course of action for patients who are dying or who are unlikely to regain consciousness. Under such circumstances, such treatments may be considered **futile** because they are unable to improve the outcome of disease or increase the patient's comfort and well-being. If patients or caregivers demand treatment that health practitioners have determined to be useless, a legal resolution may be required. Conversely, medical personnel may find it objectionable to withdraw life support when they know that the inevitable consequence will be the patient's death.

Legal Decisions

One of the landmark cases involving nutrition support concerned Nancy Cruzan, who suffered permanent and irreversible brain damage after a car crash in 1983, when she was 26 years of age.[4] After she had been in a **persistent vegetative state** for five years, her parents requested permission to discontinue tube feeding, but hospital staff refused to honor the request, and the matter was taken to court. The Missouri Supreme Court determined that Nancy had never definitively stated her "right to die" wishes and that her parents were unable to make such a request for her. The court also stated that preserving life, no matter what its quality, should take precedence over all other considerations. Nancy's parents appealed the ruling, but in 1990, the U.S. Supreme Court upheld the Missouri Supreme Court in a five-to-four decision. Three witnesses were eventually found who could testify that Nancy would not desire life-sustaining treatment under the circumstances, and the court finally granted permission to remove the feeding tube. This case illustrates the importance of having an **advance directive** (discussed in a later section) that clearly indicates one's preferences for medical treatment in the event of incapacitation.

In a more recent case that received widespread media attention, the spouse and parents of a patient in a persistent vegetative state fought a 10-year legal battle over her medical care. In 1990, at the age of 25, Terri Schiavo suffered a full cardiac arrest.[5] She initially fell into a coma, but her condition evolved into a persistent vegetative state that was considered irreversible. Despite the neurologists' diagnosis and a series of computed tomography (CT) and magnetic resonance imaging (MRI) scans showing extensive brain atrophy, her parents maintained that she was minimally conscious and could improve somewhat with rigorous treatment. Her husband, who was legally responsible for her care, insisted that she would never have wanted to be kept alive in a vegetative state. Like Nancy Cruzan, Terri had never expressed her wishes in an advance directive.

In 1998, Terri's husband filed a petition to have her feeding tube removed, and a Florida court approved the motion in February 2000. Although Terri's parents appealed, an appeals court affirmed the decision, and the Florida Supreme Court declined to review the case. In April 2001, Terri's physicians removed her feeding tube, but within a few days a federal circuit court judge ordered it to be reinserted and reopened the case. Eventually, the motions filed by the parents were dismissed and Terri's feeding tube was removed for the second time in October 2003. Within days, the Florida legislature passed a bill known as "Terri's Law" that gave the governor the authority to intervene, and Governor Jeb Bush ordered the feeding tube reinserted. A year later, Florida's Supreme Court declared Terri's Law to be unconstitutional. Although the governor appealed the decision, his appeal was rejected in January 2005. Terri's feeding tube was removed for the third time in March 2005. Despite emergency petitions by her parents and an attempt by the U.S. Congress to have her case reconsidered, the courts refused to grant a restraining order, and Terri died 13 days after her feeding tube was removed.

Religious Viewpoints

The withdrawal of nutrition support and other life-sustaining treatments may not be acceptable to persons of some religious faiths. For example, Orthodox Jews believe that the soul is present in people who are alive (even if permanently unconscious) and therefore deem it necessary to maintain life.[6] If a person's or family's religious beliefs are not in accord with medical recommendations, health practitioners are expected to consider the viewpoint and try to resolve the issue in some way. If practitioners are unable to comply with the wishes of a patient or caregiver, the care of the patient should be transferred elsewhere.

Advance Planning

Individuals are encouraged to discuss their medical preferences with family members and surrogate decision makers so that their wishes will be considered in the event that they become incapacitated. In addition, written instructions regarding one's preferences, called *advance directives*, can be incorporated into the medical record and updated when appropriate. Advance directions take effect only if a physician determines that a patient lacks the ability to understand and make decisions about available treatments. If a person's preferences are unknown, decisions are based on a patient's best interests as determined by a caregiver or family member.[7]

Advance Directives

A person can declare preferences about medical treatments in a **living will**, sometimes called a *medical directive*. Living wills can include detailed instructions about life-sustaining procedures that a person does or does not want. Another important directive is a **durable power of attorney** (sometimes called a *health care proxy*), in which another person (a **health care agent**) is appointed to act as decision maker in the event of incapacitation. The agent should understand one's medical preferences and be absolutely trustworthy. Only one person can be designated, although one or two alternates may also be listed. If an agent is given comprehensive power to supervise care, he or she may make decisions about medical staff, health care facilities, and medical procedures.

Laws regarding advance directives vary from state to state. In some states, nutrition and hydration are not considered life-sustaining treatments, and a person's instructions about them may need to be indicated separately. Some states restrict the use of advance directives to terminal illness or disallow them if a woman is pregnant. State statutes also specify characteristics of people who may serve as health care agents and witnesses. Generally, advance directives created in one state are honored in another.

The Do-Not-Resuscitate Order

A **do-not-resuscitate (DNR) order** is frequently used to withhold CPR in the event of cardiopulmonary arrest, which occurs too suddenly for deliberate decision making.[8] A DNR order is written in the medical record as other directives are, but it does not exclude the use of other life-prolonging measures. A DNR order is most often used in patients with serious illnesses or advanced age. Some institutions allow a physician to write a DNR order for a patient who has a poor prognosis, but the physician must inform the patient or surrogate if this is done.

Organ and Tissue Donation

End-of-life decisions invariably raise questions about a dying patient's preferences concerning organ and tissue donation. Even if a donor card has been signed, it is important to let family members know one's wishes, as the family may need to sign a consent form in order for donation to occur. Although organ donation is a difficult topic to bring up near the time of death, potential donors can be assured that their gift could greatly enhance or save the lives of others.

Ethical questions sometimes arise when organs are donated. A physician must alert an organ procurement team about a donor's existence and arrange to maintain organ functions until organs are retrieved. Treatments that maintain the viability of organs and tissues cannot be used if they may harm the donor. Sometimes the care of a donor and the needs of a potential recipient may appear to be in conflict, but the care of donors and recipients is always kept separate and performed by different physicians.

Ongoing Issues

Despite the availability of advance directives, only about 20 percent of people in the United States have completed one.[9] Furthermore, advance directives are often unavailable when intensive care decisions are made: one study found that only 57.5 percent of patient charts indicating the existence of an advance directive actually contained a copy.[10] In addition, advance directives are sometimes too general or vague to guide treatment decisions.

Physicians must often make treatment decisions before they have a chance to discuss them with patients or caregivers.[11] In many cases, life-sustaining treatments are begun without the prior knowledge of patients or their decision makers, or treatments continue even if patients want them stopped. Patients who are fully aware of treatment options and clearly state their preferences are more likely to be successful at obtaining the care they desire.

Medical decisions that are planned in advance and discussed with close friends and family can help to prevent decision-making dilemmas during emergency situations. Health practitioners should strive to provide the best information possible so that patients can consider all of the medical options available and make their preferences known to medical personnel.

References

1. M. A. Grippi, Ethics in critical care, in A. P. Fishman and coeditors, *Fishman's Manual of Pulmonary Diseases and Disorders* (New York: McGraw-Hill, 2002), pp. 1111–1114.
2. E. J. Emanuel, Bioethics in the practice of medicine, in L. Goldman and D. Ausiello, eds., *Cecil Medicine* (Philadelphia: Saunders, 2008), pp. 6–11.
3. Emanuel, 2008.
4. J. O. Maillet, Position of the American Dietetic Association: Ethical and legal issues in nutrition, hydration, and feeding, *Journal of the American Dietetic Association* 108 (2008): 873–882.
5. R. Cranford, Facts, lies, and videotapes: The permanent vegetative state and the sad case of Terri Schiavo, *Journal of Law, Medicine, and Ethics* 33 (2005): 363–372.
6. Maillet, 2008.
7. L. Snyder and C. Leffler, Position paper: Ethics manual, *Annals of Internal Medicine* (2005): 560–582.
8. Snyder and Leffler, 2005.
9. Emanuel, 2008.
10. Institute of Medicine, *Approaching Death: Improving Care at the End of Life* (Washington, DC: National Academy Press, 1997), pp. 202–203.
11. Emanuel, 2008.

CHAPTER

22

© Masterfile

Nutrition in the Clinical Setting

The body's dramatic response to severe stress can alter metabolism enough to threaten survival. Many patients with severe stress require life support measures and intensive monitoring. Stress also raises nutritional needs considerably—increasing the risk of malnutrition even in previously healthy individuals. Providing nutrition care for these patients is not only challenging, it is often ineffective for preventing loss of weight and muscle tissue. Despite these difficulties, the health care professional must determine the best measures to take to limit damage and promote recovery.

Throughout this chapter, the CourseMate icon indicates an opportunity for online self-study, linking you to activities to increase your understanding of chapter concepts. **www.cengagebrain.com** (search for ISBN 084006845X)

Metabolic and Respiratory Stress

CHAPTER OUTLINE

This chapter addresses the nutrition care provided to patients who undergo certain types of physiological stress. **Metabolic stress**, a disruption in the body's internal chemical environment, can result from uncontrolled infections or extensive tissue damage, such as deep, penetrating wounds or multiple broken bones. As the first part of this chapter explains, the body's stress response is an attempt to restore balance, but it can have both helpful and harmful effects. Later sections of this chapter describe **respiratory stress**, which is characterized by inadequate oxygen and excessive carbon dioxide in the blood and tissues. Both metabolic and respiratory stress can lead to **hypermetabolism** (above-normal metabolic rate), **wasting** (breakdown of muscle mass and loss of strength), and in severe circumstances, life-threatening complications. The highlight following this chapter discusses the causes and consequences of **multiple organ dysfunction syndrome**, the simultaneous dysfunction of two or more organ systems, which is often fatal.

metabolic stress: a disruption in the body's chemical environment due to the effects of disease or injury. Metabolic stress is characterized by changes in metabolic rate, heart rate, blood pressure, hormonal status, and nutrient metabolism.

respiratory stress: abnormal gas exchange between the air and blood, resulting in lower-than-normal oxygen levels and higher-than-normal carbon dioxide levels.

hypermetabolism: a higher-than-normal metabolic rate.

wasting: the breakdown of muscle tissue that results from disease or malnutrition.

multiple organ dysfunction syndrome: the progressive dysfunction of two or more organ systems that develops during intensive care; often results in death.

stress response: the chemical and physical changes that occur within the body during stress.

The Body's Responses to Stress and Injury

The **stress response** is the body's *nonspecific* response to a variety of stressors, such as infection, fractures, surgery, burns, and wounds. During stress, the metabolic processes that support immediate survival are given priority, while those of lesser consequence are delayed. Energy is of primary importance, and therefore the energy nutrients are mobilized from storage and made available in the blood. Heart rate and respiration (breathing rate) increase to deliver oxygen and nutrients to cells more quickly, and blood pressure rises. Meanwhile, energy is diverted from processes that are not life sustaining, such as growth, reproduction, and long-term immunity. If stress continues for a long period, interference with these processes begins to cause damage, possibly resulting in growth retardation and illness.

Hormonal Responses to Stress

The stress response is mediated by several hormones, which are released into the blood soon after the onset of injury (see Table 22-1).[1] The catecholamines (epinephrine and norepinephrine), often called the *fight-or-flight hormones*, stimulate heart muscle, raise blood pressure, and increase metabolic rate. Epinephrine also promotes glucagon secretion from the pancreas, prompting the release of nutrients from storage. The steroid hormone cortisol enhances protein degradation, raising amino acid levels in the

Pressure sores, wounds that develop when prolonged pressure cuts off blood circulation to the skin and underlying tissues, are a frequent source of metabolic stress in bedridden and wheelchair-bound patients.

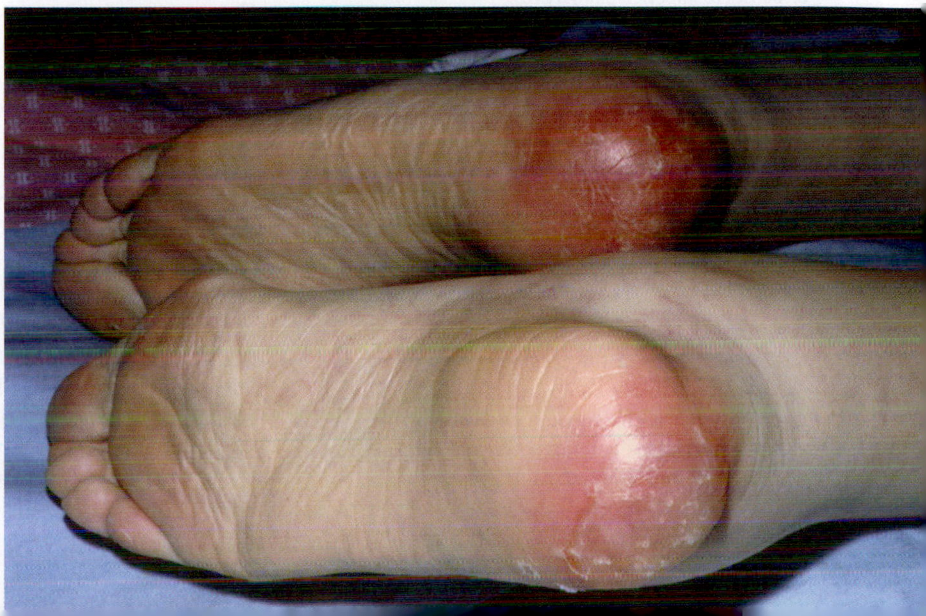

Mike Devlin / Photo Researchers, Inc.

TABLE 22-1 Metabolic Effects of Hormones Released during the Stress Response

Hormone	Metabolic Effects
Catecholamines	• Increase in metabolic rate • Glycogen breakdown in liver and muscle • Glucose production from amino acids • Release of fatty acids from adipose tissue • Glucagon secretion from pancreas
Glucagon	• Glycogen breakdown in liver • Glucose production from amino acids • Release of fatty acids from adipose tissue
Cortisol	• Protein degradation • Enhancement of glucagon's action on liver glycogen • Glucose production from amino acids • Release of fatty acids from adipose tissue
Aldosterone	• Sodium reabsorption in kidneys
Antidiuretic hormone	• Water reabsorption in kidneys

blood and making amino acids available for conversion to glucose. All of these hormones have similar effects on glucose and fat metabolism, causing the breakdown of glycogen (glycogenolysis), the production of glucose from amino acids (gluconeogenesis), and the breakdown of triglycerides in adipose tissue (lipolysis). ♦ Thus, the combined effects of these hormones contribute to hyperglycemia, which often accompanies critical illness. Two other hormones induced by stress, aldosterone and antidiuretic hormone, help to maintain blood volume by stimulating the kidneys to reabsorb more sodium and water, respectively.

♦ The catecholamines, glucagon, and cortisol have actions that oppose those of insulin and are therefore referred to as *counterregulatory hormones*.

Cortisol's effects can be detrimental when stress is prolonged. In excess, cortisol causes the depletion of protein in muscle, bone, connective tissue, and skin. It impairs wound healing, so high cortisol levels may be especially dangerous for a patient with severe injuries. Because cortisol inhibits protein synthesis, consuming more protein cannot easily reverse tissue losses. Excess cortisol also leads to insulin resistance, contributing to hyperglycemia. In addition, cortisol suppresses immune responses, increasing susceptibility to infection. Note that the pharmaceutical forms of cortisol are common anti-inflammatory medications (such as *cortisone* and *prednisone*); their long-term use can cause undesirable side effects such as muscle wasting, thinning of the skin, diabetes, and early osteoporosis.

The Inflammatory Response

Cells of the immune system mount a quick, nonspecific response to infection or tissue injury. This so-called **inflammatory response** serves to contain and destroy infectious agents (and their products) and prevent further tissue damage. The inflammatory response also triggers various events that promote healing. As in the stress response, however, there is a delicate balance between a response that protects tissues from further injury and an excessive response that can cause additional damage to tissue.

inflammatory response: a group of nonspecific immune responses to infection or injury.

phagocytes (FAG-oh-sites): white blood cells (neutrophils and macrophages) that have the ability to engulf and destroy antigens.

• **phagein** = to eat

The Inflammatory Process

The inflammatory response begins with the dilation of arterioles and capillaries at the site of injury, which increases the blood flow to the affected area. The capillaries within the damaged tissue become more permeable, allowing some blood plasma to escape into the tissue and cause local edema (see Figure 22-1). The various changes in blood vessels ♦ also encourage the entry of immune cells that can destroy foreign agents. Among the first cells to arrive are the **phagocytes**, which slip through gaps between the endothelial cells that form the vessel walls. The phagocytes engulf microorganisms and destroy them with hydrolytic enzymes and reactive forms of oxygen. When inflammation becomes chronic, these normally useful products of phagocytes can damage healthy tissue.

♦ The classic signs of inflammation that accompany altered blood flow are:

- *Redness*—due to dilation of blood vessels in the injured area
- *Heat*—due to the influx of warm arterial blood
- *Swelling*—due to the accumulation of fluid at the site of injury
- *Pain*—due to the pressure of edema within damaged tissue and the actions of certain chemical mediators on pain receptors

FIGURE 22-1 The Inflammatory Process

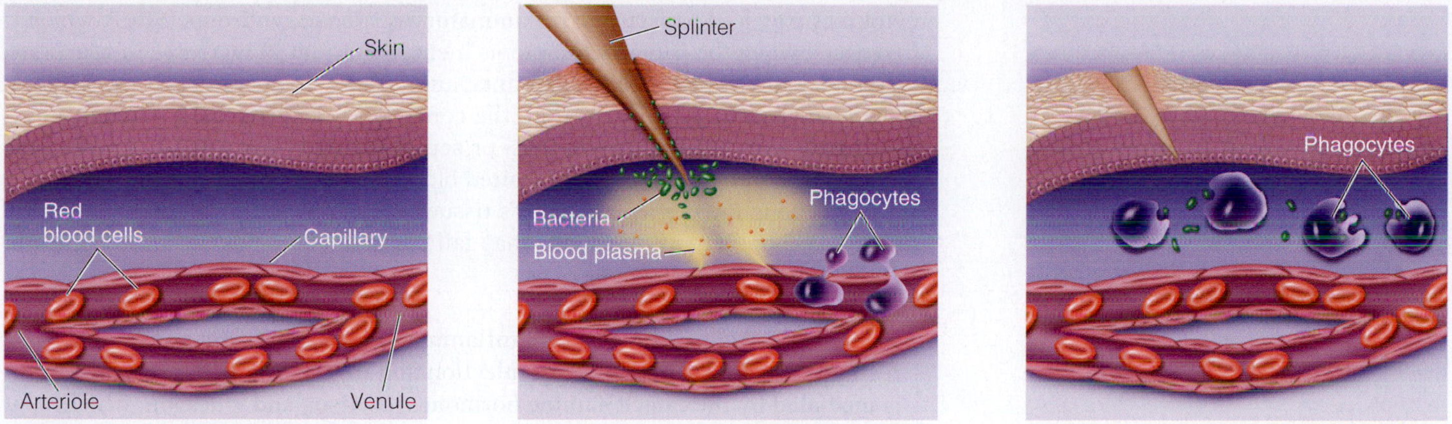

Cells lining the blood vessels lie close together, and normally do not allow the contents to cross into tissue.

When tissues are damaged, immune cells release histamine, which dilates some blood vessels, increasing blood flow to the damaged area. Fluid leaks out of capillaries (causing swelling), and phagocytes escape between the small gaps in the blood vessel walls.

Phagocytes engulf bacteria and disable them with hydrolytic enzymes and reactive forms of oxygen.

Mediators of Inflammation Numerous chemical substances control the inflammatory process. These *mediators* are released from damaged tissue, blood vessel cells, and activated immune cells. Many of them help to regulate more than one step in the process. Some of the examples that follow were introduced in Highlight 17's discussion of immunity. Histamine, a small molecule similar to an amino acid in structure, is released from granules within **mast cells**, causing vasodilation and capillary permeability. ♦ Fragments of complement proteins ♦ trigger histamine's release from mast cells and help to recruit and activate phagocytes. Other compounds that participate in the inflammatory process include several cytokines ♦ (especially interleukin-1, interleukin-6, and tumor necrosis factor-α) and various **eicosanoids** (which are derived from dietary fatty acids). Note that most anti-inflammatory medications, including steroidal drugs (such as cortisone and prednisone) and nonsteroidal anti-inflammatory drugs (such as aspirin and ibuprofen), act by blocking eicosanoid synthesis.

Changing dietary fat sources can have subtle effects on the inflammatory process.[2] The major precursor for the eicosanoids is arachidonic acid, which derives from the omega-6 fatty acids in vegetable oils. Some omega-3 fatty acids compete with arachidonic acid and inhibit the production of the most powerful inflammatory mediators. Partially replacing vegetable oils rich in omega-6 fatty acids with food sources high in omega-3 fatty acids (such as fish oil) helps to suppress inflammation, but it is not a reliable treatment.

Systemic Effects of Inflammation Cytokines released during the inflammatory process induce **systemic** effects as well as the localized effects described earlier. Within hours (or, in some cases, days) after inflammation, infection, or severe injury, the liver steps up its production of certain proteins in an effort known as the **acute-phase response**.[3] These acute-phase proteins include **C-reactive protein**, ♦ complement, blood-clotting proteins such as fibrinogen and prothrombin, and others. At the same time, plasma concentrations of albumin, iron, and zinc fall (recall from Chapter 17 that albumin levels are often measured to assess health and nutrition status). The acute-phase response is accompanied by muscle catabolism to make amino acids available for glucose production, tissue repair, and immune protein synthesis; consequently, negative nitrogen balance (and

♦ *Antihistamines* are medications taken to reduce the effects of histamine.

♦ Reminder: *Complement* is a collective term for a group of plasma proteins that assist the activities of antibodies.

♦ Reminder: *Cytokines* are hormone-like proteins that regulate immune responses.

mast cells: cells within connective tissue that produce and release histamine.

eicosanoids (eye-KO-sa-noids): 20-carbon molecules derived from dietary fatty acids that help to regulate blood pressure, blood clotting, and other body functions.

- **eicosa** = twenty

systemic (sih-STEM-ic): relating to the entire body.

acute-phase response: changes in body chemistry resulting from infection, inflammation, or injury; characterized by alterations in plasma proteins.

C-reactive protein: an acute-phase protein released from the liver during acute inflammation or stress.

♦ C-reactive protein, the best clinical indicator of the acute-phase response, becomes elevated during many chronic diseases.

wasting) frequently results. Other clinical features include an elevated metabolic rate, increased blood neutrophil levels, lethargy, anorexia, and, often, fever.

If inflammation does not resolve, the continued production of pro-inflammatory cytokines may lead to **systemic inflammatory response syndrome** (**SIRS**), which is diagnosed when the patient's symptoms include substantial increases in heart rate, respiratory rate, white blood cell counts, and/or body temperature. If these symptoms result from a severe infection, the condition is called **sepsis**. Complications associated with severe cases of SIRS or sepsis include fluid retention and tissue edema, low blood pressure, and impaired blood flow. If the reduction in blood flow is severe enough to deprive the body's tissues of oxygen and nutrients (a condition known as **shock**), multiple organs may fail simultaneously, as discussed in Highlight 22.

IN SUMMARY **The stress and inflammatory responses are nonspecific responses to stressors that cause infection and injury. The stress response is mediated by the catecholamine hormones, cortisol, and glucagon, which together raise nutrient levels in blood, stimulate heart rate, raise blood pressure, and increase metabolic rate. Aldosterone and antidiuretic hormone help to maintain adequate blood volume. The inflammatory process—mediated by compounds released from damaged tissues, immune cells, and blood vessels—results in systemic effects that alter nutrient metabolism, heart rate, blood pressure, body temperature, and immune cell functions. Signs of inflammation in injured tissues include redness, heat, swelling, and pain. Persistent, severe inflammation may result in shock and increases the risk of multiple organ dysfunction.**

Nutrition Treatment of Acute Stress

As described earlier, an excessive response to metabolic stress can worsen illness and even threaten survival. Therefore, medical personnel must manage both the acute medical condition that initiated stress and the complications that arise as a result of the stress and inflammatory responses. Immediate concerns during severe stress are to restore lost fluids and electrolytes and to remove underlying stressors. Thus, initial treatments include administering intravenous solutions to correct fluid and electrolyte imbalances, treating infections, repairing wounds, draining **abscesses** (pus), and removing dead tissue (**debridement**). After stabilization, nutrient needs can be estimated and nutrition therapy provided.

systemic inflammatory response syndrome (SIRS): a whole-body inflammatory response caused by severe illness or trauma; characterized by raised heart and respiratory rates, abnormal white blood cell counts, and elevated body temperature.

sepsis: a whole-body inflammatory response caused by infection; characterized by symptoms similar to those of SIRS.

shock: a severe reduction in blood flow that deprives the body's tissues of oxygen and nutrients; characterized by reduced blood pressure, raised heart and respiratory rates, and muscle weakness.

abscesses (AB-sess-es): accumulations of pus.

debridement: the surgical removal of dead, damaged, or contaminated tissue resulting from burns or wounds; helps to prevent infection and hasten healing.

Determining Nutritional Requirements

The most notable metabolic changes in patients undergoing metabolic stress include hypermetabolism, negative nitrogen balance, insulin resistance, and hyperglycemia. Hypermetabolism and negative nitrogen balance can lead to wasting, which may impair organ function and delay recovery. Hyperglycemia increases the risk of infection, a dangerous problem during critical illness. Thus, the principal goal of nutrition therapy is to provide a diet that preserves lean (muscle) tissue, maintains immune defenses, and promotes healing.

Feeding an acutely stressed patient is often challenging. Overfeeding increases the risks of refeeding syndrome ♦ and its associated hyperglycemia. Underfeeding may worsen negative nitrogen balance and increase lean tissue losses. Assessing nutritional needs can be complicated, however, because fluid imbalances prevent accurate weight measurements, and laboratory data may reflect the metabolic alterations of illness rather than the person's nutrition status.

♦ Reminder: *Refeeding syndrome* can develop when a severely malnourished person is aggressively fed; it is associated with fluid and electrolyte imbalances and hyperglycemia.

The amounts of protein and energy to provide during acute illness are controversial and still under investigation. Research results have been mixed, in part because various conditions can lead to metabolic stress and each patient's situation is somewhat different.[4] Moreover, protein and energy needs can vary substantially over the course of illness. The guidelines presented here are subject to change as new

TABLE 22-2 Disease-Specific Stress Factors for Estimating Energy Needs during Metabolic Stress

Method:

Step 1. Estimate the energy needed to support resting metabolic rate (RMR) using indirect calorimetry or a predictive equation (see Table 18-3, p. 599).

Step 2. Multiply the patient's RMR by an appropriate stress factor for acute illness (see the example in the "How To" box on p. 600).

Examples of stress factors:

- Advanced liver disease: 1.0 to 1.16
- Inflammatory bowel disease (active): 1.05 to 1.10
- Pancreatitis: 1.13 to 1.21
- Surgery: 1.2 to 1.4
- Mechanical ventilation: 1.32 to 1.34
- Burns (20 to 30 percent of body surface): 1.6 to 1.7
- Burns (30 to 40 percent of body surface): 1.8 to 1.9
- Burns (40 to 45 percent of body surface): 2.0

SOURCE: American Dietetic Association, *Nutrition Care Manual* (Chicago: American Dietetic Association, 2007); N. Barak, E. Wall-Alonso, and M. D. Sitrin, Evaluation of stress factors and body weight adjustments currently used to estimate energy expenditure in hospitalized patients, *Journal of Parenteral and Enteral Nutrition* 26 (2002): 231–238.

findings help to resolve the complex issues related to nutrient intakes and delivery methods. To help guide their decisions about treatment, clinicians need to closely observe patients' responses to feedings and readjust nutrient intakes as necessary.

Estimating Energy Needs for Acute Stress A common method for determining the energy needs of acutely stressed individuals is to estimate or measure the resting metabolic rate (RMR) and then multiply the result by a stress factor to account for the increased energy requirements of stress and healing. This method was introduced in Chapter 18 (see the "How To" on p. 600); Table 18-3 (p. 599) lists some common predictive equations used for estimating RMR.

Stress factors depend on the severity of the illness and the patient's overall nutrition status. Generally, energy needs are increased by fever, mechanical ventilation, and the presence of open wounds; patients with burns and infections often have the highest energy needs. For many critically ill patients, the stress factor 1.2 provides adequate energy, and higher factors may result in overfeeding.[5] Patients who are critically ill are usually bedridden and inactive, so the energy needed for physical activity is minimal. Table 22-2 reviews the use of stress factors and provides examples appropriate for critical care patients; keep in mind that the values used may vary somewhat among institutions.

Predictive equations used for determining energy needs sometimes include "built-in" stress factors to account for stress, injury, or intensive treatment. Table 22-3 shows examples of equations used in critical care populations and describes the use of the Ireton–Jones equation, which includes multipliers for the presence of trauma and burn injuries. Other equations in current use include factors for other pertinent variables, such as body temperature, heart rate, and respiratory rate.[6]

Daily energy requirements for nonobese critical care patients often fall within the range of 25 to 30 kcalories per kilogram of body weight;[7] a patient weighing 160 pounds (72.7 kilograms) may therefore require between 1818 and 2181 kcalories per day. ♦ The energy intake is sometimes started within this range and then adjusted as the patient's body weight and other determinants of nutrition status change. For critically ill obese patients (BMI $>$ 30), energy needs may range from 22 to 25 kcalories per kilogram of ideal body weight per day.[8] ♦

Protein Requirements in Acute Stress To maintain lean tissue, the protein intakes recommended during acute stress are higher than DRI values. ♦ In nonobese critically ill patients, protein needs may range from 1.2 to 2.0 grams per

♦ 72.7 kg $\times$ 25 kcal/kg = 1818 kcal
72.7 kg $\times$ 30 kcal/kg = 2181 kcal

♦ Reminder: Ideal body weights typically fall within the BMI range of 18.5 to 25.

♦ Reminder: The protein RDA for adults is 0.8 gram per kilogram body weight.

TABLE 22-3 Selected Equations for Estimating Energy Needs in Critical Care Patients

Ireton–Jones

Energy needs (kcal/day) = 1925 + [5 × wt (kg)] − [10 × age (yr)] + [281 × sex] + [292 × trauma] + [851 × burn]

where sex is male (× 1) or female (× 0), trauma is the presence of physical injury (× 1) or not (× 0), and burn is the presence of a burn injury (× 1) or not (× 0).

Penn State

Energy needs (kcal/day) = [RMR × 0.85] + [V_E × 33] + [T_{max} × 175] − 6433

where RMR is calculated using the Harris-Benedict equation (see Table 18-3, p. 599), V_E is minute ventilation in liters per minute (reading taken from the ventilator), and T_{max} is the patient's maximum body temperature (in degrees Celsius) in the preceding 24 hours.

Example (Ireton–Jones equation): Erin is a 27-year-old female patient who weighs 140 pounds (63.6 kilograms). Two days ago, she was severely injured in an automobile accident and is currently being cared for in a critical care unit where she is receiving mechanical ventilation. She did not suffer a burn injury. Using the Ireton–Jones equation shown above, her daily energy needs can be estimated as follows:

Energy needs (kcal/day) = 1925 + [5 × wt (kg)] − [10 × age (yr)] + [281 × sex] + [292 × trauma] + [851 × burn]
= 1925 + (5 × 63.6 kg) − (10 × 27 yr) + (281 × 0) + (292 × 1) + (851 × 0)
= 1925 + 318 − 270 + 0 + 292 + 0 = 2265 kcal

NOTE: The equations shown here were developed for use in ventilator-dependent patients.

kilogram body weight per day;[9] burn patients may require between 2 and 3 grams per kilogram body weight each day due to the substantial losses of protein associated with burn wounds.[10] For most obese patients, protein requirements may range from 2.0 to 2.5 grams per kilogram of ideal body weight per day.[11] Even with adequate protein, however, negative nitrogen balance cannot be prevented during acute stress because hormonal changes encourage protein catabolism. The bed rest required during critical illness also contributes substantially to muscle breakdown.

The amino acids glutamine and arginine are sometimes added to the diets of acutely stressed and immune-compromised patients. Some studies have suggested that glutamine supplementation may improve immune function, preserve muscle mass, and reduce mortality rates in critically ill patients.[12] Arginine supplementation may have beneficial effects on the immune responses and nitrogen balance of critically ill and postoperative patients.[13] Although glutamine and arginine are often added to enteral formulas promoted for wound healing and enhanced immunity, their use remains controversial.[14]

Carbohydrate and Fat Intakes in Acute Stress The bulk of energy needs are supplied from carbohydrate and fat. Carbohydrate is usually the main source of energy, providing 50 to 60 percent of total energy requirements.[15] When parenteral feedings are necessary, dextrose is provided to critically ill patients at no more than 5 milligrams per kilogram body weight per minute to prevent hyperglycemia (see p. 673). In patients with severe hyperglycemia, fat may supply up to 50 percent of kcalories, although high fat intakes may suppress immune function and increase the risks of developing infections and hypertriglyceridemia. Patients with blood triglyceride levels above 300 to 400 milligrams per deciliter may require fat restriction.[16]

Micronutrient Needs in Acute Stress Acutely stressed patients are believed to have increased micronutrient needs, but specific requirements remain unknown.[17] In hypermetabolic patients, the need for B vitamins may be higher to support the increase in energy metabolism. A number of micronutrients, such as vitamin A, vitamin C, and zinc, have critical roles in immunity and wound healing, and their supplementation may speed recovery under certain circumstances. Patients with burns and tissue injuries may have increased requirements for trace minerals due to tissue losses; in several studies, supplementation of zinc, copper, and selenium improved immune responses in severely burned patients.[18]

Plasma levels of micronutrients are often altered during critical illness. The acute-phase response causes a redistribution in the tissue content of some micronutrients that either raises or lowers their blood levels; therefore, micronutrient status is sometimes difficult to interpret. ♦ Blood concentrations of trace minerals are monitored in patients receiving parenteral nutrition support to ensure that excessive amounts are not given intravenously.

♦ During the acute-phase response, plasma levels of iron and zinc fall, whereas plasma copper levels rise.

Approaches to Nutrition Care in Acute Stress

As mentioned earlier, the initial care following acute stress focuses on maintaining fluid and electrolyte balances. Simple intravenous solutions often contain dextrose, providing minimal kcalories. Once patients are stable, nutrition support may be necessary if poor appetite, the medical condition, or a medical procedure (such as mechanical ventilation) interferes with food intake. For patients with a functional GI tract, early enteral feedings—started in the first 24 to 48 hours after hospitalization—are associated with fewer complications and shorter hospital stays compared with delayed feedings. If enteral nutrition is not possible, malnourished patients may receive parenteral nutrition support soon after admission to the hospital. In previously healthy patients, however, parenteral nutrition support may be withheld during the first seven days of hospitalization to avoid the risk of infectious complications.[19]

Once patients can tolerate oral feedings, a high-kcalorie, high-protein diet is often prescribed, although care must be taken not to overfeed patients who are at risk of developing refeeding syndrome or hyperglycemia. Because meeting protein and energy needs may be difficult, enteral formulas are often provided to supplement the diet. Many such formulas have high nutrient density, and some contain extra amounts of nutrients believed to promote healing or benefit immune function, such as the amino acids arginine and glutamine, omega-3 fatty acids, and the antioxidant nutrients. Nutrient needs should be reassessed frequently until the patient's condition improves.

Patients with Burn Injuries

Burns are among the most severe injuries that a person may experience, and they have destructive effects on growth and health that may persist long after the burns have healed. Causes of burns include flames or scalding water, chemical agents, electricity, and irradiation. Frequent complications include infection and **hypovolemia**, which can increase the risk of death.[20]

Burn Classification

Burns are classified according to how deeply they penetrate the skin and underlying tissue. First-degree burns affect only the **epidermis** and are pink or red, dry, and painful (for example, a sunburn). Second-degree burns (also called partial-thickness burns) involve both the epidermis and a portion of the **dermis**. They are red, wet, and blistery, and extremely painful because nerve endings are exposed. Third-degree burns (also known as full-thickness burns) destroy both the epidermis and dermis and may extend into the tissues below; their appearance may be waxy white, brown and leathery, or black and charred. These burns are deep enough to destroy nerves and are therefore painless.[21]

Burn size in adults is often estimated by dividing the body into 11 parts; each part represents about 9 percent of the total body surface area (TBSA).[22] The head

hypovolemia (HIGH-poe-voe-LEE-me-ah): low blood volume.

epidermis (eh-pih-DER-miss): the outer layer of the skin.

dermis: the connective tissue layer underneath the epidermis that contains the skin's blood vessels and nerves.

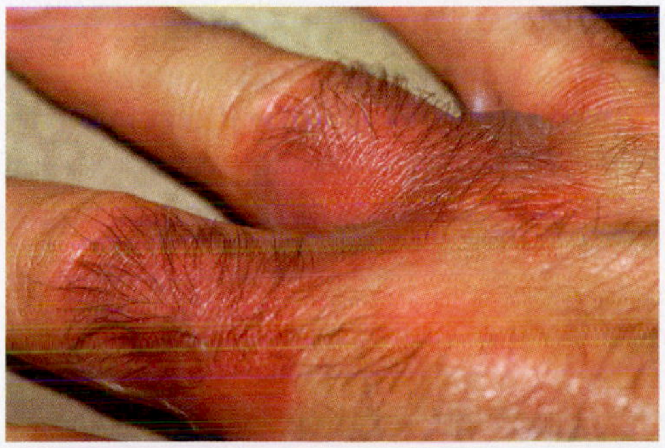

A first-degree burn injures the epidermis and is characterized by pink or reddened skin.

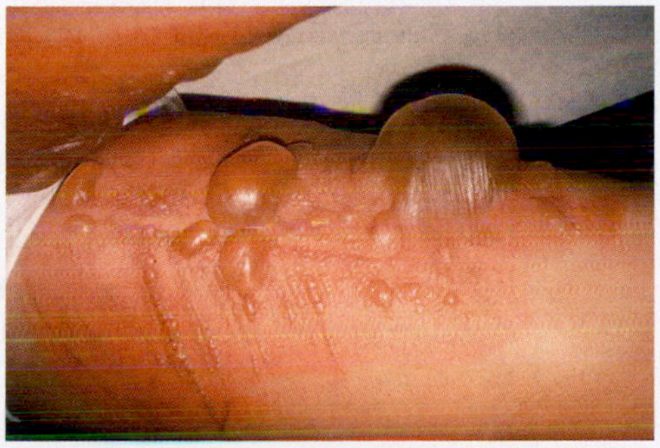

A second-degree burn damages the epidermis and a portion of the dermis and causes redness, swelling, and blistering.

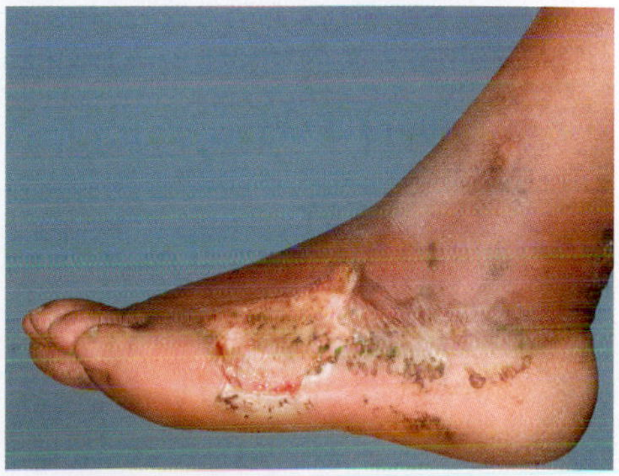

A third-degree burn destroys both the epidermis and dermis and may involve the tissues beneath the skin.

♦ Burn size can be estimated by sectioning the total body surface area (TBSA) as shown.

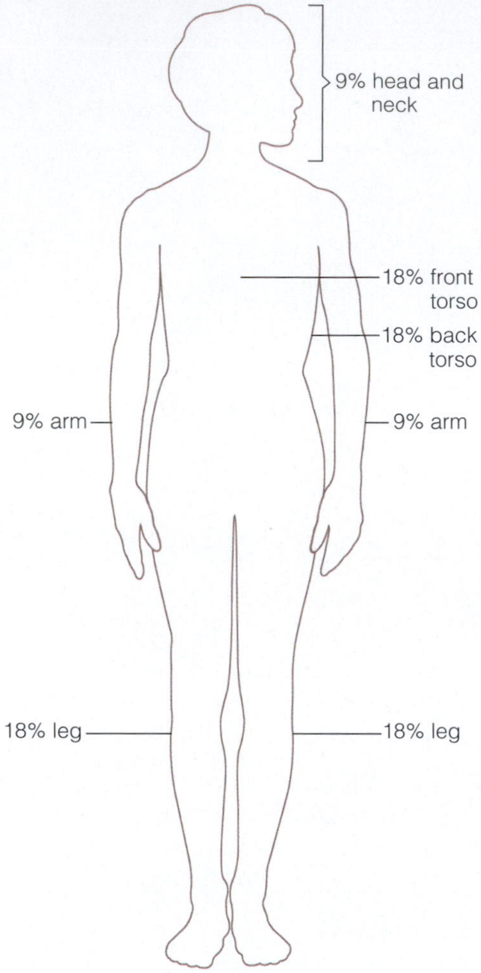

♦ Chapter 18 provides information about the high-kcalorie, high-protein diet (see p. 604).

and neck region and each arm are equivalent to about 9 percent TBSA each; the front torso, the back torso, and each leg represent approximately 18 percent TBSA each (see the margin drawing). ♦ The severity of a burn is based both on its thickness and on the amount of surface area involved.

Treatment for Burn Injuries Emergency measures after a burn include the removal of clothing and smoldering material from the skin. Burns caused by acid or chemical compounds must be flushed with copious amounts of water. Wounds are cleaned and debris removed. Blisters and dead tissue are debrided, if necessary. Finally, the surface is covered with topical antibacterial agents and sterile dressings. Immediate care also includes fluid replacement and electrolyte management, as the fluid losses through burned skin can be considerable. Some burn victims need immediate oxygen support or mechanical ventilation. Pain relief medication is also required soon after injury.

Metabolic Changes in Burn Patients Burn injuries cause severe metabolic stress and the inflammatory response, as discussed earlier in this chapter, and therefore they result in hypermetabolism, tissue breakdown, and altered nutrient metabolism. With the protective skin barrier partially destroyed, burns are accompanied by losses of evaporative water and body heat. Second- and third-degree burns can cause substantial losses of protein and micronutrients. Extensive burns can disrupt gastrointestinal function.

Nutrition Therapy for Burn Patients The objectives of nutrition care for burn patients are to achieve nitrogen balance, minimize tissue losses, and maintain appropriate body weight. The nutrition prescription is typically a high-kcalorie, high-protein diet.[23] ♦ Energy needs can be calculated as described previously (see Table 22-2 on p. 687 and Table 22-3 on p. 688); equations that consider factors such as burn severity, surgical procedures, or ventilator use may be used during peak recovery periods. The suggested protein intake is 2 to 3 grams per kilogram body weight or 20 to 25 percent of total kcalories in individuals with burns greater than 10 percent TBSA.[24] For less extensive burns, a protein intake of 1.2 grams per kilogram body weight is usually adequate. Adequacy of protein intake is assessed by monitoring nitrogen balance, serum proteins, and wound-healing ability. Micronutrient supplements are typically provided and may include high amounts of vitamin A, vitamin C, and zinc, which are thought to support immunity and promote wound healing. Fluid and electrolyte needs must be monitored carefully during the recovery period; the patient's hydration status can be evaluated by monitoring urine output and serum electrolyte levels.

Some patients may need to be evaluated for feeding ability; problems that may interfere with eating include burns on the face, hands, and arms; bulky dressings; frequent dressing changes; and pain medications that cause sedation. Patients who are able to eat are often offered small, frequent meals rather than large meals and are provided with oral supplements and nutrient-dense snacks to help them meet energy and protein needs. A combination of oral feedings and tube feedings is often necessary. Some burn patients develop gastroparesis or intestinal ileus (stomach or intestinal paralysis) and may require nasoenteric feedings (discussed in Chapter 20). Parenteral support may be required if intestinal function is lacking, if complications that interfere with enteral feedings develop, or if nutrient requirements cannot be met by tube feeding alone. The accompanying Case Study reviews the nutrition care of a burn patient.

IN SUMMARY **Severe metabolic stress can cause hypermetabolism, negative nitrogen balance, and hyperglycemia, and may result in wasting. The objective of nutrition care during acute stress is to provide a diet that preserves muscle tissue, maintains immune defenses, and promotes healing. Energy needs are often estimated by modifying RMR values with stress factors that account for the increased demands of the medical condition or treatment. Protein recommendations are higher than DRI levels to help prevent tissue losses and**

CASE STUDY

Patient with a Severe Burn

David Bray, a 42-year-old man, has been admitted to intensive care. He suffered a severe burn covering 35 percent of his body when he was trapped inside a burning building. His wife told the nurse that Mr. Bray's height is 6 feet and that he usually weighs about 175 pounds. The physician ordered lab work, including serum protein concentrations, but the results have not yet been received.

1. Identify Mr. Bray's immediate needs after the injury. Describe the initial concerns of the health care team and the measures they might take soon after Mr. Bray's arrival at the hospital.
2. Considering Mr. Bray's condition, what problems might the health care team encounter when they attempt to obtain information that can help them assess his nutrition status? What additional concerns might they have if Mr. Bray was malnourished before he experienced the burn?
3. Estimate Mr. Bray's energy and protein needs (use a protein factor of 2.5 grams per kilogram). What problems may interfere with Mr. Bray's ability to meet his nutrient needs?
4. After Mr. Bray transitions to oral feedings, he is able to obtain only 65 percent of his energy requirements. What other feeding options may be considered?

allow healing of damaged tissue. Enteral or parenteral nutrition support may be needed to meet the high nutrient requirements of acutely stressed patients. Burn patients typically require fluid replacement and electrolyte management after a burn injury, and a high-kcalorie, high-protein diet during the recovery period.

Nutrition and Respiratory Stress

Some medical problems upset the gas exchange process between the air and blood and result in respiratory stress, which is characterized by a reduction in the blood's oxygen supply and an increase in carbon dioxide levels. Excessive carbon dioxide in the blood may disturb the breathing pattern enough to interfere with food intake. Moreover, the labored breathing caused by many respiratory disorders entails a higher energy cost than normal breathing does, raising energy needs and increasing carbon dioxide production further. Lung diseases make physical activity difficult and can lead to muscle wasting. Weight loss and malnutrition therefore become dangerous outcomes of some types of respiratory illnesses.

Chronic Obstructive Pulmonary Disease

Chronic obstructive pulmonary disease (COPD) refers to a group of conditions characterized by the persistent obstruction of airflow through the lungs. Figure 22-2 illustrates the main airways (**bronchi** and **bronchioles**) and air sacs (**alveoli**) of the normal respiratory system, and Figure 22-3 shows how they are altered in COPD. The two main types of COPD are **chronic bronchitis** and **emphysema**, and many patients display features of both conditions:[25]

- *Chronic bronchitis* is characterized by persistent inflammation and excessive secretions of mucus in the main airways of the lungs, which may ultimately thicken and become too narrow for adequate mucus clearance. Chronic bronchitis is diagnosed when a chronic, productive cough persists for at least three months of the year for two consecutive years.

chronic obstructive pulmonary disease (COPD): a group of lung diseases characterized by persistent obstructed airflow through the lungs and airways; includes chronic bronchitis and emphysema.

bronchi (BRON-key), **bronchioles** (BRON-key-oles): the main airways of the lungs. The singular form of bronchi is *bronchus*.

alveoli (al-VEE-oh-lie): air sacs in the lungs. One air sac is an *alveolus*.

chronic bronchitis (bron-KYE-tis): a lung disorder characterized by persistent inflammation and excessive secretions of mucus in the main airways of the lungs.

emphysema (EM-fih-ZEE-mah): a progressive lung condition characterized by the breakdown of the lungs' elastic structure and destruction of the walls of the bronchioles and alveoli, reducing the surface area involved in respiration.

FIGURE 22-2 The Respiratory System

Inhaled air travels via the trachea to the bronchi and bronchioles, the major airways of the lungs. Oxygen and carbon dioxide are exchanged across the thin-walled alveoli, which are surrounded by capillaries.

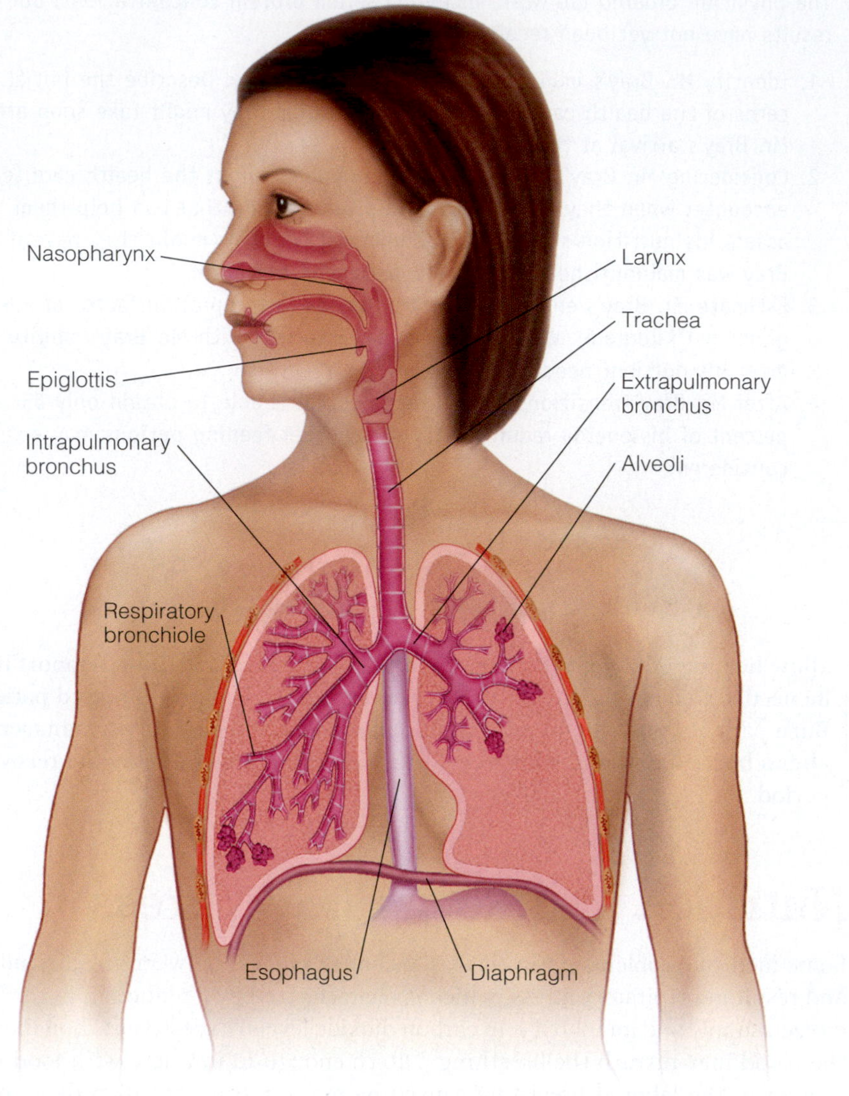

SOURCE: Based on a drawing in Carol Mattson Porth, *Pathophysiology*, 5th ed. (Lippincott Williams & Wilkins, 1998).

- *Emphysema* is characterized by the breakdown of the lungs' elastic structure and destruction of the walls of the bronchioles and alveoli, changes that significantly reduce the surface area available for respiration. Emphysema is diagnosed on the basis of clinical signs and the results of lung function tests.

Both chronic bronchitis and emphysema are associated with abnormal levels of oxygen and carbon dioxide in the blood and shortness of breath (**dyspnea**). COPD may eventually lead to respiratory or heart failure and, together with other chronic respiratory illnesses, ranks as the third leading cause of death in the United States.[26]

COPD is a debilitating condition. Generally, dyspnea worsens as the condition progresses, resulting in dramatic reductions in physical activity and quality of life. Activities of daily living such as bathing or dressing may cause exhaustion or breathlessness. Weight loss and wasting are common in the advanced stages of disease and may result from hypermetabolism, poor food intake, and the actions of

dyspnea (DISP-nee-ah): shortness of breath.

FIGURE 22-3 **Chronic Obstructive Pulmonary Disease**

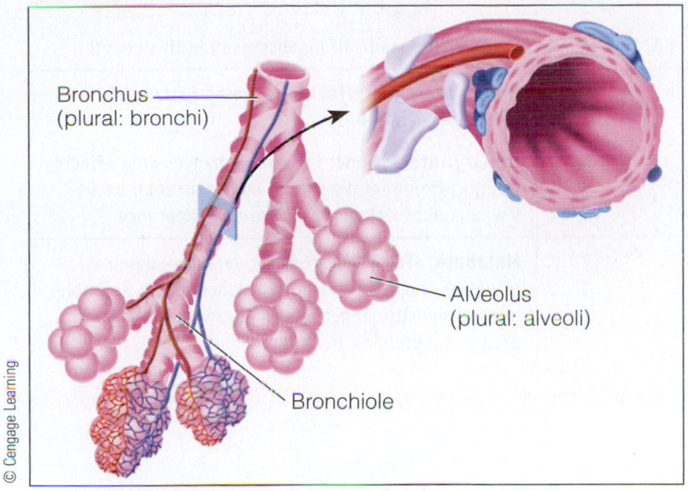

Healthy bronchi provide an open passageway for air. Healthy alveoli permit gas exchange between the air and blood.

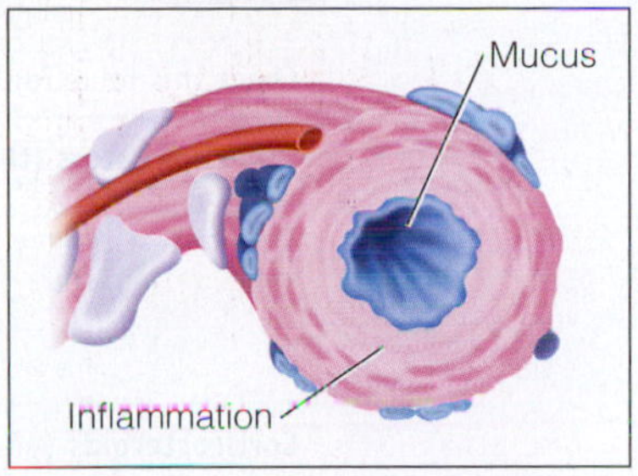

Chronic bronchitis is characterized by inflammation, excessive secretion of mucus, and narrowing of the bronchi – factors that reduce normal airflow.

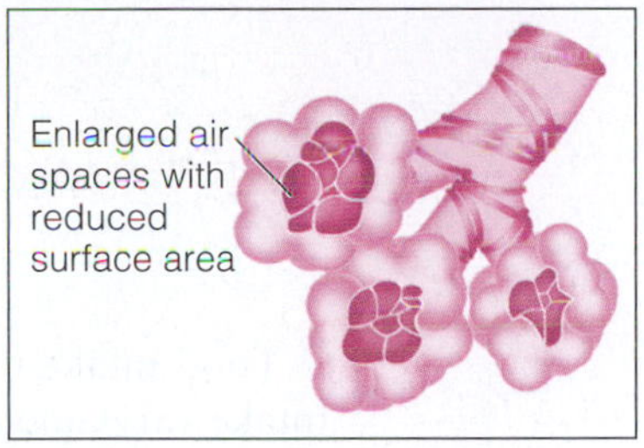

Emphysema is characterized by gradual destruction of the walls separating the alveoli and reduced lung elasticity.

various inflammatory proteins. As with other chronic illnesses, anxiety and depression are a concern, and psychological distress may reduce a COPD patient's ability to cope with the demands of treatment.

Causes of COPD Cigarette smoking is the primary risk factor in 90 percent of COPD cases[27] and is especially damaging when combined with respiratory infections or an occupational exposure to dusts or chemicals. Only a minority of smokers (about 15 percent) develops COPD, however; thus, genetic susceptibility also contributes to its development. Genetic factors are especially likely in patients with early-onset COPD. Alpha-1-antitrypsin deficiency, an inherited disorder, occurs in 1 to 2 percent of patients with COPD.[28] These individuals have inadequate blood levels of a plasma protein (alpha-1 antitrypsin) that normally inhibits enzymatic breakdown of the lungs' connective tissue.

Treatment of COPD The primary objectives of COPD treatment are to prevent the disease from progressing and relieve major symptoms (dyspnea and coughing). Individuals with COPD are encouraged to quit smoking to prevent disease progression and to get vaccinated against influenza and pneumonia to avoid complications. The most frequently prescribed medications are bronchodilators, which improve airflow, and corticosteroids (anti-inflammatory medications), which help to prevent symptom recurrence; note that corticosteroids promote catabolic processes and can exacerbate the muscle loss that often accompanies COPD. For people with severe COPD, supplemental oxygen therapy (12 hours daily) can maintain normal oxygen levels in the blood and reduce mortality risk. The Diet-Drug Interactions feature lists nutrition-related effects of the medications used to treat COPD.

Nutrition Therapy for COPD The main goals of nutrition therapy for COPD are to correct malnutrition (which affects up to 60 percent of patients with COPD[29]), promote the maintenance of a healthy body weight, and prevent muscle wasting. Research studies have found that underweight COPD patients have higher mortality rates;[30] thus, encouraging adequate food intake is typically the main focus of the nutrition care plan. Energy needs of COPD patients are usually raised due to hypermetabolism (about 20 percent above normal), which results from chronic inflammation and the increased workload of respiratory muscles.[31] Because excess body weight places an additional strain on the respiratory system, COPD patients who are overweight or obese may benefit from energy restriction and gradual weight reduction.[32]

Patients who need supplemental oxygen can use lightweight, portable equipment that allows them to move about freely.

DIET-DRUG Interactions

Check this table for notable nutrition-related effects of the medications discussed in this chapter.

Bronchodilators (theophylline, dyphylline)	**Gastrointestinal effects:** Increased gastric acid secretion, acid reflux **Dietary interactions:** Caffeine enhances drug effects (daily caffeine intake should be consistent); excessive alcohol intake may reduce drug clearance
Corticosteroids (prednisone)	**Metabolic effects:** Glucose intolerance, sodium retention, negative nitrogen balance, decreased bone mineral density, appetite stimulation, weight gain, growth suppression in children

Food intake often declines as COPD progresses, although the causes of poor intake vary among patients. Dyspnea may interfere with chewing or swallowing. Physical changes in the diaphragm and lungs may reduce abdominal volume, leading to early satiety. Appetite may be affected by medications, depression or anxiety, or altered taste perception. ♦ Some patients may become too disabled to shop or prepare food or may lack adequate support at home. The health practitioner must assess the unique needs of a COPD patient before proposing a nutrition care plan.

♦ The altered sense of taste in patients with COPD may be due to chronic mouth breathing, which dries the mouth. Taste is also affected by the use of certain medications, including some bronchodilators.

Some patients may benefit from eating frequent, small meals spaced throughout the day rather than two or three large ones. The lower energy content of small meals reduces the carbon dioxide load, and the smaller meals may produce less abdominal discomfort and dyspnea. Some individuals may eat better if they receive supplemental oxygen at mealtimes. Consuming adequate fluids should be encouraged to help prevent the secretion of overly thick mucus; however, some patients should consume liquids between meals so as not to interfere with food intake. For undernourished patients, a high-kcalorie, high-protein diet may be helpful, but excessive energy intakes increase the amount of carbon dioxide produced and can increase respiratory stress. Liquid supplements may be recommended as between-meal snacks to improve weight gain or endurance, but patients should be cautioned not to consume amounts that reduce energy intake at mealtime.

Pulmonary Formulas Enteral formulas designed for use in COPD provide more kcalories from fat and fewer from carbohydrate than standard formulas. The ratio of carbon dioxide production to oxygen consumption in cells is lower when fat is consumed, so theoretically these formulas should lower respiratory requirements. However, research studies have not confirmed that the reduced-carbohydrate formulas improve clinical outcomes more than moderate energy intakes, so the benefits of using these pulmonary formulas are uncertain.[33]

Incorporating an Exercise Program Loss of muscle can be more readily prevented or reversed if the treatment plan includes an effective exercise program. With exercise, patients are likely to see improvements in their endurance and become less fearful of their physical limitations. For some patients, the combination of an exercise plan and oral supplements may be better for maintaining weight and improving muscle mass than either component of treatment alone.[34] The accompanying Case Study allows you to review the nutrition care for a patient with COPD.

Respiratory Failure

In respiratory failure, the gas exchange between the air and circulating blood is severely impaired, causing abnormal levels of tissue gases that can be life threatening. Any of a large number of conditions that cause lung injury or impair lung function can be the underlying cause of failure. Possible causes of respiratory failure include infection (such as pneumonia or sepsis), aspira-

CASE STUDY

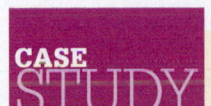

Elderly Man with Emphysema

John Norback is an 82-year-old man who has emphysema that severely affects both lungs. He is 5 feet 9 inches tall and currently weighs 150 pounds, about 20 pounds less than his weight in earlier years. He lives with a daughter and son-in-law and eats meals with their family. He becomes breathless when eating and when walking around the house, and he feels tired all the time. A medical clinic recently ordered oxygen therapy for home use, but supplies have not yet arrived. Mr. Norback's daughter is concerned about her father's recent weight loss and breathlessness.

1. Assess Mr. Norback's risk of malnutrition, using information from Table 17-8 in Chapter 17 (p. 583). What factors may have contributed to his weight loss?
2. What are possible reasons for Mr. Norback's difficulty with eating? List some dietary suggestions that may help to improve his appetite and food intake. How might the use of oxygen therapy help?
3. Based on the history given, what factors may account for Mr. Norback's tiredness? What suggestions would you give Mr. Norback and his daughter regarding physical activity?

tion of stomach contents, physical trauma, neuromuscular disorders, smoke inhalation, and airway obstruction.[35]

Severe lung damage can lead to **acute respiratory distress syndrome** (ARDS), a life-threatening condition that requires the use of mechanical ventilation to restore normal oxygen and carbon dioxide levels. In ARDS, the lungs exhibit extensive inflammation and fluid buildup (called *pulmonary edema*) that interfere with lung ventilation and gas exchange in the alveoli. Later stages are associated with a proliferation of lung cells, which causes fibrosis and disrupts lung structure. A dangerous complication of ARDS is the progression to multiple organ dysfunction syndrome, described in the highlight following this chapter.

Consequences of Respiratory Failure Respiratory failure is characterized by severe **hypoxemia** (low oxygen levels in the blood) and **hypercapnia** (excessive carbon dioxide in the blood). The low oxygen content of tissues (**hypoxia**) impedes cellular function and may lead to cell death. Severe hypercapnia can cause **acidosis**, which interferes with normal functioning of the central nervous system. To compensate for respiratory failure, a person breathes more rapidly, and the heart rate increases. The skin may become sweaty and develop a bluish cast (**cyanosis**). Headache, confusion, and drowsiness may occur. Severe cases of respiratory failure can cause heart arrhythmias and, ultimately, coma.

Treatment of Respiratory Failure The treatment of respiratory failure focuses on supporting lung function and correcting the underlying disorder. Because respiratory failure can be caused by a number of different conditions, treatment plans vary considerably. Individuals with chronic lung disorders may be provided with oxygen therapy via a face mask or nasal tubing to relieve symptoms, whereas patients with ARDS receive mechanical ventilation until they are able to breathe independently. Diuretics may be prescribed to mobilize the fluid that has accumulated in lung tissue. Medications may be given to treat infections, keep airways open, or relieve inflammation. Complications are common in ARDS and must be forestalled to prevent multiple organ dysfunction.

Nutrition Therapy for Acute Respiratory Failure Patients with lung injuries or ARDS are frequently hypermetabolic and/or catabolic and at high risk of muscle wasting. The primary concerns are therefore to provide enough energy and protein to sustain muscle mass and lung function without overtaxing the respiratory

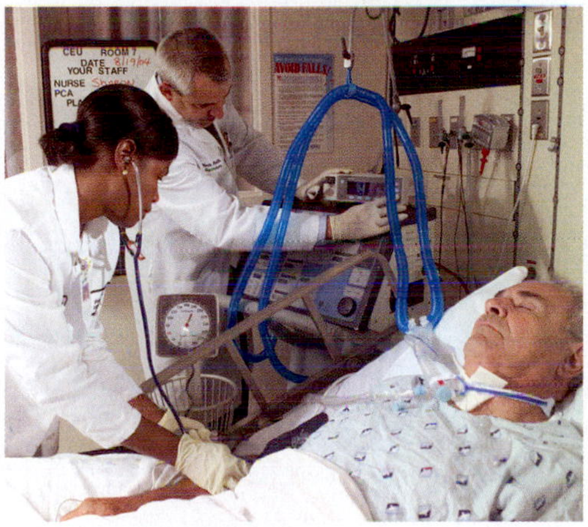

Mechanical ventilation controls the rate and amount of oxygen supplied to a person's airways.

acute respiratory distress syndrome (ARDS): respiratory failure triggered by severe lung injury; a medical emergency that causes dyspnea and pulmonary edema and usually requires mechanical ventilation.

hypoxemia (high-pock-SEE-me-ah): a low level of oxygen in the blood.

hypercapnia (high-per-CAP-nee-ah): excessive carbon dioxide in the blood.

hypoxia (high-POCK-see-ah): a low amount of oxygen in body tissues.

acidosis: acid accumulation in body tissues; depresses the central nervous system and can lead to disorientation and, eventually, coma.

cyanosis (sigh-ah-NOH-sis): a bluish cast in the skin due to the color of deoxygenated hemoglobin. Cyanosis is most evident in individuals with lighter, thinner skin; it is mostly seen on lips, cheeks, and ears and under the nails.

system. Fluid restrictions may be necessary to help correct pulmonary edema. When nutrition support is necessary, enteral nutrition is preferred over parenteral nutrition.

Energy Needs Energy needs typically range from 25 to 35 kilocalories per kilogram.[36] The body weight used to determine energy needs may need to be corrected for pulmonary edema. Overfeeding should be avoided because it can cause excessive carbon dioxide production and worsen respiratory function.

Energy needs can also be estimated using predictive equations such as those described in Tables 22-2 and 22-3. For patients using mechanical ventilation, the stress factors 1.34 and 1.32 are often appropriate for men and women, respectively.[37] Several predictive equations used in ventilated patients include respiratory rates (in breaths per minute or liters per minute) or lung volume (in liters). Note that mechanical ventilation often reduces energy expenditure, as patients using ventilators are often heavily sedated and immobile.

Protein Protein requirements are increased in patients with lung inflammation or ARDS. For mild or moderate lung injury, protein recommendations range from 1.0 to 1.5 grams of protein per kilogram body weight per day. Patients with ARDS may require 1.5 to 2.0 grams of protein per kilogram body weight daily.

Fluids Although most patients have normal fluid requirements, fluid status should be monitored daily to prevent fluid imbalance. Some patients may require fluid restriction to prevent edema in lung tissue, whereas others may become dehydrated due to diuretic therapy, an increase in bronchial secretions, or a low fluid intake. The presence of edema can make it difficult to assess whether a critically ill patient is maintaining weight.

Nutrition Support Patients with severe cases of respiratory failure may be unable to eat meals and may require nutrition support. Tube feedings are used if the intestine is functional, and intestinal feedings may be preferred over gastric feedings because they reduce the risk of aspiration. Patients with acute lung injuries or ARDS may benefit from enteral formulas designed to reduce inflammation and promote healing; these formulas are typically fortified with omega-3 fatty acids and antioxidant nutrients.[38] Nutrient-dense formulas (1.5 to 2.0 kcalories per milliliter) are prescribed for patients with fluid restrictions. If the risk of aspiration is too high to continue enteral feedings, parenteral nutrition support may be considered.

IN SUMMARY **Respiratory stress from chronic or acute disease affects body weight, muscle mass, and the normal functioning of all body tissues. Chronic obstructive pulmonary diseases (COPD) are debilitating, progressive illnesses that can lead to malnutrition, muscle wasting, and activity intolerance. The goals of nutrition therapy for COPD are to improve food intake, maintain proper weight, preserve muscle tissue, and improve exercise endurance. Respiratory failure, characterized by hypoxemia and hypercapnia, can result from conditions that cause lung injury or impair lung function. Acute respiratory distress syndrome (ARDS) is a severe form of respiratory failure that requires mechanical ventilation. Goals of nutrition therapy for respiratory failure are to supply enough energy and protein to support lung function without burdening the respiratory system. Fluid restrictions may be necessary to reverse pulmonary edema.**

Clinical Portfolio

1. Adam is a 29-year-old male who is 6 feet 2 inches tall and has a usual body weight of 180 pounds. He underwent emergency surgery following an accident and is now being cared for in the intensive care unit. Using the method outlined in Table 22-2 on p. 687, estimate Adam's energy requirement. Estimate his protein requirement, using the factor 1.5 grams per kilogram of body weight.
2. Turning again to the case described in item 1, assume that Adam requires a tube feeding and can tolerate a standard enteral formula. Check Appendix K to find at least three formulas that the nutrition support team might select for tube feeding. Determine the volume of each formula that would be needed to meet Adam's energy and protein needs. Would this volume also meet the recommendations for vitamins and minerals?
3. Ayla is a 23-year-old law student admitted to the hospital following an automobile accident in which she broke several bones and ruptured part of her small intestine. She has been in the hospital for several weeks and has just begun eating table foods. Her brother, who was driving the vehicle, was also seriously injured and nearly lost his life. Aside from the increased nutritional needs imposed by the stress of the accident, discuss how the following factors might interfere with Ayla's ability to improve her nutrition status:
   - Ayla's injuries are painful.
   - Ayla's medications cause drowsiness.
   - Ayla is depressed.
   - Ayla is often out of her room for X-rays and other diagnostic tests when the menus and food trays arrive.
   - Ayla's food intake is sometimes restricted due to the procedures she is undergoing.

   How might these problems be resolved to improve Ayla's food intake?

Nutrition on the Net

For further study of topics covered in this chapter, log on to **www.cengagebrain.com** and search for ISBN 084006845X.

- To uncover additional information relevant to critical care, visit these sites:

  American Association of Critical-Care Nurses: **www.aacn.org**

  American Society for Parenteral and Enteral Nutrition: **www.nutritioncare.org**
- These sites provide resources regarding burns for patients and their families:

  American Burn Association: **www.ameriburn.org**

  The Burn Resource Center: **www.burnsurvivor.com**
- This nonprofit group provides comprehensive, up-to-date information for burn care professionals: **www.burnsurgery.org**
- To learn more about lung diseases, visit these sites:

  American Lung Association: **www.lungusa.org**

  Canadian Lung Association: **www.lung.ca**

  National Heart, Lung, and Blood Institute: **www.nhlbi.nih.gov**

Nutrition Assessment Checklist for People Undergoing Metabolic or Respiratory Stress

Medical History

Check the medical record to determine:

- Cause of stress
- Severity of stress
- Whether any organ system is compromised
- Whether nutrition support is required

For patients with COPD, check to determine:

- Degree of breathing difficulty
- Use of oxygen therapy
- Activity tolerance

Review the medical record for complications related to underfeeding or overfeeding, such as:

- Dehydration or fluid overload
- Electrolyte imbalances
- Fatty liver
- Hyperglycemia
- Hypertriglyceridemia

Medications

Record all medications and note:

- Side effects that may alter food intake or nutrition status
- Use of theophylline in patients who may need to monitor caffeine intake

Dietary Intake

If the patient is not meeting nutrition goals:

- Monitor intakes to ensure that the patient is receiving the diet prescribed.
- Investigate appetite problems or difficulties with eating.
- Consider interventions to improve food intake.
- Consider the need for supplementation.
- In patients with COPD, consider problems that may hamper the patient's ability to prepare or consume foods.

Anthropometric Data

Measure baseline height and weight, and monitor daily weights. Remember that body weight can fluctuate in acutely ill patients who undergo fluid resuscitation. Once a patient's weight has stabilized:

- Reevaluate protein and energy needs.
- Consider the need to alter the energy prescription to meet weight goals.

Laboratory Tests

Laboratory tests that may be affected by stress and therefore require careful interpretation include:

- Albumin
- C-reactive protein
- Prealbumin
- Serum iron and zinc
- Transferrin
- White blood cell count

Monitor laboratory tests for signs of:

- Dehydration or fluid overload
- Electrolyte and acid-base imbalances
- Hyperglycemia
- Hypertriglyceridemia
- Nutrient deficiencies
- Negative nitrogen balance
- Organ dysfunction or organ function that has normalized

Physical Signs

Regularly assess vital signs, including:

- Blood pressure
- Body temperature
- Pulse
- Respiratory rate

Look for physical signs of:

- Protein-energy malnutrition
- Dehydration or fluid overload
- Nutrient deficiencies and excesses

References

1. S. F. Lowry and J. M. Perez, The hypercatabolic state, in M. E. Shils and coeditors, *Modern Nutrition in Health and Disease* (Baltimore: Lippincott Williams & Wilkins, 2006), pp. 1381–1400; M. I. T. D. Correia and C. T. de Almeida, Metabolic response to stress, in G. Cresci, ed., *Nutrition Support for the Critically Ill Patient: A Guide to Practice* (Boca Raton, FL: Taylor & Francis Group, 2005), pp. 3–13.
2. N. D. Riediger and coauthors, A systemic review of the roles of n-3 fatty acids in health and disease, *Journal of the American Dietetic Association* 109 (2009): 668–679; P. C. Calder, Polyunsaturated fatty acids, inflammatory processes and inflammatory bowel diseases, *Molecular Nutrition and Food Research* 8 (2008): 885–897.
3. V. Kumar and coauthors, Acute and chronic inflammation, in V. Kumar and coeditors, *Robbins and Cotran Pathologic Basis of Disease* (Philadelphia: Saunders, 2010), pp. 43–77; D. S. Pisetsky, Laboratory testing in the rheumatic diseases, in L. Goldman and D. Ausiello, eds., *Cecil Medicine* (Philadelphia: Saunders, 2008), pp. 1966–1970.
4. Y. Debaveye and G. Van den Berghe, Risks and benefits of nutritional support during critical illness, *Annual Review of Nutrition* 26 (2006): 513–538.
5. D. A. Schoeller, Making indirect calorimetry a gold standard for predicting energy requirements for institutionalized patients, *Journal of the American Dietetic Association* 107 (2007): 390–392; K. A. Kudsk and G. S. Sacks, Nutrition in the care of the patient with surgery, trauma, and sepsis, in M. E. Shils and coeditors, *Modern Nutrition in Health and Disease* (Baltimore: Lippincott Williams & Wilkins, 2006), pp. 1414–1435.

6. D. Frankenfield, Prediction of resting metabolic rate in critically ill adult patients: Results of a systematic review of the evidence, *Journal of the American Dietetic Association* 107 (2007): 1552–1561.
7. Kudsk and Sacks, 2006.
8. S. A. McClave and coauthors, Guidelines for the provision and assessment of nutrition support therapy in the adult critically ill patient: Society of Critical Care Medicine (SCCM) and American Society for Parenteral and Enteral Nutrition (A.S.P.E.N.), *Journal of Parenteral and Enteral Nutrition* 33 (2009): 277–316.
9. McClave and coauthors, 2009.
10. American Dietetic Association, *Nutrition Care Manual* (Chicago: American Dietetic Association, 2010).
11. McClave and coauthors, 2009.
12. McClave and coauthors, 2009; Lowry and Perez, 2006.
13. M. Zhou and R. G. Martindale, Arginine in the critical care setting, *Journal of Nutrition* 137 (2007): 1687S–1692S; P. Furst, Protein and amino acid metabolism: Comparison of stressed and nonstressed states, in G. Cresci, ed., *Nutrition Support for the Critically Ill Patient: A Guide to Practice* (Boca Raton, FL: Taylor & Francis Group, 2005), pp. 27–47.
14. Debaveye and Van den Berghe, 2006.
15. Kudsk and Sacks, 2006.
16. J. Lefton and P. P. Lopez, Macronutrient requirements: Carbohydrate, protein, and lipid, in G. Cresci, ed., *Nutrition Support for the Critically Ill Patient: A Guide to Practice* (Boca Raton, FL: Taylor & Francis Group, 2005), pp. 99–108.
17. K. Sriram and J. I. Cué, Micronutrient and antioxidant therapy in critically ill patients, in G. Cresci, ed., *Nutrition Support for the Critically Ill Patient: A Guide to Practice* (Boca Raton, FL: Taylor & Francis Group, 2005), pp. 109–123.
18. Kudsk and Sacks, 2006.
19. McClave and coauthors, 2009.
20. M. H. Beers and coeditors, *The Merck Manual of Diagnosis and Therapy* (Whitehouse Station, NJ: Merck Research Laboratories, 2006), pp. 2592–2597.
21. R. H. Demling and J. D. Gates, Medical aspects of trauma and burn care, in L. Goldman and D. Ausiello, eds., *Cecil Medicine* (Philadelphia: Saunders, 2008), pp. 790–797; Beers and coeditors, 2006, pp. 2592–2597.
22. Beers and coeditors, 2006, pp. 2592–2597.
23. American Dietetic Association, 2010.
24. American Dietetic Association, 2010.
25. A. N. Husain, The lung, in V. Kumar and coeditors, *Robbins and Cotran Pathologic Basis of Disease* (Philadelphia: Saunders, 2010), pp. 677–737; N. Anthonisen, Chronic obstructive pulmonary disease, in L. Goldman and D. Ausiello, eds., *Cecil Medicine* (Philadelphia: Saunders, 2008), pp. 619–627.
26. A. M. Miniño, J. Xu, and K. D. Kochanek, Deaths: Preliminary data for 2008, *National Vital Statistics Reports* 59, no. 2 (Hyattsville, MD: National Center for Health Statistics, 2010).
27. T. J. Prendergast and S. J. Ruoss, Pulmonary disease, in S. J. McPhee and W. F. Ganong, eds., *Pathophysiology of Disease: An Introduction to Clinical Medicine* (New York: McGraw-Hill/Lange, 2006), pp. 218–258.
28. Beers and coeditors, 2006, pp. 400–422.
29. B. Suckling, M. M. Johnson, and R. Chin, Jr., Nutrition, respiratory function, and disease, in M. E. Shils and coeditors, *Modern Nutrition in Health and Disease* (Baltimore: Lippincott Williams & Wilkins, 2006), pp. 1462–1474.
30. D. A. King, F. Cordova, and S. M. Scharf, Nutritional aspects of chronic obstructive pulmonary disease, *Proceedings of the American Thoracic Society* 5 (2008): 519–523.
31. King, Cordova, and Scharf, 2008.
32. M. Poulain and coauthors, The effect of obesity on chronic respiratory diseases: Pathophysiology and therapeutic strategies, *Canadian Medical Association Journal* 174 (2006): 1293–1299.
33. McClave and coauthors, 2009; A. Malone, Enteral formula selection, in P. Charney and A. Malone, eds., *ADA Pocket Guide to Enteral Nutrition* (Chicago: American Dietetic Association, 2006), pp. 63–122.
34. M. C. Steiner and coauthors, Nutritional enhancement of exercise performance in chronic obstructive pulmonary disease: A randomised controlled trial, *Thorax* 58 (2003): 745–751.
35. L. D. Hudson and A. S. Slutsky, Acute respiratory failure, in L. Goldman and D. Ausiello, eds., *Cecil Medicine* (Philadelphia: Saunders, 2008), pp. 723–734.
36. American Dietetic Association, 2010.
37. N. Barak, E. Wall-Alonso, and M. D. Sitrin, Evaluation of stress factors and body weight adjustments currently used to estimate energy expenditure in hospitalized patients, *Journal of Parenteral and Enteral Nutrition* 26 (2002): 231–238.
38. McClave and coauthors, 2009.

HIGHLIGHT 22

Multiple Organ Dysfunction Syndrome

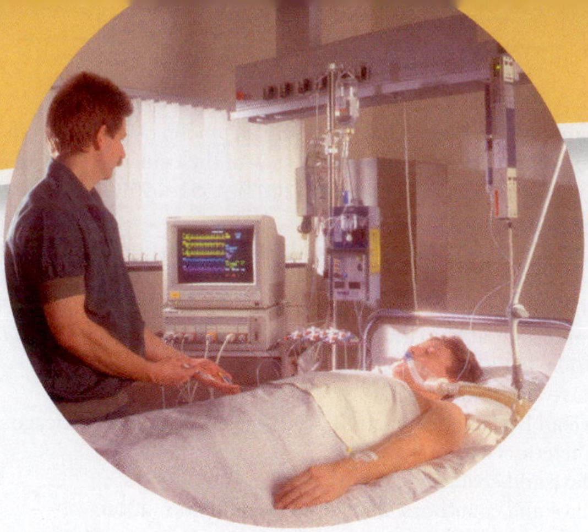
© Hein Hopmans/Phototake

Multiple organ dysfunction syndrome (MODS), also called *multiple organ failure*, is a frequent cause of death in intensive care patients. Described as the progressive dysfunction of two or more of the body's organ systems, MODS most often involves the lungs, liver, kidneys, and gastrointestinal (GI) tract. MODS is not a disease *per se*, but rather a late stage of severe illness or injury that results from a severe inflammatory response (discussed in Chapter 22).[1] MODS can be initiated by a number of very different critical illnesses and conditions, including acute respiratory failure, trauma, sepsis, burn injuries, extensive surgery, and pancreatitis. This highlight discusses how MODS develops, the manner in which it is treated, and the importance of its prevention.

Multiple organ dysfunction syndrome was recognized as a clinical entity only after World War II. Prior to the mid-20th century, patients with severe illnesses or multiple injuries frequently died of shock or circulatory failure. After fluid replacement and blood transfusions became standard treatments, the kidneys became the organs at highest risk, and kidney failure became the most common cause of death. Eventually, physicians learned to better support kidney function by providing appropriate electrolyte solutions and improving urine output. With improved kidney care, the lungs became the most vulnerable organ after severe injury. Improved treatment of respiratory failure eventually led to the current situation: advances in critical care allow patients to survive severe illnesses and injuries, but the body's defenses often overburden organs that were not originally injured.

Development of MODS

As discussed earlier in Chapter 22, injury and infection cause the release of chemical mediators that have systemic (whole-body) effects. A severe, persistent inflammatory response can lead to systemic inflammatory response syndrome (SIRS), which is associated with a constellation of symptoms including fever, raised heart and respiratory rates, and abnormal white blood cell counts. SIRS is a normal adaptive response to a severe insult, but if not reversed quickly enough it can progress to shock, which is characterized by extremely low blood pressure and an inadequate blood supply for the tissues and organs of the body.[2]

As might be expected from a systemic reduction in blood availability, shock can impair numerous organ systems. The abnormal delivery of oxygen and nutrients to tissues and insufficient removal of wastes result in irreversible injury to cells and tissues. Although each organ system is affected differently, ultimately one or more organs may begin to fail. The failure of one organ may place excessive demands on another, causing the second to fail as well. The progression of SIRS to MODS reflects the inability of the body's defenses and medical treatments to counter the detrimental effects of a sustained and potent inflammatory response.

TABLE H22-1 Physiological Effects of Organ or System Failure

Organ or System	Effects of Failure
Lungs	Inability to maintain gas exchange
Liver	Altered metabolic processes
Kidneys	Inability to regulate blood volume, maintain electrolytes, remove wastes
Heart	Low cardiac output, low blood pressure, inadequate circulation, shock
GI tract	Impaired digestion and absorption, abnormal bleeding, bacterial translocation
Immune system	Infection, sepsis
Coagulation system	Excessive bleeding or blood clotting
Central nervous system	Decreased perceptions, brain injury, coma

Although the clinical course differs substantially among patients, the sequence of organ dysfunction often follows a similar pattern: first the lungs fail, then the liver, and finally the kidneys, GI tract, or heart. Other organs or systems may also become involved, and each additional failure reduces the likelihood of survival. Table H22-1 lists the organs and systems most often involved in MODS and the potential consequences of their failure.

Factors That Influence Organ Dysfunction

The specific pathophysiology of MODS is poorly understood. Although early reports attempted to link the development of MODS directly to sepsis, sepsis is not present in all cases. Infection often results from impaired immune function and therefore is a frequent consequence of MODS, but it is not necessarily the

FIGURE H22-1 Relationships among SIRS, Sepsis, and Multiple Organ Dysfunction Syndrome

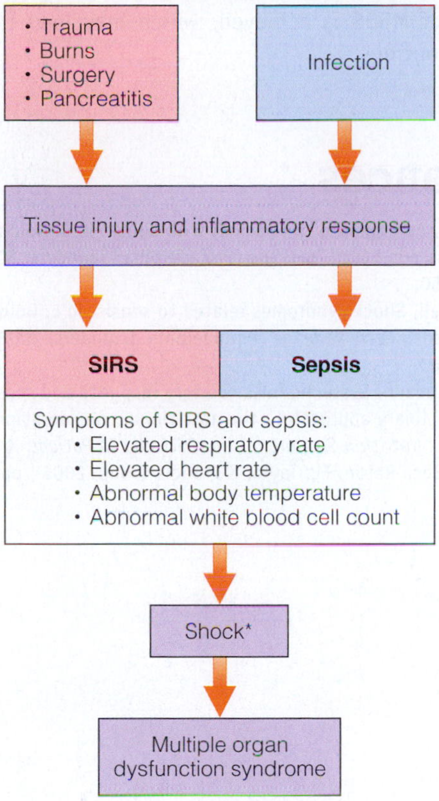

*After critical injury, shock may sometimes precede and be the cause of SIRS.

underlying trigger of organ dysfunction. Recall from this chapter that sepsis gives rise to symptoms identical to those seen in SIRS. Figure H22-1 illustrates the relationships among SIRS, infection, sepsis, and MODS.

Finding the exact cause of MODS is difficult because each patient's situation is somewhat different. Epidemiological studies have, however, identified a number of factors that increase risk. For example, people who develop MODS are often older, have multiple or severe injuries, and develop severe infections. Table H22-2 lists the major risk factors associated with MODS, some of which are discussed in the following sections.

TABLE H22-2 Factors That Influence Risk of Multiple Organ Dysfunction Syndrome

Age over 55 years
Prior chronic disease
Persistent SIRS
Major infection
Blood transfusions
Severity of tissue injury
Length of time between injury and arrival at hospital
Malnutrition

Age

Patients over 55 years old are several times more likely to develop MODS than are younger patients. In elderly patients, the increased risk may be due to the presence of chronic illnesses that directly affect organ function, such as heart disease, lung disease, diabetes, or liver damage. Aging also decreases the functional reserve of organs, thereby reducing an older patient's ability to deal with the additional stress that arises during critical illness.

Severity of SIRS

The length of time that SIRS persists is related to the development of MODS. Patients who have SIRS that persists for more than three days are more likely to develop MODS than patients who have SIRS for less than two days.

Infection

Prolonged SIRS can suppress immune function and increase the risk of developing an infection. During hospital stays, critically ill patients often contract pneumonia—the principal infection associated with MODS. The risks of infection and sepsis greatly increase with the use of invasive catheters, which are frequently needed during intensive care to provide oxygen support, intravenous fluid resuscitation, nutrition support, and urine clearance.

Blood Transfusions

Blood transfusions are immunosuppressive and may increase a patient's risks of developing infection or sepsis. Blood transfusions frequently have adverse effects that can add further stress; they may cause acute lung injury, allergic reactions, red blood cell hemolysis (breakdown), and other complications.

Treatment for MODS

Once MODS has developed, extensive medical support is needed until the inflammatory response has abated. Unfortunately, aggressive treatments can have damaging effects of their own and may cause further injury to organs that are already weakened by illness. Health practitioners must be aware of the adverse effects of aggressive therapies and remain alert to a patient's responses to treatments. Therapies that are often used to manage MODS include:[3]

- *Lung support.* Mechanical ventilation is used to assist injured lungs and sustain gas exchange.
- *Fluid resuscitation.* Fluids and electrolytes are supplied to restore blood volume and maintain electrolyte balance.
- *Support of heart and blood vessel function.* Medications help to sustain or increase cardiac output and maintain adequate blood pressure.

- *Kidney support.* Hemofiltration or dialysis helps to prevent the buildup of toxic metabolites in blood.
- *Protection against infection.* Antibiotic therapy may reverse or prevent infections.
- *Nutrition support.* Enteral and parenteral nutrition support provide nutrients, help to prevent excessive wasting, and promote recovery.

Because mortality rates for MODS are so high, prevention must be considered at the earliest stages of injury and treatment, before an excessive inflammatory response can cause further damage. Health practitioners have learned to identify the conditions that can increase organ stress whether they are due to a disease process, an inflammatory response, or an aggressive treatment that is intended to provide organ support. Although improvements in care over the past few decades have reduced some of the complications that arise during intensive care, rates of mortality from MODS have not changed. Thus, a focus on prevention is critical until a better understanding of the pathophysiology of MODS is achieved, which may lead to additional therapeutic options.

References

1. J. Parrillo, Approach to the patient with shock, in L. Goldman and D. Ausiello, eds., *Cecil Medicine* (Philadelphia: Saunders, 2008), pp. 742–750.
2. J. A. Russell, Shock syndromes related to sepsis, in L. Goldman and D. Ausiello, eds., *Cecil Medicine* (Philadelphia: Saunders, 2008), pp. 755–763.
3. M. H. Oltermann, Systemic inflammatory response and sepsis: A multidisciplinary approach to the nutritional considerations, in G. Cresci, ed., *Nutrition Support for the Critically Ill Patient: A Guide to Practice* (Boca Raton, FL: Taylor & Francis Group, 2005), pp. 565–577.

CHAPTER

23

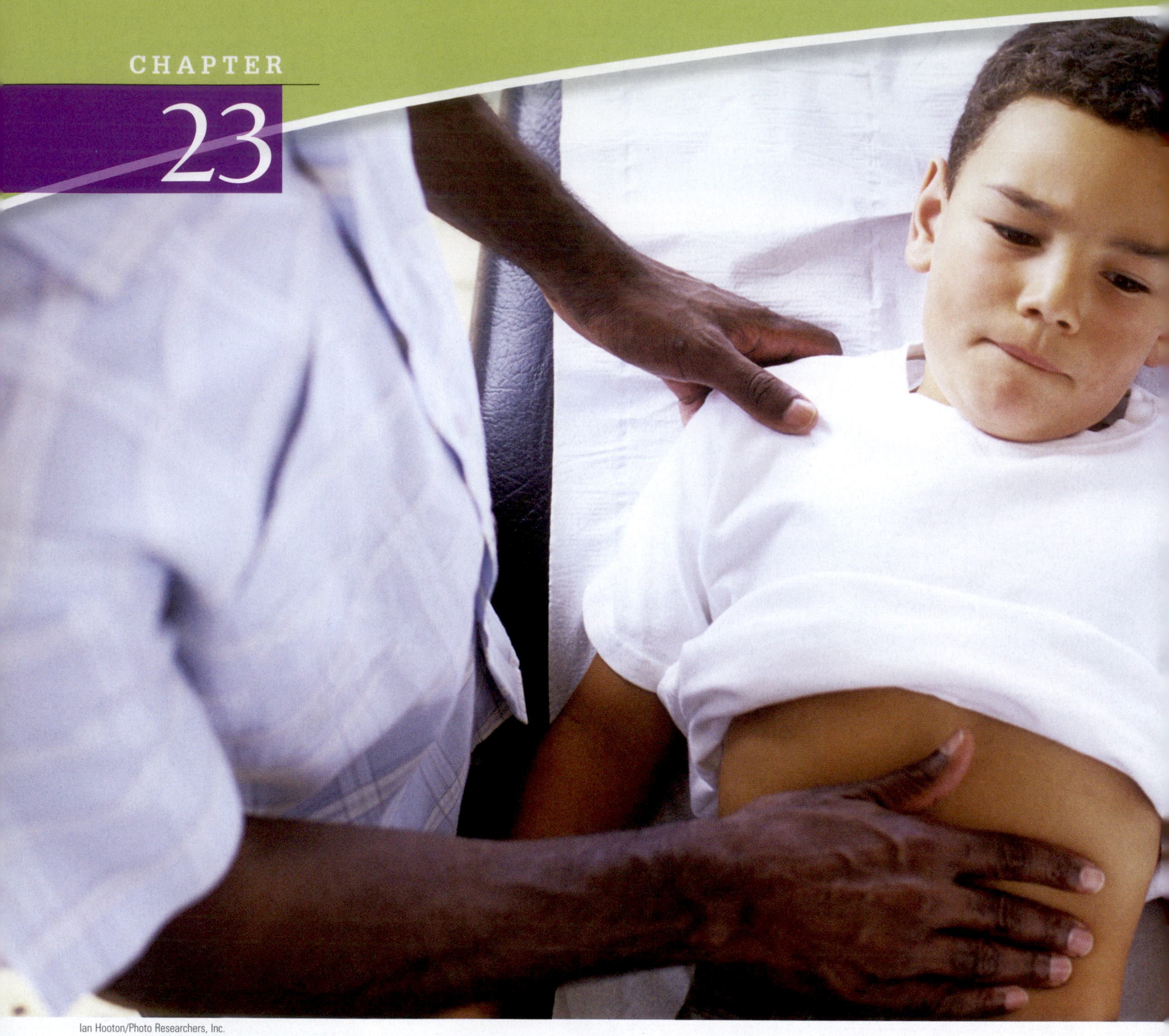

Ian Hooton/Photo Researchers, Inc.

Nutrition in the Clinical Setting

Gastrointestinal illnesses account for a significant fraction of hospital admissions and visits to health practitioners each year. Diagnosis is not always straightforward, however, because many patients with gastrointestinal complaints exhibit no physical abnormalities. Evaluation therefore requires a detailed review of a patient's symptoms and responses to dietary adjustments. Because gastrointestinal complications frequently accompany other illnesses, the medical history can sometimes uncover the underlying source of distress.

Throughout this chapter, the CourseMate icon indicates an opportunity for online self-study, linking you to activities to increase your understanding of chapter concepts. **www.cengagebrain.com** (search for ISBN 084006845X)

Upper Gastrointestinal Disorders

CHAPTER OUTLINE

The remarkable gastrointestinal (GI) tract provides a means of delivering nutrients to the body's interior. When a medical condition impairs some of the GI tract's functions, dietary adjustments may help to ease symptoms and prevent malnutrition. This chapter discusses common upper GI tract symptoms and disorders; the next chapter describes conditions that affect the lower GI tract. Highlight 23 presents several mouth and dental problems and their associations with chronic disease.

Figure 23-1 illustrates the upper GI tract and reviews its functions. ♦ In the mouth, the teeth and jaw muscles work together to break down food to a consistency that is easily swallowed. Upon swallowing, a bolus of food passes through the pharynx to the esophagus, where peristaltic contractions move the bolus toward the stomach. The lower esophageal sphincter relaxes to allow the bolus to enter the stomach and then closes to prevent reflux (backward flow) of stomach contents. The stomach adds gastric juices to liquefy the food materials and begin the process of digestion.

♦ See Chapter 3 for a complete review of the GI tract and its functions; common digestive problems and simple self-help measures were introduced in Highlight 3.

Conditions Affecting the Esophagus

This section examines the causes and treatments of the two most common disorders affecting the esophagus: dysphagia (difficulty swallowing), which was introduced in Chapter 18, and gastroesophageal reflux disease, often referred to as "heartburn."

Dysphagia

The act of swallowing involves multiple processes. In the initial, or **oropharyngeal**, phase of swallowing, muscles in the mouth and tongue propel the bolus of food through the pharynx and into the esophagus. At the same time, tissues of the soft palate prevent food from entering the nasal cavity, and the epiglottis blocks the opening to the trachea to prevent aspiration of food substances or saliva into the lungs. In the second, or **esophageal**, phase of swallowing, peristalsis forces the bolus through the esophagus, and the lower esophageal sphincter relaxes to allow passage of the bolus into the stomach. Due to the many tasks involved in swallowing, dysphagia can result from a number of different physical or neurological conditions. Table 23-1 lists some potential causes of dysphagia, which are categorized according to the phase of swallowing that is impaired.[1]

Oropharyngeal Dysphagia

A person with **oropharyngeal dysphagia** typically has a neuromuscular condition that upsets the swallowing reflex or impairs the mobility of the tongue and other oral tissues. Symptoms include an inability to initiate swallowing, coughing during or after swallowing (due to aspiration), and nasal

oropharyngeal (OR-oh-fah-ren-JEE-al): involving the mouth and pharynx.

esophageal (eh-SOF-ah-JEE-al): involving the esophagus.

oropharyngeal dysphagia: an inability to transfer food from the mouth and pharynx to the esophagus; usually caused by a neurological or muscular disorder.

FIGURE 23-1 The Upper GI Tract

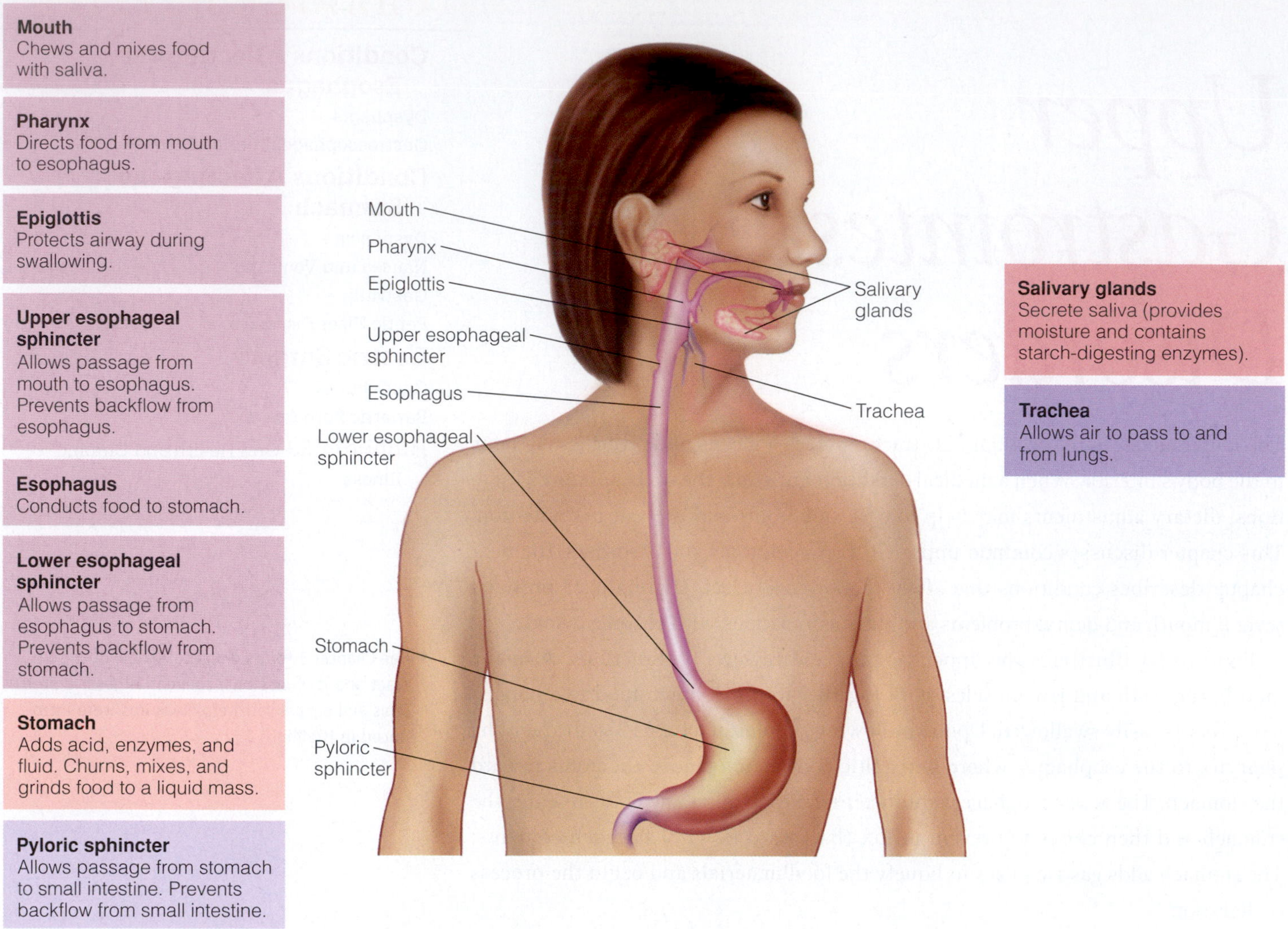
Mouth
Chews and mixes food with saliva.
Pharynx
Directs food from mouth to esophagus.
Epiglottis
Protects airway during swallowing.
Upper esophageal sphincter
Allows passage from mouth to esophagus. Prevents backflow from esophagus.
Esophagus
Conducts food to stomach.
Lower esophageal sphincter
Allows passage from esophagus to stomach. Prevents backflow from stomach.
Stomach
Adds acid, enzymes, and fluid. Churns, mixes, and grinds food to a liquid mass.
Pyloric sphincter
Allows passage from stomach to small intestine. Prevents backflow from small intestine.
Mouth
Pharynx
Epiglottis
Upper esophageal sphincter
Esophagus
Lower esophageal sphincter
Stomach
Pyloric sphincter
Salivary glands
Trachea
Salivary glands
Secrete saliva (provides moisture and contains starch-digesting enzymes).
Trachea
Allows air to pass to and from lungs.

regurgitation. Other signs include bad breath, a gurgling noise after swallowing, a hoarse or "wet" voice, or a speech disorder. Oropharyngeal dysphagia is common in elderly persons and frequently follows a stroke.[2]

Esophageal Dysphagia A person with **esophageal dysphagia** usually has an obstruction in the esophagus or a motility disorder. The main symptom is the sensation of food "sticking" in the esophagus after it is swallowed. An obstruction can be caused by a **stricture** (abnormal narrowing), tumor, or compression of the esophagus by surrounding tissues. Whereas an obstruction can prevent the passage of solid foods but may not affect liquids, a motility disorder hinders the passage of both solids and liquids. **Achalasia**, the most common motility disorder, is a degenerative nerve condition affecting the esophagus; it is characterized by impaired peristalsis and incomplete relaxation of the lower esophageal sphincter when swallowing.[3]

Complications of Dysphagia Health practitioners should be alert to the various complications that may accompany dysphagia. If the condition restricts food consumption, malnutrition and weight loss may occur. Individuals who cannot swallow liquids are at increased risk of dehydration. If aspiration occurs, it may cause choking, airway obstruction, or respiratory infections, including pneumonia. If a person does not have a normal cough reflex, aspiration is more difficult to diagnose and may go unnoticed.

esophageal dysphagia: an inability to move food through the esophagus; usually caused by an obstruction or a motility disorder.

stricture: abnormal narrowing of a passageway; often due to inflammation, scarring, or a congenital abnormality.

achalasia (ack-ah-LAY-zhah): an esophageal disorder characterized by weakened peristalsis and impaired relaxation of the lower esophageal sphincter.

- **a** = without
- **chalasia** = relaxation

TABLE 23-1 Selected Causes of Dysphagia

Oropharyngeal Dysphagia
• Alzheimer's disease (advanced stages) • Developmental disabilities • Goiter • Lou Gehrig's disease (amyotrophic lateral sclerosis) • Multiple sclerosis • Muscular dystrophy • Myasthenia gravis • Parkinson's disease • Poliomyelitis • Stroke
Esophageal Dysphagia
• Achalasia • Enlarged right atrium (upper right chamber of heart) • Esophageal cancer • Esophageal spasm • Scleroderma • Strictures (from inflammation, scarring, or a congenital abnormality) • Thoracic tumor (usually lung cancer)

Evaluation of Dysphagia Diagnosing the exact cause of dysphagia often requires a thorough examination. One assessment method is the barium swallow study, in which the swallowing process is monitored using video X-ray equipment; for this procedure, the patient must consume foods or liquids that contain barium, a metallic element visible on X-rays. Another technique, endoscopy, uses a thin, flexible tube to examine the esophageal lumen directly. Peristalsis and sphincter pressure can be measured using a manometer, a flexible catheter with multiple pressure sensors that is passed into the esophagus. A neurological examination may be needed to evaluate mental status, physical reflexes, and the cranial nerves associated with swallowing.

Nutrition Intervention for Dysphagia To compensate for swallowing difficulties, a person with dysphagia may need to consume foods and beverages that have been physically modified so that they are easier to swallow. Because a wide variety of defects can cause dysphagia, finding the best diet is often a challenge. Furthermore, a person's swallowing ability can fluctuate over time, so the dietary plan needs frequent reassessment.

The National Dysphagia Diet, developed in 2002 by a panel of dietitians, speech and language therapists, and a food scientist, has helped to standardize the nutrition care of dysphagia patients.[4] Table 23-2 presents brief descriptions of the different levels of the diet and some sample meals. After the appropriate dietary level is selected, it must be adjusted to suit the person's swallowing abilities and tolerances. In many cases, the most appropriate foods may be determined only by trial and error. A consultation with a swallowing expert, such as a speech and language therapist, is often necessary.

Food Properties and Preparation Foods included in dysphagia diets should have easy-to-manage textures and consistencies. Soft, cohesive foods are easier to handle than hard or crumbly foods. Moist foods are better tolerated than dry foods. Some foods within a category may be acceptable and others may not; for example, some cookies are soft and tender, whereas others are hard and brittle. Sticky or gummy foods, such as peanut butter and cream cheese, may be difficult to clear from the mouth and throat.

The textures of foods are typically altered to make them easier to swallow. Foods are often pureed, mashed, ground, or minced (review Table 18-5 on p. 602). Foods that have more than one texture, such as vegetable soup or cereal with milk, are

Can you tell that the foods in this photo are pureed foods shaped with food molds?

TABLE 23-2 National Dysphagia Diet

Level 1: Dysphagia Pureed

Foods should be pureed or well mashed, homogeneous, and cohesive. This diet is for patients with moderate to severe dysphagia and poor oral or chewing ability.

Sample menus:

- *Breakfast*: Cream of Wheat, slurried muffins or pancakes,[a] pureed scrambled eggs, plain or vanilla yogurt, well-mashed bananas, fruit juice without pulp (thickened as needed), coffee or tea (if thin liquids are acceptable).
- *Lunch or dinner*: Pureed tomato soup, slurried crackers, pureed meat or poultry, zucchini soufflé, mashed potatoes with gravy, pureed carrots or green beans, smooth applesauce, pureed peaches, chocolate pudding.

Foods to avoid: Dry breads and cereals, oatmeal, rice, fruit yogurt, cheese (including cottage cheese), peanut butter, nuts and seeds, raw fruits and vegetables, chunky applesauce, fruit preserves with chunks or seeds, tomato sauce with seeds, beverages with pulp, coarsely ground pepper, herbs.

Level 2: Dysphagia Mechanically Altered

Foods should be moist, cohesive, and soft textured and should easily form a bolus. This diet is for patients with mild to moderate dysphagia; some chewing ability is required.

Sample menus:

- *Breakfast*: Moist oatmeal, cornflakes or puffed rice cereal with milk (thickened as needed), moist pancakes or muffins (with butter, margarine, or jam; without nuts or seeds), soft scrambled eggs, cottage cheese, ripe bananas or cooked fruit without skin or seeds, fruit juice (thickened as needed), coffee or tea (if thin liquids are allowed).
- *Lunch or dinner*: Soup with easy-to-chew meat and vegetables; slurried bread or crackers; minced, tender-cooked meat; well-cooked pasta with moist meatballs and meat sauce; baked potato with gravy; soft, tender-cooked vegetables (not fibrous or rubbery); canned peach slices; soft fruit pie (with bottom crust only); soft, smooth chocolate bar.

Foods to avoid: Dry or coarse foods; breads and cereals with nuts, seeds, or dried fruit; frankfurters and sausages; hard-cooked eggs; corn and clam chowders; sandwiches; pizza; sliced cheese; rice; potato skins; French fries; raw vegetables; fibrous, rubbery, or nontender cooked vegetables such as asparagus, broccoli, brussels sprouts, cabbage, celery, corn, and peas; peanut butter; coconut; nuts and seeds; raw fruit (except banana); cooked fruits with skin or seeds; pineapple; mango; uncooked dried fruit; popcorn; chewy candies (such as caramel or licorice).

Level 3: Dysphagia Advanced

Foods should be moist and be in bite-sized pieces when swallowed; foods with mixed textures are included. This diet is for patients with mild dysphagia and adequate chewing ability.

Sample menus:

- *Breakfast*: Cereal with milk, moist pancakes or muffins (with butter, margarine, or jam; without nuts or seeds), poached or scrambled eggs, fruit yogurt, soft fresh fruit (peeled) or berries, coffee or tea (if thin liquids are tolerated).
- *Lunch or dinner*: Chicken noodle soup; moistened crackers or moist bread; thin-sliced tender meat; cheese; moist, soft-cooked potatoes or rice; tender-cooked vegetables; shredded lettuce with dressing; fresh, peeled peach or melon; canned fruit salad; moist chocolate chip cookie (without nuts).

Foods to avoid: Dry or coarse foods; breads and cereals with nuts, seeds, or dried fruit; corn and clam chowders; potato skins; raw vegetables (except shredded lettuce); corn; chunky peanut butter; coconut; nuts and seeds; hard fruit (such as apples or pears); fruit with skin, seeds, or stringy textures (such as mango or pineapple); uncooked dried fruit; fruit leathers; popcorn; chewy candies (such as caramel or licorice).

Liquid Consistencies (only those tolerated are allowed in the diet)

- *Thin*: Watery fluids; may include milk, coffee, tea, juices, carbonated beverages.
- *Nectarlike*: Fluids thicker than water that can be sipped through a straw; may include buttermilk, eggnog, tomato juice.
- *Honeylike*: Fluids that can be eaten with a spoon but do not hold their shape; may include honey, tomato sauce, yogurt.
- *Spoon-thick*: Thick fluids that must be eaten with a spoon and can hold their shape; may include milk pudding, thickened applesauce.

[a]Slurried foods are mixed with liquid until the consistency is appropriate; they may be gelled and shaped to improve appearance.

harder to handle, so ingredients may be blended to a single consistency with items such as nuts and seeds omitted.

Consuming foods that have a similar consistency can quickly become monotonous. By using commercial thickeners and food molds, pureed foods can be formed into attractive shapes. Including a variety of flavors and colors can also make a meal more appealing. The "How To" offers additional suggestions for improving the acceptance of pureed and other mechanically altered foods.

Properties of Liquids Thickened liquids are easier to swallow than thin liquids such as water or juice. Table 23-2 describes the four levels of liquid consistencies prescribed for dysphagia patients, referred to as thin, nectarlike, honeylike, and

HOW TO Improve Acceptance of Mechanically Altered Foods

Take a moment to think about a meal of pureed or ground foods. A typical dinner of baked chicken, potatoes, carrots, and green beans can look like mounds of differently colored mush. The foods may taste great, but a person may have little appetite before trying a first bite. To improve appetite, be creative when preparing and serving meals:

- Help to stimulate the appetite by preparing favorite foods and foods with pleasant smells. Enliven food flavors with seasonings and spices.
- Use attractive plates and silverware to improve the visual appeal of a meal. Consider colors and shapes when arranging foods on a plate; colorful garnishes can add color and eye appeal. Substitute brightly colored vegetables for white vegetables; for example, replace mashed potatoes with mashed sweet potatoes.
- Try layering ingredients so that the food looks like a fancy casserole or popular hors d'oeuvre. For example, food items can resemble lasagna, moussaka, tamales, or sushi.
- Shape pureed and ground foods to resemble traditional dishes; for example, flatten a spoonful of pureed meat to make a patty, or use small scoops so that meat resembles meatballs. Use food molds to restore slurried breads and pureed meats to their traditional shapes.

Efforts to improve the visual appearance of foods can go a long way toward helping people eat nourishing meals and maintain a healthy weight.

TRY IT Think of a favorite entrée or hors d'oeuvre that contains ingredients that can be pureed, mashed, or ground. Describe how this item can be prepared or presented so that it is appealing to a person who must consume a mechanically altered diet.

spoon-thick. To increase viscosity, commercial starch thickeners can be stirred into beverages and other liquid foods, such as soup broths. Some beverages can lose their appeal when thickened; for example, individuals may find thickened coffee and tea unacceptable. Moreover, hydration is more difficult to maintain when a patient has access to only thickened beverages, which are less acceptable for quenching thirst.

Alternative Feeding Strategies for Dysphagia Some patients may be able to learn alternative feeding techniques to help them compensate for their swallowing problem. For example, changing the position of the head and neck while eating and drinking can minimize some swallowing difficulties. (As an example, cups designed for dysphagia patients allow drinking without tilting the head back.) Individuals with oropharyngeal dysphagia can be taught exercises that strengthen the jaws, tongue, or larynx, or they can learn new methods of swallowing that allow them to consume a normal diet. Speech and language therapists are often responsible for teaching patients these techniques.

Gastroesophageal Reflux Disease

Gastroesophageal reflux disease (GERD) ♦ is a condition in which the stomach's acidic contents back up into the esophagus, causing discomfort and, sometimes, tissue damage. People who suffer from GERD often refer to these symptoms as *heartburn* or *acid indigestion*. Reflux does not necessarily cause symptoms or injury—it occurs occasionally in healthy people and is a problem only if it creates complications and requires lifestyle changes or medical treatment.

♦ Highlight 3 introduced gastroesophageal reflux disease and discussed strategies for preventing its recurrence.

Causes of GERD The lower esophageal sphincter is the main barrier to gastric reflux, so GERD can result if the sphincter muscle is weak or relaxes inappropriately. Medical conditions that either interfere with the sphincter's function or prevent rapid clearance of acid from the esophagus can predispose a person to GERD.

FIGURE 23-2 The Upper GI Tract, Acid Reflux, and Hiatal Hernia

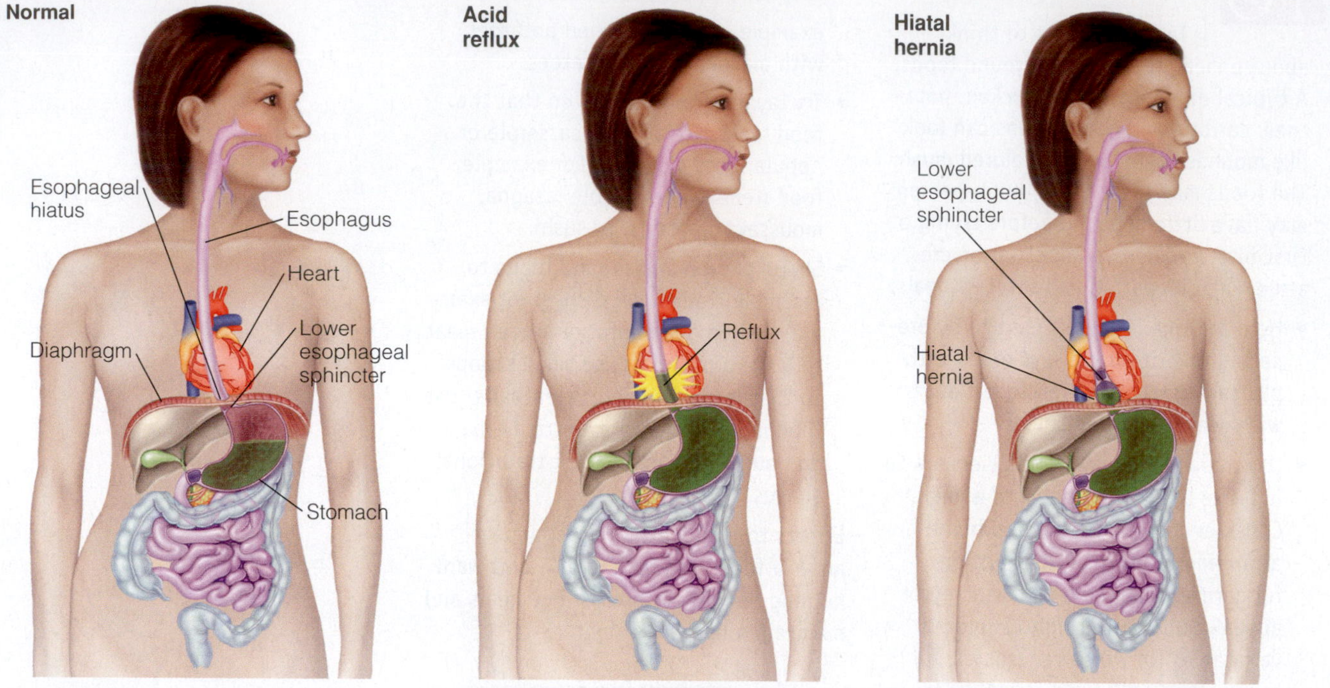

The stomach normally lies below the diaphragm, and the esophagus passes through the esophageal hiatus. The lower esophageal sphincter prevents reflux of stomach contents.

Whenever the pressure in the stomach exceeds the pressure in the esophagus, as can occur with overeating and overdrinking, the chance of reflux increases. The resulting "heartburn" is so-named because it is felt in the area of the heart.

Risk of acid reflux may increase as a consequence of a hiatal hernia. A "sliding" hiatal hernia occurs when part of the stomach, along with the lower esophageal sphincter, rises above the diaphragm.

Conditions associated with high rates of GERD include pregnancy, obesity, asthma, and **hiatal hernia**, a condition in which a portion of the stomach protrudes above the diaphragm (see Figure 23-2). Pregnancy is the most common predisposing condition—as many as two-thirds of pregnant women report heartburn, which usually worsens during the third trimester.[5] Some medications can increase the risk of reflux, as does the use of nasogastric tubes in tube feedings. Various other conditions or substances can exacerbate GERD by increasing stomach distention or weakening the lower esophageal sphincter; Table 23-3 lists examples.

Consequences of GERD If gastric acid remains in the esophagus long enough to damage the esophageal lining, the resulting inflammation is called **reflux esophagitis**. Severe and chronic inflammation may lead to esophageal ulcers, with consequent bleeding. Healing and scarring of ulcerated tissue may narrow the inner diameter of the esophagus, causing esophageal stricture. A slowly progressive dysphagia for solid foods sometimes results, and swallowing occasionally becomes painful. Pulmonary disease may develop if gastric contents are aspirated into the lungs. Chronic reflux is also associated with **Barrett's esophagus**, a condition in which damaged esophageal cells are gradually replaced by cells that resemble those in gastric or intestinal tissue; such cellular changes increase the risk of developing esophageal cancer. GERD can also damage tissues in the mouth, pharynx, and larynx, resulting in eroded tooth enamel, sore throat, and laryngitis.[6]

Treatment of GERD Treatment objectives are to alleviate symptoms and facilitate the healing of damaged tissue. Severe ulcerative disease may require immediate acid-suppressing medication, whereas a mild case may be managed with dietary and lifestyle modifications. The "How To" lists lifestyle modifications that may help to prevent the recurrence of gastrointestinal reflux.

Medications that suppress gastric acid secretion help the healing process by reducing the damaging effects of acid on esophageal tissue. **Proton-pump inhibitors**

hiatal hernia: a condition in which the upper portion of the stomach protrudes above the diaphragm; most cases are asymptomatic.

reflux esophagitis: inflammation in the esophagus related to the reflux of acidic stomach contents.

Barrett's esophagus: a condition in which esophageal cells damaged by chronic exposure to stomach acid are replaced by cells that resemble those in the stomach or small intestine, sometimes becoming cancerous.

proton-pump inhibitors: a class of drugs that inhibit the enzyme that pumps hydrogen ions (protons) into the stomach. Examples include omeprazole (Prilosec) and lansoprazole (Prevacid).

TABLE 23-3 Conditions and Substances Associated with Esophageal Reflux

Conditions That Increase Pressure within the Stomach	Substances That Weaken the Lower Esophageal Sphincter
Ascites (abdominal fluid accumulation)	Alcohol
Delayed gastric emptying	Anticholinergic drugs
Eating large meals	Antihistamines
Lying down after eating	Caffeinated beverages
Obesity	Calcium channel blockers
Pregnancy	Chocolate
Wearing tight clothing around the waist or abdomen	Cigarette smoking
	Diazepam
	Fatty foods
	Peppermint and spearmint
	Progesterone
	Theophylline
	Tricyclic antidepressants

are the most effective of the antisecretory agents and are used both for rapid healing of esophagitis and as a maintenance treatment. Other antisecretory drugs include **histamine-2 receptor blockers** (often referred to as *H2 blockers*) and antacids, which neutralize gastric acid. Antacids are frequently used to relieve occasional heartburn, but they are not necessarily appropriate for GERD because they have only short-term effects and may cause some nutrient deficiencies when used over the long term.

histamine-2 receptor blockers: a class of drugs that suppresses acid secretion by inhibiting receptors on acid-producing cells; commonly called *H2 blockers*. Examples include cimetidine (Tagamet), ranitidine (Zantac), and famotidine (Pepcid).

HOW TO Manage Gastroesophageal Reflux Disease

Management of GERD may require modifications in diet and lifestyle to minimize discomfort and reduce the recurrence of acid reflux. Recommendations typically include the following:

- Avoid eating bedtime snacks or lying down after meals. Meals should be consumed at least three hours before bedtime.
- Reduce nighttime reflux by elevating the head of the bed on 6-inch blocks, inserting a foam wedge under the mattress, or propping pillows under the head and upper torso.
- Consume only small meals, and drink liquids between meals so that the stomach does not become overly distended, which can exert pressure on the lower esophageal sphincter.
- Limit foods that weaken lower esophageal sphincter pressure or increase gastric acid secretion; these include chocolate, fried and fatty foods, spearmint and peppermint, coffee (both caffeinated and decaffeinated), and tea.
- Avoid cigarettes and alcohol; both relax the lower esophageal sphincter.
- Avoid bending over and wearing tight-fitting garments; both can cause pressure in the stomach to increase, heightening the risk of reflux.
- During periods of esophagitis, avoid foods and beverages that may irritate the esophagus, such as citrus fruits and juices, tomato products, garlic, onions, pepper, spicy foods, carbonated beverages, and very hot or very cold foods (depending on individual tolerances).
- Avoid using nonsteroidal anti-inflammatory drugs (NSAIDs) such as aspirin, naproxen, and ibuprofen, which can damage the esophageal mucosa.

Food tolerances among people with GERD can vary markedly. Health professionals can help patients pinpoint food intolerances by advising them to keep a record of the foods and beverages consumed, as well as any resulting symptoms.

TRY IT Create a one-day menu of five meals or snacks that excludes foods and beverages that are problematic for individuals with GERD.

CASE STUDY

Woman with GERD

Elyssa Rinaldi is a 39-year-old accountant who is 5 feet 4 inches tall and weighs 165 pounds. During a recent physical examination, she mentioned to her physician that she had been feeling fairly well until she began experiencing heartburn, which has progressively become more frequent and painful. The heartburn often occurs after she eats a large meal and is particularly bad after she goes to bed at night. By directly examining the esophageal lumen using an endoscope (a thin, flexible tube equipped with an optical device), the physician found evidence of reflux esophagitis and a slight narrowing throughout the length of the esophagus.

Ms. Rinaldi's medical history does not indicate any significant health problems. During her last physical exam, her physician advised her to stop smoking cigarettes and to lose 20 pounds, but she has not attempted to do either. The nutrition assessment reveals that Ms. Rinaldi is feeling stressed because it is the middle of the tax season. She usually has little time for breakfast, eats a lunch of fast foods while continuing to work at her desk, and eats a large dinner at around 8 P.M. She generally has wine with dinner and another alcoholic beverage later in the evening.

1. Explain to Ms. Rinaldi the meaning of the medical diagnoses *reflux esophagitis* and *esophageal stricture*.
2. From the brief history provided, list the factors and behaviors that increase Ms. Rinaldi's risks of experiencing reflux. What recommendations can you make to help her change these behaviors?
3. What medications might the physician prescribe, and why?

Surgery may be required in severe cases of GERD that are unresponsive to medications and lifestyle changes. In one popular procedure (called *fundoplication*), the upper section of the stomach (the fundus) is gathered up around the lower esophagus and sewn in such a way that the esophagus and sphincter are surrounded by stomach muscle; this technique increases pressure within the esophagus and fortifies the sphincter muscle. Esophageal strictures are often treated by dilating the esophagus with an inflatable balloon-like device or a fixed-size dilator, or by using surgical approaches. The Case Study will help you to review the treatments available for a patient with GERD.

IN SUMMARY **Dysphagia and gastroesophageal reflux are the most common esophageal disorders. Dysphagia may interfere with food intake and increase the risk of aspiration. Treatment may include dietary adjustments, strengthening exercises, and using different swallowing techniques. Gastroesophageal reflux disease (GERD) may lead to esophageal ulcers, inflammation, bleeding, and stricture. Treatment includes the use of acid-suppressing drugs and lifestyle changes.**

Conditions Affecting the Stomach

Stomach disorders range from occasional bouts of discomfort to severe conditions that require surgery. This section begins with a discussion of *dyspepsia* (often called "indigestion"), the sensation of pain or discomfort in the upper abdomen that occurs after food consumption. More serious stomach conditions that may benefit from dietary adjustments include *gastritis* and *peptic ulcers*, which most often result from bacterial infection or the use of medications that damage the stomach lining.

dyspepsia: a feeling of pain, bloating, or discomfort in the upper abdominal area, often called indigestion; a symptom of illness rather than a disease itself.
- **dys** = bad; impaired
- **pepsis** = digestion

Dyspepsia

Dyspepsia refers to general symptoms of indigestion in the upper abdominal region, which may include stomach pain, gnawing sensations, early satiety, nausea, vomiting, and bloating. These symptoms sometimes indicate the presence of more serious illnesses, such as GERD or peptic ulcer disease. Although about 25 percent of the population experiences dyspepsia, only one person in four seeks medical attention.[7]

Causes of Dyspepsia Abdominal symptoms don't always lead to a clear diagnosis. Various medical conditions can cause abdominal discomfort, including peptic ulcers, GERD, gastric motility disorders, gallbladder and pancreatic diseases, and tumors in the upper GI tract. Chronic diseases such as diabetes mellitus, heart disease, and hypothyroidism can sometimes be accompanied by gastric symptoms. Some medications, including aspirin (and other nonsteroidal anti-inflammatory drugs), antibiotics, digitalis, estrogens, and theophylline, can cause gastrointestinal distress. Some dietary supplements, such as iron and potassium supplements and some herbal products, may cause gastrointestinal problems. Intestinal conditions such as irritable bowel syndrome or lactose intolerance may mimic dyspepsia. Although pinpointing the cause of the symptoms can be difficult, a complete examination is in order if the individual experiences unintentional weight loss, persistent vomiting, dysphagia, anemia, or bleeding, which suggest the presence of serious illness.[8]

Potential Food Intolerances Although many people attribute their symptoms to eating certain foods or spices, controlled studies have been unable to find associations between specific foods and dyspepsia.[9] Coffee can induce symptoms in about 50 percent of dyspepsia patients, however, and also increases gastric acid production and acid reflux.[10] Spicy foods may cause some injury to the mucosal lining and exacerbate the pain from a preexisting ulcer. High-fat meals can slow gastric emptying and thereby exacerbate dyspepsia. To minimize symptoms, people with dyspepsia are typically advised to avoid consuming excessively large meals, fatty or highly spiced foods, and specific foods believed to trigger symptoms.[11]

Bloating and Stomach Gas The feeling of bloating may be caused by excessive gas in the stomach, which accumulates when air is swallowed. Air swallowing often accompanies gum chewing, smoking, rapid eating, drinking carbonated beverages, and using a straw. Omitting these practices generally helps to correct the problem.

Nausea and Vomiting

Nausea and vomiting accompany many illnesses and are common side effects of medications. Although occasional vomiting is not dangerous, prolonged vomiting can cause fluid and electrolyte imbalances and may require medical care. A chronic vomiting problem can reduce food intake and lead to malnutrition and nutrient deficiencies.

The timing of vomiting gives clues about its cause. Vomiting that occurs within an hour after a meal suggests a peptic ulcer or a psychological cause. If it occurs more than one hour after a meal, possible causes include food poisoning, an obstruction that prevents stomach emptying, or a stomach motility disorder.

Treatment of Nausea and Vomiting Most cases are short lived and require no treatment. When treatment is necessary, the main goal is to find and correct the underlying disorder. Restoring hydration and electrolyte balance may also be necessary in some individuals. If a medication is the cause, taking it with food may help. If the cause is unknown or the underlying disorder cannot be corrected, medications that suppress nausea and vomiting can be prescribed. People with **intractable vomiting**—vomiting that is not easily controlled—may require intravenous nutrition support.

Dietary Interventions Sometimes nausea can be prevented or improved with dietary measures.[12] To minimize stomach distention, patients should consume small meals and drink beverages between meals rather than during a meal. Dry, starchy foods such as toast, crackers, and pretzels may help to reduce nausea, whereas fatty or spicy foods and foods with strong odors may worsen symptoms. Foods that are cold or at room temperature may be better tolerated than hot foods. Individuals often have strong food aversions when nauseated, and tolerances vary greatly.

intractable vomiting: vomiting that is not easily managed or controlled.

♦ The suffix *-itis* refers to the presence of inflammation in an organ or tissue. Some forms of gastritis are characterized by tissue destruction rather than inflammation.

TABLE 23-4 Potential Causes of Gastritis

Infection
• Bacterial: *Helicobacter pylori, Actinomyces israelii* • Fungal: *Candida albicans* • Parasitic: Cryptosporidiosis, nematode infection • Viral: Cytomegalovirus, herpes simplex virus
Chemical Substances
• Alcohol • Cancer chemotherapy • Drugs (especially aspirin and other NSAIDs) • Ingestion of toxins or corrosive materials
Internal (Bodily) Causes
• Autoimmune disease • Bile reflux • Severe stress • Systemic illness or sepsis
Miscellaneous
• Food allergy • Foreign bodies • High salt intake • Radiation therapy

♦ The specific reasons why ulcers develop are not known; only 10 to 15 percent of individuals with chronic *H. pylori* infection develop a peptic ulcer.

♦ Reminder: *Gastrin* is a hormone that signals stomach cells to secrete hydrochloric acid.

gastritis: inflammation of stomach tissue.

***Helicobacter pylori (H. pylori)*:** a species of bacterium that colonizes gastric mucosa; a primary cause of gastritis and peptic ulcer disease.

hypochlorhydria (HIGH-poe-clor-HIGH-dree-ah): abnormally low gastric acid secretions.

achlorhydria (AY-clor-HIGH-dree-ah): absence of gastric acid secretions.

peptic ulcer: an open sore in the gastrointestinal mucosa; may develop in the esophagus, stomach, or duodenum.

• **peptic** = related to digestion

Zollinger-Ellison syndrome: a condition characterized by the presence of gastrin-secreting tumors in the duodenum or pancreas.

♦ In the United States, up to 80 percent of ulcers occur in the duodenum.

Gastritis

Gastritis is a general term that refers to inflammation of the stomach mucosa. ♦ Acute cases of gastritis typically result from irritating substances or treatments that damage the gastric mucosa, resulting in tissue erosions, ulcers, or hemorrhaging (severe bleeding). Chronic cases may be caused by long-term infections or autoimmune disease and can progress to widespread gastric inflammation and tissue atrophy. Most often, gastritis results from ***Helicobacter pylori*** infection or the use of aspirin or NSAIDs, which are primary causes of peptic ulcer disease as well. Table 23-4 lists some potential causes of gastritis.

Complications of Gastritis

The extensive tissue damage that sometimes develops in chronic gastritis can disrupt gastric secretory functions. If hydrochloric acid secretions become abnormally low (**hypochlorhydria**) or absent (**achlorhydria**), absorption of nonheme iron and vitamin B_{12} can be impaired, increasing the risk of deficiency. Pernicious anemia, a condition characterized by the autoimmune destruction of stomach cells that produce intrinsic factor, is a late complication of atrophic gastritis that can result in the macrocytic anemia of vitamin B_{12} deficiency (see p. 331).

Dietary Interventions for Gastritis

Dietary recommendations depend on an individual's symptoms. In asymptomatic cases, no dietary adjustments are needed. If pain or discomfort is present, the patient should avoid irritating foods and beverages; these often include alcohol, coffee (including decaffeinated), tea, cola beverages, spicy foods, and fried or fatty foods. If food consumption increases pain or causes nausea and vomiting, food intake should be avoided for 24 to 48 hours to rest the stomach. If hypochlorhydria or achlorhydria is present, supplementation of iron and vitamin B_{12} may be warranted.

Peptic Ulcer Disease

A **peptic ulcer** is an open sore that develops in the GI mucosa when gastric acid and pepsin overwhelm mucosal defenses and destroy mucosal tissue. A primary factor in peptic ulcer development is *H. pylori* infection, which is present in approximately 30 to 60 percent of patients with gastric ulcers and 70 to 90 percent of those with duodenal ulcers.[13] ♦ Another major factor is the use of NSAIDs, which have both topical and systemic effects that can damage the GI lining. In rare cases, ulcers may develop from disorders that cause excessive acid secretion: one such condition is **Zollinger-Ellison syndrome**, characterized by the presence of gastrin-secreting tumors in the duodenum or pancreas. ♦ Ulcer risk can be increased by cigarette smoking and psychological stress.

Effects of Emotional Stress

Although most ulcers are associated with *H. pylori* infection or NSAID use, about a quarter of ulcers develop in people for other reasons.[14] Psychological stress is not believed to cause ulcers *per se*, but it has effects on physiological processes and behaviors that may increase a person's vulnerability. The physiological effects of stress vary among individuals but may include hormonal changes that impair immune responses and wound healing, increased secretions of hydrochloric acid and pepsin, and rapid stomach emptying (which increases the acid load in the duodenum). Stress may also lead to behavioral changes, including the increased use of cigarettes, alcohol, and NSAIDs—all potential risk factors for ulcers. Thus, stress may play a contributory role in ulcer development, although its precise effects are not fully understood.

Symptoms of Peptic Ulcers

Peptic ulcer symptoms vary. Some people are asymptomatic or experience only mild discomfort. Ulcer "pain" may be experienced as a hunger pain, a sensation of gnawing, or a burning pain in the stomach region. The pain or discomfort of ulcers may be relieved by food and recur several hours after a meal, especially if the ulcer is duodenal. ♦ Gastric ulcers, however, are often aggravated by food and can cause loss of appetite and eventual weight loss. Ulcer symptoms tend to go into remission regularly and recur every few weeks or months.[15]

DIET-DRUG Interactions

Check this table for notable nutrition-related effects of the medications discussed in this chapter.

Antacids (aluminum hydroxide, magnesium hydroxide, calcium carbonate)	**Gastrointestinal effects:** Constipation (aluminum- or calcium-containing antacids), diarrhea (magnesium-containing antacids) **Dietary interactions:** May decrease iron, folate, or vitamin B_{12} absorption **Metabolic effects:** Electrolyte imbalances
Antibiotics (for *H. pylori* infection; include amoxicillin, metronidazole, tetracycline)	**Gastrointestinal effects:** Diarrhea (amoxicillin, tetracycline), nausea and vomiting (tetracycline), altered taste sensation (metronidazole) **Dietary interactions:** Avoid alcohol with metronidazole; tetracycline decreases iron absorption and binds calcium in the GI tract, reducing absorption of both the tetracycline and the calcium
Antisecretory drugs (proton-pump inhibitors, H2 blockers)	**Gastrointestinal effects:** Constipation, diarrhea, nausea and vomiting, abdominal pain (proton-pump inhibitors) **Dietary interactions:** May decrease iron, folate, and vitamin B_{12} absorption

Complications of Peptic Ulcers Peptic ulcers are a major cause of GI bleeding, which occurs in up to 15 percent of ulcer cases.[16] Bleeding is a potential cause of death and, if severe, may require surgical intervention. Severe bleeding is evidenced by black, tarry stool samples or, occasionally, vomit that resembles coffee grounds. Other serious complications of ulcers include perforations of the stomach or duodenum (sometimes leading directly into the peritoneal cavity) and **gastric outlet obstruction** due to scarring or inflammation.

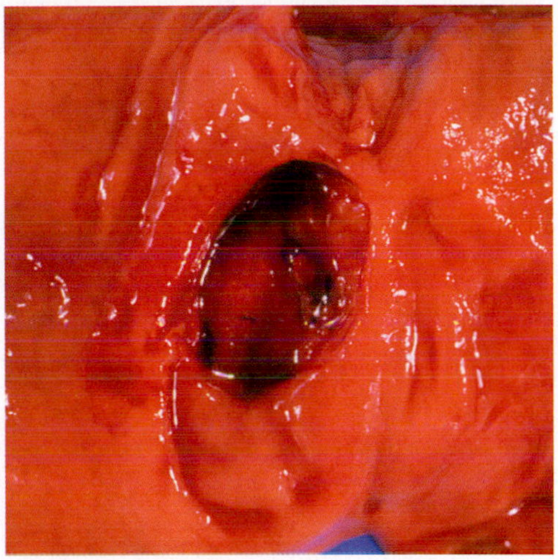

© Dr. E. Walker/Photo Researchers, Inc.

A peptic ulcer, such as the gastric ulcer shown here, damages mucosal tissue and may cause pain and bleeding.

Drug Therapy for Peptic Ulcers The goals of ulcer treatment are to relieve pain, promote healing, and prevent recurrence. In most cases, treatment requires using a combination of antibiotics to eradicate *H. pylori* infection and/or discontinuing the use of aspirin and other NSAIDs, which can irritate the gastric mucosa and delay healing. The antibiotics used to treat *H. pylori* infection most often include amoxicillin, clarithromycin, metronidazole, and tetracycline. Antisecretory drugs are prescribed to relieve pain and allow healing; these include proton-pump inhibitors, H2 blockers, and antacids (as used in GERD; see the earlier discussion on pp. 710–711). The most frequently prescribed drug regimen is a "triple therapy" that includes two antibiotics and an antisecretory drug. Bismuth preparations (such as Pepto-Bismol) and sucralfate may also help in healing ulcers by coating the mucosal lining and preventing further tissue erosion. See the Diet-Drug Interactions feature above for significant nutrition-related effects of the medications used in ulcer treatment.

Nutrition Care for Peptic Ulcers The goals of nutrition care are to correct nutrient deficiencies, if necessary, and encourage dietary and lifestyle practices that minimize symptoms.[17] Patients should avoid dietary substances that increase acid secretion or irritate the GI lining; examples include alcohol, coffee and other caffeine-containing beverages, chocolate, and pepper, although individual tolerances vary. Small meals may be better tolerated than large ones. Patients should avoid food consumption for at least two hours before bedtime. Cigarette smoking should be discouraged, as it can delay healing and increase the risk of ulcer recurrence. There is no evidence that dietary adjustments can alter the rate of healing.[18]

IN SUMMARY *Dyspepsia* refers to general symptoms of indigestion such as abdominal pain, nausea, and vomiting, which can be caused by a variety of medical conditions. Gastritis and peptic ulcer disease are most often associated with *H. pylori* infection, which can be eradicated by antibiotic therapy. In addition, NSAID use may promote gastritis and peptic ulcer disease by damaging the mucosal lining. Extensive damage to the mucosa may reduce gastric secretions and increase the risks of developing iron and vitamin B_{12} deficiencies.

gastric outlet obstruction: an obstruction that prevents the normal emptying of stomach contents into the duodenum.

> **The nutrition care for gastritis and peptic ulcer disease includes correcting any nutritional deficiencies that develop and eliminating dietary substances that can cause pain or discomfort.**

Gastric Surgery

Gastric surgery is sometimes necessary for treating stomach cancer, some ulcer complications, and ulcers that are resistant to drug therapy. In recent years, gastric surgeries have also become popular treatments for severe obesity. Because gastric surgeries can interfere with stomach function either temporarily or permanently, patients generally need to make significant dietary adjustments afterward.

Stomach cancers are often treated with a **gastrectomy**, a surgical procedure that removes the diseased areas of the stomach. To suppress gastric acid secretion, a **vagotomy** is sometimes performed; this procedure severs the **vagus nerve**, which normally stimulates the cells that produce gastric acid. Because a vagotomy may impair gastric motility, it is sometimes followed by a **pyloroplasty**, which widens the pyloric sphincter to ensure drainage from the stomach to the duodenum. **Bariatric surgery**, the type of surgery that treats severe obesity, was introduced in Chapter 9 (pp. 283–284) and is discussed later in this chapter.

Gastrectomy

Figure 23-3 illustrates some typical gastrectomy procedures. In a partial gastrectomy, only part of the stomach is removed, and the remaining portion is connected to the duodenum or jejunum. In a total gastrectomy, the surgeon removes the entire stomach and connects the esophagus directly to the small intestine.

Nutrition Care after Gastrectomy

The primary goals of nutrition care after a gastrectomy are to meet the nutrient needs of the postsurgical patient and promote the healing of stomach tissue. Another goal is to prevent discomfort or nutrient deficiencies that may arise due to altered stomach function. As the next section will describe, some gastric surgeries increase the risk of **dumping syndrome**, a group of symptoms that result when a large amount of food passes rapidly into the small intestine.

Following a gastrectomy, oral intake of fluids and foods is suspended until some healing has occurred, and fluids are supplied intravenously. Small sips of water, ice chips (melted in the mouth), and broth are usually the first fluids given orally. Once fluids are tolerated, patients are offered liquid meals (with no sugars) at first, and they usually progress to a soft food diet by the fourth or fifth day after surgery. Tube feedings may be necessary if complications prevent a normal progression to solid foods.[19]

FIGURE 23-3 Gastrectomy Procedures

In a gastrectomy, part or all of the stomach is surgically removed. The dashed lines show the removed section.

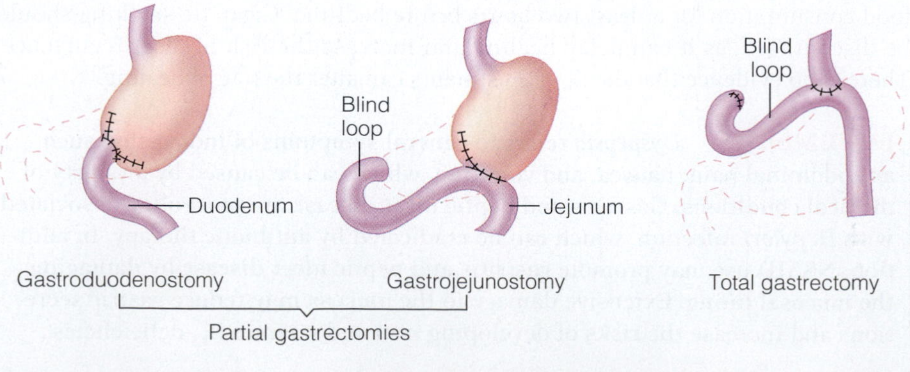

gastrectomy (gah-STREK-ta-mee): the surgical removal of part of the stomach (partial gastrectomy) or the entire stomach (total gastrectomy).

vagotomy (vay-GOT-oh-mee): surgery that severs the vagus nerve in order to suppress gastric acid secretion. This surgery may require a follow-up *pyloroplasty* procedure to allow stomach drainage.

vagus nerve: the cranial nerve that regulates hydrochloric acid secretion and peristalsis. Effects elsewhere in the body include regulation of heart rate and bronchiole constriction.

pyloroplasty (pye-LORE-oh-PLAS-tee): surgery that enlarges the pyloric sphincter.

bariatric (BAH-ree-AH-trik) **surgery:** surgery that treats severe obesity.

- **baros** = weight

dumping syndrome: a cluster of symptoms that result from the rapid emptying of an osmotic load from the stomach into the small intestine.

TABLE 23-5 Postgastrectomy Diet

Food Category	Foods Recommended (as tolerated)	Foods to Limit (unless tolerated)
Meat and meat alternates	Lean tender meats, fish, poultry, shellfish, eggs, smooth nut butters	Fried, tough, or chewy meats; frankfurters and sausages; bacon; luncheon meats; dried peas and beans
Milk and milk products	Milk, plain yogurt, mild cheeses	Milk shakes, chocolate milk, fruit yogurt
Breads and cereals	Breads, crackers, bagels, pasta, and breakfast cereals made from enriched white flour (cereals should contain no added sugars)	Breads and cereals with more than 2 grams of fiber per serving; baked goods with dried fruits, nuts, or seeds; granola; frosted cereals; pastries; doughnuts
Vegetables	Tender-cooked vegetables without peels, skins, or seeds; raw lettuce	Raw vegetables (except lettuce), beets, broccoli, brussels sprouts, cabbage, cauliflower, collard and mustard greens, corn, potato skins
Fruit	Canned fruit without added sugars, bananas, melon	Canned fruits in syrup, raw fruits (except bananas and melons), dried fruits, fruit juices
Beverages	Decaffeinated coffee and tea, beverages sweetened with artificial sweeteners	Caffeinated beverages; alcoholic beverages; beverages sweetened with sugars, corn syrup, or honey

Dietary measures after gastrectomy are influenced by the size of the remaining stomach, which influences meal size, and the stomach emptying rate, which affects food tolerances. Initially, the patient is offered small meals and snacks that include only one or two food items; these foods may contain protein (fish, lean meats, and eggs), fat, and complex carbohydrates (bread, potatoes, and vegetables). Depending on the amount of food tolerated, the patient may require as many as six to eight small meals and snacks per day. The patient should avoid sweets and sugars because they increase osmolarity in the small intestine and potentiate the dumping syndrome (discussed below). Some patients may need to avoid milk products due to lactose intolerance. Soluble fibers may be added to meals to help delay stomach emptying and reduce diarrhea. Although tolerances vary, patients may have difficulty with fatty foods, highly spiced foods, carbonated drinks, caffeine-containing beverages, alcohol, extremely hot or cold foods, peppermint, and chocolate. Liquids are restricted during meals due to limited stomach capacity and because liquids can increase the stomach emptying rate. Table 23-5 lists foods that are often permitted or limited in postgastrectomy diets.[20]

Dumping Syndrome The dumping syndrome, a common complication of gastrectomy, is characterized by a group of symptoms resulting from rapid gastric emptying. Ordinarily, the pyloric sphincter controls the rate of flow from the stomach into the duodenum. After some types of stomach surgery, the hypertonic gastric contents are no longer regulated and rush into the small intestine more quickly after meals, causing a number of unpleasant effects. Early symptoms can occur within 30 minutes and may include nausea, vomiting, abdominal cramping, diarrhea, lightheadedness, rapid heartbeat, and others (see Table 23-6). These symptoms may be due to a shift of fluid from blood vessels to the intestine that lowers

TABLE 23-6 Symptoms of Dumping Syndrome

Early Dumping Syndrome	Late Dumping Syndrome
Symptoms may begin within 30 minutes after eating.	**Symptoms may begin 1 to 3 hours after eating.**
• Abdominal pain, cramping	• Anxiety
• Diarrhea	• Confusion
• Dizziness	• Headache
• Flushing, sweating	• Hunger
• Nausea and vomiting	• Palpitations
• Rapid heartbeat	• Sweating
• Weakness, feeling faint	• Weakness, feeling faint

blood volume and increases intestinal distention, and/or the accelerated release of GI hormones that induce strong intestinal contractions. Several hours later, symptoms of hypoglycemia may occur because the unusually large spike in blood glucose following the meal (due to rapid nutrient influx and absorption) can result in an excessive insulin response.

Dietary adjustments can greatly minimize or prevent dumping syndrome. The goals are to limit the amount of food material that reaches the intestine, slow the rate of gastric emptying, and reduce foods that increase hypertonicity. Therefore, meal size is limited, fluids are restricted during meals, and sugars (including milk sugar) are restricted. In some cases, drugs that inhibit gastrointestinal motility (such as octreotide) ♦ may help. The "How To" lists practical suggestions for reducing the occurrence of dumping syndrome. The Case Study provides the opportunity to design a menu for a postgastrectomy patient who is at risk for dumping syndrome.

♦ *Octreotide* inhibits gastrointestinal motility, thereby slowing both gastric emptying and transit time in the small intestine.

Nutrition Problems following Gastrectomy After a gastrectomy, it may take time for the patient to learn the amount of food that can be consumed without discomfort. The symptoms associated with meals may lead to food avoidance, substantial weight loss, and eventually, malnutrition. Other nutrition problems that may occur after gastrectomy include the following:

- *Fat malabsorption.* Fat digestion and absorption may become impaired for a number of reasons after gastrectomy. The accelerated transit of food material may prevent the normal mixing of fat with lipase and bile. If the duodenum has been removed or bypassed, less lipase is available for fat digestion. Bacterial overgrowth, ♦ a common consequence of gastric surgeries, can lead to changes in bile acids that upset bile function. The fat malabsorption that results from these changes can eventually cause deficiencies of fat-soluble vitamins and some minerals. Supplemental pancreatic enzymes are sometimes provided to improve fat digestion. Medium-chain triglycerides, which are more easily digested and absorbed, can be used to supply additional fat calories.

♦ Bacterial overgrowth may be a consequence of reduced gastric acid secretions, altered motility of intestinal contents, or changes in intestinal anatomy due to surgical reconstruction. Chapter 24 describes bacterial overgrowth in detail.

HOW TO Alter the Diet to Reduce Symptoms of Dumping Syndrome

Dietary adjustments can greatly minimize or prevent symptoms of dumping syndrome. The following suggestions may help:

- Eat smaller meals that suit the reduced capacity of the stomach. Increase the number of meals consumed daily so that energy intake is adequate.
- Eat in a relaxed setting. Eat slowly, and chew food thoroughly.
- Limit the amount of fluid included in meals. Avoid drinking beverages within 30 minutes before and after meals, but be sure to consume adequate fluid during the day to avoid dehydration.
- Avoid juices and sweetened beverages and foods that contain high amounts of sugar.
- Avoid carbonated beverages if they cause bloating.
- Use artificial sweeteners to sweeten beverages and desserts.
- Avoid foods and beverages that are very hot or very cold, unless tolerated.
- Include fiber-rich foods in each meal. Sometimes adding soluble fibers like pectin or guar gum to meals can help to control symptoms.
- Avoid milk and most milk products, which are high in lactose. Avoid enzyme-treated milk as well, because the breakdown products of lactose (glucose and galactose) can also cause dumping symptoms. Cheese may be better tolerated than milk because its lactose content is low. Make an effort to consume nonmilk calcium sources such as green leafy vegetables, fish with bones, and tofu.
- If symptoms of hypoglycemia continue, try including a protein-rich food in each meal.
- Lie down for 20 to 30 minutes (or longer) after eating to help slow the transit of food to the small intestine. While eating a meal, sit upright.

TRY IT Based on your typical intake, list three dietary items that you would include and three items you would exclude to avoid symptoms if you were at risk of dumping syndrome.

CASE STUDY

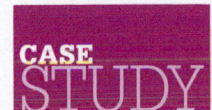

Nutrition Care for Patient after Gastric Surgery

Dirk Hanson, a 58-year-old biology teacher, was admitted to the hospital for gastric surgery after numerous medical treatments failed to manage severe complications related to his peptic ulcer disease. A gastrojejunostomy was performed, and after about 24 hours, Mr. Hanson was able to take small sips of warm water. The health care team anticipates multiple nutrition-related problems and is taking measures to prevent them.

1. Review Figure 23-3 to better understand Mr. Hanson's surgical procedure. Consider the possibilities that he might experience the following symptoms: early satiety, nausea and vomiting, weight loss, dumping syndrome, fat malabsorption, anemia, and bone disease. Explain why each of these conditions may occur.
2. What type of diet will the physician prescribe for Mr. Hanson after he begins eating solid foods? Create a day's worth of menus, using foods from Table 23-5.
3. What advice can you give Mr. Hanson that will help to prevent dumping syndrome? List several foods from each major food group that may cause symptoms of dumping syndrome.

- *Bone disease*. Osteoporosis and osteomalacia are common outcomes following a gastrectomy. The fat malabsorption described earlier can cause malabsorption of both vitamin D and calcium; ♦ furthermore, patients at risk of dumping syndrome may need to avoid milk products, which are among the best sources of these nutrients. Bone density should be monitored during the years following surgery, and supplementation of calcium and vitamin D is often recommended.
- *Anemia*. After gastrectomy, the reduction in gastric secretions impairs the absorption of both iron and vitamin B_{12}, often leading to anemia. If the duodenum has been removed or is bypassed, the risk of iron deficiency increases because the duodenum is a major site of iron absorption. Supplementation of both iron and vitamin B_{12} is usually warranted after surgery.

♦ Fat malabsorption reduces calcium absorption because the negatively charged fatty acids combine with calcium (which is positively charged) and prevent its absorption.

Bariatric Surgery Bariatric surgery is currently considered the most effective and durable treatment for severe obesity.[21] The most popular surgical options for weight reduction, the gastric bypass and gastric banding procedures, were introduced in Chapter 9 (pp. 283–284; see Figure 9-5 on p. 283). The gastric bypass operation, ♦ which accounts for about 88 percent of bariatric surgeries,[22] creates a small gastric pouch that reduces stomach capacity and thereby restricts meal size. In addition, the gastric pouch is connected directly to the jejunum, resulting in significant nutrient malabsorption because the flow of food bypasses a significant portion of the small intestine. In the gastric banding procedure, a gastric pouch is created using a fluid-filled inflatable band; adjusting the band's fluid level can tighten or loosen the band and alter the size of the opening to the rest of the stomach. A smaller opening reduces the speed at which the pouch is emptied and prolongs the sense of fullness after a meal. Whereas the gastric bypass operation is usually permanent, the gastric banding procedure is fully reversible.

♦ The gastric bypass surgery is also known as a *Roux-en-Y gastric bypass*: the Swiss surgeon *Roux* developed the procedure, which reconstructs the small intestine such that it resembles the letter *Y*.

Clinical studies indicate that the gastric bypass surgery generally results in greater weight loss than the gastric banding procedure. Although study results vary, gastric bypass patients typically lose 60 to 70 percent of their excess body weight, whereas those who undergo gastric banding lose about 50 percent of their excess weight.[23] Although bariatric surgeries are effective treatments for severe obesity, patients should have realistic expectations about the amount of weight they are likely to lose, the diet they will need to follow, and the complications that may ensue. Some types of bariatric surgery can dramatically affect health and nutrition status, and many patients require lifelong management.

Dietary Guidelines after Bariatric Surgery The main objectives of nutrition care after bariatric surgery are to maximize and maintain weight loss, ensure appropriate nutrient intakes, maintain hydration, and avoid complications such as nausea and vomiting or dumping syndrome.[24] The gastric pouch created by surgery may eventually expand to hold up to 1/2 cup of food, but its initial capacity is only a few tablespoons. Only sugar-free, noncarbonated clear liquids and low-fat broths are given during the first two days following bariatric surgery. Afterward, patients consume a liquid diet (low in sugars) at first, followed by pureed foods and then solid foods; the diet is advanced as tolerated. Once the diet progresses to solid foods, patients may consume between three and six small meals per day. Only small portions of food can be consumed at each meal because overeating can stretch the gastric pouch or result in vomiting or regurgitation. Similarly, fluids must be consumed separately from meals to avoid excessive distention. Other dietary recommendations include the following:

- *Protein intake.* Recommendations range from 1.0 to 1.5 grams of protein per kilogram of ideal body weight per day; [25] however, intakes are often lower than recommended. Patients are generally instructed to eat high-protein foods before consuming other foods in a meal and to consume liquid protein supplements regularly.
- *Vitamin and mineral deficiencies.* Bariatric patients have a high risk of developing nutrient deficiencies due to reduced food intake, reduced gastric secretions, and nutrient malabsorption. Supplemental vitamin B_{12}, vitamin D, iron, and calcium are recommended after surgery. A daily multivitamin/mineral supplement ensures that patients meet their needs for other nutrients.
- *Foods to avoid.* Some foods may obstruct the gastric outlet; these include doughy or sticky breads, pasta products, and rice; fibrous vegetables such as asparagus and celery; foods with seeds, peels, or skins; nuts; popcorn; and tough, dry meats.
- *Dumping syndrome.* To avoid symptoms of dumping syndrome, gastric bypass patients must carefully control food portions, avoid foods high in sugars, and consume liquids between meals (review the "How To" on p. 718).

After bariatric surgery, patient education and counseling are critical for weight loss and weight management, and patients also need to learn the elements of a healthy diet. The "How To" includes additional dietary suggestions for patients who have undergone bariatric surgery.

Postsurgical Concerns in Gastric Bypass Surgery Common complaints after bariatric surgery include nausea, vomiting, and constipation.[26] Although the cause of these problems varies among patients, dietary noncompliance and inadequate fluid intake are often contributing factors, and improved dietary intakes can help in resolving these conditions. The long-term complications that may develop after gastric bypass surgery are similar to those that arise after gastrectomy and include fat malabsorption, bone disease, and anemia. Rapid weight loss increases a person's risk of developing gallbladder disease; patients at especially high risk sometimes have their gallbladders removed while undergoing bariatric surgery. After weight loss, plastic surgery may be necessary to remove extra skin, especially on the abdomen, buttocks, hips, and thighs.

IN SUMMARY **Gastric surgeries, used to treat cancer, peptic ulcer complications, and obesity, require dietary adjustments after surgery and are associated with complications that may affect nutrition status. Common postsurgical complications include fat malabsorption, bone disease, anemia, and dumping syndrome. After bariatric surgery, patients must learn to consume appropriate food portions, use dietary supplements to prevent nutrient deficiencies, and**

HOW TO Alter Dietary Habits to Achieve and Maintain Weight Loss after Bariatric Surgery

Patients need to learn new dietary habits after bariatric surgery. The following recommendations may help:

- Consume only small portions of food and chew food thoroughly. Use a small spoon, and take small bites. Relax and enjoy the meal, taking at least 20 minutes to eat.
- Understand that, at first, the appropriate amount of each food served at mealtime may be only a few spoonfuls. Learn to recognize the sensations that occur when the gastric pouch is full. Signs of fullness may include pressure in the stomach region, a slight feeling of nausea, or pain in the upper chest or shoulder.
- To control vomiting, try eating smaller volumes of food, eating more slowly, and avoiding foods known to cause difficulty. Continued vomiting may be a sign that some food choices or amounts are inappropriate.
- Consume food only during designated mealtimes (usually three to six small meals per day), and avoid consuming foods at other times of the day. Snacking throughout the day can become a bad habit that causes weight to be regained.
- Learn to recognize foods that cause problems. Foods that are dry, sticky, or fibrous may be difficult to tolerate during the weeks after surgery.
- Avoid consuming liquids within 30 minutes of mealtime. Avoid high-kcalorie drinks like sweetened soda, milk shakes, and alcoholic beverages. Avoid drinking carbonated beverages or using a straw, as these practices can increase stomach gas and cause bloating.
- Sip water and other beverages throughout the day to obtain sufficient fluids. Aim to consume between 6 and 8 cups of water and other noncaloric beverages daily. Remember that most people meet a significant fraction of their fluid needs by eating food, but a person who has undergone bariatric surgery must limit food intake.
- Engage in regular physical activity. Activity is a valuable aid to weight maintenance and can help to maintain lean tissue while weight is being lost.

TRY IT Create a one-day menu of five meals or snacks that would be appropriate for patients who have recently undergone bariatric surgery. Be sure to specify portion sizes. How would you rate the micronutrient adequacy of your diet plan?

choose foods that are unlikely to cause abdominal discomfort, vomiting, or dumping syndrome.

Clinical Portfolio

1. Although some individuals require a mechanically altered diet for just a few weeks, others have medical problems that require long-term use of such diets. Consider the difference between working with a person who has had a swallowing problem for years and a person who recently had mouth surgery and is just beginning to eat again.
   - Explain how the needs of these individuals may differ. What nutrition-related problems may develop if a person has been following a restrictive dysphagia diet for several years?
   - Using Table 23-2 and the "How To" on p. 709, create a day's worth of menus for a person who requires long-term use of a pureed dysphagia diet and tolerates only liquids that have a honeylike consistency.
2. Jillian, a 35-year-old woman who is 5 feet 4 inches tall and weighs 227 pounds, has had severe hip and knee osteoarthritis for several years and was recently diagnosed with type 2 diabetes. After trying numerous diet programs without success, she eventually visits a bariatric surgeon to learn about the surgical options for treating her obesity.
   - Using information from the margin on p. 283 (Chapter 9) and the BMI table on the back inside cover of this book, determine Jillian's BMI and whether she would be a good candidate for bariatric surgery. Explain why or why not.
   - Should Jillian decide to undergo gastric bypass surgery, she will need to permanently change her dietary habits. Outline the diet progression typically recommended after bariatric surgery, and describe the dietary measures

necessary for preventing vomiting, distention of the gastric pouch, gastric outlet obstruction, and dumping syndrome.

- Explain why dehydration is a frequent complication following bariatric surgery. What tips can you give Jillian to help her avoid this problem?

Nutrition on the Net

For further study of topics covered in this chapter, log on to **www.cengagebrain.com** and search for ISBN 084006845X.

- Visit the websites of these organizations to find information that is helpful both for health practitioners and patients with gastrointestinal problems:

  American College of Gastroenterology: **www.acg.gi.org**

  American Gastroenterological Association: **www.gastro.org**

  Digestive Health Center of Excellence, University of Virginia Health System: **www.medicine.virginia.edu/clinical/departments/medicine/divisions/digestive-health**

  International Foundation for Functional Gastrointestinal Disorders: **www.iffgd.org**

  National Institute of Diabetes and Digestive and Kidney Diseases, a division of the National Institutes of Health: **www2.niddk.nih.gov**

- Find more information about dysphagia at the Dysphagia Resource Center: **www.dysphagia.com**
- Learn more about *Helicobacter pylori* from the Helicobacter Foundation: **www.helico.com**
- The Consumer Guide to Bariatric Surgery provides general information about bariatric surgeries at this website: **www.yourbariatricsurgeryguide.com**

Nutrition Assessment Checklist for People with Upper GI Tract Disorders

Medical History

Check the medical history to uncover conditions or treatments that may:

- Interfere with chewing or swallowing
- Lead to dry mouth
- Lead to dyspepsia, nausea, or vomiting

Check for a medical diagnosis of:

- Gastritis or peptic ulcer
- GERD
- Hiatal hernia
- Pernicious anemia

For a patient who has undergone gastric surgery, check for the following complications:

- Anemia
- Bone disease
- Dumping syndrome
- Fat malabsorption

Medications

Record all medications and note:

- Aspirin or NSAID use in patients with gastritis or peptic ulcer disease
- Medications that may cause dry mouth
- Medications that may cause nausea and vomiting

To help alleviate nausea, suggest that medications be taken with food, when possible.

Dietary Intake

To devise an acceptable meal plan, obtain:

- An accurate and thorough record of food intake
- A record of foods that provoke symptoms of GERD, dyspepsia, peptic ulcers, or dumping syndrome

For patients on long-term dysphagia diets, monitor:

- Appetite
- Tolerances to foods
- Variety of foods offered and regularly consumed

Anthropometric Data

Measure baseline height and weight. Address weight loss early to prevent malnutrition for patients with:

- Dumping syndrome
- Dysphagia or difficulty chewing
- Dyspepsia or long-term nausea
- Malabsorption

Laboratory Tests

Check laboratory tests for signs of dehydration for patients with:

- Constipation
- Dumping syndrome
- Persistent vomiting

Check laboratory tests for nutrition-related anemia in patients with:

- Gastritis
- Long-term use of antisecretory drugs
- Previous gastric surgeries

Physical Signs

Look for physical signs of:

- Dehydration—in patients with constipation, dumping syndrome, or persistent vomiting
- Iron and vitamin B_{12} deficiencies—in patients with hypochlorhydria or achlorhydria

References

1. R. C. Orlando, Diseases of the esophagus, in L. Goldman and D. Ausiello, eds., *Cecil Medicine* (Philadelphia: Saunders, 2008), pp. 998–1009; M. H. Beers and coeditors, *The Merck Manual of Diagnosis and Therapy* (Whitehouse Station, NJ: Merck Research Laboratories, 2006), pp. 62–183.
2. R. K. Goyal, A. Chaudhury, and H. Mashimo, Oropharyngeal and esophageal motility disorders, in N. J. Greenberger, ed., *Current Diagnosis and Treatment: Gastroenterology, Hepatology, and Endoscopy* (New York: McGraw-Hill Companies, 2009), pp. 155–171.
3. Orlando, 2008.
4. American Dietetic Association, *Nutrition Care Manual* (Chicago: American Dietetic Association, 2010).
5. J. R. Agrawal and S. Friedman, Gastrointestinal and biliary complications of pregnancy, in N. J. Greenberger, ed., *Current Diagnosis and Treatment: Gastroenterology, Hepatology, and Endoscopy* (New York: McGraw-Hill Companies, 2009), pp. 75–97.
6. Agrawal and Friedman, 2009; Orlando, 2008.
7. N. J. Talley, Functional gastrointestinal disorders: Irritable bowel syndrome, dyspepsia, and noncardiac chest pain, in L. Goldman and D. Ausiello, eds., *Cecil Medicine* (Philadelphia: Saunders, 2008), pp. 990–998.
8. V. S. Wang and R. Burakoff, Functional (nonulcer) dyspepsia, in N. J. Greenberger, ed., *Current Diagnosis and Treatment: Gastroenterology, Hepatology, and Endoscopy* (New York: McGraw-Hill Companies, 2009), pp. 189–199.
9. J. Tack, Dyspepsia, in M. Feldman and coeditors, *Sleisenger and Fordtran's Gastrointestinal and Liver Disease* (Philadelphia: Saunders, 2010), pp. 183–195.
10. Talley, 2008.
11. Wang and Burakoff, 2009; S. Escott-Stump, *Nutrition and Diagnosis-Related Care* (Baltimore: Lippincott Williams & Wilkins, 2008).
12. American Dietetic Association, 2010; Escott-Stump, 2008.
13. E. Lew, Peptic ulcer disease, in N. J. Greenberger, ed., *Current Diagnosis and Treatment: Gastroenterology, Hepatology, and Endoscopy* (New York: McGraw-Hill Companies, 2009), pp. 175–183.
14. K. Ramakrishnan and R. C. Salinas, Peptic ulcer disease, *American Family Physician* 76 (2007): 1005–1012.
15. Beers and coeditors, 2006.
16. Lew, 2009.
17. American Dietetic Association, 2010.
18. F. F. Ferri, *Ferri's Clinical Advisor 2009: Instant Diagnosis and Treatment* (Philadelphia: Mosby, 2009), pp. 693–694; W. F. Stenson, The esophagus and stomach, in M. E. Shils and coeditors, *Modern Nutrition in Health and Disease* (Baltimore: Lippincott Williams & Wilkins, 2006), pp. 1179–1188.
19. American Dietetic Association, 2010.
20. American Dietetic Association, 2010.
21. M. K. Robinson and N. J. Greenberger, Treatment of obesity: The impact of bariatric surgery, in N. J. Greenberger, ed., *Current Diagnosis and Treatment: Gastroenterology, Hepatology, and Endoscopy* (New York: McGraw-Hill Companies, 2009), pp. 210–221.
22. G. Woodard and J. Morton, Bariatric surgery, in M. Feldman and coeditors, *Sleisenger and Fordtran's Gastrointestinal and Liver Disease* (Philadelphia: Saunders, 2010), pp. 115–119.
23. Weight Management Dietetic Practice Group, *ADA Pocket Guide to Bariatric Surgery* (Chicago: American Dietetic Association, 2009), pp. 1–16.
24. American Dietetic Association, 2010; J. I. Mechanick and coauthors, American Association of Clinical Endocrinologists, the Obesity Society, and American Society for Metabolic and Bariatric Surgery medical guidelines for clinical practice for the perioperative nutrition, metabolic, and nonsurgical support of the bariatric surgery patients, *Obesity* 17 (2009): S1–S41.
25. L. Aills and coauthors, ASMBS allied health nutritional guidelines for the surgical weight loss patient, *Surgery for Obesity and Related Diseases* 4 (2008): S73–S108.
26. Robinson and Greenberger, 2009.

Oral Health and Chronic Illness

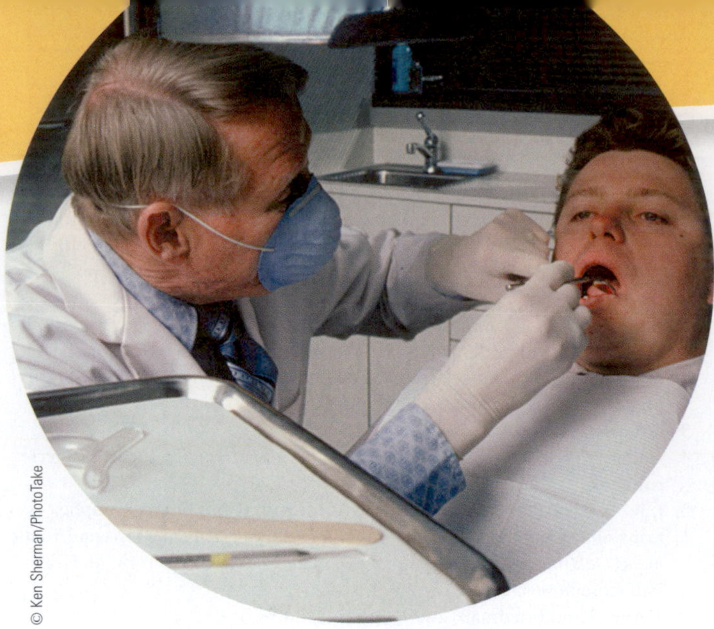

© Ken Sherman/PhotoTake

Various aspects of nutrition and oral health have been discussed in this book. Chapter 4 describes the effects of sugar and other fermentable carbohydrates on tooth decay. Chapter 13 explains how fluoride can help to prevent dental caries. Chapter 16 discusses the development of tooth decay in babies who are given bottles for prolonged periods. This highlight introduces other problems related to oral health, and describes some interactions between dental diseases and chronic illnesses. The accompanying glossary defines related terms.

Periodontal Disease

Recall from Chapter 4 that dental caries develops when the bacteria that reside in dental plaque metabolize dietary carbohydrates and produce acids that dissolve tooth enamel (review Figure 4-14 on p. 114). Deposits of plaque can thicken and lead to other dental problems as well. As plaque accumulates on the tooth surface, it fills with calcium and phosphate, eventually forming **dental calculus**. Calculus may develop either at the gum surface or in the crevice between the gum and a tooth; its presence may lead to additional plaque retention. The buildup of plaque and calculus increases the likelihood of infection and subsequent inflammation.

Periodontal disease is the name given to inflammatory conditions that involve the **periodontium**—the structures that support the tooth in its bony socket. The periodontium includes the gums (called **gingiva**), other connective tissues surrounding the tooth, and the bone underneath. Inflammation of the gums, called **gingivitis**, is characterized by redness, bleeding, and swelling of gum tissue. **Periodontitis** is an inflammation of the other tissues surrounding the tooth. As plaque invades the space below the gum line, the combination of toxic bacterial by-products and the body's immune response can damage the tissues holding a tooth in place. Left untreated, the tissues and bone of the periodontium may ultimately be destroyed, leading to permanent tooth loss.

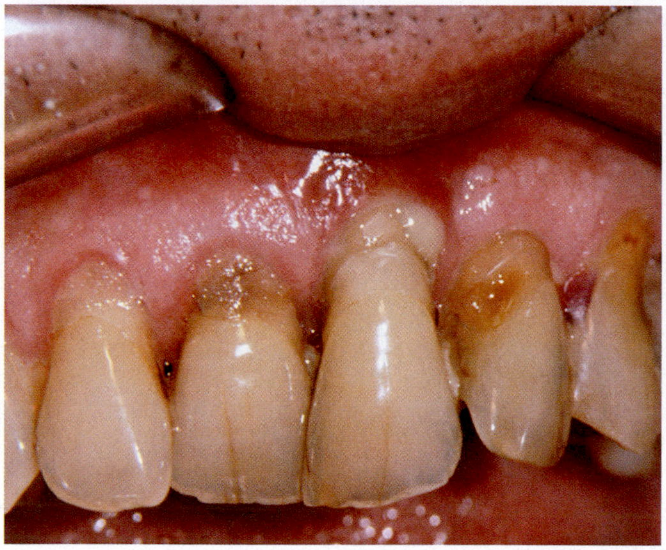

© CNRI/Photo Researchers, Inc.

Periodontal disease destroys the tissues and bones that hold teeth in place.

GLOSSARY

dental calculus: mineralized dental plaque, often associated with inflammation and bleeding.

gingiva (jin-JYE-va, JIN-jeh-va): the gums.

gingivitis (jin-jeh-VYE-tus): inflammation of the gums, characterized by redness, swelling, and bleeding.

periodontal disease: disease that affects the connective tissue that supports the teeth.

periodontitis: inflammation or degeneration of the tissues that support the teeth.

periodontium: the tissues that support the teeth, including the gums, cementum (bonelike material covering the dentin layer of the tooth), periodontal ligament, and underlying bone.

- **peri** = around, surrounding
- **odont** = tooth

Sjögren's syndrome: an autoimmune disease characterized by the destruction of secretory glands, resulting in dry mouth and dry eyes.

xerostomia: dry mouth caused by reduced salivary flow.

- **xero** = dry
- **stoma** = mouth

Reminder: *Dental caries* refers to tooth decay. *Dental plaque* is an accumulation of bacteria and their by-products that grow on teeth and can lead to dental caries and gum disease.

Risk Factors

Dental plaque is the primary risk factor associated with periodontal disease, and the severity of disease is related to the amount of plaque present. Tobacco smoking is another factor, possibly because of its destructive effects on cellular immune responses. The risk of developing periodontal disease is especially high if a person has a chronic illness that impairs immune status, such as diabetes mellitus or HIV infection. Other risk factors include stress, pregnancy, use of certain medications (including oral contraceptives, antiepileptic drugs, and anticancer drugs), and dental conditions that increase plaque accumulation, such as poorly aligned teeth or ill-fitting bridges.[1] Strategies for reducing risk focus on improving oral hygiene (proper brushing and flossing) and encouraging smoking cessation.

Signs and Symptoms

Periodontal disease typically begins with gingivitis, evidenced by tender and swollen gums that bleed readily from brushing or flossing. The gap between infected gums and teeth usually deepens, allowing food particles to get caught easily. Often, a bad taste in the mouth or persistent bad breath is the first sign of gingivitis. In severe cases, pus may surround the teeth and gums, the teeth may be sensitive, and chewing painful. If bone is destroyed, the affected gums usually recede, and teeth may loosen or change position.

Treatment

Treatment of periodontal disease depends on the extent of damage. In mild cases, deep cleaning and proper oral hygiene may reverse the condition. Antimicrobial mouth rinses and topical antibiotics are often prescribed to control infection. Surgical approaches are sometimes necessary to remove plaque or calculus deposits underneath gum tissue or to replace tissues that have been destroyed.

Dry Mouth

Secretions of the salivary glands protect the teeth and the mouth's soft tissues. Saliva rinses away the sugars and food particles that remain on teeth after meals and lubricates oral tissues. Saliva also contains antimicrobial proteins (immunoglobulins and lysozyme) that defend against bacteria and fungi. The buffers in saliva (such as bicarbonate, proteins, and phosphates) raise the mouth's pH so that tooth enamel is protected from the acid produced by caries-causing bacteria. The calcium, phosphate, and fluoride ions in saliva help to prevent dissolution of enamel and promote remineralization. Thus, saliva helps to prevent infection within the mouth, control plaque formation, and maintain tooth enamel. If salivary secretions are low or absent, the risk of developing dental caries and periodontal disease greatly increases.[2]

Dry mouth (**xerostomia**), caused by reduced salivary flow, is a side effect of many medications and is associated with a number of diseases and disease treatments. Antihistamines, antihypertensive agents, antidepressants, decongestants, and other medications can cause dry mouth. Poorly controlled diabetes mellitus is often associated with dry mouth, as are conditions that directly affect salivary gland function, such as **Sjögren's syndrome**. Radiation therapy used to treat head and neck cancers often damages salivary glands, sometimes permanently. Mouth breathing is also a common cause of dry mouth.[3]

A reduction in salivary flow can impair health in other ways as well. Dry mouth can interfere with speech and cause bad breath. Mouth infections and dental diseases are more common. Chewing and swallowing are more difficult, and taste sensation is diminished. Dentures may be uncomfortable to wear, and ulcerations may develop where they contact the mouth. Dry mouth may cause a person to reduce food intake and thereby increase malnutrition risk. Table H23-1 offers suggestions that may help to manage dry mouth.

TABLE H23-1 Suggestions for Managing Dry Mouth

- Take frequent sips of water or another sugarless beverage.
- Chew sugarless gum to help stimulate salivary flow.
- Suck on ice cubes or frozen fruit juice bars (unless their coldness causes discomfort).
- Avoid citrus juices and spicy or salty foods if they cause mouth irritation.
- Avoid dry foods like toast, chips, and crackers.
- Avoid caffeine, alcohol, and smoking, which may dry the mouth.
- Consume foods that have a high fluid content such as soups, stews, sauces and gravies, yogurt, and pureed fruits.
- Try over-the-counter saliva substitutes (available as gels, sprays, and tablets), especially just before meals and at bedtime.
- Try rinsing the mouth with small amounts of vegetable oil or softened margarine.
- Use a humidifier during the night.
- Pay strict attention to oral hygiene, brushing and flossing at least twice daily. Try to brush immediately after each meal.
- Avoid alcohol- and detergent-containing mouthwashes that may dry and irritate the mouth.
- If dry mouth is caused by a medication, ask your physician about possible alternatives.
- Ask your physician if using a medication to stimulate saliva secretion may be of benefit; examples include nicotinic acid tablets and pilocarpine.

Dental Health and Chronic Illness

Maintaining dental health is especially challenging for people who have certain chronic illnesses. Some diseases can alter the structure and function of dental tissues, impair immune responses, or cause reduced salivary flow. As mentioned earlier, some medications can reduce salivary secretions, along with the immune protection that saliva provides. This section describes how several chronic conditions may upset oral health and increase the risks of developing dental problems.

Diabetes Mellitus

For a number of reasons, periodontal disease is more prevalent among people with diabetes mellitus, especially those whose diabetes is poorly controlled. People with diabetes often have impaired immune responses and a greater susceptibility to infections. Diabetes also favors the growth of bacteria that tend to infect periodontal tissues. People with diabetes tend to have higher plaque accumulations and dry mouth. In addition, the damaging effects of hyperglycemia weaken the collagen structure of dental tissues, making them more vulnerable to destruction.[4]

Because the risk of developing dental caries and oral fungal infections is greater for people with diabetes, they must pay strict attention to oral hygiene. Smoking is discouraged because it can increase periodontal disease risk substantially in people with diabetes. Health care providers should advise patients with diabetes that glucose control and routine dental care are critical to preventing periodontal disease.

Human Immunodeficiency Virus (HIV) Infection/AIDS

HIV infection is characterized by compromised immunity, and the risk of developing periodontal disease is closely linked to the extent of HIV infection. Those at greatest risk of developing dental diseases include smokers and patients in the advanced stages of illness. HIV-infected individuals often develop dry mouth as a result of medications or salivary gland dysfunction.[5] In untreated persons, fungal and viral infections are common and may cause burning in the mouth and painful ulcerations.

Oral Cancers

The radiation treatment required for oral cancers often causes serious oral and dental complications.[6] Inflammation and tissue damage can be so severe that the radiation treatment may need to be halted or the intensity significantly reduced. Radiation can also reduce salivary flow, causing the problem of dry mouth described earlier. Other complications include fungal and viral infections, changes in taste sensation, and tissue and muscle scarring (which often reduces chewing ability). To minimize complications, dental care is often initiated before radiation therapy begins.

Dental Health and Disease Risk

Dental diseases may have adverse effects on health beyond their effects on teeth. The bacteria that reside on dental tissues can enter the bloodstream and travel to other tissues; therefore, they may be able to cause infections elsewhere in the body. Evidence supports a link between dental bacteria and other conditions, including the following:[7]

- *Systemic inflammation*. The inflammatory process induced by periodontal disease increases levels of cytokines and other mediators that have systemic effects. Systemic inflammation may contribute to the development of various chronic illnesses, including heart disease and diabetes.
- *Atherosclerosis and heart disease*. Oral bacteria are frequently found residing in the arteries of people with atherosclerosis, where they may induce the release of inflammatory mediators that promote atherosclerosis. Periodontal bacteria may also initiate plaque formation and induce blood clotting.
- *Diabetes mellitus*. The chronic inflammation caused by periodontal disease can exacerbate insulin resistance and provoke events leading to type 2 diabetes. Severe periodontal disease has also been linked to poor glycemic control in persons with diabetes.
- *Respiratory illnesses*. The teeth of hospitalized individuals can become colonized with bacteria that cause respiratory illnesses. In addition, mortality rates associated with pneumonia have been found to be four times higher in individuals with periodontal disease.[8]

Research studies are in progress to confirm cause-and-effect relationships between oral bacteria and the medical conditions described above, as well as the specific mechanisms involved.

Nutrition status has a strong influence on oral health. Developing sound eating habits and maintaining good dental hygiene are practices that can promote dental health and possibly reduce the risk of developing other medical problems. Additional studies will help to clarify the complex interactions between oral health and chronic illnesses.

References

1. J. Kim and S. Amar, Periodontal disease and systemic conditions: A bidirectional relationship, *Odontology* 94 (2006): 10–21.
2. S. K. Stookey, The effect of saliva on dental caries, *Journal of the American Dental Association* 139 (2008): 11S–17S; D. P. DePaola and coauthors, Nutrition and dental medicine, in M. E. Shils and coeditors, *Modern Nutrition in Health and Disease* (Baltimore: Lippincott Williams & Wilkins, 2006), pp. 1152–1178.
3. T. E. Daniels, Diseases of the mouth and salivary glands, in L. Goldman and D. Ausiello, eds., *Cecil Medicine* (Philadelphia: Saunders, 2008), pp. 2867–2874; DePaola and coauthors, 2006.
4. American Dietetic Association, Position of the American Dietetic Association: Oral health and nutrition, *Journal of the American Dietetic Association* (2007): 1418–1428; DePaola and coauthors, 2006.
5. J. C. C. Filho and E. M. Giovani, Xerostomy, dental caries and periodontal disease in HIV+ patients, *Brazilian Journal of Infectious Diseases* 13(2009): 13–17.
6. American Dietetic Association, 2007.
7. P. Weidlich and coauthors, Association between periodontal diseases and systemic diseases, *Brazilian Oral Research* 22 (2008): 32–43.
8. Weidlich and coauthors, 2008.

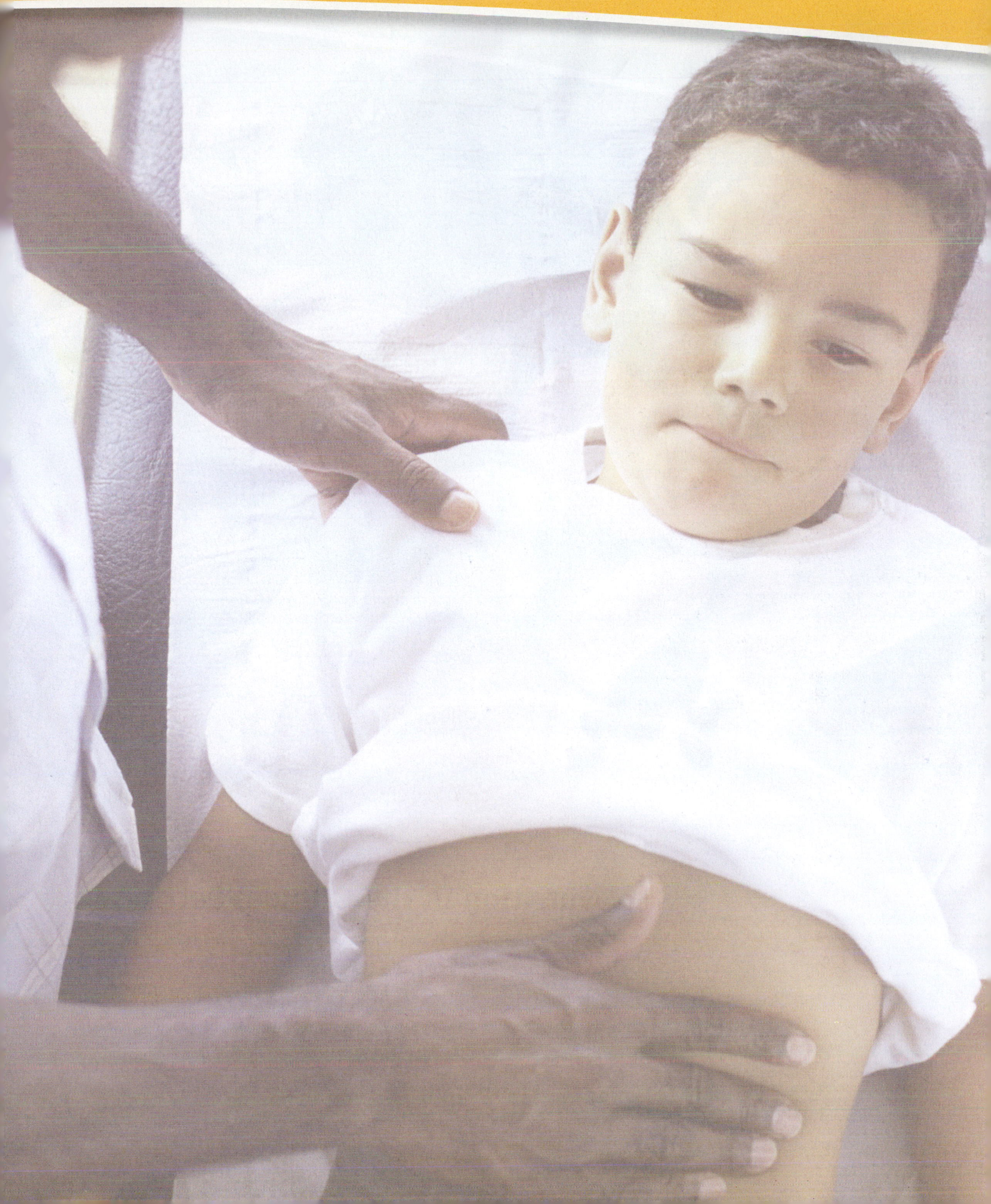

CHAPTER

24

© Masterfile

Nutrition in the Clinical Setting

Disorders affecting the lower gastrointestinal tract can interfere substantially with a patient's diet and lifestyle. Some of the diets required for these conditions are complicated and difficult to follow, and the foods that are tolerated can vary considerably. In visits with patients, health care professionals should ensure that patients understand the nutrition prescription and help to pinpoint difficult foods. They can also suggest ways to make restrictive diets more acceptable.

Throughout this chapter, the CourseMate icon indicates an opportunity for online self-study, linking you to activities to increase your understanding of chapter concepts. **www.cengagebrain.com** (search for ISBN 084006845X)

Lower Gastrointestinal Disorders

CHAPTER OUTLINE

This chapter discusses medical conditions that can upset the digestive and absorptive functions of the lower gastrointestinal (GI) tract, which consists of the small intestine (the duodenum, jejunum, and ileum) and the large intestine (the colon, rectum, and anal canal). The digestion and absorption of nutrients occur primarily in the small intestine. The pancreas and gallbladder support these complex functions by delivering digestive secretions to the duodenum, the segment of small intestine closest to the stomach. The large intestine reabsorbs water and facilitates the excretion of waste material. Figure 24-1 illustrates the lower GI tract and related organs and reviews the functions of each organ. Chapter 3 provides additional detail.

Common Intestinal Problems

Nearly all people experience occasional intestinal problems, which usually clear up without medical treatment. Intestinal discomfort can sometimes drive a person to seek medical attention, however, and the symptoms may be evidence of a serious intestinal disorder or other illness. The most common intestinal problems and their causes and treatments are discussed below.

Constipation

Diagnosis of constipation is based, in part, on a defecation frequency of fewer than three bowel movements per week. Other symptoms may include the passage of hard stool and excessive straining during defecation. In some cases, a person's perception of constipation may be due to a mistaken notion of what constitutes "normal" bowel habits, so the person's expectations about bowel function may need to be addressed. Constipation affects up to 28 percent of the population in Western countries and is particularly prevalent among women and older adults.[1]

Causes of Constipation The risk of constipation is increased in individuals with a low-fiber diet, low food intake, inadequate fluid intake, or low level of physical activity. All of these factors can extend transit time, leading to increased water reabsorption within the colon and dry, hard stools that are difficult to pass. Medical conditions often associated with constipation include diabetes mellitus and hypothyroidism. Neurological conditions such as Parkinson's disease, spinal cord lesions, and multiple sclerosis may cause motor problems that lead to constipation. During pregnancy, women often experience constipation because the enlarged uterus presses against the rectum and colon. Constipation is also a common side effect of several classes of medications and some dietary supplements, including opiate-containing analgesics, tricyclic antidepressants, anticonvulsants, calcium channel blockers, aluminum-containing antacids, and iron and calcium supplements.

FIGURE 24-1 The Lower GI Tract and Related Organs

Stomach
Adds acid, enzymes, and fluid. Churns, mixes, and grinds food to a liquid mass.

Pyloric sphincter
Allows passage from stomach to small intestine. Prevents backflow from small intestine.

Small intestine
Produces enzymes that digest energy-yielding nutrients to smaller nutrient particles. Cells absorb nutrients into blood and lymph.

Ileocecal valve (sphincter)
Allows passage from small to large intestine. Prevents backflow from large intestine and controls transit through intestine.

Large intestine (colon)
Reabsorbs water and minerals. Passes waste (fiber and some water) to rectum.

Rectum
Stores waste prior to elimination.

Anus
Holds rectum closed. Opens to allow elimination.

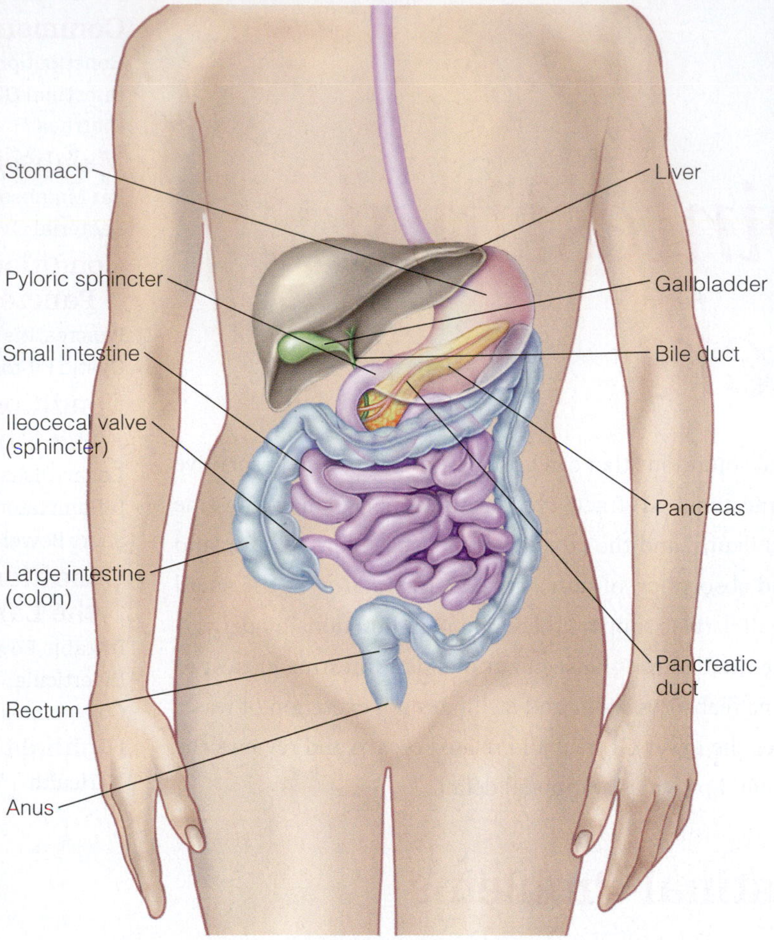

Liver
Manufactures bile salts (detergent-like substances), to help digest fats.

Gallbladder
Stores bile until needed.

Bile duct
Conducts bile from liver into small intestine.

Pancreas
Manufactures enzymes to digest energy-yielding nutrients and bicarbonate to neutralize acidic stomach contents that enter the small intestine. (Also produces insulin and glucagon.)

Pancreatic duct
Conducts pancreatic juice from pancreas into small intestine.

♦ The fiber DRI for women and men aged 19 to 50 years are 25 and 38 grams, respectively. Recommendations for increasing fiber intake are described on pp. 120–121, and Appendix H includes the fiber values of most common foods.

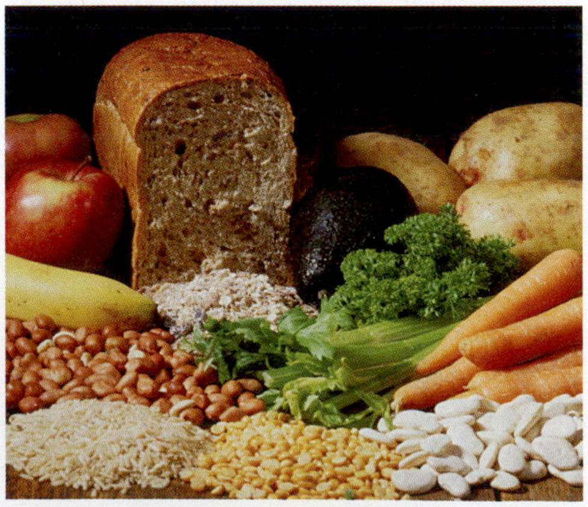

High-fiber foods promote regular bowel movements.

Treatment of Constipation The primary treatment for constipation is a gradual increase in fiber intake to about 25 grams per day.[2] ♦ High-fiber diets increase stool weight and fecal water content and promote a more rapid transit of materials through the colon. Foods that increase stool weight the most are wheat bran, fruits, and vegetables.[3] Bran intake can be increased by adding bran cereals and whole-wheat bread to the diet or by mixing bran powder with beverages or foods. The transition to a high-fiber diet may be difficult for some people because it can increase intestinal gas, so high-fiber foods should be added gradually, as tolerated. Fiber supplements such as methylcellulose (Citrucel), psyllium (Metamucil, Fiberall), and polycarbophil (a synthetic fiber) are also effective (see Table 24-1); these supplements can be mixed with beverages and taken several times daily. Unlike other fibers, methylcellulose and polycarbophil do not increase intestinal gas.

Several other measures may also help constipation. Consuming adequate fluid (usually 1.5 to 2 liters daily) helps to increase stool frequency in people who are already consuming a high-fiber diet.[4] An appropriate fluid intake prevents excessive reabsorption of water from the colon, resulting in wetter stools. Adding prunes or prune juice to the diet is often recommended because prunes contain compounds that have a mild laxative effect.

Laxatives Many laxatives can be purchased without prescription. They work by increasing stool weight, increasing the water content of the stool, or stimulating

TABLE 24-1 Laxatives and Bulk-Forming Agents

Laxative Type	Active Ingredients	Product Examples	Method of Action	Cautions
Fiber (bulk-forming agents)	Malt soup extract, methylcellulose, polycarbophil, psyllium	Citrucel, Fiberall, Fiber-Lax, Metamucil	Fiber supplements increase stool weight and aid in formation of soft, bulky stools. Similar effects are achieved by adding bran to the diet. For mild constipation. Safe for long-term use.	Some fiber supplements may increase flatulence. Psyllium may cause an allergic reaction.
Osmotic laxatives: nonabsorbable salts	Magnesium citrate, magnesium hydroxide, sodium phosphate	Epsom salts, milk of magnesia	Unabsorbed salts attract and retain water in the large intestine and stimulate contractions.	May cause bloating and watery stools or diarrhea. Should be used with caution. Avoid using in renal patients and children.
Osmotic laxatives: nonabsorbable sugars	Lactulose, polyethylene glycol, sorbitol	Cephulac, Chronulac	Unabsorbed sugars attract water to the large intestine and promote softer stools. Must be used for several days to take effect. Safe for long-term use.	May cause flatulence and cramps. Can lose effectiveness over time.
Stimulant or irritant laxatives	Aloe, bisacodyl, cascara, castor oil, senna	Correctol, Dulcolax, Ex-Lax	Act as local irritants to colonic tissue; stimulate peristalsis and mucosal secretions. For moderate to severe constipation. Long-term use is discouraged.	Usually given only after milder treatments fail. May alter fluid and electrolyte balances. May lead to laxative dependency.
Stool surfactant agents (stool softeners)	Docusate sodium	Colace, Kaopectate, Surfak	Detergent action promotes the mixing of water with stools. Prevents formation of dry, hard stools.	Do not increase stool weight. Limited effectiveness.

peristaltic contractions. Table 24-1 includes examples of common laxatives and describes their modes of action. Enemas and suppositories (chemicals introduced into the rectum) are also used to promote defecation; they work by distending and stimulating the rectum or by lubricating the stool.

Medical Interventions For patients with severe constipation who do not respond to dietary or laxative treatments, physicians may prescribe medications that stimulate colonic contractions. Physical therapy and biofeedback techniques are sometimes successful in training patients to relax their pelvic muscles more effectively. Surgical interventions are a last resort and include colonic resections and colostomy operations, which are discussed later in this chapter.

Intestinal Gas

As mentioned in the previous section, increased intestinal gas (**flatulence**) may be an unpleasant side effect of consuming a high-fiber diet. The undigested fibers pass into the colon, where they are fermented by bacteria, which produce gas as a by-product. ♦ Other incompletely digested or poorly absorbed carbohydrates have similar effects; these include fructose, sugar alcohols (sorbitol, xylitol, mannitol), the indigestible carbohydrates in beans (raffinose and stachyose), and some forms of resistant starch, found in grain products and potatoes. Table 24-2 lists examples of foods commonly associated with excessive gas production, although individual responses vary. Malabsorption disorders (discussed later in this chapter) can cause considerable flatulence because the undigested nutrients can be metabolized by colonic bacteria. Swallowed air that is not expelled by belching may travel to the intestines and be a source of intestinal gas as well (see p. 92).

♦ Reminder: The *soluble fibers* in foods are more readily fermented in the small intestine than the *insoluble fibers*.

Many people blame abdominal bloating and pain on excessive gas, but these symptoms do not correlate well with an increase in intestinal gas.[5] In fact, most people who self-diagnose a flatulence problem have no more intestinal gas than others. Some individuals who experience frequent symptoms of abdominal bloating and pain are later diagnosed with irritable bowel syndrome (see pp. 748–749) or dyspepsia (discussed in Chapter 23).

Diarrhea

Diarrhea is characterized by the passage of frequent, watery stools. In most cases, it lasts for only a day or two and subsides without complication. Severe or persistent diarrhea, however, can cause dehydration and electrolyte imbalances. If chronic, it may lead to weight loss and malnutrition. Diarrhea may be

flatulence: the condition of having excessive intestinal gas, which causes abdominal discomfort.

TABLE 24-2 Foods That May Increase Intestinal Gas

Apples
Artichokes
Asparagus
Beer
Broccoli
Brussels sprouts
Cabbage
Carbonated beverages
Carrots
Corn
Dried beans and peas
Fructose-sweetened products
Fruit juices
Green beans
Leeks
Milk products (if lactose intolerant)
Onions
Peanuts
Pears
Turnips
Wheat

accompanied by other symptoms, such as fever, abdominal cramps, dyspepsia, or bleeding, which help in diagnosing the cause.

Causes of Diarrhea Diarrhea is a complication of various GI disorders and may also be induced by infections, medications, or dietary substances. It results from inadequate fluid reabsorption in the intestines, sometimes in conjunction with an increase in intestinal secretions.[6] In *osmotic diarrhea*, unabsorbed nutrients or other substances attract water to the colon and increase fecal water content; the usual causes include high intakes of poorly absorbed sugars (such as sorbitol, mannitol, or fructose), lactase deficiency, and ingestion of laxatives that contain magnesium or phosphates. In *secretory diarrhea*, the fluid secreted by the intestines exceeds the amount that can be reabsorbed by intestinal cells. Secretory diarrhea is often due to bacterial food poisoning but can also be caused by intestinal inflammation and various chemical substances. *Motility disorders* can result in diarrhea because they accelerate the transit of colonic residue, reducing the contact time available for fluid reabsorption.

Acute cases of diarrhea start abruptly and may persist for several weeks; they are frequently caused by viral, bacterial, or protozoal infections or occur as a side effect of medications. Chronic diarrhea, which persists for three weeks or longer,[7] can result from malabsorption disorders, inflammatory diseases, motility disorders, infectious diseases, radiation treatment, and many other conditions. As mentioned in earlier chapters, diarrhea is a frequent complication of tube feedings and it may also occur when enteral feedings are resumed after a period of bowel rest (see Chapters 20 and 21).

Medical Treatment of Diarrhea Correcting the underlying medical problem is the first step in treating diarrhea. For example, antibiotics can be prescribed for treating intestinal infections. If a medication is the cause of diarrhea, a different drug may be suggested. If certain foods are responsible, they can be omitted from the diet. Bulk-forming agents such as psyllium (Metamucil) or methylcellulose (Citrucel) can help to reduce the liquidity of the stool. If chronic diarrhea does not respond to treatment, antidiarrheal drugs may be prescribed to slow GI motility or reduce intestinal secretions. Probiotics ♦ may be beneficial for certain types of diarrhea (especially diarrhea caused by infections), but standard treatment protocols have not been developed.[8] People with severe, **intractable** diarrhea sometimes require total parenteral nutrition.

♦ *Probiotics* are live bacteria provided in foods and dietary supplements for the purpose of preventing or treating disease. Highlight 24 describes the potential health benefits of probiotics.

Oral Rehydration Therapy Severe diarrhea requires the replacement of lost fluid and electrolytes. Oral rehydration solutions can be purchased or easily mixed using water, salts, and glucose or sucrose (see the recipe in the margin). ♦ The addition of carbohydrate to the solution facilitates sodium and water absorption. Commercial sports drinks are not ideal fluids for rehydration because their sodium content is too low, but they can be used if accompanied by salty snack foods.[9] When diarrhea results in extreme dehydration, intravenous solutions are used to quickly replenish fluid and electrolytes.

♦ An oral rehydration solution can be mixed from the following ingredients:
- 1/2 tsp sodium chloride (table salt)
- 1/3 tsp potassium chloride (salt substitute)
- 3/4 tsp sodium bicarbonate (baking soda)
- 1 1/3 tbs sugar
- 1 qt water

Nutrition Therapy for Diarrhea Because diarrhea can develop for numerous reasons, the nutrition prescription depends on the medical diagnosis and severity of the condition. The dietary treatment often recommended is a low-fiber, low-fat, lactose-free diet.[10] The diet limits foods that contribute to colonic residue, such as those with significant amounts of fiber, resistant starch, fructose, sugar alcohols, and lactose (in lactose-intolerant individuals). Fructose and sugar alcohols, which are poorly absorbed, retain fluids in the colon and contribute to osmotic diarrhea. Similarly, milk products may worsen osmotic diarrhea in persons who are lactose intolerant. Avoidance of fatty foods is recommended because they can sometimes aggravate diarrhea. Gas-producing foods (those with poorly digested or absorbed carbohydrates) can increase intestinal distention and cause additional discomfort. Patients should avoid caffeinated coffee and tea because caffeine stimulates GI motility and can thereby reduce water reabsorption. In the treatment of formula-fed infants, apple pectin or banana flakes are sometimes added to formulas to help

intractable: not easily managed or controlled.

TABLE 24-3 Foods That May Worsen Diarrhea[a]

Foods to Avoid	Rationale	Selected Examples
High-fiber foods	They increase colonic residue.	Breads and cereals with more than 2 g fiber per serving, fruits and vegetables with peels or skins
Foods with indigestible carbohydrates	They contribute to osmotic diarrhea.	Artichokes, asparagus, brussels sprouts, cabbage, dried beans and peas, fruit, garlic, green beans, leeks, onions, wheat
Foods that contain fructose or sugar alcohols	They contribute to osmotic diarrhea.	Dried fruits, fresh fruits (except bananas), fruit juices, fructose-sweetened soft drinks, sugar-free gums and candies
Milk products, if person is lactose intolerant	They contribute to osmotic diarrhea.	Milk and milk products
Gas-producing foods	They increase abdominal discomfort.	Foods with poorly digested or absorbed carbohydrates (including foods listed in the three rows directly above)
Caffeine-containing beverages	They increase intestinal motility.	Coffee, tea, colas, energy drinks

[a]Individual tolerances vary; the foods to avoid are best determined by trial and error.

thicken stool consistency. Table 24-3 lists examples of foods that may worsen diarrhea, although individual tolerances vary.

IN SUMMARY **The most common intestinal problems include constipation, intestinal gas, and diarrhea. Constipation accompanies many different health problems but generally correlates with low-fiber diets, low food or inadequate fluid intakes, and physical inactivity. Intestinal gas is largely produced by colonic bacteria and is usually caused by nutrient malabsorption. Diarrhea can result from malabsorption, intestinal infections, motility disorders, and dietary substances and may require oral rehydration therapy to replace fluid and electrolyte losses. Dietary modifications may help to improve bowel function and alleviate intestinal discomfort.**

Malabsorption Syndromes

To digest and absorb nutrients, we depend on normal digestive secretions and healthy intestinal mucosa. Malabsorption can therefore be caused by pancreatic disorders that lead to enzyme or bicarbonate deficiencies, conditions that lead to bile deficiency, and inflammatory conditions or medical treatments that damage intestinal tissue. In some cases, the treatment of an intestinal disorder requires surgical removal of a section (**resection**) of the small intestine, leaving minimal absorptive capacity in the portion that remains. In addition, various medications can damage the mucosa and impair the digestive and absorptive functions of the small intestine. Table 24-4 lists examples of diseases and treatments that are frequently associated with malabsorption.

Malabsorption rarely involves a single nutrient. When malabsorption is caused by pancreatic enzyme deficiencies, all macronutrients—protein, carbohydrate, and fat—may be affected. When fat is malabsorbed, fat-soluble nutrients and some minerals are usually malabsorbed as well. Malabsorption disorders and their treatments can tax nutrition status further by causing complications that alter food intake, raise nutrient needs, and incur additional nutrient losses.

Fat Malabsorption

Fat is the nutrient most frequently malabsorbed because both digestive enzymes and bile must be present for its digestion. Thus, fat malabsorption often develops when an illness interferes with the production or secretion of either pancreatic lipase or bile. For example, both pancreatitis and cystic fibrosis can reduce the secretion of pancreatic lipase, whereas severe liver disease can cause bile insufficiency. Motility disorders that accelerate gastric emptying or intestinal transit can cause fat malabsorption because they prevent the normal mixing of dietary fat with lipase and bile. Fat malabsorption may also be caused by

resection: the surgical removal of part of an organ or body structure.

TABLE 24-4 Potential Causes of Malabsorption

Genetic disorders
- Enzyme deficiencies

Intestinal disorders
- AIDS-related enteropathy
- Bacterial overgrowth
- Celiac disease
- Crohn's disease
- Radiation enteritis

Intestinal infections
- Giardiasis
- Nematode (roundworm) infections

Liver disease (bile insufficiency)

Pancreatic disorders
- Chronic pancreatitis
- Cystic fibrosis

Surgeries
- Gastric or intestinal bypass surgery
- Intestinal resection (short bowel syndrome)

FIGURE 24-2 The Consequences of Fat Malabsorption

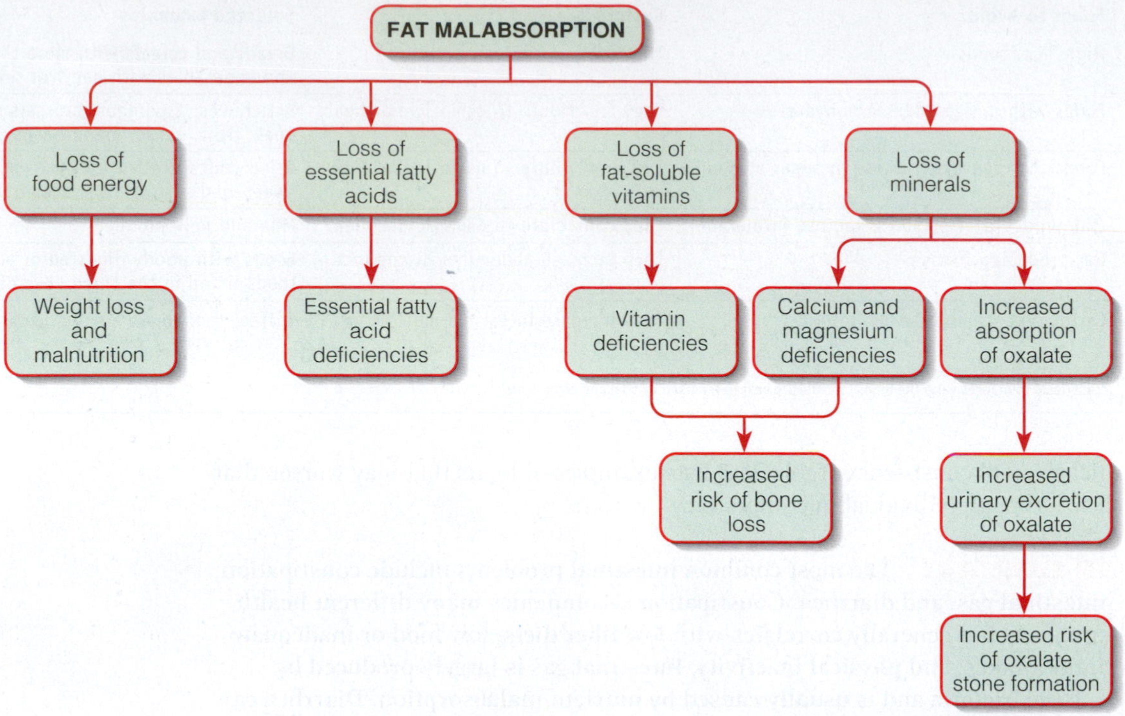

conditions or treatments that damage the intestinal mucosa, such as inflammatory bowel diseases, AIDS, and radiation treatments for cancer.

Fat malabsorption is often evidenced by **steatorrhea**, the presence of excessive fat in the stools. Steatorrhea can be evaluated by placing the patient on a high-fat diet (100 grams per day), performing a 48- to 72-hour stool collection, and measuring the stool's fat content. Healthy individuals generally excrete less than 7 grams of fat per day under these conditions.[11]

Consequences of Fat Malabsorption Fat malabsorption is associated with losses of food energy, essential fatty acids, fat-soluble vitamins, and some minerals (see Figure 24-2). Weight loss may result if the individual does not consume alternative sources of energy. Deficiencies of fat-soluble vitamins and essential fatty acids are common in chronic conditions. Malabsorption of some minerals, including calcium, magnesium, and zinc, often occurs because the minerals form **soaps** with the unabsorbed fatty acids and bile acids. Calcium deficiency may lead to bone loss, which is further aggravated by the vitamin D deficiency that may be present due to fat malabsorption.

Another consequence of fat malabsorption is an increased risk of kidney stones, which are most often composed of calcium oxalate. ♦ The oxalates in foods ordinarily bind to calcium in the small intestine and are excreted in the stool. If calcium instead binds to fatty acids or bile acids, the oxalates are free to be absorbed into the blood and are ultimately excreted in the urine. The risk of developing oxalate stones increases when urinary oxalate levels are high. Kidney stones are discussed further in Chapter 28.

Nutrition Therapy for Fat Malabsorption If steatorrhea does not improve, a fat-controlled diet may be recommended (see Table 24-5). The objectives of the diet are to relieve intestinal symptoms that are aggravated by fat intake (such as diarrhea and flatulence) and to reduce vitamin and mineral losses. Fat should not be restricted more than necessary because fat is an important source of energy. Medium-chain triglycerides (MCT), which do not require lipase or bile for digestion and absorption, can be used as an alternative source of dietary fat, although MCT

♦ Reminder: *Oxalates* are plant compounds that bind with some minerals to form complexes that the body cannot absorb. They are present in green leafy vegetables such as beet greens and spinach.

steatorrhea (stee-AT-or-REE-ah): excessive fat in the stools due to fat malabsorption; characterized by stools that are loose, frothy, and foul smelling due to a high fat content.

- **steat** = fat
- **rheo** = flow

soaps: chemical compounds formed from fatty acids and positively charged minerals.

TABLE 24-5 Fat-Controlled Diet

A fat-controlled diet includes mostly low-fat and fat-free foods. For a fat intake of 50 grams per day, limit meats and meat substitutes to 6 ounces per day, and limit fats and oils to 8 teaspoons per day.[a] Foods from other food groups should provide less than 1 gram of fat per serving.

Food Category	Foods Recommended	Foods to Avoid
Meat and meat alternates	Lean meats, fish, and skinless poultry prepared by broiling, roasting, grilling, or boiling; low-fat luncheon meats such as sliced turkey breast; meat alternates such as dried beans or peas; low-fat egg substitutes. Limit whole eggs to 2 per week.	Meat with visible fat, ground beef (unless extra lean), sausage, bacon, frankfurters, spareribs, duck, tuna packed in oil
Milk and milk products	Fat-free milk, fat-free yogurt, fat-free sour cream substitutes, fat-free half-and-half and cream substitutes, fat-free cheeses. Low-fat milk products can be used in moderation.	Milk products that are not fat free or low fat
Breads, cereals, rice, and pasta	Whole-grain and enriched breads, cooked cereals and most cold breakfast cereals, plain tortillas, bagels, English muffins, fat-free muffins, saltine crackers, graham crackers, plain rice, plain noodles and pasta.	Biscuits, pancakes, waffles, granola, snack crackers made with fat, cornbread, doughnuts, corn chips, fried rice
Vegetables	All vegetables prepared without added fat.	Buttered, creamed, breaded, or fried vegetables; vegetables prepared au gratin style; French-fried potatoes; olives
Fruits	All fruits except avocado.	Avocado; fruits prepared with fats, nuts, or coconut
Desserts	Sherbet; fruit ices; flavored gelatin; angel food cake; meringue; fat-free puddings; fat-free bakery products; fat-free ice cream or frozen yogurt; fat-free candies such as marshmallows, jelly beans, and hard candy.	Cakes, cookies, pies, and pastries made with fat; puddings made with whole milk or eggs; ice cream; candies made with fat such as caramel and chocolates
Fats	For a total fat intake of 50 grams per day, limit intake of the following foods to 8 teaspoons daily: vegetable oil, butter, margarine, mayonnaise, lard (each has 3$^1/_2$ to 4$^1/_2$ grams of fat). Each of these foods can replace 1 teaspoon of fat in the amounts specified: 1 tbs salad dressing, 2 tbs low-fat salad dressing, $^1/_2$ tbs peanut butter, 1 tbs chopped nuts, 2 tbs mashed avocado.	Dietary fat that exceeds the amount specified in the nutrition prescription
Beverages	Fruit juices, soft drinks, fat-free milk, coffee, tea, coffee substitutes.	Beverages made with milk (unless fat free) or added cream, chocolate milk, eggnog, milk shakes

Sample menu (about 50 grams of fat):

Breakfast: 6 oz orange juice, 1 c oatmeal with nonfat milk and raisins, 1 slice whole-wheat toast with 1 tsp margarine, coffee with fat-free half-and-half

Lunch: Turkey breast sandwich (2 slices whole-wheat bread, 2 oz lean turkey breast, 2 tomato slices, lettuce leaf, and 2 tsp mayonnaise), 2 c salad greens with 1 tbs salad dressing, fruit cup (1 c peaches and $^1/_2$ c berries) with $^1/_2$ c orange sherbet

Snack: 6 oz nonfat fruit yogurt, 6 saltine crackers with 1 tbs peanut butter and $^1/_2$ tbs honey

Dinner: 4 oz cod with sliced lemon, 1 slice French bread with 1 tsp butter, 1 c steamed rice with herbs and walnut oil (includes $^1/_2$ tsp oil), 1 c steamed broccoli and carrots with $^1/_2$ tsp margarine, 1 piece angel food cake with fat-free whipped cream

[a]For a fat intake of 25 grams per day, limit meats and meat substitutes to 4 ounces per day, and limit fats and oils to 2 teaspoons per day.

oil does not provide essential fatty acids. The "How To" on p. 736 offers suggestions for following the fat-controlled diet and for using MCT oil.

Bacterial Overgrowth

Ordinarily, the GI tract is protected from **bacterial overgrowth** by gastric acid, which destroys bacteria; peristalsis, which flushes bacteria through the small intestine before they multiply; and immunoglobulins secreted into the GI lumen.[12] When bacterial overgrowth does occur, it can lead to fat malabsorption because the bacteria dismantle the bile acids needed for fat emulsification. Deficiencies of the fat-soluble vitamins A, D, and E may eventually develop. The bacteria also produce enzymes and toxins that disturb the intestinal mucosa, destroying some mucosal enzymes (especially lactase) and possibly reducing the absorptive surface area. Some types of bacteria metabolize vitamin B_{12}, reducing its absorption and increasing the risk of deficiency. Although symptoms of bacterial overgrowth are often minor and nonspecific, severe cases may lead to chronic diarrhea, steatorrhea, flatulence, bloating, and weight loss.

bacterial overgrowth: excessive bacterial colonization of the stomach and small intestine; may be due to low gastric acidity, altered GI motility, mucosal damage, or contamination.

HOW TO Follow a Fat-Controlled Diet

For some individuals, fat-controlled diets may be difficult to follow. Fats add flavors, aromas, and textures to foods—characteristics that make foods more enjoyable. Unlike some diets that can be introduced gradually, a fat-controlled diet is often implemented immediately, allowing little time for adaptation. These suggestions may help:

- Fat is better tolerated if provided in small portions. Divide the day's allotment into several servings that can be consumed throughout the day.
- Use variety to enhance enjoyment of meals: vary flavors, textures, colors, and seasonings.
- Look for fat-free items when grocery shopping. Incorporate fat-free ingredients when preparing favorite recipes.
- Try fat-free and low-fat condiments to improve the diet's palatability. Experiment with herbs and spices. Instead of butter, use fruit butters on toast. Use butter-flavored granules on vegetables. Replace mayonnaise in sandwiches with spicy mustard. Replace salad dressings with flavored vinegars.
- Avoid products that contain the fat substitute olestra, which may aggravate GI symptoms.

If patients are interested in using medium-chain triglyceride (MCT) oil:

- Explain that MCT products are expensive but that the cost is sometimes covered by medical insurance.
- Advise patients to add MCT oil to the diet gradually. Diarrhea and abdominal cramps may result if too much is used at once. Tolerance to MCT oil may improve in time.
- Advise patients that MCT oil may have an unpleasant taste when used alone. Suggest using MCT oil in recipes as a substitute for regular oil. MCT oil can replace oil in salad dressings, be incorporated into sauces, and be used in cooking or baking. It can also be added to fat-free milk products to make milk shakes.
- Point out that MCT oil should not be used to fry foods because it decomposes at lower temperatures than most cooking oils.

TRY IT Write down everything you eat and drink over a 24-hour period. Then, identify foods and food preparation methods that would be inappropriate if you were on a fat-controlled diet. What changes could you make (either food substitutions or adjustments in seasonings or preparation methods) to comply with a fat-controlled diet that you would find acceptable?

Causes of Bacterial Overgrowth Conditions that impair intestinal motility and allow material to stagnate can greatly increase susceptibility to bacterial overgrowth. For example, in some types of gastric surgery, a portion of the small intestine is bypassed, preventing the flow of material in the bypassed region and allowing bacteria to flourish (see the "blind loop" shown in Figure 23-3 on p. 716). Intestinal motility can also be reduced by strictures, obstructions, and diverticula (protrusions) in the small intestine, as well as by some chronic illnesses, including diabetes mellitus, scleroderma, and chronic kidney disease.[13]

Reduced secretions of gastric acid may also lead to bacterial overgrowth. Possible causes include atrophic gastritis, use of acid-suppressing medications, and some gastrectomy procedures.

Treatment for Bacterial Overgrowth Treatment may include antibiotics to suppress bacterial growth and surgical correction of the anatomical defects that contribute to a motility disorder. Medications may be given to stimulate peristalsis. A lactose-restricted diet may reduce flatulence and diarrhea in some individuals.[14] Dietary supplements can help to correct nutrient deficiencies, especially deficiencies of the fat-soluble vitamins A, D, and E; calcium (which combines with malabsorbed fatty acids); and vitamin B_{12}.[15]

IN SUMMARY **Malabsorption syndromes can be caused by an undersupply of digestive secretions, motility disorders, and damaged intestinal mucosa. Malabsorption often affects multiple nutrients and can lead to complications that impair nutrition status further. Fat malabsorption, usually indicated by the development of steatorrhea, is associated with the loss of food energy and deficiencies of essential fatty acids, fat-soluble vitamins, and some minerals. Bacterial overgrowth can result from conditions that reduce gastric acidity or intestinal motility; it typically causes malabsorption of fat and some essential nutrients.**

Conditions Affecting the Pancreas

As mentioned previously, pancreatic disorders can lead to maldigestion and malabsorption due to the impaired secretion of digestive enzymes. ♦ This section describes several pancreatic illnesses that are characterized by widespread malabsorption.

♦ Reminder: The digestive secretions of the pancreas include bicarbonate and digestive enzymes. The bicarbonate neutralizes the acidic gastric contents that enter the duodenum, and the digestive enzymes break down protein, carbohydrate, and fat.

Pancreatitis

Pancreatitis is an inflammatory disease of the pancreas. Although mild cases may subside in a few days, other cases can persist for weeks or months. Chronic pancreatitis can lead to irreversible damage to pancreatic tissue and permanent loss of function.

Acute Pancreatitis In acute pancreatitis, the digestive enzymes within pancreatic cells become prematurely activated, causing destruction of pancreatic tissue and subsequent inflammation. About 70 to 80 percent of acute cases are caused by gallstones or alcohol abuse; less frequent causes include elevated blood triglyceride levels (greater than 1000 milligrams per deciliter) or exposure to toxins.[16]

Common symptoms of acute pancreatitis include severe abdominal pain, nausea and vomiting, and abdominal distention. Elevated serum levels of amylase and lipase—released by damaged pancreatic tissue into the blood—help to confirm the diagnosis. In most patients, the condition resolves within a week with no complications. More severe cases may lead to chronic pancreatitis, infection, the systemic inflammatory response syndrome (see p. 686), or multiple organ failure.

Nutrition Therapy for Acute Pancreatitis The initial treatment for acute pancreatitis is supportive and includes pain control, intravenous hydration, and supplementary oxygen, if necessary. Oral fluids and food are withheld until the patient is pain free and experiences no nausea or vomiting.[17] Afterward, patients may consume a liquid diet or small low-fat meals, as tolerated (fat stimulates the pancreas more than other nutrients). In severe pancreatitis, tube feedings may be necessary; either standard formulas or elemental formulas ♦ may be used, depending on patient tolerance.[18] Protein and energy needs are high in severe cases due to the catabolic and hypermetabolic effects of inflammation. Patients with acute pancreatitis require nutrient supplementation until food intake can meet nutritional needs.[19]

♦ Reminder: An *elemental formula* contains hydrolyzed nutrients that require minimal digestion and are easily absorbed.

Chronic Pancreatitis Chronic pancreatitis is characterized by progressive, permanent damage to pancreatic tissue, resulting in the impaired secretion of digestive enzymes and bicarbonate. About 70 percent of cases are caused by excessive alcohol consumption.[20] Unlike acute pancreatitis, chronic pancreatitis is not caused by gallstones.

In chronic pancreatitis, abdominal pain is often severe and unrelenting and may worsen with eating. Analgesics or opiate drugs are often needed for pain control. Fat maldigestion develops sooner than maldigestion of protein or carbohydrate, and steatorrhea is common in advanced cases. Food avoidance (due to pain associated with eating) and malabsorption may lead to weight loss and malnutrition. Long-term illness is associated with reductions in both insulin and glucagon secretions, and diabetes eventually develops in 40 to 80 percent of patients.[21]

Nutrition Therapy for Chronic Pancreatitis The objectives of nutrition therapy are to correct malnutrition, reduce malabsorption, and prevent symptom recurrence. Dietary supplements are used to correct nutrient deficiencies, which may be due to malabsorption or to the alcohol abuse that caused the disease. To improve food tolerance, patients should consume small, low-fat meals. They should also avoid alcohol completely and quit smoking cigarettes, as these substances can exacerbate illness and interfere with healing.[22]

Steatorrhea is usually treated with pancreatic enzyme replacement. Pancreatic enzymes are often **enteric coated** to resist the acidity of the stomach and do not dissolve until the pH is above 5.5. If nonenteric-coated preparations are used, acid-suppressing drugs are also required. Fecal fat concentrations must be monitored to determine if the enzyme treatment has been effective. In some cases, a

enteric coated: refers to medications or enzyme preparations that are coated to withstand gastric acidity and dissolve only at the higher pH of the small intestine.

fat-controlled diet may help to reduce symptoms, and MCT oil can be used as an alternative source of fat kcalories.

Cystic Fibrosis

Cystic fibrosis is the most common life-threatening genetic disorder among Caucasians, with an incidence of approximately 1 in 2500 to 1 in 3200 white births.[23] In the United States, about 30,000 people are affected.[24] Cystic fibrosis is caused by a genetic mutation that disturbs the transport of chloride ions in **exocrine** glands. This defect results in thickened glandular secretions and a broad range of serious complications. Until a few decades ago, few infants born with cystic fibrosis survived to adulthood. Now, with early detection and advances in medical treatment, the average life span has extended beyond 36 years of age, with many patients surviving into their 50s.

Consequences of Cystic Fibrosis

Cystic fibrosis is characterized by abnormal chloride and sodium levels in exocrine secretions. The secretions are unusually viscous and can block the ducts that normally allow their passage. The blockages that develop can disrupt tissue function and cause tissue damage. The major complications of cystic fibrosis involve the lungs, pancreas, ♦ and sweat glands:

♦ Reminder: The pancreas secretes digestive enzymes and bicarbonate into the digestive tract (*exocrine* secretions) and the hormones insulin and glucagon into the bloodstream (*endocrine* secretions).

- *Lung disease.* The abnormally thick mucus secretions cause obstructions in many of the small airways of the lungs. The obstructions lead to chronic coughing and persistent respiratory infections, both of which contribute to progressive inflammation in the bronchial tissues. The eventual lung damage results in breathing difficulties and lower exercise tolerance. As with other obstructive airway diseases, ♦ nutrition status may become impaired due to hypermetabolism, the greater energy cost of labored breathing, and anorexia (loss of appetite). The chronic respiratory infections raise energy needs further.

♦ Chapter 22 describes the nutrition problems associated with chronic obstructive lung diseases.

- *Pancreatic disease.* Approximately 85 percent of cystic fibrosis patients develop thickened pancreatic secretions that obstruct the pancreatic ducts, leading to progressive damage and scarring within pancreatic tissue.[25] Fewer pancreatic enzymes reach the small intestine, causing malabsorption of protein, fat, and fat-soluble vitamins. Other problems that may develop over time include pancreatitis and glucose intolerance or diabetes (due to destruction of the insulin-producing cells).
- *Other complications.* Because cystic fibrosis affects all exocrine secretions, complications typically develop in many other tissues or organs. Salt losses in sweat are usually excessive, increasing the risk of dehydration. Intestinal obstruction is a common symptom in newborn infants and may also occur in older patients. Gallbladder and liver diseases may result from bile duct obstructions. Abnormalities in genital tissues cause sterility in men and reduced fertility in women.

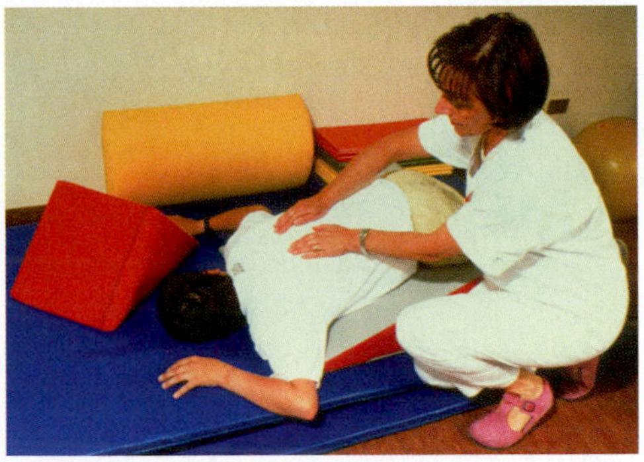

Postural drainage, a type of physical therapy used in treating cystic fibrosis, helps to clear the thick, sticky secretions that block airways and increase infection risk.

Nutrition Therapy for Cystic Fibrosis

Children with cystic fibrosis are chronically undernourished, grow poorly, and have difficulty maintaining normal body weight. Their energy and protein requirements are high due to increased requirements, nutrient malabsorption, and reduced food consumption. To achieve normal growth and appropriate weight, energy and protein needs may range from 110 to 200 percent of DRI values.[26] To compensate for fat malabsorption, about 35 to 40 percent of kcalories should come from fat. Patients with cystic fibrosis are typically encouraged to eat high-kcalorie and high-fat foods, eat frequent meals and snacks, and supplement meals with milk shakes or liquid dietary supplements. Supplemental tube feedings can help to improve nutrition status if energy intakes are inadequate.

Pancreatic enzyme replacement therapy is a central feature of cystic fibrosis treatment. Supplemental enzymes must be included with every meal or snack. For infants and small children, the contents of capsules are mixed in small amounts of liquid or a soft food (such as applesauce) and fed with a spoon. Enzyme dosages

cystic fibrosis: an inherited disease characterized by the production of abnormally viscous exocrine secretions; often leads to respiratory illness and pancreatic insufficiency.

exocrine: pertains to external secretions, such as those of the mucous membranes or the skin. Opposite of *endocrine*, which pertains to hormonal secretions into the blood.
- **exo** = outside
- **krinein** = to secrete

CASE STUDY: Child with Cystic Fibrosis

Julie is a 7-year-old girl diagnosed with cystic fibrosis. Symptoms of steatorrhea and poor growth during infancy prompted the tests that led to the diagnosis. She is currently 45 inches tall and weighs 42 pounds. Her height for age and weight for age fall near the 10th percentile (see Appendix E). Julie eats regular foods during the day and receives additional nutrients by tube feedings delivered overnight.

1. What do the height and weight percentiles tell you about Julie's nutrition status? Why is growth failure common in children with cystic fibrosis?
2. Explain why Julie's energy needs are so much higher than normal. Describe the elements of the diet that Julie should follow to improve growth. Explain why her energy requirements may change if she develops a respiratory infection.
3. Explain to Julie's parents how to use enzyme replacement therapy effectively.
4. Julie's parents are hoping to discontinue the nightly tube feedings. Do you think the tube feedings are necessary? Why or why not?

may need to be adjusted if malabsorption continues, as evidenced by poor growth or GI symptoms such as steatorrhea, intestinal gas, or abdominal pain.

The risk of nutrient deficiency depends on the degree of malabsorption. The nutrients of greatest concern include the fat-soluble vitamins, essential fatty acids, and calcium. Multivitamin and fat-soluble vitamin supplements are routinely recommended. The liberal use of table salt and salty foods is encouraged to make up for losses of sodium in sweat. The accompanying Case Study checks your understanding of the nutrition therapy for a child with cystic fibrosis.

IN SUMMARY **Chronic pancreatic disorders can lead to widespread maldigestion and malabsorption due to the impaired secretion of digestive enzymes. Acute pancreatitis is short lived and does not cause permanent damage, but it requires the withholding of food and liquids until healing has occurred. Chronic pancreatitis leads to digestive enzyme deficiencies and is treated with pancreatic enzyme replacement therapy. Cystic fibrosis, a genetic disorder associated with thickened exocrine secretions, causes obstructive lung disease and pancreatic damage. Children with cystic fibrosis have high protein and energy requirements and must use pancreatic enzyme replacement therapy and dietary supplements to reverse malnutrition.**

Conditions Affecting the Small Intestine

When the intestinal mucosa is damaged due to inflammation, infection, or other causes, malabsorption is the likely outcome. *Celiac disease* and *inflammatory bowel diseases* are intestinal conditions that can impair mucosal function, whereas *short bowel syndrome* is the malabsorption disorder that results when a significant portion of the small intestine is surgically removed.

Celiac Disease

Celiac disease is an immune disorder characterized by an abnormal immune response to a protein fraction in **wheat gluten** ♦ and to related proteins in barley and rye. The reaction to gluten causes severe damage to the intestinal mucosa and subsequent malabsorption. Celiac disease may affect approximately 1 percent of individuals in the United States.[27]

celiac (SEE-lee-ack) **disease:** immune disorder characterized by an abnormal immune response to the dietary protein gluten; also called *gluten-sensitive enteropathy* or *celiac sprue*.

wheat gluten (GLU-ten): a family of water-insoluble proteins in wheat; includes the gliadin (GLY-ah-din) fractions that are toxic to persons with celiac disease.

♦ The protein in wheat gluten that has toxic effects in celiac disease is called *gliadin*.

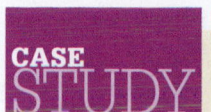

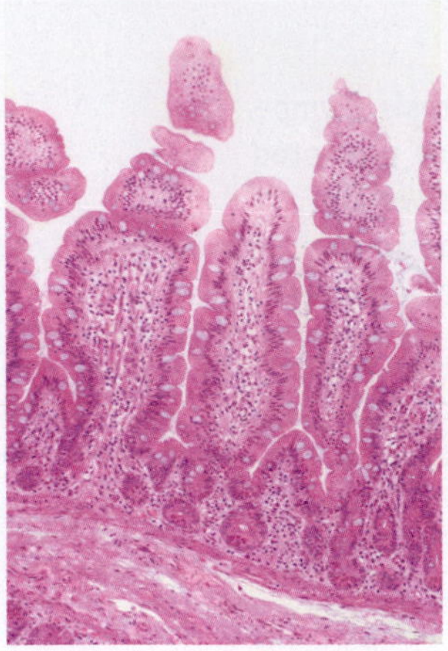

In the healthy intestine, the villi greatly increase the absorptive surface area.

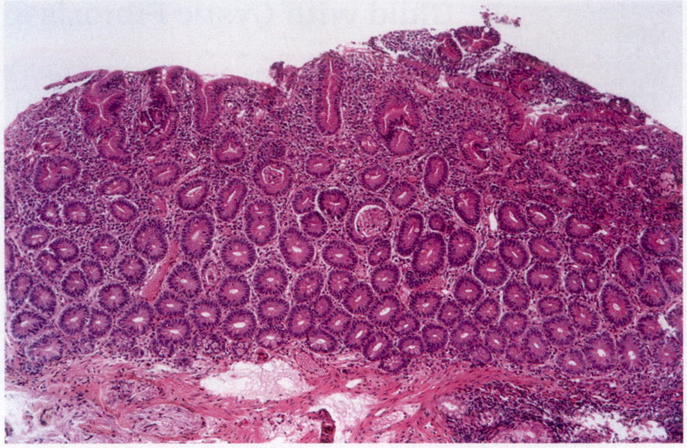

In celiac disease, the villi may be shortened or absent, resulting in substantial reductions in nutrient absorption.

Consequences of Celiac Disease The immune reaction to gluten can cause striking changes in intestinal tissue. In affected areas, the absorptive surface appears flattened due to the shortening or absence of villi and overdeveloped crypts (see the photos). ♦ The reduction in mucosal surface area (and, therefore, in intestinal digestive enzymes) can be substantial. The damage may be restricted to the duodenum or may involve the full length of the small intestine. Individuals with severe disease may malabsorb all nutrients to some degree, especially the macronutrients, calcium, iron, folate, the fat-soluble vitamins, and vitamin B_{12}.[28]

♦ Reminder: *Crypts* are tubular glands that lie between the intestinal villi and secrete intestinal juices into the small intestine.

Symptoms of celiac disease include GI disturbances such as diarrhea, steatorrhea, and flatulence. Because lactase deficiency can result from mucosal damage, milk products may exacerbate GI symptoms. Due to nutrient malabsorption, children with celiac disease often exhibit poor growth, low body weights, muscle wasting, and anemia. Adults may develop anemia, bone disorders, neurological symptoms, and fertility problems.[29] Individuals with celiac disease who do not eliminate gluten from the diet are at increased risk of developing intestinal and lymphatic cancers.[30]

Some gluten-sensitive individuals may have few GI symptoms but react to gluten by developing a severe rash. This condition is called **dermatitis herpetiformis** and requires dietary adjustments similar to those for celiac disease.

Nutrition Therapy for Celiac Disease The treatment for celiac disease is lifelong adherence to a gluten-free diet. Improvement in symptoms is often evident within several weeks, although mucosal healing can sometimes take years. If lactase deficiency is suspected, patients should avoid lactose-containing foods until the intestine has recovered. Dietary supplements can be used to meet micronutrient needs and reverse deficiencies.

Gluten-free products help people with celiac disease enjoy a wider variety of foods.

The gluten-free diet eliminates foods that contain wheat, barley, and rye (see Table 24-6). Because many foods contain ingredients derived from these grains, foods that are problematic are not always obvious. Even small amounts of gluten may cause symptoms in some people, so patients need to check ingredient lists on food labels carefully. Gluten-containing products that may be overlooked include beer, caramel coloring, coffee substitutes, communion wafers, imitation meats, malt syrup, medications, salad dressings, and soy sauce. Special gluten-free products can be purchased to replace common food items such as bread, pasta, and cereals. Although somewhat expensive, these foods increase food choices and allow celiac patients to enjoy foods that would otherwise be forbidden. Patients should also be instructed in food preparation methods that prevent cross-contamination from utensils, cutting boards, and toasters.

dermatitis herpetiformis (DERM-ah-TYE-tis HER-peh-tih-FOR-mis): a gluten-sensitive disorder characterized by a severe skin rash.

Although most people with celiac disease can safely consume moderate amounts of oats, most oats grown in the United States are contaminated with wheat, bar-

TABLE 24-6 Gluten-Free Diet

Food Category	Gluten-Free Choices	Potential Gluten Sources
Meat and meat alternates	Fresh meat, fish, or poultry; shellfish; dried peas and beans; tofu; nuts and seeds; eggs	Luncheon meats, sandwich spreads, meatloaf, meatballs, frankfurters, sausages, poultry injected with broth, imitation meat products, imitation seafood, meat extenders, miso, egg substitutes, dried egg products, dry roasted nuts, peanut butter. *Avoid*: products made with hydrolyzed vegetable protein (HVP), marinades, and soy sauce; breaded foods; foods prepared with cream sauces or gravies.
Milk and milk products	Milk, buttermilk, half-and-half, cream, plain yogurt, cheese, cottage cheese, cream cheese	Chocolate milk, milk shakes, frozen yogurt, flavored yogurt, cheese spreads, cheese sauces. *Avoid*: malted milk, malted milk powders.
Breads, cereals, rice, and pasta	Breads, bakery products, and cereals made with amaranth, arrowroot, buckwheat, corn, flax, hominy grits, millet, potato flour or potato starch, quinoa, rice, sorghum, soybean flour, tapioca, and teff; pasta and noodles made with the grains or starches listed above; corn tacos and corn tortillas	Oatmeal and oat bran (due to contamination), rice crackers, rice cakes, corn cakes. *Avoid*: breads, bakery products, cereals, tortillas, pastas, and pancake or baking mixes made with wheat, rye, barley, and triticale. *Wheat products* include bulghur, cous cous, durum flour, einkorn, emmer, farina, graham flour, kamut, semolina, spelt, wheat bran, wheat germ. *Barley products* include malt, malt flavoring, and malt extract.
Fruits and vegetables	Any fresh, frozen, or canned fruits or vegetables	French fries from fast-food restaurants, commercial salad dressings, fruit pie fillings, dried fruits (may be dusted with flour). *Avoid*: scalloped potatoes (usually made with wheat flour), creamed vegetables, vegetables dipped in batters.
Desserts	Bakery products made with gluten-free flours, most ice creams, sherbet, sorbet, Italian ices, popsicles, gelatin desserts, egg custards, most chocolate bars, chocolate chips, hard candies, whipped toppings	Some ice creams (especially if made with cookie dough, brownies, nuts, and other added ingredients), icing or frosting, candies and candy bars, marshmallows. *Avoid*: bakery products or doughnuts made with wheat, rye, or barley; puddings made with wheat flour; ice cream or sherbets that contain gluten stabilizers; ice cream cones; licorice.
Beverages	Coffee; tea; cocoa made with pure cocoa powder; soft drinks; wine; distilled alcoholic beverages such as rum, gin, whiskey, and vodka	Instant tea or coffee, coffee substitutes, chocolate drinks, hot cocoa mixes. *Avoid*: beer, ale, lager, malted beverages, cereal beverages, beverages that contain nondairy cream substitutes.

ley, or rye. Oats are usually grown in rotation with other grains and may become contaminated during harvesting or processing. However, several oat manufacturers in the United States produce oats in dedicated facilities and test the products to ensure that they are gluten free.[31] Individuals who wish to include oats in their diet should be advised to purchase only uncontaminated oats and to limit intakes to the amounts found to be safe (about 1/2 cup of dry rolled oats per day).

A gluten-free diet may become monotonous unless care is taken to diversify food choices. The diet can also be a social liability by restricting food choices when individuals eat in restaurants, visit friends, or travel. Nonadherence is common when individuals are away from home.[32] Nutrition education can help celiac patients learn how to meet their nutrient needs and expand meal options despite dietary constraints. Figure 24-3 shows an example of a menu for a gluten-free diet.

Inflammatory Bowel Diseases

Inflammatory bowel diseases are chronic inflammatory illnesses characterized by abnormal immune responses in the GI tract.[33] Both genetic and environmental factors are believed to contribute to the development of these diseases, although the exact triggers are unknown. Table 24-7 compares the two major forms of inflammatory bowel disease, **Crohn's disease** and **ulcerative colitis**. Crohn's disease usually involves the small intestine and may lead to nutrient malabsorption, whereas ulcerative colitis affects the large intestine, where little nutrient absorption occurs. ♦ Both diseases are characterized by periods of active disease interspersed with periods of remission. Nutrient losses can result from tissue damage, bleeding, and diarrhea.

Crohn's disease: an inflammatory bowel disease that usually occurs in the lower portion of the small intestine and the colon. Inflammation may pervade the entire intestinal wall.

ulcerative colitis (ko-LY-tis): an inflammatory bowel disease that involves the colon. Inflammation affects the mucosa and submucosa of the intestinal wall.

♦ Although ulcerative colitis affects the large intestine, the condition is included in this section because it is one of the major subtypes of inflammatory bowel disease.

FIGURE 24-3 Sample Menu—Gluten-Free Diet

SAMPLE MENU	
Breakfast	Orange juice
	Gluten-free pancake with maple syrup
	Plain yogurt with banana and strawberries
	Coffee with half-and-half
Lunch	Grilled chicken breast with cranberry chutney
	Baked potato topped with grated cheddar cheese
	Sliced tomato with chopped basil
	Raspberry sherbet
Snack	Tortilla chips and guacamole
	Hot cocoa (made with cocoa powder)
Dinner	Sauteed catfish with sliced lemon and dill
	Wild rice pilaf
	Collard greens and garlic sauteed in olive oil
	Green salad with oil and vinegar dressing
	Vanilla egg custard

Complications of Crohn's Disease Crohn's disease may occur in any region of the GI tract, but most cases involve the ileum and/or large intestine. Lesions may develop in different areas in the intestine, with normal tissue separating affected regions (called "skip" lesions). During exacerbations, the inflammation may extend deeply into intestinal tissue and be accompanied by ulcerations, fissures, and fistulas. ♦ Loops of intestine may become matted together. Scar tissue eventually thickens and stiffens the intestinal wall, narrowing the lumen and possibly causing strictures or obstructions. About three fourths of patients require surgery within 20 years of diagnosis.[34] Patients with Crohn's disease are also at increased risk of developing intestinal cancers.

♦ Reminder: *Fistulas* are abnormal passages between organs or tissues that allow the passage of fluids or secretions.

Malnutrition may result from malabsorption, nutrient losses (especially of protein) associated with the tissue damage, reduced food intake, and surgical resections that shorten the small intestine. If the ileum is affected, bile acids may become

TABLE 24-7 Comparison of Crohn's Disease and Ulcerative Colitis

	Crohn's Disease	Ulcerative Colitis
Location of inflammation	Approximately 40% of cases involve the ileum and cecum, 30% are in the small intestine only, and 20% are in the colon	Inflammation is confined to the rectum and colon; it begins at the rectum and spreads into the colon
Pattern of inflammation	Discrete areas separated by normal tissue ("skip" lesions)	Continuous inflammation that begins at the rectum and ends abruptly within the colon
Depth of damage	Damage throughout all layers of tissue; causes deep fissures that give intestinal tissue a "cobblestone" appearance	Damage primarily in the mucosa and submucosa (layers of intestinal tissue closest to the lumen)
Fistulas	Common	Usually do not occur
Cancer risk	Increased	Greatly increased

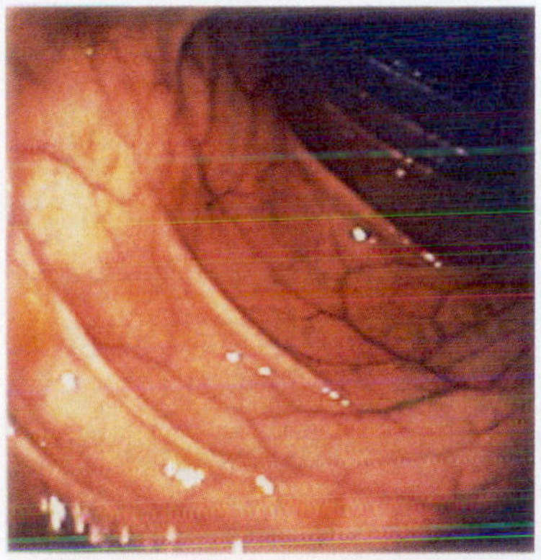

The healthy colon has a smooth surface with a visible pattern of fine blood vessels.

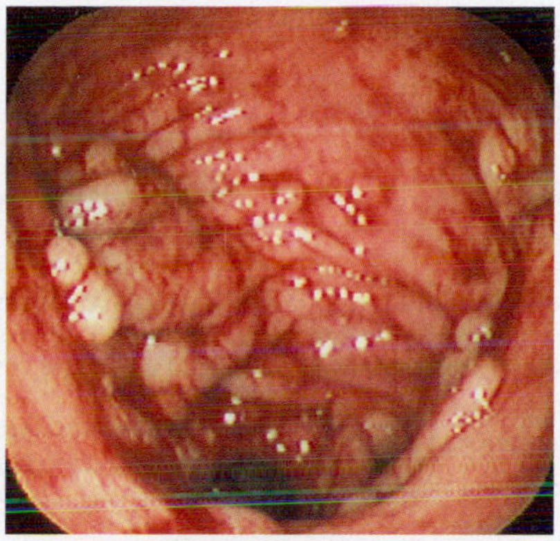

In Crohn's disease, the mucosa has a "cobblestone" appearance due to deep fissuring in the inflamed mucosal tissue.

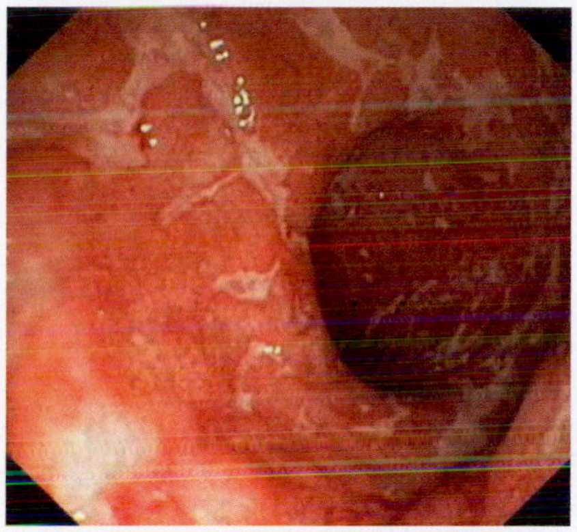

In ulcerative colitis, the colon appears inflamed and reddened, and ulcers are visible.

depleted, ♦ causing malabsorption of fat, fat-soluble vitamins, calcium, magnesium, and zinc (the minerals bind to the unabsorbed fatty acids). Because the ileum is the site of vitamin B_{12} absorption, deficiency can develop unless the patient is given vitamin B_{12} injections. Anemia may result from bleeding, inadequate absorption of the nutrients involved in blood cell formation, or the metabolic effects of chronic illness (see Highlight 25). Anorexia often develops due to abdominal discomfort and the effects of cytokines produced during the inflammatory process.[35]

♦ Reminder: Most of the bile used during digestion is eventually reabsorbed in the ileum and returned to the liver.

Complications of Ulcerative Colitis Ulcerative colitis always involves the rectum and usually extends into the colon. Inflammation is continuous along the length of intestine affected, ending abruptly at the area where healthy tissue begins. Tissue erosion or ulceration develops primarily in the mucosa and submucosa (the layers of intestinal tissue closest to the lumen). During active episodes, patients have frequent, urgent bowel movements that are small in volume. Stools are often streaked with blood and contain mucus.

Although mild disease may cause few complications, weight loss, fever, and weakness are common when most of the colon is involved. Severe disease is often associated with anemia (due to blood loss), dehydration, and electrolyte imbalances. Protein losses from the inflamed tissue can be substantial. A **colectomy** (removal of the colon) is performed in 20 to 25 percent of patients and prevents future recurrence.[36] Colon cancer risk is substantially increased in ulcerative colitis patients.

Drug Treatment of Inflammatory Bowel Diseases Medications help to control symptoms, reduce inflammation, and minimize complications. The drugs prescribed include antidiarrheal agents, immunosuppressants, anti-inflammatory drugs (usually corticosteroids and salicylates), and antibiotics. Although these medications may allow the patient to achieve and maintain remission, some may cause side effects that are detrimental to nutrition status. The Diet-Drug Interactions feature lists some nutrition-related effects of the medications used in inflammatory bowel diseases.

Nutrition Therapy for Crohn's Disease Crohn's disease often requires aggressive dietary management because it can lead to protein-energy malnutrition (PEM), nutrient deficiencies, and growth failure in children. Specific dietary measures depend on the functional status of the GI tract and the symptoms and complications that develop; thus, nutrition care varies among patients and throughout the course of illness.

During disease exacerbations, a low-fiber, low-fat diet provided in small, frequent feedings can minimize stool output and reduce symptoms of malabsorption. High-kcalorie, high-protein diets may be prescribed to prevent or treat malnutrition or promote healing; protein needs may be 50 percent higher than DRI levels. Liquid supplements may help to increase energy intake and improve weight gain. Vitamin

colectomy: removal of a portion or all of the colon.

DIET-DRUG Interactions

Check this table for notable nutrition-related effects of the medications discussed in this chapter.

Antidiarrheal drugs	**Gastrointestinal effect:** Constipation
Anti-inflammatory drugs (sulfasalazine, corticosteroids)	**Gastrointestinal effects:** Nausea, heartburn (sulfasalazine) **Dietary interactions:** Sulfasalazine may decrease folate absorption; supplementation is recommended **Metabolic effects:** Anemia (sulfasalazine); fluid retention, hyperglycemia, hypocalcemia, hypokalemia, hypophosphatemia, increased appetite, protein catabolism (corticosteroids)
Antisecretory drugs (proton-pump inhibitors, H2 blockers)	**Gastrointestinal effects:** Nausea and vomiting, diarrhea, constipation, abdominal pain (proton-pump inhibitors) **Dietary interactions:** May decrease iron, folate, and vitamin B_{12} absorption
Laxatives	**Gastrointestinal effects:** Diarrhea, flatulence, abdominal discomfort **Metabolic effects:** Dehydration, electrolyte imbalances, laxative dependency
Pancreatic enzyme replacements	**Gastrointestinal effects:** Nausea and vomiting, stomach cramping, diarrhea, constipation, irritation of GI mucosa **Metabolic effects:** Elevated serum or urinary uric acid levels (with high doses), allergic reactions (rare)

and mineral supplements are usually necessary, especially if nutrient malabsorption is present; nutrients at risk include calcium, iron, magnesium, zinc, folate, vitamin B_{12}, and vitamin D.[37] In some instances, tube feedings are used to supplement the diet or may be the sole means of providing nutrients; some patients may tolerate elemental formulas more easily than standard formulas. Table 24-8 includes examples of other dietary adjustments that may be beneficial for patients with Crohn's disease.

During periods of remission, dietary restrictions are unnecessary unless complications develop. Symptoms that may interfere with food intake include anorexia, pain, and diarrhea. A restricted intake of lactose, fructose, and sorbitol may improve symptoms of diarrhea or flatulence. Adequate fluid replacement should be encouraged in patients with diarrhea. Individuals with partial obstructions may need to restrict high-fiber foods. Although research studies suggest that supplementation with omega-3 fatty acids, glutamine, or probiotics may be helpful, more research is necessary to confirm these benefits.[38]

Nutrition Therapy for Ulcerative Colitis In most cases, the diet for ulcerative colitis requires few adjustments. As in Crohn's disease, the symptoms and complications that arise are managed with specific dietary measures (see Table 24-8). During disease exacerbations, emphasis is given to restoring fluid and electrolyte balances and correcting deficiencies that result from protein and blood losses; dietary adjustments are based on the extent of bleeding and diarrhea output. Thus, adequate protein, energy, fluid, and electrolytes need to be provided. A low-fiber diet may reduce irritation by minimizing fecal volume. If colon function becomes severely impaired, food and fluids may be withheld and fluids and electrolytes supplied intravenously until colon function is restored.[39]

Short Bowel Syndrome

The treatment of Crohn's disease, cancers of the small intestine, and other intestinal disorders may include the surgical resection

TABLE 24-8 Management of Symptoms and Complications in Crohn's Disease

Symptom or Complication	Possible Dietary Measures
Growth failure, weight loss, or muscle wasting	• High-kcalorie, high-protein diet • Liquid supplements • Elemental tube feedings
Anorexia or pain with eating	• Small, frequent meals • Liquid supplements • If long term (>5 to 7 days): elemental tube feedings
Malabsorption	• High-kcalorie diet • Nutrient supplementation
Steatorrhea (fat malabsorption)	• Fat restriction • Medium-chain triglycerides • Nutrient supplementation
Diarrhea	• Fluid and electrolyte replacement • Nutrient supplementation
Lactose intolerance	• Avoidance of lactose-containing foods
Nutrient deficiencies	• Nutrient-dense diet • Nutrient supplementation
Strictures or fistulas	• Low-fiber diet
Severe bowel obstruction, high-output fistulas, or severe exacerbations of disease	• Total parenteral nutrition

of a major portion of the small intestine. **Short bowel syndrome** is the malabsorption syndrome that results when the absorptive capacity of the remaining intestine is insufficient for meeting nutritional needs. Without appropriate dietary adjustments, short bowel syndrome can result in fluid and electrolyte imbalances and multiple nutrient deficiencies. Symptoms include diarrhea, steatorrhea, dehydration, weight loss, and growth impairment in children.

Consequences of Short Bowel Syndrome Figure 24-4 reviews nutrient absorption in the GI tract and describes how absorption is affected by surgical resections. Generally, up to 50 percent of the small intestine can be resected without serious nutritional consequences.[40] More extensive resections lead to generalized malabsorption, and patients may need lifelong parenteral nutrition to supplement oral intakes. Other problems that may develop include kidney stones (due to the effects of fat malabsorption on urinary oxalate levels; review p. 734) and gallstones (due to bile malabsorption and the subsequent imbalance between bile acid and cholesterol concentrations in bile; see Chapter 25). Furthermore, loss of the ileocecal valve (between the ileum and cecum) increases the likelihood that colonic bacteria will infiltrate the small intestine and cause bacterial overgrowth.[41]

Intestinal Adaptation After an intestinal resection, the remaining intestine undergoes **intestinal adaptation**, an adaptive response that dramatically improves the intestine's absorptive capacity. Adaptation depends on the presence of nutrients and GI secretions in the lumen, and therefore oral intakes are begun as soon as possible after surgery to stimulate the growth of intestinal tissue. Many patients can eventually return to a normal diet if adaptation compensates sufficiently for the removed length of intestine.

Adaptation begins soon after surgery and continues for several years. During this period, the remaining section of intestine develops taller villi and deeper crypts and also grows in length and diameter; these changes dramatically increase the absorptive surface area of the remaining intestine. The ileum has a greater capacity for adaptation than the jejunum; thus, removal of the ileum has more severe consequences than removal of the jejunum. Loss of the ileum permanently disrupts both

short bowel syndrome: the malabsorption syndrome that follows resection of the small intestine, resulting in inadequate absorptive capacity in the remaining intestine.

intestinal adaptation: the process of intestinal recovery following resection that leads to improved absorptive capacity.

FIGURE 24-4 Nutrient Absorption and Consequences of Intestinal Surgeries

About 90 to 95 percent of nutrient absorption takes place in the first half of the small intestine. After a resection, nutrient absorption may be reduced.

WHAT IS ABSORBED

Duodenum/jejunum
- Simple carbohydrates
- Fats
- Amino acids
- Vitamins[a]
- Minerals[a]
- Water

Ileum
- Bile salts
- Vitamin B_{12}
- Water

(Assumes absorptive function of duodenum and jejunum with adaptation)

Colon
- Water
- Electrolytes
- Short-chain fatty acids

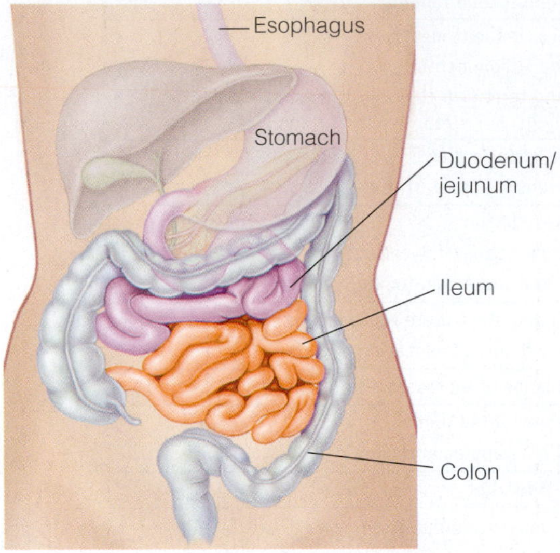

POSSIBLE CONSEQUENCES OF RESECTION

Duodenum/jejunum
- Minimal consequences if the ileum remains intact
- Calcium and iron malabsorption if duodenum resected

Ileum
- Fat malabsorption
- Protein malabsorption
- Malabsorption of fat-soluble vitamins and vitamin B_{12}
- Reduced calcium, magnesium, and zinc absorption
- Fluid losses
- Diarrhea/steatorrhea

Colon
- Fluid and electrolyte losses
- Diarrhea

(Losses are compounded if ileum is also resected)

[a]The absorption of vitamins and minerals begins in the duodenum and continues throughout the length of the small intestine.

vitamin B_{12} and bile acid absorption. Depletion of bile acids exacerbates fat malabsorption, and the unabsorbed bile acids irritate the colon walls and can worsen diarrhea. Adaptation is achieved more easily if the colon is present, because the colon's resident bacteria can metabolize unabsorbed carbohydrates and produce some usable nutrients (such as short-chain fatty acids). An intact colon also helps to reduce losses of fluids and electrolytes.

Nutrition Therapy for Short Bowel Syndrome Immediately after a resection, fluids and electrolytes must be supplied intravenously. In the first few weeks after surgery, the fluid losses from diarrhea can be substantial, so appropriate rehydration therapy is critical to recovery. Antidiarrheal medications may be needed to limit losses. The diarrhea gradually lessens as intestinal adaptation progresses.

Total parenteral nutrition meets nutrient needs after surgery and is gradually reduced as oral feedings increase. To promote intestinal adaptation, oral feedings may be started within a week after surgery, after diarrhea subsides somewhat and some bowel function is restored. Initial oral intake may consist of occasional sips of clear, sugar-free liquids, progressing to larger amounts of liquid formulas and then to solid foods, as tolerated. Very small, frequent feedings can utilize the remaining intestine most efficiently. Parenteral nutrition can be tapered and eventually discontinued once oral intakes supply adequate nourishment. Some patients require tube feedings in addition to oral feedings to successfully meet nutrient needs.

The exact diet prescribed for short bowel syndrome depends on the portion of intestine removed, the length of remaining intestine, and whether the colon is still intact; moreover, dietary readjustments may be required as intestinal adaptation progresses.[42] A high-kcalorie diet is typically recommended to compensate for malabsorption; thus, a high-fat, low-carbohydrate diet may be recommended if the fat is sufficiently tolerated. Conversely, a high-complex-carbohydrate, low-fat diet may be suggested for patients who have an intact colon, because the colon bacteria can metabolize the unabsorbed carbohydrate and produce short-chain fatty acids (contributing an additional 310 to 1170 kcalories daily[43]), and the low fat intake can

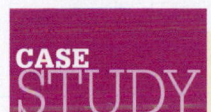

CASE STUDY: Patient with Short Bowel Syndrome

Judi Morel is a 28-year-old economist with an eight-year history of Crohn's disease. Judi is 5 feet 7 inches tall. Three years ago, she underwent a small bowel resection and remained free of active disease for two years. During that time, her symptoms subsided; she was able to tolerate most foods without any problem and gained weight. Ten months ago, Judi experienced a severe flare-up of her Crohn's disease. Since that time, she has lost 15 pounds and currently weighs 118 pounds. She has experienced severe abdominal pain and fatigue that have persisted despite aggressive medical management that included intravenous nutrition. Five days ago, Judi underwent another resection, which left her with 40 percent of healthy small intestine. Her colon is intact. She is experiencing extensive diarrhea.

1. Describe the manifestations of Crohn's disease, and explain why surgery is sometimes performed as part of the treatment. Describe the complications of disease that may affect nutrient needs.
2. Using the BMI table in the back of the book, check the ideal weight range for a person of Judi's height. What nutrition-related concerns are suggested by Judi's recent weight loss? What other nutrition problems did Judi probably experience as a consequence of Crohn's disease?
3. Discuss the complications that may follow an extensive intestinal resection. What factors may affect a person's ability to meet nutrient needs with an oral diet?
4. Discuss the dietary progression that is recommended following an intestinal resection. After Judi is able to eat solid foods, what factors may affect the type of diet that is recommended for her?

improve steatorrhea. In some cases, medium-chain triglycerides may be added as an energy source.

Dietary choices are tailored according to individual symptoms and tolerances. Patients should avoid concentrated sweets, which draw fluid into the intestines, if they cause additional diarrhea. Some patients may be lactose intolerant, but they can minimize symptoms by consuming limited amounts of milk at one time. Due to the high risk of developing kidney stones as a result of calcium malabsorption, a low-oxalate diet may be recommended (see Chapter 28).

Vitamin and mineral supplements can prevent deficiencies from developing due to malabsorption. If fat is malabsorbed, patients may need supplements of fat-soluble vitamins, calcium, magnesium, and zinc. If a large portion of the ileum has been removed, vitamin B_{12} must be injected. Iron absorption is likely to be compromised only if upper portions of the small intestine have been removed.

Individuals with fewer than 100 centimeters (about 40 inches) of jejunum and no colon usually require intravenous fluids and parenteral nutrition to survive.[44] Intestinal transplantation is an option for patients who cannot continue parenteral nutrition because of life-threatening complications. The accompanying Case Study can help you review the material on short bowel syndrome.

IN SUMMARY **Disorders of the small intestine that cause damage to mucosal tissue, such as celiac disease and Crohn's disease, often result in malabsorption. A gluten-free diet can alleviate the symptoms of celiac disease. Treatment of Crohn's disease may include medications to suppress inflammation and relieve symptoms, dietary adjustments that reduce symptoms and correct deficiencies, and intestinal resection to remove damaged tissue. Ulcerative colitis is an inflammatory condition that affects the rectum and colon only; severe cases may require colectomy. Short bowel syndrome is a possible consequence of intestinal resection; intestinal adaptation may improve absorptive capacity over time.**

Conditions Affecting the Large Intestine

In the large intestine, the colon moves undigested materials to the rectum and has a central role in maintaining fluid and electrolyte balances. Its bacterial population ferments undigested nutrients and produces short-chain fatty acids and some vitamins that our bodies can absorb and use. This section describes several conditions that may upset the normal functioning of the large intestine.

♦ Irritable bowel syndrome was introduced in Highlight 3.

Irritable Bowel Syndrome

People with **irritable bowel syndrome** ♦ experience chronic and recurring intestinal symptoms that cannot be explained by specific physical abnormalities. The symptoms usually include disturbed defecation (diarrhea and/or constipation), flatulence, and abdominal discomfort or pain; the pain is often aggravated by eating and relieved by defecation. In some patients, symptoms are mild; in others, the disturbances in colonic function can interfere with work and social activities enough to dramatically alter the person's lifestyle and sense of well-being.[45]

Although the causes of irritable bowel syndrome remain elusive, people with the disorder tend to have excessive colonic responses to meals, GI hormones, and stress. Many individuals exhibit hypersensitivity to a normal degree of intestinal distention and feel discomfort when experiencing normal meal transit or typical amounts of intestinal gas. Intestinal motility after meals may be excessive, leading to diarrhea, or be reduced, causing constipation. Some patients may have had a bacterial infection that initiated their GI problems; other show signs of low-grade intestinal inflammation of uncertain cause.[46] Symptoms often worsen during periods of psychological stress.

Diagnosing irritable bowel syndrome is often difficult because its symptoms are typical of other GI disorders and laboratory tests for the condition are nonexistent. Up to 5 percent of patients with symptoms of irritable bowel syndrome are found to have celiac disease.[47] Other GI disturbances that may cause similar symptoms include inflammatory bowel diseases, lactose intolerance, hypothyroidism or hyperthyroidism, GI infections, and intestinal cancers. Between 40 and 94 percent of patients diagnosed with irritable bowel syndrome have coexisting psychiatric illnesses, such as anxiety and depression, which can exacerbate symptoms.[48]

Treatment of Irritable Bowel Syndrome

Medical treatment of irritable bowel syndrome often includes dietary adjustments, stress management, and behavioral therapies. Medications may be prescribed to manage symptoms, although they are not always helpful. The drugs prescribed may include antidiarrheal drugs, anticholinergics (which affect GI motility), antidepressants, and laxatives.

Nutrition Therapy for Irritable Bowel Syndrome

Although dietary adjustments may reduce symptoms, responses among patients can vary considerably. The most common recommendation is to gradually increase fiber intake from food or supplements to relieve constipation and improve stool bulk. However, clinical studies suggest that additional fiber has only marginal effectiveness in improving symptoms and may worsen flatulence.[49] Psyllium supplementation may be helpful for individuals with constipation. Some individuals have fewer symptoms when they consume small, frequent meals instead of larger ones. Foods that aggravate symptoms may include fried or fatty foods, gas-producing foods (review Table 24-2), milk products (not necessarily due to lactose intolerance), wheat products, coffee (with or without caffeine), and alcoholic beverages; however, individual tolerances are best determined by trial and error. Psychological associations have a strong influence on food tolerance, so foods that patients perceive to be problematic should be discussed so that the diet is not restricted unnecessarily. A careful evaluation of the dietary patterns that exacerbate symptoms may uncover the foods and habits most closely associated with intestinal discomfort.

irritable bowel syndrome: an intestinal disorder of unknown cause that disturbs the functioning of the lower bowel; symptoms include abdominal pain, flatulence, diarrhea, and constipation.

CASE STUDY Young Adult with Irritable Bowel Syndrome

Hannah Tran is a 22-year-old recent college graduate who began her first professional job in a bank one month ago. As a college student, she occasionally experienced abdominal pain and cramping after eating. She also had frequent bouts of diarrhea and felt somewhat better after bowel movements. Once Hannah began her new job, her symptoms occurred more frequently. At first she attributed her symptoms to job stress, but when the symptoms continued for several months, she decided to see her physician. After taking a careful history and conducting tests to rule out other bowel disorders, the physician diagnosed irritable bowel syndrome. The physician prescribed bulk-forming fibers and advised Hannah to keep a record of her food intake and symptoms for one week. Hannah was then referred to a dietitian for a review of her dietary record. The dietitian noticed that Hannah routinely drank several cups of coffee in the morning and had large meals for lunch and dinner. Hannah often ate out in Mexican restaurants and favored spicy foods, refried beans, and fatty desserts. Between meals, she snacked on low-carbohydrate foods sweetened with sugar alcohols and drank several cans of soda daily. Her dietary fiber intake, however, totaled only about 13 grams daily.

1. Describe the characteristics of irritable bowel syndrome to Hannah, and indicate the role that stress might play in her illness.
2. Explain how the record of food intake and symptoms might be helpful in devising an appropriate dietary plan for Hannah. Could any of the foods that are currently in Hannah's diet be aggravating her symptoms?
3. What dietary measures might benefit individuals with irritable bowel syndrome? What problems might some of the dietary changes cause?

Treatments under investigation for irritable bowel syndrome include peppermint oil, which relaxes smooth muscle within the GI tract, and various types of probiotics (see Highlight 24). The Case Study allows you to apply your knowledge about irritable bowel syndrome to a clinical situation.

Diverticular Disease of the Colon

Diverticulosis ♦ refers to the presence of pebble-sized herniations (outpockets) in the intestinal wall, known as diverticula (see the photo). In Western societies, the diverticula occur most often in the sigmoid colon, the portion of the colon just above the rectum. The prevalence of diverticulosis increases with age, occurring in 50 to 65 percent of 80-year-old individuals.[50] Most people with diverticulosis are symptom free and remain unaware of the condition until a complication develops.

♦ Diverticulosis was introduced in Highlight 3.

Epidemiological studies suggest that the development of diverticula is strongly influenced by the amount of dietary fiber a person consumes. By increasing stool weight and bulk, high-fiber diets reduce the workload of the circular muscles that move wastes through the colon. Low-fiber diets require more vigorous muscle contractions, increasing pressure within the segments immediately adjacent to the circular muscles. This increase in pressure may induce small areas of intestinal tissue to weaken and balloon outward over time.

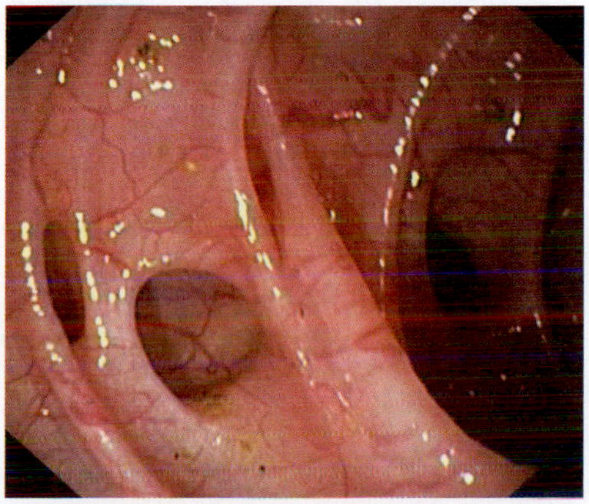

© Hans Bjorknas/Gastrolab

Diverticula are frequently seen during a colonoscopy, a procedure that uses a flexible, lighted tube to examine the inside of the colon.

Diverticulitis Inflammation or infection sometimes develops in the area around a diverticulum. This condition, called *diverticulitis,* is the most common complication of diverticulosis, affecting 10 to 25 percent of individuals with the condition.[51] It is thought to result from erosion of the diverticular wall due to high pressure within the lumen, causing inflammation and eventually a microperforation that leads to subsequent infection.[52] If the infection spreads to adjacent organs, fistulas may develop. Less frequently, the infection spreads to the peritoneal cavity, causing life-threatening illness. Symptoms of diverticulitis may include persistent abdominal pain, tenderness in the affected area, fever, constipation, and diarrhea. Anorexia, nausea, and vomiting may also occur.

Treatment for Diverticular Disease Medical treatment for diverticulosis is necessary only if symptoms develop. Because a high-fiber diet may prevent disease progression and the development of intestinal symptoms, patients are advised to increase fiber intake, with an emphasis on insoluble fiber sources. Fiber should be increased gradually to ensure tolerance. Bulk-forming agents, such as psyllium, can help to increase fiber intake if food sources are insufficient. Although avoiding nuts, seeds, and popcorn is sometimes suggested to prevent complications, no evidence is available to justify the recommendation.[53]

Patients with diverticulitis may need antibiotics to treat infections and, possibly, pain-control medications. In mild cases, a clear liquid diet may be advised initially, with progression to solid foods as symptoms resolve. In more severe cases, bowel rest is necessary (oral fluids and food are withheld), and fluids are given intravenously. Afterward, an oral diet is gradually reintroduced as the condition improves, beginning with clear liquids and progressing to a restricted-fiber diet until inflammation and bleeding subside.[54] After recovery, a high-fiber diet is recommended to prevent disease progression and symptom recurrence. Surgical interventions are sometimes necessary to treat complications of diverticulitis and may include removal of the affected portion of the colon.

IN SUMMARY **Irritable bowel syndrome and diverticular disease are common disorders affecting the large intestine. Irritable bowel syndrome is characterized by chronic, recurring intestinal symptoms such as diarrhea and/or constipation, abdominal pain, and flatulence. Although the causes are unknown, the disorder is influenced by food intake, stress, and psychological factors. Diverticulosis is often asymptomatic until complications develop; its prevalence increases with advancing age and is often associated with low fiber intakes. Patients with irritable bowel syndrome or diverticulosis may benefit from a high-fiber diet; additional measures for treating irritable bowel syndrome may include consuming small meals and omitting certain foods from the diet.**

Colostomies and Ileostomies

An *ostomy* is a surgically created opening (called a **stoma**) in the abdominal wall through which dietary wastes can be eliminated. Whereas a permanent ostomy is necessary after a partial or total colectomy, a temporary ostomy is sometimes constructed to bypass the colon after injury or extensive surgery. To create the stoma, the cut end of the remaining segment of functional intestine is routed through an opening in the abdominal wall and stitched in place so that it empties to the exterior. The stoma can be formed from a section of the colon (**colostomy**) or ileum (**ileostomy**), as shown in Figure 24-5. Conditions that may require these procedures include inflammatory bowel diseases, diverticulitis, and colorectal cancers.

To collect wastes, a disposable bag is affixed to the skin around the stoma and emptied during the day as needed. Alternatively, an interior pouch can be surgically constructed behind the stoma using intestinal tissue, and the pouch can be emptied with a catheter when convenient. Stool consistency varies according to the length of colon that is functional. If a small portion of the colon is absent or bypassed, the stools may continue to be semisolid. If the entire colon has been removed or is bypassed, absorption of fluid and electrolytes into the body is reduced substantially, and the output is liquid. Due to the difficulty in obtaining enough water to replace losses, patients with ileostomies often have low urine output and an increased risk of developing kidney stones.

Nutrition Therapy for Patients with Ostomies The nutrition care after an ostomy depends on the length of colon removed and the portion of ileum that remains, so dietary adjustments are individualized according to the surgical procedure and symptoms that develop afterward. Following surgery, the diet progresses from clear liquids that are low in sugars to a regular meal plan, as tolerated. To

stoma (STOE-ma): a surgically created opening in a body tissue or organ.

colostomy (co-LAH-stoe-me): a surgically created passage through the abdominal wall into the colon.

ileostomy (ill-ee-AH-stoe-me): a surgically created passage through the abdominal wall into the ileum.

FIGURE 24-5 Colostomy and Ileostomy

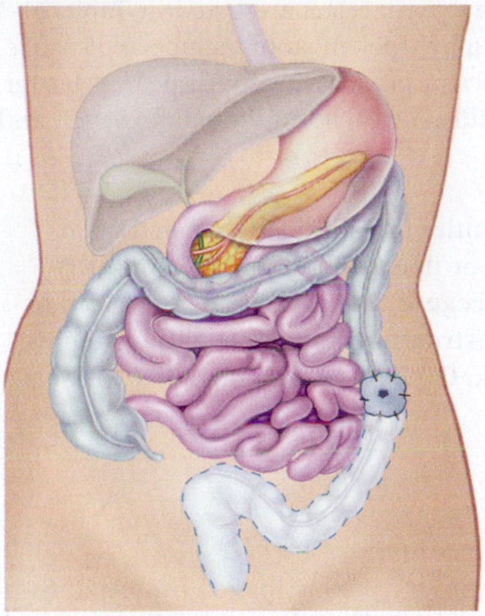

In a colostomy, a portion of the colon is removed or bypassed, and the stoma is formed from the remaining section of functional colon.

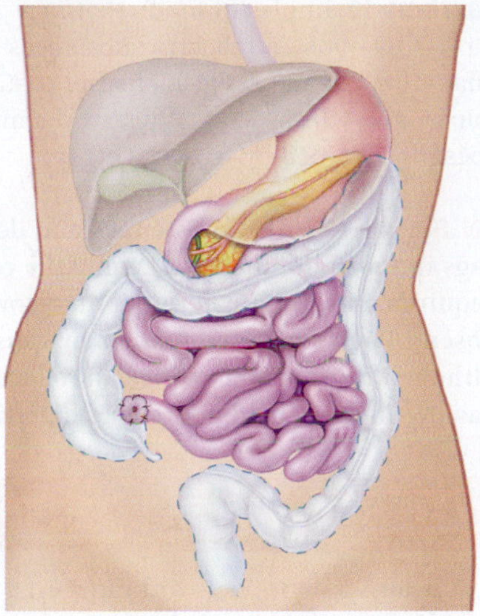

In an ileostomy, the entire colon is removed or bypassed, and the stoma is formed from the ileum.

reduce stool output, a low-fiber diet may be recommended. Small, frequent meals may be more acceptable than larger ones. To determine food tolerances, patients should try small amounts of questionable foods and assess their effects; a food that causes problems can be tried again later. Appropriate fluid and electrolyte intakes should be encouraged when a large portion of the colon has been removed.

People with ileostomies need to chew thoroughly to ensure that foods are adequately digested and to prevent obstructions, a common complication due to the small diameter of the ileal lumen. Foods high in insoluble fibers are sometimes avoided because they reduce transit time, may cause obstructions, and increase stool output. Because less fluid is reabsorbed following an ileostomy (in which the colon is removed), the diet should provide adequate liquid to prevent dehydration. To replace electrolyte losses, patients are encouraged to use salt liberally and to ingest beverages with added electrolytes (such as sports drinks and oral rehydration beverages), if necessary. If a large portion of the ileum has been removed, fat malabsorption may occur due to bile acid depletion, and vitamin B_{12} injections may be required.

Dietary concerns after colostomies depend on the length of colon remaining. Most patients have no dietary restrictions and can return to a regular diet. Patient concerns typically include stool odors, excessive gas production, and diarrhea. If a large portion of colon was removed, recommendations may be similar to those given to ileostomy patients.

Obstructions As mentioned, foods that are incompletely digested can cause obstructions, a primary concern of ileostomy patients. Although these patients can consume almost any food that is cut into small pieces and carefully chewed, the following foods may cause difficulty: celery, coconut, corn, dried fruit, grapes, nuts, popcorn, raw cabbage (for example, in coleslaw), and unpeeled apples.[55]

Reducing Gas and Odors Persons with ostomies are often concerned about foods that may increase gas production or cause strong odors. Foods that may cause excessive gas include those listed in Table 24-2 on p. 732; practices that increase gas formation include smoking, gum chewing, tobacco chewing, using drinking straws, and eating quickly. Foods that sometimes produce unpleasant odors include asparagus, beer, broccoli, brussels sprouts, cabbage, dried beans and peas, eggs, fish, garlic, and onions. Foods that may help to reduce odors include buttermilk, cranberry juice, parsley, and yogurt.[56]

Diarrhea Examples of foods that may aggravate diarrhea are listed in Table 24-3 on p. 733. Foods and dietary substances that may thicken stool include applesauce, banana (or banana flakes), cheese, pasta, pectin, potatoes, smooth peanut butter, tapioca, and white rice.[57] What works may differ for each individual, however, and is best determined by trial and error.

IN SUMMARY **Colostomies and ileostomies are surgically created openings in the abdominal wall using the colon or ileum. Fluid and electrolyte requirements are greater after an ostomy because colon function is reduced or absent. Poorly digested foods may cause obstructions in people with ostomies, although thorough chewing can reduce risk. Other concerns include excessive gas production, food odors, and diarrhea.**

Clinical Portfolio

1. A health practitioner working with a patient with a constipation problem provides him with detailed information about a high-fiber diet. At a follow-up appointment, the patient reports no change in symptoms. His food diary for that day shows that he consumed an omelet and toast for breakfast and a sandwich with juice for lunch.
   - Considering these two meals only, what additional information would help the health practitioner evaluate the man's compliance with the diet he was given?
   - Review the discussion about fiber in Chapter 4, and create a one-day menu that provides the DRI for fiber for an adult male, using the fiber values listed in Appendix H.
2. Using Table 24-5 on p. 735 as a guide, plan a day's menus for a diet containing approximately 50 grams of fat. Take care to make the meals both palatable and nutritious. How can these menus be improved using the suggestions in the "How To" on p. 736?
3. As stated in this chapter, treatment of celiac disease is deceptively simple—eliminate wheat, barley, and rye, and possibly oats. Remaining on a gluten-free diet is more challenging than it appears, however.
   - Randomly select 10 of your favorite snack and convenience foods. Take a trip to the grocery store, and check the labels of the products you selected to see if they would be allowed on a gluten-free diet. Keep in mind that the labels may not list all offending ingredients.
   - Find acceptable substitutes for the products that are not allowed, either by substituting other foods or by checking for gluten-free products in the grocery store. If you have access to the Internet, you may want to investigate websites that advertise gluten-free products to get an idea of what's available.

Nutrition on the Net

For further study of topics covered in this chapter, log on to **www.cengagebrain.com** and search for ISBN 084006845X.

- Visit the websites of these organizations to find information that is helpful both for health practitioners and patients with gastrointestinal problems:

American College of Gastroenterology: **www.acg.gi.org**

American Gastroenterological Association: **www.gastro.org**

National Institute of Diabetes and Digestive and Kidney Diseases, a division of the National Institutes of Health: **www2.niddk.nih.gov**

- Find additional information about cystic fibrosis at the website of the Cystic Fibrosis Foundation: www.cff.org/home
- Find more information about celiac disease by visiting these websites:

  Celiac Disease Foundation: www.celiac.org

  Celiac Sprue Association: www.csaceliacs.org

  Gluten Intolerance Group of North America: www.gluten.net
- Learn more about inflammatory bowel diseases at the website of the Crohn's and Colitis Foundation of America: www.ccfa.org

Nutrition Assessment Checklist for People with Lower GI Tract Disorders

Medical History

Check the medical record for diseases that:

- Cause chronic GI symptoms, such as irritable bowel syndrome or ulcerative colitis
- Interfere with pancreatic enzyme secretion, such as chronic pancreatitis or cystic fibrosis
- Interfere with nutrient absorption, such as Crohn's disease or celiac disease

Check for surgical procedures involving the lower GI tract, such as:

- Intestinal resections or bypass surgeries
- Ileostomy
- Colostomy

Check for the following symptoms or complications:

- Anemia
- Bacterial overgrowth
- Bone disease
- Constipation
- Diarrhea, dehydration
- Fistulas
- Lactose intolerance
- Nutrient deficiencies
- Obstructions
- Oxalate kidney stones
- Poor growth, in children
- Steatorrhea

Medications

Check for medications or dietary supplements that may:

- Cause constipation or diarrhea
- Interfere with food intake by causing nausea, vomiting, cramps, dry mouth, or drowsiness
- Alter appetite or nutrient needs

Dietary Intake

Note the following problems, and contact the dietitian if you suspect difficulties such as:

- Poor appetite or food intake
- Food intolerances
- Inadequate fiber intake, in patients with constipation
- Lactose intolerance, in patients with diarrhea
- Inadequate fluid intake

Anthropometric Data

Measure baseline height and weight. Address weight loss early to prevent malnutrition in patients with:

- Severe or persistent diarrhea
- Nutrient malabsorption

Laboratory Tests

Check laboratory tests for signs of dehydration, electrolyte imbalances, nutrient deficiencies, and anemia in patients with:

- Severe or persistent diarrhea
- Nutrient malabsorption
- Intestinal resections

Physical Signs

Look for physical signs of:

- Dehydration
- Essential fatty acid and fat-soluble vitamin deficiencies
- Folate and vitamin B_{12} deficiencies
- Mineral deficiencies
- Protein-energy malnutrition

References

1. A. J. Lembo and S. P. Ullman, Constipation, in M. Feldman, L. S. Friedman, and L. J. Brandt, eds., *Sleisenger and Fordtran's Gastrointestinal and Liver Disease* (Philadelphia: Saunders, 2010), pp. 259–284.
2. C. A. Ternent and coauthors, Practice parameters for the evaluation and management of constipation, *Diseases of the Colon and Rectum* 50 (2007): 2013–2022.
3. Standing Committee on the Scientific Evaluation of Dietary Reference Intakes, Food and Nutrition Board, Institute of Medicine, *Dietary Reference Intakes for Energy, Carbohydrate, Fiber, Fat, Fatty Acids, Cholesterol, Protein, and Amino Acids* (Washington, DC: National Academies Press, 2002).
4. Ternent and coauthors, 2007.
5. F. Azpiroz and M. D. Levitt, Intestinal gas, in M. Feldman, L. S. Friedman, and L. J. Brandt, eds., *Sleisenger and Fordtran's Gastrointestinal and Liver Disease* (Philadelphia: Saunders, 2010), pp. 233–240.
6. L. R. Schiller and J. H. Sellin, Diarrhea, in M. Feldman, L. S. Friedman, and L. J. Brandt, eds., *Sleisenger and Fordtran's Gastrointestinal and Liver Disease* (Philadelphia: Saunders, 2010), pp. 211–232.
7. J. S. Trier, Acute diarrheal disorders, in N. J. Greenberger, ed., *Current Diagnosis and Treatment: Gastroenterology, Hepatology, and Endoscopy* (New York: McGraw-Hill Companies, 2009), pp. 45–63.

8. M. de Vrese and P. R. Marteau, Probiotics and prebiotics: Effects on diarrhea, *American Journal of Clinical Nutrition* 137 (2007): 803S–811S.
9. Schiller and Sellin, 2010.
10. American Dietetic Association, *Nutrition Care Manual* (Chicago: American Dietetic Association, 2010).
11. C. E. Semrad and D. W. Powell, Approach to the patient with diarrhea and malabsorption, in L. Goldman and D. Ausiello, eds., *Cecil Medicine* (Philadelphia: Saunders, 2008), pp. 1019–1042.
12. J. S. Trier, Intestinal malabsorption, in N. J. Greenberger, ed., *Current Diagnosis and Treatment: Gastroenterology, Hepatology, and Endoscopy* (New York: McGraw-Hill Companies, 2009), pp. 223–242.
13. S. O'Mahony and F. Shanahan, Enteric microbiota and small intestinal bacterial overgrowth, in M. Feldman, L. S. Friedman, and L. J. Brandt, eds., *Sleisenger and Fordtran's Gastrointestinal and Liver Disease* (Philadelphia: Saunders, 2010), pp. 1769–1778.
14. J. K. DiBaise, Nutritional consequences of small intestinal bacterial overgrowth, *Practical Gastroenterology* 32 (December, 2008): 15–28.
15. Semrad and Powell, 2008.
16. V. Singh, D. L. Conwell, and P. A. Banks, Acute pancreatitis, in N. J. Greenberger, ed., *Current Diagnosis and Treatment: Gastroenterology, Hepatology, and Endoscopy* (New York: McGraw-Hill Companies, 2009), pp. 291–298.
17. S. Tenner and W. M. Steinberg, Acute pancreatitis, in M. Feldman, L. S. Friedman, and L. J. Brandt, eds., *Sleisenger and Fordtran's Gastrointestinal and Liver Disease* (Philadelphia: Saunders, 2010), pp. 959–983; Singh, Conwell, and Banks, 2009.
18. American Dietetic Association, 2010.
19. American Dietetic Association, 2010.
20. C. E. Forsmark, Chronic pancreatitis, in M. Feldman, L. S. Friedman, and L. J. Brandt, eds., *Sleisenger and Fordtran's Gastrointestinal and Liver Disease* (Philadelphia: Saunders, 2010), pp. 985–1015.
21. Forsmark, 2010.
22. Forsmark, 2010; B. Wu, D. Conwell, and P. Banks, Chronic pancreatitis, in N. J. Greenberger, ed., *Current Diagnosis and Treatment: Gastroenterology, Hepatology, and Endoscopy* (New York: McGraw-Hill Companies, 2009), pp. 299–304.
23. D. C. Whitcomb and M. E. Lowe, Hereditary, familial, and genetic disorders of the pancreas and pancreatic disorders in childhood, in M. Feldman, L. S. Friedman, and L. J. Brandt, eds., *Sleisenger and Fordtran's Gastrointestinal and Liver Disease* (Philadelphia: Saunders, 2010), pp. 931–957.
24. M. J. Welsh, Cystic fibrosis, in L. Goldman and D. Ausiello, eds., *Cecil Medicine* (Philadelphia: Saunders, 2008), pp. 627–631.
25. Welsh, 2008.
26. Whitcomb and Lowe, 2010; American Dietetic Association, 2010.
27. M. M. Niewinski, Advances in celiac disease and gluten-free diet, *Journal of the American Dietetic Association* (2008): 661–672.
28. Niewinski, 2008.
29. R. J. Farrell and C. P. Kelly, Celiac disease and refractory celiac disease, in M. Feldman, L. S. Friedman, and L. J. Brandt, eds., *Sleisenger and Fordtran's Gastrointestinal and Liver Disease* (Philadelphia: Saunders, 2010), pp. 1797–1820.
30. J. J. Heidelbaugh and coauthors, Gastroenterology, in R. E. Rakel, ed., *Textbook of Family Medicine* (Philadelphia: Saunders, 2007), pp. 1115–1171.
31. N. Raymond, J. Heap, and S. Case, The gluten-free diet: An update for health professionals, *Practical Gastroenterology* 30 (September, 2006): 67–92.
32. Raymond, Heap, and Case, 2006.
33. B. E. Sands and C. A. Siegel, Crohn's disease, in M. Feldman, L. S. Friedman, and L. J. Brandt, eds., *Sleisenger and Fordtran's Gastrointestinal and Liver Disease* (Philadelphia: Saunders, 2010), pp. 1941–1973.
34. Sands and Siegel, 2010.
35. A. M. Griffiths, Inflammatory bowel disease, in M. E. Shils and coeditors, *Modern Nutrition in Health and Disease* (Baltimore: Lippincott Williams & Wilkins, 2006), pp. 1209–1218.
36. W. F. Stenson, Inflammatory bowel disease, in L. Goldman and D. Ausiello, eds., *Cecil Medicine* (Philadelphia: Saunders, 2008), pp. 1042–1050.
37. American Dietetic Association, 2010.
38. American Dietetic Association, 2010.
39. Stenson, 2008.
40. Trier, Acute diarrheal disorders, 2009.
41. A. L. Buchman, Short bowel syndrome, in M. Feldman, L. S. Friedman, and L. J. Brandt, eds., *Sleisenger and Fordtran's Gastrointestinal and Liver Disease* (Philadelphia: Saunders, 2010), pp. 1779–1795.
42. Buchman, 2010; C. R. Parrish, The clinician's guide to short bowel syndrome, *Practical Gastroenterology* (September 2005): 67–106.
43. Buchman, 2010.
44. Buchman, 2010.
45. S. Friedman, Irritable bowel syndrome, in N. J. Greenberger, ed., *Current Diagnosis and Treatment: Gastroenterology, Hepatology, and Endoscopy* (New York: McGraw-Hill Companies, 2009), pp. 279–290.
46. N. J. Talley, Irritable bowel syndrome, in M. Feldman, L. S. Friedman, and L. J. Brandt, eds., *Sleisenger and Fordtran's Gastrointestinal and Liver Disease* (Philadelphia: Saunders, 2010), pp. 2091–2104
47. Talley, 2010.
48. Talley, 2010.
49. W. D. Heizer, S. Southern, and S. McGovern, The role of diet in symptoms of irritable bowel syndrome in adults: A narrative review, *Journal of the American Dietetic Association* 109 (2009): 1204–1214; A. Sanjeevi and D. F. Kirby, The role of food and dietary intervention in the irritable bowel syndrome, *Practical Gastroenterology* 32 (July 2008): 33–42.
50. A. C. Travis and R. S. Blumberg, Diverticular disease of the colon, in N. J. Greenberger, ed., *Current Diagnosis and Treatment: Gastroenterology, Hepatology, and Endoscopy* (New York: McGraw-Hill Companies, 2009), pp. 243–255.
51. C. Prather, Inflammatory and anatomic diseases of the intestine, peritoneum, mesentery, and omentum, in L. Goldman and D. Ausiello, eds., *Cecil Medicine* (Philadelphia: Saunders, 2008), pp. 1050–1061.
52. Travis and Blumberg, 2009.
53. Travis and Blumberg, 2009; L. L. Strate and coauthors, Nut, corn, and popcorn consumption and the incidence of diverticular disease, *Journal of the American Medical Association* 300 (2008): 907–914.
54. American Dietetic Association, 2010.
55. American Dietetic Association, 2010.
56. American Dietetic Association, 2010.
57. American Dietetic Association, 2010.

Probiotics and Intestinal Health

Soon after birth, the warm, nutrient-rich environment within the gastrointestinal tract is colonized by a wide variety of bacterial species. The approximately 10 trillion bacterial cells inhabiting our bodies (**flora**) make up more than 90 percent of all our cells. Most bacterial cells reside in our colon, which harbors over 400 different species.[1] Although the exact composition of intestinal bacteria varies among individuals, the pattern within an individual tends to remain constant over time, fluctuating somewhat due to illness, antibiotic treatment, and to some extent, dietary factors. Table H24-1 lists the predominant types of bacteria that colonize the human intestines, and Table H24-2 shows how the bacterial populations vary within different regions of the GI tract.

TABLE H24-1 Intestinal Flora

Predominant Types	Subdominant Types
Bacteroides	*Enterobacteria*
Bifidobacteria	*Enterococci*
Clostridia	*Escherichia*
Eubacteria	*Klebsiella*
Peptococci	*Lactobacilli*
Peptostreptococci	*Micrococci*
Ruminococci	*Staphylococci*

TABLE H24-2 Bacterial Populations in the Gastrointestinal Tract

Organ	Total Bacteria (per mL of contents)
Stomach, duodenum	10 to 1000
Jejunum, ileum	10^4 to 10^8
Colon	10^{10} to 10^{12}

Over the past several decades, nutritional scientists and microbiologists have tried to determine whether **probiotics**—live, **nonpathogenic** bacteria supplied in sufficient numbers to possibly benefit our health—can be useful for preventing or treating various medical conditions. Although the diseases of interest include gastrointestinal disorders, researchers have also been studying the effects of bacterial cells on cancer, immune system disorders, and other illnesses. This highlight discusses some of the research and explains some of the issues involved in selecting and consuming probiotic bacteria. The accompanying glossary defines the relevant terms.

Our Intestinal Flora

Intestinal bacteria can benefit our health in a number of different ways. First, the bacteria degrade much of our undigested or unabsorbed dietary carbohydrate, including dietary fibers, starch that is resistant to digestion, and poorly absorbed sugars and sugar alcohols. In turn, the bacteria produce some vitamins, as well as short-chain fatty acids that our colonic epithelial cells and other body cells can use as an energy source. Intestinal bacteria also assist in the development and maintenance of mucosal tissue, protect intestinal tissue from **pathogenic** bacteria, and stimulate immune defenses in mucosal cells and other body tissues.[2]

Probiotic Bacteria

For bacteria to be "probiotic"—that is, beneficial to health—they must be nonpathogenic when consumed. They must survive their transit through the digestive tract; therefore, they must be resistant to destruction by stomach acid, bile, and other digestive substances. They should be able to alter the intestinal environment in some way that is beneficial to the human host, either by producing antimicrobial substances, altering immune defenses, metabolizing undigested foodstuffs, or protecting the intestinal walls.[3]

Probiotic bacteria must be consumed in high amounts—between 100 million and 100 billion live bacteria per day—to survive in sufficient numbers to influence the bacterial populations in the large intestine; a serving of yogurt usually provides these amounts. Carefully controlled studies have not found that probiotic bacteria actually *colonize* the intestine, however, as they are no longer detected in fecal or intestinal samples once ingestion of the probiotic product stops.[4] Note that only a few different types of bacteria are used in foods, and the relatively small amounts consumed cannot compete with the huge populations that normally populate our digestive tract.

GLOSSARY

bacterial translocation: movement of bacteria across the intestinal mucosa, allowing access to body tissues.

flora: the bacteria that normally reside in a person's body.

nonpathogenic: not capable of causing disease.

pathogenic: capable of causing disease.

prebiotics: indigestible substances in foods that stimulate the growth of nonpathogenic bacteria within the large intestine.

probiotics: live bacteria provided in foods and dietary supplements for the purpose of preventing or treating disease.

Probiotic Bacteria and Disease

Although results of research studies vary, probiotic bacteria may help to prevent and treat some gastric and intestinal disorders (such as inflammatory bowel diseases and irritable bowel syndrome), alter susceptibility to food allergens and alleviate some allergy symptoms, and improve the availability and digestibility of various nutrients.[5] Other potential benefits include improved immune responses, reduced symptoms of lactose intolerance, and reduced cancer risk.[6]

Much of the research investigating probiotics and intestinal illness has focused on the prevention and treatment of infectious diarrhea. For example, controlled trials have suggested that certain strains of probiotic bacteria may shorten the duration of diarrhea caused by rotavirus infection in infants and children, decrease the incidence of traveler's diarrhea in tourists visiting high-risk areas, and prevent the recurrence of infectious diarrhea in hospitalized patients.[7] In studies of children and adults using antibiotics, some strains of probiotic bacteria have been shown to reduce the incidence and duration of antibiotic-associated diarrhea. As another example, some studies have suggested that probiotic treatment may help to reduce the recurrence of *pouchitis*, an inflammation of the surgical pouch created in patients who have had an ileostomy or colostomy.[8]

Despite promising research results thus far, there are no clear conclusions about the appropriate probiotic doses or durations of treatment for many of these conditions. Moreover, the beneficial effects of one bacterial strain cannot be extrapolated to other strains of the same species.[9] Thus, individuals who decide to consume probiotic-containing foods and supplements to benefit their health cannot be certain that the substances they use will help their condition. At best, probiotics should be considered an adjunct therapy rather than a primary treatment for an illness.

Probiotics in the Diet

Probiotics are provided mainly by fermented foods. In the United States, yogurt and acidophilus milk are produced using various species of lactobacilli and bifidobacteria, although the species are chosen for their ability to produce desirable food products rather than their potential health benefits. In Europe and Asia, food products containing probiotic bacteria include yogurt, milk, ice cream, oatmeal gruel, and soft drinks. Although lactobacilli are used to produce various other fermented food products, such as sauerkraut, pickles, brined olives, and sausages, these foods retain few, if any, live bacteria after they undergo typical food processing methods.[10]

Various species of *Lactobacillus* are used in the production of fermented food products, such as the foods shown in this photo.

A number of companies market probiotic supplements, which are available in capsules, tablets, and powders. Because probiotic bacteria are living organisms, storage conditions may affect their viability—heat, moisture, and oxygen can reduce survival times—and therefore consumers should check the expiration date before purchasing a product. When a consumer group (Consumer Lab.com) tested 13 probiotic supplements, they found that 5 of the products contained substantially fewer live bacteria than was claimed on the label.[11] Thus, there is no guarantee that a dietary supplement will contain the amount of bacteria expected.

Certain indigestible substances in food, called **prebiotics**, can stimulate the growth or activity of resident bacteria within the large intestine. Prebiotics include some of the carbohydrates found in asparagus, chicory root, garlic, Jerusalem artichokes, onions, and other foods.[12] Because the intestinal bacteria that degrade these substances produce gas as a by-product, people who consume high amounts of these foods may experience more flatulence than usual.

Safety Concerns

One major concern is the possibility that probiotic bacteria may cause infection in immune-compromised individuals. Various species of probiotic bacteria, including *Lactobacillus* species, have been isolated from the infection sites of severely ill individuals who were consuming the probiotic.[13] Risk is increased by the use of antibiotic therapy (which reduces intestinal flora populations), illnesses or medications that suppress immunity, and illnesses that increase risk of **bacterial translocation** (including inflammatory bowel illnesses and intestinal infections). Care should be taken to inquire about probiotic use in these patients.

Other safety concerns are related to the lack of industry standards for probiotics in foods and supplements: the concentrations and strains of probiotic bacteria in foods may vary substantially.[14] Thus, a consumer who wishes to try probiotics would find it difficult to determine how much of a product to consume in order to achieve the desired effect.

In recent years, the contributions of our intestinal flora to health have been increasingly recognized. Preliminary research suggests that altering our bacterial populations by consuming probiotics or prebiotics may help to improve our defenses against certain illnesses. Additional studies are needed to verify the beneficial effects of probiotics and prebiotics and to develop standard protocols that can be used for treating illness.

References

1. S. O'Mahony and F. Shanahan, Enteric microbiota and small intestinal bacterial overgrowth, in M. Feldman, L. S. Friedman, and L. J. Brandt, eds., *Sleisenger and Fordtran's Gastrointestinal and Liver Disease* (Philadelphia: Saunders, 2010), pp. 1769–1778.
2. A. M. O'Hara and F. Shanahan, The gut flora as a forgotten organ, *EMBO Reports* 7 (2006): 688–693.
3. P. Winkler and coauthors, Molecular and cellular basis of microflora-host interactions, *Journal of Nutrition* 137 (2007): 756S–772S.
4. B. Corthésy, H. R. Gaskins, and A. Mercenier, Cross-talk between probiotic bacteria and the host immune system, *Journal of Nutrition* 137 (2007): 781S–790S.
5. M. de Vrese and P. R. Marteau, Probiotics and prebiotics: Effects on diarrhea, *Journal of Nutrition* 137 (2007): 803S–811S; A. C. Ouwehand, Antiallergic effects of probiotics, *Journal of Nutrition* 137 (2007): 794S–797S.
6. S. Parvez and coauthors, Probiotics and their fermented food products are beneficial for health, *Journal of Applied Microbiology* 100 (2006): 1171–1185.
7. de Vrese and Marteau, 2007.
8. J. J. Jones and A. E. Foxx-Orenstein, Probiotics in inflammatory bowel disease, *Practical Gastroenterology* (March 2006): 44–50.
9. L. C. Douglas and M. E. Sanders, Probiotics and prebiotics in dietetics practice, *Journal of the American Dietetic Association* (2008): 510–521.
10. Douglas and Sanders, 2008.
11. ConsumerLab.com, Product review: Probiotic supplements (including *Lactobacillus acidophilus*, *Bifidobacterium*, and others), www.consumerlab.com, accessed August 19, 2010.
12. S. Kolida and G. R. Gibson, Prebiotic capacity of inulin-type fructans, *Journal of Nutrition* 137 (2007): 2503S–2506S.
13. K. Whelan and C. E. Myers, Safety of probiotics in patients receiving nutritional support: A systematic review of case reports, randomized controlled trials, and nonrandomized trials, *American Journal of Clinical Nutrition* 91 (2010): 687–703.
14. E. R. Farnworth, The evidence to support health claims for probiotics, *Journal of Nutrition* 138 (2008): 1250S–1254S.

CHAPTER

25

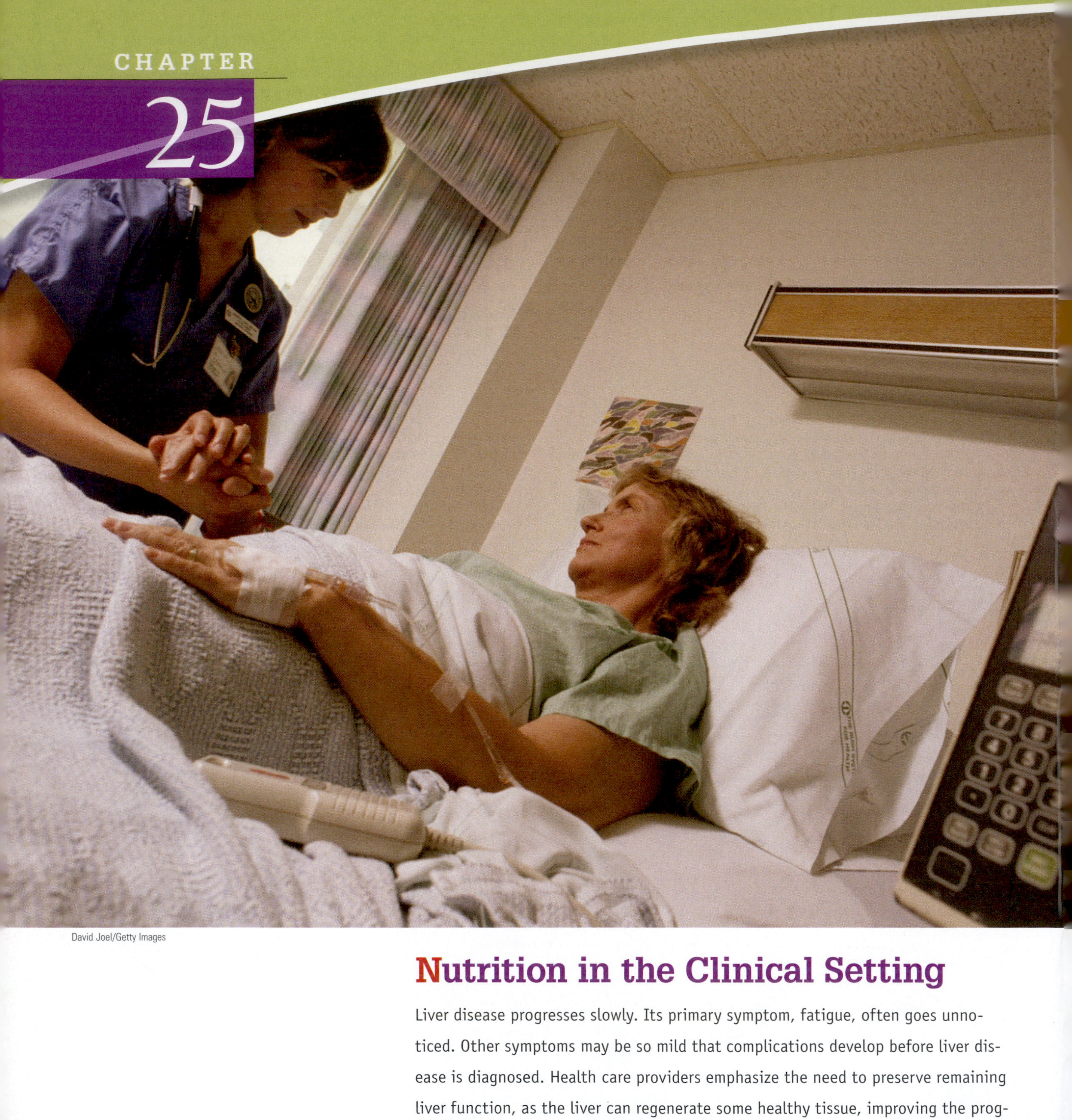

David Joel/Getty Images

Nutrition in the Clinical Setting

Liver disease progresses slowly. Its primary symptom, fatigue, often goes unnoticed. Other symptoms may be so mild that complications develop before liver disease is diagnosed. Health care providers emphasize the need to preserve remaining liver function, as the liver can regenerate some healthy tissue, improving the prognosis. Preventing additional damage is the principal means of avoiding liver failure or transplantation.

Throughout this chapter, the CourseMate icon indicates an opportunity for online self-study, linking you to activities to increase your understanding of chapter concepts. **www.cengagebrain.com** (search for ISBN 084006845X)

Liver Disease and Gallstones

CHAPTER OUTLINE

The liver is the most metabolically active organ in the body. As you may recall from Chapter 7, the liver plays a central role in processing, storing, and redistributing the nutrients provided by the foods we eat. ♦ The liver synthesizes most of the proteins that circulate in plasma, including albumin, clotting proteins, and transport proteins; it also produces the bile that emulsifies fat during digestion. In addition, the liver detoxifies drugs and alcohol and processes excess nitrogen so that it can be safely excreted as urea. If the liver's numerous roles are upset by liver damage or disease, the effects on health and nutrition status can be profound.

As Figure 25-1 shows, the liver is ideally situated for receiving and processing the nutrients absorbed by the small intestine. The portal vein's nutrient-rich blood supplies about 75 percent of the blood that enters liver tissue; the rest arrives via hepatic arteries. Blood is returned to the heart by way of the hepatic vein and then

♦ Table 7-1 on p. 207 summarizes the chemical reactions within the liver that are related to the metabolism of carbohydrates, lipids, and protein.

FIGURE 25-1 The Liver, Biliary System, and Associated Blood Vessels

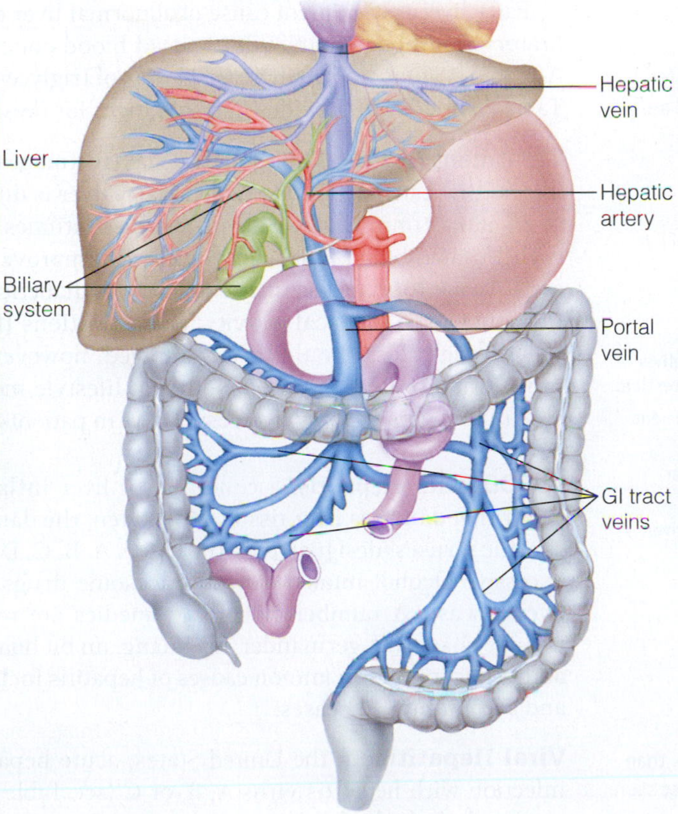

Liver
Receives nutrients from the digestive tract and processes them for distribution throughout the body.

Biliary system
Includes the gallbladder, which stores and secretes bile, and the bile ducts, which conduct bile from the liver to the gallbladder and from the gallbladder to the intestine.

Hepatic vein
Returns blood from the liver to the heart.

Hepatic artery
Supplies oxygen-rich blood from the heart to the liver.

Portal vein
Carries nutrient-rich blood from the digestive tract to the liver.

GI tract veins
Transport absorbed nutrients to the portal vein.

circulates throughout the body. The **biliary system** of channels and ducts carries bile and other substances from the liver to the duodenum while a meal is being digested. Between meals, the bile is diverted to the gallbladder, where it is stored and concentrated until needed for a subsequent meal.

Fatty Liver and Hepatitis

Fatty liver and hepatitis are the two most common disorders affecting the liver. Although both conditions may be mild and are usually reversible, each may progress to more serious illness and eventually cause liver damage.

♦ Reminder: *Fatty liver* is an accumulation of triglycerides in the liver; also called *hepatic steatosis* (STEE-ah-TOE-sis).

Fatty Liver

Fatty liver ♦ is an accumulation of fat in liver tissue. Ordinarily, the liver's excess triglycerides are packaged into very-low-density lipoproteins (VLDL) and exported to the bloodstream (see Chapter 5). Although the exact reasons why fat accumulates are often unknown, fatty liver represents an imbalance between the amount of fat synthesized in the liver or picked up from the blood and the amount exported to the blood via VLDL. Fatty liver is estimated to affect 20 percent or more of the adult population in the United States.[1]

Fatty liver is a clinical finding that is common to many conditions. It is present in the majority of patients who have alcoholic liver disease and can also result from exposure to drugs and toxic metals. It often accompanies diabetes mellitus, metabolic syndrome, obesity, and diseases of malnutrition, including kwashiorkor and marasmus. Fatty liver may also follow gastrointestinal bypass surgery and long-term total parenteral nutrition.[2]

Consequences of Fatty Liver

In many individuals, fatty liver is asymptomatic and causes no harm. In other cases, it may be associated with liver enlargement (**hepatomegaly**), inflammation (**steatohepatitis**), and fatigue. If liver damage and scarring develop, fatty liver may progress to cirrhosis (discussed in a later section), liver failure, or liver cancer.[3]

Fatty liver is a frequent cause of abnormal liver enzyme levels in the blood. Laboratory findings may include elevated blood concentrations of the liver enzymes ALT and AST, as well as increased levels of triglycerides, cholesterol, and glucose. ♦ Table 25-3 (p. 763) provides normal ranges for these liver enzymes.

♦ The liver enzymes ALT (alanine aminotransferase) and AST (aspartate aminotransferase) are involved in amino acid catabolism.

Treatment of Fatty Liver

The usual treatment for fatty liver is to eliminate the factors that cause it. For example, if fatty liver is due to alcohol abuse or drug treatment, it may improve after the patient discontinues use of the substance. In patients with elevated blood lipids, fatty liver may improve after blood lipid levels are lowered. An appropriate treatment for obese or diabetic patients might be weight reduction, increased physical activity, or medications that improve insulin sensitivity. Rapid weight loss should be discouraged, however, because it may accelerate the progression of liver disease.[4] Note that lifestyle modifications are not always successful in reversing fatty liver, especially in patients who lack the usual risk factors.

biliary system: the gallbladder and ducts that deliver bile from the liver and gallbladder to the small intestine.

hepatomegaly (HEP-ah-toe-MEG-ah-lee): enlargement of the liver.

steatohepatitis (STEE-ah-to-HEP-ah-TIE-tis): liver inflammation that is associated with fatty liver.

hepatitis (hep-ah-TYE-tis): inflammation of the liver.

Hepatitis

Hepatitis, a condition of liver inflammation, can result from any factor that damages liver tissue. Most often, the damage is caused by infection with specific viruses, designated by the letters A, B, C, D, E, and G. Other causes include excessive alcohol intake, exposure to some drugs and toxic chemicals, and fatty liver disease. A number of herbal remedies are reported to cause hepatitis; they include chaparral, germander, ma huang, jin bu huan, kava kava, kombucha, senna, and skullcap.[5] Less common causes of hepatitis include infection with other viruses and autoimmune diseases.

Viral Hepatitis

In the United States, acute hepatitis ♦ is most often caused by infection with hepatitis virus A, B, or C (see Table 25-1). Specific features of these viruses include the following:

♦ Hepatitis is considered *acute* if it lasts less than six months; *chronic* cases are those that last six months or longer.

TABLE 25-1 Features of Hepatitis Viruses

Hepatitis Virus	Major Mode of Transmission	New Cases (United States, 2007)	Chronic Disease Rate (% of cases)	Chronic Cases (United States)	Vaccination Available
A	Fecal-oral	25,000	None	0	Yes
B	Bloodborne, sexual transmission	43,000	2–7	1.25 million	Yes
C	Bloodborne	17,000	50–85	3.2 million	No

NOTE: There are fewer new cases of hepatitis C virus (HCV) infection than of hepatitis B virus (HBV) infection each year, but more HCV cases become chronic. Therefore, there are more HCV carriers than HBV carriers.

- *Hepatitis A virus* (HAV) is primarily spread via fecal-oral transmission, which usually involves the ingestion of foods or beverages that have been contaminated with fecal matter. Outbreaks of HAV infection are often associated with floods and other natural disasters, when inadequately treated sewage contaminates water supplies. Less frequently, HAV infection is contracted by consuming undercooked shellfish obtained from contaminated waters. Individuals at high risk of HAV infection include food handlers, child care workers, illicit drug users, and travelers to regions where the virus is endemic. In the United States, the incidence of HAV infection has declined 92 percent since 1995 as a result of routine vaccinations in children and high-risk individuals.[6] HAV infection usually resolves within a few months and does not cause chronic illness or permanent liver damage.
- *Hepatitis B virus* (HBV) is transmitted by infected blood or needles or by sexual contact. A major global health concern, HBV has infected one-third of the world population, although chronic illness develops in less than 10 percent of cases.[7] In the United States, HBV infections have dropped 82 percent since 1990 due to routine screenings and vaccinations for at-risk individuals.[8] Vaccinations are currently recommended for newborn infants and children, health-care workers, recipients of blood products, dialysis patients, sexually active adults, and users of injected drugs.
- *Hepatitis C virus* (HCV) is spread via infected blood or needles but is not readily spread by sexual contact. Most HCV cases progress to chronic illness, and currently HCV is the most common cause of chronic liver disease in the United States.[9] No vaccine is available to protect against HCV infection. Preventive measures include blood donor screening, viral inactivation of blood products, and infection control practices in health care settings. The incidence of HCV infection has declined markedly in recent years; newly diagnosed cases usually involve individuals with chronic infections who contracted hepatitis C decades ago.[10]

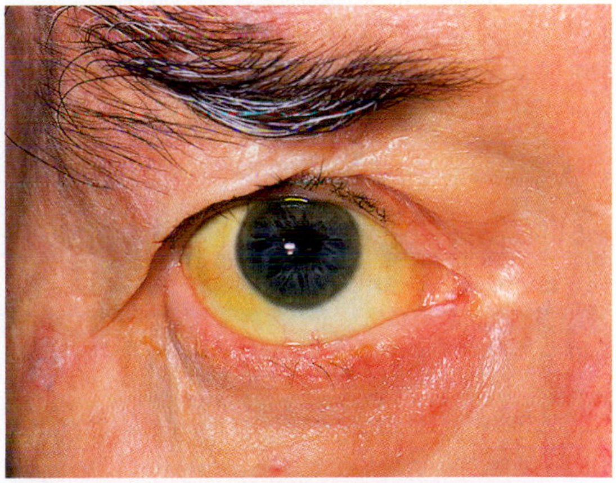

Jaundice is a yellow discoloration of the tissues that is most easily seen in the whites of the eyes.

Symptoms of Hepatitis The effects of hepatitis depend on the cause and severity of the disease. Individuals with mild or chronic cases are often asymptomatic. The onset of acute hepatitis may be accompanied by fatigue, nausea, vomiting, anorexia, and pain in the liver area. The liver is often slightly enlarged and tender. **Jaundice** (yellow discoloration of tissues) may develop, causing yellowing of the skin, urine, and the whites of the eyes. ♦ Other symptoms of hepatitis may include fever, muscle weakness, joint pain, and skin rashes. Serum levels of the liver enzymes ALT and AST are typically elevated. Chronic hepatitis can cause complications that are typical of liver cirrhosis and may lead to cirrhosis and liver cancer.

♦ Jaundice results when liver dysfunction impairs the metabolism of bilirubin, a breakdown product of hemoglobin that is normally eliminated in bile. Accumulation of bilirubin in the bloodstream leads to yellow discoloration of tissues.

jaundice (JAWN-dis): yellow discoloration of the skin and eyes due to an accumulation of bilirubin, a breakdown product of hemoglobin that normally exits the body via bile secretions.

Treatment of Hepatitis Hepatitis is treated with supportive care, such as bed rest (if necessary) and an appropriate diet. Hepatitis patients should avoid substances that irritate the liver, such as alcohol, drugs, and dietary supplements that cause liver damage. Hepatitis A infection usually resolves without the use of medications. Antiviral agents may be used to treat HBV and HCV infections; examples include lamivudine and ribavirin, which block viral replication, and interferon, which both inhibits viral replication and enhances immune responses.[11] Nonviral forms of hepatitis may be treated with anti-inflammatory and immunosuppressant drugs. Hospitalization is not required for hepatitis unless other medical conditions or complications hamper recovery.

Nutrition Therapy for Hepatitis Nutrition care varies according to a patient's symptoms and nutrition status.[12] Most individuals require no dietary changes. Those with anorexia or abdominal discomfort may find small, frequent meals easier to tolerate. Malnourished individuals need to consume adequate protein and energy to replenish nutrient stores; the diet should include about 1.0 to 1.2 grams of protein per kilogram of body weight each day. A low-fat diet, with fat limited to less than 30 percent of total kcalories, may be recommended for those with steatorrhea. Patients with persistent vomiting may require fluid and electrolyte replacement. Liquid supplements can be helpful for improving nutrient intakes.

cirrhosis (sih-ROE-sis): an advanced stage of liver disease in which extensive scarring replaces healthy liver tissue, causing impaired liver function and liver failure.

portal hypertension: elevated blood pressure in the portal vein due to obstructed blood flow through the liver.

collaterals: blood vessels that enlarge to allow an alternative pathway for diverted blood.

varices (VAH-rih-seez): abnormally dilated blood vessels (singular: *varix*).

ascites (ah-SIGH-teez): an abnormal accumulation of fluid in the abdominal cavity.

IN SUMMARY **Fatty liver can result from excessive alcohol intake, drug toxicity, and chronic disorders such as diabetes and obesity. Hepatitis is frequently caused by viral infection but can also result from alcohol abuse, drug toxicity, and other causes. Although fatty liver is often benign, hepatitis can become chronic and lead to cirrhosis and liver cancer. Treatment of hepatitis involves supportive care, such as bed rest, elimination of liver toxins, and dietary measures that maintain or improve nutrition status.**

Cirrhosis

Cirrhosis is the final phase of chronic liver disease. Long-term liver disease gradually destroys liver tissue, leading to scarring (fibrosis) in some regions and small areas of regenerated, healthy tissue in others. As the disease progresses, the scarring becomes more extensive, leaving fewer areas of healthy tissue. A cirrhotic liver is often shrunken and has an irregular, nodular appearance. Cirrhosis impairs liver function and can eventually lead to liver failure. It is the 12th leading cause of death in the United States.[13]

Table 25-2 lists some common causes of cirrhosis. In the United States, most cases are caused by alcoholic liver disease and chronic hepatitis C infection, followed by fatty liver disease and chronic hepatitis B infection.[14] Additional causes include other types of chronic hepatitis; bile duct blockages, which cause bile acids to accumulate to toxic levels in the liver; drug-induced liver injury; and inherited disorders that cause toxic substances to build up in the liver.

TABLE 25-2 Causes of Cirrhosis

Alcoholic liver disease
Autoimmune hepatitis
Bile duct obstructions • Complications of gallbladder surgery • Cystic fibrosis • Diseases that cause bile duct injury
Drug-induced liver injury
Inherited disorders • Galactosemia • Glycogen storage disease • Hemochromatosis (causes excessive liver iron) • Wilson's disease (causes excessive liver copper)
Nonalcoholic fatty liver disease
Viral hepatitis • Hepatitis B • Hepatitis C

Consequences of Cirrhosis

About 40 percent of cirrhosis patients are asymptomatic.[15] Because liver damage progresses slowly, the effects of cirrhosis may be subtle at first. Initial symptoms are usually nonspecific and may include

© Martin Rotker/Phototake

Normal liver tissue is smooth and has a regular texture.

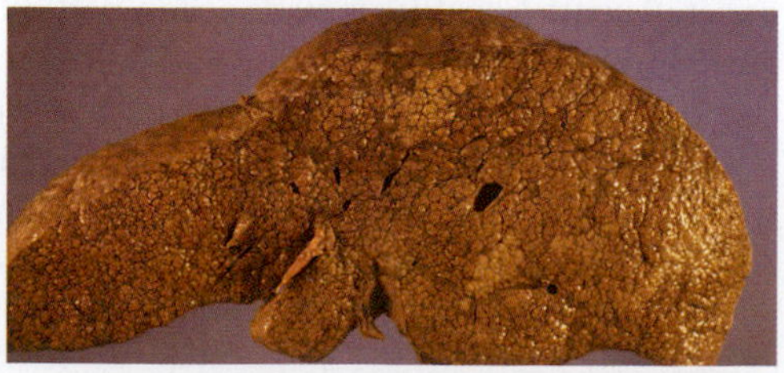

© Dr. Joseph William/Phototake

A cirrhotic liver has an irregular, nodular appearance. The nodules represent clusters of regenerating cells within the damaged liver tissue.

TABLE 25-3 Laboratory Tests for Evaluation of Liver Disease

Laboratory Test	Normal Ranges (serum)	Values in Liver Disease
Alanine aminotransferase (ALT)	Male: 10–40 U/L Female: 7–35 U/L	Elevated
Albumin	3.4–4.8 g/dL	Decreased
Alkaline phosphatase	25–100 U/L	Normal or elevated
Ammonia	15–45 μg N/dL	Elevated
Aspartate aminotransferase (AST)	10–30 U/L	Elevated
Bilirubin (total)	0.3–1.2 mg/dL	Elevated
Blood urea nitrogen (BUN)	6–20 mg/dL	Normal or decreased
Prothrombin time[a]	10–13 seconds	Prolonged

[a]The test for prothrombin time evaluates the clotting ability of blood.
NOTE: U/L = units per liter; dL = deciliter; μg = micrograms; N = nitrogen

FIGURE 25-2 Clinical Effects of Liver Cirrhosis

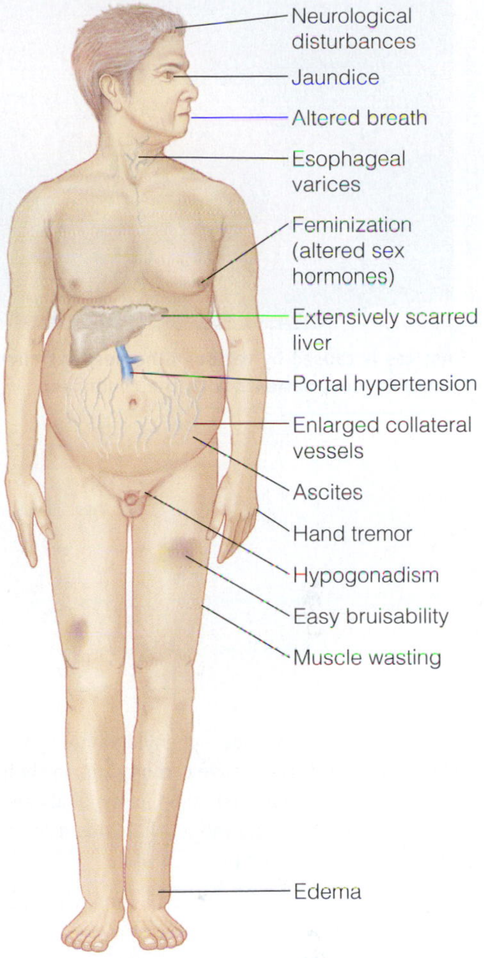

fatigue, weakness, anorexia, and weight loss. Later, the decline in liver function can lead to metabolic disturbances: patients may develop anemia, bruise easily, and be more susceptible to infections. If bile obstruction occurs, jaundice and fat malabsorption are likely. The physical changes in liver tissue may interfere with blood flow, causing fluid to accumulate in blood vessels and body tissues. Advanced cirrhosis can disrupt kidney and lung function. Figure 25-2 illustrates some of the clinical effects of liver cirrhosis, and later sections describe some of these complications in more detail.

Table 25-3 lists laboratory tests that are used to monitor the extent of liver damage. Serum liver enzyme levels are elevated in liver disease because the injured liver tissue releases the enzymes into the bloodstream. Serum levels of bilirubin may be increased if the liver is too damaged to process it or if bile ducts are blocked and prevent its excretion. Reduced synthesis of plasma proteins in the liver lowers albumin levels and extends blood-clotting time. Liver damage also impairs the conversion of ammonia to urea, causing ammonia levels in the blood to rise.

Portal Hypertension A large volume of blood normally flows through the liver. The portal vein ♦ and hepatic artery together supply approximately 1500 milliliters (about 1.5 quarts) of blood each minute to the extensive network of vessels in the liver. The scarred tissue of a cirrhotic liver impedes the flow of blood, three-fourths of which is supplied by the portal vein.[16] The resistance to blood flow within the liver causes a rise in blood pressure within the portal vein, called **portal hypertension**.

♦ Reminder: The *portal vein* is the large blood vessel that carries nutrient-rich blood from the digestive tract to the liver.

Collaterals and Gastroesophageal Varices When blood flow through the portal vein is impeded, the blood is diverted to the smaller blood vessels surrounding the liver. These **collaterals** develop throughout the gastrointestinal (GI) tract and in regions near the abdominal wall. As pressure builds, the collateral vessels become enlarged and engorged, forming abnormally dilated vessels called **varices** (see the photo). The varices that develop in the esophagus (*esophageal varices*) and stomach (*gastric varices*) are vulnerable to rupture because they have thin walls and often bulge into the lumen. If ruptured, they can cause massive bleeding that is sometimes fatal. The blood loss is exacerbated by the liver's reduced production of blood-clotting factors.

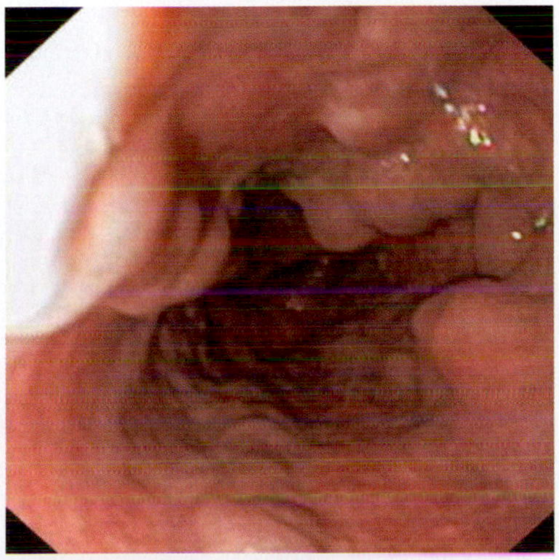

Esophageal varices, such as the one shown here, may protrude into the lumen and be vulnerable to rupture and bleeding.

Ascites Within 10 years of disease onset, about 50 percent of cirrhosis patients develop **ascites**, a large accumulation of fluid in the abdominal cavity. The development of ascites indicates that liver damage has reached a critical stage, as half of patients with ascites die within 2 years.[17] Ascites is primarily a consequence of portal hypertension, sodium and water retention in the kidneys, and reduced albumin

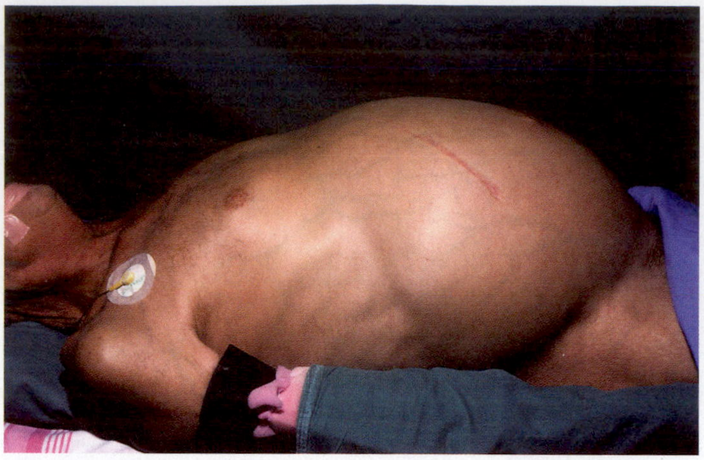

Ascites is caused by various illnesses, but cirrhosis is the underlying cause in most patients with the condition.

synthesis in the diseased liver. The increased pressure within the portal vein triggers the release of chemical factors (such as nitric oxide) that dilate local blood vessels and lower the blood volume elsewhere; these changes activate sodium and water retention in the kidneys and cause an accumulation of body fluid. Elevated pressure within the liver's small blood vessels (**sinusoids**) causes leaking of fluid into lymphatic vessels and, ultimately, the abdominal cavity. The water pooling is exacerbated by low levels of serum albumin, a protein that helps to retain fluid in blood vessels. Ascites can cause abdominal discomfort and early satiety, which contribute to malnutrition. Because ascites can raise body weight considerably, weight changes may be difficult to interpret.

Hepatic Encephalopathy Advanced liver disease sometimes leads to **hepatic encephalopathy**, a disorder characterized by abnormal neurological functioning. Symptoms of hepatic encephalopathy include changes in personality, behavior, mental abilities, and motor functions (see Table 25-4). At worst, amnesia, seizures, and **hepatic coma** may develop. Although hepatic encephalopathy is fully reversible with medical treatment, the prognosis is poor when it progresses to the advanced stages.

The exact causes of hepatic encephalopathy remain elusive, although elevated blood ammonia levels are thought to play a key role in its development due to ammonia's neurotoxicity. Other substances that may accumulate in brain tissue and disturb brain function include sulfur compounds, short-chain fatty acids, and manganese.[18] Another theory is that the brain's neurotransmitters are altered by an increased ratio of aromatic amino acids to branched-chain amino acids in brain tissue, a result of disordered amino acid metabolism in the liver. ♦ Most likely, a combination of metabolic abnormalities contributes to the disruption in neurological functioning.

♦ The aromatic amino acids—phenylalanine, tyrosine, and tryptophan—have carbon rings in their side groups. The branched-chain amino acids are leucine, isoleucine, and valine; their side groups have a branched structure.

Elevated Blood Ammonia Levels Much of the body's free ammonia is produced by bacterial action on unabsorbed dietary protein in the colon. Normally, the liver extracts this ammonia from portal blood and converts it to urea, which is then excreted by the kidneys. In advanced liver disease, the liver is unable to process the ammonia sufficiently. In addition, the ammonia-laden portal blood bypasses the liver by way of collateral vessels and reaches the general blood circulation, causing a substantial increase in the ammonia that reaches brain tissue. Although ammonia levels do not correlate well with the degree of neurological impairment in hepatic encephalopathy, ammonia-reducing medications can successfully reverse the neurological symptoms.[19]

Malnutrition and Wasting Most patients with cirrhosis develop protein-energy malnutrition (PEM) and experience some degree of wasting. Malnutrition is usually caused by a combination of factors (see Table 25-5). Patients may consume less food due to reduced appetite, GI symptoms, early satiety associated with ascites, or fatigue. If the diet is restricted in sodium (to treat ascites), foods may seem unpalatable. Fat malabsorption is common due to reduced bile flow, which may lead to ste-

TABLE 25-4 Symptoms of Hepatic Encephalopathy

Early Stages	Middle Stages	Later Stages
Short attention span	Disorientation, impaired memory	Confusion, amnesia
Depression, irritability	Anxiety, impaired judgment	Anger, paranoia
Lack of coordination, tremor	Slurred speech, abnormal reflexes	Muscular rigidity, abnormal reflexes
Sleep disorders	Lethargy	Semi-stupor, coma

sinusoids: the small, capillary-like passages that carry blood through liver tissue.

hepatic encephalopathy (en-sef-ah-LOP-ah-thie): a condition that develops in advanced liver disease that is characterized by altered neurological functioning, including changes in personality and behavior, reduced mental abilities, and disturbances in motor function.

- **encephalo** = brain
- **pathy** = disease

hepatic coma: loss of consciousness resulting from severe liver disease.

TABLE 25-5 Possible Causes of Malnutrition in Liver Disease

Mechanism	Examples
Reduced nutrient intake	Abdominal discomfort, altered mental status, anorexia, early satiety (due to ascites), effects of medications (including gastrointestinal disturbances and taste changes), fasting for medical procedures, fatigue, nausea and vomiting, restrictive diets
Malabsorption or nutrient losses	Diarrhea, effects of medications (including malabsorption and nutrient losses from diuretic use), fat malabsorption (due to reduced bile flow), gastrointestinal bleeding, vomiting
Altered metabolism or increased nutrient needs	Hypermetabolism, impaired protein synthesis, infections or inflammation, muscle catabolism, reduced nutrient storage and metabolism in the liver

atorrhea and deficiencies of the fat-soluble vitamins and some minerals. Additional nutrient losses may result from diarrhea, vomiting, and GI bleeding. If cirrhosis is a consequence of alcohol abuse, multiple nutrient deficiencies may be present.

Treatment of Cirrhosis

Medical treatment for cirrhosis aims to correct both the underlying cause of disease and any complications that develop. Supportive care, including an appropriate diet and avoidance of liver toxins, promotes recovery and helps to prevent further damage. Abstinence from alcohol is critical for preserving the remaining liver function and extending survival. Antiviral medications may be prescribed to treat viral infections. Patients should be screened and treated for life-threatening complications, such as gastroesophageal varices and liver cancer. Liver transplantation may be necessary in advanced cirrhosis.

Medications can effectively treat many of the complications that accompany cirrhosis. Individuals with portal hypertension and varices may be given propranolol (Inderal) or octreotide (Sandostatin), which reduce both portal blood pressure and bleeding risk. Diuretics can help to control portal hypertension and ascites; common examples include spironolactone (Aldactone) and furosemide (Lasix). Lactulose, a nonabsorbable disaccharide, treats hepatic encephalopathy by reducing ammonia production and absorption in the colon. The antibiotic rifaximin is an alternative treatment for elevated ammonia that works by altering bacterial populations. To stimulate the appetite and promote weight gain, megestrol acetate (Megace) or dronabinol (Marinol) may be prescribed. The Diet-Drug Interactions feature lists potential nutritional problems associated with these medications.

Nutrition Therapy for Cirrhosis

Nutrition care for cirrhosis is customized to each patient's needs, which vary considerably and depend on the accompanying complications. Common problems include PEM and muscle wasting; thus, protein and energy intakes must be sufficient for maintaining nitrogen balance. Dietary substances that may cause additional liver injury should be avoided; examples include alcohol, some herbal supplements, and vitamin or mineral megadoses. Table 25-6 lists the general dietary guidelines for cirrhosis.

Energy Energy requirements may range from 20 to 40 percent above resting metabolic rate (RMR); if possible, indirect calorimetry should be used to determine RMR. Hypermetabolism, infection, nutrient malabsorption, and recent unintentional weight loss can increase energy needs. For patients with ascites, RMR calculations should use either the patient's desirable weight or an estimated dry weight (weight without ascites). A value for dry weight can be obtained after diuretic therapy or after a medical procedure that directly removes excess abdominal fluid.

Many patients with cirrhosis have difficulty consuming enough food to achieve good nutrition status. Some individuals may find four to six small meals easier to tolerate than three large meals each day. Nutritional supplements, including liquid formulas and energy bars, can be used to improve energy intakes. The "How To" offers additional suggestions that can help a patient meet energy needs.

DIET-DRUG Interactions

Check this table for notable nutrition-related effects of the medications discussed in this chapter.

Appetite stimulants (megestrol acetate, dronabinol)	**Gastrointestinal effects:** Nausea, vomiting, diarrhea **Metabolic effects:** Hyperglycemia (megestrol acetate)
Diuretics (furosemide, spironolactone[a])	**Gastrointestinal effects:** Dry mouth, anorexia, decreased taste perception **Dietary interactions:** Furosemide's bioavailability is reduced when taken with food **Metabolic effects:** Fluid and electrolyte imbalances,[a] hyperglycemia (spironolactone), hyperlipidemia (spironolactone), thiamin and zinc deficiencies
Immunosuppressants (cyclosporine, tacrolimus)	**Gastrointestinal effects:** Nausea, vomiting, diarrhea, anorexia (tacrolimus) **Dietary interactions:** Cyclosporine potentiates the effects of alcohol. The bioavailability of tacrolimus is reduced when the drug is taken with food. Grapefruit juice can raise serum concentrations of these drugs to toxic levels. **Metabolic effects:** Electrolyte imbalances, hypertension, hyperglycemia, hyperlipidemia
Lactulose	**Gastrointestinal effects:** Diarrhea **Metabolic effects:** Fluid and electrolyte imbalances

[a]*Furosemide* is a "potassium-wasting" diuretic; patients should increase intakes of potassium-rich foods. *Spironolactone* is a "potassium-sparing" diuretic; patients should avoid supplemental potassium and potassium-containing salt substitutes.

TABLE 25-6 Nutrition Therapy for Liver Cirrhosis

Energy	• Energy needs may range from 20 to 40 percent above resting metabolic rate (RMR); if possible, use indirect calorimetry to determine RMR. • Use estimated dry body weight for RMR calculations in patients with ascites. • Energy requirements may be higher in patients with hypermetabolism, infection, malabsorption, or malnutrition. • Energy requirements may be lower in patients who would benefit from weight loss.
Meal frequency	• To improve food intake, patients should consume small meals four to six times daily.
Protein	• Provide 0.8 to 1.2 g protein per kilogram of dry body weight per day to maintain nitrogen balance and prevent wasting.
Carbohydrate	• No carbohydrate restrictions unless the patient has insulin resistance or diabetes. • For persons with insulin resistance or diabetes, monitor carbohydrate intakes and provide a diet that maintains blood glucose control.
Fat	• No fat restrictions unless fat malabsorption is present. • If fat is malabsorbed, restrict fat to 30 percent of total kcalories or as necessary to control steatorrhea; use medium-chain triglycerides (MCT) to increase kcalories.
Sodium and fluid	• Restrict sodium as necessary to control ascites; 2000 mg sodium per day is adequate restriction in most cases. • If ascites is accompanied by low serum sodium levels (less than 128 mEq/L), restrict fluids to 1200 to 1500 mL per day. In severe cases (serum sodium less than 125 mEq/L), restrict fluids to 1000 to 1200 mL per day.
Vitamins and minerals	• Ensure adequate intake from diet or supplements based on individual needs.

HOW TO Help the Cirrhosis Patient Eat Enough Food

Individuals with cirrhosis often have difficulty consuming enough food to prevent malnutrition and its consequences. Ascites and gastrointestinal symptoms such as nausea and vomiting may interfere with food intake. Fatigue may cause a lack of interest in food preparation. Sodium restrictions may make foods unpalatable. To improve food intake:

- If nutrient restrictions are necessary, make sure the patient fully understands how to modify the diet so that food intake is not restricted unnecessarily. Provide lists of acceptable foods and menus. Explain how recipes can be altered so that favorite foods can still be incorporated into the diet.
- Suggest between-meal snacks during the day and a snack at bedtime. A liquid supplement like Ensure can substitute for a snack and requires no preparation. Snacks should not be consumed within two hours of meals, or they may reduce appetite at mealtime.
- If the patient has little appetite or is quickly satiated, suggest foods that are higher in food energy, such as whole milk instead of reduced-fat milk or canned fruit that is packed in heavy syrup instead of fruit juice. Suggest that beverages be consumed separately from meals.
- Recommend energy boosters. Cream sauces and gravies can add kcalories to entrées. Fruit juices and fruit nectars can substitute for drinking water. The following additions can boost the energy content of meals:
  - Sour cream and butter—on vegetables and potatoes
  - Mayonnaise—in sandwiches and salads
  - Half-and-half and light cream—in soups and on cereals
  - Hard-cooked eggs—in casseroles and meat loaf
  - Cheese—in salads and casseroles and melted on steamed vegetables
  - Peanut butter, nut butters, and cream cheese—on crackers or celery and in milk shakes
  - Chopped nuts—in salads, cooked cereals, and bakery products

Sodium-controlled diets are recommended for treating ascites and other medical conditions, including kidney and heart disorders. The "How To" on p. 832 offers suggestions to help patients implement sodium restrictions. To improve the palatability of low-sodium meals:

- Suggest that patients replace the salt they use for cooking and seasoning with strong-flavored herbs and spices such as chili powder, coriander, cumin, curry powder, garlic, ginger, lemon, mint, and parsley.
- Advise patients to check food labels to learn the sodium content of the foods they eat. Similar products may be available that are lower in sodium. (Persons using potassium-sparing diuretics should be cautioned to avoid salt substitutes that replace sodium with potassium.)

Offer support and encouragement to the patient with cirrhosis. Significant weight loss is less likely to occur if dietary advice is provided before problems progress.

TRY IT Think of some appetizers or entrées you frequently eat. Describe how you might prepare these items to increase their protein and energy contents. If these foods are typically prepared with salt, can you think of some seasonings you can substitute instead?

Protein The protein recommendation is 0.8 to 1.2 grams of protein per kilogram of body weight per day, based on dry weight or an appropriate weight for height.[20] ♦ Patients with hepatic encephalopathy should avoid excessive protein consumption, and their protein intake should be spread throughout the day so that they consume only modest amounts at each meal. Protein restriction is not helpful, however, because an inadequate protein intake can worsen malnutrition and wasting.

♦ Reminder: The protein RDA for healthy adults is 0.8 g/kg.

In an attempt to normalize amino acid ratios in blood plasma and brain tissue and possibly improve the mental status of patients with hepatic encephalopathy, some health care providers may prescribe enteral formulas enriched with branched-chain amino acids. Clinical studies testing the use of these formulas have yielded mixed results, however, and their use is recommended only in patients who do not respond to conventional treatment.[21]

Carbohydrate Carbohydrate provides a substantial proportion of energy needs. Many patients with cirrhosis are insulin resistant, however, and require medications or insulin to manage their hyperglycemia. These individuals should follow the dietary guidelines for diabetes: monitor carbohydrate intakes and consume a diet that maintains blood glucose levels within a normal range (see Chapter 26).

Carbohydrate intakes should be fairly consistent from day to day for improved blood glucose control.

Fat Fat provides both energy and essential fatty acids. In patients with fat malabsorption, fat intake may be restricted to less than 30 percent of total kcalories or as necessary to control steatorrhea. Medium-chain triglycerides (MCT) may be used to provide additional energy, although essential fatty acids cannot be obtained from MCT oils and may need to be supplemented. Severe steatorrhea warrants supplementation of the fat-soluble vitamins, calcium, magnesium, and zinc (see Chapter 24).

Sodium and Fluid Patients with ascites are generally advised to restrict sodium. Because ascites is partly caused by sodium and water retention in the kidneys, treatment usually includes both a moderate sodium restriction (to no more than 2000 milligrams of sodium per day) and diuretic therapy to promote fluid loss. ♦ Potassium intake should be monitored if a potassium-wasting diuretic (such as furosemide) is used.

♦ Table 28-1 (p. 847) and the "How To" on p. 832 provide information about following a sodium-restricted diet.

Many patients find low-sodium diets unpalatable, so some health practitioners may allow a more liberal sodium intake and depend on diuretics to mobilize excess fluids. If patients do not respond to sodium restriction and diuretic therapy, fluid may be removed from the abdomen by surgical puncture (**paracentesis**) or may be diverted to the bloodstream using a catheter (**peritoneovenous shunt**).

Fluid restriction may be necessary when ascites is accompanied by a low concentration of serum sodium. If the sodium level falls below 128 milliequivalents per liter, the fluid intake should be limited to 1200 to 1500 milliliters daily; with a sodium level below 125 milliequivalents per liter, fluids should be restricted to 1000 to 1200 milliliters per day.[22]

Vitamins and Minerals Vitamin and mineral deficiencies are common in patients with cirrhosis due to the effects of illness, disease complications, or the alcohol abuse that may have induced the liver disease. Therefore, multivitamin supplementation is often necessary. If steatorrhea is present, fat-soluble nutrients can be provided in water-soluble forms. Patients with esophageal varices may find it easier to ingest supplements in liquid form.

Enteral and Parenteral Nutrition Support In patients who are unable to consume enough food, tube feedings may be infused overnight as a supplement to oral intakes or may replace oral feedings entirely. Although standard formulas are often appropriate, an energy-dense, moderate-protein, low-electrolyte formula may be necessary for patients with ascites or fluid restrictions. In patients with esophageal varices, the feeding tube should be as narrow and flexible as possible to prevent rupture and bleeding. Parenteral nutrition support should be considered for patients who are unable to tolerate enteral feedings due to intestinal obstruction, gastrointestinal bleeding, or uncontrollable vomiting. To avoid excessive fluid delivery, patients with ascites typically require concentrated parenteral solutions, which are infused into central veins. The Case Study allows you to apply your knowledge of cirrhosis to a clinical situation.

IN SUMMARY **Liver cirrhosis is characterized by extensive fibrosis and progressive liver dysfunction. The primary causes of cirrhosis in the United States are alcoholic liver disease and hepatitis C infection. Symptoms of cirrhosis include fatigue, GI disturbances, anorexia, and weight loss; eventually, patients may bruise easily and be more susceptible to infections. Complications of cirrhosis include portal hypertension, gastroesophageal varices, ascites, and hepatic encephalopathy. Treatment is highly individualized and depends on the accompanying symptoms and complications. Both drug therapies and dietary adjustments are usually necessary. If warranted, the diet may need to be restricted in fat, sodium, or fluids.**

paracentesis (pah-rah-sen-TEE-sis): a surgical puncture of a body cavity with an aspirator to draw out excess fluid.

peritoneovenous (PEH-rih-toe-NEE-oh-VEE-nus) **shunt:** a surgical passage created between the peritoneum and the jugular vein to divert fluid and relieve ascites. The peritoneum is the membrane that surrounds the abdominal cavity.

CASE STUDY Man with Cirrhosis

Lenny Levitt, a 49-year-old carpenter, has just been diagnosed with cirrhosis, which is a consequence of his alcohol abuse over the past 25 years. Although he understands that he has an alcohol problem and recently entered an alcohol rehabilitation program, he is still drinking. At 5 feet 8 inches tall, Mr. Levitt, who formerly weighed 160 pounds, now weighs 130 pounds. According to family members, he is showing signs of mental deterioration, such as forgetfulness and an inability to concentrate. He is jaundiced and appears thin, although his abdomen is distended with ascites. Laboratory findings indicate elevated serum concentrations of AST, ALT, and ammonia; reduced albumin levels; and hyperglycemia.

1. Do Mr. Levitt's laboratory values suggest liver disease? Compare the results of his laboratory tests with the values shown in Table 25-3.
2. From the limited information available, evaluate Mr. Levitt's nutrition status. What medical problem makes it difficult to interpret his present weight? Describe the development of that type of problem in liver disease, and explain how the diet is usually adjusted for such a patient.
3. Estimate Mr. Levitt's energy and protein needs. Describe the general diet you might recommend for him. What suggestions do you have for increasing his energy intake?
4. Explain the significance of Mr. Levitt's elevated blood ammonia levels. What are some signs that would indicate that he is undergoing mental decline?
5. Describe each of the following complications of liver disease: portal hypertension, jaundice, and gastroesophageal varices. What complication may result if the esophageal varices are not treated?

Liver Transplantation

Acute or chronic liver disease can lead to liver failure, in which case liver transplantation is the only remaining treatment option. The most common illnesses that precede liver transplantation are chronic hepatitis C infection and alcoholic liver disease, which account for about 50 percent of liver transplant cases.[23] The five-year survival rate among transplant recipients ranges from 58 to 81 percent, depending on the cause of illness.[24] Complications such as ascites and hepatic encephalopathy worsen the prognosis.

Nutrition Status of Transplant Patients As mentioned earlier, advanced liver disease is usually associated with malnutrition, which can increase the risk of complications following a liver transplant. Evaluating nutrition status in transplant candidates can be difficult, however, because liver dysfunction and malnutrition often have similar metabolic effects. If fluid retention is present, it can mask weight loss and alter anthropometric and laboratory values. Correcting malnutrition prior to transplant surgery can help speed recovery after the surgery.

Posttransplantation Concerns The immediate concerns following a transplant are organ rejection and infection. Immunosuppressive drugs, including prednisone, tacrolimus, and cyclosporine, help to reduce the immune responses that cause rejection, but they also raise the risk of infection. Infections are a potential cause of death following a liver transplant; therefore, antibiotics and antiviral medications are prescribed to reduce infection risk.

Immunosuppressive drugs can affect nutrition status in numerous ways. Gastrointestinal side effects include nausea, vomiting, diarrhea, abdominal pain, and mouth sores. Some medications may alter appetite and taste perception. Some of the drugs may cause hyperglycemia or outright diabetes, which may need to be controlled with insulin. Electrolyte and fluid imbalances are common. Other possible effects include hypertension, hyperlipidemias, protein catabolism, and increased osteoporosis risk.

Protein and energy requirements are increased after transplantation due to the stress of surgery. High-kcalorie, high-protein snacks and enteral supplements can help the transplant patient meet postsurgical needs. Vitamin and mineral supplementation is also an integral part of nutrition care. To help transplant patients avoid developing foodborne illnesses, health practitioners can provide information about food safety measures, such as cooking meats adequately, washing fresh produce, and avoiding foods that may be contaminated. Highlight 18 (pp. 609–617) provides additional information about food safety.

IN SUMMARY **Liver transplantation has improved the long-term outlook for patients with advanced liver disease. Transplant patients are usually malnourished and may have medical problems that affect transplant success. Due to the potential for organ rejection, immunosuppressive drugs are prescribed following surgery. Use of these drugs increases the risk of infection, and the drugs have side effects that can impair nutrition status and general health.**

Gallstone Disease

As described earlier in this chapter, the gallbladder concentrates and stores the bile produced by the liver until the bile is needed for fat digestion (see Figure 25-3). Disorders that obstruct the liver's release of bile can damage the liver. More commonly, disorders of the biliary system—the gallbladder and bile ducts—involve the formation of **gallstones**. Gallstones affect an estimated 20 million people in the United States, or about 12 percent of the adult population.[25]

Types of Gallstones

The formation of gallstones, or **cholelithiasis**, results from the excessive concentration and crystallization of compounds in bile. Bile is a solution of bile salts, cholesterol, phospholipids (primarily lecithin), proteins, and bile pigment (bilirubin). While stored in the gallbladder, bile's concentration increases approximately tenfold as its water content is extracted. Factors that raise bile's cholesterol concentration, promote crystal formation and development, or reduce gallbladder motility favor gallstone formation.[26]

FIGURE 25-3 The Gallbladder and Bile Ducts

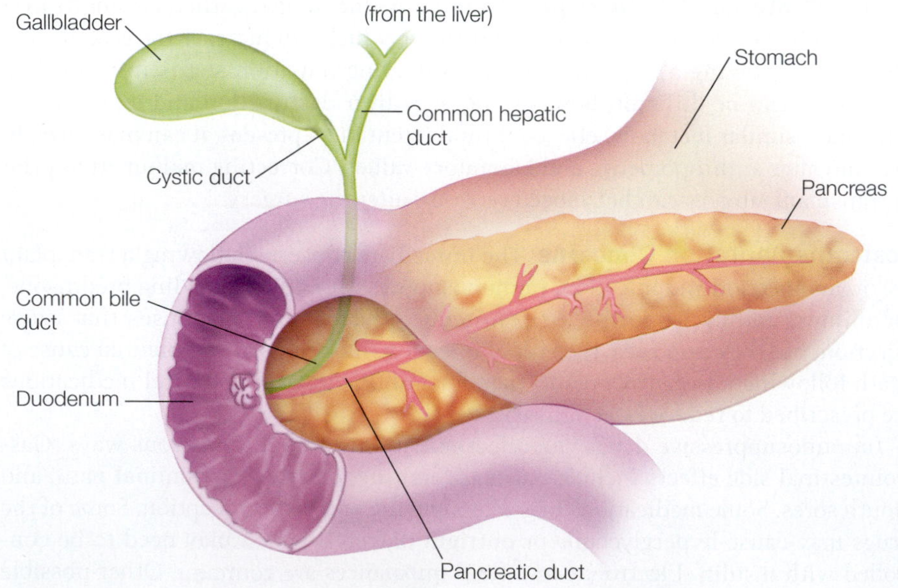

gallstones: stones that form in the gallbladder from crystalline deposits of cholesterol or bilirubin.

cholelithiasis (KOH-leh-lih-THIGH-ah-sis): formation of gallstones.

- **chole** = bile
- **lithiasis** = formation of stones

Cholesterol Gallstones In about 80 percent of cases, the gallstones are composed primarily of cholesterol, although they also contain calcium salts and bilirubin.[27] The cholesterol in bile can precipitate out of solution and form small crystals, which eventually coalesce to form stones. The stones can be as small as a pea or as large as a Ping-Pong ball. Some people tend to form many small stones, while others may form only one or two large ones.

Cholesterol gallstones often develop after the bile concentrate thickens and forms a type of **sludge** that cannot be easily expelled by gallbladder contraction. Biliary sludge may develop after rapid weight loss or fasting, gastric bypass surgery, or long-term total parenteral nutrition, and it can also occur during pregnancy.

Courtesy of Dr. David King. © Myrna Engler

Most gallstones are made primarily of cholesterol; they can be as small as a pea or as large as a Ping-Pong ball.

Pigment Gallstones Although pigment stones account for a minority of cases in the United States, they are the predominant type of gallstone in some Asian countries. Pigment stones are primarily made up of the calcium salt of bilirubin (calcium bilirubinate). They often develop as a result of bacterial infection, which alters the structure of bilirubin and causes it to precipitate out of bile and form stones. Other cases result from excessive red blood cell breakdown, leading to an abnormal accumulation of bilirubin. Conditions associated with pigment stone formation include biliary tract infections, pancreatitis, and red blood cell disorders, such as sickle-cell anemia. Pigment stones may form in either the gallbladder or a bile duct. Unlike the crystalline cholesterol stones, pigment stones are soft and easily crushed.

Consequences of Gallstones

About 80 percent of gallstones are asymptomatic and are discovered accidentally while testing for other conditions.[28] In other cases, patients may experience an aggressive course of illness with recurring symptoms.

Gallstone Symptoms Gallstone pain usually arises when a gallstone temporarily blocks the cystic duct, which leads from the gallbladder to the common bile duct (review Figure 25-3). The pain is steady and severe and may last for several minutes or several hours. Although the pain is usually located in the upper abdomen, it may radiate to the chest or to the back. Nausea, vomiting, and bloating may also be present. Symptoms usually develop after meals, especially after eating fatty foods. Pain may also occur during the night and awaken a person from sleep.

Complications of Gallstones If a gallstone remains lodged in the cystic duct, it can obstruct bile flow to the duodenum and cause **cholecystitis**—distention and inflammation of the gallbladder. Cholecystitis can lead to infection or to more severe complications, including perforation of the gallbladder, **peritonitis**, and fistulas. If gallstones obstruct the common bile duct, they can block bile flow from the liver and lead to jaundice or damage to liver tissue. An impacted stone within the bile ducts may lead to infection and the condition known as **bacterial cholangitis**, which causes severe pain, sepsis, and fever and is often a medical emergency. Gallstones can block the pancreatic duct as well—a primary cause of acute pancreatitis. Due to the potential danger of these complications, individuals should seek medical attention if gallstone pain does not resolve over time or if fever, jaundice, or persistent nausea and vomiting develop.

Risk Factors for Cholesterol Gallstones

The risk of developing cholesterol gallstones is influenced by a number of genetic and lifestyle factors. As described in the sections that follow, the risk factors may either cause an increase in bile's cholesterol concentration or a reduction in gallbladder motility, thereby promoting gallstone crystallization or subsequent stone growth.

Ethnicity Although the genetic factors related to gallstone formation are not well understood, ethnicity strongly influences gallstone formation. The Pima Indians

sludge: literally, a semisolid mass. Biliary sludge is made up of mucus, cholesterol crystals, and bilirubin granules.

cholecystitis (KOH-leh-sih-STY-tis): inflammation of the gallbladder, usually caused by obstruction of the cystic duct by gallstones.

peritonitis: inflammation of the peritoneal membrane, which lines the abdominal cavity.

bacterial cholangitis (KOH-lan-JYE-tis): bacterial infection involving the bile ducts.

are an exceptionally high-risk population; gallstones develop in about 70 percent of adult women. Other high-risk populations include Scandinavians, Chileans, Hispanic Americans, and various Native American groups in North and South America. In the United States, African Americans have a lower prevalence of gallstones than white populations.[29]

Aging Because gallstones do not dissolve spontaneously, gallstone prevalence increases with age. Moreover, bile composition tends to change with aging: the cholesterol concentration increases while bile salts decrease, leading to a greater likelihood of cholesterol crystallization.

Gender The prevalence of cholesterol gallstones is about twice as high in women as in men.[30] The reason for the gender difference is that estrogen alters cholesterol metabolism, causing an increased secretion of cholesterol into bile. The use of estrogen replacement therapy after menopause increases gallstone risk by promoting additional cholesterol secretion into bile.

Pregnancy Some women experience their first gallstone symptoms during pregnancy. Gallstone risk is increased in pregnancy due to hormonal changes: the higher serum estrogen levels raise bile's cholesterol concentration, and the increase in progesterone levels reduces gallbladder motility. The risk of gallstones worsens as the pregnancy progresses and is especially high during the third trimester.

Obesity and Weight Loss Obesity is associated with increased cholesterol synthesis in the liver, leading to higher cholesterol concentrations in bile. In a large study of disease risk in women of different weights, researchers found that the incidence of gallstones was seven times higher in women with body mass indexes (BMIs) greater than 45 as compared with nonobese women.[31] ♦

♦ Reminder: A healthy weight usually falls between a BMI of 18.5 and 25.0. Overweight and obesity are usually defined by BMIs above 25 and 30, respectively.

Gallstones frequently develop as a result of rapid weight loss, occurring in about one quarter of obese persons who undergo severe kcalorie restriction and up to half of those who undergo gastric bypass surgery.[32] Rapid weight loss increases the secretion of cholesterol into bile and may also decrease gallbladder motility. Another effect of rapid weight loss is the increased production of the gallbladder's *mucin* proteins, which are a major component of biliary sludge and also serve as a matrix for cholesterol crystals during stone growth. The oral ingestion of bile salts has been shown to reduce the risk of gallstone formation during rapid weight loss.

Other Risk Factors Long-term total parenteral nutrition usually reduces gallbladder motility, which promotes the development of biliary sludge. Some medications (such as octreotide) may have similar effects. The medication clofibrate, used for heart disease, increases the cholesterol concentration of bile, promoting gallstone crystallization. High triglyceride levels in blood are also associated with increased gallstone risk, as are spinal cord injuries and diseases affecting the ileum.

Treatment of Gallstones

Asymptomatic gallstones generally do not require treatment. Gallstones that cause symptoms or complications are usually treated with gallbladder surgery or nonsurgical procedures that dissolve or fragment the stones. To minimize symptoms before the gallbladder or gallstones are removed, a low-fat diet (with less than 30 percent of total kcalories from fat) may be prescribed; some individuals may tolerate small, frequent meals better than large meals.[33]

Surgery Gallbladder removal, or **cholecystectomy**, is the primary treatment for patients with recurring gallstones.[34] The preferred surgical approach is a **laparoscopic** method, which relies on narrow surgical telescopes (laparoscopes) to view and perform the necessary procedures via small incisions in the abdomen. The procedure takes only one or two hours, and many patients are discharged on the same day as the surgery. In patients with complications that make organ removal diffi-

cholecystectomy (KOH-leh-sis-TEK-toe-mee): surgical removal of the gallbladder.

laparoscopic: pertaining to procedures that use a laparoscope for internal examination or surgery. A laparoscope is a narrow surgical telescope that is inserted into the abdominal cavity through a small incision. A video camera is usually attached so that the procedure can be viewed on a television monitor.

cult, open cholecystectomy may be performed. In this procedure, the surgeon cuts through the abdominal muscle and exposes the abdominal cavity, allowing direct access to the gallbladder and bile ducts. An open cholecystectomy is associated with a greater risk of infection, more pain, and a lengthier recovery time than the laparoscopic procedure.

Once the gallbladder has been removed, the common bile duct collects bile between meals and releases it into the duodenum at mealtimes; thus, patients can usually tolerate a regular diet. Some individuals may experience diarrhea due to an increased amount of bile in the large intestine, which has a laxative effect. Abdominal pain is sometimes caused by the presence of residual stones within the common bile duct that were overlooked during surgery or that formed within the duct itself. Bile duct injuries occasionally result from the surgical procedure.

Nonsurgical Procedures Nonsurgical methods are used primarily in patients who have small cholesterol stones and transient conditions associated with gallstone formation. The gallstones can be treated by oral intake of ursodeoxycholic acid (ursodiol), a bile acid that reduces cholesterol secretion by the liver and eventually causes the cholesterol crystals in gallstones to dissolve. Ursodeoxycholic acid must be used for 6 to 12 months and is best suited for stones that are 5 millimeters (about 1/4 inch) in diameter or smaller. Recurrence rates after dissolution are as high as 50 percent.[35]

Cholesterol gallstones can be fragmented using **shock-wave lithotripsy**, a procedure that is also used to fragment kidney stones. This technique uses high-amplitude sound waves (called shock waves) to break gallstones into pieces that are small enough to either pass into the intestine without causing symptoms or be dissolved with ursodeoxycholic acid. Shock-wave lithotripsy can be performed only in patients with few gallstones. Success is highest in patients with solitary stones that are less than 20 millimeters (3/4 inch) in diameter. Recurrence of gallstones has been reported in up to 54 percent of patients using this procedure.[36]

IN SUMMARY **Gallstones are the most common disorder affecting the gallbladder. They are formed by the concentration of compounds in bile, especially cholesterol and the bile pigment bilirubin. Although most gallstones are asymptomatic, some gallstones can cause recurring pain and GI problems that often appear after meals and persist for several hours. The risk of gallstone disease is influenced by ethnicity, gender, pregnancy, obesity, rapid weight loss, and other factors. Treatments include gallbladder removal and gallstone dissolution or fragmentation.**

Clinical Portfolio

1. Vijaya Reddy is a college student who visited relatives near her parents' birthplace in Anantapur, India, during summer vacation. Although her relatives provided boiled or purified water at their home, they occasionally took Vijaya to local restaurants, where she drank tap water. Several weeks after Vijaya returned home, she developed flu-like symptoms and started feeling extremely tired. She also experienced upper abdominal pain and felt nauseous after meals. After her roommate told her that her eyes and skin appeared yellow, she knew something was definitely wrong. A physician at the student health center diagnosed hepatitis.
   - Which type of hepatitis does Vijaya most likely have?
   - What additional symptoms may develop? Is Vijaya's condition likely to become chronic?
   - What medical treatment is suggested for Vijaya's condition? Describe the dietary modifications that may be necessary in some cases.

shock-wave lithotripsy: a nonsurgical procedure that uses high-amplitude sound waves to fragment gallstones or kidney stones.

2. As discussed in the section on cirrhosis, many patients develop protein-energy malnutrition and wasting during the course of illness. Review Table 25-5 to find examples of problems that may lead to malnutrition. Select three nutrition or medical problems (from the "Examples" column), and discuss the complications of liver disease that may cause the problems you selected. What dietary or medical treatments can help in managing these problems?

Nutrition on the Net

For further study of topics covered in this chapter, log on to **www.cengagebrain.com** and search for ISBN 084006845X.

- To obtain additional information about liver diseases, visit the American Liver Foundation and the Canadian Liver Foundation: **www.liverfoundation.org** and **www.liver.ca**
- The websites of the following organizations include information that is helpful for both health practitioners and patients with liver diseases:

  American College of Gastroenterology: **www.acg.gi.org**

  American Gastroenterological Association: **www.gastro.org**
- Learn more about hepatitis by visiting the Hepatitis Foundation International: **www.hepfi.org**
- To uncover more information about liver transplants, search the CenterSpan Transplant News Network: **www.centerspan.org**

Nutrition Assessment Checklist for People with Disorders of the Liver and Gallbladder

Medical History

Check the medical record to determine:

- Type of liver disorder
- Cause of the liver disorder
- If the patient has received a liver transplant
- If the patient has a history of gallstones

Review the medical record for complications that may alter nutritional needs, including:

- Abdominal pain
- Anemia
- Ascites
- Esophageal varices
- Hepatic encephalopathy
- Impaired kidney or lung function
- Infections
- Insulin resistance or diabetes mellitus
- Malabsorption
- Malnutrition
- Pancreatitis

Medications

In patients with liver dysfunction, the risk of diet-drug interactions is high because most drugs are metabolized in the liver. Risk of interactions is intensified for patients with:

- Ascites (medications may take a long time to reach the liver)
- Renal failure (medications are often metabolized further in the kidneys and excreted in the urine)
- Malnutrition
- Multiple prescriptions
- Long-term medication use

Dietary Intake

For patients with fatty liver, pay special attention to:

- Energy intake, if the patient is overweight or malnourished, has diabetes, or is receiving total parenteral nutrition
- Carbohydrate intake, if the patient has diabetes or is receiving total parenteral nutrition
- Alcohol abuse

For patients with hepatitis, cirrhosis, or ascites:

- Check appetite.
- Ensure that energy and nutrient intakes are adequate.
- Determine whether alcohol is being consumed.
- Determine whether sodium or fluid restriction is warranted.
- Base energy intakes on desirable weight or an estimated dry weight to avoid overfeeding.

Anthropometric Data

Take baseline height and weight measurements, and monitor weight regularly. For patients with ascites and edema:

- Monitor weight changes to evaluate the degree of fluid retention.
- Remember that the patient may be malnourished, and weight may be deceptively high.

Laboratory Tests

Note that albumin and serum proteins are often reduced in people with liver disease and are not appropriate indicators of nutrition status. Review the following laboratory test results to assess liver function:

- Albumin
- Alkaline phosphatase
- ALT and AST
- Ammonia
- Bilirubin
- Prothrombin time

Check laboratory test results for complications associated with liver failure, including:

- Anemia
- Decreased renal function
- Fluid retention
- Hyperglycemia

Physical Signs

Look for physical signs of:

- Fluid retention (ascites and edema)
- PEM (muscle wasting and unintentional weight loss)
- Nutrient deficiencies

References

1. A. M. Diehl, Alcoholic and nonalcoholic steatohepatitis, in L. Goldman and D. Ausiello, eds., *Cecil Medicine* (Philadelphia: Saunders, 2008), pp. 1135–1139.
2. D. E. Cohen and F. A. Anania, Nonalcoholic fatty liver disease, in N. J. Greenberger, ed., *Current Diagnosis and Treatment: Gastroenterology, Hepatology, and Endoscopy* (New York: McGraw-Hill Companies, 2009), pp. 467–472; Diehl, 2008.
3. A. E. Reid, Nonalcoholic fatty liver disease, in M. Feldman, L. S. Friedman, and L. J. Brandt, eds., *Sleisenger and Fordtran's Gastrointestinal and Liver Disease* (Philadelphia: Saunders, 2010), pp. 1401–1411.
4. Reid, 2010.
5. J. H. Lewis, Liver disease caused by anesthetics, toxins, and herbal preparations, in M. Feldman, L. S. Friedman, and L. J. Brandt, eds., *Sleisenger and Fordtran's Gastrointestinal and Liver Disease* (Philadelphia: Saunders, 2010), pp. 1447–1459.
6. Centers for Disease Control and Prevention, Surveillance for acute viral hepatitis—United States, 2007, *Morbidity and Mortality Weekly Report* 58, No. SS03 (2009): 1–27.
7. J. M. Crawford and C. Liu, Liver and biliary tract, in V. Kumar and coeditors, *Robbins and Cotran Pathologic Basis of Disease* (Philadelphia: Saunders, 2010), pp. 833–890.
8. Centers for Disease Control and Prevention, 2009.
9. Crawford and Liu, 2010.
10. A. Rutherford and J. L. Dienstag, Viral hepatitis, in N. J. Greenberger, ed., *Current Diagnosis and Treatment: Gastroenterology, Hepatology, and Endoscopy* (New York: McGraw-Hill Companies, 2009), pp. 420–443.
11. S. Safrin, Antiviral agents, in B. G. Katzung, ed., *Basic and Clinical Pharmacology* (New York: McGraw-Hill/Lange, 2007), pp. 790–818.
12. American Dietetic Association, *Nutrition Care Manual* (Chicago: American Dietetic Association, 2010).
13. J. Q. Xu and coauthors, Deaths: Final data for 2007, *National Vital Statistics Reports* 58 (May 2010).
14. G. Garcia-Tsao, Cirrhosis and its sequelae, in L. Goldman and D. Ausiello, eds., *Cecil Medicine* (Philadelphia: Saunders, 2008), pp. 1140–1147.
15. J. J. Heidelbaugh and coauthors, Gastroenterology, in R. E. Rakel, ed., *Textbook of Family Medicine* (Philadelphia: Saunders, 2007), pp. 1115–1171.
16. V. H. Shah and P. S. Kamath, Portal hypertension and gastrointestinal bleeding, in M. Feldman, L. S. Friedman, and L. J. Brandt, eds., *Sleisenger and Fordtran's Gastrointestinal and Liver Disease* (Philadelphia: Saunders, 2010), pp. 1489–1516.
17. B. A. Runyon, Ascites and spontaneous bacterial peritonitis, in M. Feldman, L. S. Friedman, and L. J. Brandt, eds., *Sleisenger and Fordtran's Gastrointestinal and Liver Disease* (Philadelphia: Saunders, 2010), pp. 1517–1541.
18. N. J. Greenberger, Portal systemic encephalopathy and hepatic encephalopathy, in N. J. Greenberger, ed., *Current Diagnosis and Treatment: Gastroenterology, Hepatology, and Endoscopy* (New York: McGraw-Hill Companies, 2009), pp. 473–477.
19. Greenberger, 2009.
20. American Dietetic Association, 2010.
21. S. A. McClave and coauthors, Guidelines for the provision and assessment of nutrition support therapy in the adult critically ill patient: Society of Critical Care Medicine (SCCM) and American Society for Parenteral and Enteral Nutrition (A.S.P.E.N.), *Journal of Parenteral and Enteral Nutrition* 33 (2009): 277–316.
22. American Dietetic Association, 2010.
23. E. B. Keeffe, Hepatic failure and liver transplantation, in L. Goldman and D. Ausiello, eds., *Cecil Medicine* (Philadelphia: Saunders, 2008), pp. 1147–1152.
24. Keeffe, 2008.
25. D. Q.-H. Wang and N. H. Afdhal, Gallstone disease, in M. Feldman, L. S. Friedman, and L. J. Brandt, eds., *Sleisenger and Fordtran's Gastrointestinal and Liver Disease* (Philadelphia: Saunders, 2010), pp. 1089–1120.
26. Wang and Afdhal, 2010.
27. G. Paumgartner and N. J. Greenberger, Gallstone disease, in N. J. Greenberger, ed., *Current Diagnosis and Treatment: Gastroenterology, Hepatology, and Endoscopy* (New York: McGraw-Hill Companies, 2009), pp. 537–546.
28. Paumgartner and Greenberger, 2009.
29. Wang and Afdhal, 2010.
30. Paumgartner and Greenberger, 2009.
31. Wang and Afdhal, 2010.
32. Wang and Afdhal, 2010.
33. American Dietetic Association, 2010.
34. R. E. Glasgow and S. J. Mulvihill, Treatment of gallstone disease, in M. Feldman, L. S. Friedman, and L. J. Brandt, eds., *Sleisenger and Fordtran's Gastrointestinal and Liver Disease* (Philadelphia: Saunders, 2010), pp. 1121–1138.
35. Glasgow and Mulvihill, 2010.
36. Glasgow and Mulvihill, 2010.

HIGHLIGHT 25

Anemia in Illness

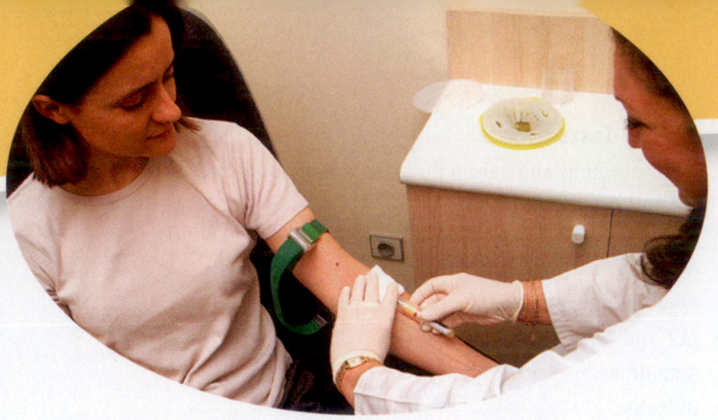

© AJPhoto/Photo Researchers, Inc.

Anemia—a reduction of red blood cells that lowers the oxygen-carrying capacity of the blood—is frequently the first sign of illness and may be the disorder that initially drives an individual to seek medical attention. Anemia is associated with a great number of diseases and is common among hospital patients: some 20 to 40 percent exhibit some degree of anemia.[1] Earlier chapters in this textbook describe some of the relationships between nutrient deficiencies and anemia. This highlight explains how and why anemia develops during the course of illness. The accompanying glossary defines the relevant terms.

Overview of Anemia

Anemia develops when red blood cells (also called *erythrocytes*) are not produced in sufficient numbers, are too quickly destroyed, or are lost due to bleeding. Because red blood cells contain the hemoglobin that supplies oxygen to tissues, their absence can result in fatigue and reduced stamina. The deficiency of oxygen in tissues is the main stimulus for the production of additional red blood cells. Table H25-1 provides an overview of some different categories of anemia and their underlying causes.

Red Blood Cell Production

The production of red blood cells (**erythropoiesis**) takes place in the bone marrow, a soft tissue found in certain types of bone. The process begins when kidney cells sense the low oxygen content of blood and release the hormone **erythropoietin** (see Figure H25-1). Erythropoietin travels to the bone marrow, where it stimulates precursor cells (stem cells) to divide and differentiate into red blood cells. The cells that are released from the bone marrow are immature red blood cells called **reticulocytes**. Reticulocytes develop into mature red blood cells over a 24- to 48-hour period while they circulate in the bloodstream.

TABLE H25-1 Types of Anemia

Type of Anemia	General Mechanism
Anemia of chronic disease	Reduced iron availability due to inflammatory processes; results in reduced red blood cell (RBC) production
Aplastic anemia	Failure of stem cells to develop into RBCs; may be due to immune disease, viruses, drugs and toxins, or genetic defects
Hemolytic anemia	Premature destruction of red blood cells; results in shortened RBC life span and fewer RBCs
Hemorrhagic anemia	Blood loss; causes reduction in circulating RBCs
Iron-deficiency anemia	Reduced iron availability due to dietary deficiency; interferes with hemoglobin production and results in small, hypochromic RBCs
Megaloblastic anemia	Reduced availability of nutrients required for DNA synthesis and cell division; results in large, immature RBCs
Sickle-cell anemia	Genetic mutation that results in altered hemoglobin molecule; causes production of abnormal, sickle-shaped RBCs
Thalassemia	Genetic mutation that reduces hemoglobin synthesis; results in reduced RBC production

Nutritional Anemias

The nutrient deficiencies that most often upset red blood cell production are those of iron, folate, and vitamin B_{12}. Iron is required for hemoglobin production, and deficiency results in **microcytic anemia**, characterized by small, hypochromic cells (see p. 429). Vitamin B_{12} and folate participate in DNA synthesis, and deficiency of either nutrient leads to **megaloblastic anemia** (also called *macrocytic anemia*), characterized by large, immature cells (see p. 329).

GLOSSARY

anemia of chronic disease: anemia that develops in persons with chronic illness; may resemble iron-deficiency anemia even though iron stores are often adequate. Also called *anemia of chronic inflammation*.

aplastic anemia: anemia characterized by the inability of bone marrow to produce adequate numbers of blood cells. Causes include drug toxicity, viruses, and genetic defects.

erythropoiesis (eh-RIH-throh-poy-EE-sis): production of red blood cells within the bone marrow.

erythropoietin (eh-RIH-throh-POY-eh-tin): a hormone produced by kidney cells that stimulates red blood cell production.

hemolytic (hee-moe-LIH-tic) **anemia:** anemia characterized by the breakdown of red blood cells.

megaloblastic anemia: anemia characterized by large, immature red blood cells, as occurs in folate and vitamin B_{12} deficiency; also called *macrocytic anemia*.

microcytic anemia: anemia characterized by small, hypochromic (pale) red blood cells, as occurs in iron deficiency.

peripheral blood smear: a blood sample spread on a glass slide and stained for analysis under a microscope. *Peripheral* refers to the use of circulating blood rather than tissue blood.

reticulocytes: immature red blood cells released into blood by bone marrow.

FIGURE H25-1 Erythropoiesis

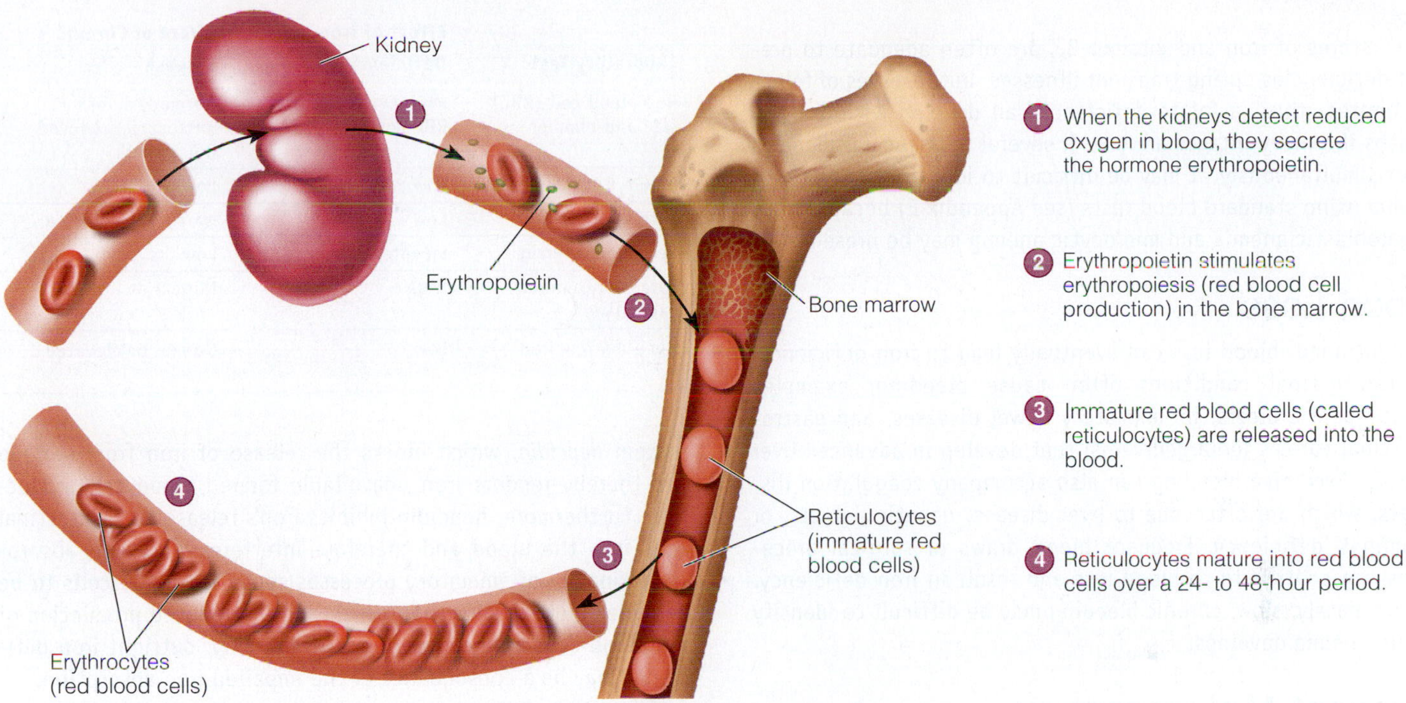

Kidney
Erythropoietin
Bone marrow
Reticulocytes (immature red blood cells)
Erythrocytes (red blood cells)
1 When the kidneys detect reduced oxygen in blood, they secrete the hormone erythropoietin.
2 Erythropoietin stimulates erythropoiesis (red blood cell production) in the bone marrow.
3 Immature red blood cells (called reticulocytes) are released into the blood.
4 Reticulocytes mature into red blood cells over a 24- to 48-hour period.

SOURCE: Reprinted with permission from L. Sherwood, *Human Physiology*, 5th ed. (Brooks/Cole, 2004), Figure 11-4, p. 395.

Other nutrient deficiencies may cause anemia, although not as frequently. Vitamin E helps to maintain cell membrane integrity, and its deficiency is associated with **hemolytic anemia** (red blood cell breakdown). Vitamin B_6 plays a role in hemoglobin production, and a deficiency may occasionally cause microcytic anemia. Vitamin C supports blood vessel integrity; fragile and bleeding capillaries may result from its deficiency. Protein-energy malnutrition leads to anemia because red blood cell development depends on protein synthesis. Although nutrient deficiencies may result from dietary inadequacy, they can also arise during the course of illness due to the effects of disease on intestinal absorption, nutrient metabolism, and nutrient losses.

Identifying Causes of Anemia

Identifying the cause of anemia is sometimes quite challenging. In some cases, anemia may be a well-known consequence of disease, as when renal failure impairs the synthesis of the hormone erythropoietin. When anemia develops rapidly, blood loss is often the cause, whereas a more gradual onset suggests malnutrition, chronic illness, or slow, chronic bleeding. The results of laboratory tests provide valuable clues, although conditions such as dehydration and inflammation can influence the values. Laboratory results are especially difficult to analyze if several disturbances are present simultaneously. A **peripheral blood smear** (see the photo) is often used to study abnormalities in red blood cell shape and may also reveal an underlying cause.

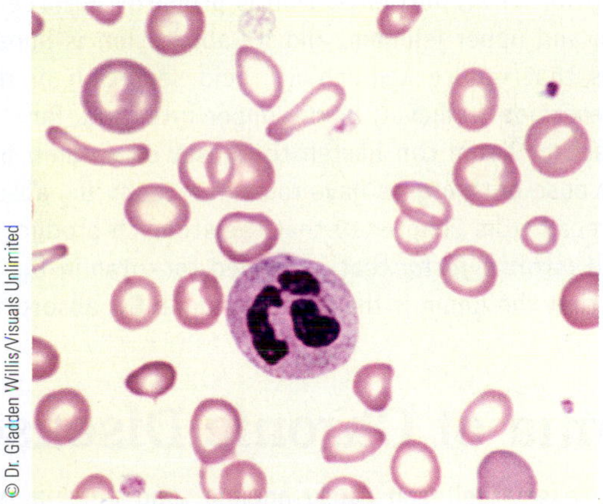

A peripheral blood smear provides information about the number and shape of blood cells.

Nutritional Anemias in Illness

There are numerous ways in which illnesses can lead to iron, folate, or vitamin B_{12} deficiencies, the main causes of the nutritional anemias. Blood loss, common to many illnesses, is a primary cause of iron deficiency. Some illnesses may result in a reduction in food intake, as discussed throughout the clinical chapters. The liver's stores of iron and vitamin B_{12} are often adequate to prevent deficiencies during transient illnesses, but reserves of folate are limited; thus, a folate deficiency can develop within a few months if dietary intakes are low. If several nutrient deficiencies occur simultaneously, it may be difficult to identify the cause of anemia using standard blood tests (see Appendix E) because both megaloblastic anemia and microcytic anemia may be present.

Blood Loss

As mentioned, blood loss can eventually lead to iron deficiency. Gastrointestinal conditions often cause bleeding; examples include peptic ulcers, inflammatory bowel diseases, and gastrointestinal varices (enlarged veins) that develop in advanced liver disease. Excessive bleeding can also accompany coagulation disorders, which are often due to liver disease, genetic defects, or vitamin K deficiency. Frequent blood draws or surgical procedures can contribute to blood loss and result in iron deficiency. Unfortunately, slow, chronic bleeding may be difficult to identify before anemia develops.

Nutrient Malabsorption

Chapter 24 explains how disorders that damage the small intestine can lead to nutrient malabsorption. Diseases like Crohn's disease and celiac disease can destroy intestinal mucosa and reduce the absorption of all nutrients. Iron is primarily absorbed in the duodenum and upper jejunum, and its absorption is impaired by conditions that reduce hydrochloric acid secretion or result in surgical resection (removal) of the upper intestine. Resection of the stomach or ileum can hasten the onset of vitamin B_{12} deficiency because both organs have roles in vitamin B_{12} absorption: you may recall from Chapter 10 that the stomach produces a protein called *intrinsic factor* that is needed for vitamin B_{12} absorption, and that the ileum is the site of vitamin B_{12} absorption.

Anemia of Chronic Disease

Chronic disease itself can cause anemia, and anemia is sometimes the initial sign that chronic disease is present. In fact, the **anemia of chronic disease** is the most common type of anemia affecting hospitalized patients and patients with chronic illnesses.[2] This type of anemia usually occurs in individuals who have inflammatory conditions, chronic infections, autoimmune disorders, or cancer. Although often a mild form of anemia, it can progress and become severe enough to require blood transfusions.

The anemia of chronic disease is characterized by alterations both in the distribution of iron among tissues and in the rates of red blood cell production and destruction.[3] During chronic illness, inflammatory mediators induce the production of the protein *hepcidin*, which blocks the release of iron from storage and thereby renders iron unavailable for red blood cell production. Furthermore, hepcidin inhibits iron's release from intestinal cells into the blood and therefore interferes with iron absorption. Finally, inflammatory processes cause red blood cells to be degraded more quickly than usual, and the reduced production of red blood cells cannot keep pace. Eventually, outright iron deficiency may be a consequence of the impaired iron absorption.

TABLE H25-2 Laboratory Tests for Evaluating Iron Deficiency and Anemia of Chronic Disease

Laboratory Test	Effect of Iron Deficiency	Effect of Chronic Disease
Red blood cell (RBC) size and number	Microcytic; reduced RBC count	Normocytic or microcytic; reduced RBC count
Serum iron	Low	Low
Serum ferritin	Low	Normal or elevated
Serum transferrin	Elevated	Low
Total iron-binding capacity	High	Normal or low
Bone marrow iron	Low	Normal or elevated

Blood tests help to distinguish between the anemia of chronic disease and iron-deficiency anemia (see Table H25-2). The combination of low serum iron and low total iron-binding capacity suggests the anemia of chronic disease rather than iron deficiency. In addition, serum ferritin levels may be normal or elevated during chronic illness, whereas they are typically low in iron deficiency. Diagnosis is more complicated if both types of anemia are present.

Medications and Anemia

Anemia is among the adverse effects that may result from medication use. Various medications may disrupt nutrient metabolism, impair blood coagulation and erythropoiesis, and increase red blood cell destruction. Because the life span of red blood cells is about 120 days, the long-term use of such medications is more likely to result in anemia than short-term use.

Drug-Nutrient Interactions

As Chapter 19 describes, there are numerous ways in which medications can alter nutrient metabolism; the most common are listed in Table 19-2 on p. 623. As an example, a number of medications are known to influence the absorption or metabolism of folate and lead to megaloblastic anemia. Sulfasalazine (used in ulcerative colitis) and some anticonvulsant drugs inhibit folate absorption, and methotrexate (an immunosuppressive) and pyrimethamine (an antimalarial) interfere with folate metabolism.[4] If a medication is known to result in deficiency, nutrient supplementation is usually recommended as an adjunct therapy.

Impaired Coagulation

Anticoagulants, which are prescribed specifically to reduce blood clotting, sometimes lead to excessive bleeding. These medications work by interfering with one of the steps involved in blood clotting, such as platelet function, vitamin K function, or the synthesis of clotting proteins. Other than anticoagulants, a large number of other drugs can impair coagulation, including commonly used drugs such as aspirin and other nonsteroidal anti-inflammatory drugs. The anticoagulant effects may be augmented if several of these drugs are used simultaneously. The slow, chronic bleeding that sometimes develops may go unnoticed until excessive blood loss has occurred.

Aplastic Anemia

Many classes of drugs are associated with **aplastic anemia**, the anemia that occurs when the bone marrow fails to produce adequate numbers of blood cells. The categories of drugs that can inhibit erythropoiesis include anticonvulsants, antibiotics, antidiabetic drugs, diuretics, antithyroid drugs, and anticancer agents.[5] Aplastic anemia can also be caused by viral infections, exposure to toxins, and genetic defects.

Hemolytic Anemia

Some patients may develop hemolytic anemia as a result of drug interactions with red blood cells. For example, a drug may alter the red blood cell membrane in such a way that a component of the membrane becomes an antigen and induces an antibody response that destroys the cell.[6] Several types of antibiotics, including penicillin and cephalosporin, may cause this type of response. Withdrawal of the drug can eventually reverse the anemia, and sometimes medications are given to suppress the immune response.

Anemia is a disorder associated with many different diseases, and it may also be caused by disease treatment. When it occurs during illness, its causes must be investigated before it leads to complications that worsen prognosis. The medical history, blood tests, and peripheral blood smears may all help to determine the reasons why anemia has developed.

References

1. K. S. Zuckerman, Approach to the anemias, in L. Goldman and D. Ausiello, eds., *Cecil Textbook of Medicine* (Philadelphia: Saunders, 2008), pp. 1179–1187.
2. G. D. Ginder, Microcytic and hypochromic anemias, in L. Goldman and D. Ausiello, eds., *Cecil Textbook of Medicine* (Philadelphia: Saunders, 2008), pp. 1187–1194.
3. Ginder, 2008.
4. A. C. Antony, Megaloblastic anemias, in L. Goldman and D. Ausiello, eds., *Cecil Textbook of Medicine* (Philadelphia: Saunders, 2008), pp. 1231–1241.
5. H. Castro-Malaspina and R. J. O'Reilly, Aplastic anemia and related disorders, in L. Goldman and D. Ausiello, eds., *Cecil Textbook of Medicine* (Philadelphia: Saunders, 2008), pp. 1241–1248.
6. R. S. Schwartz, Autoimmune and intravascular hemolytic anemias, in L. Goldman and D. Ausiello, eds., *Cecil Textbook of Medicine* (Philadelphia: Saunders, 2008), pp. 1194–1203.

CHAPTER

26

© Masterfile

Nutrition in the Clinical Setting

Diabetes is often a silent disease. The damaging effects of high blood glucose can take decades to develop, a characteristic that causes some people to ignore their condition and disregard treatment. When complications develop, there is no way to reverse the harm to the heart, kidneys, nerves, and eyes that has occurred. Because most diabetes care requires self-management, the challenge for health practitioners is to motivate patients to make the dietary and lifestyle changes that are necessary. The good news is that careful management allows individuals with diabetes to live long, healthy, and productive lives.

Throughout this chapter, the CourseMate icon indicates an opportunity for online self-study, linking you to activities to increase your understanding of chapter concepts. **www.cengagebrain.com** (search for ISBN 084006845X)

Diabetes Mellitus

CHAPTER OUTLINE

The incidence of **diabetes mellitus** ♦ is steadily increasing in the United States and many other countries (see Figure 26-1). It now affects an estimated 10.2 percent of adults aged 20 and older in the United States, or about 24 million people.[1] About 25 percent of persons with diabetes are unaware that they have it,[2] a danger because its damaging effects often occur before symptoms develop. Diabetes ranks seventh among the leading causes of death in the United States. It also contributes to the development of other life-threatening diseases, including heart disease and kidney failure, which are discussed in the two chapters that follow. The glossary on p. 782 defines diabetes-related symptoms and complications.

♦ An unrelated condition with a similar name is *diabetes insipidus*, a pituitary disorder that causes a deficiency of antidiuretic hormone.

FIGURE 26-1 **Prevalence of Diagnosed Diabetes among Adults in the United States**

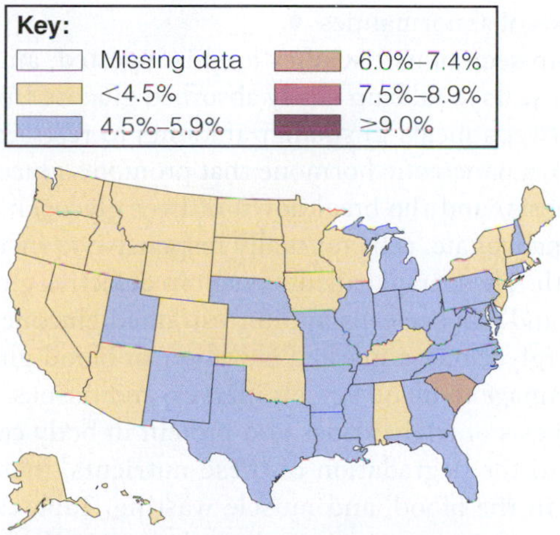

1994: Twenty-five states had a prevalence of diagnosed diabetes less than 4.5%, and only one state had a prevalence of 6% or greater.

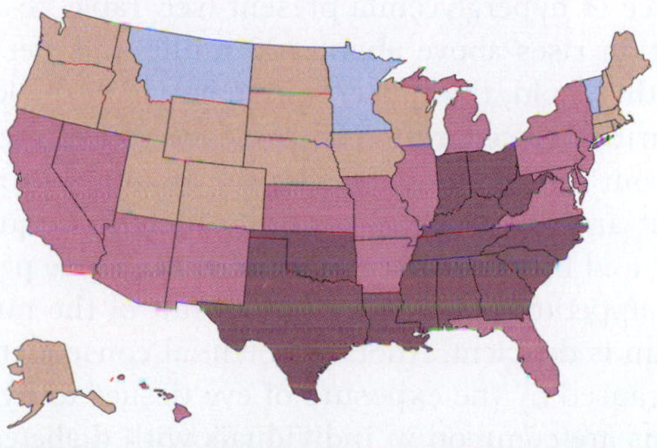

2008: No state had a prevalence of diagnosed diabetes less than 4.5%; only 3 states had a prevalence of less than 6%, and 13 states had a prevalence of 9% or greater.

diabetes (DYE-ah-BEE-teez) **mellitus:** a group of metabolic disorders characterized by hyperglycemia and disordered insulin metabolism.
- **diabetes** = siphon (in Greek), referring to the excessive passage of urine that is characteristic of untreated diabetes
- **mellitus** = sweet, honeylike

GLOSSARY OF DIABETES-RELATED SYMPTOMS AND COMPLICATIONS

acetone breath: a distinctive fruity odor on the breath of a person with ketosis.

claudication (CLAW-dih-KAY-shun): pain in the legs while walking; usually due to an inadequate supply of blood to muscles.

diabetic coma: a coma that occurs in uncontrolled diabetes; may be due to diabetic ketoacidosis, the hyperosmolar hyperglycemic syndrome, or severe hypoglycemia.

diabetic nephropathy (neh-FRAH-pah-thee): damage to the kidneys that results from long-term diabetes.

diabetic neuropathy (nur-RAH-pah-thee): complications of diabetes that cause damage to nerves.

diabetic retinopathy (REH-tih-NAH-pah-thee): retinal damage that results from long-term diabetes.

gangrene: death of tissue due to a deficient blood supply and/or infection.

gastroparesis (GAS-troe-pah-REE-sis): delayed stomach emptying caused by nerve damage in stomach tissue.

glycosuria (GLY-co-SOOR-ee-ah): the presence of glucose in the urine.

hyperglycemia: elevated blood glucose concentrations. Normal fasting plasma glucose levels are less than 100 mg/dL. Fasting plasma glucose levels between 100 and 125 mg/dL suggest prediabetes; values of 126 mg/dL and above suggest diabetes.

hyperosmolar hyperglycemic syndrome: extreme hyperglycemia associated with dehydration, hyperosmolar blood, and altered mental status; sometimes called the *hyperosmolar hyperglycemic nonketotic state*.

hypoglycemia: abnormally low concentrations of blood glucose. In diabetes, hypoglycemia is treated when plasma glucose levels fall below 70 mg/dL.

ketoacidosis (KEY-toe-ass-ih-DOE-sis): an acidosis (lowering of blood pH) that results from the excessive production of ketone bodies.

ketonuria (KEY-toe-NOOR-ee-ah): the presence of ketone bodies in the urine.

macrovascular complications: disorders that affect the large blood vessels, including the coronary arteries and arteries of the limbs.

microalbuminuria: the presence of albumin (a blood protein) in the urine, a sign of diabetic nephropathy.

microvascular complications: disorders that affect the small blood vessels and capillaries, including those in the retina and kidneys.

peripheral vascular disease: impaired blood circulation in the limbs.

polydipsia (POL-ee-DIP-see-ah): excessive thirst.

polyphagia (POL-ee-FAY-jee-ah): excessive appetite or food intake.

polyuria (POL-ee-YOOR-ree-ah): excessive urine production.

Overview of Diabetes Mellitus

The term *diabetes mellitus* refers to metabolic disorders characterized by elevated blood glucose concentrations and disordered insulin metabolism. People with diabetes may be unable to produce sufficient insulin or use insulin effectively, or they may have both types of abnormalities. ♦

♦ Reminder: *Insulin* is a pancreatic hormone that regulates blood glucose concentrations. Its actions are countered mainly by the hormone *glucagon*.

Normally, insulin secretions rise after food is ingested, and the insulin enables muscle and adipose cells to take up newly absorbed glucose from the blood. Insulin is also secreted between meals in smaller amounts to restrain the glucose-raising actions of glucagon, a pancreatic hormone that promotes glucose production in the liver (gluconeogenesis) and the breakdown of liver glycogen. In diabetes, insulin secretion may be inadequate, cells normally responsive to insulin may be resistant to its effects, or both. These impairments result in defective glucose uptake and utilization in muscle and adipose cells and unrestrained gluconeogenesis in the liver. The result is **hyperglycemia**, a marked elevation in blood glucose levels that can ultimately cause damage to blood vessels, nerves, and tissues. Because insulin also promotes the synthesis of triglycerides and protein in body cells, impaired insulin metabolism leads to the degradation of these nutrients, increased fatty acid and triglyceride levels in the blood, and muscle wasting. Table 26-1 summarizes the effects of insulin insufficiency on nutrient metabolism in the body.

Symptoms of Diabetes Mellitus

Symptoms of diabetes are usually related to the degree of hyperglycemia present (see Table 26-2). When the plasma glucose concentration rises above about 200 milligrams per deciliter (mg/dL), it exceeds the **renal threshold**, the concentration at which the kidneys begin to pass glucose into the urine (**glycosuria**). The presence of glucose in the urine draws additional water from the blood, increasing the amount of urine produced. Thus, the symptoms that arise in diabetes typically include frequent urination (**polyuria**), dehydration, and increased thirst (**polydipsia**). Some people lose weight and have an increased appetite (**polyphagia**) as a result of the nutrient depletion that occurs when insulin is deficient. Another potential consequence of hyperglycemia is blurred vision, caused by the exposure of eye tissues to hyperosmolar fluids. ♦ Increased infections are common in individuals with diabetes and may be due to weakened immune responses and impaired circulation. In some cases, constant fatigue is the only symptom and may be related to altered energy metabolism, dehydration, or other effects of the disease.

renal threshold: the blood concentration of a substance that exceeds the kidneys' capacity for reabsorption, causing the substance to be passed into the urine.

♦ Reminder: *Osmolarity* refers to the concentration of osmotically active particles in solution. Hyperglycemia causes the body's fluids to become *hyperosmolar*, meaning that they have an abnormally high osmolarity.

TABLE 26-1 Effects of Insulin Insufficiency on Nutrient Metabolism

Insulin normally promotes nutrient uptake after meals, as well as the synthesis of glycogen, triglycerides, and protein in liver, adipose, and muscle tissue. A defect in insulin metabolism inhibits these processes, leading to the effects shown in this table.

Nutrient	Effects of Insulin Insufficiency
Carbohydrate	• Decreased glucose uptake by muscle and adipose cells • Decreased glycogen synthesis in the liver and muscle • Increased glycogen breakdown in the liver and muscle • Increased gluconeogenesis in the liver • Hyperglycemia
Fat	• Decreased triglyceride synthesis in adipose tissue • Increased triglyceride breakdown in adipose tissue • Increased fatty acid and triglyceride levels in the blood • Increased production of ketone bodies in the liver
Protein	• Decreased amino acid uptake by muscle cells • Decreased synthesis of tissue protein • Increased breakdown of tissue protein • Muscle wasting and growth retardation

TABLE 26-2 Symptoms of Diabetes Mellitus

Frequent urination (polyuria)
Dehydration, dry mouth
Increased thirst (polydipsia)
Weight loss
Increased hunger (polyphagia)
Blurred vision
Increased infections
Fatigue

Diagnosis of Diabetes Mellitus

The diagnosis of diabetes is based primarily on plasma glucose levels, which can be measured under fasting conditions ♦ or at random times during the day. In some cases, an **oral glucose tolerance test** is given: the individual ingests a 75-gram glucose load, and plasma glucose is measured at one or more time intervals following glucose ingestion. **Glycated hemoglobin (HbA_{1c})** levels, which reflect hemoglobin's exposure to glucose over prolonged periods, are an indirect assessment of blood glucose levels. The following criteria are currently used to diagnose diabetes:[3]

♦ Normal fasting plasma glucose levels are approximately 75 to 100 mg/dL (published values vary).

- The plasma glucose concentration of a blood sample obtained at a random time during the day (without regard to food intake) is 200 mg/dL or higher, and classic symptoms of diabetes (such as polyuria, polydipsia, and unexplained weight loss) are present.
- The plasma glucose concentration is 126 mg/dL or higher after a fast of at least eight hours.
- The plasma glucose concentration measured two hours after a 75-gram glucose load is 200 mg/dL or higher.
- The HbA_{1c} level is 6.5 percent or higher.

Overt symptoms of hyperglycemia help to confirm the diagnosis. Otherwise, a diagnosis of diabetes is confirmed only if a subsequent test yields similar results.

The term **prediabetes** pertains to individuals who have blood glucose levels between normal and diabetic, that is, between 100 and 125 mg/dL when fasting (a condition known as *impaired fasting glucose*) or between 140 and 200 mg/dL when measured two hours after ingesting a 75-gram glucose load (a condition known as *impaired glucose tolerance*). HbA_{1c} levels between 5.7 and 6.4 percent also suggest prediabetes. Although people with prediabetes are usually asymptomatic, they are at increased risk of developing diabetes and cardiovascular diseases. Prediabetes has been estimated to affect approximately 29 percent of adults in the United States,[4] and it is especially prevalent among those who are overweight or obese.

Types of Diabetes Mellitus

Table 26-3 lists features of the two main types of diabetes, type 1 and type 2 diabetes. Pregnancy can lead to abnormal glucose tolerance and the condition known as *gestational diabetes* (discussed later in this chapter), which often resolves after pregnancy but is a risk factor for type 2

oral glucose tolerance test: a test that evaluates a person's ability to tolerate an oral glucose load.

glycated hemoglobin (HbA_{1c}): hemoglobin that has nonenzymatically attached to glucose; the level of HbA_{1c} in blood helps to diagnose diabetes and evaluate long-term glycemic control. Also called *glycosylated hemoglobin*.

prediabetes: the condition in which blood glucose levels are higher than normal but not high enough to be diagnosed as diabetes.

TABLE 26-3 Features of Type 1 and Type 2 Diabetes Mellitus

Feature	Type 1	Type 2
Prevalence in diabetic population	5 to 10 percent of cases	90 to 95 percent of cases
Age of onset	<30 years	>40 years[a]
Associated conditions	Autoimmune diseases, viral infection, inherited factors	Obesity, aging, inactivity, inherited factors
Major defect	Destruction of pancreatic beta cells; insulin deficiency	Insulin resistance; insulin deficiency relative to needs
Insulin secretion	Little or none	Varies; may be normal, increased, or decreased
Requirement for insulin therapy	Always	Sometimes
Former names	Juvenile-onset diabetes Insulin-dependent diabetes	Adult-onset diabetes Noninsulin-dependent diabetes

[a]Incidence of type 2 diabetes is increasing in children and adolescents; in more than 90 percent of these cases, it is associated with overweight or obesity and a family history of type 2 diabetes.

diabetes. ♦ Diabetes can also be caused by medical conditions that either damage the pancreas or interfere with insulin function.

♦ Gestational diabetes was introduced in Chapter 14.

Type 1 Diabetes **Type 1 diabetes** accounts for about 5 to 10 percent of diabetes cases. It is usually caused by **autoimmune** destruction of the pancreatic beta cells, which produce and secrete insulin. By the time symptoms develop, the damage to the beta cells has progressed so far that insulin must be supplied exogenously, most often by injection. Although the reason for the autoimmune attack is usually unknown, environmental toxins or infections are likely triggers. People with type 1 diabetes often have a genetic susceptibility for the disorder and are at increased risk of developing other autoimmune diseases.

Type 1 diabetes usually develops during childhood or adolescence, and symptoms may appear abruptly in previously healthy children.[5] Classic symptoms are polyuria, polydipsia, weight loss, and weakness or fatigue. **Ketoacidosis**—acidosis due to the excessive production of ketone bodies—is sometimes the first sign of disease. ♦ Disease onset tends to be more gradual in individuals who develop type 1 diabetes in later years. Blood tests that detect antibodies to insulin, pancreatic islet cells, and pancreatic enzymes can confirm the diagnosis and help to predict development of the disease in close relatives.

♦ Reminder: *Ketone bodies* are products of fat metabolism that are produced in the liver; they accumulate in tissues when fatty acids are released in abnormally high amounts from adipose tissue.

Type 2 Diabetes **Type 2 diabetes** is the most prevalent form of diabetes, accounting for 90 to 95 percent of cases, and it is often asymptomatic. The primary defect in type 2 diabetes is **insulin resistance**, the reduced sensitivity to insulin in muscle, adipose, and liver cells. To compensate, the pancreas increases its secretion of insulin, and plasma insulin concentrations may rise to abnormally high levels (**hyperinsulinemia**). Over time, the pancreas becomes less able to compensate for the cells' reduced sensitivity to insulin, and hyperglycemia worsens. The high demand for insulin can eventually exhaust the beta cells of the pancreas and lead to impaired insulin production and reduced plasma insulin concentrations. Type 2 diabetes is therefore associated with both insulin resistance and relative insulin deficiency; that is, the amount of insulin secreted is insufficient to compensate for its diminished effect in cells.

type 1 diabetes: the type of diabetes that accounts for 5 to 10 percent of diabetes cases and usually results from autoimmune destruction of pancreatic beta cells.

autoimmune: an immune response directed against the body's own tissues.

- **auto** = self

type 2 diabetes: the type of diabetes that accounts for 90 to 95 percent of diabetes cases and usually results from insulin resistance coupled with insufficient insulin secretion.

insulin resistance: reduced sensitivity to insulin in muscle, adipose, and liver cells.

hyperinsulinemia: abnormally high levels of insulin in the blood.

Although the precise causes of type 2 diabetes are unknown, risk is substantially increased by obesity (especially abdominal obesity), aging, and physical inactivity. An estimated 80 to 90 percent of individuals with type 2 diabetes are obese, and obesity itself can directly cause some degree of insulin resistance.[6] ♦ Prevalence increases with age and approaches 23 percent in persons over 60 years of age; however, many of these cases remain undiagnosed.[7] Inherited factors strongly influence

♦ Highlight 26 provides information about the relationship between obesity and insulin resistance.

risk, and type 2 diabetes is more common in certain ethnic groups, including African Americans, Asian Americans, Hispanic Americans, Mexican Americans, Native Americans, and Pacific Islanders.

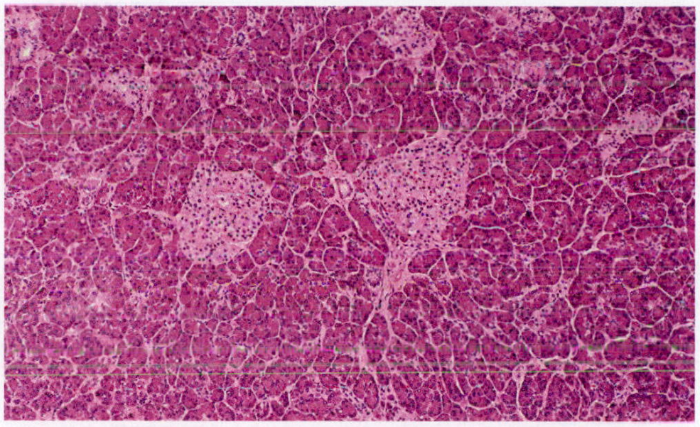

© Peter Arnold, Inc./Alamy

Cross-sections of the pancreas reveal small clusters of cells known as the islets of Langerhans; these regions contain the beta cells that produce insulin.

Type 2 Diabetes in Children and Adolescents Although most cases of type 2 diabetes are diagnosed in individuals over 40 years old, children and teenagers who are overweight or obese or have a family history of diabetes are at increased risk. Because type 2 diabetes is frequently asymptomatic, it is generally identified in youths only when high-risk groups are screened for the disease.

Increased rates of both type 1 and type 2 diabetes have been documented in children in past decades and correlate with the rise in childhood obesity. Type 1 and type 2 diabetes are sometimes difficult to distinguish in children, however, and a few studies suggest that some children diagnosed with type 1 diabetes may actually have had type 2 diabetes.[8] One multicenter study found that about half of the cases of newly diagnosed diabetes in African American and Hispanic individuals between 10 and 19 years old were classified as type 2 diabetes.[9] Although type 2 diabetes is still relatively rare in children, its increasing prevalence indicates that routine screening and diabetes prevention programs may be important safeguards for children at risk.

Prevention of Type 2 Diabetes Mellitus

Clinical studies suggest that lifestyle changes can prevent or delay the incidence of type 2 diabetes in individuals at risk. In the Diabetes Prevention Program, a multicenter trial of 3234 adults with impaired glucose tolerance, dietary changes and increased physical activity led to a 58 percent reduction in diabetes incidence.[10] Based on the results of this and similar studies, guidelines for diabetes prevention include the following strategies:[11]*

- *Weight management.* A sustained weight loss of 5 to 10 percent of body weight is recommended for overweight and obese individuals. If weight loss cannot be achieved, healthy eating behaviors should be encouraged to prevent additional weight gain.
- *Active lifestyle.* At least 150 minutes of moderate physical activity, such as brisk walking, is recommended weekly.
- *Dietary modifications.* An increased intake of whole grains and dietary fiber has been associated with a reduced risk for type 2 diabetes. Individuals who are overweight or obese should decrease their intake of dietary fat to avoid consuming excessive energy.
- *Regular monitoring.* Individuals at risk should be monitored every year to check for the possible development of type 2 diabetes. If necessary, they can be provided with additional counseling, education, or resources.

Clinical trials have found that a moderate alcohol intake (one to two drinks per day) may reduce the risk of developing type 2 diabetes, compared with either abstinence from alcohol or heavy drinking.[12] Alcohol's protective effect may be attributable to the increased secretion of adiponectin, an adipose hormone that improves insulin sensitivity. Specific recommendations regarding alcohol intake are unavailable, however, because the risk of adverse effects from alcohol ingestion must be considered on an individual basis.

Acute Complications of Diabetes Mellitus

Untreated diabetes may result in life-threatening complications. An insulin deficiency can cause significant disturbances in energy metabolism, and severe hyperglycemia can lead to dehydration and electrolyte imbalances. In treated diabetes, hypoglycemia (low blood glucose) is a possible complication of inappropriate disease management.

* The antidiabetic medication metformin may be beneficial for preventing diabetes in high-risk individuals, such as those with obesity, a sedentary lifestyle, prediabetes, or a family history of diabetes.

FIGURE 26-2 Acute Effects of Insulin Insufficiency

The effects of insulin insufficiency can be grouped according to its effects on carbohydrate, protein, and fat metabolism.

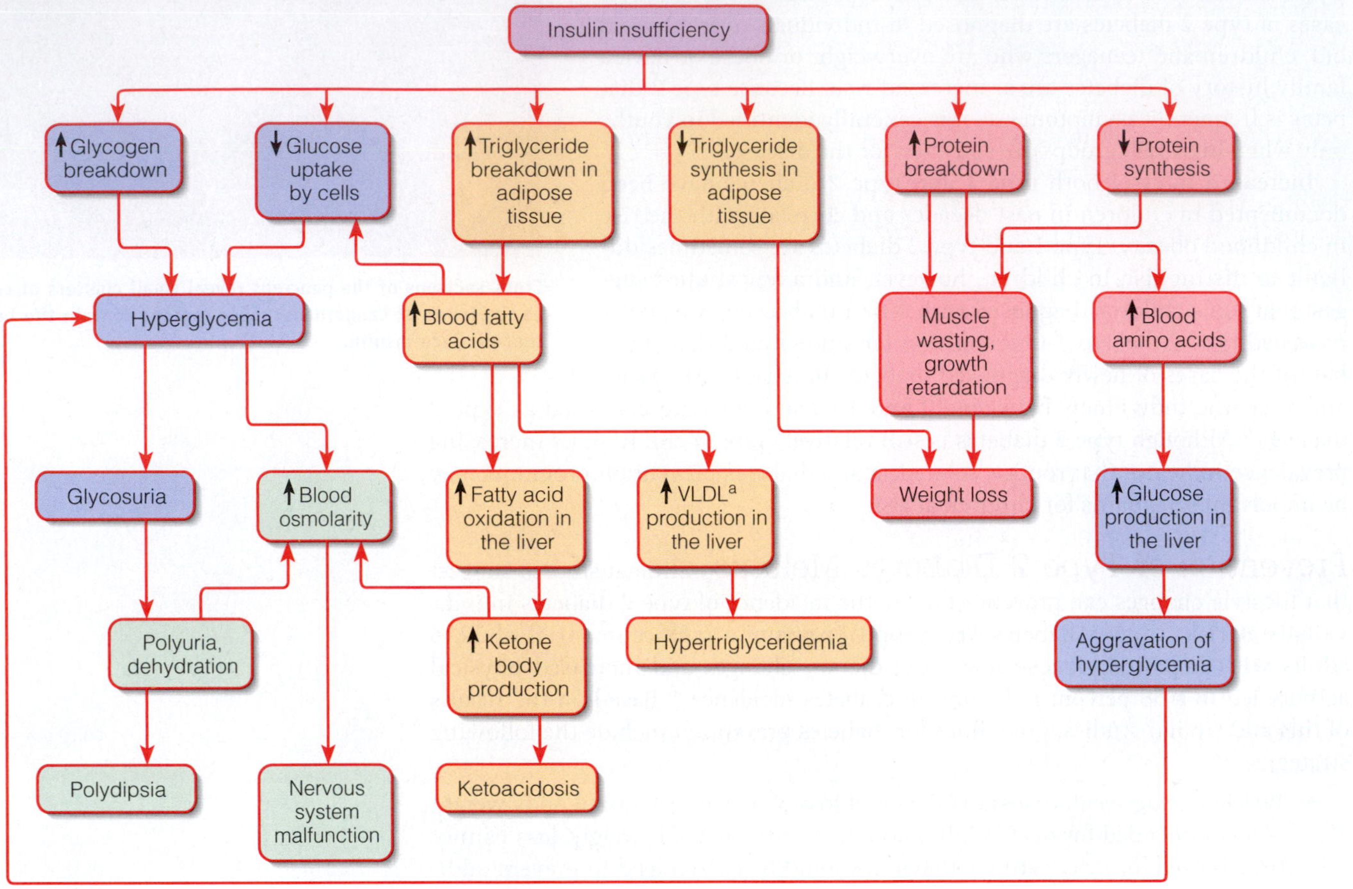

[a]Very low density lipoproteins; these lipoproteins transport triglycerides from the liver to other tissues.

Figure 26-2 presents an overview of some of the acute effects of insulin insufficiency on energy metabolism.

Diabetic Ketoacidosis in Type 1 Diabetes A severe lack of insulin causes diabetic ketoacidosis. Without insulin, glucagon's effects become more pronounced, leading to the unrestrained breakdown of the triglycerides in adipose tissue and the protein in muscle. As a result, an increased supply of fatty acids and amino acids arrives in the liver, fueling the production of ketone bodies and glucose. ♦ Ketone bodies, which are acidic, can reach dangerously high levels in the bloodstream (ketoacidosis) and spill into the urine (**ketonuria**). Blood pH typically falls below 7.30 (blood pH normally ranges between 7.35 and 7.45). Blood glucose concentrations usually exceed 250 mg/dL and may rise above 1000 mg/dL in severe cases. The main features of diabetic ketoacidosis therefore include severe ketosis, ♦ acidosis, and hyperglycemia.[13]

Patients with ketoacidosis may exhibit symptoms of both acidosis and dehydration. Acidosis is partially corrected by exhalation of carbon dioxide, so rapid or deep breathing is characteristic. ♦ Ketone accumulation is sometimes evident by a fruity odor on a person's breath (**acetone breath**). Significant fluid loss (polyuria) accompanies the hyperglycemia, lowering blood volume and blood pressure and depleting electrolytes. In response, patients may demonstrate marked fatigue, lethargy, nausea, and vomiting. Mental state may vary from alertness to comatose

♦ Chapter 7 provides details about the metabolic pathways involved in ketone body and glucose production.

♦ Reminder: *Ketosis* is a condition characterized by an abnormal increase in ketone body concentrations.

♦ Bicarbonate is a buffer in the blood that corrects acidosis. The acid (H^+) combines with bicarbonate (HCO_3^-) to form carbonic acid (H_2CO_3), which breaks down to water (H_2O) and carbon dioxide (CO_2). The carbon dioxide is then exhaled.

(**diabetic coma**). ♦ Treatment of diabetic ketoacidosis includes insulin therapy to correct hyperglycemia, intravenous fluid and electrolyte replacement, and, in some cases, bicarbonate therapy to treat acidosis.

♦ Diabetic coma was a frequent cause of death before insulin was routinely used to manage diabetes.

Diabetic ketoacidosis is sometimes the earliest sign that leads to diagnosis of type 1 diabetes, but more often it results from inappropriate diabetes treatment (such as missed insulin injections), illness or infection, alcohol abuse, or other physiological stressors.[14] The condition usually develops quickly, within one or two days. The mortality rate is nearly 5 percent in individuals under 40 years of age but exceeds 20 percent in elderly individuals.[15] Although diabetic ketoacidosis can occur in type 2 diabetes—usually due to severe stressors such as infection, trauma, or surgery—it rarely develops because even relatively low insulin concentrations are able to suppress ketone body production.

Hyperosmolar Hyperglycemic Syndrome in Type 2 Diabetes The **hyperosmolar hyperglycemic syndrome** is a condition of severe hyperglycemia and dehydration that develops in the absence of significant ketosis. As mentioned earlier, the hyperglycemia that develops in poorly controlled diabetes leads to polyuria, which results in substantial fluid and electrolyte losses. In the hyperosmolar hyperglycemic syndrome, patients are unable to recognize thirst or adequately replace fluids due to age, illness, sedation, or incapacity. The profound dehydration that eventually develops exacerbates the rise in blood glucose levels, which often exceed 600 mg/dL and may climb above 1000 mg/dL. Blood plasma may become so hyperosmolar as to cause neurological abnormalities, such as abnormal reflexes, motor impairments, reduced verbal ability, and seizures; about 10 percent of patients lapse into coma.[16] Treatment includes intravenous fluid and electrolyte replacement and insulin therapy.

The hyperosmolar hyperglycemic syndrome is sometimes the first sign of type 2 diabetes in older persons. It is usually precipitated by infection, illness, or a drug treatment that impairs insulin action or secretion. Unlike diabetic ketoacidosis, the condition often evolves slowly, over one week or longer; the absence of clinical symptoms can delay its diagnosis. The mortality rate may be as high as 20 percent, in part because the condition occurs more often in older patients who have cardiovascular disease or other major illnesses.[17]

Hypoglycemia **Hypoglycemia**, or low blood glucose, is the most frequent complication of type 1 diabetes and may occur in type 2 diabetes as well. It is due to the inappropriate management of diabetes rather than the disease itself, and is usually caused by excessive dosages of insulin or antidiabetic drugs, prolonged exercise, skipped or delayed meals, inadequate food intake, or the consumption of alcohol without food. Hypoglycemia is the most frequent cause of coma in insulin-treated patients and is believed to account for 3 to 4 percent of deaths in this population.[18]

Symptoms of hypoglycemia include dizziness, shakiness, hunger, irritability, sweating, and heart palpitations. Mental confusion may prevent a person from recognizing the problem and taking such corrective action as ingesting glucose tablets, juice, or candy. If hypoglycemia occurs during the night, patients may be completely unaware of its presence. Severe hypoglycemia or a delay in treatment can cause irreversible brain damage.

Chronic Complications of Diabetes Mellitus

Prolonged exposure to high glucose concentrations can alter cellular functions and damage cells and tissues. Glucose nonenzymatically combines with proteins, producing molecules that eventually break down to form reactive compounds known as **advanced glycation end products (AGEs)**; in diabetes, these AGEs accumulate to such high levels that they alter the structures of proteins and blood vessels. Excessive glucose also promotes the production and accumulation of sorbitol, which increases oxidative stress and alters molecular structures and functions.

Chronic complications of diabetes typically involve the large blood vessels (**macrovascular complications**), smaller blood vessels such as arterioles and capillaries (**microvascular complications**), and the nervous system (**diabetic neuropathy**).

advanced glycation end products (AGEs): reactive compounds formed after glucose combines with protein; AGEs can damage tissues and lead to diabetic complications.

Other tissues adversely affected by diabetes include the lens of the eye and the skin; cataracts, glaucoma, and various types of skin lesions sometimes develop. Infections are common in diabetes, a possible consequence of hyperglycemia, impaired circulation, or depressed immune responses. In individuals with type 2 diabetes, complications often develop before diabetes is diagnosed.

Macrovascular Complications The damage caused by diabetes accelerates the development of atherosclerosis in the coronary arteries and the arteries of the limbs. Cardiovascular disease is the leading cause of death in people with diabetes, accounting for up to 70 percent of deaths.[19] Moreover, type 2 diabetes is frequently accompanied by multiple risk factors for coronary heart disease, including hypertension, abnormal blood lipids, and obesity. ♦ People with diabetes also have increased tendencies for thrombosis (blood clot formation) and abnormal ventricle function, both of which can worsen the clinical course of heart disease.

♦ The *metabolic syndrome*, a cluster of symptoms often associated with insulin resistance (including hyperglycemia, hypertension, and abnormal blood lipids), substantially increases heart disease risk (see Highlight 26).

Peripheral vascular disease (impaired blood circulation in the limbs) increases the risk of **claudication** (pain while walking) and contributes to the development of foot ulcers (see the photo). Left untreated, foot ulcers can lead to **gangrene** (tissue death), and some patients require foot amputation, a major cause of disability in individuals with diabetes. About 15 percent of diabetic patients experience clinically significant foot ulcers during the course of illness.[20]

Microvascular Complications Long-term diabetes is associated with a thickening of the basement membrane of capillaries and small arterioles, which impairs the normal functioning of these blood vessels. The primary microvascular complications involve the capillaries in the retina of the eye and the kidneys. Diabetes is currently the leading cause of both adult blindness and kidney failure in the United States.[21]

In **diabetic retinopathy**, the weakened capillaries of the retina leak fluid, lipids, or blood, causing local edema or hemorrhaging. The defective blood flow also leads to damage and scarring within retinal tissue. New blood vessels eventually form, but they are fragile and bleed easily, releasing blood and proteins that obscure vision. The retinal changes usually occur after an individual has had diabetes for many years. Up to 80 percent of diabetes patients develop retinopathy 15 to 20 years after diabetes onset.[22] Diabetic retinopathy progresses most rapidly when diabetes is poorly controlled, and intensive management substantially reduces the risk.

In **diabetic nephropathy**, damage to the kidneys' specialized capillaries prevents adequate blood filtration, resulting in abnormal protein losses in the urine (**microalbuminuria**). As the kidney damage worsens, urine production decreases and nitrogenous wastes accumulate in the blood; eventually, the individual requires dialysis (artificial filtration of blood) to survive. ♦ Because the kidneys normally regulate blood volume and blood pressure (see pp. 387–388), inadequate kidney function leads to high blood pressure in patients with nephropathy. Kidney failure eventually develops in about 30 to 35 percent of patients with type 1 diabetes and 20 percent of those with type 2 diabetes.[23] As with diabetic retinopathy, intensive diabetes management can help slow the progression of kidney damage.

♦ Chapter 28 provides details about the progression and treatment of chronic kidney disease.

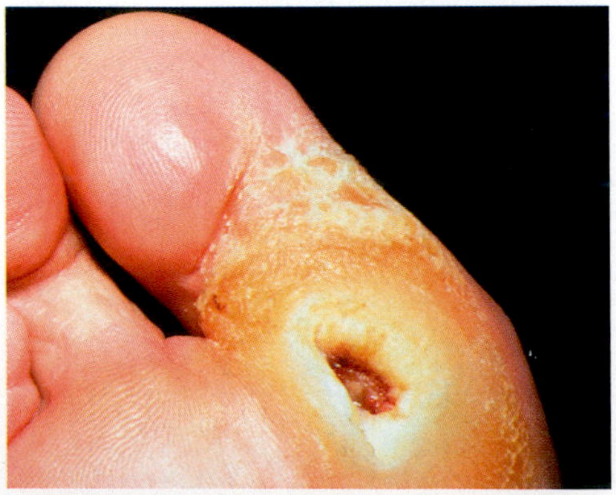

Foot ulcers are a common complication of diabetes because blood circulation is impaired, which slows healing, and nerve damage dampens foot pain, delaying recognition and treatment of cuts and bruises.

Diabetic Neuropathy The extent of neuropathy (nerve damage) that develops in diabetes depends on the severity and duration of hyperglycemia. Symptoms of neuropathy may be experienced as deep pain or burning in the legs and feet, weakness of the arms and legs, or numbness and tingling in the hands and feet. Pain and cramping, especially in the legs, are often severe during the night and may interrupt sleep. Neuropathy also contributes to the development of foot ulcers because cuts and bruises may go unnoticed until wounds are severe. Other manifestations of neuropathy include sweating abnormalities, disturbances in bladder and bowel function, sexual dysfunction, constipation, and delayed stomach emptying (**gastroparesis**). Neuropathy occurs in about 50 percent of type 1 and type 2 diabetes cases.[24]

IN SUMMARY **Diabetes mellitus is a chronic condition characterized by inadequate insulin secretion and/or impaired insulin action. Diagnosis of diabetes is based on indicators of hyperglycemia. In type 1 diabetes, the pancreas**

secretes little or no insulin, and insulin therapy is necessary for survival. Type 2 diabetes is characterized by insulin resistance coupled with relative insulin deficiency, and disease risk is increased by obesity, aging, and physical inactivity. Acute complications of diabetes include diabetic ketoacidosis, in which hyperglycemia is accompanied by ketosis and acidosis, and the hyperosmolar hyperglycemic syndrome, characterized by severe hyperglycemia, dehydration, and possible mental impairments. The development of hypoglycemia is usually due to inappropriate disease management. Chronic complications of diabetes include macrovascular disorders such as cardiovascular disease and peripheral vascular disease, microvascular conditions such as diabetic retinopathy and diabetic nephropathy, and diabetic neuropathy.

Treatment of Diabetes Mellitus

Diabetes is a chronic and progressive illness that requires lifelong treatment. Managing blood glucose levels is a delicate balancing act that involves meal planning, proper timing of medications, and physical exercise. Frequent adjustments in treatment are often necessary to establish good **glycemic** control. Individuals with type 1 diabetes require insulin therapy for survival. Type 2 diabetes is initially treated with nutrition therapy and exercise, but most patients eventually need antidiabetic medications or insulin. Diabetes management becomes even more difficult once complications develop. Although the health care team must determine the appropriate therapy, the individual with diabetes ultimately assumes much of the responsibility for treatment and therefore requires education in self-management of the disease.

Treatment Goals

The main goal of diabetes treatment is to maintain blood glucose levels within a desirable range to prevent or reduce the risk of complications. Several clinical trials have shown that *intensive* diabetes treatment, which keeps blood glucose levels tightly controlled, can greatly reduce the incidence and severity of chronic complications. Therefore, maintenance of near-normal glucose levels has become the fundamental objective of diabetes care plans.* Other goals of treatment include maintaining healthy blood lipid concentrations, controlling blood pressure, and managing weight—measures that can help to prevent or delay diabetes complications as well. Although intensive therapy is associated with some risks, including an increased risk of hypoglycemia, its benefits outweigh these disadvantages.

Benefits of Intensive Treatment

Landmark studies conducted in the 1980s and 1990s confirmed that keeping blood glucose levels as close to normal as possible offers clear advantages over less rigorous diabetes treatment. The Diabetes Control and Complications Trial was a multicenter trial that tested whether the intensive treatment of type 1 diabetes would decrease the frequency and severity of microvascular and neurological complications.[25] In this study, 1441 persons with type 1 diabetes were randomly assigned to receive either conventional or intensive therapy, as summarized in Table 26-4. The subjects were followed for an average of 6.5 years. The participants undergoing intensive therapy had delayed onset and reduced progression of retinopathy, nephropathy, and neuropathy; however, they also experienced increased incidences of severe hypoglycemia and gained more weight. A later trial, the United Kingdom Prospective Diabetes Study, found similar advantages to using intensive treatment in type 2 diabetes.[26]

Diabetes Self-Management Education

Diabetes education provides an individual with the knowledge and skills necessary to implement treatment. The

* Intensive treatment may be inappropriate for some individuals with diabetes; examples include individuals with limited life expectancies or a history of hypoglycemia and middle-aged or older adults with previous heart disease or multiple heart disease risk factors.

glycemic (gly-SEE-mic): pertaining to blood glucose.

TABLE 26-4 Comparison of Conventional and Intensive Therapies for Type 1 Diabetes

	Conventional Therapy	Intensive Therapy
Blood glucose monitoring	Monitored daily	Monitored at least three times daily
Insulin therapy	One or two daily injections; no daily adjustments	Three or more daily injections or use of an external insulin pump; dosage adjusted according to the results of blood glucose monitoring and expected carbohydrate intake
Advantages	Fewer incidences of severe hypoglycemia; less weight gain	Delayed progression of retinopathy, nephropathy, and neuropathy
Disadvantages	More rapid progression of retinopathy, nephropathy, and neuropathy	Twofold to threefold increase in severe hypoglycemia; weight gain; increased risk of becoming overweight

primary instructor is often a **Certified Diabetes Educator (CDE)**, a health care professional (often a nurse or dietitian) who has specialized knowledge about diabetes treatment and the health education process. To manage diabetes, patients need to learn about appropriate meal planning, medication administration, blood glucose monitoring, weight management, appropriate physical activity, and prevention and treatment of diabetic complications.

Evaluating Diabetes Treatment

Diabetes treatment is largely evaluated by monitoring glycemic status. Good glycemic control requires frequent home monitoring of blood glucose using a glucose meter, referred to as **self-monitoring of blood glucose**. ♦ In this procedure, a drop of blood from a finger prick is applied to a chemically treated paper strip, which is then analyzed for glucose. Glucose testing provides valuable feedback when the patient adjusts food intake, medications, and physical activity and is helpful for preventing hypoglycemia. Ideally, patients with type 1 diabetes should monitor blood glucose three or more times daily—and more frequently when therapy is adjusted.[27] Some patients may achieve better glycemic control by also using a **continuous glucose monitoring** system, which measures tissue glucose levels every few minutes using a tiny sensor placed under the skin. Although self-monitoring of blood glucose is also useful in type 2 diabetes, the recommended frequency varies according to the specific needs of individual patients.

♦ Goals for glycemic control:
- Fasting glucose: 70–130 mg/dL
- One to two hours after mealtime: <180 mg/dL
- Bedtime glucose: 90–150 mg/dL

Long-Term Glycemic Control

Health care providers periodically evaluate long-term glycemic control by measuring HbA_{1c} (glycated hemoglobin) levels. The glucose in blood freely enters red blood cells and attaches to hemoglobin molecules in direct proportion to the amount of glucose present. Because the life span of red blood cells averages 120 days, the percentage of HbA_{1c} reflects glycemic control over the preceding two to three months (the average age of circulating red blood cells).[28] The goal of diabetes treatment is an HbA_{1c} value under 7 percent,[29] ♦ but it is often markedly higher in people with diabetes, even those who are maintaining near-normal blood glucose levels. Less stringent HbA_{1c} goals may be suitable for certain patients, including those with limited life expectancy, advanced diabetic complications, or a history of severe hypoglycemia.

♦ In people without diabetes, HbA_{1c} is typically less than 6 percent of total hemoglobin.

The **fructosamine test** is sometimes conducted to determine glycemic control for the preceding two-week period. This test determines the nonenzymatic glycation of serum proteins (primarily albumin), which have a shorter half-life than hemoglobin. Most often, the fructosamine test is used to evaluate recent adjustments in diabetes treatment or glycemic control during pregnancy. The test cannot be interpreted correctly in patients with kidney or liver disease.

Monitoring for Long-Term Complications

Individuals with diabetes are routinely monitored for signs of long-term complications. Blood pressure is measured at each checkup. Annual lipid screening is suggested for adult patients. Routine checks for urinary protein (microalbuminuria) can determine if nephropathy has

Certified Diabetes Educator (CDE): a health care professional who specializes in diabetes management education. Certification is obtained from the National Certification Board for Diabetes Educators.

self-monitoring of blood glucose: home monitoring of blood glucose levels using a glucose meter.

continuous glucose monitoring: continuous monitoring of tissue glucose levels using a small sensor placed under the skin.

fructosamine test: a measurement of glycated serum proteins; used to analyze glycemic control over the preceding two weeks. Also known as the *glycated albumin test* or the *glycated serum protein test*.

developed. Physical examinations generally screen for signs of retinopathy, neuropathy, and foot problems.

Ketone Testing Ketone testing, which checks for the development of ketoacidosis, should be performed if symptoms are present or if risk has increased due to acute illness, stress, or pregnancy. Both blood and urine tests are available for home use, although the blood tests are generally more reliable. Ketone testing is most useful for patients who have type 1 diabetes or gestational diabetes. Individuals with type 2 diabetes may produce excessive ketone bodies when severely stressed by infection or trauma.

© Tony Freeman/PhotoEdit

Self-monitoring of blood glucose can help individuals with diabetes learn how to maintain blood glucose levels within a desirable range.

Body Weight Concerns

Whereas individuals with newly diagnosed type 1 diabetes are likely to be thin, most people with type 2 diabetes are overweight or obese. In children, body weight and growth patterns are monitored to evaluate whether energy intakes are appropriate.

Body Weight in Type 1 Diabetes In general, people with type 1 diabetes are less likely to be overweight than those in the general population. However, excessive weight gain is sometimes an unintentional side effect of improved glycemic control, especially in those undergoing intensive insulin therapy. Although the cause of the weight gain is unclear, it is possibly related to the insulin treatment, which may stimulate fat synthesis or induce energy intake in some way.[30] In addition, the insulin treatment eliminates urinary glucose losses and may indirectly contribute to energy excess.[31] Although patients should try to prevent excessive weight gain, concerns about weight should not discourage the use of intensive therapy, which is associated with longer life expectancy and fewer complications than occur with conventional therapy. It is also important to ensure that growing children receive sufficient energy for normal growth and development.

Body Weight in Type 2 Diabetes Because excessive body fat can worsen insulin resistance, weight loss is recommended for overweight or obese individuals who have diabetes.[32] Even moderate weight loss (10 to 20 pounds) can help to improve glycemic control, blood lipid levels, and blood pressure. Weight loss is most beneficial early in the course of diabetes, before insulin secretion has diminished.

Not all persons with type 2 diabetes are overweight or obese. Older adults and those in long-term care facilities are often underweight and may need to gain weight. Low body weight increases risks of morbidity and mortality in these individuals.

Nutrition Therapy: Nutrient Recommendations

Nutrition therapy can both improve blood glucose levels and slow the progression of diabetic complications. As always, the nutrition care plan must consider personal preferences and lifestyle habits. In addition, dietary intakes must be modified to accommodate growth, lifestyle changes, aging, and any complications that develop. Although all members of the diabetes care team should understand the principles of dietary treatment, a registered dietitian is best suited to design and implement the nutrition therapy provided to diabetes patients. This section presents the nutrient recommendations for diabetes; a later section describes meal-planning strategies.

Total Carbohydrate Intake The amount of carbohydrate consumed has the greatest influence on blood glucose levels after meals—the more grams of carbohydrate ingested, the greater the glycemic response. The carbohydrate recommendation is based in part on the person's metabolic needs (which are related to the type of diabetes, degree of glucose tolerance, and blood lipid levels), the type of insulin or other medications used to manage the diabetes, and individual preferences. Low-carbohydrate diets, which restrict carbohydrate intake to less than 130 grams per day, are not recommended.[33]

Carbohydrate Sources Different carbohydrate-containing foods have different effects on blood glucose levels; for example, consuming a portion of white rice causes blood glucose to rise more than would a similar portion of barley. A food's

♦ Chapter 4 provides additional information about the glycemic index (see pp. 110–111). The website **www.glycemicindex.com** provides glycemic index values for a wide variety of common foods.

♦ The fiber DRI for adult women and men ranges from 21 to 38 g; check the DRI table on the inside front cover of this text for specific values.

♦ The protein RDA for adults is 0.8 g/kg body weight. The recommended intake range for protein is 10 to 35 percent of kcalories.

♦ Reminder: One drink is equivalent to 12 oz of beer, 5 oz of wine, or $1^1/_2$ oz of 80 proof distilled spirits such as gin, rum, vodka, and whiskey.

glycemic effect is influenced by the type of carbohydrate in a food, the food's fiber content, the preparation method, the other foods included in a meal, and individual tolerances. For individuals with diabetes, using the glycemic index ♦ may provide some additional benefit for achieving glycemic control, compared with that obtained by considering only the amount of carbohydrate consumed.[34] In addition, high-fiber, minimally processed foods—which typically have more moderate effects on blood glucose than do highly processed, starchy foods—are among the foods frequently recommended for persons with diabetes.

Fiber Fiber recommendations for individuals with diabetes are similar to those for the general population; ♦ thus, people with diabetes are encouraged to include fiber-rich foods such as legumes, whole-grain cereals, fruits, and vegetables in their diet. Although some studies have suggested that very high intakes of fiber (50 grams or more per day) may improve glycemic control, many individuals have difficulty enjoying or tolerating such large amounts of fiber.[35]

Sugars A common misperception is that people with diabetes need to avoid sugar and sugar-containing foods. In reality, table sugar (sucrose), made up of glucose and fructose, has a lower glycemic effect than starch. Because moderate consumption of sugar has not been shown to adversely affect glycemic control,[36] sugar recommendations for people with diabetes are similar to those for the general population, which suggest minimizing foods and beverages that contain added sugars. However, sugars and sugary foods must be counted as part of the daily carbohydrate allowance.

Although fructose has a minimal glycemic effect, its use as an added sweetener is not advised because excessive dietary fructose may adversely affect blood lipid levels. (Note that it is not necessary to avoid the naturally occurring fructose in fruits and vegetables.) Sugar alcohols (such as sorbitol and maltitol) provide about half the kcalories and carbohydrate grams of sucrose and other sugars, so the amounts consumed need to be considered when determining the total carbohydrate intake. Artificial sweeteners (such as aspartame, saccharin, and sucralose) contain no digestible carbohydrate and can be safely used in place of sugar.

Dietary Fat Because people with diabetes are at high risk of developing cardiovascular diseases, guidelines for dietary fat are similar to those suggested for other persons at risk: saturated fat should be less than 7 percent of total kcalories, *trans* fat should be minimized, and cholesterol intake should be limited to less than 200 milligrams daily.[37] In addition, two or more servings of fish are recommended weekly because fish supplies omega-3 fatty acids that reduce heart disease risk and can also help to displace foods that are high in saturated fat from the diet. Dietary strategies for cardiovascular disease are discussed further in Chapter 27.

Protein Protein recommendations for people with diabetes are similar to those for the general population. ♦ In the United States, the average protein intake is about 15 percent of the energy intake. Although small, short-term studies have suggested that protein intakes above 20 percent of kcalories may improve glycemic control, increase satiety, and help with weight loss, the long-term effects of such diets on diabetes management are unknown.[38] In addition, high protein intakes are discouraged because they may be detrimental to kidney function in some individuals.

Alcohol Use in Diabetes Guidelines for alcohol intake are similar to those for the general population, which recommend that women and men limit their average daily intakes of alcohol to one drink and two drinks per day, respectively. ♦ In addition, individuals using insulin or medications that promote insulin secretion should consume food when they ingest alcoholic beverages to avoid hypoglycemia. Alcohol can cause hypoglycemia by interfering with glucose production in the liver. Conversely, an excessive alcohol intake (three or more drinks per day) can worsen hyperglycemia and raise triglyceride levels in susceptible persons. People who should avoid alcohol include pregnant women and individuals with advanced neuropathy, abnormally high triglyceride levels, or a history of alcohol abuse.

Micronutrients Micronutrient recommendations for people with diabetes are the same as for the general population. Vitamin and mineral supplementation is not recommended unless nutrient deficiencies develop; those at risk include the elderly, pregnant or lactating women, strict vegetarians, and individuals on kcalorie-restricted diets. Although some studies have suggested that supplemental chromium can improve glycemic control in type 2 diabetes, results have not been consistent.[39] At present, chromium supplementation is not recommended for those with type 2 diabetes.

Nutrition Therapy: Meal-Planning Strategies

Dietitians provide a number of meal-planning strategies to help people with diabetes maintain glycemic control. These strategies emphasize control of carbohydrate intake and portion sizes. Initial dietary instructions may include a discussion of the *Dietary Guidelines for Americans* or other recommendations designed for the general population (see Chapter 2), as well as guidelines for improving blood lipids and other cardiovascular risk factors. Sample menus that include commonly eaten foods can help to illustrate general principles. People using intensive insulin therapy must learn to coordinate insulin injections with meals and to match insulin dosages to carbohydrate intake, as discussed later.

Carbohydrate Counting Carbohydrate-counting techniques are simpler and more flexible than other menu-planning approaches and are widely used for planning diabetes diets. Carbohydrate counting works as follows: After a dietitian determines a person's nutrient and energy needs, the individual is given a daily carbohydrate allowance, often divided into a pattern of meals and snacks according to individual preferences. The carbohydrate allowance can be expressed in grams or as the number of carbohydrate portions allowed per meal (see Table 26-5). The user of the plan need only be concerned about meeting carbohydrate goals and can select from any of the carbohydrate-containing food groups when planning meals (see Table 26-6 and Figure 26-3). Although encouraged to make healthy food choices, the individual has the freedom to choose the foods desired at each meal without risking loss of glycemic control. Some people may also need guidance about non-carbohydrate foods to help them choose a healthy diet that improves blood lipids or energy intakes. The "How To" on pp. 794–795 shows how to implement carbohydrate counting in clinical practice.

Carbohydrate counting is taught at different levels of complexity depending on a person's needs and abilities. The basic carbohydrate-counting method just described can be helpful for most people, although it requires a consistent carbohydrate intake from day to day to match the medication or insulin regimen. Advanced carbohydrate counting allows more flexibility but is best suited for patients using intensive insulin therapy. With this method, a person can determine the specific dose of insulin needed to cover the amount of carbohydrate consumed at a meal. The person is then free to choose the types and portions of food desired without sacrificing glycemic control. Advanced carbohydrate counting requires some training and should be attempted only after an individual has mastered more basic methods.

Exchange Lists for Meal Planning The exchange list system is an alternative meal-planning method, although it may be more difficult for patients to learn than carbohydrate counting. This system of meal planning was introduced in Chapter 2 and is described further in Appendix G (Appendix I for Canadians). The exchange system sorts foods according to their proportions of carbohydrate, fat, and protein so that each item in a food group (or "exchange list") has a similar macronutrient and energy content (see pp. G-1 to G-2). Thus, any food on a list can be exchanged, or traded, for any other food on the same list without affecting the macronutrient balance in a day's meals. Although the exchange list system can be helpful for individuals who want a structured dietary plan that provides specific percentages of protein, carbohydrate, and fat, it offers no advantages for maintaining glycemic control and is less flexible than carbohydrate counting.

HOW TO Use Carbohydrate Counting in Clinical Practice

1. The first step in basic carbohydrate counting is to determine an appropriate carbohydrate intake and suitable distribution pattern; an example is shown in Table 26-5. A nutrition assessment can help to estimate a person's usual energy and carbohydrate intakes. The carbohydrate level should be acceptable to the person using the plan. Frequent monitoring of blood glucose levels can help determine whether additional carbohydrate restriction would be helpful.

TABLE 26-5 Sample Carbohydrate Distribution for a 2000-kCalorie Diet

	Carbohydrate Allowance	
Meals	**Grams**	**Portions[a]**
Breakfast	60	4
Lunch	60	4
Afternoon snack	30	2
Dinner	75	5
Evening snack	30	2
Totals	**255 g**	**17**

NOTE: The carbohydrate allowance in this example is approximately 50 percent of total kcalories.
[a]1 portion = 15 g carbohydrate = 1 portion of starchy food, milk, or fruit.

The example given in Table 26-5 illustrates a meal pattern for a person consuming 2000 kcalories daily with a carbohydrate allowance of 50 percent of kcalories. This is calculated as follows:

$$50\% \times 2000 \text{ kcal} = 1000 \text{ kcal of carbohydrate}$$

$$\frac{1000 \text{ kcal carbohydrate}}{4 \text{ kcal/g carbohydrate}} = 250 \text{ g carbohydrate/day}$$

$$\frac{250 \text{ g carbohydrate}}{15 \text{ g/1 carbohydrate portion}} = 16.7 \text{ carbohydrate portions/day}$$

2. The distribution of carbohydrates among meals and snacks is based on both individual preferences and metabolic needs. In type 1 diabetes, the insulin regimen must coordinate with the individual's dietary and lifestyle choices. People using conventional insulin therapy must maintain a consistent carbohydrate intake from day to day to match their particular insulin prescription, whereas those using intensive therapy can alter insulin dosages when carbohydrate intakes change. People with type 2 diabetes are encouraged to develop dietary patterns that suit their lifestyle and medication schedules. For

TABLE 26-6 Carbohydrate-Containing Food Groups and Sample Portion Sizes

Bread, cereal, rice, and pasta: 1 portion = 15 g carbohydrate
1 slice of bread or 1 tortilla
1/2 English muffin
3/4 c unsweetened, ready-to-eat cereal
1/2 c cooked oatmeal
1/3 c cooked rice or pasta
Starchy vegetables: 1 portion = 15 g carbohydrate
1 small (3 oz) potato
1/2 c canned or frozen corn
1/2 c cooked beans
1 c winter squash, cubed
Fruit: 1 portion = 15 g carbohydrate
1 medium apple, orange, or peach
1 small banana
3/4 c blueberries or chopped pineapple
1/2 c apple or orange juice
Milk products: 1 portion = 12 g carbohydrate; may be rounded up to 15 g for ease in counting carbohydrate portions
1 c milk (whole, low fat, or fat free)
1 c buttermilk
6 oz plain yogurt
Sweets and desserts: Considerable variation in carbohydrate content; portions listed contain approximately 15 g
1/2 c ice cream
2 sandwich cookies (with cream filling)
1/2 frosted cupcake
1 granola bar (1 oz)
1 tbs honey
Nonstarchy vegetables: 1 portion = 3 to 6 g carbohydrate; 3 servings are equivalent to 1 carbohydrate portion; can be disregarded if less than 3 servings are consumed
1/2 c cooked cauliflower
1/2 c cooked cabbage, collards, or kale
1/2 c cooked okra
1/2 c diced or raw tomatoes

NOTE: Unprocessed meats, fish, and poultry contain negligible amounts of carbohydrate.

(*continued*)

HOW TO Use Carbohydrate Counting in Clinical Practice (*continued*)

FIGURE 26-3 Translating Carbohydrate Portions into a Day's Meals

SAMPLE MENU

	Carbohydrate Portions
Breakfast:	
Carbohydrate goal = 4 portions or 60 g	
3/4 c unsweetened, ready-to-eat cereal	1
1/2 c low-fat milk	1/2
1 scrambled egg	—
1 slice whole-wheat toast (with margarine or butter)	1
6 oz orange juice	1 1/2
Coffee (without milk or sugar)	—
Lunch:	
Carbohydrate goal = 4 portions or 60 g	
1 tuna salad sandwich (includes 2 slices whole-grain bread, mayonnaise)	2
6 oz yogurt (plain) with 3/4 c blueberries and artificial sweetener	2
Diet cola	—

	Carbohydrate Portions
Afternoon snack:	
Carbohydrate goal = 2 portions or 30 g	
2 sandwich cookies	1
1 c low-fat millk	1
Dinner:	
Carbohydrate goal = 5 portions or 75 g	
4 oz grilled steak	—
1 small baked potato (with margarine or butter)	1
Corn on cob, 1 large ear	2
1/2 c steamed collard greens[a] and 1 c sliced, raw tomatoes[a]	1
1/2 c ice cream	1
Evening snack:	
Carbohydrate goal = 2 portions or 30 g	
1 medium apple	1
1 oz granola bar	1

[a]Three servings of nonstarchy vegetables are equivalent to 1 carbohydrate portion.

all types of diabetes, the carbohydrate recommendation may need to be altered periodically to improve blood glucose control.

3. Carbohydrate counting can be done in one of two ways:
   - Count the grams of carbohydrate provided by foods.
   - Count carbohydrate portions, expressed in terms of servings that contain approximately 15 grams each.

   Success with carbohydrate counting requires knowledge about the food sources of carbohydrates and an understanding of portion control. As shown in Table 26-6, food selections that contain about 15 grams of carbohydrate are interchangeable. The portions of foods that contain 15 grams may vary substantially, however, even among foods in a single food group. Accurate carbohydrate counting often requires instruction and practice in portion control using measuring cups, spoons, and a food scale. Food lists that indicate the carbohydrate contents of common foods are available from the American Diabetes Association and the American Dietetic Association; these are helpful resources for learning carbohydrate-counting methods.

   When using packaged foods, individuals should check the Nutrition Facts panel of food labels to find the carbohydrate content of a serving. If the fiber content is greater than 5 grams per serving, it should be subtracted from the *Total Carbohydrate* value, as fiber does not contribute to blood glucose. If the sugar alcohol content is greater than 5 grams per serving, half of the grams of sugar alcohol can be subtracted from the *Total Carbohydrate* value.

4. Once they have learned the basic carbohydrate counting method, individuals can select whatever foods they wish as long as they do not exceed their carbohydrate goals. Figure 26-3 shows a day's menu that provides the carbohydrate allowance shown in Table 26-5. Although carbohydrate counting focuses on a single macronutrient, people using this technique should be encouraged to follow a healthy eating plan that meets other dietary objectives as well.

TRY IT Using the food listings in Table 26-6, create a one-day menu of 4 meals and snacks that provides about 230 grams of carbohydrate. Make sure the carbohydrate intake is spread out fairly evenly throughout the day.

The exchange lists can be helpful resources for individuals using carbohydrate-counting methods because the portions in the exchange lists are interchangeable with the portions used in carbohydrate counting. Foods listed in the starch, fruit, and milk exchange lists, for example, are equivalent to carbohydrate "portions," as each item contains approximately 15 grams of carbohydrate (see Tables G-2, G-3, and G-4; note that the carbohydrate in the milk exchanges can be rounded up to 15 grams). In the list labeled "Sweets, Desserts, and Other Carbohydrates" (Table G-5), the number of carbohydrate portions per serving is indicated in the far-right column.

Insulin Therapy

Insulin therapy is necessary for individuals who cannot produce enough insulin to meet their metabolic needs. It is therefore required by people with type 1 diabetes and those with type 2 diabetes who cannot maintain glycemic control with medications, diet, and exercise. The pancreas normally secretes insulin in relatively low amounts between meals and during the night (called *basal insulin*) and in much higher amounts when meals are ingested. Ideally, the insulin treatment should reproduce the natural pattern of insulin secretion as closely as possible.

Insulin Preparations

The forms of insulin that are commercially available differ by their onset of action, timing of peak action, and duration of effects. Table 26-7 and Figure 26-4 show how insulin preparations are classified: they may be rapid acting (lispro, aspart, and glulisine), short acting (regular), intermediate acting (NPH), or long acting (glargine and detemir), thereby allowing substantial flexibility in establishing a suitable insulin regimen.[40] The rapid- and short-acting insulins are used at mealtimes, whereas the intermediate- and long-acting insulins provide basal insulin for the periods between meals and during the night. Thus, mixtures of several types of insulin can produce greater glycemic control than any one type alone. Several premixed formulations are also available; some examples are listed in Table 26-7.

Most insulin is produced by recombinant DNA techniques that allow the mass production of human insulin by bacteria or yeast. The different forms of insulin are made by chemically modifying insulin's amino acid sequence or by combining insulin with special buffers or peptides that alter insulin's concentration, solubility, or duration of activity in the body.

subcutaneous (sub-cue-TAY-nee-us): beneath the skin.

syringes: devices used for injecting medications. A syringe consists of a hypodermic needle attached to a hollow tube with a plunger inside.

♦ Because insulin is a protein, it would be destroyed by digestive processes if taken orally.

Insulin Delivery

Insulin is most often administered by **subcutaneous** injection, either self-administered or provided by caregivers. ♦ Disposable **syringes**, which are filled from vials that contain multiple doses of insulin, are the most common devices used for injecting insulin. Another option is to use insulin pens, injection

TABLE 26-7 Insulin Preparations

Form of Insulin	Common Preparations	Onset of Action	Peak Action	Duration of Action
Rapid acting	Lispro (Humalog) Aspart (Novolog) Glulisine (Apidra)	5 to 15 minutes	60 to 90 minutes	3 to 5 hours
Short acting	Regular	30 minutes	2 to 3 hours	5 to 8 hours
Intermediate acting	NPH	2 to 4 hours	6 to 10 hours	10 to 16 hours
Long acting	Glargine (Lantus) Detemir (Levemir)	1 to 2 hours	Steady effects	24 hours
Insulin mixtures (with sample ratios)	NPH/regular (70:30) NPH/regular (50:50)	Variable; depends on formulation	Variable; depends on formulation	Variable; depends on formulation

FIGURE 26-4 Effects of Insulin Preparations

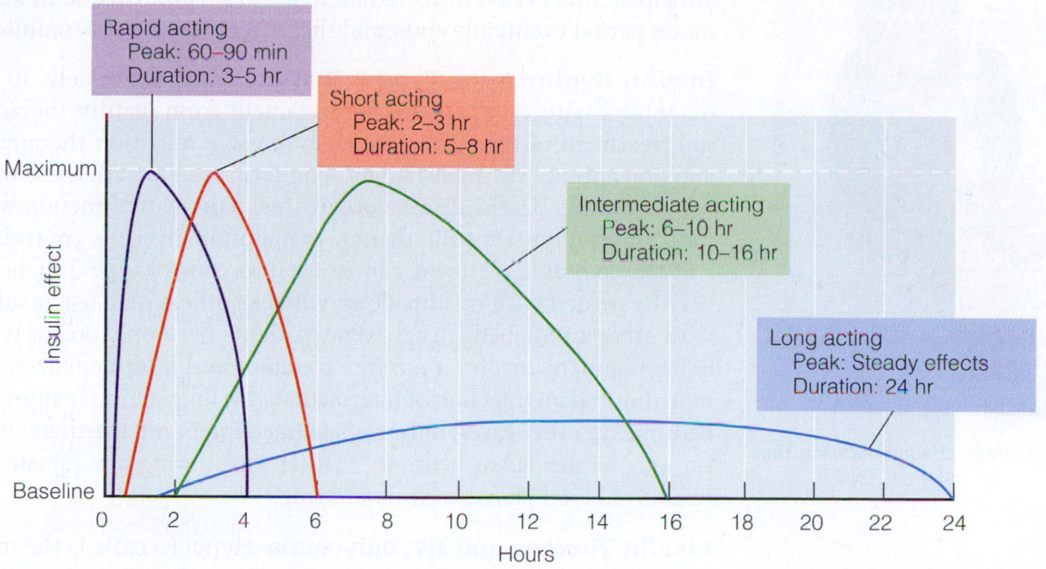

devices that resemble permanent marking pens. Disposable insulin pens are prefilled with insulin and used one time only, whereas reusable pens can be fitted with prefilled insulin cartridges and replaceable needles. To eliminate the need for multiple punctures, injection ports for insulin are sometimes inserted through the skin and left in place for several days. Some individuals use insulin pumps, computerized devices that can be programmed to deliver basal insulin continuously and bolus doses at mealtimes. These pumps infuse insulin through thin, flexible tubing that remains in the skin. An insulin pump can be worn under clothes, attached to a belt, or kept in a pocket.

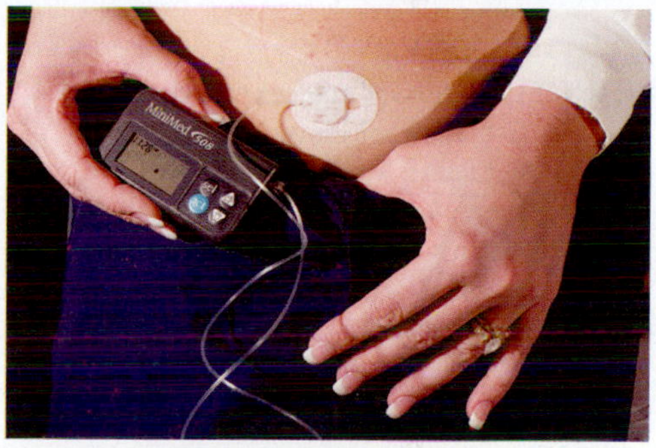

External insulin pumps deliver insulin continuously through thin, flexible tubing inserted into the skin.

An inhalation powder that supplied rapid-acting insulin was available for several years but is no longer marketed due to inadequate sales. Alternative forms of insulin that can be inhaled are currently under development.

Insulin Regimen for Type 1 Diabetes Type 1 diabetes is best managed with intensive insulin therapy, which involves multiple daily injections of several types of insulin or the use of an insulin pump. Usually, intermediate- or long-acting insulin meets basal insulin needs, and rapid- or short-acting insulin is injected before meals. ♦ Three or more daily injections are required for good glycemic control. Simpler regimens involve twice-daily injections of a mixture of intermediate- and short-acting insulin. Regimens that include three or more injections allow for greater flexibility in carbohydrate intake and meal timing. With fewer injections, the timing of both meals and injections must be similar from day to day to avoid periods of insulin deficiency or excess.

♦ Rapid-acting insulin begins working within 15 minutes, so it can be injected right before a meal. Short-acting insulin requires a half-hour wait before the meal can begin.

A person using intensive therapy must learn to accurately determine the amount of insulin to inject before each meal. The amount required depends on the premeal blood glucose level, the carbohydrate content of the meal, and the person's body weight and sensitivity to insulin. To determine insulin sensitivity, the individual keeps careful records of food intake, insulin dosages, and blood glucose levels. Eventually, these records are analyzed by medical personnel to determine the appropriate **carbohydrate-to-insulin ratio** for that individual, which assists in calculating insulin dosages at mealtime. Intensive therapy allows for substantial variation in food intake and lifestyle, but it requires frequent testing of blood glucose levels and a good understanding of carbohydrate counting.

carbohydrate-to-insulin ratio: the amount of carbohydrate that can be handled per unit of insulin. On average, every 15 grams of carbohydrate requires about 1 unit of rapid- or short-acting insulin.

After insulin therapy is initiated, persons with type 1 diabetes may experience a temporary remission of disease symptoms and a reduced need for insulin, known

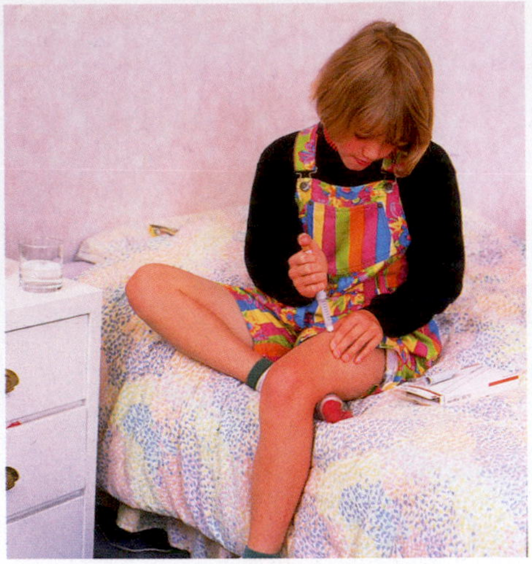
© Mark Clarke/Science Photo Library/Photo Researchers, Inc.

Children often become adept at administering the insulin they require.

as the "honeymoon period." The remission is due to a temporary improvement in pancreatic function and may last for several weeks or months.[41] It is important to anticipate this period of remission to avoid insulin excess. In all cases, the honeymoon period eventually ends, and the patient must reinstate full insulin treatment.

Insulin Regimen for Type 2 Diabetes Approximately 30 percent of people diagnosed with type 2 diabetes can benefit from insulin therapy.[42] Although initial treatment of type 2 diabetes may involve nutrition therapy, physical activity, and oral antidiabetic medications, long-term results with these treatments are often disappointing. As the disease progresses, pancreatic function worsens, and many individuals require insulin therapy to maintain glycemic control.

Many possible regimens can be used to control type 2 diabetes. Some persons may be treated with insulin alone, whereas others may use insulin in combination with other antidiabetic drugs. Many patients need only one or two daily injections. Some regimens involve a mixture of rapid- and intermediate-acting insulin in the morning and an injection of intermediate- or long-acting insulin at dinner or before bedtime. In other cases, only a single injection of intermediate- or long-acting insulin may be needed at bedtime.[43] Doses and timing are adjusted according to the results of blood glucose self-monitoring.

Insulin Therapy and Hypoglycemia Hypoglycemia is the most common complication of insulin treatment, although it may also result from the use of some oral antidiabetic drugs. It most often results from intensive insulin therapy, because the attempt to attain near-normal blood glucose levels increases the risk of overtreatment. Hypoglycemia can be corrected with the immediate intake of glucose or a carbohydrate-containing food. Usually, 15 to 20 grams of carbohydrate ♦ can relieve hypoglycemia in about 15 minutes, although patients should retest their blood glucose levels after 15 minutes in case additional treatment is necessary. Foods that provide pure glucose yield a better response than foods that contain other sugars, such as sucrose or fructose. Individuals using insulin are usually advised to carry glucose tablets or a source of carbohydrate that can be easily ingested. Those at risk of severe hypoglycemia are often given prescriptions for the hormone glucagon, which can be injected by caregivers in case of unconsciousness.

♦ Each of the following sources provides approximately 15 g of carbohydrate:
- Glucose tablets: 4 tablets
- Table sugar: 4 tsp
- Honey: 1 tbs
- Jelly beans: 15 small
- Orange juice: 1/2 c

Fasting Hyperglycemia Insulin therapy must sometimes be adjusted to prevent **fasting hyperglycemia**, which has three possible causes. The usual cause is a waning of insulin action during the night due to insufficient insulin. A second possibility, known as the **dawn phenomenon**, occurs when blood glucose levels increase in the morning due to the early morning secretion of growth hormone, which counteracts insulin's actions. Less frequently, fasting hyperglycemia develops as a result of nighttime hypoglycemia, which causes hormonal responses that stimulate glucose production; the resulting condition is known as **rebound hyperglycemia** (also called the **Somogyi effect**). Whatever the cause, fasting hyperglycemia can be treated by adjusting the dosage or formulation of insulin administered in the evening.

Antidiabetic Drugs

Treatment of type 2 diabetes often requires the use of oral medications and injectable drugs other than insulin. These drugs can improve hyperglycemia by several modes of action: they can stimulate insulin secretion, suppress glucagon secretion, decrease insulin resistance, improve glucose utilization in tissues, reduce glucose production in the liver, delay stomach emptying, or delay carbohydrate digestion and absorption. Treatment may involve the use of a single medication (monotherapy) or a combination of several medications (combination therapy). By utilizing several mechanisms at once, combination therapy achieves more rapid and sustained glycemic control than is possible with monotherapy. Table 26-8 lists examples of antidiabetic drugs, and the Diet-Drug Interactions feature lists some of their nutrition-related effects. Because medications cannot replace the benefits offered by dietary modifications and physical activity, persons with diabetes should be advised to continue both.

fasting hyperglycemia: hyperglycemia that develops after an overnight fast of at least eight hours.

dawn phenomenon: morning hyperglycemia that is caused by the early morning release of growth hormone, which counteracts insulin's glucose-lowering effects.

rebound hyperglycemia: hyperglycemia that results from the release of counterregulatory hormones following nighttime hypoglycemia; also called the **Somogyi effect**.

TABLE 26-8 Antidiabetic Drugs

Drug Category	Common Examples	Mode of Action
Alpha-glucosidase inhibitors	Acarbose (Precose) Miglitol (Glyset)	Delay carbohydrate digestion and absorption
Amylin analogs (injected)	Pramlintide (Smylin)	Suppress glucagon secretion, delay stomach emptying, suppress appetite
Biguanides	Metformin (Glucophage)	Inhibit liver glucose production, improve glucose utilization
Dipeptidyl peptidase 4 (DPP-4) inhibitors	Saxagliptin (Onglyza) Sitagliptin (Januvia)	Improve insulin secretion, suppress glucagon secretion, delay stomach emptying
Incretin mimetics (injected)	Exenatide (Byetta) Liraglutide (Victoza)	Improve insulin secretion, suppress glucagon secretion, delay stomach emptying
Meglitinides	Nateglinide (Starlix) Repaglinide (Prandin)	Stimulate insulin secretion by the pancreas
Sulfonylureas	Glipizide (Glucotrol) Glyburide (Glynase PresTab)	Stimulate insulin secretion by the pancreas
Thiazolidinediones	Pioglitazone (Actos) Rosiglitazone (Avandia)	Decrease insulin resistance

Physical Activity and Diabetes Management

Regular physical activity can improve glycemic control considerably and is therefore a central feature of disease management. A regular exercise program improves insulin sensitivity, which reduces insulin requirements. Physical activity also benefits other aspects of health, including blood lipid levels, blood pressure, body weight, and cardiovascular functioning. People with diabetes are advised to perform at least 150 minutes of moderate-intensity activity or 75 minutes of vigorous aerobic activity each week, or an equivalent combination. In addition, they should participate in a resistance exercise program at least two or three times weekly unless contraindicated by a medical condition that increases risk of injury.[44]

Medical Evaluation before Exercise

Before a person with diabetes begins a new exercise program, a medical evaluation should screen for problems that may be

DIET-DRUG Interactions

Check this table for notable nutrition-related effects of the medications discussed in this chapter.

Sulfonylureas	**Gastrointestinal effects:** Nausea, vomiting, cramps, diarrhea **Dietary interactions:** Avoid using with alcohol due to a toxic reaction that causes flushing, throbbing head and neck pain, shortness of breath, palpitations, and sweating. Avoid using with dietary supplements that contain ginseng, garlic, fenugreek, coriander, and celery, as they may increase risk of hypoglycemia. **Metabolic effects:** Hypoglycemia, weight gain, allergic skin reactions
Biguanides (metformin)	**Gastrointestinal effects:** Abdominal pain and cramps, nausea, vomiting, gas, diarrhea, metallic taste, anorexia **Dietary interactions:** Avoid using with alcohol. **Metabolic effects:** Decreased folate and vitamin B_{12} absorption
Thiazolidinediones	**Metabolic effects:** Hypoglycemia, weight gain, fluid retention, edema, anemia
Alpha-glucosidase inhibitors	**Gastrointestinal effects:** Abdominal pain and cramps, nausea, gas, diarrhea **Metabolic effects:** Elevated liver enzymes, hyperbilirubinemia

aggravated by certain activities. Complications involving the heart and blood vessels, eyes, kidneys, feet, and nervous system may limit the types of activity recommended. For individuals with a low level of fitness who have been relatively inactive, only mild or moderate exercise may be prescribed at first; for example, a short walk at a comfortable pace may be the first activity suggested. People with severe retinopathy should avoid vigorous aerobic or resistance exercise, which may lead to retinal detachment and damage to eye tissue. Peripheral neuropathy may preclude repetitive weight-bearing exercises such as jogging and step exercises because these activities may lead to foot ulcerations. ♦ To prevent dehydration, which can adversely affect blood glucose levels and heart function, proper hydration should be encouraged before and during exercise.

♦ The use of protective foot gear, such as gel soles and socks that prevent blisters, can help to prevent foot trauma during exercise.

Maintaining Glycemic Control People who do not have diabetes maintain blood glucose levels during physical activity because their normal hormonal responses—a fall in insulin levels and increased secretion of glucagon and epinephrine—promote glucose production in the liver. In people who use insulin or medications that induce insulin secretion, the natural hormonal balance is upset: blood glucose levels drop during activity because the insulin promotes rapid consumption of glucose by exercising muscles and also blocks glucose synthesis by the liver. For this reason, insulin should not be injected immediately before exercise because it can lead to hypoglycemia, and medications that promote insulin secretion may require dosage adjustments due to exercise. Conversely, a complete lack of insulin contributes to hyperglycemia, because liver glucose production is unchecked.

Individuals who use insulin or medications that induce insulin secretion should check blood glucose levels both before and after an activity. If blood glucose is below 100 mg/dL, carbohydrate should be consumed before exercise begins. Additional carbohydrate may be needed during or after prolonged activity or even several hours after the activity is completed. Strenuous exercise should be avoided if blood glucose levels exceed 250 mg/dL, and all types of physical activity should be avoided when blood glucose levels exceed 300 mg/dL or ketosis is present.[45]

Sick-Day Management

Illness, infection, or injury can cause hormonal changes that raise blood glucose levels and increase the risk of developing diabetic ketoacidosis or the hyperosmolar hyperglycemic syndrome. Hence, individuals with diabetes are counseled about sick-day management along with other self-care measures for diabetes. During illness, patients with diabetes should measure blood glucose and ketone levels in blood or urine several times daily. They should continue to take antidiabetic drugs, including insulin, as prescribed; adjustments in dosages may be necessary if hyperglycemia persists. If over-the-counter (OTC) drugs are necessary, they should use only those that are sugar- and alcohol-free. Because some OTC drugs (such as decongestants) can raise blood glucose levels and others (such as medications that contain aspirin) can interact with antidiabetic drugs, patients should check with health care providers before trying an unfamiliar drug.[46]

During illness, individuals with diabetes should consume their usual diet, if possible. If appetite is poor, they should select easy-to-manage foods and beverages that provide the prescribed amount of carbohydrate at each meal. Foods that are easily tolerated include toast, crackers, soup, yogurt, fruit, fruit juices, frozen juice bars, and carbohydrate-sweetened beverages. To prevent dehydration, especially if vomiting or diarrhea is present, patients should make sure they consume adequate amounts of liquids throughout the day.

IN SUMMARY **Diabetes treatment includes nutrition therapy, the use of insulin or other antidiabetic medications, and appropriate physical activity. Glycemic control is most often evaluated by monitoring blood glucose levels and glycated hemoglobin. The quantity of carbohydrate consumed has the greatest influence on blood glucose levels after meals. The total amount of carbohydrate ingested is more important than the type of carbohydrate consumed.**

CASE STUDY

Child with Type 1 Diabetes

Nora is a 12-year-old girl who was diagnosed with type 1 diabetes two years ago. She practices intensive therapy and has had the support of her parents and an excellent diabetes management team. With their help, Nora has been able to assume the bulk of the responsibility for her diabetes care and has managed to control her blood glucose remarkably well. In the past few months, however, Nora has been complaining bitterly about the impositions diabetes has placed on her life and her interactions with friends. Sometimes she refuses to monitor her blood glucose levels, and she has skipped insulin injections a few times. Recently, Nora was admitted to the emergency room complaining of fever, nausea, vomiting, and intense thirst. The physician noted that Nora was confused and lethargic. A urine test was positive for ketones, and her blood glucose levels were 400 mg/dL. The diagnosis was diabetic ketoacidosis.

1. Describe the metabolic events that can lead to ketoacidosis. Were Nora's symptoms and laboratory tests consistent with the diagnosis?
2. Review Table 26-4, and consider the advantages and disadvantages that intensive therapy might have for Nora.
3. Discuss how Nora's age might influence her ability to cope with and manage her diabetes. Why might she feel that diabetes is disrupting her life? What suggestions may help?
4. Review the complications associated with long-term diabetes. How might you explain the importance of glycemic control to a 12-year-old girl?

Carbohydrate counting is widely used in menu planning and can be taught at different levels of complexity, depending on individual needs and abilities. Insulin therapy is required for patients who are unable to produce sufficient insulin and may be used in both type 1 and type 2 diabetes. Antidiabetic drugs can improve insulin secretion and effectiveness, suppress glucagon secretion, reduce glucose production by the liver, and delay carbohydrate absorption. Physical activity can improve glycemic status and enhance various aspects of general health. Illness can worsen glycemic control and may necessitate adjustments in medications and careful attention to dietary and fluid requirements. The Case Study provides an opportunity to review the factors that influence treatment of type 1 diabetes.

Diabetes Management in Pregnancy

Women with diabetes face new challenges during pregnancy. Due to hormonal changes, pregnancy increases insulin resistance and the need for insulin, so maintaining glycemic control may be more difficult. In addition, about 7 percent of nondiabetic women develop gestational diabetes and require treatment during pregnancy.[47] Women with gestational diabetes are at greater risk of developing type 2 diabetes later in life, and their children are at increased risk of developing obesity and type 2 diabetes as they enter adulthood.

A pregnancy complicated by diabetes increases health risks for both mother and fetus. Uncontrolled diabetes is linked with increased incidences of miscarriage, birth defects, and fetal deaths. Newborns are more likely to suffer from respiratory distress and to develop metabolic problems such as hypoglycemia, jaundice, and hypocalcemia. Women with diabetes often deliver babies with **macrosomia** (abnormally large bodies), which makes delivery more difficult and can result in birth trauma or the need for a cesarean section. Macrosomia results because

macrosomia (MAK-roh-SOH-mee-ah): the condition of having an abnormally large body; in infants, refers to birth weights of 4000 grams (8 pounds 13 ounces) and above.

Glycemic control during pregnancy offers the best chance of a safe delivery and a healthy infant.

maternal hyperglycemia induces excessive insulin production by the fetal pancreas, which stimulates growth and fat deposition.[48]

Pregnancy in Type 1 or Type 2 Diabetes Women with diabetes who achieve glycemic control at conception and during the first trimester of their pregnancy substantially reduce the risks of birth defects and spontaneous abortion. For this reason, it is recommended that women contemplating pregnancy receive preconception care to avoid the complications associated with poorly controlled diabetes. Maintaining glycemic control during the second and third trimesters can minimize the risks of macrosomia and morbidity in newborn infants.

Nutrient requirements during pregnancy are generally similar for women with and without diabetes. The dietary adjustments suggested for improving glycemic control should be based on a woman's dietary habits and the results of blood glucose monitoring. Regular meals and snacks help to avoid hypoglycemia, which is more likely to occur during pregnancy because glucose is continuously supplied to the fetus. An evening snack is usually required to prevent overnight hypoglycemia and ketosis. Insulin and medication changes are often needed during pregnancy, and the woman may have to adjust her dietary habits further as a result.

Gestational Diabetes Risk of gestational diabetes is highest in women who have a family history of diabetes, are obese, are in a high-risk ethnic group (Hispanic American, Native American, Asian American, African American, or Pacific Islander), or have previously given birth to an infant weighing over 9 pounds. To ensure that problems are dealt with promptly, physicians routinely test all women for gestational diabetes between 24 and 28 weeks of gestation. In high-risk women, screening should begin prior to pregnancy or soon after conception. Even mild hyperglycemia can have adverse effects on a developing fetus and may lead to complications during pregnancy.[49]

Women with gestational diabetes who are overweight or obese may need to adjust their energy intakes during pregnancy. Although adequate energy is needed for fetal development, a modest kcaloric reduction (about 30 percent less than total energy needs) may improve glycemic control without increasing the risk of ketosis.[50] Restricting carbohydrate to 40 to 45 percent of total energy intake may improve blood glucose levels after meals. Carbohydrate is usually poorly tolerated in the morning; therefore, restricting carbohydrate (to about 30 grams) at breakfast is often necessary.[51] The remaining carbohydrate intake should be spaced throughout the day in several meals and snacks, including an evening snack to prevent ketosis during the night. Regular aerobic activity is often recommended because it can help to improve glycemic control. Women who fail to achieve glycemic goals by diet and exercise alone may need to use insulin or an antidiabetic drug that is safe during pregnancy.[52] The Case Study reviews the connections between gestational diabetes and type 2 diabetes.

IN SUMMARY **Careful management of blood glucose levels before and during pregnancy may reduce complications in mother and infant. Most nutrient requirements during pregnancy are similar for women with and without diabetes. Carbohydrate intake should be distributed into several meals and snacks, including an evening snack to prevent overnight ketosis. Carbohydrate restriction may be recommended, especially in women with gestational diabetes. Moderate energy restriction may help to improve glycemic control in overweight and obese women with gestational diabetes.**

CASE STUDY

Woman with Type 2 Diabetes

Alicia Cordova is a 41-year-old Mexican American woman recently diagnosed with type 2 diabetes. Mrs. Cordova developed gestational diabetes while she was pregnant with her second child. Her blood glucose levels returned to normal following pregnancy, and she was advised to get regular checkups, maintain a desirable weight, and engage in regular physical activity. Although she reports that she does not overeat and that she exercises regularly, she has been unable to maintain a healthy weight. At 5 feet 3 inches tall, Mrs. Cordova currently weighs 155 pounds. She has decided to lose weight and join a gym because she is concerned about the long-term effects of diabetes and the possibility that she may need insulin injections. She is also concerned about her husband and children because they are overweight and not very active. The physician refers Mrs. Cordova to a dietitian to help her plan a diet.

1. What factors in Mrs. Cordova's medical history increase her risk for diabetes? Are her husband and children also at risk?
2. Describe the general characteristics of a diet and exercise program that would be appropriate for Mrs. Cordova. How might weight loss and physical activity benefit her diabetes?
3. If Mrs. Cordova is unable to control her blood glucose with diet and physical activity, what treatment might be suggested? Can you explain to Mrs. Cordova why she would probably not require insulin at this time?
4. What dietary and lifestyle changes may help to prevent diabetes in Mrs. Cordova's husband and children?

Clinical Portfolio

1. Using the carbohydrate-counting method described in the "How To" on pp. 794–795, determine an appropriate carbohydrate intake (in both grams and portions) for a man with type 2 diabetes who requires approximately 2600 kcalories daily. Assume he would benefit from a carbohydrate allowance that is 50 percent of his energy intake. Using information from Tables 26-5 and 26-6, develop a one-day sample menu that is likely to meet his carbohydrate goals. Use the exchange lists in Appendix G to find additional examples of foods to include in your menu.
2. Take a trip to a pharmacy or use information from an online drugstore to price these items: blood glucose meter, test strips for the glucose meter selected, lancets, insulin, and syringes. Determine the approximate cost of insulin and syringes for a person who uses 12 units of short-acting insulin (regular) and 18 units of intermediate-acting insulin (NPH) taken in three injections daily (thus, the insulin requirement is 30 units per day). Also estimate the cost of testing blood glucose three times daily. Approximately how much would these supplies cost per month?

Nutrition on the Net

For further study of topics covered in this chapter, log on to **www.cengagebrain.com** and search for ISBN 084006845X.

- Visit the American Diabetes Association and the Joslin Diabetes Center to find information on a wide range of topics related to diabetes: **www.diabetes.org** and **www.joslin.org**
- Comprehensive and reliable information about diabetes for both health practitioners and consumers is available from the National Institute of Diabetes and Digestive and Kidney Diseases and the Centers for Disease Control and Prevention: **www2.niddk.nih.gov** and **www.cdc.gov/diabetes**
- Find out how to become a diabetes educator by visiting the website of the American Association of Diabetes Educators: **www.diabeteseducator.org**

Nutrition Assessment Checklist for People with Diabetes

Medical History

Check the medical record to determine:

- Type of diabetes
- Duration of diabetes
- Acute and chronic complications
- Conditions, including pregnancy, that may alter treatment

Medications

For people with preexisting diabetes who use antidiabetic drugs (including insulin), note:

- Type of medication
- Administration schedule

Check for use of other medications, including:

- Medications that affect blood glucose levels
- Cholesterol- and triglyceride-lowering medications
- Antihypertensive medications

Dietary Intake

To devise an acceptable meal plan and coordinate medications, obtain:

- An accurate and thorough record of food intake and meal patterns
- An account of usual physical activities

At medical checkups, reassess the person's ability to:

- Maintain an appropriate carbohydrate intake
- Maintain an appropriate energy intake
- Monitor blood glucose levels at home
- Adjust insulin and diet to accommodate sick days
- Use appropriate foods to treat hypoglycemia

Anthropometric Data

Take accurate baseline height and weight measurements as a basis for:

- Appropriate energy intake
- Initial insulin therapy

Periodically reassess height and weight for children and weight for adults and pregnant women to ensure that the meal plan provides an appropriate energy intake.

Laboratory Tests

Monitor the success of diabetes treatment using these tests:

- Blood lipid concentrations
- Blood or urinary ketones
- Fructosamine, if necessary
- Glycated hemoglobin
- Urinary protein (microalbuminuria)

Physical Signs

Look for physical signs of:

- Dehydration, especially in older adults
- Foot ulcers
- Nerve damage
- Vision problems

References

1. National Center for Health Statistics, *Health, United States, 2009: With Special Feature on Medical Technology* (Hyattsville, MD: 2010), p. 258.
2. National Center for Health Statistics, 2010.
3. American Diabetes Association, Diagnosis and classification of diabetes mellitus, *Diabetes Care* 34 (2011): S62–S69.
4. D. Lloyd-Jones and coauthors, Heart disease and stroke statistics 2010 update: A report from the American Heart Association, *Circulation* 121 (2010): e46–e215.
5. S. E. Inzucchi and R. S. Sherwin, Type 1 diabetes mellitus, in L. Goldman and D. Ausiello, eds., *Cecil Medicine* (Philadelphia: Saunders, 2008), pp. 1727–1747.
6. A. Maitra, The endocrine system, in V. Kumar and coeditors, *Robbins and Cotran Pathologic Basis of Disease* (Philadelphia: Saunders, 2010), pp. 1097–1164.
7. National Center for Health Statistics, 2010.
8. R. B. Lipton, Incidence of diabetes in children and youth—tracking a moving target, *Journal of the American Medical Association* 297 (2007): 2760–2762.
9. The Writing Group for the SEARCH for Diabetes in Youth Study Group, Incidence of diabetes in youth in the United States, *Journal of the American Medical Association* 297 (2007): 2716–2724.

10. Diabetes Mellitus: A Guide to Patient Care (Philadelphia: Lippincott Williams & Wilkins, 2007), pp. 37–49; J. Wylie-Rosett and L. M. Delahanty, Diabetes prevention, in T. A. Ross, J. L. Boucher, and B. S. O'Connell, eds., *American Dietetic Association Guide to Diabetes: Medical Nutrition Therapy and Education* (Chicago: American Dietetic Association, 2005), pp. 49–58.
11. American Diabetes Association, Standards of medical care in diabetes—2011, *Diabetes Care* 34 (2011): S11–S61; *Diabetes Mellitus: A Guide to Patient Care*, 2007, pp. 37–49.
12. A. Pietraszek, S. Gregersen, and K. Hermansen, Alcohol and type 2 diabetes: A review, *Nutrition, Metabolism, and Cardiovascular Diseases* 20 (2010): 366–375; M. M. Joosten and coauthors, Moderate alcohol consumption increases insulin sensitivity and ADIPOQ expression in postmenopausal women: A randomised, crossover trial, *Diabetologia* 51 (2008): 1375–1381.
13. U. Masharani and M. S. German, Pancreatic hormones and diabetes mellitus, in D. G. Gardner and D. Shoback, eds., *Greenspan's Basic and Clinical Endocrinology* (New York: McGraw-Hill/Lange, 2007), pp. 661–747.
14. Inzucchi and Sherwin, Type 1 diabetes mellitus, 2008.
15. Masharani and German, 2007.
16. S. E. Inzucchi and R. S. Sherwin, Type 2 diabetes mellitus, in L. Goldman and D. Ausiello, eds., *Cecil Medicine* (Philadelphia: Saunders, 2008), pp. 1748–1760.
17. D. G. Gardner, Endocrine emergencies, in D. G. Gardner and D. Shoback, eds., *Greenspan's Basic and Clinical Endocrinology* (New York: McGraw-Hill/Lange, 2007), pp. 868–893.
18. Inzucchi and Sherwin, Type 1 diabetes mellitus, 2008.
19. Inzucchi and Sherwin, Type 1 diabetes mellitus, 2008.
20. Inzucchi and Sherwin, Type 1 diabetes mellitus, 2008.
21. Inzucchi and Sherwin, Type 1 diabetes mellitus, 2008.
22. Maitra, The endocrine system, 2010.
23. Inzucchi and Sherwin, Type 1 diabetes mellitus, 2008.
24. Inzucchi and Sherwin, Type 1 diabetes mellitus, 2008.
25. Diabetes Control and Complications Trial Research Group, The effect of intensive treatment of diabetes on the development and progression of long-term complications in insulin-dependent diabetes mellitus, *New England Journal of Medicine* 329 (1993): 977–986.
26. American Diabetes Association, Implications of the United Kingdom Prospective Diabetes Study, *Diabetes Care* 21 (1998): 2180–2184.
27. American Diabetes Association, Standards of medical care in diabetes—2011, 2011.
28. Masharani and German, 2007.
29. American Diabetes Association, Standards of medical care in diabetes—2011, 2011.
30. M. Ryan and coauthors, Is a failure to recognize an increase in food intake a key to understanding insulin-induced weight gain? *Diabetes Care* (17 December 2007) epub, available at doi:10.2337/dc07-1171; A. N. Jacob and coauthors, Potential causes of weight gain in type 1 diabetes mellitus, *Diabetes, Obesity and Metabolism* 8 (2006): 404–411.
31. A. Daly, Use of insulin and weight gain: Optimizing diabetes nutrition therapy, *Journal of the American Dietetic Association* 107 (2007): 1386–1393.
32. American Diabetes Association, Standards of medical care in diabetes—2011, 2011; M. J. Franz and coauthors, The evidence for medical nutrition therapy for type 1 and type 2 diabetes in adults, *Journal of the American Dietetic Association* 110 (2010): 1852–1889.
33. American Diabetes Association, Standards of medical care in diabetes—2011, 2011; American Diabetes Association, Nutrition recommendations and interventions for diabetes, *Diabetes Care* 31 (2008): S61–S78.
34. American Diabetes Association, Standards of medical care in diabetes—2011, 2011.
35. American Diabetes Association, Nutrition recommendations and interventions for diabetes, 2008.
36. Franz and coauthors, 2010; *American* Diabetes Association, Nutrition recommendations and interventions for diabetes, 2008.
37. American Diabetes Association, Nutrition recommendations and interventions for diabetes, 2008.
38. Franz and coauthors, 2010; American Diabetes Association, Nutrition recommendations and interventions for diabetes, 2008.
39. American Diabetes Association, Nutrition recommendations and interventions for diabetes, 2008.
40. M. S. Nolte and J. H. Karam, Pancreatic hormones and antidiabetic drugs, in B. G. Katzung, ed., *Basic and Clinical Pharmacology* (New York: McGraw-Hill/Lange, 2007), pp. 683–705; Masharani and German, 2007.
41. Inzucchi and Sherwin, Type 1 diabetes mellitus, 2008.
42. Nolte and Karam, 2007.
43. Inzucchi and Sherwin, Type 2 diabetes mellitus, 2008.
44. American Diabetes Association, Standards of medical care in diabetes—2011, 2011.
45. Inzucchi and Sherwin, Type 1 diabetes mellitus, 2008; *Diabetes Mellitus: A Guide to Patient Care* (Philadelphia: Lippincott Williams & Wilkins, 2007), p. 74.
46. Diabetes Mellitus: A Guide to Patient Care, 2007, pp. 248–249.
47. American Diabetes Association, Diagnosis and classification of diabetes mellitus, 2011.
48. A. Maitra, Diseases of infancy and childhood, in V. Kumar and coeditors, *Robbins and Cotran Pathologic Basis of Disease* (Philadelphia: Saunders, 2010), pp. 447–483.
49. Z. Hussain and L. Jovanovic, Nutritional strategies in pregestational, gestational, and postpartum diabetic patients, in J. I. Mechanick and E. M. Brett, eds., *Nutritional Strategies for the Diabetic and Prediabetic Patient* (Boca Raton, FL: CRC Press, 2006), pp. 133–148.
50. American Diabetes Association, Nutrition recommendations and interventions for diabetes, 2008.
51. American Dietetic Association, *Nutrition Care Manual* (Chicago: American Dietetic Association, 2011).
52. O. Langer, Oral antidiabetic drugs in pregnancy: The other alternative, *Diabetes Spectrum* 20 (2007): 101–105.

The Metabolic Syndrome

Chapter 26 describes how insulin resistance—a reduced sensitivity to insulin in muscle, adipose, and liver cells—can contribute to hyperglycemia and hyperinsulinemia and, eventually, to type 2 diabetes. Insulin resistance is also a central feature of several other conditions, including the **metabolic syndrome**, a cluster of metabolic abnormalities that raise the risk of developing cardiovascular diseases (CVD) and type 2 diabetes. This highlight describes how the metabolic syndrome is diagnosed, how and why it might develop, its consequences, and current treatment approaches. The accompanying glossary defines the relevant terms.

Prevalence of the Metabolic Syndrome

Table H26-1 lists the laboratory values used to identify the metabolic syndrome, which is diagnosed when at least three of the following disorders are present: hyperglycemia, obesity, **hypertriglyceridemia** (high blood triglyceride levels), low HDL cholesterol levels, and hypertension (high blood pressure). An estimated 34 percent of adults in the United States may meet the criteria for the metabolic syndrome.[1] As Figure H26-1 shows, prevalence increases with age. Risk also varies among ethnic groups: Mexican Americans have the highest incidence of the metabolic syndrome in the United States, with an overall prevalence of about 37 percent.[2] Although the precise cause of the metabolic syndrome is not known, the close relationship between abdominal obesity and insulin resistance suggests that the current obesity crisis in the United States may be partly responsible for the high prevalence of the condition.

TABLE H26-1 Features of the Metabolic Syndrome

Metabolic syndrome is diagnosed when three or more of the following abnormalities are present.

Measure	Diagnostic Cut Point
Hyperglycemia	Fasting plasma glucose ≥100 mg/dL
Abdominal obesity	Waist circumference >40" in men, >35" in women
Hypertriglyceridemia	≥150 mg/dL
Reduced HDL cholesterol	<40 mg/dL in men, <50 mg/dL in women
Hypertension	≥130/85 mm Hg

FIGURE H26-1 Prevalence of the Metabolic Syndrome in the U.S. Population

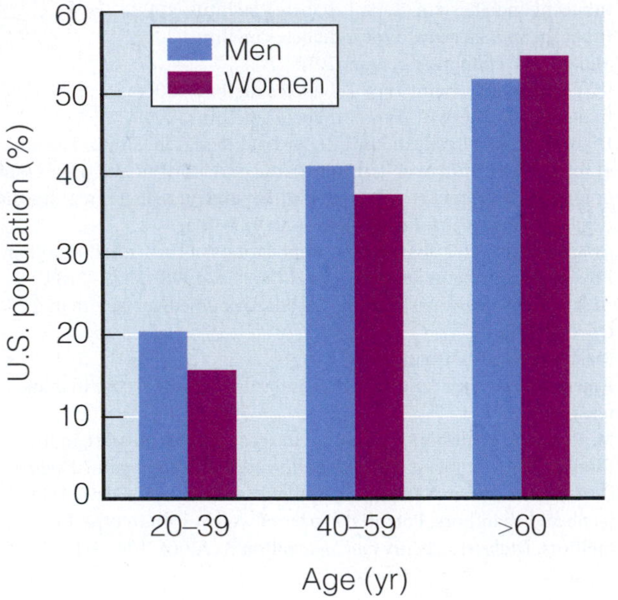

SOURCE: R. B. Ervin, Prevalence of metabolic syndrome among adults 20 years of age and over, by sex, age, race and ethnicity, and body mass index: United States, 2003–2006, *National Health Statistics Reports*, no. 13 (Hyattsville, MD: National Center for Health Statistics, 2009).

GLOSSARY

adiponectin (AH-dih-poe-NECK-tin): a hormone produced by adipose cells that improves insulin sensitivity.

fibrinogen (fye-BRIN-oh-jen): a liver protein that promotes blood clot formation.

hypertriglyceridemia (HYE-per-try-gliss-er-rye-DEE-me-ah): high blood triglyceride levels.

metabolic syndrome: a cluster of interrelated clinical disorders, including obesity, insulin resistance, high blood pressure, and abnormal blood lipids, which together increase risk of diabetes and cardiovascular disease; also known as *insulin resistance syndrome* or *syndrome X*.

nitric oxide: a compound produced by blood vessel cells that helps to regulate blood vessel function, including dilation and constriction.

plasminogen activator inhibitor-1: a protein that promotes blood clotting by inhibiting blood clot degradation within blood vessels.

resistin (re-ZIST-in): a hormone produced by adipose cells that induces insulin resistance.

Reminder: Cytokines are signaling proteins produced by the body's cells.

Obesity and the Metabolic Syndrome

Excessive body fat induces a number of metabolic changes that lead to insulin resistance, which then leads to hyperglycemia and other abnormalities. Overweight individuals (BMI from 25 to 29.9) have about a sixfold increased risk of developing metabolic syndrome compared with underweight and normal-weight persons (BMI less than 25).[3] The following sections explore several of these relationships.

Obesity and Insulin Resistance

Various theories have been proposed to explain the relationship between obesity and insulin resistance. Much of the research suggests that obesity leads to an increase in fatty acid concentrations in the blood, resulting in the abnormal deposition of triglycerides in the muscle, liver, and abdominal region.[4] The abnormally high fat content of these tissues may perturb cellular responses to insulin and result in insulin resistance. Unless the pancreas can secrete enough insulin to compensate, glucose uptake from the blood is reduced, contributing to hyperglycemia.

Obesity can also alter production of the hormones and various other proteins made in adipose cells, which influence metabolic processes and fuel use in the body.[5] For example, obesity is associated with the reduced secretion of **adiponectin**, a hormone involved in sensitivity to insulin. Conversely, the hormone **resistin**, which promotes insulin resistance, is released in greater amounts. The lipid overload in adipose cells induces a chronic inflammatory state, which draws macrophages (white blood cells) to the adipose tissue.[6] Together, the adipose cells and macrophages secrete a variety of cytokines (such as TNF-alpha) that interfere with insulin responsiveness.

Obesity and Hypertension

Obesity increases the risk of developing high blood pressure, a common component of the metabolic syndrome. Both insulin resistance and hyperinsulinemia may be implicated in raising blood pressure. Insulin resistance interferes with the normal relaxation and dilation of blood vessels. Hyperinsulinemia promotes reabsorption of sodium in the kidneys, resulting in fluid retention and increased blood volume.[7] These effects contribute to an increase in blood pressure.

Obesity and Hypertriglyceridemia

Abdominal obesity is frequently associated with blood lipid abnormalities.[8] Increases in body weight are linked with higher triglyceride and LDL cholesterol levels and lower HDL cholesterol levels. As a result of obesity, the adipose cells are less responsive to insulin and release more fatty acids into the blood. At the same time, they are less able to extract and store triglycerides from chylomicrons and VLDL. To keep up with the greater influx of fatty acids from the bloodstream, the liver must accelerate its production of VLDL, and hypertriglyceridemia develops.

Metabolic Syndrome and CVD Risk

As mentioned, individuals with the metabolic syndrome are at increased risk of developing cardiovascular diseases. The disorders that characterize the metabolic syndrome—obesity, lipid abnormalities, and hypertension—are all independent risk factors for CVD. In addition, both insulin resistance and elevated lipoprotein levels can cause damage to blood vessels, promoting inflammation and accelerating the progression of atherosclerosis.[9] Blood vessel inflammation induces liver secretion of **fibrinogen**, a protein that promotes blood clot formation (increasing risk of a heart attack or stroke). C-reactive protein, which is elevated by both inflammation and obesity, inhibits **nitric oxide** production by blood vessel cells, an effect that impairs blood vessel activity and also promotes blood clotting.[10] Another procoagulant factor—**plasminogen activator inhibitor-1**—is overproduced as a consequence of both obesity and hyperinsulinemia. The combined effect of these multiple abnormalities can worsen atherosclerosis and increase the risks of developing heart attack or stroke. Individuals with the metabolic syndrome are also at increased risk of developing diabetes, which is another major risk factor for CVD.

Treatment of the Metabolic Syndrome

The metabolic syndrome is primarily treated with dietary and lifestyle changes, with the goal of correcting abnormalities that increase CVD risk.[11] In most individuals, a combination of weight loss and physical activity can improve insulin resistance, blood pressure, and blood lipid levels. Additional dietary strategies depend on a patient's specific symptoms. If dietary and lifestyle changes are not successful, medications may be prescribed. Because effective treatment requires lifelong commitment, health care providers should work with patients to develop a treatment plan that they are willing to adopt.

Dietary Management

Weight reduction is often recommended for obese individuals, and even a small weight loss (10 to 20 pounds) can improve symptoms. Many people find it difficult to achieve and maintain weight loss, however, so they should be encouraged to make other dietary changes that can improve their health. In individuals with hypertriglyceridemia, the general recommendation is to reduce intake of added sugars and refined grain products (soda, juices, white bread, sweetened cereal, and desserts) and increase servings of whole grains and foods high in fiber (whole-wheat bread, oatmeal, legumes, fruits, and vegetables). In some people, carbohydrate restriction may help to reduce blood triglyceride levels and improve hyperglycemia.[12] Including fish in the diet each week may also improve triglyceride levels. Individuals with hypertension are encouraged to reduce sodium intake and increase consumption of fruits and vegetables and low-fat milk products. A diet low in saturated fat, *trans* fats, and cholesterol can help to reduce LDL cholesterol levels. Chapter 27 describes additional dietary modifications that can reduce CVD risk.

Physical Activity

Regular physical activity helps with weight management and may also improve blood lipid concentrations, hypertension, and insulin resistance—all changes that can reduce the risk of developing CVD. A regular exercise program can also prevent or delay the onset of diabetes in persons at risk.[13] A program that includes both aerobic exercise and strength training is best. A minimum of 30 minutes of moderate aerobic activity (brisk walking, jogging, or cycling) daily is suggested, although longer periods (one hour daily) are recommended for weight control. A sedentary lifestyle can worsen the progression of metabolic syndrome and should be discouraged.

© Rolf Bruderer/Corbis

Regular exercise can reduce the risks of developing the metabolic syndrome, cardiovascular diseases, and type 2 diabetes.

Drug Therapy

If dietary and lifestyle changes are unsuccessful, medications may be prescribed to correct hypertriglyceridemia and hypertension (Chapter 27 provides details). At present, antidiabetic drugs are not routinely used to treat insulin resistance in patients with the metabolic syndrome due to insufficient evidence that the drugs can improve long-term outcomes better than lifestyle changes.

As explained in this highlight, the metabolic syndrome consists of a cluster of interrelated disorders that increase the risk for developing CVD and type 2 diabetes. Whereas individually the features of the metabolic syndrome are risk factors for CVD, in combination they may raise risk twofold to threefold. Treatment of the metabolic syndrome emphasizes dietary and lifestyle changes. The following chapter provides additional information about the dietary and lifestyle changes that can reduce CVD risk.

References

1. R. B. Ervin, Prevalence of metabolic syndrome among adults 20 years of age and over, by sex, age, race and ethnicity, and body mass index: United States, 2003–2006, *National Health Statistics Reports*, no. 13 (Hyattsville, MD: National Center for Health Statistics, 2009).
2. Ervin, 2009.
3. Ervin, 2009.
4. J.-P. Després and I. Lemieux, Abdominal obesity and metabolic syndrome, *Nature* 444 (2006): 881–887.
5. A. Maitra, The endocrine system, in V. Kumar and coeditors, *Robbins and Cotran Pathologic Basis of Disease* (Philadelphia: Saunders, 2010), pp. 1097–1164; M. D. Jensen, Obesity, in L. Goldman and D. Ausiello, eds., *Cecil Medicine* (Philadelphia: Saunders, 2008), pp. 1643–1652.
6. A. Guilherme and coauthors, Adipocyte dysfunctions linking obesity to insulin resistance and type 2 diabetes, *Nature Reviews* 9 (May 2008): 367–377.
7. V. Kumar and coauthors, Environmental and nutritional diseases, in V. Kumar and coeditors, *Robbins and Cotran Pathologic Basis of Disease* (Philadelphia: Saunders, 2010), pp. 399–445.
8. Jensen, 2008.
9. R. F. Kushner and J. L. Roth, Nutritional strategies for patients with obesity and the metabolic syndrome, in J. I. Mechanick and E. M. Brett, eds., *Nutritional Strategies for the Diabetic and Prediabetic Patient* (Boca Raton, FL: CRC Press, 2006), pp. 55–80.
10. S. M. Grundy and coauthors, Clinical management of metabolic syndrome: Report of the American Heart Association/National Heart, Lung, and Blood Institute/American Diabetes Association Conference on Scientific Issues Related to Management, *Circulation* 109 (2004): 551–556.
11. F. Magkos and coauthors, Management of the metabolic syndrome and type 2 diabetes through lifestyle modification, *Annual Review of Nutrition* 29 (2009): 223–256; Kushner and Roth, 2006.
12. Kushner and Roth, 2006.
13. Magkos and coauthors, 2009; *Diabetes Mellitus: A Guide to Patient Care* (Philadelphia: Lippincott Williams & Wilkins, 2007), pp. 37–49.

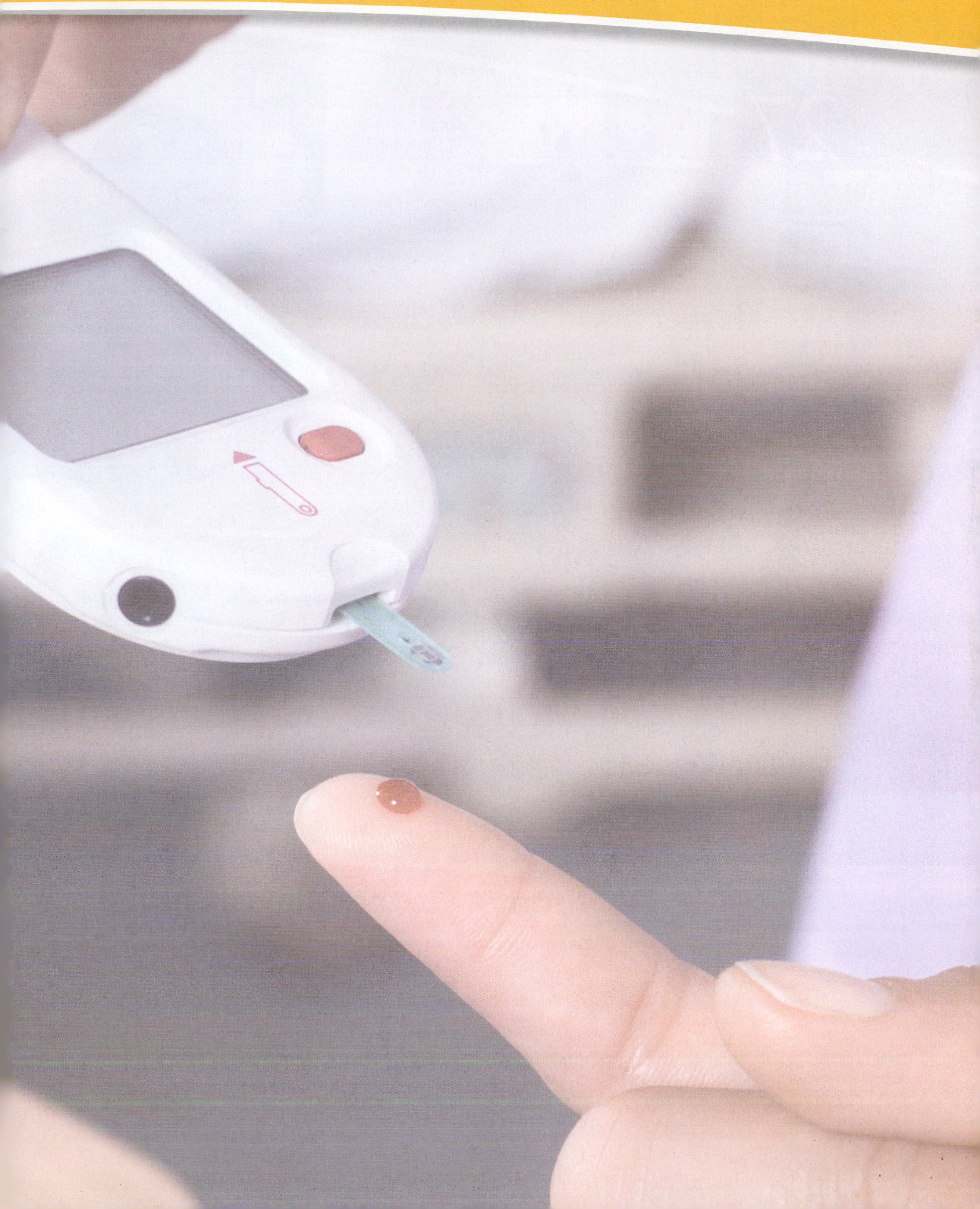

CHAPTER

27

Nutrition in the Clinical Setting

Each heartbeat sends oxygen-rich blood to the body's tissues. When the functions of the heart and blood vessels are disturbed, as is common in cardiovascular diseases, the disrupted blood supply hinders the ability of cells to carry out their metabolic functions. At first, people with cardiovascular disease may not realize that their weakness, fatigue, or shortness of breath are symptoms of a cardiovascular illness. When their condition worsens, however, the complications can be disabling and interfere with many aspects of daily life.

Throughout this chapter, the CourseMate icon indicates an opportunity for online self-study, linking you to activities to increase your understanding of chapter concepts. **www.cengagebrain.com** (search for ISBN 084006845X)

Cardiovascular Diseases

CHAPTER OUTLINE

Cardiovascular disease (CVD) is a general term describing diseases of the heart and blood vessels. ♦ **Coronary heart disease (CHD)**, the most common form of CVD, is usually caused by atherosclerosis in the coronary arteries that supply blood to the heart muscle. If atherosclerosis restricts blood flow in these arteries, the resulting deprivation of oxygen and nutrients can destroy heart tissue and cause a **myocardial infarction (MI)—a heart attack**. When the blood supply to brain tissue is blocked, a **stroke** occurs. Both heart attack and stroke may result in disablement or death. This chapter describes these and other cardiovascular disorders. Figure 27-1 shows the percentages of deaths resulting from all types of CVD. The glossary on p. 812 defines common terms related to CVD.

Cardiovascular disease is responsible for nearly 34 percent of deaths in the United States.[1] Although many people assume that heart conditions are men's diseases, more women than men die each year from the various types of CVD. Furthermore, CVD is a global health issue; it is the leading cause of death in Europe and contributes to 30 percent of deaths worldwide.[2]

♦ Cardiovascular disease is described in Chapter 5 (see p. 151) and discussed further in Highlight 15.

Atherosclerosis

In atherosclerosis, ♦ sometimes called "hardening of the arteries," the artery walls become progressively thickened due to an accumulation of fatty deposits, smooth muscle cells, and fibrous connective tissue, collectively known as **plaque**. Plaque can exist in a stable form that does not cause complications or an unstable form called **vulnerable plaque**. Vulnerable plaque has only a thin, fibrous barrier between

♦ Atherosclerosis is the most common form of *arteriosclerosis*, a more general term for arterial diseases that are characterized by abnormally thickened walls and lost elasticity.

FIGURE 27-1 Percentage Breakdown of Deaths from Cardiovascular Diseases in the United States

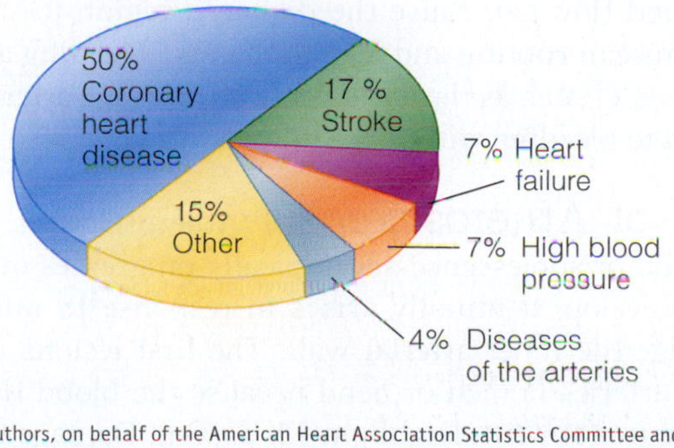

SOURCE: V. L. Roger and coauthors, on behalf of the American Heart Association Statistics Committee and Stroke Statistics Subcommittee, Heart disease and stroke statistics—2011 update: A report from the American Heart Association, *Circulation* 123 (2011), available at doi:10.1161/CIR.0b013e3182009701.

GLOSSARY

OF TERMS RELATED TO CARDIOVASCULAR DISEASES

aneurysm (AN-you-rih-zum): an abnormal enlargement or bulging of a blood vessel (usually an artery) caused by weakness in the blood vessel wall.

angina (an-JYE-nah or AN-ji-nah) **pectoris:** a condition caused by ischemia in the heart muscle that results in discomfort or dull pain in the chest region. The pain often radiates to the left shoulder and arm or to the back, neck, and lower jaw.

cardiovascular disease (CVD): a general term describing diseases of the heart and blood vessels.
- **cardio** = heart
- **vascular** = blood vessels

coronary heart disease (CHD): a chronic, progressive disease characterized by obstructed blood flow in the coronary arteries; also called *coronary artery disease*.

embolism (EM-boh-lizm): the obstruction of a blood vessel by an embolus, causing sudden tissue death.
- **embol** = to insert, plug

embolus (EM-boh-lus): an abnormal particle, such as a blood clot or air bubble, that travels in the blood.

fatty streaks: initial lesions of atherosclerosis that form on the artery wall, characterized by accumulations of foam cells, lipid material, and connective tissue.

foam cells: swollen cells in the artery wall that accumulate lipids.

intermittent claudication (claw-dih-KAY-shun): severe pain and weakness in the legs (especially the calves) caused by inadequate blood supply to the muscles; usually occurs with walking and subsides during rest.

ischemia (iss-KEE-mee-a): inadequate blood supply within tissues due to obstructed blood flow in the arteries.

myocardial (MY-oh-CAR-dee-al) **infarction** (in-FARK-shun), or **MI:** death of heart muscle caused by a sudden reduction in coronary blood flow; also called a **heart attack** or *cardiac arrest*.
- **myo** = muscle
- **cardial** = heart
- **infarct** = tissue death

plaque (PLACK): an accumulation of fatty deposits, smooth muscle cells, and fibrous connective tissue in blood vessels.

stroke: a sudden injury to brain tissue resulting from impaired blood flow through an artery that supplies blood to the brain; also called a *cerebrovascular accident*.
- **cerebro** = brain

thrombosis (throm-BOH-sis): the formation or presence of a blood clot in blood vessels. A *coronary thrombosis* occurs in a coronary artery, and a *cerebral thrombosis* occurs in an artery that supplies blood to the brain.
- **thrombo** = clot

thrombus: a blood clot formed within a blood vessel that remains attached to its place of origin.

vulnerable plaque: a form of plaque, susceptible to rupture, that is lipid-rich and has only a thin, fibrous barrier between the arterial lumen and the plaque's lipid core.

its lipid-rich core and the arterial lumen.[3] It is highly susceptible to rupture, which then promotes blood clot formation (**thrombosis**) within the artery.

Consequences of Atherosclerosis

As atherosclerosis worsens, it can eventually narrow the lumen of an artery and interfere with blood flow. If the plaque ruptures and results in thrombosis, the blood clot (**thrombus**) may enlarge in time and ultimately obstruct blood flow. A portion of a clot can also break free (**embolus**) and travel through the circulatory system until it lodges in a narrowed artery and shuts off blood flow to the surrounding tissue (**embolism**). Most complications of atherosclerosis result from the deficiency of blood and oxygen within the tissue served by an artery (**ischemia**).

Atherosclerosis can affect almost any organ or tissue in the body and, accordingly, is a major cause of disablement or death. Obstructed blood flow in the coronary arteries can cause pain or discomfort in the chest and surrounding regions (**angina pectoris**) or lead to a heart attack. As mentioned earlier, obstructed blood flow to the brain can cause injury or destruction to brain tissue, or a stroke. Impaired blood circulation in the legs can cause fatigue and pain while walking, known as **intermittent claudication**. Blockage of the arteries that supply the kidneys can result in kidney disease or even acute kidney failure.

Atherosclerosis is the most common cause of an **aneurysm**—the abnormal dilation of a blood vessel. Plaque can weaken the blood vessel wall, and eventually the pressure of blood flow can cause the damaged region to stretch and balloon outward. Aneurysms can rupture and lead to massive bleeding and death, particularly when a large vessel such as the aorta is affected. In the arteries of the brain, an aneurysm may lead to bleeding within the brain, a coma, or a stroke.

Development of Atherosclerosis

Atherosclerosis begins to develop as early as childhood or adolescence and typically progresses over several decades before symptoms develop. It initially arises in response to minimal but chronic injuries that damage the inner arterial wall. The first lesions tend to develop in regions where the arteries branch or bend because the blood flow is disturbed in those areas (see Figure H15-1 in Highlight 15, p. 534). Damage to the artery elicits an inflammatory response, ♦ attracting immune cells and increasing the permeability of artery walls. Low-density lipoproteins (LDL) slip under the artery's thin layer of **endothelial cells**, become oxidized by local enzymes, and accumulate. Arterial

♦ C-reactive protein, an acute-phase protein secreted during the inflammatory response, is associated with increased heart disease risk (see Chapter 22).

endothelial cells: the cells that line the inner surfaces of blood vessels, lymphatic vessels, and body cavities.

FIGURE 27-2 **Stages of Plaque Progression**

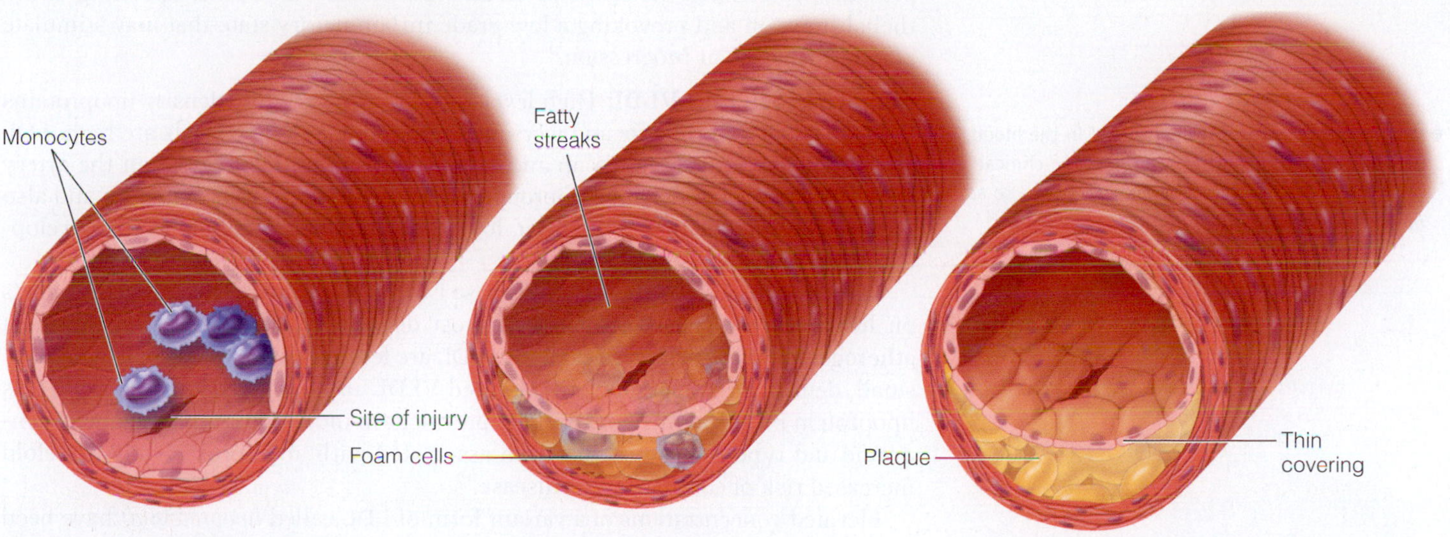

Monocytes—phagocytic white blood cells—circulate in the bloodstream and respond to injury on the artery wall.

Monocytes slip under blood vessel cells and engulf LDL cholesterol, becoming foam cells. The thin layers of foam cells that develop on artery walls are known as *fatty streaks*.

A fatty streak thickens and forms plaque as it accumulates additional lipids, smooth muscle cells, connective tissue, and cellular debris.

The artery may expand to accommodate plaque. When this occurs, the plaque that develops often contains a large lipid core with a thin, fibrous covering and is vulnerable to rupture and thrombosis.

macrophages ♦ engulf this altered LDL and become **foam cells**; these fat-laden cells are visible as fatty deposits along artery walls, known as **fatty streaks** (see Figure 27-2). To repair the damage, the artery's smooth muscle cells divide and produce fibrous proteins, which form a scar-like cap to wall off the fatty lesion. The lipid core accumulates calcium and cellular debris, and the accumulated cholesterol may crystallize and harden. The numerous processes involved in atherosclerosis occur in response to cytokines and other signaling molecules produced by the endothelial and immune cells.[4]

♦ Reminder: *Macrophages* are immune cells that engulf pathogens and cellular debris; they are derived from white blood cells called *monocytes* (see Highlight 17).

As atherosclerosis progresses, the artery may expand outward to accommodate the plaque, such that a decrease in the lumen diameter does not occur.[5] In other cases, the accumulating plaque causes narrowing, rather than expansion, of the artery. Although arteries that expand are less likely to interfere with blood flow, they are usually associated with unstable vulnerable plaque, which is more likely to rupture, induce clotting, and increase the risk of heart attack or stroke. The arteries that accommodate plaque only by narrowing may impede blood flow, but they generally have a more stable plaque structure, with a lower lipid content and a thicker barrier between the plaque and arterial lumen. The discovery of variations in plaque anatomy may help to explain why some CVD treatments can dramatically reduce the risk of heart attack and stroke even when plaque volume and the lumen diameter do not change.[6]

Causes of Atherosclerosis

The reasons why atherosclerosis develops and progresses are complex. Generally, the factors that initiate atherosclerosis either cause direct damage to the artery wall or allow lipid materials to penetrate its surface. The factors that worsen atherosclerosis or lead to complications are those that promote plaque rupture or blood coagulation. The development of advanced atherosclerosis is a long-term process that involves recurrent plaque rupture, thrombosis, and healing at sites in the artery wall.[7]

Shear Stress/Hypertension The stress of blood flow along artery walls—called *shear stress*—can cause physical damage to arteries.[8] Hypertension (high blood pressure) intensifies the stress of blood flow on arterial tissue, upsetting endothelial function and provoking a low-grade inflammatory state that may stimulate plaque formation or progression.[9]

Elevated LDL and VLDL High levels of LDL and very-low-density lipoproteins (VLDL) ♦ can both initiate and worsen atherosclerosis. When levels are high, LDL and VLDL are actively taken up and retained in susceptible regions in the artery wall.[10] Because high-density lipoproteins (HDL) help to prevent oxidation and also remove cholesterol from circulation, low levels of HDL contribute to the development of atherosclerosis as well.

♦ Reminder: LDL transport cholesterol in the blood, whereas VLDL transport triglycerides. In clinical practice, VLDL levels are commonly referred to as *blood triglycerides*.

LDL vary in size and density, and these LDL subtypes may have differing effects on heart disease risk. The smallest, most dense LDL are considered the most **atherogenic**, whereas larger, less dense LDL are less atherogenic.[11] People who have small, dense LDL frequently have elevated VLDL and low HDL levels as well. This lipoprotein profile, which is especially prevalent in individuals with metabolic syndrome and type 2 diabetes, has been associated with an approximately threefold increased risk of coronary heart disease.[12]

Elevated concentrations of a variant form of LDL called *lipoprotein(a)* have been found to speed the progression of atherosclerosis and to raise the risk of various types of CVD.[13] Lipoprotein(a) levels are primarily genetically determined and are influenced to only a minor degree by age and environmental factors.

Cigarette Smoking Chemicals in cigarette smoke (including nicotine) are toxic to endothelial cells, and the resulting damage contributes to arterial injury. Other effects of smoking include chronic inflammation, enhanced blood coagulation, increased LDL cholesterol, and decreased HDL cholesterol—all effects that can promote the progression of atherosclerosis.[14]

Diabetes Mellitus Diabetes can initiate and accelerate the development of atherosclerosis in multiple ways. Chronic hyperglycemia leads to the accumulation of advanced glycation end products (AGEs), ♦ which can directly damage endothelial cells and disturb blood vessel function.[15] AGEs also promote inflammation and oxidative stress, which contribute to plaque's progression. By other mechanisms, diabetes increases tendencies for vasoconstriction, plaque rupture, and blood clotting.

♦ Reminder: *Advanced glycation end products* are reactive compounds formed after glucose nonenzymatically attaches to proteins.

Age and Gender Advancing age promotes atherosclerosis due to the cumulative exposure to risk factors and the degeneration of arterial cells with age. Risk of atherosclerosis increases substantially in men aged 45 or older and women aged 55 or older. After menopause, women's risk increases, in part, because the reduction in estrogen levels has unfavorable effects on lipoprotein levels and arterial function.[16] ♦ Levels of the amino acid homocysteine, which may damage artery walls and promote blood clotting, rise with age and are generally higher in men; however, researchers have not determined whether homocysteine is a cause or an effect of the disease process.[17] ♦

♦ Estrogen replacement therapy after menopause has mixed effects on heart disease risk; it can improve endothelial function, lower LDL, and raise HDL, but it also promotes blood clotting.

♦ Reminder: Blood homocysteine levels are influenced by intakes of folate, vitamin B_{12}, and vitamin B_6 (see Chapter 10).

IN SUMMARY **Atherosclerosis, characterized by plaque buildup in artery walls, can lead to complications such as angina pectoris, heart attack, stroke, intermittent claudication, kidney disease, and aneurysms. Plaque develops in response to long-term, chronic inflammation at susceptible sites in artery walls. Leading causes of plaque formation and progression include shear stress, hypertension, elevated LDL and VLDL levels, cigarette smoking, diabetes, and aging.**

Coronary Heart Disease (CHD)

Coronary heart disease (CHD), also called *coronary artery disease*, is the most common type of cardiovascular disease and the leading cause of death in the United States.[18] As discussed earlier, CHD is characterized by impaired blood flow through the coronary arteries, which may lead to angina pectoris, heart attack, or even sud-

atherogenic: able to initiate or promote atherosclerosis.

den death. CHD is most often caused by atherosclerosis but occasionally results from a **spasm** or inflammatory condition that causes narrowing of the coronary arteries. Although atherosclerosis can advance enough to fully block an artery, most heart attacks occur with less than 50 percent blockage.[19]

The lifetime risk of developing CHD is 49 percent for men and 32 percent for women.[20] Women typically develop CHD about 10 years later in life than men do, and their incidences of serious complications like heart attack and sudden death lag behind men's by about 20 years. In both genders, CHD can lie dormant for years: at least half of sudden deaths from CHD occur without prior symptoms in both men and women.[21]

Symptoms of Coronary Heart Disease Symptoms of CHD usually arise only after many years of disease progression. In angina pectoris and heart attacks, pain or discomfort most often occurs in the chest region and may be perceived as a feeling of heaviness, pressure, or squeezing; the pain may radiate to the shoulders, left arm, back, neck, jaw, or teeth. In angina pectoris, the symptoms are often triggered by exertion and subside with rest; in a heart attack, the pain may be severe, last longer, and occur without exertion. Other symptoms of CHD include shortness of breath, unusual weakness or fatigue, lightheadedness or dizziness, sweating, nausea, vomiting, and lower abdominal discomfort.[22] Women are more likely than men to have a heart condition (or even a heart attack) that is unaccompanied by chest pain or acute symptoms.

Evaluating Risk for Coronary Heart Disease Because CHD develops over many years, prevention should begin well before symptoms appear. Population studies have suggested that about 90 percent of people with CHD have at least one of the four classic risk factors: cigarette smoking, high LDL cholesterol, hypertension, or diabetes.[23] These and other major risk factors for CHD are listed in Table 27-1; most of the risk factors listed can be modified by changes in diet and lifestyle.

CHD Risk Assessment Risk assessment requires several key laboratory measures (see Table 27-2) and a thorough medical history. A complete lipoprotein profile (also called a *blood lipid profile*), which includes measures of total cholesterol, LDL and HDL cholesterol, and blood triglycerides (VLDLs), should be obtained every 5 years starting at 20 years of age. Sometimes the ratio of total cholesterol to HDL cholesterol is used to predict CHD risk: a high total cholesterol value suggests elevated LDL cholesterol, and a low HDL value is often linked with other lipid abnormalities, such as small, dense LDL and elevated VLDL. Overweight and obesity predispose to CHD, particularly when abdominal obesity ♦ is present. Hypertension is a major risk factor; for people over 50 years of age, a high systolic blood pressure is

TABLE 27-1 Risk Factors for CHD

Major Risk Factors for CHD (not modifiable)

- Increasing age
- Male gender
- Family history of heart disease

Major Risk Factors for CHD (modifiable)

- High LDL cholesterol
- High blood triglyceride (VLDL) levels
- Low HDL cholesterol
- Hypertension (high blood pressure)
- Diabetes
- Obesity (especially abdominal obesity)
- Physical inactivity
- Cigarette smoking
- Alcohol overconsumption (≥3 drinks per day)
- An "atherogenic" diet (includes high saturated fat, cholesterol, and *trans* fat intakes; low fruit and vegetable intakes)

NOTE: Risk factors highlighted in yellow have relationships with diet.

SOURCES: V. L. Roger and coauthors, on behalf of the American Heart Association Statistics Committee and Stroke Statistics Subcommittee, Heart disease and stroke statistics—2011 update: A report from the American Heart Association, *Circulation* 123 (2011), available at doi:10.1161/CIR.0b013e3182009701; M. H. Criqui, Epidemiology of cardiovascular disease, in L. Goldman and D. Ausiello, eds., *Cecil Medicine* (Philadelphia: Saunders, 2008), pp. 301–305.

♦ Abdominal obesity is suggested by a waist circumference of more than 40 inches for men and more than 35 inches for women.

TABLE 27-2 Laboratory Measures for CHD Risk Assessment

Clinical Measures	Desirable	Borderline Risk	High Risk
Total blood cholesterol (mg/dL)	<200	200–239	≥240
LDL cholesterol (mg/dL)	<100[a]	130–159	160–189[b]
HDL cholesterol (mg/dL)	≥60	40–59	<40
Triglycerides, fasting (mg/dL)	<150	150–199	200–499[c]
Body mass index (BMI)[d]	18.5–24.9	25–29.9	≥30
Blood pressure (systolic and diastolic pressure)	<120/<80	120–139/80–89[e]	≥140/≥90[f]

[a]LDL levels of 100–129 mg/dL indicate a near or above optimal level; <70 mg/dL is a desirable goal for very high-risk persons.
[b]LDL levels ≥190 mg/dL indicate a very high risk.
[c]Triglyceride levels ≥500 mg/dL indicate a very high risk.
[d]Body mass index (BMI) is defined in Chapter 8; BMI standards are found on the inside back cover.
[e]These values indicate prehypertension.
[f]These values indicate stage one hypertension; ≥160/≥100 indicates stage two hypertension. Physicians use these classifications to determine medical treatment.

spasm: a sudden, forceful, and involuntary muscle contraction.

HOW TO Assess a Person's Risk of Heart Disease

This assessment estimates a person's 10-year risk for experiencing a major coronary event associated with CHD, such as a heart attack.[a] A high score does not mean that the person *will* have a heart attack, but it warns of the possibility and suggests the need to consult a physician. To use this tool, you need to know a person's age, total and HDL cholesterol levels, and blood pressure.

Age (years):

	Men	Women
20–34	−9	−7
35–39	−4	−3
40–44	0	0
45–49	3	3
50–54	6	6
55–59	8	8
60–64	10	10
65–69	11	12
70–74	12	14
75–79	13	16

HDL (mg/dL):

	Men	Women
≥60	−1	−1
50–59	0	0
40–49	1	1
<40	2	2

Systolic Blood Pressure (mm Hg):

	Untreated		Treated	
	Men	Women	Men	Women
<120	0	0	0	0
120–129	0	1	1	3
130–139	1	2	2	4
140–159	1	3	2	5
≥160	2	4	3	6

Total Cholesterol (mg/dL):

	Age 20–39		Age 40–49		Age 50–59		Age 60–69		Age 70–79	
	Men	Women	Men	Women	Men	Women	Men	Women	Men	Women
<160	0	0	0	0	0	0	0	0	0	0
160–199	4	4	3	3	2	2	1	1	0	1
200–239	7	8	5	6	3	4	1	2	0	1
240–279	9	11	6	8	4	5	2	3	1	2
≥280	11	13	8	10	5	7	3	4	1	2
Smoking (any cigarette smoking in the past month):										
Smoker	8	9	5	7	3	4	1	2	1	1
Nonsmoker	0	0	0	0	0	0	0	0	0	0

Scoring Heart Disease Risk

Add up the total points: ___. Using the table at the right, find the total in the first column for the appropriate gender, and then check the second column to learn the percentage risk of developing severe CHD within the next 10 years. A person's risk for an acute coronary event (such as a heart attack) may be identified as low (<10%), moderate (10–20%), or high (>20%). Treatment strategies vary according to a person's risk category.

Men		Women	
Total	Risk (%)	Total	Risk (%)
<0	<1	<9	<1
0–4	1	9–12	1
5–6	2	13–14	2
7	3	15	3
8	4	16	4
9	5	17	5
10	6	18	6
11	8	19	8
12	10	20	11
13	12	21	14
14	16	22	17
15	20	23	22
16	25	24	27
≥17	≥30	≥25	≥30

[a]A link to the electronic version of this assessment is available on the ATP III page of the National Heart, Lung, and Blood Institute's website (**www.nhlbi.nih.gov/guidelines/cholesterol**).
SOURCE: Adapted from Expert Panel on Detection, Evaluation, and Treatment of High Blood Cholesterol in Adults (Adult Treatment Panel III), *Third Report of the National Cholesterol Education Program (NCEP)*, NIH publication no. 02-5216 (Bethesda, MD: National Heart, Lung, and Blood Institute, 2002), section III.

TRY IT Estimate your 10-year risk for CHD using the tables above or the NHLBI's online risk calculator (see note *a*).

more predictive of CHD risk than is high diastolic blood pressure. Finally, cigarette smoking and the presence of diabetes strongly contribute to CHD risk. The "How To" on p. 816 presents a screening method for assessing a person's 10-year risk of developing CHD that includes some of these risk factors.

Blood Cholesterol Levels and CHD Risk Once a person's level of risk has been identified, much of the treatment focuses on lowering LDL cholesterol. Elevated LDL levels are directly related to the development of atherosclerosis, and clinical studies have confirmed that LDL-lowering treatments can successfully reduce CHD mortality rates. CHD is seldom seen in populations that maintain desirable LDL levels.[24]

As mentioned earlier, HDL help to protect against atherosclerosis, and low HDL levels often coexist with other lipid abnormalities; thus, a low HDL level is highly predictive of CHD risk. In addition, some risk factors for CHD—obesity, smoking, inactivity, and male gender—are associated with reduced HDL levels. It is not known whether raising HDL will help to reduce CHD risk, but weight loss, smoking cessation, and regular physical activity can all independently help to lower risk.

Treatment Strategies for Lowering LDL Cholesterol The aggressiveness of the LDL cholesterol–lowering treatment depends on a person's level of risk for CHD. The National Cholesterol Education Program, developed by a division of the National Institutes of Health, periodically issues treatment guidelines that consider an individual's risk status. ♦ The most current guidelines, known as the Adult Treatment Panel III (ATP III), advise aggressive treatment for patients who already have CHD or a CHD risk equivalent; ♦ conversely, individuals at mild or moderate risk for CHD may be able to manage their cholesterol levels with changes in diet and lifestyle. The "How To" on pp. 818–819 summarizes the use of the ATP III guidelines in clinical practice.

♦ For updated guidelines related to CHD risk reduction, access this website: **www.nhlbi.nih.gov/guidelines**

♦ A *CHD risk equivalent* carries the same risk for a major coronary event as established CHD; examples include type 2 diabetes mellitus and diseases caused by atherosclerosis, such as stroke or aortic aneurysm.

Therapeutic Lifestyle Changes for Lowering CHD Risk

People who have CHD or multiple risk factors for CHD are often advised to make dietary and lifestyle changes before considering drug treatment. An approach to risk reduction promoted by the National Cholesterol Education Program, known as Therapeutic Lifestyle Changes (TLC), is summarized in Table 27-3.[25] The main features of the TLC plan include a cholesterol-lowering diet, regular physical activity,

TABLE 27-3 Reducing Risk of CHD with Therapeutic Lifestyle Changes

Dietary Strategies

- Limit saturated fat to less than 7 percent of total kcalories and cholesterol to less than 200 milligrams per day. Maintaining a fat intake that is 25 to 35 percent of total kcalories may help with this goal.
- Replace saturated fats with unsaturated fats from fish, vegetable oils, and nuts or with carbohydrates from whole grains, legumes, fruits, and vegetables.
- Avoid food products that contain *trans*-fatty acids. The *trans* fat content in packaged foods is shown on the Nutrition Facts panel.
- Choose foods high in soluble fibers, including oats, barley, legumes, and fruit. Food supplements that contain psyllium seed husks can be used to help lower LDL cholesterol levels.
- Regularly consume food products that contain added plant sterols or stanols.
- To reduce blood pressure, limit sodium intake to 2400 milligrams per day,[a] and choose a diet that is high in fruits, vegetables, and whole grains and includes low-fat milk products and nuts.
- Fish can be consumed regularly as part of a CHD risk-reduction diet.
- If alcohol is consumed, it should be limited to one drink daily for women and two drinks daily for men.

Lifestyle Choices

- Physical activity: At least 30 minutes of moderate-intensity endurance activity should be undertaken on most days of the week. The eventual goal should be an expenditure of at least 2000 kcalories weekly.
- Smoking cessation: Exposure to any form of tobacco smoke should be minimized.

Weight Reduction

- Weight reduction may improve other CHD risk factors. The general goals of a weight-management program should be to prevent weight gain, reduce body weight, and maintain a lower body weight over the long term. The initial goal of a weight-loss program should be to lose no more than 10 percent of original body weight.

[a]According to DRI recommendations, sodium intake should be limited to 2300 milligrams daily.

HOW TO Detect, Evaluate, and Treat High Blood Cholesterol

The Adult Treatment Panel III (ATP III) guidelines identify three levels of risk for CHD and set treatment goals for each level. First, a brief risk assessment tool identifies a person's risk for an acute coronary event in the next 10 years as being less than 10% (low risk), 10 to 20% (moderate risk), or over 20% (high risk). After risk is determined, treatment strategies are addressed. Note that these are guidelines only and should not override the judgment of an attending physician.

Step 1. Obtain a complete lipoprotein profile from blood samples taken after a 9- to 12-hour fast. (Desirable blood lipid levels are shown in Table 27-2.)

Step 2. Identify the presence of diseases that confer high risk for acute CHD events; such conditions are considered CHD risk equivalents and include:

- Symptoms of CHD
- Symptoms of stroke
- Aortic aneurysm
- Intermittent claudication
- Diabetes mellitus

Step 3. Identify major risk factors other than LDL cholesterol:

- Cigarette smoking
- Hypertension (blood pressure ≥140/≥90 mm Hg) or use of antihypertensive drugs
- Low HDL cholesterol (<40 mg/dL). If HDL levels are ≥60 mg/dL, subtract one risk factor from the total count.
- Family history of premature CHD (CHD in father or brother <55 years; mother or sister <65 years)
- Age (if male, ≥45 years; if female, ≥55 years)

Step 4. Assign the 10-year risk of an acute coronary event as being high (>20%), moderate (10–20%), or low (<10%):

- If any CHD risk equivalent is present (see Step 2), the person's risk level is *high*.
- If two or more risk factors other than LDL are present (see Step 3), assess risk according to age, total cholesterol level, HDL level, systolic blood pressure, and smoking habit (use the "How To" on p. 816).
- If one risk factor (or less) is present, the person's risk level is *low*.

Step 5. Determine LDL goals and treatment options for each risk category:

Risk Category	LDL Goal	LDL Level at Which to Initiate Therapeutic Lifestyle Changes (TLC)	LDL Level at Which to Initiate Drug Therapy
High	<100 mg/dL[a]	≥100 mg/dL	≥130 mg/dL[b]
Moderate	<130 mg/dL[c]	≥130 mg/dL	≥130 mg/dL[b] if 10-year risk 10–20% ≥160 mg/dL if 10-year risk <10%
Low	<160 mg/dL	≥160 mg/dL	≥190 mg/dL

[a]<70 mg/dL is a goal for very high-risk patients.
[b]Drug therapy is sometimes considered for LDL >100 mg/dL.
[c]<100 mg/dL is a goal for some patients.

(continued)

and weight reduction. If the recommendations are followed carefully, substantial progress may be seen after six weeks. People with a high risk of CHD should try to lower LDL cholesterol with at least a three-month trial of TLC before starting drug therapy. This section describes the elements of the TLC plan in detail.

Saturated Fat Of the dietary lipids, saturated fat has the strongest effect on blood cholesterol levels, and replacing saturated fat with monounsaturated and polyunsaturated fats can generally lower LDL levels. ♦ The TLC recommendation is to consume less than 7 percent of total kcalories as saturated fat. The average saturated fat intake in the United States is about 11 percent of total kcalories consumed.[26]

♦ Reminder: Although not all saturated fatty acids have the same cholesterol-raising effect, foods high in saturated fat typically contain a mixture of fatty acids (see p. 151).

For most people, cutting down on saturated fat involves more than just switching from butter to vegetable oil, as the main sources of saturated fat in the United States are whole-milk products, high-fat meats, and baked goods. Choosing lean meats or fish, using fat-free or low-fat milk products, and avoiding certain types of bakery products are usually more effective ways of reducing saturated fat. Chapter 5 and Highlight 5 provide additional information about food selections that are low in saturated fat.

HOW TO Detect, Evaluate, and Treat High Blood Cholesterol (*continued*)

Step 6. Initiate Therapeutic Lifestyle Changes (TLC) if the patient's LDL level is above the goal. The main features of TLC include:

- Restricted intake of saturated fat (<7% of total kcal) and cholesterol (<200 mg per day).
- Increased intake of soluble fiber (10–25 g per day) and plant sterols or stanols (2 g per day).
- Moderate physical activity (about 200 kcal expended per day).
- Weight reduction, if necessary.

If there is no improvement in LDL after three months, consider drug therapy.

Step 7. Consider adding drug therapy if LDL levels exceed recommendations shown in Step 5. The drugs in the following table are typically prescribed to improve LDL, VLDL, and HDL levels; nicotinic acid can also reduce lipoprotein(a) levels.

Type of Drug	Examples	Mode of Action
Statins	Atorvastatin (Lipitor) Simvastatin (Zocor)	Reduce cholesterol synthesis in the liver
Bile acid sequestrants	Cholestyramine (Questran) Colestipol (Colestid)	Bind bile acids in the small intestine, reducing reabsorption
Fibric acid derivatives	Fenofibrate (Tricor) Gemfibrozil (Lopid)	Modulate production of proteins that regulate lipoprotein synthesis and breakdown
Nicotinic acid (a form of niacin)	Nicotinic acid (Niaspan)	Reduces triglyceride breakdown in adipose tissue and subsequent VLDL/LDL production

Step 8. Identify metabolic syndrome based on the presence of three or more of the following risk determinants:

- Abdominal obesity (waist measurement >40 inches in men; >35 inches in women).
- Elevated blood triglycerides (≥150 mg/dL).
- Low HDL cholesterol (<40 mg/dL in men; <50 mg/dL in women).
- Elevated blood pressure (≥130/≥85 mm Hg).
- Elevated fasting glucose (≥100 mg/dL).

If metabolic syndrome is present, treatment should begin with weight management and moderate physical activity. If there is no improvement in blood lipids or blood pressure, use of medications is advised.

Step 9. In most people, elevated blood triglycerides (≥150 mg/dL) are initially treated with weight management and moderate physical activity. If they remain above 200 mg/dL, drug therapy (nicotinic acid or fibric acids) should be considered. For blood triglycerides above 500 mg/dL, a very low-fat diet (<15% kcal from fat) may be necessary to prevent complications.

TRY IT Identify the CHD risk level for a 50-year-old male patient with diabetes. What treatment would be suggested if his LDL level were 110 mg/dL?

Replacing saturated fat with carbohydrate can also lower LDL cholesterol, but such a change may lower HDL cholesterol as well and also raise blood triglyceride levels. ♦ The effect on triglyceride levels can be minimized by limiting added sugars and including fiber-rich foods; ideally, the diet should include generous amounts of whole grains, legumes, fruits, and vegetables. The TLC diet recommends a carbohydrate intake in the range of 50 to 60 percent of total kcalories.

♦ Diets high in carbohydrate—especially those high in added sugars—can raise blood triglyceride levels in some people.

Polyunsaturated and Monounsaturated Fat As described in the previous section, replacing saturated fat with either monounsaturated or polyunsaturated fat helps to lower LDL levels. A switch to polyunsaturated fat tends to have the greater effect, but it also promotes a slight reduction in HDL cholesterol.[27] Other concerns are that high intakes of polyunsaturated fat may contribute to oxidative stress ♦ or increase inflammation within the body. ♦ Therefore, TLC guidelines limit polyunsaturated fat to 10 percent of total kcalories, whereas up to 20 percent of kcalories from monounsaturated fat are allowed. Note that most polyunsaturated fat in the diet consists of omega-6 fatty acids, such as linoleic acid; omega-3 fatty acids may have beneficial effects on heart disease risk, as described in a later section.

♦ Reminder: Polyunsaturated fats are more susceptible to oxidation than saturated fats, although the clinical significance of consuming high amounts is still unknown.

♦ Reminder: Polyunsaturated fatty acids are precursors for the eicosanoids, which mediate inflammation (see p. 685).

Total Fat For people whose fat intake includes substantial saturated fat, limiting total fat may indirectly reduce saturated fat. Therefore, the TLC recommendation for total fat is an intake of 25 to 35 percent of kcalories. Individuals with elevated blood triglycerides may benefit from a fat intake at the upper end of this range (30 to 35 percent) so that their carbohydrate intakes are not excessive. Fat intakes higher than 35 percent of kcalories are discouraged because they may promote weight gain in some people.

***Trans* Fat** *Trans*-fatty acids can raise LDL cholesterol levels, and when they replace saturated fats in the diet (as when stick margarine replaces butter), they may also cause a reduction in HDL cholesterol. *Trans* fats may also raise CHD risk by altering blood vessel function, promoting inflammation, and reducing LDL size.[28] The TLC recommendation is to keep *trans* fat intake as low as possible.

Most sources of *trans* fats are products made with partially hydrogenated vegetable oils; examples include baked goods such as crackers, cookies, and doughnuts; snack foods such as potato chips and corn chips; and fried foods such as French fries and fried chicken. In the past few years, food manufacturers have reformulated many food products so that they contain little or no *trans* fat. In some cases, unfortunately, the *trans* fats have been replaced with saturated fat sources, so consumers should read labels carefully to avoid both types of cholesterol-raising fats.

Dietary Cholesterol A high cholesterol intake can raise LDL levels, and reducing dietary cholesterol lowers LDL cholesterol in most people. The TLC recommendation is a cholesterol intake of less than 200 milligrams per day. Currently, the cholesterol intakes of women and men in the United States average about 230 and 362 milligrams per day, respectively.[29] Eggs contribute about one-quarter of the cholesterol in the American diet, followed by chicken, beef, and cheese.[30]

Soluble Fibers As Chapter 4 explains, a diet rich in soluble, viscous fibers can reduce LDL cholesterol levels by inhibiting bile reabsorption in the small intestine and reducing cholesterol synthesis in the liver (see p. 118). An extra 5 to 10 grams of soluble fiber daily is associated with a 3 to 5 percent reduction in total cholesterol levels. Dietary sources of soluble fibers include oats, barley, legumes, and fruits. The soluble fiber in psyllium seed husks, frequently used to treat constipation, is effective for lowering cholesterol levels when used as a dietary supplement.[31]

Plant Sterols and Stanols Foods or supplements that contain significant amounts of plant sterols or plant stanols can help to lower LDL cholesterol levels. ♦ Plant sterols or stanols are added to various food products, such as margarine and orange juice, or supplied in dietary supplements. These plant compounds work by interfering with cholesterol and bile absorption. About 2 grams of plant sterols daily (provided by 2 to 2½ tablespoons of sterol-enriched margarines) can lower LDL cholesterol by up to 15 percent without lowering HDL cholesterol.[32]

♦ Plant sterols are extracted from soybeans and pine tree oils and are hydrogenated to produce the plant stanols that are sometimes added to commercial products.

Sodium and Potassium Intakes Excessive dietary sodium may raise blood pressure, ♦ whereas potassium can help to lower blood pressure. A low-sodium diet that contains generous amounts of fruits and vegetables, low-fat milk products, nuts, and whole grains has been found to substantially reduce blood pressure, largely due to the diet's content of potassium and several other minerals that have blood pressure–lowering effects. This diet (the *DASH Eating Plan*) and other factors that influence blood pressure are discussed in a later section (see pp. 829–831).

♦ Some people are more sensitive to sodium intakes than others, as discussed in a later section.

Fish and Omega-3 Fatty Acids The omega-3 fatty acids in fatty fish, known as EPA and DHA, ♦ may benefit people who have had a heart attack by suppressing inflammation, lowering blood triglyceride levels, reducing blood clotting, and stabilizing heart rhythm. In addition, including fish in the diet can reduce CHD risk because fish is low in saturated fat and typically replaces entrées that contain animal fat (a source of saturated fat). The American Heart Association recommends consuming two or more servings of fish per week, with an emphasis on fatty fish.[33] Fish oil supplements may be helpful for some individuals with documented CHD, and they are particularly useful for lowering triglyceride levels (see p. 824); however,

♦ Reminder: *EPA* and *DHA* are abbreviations for eicosapentaenoic acid and docosahexaenoic acid, respectively, the 20- and 22-carbon polyunsaturated fatty acids found in fish. See p. 149 for a review of these fatty acids.

they should be used under physician supervision due to the potential for excessive bleeding and other harmful effects. Chapter 5 and Highlight 5 provide additional information about omega-3 fatty acids and the use of fish oil supplements.

The 18-carbon omega-3 fatty acids found in flaxseed and other land plants have lesser or different effects than the omega-3 fatty acids from marine sources. Although some evidence suggests that moderate increases in these plant sources of omega-3 fatty acids may improve CHD risk, additional research is needed to confirm their benefits.[34]

Alcohol Moderate consumption of alcohol—from beer, wine, or liquor—has favorable effects on atherosclerosis, blood-clotting activity, HDL cholesterol levels, inflammation, and insulin resistance.[35] Alcohol use is also inversely related to the incidence of heart attack. These benefits are most apparent in men and women who are at least 45 and 55 years old, respectively. Of note, only low or moderate amounts of alcohol—no more than one drink daily for women and two for men—have been found to lower CHD risk, and higher intakes are associated with higher mortality rates. One "drink" is equivalent to 12 ounces of beer, 5 ounces of wine, 10 ounces of wine cooler, or 1½ ounces of 80 proof distilled spirits such as gin, rum, vodka, and whiskey.

For some people, alcohol's negative effects can offset its health advantages. Alcohol consumption is associated with cancers of the gastrointestinal (GI) tract and several other cancers, including liver cancer and breast cancer.[36] Alcohol is destructive to the liver and male reproductive system, and high alcohol intakes can elevate blood pressure and triglyceride levels. Moreover, up to 10 percent of adults misuse alcohol.[37] For these reasons, nondrinkers are not encouraged to start drinking in an effort to decrease their risk for CHD.

Regular aerobic exercise can strengthen the cardiovascular system, promote weight loss, reduce blood pressure, and improve blood glucose and lipid levels.

Regular Physical Activity Regular aerobic activity reverses a number of risk factors for CHD: it can lower blood triglycerides, raise HDL levels, lower blood pressure, promote weight loss, improve insulin sensitivity, strengthen heart muscle, and increase coronary artery size and tone. Activities that use large muscle groups have the greatest benefits; such activities include brisk walking, running, swimming, cycling, stair-stepping, and cross-country skiing. ♦ Alternatives for busy people include heavy house cleaning, lawn mowing, raking leaves, and walking to and from work.

♦ Resistance exercise can also help to reverse some CHD risk factors, but its overall effect on CHD risk is uncertain.

Research studies have found that the most active persons have CHD rates that are about half the rates of those who are the least active.[38] The American Heart Association recommends that all adults participate in at least 30 minutes of moderate-intensity physical activity on most (preferably all) days of the week, whereas 60 minutes of physical activity is suggested for adults who are attempting or maintaining weight loss.[39] If preferred, physical activity can be divided into several sessions during the day. Note that vigorous activity increases the risk of heart attack and sudden death in individuals with diagnosed heart disease, so sedentary adults are advised to increase their activity levels gradually.

Smoking Cessation Cigarette smoking is a major risk factor for CHD as well as for other types of cardiovascular disease. ♦ In addition to promoting atherosclerosis, cigarette smoking decreases the oxygen-carrying capacity of the blood, raises the heart rate, inhibits vasodilation, reduces exercise tolerance, and promotes blood clotting, among other effects.[40] As mentioned previously, smokers tend to have higher levels of LDL and lower HDL than nonsmokers. Secondhand smoke can cause some of these effects as well.

♦ Cigar and pipe smoking can also increase the risk of CHD, but the risk may be lower because the smoke is less likely to be inhaled.

The risk from smoking depends on the amount and duration of exposure: it is related to the age when smoking started, the number of cigarettes smoked daily, and the degree of inhalation. Even smoking just one or two cigarettes daily increases CHD risk, and cigarettes that have low tar and nicotine do not lower the risk. Quitting smoking improves risk quickly; the incidence of CHD drops to levels near those of nonsmokers in just 2 years.[41] Currently, about 23 percent of men and 18 percent of women in the United States are cigarette smokers.[42]

HOW TO Implement a Heart-Healthy Diet

For some people, following a heart-healthy diet may require significant changes in food choices. It is often easier to adopt a new diet if only a few changes are made at a time. Discussing positive choices (what to eat) first, rather than negative ones (what not to eat), may improve compliance. These suggestions can help patients implement their diet:

Breads, Cereals, and Pasta

- Choose whole-grain breads and cereals. Make sure the first ingredient on bread and cereal labels is "whole wheat" rather than "enriched wheat flour." Consume oats and barley regularly, as they are good sources of soluble fibers.
- Bakery products and snack foods often contain *trans*-fatty acids. Choose products whose labels list 0 grams of *trans* fat on the Nutrition Facts panel; the ingredient lists should not include any "partially hydrogenated vegetable oil," the main source of *trans* fatty acids.
- Avoid products that contain tropical oils (coconut, palm, or palm kernel oil), which are high in saturated fat.

Fruits and Vegetables

- Incorporate at least one or two servings of fruits and vegetables into each meal. Keep the refrigerator stocked with a variety of colorful fruits and vegetables (baby carrots, blueberries, grapes) that can be eaten when the urge to nibble arises.
- Check food labels on canned products carefully. Canned vegetables (especially tomato-based products) are often high in sodium. Fruits that are canned in juice are higher in nutrient density than those canned in syrup.
- Avoid French fries from fast-food restaurants, which are often prepared with *trans* fats. Restrict high-sodium foods such as pickles, olives, sauerkraut, and kimchee.

Lunch and Dinner Entrées

- Limit meat, fish, and poultry intake to 5 ounces per day. Plan to eat fish twice a week, preferably fatty fish such as salmon, tuna, and mackerel.
- Select lean cuts of beef, such as sirloin tip and round steak; lean cuts of pork, such as loin chops and tenderloin; and skinless poultry pieces. Trim visible fat before cooking.
- Select extra-lean ground meat and drain well after cooking. Use lean ground turkey, without skin added, in place of ground beef.
- Prepare pasta and vegetable stir-fry dishes several times weekly to help reduce meat intake and increase vegetable intake. Use soybean products and other legumes as sources of protein.
- Limit cholesterol-rich organ meats (liver, brain, sweetbreads) and shrimp. Limit intake of whole eggs to two per week, as the yolks are high in cholesterol (about 210 milligrams per yolk). Replace whole eggs in recipes with egg whites or commercial egg substitutes or similar reduced-cholesterol products.
- Restrict these high-sodium foods: cured or smoked meats such as beef jerky, bologna, corned beef, frankfurters, ham, luncheon meats, salt pork, and sausage; salty or smoked fish such as anchovies, caviar, salted or dried cod, herring, and smoked salmon; and canned, frozen, or packaged soups, sauces, and entrées.

(*continued*)

Weight Reduction Obesity, especially abdominal obesity, is an independent risk factor for CHD.[43] In obesity, the enlarged adipose cells increase their production and release of inflammatory mediators and blood clotting factors; these compounds raise the risks of both atherosclerosis and heart attack. Obesity also strains the heart and blood vessels because **cardiac output** is greater, thus increasing the workload of the left ventricle, which pumps blood to the major arteries. Moreover, obesity leads to metabolic abnormalities that increase CHD risk, such as insulin resistance, hypertension, elevated triglycerides, low HDL levels, and a reduction in LDL size. ♦

♦ The pattern of metabolic complications associated with obesity characterizes the metabolic syndrome, which is described in Highlight 26.

Weight reduction can improve such CHD risk factors as hypertension, elevated blood triglycerides, low HDL cholesterol, and insulin resistance. However, individuals should focus on weight reduction only *after* they have adopted other dietary measures to lower LDL.[44] This approach ensures that LDL reduction is given priority and that the individual does not receive a multitude of dietary suggestions at one time. The initial goal of a weight-loss program should be no more than 10 percent of a person's original body weight. For some, avoiding additional weight gain may be a desirable starting point.

Successful Adherence to Lifestyle Changes Adopting many lifestyle changes at once can be challenging, and instruction and counseling are critical for success. Health practitioners can help to motivate patients by explaining the reasons for each change, setting obtainable goals, and providing practical suggestions. An initial diet history can offer clues about a person's behaviors and preferences, and fol-

cardiac output: the volume of blood pumped by the heart within a specified period of time.

HOW TO Implement a Heart-Healthy Diet (*continued*)

Milk Products

- Select fat-free or low-fat milk products only. Use yogurt or fat-free sour cream to make dips or salad dressings. Substitute evaporated fat-free milk for heavy cream.
- Restrict foods high in saturated fat or sodium, such as cheese, processed cheeses, and ice cream or other milk-based desserts.

Fats and Oils

- Prepare salad dressings and other foods with vegetable oils rich in omega-3 fatty acids, such as canola, soybean, flaxseed, and walnut oils. Select other unsaturated vegetable oils, such as corn, olive, peanut, safflower, sesame, and sunflower oils, instead of saturated fat sources such as butter and lard.
- Select margarines that indicate 0 grams of *trans* fat on the Nutrition Facts panel; these products should contain little or no "partially hydrogenated vegetable oil." Tub margarines are less likely to contain *trans* fat than stick margarines. To help lower LDL cholesterol levels, use margarines with added plant sterols or stanols.
- Add unsalted nuts or avocados to meals to make them more appetizing; these foods are good sources of unsaturated fats.
- Avoid tropical oils (coconut, palm, and palm kernel oil), which are high in saturated fat.

Spices and Seasonings

- Use salt only at the end of cooking, and you will need to add much less. Use salt substitutes at the table.
- Spices and herbs can improve food flavor without adding sodium. Try using more garlic, ginger, basil, curry or chili powder, cumin, pepper, lemon, mint, oregano, rosemary, and thyme.
- Check the sodium content on food labels. Flavorings and sauces that are usually high in sodium include bouillon cubes, soy sauce, steak and barbecue sauces, relishes, mustard, and catsup.

Snacks and Desserts

- Select low-sodium and low-saturated fat snacks such as unsalted pretzels and nuts, plain popcorn, and unsalted chips and crackers. Check labels on snack foods and desserts to ensure that they do not include *trans* fats.
- Enjoy angel food cake, which is made without egg yolks and added fat. Select low-fat frozen desserts such as sherbet, sorbet, fruit bars, and some low-fat ice creams.
- Choose canned or dried fruits and crunchy raw vegetables to boost fruit and vegetable intake.

TRY IT Plan heart-healthy meals for a day. Compare your one-day menu with the strategies listed in Table 27-3 (p. 817), and explain how you could improve any shortcomings.

low-up visits allow an opportunity to determine compliance. In some individuals, high LDL cholesterol levels may persist despite adherence to a TLC program, and drug therapy may be the only effective treatment for such people. Review Table 27-3 (p. 817) for a summary of the suggestions discussed in this section. The "How To" above offers suggestions for implementing a heart-healthy diet.

Lifestyle Changes for Hypertriglyceridemia

Hypertriglyceridemia (elevated blood triglycerides) ♦ affects nearly one-third of adults in the United States.[45] It is common in people with diabetes mellitus, obesity, and the metabolic syndrome and may also result from other disorders. Elevated blood triglycerides may coexist with elevated LDL cholesterol or occur separately. Whereas mild or moderate hypertriglyceridemia is often associated with increased risk of CHD, severe hypertriglyceridemia can cause additional complications, including fatty deposits in the skin and soft tissues and acute pancreatitis.[46]

♦ Blood triglyceride levels:
- Borderline high: 150–199 mg/dL
- High: 200–499 mg/dL
- Very high: ≥500 mg/dL

Nutrition Therapy for Hypertriglyceridemia Dietary and lifestyle changes can improve most cases of mild hypertriglyceridemia.[47] Overweight and obesity, a sedentary lifestyle, and cigarette smoking all can raise triglyceride levels. Dietary factors that increase triglyceride levels include high intakes of alcohol and refined carbohydrates; sucrose and fructose are the carbohydrates with the strongest effect. Thus, controlling body weight, being physically active, quitting smoking, restricting alcohol, and limiting intakes of refined carbohydrates (especially foods made with white flour or added sugars) are basic treatments for hypertriglyceridemia. As mentioned earlier, high triglyceride levels are often associated with low HDL, and the lifestyle changes listed here are likely to improve HDL levels as well.

Severe Hypertriglyceridemia Extreme elevations in blood triglycerides are usually caused by genetic mutations that upset lipoprotein metabolism. In addition to dietary and lifestyle changes, medications are usually necessary for lowering blood triglyceride levels above 500 milligrams per deciliter. If blood triglycerides exceed 1000 milligrams per deciliter, a very low-fat diet, providing less than 15 percent of kcalories from fat, may be required.[48] Patients must also eliminate consumption of alcoholic beverages.

Fish Oil Supplements and Hypertriglyceridemia Fish oil supplements are sometimes recommended for treating hypertriglyceridemia. Clinical trials suggest that a daily intake of 2 to 4 grams of EPA and DHA (combined) can lower triglyceride levels by 30 to 50 percent.[49] As mentioned previously, fish oil therapy should be monitored by a physician due to the potential for adverse effects. Although over-the-counter supplements are available, most provide only small amounts of EPA/DHA (about 300 milligrams per capsule), requiring the use of 7 to 13 capsules daily. A prescription form of fish oil (Lovaza) is available and contains about 840 milligrams of EPA/DHA per capsule.

Vitamin Supplementation and CHD Risk

People often ask about the potential benefits of using certain types of vitamin supplements for reducing CHD risk, particularly B vitamin and antioxidant supplements. Most clinical trials have not been able to confirm any benefits from using these supplements, as described in this section.

B Vitamin Supplements and Homocysteine Although elevated blood homocysteine levels are a known risk factor for CHD, it is unclear whether homocysteine itself is directly damaging or is simply an indicator of other abnormalities. Possibly, homocysteine has harmful effects on the artery wall, causes oxidative stress, or heightens blood-clotting activity that worsens atherosclerosis.[50] Although increased intakes of folate, vitamin B_6, and vitamin B_{12} can lower homocysteine levels, clinical trials have not demonstrated that supplementation with these vitamins can reduce the incidence of heart attacks in those at risk.[51] Hence, B vitamin supplements are not currently recommended for patients at risk for CHD.

Antioxidant Vitamin Supplements Because oxidized LDL promote atherosclerosis, researchers have hypothesized that antioxidant supplements may help to prevent atherosclerosis and reduce CHD risk. Several epidemiological studies suggested that antioxidant-rich diets protect against CHD, but because persons who consume such diets usually maintain a healthy lifestyle and body weight as well, it has been difficult to determine whether the antioxidants in the diets were responsible for the effect. Most studies that have tested supplementation with single antioxidants (such as vitamins C or E), combinations of antioxidants, or multivitamins have produced weak or inconsistent results, and several studies suggested possible harm.[52] Until more data are available, the use of antioxidant supplements is not recommended for heart disease prevention.[53]

Drug Therapies for CHD Prevention

Physicians usually recommend a trial of dietary and lifestyle changes before considering drug therapies for CHD prevention. If an individual cannot reach LDL goals with lifestyle changes alone, one or more medications may be prescribed (review Step 7 in the "How To" on p. 819). Individuals using medications for CHD prevention should continue the TLC program so that they can use the lowest possible doses of the drugs.

In addition to lipid-lowering medications, some people may require drugs that suppress blood clotting (anticoagulants and aspirin) or reduce blood pressure. Nitroglycerin (a vasodilator) may be given to alleviate angina as needed. Some med-

DIET-DRUG Interactions

Check this table for notable nutrition-related effects of the medications discussed in this chapter.

Anticoagulants (warfarin)	**Dietary interactions:** Require consistent vitamin K intake to maintain effectiveness. Drug effects are enhanced with supplementation of vitamin E, dong quai, danshen, fish oils, garlic, and ginkgo. Drug effects are reduced with coenzyme Q, ginseng, and green tea. Avoid alcohol.
Antihypertensives	
Beta-blockers	**Metabolic effects:** Elevated serum potassium levels, hypoglycemia
Calcium channel blockers	**Gastrointestinal effects:** Nausea, constipation **Dietary interactions:** Avoid herbal supplements that contain natural licorice. Avoid grapefruit juice, which may enhance drug effects.
ACE inhibitors	**Gastrointestinal effects:** Reduced taste sensation **Dietary interactions:** Avoid herbal supplements that contain natural licorice. Avoid potassium supplements and salt substitutes that contain potassium. **Metabolic effects:** Elevated serum potassium levels
Antilipimics	
Statins	**Gastrointestinal effects:** Constipation, flatulence, GI discomfort **Dietary interactions:** Avoid grapefruit juice, which may enhance drug effects. **Metabolic effects:** Elevated serum liver enzymes
Bile acid sequestrants	**Gastrointestinal effects:** Constipation, flatulence, diarrhea **Dietary interactions:** Cause reduced absorption of fat-soluble vitamins **Metabolic effects:** Electrolyte imbalances, iron deficiency
Nicotinic acid	**Gastrointestinal effects:** GI discomfort **Dietary interactions:** Avoid alcoholic beverages, coffee, and tea, which may increase side effects. **Metabolic effects:** Elevated serum liver enzymes, elevated uric acid levels, hyperglycemia, low blood pressure
Digoxin	**Gastrointestinal effects:** Anorexia, nausea, stomach cramps, diarrhea **Dietary interactions:** High-fiber foods and magnesium supplements can reduce drug absorption. St. John's wort may reduce drug efficacy. **Metabolic effects:** Elevated serum potassium and reduced serum magnesium levels. Drug toxicity can develop if body potassium levels are low.
Diuretics (furosemide, spironolactone[a])	**Gastrointestinal effects:** Dry mouth, anorexia, decreased taste perception **Dietary interactions:** Furosemide bioavailability is reduced when the drug is taken with food. **Metabolic effects:** Fluid and electrolyte imbalances,[a] hyperglycemia (spironolactone), hyperlipidemia (spironolactone), thiamin and zinc deficiencies
Nitroglycerin	**Gastrointestinal effects:** Decreased taste perception **Dietary interactions:** Increases effects of alcohol.

[a]*Furosemide* is a "potassium-wasting" diuretic; patients taking furosemide should increase intakes of potassium-rich foods. *Spironolactone* is a "potassium-sparing" diuretic; patients taking spironolactone should avoid supplemental potassium and salt substitutes that contain potassium.

ications may affect nutrition status or food intake (see the Diet-Drug Interactions feature); the interactions can be even more complicated when multiple medications are used.

Treatment of Heart Attack As explained earlier, a heart attack occurs when the blood supply to heart muscle is blocked, causing damage or death to

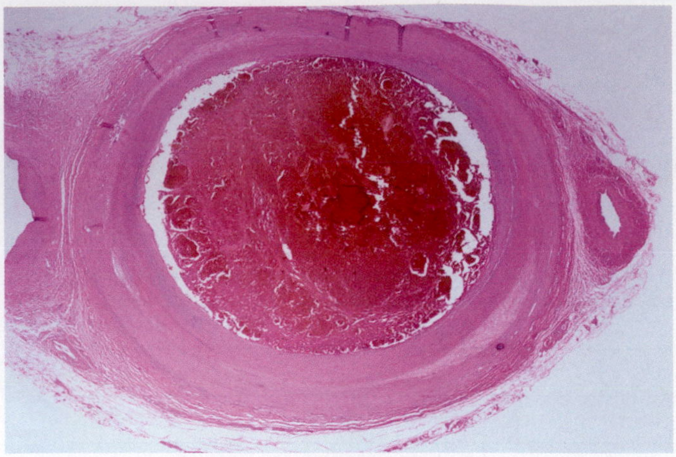

A heart attack occurs when a blood clot blocks the flow of blood in a coronary artery narrowed by atherosclerosis.

heart tissue. If the blood flow within the artery is restored quickly, the heart muscle may be saved; if not, the muscle tissue dies. Drug therapies given immediately after a heart attack may include thrombolytic drugs (clot-busting drugs), anticoagulants, aspirin, painkillers, and medications that regulate heart rhythm and reduce blood pressure.[54] Patients are not given food or beverages, except for sips of water or clear liquids, until their condition stabilizes. Once able to eat, they are initially offered small portions of foods that are low in sodium, saturated fat, and cholesterol. The sodium restriction helps to limit fluid retention but may be lifted after several days if the patient shows no signs of heart failure.

A heart attack patient needs to regain strength and learn strategies that can reduce the risk of a future heart attack; such strategies are similar to the lifestyle changes described earlier. Thus, the cardiac rehabilitation programs in hospitals and outpatient clinics include exercise therapy, instruction about heart-healthy food choices, help with smoking cessation, and medication counseling. The programs often last several months. Home-based rehabilitation programs are also beneficial, but they are more limited in scope and lack the benefit of group interaction.

IN SUMMARY **Long-term CHD management emphasizes risk reduction. Modifiable risk factors include elevated LDL and triglyceride levels, low HDL levels, hypertension, diabetes, obesity, a sedentary lifestyle, cigarette smoking, and various dietary factors. To help in reducing LDL levels and eliminating other risk factors, the Therapeutic Lifestyle Changes (TLC) approach is often suggested. Dietary recommendations are to reduce saturated fat, *trans* fats, and cholesterol; increase soluble fiber; and incorporate plant sterols or stanols and fish into the diet. Other recommendations include regular physical activity, smoking cessation, and weight reduction. Treatment for mild hypertriglyceridemia emphasizes weight control, regular physical activity, smoking cessation, avoiding a high carbohydrate intake, and alcohol restriction. Severe hypertriglyceridemia requires drug therapies and dietary fat restriction. Dietary supplements are not recommended for heart disease prevention. Medications given after a heart attack suppress blood clotting, regulate heart rhythm, and reduce blood pressure, and patients are offered small portions of heart-healthy foods. To reduce the risk of a future heart attack, patients must learn strategies similar to the TLC approach.**

Stroke

Stroke is the fourth most common cause of death in the United States[55] and a leading cause of serious long-term disability in adults. About 87 percent of strokes are **ischemic strokes**,[56] caused by the obstruction of blood flow to brain tissue. **Hemorrhagic strokes** occur in 13 percent of cases and result from bleeding within the brain, which damages brain tissue. Most ischemic strokes are a result of ruptured atherosclerotic plaque and subsequent blood clot formation, but an embolism ♦ may also cause a stroke. Hemorrhagic strokes often result from rupture of a blood vessel that has been weakened by atherosclerosis and chronic hypertension. Hemorrhagic strokes are generally more deadly: about 38 percent result in death within 30 days.[57]

♦ Reminder: An *embolism* is the obstruction of a blood vessel by a traveling blood clot or air bubble (an *embolus*).

Strokes that occur suddenly and are short-lived (lasting several minutes to several hours) are called **transient ischemic attacks (TIAs)**. These brief strokes are a warning sign that a heart attack or a more severe stroke may follow, and they need to be evaluated and treated quickly.[58] TIAs typically cause short-term neurological symptoms, such as confusion, slurred speech, numbness, paralysis, or difficulty speaking. Treatment includes the use of aspirin and other drugs that inhibit blood clotting.

ischemic strokes: strokes caused by the obstruction of blood flow to brain tissue.

hemorrhagic strokes: strokes caused by bleeding within the brain, which destroys or compresses brain tissue.

transient ischemic attacks (TIAs): brief ischemic strokes that cause short-term neurological symptoms.

Stroke Prevention Stroke is largely preventable by recognizing its risk factors and making lifestyle choices that reduce risk. Many of the risk factors are similar to those for heart disease; they include hypertension, elevated LDL cholesterol, diabetes mellitus, cigarette smoking, and a history of cardiovascular disease. Medications that suppress blood clotting reduce the risk of ischemic stroke, especially in people who have suffered a first stroke or a transient ischemic attack. The drugs typically prescribed include antiplatelet drugs (including aspirin) or anticoagulants such as warfarin (Coumadin). ♦ Anticoagulant therapy requires regular follow-up and occasional adjustments in dosage to prevent excessive bleeding.

♦ Warfarin acts by interfering with vitamin K's blood-clotting function (see Chapter 19, p. 625).

Stroke Management The effects of a stroke vary according to the area of the brain that has been injured. Body movements, senses, and speech are often impaired, and one side of the body may be weakened or paralyzed. Early diagnosis and treatment are necessary to preserve brain tissue and minimize long-term disability. Ideally, thrombolytic (clot-busting) drugs are used within the first few hours following an ischemic stroke to restore blood flow and prevent further brain damage.[59] After patients have stabilized, they are usually started on medications that help to prevent stroke recurrence or complications, including anticoagulants or antiplatelet drugs, antihypertensives, and blood-lipid lowering drugs.

Rehabilitation programs typically start as soon as possible after stabilization. Patients must be evaluated for neurological deficits, sensory loss, mobility impairments, bowel and bladder function, communication ability, and psychological problems. Rehabilitation services often include physical therapy, occupational therapy, speech and language pathology, and kinesiotherapy (training to improve strength and mobility).

The focus of nutrition care is to help patients maintain nutrition status and overall health despite the disabilities caused by the stroke. In addition, some patients may need to learn about dietary treatments that improve blood lipid levels and blood pressure. The initial assessment should determine the nature of the patient's self-feeding difficulty (if any) and the adjustments required for appropriate food intake. Dysphagia (difficulty swallowing) is a frequent complication of stroke and is associated with a poorer prognosis. ♦ Difficulty with speech prevents patients from communicating food preferences or describing the problems they may be having with eating. Coordination problems can make it hard for patients to grasp utensils or bring food from table to mouth. In some cases, tube feedings may be necessary until the patient has regained these skills. Highlight 27 describes additional options for feeding people with disabilities, such as those that follow stroke.

♦ Reminder: Chapter 23 describes the nutrition care of patients with dysphagia.

IN SUMMARY **The two major types of strokes, ischemic and hemorrhagic stroke, may be a consequence of atherosclerosis, hypertension, or both. Transient ischemic attacks, which are short-lived ischemic strokes, are a warning sign that a heart attack or a more severe stroke may follow. Strokes are largely preventable by reversing modifiable risk factors, such as hypertension, cigarette smoking, diabetes mellitus, and elevated LDL cholesterol. Treatment of a stroke includes the use of anticlotting drugs such as antiplatelet drugs and anticoagulants. Rehabilitation services evaluate the extent of neurological and functional impairment caused by a stroke and provide the therapy patients need to regain lost function. A patient who has had a major stroke may have problems eating normally due to lack of coordination and difficulty swallowing.**

Hypertension

Hypertension (high blood pressure) affects about one-third of adults in the United States.[60] Prevalence is especially high in African Americans, who develop hypertension earlier in life and sustain higher average blood pressures throughout their lives than other ethnic groups. An estimated 20 percent of people with hypertension are unaware that they have it.[61] ♦

♦ Blood pressure is measured both when heart muscle contracts (*systolic* blood pressure) and when it relaxes (*diastolic* blood pressure). Measurements are expressed as millimeters of mercury (mm Hg).

	Systolic	Diastolic
• Desirable blood pressure	<120	<80
• Prehypertension	120–139	80–89
• Hypertension	≥ 140	≥ 90

Although people cannot feel the physical effects of hypertension, it is a primary risk factor for atherosclerosis and cardiovascular diseases. For each 20/10 mm Hg increase above normal blood pressure (an increase of 20 mm Hg in systolic blood pressure and 10 mm Hg in diastolic blood pressure), the risk of death from CVD doubles.[62] Elevated blood pressure forces the heart to work harder to eject blood into the arteries; this effort weakens heart muscle and increases the risk of developing heart arrhythmias, heart failure, and even sudden death. Hypertension is also a primary cause of stroke and kidney failure; reducing blood pressure can dramatically reduce the incidence of these diseases.

Factors That Influence Blood Pressure Although the underlying causes of most cases of hypertension are not fully understood, much is known about the physiological factors that affect blood pressure, the force exerted by the blood on artery walls. As shown in Figure 27-3, blood pressure depends on the volume of blood pumped by the heart (cardiac output) and the resistance the blood encounters in the arterioles (peripheral resistance). ♦ When either cardiac output or peripheral resistance increases, blood pressure rises. Cardiac output is raised when heart rate or blood volume increases; peripheral resistance is affected mostly by the

♦ The equation describing this relationship is blood pressure (BP) = cardiac output (CO) × peripheral resistance (PR)

FIGURE 27-3 Determinants of Blood Pressure

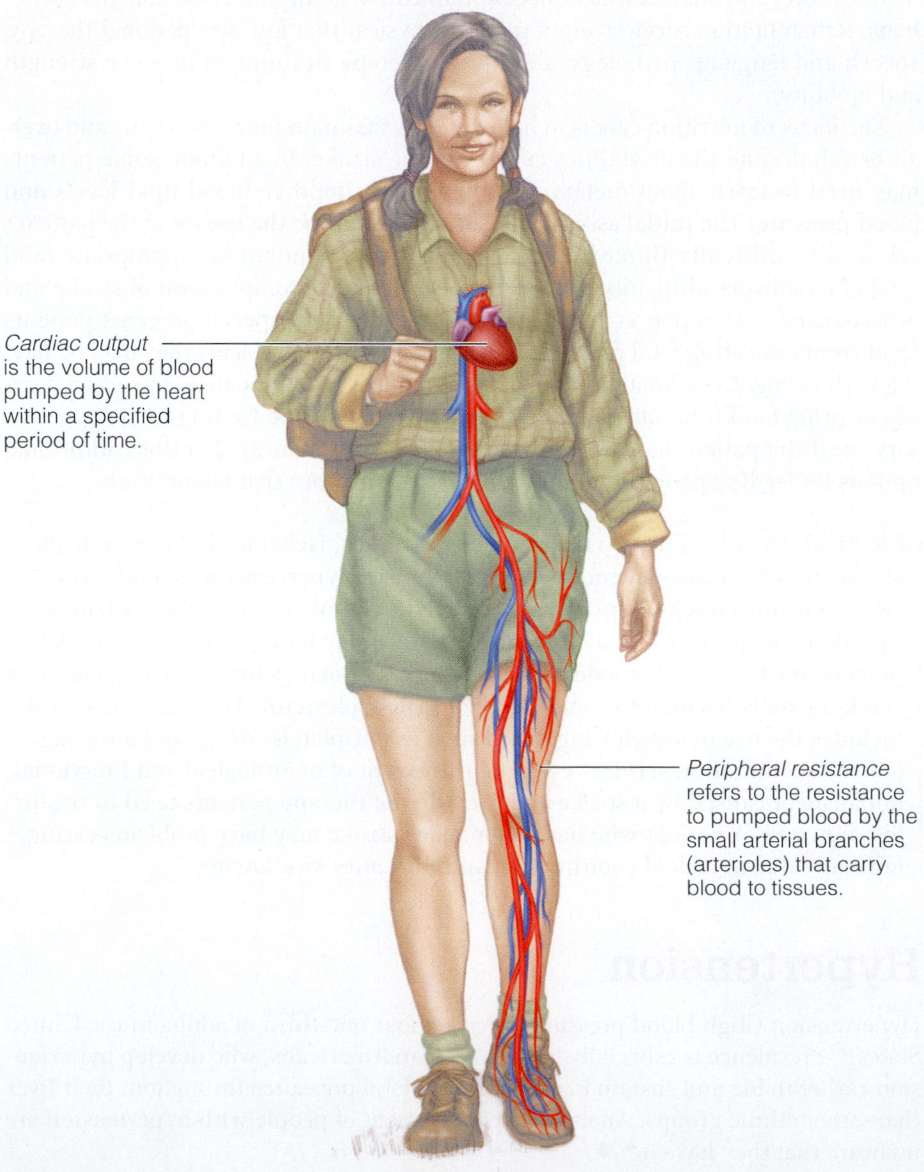
Cardiac output is the volume of blood pumped by the heart within a specified period of time.
Peripheral resistance refers to the resistance to pumped blood by the small arterial branches (arterioles) that carry blood to tissues.

diameters of the arterioles and blood viscosity. Blood pressure is therefore influenced by the nervous system, which regulates heart muscle contractions and arteriole diameters, and hormonal signals, which may cause fluid retention or blood vessel constriction. The kidneys also play a role in regulating blood pressure by controlling the secretion of the hormones involved in vasoconstriction and retention of sodium and water.

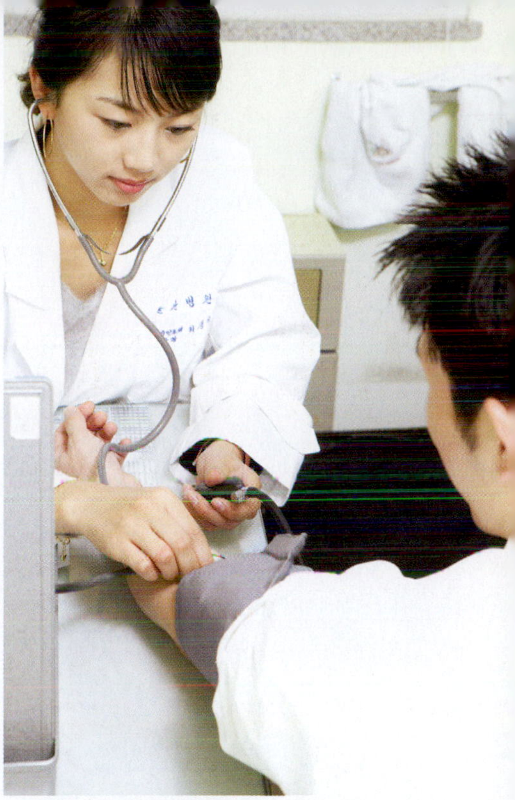

© TongRo Image Stock/Jupiter Images

Screening people for hypertension is a first step toward early detection and prevention of complications.

Factors That Contribute to Hypertension

In 90 to 95 percent of hypertension cases, the cause is unknown (called **primary** or **essential hypertension**).[63] In other cases, hypertension is caused by a known physical or metabolic disorder (**secondary hypertension**), such as an abnormality in an organ or hormone involved in blood pressure regulation. For example, conditions characterized by the narrowing of renal arteries often result in the increased production of proteins and hormones that stimulate water retention and vasoconstriction, thereby raising blood pressure. A number of hormonal disorders and medications may also cause secondary hypertension.

Various risk factors for hypertension have been identified. These include the following:

- *Aging.* Hypertension risk increases with age. More than two-thirds of persons older than 65 years have hypertension.[64] Moreover, individuals who have normal blood pressure at age 55 still have a 90 percent risk of developing high blood pressure during their lifetimes.[65]
- *Genetic factors.* Risk of hypertension is similar among family members. It is also more prevalent and severe in certain ethnic groups; for example, the prevalence in African American adults is about 44 percent, compared with a prevalence of about 33 percent in whites and 28 percent in Mexican Americans.[66]
- *Obesity.* Most people with hypertension—an estimated 60 percent—are obese.[67] Obesity raises blood pressure, in part, by increasing blood volume, promoting vasoconstriction, and increasing activity of the sympathetic nervous system.[68]
- *Salt sensitivity.* ♦ Approximately 50 percent of those with hypertension have blood pressure that is sensitive to salt.[69] These people can improve their blood pressure by reducing salt in their diets.
- *Alcohol.* Heavy drinking (three or more drinks daily) increases the incidence and severity of hypertension by stimulating the sympathetic nervous system.[70] Reducing alcohol consumption reverses this effect.
- *Dietary factors.* A person's diet may affect hypertension risk. As explained later, dietary modifications that increase intakes of potassium, calcium, and magnesium have been shown to reduce blood pressure.

♦ *Salt sensitivity* is also known as *sodium sensitivity*. Salt is the main source of sodium in the diet.

Treatment of Hypertension

Controlling hypertension improves CVD risk considerably: a blood pressure reduction of 10/5 mm Hg lowers the risks of death from CHD and stroke by about 45 and 55 percent, respectively.[71] Both lifestyle modifications and medications are used to treat hypertension. ♦ For people with prehypertension, ♦ changes in diet and lifestyle alone may lower blood pressure to a normal level.

♦ The goal of hypertension treatment is to reduce blood pressure to $<140/<90$ mm Hg. For people with diagnosed CHD, diabetes, or kidney disease, the blood pressure goal is $<130/<80$ mm Hg.

♦ Prehypertension: 120–139/80–89 mm Hg.

Table 27-4 lists lifestyle modifications that can reduce blood pressure and the expected reduction in systolic blood pressure for each change. The recommendations include weight reduction if overweight or obese; a diet low in sodium and rich in potassium, calcium, and magnesium; regular physical activity; and a moderate alcohol intake, if one chooses to drink.[72] ♦ Combining two or more of these modifications can enhance results. As Table 27-4 shows, weight reduction and dietary adjustments generally have the most dramatic effects on blood pressure.

♦ For updated guidelines related to hypertension treatment, access this website: **www.nhlbi.nih.gov/guidelines**

Weight Reduction

For obese individuals, weight reduction may reduce blood pressure significantly. Clinical studies suggest that systolic blood pressure can be reduced by about 1 mm Hg for each kilogram of weight loss and that the blood

primary hypertension: hypertension with an unknown cause; also known as **essential hypertension**.

secondary hypertension: hypertension that results from a known physiological abnormality.

TABLE 27-4 Lifestyle Modifications for Blood Pressure Reduction

Modification	Recommendation	Expected Reduction in Systolic Blood Pressure
Weight reduction	Maintain healthy body weight (BMI below 25).	5–20 mm Hg/10 kg weight loss
DASH Eating Plan[a]	Adopt a diet rich in fruits, vegetables, and low-fat milk products with reduced saturated fat intake.	8–14 mm Hg
Sodium restriction	Reduce dietary sodium intake to less than 2400 milligrams sodium (less than 6 grams salt) per day.[b]	2–8 mm Hg
Physical activity	Perform aerobic physical activity for at least 30 minutes per day, most days of the week.	4–9 mm Hg
Moderate alcohol consumption	Men: Limit to two drinks per day. Women and lighter-weight men: Limit to one drink per day.	2–4 mm Hg

[a]The DASH Eating Plan was tested in a study called Dietary Approaches to Stop Hypertension.
[b]According to DRI recommendations, sodium intake should be limited to 2300 milligrams daily.
SOURCE: Adapted from *Reference Card from the Seventh Report of the Joint National Committee on Prevention, Detection, Evaluation, and Treatment of High Blood Pressure (JNC 7)*, NIH publication no. 03-5231 (Bethesda, MD: National Institutes of Health, National Heart, Lung, and Blood Institute, and National High Blood Pressure Education Program, May 2003).

pressure reduction may be sustained for several years.[73] In the long term, however (more than three years), blood pressure tends to revert to initial levels, even when weight loss is partially maintained. Weight reduction is most beneficial for blood pressure control during periods of weight loss and weight maintenance.[74]

Dietary Approaches for Reducing Blood Pressure Several research studies have shown that a significant reduction in blood pressure can be achieved by following a diet that emphasizes fruits, vegetables, and low-fat dairy products and includes whole grains, poultry, fish, and nuts.[75] The diet tested in these studies, now known as the *DASH Eating Plan*, ♦ provides more fiber, potassium, magnesium, and calcium than the typical American diet. The diet also limits red meat, sweets, sugar-containing beverages, saturated fat (to 7 percent of kcalories), and cholesterol (to 150 milligrams per day), so it is beneficial for reducing CHD risk as well. During the eight-week study period when hypertensive subjects consumed the DASH diet, their systolic blood pressures fell by 11.4 mm Hg more than the blood pressures of subjects who remained on the standard American control diet.[76] The DASH Eating Plan, shown in Table 27-5, is a dietary pattern that meets the goals specified in the *Dietary Guidelines for Americans* (see Chapter 2).

♦ The DASH Eating Plan is based on the test diet used in a study called *D*ietary *A*pproaches to *S*top *H*ypertension.

The DASH Eating Plan is even more effective when accompanied by a low sodium intake. In a research study that tested the blood pressure–lowering effects of the DASH dietary pattern in combination with sodium restriction, the best results were achieved when sodium was reduced to 1500 milligrams daily—a lower level than the maximum intake of 2400 milligrams ♦ recommended for people with hypertension.[77] These results suggest that the optimal sodium intake for people with hypertension may be far lower than is typically recommended.

♦ The maximum sodium intake recommended for people with hypertension (2400 mg) is similar to the Tolerable Upper Intake Level (UL) for sodium, which was set at 2300 mg to prevent the adverse effects of high sodium intakes on blood pressure in the general population.

Sodium restriction by itself can have a modest blood pressure–lowering effect (review Table 27-4), but some people are more responsive than others. Although a low-sodium diet may improve blood pressure to some extent, it should be combined with other lifestyle modifications for greater effect. The "How To" (p. 832) lists practical suggestions for restricting sodium intake; additional detail is provided in Table 28-1 on p. 847.

Physical Activity Regular aerobic exercise improves both systolic and diastolic blood pressure. Of note, low- to moderate-intensity exercise (such as brisk walking)

TABLE 27-5 The DASH Eating Plan

Food Group	Recommended Servings for Different Energy Intakes (servings per day except as noted) 1600 kcal	2000 kcal	2600 kcal	3100 kcal
Grains and grain products[a] (1 serving = 1 slice bread, 1 oz dry cereal,[b] or 1/2 c cooked rice, pasta, or cereal)	6	6–8	10–11	12–13
Vegetables (1 serving = 1/2 c cooked vegetables, 1 c raw leafy vegetables, or 1/2 c vegetable juice)	3–4	4–5	5–6	6
Fruits (1 serving = 1 medium fruit; 1/2 c fresh, frozen, or canned fruit; 1/4 c dried fruit; or 1/2 c fruit juice)	4	4–5	5–6	6
Milk products (low fat or fat free) (1 serving = 1 c milk or yogurt, or 1 1/2 oz cheese)	2–3	2–3	3	3–4
Meat, poultry, and fish (1 serving = 3 oz cooked lean meat, poultry, or fish)	1–2	1–2	2	2–3
Nuts, seeds, and legumes (1 serving = 1/3 c nuts, 2 tbs peanut butter, 2 tbs seeds, or 1/2 c cooked dry beans or peas)	3 per week	4–5 per week	1	1
Fats and oils (1 serving = 1 tsp vegetable oil or soft margarine, 1/2 tbs mayonnaise, or 1 tbs salad dressing)	2	2–3	3	4
Sweets (1 serving = 1 tbs sugar, jelly, or jam; 1/2 c sorbet; or 1 c lemonade)	0	up to 5 per week	≤ 2	≤ 2

[a]Whole grains are recommended for most servings consumed.
[b]One ounce of dry cereal may be equivalent to 1/2 to 1 1/4 cups, depending on the cereal. Check the food label for the portion size.
SOURCE: National Heart, Lung, and Blood Institute, National Institutes of Health, *Your Guide to Lowering Your Blood Pressure with DASH*, NIH publication no. 06-4082 (Bethesda, MD: National Heart, Lung, and Blood Institute, 2006).

is more effective for lowering blood pressure than high-intensity exercise.[78] Intensive resistance exercise (such as heavy weight listing) can raise blood pressure to some extent and should be avoided.[79]

Drug Therapies for Reducing Blood Pressure People with hypertension usually require two or more medications to meet their blood pressure goals. Using a combination of drugs with different modes of action can reduce the doses of each drug needed and minimize side effects. Most treatments include diuretics, which lower blood pressure by reducing blood volume. Other medications prescribed include ACE inhibitors, ♦ beta-blockers, and calcium channel blockers; these drugs are also used to treat various heart conditions. Drug dosages may need regular adjustment until the blood pressure goal is reached. The Diet-Drug Interactions feature lists nutrition-related side effects of these medications.

♦ • *ACE inhibitors* interfere with the production of angiotensin II, a peptide that regulates blood pressure.
• *Beta-blockers* reduce heart rate and cardiac output.
• *Calcium channel blockers* inhibit vasoconstriction.

IN SUMMARY **About one in three persons in the United States has hypertension, which increases the risk of developing CHD, stroke, heart failure, and kidney failure. Blood pressure is elevated by factors that increase blood volume, heart rate, or resistance to blood flow. Although the underlying cause of most hypertension cases is unknown, risk factors include aging, family history, ethnicity, obesity, and various dietary factors. Treatment usually includes a combination of lifestyle modifications and drug therapies.**

The Case Study (p. 833) provides an opportunity to review the risk factors and treatments for CHD and hypertension.

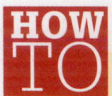

HOW TO Reduce Sodium Intake

- Select fresh, unprocessed foods. Packaged foods, canned goods, and frozen meals are often high in sodium.
- Do not use salt at the table or while cooking. Salt substitutes may be useful for some people. Salt substitutes often contain potassium, however, and are not appropriate for people using diuretics that promote potassium retention in the blood.
- Avoid eating in fast-food restaurants; most menu choices are very high in sodium.
- Check food labels. The labeling term *low sodium* is a better guide than the terms *reduced sodium* (contains 25 percent less sodium than the regular product) or *light in sodium* (contains 50 percent less sodium). To be labeled *low sodium*, a food product must contain less than 140 milligrams of sodium per serving. Keep your goal sodium level in mind when you read labels.
- Recognize the high-sodium foods in each food category, and purchase only unsalted or low-sodium varieties of these products if they are available. High-sodium foods include the following:
  - Snack foods made with added salt, such as tortilla chips, popcorn, and nuts.
  - Processed meats, such as ham, corned beef, bologna, salami, sausage, bacon, frankfurters, and pastrami.
  - Processed fish, such as salted fish and canned fish.
  - Tomato-based products, such as tomato sauce, tomato juice, pizza, canned tomatoes, and catsup.
  - Canned soups and broths; note that even reduced-sodium varieties may contain excessive sodium.
  - Cheese, such as cottage cheese, American cheese, and Parmesan and most other hard cheeses.
  - Bakery products made with baking powder or baking soda (sodium bicarbonate), such as cakes, cookies, doughnuts, and muffins.
  - Condiments and relishes, such as bouillon cubes, olives, and pickled vegetables.
  - Flavoring sauces, such as soy sauce, barbecue sauce, and steak sauce.
- Check for the word *sodium* on medication labels. Sodium is often an ingredient in some types of antacids and laxatives.

TRY IT List five high-sodium foods you often include in your diet. Using a food composition table (such as the one in Appendix H), list the sodium content for the portions you typically consume. Can you think of a low-sodium alternative for each of these foods?

Heart Failure

Heart failure, also called *congestive heart failure*, is characterized by the heart's inability to pump adequate blood, resulting in inadequate blood delivery and a buildup of fluids in the veins and tissues. Heart failure has various causes, but it is often a consequence of chronic disorders that create extra work for the heart muscle, such as hypertension or CHD. To accommodate the extra workload, the heart enlarges or pumps faster or harder, but it eventually may weaken enough to fail completely. Heart failure develops mostly in older adults and the elderly: approximately 75 percent of persons with heart failure in the United States are age 65 or older.[80]

heart failure: a condition with various causes that is characterized by the heart's inability to pump adequate blood to the body's cells, resulting in fluid accumulation in the tissues; also called *congestive heart failure*.

Consequences of Heart Failure

The symptoms and consequences of heart failure depend on the side of the heart that fails. The right side of the heart normally receives blood from the peripheral tissues and pumps the blood to the lungs. With impaired pumping, blood backs up in the peripheral tissues and abdominal organs. Fluid may accumulate in the lower extremities and in the liver and abdomen, causing chest pain, difficulty with digestion and absorption, and swelling in the legs, ankles, and feet. In contrast, the left side of the heart receives blood from the lungs and pumps it to the peripheral tissues. A weakened left heart can cause a buildup of fluid in the lungs (called *pulmonary edema*), resulting in extreme shortness of breath and limited oxygen for activity; in severe cases, it can lead to respiratory failure. ♦ With inadequate blood flow, the functions of various organs, such as the liver and kidneys, may become impaired. The effects of heart failure also depend on the severity of illness: mild cases may be asymptomatic, but severe cases may cause considerable damage to health.

♦ Left-sided heart failure burdens the right side of the heart and eventually leads to right-sided failure; in time, many patients have symptoms of both types of failure.

CASE STUDY Patient with Cardiovascular Disease

Robert Reid, a 48-year-old African American computer programmer, is 5 feet 9 inches tall and weighs 240 pounds. He sits for long hours at work and is too tired to exercise when he gets home at night. His meals usually include fatty meats, eggs, and cheese, and he likes dairy desserts such as pudding and ice cream. He has a family history of CHD and hypertension. His recent laboratory tests show that his blood pressure is 160/100 mm Hg, and his LDL and HDL levels are 160 mg/dL and 35 mg/dL, respectively. He smokes a pack of cigarettes each day and usually has two glasses of wine at both lunch and dinner.

1. Identify Mr. Reid's major risk factors for CHD and hypertension. Which can be modified? What complications might occur if he delays treatment for his blood lipids and blood pressure?
2. What dietary changes would you recommend that could help to improve Mr. Reid's blood pressure and his LDL and HDL cholesterol? Explain the rationale for each dietary change. Prepare a day's menu for Mr. Reid using the DASH Eating Plan as an outline for your choices.
3. What other laboratory tests or measurements would you need to better assess Mr. Reid's condition? Why?
4. Describe several benefits that Mr. Reid might obtain from a program that includes weight reduction and regular physical activity. Explain why the use of alcohol can be both a protective and a damaging lifestyle habit.
5. Assuming that Mr. Reid does not make any changes in his diet and lifestyle and suffers a heart attack, identify the elements of a cardiac rehabilitation program that would be critical for his long-term survival.

Heart failure often affects a person's food intake and level of physical activity. In persons with abdominal bloating and liver enlargement, pain and discomfort may worsen with meals. Limb weakness and fatigue can limit physical activity. End-stage heart failure is often accompanied by **cardiac cachexia**, a condition of severe malnutrition characterized by severe weight loss and tissue wasting. Cardiac cachexia may develop due to increased levels of pro-inflammatory cytokines (which promote catabolism), elevated metabolic rate, loss of appetite, and reduced food intake.[81] The resultant weakness further lowers the person's strength, functional capacity, and activity levels.

cardiac cachexia: a condition of severe malnutrition that develops in heart failure patients; characterized by weight loss and tissue wasting.

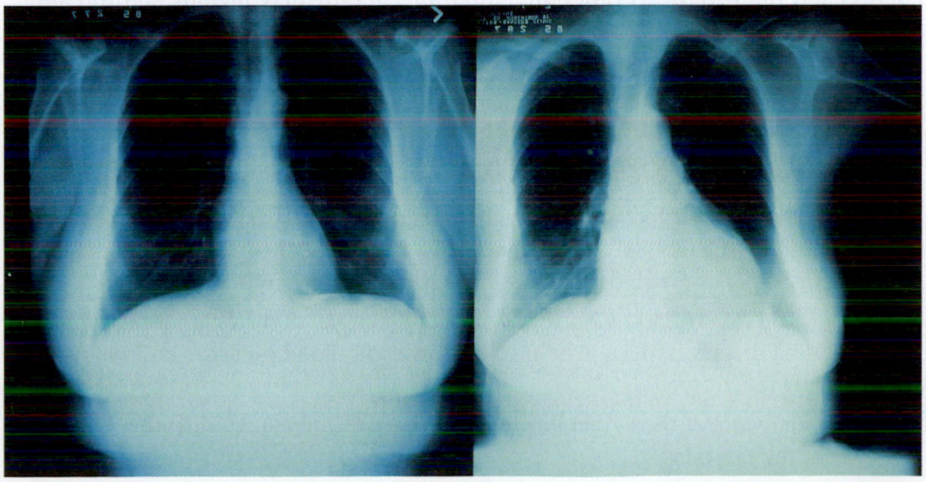

An overburdened heart enlarges in an effort to supply blood to the body's tissues.

Medical Management of Heart Failure

Heart failure is a chronic, progressive illness that may require frequent hospitalizations. Many patients face a combination of debilitating symptoms, complex treatments, and an uncertain outcome. Important goals of medical therapy are to slow disease progression and enhance the patient's quality of life.

The specific treatment for heart failure depends on the nature and severity of the illness. Medications help to manage fluid retention and improve heart function. Dietary sodium and fluid restrictions can help to prevent fluid accumulation. Vaccinations for influenza and pneumonia reduce the risk of developing respiratory infections. Treatment of CHD risk factors, such as hypertension and lipid disorders, can help to slow disease progression. Heart failure patients are also encouraged to participate in exercise programs to avoid becoming physically disabled and to improve endurance.

Drug Therapies for Heart Failure The medications prescribed for heart failure include diuretics, ACE inhibitors, angiotensin receptor blockers, beta-blockers, vasodilators, and digitalis.[82] The diuretics are given to reverse or prevent fluid retention. The patient must monitor fluid fluctuations with daily weight measurements and can make small adjustments in the diuretic dose as needed. The other drugs listed help to improve heart and blood vessel functioning and blood flow.

Nutrition Therapy for Heart Failure The main dietary recommendation for heart failure is a sodium restriction of 2000 milligrams or less daily to reduce the likelihood of fluid retention.[83] ♦ In patients with persistent or recurrent fluid retention, fluid intakes may be restricted to 2 liters per day or less. Individuals who have difficulty eating due to abdominal or chest pain may tolerate small, frequent meals better than large meals.

♦ The recommendation to limit sodium intake to 2000 mg is not very restrictive. The current DRI recommendation is to limit sodium intake to 2300 mg daily.

Other Dietary Recommendations Patients with heart failure may be prone to constipation due to diuretic use and reduced physical activity. Maintaining an adequate fiber intake can help to minimize constipation problems. Because of alcohol's deleterious effects on blood pressure and heart function, most patients need to restrict or avoid alcoholic beverages.

Cardiac Cachexia No known therapies can reverse cardiac cachexia, and prognosis is poor. For some patients, liquid supplements, tube feedings, or parenteral nutrition support can be supportive additions to treatment.

IN SUMMARY **Heart failure is usually a chronic, progressive condition that results from other cardiovascular illnesses. In heart failure, the heart is unable to pump adequate blood to tissues. Consequences may include fluid accumulation in the veins, lungs, and other organs and impaired organ function. Treatment of heart failure usually includes drug therapies that reduce fluid accumulation and improve heart function. Nutrition therapy may include sodium and fluid restrictions.**

Clinical Portfolio

1. List risk factors for coronary heart disease, and identify possible interrelationships among the factors. For example, a woman over age 55 is also at risk for diabetes; a person with diabetes is more likely to have hypertension.
2. Review the DASH Eating Plan shown in Table 27-5. As the chapter describes, the DASH dietary pattern is helpful for lowering blood pressure and for reducing CHD risk as well.
   - List elements of the DASH Eating Plan that are consistent with the TLC recommendations.

- Suggest ways in which a person following the DASH Eating Plan might accomplish the following additional dietary modifications: consume a higher percentage of fat from monounsaturated sources, reduce intake of *trans*-fatty acids, and include EPA/DHA and plant sterols in the diet.

Nutrition on the Net

For further study of topics covered in this chapter, log on to **www.cengagebrain.com** and search for ISBN 084006845X.

- To search for additional information about cardiovascular diseases, obtain the American Heart Association dietary recommendations, or find links to other relevant materials, visit the website of the American Heart Association: **www.heart.org**
- Information about cardiovascular diseases, the DASH Eating Plan, and implementation of heart-healthy diets is available at the websites of the National Heart, Lung, and Blood Institute and the Heart and Stroke Foundation of Canada: **www.nhlbi.nih.gov** and **www.heartandstroke.ca**
- To learn about improving health care and life expectancy of ethnic minority populations at high risk for cardiovascular diseases, visit the website of the International Society on Hypertension in Blacks: **www.ishib.org**
- Learn more about the prevention and treatment of stroke at the website of the National Stroke Association: **www.stroke.org**

Nutrition Assessment Checklist for People with Cardiovascular Diseases

Medical History

Check the medical record for a diagnosis of:

- Coronary heart disease
- Stroke
- Hypertension
- Heart failure

Review the medical record for complications related to cardiovascular diseases:

- Heart attack
- Transient ischemic attack
- Cardiac cachexia

Note CVD risk factors that are related to diet, including:

- Elevated LDL or triglyceride levels
- Obesity or overweight
- Diabetes
- Hypertension

Medications

For patients using drug treatments for cardiovascular diseases, note:

- Side effects that may alter food intake
- Medications that may interact with grapefruit juice
- Use of warfarin, which requires a consistent vitamin K intake
- Use of diuretics or other drugs associated with potassium imbalances
- Potential diet-drug or herb-drug interactions

Dietary Intake

For patients with CHD, a previous stroke, or hypertension, assess the diet for:

- Energy intake
- Saturated fat, *trans* fat, cholesterol, and sodium content
- Soluble fiber and plant sterol or plant stanol content
- Intake of whole grains, fruits, vegetables, legumes, and nuts
- Alcohol content

For patients with complications resulting from cardiovascular diseases:

- Check physical disabilities that may interfere with food preparation or intake following a stroke.
- Check adequacy of food intake in patients with heart failure.

Anthropometric Data

Measure baseline height and weight, and reassess weight at each medical checkup. Note whether patients are meeting weight goals, including:

- Weight loss or maintenance in patients who are overweight
- Weight maintenance in patients with heart failure

Remember that weight may be deceptively high in people who are retaining fluids, especially individuals with heart failure.

Laboratory Tests

Monitor the following laboratory tests in people with cardiovascular diseases:

- LDL cholesterol, blood triglycerides, and HDL cholesterol
- Blood glucose in patients with diabetes
- Serum potassium in patients using diuretics, antihypertensive medications, or digoxin
- Blood-clotting times in patients using anticoagulants
- Indicators of fluid retention in patients with heart failure

Physical Signs

Blood pressure measurement is routine in physical exams but is especially important for people who:

- Have cardiovascular diseases
- Have experienced a heart attack or stroke
- Have risk factors for CHD or hypertension

Look for signs of:

- Potassium imbalances (muscle weakness, numbness and tingling, irregular heartbeat) in those using diuretics, antihypertensive medications, or digoxin
- Fluid overload in patients with heart failure

References

1. V. L. Roger and coauthors, on behalf of the American Heart Association Statistics Committee and Stroke Statistics Subcommittee, Heart disease and stroke statistics––2011 update: A report from the American Heart Association, *Circulation* 123 (2011), available at doi:10.1161/CIR.0b013e3182009701.
2. American Heart Association, International Cardiovascular Disease Statistics (2009), http://www.americanheart.org/presenter.jhtml?identifier=3001008, accessed November 26, 2010.
3. V. Fuster, Atherosclerosis, thrombosis, and vascular biology, in L. Goldman and D. Ausiello, eds., *Cecil Medicine* (Philadelphia: Saunders, 2008), pp. 472–477.
4. Fuster, 2008.
5. C. L. Jackson, Is there life after plaque rupture? *Biochemical Society Transactions* 35 (2007): 887–889; P. Libby and P. Theroux, Pathophysiology of coronary artery disease, *Circulation* 111 (2005): 3481–3488.
6. Libby and Theroux, 2005.
7. R. N. Mitchell and F. J. Schoen, Blood vessels, in V. Kumar and coeditors, *Robbins and Cotran Pathologic Basis of Disease* (Philadelphia: Saunders, 2010), pp. 487–528; W. Insull, The pathology of atherosclerosis: Plaque development and plaque responses to medical treatment, *American Journal of Medicine* 122 (2009): S3–S14.
8. C. Cheng and coauthors, Atherosclerotic lesion size and vulnerability are determined by patterns of fluid shear stress, *Circulation* 113 (2006): 2744–2753.
9. V. Cachofeiro and coauthors, Inflammation: A link between hypertension and atherosclerosis, *Current Hypertension Reviews* 5 (2009): 40–48.
10. Insull, 2009; I. Tabas, K. J. Williams, and J. Boren, Subendothelial lipoprotein retention as the initiating process in atherosclerosis: Update and therapeutic implications, *Circulation* 116 (2007): 1832–1844.
11. K. Musunuru, Atherogenic dyslipidemia: Cardiovascular risk and dietary intervention, *Lipids* 45 (2010): 907–914; M. J. Chapman, Metabolic syndrome and type 2 diabetes: Lipid and physiological consequences, *Diabetes and Vascular Disease Research* 4 (2007): S5–S8.
12. T. B. Twickler and coauthors, Elevated remnant-like particle cholesterol concentration: A characteristic feature of the atherogenic lipoprotein phenotype, *Circulation* 109 (2004): 1918–1925.
13. G. T. Jones and coauthors, Plasma lipoprotein(a) indicates risk for 4 distinct forms of vascular disease, *Clinical Chemistry* 53 (2007): 679–685.
14. N. L. Benowitz, Tobacco, in L. Goldman and D. Ausiello, eds., *Cecil Medicine* (Philadelphia: Saunders, 2008), pp. 162–166.
15. S. Vasdev, V. Gill, and P. Singal, Role of advanced glycation end products in hypertension and atherosclerosis: Therapeutic implications, *Cell Biochemistry and Biophysics* 49 (2007): 48–63; M. H. Beers and coeditors, *The Merck Manual of Diagnosis and Therapy* (Whitehouse Station, NJ: Merck Research Laboratories, 2006), pp. 570–772.
16. M. A. Maturana, M. C. Irigoyen, and P. M. Spritzer, Menopause, estrogens, and endothelial dysfunction: Current concepts, *Clinics* 62 (2007): 77–86.
17. K. S. McCully, Homocysteine, vitamins, and vascular disease prevention, *American Journal of Clinical Nutrition* 86 (2007): 1563S–1568S.
18. Roger and coauthors, 2011.
19. P. P. Toth and coauthors, Cardiovascular disease, in R. E. Rakel, ed., *Textbook of Family Medicine* (Philadelphia: Saunders, 2007), pp. 735–805; Beers and coeditors, 2006.
20. Roger and coauthors, 2011.
21. Roger and coauthors, 2011.
22. P. P. Toth and coauthors, Cardiovascular disease, in R. E. Rakel, ed., *Textbook of Family Medicine* (Philadelphia: Saunders, 2007), pp. 735–805; Beers and coeditors, 2006.
23. Roger and coauthors, 2011.
24. Tabas, Williams, and Boren, 2007.
25. National Cholesterol Education Program, *Third Report of the National Cholesterol Education Program (NCEP) Expert Panel on Detection, Evaluation, and Treatment of High Blood Cholesterol in Adults (Adult Treatment Panel III): Final Report*, NIH publication no. 02-5215 (Bethesda, MD: National Heart, Lung, and Blood Institute, 2002).
26. U.S. Department of Agriculture, Agricultural Research Service, Nutrient intakes from food: Mean amounts consumed per individual, by gender and age, *What We Eat in America, NHANES 2007–2008* (2010), www.ars.usda.gov/ba/bhnrc/fsrg, accessed December 12, 2010.
27. American Dietetic Association, Position of the American Dietetic Association and Dietitians of Canada: Dietary fatty acids, *Journal of the American Dietetic Association* 107 (2007): 1599–1611.
28. American Dietetic Association, Position of the American Dietetic Association and Dietitians of Canada, 2007.
29. U.S. Department of Agriculture, Agricultural Research Service, 2010.
30. National Cancer Institute, Applied Research Program, *Sources of Cholesterol among the U.S. Population, 2005–06* (updated July 21, 2010), http://riskfactor.cancer.gov/diet/foodsources/cholesterol, accessed December 4, 2010.
31. L. Van Horn and coauthors, The evidence for dietary prevention and treatment of cardiovascular disease, *Journal of the American Dietetic Association* 108 (2008): 287–331.
32. A. H. Lichtenstein and coauthors, Diet and lifestyle recommendations revision 2006: A scientific statement from the American Heart Association Nutrition Committee, *Circulation* 114 (2006): 82–96.
33. Lichtenstein and coauthors, 2006.
34. Van Horn and coauthors, 2008; C. Wang and coauthors, n-3 Fatty acids from fish or fish-oil supplements, but not alpha-linolenic acid, benefit cardiovascular disease outcomes in primary- and secondary-prevention studies: A systematic review, *American Journal of Clinical Nutrition* 84 (2006): 5–17.
35. Lichtenstein and coauthors, 2006; I. Gigleux and coauthors, Moderate alcohol consumption is more cardioprotective in men with the metabolic syndrome, *Journal of Nutrition* 136 (2006): 3027–3032.
36. World Cancer Research Fund/American Institute for Cancer Research, *Food, Nutrition, Physical Activity, and the Prevention of Cancer: A Global Perspective* (Washington, DC: American Institute for Cancer Research, 2007), pp. 157–171.
37. P. G. O'Connor, Alcohol abuse and dependence, in L. Goldman and D. Ausiello, eds., *Cecil Medicine* (Philadelphia: Saunders, 2008), pp. 167–174.
38. P. D. Thompson and coauthors, Exercise and physical activity in the prevention and treatment of atherosclerotic cardiovascular disease, *Circulation* 107 (2003): 3109–3116.
39. Lichtenstein and coauthors, 2006.
40. Benowitz, 2008.
41. M. H. Criqui, Epidemiology of cardiovascular disease, in L. Goldman and D. Ausiello, eds., *Cecil Medicine* (Philadelphia: Saunders, 2008), pp. 301–305.
42. Roger and coauthors, 2011.
43. Van Horn and coauthors, 2008; P. Poirier and coauthors, Obesity and cardiovascular disease: Pathophysiology, evaluation, and effect of weight loss, *Circulation* 113 (2006): 898–918.
44. National Cholesterol Education Program, 2002.
45. E. S. Ford and coauthors, Hypertriglyceridemia and its pharmacologic treatment among US Adults, *Archives of Internal Medicine* 169 (2009): 572–578.
46. G. Yuan, K. Z. Al-Shali, and R. A. Hegele, Hypertriglyceridemia: Its etiology, effects, and treatment, *Canadian Medical Association Journal* 176 (2007): 1113–1120; M. J. Malloy and J. P. Kane, Disorders of lipoprotein metabolism, in D. G. Gardner and D. Shoback, eds., *Greenspan's Basic and Clinical Endocrinology* (New York: McGraw-Hill/Lange, 2007), pp. 770–795.
47. American Dietetic Association, *Nutrition Care Manual* (Chicago: American Dietetic Association, 2011); Yuan, Al-Shali, and Hegele, 2007.
48. R. C. Oh and J. B. Lanier, Management of hypertriglyceridemia, *American Family Physician* 75 (2007): 1365–1371.
49. Oh and Lanier, 2007.
50. McCully, 2007; E. Lonn and coauthors, Homocysteine lowering with folic acid and B vitamins in vascular disease, *New England Journal of Medicine* 354 (2006): 1567–1577.

51. J. M. Armitage and coauthors, Study of the Effectiveness of Additional Reductions in Cholesterol and Homocysteine (SEARCH) Collaborative Group, Effects of homocysteine-lowering with folic acid plus vitamin B_{12} vs. placebo on mortality and major morbidity in myocardial infarction survivors: A randomized trial, *Journal of the American Medical Association* 303 (2010): 2486–2494; A. J. Martí-Carvajal and coauthors, Homocysteine lowering interventions for preventing cardiovascular events, *Cochrane Database of Systematic Reviews* 4, no. CD006612 (2009), available at doi:10.1002/14651858.CD006612.pub2; Van Horn and coauthors, 2008.
52. C. Hatzigeorgiou and coauthors, Antioxidant vitamin intake and subclinical coronary atherosclerosis, *Preventive Cardiology* 9 (2006): 75–81; D. H. Lee and coauthors, Does supplemental vitamin C increase cardiovascular disease risk in women with diabetes? *American Journal of Clinical Nutrition* 80 (2004): 1194–1200.
53. Van Horn and coauthors, 2008; L. Mosca and coauthors, Evidence-based guidelines for cardiovascular disease prevention in women: 2007 update, *Circulation* 115 (2007): 1481–1501; Lichtenstein and coauthors, 2006.
54. J. L. Anderson, ST segment elevation acute myocardial infarction and complications of myocardial infarction, in L. Goldman and D. Ausiello, eds., *Cecil Medicine* (Philadelphia: Saunders, 2008), pp. 500–518.
55. A. M. Miniño, J. Xu, and K. D. Kochanek, Deaths: Preliminary data for 2008, *National Vital Statistics Reports* 59, no. 2 (Hyattsville, MD: National Center for Health Statistics, 2010).
56. Roger and coauthors, 2011.
57. Roger and coauthors, 2011.
58. B. T. Vanderhoff and W. Carroll, Neurology, in R. E. Rakel, ed., *Textbook of Family Medicine* (Philadelphia: Saunders, 2007), pp. 1283–1334.
59. J. A. Zivin, Ischemic cerebrovascular disease, in L. Goldman and D. Ausiello, eds., *Cecil Medicine* (Philadelphia: Saunders, 2008), pp. 2708–2719.
60. Roger and coauthors, 2011.
61. Roger and coauthors, 2011.
62. C. Rosendorff and coauthors, Treatment of hypertension in the prevention and management of ischemic heart disease: A scientific statement from the American Heart Association Council for High Blood Pressure Research and the Councils on Clinical Cardiology and Epidemiology and Prevention, *Circulation* 115 (2007): 2761–2788.
63. R. G. Victor, Arterial hypertension, in L. Goldman and D. Ausiello, eds., *Cecil Medicine* (Philadelphia: Saunders, 2008), pp. 430–450.
64. Roger and coauthors, 2011.
65. Beers and coeditors, 2006.
66. Roger and coauthors, 2011.
67. T. A. Kotchen and J. M. Kotchen, Nutrition, diet, and hypertension, in M. E. Shils and coeditors, *Modern Nutrition in Health and Disease* (Philadelphia: Lippincott Williams & Wilkins, 2006), pp. 1095–1107.
68. M. D. Jensen, Obesity, in L. Goldman and D. Ausiello, eds., *Cecil Medicine* (Philadelphia: Saunders, 2008), pp.1643–1652.
69. B. Rodriguez-Iturbe and N. D. Vaziri, Salt-sensitive hypertension—Update on novel findings, *Nephrology Dialysis Transplantation* 22 (2007): 992–995.
70. Victor, 2008.
71. Rosendorff and coauthors, 2007.
72. National High Blood Pressure Education Program/National Institutes of Health, *The Seventh Report of the Joint National Committee on Prevention, Detection, Evaluation, and Treatment of High Blood Pressure (JNC 7)*, NIH publication no. 03-5233 (Bethesda, MD: National Heart, Lung, and Blood Institute, 2003).
73. L. H. Kuller, Weight loss and reduction of blood pressure and hypertension, *Hypertension* 54 (2009): 700–701; L. Aucott and coauthors, Long-term weight loss from lifestyle intervention benefits blood pressure? A systematic review, *Hypertension* 54 (2009): 756–762.
74. Kuller, 2009; Aucott and coauthors, 2009.
75. F. M. Sacks and coauthors, Effects on blood pressure of reduced dietary sodium and the Dietary Approaches to Stop Hypertension (DASH) Diet, *New England Journal of Medicine* 344 (2001): 3–10; L. J. Appel and coauthors, A clinical trial on the effects of dietary patterns on blood pressure, *New England Journal of Medicine* 336 (1997): 1117–1124.
76. Appel and coauthors, 1997.
77. Sacks and coauthors, 2001.
78. P. F. Kokkinos and coauthors, Physical activity in the prevention and management of high blood pressure, *Hellenic Journal of Cardiology* 50 (2009): 52–59.
79. G. Mancia and coauthors, 2007 Guidelines for the management of arterial hypertension: The Task Force for the Management of Arterial Hypertension of the European Society of Hypertension (ESH) and of the European Society of Cardiology (ESC), *European Heart Journal* 28 (2007): 1462–1536.
80. B. M. Massie, Heart failure: Pathophysiology and diagnosis, in L. Goldman and D. Ausiello, eds., *Cecil Medicine* (Philadelphia: Saunders, 2008), pp. 345–354.
81. Massie, 2008.
82. B. G. Katzung and W. W. Parmley, Drugs used in heart failure, in B. G. Katzung, ed., *Basic and Clinical Pharmacology* (New York: McGraw-Hill/Lange, 2007), pp. 198–210.
83. American Dietetic Association, *Nutrition Care Manual*, 2011.

Feeding Disabilities

Chapter 27 refers to difficulties following a stroke that can interfere with the ability to eat independently. This highlight discusses the problems faced by individuals who must cope with disabilities that interfere with the process of eating, including those that interfere with chewing, swallowing, or bringing food to mouth. These obstacles can arise at any time during a person's life and from any number of causes. An infant may be born with a physical impairment such as cleft palate; an adolescent may lose motor control following injuries sustained in an automobile accident; an older adult may struggle with the pain of arthritis or the mental deterioration of dementia. Table H27-1 lists some of the conditions that may lead to feeding problems.

Effects of Disabilities on Nutrition Status

Eating and drinking require a considerable number of individual coordinated motions. Consider an infant learning the skills required for feeding: each step—sitting, grasping cups and utensils, bringing food to the mouth, biting, chewing, and swallowing—requires coordinated movements. An injury or disability that interferes with any of these movements can lead to feeding problems and inadequate food intake. Total food intake is often significantly reduced when individuals with inefficient motor function take a long time to eat.[1] Difficulties that affect procurement of food, such as the inability to drive, walk, or carry groceries, can also lower food intake and lead to malnutrition and weight loss.

TABLE H27-1 Conditions That May Lead to Feeding Problems

The following conditions may lead to feeding problems by interfering with a person's ability to suck, bite, chew, swallow, or coordinate hand-to-mouth movements.

• Accidents	• Language, visual, or hearing impairments
• Amputations	• Multiple sclerosis
• Arthritis	• Muscle weakness
• Birth defects	• Muscular dystrophy
• Cerebral palsy	• Neuromotor dysfunction
• Cleft palate	• Parkinson's disease
• Down syndrome	• Polio
• Head injuries	• Spinal cord injuries
• Huntington's chorea	• Stroke

Energy Requirements

Certain disabilities can either increase or decrease energy requirements.[2] Disabilities that affect muscle tension or mobility can reduce physical activity and, consequently, energy requirements. Other disabilities, such as certain forms of cerebral palsy, cause involuntary muscle activity that raises energy requirements. Loss of a limb due to amputation reduces energy needs in proportion to the weight and metabolism represented by the missing limb, but energy needs may be greater if an individual increases activity to compensate for the loss, such as by propelling a wheelchair. Because the effects of disabilities are often unpredictable, the health care practitioner may find it difficult to assess energy requirements until weight gain or loss has occurred.

Overweight and obesity often accompany conditions that limit mobility or result in short stature; examples include Down syndrome and spina bifida. Obesity may also develop because the family or caregiver provides an inappropriate amount of food, sometimes out of sympathy for the individual who has a disability. In these cases, the health practitioner may need to counsel the family or caregiver about appropriate food choices and portion sizes.

Effects of Disease Symptoms and Medications

Physical symptoms of disease sometimes interfere with eating and nutrition status. Examples include nausea, frequent coughing or choking, difficulty breathing, and gastroesophageal reflux. Individuals with speech or hearing problems may have difficulty communicating with caregivers about thirst and hunger. Mobility problems can influence nutrition status by leading to bone demineralization and pressure sores.

Conditions that require the use of multiple medications can also have a significant impact on nutrition status (see Chapter 19). Medications may increase or decrease appetite, interfere with nutrient metabolism, or have gastrointestinal effects that cause pain or discomfort with eating.

Social Concerns

Mealtimes are a critical time for social interaction, and therefore individuals with feeding problems may encounter emotional and social problems if they are unable to participate. Children may fail to develop social skills, whereas adults may miss the social stimulation that mealtimes provide. Individuals should be encouraged to sit with family and friends during meals so that they are not deprived of the social and cultural aspects of eating.

Independent Eating for People with Disabilities

The evaluation and treatment of feeding problems often involve the joint efforts of health care professionals from a variety of disciplines, including dietitians, nurses, occupational and physical

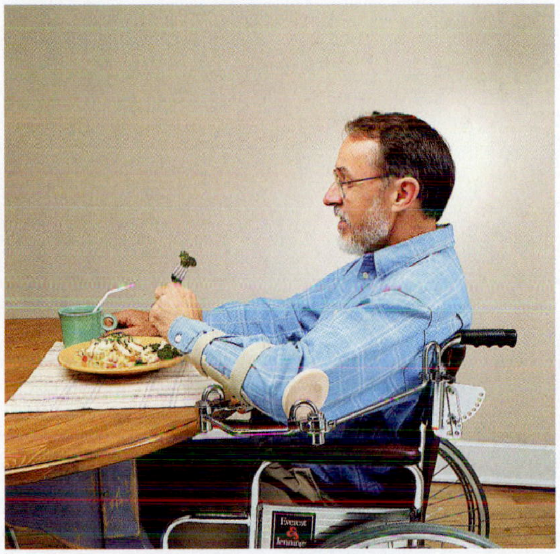

Adaptive feeding equipment can help patients with feeding disabilities gain independence.

TABLE H27-2 Interventions for Feeding-Related Problems

Inability to Suck

- Use squeeze bottles, which do not require sucking, to express liquids into the mouth.
- Place a spoon on the center of the tongue, and apply downward pressure to stimulate sucking.
- Apply rhythmic, slow strokes on the tongue to alter tongue position and improve the sucking response.

Inability to Chew

- Place foods between gums and teeth to promote chewing.
- Improve chewing skills with foods of different textures; for example, fruit leathers stimulate jaw movements but dissolve quickly enough to minimize choking.
- Provide soft foods that require minimal chewing or are easily chewed.

Inability to Swallow

- Provide thickened liquids, pureed foods, and moist foods that form boluses easily.
- Provide cold formulas, frozen fruit juice bars, and ice; cold substances promote swallowing movements by the tongue and soft palate.
- Make sure the patient's jaw and lips are closed to facilitate swallowing action.
- Correct posture and head position if they interfere with swallowing ability.

Inability to Grasp or Coordinate Movements

- Provide utensils that have modified handles, or are smaller or larger as necessary.
- Encourage the use of hands for feeding if utensils are difficult to maneuver.
- Provide plates with food guards to prevent spilling.
- Supply clothing protection.

Impaired Vision

- Place foods (meats, vegetables) in similar locations on the plate at each meal.
- Provide plates with food guards to prevent spilling.

therapists, speech-language pathologists, and dentists. Together, these professionals evaluate each patient's dietary needs and assess abilities to chew, sip, swallow, grasp utensils, use utensils to pick up foods, and bring foods from the plate to the mouth. A speech-language pathologist most often evaluates chewing and swallowing abilities and trains patients to use lips, tongue, and throat for eating and speaking. An occupational therapist can demonstrate alternative feeding strategies, including changes in body position that improve feeding, techniques for handling utensils and food, and use of special feeding devices.

Feeding Strategies

Direct observation of a patient during mealtimes allows health care professionals to assess current eating behaviors, demonstrate feeding techniques, monitor the patient's and caregiver's understanding of the techniques, and evaluate how well the care plan is working. To illustrate, consider a child with a feeding problem caused by hypersensitivity to oral stimulation. The health care professional may start by teaching the caregiver to gently stroke the child's face with a hand, washcloth, or soft toy. Once the child tolerates touch on less sensitive areas of the face, the health care professional may encourage the caregiver to slowly begin to rub the child's lips, gums, palate, and tongue. With time, the child may be better able to tolerate the presence of food in the mouth. Examples of other strategies that can help feeding problems are listed in Table H27-2.

Adaptive Feeding Equipment

Adaptive feeding devices can make a remarkable difference in a person's ability to eat independently. Figure H27-1 shows a few of the many special feeding devices that are available and describes their uses. Other examples of adaptive equipment include specialized chairs to improve posture, bolsters inserted under arms to improve elbow stability, and raised trays or eating surfaces to simplify hand-to-mouth movements.

Sometimes, despite the best efforts of all involved, a patient is unable to consume enough food by mouth. In these cases, tube feedings can help to improve nutrition status. Tube feedings are typically recommended for patients with severe dysphagia (difficulty swallowing), aspiration pneumonia, recurrent malnutrition, or failure to thrive.[3]

A Note for Caregivers

The responsibility of caring for a person with a feeding problem can frequently overwhelm a caregiver.[4] Caring for a person with disabilities requires time and patience—and many new therapies to be learned and administered. The caregiver may spend many hours preparing special foods, monitoring the use of adaptive feeding equipment, and helping with feedings. Moreover, a person with disabilities may need help with other tasks as well, and

HIGHLIGHT 27

FIGURE H27-1 **Examples of Adaptive Feeding Devices**

Utensils

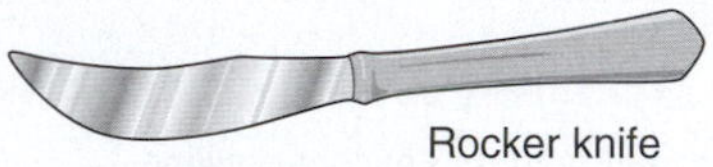

Rocker knife

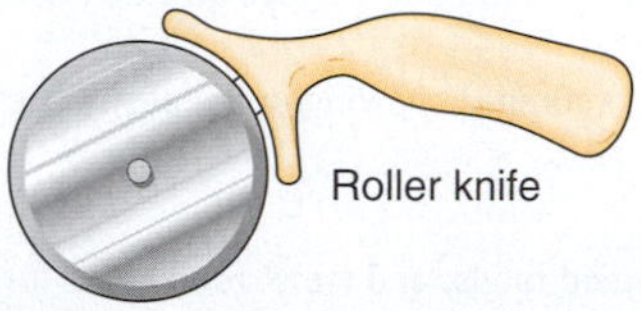

Roller knife

People with only one arm or hand may have difficulty cutting foods and may appreciate using a *rocker knife* or a *roller knife.*

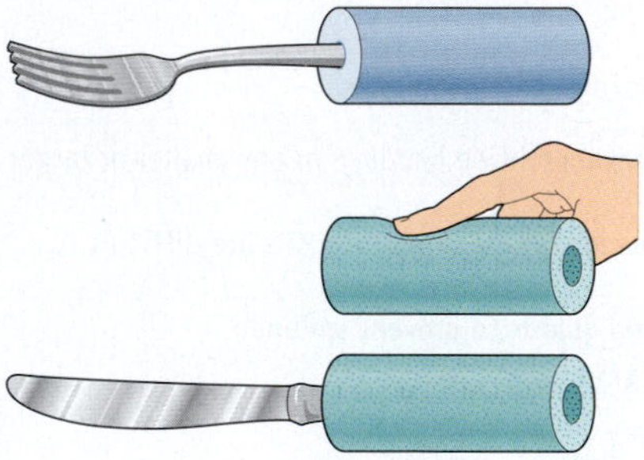

People with a limited range of motion can feed themselves better when they use *flatware with built-up handles.*

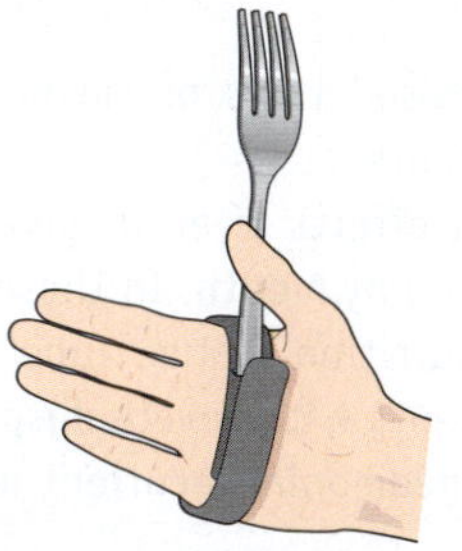

People with extreme muscle weakness may be able to eat with a *utensil holder.*

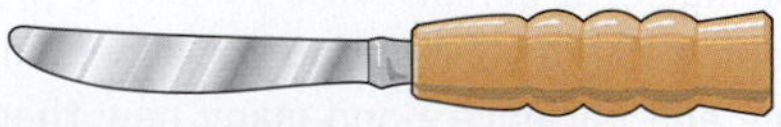

For people with tremors, spasticity, and uneven jerky movements, *weighted utensils* can aid the feeding process.

Battery-powered feeding machines enable people with severe limitations to eat with less assistance from others.

Plates

People who have limited dexterity and difficulty maneuvering food find *scoop dishes* or *food guards* useful.

People with uncontrolled or excessive movements might move dishes around while eating and may benefit from using *unbreakable dishes with suction cups.*

Cups

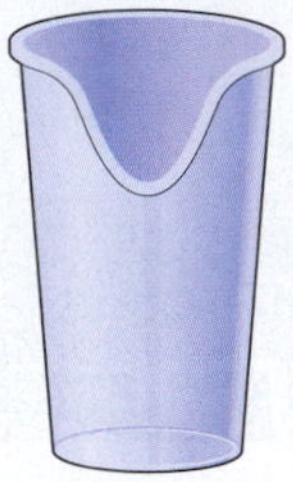

People with limited neck motion can use a *cutout plastic cup.*

Two-handed cups enable people with moderate muscle weakness to lift a cup with two hands.

People with uncontrolled or excessive movements might prefer to drink liquids from a *covered cup* or glass with a *slotted opening* or *spout.*

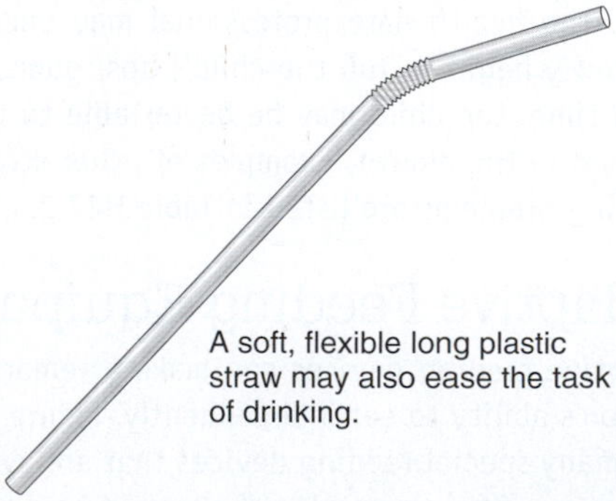

A soft, flexible long plastic straw may also ease the task of drinking.

all may require a considerable amount of time. In many cases, a caregiver receives little or no assistance. These conditions may lead to strained interactions between caregiver and patient and cause stress and frustration. Psychologists can offer counseling to patients or caregivers to help them adjust, and all members of the health care team can offer emotional support and practical suggestions to ease caregivers' responsibilities and frustrations.

Successful therapy for people with feeding disabilities requires the involvement of many health care professionals and depends on accurate identification of impaired feeding skills and determination of appropriate interventions. Ideally, with training, people with disabilities attain total independence—they are able to prepare, serve, and eat nutritionally adequate food daily without help. In some cases, these goals can be met with the help of caregivers. The combined efforts of the health care team can support both patients and caregivers in enhancing quality of life and in achieving independence to the greatest degree possible.

References

1. Position of the Canadian Paediatric Society: Nutrition in neurologically impaired children, *Paediatrics and Child Health* 14 (2009): 395–401.
2. Position of the American Dietetic Association: Providing nutrition services for people with developmental disabilities and special health care needs, *Journal of the American Dietetic Association* 110 (2010): 296–307.
3. Position of the American Dietetic Association, 2010.
4. P. Raina and coauthors, The health and well-being of caregivers of children with cerebral palsy, *Pediatrics* 115 (2005): e626–e636.

CHAPTER

28

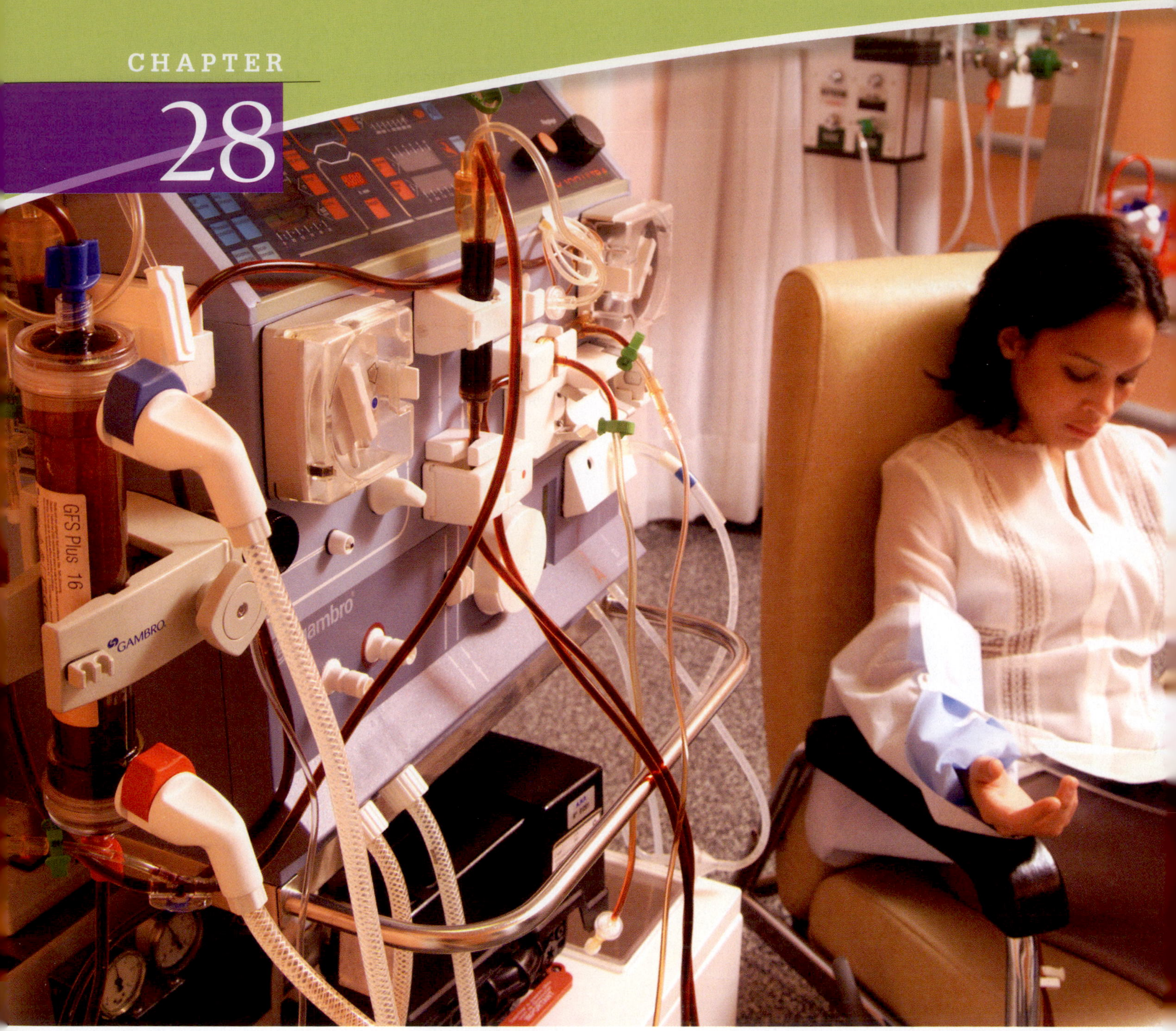

AJPhoto/Photo Researchers, Inc.

Nutrition in the Clinical Setting

Each bean-shaped kidney is only about the size of a fist, yet the kidneys carry out many critical functions. Among other tasks, the kidneys shoulder much of the responsibility for maintaining the body's chemical balance. If the kidneys fail to function, toxic compounds build up in the blood, causing a wide range of symptoms and life-threatening complications. Unfortunately, acute kidney diseases have high mortality rates, and chronic kidney disease is underdiagnosed and undertreated, as symptoms do not arise until the later stages. Health practitioners must recognize and treat renal diseases early before kidney damage progresses and causes irreversible illness.

Throughout this chapter, the CourseMate icon indicates an opportunity for online self-study, linking you to activities to increase your understanding of chapter concepts. **www.cengagebrain.com** (search for ISBN 084006845X)

Kidney Diseases

The two kidneys sit just above the waist on each side of the spinal column. As part of the urinary system (see Figure 28-1), they are responsible for filtering the blood and removing excess fluid and wastes for elimination in urine. Because the kidneys are so proficient at this task, disturbances in body fluids that result from food intake, physical activity, and metabolism are normally corrected within hours. The kidneys also perform a number of other metabolic roles, as discussed in the first section of this chapter. Thus, kidney disorders not only result in fluid and electrolyte imbalances, but can have widespread effects on health.

CHAPTER OUTLINE

Functions of the Kidneys

The functional unit of the kidneys is the **nephron**, introduced on p. 387 (see Figure 12-2). Within each nephron, the **glomerulus**, a ball-shaped tuft of capillaries, serves as a gateway through which the components of blood must pass to form **filtrate**. ♦ The glomerulus and surrounding **Bowman's capsule** function like a sieve, retaining blood cells and most plasma proteins in the blood while allowing fluid and small solutes to enter the nephron's system of **tubules**. As the filtrate passes through the tubules, its composition continuously changes as some of its components are reabsorbed and returned to the blood via capillaries surrounding each tubule. Eventually, the remaining filtrate enters a **collecting duct** shared by several nephrons, and additional water is reabsorbed to form the final urine product. ♦ The urine travels through the ureters to the bladder for temporary storage. By filtering the blood and forming urine, the kidneys regulate the extracellular fluid volume and osmolarity, electrolyte concentrations, and acid-base balance. They also excrete metabolic waste products such as urea and creatinine, as well as various drugs and toxins.

♦ The rate at which the kidneys form filtrate is known as the *glomerular filtration rate*, discussed later in this chapter.

♦ About 99 percent of the substances in filtrate, including water, are reabsorbed, leaving only 1 to 2 liters of urine to be excreted daily.

In addition to their remarkable role in maintaining homeostasis, the kidneys have other vital roles:

- The kidneys help to regulate blood pressure by secreting the enzyme *renin*. Renin catalyzes the formation of angiotensin I from the plasma protein angiotensinogen. In the lungs and elsewhere, angiotensin I is converted to angiotensin II, a potent vasoconstrictor that narrows the diameters of arterioles and thereby raises blood pressure. Angiotensin II also stimulates the release of aldosterone, an adrenal hormone that induces the kidneys to increase reabsorption of sodium and water; this increases plasma volume, which raises blood pressure. (For more details, review Figure 12-3 on p. 388.)
- The kidneys produce the hormone **erythropoietin**, which stimulates the production of red blood cells in the bone marrow (see Highlight 25 for details).
- The kidneys convert vitamin D to its active form, 1,25-dihydroxyvitamin D_3, thereby playing a central role in regulating calcium balance and maintaining bone tissue (see Figure 11-9 on p. 363).

Subsequent sections of this chapter explain how **renal** diseases can interfere with the kidneys' various functions and severely disrupt health.

IN SUMMARY **The kidneys are responsible for filtering the blood and removing wastes for excretion in urine. By adjusting the blood's volume and composition, the kidneys help to maintain homeostasis within the body. Other**

nephron (NEF-ron): the functional unit of the kidneys, consisting of a glomerulus and tubules.

- **nephros** = kidney

glomerulus (gloh-MEHR-yoo-lus): a tuft of capillaries within the nephron that filters water and solutes from the blood as urine production begins (plural: *glomeruli*).

filtrate: the substances that pass through the glomerulus and travel through the nephron's tubules, eventually forming urine.

Bowman's (BOE-minz) **capsule:** a cuplike component of the nephron that surrounds the glomerulus and collects the filtrate that is passed to the tubules.

tubules: tubelike structures of the nephron that process filtrate during urine production. The tubules are surrounded by capillaries that reabsorb substances retained by tubule cells.

collecting duct: the last portion of a nephron's tubule, where the final concentration of urine occurs. One collecting duct is shared by several nephrons.

erythropoietin (eh-RITH-ro-POY-eh-tin): a hormone made by the kidneys that stimulates red blood cell production.

renal (REE-nal): pertaining to the kidneys.

FIGURE 28-1 The Kidneys and Urinary Tract

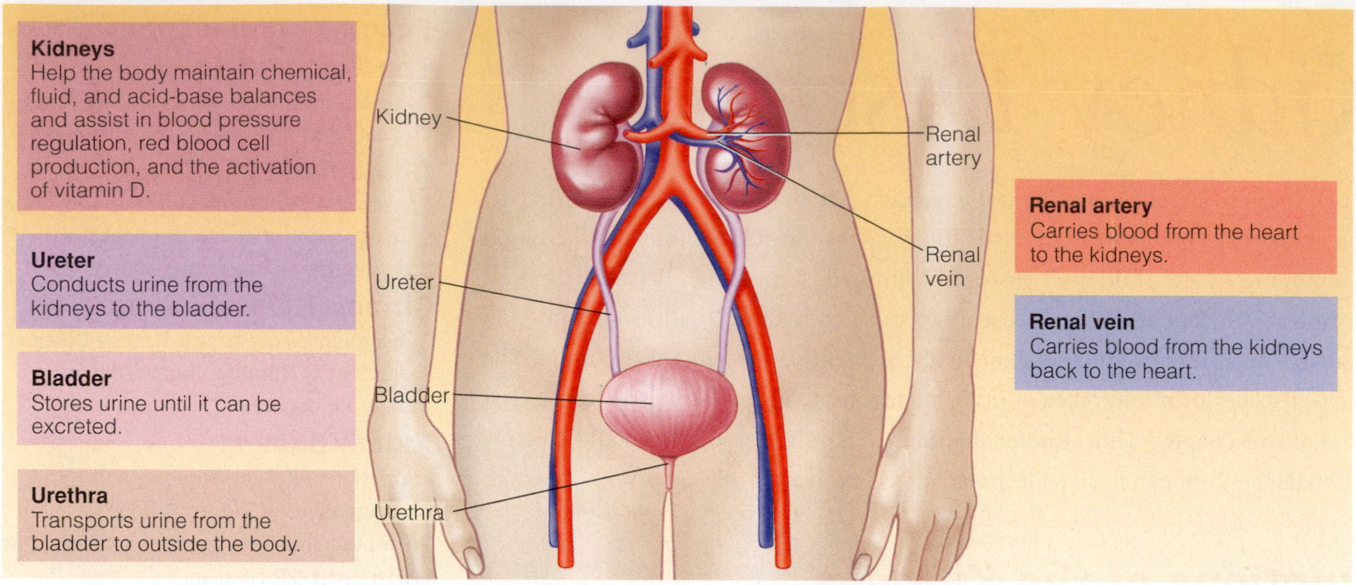
Kidneys
Help the body maintain chemical, fluid, and acid-base balances and assist in blood pressure regulation, red blood cell production, and the activation of vitamin D.
Ureter
Conducts urine from the kidneys to the bladder.
Bladder
Stores urine until it can be excreted.
Urethra
Transports urine from the bladder to outside the body.
Kidney
Ureter
Bladder
Urethra
Renal artery
Renal vein
Renal artery
Carries blood from the heart to the kidneys.
Renal vein
Carries blood from the kidneys back to the heart.

kidney functions include the production of enzymes and hormones that regulate blood pressure, stimulate red blood cell production, and activate vitamin D.

The Nephrotic Syndrome

The **nephrotic syndrome** is not a specific disease; rather, the term refers to kidney disorders that result in significant urinary protein losses (**proteinuria**) due to severe glomerular damage. The condition arises because damage to the glomeruli increases their permeability to plasma proteins, allowing the proteins to escape into the urine. The loss of plasma proteins (typically more than 3½ grams per day) causes serious consequences, including edema, blood lipid abnormalities, blood coagulation disorders, and infections. In some cases, the nephrotic syndrome can progress to renal failure.

Causes of the nephrotic syndrome include glomerular disorders, diabetic nephropathy, immunological and hereditary diseases, infections (involving the kidneys or elsewhere in the body), chemical damage (from medications or illicit drugs), and some cancers.[1] Depending on the underlying condition, some patients may experience one or more relapses and require additional treatment to prevent the proteinuria from recurring. Note that the causes of the nephrotic syndrome and course of illness tend to differ somewhat between children and adults.

nephrotic (neh-FROT-ik) **syndrome:** a syndrome associated with kidney disorders that damage the glomerulus and cause urinary protein losses exceeding 3½ g/day.

proteinuria (PRO-teen-NOO-ree-ah): loss of protein, mostly albumin, in the urine; also known as *albuminuria*.

hypoalbuminemia: low plasma albumin concentrations.

♦ Reminder: Plasma proteins, such as albumin, help to maintain fluid balance within the blood.

Consequences of the Nephrotic Syndrome

In the nephrotic syndrome, urinary protein losses generally average about 8 grams daily.[2] The liver attempts to compensate for these losses by increasing its synthesis of various plasma proteins, but some of the proteins are produced in excessive amounts.

Edema Albumin is the most abundant plasma protein, and it is the protein with the most significant urinary losses as well. The **hypoalbuminemia** characteristic of the nephrotic syndrome contributes to a fluid shift from blood plasma to the interstitial spaces and, thus, edema. ♦ Impaired sodium excretion also contributes to edema: the nephrotic kidney tends to reabsorb sodium in greater amounts than usual, causing sodium and water retention within the body.[3]

Blood Lipid and Blood Clotting Abnormalities Individuals with the nephrotic syndrome frequently have elevated levels of low-density lipoproteins (LDL), very-low-density lipoproteins (VLDL), and the more atherogenic LDL variant known as

FIGURE 28-2 **Effects of Urinary Protein Losses in the Nephrotic Syndrome**

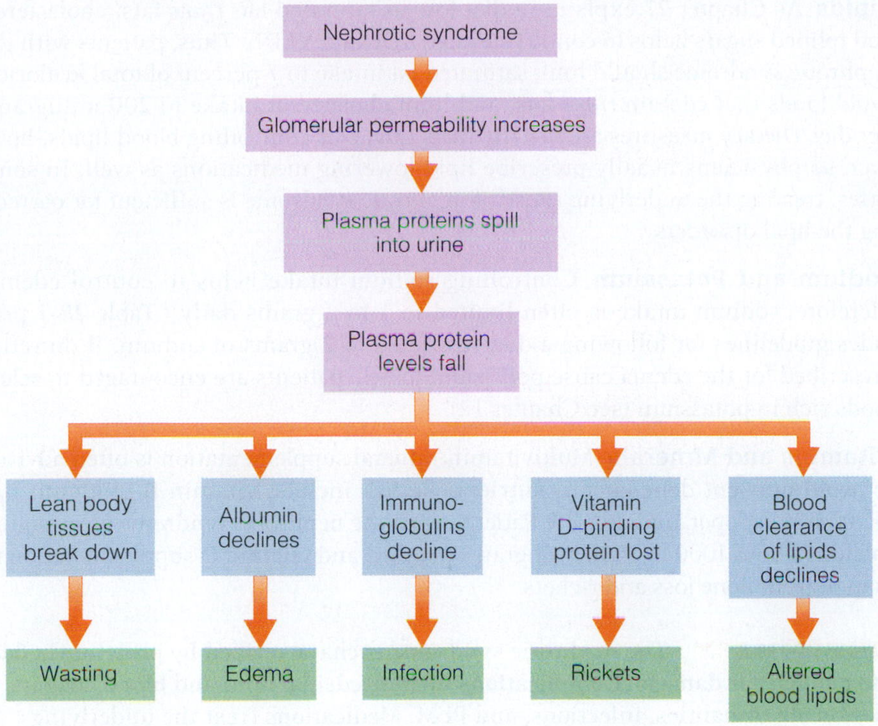

lipoprotein(a). Furthermore, the risk of blood clotting is increased due to urinary losses of proteins that inhibit blood clotting and elevated levels of plasma proteins that favor clotting. The blood clotting abnormalities increase the risk of **deep vein thrombosis** and similar disorders. Clinical studies have found that the nephrotic syndrome is associated with accelerated atherosclerosis and sharply increased risks of heart disease and stroke.[4]

Other Effects of the Nephrotic Syndrome The proteins lost in urine include immunoglobulins (antibodies) and vitamin D–binding protein. Depletion of immunoglobulins increases susceptibility to infection. Loss of vitamin D–binding protein lowers vitamin D and calcium levels and increases the risk of rickets in children. Patients with the nephrotic syndrome frequently develop protein-energy malnutrition (PEM) and muscle wasting from the continued proteinuria. Figure 28-2 summarizes the effects of urinary protein losses in the nephrotic syndrome.

Treatment of the Nephrotic Syndrome

Medical treatment of the nephrotic syndrome requires diagnosis and management of the underlying disorder responsible for the proteinuria. Complications are managed with medications and nutrition therapy. The drugs prescribed may include diuretics, ACE inhibitors (which reduce protein losses), lipid-lowering drugs, anticoagulants, anti-inflammatory drugs (usually corticosteroids), and immunosuppressants (such as cyclosporine).[5] Nutrition therapy can help to prevent PEM, correct lipid abnormalities, and alleviate edema.

Protein and Energy Meeting protein and energy needs helps to minimize losses of muscle tissue. High-protein diets are not advised, however, because they can exacerbate urinary protein losses and result in further damage to the kidneys. Instead, the protein intake should fall between 0.8 and 1.0 gram per kilogram of body weight per day; at least half of the protein consumed should be from high-quality sources such as milk products, meat, fish, poultry, eggs, and soy products.[6]

deep vein thrombosis: formation of a stationary blood clot (thrombus) in a deep vein, usually in the leg, which causes inflammation, pain, and swelling, and is potentially fatal.

An adequate energy intake (about 35 kcalories per kilogram of body weight daily) sustains weight and spares protein. Weight loss or infections suggest the need for additional energy.

Lipids As Chapter 27 explains, a diet low in saturated fat, *trans* fats, cholesterol, and refined sugars helps to control elevated LDL and VLDL. Thus, patients with the nephrotic syndrome should limit saturated fat intake to 7 percent of total kcalories, avoid foods that contain *trans* fats, and limit cholesterol intake to 200 milligrams per day. Dietary measures are usually inadequate for controlling blood lipids, however, so physicians usually prescribe lipid-lowering medications as well. In some cases, treating the underlying cause of nephrotic syndrome is sufficient for correcting the lipid disorders.[7]

Sodium and Potassium Controlling sodium intake helps to control edema; therefore, sodium intake is often limited to 1 to 2 grams daily.[8] Table 28-1 provides guidelines for following a diet restricted to 2 grams of sodium. If diuretics prescribed for the edema cause potassium losses, patients are encouraged to select foods rich in potassium (see Chapter 12).

Vitamins and Minerals Multivitamin/mineral supplementation is often advised to avoid nutrient deficiencies; nutrients at risk include vitamin B_6, vitamin B_{12}, folate, iron, copper, and zinc.[9] ♦ Patients with the nephrotic syndrome also require calcium (about 1000 to 1500 milligrams per day) and vitamin D supplementation to help prevent bone loss and rickets.

♦ Nutrient deficiencies may develop if the carrier proteins for nutrients are lost in the urine.

IN SUMMARY **The nephrotic syndrome is characterized by proteinuria due to glomerular damage. Complications include edema, lipid and blood coagulation abnormalities, infections, and PEM. Medications treat the underlying cause of proteinuria and help to manage complications. The diet should provide sufficient protein and energy to maintain health, but patients should avoid consuming excess protein. Other dietary adjustments can help to correct lipid disorders, edema, and nutrient deficiencies.**

Acute Kidney Injury

In **acute kidney injury**, kidney function deteriorates rapidly, over hours or days. The loss of kidney function reduces urine output and allows nitrogenous wastes to build up in the blood. The degree of renal dysfunction varies from mild to severe. With prompt treatment, acute kidney injury is often reversible, although mortality rates are high, ranging from 36 to 86 percent.[10] Up to 7 percent of hospital patients develop acute kidney injury.[11]

Causes of Acute Kidney Injury

Many different disorders can lead to acute kidney injury, and it often develops as a consequence of severe illness, sepsis, or injury. To aid in diagnosis and treatment, its causes are commonly classified as prerenal, intrarenal, or postrenal. *Prerenal* factors are those that cause a sudden reduction in blood flow to the kidneys; they often involve a severe stressor such as heart failure, shock, or blood loss. Factors that damage kidney tissue, such as infections, toxicants, drugs, or direct trauma, are classified as *intrarenal* causes of acute kidney injury. *Postrenal* factors are those that prevent excretion of urine due to urinary tract obstructions. Table 28-2 provides examples of specific disorders that may cause acute kidney injury.

acute kidney injury: a condition associated with the abrupt decline of kidney function over a period of hours or days; potentially a cause of acute renal failure.

oliguria (OL-lih-GOO-ree-ah): an abnormally low amount of urine, often less than 400 mL/day.

anuria (ah-NOO-ree-ah): the absence of urine; clinically identified as urine output less than 50 mL/day.

Consequences of Acute Kidney Injury

A decline in renal function alters the composition of blood and urine. The kidneys become unable to regulate the levels of electrolytes, acid, and nitrogenous wastes ♦ in the blood. Urine may be diminished in quantity (**oliguria**) or absent (**anuria**), leading to fluid retention.

♦ A sudden rise in blood urea nitrogen (BUN) and/or serum creatinine suggests the presence of acute kidney injury. Refer to Table 17-9 on p. 584 for normal laboratory values.

TABLE 28-1 Sodium-Controlled Diet

General Guidelines

About 75 percent of the sodium in a typical diet comes from processed foods, about 10 percent from unprocessed natural foods, and about 15 percent from table salt. With this in mind:

- Whenever possible, select fresh foods, which are usually low in sodium.
- Select frozen and canned food products that have been prepared without added salt.
- Avoid adding salt to foods while cooking or at the table.
- When dining in restaurants, ask that meals be prepared without salt.

Sodium in Foods

All foods contain sodium, but some contain more than others. Use the following information to plan meals that are low in sodium.

Food Group	Serving Size	Sodium per Serving (mg)
Milk products	1 cup milk or yogurt; 1 oz hard cheese (cheddar, Swiss, jack)	150–200
	Avoid: buttermilk, cottage cheese, cheese spreads, prepared cheeses (such as American cheese)	
Meat, fish, poultry, and eggs	3 oz fresh meat, fish, or poultry; 1 large egg	60
	Avoid: luncheon meats, corned beef, salt pork, sausage, frankfurters, bacon, canned meats or fish, fresh meats prepared with injected broth	
Fruits and vegetables	$^1/_2$ cup fresh vegetables, $^1/_2$ cup fresh or frozen fruit, 6 oz fruit juice; 6 oz tomato or vegetable juice without added salt	10–20
	Avoid: pickled vegetables, olives, tomato or vegetable juices with added salt; dried fruits with added sodium sulfite	
Breads and cereals	$^1/_2$–$^2/_3$ cup dry or cooked cereal without added salt, $^1/_2$ cup cooked rice or pasta	0–10
	1 slice bread, 1 roll or tortilla, $^1/_2$–$^2/_3$ cup dry or cooked cereal prepared with salt	150
	Avoid: pancakes, waffles, muffins, biscuits, and quick breads made with baking powder or baking soda; instant and ready-to-eat cereals with >175 mg sodium; salted snack foods	
Condiments	Unsalted butter; low-sodium salad dressings, mayonnaise, sauces, and gravies; low-sodium catsup, mustard, and hot sauces; garlic or onion powders without added salt; lemon or lime juice, vinegar	Varies; check labels
	Avoid: commercial salad dressings, gravy or soup mixes, barbeque sauces, soy sauce, steak sauces, and spices or herb products made with salt; bouillon, meat tenderizers, monosodium glutamate	

A Sample Diet Restricted to 2 Grams (2000 mg) of Sodium

Using the guidelines provided here, an individual can develop a variety of sample menus. A possible plan for a day might look like this:

Food Group	Sodium (mg)
Meat, 6 oz (2 servings × 60 mg)	120
Milk, 3 c (3 servings × 150 mg)	450
Fruit, 2 servings	negligible
Vegetables, 3 servings	45
Whole-grain bread, 4 slices (4 × 150 mg)	600
Salt, $^1/_4$ tsp (used lightly at meals)	600
Total	**1815**

Individuals can use the remainder of the sodium allowance for whatever foods they desire. The sodium content of most foods can be determined by reading food labels or using food composition tables (such as Appendix H). See additional information about reducing sodium intake in Chapter 27 (p. 832).

TABLE 28-2 Causes of Acute Kidney Injury

Prerenal Factors (60 to 70% of cases)	Intrarenal Factors (25 to 40% of cases)	Postrenal Factors (5 to 10% of cases)
• **Low blood volume or pressure:** hemorrhage, burns, sepsis or shock, anaphylactic reactions, nephrotic syndrome, gastrointestinal losses, diuretics, antihypertensive medications • **Renal artery disorders:** blood clots or emboli, stenosis, aneurysm, trauma • **Heart disorders:** heart failure, heart attack, arrhythmias	• **Vascular disorders:** sickle-cell disease, diabetes mellitus, transfusion reactions • **Obstructions (within kidney):** inflammation, tumors, stones, scar tissue • **Renal injury:** infections, environmental contaminants, drugs, medications, *E. coli* food poisoning	• **Obstructions (ureter or bladder):** strictures, tumors, stones, trauma • **Prostate disorders:** cancer or hyperplasia • **Renal vein thrombosis** • **Bladder disorders:** neurological conditions, bladder rupture • **Pregnancy**

Diagnosis is often a complex task because the clinical effects can be subtle and vary according to the underlying cause of disease.

Fluid and Electrolyte Imbalances About half of patients with acute kidney injury experience oliguria, producing less than 400 milliliters of urine per day.[12] ♦ The reduced excretion of fluids and electrolytes results in sodium retention and elevated levels of potassium, phosphate, and magnesium in the blood. Elevated potassium (**hyperkalemia**) is of particular concern, because potassium imbalances can alter heart rhythm and lead to heart failure. Elevated serum phosphate levels (**hyperphosphatemia**) ♦ promote excessive secretion of parathyroid hormone, which leads to losses of bone calcium. Due to the sodium retention and reduced urine production, edema is a common symptom of acute kidney injury and may be apparent as puffiness in the face and hands and swelling of the feet and ankles.

♦ Normal urine volume typically exceeds 800 milliliters per day, or about 3½ cups.

♦ To measure serum phosphate levels, the phosphorus content of the blood is analyzed; thus, the terms *serum phosphate* and *serum phosphorus* are often used interchangeably.

Uremia As a result of impaired kidney function, nitrogen-containing compounds and various other waste products may accumulate in the blood; a condition often referred to as **uremia**. ♦ The clinical outcome, called the *uremic syndrome*, includes a cluster of symptoms caused by impairments in multiple body systems. Although the clinical signs vary considerably among patients, the symptoms may reflect hormonal imbalances, electrolyte and acid-base imbalances, disturbed heart and gastrointestinal functioning, neuromuscular disturbances, and depressed immunity, among other conditions. The uremic syndrome is described in more detail later in this chapter.

♦ The related term *azotemia* refers specifically to the accumulation of nitrogenous wastes in the blood.

Treatment of Acute Kidney Injury

Treatment of acute kidney injury involves a combination of drug therapy, **dialysis**, ♦ and nutrition therapy to restore fluid and electrolyte balances and minimize blood concentrations of toxic waste products. Both medical care and dietary measures are highly individualized to suit each patient's needs. Correcting the underlying illness is necessary to prevent further damage to the kidneys.

In oliguric patients (those with reduced urine output), recovery from kidney injury sometimes begins with a period of **diuresis**, in which large amounts of fluid (up to 3 liters daily) are excreted.[13] Because tubular function is minimal at this stage, electrolytes may not be sufficiently reabsorbed; consequently, both fluid depletion and electrolyte losses become a concern. Patients with this pattern of recovery (generally those with tubular injury) require close monitoring in case they need fluid and electrolyte replacement.

♦ Highlight 28 describes common dialysis procedures, including *continuous renal replacement therapy*, the approach usually used for treating acute kidney injury.

hyperkalemia (HIGH-per-ka-LEE-me-ah): elevated serum potassium levels.

hyperphosphatemia (HIGH-per-fos-fa-TEE-me-ah): elevated serum phosphate levels.

uremia (you-REE-me-ah): the accumulation of nitrogenous and various other waste products in the blood; often associated with symptoms that reflect impairments in multiple body systems (literally, "urine in the blood"). The term may also be used to indicate the toxic state that results when wastes are retained in the blood.

dialysis (dye-AH-lih-sis): a treatment that removes wastes and excess fluid from the blood after the kidneys have stopped functioning.

diuresis (DYE-uh-REE-sis): increased urine production.

Drug Treatment in Acute Kidney Injury Because kidney function is required for drug excretion, patients may need to use lower doses of their usual medications to compensate for limited urine output. Conversely, dialysis treatment may increase losses of some drugs, and doses may need to be increased. Drugs that are **nephrotoxic** (including some antibiotics and nonsteroidal anti-inflammatory drugs) must be avoided until kidney function improves.

The medications prescribed for acute kidney injury depend on the underlying cause of illness and the complications that develop. Inflammatory conditions may require treatment with immunosuppressants. Edema is treated with diuretics to mobilize fluids; furosemide (Lasix) is the usual choice. Patients with hyperkalemia are given potassium-exchange resins that bind potassium ions in the gastrointestinal (GI) tract, ensuring the elimination of potassium in the stool. Rapid correction of hyperkalemia requires the use of insulin, which causes a temporary shift of extracellular potassium into the cells. (Glucose must be supplied along with insulin to prevent hypoglycemia.) If acidosis is present, bicarbonate may be administered orally or intravenously.

Energy and Protein Although its effects are highly variable, acute kidney injury is often a catabolic condition associated with hypermetabolism and muscle wasting. Therefore, sufficient energy and protein must be ingested to preserve muscle mass. Initially, the patient may be provided with 25 to 35 kcalories per kilogram of body weight per day, while body weight is monitored to ensure that energy intake is adequate.[14] If available, indirect calorimetry provides the best estimate of energy needs.

Protein contributes nitrogen, increasing the kidneys' workload, but intake should be sufficient to maintain nitrogen balance (to the extent possible) and prevent additional wasting. Protein recommendations are influenced by kidney function, the degree of catabolism, and the use of dialysis (dialysis removes nitrogenous wastes). For patients who are not treated with dialysis, protein intakes should be limited to 0.8 to 1.2 grams per kilogram body weight per day.[15] Higher intakes (1.2 to 1.5 grams per kilogram daily) may be recommended if kidney function improves or the treatment includes dialysis. Patients who are catabolic or septic typically have high protein needs but they require dialysis to accommodate the additional nitrogen load.

Fluids Health practitioners can assess fluid status by monitoring weight fluctuations, blood pressure, pulse rates, and appearance of the skin and mucous membranes. Another method is to measure serum sodium concentrations: a low level of sodium often indicates excessive fluid intake, and a high level suggests inadequate intake.

Fluid balance must be restored in patients who are either overhydrated or dehydrated. Thereafter, fluid needs can be estimated by measuring urine output and adding about 500 milliliters to account for the water lost from skin, lungs, and perspiration. An individual with fever, vomiting, or diarrhea requires additional fluid. Patients undergoing dialysis can ingest fluids more freely.

Electrolytes Serum electrolyte levels are monitored closely to determine appropriate electrolyte intakes. Depending on the results of laboratory tests and the clinical assessment, restrictions may be necessary for potassium (2000 to 3000 milligrams per day), phosphorus (8 to 15 milligrams per kilogram body weight per day), and sodium (2000 to 3000 milligrams per day).[16] Patients undergoing dialysis may be allowed more liberal intakes. As mentioned previously, oliguric patients who experience diuresis at the beginning of the recovery period may need electrolyte replacement to compensate for urinary losses.

Enteral and Parenteral Nutrition Some patients need nutrition support to obtain adequate energy. Enteral support (tube feeding) is preferred over parenteral nutrition because it is less likely to cause infection and sepsis. Enteral formulas for patients with acute kidney injury are more kcalorically dense and may have lower protein and electrolyte concentrations than standard formulas. Total parenteral nutrition is necessary only if patients are severely malnourished or cannot consume food or tolerate tube feedings for more than 14 days.[17]

nephrotoxic: toxic to the kidneys.

CASE STUDY: Woman with Acute Kidney Injury

Catherine Garber is a 42-year-old office manager admitted to the hospital's intensive care unit. She was first seen in the emergency room with severe edema, headache, nausea and vomiting, and a rapid heart rate. She reported an inability to pass more than minimal amounts of urine in the past two days. Her son, who drove her to the emergency room, reported that she had missed work for several days and seemed confused and unusually tired. Laboratory tests revealed elevated serum creatinine, BUN, and potassium levels. After learning from her medical history that Mrs. Garber had begun taking penicillin earlier in the week, the physician diagnosed acute kidney injury, probably caused by a reaction to the medication. Mrs. Garber is 5 feet 3 inches tall and weighs 125 pounds.

1. Describe the probable reason for Mrs. Garber's inability to produce urine. Is her reaction to penicillin considered a prerenal, intrarenal, or postrenal cause of kidney injury? Give examples of other medical problems that can cause acute kidney injury.
2. What medications can the physician prescribe to treat Mrs. Garber's edema and hyperkalemia? What recommendation is likely regarding her continued use of penicillin?
3. What concerns should be kept in mind when determining Mrs. Garber's energy, protein, fluid, and electrolyte needs during acute kidney injury? How would dialysis treatment alter recommendations?
4. After treatment begins, Mrs. Garber suddenly begins producing copious amounts of urine. How should this development alter dietary treatment?

As you read through the discussion of chronic kidney disease, consider how Mrs. Garber's diet should change if her kidney problems become chronic.

IN SUMMARY **Acute kidney injury is characterized by a rapid loss in kidney function, causing a buildup of waste products in the blood. Causes of acute kidney injury include prerenal, intrarenal, or postrenal factors. Consequences may include fluid and electrolyte imbalances and uremia. If hyperkalemia develops, it can alter heart rhythm and lead to heart failure. Acute kidney injury is treated with medications, dialysis, and dietary modifications.**

The accompanying Case Study checks your understanding of acute kidney injury.

Chronic Kidney Disease

Unlike acute kidney injury, in which kidney function declines suddenly and rapidly, **chronic kidney disease** is characterized by gradual, irreversible deterioration. Because the kidneys have a large functional reserve, ♦ the disease typically progresses over many years without causing symptoms. Patients are typically diagnosed late in the course of illness after most kidney function has been lost.[18]

♦ The kidneys' ability to function despite loss of nephrons is referred to as *renal reserve*.

The most common causes of chronic kidney disease are diabetes mellitus and hypertension, which are estimated to cause 45 and 27 percent of cases, respectively.[19] Other conditions that lead to chronic kidney disease include inflammatory, immunological, and hereditary diseases that directly involve the kidneys. Chronic kidney disease affects approximately 13 percent of the U.S. population.[20]

Consequences of Chronic Kidney Disease

In the early stages of chronic kidney disease, the nephrons compensate by enlarging so that they can handle the extra workload. As the nephrons deteriorate, however, there is addi-

chronic kidney disease: kidney disease characterized by gradual, irreversible deterioration of the kidneys; also called *chronic renal failure*.

TABLE 28-3 Clinical Effects of Chronic Kidney Disease

Early Stages

- Anorexia
- Exercise intolerance
- Fatigue
- Headache
- Hypercoagulation
- Hypertension
- Proteinuria, hematuria (blood in urine)

Advanced Stages

- Anemia, bleeding tendency
- Cardiovascular disease
- Confusion, mental impairments
- Electrolyte imbalances
- Fluid retention, edema
- Hormonal abnormalities
- Itching
- Metabolic acidosis
- Nausea and vomiting
- Peripheral neuropathy
- Protein-energy malnutrition
- Reduced immunity
- Renal osteodystrophy

TABLE 28-4 Evaluation of Chronic Kidney Disease

Stage of Disease	Description	GFR[a] (mL/min per 1.73 m^2)
1	Kidney damage with normal or increased GFR	≥90
2	Kidney damage with mildly decreased GFR	60–89
3	Moderately decreased GFR	30–59
4	Severely decreased GFR	15–29
5	Kidney failure	<15 (or undergoing dialysis)

[a]Glomerular filtration rate, or GFR, is estimated from the Modification of Diet in Renal Disease study equation and is based on age, gender, race, and calibration for serum creatinine. Normal GFR is approximately 125 milliliters per minute.
SOURCE: A. S. Levey and coauthors, National Kidney Foundation practice guidelines for chronic kidney disease: Evaluation, classification, and stratification, *Annals of Internal Medicine* 139 (2003): 137–147. Used by permission.

tional work for the remaining nephrons. The overburdened nephrons continue to degenerate until finally the kidneys are unable to function adequately, resulting in kidney failure. Once the extent of kidney damage necessitates active treatment—either dialysis or a kidney transplant—the condition is classified as **end-stage renal disease** (ESRD).[21] Without intervention at this stage, an individual cannot survive. Table 28-3 lists common clinical effects of the early and advanced stages of chronic kidney disease. Many symptoms of chronic kidney disease are nonspecific, which may delay diagnosis of the condition.

Renal disease is evaluated using the **glomerular filtration rate** (GFR), the rate at which the kidneys form filtrate. The GFR can be estimated using predictive equations that are based on serum creatinine levels, ♦ age, gender, race, and body size. Table 28-4 shows how chronic kidney disease is classified according to estimated GFR. Other laboratory measures used to assess kidney function include urinary protein levels, BUN, and the ratio of albumin to creatinine in a urine sample.[22]

Altered Electrolytes and Hormones As the GFR falls, the increased activity by the remaining nephrons is often sufficient to maintain electrolyte excretion. Thus, fluid and electrolyte disturbances may not develop until the third or fourth stage of chronic kidney disease. A number of hormonal adaptations also help to regulate electrolyte levels, but these changes may cause complications of their own. The increased secretion of aldosterone ♦ helps to prevent increases in serum potassium but contributes to fluid overload and the development of hypertension (in patients who were not previously hypertensive). Increased secretion of parathyroid hormone ♦ helps to prevent elevations in serum phosphorus but contributes to bone loss and the development of **renal osteodystrophy**, a bone disorder common in renal patients. Electrolyte imbalances are likely when the GFR becomes extremely low (less than 5 milliliters per minute), when hormonal adaptations are inadequate, or when intakes of water and electrolytes are either very restricted or excessive.

Because the kidneys are responsible for maintaining acid-base balance, acidosis often develops in chronic kidney disease. Although usually mild, the acidosis

end-stage renal disease (ESRD): an advanced stage of chronic kidney disease in which dialysis or a kidney transplant is necessary to sustain life.

glomerular filtration rate (GFR): the rate at which filtrate is formed within the kidneys, normally about 125 mL/min.

renal osteodystrophy: a bone disorder that develops in patients with chronic kidney disease as a consequence of the increased secretion of parathyroid hormone, reduced serum calcium, acidosis, and impaired vitamin D activation by the kidneys.

♦ Reminder: *Creatinine* is a waste product of creatine, a nitrogen-containing compound in muscle cells.

♦ Reminder: *Aldosterone* promotes sodium (and therefore water) retention and potassium excretion.

♦ Reminder: *Parathyroid hormone* helps to regulate serum concentrations of calcium and phosphorus. Elevated parathyroid hormone stimulates bone turnover and the release of calcium from bone into blood.

exacerbates renal bone disease because compounds in bone (for example, protein and phosphates) are released to buffer the acid in blood.

Uremic Syndrome Uremia usually develops during the final stages of chronic renal failure, when the GFR falls below about 15 milliliters per minute.[23] As mentioned previously, the numerous symptoms and complications that result from uremia are collectively known as the **uremic syndrome**. Clinical effects may include the following:[24]

- *Hormonal imbalances.* Diseased kidneys are unable to produce erythropoietin, causing anemia. Reduced production of active vitamin D contributes to bone disease. Altered levels of various other hormones may upset growth, reproductive function (menstruation, sperm production), and blood glucose regulation.
- *Altered heart function/increased heart disease risk.* Fluid and electrolyte imbalances result in hypertension, arrhythmias, and eventual heart muscle enlargement. Excessive parathyroid hormone secretion leads to calcification of arteries and heart tissue. Patients with uremia are at increased risk of stroke, heart attack, and heart failure.
- *Neuromuscular disturbances.* Initial symptoms may be mild, and include malaise, irritability, and altered thought processes. Later effects include muscle cramping, restless leg syndrome, sensory deficits, tremor, and seizures.
- *Other effects.* Defects in platelet function and clotting factors prolong bleeding time and contribute to bruising, gastrointestinal bleeding, and anemia. Skin changes include increased pigmentation and severe pruritus (itchiness). Patients with uremia typically have suppressed immune responses and are at high risk of developing infections.

Protein-Energy Malnutrition Patients with chronic kidney disease often develop PEM and wasting. ♦ Anorexia is thought to contribute to the poor food intake of kidney patients and may result from hormonal disturbances, nausea and vomiting, restrictive diets, uremia, and medications. Nutrient losses also contribute to malnutrition and may be a consequence of vomiting, diarrhea, gastrointestinal bleeding, and dialysis. In addition, many of the illnesses that lead to chronic kidney disease induce a catabolic state that contributes to protein losses.[25]

♦ Reminder: A screening method sometimes used for assessing PEM risk is the *Subjective Global Assessment*, described in Chapter 17 (see Table 17-3 on p. 572).

Treatment of Chronic Kidney Disease

The goals of treatment for patients with chronic kidney disease are to slow disease progression and prevent or alleviate symptoms. Nutrition therapy helps to prevent PEM and weight loss. Once kidney disease reaches the final stages, dialysis or a kidney transplant is necessary to sustain life.

Drug Therapy for Chronic Kidney Disease Medications help to control some of the complications associated with chronic kidney disease. Treatment of hypertension is critical for slowing disease progression and reducing cardiovascular disease risk; thus, antihypertensive drugs are usually prescribed (see Chapter 27). Some antihypertensive drugs (such as ACE inhibitors) can reduce proteinuria, helping to prevent additional kidney damage. Anemia is treated by injection or intravenous administration of erythropoietin (epoetin). Other drug treatments include phosphate binders (taken with food) to reduce serum phosphorus levels, sodium bicarbonate to reverse acidosis, and cholesterol-lowering medications. Supplementation with active vitamin D (called *calcitriol*) helps to raise serum calcium and reduce parathyroid hormone levels.

Dialysis Dialysis replaces kidney function by removing excess fluid and wastes from the blood. In **hemodialysis**, the blood is circulated through a **dialyzer** (artificial kidney), where it is bathed by a **dialysate**, a solution that selectively removes fluid and wastes. In **peritoneal dialysis**, the dialysate is infused into a person's peritoneal cavity, and blood is filtered by the peritoneum (the membrane that sur-

uremic syndrome: the cluster of symptoms associated with inadequate kidney function; the symptoms reflect fluid, electrolyte, and hormonal imbalances; altered heart function; neuromuscular disturbances; and other metabolic derangements.

hemodialysis (HE-moe-dye-AL-ih-sis): a treatment that removes fluids and wastes from the blood by passing the blood through a dialyzer.

dialyzer (DYE-ah-LYE-zer): a machine used in hemodialysis to filter the blood; also called an *artificial kidney*.

dialysate (dye-AL-ih-sate): the solution used in dialysis to draw wastes and fluids from the blood.

peritoneal (PEH-rih-toe-NEE-al) **dialysis:** a treatment that removes fluids and wastes from the blood by using the body's peritoneal membrane as a filter.

rounds the abdominal cavity). After several hours, the dialysate is drained, removing unneeded fluid and wastes. Highlight 28 provides additional information about dialysis.

Nutrition Therapy for Chronic Kidney Disease The patient's diet strongly influences disease progression, the development of complications, and serum levels of nitrogenous wastes and electrolytes. Because the dietary measures for chronic kidney disease are complex and nutrient needs change frequently during the course of illness, a dietitian who specializes in renal disease is best suited to provide nutrition therapy. Table 28-5 summarizes the general dietary guidelines for patients in the different stages of chronic kidney disease. The predialysis guidelines apply to patients in stages 1 through 4; by stage 5, either hemodialysis or peritoneal dialysis is necessary. Because patients' needs vary considerably, actual recommendations should be based on the results of a careful and complete nutrition assessment.

Energy The energy intake should be high enough to allow patients to maintain a healthy weight and to prevent wasting. Foods and beverages with a high energy density are typically recommended. ♦ Malnourished patients may require oral supplements or tube feedings to maintain weight. (The "How To" on p. 876 in Chapter 29 includes suggestions for increasing the energy content of meals.)

The dialysate used in peritoneal dialysis contains glucose in order to draw fluid from the blood to the peritoneal cavity by osmosis; about 60 percent of this glucose is absorbed.[26] The kcalories from glucose (as many as 800 kcalories daily) must be included in estimates of energy intake. Weight gain is sometimes a problem when peritoneal dialysis continues for a long period.[27]

Protein A low-protein diet is usually prescribed to reduce the amount of nitrogenous waste produced. In addition, low-protein diets supply less phosphorus than high-protein diets, reducing the risks associated with hyperphosphatemia. Because renal patients often develop PEM, however, their diet must provide enough protein to meet needs and prevent wasting. During the predialysis period, the recommended protein intake is 0.60 to 0.75 grams per kilogram of body weight per day.[28] ♦ At least 50 percent of the protein consumed should come from high-quality protein sources (such as eggs, milk products, meat, poultry, fish, and soybeans) to ensure that the patient consumes adequate amounts of the essential amino acids. Low-protein breads, pastas, and other grain-based products are commercially

♦ Reminder: Foods with a high energy density contain a high number of kcalories per unit weight; these foods are generally high in fat and low in water content.

♦ The protein RDA for adults is 0.8 g/kg body weight.

TABLE 28-5 Dietary Recommendations for Chronic Kidney Disease

Nutrient	Predialysis	Hemodialysis	Peritoneal Dialysis
Energy[a] (kcal/kg body weight)	35 for <60 years old 30–35 for ≥60 years old	35 for <60 years old 30–35 for ≥60 years old	35 for <60 years old 30–35 for ≥60 years old (total energy intake includes kcalories absorbed from the dialysate)
Protein (g/kg body weight)	0.60–0.75 (≥50% high-quality proteins)	≥1.2 (≥50% high-quality proteins)	≥1.2–1.3 (≥50% high-quality proteins)
Fat	As necessary to maintain a healthy lipid profile	As necessary to maintain a healthy lipid profile	As necessary to maintain a healthy lipid profile
Fluid (mL/day)	Unrestricted if urine output is normal	500–1000 plus daily urine output	As necessary to maintain fluid balance
Sodium (mg/day)	1000–3000	1000–3000	2000–4000
Potassium (mg/day)	Unrestricted unless hyperkalemia is present	2000–3000; adjust according to serum potassium levels	3000–4000; adjust according to serum potassium levels
Calcium (mg/day)	1000–1500	≤2000 from diet and medications	≤2000 from diet and medications
Phosphorus (mg/day)	800–1000 if serum phosphorus or parathyroid hormone is elevated	800–1000 if serum phosphorus or parathyroid hormone is elevated	800–1000 if serum phosphorus or parathyroid hormone is elevated

[a]Values listed apply to adults; recommendations for children should not fall below DRI levels.

SOURCES: American Dietetic Association, *Nutrition Care Manual* (Chicago: American Dietetic Association, 2011); D. J. Goldstein-Fuchs and C. M. Goeddeke-Merickel, Nutrition and kidney disease, in A. Greenberg, ed., *Primer on Kidney Diseases* (Philadelphia: Saunders, 2009), pp. 479–486.

© Craig M. Moore

In a renal diet, at least half of the protein consumed should be from high-quality protein sources such as eggs, milk products, meat, poultry, and fish.

♦ Reminder: Most salad dressings and mayonnaise products are made with polyunsaturated or monounsaturated vegetable oils.

available to help renal patients improve their energy intakes without increasing protein consumption.

Because of the high risk of wasting and compliance difficulties associated with low-protein diets, some dietitians suggest that patients consume higher amounts of protein to preserve health.[29] Once dialysis has begun, protein restrictions can be relaxed, because dialysis removes nitrogenous wastes and results in some amino acid losses as well.

Lipids To control elevated blood lipids and reduce heart disease risk, patients with chronic kidney disease are advised to restrict their intakes of saturated fat, *trans* fat, and cholesterol. Although patients are often encouraged to consume high-fat foods to improve their energy intakes, the foods they select should provide mostly unsaturated fats. Good choices include nuts and seeds, oil-based salad dressings, mayonnaise, ♦ avocados, and soybean products (Highlight 5 provides additional suggestions).

Sodium and Fluids As kidney disease progresses, patients excrete less urine and cannot handle normal intakes of sodium and fluids. Recommendations depend on the total urine output, changes in body weight and blood pressure, and serum sodium levels. A rise in body weight and blood pressure suggests that the person is retaining sodium and fluid; conversely, declines in these measurements indicate fluid loss. Most persons with kidney disease tend to retain sodium and may benefit from mild restriction; less frequently, a patient may have a salt-wasting condition that requires additional dietary sodium.

Fluids are not restricted until urine output decreases. For a person who is neither dehydrated nor overhydrated, the daily fluid intake should match the daily urine output. (Obligatory water losses—from skin and lungs—are replaced by the water contained in the solid foods that are consumed.) Once a person is on dialysis, sodium and fluid intakes should be controlled so that only about 2 pounds of water weight are gained daily—this excess fluid is then removed during the next dialysis treatment. Patients on fluid-restricted diets should be advised that foods such as flavored gelatin, soups, fruit ices, frozen fruit juice bars, and ice milk contribute to the fluid allowance.

Potassium Before dialysis treatments begin, most renal patients can handle typical intakes of potassium; restrictions are usually necessary only in those with elevated potassium levels. Individuals with diabetic nephropathy are at high risk of hyperkalemia and may also need to limit dietary potassium during the early stages of disease. Conversely, potassium supplementation may be necessary for persons using potassium-wasting diuretics.

Dialysis patients must control potassium intakes to prevent hyperkalemia or, more rarely, **hypokalemia**. Restriction is necessary for people treated with hemodialysis, whereas those undergoing peritoneal dialysis can consume potassium more freely. Recommended intakes are based on serum potassium levels, renal function, medications, and the dialysis procedure used.

hypokalemia (HIGH-po-ka-LEE-me-ah): low serum potassium levels.

hypercalcemia (HIGH-per-kal-SEE-me-ah): elevated serum calcium levels.

All fresh foods provide potassium, but some fruits and vegetables contain such high amounts that some patients must restrict intakes. Table 28-6 shows the potassium content of some common fruits and vegetables. Foods in other food groups may be high in potassium as well; examples include dried beans, fish, milk and milk products, molasses, nuts and nut butters, and wheat bran. Patients should be cautioned that salt substitutes and other low-sodium products often contain potassium chloride, which people on a potassium-restricted diet should avoid. Appendix H provides additional information about the potassium content of common foods.

© Craig M. Moore

People on a renal diet can consume most fruits and vegetables in limited amounts.

Calcium, Phosphorus, and Vitamin D To prevent bone disease, calcium and phosphorus intakes may need adjustment, even during the early stages of kidney disease. Laboratory values usually help to guide dietary recommendations for these nutrients. Serum calcium levels must be monitored to guard against **hypercalcemia**, which can develop in response to simultaneous calcium and vitamin D supplementation. Elevated serum phosphorus levels indicate the need for dietary phospho-

TABLE 28-6 Potassium Guide—Fruits and Vegetables

This table lists common fruits and vegetables according to their potassium content. One serving is 1/2 cup raw fruit or cooked vegetable unless otherwise noted. Keep in mind that the portion size may determine how a food is categorized. Check Appendix H for additional information about the potassium content of foods.

High Potassium (>250 mg per serving)	Medium Potassium (150–250 mg per serving)	Low Potassium (<150 mg per serving)
Avocado	Apple (1 medium)	Blueberries
Banana	Apricots (2 whole)	Cabbage
Beets	Asparagus	Carrots (1 medium)
Chard	Broccoli	Cauliflower
Dates (3 whole)	Cantaloupe	Cucumbers
Nectarine (1 small)	Celery	Eggplant
Orange (1 medium)	Corn	Grapes
Parsnips	Grapefruit (1/2 fruit)	Green beans
Potatoes	Honeydew melon	Green pepper
Pumpkin	Kale	Lettuce (4 leaves, raw)
Raisins	Peach (1 small)	Onions (1 small)
Spinach	Pear (1 medium)	Plum (1 small)
Sweet potatoes	Peas	Strawberries
Tomato	Zucchini	Watermelon

rus restriction. Vitamin D supplementation is standard treatment for many renal patients, ♦ but the amount prescribed depends on the serum levels of calcium, phosphorus, and parathyroid hormone.

♦ Reminder: Diseased kidneys are unable to produce activated vitamin D, which normally regulates calcium absorption and helps to maintain serum calcium levels.

High-protein foods are also high in phosphorus, so the protein-restricted diets consumed by predialysis patients curb phosphorus intakes as well. After dialysis treatments begin and protein intakes are liberalized, phosphate binders (taken with meals) become essential for phosphorus control. Because foods that are rich in calcium (such as milk and milk products) are usually high in phosphorus and are therefore restricted, patients must rely on calcium supplements to meet their calcium needs. Table 28-7 lists examples of foods that are high in phosphorus.

TABLE 28-7 Foods High in Phosphorus[a]

- Barley
- Bran (oat, wheat)
- Buckwheat groats
- Bulgur
- Canned iced teas
- Canned lemonade
- Coconut
- Cola beverages
- Cornmeal
- Couscous
- Dried peas and beans
- Fish
- Milk products
- Nuts and seeds
- Organ meats
- Peanut butter
- Processed meats
- Soybeans, tofu

[a]For a complete list, visit the USDA's Nutrient Database at **www.nal.usda.gov/fnic/foodcomp/search**. Click on "Nutrient Lists," and then find the list of foods sorted in descending order by phosphorus content (click on the letter "W" to the right of the word "Phosphorus").

Vitamins and Minerals The restrictive renal diet interferes with vitamin and mineral intakes, increasing the risk of deficiencies. In addition, patients treated with dialysis lose water-soluble vitamins and some trace minerals into the dialysate. Dietary supplements for dialysis patients typically supply generous amounts of folic acid and vitamin B_6—0.8 to 1.0 milligram and 5 to 10 milligrams per day, respectively—along with recommended amounts of the other water-soluble vitamins.[30] Supplemental vitamin C should be limited to 100 milligrams per day, because excessive intakes can contribute to kidney stone formation in individuals at risk (see p. 859). Vitamin A supplements are not recommended because vitamin A levels tend to rise as kidney function worsens.

Iron deficiency is common in hemodialysis patients and may be due to inadequate erythropoietin, gastrointestinal bleeding, reduced iron absorption, or blood losses associated with the dialysis treatment.[31] Intravenous administration of iron, in conjunction with erythropoietin therapy, is more effective than oral iron supplementation for improving iron status.

Enteral and Parenteral Nutrition Nutrition support is sometimes necessary for renal patients who cannot consume adequate amounts of food. The enteral formulas suitable for patients with chronic kidney disease are more kcalorically dense and have lower protein and electrolyte concentrations than standard formulas. **Intradialytic parenteral nutrition** is an option for supplying supplemental nutrients to dialysis patients; this technique combines parenteral infusions with hemodialysis treatments. An advantage of this approach is that the volume of parenteral

intradialytic parenteral nutrition: the infusion of nutrients during hemodialysis, often providing amino acids, dextrose, lipids, and some trace minerals.

HOW TO Help Patients Comply with a Renal Diet

Patients with renal disease and their caregivers face considerable challenges as they learn to manage a renal diet. The following suggestions may help:

1. *To keep track of fluid intake*:
   - Fill a container with an amount of water equal to your total fluid allowance. Each time you consume a liquid food or beverage, discard an equivalent amount of water from the container. The amount remaining in the container will show you how much fluid you have left for the day.
   - Be sure to save enough fluid to take medications.
2. *To help control thirst*:
   - Chew gum or suck hard candy.
   - Suck on frozen grapes.
   - Freeze beverages to a semisolid state so that they take longer to consume.
   - Add lemon juice or crumpled mint leaves to water to make it more refreshing.
   - Gargle with refrigerated mouthwash.
3. *To increase the energy content of meals*:
   - Add extra margarine or a flavored oil to rice, noodles, breads, crackers, and cooked vegetables. Add extra salad dressing or mayonnaise to salads.
   - Add nondairy whipped toppings to desserts.
   - Include fried foods in your diet.
4. *To include more of your favorite vegetables in meals*:
   - Consult your nurse or dietitian to learn whether you can safely use the process of leaching to remove some of the potassium from vegetables.
   - To leach potassium from vegetables: Cut the vegetables into $^{1}/_{8}$-inch slices and rinse. Soak the vegetables in a large amount of warm water for two hours—about 10 parts of water to 1 part of vegetables. Rinse vegetables well. Boil vegetables using 5 parts of water to 1 part of vegetables.
5. *To prevent the diet from becoming monotonous*:
   - Experiment with new combinations of allowed foods.
   - Substitute nondairy products for milk products. Nondairy products, which are lower in protein, phosphorus, and potassium, can substitute for milk and add energy to the diet.
   - Add flavor to foods by seasoning with garlic, onion, chili powder, curry powder, oregano, mint, basil, parsley, pepper, or lemon juice.
   - Consult a nurse or dietitian when you want to eat restricted foods. Many restricted foods can be used occasionally and in small amounts if the menu is carefully adjusted.

TRY IT Individuals often find unique ways to comply with difficult dietary modifications. For three of the five categories listed, suggest a novel technique that would be helpful for you if you needed to deal with the restrictions of a renal diet.

solution infused can be simultaneously removed (recall that fluid intake is controlled in dialysis patients). However, clinical studies have not shown intradialytic parenteral nutrition to be more successful than oral supplementation in improving nutrition status or mortality rates in malnourished dialysis patients.[32]

Dietary Compliance Adhering to a renal diet is probably the most challenging aspect of treatment for patients with renal disease. These patients often require extensive counseling once multiple dietary restrictions become necessary. Depending on the stage of illness and the patient's laboratory values, the renal diet may limit protein, fluids, sodium, potassium, and phosphorus, thereby affecting food selections from all major food groups. In addition, adjustments in nutrient intake are required as the disease progresses. Because these diets have so many restrictions, patient compliance is often a problem. The "How To" provides suggestions to help patients comply with renal diets, and Table 28-8 shows an example of a one-day menu that includes some of these restrictions. The accompanying Case Study allows you to apply your knowledge about chronic kidney disease and hemodialysis.

Kidney Transplants

A preferred alternative to dialysis in patients with end-stage renal disease is kidney transplantation.[33] A successful kidney transplant restores kidney function, allows a more liberal diet, and frees the patient from routine dialysis. Given the choice, many patients would prefer transplants, but the demand for suitable kidneys far exceeds the supply. Other barriers to transplantation include advanced age, poor health, and financial difficulties. Approximately 30 percent of patients who develop end-stage renal disease receive a kidney transplant.[34]

TABLE 28-8 Chronic Kidney Disease—One-Day Menu

The following menu provides 2028 kcalories, 46 g protein, 784 mg phosphorus, 2190 mg potassium, and 1510 mg sodium.[a] The energy and protein content would be appropriate for a 135-pound predialysis patient.

Breakfast

- Corn flakes with milk (1 cup cereal, 1/2 cup whole milk)
- Apricot nectar (1 cup)
- Caffé latte (brewed coffee, 2 tsp sugar, 1/2 cup cream substitute)

Lunch

- Turkey sandwich (2 slices white bread, 1 1/2 oz dark meat, 5 slices cucumber, 1 tbs mayonnaise)
- Grape juice (1 cup)
- Orange sherbet (1/2 cup)

Dinner

- Spaghetti with tomato sauce (1 cup cooked spaghetti, 1/2 cup bottled tomato sauce, 1/2 tbs grated cheese)
- Green beans with olive oil (1 cup cooked green beans, 1 tbs olive oil)
- Biscuit with margarine (2 1/2-inch biscuit, 1/2 tbs margarine)
- Baked apple with nondairy sour cream (1 large apple, 1/4 cup nondairy sour cream)

[a]Energy and nutrient values were obtained from the USDA National Nutrient Database for Standard Reference: **www.nal.usda.gov/fnic/foodcomp/search**

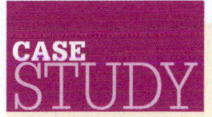

CASE STUDY Man with Chronic Kidney Disease

Thomas Stone is a 55-year-old banker who developed chronic kidney disease as a result of hypertension. His condition was discovered several years ago, when routine laboratory tests revealed elevated serum creatinine and BUN levels. Since then, he has been taking antihypertensive medications and restricting dietary sodium; he reported difficulty following the low-protein diet that was also prescribed. Mr. Stone recently visited his doctor with complaints of low urine output and reduced sensation in his hands and feet. He also reported feeling drowsy at work and mentioned that he was bruising more than usual. The examination revealed a 9-pound weight gain since his last visit and swelling in his ankles and feet. Tests revealed that his GFR had fallen to 10 milliliters per minute. Mr. Stone is 5 feet 8 inches tall and normally weighs 160 pounds.

1. Explain how chronic kidney disease progresses. What happens to GFR, serum creatinine levels, and BUN as renal function declines?
2. Describe the clinical effects you would expect during the final stage of disease, when kidney failure develops. Explain the significance of each of Mr. Stone's physical complaints.
3. Explain why a low-sodium, low-protein diet was prescribed for Mr. Stone at a former visit. What energy and protein intakes were probably recommended at that time?
4. The physician determines that Mr. Stone's kidney disease has reached the final stage and prescribes hemodialysis. How will dialysis alter Mr. Stone's diet? Calculate his new protein recommendation, and compare it to the amount of protein recommended before dialysis. What other changes in nutrient intake may be necessary?

DIET-DRUG Interactions

Check this table for notable nutrition-related effects of the medications discussed in this chapter.

Immunosuppressants	
Cyclosporine, tacrolimus	**Gastrointestinal effects:** Nausea, vomiting, diarrhea, anorexia (tacrolimus) **Dietary interactions:** Cyclosporine potentiates the effects of alcohol. The bioavailability of tacrolimus is reduced when the drug is taken with food. Grapefruit juice can raise serum concentrations of these drugs to toxic levels. **Metabolic effects:** Electrolyte imbalances, hypertension, hyperglycemia, hyperlipidemia
Corticosteroids	**Metabolic effects:** Glucose intolerance, sodium retention, negative nitrogen balance, appetite stimulation, weight gain, growth suppression in children
Phosphate binders (calcium-containing)	**Gastrointestinal effect:** Constipation **Metabolic effects:** Electrolyte imbalances
Potassium-exchange resins (sodium polystyrene sulfonate)	**Metabolic effects:** Fluid retention, hypokalemia, hypocalcemia
Potassium citrate	**Gastrointestinal effects:** Nausea, vomiting, stomach pain, diarrhea **Metabolic effect:** Hyperkalemia

Immunosuppressive Drug Therapy To prevent tissue rejection following transplant surgery, patients require high doses of immunosuppressive drugs. ♦ These drugs have multiple effects that can alter nutrition status, including nausea, vomiting, diarrhea, glucose intolerance, altered blood lipids, fluid retention, hypertension, and increased risk of infection. Because immunosuppressive drugs increase the risk of foodborne infection, food safety guidelines should be provided to patients and caregivers. The Diet-Drug Interactions feature summarizes the nutrition-related effects of the drugs mentioned in this chapter.

♦ Examples of immunosuppressive drugs used after a kidney transplant:
- Azathioprine (Imuran)
- Corticosteroids (prednisone)
- Cyclosporine (Sandimmune)
- Tacrolimus (Prograf)

Nutrition Therapy after a Kidney Transplant Energy and protein requirements increase after surgery due to stress and the catabolic effects of drug therapy. Once recovery is under way, recommendations for energy and most nutrients are similar to those suggested for the general population. Patients should attempt to maintain a healthy body weight and consume a diet that reduces their risk for cardiovascular diseases.

For most transplant patients, the side effects of drugs are the primary reason that dietary adjustments may be required. Although sodium, potassium, phosphorus, and fluid intakes are usually liberalized following a transplant, serum electrolyte levels must be monitored because some drug therapies can cause electrolyte imbalances or fluid retention. If corticosteroids are used as immunosuppressants, calcium supplementation is recommended because the medication increases urinary calcium losses. If drug treatment leads to hyperglycemia, patients should limit intakes of refined carbohydrates and concentrated sweets; for some individuals, oral medications or insulin therapy may be necessary. As noted earlier, patients must carefully follow food safety guidelines to avoid foodborne illness (see Highlight 18).

IN SUMMARY Chronic kidney disease causes gradual loss of kidney function and often results from long-standing diabetes mellitus or hypertension. Depending on the stage of illness, complications of chronic kidney disease may include fluid and electrolyte imbalances, hypertension, renal osteodystrophy, mental impairments, bleeding abnormalities, anemia, increased risk for cardiovascular disease, and reduced immunity. Treatment can slow disease progression and correct complications; it includes drug therapies, dialysis, and nutrition therapy. Dietary measures usually feature a low-protein diet, controlled fluid and sodium intakes, phosphorus restrictions, and calcium and vitamin D supplementation; potassium restrictions are usually necessary after dialysis treatment begins. Kidney transplantation can restore renal function and liberalize dietary restrictions.

Kidney Stones

Approximately 12 percent of men and 6 percent of women in the United States develop one or more **kidney stones** during their lifetimes.[35] A kidney stone is a crystalline mass that forms within the urinary tract. Although stones are often asymptomatic, their passage can cause severe pain or block the urinary tract. Stones tend to recur but can be prevented with dietary measures and medical treatment.

© Nathan Griffith/Corbis

The most common type of kidney stone is composed of calcium oxalate crystals, as shown here. Kidney stones may be as small as a bread crumb or as large as a golf ball.

Formation of Kidney Stones

Kidney stones develop when stone constituents become concentrated in urine, allowing crystals to form and grow. About 70 percent of kidney stones are made up primarily of calcium oxalate. Less commonly, stones are composed of calcium phosphate, uric acid, the amino acid cystine, or magnesium ammonium phosphate (the latter are known as *struvite* stones). Factors that predispose an individual to stone formation include the following:

- *Dehydration* or *low urine volume*, which promotes the crystallization of minerals and other compounds in urine.
- *Obstruction*, which prevents the flow of urine and encourages salt precipitation.
- *Urine acidity*, which affects the dissolution of urinary constituents. Some stones form more readily in acidic urine, whereas others form in alkaline urine.
- *Metabolic factors*, which affect the presence of compounds that either promote or inhibit crystal growth.
- *Renal disease*, which is associated with calcification of tissues and phosphate accumulation.

The most common types of kidney stones are described in this section.

Calcium Oxalate Stones The most common abnormality in people with calcium oxalate stones is **hypercalciuria** (elevated urinary calcium levels). Hypercalciuria can result from excessive calcium absorption, impaired calcium reabsorption in kidney tubules, or elevated serum levels of parathyroid hormone or vitamin D. However, some people with calcium oxalate stones excrete normal amounts of calcium in the urine, and the reason they form stones is unknown.

Elevated urinary oxalate levels, or **hyperoxaluria**, also promote the formation of calcium oxalate crystals. Oxalate is a normal product of metabolism that readily binds to calcium. Hyperoxaluria may reflect an increase in the body's synthesis of oxalate or increased absorption from dietary sources. ♦ People who form calcium oxalate stones are often advised to reduce their dietary intake of oxalate (see Table 28-9) and to avoid supplementation with vitamin C, which degrades to oxalate in the body.[36]

♦ Reminder: Fat malabsorption promotes oxalate absorption, thereby increasing the risk of forming calcium oxalate stones (see Chapter 24).

Calcium Phosphate Stones Calcium phosphate is often a minor constituent of calcium oxalate stones, but some individuals form kidney stones in which calcium phosphate is the main constituent. Although uncommon, calcium phosphate stones sometimes form in people with hypercalciuria who produce alkaline urine.

Uric Acid Stones Uric acid stones develop when the urine is abnormally acidic, contains excessive uric acid, or both. These stones are frequently associated with **gout**, a metabolic disorder characterized by elevated uric acid levels in the blood and urine. A diet rich in **purines** also contributes to high uric acid levels; purines

kidney stones: crystalline masses that form in the urinary tract; also called *renal calculi* and *nephrolithiasis*.

hypercalciuria (HIGH-per-kal-see-YOO-ree-ah): elevated urinary calcium levels.

hyperoxaluria (HIGH-per-ox-ah-LOO-ree-ah): elevated urinary oxalate levels.

gout (GOWT): a metabolic disorder characterized by elevated uric acid levels in the blood and urine and the deposition of uric acid in and around the joints, causing acute joint inflammation.

purines (PYOO-reens): products of nucleotide metabolism that degrade to uric acid.

TABLE 28-9 Foods High in Oxalate

Vegetables	Fruits	Other
Beets*	Apricots, dried	Barley
Chard	Blackberries	Buckwheat
Collard greens	Blueberries	Chocolate*
Dried beans	Currants, red	Cocoa
Eggplant	Figs	Cornmeal, grits
Escarole	Grapes, Concord	Miso
Green beans	Kiwi	Nuts, nut butters*
Kale	Lemon peel	Peanut butter*
Leeks	Oranges, orange peel	Sesame seeds, tahini
Mustard greens	Raspberries	Soybean products
Parsley	Rhubarb*	Tea*
Spinach*	Strawberries*	Wheat bran*
Sweet potatoes		Whole-wheat flour

NOTE: The oxalate content of many foods has not been analyzed, and few studies have been conducted to determine which foods raise urinary oxalate levels.
*The foods marked with an asterisk have been documented to raise urinary oxalate levels and should be avoided by people who form calcium oxalate stones.

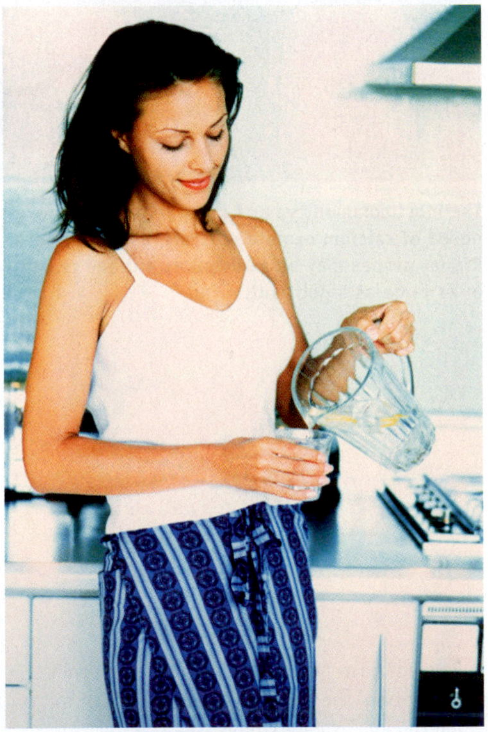
© James Darell/Getty Images

Drinking plenty of water throughout the day is the most important measure for preventing kidney stones.

are abundant in animal proteins (meat, poultry, seafood) and degrade to uric acid in the body. In addition, a high intake of animal protein increases urine acidity, which promotes the crystallization of uric acid.

Cystine and Struvite Stones

Cystine stones can form in people with the inherited disorder **cystinuria**, in which the renal tubules are unable to reabsorb the amino acid cystine. This abnormality results in abnormally high concentrations of cystine in the urine, leading to subsequent crystallization and stone formation. **Struvite** stones, composed primarily of magnesium ammonium phosphate, form in alkaline urine; the urinary pH is elevated due to the bacterial degradation of urea to ammonia. Struvite stones can accompany chronic urinary infections or disorders that interfere with urinary flow.

Consequences of Kidney Stones

In most cases, kidney stones do not pose serious medical problems. Small stones can readily pass through the ureters and out of the body with minimal treatment.

Renal Colic

A stone passing through the ureter can produce severe, stabbing pain, called **renal colic**. Generally, the pain begins in the back and intensifies as the stone travels toward the bladder (review Figure 28-1 on p. 844). The pain can be severe enough to cause nausea and vomiting and sometimes requires medication. When the stone reaches the bladder, the pain abruptly stops. Blood may appear in the urine (**hematuria**) as a result of damage to the kidney or ureter lining.

Urinary Tract Complications

Depending on the location of the stone, symptoms may include urination urgency, frequent urination, or inability to urinate. Stones that are unable to pass through the ureter can cause a urinary tract obstruction and possibly lead to infection or acute kidney injury.

Prevention and Treatment of Kidney Stones

Solutes are less likely to crystallize and form stones in dilute urine. Therefore, people who form kidney stones are advised to drink 12 to 16 cups of fluids throughout the day in order to maintain urine volumes of least 2 to 2½ liters per day.[37] Additional fluid may

cystinuria (SIS-tin-NOO-ree-ah): an inherited disorder characterized by the elevated urinary excretion of several amino acids, including cystine.

struvite (STROO-vite): crystals of magnesium ammonium phosphate.

renal colic: the intense pain that occurs when a kidney stone passes through the ureter.

hematuria (HE-mah-TOO-ree-ah): blood in the urine.

be needed in hot weather or if an individual is extremely active. For some patients, dietary modifications, medications, or surgical stone removal may be necessary.

Calcium Oxalate Stones Most dietary strategies and drug treatments for calcium oxalate stones aim to reduce urinary calcium and oxalate levels. Dietary measures may include adjustments in calcium, oxalate, protein, and sodium intakes.[38] Patients should consume adequate calcium from food sources (about 800 and 1200 milligrams per day for men and women, respectively) because dietary calcium combines with oxalate in the intestines, reducing oxalate absorption and helping to control hyperoxaluria. ♦ Conversely, low-calcium diets promote oxalate absorption and higher urinary oxalate levels. Some individuals with hyperoxaluria may benefit from dietary oxalate restriction (review Table 28-9). High protein and sodium intakes increase urinary calcium excretion, so moderate protein consumption (0.8 to 1.0 gram per kilogram of body weight per day) and a controlled sodium intake (no more than 3450 milligrams daily) are also advised. Vitamin C intakes should not exceed the RDA (90 and 75 milligrams for men and women, respectively) because vitamin C degrades to oxalate.

♦ Because calcium supplements can elevate urinary calcium levels, they are not as helpful as food sources of calcium.

Thiazide diuretics (such as hydrochlorothiazide) are a mainstay of drug therapy and help to reduce urinary calcium by enhancing calcium reabsorption in the kidney tubules. Other medications that may be prescribed include cholestyramine (Questran), which reduces oxalate absorption; potassium citrate (a base), which inhibits crystal formation; and allopurinol (Zyloprim), which reduces uric acid production in the body and may have other effects.[39]

Uric Acid Stones Diets restricted in purines may help to control urinary uric acid levels. Because all animal proteins contain purines, strict dietary control over a long period may be difficult to achieve. In addition, the benefits of purine restriction are unclear. The drug treatments used for uric acid stones include allopurinol to reduce uric acid levels and potassium citrate to reduce urine acidity.

Cystine and Struvite Stones High fluid intakes may prevent the formation of cystine stones in some patients, whereas other individuals require drug therapy to reduce cystine production in the body. Medications frequently prescribed include penicillamine (Cuprimine) and tiopronin (Thiola), which increase the solubility of cystine, and potassium citrate, which reduces urine acidity.

Preventing urinary tract infections is an important strategy for preventing struvite stones. Patients with these stones may require antibiotic therapy to prevent further stone formation.

Medical Treatment for Kidney Stones Medical treatment may be necessary for a kidney stone that is too large to pass, blocks urine flow, or causes severe pain or bleeding. Medications that may be given to facilitate stone passage include alpha-blockers and calcium channel blockers, which relax the ureter and increase urine flow. Sometimes a *stent* (a thin, flexible tube) is placed in the ureter to promote stone passage, although the stent may be uncomfortable and cause excessive bleeding. Some kidney stones can be fragmented into pieces that are small enough to pass in the urine; the most common method is *extracorporeal shock wave lithotripsy*, a procedure that uses high-amplitude sound waves to break down the kidney stone. Various other surgical methods—which involve physical removal of the stone—have a higher success rate but are also more invasive.

IN SUMMARY **Kidney stones form when stone constituents—calcium oxalate, calcium phosphate, uric acid, cystine, or magnesium ammonium phosphate—crystallize in urine. Complications include renal colic, difficulty with urination, and obstruction. Kidney stones may be prevented by maintaining urine volumes of at least 2½ liters daily. Other dietary measures include the consumption of appropriate amounts of calcium, oxalates, protein, sodium, and purines. Symptomatic kidney stones are sometimes treated with medications or treatments that facilitate stone passage or surgeries that fragment or remove stones.**

Clinical Portfolio

1. A person with chronic kidney disease may need multiple medications to control disease progression and treat symptoms and complications. For people with diabetes and hyperlipidemias who develop chronic kidney disease, medications might include insulin, oral hypoglycemic drugs, antihypertensives, diuretics, lipid-lowering medications, and phosphate binders. Review the nutrition-related side effects of these medications. Describe the ways in which these medications may make it harder for people to maintain nutrition status.
2. Identify the recommended energy, protein, and sodium intakes for a 65-year-old hemodialysis patient who weighs 60 kilograms (use the guidelines shown in Table 28-5). Then, consider the type of diet that would be appropriate for this patient by following these steps:
   - Create a one-day menu that provides appropriate amounts of energy, protein, and sodium for this patient (use the energy and nutrient values in Appendix H or another food composition table).
   - Assuming that the patient's laboratory test results suggest that potassium and phosphorus restrictions are necessary, would the day's intake of these nutrients be within the ranges suggested in Table 28-5? If not, adjust the food list to better match the guidelines.
   - If this patient were to begin peritoneal dialysis, which nutrient recommendations would change? Explain why the diet would be easier to follow than the diet required during hemodialysis.

Nutrition on the Net

For further study of topics covered in this chapter, log on to **www.cengagebrain.com** and search for ISBN 084006845X.

- To search for specific topics related to kidney diseases, dialysis, and kidney transplants, visit these sites:

  Kidney Foundation of Canada: **www.kidney.ca**

  National Institute of Diabetes and Digestive and Kidney Diseases: **www2.niddk.nih.gov**

  National Kidney Foundation: **www.kidney.org**
- To find materials for patients with kidney diseases, visit the American Association of Kidney Patients: **www.aakp.org**
- To find more information about kidney stones, visit the Oxalosis and Hyperoxaluria Foundation: **www.ohf.org**
- To see photographs of kidney stones, visit the website of the Louis C. Herring and Company Laboratory: **www.herringlab.com**

Nutrition Assessment Checklist for People with Kidney Diseases

Medical History

Check the medical record to determine:

- Degree of kidney function
- Cause of the nephrotic syndrome or kidney disease
- Type of dialysis, if appropriate
- Whether the patient has received a kidney transplant
- Type of kidney stone

Review the medical record for complications that may alter nutritional needs:

- Anemia
- Diabetes mellitus
- Edema or oliguria
- Hyperlipidemia
- Hypertension
- Metabolic stress or infection
- Protein-energy malnutrition

Medications

Assess risks for medication-related malnutrition related to:

- Long-term use of medications
- Multiple medication use, especially if medications affect nutrition status

For all patients with kidney diseases, note:

- Whether medications or supplements contain electrolytes that must be controlled
- Use of drugs or herbs that may be toxic to the kidneys

Dietary Intake

For patients with the nephrotic syndrome, kidney disease, or kidney transplant, assess intakes of:

- Protein and energy
- Fluid
- Vitamins, especially vitamin D
- Minerals, especially calcium, phosphorus, iron, and electrolytes

For patients with kidney stones or a history of kidney stones:

- Stress the need to drink plenty of fluids throughout the day.
- Assess intake of calcium, oxalate, sodium, protein, purines, or vitamin C, as appropriate for the type of stone.

Anthropometric Data

Take accurate baseline height and weight measurements. Keep in mind that:

- Fluid retention due to the nephrotic syndrome or kidney failure can mask malnutrition.
- For dialysis patients, the weight measured immediately after the dialysis treatment (called the *dry weight*) most accurately reflects the person's true weight. Rapid weight gain between dialysis treatments reflects fluid retention. If fluid retention is excessive, review fluid intake to determine if the patient understands and is complying with diet recommendations.

Laboratory Tests

Note that serum protein levels are often low in patients with nephrotic syndrome or advanced kidney disease. Review the following laboratory test results to assess the degree of kidney function and response to treatments:

- Blood urea nitrogen (BUN)
- Creatinine
- Glomerular filtration rate (GFR)
- Serum electrolytes
- Urinary protein

Check laboratory test results for complications associated with kidney disease, including:

- Anemia
- Hyperglycemia
- Hyperlipidemia
- Hyperparathyroidism (related to bone disease)

Physical Signs

For patients with nephrotic syndrome or kidney disease, look for physical signs of:

- Bone disease
- Dehydration or fluid retention
- Hyperkalemia
- Iron deficiency
- Uremia

References

1. G. B. Appel, Glomerular disorders and nephrotic syndromes, in L. Goldman and D. Ausiello, eds., *Cecil Medicine* (Philadelphia: Saunders, 2008), pp. 866–876.
2. B. R. Don and G. A. Kaysen, Proteinuria and nephrotic syndrome, in R. W. Schrier, ed., *Renal and Electrolyte Disorders* (Philadelphia: Lippincott Williams & Wilkins, 2010), pp. 519–558.
3. Don and Kaysen, 2010.
4. Don and Kaysen, 2010.
5. Appel, 2008; J. A. Charlesworth, D. M. Gracey, and B. A. Pussell, Adult nephrotic syndrome: Non-specific strategies for treatment, *Nephrology* 13 (2008): 45–50.
6. American Dietetic Association, *Nutrition Care Manual* (Chicago: American Dietetic Association, 2011); Don and Kaysen, 2010.
7. Don and Kaysen, 2010.
8. American Dietetic Association, 2011.
9. American Dietetic Association, 2011; G. M. Podda and coauthors, Abnormalities of homocysteine and B vitamins in the nephrotic syndrome, *Thrombosis Research* 120 (2007): 647–652.
10. B. A. Molitoris, Acute kidney injury, in L. Goldman and D. Ausiello, eds., *Cecil Medicine* (Philadelphia: Saunders, 2008), pp. 862–866.
11. R. W. Schrier and C. L. Edelstein, Acute kidney injury: Pathogenesis, diagnosis, and management, in R. W. Schrier, ed., *Renal and Electrolyte Disorders* (Philadelphia: Lippincott Williams & Wilkins, 2010), pp. 325–388.
12. Schrier and Edelstein, 2010.
13. C. E. Alpers, The kidney, in V. Kumar and coeditors, *Robbins and Cotran Pathologic Basis of Disease* (Philadelphia: Saunders, 2010), pp 905–969.
14. American Dietetic Association, 2011.
15. American Dietetic Association, 2011.
16. American Dietetic Association, 2011.
17. Schrier and Edelstein, 2010.
18. M. Chonchol and L. Chan, Chronic kidney disease: Manifestations and pathogenesis, in R. W. Schrier, ed., *Renal and Electrolyte Disorders* (Philadelphia: Lippincott Williams & Wilkins, 2010), pp. 389–425.
19. W. E. Mitch, Chronic kidney disease, in L. Goldman and D. Ausiello, eds., *Cecil Medicine* (Philadelphia: Saunders, 2008), pp. 921–930.
20. J. Coresh and coauthors, Prevalence of chronic kidney disease in the United States, *Journal of the American Medical Association* 298 (2007): 2038–2047.
21. L. A. Stevens, N. Stoycheff, and A. S. Levey, Staging and management of chronic kidney disease, in A. Greenberg, ed., *Primer on Kidney Diseases* (Philadelphia: Saunders, 2009), pp. 436–445.
22. Mitch, 2008.
23. G. T. Obrador, Chronic renal failure and the uremic syndrome, in E. V. Lerma and coeditors, *Current Diagnosis and Treatment: Nephrology and Hypertension* (New York: McGraw-Hill/Lange, 2009), pp. 149–154.
24. Chonchol and Chan, 2010; A. Schieppati, R. Pisoni, and G. Remuzzi, Pathophysiology of chronic kidney disease, in A. Greenberg, ed., *Primer on Kidney Diseases* (Philadelphia: Saunders, 2009), pp. 422–445.
25. J. D. Kopple, Nutrition, diet, and the kidney, in M. E. Shils and coeditors, *Modern Nutrition in Health and Disease* (Baltimore, MD: Lippincott Williams & Wilkins, 2006), pp. 1475–1511.
26. American Dietetic Association, 2011.
27. A. J. Hutchison and A. Vardhan, Peritoneal dialysis, in A. Greenberg, ed., *Primer on Kidney Diseases* (Philadelphia: Saunders, 2009), pp. 459–471.
28. American Dietetic Association, 2011.
29. J. A. Beto and V. K. Bansal, Medical nutrition therapy in chronic kidney failure: Integrating clinical practice guidelines, *Journal of the American Dietetic Association* 104 (2004): 404–409.
30. D. J. Goldstein-Fuchs and C. M. Goeddeke-Merickel, Nutrition and kidney disease, in A. Greenberg, ed., *Primer on Kidney Diseases* (Philadelphia: Saunders, 2009), pp. 478–486.

31. N. Tolkoff-Rubin, Treatment of irreversible renal failure, in L. Goldman and D. Ausiello, eds., *Cecil Medicine* (Philadelphia: Saunders, 2008), pp. 936–947.
32. N. J. Cano and coauthors, Intradialytic parenteral nutrition does not improve survival in malnourished hemodialysis patients: A 2-year multicenter, prospective, randomized study, *Journal of the American Society of Nephrology* 18 (2007): 2583–2591; L. B. Pupim and coauthors, Intradialytic oral nutrition improves protein homeostasis in chronic hemodialysis patients with deranged nutritional status, *Journal of the American Society of Nephrology* 17 (2006): 149–157.
33. Tolkoff-Rubin, 2008.
34. S. Beddhu, Outcome of end-stage renal disease therapies, in A. Greenberg, ed., *Primer on Kidney Diseases* (Philadelphia: Saunders, 2009), pp. 472–477.
35. G. C. Curhan, Nephrolithiasis, in A. Greenberg, ed., *Primer on Kidney Diseases* (Philadelphia: Saunders, 2009), pp. 382–388.
36. L. K. Massey, M. Liebman, and S. A. Kynast-Gales, Ascorbate increases human oxaluria and kidney stone risk, *Journal of Nutrition* 135 (2005): 1673–1677.
37. American Dietetic Association, 2011; Curhan, 2009.
38. American Dietetic Association, 2011.
39. Curhan, 2009.

Dialysis

Although there is no perfect substitute for one's own kidneys, dialysis offers a life-sustaining treatment option for people with chronic kidney disease who develop renal failure. Dialysis can serve as a permanent treatment or as a temporary measure to sustain life until a suitable kidney donor can be found. Dialysis can also restore fluid and electrolyte balances in patients with acute renal failure. Clinicians who routinely work with renal patients should understand how dialysis procedures work. This highlight describes the process of dialysis and outlines the different types of procedures that are available. The accompanying glossary defines the relevant terms.

The Basics of Dialysis

As described in this section, dialysis removes excess fluids and wastes from the blood by employing the processes of **diffusion**, **osmosis**, and **ultrafiltration** (see Figure H28-1). The dialysate, a solution similar in composition to normal blood plasma, is delivered to a compartment beside a **semipermeable membrane**; the person's blood flows along the other side of the membrane. The semipermeable membrane acts like a filter: small molecules such as urea and glucose can pass through microscopic pores in the membrane, whereas large molecules are unable to cross.

In *hemodialysis*, the tiny tubes that carry blood through the dialyzer are made of materials that serve as semipermeable membranes. In *peritoneal dialysis*, the body's peritoneal membrane, rich with blood vessels, is used to filter the blood.

Removal of Solutes

The chemical composition of the dialysate affects the movement of solutes across the semipermeable membrane. When the concentration of a substance is lower in the dialysate than in the blood, the substance—provided it can cross the membrane—will diffuse out of the blood. For example, the goal is to remove as much as possible of the waste product urea from the blood, so the dialysate contains no urea. For many other solutes, the dialysate is adjusted so that only excesses will be removed. Potassium can be removed from the blood, for example, by providing a dialysate that has a lower concentration of potassium than is found in the person's blood. The dialysate must contain some potassium, however; otherwise the blood potassium would fall too low.

The dialysate can also be used to add needed components back into the blood. For a person with acidosis, for example, bases such as bicarbonate are added to the dialysate; the bases then move by diffusion into the blood to alleviate the acidosis.

Removal of Fluid

Because albumin and other plasma proteins are so adept at retaining fluids in blood, osmosis alone is not an efficient process for removing fluid. In hemodialysis, a **pressure gradient** is created between the blood and the dialysate. Most modern dialyzers produce *positive* pressure in the blood compartment and *negative* pressure in the dialysate compartment, establishing a pressure

FIGURE H28-1 Diffusion, Osmosis, and Ultrafiltration

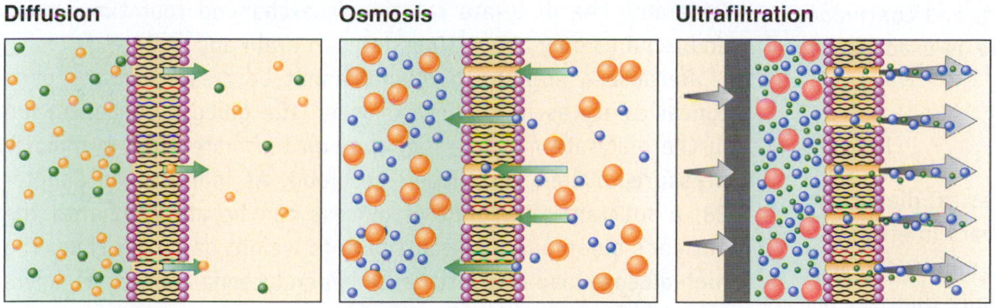

Small molecules (electrolytes and waste products) move from an area of high concentration to an area of low concentration by diffusion.

Water moves from an area of high water concentration to an area of low water concentration. In other words, water moves toward the side where solutes are more concentrated.

Pressure squeezes water and small molecules through the pores of a semipermeable membrane during ultrafiltration.

GLOSSARY

continuous ambulatory peritoneal dialysis (CAPD): the most common method of peritoneal dialysis; involves frequent exchanges of dialysate, which remains in the peritoneal cavity throughout the day.

continuous renal replacement therapy (CRRT): a slow, continuous method of removing solutes and/or fluids from blood by gently pumping blood across a filtration membrane over a prolonged time period.

diffusion: movement of solutes from an area of high concentration to one of low concentration.

hemofiltration: removal of fluid and solutes by pumping blood across a membrane; no osmotic gradients are created during the process.

oncotic pressure: the pressure exerted by fluid on one side of a membrane as a result of osmosis.

osmosis: movement of water across a membrane toward the side where solutes are more concentrated.

peritonitis: inflammation of the peritoneal membrane.

pressure gradient: the change in pressure over a given distance. In dialysis, a pressure gradient is created between the blood and the dialysate.

semipermeable membrane: a membrane that allows some particles to pass through, but not others.

ultrafiltration: removal of fluids and solutes from blood by using pressure to transfer the blood across a semipermeable membrane.

urea kinetic modeling: a method of determining the adequacy of dialysis treatment by calculating the urea clearance from blood.

gradient that "pushes" water (and accompanying solutes) through the pores of the membrane.[1] This process, called ultrafiltration, relies on pumps to establish an appropriate flow rate between the blood and the dialysate.

Evaluation of Dialysis Treatment

A number of methods have been devised for gauging the adequacy of dialysis treatment. The most common method is **urea kinetic modeling**, a technique that evaluates the amount of urea cleared from the blood. The formula used most often is Kt/V, where K is the amount of urea cleared, t is the time spent on dialysis, and V is the blood volume. The value obtained indicates whether the patient has undergone sufficient dialysis; the goal is a Kt/V result of approximately 1.2. Because technical data (such as dialyzer clearance data, blood flow rate, and dialysate flow rate) need to be incorporated into the calculation, the computation is usually done by computer analysis. Current treatment guidelines recommend that hemodialysis adequacy be evaluated at least monthly, or more often if problems develop or patients are noncompliant.[2]

Types of Dialysis

Three approaches are currently used to remove fluids and wastes from the body: hemodialysis, peritoneal dialysis, and continuous renal replacement therapy. The latter procedure is used only to treat acute renal failure.

Hemodialysis

As described previously, hemodialysis utilizes a dialyzer to cleanse the patient's blood. Although dialyzers vary in efficiency, the treatment usually lasts 3 to 4 hours and is required at least 3 times weekly. Other options include short daily dialysis, performed for about 2 hours per day, and daily nocturnal hemodialysis, in which dialysis is done at home overnight while the patient is sleeping. Although some studies have reported improved outcomes in patients who undergo daily dialysis, these approaches have not been widely adopted.[3] Note that most patients must visit dialysis centers to obtain treatment; few patients have access to a dialysis machine at home.

Although lifesaving, hemodialysis is associated with a substantial number of complications.[4] Problems at the vascular access site include infections and blood clotting. Hypotension can develop while blood is circulated through the dialyzer. Muscle cramping often occurs during the procedure, especially in the hands, legs, and feet. Blood losses can worsen anemia, which is already severe in two-thirds of patients beginning hemodialysis treatment.[5] Patients may also experience headaches, weakness, nausea, vomiting, restlessness, and agitation. Many patients experience extreme fatigue after a hemodialysis treatment, and some may require rest or sleep.

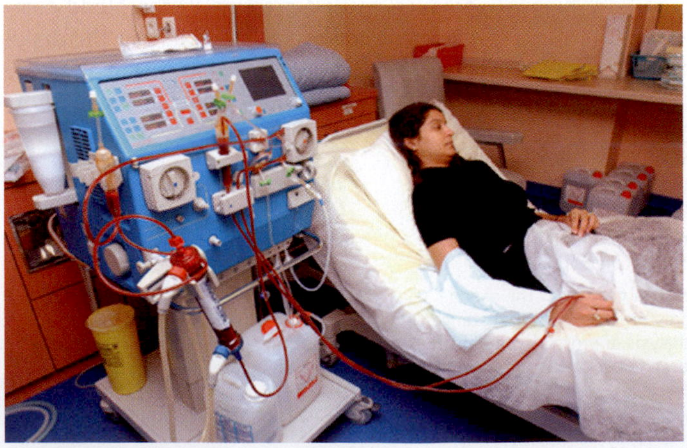

Hpa-Voisin/Photo Researchers

During hemodialysis, blood passes through a dialyzer where wastes are extracted, and the cleansed blood is returned to the body.

Peritoneal Dialysis

In peritoneal dialysis, the peritoneal membrane surrounding the abdominal organs serves as a semipermeable membrane. The dialysate is infused into a catheter that empties into the peritoneal space—the space within the abdomen near the intestines (see Figure H28-2). In the most common procedure, **continuous ambulatory peritoneal dialysis (CAPD)**, the dialysate remains in the peritoneal cavity for 4 to 6 hours, after which it is drained and replaced with fresh dialysate (about 2 to 3 liters in adults). Generally, the dialysate solution is exchanged four times daily and requires only about 30 minutes to drain and replace.

Because a pressure gradient cannot be created in the peritoneal cavity as it can in a dialyzer, the glucose concentration in the dialysate must be high enough to create enough **oncotic pressure** to draw fluid from the blood. As indicated in Chapter 28, a substantial amount of glucose can be absorbed into the patient's blood and may contribute to weight gain over time. The high glucose load may also cause hyperglycemia and hypertriglyceridemia in some patients.

Peritoneal dialysis offers a number of advantages over hemodialysis: vascular access is not required, dietary restrictions are fewer, and the procedure can be scheduled when convenient. The most common complication is infection, which can occur at the catheter site or within the peritoneal cavity (**peritonitis**). Other problems that may arise include blood clotting in the catheter, catheter migration, and abdominal hernia due to the dialysate volume.

Continuous Renal Replacement Therapy

In people with acute kidney injury, **continuous renal replacement therapy (CRRT)** removes fluids and wastes. CRRT utilizes the process of **hemofiltration**, in which blood is gently pumped across a filtration membrane over a prolonged time period. (This process differs from dialysis treatments that rely on the diffusion of wastes across a membrane into the dialysate.) Either a pump

FIGURE H28-2 Peritoneal Dialysis

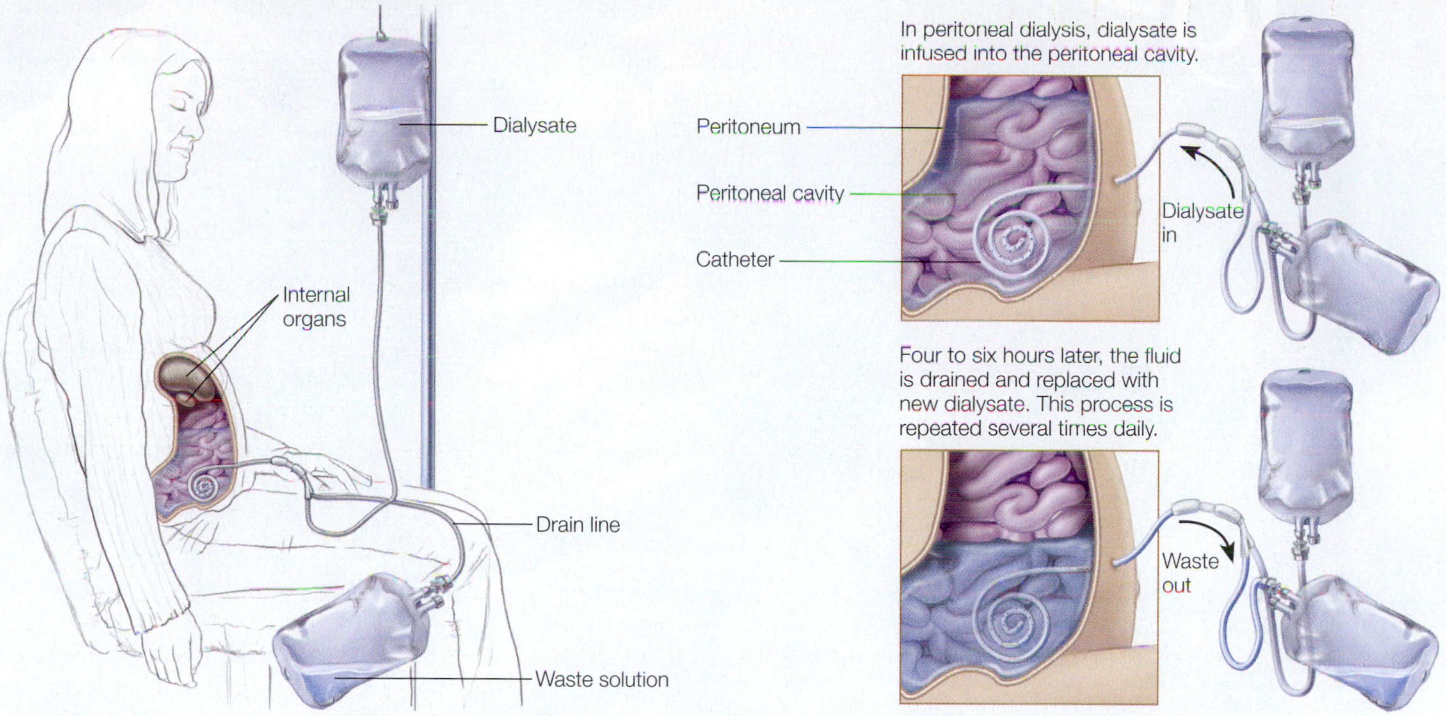
Dialysate
Internal organs
Drain line
Waste solution
In peritoneal dialysis, dialysate is infused into the peritoneal cavity.
Peritoneum
Peritoneal cavity
Catheter
Dialysate in
Four to six hours later, the fluid is drained and replaced with new dialysate. This process is repeated several times daily.
Waste out

or the patient's own blood pressure moves the blood across the membrane. The procedure can be used to remove fluids, solutes, or both. Some patients require fluid replacement during the procedure to maintain adequate blood volume, so hydration status must be closely monitored.

The use of CRRT is advantageous in acute care situations because it corrects imbalances without causing sudden shifts in blood volume, which are poorly tolerated in acute care patients. In addition, replacement fluids can include parenteral feedings without upsetting fluid balance. Complications include clotting problems, damage to arteries, and inadequate blood flow rates in hypotensive patients.

Dialysis and CRRT help to remove the wastes and fluids that are normally removed by healthy kidneys. Although these procedures cannot restore the kidneys' hormonal functions, they provide a lifesaving means of alleviating symptoms of uremia, hypertension, and edema.

References

1. J. Z. Kallenbach and coauthors, *Review of Hemodialysis for Nurses and Dialysis Personnel* (St. Louis: Elsevier/Mosby, 2005), pp. 61–70.
2. National Kidney Foundation, K/DOQI clinical practice guidelines for hemodialysis adequacy: Update 2006, www.kidney.org/PROFESSIONALS/kdoqi/guideline_upHD_PD_VA/hd_guide2.htm, accessed December 23, 2010.
3. K. L. Johansen and coauthors, Survival and hospitalization among patients using nocturnal and short daily compared to conventional hemodialysis: A USRDS study, *Kidney International* 76 (2009): 984–990.
4. N. Tolkoff-Rubin, Treatment of irreversible renal failure, in L. Goldman and D. Ausiello, eds., *Cecil Medicine* (Philadelphia: Saunders, 2008), pp. 936–947.
5. Tolkoff-Rubin, 2008.

CHAPTER

29

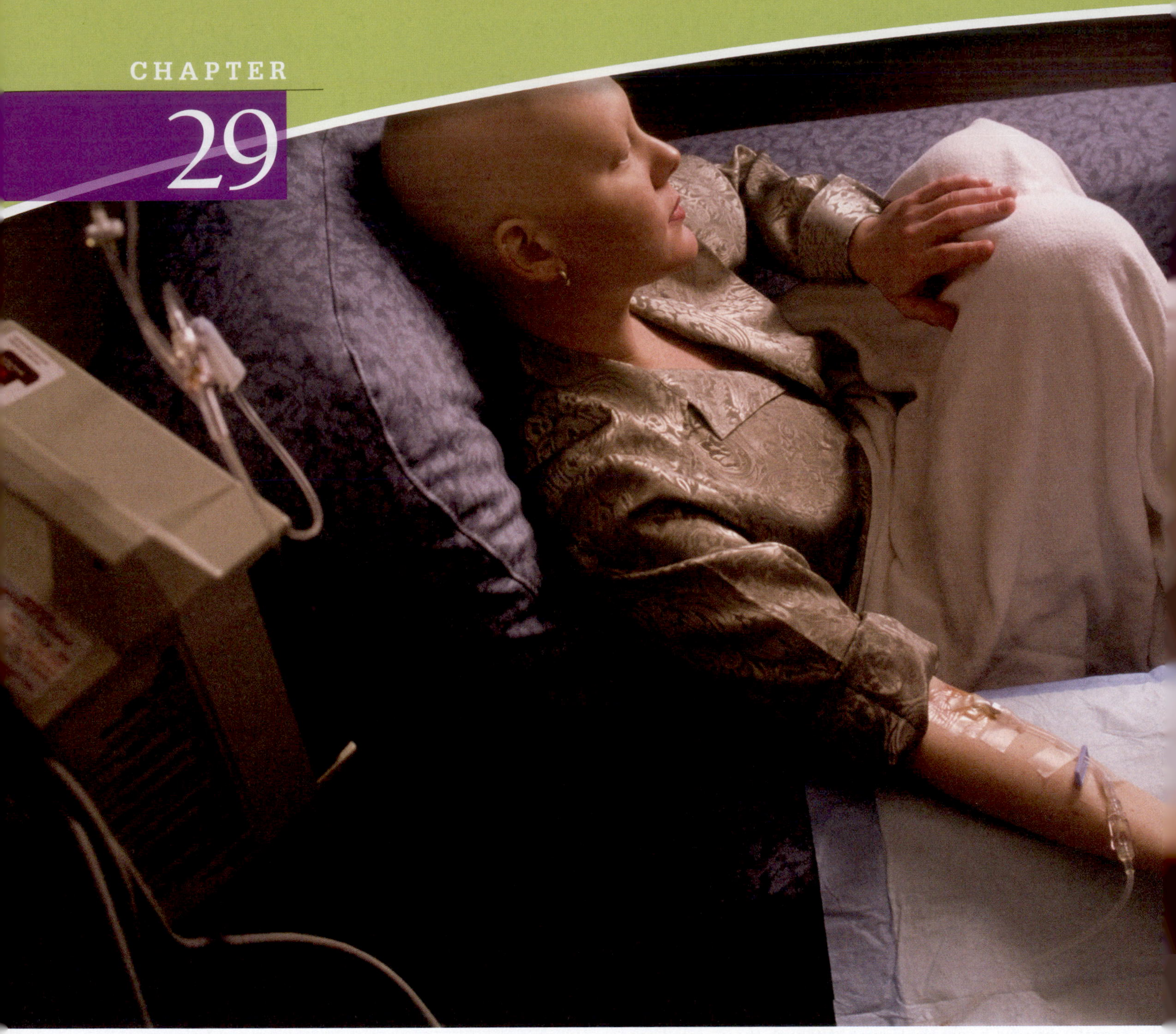

Kevin Laubacher/Getty Images

Nutrition in the Clinical Setting

A diagnosis of cancer or HIV infection can be devastating. Patients will likely expect an ever-worsening course of illness and, possibly, death. Medical management soon becomes an ever-present burden, and treatments are often unpleasant. For both illnesses, however, extraordinary therapeutic advances have been made. Treatment options have expanded, and patients have benefited from vast improvements in quality of life. The health practitioner's knowledge and empathy are the patient's most important resources—and an important source of hope.

Throughout this chapter, the CourseMate icon indicates an opportunity for online self-study, linking you to activities to increase your understanding of chapter concepts. **www.cengagebrain.com** (search for ISBN 084006845X)

Cancer and HIV Infection

CHAPTER OUTLINE

Although **cancers** and **human immunodeficiency virus** (HIV) infections are distinct disorders, from a nutritional standpoint, they share some similarities. Both disorders have debilitating effects that influence nutritional needs, and both can lead to severe wasting in advanced cases. These illnesses require nutrition therapy that is highly individualized based on the symptoms manifested and the organ systems involved.

Cancer

Cancer, the growth of **malignant** tissue, is the second most common cause of death in the United States, ranking just below cardiovascular disease. Cancer is not a single disorder, however; there are many different kinds of malignant growths. The different types of cancer have different characteristics, occur in different locations in the body, take different courses, and require different treatments. ♦ Whereas an isolated, nonspreading type of skin cancer may be removed in a physician's office with no effect on nutrition status, advanced cancers—especially those of the gastrointestinal (GI) tract and pancreas—can seriously impair nutrition status. In the United States, the most common cancers are breast cancer (in women), prostate cancer (in men), lung cancer, and colorectal cancers.[1]

How Cancer Develops

The development of cancer, called **carcinogenesis**, often proceeds slowly and continues for several decades. A cancer arises from mutations in the genes that control cell division in a single cell.[2] These mutations may promote cellular growth, interfere with growth restraint, or prevent cellular death. The affected cell thereby loses its built-in capacity for halting cell division and produces daughter cells with the same genetic defects. As the abnormal mass of cells, called a **tumor** (or *neoplasm*), grows, ♦ a network of blood vessels forms to supply the tumor with the nutrients it needs to support its growth. The tumor can disrupt the functioning of the normal tissue around it, and some tumor cells may **metastasize**, spreading to other regions in the body. Figure 29-1 illustrates the steps in cancer development. In leukemia (cancer of the white blood cells), the abnormal cells do not form a tumor; they accumulate in the blood and other tissues.

The reasons cancers develop are numerous and varied. Vulnerability to cancer is sometimes inherited, as when a person is born with a genetic defect that alters DNA structure, function, or repair. Certain metabolic processes may initiate carcinogenesis, as when phagocytes (immune cells) produce oxidants that cause DNA damage, or when chronic inflammation increases the rate of cell division and the risk of a damaging mutation. More often, cancers are caused by interactions between a person's genes and the environment. Exposure to cancer-causing substances, or **carcinogens**, may induce genetic mutations that lead to cancer; other substances may stimulate division or proliferation of the altered cells. Table 29-1 provides examples of environmental factors that increase cancer risk.

♦ Cancers are classified by the tissues or cells from which they develop:
- *Adenocarcinomas* (ADD-eh-no-CAR-sih-NO-muz) arise from glandular tissues.
- *Carcinomas* (CAR-sih-NO-muz) arise from epithelial tissues.
- *Leukemias* (loo-KEY-mee-uz) arise from white blood cell precursors.
- *Lymphomas* (lim-FOE-muz) arise from lymphoid tissue.
- *Melanomas* (MEL-ah-NO-muz) arise from pigmented skin cells.
- *Myelomas* (MY-ah-LOE-muz) arise from plasma cells in the bone marrow.
- *Sarcomas* (sar-KO-muz) arise from connective tissues, such as muscle or bone.

♦ An abnormal mass of cells that is noncancerous is called a *benign* tumor.

cancers: diseases characterized by the uncontrolled growth of a group of cells, which can destroy adjacent tissues and spread to other areas of the body via lymph or blood.

human immunodeficiency virus (HIV): the virus that causes acquired immune deficiency syndrome (AIDS). HIV destroys immune cells and progressively impedes the body's ability to fight infections and certain cancers.

malignant (ma-LIG-nent): describes a cancerous cell or tumor, which can injure healthy tissue and spread cancer to other regions of the body.

carcinogenesis (CAR-sin-oh-JEN-eh-sis): the process of cancer development.

tumor: an abnormal tissue mass that has no physiological function; also called a *neoplasm* (NEE-oh-plazm).

metastasize (meh-TAS-tah-size): to spread from one part of the body to another; refers to cancer cells.

carcinogens (CAR-sin-oh-jenz or car-SIN-oh-jenz): substances that can cause cancer (the adjective is *carcinogenic*).

FIGURE 29-1 Cancer Development

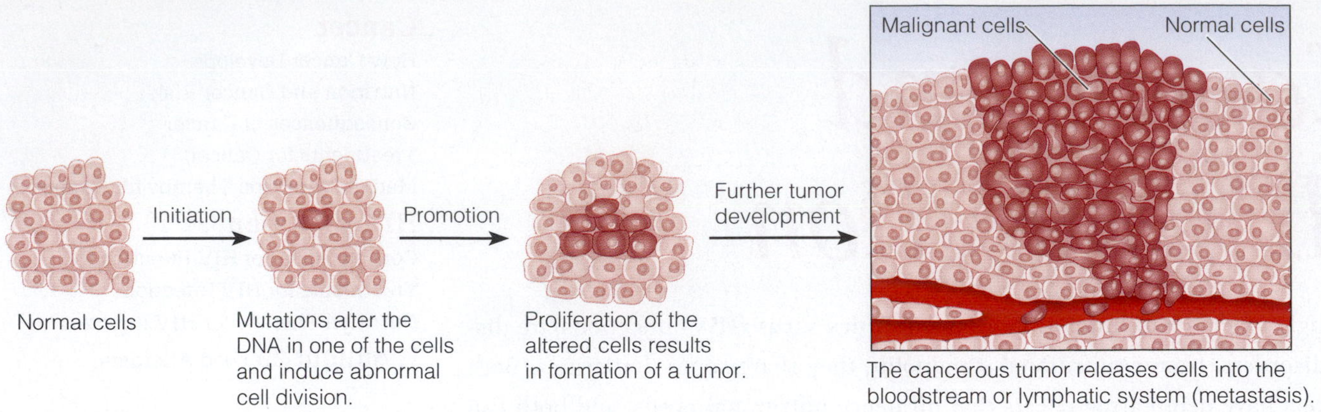
Malignant cells
Normal cells
Normal cells
Initiation
Mutations alter the DNA in one of the cells and induce abnormal cell division.
Promotion
Proliferation of the altered cells results in formation of a tumor.
Further tumor development
The cancerous tumor releases cells into the bloodstream or lymphatic system (metastasis).

Nutrition and Cancer Risk

Like other environmental factors, diet and lifestyle strongly influence cancer risk. Certain food components may directly damage DNA, alter the metabolism of carcinogens by liver enzymes, or inhibit the formation of carcinogens in the body.[3] In addition, energy balance and growth rates affect the rate of cell division and consequently influence the rates at which mutations form and are replicated. Table 29-2 lists examples of nutrition-related factors that may increase or decrease the risk of developing cancer.

Nutrition and Increased Cancer Risk

As shown in Table 29-2, obesity is a risk factor for a number of different cancers, including some relatively common cancers such as colon cancer and postmenopausal breast cancer. Obesity increases cancer risk, in part, by altering levels of hormones that influence cell growth, such as the sex hormones, insulin, and several kinds of growth factors.[4] For example, in the case of breast cancer in postmenopausal women, the hormone estrogen is likely involved: obese women have higher estrogen levels than lean women do because

TABLE 29-1 Environmental Factors That Increase Cancer Risk

Environmental Factors	Cancer Sites
Aflatoxins (toxins in moldy peanuts or grains)	Liver
Asbestos[a]	Lung, pleura, peritoneum
Chromium (hexavalent) compounds	Nasal cavity, lung
Estrogen-progesterone replacement therapy	Breast
Immunosuppressive medications	Lymphoid tissues, liver
Infection with *Helicobacter pylori*	Stomach
Infection with hepatitis B and hepatitis C viruses	Liver
Infection with human papillomavirus (HPV)	Cervix
Ionizing radiation (X-rays, radioactive isotopes, and other sources)	White blood cells (leukemia), esophagus, stomach, colon, thyroid, lung, bladder, breast
Tobacco[b]	Nasal cavity, lung, mouth, pharynx, larynx, esophagus, stomach, colon, rectum, liver, pancreas, kidney, renal pelvis, bladder
Ultraviolet radiation (sun exposure)	Skin

[a]Risk is greatly increased in cigarette smokers.
[b]A combined exposure to tobacco and alcohol multiplies the risks of developing cancers of the oral cavity, pharynx, larynx, and esophagus.
SOURCES: M. J. Thun, Epidemiology of cancer, in L. Goldman and D. Ausiello, eds., *Cecil Medicine* (Philadelphia: Saunders, 2008), pp. 1335–1340; World Cancer Research Fund/American Institute for Cancer Research, *Food, Nutrition, Physical Activity, and the Prevention of Cancer: A Global Perspective* (Washington, DC: American Institute for Cancer Research, 2007), pp. 157–171.

TABLE 29-2 Nutrition-Related Factors That Influence Cancer Risk

Nutrition-Related Factors[a]	Cancer Sites
Factors that may increase cancer risk:	
Obesity	Esophagus, colon, rectum, pancreas, gallbladder, kidney, breast (postmenopausal), endometrium
Red meat, processed meats	Colon, rectum
Salted and salt-preserved foods	Stomach
Beta-carotene supplements	Lung[b]
High-calcium diets (over 1500 mg daily)	Prostate
Alcohol[c]	Mouth, pharynx, larynx, esophagus, colon, rectum, liver, breast
Low level of physical activity[d]	Colon, breast (postmenopausal), endometrium
Factors that may decrease cancer risk:	
Fruits and nonstarchy vegetables	Lung, mouth, pharynx, larynx, esophagus, stomach
Carotenoid-containing foods	Lung, mouth, pharynx, larynx, esophagus
Tomato products	Prostate
Allium vegetables (onion, garlic)	Stomach, colon, rectum
Vitamin C–containing foods	Esophagus
Folate-containing foods	Pancreas
Fiber-containing foods	Colon, rectum
Milk and calcium supplements	Colon, rectum
High level of physical activity[d]	Colon, breast (postmenopausal), endometrium

[a]Altered cancer risk is associated with high intakes of the dietary substances listed.
[b]Cancer risk is increased in tobacco smokers and may not apply to other groups.
[c]A combined exposure to alcohol and tobacco multiplies the risks of developing cancers of the oral cavity, pharynx, larynx, and esophagus.
[d]Physical activity may influence cancer risk by altering body fatness, intestinal transit time, insulin sensitivity, hormone levels, enzyme activities, and immune responses.
SOURCE: World Cancer Research Fund/American Institute for Cancer Research, *Food, Nutrition, Physical Activity, and the Prevention of Cancer: A Global Perspective* (Washington, DC: American Institute for Cancer Research, 2007).

adipose tissue is the primary source of estrogen after menopause. The increase in circulating estrogen may create an environment that encourages carcinogenesis in breast tissue.

Alcohol consumption correlates strongly with cancers of the head and neck, colon, rectum, and breast. For head and neck cancers, the risk is multiplied when alcohol drinkers also smoke tobacco.[5] In addition, alcohol abuse may damage the liver and instigate the development of liver cancer. These findings illustrate why the potential benefits of moderate alcohol consumption on cardiovascular disease risk must be weighed against the potential dangers.

Food preparation methods are responsible for producing certain types of carcinogens. Cooking meat, poultry, and fish at high temperatures (frying, broiling) causes the amino acids and creatine in these foods to react together and form carcinogens.[6]* Carcinogens also accompany the smoke that adheres to food during grilling, and they are present in the charred surfaces of grilled meat and fish.** However, the cancer risk from eating such foods is unclear because the biological actions of these carcinogens are modulated by other dietary components, including compounds in vegetables and other plant foods. In several population studies, consumption of well-cooked meats was linked to cancers of the stomach, colon, breast, and prostate.[7] ♦

Nutrition and Decreased Cancer Risk The consumption of fruits and vegetables may provide some benefits in protecting against the development of cancer

*These carcinogens are *heterocyclic amines*.
**These carcinogens are *polycyclic aromatic hydrocarbons*.

♦ To minimize carcinogen formation during cooking:
- Marinate meats before cooking.
- Use lower-heat options such as roasting, stewing, or microwaving.
- Choose lean meats for grilling, and take care not to blacken surfaces.
- To reduce smoke formation, prevent fat from dripping onto the heat source.

Cruciferous vegetables, such as cauliflower, broccoli, and brussels sprouts, contain nutrients and phytochemicals that inhibit cancer development.

(see Table 29-2). Fruits and vegetables contain both nutrients and phytochemicals with antioxidant activity, and these substances may prevent or reduce the oxidative reactions in cells that cause DNA damage. Phytochemicals may also help to inhibit carcinogen production in the body, enhance immune functions that protect against cancer development, or promote enzyme reactions that inactivate carcinogens.[8] The B vitamin folate, which is provided by certain fruits and vegetables, plays roles in DNA synthesis and repair; thus, inadequate folate intakes may allow DNA damage to accumulate. Fruits and vegetables also contribute dietary fiber, which may help to protect against colon and rectal cancers by diluting potential carcinogens in fecal matter and accelerating their removal from the GI tract. Table 29-3 summarizes the dietary and lifestyle practices that may help to reduce the risk of developing cancer.

Consequences of Cancer

Once cancer develops, its consequences depend on the location of the tumor, its severity, and the treatment. The complications that develop are often due to the tumor's impingement on surrounding tissues. Nonspecific effects of cancer include **anorexia**, lethargy, weight loss, night sweats, and fever.[9] During the early stages, many cancers produce no symptoms, and the person may be unaware of the threat to health.

Wasting Associated with Cancer

Anorexia, muscle wasting, weight loss, anemia, and fatigue typify **cancer cachexia**, a condition of severe malnutrition that develops in up to 50 percent of cancer patients.[10] Weight loss is often evident at the time that cancer is diagnosed, and severe malnutrition, typically seen in the later

TABLE 29-3 Recommendations for Reducing Cancer Risk

Maintain a healthy body weight throughout life.
• Be as lean as possible within the normal range of body weight for your height. • Avoid weight gain and increases in waist circumference throughout adulthood.
Be physically active as part of everyday life.
• For adults: engage in moderate to vigorous physical activity for at least 30 minutes on at least 5 days per week; 45 to 60 minutes is preferable. • For children and adolescents: engage in moderate to vigorous activity for 60 minutes on at least 5 days of the week. • Limit sedentary habits such as watching television.
Choose a healthy diet that emphasizes plant sources.
• Limit consumption of energy-dense foods (>225 kcal per 100 g food) and sugary drinks that contribute to weight gain. • Consume relatively unprocessed grains and/or legumes with every meal. Choose whole-grain products instead of processed (refined) grains. • Consume five or more servings of nonstarchy vegetables and fruits every day. • Limit consumption of red meats (such as beef, pork, and lamb) and processed meats (such as those preserved by smoking, curing, or salting). • Avoid salt-preserved, salted, and salty foods. • Avoid moldy grains and legumes.
Limit consumption of alcoholic beverages.
• For women: drink no more than one drink daily. • For men: drink no more than two drinks daily.
Aim to meet nutritional needs through the diet.
• Obtain necessary nutrients from the diet. Dietary supplements are not recommended for cancer prevention, and they may have unexpected adverse effects.
Avoid using tobacco in any form.

SOURCES: World Cancer Research Fund/American Institute for Cancer Research, *Food, Nutrition, Physical Activity, and the Prevention of Cancer: A Global Perspective* (Washington, DC: American Institute for Cancer Research, 2007); L. H. Kushi and coauthors, American Cancer Society guidelines on nutrition and physical activity for cancer prevention: Reducing the risk of cancer with healthy food choices and physical activity, *CA: A Cancer Journal for Clinicians* 56 (2006): 254–281.

anorexia: lack of appetite.

cancer cachexia (ka-KEK-see-ah): a wasting syndrome associated with cancer that is characterized by anorexia, muscle wasting, weight loss, and fatigue.

stages of cancer, is the ultimate cause of death in many cases. Without adequate energy and nutrients, the body is poorly equipped to maintain organ function, support immune defenses, and mend damaged tissues. An involuntary weight loss of more than 10 percent, which indicates significant malnutrition, is cause for concern.[11] Care must be taken not to overlook unintentional weight loss in patients who are overweight or obese.

Many factors play a role in the wasting associated with cancer. Cytokines, ♦ released by both tumor cells and immune cells, induce a catabolic state. The combined effects of a poor appetite, accelerated and abnormal metabolism, and the diversion of nutrients to support tumor growth result in a lower supply of energy and nutrients at a time when demands are high. Appetite and food intake are further disturbed by the effects of treatments and medications prescribed for cancer patients. Unlike in starvation, nutrition intervention alone is unable to reverse cachexia.[12]

♦ The cytokines that induce cachexia include tumor necrosis factor-α, interleukin-1, interleukin-6, and γ-interferon.

Metabolic Changes The metabolic changes that arise in cancer exacerbate the wasting described in the previous section.[13] Cancer patients exhibit an increased rate of protein turnover, but reduced muscle protein synthesis. ♦ Muscle contributes amino acids for gluconeogenesis (glucose production), further depleting the body's supply of protein. Triglyceride breakdown increases, elevating serum lipids. Many patients develop insulin resistance. These metabolic abnormalities help to explain why people with cancer fail to regain lean tissue or maintain healthy body weights even when they are consuming adequate energy and nutrients.

♦ Reminder: *Protein turnover* refers to the continuous degradation and synthesis of the body's proteins.

Anorexia and Reduced Food Intake Anorexia is a major contributor to the wasting associated with cancer. Some factors that contribute to anorexia or otherwise reduce food intake include:

- *Chronic nausea and early satiety.* People with cancer frequently experience nausea and a premature feeling of fullness after eating small amounts of food.
- *Fatigue.* People with cancer often tire easily and lack the energy to prepare and eat meals. Once cachexia develops, these tasks become even more difficult.
- *Pain.* People in pain may have little interest in eating, particularly if eating makes the pain worse.
- *Mental stress.* A cancer diagnosis can cause distress, anxiety, and depression, all of which may reduce appetite. Facing and undergoing cancer treatments induces additional psychological stress.
- *Gastrointestinal obstructions.* A tumor may partially or completely obstruct a portion of the GI tract, causing complications such as nausea and vomiting, early satiety, delayed gastric emptying, and bacterial overgrowth. Some patients with obstructions are unable to tolerate oral diets.
- *Effects of cancer therapies.* Chemotherapy and radiation treatments for cancer frequently have side effects that make food consumption difficult, such as nausea, vomiting, dry mouth, altered taste perceptions, food aversions, mouth sores, inflammation of the mouth and esophagus, difficulty swallowing, abdominal pain or discomfort, diarrhea, and constipation.

Treatments for Cancer

The primary medical treatments for cancer—surgery, chemotherapy, radiation therapy, or any combination of the three—aim to remove cancer cells, prevent further tumor growth, and alleviate symptoms.[14] The likelihood of effective treatment is highest with early detection and intervention. Because treatment decisions are difficult and cancer therapies have considerable side effects, patients rely on health care providers to help them make informed decisions.

Surgery Surgery is performed to remove tumors, determine the extent of cancer, and protect nearby tissues. Often, surgery must be followed by other cancer treatments to prevent growth of new tumors. The acute metabolic stress caused by surgery raises protein and energy needs and can exacerbate wasting. Surgery

♦ One drug that inhibits cell division is *methotrexate*, which closely resembles the B vitamin folate (see Figure 19-1 on p. 625). Folate is required for cell division because it is needed for DNA synthesis. Methotrexate works by blocking activity of the enzyme that converts folate to its active form.

chemotherapy: the use of drugs to arrest or destroy cancer cells; these drugs are called *antineoplastic agents*.

neutropenia: a low white blood cell (neutrophil) count, which increases susceptibility to infection.

radiation therapy: the use of X-rays, gamma rays, or atomic particles to destroy cancer cells.

radiation enteritis: inflammation of intestinal tissue caused by radiation therapy.

hematopoietic stem cell transplantation: transplantation of the stem cells that produce red blood cells and white blood cells; the stem cells are obtained from the bone marrow (*bone marrow transplantation*) or circulating blood.

tissue rejection: destruction of donor tissue by the recipient's immune system, which recognizes the donor cells as foreign.

♦ Reminder: Reactive oxygen species and their effects on cells were described in Highlight 11.

TABLE 29-4 Nutrition-Related Side Effects of Cancer Surgeries

Head and Neck Surgeries
Aspiration
Dry or sore mouth
Reduced chewing or swallowing ability
Reduced sense of taste or smell
Esophageal Resection
Acid reflux
Altered gastric motility
Reduced swallowing ability
Gastric Resection
Dumping syndrome
Early satiety
Inadequate gastric acid secretion
Malabsorption of iron, folate, and vitamin B_{12}
Intestinal Resection
Bile insufficiency
Diarrhea
Fluid and electrolyte imbalances
General malabsorption
Pancreatic Resection
Diabetes mellitus
General malabsorption

also contributes to pain, fatigue, and anorexia, all of which can reduce food intake at a time when nutritional needs are substantial. Blood loss contributes to nutrient losses and further exacerbates malnutrition. Some surgeries can have long-term effects on nutrition status (see Table 29-4).

Chemotherapy **Chemotherapy** relies on the use of drugs to treat cancer, and is used to inhibit tumor growth, shrink tumors before surgery, and prevent or eradicate metastasis. Some cancer drugs interfere with the process of cell division; ♦ others sterilize cells that are in a resting phase and are not actively dividing. Ideally, chemotherapy would wipe out cancer cells without destroying healthy ones. Unfortunately, most of these drugs have toxic effects on normal cells as well and are especially damaging to rapidly dividing cells, such as those of the GI tract, skin, and bone marrow. The bone marrow damage can suppress the production of red blood cells (causing anemia) and white blood cells (causing **neutropenia**). Some of the newer drugs are able to target properties specific to cancer cells and are better tolerated by the body's tissues. Table 29-5 describes some nutrition-related side effects that may result from chemotherapy.

Radiation Therapy **Radiation therapy** treats cancer by bombarding cancer cells with X-rays, gamma rays, or various atomic particles. These treatments generate reactive forms of oxygen, such as superoxide and hydroxyl radicals, ♦ which can damage cellular DNA and cause cell death. Newer techniques are able to focus radiation directly at tumors and minimize damage to nearby tissues. An advantage of radiation therapy over surgery is that it can shrink tumors while preserving organ structure and function. Compared with chemotherapy, radiation therapy is better able to target specific regions of the body, rather than involving all body cells. Nonetheless, radiation therapy can damage healthy tissues and sometimes has long-term detrimental effects on nutrition status. Radiation to the head and neck area can damage the salivary glands and taste buds, causing inflammation, dry mouth, and a reduced sense of taste; in severe cases, the damage may be permanent. Radiation treatment in the lower abdominal area can cause **radiation enteritis**, an inflammatory condition of the small intestine that causes nausea, vomiting, and diarrhea; the condition may persist for months or years and lead to chronic malabsorption in some individuals. Table 29-5 includes additional nutrition-related side effects of radiation treatment.

Hematopoietic Stem Cell Transplantation **Hematopoietic stem cell transplantation** replaces the blood-forming stem cells that have been destroyed by high-dose chemotherapy or radiation therapy. These procedures may be used to treat leukemia, lymphomas, and multiple myeloma.[15] If possible, stem cells are collected from the patient before chemotherapy or radiation treatment begins so that it is not necessary to find a separate donor. If another person's cells are used, the patient must take immunosuppressant drugs to prevent **tissue rejection**.

The treatments required for stem cell transplantation can have a substantial impact on food intake and nutrition status. The high-dose chemotherapy or radiation therapy preceding the transplant and the immunosuppressant drugs often required afterward can impair immune function substantially and increase the risk of foodborne illness. Other common complications include anorexia, nausea, vomiting, dry mouth, altered taste sensations, inflamed mucous membranes, malabsorption, and diarrhea. Patients are often unable to consume adequate food during or after the procedures and usually require nutrition support.

Biological Therapies Newer therapies for cancer include the use of biological molecules that stimulate immune responses against cancer cells (also called *immunotherapy*). These substances include antibodies, cytokines, and other proteins that strengthen the body's immune defenses, enable the destruction of cancer cells, or interfere with cancer development in some way. Although side effects vary, many of these treatments can cause anorexia, GI symptoms, and general discomfort, reducing a person's ability or desire to consume adequate amounts of food.

TABLE 29-5 Nutrition-Related Side Effects of Chemotherapy and Radiation Therapy

	Reduced Nutrient Intake	Accelerated Nutrient Losses	Altered Metabolism
Chemotherapy	Abdominal pain Anorexia Mouth ulcers Nausea and vomiting Reduced taste sensation	Diarrhea Gastrointestinal inflammation Malabsorption Vomiting	Anemia, neutropenia Fluid and electrolyte imbalances as a consequence of vomiting, diarrhea, or malabsorption Hyperglycemia Interference with vitamins or body compounds Negative nitrogen and micronutrient balances Secondary effects of malnutrition, infection, or inflammation
Radiation therapy	Anorexia Damage to teeth, jaws, or salivary glands Dysphagia Esophagitis Mouth sores Nausea and vomiting Reduced salivary secretions Reduced taste sensation	Blood loss from intestine and bladder Diarrhea Fistulas Intestinal obstructions Malabsorption Radiation enteritis Vomiting	Fluid and electrolyte imbalances as a consequence of vomiting, diarrhea, or malabsorption Secondary effects of malnutrition, infection, or inflammation

Medications to Combat Anorexia and Wasting To help cancer patients combat anorexia, medications may be prescribed to stimulate the appetite and promote weight gain. Examples include megestrol acetate (Megace), a synthetic compound similar in structure to the hormone progesterone, and dronabinol (Marinol), which resembles the psychoactive ingredient in marijuana and stimulates the appetite at doses that have minimal mental effects. Under investigation are medications that promote muscle protein synthesis, induce the secretion of growth hormone or growth factors, or inhibit some pro-inflammatory cytokines (which have catabolic effects).[16]

Alternative Therapies Many patients turn to *complementary and alternative medicine (CAM)* ♦ to assist them in their fight against cancer. Patients may turn to CAM because they wish to gain more control over treatment or because they are concerned about the effectiveness of conventional approaches. Although few abandon conventional medicine, up to 80 percent of cancer patients combine one or more CAM approaches with standard treatment.[17] Many patients do not discuss their use of CAM with physicians.

♦ Reminder: *Complementary and alternative medicine (CAM)* refers to health care practices that have not been proved to be effective and consequently are not included as part of conventional treatment (see Highlight 19).

Multivitamin and herbal supplements are among the most frequently used CAM therapies. Although many supplements can be used without risk, some may have adverse effects or interfere with conventional treatments. Use of the herb St. John's wort, for example, can reduce the effectiveness of some anticancer drugs.[18] As another example, some studies suggest that antioxidant supplements interfere with chemotherapy and radiation treatments.[19] Clinical trials of several popular supplements are in progress to learn more about their potential effects and interactions with cancer treatments.

Medical Nutrition Therapy for Cancer

The objectives of nutrition therapy for cancer patients are to minimize loss of weight and muscle tissue, correct nutrient deficiencies, and provide a diet that patients can tolerate and enjoy despite the complications of illness. Appropriate nutrition care helps patients preserve their strength and improves recovery after stressful cancer treatments. Moreover, malnourished cancer patients develop more complications and have shorter survival times than patients who maintain good nutrition status.[20]

Because there are many forms of cancer and a variety of potential treatments, nutritional needs among cancer patients vary considerably. Furthermore, a person's needs may change at different stages of illness. Patients should be screened for malnutrition when cancer is diagnosed and reassessed during the treatment and recovery periods.

Protein and Energy For patients at risk of weight loss and wasting, the focus of nutrition care is to ensure appropriate intakes of protein and energy. To maintain muscle tissue, protein needs may range from 1.0 to 1.5 grams of protein per kilogram of body weight per day; to restore lean tissue, recommended intakes may be 1.5 to 2.0 grams per kilogram body weight per day.[21] Daily energy needs may be 25 to 35 kcalories per kilogram of body weight for weight maintenance, and 35 to 45 kcalories per kilogram for weight regain. Health practitioners should monitor patients' weight changes and adjust intake recommendations as necessary. Patients who cannot eat adequate food may be able to meet their needs by supplementing the diet with nutrient-dense formulas. The "How To" provides suggestions that can help to increase the energy and protein content of meals. ♦

♦ Reminder: The high-kcalorie, high-protein diet, which is appropriate for some individuals with cancer, is described in Chapter 18 (see p. 604).

Although weight loss is a problem for many cancer patients, breast cancer patients often gain weight.[22] The weight gain occurs during the first two years after breast cancer diagnosis and is associated with an increase in total body fat. By dis-

HOW TO Increase kCalories and Protein in Meals

To increase the energy content of a meal, try these suggestions:

- *Meats*. Choose high-fat meats instead of lean meats. Sauté or pan-fry meat instead of baking or roasting it, and use sauces or gravies liberally. Sprinkle bacon bits or sausage pieces on vegetable dishes.
- *Cheese*. Include cheese slices or cream cheese in sandwiches made with luncheon meats. Spread cream cheese on raw vegetables, toast, and crackers or mix into dishes that contain chopped fruit.
- *Half-and-half and cream*. Replace milk or water with half-and-half or cream in soups, sauces, hot chocolate, desserts, mashed potatoes, and cold cereals. Use sour cream or cream sauces on potato dishes, vegetable dishes, and soups. Add whipped cream to fruit salads and desserts.
- *Breads and cereals*. Choose high-fat grain products such as granola, pancakes, waffles, French toast, and biscuits. Prepare hot cereals with whole milk or cream, or added fat.
- *Fruits*. Mash avocados to make guacamole, or use mashed avocado as a sandwich spread. Add chopped dried fruits to salads and baked goods. Snack on dried fruits between meals.
- *Nuts*. Add chopped nuts to pasta dishes, stir-fried vegetables, fruit salads, and green salads. Use nut meats in baked products. Spread nut butters on bread and crackers.
- *Butter or margarine*. Melt on pasta, potatoes, rice, and cooked vegetables. Add to hot cereals, casseroles, and soups. Spread liberally on bread, crackers, and rolls.
- *Mayonnaise or salad dressings*. Add to pasta, tuna, and potato salads. Use as dressings for raw or cooked vegetables.
- *Beverages*. Replace water and non-kcaloric beverages with sweetened drinks, fruit juices, and milk shakes. Drink whole milk instead of low-fat or nonfat milk. Add strawberry or chocolate syrup to plain milk to boost kcalories.

These suggestions can help to add protein to a meal:

- *Meats*. Add small chunks of meat to soups, egg dishes, casseroles, bean dishes, and pasta sauces. Add minced meats to vegetable dishes. Add chunks of cooked chicken or turkey to salads.
- *Eggs*. Add raw eggs when preparing casseroles, meatballs, and hamburgers. Add chopped hard-cooked eggs to salads, vegetable dishes, sandwich fillings, and pasta and potato salads.
- *Cheese*. Melt on burgers, meat loaf, cooked vegetables, scrambled eggs, casseroles, and potatoes. Add cottage cheese to casseroles, egg dishes, pasta recipes, and salad dressings. Grate hard cheeses and sprinkle on soups, salads, and cooked vegetable dishes. Avoid using reduced-fat cheeses.
- *Milk*. Use in place of water when preparing cereals and soups. Use cream sauces (which are made with milk) to flavor vegetable and pasta dishes.
- *Powdered milk (use full-fat milk powder if available)*. Add to recipes that include milk. Dissolve extra milk powder into milk-containing beverages. Stir into hot cereals, potato dishes, casseroles, and sauces. Add to scrambled eggs, hamburgers, and meat loaf.
- *Protein supplements*. Snack on protein bars between meals. Add protein powders to beverages and shakes. Drink meal replacement formulas, such as Ensure or Boost, instead of juices or soda.

TRY IT Make a list of the foods and beverages you consumed in the past 24 hours. Then, describe five ways you could have increased your energy intake and five ways you could have increased your protein intake during this period.

TABLE 29-6 Dietary Considerations for Specific Cancers

Cancer Sites	Common Complications[a]	Possible Dietary Measures
Brain and nervous system	Chewing or swallowing difficulty, headache, altered taste or smell sensation, difficulty feeding oneself	Mechanically altered diet, use of adaptive feeding devices (see Highlight 27)
Head and neck[b]	Chewing or swallowing difficulty, aspiration, inflamed mucosa, dry mouth, altered taste or smell sensation	Tube feeding, mechanically altered diet
Esophagus	Swallowing difficulty, aspiration, obstruction, acid reflux, inflamed mucosa	Tube feeding, mechanically altered diet
Stomach	Anorexia, early satiety, reduced secretion of gastric acid and intrinsic factor, delayed stomach emptying, dumping syndrome, malabsorption, nutrient deficiencies	Tube feeding (for obstruction or unmanageable dumping syndrome); postgastrectomy diet; small, frequent meals; limited sugars and insoluble fibers (see Chapter 23); nutrient supplementation
Intestine	Inflamed mucosa, bacterial overgrowth, obstruction, lactose intolerance, general malabsorption, bile insufficiency, nutrient deficiencies, short bowel syndrome (if resected), altered bowel function, fluid and electrolyte imbalances	Tube feeding or total parenteral nutrition for obstruction, enteritis, or short bowel syndrome; fat- and lactose-restricted diet (see Chapter 24); nutrient supplementation
Pancreas	Reduced secretion of digestive enzymes, bile insufficiency, general malabsorption, nutrient deficiencies, hyperglycemia	Fat-restricted diet; enzyme replacement (see Chapter 24); small, frequent meals; carbohydrate-controlled diet (Chapter 26); nutrient supplementation

[a]Actual complications depend on the exact location of the cancer and the specific methods used for treating the cancer.
[b]Includes cancers of the salivary glands, oral and nasal cavities, pharynx, and larynx.

cussing weight maintenance soon after diagnosis and encouraging physical activity, health practitioners can help patients avoid unnecessary weight gain.

Managing Symptoms and Complications A thorough nutrition assessment often uncovers specific problems or symptoms that interfere with food consumption. Table 29-6 lists dietary considerations related to cancers affecting different sites in the body. The "How To" on pp. 878–879 describes dietary strategies that may alleviate symptoms and improve food intake. Patients' responses to these strategies can vary considerably, and in some cases a number of adjustments may be necessary.

Low-Microbial Diet Patients with suppressed immunity or neutropenia may be prescribed a **low-microbial diet** (also called a *neutropenic diet*), which includes only foods that are unlikely to be contaminated with bacteria or other microbes.[23] Generally, patients should consume only well-cooked meats and eggs, pasteurized milk products, well-washed fruits and vegetables, and shelf-stable packaged foods. Foods that must be avoided include unwashed raw fruits and vegetables; unpasteurized juices and milk products; undercooked meat, poultry, and eggs; leftover luncheon meats and meat spreads; leftover foods that have not been adequately reheated; and foods from salad bars or street vendors. In addition, patients should be instructed to follow safe food-handling practices to minimize the risk of foodborne illness (see Highlight 18).

Enteral and Parenteral Nutrition Support Nutrition support is used in limited situations during cancer treatment. Generally, tube feedings and parenteral nutrition are provided to patients who have long-term or permanent gastrointestinal impairment or are experiencing complications that interfere with food intake.[24] For example, many patients undergoing radiation therapy for head and neck cancers

low-microbial diet: a diet that contains foods that are unlikely to be contaminated with bacteria and other microbes.

HOW TO Help Patients Handle Food-Related Problems

In people with cancer or HIV infection, various complications can interfere with food intake. Health care providers can try to identify a patient's specific problems and offer appropriate solutions. Not every suggestion will work for each person; encourage patients to experiment and find strategies that work best.

I just don't have an appetite.

- Eat small meals and snacks at regular times each day.
- Eat the largest meal at the time of day when you feel the best.
- Include nutrient-dense foods in meals, and consume them before other foods.
- Indulge in favorite foods throughout the day. Serve foods attractively.
- Avoid drinking large amounts of liquids before or with meals.
- Eat in a pleasant and relaxed environment. Eat with family and friends when possible.
- Listen to your favorite music or enjoy a TV or radio program while you eat.
- Ask your doctor about appetite-enhancing medications.

I am too tired to fix meals and eat.

- Let family members and friends prepare food for you.
- Obtain foods that are easy to prepare and easy to eat, such as sandwiches, frozen dinners, take-out meals from restaurants, instant breakfast drinks, liquid formulas, and energy bars.
- Find time to rest before you attempt to prepare a large meal.
- Prepare soups, stews, and casserole dishes in sufficient quantity to provide enough for several meals, so that you will have enough to eat at times when you are too tired to cook.

Foods just don't taste right.

- Brush your teeth or use mouthwash before you eat.
- Consume foods chilled or at room temperature. Use plastic, rather than metal, eating utensils.
- Choose eggs, fish, poultry, and milk products instead of meats.
- Experiment with sauces, seasonings, herbs, spices, and sweeteners to improve food taste and flavor.
- Save your favorite foods for times when you are not feeling nauseated.

I am nauseated a lot of the time, and sometimes I need to vomit.

- Consume liquids throughout the day to replace fluids.
- If you become nauseated from chemotherapy treatments, avoid eating for at least two hours before treatments.
- Consume your largest meal at a time when you are least likely to feel nauseous.
- Try consuming smaller meals, and eat slowly. Experiment with foods to see if some foods cause nausea more than others.
- Avoid foods and meals that have strong odors or are fatty, greasy, or gas forming.

I have problems chewing and swallowing food.

- Experiment with food consistencies to find the ones you can manage best. Thin liquids, dry foods, and sticky foods (such as peanut butter) are often difficult to swallow.
- Add sauces and gravies to dry foods.
- Drink fluids during meals to ease chewing and swallowing.
- Try using a straw to drink liquids. Experiment with beverage thickeners if you cannot tolerate thin beverages.
- Tilt your head forward and backward to see if you can swallow more easily when your head is positioned differently.

(*continued*)

◆ Irradiation to the head and neck regions often causes dysphagia and mouth sores.

require long-term tube feeding and may need to continue tube feedings at home. ◆ Parenteral nutrition is reserved for patients who have inadequate GI function, such as individuals with chronic radiation enteritis. Whenever possible, enteral nutrition is strongly preferred over parenteral nutrition, to preserve GI function and avoid infection.

IN SUMMARY Cancer arises from mutations in the genes that control cell division. Some dietary substances promote carcinogenesis, while others may help to prevent cancer. Cancer's effects on nutrition status depend on the type of cancer a person has, its severity, and the methods used to treat the cancer. Cancer cachexia is a frequent complication of cancer and may be related to anorexia, altered metabolism, and responses to treatment. Medical treatments for cancer include surgery, chemotherapy, radiation therapy, and biological

HOW TO Help Patients Handle Food-Related Problems (*continued*)

I have sores in my mouth, and they hurt when I eat.

- Try eating chilled or frozen foods; they are often soothing.
- Try soft foods such as ice cream, milk shakes, bananas, applesauce, mashed potatoes, cottage cheese, and macaroni and cheese. Mix dry foods with sauces or gravies.
- Cut foods into smaller pieces, so they are less likely to irritate the mouth.
- Avoid foods that irritate mouth sores, such as citrus fruits and juices, tomatoes and tomato-based products, spicy foods, foods that are very salty, foods with seeds (such as poppy seeds and sesame seeds) that can scrape the sores, and coarse foods such as raw vegetables, crackers, corn chips, and toast.
- Ask your doctor about using a local anesthetic solution such as lidocaine before eating to reduce pain.
- Use a straw for drinking liquids, in order to bypass the sores.

My mouth is really dry.

- Rinse your mouth with warm salt water or mouthwash frequently. Avoid using mouthwash that contains alcohol.
- Drink small amounts of liquid frequently between meals.
- Ask your doctor or pharmacist about medications or saliva substitutes that can help a dry mouth condition.
- Use sour candy or chewing gum to stimulate the flow of saliva.
- Sip fluids frequently while eating. Add broth, sauces, gravies, mayonnaise, butter, or margarine to dry foods.
- Make sure you brush your teeth and floss regularly to prevent tooth decay and oral infections.

I am having trouble with constipation.

- Drink plenty of fluids. Try warm fluids, especially in the morning.
- Eat whole-grain breads and cereals, nuts, fresh fruits and vegetables, prunes, and prune juice. Avoid refined carbohydrate foods such as white bread, white rice, and pasta.
- Engage in physical activity regularly.
- Try an over-the-counter bulk-forming agent, such as methylcellulose (Citrucel), psyllium (Metamucil or Fiberall), or polycarbophil (Fiber-Lax).

I am having trouble with diarrhea.

- To avoid dehydration, drink plenty of fluids throughout the day. Salty broths and soups, diluted fruit juices, and sports drinks are good choices. Avoid caffeine- and alcohol-containing beverages. For severe diarrhea, try oral rehydration formulas that are commercially prepared.
- Avoid foods and beverages that increase gas, such as legumes, onions, vegetables of the cabbage family, foods that contain sorbitol or mannitol, and carbonated beverages.
- Try using lactase enzyme replacements when you use milk products in case you are experiencing lactose intolerance. Yogurt and aged cheeses may be easier to tolerate than milk and fresh cheeses.
- Avoid fatty foods if you are fat intolerant. Try reducing your intake of whole-grain breads and cereals if they worsen the diarrhea.
- Eat small, frequent meals instead of large ones. Try consuming cool or lukewarm foods instead of very cold or hot foods.
- Ask your doctor about using a bulk-forming agent or antidiarrhea medication.

TRY IT Identify two or three food-related problems that you have experienced in the past. Based on your experiences, which suggestions listed above for each of these problems seem the most and least helpful? Why?

therapies, which remove cancer cells, prevent tumor growth, and alleviate symptoms. Nutrition therapy aims to minimize weight loss and wasting, correct deficiencies, and manage complications that impair food intake. The Case Study allows you to apply information about nutrition and cancer to a clinical situation.

HIV Infection

Possibly the most infamous infectious disease today is **acquired immune deficiency syndrome (AIDS)**. AIDS develops from infection with human immunodeficiency virus (HIV), which attacks the immune system and disables a person's defenses against other diseases, including infections and certain cancers. Then these

acquired immune deficiency syndrome (AIDS): the late stage of illness caused by infection with the human immunodeficiency virus (HIV); characterized by severe damage to immune function.

CASE STUDY Woman with Cancer

Shannon Miraglia is a 58-year-old public relations consultant who was recently diagnosed with colon cancer after a routine colonoscopy, a procedure in which the colon is examined using a flexible tube attached to an optical device. Mrs. Miraglia is scheduled to have surgery to remove the segment of colon that contains the tumor and to determine if the cancer has spread to the surrounding lymph nodes and, possibly, other organs. The nurse completing the nutrition assessment finds that Mrs. Miraglia is 5 feet 5 inches tall and weighs 178 pounds. Mrs. Miraglia usually spends most of the day sitting and has little time to engage in recreational exercise. Her diet typically includes red meat at both lunch and dinner, and she consumes one or two glasses of wine with both meals. She eats two or three servings of fruits and vegetables each day, although she does not like green leafy vegetables very much. She rarely drinks milk or consumes milk products.

1. Review Table 29-2 on p. 871, and describe the factors in Mrs. Miraglia's diet and lifestyle that may have contributed to the development of colon cancer.
2. What symptoms and complications may arise after colon surgery and impair nutrition status? If the cancer team decides that Mrs. Miraglia needs follow-up chemotherapy, how might the chemotherapy affect her nutrition status?
3. If Mrs. Miraglia is unresponsive to treatment and her cancer progresses, she may develop cancer cachexia. Describe this syndrome, its causes, and its consequences.
4. Provide suggestions that may help Mrs. Miraglia handle the following problems should they develop: poor appetite, fatigue, taste alterations, nausea and vomiting, chewing and swallowing difficulties, mouth sores, dry mouth, diarrhea, constipation, and weight loss.

diseases—which would cause few, if any, symptoms in people with healthy immune systems—destroy health and life.

The HIV/AIDS epidemic continues to sweep across countries, especially in sub-Saharan Africa (see Table 29-7). For many years, the destructive effects of HIV infection seemed unstoppable, but in the mid- to late 1990s the death rate from AIDS began to decline in the United States, and the progression from HIV infection to AIDS slowed considerably. Although AIDS is still incurable, remarkable progress has been made in understanding and treating HIV infection.

Without a cure for AIDS, the best course is prevention. HIV is most often sexually transmitted and can be spread by direct contact with contaminated body fluids, such as blood, semen, vaginal secretions, and breast milk. Because many people remain symptom-free during the early stages of infection, they may not realize that they can pass the infection to others. To reduce the spread of HIV infection, individuals at risk (see Table 29-8) are encouraged to undergo testing. A blood test

TABLE 29-7 The HIV and AIDS Epidemic at a Glance, 2009

Stage of Epidemic	World	Sub-Saharan Africa	North America
Individuals living with HIV infection	33,300,000	22,500,000	1,500,000
Individuals newly infected with HIV	2,600,000	1,800,000	70,000
AIDS-related deaths	1,800,000	1,300,000	26,000

SOURCE: Joint United Nations Programme on HIV/AIDS (UNAIDS), *Global report: UNAIDS report on the global AIDS epidemic 2010*, www.unaids.org/documents/20101123_GlobalReport_em.pdf; accessed December 21, 2010.

TABLE 29-8 Risk Factors for HIV Infection

- History of receiving blood transfusions or blood components before 1985
- Infant born to mother with HIV infection
- Intravenous drug use in which syringes are shared among users
- Sexual contact with intravenous drug users, prostitutes, or individuals with a history of HIV or other sexually transmitted diseases
- Sexual contact with multiple partners
- Unsafe sexual practices

can usually detect HIV antibodies within several months after exposure and, often, after two or three weeks. An estimated 21 percent of persons in the United States who have HIV infection are unaware that they are infected.[25]

Consequences of HIV Infection

HIV infection destroys immune cells that have a protein called CD4 on their surfaces. The cells most affected are the **helper T cells**, ♦ also called *CD4+ T cells* because the presence of CD4 is a primary characteristic. HIV is able to enter the helper T cells and induce them to produce additional copies of the virus, thus perpetuating and exacerbating the infection. Other cells that have the CD4 protein (and are infected by HIV) include tissue macrophages, blood monocytes, and certain cells of the central nervous system.[26] Early symptoms of HIV infection are nonspecific and may include fever, sore throat, malaise, swollen lymph nodes, skin rashes, muscle and joint pain, and diarrhea. After these symptoms subside, many people remain symptom-free for 5 to 10 years or even longer. If the HIV infection is not treated, however, the depletion of T cells eventually increases the person's susceptibility to **opportunistic infections**—that is, infections caused by microorganisms that normally do not cause disease in healthy individuals.

The term *AIDS* applies to the advanced stages of HIV infection, in which the inability to fight illness allows a number of serious diseases and complications to develop; such **AIDS-defining illnesses** include severe infections, certain cancers, and wasting of muscle tissue. Without treatment, AIDS develops in 26 to 36 percent of HIV-infected persons within 7 years.[27] Health practitioners evaluate disease progression by measuring the concentrations of helper T cells and circulating virus (called the *viral load*) and by monitoring clinical symptoms. Although current drug therapies can dramatically slow the progression of HIV infection, the drugs' side effects may make it difficult for patients to adhere to treatments, as discussed in several of the following sections.

♦ *T cells* are lymphocytes that develop in the thymus gland. The other lymphocytes are the *B cells* (which develop in bone marrow) and *natural killer cells*.

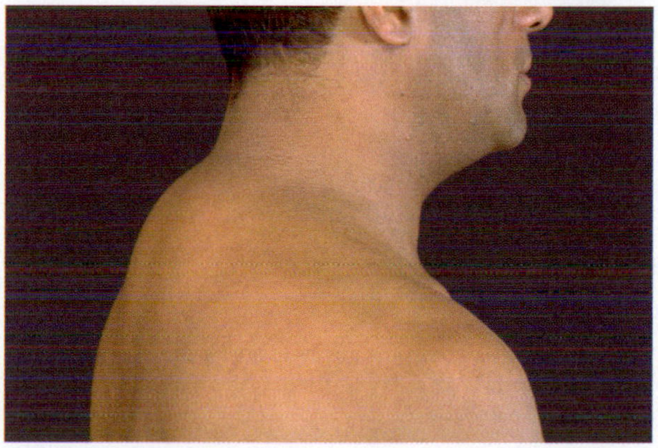

© Medical-on-Line/Alamy

HIV lipodystrophy is sometimes evident by the accumulation of fatty tissue at the base of the neck, referred to as *buffalo hump*.

Lipodystrophy

About one-third or more of patients using drug therapies to suppress HIV infection develop abnormalities in glucose and fat metabolism.[28] These complications, collectively known as the **HIV-lipodystrophy syndrome**, include body fat redistribution, abnormal blood lipid levels, and insulin resistance. Patients may accumulate abdominal fat, lose fat from the face and extremities, or both.[29] Also observed are breast enlargement (in both men and women), fat accumulation at the base of the neck (called a **buffalo hump**), and benign growths composed of fat tissue (called **lipomas**). The changes in body composition are often disfiguring and may cause physical discomfort; moreover, patients often develop hypertriglyceridemia, elevated low-density lipoprotein (LDL) cholesterol levels, low high-density lipoprotein (HDL) cholesterol levels, glucose intolerance, and hyperinsulinemia. The reasons for the development of lipodystrophy are unknown.

Weight Loss and Wasting

Even with effective treatment of HIV infection, weight loss and wasting are ongoing problems for many HIV-infected patients. *HIV-associated wasting* is diagnosed in patients who unintentionally lose 7.5 percent of body weight within 6 months, or 10 percent of body weight within 12 months.[30] The

helper T cells: lymphocytes that have a specific protein called CD4 on their surfaces and therefore are also known as *CD4+ T cells*; these are the cells most affected in HIV infection.

opportunistic infections: infections from microorganisms that normally do not cause disease in healthy people but are damaging to persons with compromised immune function.

AIDS-defining illnesses: diseases and complications associated with the later stages of an HIV infection, including wasting, recurrent bacterial pneumonia, opportunistic infections, and certain cancers.

HIV-lipodystrophy (LIP-oh-DIS-tro-fee) **syndrome:** a collection of abnormalities in fat and glucose metabolism that may result from drug treatments for HIV infection; changes include body fat redistribution, abnormal blood lipid levels, and insulin resistance. The accumulation of abdominal fat is sometimes called *protease paunch*.

buffalo hump: the accumulation of fatty tissue at the base of the neck.

lipomas (lih-POE-muz): benign tumors composed of fatty tissue.

wasting has been linked with accelerated disease progression, reduced strength, and fatigue. In the later stages of AIDS, the wasting is severe and increases the risk of death. Much as in cancer, wasting associated with HIV infection has many causes: anorexia and inadequate food intake, altered metabolism, malabsorption, chronic diarrhea, and diet-drug interactions.

Anorexia and Reduced Food Intake Inadequate food intake is a key factor in the development of wasting. Poor food intake may result from various factors, including the following:

- *Emotional distress, pain, and fatigue.* The physical and social problems that accompany chronic illness may cause fear, anxiety, and depression, which contribute to anorexia. Pain and fatigue, which may be associated with some disease complications, can cause anorexia and difficulty with eating.
- *Oral infections.* The oral infections associated with HIV infection can cause discomfort and interfere with food consumption. Common infections include **candidiasis** and **herpes simplex virus** infection. Candidiasis (commonly called *thrush*) can cause mouth pain, dysphagia (difficulty swallowing), and altered taste sensation; infection with herpes simplex virus may cause painful lesions around the lips and in the mouth.
- *Respiratory disorders.* Respiratory infections, including pneumonia and tuberculosis, are common in people with HIV infection. Symptoms may include chest pain, shortness of breath, and cough, which interfere with eating and contribute to anorexia.
- *Cancer.* As described earlier in this chapter, cancer leads to anorexia for numerous reasons. In addition, **Kaposi's sarcoma**, a type of cancer frequently associated with HIV infection, can cause lesions in the mouth and throat that make eating painful.
- *Medications.* The medications given to treat HIV infection, other infections, and cancer often cause anorexia, nausea and vomiting, altered taste sensation, food aversions, and diarrhea.

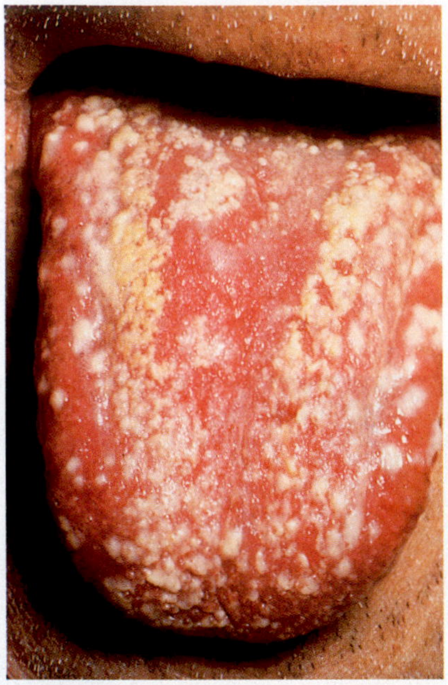

The oral infection *thrush* is easily identified by the characteristic milky white patches that appear on the tongue.

GI Tract Complications Complications of HIV infection involving the GI tract may result from opportunistic infections, the HIV infection itself, or medications.[31] In addition to the oral infections described previously, infections may develop in the esophagus, stomach, and intestines. Furthermore, advanced AIDS is often accompanied by characteristic changes in the small intestinal lining: the villi appear shortened and flattened, and the absorptive area is substantially reduced; ♦ these changes can cause malabsorption, steatorrhea, and diarrhea.

♦ The AIDS-related abnormalities in the intestinal mucosa are sometimes referred to as *AIDS enteropathy* (EN-ter-OP-ah-thy).

As described earlier, many patients are unable to tolerate the medications used to suppress HIV and develop nausea, vomiting, and diarrhea. Furthermore, the medications that treat the viral, parasitic, and fungal infections in the GI tract contribute to bacterial overgrowth. Thus, HIV-infected patients using standard treatments face an extremely high risk of malnutrition due to the combination of intestinal discomfort, bacterial overgrowth, malabsorption, and nutrient losses from vomiting, steatorrhea, and diarrhea.

candidiasis: a fungal infection on the mucous membranes of the oral cavity and elsewhere; usually caused by *Candida albicans*.

herpes simplex virus: a common virus that can cause blisterlike lesions on the lips and in the mouth.

Kaposi's (kah-POH-seez) **sarcoma:** a common cancer in HIV-infected persons that is characterized by lesions in the skin, lungs, and GI tract.

Neurological Complications Neurological complications may be a consequence of HIV infection, immune suppression, or cancers and infections that target brain tissue.[32] Clinical features include mild to severe dementia; muscle weakness and gait disturbances; and pain, numbness, and tingling in the legs and feet. Neurological impairments are usually more pronounced in the advanced stages of AIDS.

Other Complications Patients with HIV infection can develop anemia due to nutrient malabsorption, blood loss, disturbed bone marrow function, medication side effects, or the chronic illness itself. ♦ HIV infection may also lead to skin disorders (rashes, infections, cancers), kidney diseases (nephrotic syndrome, chronic kidney disease), eye disorders (retinal infection or detachment), and coronary heart disease.[33]

♦ Reminder: The *anemia of chronic disease* often develops during chronic illness and is characterized by altered iron distribution in tissues and reduced red blood cell synthesis, among other abnormalities (see Highlight 25).

TABLE 29-9 Antiretroviral Drugs for Treatment of HIV Infection

Category	Examples	Mode of Action
CCR5 antagonist	Maraviroc	CCR5 antagonists prevent HIV from entering cells by blocking a membrane receptor on the host cell.
Fusion inhibitor	Enfuvirtide	Fusion inhibitors prevent HIV from entering cells by binding a viral protein needed for its entry.
Integrase inhibitor	Raltegravir	Integrase inhibitors impair the function of HIV's integrase enzyme, which incorporates viral DNA into the host cell's genome.
Non-nucleoside reverse transcriptase inhibitor (NNRTI)	Delavirdine, efavirenz, nevirapine	NNRTI bind active sites on HIV's reverse transcriptase enzyme, blocking the ability of HIV to produce DNA copies of its genetic material.
Nucleoside reverse transcriptase inhibitor (NRTI)	Didanosine, lamivudine, zidovudine	As analogs of the nucleosides needed for DNA synthesis, NRTI impair the ability of HIV's reverse transcriptase enzyme to produce usable copies of DNA.
Protease inhibitor (PI)	Saquinavir, ritonavir, indinavir	PI inhibit HIV's protease enzyme, which cleaves HIV's gene products into usable structural proteins.

SOURCES: U.S. Department of Health and Human Services, Panel on Antiretroviral Guidelines for Adults and Adolescents, *Guidelines for the Use of Antiretroviral Agents in HIV-1-Infected Adults and Adolescents*, December 1, 2009, pp. 1–168, http://aidsinfo.nih.gov/contentfiles/AdultandAdolescentGL.pdf, accessed November 21, 2010; S. Safrin, Antiviral agents, in B. G. Katzung, ed., *Basic and Clinical Pharmacology* (New York: McGraw-Hill/Lange, 2007), pp. 790–818.

Treatments for HIV Infection

Although there is no cure for HIV infection, treatments can help to slow its progression, reduce complications, and alleviate pain. The standard treatment for suppressing HIV infection, called *highly active antiretroviral therapy* (*HAART*), combines three or more antiretroviral drugs.[34] Table 29-9 lists the major drug categories included in antiretroviral therapy and describes the drugs' modes of action. These antiretroviral agents have multiple adverse effects that make their long-term use difficult to tolerate. In addition to the GI effects discussed previously, side effects include skin rashes, headache, anemia, tingling and numbness, hepatitis, pancreatitis, and kidney stones. Thus, although HAART has improved life span and quality of life for many patients, the drug regimens are difficult to adhere to and cause complications that require continual management. The Diet-Drug Interactions feature summarizes the nutrition-related effects of some of the antiretroviral agents and other drugs mentioned in this chapter.

Control of Anorexia and Wasting Anabolic hormones, appetite stimulants, and regular physical activity have been successful in reversing weight loss and increasing muscle mass in HIV-infected patients. Testosterone and human growth hormone have demonstrated positive effects on nitrogen balance and lean tissue content, especially in combination with resistance training. A regular program of resistance exercise improves muscle mass and strength and corrects some of the metabolic abnormalities (altered blood lipids and insulin resistance) that are common in HIV-infected patients. The medications megestrol acetate and dronabinol (described on p. 875) are sometimes prescribed to stimulate appetite and improve weight gain, although much of the weight increase is attributable to a gain of fat rather than lean tissue.[35]

Control of Lipodystrophy Treatment strategies for lipodystrophy are under investigation. Both aerobic activity and resistance training may help to reduce abdominal fat, although some patients opt for cosmetic surgery. Patients may be given alternative antiretroviral drugs to alleviate symptoms. Medications may be prescribed to treat abnormal blood lipid levels and insulin resistance.

Alternative Therapies Like cancer patients, people with HIV infection and AIDS are frequently tempted to try unconventional methods of treatment. Although many alternative therapies are harmless, some may have side effects that worsen

Resistance training can help a person with HIV infection maintain muscle mass and strength.

Andresr/Shutterstock.com

DIET-DRUG Interactions

Check this table for notable nutrition-related effects of the medications discussed in this chapter.

Appetite stimulants (megestrol acetate, dronabinol)	**Gastrointestinal effects:** Nausea, vomiting, diarrhea **Metabolic effects:** Hyperglycemia (megestrol acetate)
Didanosine	**Gastrointestinal effects:** Nausea, vomiting, dry mouth, altered taste perception, anorexia, diarrhea, constipation **Dietary interactions:** Take medication either one-half hour before or two hours after a meal; avoid alcohol and aluminum- and magnesium-containing antacids. **Metabolic effects:** Pancreatitis; anemia; reduced serum copper, zinc, and vitamin B_{12} levels
Methotrexate	**Gastrointestinal effects:** Nausea, vomiting, diarrhea, reduced absorption of vitamin B_{12} and calcium **Dietary interactions:** Milk may reduce methotrexate absorption if the milk and methotrexate are ingested together. **Metabolic effects:** Increased serum uric acid levels, anemia, liver toxicity
Ritonavir	**Gastrointestinal effects:** Nausea, vomiting, altered taste perception, anorexia, diarrhea **Metabolic effects:** Pancreatitis, hyperglycemia, reduced serum copper and zinc levels; increased levels of blood triglycerides, LDL cholesterol, uric acid, liver enzymes, and creatine kinase
Zidovudine	**Gastrointestinal effects:** Nausea, vomiting, altered taste perception, anorexia, mouth sores, constipation **Dietary interactions:** Do not take medication with a high-fat meal. **Metabolic effects:** Anemia; reduced serum copper, zinc, and vitamin B_{12} levels

NOTE: Most antiretroviral drugs that treat HIV infection have gastrointestinal and metabolic side effects; only a few are listed here as examples.

complications or interfere with treatment. The herb St. John's wort, for example, can reduce the effectiveness of some antiretroviral drugs.[36] Monitoring patients' use of dietary supplements is essential to reduce the possibility of nutrient-drug and herb-drug interactions.

Nutrition Therapy for HIV Infection

HIV-infected individuals must learn how to maintain body weight and muscle mass, prevent malnutrition, and cope with nutrition-related side effects of medications. Therefore, nutrition assessment and counseling should begin as soon as a patient is diagnosed with HIV infection. The initial assessment should include an evaluation of body weight and body composition. Follow-up measurements may indicate the need to adjust dietary recommendations and drug therapies.

Weight Management Since the development of successful drug therapies for HIV infection, obesity and overweight have become more prevalent than wasting among HIV-infected individuals in the United States.[37] Because excessive body weight can increase risks for cardiovascular disease and diabetes, moderate weight loss is recommended for patients with HIV infection who are overweight or obese.

Individuals who experience weight loss and wasting may benefit from a high-kcalorie, high-protein diet. Daily energy needs may be 30 to 40 kcalories per kilogram body weight, and protein requirements may be as high as 1.2 to 2.0 grams per kilogram body weight.[38] If food consumption is difficult, small, frequent feedings may be better tolerated than several large meals. The addition of nutrient-dense snacks, protein or energy bars, and oral supplements can improve intakes. Liquid formulas may be useful for the person who is too tired to eat or prepare meals.

Review the "How To" on p. 876 for additional suggestions for adding energy and protein to the diet.

Metabolic Complications As mentioned, individuals using antiretroviral drugs frequently develop insulin resistance and elevated triglyceride and LDL cholesterol levels. Treating these problems often requires both medications and dietary adjustments. Patients should be advised to achieve or maintain a desirable weight, replace saturated fats with monounsaturated and polyunsaturated fats, increase fiber intake, and limit intakes of *trans*-fatty acids, cholesterol, added sugars, and alcohol. Regular physical activity can improve both insulin resistance and blood lipid levels. ♦ If problems persist, alternative antiretroviral medications may be prescribed in an attempt to improve the metabolic abnormalities.

♦ Additional suggestions for managing insulin resistance and hyperlipidemias are available in Chapters 26 and 27, respectively.

Vitamins and Minerals Vitamin and mineral needs of people with HIV infections are highly variable, and little information is available concerning specific needs. Because nutrient deficiencies are likely to result from reduced food intake, malabsorption, diet-drug interactions, and nutrient losses, multivitamin-mineral supplements are typically recommended. Patients should be cautioned to maintain intakes that are close to DRI recommendations, however, due to the risk of adverse interactions between excessive amounts of vitamins and minerals, and antiretroviral drugs.[39]

Symptom Management The discomfort associated with antiretroviral therapy, opportunistic GI infections, and malabsorption symptoms can make food consumption difficult, and problems such as vomiting and diarrhea contribute to fluid and electrolyte losses. The "How To" on pp. 878–879 describes measures for improving food and fluid intakes in individuals with these problems.

Food Safety The depressed immunity of people with HIV infections places them at extremely high risk of developing foodborne illnesses. Health practitioners should caution patients about their high susceptibility to foodborne illness and provide detailed instructions about the safe handling and preparation of foods; for some individuals, a low-microbial diet may be suggested (see p. 877). Water can also be a source of foodborne illness and is a common cause of **cryptosporidiosis** in HIV-infected individuals. In places where water quality is questionable, patients should consult their local health departments to determine whether the tap water is safe to drink. If not, or to take additional safety measures, water used for drinking and making ice cubes should be boiled for one minute.

Enteral and Parenteral Nutrition Support In later stages of illness, people with HIV infections may be unable to consume enough food and may need aggressive nutrition support. Tube feedings are preferred whenever the GI tract is functional; they can be provided at night to supplement oral diets consumed during the day. Parenteral nutrition is reserved for patients who are unable to tolerate enteral nutrition, such as those with GI obstructions that prevent food intake. For individuals with severe malabsorption, orally administered hydrolyzed formulas containing medium-chain triglycerides may be as effective as parenteral nutrition for reversing weight loss and wasting. For either type of nutrition support, careful measures are necessary to avoid bacterial contamination of nutrient formulas and feeding equipment.

IN SUMMARY **By attacking immune cells, HIV causes progressive damage to immune function and may eventually lead to AIDS. Improved drug therapies have slowed the progression of HIV infection; however, these drugs may promote the HIV-lipodystrophy syndrome, characterized by body fat redistribution, abnormal lipid levels, and insulin resistance. HIV infection may lead to weight loss and wasting, anorexia, and various complications that affect**

cryptosporidiosis (KRIP-toe-spor-ih-dee-OH-sis): a foodborne illness caused by the parasite *Cryptosporidium parvum*.

CASE STUDY

Man with HIV Infection

Three years ago, Brian Wolfe, a 37-year-old financial planner, sought medical help when he began feeling run-down and developed a painful white fungal infection over his mouth and tongue. The presence of thrush, recent weight loss, and anemia alerted Mr. Wolfe's physician to the possibility of an HIV infection. When Mr. Wolfe tested positive for HIV, he and his family and friends were devastated by the news, but those close to him have remained supportive. During the three years since Mr. Wolfe began antiretroviral drug therapy, he has maintained his weight but has also developed lipodystrophy and hypertriglyceridemia. Mr. Wolfe is 6 feet tall and currently weighs 185 pounds. He occasionally develops diarrhea and sometimes anorexia.

1. Describe lipodystrophy, and discuss its typical pattern in people who have an HIV infection. What adjustments in treatment and lifestyle may be helpful for Mr. Wolfe?
2. Describe an appropriate diet for Mr. Wolfe. What strategies may improve his problems with diarrhea and anorexia? Suggest reasons why diarrhea and anorexia may develop in people with HIV infections.
3. Explain why an HIV infection can lead to wasting as the disease progresses to the later stages. What recommendations may be helpful for maintaining weight and health if wasting becomes a problem?

food intake. Dietary adjustments, resistance training, and medications can help patients maintain their weight and prevent wasting. Following food safety guidelines can help patients avoid foodborne illnesses.

The Case Study provides an opportunity to review the nutrition concerns of a person with HIV infection.

Clinical Portfolio

1. Consider the nutrition problems that may develop in a 36-year-old woman with a malignant brain tumor that affects her ability to move the right side of her body (including the tongue) and to speak coherently. She is taking a pain medication that makes her nauseated and sleepy. Her expected survival time is only about six months.
    - If she is right-handed, how might her impairment interfere with eating? What suggestions do you have for overcoming this problem?
    - How might her nutrition status be affected by her inability to communicate effectively? What suggestions may help?
    - In what ways might the pain medication she is taking affect her nutrition status?
2. Various types of chronic conditions can lead to weight loss and wasting. For some of these conditions, such as Crohn's disease and celiac disease (Chapter 24), diet is a cornerstone of treatment. For others, such as cancer and HIV infection, nutrition plays a supportive role. What determines whether nutrition plays a primary role or a supportive role in the treatment of disease?

Nutrition on the Net

For further study of topics covered in this chapter, log on to **www.cengagebrain.com** and search for ISBN 084006845X.

- To learn more about cancer, including risk factors, prevention, screening, detection, treatments (including nutrition), and support networks, visit these sites:

  American Association for Cancer Research: **www.aacr.org**

  American Cancer Society: **www.cancer.org**

  American Institute for Cancer Research: **www.aicr.org**

  National Cancer Institute: **www.cancer.gov**

- To find additional information about HIV infection and AIDS, visit these sites:

  AIDS Education Global Information System: **www.aegis.com**

  AIDSinfo, an information service provided by the U.S. Department of Health and Human Services: **aidsinfo.nih.gov**

  The Body: **www.thebody.com**

  UCSF Center for HIV Information: **hivinsite.ucsf.edu**

- To review information about safe food handling provided by U.S. government agencies, visit this site: **www.foodsafety.gov**

Nutrition Assessment Checklist for People with Cancer or HIV Infections

Medical History

Check the medical record to determine:

- Type and stage of cancer
- Stage of HIV infection

Review the medical record for complications that may alter nutrition therapy, including:

- Altered organ function
- Altered taste perception
- Anorexia
- Dry mouth and oral infections
- GI symptoms and infections
- Hyperlipidemias
- Insulin resistance
- Malnutrition and wasting

Medications

For patients with cancer or HIV infections:

- Check medications to identify potential diet-drug interactions.
- Recommend the use of antinauseants at mealtime, if needed.
- Ask about the use of dietary supplements, including herbal products.

For cancer patients who require chemotherapy:

- Recommend strategies to prevent food aversions.
- Offer suggestions for managing drug-related complications.

For HIV-infected patients using antiretroviral drug therapy:

- Remind patients that some drugs are better absorbed with foods and that others must be taken on an empty stomach.
- Help patients work out a medication schedule that suits their lifestyle and is timed appropriately in regard to food intake.
- Offer suggestions for managing drug-related complications.

Dietary Intake

For patients with poor food intakes and weight loss:

- Determine the reasons for reduced food intake.
- Offer appropriate suggestions to improve food intake.
- Provide interventions before weight loss progresses too far.

For patients with HIV infections who experience weight gain, elevated triglyceride or LDL cholesterol levels, or hyperglycemia:

- Assess the diet for energy, total fat, types of fat, carbohydrates, fiber, and sugars.
- For patients with hyperlipidemias, recommend a diet low in saturated fat, *trans*-fatty acids, cholesterol, and sugars.
- For patients with hyperglycemia, recommend a consistent carbohydrate intake at meals and snacks that emphasizes complex carbohydrates and limits concentrated sweets.
- Recommend regular physical activity for weight control and for improving blood lipid levels and insulin resistance.

Anthropometric Data

Take baseline height and weight measurements, monitor weight regularly, and suggest dietary adjustments for weight maintenance, if necessary. Remember that body composition may change without affecting body weight. Perform baseline and periodic body composition measurements in HIV-infected patients who are using antiretroviral drug therapy.

Laboratory Tests

Note that albumin and other serum proteins may be reduced in patients with cancer or HIV infections, especially in those experiencing wasting. Check laboratory tests for indications of:

- Anemia
- Dehydration
- Elevated LDL cholesterol levels
- Elevated triglyceride levels
- Hyperglycemia

For patients with HIV infections, evaluate disease progression by checking:

- Helper T cell counts
- Viral load

Physical Signs

Look for physical signs of:

- Dehydration (especially for patients with fever, vomiting, or diarrhea)
- Kaposi's sarcoma
- Oral infections
- Protein-energy malnutrition and wasting

References

1. National Center for Health Statistics, *Health, United States, 2009: With Special Feature on Medical Technology* (Hyattsville, MD: U.S. Department of Health and Human Services, 2009), pp. 254–257.
2. R. S. K. Chaganti, Genetics of cancer, in L. Goldman and D. Ausiello, eds., *Cecil Medicine* (Philadelphia: Saunders, 2008), pp. 1340–1344.
3. W. C. Willett and E. Giovannucci, Epidemiology of diet and cancer risk, in M. E. Shils and coeditors, *Modern Nutrition in Health and Disease* (Philadelphia: Lippincott Williams & Wilkins, 2006), pp. 1267–1279.
4. World Cancer Research Fund/American Institute for Cancer Research, *Food, Nutrition, Physical Activity, and the Prevention of Cancer: A Global Perspective* (Washington, DC: American Institute for Cancer Research, 2007).
5. World Cancer Research Fund/American Institute for Cancer Research, 2007.
6. R. J. Turesky, Formation and biochemistry of carcinogenic heterocyclic aromatic amines in cooked meats, *Toxicology Letters* 168 (2007): 219–227; T. Sugimura and coauthors, Heterocyclic amines: Mutagens/carcinogens produced during cooking of meat and fish, *Cancer Science* 95 (2004): 290–299.
7. S. Koutros and coauthors, Meat and meat mutagens and risk of prostate cancer in the agricultural health study, *Cancer Epidemiology, Biomarkers, and Prevention* 17 (2008): 80–87; World Cancer Research Fund/American Institute for Cancer Research, 2007; Turesky, 2007.
8. World Cancer Research Fund/American Institute for Cancer Research, 2007.
9. H. S. Rugo, Paraneoplastic syndromes and other non-neoplastic effects of cancer, in L. Goldman and D. Ausiello, eds., *Cecil Medicine* (Philadelphia: Saunders, 2008), pp. 1353–1362.
10. M. J. Tisdale, Mechanisms of cancer cachexia, *Physiological Reviews* 89 (2009): 381–410.
11. M. Schattner and M. Shike, Nutrition support of the patient with cancer, in M. E. Shils and coeditors, *Modern Nutrition in Health and Disease* (Philadelphia: Lippincott Williams & Wilkins, 2006), pp. 1290–1313.
12. S. Dodson and coauthors, Muscle wasting in cancer cachexia: Clinical implications, diagnosis, and emerging treatment strategies, *Annual Review of Medicine* 62 (2011): 8.1–8.15.
13. T. Agustsson and coauthors, Mechanism of increased lipolysis in cancer cachexia, *Cancer Research* 67 (2007): 5531–5537; L. G. Melstrom and coauthors, Mechanisms of skeletal muscle degradation and its therapy in cancer cachexia, *Histology and Histopathology* 22 (2007): 805–814; A. Lelbach, G. Muzes, and J. Feher, Current perspectives of catabolic mediators of cancer cachexia, *Medical Science Monitor* 13 (2007): RA168–173.
14. M. C. Perry, Principles of cancer therapy, in L. Goldman and D. Ausiello, eds., *Cecil Medicine* (Philadelphia: Saunders, 2008), pp. 1370–1387.
15. J. M. Vose and S. Z. Pavletic, Hematopoietic stem cell transplantation, in L. Goldman and D. Ausiello, eds., *Cecil Medicine* (Philadelphia: Saunders, 2008), pp. 1328–1332.
16. Dodson and coauthors, 2011.
17. C. S. Roberts, Patient-physician communication regarding use of complementary therapies during cancer treatment, *Journal of Psychosocial Oncology* 23 (2005): 35–60.
18. C. E. Dennehy and C. Tsourounis, Botanicals ("herbal medications") and nutritional supplements, in B. G. Katzung, ed., *Basic and Clinical Pharmacology* (New York: McGraw-Hill/Lange, 2007), pp. 1050–1062.
19. M. L. Heaney and coauthors, Vitamin C antagonizes the cytotoxic effects of antineoplastic drugs, *Cancer Research* 68 (2008): 8031–8038; B. Bruemmer and coauthors, The association between vitamin C and vitamin E supplement use before hematopoietic stem cell transplant and outcomes to two years, *Journal of the American Dietetic Association* 103 (2003): 982–990.
20. Schattner and Shike, 2006.
21. S. Escott-Stump, *Nutrition and Diagnosis-Related Care* (Baltimore: Lippincott Williams and Wilkins, 2008), pp. 672–686.
22. N. Saquib and coauthors, Weight gain and recovery of pre-cancer weight after breast cancer treatments: Evidence from the women's healthy eating and living (WHEL) study, *Breast Cancer Research and Treatment* 105 (2007): 177–186; G. Makari-Judson, C. H. Judson, and W. C. Mertens, Longitudinal patterns of weight gain after breast cancer diagnosis: Observations beyond the first year, *Breast Journal* 13 (2007): 258–265.
23. American Dietetic Association, *Nutrition Care Manual* (Chicago: American Dietetic Association, 2011.
24. Schattner and Shike, 2006.
25. UNAIDS/WHO, *AIDS Epidemic Update: December 2009*, http://data.unaids.org/pub/Report/2009/2009_epidemic_update_en.pdf; accessed November 21, 2010.
26. G. M. Shaw, Biology of human immunodeficiency viruses, in L. Goldman and D. Ausiello, eds., *Cecil Medicine* (Philadelphia: Saunders, 2008), pp. 2557–2561.
27. Shaw, 2008.
28. H. Masur, L. Healey, and C. Hadigan, Treatment of human immunodeficiency virus infection and acquired immunodeficiency syndrome, in L. Goldman and D. Ausiello, eds., *Cecil Medicine* (Philadelphia: Saunders, 2008), pp. 2571–2582.
29. K. R. Dong, HIV-associated fat deposition, in K. M. Hendricks, K. R. Dong, and J. L. Gerrior, eds., *Nutrition Management of HIV and AIDS* (Chicago: American Dietetic Association, 2009), pp. 57–65; Masur, Healey, and Hadigan, 2008.
30. J. L. Gerrior, Unintentional weight loss and wasting in HIV infection, in K. M. Hendricks, K. R. Dong, and J. L. Gerrior, eds., *Nutrition Management of HIV and AIDS* (Chicago: American Dietetic Association, 2009), pp. 41–55.
31. J. G. Bartlett, Gastrointestinal manifestations of human immunodeficiency virus and acquired immunodeficiency syndrome, in L. Goldman and D. Ausiello, eds., *Cecil Medicine* (Philadelphia: Saunders, 2008), pp. 2582–2585.
32. J. R. Berger and A. Nath, Neurologic complications of human immunodeficiency virus infection, in L. Goldman and D. Ausiello, eds., *Cecil Medicine* (Philadelphia: Saunders, 2008), pp. 2607–2611.
33. E. G. L. Wilkins, Human immunodeficiency virus infection and the human acquired immunodeficiency syndrome, in N. A. Boon, N. R. Colledge, and B. R. Walker, eds., *Davidson's Principles and Practice of Medicine* (Philadelphia: Churchill Livingstone/Elsevier, 2006), pp. 377–402.
34. U.S. Department of Health and Human Services, Panel on Antiretroviral Guidelines for Adults and Adolescents, *Guidelines for the Use of Antiretroviral Agents in HIV-1-Infected Adults and Adolescents*, December 1, 2009, pp. 1–168, http://aidsinfo.nih.gov/contentfiles/AdultandAdolescentGL.pdf, accessed November 21, 2010; S. Safrin, Antiviral agents, in B. G. Katzung, ed., *Basic and Clinical Pharmacology* (New York: McGraw-Hill/Lange, 2007), pp. 790–818.
35. K. Mulligan and coauthors, Testosterone supplementation of megestrol therapy does not enhance lean tissue accrual in men with human immunodeficiency virus-associated weight loss: A randomized, double-blind, placebo-controlled, multicenter trial, *Journal of Clinical Endocrinology and Metabolism* 92 (2007): 563–570.
36. Dennehy and Tsourounis, 2007.
37. L. Vining, General nutrition issues for healthy living with HIV infection, in K. M. Hendricks, K. R. Dong, and J. L. Gerrior, eds., *Nutrition Management of HIV and AIDS* (Chicago: American Dietetic Association, 2009), pp. 23–40.
38. Gerrior, 2009.
39. A. Howard, K. M. Hendricks, and J. Dwyer, Dietary supplement use in HIV infection, in K. M. Hendricks, K. R. Dong, and J. L. Gerrior, eds., *Nutrition Management of HIV and AIDS* (Chicago: American Dietetic Association, 2009), pp. 109–127; P. K. Drain and coauthors, Micronutrients in HIV-positive persons receiving highly active antiretroviral therapy, *American Journal of Clinical Nutrition* 85 (2007).

HIGHLIGHT 29

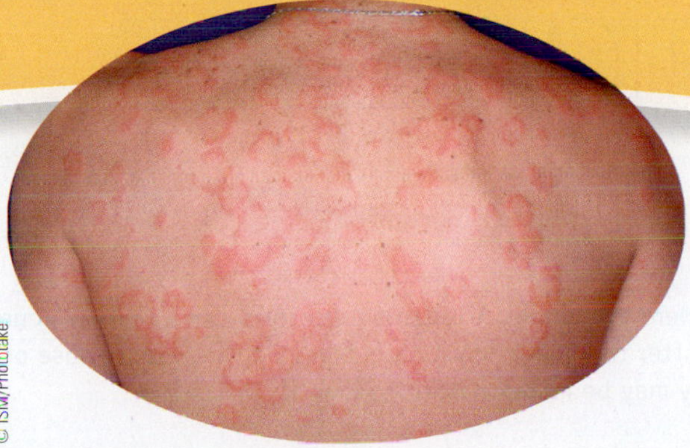

© ISM/Phototake

Food Allergies

Some of the diseases discussed in this book involve adverse reactions to specific foods. Chapter 15 explains that such responses can be categorized either as *food allergies*, which elicit an immune response, or *food intolerances*, which are caused by other physiological processes. Celiac disease and dermatitis herpetiformis, for example, are characterized by allergic reactions to gluten, whereas lactose intolerance, a result of lactase deficiency, is a type of food intolerance. This highlight focuses on the diagnosis and treatment of food allergies, beginning with a brief review of the body's reactions to an **allergen**. The accompanying glossary defines the relevant terms.

A Review of Food Allergy

A food allergy occurs when a food component, usually an incompletely digested protein fragment, is absorbed into the blood and elicits an immune response.[1] The allergen is treated as a foreign particle that needs to be neutralized, and allergen-specific antibodies are produced to mount a defense. These antibodies bind to specialized cells (mast cells and basophils) that release histamine and other inflammatory mediators when the antibodies encounter the allergens. The mediators circulate in the blood and may trigger symptoms in the GI tract, skin, respiratory system, and circulatory system. (In some cases of food allergy, the immune response is controlled by certain immune cells rather than antibodies.) The foods most likely to cause an allergy include eggs, fish, milk, peanuts, shellfish, soybeans, tree nuts, and wheat.[2]

Common symptoms of food allergies include skin rashes, itching, abdominal pain, vomiting, and diarrhea. **Hives** occur frequently; these raised, swollen patches of skin or mucous membranes are associated with intense itching. The most dangerous effect of allergy is **anaphylaxis**, a systemic (whole-body) reaction that can cause breathing difficulty and a dangerous fall in blood pressure, potentially leading to shock. People whose food allergies are intense enough to cause anaphylaxis are often prescribed epinephrine (to counteract the actions of histamine), which they can self-inject in an emergency.

Contact dermatitis or hives can also develop on skin after physical contact with food. In the condition known as **oral allergy syndrome**, hives, swelling, and itching are mostly confined to the lips, tongue, mouth, and throat. These symptoms usually develop following the consumption of raw fruits and vegetables.[3]

Diagnosis of Food Allergy

If a food allergy is suspected, an accurate diagnosis can help a person avoid unnecessary dietary restrictions. Parents who believe a food allergy is causing health or behavioral problems often limit their children's food intakes, which can adversely affect growth and nutrition status.[4] A timely diagnosis can also help a person avoid accidental exposure to a food allergen.

Diagnosis of food allergy requires a thorough medical history, physical examination, and certain types of laboratory tests. The medical history can help to establish whether the symptoms are a response to a true food allergy rather than a food intolerance, foodborne illness, or food toxicity. To help pinpoint the foods that cause symptoms, patients are generally advised to keep a **food and symptom diary**, which provides a record of the foods consumed, the amounts, the symptoms that develop, and the timing between food consumption and symptom onset. Other helpful data include the ingredient lists of prepared or packaged foods

GLOSSARY

allergen: a substance that triggers an allergic response.

anaphylaxis: a severe allergic reaction that may include gastrointestinal upset, skin inflammation, breathing difficulty, and low blood pressure, potentially leading to shock.

cross-reactivity: the ability of an antibody to react to an antigen that is similar, but not identical, to the one that induced the antibody's formation.

food and symptom diary: a food record kept by a patient to determine the cause of an adverse reaction; includes the specific foods and beverages consumed, symptoms experienced, and the timing of meals and symptom onset.

hives: an allergic reaction characterized by raised, swollen patches of skin or mucous membranes that are associated with intense itching; also called *urticaria*.

oral allergy syndrome: an allergic response in which symptoms of hives, swelling, or itching occur only in the mouth and throat; usually a short-lived response that resolves quickly.

Reminder: A *food allergy* is an adverse reaction to food that involves an immune response; also called *food hypersensitivity*. A *food intolerance* is an adverse reaction to food that does not involve an immune response.

that were consumed. If allergy symptoms arise several hours or days after the offending food is ingested, the exact cause of the allergy may be more difficult to identify.

Skin-Prick Testing

The skin-prick test evaluates the patient's responses to commercially prepared food extracts that are introduced into the skin (see the photo). Substances that cause areas of redness and swelling greater than 3 millimeters in diameter are considered possible allergens, and larger responses suggest a greater potential for allergy. Although a positive skin-prick result correctly identifies an allergen in only 50 percent or fewer cases, a negative result is fairly good evidence that the test substance is not the cause of an allergy.[5]

Antibody Blood Testing

Measures of food-specific serum antibodies are useful for assessing the presence of food allergies; generally, a high antibody level suggests an increased risk of an allergic response to a food. Because a person with low antibody levels may still experience an allergic reaction, however, antibody test results need to be considered along with other methods of diagnosis.

Elimination Diets

In an elimination diet, the patient omits all suspected food allergens from the diet until symptoms subside, and then reintroduces individual foods one by one. Although foods that cause symptoms are sometimes easily identified using this method, it may be difficult to identify allergens when they are ingredients in packaged foods. Also, allergic reactions sometimes persist for some time after the allergens are removed from the diet; in these cases, an elemental formula diet (which contains no intact proteins) may be needed to stabilize the patient before foods are reintroduced.

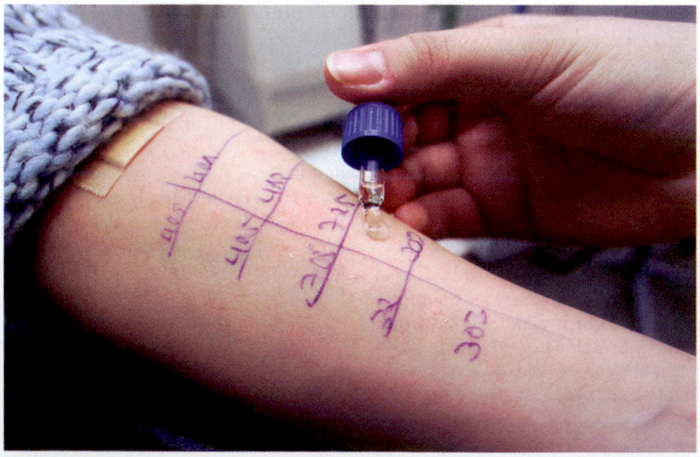

CNImaging/Das Fotoarchiv/Photolibrary

In a skin-prick test, extracts containing food allergens are placed on the skin, and the skin is pricked using a lancet or needle. This technique introduces small amounts of the allergens into the skin.

Oral Food Challenges

When performed properly, oral food challenges are considered the gold standard for diagnosing or confirming a particular food allergy.[6] In an oral challenge, the food suspected of causing allergy is presented to the patient in a dose suggested by the medical history. If the test substance does not cause symptoms, the challenge is repeated to rule out a false-negative result.* Ideally, oral food challenges are double blinded and placebo controlled: the test foods are mixed into other foods or provided in capsules, and placebos are identical in appearance, taste, and texture. Oral challenges can be labor intensive and cannot be performed if a patient has a history of severe anaphylaxis.

Management of Food Allergy

Food allergies are usually managed by eliminating all dietary sources of an allergen. Successful management depends in part on the patient's ability to identify hidden sources of allergens in foods with multiple ingredients. Problem foods may also be consumed at restaurants, schools, and other public places, where the foods' ingredients are not always obvious. In addition, inadvertent ingestion of allergens may occur when foods become contaminated during meal preparation or food processing. The foods that account for most allergic reactions in infants and children are cow's milk, eggs, and peanuts;[7] Table H29-1 lists foods that should be avoided by persons with these allergies.

Although the presence of major food allergens must be listed on food labels (see Chapter 15), labeling requirements do not apply to nonfood items, such as cosmetics, soaps, lotions, shampoos, and medications. Therefore, ingredient lists on these products should be checked carefully to avoid exposure.

Milk Allergy

Milk and the proteins derived from milk are common ingredients in many prepared and packaged foods. In addition, individuals with milk allergies need to avoid milk from all animals due to the potential for **cross-reactivity**. Obtaining sufficient calcium and vitamin D from nonmilk sources may be difficult, and supplementation is often warranted. A milk allergy may be difficult to differentiate from lactose intolerance because both conditions can produce gastrointestinal symptoms.

Egg Allergy

Eggs and egg proteins are common ingredients in many recipes and processed foods. People with egg allergy should avoid eggs from all birds to prevent cross-reactivity. Because flu vaccines

*A *false-negative* test result indicates that a condition is not present (a negative result) when in fact it is (therefore, it is a false result). Conversely, a *false-positive* test result indicates that a condition is present (a positive result) when in fact it is not (therefore, it is a false result).

TABLE H29-1 Milk, Egg, and Peanut Allergies: Foods to Avoid

Food Allergy	Food Ingredients to Exclude	Hidden Sources
Milk allergy	Milk (including dried, evaporated, and condensed milks), milk solids, buttermilk, yogurt, cheese, butter, ghee, artificial butter flavor, half-and-half, cream, whipped cream, custard, pudding, ice cream, casein (or caseinates), whey, protein hydrolysates, lactalbumin, lactoferrin, lactoglobulin, lactulose.	Margarine, luncheon meats, frankfurters and sausages, high-protein products (including bars, flours, and beverages), nougat candy, chocolate bars, caramel color or flavorings, coffee whiteners, bakery glazes, salad dressings, sauces. Meats sliced at a delicatessen are subject to cross-contamination from sliced cheeses.
Egg allergy	Eggs (including powdered eggs and egg substitutes), eggnog, egg white, meringue, albumin, globulin, lysozyme, ovalbumin, ovoglobulin, ovomucin, ovomucoid, ovotransferrin, ovovitellin, lecithin (some food labels may indicate that a "binder" or "emulsifier" was added).	Many baked products and baking mixes, noodles and pastas, mayonnaise, béarnaise and hollandaise sauces, breaded meats and vegetables, candies, fondants, marshmallows, frozen desserts, ice cream, custard, pudding, frankfurters and sausages, processed meats, surimi, cocoa drinks, salad dressings, bakery glazes.
Peanut allergy	Peanuts (also called ground nuts), peanut butter, peanut flour, nut pieces, mixed nuts, beer nuts, artificial nuts, mandalona nuts, peanut sauces (common in Asian cuisine), hydrolyzed vegetable protein (HVP), cold-pressed or gourmet peanut oils (may contain peanut residue), lupin flour.	Chocolate and candy bars, power bars, marzipan, nougat, breakfast cereals, egg rolls, satay sauce, curries, salad dressings. Cross-contamination is possible from food-processing equipment; caution is required when purchasing baked products, ice creams, candies, nut butters, and sunflower seeds.

are prepared using egg embryos, people with egg allergies need to check with their physicians before being vaccinated.

Peanut Allergy

Some people with peanut allergies have severe reactions, including anaphylaxis, to even the smallest quantities of peanuts. Although peanut allergy is not ordinarily associated with other nut allergies, patients may be advised to avoid all nuts due to potential contamination from food-processing equipment (see Table H29-1). People with peanut allergies may also react to lupin flour (produced from seeds of the lupin plant), which is sometimes used as a wheat flour additive in Europe and Australia.

Reevaluation of Food Allergy

Due to the stringent dietary restrictions required for some food allergies, health providers advise that patients with these allergies be reevaluated periodically so that they do not continue the restrictions unnecessarily. Most young children outgrow food allergies within three to five years, and many older children and adults also lose their allergies in time.[8] Individuals with allergies to peanuts, tree nuts, and seafood are least likely to develop tolerance. Reevaluation may require skin-prick tests and oral food challenges, although substantial caution is necessary in patients who experienced severe allergic reactions after consuming certain foods.

References

1. H. A. Sampson, Food allergies, in M. Feldman, L. S. Friedman, and L. J. Brandt, eds., *Sleisenger and Fordtran's Gastrointestinal and Liver Disease* (Philadelphia: Saunders, 2010), pp. 139–148.
2. S. L. Taylor and S. L. Hefle, Food allergies and intolerances, in M. E. Shils and coeditors, *Modern Nutrition in Health and Disease* (Baltimore: Lippincott Williams & Wilkins, 2006), pp. 1512–1530.
3. Sampson, 2010.
4. Taylor and Hefle, 2006.
5. J. A. Chapman and coauthors, Food allergy: A practice parameter, *Annals of Allergy, Asthma, and Immunology* 96 (2006): S1–S68.
6. S. H. Sicherer and H. A. Sampson, Food allergy: Recent advances in pathophysiology and treatment, *Annual Review of Medicine* 60 (2009): 261–277.
7. Sampson, 2010.
8. Sampson, 2010; L. B. Schwartz, Systemic anaphylaxis, food allergy, and insect sting allergy, in L. Goldman and D. Ausiello, eds., *Cecil Medicine* (Philadelphia: Saunders, 2008), pp. 1949–1950.

Appendixes

APPENDIX A

CONTENTS

Cells, Hormones, and Nerves

This appendix is offered as an optional chapter for readers who want to enhance their understanding of how the body coordinates its activities. It presents a brief summary of the structure and function of the body's basic working unit (the cell) and of the body's two major regulatory systems (the hormonal system and the nervous system).

Cells

The body's organs are made up of millions of cells and of materials produced by them. Each **cell** is specialized to perform its organ's functions, but all cells have common structures (see the accompanying glossary and Figure A-1). Every cell is contained within a **cell membrane**. The cell membrane assists in moving materials into and out of the cell, and some of its special proteins act as "pumps" (described in Chapter 6). Some features of cell membranes, such as microvilli (Chapter 3), permit cells to interact with other cells and with their environments in highly specific ways.

Inside the membrane lies the **cytoplasm**, which is filled with **cytosol**, a jelly-like fluid. The cytoplasm contains much more than just cytosol, though. It is a highly organized system of fibers, tubes, membranes, particles, and subcellular **organelles** as complex as a city. These parts intercommunicate, manufacture and exchange materials, package and prepare materials for export, and maintain and repair themselves.

Within each cell is another membrane-enclosed body, the **nucleus**. Inside the nucleus are the **chromosomes**, which contain the genetic material, DNA. The DNA encodes all the instructions for carrying out the cell's activities. The role of DNA in coding for cell proteins is summarized in Figure 6-7 on p. 179. Chapter 6

GLOSSARY OF CELL STRUCTURES

cell: the basic structural unit of all living things.

cell membrane: the thin layer of tissue that surrounds the cell and encloses its contents; made primarily of lipid and protein.

chromosomes: a set of structures within the nucleus of every cell that contains the cell's genetic material, DNA, associated with other materials (primarily proteins).

cytoplasm (SIGH-toh-plazm): the cell contents, except for the nucleus.
- **cyto** = cell
- **plasm** = a form

cytosol: the fluid of cytoplasm; contains water, ions, nutrients, and enzymes.

endoplasmic reticulum (en-doh-PLAZ-mic reh-TIC-you-lum): a complex network of intracellular membranes. The *rough endoplasmic reticulum* is dotted with ribosomes, where protein synthesis takes place. The *smooth endoplasmic reticulum* bears no ribosomes.
- **endo** = inside
- **plasm** = the cytoplasm

Golgi (GOAL-gee) **apparatus:** a set of membranes within the cell where secretory materials are packaged for export.

lysosomes (LYE-so-zomes): cellular organelles; membrane-enclosed sacs of degradative enzymes.
- **lysis** = dissolution

mitochondria (my-toh-KON-dree-uh); singular *mitochondrion:* the cellular organelles responsible for producing ATP aerobically; made of membranes (lipid and protein) with enzymes mounted on them.
- **mitos** = thread (referring to their slender shape)
- **chondros** = cartilage (referring to their external appearance)

nucleus: a major membrane-enclosed body within every cell, which contains the cell's genetic material, DNA, embedded in chromosomes.
- **nucleus** = a kernel

organelles: subcellular structures such as ribosomes, mitochondria, and lysosomes.
- **organelle** = little organ

ribosomes (RYE-boh-zomes): protein-making organelles in cells; composed of RNA and protein.
- **ribo** = containing the sugar ribose (in RNA)
- **some** = body

FIGURE A-1 The Structure of a Typical Cell

The cell shown might be one in a gland (such as the pancreas) that produces secretory products (enzymes) for export (to the intestine). The rough endoplasmic reticulum with its ribosomes produces the enzymes; the smooth reticulum conducts them to the Golgi region; the Golgi membranes merge with the cell membrane, where the enzymes can be released into the extracellular fluid.

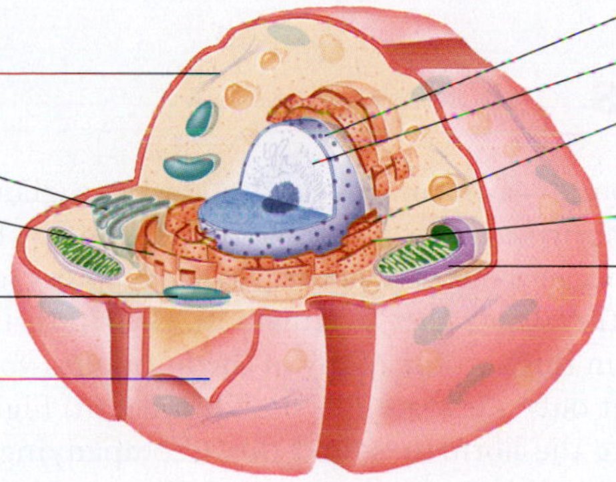

also describes the variety of proteins produced by cells and the ways they perform the body's work.

Among the organelles within a cell are ribosomes, mitochondria, and lysosomes. Figure 6-7 briefly refers to the **ribosomes**; they assemble amino acids into proteins, following directions conveyed to them by RNA.

The **mitochondria** are made of intricately folded membranes that bear thousands of highly organized sets of enzymes on their inner and outer surfaces. Mitochondria are crucial to energy metabolism (described in Chapter 7) and muscles conditioned to work aerobically are packed with them. Their presence is implied whenever the TCA cycle and electron transport chain are mentioned because the mitochondria house the needed enzymes.*

The **lysosomes** are membranes that enclose degradative enzymes. When a cell needs to self-destruct or to digest materials in its surroundings, its lysosomes free their enzymes. Lysosomes are active when tissue repair or remodeling is taking place—for example, in cleaning up infections, healing wounds, shaping embryonic organs, and remodeling bones.

Besides these and other cellular organelles, the cell's cytoplasm contains a highly organized system of membranes, the **endoplasmic reticulum**. The ribosomes may either float free in the cytoplasm or be mounted on these membranes. A membranous surface dotted with ribosomes looks speckled under the microscope and is called "rough" endoplasmic reticulum; such a surface without ribosomes is called "smooth." Some intracellular membranes are organized into tubules that collect cellular materials, merge with the cell membrane, and discharge their contents to the outside of the cell; these membrane systems are named the **Golgi apparatus**, after the scientist who first described them. The rough and smooth endoplasmic reticula and the Golgi apparatus are continuous with one another, so secretions produced deep in the interior of the cell can be efficiently transported to the outside and released. These and other cell structures enable cells to perform the multitudes of functions for which they are specialized.

*For the reactions of glycolysis, the TCA cycle, and the electron transport chain, see Chapter 7 and Appendix C. The reactions of glycolysis take place in the cytoplasm; the conversion of pyruvate to acetyl CoA takes place in the mitochondria, as do the TCA cycle and electron transport chain reactions. The mitochondria then release carbon dioxide, water, and ATP as their end products.

The actions of cells are coordinated by both hormones and nerves, as the next sections show. Among the types of cellular organelles are receptors for the hormones delivering instructions that originate elsewhere in the body. Some hormones penetrate the cell and its nucleus and attach to receptors on chromosomes, where they activate certain genes to initiate, stop, speed up, or slow down synthesis of certain proteins as needed. Other hormones attach to receptors on the cell surface and transmit their messages from there. The hormones ♦ are described in the next section; the nerves, in the one following.

♦ The study of hormones and their effects is **endocrinology.**

Hormones

A chemical compound—a **hormone**—originates in a gland and travels in the bloodstream. The hormone flows everywhere in the body, but only its target organs respond to it because only they possess the receptors to receive it.

The hormones, the glands they originate in, their target organs, and their effects are described in this section. Many of the hormones you might be interested in are included, but only a few are discussed in detail. Figure A-2 identifies the glands that produce the hormones, and the accompanying glossary defines the hormones discussed in this section.

Hormones of the Pituitary Gland and Hypothalamus

The anterior pituitary gland ♦ produces the following hormones, each of which acts on one or more target organs and elicits a characteristic response:

- **Adrenocorticotropin** (ACTH) acts on the adrenal cortex, promoting the production and release of its hormones.
- **Thyroid-stimulating hormone** (TSH) acts on the thyroid gland, promoting the production and release of thyroid hormones.

♦ The **pituitary gland** in the brain has two parts—the **anterior** (front) and the **posterior** (hind).

hormone: a chemical messenger. Hormones are secreted by a variety of endocrine glands in response to altered conditions in the body. Each hormone travels to one or more specific target tissues or organs, where it elicits a specific response to maintain homeostasis.

GLOSSARY OF HORMONES

adrenocorticotropin (ad-REE-noh-KORE-tee-koh-TROP-in) or **ACTH:** a hormone, so named because it stimulates *(trope)* the adrenal cortex. The adrenal gland, like the pituitary, has two parts, in this case an outer portion *(cortex)* and an inner core *(medulla)*. The realease of ACTH is mediated by *corticotropin-releasing hormone (CRH)*.

aldosterone: a hormone from the adrenal gland involved in blood pressure regulation.
- **aldo** = aldehyde

angiotensin: a hormone involved in blood pressure regulation that is activated by *renin* (REN-in), an enzyme from the kidneys.
- **angio** = blood vessels
- **tensin** = pressure
- **ren** = kidneys

antidiuretic hormone (ADH): the hormone that prevents water loss in urine (also called **vasopressin**).
- **anti** = against
- **di** = through
- **ure** = urine
- **vaso** = blood vessels
- **pressin** = pressure

calcitonin (KAL-see-TOH-nin): a hormone secreted by the thyroid gland that regulates (tones) calcium metabolism.

erythropoietin (eh-RITH-ro-POY-eh-tin): a hormone that stimulates red blood cell production.
- **erythro** = red (blood cell)
- **poiesis** = creating (like poetry)

estrogens: hormones responsible for the menstrual cycle and other female characteristics.
- **oestrus** = the egg-making cycle
- **gen** = gives rise to

follicle-stimulating hormone (FSH): a hormone that stimulates maturation of the ovarian follicles in females and the production of sperm in males. (The ovarian follicles are part of the female reproductive system where the eggs are produced.) The release of FSH is mediated by **follicle-stimulating hormone releasing hormone (FSH–RH).**

glucocorticoids: hormones from the adrenal cortex that affect the body's management of glucose.
- **gluco** = glucose
- **corticoid** = from the cortex

growth hormone (GH): a hormone secreted by the pituitary that regulates the cell division and protein synthesis needed for normal growth (also called **somatotropin**). The release of GH is mediated by **GH-releasing hormone (GHRH)** and **GH-inhibiting hormone (GHIH).**

luteinizing (LOO-tee-in-EYE-zing) **hormone (LH):** a hormone that stimulates ovulation and the development of the corpus luteum (the small tissue that develops from a ruptured ovarian follicle and secretes hormones); so called because the follicle turns yellow as it matures. In men, LH stimulates testosterone secretion. The release of LH is mediated by **luteinizing hormone–releasing hormone (LH–RH).**
- **lutein** = a yellow pigment

oxytocin (OCK-see-TOH-sin): a hormone that stimulates the mammary glands to eject milk during lactation and the uterus to contract during childbirth.
- **oxy** = quick
- **tocin** = childbirth

progesterone: the hormone of gestation (pregnancy).
- **pro** = promoting
- **gest** = gestation (pregnancy)
- **sterone** = a steroid hormone

prolactin (proh-LAK-tin): a hormone so named because it promotes *(pro)* the production of milk *(lacto)*. The release of prolactin is mediated by **prolactin-inhibiting hormone (PIH).**

relaxin: the hormone of late pregnancy.

somatostatin (GHIH): a hormone that inhibits the release of growth hormone; the opposite of **somatotropin (GH).**
- **somato** = body
- **stat** = keep the same
- **tropin** = make more

testosterone: a steroid hormone from the testicles, or testes. The steroids, as explained in Chapter 5, are chemically related to, and some are derived from, the lipid cholesterol.
- **sterone** = a steroid hormone

thyroid-stimulating hormone (TSH): a hormone secreted by the pituitary that stimulates the thyroid gland to secrete its hormones—thyroxine and triiodothyronine. The release of TSH is mediated by **TSH-releasing hormone (TRH).**

FIGURE A-2 The Endocrine System

These organs and glands release hormones that regulate body processes. An *endocrine gland* secretes its product directly into *(endo)* the blood; for example, the pancreas cells that produce insulin. An *exocrine gland* secretes its product(s) out *(exo)* to an epithelial surface either directly or through a duct; the sweat glands of the skin and the enzyme-producing glands of the pancreas are both examples. The pancreas is therefore both an endocrine and an exocrine gland.

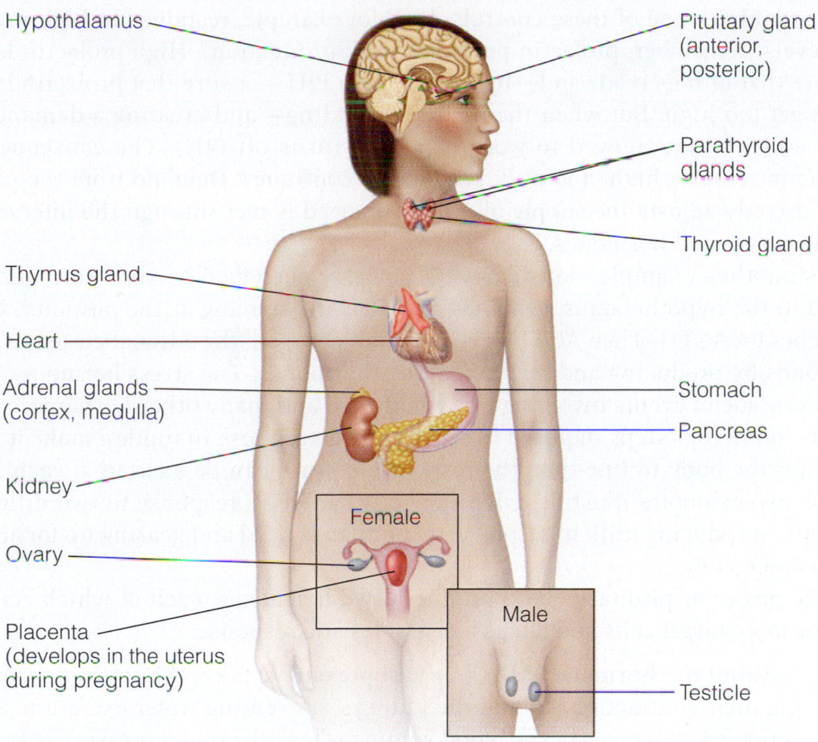

- **Growth hormone** (GH) or **somatotropin** acts on all tissues, promoting growth, fat breakdown, and the formation of antibodies.
- **Follicle-stimulating hormone** (FSH) acts on the ovaries in the female, promoting their maturation, and on the testicles in the male, promoting sperm formation.
- **Luteinizing hormone** (LH) also acts on the ovaries, stimulating their maturation, the production and release of progesterone and estrogens, and ovulation; and on the testicles, promoting the production and release of testosterone.
- **Prolactin**, secreted in the female during pregnancy and lactation, acts on the mammary glands to stimulate their growth and the production of milk.

Each of these hormones has one or more signals that turn it on and another (or others) that turns it off. ♦ Among the controlling signals are several hormones from the hypothalamus:

- **Corticotropin-releasing hormone** (CRH), which promotes release of ACTH, is turned on by stress and turned off by ACTH when enough has been released.
- **TSH-releasing hormone** (TRH), which promotes release of TSH, is turned on by large meals or low body temperature.
- **GH-releasing hormone** (GHRH), which stimulates the release of growth hormone, is turned on by insulin.

♦ Hormones that are turned off by their own effects are said to be regulated by **negative feedback** (see Figure 3-13 on p. 84).

- **GH-inhibiting hormone (GHIH** or **somatostatin)**, which inhibits the release of GH and interferes with the release of TSH, is turned on by hypoglycemia and/or physical activity and is rapidly destroyed by body tissues so that it does not accumulate.
- **FSH/LH–releasing hormone (FSH/LH–RH)** is turned on in the female by nerve messages or low estrogen and in the male by low testosterone.
- **Prolactin-inhibiting hormone (PIH)** is turned on by high prolactin levels and off by estrogen, testosterone, and suckling (by way of nerve messages).

Let's examine some of these controls. PIH, for example, responds to high prolactin levels (remember, prolactin promotes milk production). High prolactin levels ensure that milk is made and—by calling forth PIH—ensure that prolactin levels don't get too high. But when the infant is suckling—and creating a demand for milk—PIH is not allowed to work (suckling turns off PIH). The consequence: prolactin remains high, and milk production continues. Demand from the infant thus directly adjusts the supply of milk. The need is met through the interaction of the nerves and hormones.

As another example, consider CRH. Stress, perceived in the brain and relayed to the hypothalamus, switches on CRH. On arriving at the pituitary, CRH switches on ACTH. Then ACTH acts on its target organ, the adrenal cortex, which responds by producing and releasing stress hormones. The stress hormones trigger a cascade of events involving every body cell and many other hormones.

The numerous steps required to set the stress response in motion make it possible for the body to fine-tune the response; control can be exerted at each step. These two examples illustrate what the body can do in response to two different stimuli—producing milk in response to an infant's need and gearing up for action in an emergency.

The posterior pituitary gland produces two hormones, each of which acts on one or more target cells and elicits a characteristic response:

- **Antidiuretic hormone (ADH)**, or **vasopressin**, acts on the arteries, promoting their contraction, and on the kidneys, preventing water excretion. ADH is turned on whenever the blood volume is low, the blood pressure is low, or the salt concentration of the blood is high (see Chapter 12). It is turned off by the return of these conditions to normal.
- **Oxytocin** acts during late pregnancy on the uterus, inducing contractions, and during lactation on the mammary glands, causing milk ejection. Oxytocin is produced in response to reduced progesterone levels, suckling, or the stretching of the cervix.

Hormones that Regulate Energy Metabolism

Hormones produced by a number of different glands have effects on energy metabolism:

- Insulin from the pancreas beta cells is turned on by many stimuli, including high blood glucose. It acts on cells to increase glucose and amino acid uptake into them and to promote the secretion of GHRH.
- Glucagon from the pancreas alpha cells responds to low blood glucose and acts on the liver to promote the breakdown of glycogen to glucose, the conversion of amino acids to glucose, and the release of glucose into the blood.
- Thyroxine from the thyroid gland responds to TSH and acts on many cells to increase their metabolic rate, growth, and heat production.
- Norepinephrine and epinephrine ♦ from the adrenal medulla respond to stimulation by sympathetic nerves and produce reactions in many cells that facilitate the body's readiness for fight or flight: increased heart activity, blood vessel constriction, breakdown of glycogen and glucose, raised blood glucose levels, and fat breakdown. Norepinephrine and epinephrine also influence the secretion of the many hormones from the hypothalamus that exert control on the body's other systems.

♦ Norepinephrine and epinephrine were formerly called **noradrenalin** and **adrenalin,** respectively.

- Growth hormone (GH) from the anterior pituitary (already mentioned).
- **Glucocorticoids** from the adrenal cortex become active during times of stress and carbohydrate metabolism.

Every body part is affected by these hormones. Each different hormone has unique effects; and hormones that oppose each other are produced in carefully regulated amounts, so each can respond to the exact degree that is appropriate to the condition.

Hormones that Adjust Other Body Balances Hormones are involved in moving calcium into and out of the body's storage deposits in the bones:

- **Calcitonin** from the thyroid gland acts on the bones, which respond by storing calcium from the bloodstream whenever blood calcium rises above the normal range. It also acts on the kidneys to increase excretion of both calcium and phosphorus in the urine. Calcitonin plays a major role in infants and young children, but is less active in adults.
- Parathyroid hormone (parathormone or PTH) from the parathyroid gland responds to the opposite condition—lowered blood calcium—and acts on three targets: the bones, which release stored calcium into the blood; the kidneys, which slow the excretion of calcium; and the intestine, which increases calcium absorption.
- Vitamin D from the skin and activated in the kidneys acts with parathyroid hormone and is essential for the absorption of calcium in the intestine.

Figure 12-12 on p. 401 diagrams the ways vitamin D and the hormones calcitonin and parathyroid hormone regulate calcium homeostasis.

Another hormone has effects on blood-making activity:

- **Erythropoietin** from the kidneys is responsive to oxygen depletion of the blood and to anemia. It acts on the bone marrow to stimulate the making of red blood cells.

Another hormone is special for pregnancy:

- **Relaxin** from the ovaries is secreted in response to the raised progesterone and estrogen levels of late pregnancy. This hormone acts on the cervix and pelvic ligaments to allow them to stretch so that they can accommodate the birth process without strain.

Other agents help regulate blood pressure:

- **Renin** (an enzyme), from the kidneys, in cooperation with **angiotensin** in the blood responds to a reduced blood supply experienced by the kidneys and acts in several ways to increase blood pressure. Renin and angiotensin also stimulate the adrenal cortex to secrete the hormone aldosterone.
- **Aldosterone**, a hormone from the adrenal cortex, targets the kidneys, which respond by reabsorbing sodium. The effect is to retain more water in the bloodstream—thus, again, raising the blood pressure. Figure 12-3 (on p. 388) in Chapter 12 provides more details.

The Gastrointestinal Hormones Several hormones are produced in the stomach and intestines in response to the presence of food or the components of food:

- Gastrin from the stomach and duodenum stimulates the production and release of gastric acid and other digestive juices and the movement of the GI contents through the system.
- Cholecystokinin from the duodenum signals the gallbladder and pancreas to release their contents into the intestine to aid in digestion.
- Secretin from the duodenum calls forth acid-neutralizing bicarbonate from the pancreas into the intestine and slows the action of the stomach and its secretion of acid and digestive juices.

- Gastric-inhibitory peptide from the duodenum and jejunum inhibits the secretion of gastric acid and slows the process of digestion.

These hormones are defined and discussed in Chapter 3.

The Sex Hormones There are three major sex hormones:

- **Testosterone** from the testicles is released in response to LH (described earlier) and acts on all the tissues that are involved in male sexuality, promoting their development and maintenance.
- **Estrogens** from the ovaries are released in response to both FSH and LH and act similarly in females.
- **Progesterone** from the ovaries' corpus luteum and from the placenta acts on the uterus and mammary glands, preparing them for pregnancy and lactation.

This brief description of the hormones and their functions should suffice to provide an awareness of the enormous impact these compounds have on body processes. The other overall regulating agency is the nervous system.

Nerves

The nervous system has a central control system that can evaluate information about conditions within and outside the body, and a vast system of wiring that receives information and sends instructions. The control unit is the brain and spinal cord, called the **central nervous system**; and the vast complex of wiring between the center and the parts is the **peripheral nervous system**. The smooth functioning that results from the systems' adjustments to changing conditions is homeostasis.

The nervous system has two general functions: it controls voluntary muscles in response to sensory stimuli from them, and it controls involuntary, internal muscles and glands in response to nerve-borne and chemical signals about their status. In fact, the nervous system is best understood as two systems that use the same or similar pathways to receive and transmit their messages. The **somatic nervous system** controls the voluntary muscles; the **autonomic nervous system** controls the internal organs.

When scientists were first studying the autonomic nervous system, they noticed that when something hurt one organ of the body, some of the other organs reacted as if in sympathy for the afflicted one. They therefore named the nerve network they were studying the sympathetic nervous system. The term is still used today to refer to that branch of the autonomic nervous system that responds to pain and stress. The other branch is called the parasympathetic nervous system. (Think of the sympathetic branch as the responder when homeostasis needs restoring and the parasympathetic branch as the commander of function during normal times.) Both systems transmit their messages through the brain and spinal cord. Nerves of the two branches travel side by side along the same pathways to transmit their messages, but they oppose each other's actions (see Figure A-3).

An example will show how the sympathetic and parasympathetic nervous systems work to maintain homeostasis. When you go outside in cold weather, your skin's temperature receptors send "cold" messages to the spinal cord and brain.

GLOSSARY OF NERVOUS SYSTEM

autonomic nervous system: the division of the nervous system that controls the body's automatic responses. Its two branches are the **sympathetic** branch, which helps the body respond to stressors from the outside environment, and the **parasympathetic** branch, which regulates normal body activities between stressful times.
- **autonomos** = self-governing

central nervous system: the central part of the nervous system; the brain and spinal cord.

peripheral (puh-RIFF-er-ul) **nervous system:** the peripheral (outermost) part of the nervous system; the vast complex of wiring that extends from the central nervous system to the body's outermost areas. It contains both somatic and autonomic components.

somatic (so-MAT-ick) **nervous system:** the division of the nervous system that controls the voluntary muscles, as distinguished from the autonomic nervous system, which controls involuntary functions.
- **soma** = body

FIGURE A-3 The Organization of the Nervous System

The brain and spinal cord evaluate information about conditions within and outside the body, and the peripheral nerves receive information and send instructions.

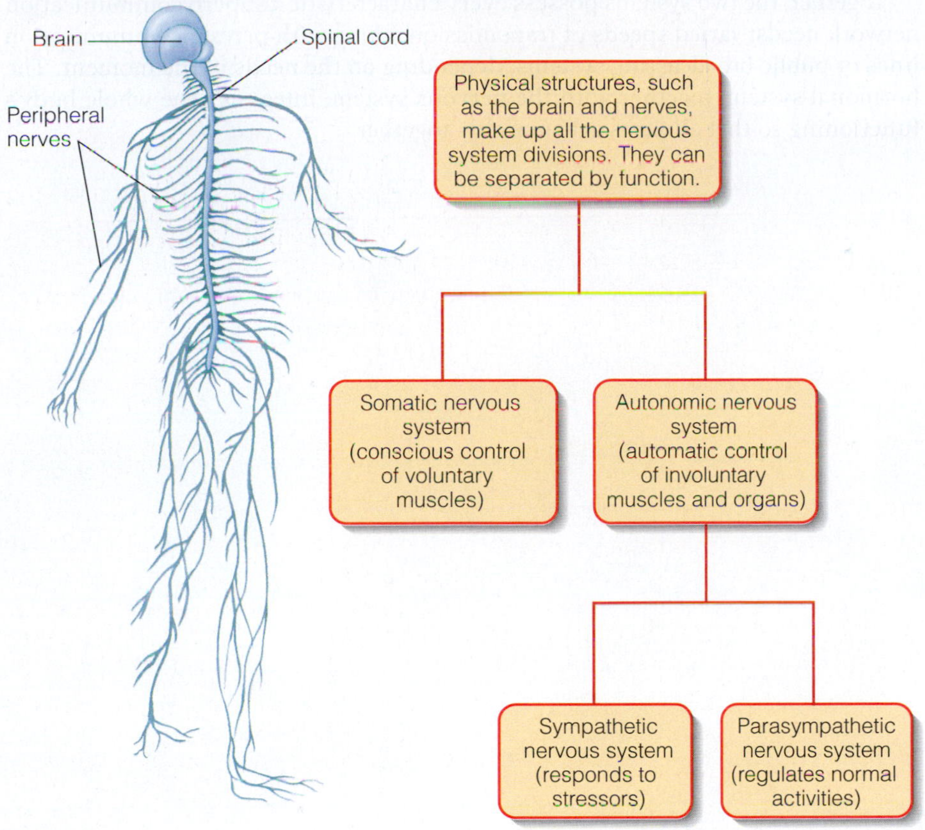

Your conscious mind may intervene at this point to tell you to zip your jacket, but let's say you have no jacket. Your sympathetic nervous system reacts to the external stressor, the cold. It signals your skin-surface capillaries to shut down so that your blood will circulate deeper in your tissues, where it will conserve heat. Your sympathetic nervous system also signals involuntary contractions of the small muscles just under the skin surface. The product of these muscle contractions is heat, and the visible result is goose bumps. If these measures do not raise your body temperature enough, then the sympathetic nerves signal your large muscle groups to shiver; the contractions of these large muscles produce still more heat. All of this activity helps to maintain your homeostasis (with respect to temperature) under conditions of external extremes (cold) that would throw it off balance. The cold was a stressor; the body's response was resistance.

Now let's say you come in and sit by a fire and drink hot cocoa. You are warm and no longer need all that sympathetic activity. At this point, your parasympathetic nerves take over; they signal your skin-surface capillaries to dilate again, your goose bumps to subside, and your muscles to relax. Your body is back to normal. This is recovery.

Putting It Together

The hormonal and nervous systems coordinate body functions by transmitting and receiving messages. The point-to-point messages of the nervous system travel through a central switchboard (the spinal cord and brain), whereas the messages of the hormonal system are broadcast over the airways (the bloodstream), and any organ with the appropriate receptors can pick them up. Nerve impulses travel faster than hormonal messages do—although both are remarkably swift. Whereas your brain's command to wiggle your toes reaches the toes within a fraction of a second and stops as quickly, a gland's message to alter a body condition may take several seconds or minutes to get started and may fade away equally slowly.

Together, the two systems possess every characteristic a superb communication network needs: varied speeds of transmission, along with private communication lines or public broadcasting systems, depending on the needs of the moment. The hormonal system, together with the nervous system, integrates the whole body's functioning so that all parts act smoothly together.

Basic Chemistry Concepts

APPENDIX B

CONTENTS

This appendix is intended to provide the background in basic chemistry you need to understand the nutrition concepts presented in this book. Chemistry is the branch of natural science that is concerned with the description and classification of **matter**, the changes that matter undergoes, and the **energy** associated with these changes. The accompanying glossary defines matter, energy, and other related terms.

GLOSSARY

atoms: the smallest components of an element that have all of the properties of the element.

compound: a substance composed of two or more different atoms—for example, water (H_2O).

element: a substance composed of atoms that are alike—for example, iron (Fe).

energy: the capacity to do work.

matter: anything that takes up space and has mass.

molecule: two or more atoms of the same or different elements joined by chemical bonds. Examples are molecules of the element oxygen, composed of two oxygen atoms (O_2), and molecules of the compound water, composed of two hydrogen atoms and one oxygen atom (H_2O).

Matter: The Properties of Atoms

Every substance has physical and chemical properties that distinguish it from all other substances and thus give it a unique identity. The physical properties include such characteristics as color, taste, texture, and odor, as well as the temperatures at which a substance changes its state (from a solid to a liquid or from a liquid to a gas) and the weight of a unit volume (its density). The chemical properties of a substance have to do with how it reacts with other substances or responds to a change in its environment so that new substances with different sets of properties are produced.

A physical change does not change a substance's chemical composition. The three physical states—ice, water, and steam—all consist of two hydrogen atoms and one oxygen atom bound together. In contrast, a chemical change occurs when an electric current passes through water. The water disappears, and two different substances are formed: hydrogen gas, which is flammable, and oxygen gas, which supports life.

Substances: Elements and Compounds

The smallest part of a substance that can exist separately without losing its physical and chemical properties is a **molecule.** If a molecule is composed of **atoms** that are alike, the substance is an **element** (for example, O_2). If a molecule is composed of two or more different kinds of atoms, the substance is a **compound** (for example, H_2O).

Just over 100 elements are known, and these are listed in Table B-1 (p. B-2). A familiar example is hydrogen, whose molecules are composed only of hydrogen atoms linked together in pairs (H_2). On the other hand, more than a million compounds are known. An example is the sugar glucose. Each of its molecules is composed of 6 carbon, 6 oxygen, and 12 hydrogen atoms linked together in a specific arrangement (as described in Chapter 4).

The Nature of Atoms

Atoms themselves are made of smaller particles. Within the atomic nucleus are protons (positively charged particles), and surrounding the nucleus are electrons (negatively charged particles). The number of protons (+) in the nucleus of an atom determines the number of electrons (−) around it. The positive charge on a proton is equal to the negative charge on an electron, so the charges cancel each other out and leave the atom neutral to its surroundings.

The nucleus may also include neutrons, subatomic particles that have no charge. Protons and neutrons are of equal mass, and together they give an atom its weight. Electrons bond atoms together to make molecules, and they are involved in chemical reactions.

Each type of atom has a characteristic number of protons in its nucleus. The hydrogen atom is the simplest of all. It possesses a single proton, with a single electron associated with it:

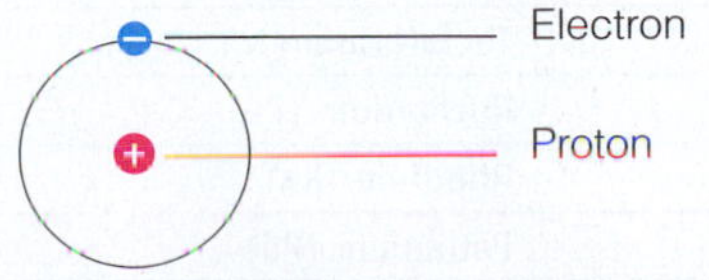

Hydrogen atom (H), atomic number 1.

Just as hydrogen always has one proton, helium always has two, lithium three, and so on. The atomic number of each element is the number of protons in the nucleus of that atom, and

APPENDIX B

TABLE B-1 **Chemical Symbols for the Elements**

Number of Protons (Atomic Number)	Element	Number of Electrons in Outer Shell	Number of Protons (Atomic Number)	Element	Number of Electrons in Outer Shell
1	Hydrogen (H)	1	50	Tin (Sn)	4
2	Helium (He)	2	51	Antimony (Sb)	5
3	Lithium (Li)	1	52	Tellurium (Te)	6
4	Beryllium (Be)	2	53	Iodine (I)	7
5	Boron (B)	3	54	Xenon (Xe)	8
6	Carbon (C)	4	55	Cesium (Cs)	1
7	Nitrogen (N)	5	56	Barium (Ba)	2
8	Oxygen (O)	6	57	Lanthanum (La)	2
9	Fluorine (F)	7	58	Cerium (Ce)	2
10	Neon (Ne)	8	59	Praseodymium (Pr)	2
11	Sodium (Na)	1	60	Neodymium (Nd)	2
12	Magnesium (Mg)	2	61	Promethium (Pm)	2
13	Aluminum (Al)	3	62	Samarium (Sm)	2
14	Silicon (Si)	4	63	Europium (Eu)	2
15	Phosphorus (P)	5	64	Gadolinium (Gd)	2
16	Sulfur (S)	6	65	Terbium (Tb)	2
17	Chlorine (Cl)	7	66	Dysprosium (Dy)	2
18	Argon (Ar)	8	67	Holmium (Ho)	2
19	Potassium (K)	1	68	Erbium (Er)	2
20	Calcium (Ca)	2	69	Thulium (Tm)	2
21	Scandium (Sc)	2	70	Ytterbium (Yb)	2
22	Titanium (Ti)	2	71	Lutetium (Lu)	2
23	Vanadium (V)	2	72	Hafnium (Hf)	2
24	Chromium (Cr)	1	73	Tantalum (Ta)	2
25	Manganese (Mn)	2	74	Tungsten (W)	2
26	Iron (Fe)	2	75	Rhenium (Re)	2
27	Cobalt (Co)	2	76	Osmium (Os)	2
28	Nickel (Ni)	2	77	Iridium (Ir)	2
29	Copper (Cu)	1	78	Platinum (Pt)	1
30	Zinc (Zn)	2	79	Gold (Au)	1
31	Gallium (Ga)	3	80	Mercury (Hg)	2
32	Germanium (Ge)	4	81	Thallium (Tl)	3
33	Arsenic (As)	5	82	Lead (Pb)	4
34	Selenium (Se)	6	83	Bismuth (Bi)	5
35	Bromine (Br)	7	84	Polonium (Po)	6
36	Krypton (Kr)	8	85	Astatine (At)	7
37	Rubidium (Rb)	1	86	Radon (Rn)	8
38	Strontium (Sr)	2	87	Francium (Fr)	1
39	Yttrium (Y)	2	88	Radium (Ra)	2
40	Zirconium (Zr)	2	89	Actinium (Ac)	2
41	Niobium (Nb)	1	90	Thorium (Th)	2
42	Molybdenum (Mo)	1	91	Protactinium (Pa)	2
43	Technetium (Tc)	1	92	Uranium (U)	2
44	Ruthenium (Ru)	1	93	Neptunium (Np)	2
45	Rhodium (Rh)	1	94	Plutonium (Pu)	2
46	Palladium (Pd)	—	95	Americium (Am)	2
47	Silver (Ag)	1	96	Curium (Cm)	2
48	Cadmium (Cd)	2	97	Berkelium (Bk)	2
49	Indium (In)	3	98	Californium (Cf)	2

(continued)

TABLE B-1 Chemical Symbols for the Elements (*continued*)

Number of Protons (Atomic Number)	Element	Number of Electrons in Outer Shell	Number of Protons (Atomic Number)	Element	Number of Electrons in Outer Shell
99	Einsteinium (Es)	2	105	Dubnium (Db)	2
100	Fermium (Fm)	2	106	Seaborgium (Sg)	2
101	Mendelevium (Md)	2	107	Bohrium (Bh)	2
102	Nobelium (No)	2	108	Hassium (Hs)	2
103	Lawrencium (Lr)	2	109	Meitnerium (Mt)	2
104	Rutherfordium (Rf)	2	110	Darmstadtium (Ds)	2

Key
- Elements found in energy-yielding nutrients, vitamins, and water
- Major minerals
- Trace minerals

this never changes in a chemical reaction; it gives the atom its identity. The atomic numbers for the known elements are listed in Table B-1.

Besides hydrogen, the atoms most common in living things are carbon (C), nitrogen (N), and oxygen (O), whose atomic numbers are 6, 7, and 8, respectively. Their structures are more complicated than that of hydrogen, but each of them possesses the same number of electrons as there are protons in the nucleus. These electrons are found in orbits, or shells (shown below).

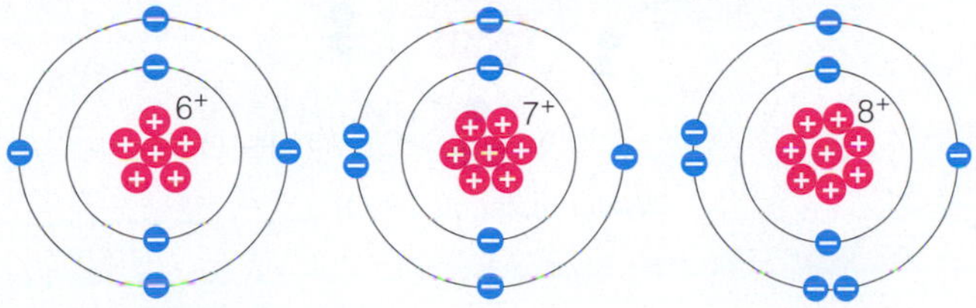

In these and all diagrams of atoms that follow, only the protons and electrons are shown. The neutrons, which contribute only to atomic weight, not to charge, are omitted.

The most important structural feature of an atom for determining its chemical behavior is the number of electrons in its outermost shell. The first, or innermost, shell is full when it is occupied by two electrons; so an atom with two or more electrons has a filled first shell. When the first shell is full, electrons begin to fill the second shell.

The second shell is completely full when it has eight electrons. A substance that has a full outer shell tends not to enter into chemical reactions. Atomic number 10, neon, is a chemically inert substance because its outer shell is complete. Fluorine, atomic number 9, has a great tendency to draw an electron from other substances to complete its outer shell, and thus it is highly reactive. Carbon has a half-full outer shell, which helps explain its great versatility; it can combine with other elements in a variety of ways to form a large number of compounds.

Atoms seek to reach a state of maximum stability or of lowest energy in the same way that a ball will roll down a hill until it reaches the lowest place. An atom achieves a state of maximum stability:

- By gaining or losing electrons to either fill or empty its outer shell.
- By sharing its electrons with other atoms and thereby completing its outer shell.

The number of electrons determines how the atom will chemically react with other atoms. The atomic number, not the weight, is what gives an atom its chemical nature.

Chemical Bonding

Atoms often complete their outer shells by sharing electrons with other atoms. In order to complete its outer shell, a carbon atom requires four electrons. A hydrogen atom requires one. Thus, when a carbon atom shares electrons with four hydrogen atoms, each completes its outer shell (as shown below). Electron sharing binds the atoms together and satisfies the conditions of maximum stability for the molecule. The outer shell of each atom is complete because hydrogen effectively has the required 2 electrons in its first (outer) shell, and carbon has 8 electrons in its second (outer) shell; and the molecule is electrically neutral, with a total of 10 protons and 10 electrons.

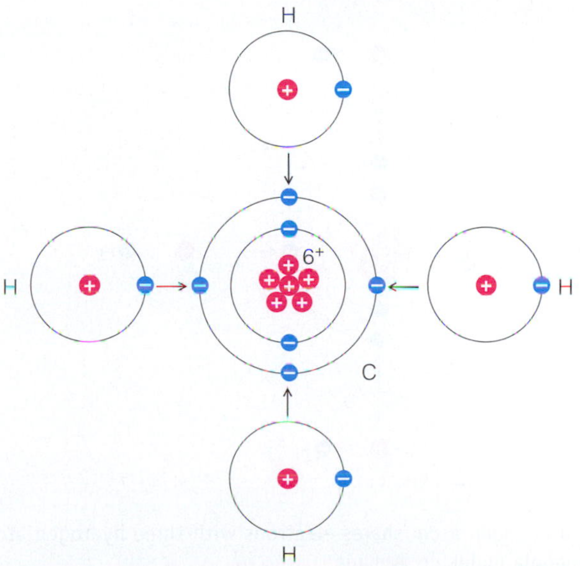

When a carbon atom shares electrons with four hydrogen atoms, a methane molecule is made.

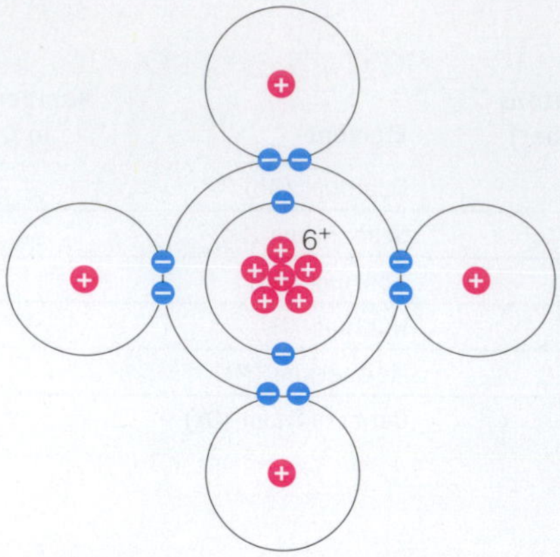

The chemical formula for methane is CH_4. Note that by sharing electrons, every atom achieves a filled outer shell.

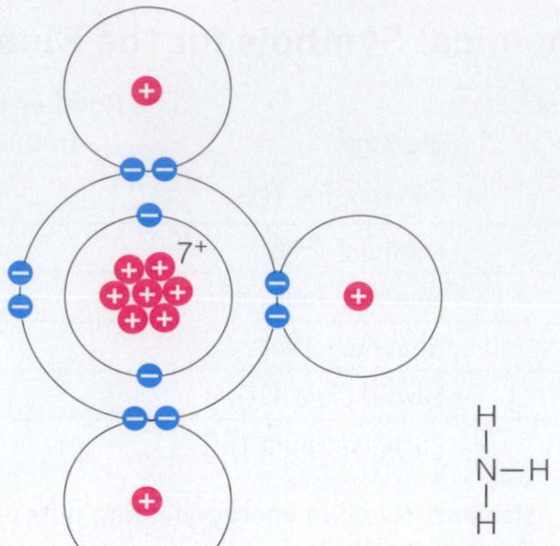

Ammonia (NH_3)

The chemical formula for ammonia is NH_3. Count the electrons in each atom's outer shell to confirm that it is filled.

Bonds that involve the sharing of electrons, like the bonds between carbon and the four hydrogens, are the most stable kind of association that atoms can form with one another. These bonds are called covalent bonds, and the resulting combination of atoms is called a molecule. A single pair of shared electrons forms a single bond. A simplified way to represent a single bond is with a single line. Thus the structure of methane (CH_4) could be represented like this:

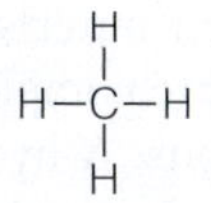

Methane (CH_4)

Similarly, one nitrogen atom and three hydrogen atoms can share electrons to form one molecule of ammonia (NH_3):

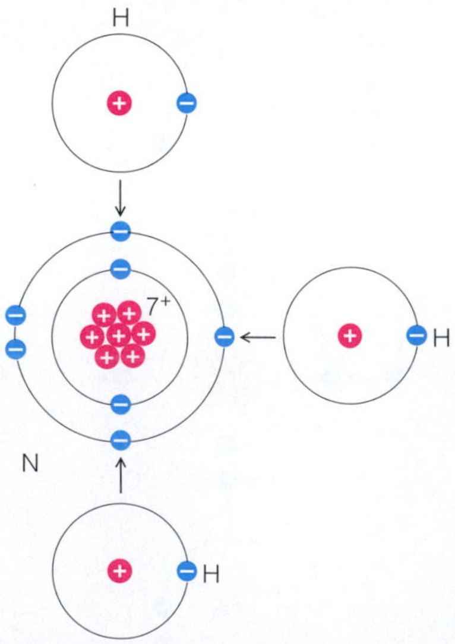

When a nitrogen atom shares electrons with three hydrogen atoms, an ammonia molecule is made.

One oxygen atom may be bonded to two hydrogen atoms to form one molecule of water (H_2O):

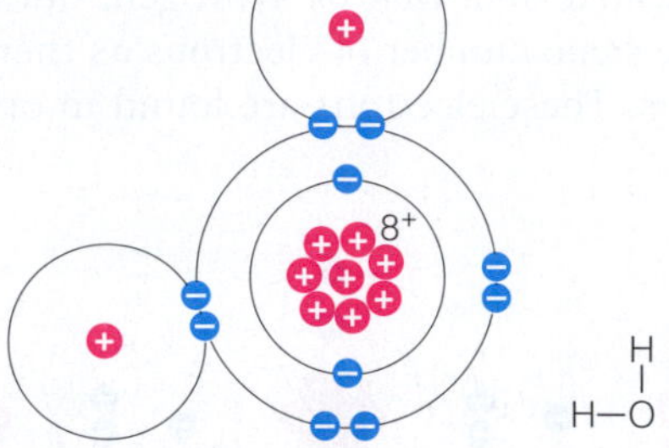

Water molecule (H_2O)

When two oxygen atoms form a molecule of oxygen, they must share two pairs of electrons. This double bond may be represented as two single lines:

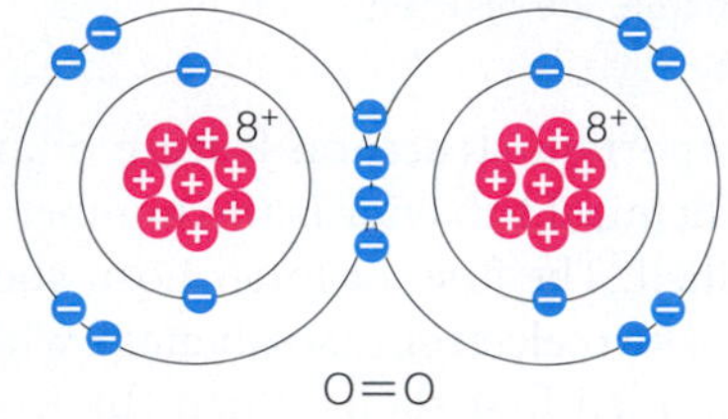

Oxygen molecule (O_2)

Small atoms form the tightest, most stable bonds. H, O, N, and C are the smallest atoms capable of forming one, two, three, and four electron-pair bonds, respectively. This is the basis for the statement in Chapter 4 that in drawings of compounds containing these atoms, hydrogen must always have one, oxygen two, nitrogen three, and carbon four bonds radiating to other atoms:

TABLE B-2 Elemental Composition of the Human Body

Element	Chemical Symbol	By Weight (%)
Oxygen	O	65.0
Carbon	C	18.0
Hydrogen	H	10.0
Nitrogen	N	3.0
Calcium	Ca	1.5
Phosphorus	P	1.0
Potassium	K	0.4
Sulfur	S	0.3
Sodium	Na	0.2
Chloride	Cl	0.1
Magnesium	Mg	0.1
Total		99.6[a]

[a]The remaining 0.4 percent by weight is contributed by the trace elements: chromium (Cr), copper (Cu), zinc (Zn), selenium (Se), molybdenum (Mo), fluorine (F), iodine (I), manganese (Mn), and iron (Fe). Cells may also contain variable traces of some of the following: boron (B), cobalt (Co), lithium (Li), strontium (Sr), aluminum (Al), silicon (Si), lead (Pb), vanadium (V), arsenic (As), bromine (Br), and others.

The stability of the associations between these small atoms and the versatility with which they can combine make them very common in living things. Interestingly, all cells, whether they come from animals, plants, or bacteria, contain the same elements in very nearly the same proportions. The elements commonly found in living things are shown in Table B-2.

Formation of Ions

An atom such as sodium (Na, atomic number 11) cannot easily fill its outer shell by sharing. Sodium possesses a filled first shell of two electrons and a filled second shell of eight; there is only one electron in its outermost shell:

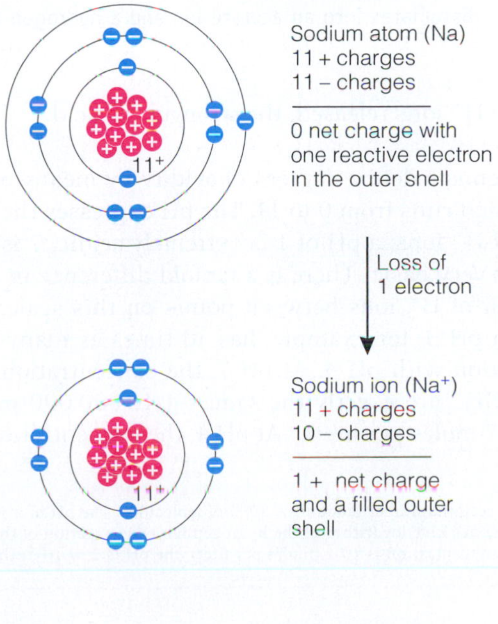

If sodium loses this electron, it satisfies one condition for stability: a filled outer shell (now its second shell counts as the outer shell). However, it is not electrically neutral. It has 11 protons (positive) and only 10 electrons (negative). It therefore has a net positive charge. An atom or molecule that has lost or gained one or more electrons and so is electrically charged is called an ion.

An atom such as chlorine (Cl, atomic number 17), with seven electrons in its outermost shell, can share electrons to fill its outer shell, or it can gain one electron to complete its outer shell and thus give it a negative charge:

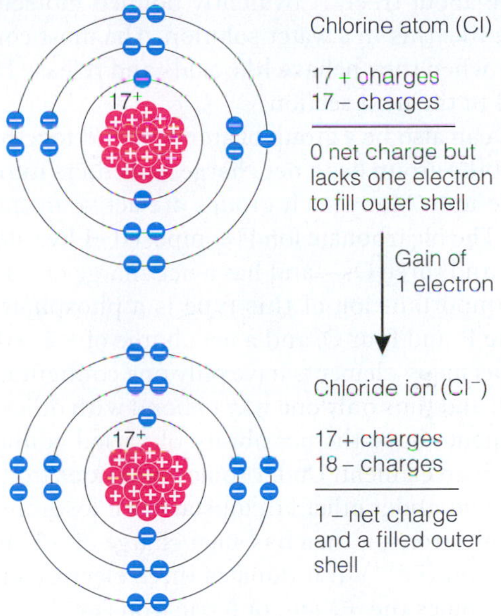

A positively charged ion such as sodium ion (Na^+) is called a cation; a negatively charged ion such as a chloride ion (Cl^-) is called an anion. Cations and anions attract one another to form salts:

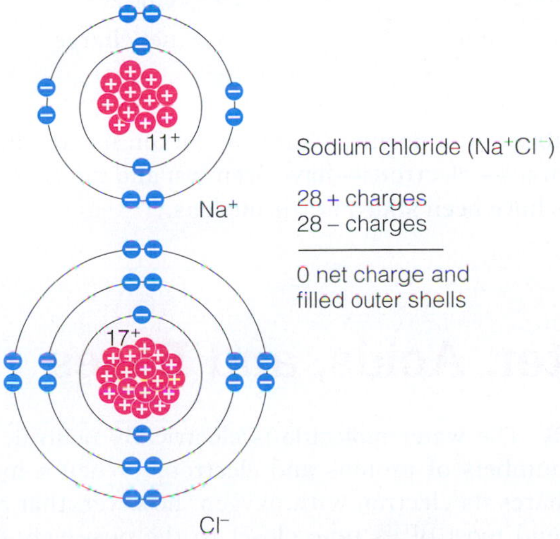

With all its electrons, sodium is a shiny, highly reactive metal; chlorine is the poisonous greenish yellow gas that was used in

World War I. But after sodium and chlorine have transferred electrons, they form the stable white salt familiar to you as table salt, or sodium chloride (Na^+Cl^-). The dramatic difference illustrates how profoundly the electron arrangement can influence the nature of a substance. The wide distribution of salt in nature attests to the stability of the union between the ions. Each meets the other's needs (a good marriage).

When dry, salt exists as crystals; its ions are stacked very regularly into a lattice, with positive and negative ions alternating in a three-dimensional checkerboard structure. In water, however, the salt quickly dissolves, and its ions separate from one another, forming an electrolyte solution in which they move about freely. Covalently bonded molecules rarely dissociate like this in a water solution. The most common exception is when they behave like acids and release H^+ ions, as discussed in the next section.

An ion can also be a group of atoms bound together in such a way that the group has a net charge and enters into reactions as a single unit. Many such groups are active in the fluids of the body. The bicarbonate ion is composed of five atoms—one H, one C, and three Os—and has a net charge of −1 (HCO_3^-). Another important ion of this type is a phosphate ion with one H, one P, and four O, and a net charge of −2 (HPO_4^{-2}).

Whereas many elements have only one configuration in the outer shell and thus only one way to bond with other elements, some elements have the possibility of varied configurations. Iron is such an element. Under some conditions iron loses two electrons, and under other circumstances it loses three. If iron loses two electrons, it then has a net charge of +2, and we call it ferrous iron (Fe^{++}). If it donates three electrons to another atom, it becomes the +3 ion, or ferric iron (Fe^{+++}).

Ferrous iron (Fe^{++})	Ferric iron (Fe^{+++})
(had 2 outer-shell electrons but has lost them)	(had 3 outer-shell electrons but has lost them)
26 + charges	26 + charges
24 − charges	23 − charges
2 + net charge	3 + net charge

Remember that a positive charge on an ion means that negative charges—electrons—have been lost and not that positive charges have been added to the nucleus.

Water, Acids, and Bases

Water The water molecule is electrically neutral, having equal numbers of protons and electrons. When a hydrogen atom shares its electron with oxygen, however, that electron will spend most of its time closer to the positively charged oxygen nucleus. This leaves the positive proton (nucleus of the hydrogen atom) exposed on the outer part of the water molecule. We know, too, that the two hydrogens both bond toward the same side of the oxygen. These two facts explain why water molecules are polar: they have regions of more positive and more negative charge.

Polar molecules like water are drawn to one another by the attractive forces between the positive polar areas of one and the negative poles of another. These attractive forces, sometimes known as polar bonds or hydrogen bonds, occur among many molecules and also within the different parts of single large molecules. Although very weak in comparison with covalent bonds, polar bonds may occur in such abundance that they become exceedingly important in determining the structure of such large molecules as proteins and DNA.

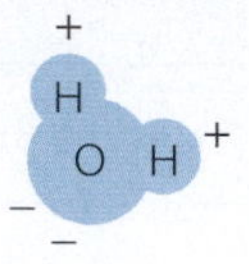

This diagram of the polar water molecule shows displacement of electrons toward the O nucleus; thus the negative region is near the O and the positive regions are near the H atoms.

Water molecules have a slight tendency to ionize, separating into positive (H^+) and negative (OH^-) ions. In pure water, a small but constant number of these ions is present, and the number of positive ions exactly equals the number of negative ions.

Acid An acid is a substance that releases H^+ ions (protons) in a water solution. Hydrochloric acid (HCl^-) is such a substance because it dissociates in a water solution into H^+ and Cl^- ions. Acetic acid is also an acid because it dissociates in water to acetate ions and free H^+:

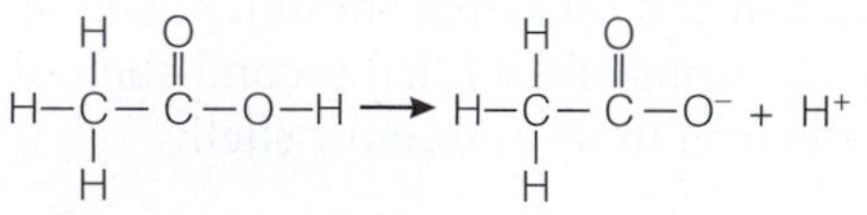

Acetic acid dissociates into an acetate ion and a hydrogen ion.

The more H^+ ions released, the stronger the acid.

pH Chemists define degrees of acidity by means of the pH scale, which runs from 0 to 14. The pH expresses the concentration of H^+ ions: a pH of 1 is extremely acidic, 7 is neutral, and 13 is very basic. There is a tenfold difference in the concentration of H^+ ions between points on this scale. A solution with pH 3, for example, has 10 times as many H^+ ions as a solution with pH 4. At pH 7, the concentrations of free H^+ and OH^- are exactly the same—1/10,000,000 moles per liter (1027 moles per liter).* At pH 4, the concentration of free

*A mole is a certain number (about 6×10^{23}) of molecules. The pH of a solution is defined as the negative logarithm of the hydrogen ion concentration of the solution. Thus, if the concentration is 10^{-2} (moles per liter), the pH is 2; if 10^{-8}, the pH is 8; and so on.

H^+ ions is 1/10,000 (1024) moles per liter. This is a higher concentration of H^+ ions, and the solution is therefore acidic. Figure 3-7 on p. 76 presents the pH scale.

Bases A base is a substance that can combine with H^+ ions, thus reducing the acidity of a solution. The compound ammonia is such a substance. The ammonia molecule has two electrons that are not shared with any other atom; a hydrogen ion (H^+) is just a naked proton with no shell of electrons at all. The proton readily combines with the ammonia molecule to form an ammonium ion; thus a free proton is withdrawn from the solution and no longer contributes to its acidity. Many compounds containing nitrogen are important bases in living systems. Acids and bases neutralize each other to produce substances that are neither acid nor base.

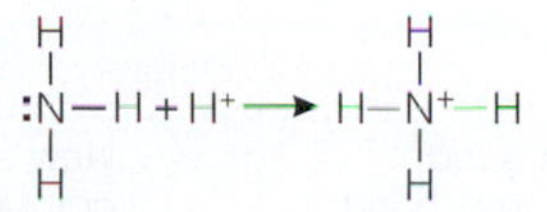

Ammonia captures a hydrogen ion from water. The two dots here represent the two electrons not shared with another atom. These dots are ordinarily not shown in chemical structure drawings. Compare this drawing with the earlier diagram of an ammonia molecule (p. B-4).

Chemical Reactions

A chemical reaction, or chemical change, results in the breakdown of substances and the formation of new ones. Almost all such reactions involve a change in the bonding of atoms. Old bonds are broken, and new ones are formed. The nuclei of atoms are never involved in chemical reactions—only their outer-shell electrons take part. At the end of a chemical reaction, the number of atoms of each type is always the same as at the beginning. For example, two hydrogen molecules ($2H_2$) can react with one oxygen molecule (O_2) to form two water molecules ($2H_2O$). In this reaction two substances (hydrogen and oxygen) disappear, and a new one (water) is formed, but at the end of the reaction there are still four H atoms and two O atoms, just as there were at the beginning. Because the atoms are now linked in a different way, their characteristics or properties have changed.

In many instances chemical reactions involve not the relinking of molecules but the exchanging of electrons or protons among them. In such reactions the molecule that gains one or more electrons (or loses one or more hydrogen ions) is said to be reduced; the molecule that loses electrons (or gains protons) is oxidized. A hydrogen ion is equivalent to a proton. Oxidation and reduction reactions take place simultaneously because an electron or proton that is lost by one molecule is accepted by another. The addition of an atom of oxygen is also oxidation because oxygen (with six electrons in the outer shell) accepts two electrons in becoming bonded. Oxidation, then, is loss of electrons, gain of protons, or addition of oxy-

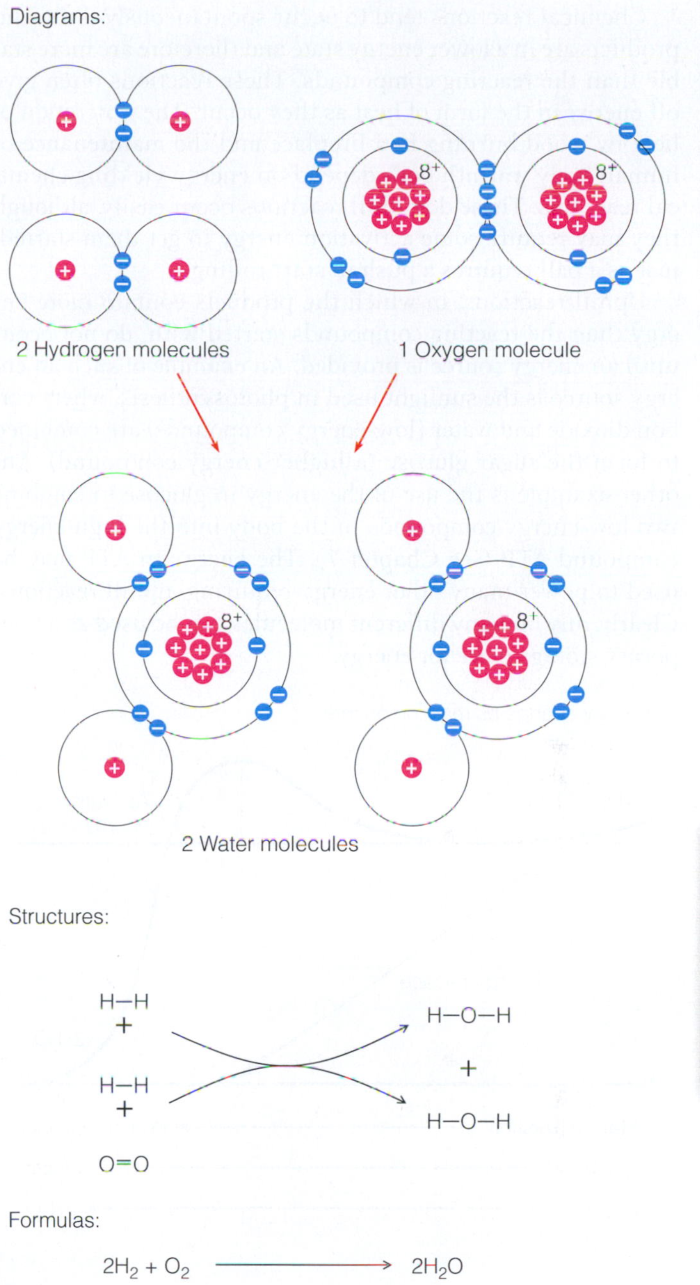

Hydrogen and oxygen react to form water.

gen (with six electrons); reduction is the opposite—gain of electrons, loss of protons, or loss of oxygen. The addition of hydrogen atoms to oxygen to form water can thus be described as the reduction of oxygen *or* the oxidation of hydrogen.

If a reaction results in a net increase in the energy of a compound, it is called an endergonic, or "uphill," reaction (energy, *erg,* is added into, *endo,* the compound). An example is the chief result of photosynthesis, the making of sugar in a plant from carbon dioxide and water using the energy of sunlight. Conversely, the oxidation of sugar to carbon dioxide and water is an exergonic, or "downhill," reaction because the end products have less energy than the starting products. Oftentimes, but not always, reduction reactions are endergonic, resulting in an increase in the energy of the products. Oxidation reactions often, but not always, are exergonic.

Chemical reactions tend to occur spontaneously if the end products are in a lower energy state and therefore are more stable than the reacting compounds. These reactions often give off energy in the form of heat as they occur. The generation of heat by wood burning in a fireplace and the maintenance of human body warmth both depend on energy-yielding chemical reactions. These downhill reactions occur easily, although they may require some activation energy to get them started, just as a ball requires a push to start rolling.

Uphill reactions, in which the products contain more energy than the reacting compounds started with, do not occur until an energy source is provided. An example of such an energy source is the sunlight used in photosynthesis, where carbon dioxide and water (low-energy compounds) are combined to form the sugar glucose (a higher-energy compound). Another example is the use of the energy in glucose to combine two low-energy compounds in the body into the high-energy compound ATP (see Chapter 7). The energy in ATP may be used to power many other energy-requiring, uphill reactions. Clearly, any of many different molecules can be used as a temporary storage place for energy.

Energy change as reaction occurs

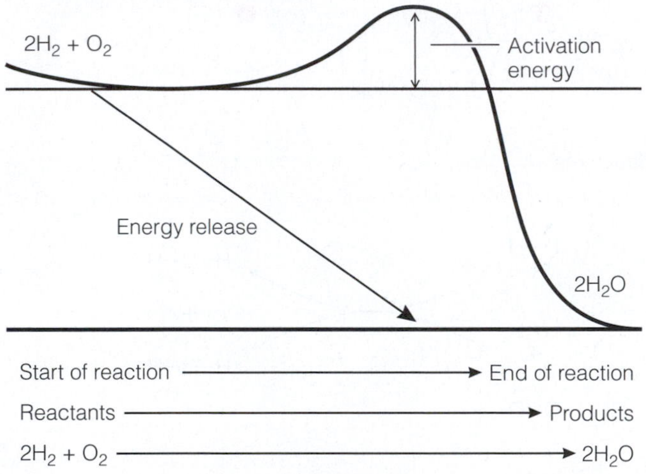

Formation of Free Radicals

Normally, when a chemical reaction takes place, bonds break and re-form with some redistribution of atoms and rearrangement of bonds to form new, stable compounds. Normally, bonds don't split in such a way as to leave a molecule with an odd, unpaired electron. When they do, free radicals are formed. Free radicals are highly unstable and quickly react with other compounds, forming more free radicals in a chain reaction. A cascade may ensue in which many highly reactive radicals are generated, resulting finally in the disruption of a living structure such as a cell membrane.

H—O—O—H or R—O—O—H —(Heat or light)→ H—O· + ·O—H or R—O· + ·O—H

Hydrogen peroxide or any hydroperoxide (R is any carbon chain with appropriate numbers of H) — Free radical

Free radicals are formed. The dots represent single electrons that are available for sharing (the atom needs another electron to fill its outer shell).

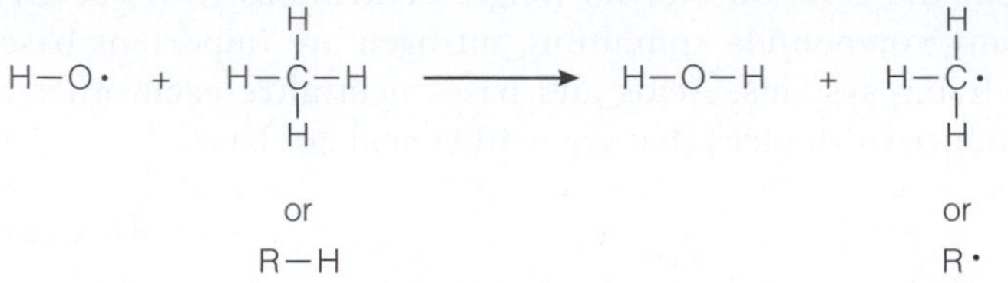

Free radical — Compound with weak bond (perhaps an unsaturated fatty acid) — New stable compound (water or an alcohol) — Free radical

Free radicals destroy biological compounds. The free radical attacks a weak bond in a biological compound, disrupting it and forming a new stable molecule and another free radical. This free radical can attack another biological compound, and so on.

Oxidation of some compounds can be induced by air at room temperature in the presence of light. Such reactions are thought to take place through the formation of compounds called peroxides:

Peroxides:

H—O—O—H	Hydrogen peroxide
R—O—O—H	Hydroperoxides (R is any carbon chain with appropriate numbers of H)
R—O—O—R	Peroxide

Some peroxides readily disintegrate into free radicals, initiating chain reactions like those just described.

Free radicals are of special interest in nutrition because the antioxidant properties of vitamins C and E as well as beta-carotene and the mineral selenium are thought to protect against the destructive effects of these free radicals (see Highlight 11). For example, vitamin E on the surface of the lungs reacts with, and is destroyed by, free radicals, thus preventing the radicals from reaching underlying cells and oxidizing the lipids in their membranes.

Biochemical Structures and Pathways

APPENDIX C

CONTENTS

The diagrams of nutrients presented here are meant to enhance your understanding of the most important organic molecules in the human diet. Following the diagrams of nutrients are sections on the major metabolic pathways mentioned in Chapter 7—glycolysis, fatty acid oxidation, amino acid degradation, the TCA cycle, and the electron transport chain—and a description of how alcohol interferes with these pathways. Discussions of the urea cycle and the formation of ketone bodies complete the appendix.

Carbohydrates

Monosaccharides

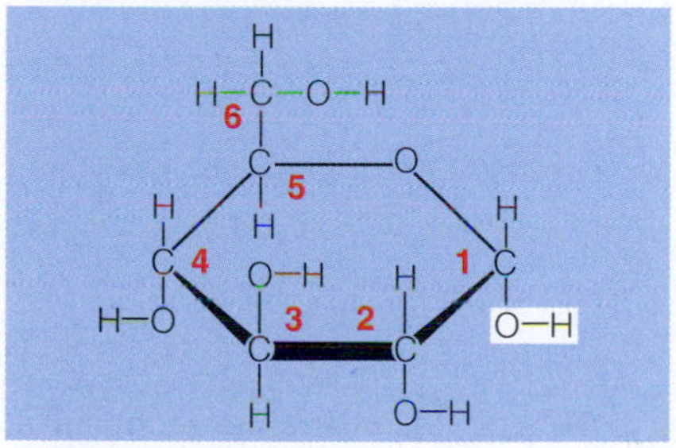

Glucose (alpha form). The ring would be at right angles to the plane of the paper. The bonds directed upward are above the plane; those directed downward are below the plane. This molecule is considered an alpha form because the OH on carbon 1 points downward.

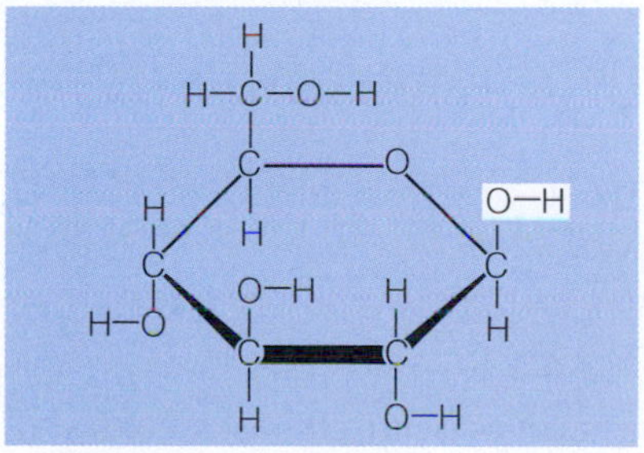

Glucose (beta form). The OH on carbon 1 points upward.
Fructose, galactose: see Chapter 4.

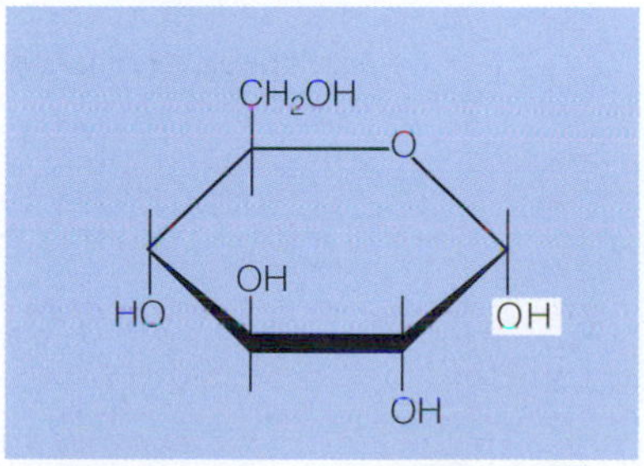

Glucose (alpha form) shorthand notation. This notation, in which the carbons in the ring and single hydrogens have been eliminated, will be used throughout this appendix.

Disaccharides

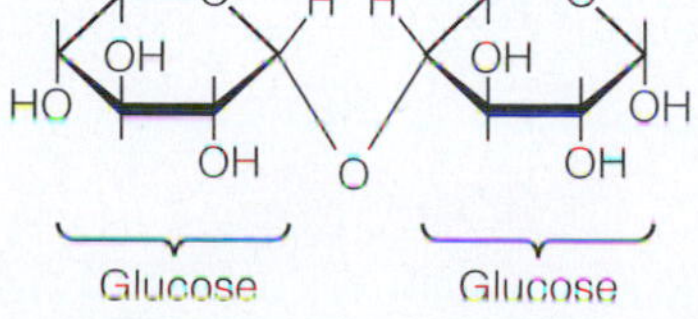

Maltose.

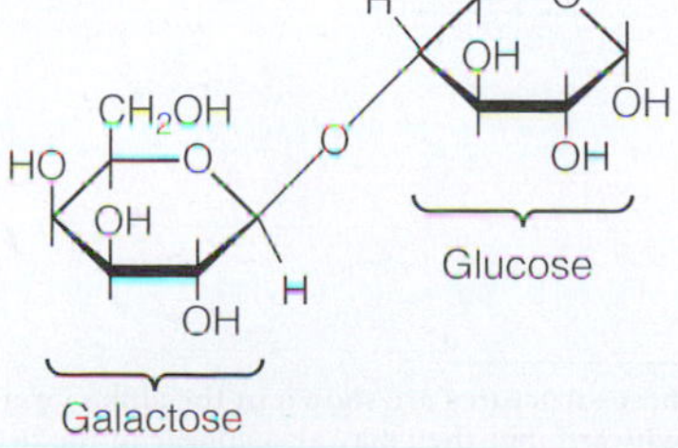

Lactose (alpha form).

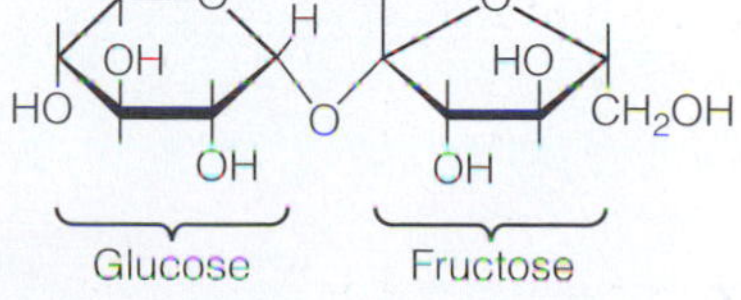

Sucrose.

Polysaccharides

As described in Chapter 4, starch, glycogen, and cellulose are all long chains of glucose molecules covalently linked together.

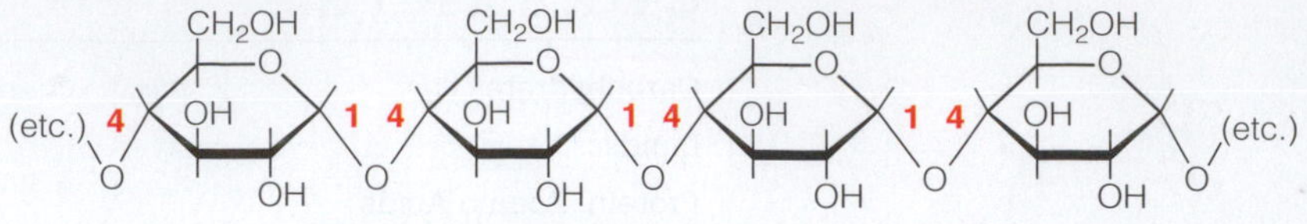

Amylose (unbranched starch)

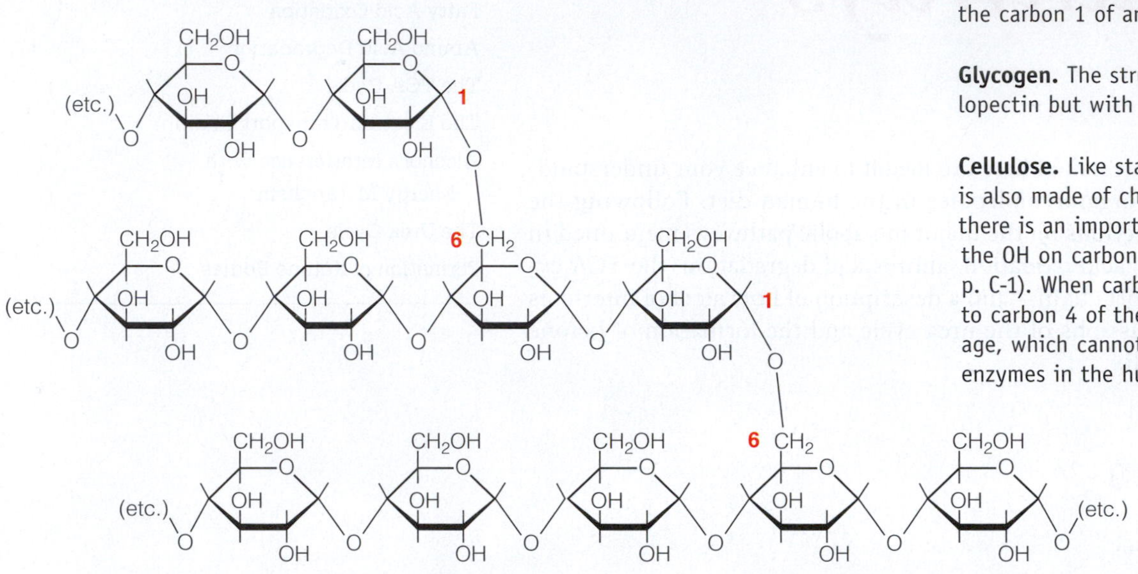

Amylopectin (branched starch)

Starch. Two kinds of covalent linkages occur between glucose molecules in starch, giving rise to two kinds of chains. Amylose is composed of straight chains, with carbon 1 of one glucose linked to carbon 4 of the next (α-1,4 linkage). Amylopectin is made up of straight chains like amylose but has occasional branches arising where the carbon 6 of a glucose is also linked to the carbon 1 of another glucose (α-1,6 linkage).

Glycogen. The structure of glycogen is like amylopectin but with many more branches.

Cellulose. Like starch and glycogen, cellulose is also made of chains of glucose units, but there is an important difference: in cellulose, the OH on carbon 1 is in the beta position (see p. C-1). When carbon 1 of one glucose is linked to carbon 4 of the next, it forms a β-1,4 linkage, which cannot be broken by digestive enzymes in the human GI tract.

Fibers, such as hemicelluloses, consist of long chains of various monosaccharides.

Monosaccharides common in the backbone chain of hemicelluloses:

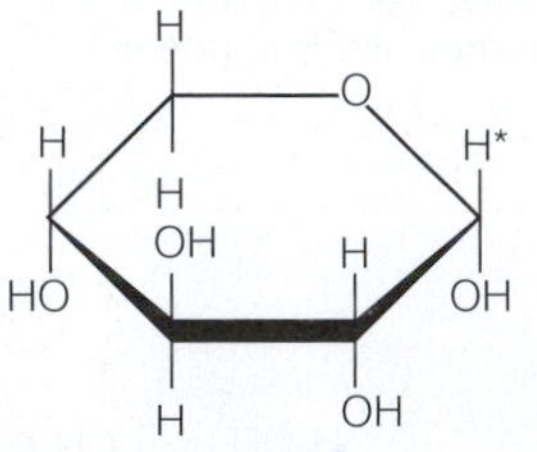

Xylose

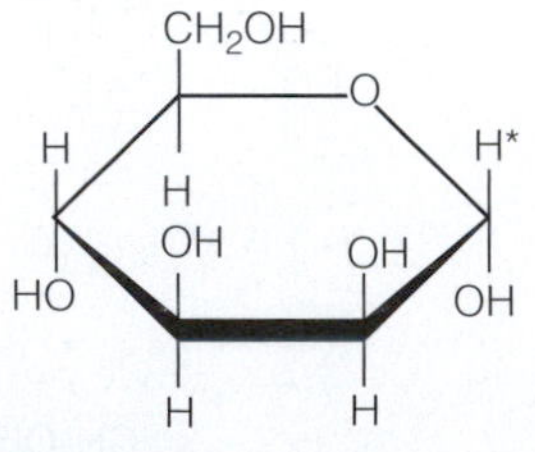

Mannose

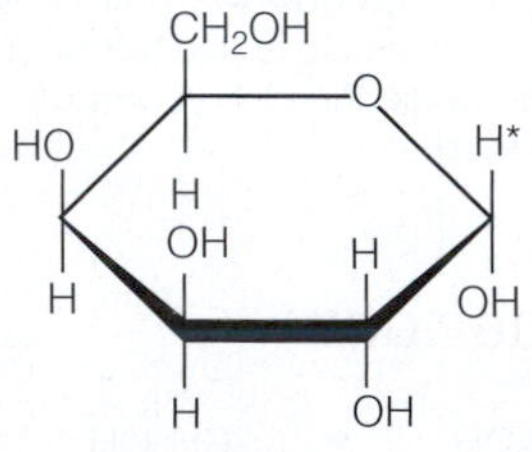

Galactose

*These structures are shown in the alpha form with the H on the carbon pointing upward and the OH pointing downward, but they may also appear in the beta form with the H pointing downward and the OH upward.

Monosaccharides common in the side chains of hemicelluloses:

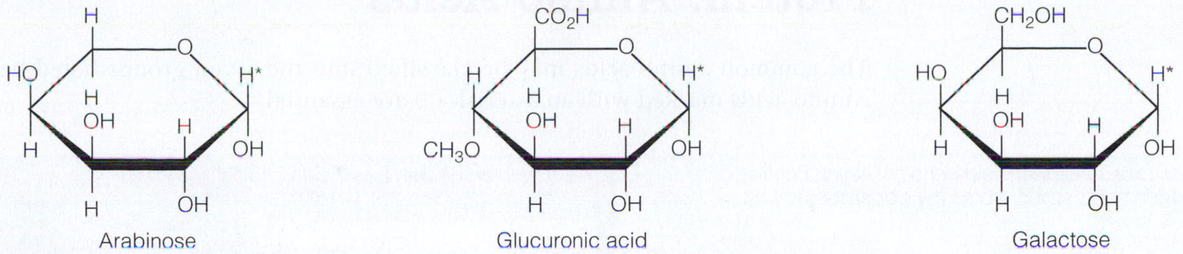

Hemicelluloses. The most common hemicelluloses are composed of a backbone chain of xylose, mannose, and galactose, with branching side chains of arabinose, glucuronic acid, and galactose.

Lipids

TABLE C-1 Saturated Fatty Acids Found in Natural Fats

Saturated Fatty Acids	Chemical Formulas	Number of Carbons	Major Food Sources
Butyric	C_3H_7COOH	4	Butterfat
Caproic	$C_5H_{11}COOH$	6	Butterfat
Caprylic	$C_7H_{15}COOH$	8	Coconut oil
Capric	$C_9H_{19}COOH$	10	Palm oil
Lauric	$C_{11}H_{23}COOH$	12	Coconut oil, palm oil
Myristic[a]	$C_{13}H_{27}COOH$	14	Coconut oil, palm oil
Palmitic[a]	$C_{15}H_{31}COOH$	16	Palm oil
Stearic[a]	$C_{17}H_{35}COOH$	18	Most animal fats
Arachidic	$C_{19}H_{39}COOH$	20	Peanut oil
Behenic	$C_{21}H_{43}COOH$	22	Seeds
Lignoceric	$C_{23}H_{47}COOH$	24	Peanut oil

[a]Most common saturated fatty acids.

TABLE C-2 Unsaturated Fatty Acids Found in Natural Fats

Unsaturated Fatty Acids	Chemical Formulas	Number of Carbons	Number of Double Bonds	Standard Notation[a]	Omega Notation[b]	Major Food Sources
Palmitoleic	$C_{15}H_{29}COOH$	16	1	16:1;9	16:1ω7	Seafood, beef
Oleic	$C_{17}H_{33}COOH$	18	1	18:1;9	18:1ω9	Olive oil, canola oil
Linoleic	$C_{17}H_{31}COOH$	18	2	18:2;9,12	18:2ω6	Sunflower oil, safflower oil
Linolenic	$C_{17}H_{29}COOH$	18	3	18:3;9,12,15	18:3ω3	Soybean oil, canola oil
Arachidonic	$C_{19}H_{31}COOH$	20	4	20:4;5,8,11,14	20:4ω6	Eggs, most animal fats
Eicosapentaenoic	$C_{19}H_{29}COOH$	20	5	20:5;5,8,11,14,17	20:5ω3	Seafood
Docosahexaenoic	$C_{21}H_{31}COOH$	22	6	22:6;4,7,10,13,16,19	22:6ω3	Seafood

NOTE: A fatty acid has two ends; designated the methyl (CH_3) end and the carboxyl, or acid (COOH), end.

[a]Standard chemistry notation begins counting carbons at the acid end. The number of carbons the fatty acid contains comes first, followed by a colon and another number that indicates the number of double bonds; next comes a semicolon followed by a number or numbers indicating the positions of the double bonds. Thus the notation for linoleic acid, an 18-carbon fatty acid with two double bonds between carbons 9 and 10 and between carbons 12 and 13, is 18:2;9,12.

[b]Because fatty acid chains are lengthened by adding carbons at the acid end of the chain, chemists use the omega system of notation to ease the task of identifying them. The omega system begins counting carbons at the methyl end. The number of carbons the fatty acid contains comes first, followed by a colon and the number of double bonds; next come the omega symbol (ω) and a number indicating the position of the double bond nearest the methyl end. Thus linoleic acid with its first double bond at the sixth carbon from the methyl end would be noted 18:2ω6 in the omega system.

Protein: Amino Acids

The common amino acids may be classified into the seven groups listed below. Amino acids marked with an asterisk (*) are essential.

1. Amino acids with aliphatic side chains, which consist of hydrogen and carbon atoms (hydrocarbons):

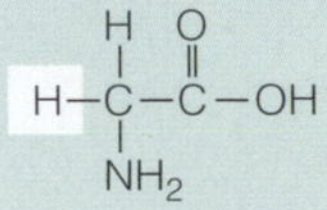

Glycine (Gly)

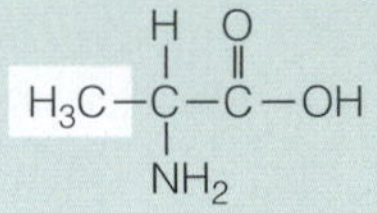

Alanine (Ala)

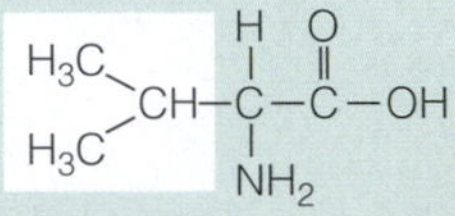

Valine* (Val)

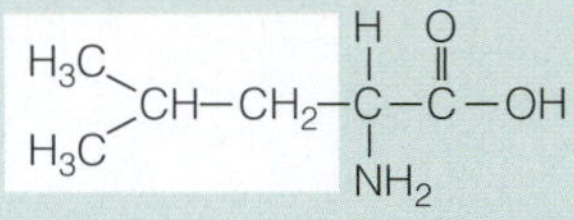

Leucine* (Leu)

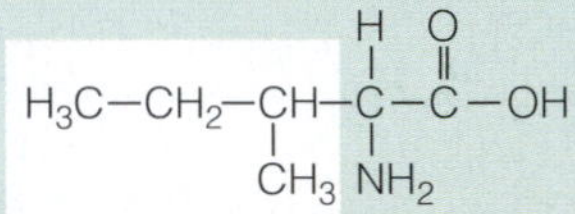

Isoleucine* (Ile)

2. Amino acids with hydroxyl (OH) side chains:

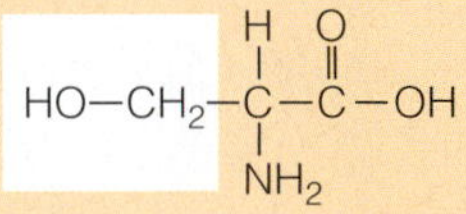

Serine (Ser)

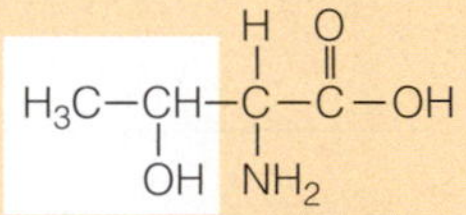

Threonine* (Thr)

3. Amino acids with side chains containing acidic groups or their amides, which contain the group NH_2:

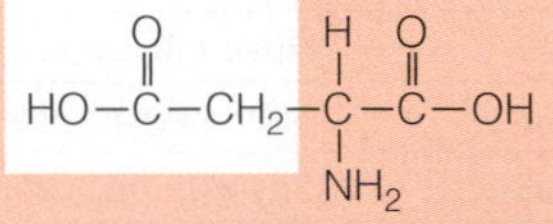

Aspartic acid (Asp)

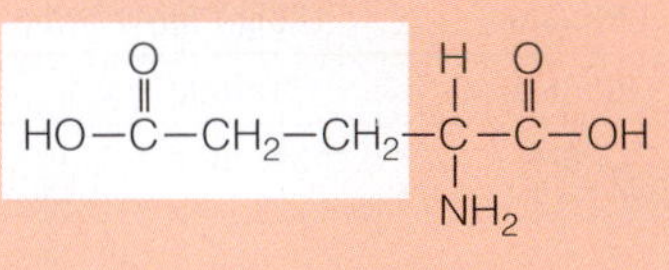

Glutamic acid (Glu)

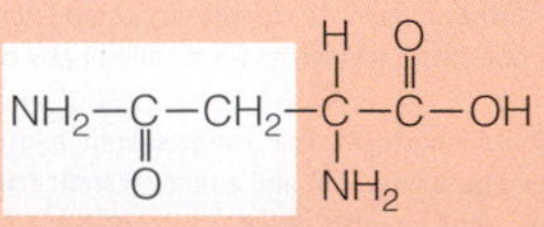

Asparagine (Asn)

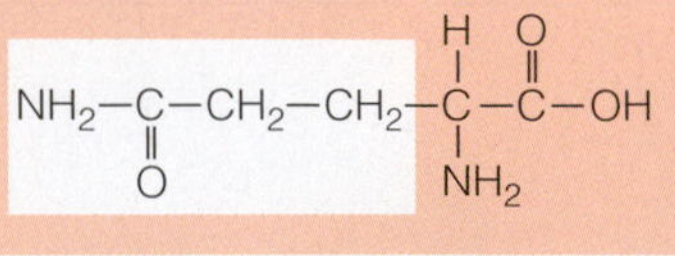

Glutamine (Gln)

4. Amino acids with basic side chains:

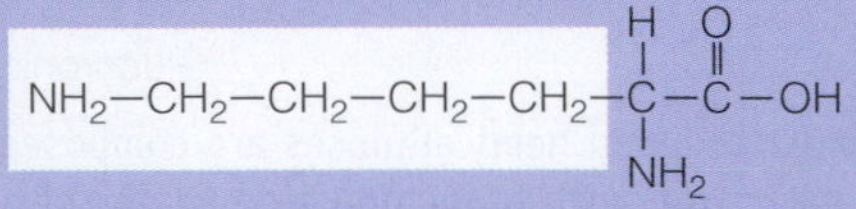

Lysine* (Lys)

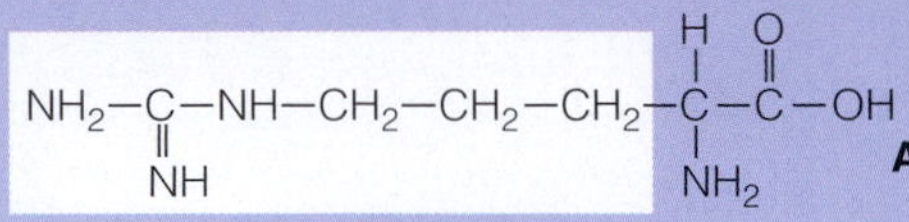

Arginine (Arg)

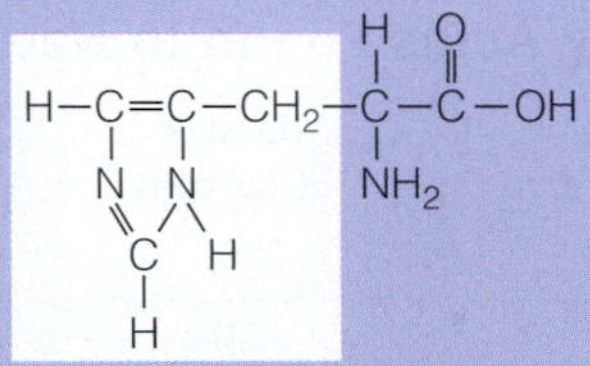

Histidine* (His)

5. Amino acids with aromatic side chains, which are characterized by the presence of at least one ring structure:

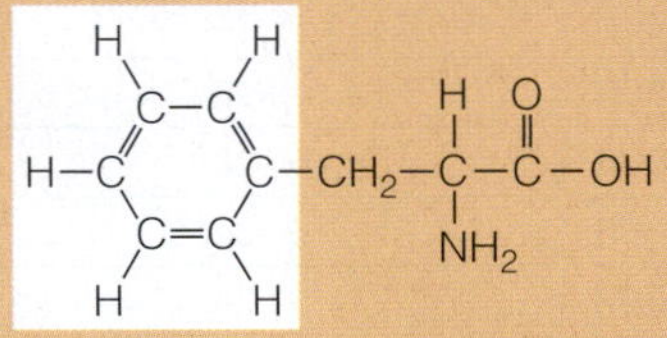

Phenylalanine* (Phe)

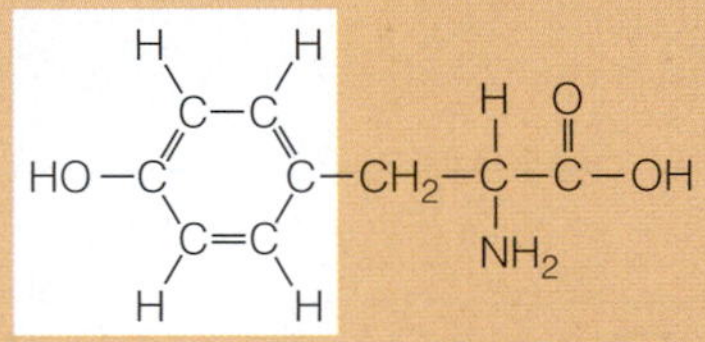

Tyrosine (Tyr)

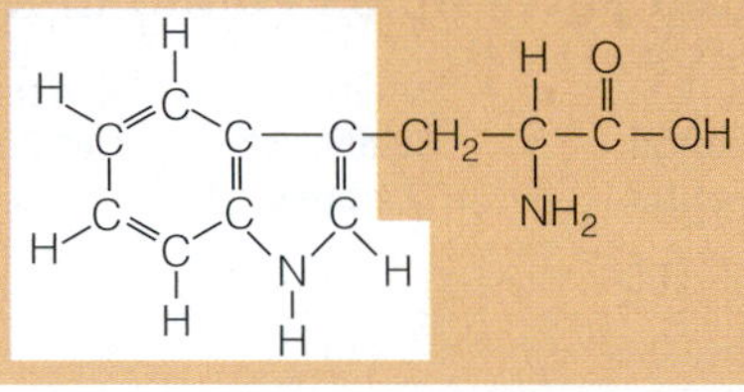

Tryptophan* (Trp)

6. Amino acids with side chains containing sulfur atoms:

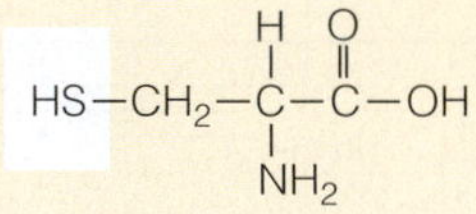

Cysteine (Cys)

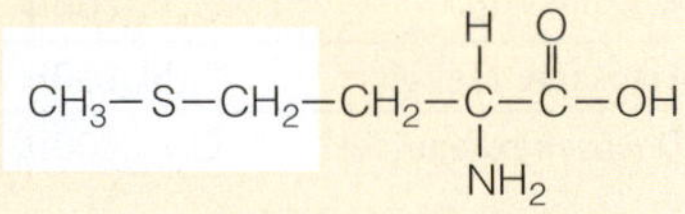

Methionine* (Met)

7. Imino acid:

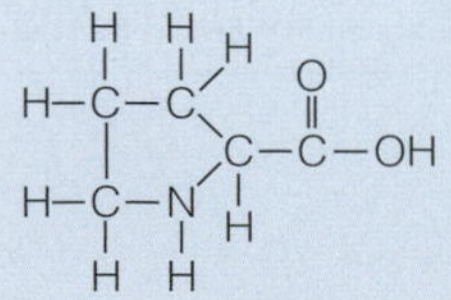

Proline (Pro)

Proline has the same chemical structure as the other amino acids, but its amino group has given up a hydrogen to form a ring.

Vitamins and Coenzymes

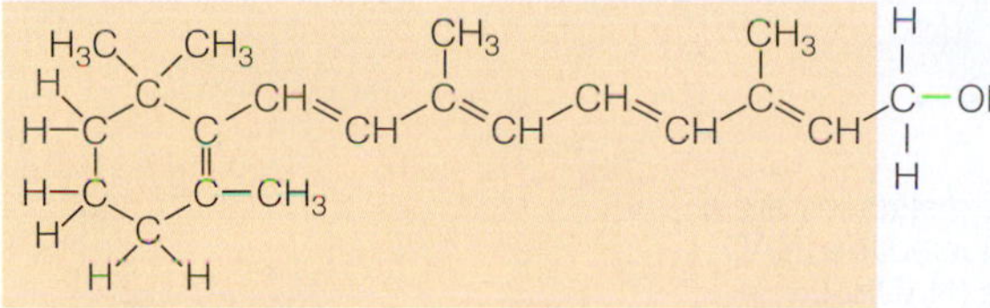

Vitamin A: retinol. This molecule is the alcohol form of vitamin A.

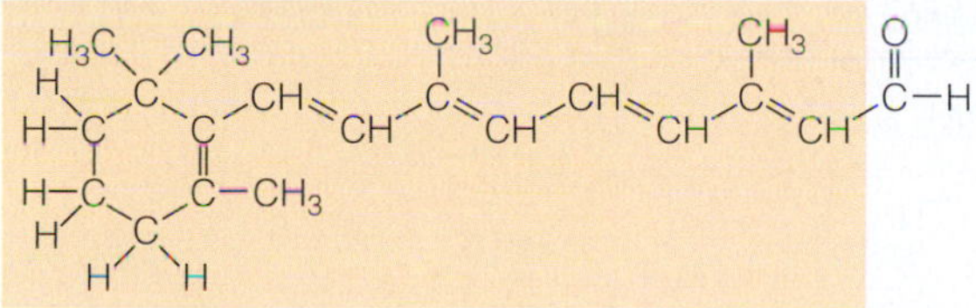

Vitamin A: retinal. This molecule is the aldehyde form of vitamin A.

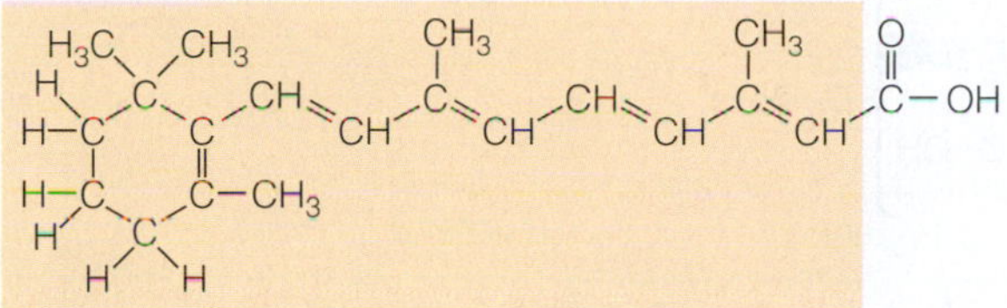

Vitamin A: retinoic acid. This molecule is the acid form of vitamin A.

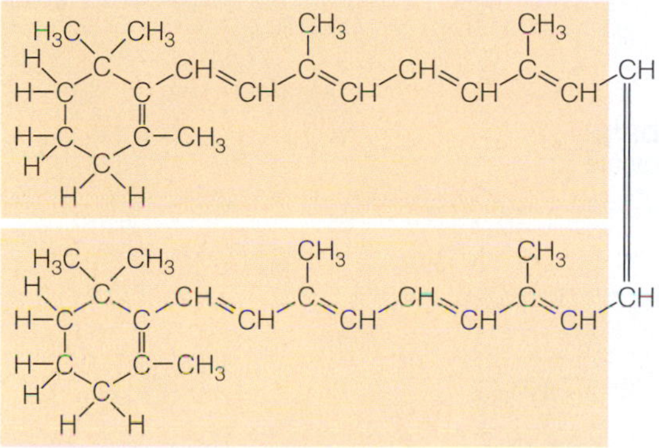

Vitamin A precursor: beta-carotene. This molecule is the carotenoid with the most vitamin A activity.

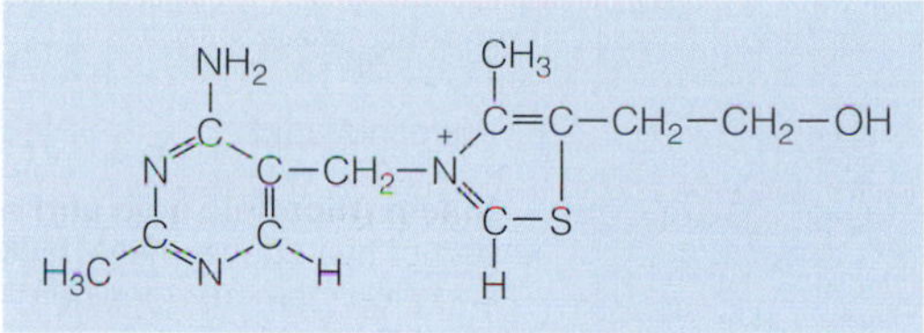

Thiamin. This molecule is part of the coenzyme thiamin pyrophosphate (TPP).

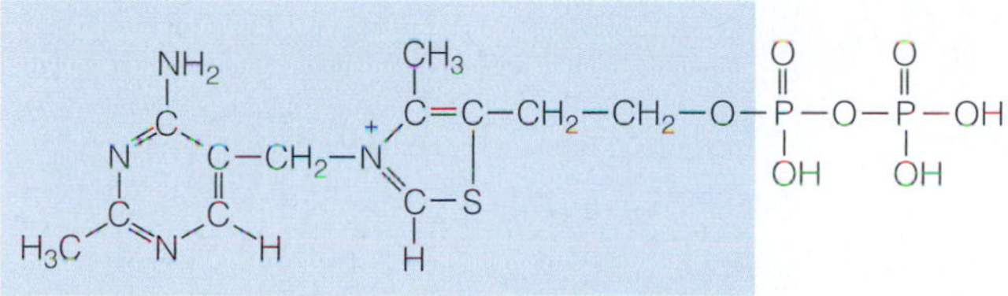

Thiamin pyrophosphate (TPP). TPP is a coenzyme that includes the thiamin molecule as part of its structure.

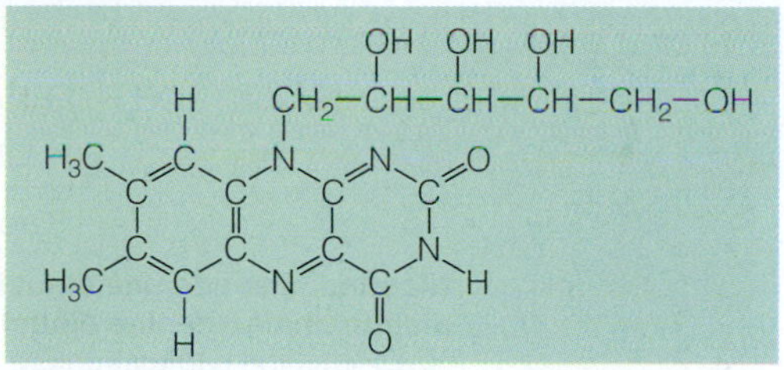

Riboflavin. This molecule is a part of two coenzymes—flavin mononucleotide (FMN) and flavin adenine dinucleotide (FAD).

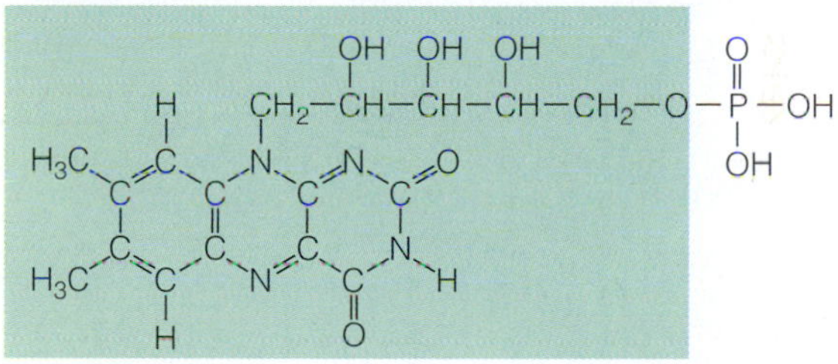

Flavin mononucleotide (FMN). FMN is a coenzyme that includes the riboflavin molecule as part of its structure.

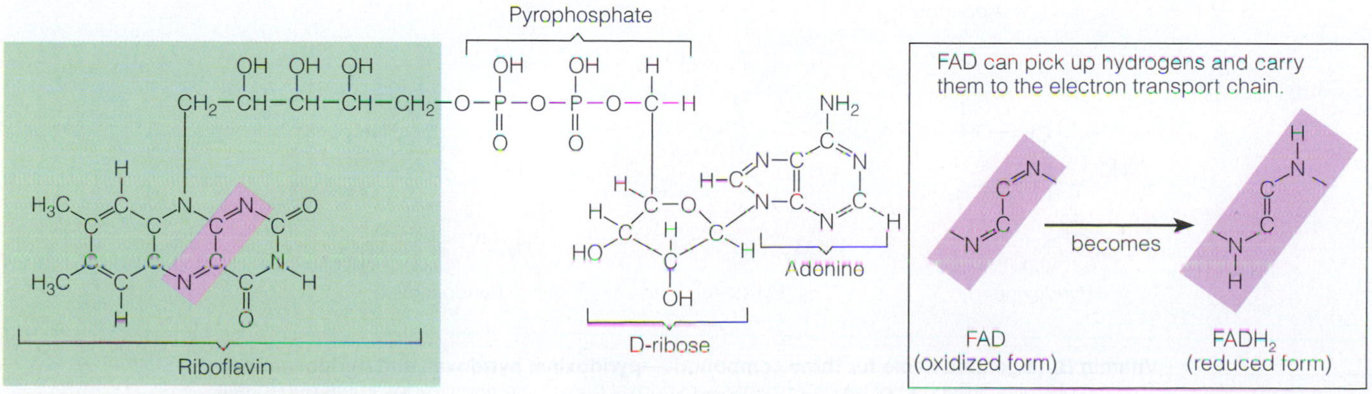

Flavin adenine dinucleotide (FAD). FAD is a coenzyme that includes the riboflavin molecule as part of its structure.

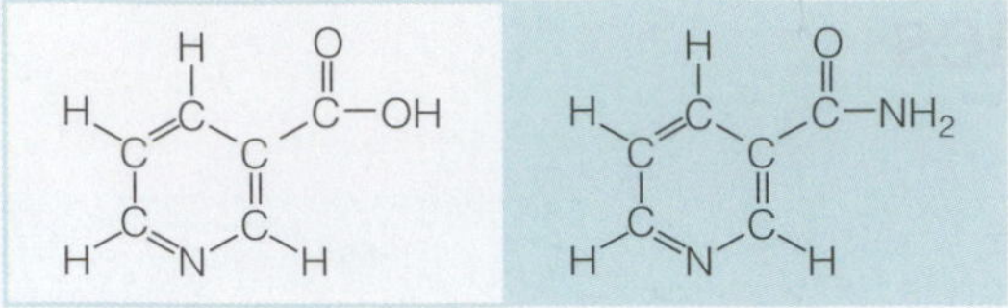

Nicotinic acid　　Nicotinamide

Niacin (nicotinic acid and nicotinamide). These molecules are a part of two coenzymes—nicotinamide adenine dinucleotide (NAD^+) and nicotinamide adenine dinucleotide phosphate ($NADP^+$).

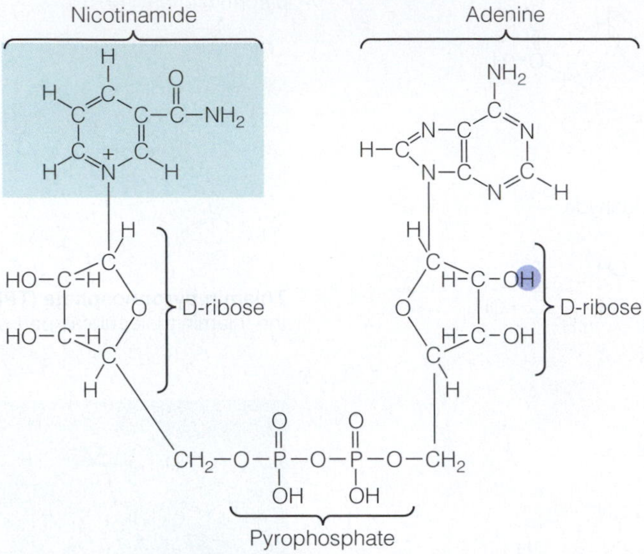

Nicotinamide adenine dinucleotide (NAD^+) and nicotinamide adenine dinucleotide phosphate ($NADP^+$). NADP has the same structure as NAD but with a phosphate group attached to the O instead of the H.

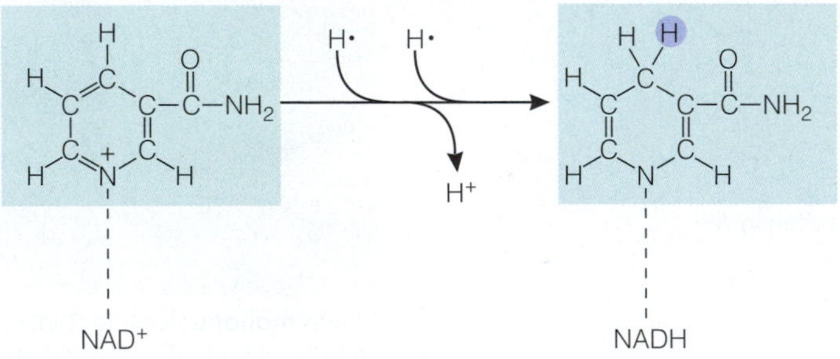

Reduced NAD^+ (NADH). When NAD^+ is reduced by the addition of H^+ and two electrons, it becomes the coenzyme NADH. (The dots on the H entering this reaction represent electrons—see Appendix B.)

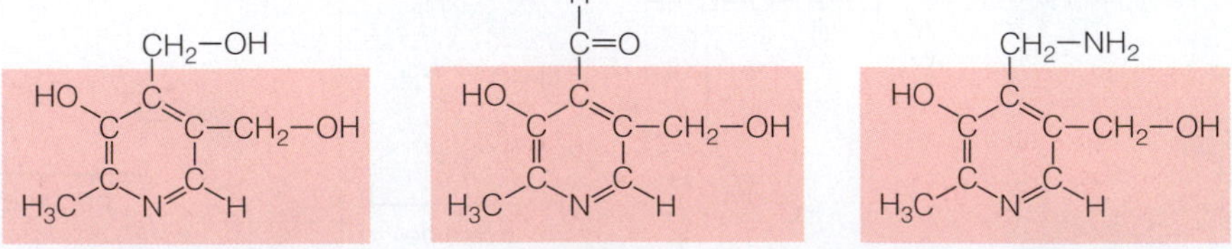

Pyridoxine　　Pyridoxal　　Pyridoxamine

Vitamin B_6 (a general name for three compounds—pyridoxine, pyridoxal, and pyridoxamine). These molecules are a part of two coenzymes—pyridoxal phosphate and pyridoxamine phosphate.

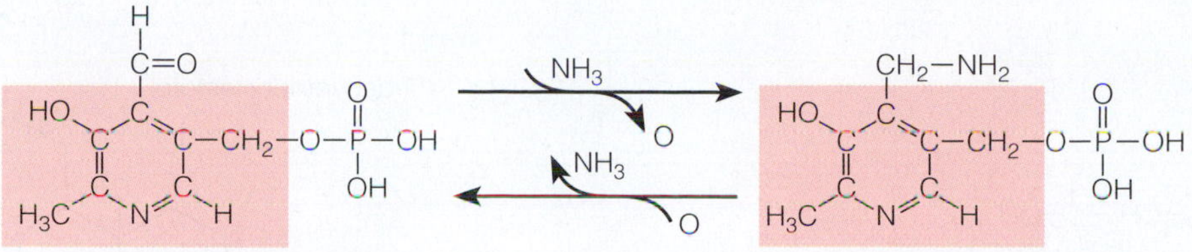

Pyridoxal phosphate   Pyridoxamine phosphate

Pyridoxal phosphate (PLP) and pyridoxamine phosphate. These coenzymes include vitamin B_6 as part of their structures.

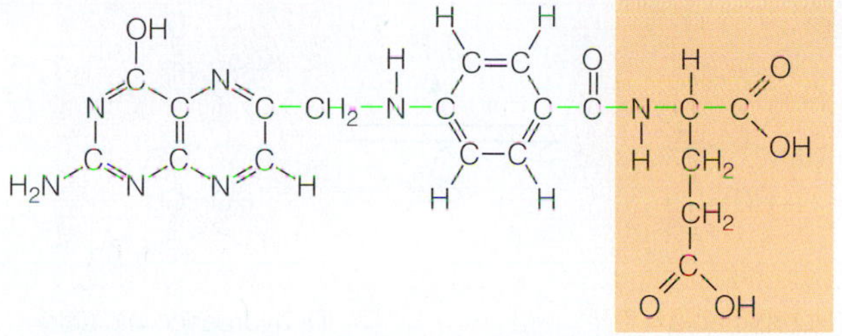

Folate (folacin or folic acid). This molecule consists of a double ring combined with a single ring and at least one glutamate (a nonessential amino acid marked in the box). Folate's biologically active form is tetrahydrofolate.

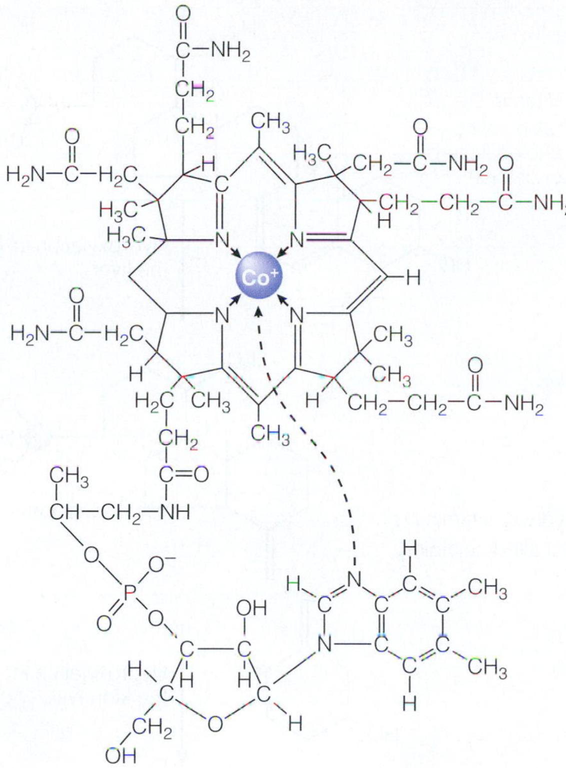

Vitamin B_{12} (cyanocobalamin). The arrows in this diagram indicate that the spare electron pairs on the nitrogens attract them to the cobalt.

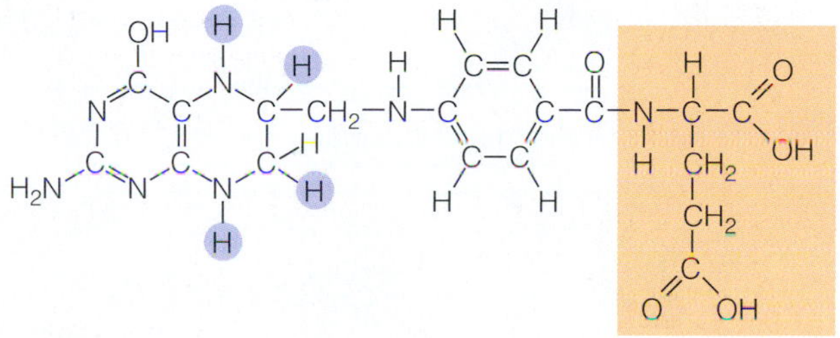

Tetrahydrofolate. This active coenzyme form of folate has four added hydrogens. An intermediate form, dihydrofolate, has two added hydrogens.

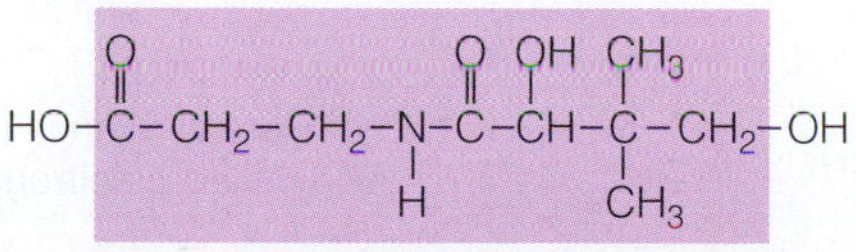

Pantothenic acid. This molecule is part of coenzyme A (CoA).

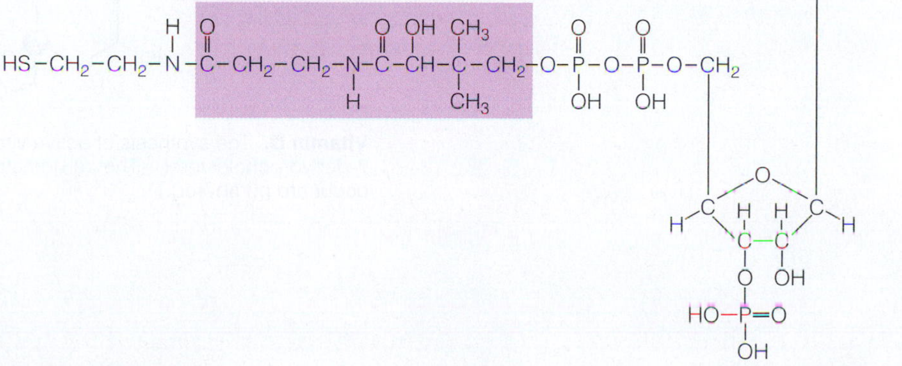

Coenzyme A (CoA). Coenzyme A is a coenzyme that includes pantothenic acid as part of its structure.

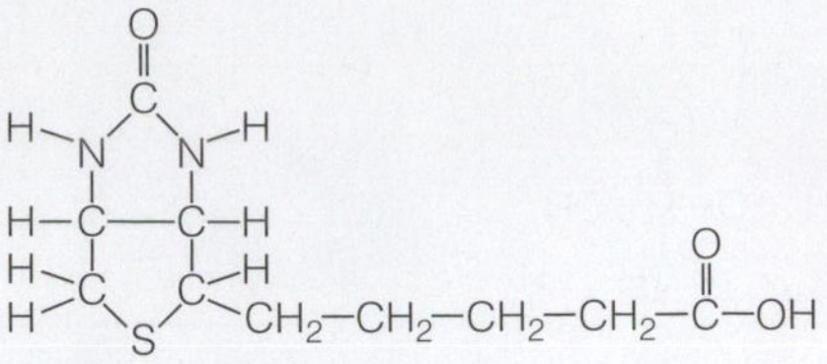

Biotin.

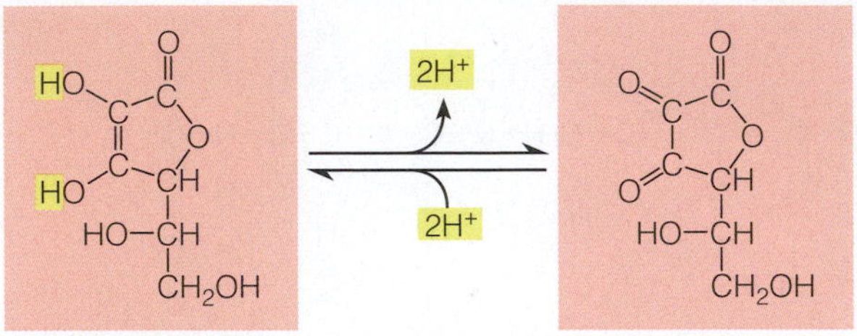

Ascorbic acid (reduced form)

Dehydroascorbic acid (oxidized form)

Vitamin C. Two hydrogen atoms with their electrons are lost when ascorbic acid is oxidized and gained when it is reduced again.

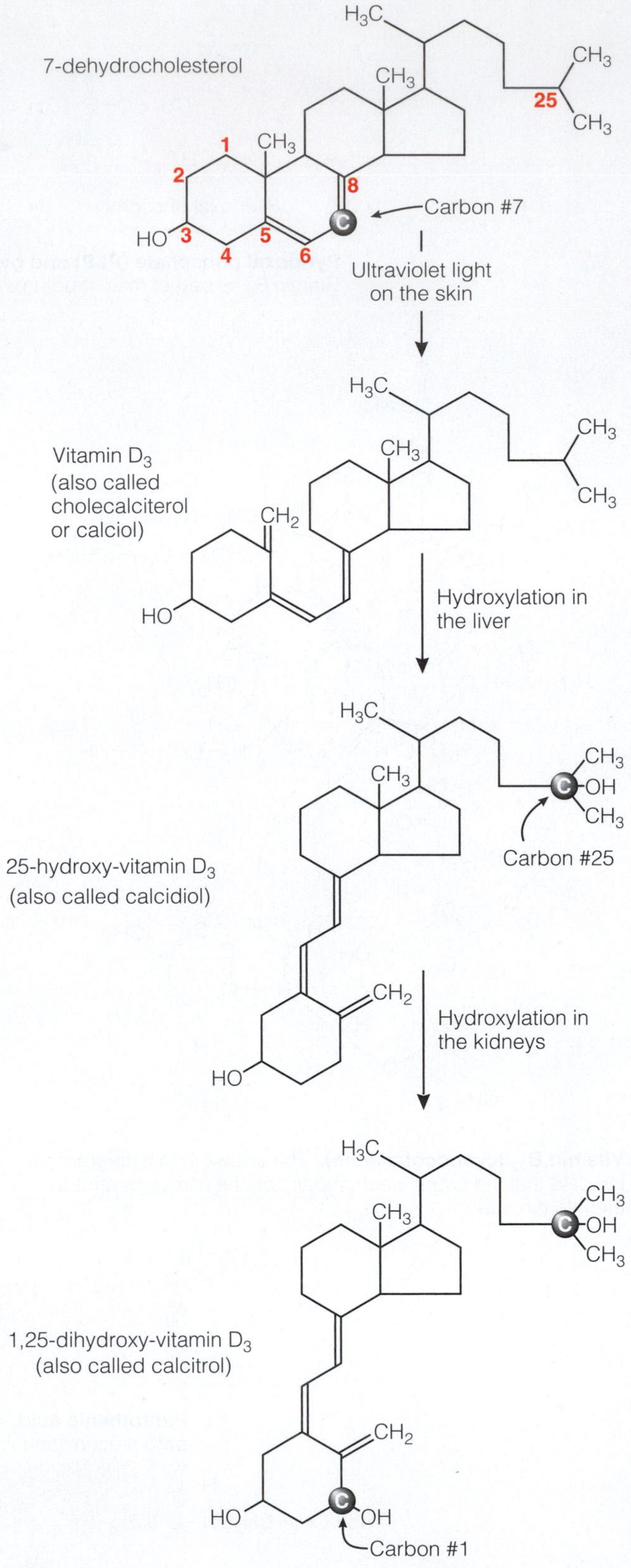

Vitamin D. The synthesis of active vitamin D begins with 7-dehydrocholesterol. (The carbon atoms at which changes occur are numbered.)

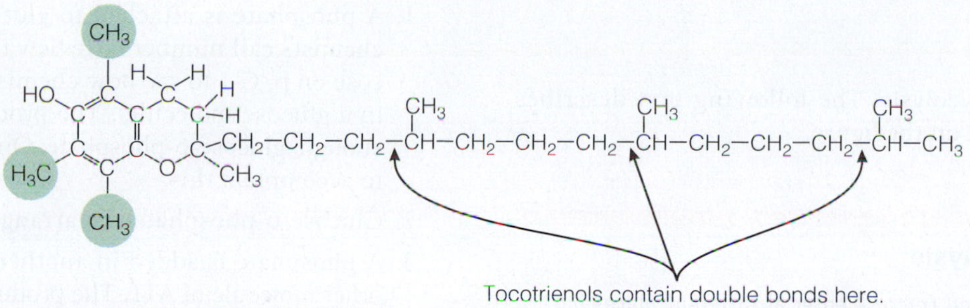

Vitamin E (alpha-tocopherol). The number and position of the methyl groups (CH_3) bonded to the ring structure differentiate among the tocopherols.

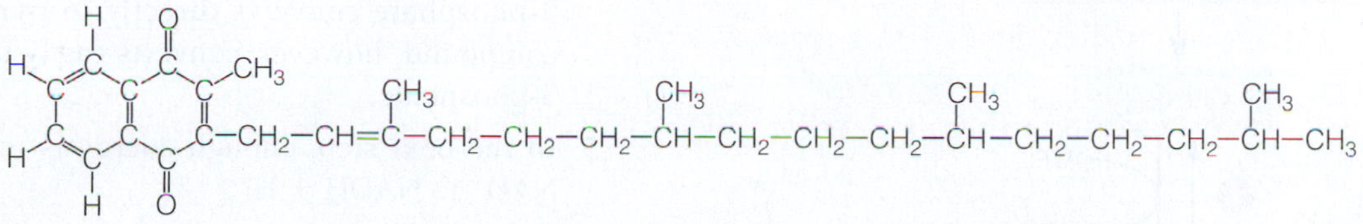

Vitamin K. Naturally occurring compounds with vitamin K activity include phylloquinones (from plants) and menaquinones (from bacteria).

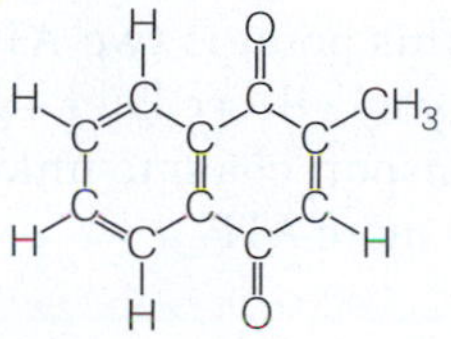

Menadione. This synthetic compound has the same activity as natural vitamin K.

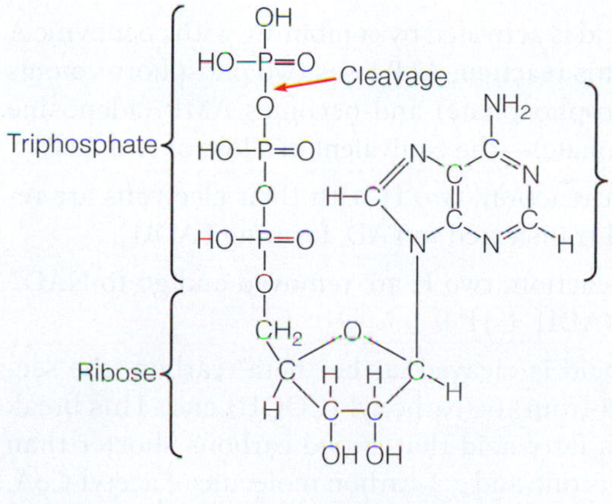

Adenosine triphosphate (ATP), the energy carrier. The cleavage point marks the bond that is broken when ATP splits to become ADP + P.

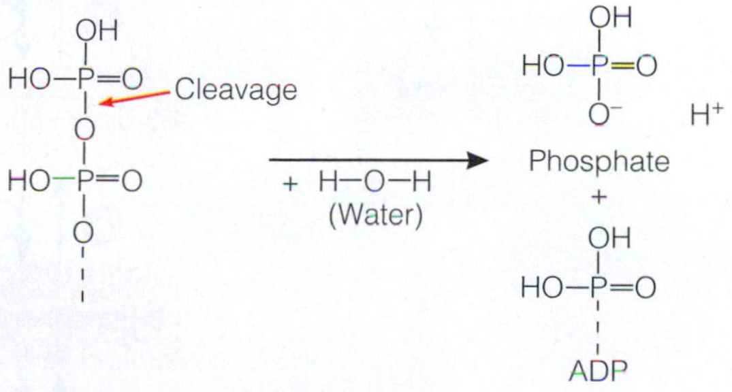

Adenosine diphosphate (ADP).

Glycolysis

Figure C-1 depicts glycolysis. The following text describes key steps as numbered on the figure.

FIGURE C-1 Glycolysis

Notice that galactose and fructose enter at different places but continue on the same pathway.

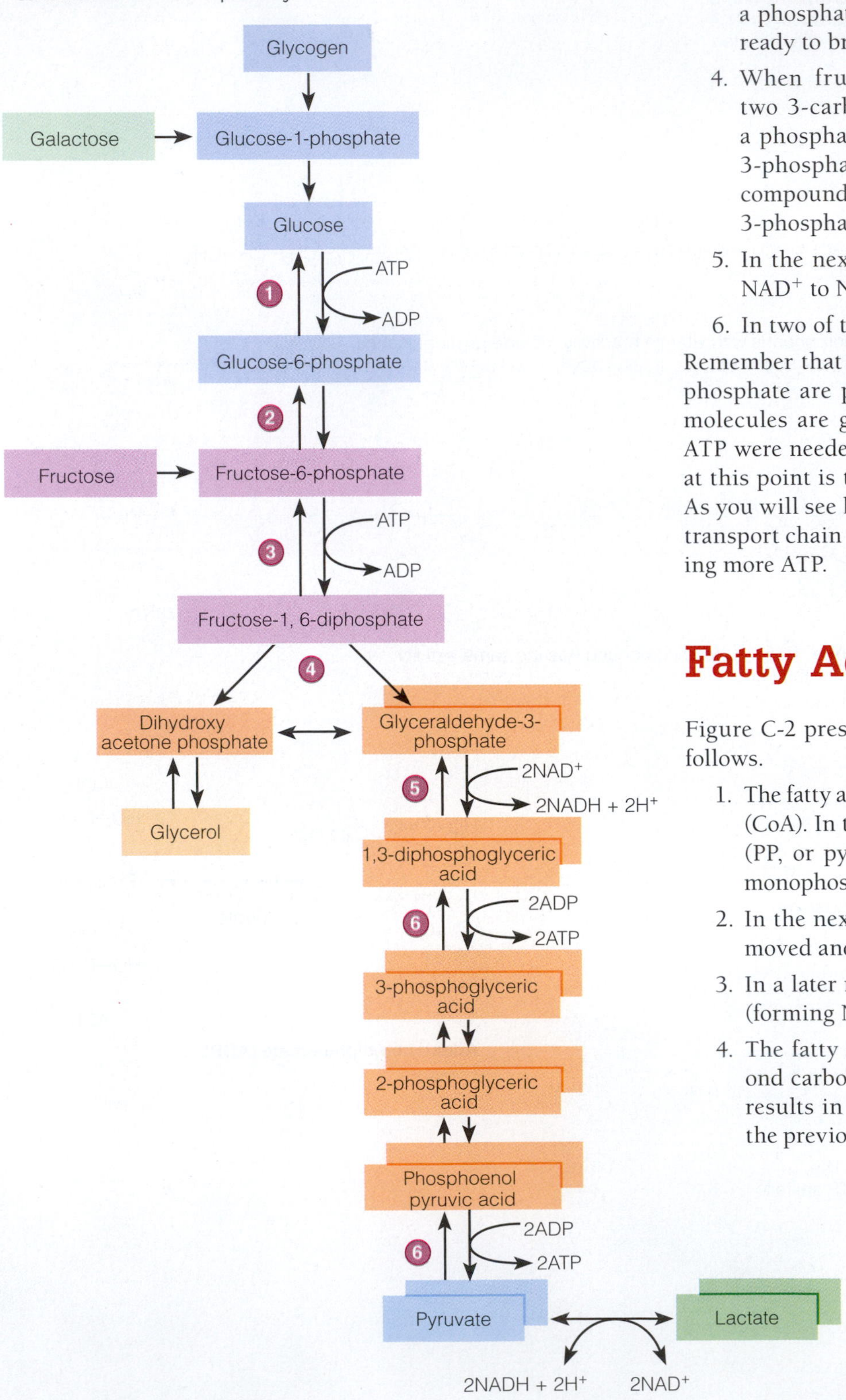

1. A phosphate is attached to glucose at the carbon that chemists call number 6 (review the first diagram of glucose on p. C-1 to see how chemists number the carbons in a glucose molecule). The product is called, logically enough, glucose-6-phosphate. One ATP molecule is used to accomplish this.
2. Glucose-6-phosphate is rearranged by an enzyme.
3. A phosphate is added in another reaction that uses another molecule of ATP. The product this time is fructose-1,6-diphosphate. At this point the 6-carbon sugar has a phosphate group on its first and sixth carbons and is ready to break apart.
4. When fructose-1,6-diphosphate breaks in half, the two 3-carbon compounds are not identical. Each has a phosphate group attached, but only glyceraldehyde-3-phosphate converts directly to pyruvate. The other compound, however, converts easily to glyceraldehyde-3-phosphate.
5. In the next step, enough energy is released to convert NAD^+ to $NADH + H^+$.
6. In two of the following steps ATP is regenerated.

Remember that in effect two molecules of glyceraldehyde-3-phosphate are produced from glucose; therefore, four ATP molecules are generated from each glucose molecule. Two ATP were needed to get the sequence started, so the net gain at this point is two ATP and two molecules of $NADH + H^+$. As you will see later, each $NADH + H^+$ moves to the electron transport chain to unload its hydrogens onto oxygen, producing more ATP.

Fatty Acid Oxidation

Figure C-2 presents fatty acid oxidation. The sequence is as follows.

1. The fatty acid is activated by combining with coenzyme A (CoA). In this reaction, ATP loses two phosphorus atoms (PP, or pyrophosphate) and becomes AMP (adenosine monophosphate)—the equivalent of a loss of two ATP.
2. In the next reaction, two H with their electrons are removed and transferred to FAD, forming $FADH_2$.
3. In a later reaction, two H are removed and go to NAD^+ (forming $NADH + H^+$).
4. The fatty acid is cleaved at the "beta" carbon, the second carbon from the carboxyl (COOH) end. This break results in a fatty acid that is two carbons shorter than the previous one and a 2-carbon molecule of acetyl CoA.

FIGURE C-2 Fatty Acid Oxidation

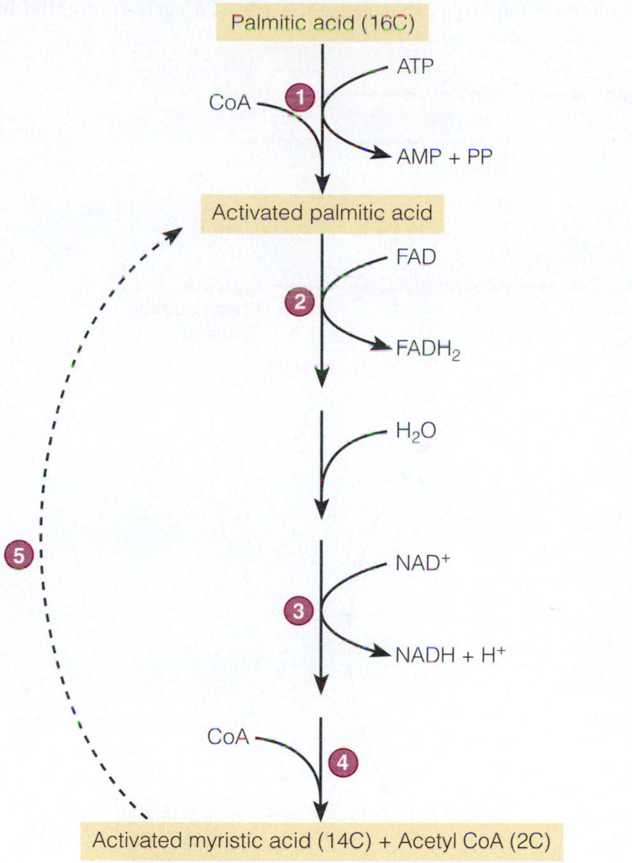

At the same time, another CoA is attached to the fatty acid, thus activating it for its turn through the series of reactions.

5. The sequence is repeated with each cycle producing an acetyl CoA and a shorter fatty acid until only a 2-carbon fatty acid remains—acetyl CoA.

In the example shown in Figure C-2, palmitic acid (a 16-carbon fatty acid) will go through this series of reactions seven times, using the equivalent of two ATP for the initial activation and generating seven $FADH_2$, seven $NADH + H^+$, and eight acetyl CoA. As you will see later, each of the seven $FADH_2$ will enter the electron transport chain to unload its hydrogens onto oxygen, yielding two ATP (for a total of 14). Similarly, each $NADH + H^+$ will enter the electron transport chain to unload its hydrogens onto oxygen, yielding three ATP (for a total of 21). Thus the oxidation of a 16-carbon fatty acid uses 2 ATP and generates 35 ATP. When the eight acetyl CoA enter the TCA cycle, even more ATP will be generated, as a later section describes.

Amino Acid Degradation

The first step in amino acid degradation is the removal of the nitrogen-containing amino group through either deamination (Figure 6-11 on p. 186) or transamination (Figure 6-12 on p. 186) reactions. Then the remaining carbon skeletons may enter the metabolic pathways at different places, as shown in Figure C-3 (p. C-12).

The TCA Cycle

The tricarboxylic acid, or TCA, cycle is the set of reactions that break down acetyl CoA to carbon dioxide and hydrogens. To link glycolysis to the TCA cycle, pyruvate enters the mitochondrion, loses a carbon group, and bonds with a molecule of CoA to become acetyl CoA. The TCA cycle uses any substance that can be converted to acetyl CoA directly or indirectly through pyruvate.

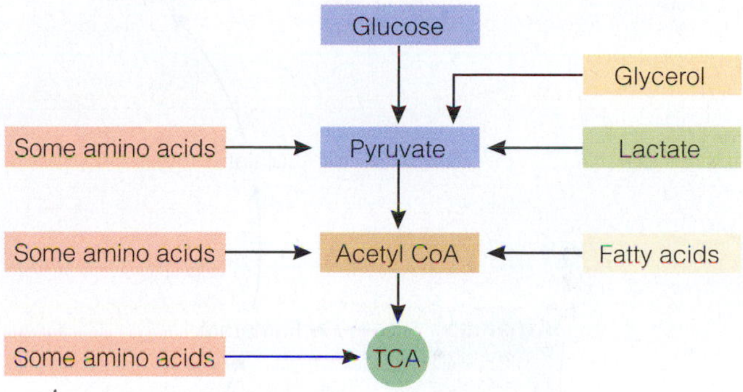

Any substance that can be converted to acetyl CoA directly, or indirectly through pyruvate, may enter the TCA cycle.

The step from pyruvate to acetyl CoA is complex. We have included only those substances that will help you understand the transfer of energy from the nutrients. Pyruvate loses a carbon to carbon dioxide and is attached to a molecule of CoA. In the process, NAD^+ picks up two hydrogens with their associated electrons, becoming $NADH + H^+$.

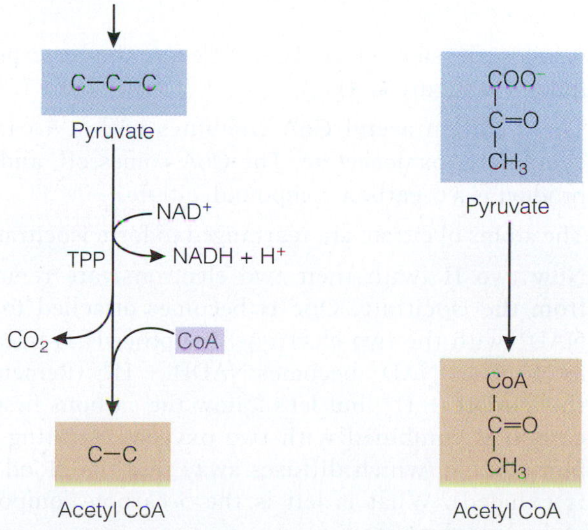

The step from pyruvate to acetyl CoA. (TPP and NAD are coenzymes containing the B vitamins thiamin and niacin, respectively.)

FIGURE C-3 Amino Acids Enter the Metabolic Pathways

After losing their amino groups, carbon skeletons can be converted to one of seven molecules that can enter the TCA cycle (presented in Figure C-4).

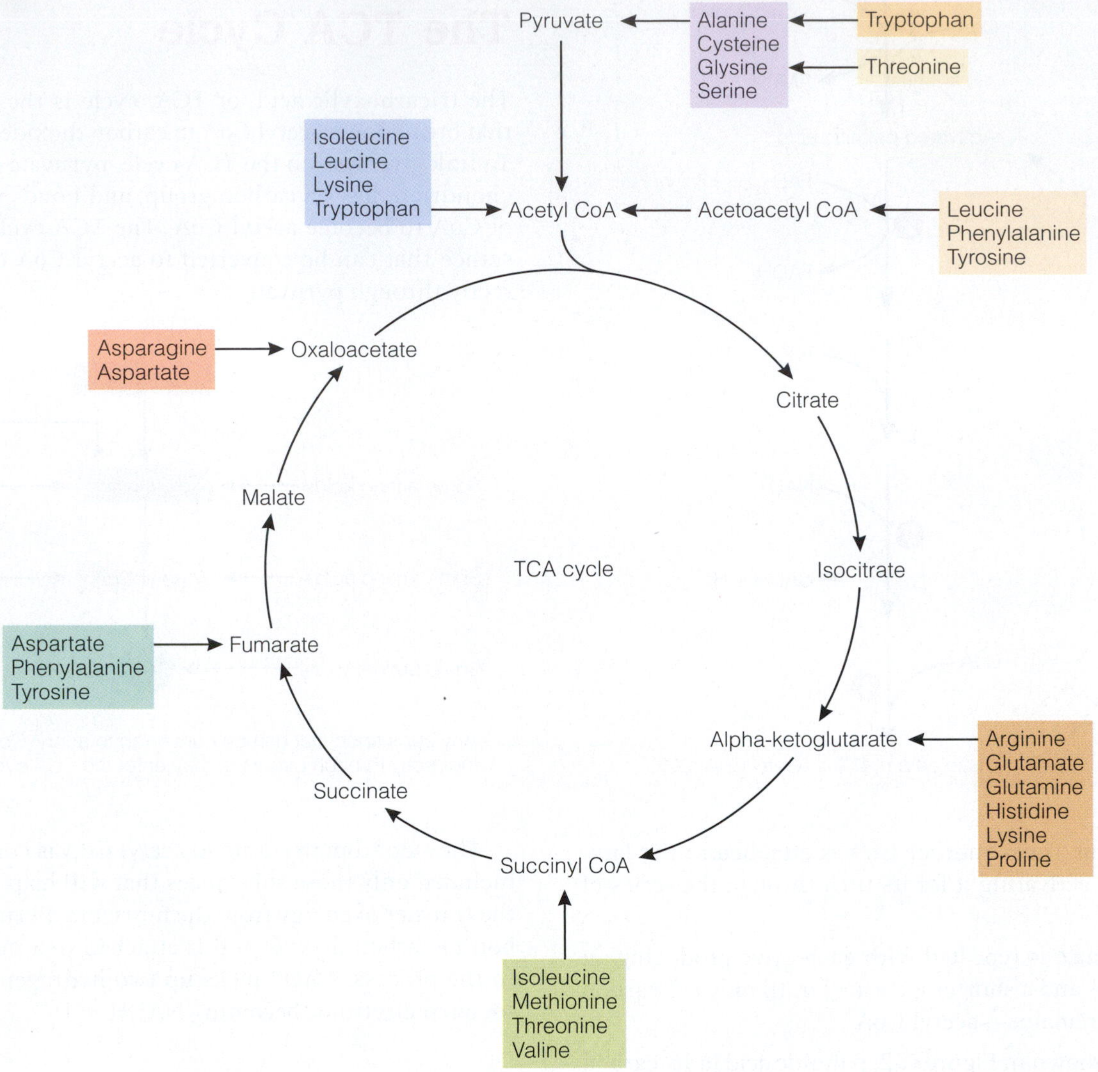

Let's follow the steps of the TCA cycle (see the corresponding numbers in Figure C-4).

1. The 2-carbon acetyl CoA combines with a 4-carbon compound, oxaloacetate. The CoA comes off, and the product is a 6-carbon compound, citrate.
2. The atoms of citrate are rearranged to form isocitrate.
3. Now two H (with their two electrons) are removed from the isocitrate. One H becomes attached to the NAD^+ with the two electrons; the other H is released as H^+. Thus NAD^+ becomes NADH + H^+. (Remember this NADH + H^+, but let's follow the carbons first.) A carbon is combined with two oxygens, forming carbon dioxide (which diffuses away into the blood and is exhaled). What is left is the 5-carbon compound alpha-ketoglutarate.
4. Now two compounds interact with alpha-ketoglutarate — a molecule of CoA and a molecule of NAD^+. In this complex reaction, a carbon and two oxygens are removed (forming carbon dioxide); two hydrogens are removed and go to NAD^+ (forming NADH + H^+); and the remaining 4-carbon compound is attached to the CoA, forming succinyl CoA. (Remember this NADH + H^+ also. You will see later what happens to it.)
5. Now two molecules react with succinyl CoA—a molecule called GDP and one of phosphate (P). The CoA comes off, the GDP and P combine to form the high-energy compound GTP (similar to ATP), and succinate remains. (Remember this GTP.)
6. In the next reaction, two H with their electrons are removed from succinate and are transferred to a molecule of FAD (a coenzyme like NAD^+) to form $FADH_2$. The product that remains is fumarate. (Remember this $FADH_2$.)
7. Next a molecule of water is added to fumarate, forming malate.

FIGURE C-4 **The TCA Cycle**

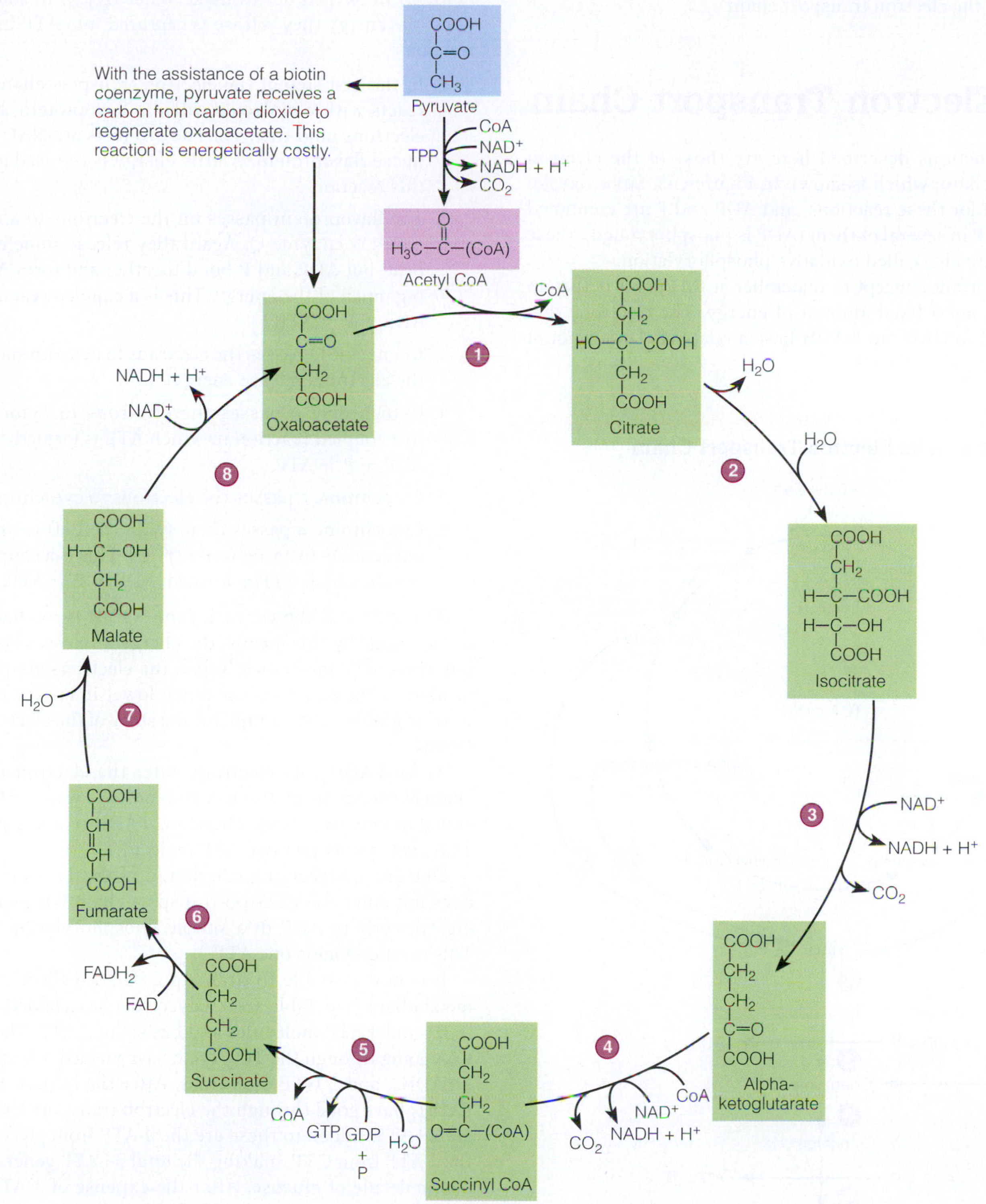

8. A molecule of NAD^+ reacts with the malate; two H with their associated electrons are removed from the malate and form $NADH + H^+$. The product that remains is the 4-carbon compound oxaloacetate. (Remember this $NADH + H^+$.)

We are back where we started. The oxaloacetate formed in this process can combine with another molecule of acetyl CoA (step 1), and the cycle can begin again, as shown in Figure C-4.

So far, we have seen two carbons brought in with acetyl CoA and two carbons ending up in carbon dioxide. But where are the energy and the ATP we promised?

A review of the eight steps of the TCA cycle shows that the compounds $NADH + H^+$ (three molecules), $FADH_2$, and GTP capture energy originally found in acetyl CoA. To see how this energy ends up in ATP, we must follow the electrons further—into the electron transport chain.

The Electron Transport Chain

The six reactions described here are those of the electron transport chain, which is shown in Figure C-5. Since oxygen is required for these reactions, and ADP and P are combined to form ATP in several of them (ADP is phosphorylated), these reactions are also called oxidative phosphorylation.

An important concept to remember at this point is that an electron is not a fixed amount of energy. The electrons that bond the H to NAD^+ in NADH have a relatively large amount of energy. In the series of reactions that follow, they release this energy in small amounts, until at the end they are attached (with H) to oxygen (O) to make water (H_2O). In some of the steps, the energy they release is captured into ATP in coupled reactions.

1. In the first step of the electron transport chain, NADH reacts with a molecule called a flavoprotein, losing its electrons (and their H). The products are NAD^+ and reduced flavoprotein. A little energy is released as heat in this reaction.
2. The flavoprotein passes on the electrons to a molecule called coenzyme Q. Again they release some energy as heat, but ADP and P bond together and form ATP, storing much of the energy. This is a coupled reaction: ADP + P → ATP.
3. Coenzyme Q passes the electrons to cytochrome *b*. Again the electrons release energy.
4. Cytochrome *b* passes the electrons to cytochrome *c* in a coupled reaction in which ATP is formed: ADP + P → ATP.
5. Cytochrome *c* passes the electrons to cytochrome *a*.
6. Cytochrome *a* passes them (with their H) to an atom of oxygen (O), forming water (H_2O). This is a coupled reaction in which ATP is formed: ADP + P → ATP.

FIGURE C-5 **The Electron Transport Chain**

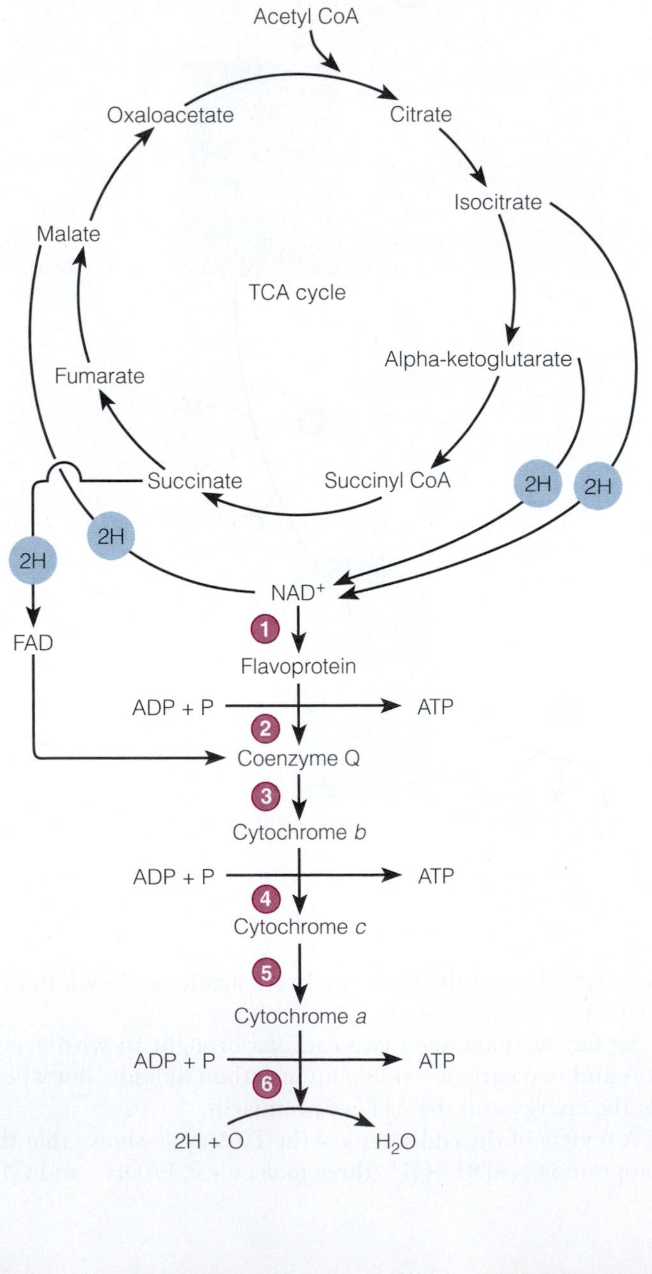

As Figure C-5 shows, each time NADH is oxidized (loses its electrons) by this means, the energy it releases is captured into three ATP molecules. When the electrons are passed on to water at the end, they are much lower in energy than they were originally. This completes the story of the electrons from NADH.

As for $FADH_2$, its electrons enter the electron transport chain at coenzyme Q. From coenzyme Q to water, ATP is generated in only two steps. Therefore, $FADH_2$ coming out of the TCA cycle yields just two ATP molecules.

One energy-receiving compound of the TCA cycle (GTP) does not enter the electron transport chain but gives its energy directly to ADP in a simple phosphorylation reaction. This reaction yields one ATP.

It is now possible to draw up a balance sheet of glucose metabolism (see Table C-3). Glycolysis has yielded 4 NADH + H^+ and 4 ATP molecules and has spent 2 ATP. The 2 acetyl CoA going through the TCA cycle have yielded 6 NADH + H^+, 2 $FADH_2$, and 2 GTP molecules. After the NADH + H^+ and $FADH_2$ have gone through the electron transport chain, there are 28 ATP. Added to these are the 4 ATP from glycolysis and the 2 ATP from GTP, making the total 34 ATP generated from one molecule of glucose. After the expense of 2 ATP is subtracted, there is a net gain of 32 ATP.*

*The total may sometimes be 30 ATP. The NADH + H^+ generated in the cytoplasm during glycolysis pass their electrons on to shuttle molecules, which move them into the mitochondria. One shuttle, malate, contributes its electrons to the electron transport chain before the first site of ATP synthesis, yielding 5 ATP. Another, glycerol phosphate, adds its electrons into the chain beyond that first site, yielding 3 ATP. Thus sometimes 5, and sometimes 3, ATP result from the NADH + H^+ that arise from glycolysis. The amount depends on the cell.

TABLE C-3 Balance Sheet for Glucose Metabolism

		ATP
Glycolysis:	4 ATP − 2 ATP	2
1 glucose to 2 pyruvate	2 NADH + H^+	3–5[a]
2 pyruvate to 2 acetyl CoA	2 NADH + H^+	5
TCA cycle and electron transport chain:		
2 isocitrate	2 NADH + H^+	5
2 alpha-ketoglutarate	2 NADH + H^+	5
2 succinyl CoA	2 GTP	2
2 succinate	2 $FADH_2$	3
2 malate	2 NADH + H^+	5
Total ATP collected from one molecule glucose:		30–32

[a]Each NADH + H^+ from glycolysis can yield 1.5 or 2.5 ATP. See the accompanying text.

A similar balance sheet from the complete breakdown of one 16-carbon fatty acid would show a net gain of 129 ATP. As mentioned earlier, 35 ATP were generated from the 7 $FADH_2$ and 7 NADH + H^+ produced during fatty acid oxidation. The 8 acetyl CoA produced will each generate 12 ATP as they go through the TCA cycle and the electron transport chain, for a total of 96 more ATP. After subtracting the 2 ATP needed to activate the fatty acid initially, the net yield from one 16-carbon fatty acid: 35 + 96 − 2 = 129 ATP.

These calculations help explain why fat yields more energy (measured as kcalories) per gram than carbohydrate or protein. The more hydrogen atoms a fuel contains, the more ATP will be generated during oxidation. The 16-carbon fatty acid molecule, with its 32 hydrogen atoms, generates 129 ATP, whereas glucose, with its 12 hydrogen atoms, yields only 32 ATP.

The TCA cycle and the electron transport chain are the body's major means of capturing the energy from nutrients in ATP molecules. Other means, such as anaerobic glycolysis, contribute energy quickly, but the aerobic processes are the most efficient. Biologists and chemists understand much more about these processes than has been presented here.

Alcohol's Interference with Energy Metabolism

Highlight 7 provides an overview of how alcohol interferes with energy metabolism. With an understanding of the TCA cycle, a few more details may be appreciated. During alcohol metabolism, the enzyme alcohol dehydrogenase oxidizes alcohol to acetaldehyde while it simultaneously reduces a molecule of NAD^+ to NADH + H^+. The related enzyme acetaldehyde dehydrogenase reduces another NAD^+ to NADH + H^+ while it oxidizes acetaldehyde to acetyl CoA, the compound that enters the TCA cycle to generate energy. Thus, whenever alcohol is being metabolized in the body, NAD^+ diminishes, and NADH + H^+ accumulates. Chemists say that the body's "redox state" is altered, because NAD^+ can oxidize, and NADH + H^+ can reduce, many other body compounds. During alcohol metabolism, NAD^+ becomes unavailable for the multitude of reactions for which it is required.

As the previous sections just explained, for glucose to be completely metabolized, the TCA cycle must be operating, and NAD^+ must be present. If these conditions are not met (and when alcohol is present, they may not be), the pathway will be blocked, and traffic will back up—or an alternate route will be taken. Think about this as you follow the pathway shown in Figure C-6.

In each step of alcohol metabolism in which NAD^+ is converted to NADH + H^+, hydrogen ions accumulate, resulting in a dangerous shift of the acid-base balance toward acid (Chapter 12 explains acid-base balance). The accumulation of NADH + H^+ slows TCA cycle activity, so pyruvate and acetyl CoA build

FIGURE C-6 Ethanol Enters the Metabolic Pathways

This is a simplified version of the glucose-to-energy pathway showing the entry of ethanol. The coenzyme NAD (which is the active form of the B vitamin niacin) is the only one shown here; however, many others are involved.

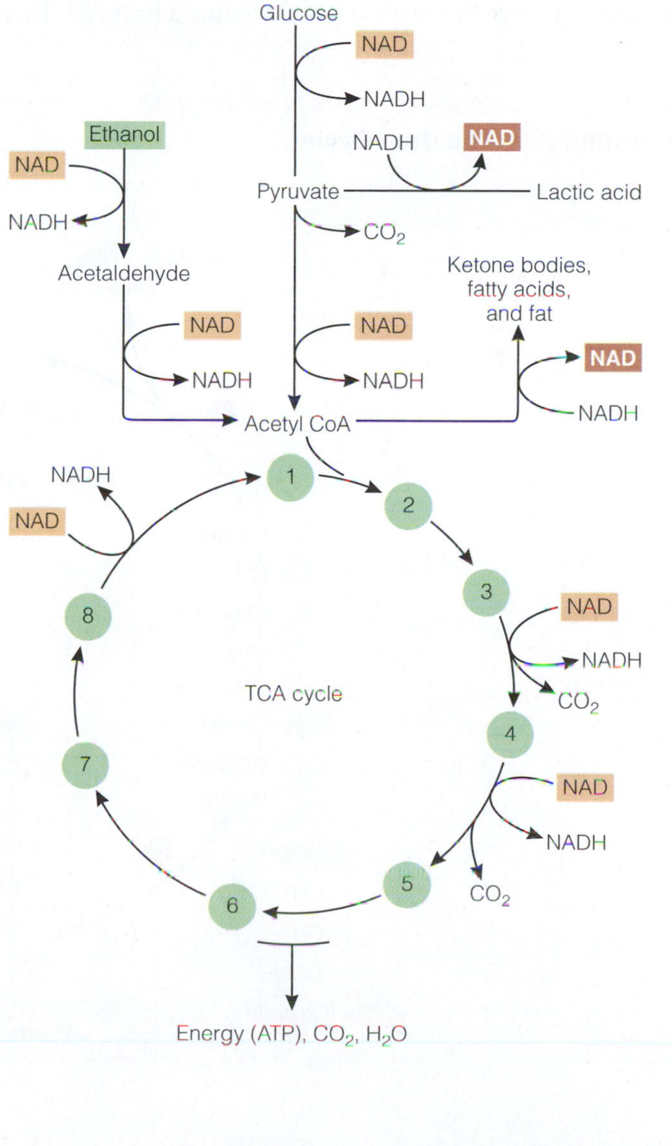

up. This condition favors the conversion of pyruvate to lactate, which serves as a temporary storage place for hydrogens from NADH + H^+. The conversion of pyruvate to lactate restores some NAD^+, but a lactate buildup has serious consequences of its own. It adds to the body's acid burden and interferes with the excretion of uric acid, causing goutlike symptoms. Molecules of acetyl CoA become building blocks for fatty acids or ketone bodies. The making of ketone bodies consumes acetyl CoA and generates NAD^+ ; but some ketone bodies are acids, so they push the acid-base balance further toward acid.

Thus alcohol cascades through the metabolic pathways, wreaking havoc along the way. These consequences have physical effects, which Highlight 7 describes.

The Urea Cycle

Chapter 6 sums up the process by which waste nitrogen is eliminated from the body by stating that ammonia molecules combine with carbon dioxide to produce urea. This is true, but it is not the whole story. Urea is produced in a multistep process within the cells of the liver.

Ammonia, freed from an amino acid or other compound during metabolism anywhere in the body, arrives at the liver by way of the bloodstream and is taken into a liver cell. There, it is first combined with carbon dioxide and a phosphate group from ATP to form carbamyl phosphate:

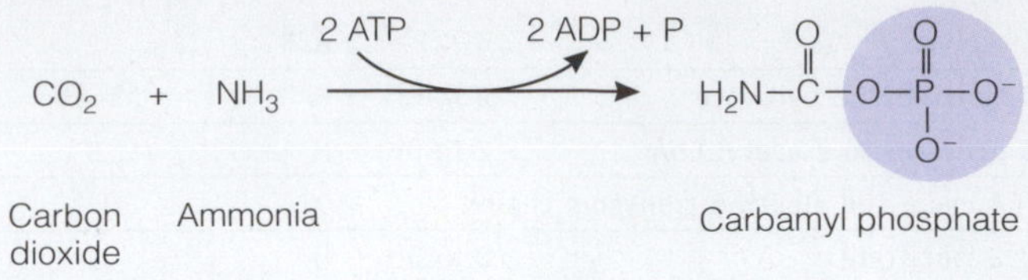

Figure C-7 shows the cycle of four reactions that follow.

1. Carbamyl phosphate combines with the amino acid ornithine, losing its phosphate group. The compound formed is citrulline.
2. Citrulline combines with the amino acid aspartic acid, to form argininosuccinate. The reaction requires energy from ATP. (ATP was shown earlier losing one phosphorus atom in a phosphate group, P, to become ADP. In this reaction, it loses two phosphorus atoms joined together, PP, and becomes adenosine monophosphate, AMP.)
3. Argininosuccinate is split, forming another acid, fumarate, and the amino acid arginine.
4. Arginine loses its terminal carbon with two attached amino groups and picks up an oxygen from water. The end product is urea, which the kidneys excrete in the

FIGURE C-7 **The Urea Cycle**

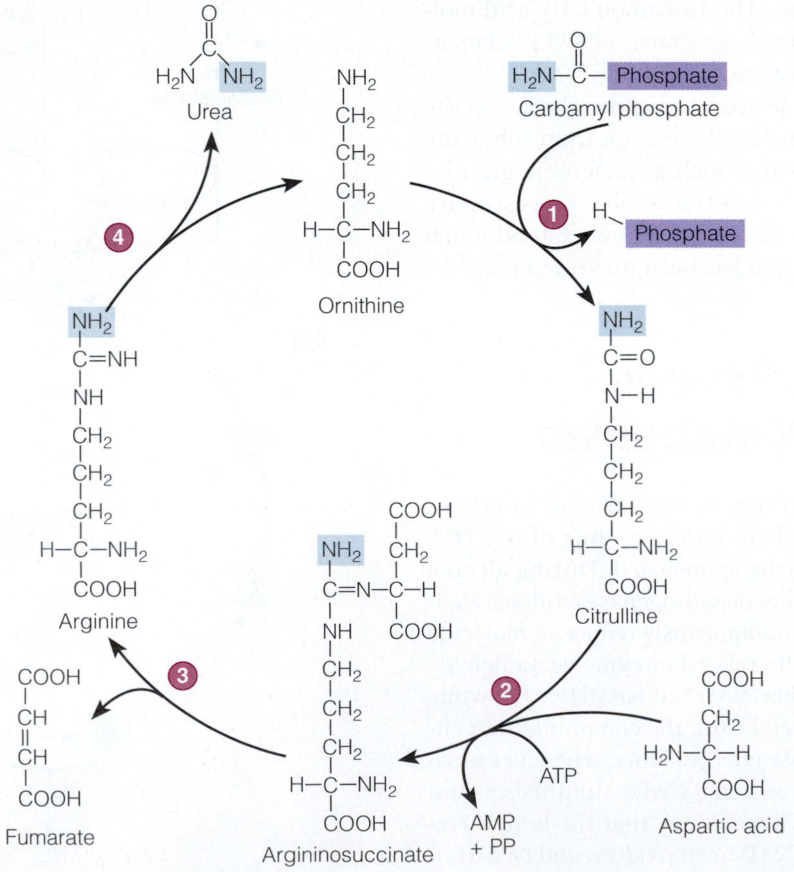

urine. The compound that remains is ornithine, identical to the ornithine with which this series of reactions began, and ready to react with another molecule of carbamyl phosphate and turn the cycle again.

Formation of Ketone Bodies

Normally, fatty acid oxidation proceeds all the way to carbon dioxide and water. However, in ketosis (discussed in Chapter 7), an intermediate is formed from the condensation of two molecules of acetyl CoA: acetoacetyl CoA. Figure C-8 shows the formation of ketone bodies from that intermediate.

1. Acetoacetyl CoA condenses with acetyl CoA to form a 6-carbon intermediate, beta-hydroxy-beta-methylglutaryl CoA.
2. This intermediate is cleaved to acetyl CoA and acetoacetate.
3. Acetoactate can be metabolized either to beta-hydroxybutyrate acid (step 3a) or to acetone (3b).

Acetoacetate, beta-hydroxybutyrate, and acetone are the ketone bodies of ketosis. Two are real ketones (they have a C=O group between two carbons); the other is an alcohol that has been produced during ketone formation—hence the term *ketone bodies,* rather than ketones, to describe the three of them. There are many other ketones in nature; these three are characteristic of ketosis in the body.

FIGURE C-8 **The Formation of Ketone Bodies**

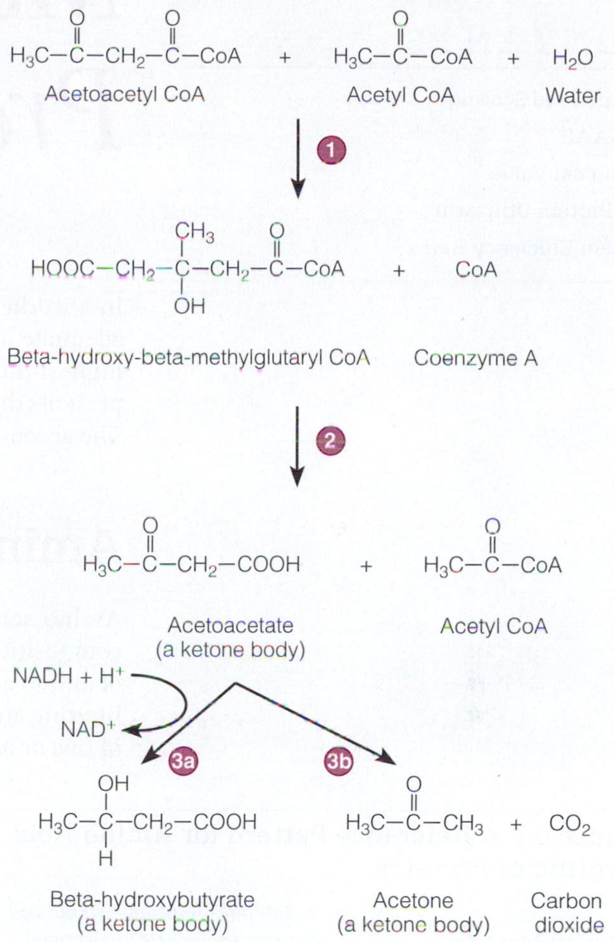

APPENDIX D

Measures of Protein Quality

CONTENTS

In a world where food is scarce and many people's diets contain marginal or inadequate amounts of protein, it is important to know which foods contain the highest-quality protein. Chapter 6 describes protein quality, and this appendix presents different measures researchers use to assess the quality of a food protein. The accompanying glossary defines related terms.

Amino Acid Scoring

Amino acid scoring evaluates a protein's quality by determining its amino acid composition and comparing it with that of a reference protein. The advantages of amino acid scoring are that it is simple and inexpensive, it easily identifies the limiting amino acid, and it can be used to score mixtures of different proportions of two or more proteins mathematically without having to make up a mixture and test it. Its chief weaknesses are that it fails to estimate the digestibility of a protein, which may strongly affect the protein's quality; it relies on a chemical procedure in which certain amino acids may be destroyed, making the pattern that is analyzed inaccurate; and it is blind to other features of the protein (such as the presence of substances that may inhibit the digestion or utilization of the protein) that would only be revealed by a test in living animals.

Table D-1 shows the reference pattern for the nine essential amino acids. To interpret the table, read, "For every 3210 units of essential amino acids, 145 must be histidine, 340 must be isoleucine, 540 must be leucine," and so on. To compare a test protein with the reference protein, the experimenter first obtains a chemical analysis of the test protein's amino acids. Then, taking 3210 units of the amino acids, the experimenter compares the amount of each amino acid to the amount found in 3210 units of essential amino acids in egg protein. For example, suppose the test protein contained (per 3210 units) 360 units of isoleucine; 500 units of leucine; 350 of lysine; and for each of the other amino acids, more units than egg protein contains. The two amino acids that are low are leucine (500 as compared with 540 in egg) and lysine (350 versus 440 in egg). The ratio, amino acid in the test protein divided by amino acid in

TABLE D-1 A Reference Pattern for Amino Acid Scoring of Proteins

Essential Amino Acids	Reference Protein—Whole Egg (mg amino acid/g nitrogen)
Histidine	145
Isoleucine	340
Leucine	540
Lysine	440
Methionine + cystine[a]	355
Phenylalanine + tyrosine[b]	580
Threonine	294
Tryptophan	106
Valine	410
Total	3210

[a]Methionine is essential and is also used to make cystine. Thus the methionine requirement is lower if cystine is supplied.
[b]Phenylalanine is essential and is also used to make tyrosine if not enough of the latter is available. Thus the phenylalanine requirement is lower if tyrosine is also supplied.

GLOSSARY

amino acid scoring: a measure of protein quality assessed by comparing a protein's amino acid pattern with that of a reference protein; sometimes called **chemical scoring.**

biological value (BV): a measure of protein quality assessed by measuring the amount of protein nitrogen that is retained from a given amount of protein nitrogen absorbed.

net protein utilization (NPU): a measure of protein quality assessed by measuring the amount of protein nitrogen that is retained from a given amount of protein nitrogen eaten.

PDCAAS (protein digestibility–corrected amino acid score): a measure of protein quality assessed by comparing the amino acid score of a food protein with the amino acid requirements of preschool-age children and then correcting for the true digestibility of the protein; recommended by the FAO/WHO and used to establish protein quality of foods for Daily Value percentages on food labels.

protein efficiency ratio (PER): a measure of protein quality assessed by determining how well a given protein supports weight gain in growing rats; used to establish the protein quality for infant formulas and baby foods.

egg, is 500/540 (or about 0.93) for leucine and 350/440 (or about 0.80) for lysine. Lysine is the limiting amino acid (the one that falls shortest compared with egg). If the protein's limiting amino acid is 80 percent of the amount found in the reference protein, it receives a score of 80.

PDCAAS

The **protein digestibility–corrected amino acid score, or PDCAAS**, compares the amino acid composition of a protein with human amino acid requirements and corrects for digestibility. First the protein's amino acid composition is determined, and then it is compared against the amino acid requirements of preschool-aged children. This comparison reveals the most limiting amino acid—the one that falls shortest compared with the reference. If a food protein's limiting amino acid is 70 percent of the amount found in the reference protein, it receives a score of 70. The amino acid score is multiplied by the food's protein digestibility percentage to determine the PDCAAS. The accompanying "How To" provides an example of how to calculate the PDCAAS, and Table D-2 lists the PDCAAS values of selected foods.

TABLE D-2 PDCAAS Values of Selected Foods

Casein (milk protein)	1.00
Egg white	1.00
Soybean (isolate)	.99
Beef	.92
Pea flour	.69
Kidney beans (canned)	.68
Chickpeas (canned)	.66
Pinto beans (canned)	.66
Rolled oats	.57
Lentils (canned)	.52
Peanut meal	.52
Whole wheat	.40

NOTE: 1.0 is the maximum PDCAAS a food protein can receive.

HOW TO Measure Protein Quality Using PDCAAS

To calculate the PDCAAS (protein digestibility–corrected amino acid score), researchers first determine the amino acid profile of the test protein (in this example, pinto beans). The second column of the table below presents the essential amino acid profile for pinto beans. The third column presents the amino acid reference pattern.

To determine how well the food protein meets human needs, researchers calculate the ratio by dividing the second column by the third column (for example, 30 ÷ 18 = 1.67). The amino acid with the lowest ratio is the most limiting amino acid—in this case, methionine. Its ratio is the amino acid score for the protein—in this case, 0.84.

The amino acid score alone, however, does not account for digestibility. Protein digestibility, as determined by rat studies, yields a value of 79 percent for pinto beans. Together, the amino acid score and the digestibility value determine the PDCAAS:

$$\text{PDCAAS} = \text{protein digestibility} \times \text{amino acid score}$$

$$\text{PDCAAS for pinto beans} = 0.79 \times 0.84 = 0.66$$

Thus the PDCAAS for pinto beans is 0.66. Table D-2 lists the PDCAAS values of selected foods.

The PDCAAS is used to determine the % Daily Value on food labels. To calculate the % Daily Value for protein for canned pinto beans, multiply the number of grams of protein in a standard serving (in the case of pinto beans, 7 grams per ½ cup) by the PDCAAS:

$$7\text{ g} \times 0.66 = 4.62$$

This value is then divided by the recommended standard for protein (for children over age four and adults, 50 grams):

$$4.62 \div 50 = 0.09 \text{ (or 9\%)}$$

The food label for this can of pinto beans would declare that one serving provides 7 grams protein, and if the label included a % Daily Value for protein (which is optional), the value would be 9 percent.

Essential Amino Acids	Amino Acid Profile of Pinto Beans (mg/g protein)	Amino Acid Reference Pattern (mg/g protein)	Amino Acid Score
Histidine	30.0	18	1.67
Isoleucine	42.5	25	1.70
Leucine	80.4	55	1.46
Lysine	69.0	51	1.35
Methionine (+ cystine)	21.1	25	0.84
Phenylalanine (+ tyrosine)	90.5	47	1.93
Threonine	43.7	27	1.62
Tryptophan	8.8	7	1.26
Valine	50.1	32	1.57

Biological Value

The **biological value (BV)** of a protein measures its efficiency in supporting the body's needs. In a test of biological value, two nitrogen balance studies are done. In the first, no protein is fed, and nitrogen (N) excretions in the urine and feces are measured. It is assumed that under these conditions, N lost in the urine is the amount the body always necessarily loses by filtration into the urine each day, regardless of what protein is fed (endogenous N). The N lost in the feces (called metabolic N) is the amount the body invariably loses into the intestine each day, whether or not food protein is fed. (To help you remember the terms: endogenous N is "urinary N on a zero-protein diet"; metabolic N is "fecal N on a zero-protein diet.")

In the second study, an amount of protein slightly below the requirement is fed. Intake and losses are measured; then the BV is derived using this formula:

$$\text{BV} = \frac{\text{N retained}}{\text{N absorbed}} \times 100$$

The denominator of this equation expresses the amount of nitrogen *absorbed:* food N minus fecal N (excluding the metabolic N the body would lose in the feces anyway, even without food). The numerator expresses the amount of N *retained* from the N absorbed: absorbed N (as in the denominator) minus the N excreted in the urine (excluding the endogenous N the body would lose in the urine anyway, even without food). The more nitrogen retained, the higher the protein quality. (Recall that when an essential amino acid is missing, protein synthesis stops, and the remaining amino acids are deaminated and the nitrogen excreted.)

Egg protein has a BV of 100, indicating that 100 percent of the nitrogen absorbed is retained. Supplied in adequate quantity, a protein with a BV of 70 or greater can support human growth as long as energy intake is adequate. Table D-3 presents the BV for selected foods.

TABLE D-3 Biological Values (BV) of Selected Foods

Egg	100
Milk	93
Beef	75
Fish	75
Corn	72

NOTE: 100 is the maximum BV a food protein can receive.

This method has the advantages of being based on experiments with human beings (it can be done with animals, too, of course) and of measuring actual nitrogen retention. But it is also cumbersome, expensive, and often impractical, and it is based on several assumptions that may not be valid. For example, the physiology, normal environment, or typical food intake of the subjects used for testing may not be similar to those for whom the test protein may ultimately be used. For another example, the retention of protein in the body does not necessarily mean that it is being well utilized. Considerable exchange of protein among tissues (protein turnover) occurs, but is hidden from view when only N intake and output are measured. The test of biological value wouldn't detect if one tissue were shorted.

Net Protein Utilization

Like BV, **net protein utilization (NPU)** measures how efficiently a protein is used by the body and involves two balance studies. The difference is that NPU measures retention of food nitrogen rather than food nitrogen absorbed (as in BV). The formula for NPU is:

$$\text{NPU} = \frac{\text{N retained}}{\text{N intake}} \times 100$$

The numerator is the same as for BV, but the denominator represents food N intake only—not N absorbed.

This method offers advantages similar to those of BV determinations and is used more frequently, with animals as the test subjects. A drawback is that if a low NPU is obtained, the test results offer no help in distinguishing between two possible causes: a poor amino acid composition of the test protein or poor digestibility. There is also a limit to the extent to which animal test results can be assumed to be applicable to human beings.

Protein Efficiency Ratio

The **protein efficiency ratio (PER)** measures the weight gain of a growing animal and compares it to the animal's protein intake. Until recently, the PER was generally accepted in the United States and Canada as the official method for assessing protein quality, and it is still used to evaluate proteins for infants.

Young rats are fed a measured amount of protein and weighed periodically as they grow. The PER is expressed as:

$$\text{PER} = \frac{\text{weight gain (g)}}{\text{protein intake (g)}}$$

This method has the virtues of economy and simplicity, but it also has many drawbacks. The experiments are time-consuming; the amino acid needs of rats are not the same as those of human beings; and the amino acid needs for growth are not the same as for the maintenance of adult animals (growing animals need more lysine, for example). Table D-4 presents PER values for selected foods.

TABLE D-4 Protein Efficiency Ratio (PER) Values of Selected Proteins

Casein (milk)	2.8
Soy	2.4
Glutein (wheat)	0.4

APPENDIX E

Nutrition Assessment: Supplemental Information

CONTENTS

Chapter 17 describes data from nutrition assessments that help health professionals evaluate patients' nutrition status and nutrient needs. This appendix provides additional information that may be useful for complete assessments.

Growth Charts

♦ Reminder: The *body mass index (BMI)* is an index of a person's weight in relation to height, determined by dividing the weight in kilograms by the square of the height in meters:

$$\text{BMI} = \frac{\text{Weight (kg)}}{\text{Height (m)}^2}$$

♦ Additional growth charts are available at **www.cdc.gov/growthcharts**

Health professionals evaluate physical development by monitoring growth rates of children and comparing these rates with those on standard growth charts. Standard charts compare length or height to age, weight to age, weight to length, head circumference to age, and body mass index (BMI) ♦ to age. Although individual growth patterns vary, a child's growth curve will generally stay at about the same percentile throughout childhood. In children whose growth has been retarded, nutrition rehabilitation will ideally cause height and weight to increase to higher percentiles. In overweight children, the goal is for weight to remain stable as height increases, until weight becomes appropriate for height.

To evaluate growth in infants, an assessor uses a chart such as those in Figures E-1 through E-3. ♦ For example, the assessor follows these steps to determine the weight percentile:

- Select the appropriate chart based on age and gender.
- Locate the child's age along the horizontal axis on the bottom of the chart.
- Locate the child's weight in pounds or kilograms along the vertical axis.
- Mark the chart where the age and weight lines intersect, and read off the percentile.

For other measures, the assessor follows a similar procedure, using the appropriate chart. (When length is measured, use the chart for birth to 36 months; when height is measured, use the chart for 2 to 20 years.) Once all of the measures are plotted on growth charts, a skilled clinician can begin to interpret the data. Ideally, the height, weight, and head circumference should be in roughly the same percentile.

Head circumference is generally measured in children under two years of age. Since the brain grows rapidly before birth and during early infancy, extreme and chronic malnutrition during these times can impair brain development, curtailing the number of brain cells and the size of head circumference. Nonnutritional factors, such as certain disorders and genetic variation, can also influence head circumference.

FIGURE E-1

Length-for-Age and Weight-for-Age Percentiles: NAME ____________
Boys, Birth to 36 Months RECORD # ________

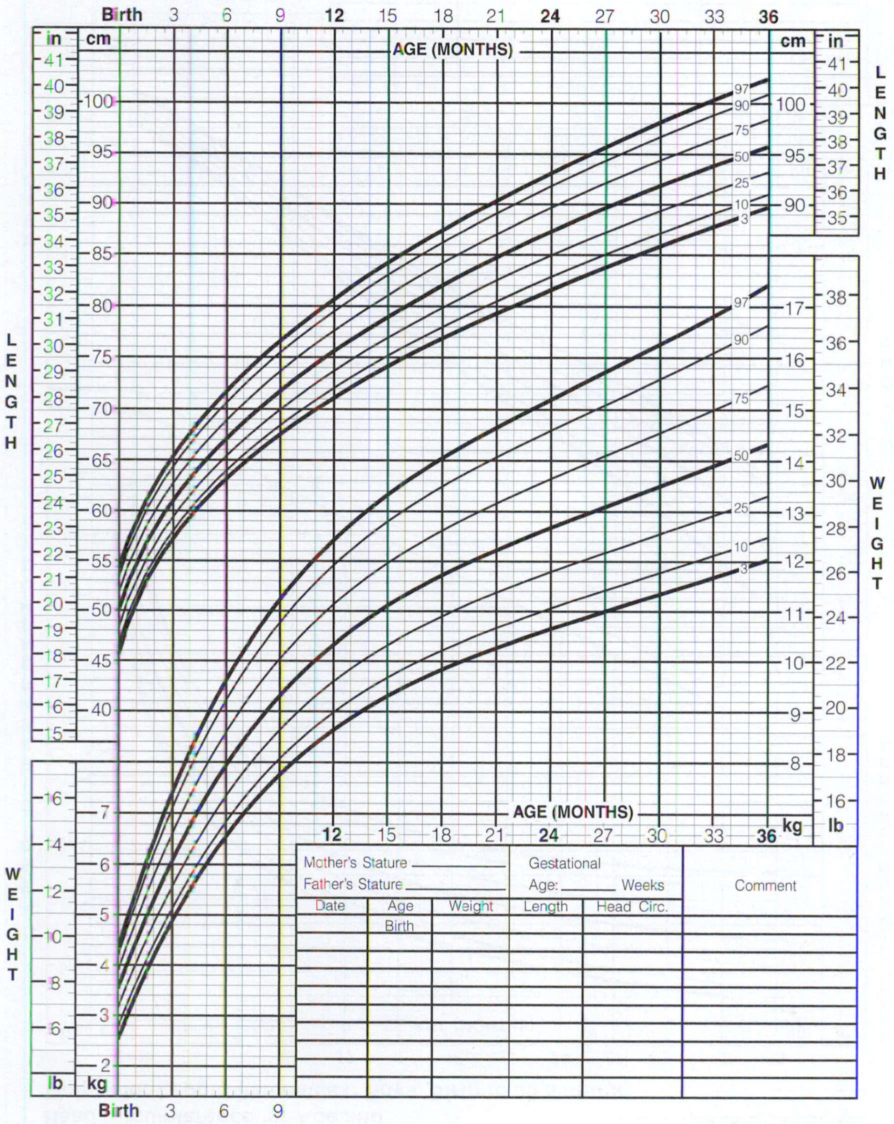

Published May 30, 2000 (modified 4/20/01).
SOURCE: Developed by the National Center for Health Statistics in collaboration with the National Center for Chronic Disease Prevention and Health Promotion (2000).
www.cdc.gov/growthcharts.

CDC SAFER • HEALTHIER • PEOPLE™

Length-for-Age and Weight-for-Age Percentiles: NAME ____________
Girls, Birth to 36 Months RECORD # ________

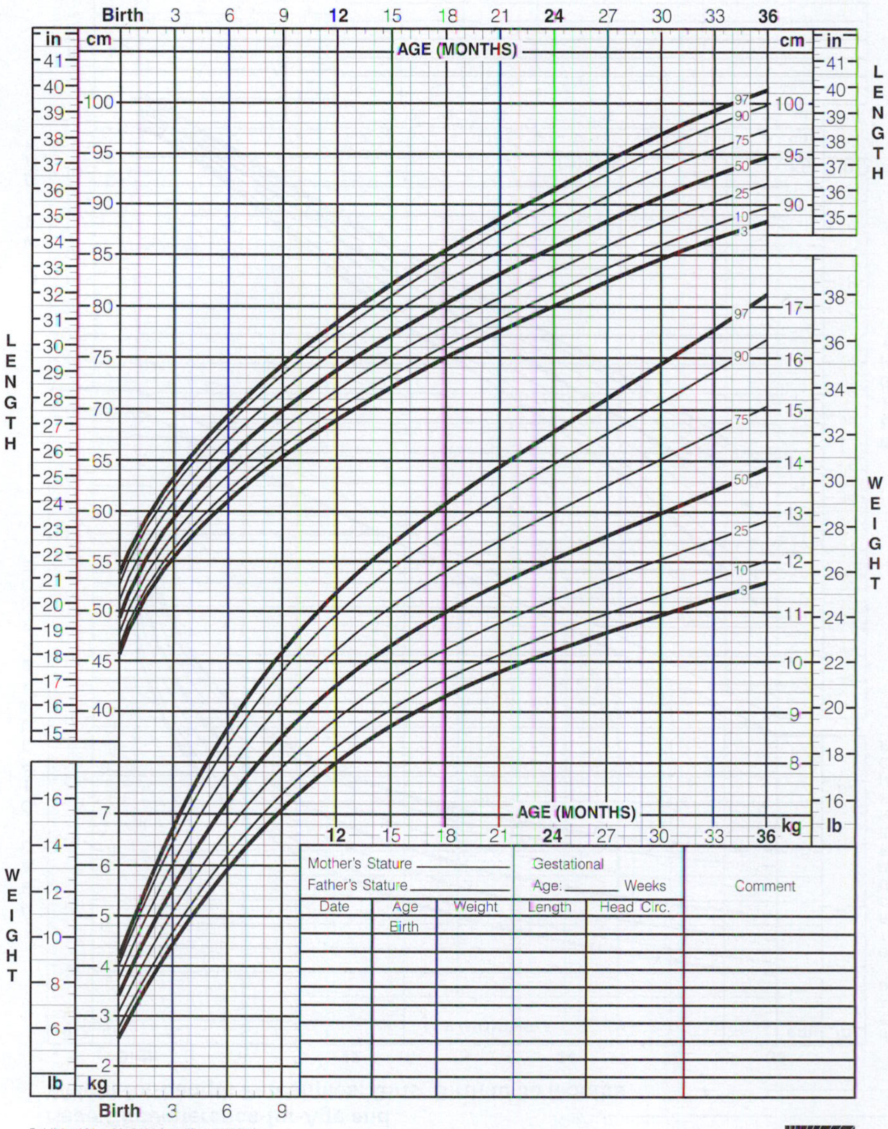

Published May 30, 2000 (modified 4/20/01).
SOURCE: Developed by the National Center for Health Statistics in collaboration with the National Center for Chronic Disease Prevention and Health Promotion (2000).
www.cdc.gov/growthcharts.

CDC SAFER • HEALTHIER • PEOPLE™

FIGURE E-2

Head Circumference-for-Age and Weight-for-Length Percentiles: Boys, Birth to 36 Months

NAME ____________
RECORD # ____________

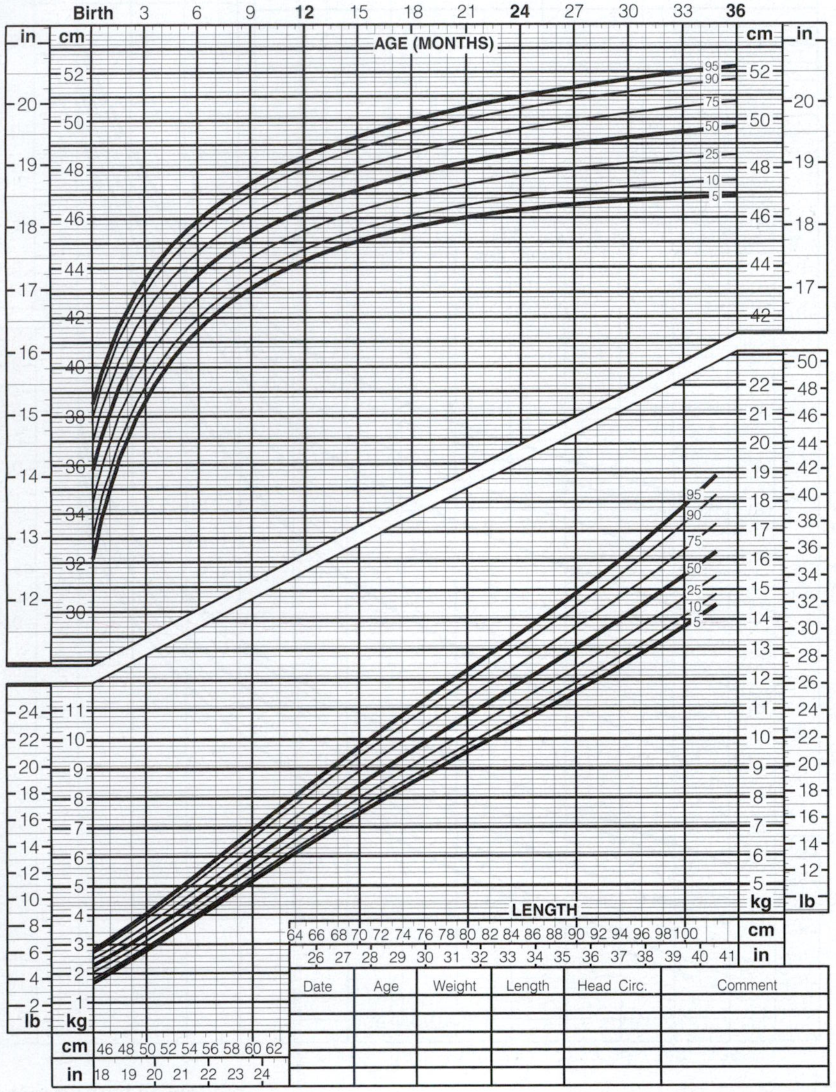

Published May 30, 2000 (modified 10/16/00).
SOURCE: Developed by the National Center for Health Statistics in collaboration with the National Center for Chronic Disease Prevention and Health Promotion (2000).
www.cdc.gov/growthcharts.

CDC SAFER • HEALTHIER • PEOPLE™

Head Circumference-for-Age and Weight-for-Length Percentiles: Girls, Birth to 36 Months

NAME ____________
RECORD # ____________

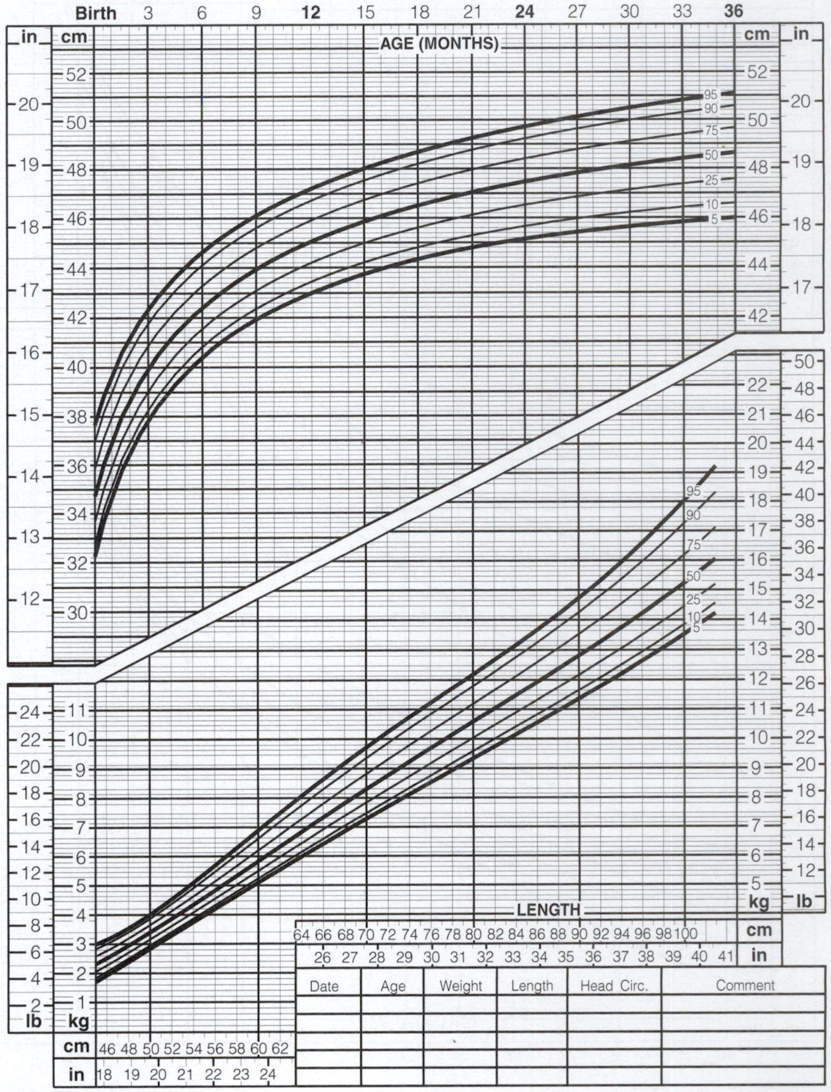

Published May 30, 2000 (modified 10/16/00).
SOURCE: Developed by the National Center for Health Statistics in collaboration with the National Center for Chronic Disease Prevention and Health Promotion (2000).
www.cdc.gov/growthcharts.

CDC SAFER • HEALTHIER • PEOPLE™

FIGURE E-3

Stature-for-Age and Weight-for-Age Percentiles: Boys, 2 to 20 Years

NAME ______________________

RECORD # ____________

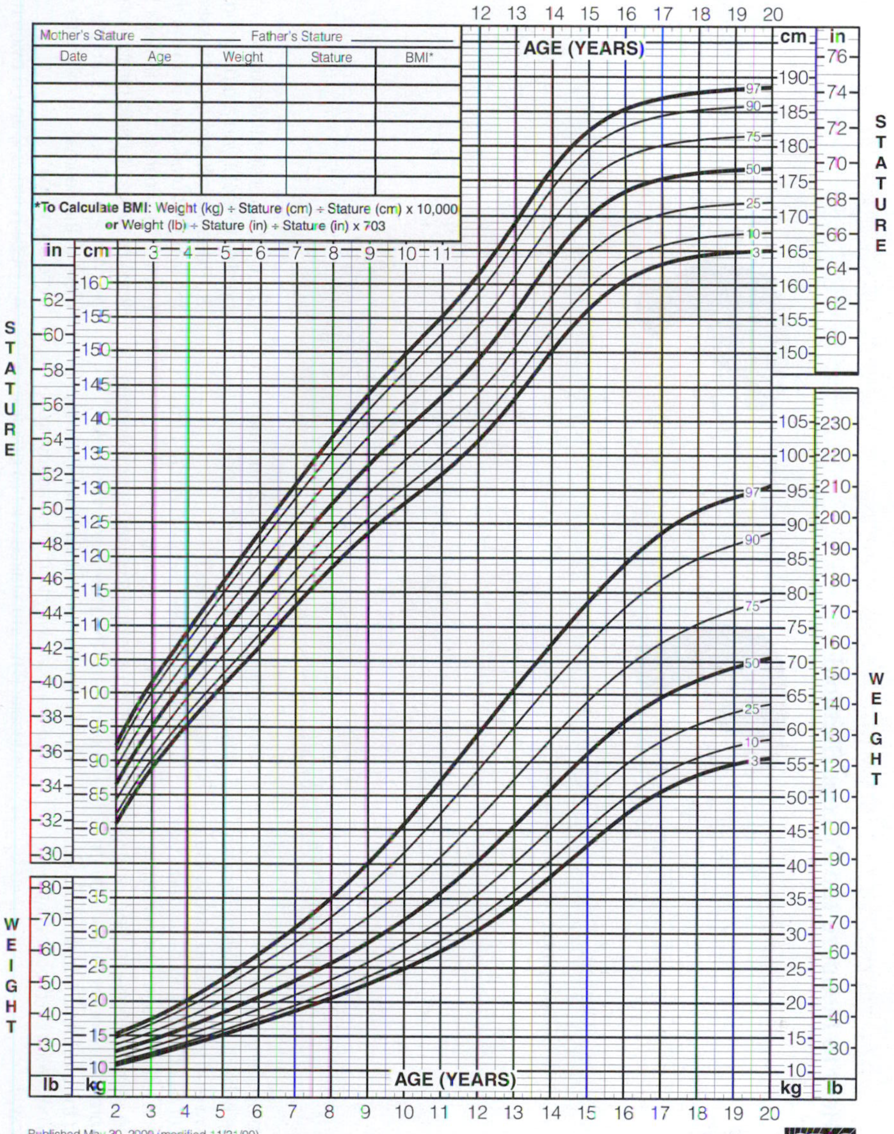

Published May 30, 2000 (modified 11/21/00).
SOURCE: Developed by the National Center for Health Statistics in collaboration with the National Center for Chronic Disease Prevention and Health Promotion (2000).
www.cdc.gov/growthcharts.

CDC SAFER • HEALTHIER • PEOPLE™

Stature-for-Age and Weight-for-Age Percentiles: Girls, 2 to 20 Years

NAME ______________________

RECORD # ____________

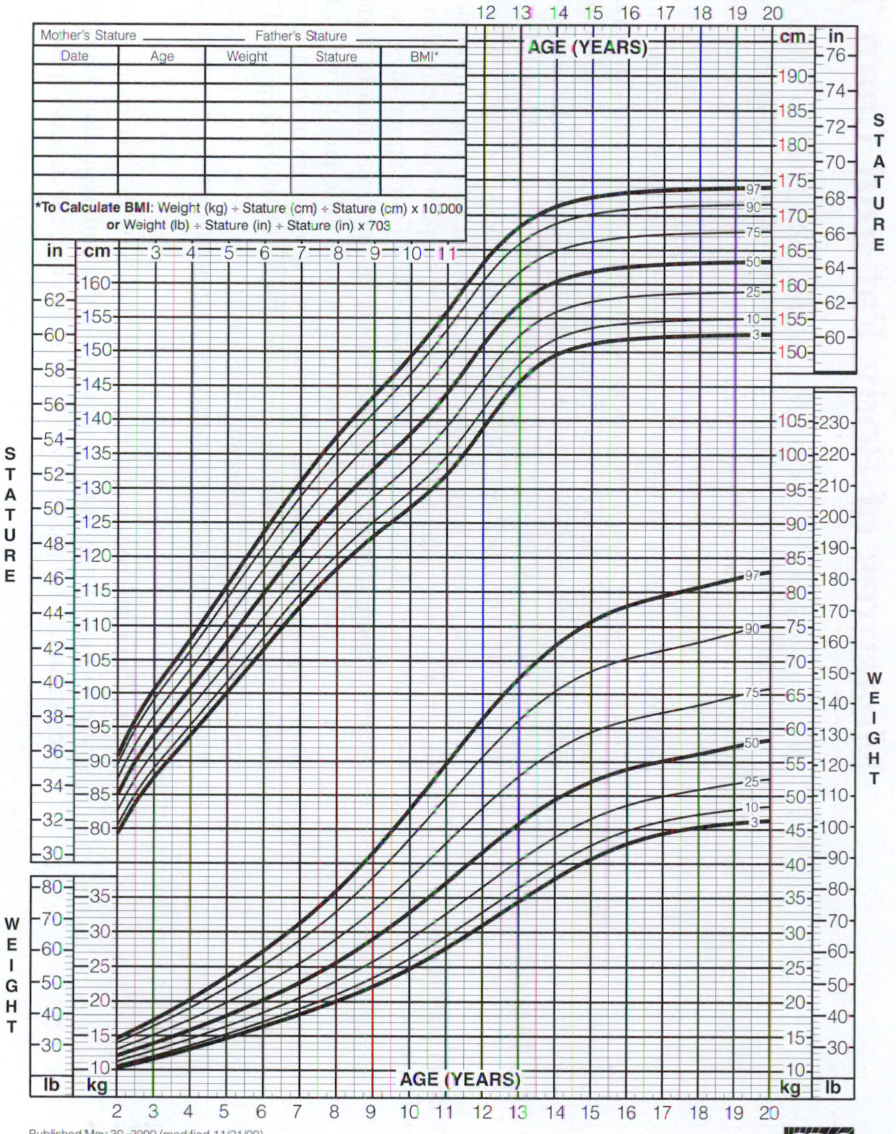

Published May 30, 2000 (modified 11/21/00).
SOURCE: Developed by the National Center for Health Statistics in collaboration with the National Center for Chronic Disease Prevention and Health Promotion (2000).
www.cdc.gov/growthcharts.

CDC SAFER • HEALTHIER • PEOPLE™

Measures of Body Fat and Lean Tissue

Significant weight changes in both children and adults can reflect overnutrition or undernutrition with respect to energy and protein. To estimate the degree to which fat stores or lean tissues are affected by malnutrition, several anthropometric measurements are useful.

Skinfold Measures Skinfold measures provide a good estimate of total body fat and a fair assessment of the fat's location. Most body fat lies directly beneath the skin, and the thickness of this subcutaneous fat correlates with total body fat. In some parts of the body, such as the back and the back of the arm over the triceps muscle, this fat is loosely attached. ♦ As illustrated in Figure E-4, an assessor can measure the thickness of the fat with calipers that apply a fixed amount of pressure. If a person gains body fat, the skinfold increases proportionately; if the person loses fat, it decreases. Measurements taken from central-body sites better reflect changes in fatness than those taken from upper sites (arm and back). Because subcutaneous fat may be thicker in one area than in another, skinfold measurements are often taken at three or four different places on the body (including upper-, central-, and lower-body sites); the sum of these measures is then compared to standard values. In some situations, the triceps skinfold measurement alone may be used because it is easily accessible. Triceps skinfold measures greater than 15 millimeters in men or 25 millimeters in women suggest excessive body fat.

♦ Common sites for skinfold measures:
- Triceps
- Biceps
- Subscapular (below shoulder blade)
- Suprailiac (above hip bone)
- Abdomen
- Upper thigh

Waist Circumference Chapter 8 explains how fat distribution correlates with health risks and mentions that the waist circumference is a valuable indicator of abdominal fat. To measure waist circumference, the assessor places a nonstretch-

FIGURE E-4 How to Measure the Triceps Skinfold

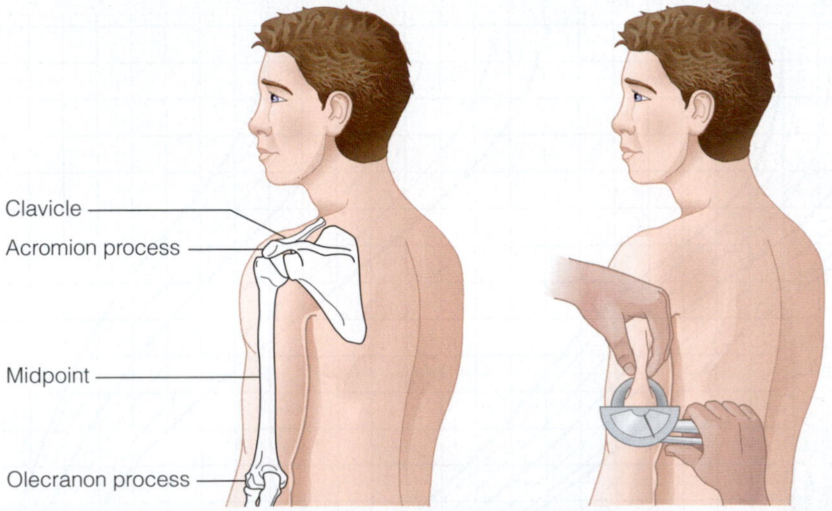

A. Find the midpoint of the arm:
1. Ask the subject to bend his or her arm at the elbow and lay the hand across the stomach. (If he or she is right-handed, measure the left arm, and vice versa.)
2. Feel the shoulder to locate the acromion process. It helps to slide your fingers along the clavicle to find the acromion process. The olecranon process is the tip of the elbow.
3. Place a measuring tape from the acromion process to the tip of the elbow. Divide this measurement by 2 and mark the midpoint of the arm with a pen.

B. Measure the skinfold:
1. Ask the subject to let his or her arm hang loosely to the side.
2. Grasp a fold of skin and subcutaneous fat between the thumb and forefinger slightly above the midpoint mark. Gently pull the skin away from the underlying muscle. (This step takes a lot of practice. If you want to be sure you don't have muscle as well as fat, ask the subject to contract and relax the muscle. You should be able to feel if you are pinching muscle.)
3. Place the calipers over the skinfold at the midpoint mark, and read the measurement to the nearest 1.0 millimeter in two to three seconds. (If using plastic calipers, align pressure lines, and read the measurement to the nearest 1.0 millimeter in two to three seconds.)
4. Repeat steps 2 and 3 twice more. Add the three readings, and then divide by 3 to find the average.

able tape around the person's body, crossing just above the upper hip bones and making sure that the tape remains on a level horizontal plane on all sides (see Figure E-5). The tape is tightened slightly, but without compressing the skin.

Waist-to-Hip Ratio The waist-to-hip ratio assesses abdominal obesity, but it offers no advantage over the waist circumference alone. To calculate the waist-to-hip ratio, divide the waistline measurement by the hip measurement. ♦ In general, women with a waist-to-hip ratio of 0.8 or greater and men with a waist-to-hip ratio of 0.9 or greater have an increased risk of developing diabetes and cardiovascular diseases.

♦ The calculation of waist-to-hip ratio in a woman with a 28-inch waist and 38-inch hips is 28 ÷ 38 = 0.74.

Hydrodensitometry To estimate body density using hydrodensitometry, the person is weighed twice—first on land and then again when submerged in water. Underwater weighing usually generates a good estimate of body fat and is useful in research, although the technique has drawbacks: it requires bulky, expensive, and nonportable equipment. Furthermore, submerging some people in water (especially those who are very young, very old, ill, or fearful) is difficult and not well tolerated.

Bioelectrical Impedance To measure body fat using the bioelectrical impedance method, a very-low-intensity electrical current is briefly sent through the body by way of electrodes placed on the wrist and ankle. Fat impedes the flow of electricity; thus, the magnitude of the current is influenced by the body fat content. Recent food intake and hydration status can influence results. As with other anthropometric methods, bioelectrical impedance requires standardized procedures and calibrated instruments.

A number of other methods are sometimes used to estimate the body's content of body fat. Table E-1 describes common techniques often used in the clinical or research setting.

FIGURE E-5 How to Measure Waist Circumference

Place the measuring tape around the waist just above the bony crest of the hip. The tape runs parallel to the floor and is snug (but does not compress the skin). The measurement is taken at the end of normal expiration.

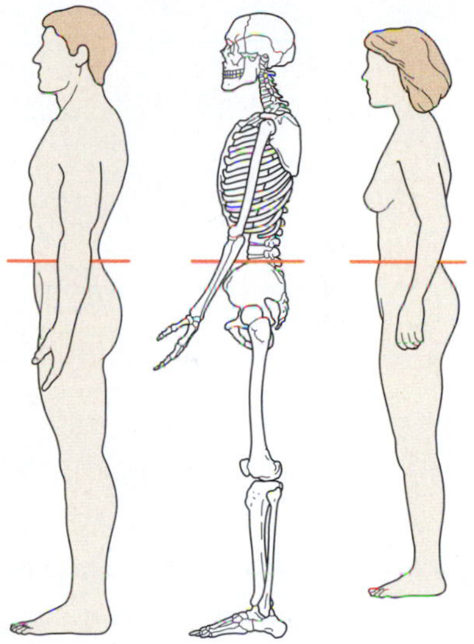

Source: National Institutes of Health Obesity Education Initiative, Clinical Guidelines on the Identification, Evaluation, and Treatment of Overweight and Obesity in Adults (Washington, D.C.: U.S. Department of Health and Human Services, 1998), p. 59.

Nutritional Anemias

Anemia, a symptom of a wide variety of nutrition- and nonnutrition-related disorders, is characterized by the reduced oxygen-carrying capacity of blood. Iron, folate, and vitamin B_{12} deficiencies—caused by inadequate intake, poor absorption, or abnormal metabolism of these nutrients—are the most common causes of nutritional anemias. Table E-2 lists laboratory tests that distinguish among the various nutrition-related anemias. Some nonnutrition-related causes of anemia

TABLE E-1 Selected Methods for Estimating Body Fat Content

Method	Description
Air-displacement plethysmography (BodPod®)	Estimates body density by measuring the body's volume (density = mass/volume); the density value allows derivation of the body's fat and lean tissue contents
Bioelectrical impedance assay	Measures the magnitude of an electrical current passed through the body; electrical conductivity is higher in lean tissues than in fat tissue
Dual energy X-ray absorptiometry	Analyzes the change in X-rays after they contact body tissues; fat and lean tissues have different effects on X-rays, allowing quantification of the various tissues
Hydrodensitometry (underwater weighing)	Estimates body density by comparing the body's weight on land and in water or by measuring the body's volume (density = mass/volume); the density value allows derivation of the body's fat and lean tissue contents
Isotope dilution—deuterated water	Measures total body water content by analyzing the dilution of heavy water (water with a heavy form of hydrogen) in body tissues; allows an estimate of lean tissue and body fat content
Skinfolds	Estimates subcutaneous fat in several regions of the body by using calipers to measure skinfold thicknesses
Ultrasound	Estimates subcutaneous fat in several regions of the body by using ultrasound to measure skinfold thicknesses

TABLE E-2 Laboratory Tests Useful in Evaluating Nutrition-Related Anemias

Test or Test Result	What It Reflects
For Anemia (general)	
Hemoglobin (Hg)	Total amount of hemoglobin in the red blood cells (RBC)
Hematocrit (Hct)	Percentage of RBC in the total blood volume
Red blood cell (RBC) count	Number of RBC
Mean corpuscular volume (MCV)	RBC size; helps determine if anemia is microcytic (iron deficiency) or macrocytic (folate or vitamin B_{12} deficiency)
Mean corpuscular hemoglobin concentration (MCHC)	Hemoglobin concentration within the average RBC; helps determine if anemia is hypochromic (iron deficiency) or normochromic (folate or vitamin B_{12} deficiency)
Bone marrow aspiration	The manufacture of blood cells in different developmental states
For Iron-Deficiency Anemia	
↓ Serum ferritin	Early deficiency state with depleted iron stores
↓ Transferrin saturation	Progressing deficiency state with diminished transport iron
↑ Erythrocyte protoporphyrin	Later deficiency state with limited hemoglobin production
For Folate-Deficiency Anemia	
↓ Serum folate	Progressing deficiency state
↓ RBC folate	Later deficiency state
For Vitamin B_{12}-Deficiency Anemia	
↓ Serum vitamin B_{12}	Progressing deficiency state
↑ Serum methylmalonic acid	Vitamin B_{12} deficiency
Schilling test	Adequacy of vitamin B_{12} absorption

include massive blood loss, infections, hereditary blood disorders such as sickle-cell anemia, and chronic liver or kidney disease.

◆ Reminder: Iron deficiency progresses as follows:
1. Iron stores diminish
2. Transport iron decreases
3. Hemoglobin production falls

Assessment of Iron Status Chapter 13 describes the progression of iron deficiency in detail, ◆ as well as the roles of some of the proteins involved in iron metabolism. This section describes the various tests that assess iron status, and Table E-3 provides acceptable values. Although other tests are more specific for detecting the early stages of iron deficiency, hemoglobin and hematocrit are most often used to detect iron-deficiency anemia because they are inexpensive and easily measured.

Serum Ferritin In the initial stage of iron deficiency, iron stores diminish. Iron is stored in the protein ferritin, which is located in the liver, spleen, and bone marrow. Serum ferritin values provide a noninvasive estimate of iron stores, because the ferritin levels in blood reflect the amounts stored in the tissues. Serum ferritin is not a reliable indicator of iron deficiency, however, because its concentrations are increased by infection, inflammation, alcohol consumption, and liver disease.

Serum Iron and Total Iron-Binding Capacity (TIBC) Early stages of iron deficiency are characterized by reduced levels of serum iron, which represent the amount of iron bound to transferrin, the iron transport protein. Total iron-binding capacity (TIBC) is a measure of the total amount of iron that the transferrin in blood can carry; thus, it is an indirect measure of the transferrin content of blood. During iron deficiency, the liver produces more transferrin in an effort to increase iron transport capacity, and therefore iron depletion is characterized by an increase in TIBC. TIBC reflects liver function as well as changes in iron metabolism.

TABLE E-3 Criteria for Assessing Iron Status

Laboratory Test	Acceptable Values	Effect of Iron Deficiency
Serum ferritin	Male: 20–250 ng/mL Female: 10–120 ng/mL	Lower than normal
Serum iron	Male: 60–175 μg/dL Female: 50–170 μg/dL	Lower than normal
Total iron-binding capacity	250–450 μg/dL	Higher than normal
Transferrin saturation	Male: 20–50% Female: 15–50%	Lower than normal
Erythrocyte protoporphyrin	$<$ 70 μg/dL red blood cells	Higher than normal
Hemoglobin (Hb)	Male: 13.5–17.5 g/dL Female: 12.0–16.0 g/dL	Lower than normal
Hematocrit (Hct)	Male: 39–49% Female: 35–45%	Lower than normal
Mean corpuscular volume (MCV)	80–100 fL	Lower than normal

NOTE: ng = nanogram, μg = microgram, dL = deciliter, nmol = nanomole, fL=femtoliter
SOURCES: L. Goldman and D. Ausiello, coeditors, *Cecil Medicine* (Philadelphia: Saunders, 2008), pp. 2983–2991; R. J. Wood and A. G. Ronnenberg, Iron, in M. E. Shils and coeditors, *Modern Nutrition in Health and Disease* (Baltimore: Lippincott Williams & Wilkins, 2006), pp. 248–270.

Transferrin Saturation The percentage of transferrin that is saturated with iron is an indirect measure derived from the serum iron and total iron-binding capacity measures, as follows:

$$\%\text{Transferrin} = \frac{\text{serum iron}}{\text{total iron-binding capacity}} \times 100$$

During iron deficiency, transferrin saturation decreases. The transferrin saturation value is a useful indicator of iron status because it includes information about both the iron and transferrin content of blood.

Erythrocyte Protoporphyrin The iron-containing molecule in hemoglobin is heme, which is formed from iron and protoporphyrin. Protoporphyrin accumulates in the blood when iron supplies are inadequate for the formation of heme. However, levels of protoporphyrin may increase when hemoglobin synthesis is impaired for other reasons, such as lead poisoning or inflammation.

Hemoglobin When iron stores become depleted, hemoglobin production is impaired, and symptoms of anemia may eventually develop. Hemoglobin's usefulness in evaluating iron status is limited, however, because hemoglobin concentrations drop fairly late in the development of iron deficiency, and other nutrient deficiencies and medical conditions can also alter hemoglobin concentrations.

Hematocrit The hematocrit is the percentage of the total blood volume occupied by red blood cells. To measure the hematocrit, a clinician spins the blood samples in a centrifuge to separate the red blood cells from the plasma. Low values indicate a reduced number or size of red blood cells. Although this test is not specific for iron status, it can help detect the presence of iron-deficiency anemia.

Mean Corpuscular Volume (MCV) The hematocrit value divided by the red blood cell count provides a measure of the average size of a red blood cell, referred to as the mean corpuscular volume (MCV). Such a measure helps classify the type of anemia that is present. In iron deficiency, the red blood cells are smaller than average (microcytic cells).

Assessment of Folate and Vitamin B_{12} Status Folate deficiency and vitamin B_{12} deficiency present a similar clinical picture—an anemia characterized by abnormally large, misshapen, and immature red blood cells (megaloblastic cells). Distinguishing between folate and vitamin B_{12} deficiency is essential, however, because their treatments differ. Giving folate to a person with vitamin B_{12} deficiency improves many of the test results indicative of vitamin B_{12} deficiency, but this would be a dangerous treatment because vitamin B_{12} deficiency causes nerve damage that folate cannot correct. Thus, inappropriate folate administration masks vitamin B_{12}-deficiency anemia, and nerve damage worsens. For this reason, it is critical to determine whether the anemia results from a folate deficiency or from a vitamin B_{12} deficiency. Several of the following assessment measures help make this distinction.

Mean Corpuscular Volume (MCV) As previously mentioned, MCV is a measure of red blood cell size. In folate and vitamin B_{12} deficiencies, the red blood cells are larger than average, or macrocytic. Macrocytic cells are not necessarily indicative of nutrient deficiency, however, as they may also result from a high alcohol intake, liver disease, and various medications.

Serum Folate and Vitamin B_{12} Levels Analyses of serum folate and vitamin B_{12} levels are usually among the first tests conducted to determine the cause of macrocytic red blood cells. The presence of low serum levels of either nutrient is consistent with a deficiency of that nutrient, whereas adequate levels can help rule out deficiency. Folate levels are not a specific measure of folate status, however; they may increase after folate consumption and decrease due to alcohol consumption, pregnancy, or use of anticonvulsants. The folate levels in red blood cells (called erythrocyte folate) correlate well with folate stores and can help diagnose folate deficiency, but the more reliable testing methods are not widely available. Table E-4 shows the acceptable ranges for these tests.

Methylmalonic Acid and Homocysteine Levels To determine whether a nutrient deficiency is present, clinicians can measure the levels of substances that accumulate when the functions of that nutrient are impaired. For example, blood levels of the amino acid homocysteine are usually increased by both folate and vitamin B_{12} deficiency, because both nutrients are needed for its metabolism. Methylmalonic acid, a breakdown product of several amino acids, requires vitamin B_{12} for its metabolism; hence, serum levels increase as a result of vitamin B_{12} deficiency. Because methylmalonic acid levels are not influenced by folate status, this measure is useful in distinguishing between folate and vitamin B_{12} deficiency.

Schilling Test As Chapter 10 explains, vitamin B_{12} deficiency most often results from malabsorption, not poor intake. The Schilling test can help diagnose malabsorption of vitamin B_{12}: after the patient takes an oral dose of radioactive vitamin

TABLE E-4 Criteria for Assessing Folate and Vitamin B_{12} Status

Laboratory Test	Acceptable Range	Effect of Folate or Vitamin B_{12} Deficiency
Serum folate	3–16 ng/mL	Reduced in folate deficiency
Erythrocyte folate	140–628 ng/mL packed cells	Reduced in folate deficiency
Serum vitamin B_{12}	200–835 pg/mL	Reduced in vitamin B_{12} deficiency
Serum methylmalonic acid	70–270 nmol/L	Increased in vitamin B_{12} deficiency
Serum homocysteine	5–14 µmol/L	Increased in folate or vitamin B_{12} deficiency

NOTE: ng = nanogram, pg = picogram, nmol = nanomole, µmol = micromole
SOURCE: L. Goldman and D. Ausiello, coeditors, *Cecil Medicine* (Philadelphia: Saunders, 2008), p. 1238 and pp. 2983–2991.

B_{12}, a urine test determines whether the vitamin B_{12} was absorbed. The Schilling test is rarely performed at present due to the difficulty in obtaining the chemical reagents needed for the test.

Antibodies to Intrinsic Factor The presence of serum antibodies for intrinsic factor can help confirm a diagnosis of pernicious anemia, an autoimmune disease characterized by destruction of the cells that produce intrinsic factor (a protein required for vitamin B_{12} absorption; see Chapter 10). Serum antibodies to the parietal cells that produce and release intrinsic factor may also indicate pernicious anemia, but these antibodies may be present in various other conditions as well.

Cautions about Nutrition Assessment

The tests outlined in this appendix yield information that becomes meaningful only when they are conducted and interpreted by a skilled clinician. Potential sources of error may be introduced at any step, from the collection of samples to the analysis and reporting of data. Equipment must be regularly calibrated to ensure accuracy of measurements. In addition, the assessor must keep in mind that few tests may be specific to the nutrient of interest alone, and lab results may reflect physiological processes other than the ones being tested. Furthermore, because many tests are not sensitive enough to detect the early stages of deficiency, follow-up testing is often necessary to identify a nutrition problem.

B_{12}, a urine test determines whether the vitamin B_{12} was absorbed. The Schilling test is rarely performed at present due to the difficulty in obtaining the chemical reagents needed for the test.

Antibodies to Intrinsic Factor. The presence of serum antibodies to intrinsic factor can help confirm a diagnosis of pernicious anemia, an autoimmune disease characterized by destruction of the cells that produce intrinsic factor (a protein required for vitamin B_{12} absorption; see Chapter 10). Serum antibodies to the parietal cells that produce and release intrinsic factor may also indicate pernicious anemia, but these antibodies may be present in various other conditions as well.

Cautions about Nutrition Assessment

The tests outlined in this appendix yield information that becomes meaningful only when they are conducted and interpreted by a skilled clinician. Potential sources of error may be introduced at any step, from the collection of samples to the analysis and reporting of data. Equipment must be regularly calibrated to ensure accuracy of measurements. In addition, the assessor must keep in mind that few tests may be specific to the nutrient of interest alone, and lab results may reflect physiological processes other than the ones being tested. Furthermore, because many tests are not sensitive enough to detect the early stages of deficiency, follow-up testing is often necessary to identify a nutritional problem.

Physical Activity and Energy Requirements

APPENDIX F

CONTENTS

Chapter 8 describes how to calculate estimated energy requirements (EER) for adults by using an equation that accounts for gender, age, weight, height, and physical activity level. Table F-1 (p. F-2) presents additional equations to determine the EER for infants, children, adolescents, and pregnant and lactating women.

This appendix helps you determine the correct physical activity (PA) factor to use in the equations, either by calculating the physical activity level or by estimating it. For those who prefer to bypass these steps, the appendix presents tables that provide a shortcut to estimating total energy expenditure.*

Calculating Physical Activity Level

To calculate your physical activity level, record all of your activities for a typical 24-hour day, noting the type of activity, the level of intensity, and the duration. Then, using a copy of Table F-2 (p. F-2), find your activity in the first column (or an activity that is reasonably similar) and multiply the number of minutes spent on that activity by the factor in the third column. Put your answer in the last column and total the accumulated values for the day. Now add the subtotal of the last column to 1.1 (to account for basal energy and the thermic effect of food) as shown. This score indicates your physical activity level. Using Table F-3 (p. F-3), find the PA factor for your age and gender that correlates with your physical activity level and use it in the energy equations presented in Table F-1.

Estimating Physical Activity Level

As an alternative to recording your activities for a day, you can use the third column of Table F-3 to decide if your daily activity is sedentary, low active, active, or very active. Find the PA factor for your age and gender that correlates with your typical physical activity level and use it in the energy equations presented in Table F-1.

Using a Shortcut to Estimate Total Energy Expenditure

The DRI Committee has developed estimates of total energy expenditure based on the equations for adults presented in Table F-1. These estimates are presented in Table F-4 (p. F-4) for women and Table F-5 (p. F-5) for men. You can use these tables to estimate your energy requirement—that is, the number of kcalories needed to maintain your current body weight. On the table appropriate for your gender, find your height in meters (or inches) in the left-hand column. Then follow the row

*This appendix, including the tables, is adapted from Committee on Dietary Reference Intakes, *Dietary Reference Intakes for Energy, Carbohydrate, Fiber, Fat, Fatty Acids, Cholesterol, Protein, and Amino Acids* (Washington, D.C.: National Academies Press, 2005).

APPENDIX F

across to find your weight in kilograms (or pounds). (If you can't find your exact height and weight, choose a value between the two closest ones.) Look down the column to find the number of kcalories that corresponds to your activity level.

Importantly, the values given in the tables are for 30-year-old people. Women 19 to 29 should add 7 kcalories per day for each year younger than age 30; older women should subtract 7 kcalories per day for each year older than age 30. Similarly, men 19 to 29 should add 10 kcalories per day for each year younger than age 30; older men should subtract 10 kcalories per day for each year older than age 30.

TABLE F-1 Equations to Determine Estimated Energy Requirement (EER)

Infants	
0–3 months	EER = (89 × weight − 100) + 175
4–6 months	EER = (89 × weight − 100) + 56
7–12 months	EER = (89 × weight − 100) + 22
13–15 months	EER = (89 × weight − 100) + 20
Children and Adolescents	
Boys	
3–8 years	EER = 88.5 − (61.9 × age) + PA × [(26.7 × weight) + (903 × height)] + 20
9–18 years	EER = 88.5 − (61.9 × age) + PA × [(26.7 × weight) + (903 × height)] + 25
Girls	
3–8 years	EER = 135.3 − (30.8 × age) + PA × [(10.0 × weight) + (934 × height)] + 20
9–18 years	EER = 135.3 − (30.8 × age) + PA × [(10.0 × weight) + (934 × height)] + 25
Adults	
Men	EER = 662 − (9.53 × age) + PA × [(15.91 × weight) + (539.6 × height)]
Women	EER = 354 − (6.91 × age) + PA × [(9.36 × weight) + (726 × height)]
Pregnancy	
1st trimester	EER = nonpregnant EER + 0
2nd trimester	EER = nonpregnant EER + 340
3rd trimester	EER = nonpregnant EER + 452
Lactation	
0–6 months postpartum	EER = nonpregnant EER + 500 − 170
7–12 months postpartum	EER = nonpregnant EER + 400 − 0

NOTE: Select the appropriate equation for gender and age and insert weight in kilograms, height in meters, and age in years. See the text and Table F-3 to determine PA.

TABLE F-2 Physical Activities and Their Scores

If your activity was equivalent to this . . .	Then list the number of minutes here and . . .	Multiply by this factor . . .	Add this column to get your physical activity level score:
Activities of Daily Living			
Gardening (no lifting)		0.0032	
Household tasks (moderate effort)		0.0024	
Lifting items continuously		0.0029	
Loading/unloading car		0.0019	
Lying quietly		0.0000	
Mopping		0.0024	
Mowing lawn (power mower)		0.0033	
Raking lawn		0.0029	
Riding in a vehicle		0.0000	
Sitting (idle)		0.0000	
Sitting (doing light activity)		0.0005	
Taking out trash		0.0019	
Vacuuming		0.0024	
Walking the dog		0.0019	
Walking from house to car or bus		0.0014	
Watering plants		0.0014	*(continued)*

TABLE F-2 Physical Activities and Their Scores (*continued*)

If your activity was equivalent to this . . .	Then list the number of minutes here and . . .	Multiply by this factor . . .	Add this column to get your physical activity level score:
Additional Activities			
Billiards		0.0013	
Calisthenics (no weight)		0.0029	
Canoeing (leisurely)		0.0014	
Chopping wood		0.0037	
Climbing hills (carrying 11 lb load)		0.0061	
Climbing hills (no load)		0.0056	
Cycling (leisurely)		0.0024	
Cycling (moderately)		0.0045	
Dancing (aerobic or ballet)		0.0048	
Dancing (ballroom, leisurely)		0.0018	
Dancing (fast ballroom or square)		0.0043	
Golf (with cart)		0.0014	
Golf (without cart)		0.0032	
Horseback riding (walking)		0.0012	
Horseback riding (trotting)		0.0053	
Jogging (6 mph)		0.0088	
Music (playing accordion)		0.0008	
Music (playing cello)		0.0012	
Music (playing flute)		0.0010	
Music (playing piano)		0.0012	
Music (playing violin)		0.0014	
Rope skipping		0.0105	
Skating (ice)		0.0043	
Skating (roller)		0.0052	
Skiing (water or downhill)		0.0055	
Squash		0.0106	
Surfing		0.0048	
Swimming (slow)		0.0033	
Swimming (fast)		0.0057	
Tennis (doubles)		0.0038	
Tennis (singles)		0.0057	
Volleyball (noncompetitive)		0.0018	
Walking (2 mph)		0.0014	
Walking (3 mph)		0.0022	
Walking (4 mph)		0.0033	
Walking (5 mph)		0.0067	
Subtotal			
Factor for basal energy and the thermic effect of food			**1.1**
Your physical activity level score			

TABLE F-3 Physical Activity Equivalents and Their PA Factors

Physical Activity Level	Description	Physical Activity Equivalents	Men, 19+ yr PA Factor	Women, 19+ yr PA Factor	Boys, 3–18 yr PA Factor	Girls, 3–18 yr PA Factor
1.0 to 1.39	Sedentary	Only those physical activities required for typical daily living	1.0	1.0	1.0	1.0
1.4 to 1.59	Low active	Daily living + 30–60 min moderate activity[a]	1.11	1.12	1.13	1.16
1.6 to 1.89	Active	Daily living + ≥ 60 min moderate activity	1.25	1.27	1.26	1.31
1.9 and above	Very active	Daily living + ≥ 60 min moderate activity *and* ≥ 60 min vigorous activity *or* ≥ 120 min moderate activity	1.48	1.45	1.42	1.56

[a]Moderate activity is equivalent to walking at a pace of 3 to 4½ miles per hour.

TABLE F-4 Total Energy Expenditure (TEE in kCalories per Day) for Women 30 Years of Age[a] at Various Levels of Activity and Various Heights and Weights

Heights m (in)	Physical Activity Level	Weight[b] kg (lb)					
1.45 (57)		38.9 (86)	45.2 (100)	52.6 (116)	63.1 (139)	73.6 (162)	84.1 (185)
		kCalories					
	Sedentary	1564	1623	1698	1813	1927	2042
	Low active	1734	1800	1912	2043	2174	2304
	Active	1946	2021	2112	2257	2403	2548
	Very active	2201	2287	2387	2553	2719	2886
1.50 (59)		41.6 (92)	48.4 (107)	56.3 (124)	67.5 (149)	78.8 (174)	90.0 (198)
		kCalories					
	Sedentary	1625	1689	1771	1894	2017	2139
	Low active	1803	1874	1996	2136	2276	2415
	Active	2025	2105	2205	2360	2516	2672
	Very active	2291	2382	2493	2671	2849	3027
1.55 (61)		44.4 (98)	51.7 (114)	60.1 (132)	72.1 (159)	84.1 (185)	96.1 (212)
		kCalories					
	Sedentary	1688	1756	1846	1977	2108	2239
	Low active	1873	1949	2081	2230	2380	2529
	Active	2104	2190	2299	2466	2632	2798
	Very active	2382	2480	2601	2791	2981	3171
1.60 (63)		47.4 (104)	55.0 (121)	64.0 (141)	76.8 (169)	89.6 (197)	102.4 (226)
		kCalories					
	Sedentary	1752	1824	1922	2061	2201	2340
	Low active	1944	2025	2168	2327	2486	2645
	Active	2185	2276	2396	2573	2750	2927
	Very active	2474	2578	2712	2914	3116	3318
1.65 (65)		50.4 (111)	58.5 (129)	68.1 (150)	81.7 (180)	95.3 (210)	108.9 (240)
		kCalories					
	Sedentary	1816	1893	1999	2148	2296	2444
	Low active	2016	2102	2556	2425	2594	2763
	Active	2267	2364	2494	2682	2871	3059
	Very active	2567	2678	2824	3039	3254	3469
1.70 (67)		53.5 (118)	62.1 (137)	72.3 (159)	86.7 (191)	101.2 (223)	115.6 (255)
		kCalories					
	Sedentary	1881	1963	2078	2235	2393	2550
	Low active	2090	2180	2345	2525	2705	2884
	Active	2350	2453	2594	2794	2994	3194
	Very active	2662	2780	2938	3166	3395	3623
1.75 (69)		56.7 (125)	65.8 (145)	76.6 (169)	91.9 (202)	107.2 (236)	122.5 (270)
		kCalories					
	Sedentary	1948	2034	2158	2325	2492	2659
	Low active	2164	2260	2437	2627	2817	3007
	Active	2434	2543	2695	2907	3119	3331
	Very active	2758	2883	3054	3296	3538	3780
1.80 (71)		59.9 (132)	69.7 (154)	81.0 (178)	97.2 (214)	113.4 (250)	129.6 (285)
		kCalories					
	Sedentary	2015	2106	2239	2416	2593	2769
	Low active	2239	2341	2529	2731	2932	3133
	Active	2519	2634	2799	3023	3247	3472
	Very active	2855	2987	3172	3428	3684	3940

(*continued*)

[a]For each year younger than 30, add 10 kcalories/day to TEE. For each year older than 30, subtract 10 kcalories/day from TEE.
[b]These columns represent a BMI of 18.5, 22.5, 25, 30, 35, and 40, respectively.

TABLE F-4 Total Energy Expenditure (TEE in kCalories per Day) for Women 30 Years of Age[a] at Various Levels of Activity and Various Heights and Weights (*continued*)

Heights m (in)	Physical Activity Level	Weight[b] kg (lb)					
1.85 (73)		63.3 (139)	73.6 (162)	85.6 (189)	102.7 (226)	119.8 (264)	136.9 (302)
		kCalories					
	Sedentary	2083	2179	2322	2509	2695	2882
	Low active	2315	2422	2624	2836	3049	3262
	Active	2605	2727	2904	3141	3378	3615
	Very active	2954	3093	3292	3562	3833	4103
1.90 (75)		66.8 (147)	77.6 (171)	90.3 (199)	108.3 (239)	126.4 (278)	144.4 (318)
		kCalories					
	Sedentary	2151	2253	2406	2603	2800	2996
	Low active	2392	2505	2720	2944	3168	3393
	Active	2693	2821	3011	3261	3511	3760
	Very active	3053	3200	3414	3699	3984	4270
1.95 (77)		70.3 (155)	81.8 (180)	95.1 (209)	114.1 (251)	133.1 (293)	152.1 (335)
		kCalories					
	Sedentary	2221	2328	2492	2699	2906	3113
	Low active	2470	2589	2817	3053	3290	3526
	Active	2781	2917	3119	3383	3646	3909
	Very active	3154	3309	3538	3838	4139	4439

[a]For each year younger than 30, add 7 kcalories/day to TEE. For each year older than 30, subtract 7 kcalories/day from TEE.
[b]These columns represent a BMI of 18.5, 22.5, 25, 30, 35, and 40, respectively.

TABLE F-5 Total Energy Expenditure (TEE in kCalories per Day) for Men 30 Years of Age[a] at Various Levels of Activity and Various Heights and Weights

Heights m (in)	Physical Activity Level	Weight[b] kg (lb)					
1.45 (57)		38.9 (86)	47.3 (100)	52.6 (116)	63.1 (139)	73.6 (163)	84.1 (185)
		kCalories					
	Sedentary	1777	1911	2048	2198	2347	2496
	Low active	1931	2080	2225	2393	2560	2727
	Active	2127	2295	2447	2636	2826	3015
	Very active	2450	2648	2845	3075	3305	3535
1.50 (59)		41.6 (92)	50.6 (107)	56.3 (124)	67.5 (149)	78.8 (174)	90.0 (198)
		kCalories					
	Sedentary	1848	1991	2126	2286	2445	2605
	Low active	2009	2168	2312	2491	2670	2849
	Active	2215	2394	2545	2748	2951	3154
	Very active	2554	2766	2965	3211	3457	3703
1.55 (61)		44.4 (98)	54.1 (114)	60.1 (132)	72.1 (159)	84.1 (185)	96.1 (212)
		kCalories					
	Sedentary	1919	2072	2205	2376	2546	2717
	Low active	2089	2259	2401	2592	2783	2974
	Active	2305	2496	2646	2862	3079	3296
	Very active	2660	2887	3087	3349	3612	3875

(*continued*)

[a]For each year younger than 30, add 10 kcalories/day to TEE. For each year older than 30, subtract 10 kcalories/day from TEE.
[b]These columns represent a BMI of 18.5, 22.5, 25, 30, 35, and 40, respectively.

TABLE F-5 Total Energy Expenditure (TEE in kCalories per Day) for Men 30 Years of Age[a] at Various Levels of Activity and Various Heights and Weights (*continued*)

Heights m (in)	Physical Activity Level	Weight[b] kg (lb)					
1.60 (63)		47.4 (104)	57.6 (121)	64.0 (141)	76.8 (169)	89.6 (197)	102.4 (226)
		kCalories					
	Sedentary	1993	2156	2286	2468	2650	2831
	Low active	2171	2351	2492	2695	2899	3102
	Active	2397	2601	2749	2980	3210	3441
	Very active	2769	3010	3211	3491	3771	4051
1.65 (65)		50.4 (111)	61.3 (129)	68.1 (150)	81.7 (180)	95.3 (210)	108.9 (240)
		kCalories					
	Sedentary	2068	2241	2369	2562	2756	2949
	Low active	2254	2446	2585	2801	3017	3234
	Active	2490	2707	2854	3099	3345	3590
	Very active	2880	3136	3339	3637	3934	4232
1.70 (67)		53.5 (118)	65.0 (137)	72.3 (159)	86.7 (191)	101.2 (223)	115.6 (255)
		kCalories					
	Sedentary	2144	2328	2454	2659	2864	3069
	Low active	2338	2542	2679	2909	3139	3369
	Active	2586	2816	2961	3222	3483	3743
	Very active	2992	3265	3469	3785	4101	4417
1.75 (69)		56.7 (125)	68.9 (145)	76.6 (169)	91.9 (202)	107.2 (236)	122.5 (270)
		kCalories					
	Sedentary	2222	2416	2540	2757	2975	3192
	Low active	2425	2641	2776	3020	3263	3507
	Active	2683	2927	3071	3347	3623	3900
	Very active	3108	3396	3602	3937	4272	4607
1.80 (71)		59.9 (132)	72.9 (154)	81.0 (178)	97.2 (214)	113.4 (250)	129.6 (285)
		kCalories					
	Sedentary	2301	2507	2628	2858	3088	3318
	Low active	2513	2741	2875	3132	3390	3648
	Active	2782	3040	3183	3475	3767	4060
	Very active	3225	3530	3738	4092	4447	4801
1.85 (73)		63.3 (139)	77.0 (162)	85.6 (189)	102.7 (226)	119.8 (264)	136.9 (302)
		kCalories					
	Sedentary	2382	2599	2718	2961	3204	3447
	Low active	2602	2844	2976	3248	3520	3792
	Active	2883	3155	3297	3606	3915	4223
	Very active	3344	3667	3877	4251	4625	4999
1.90 (75)		66.8 (147)	81.2 (171)	90.3 (199)	108.3 (239)	126.4 (278)	144.4 (318)
		kCalories					
	Sedentary	2464	2693	2810	3066	3322	3579
	Low active	2693	2948	3078	3365	3652	3939
	Active	2986	3273	3414	3739	4065	4390
	Very active	3466	3806	4018	4413	4807	5202
1.95 (77)		70.3 (155)	85.6 (180)	95.1 (209)	114.1 (251)	133.1 (293)	152.1 (335)
		kCalories					
	Sedentary	2547	2789	2903	3173	3443	3713
	Low active	2786	3055	3183	3485	3788	4090
	Active	3090	3393	3533	3875	4218	4561
	Very active	3590	3948	4162	4578	4993	5409

[a]For each year younger than 30, add 10 kcalories/day to TEE. For each year older than 30, subtract 10 kcalories/day from TEE.
[b]These columns represent a BMI of 18.5, 22.5, 25, 30, 35, and 40, respectively.

Exchange Lists for Diabetes

APPENDIX G

CONTENTS

Chapter 2 introduces the exchange system, and this appendix provides details from the *2008 Choose Your Foods: Exchange Lists for Diabetes*. Appendix I presents Canada's meal-planning system.

Exchange lists can help people with diabetes to manage their blood glucose levels by controlling the amount and kinds of carbohydrates they consume. These lists can also help in planning diets for weight management by controlling kcalorie and fat intake.

The Exchange System

The exchange system sorts foods into groups by their proportions of carbohydrate, fat, and protein (Table G-1 on p. G-2). These groups may be organized into several exchange lists of foods (Tables G-2 through G-12 on pp. G-3–G-16). For example, the carbohydrate group includes these exchange lists:

- Starch
- Fruits
- Milk (fat-free, reduced-fat, and whole)
- Sweets, Desserts, and Other Carbohydrates
- Nonstarchy Vegetables

Then any food on a list can be "exchanged" for any other on that same list. Another group for alcohol has been included as a reminder that these beverages often deliver substantial carbohydrate and kcalories, and therefore warrant their own list.

Serving Sizes

The serving sizes have been carefully adjusted and defined so that a serving of any food on a given list provides roughly the same amount of carbohydrate, fat, and protein, and, therefore, total energy. Any food on a list can thus be exchanged, or traded, for any other food on the same list without significantly affecting the diet's energy-nutrient balance or total kcalories. For example, a person may select 17 small grapes or ½ large grapefruit as one fruit exchange, and either choice would provide roughly 15 grams of carbohydrate and 60 kcalories. A whole grapefruit, however, would count as 2 fruit exchanges.

To apply the system successfully, users must become familiar with the specified serving sizes. A convenient way to remember the serving sizes and energy values is to keep in mind a typical item from each list (review Table G-1).

The Foods on the Lists

Foods do not always appear on the exchange list where you might first expect to find them. They are grouped according to their energy-nutrient contents rather than by their source (such as milks), their outward appearance, or their vitamin and mineral contents. For example, cheeses are grouped with meats (not milk) because, like meats, cheeses contribute energy from protein and fat but provide negligible carbohydrate.

For similar reasons, starchy vegetables such as corn, green peas, and potatoes are found on the Starch list with breads and cereals, not with the vegetables. Likewise, bacon is grouped with the fats and oils, not with the meats.

Diet planners learn to view mixtures of foods, such as casseroles and soups, as combinations of foods from different exchange lists. They also learn to interpret food labels with the exchange system in mind.

Controlling Energy, Fat, and Sodium

The exchange lists help people control their energy intakes by paying close attention to serving sizes. People wanting to lose weight can limit foods from the Sweets, Desserts, and Other Carbohydrates and Fats lists, and they might choose to avoid the Alcohol list altogether. The Free Foods list provide low-kcalorie choices.

By assigning items like bacon to the Fats list, the exchange lists alert consumers to foods that are unexpectedly high in fat. Even the Starch list specifies which grain products contain added fat (such as biscuits, cornbread, and waffles) by marking them with a symbol to indicate added fat (the symbols are explained in the table keys). In addition, the exchange lists encourage users to think of fat-free milk as milk and of whole milk as milk with added fat, and to think of lean meats as meats and of medium-fat and high-fat meats as meats with added fat. To that end, foods on the milk and meat lists are separated into categories based on their fat contents (review Table G-1). The Milk list is subdivided for fat-free, reduced fat, and whole; the meat list is subdivided for lean, medium fat, and high fat. The meat list also includes plant-based proteins, which tend to be rich in fiber. Notice that many of these foods (p. G-11) bear the symbol for "high fiber."

People wanting to control the sodium in their diets can begin by eliminating any foods bearing the "high sodium" symbol. In most cases, the symbol identifies foods that, in one serving, provide 480 milligrams or more of sodium. Foods

TABLE G-1 The Food Lists

Lists	Typical Item/Portion Size	Carbohydrate (g)	Protein (g)	Fat (g)	Energy[a] (kcal)
Carbohydrates					
Starch[b]	1 slice bread	15	0–3	0–1	80
Fruits	1 small apple	15	—	—	60
Milk					
Fat-free, low-fat, 1%	1 c fat-free milk	12	8	0–3	100
Reduced-fat, 2%	1 c reduced-fat milk	12	8	5	120
Whole	1 c whole milk	12	8	8	160
Sweets, desserts, and other carbohydrates[c]	2 small cookies	15	varies	varies	varies
Nonstarchy vegetables	½ c cooked carrots	5	2	—	25
Meat and Meat Substitutes					
Lean	1 oz chicken (no skin)	—	7	0–3	45
Medium-fat	1 oz ground beef	—	7	4–7	75
High-fat	1 oz pork sausage	—	7	8+	100
Plant-based proteins	½ c tofu	varies	7	varies	varies
Fats	1 tsp butter	—	—	5	45
Alcohol	12 oz beer	varies	—	—	100

[a]The energy value for each exchange list represents an approximate average for the group and does not reflect the precise number of grams of carbohydrate, protein, and fat. For example, a slice of bread contains 15 grams of carbohydrate (60 kcalories), 3 grams protein (12 kcalories), and a little fat—rounded to 80 kcalories for ease in calculating. A ½ cup of vegetables (not including starchy vegetables) contains 5 grams carbohydrate (20 kcalories) and 2 grams protein (8 more), which has been rounded down to 25 kcalories.

[b]The Starch list includes cereals, grains, breads, crackers, snacks, starchy vegetables (such as corn, peas, and potatoes), and legumes (dried beans, peas, and lentils).

[c]The Sweets, Desserts, and Other Carbohydrates list includes foods that contain added sugars and fats such as sodas, candy, cakes, cookies, doughnuts, ice cream, pudding, syrup, and frozen yogurt.

on the Combination Foods or Fast Foods lists that bear the symbol provide more than 600 milligrams of sodium. Other foods may also contribute substantially to sodium (consult Chapter 12 for details).

Planning a Healthy Diet

To obtain a daily variety of foods that provide healthful amounts of carbohydrate, protein, and fat, as well as vitamins, minerals, and fiber, the meal plan for adults and teenagers should include at least:

- Two to three servings of nonstarchy vegetables
- Two servings of fruits
- Six servings of grains (at least three of whole grains), beans, and starchy vegetables
- Two servings of low-fat or fat-free milk
- About 6 ounces of meat or meat substitutes
- *Small* amounts of fat and sugar

The actual amounts are determined by age, gender, activity levels, and other factors that influence energy needs. Refer to Chapter 8 as you read through these sections to get an idea of how exchange lists can be useful in planning a diet.

TABLE G-2 Starch

The Starch list includes bread, cereals and grains, starchy vegetables, crackers and snacks, and legumes (dried beans, peas, and lentils).
1 starch choice = 15 grams carbohydrate, 0–3 grams protein, 0–1 grams fat, and 80 kcalories.

NOTE: In general, one starch exchange is ½ cup cooked cereal, grain, or starchy vegetable; ⅓ cup cooked rice or pasta; 1 ounce of bread product; ¾ ounce to 1 ounce of most snack foods.

Bread

Food	Serving Size
Bagel, large (about 4 oz)	¼ (1 oz)
⚠ Biscuit, 2½ inches across	1
Bread	
☺ reduced-kcalorie	2 slices (1½ oz)
white, whole-grain, pumpernickel, rye, unfrosted raisin	1 slice (1 oz)
Chapatti, small, 6 inches across	1
⚠ Cornbread, 1¾ inch cube	1 (1½ oz)
English muffin	½
Hot dog bun or hamburger bun	½ (1 oz)
Naan, 8 inches by 2 inches	¼
Pancake, 4 inches across, ¼ inch thick	1
Pita, 6 inches across	½
Roll, plain, small	1 (1 oz)
⚠ Stuffing, bread	⅓ cup
⚠ Taco shell, 5 inches across	2
Tortilla, corn, 6 inches across	1
Tortilla, flour, 6 inches across	1
Tortilla, flour, 10 inches across	⅓
⚠ Waffle, 4-inch square or 4 inches across	1

Cereals and Grains

Food	Serving Size
Barley, cooked	⅓ cup
Bran, dry	
☺ oat	¼ cup
☺ wheat	½ cup
☺ Bulgur (cooked)	½ cup
Cereals	
☺ bran	½ cup
cooked (oats, oatmeal)	½ cup
puffed	1½ cups
shredded wheat, plain	½ cup
sugar-coated	½ cup
unsweetened, ready-to-eat	¾ cup
Couscous	⅓ cup
Granola	
low-fat	¼ cup
⚠ regular	¼ cup
Grits, cooked	½ cup
Kasha	½ cup
Millet, cooked	⅓ cup

(continued)

KEY

☺ = More than 3 grams of dietary fiber per serving.

⚠ = Extra fat, or prepared with added fat. (Count as 1 starch + 1 fat.)

🧂 = 480 milligrams or more of sodium per serving.

TABLE G-2 Starch (*continued*)

Cereals and Grains—continued

Food	Serving Size
Muesli	¼ cup
Pasta, cooked	⅓ cup
Polenta, cooked	⅓ cup
Quinoa, cooked	⅓ cup
Rice, white or brown, cooked	⅓ cup
Tabbouleh (tabouli), prepared	½ cup
Wheat germ, dry	3 Tbsp
Wild rice, cooked	½ cup

Starchy Vegetables

Food	Serving Size
Cassava	⅓ cup
Corn	½ cup
on cob, large	½ cob (5 oz)
☺ Hominy, canned	¾ cup
☺ Mixed vegetables with corn, peas, or pasta	1 cup
☺ Parsnips	½ cup
☺ Peas, green	½ cup
Plantain, ripe	⅓ cup
Potato	
baked with skin	¼ large (3 oz)
boiled, all kinds	½ cup or ½ medium (3 oz)
⚠ mashed, with milk and fat	½ cup
french fried (oven-baked)[a]	1 cup (2 oz)
☺ Pumpkin, canned, no sugar added	1 cup
Spaghetti/pasta sauce	½ cup
☺ Squash, winter (acorn, butternut)	1 cup
☺ Succotash	½ cup
Yam, sweet potato, plain	½ cup

Crackers and Snacks[b]

Food	Serving Size
Animal crackers	8
Crackers	
⚠ round-butter type	6
saltine-type	6
⚠ sandwich-style, cheese or peanut butter filling	3
⚠ whole-wheat regular	2–5 (¾ oz)
☺ whole-wheat lower fat or crispbreads	2–5 (¾ oz)
Graham cracker, 2½-inch square	3
Matzoh	¾ oz
Melba toast, about 2-inch by 4-inch piece	4
Oyster crackers	20
Popcorn	3 cups
⚠ ☺ with butter	3 cups
☺ no fat added	3 cups
☺ lower fat	3 cups
Pretzels	¾ oz
Rice cakes, 4 inches across	2
Snack chips	
fat-free or baked (tortilla, potato), baked pita chips	15–20 (¾ oz)
⚠ regular (tortilla, potato)	9–13 (¾ oz)

Beans, Peas, and Lentils[c]

The choices on this list count as 1 starch + 1 lean meat.

Food	Serving Size
☺ Baked beans	⅓ cup
☺ Beans, cooked (black, garbanzo, kidney, lima, navy, pinto, white)	½ cup
☺ Lentils, cooked (brown, green, yellow)	½ cup
☺ Peas, cooked (black-eyed, split)	½ cup
🧂 ☺ Refried beans, canned	½ cup

KEY

☺ = More than 3 grams of dietary fiber per serving.

⚠ = Extra fat, or prepared with added fat. (Count as 1 starch + 1 fat.)

🧂 = 480 milligrams or more of sodium per serving.

[a]Restaurant-style french fries are on the Fast Foods list.

[b]For other snacks, see the Sweets, Desserts, and Other Carbohydrates list. For a quick estimate of serving size, an open handful is equal to about 1 cup or 1 to 2 ounces of snack food.

[c]Beans, peas, and lentils are also found on the Meat and Meat Substitutes list.

TABLE G-3 Fruits

Fruit[a]

The Fruits list includes fresh, frozen, canned, and dried fruits and fruit juices. 1 fruit choice = 15 grams carbohydrate, 0 grams protein, 0 grams fat, and 60 kcalories.

NOTE: In general, one fruit exchange is ½ cup canned or fresh fruit or unsweetened fruit juice; 1 small fresh fruit (4 ounces); 2 tablespoons dried fruit.

Food	Serving Size
Apple, unpeeled, small	1 (4 oz)
Apples, dried	4 rings
Applesauce, unsweetened	½ cup
Apricots	
canned	½ cup
dried	8 halves
☺ fresh	4 whole (5½ oz)
Banana, extra small	1 (4 oz)
☺ Blackberries	¾ cup
Blueberries	¾ cup
Cantaloupe, small	⅓ melon or 1 cup cubed (11 oz)
Cherries	
sweet, canned	½ cup
sweet fresh	12 (3 oz)
Dates	3
Dried fruits (blueberries, cherries, cranberries, mixed fruit, raisins)	2 Tbsp
Figs	
dried	1½
☺ fresh	1½ large or 2 medium (3½ oz)
Fruit cocktail	½ cup
Grapefruit	
large	½ (11 oz)
sections, canned	¾ cup
Grapes, small	17 (3 oz)
Honeydew melon	1 slice or 1 cup cubed (10 oz)
☺ Kiwi	1 (3½ oz)
Mandarin oranges, canned	¾ cup
Mango, small	½ (5½ oz) or ½ cup

Food	Serving Size
Nectarine, small	1 (5 oz)
☺ Orange, small	1 (6½ oz)
Papaya	½ or 1 cup cubed (8 oz)
Peaches	
canned	½ cup
fresh, medium	1 (6 oz)
Pears	
canned	½ cup
fresh, large	½ (4 oz)
Pineapple	
canned	½ cup
fresh	¾ cup
Plums	
canned	½ cup
dried (prunes)	3
small	2 (5 oz)
☺ Raspberries	1 cup
☺ Strawberries	1¼ cup whole berries
☺ Tangerines, small	2 (8 oz)
Watermelon	1 slice or 1¼ cups cubes (13½ oz)

Fruit Juice

Food	Serving Size
Apple juice/cider	½ cup
Fruit juice blends, 100% juice	⅓ cup
Grape juice	⅓ cup
Grapefruit juice	½ cup
Orange juice	½ cup
Pineapple juice	½ cup
Prune juice	⅓ cup

KEY

☺ = More than 3 grams of dietary fiber per serving.

▽ = Extra fat, or prepared with added fat. (Count as 1 starch + 1 fat.)

🧂 = 480 milligrams or more of sodium per serving.

[a]The weight listed includes skin, core, seeds, and rind.

TABLE G-4 Milk

The Milk list groups milks and yogurts based on the amount of fat they have (fat-free/low fat, reduced fat, and whole). Cheeses are found on the Meat and Meat Substitutes list and cream and other dairy fats are found on the Fats list.

NOTE: In general, one milk choice is 1 cup (8 fluid ounces or ½ pint) milk or yogurt.

Milk and Yogurts

Food	Serving Size
Fat-free or low-fat (1%)	
1 fat-free/low-fat milk choice = 12 g carbohydrate, 8 g protein, 0–3 g fat, and 100 kcal.	
Milk, buttermilk, acidophilus milk, Lactaid	1 cup
Evaporated milk	½ cup
Yogurt, plain or flavored with an artificial sweetener	⅔ cup (6 oz)
Reduced-fat (2%)	
1 reduced-fat milk choice = 12 g carbohydrate, 8 g protein, 5 g fat, and 120 kcal.	
Milk, acidophilus milk, kefir, Lactaid	1 cup
Yogurt, plain	⅔ cup (6 oz)
Whole	
1 whole milk choice = 12 g carbohydrate, 8 g protein, 8 g fat, and 160 kcal.	
Milk, buttermilk, goat's milk	1 cup
Evaporated milk	½ cup
Yogurt, plain	8 oz

Dairy-Like Foods

Food	Serving Size	Count as
Chocolate milk		
fat-free	1 cup	1 fat-free milk + 1 carbohydrate
whole	1 cup	1 whole milk + 1 carbohydrate
Eggnog, whole milk	½ cup	1 carbohydrate + 2 fats
Rice drink		
flavored, low fat	1 cup	2 carbohydrates
plain, fat-free	1 cup	1 carbohydrate
Smoothies, flavored, regular	10 oz	1 fat-free milk + 2½ carbohydrates
Soy milk		
light	1 cup	1 carbohydrate + ½ fat
regular, plain	1 cup	1 carbohydrate + 1 fat
Yogurt		
and juice blends	1 cup	1 fat-free milk + 1 carbohydrate
low carbohydrate (less than 6 grams carbohydrate per choice)	⅔ cup (6 oz)	½ fat-free milk
with fruit, low-fat	⅔ cup (6 oz)	1 fat-free milk + 1 carbohydrate

TABLE G-5 Sweets, Desserts, and Other Carbohydrates

1 other carbohydrate choice = 15 grams carbohydrate, variable grams protein, variable grams fat, and variable kcalories.

NOTE: In general, one choice from this list can substitute for foods on the Starch, Fruits, or Milk lists.

Beverages, Soda, and Energy/Sports Drinks

Food	Serving Size	Count as
Cranberry juice cocktail	½ cup	1 carbohydrate
Energy drink	1 can (8.3 oz)	2 carbohydrates
Fruit drink or lemonade	1 cup (8 oz)	2 carbohydrates

(*continued*)

TABLE G-5 Sweets, Desserts, and Other Carbohydrates (*continued*)

Beverages, Soda, and Energy/Sports Drinks—continued

Food	Serving Size	Count as
Hot chocolate		
regular	1 envelope added to 8 oz water	1 carbohydrate + 1 fat
sugar-free or light	1 envelope added to 8 oz water	1 carbohydrate
Soft drink (soda), regular	1 can (12 oz)	2½ carbohydrates
Sports drink	1 cup (8 oz)	1 carbohydrate

Brownies, Cake, Cookies, Gelatin, Pie, and Pudding

Food	Serving Size	Count as
Brownie, small, unfrosted	1¼-inch square, ⅞ inch high (about 1 oz)	1 carbohydrate + 1 fat
Cake		
angel food, unfrosted	1⁄12 of cake (about 2 oz)	2 carbohydrates
frosted	2-inch square (about 2 oz)	2 carbohydrates + 1 fat
unfrosted	2-inch square (about 2 oz)	1 carbohydrate + 1 fat
Cookies		
chocolate chip	2 cookies (2¼ inches across)	1 carbohydrate + 2 fats
gingersnap	3 cookies	1 carbohydrate
sandwich, with crème filling	2 small (about ⅔ oz)	1 carbohydrate + 1 fat
sugar-free	3 small or 1 large (¾–1 oz)	1 carbohydrate + 1–2 fats
vanilla wafer	5 cookies	1 carbohydrate + 1 fat
Cupcake, frosted	1 small (about 1¾ oz)	2 carbohydrates + 1–1½ fats
Fruit cobbler	½ cup (3½ oz)	3 carbohydrates + 1 fat
Gelatin, regular	½ cup	1 carbohydrate
Pie		
commercially prepared fruit, 2 crusts	⅙ of 8-inch pie	3 carbohydrates + 2 fats
pumpkin or custard	⅛ of 8-inch pie	1½ carbohydrates + 1½ fats
Pudding		
regular (made with reduced-fat milk)	½ cup	2 carbohydrates
sugar-free or sugar- and fat-free (made with fat-free milk)	½ cup	1 carbohydrate

Candy, Spreads, Sweets, Sweeteners, Syrups, and Toppings

Food	Serving Size	Count as
Candy bar, chocolate/peanut	2 "fun size" bars (1 oz)	1½ carbohydrates + 1½ fats
Candy, hard	3 pieces	1 carbohydrate
Chocolate "kisses"	5 pieces	1 carbohydrate + 1 fat
Coffee creamer		
dry, flavored	4 tsp	½ carbohydrate + ½ fat
liquid, flavored	2 Tbsp	1 carbohydrate
Fruit snacks, chewy (pureed fruit concentrate)	1 roll (¾ oz)	1 carbohydrate
Fruit spreads, 100% fruit	1½ Tbsp	1 carbohydrate
Honey	1 Tbsp	1 carbohydrate
Jam or jelly, regular	1 Tbsp	1 carbohydrate
Sugar	1 Tbsp	1 carbohydrate
Syrup		
chocolate	2 Tbsp	2 carbohydrates
light (pancake type)	2 Tbsp	1 carbohydrate
regular (pancake type)	1 Tbsp	1 carbohydrate

(*continued*)

TABLE G-5 Sweets, Desserts, and Other Carbohydrates (*continued*)

Condiments and Sauces[a]

Food	Serving Size	Count as
Barbeque sauce	3 Tbsp	1 carbohydrate
Cranberry sauce, jellied	¼ cup	1½ carbohydrates
Gravy, canned or bottled	½ cup	½ carbohydrate + ½ fat
Salad dressing, fat-free, low-fat, cream-based	3 Tbsp	1 carbohydrate
Sweet and sour sauce	3 Tbsp	1 carbohydrate

Doughnuts, Muffins, Pastries, and Sweet Breads

Food	Serving Size	Count as
Banana nut bread	1-inch slice (1 oz)	2 carbohydrates + 1 fat
Doughnut		
cake, plain	1 medium (1½ oz)	1½ carbohydrates + 2 fats
yeast type, glazed	3¾ inches across (2 oz)	2 carbohydrates + 2 fats
Muffin (4 oz)	¼ muffin (1 oz)	1 carbohydrate + ½ fat
Sweet roll or Danish	1 (2½ oz)	2½ carbohydrates + 2 fats

Frozen Bars, Frozen Desserts, Frozen Yogurt, and Ice Cream

Food	Serving Size	Count as
Frozen pops	1	½ carbohydrate
Fruit juice bars, frozen, 100% juice	1 bar (3 oz)	1 carbohydrate
Ice cream		
fat-free	½ cup	1½ carbohydrates
light	½ cup	1 carbohydrate + 1 fat
no sugar added	½ cup	1 carbohydrate + 1 fat
regular	½ cup	1 carbohydrate + 2 fats
Sherbet, sorbet	½ cup	2 carbohydrates
Yogurt, frozen		
fat-free	⅓ cup	1 carbohydrate
regular	½ cup	1 carbohydrate + 0–1 fat

Granola Bars, Meal Replacement Bars/Shakes, and Trail Mix

Food	Serving Size	Count as
Granola or snack bar, regular or low-fat	1 bar (1 oz)	1½ carbohydrates
Meal replacement bar	1 bar (1⅓ oz)	1½ carbohydrates + 0–1 fat
Meal replacement bar	1 bar (2 oz)	2 carbohydrates + 1 fat
Meal replacement shake, reduced kcalorie	1 can (10–11 oz)	1½ carbohydrates + 0–1 fat
Trail mix		
candy/nut-based	1 oz	1 carbohydrate + 2 fats
dried fruit-based	1 oz	1 carbohydrate + 1 fat

KEY

= 480 milligrams or more of sodium per serving.

[a] You can also check the Fats list and Free Foods list for other condiments.

TABLE G-6 Nonstarchy Vegetables

The Nonstarchy Vegetables list includes vegetables that have few grams of carbohydrates or kcalories; starchy vegetables are found on the Starch list. 1 nonstarchy vegetable choice = 5 grams carbohydrate, 2 grams protein, 0 grams fat, and 25 kcalories.

NOTE: In general, one nonstarchy vegetable choice is ½ cup cooked vegetables or vegetable juice or 1 cup raw vegetables. Count 3 cups of raw vegetables or 1½ cups of cooked vegetables as one carbohydrate choice.

Nonstarchy Vegetables[a]

Amaranth or Chinese spinach	Kohlrabi
Artichoke	Leeks
Artichoke hearts	Mixed vegetables (without corn, peas, or pasta)
Asparagus	Mung bean sprouts
Baby corn	Mushrooms, all kinds, fresh
Bamboo shoots	Okra
Beans (green, wax, Italian)	Onions
Bean sprouts	Oriental radish or daikon
Beets	Pea pods
🧂 Borscht	☺ Peppers (all varieties)
Broccoli	Radishes
☺ Brussels sprouts	Rutabaga
Cabbage (green, bok choy, Chinese)	🧂 Sauerkraut
☺ Carrots	Soybean sprouts
Cauliflower	Spinach
Celery	Squash (summer, crookneck, zucchini)
☺ Chayote	Sugar pea snaps
Coleslaw, packaged, no dressing	☺ Swiss chard
Cucumber	Tomato
Eggplant	Tomatoes, canned
Gourds (bitter, bottle, luffa, bitter melon)	🧂 Tomato sauce
Green onions or scallions	🧂 Tomato/vegetable juice
Greens (collard, kale, mustard, turnip)	Turnips
Hearts of palm	Water chestnuts
Jicama	Yard-long beans

KEY

☺ = More than 3 grams of dietary fiber per serving.

🧂 = 480 milligrams or more of sodium per serving.

[a]Salad greens (like chicory, endive, escarole, lettuce, romaine, spinach, arugula, radicchio, watercress) are on the Free Foods list.

TABLE G-7 Meat and Meat Substitutes

The Meat and Meat Substitutes list groups foods based on the amount of fat they have (lean meat, medium-fat meat, high-fat meat, and plant-based proteins).

Lean Meats and Meat Substitutes

1 lean meat choice = 0 grams carbohydrate, 7 grams protein, 0–3 grams fat, and 100 kcalories.

Food	Amount
Beef: Select or Choice grades trimmed of fat: ground round, roast (chuck, rib, rump), round, sirloin, steak (cubed, flank, porterhouse, T-bone), tenderloin	1 oz
🧂 Beef jerky	1 oz
Cheeses with 3 grams of fat or less per oz	1 oz
Cottage cheese	¼ cup
Egg substitutes, plain	¼ cup
Egg whites	2
Fish, fresh or frozen, plain: catfish, cod, flounder, haddock, halibut, orange roughy, salmon, tilapia, trout, tuna	1 oz
🧂 Fish, smoked: herring or salmon (lox)	1 oz
Game: buffalo, ostrich, rabbit, venison	1 oz
🧂 Hot dog with 3 grams of fat or less per oz (8 dogs per 14 oz package) *Note: May be high in carbohydrate.*	1
Lamb: chop, leg, or roast	1 oz
Organ meats: heart, kidney, liver *Note: May be high in cholesterol.*	1 oz
Oysters, fresh or frozen	6 medium
Pork, lean	
🧂 Canadian bacon	1 oz
rib or loin chop/roast, ham, tenderloin	1 oz
Poultry, without skin: Cornish hen, chicken, domestic duck or goose (well-drained of fat), turkey	1 oz
Processed sandwich meats with 3 grams of fat or less per oz: chipped beef, deli thin-sliced meats, turkey ham, turkey kielbasa, turkey pastrami	1 oz
Salmon, canned	1 oz
Sardines, canned	2 medium
🧂 Sausage with 3 grams of fat or less per oz	1 oz
Shellfish: clams, crab, imitation shellfish, lobster, scallops, shrimp	1 oz
Tuna, canned in water or oil, drained	1 oz
Veal, lean chop, roast	1 oz

Medium-Fat Meat and Meat Substitutes

1 medium-fat meat choice = 0 grams carbohydrate, 7 grams protein, 4–7 grams fat, and 130 kcalories.

Food	Amount
Beef: corned beef, ground beef, meatloaf, Prime grades trimmed of fat (prime rib), short ribs, tongue	1 oz

Medium-Fat Meat and Meat Substitutes—continued

Food	Amount
Cheeses with 4–7 grams of fat per oz: feta, mozzarella, pasteurized processed cheese spread, reduced-fat cheeses, ' string	1 oz
Egg *Note: High in cholesterol, so limit to 3 per week.*	1
Fish, any fried product	1 oz
Lamb: ground, rib roast	1 oz
Pork: cutlet, shoulder roast	1 oz
Poultry: chicken with skin; dove, pheasant, wild duck, or goose; fried chicken; ground turkey	1 oz
Ricotta cheese	2 oz or ¼ cup
🧂 Sausage with 4–7 grams of fat per oz	1 oz
Veal, cutlet (no breading)	1 oz

High-Fat Meat and Meat Substitutes

1 high-fat meat choice = 0 grams carbohydrate, 7 grams protein, 8+ grams fat, and 150 kcalories. These foods are high in saturated fat, cholesterol, and kcalories and may raise blood cholesterol levels if eaten on a regular basis. Try to eat 3 or fewer servings from this group per week.

Food	Amount
Bacon	
🧂 pork	2 slices (16 slices per lb or 1 oz each, before cooking)
🧂 turkey	3 slices (½ oz each before cooking)
Cheese, regular: American, bleu, brie, cheddar, hard goat, Monterey jack, queso, and Swiss	1 oz
⚠ 🧂 Hot dog: beef, pork, or combination (10 per lb-sized package)	1
🧂 Hot dog: turkey or chicken (10 per lb-sized package)	1
Pork: ground, sausage, spareribs	1 oz
Processed sandwich meats with 8 grams of fat or more per oz: bologna, pastrami, hard salami	1 oz
🧂 Sausage with 8 grams fat or more per oz: bratwurst, chorizo, Italian, knockwurst, Polish, smoked, summer	1 oz

(continued)

KEY

☺ = More than 3 grams of dietary fiber per serving.

⚠ = Extra fat, or prepared with added fat. (Count as 1 starch + 1 fat.)

🧂 = 480 milligrams or more of sodium per serving.

TABLE G-7 Meat and Meat Substitutes (*continued*)

Plant-Based Proteins

1 plant-based protein choice = variable grams carbohydrate, 7 grams protein, variable grams fat, and variable kcalories.
Because carbohydrate content varies among plant-based proteins, you should read the food label.

Food	Serving Size	Count as
"Bacon" strips, soy-based	3 strips	1 medium-fat meat
☺ Baked beans	⅓ cup	1 starch + 1 lean meat
☺ Beans, cooked: black, garbanzo, kidney, lima, navy, pinto, white[a]	½ cup	1 starch + 1 lean meat
☺ "Beef" or "sausage" crumbles, soy-based	2 oz	½ carbohydrate + 1 lean meat
"Chicken" nuggets, soy-based	2 nuggets (1½ oz)	½ carbohydrate + 1 medium-fat meat
☺ Edamame	½ cup	½ carbohydrate + 1 lean meat
Falafel (spiced chickpea and wheat patties)	3 patties (about 2 inches across)	1 carbohydrate + 1 high-fat meat
Hot dog, soy-based	1 (1½ oz)	½ carbohydrate + 1 lean meat
☺ Hummus	⅓ cup	1 carbohydrate + 1 high-fat meat
☺ Lentils, brown, green, or yellow	½ cup	1 carbohydrate + 1 lean meat
☺ Meatless burger, soy-based	3 oz	½ carbohydrate + 2 lean meats
☺ Meatless burger, vegetable- and starch-based	1 patty (about 2½ oz)	1 carbohydrate + 2 lean meats
Nut spreads: almond butter, cashew butter, peanut butter, soy nut butter	1 Tbsp	1 high-fat meat
☺ Peas, cooked: black-eyed and split peas	½ cup	1 starch + 1 lean meat
🧂 ☺ Refried beans, canned	½ cup	1 starch + 1 lean meat
"Sausage" patties, soy-based	1 (1½ oz)	1 medium-fat meat
Soy nuts, unsalted	¾ oz	½ carbohydrate + 1 medium-fat meat
Tempeh	¼ cup	1 medium-fat meat
Tofu	4 oz (½ cup)	1 medium-fat meat
Tofu, light	4 oz (½ cup)	1 lean meat

KEY

☺ = More than 3 grams of dietary fiber per serving.

⚠ = Extra fat, or prepared with added fat. (Add an additional fat choice to this food.)

🧂 = 480 milligrams or more of sodium per serving (based on the sodium content of a typical 3-oz serving of meat, unless 1 or 2 oz is the normal serving size).

[a]Beans, peas, and lentils are also found on the Starch list; nut butters in smaller amounts are found in the Fats list.

TABLE G-8 Fats

Fats and oils have mixtures of unsaturated (polyunsaturated and monounsaturated) and saturated fats. Foods on the Fats list are grouped together based on the major type of fat they contain. 1 fat choice = 0 grams carbohydrate, 0 grams protein, 5 grams fat, and 45 kcalories.

NOTE: In general, one fat exchange is 1 teaspoon of regular margarine, vegetable oil, or butter; 1 tablespoon of regular salad dressing.

When used in large amounts, bacon and peanut butter are counted as high-fat meat choices (see Meat and Meat Substitutes list). Fat-free salad dressings are found on the Sweets, Desserts, and Other Carbohydrates list. Fat-free products such as margarines, salad dressings, mayonnaise, sour cream, and cream cheese are found on the Free Foods list.

Monounsaturated Fats

Food	Serving Size
Avocado, medium	2 Tbsp (1 oz)
Nut butters (*trans* fat-free): almond butter, cashew butter, peanut butter (smooth or crunchy)	1½ tsp
Nuts	
almonds	6 nuts
Brazil	2 nuts
cashews	6 nuts
filberts (hazelnuts)	5 nuts
macadamia	3 nuts
mixed (50% peanuts)	6 nuts
peanuts	10 nuts
pecans	4 halves
pistachios	16 nuts
Oil: canola, olive, peanut	1 tsp
Olives	
black (ripe)	8 large
green, stuffed	10 large

Polyunsaturated Fats

Food	Serving Size
Margarine: lower-fat spread (30%–50% vegetable oil, *trans* fat-free)	1 Tbsp
Margarine: stick, tub (*trans* fat-free) or squeeze (*trans* fat-free)	1 tsp
Mayonnaise	
reduced-fat	1 Tbsp
regular	1 tsp
Mayonnaise-style salad dressing	
reduced-fat	1 Tbsp
regular	2 tsp
Nuts	
Pignolia (pine nuts)	1 Tbsp
walnuts, English	4 halves
Oil: corn, cottonseed, flaxseed, grape seed, safflower, soybean, sunflower	1 tsp
Oil: made from soybean and canola oil—Enova	1 tsp
Plant stanol esters	
light	1 Tbsp
regular	2 tsp

Polyunsaturated Fats—continued

Food	Serving Size
Salad dressing	
🧂 reduced-fat *Note: May be high in carbohydrate.*	2 Tbsp
🧂 regular	1 Tbsp
Seeds	
flaxseed, whole	1 Tbsp
pumpkin, sunflower	1 Tbsp
sesame seeds	1 Tbsp
Tahini or sesame paste	2 tsp

Saturated Fats

Food	Serving Size
Bacon, cooked, regular or turkey	1 slice
Butter	
reduced-fat	1 Tbsp
stick	1 tsp
whipped	2 tsp
Butter blends made with oil	
reduced-fat or light	1 Tbsp
regular	1½ tsp
Chitterlings, boiled	2 Tbsp (½ oz)
Coconut, sweetened, shredded	2 Tbsp
Coconut milk	
light	⅓ cup
regular	1½ Tbsp
Cream	
half and half	2 Tbsp
heavy	1 Tbsp
light	1½ Tbsp
whipped	2 Tbsp
whipped, pressurized	¼ cup
Cream cheese	
reduced-fat	1½ Tbsp (¾ oz)
regular	1 Tbsp (½ oz)
Lard	1 tsp
Oil: coconut, palm, palm kernel	1 tsp
Salt pork	¼ oz
Shortening, solid	1 tsp
Sour cream	
reduced-fat or light	3 Tbsp
regular	2 Tbsp

KEY

🧂 = 480 milligrams or more of sodium per serving.

TABLE G-9 Free Foods

A "free" food is any food or drink choice that has less than 20 kcalories and 5 grams or less of carbohydrate per serving.

- Most foods on this list should be limited to 3 servings (as listed here) per day. Spread out the servings throughout the day. If you eat all 3 servings at once, it could raise your blood glucose level.
- Food and drink choices listed here without a serving size can be eaten whenever you like.

Low Carbohydrate Foods

Food	Serving Size
Cabbage, raw	½ cup
Candy, hard (regular or sugar-free)	1 piece
Carrots, cauliflower, or green beans, cooked	¼ cup
Cranberries, sweetened with sugar substitute	½ cup
Cucumber, sliced	½ cup
Gelatin	
dessert, sugar-free	
unflavored	
Gum	
Jam or jelly, light or no sugar added	2 tsp
Rhubarb, sweetened with sugar substitute	½ cup
Salad greens	
Sugar substitutes (artificial sweeteners)	
Syrup, sugar-free	2 Tbsp

Modified Fat Foods with Carbohydrate

Food	Serving Size
Cream cheese, fat-free	1 Tbsp (½ oz)
Creamers	
nondairy, liquid	1 Tbsp
nondairy, powdered	2 tsp
Margarine spread	
fat-free	1 Tbsp
reduced-fat	1 tsp
Mayonnaise	
fat-free	1 Tbsp
reduced-fat	1 tsp
Mayonnaise-style salad dressing	
fat-free	1 Tbsp
reduced-fat	1 tsp
Salad dressing	
fat-free or low-fat	1 Tbsp
fat-free, Italian	2 Tbsp
Sour cream, fat-free or reduced-fat	1 Tbsp
Whipped topping	
light or fat-free	2 Tbsp
regular	1 Tbsp

Condiments

Food	Serving Size
Barbecue sauce	2 tsp
Catsup (ketchup)	1 Tbsp
Honey mustard	1 Tbsp
Horseradish	
Lemon juice	
Miso	1½ tsp
Mustard	
Parmesan cheese, freshly grated	1 Tbsp
Pickle relish	1 Tbsp
Pickles	
🧂 dill	1½ medium
sweet, bread and butter	2 slices
sweet, gherkin	¾ oz
Salsa	¼ cup
🧂 Soy sauce, light or regular	1 Tbsp
Sweet and sour sauce	2 tsp
Sweet chili sauce	2 tsp
Taco sauce	1 Tbsp
Vinegar	
Yogurt, any type	2 Tbsp

Drinks/Mixes

Any food on the list—without a serving size listed—can be consumed in any moderate amount.
🧂 Bouillon, broth, consommé
Bouillon or broth, low-sodium
Carbonated or mineral water
Club soda
Cocoa powder, unsweetened (1 Tbsp)
Coffee, unsweetened or with sugar substitute
Diet soft drinks, sugar-free
Drink mixes, sugar-free
Tea, unsweetened or with sugar substitute
Tonic water, diet
Water
Water, flavored, carbohydrate free

Seasonings

Any food on this list can be consumed in any moderate amount.
Flavoring extracts (for example, vanilla, almond, peppermint)
Garlic
Herbs, fresh or dried
Nonstick cooking spray
Pimento
Spices
Hot pepper sauce
Wine, used in cooking
Worcestershire sauce

KEY

🧂 = 480 milligrams or more of sodium per serving.

TABLE G-10 Combination Foods

Many foods are eaten in various combinations, such as casseroles. Because "combination" foods do not fit into any one choice list, this list of choices provides some typical combination foods.

Food	Serving Size	Count as
Entrees		
Casserole type (tuna noodle, lasagna, spaghetti with meatballs, chili with beans, macaroni and cheese)	1 cup (8 oz)	2 carbohydrates + 2 medium-fat meats
Stews (beef/other meats and vegetables)	1 cup (8 oz)	1 carbohydrate + 1 medium-fat meat + 0–3 fats
Tuna salad or chicken salad	½ cup (3½ oz)	½ carbohydrate + 2 lean meats + 1 fat
Frozen Meals/Entrees		
Burrito (beef and bean)	1 (5 oz)	3 carbohydrates + 1 lean meat + 2 fats
Dinner-type meal	generally 14–17 oz	3 carbohydrates + 3 medium-fat meats + 3 fats
Entrée or meal with less than 340 kcalories	about 8–11 oz	2–3 carbohydrates + 1–2 lean meats
Pizza		
cheese/vegetarian, thin crust	¼ of a 12 inch (4½–5 oz)	2 carbohydrates + 2 medium-fat meats
meat topping, thin crust	¼ of a 12 inch (5 oz)	2 carbohydrates + 2 medium-fat meats + 1½ fats
Pocket sandwich	1 (4½ oz)	3 carbohydrates + 1 lean meat + 1–2 fats
Pot pie	1 (7 oz)	2½ carbohydrates + 1 medium-fat meat + 3 fats
Salads (Deli-Style)		
Coleslaw	½ cup	1 carbohydrate + 1½ fats
Macaroni/pasta salad	½ cup	2 carbohydrates + 3 fats
Potato salad	½ cup	1½–2 carbohydrates + 1–2 fats
Soups		
Bean, lentil, or split pea	1 cup	1 carbohydrate + 1 lean meat
Chowder (made with milk)	1 cup (8 oz)	1 carbohydrate + 1 lean meat + 1½ fats
Cream (made with water)	1 cup (8 oz)	1 carbohydrate + 1 fat
Instant	6 oz prepared	1 carbohydrate
with beans or lentils	8 oz prepared	2½ carbohydrates + 1 lean meat
Miso soup	1 cup	½ carbohydrate + 1 fat
Oriental noodle	1 cup	2 carbohydrates + 2 fats
Rice (congee)	1 cup	1 carbohydrate
Tomato (made with water)	1 cup (8 oz)	1 carbohydrate
Vegetable beef, chicken noodle, or other broth-type	1 cup (8 oz)	1 carbohydrate

KEY

= More than 3 grams of dietary fiber per serving.

= Extra fat, or prepared with added fat.

= 600 milligrams or more of sodium per serving (for combination food main dishes/meals).

TABLE G-11 Fast Foods

The choices in the Fast Foods list are not specific fast-food meals or items, but are estimates based on popular foods. Ask the restaurant or check its website for nutrition information about your favorite fast foods.

Breakfast Sandwiches

Food	Serving Size	Count as
🧂 Egg, cheese, meat, English muffin	1 sandwich	2 carbohydrates + 2 medium-fat meats
🧂 Sausage biscuit sandwich	1 sandwich	2 carbohydrates + 2 high-fat meats + 3½ fats

Main Dishes/Entrees

Food	Serving Size	Count as
🧂 ☺ Burrito (beef and beans)	1 (about 8 oz)	3 carbohydrates + 3 medium-fat meats + 3 fats
🧂 Chicken breast, breaded and fried	1 (about 5 oz)	1 carbohydrate + 4 medium-fat meats
Chicken drumstick, breaded and fried	1 (about 2 oz)	2 medium-fat meats
🧂 Chicken nuggets	6 (about 3½ oz)	1 carbohydrate + 2 medium-fat meats + 1 fat
🧂 Chicken thigh, breaded and fried	1 (about 4 oz)	½ carbohydrate + 3 medium-fat meats + 1½ fats
🧂 Chicken wings, hot	6 (5 oz)	5 medium-fat meats + 1½ fats

Oriental

Food	Serving Size	Count as
🧂 Beef/chicken/shrimp with vegetables in sauce	1 cup (about 5 oz)	1 carbohydrate + 1 lean meat + 1 fat
🧂 Egg roll, meat	1 (about 3 oz)	1 carbohydrate + 1 lean meat + 1 fat
Fried rice, meatless	½ cup	1½ carbohydrates + 1½ fats
🧂 Meat and sweet sauce (orange chicken)	1 cup	3 carbohydrates + 3 medium-fat meats + 2 fats
🧂 ☺ Noodles and vegetables in sauce (chow mein, lo mein)	1 cup	2 carbohydrates + 1 fat

Pizza

Food	Serving Size	Count as
Pizza		
🧂 cheese, pepperoni, regular crust	⅛ of a 14 inch (about 4 oz)	2½ carbohydrates + 1 medium-fat meat + 1½ fats
🧂 cheese/vegetarian, thin crust	¼ of a 12 inch (about 6 oz)	2½ carbohydrates + 2 medium-fat meats + 1½ fats

Sandwiches

Food	Serving Size	Count as
🧂 Chicken sandwich, grilled	1	3 carbohydrates + 4 lean meats
🧂 Chicken sandwich, crispy	1	3½ carbohydrates + 3 medium-fat meats + 1 fat
Fish sandwich with tartar sauce	1	2½ carbohydrates + 2 medium-fat meats + 2 fats
Hamburger		
🧂 large with cheese	1	2½ carbohydrates + 4 medium-fat meats + 1 fat
regular	1	2 carbohydrates + 1 medium-fat meat + 1 fat
🧂 Hot dog with bun	1	1 carbohydrate + 1 high-fat meat + 1 fat
Submarine sandwich		
🧂 less than 6 grams fat	6-inch sub	3 carbohydrates + 2 lean meats
🧂 regular	6-inch sub	3½ carbohydrates + 2 medium-fat meats + 1 fat
Taco, hard or soft shell (meat and cheese)	1 small	1 carbohydrate + 1 medium-fat meat + 1½ fats

(continued)

KEY

☺ = More than 3 grams of dietary fiber per serving.

⚠ = Extra fat, or prepared with added fat.

🧂 = 600 milligrams or more of sodium per serving (for fast-food main dishes/meals).

TABLE G-11 Fast Foods (*continued*)

Salads		
Food	**Serving Size**	**Count as**
🧂 ☺ Salad, main dish (grilled chicken type, no dressing or croutons)		1 carbohydrate + 4 lean meats
Salad, side, no dressing or cheese	Small (about 5 oz)	1 vegetable
Sides/Appetizers		
Food	**Serving Size**	**Count as**
▽ French fries, restaurant style	small	3 carbohydrates + 3 fats
	medium	4 carbohydrates + 4 fats
	large	5 carbohydrates + 6 fats
🧂 Nachos with cheese	small (about 4½ oz)	2½ carbohydrates + 4 fats
🧂 Onion rings	1 serving (about 3 oz)	2½ carbohydrates + 3 fats
Desserts		
Food	**Serving Size**	**Count as**
Milkshake, any flavor	12 oz	6 carbohydrates + 2 fats
Soft-serve ice cream cone	1 small	2½ carbohydrates + 1 fat

KEY

☺ = More than 3 grams of dietary fiber per serving.

▽ = Extra fat, or prepared with added fat.

🧂 = 600 milligrams or more of sodium per serving (for fast-food main dishes/meals).

TABLE G-12 Alcohol

1 alcohol equivalent = variable grams carbohydrate, 0 grams protein, 0 grams fat, and 100 kcalories.

NOTE: In general, one alcohol choice (½ ounce absolute alcohol) has about 100 kcalories. For those who choose to drink alcohol, guidelines suggest limiting alcohol intake to 1 drink or less per day for women, and 2 drinks or less per day for men. To reduce your risk of low blood glucose (hypoglycemia), especially if you take insulin or a diabetes pill that increases insulin, always drink alcohol with food. While alcohol, by itself, does not directly affect blood glucose, be aware of the carbohydrate (for example, in mixed drinks, beer, and wine) that may raise your blood glucose.

Alcoholic Beverage	Serving Size	Count as
Beer		
light (4.2%)	12 fl oz	1 alcohol equivalent + ½ carbohydrate
regular (4.9%)	12 fl oz	1 alcohol equivalent + 1 carbohydrate
Distilled spirits: vodka, rum, gin, whiskey, 80 or 86 proof	1½ fl oz	1 alcohol equivalent
Liqueur, coffee (53 proof)	1 fl oz	1 alcohol equivalent + 1 carbohydrate
Sake	1 fl oz	½ alcohol equivalent
Wine		
dessert (sherry)	3½ fl oz	1 alcohol equivalent + 1 carbohydrate
dry, red or white (10%)	5 fl oz	1 alcohol equivalent

The Exchange Lists are the basis of a meal planning system designed by a committee of the American Diabetes Association and The American Dietetic Association. While designed primarily for people with diabetes and others who must follow special diets, the Exchange Lists are based on principles of good nutrition that apply to everyone.

Table of Food Composition

APPENDIX H

CONTENTS

This edition of the table of food composition includes a wide variety of foods. It is updated with each edition to reflect current nutrient data for foods, to remove outdated foods, and to add foods that are new to the marketplace.* The nutrient database for this appendix is compiled from a variety of sources, including the USDA Nutrient Database and manufacturers' data. The USDA database provides data for a wider variety of foods and nutrients than other sources. Because laboratory analysis for each nutrient can be quite costly, manufacturers tend to provide data only for those nutrients mandated on food labels. Consequently, data for their foods are often incomplete; any missing information on this table is designated as a dash. Keep in mind that a dash means only that the information is unknown and should not be interpreted as a zero. A zero means that the nutrient is not present in the food.

Whenever using nutrient data, remember that many factors influence the nutrient contents of foods. These factors include the mineral content of the soil, the diet fed to the animal or the fertilizer used on the plant, the season of harvest, the method of processing, the length and method of storage, the method of cooking, the method of analysis, and the moisture content of the sample analyzed. With so many influencing factors, users should view nutrient data as a close approximation of the actual amount.

For updates, corrections, and a list of more than 8000 foods and codes found in the diet analysis software that accompanies this text, visit www.cengage.com/nutrition and click on Diet Analysis Plus.

- ***Fats*** Total fats, as well as the breakdown of total fats to saturated, monounsaturated, polyunsaturated, and *trans* fats, are listed in the table. The fatty acids seldom add up to the total in part due to rounding but also because values may include some non-fatty acids, such as glycerol, phosphate, or sterols.
- **Trans *Fats*** *Trans* fat data have been listed in the table. Because food manufacturers have only been required to report *trans* fats on food labels since January 2006, much of the data is incomplete. Missing *trans* fat data are designated with a dash. As additional *trans* fat data become available, the table will be updated.
- ***Vitamin A and Vitamin E*** In keeping with the 2001 RDA for vitamin A, this appendix presents data for vitamin A in micrograms (μg) RAE. Similarly, because the 2000 RDA for vitamin E is based only on the alpha-tocopherol form of vitamin E, this appendix reports vitamin E data in milligrams (mg) alpha-tocopherol, listed on the table as Vit E (mg α).
- ***Bioavailability*** Keep in mind that the availability of nutrients from foods depends not only on the quantity provided by a food, but also on the amount absorbed and used by the body—the bioavailability. The bioavailability of folate from fortified foods, for example, is greater than from naturally occurring sources. Similarly, the body can make niacin from the amino acid tryptophan, but niacin values in this table (and most databases) report preformed niacin only. Chapter 10 provides conversion factors and additional details.

*This food composition table has been prepared by Cengage Learning. The nutritional data are supplied by Axxya Systems.

- *Using the Table* The foods and beverages in this table are organized into several categories, which are listed at the head of each right-hand page. Page numbers are provided, and each group is color-coded to make it easier to find individual foods.
- *Caffeine Sources* Caffeine occurs in several plants, including the familiar coffee bean, the tea leaf, and the cocoa bean from which chocolate is made. Most human societies use caffeine regularly, most often in beverages, for its stimulant effect and flavor. Caffeine contents of beverages vary depending on the plants they are made from, the climates and soils where the plants are grown, the grind or cut size, the method and duration of brewing, and the amounts served. The accompanying table shows that, in general, a cup of coffee contains the most caffeine; a cup of tea, less than half as much; and cocoa or chocolate, less still. As for cola beverages, they are made from kola nuts, which contain caffeine, but most of their caffeine is added, using the purified compound obtained from decaffeinated coffee beans. The FDA lists caffeine as a multipurpose GRAS substance ♦ that may be added to foods and beverages. Drug manufacturers use caffeine in many products.

♦ A GRAS substance is one that is "generally recognized as safe."

TABLE Caffeine Content of Selected Beverages, Foods, and Medications

Beverages and Foods	Serving Size	Average (mg)
Coffee		
Brewed	8 oz	95
Decaffeinated	8 oz	2
Instant	8 oz	64
Tea		
Brewed, green	8 oz	30
Brewed, herbal	8 oz	0
Brewed, leaf or bag	8 oz	47
Instant	8 oz	26
Lipton Brisk iced tea	12 oz	7
Nestea Cool iced tea	12 oz	12
Snapple iced tea (all flavors)	16 oz	42

Beverages and Foods	Serving Size	Average (mg)
Soft Drinks		
A&W Creme Soda	12 oz	29
Barq's Root Beer	12 oz	18
Coca-Cola	12 oz	30
Dr. Pepper, Mr. Pibb, Sunkist Orange	12 oz	36
A&W Root Beer, club soda, Fresca, ginger ale, 7-Up, Sierra Mist, Sprite, Squirt, tonic water, caffeine-free soft drinks	12 oz	0
Mello Yello	12 oz	51
Mountain Dew	12 oz	45
Pepsi	12 oz	32

TABLE Caffeine Content of Selected Beverages, Foods, and Medications (*continued*)

Beverages and Foods	Serving Size	Average (mg)
Energy Drinks		
Amp	8.4 oz	70
Aqua Blast	.5 L	90
Aqua Java	.5 L	55
E Maxx	8.4 oz	74
Java Water	.5 L	125
KMX	8.4 oz	33
Krank	.5 L	100
Red Bull	8.3 oz	67
Red Devil	8.4 oz	42
Sobe Adrenaline Rush	8.3 oz	77
Sobe No Fear	16 oz	141
Water Joe	.5 L	65
Other Beverages		
Chocolate milk or hot cocoa	8 oz	5
Starbucks Frappuccino Mocha	9.5 oz	72
Starbucks Frappuccino Vanilla	9.5 oz	64
Yoohoo chocolate drink	9 oz	3
Candies		
Baker's chocolate	1 oz	26
Dark chocolate covered coffee beans	1 oz	235
Dark chocolate, semisweet	1 oz	18
Milk chocolate	1 oz	6
Milk chocolate covered coffee beans	1 oz	224
White chocolate	1 oz	0
Foods		
Frozen yogurt, Ben & Jerry's coffee fudge	1 cup	85
Frozen yogurt, Häagen-Dazs coffee	1 cup	40
Ice cream, Starbucks coffee	1 cup	50
Ice cream, Starbucks Frappuccino bar	1 bar	15
Yogurt, Dannon coffee flavored	1 cup	45

Drugs[a]	Serving Size	Average (mg)
Cold Remedies		
Coryban-D, Dristan	1 tablet	30
Diuretics		
Aqua-Ban	1 tablet	100
Pre-Mens Forte	1 tablet	100
Pain Relievers		
Anacin, BC Fast Pain Reliever	1 tablet	32
Excedrin, Midol, Midol Max Strength	1 tablet	65
Stimulants		
Awake, NoDoz	1 tablet	100
Awake Maximum Strength, Caffedrine, NoDoz Maximum Strength, Stay Awake, Vivarin	1 tablet	200
Weight-Control Aids		
Dexatrim	1 tablet	200

[a]A pharmacologically active dose of caffiene is defined as 200 milligrams.

NOTE: The FDA suggests a maximum of 65 milligrams per 12-ounce cola beverage but does not regulate the caffeine contents of other beverages. Because products change, contact the manufacturer for an update on products you use regularly.

SOURCE: Adapted from USDA database Release 18 (www.nal.usda.gov/fnic/foodcomp/Data/), Caffeine content of foods and drugs, Center for Science and the Public Interest (www.cspinet.org/new/cafchart.htm), and R. R. McCusker, B. A. Goldberger, and E. J. Cone, Caffeine content of energy drinks, carbonated sodas, and other beverages, *Journal of Analytical Toxicology* 30 (2006): 112–114.

TABLE H-1 Table of Food Composition (Computer code is for Cengage Diet Analysis program) (For purposes of calculations, use "0" for t, <1, <.1, <.01, etc.)

DA+ Code	Food Description	Quantity	Measure	Wt (g)	H_2O (g)	Ener (kcal)	Prot (g)	Carb (g)	Fiber (g)	Fat (g)	Fat Breakdown (g) Sat	Mono	Poly	*Trans*
Breads, Baked Goods, Cakes, Cookies, Crackers, Chips, Pies														
	Bagels													
8534	Cinnamon and raisin	1	item(s)	71	22.7	194	7.0	39.2	1.6	1.2	0.2	0.1	0.5	—
14395	Multi-grain	1	item(s)	61	—	170	6.0	35.0	1.0	1.5	0.5	0.1	0.4	—
8538	Oat bran	1	item(s)	71	23.4	181	7.6	37.8	2.6	0.9	0.1	0.2	0.3	—
4910	Plain, enriched	1	item(s)	71	25.8	182	7.1	35.9	1.6	1.2	0.3	0.4	0.5	0
4911	Plain, enriched, toasted	1	item(s)	66	18.7	190	7.4	37.7	1.7	1.1	0.2	0.3	0.6	0
	Biscuits													
25008	Biscuits	1	item(s)	41	15.8	121	2.6	16.4	0.5	4.9	1.4	1.4	1.8	—
16729	Scone	1	item(s)	42	11.5	148	3.8	19.1	0.6	6.2	2.0	2.5	1.3	—
25166	Wheat biscuits	1	item(s)	55	21.0	162	3.6	21.9	1.4	6.7	1.9	1.9	2.5	—
	Bread													
325	Boston brown, canned	1	slice(s)	45	21.2	88	2.3	19.5	2.1	0.7	0.1	0.1	0.3	—
8716	Bread sticks, plain	4	item(s)	24	1.5	99	2.9	16.4	0.7	2.3	0.3	0.9	0.9	—
25176	Cornbread	1	piece(s)	55	25.9	141	4.7	18.3	0.9	5.4	2.1	1.4	1.5	0
327	Cracked wheat	1	slice(s)	25	9.0	65	2.2	12.4	1.4	1.0	0.2	0.5	0.2	—
9079	Croutons, plain	¼	cup(s)	8	0.4	31	0.9	5.5	0.4	0.5	0.1	0.2	0.1	—
8582	Egg	1	slice(s)	40	13.9	113	3.8	19.1	0.9	2.4	0.6	0.9	0.4	—
8585	Egg, toasted	1	slice(s)	37	10.5	117	3.9	19.5	0.9	2.4	0.6	1.1	0.4	—
329	French	1	slice(s)	32	8.9	92	3.8	18.1	0.8	0.6	0.2	0.1	0.3	—
8591	French, toasted	1	slice(s)	23	4.7	73	3.0	14.2	0.7	0.5	0.1	0.1	0.2	—
42096	Indian fry, made with lard (Navajo)	3	ounce(s)	85	26.9	281	5.7	41.0	—	10.4	3.9	3.8	0.9	—
332	Italian	1	slice(s)	30	10.7	81	2.6	15.0	0.8	1.1	0.3	0.2	0.4	—
1393	Mixed grain	1	slice(s)	26	9.6	69	3.5	11.3	1.9	1.1	0.2	0.2	0.5	0
8604	Mixed grain, toasted	1	slice(s)	24	7.6	69	3.5	11.3	1.9	1.1	0.2	0.2	0.5	0
8605	Oat bran	1	slice(s)	30	13.2	71	3.1	11.9	1.4	1.3	0.2	0.5	0.5	—
8608	Oat bran, toasted	1	slice(s)	27	10.4	70	3.1	11.8	1.3	1.3	0.2	0.5	0.5	—
8609	Oatmeal	1	slice(s)	27	9.9	73	2.3	13.1	1.1	1.2	0.2	0.4	0.5	—
8613	Oatmeal, toasted	1	slice(s)	25	7.8	73	2.3	13.2	1.1	1.2	0.2	0.4	0.5	—
1409	Pita	1	item(s)	60	19.3	165	5.5	33.4	1.3	0.7	0.1	0.1	0.3	—
7905	Pita, whole wheat	1	item(s)	64	19.6	170	6.3	35.2	4.7	1.7	0.3	0.2	0.7	—
338	Pumpernickel	1	slice(s)	32	12.1	80	2.8	15.2	2.1	1.0	0.1	0.3	0.4	—
334	Raisin, enriched	1	slice(s)	26	8.7	71	2.1	13.6	1.1	1.1	0.3	0.6	0.2	—
8625	Raisin, toasted	1	slice(s)	24	6.7	71	2.1	13.7	1.1	1.2	0.3	0.6	0.2	—
10168	Rice, white, gluten free, wheat free	1	slice(s)	38	—	130	1.0	18.0	0.5	6.0	0	—	—	0
8653	Rye	1	slice(s)	32	11.9	83	2.7	15.5	1.9	1.1	0.2	0.4	0.3	—
8654	Rye, toasted	1	slice(s)	29	9.0	82	2.7	15.4	1.9	1.0	0.2	0.4	0.3	—
336	Rye, light	1	slice(s)	25	9.3	65	2.0	12.0	1.6	1.0	0.2	0.3	0.3	—
8588	Sourdough	1	slice(s)	25	7.0	72	2.9	14.1	0.6	0.5	0.1	0.1	0.2	—
8592	Sourdough, toasted	1	slice(s)	23	4.7	73	3.0	14.2	0.7	0.5	0.1	0.1	0.2	—
491	Submarine or hoagie roll	1	item(s)	135	40.6	400	11.0	72.0	3.8	8.0	1.8	3.0	2.2	—
8596	Vienna, toasted	1	slice(s)	23	4.7	73	3.0	14.2	0.7	0.5	0.1	0.1	0.2	—
8670	Wheat	1	slice(s)	25	8.9	67	2.7	11.9	0.9	0.9	0.2	0.2	0.4	—
8671	Wheat, toasted	1	slice(s)	23	5.6	72	3.0	12.8	1.1	1.0	0.2	0.2	0.4	—
340	White	1	slice(s)	25	9.1	67	1.9	12.7	0.6	0.8	0.2	0.2	0.3	—
1395	Whole wheat	1	slice(s)	46	15.0	128	3.9	23.6	2.8	2.5	0.4	0.5	1.4	—
	Cakes													
386	Angel food, prepared from mix	1	piece(s)	50	16.5	129	3.1	29.4	0.1	0.2	0	0	0.1	—
8772	Butter pound, ready to eat, commercially prepared	1	slice(s)	75	18.5	291	4.1	36.6	0.4	14.9	8.7	4.4	0.8	—
28517	Carrot	1	slice(s)	131	56.6	339	4.8	56.5	1.9	11.1	1.0	5.7	3.8	—
4931	Chocolate with chocolate icing, commercially prepared	1	slice(s)	64	14.7	235	2.6	34.9	1.8	10.5	3.1	5.6	1.2	—
8756	Chocolate, prepared from mix	1	slice(s)	95	23.2	352	5.0	50.7	1.5	14.3	5.2	5.7	2.6	—
393	Devil's food cupcake with chocolate frosting	1	item(s)	35	8.4	120	2.0	20.0	0.7	4.0	1.8	1.6	0.6	—
8757	Fruitcake, ready to eat, commercially prepared	1	piece(s)	43	10.9	139	1.2	26.5	1.6	3.9	0.5	1.8	1.4	—
1397	Pineapple upside down, prepared from mix	1	slice(s)	115	37.1	367	4.0	58.1	0.9	13.9	3.4	6.0	3.8	—
411	Sponge, prepared from mix	1	slice(s)	63	18.5	187	4.6	36.4	0.3	2.7	0.8	1.0	0.4	—
8817	White with coconut frosting, prepared from mix	1	slice(s)	112	23.2	399	4.9	70.8	1.1	11.5	4.4	4.1	2.4	—
8819	Yellow with chocolate frosting, ready to eat, commercially prepared	1	slice(s)	64	14.0	243	2.4	35.5	1.2	11.1	3.0	6.1	1.4	—

PAGE KEY: H-4 = Breads/Baked Goods H-10 = Cereal/Rice/Pasta H-14 = Fruit H-18 = Vegetables/Legumes H-28 = Nuts/Seeds H-30 = Vegetarian H-32 = Dairy H-40 = Eggs H-40 = Seafood H-42 = Meats H-46 = Poultry H-46 = Processed Meats H-48 = Beverages H-52 = Fats/Oils H-54 = Sweets H-56 = Spices/Condiments/Sauces H-60 = Mixed Foods/Soups/Sandwiches H-64 = Fast Food H-84 = Convenience H-86 = Baby Foods

Chol (mg)	Calc (mg)	Iron (mg)	Magn (mg)	Pota (mg)	Sodi (mg)	Zinc (mg)	Vit A (µg)	Thia (mg)	Vit E (mg α)	Ribo (mg)	Niac (mg)	Vit B_6 (mg)	Fola (µg)	Vit C (mg)	Vit B_{12} (µg)	Sele (µg)
0	13	2.69	19.9	105.1	228.6	0.80	14.9	0.27	0.22	0.19	2.18	0.04	78.8	0.5	0	22.0
0	60	1.08	—	—	310.0	—	0	—	—	—	—	—	—	0	—	—
0	9	2.18	22.0	81.7	360.0	0.63	0.7	0.23	0.23	0.24	2.10	0.03	69.6	0.1	0	24.3
0	63	4.29	15.6	53.3	318.1	1.34	0	0.42	0.07	0.18	2.82	0.04	103.0	0.7	0	16.2
0	65	2.97	15.8	56.1	316.8	0.85	0	0.39	0.07	0.17	2.88	0.04	86.5	0	0	16.6
0	38	0.94	6.0	47.4	206.0	0.20	—	0.16	0.01	0.12	1.20	0.01	31.7	0.1	0.1	7.1
49	79	1.35	7.1	48.7	277.2	0.29	64.7	0.14	0.42	0.15	1.19	0.02	32.3	0	0.1	10.9
0	57	1.21	16.1	81.0	321.1	0.42	—	0.19	0.01	0.14	1.65	0.03	35.3	0.1	0.1	0
0	32	0.94	28.4	143.1	284.0	0.22	11.3	0.01	0.14	0.05	0.50	0.03	5.0	0	0	9.9
0	5	1.02	7.7	29.8	157.7	0.21	0	0.14	0.24	0.13	1.26	0.01	38.9	0	0	9.0
21	94	0.91	10.5	71.5	209.8	0.48	—	0.14	0.32	0.15	1.03	0.04	34.6	1.7	0.2	6.2
0	11	0.70	13.0	44.3	134.5	0.31	0	0.09	—	0.06	0.92	0.08	15.3	0	0	6.3
0	6	0.30	2.3	9.3	52.4	0.06	0	0.04	—	0.02	0.40	0.00	9.9	0	0	2.8
20	37	1.21	7.6	46.0	196.8	0.31	25.2	0.17	0.10	0.17	1.93	0.02	42.0	0	0	12.0
21	38	1.23	7.8	46.6	199.8	0.31	25.5	0.14	0.10	0.16	1.77	0.02	36.3	0	0	12.2
0	14	1.16	9.0	41.0	208.0	0.29	0	0.13	0.05	0.09	1.52	0.03	47.4	0.1	0	8.7
0	11	0.89	7.1	32.2	165.6	0.24	0	0.10	0.04	0.09	1.24	0.02	32.2	0	0	6.8
6	48	3.43	15.3	65.5	279.8	0.29	0	0.36	0.00	0.18	3.91	0.03	103.8	—	0	15.8
0	23	0.88	8.1	33.0	175.2	0.25	0	0.14	0.08	0.08	1.31	0.01	57.3	0	0	8.2
0	27	0.65	20.3	59.8	109.2	0.44	0	0.07	0.09	0.03	1.05	0.06	19.5	0	0	8.6
0	27	0.65	20.4	60.0	109.7	0.44	0	0.06	0.10	0.03	1.05	0.07	16.8	0	0	8.6
0	20	0.93	10.5	44.1	122.1	0.26	0.6	0.15	0.13	0.10	1.44	0.02	24.3	0	0	9.0
0	19	0.92	9.2	33.2	121.0	0.28	0.5	0.12	0.13	0.09	1.29	0.01	18.6	0	0	8.9
0	18	0.72	10.0	38.3	161.7	0.27	1.4	0.10	0.13	0.06	0.84	0.01	16.7	0	0	6.6
0	18	0.74	10.3	38.5	162.8	0.28	1.3	0.09	0.13	0.06	0.77	0.02	13.3	0.1	0	6.7
0	52	1.57	15.6	72.0	321.6	0.50	0	0.35	0.18	0.19	2.77	0.02	64.2	0	0	16.3
0	10	1.95	44.2	108.8	340.5	0.97	0	0.21	0.39	0.05	1.81	0.17	22.4	0	0	28.2
0	22	0.91	17.3	66.6	214.7	0.47	0	0.10	0.13	0.09	0.98	0.04	29.8	0	0	7.8
0	17	0.75	6.8	59.0	101.4	0.18	0	0.08	0.07	0.10	0.90	0.01	27.6	0	0	5.2
0	17	0.76	6.7	59.0	101.8	0.19	0	0.07	0.07	0.09	0.81	0.02	23.5	0.1	0	5.2
0	100	1.08	—	—	140	—	—	0.15	—	0.10	1.20	—	32.0	0	—	—
0	23	0.90	12.8	53.1	211.2	0.36	0	0.13	0.10	0.10	1.21	0.02	35.2	0.1	0	9.9
0	23	0.89	12.5	53.1	210.3	0.36	0	0.11	0.10	0.09	1.09	0.02	29.9	0.1	0	9.9
0	20	0.70	3.9	51.0	175.0	0.18	0	0.10	—	0.08	0.80	0.01	5.3	0	0	8.0
0	11	0.91	7.0	32.0	162.5	0.23	0	0.11	0.05	0.07	1.19	0.03	37.0	0.1	0	6.8
0	11	0.89	7.1	32.2	165.6	0.24	0	0.10	0.04	0.09	1.24	0.02	32.2	0	0	6.8
0	100	3.80	—	128.0	683.0	—	0	0.54	—	0.33	4.50	0.04	—	0	—	42.0
0	11	0.89	7.1	32.2	165.6	0.24	0	0.10	0.04	0.09	1.24	0.02	32.2	0	0	6.8
0	36	0.87	12.0	46.0	130.3	0.30	0	0.09	0.05	0.08	1.30	0.03	21.3	0.1	0	7.2
0	38	0.94	13.6	51.3	140.5	0.34	0	0.10	0.06	0.09	1.44	0.04	19.8	0	0	7.7
0	38	0.94	5.8	25.0	170.3	0.19	0	0.11	0.06	0.08	1.10	0.02	27.8	0	0	4.3
0	15	1.42	37.3	144.4	159.2	0.69	0	0.13	0.35	0.10	1.83	0.09	29.9	0	0	17.8
0	42	0.11	4.0	67.5	254.5	0.06	0	0.04	0.01	0.10	0.08	0.00	9.5	0	0	7.7
166	26	1.03	8.3	89.3	298.5	0.34	111.8	0.10	—	0.17	0.98	0.03	30.8	0	0.2	6.6
0	65	2.18	23.0	279.6	367.7	0.44	—	0.25	0.01	0.19	1.73	0.10	43.4	4.6	0	14.7
27	28	1.40	21.8	128.0	213.8	0.44	16.6	0.01	0.62	0.08	0.36	0.02	10.9	0.1	0.1	2.1
55	57	1.53	30.4	133.0	299.3	0.65	38.0	0.13	—	0.20	1.08	0.03	25.7	0.2	0.2	11.3
19	21	0.70	—	46.0	92.0	—	—	0.04	—	0.05	0.30	—	2.1	0	—	2.0
2	14	0.89	6.9	65.8	116.1	0.11	3.0	0.02	0.38	0.04	0.34	0.02	8.6	0.2	0	0.9
25	138	1.70	15.0	128.8	366.9	0.35	71.3	0.17	—	0.17	1.36	0.03	29.9	1.4	0.1	10.8
107	26	0.99	5.7	88.8	143.6	0.37	48.5	0.10	—	0.19	0.75	0.03	24.6	0	0.2	11.7
1	101	1.29	13.4	110.9	318.1	0.37	13.4	0.14	0.13	0.21	1.19	0.03	34.7	0.1	0.1	12.0
35	24	1.33	19.2	113.9	215.7	0.39	21.1	0.07	—	0.10	0.79	0.02	14.1	0	0.1	2.2

TABLE H-1 Table of Food Composition (*continued*) (Computer code is for Cengage Diet Analysis program) (For purposes of calculations, use "0" for t, <1, <.1, <.01, etc.)

DA+ Code	Food Description	Quantity	Measure	Wt (g)	H_2O (g)	Ener (kcal)	Prot (g)	Carb (g)	Fiber (g)	Fat (g)	Fat Breakdown (g)			
											Sat	Mono	Poly	*Trans*
Breads, Baked Goods, Cakes, Cookies, Crackers, Chips, Pies—*continued*														
8822	Yellow with vanilla frosting, ready to eat, commercially prepared	1	slice(s)	64	14.1	239	2.2	37.6	0.2	9.3	1.5	3.9	3.3	—
	Snack cakes													
8791	Chocolate snack cake, creme filled, with frosting	1	item(s)	50	9.3	200	1.8	30.2	1.6	8.0	2.4	4.3	0.9	—
25010	Cinnamon coffee cake	1	piece(s)	72	22.6	231	3.6	35.8	0.7	8.3	2.2	2.6	3.0	—
16777	Funnel cake	1	item(s)	90	37.6	276	7.3	29.1	0.9	14.4	2.7	4.7	6.1	—
8794	Sponge snack cake, creme filled	1	item(s)	43	8.6	155	1.3	27.2	0.2	4.8	1.1	1.7	1.4	—
	Snacks, chips, pretzels													
29428	Bagel chips, plain	3	item(s)	29	—	130	3.0	19.0	1.0	4.5	0.5	—	—	—
29429	Bagel chips, toasted onion	3	item(s)	29	—	130	4.0	20.0	1.0	4.5	0.5	—	—	—
38192	Chex traditional snack mix	1	cup(s)	45	—	197	3.0	33.3	1.5	6.1	0.8	—	—	—
654	Potato chips, salted	1	ounce(s)	28	0.6	155	1.9	14.1	1.2	10.6	3.1	2.8	3.5	—
8816	Potato chips, unsalted	1	ounce(s)	28	0.5	152	2.0	15.0	1.4	9.8	3.1	2.8	3.5	—
5096	Pretzels, plain, hard, twists	5	item(s)	30	1.0	114	2.7	23.8	1.0	1.1	0.2	0.4	0.4	—
4632	Pretzels, whole wheat	1	ounce(s)	28	1.1	103	3.1	23.0	2.2	0.7	0.2	0.3	0.2	—
4641	Tortilla chips, plain	6	item(s)	11	0.2	53	0.8	7.1	0.6	2.5	0.3	0.8	0.5	0.3
	Cookies													
8859	Animal crackers	12	item(s)	30	1.2	134	2.1	22.2	0.3	4.1	1.0	2.3	0.6	—
8876	Brownie, prepared from mix	1	item(s)	24	3.0	112	1.5	12.0	0.5	7.0	1.8	2.6	2.3	—
25207	Chocolate chip cookies	1	item(s)	30	3.7	140	2.0	16.2	0.6	7.9	2.1	3.3	2.1	—
8915	Chocolate sandwich cookie with extra creme filling	1	item(s)	13	0.2	65	0.6	8.9	0.4	3.2	0.7	2.1	0.3	1.1
14145	Fig Newtons cookies	1	item(s)	16	—	55	0.5	11.0	0.5	1.3	0	—	—	0
8920	Fortune cookie	1	item(s)	8	0.6	30	0.3	6.7	0.1	0.2	0.1	0.1	0	—
25208	Oatmeal cookies	1	item(s)	69	12.3	234	5.7	45.1	3.1	4.2	0.7	1.3	1.8	—
25213	Peanut butter cookies	1	item(s)	35	4.1	163	4.2	16.9	0.9	9.2	1.7	4.7	2.3	—
33095	Sugar cookies	1	item(s)	16	4.1	61	1.1	7.4	0.1	3.0	0.6	1.3	0.9	—
9002	Vanilla sandwich cookie with creme filling	1	item(s)	10	0.2	48	0.5	7.2	0.2	2.0	0.3	0.8	0.8	—
	Crackers													
9012	Cheese cracker sandwich with peanut butter	4	item(s)	28	0.9	139	3.5	15.9	1.0	7.0	1.2	3.6	1.4	—
9008	Cheese crackers (mini)	30	item(s)	30	0.9	151	3.0	17.5	0.7	7.6	2.8	3.6	0.7	—
33362	Cheese crackers, low sodium	1	serving(s)	30	0.9	151	3.0	17.5	0.7	7.6	2.9	3.6	0.7	—
8928	Honey graham crackers	4	item(s)	28	1.2	118	1.9	21.5	0.8	2.8	0.4	1.1	1.1	—
9016	Matzo crackers, plain	1	item(s)	28	1.2	112	2.8	23.8	0.9	0.4	0.1	0	0.2	—
9024	Melba toast	3	item(s)	15	0.8	59	1.8	11.5	0.9	0.5	0.1	0.1	0.2	—
9028	Melba toast, rye	3	item(s)	15	0.7	58	1.7	11.6	1.2	0.5	0.1	0.1	0.2	—
14189	Ritz crackers	5	item(s)	16	0.5	80	1.0	10.0	0	4.0	1.0	—	—	0
9014	Rye crispbread crackers	1	item(s)	10	0.6	37	0.8	8.2	1.7	0.1	0	0	0.1	—
9040	Rye wafer	1	item(s)	11	0.6	37	1.1	8.8	2.5	0.1	0	0	0	—
432	Saltine crackers	5	item(s)	15	0.8	64	1.4	10.6	0.5	1.7	0.2	1.1	0.2	0.5
9046	Saltine crackers, low salt	5	item(s)	15	0.6	65	1.4	10.7	0.5	1.8	0.4	1.0	0.3	—
9052	Snack cracker sandwich with cheese filling	4	item(s)	28	1.1	134	2.6	17.3	0.5	5.9	1.7	3.2	0.7	—
9054	Snack cracker sandwich with peanut butter filling	4	item(s)	28	0.8	138	3.2	16.3	0.6	6.9	1.4	3.9	1.3	—
9048	Snack crackers, round	10	item(s)	30	1.1	151	2.2	18.3	0.5	7.6	1.1	3.2	2.9	—
9050	Snack crackers, round, low salt	10	item(s)	30	1.1	151	2.2	18.3	0.5	7.6	1.1	3.2	2.9	—
9044	Soda crackers	5	tem(s)	15	0.8	64	1.4	10.6	0.5	1.7	0.2	1.1	0.2	0.5
9059	Wheat cracker sandwich with cheese filling	4	item(s)	28	0.9	139	2.7	16.3	0.9	7.0	1.2	2.9	2.6	—
9061	Wheat cracker sandwich with peanut butter filling	4	item(s)	28	1.0	139	3.8	15.1	1.2	7.5	1.3	3.3	2.5	—
9055	Wheat crackers	10	item(s)	30	0.9	142	2.6	19.5	1.4	6.2	1.6	3.4	0.8	—
9057	Wheat crackers, low salt	10	item(s)	30	0.9	142	2.6	19.5	1.4	6.2	1.6	3.4	0.8	—
9022	Whole wheat crackers	7	item(s)	28	0.8	124	2.5	19.2	2.9	4.8	1.0	1.6	1.8	—
	Pastry													
16754	Apple fritter	1	item(s)	17	6.4	61	1.0	5.5	0.2	3.9	0.9	1.7	1.1	—
41565	Cinnamon rolls with icing, refrigerated dough	1	serving(s)	44	12.3	145	2.0	23.0	0.5	5.0	1.5	—	—	2.0
4945	Croissant, butter	1	item(s)	57	13.2	231	4.7	26.1	1.5	12.0	6.6	3.1	0.6	—
9096	Danish, nut	1	item(s)	65	13.3	280	4.6	29.7	1.3	16.4	3.8	8.9	2.8	—
9115	Doughnut with creme filling	1	item(s)	85	32.5	307	5.4	25.5	0.7	20.8	4.6	10.3	2.6	—

PAGE KEY: H-4 = Breads/Baked Goods H-10 = Cereal/Rice/Pasta H-14 = Fruit H-18 = Vegetables/Legumes H-28 = Nuts/Seeds H-30 = Vegetarian H-32 = Dairy H-40 = Eggs H-40 = Seafood H-42 = Meats H-46 = Poultry H-46 = Processed Meats H-48 = Beverages H-52 = Fats/Oils H-54 = Sweets H-56 = Spices/Condiments/Sauces H-60 = Mixed Foods/Soups/Sandwiches H-64 = Fast Food H-84 = Convenience H-86 = Baby Foods

Chol (mg)	Calc (mg)	Iron (mg)	Magn (mg)	Pota (mg)	Sodi (mg)	Zinc (mg)	Vit A (µg)	Thia (mg)	Vit E (mg α)	Ribo (mg)	Niac (mg)	Vit B_6 (mg)	Fola (µg)	Vit C (mg)	Vit B_{12} (µg)	Sele (µg)
35	40	0.68	3.8	33.9	220.2	0.16	12.2	0.06	—	0.04	0.32	0.01	17.3	0	0.1	3.5
0	58	1.80	18.0	88.0	194.5	0.52	0.5	0.01	0.54	0.03	0.46	0.07	13.0	1.0	0	1.7
26	55	1.36	9.9	91.9	277.6	0.30	—	0.17	0.23	0.16	1.29	0.02	36.1	0.3	0.1	9.6
62	126	1.90	16.2	152.1	269.1	0.65	49.5	0.23	1.54	0.32	1.86	0.04	50.4	0	0.3	17.7
7	19	0.54	3.4	37.0	155.1	0.12	2.1	0.06	0.50	0.05	0.52	0.01	17.0	0	0	1.3
0	0	0.72	—	45.0	70.0	—	0	—	—	—	—	—	—	0	0	—
0	0	0.72	—	50.0	300.0	—	0	—	—	—	—	—	—	0	0	—
0	0	0.55	—	75.8	621.2	—	0	0.09	—	0.05	1.21	—	12.1	0	—	—
0	7	0.45	19.8	465.5	148.8	0.67	0	0.01	1.91	0.06	1.18	0.20	21.3	5.3	0	2.3
0	7	0.46	19.0	361.5	2.3	0.30	0	0.04	2.58	0.05	1.08	0.18	12.8	8.8	0	2.3
0	11	1.29	10.5	43.8	514.5	0.25	0	0.13	0.10	0.18	1.57	0.03	51.3	0	0	1.7
0	8	0.76	8.5	121.9	57.6	0.17	0	0.12	—	0.08	1.85	0.07	15.3	0.3	0	—
0	19	0.25	15.8	23.2	45.5	0.26	0	0.00	0.46	0.01	0.13	0.02	2.2	0	0	0.7
0	13	0.82	5.4	30.0	117.9	0.19	0	0.10	0.03	0.09	1.04	0.01	30.9	0	0	2.1
18	14	0.44	12.7	42.2	82.3	0.23	42.2	0.03	—	0.05	0.24	0.02	7.0	0.1	0	2.8
13	11	0.69	12.4	62.1	108.8	0.24	—	0.08	0.54	0.06	0.87	0.01	17.8	0	0	4.1
0	2	1.01	4.7	17.8	45.6	0.10	0	0.02	0.25	0.02	0.25	0.00	6.0	0	0	1.1
0	10	0.36	—	—	57.5	—	0	—	—	—	—	—	—	0	—	—
0	1	0.12	0.6	3.3	21.9	0.01	0.1	0.01	0.00	0.01	0.15	0.00	5.3	0	0	0.2
0	26	1.93	48.8	176.7	311.1	1.42	—	0.26	0.23	0.13	1.35	0.09	34.9	0.3	0	17.4
13	27	0.65	21.1	112.8	154.1	0.46	—	0.08	0.73	0.09	1.85	0.05	23.5	0.1	0.1	4.8
18	5	0.30	1.7	12.2	49.4	0.08	—	0.04	0.28	0.05	0.31	0.01	9.5	0	0	3.1
0	3	0.22	1.4	9.1	34.9	0.04	0	0.02	0.16	0.02	0.27	0.00	5.0	0	0	0.3
0	14	0.76	15.7	61.0	198.8	0.29	0.3	0.15	0.66	0.08	1.63	0.04	26.3	0	0.1	2.3
4	45	1.43	10.8	43.5	298.5	0.33	8.7	0.17	0.01	0.12	1.40	0.16	45.6	0	0.1	2.6
4	45	1.43	10.8	31.8	137.4	0.33	5.1	0.17	0.09	0.12	1.40	0.16	26.7	0	0.1	2.6
0	7	1.04	8.4	37.8	169.4	0.22	0	0.06	0.09	0.08	1.15	0.01	12.9	0	0	2.9
0	4	0.89	7.1	31.8	0.6	0.19	0	0.11	0.01	0.08	1.10	0.03	4.8	0	0	10.5
0	14	0.55	8.9	30.3	124.4	0.30	0	0.06	0.06	0.04	0.61	0.01	18.6	0	0	5.2
0	12	0.55	5.9	29.0	134.9	0.20	0	0.07	—	0.04	0.70	0.01	12.8	0	0	5.8
0	20	0.72	—	10.0	135.0	—	—	—	—	—	—	—	—	0	—	—
0	3	0.24	7.8	31.9	26.4	0.23	0	0.02	0.08	0.01	0.10	0.02	4.7	0	0	3.7
0	4	0.65	13.3	54.5	87.3	0.30	0	0.04	0.08	0.03	0.17	0.03	5.0	0	0	2.6
0	10	0.84	3.3	23.1	160.8	0.12	0	0.01	0.14	0.06	0.78	0.01	20.9	0	0	1.5
0	18	0.81	4.1	108.6	95.4	0.11	0	0.08	0.01	0.06	0.78	0.01	18.6	0	0	2.9
1	72	0.66	10.1	120.1	392.3	0.17	4.8	0.12	0.06	0.19	1.05	0.01	28.0	0	0	6.0
0	23	0.77	15.4	60.2	201.0	0.31	0.3	0.13	0.57	0.07	1.71	0.04	24.1	0	0	3.0
0	36	1.08	8.1	39.9	254.1	0.20	0	0.12	0.60	0.10	1.21	0.01	27.0	0	0	2.0
0	36	1.08	8.1	106.5	111.9	0.20	0	0.12	0.60	0.10	1.21	0.01	27.0	0	0	2.0
0	10	0.84	3.3	23.1	160.8	0.12	0	0.01	0.14	0.06	0.78	0.01	20.9	0	0	1.5
2	57	0.73	15.1	85.7	255.6	0.24	4.8	0.10	—	0.12	0.89	0.07	17.9	0.4	0	6.8
0	48	0.74	10.6	83.2	226.0	0.23	0	0.10	—	0.08	1.64	0.03	19.6	0	0	6.1
0	15	1.32	18.6	54.9	238.5	0.48	0	0.15	0.15	0.09	1.48	0.04	35.1	0	0	1.9
0	15	1.32	18.6	60.9	84.9	0.48	0	0.15	0.15	0.09	1.48	0.04	15.0	0	0	10.1
0	14	0.86	27.7	83.2	184.5	0.60	0	0.05	0.24	0.02	1.26	0.05	7.8	0	0	4.1
14	9	0.26	2.2	22.4	6.8	0.09	7.1	0.03	0.07	0.04	0.23	0.01	6.3	0.2	0.1	2.6
0	—	0.72	—	—	340.1	—	0	—	—	—	—	—	—	—	—	—
38	21	1.15	9.1	67.3	424.1	0.42	117.4	0.22	0.47	0.13	1.24	0.03	50.2	0.1	0.1	12.9
30	61	1.17	20.8	61.8	236.0	0.56	5.9	0.14	0.53	0.15	1.49	0.06	54.0	1.1	0.1	9.2
20	21	1.55	17.0	68.0	262.7	0.68	9.4	0.28	0.24	0.12	1.90	0.05	59.5	0	0.1	9.2

TABLE H-1 Table of Food Composition (*continued*) (Computer code is for Cengage Diet Analysis program) (For purposes of calculations, use "0" for t, <1, <.1, <.01, etc

DA+ Code	Food Description	Quantity	Measure	Wt (g)	H_2O (g)	Ener (kcal)	Prot (g)	Carb (g)	Fiber (g)	Fat (g)	Fat Breakdown (g) Sat	Mono	Poly	*Trans*
Breads, Baked Goods, Cakes, Cookies, Crackers, Chips, Pies—*continued*														
9117	Doughnut with jelly filling	1	item(s)	85	30.3	289	5.0	33.2	0.8	15.9	4.1	8.7	2.0	—
4947	Doughnut, cake	1	item(s)	47	9.8	198	2.4	23.4	0.7	10.8	1.7	4.4	3.7	—
9105	Doughnut, cake, chocolate glazed	1	item(s)	42	6.8	175	1.9	24.1	0.9	8.4	2.2	4.7	1.0	—
437	Doughnut, glazed	1	item(s)	60	15.2	242	3.8	26.6	0.7	13.7	3.5	7.7	1.7	—
10617	Toaster pastry, brown sugar cinnamon	1	item(s)	50	5.3	210	3.0	35.0	1.0	6.0	1.0	4.0	1.0	—
30928	Toaster pastry, cream cheese	1	item(s)	54	—	200	3.0	23.0	0	11.0	4.5	—	—	1.5
	Muffins													
25015	Blueberry	1	item(s)	63	29.7	160	3.4	23.0	0.8	6.0	0.9	1.5	3.3	—
9189	Corn, ready to eat	1	item(s)	57	18.6	174	3.4	29.0	1.9	4.8	0.8	1.2	1.8	—
9121	English muffin, plain, enriched	1	item(s)	57	24.0	134	4.4	26.2	1.5	1.0	0.1	0.2	0.5	—
29582	English muffin, toasted	1	item(s)	50	18.6	128	4.2	25.0	1.5	1.0	0.1	0.2	0.5	—
9145	English muffin, wheat	1	item(s)	57	24.1	127	5.0	25.5	2.6	1.1	0.2	0.2	0.5	—
8894	Oat bran	1	item(s)	57	20.0	154	4.0	27.5	2.6	4.2	0.6	1.0	2.4	—
	Granola bars													
38161	Kudos milk chocolate granola bars w/fruit and nuts	1	item(s)	28	—	90	2.0	15.0	1.0	3.0	1.0	—	—	—
38196	Nature Valley banana nut crunchy granola bars	2	item(s)	42	—	190	4.0	28.0	2.0	7.0	1.0	—	—	—
38187	Nature Valley fruit 'n' nut trail mix bar	1	item(s)	35	—	140	3.0	25.0	2.0	4.0	0.5	—	—	—
1383	Plain, hard	1	item(s)	25	1.0	115	2.5	15.8	1.3	4.9	0.6	1.1	3.0	—
4606	Plain, soft	1	item(s)	28	1.8	126	2.1	19.1	1.3	4.9	2.1	1.1	1.5	—
	Pies													
454	Apple pie, prepared from home recipe	1	slice(s)	155	73.3	411	3.7	57.5	2.3	19.4	4.7	8.4	5.2	—
470	Pecan pie, prepared from home recipe	1	slice(s)	122	23.8	503	6.0	63.7	—	27.1	4.9	13.6	7.0	—
33356	Pie crust mix, prepared, baked	1	slice(s)	20	2.1	100	1.3	10.1	0.4	6.1	1.5	3.5	0.8	—
9007	Pie crust, ready to bake, frozen, enriched, baked	1	slice(s)	16	1.8	82	0.7	7.9	0.2	5.2	1.7	2.5	0.6	—
472	Pumpkin pie, prepared from home recipe	1	slice(s)	155	90.7	316	7.0	40.9	—	14.4	4.9	5.7	2.8	—
	Rolls													
8555	Crescent dinner roll	1	item(s)	28	9.7	78	2.7	13.8	0.6	1.2	0.3	0.3	0.6	—
489	Hamburger roll or bun, plain	1	item(s)	43	14.9	120	4.1	21.3	0.9	1.9	0.5	0.5	0.8	—
490	Hard roll	1	item(s)	57	17.7	167	5.6	30.0	1.3	2.5	0.3	0.6	1.0	—
5127	Kaiser roll	1	item(s)	57	17.7	167	5.6	30.0	1.3	2.5	0.3	0.6	1.0	—
5130	Whole wheat roll or bun	1	item(s)	28	9.4	75	2.5	14.5	2.1	1.3	0.2	0.3	0.6	—
	Sport bars													
37026	Balance original chocolate bar	1	item(s)	50	—	200	14.0	22.0	0.5	6.0	3.5	—	—	—
37024	Balance original peanut butter bar	1	item(s)	50	—	200	14.0	22.0	1.0	6.0	2.5	—	—	—
36580	Clif Bar chocolate brownie energy bar	1	item(s)	68	—	240	10.0	45.0	5.0	4.5	1.5	—	—	0
36583	Clif Bar crunchy peanut butter energy bar	1	item(s)	68	—	250	12.0	40.0	5.0	6.0	1.5	—	—	0
36589	Clif Luna Nutz over Chocolate energy bar	1	item(s)	48	—	180	10.0	25.0	3.0	4.5	2.5	—	—	0
12005	PowerBar apple cinnamon	1	item(s)	65	—	230	9.0	45.0	3.0	2.5	0.5	1.5	0.5	0
16078	PowerBar banana	1	item(s)	65	—	230	9.0	45.0	3.0	2.5	0.5	1.0	0.5	0
16080	PowerBar chocolate	1	item(s)	65	6.4	230	10.0	45.0	3.0	2.0	0.5	0.5	1.0	0
29092	PowerBar peanut butter	1	item(s)	65	—	240	10.0	45.0	3.0	3.5	0.5	—	—	0
	Tortillas													
1391	Corn tortillas, soft	1	item(s)	26	11.9	57	1.5	11.6	1.6	0.7	0.1	0.2	0.4	—
1669	Flour tortilla	1	item(s)	32	9.7	100	2.7	16.4	1.0	2.5	0.6	1.2	0.5	—
	Pancakes, waffles													
8926	Pancakes, blueberry, prepared from recipe	3	item(s)	114	60.6	253	7.0	33.1	0.8	10.5	2.3	2.6	4.7	—
5037	Pancakes, prepared from mix with egg and milk	3	item(s)	114	60.3	249	8.9	32.9	2.1	8.8	2.3	2.4	3.3	—
1390	Taco shells, hard	1	item(s)	13	1.0	62	0.9	8.3	0.6	2.8	0.6	1.6	0.5	0.6
30311	Waffle, 100% whole grain	1	item(s)	75	32.3	200	6.9	25.0	1.9	8.4	2.3	3.3	2.1	—
9219	Waffle, plain, frozen, toasted	2	item(s)	66	20.2	206	4.7	32.5	1.6	6.3	1.1	3.2	1.5	—
500	Waffle, plain, prepared from recipe	1	item(s)	75	31.5	218	5.9	24.7	1.7	10.6	2.1	2.6	5.1	—

PAGE KEY: H-4 = Breads/Baked Goods H-10 = Cereal/Rice/Pasta H-14 = Fruit H-18 = Vegetables/Legumes H-28 = Nuts/Seeds H-30 = Vegetarian H-32 = Dairy H-40 = Eggs H-40 = Seafood H-42 = Meats H-46 = Poultry H-46 = Processed Meats H-48 = Beverages H-52 = Fats/Oils H-54 = Sweets H-56 = Spices/Condiments/Sauces H-60 = Mixed Foods/Soups/Sandwiches H-64 = Fast Food H-84 = Convenience H-86 = Baby Foods

Chol (mg)	Calc (mg)	Iron (mg)	Magn (mg)	Pota (mg)	Sodi (mg)	Zinc (mg)	Vit A (μg)	Thia (mg)	Vit E (mg α)	Ribo (mg)	Niac (mg)	Vit B_6 (mg)	Fola (μg)	Vit C (mg)	Vit B_{12} (μg)	Sele (μg)
22	21	1.49	17.0	67.2	249.1	0.63	14.5	0.26	0.36	0.12	1.81	0.08	57.8	0	0.2	10.6
17	21	0.91	9.4	59.7	256.6	0.25	17.9	0.10	0.90	0.11	0.87	0.02	24.4	0.1	0.1	4.4
24	89	0.95	14.3	44.5	142.8	0.23	5.0	0.01	0.08	0.02	0.19	0.01	18.9	0	0	1.7
4	26	0.36	13.2	64.8	205.2	0.46	2.4	0.53	—	0.04	0.39	0.03	13.2	0.1	0.1	5.0
0	0	1.80	—	70.0	190.0	—	—	0.15	—	0.17	2.00	0.20	40.0	0	0	—
10	100	1.80	—	—	220.0	—	—	0.15	—	0.17	2.00	—	40.0	0	0.6	—
20	56	1.02	7.8	70.2	289.4	0.28	—	0.17	0.75	0.15	1.25	0.02	34.0	0.4	0.1	8.8
15	42	1.60	18.2	39.3	297.0	0.30	29.6	0.15	0.45	0.18	1.16	0.04	45.6	0	0.1	8.7
0	30	1.42	12.0	74.7	264.5	0.39	0	0.25	—	0.16	2.21	0.02	42.2	0	0	—
0	95	1.36	11.0	71.5	252.0	0.38	0	0.19	0.16	0.14	1.90	0.02	43.5	0.1	0	13.5
0	101	1.63	21.1	106.0	217.7	0.61	0	0.24	0.25	0.16	1.91	0.05	36.5	0	0	16.6
0	36	2.39	89.5	289.0	224.0	1.04	0	0.14	0.37	0.05	0.23	0.09	50.7	0	0	6.3
0	200	0.36	—	—	60.0	—	0	—	—	—	—	—	—	0	0	—
0	20	1.08	—	120.0	160.0	—	0	—	—	—	—	—	—	0	—	—
0	0	0.00	—	—	95.0	—	0	—	—	—	—	—	—	0	—	—
0	15	0.72	23.8	82.3	72.0	0.50	0	0.06	—	0.03	0.39	0.02	5.6	0.2	0	4.0
0	30	0.72	21.0	92.3	79.0	0.42	0	0.08	—	0.04	0.14	0.02	6.8	0	0.1	4.6
0	11	1.73	10.9	122.5	327.1	0.29	17.1	0.22	—	0.16	1.90	0.05	37.2	2.6	0	12.1
106	39	1.80	31.7	162.3	319.6	1.24	100.0	0.22	—	0.22	1.03	0.07	31.7	0.2	0.2	14.6
0	12	0.43	3.0	12.4	145.8	0.07	0	0.06	—	0.03	0.47	0.01	14.0	0	0	4.4
0	3	0.36	2.9	17.6	103.5	0.05	0	0.04	0.42	0.06	0.39	0.01	8.8	0	0	0.5
65	146	1.96	29.5	288.3	348.8	0.71	660.3	0.14	—	0.31	1.21	0.07	32.6	2.6	0.1	11.0
0	39	0.93	5.9	26.3	134.1	0.18	0	0.11	0.02	0.09	1.16	0.02	31.1	0	0.1	5.5
0	59	1.42	9.0	40.4	206.0	0.28	0	0.17	0.03	0.13	1.78	0.03	47.7	0	0.1	8.4
0	54	1.87	15.4	61.6	310.1	0.53	0	0.27	0.23	0.19	2.41	0.02	54.2	0	0	22.3
0	54	1.86	15.4	61.6	310.1	0.53	0	0.27	0.23	0.19	2.41	0.01	54.2	0	0	22.3
0	30	0.69	24.1	77.1	135.5	0.57	0	0.07	0.26	0.04	1.04	0.06	8.5	0	0	14.0
3	100	4.50	40.0	160.0	180.0	3.75	—	0.37	—	0.42	5.00	0.50	100.0	60.0	1.5	17.5
3	100	4.50	40.0	130.0	230.0	3.75	—	0.37	—	0.42	5.00	0.50	100.0	60.0	1.5	17.5
0	250	4.50	100.0	370.0	150.0	3.00	—	0.37	—	0.25	3.00	0.40	80.0	60.0	0.9	14.0
0	250	4.50	100.0	230.0	250.0	3.00	—	0.37	—	0.25	3.00	0.40	80.0	60.0	0.9	14.0
0	350	5.40	80.0	190.0	190.0	5.25	—	1.20	—	1.36	16.00	2.00	400.0	60.0	6.0	24.5
0	300	6.30	140.0	125.0	100.0	5.25	—	1.50	—	1.70	20.00	2.00	400.0	60.0	6.0	—
0	300	6.30	140.0	190.0	100.0	5.25	0	1.50	—	1.70	20.00	2.00	400.0	60.0	6.0	—
0	300	6.30	140.0	200.0	95.0	5.25	0	1.50	—	1.70	20.00	2.00	400.0	60.0	6.0	5.1
0	300	6.30	140.0	130.0	120.0	5.25	0	1.50	—	1.70	20.00	2.00	400.0	60.0	6.0	—
0	21	0.32	18.7	48.4	11.7	0.34	0	0.02	0.07	0.02	0.39	0.06	1.3	0	0	1.6
0	41	1.06	7.0	49.6	203.5	0.17	0	0.17	0.06	0.08	1.14	0.01	33.3	0	0	7.1
64	235	1.96	18.2	157.3	469.7	0.61	57.0	0.22	—	0.31	1.73	0.05	41.0	2.5	0.2	16.0
81	245	1.48	25.1	226.9	575.7	0.85	82.1	0.22	—	0.35	1.40	0.12	104.6	0.7	0.4	—
0	13	0.25	11.3	29.7	51.7	0.21	0.1	0.03	0.09	0.01	0.25	0.03	9.2	0	0	0.6
71	194	1.60	28.5	171.0	371.3	0.87	48.8	0.15	0.32	0.25	1.47	0.08	28.5	0	0.4	20.0
10	203	4.56	15.8	95.0	481.8	0.35	262.7	0.34	0.64	0.46	5.86	0.68	49.5	0	1.9	8.3
52	191	1.73	14.3	119.3	383.3	0.51	48.8	0.19	—	0.26	1.55	0.04	34.5	0.3	0.2	34.7

TABLE H-1 Table of Food Composition (*continued*) (Computer code is for Cengage Diet Analysis program) (For purposes of calculations, use "0" for t, <1, <.1, <.01, etc.)

DA+ Code	Food Description	Quantity	Measure	Wt (g)	H_2O (g)	Ener (kcal)	Prot (g)	Carb (g)	Fiber (g)	Fat (g)	Fat Breakdown (g) Sat	Mono	Poly	*Trans*
	Cereal, Flour, Grain, Pasta, Noodles, Popcorn													
	Grain													
2861	Amaranth, dry	½	cup(s)	98	9.6	365	14.1	64.5	9.1	6.3	1.6	1.4	2.8	—
1953	Barley, pearled, cooked	½	cup(s)	79	54.0	97	1.8	22.2	3.0	0.3	0.1	0	0.2	—
1956	Buckwheat groats, cooked, roasted	½	cup(s)	84	63.5	77	2.8	16.8	2.3	0.5	0.1	0.2	0.2	—
1957	Bulgur, cooked	½	cup(s)	91	70.8	76	2.8	16.9	4.1	0.2	0	0	0.1	—
1963	Couscous, cooked	½	cup(s)	79	57.0	88	3.0	18.2	1.1	0.1	0	0	0.1	—
1967	Millet, cooked	½	cup(s)	120	85.7	143	4.2	28.4	1.6	1.2	0.2	0.2	0.6	—
1969	Oat bran, dry	½	cup(s)	47	3.1	116	8.1	31.1	7.2	3.3	0.6	1.1	1.3	—
1972	Quinoa, dry	½	cup(s)	85	11.3	313	12.0	54.5	5.9	5.2	0.6	1.4	2.8	—
	Rice													
129	Brown, long grain, cooked	½	cup(s)	98	71.3	108	2.5	22.4	1.8	0.9	0.2	0.3	0.3	—
2863	Brown, medium grain, cooked	½	cup(s)	98	71.1	109	2.3	22.9	1.8	0.8	0.2	0.3	0.3	—
37488	Jasmine, saffroned, cooked	½	cup(s)	280	—	340	8.0	78.0	0	0	0	0	0	0
30280	Pilaf, cooked	½	cup(s)	103	74.0	129	2.1	22.2	0.6	3.3	0.6	1.5	1.0	—
28066	Spanish, cooked	½	cup(s)	244	184.2	241	5.7	50.2	3.3	1.9	0.4	0.6	0.7	0
2867	White glutinous, cooked	½	cup(s)	87	66.7	84	1.8	18.3	0.9	0.2	0	0.1	0.1	—
484	White, long grain, boiled	½	cup(s)	79	54.1	103	2.1	22.3	0.3	0.2	0.1	0.1	0.1	—
482	White, long grain, enriched, instant, boiled	½	cup(s)	83	59.4	97	1.8	20.7	0.5	0.4	0	0.1	0	—
486	White, long grain, enriched, parboiled, cooked	½	cup(s)	79	55.6	97	2.3	20.6	0.7	0.3	0.1	0.1	0.1	—
1194	Wild brown, cooked	½	cup(s)	82	60.6	83	3.3	17.5	1.5	0.3	0	0	0.2	—
	Flour and grain fractions													
505	All purpose flour, self-rising, enriched	½	cup(s)	63	6.6	221	6.2	46.4	1.7	0.6	0.1	0	0.2	—
503	All purpose flour, white, bleached, enriched	½	cup(s)	63	7.4	228	6.4	47.7	1.7	0.6	0.1	0	0.2	—
1643	Barley flour	½	cup(s)	56	5.5	198	4.2	44.7	2.1	0.8	0.2	0.1	0.4	—
383	Buckwheat flour, whole groat	½	cup(s)	60	6.7	201	7.6	42.3	6.0	1.9	0.4	0.6	0.6	—
504	Cake wheat flour, enriched	½	cup(s)	69	8.6	248	5.6	53.5	1.2	0.6	0.1	0.1	0.3	—
426	Cornmeal, degermed, enriched	½	cup(s)	69	7.8	255	5.0	54.6	2.8	1.2	0.1	0.2	0.5	0
424	Cornmeal, yellow whole grain	½	cup(s)	61	6.2	221	4.9	46.9	4.4	2.2	0.3	0.6	1.0	—
1978	Dark rye flour	½	cup(s)	64	7.1	207	9.0	44.0	14.5	1.7	0.2	0.2	0.8	—
1644	Masa corn flour, enriched	½	cup(s)	57	5.1	208	5.3	43.5	5.5	2.1	0.3	0.6	1.0	—
1976	Rice flour, brown	½	cup(s)	79	9.4	287	5.7	60.4	3.6	2.2	0.4	0.8	0.8	—
1645	Rice flour, white	½	cup(s)	79	9.4	289	4.7	63.3	1.9	1.1	0.3	0.3	0.3	—
1980	Semolina, enriched	½	cup(s)	84	10.6	301	10.6	60.8	3.2	0.9	0.1	0.1	0.4	—
2827	Soy flour, raw	½	cup(s)	42	2.2	185	14.7	14.9	4.1	8.8	1.3	1.9	4.9	—
1990	Wheat germ, crude	2	tablespoon(s)	14	1.6	52	3.3	7.4	1.9	1.4	0.2	0.2	0.9	—
506	Whole wheat flour	½	cup(s)	60	6.2	203	8.2	43.5	7.3	1.1	0.2	0.1	0.5	—
	Breakfast bars													
39230	Atkins Morning Start apple crisp breakfast bar	1	item(s)	37	—	170	11.0	12.0	6.0	9.0	4.0	—	—	—
10571	Nutri-Grain apple cinnamon cereal bar	1	item(s)	37	—	140	2.0	27.0	1.0	3.0	0.5	2.0	0.5	—
10647	Nutri-Grain blueberry cereal bar	1	item(s)	37	5.4	140	2.0	27.0	1.0	3.0	0.5	2.0	0.5	—
10648	Nutri-Grain raspberry cereal bar	1	item(s)	37	5.4	140	2.0	27.0	1.0	3.0	0.5	2.0	0.5	—
10649	Nutri-Grain strawberry cereal bar	1	item(s)	37	5.4	140	2.0	27.0	1.0	3.0	0.5	2.0	0.5	—
	Breakfast cereals, hot													
1260	Cream of Wheat, instant, prepared	½	cup(s)	121	—	388	12.9	73.3	4.3	0	0	0	0	0
365	Farina, enriched, cooked w/water and salt	½	cup(s)	117	102.4	56	1.7	12.2	0.3	0.1	0	0	0	—
363	Grits, white corn, regular and quick, enriched, cooked w/water and salt	½	cup(s)	121	103.3	71	1.7	15.6	0.4	0.2	0	0.1	0.1	—
8636	Grits, yellow corn, regular and quick, enriched, cooked w/salt	½	cup(s)	121	103.3	71	1.7	15.6	0.4	0.2	0	0.1	0.1	—
8657	Oatmeal, cooked w/water	½	cup(s)	117	97.8	83	3.0	14.0	2.0	1.8	0.4	0.5	0.7	0
5500	Oatmeal, maple and brown sugar, instant, prepared	1	item(s)	198	150.2	200	4.8	40.4	2.4	2.2	0.4	0.7	0.8	—
5510	Oatmeal, ready to serve, packet, prepared	1	item(s)	186	158.7	112	4.1	19.8	2.7	2.0	0.4	0.7	0.8	—
	Breakfast cereals, ready to eat													
1197	All-Bran	1	cup(s)	62	1.3	160	8.1	46.0	18.2	2.0	0.4	0.4	1.3	0
1200	All-Bran Buds	1	cup(s)	91	2.7	212	6.4	72.7	39.1	1.9	0.4	0.5	1.2	0

PAGE KEY: H-4 = Breads/Baked Goods H-10 = Cereal/Rice/Pasta H-14 = Fruit H-18 = Vegetables/Legumes H-28 = Nuts/Seeds H-30 = Vegetarian H-32 = Dairy H-40 = Eggs H-40 = Seafood H-42 = Meats H-46 = Poultry H-46 = Processed Meats H-48 = Beverages H-52 = Fats/Oils H-54 = Sweets H-56 = Spices/Condiments/Sauces H-60 = Mixed Foods/Soups/Sandwiches H-64 = Fast Food H-84 = Convenience H-86 = Baby Foods

Chol (mg)	Calc (mg)	Iron (mg)	Magn (mg)	Pota (mg)	Sodi (mg)	Zinc (mg)	Vit A (µg)	Thia (mg)	Vit E (mg α)	Ribo (mg)	Niac (mg)	Vit B_6 (mg)	Fola (µg)	Vit C (mg)	Vit B_{12} (µg)	Sele (µg)
0	149	7.40	259.3	356.8	20.5	3.10	0	0.06	—	0.20	1.24	0.20	47.8	4.1	0	—
0	9	1.04	17.3	73.0	2.4	0.64	0	0.06	0.01	0.04	1.61	0.09	12.6	0	0	6.8
0	6	0.67	42.8	73.9	3.4	0.51	0	0.03	0.07	0.03	0.79	0.06	11.8	0	0	1.8
0	9	0.87	29.1	61.9	4.6	0.51	0	0.05	0.01	0.02	0.91	0.07	16.4	0	0	0.5
0	6	0.30	6.3	45.5	3.9	0.20	0	0.05	0.10	0.02	0.77	0.04	11.8	0	0	21.6
0	4	0.75	52.8	74.4	2.4	1.09	0	0.12	0.02	0.09	1.59	0.13	22.8	0	0	1.1
0	27	2.54	110.5	266.0	1.9	1.46	0	0.55	0.47	0.10	0.43	0.07	24.4	0	0	21.2
0	40	3.88	167.4	478.5	4.2	2.62	0.8	0.30	2.06	0.26	1.28	0.40	156.4	0	0	7.2
0	10	0.41	41.9	41.9	4.9	0.61	0	0.09	0.02	0.02	1.49	0.14	3.9	0	0	9.6
0	10	0.51	42.9	77.0	1.0	0.60	0	0.09	—	0.01	1.29	0.14	3.9	0	0	38.0
0	—	2.16	—	—	780.0	—	—	—	—	—	—	—	—	—	—	—
0	11	1.16	9.3	54.6	390.4	0.37	33.0	0.13	0.28	0.02	1.23	0.06	44.3	0.4	0	4.3
0	37	1.52	95.4	330.5	97.1	1.40	—	0.27	0.12	0.05	3.24	0.38	19.0	22.6	0	14.3
0	2	0.12	4.4	8.7	4.4	0.35	0	0.01	0.03	0.01	0.25	0.02	0.9	0	0	4.9
0	8	0.94	9.5	27.7	0.8	0.38	0	0.12	0.03	0.01	1.16	0.07	45.8	0	0	5.9
0	7	1.46	4.1	7.4	3.3	0.40	0	0.06	0.01	0.01	1.43	0.04	57.8	0	0	4.0
0	15	1.43	7.1	44.2	1.6	0.29	0	0.16	0.01	0.01	1.82	0.12	64.0	0	0	7.3
0	2	0.49	26.2	82.8	2.5	1.09	0	0.04	0.19	0.07	1.05	0.11	21.3	0	0	0.7
0	211	2.90	11.9	77.5	793.7	0.38	0	0.42	0.02	0.24	3.64	0.02	122.5	0	0	21.5
0	9	2.90	13.7	66.9	1.2	0.42	0	0.48	0.02	0.30	3.68	0.02	114.4	0	0	21.2
0	16	0.70	45.4	185.9	4.5	1.04	0	0.06	—	0.02	2.56	0.14	12.9	0	0	2.0
0	25	2.42	150.6	346.2	6.6	1.86	0	0.24	0.18	0.10	3.68	0.34	32.4	0	0	3.4
0	10	5.01	11.0	71.9	1.4	0.42	0	0.61	0.01	0.29	4.65	0.02	127.4	0	0	3.4
0	2	2.98	24.1	104.9	4.8	0.48	7.6	0.42	0.10	0.28	3.66	0.12	148.3	0	0	8.0
0	4	2.10	77.5	175.1	21.3	1.10	6.7	0.22	0.24	0.12	2.20	0.18	15.2	0	0	9.4
0	36	4.12	158.7	467.2	0.6	3.58	0.6	0.20	0.90	0.16	2.72	0.28	38.4	0	0	22.8
0	80	4.10	62.7	169.9	2.8	1.00	0	0.80	0.08	0.42	5.60	0.20	132.8	0	0	8.5
0	9	1.56	88.5	228.3	6.3	1.92	0	0.34	0.94	0.06	5.00	0.58	12.6	0	0	—
0	8	0.26	27.6	60.0	0	0.62	0	0.10	0.08	0.02	2.04	0.34	3.2	0	0	11.9
0	14	3.64	39.2	155.3	0.8	0.86	0	0.66	0.20	0.46	5.00	0.08	152.8	0	0	74.6
0	87	2.70	182.0	1067.0	5.5	1.65	2.5	0.24	0.82	0.48	1.83	0.18	146.4	0	0	3.2
0	6	0.90	34.4	128.2	1.7	1.76	0	0.27	—	0.07	0.97	0.18	40.4	0	0	11.4
0	20	2.32	82.8	243.0	3.0	1.74	0	0.26	0.48	0.12	3.82	0.20	26.4	0	0	42.4
0	200	—	—	90.0	70.0	—	—	0.22	—	0.25	3.00	—	—	9.0	—	—
0	200	1.80	8.0	75.0	110.0	1.50	—	0.37	—	0.42	5.00	0.50	40.0	0	—	—
0	200	1.80	8.0	75.0	110.0	1.50	—	0.37	—	0.42	5.00	0.50	40.0	0	0	—
0	200	1.80	8.0	70.0	110.0	1.50	—	0.37	—	0.42	5.00	0.50	40.0	0	0	—
0	200	1.80	8.0	55.0	110.0	1.50	—	0.37	—	0.42	5.00	0.50	40.0	0	0	—
0	862	34.91	21.4	150.8	732.6	0.86	—	1.59	—	1.47	21.55	2.15	431.0	0	0	—
0	5	0.58	2.3	15.1	383.3	0.09	0	0.07	0.01	0.05	0.57	0.01	39.6	0	0	10.6
0	4	0.73	6.1	25.4	269.8	0.08	0	0.10	0.02	0.07	0.87	0.03	39.9	0	0	3.8
0	4	0.73	6.1	25.4	269.8	0.08	2.4	0.10	0.02	0.07	0.87	0.03	39.9	0	0	3.3
0	11	1.05	31.6	81.9	4.7	1.17	0	0.09	0.09	0.02	0.26	0.01	7.0	0	0	6.3
0	26	6.83	49.9	126.4	403.5	1.03	0	1.02	—	0.05	1.56	0.30	42.2	0	0	11.1
0	21	3.96	44.7	112.4	240.9	0.92	0	0.60	—	0.04	0.77	0.18	18.7	0	0	3.8
0	241	10.90	224.4	632.4	150.0	3.00	300.1	1.40	—	1.68	9.16	7.44	800.0	12.4	12.0	5.8
0	57	13.64	186.4	909.1	614.5	4.55	464.5	1.09	1.42	1.27	15.45	6.09	1222.7	18.2	18.2	26.3

TABLE H-1 Table of Food Composition (*continued*) (Computer code is for Cengage Diet Analysis program) (For purposes of calculations, use "0" for t, <1, <.1, <.01, etc.)

DA+ Code	Food Description	Quantity	Measure	Wt (g)	H_2O (g)	Ener (kcal)	Prot (g)	Carb (g)	Fiber (g)	Fat (g)	Fat Breakdown (g)			
											Sat	Mono	Poly	*Trans*
Cereal, Flour, Grain, Pasta, Noodles, Popcorn—*continued*														
1199	Apple Jacks	1	cup(s)	33	0.9	130	1.0	30.0	0.5	0.5	0	—	—	0
1204	Cap'n Crunch	1	cup(s)	36	0.9	147	1.3	30.7	1.3	2.0	0.5	0.4	0.3	—
1205	Cap'n Crunch Crunchberries	1	cup(s)	35	0.9	133	1.3	29.3	1.3	2.0	0.5	0.4	0.3	—
1206	Cheerios	1	cup(s)	30	1.0	110	3.0	22.0	3.0	2.0	0	0.5	0.5	—
3415	Cocoa Puffs	1	cup(s)	30	0.6	120	1.0	26.0	0.2	1.0	—	—	—	—
1207	Cocoa Rice Krispies	1	cup(s)	41	1.0	160	1.3	36.0	1.3	1.3	0.7	0	0	—
5522	Complete wheat bran flakes	1	cup(s)	39	1.4	120	4.0	30.7	6.7	0.7	—	—	—	0
1211	Corn Flakes	1	cup(s)	28	0.9	100	2.0	24.0	1.0	0	0	0	0	0
1247	Corn Pops	1	cup(s)	31	0.9	120	1.0	28.0	0.3	0	0	0	0	0
1937	Cracklin' Oat Bran	1	cup(s)	65	2.3	267	5.3	46.7	8.0	9.3	4.0	4.7	1.3	0
1220	Froot Loops	1	cup(s)	32	0.8	120	1.0	28.0	1.0	1.0	0.5	0	0	—
38214	Frosted Cheerios	1	cup(s)	37	—	149	2.5	31.1	1.2	1.2	—	—	—	—
372	Frosted Flakes	1	cup(s)	41	1.1	160	1.3	37.3	1.3	0	0	0	0	0
38215	Frosted Mini Chex	1	cup(s)	40	—	147	1.3	36.0	0	0	0	0	0	0
10268	Frosted Mini-Wheats	1	cup(s)	59	3.1	208	5.8	47.4	5.8	1.2	0	0	0.6	0
38216	Frosted Wheaties	1	cup(s)	40	—	147	1.3	36.0	0.3	0	0	0	0	0
1223	Granola, prepared	½	cup(s)	61	3.3	298	9.1	32.5	5.5	14.7	2.5	5.8	5.6	0
2415	Honey Bunches of Oats honey roasted	1	cup(s)	40	0.9	160	2.7	33.3	1.3	2.0	0.7	1.2	0.1	—
1227	Honey Nut Cheerios	1	cup(s)	37	0.9	149	3.7	29.9	2.5	1.9	0	0.6	0.6	—
2424	Honeycomb	1	cup(s)	22	0.3	83	1.5	19.5	0.8	0.4	0	—	—	—
10286	Kashi whole grain puffs	1	cup(s)	19	—	70	2.0	15.0	1.0	0.5	0	—	—	0
41142	Kellogg's Mueslix	1	cup(s)	83	7.2	298	7.6	60.8	6.1	4.6	0.7	2.4	1.5	0
1231	Kix	1	cup(s)	24	0.5	96	1.6	20.8	0.8	0.4	—	—	—	—
30569	Life	1	cup(s)	43	1.7	160	4.0	33.3	2.7	2.0	0.3	0.6	0.6	—
1233	Lucky Charms	1	cup(s)	24	0.6	96	1.6	20.0	0.8	0.8	—	—	—	—
38220	Multi Grain Cheerios	1	cup(s)	30	—	110	3.0	24.0	3.0	1.0	—	—	—	—
1201	Multi-Bran Chex	1	cup(s)	63	1.3	216	4.3	52.9	8.6	1.6	0	0	0.5	0
13633	Post Bran Flakes	1	cup(s)	40	1.5	133	4.0	32.0	6.7	0.7	0	—	—	—
1241	Product 19	1	cup(s)	30	1.0	100	2.0	25.0	1.0	0	0	0	0	0
32432	Puffed rice, fortified	1	cup(s)	14	0.4	56	0.9	12.6	0.2	0.1	0	—	—	—
32433	Puffed wheat, fortified	1	cup(s)	12	0.4	44	1.8	9.6	0.5	0.1	0	—	—	—
13334	Quaker 100% natural granola oats and honey	½	cup(s)	48	—	220	5.0	31.0	3.0	9.0	3.8	4.1	1.2	—
13335	Quaker 100% natural granola oats, honey, and raisins	½	cup(s)	51	—	230	5.0	34.0	3.0	9.0	3.6	3.8	1.1	—
2420	Raisin Bran	1	cup(s)	59	5.0	190	4.0	46.0	8.0	1.0	0	0.1	0.4	—
1244	Rice Chex	1	cup(s)	31	0.8	120	2.0	27.0	0.3	0	0	0	0	0
1245	Rice Krispies	1	cup(s)	26	0.8	96	1.6	23.2	0	0	0	0	0	0
5593	Shredded Wheat	1	cup(s)	49	0.4	177	5.8	40.9	6.9	1.1	0.1	0	0.2	0
1248	Smacks	1	cup(s)	36	1.1	133	2.7	32.0	1.3	0.7	—	—	—	—
1246	Special K	1	cup(s)	31	0.9	110	7.0	22.0	0.5	0	0	0	0	0
3428	Total corn flakes	1	cup(s)	23	0.6	83	1.5	18.0	0.6	0	0	0	0	0
1253	Total whole grain	1	cup(s)	40	1.1	147	2.7	30.7	4.0	1.3	—	—	—	—
1254	Trix	1	cup(s)	30	0.6	120	1.0	27.0	1.0	1.0	—	—	—	—
382	Wheat germ, toasted	2	tablespoon(s)	14	0.8	54	4.1	7.0	2.1	1.5	0.3	0.2	0.9	—
1257	Wheaties	1	cup(s)	36	1.2	132	3.6	28.8	3.6	1.2	—	—	—	—
	Pasta, noodles													
449	Chinese chow mein noodles, cooked	½	cup(s)	23	0.2	119	1.9	12.9	0.9	6.9	1.0	1.7	3.9	—
1995	Corn pasta, cooked	½	cup(s)	70	47.8	88	1.8	19.5	3.4	0.5	0.1	0.1	0.2	—
448	Egg noodles, enriched, cooked	½	cup(s)	80	54.2	110	3.6	20.1	1.0	1.7	0.3	0.5	0.4	0
1563	Egg noodles, spinach, enriched, cooked	½	cup(s)	80	54.8	106	4.0	19.4	1.8	1.3	0.3	0.4	0.3	—
440	Macaroni, enriched, cooked	½	cup(s)	70	43.5	111	4.1	21.6	1.3	0.7	0.1	0.1	0.2	0
2000	Macaroni, tricolor vegetable, enriched, cooked	½	cup(s)	67	45.8	86	3.0	17.8	2.9	0.1	0	0	0	—
1996	Plain pasta, fresh-refrigerated, cooked	½	cup(s)	64	43.9	84	3.3	16.0	—	0.7	0.1	0.1	0.3	—
1725	Ramen noodles, cooked	½	cup(s)	114	94.5	104	3.0	15.4	1.0	4.3	0.2	0.2	0.2	—
2878	Soba noodles, cooked	½	cup(s)	95	69.4	94	4.8	20.4	—	0.1	0	0	0	—
2879	Somen noodles, cooked	½	cup(s)	88	59.8	115	3.5	24.2	—	0.2	0	0	0.1	—
493	Spaghetti, al dente, cooked	½	cup(s)	65	41.6	95	3.5	19.5	1.0	0.5	0.1	0.1	0.2	—
2884	Spaghetti, whole wheat, cooked	½	cup(s)	70	47.0	87	3.7	18.6	3.2	0.4	0.1	0.1	0.1	—
	Popcorn													
476	Air popped	1	cup(s)	8	0.3	31	1.0	6.2	1.2	0.4	0	0.1	0.2	—
4619	Caramel	1	cup(s)	35	1.0	152	1.3	27.8	1.8	4.5	1.3	1.0	1.6	—

PAGE KEY: H-4 = Breads/Baked Goods H-10 = Cereal/Rice/Pasta H-14 = Fruit H-18 = Vegetables/Legumes H-28 = Nuts/Seeds H-30 = Vegetarian H-32 = Dairy H-40 = Eggs H-40 = Seafood H-42 = Meats H-46 = Poultry H-46 = Processed Meats H-48 = Beverages H-52 = Fats/Oils H-54 = Sweets H-56 = Spices/Condiments/Sauces H-60 = Mixed Foods/Soups/Sandwiches H-64 = Fast Food H-84 = Convenience H-86 = Baby Foods

Chol (mg)	Calc (mg)	Iron (mg)	Magn (mg)	Pota (mg)	Sodi (mg)	Zinc (mg)	Vit A (µg)	Thia (mg)	Vit E (mg α)	Ribo (mg)	Niac (mg)	Vit B_6 (mg)	Fola (µg)	Vit C (mg)	Vit B_{12} (µg)	Sele (µg)
0	0	4.50	8.0	30.0	130.0	1.50	150.2	0.37	—	0.42	5.00	0.50	100.0	15.0	1.5	2.4
0	5	6.80	20.0	73.3	266.7	5.00	2.5	0.51	—	0.57	6.68	0.67	133.5	0	0	6.7
0	7	6.53	18.7	73.3	240.0	5.13	2.4	0.51	—	0.57	6.68	0.67	133.7	0	0	6.7
0	100	8.10	40.0	95.0	280.0	3.75	150.3	0.37	—	0.42	5.00	0.50	200.0	6.0	1.5	11.3
0	100	4.50	8.0	50.0	170.0	3.75	0	0.37	—	0.42	5.00	0.50	100.0	6.0	1.5	2.0
0	53	6.00	10.7	66.7	253.3	2.00	200.1	0.49	—	0.56	6.67	0.67	133.3	20.0	2.0	5.8
0	0	24.00	53.3	226.7	280.0	20.00	300.1	2.00	—	2.27	26.67	2.67	533.3	80.0	8.0	4.1
0	0	8.10	3.4	25.0	200.0	0.16	149.8	0.37	—	0.42	5.00	0.50	100.0	6.0	1.5	1.4
0	0	1.80	2.5	25.0	120.0	1.50	150.0	0.37	—	0.42	5.00	0.50	100.0	6.0	1.5	2.0
0	27	2.40	80.0	293.3	200.0	2.00	299.9	0.49	—	0.56	6.67	0.67	133.3	20.0	2.0	14.4
0	0	4.50	8.0	35.0	150.0	1.50	150.1	0.37	—	0.42	5.00	0.50	100.0	15.0	1.5	2.3
0	124	5.60	19.9	68.4	261.3	4.67	—	0.46	—	0.52	6.22	0.62	124.4	7.5	1.9	—
0	0	6.00	3.7	26.7	200.0	0.20	200.1	0.49	—	0.56	6.67	0.67	133.3	8.0	2.0	1.8
0	133	12.00	—	33.3	266.7	4.00	—	0.49	—	0.56	6.67	0.67	266.7	8.0	2.0	—
0	0	16.66	69.4	196.7	5.8	1.74	0	0.43	—	0.49	5.78	0.58	115.7	0	1.7	2.4
0	133	10.80	0	46.7	266.7	10.00	—	1.00	—	1.13	13.33	1.33	533.3	8.0	4.0	—
0	48	2.58	106.8	329.4	15.3	2.45	0.6	0.44	6.77	0.17	1.30	0.17	50.0	0.7	0	17.0
0	0	10.80	21.3	0	253.3	0.40	—	0.49	—	0.56	6.67	0.67	133.0	0	2.0	—
0	124	5.60	39.8	112.0	336.0	4.67	—	0.46	—	0.52	6.22	0.62	248.9	7.5	1.9	8.8
0	0	2.03	6.0	26.3	165.4	1.13	—	0.28	—	0.32	3.74	0.37	75.0	0	1.1	—
0	0	0.36	—	60.0	0	—	0	0.03	—	0.03	0.80	0.00	—	0	—	—
0	48	6.83	74.2	363.3	257.5	5.67	136.7	0.67	6.00	0.67	8.33	3.08	615.0	0.3	9.2	14.4
0	120	6.48	6.4	28.0	216.0	3.00	120.2	0.30	—	0.34	4.00	0.40	160.0	4.8	1.2	4.8
0	149	11.87	41.3	120.0	213.3	5.33	0.9	0.53	—	0.60	7.12	0.71	142.4	0	0	10.7
0	80	3.60	12.8	48.0	168.0	3.00	—	0.30	—	0.34	4.00	0.40	160.0	4.8	1.2	4.8
0	100	18.00	24.0	85.0	200.0	15.00	—	1.50	—	1.70	20.00	2.00	400.0	15.0	6.0	—
0	108	17.50	64.8	237.7	410.6	4.05	171.1	0.40	—	0.45	5.40	0.54	432.2	6.5	1.6	4.9
0	0	10.80	80.0	266.7	280.0	2.00	—	0.49	—	0.56	6.67	0.67	133.3	0	2.0	—
0	0	18.00	16.0	50.0	210.0	15.00	225.3	1.50	—	1.70	20.00	2.00	400.0	60.0	6.0	3.6
0	1	4.43	3.5	15.8	0.4	0.14	0	0.36	—	0.25	4.94	0.01	2.7	0	0	1.5
0	3	3.80	17.4	41.8	0.5	0.28	0	0.31	—	0.21	4.23	0.02	3.8	0	0	14.8
0	61	1.20	51.0	220.0	20.0	1.05	0.5	0.13	—	0.12	0.82	0.07	15.0	0.2	0.1	8.3
0	59	1.20	49.0	250.0	20.0	0.99	0.5	0.13	—	0.12	0.80	0.08	14.1	0.4	0.1	8.8
0	20	10.80	80.0	360.0	360.0	2.25	—	0.37	—	0.42	5.00	0.50	100.0	0	2.1	—
0	100	9.00	9.3	35.0	290.0	3.75	—	0.37	—	0.42	5.00	0.50	200.0	6.0	1.5	1.2
0	0	1.44	12.8	32.0	256.0	0.48	120.1	0.30	—	0.34	4.80	0.40	80.0	4.8	1.2	4.1
0	18	2.90	60.3	179.3	1.1	1.37	0	0.14	—	0.12	3.47	0.18	21.1	0	0	2.0
0	0	0.48	10.7	53.3	66.7	0.40	200.2	0.49	—	0.56	6.67	0.67	133.3	8.0	2.0	17.5
0	0	8.10	16.0	60.0	220.0	0.90	225.1	0.52	—	0.59	7.00	2.00	400.0	21.0	6.0	7.0
0	752	13.53	0	22.6	157.9	11.28	112.8	1.13	22.56	1.28	15.04	1.50	300.8	45.1	4.5	1.2
0	1333	24.00	32.0	120.0	253.3	20.00	200.4	2.00	31.32	2.27	26.67	2.67	533.3	80.0	8.0	1.9
0	100	4.50	0	15.0	190.0	3.75	150.3	0.37	—	0.42	5.00	0.50	100.0	6.0	1.5	6.0
0	6	1.28	45.2	133.8	0.6	2.35	0.7	0.23	2.25	0.11	0.79	0.13	49.7	0.8	0	9.2
0	24	9.72	38.4	126.0	264.0	9.00	180.4	0.90	—	1.02	12.00	1.20	240.0	7.2	3.6	1.7
0	5	1.06	11.7	27.0	98.8	0.31	0	0.13	0.78	0.09	1.33	0.02	20.3	0	0	9.7
0	1	0.18	25.2	21.7	0	0.44	2.1	0.04	—	0.02	0.39	0.04	4.2	0	0	2.0
23	10	1.17	16.8	30.4	4.0	0.52	4.8	0.23	0.13	0.11	1.66	0.03	67.2	0	0.1	19.1
26	15	0.87	19.2	29.6	9.6	0.50	8.0	0.19	0.46	0.09	1.17	0.09	51.2	0	0.1	17.4
0	5	0.90	12.6	30.8	0.7	0.36	0	0.19	0.04	0.10	1.18	0.03	51.1	0	0	18.5
0	7	0.33	12.7	20.8	4.0	0.30	3.4	0.08	0.14	0.04	0.72	0.02	43.6	0	0	13.3
21	4	0.73	11.5	15.4	3.8	0.36	3.8	0.13	—	0.10	0.64	0.02	41.0	0	0.1	—
18	9	0.89	8.5	34.5	414.5	0.30	—	0.08	—	0.04	0.71	0.03	4.0	0.1	0	—
0	4	0.45	8.5	33.2	57.0	0.11	0	0.09	—	0.02	0.48	0.03	6.6	0	0	—
0	7	0.45	1.8	25.5	141.7	0.19	0	0.01	—	0.03	0.08	0.01	1.8	0	0	—
0	7	1.00	12.4	51.5	0.5	0.35	0	0.12	0.04	0.07	0.90	0.04	7.8	0	0	40.0
0	11	0.74	21.0	30.8	2.1	0.57	0	0.08	0.21	0.03	0.50	0.06	3.5	0	0	18.1
0	1	0.25	11.5	26.3	0.6	0.25	0.8	0.01	0.02	0.01	0.18	0.01	2.5	0	0	0
2	15	0.61	12.3	38.4	72.5	0.20	0.7	0.02	0.42	0.02	0.77	0.01	1.8	0	0	1.3

TABLE H-1 Table of Food Composition (*continued*) (Computer code is for Cengage Diet Analysis program) (For purposes of calculations, use "0" for t, <1, <.1, <.01, etc.

DA+ Code	Food Description	Quantity	Measure	Wt (g)	H_2O (g)	Ener (kcal)	Prot (g)	Carb (g)	Fiber (g)	Fat (g)	Fat Breakdown (g) Sat	Mono	Poly	*Trans*
	Cereal, Flour, Grain, Pasta, Noodles, Popcorn—*continued*													
4620	Cheese flavored	1	cup(s)	36	0.9	188	3.3	18.4	3.5	11.8	2.3	3.5	5.5	—
477	Popped in oil	1	cup(s)	11	0.1	64	0.8	5.0	0.9	4.8	0.8	1.1	2.6	—
	Fruit and Fruit Juices													
	Apples													
952	Juice, prepared from frozen concentrate	½	cup(s)	120	105.0	56	0.2	13.8	0.1	0.1	0	0	0	—
225	Juice, unsweetened, canned	½	cup(s)	124	109.0	58	0.1	14.5	0.1	0.1	0	0	0	—
224	Slices	½	cup(s)	55	47.1	29	0.1	7.6	1.3	0.1	0	0	0	—
946	Slices without skin, boiled	½	cup(s)	86	73.1	45	0.2	11.7	2.1	0.3	0	0	0.1	—
223	Raw medium, with peel	1	item(s)	138	118.1	72	0.4	19.1	3.3	0.2	0	0	0.1	—
948	Dried, sulfured	¼	cup(s)	22	6.8	52	0.2	14.2	1.9	0.1	0	0	0	—
226	Applesauce, sweetened, canned	½	cup(s)	128	101.5	97	0.2	25.4	1.5	0.2	0	0	0.1	—
227	Applesauce, unsweetened, canned	½	cup(s)	122	107.8	52	0.2	13.8	1.5	0.1	0	0	0	—
38492	Crabapples	1	item(s)	35	27.6	27	0.1	7.0	0.9	0.1	0	0	0	—
	Apricot													
228	Fresh without pits	4	item(s)	140	120.9	67	2.0	15.6	2.8	0.5	0	0.2	0.1	—
229	Halves with skin, canned in heavy syrup	½	cup(s)	129	100.1	107	0.7	27.7	2.1	0.1	0	0	0	—
230	Halves, dried, sulfured	¼	cup(s)	33	10.1	79	1.1	20.6	2.4	0.2	0	0	0	—
	Avocado													
233	California, whole, without skin or pit	½	cup(s)	115	83.2	192	2.2	9.9	7.8	17.7	2.4	11.3	2.1	—
234	Florida, whole, without skin or pit	½	cup(s)	115	90.6	138	2.5	9.0	6.4	11.5	2.2	6.3	1.9	—
2998	Pureed	⅛	cup(s)	28	20.2	44	0.5	2.4	1.8	4.0	0.6	2.7	0.5	—
	Banana													
4580	Dried chips	¼	cup(s)	55	2.4	285	1.3	32.1	4.2	18.5	15.9	1.1	0.3	—
235	Fresh whole, without peel	1	item(s)	118	88.4	105	1.3	27.0	3.1	0.4	0.1	0	0.1	—
	Blackberries													
237	Raw	½	cup(s)	72	63.5	31	1.0	6.9	3.8	0.4	0	0	0.2	—
958	Unsweetened, frozen	½	cup(s)	76	62.1	48	0.9	11.8	3.8	0.3	0	0	0.2	—
	Blueberries													
959	Canned in heavy syrup	½	cup(s)	128	98.3	113	0.8	28.2	2.0	0.4	0	0.1	0.2	—
238	Raw	½	cup(s)	73	61.1	41	0.5	10.5	1.7	0.2	0	0	0.1	—
960	Unsweetened, frozen	½	cup(s)	78	67.1	40	0.3	9.4	2.1	0.5	0	0.1	0.2	—
	Boysenberries													
961	Canned in heavy syrup	½	cup(s)	128	97.6	113	1.3	28.6	3.3	0.2	0	0	0.1	—
962	Unsweetened, frozen	½	cup(s)	66	56.7	33	0.7	8.0	3.5	0.2	0	0	0.1	—
35576	**Breadfruit**	1	item(s)	384	271.3	396	4.1	104.1	18.8	0.9	0.2	0.1	0.3	—
	Cherries													
967	Sour red, canned in water	½	cup(s)	122	109.7	44	0.9	10.9	1.3	0.1	0	0	0	—
3000	Sour red, raw	½	cup(s)	78	66.8	39	0.8	9.4	1.2	0.2	0.1	0.1	0.1	—
3004	Sweet, canned in heavy syrup	½	cup(s)	127	98.2	105	0.8	26.9	1.9	0.2	0	0.1	0.1	—
969	Sweet, canned in water	½	cup(s)	124	107.9	57	1.0	14.6	1.9	0.2	0	0	0	—
240	Sweet, raw	½	cup(s)	73	59.6	46	0.8	11.6	1.5	0.1	0	0	0	—
	Cranberries													
3007	Chopped, raw	½	cup(s)	55	47.9	25	0.2	6.7	2.5	0.1	0	0	0	—
1717	Cranberry apple juice drink	½	cup(s)	123	102.6	77	0	19.4	0	0.1	0	0	0.1	—
1638	Cranberry juice cocktail	½	cup(s)	127	109.0	68	0	17.1	0	0.1	0	0	0.1	—
241	Cranberry juice cocktail, low calorie, with saccharin	½	cup(s)	119	112.8	23	0	5.5	0	0	0	0	0	—
242	Cranberry sauce, sweetened, canned	¼	cup(s)	69	42.0	105	0.1	26.9	0.7	0.1	0	0	0	—
	Dates													
244	Domestic, chopped	¼	cup(s)	45	9.1	125	1.1	33.4	3.6	0.2	0	0	0	—
243	Domestic, whole	¼	cup(s)	45	9.1	125	1.1	33.4	3.6	0.2	0	0	0	—
	Figs													
975	Canned in heavy syrup	½	cup(s)	130	98.8	114	0.5	29.7	2.8	0.1	0	0	0.1	—
974	Canned in water	½	cup(s)	124	105.7	66	0.5	17.3	2.7	0.1	0	0	0.1	—
973	Raw, medium	2	item(s)	100	79.1	74	0.7	19.2	2.9	0.3	0.1	0.1	0.1	—
	Fruit cocktail and salad													
245	Fruit cocktail, canned in heavy syrup	½	cup(s)	124	99.7	91	0.5	23.4	1.2	0.1	0	0	0	—
978	Fruit cocktail, canned in juice	½	cup(s)	119	103.6	55	0.5	14.1	1.2	0	0	0	0	—
977	Fruit cocktail, canned in water	½	cup(s)	119	107.6	38	0.5	10.1	1.2	0.1	0	0	0	—
979	Fruit salad, canned in water	½	cup(s)	123	112.1	37	0.4	9.6	1.2	0.1	0	0	0	—

PAGE KEY: H-4 = Breads/Baked Goods H-10 = Cereal/Rice/Pasta H-14 = Fruit H-18 = Vegetables/Legumes H-28 = Nuts/Seeds H-30 = Vegetarian H-32 = Dairy H-40 = Eggs H-40 = Seafood H-42 = Meats H-46 = Poultry H-46 = Processed Meats H-48 = Beverages H-52 = Fats/Oils H-54 = Sweets H-56 = Spices/Condiments/Sauces H-60 = Mixed Foods/Soups/Sandwiches H-64 = Fast Food H-84 = Convenience H-86 = Baby Foods

Chol (mg)	Calc (mg)	Iron (mg)	Magn (mg)	Pota (mg)	Sodi (mg)	Zinc (mg)	Vit A (µg)	Thia (mg)	Vit E (mg α)	Ribo (mg)	Niac (mg)	Vit B_6 (mg)	Fola (µg)	Vit C (mg)	Vit B_{12} (µg)	Sele (µg)
4	40	0.79	32.5	93.2	317.4	0.71	13.6	0.04	—	0.08	0.52	0.08	3.9	0.2	0.2	4.3
0	0	0.22	8.7	20.0	116.4	0.34	0.9	0.01	0.27	0.00	0.13	0.01	2.8	0	0	0.2
0	7	0.31	6.0	150.6	8.4	0.05	0	0.00	0.01	0.02	0.05	0.04	0	0.7	0	0.1
0	9	0.46	3.7	147.6	3.7	0.04	0	0.03	0.01	0.02	0.12	0.04	0	1.1	0	0.1
0	3	0.06	2.7	58.8	0.5	0.02	1.6	0.01	0.10	0.01	0.05	0.02	1.6	2.5	0	0
0	4	0.16	2.6	75.2	0.9	0.03	1.7	0.01	0.04	0.01	0.08	0.04	0.9	0.2	0	0.3
0	8	0.16	6.9	147.7	1.4	0.05	4.1	0.02	0.24	0.03	0.12	0.05	4.1	6.3	0	0
0	3	0.30	3.4	96.8	18.7	0.04	0	0.00	0.11	0.03	0.20	0.03	0	0.8	0	0.3
0	5	0.44	3.8	77.8	3.8	0.05	1.3	0.01	0.26	0.03	0.24	0.03	1.3	2.2	0	0.4
0	4	0.14	3.7	91.5	2.4	0.03	1.2	0.01	0.25	0.03	0.22	0.03	1.2	1.5	0	0.4
0	6	0.12	2.5	67.9	0.4	—	0.7	0.01	0.20	0.01	0.03	—	2.0	2.8	0	—
0	18	0.54	14.0	362.6	1.4	0.28	134.4	0.04	1.24	0.05	0.84	0.07	12.6	14.0	0	0.1
0	12	0.38	9.0	180.6	5.2	0.14	80.0	0.02	0.77	0.02	0.48	0.07	2.6	4.0	0	0.1
0	18	0.87	10.5	381.5	3.3	0.12	59.1	0.00	1.42	0.02	0.85	0.05	3.3	0.3	0	0.7
0	15	0.66	33.3	583.0	9.2	0.78	8.0	0.08	2.23	0.16	2.19	0.31	102.3	10.1	0	0.4
0	12	0.19	27.6	403.6	2.3	0.45	8.0	0.02	3.03	0.04	0.76	0.08	40.3	20.0	0	—
0	3	0.14	8.0	134.1	1.9	0.17	1.9	0.01	0.57	0.03	0.48	0.07	22.4	2.8	0	0.1
0	10	0.69	41.8	294.8	3.3	0.40	2.2	0.04	0.13	0.01	0.39	0.14	7.7	3.5	0	0.8
0	6	0.30	31.9	422.4	1.2	0.17	3.5	0.03	0.11	0.08	0.78	0.43	23.6	10.3	0	1.2
0	21	0.45	14.4	116.6	0.7	0.38	7.9	0.01	0.84	0.02	0.47	0.02	18.0	15.1	0	0.3
0	22	0.60	16.6	105.7	0.8	0.19	4.5	0.02	0.88	0.03	0.91	0.05	25.7	2.3	0	0.3
0	6	0.42	5.1	51.2	3.8	0.09	2.6	0.04	0.49	0.07	0.14	0.05	2.6	1.4	0	0.1
0	4	0.20	4.4	55.8	0.7	0.12	2.2	0.03	0.41	0.03	0.30	0.04	4.4	7.0	0	0.1
0	6	0.14	3.9	41.9	0.8	0.05	1.6	0.03	0.37	0.03	0.40	0.05	5.4	1.9	0	0.1
0	23	0.55	14.1	115.2	3.8	0.24	2.6	0.03	—	0.04	0.29	0.05	43.5	7.9	0	0.5
0	18	0.56	10.6	91.7	0.7	0.15	2.0	0.04	0.57	0.02	0.51	0.04	41.6	2.0	0	0.1
0	65	2.07	96.0	1881.6	7.7	0.46	0	0.42	0.38	0.11	3.45	0.38	53.8	111.4	0	2.3
0	13	1.67	7.3	119.6	8.5	0.09	46.4	0.02	0.28	0.05	0.22	0.05	9.8	2.6	0	0
0	12	0.25	7.0	134.1	2.3	0.08	49.6	0.02	0.05	0.03	0.31	0.03	6.2	7.8	0	0
0	11	0.44	11.4	183.4	3.8	0.12	10.1	0.02	0.29	0.05	0.50	0.03	5.1	4.6	0	0
0	14	0.45	11.2	162.4	1.2	0.10	9.9	0.03	0.29	0.05	0.51	0.04	5.0	2.7	0	0
0	9	0.26	8.0	161.0	0	0.05	2.2	0.02	0.05	0.02	0.11	0.04	2.9	5.1	0	0
0	4	0.13	3.3	46.8	1.1	0.05	1.7	0.01	0.66	0.01	0.05	0.03	0.6	7.3	0	0.1
0	4	0.09	1.2	20.8	2.5	0.02	0	0.00	0.15	0.00	0.00	0.00	0	48.4	0	0
0	4	0.13	1.3	17.7	2.5	0.04	0	0.00	0.28	0.00	0.05	0.00	0	53.5	0	0.3
0	11	0.05	2.4	29.6	3.6	0.02	0	0.00	0.06	0.00	0.00	0.00	0	38.2	0	0
0	3	0.15	2.1	18.0	20.1	0.03	1.4	0.01	0.57	0.01	0.06	0.01	0.7	1.4	0	0.2
0	17	0.45	19.1	291.9	0.9	0.12	0	0.02	0.02	0.02	0.56	0.07	8.5	0.2	0	1.3
0	17	0.45	19.1	291.9	0.9	0.12	0	0.02	0.02	0.02	0.56	0.07	8.5	0.2	0	1.3
0	35	0.36	13.0	128.2	1.3	0.14	2.6	0.03	0.16	0.05	0.55	0.09	2.6	1.3	0	0.3
0	35	0.36	12.4	127.7	1.2	0.15	2.5	0.03	0.10	0.05	0.55	0.09	2.5	1.2	0	0.1
0	35	0.36	17.0	232.0	1.0	0.14	7.0	0.06	0.10	0.04	0.40	0.10	6.0	2.0	0	0.2
0	7	0.36	6.2	109.1	7.4	0.09	12.4	0.02	0.49	0.02	0.46	0.06	3.7	2.4	0	0.6
0	9	0.25	8.3	112.6	4.7	0.11	17.8	0.01	0.47	0.02	0.48	0.06	3.6	3.2	0	0.6
0	6	0.30	8.3	111.4	4.7	0.11	15.4	0.02	0.47	0.01	0.43	0.06	3.6	2.5	0	0.6
0	9	0.37	6.1	95.6	3.7	0.10	27.0	0.02	—	0.03	0.46	0.04	3.7	2.3	0	1.0

TABLE H-1 Table of Food Composition (*continued*) (Computer code is for Cengage Diet Analysis program) (For purposes of calculations, use "0" for t, <1, <.1, <.01, etc

DA+ Code	Food Description	Quantity	Measure	Wt (g)	H_2O (g)	Ener (kcal)	Prot (g)	Carb (g)	Fiber (g)	Fat (g)	Fat Breakdown (g) Sat	Mono	Poly	*Trans*
Fruit and Fruit Juices—*continued*														
	Gooseberries													
982	Canned in light syrup	½	cup(s)	126	100.9	92	0.8	23.6	3.0	0.3	0	0	0.1	—
981	Raw	½	cup(s)	75	65.9	33	0.7	7.6	3.2	0.4	0	0	0.2	—
	Grapefruit													
251	Juice, pink, sweetened, canned	½	cup(s)	125	109.1	57	0.7	13.9	0.1	0.1	0	0	0	—
249	Juice, white	½	cup(s)	124	111.2	48	0.6	11.4	0.1	0.1	0	0	0	—
3022	Pink or red, raw	½	cup(s)	114	100.8	48	0.9	12.2	1.8	0.2	0	0	0	—
248	Sections, canned in light syrup	½	cup(s)	127	106.2	76	0.7	19.6	0.5	0.1	0	0	0	—
983	Sections, canned in water	½	cup(s)	122	109.6	44	0.7	11.2	0.5	0.1	0	0	0	—
247	White, raw	½	cup(s)	115	104.0	38	0.8	9.7	1.3	0.1	0	0	0	—
	Grapes													
255	American, slip skin	½	cup(s)	46	37.4	31	0.3	7.9	0.4	0.2	0.1	0	0	—
256	European, red or green, adherent skin	½	cup(s)	76	60.8	52	0.5	13.7	0.7	0.1	0	0	0	—
3159	Grape juice drink, canned	½	cup(s)	125	106.6	71	0	18.2	0.1	0	0	0	0	0
259	Grape juice, sweetened, with added vitamin C, prepared from frozen concentrate	½	cup(s)	125	108.6	64	0.2	15.9	0.1	0.1	0	0	0	—
3060	Raisins, seeded, packed	¼	cup(s)	41	6.8	122	1.0	32.4	2.8	0.2	0.1	0	0.1	—
987	**Guava, raw**	1	item(s)	55	44.4	37	1.4	7.9	3.0	0.5	0.2	0	0.2	—
35593	**Guavas, strawberry**	1	item(s)	6	4.8	4	0	1.0	0.3	0	0	0	0	—
3027	**Jackfruit**	½	cup(s)	83	60.4	78	1.2	19.8	1.3	0.2	0	0	0.1	—
990	**Kiwi fruit or Chinese gooseberries**	1	item(s)	76	63.1	46	0.9	11.1	2.3	0.4	0	0	0.2	—
	Lemon													
262	Juice	1	tablespoon(s)	15	13.8	4	0.1	1.3	0.1	0	0	0	0	0
993	Peel	1	teaspoon(s)	2	1.6	1	0	0.3	0.2	0	0	0	0	—
992	Raw	1	item(s)	108	94.4	22	1.3	11.6	5.1	0.3	0	0	0.1	—
	Lime													
269	Juice	1	tablespoon(s)	15	14.0	4	0.1	1.3	0.1	0	0	0	0	—
994	Raw	1	item(s)	67	59.1	20	0.5	7.1	1.9	0.1	0	0	0	—
995	**Loganberries, frozen**	½	cup(s)	74	62.2	40	1.1	9.6	3.9	0.2	0	0	0.1	—
	Mandarin orange													
1038	Canned in juice	½	cup(s)	125	111.4	46	0.8	11.9	0.9	0	0	0	0	—
1039	Canned in light syrup	½	cup(s)	126	104.7	77	0.6	20.4	0.9	0.1	0	0	0	—
999	**Mango**	½	cup(s)	83	67.4	54	0.4	14.0	1.5	0.2	0.1	0.1	0	—
1005	**Nectarine, raw, sliced**	½	cup(s)	69	60.4	30	0.7	7.3	1.2	0.2	0	0.1	0.1	—
	Melons													
271	Cantaloupe	½	cup(s)	80	72.1	27	0.7	6.5	0.7	0.1	0	0	0.1	—
1000	Casaba melon	½	cup(s)	85	78.1	24	0.9	5.6	0.8	0.1	0	0	0	—
272	Honeydew	½	cup(s)	89	79.5	32	0.5	8.0	0.7	0.1	0	0	0	—
318	Watermelon	½	cup(s)	76	69.5	23	0.5	5.7	0.3	0.1	0	0	0	—
	Orange													
14412	Juice with calcium and vitamin D	½	cup(s)	120	—	55	1.0	13.0	0	0	0	0	0	0
29630	Juice, fresh squeezed	½	cup(s)	124	109.5	56	0.9	12.9	0.2	0.2	0	0	0	—
14411	Juice, not from concentrate	½	cup(s)	120	—	55	1.0	13.0	0	0	0	0	0	0
278	Juice, unsweetened, prepared from frozen concentrate	½	cup(s)	125	109.7	56	0.8	13.4	0.2	0.1	0	0	0	—
3040	Peel	1	teaspoon(s)	2	1.5	2	0	0.5	0.2	0	0	0	0	—
273	Raw	1	item(s)	131	113.6	62	1.2	15.4	3.1	0.2	0	0	0	—
274	Sections	½	cup(s)	90	78.1	42	0.8	10.6	2.2	0.1	0	0	0	—
	Papaya, raw													
16830	Dried, strips	2	item(s)	46	12.0	119	1.9	29.9	5.5	0.4	0.1	0.1	0.1	—
282	Papaya	½	cup(s)	70	62.2	27	0.4	6.9	1.3	0.1	0	0	0	—
35640	**Passion fruit, purple**	1	item(s)	18	13.1	17	0.4	4.2	1.9	0.1	0	0	0.1	—
	Peach													
285	Halves, canned in heavy syrup	½	cup(s)	131	103.9	97	0.6	26.1	1.7	0.1	0	0	0.1	—
286	Halves, canned in water	½	cup(s)	122	113.6	29	0.5	7.5	1.6	0.1	0	0	0	—
290	Slices, sweetened, frozen	½	cup(s)	125	93.4	118	0.8	30.0	2.3	0.2	0	0.1	0.1	—
283	Raw, medium	1	item(s)	150	133.3	59	1.4	14.3	2.3	0.4	0	0.1	0.1	—
	Pear													
8672	Asian	1	item(s)	122	107.7	51	0.6	13.0	4.4	0.3	0	0.1	0.1	—
293	D'Anjou	1	item(s)	200	168.0	120	1.0	30.0	5.2	1.0	0	0.2	0.2	—
294	Halves, canned in heavy syrup	½	cup(s)	133	106.9	98	0.3	25.5	2.1	0.2	0	0	0	—
1012	Halves, canned in juice	½	cup(s)	124	107.2	62	0.4	16.0	2.0	0.1	0	0	0	—
291	Raw	1	item(s)	166	139.0	96	0.6	25.7	5.1	0.2	0	0	0	—

PAGE KEY: H-4 = Breads/Baked Goods H-10 = Cereal/Rice/Pasta H-14 = Fruit H-18 = Vegetables/Legumes H-28 = Nuts/Seeds H-30 = Vegetarian H-32 = Dairy H-40 = Eggs H-40 = Seafood H-42 = Meats H-46 = Poultry H-46 = Processed Meats H-48 = Beverages H-52 = Fats/Oils H-54 = Sweets H-56 = Spices/Condiments/Sauces H-60 = Mixed Foods/Soups/Sandwiches H-64 = Fast Food H-84 = Convenience H-86 = Baby Foods

Chol (mg)	Calc (mg)	Iron (mg)	Magn (mg)	Pota (mg)	Sodi (mg)	Zinc (mg)	Vit A (µg)	Thia (mg)	Vit E (mg α)	Ribo (mg)	Niac (mg)	Vit B_6 (mg)	Fola (µg)	Vit C (mg)	Vit B_{12} (µg)	Sele (µg)
0	20	0.42	7.6	97.0	2.5	0.14	8.8	0.03	—	0.07	0.19	0.02	3.8	12.6	0	0.5
0	19	0.23	7.5	148.5	0.8	0.09	11.3	0.03	0.28	0.02	0.23	0.06	4.5	20.8	0	0.5
0	10	0.45	12.5	202.2	2.5	0.08	0	0.05	0.05	0.03	0.40	0.03	12.5	33.6	0	0.1
0	11	0.25	14.8	200.1	1.2	0.06	1.2	0.05	0.27	0.02	0.25	0.05	12.4	46.9	0	0.1
0	25	0.09	10.3	154.5	0	0.07	66.4	0.04	0.14	0.03	0.23	0.06	14.9	35.7	0	0.1
0	18	0.50	12.7	163.8	2.5	0.10	0	0.04	0.11	0.02	0.30	0.02	11.4	27.1	0	1.1
0	18	0.50	12.2	161.0	2.4	0.11	0	0.05	0.11	0.03	0.30	0.02	11.0	26.6	0	1.1
0	14	0.07	10.4	170.2	0	0.08	2.3	0.04	0.15	0.02	0.30	0.05	11.5	38.3	0	1.6
0	6	0.13	2.3	87.9	0.9	0.02	2.3	0.04	0.09	0.02	0.14	0.05	1.8	1.8	0	0
0	8	0.27	5.3	144.2	1.5	0.05	2.3	0.05	0.14	0.05	0.14	0.07	1.5	8.2	0	0.1
0	9	0.16	7.5	41.3	11.3	0.04	0	0.28	0.00	0.44	0.18	0.04	1.3	33.1	0	0.1
0	5	0.13	5.0	26.3	2.5	0.05	0	0.02	0.00	0.03	0.16	0.05	1.3	29.9	0	0.1
0	12	1.06	12.4	340.3	11.6	0.07	0	0.04	—	0.07	0.46	0.07	1.2	2.2	0	0.2
0	10	0.14	12.1	229.4	1.1	0.12	17.1	0.03	0.40	0.02	0.59	0.06	27.0	125.6	0	0.3
0	1	0.01	1.0	17.5	2.2	—	0.3	0.00	—	0.00	0.03	0.00	—	2.2	0	
0	28	0.49	30.5	250.0	2.5	0.35	12.4	0.02	—	0.09	0.33	0.09	11.5	5.5	0	0.5
0	26	0.23	12.9	237.1	2.3	0.10	3.0	0.02	1.11	0.01	0.25	0.04	19.0	70.5	0	0.2
0	1	0.00	0.9	18.9	0.2	0.01	0.2	0.00	0.02	0.00	0.02	0.01	2.0	7.0	0	0
0	3	0.01	0.3	3.2	0.1	0.01	0.1	0.00	0.01	0.00	0.01	0.00	0.3	2.6	0	0
0	66	0.75	13.0	156.6	3.2	0.10	2.2	0.05	—	0.04	0.21	0.11	—	83.2	0	1.0
0	2	0.02	1.2	18.0	0.3	0.01	0.3	0.00	0.03	0.00	0.02	0.01	1.5	4.6	0	0
0	22	0.40	4.0	68.3	1.3	0.07	1.3	0.02	0.14	0.01	0.13	0.02	5.4	19.5	0	0.3
0	19	0.47	15.4	106.6	0.7	0.25	1.5	0.04	0.64	0.03	0.62	0.05	19.1	11.2	0	0.1
0	14	0.34	13.7	165.6	6.2	0.64	53.5	0.10	0.12	0.04	0.55	0.05	6.2	42.6	0	0.5
0	9	0.47	10.1	98.3	7.6	0.30	52.9	0.07	0.13	0.06	0.56	0.05	6.3	24.9	0	0.5
0	8	0.10	7.4	128.7	1.7	0.03	31.4	0.05	0.92	0.04	0.48	0.11	11.6	22.8	0	0.5
0	4	0.19	6.2	138.7	0	0.12	11.7	0.02	0.53	0.02	0.78	0.02	3.5	3.7	0	0
0	7	0.17	9.6	213.6	12.8	0.14	135.2	0.03	0.04	0.01	0.59	0.05	16.8	29.4	0	0.3
0	9	0.29	9.4	154.7	7.7	0.06	0	0.01	0.04	0.03	0.20	0.14	6.8	18.5	0	0.3
0	5	0.15	8.8	201.8	15.9	0.07	2.7	0.03	0.01	0.01	0.37	0.07	16.8	15.9	0	0.6
0	5	0.18	7.6	85.1	0.8	0.07	21.3	0.02	0.04	0.01	0.13	0.03	2.3	6.2	0	0.3
0	175	0.00	12.0	225.0	0	—	0	0.08	—	0.03	0.40	0.06	30.0	36.0	0	—
0	14	0.25	13.6	248.0	1.2	0.06	12.4	0.11	0.05	0.04	0.50	0.05	37.2	62.0	0	0.1
0	10	0.00	12.5	225.0	0	0.06	0	0.08	—	0.03	0.40	0.06	30.0	36.0	0	0.1
0	11	0.12	12.5	236.6	1.2	0.06	6.2	0.10	0.25	0.02	0.25	0.06	54.8	48.4	0	0.1
0	3	0.01	0.4	4.2	0.1	0.01	0.4	0.00	0.01	0.00	0.01	0.00	0.6	2.7	0	0
0	52	0.13	13.1	237.1	0	0.09	14.4	0.11	0.23	0.05	0.36	0.07	39.3	69.7	0	0.7
0	36	0.09	9.0	162.9	0	0.06	9.9	0.07	0.16	0.03	0.25	0.05	27.0	47.9	0	0.4
0	73	0.30	30.4	782.9	9.2	0.21	83.7	0.06	2.22	0.08	0.93	0.05	58.0	37.7	0	1.8
0	17	0.07	7.0	179.9	2.1	0.05	38.5	0.02	0.51	0.02	0.24	0.01	26.6	43.3	0	0.4
0	2	0.28	5.2	62.6	5.0	0.01	11.5	0.00	0.00	0.02	0.27	0.01	2.5	5.4	0	0.1
0	4	0.35	6.6	120.5	7.9	0.11	22.3	0.01	0.64	0.03	0.80	0.02	3.9	3.7	0	0.4
0	2	0.39	6.1	120.8	3.7	0.11	32.9	0.01	0.59	0.02	0.63	0.02	3.7	3.5	0	0.4
0	4	0.46	6.3	162.5	7.5	0.06	17.5	0.01	0.77	0.04	0.81	0.02	3.8	117.8	0	0.5
0	9	0.37	13.5	285.0	0	0.25	24.0	0.03	1.09	0.04	1.20	0.03	6.0	9.9	0	0.2
0	5	0.00	9.8	147.6	0	0.02	0	0.01	0.14	0.01	0.26	0.02	9.8	4.6	0	0.1
0	22	0.50	12.0	250.0	0	0.24	—	0.04	1.00	0.08	0.20	0.03	14.6	8.0	0	1.0
0	7	0.29	5.3	86.5	6.7	0.10	0	0.01	0.10	0.02	0.32	0.01	1.3	1.5	0	0
0	11	0.36	8.7	119.0	5.0	0.11	0	0.01	0.10	0.01	0.25	0.02	1.2	2.0	0	0
0	15	0.28	11.6	197.5	1.7	0.16	1.7	0.02	0.19	0.04	0.26	0.04	11.6	7.0	0	0.2

TABLE H-1 Table of Food Composition (*continued*) (Computer code is for Cengage Diet Analysis program) (For purposes of calculations, use "0" for t, <1, <.1, <.01, et

DA+ Code	Food Description	Quantity	Measure	Wt (g)	H_2O (g)	Ener (kcal)	Prot (g)	Carb (g)	Fiber (g)	Fat (g)	Fat Breakdown (g) Sat	Mono	Poly	*Trans*
	Fruit and Fruit Juices—*continued*													
1017	**Persimmon**	1	item(s)	25	16.1	32	0.2	8.4	—	0.1	0	0	0	—
	Pineapple													
3053	Canned in extra heavy syrup	½	cup(s)	130	101.0	108	0.4	28.0	1.0	0.1	0	0	0	—
1019	Canned in juice	½	cup(s)	125	104.0	75	0.5	19.5	1.0	0.1	0	0	0	—
296	Canned in light syrup	½	cup(s)	126	108.0	66	0.5	16.9	1.0	0.2	0	0	0.1	—
1018	Canned in water	½	cup(s)	123	111.7	39	0.5	10.2	1.0	0.1	0	0	0	—
299	Juice, unsweetened, canned	½	cup(s)	125	108.0	66	0.5	16.1	0.3	0.2	0	0	0.1	—
295	Raw, diced	½	cup(s)	78	66.7	39	0.4	10.2	1.1	0.1	0	0	0	—
1024	**Plantain, cooked**	½	cup(s)	77	51.8	89	0.6	24.0	1.8	0.1	0.1	0	0	—
300	**Plum, raw, large**	1	item(s)	66	57.6	30	0.5	7.5	0.9	0.2	0	0.1	0	—
1027	**Pomegranate**	1	item(s)	154	124.7	105	1.5	26.4	0.9	0.5	0.1	0.1	0.1	—
	Prunes													
5644	Dried	2	item(s)	17	5.2	40	0.4	10.7	1.2	0.1	0	0	0	—
305	Dried, stewed	½	cup(s)	124	86.5	133	1.2	34.8	3.8	0.2	0	0.1	0	—
306	Juice, canned	1	cup(s)	256	208.0	182	1.6	44.7	2.6	0.1	0	0.1	0	—
	Raspberries													
309	Raw	½	cup(s)	62	52.7	32	0.7	7.3	4.0	0.4	0	0	0.2	—
310	Red, sweetened, frozen	½	cup(s)	125	90.9	129	0.9	32.7	5.5	0.2	0	0	0.1	—
311	**Rhubarb, cooked with sugar**	½	cup(s)	120	81.5	140	0.5	37.5	2.7	0.1	0	0	0.1	—
	Strawberries													
313	Raw	½	cup(s)	72	65.5	23	0.5	5.5	1.4	0.2	0	0	0.1	—
315	Sweetened, frozen, thawed	½	cup(s)	128	99.5	99	0.7	26.8	2.4	0.2	0	0	0.1	—
16828	**Tangelo**	1	item(s)	95	82.4	45	0.9	11.2	2.3	0.1	0	0	0	—
	Tangerine													
1040	Juice	½	cup(s)	124	109.8	53	0.6	12.5	0.2	0.2	0	0	0	—
316	Raw	1	item(s)	88	74.9	47	0.7	11.7	1.6	0.3	0	0.1	0.1	—
	Vegetables, Legumes													
	Amaranth													
1043	Leaves, boiled, drained	½	cup(s)	66	60.4	14	1.4	2.7	—	0.1	0	0	0.1	—
1042	Leaves, raw	1	cup(s)	28	25.7	6	0.7	1.1	—	0.1	0	0	0	—
8683	**Arugula leaves, raw**	1	cup(s)	20	18.3	5	0.5	0.7	0.3	0.1	0	0	0.1	—
	Artichoke													
1044	Boiled, drained	1	item(s)	120	100.9	64	3.5	14.3	10.3	0.4	0.1	0	0.2	—
2885	Hearts, boiled, drained	½	cup(s)	84	70.6	45	2.4	10.0	7.2	0.3	0.1	0	0.1	—
	Asparagus													
566	Boiled, drained	½	cup(s)	90	83.4	20	2.2	3.7	1.8	0.2	0	0	0.1	—
568	Canned, drained	½	cup(s)	121	113.7	23	2.6	3.0	1.9	0.8	0.2	0	0.3	—
565	Tips, frozen, boiled, drained	½	cup(s)	90	84.7	16	2.7	1.7	1.4	0.4	0.1	0	0.2	—
	Bamboo shoots													
1048	Boiled, drained	½	cup(s)	60	57.6	7	0.9	1.2	0.6	0.1	0	0	0.1	—
1049	Canned, drained	½	cup(s)	66	61.8	12	1.1	2.1	0.9	0.3	0.1	0	0.1	—
	Beans													
1801	Adzuki beans, boiled	½	cup(s)	115	76.2	147	8.6	28.5	8.4	0.1	0	—	—	—
511	Baked beans with franks, canned	½	cup(s)	130	89.8	184	8.7	19.9	8.9	8.5	3.0	3.7	1.1	—
513	Baked beans with pork in sweet sauce, canned	½	cup(s)	127	89.3	142	6.7	26.7	5.3	1.8	0.6	0.6	0.5	0
512	Baked beans with pork in tomato sauce, canned	½	cup(s)	127	93.0	119	6.5	23.6	5.1	1.2	0.5	0.7	0.3	—
1805	Black beans, boiled	½	cup(s)	86	56.5	114	7.6	20.4	7.5	0.5	0.1	0	0.2	—
14597	Chickpeas, garbanzo beans or bengal gram, boiled	½	cup(s)	82	49.4	134	7.3	22.5	6.2	2.1	0.2	0.5	0.9	—
569	Fordhook lima beans, frozen, boiled, drained	½	cup(s)	85	62.0	88	5.2	16.4	4.9	0.3	0.1	0	0.1	—
1806	French beans, boiled	½	cup(s)	89	58.9	114	6.2	21.3	8.3	0.7	0.1	0	0.4	—
2773	Great northern beans, boiled	½	cup(s)	89	61.1	104	7.4	18.7	6.2	0.4	0.1	0	0.2	—
2736	Hyacinth beans, boiled, drained	½	cup(s)	44	37.8	22	1.3	4.0	—	0.1	0.1	0.1	0	—
570	Lima beans, baby, frozen, boiled, drained	½	cup(s)	90	65.1	95	6.0	17.5	5.4	0.3	0.1	0	0.1	—
515	Lima beans, boiled, drained	½	cup(s)	85	57.1	105	5.8	20.1	4.5	0.3	0.1	0	0.1	—
579	Mung beans, sprouted, boiled, drained	½	cup(s)	62	57.9	13	1.3	2.6	0.5	0.1	0	0	0	—
510	Navy beans, boiled	½	cup(s)	91	58.1	127	7.5	23.7	9.6	0.6	0.1	0.1	0.4	0
32816	Pinto beans, boiled, drained, no salt added	½	cup(s)	85	53.8	122	7.7	22.4	7.7	0.6	0.1	0.1	0.2	—

PAGE KEY: H-4 = Breads/Baked Goods H-10 = Cereal/Rice/Pasta H-14 = Fruit H-18 = Vegetables/Legumes H-28 = Nuts/Seeds H-30 = Vegetarian H-32 = Dairy H-40 = Eggs H-40 = Seafood H-42 = Meats H-46 = Poultry H-46 = Processed Meats H-48 = Beverages H-52 = Fats/Oils H-54 = Sweets H-56 = Spices/Condiments/Sauces H-60 = Mixed Foods/Soups/Sandwiches H-64 = Fast Food H-84 = Convenience H-86 = Baby Foods

Chol (mg)	Calc (mg)	Iron (mg)	Magn (mg)	Pota (mg)	Sodi (mg)	Zinc (mg)	Vit A (μg)	Thia (mg)	Vit E (mg α)	Ribo (mg)	Niac (mg)	Vit B_6 (mg)	Fola (μg)	Vit C (mg)	Vit B_{12} (μg)	Sele (μg)
0	7	0.62	—	77.5	0.3	—	0	—	—	—	—	—	—	16.5	0	—
0	18	0.49	19.5	132.6	1.3	0.14	1.3	0.11	—	0.03	0.36	0.09	6.5	9.5	0	—
0	17	0.35	17.4	151.9	1.2	0.12	2.5	0.12	0.01	0.02	0.35	0.09	6.2	11.8	0	0.5
0	18	0.49	20.2	132.3	1.3	0.15	2.5	0.11	0.01	0.03	0.36	0.09	6.3	9.5	0	0.5
0	18	0.49	22.1	156.2	1.2	0.15	2.5	0.11	0.01	0.03	0.37	0.09	6.2	9.5	0	0.5
0	16	0.39	15.0	162.5	2.5	0.14	0	0.07	0.03	0.03	0.25	0.13	22.5	12.5	0	0.1
0	10	0.22	9.3	84.5	0.8	0.09	2.3	0.06	0.02	0.03	0.39	0.09	14.0	37.0	0	0.1
0	2	0.45	24.6	358.1	3.9	0.10	34.7	0.04	0.10	0.04	0.58	0.19	20.0	8.4	0	1.1
0	4	0.11	4.6	103.6	0	0.06	11.2	0.02	0.17	0.02	0.27	0.02	3.3	6.3	0	0
0	5	0.46	4.6	398.9	4.6	0.18	7.7	0.04	0.92	0.04	0.46	0.16	9.2	9.4	0	0.9
0	7	0.16	6.9	123.0	0.3	0.07	6.6	0.01	0.07	0.03	0.32	0.03	0.7	0.1	0	0
0	24	0.51	22.3	398.0	1.2	0.24	21.1	0.03	0.24	0.12	0.90	0.27	0	3.6	0	0.1
0	31	3.02	35.8	706.6	10.2	0.53	0	0.04	0.30	0.17	2.01	0.55	0	10.5	0	1.5
0	15	0.42	13.5	92.9	0.6	0.26	1.2	0.02	0.54	0.02	0.37	0.03	12.9	16.1	0	0.1
0	19	0.81	16.3	142.5	1.3	0.22	3.8	0.02	0.90	0.05	0.28	0.04	32.5	20.6	0	0.4
0	174	0.25	16.2	115.0	1.0	—	—	0.02	—	0.03	0.25	—	—	4.0	0	—
0	12	0.30	9.4	110.2	0.7	0.10	0.7	0.02	0.21	0.02	0.28	0.03	17.3	42.3	0	0.3
0	14	0.59	7.7	125.0	1.3	0.06	1.3	0.01	0.30	0.09	0.37	0.03	5.1	50.4	0	0.9
0	38	0.09	9.5	172.0	0	0.06	10.5	0.08	0.17	0.03	0.26	0.05	28.5	50.5	0	0.5
0	22	0.25	9.9	219.8	1.2	0.04	16.1	0.07	0.16	0.02	0.12	0.05	6.2	38.3	0	0.1
0	33	0.13	10.6	146.1	1.8	0.06	29.9	0.05	0.18	0.03	0.33	0.07	14.1	23.5	0	0.1
0	138	1.49	36.3	423.1	13.9	0.58	91.7	0.01	—	0.09	0.37	0.12	37.6	27.1	0	0.6
0	60	0.65	15.4	171.1	5.6	0.25	40.9	0.01	—	0.04	0.18	0.05	23.8	12.1	0	0.3
0	32	0.29	9.4	73.8	5.4	0.09	23.8	0.01	0.09	0.02	0.06	0.01	19.4	3.0	0	0.1
0	25	0.73	50.4	343.2	72.0	0.48	1.2	0.06	0.22	0.10	1.33	0.09	106.8	8.9	0	0.2
0	18	0.51	35.3	240.2	50.4	0.33	0.8	0.04	0.16	0.07	0.93	0.06	74.8	6.2	0	0.2
0	21	0.81	12.6	201.6	12.6	0.54	45.0	0.14	1.35	0.12	0.97	0.07	134.1	6.9	0	5.5
0	19	2.21	12.1	208.1	347.3	0.48	49.6	0.07	1.47	0.12	1.15	0.13	116.2	22.3	0	2.1
0	16	0.50	9.0	154.8	2.7	0.36	36.0	0.05	1.08	0.09	0.93	0.01	121.5	22.0	0	3.5
0	7	0.14	1.8	319.8	2.4	0.28	0	0.01	—	0.03	0.18	0.06	1.2	0	0	0.2
0	5	0.21	2.6	52.4	4.6	0.43	0.7	0.02	0.41	0.02	0.09	0.09	2.0	0.7	0	0.3
0	32	2.30	59.8	611.8	9.2	2.03	0	0.13	—	0.07	0.82	0.11	139.2	0	0	1.4
8	62	2.24	36.3	304.3	556.9	2.42	5.2	0.08	0.21	0.07	1.17	0.06	38.9	3.0	0.4	8.4
9	75	2.08	41.7	326.4	422.5	1.73	0	0.05	0.03	0.07	0.44	0.07	10.1	3.5	0	6.3
9	71	4.09	43.0	373.2	552.8	6.93	5.1	0.06	0.12	0.05	0.62	0.08	19.0	3.8	0	5.9
0	23	1.80	60.2	305.3	0.9	0.96	0	0.21	—	0.05	0.43	0.05	128.1	0	0	1.0
0	40	2.36	39.4	238.6	5.7	1.25	0.8	0.09	0.28	0.05	0.43	0.11	141.0	1.1	0	3.0
0	26	1.54	35.7	258.4	58.7	0.62	8.5	0.06	0.24	0.05	0.90	0.10	17.9	10.9	0	0.5
0	56	0.95	49.6	327.5	5.3	0.56	0	0.11	—	0.05	0.48	0.09	66.4	1.1	0	1.1
0	60	1.88	44.3	346.0	1.8	0.77	0	0.14	—	0.05	0.60	0.10	90.3	1.2	0	3.6
0	18	0.33	18.3	114.0	0.9	0.16	3.0	0.02	—	0.03	0.20	0.01	20.4	2.2	0	0.7
0	25	1.76	50.4	369.9	26.1	0.49	7.2	0.06	0.57	0.04	0.69	0.10	14.4	5.2	0	1.5
0	27	2.08	62.9	484.5	14.5	0.67	12.8	0.11	0.11	0.08	0.88	0.16	22.1	0.6	0	1.7
0	7	0.40	8.7	62.6	6.2	0.29	0.6	0.03	0.04	0.06	0.51	0.03	18.0	7.1	0	0.4
—	63	2.14	48.2	354.0	0	0.93	0	0.21	0.01	0.06	0.59	0.12	127.4	0.8	0	2.6
0	39	1.79	43.0	373.0	1.0	0.84	0	0.16	0.80	0.05	0.27	0.19	147.0	0.7	0	5.3

TABLE H-1 Table of Food Composition (*continued*) (Computer code is for Cengage Diet Analysis program) (For purposes of calculations, use "0" for t, <1, <.1, <.01, etc.)

DA+ Code	Food Description	Quantity	Measure	Wt (g)	H_2O (g)	Ener (kcal)	Prot (g)	Carb (g)	Fiber (g)	Fat (g)	Fat Breakdown (g) Sat	Mono	Poly	*Trans*
Vegetables, Legumes—*continued*														
1052	Pinto beans, frozen, boiled, drained	½	cup(s)	47	27.3	76	4.4	14.5	4.0	0.2	0	0	0.1	—
514	Red kidney beans, canned	½	cup(s)	128	99.0	108	6.7	19.9	6.9	0.5	0.1	0.2	0.2	—
1810	Refried beans, canned	½	cup(s)	127	96.1	119	6.9	19.6	6.7	1.6	0.6	0.7	0.2	—
1053	Shell beans, canned	½	cup(s)	123	111.1	37	2.2	7.6	4.2	0.2	0	0	0.1	—
1670	Soybeans, boiled	½	cup(s)	86	53.8	149	14.3	8.5	5.2	7.7	1.1	1.7	4.4	—
1108	Soybeans, green, boiled, drained	½	cup(s)	90	61.7	127	11.1	9.9	3.8	5.8	0.7	1.1	2.7	—
1807	White beans, small, boiled	½	cup(s)	90	56.6	127	8.0	23.1	9.3	0.6	0.1	0.1	0.2	—
575	Yellow snap, string or wax beans, boiled, drained	½	cup(s)	63	55.8	22	1.2	4.9	2.1	0.2	0	0	0.1	—
576	Yellow snap, string or wax beans, frozen, boiled, drained	½	cup(s)	68	61.7	19	1.0	4.4	2.0	0.1	0	0	0.1	—
	Beets													
584	Beet greens, boiled, drained	½	cup(s)	72	64.2	19	1.9	3.9	2.1	0.1	0	0	0.1	—
2730	Pickled, canned with liquid	½	cup(s)	114	92.9	74	0.9	18.5	3.0	0.1	0	0	0	—
581	Sliced, boiled, drained	½	cup(s)	85	74.0	37	1.4	8.5	1.7	0.2	0	0	0.1	—
583	Sliced, canned, drained	½	cup(s)	85	77.3	26	0.8	6.1	1.5	0.1	0	0	0	—
580	Whole, boiled, drained	2	item(s)	100	87.1	44	1.7	10.0	2.0	0.2	0	0	0.1	—
585	**Cowpeas or black-eyed peas, boiled, drained**	½	cup(s)	83	62.3	80	2.6	16.8	4.1	0.3	0.1	0	0.1	—
	Broccoli													
588	Chopped, boiled, drained	½	cup(s)	78	69.6	27	1.9	5.6	2.6	0.3	0.1	0	0.1	—
590	Frozen, chopped, boiled, drained	½	cup(s)	92	83.5	26	2.9	4.9	2.8	0.1	0	0	0.1	—
587	Raw, chopped	½	cup(s)	46	40.6	15	1.3	3.0	1.2	0.2	0	0	0	—
16848	**Broccoflower, raw, chopped**	½	cup(s)	32	28.7	10	0.9	1.9	1.0	0.1	0	0	0	—
	Brussels sprouts													
591	Boiled, drained	½	cup(s)	78	69.3	28	2.0	5.5	2.0	0.4	0.1	0	0.2	—
592	Frozen, boiled, drained	½	cup(s)	78	67.2	33	2.8	6.4	3.2	0.3	0.1	0	0.2	—
	Cabbage													
595	Boiled, drained, no salt added	1	cup(s)	150	138.8	35	1.9	8.3	2.8	0.1	0	0	0	—
35611	Chinese (pak choi or bok choy), boiled with salt, drained	1	cup(s)	170	162.4	20	2.6	3.0	1.7	0.3	0	0	0.1	—
16869	Kim chee	1	cup(s)	150	137.5	32	2.5	6.1	1.8	0.3	0	0	0.2	—
594	Raw, shredded	1	cup(s)	70	64.5	17	0.9	4.1	1.7	0.1	0	0	0	—
596	Red, shredded, raw	1	cup(s)	70	63.3	22	1.0	5.2	1.5	0.1	0	0	0.1	—
597	Savoy, shredded, raw	1	cup(s)	70	63.7	19	1.4	4.3	2.2	0.1	0	0	0	—
35417	**Capers**	1	teaspoon(s)	4	—	2	0	0	0	0	0	0	0	0
	Carrots													
8691	Baby, raw	8	item(s)	80	72.3	28	0.5	6.6	2.3	0.1	0	0	0.1	—
601	Grated	½	cup(s)	55	48.6	23	0.5	5.3	1.5	0.1	0	0	0.1	0
1055	Juice, canned	½	cup(s)	118	104.9	47	1.1	11.0	0.9	0.2	0	0	0.1	—
600	Raw	½	cup(s)	61	53.9	25	0.6	5.8	1.7	0.1	0	0	0.1	0
602	Sliced, boiled, drained	½	cup(s)	78	70.3	27	0.6	6.4	2.3	0.1	0	0	0.1	—
32725	**Cassava or manioc**	½	cup(s)	103	61.5	165	1.4	39.2	1.9	0.3	0.1	0.1	0	—
	Cauliflower													
606	Boiled, drained	½	cup(s)	62	57.7	14	1.1	2.5	1.4	0.3	0	0	0.1	—
607	Frozen, boiled, drained	½	cup(s)	90	84.6	17	1.4	3.4	2.4	0.2	0	0	0.1	—
605	Raw, chopped	½	cup(s)	50	46.0	13	1.0	2.6	1.2	0	0	0	0	—
	Celery													
609	Diced	½	cup(s)	51	48.2	8	0.3	1.5	0.8	0.1	0	0	0	—
608	Stalk	2	item(s)	80	76.3	13	0.6	2.4	1.3	0.1	0	0	0.1	—
	Chard													
1057	Swiss chard, boiled, drained	½	cup(s)	88	81.1	18	1.6	3.6	1.8	0.1	0	0	0	—
1056	Swiss chard, raw	1	cup(s)	36	33.4	7	0.6	1.3	0.6	0.1	0	0	0	—
	Collard greens													
610	Boiled, drained	½	cup(s)	95	87.3	25	2.0	4.7	2.7	0.3	0	0	0.2	—
611	Frozen, chopped, boiled, drained	½	cup(s)	85	75.2	31	2.5	6.0	2.4	0.3	0.1	0	0.2	—
	Corn													
29614	Yellow corn, fresh, cooked	1	item(s)	100	69.2	107	3.3	25.0	2.8	1.3	0.2	0.4	0.6	—
615	Yellow creamed sweet corn, canned	½	cup(s)	128	100.8	92	2.2	23.2	1.5	0.5	0.1	0.2	0.3	—
612	Yellow sweet corn, boiled, drained	½	cup(s)	82	57.0	89	2.7	20.6	2.3	1.1	0.2	0.3	0.5	—

PAGE KEY: H-4 = Breads/Baked Goods H-10 = Cereal/Rice/Pasta H-14 = Fruit H-18 = Vegetables/Legumes H-28 = Nuts/Seeds H-30 = Vegetarian H-32 = Dairy H-40 = Eggs H-40 = Seafood H-42 = Meats H-46 = Poultry H-46 = Processed Meats H-48 = Beverages H-52 = Fats/Oils H-54 = Sweets H-56 = Spices/Condiments/Sauces H-60 = Mixed Foods/Soups/Sandwiches H-64 = Fast Food H-84 = Convenience H-86 = Baby Foods

Chol (mg)	Calc (mg)	Iron (mg)	Magn (mg)	Pota (mg)	Sodi (mg)	Zinc (mg)	Vit A (μg)	Thia (mg)	Vit E (mg α)	Ribo (mg)	Niac (mg)	Vit B_6 (mg)	Fola (μg)	Vit C (mg)	Vit B_{12} (μg)	Sele (μg)
0	24	1.27	25.4	303.6	39.0	0.32	0	0.12	—	0.05	0.29	0.09	16.0	0.3	0	0.7
0	32	1.62	35.8	327.7	330.2	2.09	0	0.13	0.02	0.11	0.57	0.10	25.6	1.4	0	0.6
10	44	2.10	41.7	337.8	378.2	1.48	0	0.03	0.00	0.02	0.39	0.18	13.9	7.6	0	1.6
0	36	1.21	18.4	133.5	409.2	0.33	13.5	0.04	0.04	0.07	0.25	0.06	22.1	3.8	0	2.6
0	88	4.42	74.0	442.9	0.9	0.98	0	0.13	0.30	0.24	0.34	0.20	46.4	1.5	0	6.3
0	131	2.25	54.0	485.1	12.6	0.82	7.2	0.23	—	0.14	1.13	0.05	99.9	15.3	0	1.3
0	65	2.54	60.9	414.4	1.8	0.97	0	0.21	—	0.05	0.24	0.11	122.6	0	0	1.2
0	29	0.80	15.6	186.9	1.9	0.23	2.5	0.05	0.28	0.06	0.38	0.04	20.6	6.1	0	0.3
0	33	0.59	16.2	85.1	6.1	0.32	4.1	0.02	0.03	0.06	0.26	0.04	15.5	2.8	0	0.3
0	82	1.36	49.0	654.5	173.5	0.36	275.8	0.08	1.30	0.20	0.35	0.09	10.1	17.9	0	0.6
0	12	0.46	17.0	168.0	299.6	0.29	1.1	0.01	—	0.05	0.28	0.05	30.6	2.6	0	1.1
0	14	0.67	19.6	259.3	65.5	0.30	1.7	0.02	0.03	0.03	0.28	0.05	68.0	3.1	0	0.6
0	13	1.54	14.5	125.8	164.9	0.17	0.9	0.01	0.02	0.03	0.13	0.04	25.5	3.5	0	0.4
0	16	0.79	23.0	305.0	77.0	0.35	2.0	0.02	0.04	0.04	0.33	0.06	80.0	3.6	0	0.7
0	106	0.92	42.9	344.9	3.3	0.85	33.0	0.08	0.18	0.12	1.15	0.05	104.8	1.8	0	2.1
0	31	0.52	16.4	228.5	32.0	0.35	60.1	0.04	1.13	0.09	0.43	0.15	84.2	50.6	0	1.2
0	30	0.56	12.0	130.6	10.1	0.25	46.9	0.05	1.21	0.07	0.42	0.12	51.5	36.9	0	0.6
0	21	0.33	9.6	143.8	15.0	0.19	14.1	0.03	0.36	0.05	0.29	0.08	28.7	40.6	0	1.1
0	11	0.23	6.4	96.0	7.4	0.20	2.6	0.02	0.01	0.03	0.23	0.07	18.2	28.2	0	0.2
0	28	0.93	15.6	247.3	16.4	0.25	30.4	0.08	0.33	0.06	0.47	0.13	46.8	48.4	0	1.2
0	20	0.37	14.0	224.8	11.6	0.18	35.7	0.08	0.39	0.08	0.41	0.22	78.3	35.4	0	0.5
0	72	0.24	22.5	294.0	12.0	0.30	6.0	0.08	0.20	0.04	0.36	0.16	45.0	56.2	0	0.9
0	158	1.76	18.7	630.7	459.0	0.28	360.4	0.04	0.14	0.10	0.72	0.28	69.7	44.2	0	0.7
0	144	1.26	27.0	379.5	996.0	0.36	288.0	0.06	0.36	0.10	0.80	0.32	88.5	79.6	0	1.5
0	28	0.33	8.4	119.0	12.6	0.12	3.5	0.04	0.10	0.02	0.16	0.08	30.1	25.6	0	0.2
0	31	0.56	11.2	170.1	18.9	0.15	39.2	0.04	0.07	0.05	0.29	0.14	12.6	39.9	0	0.4
0	24	0.28	19.6	161.0	19.6	0.18	35.0	0.05	0.11	0.02	0.21	0.13	56.0	21.7	0	0.6
0	0	0.00	—	—	140	—	0	—	—	—	—	—	—	0	—	—
0	26	0.71	8.0	189.6	62.4	0.13	552.0	0.02	—	0.02	0.44	0.08	21.6	2.1	0	0.7
0	18	0.16	6.6	176.0	37.9	0.13	459.2	0.03	0.36	0.03	0.54	0.07	10.4	3.2	0	0.1
0	28	0.54	16.5	344.6	34.2	0.21	1128.1	0.11	1.37	0.07	0.46	0.26	4.7	10.0	0	0.7
0	20	0.18	7.3	195.2	42.1	0.15	509.4	0.04	0.40	0.04	0.60	0.08	11.6	3.6	0	0.1
0	23	0.26	7.8	183.3	45.2	0.15	664.6	0.05	0.80	0.03	0.50	0.11	10.9	2.8	0	0.5
0	16	0.27	21.6	279.1	14.4	0.35	1.0	0.08	0.19	0.04	0.87	0.09	27.8	21.2	0	0.7
0	10	0.19	5.6	88.0	9.3	0.10	0.6	0.02	0.04	0.03	0.25	0.10	27.3	27.5	0	0.4
0	15	0.36	8.1	125.1	16.2	0.11	0	0.03	0.05	0.04	0.27	0.07	36.9	28.2	0	0.5
0	11	0.22	7.5	151.5	15.0	0.14	0.5	0.03	0.04	0.03	0.26	0.11	28.5	23.2	0	0.3
0	20	0.10	5.6	131.3	40.4	0.07	11.1	0.01	0.14	0.03	0.16	0.04	18.2	1.6	0	0.2
0	32	0.16	8.8	208.0	64.0	0.10	17.6	0.01	0.21	0.04	0.25	0.05	28.8	2.5	0	0.3
0	51	1.98	75.3	480.4	156.6	0.29	267.8	0.03	1.65	0.08	0.32	0.07	7.9	15.8	0	0.8
0	18	0.64	29.2	136.4	76.7	0.13	110.2	0.01	0.68	0.03	0.14	0.03	5.0	10.8	0	0.3
0	133	1.10	19.0	110.2	15.2	0.21	385.7	0.03	0.83	0.10	0.54	0.12	88.4	17.3	0	0.5
0	179	0.95	25.5	213.4	42.5	0.22	488.8	0.04	1.06	0.09	0.54	0.09	64.6	22.4	0	1.3
0	2	0.61	32.0	248.0	242.0	0.48	13.0	0.20	0.09	0.07	1.60	0.06	46.0	6.2	0	0.2
0	4	0.48	21.8	171.5	364.8	0.67	5.1	0.03	0.09	0.06	1.22	0.08	57.6	5.9	0	0.5
0	2	0.36	21.3	173.8	0	0.50	10.7	0.17	0.07	0.05	1.32	0.04	37.7	5.1	0	0.2

TABLE H-1 Table of Food Composition (*continued*) (Computer code is for Cengage Diet Analysis program) (For purposes of calculations, use "0" for t, <1, <.1, <.01, etc.)

DA+ Code	Food Description	Quantity	Measure	Wt (g)	H_2O (g)	Ener (kcal)	Prot (g)	Carb (g)	Fiber (g)	Fat (g)	Fat Breakdown (g) Sat	Mono	Poly	*Trans*
	Vegetables, Legumes—*continued*													
614	Yellow sweet corn, frozen, boiled, drained	½	cup(s)	82	63.2	66	2.1	15.8	2.0	0.5	0.1	0.2	0.3	—
618	**Cucumber**	¼	item(s)	75	71.7	11	0.5	2.7	0.4	0.1	0	0	0	—
16870	**Cucumber, kim chee**	½	cup(s)	75	68.1	16	0.8	3.6	1.1	0.1	0	0	0	—
	Dandelion greens													
620	Chopped, boiled, drained	½	cup(s)	53	47.1	17	1.1	3.4	1.5	0.3	0.1	0	0.1	—
2734	Raw	1	cup(s)	55	47.1	25	1.5	5.1	1.9	0.4	0.1	0	0.2	—
1066	**Eggplant, boiled, drained**	½	cup(s)	50	44.4	17	0.4	4.3	1.2	0.1	0	0	0	—
621	**Endive or escarole, chopped, raw**	1	cup(s)	50	46.9	8	0.6	1.7	1.5	0.1	0	0	0	—
8784	**Jicama or yambean**	½	cup(s)	65	116.5	49	0.9	11.4	6.3	0.1	0	0	0.1	—
	Kale													
623	Frozen, chopped, boiled, drained	½	cup(s)	65	58.8	20	1.8	3.4	1.3	0.3	0	0	0.2	—
29313	Raw	1	cup(s)	67	56.6	33	2.2	6.7	1.3	0.5	0.1	0	0.2	—
	Kohlrabi													
1072	Boiled, drained	½	cup(s)	83	74.5	24	1.5	5.5	0.9	0.1	0	0	0	—
1071	Raw	1	cup(s)	135	122.9	36	2.3	8.4	4.9	0.1	0	0	0.1	—
	Leeks													
1074	Boiled, drained	½	cup(s)	52	47.2	16	0.4	4.0	0.5	0.1	0	0	0	—
1073	Raw	1	cup(s)	89	73.9	54	1.3	12.6	1.6	0.3	0	0	0.1	—
	Lentils													
522	Boiled	¼	cup(s)	50	34.5	57	4.5	10.0	3.9	0.2	0	0	0.1	—
1075	Sprouted	1	cup(s)	77	51.9	82	6.9	17.0	—	0.4	0	0.1	0.2	—
	Lettuce													
625	Butterhead leaves	11	piece(s)	83	78.9	11	1.1	1.8	0.9	0.2	0	0	0.1	—
624	Butterhead, Boston or Bibb	1	cup(s)	55	52.6	7	0.7	1.2	0.6	0.1	0	0	0.1	—
626	Iceberg	1	cup(s)	55	52.6	8	0.5	1.6	0.7	0.1	0	0	0	—
628	Iceberg, chopped	1	cup(s)	55	52.6	8	0.5	1.6	0.7	0.1	0	0	0	—
629	Looseleaf	1	cup(s)	36	34.2	5	0.5	1.0	0.5	0.1	0	0	0	—
1665	Romaine, shredded	1	cup(s)	56	53.0	10	0.7	1.8	1.2	0.2	0	0	0.1	—
	Mushrooms													
15585	Crimini (about 6)	3	ounce(s)	85	—	28	3.7	2.8	1.9	0	0	0	0	0
8700	Enoki	30	item(s)	90	79.7	40	2.3	6.9	2.4	0.3	0	0	0.1	—
1079	Mushrooms, boiled, drained	½	cup(s)	78	71.0	22	1.7	4.1	1.7	0.4	0	0	0.1	—
1080	Mushrooms, canned, drained	½	cup(s)	78	71.0	20	1.5	4.0	1.9	0.2	0	0	0.1	—
630	Mushrooms, raw	½	cup(s)	48	44.4	11	1.5	1.6	0.5	0.2	0	0	0.1	—
15587	Portabella, raw	1	item(s)	84	—	30	3.0	3.9	3.0	0	0	0	0	0
2743	Shiitake, cooked	½	cup(s)	73	60.5	41	1.1	10.4	1.5	0.2	0	0.1	0	—
	Mustard greens													
2744	Frozen, boiled, drained	½	cup(s)	75	70.4	14	1.7	2.3	2.1	0.2	0	0.1	0	—
29319	Raw	1	cup(s)	56	50.8	15	1.5	2.7	1.8	0.1	0	0	0	—
	Okra													
16866	Batter coated, fried	11	piece(s)	83	55.6	156	2.1	12.7	2.0	11.2	1.5	3.7	5.5	—
32742	Frozen, boiled, drained, no salt added	½	cup(s)	92	83.8	26	1.9	5.3	2.6	0.3	0.1	0	0.1	—
632	Sliced, boiled, drained	½	cup(s)	80	74.1	18	1.5	3.6	2.0	0.2	0	0	0	—
	Onions													
635	Chopped, boiled, drained	½	cup(s)	105	92.2	46	1.4	10.7	1.5	0.2	0	0	0.1	—
2748	Frozen, boiled, drained	½	cup(s)	106	97.8	30	0.8	7.0	1.9	0.1	0	0	0	—
1081	Onion rings, breaded and pan fried, frozen, heated	10	piece(s)	71	20.2	289	3.8	27.1	0.9	19.0	6.1	7.7	3.6	—
633	Raw, chopped	½	cup(s)	80	71.3	32	0.9	7.5	1.4	0.1	0	0	0	—
16850	Red onions, sliced, raw	½	cup(s)	57	50.7	24	0.5	5.8	0.8	0	0	0	0	—
636	Scallions, green or spring onions	2	item(s)	30	26.9	10	0.5	2.2	0.8	0.1	0	0	0	—
16860	**Palm hearts, cooked**	½	cup(s)	73	50.7	84	2.0	18.7	1.1	0.1	0	0	0.1	—
637	**Parsley, chopped**	1	tablespoon(s)	4	3.3	1	0.1	0.2	0.1	0	0	0	0	—
638	**Parsnips, sliced, boiled, drained**	½	cup(s)	78	62.6	55	1.0	13.3	2.8	0.2	0	0.1	0	—
	Peas													
639	Green peas, canned, drained	½	cup(s)	85	69.4	59	3.8	10.7	3.5	0.3	0.1	0	0.1	—
641	Green peas, frozen, boiled, drained	½	cup(s)	80	63.6	62	4.1	11.4	4.4	0.2	0	0	0.1	—
35694	Pea pods, boiled with salt, drained	½	cup(s)	80	71.1	32	2.6	5.2	2.2	0.2	0	0	0.1	—

PAGE KEY: H-4 = Breads/Baked Goods H-10 = Cereal/Rice/Pasta H-14 = Fruit H-18 = Vegetables/Legumes H-28 = Nuts/Seeds H-30 = Vegetarian H-32 = Dairy H-40 = Eggs H-40 = Seafood H-42 = Meats H-46 = Poultry H-46 = Processed Meats H-48 = Beverages H-52 = Fats/Oils H-54 = Sweets H-56 = Spices/Condiments/Sauces H-60 = Mixed Foods/Soups/Sandwiches H-64 = Fast Food H-84 = Convenience H-86 = Baby Foods

Chol (mg)	Calc (mg)	Iron (mg)	Magn (mg)	Pota (mg)	Sodi (mg)	Zinc (mg)	Vit A (μg)	Thia (mg)	Vit E (mg α)	Ribo (mg)	Niac (mg)	Vit B_6 (mg)	Fola (μg)	Vit C (mg)	Vit B_{12} (μg)	Sele (μg)
0	2	0.38	23.0	191.1	0.8	0.51	8.2	0.02	0.05	0.05	1.07	0.08	28.7	2.9	0	0.6
0	12	0.20	9.8	110.6	1.5	0.14	3.8	0.01	0.01	0.01	0.07	0.03	5.3	2.1	0	0.2
0	7	3.61	6.0	87.8	765.8	0.38	—	0.02	—	0.02	0.34	0.08	17.3	2.6	0	—
0	74	0.95	12.6	121.8	23.1	0.15	179.6	0.07	1.28	0.09	0.27	0.08	6.8	9.5	0	0.2
0	103	1.70	19.8	218.3	41.8	0.22	279.4	0.10	1.89	0.14	0.44	0.13	14.8	19.2	0	0.3
0	3	0.12	5.4	60.9	0.5	0.06	1.0	0.04	0.20	0.01	0.30	0.04	6.9	0.6	0	0
0	26	0.41	7.5	157.0	11.0	0.39	54.0	0.04	0.22	0.03	0.20	0.01	71.0	3.2	0	0.1
0	16	0.78	15.5	194.0	5.2	0.20	1.3	0.02	0.59	0.04	0.25	0.05	15.5	26.1	0	0.9
0	90	0.61	11.7	208.7	9.8	0.11	477.8	0.02	0.59	0.07	0.43	0.05	9.1	16.4	0	0.6
0	90	1.14	22.8	299.5	28.8	0.29	515.2	0.07	—	0.08	0.66	0.18	19.4	80.4	0	0.6
0	21	0.33	15.7	280.5	17.3	0.26	1.7	0.03	0.43	0.02	0.32	0.13	9.9	44.6	0	0.7
0	32	0.54	25.7	472.5	27.0	0.04	2.7	0.06	0.64	0.02	0.54	0.20	21.6	83.7	0	0.9
0	16	0.56	7.3	45.2	5.2	0.02	1.0	0.01	—	0.01	0.10	0.04	12.5	2.2	0	0.3
0	53	1.86	24.9	160.2	17.8	0.10	73.9	0.05	0.81	0.02	0.35	0.20	57.0	10.7	0	0.9
0	9	1.65	17.8	182.7	1.0	0.63	0	0.08	0.05	0.04	0.52	0.09	89.6	0.7	0	1.4
0	19	2.47	28.5	247.9	8.5	1.16	1.5	0.17	—	0.09	0.86	0.14	77.0	12.7	0	0.5
0	29	1.02	10.7	196.4	4.1	0.16	137.0	0.04	0.14	0.05	0.29	0.06	60.2	3.1	0	0.5
0	19	0.68	7.1	130.9	2.7	0.11	91.3	0.03	0.09	0.03	0.19	0.04	40.1	2.0	0	0.3
0	10	0.22	3.8	77.5	5.5	0.08	13.7	0.02	0.09	0.01	0.07	0.02	15.9	1.5	0	0.1
0	10	0.22	3.8	77.5	5.5	0.08	13.7	0.02	0.09	0.01	0.07	0.02	15.9	1.5	0	0.1
0	13	0.31	4.7	69.8	10.1	0.06	133.2	0.02	0.10	0.02	0.13	0.03	13.7	6.5	0	0.2
0	18	0.54	7.8	138.3	4.5	0.13	162.4	0.04	0.07	0.03	0.17	0.04	76.2	13.4	0	0.2
0	0	0.67	—	—	32.6	—	0	—	—	—	—	—	—	0	0	—
0	1	0.98	14.4	331.2	2.7	0.54	0	0.16	0.01	0.14	5.31	0.07	46.8	0	0	2.0
0	5	1.35	9.4	277.7	1.6	0.67	0	0.05	0.01	0.23	3.47	0.07	14.0	3.1	0	9.3
0	9	0.61	11.7	100.6	331.5	0.56	0	0.06	0.01	0.01	1.24	0.04	9.4	0	0	3.2
0	1	0.24	4.3	152.6	2.4	0.25	0	0.04	0.01	0.10	1.73	0.05	7.7	1.0	0	4.5
0	39	0.35	—	—	9.9	—	0	—	—	—	—	—	—	0	0	—
0	2	0.31	10.2	84.8	2.9	0.96	0	0.02	0.02	0.12	1.08	0.11	15.2	0.2	0	18.0
0	76	0.84	9.8	104.3	18.8	0.15	265.5	0.03	1.01	0.04	0.19	0.08	52.5	10.4	0	0.5
0	58	0.81	17.9	198.2	14.0	0.11	294.0	0.04	1.12	0.06	0.45	0.10	104.7	39.2	0	0.5
2	54	1.13	32.2	170.8	109.7	0.44	14.0	0.16	1.50	0.12	1.29	0.11	39.6	9.2	0	3.6
0	88	0.61	46.9	215.3	2.8	0.57	15.6	0.09	0.29	0.11	0.72	0.04	134.3	11.2	0	0.6
0	62	0.22	28.8	108.0	4.8	0.34	11.2	0.10	0.21	0.04	0.69	0.15	36.8	13.0	0	0.3
0	23	0.24	11.5	174.3	3.1	0.21	0	0.03	0.02	0.02	0.17	0.12	15.7	5.5	0	0.6
0	17	0.32	6.4	114.5	12.7	0.06	0	0.02	0.01	0.02	0.14	0.06	13.8	2.8	0	0.4
0	22	1.20	13.5	91.6	266.3	0.29	7.8	0.19	—	0.09	2.56	0.05	46.9	1.0	0	2.5
0	18	0.16	8.0	116.8	3.2	0.13	0	0.03	0.01	0.02	0.09	0.09	15.2	5.9	0	0.4
0	13	0.10	5.7	82.4	1.7	0.09	0	0.02	0.01	0.01	0.04	0.08	10.9	3.7	0	0.3
0	22	0.44	6.0	82.8	4.8	0.11	15.0	0.01	0.16	0.02	0.15	0.01	19.2	5.6	0	0.2
0	13	1.23	7.3	1318.4	10.2	2.72	2.2	0.03	0.36	0.12	0.62	0.53	14.6	5.0	0	0.5
0	5	0.23	1.9	21.1	2.1	0.04	16.0	0.00	0.02	0.00	0.05	0.00	5.8	5.1	0	0
0	29	0.45	22.6	286.3	7.8	0.20	0	0.06	0.78	0.04	0.56	0.07	45.2	10.1	0	1.3
0	17	0.80	14.5	147.1	214.2	0.60	23.0	0.10	0.02	0.06	0.62	0.05	37.4	8.2	0	1.4
0	19	1.21	17.6	88.0	57.6	0.53	84.0	0.22	0.02	0.08	1.18	0.09	47.2	7.9	0	0.8
0	34	1.57	20.8	192.0	192.0	0.29	41.6	0.10	0.31	0.06	0.43	0.11	23.2	38.3	0	0.6

TABLE H-1 Table of Food Composition (*continued*) (Computer code is for Cengage Diet Analysis program) (For purposes of calculations, use "0" for t, <1, <.1, <.01, etc.

DA+ Code	Food Description	Quantity	Measure	Wt (g)	H_2O (g)	Ener (kcal)	Prot (g)	Carb (g)	Fiber (g)	Fat (g)	Fat Breakdown (g)			
											Sat	Mono	Poly	*Trans*
	Vegetables, Legumes—*continued*													
1082	Peas and carrots, canned with liquid	½	cup(s)	128	112.4	48	2.8	10.8	2.6	0.3	0.1	0	0.2	—
1083	Peas and carrots, frozen, boiled, drained	½	cup(s)	80	68.6	38	2.5	8.1	2.5	0.3	0.1	0	0.2	—
2750	Snow or sugar peas, frozen, boiled, drained	½	cup(s)	80	69.3	42	2.8	7.2	2.5	0.3	0.1	0	0.1	—
640	Snow or sugar peas, raw	½	cup(s)	32	28.0	13	0.9	2.4	0.8	0.1	0	0	0	—
29324	Split peas, sprouted	½	cup(s)	60	37.4	77	5.3	16.9	—	0.4	0.1	0	0.2	—
	Peppers													
644	Green bell or sweet, boiled, drained	½	cup(s)	68	62.5	19	0.6	4.6	0.8	0.1	0	0	0.1	—
643	Green bell or sweet, raw	½	cup(s)	75	69.9	15	0.6	3.5	1.3	0.1	0	0	0	—
1664	Green hot chili	1	item(s)	45	39.5	18	0.9	4.3	0.7	0.1	0	0	0	—
1663	Green hot chili, canned with liquid	½	cup(s)	68	62.9	14	0.6	3.5	0.9	0.1	0	0	0	—
1086	Jalapeno, canned with liquid	½	cup(s)	68	60.4	18	0.6	3.2	1.8	0.6	0.1	0	0.3	—
8703	Yellow bell or sweet	1	item(s)	186	171.2	50	1.9	11.8	1.7	0.4	0.1	0	0.2	—
1087	**Poi**	½	cup(s)	120	86.0	134	0.5	32.7	0.5	0.2	0	0	0.1	—
	Potatoes													
1090	Au gratin mix, prepared with water, whole milk and butter	½	cup(s)	124	97.7	115	2.8	15.9	1.1	5.1	3.2	1.5	0.2	—
1089	Au gratin, prepared with butter	½	cup(s)	123	90.7	162	6.2	13.8	2.2	9.3	5.8	2.6	0.3	—
5791	Baked, flesh and skin	1	item(s)	202	151.3	188	5.1	42.7	4.4	0.3	0.1	0	0.1	—
645	Baked, flesh only	½	cup(s)	61	46.0	57	1.2	13.1	0.9	0.1	0	0	0	—
1088	Baked, skin only	1	item(s)	58	27.4	115	2.5	26.7	4.6	0.1	0	0	0	—
5795	Boiled in skin, flesh only, drained	1	item(s)	136	104.7	118	2.5	27.4	2.1	0.1	0	0	0.1	—
5794	Boiled, drained, skin and flesh	1	item(s)	150	115.9	129	2.9	29.8	2.5	0.2	0	0	0.1	—
647	Boiled, flesh only	½	cup(s)	78	60.4	67	1.3	15.6	1.4	0.1	0	0	0	—
648	French fried, deep fried, prepared from raw	14	item(s)	70	32.8	187	2.7	23.5	2.9	9.5	1.9	4.2	3.0	—
649	French fried, frozen, heated	14	item(s)	70	43.7	94	1.9	19.4	2.0	3.7	0.7	2.3	0.2	—
1091	Hashed brown	½	cup(s)	78	36.9	207	2.3	27.4	2.5	9.8	1.5	4.1	3.7	—
652	Mashed with margarine and whole milk	½	cup(s)	105	79.0	119	2.1	17.7	1.6	4.4	1.0	2.0	1.2	0.7
653	Mashed, prepared from dehydrated granules with milk, water, and margarine	½	cup(s)	105	79.8	122	2.3	16.9	1.4	5.0	1.3	2.1	1.4	—
2759	Microwaved	1	item(s)	202	145.5	212	4.9	49.0	4.6	0.2	0.1	0	0.1	—
2760	Microwaved in skin, flesh only	½	cup(s)	78	57.1	78	1.6	18.1	1.2	0.1	0	0	0	—
5804	Microwaved, skin only	1	item(s)	58	36.8	77	2.5	17.2	4.2	0.1	0	0	0	—
1097	Potato puffs, frozen, heated	½	cup(s)	64	38.2	122	1.3	17.8	1.6	5.5	1.2	3.9	0.3	—
1094	Scalloped mix, prepared with water, whole milk and butter	½	cup(s)	124	98.4	116	2.6	15.9	1.4	5.3	3.3	1.5	0.2	—
1093	Scalloped, prepared with butter	½	cup(s)	123	99.2	108	3.5	13.2	2.3	4.5	2.8	1.3	0.2	—
	Pumpkin													
1773	Boiled, drained	½	cup(s)	123	114.8	25	0.9	6.0	1.3	0.1	0	0	0	—
656	Canned	½	cup(s)	123	110.2	42	1.3	9.9	3.6	0.3	0.2	0	0	—
	Radicchio													
8731	Leaves, raw	1	cup(s)	40	37.3	9	0.6	1.8	0.4	0.1	0	0	0	—
2498	Raw	1	cup(s)	40	37.3	9	0.6	1.8	0.4	0.1	0	0	0	—
657	**Radishes**	6	item(s)	27	25.7	4	0.2	0.9	0.4	0	0	0	0	—
1099	**Rutabaga, boiled, drained**	½	cup(s)	85	75.5	33	1.1	7.4	1.5	0.2	0	0	0.1	—
658	**Sauerkraut, canned**	½	cup(s)	118	109.2	22	1.1	5.1	3.4	0.2	0	0	0.1	—
	Seaweed													
1102	Kelp	½	cup(s)	40	32.6	17	0.6	3.8	0.5	0.2	0.1	0	0	—
1104	Spirulina, dried	½	cup(s)	8	0.4	22	4.3	1.8	0.3	0.6	0.2	0.1	0.2	—
1106	**Shallots**	3	tablespoon(s)	30	23.9	22	0.8	5.0	—	0	0	0	0	—
	Soybeans													
1670	Boiled	½	cup(s)	86	53.8	149	14.3	8.5	5.2	7.7	1.1	1.7	4.4	—
2825	Dry roasted	½	cup(s)	86	0.7	388	34.0	28.1	7.0	18.6	2.7	4.1	10.5	—
2824	Roasted, salted	½	cup(s)	86	1.7	405	30.3	28.9	15.2	21.8	3.2	4.8	12.3	—
8739	Sprouted, stir fried	½	cup(s)	63	42.3	79	8.2	5.9	0.5	4.5	0.6	1.0	2.5	0
	Soy products													
1813	Soy milk	1	cup(s)	240	211.3	130	7.8	15.1	1.4	4.2	0.5	1.0	2.3	0
2838	Tofu, dried, frozen (koyadofu)	3	ounce(s)	85	4.9	408	40.8	12.4	6.1	25.8	3.7	5.7	14.6	—

PAGE KEY: H-4 = Breads/Baked Goods H-10 = Cereal/Rice/Pasta H-14 = Fruit H-18 = Vegetables/Legumes H-28 = Nuts/Seeds H-30 = Vegetarian H-32 = Dairy H-40 = Eggs H-40 = Seafood H-42 = Meats H-46 = Poultry H-46 = Processed Meats H-48 = Beverages H-52 = Fats/Oils H-54 = Sweets H-56 = Spices/Condiments/Sauces H-60 = Mixed Foods/Soups/Sandwiches H-64 = Fast Food H-84 = Convenience H-86 = Baby Foods

Chol (mg)	Calc (mg)	Iron (mg)	Magn (mg)	Pota (mg)	Sodi (mg)	Zinc (mg)	Vit A (µg)	Thia (mg)	Vit E (mg α)	Ribo (mg)	Niac (mg)	Vit B_6 (mg)	Fola (µg)	Vit C (mg)	Vit B_{12} (µg)	Sele (µg)
0	29	0.96	17.9	127.5	331.5	0.74	368.5	0.09	—	0.07	0.74	0.11	23.0	8.4	0	1.1
0	18	0.75	12.8	126.4	54.4	0.36	380.8	0.18	0.41	0.05	0.92	0.07	20.8	6.5	0	0.9
0	47	1.92	22.4	173.6	4.0	0.39	52.8	0.05	0.37	0.09	0.45	0.13	28.0	17.6	0	0.6
0	14	0.65	7.6	63.0	1.3	0.08	17.0	0.04	0.12	0.02	0.19	0.05	13.2	18.9	0	0.2
0	22	1.34	33.6	228.6	12.0	0.62	4.8	0.12	—	0.08	1.84	0.14	86.4	6.2	0	0.4
0	6	0.31	6.8	112.9	1.4	0.08	15.6	0.04	0.34	0.02	0.32	0.15	10.9	50.6	0	0.2
0	7	0.25	7.5	130.4	2.2	0.09	13.4	0.04	0.27	0.02	0.35	0.16	7.5	59.9	0	0
0	8	0.54	11.3	153.0	3.2	0.13	26.6	0.04	0.31	0.04	0.42	0.12	10.4	109.1	0	0.2
0	5	0.34	9.5	127.2	797.6	0.10	24.5	0.01	0.46	0.02	0.54	0.10	6.8	46.2	0	0.2
0	16	1.28	10.2	131.2	1136.3	0.23	57.8	0.03	0.47	0.03	0.27	0.13	9.5	6.8	0	0.3
0	20	0.85	22.3	394.3	3.7	0.31	18.6	0.05	—	0.04	1.65	0.31	48.4	341.3	0	0.6
0	19	1.06	28.8	219.6	14.4	0.26	3.6	0.16	2.76	0.05	1.32	0.33	25.2	4.8	0	0.8
19	103	0.39	18.6	271.0	543.3	0.29	64.4	0.02	—	0.10	1.16	0.05	8.7	3.8	0	3.3
28	146	0.78	24.5	485.1	530.4	0.85	78.4	0.08	—	0.14	1.22	0.21	13.5	12.1	0	3.3
0	30	2.18	56.6	1080.7	20.2	0.72	2.0	0.12	0.08	0.09	2.84	0.62	56.6	19.4	0	0.8
0	3	0.21	15.3	238.5	3.1	0.18	0	0.06	0.02	0.01	0.85	0.18	5.5	7.8	0	0.2
0	20	4.08	24.9	332.3	12.2	0.28	0.6	0.07	0.02	0.06	1.77	0.35	12.8	7.8	0	0.4
0	7	0.42	29.9	515.4	5.4	0.40	0	0.14	0.01	0.02	1.95	0.40	13.6	17.7	0	0.4
0	13	1.27	34.1	572.0	7.4	0.46	0	0.14	0.01	0.03	2.13	0.44	15.0	18.4	0	—
0	6	0.24	15.6	255.8	3.9	0.21	0	0.07	0.01	0.01	1.02	0.21	7.0	5.8	0	0.2
0	16	1.05	30.8	567.0	8.4	0.39	0	0.08	0.09	0.03	1.34	0.37	16.1	21.2	0	0.4
0	8	0.51	18.2	315.7	271.6	0.26	0	0.09	0.07	0.02	1.55	0.12	19.6	9.3	0	0.1
0	11	0.43	27.3	449.3	266.8	0.37	0	0.13	0.01	0.03	1.80	0.37	12.5	10.1	0	0.4
1	23	0.27	19.9	344.4	349.6	0.31	43.0	0.09	0.44	0.04	1.23	0.25	9.4	11.0	0.1	0.8
2	36	0.21	21.0	164.8	179.5	0.26	49.3	0.09	0.53	0.09	0.90	0.16	8.4	6.8	0.1	5.9
0	22	2.50	54.5	902.9	16.2	0.72	0	0.24	—	0.06	3.46	0.69	24.2	30.5	0	0.8
0	4	0.31	19.4	319.0	5.4	0.25	0	0.10	—	0.01	1.26	0.25	9.3	11.7	0	0.3
0	27	3.44	21.5	377.0	9.3	0.29	0	0.04	0.01	0.04	1.28	0.28	9.9	8.9	0	0.3
0	9	0.41	10.9	199.7	307.2	0.21	0	0.08	0.15	0.02	0.97	0.08	9.0	4.0	0	0.4
14	45	0.47	17.4	252.2	423.7	0.31	43.5	0.02	—	0.06	1.28	0.05	12.4	4.1	0	2.0
15	70	0.70	23.3	463.1	410.4	0.49	0	0.08	—	0.11	1.29	0.22	13.5	13.0	0	2.0
0	18	0.69	11.0	281.8	1.2	0.28	306.3	0.03	0.98	0.09	0.50	0.05	11.0	5.8	0	0.2
0	32	1.70	28.2	252.4	6.1	0.20	953.1	0.02	1.29	0.06	0.45	0.06	14.7	5.1	0	0.5
0	8	0.23	5.2	120.8	8.8	0.25	0.4	0.01	0.90	0.01	0.10	0.02	24.0	3.2	0	0.4
0	8	0.23	5.2	120.8	8.8	0.25	0.4	0.01	0.90	0.01	0.10	0.02	24.0	3.2	0	0.4
0	7	0.09	2.7	62.9	10.5	0.07	0	0.00	0.00	0.01	0.06	0.01	6.8	4.0	0	0.2
0	41	0.45	19.6	277.1	17.0	0.30	0	0.07	0.27	0.04	0.61	0.09	12.8	16.0	0	0.6
0	35	1.73	15.3	200.6	780.0	0.22	1.2	0.03	0.17	0.03	0.17	0.15	28.3	17.3	0	0.7
0	67	1.12	48.4	35.6	93.2	0.48	2.4	0.02	0.32	0.04	0.16	0.00	72.0	1.2	0	0.3
0	9	2.14	14.6	102.2	78.6	0.15	2.2	0.18	0.38	0.28	0.96	0.03	7.1	0.8	0	0.5
0	11	0.36	6.3	100.2	3.6	0.12	18.0	0.02	—	0.01	0.06	0.09	10.2	2.4	—	0.4
0	88	4.42	74.0	442.9	0.9	0.98	0	0.13	0.30	0.24	0.34	0.20	46.4	1.5	0	6.3
0	120	3.39	196.1	1173.0	1.7	4.10	0	0.36	—	0.64	0.90	0.19	176.3	4.0	0	16.6
0	119	3.35	124.7	1264.2	140.2	2.70	8.6	0.08	0.78	0.12	1.21	0.17	181.5	1.9	0	16.4
0	52	0.25	60.4	356.6	8.8	1.32	0.6	0.26	—	0.12	0.69	0.10	79.9	7.5	0	0.4
0	60	1.53	60.0	283.2	122.4	0.28	0	0.14	0.26	0.16	1.23	0.18	43.2	0	0	11.5
0	310	8.27	50.2	17.0	5.1	4.16	22.1	0.42	—	0.27	1.01	0.24	78.2	0.6	0	46.2

TABLE H-1 Table of Food Composition (*continued*) (Computer code is for Cengage Diet Analysis program) (For purposes of calculations, use "0" for t, <1, <.1, <.01, etc.)

DA+ Code	Food Description	Quantity	Measure	Wt (g)	H_2O (g)	Ener (kcal)	Prot (g)	Carb (g)	Fiber (g)	Fat (g)	Fat Breakdown (g) Sat	Mono	Poly	*Trans*
Vegetables, Legumes—*continued*														
13844	Tofu, extra firm	3	ounce(s)	85	—	86	8.6	2.2	1.1	4.3	0.5	0.9	2.8	—
13843	Tofu, firm	3	ounce(s)	85	—	75	7.5	2.2	0.5	3.2	0	0.9	2.3	—
1816	Tofu, firm, with calcium sulfate and magnesium chloride (nigari)	3	ounce(s)	85	72.2	60	7.0	1.4	0.8	3.5	0.7	1.0	1.5	—
1817	Tofu, fried	3	ounce(s)	85	43.0	230	14.6	8.9	3.3	17.2	2.5	3.8	9.7	—
13841	Tofu, silken	3	ounce(s)	85	—	42	3.7	1.9	0	2.3	0.5	—	—	—
13842	Tofu, soft	3	ounce(s)	85	—	65	6.5	1.1	0.5	3.2	0.5	1.1	2.2	—
1671	Tofu, soft, with calcium sulfate and magnesium chloride (nigari)	3	ounce(s)	85	74.2	52	5.6	1.5	0.2	3.1	0.5	0.7	1.8	—
	Spinach													
663	Canned, drained	½	cup(s)	107	98.2	25	3.0	3.6	2.6	0.5	0.1	0	0.2	—
660	Chopped, boiled, drained	½	cup(s)	90	82.1	21	2.7	3.4	2.2	0.2	0	0	0.1	—
661	Chopped, frozen, boiled, drained	½	cup(s)	95	84.5	32	3.8	4.6	3.5	0.8	0.1	0	0.4	—
662	Leaf, frozen, boiled, drained	½	cup(s)	95	84.5	32	3.8	4.6	3.5	0.8	0.1	0	0.4	—
659	Raw, chopped	1	cup(s)	30	27.4	7	0.9	1.1	0.7	0.1	0	0	0	—
8470	Trimmed leaves	1	cup(s)	32	27.5	3	0.9	0	2.8	0.1	—	—	—	—
	Squash													
1662	Acorn winter, baked	½	cup(s)	103	85.0	57	1.1	14.9	4.5	0.1	0	0	0.1	—
29702	Acorn winter, boiled, mashed	½	cup(s)	123	109.9	42	0.8	10.8	3.2	0.1	0	0	0	—
29451	Butternut, frozen, boiled	½	cup(s)	122	106.9	47	1.5	12.2	1.8	0.1	0	0	0	—
1661	Butternut winter, baked	½	cup(s)	102	89.5	41	0.9	10.7	3.4	0.1	0	0	0	—
32773	Butternut winter, frozen, boiled, mashed, no salt added	½	cup(s)	121	106.4	47	1.5	12.2	—	0.1	0	0	0	—
29700	Crookneck and straightneck summer, boiled, drained	½	cup(s)	65	60.9	12	0.6	2.6	1.2	0.1	0	0	0.1	—
29703	Hubbard winter, baked	½	cup(s)	102	86.8	51	2.5	11.0	—	0.6	0.1	0	0.3	—
1660	Hubbard winter, boiled, mashed	½	cup(s)	118	107.5	35	1.7	7.6	3.4	0.4	0.1	0	0.2	—
29704	Spaghetti winter, boiled, drained, or baked	½	cup(s)	78	71.5	21	0.5	5.0	1.1	0.2	0	0	0.1	—
664	Summer, all varieties, sliced, boiled, drained	½	cup(s)	90	84.3	18	0.8	3.9	1.3	0.3	0.1	0	0.1	—
665	Winter, all varieties, baked, mashed	½	cup(s)	103	91.4	38	0.9	9.1	2.9	0.4	0.1	0	0.2	—
1112	Zucchini summer, boiled, drained	½	cup(s)	90	85.3	14	0.6	3.5	1.3	0	0	0	0	—
1113	Zucchini summer, frozen, boiled, drained	½	cup(s)	112	105.6	19	1.3	4.0	1.4	0.1	0	0	0.1	—
	Sweet potatoes													
666	Baked, peeled	½	cup(s)	100	75.8	90	2.0	20.7	3.3	0.2	0	0	0.1	—
667	Boiled, mashed	½	cup(s)	164	131.4	125	2.2	29.1	4.1	0.2	0.1	0	0.1	—
668	Candied, home recipe	½	cup(s)	91	61.1	132	0.8	25.4	2.2	3.0	1.2	0.6	0.1	—
670	Canned, vacuum pack	½	cup(s)	100	76.0	91	1.7	21.1	1.8	0.2	0	0	0.1	—
2765	Frozen, baked	½	cup(s)	88	64.5	88	1.5	20.5	1.6	0.1	0	0	0	—
1136	Yams, baked or boiled, drained	½	cup(s)	68	47.7	79	1.0	18.7	2.7	0.1	0	0	0	—
32785	**Taro shoots, cooked, no salt added**	½	cup(s)	70	66.7	10	0.5	2.2	—	0.1	0	0	0	—
	Tomatillo													
8774	Raw	2	item(s)	68	62.3	22	0.7	4.0	1.3	0.7	0.1	0.1	0.3	—
8777	Raw, chopped	½	cup(s)	66	60.5	21	0.6	3.9	1.3	0.7	0.1	0.1	0.3	—
	Tomato													
16846	Cherry, fresh	5	item(s)	85	80.3	15	0.7	3.3	1.0	0.2	0	0	0.1	—
671	Fresh, ripe, red	1	item(s)	123	116.2	22	1.1	4.8	1.5	0.2	0	0	0.1	—
675	Juice, canned	½	cup(s)	122	114.1	21	0.9	5.2	0.5	0.1	0	0	0	—
75	Juice, no salt added	½	cup(s)	122	114.1	21	0.9	5.2	0.5	0.1	0	0	0	—
1699	Paste, canned	2	tablespoon(s)	33	24.1	27	1.4	6.2	1.3	0.2	0	0	0.1	—
1700	Puree, canned	¼	cup(s)	63	54.9	24	1.0	5.6	1.2	0.1	0	0	0.1	—
1118	Red, boiled	½	cup(s)	120	113.2	22	1.1	4.8	0.8	0.1	0	0	0.1	—
3952	Red, diced	½	cup(s)	90	85.1	16	0.8	3.5	1.1	0.2	0	0	0.1	—
1120	Red, stewed, canned	½	cup(s)	128	116.7	33	1.2	7.9	1.3	0.2	0	0	0.1	—
1125	Sauce, canned	¼	cup(s)	61	55.6	15	0.8	3.3	0.9	0.1	0	0	0	—
8778	Sun dried	½	cup(s)	27	3.9	70	3.8	15.1	3.3	0.8	0.1	0.1	0.3	—
8783	Sun dried in oil, drained	¼	cup(s)	28	14.8	59	1.4	6.4	1.6	3.9	0.5	2.4	0.6	—
	Turnips													
678	Turnip greens, chopped, boiled, drained	½	cup(s)	72	67.1	14	0.8	3.1	2.5	0.2	0	0	0.1	—

PAGE KEY: H-4 = Breads/Baked Goods H-10 = Cereal/Rice/Pasta H-14 = Fruit H-18 = Vegetables/Legumes H-28 = Nuts/Seeds H-30 = Vegetarian H-32 = Dairy H-40 = Eggs H-40 = Seafood H-42 = Meats H-46 = Poultry H-46 = Processed Meats H-48 = Beverages H-52 = Fats/Oils H-54 = Sweets H-56 = Spices/Condiments/Sauces H-60 = Mixed Foods/Soups/Sandwiches H-64 = Fast Food H-84 = Convenience H-86 = Baby Foods

Chol (mg)	Calc (mg)	Iron (mg)	Magn (mg)	Pota (mg)	Sodi (mg)	Zinc (mg)	Vit A (µg)	Thia (mg)	Vit E (mg α)	Ribo (mg)	Niac (mg)	Vit B_6 (mg)	Fola (µg)	Vit C (mg)	Vit B_{12} (µg)	Sele (µg)
0	65	1.16	84.1	—	0	—	0	—	—	—	—	—	—	0	0	—
0	108	1.16	56.1	—	0	—	0	—	—	—	—	—	—	0	0	—
0	171	1.36	31.5	125.9	10.2	0.70	0	0.05	0.01	0.05	0.08	0.06	16.2	0.2	0	8.4
0	316	4.14	51.0	124.2	13.6	1.69	0.9	0.14	0.03	0.04	0.08	0.08	23.0	0	0	24.2
0	56	0.34	33.1	—	4.7	—	0	—	—	—	—	—	—	0	1.7	—
0	108	1.16	35.5	—	0	—	0	—	—	—	—	—	—	0	1.9	—
0	94	0.94	23.0	102.1	6.8	0.54	0	0.04	0.01	0.03	0.45	0.04	37.4	0.2	0	7.6
0	136	2.45	81.3	370.2	28.9	0.48	524.3	0.02	2.08	0.14	0.41	0.11	104.8	15.3	0	1.5
0	122	3.21	78.3	419.4	63.0	0.68	471.6	0.08	1.87	0.21	0.44	0.21	131.4	8.8	0	1.4
0	145	1.86	77.9	286.9	92.2	0.46	572.9	0.07	3.36	0.16	0.41	0.12	115.0	2.1	0	5.2
0	145	1.86	77.9	286.9	92.2	0.46	572.9	0.07	3.36	0.16	0.41	0.12	115.0	2.1	0	5.2
0	30	0.81	23.7	167.4	23.7	0.16	140.7	0.02	0.61	0.06	0.22	0.06	58.2	8.4	0	0.3
0	25	2.13	25.5	134.1	38.0	0.18	—	0.03	—	0.05	0.18	0.07	0	7.5	0	—
0	45	0.95	44.1	447.9	4.1	0.17	21.5	0.17	—	0.01	0.90	0.19	19.5	11.1	0	0.7
0	32	0.68	31.9	322.2	3.7	0.13	50.2	0.12	—	0.01	0.65	0.14	13.5	8.0	0	0.5
0	23	0.70	10.9	161.9	2.4	0.14	203.3	0.06	0.14	0.05	0.56	0.08	19.5	4.3	0	0.6
0	42	0.61	29.6	289.6	4.1	0.13	569.1	0.07	1.31	0.01	0.99	0.12	19.4	15.4	0	0.5
0	23	0.70	10.9	161.2	2.4	0.14	202.4	0.06	—	0.05	0.56	0.08	19.4	4.2	0	0.6
0	14	0.31	13.6	137.1	1.3	0.19	5.2	0.03	—	0.02	0.29	0.07	14.9	5.4	0	0.1
0	17	0.48	22.4	365.1	8.2	0.15	308.0	0.07	—	0.04	0.57	0.17	16.3	9.7	0	0.6
0	12	0.33	15.3	252.5	5.9	0.11	236.0	0.05	0.14	0.03	0.39	0.12	11.8	7.7	0	0.4
0	16	0.26	8.5	90.7	14.0	0.15	4.7	0.02	0.09	0.01	0.62	0.07	6.2	2.7	0	0.2
0	24	0.32	21.6	172.8	0.9	0.35	9.9	0.04	0.12	0.03	0.46	0.05	18.0	5.0	0	0.2
0	23	0.45	13.3	247.0	1.0	0.23	267.5	0.02	0.12	0.07	0.51	0.17	20.5	9.8	0	0.4
0	12	0.32	19.8	227.7	2.7	0.16	50.4	0.04	0.11	0.04	0.39	0.07	15.3	4.1	0	0.2
0	19	0.54	14.5	216.3	2.2	0.22	10	0.05	0.13	0.04	0.43	0.05	8.9	4.1	0	0.2
0	38	0.69	27.0	475.0	36.0	0.32	961.0	0.10	0.71	0.10	1.48	0.28	6.0	19.6	0	0.2
0	44	1.18	29.5	377.2	44.3	0.33	1290.7	0.09	1.54	0.08	0.88	0.27	9.8	21.0	0	0.3
7	24	1.03	10.0	172.6	63.9	0.13	0	0.01	—	0.03	0.36	0.03	10.0	6.1	0	0.7
0	22	0.89	22.0	312.0	53.0	0.18	399.0	0.04	1.00	0.06	0.74	0.19	17.0	26.4	0	0.7
0	31	0.47	18.4	330.1	7.0	0.26	913.3	0.05	0.67	0.04	0.49	0.16	19.3	8.0	0	0.5
0	10	0.35	12.2	455.6	5.4	0.13	4.1	0.06	0.23	0.01	0.37	0.15	10.9	8.2	0	0.5
0	10	0.28	5.6	240.8	1.4	0.37	2.1	0.02	—	0.03	0.56	0.07	2.1	13.2	0	0.7
0	5	0.42	13.6	182.2	0.7	0.15	4.1	0.03	0.25	0.02	1.25	0.03	4.8	8.0	0	0.3
0	5	0.41	13.2	176.9	0.7	0.15	4.0	0.03	0.25	0.02	1.22	0.04	4.6	7.7	0	0.3
0	9	0.22	9.4	201.5	4.3	0.14	35.7	0.03	0.45	0.01	0.50	0.06	12.8	10.8	0	0
0	12	0.33	13.5	291.5	6.2	0.20	51.7	0.04	0.66	0.02	0.73	0.09	18.5	15.6	0	0
0	12	0.52	13.4	278.2	326.8	0.18	27.9	0.06	0.39	0.04	0.82	0.14	24.3	22.2	0	0.4
0	12	0.52	13.4	278.2	12.2	0.18	27.9	0.06	0.39	0.04	0.82	0.14	24.3	22.2	0	0.4
0	12	0.97	13.8	332.6	259.1	0.20	24.9	0.02	1.41	0.05	1.00	0.07	3.9	7.2	0	1.7
0	11	1.11	14.4	274.4	249.4	0.22	16.3	0.01	1.23	0.05	0.91	0.07	6.9	6.6	0	0.4
0	13	0.82	10.8	261.6	13.2	0.17	28.8	0.04	0.67	0.03	0.64	0.10	15.6	27.4	0	0.6
0	9	0.24	9.9	213.3	4.5	0.15	37.8	0.03	0.48	0.01	0.53	0.07	13.5	11.4	0	0
0	43	1.70	15.3	263.9	281.8	0.22	11.5	0.06	1.06	0.04	0.91	0.02	6.4	10.1	0	0.8
0	8	0.62	9.8	201.9	319.6	0.12	10.4	0.01	0.87	0.04	0.59	0.06	6.7	4.3	0	0.1
0	30	2.45	52.4	925.3	565.7	0.53	11.9	0.14	0.00	0.13	2.44	0.09	18.4	10.6	0	1.5
0	13	0.73	22.3	430.4	73.2	0.21	17.6	0.05	—	0.10	0.99	0.08	6.3	28.0	0	0.8
0	99	0.58	15.8	146.2	20.9	0.10	274.3	0.03	1.35	0.05	0.30	0.13	85.0	19.7	0	0.6

TABLE H-1 Table of Food Composition (*continued*) (Computer code is for Cengage Diet Analysis program) (For purposes of calculations, use "0" for t, <1, <.1, <.01, et

DA+ Code	Food Description	Quantity	Measure	Wt (g)	H_2O (g)	Ener (kcal)	Prot (g)	Carb (g)	Fiber (g)	Fat (g)	Fat Breakdown (g) Sat	Mono	Poly	*Trans*
	Vegetables, Legumes—*continued*													
679	Turnip greens, frozen, chopped, boiled, drained	½	cup(s)	82	74.1	24	2.7	4.1	2.8	0.3	0.1	0	0.1	—
677	Turnips, cubed, boiled, drained	½	cup(s)	78	73.0	17	0.6	3.9	1.6	0.1	0	0	0	—
	Vegetables, mixed													
1132	Canned, drained	½	cup(s)	82	70.9	40	2.1	7.5	2.4	0.2	0	0	0.1	—
680	Frozen, boiled, drained	½	cup(s)	91	75.7	59	2.6	11.9	4.0	0.1	0	0	0.1	—
7489	V8 100% vegetable juice	½	cup(s)	120	—	25	1.0	5.0	1.0	0	0	0	0	0
7490	V8 low sodium vegetable juice	½	cup(s)	120	—	25	0	6.5	1.0	0	0	0	0	0
7491	V8 spicy hot vegetable juice	½	cup(s)	120	—	25	1.0	5.0	0.5	0	0	0	0	0
	Water chestnuts													
31073	Sliced, drained	½	cup(s)	75	70.0	20	0	5.0	1.0	0	0	0	0	0
31087	Whole	½	cup(s)	75	70.0	20	0	5.0	1.0	0	0	0	0	0
1135	**Watercress**	1	cup(s)	34	32.3	4	0.8	0.4	0.2	0	0	0	0	—
	Nuts, Seeds, and Products													
	Almonds													
32940	Almond butter with salt added	1	tablespoon(s)	16	0.2	101	2.4	3.4	0.6	9.5	0.9	6.1	2.0	—
1137	Almond butter, no salt added	1	tablespoon(s)	16	0.2	101	2.4	3.4	0.6	9.5	0.9	6.1	2.0	—
32886	Blanched	¼	cup(s)	36	1.6	211	8.0	7.2	3.8	18.3	1.4	11.7	4.4	—
32887	Dry roasted, no salt added	¼	cup(s)	35	0.9	206	7.6	6.7	4.1	18.2	1.4	11.6	4.4	—
29724	Dry roasted, salted	¼	cup(s)	35	0.9	206	7.6	6.7	4.1	18.2	1.4	11.6	4.4	—
29725	Oil roasted, salted	¼	cup(s)	39	1.1	238	8.3	6.9	4.1	21.7	1.7	13.7	5.3	—
508	Slivered	¼	cup(s)	27	1.3	155	5.7	5.9	3.3	13.3	1.0	8.3	3.3	0
1138	**Beechnuts, dried**	¼	cup(s)	57	3.8	328	3.5	19.1	5.3	28.5	3.3	12.5	11.4	—
517	**Brazil nuts, dried, unblanched**	¼	cup(s)	35	1.2	230	5.0	4.3	2.6	23.3	5.3	8.6	7.2	—
1166	**Breadfruit seeds, roasted**	¼	cup(s)	57	28.3	118	3.5	22.8	3.4	1.5	0.4	0.2	0.8	—
1139	**Butternuts, dried**	¼	cup(s)	30	1.0	184	7.5	3.6	1.4	17.1	0.4	3.1	12.8	—
	Cashews													
32931	Cashew butter with salt added	1	tablespoon(s)	16	0.5	94	2.8	4.4	0.3	7.9	1.6	4.7	1.3	—
32889	Cashew butter, no salt added	1	tablespoon(s)	16	0.5	94	2.8	4.4	0.3	7.9	1.6	4.7	1.3	—
1140	Dry roasted	¼	cup(s)	34	0.6	197	5.2	11.2	1.0	15.9	3.1	9.4	2.7	—
518	Oil roasted	¼	cup(s)	32	1.1	187	5.4	9.6	1.1	15.4	2.7	8.4	2.8	—
	Coconut, shredded													
32896	Dried, not sweetened	¼	cup(s)	23	0.7	152	1.6	5.4	3.8	14.9	13.2	0.6	0.2	—
1153	Dried, shredded, sweetened	¼	cup(s)	23	2.9	116	0.7	11.1	1.0	8.3	7.3	0.4	0.1	—
520	Shredded	¼	cup(s)	20	9.4	71	0.7	3.0	1.8	6.7	5.9	0.3	0.1	—
	Chestnuts													
1152	Chinese, roasted	¼	cup(s)	36	14.6	87	1.6	19.0	—	0.4	0.1	0.2	0.1	—
32895	European, boiled and steamed	¼	cup(s)	46	31.3	60	0.9	12.8	—	0.6	0.1	0.2	0.2	—
32911	European, roasted	¼	cup(s)	36	14.5	88	1.1	18.9	1.8	0.8	0.1	0.3	0.3	—
32922	Japanese, boiled and steamed	¼	cup(s)	36	31.0	20	0.3	4.5	—	0.1	0	0	0	—
32923	Japanese, roasted	¼	cup(s)	36	18.1	73	1.1	16.4	—	0.3	0	0.1	0.1	—
4958	**Flax seeds or linseeds**	¼	cup(s)	43	3.3	225	8.4	12.3	11.9	17.7	1.7	3.2	12.6	0
32904	**Ginkgo nuts, dried**	¼	cup(s)	39	4.8	136	4.0	28.3	—	0.8	0.1	0.3	0.3	—
	Hazelnuts or filberts													
32901	Blanched	¼	cup(s)	30	1.7	189	4.1	5.1	3.3	18.3	1.4	14.5	1.7	—
32902	Dry roasted, no salt added	¼	cup(s)	30	0.8	194	4.5	5.3	2.8	18.7	1.3	14.0	2.5	—
1156	**Hickory nuts, dried**	¼	cup(s)	30	0.8	197	3.8	5.5	1.9	19.3	2.1	9.8	6.6	—
	Macadamias													
32905	Dry roasted, no salt added	¼	cup(s)	34	0.5	241	2.6	4.5	2.7	25.5	4.0	19.9	0.5	—
32932	Dry roasted, with salt added	¼	cup(s)	34	0.5	240	2.6	4.3	2.7	25.5	4.0	19.9	0.5	—
1157	Raw	¼	cup(s)	34	0.5	241	2.6	4.6	2.9	25.4	4.0	19.7	0.5	—
	Mixed nuts													
1159	With peanuts, dry roasted	¼	cup(s)	34	0.6	203	5.9	8.7	3.1	17.6	2.4	10.8	3.7	—
32933	With peanuts, dry roasted, with salt added	¼	cup(s)	34	0.6	203	5.9	8.7	3.1	17.6	2.4	10.8	3.7	—
32906	Without peanuts, oil roasted, no salt added	¼	cup(s)	36	1.1	221	5.6	8.0	2.0	20.2	3.3	11.9	4.1	—
	Peanuts													
2807	Dry roasted	¼	cup(s)	37	0.6	214	8.6	7.9	2.9	18.1	2.5	9.0	5.7	—
2806	Dry roasted, salted	¼	cup(s)	37	0.6	214	8.6	7.9	2.9	18.1	2.5	9.0	5.7	—
1763	Oil roasted, salted	¼	cup(s)	36	0.5	216	10.1	5.5	3.4	18.9	3.1	9.4	5.5	—
1884	Peanut butter, chunky	1	tablespoon(s)	16	0.2	94	3.8	3.5	1.3	8.0	1.3	3.9	2.4	—
30303	Peanut butter, low sodium	1	tablespoon(s)	16	0.2	95	4.0	3.1	0.9	8.2	1.8	3.9	2.2	—
30305	Peanut butter, reduced fat	1	tablespoon(s)	18	0.2	94	4.7	6.4	0.9	6.1	1.3	2.9	1.8	—

PAGE KEY: H-4 = Breads/Baked Goods H-10 = Cereal/Rice/Pasta H-14 = Fruit H-18 = Vegetables/Legumes H-28 = Nuts/Seeds H-30 = Vegetarian H-32 = Dairy H-40 = Eggs H-40 = Seafood H-42 = Meats H-46 = Poultry H-46 = Processed Meats H-48 = Beverages H-52 = Fats/Oils H-54 = Sweets H-56 = Spices/Condiments/Sauces H-60 = Mixed Foods/Soups/Sandwiches H-64 = Fast Food H-84 = Convenience H-86 = Baby Foods

Chol (mg)	Calc (mg)	Iron (mg)	Magn (mg)	Pota (mg)	Sodi (mg)	Zinc (mg)	Vit A (μg)	Thia (mg)	Vit E (mg α)	Ribo (mg)	Niac (mg)	Vit B_6 (mg)	Fola (μg)	Vit C (mg)	Vit B_{12} (μg)	Sele (μg)
0	125	1.59	21.3	183.7	12.3	0.34	441.2	0.04	2.18	0.06	0.38	0.06	32.0	17.9	0	1.0
0	26	0.14	7.0	138.1	12.5	0.09	0	0.02	0.02	0.02	0.23	0.05	7.0	9.0	0	0.2
0	22	0.86	13.0	237.2	121.4	0.33	475.1	0.04	0.24	0.04	0.47	0.06	19.6	4.1	0	0.2
0	23	0.74	20.0	153.8	31.9	0.44	194.7	0.06	0.34	0.10	0.77	0.06	17.3	2.9	0	0.3
0	20	0.36	12.9	260.0	310.0	0.24	100.0	0.05	—	0.03	0.87	0.17	—	30.0	0	—
0	20	0.36	—	450.0	70.0	—	100.0	0.02	—	0.02	0.75	—	—	30.0	0	—
0	20	0.36	12.9	240.0	360.0	0.24	50.0	0.05	—	0.03	0.88	0.17	—	15.0	0	—
0	7	0.00	—	—	5.0	—	0	—	—	—	—	—	—	2.0	—	—
0	7	0.00	—	—	5.0	—	0	—	—	—	—	—	—	2.0	—	—
0	41	0.06	7.1	112.2	13.9	0.03	54.4	0.03	0.34	0.04	0.06	0.04	3.1	14.6	0	0.3
0	43	0.59	48.5	121.3	72.0	0.49	0	0.02	4.16	0.10	0.46	0.01	10.4	0.1	0	0.8
0	43	0.59	48.5	121.3	1.8	0.48	0	0.02	—	0.09	0.46	0.01	10.4	0.1	0	—
0	78	1.34	99.7	249.0	10.2	1.13	0	0.07	8.95	0.20	1.32	0.04	10.9	0	0	1.0
0	92	1.55	98.7	257.4	0.3	1.22	0	0.02	8.97	0.29	1.32	0.04	11.4	0	0	1.0
0	92	1.55	98.7	257.4	117.0	1.22	0	0.02	8.97	0.29	1.32	0.04	11.4	0	0	1.0
0	114	1.44	107.5	274.4	133.1	1.20	0	0.03	10.19	0.30	1.43	0.04	10.6	0	0	1.1
0	71	1.00	72.4	190.4	0.3	0.83	0	0.05	7.07	0.27	0.91	0.03	13.5	0	0	0.7
0	1	1.39	0	579.7	21.7	0.20	0	0.16	—	0.20	0.48	0.38	64.4	8.8	0	4.0
0	56	0.85	131.6	230.7	1.1	1.42	0	0.21	2.00	0.01	0.10	0.03	7.7	0.2	0	671.0
0	49	0.50	35.3	616.7	15.9	0.58	8.5	0.22	—	0.12	4.20	0.22	33.6	4.3	0	8.0
0	16	1.21	71.1	126.3	0.3	0.94	1.8	0.12	—	0.04	0.31	0.17	19.8	1.0	0	5.2
0	7	0.81	41.3	87.4	98.2	0.83	0	0.05	0.15	0.03	0.26	0.04	10.9	0	0	1.8
0	7	0.81	41.3	87.4	2.4	0.83	0	0.05	—	0.03	0.26	0.04	10.9	0	0	1.8
0	15	2.06	89.1	193.5	5.5	1.92	0	0.07	0.32	0.07	0.48	0.09	23.6	0	0	4.0
0	14	1.95	88.0	203.8	4.2	1.73	0	0.12	0.30	0.07	0.56	0.10	8.1	0.1	0	6.5
0	6	0.76	20.7	125.2	8.5	0.46	0	0.01	0.10	0.02	0.13	0.07	2.1	0.3	0	4.3
0	3	0.45	11.6	78.4	60.9	0.42	0	0.01	0.09	0.00	0.11	0.06	1.9	0.2	0	3.9
0	3	0.48	6.4	71.2	4.0	0.21	0	0.01	0.04	0.00	0.11	0.01	5.2	0.7	0	2.0
0	7	0.54	32.6	173.0	1.4	0.33	0	0.05	—	0.03	0.54	0.15	26.1	13.9	0	2.6
0	21	0.80	24.8	328.9	12.4	0.11	0.5	0.06	—	0.03	0.32	0.10	17.5	12.3	0	—
0	10	0.32	11.8	211.6	0.7	0.20	0.4	0.08	0.18	0.05	0.48	0.18	25.0	9.3	0	0.4
0	4	0.19	6.5	42.8	1.8	0.14	0.4	0.04	—	0.01	0.19	0.03	6.1	3.4	0	—
0	13	0.75	23.2	154.8	6.9	0.51	1.4	0.15	—	—	0.24	0.14	21.4	10.1	0	—
0	142	2.13	156.1	354.0	11.9	1.83	0	0.06	0.14	0.06	0.59	0.39	118.4	0.5	0	2.3
0	8	0.62	20.7	390.2	5.1	0.26	21.5	0.17	—	0.07	4.58	0.25	41.4	11.4	0	—
0	45	0.98	48.0	197.4	0	0.66	0.6	0.14	5.25	0.03	0.46	0.17	23.4	0.6	0	1.2
0	37	1.31	51.9	226.5	0	0.74	0.9	0.10	4.58	0.03	0.61	0.18	26.4	1.1	0	1.2
0	18	0.64	51.9	130.8	0.3	1.29	2.1	0.26	—	0.04	0.27	0.06	12.0	0.6	0	2.4
0	23	0.88	39.5	121.6	1.3	0.43	0	0.23	0.19	0.02	0.76	0.12	3.4	0.2	0	3.9
0	23	0.88	39.5	121.6	88.8	0.43	0	0.23	0.19	0.02	0.76	0.12	3.4	0.2	0	3.9
0	28	1.24	43.6	123.3	1.7	0.44	0	0.40	0.18	0.05	0.83	0.09	3.7	0.4	0	1.2
0	24	1.27	77.1	204.5	4.1	1.30	0.3	0.07	—	0.07	1.61	0.10	17.1	0.1	0	1.0
0	24	1.26	77.1	204.5	229.1	1.30	0	0.06	3.74	0.06	1.61	0.10	17.1	0.1	0	2.6
0	38	0.92	90.4	195.8	4.0	1.67	0.4	0.18	—	0.17	0.70	0.06	20.2	0.2	0	—
0	20	0.82	64.2	240.2	2.2	1.20	0	0.16	2.52	0.03	4.93	0.09	52.9	0	0	2.7
0	20	0.82	64.2	240.2	296.7	1.20	0	0.16	2.84	0.03	4.93	0.09	52.9	0	0	2.7
0	22	0.54	63.4	261.4	115.2	1.18	0	0.03	2.49	0.03	4.97	0.16	43.2	0.3	0	1.2
0	7	0.30	25.6	119.2	77.8	0.45	0	0.02	1.01	0.02	2.19	0.07	14.7	0	0	1.3
0	6	0.29	25.4	107.0	2.7	0.47	0	0.01	1.23	0.02	2.14	0.07	11.8	0	0	1.2
0	6	0.34	30.6	120.4	97.2	0.50	0	0.05	1.20	0.01	2.63	0.06	10.8	0	0	1.4

TABLE H-1 Table of Food Composition (*continued*) (Computer code is for Cengage Diet Analysis program) (For purposes of calculations, use "0" for t, <1, <.1, <.01, etc.)

DA+ Code	Food Description	Quantity	Measure	Wt (g)	H_2O (g)	Ener (kcal)	Prot (g)	Carb (g)	Fiber (g)	Fat (g)	Fat Breakdown (g) Sat	Mono	Poly	*Trans*
	Nuts, Seeds, and Products—*continued*													
524	Peanut butter, smooth	1	tablespoon(s)	16	0.3	94	4.0	3.1	1.0	8.1	1.7	3.9	2.3	—
2804	Raw	¼	cup(s)	37	2.4	207	9.4	5.9	3.1	18.0	2.5	8.9	5.7	—
	Pecans													
32907	Dry roasted, no salt added	¼	cup(s)	28	0.3	198	2.6	3.8	2.6	20.7	1.8	12.3	5.7	—
32936	Dry roasted, with salt added	¼	cup(s)	27	0.3	192	2.6	3.7	2.5	20.0	1.7	11.9	5.6	—
1162	Oil roasted	¼	cup(s)	28	0.3	197	2.5	3.6	2.6	20.7	2.0	11.3	6.5	—
526	Raw	¼	cup(s)	27	1.0	188	2.5	3.8	2.6	19.6	1.7	11.1	5.9	—
12973	**Pine nuts or pignolia, dried**	1	tablespoon(s)	9	0.2	58	1.2	1.1	0.3	5.9	0.4	1.6	2.9	—
	Pistachios													
1164	Dry roasted	¼	cup(s)	31	0.6	176	6.6	8.5	3.2	14.1	1.7	7.4	4.3	—
32938	Dry roasted, with salt added	¼	cup(s)	32	0.6	182	6.8	8.6	3.3	14.7	1.8	7.7	4.4	—
1167	**Pumpkin or squash seeds, roasted**	¼	cup(s)	57	4.0	296	18.7	7.6	2.2	23.9	4.5	7.4	10.9	—
	Sesame													
32912	Sesame butter paste	1	tablespoon(s)	16	0.3	94	2.9	3.8	0.9	8.1	1.1	3.1	3.6	—
32941	Tahini or sesame butter	1	tablespoon(s)	15	0.5	89	2.6	3.2	0.7	8.0	1.1	3.0	3.5	—
1169	Whole, roasted, toasted	3	tablespoon(s)	10	0.3	54	1.6	2.4	1.3	4.6	0.6	1.7	2.0	—
	Soy nuts													
34173	Deep sea salted	¼	cup(s)	28	—	119	11.9	8.9	4.9	4.0	1.0	—	—	—
34174	Unsalted	¼	cup(s)	28	—	119	11.9	8.9	4.9	4.0	0	—	—	—
	Sunflower seeds													
528	Kernels, dried	1	tablespoon(s)	9	0.4	53	1.9	1.8	0.8	4.6	0.4	1.7	2.1	—
29721	Kernels, dry roasted, salted	1	tablespoon(s)	8	0.1	47	1.5	1.9	0.7	4.0	0.4	0.8	2.6	—
29723	Kernels, toasted, salted	1	tablespoon(s)	8	0.1	52	1.4	1.7	1.0	4.8	0.5	0.9	3.1	—
32928	Sunflower seed butter with salt added	1	tablespoon(s)	16	0.2	93	3.1	4.4	—	7.6	0.8	1.5	5.0	—
	Trail mix													
4646	Trail mix	¼	cup(s)	38	3.5	173	5.2	16.8	2.0	11.0	2.1	4.7	3.6	—
4647	Trail mix with chocolate chips	¼	cup(s)	38	2.5	182	5.3	16.8	—	12.0	2.3	5.1	4.2	—
4648	Tropical trail mix	¼	cup(s)	35	3.2	142	2.2	23.0	—	6.0	3.0	0.9	1.8	—
	Walnuts													
529	Dried black, chopped	¼	cup(s)	31	1.4	193	7.5	3.1	2.1	18.4	1.1	4.7	11.0	—
531	English or Persian	¼	cup(s)	29	1.2	191	4.5	4.0	2.0	19.1	1.8	2.6	13.8	—
	Vegetarian Foods													
	Prepared													
34222	Brown rice and tofu stir-fry (vegan)	8	ounce(s)	227	244.4	302	16.5	18.0	3.2	21.0	1.7	4.7	13.4	0
34368	Cheese enchilada casserole (lacto)	8	ounce(s)	227	80.3	385	16.6	38.4	4.1	17.8	9.5	6.1	1.1	—
34247	Five bean casserole (vegan)	8	ounce(s)	227	175.8	178	5.9	26.6	6.0	5.8	1.1	2.5	1.9	0
34261	Lentil stew (vegan)	8	ounce(s)	227	227.9	188	11.5	35.9	11.0	0.7	0.1	0.1	0.3	0
34397	Macaroni and cheese (lacto)	8	ounce(s)	227	352.1	391	18.1	37.1	1.0	18.7	9.8	6.0	1.8	0
34238	Steamed rice and vegetables (vegan)	8	ounce(s)	227	222.9	587	11.2	87.9	5.8	23.1	4.1	8.7	9.1	0
34308	Tofu rice burgers (ovo-lacto)	1	piece(s)	218	77.6	435	22.4	68.6	5.6	8.4	1.7	2.4	3.5	—
34276	Vegan spinach enchiladas (vegan)	1	piece(s)	82	59.2	93	4.9	14.5	1.8	2.4	0.3	0.6	1.3	—
34243	Vegetable chow mein (vegan)	8	ounce(s)	227	163.3	166	6.5	22.1	2.0	6.4	0.7	2.7	2.5	0
34454	Vegetable lasagna (lacto)	8	ounce(s)	227	178.9	208	13.7	29.9	2.6	4.1	2.3	1.1	0.3	—
34339	Vegetable marinara (vegan)	8	ounce(s)	252	200.7	104	3.0	16.7	1.4	3.1	0.4	1.4	1.0	0
34356	Vegetable rice casserole (lacto)	8	ounce(s)	227	178.9	238	9.7	24.4	4.0	12.5	4.9	3.5	3.1	—
34311	Vegetable strudel (ovo-lacto)	8	ounce(s)	227	63.1	478	12.0	32.4	2.5	33.8	11.5	16.7	3.9	0
34371	Vegetable taco (lacto)	1	item(s)	85	46.5	117	4.2	13.6	2.9	5.6	2.1	1.9	1.3	—
34282	Vegetarian chili (vegan)	8	ounce(s)	227	191.4	115	5.6	21.4	7.1	1.5	0.2	0.3	0.7	0
34367	Vegetarian vegetable soup (vegan)	8	ounce(s)	227	257.9	111	3.2	16.0	3.2	5.0	1.0	2.1	1.6	0
	Boca burger													
32067	All American flamed grilled patty	1	item(s)	71	—	90	14.0	4.0	3.0	3.0	1.0	—	—	0
32074	Boca chik'n nuggets	4	item(s)	87	—	180	14.0	17.0	3.0	7.0	1.0	—	—	0
32075	Boca meatless ground burger	½	cup(s)	57	—	60	13.0	6.0	3.0	0.5	0	—	—	0
32072	Breakfast links	2	item(s)	45	—	70	8.0	5.0	2.0	3.0	0.5	—	—	0
32071	Breakfast patties	1	item(s)	38	—	60	7.0	5.0	2.0	2.5	0	—	—	0
35780	Cheeseburger meatless burger patty	1	item(s)	71	—	100	12.0	5.0	3.0	5.0	1.5	—	—	0
33958	Original meatless chik'n patties	1	item(s)	71	—	160	11.0	15.0	2.0	6.0	1.0	—	—	0

PAGE KEY: H-4 = Breads/Baked Goods H-10 = Cereal/Rice/Pasta H-14 = Fruit H-18 = Vegetables/Legumes H-28 = Nuts/Seeds H-30 = Vegetarian H-32 = Dairy H-40 = Eggs H-40 = Seafood H-42 = Meats H-46 = Poultry H-46 = Processed Meats H-48 = Beverages H-52 = Fats/Oils H-54 = Sweets H-56 = Spices/Condiments/Sauces H-60 = Mixed Foods/Soups/Sandwiches H-64 = Fast Food H-84 = Convenience H-86 = Baby Foods

Chol (mg)	Calc (mg)	Iron (mg)	Magn (mg)	Pota (mg)	Sodi (mg)	Zinc (mg)	Vit A (μg)	Thia (mg)	Vit E (mg α)	Ribo (mg)	Niac (mg)	Vit B_6 (mg)	Fola (μg)	Vit C (mg)	Vit B_{12} (μg)	Sele (μg)
0	7	0.30	24.6	103.8	73.4	0.47	0	0.01	1.44	0.02	2.14	0.09	11.8	0	0	0.9
0	34	1.67	61.3	257.3	6.6	1.19	0	0.23	3.04	0.04	4.40	0.12	87.6	0	0	2.6
0	20	0.78	36.8	118.3	0.3	1.41	1.9	0.12	0.35	0.03	0.32	0.05	4.5	0.2	0	1.1
0	19	0.75	35.6	114.5	103.4	1.36	1.9	0.11	0.34	0.03	0.31	0.05	4.3	0.2	0	1.1
0	18	0.68	33.3	107.8	0.3	1.23	1.4	0.13	0.70	0.03	0.33	0.05	4.1	0.2	0	1.7
0	19	0.69	33.0	111.7	0	1.23	0.8	0.18	0.38	0.04	0.32	0.06	6.0	0.3	0	1.0
0	1	0.47	21.6	51.3	0.2	0.55	0.1	0.03	0.80	0.02	0.37	0.01	2.9	0.1	0	0.1
0	34	1.29	36.9	320.4	3.1	0.71	4.0	0.26	0.59	0.05	0.44	0.39	15.4	0.7	0	2.9
0	35	1.34	38.4	333.4	129.6	0.73	4.2	0.26	0.61	0.05	0.45	0.40	16.0	0.7	0	3.0
0	24	8.48	303.0	457.4	10.2	4.22	10.8	0.12	0.00	0.18	0.99	0.05	32.3	1.0	0	3.2
0	154	3.07	57.9	93.1	1.9	1.17	0.5	0.04	—	0.03	1.07	0.13	16.0	0	0	0.9
0	21	0.66	14.3	68.9	5.3	0.69	0.5	0.24	—	0.02	0.85	0.02	14.7	0.6	0	0.3
0	94	1.40	33.8	45.1	1.0	0.68	0	0.07	—	0.02	0.43	0.07	9.3	0	0	0.5
0	59	1.07	—	—	148.1	—	0	—	—	—	—	—	—	0	—	—
0	59	1.07	—	—	9.9	—	0	—	—	—	—	—	—	0	—	—
0	7	0.47	29.3	58.1	0.8	0.45	0.3	0.13	2.99	0.03	0.75	0.12	20.4	0.1	0	4.8
0	6	0.30	10.3	68.0	32.8	0.42	0	0.01	2.09	0.02	0.56	0.06	19.0	0.1	0	6.3
0	5	0.57	10.8	41.1	51.3	0.44	0	0.03	—	0.02	0.35	0.07	19.9	0.1	0	5.2
0	20	0.76	59.0	11.5	83.2	0.85	0.5	0.05	—	0.05	0.85	0.13	37.9	0.4	0	—
0	29	1.14	59.3	256.9	85.9	1.20	0.4	0.17	—	0.07	1.76	0.11	26.6	0.5	0	—
2	41	1.27	60.4	243.0	45.4	1.17	0.8	0.15	—	0.08	1.65	0.09	24.4	0.5	0	—
0	20	0.92	33.6	248.2	3.5	0.41	0.7	0.15	—	0.04	0.51	0.11	14.7	2.7	0	—
0	19	0.97	62.8	163.4	0.6	1.05	0.6	0.01	0.56	0.04	0.14	0.18	9.7	0.5	0	5.3
0	29	0.85	46.2	129.0	0.6	0.90	0.3	0.10	0.20	0.04	0.32	0.15	28.7	0.4	0	1.4
0	353	6.34	118.3	501.4	142.2	2.03	—	0.23	0.07	0.14	1.49	0.36	51.8	24.8	0	14.8
39	441	2.44	34.6	191.2	1139.7	1.84	—	0.31	0.05	0.35	2.23	0.11	87.1	20.4	0.4	20.0
0	48	1.78	40.8	364.1	613.6	0.61	—	0.10	0.52	0.07	0.93	0.11	39.4	8.3	0	3.3
0	34	3.23	50.0	548.8	436.5	1.42	—	0.24	0.14	0.16	2.31	0.29	91.4	26.4	0	12.1
43	415	1.71	45.4	267.8	1641.0	2.32	—	0.32	0.27	0.48	2.18	0.13	74.2	0.9	0.8	33.3
0	91	3.31	153.1	810.1	3117.8	2.04	—	0.37	3.03	0.21	6.16	0.64	61.7	35.2	0	18.8
52	467	9.01	89.7	455.6	2449.5	2.06	—	0.27	0.12	0.26	3.43	0.29	106.6	2.0	0.1	43.0
0	117	1.13	40.4	170.5	134.2	0.68	—	0.07	—	0.07	0.53	0.10	52.1	1.8	0	5.1
0	189	3.70	28.0	310.3	372.7	0.76	—	0.13	0.05	0.11	1.43	0.14	45.6	8.0	0	6.5
10	176	1.86	41.9	470.0	759.4	1.14	—	0.26	0.05	0.25	2.49	0.22	64.7	19.0	0.4	21.8
0	17	0.94	19.1	189.9	439.6	0.42	—	0.15	0.55	0.08	1.36	0.12	40.6	23.5	0	10.8
17	190	1.28	29.3	414.2	626.0	1.24	—	0.16	0.35	0.29	2.00	0.19	72.6	56.0	0.2	5.8
29	200	2.15	24.5	181.0	512.1	1.24	—	0.28	0.20	0.31	2.88	0.11	88.3	17.4	0.2	19.7
7	77	0.88	26.3	174.1	280.7	0.59	—	0.08	0.04	0.06	0.49	0.08	38.7	4.6	0	3.0
0	65	1.98	41.0	543.1	390.7	0.74	—	0.14	0.15	0.10	1.31	0.18	59.3	20.3	0	4.4
0	46	1.87	34.9	550.3	729.5	0.56	—	0.13	0.55	0.09	1.99	0.27	47.8	29.9	0	1.4
5	150	1.80	—	—	280.0	—	0	—	—	—	—	—	—	0	—	—
0	40	1.44	—	—	500.0	—	—	—	—	—	—	—	—	0	—	—
0	60	1.80	—	—	270.0	—	0	—	—	—	—	—	—	0	—	—
0	20	1.44	—	—	330.0	—	0	—	—	—	—	—	—	0	—	—
0	20	1.08	—	—	280.0	—	0	—	—	—	—	—	—	0	—	—
5	80	1.80	—	—	360.0	—	—	—	—	—	—	—	—	0	—	—
0	40	1.80	—	—	430.0	—	—	—	—	—	—	—	—	0	—	—

TABLE H-1 Table of Food Composition (*continued*) (Computer code is for Cengage Diet Analysis program) (For purposes of calculations, use "0" for t, <1, <.1, <.01, etc.)

DA+ Code	Food Description	Quantity	Measure	Wt (g)	H_2O (g)	Ener (kcal)	Prot (g)	Carb (g)	Fiber (g)	Fat (g)	Fat Breakdown (g)			
											Sat	Mono	Poly	*Trans*
	Vegetarian Foods—*continued*													
32066	Original patty	1	item(s)	71	—	70	13.0	6.0	4.0	0.5	0	—	—	0
32068	Roasted garlic patty	1	item(s)	71	—	70	12.0	6.0	4.0	1.5	0	—	—	0
37814	Roasted onion meatless burger patty	1	item(s)	71	—	70	11.0	7.0	4.0	1.0	0	—	—	0
	Gardenburger													
37810	BBQ chik'n with sauce	1	item(s)	142	—	250	14.0	30.0	5.0	8.0	1.0	—	—	0
39661	Black bean burger	1	item(s)	71	—	80	8.0	11.0	4.0	2.0	0	—	—	0
39666	Buffalo chik'n wing	3	item(s)	95	—	180	9.0	8.0	5.0	12.0	1.5	—	—	0
39665	Country fried chicken with creamy pepper gravy	1	item(s)	142	—	190	9.0	16.0	2.0	9.0	1.0	—	—	0
37808	Flamed grilled chik'n	1	item(s)	71	—	100	13.0	5.0	3.0	2.5	0	—	—	0
37803	Garden vegan	1	item(s)	71	—	100	10.0	12.0	2.0	1.0	—	—	—	0
39663	Homestyle classic burger	1	item(s)	71	—	110	12.0	6.0	4.0	5.0	0.5	—	—	0
37807	Meatless breakfast sausage	1	item(s)	43	—	50	5.0	2.0	2.0	3.5	0	—	—	0
37809	Meatless meatballs	6	item(s)	85	—	110	12.0	8.0	4.0	4.5	1.0	—	—	0
37806	Meatless riblets with sauce	1	item(s)	142	—	160	17.0	11.0	4.0	5.0	0	—	—	0
29913	Original	1	item(s)	71	—	90	10.0	8.0	3.0	2.0	0.5	—	—	0
39662	Sun-dried tomato basil burger	1	item(s)	71	—	80	10.0	11.0	3.0	1.5	0.5	—	—	0
29915	Veggie medley	1	item(s)	71	—	90	9.0	11.0	4.0	2.0	0	—	—	0
	Loma Linda													
9311	Big franks, canned	1	item(s)	51	—	110	11.0	3.0	2.0	6.0	1.0	1.5	3.5	0
9323	Fried chik'n with gravy	2	piece(s)	80	45.9	150	12.0	5.0	2.0	10	1.5	2.5	5.0	0
9326	Linketts, canned	1	item(s)	35	21.0	70	7.0	1.0	1.0	4.0	0.5	1.0	2.5	0
9336	Redi-Burger patties, canned	1	slice(s)	85	50.5	120	18.0	7.0	4.0	2.5	0.5	0.5	1.5	0
9350	Swiss Stake pattie with gravy, frozen	1	piece(s)	92	65.7	130	9.0	9.0	3.0	6.0	1.0	1.5	3.5	0
9354	Tender Rounds meatball substitute, canned in gravy	6	piece(s)	80	53.9	120	13.0	6.0	1.0	4.5	0.5	1.5	2.5	0
	Morningstar Farms													
33707	America's Original Veggie Dog links	1	item(s)	57	—	80	11.0	6.0	1.0	0.5	0	—	—	0
9362	Better'n Eggs egg substitute	¼	cup(s)	57	50.3	20	5.0	0	0	0	0	0	0	0
9371	Breakfast bacon strips	2	item(s)	16	6.8	60	2.0	2.0	0.5	4.5	0.5	1.0	3.0	0
9368	Breakfast sausage links	2	item(s)	45	26.8	80	9.0	3.0	2.0	3.0	0.5	1.5	1.0	0
33705	Chik'n nuggets	4	piece(s)	86	—	190	12.0	18.0	2.0	7.0	1.0	2.0	4.0	0
11587	Chik patties	1	item(s)	71	36.3	150	9.0	16.0	2.0	6.0	1.0	1.5	2.5	0
2531	Garden veggie patties	1	item(s)	67	40.1	100	10.0	9.0	4.0	2.5	0.5	0.5	1.5	0
33702	Spicy black bean veggie burger	1	item(s)	78	—	140	12.0	15.0	3.0	4.0	0.5	1.0	2.5	0
9412	Vegetarian chili, canned	1	cup(s)	230	172.6	180	16.0	25.0	10.0	1.5	0.5	0.5	0.5	0
	Worthington													
9424	Chili, canned	1	cup(s)	230	167.0	280	24.0	25.0	8.0	10.0	1.5	1.5	7.0	0
9436	Diced chik, canned	¼	cup(s)	55	42.7	50	9.0	2.0	1.0	0	0	0	0	0
9440	Dinner roast, frozen	1	slice(s)	85	53.2	180	14.0	6.0	3.0	11.0	1.5	4.5	5.0	0
9420	Meatless chicken slices, frozen	3	slice(s)	57	38.9	90	9.0	2.0	0.5	4.5	1.0	1.0	2.5	0
36702	Meatless chicken style roll, frozen	1	slice(s)	55	—	90	9.0	2.0	1.0	4.5	1.0	1.0	2.5	0
9428	Meatless corned beef, sliced, frozen	3	slice(s)	57	31.2	140	10.0	5.0	0	9.0	1.0	2.0	5.0	0
9470	Meatless salami, sliced, frozen	3	slice(s)	57	32.4	120	12.0	3.0	2.0	7.0	1.0	1.0	5.0	0
9480	Meatless smoked turkey, sliced	3	slice(s)	57	—	140	10.0	4.0	0	9.0	1.5	2.0	5.0	0
9462	Prosage links	2	item(s)	45	26.8	80	9.0	3.0	2.0	3.0	0.5	0.5	2.0	0
9484	Stakelets patty beef steak substitute, frozen	1	piece(s)	71	41.5	150	14.0	7.0	2.0	7.0	1.0	2.5	3.5	0
9486	Stripples bacon substitute	2	item(s)	16	6.8	60	2.0	2.0	0.5	4.5	0.5	1.0	3.0	0
9496	Vegetable Skallops meat substitute, canned	½	cup(s)	85	—	90	17.0	4.0	3.0	1.0	0	0	0.5	0
	Dairy													
	Cheese													
1433	Blue, crumbled	1	ounce(s)	28	12.0	100	6.1	0.7	0	8.1	5.3	2.2	0.2	—
884	Brick	1	ounce(s)	28	11.7	105	6.6	0.8	0	8.4	5.3	2.4	0.2	—
885	Brie	1	ounce(s)	28	13.7	95	5.9	0.1	0	7.8	4.9	2.3	0.2	—
34821	Camembert	1	ounce(s)	28	14.7	85	5.6	0.1	0	6.9	4.3	2.0	0.2	—
5	Cheddar, shredded	¼	cup(s)	28	10.4	114	7.0	0.4	0	9.4	6.0	2.7	0.3	—
888	Cheddar or colby	1	ounce(s)	28	10.8	112	6.7	0.7	0	9.1	5.7	2.6	0.3	—
32096	Cheddar or colby, low fat	1	ounce(s)	28	17.9	49	6.9	0.5	0	2.0	1.2	0.6	0.1	—
889	Edam	1	ounce(s)	28	11.8	101	7.1	0.4	0	7.9	5.0	2.3	0.2	—

PAGE KEY: H-4 = Breads/Baked Goods H-10 = Cereal/Rice/Pasta H-14 = Fruit H-18 = Vegetables/Legumes H-28 = Nuts/Seeds H-30 = Vegetarian H-32 = Dairy H-40 = Eggs H-40 = Seafood H-42 = Meats H-46 = Poultry H-46 = Processed Meats H-48 = Beverages H-52 = Fats/Oils H-54 = Sweets H-56 = Spices/Condiments/Sauces H-60 = Mixed Foods/Soups/Sandwiches H-64 = Fast Food H-84 = Convenience H-86 = Baby Foods

Chol (mg)	Calc (mg)	Iron (mg)	Magn (mg)	Pota (mg)	Sodi (mg)	Zinc (mg)	Vit A (μg)	Thia (mg)	Vit E (mg α)	Ribo (mg)	Niac (mg)	Vit B_6 (mg)	Fola (μg)	Vit C (mg)	Vit B_{12} (μg)	Sele (μg)
0	60	1.80	—	—	280.0	—	0	—	—	—	—	—	—	0	—	—
0	60	1.80	—	—	370.0	—	0	—	—	—	—	—	—	0	—	—
0	100	2.70	—	—	300.0	—	—	—	—	—	—	—	—	0	—	—
0	150	1.08	—	—	890.0	—	—	—	—	—	—	—	—	0	—	—
0	40	1.44	—	—	330.0	—	—	—	—	—	—	—	—	0	—	—
0	40	0.72	—	—	1000.0	—	—	—	—	—	—	—	—	0	—	—
5	40	1.44	—	—	550.0	—	—	—	—	—	—	—	—	0	—	—
0	60	3.60	—	—	360.0	—	—	—	—	—	—	—	—	0	—	—
0	40	4.50	—	—	230.0	—	—	—	—	—	—	—	—	0	—	—
0	80	1.44	—	—	380.0	—	—	—	—	—	—	—	—	0	—	—
0	20	0.72	—	—	120.0	—	—	—	—	—	—	—	—	0	—	—
0	60	1.80	—	—	400.0	—	—	—	—	—	—	—	—	0	—	—
0	60	1.80	—	—	720.0	—	—	—	—	—	—	—	—	3.6	—	—
0	80	1.08	30.4	193.4	490.0	0.89	—	0.10	—	0.15	1.08	0.08	10.1	1.2	0.1	7.0
5	60	1.44	—	—	260.0	—	—	—	—	—	—	—	—	3.6	—	—
0	40	1.44	27.0	182.0	290.0	0.46	—	0.07	—	0.08	0.90	0.09	10.6	9.0	0	4.0
0	0	0.77	—	50.0	220.0	—	0	0.22	—	0.10	2.00	0.70	—	0	2.4	—
0	20	1.80	—	70.0	430.0	0.33	0	1.05	—	0.34	4.00	0.30	—	0	2.4	—
0	0	0.36	—	20.0	160.0	0.46	0	0.12	—	0.20	0.80	0.16	—	0	0.9	—
0	0	1.06	—	140.0	450.0	—	0	0.15	—	0.25	4.00	0.40	—	0	1.2	—
0	0	0.72	—	200.0	430.0	—	0	0.45	—	0.25	10.00	1.00	—	0	5.4	—
0	20	1.08	—	80.0	340.0	0.66	0	0.75	—	0.17	2.00	0.16	—	0	1.2	—
0	0	0.72	—	60.0	580.0	—	0	—	—	—	—	—	—	0	—	—
0	20	0.72	—	75.0	90.0	0.60	37.5	0.03	—	0.34	0.00	0.08	24.0	—	0.6	—
0	0	0.36	—	15.0	220.0	0.05	0	0.75	—	0.04	0.40	0.07	—	0	0.2	—
0	0	1.80	—	50.0	300.0	—	0	0.37	—	0.17	7.00	0.50	—	0	3.0	—
0	20	2.70	—	320.0	490.0	—	0	0.52	—	0.25	5.00	0.30	—	0	1.5	—
0	0	1.80	—	210.0	540.0	—	0	1.80	—	0.17	2.00	0.20	—	0	1.2	—
0	40	0.72	—	180.0	350.0	—	—	—	—	—	—	—	—	0	—	—
0	40	1.80	—	320.0	470.0	—	0	—	—	—	0.00	—	—	0	—	—
0	40	3.60	—	660.0	900.0	—	—	—	—	—	—	—	—	0	—	—
0	40	3.60	—	330.0	1130.0	—	0	0.30	—	0.13	2.00	0.70	—	0	1.5	—
0	0	1.08	—	100.0	220.0	0.24	0	0.06	—	0.10	4.00	0.08	—	0	0.2	—
0	20	1.80	—	120.0	580.0	0.64	0	1.80	—	0.25	6.00	0.60	—	0	1.5	—
0	250	1.80	—	250.0	250.0	0.26	0	0.37	—	0.13	4.00	0.30	—	0	1.8	—
0	100	1.08	—	240.0	240.0	—	0	0.37	—	0.13	4.00	0.30	—	0	1.8	—
0	0	1.80	—	130.0	460.0	0.26	0	0.45	—	0.17	5.00	0.30	—	0	1.8	—
0	0	1.08	—	95.0	800.0	0.30	0	0.75	—	0.17	4.00	0.20	—	0	0.6	—
0	60	2.70	—	60.0	450.0	0.23	0	1.80	—	0.17	6.00	0.40	—	0	3.0	—
0	0	1.44	—	50.0	320.0	0.36	0	1.80	—	0.17	2.00	0.30	—	0	3.0	—
0	40	1.08	—	130.0	480.0	0.50	0	1.20	—	0.13	3.00	0.30	—	0	1.5	—
0	0	0.36	—	15.0	220.0	0.05	0	0.75	—	0.03	0.40	0.08	—	0	0.2	—
0	0	0.36	—	10.0	390.0	0.67	0	0.03	—	0.03	0.00	0.01	—	0	0	—
21	150	0.08	6.5	72.6	395.5	0.75	56.1	0.01	0.07	0.10	0.28	0.04	10.2	0	0.3	4.1
27	191	0.12	6.8	38.6	158.8	0.73	82.8	0.00	0.07	0.10	0.03	0.01	5.7	0	0.4	4.1
28	52	0.14	5.7	43.1	178.3	0.67	49.3	0.02	0.06	0.14	0.10	0.06	18.4	0	0.5	4.1
20	110	0.09	5.7	53.0	238.7	0.67	68.3	0.01	0.06	0.14	0.18	0.06	17.6	0	0.4	4.1
30	204	0.19	7.9	27.7	175.4	0.87	74.9	0.01	0.08	0.10	0.02	0.02	5.1	0	0.2	3.9
27	194	0.21	7.4	36.0	171.2	0.87	74.8	0.00	0.07	0.10	0.02	0.02	5.1	0	0.2	4.1
6	118	0.11	4.5	18.7	173.5	0.51	17.0	0.00	0.01	0.06	0.01	0.01	3.1	0	0.1	4.1
25	207	0.12	8.5	53.3	273.6	1.06	68.9	0.01	0.06	0.11	0.02	0.02	4.5	0	0.4	4.1

TABLE H-1 Table of Food Composition (*continued*) (Computer code is for Cengage Diet Analysis program) (For purposes of calculations, use "0" for t, <1, <.1, <.01, etc.)

DA+ Code	Food Description	Quantity	Measure	Wt (g)	H_2O (g)	Ener (kcal)	Prot (g)	Carb (g)	Fiber (g)	Fat (g)	Fat Breakdown (g) Sat	Mono	Poly	*Trans*
Dairy—*continued*														
890	Feta	1	ounce(s)	28	15.7	75	4.0	1.2	0	6.0	4.2	1.3	0.2	—
891	Fontina	1	ounce(s)	28	10.8	110	7.3	0.4	0	8.8	5.4	2.5	0.5	—
8527	Goat cheese, soft	1	ounce(s)	28	17.2	76	5.3	0.3	0	6.0	4.1	1.4	0.1	—
893	Gouda	1	ounce(s)	28	11.8	101	7.1	0.6	0	7.8	5.0	2.2	0.2	—
894	Gruyere	1	ounce(s)	28	9.4	117	8.5	0.1	0	9.2	5.4	2.8	0.5	—
895	Limburger	1	ounce(s)	28	13.7	93	5.7	0.1	0	7.7	4.7	2.4	0.1	—
896	Monterey jack	1	ounce(s)	28	11.6	106	6.9	0.2	0	8.6	5.4	2.5	0.3	—
13	Mozzarella, part skim milk	1	ounce(s)	28	15.2	72	6.9	0.8	0	4.5	2.9	1.3	0.1	—
12	Mozzarella, whole milk	1	ounce(s)	28	14.2	85	6.3	0.6	0	6.3	3.7	1.9	0.2	—
897	Muenster	1	ounce(s)	28	11.8	104	6.6	0.3	0	8.5	5.4	2.5	0.2	—
898	Neufchatel	1	ounce(s)	28	17.6	74	2.8	0.8	0	6.6	4.2	1.9	0.2	—
14	Parmesan, grated	1	tablespoon(s)	5	1.0	22	1.9	0.2	0	1.4	0.9	0.4	0.1	—
17	Provolone	1	ounce(s)	28	11.6	100	7.3	0.6	0	7.5	4.8	2.1	0.2	—
19	Ricotta, part skim milk	¼	cup(s)	62	45.8	85	7.0	3.2	0	4.9	3.0	1.4	0.2	—
18	Ricotta, whole milk	¼	cup(s)	62	44.1	107	6.9	1.9	0	8.0	5.1	2.2	0.2	—
20	Romano	1	tablespoon(s)	5	1.5	19	1.6	0.2	0	1.3	0.9	0.4	0	—
900	Roquefort	1	ounce(s)	28	11.2	105	6.1	0.6	0	8.7	5.5	2.4	0.4	—
21	Swiss	1	ounce(s)	28	10.5	108	7.6	1.5	0	7.9	5.0	2.1	0.3	—
	Imitation cheese													
42245	Imitation American cheddar cheese	1	ounce(s)	28	15.1	68	4.7	3.3	0	4.0	2.5	1.2	0.1	—
53914	Imitation cheddar	1	ounce(s)	28	15.1	68	4.7	3.3	0	4.0	2.5	1.2	0.1	—
	Cottage cheese													
9	Low fat, 1% fat	½	cup(s)	113	93.2	81	14.0	3.1	0	1.2	0.7	0.3	0	—
8	Low fat, 2% fat	½	cup(s)	113	89.6	102	15.5	4.1	0	2.2	1.4	0.6	0.1	—
	Cream cheese													
11	Cream cheese	2	tablespoon(s)	29	15.6	101	2.2	0.8	0	10.1	6.4	2.9	0.4	—
17366	Fat-free cream cheese	2	tablespoon(s)	30	22.7	29	4.3	1.7	0	0.4	0.3	0.1	0	—
10438	Tofutti Better than Cream Cheese	2	tablespoon(s)	30	—	80	1.0	1.0	0	8.0	2.0	—	6.0	—
	Processed cheese													
24	American cheese food, processed	1	ounce(s)	28	12.3	94	5.2	2.2	0	7.1	4.2	2.0	0.3	—
25	American cheese spread, processed	1	ounce(s)	28	13.5	82	4.7	2.5	0	6.0	3.8	1.8	0.2	—
22	American cheese, processed	1	ounce(s)	28	11.1	106	6.3	0.5	0	8.9	5.6	2.5	0.3	—
9110	Kraft deluxe singles pasteurized process American cheese	1	ounce(s)	28	—	108	5.4	0	0	9.5	5.4	—	—	—
23	Swiss cheese, processed	1	ounce(s)	28	12.0	95	7.0	0.6	0	7.1	4.5	2.0	0.2	—
	Soy cheese													
10437	Galaxy Foods vegan grated parmesan cheese alternative	1	tablespoon(s)	8	—	23	3.0	1.5	0	0	0	0	0	0
10430	Nu Tofu cheddar flavored cheese alternative	1	ounce(s)	28	—	70	6.0	1.0	0	4.0	0.5	2.5	1.0	—
	Cream													
26	Half and half cream	1	tablespoon(s)	15	12.1	20	0.4	0.6	0	1.7	1.1	0.5	0.1	—
32	Heavy whipping cream, liquid	1	tablespoon(s)	15	8.7	52	0.3	0.4	0	5.6	3.5	1.6	0.2	—
28	Light coffee or table cream, liquid	1	tablespoon(s)	15	11.1	29	0.4	0.5	0	2.9	1.8	0.8	0.1	—
30	Light whipping cream, liquid	1	tablespoon(s)	15	9.5	44	0.3	0.4	0	4.6	2.9	1.4	0.1	—
34	Whipped cream topping, pressurized	1	tablespoon(s)	3	1.8	8	0.1	0.4	0	0.7	0.4	0.2	0	—
	Sour cream													
30556	Fat-free sour cream	2	tablespoon(s)	32	25.8	24	1.0	5.0	0	0	0	0	0	0
36	Sour cream	2	tablespoon(s)	24	17.0	51	0.8	1.0	0	5.0	3.1	1.5	0.2	—
	Imitation cream													
3659	Coffeemate nondairy creamer, liquid	1	tablespoon(s)	15	—	20	0	2.0	0	1.0	0	0.5	0	—
40	Cream substitute, powder	1	teaspoon(s)	2	0	11	0.1	1.1	0	0.7	0.7	0	0	—
904	Imitation sour cream	2	tablespoon(s)	29	20.5	60	0.7	1.9	0	5.6	5.1	0.2	0	—
35972	Nondairy coffee whitener, liquid, frozen	1	tablespoon(s)	15	11.7	21	0.2	1.7	0	1.5	0.3	1.1	0	—
35976	Nondairy dessert topping, frozen	1	tablespoon(s)	5	2.4	15	0.1	1.1	0	1.2	1.0	0.1	0	—
35975	Nondairy dessert topping, pressurized	1	tablespoon(s)	4	2.7	12	0	0.7	0	1.0	0.8	0.1	0	—

PAGE KEY: H-4 = Breads/Baked Goods H-10 = Cereal/Rice/Pasta H-14 = Fruit H-18 = Vegetables/Legumes H-28 = Nuts/Seeds H-30 = Vegetarian H-32 = Dairy H-40 = Eggs H-40 = Seafood H-42 = Meats H-46 = Poultry H-46 = Processed Meats H-48 = Beverages H-52 = Fats/Oils H-54 = Sweets H-56 = Spices/Condiments/Sauces H-60 = Mixed Foods/Soups/Sandwiches H-64 = Fast Food H-84 = Convenience H-86 = Baby Foods

Chol (mg)	Calc (mg)	Iron (mg)	Magn (mg)	Pota (mg)	Sodi (mg)	Zinc (mg)	Vit A (μg)	Thia (mg)	Vit E (mg α)	Ribo (mg)	Niac (mg)	Vit B_6 (mg)	Fola (μg)	Vit C (mg)	Vit B_{12} (μg)	Sele (μg)
25	140	0.18	5.4	17.6	316.4	0.81	35.4	0.04	0.05	0.23	0.28	0.12	9.1	0	0.5	4.3
33	156	0.06	4.0	18.1	226.8	0.99	74.0	0.01	0.07	0.05	0.04	0.02	1.7	0	0.5	4.1
13	40	0.53	4.5	7.4	104.3	0.26	81.6	0.02	0.05	0.10	0.12	0.07	3.4	0	0.1	0.8
32	198	0.06	8.2	34.3	232.2	1.10	46.8	0.01	0.06	0.09	0.01	0.02	6.0	0	0.4	4.1
31	287	0.04	10.2	23.0	95.3	1.10	76.8	0.01	0.07	0.07	0.03	0.02	2.8	0	0.5	4.1
26	141	0.03	6.0	36.3	226.8	0.59	96.4	0.02	0.06	0.14	0.04	0.02	16.4	0	0.3	4.1
25	211	0.20	7.7	23.0	152.0	0.85	56.1	0.00	0.07	0.11	0.02	0.02	5.1	0	0.2	4.1
18	222	0.06	6.5	23.8	175.5	0.78	36.0	0.01	0.04	0.08	0.03	0.02	2.6	0	0.2	4.1
22	143	0.12	5.7	21.5	177.8	0.82	50.7	0.01	0.05	0.08	0.02	0.01	2.0	0	0.6	4.8
27	203	0.11	7.7	38.0	178.0	0.79	84.5	0.00	0.07	0.09	0.02	0.01	3.4	0	0.4	4.1
22	21	0.07	2.3	32.3	113.1	0.14	84.5	0.00	—	0.05	0.03	0.01	3.1	0	0.1	0.9
4	55	0.04	1.9	6.3	76.5	0.19	6.0	0.00	0.01	0.02	0.01	0.00	0.5	0	0.1	0.9
20	214	0.14	7.9	39.1	248.3	0.91	66.9	0.01	0.06	0.09	0.04	0.02	2.8	0	0.4	4.1
19	167	0.27	9.2	76.9	76.9	0.82	65.8	0.01	0.04	0.11	0.04	0.01	8.0	0	0.2	10.3
31	127	0.23	6.8	64.6	51.7	0.71	73.8	0.01	0.06	0.12	0.06	0.02	7.4	0	0.2	8.9
5	53	0.03	2.1	4.3	60	0.12	4.8	0.00	0.01	0.01	0.00	0.00	0.4	0	0.1	0.7
26	188	0.15	8.5	25.8	512.9	0.59	83.3	0.01	—	0.16	0.20	0.03	13.9	0	0.2	4.1
26	224	0.05	10.8	21.8	54.4	1.23	62.4	0.01	0.10	0.08	0.02	0.02	1.7	0	0.9	5.2
10	159	0.08	8.2	68.6	381.3	0.73	32.3	0.01	0.07	0.12	0.03	0.03	2.0	0	0.1	4.3
10	159	0.09	8.2	68.6	381.3	0.73	32.3	0.01	0.07	0.12	0.04	0.03	2.0	0	0.1	4.3
5	69	0.15	5.7	97.2	458.8	0.42	12.4	0.02	0.01	0.18	0.14	0.07	13.6	0	0.7	10.2
9	78	0.18	6.8	108.5	458.8	0.47	23.7	0.02	0.02	0.20	0.16	0.08	14.7	0	0.8	11.5
32	23	0.34	1.7	34.5	85.8	0.15	106.1	0.01	0.08	0.05	0.02	0.01	3.8	0	0.1	0.7
2	56	0.05	4.2	48.9	163.5	0.26	83.7	0.01	0.00	0.05	0.04	0.01	11.1	0	0.2	1.5
0	0	0.00	—	—	135.0	—	0	—	—	—	—	—	—	0	—	—
23	162	0.16	8.8	82.5	358.6	0.90	57.0	0.01	0.06	0.14	0.04	0.02	2.0	0	0.4	4.6
16	159	0.09	8.2	68.6	381.3	0.73	49.0	0.01	0.05	0.12	0.03	0.03	2.0	0	0.1	3.2
27	156	0.05	7.7	47.9	422.1	0.80	72.0	0.01	0.07	0.10	0.02	0.02	2.3	0	0.2	4.1
27	338	0.00	0	33.8	459.0	1.22	114.0	—	—	0.14	—	—	—	0	0.2	—
24	219	0.17	8.2	61.2	388.4	1.02	56.1	0.00	0.09	0.07	0.01	0.01	1.7	0	0.3	4.5
0	60	0.00	—	75.0	97.5	—	—	—	—	—	—	—	—	—	—	—
0	200	0.36	—	—	190.0	—	—	—	—	—	—	—	—	0	—	—
6	16	0.01	1.5	19.5	6.2	0.08	14.6	0.01	0.05	0.02	0.01	0.01	0.5	0.1	0	0.3
21	10	0.00	1.1	11.3	5.7	0.03	61.7	0.00	0.15	0.01	0.01	0.00	0.6	0.1	0	0.1
10	14	0.01	1.4	18.3	6.0	0.04	27.2	0.01	0.08	0.02	0.01	0.01	0.3	0.1	0	0.1
17	10	0.00	1.1	14.6	5.1	0.03	41.9	0.00	0.13	0.01	0.01	0.00	0.6	0.1	0	0.1
2	3	0.00	0.3	4.4	3.9	0.01	5.6	0.00	0.01	0.00	0.00	0.00	0.1	0	0	0
3	40	0.00	3.2	41.3	45.1	0.16	23.4	0.01	0.00	0.04	0.02	0.01	3.5	0	0.1	1.7
11	28	0.01	2.6	34.6	12.7	0.06	42.5	0.01	0.14	0.03	0.01	0.00	2.6	0.2	0.1	0.5
0	0	0.00	—	30.0	0	—	0	0.01	—	0.01	0.20	—	—	0	—	—
0	0	0.02	0.1	16.2	3.6	0.01	0	0.00	0.01	0.00	0.00	0.00	0	0	0	0
0	1	0.11	1.7	46.3	29.3	0.34	0	0.00	0.21	0.00	0.00	0.00	0	0	0	0.7
0	1	0.00	0	28.9	12.0	0.00	0.2	0.00	0.12	0.00	0.00	0.00	0	0	0	0.2
0	0	0.00	0.1	0.9	1.2	0.00	0.3	0.00	0.05	0.00	0.00	0.00	0	0	0	0.1
0	0	0.00	0	0.8	2.8	0.00	0.2	0.00	0.04	0.00	0.00	0.00	0	0	0	0.1

TABLE H-1 Table of Food Composition (*continued*) (Computer code is for Cengage Diet Analysis program) (For purposes of calculations, use "0" for t, <1, <.1, <.01, e

DA+ Code	Food Description	Quantity	Measure	Wt (g)	H_2O (g)	Ener (kcal)	Prot (g)	Carb (g)	Fiber (g)	Fat (g)	Fat Breakdown (g) Sat	Mono	Poly	*Trans*
	Dairy—*continued*													
	Fluid milk													
60	Buttermilk, low fat	1	cup(s)	245	220.8	98	8.1	11.7	0	2.2	1.3	0.6	0.1	—
54	Low fat, 1%	1	cup(s)	244	219.4	102	8.2	12.2	0	2.4	1.5	0.7	0.1	—
55	Low fat, 1%, with nonfat milk solids	1	cup(s)	245	220.0	105	8.5	12.2	0	2.4	1.5	0.7	0.1	—
57	Nonfat, skim or fat free	1	cup(s)	245	222.6	83	8.3	12.2	0	0.2	0.1	0.1	0	—
58	Nonfat, skim or fat free with nonfat milk solids	1	cup(s)	245	221.4	91	8.7	12.3	0	0.6	0.4	0.2	0	—
51	Reduced fat, 2%	1	cup(s)	244	218.0	122	8.1	11.4	0	4.8	3.1	1.4	0.2	—
52	Reduced fat, 2%, with nonfat milk solids	1	cup(s)	245	217.7	125	8.5	12.2	0	4.7	2.9	1.4	0.2	—
50	Whole, 3.3%	1	cup(s)	244	215.5	146	7.9	11.0	0	7.9	4.6	2.0	0.5	—
	Canned milk													
62	Nonfat or skim evaporated	2	tablespoon(s)	32	25.3	25	2.4	3.6	0	0.1	0	0	0	—
63	Sweetened condensed	2	tablespoon(s)	38	10.4	123	3.0	20.8	0	3.3	2.1	0.9	0.1	—
61	Whole evaporated	2	tablespoon(s)	32	23.3	42	2.1	3.2	0	2.4	1.4	0.7	0.1	—
	Dried milk													
64	Buttermilk	¼	cup(s)	30	0.9	117	10.4	14.9	0	1.8	1.1	0.5	0.1	—
65	Instant nonfat with added vitamin A	¼	cup(s)	17	0.7	61	6.0	8.9	0	0.1	0.1	0	0	—
5234	Skim milk powder	¼	cup(s)	17	0.7	62	6.1	9.1	0	0.1	0.1	0	0	—
907	Whole dry milk	¼	cup(s)	32	0.8	159	8.4	12.3	0	8.5	5.4	2.5	0.2	—
909	**Goat milk**	1	cup(s)	244	212.4	168	8.7	10.9	0	10.1	6.5	2.7	0.4	—
	Chocolate milk													
33155	Chocolate syrup, prepared with milk	1	cup(s)	282	227.0	254	8.7	36.0	0.8	8.3	4.7	2.1	0.5	—
33184	Cocoa mix with aspartame, added sodium and vitamin A, no added calcium or phosphorus, prepared with water	1	cup(s)	192	177.4	56	2.3	10.8	1.2	0.4	0.3	0.1	0	—
908	Hot cocoa, prepared with milk	1	cup(s)	250	206.4	193	8.8	26.6	2.5	5.8	3.6	1.7	0.1	0.2
69	Low fat	1	cup(s)	250	211.3	158	8.1	26.1	1.3	2.5	1.5	0.8	0.1	—
68	Reduced fat	1	cup(s)	250	205.4	190	7.5	30.3	1.8	4.8	2.9	1.1	0.2	—
67	Whole	1	cup(s)	250	205.8	208	7.9	25.9	2.0	8.5	5.3	2.5	0.3	—
70	**Eggnog**	1	cup(s)	254	188.9	343	9.7	34.4	0	19.0	11.3	5.7	0.9	—
	Breakfast drinks													
10093	Carnation Instant Breakfast classic chocolate malt, prepared with skim milk, no sugar added	1	cup(s)	243	—	142	11.1	21.3	0.7	1.3	0.7	—	—	—
10092	Carnation Instant Breakfast classic French vanilla, prepared with skim milk, no sugar added	1	cup(s)	273	—	150	12.9	24.0	0	0.4	0.4	—	—	—
10094	Carnation Instant Breakfast stawberry sensation, prepared with skim milk, no sugar added	1	cup(s)	243	—	142	11.1	21.3	0	0.4	0.4	—	—	—
10091	Carnation Instant Breakfast strawberry sensation, prepared with skim milk	1	cup(s)	273	—	220	12.5	38.8	0	0.4	0.4	—	—	—
1417	Ovaltine rich chocolate flavor, prepared with skim milk	1	cup(s)	258	—	170	8.5	31.0	0	0	0	0	0	0
8539	**Malted milk, chocolate mix, fortified, prepared with milk**	1	cup(s)	265	215.8	223	8.9	28.9	1.1	8.6	5.0	2.2	0.5	—
	Milkshakes													
73	Chocolate	1	cup(s)	227	164.0	270	6.9	48.1	0.7	6.1	3.8	1.8	0.2	—
3163	Strawberry	1	cup(s)	226	167.8	256	7.7	42.8	0.9	6.3	3.9	—	—	—
74	Vanilla	1	cup(s)	227	169.2	254	8.8	40.3	0	6.9	4.3	2.0	0.3	—
	Ice cream													
4776	Chocolate	½	cup(s)	66	36.8	143	2.5	18.6	0.8	7.3	4.5	2.1	0.3	—
12137	Chocolate fudge, no sugar added	½	cup(s)	71	—	100	3.0	16.0	2.0	3.0	1.5	—	—	0
16514	Chocolate, soft serve	½	cup(s)	87	49.9	177	3.2	24.1	0.7	8.4	5.2	2.4	0.3	—
16523	Sherbet, all flavors	½	cup(s)	97	63.8	139	1.1	29.3	3.2	1.9	1.1	0.5	0.1	—
4778	Strawberry	½	cup(s)	66	39.6	127	2.1	18.2	0.6	5.5	3.4	—	—	—
76	Vanilla	½	cup(s)	72	43.9	145	2.5	17.0	0.5	7.9	4.9	2.1	0.3	—
12146	Vanilla chocolate swirl, fat-free, no sugar added	½	cup(s)	71	—	100	3.0	14.0	2.0	3.0	2.0	—	—	0
82	Vanilla, light	½	cup(s)	76	48.3	125	3.6	19.6	0.2	3.7	2.2	1.0	0.2	—
78	Vanilla, light, soft serve	½	cup(s)	88	61.2	111	4.3	19.2	0	2.3	1.4	0.7	0.1	—

PAGE KEY: H-4 = Breads/Baked Goods H-10 = Cereal/Rice/Pasta H-14 = Fruit H-18 = Vegetables/Legumes H-28 = Nuts/Seeds H-30 = Vegetarian H-32 = Dairy H-40 = Eggs H-40 = Seafood H-42 = Meats H-46 = Poultry H-46 = Processed Meats H-48 = Beverages H-52 = Fats/Oils H-54 = Sweets H-56 = Spices/Condiments/Sauces H-60 = Mixed Foods/Soups/Sandwiches H-64 = Fast Food H-84 = Convenience H-86 = Baby Foods

Chol (mg)	Calc (mg)	Iron (mg)	Magn (mg)	Pota (mg)	Sodi (mg)	Zinc (mg)	Vit A (µg)	Thia (mg)	Vit E (mg α)	Ribo (mg)	Niac (mg)	Vit B_6 (mg)	Fola (µg)	Vit C (mg)	Vit B_{12} (µg)	Sele (µg)
10	284	0.12	27.0	370.0	257.3	1.02	17.2	0.08	0.12	0.37	0.14	0.08	12.3	2.5	0.5	4.9
12	290	0.07	26.8	366.0	107.4	1.02	141.5	0.04	0.02	0.45	0.22	0.09	12.2	0	1.1	8.1
10	314	0.12	34.3	396.9	127.4	0.98	144.6	0.09	—	0.42	0.22	0.11	12.3	2.5	0.9	5.6
5	306	0.07	27.0	382.2	102.9	1.02	149.5	0.11	0.02	0.44	0.23	0.09	12.3	0	1.3	7.6
5	316	0.12	36.8	419.0	129.9	1.00	149.5	0.10	0.00	0.42	0.22	0.11	12.3	2.5	1.0	5.4
20	285	0.07	26.8	366.0	100.0	1.04	134.2	0.09	0.07	0.45	0.22	0.09	12.2	0.5	1.1	6.1
20	314	0.12	34.3	396.9	127.4	0.98	137.2	0.09	—	0.42	0.22	0.11	12.3	2.5	0.9	5.6
24	276	0.07	24.4	348.9	97.6	0.97	68.3	0.10	0.14	0.44	0.26	0.08	12.2	0	1.1	9.0
1	93	0.09	8.6	105.9	36.7	0.28	37.6	0.01	0.00	0.09	0.05	0.01	2.9	0.4	0.1	0.8
13	109	0.07	9.9	141.9	48.6	0.36	28.3	0.03	0.06	0.16	0.08	0.02	4.2	1.0	0.2	5.7
9	82	0.06	7.6	95.4	33.4	0.24	20.5	0.01	0.04	0.10	0.06	0.01	2.5	0.6	0.1	0.7
21	359	0.09	33.3	482.5	156.7	1.21	14.9	0.11	0.03	0.48	0.27	0.10	14.2	1.7	1.2	6.2
3	209	0.05	19.9	289.9	93.3	0.75	120.5	0.07	0.00	0.30	0.15	0.06	8.5	1.0	0.7	4.6
3	214	0.05	20.3	296.0	95.3	0.76	123.1	0.07	0.00	0.30	0.15	0.06	8.7	1.0	0.7	4.7
31	292	0.15	27.2	425.6	118.7	1.06	82.2	0.09	0.15	0.38	0.20	0.09	11.8	2.8	1.0	5.2
27	327	0.12	34.2	497.8	122.0	0.73	139.1	0.11	0.17	0.33	0.67	0.11	2.4	3.2	0.2	3.4
25	251	0.90	50.8	408.9	132.5	1.21	70.5	0.11	0.14	0.46	0.38	0.09	14.1	0	1.1	9.6
0	92	0.74	32.6	405.1	138.2	0.51	0	0.04	0.00	0.20	0.16	0.04	1.9	0	0.2	2.5
20	263	1.20	57.5	492.5	110.0	1.57	127.5	0.09	0.07	0.45	0.33	0.10	12.5	0.5	1.1	6.8
8	288	0.60	32.5	425.0	152.5	1.02	145.0	0.09	0.05	0.41	0.31	0.10	12.5	2.3	0.9	4.8
20	273	0.60	35.0	422.5	165.0	0.97	160.0	0.11	0.10	0.45	0.41	0.06	5.0	0	0.8	8.5
30	280	0.60	32.5	417.5	150.0	1.02	65.0	0.09	0.15	0.40	0.31	0.10	12.5	2.3	0.8	4.8
150	330	0.50	48.3	419.1	137.2	1.16	116.8	0.08	0.50	0.48	0.26	0.12	2.5	3.8	1.1	10.7
9	444	4.00	88.9	631.1	195.6	3.38	—	0.33	—	0.45	4.44	0.44	4.0	26.7	1.3	8.0
9	500	4.50	100.0	665.0	192.0	3.75	—	0.37	—	0.51	5.00	0.49	100.0	30.0	1.5	9.0
9	444	4.00	88.9	568.9	186.7	3.38	—	0.33	—	0.45	4.44	0.44	88.9	26.7	1.3	8.0
9	500	4.47	100.0	665.0	288.0	3.75	—	0.37	—	0.51	5.07	0.50	100.0	30.0	1.5	8.8
5	350	3.60	100.0	—	270.0	3.75	—	0.37	—	—	4.00	0.40	—	12.0	1.2	—
27	339	3.76	45.1	577.7	230.6	1.16	903.7	0.75	0.15	1.31	11.08	1.01	18.6	31.8	1.1	12.5
25	300	0.70	36.4	508.9	252.2	1.09	40.9	0.10	0.11	0.50	0.28	0.05	11.4	0	0.7	4.3
25	256	0.24	29.4	412.0	187.9	0.81	58.9	0.10	—	0.44	0.39	0.10	6.8	1.8	0.7	4.8
27	332	0.22	27.3	415.8	215.8	0.88	56.8	0.06	0.11	0.44	0.33	0.09	15.9	0	1.2	5.2
22	72	0.61	19.1	164.3	50.2	0.38	77.9	0.02	0.19	0.12	0.14	0.03	10.6	0.5	0.2	1.7
10	100	0.36	—	—	65.0	—	—	—	—	—	—	—	—	0	—	—
22	103	0.32	19.0	192.0	43.3	0.45	66.6	0.03	0.22	0.13	0.11	0.03	4.3	0.5	0.3	2.5
0	52	0.13	7.7	92.6	44.4	0.46	9.7	0.02	0.02	0.08	0.07	0.02	6.8	5.6	0.1	1.3
19	79	0.13	9.2	124.1	39.6	0.22	63.4	0.03	—	0.16	0.11	0.03	7.9	5.1	0.2	1.3
32	92	0.06	10.1	143.3	57.6	0.49	85.0	0.03	0.21	0.17	0.08	0.03	3.6	0.4	0.3	1.3
10	100	0.00	—	—	65.0	—	—	—	—	—	—	—	—	0	—	—
21	122	0.14	10.6	158.1	56.2	0.55	97.3	0.04	0.09	0.19	0.10	0.03	4.6	0.9	0.4	1.5
11	138	0.05	12.3	194.5	61.6	0.46	25.5	0.04	0.05	0.17	0.10	0.04	4.4	0.8	0.4	3.2

TABLE H-1 Table of Food Composition (*continued*) (Computer code is for Cengage Diet Analysis program) (For purposes of calculations, use "0" for t, <1, <.1, <.01, et

DA+ Code	Food Description	Quantity	Measure	Wt (g)	H_2O (g)	Ener (kcal)	Prot (g)	Carb (g)	Fiber (g)	Fat (g)	Fat Breakdown (g)			
											Sat	Mono	Poly	*Trans*
Dairy—*continued*														
	Soy desserts													
10694	Tofutti low fat vanilla fudge nondairy frozen dessert	½	cup(s)	70	—	140	2.0	24.0	0	4.0	1.0	—	—	—
15721	Tofutti premium chocolate supreme nondairy frozen dessert	½	cup(s)	70	—	180	3.0	18.0	0	11.0	2.0	—	—	—
15720	Tofutti premium vanilla non-dairy frozen dessert	½	cup(s)	70	—	190	2.0	20.0	0	11.0	2.0	—	—	—
	Ice milk													
16517	Chocolate	½	cup(s)	66	42.9	94	2.8	16.9	0.3	2.1	1.3	0.6	0.1	—
16516	Flavored, not chocolate	½	cup(s)	66	41.4	108	3.5	17.5	0.2	2.6	1.7	0.6	0.1	—
	Pudding													
25032	Chocolate	½	cup(s)	144	109.7	155	5.1	22.7	0.7	5.4	3.1	1.7	0.2	0
1923	Chocolate, sugar free, prepared with 2% milk	½	cup(s)	133	—	100	5.0	14.0	0.3	3.0	1.5	—	—	—
1722	Rice	½	cup(s)	113	75.6	151	4.1	29.9	0.5	1.9	1.1	0.5	0.1	—
4747	Tapioca, ready to eat	1	item(s)	142	102.0	185	2.8	30.8	0	5.5	1.4	3.6	0.1	—
25031	Vanilla	½	cup(s)	136	109.7	116	4.7	17.6	0	2.8	1.6	0.9	0.2	0
1924	Vanilla, sugar free, prepared with 2% milk	½	cup(s)	133	—	90	4.0	12.0	0.2	2.0	1.5	—	—	—
	Frozen yogurt													
4785	Chocolate, soft serve	½	cup(s)	72	45.9	115	2.9	17.9	1.6	4.3	2.6	1.3	0.2	—
1747	Fruit varieties	½	cup(s)	113	80.5	144	3.4	24.4	0	4.1	2.6	1.1	0.1	—
4786	Vanilla, soft serve	½	cup(s)	72	47.0	117	2.9	17.4	0	4.0	2.5	1.1	0.2	—
	Milk substitutes													
	Lactose free													
16081	Fat-free, calcium fortified [milk]	1	cup(s)	240	—	80	8.0	13.0	0	0	0	0	0	0
36486	Low fat milk	1	cup(s)	240	—	110	8.0	13.0	0	2.5	1.5	—	—	—
36487	Reduced fat milk	1	cup(s)	240	—	130	8.0	12.0	0	5.0	3.0	—	—	—
36488	Whole milk	1	cup(s)	240	—	150	8.0	12.0	0	8.0	5.0	—	—	—
	Rice													
10083	Rice Dream carob rice beverage	1	cup(s)	240	—	150	1.0	32.0	0	2.5	0	—	—	—
17089	Rice Dream original rice beverage, enriched	1	cup(s)	240	—	120	1.0	25.0	0	2.0	0	—	—	—
10087	Rice Dream vanilla enriched rice beverage	1	cup(s)	240	—	130	1.0	28.0	0	2.0	0	—	—	—
	Soy													
34750	Soy Dream chocolate enriched soy beverage	1	cup(s)	240	—	210	7.0	37.0	1.0	3.5	0.5	—	—	—
34749	Soy Dream vanilla enriched soy beverage	1	cup(s)	240	—	150	7.0	22.0	0	4.0	0.5	—	—	—
13840	Vitasoy light chocolate soymilk	1	cup(s)	240	—	100	4.0	17.0	0	2.0	0.5	0.5	1.0	—
13839	Vitasoy light vanilla soymilk	1	cup(s)	240	—	70	4.0	10.0	0	2.0	0.5	0.5	1.0	—
13836	Vitasoy rich chocolate soymilk	1	cup(s)	240	—	160	7.0	24.0	1.0	4.0	0.5	1.0	2.5	—
13835	Vitasoy vanilla delite soymilk	1	cup(s)	240	—	120	7.0	13.0	1.0	4.0	0.5	1.0	2.5	—
	Yogurt													
3615	Custard style, fruit flavors	6	ounce(s)	170	127.1	190	7.0	32.0	0	3.5	2.0	—	—	—
3617	Custard style, vanilla	6	ounce(s)	170	134.1	190	7.0	32.0	0	3.5	2.0	0.9	0.1	—
32101	Fruit, low fat	1	cup(s)	245	184.5	243	9.8	45.7	0	2.8	1.8	0.8	0.1	—
29638	Fruit, nonfat, sweetened with low-calorie sweetener	1	cup(s)	241	208.3	123	10.6	19.4	1.2	0.4	0.2	0.1	0	—
93	Plain, low fat	1	cup(s)	245	208.4	154	12.9	17.2	0	3.8	2.5	1.0	0.1	—
94	Plain, nonfat	1	cup(s)	245	208.8	137	14.0	18.8	0	0.4	0.3	0.1	0	—
32100	Vanilla, low fat	1	cup(s)	245	193.6	208	12.1	33.8	0	3.1	2.0	0.8	0.1	—
5242	Yogurt beverage	1	cup(s)	245	199.8	172	6.2	32.8	0	2.2	1.4	0.6	0.1	—
38202	Yogurt smoothie, nonfat, all flavors	1	item(s)	325	—	290	10.0	60.0	6.0	0	0	0	0	0
	Soy yogurt													
34617	Stonyfield Farm O'Soy strawberry-peach pack organic cultured soy yogurt	1	item(s)	113	—	100	5.0	16.0	3.0	2.0	0	—	—	0
34616	Stonyfield Farm O'Soy vanilla organic cultured soy yogurt	1	item(s)	170	—	150	7.0	26.0	4.0	2.0	0	—	—	0
10453	White Wave plain silk cultured soy yogurt	8	ounce(s)	227	—	140	5.0	22.0	1.0	3.0	0.5	—	—	0

PAGE KEY: H-4 = Breads/Baked Goods H-10 = Cereal/Rice/Pasta H-14 = Fruit H-18 = Vegetables/Legumes H-28 = Nuts/Seeds H-30 = Vegetarian H-32 = Dairy H-40 = Eggs H-40 = Seafood H-42 = Meats H-46 = Poultry H-46 = Processed Meats H-48 = Beverages H-52 = Fats/Oils H-54 = Sweets H-56 = Spices/Condiments/Sauces H-60 = Mixed Foods/Soups/Sandwiches H-64 = Fast Food H-84 = Convenience H-86 = Baby Foods

Chol (mg)	Calc (mg)	Iron (mg)	Magn (mg)	Pota (mg)	Sodi (mg)	Zinc (mg)	Vit A (µg)	Thia (mg)	Vit E (mg α)	Ribo (mg)	Niac (mg)	Vit B_6 (mg)	Fola (µg)	Vit C (mg)	Vit B_{12} (µg)	Sele (µg)
0	0	0.00	—	8.0	90.0	—	0	—	—	—	—	—	—	0	—	—
0	0	0.00	—	7.0	180.0	—	0	—	—	—	—	—	—	0	—	—
0	0	0.00	—	2.0	210.0	—	0	—	—	—	—	—	—	0	—	—
6	94	0.15	13.1	155.2	40.6	0.36	15.7	0.03	0.05	0.11	0.08	0.02	3.9	0.5	0.3	2.2
16	76	0.05	9.2	136.2	48.5	0.47	90.4	0.02	0.05	0.11	0.06	0.01	3.3	0.1	0.2	1.3
35	149	0.46	31.3	226.7	137.0	0.71	—	0.05	0.00	0.22	0.15	0.06	8.3	1.2	0.5	4.9
10	150	0.72	—	330.0	310.0	—	—	0.06	—	0.26	—	—	—	0	—	—
7	113	0.28	15.8	201.4	66.4	0.52	41.6	0.03	0.05	0.17	0.34	0.06	4.5	0.2	0.2	4.8
1	101	0.15	8.5	130.6	205.9	0.31	0	0.03	0.21	0.13	0.09	0.03	4.3	0.4	0.3	0
35	146	0.17	17.2	188.9	136.4	0.52	—	0.04	0.00	0.22	0.10	0.05	8.0	1.2	0.5	4.6
10	150	0.00	—	190.0	380.0	—	—	0.03	—	0.17	—	—	—	0	—	—
4	106	0.90	19.4	187.9	70.6	0.35	31.7	0.02	—	0.15	0.22	0.05	7.9	0.2	0.2	1.7
15	113	0.52	11.3	176.3	71.2	0.31	55.4	0.04	0.10	0.20	0.07	0.04	4.5	0.8	0.1	2.1
1	103	0.21	10.1	151.9	62.6	0.30	42.5	0.02	0.07	0.16	0.20	0.05	4.3	0.6	0.2	2.4
3	500	0.00	—	—	125.0	—	100.0	—	—	—	—	—	—	0	0	—
10	300	0.00	—	—	125.0	—	100.0	—	—	—	—	—	—	0	—	—
20	300	0.00	—	—	125.0	—	98.2	—	—	—	—	—	—	0	—	—
35	300	0.00	—	—	125.0	—	58.1	—	—	—	—	—	—	0	—	—
0	20	0.72	—	82.5	100.0	—	—	—	—	—	—	—	—	1.2	—	—
0	300	0.00	13.3	60.0	90.0	0.24	—	0.06	—	0.00	0.84	0.07	—	0	1.5	—
0	300	0.00	—	53.0	90.0	—	—	—	—	—	—	—	—	0	1.5	—
0	300	1.80	60.0	350.0	160.0	0.60	33.3	0.15	—	0.06	0.80	0.12	60.0	0	3.0	—
0	300	1.80	40.0	260.0	140.0	0.60	33.3	0.15	—	0.06	0.80	0.12	60.0	0	3.0	—
0	300	0.72	24.0	200.0	140.0	0.90	—	0.09	—	0.34	—	—	24.0	0	0.9	—
0	300	0.72	24.0	200.0	120.0	0.90	—	0.09	—	0.34	—	—	24.0	0	0.9	—
0	300	1.08	40.0	320.0	150.0	0.90	—	0.15	—	0.34	—	—	60.0	0	0.9	—
0	40	0.72	—	320.0	115.0	—	0	—	—	—	—	—	—	0	—	—
15	300	0.00	16.0	310.0	100.0	—	—	—	—	0.25	—	—	—	0	—	—
15	300	0.00	16.0	310.0	100.0	—	—	—	—	0.25	—	—	—	0	—	—
12	338	0.14	31.9	433.7	129.9	1.64	27.0	0.08	0.04	0.39	0.21	0.09	22.1	1.5	1.1	6.9
5	369	0.62	41.0	549.5	139.8	1.83	4.8	0.10	0.16	0.44	0.49	0.10	31.3	26.5	1.1	7.0
15	448	0.19	41.7	573.3	171.5	2.18	34.3	0.10	0.07	0.52	0.27	0.12	27.0	2.0	1.4	8.1
5	488	0.22	46.6	624.8	188.7	2.37	4.9	0.11	0.00	0.57	0.30	0.13	29.4	2.2	1.5	8.8
12	419	0.17	39.2	536.6	161.7	2.03	29.4	0.10	0.04	0.49	0.26	0.11	27.0	2.0	1.3	12.0
13	260	0.22	39.2	399.4	98.0	1.10	14.7	0.11	0.00	0.51	0.30	0.14	29.4	2.1	1.5	—
5	300	2.70	100.0	580.0	290.0	2.25	—	0.37	—	0.42	5.00	0.50	100.0	15.0	1.5	—
0	100	1.08	24.0	5.0	20.0	—	0	0.22	—	0.10	—	0.04	—	0	0	—
0	150	1.44	40.0	15.0	40.0	—	—	0.30	—	0.13	—	0.08	—	0	0	—
0	400	1.44	—	0	30.0	—	0	—	—	—	—	—	—	0	—	—

TABLE H-1 Table of Food Composition (*continued*) (Computer code is for Cengage Diet Analysis program) (For purposes of calculations, use "0" for t, <1, <.1, <.01, et

DA+ Code	Food Description	Quantity	Measure	Wt (g)	H_2O (g)	Ener (kcal)	Prot (g)	Carb (g)	Fiber (g)	Fat (g)	Fat Breakdown (g)			
											Sat	Mono	Poly	*Trans*
Eggs														
	Eggs													
99	Fried	1	item(s)	46	31.8	90	6.3	0.4	0	7.0	2.0	2.9	1.2	—
100	Hard boiled	1	item(s)	50	37.3	78	6.3	0.6	0	5.3	1.6	2.0	0.7	—
101	Poached	1	item(s)	50	37.8	71	6.3	0.4	0	5.0	1.5	1.9	0.7	—
97	Raw, white	1	item(s)	33	28.9	16	3.6	0.2	0	0.1	0	0	0	—
96	Raw, whole	1	item(s)	50	37.9	72	6.3	0.4	0	5.0	1.5	1.9	0.7	—
98	Raw, yolk	1	item(s)	17	8.9	54	2.7	0.6	0	4.5	1.6	2.0	0.7	—
102	Scrambled, prepared with milk and butter	2	item(s)	122	89.2	204	13.5	2.7	0	14.9	4.5	5.8	2.6	0.7
	Egg substitute													
4028	Egg Beaters	¼	cup(s)	61	—	30	6.0	1.0	0	0	0	0	0	0
920	Frozen	¼	cup(s)	60	43.9	96	6.8	1.9	0	6.7	1.2	1.5	3.7	—
918	Liquid	¼	cup(s)	63	51.9	53	7.5	0.4	0	2.1	0.4	0.6	1.0	—
Seafood														
	Cod													
6040	Atlantic cod or scrod, baked or broiled	3	ounce(s)	85	64.6	89	19.4	0	0	0.7	0.1	0.1	0.2	—
1573	Atlantic cod, cooked, dry heat	3	ounce(s)	85	64.6	89	19.4	0	0	0.7	0.1	0.1	0.2	—
2905	**Eel, raw**	3	ounce(s)	85	58.0	156	15.7	0	0	9.9	2.0	6.1	0.8	—
	Fish fillets													
25079	Baked	3	ounce(s)	84	79.9	99	21.7	0	0	0.7	0.1	0.1	0.3	—
8615	Batter coated or breaded, fried	3	ounce(s)	85	45.6	197	12.5	14.4	0.4	10.5	2.4	2.2	5.3	—
25082	Broiled fish steaks	3	ounce(s)	85	68.1	128	24.2	0	0	2.6	0.4	0.9	0.8	—
25083	Poached fish steaks	3	ounce(s)	85	67.1	111	21.1	0	0	2.3	0.3	0.8	0.7	—
25084	Steamed	3	ounce(s)	85	72.2	79	17.2	0	0	0.6	0.1	0.1	0.2	—
25089	**Flounder, baked**	3	ounce(s)	85	64.4	113	14.8	0.4	0.1	5.5	1.1	2.2	1.4	—
1825	**Grouper, cooked, dry heat**	3	ounce(s)	85	62.4	100	21.1	0	0	1.1	0.3	0.2	0.3	—
	Haddock													
6049	Baked or broiled	3	ounce(s)	85	63.2	95	20.6	0	0	0.8	0.1	0.1	0.3	—
1578	Cooked, dry heat	3	ounce(s)	85	63.1	95	20.6	0	0	0.8	0.1	0.1	0.3	—
1886	**Halibut, Atlantic and Pacific, cooked, dry heat**	3	ounce(s)	85	61.0	119	22.7	0	0	2.5	0.4	0.8	0.8	—
1582	**Herring, Atlantic, pickled**	4	piece(s)	60	33.1	157	8.5	5.8	0	10.8	1.4	7.2	1.0	—
1587	**Jack mackerel, solids, canned, drained**	2	ounce(s)	57	39.2	88	13.1	0	0	3.6	1.1	1.3	0.9	—
8580	**Octopus, common, cooked, moist heat**	3	ounce(s)	85	51.5	139	25.4	3.7	0	1.8	0.4	0.3	0.4	—
1831	**Perch, mixed species, cooked, dry heat**	3	ounce(s)	85	62.3	100	21.1	0	0	1.0	0.2	0.2	0.4	—
1592	**Pacific rockfish, cooked, dry heat**	3	ounce(s)	85	62.4	103	20.4	0	0	1.7	0.4	0.4	0.5	—
	Salmon													
2938	Coho, farmed, raw	3	ounce(s)	85	59.9	136	18.1	0	0	6.5	1.5	2.8	1.6	—
1594	Broiled or baked with butter	3	ounce(s)	85	53.9	155	23.0	0	0	6.3	1.2	2.3	2.3	—
29727	Smoked chinook (lox)	2	ounce(s)	57	40.8	66	10.4	0	0	2.4	0.5	1.1	0.6	—
154	**Sardine, Atlantic with bones, canned in oil**	3	ounce(s)	85	50.7	177	20.9	0	0	9.7	1.3	3.3	4.4	—
	Scallops													
155	Mixed species, breaded, fried	3	item(s)	47	27.2	100	8.4	4.7	—	5.1	1.2	2.1	1.3	—
1599	Steamed	3	ounce(s)	85	64.8	90	13.8	2.0	0	2.6	0.4	1.0	0.8	—
1839	**Snapper, mixed species, cooked, dry heat**	3	ounce(s)	85	59.8	109	22.4	0	0	1.5	0.3	0.3	0.5	—
	Squid													
1868	Mixed species, fried	3	ounce(s)	85	54.9	149	15.3	6.6	0	6.4	1.6	2.3	1.8	—
16617	Steamed or boiled	3	ounce(s)	85	63.3	89	15.2	3.0	0	1.3	0.4	0.1	0.5	—
1570	**Striped bass, cooked, dry heat**	3	ounce(s)	85	62.4	105	19.3	0	0	2.5	0.6	0.7	0.9	—
1601	**Sturgeon, steamed**	3	ounce(s)	85	59.4	111	17.0	0	0	4.3	1.0	2.0	0.7	—
1840	**Surimi, formed**	3	ounce(s)	85	64.9	84	12.9	5.8	0	0.8	0.2	0.1	0.4	—
1842	**Swordfish, cooked, dry heat**	3	ounce(s)	85	58.5	132	21.6	0	0	4.4	1.2	1.7	1.0	—
1846	**Tuna, yellowfin or ahi, raw**	3	ounce(s)	85	60.4	92	19.9	0	0	0.8	0.2	0.1	0.2	—
	Tuna, canned													
159	Light, canned in oil, drained	2	ounce(s)	57	33.9	112	16.5	0	0	4.6	0.9	1.7	1.6	—
355	Light, canned in water, drained	2	ounce(s)	57	42.2	66	14.5	0	0	0.5	0.1	0.1	0.2	—
33211	Light, no salt, canned in oil, drained	2	ounce(s)	57	33.9	112	16.5	0	0	4.7	0.9	1.7	1.6	—
33212	Light, no salt, canned in water, drained	2	ounce(s)	57	42.6	66	14.5	0	0	0.5	0.1	0.1	0.2	—

PAGE KEY: H-4 = Breads/Baked Goods H-10 = Cereal/Rice/Pasta H-14 = Fruit H-18 = Vegetables/Legumes H-28 = Nuts/Seeds H-30 = Vegetarian H-32 = Dairy H-40 = Eggs H-40 = Seafood H-42 = Meats H-46 = Poultry H-46 = Processed Meats H-48 = Beverages H-52 = Fats/Oils H-54 = Sweets H-56 = Spices/Condiments/Sauces H-60 = Mixed Foods/Soups/Sandwiches H-64 = Fast Food H-84 = Convenience H-86 = Baby Foods

Chol (mg)	Calc (mg)	Iron (mg)	Magn (mg)	Pota (mg)	Sodi (mg)	Zinc (mg)	Vit A (μg)	Thia (mg)	Vit E (mg α)	Ribo (mg)	Niac (mg)	Vit B_6 (mg)	Fola (μg)	Vit C (mg)	Vit B_{12} (μg)	Sele (μg)
210	27	0.91	6.0	67.6	93.8	0.55	91.1	0.03	0.56	0.23	0.03	0.07	23.5	0	0.6	15.7
212	25	0.59	5.0	63.0	62.0	0.52	84.5	0.03	0.51	0.25	0.03	0.06	22.0	0	0.6	15.4
211	27	0.91	6.0	66.5	147.0	0.55	69.5	0.02	0.48	0.20	0.03	0.06	17.5	0	0.6	15.8
0	2	0.02	3.6	53.8	54.8	0.01	0	0.00	0.00	0.14	0.03	0.00	1.3	0	0	6.6
212	27	0.91	6.0	67.0	70.0	0.55	70.0	0.03	0.48	0.23	0.03	0.07	23.5	0	0.6	15.9
210	22	0.46	0.9	18.5	8.2	0.39	64.8	0.03	0.43	0.09	0.00	0.06	24.8	0	0.3	9.5
429	87	1.46	14.6	168.4	341.6	1.22	174.5	0.06	1.33	0.53	0.09	0.14	36.6	0.2	0.9	27.5
0	20	1.08	4.0	85.0	115.0	0.60	112.5	0.15	—	0.85	0.20	0.08	60.0	0	1.2	—
1	44	1.18	9.0	127.8	119.4	0.58	6.6	0.07	0.95	0.23	0.08	0.08	9.6	0.3	0.2	24.8
1	33	1.32	5.6	207.1	111.1	0.82	11.3	0.07	0.17	0.19	0.07	0.00	9.4	0	0.2	15.6
47	12	0.41	35.7	207.5	66.3	0.49	11.9	0.07	0.68	0.06	2.13	0.24	6.8	0.8	0.9	32.0
47	12	0.41	35.7	207.5	66.3	0.49	11.9	0.07	0.68	0.06	2.13	0.24	6.8	0.9	0.9	32.0
107	17	0.42	17.0	231.3	43.4	1.37	887.0	0.13	3.40	0.03	2.97	0.05	12.8	1.5	2.6	5.5
44	8	0.31	29.1	489.0	86.1	0.48	—	0.02	—	0.05	2.47	0.46	8.1	3.0	1.0	44.3
29	15	1.79	20.4	272.2	452.5	0.37	9.4	0.09	—	0.09	1.78	0.08	14.5	0	0.9	7.7
37	55	0.97	96.7	524.3	62.9	0.49	—	0.05	—	0.08	6.47	0.36	12.6	0	1.2	42.5
32	48	0.85	84.0	455.6	54.7	0.42	—	0.05	—	0.07	5.92	0.33	11.5	0	1.1	37.0
41	12	0.29	24.7	319.3	41.7	0.34	—	0.06	—	0.06	1.89	0.21	6.1	0.8	0.8	32.0
44	19	0.34	47.3	224.7	280.2	0.20	—	0.06	0.40	0.07	2.02	0.18	7.4	2.8	1.6	33.5
40	18	0.96	31.5	404.0	45.1	0.43	42.5	0.06	—	0.01	0.32	0.29	8.5	0	0.6	39.8
63	36	1.15	42.5	339.4	74.0	0.40	16.2	0.03	0.42	0.03	3.94	0.29	6.8	0	1.2	34.4
63	36	1.14	42.5	339.3	74.0	0.40	16.2	0.03	—	0.03	3.93	0.29	11.1	0	1.2	34.4
35	51	0.91	91.0	489.9	58.7	0.45	45.9	0.05	—	0.07	6.05	0.33	11.9	0	1.2	39.8
8	46	0.73	4.8	41.4	522.0	0.31	154.8	0.02	1.02	0.08	1.98	0.10	1.2	0	2.6	35.1
45	137	1.15	21.0	110.0	214.9	0.57	73.7	0.02	0.58	0.12	3.50	0.11	2.8	0.5	3.9	21.4
82	90	8.11	51.0	535.8	391.2	2.85	76.5	0.04	1.02	0.06	3.21	0.55	20.4	6.8	30.6	76.2
98	87	0.98	32.3	292.6	67.2	1.21	8.5	0.06	—	0.10	1.61	0.11	5.1	1.4	1.9	13.7
37	10	0.45	28.9	442.3	65.5	0.45	60.4	0.03	1.32	0.07	3.33	0.22	8.5	0	1.0	39.8
43	10	0.29	26.4	382.7	40.0	0.36	47.6	0.08	—	0.09	5.79	0.56	11.1	0.9	2.3	10.7
40	15	1.02	26.9	376.6	98.6	0.56	—	0.13	1.14	0.05	8.33	0.18	4.2	1.8	2.3	41.0
13	6	0.48	10.2	99.2	1134.0	0.17	14.7	0.01	—	0.05	2.67	0.15	1.1	0	1.8	21.6
121	325	2.48	33.2	337.6	429.5	1.10	27.2	0.04	1.70	0.18	4.43	0.14	10.2	0	7.6	44.8
28	20	0.38	27.4	154.8	215.8	0.49	10.7	0.02	—	0.05	0.70	0.06	17.2	1.1	0.6	12.5
27	20	0.22	45.9	238.0	358.7	0.78	32.3	0.01	0.16	0.05	0.84	0.11	10.2	2.0	1.1	18.2
40	34	0.20	31.5	444.0	48.5	0.37	29.8	0.04	—	0.00	0.29	0.39	5.1	1.4	3.0	41.7
221	33	0.85	32.3	237.3	260.3	1.48	9.4	0.04	—	0.39	2.21	0.04	11.9	3.6	1.0	44.1
227	31	0.62	28.9	192.1	356.2	1.49	8.5	0.01	1.17	0.32	1.69	0.04	3.4	3.2	1.0	43.7
88	16	0.91	43.4	279.0	74.8	0.43	26.4	0.09	—	0.03	2.17	0.29	8.5	0	3.8	39.8
63	11	0.59	29.8	239.7	388.5	0.35	198.9	0.06	0.52	0.07	8.30	0.19	14.5	0	2.2	13.3
26	8	0.22	36.6	95.3	121.6	0.28	17.0	0.01	0.53	0.01	0.18	0.02	1.7	0	1.4	23.9
43	5	0.88	28.9	313.8	97.8	1.25	34.9	0.03	—	0.09	10.02	0.32	1.7	0.9	1.7	52.5
38	14	0.62	42.5	377.6	31.5	0.44	15.3	0.37	0.42	0.04	8.33	0.77	1.7	0.8	0.4	31.0
10	7	0.79	17.6	117.3	200.6	0.51	13.0	0.02	0.49	0.07	7.03	0.06	2.8	0	1.2	43.1
17	6	0.87	15.3	134.3	191.5	0.43	9.6	0.01	0.19	0.04	7.52	0.19	2.3	0	1.7	45.6
10	7	0.78	17.6	117.4	28.3	0.51	0	0.02	—	0.06	7.03	0.06	2.8	0	1.2	43.1
17	6	0.86	15.3	134.4	28.3	0.43	0	0.01	—	0.04	7.52	0.19	2.3	0	1.7	45.6

TABLE H-1 Table of Food Composition (*continued*) (Computer code is for Cengage Diet Analysis program) (For purposes of calculations, use "0" for t, <1, <.1, <.01, et

DA+ Code	Food Description	Quantity	Measure	Wt (g)	H_2O (g)	Ener (kcal)	Prot (g)	Carb (g)	Fiber (g)	Fat (g)	Fat Breakdown (g) Sat	Mono	Poly	*Trans*
	Seafood—*continued*													
2961	White, canned in oil, drained	2	ounce(s)	57	36.3	105	15.0	0	0	4.6	0.7	1.8	1.7	—
351	White, canned in water, drained	2	ounce(s)	57	41.5	73	13.4	0	0	1.7	0.4	0.4	0.6	—
33213	White, no salt, canned in oil, drained	2	ounce(s)	57	36.3	105	15.0	0	0	4.6	0.9	1.4	1.9	—
33214	White, no salt, canned in water, drained	2	ounce(s)	57	42.0	73	13.4	0	0	1.7	0.4	0.4	0.6	—
	Yellowtail													
8548	Mixed species, cooked, dry heat	3	ounce(s)	85	57.3	159	25.2	0	0	5.7	1.4	2.2	1.5	—
2970	Mixed species, raw	2	ounce(s)	57	42.2	83	13.1	0	0	3.0	0.7	1.1	0.8	—
	Shellfish, meat only													
1857	Abalone, mixed species, fried	3	ounce(s)	85	51.1	161	16.7	9.4	0	5.8	1.4	2.3	1.4	—
16618	Abalone, steamed or poached	3	ounce(s)	85	40.7	177	28.8	10.1	0	1.3	0.3	0.2	0.2	—
	Crab													
1851	Blue crab, canned	2	ounce(s)	57	43.2	56	11.6	0	0	0.7	0.1	0.1	0.2	—
1852	Blue crab, cooked, moist heat	3	ounce(s)	85	65.9	87	17.2	0	0	1.5	0.2	0.2	0.6	—
8562	Dungeness crab, cooked, moist heat	3	ounce(s)	85	62.3	94	19.0	0.8	0	1.1	0.1	0.2	0.3	—
1860	**Clams, cooked, moist heat**	3	ounce(s)	85	54.1	126	21.7	4.4	0	1.7	0.2	0.1	0.5	—
1853	**Crayfish, farmed, cooked, moist heat**	3	ounce(s)	85	68.7	74	14.9	0	0	1.1	0.2	0.2	0.4	—
	Oysters													
8720	Baked or broiled	3	ounce(s)	85	68.6	89	5.6	3.2	0	5.8	1.3	2.1	1.9	—
152	Eastern, farmed, raw	3	ounce(s)	85	73.3	50	4.4	4.7	0	1.3	0.4	0.1	0.5	—
8715	Eastern, wild, cooked, moist heat	3	ounce(s)	85	59.8	117	12.0	6.7	0	4.2	1.3	0.5	1.6	—
8584	Pacific, cooked, moist heat	3	ounce(s)	85	54.5	139	16.1	8.4	0	3.9	0.9	0.7	1.5	—
1865	Pacific, raw	3	ounce(s)	85	69.8	69	8.0	4.2	0	2.0	0.4	0.3	0.8	—
1854	**Lobster, northern, cooked, moist heat**	3	ounce(s)	85	64.7	83	17.4	1.1	0	0.5	0.1	0.1	0.1	—
1862	**Mussel, blue, cooked, moist heat**	3	ounce(s)	85	52.0	146	20.2	6.3	0	3.8	0.7	0.9	1.0	—
	Shrimp													
158	Mixed species, breaded, fried	3	ounce(s)	85	44.9	206	18.2	9.8	0.3	10.4	1.8	3.2	4.3	—
1855	Mixed species, cooked, moist heat	3	ounce(s)	85	65.7	84	17.8	0	0	0.9	0.2	0.2	0.4	—
	Beef, Lamb, Pork													
	Beef													
4450	Breakfast strips, cooked	2	slice(s)	23	5.9	101	7.1	0.3	0	7.8	3.2	3.8	0.4	—
174	Corned beef, canned	3	ounce(s)	85	49.1	213	23.0	0	0	12.7	5.3	5.1	0.5	—
33147	Cured, thin siced	2	ounce(s)	57	32.9	100	15.9	3.2	0	2.2	0.9	1.0	0.1	—
4581	Jerky	1	ounce(s)	28	6.6	116	9.4	3.1	0.5	7.3	3.1	3.2	0.3	—
	Ground beef													
5898	Lean, broiled, medium	3	ounce(s)	85	50.4	202	21.6	0	0	12.2	4.8	5.3	0.4	—
5899	Lean, broiled, well done	3	ounce(s)	85	48.4	214	23.8	0	0	12.5	5.0	5.7	0.3	0.4
5914	Regular, broiled, medium	3	ounce(s)	85	46.1	246	20.5	0	0	17.6	6.9	7.7	0.6	—
5915	Regular, broiled, well done	3	ounce(s)	85	43.8	259	21.6	0	0	18.4	7.5	8.5	0.5	0.6
	Beef rib													
4241	Rib, small end, separable lean, 0″ fat, broiled	3	ounce(s)	85	53.2	164	25.0	0	0	6.4	2.4	2.6	0.2	—
4183	Rib, whole, lean and fat, 1/4″ fat, roasted	3	ounce(s)	85	39.0	320	18.9	0	0	26.6	10.7	11.4	0.9	—
	Beef roast													
16981	Bottom round, choice, separable lean and fat, 1/8″ fat, braised	3	ounce(s)	85	46.2	216	27.9	0	0	10.7	4.1	4.6	0.4	—
16979	Bottom round, separable lean and fat, 1/8″ fat, roasted	3	ounce(s)	85	52.4	185	22.5	0	0	9.9	3.8	4.2	0.4	—
16924	Chuck, arm pot roast, separable lean and fat, 1/8″ fat, braised	3	ounce(s)	85	42.9	257	25.6	0	0	16.3	6.5	7.0	0.6	—
16930	Chuck, blade roast, separable lean and fat, 1/8″ fat, braised	3	ounce(s)	85	40.5	290	22.8	0	0	21.4	8.5	9.2	0.8	—
5853	Chuck, blade roast, separable lean, 0″ trim, pot roasted	3	ounce(s)	85	47.4	202	26.4	0	0	9.9	3.9	4.3	0.3	—
4296	Eye of round, choice, separable lean, 0″ fat, roasted	3	ounce(s)	85	56.5	138	24.4	0	0	3.7	1.3	1.5	0.1	—
16989	Eye of round, separable lean and fat, 1/8″ fat, roasted	3	ounce(s)	85	52.2	180	24.2	0	0	8.5	3.2	3.6	0.3	—

PAGE KEY: H-4 = Breads/Baked Goods H-10 = Cereal/Rice/Pasta H-14 = Fruit H-18 = Vegetables/Legumes H-28 = Nuts/Seeds H-30 = Vegetarian H-32 = Dairy H-40 = Eggs H-40 = Seafood H-42 = Meats H-46 = Poultry H-46 = Processed Meats H-48 = Beverages H-52 = Fats/Oils H-54 = Sweets H-56 = Spices/Condiments/Sauces H-60 = Mixed Foods/Soups/Sandwiches H-64 = Fast Food H-84 = Convenience H-86 = Baby Foods

Chol (mg)	Calc (mg)	Iron (mg)	Magn (mg)	Pota (mg)	Sodi (mg)	Zinc (mg)	Vit A (µg)	Thia (mg)	Vit E (mg α)	Ribo (mg)	Niac (mg)	Vit B_6 (mg)	Fola (µg)	Vit C (mg)	Vit B_{12} (µg)	Sele (µg)
18	2	0.36	19.3	188.8	224.5	0.26	2.8	0.01	1.30	0.04	6.63	0.24	2.8	0	1.2	34.1
24	8	0.55	18.7	134.3	213.6	0.27	3.4	0.00	0.48	0.02	3.28	0.12	1.1	0	0.7	37.2
18	2	0.36	19.3	188.8	28.3	0.26	0	0.01	—	0.04	6.63	0.24	2.8	0	1.2	34.1
24	8	0.54	18.7	134.4	28.3	0.27	3.4	0.00	—	0.02	3.28	0.12	1.1	0	0.7	37.3
60	25	0.53	32.3	457.6	42.5	0.56	26.4	0.14	—	0.04	7.41	0.15	3.4	2.5	1.1	39.8
31	13	0.28	17.0	238.1	22.1	0.29	16.4	0.08	—	0.02	3.86	0.09	2.3	1.6	0.7	20.7
80	31	3.23	47.6	241.5	502.6	0.80	1.7	0.18	—	0.11	1.61	0.12	11.9	1.5	0.6	44.1
144	50	4.84	68.9	295.0	980.1	1.38	3.4	0.28	6.74	0.12	1.89	0.21	6.0	2.6	0.7	75.6
50	57	0.47	22.1	212.1	188.8	2.27	1.1	0.04	1.04	0.04	0.77	0.08	24.4	1.5	0.3	18.0
85	88	0.77	28.1	275.6	237.3	3.58	1.7	0.08	1.56	0.04	2.80	0.15	43.4	2.8	6.2	34.2
65	50	0.36	49.3	347.0	321.5	4.65	26.4	0.04	—	0.17	3.08	0.14	35.7	3.1	8.8	40.5
57	78	23.78	15.3	534.1	95.3	2.32	145.4	0.12	—	0.36	2.85	0.09	24.7	18.8	84.1	54.4
117	43	0.94	28.1	202.4	82.5	1.25	12.8	0.03	—	0.06	1.41	0.11	9.4	0.4	2.6	29.1
43	36	5.30	37.4	125.0	403.8	72.22	60.4	0.07	0.98	0.06	1.04	0.04	7.7	2.8	14.7	50.7
21	37	4.91	28.1	105.4	151.3	32.23	6.8	0.08	—	0.05	1.07	0.05	15.3	4.0	13.8	54.1
89	77	10.19	80.8	239.0	358.9	154.45	45.9	0.16	—	0.15	2.11	0.10	11.9	5.1	29.8	60.9
85	14	7.82	37.4	256.8	180.3	28.27	124.2	0.10	0.72	0.37	3.07	0.07	12.8	10.9	24.5	131.0
43	7	4.34	18.7	142.9	90.1	14.13	68.9	0.05	—	0.20	1.70	0.04	8.5	6.8	13.6	65.5
61	52	0.33	29.8	299.4	323.2	2.48	22.1	0.01	0.85	0.05	0.91	0.06	9.4	0	2.6	36.3
48	28	5.71	31.5	227.9	313.8	2.27	77.4	0.25	—	0.35	2.55	0.08	64.6	11.6	20.4	76.2
150	57	1.07	34.0	191.3	292.4	1.17	0	0.11	—	0.11	2.60	0.08	15.3	1.3	1.6	35.4
166	33	2.62	28.9	154.8	190.5	1.32	57.8	0.02	1.17	0.02	2.20	0.10	3.4	1.9	1.3	33.7
27	2	0.71	6.1	93.1	509.2	1.44	0	0.02	0.06	0.05	1.46	0.07	1.8	0	0.8	6.1
73	10	1.76	11.9	115.7	855.6	3.03	0	0.01	0.12	0.12	2.06	0.11	7.7	0	1.4	36.5
23	6	1.53	10.8	243.2	815.9	2.25	0	0.04	0.00	0.10	2.98	0.19	6.2	0	1.5	16.0
14	6	1.53	14.5	169.2	627.4	2.29	0	0.04	0.13	0.04	0.49	0.05	38.0	0	0.3	3.0
58	6	2.00	17.9	266.2	59.5	4.63	0	0.05	—	0.23	4.21	0.23	7.6	0	1.8	16.0
69	12	2.21	18.4	250.0	62.4	5.86	0	0.08	—	0.23	5.10	0.16	9.4	0	1.7	19.0
62	9	2.07	17.0	248.3	70.6	4.40	0	0.02	—	0.16	4.90	0.23	7.6	0	2.5	16.2
71	12	2.30	18.5	242.4	72.4	5.18	0	0.08	—	0.23	4.93	0.17	8.5	0	1.6	18.0
65	16	1.59	21.3	319.8	51.9	4.64	0	0.06	0.34	0.12	7.15	0.53	8.5	0	1.4	29.2
72	9	1.96	16.2	251.7	53.6	4.45	0	0.06	—	0.14	2.85	0.19	6.0	0	2.1	18.7
68	6	2.29	17.9	223.7	35.7	4.59	0	0.05	0.41	0.15	5.05	0.36	8.5	0	1.7	29.3
64	5	1.83	14.5	182.0	29.8	3.76	0	0.05	0.34	0.12	3.92	0.29	6.8	0	1.3	23.0
67	14	2.15	17.0	205.8	42.5	5.93	0	0.05	0.45	0.15	3.63	0.25	7.7	0	1.9	24.1
88	11	2.66	16.2	198.2	55.3	7.15	0	0.06	0.17	0.20	2.06	0.22	4.3	0	1.9	20.9
73	11	3.12	19.6	223.7	60.4	8.73	0	0.06	—	0.23	2.27	0.24	5.1	0	2.1	22.7
49	5	2.16	16.2	200.7	32.3	4.28	0	0.05	0.30	0.15	4.69	0.34	8.5	0	1.4	28.0
54	5	1.98	15.3	193.1	31.5	3.95	0	0.05	0.34	0.13	4.37	0.31	7.7	0	1.5	25.2

TABLE H-1 Table of Food Composition (*continued*) (Computer code is for Cengage Diet Analysis program) (For purposes of calculations, use "0" for t, <1, <.1, <.01, etc.)

DA+ Code	Food Description	Quantity	Measure	Wt (g)	H_2O (g)	Ener (kcal)	Prot (g)	Carb (g)	Fiber (g)	Fat (g)	Fat Breakdown (g) Sat	Mono	Poly	*Trans*
Beef, Lamb, Pork—*continued*														
	Beef steak													
4348	Short loin, t-bone steak, lean and fat, ¼″ fat, broiled	3	ounce(s)	85	43.2	274	19.4	0	0	21.2	8.3	9.6	0.8	—
4349	Short loin, t-bone steak, lean, ¼″ fat, broiled	3	ounce(s)	85	52.3	174	22.8	0	0	8.5	3.1	4.2	0.3	—
4360	Top loin, prime, lean and fat, ¼″ fat, broiled	3	ounce(s)	85	42.7	275	21.6	0	0	20.3	8.2	8.6	0.7	—
	Beef variety													
188	Liver, pan fried	3	ounce(s)	85	52.7	149	22.6	4.4	0	4.0	1.3	0.5	0.5	0.2
4447	Tongue, simmered	3	ounce(s)	85	49.2	242	16.4	0	0	19.0	6.9	8.6	0.6	0.7
	Lamb chop													
3275	Loin, domestic, lean and fat, ¼″ fat, broiled	3	ounce(s)	85	43.9	269	21.4	0	0	19.6	8.4	8.3	1.4	—
	Lamb leg													
3264	Domestic, lean and fat, ¼″ fat, cooked	3	ounce(s)	85	45.7	250	20.9	0	0	17.8	7.5	7.5	1.3	—
	Lamb rib													
182	Domestic, lean and fat, ¼″ fat, broiled	3	ounce(s)	85	40.0	307	18.8	0	0	25.2	10.8	10.3	2.0	—
183	Domestic, lean, ¼″ fat, broiled	3	ounce(s)	85	50.0	200	23.6	0	0	11.0	4.0	4.4	1.0	—
	Lamb shoulder													
186	Shoulder, arm and blade, domestic, choice, lean and fat, ¼″ fat, roasted	3	ounce(s)	85	47.8	235	19.1	0	0	17.0	7.2	6.9	1.4	—
187	Shoulder, arm and blade, domestic, choice, lean, ¼″ fat, roasted	3	ounce(s)	85	53.8	173	21.2	0	0	9.2	3.5	3.7	0.8	—
3287	Shoulder, arm, domestic, lean and fat, ¼″ fat, braised	3	ounce(s)	85	37.6	294	25.8	0	0	20.4	8.4	8.7	1.5	—
3290	Shoulder, arm, domestic, lean, ¼″ fat, braised	3	ounce(s)	85	41.9	237	30.2	0	0	12.0	4.3	5.2	0.8	—
	Lamb variety													
3375	Brain, pan fried	3	ounce(s)	85	51.6	232	14.4	0	0	18.9	4.8	3.4	1.9	—
3406	Tongue, braised	3	ounce(s)	85	49.2	234	18.3	0	0	17.2	6.7	8.5	1.1	—
	Pork, cured													
29229	Bacon, Canadian style, cured	2	ounce(s)	57	37.9	89	11.7	1.0	0	4.0	1.3	1.8	0.4	—
161	Bacon, cured, broiled, pan fried or roasted	2	slice(s)	16	2.0	87	5.9	0.2	0	6.7	2.2	3.0	0.7	0
35422	Breakfast strips, cured, cooked	3	slice(s)	34	9.2	156	9.8	0.4	0	12.5	4.3	5.6	1.9	—
189	Ham, cured, boneless, 11% fat, roasted	3	ounce(s)	85	54.9	151	19.2	0	0	7.7	2.7	3.8	1.2	—
29215	Ham, cured, extra lean, 4% fat, canned	2	2 ounce(s)	57	41.7	68	10.5	0	0	2.6	0.9	1.3	0.2	—
1316	Ham, cured, extra lean, 5% fat, roasted	3	ounce(s)	85	57.6	123	17.8	1.3	0	4.7	1.5	2.2	0.5	—
16561	Ham, smoked or cured, lean, cooked	1	slice(s)	42	27.6	66	10.5	0	0	2.3	0.8	1.1	0.3	—
	Pork chop													
32671	Loin, blade, chops, lean and fat, pan fried	3	ounce(s)	85	42.5	291	18.3	0	0	23.6	8.6	10	2.6	—
32672	Loin, center cut, chops, lean and fat, pan fried	3	ounce(s)	85	45.1	236	25.4	0	0	14.1	5.1	6.0	1.6	—
32682	Loin, center rib, chops, boneless, lean and fat, braised	3	ounce(s)	85	49.5	217	22.4	0	0	13.4	5.2	6.1	1.1	—
32603	Loin, center rib, chops, lean, broiled	3	ounce(s)	85	55.4	158	21.9	0	0	7.1	2.4	3.0	0.8	0.1
32478	Loin, whole, lean and fat, braised	3	ounce(s)	85	49.6	203	23.2	0	0	11.6	4.3	5.2	1.0	—
32481	Loin, whole, lean, braised	3	ounce(s)	85	52.2	174	24.3	0	0	7.8	2.9	3.5	0.6	—
	Pork leg or ham													
32471	Pork leg or ham, rump portion, lean and fat, roasted	3	ounce(s)	85	48.3	214	24.6	0	0	12.1	4.5	5.4	1.2	—
32468	Pork leg or ham, whole, lean and fat, roasted	3	ounce(s)	85	46.8	232	22.8	0	0	15.0	5.5	6.7	1.4	—
	Pork ribs													
32693	Loin, country style, lean and fat, roasted	3	ounce(s)	85	43.3	279	19.9	0	0	21.6	7.8	9.4	1.7	—
32696	Loin, country style, lean, roasted	3	ounce(s)	85	49.5	210	22.6	0	0	12.6	4.5	5.5	0.9	—

PAGE KEY: H-4 = Breads/Baked Goods H-10 = Cereal/Rice/Pasta H-14 = Fruit H-18 = Vegetables/Legumes H-28 = Nuts/Seeds H-30 = Vegetarian H-32 = Dairy H-40 = Eggs H-40 = Seafood H-42 = Meats H-46 = Poultry H-46 = Processed Meats H-48 = Beverages H-52 = Fats/Oils H-54 = Sweets H-56 = Spices/Condiments/Sauces H-60 = Mixed Foods/Soups/Sandwiches H-64 = Fast Food H-84 = Convenience H-86 = Baby Foods

Chol (mg)	Calc (mg)	Iron (mg)	Magn (mg)	Pota (mg)	Sodi (mg)	Zinc (mg)	Vit A (µg)	Thia (mg)	Vit E (mg α)	Ribo (mg)	Niac (mg)	Vit B_6 (mg)	Fola (µg)	Vit C (mg)	Vit B_{12} (µg)	Sele (µg)
58	7	2.56	17.9	233.9	57.8	3.56	0	0.07	0.18	0.17	3.29	0.27	6.0	0	1.8	10.0
50	5	3.11	22.1	278.1	65.5	4.34	0	0.09	0.11	0.21	3.93	0.33	6.8	0	1.9	8.5
67	8	1.88	19.6	294.3	53.6	3.85	0	0.06	—	0.15	3.96	0.31	6.0	0	1.6	19.5
324	5	5.24	18.7	298.5	65.5	4.44	6586.3	0.15	0.39	2.91	14.86	0.87	221.1	0.6	70.7	27.9
112	4	2.22	12.8	156.5	55.3	3.47	0	0.01	0.25	0.25	2.96	0.13	6.0	1.1	2.7	11.2
85	17	1.53	20.4	278.1	65.5	2.96	0	0.08	0.11	0.21	6.03	0.11	15.3	0	2.1	23.3
82	14	1.59	19.6	263.7	61.2	3.79	0	0.08	0.11	0.21	5.66	0.11	15.3	0	2.2	22.5
84	16	1.59	19.6	229.5	64.6	3.40	0	0.07	0.10	0.18	5.95	0.09	11.9	0	2.2	20.3
77	14	1.87	24.7	266.1	72.3	4.47	0	0.08	0.15	0.21	5.56	0.12	17.9	0	2.2	26.4
78	17	1.67	19.6	213.4	56.1	4.44	0	0.07	0.11	0.20	5.22	0.11	17.9	0	2.2	22.3
74	16	1.81	21.3	225.3	57.8	5.13	0	0.07	0.15	0.22	4.89	0.12	21.3	0	2.3	24.2
102	21	2.03	22.1	260.3	61.2	5.17	0	0.06	0.12	0.21	5.66	0.09	15.3	0	2.2	31.6
103	22	2.29	24.7	287.5	64.6	6.20	0	0.06	0.15	0.23	5.38	0.11	18.7	0	2.3	32.1
2130	18	1.73	18.7	304.5	133.5	1.70	0	0.14	—	0.31	3.87	0.19	6.0	19.6	20.5	10.2
161	9	2.23	13.6	134.4	57.0	2.54	0	0.06	—	0.35	3.13	0.14	2.6	6.0	5.4	23.8
28	5	0.38	9.6	195.0	798.9	0.78	0	0.42	0.11	0.09	3.53	0.22	2.3	0	0.4	14.2
18	2	0.22	5.3	90.4	369.6	0.56	1.8	0.06	0.04	0.04	1.76	0.04	0.3	0	0.2	9.9
36	5	0.67	8.8	158.4	713.7	1.25	0	0.25	0.08	0.12	2.58	0.11	1.4	0	0.6	8.4
50	7	1.13	18.7	347.7	1275.0	2.09	0	0.62	0.26	0.28	5.22	0.26	2.6	0	0.6	16.8
22	3	0.53	9.6	206.4	711.6	1.09	0	0.47	0.09	0.13	3.00	0.25	3.4	0	0.5	8.2
45	7	1.25	11.9	244.1	1023.1	2.44	0	0.64	0.21	0.17	3.42	0.34	2.6	0	0.6	16.6
23	3	0.39	9.2	132.7	557.3	1.07	0	0.28	0.10	0.10	2.10	0.19	1.7	0	0.3	10.7
72	26	0.74	17.9	282.4	57.0	2.71	1.7	0.52	0.17	0.25	3.35	0.28	3.4	0.5	0.7	29.7
78	23	0.77	24.7	361.5	68.0	1.96	1.7	0.96	0.21	0.25	4.76	0.39	5.1	0.9	0.6	33.2
62	4	0.78	14.5	329.1	34.0	1.76	1.7	0.44	—	0.20	3.66	0.26	3.4	0.3	0.4	28.4
56	22	0.57	21.3	291.7	48.5	1.91	0	0.48	0.08	0.18	6.68	0.57	0	0	0.4	38.6
68	18	0.91	16.2	318.1	40.8	2.02	1.7	0.53	0.20	0.21	3.75	0.31	2.6	0.5	0.5	38.5
67	15	0.96	17.0	329.1	42.5	2.10	1.7	0.56	0.17	0.22	3.90	0.32	3.4	0.5	0.5	41.0
82	10	0.89	23.0	318.1	52.7	2.39	2.6	0.63	0.18	0.28	3.95	0.26	2.6	0.2	0.6	39.8
80	12	0.85	18.7	299.4	51.0	2.51	2.6	0.54	0.18	0.26	3.89	0.34	8.5	0.3	0.6	38.5
78	21	0.90	19.6	292.6	44.2	2.00	2.6	0.75	—	0.29	3.67	0.37	4.3	0.3	0.7	31.6
79	25	1.09	20.4	296.8	24.7	3.24	1.7	0.48	—	0.29	3.96	0.37	4.3	0.3	0.7	36.0

TABLE H-1 Table of Food Composition (*continued*) (Computer code is for Cengage Diet Analysis program) (For purposes of calculations, use "0" for t, <1, <.1, <.01, et

DA+ Code	Food Description	Quantity	Measure	Wt (g)	H_2O (g)	Ener (kcal)	Prot (g)	Carb (g)	Fiber (g)	Fat (g)	Fat Breakdown (g) Sat	Mono	Poly	*Trans*
Beef, Lamb, Pork—*continued*														
	Pork shoulder													
32626	Shoulder, arm picnic, lean and fat, roasted	3	ounce(s)	85	44.3	270	20.0	0	0	20.4	7.5	9.1	2.0	—
32629	Shoulder, arm picnic, lean, roasted	3	ounce(s)	85	51.3	194	22.7	0	0	10.7	3.7	5.1	1.0	—
	Rabbit													
3366	Domesticated, roasted	3	ounce(s)	85	51.5	168	24.7	0	0	6.8	2.0	1.8	1.3	—
3367	Domesticated, stewed	3	ounce(s)	85	50.0	175	25.8	0	0	7.2	2.1	1.9	1.4	—
	Veal													
3391	Liver, braised	3	ounce(s)	85	50.9	163	24.2	3.2	0	5.3	1.7	1.0	0.9	0.3
3319	Rib, lean only, roasted	3	ounce(s)	85	55.0	151	21.9	0	0	6.3	1.8	2.3	0.6	—
1732	Deer or venison, roasted	3	ounce(s)	85	55.5	134	25.7	0	0	2.7	1.1	0.7	0.5	—
Poultry														
	Chicken													
29562	Flaked, canned	2	ounce(s)	57	39.3	97	10.3	0.1	0	5.8	1.6	2.3	1.3	—
	Chicken, fried													
29632	Breast, meat only, breaded, baked or fried	3	ounce(s)	85	44.3	193	25.3	6.9	0.2	6.6	1.6	2.7	1.7	—
35327	Broiler breast, meat only, fried	3	ounce(s)	85	51.2	159	28.4	0.4	0	4.0	1.1	1.5	0.9	—
36413	Broiler breast, meat and skin, flour coated, fried	3	ounce(s)	85	48.1	189	27.1	1.4	0.1	7.5	2.1	3.0	1.7	—
36414	Broiler drumstick, meat and skin, flour coated, fried	3	ounce(s)	85	48.2	208	22.9	1.4	0.1	11.7	3.1	4.6	2.7	—
35389	Broiler drumstick, meat only, fried	3	ounce(s)	85	52.9	166	24.3	0	0	6.9	1.8	2.5	1.7	—
35406	Broiler leg, meat only, fried	3	ounce(s)	85	51.5	177	24.1	0.6	0	7.9	2.1	2.9	1.9	—
35484	Broiler wing, meat only, fried	3	ounce(s)	85	50.9	179	25.6	0	0	7.8	2.1	2.6	1.8	—
29580	Patty, fillet or tenders, breaded, cooked	3	ounce(s)	85	40.2	256	14.5	12.2	0	16.5	3.7	8.4	3.7	—
	Chicken, roasted, meat only													
35409	Broiler leg, meat only, roasted	3	ounce(s)	85	55.0	162	23.0	0	0	7.2	1.9	2.6	1.7	—
35486	Broiler wing, meat only, roasted	3	ounce(s)	85	53.4	173	25.9	0	0	6.9	1.9	2.2	1.5	—
35138	Roasting chicken, dark meat, meat only, roasted	3	ounce(s)	85	57.0	151	19.8	0	0	7.4	2.1	2.8	1.7	—
35136	Roasting chicken, light meat, meat only, roasted	3	ounce(s)	85	57.7	130	23.1	0	0	3.5	0.9	1.3	0.8	—
35132	Roasting chicken, meat only, roasted	3	ounce(s)	85	57.3	142	21.3	0	0	5.6	1.5	2.1	1.3	—
	Chicken, stewed													
1268	Gizzard, simmered	3	ounce(s)	85	57.8	124	25.8	0	0	2.3	0.6	0.4	0.3	0.1
1270	Liver, simmered	3	ounce(s)	85	56.8	142	20.8	0.7	0	5.5	1.8	1.2	1.7	0.1
3174	Meat only, stewed	3	ounce(s)	85	56.8	151	23.2	0	0	5.7	1.6	2.0	1.3	—
	Duck													
1286	Domesticated, meat and skin, roasted	3	ounce(s)	85	44.1	287	16.2	0	0	24.1	8.2	11.0	3.1	—
1287	Domesticated, meat only, roasted	3	ounce(s)	85	54.6	171	20.0	0	0	9.5	3.5	3.1	1.2	—
	Goose													
35507	Domesticated, meat and skin, roasted	3	ounce(s)	85	44.2	259	21.4	0	0	18.6	5.8	8.7	2.1	—
35524	Domesticated, meat only, roasted	3	ounce(s)	85	48.7	202	24.6	0	0	10.8	3.9	3.7	1.3	—
1297	Liver pate, smoked, canned	4	tablespoon(s)	52	19.3	240	5.9	2.4	0	22.8	7.5	13.3	0.4	—
	Turkey													
3256	Ground turkey, cooked	3	ounce(s)	85	50.5	200	23.3	0	0	11.2	2.9	4.2	2.7	—
3263	Patty, batter coated, breaded, fried	1	item(s)	94	46.7	266	13.2	14.8	0.5	16.9	4.4	7.0	4.4	—
219	Roasted, dark meat, meat only	3	ounce(s)	85	53.7	159	24.3	0	0	6.1	2.1	1.4	1.8	—
222	Roasted, fryer roaster breast, meat only	3	ounce(s)	85	58.2	115	25.6	0	0	0.6	0.2	0.1	0.2	—
220	Roasted, light meat, meat only	3	ounce(s)	85	56.4	134	25.4	0	0	2.7	0.9	0.5	0.7	—
1303	Turkey roll, light and dark meat	2	slice(s)	57	39.8	84	10.3	1.2	0	4.0	1.2	1.3	1.0	—
1302	Turkey roll, light meat	2	slice(s)	57	42.5	56	8.4	2.9	0	0.9	0.2	0.2	0.1	0
Processed Meats														
	Beef													
1331	Corned beef loaf, jellied, sliced	2	slice(s)	57	39.2	87	13.0	0	0	3.5	1.5	1.5	0.2	—

PAGE KEY: H-4 = Breads/Baked Goods H-10 = Cereal/Rice/Pasta H-14 = Fruit H-18 = Vegetables/Legumes H-28 = Nuts/Seeds H-30 = Vegetarian H-32 = Dairy H-40 = Eggs H-40 = Seafood H-42 = Meats H-46 = Poultry H-46 = Processed Meats H-48 = Beverages H-52 = Fats/Oils H-54 = Sweets H-56 = Spices/Condiments/Sauces H-60 = Mixed Foods/Soups/Sandwiches H-64 = Fast Food H-84 = Convenience H-86 = Baby Foods

Chol (mg)	Calc (mg)	Iron (mg)	Magn (mg)	Pota (mg)	Sodi (mg)	Zinc (mg)	Vit A (µg)	Thia (mg)	Vit E (mg α)	Ribo (mg)	Niac (mg)	Vit B_6 (mg)	Fola (µg)	Vit C (mg)	Vit B_{12} (µg)	Sele (µg)
80	16	1.00	14.5	276.4	59.5	2.93	1.7	0.44	—	0.25	3.33	0.29	3.4	0.2	0.6	28.6
81	8	1.20	17.0	298.5	68.0	3.46	1.7	0.49	—	0.30	3.66	0.34	4.3	0.3	0.7	32.7
70	16	1.93	17.9	325.7	40.0	1.93	0	0.07	—	0.17	7.17	0.40	9.4	0	7.1	32.7
73	17	2.01	17.0	255.1	31.5	2.01	0	0.05	0.37	0.14	6.09	0.28	7.7	0	5.5	32.7
435	5	4.34	17.0	279.8	66.3	9.55	8026	0.15	0.57	2.43	11.18	0.78	281.5	0.9	72.0	16.4
98	10	0.81	20.4	264.5	82.5	3.81	0	0.05	0.30	0.24	6.37	0.23	11.9	0	1.3	9.4
95	6	3.80	20.4	284.9	45.9	2.33	0	0.15	—	0.51	5.70	—	—	0	—	11.0
35	8	0.89	6.8	147.4	408.2	0.79	19.3	0.01	—	0.07	3.58	0.19	2.3	0	0.2	—
67	19	1.05	24.7	222.6	450.2	0.84	—	0.08	—	0.09	10.97	0.46	4.3	0	0.3	—
77	14	0.96	26.4	234.7	67.2	0.91	6.0	0.06	0.35	0.10	12.57	0.54	3.4	0	0.3	22.3
76	14	1.01	25.5	220.3	64.6	0.93	12.8	0.06	0.39	0.11	11.68	0.49	5.1	0	0.3	20.3
77	10	1.13	19.6	194.8	75.7	2.45	21.3	0.06	0.65	0.19	5.13	0.29	8.5	0	0.3	15.6
80	10	1.12	20.4	211.8	81.6	2.73	15.3	0.06	—	0.20	5.22	0.33	7.7	0	0.3	16.7
84	11	1.19	21.3	216.0	81.6	2.53	17.0	0.07	0.38	0.21	5.68	0.33	7.7	0	0.3	16.0
71	13	0.96	17.9	176.9	77.4	1.80	15.3	0.03	0.40	0.10	6.15	0.50	3.4	0	0.3	21.6
49	11	0.75	19.6	244.8	411.4	0.79	4.3	0.09	1.04	0.12	5.99	0.24	24.7	0	0.2	13.9
80	10	1.11	20.4	205.8	77.4	2.43	16.2	0.06	0.22	0.19	5.37	0.31	6.8	0	0.3	18.8
72	14	0.98	17.9	178.6	78.2	1.82	15.3	0.03	0.22	0.10	6.21	0.50	3.4	0	0.3	21.0
64	9	1.13	17.0	190.5	80.8	1.81	13.6	0.05	—	0.16	4.87	0.26	6.0	0	0.2	16.7
64	11	0.91	19.6	200.7	43.4	0.66	6.8	0.05	0.22	0.07	8.90	0.45	2.6	0	0.3	21.9
64	10	1.02	17.9	194.8	63.8	1.29	10.2	0.05	—	0.12	6.70	0.34	4.3	0	0.2	20.9
315	14	2.71	2.6	152.2	47.6	3.75	0	0.02	0.17	0.17	2.65	0.06	4.3	0	0.9	35.0
479	9	9.89	21.3	223.7	64.6	3.38	3385.8	0.24	0.69	1.69	9.39	0.64	491.6	23.7	14.3	70.1
71	12	0.99	17.9	153.1	59.5	1.69	12.8	0.04	0.22	0.13	5.20	0.22	5.1	0	0.2	17.8
71	9	2.29	13.6	173.5	50.2	1.58	53.6	0.14	0.59	0.22	4.10	0.15	5.1	0	0.3	17.0
76	10	2.29	17.0	214.3	55.3	2.21	19.6	0.22	0.59	0.39	4.33	0.21	8.5	0	0.3	19.1
77	11	2.40	18.7	279.8	59.5	2.22	17.9	0.06	1.47	0.27	3.54	0.31	1.7	0	0.3	18.5
82	12	2.44	21.3	330.0	64.6	2.69	10.2	0.07	—	0.33	3.47	0.39	10.2	0	0.4	21.7
78	36	2.86	6.8	71.8	362.4	0.47	520.5	0.04	—	0.15	1.30	0.03	31.2	0	4.9	22.9
87	21	1.64	20.4	229.6	91.0	2.43	0	0.04	0.28	0.14	4.09	0.33	6.0	0	0.3	31.6
71	13	2.06	14.1	258.5	752.0	1.35	9.4	0.09	0.87	0.17	2.16	0.18	38.5	0	0.2	20.8
72	27	1.98	20.4	246.6	67.2	3.79	0	0.05	0.54	0.21	3.10	0.30	7.7	0	0.3	34.8
71	10	1.30	24.7	248.3	44.2	1.48	0	0.03	0.07	0.11	6.37	0.47	5.1	0	0.3	27.3
59	16	1.14	23.8	259.4	54.4	1.73	0	0.05	0.07	0.11	5.81	0.45	5.1	0	0.3	27.3
31	18	0.76	10.2	153.1	332.3	1.13	0	0.05	0.19	0.16	2.72	0.15	2.8	0	0.1	16.6
19	4	0.21	10.8	242.1	590.8	0.50	0	0.01	0.07	0.08	4.05	0.23	2.3	0	0.2	7.4
27	6	1.15	6.2	57.3	540.4	2.31	0	0.00	—	0.06	0.99	0.06	4.5	0	0.7	9.8

TABLE H-1 Table of Food Composition (*continued*) (Computer code is for Cengage Diet Analysis program) (For purposes of calculations, use "0" for t, <1, <.1, <.01, et

DA+ Code	Food Description	Quantity	Measure	Wt (g)	H_2O (g)	Ener (kcal)	Prot (g)	Carb (g)	Fiber (g)	Fat (g)	Fat Breakdown (g) Sat	Mono	Poly	*Trans*
Processed Meats—*continued*														
	Bologna													
13459	Beef	1	slice(s)	28	15.1	90	3.0	1.0	0	8.0	3.5	4.3	0.3	—
13461	Light, made with pork and chicken	1	slice(s)	28	18.2	60	3.0	2.0	0	4.0	1.0	2.0	0.4	—
13458	Made with chicken and pork	1	slice(s)	28	15.0	90	3.0	1.0	0	8.0	3.0	4.1	1.1	—
13565	Turkey bologna	1	slice(s)	28	19.0	50	3.0	1.0	0	4.0	1.0	1.1	1.0	0
	Chicken													
7125	Breast, smoked	1	slice(s)	10	—	10	1.8	0.3	0	0.2	0	—	—	—
	Ham													
7127	Deli-sliced, honey	1	slice(s)	10	—	10	1.7	0.3	0	0.3	0.1	—	—	—
7126	Deli-sliced, smoked	1	slice(s)	10	—	10	1.7	0.2	0	0.3	0.1	—	—	—
8614	**Beef and pork mortadella, sliced**	2	slice(s)	46	24.1	143	7.5	1.4	0	11.7	4.4	5.2	1.4	—
1323	**Pork olive loaf**	2	slice(s)	57	33.1	133	6.7	5.2	0	9.4	3.3	4.5	1.1	—
1324	**Pork pickle and pimento loaf**	2	slice(s)	57	34.2	128	6.4	4.8	0.9	9.1	3.0	4.0	1.6	—
	Sausages and frankfurters													
37296	Beerwurst beef, beer salami (bierwurst)	1	slice(s)	29	16.6	74	4.1	1.2	0	5.7	2.5	2.7	0.2	—
37257	Beerwurst pork, beer salami	1	slice(s)	21	12.9	50	3.0	0.4	0	4.0	1.3	1.9	0.5	—
35338	Berliner, pork and beef	1	ounce(s)	28	17.3	65	4.3	0.7	0	4.9	1.7	2.3	0.4	—
37298	Bratwurst pork, cooked	1	piece(s)	74	42.3	181	10.4	1.9	0	14.3	5.1	6.7	1.5	—
37299	Braunschweiger pork liver sausage	1	slice(s)	15	8.2	51	2.0	0.3	0	4.5	1.5	2.1	0.5	—
1329	Cheesefurter or cheese smokie, beef and pork	1	item(s)	43	22.6	141	6.1	0.6	0	12.5	4.5	5.9	1.3	—
1330	Chorizo, beef and pork	2	ounce(s)	57	18.1	258	13.7	1.1	0	21.7	8.2	10.4	2.0	—
8600	Frankfurter, beef	1	item(s)	45	23.4	149	5.1	1.8	0	13.3	5.3	6.4	0.5	—
202	Frankfurter, beef and pork	1	item(s)	45	25.2	137	5.2	0.8	0	12.4	4.8	6.2	1.2	—
1293	Frankfurter, chicken	1	item(s)	45	28.1	100	7.0	1.2	0.2	7.3	1.7	2.7	1.7	0.1
3261	Frankfurter, turkey	1	item(s)	45	28.3	100	5.5	1.7	0	7.8	1.8	2.6	1.8	0.4
37275	Italian sausage, pork, cooked	1	item(s)	68	32.0	234	13.0	2.9	0.1	18.6	6.5	8.1	2.2	—
37307	Kielbasa or kolbassa, pork and beef	1	slice(s)	30	18.5	67	5.0	1.0	0	4.7	1.7	2.2	0.5	—
1333	Knockwurst or knackwurst, beef and pork	2	ounce(s)	57	31.4	174	6.3	1.8	0	15.7	5.8	7.3	1.7	—
37285	Pepperoni, beef and pork	1	slice(s)	11	3.4	51	2.2	0.4	0.2	4.4	1.8	2.1	0.3	—
37313	Polish sausage, pork	1	slice(s)	21	11.4	60	2.8	0.7	0	5.0	1.8	2.3	0.5	—
206	Salami, beef, cooked, sliced	2	slice(s)	52	31.2	136	6.5	1.0	0	11.5	5.1	5.5	0.5	—
37272	Salami, pork, dry or hard	1	slice(s)	13	4.6	52	2.9	0.2	0	4.3	1.5	2.0	0.5	—
40987	Sausage, turkey, cooked	2	ounce(s)	57	36.9	111	13.5	0	0	5.9	1.3	1.7	1.5	0.2
8620	Smoked sausage, beef and pork	2	ounce(s)	57	30.6	181	6.8	1.4	0	16.3	5.5	6.9	2.2	0
8619	Smoked sausage, pork	2	ounce(s)	57	32.0	178	6.8	1.2	0	16.0	5.3	6.4	2.1	0.1
37273	Smoked sausage, pork link	1	piece(s)	76	29.8	295	16.8	1.6	0	24.0	8.6	11.1	2.8	—
1336	Summer sausage, thuringer, or cervelat, beef and pork	2	ounce(s)	57	25.6	205	9.9	1.9	0	17.3	6.5	7.4	0.7	—
37294	Vienna sausage, cocktail, beef and pork, canned	1	piece(s)	16	10.4	37	1.7	0.4	0	3.1	1.1	1.5	0.2	—
	Spreads													
1318	Ham salad spread	¼	cup(s)	60	37.6	130	5.2	6.4	0	9.3	3.0	4.3	1.6	—
32419	Pork and beef sandwich spread	4	tablespoon(s)	60	36.2	141	4.6	7.2	0.1	10.4	3.6	4.6	1.5	—
	Turkey													
13604	Breast, fat free, oven roasted	1	slice(s)	28	—	25	4.0	1.0	0	0	0	0	0	0
13606	Breast, hickory smoked fat free	1	slice(s)	28	—	25	4.0	1.0	0	0	0	0	0	0
16049	Breast, hickory smoked slices	1	slice(s)	56	—	50	11.0	1.0	0	0	0	0	0	0
16047	Breast, honey roasted slices	1	slice(s)	56	—	60	11.0	3.0	0	0	0	0	0	0
16048	Breast, oven roasted slices	1	slice(s)	56	—	50	11.0	1.0	0	0	0	0	0	0
7124	Breast, oven roasted	1	slice(s)	10	—	10	1.8	0.3	0	0.1	0	0	0	—
13567	Turkey ham, 10% water added	2	slice(s)	56	40.9	70	10.0	2.0	0	3.0	0	0.4	0.6	0
37270	Turkey pastrami	1	slice(s)	28	20.3	35	4.6	1.0	0	1.2	0.3	0.4	0.3	—
3262	Turkey salami	2	slice(s)	57	39.1	98	10.9	0.9	0.1	5.2	1.6	1.8	1.4	0
37318	Turkey salami, cooked	1	slice(s)	28	20.4	43	4.3	0.1	0	2.7	0.8	0.9	0.7	—
Beverages														
	Beer													
866	Ale, mild	12	fluid ounce(s)	360	332.3	148	1.1	13.3	0.4	0	0	0	0	0
686	Beer	12	fluid ounce(s)	356	327.7	153	1.6	12.7	0	0	0	0	0	0
16886	Beer, non alcoholic	12	fluid ounce(s)	360	328.1	133	0.8	29.0	0	0.4	0.1	0	0.2	0

PAGE KEY: H-4 = Breads/Baked Goods H-10 = Cereal/Rice/Pasta H-14 = Fruit H-18 = Vegetables/Legumes H-28 = Nuts/Seeds H-30 = Vegetarian H-32 = Dairy H-40 = Eggs H-40 = Seafood H-42 = Meats H-46 = Poultry H-46 = Processed Meats H-48 = Beverages H-52 = Fats/Oils H-54 = Sweets H-56 = Spices/Condiments/Sauces H-60 = Mixed Foods/Soups/Sandwiches H-64 = Fast Food H-84 = Convenience H-86 = Baby Foods

Chol (mg)	Calc (mg)	Iron (mg)	Magn (mg)	Pota (mg)	Sodi (mg)	Zinc (mg)	Vit A (μg)	Thia (mg)	Vit E (mg α)	Ribo (mg)	Niac (mg)	Vit B_6 (mg)	Fola (μg)	Vit C (mg)	Vit B_{12} (μg)	Sele (μg)
20	0	0.36	3.9	47.0	310.0	0.56	0	0.01	—	0.03	0.67	0.04	3.6	0	0.4	—
20	40	0.36	5.6	45.6	300.0	0.45	0	—	—	—	—	—	—	0	—	—
30	20	0.36	5.9	43.1	300.0	0.39	0	—	—	—	—	—	—	0	—	—
20	40	0.36	6.2	42.6	270.0	0.51	0	—	—	—	—	—	—	0	—	—
4	0	0.00	—	—	100.0	—	0	—	—	—	—	—	—	0	—	—
4	0	0.12	—	—	100.0	—	0	—	—	—	—	—	—	0.6	—	—
4	0	0.12	—	—	103.3	—	0	—	—	—	—	—	—	0.6	—	—
26	8	0.64	5.1	75.0	573.2	0.96	0	0.05	0.10	0.07	1.23	0.06	1.4	0	0.7	10.4
22	62	0.30	10.8	168.7	842.9	0.78	34.1	0.16	0.14	0.14	1.04	0.13	1.1	0	0.7	9.3
33	62	0.75	19.3	210.7	740.7	0.95	44.3	0.22	0.22	0.06	1.41	0.23	21.0	4.4	0.3	4.5
18	3	0.44	3.5	66.5	264.9	0.71	0	0.02	0.05	0.03	0.98	0.04	0.9	0	0.6	4.7
12	2	0.15	2.7	53.3	261.0	0.36	0	0.11	0.03	0.04	0.68	0.07	0.6	0	0.2	4.4
13	3	0.32	4.3	80.2	367.7	0.70	0	0.10	—	0.06	0.88	0.05	1.4	0	0.8	4.0
44	33	0.95	11.1	156.9	412.2	1.70	0	0.37	0.01	0.13	2.36	0.15	1.5	0.7	0.7	15.7
24	1	1.42	1.7	27.5	131.5	0.42	641.0	0.03	0.05	0.23	1.27	0.05	6.7	0	3.1	8.8
29	25	0.46	5.6	88.6	465.3	0.96	20.2	0.10	0.10	0.06	1.24	0.05	1.3	0	0.7	6.8
50	5	0.90	10.2	225.7	700.2	1.93	0	0.35	0.12	0.17	2.90	0.30	1.1	0	1.1	12.0
24	6	0.67	6.3	70.2	513.0	1.10	0	0.01	0.09	0.06	1.06	0.04	2.3	0	0.8	3.7
23	5	0.51	4.5	75.2	504.0	0.82	8.1	0.09	0.11	0.05	1.18	0.05	1.8	0	0.6	6.2
43	33	0.52	9.0	90.9	379.8	0.50	0	0.02	0.09	0.11	2.10	0.14	3.2	0	0.2	10.4
35	67	0.66	6.3	176.4	485.1	0.82	0	0.01	0.27	0.08	1.65	0.06	4.1	0	0.4	6.8
39	14	0.97	12.2	206.7	820.8	1.62	6.8	0.42	0.17	0.15	2.83	0.22	3.4	0.1	0.9	15.0
20	13	0.44	4.9	84.4	283.0	0.61	0	0.06	0.06	0.06	0.87	0.05	1.5	0	0.5	5.4
34	6	0.37	6.2	112.8	527.3	0.94	0	0.19	0.32	0.07	1.55	0.09	1.1	0	0.7	7.7
13	2	0.15	2.0	34.7	196.7	0.30	0	0.05	0.00	0.02	0.59	0.04	0.7	0.1	0.2	2.4
15	2	0.29	2.9	37.3	199.3	0.40	0	0.10	0.04	0.03	0.71	0.03	0.4	0.2	0.2	3.7
37	3	1.14	6.8	97.8	592.8	0.92	0	0.04	0.08	0.08	1.68	0.08	1.0	0	1.6	7.6
10	2	0.16	2.8	48.4	289.3	0.53	0	0.11	0.02	0.04	0.71	0.07	0.3	0	0.4	3.3
52	12	0.84	11.9	169.0	377.1	2.19	7.4	0.04	0.10	0.14	3.24	0.18	3.4	0.4	0.7	0
33	7	0.42	7.4	101.5	516.5	0.71	7.4	0.10	0.07	0.06	1.66	0.09	1.1	0	0.3	0
35	6	0.33	6.2	273.9	468.9	0.74	0	0.12	0.14	0.10	1.59	0.10	0.6	0	0.4	10.4
52	23	0.87	14.4	254.6	1136.6	2.13	0	0.53	0.18	0.19	3.43	0.26	3.8	1.5	1.2	16.4
42	5	1.15	7.9	147.4	737.1	1.45	0	0.08	0.12	0.18	2.44	0.14	1.1	9.4	3.1	11.5
14	2	0.14	1.1	16.2	155.0	0.25	0	0.01	0.03	0.01	0.25	0.01	0.6	0	0.2	2.7
22	5	0.35	6.0	90.0	547.2	0.66	0	0.26	1.04	0.07	1.25	0.09	0.6	0	0.5	10.7
23	7	0.47	4.8	66.0	607.8	0.61	15.6	0.10	1.04	0.08	1.03	0.07	1.2	0	0.7	5.8
10	0	0.00	—	—	340.0	—	0	—	—	—	—	—	—	0	—	—
10	0	0.00	—	—	300.0	—	0	—	—	—	—	—	—	0	—	—
25	0	0.72	—	—	720.0	—	0	—	—	—	—	—	—	0	—	—
20	0	0.72	—	—	660.0	—	0	—	—	—	—	—	—	0	—	—
20	0	0.72	—	—	660.0	—	0	—	—	—	—	—	—	0	—	—
4	0	0.06	—	—	103.3	—	0	—	—	—	—	—	—	0	—	—
40	0	0.72	12.3	162.4	700.0	1.44	0	—	—	—	—	—	—	0	—	—
19	3	1.19	4.0	97.8	278.1	0.61	1.1	0.01	0.06	0.07	1.00	0.07	1.4	4.6	0.1	4.6
43	23	0.70	12.5	122.5	569.3	1.31	1.1	0.24	0.13	0.17	2.25	0.24	5.7	0	0.6	15.0
22	11	0.35	6.2	61.2	284.6	0.65	0.6	0.12	0.06	0.08	1.12	0.12	2.8	0	0.3	7.5
0	18	0.07	21.6	90.0	14.4	0.03	0	0.03	0.00	0.10	1.62	0.18	21.6	0	0.1	2.5
0	14	0.07	21.4	96.2	14.3	0.03	0	0.01	0.00	0.08	1.82	0.16	21.4	0	0.1	2.1
0	25	0.21	25.2	28.8	46.8	0.07	—	0.07	0.00	0.18	3.99	0.10	50.4	1.8	0.1	4.3

TABLE H-1 Table of Food Composition (*continued*) (Computer code is for Cengage Diet Analysis program) (For purposes of calculations, use "0" for t, <1, <.1, <.01, et

DA+ Code	Food Description	Quantity	Measure	Wt (g)	H_2O (g)	Ener (kcal)	Prot (g)	Carb (g)	Fiber (g)	Fat (g)	Fat Breakdown (g) Sat	Mono	Poly	*Trans*
	Beverages—*continued*													
31609	Bud Light beer	12	fluid ounce(s)	355	335.5	110	0.9	6.6	0	0	0	0	0	0
31608	Budweiser beer	12	fluid ounce(s)	355	327.7	145	1.3	10.6	0	0	0	0	0	0
869	Light beer	12	fluid ounce(s)	354	335.9	103	0.9	5.8	0	0	0	0	0	0
31613	Michelob beer	12	fluid ounce(s)	355	323.4	155	1.3	13.3	0	0	0	0	0	0
31614	Michelob Light beer	12	fluid ounce(s)	355	329.8	134	1.1	11.7	0	0	0	0	0	0
	Gin, rum, vodka, whiskey													
857	Distilled alcohol, 100 proof	1	fluid ounce(s)	28	16.0	82	0	0	0	0	0	0	0	0
687	Distilled alcohol, 80 proof	1	fluid ounce(s)	28	18.5	64	0	0	0	0	0	0	0	0
688	Distilled alcohol, 86 proof	1	fluid ounce(s)	28	17.8	70	0	0	0	0	0	0	0	0
689	Distilled alcohol, 90 proof	1	fluid ounce(s)	28	17.3	73	0	0	0	0	0	0	0	0
856	Distilled alcohol, 94 proof	1	fluid ounce(s)	28	16.8	76	0	0	0	0	0	0	0	0
	Liqueurs													
33187	Coffee liqueur, 53 proof	1	fluid ounce(s)	35	10.8	113	0	16.3	0	0.1	0	0	0	—
3142	Coffee liqueur, 63 proof	1	fluid ounce(s)	35	14.4	107	0	11.2	0	0.1	0	0	0	—
736	Cordials, 54 proof	1	fluid ounce(s)	30	8.9	106	0	13.3	0	0.1	0	0	0	—
	Wine													
861	California red wine	5	fluid ounce(s)	150	133.4	125	0.3	3.7	0	0	0	0	0	0
858	Domestic champagne	5	fluid ounce(s)	150	—	105	0.3	3.8	0	0	0	0	0	0
690	Sweet dessert wine	5	fluid ounce(s)	147	103.7	235	0.3	20.1	0	0	0	0	0	0
1481	White wine	5	fluid ounce(s)	148	128.1	121	0.1	3.8	0	0	0	0	0	0
1811	Wine cooler	10	fluid ounce(s)	300	267.4	159	0.3	20.2	0	0.1	0	0	0	—
	Carbonated													
31898	7 Up	12	fluid ounce(s)	360	321.0	140	0	39.0	0	0	0	0	0	0
692	Club soda	12	fluid ounce(s)	355	354.8	0	0	0	0	0	0	0	0	0
12010	Coca-Cola Classic cola soda	12	fluid ounce(s)	360	319.4	146	0	40.5	0	0	0	0	0	0
693	Cola	12	fluid ounce(s)	368	332.7	136	0.3	35.2	0	0.1	0	0	0	—
2391	Cola or pepper-type soda, low calorie with saccharin	12	fluid ounce(s)	355	354.5	0	0	0.3	0	0	0	0	0	0
9522	Cola soda, decaffeinated	12	fluid ounce(s)	372	333.4	153	0	39.3	0	0	0	0	0	0
9524	Cola, decaffeinated, low calorie with aspartame	12	fluid ounce(s)	355	354.3	4	0.4	0.5	0	0	0	0	0	0
1415	Cola, low calorie with aspartame	12	fluid ounce(s)	355	353.6	7	0.4	1.0	0	0.1	0	0	0	—
1412	Cream soda	12	fluid ounce(s)	371	321.5	189	0	49.3	0	0	0	0	0	0
31899	Diet 7 Up	12	fluid ounce(s)	360	—	0	0	0	0	0	0	0	0	0
12031	Diet Coke cola soda	12	fluid ounce(s)	360	—	2	0	0.2	0	0	0	0	0	0
29392	Diet Mountain Dew soda	12	fluid ounce(s)	360	—	0	0	0	0	0	0	0	0	0
29389	Diet Pepsi cola soda	12	fluid ounce(s)	360	—	0	0	0	0	0	0	0	0	0
12034	Diet Sprite soda	12	fluid ounce(s)	360	—	4	0	0	0	0	0	0	0	0
695	Ginger ale	12	fluid ounce(s)	366	333.9	124	0	32.1	0	0	0	0	0	0
694	Grape soda	12	fluid ounce(s)	372	330.3	160	0	41.7	0	0	0	0	0	0
1876	Lemon lime soda	12	fluid ounce(s)	368	330.8	147	0.2	37.4	0	0.1	0	0	0	—
29391	Mountain Dew soda	12	fluid ounce(s)	360	314.0	170	0	46.0	0	0	0	0	0	0
3145	Orange soda	12	fluid ounce(s)	372	325.9	179	0	45.8	0	0	0	0	0	0
1414	Pepper-type soda	12	fluid ounce(s)	368	329.3	151	0	38.3	0	0.4	0.3	0	0	—
29388	Pepsi regular cola soda	12	fluid ounce(s)	360	318.9	150	0	41.0	0	0	0	0	0	0
696	Root beer	12	fluid ounce(s)	370	330.0	152	0	39.2	0	0	0	0	0	0
12044	Sprite soda	12	fluid ounce(s)	360	321.0	144	0	39.0	0	0	0	0	0	0
	Coffee													
731	Brewed	8	fluid ounce(s)	237	235.6	2	0.3	0	0	0	0	0	0	0
9520	Brewed, decaffeinated	8	fluid ounce(s)	237	234.3	5	0.3	1.0	0	0	0	0	0	0
16882	Cappuccino	8	fluid ounce(s)	240	224.8	79	4.1	5.8	0.2	4.9	2.3	1.0	0.2	—
16883	Cappuccino, decaffeinated	8	fluid ounce(s)	240	224.8	79	4.1	5.8	0.2	4.9	2.3	1.0	0.2	—
16880	Espresso	8	fluid ounce(s)	237	231.8	21	0	3.6	0	0.4	0.2	0	0.2	0
16881	Espresso, decaffeinated	8	fluid ounce(s)	237	231.8	21	0	3.6	0	0.4	0.2	0	0.2	0
732	Instant, prepared	8	fluid ounce(s)	239	236.5	5	0.2	0.8	0	0	0	0	0	0
	Fruit drinks													
29357	Crystal Light sugar-free lemonade drink	8	fluid ounce(s)	240	—	5	0	0	0	0	0	0	0	0
6012	Fruit punch drink with added vitamin C, canned	8	fluid ounce(s)	248	218.2	117	0	29.7	0.5	0	0	0	0	0
31143	Gatorade Thirst Quencher, all flavors	8	fluid ounce(s)	240	—	50	0	14.0	0	0	0	0	0	0
260	Grape drink, canned	8	fluid ounce(s)	250	210.5	153	0	39.4	0	0	0	0	0	0
17372	Kool-Aid (lemonade/punch/fruit drink)	8	fluid ounce(s)	248	220.0	108	0.1	27.8	0.2	0	0	0	0	—

PAGE KEY: H-4 = Breads/Baked Goods H-10 = Cereal/Rice/Pasta H-14 = Fruit H-18 = Vegetables/Legumes H-28 = Nuts/Seeds H-30 = Vegetarian H-32 = Dairy H-40 = Eggs H-40 = Seafood H-42 = Meats H-46 = Poultry H-46 = Processed Meats H-48 = Beverages H-52 = Fats/Oils H-54 = Sweets H-56 = Spices/Condiments/Sauces H-60 = Mixed Foods/Soups/Sandwiches H-64 = Fast Food H-84 = Convenience H-86 = Baby Foods

Chol (mg)	Calc (mg)	Iron (mg)	Magn (mg)	Pota (mg)	Sodi (mg)	Zinc (mg)	Vit A (µg)	Thia (mg)	Vit E (mg α)	Ribo (mg)	Niac (mg)	Vit B_6 (mg)	Fola (µg)	Vit C (mg)	Vit B_{12} (µg)	Sele (µg)
0	18	0.14	17.8	63.9	9.0	0.10	0	0.03	—	0.10	1.39	0.12	14.6	0	0	4.0
0	18	0.10	21.3	88.8	9.0	0.07	0	0.02	—	0.09	1.60	0.17	21.3	0	0.1	4.0
0	14	0.10	17.7	74.3	14.2	0.03	0	0.01	0.00	0.05	1.38	0.12	21.2	0	0.1	1.4
0	18	0.10	21.3	88.8	9.0	0.07	0	0.02	—	0.09	1.60	0.17	21.3	0	0.1	4.0
0	18	0.14	17.8	63.9	9.0	0.10	0	0.03	—	0.10	1.39	0.12	14.6	0	0	4.0
0	0	0.01	0	0.6	0.3	0.01	0	0.00	—	0.00	0.00	0.00	0	0	0	0
0	0	0.01	0	0.6	0.3	0.01	0	0.00	0.00	0.00	0.00	0.00	0	0	0	0
0	0	0.01	0	0.6	0.3	0.01	0	0.00	0.00	0.00	0.00	0.00	0	0	0	0
0	0	0.01	0	0.6	0.3	0.01	0	0.00	0.00	0.00	0.00	0.00	0	0	0	0
0	0	0.01	0	0.6	0.3	0.01	0	0.00	—	0.00	0.00	0.00	0	0	0	0
0	0	0.02	1.0	10.4	2.8	0.01	0	0.00	0.00	0.00	0.05	0.00	0	0	0	0.1
0	0	0.02	1.0	10.4	2.8	0.01	0	0.00	—	0.00	0.05	0.00	0	0	0	0.1
0	0	0.02	0.6	4.5	2.1	0.01	0	0.00	0.00	0.00	0.02	0.00	0	0	0	0.1
0	12	1.43	16.2	170.6	15.0	0.14	0	0.01	0.00	0.04	0.11	0.05	1.5	0	0	—
0	—	—	—	—	—	—	—	—	—	—	—	—	—	—	0	—
0	12	0.34	13.2	135.4	13.2	0.10	0	0.01	0.00	0.01	0.30	0.00	0	0	0	0.7
0	13	0.39	14.8	104.7	7.4	0.18	0	0.01	0.00	0.01	0.15	0.06	1.5	0	0	0.1
0	18	0.75	15.0	129.0	24.0	0.18	—	0.01	0.03	0.03	0.13	0.03	3.0	5.4	0	0.6
0	—	—	—	0.6	75.0	—	—	—	—	—	—	—	—	—	—	—
0	18	0.03	3.5	7.1	74.6	0.35	0	0.00	0.00	0.00	0.00	0.00	0	0	0	0
0	—	—	—	0	49.5	—	0	—	—	—	—	—	—	0	—	—
0	7	0.41	0	7.4	14.7	0.06	0	0.00	0.00	0.00	0.00	0.00	0	0	0	0.4
0	14	0.06	3.5	14.2	56.8	0.11	0	0.00	0.00	0.00	0.00	0.00	0	0	0	0.3
0	7	0.08	0	11.2	14.9	0.03	0	0.00	0.00	0.00	0.00	0.00	0	0	0	0.4
0	11	0.06	0	24.9	14.2	0.03	0	0.02	0.00	0.08	0.00	0.00	0	0	0	0.3
0	11	0.39	3.5	28.4	28.4	0.03	0	0.02	0.00	0.08	0.00	0.00	0	0	0	0
0	19	0.18	3.7	3.7	44.5	0.26	0	0.00	0.00	0.00	0.00	0.00	0	0	0	0
0	—	—	—	77.0	45.0	—	—	—	—	—	—	—	—	—	—	—
0	—	—	—	18.0	42.0	—	0	—	—	—	—	—	—	0	—	—
0	—	—	—	70.0	35.0	—	—	—	—	—	—	—	—	—	—	—
0	—	—	—	30.0	35.0	—	—	—	—	—	—	—	—	—	—	—
0	—	—	—	109.5	36.0	—	0	—	—	—	—	—	—	0	—	—
0	11	0.65	3.7	3.7	25.6	0.18	0	0.00	0.00	0.00	0.00	0.00	0	0	0	0.4
0	11	0.29	3.7	3.7	55.8	0.26	0	0.00	—	0.00	0.00	0.00	0	0	0	0
0	7	0.41	3.7	3.7	33.2	0.14	0	0.00	0.00	0.00	0.05	0.00	0	0	0	0
0	—	—	—	0	70.0	—	—	—	—	—	—	—	—	—	—	—
0	19	0.21	3.7	7.4	44.6	0.36	0	0.00	—	0.00	0.00	0.00	0	0	0	0
0	11	0.14	0	3.7	36.8	0.14	0	0.00	—	0.00	0.00	0.00	0	0	0	0.4
0	—	—	—	0	35.0	—	—	—	—	—	—	—	—	—	—	—
0	18	0.18	3.7	3.7	48.0	0.26	0	0.00	0.00	0.00	0.00	0.00	0	0	0	0.4
0	—	—	—	0	70.5	—	0	—	—	—	—	—	—	0	—	—
0	5	0.02	7.1	116.1	4.7	0.04	0	0.03	0.02	0.18	0.45	0.00	4.7	0	0	0
0	7	0.14	11.8	108.9	4.7	0.00	0	0.00	0.00	0.03	0.66	0.00	0	0	0	0.5
12	144	0.19	14.4	232.8	50.4	0.50	33.6	0.04	0.09	0.27	0.13	0.04	7.2	0	0.4	4.6
12	144	0.19	14.4	232.8	50.4	0.50	33.6	0.04	0.09	0.27	0.13	0.04	7.2	0	0.4	4.6
0	5	0.30	189.6	272.6	33.2	0.11	0	0.00	0.04	0.42	12.34	0.00	2.4	0.5	0	0
0	5	0.30	189.6	272.6	33.2	0.11	0	0.00	0.04	0.42	12.34	0.00	2.4	0.5	0	0
0	10	0.09	9.5	71.6	9.5	0.01	0	0.00	0.00	0.00	0.56	0.00	0	0	0	0.2
0	0	0.00	—	160.0	40.0	—	0	—	—	—	—	—	—	0	—	—
0	20	0.22	7.4	62.0	94.2	0.02	5.0	0.05	0.04	0.05	0.05	0.02	9.9	89.3	0	0.5
0	0	0.00	—	30.0	110.0	—	0	—	—	—	—	—	—	0	—	—
0	130	0.17	2.5	30.0	40.0	0.30	0	0.00	0.00	0.01	0.02	0.01	0	78.5	0	0.3
0	14	0.45	5.0	49.6	31.0	0.19	—	0.03	—	0.05	0.04	0.01	4.3	41.6	0	1.0

TABLE H-1 Table of Food Composition (*continued*) (Computer code is for Cengage Diet Analysis program) (For purposes of calculations, use "0" for t, <1, <.1, <.01, etc

DA+ Code	Food Description	Quantity	Measure	Wt (g)	H_2O (g)	Ener (kcal)	Prot (g)	Carb (g)	Fiber (g)	Fat (g)	Fat Breakdown (g) Sat	Mono	Poly	*Trans*
	Beverages—*continued*													
17225	Kool-Aid sugar free, low calorie tropical punch drink mix, prepared	8	fluid ounce(s)	240	—	5	0	0	0	0	0	0	0	0
266	Lemonade, prepared from frozen concentrate	8	fluid ounce(s)	248	221.6	99	0.2	25.8	0	0.1	0	0	0	—
268	Limeade, prepared from frozen concentrate	8	fluid ounce(s)	247	212.6	128	0	34.1	0	0	0	0	0	—
14266	Odwalla strawberry C monster smoothie blend	8	fluid ounce(s)	240	—	160	2.0	38.0	0	0	0	0	0	0
10080	Odwalla strawberry lemonade quencher	8	fluid ounce(s)	240	—	110	0	28.0	0	0	0	0	0	0
10099	Snapple fruit punch fruit drink	8	fluid ounce(s)	240	—	110	0	29.0	0	0	0	0	0	0
10096	Snapple kiwi strawberry fruit drink	8	fluid ounce(s)	240	211.2	110	0	28.0	0	0	0	0	0	0
	Slim Fast ready-to-drink shake													
16054	French vanilla ready to drink shake	11	fluid ounce(s)	325	—	220	10.0	40.0	5.0	2.5	0.5	1.5	0.5	—
40447	Optima rich chocolate royal ready-to-drink shake	11	fluid ounce(s)	330	—	180	10.0	24.0	5.0	5.0	1.0	3.5	0.5	0
16055	Strawberries n cream ready to drink shake	11	fluid ounce(s)	325	—	220	10.0	40.0	5.0	2.5	0.5	1.5	0.5	—
	Tea													
33179	Decaffeinated, prepared	8	fluid ounce(s)	237	236.3	2	0	0.7	0	0	0	0	0	0
1877	Herbal, prepared	8	fluid ounce(s)	237	236.1	2	0	0.5	0	0	0	0	0	0
735	Instant tea mix, lemon flavored with sugar, prepared	8	fluid ounce(s)	259	236.2	91	0	22.3	0.3	0.2	0	0	0	—
734	Instant tea mix, unsweetened, prepared	8	fluid ounce(s)	237	236.1	2	0.1	0.4	0	0	0	0	0	0
733	Tea, prepared	8	fluid ounce(s)	237	236.3	2	0	0.7	0	0	0	0	0	0
	Water													
1413	Mineral water, carbonated	8	fluid ounce(s)	237	236.8	0	0	0	0	0	0	0	0	0
33183	Poland spring water, bottled	8	fluid ounce(s)	237	237.0	0	0	0	0	0	0	0	0	0
1821	Tap water	8	fluid ounce(s)	237	236.8	0	0	0	0	0	0	0	0	0
1879	Tonic water	8	fluid ounce(s)	244	222.3	83	0	21.5	0	0	0	0	0	0
	Fats and Oils													
	Butter													
104	Butter	1	tablespoon(s)	14	2.3	102	0.1	0	0	11.5	7.3	3.0	0.4	—
2522	Butter Buds, dry butter substitute	1	teaspoon(s)	2	—	5	0	2.0	0	0	0	0	0	0
921	Unsalted	1	tablespoon(s)	14	2.5	102	0.1	0	0	11.5	7.3	3.0	0.4	—
107	Whipped	1	tablespoon(s)	9	1.5	67	0.1	0	0	7.6	4.7	2.2	0.3	—
944	Whipped, unsalted	1	tablespoon(s)	11	2.0	82	0.1	0	0	9.2	5.9	2.4	0.3	—
	Fats, cooking													
2671	Beef tallow, semisolid	1	tablespoon(s)	13	0	115	0	0	0	12.8	6.4	5.4	0.5	—
922	Chicken fat	1	tablespoon(s)	13	0	115	0	0	0	12.8	3.8	5.7	2.7	—
5454	Household shortening with vegetable oil	1	tablespoon(s)	13	0	115	0	0	0	13.0	3.4	5.5	2.7	2.2
111	Lard	1	tablespoon(s)	13	0	115	0	0	0	12.8	5.0	5.8	1.4	—
	Margarine													
114	Margarine	1	tablespoon(s)	14	2.3	101	0	0.1	0	11.4	2.1	5.5	3.4	2.1
5439	Soft	1	tablespoon(s)	14	2.3	103	0.1	0.1	0	11.6	1.7	4.4	2.1	3.0
32329	Soft, unsalted, with hydrogenated soybean and cottonseed oils	1	tablespoon(s)	14	2.5	101	0.1	0.1	0	11.3	2.0	5.4	3.5	—
928	Unsalted	1	tablespoon(s)	14	2.6	101	0.1	0.1	0	11.3	2.1	5.2	3.5	—
119	Whipped	1	tablespoon(s)	9	1.5	64	0.1	0.1	0	7.2	1.2	3.2	2.5	—
	Spreads													
54657	I Can't Believe It's Not Butter!, tub, soya oil (non-hydrogenated)	1	tablespoon(s)	14	2.3	103	0.1	0.1	0	11.6	2.8	2.0	5.1	0.1
2708	Mayonnaise with soybean and safflower oils	1	tablespoon(s)	14	2.1	99	0.2	0.4	0	11.0	1.2	1.8	7.6	—
16157	Promise vegetable oil spread, stick	1	tablespoon(s)	14	4.2	90	0	0	0	10.0	2.5	2.0	4.0	—
	Oils													
2681	Canola	1	tablespoon(s)	14	0	120	0	0	0	13.6	1.0	8.6	3.8	0.1
120	Corn	1	tablespoon(s)	14	0	120	0	0	0	13.6	1.8	3.8	7.4	0
122	Olive	1	tablespoon(s)	14	0	119	0	0	0	13.5	1.9	9.9	1.4	—

PAGE KEY: H-4 = Breads/Baked Goods H-10 = Cereal/Rice/Pasta H-14 = Fruit H-18 = Vegetables/Legumes H-28 = Nuts/Seeds H-30 = Vegetarian H-32 = Dairy H-40 = Eggs H-40 = Seafood H-42 = Meats H-46 = Poultry H-46 = Processed Meats H-48 = Beverages H-52 = Fats/Oils H-54 = Sweets H-56 = Spices/Condiments/Sauces H-60 = Mixed Foods/Soups/Sandwiches H-64 = Fast Food H-84 = Convenience H-86 = Baby Foods

Chol (mg)	Calc (mg)	Iron (mg)	Magn (mg)	Pota (mg)	Sodi (mg)	Zinc (mg)	Vit A (µg)	Thia (mg)	Vit E (mg α)	Ribo (mg)	Niac (mg)	Vit B_6 (mg)	Fola (µg)	Vit C (mg)	Vit B_{12} (µg)	Sele (µg)
0	0	0.00	—	10.1	10.1	—	0	—	—	—	—	—	—	6.0	—	—
0	10	0.39	5.0	37.2	9.9	0.05	0	0.01	0.02	0.05	0.04	0.01	2.5	9.7	0	0.2
0	5	0.00	4.9	24.7	7.4	0.02	0	0.01	0.00	0.01	0.02	0.01	2.5	7.7	0	0.2
0	20	0.72	—	0	20.0	—	0	—	—	—	—	—	—	600.0	0	—
0	0	0.00	—	70.0	10.0	—	0	—	—	—	—	—	—	54.0	0	—
0	0	0.00	—	20.0	10.0	—	0	—	—	—	—	—	—	0	0	—
0	0	0.00	—	40.0	10.0	—	0	—	—	—	—	—	—	0	0	—
5	400	2.70	140.0	600.0	220.0	2.25	—	0.52	—	0.59	7.00	0.70	120.0	60.0	2.1	17.5
5	1000	2.70	140.0	600.0	220.0	2.25	—	0.52	—	0.59	7.00	0.70	120.0	30.0	2.1	17.5
5	400	2.70	140.0	600.0	220.0	2.25	—	0.52	—	0.59	7.00	0.70	120.0	60.0	2.1	17.5
0	0	0.04	7.1	87.7	7.1	0.04	0	0.00	0.00	0.03	0.00	0.00	11.9	0	0	0
0	5	0.18	2.4	21.3	2.4	0.09	0	0.02	0.00	0.01	0.00	0.00	2.4	0	0	0
0	5	0.05	2.6	38.9	5.2	0.02	0	0.00	0.00	0.00	0.02	0.00	0	0	0	0.3
0	7	0.02	4.7	42.7	9.5	0.02	0	0.00	0.00	0.01	0.07	0.00	0	0	0	0
0	0	0.04	7.1	87.7	7.1	0.04	0	0.00	0.00	0.03	0.00	0.00	11.9	0	0	0
0	33	0.00	0	0	2.4	0.00	0	0.00	—	0.00	0.00	0.00	0	0	0	0
0	2	0.02	2.4	0	2.4	0.00	0	0.00	—	0.00	0.00	0.00	0	0	0	0
0	7	0.00	2.4	2.4	7.1	0.00	0	0.00	0.00	0.00	0.00	0.00	0	0	0	0
0	2	0.02	0	0	29.3	0.24	0	0.00	0.00	0.00	0.00	0.00	0	0	0	0
31	3	0.00	0.3	3.4	81.8	0.01	97.1	0.00	0.32	0.01	0.01	0.00	0.4	0	0	0.1
0	0	0.00	0	1.6	120.0	0.00	0	0.00	0.00	0.00	0.00	0.00	0	0	0	—
31	3	0.00	0.3	3.4	1.6	0.01	97.1	0.00	0.32	0.01	0.01	0.00	0.4	0	0	0.1
21	2	0.01	0.2	2.4	77.7	0.01	64.3	0.00	0.21	0.00	0.00	0.00	0.3	0	0	0.1
25	3	0.00	0.2	2.7	1.3	0.01	78.0	0.00	0.26	0.00	0.00	0.00	0.3	0	0	0.1
14	0	0.00	0	0	0	0.00	0	0.00	0.34	0.00	0.00	0.00	0	0	0	0
11	0	0.00	0	0	0	0.00	0	0.00	0.34	0.00	0.00	0.00	0	0	0	0
0	0	0.00	0	0	0	0.00	0	0.00	—	0.00	0.00	0.00	0	0	0	—
12	0	0.00	0	0	0	0.01	0	0.00	0.07	0.00	0.00	0.00	0	0	0	0
0	4	0.01	0.4	5.9	133.0	0.00	115.5	0.00	1.26	0.01	0.00	0.00	0.1	0	0	0
0	4	0.00	0.3	5.5	155.4	0.00	142.7	0.00	1.00	0.00	0.00	0.00	0.1	0	0	0
0	4	0.00	0.3	5.4	3.9	0.00	103.1	0.00	0.98	0.00	0.00	0.00	0.1	0	0	0
0	2	0.00	0.3	3.5	0.3	0.00	115.5	0.00	1.80	0.00	0.00	0.00	0.1	0	0	0
0	2	0.00	0.2	3.4	97.1	0.00	73.7	0.00	0.45	0.00	0.00	0.00	0.1	0	0	0
0	4	0.00	0.3	5.5	155.3	0.00	142.6	0.00	0.72	0.00	0.00	0.00	0.1	0	0	0
8	2	0.06	0.1	4.7	78.4	0.01	11.6	0.00	3.03	0.00	0.00	0.08	1.1	0	0	0.2
0	10	0.18	—	8.7	90.0	—	—	0.00	—	0.00	0.00	—	—	0.6	—	
0	0	0.00	0	0	0	0.00	0	0.00	2.37	0.00	0.00	0.00	0	0	0	0
0	0	0.00	0	0	0	0.00	0	0.00	1.94	0.00	0.00	0.00	0	0	0	0
0	0	0.07	0	0.1	0.3	0.00	0	0.00	1.93	0.00	0.00	0.00	0	0	0	0

TABLE H-1 Table of Food Composition (*continued*) (Computer code is for Cengage Diet Analysis program) (For purposes of calculations, use "0" for t, <1, <.1, <.01, et

DA+ Code	Food Description	Quantity	Measure	Wt (g)	H_2O (g)	Ener (kcal)	Prot (g)	Carb (g)	Fiber (g)	Fat (g)	Fat Breakdown (g) Sat	Mono	Poly	*Trans*
	Fats and Oils—*continued*													
124	Peanut	1	tablespoon(s)	14	0	119	0	0	0	13.5	2.3	6.2	4.3	—
2693	Safflower	1	tablespoon(s)	14	0	120	0	0	0	13.6	0.8	10.2	2.0	—
923	Sesame	1	tablespoon(s)	14	0	120	0	0	0	13.6	1.9	5.4	5.7	—
128	Soybean, hydrogenated	1	tablespoon(s)	14	0	120	0	0	0	13.6	2.0	5.8	5.1	—
130	Soybean, with soybean and cottonseed oil	1	tablespoon(s)	14	0	120	0	0	0	13.6	2.4	4.0	6.5	—
2700	Sunflower	1	tablespoon(s)	14	0	120	0	0	0	13.6	1.8	6.3	5.0	—
357	**Pam original no stick cooking spray**	1	serving(s)	0	0.2	0	0	0	0	0	0	0	0	—
	Salad dressing													
132	Blue cheese	2	tablespoon(s)	30	9.7	151	1.4	2.2	0	15.7	3.0	3.7	8.3	—
133	Blue cheese, low calorie	2	tablespoon(s)	32	25.4	32	1.6	0.9	0	2.3	0.8	0.6	0.8	—
1764	Caesar	2	tablespoon(s)	30	10.3	158	0.4	0.9	0	17.3	2.6	4.1	9.9	—
29654	Creamy, reduced calorie, fat-free, cholesterol-free, sour cream and/or buttermilk and oil	2	tablespoon(s)	32	23.9	34	0.4	6.4	0	0.9	0.2	0.2	0.5	—
29617	Creamy, reduced calorie, sour cream and/or buttermilk and oil	2	tablespoon(s)	30	22.2	48	0.5	2.1	0	4.2	0.6	1.0	2.4	—
134	French	2	tablespoon(s)	32	11.7	146	0.2	5.0	0	14.3	1.8	2.7	6.7	—
135	French, low fat	2	tablespoon(s)	32	17.4	74	0.2	9.4	0.4	4.3	0.4	1.9	1.6	—
136	Italian	2	tablespoon(s)	29	16.6	86	0.1	3.1	0	8.3	1.3	1.9	3.8	—
137	Italian, diet	2	tablespoon(s)	30	25.4	23	0.1	1.4	0	1.9	0.1	0.7	0.5	—
139	Mayonnaise-type	2	tablespoon(s)	29	11.7	115	0.3	7.0	0	9.8	1.4	2.6	5.3	—
942	Oil and vinegar	2	tablespoon(s)	32	15.2	144	0	0.8	0	16.0	2.9	4.7	7.7	—
1765	Ranch	2	tablespoon(s)	30	11.6	146	0.1	1.6	0	15.8	2.3	5.2	7.6	—
3666	Ranch, reduced calorie	2	tablespoon(s)	30	20.5	62	0.1	2.2	0	6.1	1.1	1.8	2.9	—
940	Russian	2	tablespoon(s)	30	11.6	107	0.5	9.3	0.7	7.8	1.2	1.8	4.4	—
939	Russian, low calorie	2	tablespoon(s)	32	20.8	45	0.2	8.8	0.1	1.3	0.2	0.3	0.7	—
941	Sesame seed	2	tablespoon(s)	30	11.8	133	0.9	2.6	0.3	13.6	1.9	3.6	7.5	—
142	Thousand Island	2	tablespoon(s)	32	14.9	118	0.3	4.7	0.3	11.2	1.6	2.5	5.8	—
143	Thousand Island, low calorie	2	tablespoon(s)	30	18.2	61	0.3	6.7	0.4	3.9	0.2	1.9	0.8	—
	Sandwich spreads													
138	Mayonnaise with soybean oil	1	tablespoon(s)	14	2.1	99	0.1	0.4	0	11.0	1.6	2.7	5.8	0
140	Mayonnaise, low calorie	1	tablespoon(s)	16	10.0	37	0	2.6	0	3.1	0.5	0.7	1.7	—
141	Tartar sauce	2	tablespoon(s)	28	8.7	144	0.3	4.1	0.1	14.4	2.2	3.8	7.7	—
	Sweets													
4799	**Butterscotch or caramel topping**	2	tablespoon(s)	41	13.1	103	0.6	27.0	0.4	0	0	0	0	—
	Candy													
1786	Almond Joy candy bar	1	item(s)	45	4.3	220	2.0	27.0	2.0	12.0	8.0	3.3	0.7	0
1785	Bit-O-Honey candy	6	item(s)	40	—	190	1.0	39.0	0	3.5	2.5	—	—	—
33375	Butterscotch candy	2	piece(s)	12	0.6	47	0	10.8	0	0.4	0.2	0.1	0	—
1701	Chewing gum, stick	1	item(s)	3	0.1	7	0	2.0	0.1	0	0	0	0	—
33378	Chocolate fudge with nuts, prepared	2	piece(s)	38	2.9	175	1.7	25.8	1.0	7.2	2.5	1.5	2.9	0.1
1787	Jelly beans	15	item(s)	43	2.7	159	0	39.8	0.1	0	0	0	0	—
1784	Kit Kat wafer bar	1	item(s)	42	0.8	210	3.0	27.0	0.5	11.0	7.0	3.5	0.3	0
4674	Krackel candy bar	1	item(s)	41	0.6	210	2.0	28.0	0.5	10.0	6.0	3.9	0.4	0
4934	Licorice	4	piece(s)	44	7.3	154	1.1	35.1	0	1.0	0	0.1	0	—
1780	Life Savers candy	1	item(s)	2	—	8	0	2.0	0	0	0	0	0	0
1790	Lollipop	1	item(s)	28	—	108	0	28.0	0	0	0	0	0	0
4679	M & Ms peanut chocolate candy, small bag	1	item(s)	49	0.9	250	5.0	30.0	2.0	13.0	5.0	5.4	2.1	—
1781	M & Ms plain chocolate candy, small bag	1	item(s)	48	0.8	240	2.0	34.0	1.0	10.0	6.0	3.3	0.3	—
4673	Milk chocolate bar, Symphony	1	item(s)	91	0.9	483	7.7	52.8	1.5	27.8	16.7	7.2	0.6	—
1783	Milky Way bar	1	item(s)	58	3.7	270	2.0	41.0	1.0	10.0	5.0	3.5	0.3	—
1788	Peanut brittle	1½	ounce(s)	43	0.3	207	3.2	30.3	1.1	8.1	1.8	3.4	1.9	—
1789	Reese's peanut butter cups	2	piece(s)	51	0.8	280	6.0	19.0	2.0	15.5	6.0	7.2	2.7	0
4689	Reese's pieces candy, small bag	1	item(s)	43	1.1	220	5.0	26.0	1.0	11.0	7.0	0.9	0.4	0
33399	Semisweet chocolate candy, made with butter	½	ounce(s)	14	0.1	68	0.6	9.0	0.8	4.2	2.5	1.4	0.1	—
1782	Snickers bar	1	item(s)	59	3.2	280	4.0	35.0	1.0	14.0	5.0	6.1	2.9	—
4694	Special Dark chocolate bar	1	item(s)	41	0.4	220	2.0	25.0	3.0	12.0	8.0	4.6	0.4	0
4695	Starburst fruit chews, original fruits	1	package(s)	59	3.9	240	0	48.0	0	5.0	1.0	2.1	1.8	—

PAGE KEY: H-4 = Breads/Baked Goods H-10 = Cereal/Rice/Pasta H-14 = Fruit H-18 = Vegetables/Legumes H-28 = Nuts/Seeds H-30 = Vegetarian H-32 = Dairy H-40 = Eggs H-40 = Seafood H-42 = Meats H-46 = Poultry H-46 = Processed Meats H-48 = Beverages H-52 = Fats/Oils H-54 = Sweets H-56 = Spices/Condiments/Sauces H-60 = Mixed Foods/Soups/Sandwiches H-64 = Fast Food H-84 = Convenience H-86 = Baby Foods

Chol (mg)	Calc (mg)	Iron (mg)	Magn (mg)	Pota (mg)	Sodi (mg)	Zinc (mg)	Vit A (µg)	Thia (mg)	Vit E (mg α)	Ribo (mg)	Niac (mg)	Vit B_6 (mg)	Fola (µg)	Vit C (mg)	Vit B_{12} (µg)	Sele (µg)
0	0	0.00	0	0	0	0.00	0	0.00	2.11	0.00	0.00	0.00	0	0	0	0
0	0	0.00	0	0	0	0.00	0	0.00	4.63	0.00	0.00	0.00	0	0	0	0
0	0	0.00	0	0	0	0.00	0	0.00	0.19	0.00	0.00	0.00	0	0	0	0
0	0	0.00	0	0	0	0.00	0	0.00	1.10	0.00	0.00	0.00	0	0	0	0
0	0	0.00	0	0	0	0.00	0	0.00	1.64	0.00	0.00	0.00	0	0	0	0
0	0	0.00	0	0	0	0.00	0	0.00	5.58	0.00	0.00	0.00	0	0	0	0
0	0	0.00	0	0.3	1.5	0.01	0.1	0.00	0.00	0.00	0.00	0.00	0	0	0	0
5	24	0.06	0	11.1	328.2	0.08	20.1	0.00	1.80	0.03	0.03	0.01	7.8	0.6	0.1	0.3
0	28	0.16	2.2	1.6	384.0	0.08	—	0.01	0.08	0.03	0.01	0.01	1.0	0.1	0.1	0.5
1	7	0.05	0.6	8.7	323.4	0.03	0.6	0.00	1.56	0.00	0.01	0.00	0.9	0	0	0.5
0	12	0.08	1.6	42.6	320.0	0.05	0.3	0.00	0.21	0.01	0.01	0.01	1.9	0	0	0.5
0	2	0.03	0.6	10.8	306.9	0.01	—	0.00	0.71	0.00	0.01	0.01	0	0.1	0	0.5
0	8	0.25	1.6	21.4	267.5	0.09	7.4	0.01	1.60	0.01	0.06	0.00	0	0	0	0
0	4	0.27	2.6	34.2	257.3	0.06	8.6	0.01	0.09	0.01	0.14	0.01	0.6	0	0	0.5
0	2	0.18	0.9	14.1	486.3	0.03	0.6	0.00	1.47	0.01	0.00	0.01	0	0	0	0.6
2	3	0.19	1.2	25.5	409.8	0.05	0.3	0.00	0.06	0.00	0.00	0.02	0	0	0	2.4
8	4	0.05	0.6	2.6	209.0	0.05	6.2	0.00	0.60	0.01	0.00	0.01	1.8	0	0.1	0.5
0	0	0.00	0	2.6	0.3	0.00	0	0.00	1.46	0.00	0.00	0.00	0	0	0	0.5
1	4	0.03	1.2	8.4	354.0	0.01	5.4	0.00	1.84	0.01	0.00	0.00	0.3	0.1	0	0.1
0	5	0.01	1.5	8.4	413.7	0.01	0.9	0.00	0.72	0.01	0.00	0.00	0.3	0.1	0	0.1
0	6	0.20	3.0	51.9	282.3	0.06	13.2	0.01	0.98	0.01	0.16	0.02	1.5	1.4	0	0.5
2	6	0.18	0	50.2	277.8	0.02	0.6	0.00	0.12	0.00	0.00	0.00	1.0	1.9	0	0.5
0	6	0.18	0	47.1	300.0	0.02	0.6	0.00	1.50	0.00	0.00	0.00	0	0	0	0.5
8	5	0.37	2.6	34.2	276.2	0.08	4.5	0.46	1.28	0.01	0.13	0.00	0	0	0	0.5
0	5	0.27	2.1	60.6	249.3	0.05	4.8	0.01	0.30	0.01	0.13	0.00	0	0	0	0
5	1	0.03	0.1	1.7	78.4	0.02	11.2	0.01	0.72	0.01	0.00	0.08	0.7	0	0	0.2
4	0	0.00	0	1.6	79.5	0.01	0	0.00	0.32	0.00	0.00	0.00	0	0	0	0.3
8	6	0.20	0.8	10.1	191.5	0.05	20.2	0.00	0.97	0.00	0.01	0.07	2.0	0.1	0.1	0.5
0	22	0.08	2.9	34.4	143.1	0.07	11.1	0.01	—	0.03	0.01	0.01	0.8	0.1	0	0
0	18	0.33	30.3	126.5	65.0	0.36	0	0.01	—	0.06	0.21	—	—	0	—	—
0	20	0.00	—	—	150.0	—	0	—	—	—	—	—	—	0	—	—
1	0	0.00	0	0.4	46.9	0.01	3.4	0.00	0.01	0.00	0.00	0.00	0	0	0	0.1
0	0	0.00	0	0.1	0	0.00	0	0.00	0.00	0.00	0.00	0.00	0	0	0	0
5	22	0.74	20.9	69.5	14.8	0.54	14.4	0.02	0.09	0.03	0.12	0.03	6.1	0.1	0	1.1
0	1	0.05	0.9	15.7	21.3	0.02	0	0.00	0.00	0.01	0.00	0.00	0	0	0	0.5
3	60	0.36	16.4	126.0	30.0	0.51	0	0.07	—	0.22	1.07	0.05	59.6	0	0.1	2.0
3	40	0.36	—	168.8	50.0	—	0	—	—	—	—	—	—	0	—	—
0	0	0.22	2.6	28.2	126.3	0.07	0	0.01	0.07	0.01	0.04	0.00	0	0	0	—
0	0	0.00	—	0	0	—	0	0.00	—	0.00	0.00	—	—	0	—	0
0	0	0.00	—	—	10.8	—	0	0.00	—	0.00	0.00	—	—	0	—	1.0
5	40	0.36	36.5	170.6	25.0	1.13	14.8	0.03	—	0.06	1.60	0.04	17.3	0.6	0.1	1.9
5	40	0.36	19.6	127.4	30.0	0.46	14.8	0.02	—	0.06	0.10	0.01	2.9	0.6	0.1	1.4
22	228	0.82	61.0	398.6	91.9	1.00	0	0.06	—	0.25	0.14	0.10	10.9	2.0	0.4	—
5	60	0.18	19.8	140.1	95.0	0.41	15.1	0.02	—	0.06	0.20	0.02	5.8	0.6	0.2	3.3
5	11	0.51	17.9	71.4	189.2	0.37	16.6	0.05	1.08	0.01	1.12	0.03	19.6	0	0	1.1
3	40	0.72	45.4	217.4	180.0	0.93	0	0.12	—	0.08	2.35	0.07	28.1	0	0.1	2.3
0	20	0.00	18.9	169.9	80.0	0.32	0	0.04	—	0.06	1.22	0.03	12.0	0	0.1	0.8
3	5	0.44	16.3	51.7	1.6	0.23	0.4	0.01	—	0.01	0.06	0.01	0.4	0	0	0.5
5	40	0.36	42.3	—	140.0	1.37	15.3	0.03	—	0.06	1.60	0.05	23.5	0.6	0.1	2.7
0	0	1.80	45.5	136.0	50.0	0.59	0	0.01	—	0.02	0.16	0.01	0.8	0	0	1.2
0	10	0.18	0.6	1.2	0	0.00	—	0.00	—	0.00	0.00	0.00	0	30	0	0.5

TABLE H-1 Table of Food Composition (*continued*) (Computer code is for Cengage Diet Analysis program) (For purposes of calculations, use "0" for t, <1, <.1, <.01, etc

DA+ Code	Food Description	Quantity	Measure	Wt (g)	H_2O (g)	Ener (kcal)	Prot (g)	Carb (g)	Fiber (g)	Fat (g)	Fat Breakdown (g) Sat	Mono	Poly	*Trans*
Sweets—*continued*														
4698	Taffy	3	piece(s)	45	2.2	179	0	41.2	0	1.5	0.9	0.4	0.1	0.1
4699	Three Musketeers bar	1	item(s)	60	3.5	260	2.0	46.0	1.0	8.0	4.5	2.6	0.3	—
4702	Twix caramel cookie bars	2	item(s)	58	2.4	280	3.0	37.0	1.0	14.0	5.0	7.7	0.5	—
4705	York peppermint pattie	1	item(s)	39	3.9	160	0.5	32.0	0.5	3.0	1.5	1.2	0.1	0
	Frosting, icing													
4760	Chocolate frosting, ready to eat	2	tablespoon(s)	31	5.2	122	0.3	19.4	0.3	5.4	1.7	2.8	0.6	—
4771	Creamy vanilla frosting, ready to eat	2	tablespoon(s)	28	4.2	117	0	19.0	0	4.5	0.8	1.4	2.2	0
17291	Dec-A-Cake variety pack candy decoration	1	teaspoon(s)	4	—	15	0	3.0	0	0.5	0	—	—	—
536	White icing	2	tablespoon(s)	40	3.6	162	0.1	31.8	0	4.2	0.8	2.0	1.2	—
	Gelatin													
13697	Gelatin snack, all flavors	1	item(s)	99	96.8	70	1.0	17.0	0	0	0	0	0	0
2616	Sugar free, low calorie mixed fruit gelatin mix, prepared	½	cup(s)	121	—	10	1.0	0	0	0	0	0	0	0
548	**Honey**	1	tablespoon(s)	21	3.6	64	0.1	17.3	0	0	0	0	0	0
	Jams, jellies													
550	Jam or preserves	1	tablespoon(s)	20	6.1	56	0.1	13.8	0.2	0	0	0	0	—
42199	Jams, preserves, dietetic, all flavors, w/sodium saccharin	1	tablespoon(s)	14	6.4	18	0	7.5	0.4	0	0	0	0	—
552	Jelly	1	tablespoon(s)	21	6.3	56	0	14.7	0.2	0	0	0	0	—
545	**Marshmallows**	4	item(s)	29	4.7	92	0.5	23.4	0	0.1	0	0	0	—
4800	**Marshmallow cream topping**	2	tablespoon(s)	40	7.9	129	0.3	31.6	0	0.1	0	0	0	—
555	**Molasses**	1	tablespoon(s)	20	4.4	58	0	14.9	0	0	0	0	0	—
4780	**Popsicle or ice pop**	1	item(s)	59	47.5	47	0	11.3	0	0.1	0	0	0	—
	Sugar													
559	Brown sugar, packed	1	teaspoon(s)	5	0.1	17	0	4.5	0	0	0	0	0	0
563	Powdered sugar, sifted	⅓	cup(s)	33	0.1	130	0	33.2	0	0	0	0	0	—
561	White granulated sugar	1	teaspoon(s)	4	0	16	0	4.2	0	0	0	0	0	0
	Sugar substitute													
1760	Equal sweetener, packet size	1	item(s)	1	—	0	0	0.9	0	0	0	0	0	0
13029	Splenda granular no calorie sweetener	1	teaspoon(s)	1	—	0	0	0.5	0	0	0	0	0	0
1759	Sweet N Low sugar substitute, packet	1	item(s)	1	0.1	4	0	0.5	0	0	0	0	0	0
	Syrup													
3148	Chocolate syrup	2	tablespoon(s)	38	11.6	105	0.8	24.4	1.0	0.4	0.2	0.1	0	—
29676	Maple syrup	¼	cup(s)	80	25.7	209	0	53.7	0	0.2	0	0.1	0.1	—
4795	Pancake syrup	¼	cup(s)	80	30.4	187	0	49.2	0	0	0	0	0	0
Spices, Condiments, Sauces														
	Spices													
807	Allspice, ground	1	teaspoon(s)	2	0.2	5	0.1	1.4	0.4	0.2	0	0	0	—
1171	Anise seeds	1	teaspoon(s)	2	0.2	7	0.4	1.1	0.3	0.3	0	0.2	0.1	—
729	Bakers' yeast, active	1	teaspoon(s)	4	0.3	12	1.5	1.5	0.8	0.2	0	0.1	0	—
683	Baking powder, double acting with phosphate	1	teaspoon(s)	5	0.2	2	0	1.1	0	0	0	0	0	0
1611	Baking soda	1	teaspoon(s)	5	0	0	0	0	0	0	0	0	0	0
8552	Basil	1	teaspoon(s)	1	0.8	0	0	0	0	0	0	0	0	—
34959	Basil, fresh	1	piece(s)	1	0.5	0	0	0	0	0	0	0	0	—
808	Basil, ground	1	teaspoon(s)	1	0.1	4	0.2	0.9	0.6	0.1	0	0	0	—
809	Bay leaf	1	teaspoon(s)	1	0	2	0	0.5	0.2	0.1	0	0	0	—
11720	Betel leaves	1	ounce(s)	28	—	17	1.8	2.4	0	0	—	—	—	—
730	Brewers' yeast	1	teaspoon(s)	3	0.1	8	1.0	1.0	0.8	0	0	0	0	0
11710	Capers	1	teaspoon(s)	5	—	0	0	0	0	0	0	0	0	0
1172	Caraway seeds	1	teaspoon(s)	2	0.2	7	0.4	1.0	0.8	0.3	0	0.2	0.1	—
1173	Celery seeds	1	teaspoon(s)	2	0.1	8	0.4	0.8	0.2	0.5	0	0.3	0.1	—
1174	Chervil, dried	1	teaspoon(s)	1	0	1	0.1	0.3	0.1	0	0	0	0	—
810	Chili powder	1	teaspoon(s)	3	0.2	8	0.3	1.4	0.9	0.4	0.1	0.1	0.2	—
8553	Chives, chopped	1	teaspoon(s)	1	0.9	0	0	0	0	0	0	0	0	—
51420	Cilantro (coriander)	1	teaspoon(s)	0	0.3	0	0	0	0	0	0	0	0	—
811	Cinnamon, ground	1	teaspoon(s)	2	0.2	6	0.1	1.9	1.2	0	0	0	0	0
812	Cloves, ground	1	teaspoon(s)	2	0.1	7	0.1	1.3	0.7	0.4	0.1	0	0.1	—
1175	Coriander leaf, dried	1	teaspoon(s)	1	0	2	0.1	0.3	0.1	0	0	0	0	—
1176	Coriander seeds	1	teaspoon(s)	2	0.2	5	0.2	1.0	0.8	0.3	0	0.2	0	—
1706	Cornstarch	1	tablespoon(s)	8	0.7	30	0	7.3	0.1	0	0	0	0	—

PAGE KEY: H-4 = Breads/Baked Goods H-10 = Cereal/Rice/Pasta H-14 = Fruit H-18 = Vegetables/Legumes H-28 = Nuts/Seeds H-30 = Vegetarian H-32 = Dairy H-40 = Eggs H-40 = Seafood H-42 = Meats H-46 = Poultry H-46 = Processed Meats H-48 = Beverages H-52 = Fats/Oils H-54 = Sweets H-56 = Spices/Condiments/Sauces H-60 = Mixed Foods/Soups/Sandwiches H-64 = Fast Food H-84 = Convenience H-86 = Baby Foods

Chol (mg)	Calc (mg)	Iron (mg)	Magn (mg)	Pota (mg)	Sodi (mg)	Zinc (mg)	Vit A (µg)	Thia (mg)	Vit E (mg α)	Ribo (mg)	Niac (mg)	Vit B_6 (mg)	Fola (µg)	Vit C (mg)	Vit B_{12} (µg)	Sele (µg)
4	4	0.00	0	1.4	23.4	0.09	12.2	0.01	0.04	0.01	0.00	0.00	0	0	0	0.3
5	20	0.36	17.5	80.3	110.0	0.33	14.5	0.01	—	0.03	0.20	0.01	0	0.6	0.1	1.5
5	40	0.36	18.5	116.8	115.0	0.45	15.0	0.09	—	0.13	0.69	0.01	13.9	0.6	0.1	1.2
0	0	0.33	23.4	66.1	10.0	0.28	0	0.01	—	0.03	0.31	0.01	1.5	0	0	—
0	2	0.44	6.4	60.0	56.1	0.09	0	0.00	0.48	0.00	0.03	0.00	0.3	0	0	0.2
0	1	0.04	0.3	9.5	51.5	0.01	0	0.00	0.43	0.08	0.06	0.00	2.2	0	0	0
0	0	0.00	—	—	15.0	—	0	—	—	—	—	—	—	0	—	—
0	4	0.01	0.4	5.6	76.4	0.01	44.4	0.00	0.32	0.01	0.00	0.00	0	0	0	0.3
0	0	0.00	—	0	40.0	—	0	—	—	—	—	—	—	0	—	—
0	0	0.00	0	0	50.0	0.00	0	0.00	0.00	0.00	0.00	0.00	0	0	0	—
0	1	0.08	0.4	10.9	0.8	0.04	0	0.00	0.00	0.01	0.02	0.01	0.4	0.1	0	0.2
0	4	0.10	0.8	15.4	6.4	0.01	0	0.00	0.02	0.02	0.01	0.00	2.2	1.8	0	0.4
0	1	0.56	0.7	9.7	0	0.01	0	0.00	0.01	0.00	0.00	0.00	1.3	0	0	0.2
0	1	0.04	1.3	11.3	6.3	0.01	0	0.00	0.00	0.01	0.01	0.00	0.4	0.2	0	0.1
0	1	0.06	0.6	1.4	23.0	0.01	0	0.00	0.00	0.00	0.02	0.00	0.3	0	0	0.5
0	1	0.08	0.8	2.0	32.0	0.01	0	0.00	0.00	0.00	0.03	0.00	0.4	0	0	0.7
0	41	0.94	48.4	292.8	7.4	0.05	0	0.01	0.00	0.00	0.18	0.13	0	0	0	3.6
0	0	0.31	0.6	8.9	4.1	0.08	0	0.00	0.00	0.00	0.00	0.00	0	0.4	0	0.1
0	4	0.03	0.4	6.1	1.3	0.00	0	0.00	0.00	0.00	0.01	0.00	0	0	0	0.1
0	0	0.01	0	0.7	0.3	0.00	0	0.00	0.00	0.00	0.00	0.00	0	0	0	0.2
0	0	0.00	0	0.1	0	0.00	0	0.00	0.00	0.00	0.00	0.00	0	0	0	0
0	0	0.00	0	0	0	0.00	0	0.00	0.00	0.00	0.00	0.00	0	0	0	0
0	0	0.00	—	—	0	—	—	0.00	—	0.00	0.00	—	—	0	0	—
0	0	0.00	—	—	0	—	0	—	0.00	—	—	—	—	0	—	—
0	5	0.79	24.4	84.0	27.0	0.27	0	0.00	0.01	0.01	0.12	0.00	0.8	0.1	0	0.5
0	54	0.96	11.2	163.2	7.2	3.32	0	0.01	0.00	0.01	0.02	0.00	0	0	0	0.5
0	2	0.02	1.6	12.0	65.6	0.06	0	0.01	0.00	0.01	0.00	0.00	0	0	0	0
0	13	0.13	2.6	19.8	1.5	0.01	0.5	0.00	—	0.00	0.05	0.00	0.7	0.7	0	0.1
0	14	0.77	3.6	30.3	0.3	0.11	0.3	0.01	—	0.01	0.06	0.01	0.2	0.4	0	0.1
0	3	0.66	3.9	80.0	2.0	0.25	0	0.09	0.00	0.21	1.59	0.06	93.6	0	0	1.0
0	339	0.51	1.8	0.2	363.1	0.00	0	0.00	0.00	0.00	0.00	0.00	0	0	0	0
0	0	0.00	0	0	1258.6	0.00	0	0.00	0.00	0.00	0.00	0.00	0	0	0	0
0	2	0.02	0.6	2.6	0	0.01	2.3	0.00	0.01	0.00	0.01	0.00	0.6	0.2	0	0
0	1	0.01	0.4	2.3	0	0.00	1.3	0.00	—	0.00	0.00	0.00	0.3	0.1	0	0
0	30	0.58	5.9	48.1	0.5	0.08	6.6	0.00	0.10	0.00	0.09	0.03	3.8	0.9	0	0
0	5	0.25	0.7	3.2	0.1	0.02	1.9	0.00	—	0.00	0.01	0.01	1.1	0.3	0	0
0	110	2.29	—	155.9	2.0	—	—	0.04	—	0.07	0.19	—	—	0.9	0	—
0	6	0.46	6.1	50.7	3.3	0.21	0	0.41	—	0.11	1.00	0.06	104.3	0	0	0
0	—	—	—	—	105.0	—	—	—	—	—	—	—	—	—	0	—
0	14	0.34	5.4	28.4	0.4	0.11	0.4	0.01	0.05	0.01	0.07	0.01	0.2	0.4	0	0.3
0	35	0.89	8.8	28.0	3.2	0.13	0.1	0.01	0.02	0.01	0.06	0.01	0.2	0.3	0	0.2
0	8	0.19	0.8	28.4	0.5	0.05	1.8	0.00	—	0.00	0.03	0.01	1.6	0.3	0	0.2
0	7	0.37	4.4	49.8	26.3	0.07	38.6	0.01	0.75	0.02	0.20	0.09	2.6	1.7	0	0.2
0	1	0.01	0.4	3.0	0	0.01	2.2	0.00	0.00	0.00	0.01	0.00	1.1	0.6	0	0
0	0	0.01	0.1	1.7	0.2	0.00	1.1	0.00	0.01	0.00	0.00	0.00	0.2	0.1	0	0
0	23	0.19	1.4	9.9	0.2	0.04	0.3	0.00	0.05	0.00	0.03	0.00	0.1	0.1	0	0.1
0	14	0.18	5.5	23.1	5.1	0.02	0.6	0.00	0.17	0.01	0.03	0.01	2.0	1.7	0	0.1
0	7	0.25	4.2	26.8	1.3	0.02	1.8	0.01	0.01	0.01	0.06	0.00	1.6	3.4	0	0.2
0	13	0.29	5.9	22.8	0.6	0.08	0	0.00	—	0.01	0.03	—	0	0.4	0	0.5
0	0	0.03	0.2	0.2	0.7	0.01	0	0.00	0.00	0.00	0.00	0.00	0	0	0	0.2

TABLE H-1 Table of Food Composition (*continued*) (Computer code is for Cengage Diet Analysis program) (For purposes of calculations, use "0" for t, <1, <.1, <.01, etc

DA+ Code	Food Description	Quantity	Measure	Wt (g)	H_2O (g)	Ener (kcal)	Prot (g)	Carb (g)	Fiber (g)	Fat (g)	Fat Breakdown (g)			
											Sat	Mono	Poly	*Trans*
Spices, Condiments, Sauces—*continued*														
1177	Cumin seeds	1	teaspoon(s)	2	0.2	8	0.4	0.9	0.2	0.5	0	0.3	0.1	—
11729	Cumin, ground	1	teaspoon(s)	5	—	11	0.4	0.8	0.8	0.4	—	—	—	—
1178	Curry powder	1	teaspoon(s)	2	0.2	7	0.3	1.2	0.7	0.3	0	0.1	0.1	—
1179	Dill seeds	1	teaspoon(s)	2	0.2	6	0.3	1.2	0.4	0.3	0	0.2	0	—
1180	Dill weed, dried	1	teaspoon(s)	1	0.1	3	0.2	0.6	0.1	0	0	0	0	—
34949	Dill weed, fresh	5	piece(s)	1	0.9	0	0	0.1	0	0	0	0	0	—
4949	Fennel leaves, fresh	1	teaspoon(s)	1	0.9	0	0	0.1	0	0	—	—	—	—
1181	Fennel seeds	1	teaspoon(s)	2	0.2	7	0.3	1.0	0.8	0.3	0	0.2	0	—
1182	Fenugreek seeds	1	teaspoon(s)	4	0.3	12	0.9	2.2	0.9	0.2	0.1	—	—	—
11733	Garam masala, powder	1	ounce(s)	28	—	107	4.4	12.8	0	4.3	—	—	—	—
1067	Garlic clove	1	item(s)	3	1.8	4	0.2	1.0	0.1	0	0	0	0	—
813	Garlic powder	1	teaspoon(s)	3	0.2	9	0.5	2.0	0.3	0	0	0	0	—
1068	Ginger root	2	teaspoon(s)	4	3.1	3	0.1	0.7	0.1	0	0	0	0	—
1183	Ginger, ground	1	teaspoon(s)	2	0.2	6	0.2	1.3	0.2	0.1	0	0	0	—
35497	Leeks, bulb and lower-leaf, freeze-dried	¼	cup(s)	1	0	3	0.1	0.6	0.1	0	0	0	0	—
1184	Mace, ground	1	teaspoon(s)	2	0.1	8	0.1	0.9	0.3	0.6	0.2	0.2	0.1	—
1185	Marjoram, dried	1	teaspoon(s)	1	0	2	0.1	0.4	0.2	0	0	0	0	—
1186	Mustard seeds, yellow	1	teaspoon(s)	3	0.2	15	0.8	1.2	0.5	0.9	0	0.7	0.2	—
814	Nutmeg, ground	1	teaspoon(s)	2	0.1	12	0.1	1.1	0.5	0.8	0.6	0.1	0	—
2747	Onion flakes, dehydrated	1	teaspoon(s)	2	0.1	6	0.1	1.4	0.2	0	0	0	0	—
1187	Onion powder	1	teaspoon(s)	2	0.1	7	0.2	1.7	0.1	0	0	0	0	—
815	Oregano, ground	1	teaspoon(s)	2	0.1	5	0.2	1.0	0.6	0.2	0	0	0.1	—
816	Paprika	1	teaspoon(s)	2	0.2	6	0.3	1.2	0.8	0.3	0	0	0.2	—
817	Parsley, dried	1	teaspoon(s)	0	0	1	0.1	0.2	0.1	0	0	0	0	—
818	Pepper, black	1	teaspoon(s)	2	0.2	5	0.2	1.4	0.6	0.1	0	0	0	—
819	Pepper, cayenne	1	teaspoon(s)	2	0.1	6	0.2	1.0	0.5	0.3	0.1	0	0.2	—
1188	Pepper, white	1	teaspoon(s)	2	0.3	7	0.3	1.6	0.6	0.1	0	0	0	—
1189	Poppy seeds	1	teaspoon(s)	3	0.2	15	0.5	0.7	0.3	1.3	0.1	0.2	0.9	—
1190	Poultry seasoning	1	teaspoon(s)	2	0.1	5	0.1	1.0	0.2	0.1	0	0	0	—
1191	Pumpkin pie spice, powder	1	teaspoon(s)	2	0.1	6	0.1	1.2	0.3	0.2	0.1	0	0	—
1192	Rosemary, dried	1	teaspoon(s)	1	0.1	4	0.1	0.8	0.5	0.2	0.1	0	0	—
11723	Rosemary, fresh	1	teaspoon(s)	1	0.5	1	0	0.1	0.1	0	0	0	0	—
2722	Saffron powder	1	teaspoon(s)	1	0.1	2	0.1	0.5	0	0	0	0	0	—
11724	Sage	1	teaspoon(s)	1	—	1	0	0.1	0	0	—	—	—	—
1193	Sage, ground	1	teaspoon(s)	1	0.1	2	0.1	0.4	0.3	0.1	0	0	0	—
30189	Salt substitute	¼	teaspoon(s)	1	—	0	0	0	0	0	0	0	0	0
30190	Salt substitute, seasoned	¼	teaspoon(s)	1	—	1	0	0.1	0	0	0	—	—	—
822	Salt, table	¼	teaspoon(s)	2	0	0	0	0	0	0	0	0	0	0
1194	Savory, ground	1	teaspoon(s)	1	0.1	4	0.1	1.0	0.6	0.1	0	—	—	—
820	Sesame seed kernels, toasted	1	teaspoon(s)	3	0.1	15	0.5	0.7	0.5	1.3	0.2	0.5	0.6	—
11725	Sorrel	1	teaspoon(s)	3	—	1	0.1	0.1	0	0	0	—	—	—
11721	Spearmint	1	teaspoon(s)	2	1.6	1	0.1	0.2	0.1	0	0	0	0	—
35498	Sweet green peppers, freeze-dried	¼	cup(s)	2	0	5	0.3	1.1	0.3	0	0	0	0	—
11726	Tamarind leaves	1	ounce(s)	28	—	33	1.6	5.2	0	0.6	—	—	—	—
11727	Tarragon	1	ounce(s)	28	—	14	1.0	1.8	0	0.3	—	—	—	—
1195	Tarragon, ground	1	teaspoon(s)	2	0.1	5	0.4	0.8	0.1	0.1	0	0	0.1	—
11728	Thyme, fresh	1	teaspoon(s)	1	0.5	1	0	0.2	0.1	0	0	0	0	—
821	Thyme, ground	1	teaspoon(s)	1	0.1	4	0.1	0.9	0.5	0.1	0	0	0	—
1196	Turmeric, ground	1	teaspoon(s)	2	0.3	8	0.2	1.4	0.5	0.2	0.1	0	0	—
11995	Wasabi	1	tablespoon(s)	14	10.7	10	0.7	2.3	0.2	0	—	—	—	—
	Condiments													
674	Catsup or ketchup	1	tablespoon(s)	15	10.4	15	0.3	3.8	0	0	0	0	0	—
703	Dill pickle	1	ounce(s)	28	26.7	3	0.2	0.7	0.3	0	0	0	0	—
138	Mayonnaise with soybean oil	1	tablespoon(s)	14	2.1	99	0.1	0.4	0	11.0	1.6	2.7	5.8	0
140	Mayonnaise, low calorie	1	tablespoon(s)	16	10.0	37	0	2.6	0	3.1	0.5	0.7	1.7	—
1682	Mustard, brown	1	teaspoon(s)	5	4.1	5	0.3	0.3	0	0.3	—	—	—	—
700	Mustard, yellow	1	teaspoon(s)	5	4.1	3	0.2	0.3	0.2	0.2	0	0.1	0	0
706	Sweet pickle relish	1	tablespoon(s)	15	9.3	20	0.1	5.3	0.2	0.1	0	0	0	—
141	Tartar sauce	2	tablespoon(s)	28	8.7	144	0.3	4.1	0.1	14.4	2.2	3.8	7.7	—
	Sauces													
685	Barbecue sauce	2	tablespoon(s)	31	18.9	47	0	11.3	0.2	0.1	0	0	0.1	0
834	Cheese sauce	¼	cup(s)	63	44.4	110	4.2	4.3	0.3	8.4	3.8	2.4	1.6	—
32123	Chili enchilada sauce, green	2	tablespoon(s)	57	53.0	15	0.6	3.1	0.7	0.3	0	0	0.1	0
32122	Chili enchilada sauce, red	2	tablespoon(s)	32	24.5	27	1.1	5.0	2.1	0.8	0.1	0	0.4	0

PAGE KEY: H-4 = Breads/Baked Goods H-10 = Cereal/Rice/Pasta H-14 = Fruit H-18 = Vegetables/Legumes H-28 = Nuts/Seeds H-30 = Vegetarian H-32 = Dairy H-40 = Eggs H-40 = Seafood H-42 = Meats H-46 = Poultry H-46 = Processed Meats H-48 = Beverages H-52 = Fats/Oils H-54 = Sweets H-56 = Spices/Condiments/Sauces H-60 = Mixed Foods/Soups/Sandwiches H-64 = Fast Food H-84 = Convenience H-86 = Baby Foods

Chol (mg)	Calc (mg)	Iron (mg)	Magn (mg)	Pota (mg)	Sodi (mg)	Zinc (mg)	Vit A (µg)	Thia (mg)	Vit E (mg α)	Ribo (mg)	Niac (mg)	Vit B_6 (mg)	Fola (µg)	Vit C (mg)	Vit B_{12} (µg)	Sele (µg)
0	20	1.39	7.7	37.5	3.5	0.10	1.3	0.01	0.07	0.01	0.09	0.01	0.2	0.2	0	0.1
0	20	—	—	43.6	4.8	—	—	—	—	—	—	—	—	—	—	—
0	10	0.59	5.1	30.9	1.0	0.08	1.0	0.01	0.44	0.01	0.06	0.02	3.1	0.2	0	0.3
0	32	0.34	5.4	24.9	0.4	0.10	0.1	0.01	—	0.01	0.05	0.01	0.2	0.4	0	0.3
0	18	0.48	4.5	33.1	2.1	0.03	2.9	0.00	—	0.00	0.02	0.01	1.5	0.5	0	—
0	2	0.06	0.6	7.4	0.6	0.01	3.9	0.00	0.01	0.00	0.01	0.00	1.5	0.9	0	—
0	1	0.02	—	4.0	0.1	—	—	0.00	—	0.00	0.01	0.00	—	0.3	0	—
0	24	0.37	7.7	33.9	1.8	0.07	0.1	0.01	—	0.01	0.12	0.01	—	0.4	0	—
0	7	1.24	7.1	28.5	2.5	0.09	0.1	0.01	—	0.01	0.06	0.02	2.1	0.1	0	0.2
0	215	9.24	93.6	411.1	27.5	1.07	—	0.09	—	0.09	0.70	—	0	0	0	—
0	5	0.05	0.8	12.0	0.5	0.03	0	0.01	0.00	0.00	0.02	0.03	0.1	0.9	0	0.4
0	2	0.07	1.6	30.8	0.7	0.07	0	0.01	0.01	0.00	0.01	0.08	0.1	0.5	0	1.1
0	1	0.02	1.7	16.6	0.5	0.01	0	0.00	0.01	0.00	0.02	0.01	0.4	0.2	0	0
0	2	0.20	3.3	24.2	0.6	0.08	0.1	0.00	0.32	0.00	0.09	0.01	0.7	0.1	0	0.7
0	3	0.06	1.3	19.2	0.3	0.01	0.1	0.01	—	0.00	0.02	0.01	2.9	0.9	0	0
0	4	0.23	2.8	7.9	1.4	0.03	0.7	0.01	—	0.01	0.02	0.00	1.3	0.4	0	0
0	12	0.49	2.1	9.1	0.5	0.02	2.4	0.00	0.01	0.00	0.02	0.01	1.6	0.3	0	0
0	17	0.32	9.8	22.5	0.2	0.18	0.1	0.01	0.09	0.01	0.26	0.01	2.5	0.1	0	4.4
0	4	0.06	4.0	7.7	0.4	0.04	0.1	0.01	0.00	0.00	0.02	0.00	1.7	0.1	0	0
0	4	0.02	1.5	27.1	0.4	0.03	0	0.01	0.00	0.00	0.01	0.02	2.8	1.3	0	0.1
0	8	0.05	2.6	19.8	1.1	0.04	0	0.01	0.01	0.00	0.01	0.02	3.5	0.3	0	0
0	24	0.66	4.1	25.0	0.2	0.06	5.2	0.01	0.28	0.01	0.09	0.01	4.1	0.8	0	0.1
0	4	0.49	3.9	49.2	0.7	0.08	55.4	0.01	0.62	0.03	0.32	0.08	2.2	1.5	0	0.1
0	4	0.29	0.7	11.4	1.4	0.01	1.5	0.00	0.02	0.00	0.02	0.00	0.5	0.4	0	0.1
0	9	0.60	4.1	26.4	0.9	0.03	0.3	0.00	0.01	0.01	0.02	0.01	0.2	0.4	0	0.1
0	3	0.14	2.7	36.3	0.5	0.04	37.5	0.01	0.53	0.01	0.15	0.04	1.9	1.4	0	0.2
0	6	0.34	2.2	1.8	0.1	0.02	0	0.00	—	0.00	0.01	0.00	0.2	0.5	0	0.1
0	41	0.26	9.3	19.6	0.6	0.28	0	0.02	0.03	0.01	0.02	0.01	1.6	0.1	0	0
0	15	0.53	3.4	10.3	0.4	0.04	2.0	0.00	0.02	0.00	0.04	0.02	2.1	0.2	0	0.1
0	12	0.33	2.3	11.3	0.9	0.04	0.2	0.00	0.01	0.00	0.03	0.01	0.9	0.4	0	0.2
0	15	0.35	2.6	11.5	0.6	0.03	1.9	0.01	—	0.01	0.01	0.02	3.7	0.7	0	0.1
0	2	0.04	0.6	4.7	0.2	0.01	1.0	0.00	—	0.00	0.01	0.00	0.8	0.2	0	—
0	1	0.07	1.8	12.1	1.0	0.01	0.2	0.00	—	0.00	0.01	0.01	0.7	0.6	0	0
0	4	—	1.1	2.7	0	0.01	—	0.00	—	—	—	—	—	—	0	—
0	12	0.19	3.0	7.5	0.1	0.03	2.1	0.01	0.05	0.00	0.04	0.01	1.9	0.2	0	0
0	7	0.00	0	603.6	0.1	—	0	—	—	—	—	—	—	0	—	—
0	0	0.00	—	476.3	0.1	—	0	—	—	—	—	—	—	0	—	—
0	0	0.01	0	0.1	581.4	0.00	0	0.00	0.00	0.00	0.00	0.00	0	0	0	0
0	30	0.53	5.3	14.7	0.3	0.06	3.6	0.01	—	—	0.05	0.02	—	0.7	0	0.1
0	3	0.21	9.2	10.8	1.0	0.27	0.1	0.03	0.01	0.01	0.15	0.00	2.6	0	0	0
0	—	—	—	—	0.1	—	—	—	—	—	—	—	—	—	—	—
0	4	0.22	1.2	8.7	0.6	0.02	3.9	0.00	—	0.00	0.01	0.00	2.0	0.3	0	—
0	2	0.16	3.0	50.7	3.1	0.03	4.5	0.01	0.06	0.01	0.11	0.03	3.7	30.4	0	0.1
0	85	1.48	20.2	—	—	—	—	0.06	—	0.02	1.16	—	—	0.9	0	—
0	48	—	14.5	128.1	2.6	0.17	—	0.04	—	—	—	—	—	0.6	0	—
0	18	0.51	5.6	48.3	1.0	0.06	3.4	0.00	—	0.02	0.14	0.03	4.4	0.8	0	0.1
0	3	0.14	1.3	4.9	0.1	0.01	1.9	0.00	—	0.00	0.01	0.00	0.4	1.3	0	—
0	26	1.73	3.1	11.4	0.8	0.08	2.7	0.01	0.10	0.01	0.06	0.01	3.8	0.7	0	0.1
0	4	0.91	4.2	55.6	0.8	0.09	0	0.00	0.06	0.01	0.11	0.04	0.9	0.6	0	0.1
0	13	0.11	—	—	—	—	—	0.02	—	0.01	0.07	—	—	11.2	0	—
0	3	0.07	2.9	57.3	167.1	0.03	7.1	0.00	0.21	0.02	0.21	0.02	1.5	2.3	0	0
0	12	0.10	2.0	26.1	248.1	0.03	2.6	0.01	0.02	0.01	0.03	0.01	0.3	0.2	0	0
5	1	0.03	0.1	1.7	78.4	0.02	11.2	0.01	0.72	0.01	0.00	0.08	0.7	0	0	0.2
4	0	0.00	0	1.6	79.5	0.01	0	0.00	0.32	0.00	0.00	0.00	0	0	0	0.3
0	6	0.09	1.0	6.8	68.1	0.01	0	0.00	0.09	0.00	0.01	0.00	0.2	0.1	0	—
0	3	0.07	2.5	6.9	56.8	0.03	0.2	0.01	0.01	0.00	0.02	0.00	0.4	0.1	0	1.6
0	0	0.13	0.8	3.8	121.7	0.02	9.2	0.00	0.08	0.01	0.03	0.00	0.2	0.2	0	0
8	6	0.20	0.8	10.1	191.5	0.05	20.2	0.00	0.97	0.00	0.01	0.07	2.0	0.1	0.1	0.5
0	4	0.06	3.8	65.0	349.7	0.04	3.8	0.00	0.20	0.01	0.15	0.01	0.6	0.2	0	0.4
18	116	0.13	5.7	18.9	521.6	0.61	50.4	0.00	—	0.07	0.01	0.01	2.5	0.3	0.1	2.0
0	5	0.36	9.5	125.7	61.9	0.11	—	0.02	0.00	0.02	0.63	0.06	5.7	43.9	0	0
0	7	1.05	11.1	231.3	113.8	0.14	—	0.01	0.00	0.21	0.61	0.34	6.6	0.3	0	0.3

TABLE H-1 Table of Food Composition (*continued*) (Computer code is for Cengage Diet Analysis program) (For purposes of calculations, use "0" for t, <1, <.1, <.01, etc.)

DA+ Code	Food Description	Quantity	Measure	Wt (g)	H_2O (g)	Ener (kcal)	Prot (g)	Carb (g)	Fiber (g)	Fat (g)	Fat Breakdown (g) Sat	Mono	Poly	*Trans*
Spices, Condiments, Sauces—*continued*														
29688	Hoisin sauce	1	tablespoon(s)	16	7.1	35	0.5	7.1	0.4	0.5	0.1	0.2	0.3	—
1641	Horseradish sauce, prepared	1	teaspoon(s)	5	3.3	10	0.1	0.2	0	1.0	0.6	0.3	0	—
16670	Mole poblano sauce	½	cup(s)	133	102.7	156	5.3	11.4	2.7	11.3	2.6	5.1	3.0	—
29689	Oyster sauce	1	tablespoon(s)	16	12.8	8	0.2	1.7	0	0	0	0	0	—
1655	Pepper sauce or Tabasco	1	teaspoon(s)	5	4.8	1	0.1	0	0	0	0	0	0	—
347	Salsa	2	tablespoon(s)	32	28.8	9	0.5	2.0	0.5	0.1	0	0	0	—
52206	Soy sauce, tamari	1	tablespoon(s)	18	12.0	11	1.9	1.0	0.1	0	0	0	0	—
839	Sweet and sour sauce	2	tablespoon(s)	39	29.8	37	0.1	9.1	0.1	0	0	0	0	—
1613	Teriyaki sauce	1	tablespoon(s)	18	12.2	16	1.1	2.8	0	0	0	0	0	0
25294	Tomato sauce	½	cup(s)	150	132.8	63	2.2	11.9	2.6	1.8	0.2	0.4	0.9	0
728	White sauce, medium	¼	cup(s)	63	46.8	92	2.4	5.7	0.1	6.7	1.8	2.8	1.8	—
1654	Worcestershire sauce	1	teaspoon(s)	6	4.5	4	0	1.1	0	0	0	0	0	0
	Vinegar													
30853	Balsamic	1	tablespoon(s)	15	—	10	0	2.0	0	0	0	0	0	0
727	Cider	1	tablespoon(s)	15	14.0	3	0	0.1	0	0	0	0	0	0
1673	Distilled	1	tablespoon(s)	15	14.3	2	0	0.8	0	0	0	0	0	0
12948	Tarragon	1	tablespoon(s)	15	13.8	2	0	0.1	0	0	0	0	0	0
Mixed Foods, Soups, Sandwiches														
	Mixed dishes													
16652	Almond chicken	1	cup(s)	242	186.8	281	21.8	15.8	3.4	14.7	1.8	6.3	5.6	—
25224	Barbecued chicken	1	serving(s)	177	99.3	327	27.1	15.7	0.5	17.1	4.8	6.8	3.8	0
25227	Bean burrito	1	item(s)	149	81.8	326	16.1	33.0	5.6	14.8	8.3	4.7	0.9	—
9516	Beef and vegetable fajita	1	item(s)	223	143.9	397	22.4	35.3	3.1	18.0	5.9	8.0	2.5	—
16796	Beef or pork egg roll	2	item(s)	128	85.2	225	9.9	18.4	1.4	12.4	2.9	6.0	2.6	—
177	Beef stew with vegetables, prepared	1	cup(s)	245	201.0	220	16.0	15.0	3.2	11.0	4.4	4.5	0.5	—
30233	Beef stroganoff with noodles	1	cup(s)	256	190.1	343	19.7	22.8	1.5	19.1	7.4	5.7	4.4	—
16651	Cashew chicken	1	cup(s)	242	186.8	281	21.8	15.8	3.4	14.7	1.8	6.3	5.6	—
30274	Cheese pizza with vegetables, thin crust	2	slice(s)	140	76.6	298	12.7	35.4	2.5	12.0	4.9	4.7	1.6	—
30330	Cheese quesadilla	1	item(s)	54	18.3	190	7.7	15.3	1.0	10.8	5.2	3.6	1.3	—
215	Chicken and noodles, prepared	1	cup(s)	240	170.0	365	22.0	26.0	1.3	18.0	5.1	7.1	3.9	—
30239	Chicken and vegetables with broccoli, onion, bamboo shoots in soy based sauce	1	cup(s)	162	125.5	180	15.8	9.3	1.8	8.6	1.7	3.0	3.1	—
25093	Chicken cacciatore	1	cup(s)	244	175.7	284	29.9	5.7	1.3	15.3	4.3	6.2	3.3	0
28020	Chicken fried turkey steak	3	ounce(s)	492	276.2	706	77.1	68.7	3.6	12.0	3.4	2.9	3.9	—
218	Chicken pot pie	1	cup(s)	252	154.6	542	22.6	41.4	3.5	31.3	9.8	12.5	7.1	—
30240	Chicken teriyaki	1	cup(s)	244	158.3	364	51.0	15.2	0.7	7.0	1.8	2.0	1.7	—
25119	Chicken waldorf salad	½	cup(s)	100	67.2	179	14.0	6.8	1.0	10.8	1.8	3.1	5.2	—
25099	Chili con carne	¾	cup(s)	215	174.4	198	13.7	21.4	7.5	6.9	2.5	2.8	0.5	0
1062	Coleslaw	¾	cup(s)	90	73.4	70	1.2	11.2	1.4	2.3	0.3	0.6	1.2	—
1574	Crab cakes, from blue crab	1	item(s)	60	42.6	93	12.1	0.3	0	4.5	0.9	1.7	1.4	—
32144	Enchiladas with green chili sauce (enchiladas verdes)	1	item(s)	144	103.8	207	9.3	17.6	2.6	11.7	6.4	3.6	1.0	0
2793	Falafel patty	3	item(s)	51	17.7	170	6.8	16.2	—	9.1	1.2	5.2	2.1	—
28546	Fettuccine alfredo	1	cup(s)	244	88.7	279	13.1	46.1	1.4	4.2	2.2	1.0	0.4	0
32146	Flautas	3	item(s)	162	78.0	438	24.9	36.3	4.1	21.6	8.2	8.8	2.3	—
29629	Fried rice with meat or poultry	1	cup(s)	198	128.5	333	12.3	41.8	1.4	12.3	2.2	3.5	5.7	—
16649	General Tso chicken	1	cup(s)	146	91.0	296	18.7	16.4	0.9	17.0	4.0	6.3	5.3	—
1826	Green salad	¾	cup(s)	104	98.9	17	1.3	3.3	2.2	0.1	0	0	0	—
1814	Hummus	½	cup(s)	123	79.8	218	6.0	24.7	4.9	10.6	1.4	6.0	2.6	—
16650	Kung pao chicken	1	cup(s)	162	87.2	434	28.8	11.7	2.3	30.6	5.2	13.9	9.7	—
16622	Lamb curry	1	cup(s)	236	187.9	257	28.2	3.7	0.9	13.8	3.9	4.9	3.3	—
25253	Lasagna with ground beef	1	cup(s)	237	158.4	284	16.9	22.3	2.4	14.5	7.5	4.9	0.8	—
442	Macaroni and cheese, prepared	1	cup(s)	200	122.3	390	14.9	40.6	1.6	18.6	7.9	6.4	2.9	—
29637	Meat filled ravioli with tomato or meat sauce, canned	1	cup(s)	251	198.7	208	7.8	36.5	1.3	3.7	1.5	1.4	0.3	—
25105	Meat loaf	1	slice(s)	115	84.5	245	17.0	6.6	0.4	16.0	6.1	6.9	0.9	0
16646	Moo shi pork	1	cup(s)	151	76.8	512	18.9	5.3	0.6	46.4	6.9	15.8	21.2	—
16788	Nachos with beef, beans, cheese, tomatoes and onions	1	serving(s)	551	253.5	1576	59.1	137.5	20.4	90.8	32.6	41.9	9.4	—
6116	Pepperoni pizza	2	slice(s)	142	66.1	362	20.2	39.7	2.9	13.9	4.5	6.3	2.3	—
29601	Pizza with meat and vegetables, thin crust	2	slice(s)	158	81.4	386	16.5	36.8	2.7	19.1	7.7	8.1	2.2	—
655	Potato salad	½	cup(s)	125	95.0	179	3.4	14.0	1.6	10.3	1.8	3.1	4.7	—

PAGE KEY: H-4 = Breads/Baked Goods H-10 = Cereal/Rice/Pasta H-14 = Fruit H-18 = Vegetables/Legumes H-28 = Nuts/Seeds H-30 = Vegetarian H-32 = Dairy H-40 = Eggs H-40 = Seafood H-42 = Meats H-46 = Poultry H-46 = Processed Meats H-48 = Beverages H-52 = Fats/Oils H-54 = Sweets H-56 = Spices/Condiments/Sauces H-60 = Mixed Foods/Soups/Sandwiches H-64 = Fast Food H-84 = Convenience H-86 = Baby Foods

Chol (mg)	Calc (mg)	Iron (mg)	Magn (mg)	Pota (mg)	Sodi (mg)	Zinc (mg)	Vit A (µg)	Thia (mg)	Vit E (mg α)	Ribo (mg)	Niac (mg)	Vit B_6 (mg)	Fola (µg)	Vit C (mg)	Vit B_{12} (µg)	Sele (µg)
0	5	0.16	3.8	19.0	258.4	0.05	0	0.00	0.04	0.03	0.18	0.01	3.7	0.1	0	0.3
2	5	0.00	0.5	6.7	14.6	0.01	8.0	0.00	0.02	0.01	0.00	0.00	0.5	0.1	0	0.1
1	38	1.81	58.3	280.9	304.8	1.15	13.3	0.06	1.72	0.08	1.84	0.09	15.9	3.4	0.1	1.1
0	5	0.02	0.6	8.6	437.3	0.01	0	0.00	0.00	0.02	0.23	0.00	2.4	0	0.1	0.7
0	1	0.05	0.6	6.4	31.7	0.01	4.1	0.00	0.00	0.00	0.01	0.01	0.1	0.2	0	0
0	9	0.14	4.8	95.0	192.0	0.11	4.8	0.01	0.37	0.01	0.02	0.05	1.3	0.6	0	0.3
0	4	0.43	7.3	38.7	1018.9	0.08	0	0.01	0.00	0.03	0.72	0.04	3.3	0	0	0.1
0	5	0.20	1.2	8.2	97.5	0.01	0	0.00	—	0.01	0.11	0.03	0.2	0	0	—
0	5	0.30	11.0	40.5	689.9	0.01	0	0.01	0.00	0.01	0.22	0.01	1.4	0	0	0.2
0	23	1.24	28.9	536.8	268.6	0.36	—	0.08	0.52	0.08	1.64	0.20	23.2	32.0	0	1.0
4	74	0.20	8.8	97.5	221.3	0.25	—	0.04	—	0.11	0.25	0.02	3.1	0.5	0.2	—
0	6	0.30	0.7	45.4	55.6	0.01	0.3	0.00	0.00	0.01	0.03	0.00	0.5	0.7	0	0
0	0	0.00	—	—	0	—	0	—	—	—	—	—	—	0	—	—
0	1	0.03	0.7	10.9	0.7	0.01	0	0.00	0.00	0.00	0.00	0.00	0	0	0	0
0	1	0.09	0	2.3	0.1	0.00	0	0.00	0.00	0.00	0.00	0.00	0	0	0	5.0
0	0	0.07	—	2.3	0.7	—	—	0.07	—	0.07	0.07	—	—	0.3	0	—
41	68	1.86	58.1	539.7	510.6	1.50	31.5	0.07	4.11	0.22	9.57	0.43	26.6	5.1	0.3	13.6
120	26	1.70	32.4	419.7	500.9	2.67	—	0.09	0.01	0.24	6.87	0.40	15.0	7.9	0.3	19.5
38	333	3.01	52.5	447.6	510.6	1.98	—	0.28	0.01	0.30	1.92	0.19	122.9	8.2	0.3	15.9
45	85	3.65	37.9	475.0	756.0	3.52	17.8	0.38	0.80	0.29	5.33	0.39	69.1	23.4	2.1	28.3
74	31	1.68	20.5	248.3	547.8	0.89	25.6	0.32	1.28	0.24	2.55	0.18	38.4	4.0	0.3	17.5
71	29	2.90	—	613.0	292.0	—	—	0.15	0.51	0.17	4.70	—	—	17.0	0	15.0
74	69	3.25	35.8	391.7	816.6	3.63	69.1	0.21	1.25	0.30	3.80	0.21	48.6	1.3	1.8	27.9
41	68	1.86	58.1	539.7	510.6	1.50	31.5	0.07	4.11	0.22	9.57	0.43	26.6	5.1	0.3	13.6
17	249	2.78	28.0	294.0	739.2	1.42	47.6	0.29	1.05	0.33	2.84	0.14	61.6	15.3	0.4	18.6
23	190	1.04	13.5	75.6	469.3	0.86	58.3	0.11	0.43	0.15	0.89	0.02	21.6	2.4	0.1	9.2
103	26	2.20	—	149.0	600.0	—	—	0.05	—	0.17	4.30	—	—	0	—	29.0
42	28	1.19	22.7	299.7	620.5	1.32	81.0	0.07	1.11	0.14	5.28	0.36	16.2	22.5	0.2	12.0
109	47	1.97	40.0	489.3	492.1	2.13	—	0.11	0.00	0.20	9.81	0.57	16.1	14.0	0.3	22.6
156	423	8.79	110.3	1182.9	880.4	6.18	—	0.72	0.00	1.05	20.16	1.20	113.9	2.7	1.3	97.6
68	66	3.32	37.8	390.6	652.7	1.94	259.6	0.39	1.05	0.39	7.25	0.23	80.6	10.3	0.2	27.0
156	51	3.26	68.3	588.0	3208.6	3.75	31.7	0.15	0.58	0.36	16.68	0.88	24.4	2.0	0.5	36.1
42	20	0.82	23.9	202.5	246.5	1.13	—	0.05	0.62	0.09	4.06	0.25	15.8	2.5	0.2	10.7
27	42	2.83	50.6	636.8	864.8	2.36	—	0.15	0.01	0.22	3.18	0.19	58.1	10.3	0.6	7.3
7	41	0.53	9.0	162.9	20.7	0.18	47.7	0.06	—	0.05	0.24	0.11	24.3	29.4	0	0.6
90	63	0.64	19.8	194.4	198.0	2.45	34.2	0.05	—	0.04	1.74	0.10	31.8	1.7	3.6	24.4
27	266	1.07	38.5	251.4	276.3	1.26	—	0.07	0.02	0.16	1.27	0.17	44.6	59.3	0.2	6.0
0	28	1.74	41.8	298.4	149.9	0.76	0.5	0.07	—	0.08	0.53	0.06	47.4	0.8	0	0.5
9	218	1.83	38.4	163.9	472.9	1.24	—	0.41	0.00	0.35	2.85	0.09	96.2	1.6	0.4	38.4
73	146	2.66	61.3	222.9	885.7	3.43	0	0.10	0.10	0.16	3.00	0.26	95.7	0	1.2	36.7
103	38	2.77	33.7	196.0	833.6	1.34	41.6	0.33	1.60	0.18	4.17	0.27	97.0	3.4	0.3	22.0
66	26	1.46	23.4	248.2	849.7	1.40	29.2	0.10	1.62	0.18	6.28	0.28	23.4	12.0	0.2	19.9
0	13	0.65	11.4	178.0	26.9	0.21	59.0	0.03	—	0.05	0.56	0.08	38.3	24.0	0	0.4
0	60	1.91	35.7	212.8	297.7	1.34	0	0.10	0.92	0.06	0.49	0.49	72.6	9.7	0	3.0
65	50	1.96	63.2	427.7	907.2	1.50	38.9	0.15	4.32	0.14	13.22	0.58	42.1	7.5	0.3	23.0
90	38	2.95	40.1	493.2	495.6	6.60	—	0.08	1.29	0.28	8.03	0.21	28.3	1.4	2.9	30.4
68	233	2.22	40.1	420.1	433.6	2.70	—	0.21	0.21	0.29	3.06	0.22	50.4	15.0	0.8	21.4
34	310	2.06	40.0	258.0	784.0	2.06	180.0	0.27	0.72	0.43	2.18	0.08	64.0	0	0.5	30.6
15	35	2.10	20.1	283.6	1352.9	1.28	27.6	0.19	0.70	0.16	2.77	0.14	42.7	21.6	0.4	13.3
85	59	1.87	21.8	300.8	411.7	3.40	—	0.08	0.00	0.27	3.72	0.13	18.7	0.9	1.6	17.9
172	32	1.57	25.7	333.7	1052.5	1.82	49.8	0.49	5.39	0.36	2.88	0.31	21.1	8.0	0.8	30.0
154	948	7.32	242.4	1201.2	1862.4	10.68	259.0	0.29	7.71	0.81	6.39	1.09	148.8	16.0	2.6	44.1
28	129	1.87	17.0	305.3	533.9	1.03	105.1	0.26	—	0.46	6.09	0.11	73.8	3.3	0.4	26.1
36	258	3.14	31.6	352.3	971.7	2.02	49.0	0.37	1.13	0.37	3.77	0.19	64.8	15.6	0.6	22.8
85	24	0.81	18.8	317.5	661.3	0.38	40.0	0.09	—	0.07	1.11	0.17	8.8	12.5	0	5.1

TABLE H-1 Table of Food Composition (*continued*) (Computer code is for Cengage Diet Analysis program) (For purposes of calculations, use "0" for t, <1, <.1, <.01, etc.)

DA+ Code	Food Description	Quantity	Measure	Wt (g)	H_2O (g)	Ener (kcal)	Prot (g)	Carb (g)	Fiber (g)	Fat (g)	Fat Breakdown (g)			
											Sat	Mono	Poly	*Trans*
Mixed Foods, Soups, Sandwiches—*continued*														
25109	Salisbury steaks with mushroom sauce	1	serving(s)	135	101.8	251	17.1	9.3	0.5	15.5	6.0	6.7	0.8	0
16637	Shrimp creole with rice	1	cup(s)	243	176.6	309	27.0	27.7	1.2	9.2	1.7	3.6	2.9	—
497	Spaghetti and meatballs with tomato sauce, prepared	1	cup(s)	248	174.0	330	19.0	39.0	2.7	12.0	3.9	4.4	2.2	—
28585	Spicy thai noodles (pad thai)	8	ounce(s)	227	73.3	221	8.9	35.7	3.0	6.4	0.8	3.3	1.8	—
33073	Stir fried pork and vegetables with rice	1	cup(s)	235	173.6	348	15.4	33.5	1.9	16.3	5.6	6.9	2.6	0
28588	Stuffed shells	2½	item(s)	249	157.5	243	15.0	28.0	2.5	8.1	3.1	3.0	1.3	—
16821	Sushi with egg in seaweed	6	piece(s)	156	116.5	190	8.9	20.5	0.3	7.9	2.2	3.2	1.5	—
16819	Sushi with vegetables and fish	6	piece(s)	156	101.6	218	8.4	43.7	1.7	0.6	0.2	0.1	0.2	—
16820	Sushi with vegetables in seaweed	6	piece(s)	156	110.3	183	3.4	40.6	0.8	0.4	0.1	0.1	0.1	—
25266	Sweet and sour pork	¾	cup(s)	249	205.9	265	29.2	17.1	1.0	8.1	2.6	3.5	1.5	0
16824	Tabouli, tabbouleh or tabuli	1	cup(s)	160	123.7	198	2.6	15.9	3.7	14.9	2.0	10.9	1.6	—
25276	Three bean salad	½	cup(s)	99	82.2	95	1.9	9.7	2.6	5.9	0.8	1.4	3.5	0
160	Tuna salad	½	cup(s)	103	64.7	192	16.4	9.6	0	9.5	1.6	3.0	4.2	—
25241	Turkey and noodles	1	cup(s)	319	228.5	270	24.0	21.2	1.0	9.2	2.4	3.5	2.3	—
16794	Vegetable egg roll	2	item(s)	128	89.8	201	5.1	19.5	1.7	11.6	2.5	5.7	2.6	—
16818	Vegetable sushi, no fish	6	piece(s)	156	99.0	226	4.8	49.9	2.0	0.4	0.1	0.1	0.1	—
	Sandwiches													
1744	Bacon, lettuce and tomato with mayonnaise	1	item(s)	164	97.2	341	11.6	34.2	2.3	17.6	3.8	5.5	6.7	—
30287	Bologna and cheese with margarine	1	item(s)	111	45.6	345	13.4	29.3	1.2	19.3	8.1	7.0	2.4	—
30286	Bologna with margarine	1	item(s)	83	33.6	251	8.1	27.3	1.2	12.1	3.7	5.0	2.1	—
16546	Cheese	1	item(s)	83	31.0	261	9.1	27.6	1.2	12.7	5.4	4.2	2.1	—
8789	Cheeseburger, large, plain	1	item(s)	185	78.9	564	32.0	38.5	2.6	31.5	12.5	10.2	1.0	1.8
8624	Cheeseburger, large, with bacon, vegetables, and condiments	1	item(s)	195	91.4	550	30.8	36.8	2.5	30.9	11.9	10.6	1.3	1.5
1745	Club with bacon, chicken, tomato, lettuce, and mayonnaise	1	item(s)	246	137.5	546	31.0	48.9	3.0	24.5	5.3	7.5	9.4	—
1908	Cold cut submarine with cheese and vegetables	1	item(s)	228	131.8	456	21.8	51.0	2.0	18.6	6.8	8.2	2.3	—
30247	Corned beef	1	item(s)	130	74.9	265	18.2	25.3	1.6	9.6	3.6	3.5	1.0	—
25283	Egg salad	1	item(s)	126	72.1	278	10.7	28.0	1.4	13.5	2.9	4.2	5.0	—
16686	Fried egg	1	item(s)	96	49.7	226	10.0	26.2	1.2	8.6	2.3	3.2	1.9	—
16547	Grilled cheese	1	item(s)	83	27.5	291	9.2	27.9	1.2	15.8	6.0	5.7	3.0	—
16659	Gyro with onion and tomato	1	item(s)	105	68.3	163	12.0	20.0	1.1	3.5	1.3	1.3	0.5	—
1906	Ham and cheese	1	item(s)	146	74.2	352	20.7	33.3	2.0	15.5	6.4	6.7	1.4	—
31890	Ham with mayonnaise	1	item(s)	112	56.3	271	13.0	27.9	1.9	11.6	2.8	4.0	4.0	—
756	Hamburger, double patty, large, with condiments and vegetables	1	item(s)	226	121.5	540	34.3	40.3	—	26.6	10.5	10.3	2.8	—
8793	Hamburger, large, plain	1	item(s)	137	57.7	426	22.6	31.7	1.5	22.9	8.4	9.9	2.1	—
8795	Hamburger, large, with vegetables and condiments	1	item(s)	218	121.4	512	25.8	40.0	3.1	27.4	10.4	11.4	2.2	—
25134	Hot chicken salad	1	item(s)	98	48.4	242	15.2	23.8	1.3	9.2	2.9	2.5	3.0	—
25133	Hot turkey salad	1	item(s)	98	50.1	224	15.6	23.8	1.3	6.9	2.3	1.6	2.5	—
1411	Hotdog with bun, plain	1	item(s)	98	52.9	242	10.4	18.0	1.6	14.5	5.1	6.9	1.7	—
30249	Pastrami	1	item(s)	134	71.2	328	13.4	27.8	1.6	17.7	6.1	8.3	1.2	—
16701	Peanut butter	1	item(s)	93	23.6	345	12.2	37.6	3.3	17.4	3.4	7.7	5.2	—
30306	Peanut butter and jelly	1	item(s)	93	24.2	330	10.3	41.9	2.9	14.7	2.9	6.5	4.4	—
1909	Roast beef submarine with mayonnaise and vegetables	1	item(s)	216	127.4	410	28.6	44.3	—	13.0	7.1	1.8	2.6	—
1910	Roast beef, plain	1	item(s)	139	67.6	346	21.5	33.4	1.2	13.8	3.6	6.8	1.7	—
1907	Steak with mayonnaise and vegetables	1	item(s)	204	104.2	459	30.3	52.0	2.3	14.1	3.8	5.3	3.3	—
25288	Tuna salad	1	item(s)	179	102.2	415	24.5	28.4	1.6	22.4	3.5	6.2	11.4	—
30283	Turkey submarine with cheese, lettuce, tomato, and mayonnaise	1	item(s)	277	168.0	529	30.4	49.4	3.0	22.8	6.8	6.0	8.6	—
31891	Turkey with mayonnaise	1	item(s)	143	74.5	329	28.7	26.4	1.3	11.2	2.6	2.6	4.8	—
	Soups													
25296	Bean	1	cup(s)	301	253.1	191	13.8	29.0	6.5	2.3	0.7	0.8	0.5	0
711	Bean with pork, condensed, prepared with water	1	cup(s)	253	215.9	159	7.3	21.0	7.3	5.5	1.4	2.0	1.7	—

PAGE KEY: H-4 = Breads/Baked Goods H-10 = Cereal/Rice/Pasta H-14 = Fruit H-18 = Vegetables/Legumes H-28 = Nuts/Seeds H-30 = Vegetarian H-32 = Dairy H-40 = Eggs H-40 = Seafood H-42 = Meats H-46 = Poultry H-46 = Processed Meats H-48 = Beverages H-52 = Fats/Oils H-54 = Sweets H-56 = Spices/Condiments/Sauces H-60 = Mixed Foods/Soups/Sandwiches H-64 = Fast Food H-84 = Convenience H-86 = Baby Foods

Chol (mg)	Calc (mg)	Iron (mg)	Magn (mg)	Pota (mg)	Sodi (mg)	Zinc (mg)	Vit A (µg)	Thia (mg)	Vit E (mg α)	Ribo (mg)	Niac (mg)	Vit B_6 (mg)	Fola (µg)	Vit C (mg)	Vit B_{12} (µg)	Sele (µg)
60	74	1.94	23.8	314.5	360.5	3.45	—	0.10	0.00	0.27	3.95	0.13	20.8	0.7	1.6	17.4
180	102	4.68	63.2	413.1	330.5	1.72	94.8	0.29	2.06	0.11	4.75	0.21	75.3	12.9	1.2	49.3
89	124	3.70	—	665.0	1009.0	—	81.5	0.25	—	0.30	4.00	—	—	22.0	—	22.0
37	31	1.56	49.4	181.3	591.6	1.05	—	0.18	0.35	0.13	1.82	0.17	44.6	22.6	0.1	3.2
46	38	2.71	33.0	396.9	569.5	2.08	—	0.51	0.38	0.20	5.07	0.30	103.0	18.8	0.4	22.8
30	188	2.26	49.4	403.0	471.5	1.41	—	0.26	0.00	0.26	3.83	0.24	80.0	17.9	0.2	28.9
214	45	1.84	18.7	135.7	463.3	0.98	106.1	0.13	0.67	0.28	1.35	0.13	62.4	1.9	0.7	20.3
11	23	2.15	25.0	202.8	340.1	0.78	45.2	0.26	0.24	0.07	2.76	0.14	76.4	3.6	0.3	13.9
0	20	1.54	18.7	96.7	152.9	0.68	25.0	0.19	0.12	0.03	1.86	0.13	73.3	2.3	0	9.8
74	40	1.76	35.6	619.9	621.8	2.53	—	0.81	0.20	0.37	6.69	0.66	14.7	11.9	0.7	49.6
0	30	1.21	35.2	249.6	796.8	0.48	54.4	0.07	2.43	0.04	1.11	0.11	30.4	26.1	0	0.5
0	26	0.96	15.5	144.8	224.2	0.30	—	0.02	0.88	0.04	0.26	0.04	32.2	10.0	0	2.7
13	17	1.02	19.5	182.5	412.1	0.57	24.6	0.03	—	0.07	6.86	0.08	8.2	2.3	1.2	42.2
77	69	2.56	33.2	400.8	577.1	2.51	—	0.23	0.28	0.30	6.41	0.29	61.4	1.4	1.1	33.4
60	29	1.65	17.9	193.3	549.1	0.48	25.6	0.15	1.28	0.20	1.59	0.09	46.1	5.5	0.2	11.3
0	23	2.38	21.8	157.6	369.7	0.82	48.4	0.28	0.15	0.05	2.44	0.12	85.8	3.7	0	8.1
21	79	2.36	27.9	351.0	944.6	1.08	44.3	0.32	1.16	0.24	4.36	0.21	73.8	9.7	0.2	27.1
40	258	2.38	24.4	215.3	941.3	1.88	102.1	0.31	0.55	0.35	2.97	0.14	59.9	0.2	0.8	20.4
17	100	2.24	16.6	138.6	579.3	1.02	44.8	0.29	0.49	0.21	2.92	0.12	58.1	0.2	0.5	16.0
22	233	2.04	19.9	127.0	733.7	1.24	97.1	0.25	0.47	0.30	2.25	0.05	58.1	0	0.3	13.1
104	309	4.47	44.4	401.5	986.1	5.75	0	0.35	—	0.77	8.26	0.49	109.2	0	2.8	38.9
98	267	4.03	44.9	464.1	1314.3	5.20	0	0.33	—	0.67	8.25	0.47	95.6	1.4	2.4	6.6
71	157	4.57	46.7	464.9	1087.3	1.82	41.8	0.54	1.52	0.40	12.82	0.61	118.1	6.4	0.4	42.3
36	189	2.50	68.4	394.4	1650.7	2.57	70.7	1.00	—	0.79	5.49	0.13	86.6	12.3	1.1	30.8
46	81	3.04	19.5	127.4	1206.4	2.26	2.6	0.23	0.20	0.24	3.42	0.11	59.8	0.3	0.9	31.2
219	85	2.25	18.8	159.0	423.1	0.87	—	0.27	0.12	0.43	2.06	0.15	73.6	0.9	0.6	29.9
206	104	2.79	17.3	117.1	438.7	0.92	89.3	0.26	0.66	0.40	2.26	0.10	79.7	0	0.6	24.2
22	235	2.05	19.9	128.7	763.6	1.26	129.5	0.19	0.72	0.28	2.05	0.05	38.2	0	0.2	13.2
28	47	1.77	22.1	218.4	235.2	2.33	9.5	0.23	0.26	0.20	3.12	0.13	44.1	3.2	0.9	18.3
58	130	3.24	16.1	290.5	770.9	1.37	96.4	0.30	0.29	0.48	2.68	0.20	75.9	2.8	0.5	23.1
34	91	2.47	23.5	210.6	1097.6	1.13	5.6	0.57	0.50	0.26	3.80	0.25	60.5	2.2	0.2	20.2
122	102	5.85	49.7	569.5	791.0	5.67	0	0.36	—	0.38	7.57	0.54	76.8	1.1	4.1	25.5
71	74	3.57	27.4	267.2	474.0	4.11	0	0.28	—	0.28	6.24	0.23	60.3	0	2.1	27.1
87	96	4.92	43.6	479.6	824.0	4.88	0	0.41	—	0.37	7.28	0.32	82.8	2.6	2.4	33.6
39	115	1.88	20.0	172.4	505.1	1.19	—	0.23	0.28	0.22	4.84	0.19	46.4	0.5	0.2	19.9
37	114	1.99	21.3	189.0	494.5	1.07	—	0.22	0.28	0.20	4.26	0.22	46.4	0.5	0.2	23.4
44	24	2.31	12.7	143.1	670.3	1.98	0	0.23	—	0.27	3.64	0.04	48.0	0.1	0.5	26.0
51	80	3.02	22.8	182.2	1364.1	2.70	2.7	0.28	0.26	0.26	4.97	0.14	60.3	0.3	1.0	14.3
0	110	2.92	66.0	226.0	580.3	1.32	0	0.31	2.39	0.23	6.72	0.18	92.1	0	0	13.2
0	94	2.50	56.7	198.1	492.9	1.12	—	0.26	2.01	0.20	5.66	0.15	78.1	0.1	0	11.2
73	41	2.80	67.0	330.5	844.6	4.38	30.2	0.41	—	0.41	5.96	0.32	71.3	5.6	1.8	25.7
51	54	4.22	30.6	315.5	792.3	3.39	11.1	0.37	—	0.30	5.86	0.26	57.0	2.1	1.2	29.2
73	92	5.16	49.0	524.3	797.6	4.52	0	0.40	—	0.36	7.30	0.36	89.8	5.5	1.6	42.0
59	78	2.97	35.9	316.3	724.7	1.02	—	0.27	0.34	0.26	12.07	0.46	60.8	1.8	2.4	76.9
64	307	4.59	49.9	534.6	1759.0	2.74	74.8	0.49	1.19	0.65	3.91	0.31	121.9	10.5	0.6	42.1
67	100	3.46	34.3	304.6	564.9	3.00	5.7	0.28	0.74	0.32	6.84	0.46	64.4	0	0.3	40
5	79	3.05	61.8	588.8	689.0	1.41	—	0.27	0.02	0.15	3.63	0.23	140.1	3.6	0.2	7.9
3	78	1.89	43.0	371.9	883.0	0.96	43.0	0.08	1.08	0.03	0.52	0.03	30.4	1.5	0	7.8

TABLE H-1 Table of Food Composition (*continued*) (Computer code is for Cengage Diet Analysis program) (For purposes of calculations, use "0" for t, <1, <.1, <.01, etc.)

DA+ Code	Food Description	Quantity	Measure	Wt (g)	H_2O (g)	Ener (kcal)	Prot (g)	Carb (g)	Fiber (g)	Fat (g)	Fat Breakdown (g)			
											Sat	Mono	Poly	*Trans*
	Mixed Foods, Soups, Sandwiches—*continued*													
713	Beef noodle, condensed, prepared with water	1	cup(s)	244	224.9	83	4.7	8.7	0.7	3.0	1.1	1.2	0.5	—
825	Cheese, condensed, prepared with milk	1	cup(s)	251	206.9	231	9.5	16.2	1.0	14.6	9.1	4.1	0.5	—
826	Chicken broth, condensed, prepared with water	1	cup(s)	244	234.1	39	4.9	0.9	0	1.4	0.4	0.6	0.3	—
25297	Chicken noodle soup	1	cup(s)	286	258.4	117	10.8	10.9	0.9	2.9	0.8	1.1	0.7	—
827	Chicken noodle, condensed, prepared with water	1	cup(s)	241	226.1	60	3.1	7.1	0.5	2.3	0.6	1.0	0.6	0
724	Chicken noodle, dehydrated, prepared with water	1	cup(s)	252	237.3	58	2.1	9.2	0.3	1.4	0.3	0.5	0.4	—
823	Cream of asparagus, condensed, prepared with milk	1	cup(s)	248	213.3	161	6.3	16.4	0.7	8.2	3.3	2.1	2.2	—
824	Cream of celery, condensed, prepared with milk	1	cup(s)	248	214.4	164	5.7	14.5	0.7	9.7	3.9	2.5	2.7	—
708	Cream of chicken, condensed, prepared with milk	1	cup(s)	248	210.4	191	7.5	15.0	0.2	11.5	4.6	4.5	1.6	—
715	Cream of chicken, condensed, prepared with water	1	cup(s)	244	221.1	117	3.4	9.3	0.2	7.4	2.1	3.3	1.5	—
709	Cream of mushroom, condensed, prepared with milk	1	cup(s)	248	215.0	166	6.2	14.0	0	9.6	3.3	2.0	1.8	0.1
716	Cream of mushroom, condensed, prepared with water	1	cup(s)	244	224.6	102	1.9	8.0	0	7.0	1.6	1.3	1.7	—
25298	Cream of vegetable	1	cup(s)	285	250.7	165	7.2	15.2	1.9	8.6	1.6	4.6	1.9	—
16689	Egg drop	1	cup(s)	244	228.9	73	7.5	1.1	0	3.8	1.1	1.5	0.6	—
25138	Golden squash	1	cup(s)	258	223.9	145	7.6	20.4	0.4	4.1	0.8	2.2	0.9	—
16663	Hot and sour	1	cup(s)	244	209.7	161	15.0	5.4	0.5	7.9	2.7	3.4	1.1	—
28054	Lentil chowder	1	cup(s)	244	202.8	153	11.4	27.7	12.6	0.5	0.1	0.1	0.2	0
28560	Macaroni and bean	1	cup(s)	246	138.8	146	5.8	22.9	5.1	3.7	0.5	2.2	0.6	0
714	Manhattan clam chowder, condensed, prepared with water	1	cup(s)	244	225.1	73	2.1	11.6	1.5	2.1	0.4	0.4	1.2	—
28561	Minestrone	1	cup(s)	241	185.4	103	4.5	16.8	4.8	2.3	0.3	1.4	0.4	0
717	Minestrone, condensed, prepared with water	1	cup(s)	241	220.1	82	4.3	11.2	1.0	2.5	0.6	0.7	1.1	—
28038	Mushroom and wild rice	1	cup(s)	244	199.7	86	4.7	13.2	1.7	0.3	0	0	0.2	0
828	New England clam chowder, condensed, prepared with milk	1	cup(s)	248	212.2	151	8.0	18.4	0.7	5.0	2.1	0.7	0.6	0
28036	New England style clam chowder	1	cup(s)	244	227.5	61	3.8	8.8	1.8	0.2	0.1	0	0	0
28566	Old country pasta	1	cup(s)	252	183.3	146	6.5	18.3	3.6	4.5	2.0	2.4	0.9	0
725	Onion, dehydrated, prepared with water	1	cup(s)	246	235.7	30	0.8	6.8	0.7	0	0	0	0	—
16667	Shrimp gumbo	1	cup(s)	244	207.2	166	9.5	18.2	2.4	6.7	1.3	2.9	2.0	—
28037	Southwestern corn chowder	1	cup(s)	244	217.8	98	4.9	17.0	2.4	0.5	0.1	0.1	0.2	0
30282	Soybean (miso)	1	cup(s)	240	218.6	84	6.0	8.0	1.9	3.4	0.6	1.1	1.4	—
25140	Split pea	1	cup(s)	165	119.7	72	4.5	16.0	1.6	0.3	0.1	0	0.2	0
718	Split pea with ham, condensed, prepared with water	1	cup(s)	253	206.9	190	10.3	28.0	2.3	4.4	1.8	1.8	0.6	—
726	Tomato vegetable, dehydrated, prepared with water	1	cup(s)	253	238.4	56	2.0	10.2	0.8	0.9	0.4	0.3	0.1	0
710	Tomato, condensed, prepared with milk	1	cup(s)	248	213.4	136	6.2	22.0	1.5	3.2	1.8	0.9	0.3	—
719	Tomato, condensed, prepared with water	1	cup(s)	244	223.0	73	1.9	16.0	1.5	0.7	0.2	0.2	0.2	—
28595	Turkey noodle	1	cup(s)	244	216.9	114	8.1	15.1	1.9	2.4	0.3	1.1	0.7	—
28051	Turkey vegetable	1	cup(s)	244	220.8	96	12.2	6.6	2.0	1.1	0.3	0.2	0.3	0
25141	Vegetable	1	cup(s)	252	228.1	82	5.2	16.5	4.5	0.3	0	0	0.1	0
720	Vegetable beef, condensed, prepared with water	1	cup(s)	244	224.0	76	5.4	9.9	2.0	1.9	0.8	0.8	0.1	—
28598	Vegetable gumbo	1	cup(s)	252	184.5	170	4.4	28.9	3.6	4.7	0.7	3.2	0.5	0
721	Vegetarian vegetable, condensed, prepared with water	1	cup(s)	241	222.7	67	2.1	11.8	0.7	1.9	0.3	0.8	0.7	—
	Fast Food													
	Arby's													
36094	Au jus sauce	1	serving(s)	85	—	43	1.0	7.0	0	1.3	0.4	—	—	0.4
751	Beef 'n cheddar sandwich	1	item(s)	195	—	445	22.0	44.0	2.0	21.0	6.0	—	—	1.0
9279	Cheddar curly fries	1	serving(s)	198	—	631	8.0	73.0	7.0	37.4	6.8	—	—	5.7

PAGE KEY: H-4 = Breads/Baked Goods H-10 = Cereal/Rice/Pasta H-14 = Fruit H-18 = Vegetables/Legumes H-28 = Nuts/Seeds H-30 = Vegetarian H-32 = Dairy H-40 = Eggs H-40 = Seafood H-42 = Meats H-46 = Poultry H-46 = Processed Meats H-48 = Beverages H-52 = Fats/Oils H-54 = Sweets H-56 = Spices/Condiments/Sauces H-60 = Mixed Foods/Soups/Sandwiches H-64 = Fast Food H-84 = Convenience H-86 = Baby Foods

Chol (mg)	Calc (mg)	Iron (mg)	Magn (mg)	Pota (mg)	Sodi (mg)	Zinc (mg)	Vit A (µg)	Thia (mg)	Vit E (mg α)	Ribo (mg)	Niac (mg)	Vit B_6 (mg)	Fola (µg)	Vit C (mg)	Vit B_{12} (µg)	Sele (µg)
5	20	1.07	7.3	97.6	929.6	1.51	12.2	0.06	1.22	0.05	1.03	0.03	19.5	0.5	0.2	7.3
48	289	0.80	20.1	341.4	1019.1	0.67	358.9	0.06	—	0.33	0.50	0.07	10.0	1.3	0.4	7.0
0	10	0.51	2.4	209.8	775.9	0.24	0	0.01	0.04	0.07	3.34	0.02	4.9	0	0.2	0
24	25	1.38	16.4	340.1	774.5	0.77	—	0.15	0.02	0.16	5.57	0.13	37.2	1.8	0.3	10.2
12	14	1.59	9.6	53.0	638.7	0.38	26.5	0.13	0.07	0.10	1.30	0.04	19.3	0	0	11.6
10	5	0.50	7.6	32.8	577.1	0.20	2.5	0.20	0.12	0.07	1.08	0.02	17.6	0	0.1	9.6
22	174	0.86	19.8	359.6	1041.6	0.91	62.0	0.10	—	0.27	0.88	0.06	29.8	4.0	0.5	8.0
32	186	0.69	22.3	310.0	1009.4	0.19	114.1	0.07	—	0.24	0.43	0.06	7.4	1.5	0.5	4.7
27	181	0.67	17.4	272.8	1046.6	0.67	178.6	0.07	—	0.25	0.92	0.06	7.4	1.2	0.5	8.0
10	34	0.61	2.4	87.8	985.8	0.63	163.5	0.02	—	0.06	0.82	0.01	2.4	0.2	0.1	7.0
10	164	1.36	19.8	267.8	823.4	0.79	81.8	0.10	1.01	0.29	0.62	0.05	7.4	0.2	0.6	6.0
0	17	1.31	4.9	73.2	775.9	0.24	9.8	0.05	0.97	0.05	0.50	0.00	2.4	0	0	2.9
1	80	1.20	17.5	340.7	787.9	0.56	—	0.12	1.05	0.18	3.32	0.12	39.5	10.7	0.3	4.3
102	22	0.75	4.9	219.6	729.6	0.48	41.5	0.02	0.29	0.19	3.02	0.05	14.6	0	0.5	7.6
4	262	0.78	42.4	542.2	515.6	0.88	—	0.16	0.52	0.30	1.14	0.16	32.7	12.5	0.7	6.0
34	29	1.24	19.5	373.3	1561.6	1.43	—	0.26	0.12	0.24	4.96	0.20	14.6	0.5	0.4	19.3
0	48	4.38	59.3	626.2	26.7	1.57	—	0.24	0.06	0.12	1.87	0.32	176.3	16.1	0	3.5
0	59	1.90	32.4	275.9	531.0	0.51	—	0.16	0.37	0.12	1.44	0.10	58.2	9.1	0	8.8
2	27	1.56	9.8	180.6	551.4	0.87	48.8	0.02	1.22	0.03	0.77	0.09	9.8	3.9	3.9	9.0
0	62	1.70	29.9	287.5	442.7	0.42	—	0.09	0.23	0.09	0.70	0.07	47.3	13.3	0	3.5
2	34	0.91	7.2	313.3	911.0	0.74	118.1	0.05	—	0.04	0.94	0.09	36.2	1.2	0	8.0
0	30	1.38	27.3	376.9	283.8	1.00	—	0.06	0.07	0.23	3.21	0.14	18.1	3.7	0.1	4.7
17	169	3.00	29.8	456.3	887.8	0.99	91.8	0.20	0.54	0.43	1.96	0.17	22.3	5.2	11.9	10.9
3	89	1.30	29.2	503.7	256.8	0.52	—	0.06	0.02	0.10	1.30	0.15	24.2	11.8	3.0	3.8
5	57	2.43	51.9	500.4	355.7	0.78	—	0.21	0.01	0.14	2.64	0.20	80.0	22.0	0.1	10.3
0	22	0.12	9.8	76.3	851.2	0.12	0	0.03	0.02	0.03	0.15	0.06	0	0.2	0	0.5
51	105	2.85	48.8	461.2	441.6	0.90	80.5	0.18	1.90	0.12	2.52	0.19	85.4	17.6	0.3	14.9
1	83	1.03	26.3	434.4	217.4	0.57	—	0.08	0.09	0.13	1.81	0.21	32.1	39.6	0.2	1.7
0	65	1.87	36.0	362.4	988.8	0.86	232.8	0.06	0.96	0.16	2.61	0.15	57.6	4.6	0.2	1.0
0	28	1.26	29.8	328.1	602.4	0.52	—	0.10	0.00	0.07	1.50	0.16	49.5	8.1	0	0.7
8	23	2.27	48.1	399.7	1006.9	1.31	22.8	0.14	—	0.07	1.47	0.06	2.5	1.5	0.3	8.0
0	20	0.60	10.1	169.5	334.0	0.20	10.1	0.06	0.43	0.09	1.26	0.06	12.7	3.0	0.1	2.0
10	166	1.36	29.8	466.2	711.8	0.84	94.2	0.09	0.44	0.31	1.35	0.15	5.0	15.6	0.6	9.2
0	20	1.31	17.1	273.3	663.7	0.29	24.4	0.04	0.41	0.07	1.23	0.10	0	15.4	0	6.1
26	28	1.40	24.1	223.8	395.9	0.75	—	0.21	0.01	0.12	2.85	0.15	45.0	7.1	0.1	13.7
21	38	1.48	25.0	423.4	348.7	0.99	—	0.09	0.01	0.09	3.64	0.27	23.9	10.6	0.2	10.3
0	40	2.08	39.1	681.5	670.3	0.67	—	0.16	0.00	0.09	2.70	0.26	36.3	22.4	0	2.2
5	20	1.09	7.3	168.4	773.5	1.51	190.3	0.03	0.58	0.04	1.00	0.07	9.8	2.4	0.3	2.7
0	56	1.85	38.9	360.2	518.4	0.64	—	0.18	0.64	0.08	1.77	0.17	57.6	21.9	0	4.1
0	24	1.06	7.2	207.3	814.6	0.45	171.1	0.05	1.39	0.04	0.90	0.05	9.6	1.4	0	4.3
0	0	—	—	—	1510.0	—	—	—	—	—	—	—	—	—	—	—
51	80	3.96	—	—	1274.0	—	—	—	—	—	—	—	—	1.8	—	—
0	80	3.24	—	—	1476.0	—	—	—	—	—	—	—	—	9.6	—	—

TABLE H-1 Table of Food Composition (*continued*) (Computer code is for Cengage Diet Analysis program) (For purposes of calculations, use "0" for t, <1, <.1, <.01, etc

DA+ Code	Food Description	Quantity	Measure	Wt (g)	H_2O (g)	Ener (kcal)	Prot (g)	Carb (g)	Fiber (g)	Fat (g)	Fat Breakdown (g) Sat	Mono	Poly	*Trans*
Fast Food—*continued*														
34770	Chicken breast fillet sandwich, grilled	1	item(s)	233	—	414	32.0	36.0	3.0	17.0	3.0	—	—	0
36131	Chocolate shake, regular	1	serving(s)	397	—	507	13.0	83.0	0	13.0	8.0	—	—	0
36045	Curly fries, large size	1	serving(s)	198	—	631	8.0	73.0	7.0	37.0	7.0	—	—	6.0
36044	Curly fries, medium size	1	serving(s)	128	—	406	5.0	47.0	5.0	24.0	4.0	—	—	4.0
752	Ham 'n cheese sandwich	1	item(s)	167	—	304	23.0	35.0	1.0	7.0	2.0	—	—	0
36048	Homestyle fries, large size	1	serving(s)	213	—	566	6.0	82.0	6.0	37.0	7.0	—	—	5.0
36047	Homestyle fries, medium size	1	serving(s)	142	—	377	4.0	55.0	4.0	25.0	4.0	—	—	4.0
33465	Homestyle fries, small size	1	serving(s)	113	—	302	3.0	44.0	3.0	20.0	4.0	—	—	3.0
9249	Junior roast beef sandwich	1	item(s)	125	—	272	16.0	34.0	2.0	10.0	4.0	—	—	0
9251	Large roast beef sandwich	1	item(s)	281	—	547	42.0	41.0	3.0	28.0	12.0	—	—	2.0
39640	Market Fresh chicken salad with pecans sandwich	1	item(s)	322	—	769	30.0	79.0	9.0	39.0	10.0	—	—	0
39641	Market Fresh Martha's Vineyard salad, without dressing	1	serving(s)	330	—	277	26.0	24.0	5.0	8.0	4.0	—	—	0
34769	Market Fresh roast turkey and Swiss sandwich	1	serving(s)	359	—	725	45.0	75.0	5.0	30.0	8.0	—	—	1.0
9267	Market Fresh roast turkey ranch and bacon sandwich	1	serving(s)	382	—	834	49.0	75.0	5.0	38.0	11.0	—	—	1.0
39642	Market Fresh Santa Fe salad, without dressing	1	serving(s)	372	—	499	30.0	42.0	7.0	23.0	8.0	—	—	2.0
39650	Market Fresh Southwest chicken wrap	1	serving(s)	251	—	567	36.0	42.0	4.0	29.0	9.0	—	—	1.0
37021	Market Fresh Ultimate BLT sandwich	1	item(s)	294	—	779	23.0	75.0	6.0	45.0	11.0	—	—	1.0
750	Roast beef sandwich, regular	1	item(s)	154	—	320	21.0	34.0	2.0	14.0	5.0	—	—	1.0
36132	Strawberry shake, regular	1	serving(s)	397	—	498	13.0	81.0	0	13.0	8.0	—	—	0
2009	Super roast beef sandwich	1	item(s)	198	—	398	21.0	40.0	2.0	19.0	6.0	—	—	1.0
36130	Vanilla shake, regular	1	serving(s)	369	—	437	13.0	66.0	0	13.0	8.0	—	—	0
	Auntie Anne's													
35371	Cheese dipping sauce	1	serving(s)	35	—	100	3.0	4.0	0	8.0	4.0	—	—	0
35353	Cinnamon sugar soft pretzel	1	item(s)	120	—	350	9.0	74.0	2.0	2.0	0	—	—	0
35354	Cinnamon sugar soft pretzel with butter	1	item(s)	120	—	450	8.0	83.0	3.0	9.0	5.0	—	—	0
35372	Marinara dipping sauce	1	serving(s)	35	—	10	0	4.0	0	0	0	0	0	0
35357	Original soft pretzel	1	serving(s)	120	—	340	10.0	72.0	3.0	1.0	0	—	—	0
35358	Original soft pretzel with butter	1	item(s)	120	—	370	10.0	72.0	3.0	4.0	2.0	—	—	0
35359	Parmesan herb soft pretzel	1	item(s)	120	—	390	11.0	74.0	4.0	5.0	2.5	—	—	—
35360	Parmesan herb soft pretzel with butter	1	item(s)	120	—	440	10.0	72.0	9.0	13.0	7.0	—	—	—
35361	Sesame soft pretzel	1	item(s)	120	—	350	11.0	63.0	3.0	6.0	1.0	—	—	0
35362	Sesame soft pretzel with butter	1	item(s)	120	—	410	12.0	64.0	7.0	12.0	4.0	—	—	0
35364	Sour cream and onion soft pretzel	1	item(s)	120	—	310	9.0	66.0	2.0	1.0	0	—	—	0
35366	Sour cream and onion soft pretzel with butter	1	item(s)	120	—	340	9.0	66.0	2.0	5.0	3.0	—	—	0
35373	Sweet mustard dipping sauce	1	serving(s)	35	—	60	0.5	8.0	0	1.5	1.0	—	—	0
35367	Whole wheat soft pretzel	1	item(s)	120	—	350	11.0	72.0	7.0	1.5	0	—	—	0
35368	Whole wheat soft pretzel with butter	1	item(s)	120	—	370	11.0	72.0	7.0	4.5	1.5	—	—	0
	Boston Market													
34978	Butternut squash	¾	cup(s)	143	—	140	2.0	25.0	2.0	4.5	3.0	—	—	0
35006	Caesar side salad	1	serving(s)	71	—	40	3.0	3.0	1.0	20.0	2.0	—	—	1.5
35013	Chicken Carver sandwich with cheese and sauce	1	item(s)	321	—	700	44.0	68.0	3.0	29.0	7.0	—	—	0
34979	Chicken gravy	4	ounce(s)	113	—	15	1.0	4.0	0	0.5	0	—	—	0
35053	Chicken noodle soup	¾	cup(s)	283	—	180	13.0	16.0	1.0	7.0	2.0	—	—	0
34973	Chicken pot pie	1	item(s)	425	—	800	29.0	59.0	4.0	49.0	18.0	—	—	7.0
35054	Chicken tortilla soup with toppings	¾	cup(s)	227	—	340	12.0	24.0	1.0	22.0	7.0	—	—	0
35007	Cole slaw	¾	cup(s)	125	—	170	2.0	21.0	2.0	9.0	2.0	—	—	0
35057	Cornbread	1	item(s)	45	—	130	1.0	21.0	0	3.5	1.0	—	—	1.0
34980	Creamed spinach	¾	cup(s)	191	—	280	9.0	12.0	4.0	23.0	15.0	—	—	0
34998	Fresh vegetable stuffing	1	cup(s)	136	—	190	3.0	25.0	2.0	8.0	1.0	—	—	0
34991	Garlic dill new potatoes	¾	cup(s)	156	—	140	3.0	24.0	3.0	3.0	1.0	—	—	0
34983	Green bean casserole	¾	cup(s)	170	—	60	2.0	9.0	2.0	2.0	1.0	—	—	0
34982	Green beans	¾	cup(s)	91	—	60	2.0	7.0	3.0	3.5	1.5	—	—	0

PAGE KEY: H-4 = Breads/Baked Goods H-10 = Cereal/Rice/Pasta H-14 = Fruit H-18 = Vegetables/Legumes H-28 = Nuts/Seeds H-30 = Vegetarian H-32 = Dairy H-40 = Eggs H-40 = Seafood H-42 = Meats H-46 = Poultry H-46 = Processed Meats H-48 = Beverages H-52 = Fats/Oils H-54 = Sweets H-56 = Spices/Condiments/Sauces H-60 = Mixed Foods/Soups/Sandwiches H-64 = Fast Food H-84 = Convenience H-86 = Baby Foods

Chol (mg)	Calc (mg)	Iron (mg)	Magn (mg)	Pota (mg)	Sodi (mg)	Zinc (mg)	Vit A (µg)	Thia (mg)	Vit E (mg α)	Ribo (mg)	Niac (mg)	Vit B_6 (mg)	Fola (µg)	Vit C (mg)	Vit B_{12} (µg)	Sele (µg)
9	90	3.06	—	—	913.0	—	—	—	—	—	—	—	—	10.8	—	—
34	510	0.54	—	—	357.0	—	—	—	—	—	—	—	—	5.4	—	—
0	80	3.24	—	—	1476.0	—	—	—	—	—	—	—	—	9.6	—	—
0	50	1.98	—	—	949.0	—	—	—	—	—	—	—	—	6.0	—	—
35	160	2.70	—	—	1420.0	—	—	—	—	—	—	—	—	1.2	—	—
0	50	1.62	—	—	1029.0	—	—	—	—	—	—	—	—	12.6	—	—
0	30	1.08	—	—	686.0	—	—	—	—	—	—	—	—	8.4	—	—
0	30	0.90	—	—	549.0	—	—	—	—	—	—	—	—	6.6	—	—
29	60	3.06	—	—	740.0	—	0	—	—	—	—	—	—	0	—	—
102	70	6.30	—	—	1869.0	—	0	—	—	—	—	—	—	0.6	—	—
74	180	4.32	—	—	1240.0	—	—	—	—	—	—	—	—	30.0	—	—
72	200	1.62	—	—	454.0	—	—	—	—	—	—	—	—	33.6	—	—
91	360	5.22	—	—	1788.0	—	—	—	—	—	—	—	—	10.2	—	—
109	330	5.40	—	—	2258.0	—	—	—	—	—	—	—	—	11.4	—	—
59	420	3.60	—	—	1231.0	—	—	—	—	—	—	—	—	36.6	—	—
88	240	4.50	—	—	1451.0	—	—	—	—	—	—	—	—	7.8	—	—
51	170	4.68	—	—	1571.0	—	—	—	—	—	—	—	—	16.8	—	—
44	60	3.60	—	—	953.0	—	0	—	—	—	—	—	—	0	—	—
34	510	0.72	—	—	363.0	—	—	—	—	—	—	—	—	6.6	—	—
44	70	3.78	—	—	1060.0	—	—	—	—	—	—	—	—	6.0	—	—
34	510	0.36	—	—	350.0	—	—	—	—	—	—	—	—	5.4	—	—
10	100	0.00	—	—	510.0	—	—	—	—	—	—	—	—	0	—	—
0	20	1.98	—	—	410.0	—	0	—	—	—	—	—	—	0	—	—
25	30	2.34	—	—	430.0	—	—	—	—	—	—	—	—	0	—	—
0	0	0.00	—	—	180.0	—	0	—	—	—	—	—	—	0	—	—
0	30	2.34	—	—	900.0	—	0	—	—	—	—	—	—	0	—	—
10	30	2.16	—	—	930.0	—	—	—	—	—	—	—	—	0	—	—
10	80	1.80	—	—	780.0	—	—	—	—	—	—	—	—	1.2	—	—
30	60	1.80	—	—	660.0	—	—	—	—	—	—	—	—	1.2	—	—
0	20	2.88	—	—	840.0	—	0	—	—	—	—	—	—	0	—	—
15	20	2.70	—	—	860.0	—	—	—	—	—	—	—	—	0	—	—
0	30	1.98	—	—	920.0	—	—	—	—	—	—	—	—	0	—	—
10	40	2.16	—	—	930.0	—	—	—	—	—	—	—	—	0	—	—
40	0	0.00	—	—	120.0	—	0	—	—	—	—	—	—	0	—	—
0	30	1.98	—	—	1100.0	—	0	—	—	—	—	—	—	0	—	—
10	30	2.34	—	—	1120.0	—	—	—	—	—	—	—	—	0	—	—
10	59	0.80	—	—	35.0	—	—	—	—	—	—	—	—	22.2	—	—
0	60	0.43	—	—	75.0	—	—	—	—	—	—	—	—	5.4	—	—
90	211	2.85	—	—	1560.0	—	—	—	—	—	—	—	—	15.8	—	—
0	0	0.00	—	—	570.0	—	0	—	—	—	—	—	—	0	—	—
55	0	1.07	—	—	220.0	—	—	—	—	—	—	—	—	1.8	—	—
115	40	4.50	—	—	800.0	—	—	—	—	—	—	—	—	1.2	—	—
45	123	1.32	—	—	1310.0	—	—	—	—	—	—	—	—	18.4	—	—
10	41	0.48	—	—	270.0	—	—	—	—	—	—	—	—	24.5	—	—
5	0	0.71	—	—	220.0	—	0	—	—	—	—	—	—	0	—	—
70	264	2.84	—	—	580.0	—	—	—	—	—	—	—	—	9.5	—	—
0	41	1.48	—	—	580.0	—	—	—	—	—	—	—	—	2.5	—	—
0	0	0.85	—	—	120.0	—	0	—	—	—	—	—	—	14.3	—	—
5	20	0.72	—	—	620.0	—	—	—	—	—	—	—	—	2.4	—	—
0	43	0.38	—	—	180.0	—	—	—	—	—	—	—	—	5.1	—	—

TABLE H-1 Table of Food Composition (*continued*) (Computer code is for Cengage Diet Analysis program) (For purposes of calculations, use "0" for t, <1, <.1, <.01, etc

DA+ Code	Food Description	Quantity	Measure	Wt (g)	H_2O (g)	Ener (kcal)	Prot (g)	Carb (g)	Fiber (g)	Fat (g)	Fat Breakdown (g)			
											Sat	Mono	Poly	*Trans*
Fast Food—*continued*														
34984	Homestyle mashed potatoes	¾	cup(s)	221	—	210	4.0	29.0	3.0	9.0	6.0	—	—	0
34985	Homestyle mashed potatoes and gravy	1	cup(s)	334	—	225	5.0	33.0	3.0	9.5	6.0	—	—	0
34988	Hot cinnamon apples	¾	cup(s)	145	—	210	0	47.0	3.0	3.0	0	—	—	0
34989	Macaroni and cheese	¾	cup(s)	221	—	330	14.0	39.0	1.0	12.0	7.0	—	—	0.5
51193	Market chopped salad with dressing	1	item(s)	563	—	580	11.0	31.0	9.0	48.0	9.0	—	—	1.0
34970	Meatloaf	1	serving(s)	218	—	480	29.0	23.0	2.0	33.0	13.0	—	—	0
39383	Nestle Toll House chocolate chip cookie	1	item(s)	78	—	370	4.0	49.0	2.0	19.0	9.0	—	—	0
34965	Quarter chicken, dark meat, no skin	1	item(s)	134	—	260	30.0	2.0	0	13.0	4.0	—	—	0
34966	Quarter chicken, dark meat, with skin	1	item(s)	149	—	280	31.0	3.0	0	15.0	4.5	—	—	0
34963	Quarter chicken, white meat, no skin or wing	1	item(s)	173	—	250	41.0	4.0	0	8.0	2.5	—	—	0
34964	Quarter chicken, white meat, with skin and wing	1	item(s)	110	—	330	50.0	3.0	0	12.0	4.0	—	—	0
34968	Roasted turkey breast	5	ounce(s)	142	—	180	38.0	0	0	3.0	1.0	—	—	0
35011	Seasonal fresh fruit salad	1	serving(s)	142	—	60	1.0	15.0	1.0	0	0	0	0	0
51192	Spinach with garlic butter sauce	1	serving(s)	170	—	130	5.0	9.0	5.0	9.0	6.0	—	—	0
34969	Spiral sliced holiday ham	8	ounce(s)	227	—	450	40.0	13.0	0	26.0	10.0	—	—	0
35003	Steamed vegetables	1	cup(s)	136	—	50	2.0	8.0	3.0	2.0	0	—	—	0
35005	Sweet corn	¾	cup(s)	176	—	170	6.0	37.0	2.0	4.0	1.0	—	—	0
35004	Sweet potato casserole	¾	cup(s)	198	—	460	4.0	77.0	3.0	17.0	6.0	—	—	0
	Burger King													
29731	Biscuit with sausage, egg, and cheese	1	item(s)	191	—	610	20.0	33.0	1.0	45.0	15.0	—	—	1.0
14249	Cheeseburger	1	item(s)	133	—	330	17.0	31.0	1.0	16.0	7.0	—	—	0.5
14251	Chicken sandwich	1	item(s)	219	—	660	24.0	52.0	4.0	40.0	8.0	—	—	2.5
3808	Chicken Tenders, 8 pieces	1	serving(s)	123	—	340	19.0	21.0	0.5	20.0	5.0	—	—	3.0
14259	Chocolate shake, small	1	item(s)	315	—	470	8.0	75.0	1.0	14.0	9.0	—	—	0
29732	Croissanwich with sausage and cheese	1	item(s)	106	37.2	370	14.0	23.0	0.5	25.0	9.0	12.7	3.3	2.0
14261	Croissanwich with sausage, egg, and cheese	1	item(s)	159	71.4	470	19.0	26.0	0.5	32.0	11.0	15.8	6.1	2.5
3809	Double cheeseburger	1	item(s)	189	—	500	30.0	31.0	1.0	29.0	14.0	—	—	1.5
14244	Double Whopper sandwich	1	item(s)	373	—	900	47.0	51.0	3.0	57.0	19.0	—	—	2.0
14245	Double Whopper with cheese sandwich	1	item(s)	398	—	990	52.0	52.0	3.0	64.0	24.0	—	—	2.5
14250	Fish Filet sandwich	1	item(s)	250	—	630	24.0	67.0	4.0	30.0	6.0	—	—	2.5
14255	French fries, medium, salted	1	serving(s)	116	—	360	4.0	41.0	4.0	20.0	4.5	—	—	4.5
14262	French toast sticks, 5 pieces	1	serving(s)	112	37.6	390	6.0	46.0	2.0	20.0	4.5	10.6	2.9	4.5
14248	Hamburger	1	item(s)	121	—	290	15.0	30.0	1.0	12.0	4.5	—	—	0
14263	Hash brown rounds, small	1	serving(s)	75	27.1	230	2.0	23.0	2.0	15.0	4.0	—	—	5.0
14256	Onion rings, medium	1	serving(s)	91	—	320	4.0	40.0	3.0	16.0	4.0	—	—	3.5
39000	Tendercrisp chicken sandwich	1	item(s)	286	—	780	25.0	73.0	4.0	43.0	8.0	—	—	4.0
37514	TenderGrill chicken sandwich	1	item(s)	258	—	450	37.0	53.0	4.0	10.0	2.0	—	—	0
14258	Vanilla shake, small	1	item(s)	296	—	400	8.0	57.0	0	15.0	9.0	—	—	0
1736	Whopper sandwich	1	item(s)	290	—	670	28.0	51.0	3.0	39.0	11.0	—	—	1.5
14243	Whopper with cheese sandwich	1	item(s)	315	—	760	33.0	52.0	3.0	47.0	16.0	—	—	1.5
	Carl's Jr													
33962	Carl's bacon Swiss crispy chicken sandwich	1	item(s)	268	—	750	31.0	91.0	—	28.0	28.0	—	—	—
10801	Carl's Catch fish sandwich	1	item(s)	215	—	560	19.0	58.0	2.0	27.0	7.0	—	1.9	—
10862	Carl's Famous Star hamburger	1	item(s)	254	—	590	24.0	50.0	3.0	32.0	9.0	—	—	—
10785	Charbroiled chicken club sandwich	1	item(s)	270	—	550	42.0	43.0	4.0	23.0	7.0	—	2.9	—
10866	Charbroiled chicken salad	1	item(s)	437	—	330	34.0	17.0	5.0	7.0	4.0	—	1.0	—
10855	Charbroiled Santa Fe chicken sandwich	1	item(s)	266	—	610	38.0	43.0	4.0	32.0	8.0	—	—	—
10790	Chicken stars, 6 pieces	1	serving(s)	85	—	260	13.0	14.0	1.0	16.0	4.0	—	1.6	—
34864	Chocolate shake, small	1	serving(s)	595	—	540	15.0	98.0	0	11.0	7.0	—	—	—
10797	Crisscut fries	1	serving(s)	139	—	410	5.0	43.0	4.0	24.0	5.0	—	—	—
10799	Double Western Bacon cheeseburger	1	item(s)	308	—	920	51.0	65.0	2.0	50	21.0	—	6.6	—

PAGE KEY: H-4 = Breads/Baked Goods H-10 = Cereal/Rice/Pasta H-14 = Fruit H-18 = Vegetables/Legumes H-28 = Nuts/Seeds H-30 = Vegetarian H-32 = Dairy H-40 = Eggs H-40 = Seafood H-42 = Meats H-46 = Poultry H-46 = Processed Meats H-48 = Beverages H-52 = Fats/Oils H-54 = Sweets H-56 = Spices/Condiments/Sauces H-60 = Mixed Foods/Soups/Sandwiches H-64 = Fast Food H-84 = Convenience H-86 = Baby Foods

Chol (mg)	Calc (mg)	Iron (mg)	Magn (mg)	Pota (mg)	Sodi (mg)	Zinc (mg)	Vit A (µg)	Thia (mg)	Vit E (mg α)	Ribo (mg)	Niac (mg)	Vit B_6 (mg)	Fola (µg)	Vit C (mg)	Vit B_{12} (µg)	Sele (µg)
25	51	0.46	—	—	660.0	—	—	—	—	—	—	—	—	19.2	—	—
25	100	0.59	—	—	1230.0	—	—	—	—	—	—	—	—	24.9	—	—
0	16	0.28	—	—	15.0	—	—	—	—	—	—	—	—	0	—	—
30	345	1.65	—	—	1290.0	—	—	—	—	—	—	—	—	0	—	—
10	—	—	—	—	2010.0	—	—	—	—	—	—	—	—	—	—	—
125	140	3.77	—	—	970.0	—	—	—	—	—	—	—	—	1.8	—	—
20	0	1.32	—	—	340.0	—	—	—	—	—	—	—	—	0	—	—
155	0	1.52	—	—	260.0	—	0	—	—	—	—	—	—	0	—	—
155	0	2.14	—	—	660.0	—	0	—	—	—	—	—	—	0	—	—
125	0	0.89	—	—	480.0	—	0	—	—	—	—	—	—	0	—	—
165	0	0.78	—	—	960.0	—	0	—	—	—	—	—	—	0	—	—
70	20	1.80	—	—	620.0	—	0	—	—	—	—	—	—	0	—	—
0	16	0.29	—	—	20.0	—	—	—	—	—	—	—	—	29.5	—	—
20	—	—	—	—	200.0	—	—	—	—	—	—	—	—	—	—	—
140	0	1.73	—	—	2230.0	—	0	—	—	—	—	—	—	0	—	—
0	53	0.46	—	—	45.0	—	—	—	—	—	—	—	—	24.0	—	—
0	0	0.43	—	—	95.0	—	—	—	—	—	—	—	—	5.8	—	—
20	44	1.18	—	—	210.0	—	—	—	—	—	—	—	—	9.8	—	—
210	250	2.70	—	—	1620.0	—	89.9	—	—	—	—	—	—	0	—	—
55	150	2.70	—	—	780.0	—	—	0.24	—	0.31	4.17	—	—	1.2	—	—
70	64	2.89	—	—	1440.0	—	—	0.50	—	0.32	10.29	—	—	0	—	—
55	20	0.72	—	—	960.0	—	—	0.14	—	0.11	10.93	—	—	0	—	—
55	333	0.79	—	—	350.0	—	—	0.11	—	0.61	0.26	—	—	2.7	0	—
50	99	1.78	20.1	217.3	810.0	1.51	—	0.34	1.03	0.33	4.33	—	—	0	0.6	22.2
180	146	2.63	28.6	313.2	1060.0	2.08	—	0.38	1.66	0.51	4.72	0.28	—	0	1.1	38.0
105	250	4.50	—	—	1030.0	—	—	0.26	—	0.44	6.37	—	—	1.2	—	—
175	150	8.07	—	—	1090.0	—	—	0.39	—	0.59	11.05	—	—	9.0	—	—
195	299	8.08	—	—	1520.0	—	—	0.39	—	0.66	11.03	—	—	9.0	—	—
60	101	3.62	—	—	1380.0	—	—	—	—	—	—	—	—	3.6	—	—
0	20	0.71	—	—	590.0	—	0	0.15	—	0.48	2.30	—	—	8.9	—	—
0	60	1.80	21.3	124.3	440.0	0.57	—	0.31	0.98	0.19	2.88	0.05	—	0	0	13.7
40	80	2.70	—	—	560.0	—	—	0.25	—	0.28	4.25	—	—	1.2	—	—
0	0	0.36	—	—	450.0	—	0	0.11	0.83	0.06	1.35	0.17	—	1.2	—	—
0	100	0.00	—	—	460.0	—	0	0.14	—	0.09	2.32	—	—	0	—	—
75	79	4.43	—	—	1730.0	—	—	—	—	—	—	—	—	8.9	—	—
75	57	6.82	—	—	1210.0	—	—	—	—	—	—	—	—	5.7	—	—
60	348	0.00	—	—	240.0	—	—	0.11	—	0.63	0.21	—	—	2.4	0	—
51	100	5.38	—	—	1020.0	—	—	0.38	—	0.43	7.30	—	—	9.0	—	—
115	249	5.38	—	—	1450.0	—	—	0.38	—	0.51	7.28	—	—	9.0	—	—
80	200	5.40	—	—	1900.0	—	—	—	—	—	—	—	—	2.4	—	—
80	150	2.70	—	—	990.0	—	60.0	—	—	—	—	—	—	2.4	—	—
70	100	4.50	—	—	910.0	—	—	—	—	—	—	—	—	6.0	—	—
95	200	3.60	—	—	1330.0	—	—	—	—	—	—	—	—	9.0	—	—
75	200	1.80	—	—	880.0	—	—	—	—	—	—	—	—	30.0	—	—
100	200	3.60	—	—	1440.0	—	—	—	—	—	—	—	—	9.0	—	—
35	19	1.02	—	—	470.0	—	0	—	—	—	—	—	—	0	—	—
45	600	1.08	—	—	360.0	—	0	—	—	—	—	—	—	0	—	—
0	20	1.80	—	—	950.0	—	0	—	—	—	—	—	—	12.0	—	—
155	300	7.20	—	—	1730.0	—	—	—	—	—	—	—	—	1.2	—	—

APPENDIX H

TABLE H-1 Table of Food Composition (*continued*) (Computer code is for Cengage Diet Analysis program) (For purposes of calculations, use "0" for t, <1, <.1, <.01, etc

DA+ Code	Food Description	Quantity	Measure	Wt (g)	H_2O (g)	Ener (kcal)	Prot (g)	Carb (g)	Fiber (g)	Fat (g)	Fat Breakdown (g) Sat	Mono	Poly	*Trans*
	Fast Food—*continued*													
14238	French fries, small	1	serving(s)	92	—	290	5.0	37.0	3.0	14.0	3.0	—	—	—
10798	French toast dips without syrup, 5 pieces	1	serving(s)	155	—	370	8.0	49.0	0	17.0	5.0	—	1.4	—
10802	Onion rings	1	serving(s)	128	—	440	7.0	53.0	3.0	22.0	5.0	—	0.8	—
34858	Spicy chicken sandwich	1	item(s)	198	—	480	14.0	48.0	2.0	26.0	5.0	—	—	—
34867	Strawberry shake, small	1	serving(s)	595	—	520	14.0	93.0	0	11.0	7.0	—	—	—
10865	Super Star hamburger	1	item(s)	348	—	790	41.0	52.0	3.0	47.0	14.0	—	—	—
38925	The Six Dollar burger	1	item(s)	429	—	1010	40.0	60.0	3.0	66.0	26.0	—	—	—
10818	Vanilla shake, small	1	item(s)	398	—	314	10.0	51.5	0	7.4	4.7	—	—	—
10770	Western Bacon cheeseburger	1	item(s)	225	—	660	32.0	64.0	2.0	30.0	12.0	—	4.8	—
	Chick Fil-A													
38746	Biscuit with bacon, egg, and cheese	1	item(s)	163	—	470	18.0	39.0	1.0	26.0	9.0	—	—	3.0
38747	Biscuit with egg	1	item(s)	135	—	350	11.0	38.0	1.0	16.0	4.5	—	—	3.0
38748	Biscuit with egg and cheese	1	item(s)	149	—	400	14.0	38.0	1.0	21.0	7.0	—	—	3.0
38753	Biscuit with gravy	1	item(s)	192	—	330	5.0	43.0	1.0	15.0	4.0	—	—	4.0
38752	Biscuit with sausage, egg, and cheese	1	item(s)	212	—	620	22.0	39.0	2.0	42.0	14.0	—	—	3.0
38771	Carrot and raisin salad	1	item(s)	113	—	170	1.0	28.0	2.0	6.0	1.0	—	—	0
38761	Chargrilled chicken Cool Wrap	1	item(s)	245	—	390	29.0	54.0	3.0	7.0	3.0	—	—	0
38766	Chargrilled chicken garden salad	1	item(s)	275	—	180	22.0	9.0	3.0	6.0	3.0	—	—	0
38758	Chargrilled chicken sandwich	1	item(s)	193	—	270	28.0	33.0	3.0	3.5	1.0	—	—	0
38742	Chicken biscuit	1	item(s)	145	—	420	18.0	44.0	2.0	19.0	4.5	—	—	3.0
38743	Chicken biscuit with cheese	1	item(s)	159	—	470	21.0	45.0	2.0	23.0	8.0	—	—	3.0
38762	Chicken Caesar Cool Wrap	1	item(s)	227	—	460	36.0	52.0	3.0	10.0	6.0	—	—	0
38757	Chicken deluxe sandwich	1	item(s)	208	—	420	28.0	39.0	2.0	16.0	3.5	—	—	0
38764	Chicken salad sandwich on wheat bun	1	item(s)	153	—	350	20.0	32.0	5.0	15.0	3.0	—	—	0
38756	Chicken sandwich	1	item(s)	170	—	410	28.0	38.0	1.0	16.0	3.5	—	—	0
38768	Chick-n-Strip salad	1	item(s)	327	—	400	34.0	21.0	4.0	20.0	6.0	—	—	0
38763	Chick-n-Strips	4	item(s)	127	—	300	28.0	14.0	1.0	15.0	2.5	—	—	0
38770	Cole slaw	1	item(s)	128	—	260	2.0	17.0	2.0	21.0	3.5	—	—	0
38776	Diet lemonade, small	1	cup(s)	255	—	25	0	5.0	0	0	0	0	0	0
38755	Hashbrowns	1	serving(s)	84	—	260	2.0	25.0	3.0	17.0	3.5	—	—	1.0
38765	Hearty breast of chicken soup	1	cup(s)	241	—	140	8.0	18.0	1.0	3.5	1.0	—	—	0
38741	Hot buttered biscuit	1	item(s)	79	—	270	4.0	38.0	1.0	12.0	3.0	—	—	3.0
38778	IceDream, small cone	1	item(s)	135	—	160	4.0	28.0	0	4.0	2.0	—	—	0
38774	IceDream, small cup	1	serving(s)	227	—	240	6.0	41.0	0	6.0	3.5	—	—	0
38775	Lemonade, small	1	cup(s)	255	—	170	0	41.0	0	0.5	0	—	—	0
38777	Nuggets	8	item(s)	113	—	260	26.0	12.0	0.5	12.0	2.5	—	—	0
38769	Side salad	1	item(s)	108	—	60	3.0	4.0	2.0	3.0	1.5	—	—	0
38767	Southwest chargrilled salad	1	item(s)	303	—	240	25.0	17.0	5.0	8.0	3.5	—	—	0
40481	Spicy chicken cool wrap	1	serving(s)	230	—	380	30.0	52.0	3.0	6.0	3.0	—	—	0
38772	Waffle potato fries, small, salted	1	serving(s)	85	—	270	3.0	34.0	4.0	13.0	3.0	—	—	1.5
	Cinnabon													
39572	Caramellata Chill w/whipped cream	16	fluid ounce(s)	480	—	406	10.0	61.0	0	14.0	8.0	—	—	0
39571	Cinnabon Bites	1	serving(s)	149	—	510	8.0	77.0	2.0	19.0	5.0	—	—	5.0
39570	Cinnabon Stix	5	item(s)	85	—	379	6.0	41.0	1.0	21.0	6.0	—	—	4.0
39567	Classic roll	1	item(s)	221	—	813	15.0	117.0	4.0	32.0	8.0	—	—	5.0
39568	Minibon	1	item(s)	92	—	339	6.0	49.0	2.0	13.0	3.0	—	—	2.0
39573	Mochalatta Chill w/whipped cream	16	fluid ounce(s)	480	—	362	9.0	55.0	0	13.0	8.0	—	—	0
39569	Pecanbon	1	item(s)	272	—	1100	16.0	141.0	8.0	56.0	10.0	—	—	5.0
	Dairy Queen													
1466	Banana split	1	item(s)	369	—	510	8.0	96.0	3.0	12.0	8.0	—	—	0
38552	Brownie Earthquake®	1	serving(s)	304	—	740	10.0	112.0	0	27.0	16.0	—	—	0.5
38561	Chocolate chip cookie dough blizzard,® small	1	item(s)	319	—	720	12.0	105.0	0	28.0	14.0	—	—	2.5
1464	Chocolate malt, small	1	item(s)	418	—	640	15.0	111.0	1.0	16.0	11.0	—	—	0.5
38541	Chocolate shake, small	1	item(s)	397	—	560	13.0	93.0	1.0	15.0	10.0	—	—	0.5
17257	Chocolate soft serve	½	cup(s)	94	—	150	4.0	22.0	0	5.0	3.5	—	—	0
1463	Chocolate sundae, small	1	item(s)	163	—	280	5.0	49.0	0	7.0	4.5	—	—	0

PAGE KEY: H-4 = Breads/Baked Goods H-10 = Cereal/Rice/Pasta H-14 = Fruit H-18 = Vegetables/Legumes H-28 = Nuts/Seeds H-30 = Vegetarian H-32 = Dairy H-40 = Eggs H-40 = Seafood H-42 = Meats H-46 = Poultry H-46 = Processed Meats H-48 = Beverages H-52 = Fats/Oils H-54 = Sweets H-56 = Spices/Condiments/Sauces H-60 = Mixed Foods/Soups/Sandwiches H-64 = Fast Food H-84 = Convenience H-86 = Baby Foods

Chol (mg)	Calc (mg)	Iron (mg)	Magn (mg)	Pota (mg)	Sodi (mg)	Zinc (mg)	Vit A (µg)	Thia (mg)	Vit E (mg α)	Ribo (mg)	Niac (mg)	Vit B_6 (mg)	Fola (µg)	Vit C (mg)	Vit B_{12} (µg)	Sele (µg)
0	0	1.08	—	—	170.0	—	0	—	—	—	—	—	—	21.0	—	—
3	0	0.00	—	—	470.0	—	0	0.25	—	0.23	2.00	—	—	0	—	—
0	20	0.72	—	—	700.0	—	0	—	—	—	—	—	—	3.6	—	—
40	100	3.60	—	—	1220.0	—	—	—	—	—	—	—	—	6.0	—	—
45	600	0.00	—	—	340.0	—	0	—	—	—	—	—	—	0	—	—
130	100	7.20	—	—	980.0	—	—	—	—	—	—	—	—	9.0	—	—
145	279	4.29	—	—	1960.0	—	—	—	—	—	—	—	—	16.7	—	—
30	401	0.00	—	—	234.0	—	0	—	—	—	—	—	—	0	—	—
85	200	5.40	—	—	1410.0	—	60.0	—	—	—	—	—	—	1.2	—	—
270	150	2.70	—	—	1190.0	—	—	—	—	—	—	—	—	0	—	—
240	80	2.70	—	—	740.0	—	—	—	—	—	—	—	—	0	—	—
255	150	2.70	—	—	970.0	—	—	—	—	—	—	—	—	0	—	—
5	60	1.80	—	—	930.0	—	0	—	—	—	—	—	—	0	—	—
300	200	3.60	—	—	1360.0	—	—	—	—	—	—	—	—	0	—	—
10	40	0.36	—	—	110.0	—	—	—	—	—	—	—	—	4.8	—	—
65	200	3.60	—	—	1020.0	—	—	—	—	—	—	—	—	6.0	—	—
65	150	0.72	—	—	620.0	—	—	—	—	—	—	—	—	30	—	—
65	80	2.70	—	—	940.0	—	—	—	—	—	—	—	—	6.0	—	—
35	60	2.70	—	—	1270.0	—	0	—	—	—	—	—	—	0	—	—
50	150	2.70	—	—	1500.0	—	—	—	—	—	—	—	—	0	—	—
80	500	3.60	—	—	1350.0	—	—	—	—	—	—	—	—	1.2	—	—
60	100	2.70	—	—	1300.0	—	—	—	—	—	—	—	—	2.4	—	—
65	150	1.80	—	—	880.0	—	—	—	—	—	—	—	—	0	—	—
60	100	2.70	—	—	1300.0	—	—	—	—	—	—	—	—	0	—	—
80	150	1.44	—	—	1070.0	—	—	—	—	—	—	—	—	6.0	—	—
65	40	1.44	—	—	940.0	—	—	—	—	—	—	—	—	0	—	—
25	60	0.36	—	—	220.0	—	—	—	—	—	—	—	—	36.0	—	—
0	0	0.36	—	—	5.0	—	0	—	—	—	—	—	—	15.0	—	—
5	20	0.72	—	—	380.0	—	—	—	—	—	—	—	—	0	—	—
25	40	1.08	—	—	900.0	—	—	—	—	—	—	—	—	0	—	—
0	60	1.80	—	—	660.0	—	0	—	—	—	—	—	—	0	—	—
15	100	0.36	—	—	80.0	—	—	—	—	—	—	—	—	0	—	—
25	200	0.36	—	—	105.0	—	—	—	—	—	—	—	—	0	—	—
0	0	0.36	—	—	10.0	—	0	—	—	—	—	—	—	15.0	—	—
70	40	1.08	—	—	1090.0	—	0	—	—	—	—	—	—	0	—	—
10	100	0.00	—	—	75.0	—	—	—	—	—	—	—	—	15.0	—	—
60	200	1.08	—	—	770.0	—	—	—	—	—	—	—	—	24.0	—	—
60	200	3.60	—	—	1090.0	—	—	—	—	—	—	—	—	3.6	—	—
0	20	1.08	—	—	115.0	—	0	—	—	—	—	—	—	1.2	—	—
46	—	—	—	—	187.0	—	—	—	—	—	—	—	—	—	—	—
35	—	—	—	—	530.0	—	—	—	—	—	—	—	—	—	—	—
16	—	—	—	—	413.0	—	—	—	—	—	—	—	—	—	—	—
67	—	—	—	—	801.0	—	—	—	—	—	—	—	—	—	—	—
27	—	—	—	—	337.0	—	—	—	—	—	—	—	—	—	—	—
46	—	—	—	—	252.0	—	—	—	—	—	—	—	—	—	—	—
63	—	—	—	—	600.0	—	—	—	—	—	—	—	—	—	—	—
30	250	1.80	—	—	180.0	—	—	—	—	—	—	—	—	15.0	—	—
50	250	1.80	—	—	350.0	—	—	—	—	—	—	—	—	0	—	—
50	350	2.70	—	—	370.0	—	—	—	—	—	—	—	—	1.2	—	—
55	450	1.80	—	—	340.0	—	—	—	—	—	—	—	—	2.4	—	—
50	450	1.44	—	—	280.0	—	—	—	—	—	—	—	—	2.4	—	—
15	100	0.72	—	—	75.0	—	—	—	—	—	—	—	—	0	—	—
20	200	1.08	—	—	140.0	—	—	—	—	—	—	—	—	0	—	—

TABLE H-1 Table of Food Composition (*continued*) (Computer code is for Cengage Diet Analysis program) (For purposes of calculations, use "0" for t, <1, <.1, <.01, etc

DA+ Code	Food Description	Quantity	Measure	Wt (g)	H_2O (g)	Ener (kcal)	Prot (g)	Carb (g)	Fiber (g)	Fat (g)	Fat Breakdown (g)			
											Sat	Mono	Poly	*Trans*
Fast Food—*continued*														
1462	Dipped cone, small	1	item(s)	156	—	340	6.0	42.0	1.0	17.0	9.0	4.0	3.0	1.0
38555	Oreo cookies blizzard, small	1	item(s)	283	—	570	11.0	83.0	0.5	21.0	10.0	—	—	2.5
38547	Royal Treats Peanut Buster® Parfait	1	item(s)	305	—	730	16.0	99.0	2.0	31.0	17.0	—	—	0
17256	Vanilla soft serve	½	cup(s)	94	—	140	3.0	22.0	0	4.5	3.0	—	—	0
	Domino's													
31606	Barbeque buffalo wings	1	item(s)	25	—	50	6.0	2.0	0	2.5	0.5	—	—	—
31604	Breadsticks	1	item(s)	30	—	115	2.0	12.0	0	6.3	1.1	—	—	—
37551	Buffalo Chicken Kickers	1	item(s)	24	—	47	4.0	3.0	0	2.0	0.5	—	—	—
37548	CinnaStix	1	item(s)	30	—	123	2.0	15.0	1.0	6.1	1.1	—	—	—
37549	Dot, cinnamon	1	item(s)	28	7.6	99	1.9	14.9	0.7	3.7	0.7	—	—	—
31605	Double cheesy bread	1	item(s)	35	—	123	4.0	13.0	0	6.5	1.9	—	—	—
31607	Hot buffalo wings	1	item(s)	25	—	45	5.0	1.0	0	2.5	0.5	—	—	—
	Domino's Classic hand tossed pizza													
31573	America's favorite feast, 12″	1	slice(s)	102	—	257	10.0	29.0	2.0	11.5	4.5	—	—	—
31574	America's favorite feast, 14″	1	slice(s)	141	—	353	14.0	39.0	2.0	16.0	6.0	—	—	—
37543	Bacon cheeseburger feast, 12″	1	slice(s)	99	—	273	12.0	28.0	2.0	13.0	5.5	—	—	—
37545	Bacon cheeseburger feast, 14″	1	slice(s)	137	—	379	17.0	38.0	2.0	18.0	8.0	—	—	—
37546	Barbeque feast, 12″	1	slice(s)	96	—	252	11.0	31.0	1.0	10.0	4.5	—	—	—
37547	Barbeque feast, 14″	1	slice(s)	131	—	344	14.0	43.0	2.0	13.5	6.0	—	—	—
31569	Cheese, 12″	1	slice(s)	55	—	160	6.0	28.0	1.0	3.0	1.0	—	—	0
31570	Cheese, 14″	1	slice(s)	75	—	220	8.0	38.0	2.0	4.0	1.0	—	—	0
37538	Deluxe feast, 12″	1	slice(s)	201	101.8	465	19.5	57.4	3.5	18.2	7.7	—	—	—
37540	Deluxe feast, 14″	1	slice(s)	273	138.4	627	26.4	78.3	4.7	24.1	10.2	—	—	—
31685	Deluxe, 12″	1	slice(s)	100	—	234	9.0	29.0	2.0	9.5	3.5	—	—	—
31694	Deluxe, 14″	1	slice(s)	136	—	316	13.0	39.0	2.0	12.5	5.0	—	—	—
31686	Extravaganzza, 12″	1	slice(s)	122	—	289	13.0	30.0	2.0	14.0	5.5	—	—	—
31695	Extravaganzza, 14″	1	slice(s)	165	—	388	17.0	40.0	3.0	18.5	7.5	—	—	—
31575	Hawaiian feast, 12″	1	slice(s)	102	—	223	10.0	30.0	2.0	8.0	3.5	—	—	—
31576	Hawaiian feast, 14″	1	slice(s)	141	—	309	14.0	41.0	2.0	11.0	4.5	—	—	—
31687	Meatzza, 12″	1	slice(s)	108	—	281	13.0	29.0	2.0	13.5	5.5	—	—	—
31696	Meatzza, 14″	1	slice(s)	146	—	378	17.0	39.0	2.0	18.0	7.5	—	—	—
31571	Pepperoni feast, extra pepperoni and cheese, 12″	1	slice(s)	98	—	265	11.0	28.0	2.0	12.5	5.0	—	—	—
31572	Pepperoni feast, extra pepperoni and cheese, 14″	1	slice(s)	135	—	363	16.0	39.0	2.0	17.0	7.0	—	—	—
31577	Vegi feast, 12″	1	slice(s)	102	—	218	9.0	29.0	2.0	8.0	3.5	—	—	—
31578	Vegi feast, 14″	1	slice(s)	139	—	300	13.0	40.0	3.0	11.0	4.5	—	—	—
	Domino's thin crust pizza													
31583	America's favorite, 12″	1	slice(s)	72	—	208	8.0	15.0	1.0	13.5	5.0	—	—	—
31584	America's favorite, 14″	1	slice(s)	100	—	285	11.0	20.0	2.0	18.5	7.0	—	—	—
31579	Cheese, 12″	1	slice(s)	49	—	137	5.0	14.0	1.0	7.0	2.5	—	—	—
31580	Cheese, 14″	1	slice(s)	68	27.0	214	8.8	19.0	1.4	11.4	4.6	2.9	2.5	—
31688	Deluxe, 12″	1	slice(s)	70	—	185	7.0	15.0	1.0	11.5	4.0	—	—	—
31697	Deluxe, 14″	1	slice(s)	94	—	248	10.0	20.0	2.0	15.0	5.5	—	—	—
31689	Extravaganzza, 12″	1	slice(s)	92	—	240	11.0	16.0	1.0	15.5	6.0	—	—	—
31698	Extravaganzza, 14″	1	slice(s)	123	—	320	14.0	21.0	2.0	20.5	8.0	—	—	—
31585	Hawaiian, 12″	1	slice(s)	71	—	174	8.0	16.0	1.0	9.5	3.5	—	—	—
31586	Hawaiian, 14″	1	slice(s)	100	—	240	11.0	21.0	2.0	13.0	5.0	—	—	—
31690	Meatzza, 12″	1	slice(s)	78	—	232	11.0	15.0	1.0	15.0	6.0	—	—	—
31699	Meatzza, 14″	1	slice(s)	104	—	310	14.0	20.0	2.0	20	8.0	—	—	—
31581	Pepperoni, extra pepperoni and cheese, 12″	1	slice(s)	68	—	216	9.0	14.0	1.0	14.0	5.5	—	—	—
31582	Pepperoni, extra pepperoni and cheese, 14″	1	slice(s)	93	—	295	13.0	20.0	1.0	19.0	7.5	—	—	—
31587	Vegi, 12″	1	slice(s)	71	—	168	7.0	15.0	1.0	9.5	3.5	—	—	—
31588	Vegi, 14″	1	slice(s)	97	—	231	10.0	21.0	2.0	13.5	5.0	—	—	—
	Domino's Ultimate deep dish pizza													
31596	America's favorite, 12″	1	slice(s)	115	—	309	12.0	29.0	2.0	17.0	6.0	—	—	—
31702	America's favorite, 14″	1	slice(s)	162	—	433	17.0	42.0	3.0	23.5	8.0	—	—	—
31590	Cheese, 12″	1	slice(s)	90	—	238	9.0	28.0	2.0	11.0	3.5	—	—	—
31591	Cheese, 14″	1	slice(s)	128	53.9	351	14.5	41.0	2.9	13.2	5.2	3.8	2.5	—
31589	Cheese, 6″	1	item(s)	215	—	598	22.9	68.4	3.9	27.6	9.9	—	—	—
31691	Deluxe, 12″	1	slice(s)	122	—	287	11.0	29.0	2.0	15.0	5.0	—	—	—

PAGE KEY: H-4 = Breads/Baked Goods H-10 = Cereal/Rice/Pasta H-14 = Fruit H-18 = Vegetables/Legumes H-28 = Nuts/Seeds H-30 = Vegetarian H-32 = Dairy H-40 = Eggs H-40 = Seafood H-42 = Meats H-46 = Poultry H-46 = Processed Meats H-48 = Beverages H-52 = Fats/Oils H-54 = Sweets H-56 = Spices/Condiments/Sauces H-60 = Mixed Foods/Soups/Sandwiches H-64 = Fast Food H-84 = Convenience H-86 = Baby Foods

Chol (mg)	Calc (mg)	Iron (mg)	Magn (mg)	Pota (mg)	Sodi (mg)	Zinc (mg)	Vit A (µg)	Thia (mg)	Vit E (mg α)	Ribo (mg)	Niac (mg)	Vit B_6 (mg)	Fola (µg)	Vit C (mg)	Vit B_{12} (µg)	Sele (µg)
20	200	1.08	—	—	130.0	—	—	—	—	—	—	—	—	1.2	—	—
40	350	2.70	—	—	430.0	—	—	—	—	—	—	—	—	1.2	—	—
35	300	1.80	—	—	400.0	—	—	—	—	—	—	—	—	1.2	—	—
15	150	0.72	—	—	70.0	—	—	—	—	—	—	—	—	0	—	—
26	10	0.36	—	—	175.5	—	—	—	—	—	—	—	—	0	—	—
0	0	0.72	—	—	122.1	—	—	—	—	—	—	—	—	0	—	—
9	0	0.00	—	—	162.5	—	—	—	—	—	—	—	—	0	—	—
0	0	0.72	—	—	111.4	—	—	—	—	—	—	—	—	0	—	—
0	6	0.59	—	—	85.7	—	—	—	—	—	—	—	—	0	—	—
6	40	0.72	—	—	162.3	—	—	—	—	—	—	—	—	0	—	—
26	10	0.36	—	—	254.5	—	—	—	—	—	—	—	—	1.2	—	—
22	100	1.80	—	—	625.5	—	—	—	—	—	—	—	—	0.6	—	—
31	140	2.52	—	—	865.5	—	—	—	—	—	—	—	—	0.6	—	—
27	140	1.80	—	—	634.0	—	—	—	—	—	—	—	—	0	—	—
38	190	2.52	—	—	900.0	—	—	—	—	—	—	—	—	0	—	—
20	140	1.62	—	—	600.0	—	—	—	—	—	—	—	—	0.6	—	—
27	190	2.16	—	—	831.5	—	—	—	—	—	—	—	—	0.6	—	—
0	0	1.80	—	—	110.0	—	0	—	—	—	—	—	—	0	—	—
0	0	2.70	—	—	150.0	—	0	—	—	—	—	—	—	0	—	—
40	199	3.56	—	—	1063.1	—	—	—	—	—	—	—	—	1.4	—	—
53	276	4.84	—	—	1432.2	—	—	—	—	—	—	—	—	1.8	—	—
17	100	1.80	—	—	541.5	—	—	—	—	—	—	—	—	0.6	—	—
23	130	2.34	—	—	728.5	—	—	—	—	—	—	—	—	1.2	—	—
28	140	1.98	—	—	764.0	—	—	—	—	—	—	—	—	0.6	—	—
37	190	2.70	—	—	1014.0	—	—	—	—	—	—	—	—	1.2	—	—
16	130	1.62	—	—	546.5	—	—	—	—	—	—	—	—	1.2	—	—
23	180	2.34	—	—	765.0	—	—	—	—	—	—	—	—	1.2	—	—
28	130	1.80	—	—	739.5	—	—	—	—	—	—	—	—	0	—	—
37	190	2.52	—	—	983.5	—	—	—	—	—	—	—	—	0	—	—
24	130	1.62	—	—	670.0	—	70.9	—	—	—	—	—	—	0	—	—
33	180	2.34	—	—	920.0	—	104.7	—	—	—	—	—	—	0	—	—
13	130	1.62	—	—	489.0	—	—	—	—	—	—	—	—	0.6	—	—
18	180	2.34	—	—	678.0	—	—	—	—	—	—	—	—	0.6	—	—
23	100	0.90	—	—	533.0	—	—	—	—	—	—	—	—	2.4	—	—
32	140	1.26	—	—	736.5	—	—	—	—	—	—	—	—	3.0	—	—
10	90	0.54	—	—	292.5	—	60.0	—	—	—	—	—	—	1.8	—	—
14	151	0.48	17.7	125.1	338.0	0.02	64.6	0.05	1.01	0.07	0.69	—	—	2.4	0.5	24.1
19	100	0.90	—	—	449.0	—	—	—	—	—	—	—	—	2.4	—	—
24	130	1.08	—	—	601.0	—	—	—	—	—	—	—	—	3.6	—	—
29	140	1.08	—	—	671.5	—	—	—	—	—	—	—	—	2.4	—	—
38	190	1.44	—	—	886.5	—	—	—	—	—	—	—	—	3.6	—	—
17	130	0.72	—	—	454.0	—	—	—	—	—	—	—	—	3.0	—	—
24	180	0.90	—	—	637.5	—	—	—	—	—	—	—	—	3.6	—	—
29	140	0.90	—	—	647.0	—	—	—	—	—	—	—	—	1.8	—	—
38	190	1.26	—	—	865.5	—	—	—	—	—	—	—	—	2.4	—	—
26	130	0.72	—	—	577.0	—	80.0	—	—	—	—	—	—	1.8	—	—
35	80	1.08	—	—	792.5	—	105.8	—	—	—	—	—	—	2.4	—	—
14	130	0.72	—	—	396.5	—	—	—	—	—	—	—	—	2.4	—	—
19	180	1.08	—	—	550.5	—	—	—	—	—	—	—	—	3.0	—	—
25	120	2.34	—	—	796.5	—	—	—	—	—	—	—	—	0.6	—	—
34	170	3.24	—	—	1110.0	—	—	—	—	—	—	—	—	0.6	—	—
11	110	1.98	—	—	555.5	—	70.0	—	—	—	—	—	—	0	—	—
18	189	3.78	32.0	209.9	718.1	1.75	99.8	0.29	1.13	0.31	5.44	—	—	0	0.6	45.6
36	295	4.67	—	—	1341.4	—	174.0	—	—	—	—	—	—	0.5	—	—
20	120	2.16	—	—	712.0	—	—	—	—	—	—	—	—	1.2	—	—

TABLE H-1 Table of Food Composition (*continued*) (Computer code is for Cengage Diet Analysis program) (For purposes of calculations, use "0" for t, <1, <.1, <.01, etc.)

DA+ Code	Food Description	Quantity	Measure	Wt (g)	H_2O (g)	Ener (kcal)	Prot (g)	Carb (g)	Fiber (g)	Fat (g)	Fat Breakdown (g) Sat	Mono	Poly	*Trans*
Fast Food—*continued*														
31700	Deluxe, 14″	1	slice(s)	156	—	396	15.0	42.0	3.0	20.0	7.0	—	—	—
31692	Extravaganzza, 12″	1	slice(s)	136	—	341	14.0	30.0	2.0	19.0	7.0	—	—	—
31701	Extravaganzza, 14″	1	slice(s)	186	—	468	20.0	43.0	3.0	25.5	9.5	—	—	—
31599	Hawaiian, 12″	1	slice(s)	114	—	275	12.0	30.0	2.0	13.0	5.0	—	—	—
31600	Hawaiian, 14″	1	slice(s)	162	—	389	17.0	43.0	3.0	18.0	6.5	—	—	—
31693	Meatzza, 12″	1	slice(s)	121	—	333	14.0	29.0	2.0	19.0	7.0	—	—	—
31703	Meatzza, 14″	1	slice(s)	167	—	458	19.0	42.0	3.0	25.0	9.5	—	—	—
31593	Pepperoni, extra pepperoni and cheese, 12″	1	slice(s)	110	—	317	13.0	29.0	2.0	17.5	6.5	—	—	—
31594	Pepperoni, extra pepperoni and cheese, 14″	1	slice(s)	155	—	443	18.0	42.0	3.0	24.0	9.0	—	—	—
31602	Vegi, 12″	1	slice(s)	114	—	270	11.0	30.0	2.0	13.5	5.0	—	—	—
31603	Vegi, 14″	1	slice(s)	159	—	380	15.0	43.0	3.0	18.0	6.5	—	—	—
31598	With ham and pineapple tidbits, 6″	1	item(s)	430	—	619	25.2	69.9	4.0	28.3	10.2	—	—	—
31595	With Italian sausage, 6″	1	item(s)	430	—	642	24.8	69.6	4.2	31.1	11.3	—	—	—
31592	With pepperoni, 6″	1	item(s)	430	—	647	25.1	68.5	3.9	32.0	11.7	—	—	—
31601	With vegetables, 6″	1	item(s)	430	—	619	23.4	70.8	4.6	28.7	10.1	—	—	—
	In-n-Out Burger													
34391	Cheeseburger with mustard and ketchup	1	serving(s)	268	—	400	22.0	41.0	3.0	18.0	9.0	—	—	0.5
34374	Cheeseburger	1	serving(s)	268	—	480	22.0	39.0	3.0	27.0	10.0	—	—	0.5
34390	Cheeseburger, lettuce leaves instead of buns	1	serving(s)	300	—	330	18.0	11.0	3.0	25.0	9.0	—	—	0
34377	Chocolate shake	1	serving(s)	425	—	690	9.0	83.0	0	36.0	24.0	—	—	1.0
34375	Double-Double cheeseburger	1	serving(s)	330	—	670	37.0	39.0	3.0	41.0	18.0	—	—	1.0
34393	Double-Double cheeseburger with mustard and ketchup	1	serving(s)	330	—	590	37.0	41.0	3.0	32.0	17.0	—	—	1.0
34392	Double-Double cheeseburger, lettuce leaves instead of buns	1	serving(s)	362	—	520	33.0	11.0	3.0	39.0	17.0	—	—	1.0
34376	French fries	1	serving(s)	125	—	400	7.0	54.0	2.0	18.0	5.0	—	—	0
34373	Hamburger	1	item(s)	243	—	390	16.0	39.0	3.0	19.0	5.0	—	—	0
34389	Hamburger with mustard and ketchup	1	serving(s)	243	—	310	16.0	41.0	3.0	10.0	4.0	—	—	0
34388	Hamburger, lettuce leaves instead of buns	1	serving(s)	275	—	240	13.0	11.0	3.0	17.0	4.0	—	—	0
34379	Strawberry shake	1	serving(s)	425	—	690	9.0	91.0	0	33.0	22.0	—	—	0.5
34378	Vanilla shake	1	serving(s)	425	—	680	9.0	78.0	0	37.0	25.0	—	—	1.0
	Jack in the Box													
30392	Bacon ultimate cheeseburger	1	item(s)	338	—	1090	46.0	53.0	2.0	77.0	30.0	—	—	3.0
1740	Breakfast Jack	1	item(s)	125	—	290	17.0	29.0	1.0	12.0	4.5	—	—	0
14074	Cheeseburger	1	item(s)	131	—	350	18.0	31.0	1.0	17.0	8.0	—	—	1.0
14106	Chicken breast strips, 4 pieces	1	serving(s)	201	—	500	35.0	36.0	3.0	25.0	6.0	—	—	6.0
37241	Chicken club salad, plain, without salad dressing	1	serving(s)	431	—	300	27.0	13.0	4.0	15.0	6.0	—	—	0
14064	Chicken sandwich	1	item(s)	145	—	400	15.0	38.0	2.0	21.0	4.5	—	—	2.5
14111	Chocolate ice cream shake, small	1	serving(s)	414	—	880	14.0	107.0	1.0	45.0	31.0	—	—	2.0
14073	Hamburger	1	item(s)	118	—	310	16.0	30.0	1.0	14.0	6.0	—	—	1.0
14090	Hash browns	1	serving(s)	57	—	150	1.0	13.0	2.0	10.0	2.5	—	—	3.0
14072	Jack's Spicy Chicken sandwich	1	item(s)	270	—	620	25.0	61.0	4.0	31.0	6.0	—	—	3.0
1468	Jumbo Jack hamburger	1	item(s)	261	—	600	21.0	51.0	3.0	35.0	12.0	—	—	1.5
1469	Jumbo Jack hamburger with cheese	1	item(s)	286	—	690	25.0	54.0	3.0	42.0	16.0	—	—	1.5
14099	Natural cut french fries, large	1	serving(s)	196	—	530	8.0	69.0	5.0	25.0	6.0	—	—	7.0
14098	Natural cut french fries, medium	1	serving(s)	133	—	360	5.0	47.0	4.0	17.0	4.0	—	—	5.0
1470	Onion rings	1	serving(s)	119	—	500	6.0	51.0	3.0	30.0	6.0	—	—	10
33141	Sausage, egg, and cheese biscuit	1	item(s)	234	—	740	27.0	35.0	2.0	55.0	17.0	—	—	6.0
14095	Seasoned curly fries, medium	1	serving(s)	125	—	400	6.0	45.0	5.0	23.0	5.0	—	—	7.0
14077	Sourdough Jack	1	item(s)	245	—	710	27.0	36.0	3.0	51.0	18.0	—	—	3.0
37249	Southwest chicken salad, plain, without salad dressing	1	serving(s)	488	—	300	24.0	29.0	7.0	11.0	5.0	—	—	0
14112	Strawberry ice cream shake, small	1	serving(s)	417	—	880	13.0	105.0	0	44.0	31.0	—	—	2.0

PAGE KEY: H-4 = Breads/Baked Goods H-10 = Cereal/Rice/Pasta H-14 = Fruit H-18 = Vegetables/Legumes H-28 = Nuts/Seeds H-30 = Vegetarian H-32 = Dairy H-40 = Eggs H-40 = Seafood H-42 = Meats H-46 = Poultry H-46 = Processed Meats H-48 = Beverages H-52 = Fats/Oils H-54 = Sweets H-56 = Spices/Condiments/Sauces H-60 = Mixed Foods/Soups/Sandwiches H-64 = Fast Food H-84 = Convenience H-86 = Baby Foods

Chol (mg)	Calc (mg)	Iron (mg)	Magn (mg)	Pota (mg)	Sodi (mg)	Zinc (mg)	Vit A (µg)	Thia (mg)	Vit E (mg α)	Ribo (mg)	Niac (mg)	Vit B_6 (mg)	Fola (µg)	Vit C (mg)	Vit B_{12} (µg)	Sele (µg)
26	170	3.06	—	—	974.5	—	—	—	—	—	—	—	—	1.2	—	—
31	160	2.52	—	—	934.5	—	—	—	—	—	—	—	—	1.2	—	—
40	220	3.42	—	—	1260.0	—	—	—	—	—	—	—	—	1.2	—	—
19	150	1.98	—	—	717.0	—	—	—	—	—	—	—	—	1.2	—	—
26	210	2.88	—	—	1011.0	—	—	—	—	—	—	—	—	1.8	—	—
31	160	2.34	—	—	910.5	—	—	—	—	—	—	—	—	0	—	—
40	220	3.24	—	—	1230.0	—	—	—	—	—	—	—	—	0.6	—	—
27	150	2.16	—	—	840.5	—	86.5	—	—	—	—	—	—	0	—	—
37	220	3.06	—	—	1166.0	—	115.4	—	—	—	—	—	—	0.6	—	—
15	150	2.16	—	—	659.5	—	—	—	—	—	—	—	—	0.6	—	—
21	220	3.06	—	—	924.0	—	—	—	—	—	—	—	—	1.2	—	—
43	298	4.84	—	—	1497.8	—	—	—	—	—	—	—	—	1.5	—	—
45	302	4.89	—	—	1478.1	—	—	—	—	—	—	—	—	0.6	—	—
47	299	4.81	—	—	1523.7	—	167.9	—	—	—	—	—	—	0.6	—	—
36	307	5.10	—	—	1472.5	—	—	—	—	—	—	—	—	4.7	—	—
60	200	3.60	—	—	1080.0	—	—	—	—	—	—	—	—	12.0	—	—
60	200	3.60	—	—	1000.0	—	—	—	—	—	—	—	—	9.0	—	—
60	200	2.70	—	—	720.0	—	—	—	—	—	—	—	—	12.0	—	—
95	300	0.72	—	—	350.0	—	—	—	—	—	—	—	—	0	—	—
120	350	5.40	—	—	1440.0	—	—	—	—	—	—	—	—	9.0	—	—
115	350	5.40	—	—	1520.0	—	—	—	—	—	—	—	—	12.0	—	—
120	350	4.50	—	—	1160.0	—	—	—	—	—	—	—	—	12.0	—	—
0	20	1.80	—	—	245.0	—	0	—	—	—	—	—	—	0	—	—
40	40	3.60	—	—	650.0	—	—	—	—	—	—	—	—	9.0	—	—
35	40	3.60	—	—	730.0	—	—	—	—	—	—	—	—	12.0	—	—
40	40	2.70	—	—	370.0	—	—	—	—	—	—	—	—	12.0	—	—
85	300	0.00	—	—	280.0	—	—	—	—	—	—	—	—	0	—	—
90	300	0.00	—	—	390.0	—	—	—	—	—	—	—	—	0	—	—
140	308	7.38	—	540.0	2040.0	—	—	—	—	—	—	—	—	0.6	—	—
220	145	3.48	—	210.0	760.0	—	—	—	—	—	—	—	—	3.5	—	—
50	151	3.61	—	270.0	790.0	—	40.2	—	—	—	—	—	—	0	—	—
80	18	1.60	—	530.0	1260.0	—	—	—	—	—	—	—	—	1.1	—	—
65	280	3.35	—	560.0	880.0	—	—	—	—	—	—	—	—	50.4	—	—
35	100	2.70	—	240.0	730.0	—	—	—	—	—	—	—	—	4.8	—	—
135	460	0.47	—	840.0	330.0	—	—	—	—	—	—	—	—	0	—	—
40	100	3.60	—	250.0	600.0	—	0	—	—	—	—	—	—	0	—	—
0	10	0.18	—	190.0	230.0	—	0	—	—	—	—	—	—	0	—	—
50	150	1.80	—	450.0	1100.0	—	—	—	—	—	—	—	—	9.0	—	—
45	164	4.92	—	380.0	940.0	—	—	—	—	—	—	—	—	9.8	—	—
70	234	4.20	—	410.0	1310.0	—	—	—	—	—	—	—	—	8.4	—	—
0	20	1.42	—	1240.0	870.0	—	0	—	—	—	—	—	—	8.9	—	—
0	19	1.01	—	840.0	590.0	—	0	—	—	—	—	—	—	5.6	—	—
0	40	2.70	—	140.0	420.0	—	40.0	—	—	—	—	—	—	18.0	—	—
280	88	2.36	—	310.0	1430.0	—	—	—	—	—	—	—	—	0	—	—
0	40	1.80	—	580.0	890.0	—	—	—	—	—	—	—	—	0	—	—
75	200	4.50	—	430.0	1230.0	—	—	—	—	—	—	—	—	9.0	—	—
55	274	4.10	—	670.0	860.0	—	—	—	—	—	—	—	—	43.8	—	—
135	466	0.00	—	750.0	290.0	—	—	—	—	—	—	—	—	0	—	—

TABLE H-1 Table of Food Composition (*continued*) (Computer code is for Cengage Diet Analysis program) (For purposes of calculations, use "0" for t, <1, <.1, <.01, etc

DA+ Code	Food Description	Quantity	Measure	Wt (g)	H_2O (g)	Ener (kcal)	Prot (g)	Carb (g)	Fiber (g)	Fat (g)	Fat Breakdown (g) Sat	Mono	Poly	*Trans*
Fast Food—*continued*														
14078	Ultimate cheeseburger	1	item(s)	323	—	1010	40.0	53.0	2.0	71.0	28.0	—	—	3.0
14110	Vanilla ice cream shake, small	1	serving(s)	379	—	790	13.0	83.0	0	44.0	31.0	—	—	2.0
	Jamba Juice													
31645	Aloha Pineapple smoothie	24	fluid ounce(s)	730	—	500	8.0	117.0	4.0	1.5	1.0	—	—	—
31646	Banana Berry smoothie	24	fluid ounce(s)	719	—	480	5.0	112.0	4.0	1.0	0	—	—	—
31656	Berry Lime Sublime smoothie	24	fluid ounce(s)	728	—	460	3.0	106.0	5.0	2.0	1.0	—	—	—
31647	Carribean Passion smoothie	24	fluid ounce(s)	730	—	440	4.0	102.0	4.0	2.0	1.0	—	—	—
38422	Carrot juice	16	fluid ounce(s)	472	—	100	3.0	23.0	0	0.5	0	—	—	—
31648	Chocolate Moo'd smoothie	24	fluid ounce(s)	634	—	720	17.0	148.0	3.0	8.0	5.0	—	—	—
31649	Citrus Squeeze smoothie	24	fluid ounce(s)	727	—	470	5.0	110.0	4.0	2.0	1.0	—	—	—
31651	Coldbuster smoothie	24	fluid ounce(s)	724	—	430	5.0	100.0	5.0	2.5	1.0	—	—	—
31652	Cranberry Craze smoothie	24	fluid ounce(s)	793	—	460	6.0	104.0	4.0	0.5	0	—	—	—
31654	Jamba Powerboost smoothie	24	fluid ounce(s)	738	—	440	6.0	105.0	6.0	1.0	0	—	—	—
38423	Lemonade	16	fluid ounce(s)	483	—	300	1.0	75.0	0	0	0	0	0	0
31657	Mango-a-go-go smoothie	24	fluid ounce(s)	690	—	440	3.0	104.0	4.0	1.5	0.5	—	—	—
38424	Orange juice, freshly squeezed	16	fluid ounce(s)	496	—	220	3.0	52.0	0.5	1.0	0	—	—	—
38426	Orange/carrot juice	16	fluid ounce(s)	484	—	160	3.0	37.0	0	1.0	0	—	—	—
31660	Orange-a-peel smoothie	24	fluid ounce(s)	726	—	440	8.0	102.0	5.0	1.5	0	—	—	—
31662	Peach Pleasure smoothie	24	fluid ounce(s)	720	—	460	4.0	108.0	4.0	2.0	1.0	—	—	—
31665	Protein Berry Pizzaz smoothie	24	fluid ounce(s)	710	—	440	20.0	92.0	5.0	1.5	0	—	—	—
31668	Razzmatazz smoothie	24	fluid ounce(s)	730	—	480	3.0	112.0	4.0	2.0	1.0	—	—	—
31669	Strawberries Wild smoothie	24	fluid ounce(s)	725	—	450	6.0	105.0	4.0	0.5	0	—	—	—
38421	Strawberry Tsunami smoothie	24	fluid ounce(s)	740	—	530	4.0	128.0	4.0	2.0	1.0	—	—	—
38427	Vibrant C juice	16	fluid ounce(s)	448	—	210	2.0	50.0	1.0	0	0	0	0	0
38428	Wheatgrass juice, freshly squeezed	1	ounce(s)	28	—	5	0.5	1.0	0	0	0	0	0	0
	Kentucky Fried Chicken (KFC)													
31850	BBQ baked beans	1	serving(s)	136	—	220	8.0	45.0	7.0	1.0	0	—	—	0
31853	Biscuit	1	item(s)	57	—	220	4.0	24.0	1.0	11.0	2.5	—	—	3.5
51223	Boneless Fiery Buffalo Wings	6	item(s)	211	—	530	30.0	44.0	3.0	26.0	5.0	—	—	2.5
39386	Boneless Honey BBQ Wings	6	item(s)	213	—	570	30.0	54.0	5.0	26.0	5.0	—	—	2.5
51224	Boneless Sweet & Spicy Wings	6	item(s)	203	—	550	30.0	50.0	3.0	26.0	5.0	—	—	2.5
31851	Cole slaw	1	serving(s)	130	—	180	1.0	22.0	3.0	10.0	1.5	—	—	0
31842	Colonel's Crispy Strips	3	item(s)	151	—	370	28.0	17.0	1.0	20.0	4.0	—	—	2.5
31849	Corn on the cob	1	item(s)	162	—	150	5.0	26.0	7.0	3.0	1.0	—	—	0
51221	Double Crunch sandwich	1	item(s)	213	—	520	27.0	39.0	3.0	29.0	5.0	—	—	1.5
3761	Extra Crispy chicken, breast	1	item(s)	162	—	370	33.0	10.0	2.0	22.0	5.0	—	—	1.5
3762	Extra Crispy chicken, drumstick	1	item(s)	60	—	150	12.0	4.0	0	10.0	2.5	—	—	1.0
3763	Extra Crispy chicken, thigh	1	item(s)	114	—	290	17.0	16.0	1.0	18.0	4.0	—	—	1.5
3764	Extra Crispy chicken, whole wing	1	item(s)	52	—	150	11.0	11.0	1.0	7.0	1.5	—	—	0
51218	Famous Bowls mashed potatoes with gravy	1	serving(s)	531	—	720	26.0	79.0	6.0	34.0	9.0	—	—	3.5
51219	Famous Bowls rice with gravy	1	serving(s)	384	—	610	25.0	67.0	5.0	27.0	8.0	—	—	2.5
31841	Honey BBQ chicken sandwich	1	item(s)	147	—	290	23.0	40.0	2.0	4.0	1.0	—	—	0
31833	Honey BBQ wing pieces	6	item(s)	157	—	460	27.0	26.0	3.0	27.0	6.0	—	—	2.0
10859	Hot wings pieces	6	piece(s)	134	—	450	26.0	19.0	2.0	30.0	7.0	—	—	2.0
42382	KFC Snacker sandwich	1	serving(s)	119	—	320	14.0	29.0	2.0	17.0	3.0	—	—	1.0
31848	Macaroni and cheese	1	serving(s)	136	—	180	8.0	18.0	0	8.0	3.5	—	—	1.0
31847	Mashed potatoes with gravy	1	serving(s)	151	—	140	2.0	20.0	1.0	5.0	1.0	—	—	0.5
10825	Original Recipe chicken, breast	1	item(s)	161	—	340	38.0	9.0	2.0	17.0	4.0	—	—	1.0
10826	Original Recipe chicken, drumstick	1	item(s)	59	—	140	13.0	3.0	0	8.0	2.0	—	—	0.5
10827	Original Recipe chicken, thigh	1	item(s)	126	—	350	19.0	7.0	1.0	27.0	7.0	—	—	1.0
10828	Original Recipe chicken, whole wing	1	item(s)	47	—	140	10.0	4.0	0	9.0	2.0	—	—	0.5
51222	Oven roasted Twister chicken wrap	1	item(s)	269	—	520	30.0	46.0	4.0	23.0	3.5	—	—	0
31844	Popcorn chicken, small or individual	1	item(s)	114	—	370	19.0	21.0	2.0	24.0	4.5	—	—	2.5
31852	Potato salad	1	serving(s)	128	—	180	2.0	22.0	2.0	9.0	1.5	—	—	0
10845	Potato wedges, small	1	serving(s)	102	—	250	4.0	32.0	3.0	12.0	2.0	—	—	1.5
31839	Tender Roast chicken sandwich with sauce	1	item(s)	236	—	430	37.0	29.0	2.0	18.0	3.5	—	—	0

PAGE KEY: H-4 = Breads/Baked Goods H-10 = Cereal/Rice/Pasta H-14 = Fruit H-18 = Vegetables/Legumes H-28 = Nuts/Seeds H-30 = Vegetarian H-32 = Dairy H-40 = Eggs H-40 = Seafood H-42 = Meats H-46 = Poultry H-46 = Processed Meats H-48 = Beverages H-52 = Fats/Oils H-54 = Sweets H-56 = Spices/Condiments/Sauces H-60 = Mixed Foods/Soups/Sandwiches H-64 = Fast Food H-84 = Convenience H-86 = Baby Foods

Chol (mg)	Calc (mg)	Iron (mg)	Magn (mg)	Pota (mg)	Sodi (mg)	Zinc (mg)	Vit A (μg)	Thia (mg)	Vit E (mg α)	Ribo (mg)	Niac (mg)	Vit B_6 (mg)	Fola (μg)	Vit C (mg)	Vit B_{12} (μg)	Sele (μg)
125	308	7.39	—	480.0	1580.0	—	—	—	—	—	—	—	—	0.6	—	—
135	532	0.00	—	750.0	280.0	—	—	—	—	—	—	—	—	0	—	—
5	200	1.80	60.0	1000.0	30.0	0.30	—	0.37	—	0.34	2.00	0.60	60.0	102.0	0	1.4
0	200	1.44	40.0	1010.0	115.0	0.60	—	0.09	—	0.25	0.80	0.70	24.0	15.0	0.2	1.4
5	200	1.80	16.0	510.0	35.0	0.30	—	0.06	—	0.25	6.00	0.70	140.0	54.0	0	1.4
5	100	1.80	24.0	810.0	60.0	0.30	—	0.09	—	0.25	5.00	0.50	100.0	78.0	0	1.4
0	150	2.70	80.0	1030.0	250.0	0.90	—	0.52	—	0.25	5.00	0.70	80.0	18.0	0	5.6
30	500	1.08	60.0	810.0	380.0	1.50	—	0.22	—	0.76	0.40	0.16	16.0	6.0	1.5	4.2
5	100	1.80	80.0	1170.0	35.0	0.30	—	0.37	—	0.34	1.90	0.60	100.0	180.0	0	1.4
5	100	1.08	60.0	1260.0	35.0	16.50	—	0.37	—	0.34	3.00	0.40	121.5	1302.0	0	1.4
0	250	1.44	16.0	500.0	50.0	0.30	—	0.03	—	0.25	5.00	0.60	120.0	54.0	0	1.4
0	1200	1.80	480.0	1070.0	45.0	16.50	—	5.55	—	6.12	68.00	7.40	640.0	288.0	10.8	77.0
0	20	0.00	8.0	200.0	10.0	0.00	—	0.03	—	0.17	14.00	1.80	320.0	36.0	0	0
5	100	1.08	24.0	780.0	50.0	0.30	—	0.15	—	0.25	5.00	0.70	120.0	72.0	0	1.4
0	60	1.08	60.0	990.0	0	0.30	—	0.45	—	0.13	2.00	0.20	160.0	246.0	0	0
0	100	1.80	60.0	1010.0	125.0	0.60	—	0.45	—	0.25	3.00	0.50	120.0	132.0	0	2.8
0	250	1.80	80.0	1380.0	160.0	0.90	—	0.45	—	0.42	2.00	0.50	140.0	240.0	0.6	1.4
5	100	0.72	32.0	740.0	60.0	0.30	—	0.06	—	0.25	4.00	0.60	80.0	18.0	0	1.4
0	1100	2.62	60.0	650.0	240.0	0.58	—	0.08	—	0.17	1.20	0.70	58.3	60.0	0	5.6
5	150	1.80	32.0	810.0	70.0	0.30	—	0.09	—	0.34	6.00	1.00	160.0	60.0	0	1.4
5	250	1.80	40.0	1050.0	180.0	0.90	—	0.12	—	0.34	0.80	0.40	40.0	60.0	0.6	1.4
5	100	1.08	24.0	480.0	10.0	0.30	—	0.06	—	0.34	14.00	1.80	320.0	90.0	0	1.4
0	20	1.08	40.0	720.0	0	0.30	—	0.30	—	0.10	1.60	0.40	80.0	678.0	0	0
0	0	1.80	8.0	80.0	0	0.00	0	0.03	—	0.03	0.40	0.04	16.0	3.6	0	2.8
0	100	2.70	—	—	730.0	—	—	—	—	—	—	—	—	1.2	—	—
0	40	1.80	—	—	640.0	—	—	—	—	—	—	—	—	0	—	—
65	40	1.80	—	—	2670.0	—	—	—	—	—	—	—	—	1.2	—	—
65	40	1.80	—	—	2210.0	—	—	—	—	—	—	—	—	1.2	—	—
65	60	1.80	—	—	2000.0	—	—	—	—	—	—	—	—	1.2	—	—
5	40	0.72	—	—	270.0	—	—	—	—	—	—	—	—	12.0	—	—
65	40	1.44	—	—	1220.0	—	0	—	—	—	—	—	—	1.2	—	—
0	60	1.08	—	—	10.0	—	—	—	—	—	—	—	—	6.0	—	—
55	100	2.70	—	—	1220.0	—	—	—	—	—	—	—	—	6.0	—	—
85	20	2.70	—	—	1020.0	—	—	—	—	—	—	—	—	1.2	—	—
55	0	1.44	—	—	300.0	—	0	—	—	—	—	—	—	0	—	—
95	20	2.70	—	—	700.0	—	—	—	—	—	—	—	—	—	—	—
45	20	1.08	—	—	340.0	—	—	—	—	—	—	—	—	0	—	—
35	200	5.40	—	—	2330.0	—	—	—	—	—	—	—	—	6.0	—	—
35	200	4.50	—	—	2130.0	—	—	—	—	—	—	—	—	6.0	—	—
60	80	2.70	—	—	710.0	—	—	—	—	—	—	—	—	2.4	—	—
140	40	1.80	—	—	970.0	—	—	—	—	—	—	—	—	21.0	—	—
115	40	1.44	—	—	990.0	—	—	—	—	—	—	—	—	1.2	—	—
25	60	2.70	—	—	690.0	—	—	—	—	—	—	—	—	2.4	—	—
15	150	0.72	—	—	800.0	—	—	—	—	—	—	—	—	1.2	—	—
0	40	1.44	—	—	560.0	—	—	—	—	—	—	—	—	1.2	—	—
135	20	2.70	—	—	960.0	—	—	—	—	—	—	—	—	6.0	—	—
70	20	1.08	—	—	340.0	—	—	—	—	—	—	—	—	0	—	—
110	20	2.70	—	—	870.0	—	—	—	—	—	—	—	—	1.2	—	—
50	20	1.44	—	—	350.0	—	0	—	—	—	—	—	—	1.2	—	—
60	40	6.30	—	—	1380.0	—	—	—	—	—	—	—	—	15.0	—	—
25	40	1.80	—	—	1110.0	—	0	—	—	—	—	—	—	0	—	—
5	0	0.36	—	—	470.0	—	—	—	—	—	—	—	—	6.0	—	—
0	20	1.08	—	—	700.0	—	0	—	—	—	—	—	—	0	—	—
80	80	2.70	—	—	1180.0	—	—	—	—	—	—	—	—	9.0	—	—

TABLE H-1 Table of Food Composition (*continued*) (Computer code is for Cengage Diet Analysis program) (For purposes of calculations, use "0" for t, <1, <.1, <.01, etc.)

DA+ Code	Food Description	Quantity	Measure	Wt (g)	H_2O (g)	Ener (kcal)	Prot (g)	Carb (g)	Fiber (g)	Fat (g)	Fat Breakdown (g) Sat	Mono	Poly	*Trans*
Fast Food—*continued*														
	Long John Silver													
39392	Baked cod	1	serving(s)	101	—	120	22.0	1.0	0	4.5	1.0	—	—	0
3777	Batter dipped fish sandwich	1	item(s)	177	—	470	18.0	48.0	3.0	23.0	5.0	—	—	4.5
37568	Battered fish	1	item(s)	92	—	260	12.0	17.0	0.5	16.0	4.0	—	—	4.5
37569	Breaded clams	1	serving(s)	85	—	240	8.0	22.0	1.0	13.0	2.0	—	—	2.5
37566	Chicken plank	1	item(s)	52	—	140	8.0	9.0	0.5	8.0	2.0	—	—	2.5
39404	Clam chowder	1	item(s)	227	—	220	9.0	23.0	0	10.0	4.0	—	—	1.0
39398	Cocktail sauce	1	ounce(s)	28	—	25	0	6.0	0	0	0	0	0	0
3770	Coleslaw	1	serving(s)	113	—	200	1.0	15.0	3.0	15.0	2.5	1.8	4.1	0
39400	French fries, large	1	item(s)	142	—	390	4.0	56.0	5.0	17.0	4.0	—	—	5.0
3774	Fries, regular	1	serving(s)	85	—	230	3.0	34.0	3.0	10.0	2.5	—	—	3.0
3779	Hushpuppy	1	piece(s)	23	—	60	1.0	9.0	1.0	2.5	0.5	—	—	1.0
3781	Shrimp, batter-dipped, 1 piece	1	piece(s)	14	—	45	2.0	3.0	0	3.0	1.0	—	—	1.0
39399	Tartar sauce	1	ounce(s)	28	—	100	0	4.0	0	9.0	1.5	—	—	—
39395	Ultimate Fish sandwich	1	item(s)	199	—	530	21.0	49.0	3.0	28.0	8.0	—	—	5.0
	McDonald's													
50828	Asian salad with grilled chicken	1	item(s)	362	—	290	31.0	23.0	6.0	10.0	1.0	—	—	0
2247	Barbecue sauce	1	item(s)	28	—	45	0	11.0	0	0	0	0	0	0
737	Big Mac hamburger	1	item(s)	219	—	560	25.0	47.0	3.0	30.0	10.0	—	—	1.5
29777	Caesar salad dressing	1	package(s)	44	—	150	1.0	5.0	0	13.0	2.5	—	—	—
38391	Caesar salad with grilled chicken, no dressing	1	serving(s)	278	230.6	181	26.4	10.5	3.1	6.0	2.9	1.7	0.8	0.2
38393	Caesar salad without chicken, no dressing	1	serving(s)	190	170.4	84	6.0	8.1	3.0	3.9	2.2	0.9	0.3	0.1
738	Cheeseburger	1	item(s)	119	—	310	15.0	35.0	1.0	12.0	6.0	—	—	1.0
29775	Chicken McGrill sandwich	1	item(s)	213	—	400	27.0	38.0	3.0	16.0	3.0	—	—	0
1873	Chicken McNuggets, 6 piece	1	serving(s)	96	—	250	15.0	15.0	0	15.0	3.0	—	—	1.5
3792	Chicken McNuggets, 4 piece	1	serving(s)	64	—	170	10.0	10.0	0	10.0	2.0	—	—	1.0
29774	Crispy chicken sandwich	1	item(s)	232	121.8	500	27.0	63.0	3.0	16.0	3.0	5.7	7.4	1.5
743	Egg McMuffin	1	item(s)	139	76.8	300	17.0	30.0	2.0	12.0	4.5	3.8	2.5	0
742	Filet-O-Fish sandwich	1	item(s)	141	—	400	14.0	42.0	1.0	18.0	4.0	—	—	1.0
2257	French fries, large	1	serving(s)	170	—	570	6.0	70.0	7.0	30.0	6.0	—	—	8.0
1872	French fries, small	1	serving(s)	74	—	250	2.0	30.0	3.0	13.0	2.5	—	—	3.5
33822	Fruit 'n Yogurt Parfait	1	item(s)	149	111.2	160	4.0	31.0	1.0	2.0	1.0	0.2	0.1	0
739	Hamburger	1	item(s)	105	—	260	13.0	33.0	1.0	9.0	3.5	—	—	0.5
2003	Hash browns	1	item(s)	53	—	140	1.0	15.0	2.0	8.0	1.5	—	—	2.0
2249	Honey sauce	1	item(s)	14	—	50	0	12.0	0	0	0	0	0	0
38397	Newman's Own creamy caesar salad dressing	1	item(s)	59	32.5	190	2.0	4.0	0	18.0	3.5	4.6	9.6	0
38398	Newman's Own low fat balsamic vinaigrette salad dressing	1	item(s)	44	29.1	40	0	4.0	0	3.0	0	1.0	1.2	0
38399	Newman's Own ranch salad dressing	1	item(s)	59	30.1	170	1.0	9.0	0	15.0	2.5	9.0	3.7	0
1874	Plain Hotcakes with syrup and margarine	3	item(s)	221	—	600	9.0	102.0	2.0	17.0	4.0	—	—	4.0
740	Quarter Pounder hamburger	1	item(s)	171	—	420	24.0	40.0	3.0	18.0	7.0	—	—	1.0
741	Quarter Pounder hamburger with cheese	1	item(s)	199	—	510	29.0	43.0	3.0	25.0	12.0	—	—	1.5
2005	Sausage McMuffin with egg	1	item(s)	165	82.4	450	20.0	31.0	2.0	27.0	10.0	10.9	4.6	0.5
50831	Side salad	1	item(s)	87	—	20	1.0	4.0	1.0	0	0	0	0	0
	Pizza Hut													
39009	Hot chicken wings	2	item(s)	57	—	110	11.0	1.0	0	6.0	2.0	—	—	0.3
14025	Meat Lovers hand tossed pizza	1	slice(s)	118	—	300	15.0	29.0	2.0	13.0	6.0	—	—	0.5
14026	Meat Lovers pan pizza	1	slice(s)	123	—	340	15.0	29.0	2.0	19.0	7.0	—	—	0.5
31009	Meat Lovers stuffed crust pizza	1	slice(s)	169	—	450	21.0	43.0	3.0	21.0	10.0	—	—	1.0
14024	Meat Lovers thin 'n crispy pizza	1	slice(s)	98	—	270	13.0	21.0	2.0	14.0	6.0	—	—	0.5
14031	Pepperoni Lovers hand tossed pizza	1	slice(s)	113	—	300	15.0	30.0	2.0	13.0	7.0	—	—	0.5
14032	Pepperoni Lovers pan pizza	1	slice(s)	118	—	340	15.0	29.0	2.0	19.0	7.0	—	—	0.5
31011	Pepperoni Lovers stuffed crust pizza	1	slice(s)	163	—	420	21.0	43.0	3.0	19.0	10.0	—	—	1.0
14030	Pepperoni Lovers thin 'n crispy pizza	1	slice(s)	92	—	260	13.0	21.0	2.0	14.0	7.0	—	—	0.5
10834	Personal Pan pepperoni pizza	1	slice(s)	61	—	170	7.0	18.0	0.5	8.0	3.0	—	—	1.0
10842	Personal Pan supreme pizza	1	slice(s)	77	—	190	8.0	19.0	1.0	9.0	3.5	—	—	1.0

PAGE KEY: H-4 = Breads/Baked Goods H-10 = Cereal/Rice/Pasta H-14 = Fruit H-18 = Vegetables/Legumes H-28 = Nuts/Seeds H-30 = Vegetarian H-32 = Dairy H-40 = Eggs H-40 = Seafood H-42 = Meats H-46 = Poultry H-46 = Processed Meats H-48 = Beverages H-52 = Fats/Oils H-54 = Sweets H-56 = Spices/Condiments/Sauces H-60 = Mixed Foods/Soups/Sandwiches H-64 = Fast Food H-84 = Convenience H-86 = Baby Foods

Chol (mg)	Calc (mg)	Iron (mg)	Magn (mg)	Pota (mg)	Sodi (mg)	Zinc (mg)	Vit A (µg)	Thia (mg)	Vit E (mg α)	Ribo (mg)	Niac (mg)	Vit B_6 (mg)	Fola (µg)	Vit C (mg)	Vit B_{12} (µg)	Sele (µg)
90	20	0.72	—	—	240.0	—	—	—	—	—	—	—	—	0	—	—
45	60	2.70	—	—	1210.0	—	—	—	—	—	—	—	—	2.4	—	—
35	20	0.72	—	—	790.0	—	—	—	—	—	—	—	—	4.8	—	—
10	20	1.08	—	—	1110.0	—	0	—	—	—	—	—	—	0	—	—
20	0	0.72	—	—	480.0	—	0	—	—	—	—	—	—	2.4	—	—
25	150	0.72	—	—	810.0	—	—	—	—	—	—	—	—	0	—	—
0	0	0.00	—	—	250.0	—	—	—	—	—	—	—	—	0	—	—
20	40	0.36	—	222.7	340.0	0.70	—	0.07	—	0.08	2.34	—	—	18.0	—	—
0	0	0.00	—	—	580.0	—	0	—	—	—	—	—	—	24.0	—	—
0	0	0.00	—	370.0	350.0	0.30	0	0.09	—	0.01	1.60	—	—	15.0	—	—
0	20	0.36	—	—	200.0	—	0	—	—	—	—	—	—	0	—	—
15	0	0.00	—	—	160.0	—	0	—	—	—	—	—	—	1.2	—	—
15	0	0.00	—	—	250.0	—	0	—	—	—	—	—	—	0	—	—
60	150	2.70	—	—	1400.0	—	—	—	—	—	—	—	—	4.8	—	—
65	150	3.60	—	—	890.0	—	—	—	—	—	—	—	—	54.0	—	—
0	0	0.00	—	55.0	260.0	—	—	—	—	—	—	—	—	0	—	—
80	250	4.50	—	400.0	1010.0	—	—	—	—	—	—	—	—	1.2	—	—
10	40	0.18	—	30.0	400.0	—	—	—	—	—	—	—	—	0.6	—	—
67	178	1.77	—	708.9	767.3	—	—	0.15	—	0.19	10.62	—	127.9	29.2	0.2	—
10	163	1.15	17.1	410.4	157.7	—	—	0.08	—	0.07	0.40	—	102.6	26.8	0	0.4
40	200	2.70	—	240.0	740.0	—	60.0	—	—	—	—	—	—	1.2	—	—
70	150	2.70	—	510.0	1010.0	—	—	—	—	—	—	—	—	6.0	—	—
35	20	0.72	—	240.0	670.0	—	—	—	—	—	—	—	—	1.2	—	—
25	0	0.36	—	160.0	450.0	—	—	—	—	—	—	—	—	1.2	—	—
60	80	3.60	62.6	526.6	1380.0	1.53	41.8	0.46	2.27	0.39	12.85	—	104.4	6.0	0.4	—
230	300	2.70	26.4	218.2	860.0	1.59	—	0.36	0.82	0.51	4.31	0.20	109.8	1.2	0.9	—
40	150	1.80	—	250.0	640.0	—	36.2	—	—	—	—	—	—	0	—	—
0	20	1.80	—	—	330.0	—	0	—	—	—	—	—	—	9.0	—	—
0	20	0.72	—	—	140.0	—	0	—	—	—	—	—	—	3.6	—	—
5	150	0.67	20.9	248.8	85.0	0.53	0	0.06	—	0.17	0.35	—	19.4	9.0	0.3	—
30	150	2.70	—	210.0	530.0	—	5.0	—	—	—	—	—	—	1.2	—	—
0	0	0.36	—	210.0	290.0	—	0	—	—	—	—	—	—	1.2	—	—
0	0	0.00	—	0	0	—	0	—	—	—	—	—	—	0	—	—
20	61	0.00	3.0	16.0	500.0	0.20	—	0.01	15.43	0.02	0.01	0.64	2.4	0	0.1	0.1
0	4	0.00	1.3	8.8	730.0	0.01	—	0.00	0.00	0.00	0.00	0.00	0	2.4	0	0
0	40	0.00	1.8	70.4	530.0	0.03	0	0.01	—	0.08	0.01	0.02	0.6	0	0	0.2
20	150	2.70	—	280.0	620.0	—	—	—	—	—	—	—	—	0	—	—
70	150	4.50	—	390.0	730.0	—	10.0	—	—	—	—	—	—	1.2	—	—
95	300	4.50	—	440.0	1150.0	—	100.0	—	—	—	—	—	—	1.2	—	—
255	300	3.60	29.7	282.2	950.0	2.01	—	0.43	0.82	0.56	4.83	0.24	—	0	1.2	—
0	20	0.72	—	—	10.0	—	—	—	—	—	—	—	—	15.0	—	—
70	0	0.36	—	—	450.0	—	—	—	—	—	—	—	—	0	—	—
35	150	1.80	—	—	760.0	—	—	—	—	—	—	—	—	6.0	—	—
35	150	2.70	—	—	750.0	—	—	—	—	—	—	—	—	6.0	—	—
55	250	2.70	—	—	1250.0	—	—	—	—	—	—	—	—	9.0	—	—
35	150	1.44	—	—	740.0	—	—	—	—	—	—	—	—	6.0	—	—
40	200	1.80	—	—	710.0	—	57.7	—	—	—	—	—	—	2.4	—	—
40	200	2.70	—	—	700.0	—	57.7	—	—	—	—	—	—	2.4	—	—
55	300	2.70	—	—	1120.0	—	—	—	—	—	—	—	—	3.6	—	—
40	200	1.44	—	—	690.0	—	58.0	—	—	—	—	—	—	2.4	—	—
15	80	1.44	—	—	340.0	—	38.5	—	—	—	—	—	—	1.4	—	—
20	80	1.86	—	—	420.0	—	—	—	—	—	—	—	—	3.6	—	—

TABLE H-1 Table of Food Composition (*continued*) (Computer code is for Cengage Diet Analysis program) (For purposes of calculations, use "0" for t, <1, <.1, <.01, etc

DA+ Code	Food Description	Quantity	Measure	Wt (g)	H_2O (g)	Ener (kcal)	Prot (g)	Carb (g)	Fiber (g)	Fat (g)	Fat Breakdown (g) Sat	Mono	Poly	*Trans*
Fast Food—*continued*														
39013	Personal Pan Veggie Lovers pizza	1	slice(s)	69	—	150	6.0	19.0	1.0	6.0	2.0	—	—	0.5
14028	Veggie Lovers hand tossed pizza	1	slice(s)	118	—	220	10.0	31.0	2.0	6.0	3.0	—	—	0.3
14029	Veggie Lovers pan pizza	1	slice(s)	119	—	260	10.0	30.0	2.0	12.0	4.0	—	—	0.3
31010	Veggie Lovers stuffed crust pizza	1	slice(s)	172	—	360	16.0	45.0	3.0	14.0	7.0	—	—	0.5
14027	Veggie Lovers thin 'n crispy pizza	1	slice(s)	101	—	180	8.0	23.0	2.0	7.0	3.0	—	—	0.5
39012	Wing blue cheese dipping sauce	1	item(s)	43	—	230	2.0	2.0	0	24.0	5.0	—	—	1.0
39011	Wing ranch dipping sauce	1	item(s)	43	—	210	0.5	4.0	0	22.0	3.5	—	—	0.5
	Starbucks													
38052	Cappuccino, tall	12	fluid ounce(s)	360	—	120	7.0	10.0	0	6.0	4.0	—	—	—
38053	Cappuccino, tall nonfat	12	fluid ounce(s)	360	—	80	7.0	11.0	0	0	0	0	0	0
38054	Cappuccino, tall soymilk	12	fluid ounce(s)	360	—	100	5.0	13.0	0.5	2.5	0	—	—	—
38059	Cinnamon spice mocha, tall nonfat w/o whipped cream	12	fluid ounce(s)	360	—	170	11.0	32.0	0	0.5	—	—	—	—
38057	Cinnamon spice mocha, tall w/whipped cream	12	fluid ounce(s)	360	—	320	10.0	31.0	0	17.0	11.0	—	—	—
38051	Espresso, single shot	1	fluid ounce(s)	30	—	5	0	1.0	0	0	0	0	0	0
38088	Flavored syrup, 1 pump	1	serving(s)	10	—	20	0	5.0	0	0	0	0	0	0
32562	Frappuccino bottled coffee drink, mocha	9½	fluid ounce(s)	298	—	190	6.0	39.0	3.0	3.0	2.0	—	—	—
32561	Frappuccino coffee drink, all bottled flavors	9½	fluid ounce(s)	281	—	190	7.0	35.0	0	3.5	2.5	—	—	—
38073	Frappuccino, mocha	12	fluid ounce(s)	360	—	220	5.0	44.0	0	3.0	1.5	—	—	—
38067	Frappuccino, tall caramel w/o whipped cream	12	fluid ounce(s)	360	—	210	4.0	43.0	0	2.5	1.5	—	—	—
38070	Frappuccino, tall coffee	12	fluid ounce(s)	360	—	190	4.0	38.0	0	2.5	1.5	—	—	—
39894	Frappuccino, tall coffee, light blend	12	fluid ounce(s)	360	—	110	5.0	22.0	2.0	1.0	0	—	—	—
38071	Frappuccino, tall espresso	12	fluid ounce(s)	360	—	160	4.0	33.0	0	2.0	1.5	—	—	—
39897	Frappuccino, tall mocha, light blend	12	fluid ounce(s)	360	—	140	5.0	28.0	3.0	1.5	0	—	—	—
39887	Frappuccino, tall Strawberries and Creme, w/o whipped cream	12	fluid ounce(s)	360	—	330	10.0	65.0	0	3.5	1.0	—	—	—
38063	Frappuccino, tall Tazo chai creme w/o whipped cream	12	fluid ounce(s)	360	—	280	10.0	52.0	0	3.5	1.0	—	—	—
38066	Frappuccino, tall Tazoberry	12	fluid ounce(s)	360	—	140	0.5	36.0	0.5	0	0	0	0	0
38065	Frappuccino, tall Tazoberry Crème	12	fluid ounce(s)	360	—	240	4.0	54.0	0.5	1.0	0	—	—	—
38080	Frappuccino, tall vanilla w/o whipped cream	12	fluid ounce(s)	360	—	270	10.0	51.0	0	3.5	1.0	—	—	—
39898	Frappuccino, tall white chocolate mocha, light blend	12	fluid ounce(s)	360	—	160	6.0	32.0	2.0	2.0	1.0	—	—	—
38074	Frappuccino, tall white chocolate w/o whipped cream	12	fluid ounce(s)	360	—	240	5.0	48.0	0	3.5	2.5	—	—	—
39883	Java Chip Frappuccino, tall w/o whipped cream	12	fluid ounce(s)	360	—	270	5.0	51.0	1.0	7.0	4.5	—	—	—
33111	Latte, tall w/nonfat milk	12	fluid ounce(s)	360	335.3	120	12.0	18.0	0	0	0	0	0	0
33112	Latte, tall w/whole milk	12	fluid ounce(s)	360	—	200	11.0	16.0	0	11.0	7.0	—	—	—
33109	Macchiato, tall caramel w/nonfat milk	12	fluid ounce(s)	360	—	170	11.0	30.0	0	1.0	0	—	—	—
33110	Macchiato, tall caramel w/whole milk	12	fluid ounce(s)	360	—	240	10.0	28.0	0	10.0	6.0	—	—	—
33107	Mocha coffee drink, tall nonfat, w/o whipped cream	12	fluid ounce(s)	360	—	170	11.0	33.0	1.0	1.5	0	—	—	—
38089	Mocha syrup	1	serving(s)	17	—	25	1.0	6.0	0	0.5	0	—	—	—
33108	Mocha, tall mocha w/whole milk	12	fluid ounce(s)	360	—	310	10.0	32.0	1.0	17.0	10.0	—	—	—
38042	Steamed apple cider, tall	12	fluid ounce(s)	360	—	180	0	45.0	0	0	0	0	0	0
38087	Tazo chai black tea, soymilk, tall	12	fluid ounce(s)	360	—	190	4.0	39.0	0.5	2.0	0	—	—	—
38084	Tazo chai black tea, tall	12	fluid ounce(s)	360	—	210	6.0	36.0	0	5.0	3.5	—	—	—
38083	Tazo chai black tea, tall nonfat	12	fluid ounce(s)	360	—	170	6.0	37.0	0	0	0	0	0	0
38076	Tazo iced tea, tall	12	fluid ounce(s)	360	—	60	0	16.0	0	0	0	0	0	0
38077	Tazo tea, grande lemonade	16	fluid ounce(s)	480	—	120	0	31.0	0	0	0	0	0	0

PAGE KEY: H-4 = Breads/Baked Goods H-10 = Cereal/Rice/Pasta H-14 = Fruit H-18 = Vegetables/Legumes H-28 = Nuts/Seeds H-30 = Vegetarian H-32 = Dairy H-40 = Eggs H-40 = Seafood H-42 = Meats H-46 = Poultry H-46 = Processed Meats H-48 = Beverages H-52 = Fats/Oils H-54 = Sweets H-56 = Spices/Condiments/Sauces H-60 = Mixed Foods/Soups/Sandwiches H-64 = Fast Food H-84 = Convenience H-86 = Baby Foods

Chol (mg)	Calc (mg)	Iron (mg)	Magn (mg)	Pota (mg)	Sodi (mg)	Zinc (mg)	Vit A (µg)	Thia (mg)	Vit E (mg α)	Ribo (mg)	Niac (mg)	Vit B_6 (mg)	Fola (µg)	Vit C (mg)	Vit B_{12} (µg)	Sele (µg)
10	80	1.80	—	—	280.0	—	—	—	—	—	—	—	—	3.6	—	—
15	150	1.80	—	—	490.0	—	—	—	—	—	—	—	—	9.0	—	—
15	150	2.70	—	—	470.0	—	—	—	—	—	—	—	—	9.0	—	—
35	250	2.70	—	—	980.0	—	—	—	—	—	—	—	—	9.0	—	—
15	150	1.44	—	—	480.0	—	—	—	—	—	—	—	—	9.0	—	—
25	20	0.00	—	—	550.0	—	0	—	—	—	—	—	—	0	—	—
10	0	0.00	—	—	340.0	—	0	—	—	—	—	—	—	0	—	—
25	250	0.00	—	—	95.0	—	—	—	—	—	—	—	—	1.2	0	—
3	200	0.00	—	—	100.0	—	—	—	—	—	—	—	—	0	0	—
0	250	0.72	—	—	75.0	—	—	—	—	—	—	—	—	0	0	—
5	300	0.72	—	—	150.0	—	—	—	—	—	—	—	—	0	0	—
70	350	1.08	—	—	140.0	—	—	—	—	—	—	—	—	2.4	0	—
0	0	0.00	—	—	0	—	0	—	—	—	—	—	—	0	0	—
0	0	0.00	—	—	0	—	0	—	—	—	—	—	—	0	0	—
12	219	1.08	—	530.0	110.0	—	—	—	—	—	—	—	—	0	—	—
15	250	0.36	—	510.0	105.0	—	—	—	—	—	—	—	—	0	—	—
10	150	0.72	—	—	180.0	—	—	—	—	—	—	—	—	0	0	—
10	150	0.00	—	—	180.0	—	—	—	—	—	—	—	—	0	0	—
10	150	0.00	—	—	180.0	—	—	—	—	—	—	—	—	0	0	—
0	150	0.00	—	—	220.0	—	—	—	—	—	—	—	—	0	—	—
10	100	0.00	—	—	160.0	—	—	—	—	—	—	—	—	0	0	—
0	150	0.72	—	—	220.0	—	—	—	—	—	—	—	—	0	—	—
3	350	0.00	—	—	270.0	—	—	—	—	—	—	—	—	21.0	—	—
3	350	0.00	—	—	270.0	—	—	—	—	—	—	—	—	3.6	0	—
0	0	0.00	—	—	30.0	—	0	—	—	—	—	—	—	0	0	—
0	150	0.00	—	—	125.0	—	0	—	—	—	—	—	—	1.2	0	—
3	350	0.00	—	—	370.0	—	—	—	—	—	—	—	—	3.6	0	—
3	150	0.00	—	—	250.0	—	—	—	—	—	—	—	—	0	—	—
10	150	0.00	—	—	210.0	—	—	—	—	—	—	—	—	0	0	—
10	150	1.44	—	—	220.0	—	—	—	—	—	—	—	—	0	—	—
5	350	0.00	39.8	—	170.0	1.35	—	0.12	—	0.47	0.36	0.13	17.5	0	1.3	—
45	400	0.00	46.6	—	160.0	1.28	—	0.12	—	0.54	0.34	0.14	16.8	2.4	1.2	—
5	300	0.00	—	—	160.0	—	—	—	—	—	—	—	—	1.2	—	—
30	300	0.00	—	—	135.0	—	—	—	—	—	—	—	—	2.4	—	—
5	300	2.70	—	—	135.0	—	—	—	—	—	—	—	—	0	—	—
0	0	0.72	—	—	0	—	0	—	—	—	—	—	—	0	0	—
55	300	2.70	—	—	115.0	—	—	—	—	—	—	—	—	0	—	—
0	0	1.08	—	—	15.0	—	0	—	—	—	—	—	—	0	0	—
0	200	0.72	—	—	70.0	—	—	—	—	—	—	—	—	0	0	—
20	200	0.36	—	—	85.0	—	—	—	—	—	—	—	—	1.2	0	—
5	200	0.36	—	—	95.0	—	—	—	—	—	—	—	—	0	0	—
0	0	0.00	—	—	0	—	0	—	—	—	—	—	—	0	0	—
0	0	0.00	—	—	15.0	—	0	—	—	—	—	—	—	4.8	0	—

TABLE H-1 Table of Food Composition (*continued*) (Computer code is for Cengage Diet Analysis program) (For purposes of calculations, use "0" for t, <1, <.1, <.01, etc

DA+ Code	Food Description	Quantity	Measure	Wt (g)	H_2O (g)	Ener (kcal)	Prot (g)	Carb (g)	Fiber (g)	Fat (g)	Fat Breakdown (g) Sat	Mono	Poly	*Trans*
	Fast Food—*continued*													
38045	Vanilla crème steamed nonfat milk, tall w/whipped cream	12	fluid ounce(s)	360	—	260	11.0	33.0	0	8.0	5.0	—	—	—
38046	Vanilla crème steamed soymilk, tall w/whipped cream	12	fluid ounce(s)	360	—	300	8.0	37.0	1.0	12.0	6.0	—	—	—
38044	Vanilla crème steamed whole milk, tall w/whipped cream	12	fluid ounce(s)	360	—	330	10.0	31.0	0	18.0	11.0	—	—	—
38090	Whipped cream	1	serving(s)	27	—	100	0	2.0	0	9.0	6.0	—	—	—
38062	White chocolate mocha, tall nonfat w/o whipped cream	12	fluid ounce(s)	360	—	260	12.0	45.0	0	4.0	3.0	—	—	—
38061	White chocolate mocha, tall w/whipped cream	12	fluid ounce(s)	360	—	410	11.0	44.0	0	20.0	13.0	—	—	—
38048	White hot chocolate, tall non-fat w/o whipped cream	12	fluid ounce(s)	360	—	300	15.0	51.0	0	4.5	3.5	—	—	—
38050	White hot chocolate, tall soymilk w/whipped cream	12	fluid ounce(s)	360	—	420	11.0	56.0	1.0	16.0	9.0	—	—	—
38047	White hot chocolate, tall w/whipped cream	12	fluid ounce(s)	360	—	460	13.0	50.0	0	22.0	15.0	—	—	—
	Subway													
15842	Cheese steak sandwich, 6″, wheat bread	1	item(s)	250	—	360	24.0	47.0	5.0	10.0	4.5	—	—	0
40478	Chicken and bacon ranch sand-wich, 6″, white or wheat bread	1	serving(s)	297	—	540	36.0	47.0	5.0	25.0	10.0	—	—	0.5
38622	Chicken and bacon ranch wrap with cheese	1	item(s)	257	—	440	41.0	18.0	9.0	27.0	10.0	—	—	0.5
32045	Chocolate chip cookie	1	item(s)	45	—	210	2.0	30.0	1.0	10.0	6.0	—	—	0
32048	Chocolate chip M&M cookie	1	item(s)	45	—	210	2.0	32.0	0.5	10.0	5.0	—	—	0
32049	Chocolate chunk cookie	1	item(s)	45	—	220	2.0	30.0	0.5	10.0	5.0	—	—	0
4024	Classic Italian B.M.T. sandwich, 6″, white bread	1	item(s)	236	—	440	22.0	45.0	2.0	21.0	8.5	—	—	0
15838	Classic tuna sandwich, 6″, wheat bread	1	item(s)	250	—	530	22.0	45.0	4.0	31.0	7.0	—	—	0.5
15837	Classic tuna sandwich, 6″, white bread	1	item(s)	243	—	520	21.0	43.0	2.0	31.0	7.5	—	—	0.5
16397	Club salad, no dressing and croutons	1	item(s)	412	—	160	18.0	15.0	4.0	4.0	1.5	—	—	0
3422	Club sandwich, 6″, white bread	1	item(s)	250	—	310	23.0	45.0	2.0	6.0	2.5	—	—	0
4030	Cold cut combo sandwich, 6″, white bread	1	item(s)	242	—	400	20.0	45.0	2.0	17.0	7.5	—	—	0.5
34030	Ham and egg breakfast sandwich	1	item(s)	142	—	310	16.0	35.0	3.0	13.0	3.5	—	—	0
3885	Ham sandwich, 6″, white bread	1	item(s)	238	—	310	17.0	52.0	2.0	5.0	2.0	—	—	0
3888	Meatball marinara sandwich, 6″, wheat bread	1	item(s)	377	—	560	24.0	63.0	7.0	24.0	11.0	—	—	1.0
4651	Meatball sandwich, 6″, white bread	1	item(s)	370	—	550	23.0	61.0	5.0	24.0	11.5	—	—	1.0
15839	Melt sandwich, 6″, white bread	1	item(s)	260	—	410	25.0	47.0	4.0	15.0	5.0	—	—	—
32046	Oatmeal raisin cookie	1	item(s)	45	—	200	3.0	30.0	1.0	8.0	4.0	—	—	0
16379	Oven-roasted chicken breast sandwich, 6″, wheat bread	1	item(s)	238	—	330	24.0	48.0	5.0	5.0	1.5	—	—	0
32047	Peanut butter cookie	1	item(s)	45	—	220	4.0	26.0	1.0	12.0	5.0	—	—	0
4655	Roast beef sandwich, 6″, wheat bread	1	item(s)	224	—	290	19.0	45.0	4.0	5.0	2.0	—	—	0
3957	Roast beef sandwich, 6″, white bread	1	item(s)	217	—	280	18.0	43.0	2.0	5.0	2.5	—	—	0
16378	Roasted chicken breast, 6″, white bread	1	item(s)	231	—	320	23.0	46.0	3.0	5.0	2.0	—	—	0
34028	Southwest steak and cheese sandwich, 6″, Italian bread	1	item(s)	271	—	450	24.0	48.0	6.0	20.0	6.0	—	—	0
4032	Spicy Italian sandwich, 6″, white bread	1	item(s)	220	—	470	20.0	43.0	2.0	25.0	9.5	—	—	0
4031	Steak and cheese sandwich, 6″, white bread	1	item(s)	243	—	350	23.0	45.0	3.0	10.0	5.0	—	—	0
32050	Sugar cookie	1	item(s)	45	—	220	2.0	28.0	0.5	12.0	6.0	—	—	0
40477	Sweet onion chicken teriyaki sandwich, 6″, white or wheat bread	1	serving(s)	281	—	370	26.0	59.0	4.0	5.0	1.5	—	—	0
38623	Turkey breast and bacon melt wrap with chipotle sauce	1	item(s)	228	—	380	31.0	20.0	9.0	24.0	7.0	—	—	0

PAGE KEY: H-4 = Breads/Baked Goods H-10 = Cereal/Rice/Pasta H-14 = Fruit H-18 = Vegetables/Legumes H-28 = Nuts/Seeds H-30 = Vegetarian H-32 = Dairy H-40 = Eggs H-40 = Seafood H-42 = Meats H-46 = Poultry H-46 = Processed Meats H-48 = Beverages H-52 = Fats/Oils H-54 = Sweets H-56 = Spices/Condiments/Sauces H-60 = Mixed Foods/Soups/Sandwiches H-64 = Fast Food H-84 = Convenience H-86 = Baby Foods

Chol (mg)	Calc (mg)	Iron (mg)	Magn (mg)	Pota (mg)	Sodi (mg)	Zinc (mg)	Vit A (µg)	Thia (mg)	Vit E (mg α)	Ribo (mg)	Niac (mg)	Vit B_6 (mg)	Fola (µg)	Vit C (mg)	Vit B_{12} (µg)	Sele (µg)
35	350	0.00	—	—	170.0	—	—	—	—	—	—	—	—	0	0	—
30	400	1.44	—	—	130.0	—	—	—	—	—	—	—	—	0	0	—
65	350	0.00	—	—	140.0	—	—	—	—	—	—	—	—	0	0	—
40	0	0.00	—	—	10.0	—	—	—	—	—	—	—	—	0	0	—
5	400	0.00	—	—	210.0	—	—	—	—	—	—	—	—	0	0	—
70	400	0.00	—	—	210.0	—	—	—	—	—	—	—	—	2.4	0	—
10	450	0.00	—	—	250.0	—	—	—	—	—	—	—	—	0	0	—
35	500	1.44	—	—	210.0	—	—	—	—	—	—	—	—	0	0	—
75	500	0.00	—	—	250.0	—	—	—	—	—	—	—	—	3.6	0	—
35	150	8.10	—	—	1090.0	—	—	—	—	—	—	—	—	18.0	—	—
90	250	4.50	—	—	1400.0	—	—	—	—	—	—	—	—	21.0	—	—
90	300	2.70	—	—	1680.0	—	—	—	—	—	—	—	—	9.0	—	—
15	0	1.08	—	—	150.0	—	—	—	—	—	—	—	—	0	—	—
10	20	1.00	—	—	100.0	—	—	—	—	—	—	—	—	0	—	—
10	0	1.00	—	—	100.0	—	—	—	—	—	—	—	—	0	—	—
55	150	2.70	—	—	1770.0	—	—	—	—	—	—	—	—	16.8	—	—
45	100	5.40	—	—	1030.0	—	—	—	—	—	—	—	—	21.0	—	—
45	100	3.60	—	—	1010.0	—	—	—	—	—	—	—	—	16.8	—	—
35	60	3.60	—	—	880.0	—	—	—	—	—	—	—	—	30.0	—	—
35	60	3.60	—	—	1290.0	—	—	—	—	—	—	—	—	13.8	—	—
60	150	3.60	—	—	1530.0	—	—	—	—	—	—	—	—	16.8	—	—
190	80	4.50	—	—	720.0	—	66.7	—	—	—	—	—	—	3.6	—	—
25	60	2.70	—	—	1375.0	—	—	—	—	—	—	—	—	13.8	—	—
45	200	7.20	—	—	1610.0	—	—	—	—	—	—	—	—	36.0	—	—
45	200	5.40	—	—	1590.0	—	—	—	—	—	—	—	—	31.8	—	—
45	150	5.40	—	—	1720.0	—	—	—	—	—	—	—	—	24.0	—	—
15	20	1.08	—	—	170.0	—	—	—	—	—	—	—	—	0	—	—
45	60	4.50	—	—	1020.0	—	—	—	—	—	—	—	—	18.0	—	—
15	20	0.72	—	—	200.0	—	—	—	—	—	—	—	—	0	—	—
20	60	6.30	—	—	920.0	—	—	—	—	—	—	—	—	18.0	—	—
20	60	4.50	—	—	900.0	—	—	—	—	—	—	—	—	13.8	—	—
45	60	2.70	—	—	1000.0	—	—	—	—	—	—	—	—	13.8	—	—
45	150	8.10	—	—	1310.0	—	—	—	—	—	—	—	—	21.0	—	—
55	60	2.70	—	—	1650.0	—	—	—	—	—	—	—	—	16.8	—	—
35	150	6.30	—	—	1070.0	—	—	—	—	—	—	—	—	13.8	—	—
15	0	0.72	—	—	140.0	—	—	—	—	—	—	—	—	0	—	—
50	80	4.50	—	—	1220.0	—	—	—	—	—	—	—	—	24.0	—	—
50	200	2.70	—	—	1780.0	—	—	—	—	—	—	—	—	6.0	—	—

TABLE H-1 Table of Food Composition (*continued*) (Computer code is for Cengage Diet Analysis program) (For purposes of calculations, use "0" for t, <1, <.1, <.01, etc

DA+ Code	Food Description	Quantity	Measure	Wt (g)	H_2O (g)	Ener (kcal)	Prot (g)	Carb (g)	Fiber (g)	Fat (g)	Fat Breakdown (g)			
											Sat	Mono	Poly	*Trans*
Fast Food—*continued*														
15834	Turkey breast and ham sandwich, 6″, white bread	1	item(s)	227	—	280	19.0	45.0	2.0	5.0	2.0	—	—	0
16376	Turkey breast sandwich, 6″, white bread	1	item(s)	217	—	270	17.0	44.0	2.0	4.5	2.0	—	—	0
15841	Veggie Delite sandwich, 6″, wheat bread	1	item(s)	167	—	230	9.0	44.0	4.0	3.0	1.0	—	—	0
16375	Veggie Delite, 6″, white bread	1	item(s)	160	—	220	8.0	42.0	2.0	3.0	1.5	—	—	0
32051	White chip macadamia nut cookie	1	item(s)	45	—	220	2.0	29.0	0.5	11.0	5.0	—	—	0
	Taco Bell													
29906	7-Layer burrito	1	item(s)	283	—	490	17.0	65.0	9.0	18.0	7.0	—	—	1.0
744	Bean burrito	1	item(s)	198	—	340	13.0	54.0	8.0	9.0	3.5	—	—	0.5
749	Beef burrito supreme	1	item(s)	248	—	410	17.0	51.0	7.0	17.0	8.0	—	—	1.0
33417	Beef Chalupa Supreme	1	item(s)	153	—	380	14.0	30.0	3.0	23.0	7.0	—	—	0.5
34474	Beef Gordita Baja	1	item(s)	153	—	340	13.0	29.0	4.0	19.0	5.0	—	—	0
29910	Beef Gordita Supreme	1	item(s)	153	—	310	14.0	29.0	3.0	16.0	6.0	—	—	0.5
2014	Beef soft taco	1	item(s)	99	—	200	10.0	21.0	3.0	9.0	4.0	—	—	0
10860	Beef soft taco supreme	1	item(s)	135	—	250	11.0	23.0	3.0	13.0	6.0	—	—	0.5
34472	Chicken burrito supreme	1	item(s)	248	—	390	20.0	49.0	6.0	13.0	6.0	—	—	0.5
33418	Chicken Chalupa Supreme	1	item(s)	153	—	360	17.0	29.0	2.0	20.0	5.0	—	—	0
34475	Chicken Gordita Baja	1	item(s)	153	—	320	17.0	28.0	3.0	16.0	3.5	—	—	0
29909	Chicken quesadilla	1	item(s)	184	—	520	28.0	40.0	3.0	28.0	12.0	—	—	0.5
29907	Chili cheese burrito	1	item(s)	156	—	390	16.0	40.0	3.0	18.0	9.0	—	—	1.5
10794	Cinnamon twists	1	serving(s)	35	—	170	1.0	26.0	1.0	7.0	0	—	—	0
29911	Grilled chicken Gordita Supreme	1	item(s)	153	—	290	17.0	28.0	2.0	12.0	5.0	—	—	0
14463	Grilled chicken soft taco	1	item(s)	99	—	190	14.0	19.0	1.0	6.0	2.5	—	—	—
29912	Grilled Steak Gordita Supreme	1	item(s)	153	—	290	15.0	28.0	2.0	13.0	5.0	—	—	0
29904	Grilled steak soft taco	1	item(s)	128	—	270	12.0	20.0	2.0	16.0	4.5	—	—	0
29905	Grilled steak soft taco supreme	1	item(s)	135	—	235	13.0	21.0	1.0	11.0	6.0	—	—	—
2021	Mexican pizza	1	serving(s)	216	—	530	20.0	42.0	7.0	30.0	8.0	—	—	1.0
29894	Mexican rice	1	serving(s)	131	—	170	6.0	23.0	1.0	11.0	3.0	—	—	0
10772	Meximelt	1	serving(s)	128	—	280	15.0	22.0	3.0	14.0	7.0	—	—	0.5
2011	Nachos	1	serving(s)	99	—	330	4.0	32.0	2.0	21.0	3.5	—	—	2.0
2012	Nachos Bellgrande	1	serving(s)	308	—	770	19.0	77.0	12.0	44.0	9.0	—	—	3.0
2023	Pintos 'n cheese	1	serving(s)	128	—	150	9.0	19.0	7.0	6.0	3.0	—	—	0.5
34473	Steak burrito supreme	1	item(s)	248	—	380	18.0	49.0	6.0	14.0	7.0	—	—	0.5
33419	Steak Chalupa Supreme	1	item(s)	153	—	360	15.0	28.0	2.0	21.0	6.0	—	—	0
747	Taco	1	item(s)	78	—	170	8.0	13.0	3.0	10.0	3.5	—	—	0
2015	Taco salad with salsa, with shell	1	serving(s)	548	—	840	30.0	80.0	15.0	45.0	11.0	—	—	1.5
14459	Taco supreme	1	item(s)	113	—	210	9.0	15.0	3.0	13.0	6.0	—	—	0
748	Tostada	1	item(s)	170	—	230	11.0	27.0	7.0	10.0	3.5	—	—	0.5
Convenience Meals														
	Banquet													
29961	Barbeque chicken meal	1	item(s)	281	—	330	16.0	37.0	2.0	13.0	3.0	—	—	—
14788	Boneless white fried chicken meal	1	item(s)	286	—	310	10.0	21.0	4.0	20.0	5.0	—	—	—
29960	Fish sticks meal	1	item(s)	207	—	470	13.0	58.0	1.0	20.0	3.5	—	—	—
29957	Lasagna with meat sauce meal	1	item(s)	312	—	320	15.0	46.0	7.0	9.0	4.0	—	—	—
14777	Macaroni and cheese meal	1	item(s)	340	—	420	15.0	57.0	5.0	14.0	8.0	—	—	—
1741	Meatloaf meal	1	item(s)	269	—	240	14.0	20.0	4.0	11.0	4.0	—	—	—
39418	Pepperoni pizza meal	1	item(s)	191	—	480	11.0	56.0	5.0	23.0	8.0	—	—	—
33759	Roasted white turkey meal	1	item(s)	255	—	230	14.0	30.0	5.0	6.0	2.0	—	—	—
1743	Salisbury steak meal	1	item(s)	269	196.9	380	12.0	28.0	3.0	24.0	12.0	—	—	—
	Budget Gourmet													
1914	Cheese manicotti with meat sauce entrée	1	item(s)	284	194.0	420	18.0	38.0	4.0	22.0	11.0	6.0	1.3	—
1915	Chicken with fettucini entrée	1	item(s)	284	—	380	20.0	33.0	3.0	19.0	10.0	—	—	—
3986	Light beef stroganoff entrée	1	item(s)	248	177.0	290	20.0	32.0	3.0	7.0	4.0	—	—	—
3996	Light sirloin of beef in herb sauce entrée	1	item(s)	269	214.0	260	19.0	30.0	5.0	7.0	4.0	2.3	0.3	—
3987	Light vegetable lasagna entrée	1	item(s)	298	227.0	290	15.0	36.0	4.8	9.0	1.8	0.9	0.6	—

PAGE KEY: H-4 = Breads/Baked Goods H-10 = Cereal/Rice/Pasta H-14 = Fruit H-18 = Vegetables/Legumes H-28 = Nuts/Seeds H-30 = Vegetarian H-32 = Dairy H-40 = Eggs H-40 = Seafood H-42 = Meats H-46 = Poultry H-46 = Processed Meats H-48 = Beverages H-52 = Fats/Oils H-54 = Sweets H-56 = Spices/Condiments/Sauces H-60 = Mixed Foods/Soups/Sandwiches H-64 = Fast Food H-84 = Convenience H-86 = Baby Foods

Chol (mg)	Calc (mg)	Iron (mg)	Magn (mg)	Pota (mg)	Sodi (mg)	Zinc (mg)	Vit A (µg)	Thia (mg)	Vit E (mg α)	Ribo (mg)	Niac (mg)	Vit B_6 (mg)	Fola (µg)	Vit C (mg)	Vit B_{12} (µg)	Sele (µg)
25	60	2.70	—	—	1210.0	—	—	—	—	—	—	—	—	13.8	—	—
20	60	2.70	—	—	1000.0	—	—	—	—	—	—	—	—	13.8	—	—
0	60	4.50	—	—	520.0	—	—	—	—	—	—	—	—	18.0	—	—
0	60	2.70	—	—	500.0	—	—	—	—	—	—	—	—	13.8	—	—
15	20	0.72	—	—	160.0	—	—	—	—	—	—	—	—	0	—	—
25	250	5.40	—	—	1350.0	—	—	—	—	—	—	—	—	15.0	—	—
5	200	4.50	—	—	1190.0	—	5.9	—	—	—	—	—	—	4.8	—	—
40	200	4.50	—	—	1340.0	—	9.9	—	—	—	—	—	—	6.0	—	—
40	150	2.70	—	—	620.0	—	—	—	—	—	—	—	—	3.6	—	—
35	100	2.70	—	—	780.0	—	—	—	—	—	—	—	—	2.4	—	—
40	150	2.70	—	—	620.0	—	—	—	—	—	—	—	—	3.6	—	—
25	100	1.80	—	—	630.0	—	—	—	—	—	—	—	—	1.2	—	—
40	150	2.70	—	—	650.0	—	—	—	—	—	—	—	—	3.6	—	—
45	200	4.50	—	—	1360.0	—	—	—	—	—	—	—	—	9.0	—	—
45	100	2.70	—	—	650.0	—	—	—	—	—	—	—	—	4.8	—	—
40	100	1.80	—	—	800.0	—	—	—	—	—	—	—	—	3.6	—	—
75	450	3.60	—	—	1420.0	—	—	—	—	—	—	—	—	1.2	—	—
40	300	1.80	—	—	1080.0	—	—	—	—	—	—	—	—	0	—	—
0	0	0.37	—	—	200.0	—	0	—	—	—	—	—	—	0	—	—
45	150	1.80	—	—	650.0	—	—	—	—	—	—	—	—	4.8	—	—
30	100	1.08	—	—	550.0	—	14.6	—	—	—	—	—	—	1.2	—	—
40	100	2.70	—	—	530.0	—	—	—	—	—	—	—	—	3.6	—	—
35	100	2.70	—	—	660.0	—	—	—	—	—	—	—	—	3.6	—	—
35	120	1.44	—	—	565.0	—	29.2	—	—	—	—	—	—	3.6	—	—
40	350	3.60	—	—	1000.0	—	—	—	—	—	—	—	—	4.8	—	—
15	100	1.44	—	—	790.0	—	—	—	—	—	—	—	—	3.6	—	—
40	250	2.70	—	—	880.0	—	—	—	—	—	—	—	—	2.4	—	—
3	80	0.71	—	—	530.0	—	0	—	—	—	—	—	—	0	—	—
35	200	3.60	—	—	1280.0	—	—	—	—	—	—	—	—	4.8	—	—
15	150	1.44	—	—	670.0	—	—	—	—	—	—	—	—	3.6	—	—
35	200	4.50	—	—	1250.0	—	9.9	—	—	—	—	—	—	9.0	—	—
40	100	2.70	—	—	530.0	—	—	—	—	—	—	—	—	3.6	—	—
25	80	1.08	—	—	350.0	—	—	—	—	—	—	—	—	1.2	—	—
65	450	7.20	—	—	1780.0	—	—	—	—	—	—	—	—	12.0	—	—
40	100	1.08	—	—	370.0	—	—	—	—	—	—	—	—	3.6	—	—
15	200	1.80	—	—	730.0	—	—	—	—	—	—	—	—	4.8	—	—
																—
50	40	1.08	—	—	1210.0	—	0	—	—	—	—	—	—	4.8	—	—
45	80	1.44	—	—	1200.0	—	—	—	—	—	—	—	—	18.0	—	—
55	20	1.44	—	—	710.0	—	—	—	—	—	—	—	—	0	—	—
20	100	2.70	—	—	1170.0	—	—	—	—	—	—	—	—	0	—	—
20	150	1.44	—	—	1330.0	—	0	—	—	—	—	—	—	0	—	—
30	0	1.80	—	—	1040.0	—	0	—	—	—	—	—	—	0	—	—
35	150	1.80	—	—	870.0	—	0	—	—	—	—	—	—	0	—	—
25	60	1.80	—	—	1070.0	—	—	—	—	—	—	—	—	3.6	—	—
60	40	1.44	—	—	1140.0	—	0	—	—	—	—	—	—	0	—	—
85	300	2.70	45.4	484.0	810.0	2.29	—	0.45	—	0.51	4.00	0.22	30.7	0	0.7	—
85	100	2.70	—	—	810.0	—	—	0.15	—	0.42	6.00	—	—	0	—	—
35	40	1.80	38.9	280.0	580.0	4.71	—	0.17	—	0.36	4.28	0.27	18.9	2.4	2.5	—
30	40	1.80	57.7	540.0	850.0	4.81	—	0.15	—	0.29	5.53	0.37	38.4	6.0	1.6	—
15	283	3.03	78.5	420.0	780.0	1.39	—	0.22	—	0.45	3.13	0.32	74.8	59.1	0.2	—

TABLE H-1 Table of Food Composition (*continued*) (Computer code is for Cengage Diet Analysis program) (For purposes of calculations, use "0" for t, <1, <.1, <.01, etc

DA+ Code	Food Description	Quantity	Measure	Wt (g)	H_2O (g)	Ener (kcal)	Prot (g)	Carb (g)	Fiber (g)	Fat (g)	Fat Breakdown (g) Sat	Mono	Poly	*Trans*
Convenience Meals—*continued*														
	Healthy Choice													
9425	Cheese French bread pizza	1	item(s)	170	—	340	22.0	51.0	5.0	5.0	1.5	—	—	—
9306	Chicken enchilada suprema meal	1	item(s)	320	251.5	360	13.0	59.0	8.0	7.0	3.0	2.0	2.0	—
3821	Familiar Favorites lasagna bake with meat sauce entrée	1	item(s)	255	—	270	13.0	38.0	4.0	7.0	2.5	—	—	—
13744	Familiar Favorites sesame chicken with vegetables and rice entrée	1	item(s)	255	—	260	17.0	34.0	4.0	6.0	2.0	2.0	2.0	—
9316	Lemon pepper fish meal	1	item(s)	303	—	280	11.0	49.0	5.0	5.0	2.0	1.0	2.0	—
9322	Traditional salisbury steak meal	1	item(s)	354	250.3	360	23.0	45.0	5.0	9.0	3.5	4.0	1.0	—
9359	Traditional turkey breasts meal	1	item(s)	298	—	330	21.0	50.0	4.0	5.0	2.0	1.5	1.5	—
	Stouffers													
2313	Cheese French bread pizza	1	serving(s)	294	—	380	15.0	43.0	3.0	16.0	6.0	—	—	—
11138	Cheese manicotti with tomato sauce entrée	1	item(s)	255	—	360	18.0	41.0	2.0	14.0	6.0	—	—	—
2366	Chicken pot pie entrée	1	item(s)	284	—	740	23.0	56.0	4.0	47.0	18.0	12.4	10.5	—
11116	Homestyle baked chicken breast with mashed potatoes and gravy entrée	1	item(s)	252	—	270	21.0	21.0	2.0	11.0	3.5	—	—	—
11146	Homestyle beef pot roast and potatoes entrée	1	item(s)	252	—	260	16.0	24.0	3.0	11.0	4.0	—	—	—
11152	Homestyle roast turkey breast with stuffing and mashed potatoes entrée	1	item(s)	273	—	290	16.0	30.0	2.0	12.0	3.5	—	—	—
11043	Lean Cuisine Comfort Classics baked chicken and whipped potatoes and stuffing entrée	1	item(s)	245	—	240	15.0	34.0	3.0	4.5	1.0	2.0	1.0	0
11046	Lean Cuisine Comfort Classics honey mustard chicken with rice pilaf entrée	1	item(s)	227	—	250	17.0	37.0	1.0	4.0	1.0	1.0	1.0	0
9479	Lean Cuisine Deluxe French bread pizza	1	item(s)	174	—	310	16.0	44.0	3.0	9.0	3.5	0.5	0.5	0
360	Lean Cuisine One Dish Favorites chicken chow mein with rice	1	item(s)	255	—	190	13.0	29.0	2.0	2.5	0.5	1.0	0.5	0
11054	Lean Cuisine One Dish Favorites chicken enchilada Suiza with Mexican-style rice	1	serving(s)	255	—	270	10.0	47.0	3.0	4.5	2.0	1.5	1.0	0
9467	Lean Cuisine One Dish Favorites fettucini alfredo entrée	1	item(s)	262	—	270	13.0	39.0	2.0	7.0	3.5	2.0	1.0	0
11055	Lean Cuisine One Dish Favorites lasagna with meat sauce entrée	1	item(s)	298	—	320	19.0	44.0	4.0	7.0	3.0	2.0	0.5	0
	Weight Watchers													
11164	Smart Ones chicken enchiladas suiza entrée	1	item(s)	255	—	340	12.0	38.0	3.0	10.0	4.5	—	—	—
39763	Smart Ones chicken oriental entrée	1	item(s)	255	—	230	15.0	34.0	3.0	4.5	1.0	—	—	—
11187	Smart Ones pepperoni pizza	1	item(s)	198	—	400	22.0	58.0	4.0	9.0	3.0	—	—	—
39765	Smart Ones spaghetti bolognese entrée	1	item(s)	326	—	280	17.0	43.0	5.0	5.0	2.0	—	—	—
31512	Smart Ones spicy szechuan style vegetables and chicken	1	item(s)	255	—	220	11.0	34.0	4.0	5.0	1.0	—	—	—
Baby Foods														
787	Apple juice	4	fluid ounce(s)	127	111.6	60	0	14.8	0.1	0.1	0	0	0	—
778	Applesauce, strained	4	tablespoon(s)	64	56.7	26	0.1	6.9	1.1	0.1	0	0	0	—
779	Bananas with tapioca, strained	4	tablespoon(s)	60	50.4	34	0.2	9.2	1.0	0	0	0	0	—
604	Carrots, strained	4	tablespoon(s)	56	51.7	15	0.4	3.4	1.0	0.1	0	0	0	—
770	Chicken noodle dinner, strained	4	tablespoon(s)	64	54.8	42	1.7	5.8	1.3	1.3	0.4	0.5	0.3	—
801	Green beans, strained	4	tablespoon(s)	60	55.1	16	0.7	3.8	1.3	0.1	0	0	0	—
910	Human milk, mature	2	fluid ounce(s)	62	53.9	43	0.6	4.2	0	2.7	1.2	1.0	0.3	—
760	Mixed cereal, prepared with whole milk	4	ounce(s)	113	84.6	128	5.4	18.0	1.5	4.0	2.2	1.2	0.4	—
772	Mixed vegetable dinner, strained	2	ounce(s)	57	50.3	23	0.7	5.4	0.8	0	—	—	0	—
762	Rice cereal, prepared with whole milk	4	ounce(s)	113	84.6	130	4.4	18.9	0.1	4.1	2.6	1.0	0.2	—
758	Teething biscuits	1	item(s)	11	0.7	44	1.0	8.6	0.2	0.6	0.2	0.2	0.1	—

PAGE KEY: H-4 = Breads/Baked Goods H-10 = Cereal/Rice/Pasta H-14 = Fruit H-18 = Vegetables/Legumes H-28 = Nuts/Seeds H-30 = Vegetarian H-32 = Dairy H-40 = Eggs H-40 = Seafood H-42 = Meats H-46 = Poultry H-46 = Processed Meats H-48 = Beverages H-52 = Fats/Oils H-54 = Sweets H-56 = Spices/Condiments/Sauces H-60 = Mixed Foods/Soups/Sandwiches H-64 = Fast Food H-84 = Convenience H-86 = Baby Foods

Chol (mg)	Calc (mg)	Iron (mg)	Magn (mg)	Pota (mg)	Sodi (mg)	Zinc (mg)	Vit A (µg)	Thia (mg)	Vit E (mg α)	Ribo (mg)	Niac (mg)	Vit B_6 (mg)	Fola (µg)	Vit C (mg)	Vit B_{12} (µg)	Sele (µg)
10	350	3.60	—	—	600.0	—	—	—	—	—	—	—	—	0	—	—
30	40	1.44	—	—	580.0	—	—	—	—	—	—	—	—	3.6	—	—
20	100	1.80	—	—	600.0	—	—	—	—	—	—	—	—	0	—	—
35	18	0.72	—	—	580.0	—	—	—	—	—	—	—	—	12.0	—	—
35	20	0.36	—	—	580.0	—	—	—	—	—	—	—	—	30.0	—	—
45	80	2.70	—	—	580.0	—	—	—	—	—	—	—	—	21.0	—	—
35	40	1.80	—	—	600.0	—	—	—	—	—	—	—	—	0	—	—
30	200	1.80	—	230.0	660.0	—	—	—	—	—	—	—	—	2.4	—	—
70	250	1.44	—	550.0	920.0	—	—	—	—	—	—	—	—	6.0	—	—
65	150	2.70	—	—	1170.0	—	—	—	—	—	—	—	—	2.4	—	—
55	20	0.72	—	490.0	770.0	—	0	—	—	—	—	—	—	0	—	—
35	20	1.80	—	800.0	960.0	—	—	—	—	—	—	—	—	6.0	—	—
45	40	1.08	—	490.0	970.0	—	—	—	—	—	—	—	—	3.6	—	—
25	40	1.16	—	500.0	650.0	—	—	—	—	—	—	—	—	3.6	—	—
30	64	0.38	—	370.0	650.0	—	—	—	—	—	—	—	—	0	—	—
20	150	2.70	—	300.0	700.0	—	—	—	—	—	—	—	—	15.0	—	—
25	40	0.72	—	380.0	650.0	—	—	—	—	—	—	—	—	2.4	—	—
20	150	0.72	—	350.0	510.0	—	—	—	—	—	—	—	—	2.4	—	—
15	200	0.72	—	290.0	690.0	—	0	—	—	—	—	—	—	0	—	—
30	250	1.47	—	610.0	690.0	—	—	—	—	—	—	—	—	2.4	—	—
40	200	0.72	—	—	800.0	—	—	—	—	—	—	—	—	2.4	—	—
35	40	0.72	—	—	790.0	—	—	—	—	—	—	—	—	6.0	—	—
15	200	1.08	—	401.0	700.0	—	69.1	—	—	—	—	—	—	4.8	—	—
15	150	3.60	—	—	670.0	—	—	—	—	—	—	—	—	9.0	—	—
10	40	1.44	—	—	890.0	—	—	—	—	—	—	—	—	0	—	—
0	5	0.72	3.8	115.4	3.8	0.03	1.3	0.01	0.76	0.02	0.10	0.03	0	73.4	0	0.1
0	3	0.12	1.9	45.4	1.3	0.01	0.6	0.01	0.36	0.02	0.04	0.02	1.3	24.5	0	0.2
0	3	0.12	6.0	52.8	5.4	0.04	1.2	0.01	0.36	0.02	0.08	0.04	3.6	10.0	0	0.4
0	12	0.20	5.0	109.8	20.7	0.08	320.9	0.01	0.29	0.02	0.25	0.04	8.4	3.2	0	0.1
10	17	0.40	9.0	89.0	14.7	0.32	70.4	0.03	0.12	0.04	0.44	0.04	7.0	0	0	2.4
0	23	0.40	12.0	87.6	3.0	0.12	10.8	0.02	0.04	0.04	0.20	0.02	14.4	0.2	0	0
9	20	0.02	1.8	31.4	10.5	0.10	37.6	0.01	0.04	0.02	0.10	0.01	3.1	3.1	0	1.1
12	249	11.82	30.6	225.7	53.3	0.80	28.4	0.49	—	0.65	6.54	0.07	12.5	1.4	0.3	
—	12	0.18	6.2	68.6	4.5	0.08	77.1	0.01	—	0.02	0.28	0.04	4.5	1.6	0	0.4
12	271	13.82	51.0	215.5	52.2	0.72	24.9	0.52	—	0.56	5.90	0.12	9.1	1.4	0.3	4.0
0	11	0.39	3.9	35.5	28.4	0.10	3.1	0.02	0.02	0.05	0.47	0.01	5.4	1.0	0	2.6

APPENDIX H

APPENDIX I

CONTENTS

WHO: Nutrition Recommendations Canada: Guidelines and Meal Planning

This appendix presents nutrition recommendations from the World Health Organization (WHO) and details for Canadians on the *Eating Well with Canada's Food Guide* and the *Beyond the Basics* meal-planning system.

Nutrition Recommendations from WHO

The World Health Organization (WHO) has assessed the relationships between diet and the development of chronic diseases. Its recommendations include:

- Energy: sufficient to support growth, physical activity, and a healthy body weight (BMI between 18.5 and 24.9) and to avoid weight gain greater than 11 pounds (5 kilograms) during adult life
- Total fat: 15 to 30 percent of total energy
- Saturated fatty acids: <10 percent of total energy
- Polyunsaturated fatty acids: 6 to 10 percent of total energy
- Omega-6 polyunsaturated fatty acids: 5 to 8 percent of total energy
- Omega-3 polyunsaturated fatty acids: 1 to 2 percent of total energy
- *Trans*-fatty acids: <1 percent of total energy
- Total carbohydrate: 55 to 75 percent of total energy
- Sugars: <10 percent of total energy
- Protein: 10 to 15 percent of total energy
- Cholesterol: <300 mg per day
- Salt (sodium): <5 g salt per day (<2 g sodium per day), appropriately iodized
- Fruits and vegetables: ≥400 g per day (about 1 pound)
- Total dietary fiber: >25 g per day from foods
- Physical activity: one hour of moderate-intensity activity, such as walking, on most days of the week

Eating Well with Canada's Food Guide

Figure I-1 presents the 2007 *Eating Well with Canada's Food Guide*. Additional publications, which are available from Health Canada ♦ through its website, provide many more details.

♦ Search for "Canada's food guide" at Health Canada: www.hc-sc.gc.ca

FIGURE I-1 *Eating Well with Canada's Food Guide*

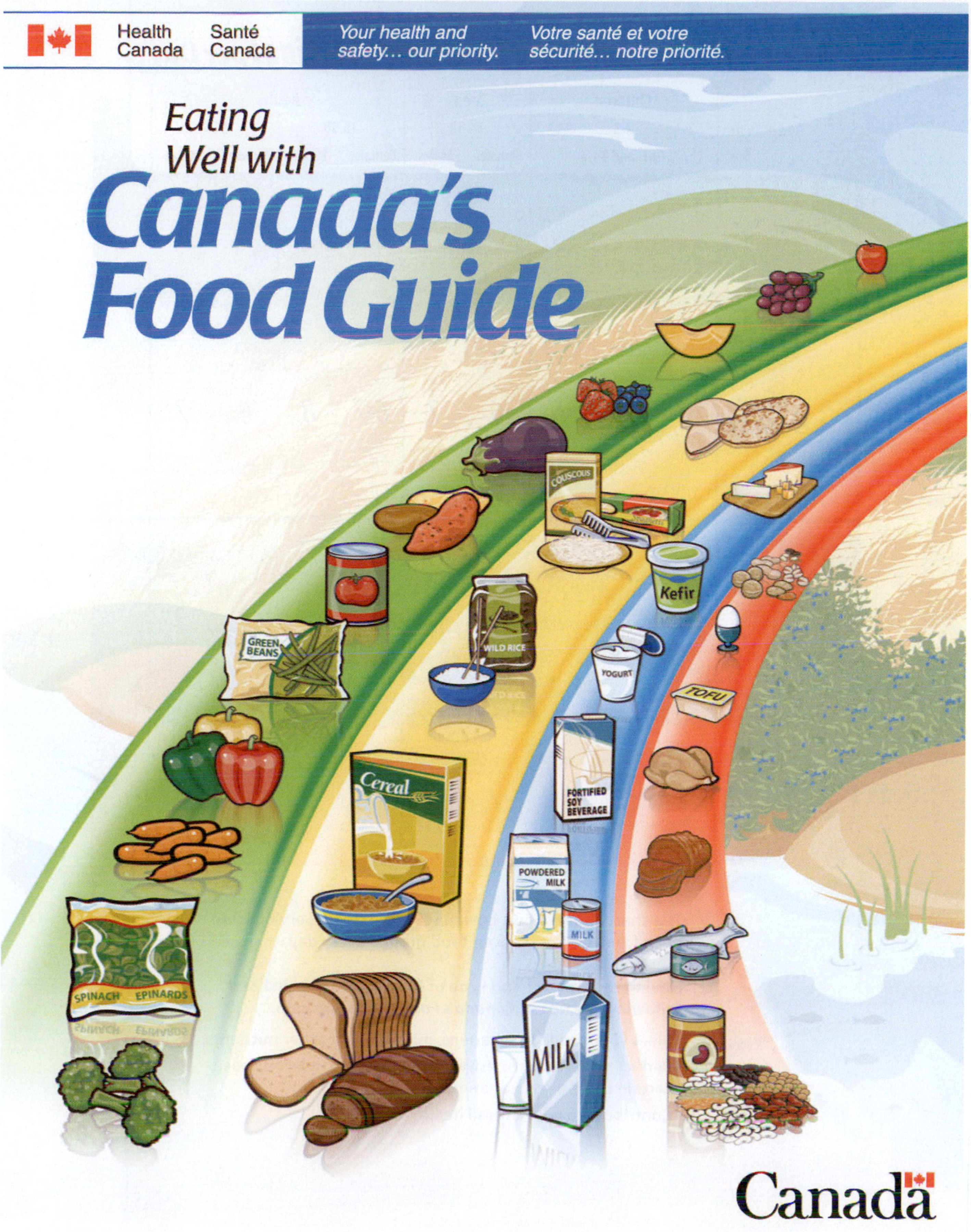

Health Canada
Santé Canada
Your health and safety… our priority.
Votre santé et votre sécurité… notre priorité.
Eating Well with Canada's Food Guide
COUSCOUS
Kefir
GREEN BEANS
WILD RICE
YOGURT
TOFU
FORTIFIED SOY BEVERAGE
Cereal
POWDERED MILK
MILK
SPINACH EPINARDS
MILK
Canada

FIGURE I-1 *Eating Well with Canada's Food Guide—continued*

Recommended Number of Food Guide Servings per Day

	Children			Teens		Adults			
Age in Years	2-3	4-8	9-13	14-18		19-50		51+	
Sex	Girls and Boys			Females	Males	Females	Males	Females	Males
Vegetables and Fruit	4	5	6	7	8	7-8	8-10	7	7
Grain Products	3	4	6	6	7	6-7	8	6	7
Milk and Alternatives	2	2	3-4	3-4	3-4	2	2	3	3
Meat and Alternatives	1	1	1-2	2	3	2	3	2	3

The chart above shows how many Food Guide Servings you need from each of the four food groups every day.

Having the amount and type of food recommended and following the tips in *Canada's Food Guide* will help:

- Meet your needs for vitamins, minerals and other nutrients.
- Reduce your risk of obesity, type 2 diabetes, heart disease, certain types of cancer and osteoporosis.
- Contribute to your overall health and vitality.

FIGURE I-1 *Eating Well with Canada's Food Guide—continued*

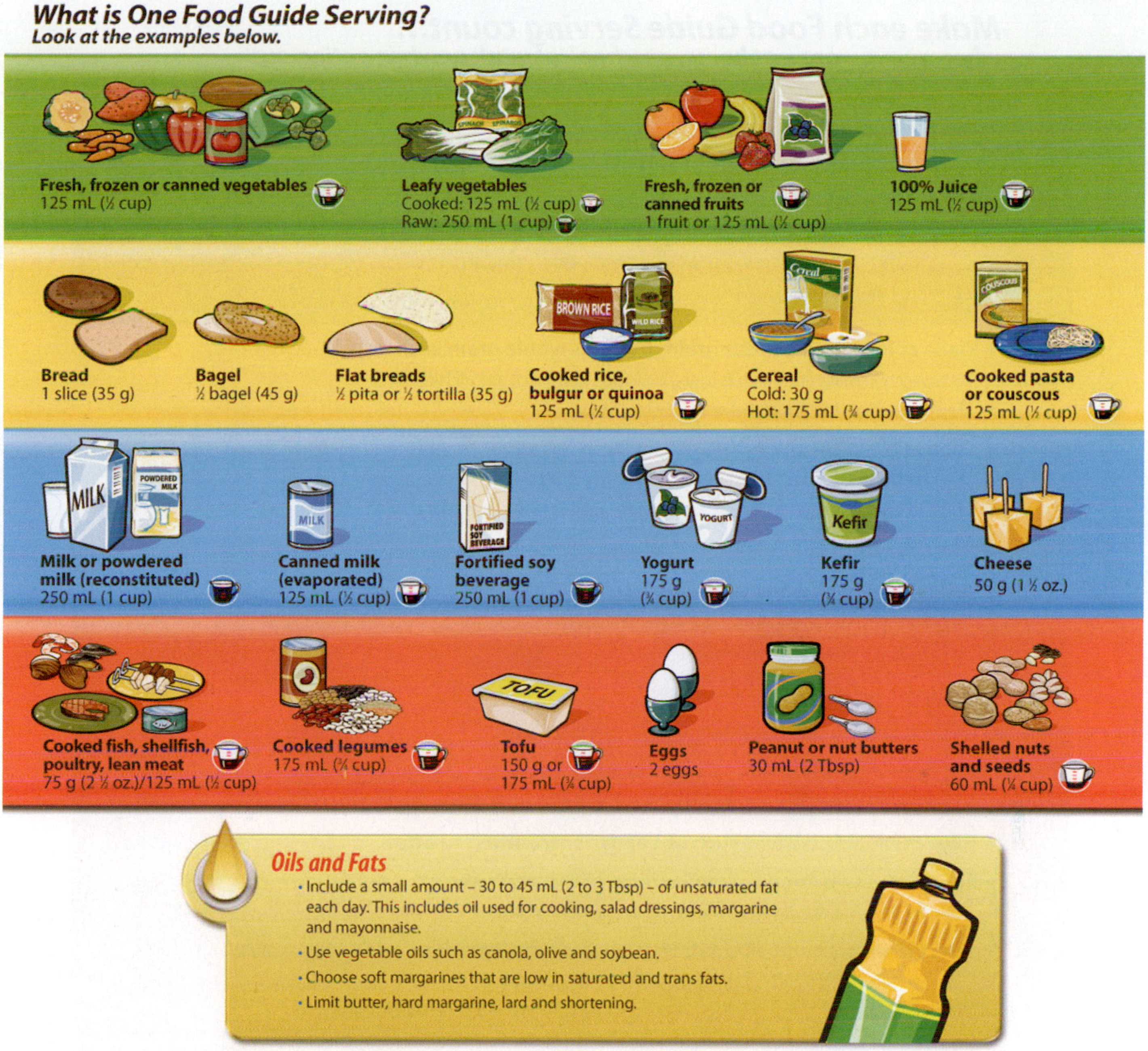
What is One Food Guide Serving?
Look at the examples below.
Fresh, frozen or canned vegetables
125 mL (½ cup)
Leafy vegetables
Cooked: 125 mL (½ cup)
Raw: 250 mL (1 cup)
Fresh, frozen or canned fruits
1 fruit or 125 mL (½ cup)
100% Juice
125 mL (½ cup)
Bread
1 slice (35 g)
Bagel
½ bagel (45 g)
Flat breads
½ pita or ½ tortilla (35 g)
Cooked rice, bulgur or quinoa
125 mL (½ cup)
Cereal
Cold: 30 g
Hot: 175 mL (¾ cup)
Cooked pasta or couscous
125 mL (½ cup)
Milk or powdered milk (reconstituted)
250 mL (1 cup)
Canned milk (evaporated)
125 mL (½ cup)
Fortified soy beverage
250 mL (1 cup)
Yogurt
175 g
(¾ cup)
Kefir
175 g
(¾ cup)
Cheese
50 g (1 ½ oz.)
Cooked fish, shellfish, poultry, lean meat
75 g (2 ½ oz.)/125 mL (½ cup)
Cooked legumes
175 mL (¾ cup)
Tofu
150 g or
175 mL (¾ cup)
Eggs
2 eggs
Peanut or nut butters
30 mL (2 Tbsp)
Shelled nuts and seeds
60 mL (¼ cup)
Oils and Fats
• Include a small amount – 30 to 45 mL (2 to 3 Tbsp) – of unsaturated fat each day. This includes oil used for cooking, salad dressings, margarine and mayonnaise.
• Use vegetable oils such as canola, olive and soybean.
• Choose soft margarines that are low in saturated and trans fats.
• Limit butter, hard margarine, lard and shortening.

FIGURE I-1 *Eating Well with Canada's Food Guide—continued*

Make each Food Guide Serving count…
wherever you are – at home, at school, at work or when eating out!

- **Eat at least one dark green and one orange vegetable each day.**
  - Go for dark green vegetables such as broccoli, romaine lettuce and spinach.
  - Go for orange vegetables such as carrots, sweet potatoes and winter squash.
- **Choose vegetables and fruit prepared with little or no added fat, sugar or salt.**
  - Enjoy vegetables steamed, baked or stir-fried instead of deep-fried.
- **Have vegetables and fruit more often than juice.**

- **Make at least half of your grain products whole grain each day.**
  - Eat a variety of whole grains such as barley, brown rice, oats, quinoa and wild rice.
  - Enjoy whole grain breads, oatmeal or whole wheat pasta.
- **Choose grain products that are lower in fat, sugar or salt.**
  - Compare the Nutrition Facts table on labels to make wise choices.
  - Enjoy the true taste of grain products. When adding sauces or spreads, use small amounts.

- **Drink skim, 1%, or 2% milk each day.**
  - Have 500 mL (2 cups) of milk every day for adequate vitamin D.
  - Drink fortified soy beverages if you do not drink milk.
- **Select lower fat milk alternatives.**
  - Compare the Nutrition Facts table on yogurts or cheeses to make wise choices.

- **Have meat alternatives such as beans, lentils and tofu often.**
- **Eat at least two Food Guide Servings of fish each week.***
  - Choose fish such as char, herring, mackerel, salmon, sardines and trout.
- **Select lean meat and alternatives prepared with little or no added fat or salt.**
  - Trim the visible fat from meats. Remove the skin on poultry.
  - Use cooking methods such as roasting, baking or poaching that require little or no added fat.
  - If you eat luncheon meats, sausages or prepackaged meats, choose those lower in salt (sodium) and fat.

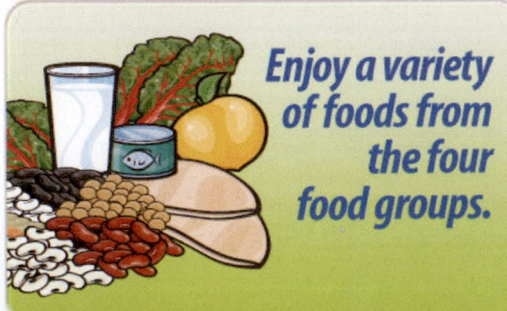

Enjoy a variety of foods from the four food groups.

Satisfy your thirst with water!

Drink water regularly. It's a calorie-free way to quench your thirst. Drink more water in hot weather or when you are very active.

* Health Canada provides advice for limiting exposure to mercury from certain types of fish. Refer to www.hc-sc.gc.ca for the latest information.

FIGURE I-1 *Eating Well with Canada's Food Guide—continued*

Advice for different ages and stages...

Children

Following *Canada's Food Guide* helps children grow and thrive.

Young children have small appetites and need calories for growth and development.

- Serve small nutritious meals and snacks each day.
- Do not restrict nutritious foods because of their fat content. Offer a variety of foods from the four food groups.
- Most of all... be a good role model.

Women of childbearing age

All women who could become pregnant and those who are pregnant or breastfeeding need a multivitamin containing **folic acid** every day. Pregnant women need to ensure that their multivitamin also contains **iron**. A health care professional can help you find the multivitamin that's right for you.

Pregnant and breastfeeding women need more calories. Include an extra 2 to 3 Food Guide Servings each day.

Here are two examples:

- Have fruit and yogurt for a snack, or
- Have an extra slice of toast at breakfast and an extra glass of milk at supper.

Men and women over 50

The need for **vitamin D** increases after the age of 50.

In addition to following *Canada's Food Guide*, everyone over the age of 50 should take a daily vitamin D supplement of 10 µg (400 IU).

How do I count Food Guide Servings in a meal?

Here is an example:

Vegetable and beef stir-fry with rice, a glass of milk and an apple for dessert		
250 mL (1 cup) mixed broccoli, carrot and sweet red pepper	=	2 **Vegetables and Fruit** Food Guide Servings
75 g (2 ½ oz.) lean beef	=	1 **Meat and Alternatives** Food Guide Serving
250 mL (1 cup) brown rice	=	2 **Grain Products** Food Guide Servings
5 mL (1 tsp) canola oil	=	part of your **Oils and Fats** intake for the day
250 mL (1 cup) 1% milk	=	1 **Milk and Alternatives** Food Guide Serving
1 apple	=	1 **Vegetables and Fruit** Food Guide Serving

FIGURE I-1 *Eating Well with Canada's Food Guide—continued*

Eat well and be active today and every day!

The benefits of eating well and being active include:

- Better overall health.
- Lower risk of disease.
- A healthy body weight.
- Feeling and looking better.
- More energy.
- Stronger muscles and bones.

Be active

To be active every day is a step towards better health and a healthy body weight.

Canada's Physical Activity Guide recommends building 30 to 60 minutes of moderate physical activity into daily life for adults and at least 90 minutes a day for children and youth. You don't have to do it all at once. Add it up in periods of at least 10 minutes at a time for adults and five minutes at a time for children and youth.

Start slowly and build up.

Eat well

Another important step towards better health and a healthy body weight is to follow *Canada's Food Guide* by:

- Eating the recommended amount and type of food each day.
- Limiting foods and beverages high in calories, fat, sugar or salt (sodium) such as cakes and pastries, chocolate and candies, cookies and granola bars, doughnuts and muffins, ice cream and frozen desserts, french fries, potato chips, nachos and other salty snacks, alcohol, fruit flavoured drinks, soft drinks, sports and energy drinks, and sweetened hot or cold drinks.

Read the label

- Compare the Nutrition Facts table on food labels to choose products that contain less fat, saturated fat, trans fat, sugar and sodium.
- Keep in mind that the calories and nutrients listed are for the amount of food found at the top of the Nutrition Facts table.

Nutrition Facts
Per 0 mL (0 g)

Amount			% Daily Value
Calories 0			
Fat 0 g			0 %
Saturates 0 g + Trans 0 g			0 %
Cholesterol 0 mg			
Sodium 0 mg			0 %
Carbohydrate 0 g			0 %
Fibre 0 g			0 %
Sugars 0 g			
Protein 0 g			
Vitamin A	0 %	Vitamin C	0 %
Calcium	0 %	Iron	0 %

Limit trans fat

When a Nutrition Facts table is not available, ask for nutrition information to choose foods lower in trans and saturated fats.

Take a step today...

- ✓ Have breakfast every day. It may help control your hunger later in the day.
- ✓ Walk wherever you can – get off the bus early, use the stairs.
- ✓ Benefit from eating vegetables and fruit at all meals and as snacks.
- ✓ Spend less time being inactive such as watching TV or playing computer games.
- ✓ Request nutrition information about menu items when eating out to help you make healthier choices.
- ✓ Enjoy eating with family and friends!
- ✓ Take time to eat and savour every bite!

For more information, interactive tools, or additional copies visit Canada's Food Guide on-line at:
www.hc-sc.gc.ca

or contact:

Publications
Health Canada
Ottawa, Ontario K1A 0K9
E-Mail: publications@hc-sc.gc.ca
Tel.: 1-866-225-0709
Fax: (613) 941-5366
TTY: 1-800-267-1245

Également disponible en français sous le titre : Bien manger avec le Guide alimentaire canadien

This publication can be made available on request on diskette, large print, audio-cassette and braille.

 HC Pub.: 4651 Cat.: H164-38/1-2007E ISBN: 0-662-44467-1

Beyond the Basics: Meal Planning for Healthy Eating, Diabetes Prevention and Management

Beyond the Basics: Meal Planning for Healthy Eating, Diabetes Prevention and Management is Canada's system of meal planning.[1] Similar to the U.S. exchange system, *Beyond the Basics* sorts foods into groups and defines portion sizes to help people manage their blood glucose and maintain a healthy weight. Because foods that contain carbohydrate raise blood glucose, the food groups are organized into two sections—those that contain carbohydrate (presented in Table I-1) and those that contain little or no carbohydrate (shown in Table I-2). One portion from any of the food groups listed in Table I-1 provides about 15 grams of available carbohydrate (total carbohydrate minus fiber) and counts as one carbohydrate choice. Within each group, foods are identified as those to "choose more often" (generally higher in vitamins, minerals, and fiber) and those to "choose less often" (generally higher in sugar, saturated fat, or *trans* fat).

[1]The tables for the Canadian meal planning system are adapted from *Beyond the Basics: Meal Planning for Healthy Eating, Diabetes Prevention and Management,* copyright 2005, with permission of the Canadian Diabetes Association. Additional information is available from www.diabetes.ca.

TABLE I-1 Food Groups that Contain Carbohydrate

1 serving = 15 g carbohydrate or 1 carbohydrate choice

Food	Measure
Grains and starches: 15 g carbohydrate, 3 g protein, 0 g fat, 286 kJ (68 kcal)	
▲ Bagel, large (11 cm, 4.5″)	¼
▲ Bagel, small (8 cm, 3″)	½
▲ Bannock, fried	1.5″ × 2.5″
● Bannock, whole grain baked	1.5″ × 2.5″
● Barley, pearled, cooked	125 mL (½ c)
▲ Bread, white	1 slice (30 g)
● Bread, whole grain	1 slice (30 g)
● Bulghur, cooked	125 mL (½ c)
▲ Bun, hamburger or hotdog	½
▲ Cereal, flaked unsweetened	125 mL (½ c)
● Cream of Wheat, cooked	175 mL (¾ c)
● Red River, cooked	125 mL (½ c)
● Chapati (15 cm, 6″)	1 (44 g)
● Corn	125 mL (½ c)
● Couscous, cooked	125 mL (½ c)
▲ Crackers, soda type	7
▲ Croutons	175 mL (¾ c)
● English muffin, whole grain	½ (28 g)
▲ French fries	10 (50 g)
● Millet, cooked	⅓ c (75 mL)
▲ Naan bread (15 cm, 6″)	¼
▲ Pancake (10 cm, 4″)	1
● Pasta, cooked	125 mL (½ c)
▲ Pita bread, white (15 cm, 6″)	½
● Pita bread, whole wheat (15 cm, 6″)	½
▲ Pizza crust (30 cm, 12″)	1/12 (90 g)

(continued)

KEY

● Choose more often
▲ Choose less often

KEY

- ● Choose more often
- ▲ Choose less often

TABLE I-1 Food Groups that Contain Carbohydrate (*continued*)

Food	Measure
Grains and starches: 15 g carbohydrate, 3 g protein, 0 g fat, 286 kJ (68 kcal)—continued	
● Plantain, cooked, mashed	⅓ c (75 mL)
● Potatoes, boiled or baked	½ medium (84 g)
● Rice, white or brown, cooked	⅓ c (75 mL)
● Roti (15 cm, 6″)	1 (44 g)
● Soup, thick, chunky type	250 mL (1 c)
● Sweet potato, mashed	⅓ c (75 mL)
▲ Taco shells (13 cm, 5″)	2 (17 g)
● Tortilla, wheat flour (25 cm, 10″)	½
▲ Waffle (medium)	1 (39 g)
Fruits: 15 g carbohydrate, 1–2 g protein, 0 g fat, 269 kJ (64 kcal)	
● Apple	1 medium (138 g)
● Applesauce, unsweetened	125 mL (½ c)
● Banana	1 small or ½ large
● Blackberries	500 mL (2 c)
● Cherries	15 (102 g)
● Fruit, canned in juice	125 mL (½ c)
▲ Fruit, dried	60 mL (¼ c)
● Grapefruit	1 small
● Grapes	15 or ½ c (80 g)
● Kiwi	2 medium (150 g)
▲ Juice	125 mL (½ c)
● Mango	½ medium
● Melon	250 mL (1 c)
● Orange	1 medium
● Other berries	250 mL (1 c)
● Pear	1 medium
● Pineapple	175 mL (¾ c)
● Plum	2 medium
● Raspberries	500 mL (2 c)
● Strawberries	500 mL (2 c)
Milk and alternatives: 15 g carbohydrate, 7 g protein, variable fat, 386–651 kJ (92–155 kcal)	
● Evaporated milk, canned	125 mL (½ c)
● Milk, fluid	250 mL (1 c)
● Milk powder, skim	60 mL (4 tbs)
● Soy beverage, flavoured	125 mL (½ c)
● Soy beverage, plain	250 mL (1 c)
● Soy yogourt, flavoured	75 mL (⅓ c)
● Yogourt, nonfat, plain	175 mL (¾ c)
● Yogourt, skim, artificially sweetened	250 mL (1 c)
Other choices (sweet foods and snacks): 15 g carbohydrate, variable protein and fat	
▲ Brownies, unfrosted	5 cm × 5 cm (2″ × 2″)
▲ Cake, unfrosted	5 cm × 5 cm (2″ × 2″)
▲ Cookies, arrowroot or gingersnap	3–4
▲ Jam, jelly, marmalade	15 mL (1 tbs)

(*continued*)

TABLE I-1 Food Groups that Contain Carbohydrate (*continued*)

Food	Measure
Other choices (sweet foods and snacks): 15 g carbohydrate, variable protein and fat—continued	
● Milk pudding, skim, no sugar added	125 ml (½ c)
▲ Muffin, plain	1 small (45 g)
▲ Oatmeal granola bar	1 (28 g)
● Popcorn, low fat, air popped	750 mL (3 c)
▲ Pretzels, low fat, large	7
▲ Pretzels, low fat, sticks	30
▲ Sugar, white	15 mL (1 tbs, 3 tsp, or 3 packets)
▲ Syrup, honey, molasses	1 tbs (15 mL)

KEY

● Choose more often
▲ Choose less often

TABLE I-2 Food Groups that Contain Little or No Carbohydrate

Food	Measure
Vegetables: To encourage consumption, most vegetables are considered "free"	
▲ Artichokes, Jerusalem[a]	
● Asparagus	
● Beans, yellow or green	
● Bean sprouts	
● Beets	
● Broccoli	
● Cabbage	
● Carrots	
● Cauliflower	
● Celery	
● Cucumber	
● Eggplant	
● Kale	
● Leeks	
● Mushrooms	
● Okra	
● Onions	
▲ Parsnips[a]	
▲ Peas[a]	
● Peppers	
● Rutabagas	
● Salad vegetables	
▲ Squash, Hubbard, pumpkin, spaghetti	
▲ Squash, acorn[a], butternut[a]	
● Tomatoes, fresh	
● Tomatoes, canned, regular	
▲ Tomatoes, canned, stewed[a]	
● Turnips	

(*continued*)

[a]These vegetables contain enough carbohydrate to be counted as one carbohydrate choice (15 g of available carbohydrate) when the portion size eaten is 1 cup (250 mL) or more.

APPENDIX I

KEY

- ● Choose more often
- ▲ Choose less often

TABLE I-2 Food Groups that Contain Little or No Carbohydrate (*continued*)

Food	Measure
Meat and alternatives: 0 g carbohydrate, 7 g protein, 3–5 g fat, 307 kJ (73 kcal)	
● Cheese, skim (<7% milk fat)	2.5 cm × 2.5 cm × 2.5 cm (1″ × 1″ × 1″)
● Cheese, light (<20% milk fat)	2.5 cm × 2.5 cm × 2.5 cm (1″ × 1″ × 1″)
▲ Cheese, regular (≥21% milk fat)	2.5 cm × 2.5 cm × 2.5 cm (1″ × 1″ × 1″)
● Cottage cheese (1–2% milk fat)	60 mL (¼ c)
● Egg	1 medium-large
● Fish, canned in oil or water	60 mL (¼ c)
● Fish, fresh or frozen, cooked	30 g (1 oz)
● Hummus	90 g (⅓ c)
● Legumes, cooked	125 mL (½ c)
● Meat, game, cooked	30 g (1 slice)
● Meat, ground, lean or extra lean, cooked	30 g (2 tbs)
● Meat, lean, cooked	30 g (1 slice)
● Meat, organ or tripe, cooked	30 g (1 slice)
● Meat, prepared, low fat	30 g (1–3 slices)
▲ Meat, prepared, regular fat	30 g (1–3 slices)
▲ Meat, regular, cooked	30 g (1–3 slices)
● Peameal/back bacon, cooked	30 g (1–2 slices)
● Poultry, ground, lean, cooked	30 g (2 tbs)
● Poultry, skinless, cooked	30 g (1 slice)
▲ Poultry/wings, skin on, cooked	45 g (2)
● Shellfish, cooked	30 g (3 medium)
● Tofu (soybean)	½ block (100 g)
● Vegetarian meat alternatives	30 g (1 oz)
Fats: 0 g carbohydrate, 0 g protein, 5 g fat, 189 kJ (45 kcal)	
▲ Avocado	34 g (⅙)
▲ Bacon	30 g (1 slice)
▲ Butter	5 mL (1 tsp)
▲ Cheese, spreadable	15 mL (1 tbs)
▲ Margarine, non-hydrogenated, regular	5 mL (1 tsp)
▲ Mayonnaise, light	15 mL (1 tbs)
▲ Nuts	15 mL (1 tbs)
▲ Oil, canola or olive	5 mL (1 tsp)
▲ Salad dressing, regular	5 mL (1 tsp)
▲ Seeds	15 g (1 tbs)
▲ Tahini	8 mL (½ tbs)
Extras: <5 g carbohydrate, 84 kJ (20 kcal)	
Broth	
Coffee	
Herbs and spices	
Ketchup	
Mustard	
Relish	
Sugar-free gelatin	
Sugar-free soft drinks	
Tea	

Healthy People 2020

Table 1-5 (p. 24) lists the objectives from the Nutrition and Weight Status section of the Healthy People 2020 initiative. This table presents additional nutrition-related objectives from other topic areas.

TABLE J-1 Nutrition-Related Objectives from Other Topic Areas

Access to Health Services

- Increase the proportion of persons who receive appropriate evidence-based clinical preventive services.

Adolescent Health

- Increase the proportion of schools with a school breakfast program.

Arthritis, Osteoporosis, and Chronic Back Conditions

- Reduce hip fractures among older adults.
- Reduce the proportion of adults with osteoporosis.

Cancer

- Reduce the cancer death rate.
- Increase the mental and physical health-related quality of life of cancer survivors.

Diabetes

- Reduce the annual number of new cases of diagnosed diabetes in the population.
- Reduce the death rate among the population with diabetes.
- Reduce the diabetes death rate.
- Improve glycemic control among the population with diagnosed diabetes.
- Improve lipid control among persons with diagnosed diabetes.
- Increase the proportion of the population with diagnosed diabetes whose blood pressure is under control.
- Increase the proportion of persons with diagnosed diabetes who receive formal diabetes education.
- Increase prevention behaviors in persons at high risk for diabetes with pre-diabetes.

Early and Middle Childhood

- Increase the proportion of elementary, middle, and senior high schools that require school health education.

Educational and Community-Based Programs

- Increase the proportion of preschool Early Head Start and Head Start programs that provide health education to prevent health problems in the following areas: unintentional injury; violence; tobacco use and addiction; alcohol and drug use, unhealthy dietary patterns; and inadequate physical activity, dental health, and safety.
- Increase the proportion of elementary, middle, and senior high schools that provide comprehensive school health education to prevent health problems in the following areas: unintentional injury; violence; suicide; tobacco use and addiction; alcohol or other drug use; unintended pregnancy, HIV/AIDS, and STD infection; unhealthy dietary patterns; and inadequate physical activity.

(continued)

TABLE J-1 Nutrition-Related Objectives from Other Topic Areas (*continued*)

- Increase the proportion of college and university students who receive information from their institution on each of the priority health risk behavior areas (all priority areas; unintentional injury; violence; suicide; tobacco use and addiction; alcohol and other drug use; unintended pregnancy, HIV/AIDS, and STD infection; unhealthy dietary patterns; and inadequate physical activity).
- Increase the proportion of worksites that offer an employee health promotion program to their employees.
- Increase the number of community-based organizations (including local health departments, tribal health services, nongovernmental organizations, and state agencies) providing population-based primary prevention services.

Environmental Health

- Reduce blood lead levels in children.
- Reduce the number of U.S. homes that are found to have lead-based paint or related hazards.

Food Safety

- Reduce infections caused by key pathogens transmitted commonly through food.
- Reduce the number of outbreak-associated infections due to Shiga toxin-producing *E. coli* 0157, or *Campylobacter*, *Listeria*, or *Salmonella* species associated with food commodity groups.
- Reduce severe allergic reactions to food among adults with a food allergy diagnosis.
- Increase the proportion of consumers who follow key food safety practices.
- Improve food safety practices associated with foodborne illness in foodservice and retail establishments.

Heart Disease and Stroke

- Increase overall cardiovascular health in the U.S. population.
- Reduce coronary heart disease deaths.
- Reduce stroke deaths.
- Reduce the proportion of persons in the population with hypertension.
- Reduce the proportion of adults with high total blood cholesterol levels.
- Reduce the mean total blood cholesterol levels among adults.
- Increase the proportion of adults with prehypertension who meet the recommended guidelines.
- Increase the proportion of adults with hypertension who meet the recommended guidelines.
- Increase the proportion of adults with elevated LDL cholesterol who have been advised by a health care provider regarding cholesterol lowering management including lifestyle changes and, if indicated, medication.
- Increase the proportion of adults with elevated LDL-cholesterol who adhere to the prescribed LDL-cholesterol lowering management lifestyle changes and, if indicated, medication.

Maternal, Infant, and Child Health

- Reduce low birth weight (LBW) and very low birth weight (VLBW).
- Reduce preterm births.
- Increase the proportion of pregnant women who receive early and adequate prenatal care.
- Increase the proportion of mothers who achieve a recommended weight gain during their pregnancies.
- Increase the proportion of women of childbearing potential with intake of at least 400 micrograms of folic acid from fortified foods or dietary supplements.
- Reduce the proportion of women of childbearing potential who have low red blood cell folate concentrations.
- Increase the proportion of women delivering a live birth who received preconception care services and practiced key recommended preconception health behaviors.
- Increase the proportion of infants who are breastfed.
- Increase the proportion of employers that have worksite lactation support programs.
- Reduce the proportion of breastfed newborns who receive formula supplementation within the first 2 days of life.
- Reduce the occurrence of fetal alcohol syndrome (FAS).
- Reduce occurrence of neural tube defects.

Mental Health and Mental Disorders

- Reduce the proportion of adolescents who engage in disordered eating behaviors in an attempt to control their weight.

TABLE J-1 Nutrition-Related Objectives from Other Topic Areas (*continued*)

Older Adults

- Increase the proportion of the health care workforce (including dietitians) with geriatric certification.

Oral Health

- Increase the proportion of the U.S. population served by community water systems with optimally fluoridated water.

Physical Activity

- Reduce the proportion of adults who engage in no leisure-time physical activity.
- Increase the proportion of adolescents and adults who meet current federal physical activity guidelines for aerobic physical activity and for muscle-strengthening activity.
- Increase the proportion of the nation's public and private schools that require daily physical education for all students.
- Increase the proportion of adolescents who participate in daily school physical education.
- Increase regularly scheduled elementary school recess in the United States.
- Increase the proportion of children and adolescents who do not exceed recommended limits for screen time—watching television, playing electronic games, or using the computer (other than for homework).

SOURCE: Healthy People 2020, available at www.healthypeople.gov/2020.

Enteral Formulas

The large number of enteral formulas available allows patients to meet a wide variety of medical needs. The first step in narrowing the choice of formulas is to determine the patient's ability to digest and absorb nutrients. Table K-1 on p. K-1 lists examples of standard formulas for patients who can adequately digest and absorb nutrients, and Table K-2 on p. K-2 provides examples of elemental formulas for patients with a limited ability to digest or absorb nutrients. Each formula is listed only once, although a formula may have more than one use. A high-protein formula, for example, may also be a fiber-containing formula. Tables K-3 and K-4 on p. K-2 list modules that can be used to prepare modular formulas or enhance enteral formulas.

The information shown in this appendix reflects the literature provided by manufacturers and does not suggest endorsement by the authors. Manufacturers frequently add new formulas, discontinue old ones, and change formula composition. Consult the manufacturers' literature and websites for updates and additional examples of enteral formulas.[a] The following products are listed in this appendix:

- *Abbott Nutrition:* Glucerna 1.0 Cal, Jevity 1 Cal, Jevity 1.5 Cal, Nepro with Carb Steady, Optimental, Osmolite 1 Cal, Oxepa, Pivot 1.5 Cal, Polycose Powder, Promote, Promote with Fiber, Pulmocare, Suplena with Carb Steady
- *Nestlé Nutrition:* Compleat, Compleat Pediatric, Crucial, Diabetisource AC, Fibersource HN, Impact, Impact 1.5, Impact Glutamine, Isosource HN, MCT Oil, Microlipid, Novasource Renal, Nutren 1.0, Nutren 1.0 Fiber, Nutren 1.5, Nutren 2.0, Nutren Glytrol, Nutren Junior, Nutren Pulmonary, Nutren Replete, Nutren Replete Fiber, NutriHep, Peptamen, Peptamen Junior, Resource Beneprotein Instant Protein Powder, Vivonex Pediatric, Vivonex T.E.N.

[a]Sources for the information in this appendix: Abbott Nutrition, **www.abbottnutrition.com**, Nestlé Nutrition, **www.nestle-nutrition.com**.

TABLE K-1 Standard Formulas

Product	Volume to Meet 100% RDI[a] (mL)	Energy (kcal/mL)	Protein or Amino Acids (g/L)	Carbohydrate (g/L)	Fat (g/L)	Notes
Lactose-Free, Standard Formulas						
Compleat	1313	1.07	48	128	40	Blenderized formula, 6 g fiber/L
Nutren 1.0	1500	1.00	40	127	38	25% fat from MCT
Osmolite 1 Cal	1321	1.06	44	144	35	20% fat from MCT
Lactose-Free, Fiber-Enhanced Formulas						
Jevity 1 Cal	1321	1.06	44	155	35	14 g fiber/L
Nutren 1.0 Fiber	1500	1.00	40	127	38	14 g fiber/L
Promote with Fiber	1000	1.00	63	138	28	14 g fiber/L
Lactose-Free, High-kCalorie Formulas						
Jevity 1.5 Cal	1500	1.50	64	216	50	22 g fiber/L
Nutren 1.5	1000	1.50	60	169	68	50% fat from MCT
Nutren 2.0	750	2.00	80	196	104	75% fat from MCT
Lactose-Free, High-Protein Formulas						
Fibersource HN	1165	1.20	53	160	39	20% fat from MCT, 10 g fiber/L
Isosource HN	1165	1.20	53	160	39	20% fat from MCT, low residue
Promote	1000	1.00	63	130	26	20% fat from MCT, low residue
Specialized Formulas: Pediatric (1 to 10 years)						
Compleat Pediatric	Varies[b]	1.00	38	130	39	Blenderized formula, 7 g fiber/L
Nutren Junior	Varies[b]	1.00	30	110	50	21% fat from MCT
Specialized Formulas: Glucose Intolerance						
Diabetisource AC	1250	1.20	60	100	59	36% kcal from carbohydrate, 15 g fiber/L
Glucerna 1.0 Cal	1420	1.00	42	96	54	34% kcal from carbohydrate, 14 g fiber/L
Nutren Glytrol	1400	1.00	45	100	48	40% kcal from carbohydrate, 15 g fiber/L
Specialized Formulas: Immune System Support						
Impact	1500	1.00	56	130	28	Enriched with arginine, nucleic acids, and omega-3 fatty acids
Impact 1.5	1250	1.50	84	140	69	Same as above
Impact Glutamine	1000	1.30	78	150	43	Same as above, and enriched with glutamine
Specialized Formulas: Chronic Kidney Disease (CKD)						
Nepro with Carb Steady	948	1.80	81	167	96	Low potassium, low phosphorus; to be used after dialysis has been instituted
Novasource Renal	1000	2.00	74	200	100	Low in electrolytes; to be used after dialysis has been instituted
Suplena with Carb Steady	948	1.80	45	205	96	Low in protein and electrolytes; for patients with CKD (stage 3 or 4)
Specialized Formulas: Respiratory Insufficiency						
Nutren Pulmonary	1000	1.50	68	100	95	55% kcal from fat, 40% fat from MCT
Oxepa	946	1.50	63	105	94	Enriched with omega-3 fatty acids and antioxidants; for mechanically ventilated patients
Pulmocare	947	1.50	63	106	93	55% kcal from fat, 20% fat from MCT, enriched with antioxidant nutrients
Specialized Formulas: Wound Healing						
Nutren Replete	1000	1.00	62	113	34	Enhanced with vitamins and minerals; for patients recovering from surgery, burns, and pressure ulcers
Nutren Replete Fiber	1000	1.00	62	113	34	Same as above; 14 g fiber/L

NOTE: MCT = Medium-chain triglycerides
[a] RDI = Reference Daily Intakes, which are labeling standards for vitamins, minerals, and protein. Consuming 100 percent of the RDI will meet the nutrient needs of most people using the product.
[b] Depends on age of child

TABLE K-2 Elemental Formulas

Product	Volume to Meet 100% RDI[a] (mL)	Energy (kcal/mL)	Protein or Amino Acids (g/L)	Carbohydrate (g/L)	Fat (g/L)	Notes
Specialized Elemental Formula: Hepatic Insufficiency						
NutriHep	1000	1.50	40	290	21	Free amino acids, high in branched-chain amino acids and low in aromatic amino acids
Specialized Elemental Formulas: Immune System Support						
Crucial	1000	1.50	94	134	68	Enriched with arginine, omega-3 fatty acids, antioxidant nutrients, and zinc
Pivot 1.5 Cal	1500	1.50	94	172	51	Enriched with arginine, glutamine, omega-3 fatty acids, and antioxidant nutrients
Specialized Elemental Formulas: Malabsorption						
Optimental	1422	1.00	51	139	28	Contains MCT and arginine, enriched with antioxidants and omega-3 fatty acids
Peptamen	1500	1.00	40	127	39	70% fat from MCT
Vivonex T.E.N.	2000	1.00	38	210	3	Powder form, 100% free amino acids, enriched with glutamine
Specialized Elemental Formulas: Pediatric (1 to 10 years)						
Peptamen Junior	Varies[b]	1.00	30	138	38	60% fat from MCT
Vivonex Pediatric	Varies[b]	0.80	24	130	24	Powder form, 100% free amino acids

NOTE: MCT = Medium-chain triglycerides
[a] RDI = Reference Daily Intakes, which are labeling standards for vitamins, minerals, and protein. Consuming 100 percent of the RDI will meet the nutrient needs of most people using the product.
[b] Depends on age of child

TABLE K-3 Protein and Carbohydrate Modules

Product	Major Ingredient	Energy (kcal/g)	Nutrient Content (g/100 g)
Resource Beneprotein Instant Protein Powder	Whey protein	3.6	86 g protein
Polycose Powder	Hydrolyzed cornstarch	3.8	94 g carbohydrate

TABLE K-4 Fat Modules

Product	Major Ingredient	Energy (kcal/mL)	Fat Content (g/100 mL)
MCT Oil	Coconut oil	7.7	93
Microlipid	Safflower oil	4.5	50

Glossary

Many medical terms have their origins in Latin or Greek. By learning a few common derivations, you can glean the meaning of words you have never heard of before. For example, once you know that "hyper" means above normal, "glyc" means glucose, and "emia" means blood, you can easily determine that "hyperglycemia" means high blood glucose. The derivations at left will help you to learn many terms presented in this glossary.

GENERAL

a- or *an-* = not or without
ana- = up
ant- or *anti-* = against
ante- or *pre-* or *pro-* = before
bi- or *di-* = two, twice
cata- = down
co- = with or together
dys- or *mal-* = bad, difficult, painful
endo- = inner or within
epi- = upon
exo- = outside of or without
extra- = outside of, beyond, or in addition
gen- or *-gen* = gives rise to, producing
homeo- = like, similar, constant unchanging state
hyper- = over, above, excessive
hypo- = below, under, beneath
in- = not
inter- = between, in the midst
intra- = within
-itis = infection or inflammation
-lysis = break
macro- = large or long
micro- = small
mono- = one, single
neo- = new, recent
oligo- = few or small
-osis or *-asis* = condition
para- = near
peri- = around, about
poly- = many or much
semi- = half
-stat or *-stasis-* = stationary
tri- = three

BODY

angi- or *vaso-* = vessel
arterio- = artery
cardiac or *cardio-* = heart
-cyte = cell
enteron = intestine
gastro- = stomach
hema- or *-emia* = blood
hepatic = liver
myo- or *sarco-* = muscle
nephr- or *renal* = kidney
neuro- = nerve
osteo- = bone
pulmo- = lung
ure- or *-uria* = urine
vena = vein

CHEMISTRY

-al = aldehyde
-ase = enzyme
-ate = salt
glyc- or *gluc-* = sweet (glucose)
hydro- or *hydrate* = water
lipo- = lipid
-ol = alcohol
-ose = carbohydrate
saccha- = sugar

1,25-dihydroxyvitamin D: vitamin D that is made from the hydroxylation of calcidiol in the kidneys; the biologically active hormone; also called **calcitriol** or **active vitamin D**.

2-in-1 solution: a parenteral solution that contains dextrose and amino acids, but excludes lipids.

24-hour recall: a record of foods consumed in the previous 24 hours; sometimes modified to include foods consumed in a typical day.

25-hydroxyvitamin D: vitamin D found in the blood that is made from the hydroxylation of cholecalciferol in the liver; also called **calcidiol**.

A

abscesses (AB-sess-es): accumulations of pus.

absorption: the uptake of nutrients by the cells of the small intestine for transport into either the blood or the lymph.

Acceptable Macronutrient Distribution Ranges (AMDR): ranges of intakes for the energy nutrients that provide adequate energy and nutrients and reduce the risk of chronic diseases.

accredited: approved; in the case of medical centers or universities, certified by an agency recognized by the U.S. Department of Education.

acesulfame (AY-sul-fame) **potassium:** an artificial sweetener composed of an organic salt that has been approved for use in both the United States and Canada; also known as **acesulfame-K** because K is the chemical symbol for potassium.

acetaldehyde (ass-et-AL-duh-hide): an intermediate in alcohol metabolism.

acetone breath: a distinctive fruity odor on the breath of a person with ketosis.

acetyl CoA (ASS-eh-teel, or ah-SEET-il, coh-AY): a 2-carbon compound (*acetate*, or *acetic acid*) to which a molecule of CoA is attached.

achalasia (ack-ah-LAY-zhah): an esophageal disorder characterized by weakened peristalsis and impaired relaxation of the lower esophageal sphincter.

achlorhydria (AY-clor-HIGH-dree-ah): absence of gastric acid secretions.

acid-base balance: the equilibrium in the body between acid and base concentrations (see Chapter 12).

acid controllers: medications used to prevent or relieve indigestion by suppressing production of acid in the stomach; also called *H2 blockers*.

acidosis (assi-DOE-sis): above-normal acidity (acid accumulation) in the blood and body fluids; depresses the central nervous system and can lead to disorientation and, eventually, coma.

acids: compounds that release hydrogen ions in a solution.

acne: a chronic inflammation of the skin's follicles and oil-producing glands, which leads to an accumulation of oils inside the ducts that surround hairs; usually associated with the maturation of young adults.

acquired immune deficiency syndrome (AIDS): the late stage of illness caused by infection with the human immunodeficiency virus (HIV); characterized by severe damage to immune function.

active vitamin D: vitamin D that is made from the hydroxylation of calcidiol in the kidneys; the biologically active hormone; also called **1,25-dihydroxyvitamin D** or **calcitriol**.

acupuncture (AK-you-PUNK-chur): a therapy that involves inserting thin needles into the skin at specific anatomical points, allegedly to correct disruptions in the flow of energy within the body.

acute kidney injury: a condition associated with the abrupt decline of kidney function over a period of hours or days; potentially a cause of acute renal failure.

acute PEM: protein-energy malnutrition caused by recent severe food restriction; characterized in children by thinness for height (wasting).

acute-phase proteins: plasma proteins released from the liver at the onset of acute infection. An example is **C-reactive protein**, which is considered one of the main indicators of severe infection and has antimicrobial effects.

acute-phase response: changes in body chemistry resulting from infection, inflammation, or injury; characterized by alterations in plasma proteins.

acute respiratory distress syndrome (ARDS): respiratory failure triggered by severe lung injury; a medical emergency that causes dyspnea and pulmonary edema and usually requires mechanical ventilation.

adaptive immunity: immunity that is specific for particular antigens; it adapts to antigens in an individual's environment and is characterized by "memory" for particular antigens. Also called **acquired immunity**.

adaptive thermogenesis: adjustments in energy expenditure related to changes in environment such as extreme cold and to physiological events such as overfeeding, trauma, and changes in hormone status.

added sugars: sugars and syrups used as an ingredient in the processing and preparation of foods such as breads, cakes, beverages, jellies, and ice cream as well as sugars eaten separately or added to foods at the table.

adequacy (dietary): providing all the essential nutrients, fiber, and energy in amounts sufficient to maintain health.

Adequate Intake (AI): the average daily amount of a nutrient that appears sufficient to maintain a specified criterion; a value used as a guide for nutrient intake when an RDA cannot be determined.

adipokines: proteins synthesized and secreted by adipose cells.

adiponectin (AH-dih-poe-NECK-tin): a hormone produced by adipose cells that improves insulin sensitivity; it inhibits inflammation and protects against insulin resistance, type 2 diabetes, and cardiovascular disease.

adipose (ADD-ih-poce) **tissue:** the body's fat tissue; consists of masses of triglyceride-storing cells.

adipose tissue lipase: an enzyme involved in the breakdown of body fat.

adjustment: the manipulative therapy practiced by chiropractors.

adolescence: the period from the beginning of puberty until maturity.

adrenal glands: glands adjacent to, and just above, each kidney.

adrenocorticotropin (ad-REE-noh-KORE-tee-koh-TROP-in) **or ATCH:** a hormone, so named because it stimulates (*trope*) the adrenal cortex. The adrenal gland, like the pituitary, has two parts, in this case the outer portion (*cortex*) and an inner core (*medulla*). The release of ACTH is mediated by *corticotropin-releasing hormone (CRH)*.

advance directive: written or oral instruction regarding one's preferences for medical treatment to be used in the event of becoming incapacitated.

advanced glycation end products (AGEs): reactive compounds formed after glucose combines with protein; AGEs can damage tissues and lead to diabetic complications.

adverse reactions: unusual responses to food (including intolerances and allergies).

aerobic (air-ROE-bic): requiring oxygen.

AIDS-defining illnesses: diseases and complications associated with the later stages of an HIV infection, including wasting, recurrent bacterial pneumonia, opportunistic infections, and certain cancers.

alcohol: a class of organic compounds containing hydroxyl (OH) groups.

alcohol abuse: a pattern of drinking that includes failure to fulfill work, school, or home responsibilities; drinking in situations that are physically dangerous (as in driving while intoxicated); recurring alcohol-related legal problems (as in aggravated assault charges); or continued drinking despite ongoing social problems that are caused by or worsened by alcohol.

alcohol dehydrogenase (dee-high-DROJ-ehnayz): an enzyme active in the stomach and the liver that converts ethanol to acetaldehyde.

alcohol-related birth defects (ARBD): malformations in the skeletal and organ systems (heart, kidneys, eyes, ears) associated with prenatal alcohol exposure.

alcohol-related neurodevelopmental disorder (ARND): abnormalities in the central nervous system and cognitive development associated with prenatal alcohol exposure.

alcoholism: a pattern of drinking that includes a strong craving for alcohol, a loss of control and an inability to stop drinking once begun, withdrawal symptoms (nausea, sweating, shakiness, and anxiety) after heavy drinking, and the need for increasing amounts of alcohol to feel "high."

aldosterone (al-DOS-ter-own): a hormone secreted by the adrenal glands that regulates blood pressure by increasing the reabsorption of sodium by the kidneys. Aldosterone also regulates chloride and potassium concentrations.

alitame (AL-ih-tame): an artificial sweetener composed of two amino acids (alanine and aspartic acid); FDA approval pending.

alkalosis (alka-LOE-sis): above-normal alkalinity (base) in the blood and body fluids.

allergen: any substance that triggers an inappropriate immune response.

allergy: an excessive and inappropriate immune reaction to a harmless substance.

alpha-lactalbumin (lact-AL-byoo-min): a major protein in human breast milk, as opposed to *casein* (CAY-seen), a major protein in cow's milk.

alpha-tocopherol: the active vitamin E compound.

alveoli (al-VEE-oh-lie): air sacs in the lungs. One air sac is an *alveolus*.

Alzheimer's disease: a degenerative disease of the brain involving memory loss and major structural changes in neuron networks; also known as *senile dementia of the Alzheimer's type (SDAT)*, *primary degenerative dementia of senile onset*, or *chronic brain syndrome*.

amenorrhea (ay-MEN-oh-REE-ah): the absence of or cessation of menstruation. *Primary amenorrhea* is menarche delayed beyond 16 years of age. *Secondary amenorrhea* is the absence of three to six consecutive menstrual cycles.

American Dietetic Association (ADA): the professional organization of dietitians in the United States. The Canadian equivalent is Dietitians of Canada, which operates similarly.

amino (a-MEEN-oh) **acid pool:** the supply of amino acids derived from either food proteins or body proteins that collect in the cells and circulating blood and stand ready to be incorporated in proteins and other compounds or used for energy.

amino acid scoring: a measure of protein quality assessed by comparing a protein's amino acid pattern with that of a reference protein; sometimes called *chemical scoring*.

amino acids: building blocks of proteins. Each contains an amino group, an acid group, a hydrogen atom, and a distinctive side group, all attached to a central carbon atom.

ammonia: a compound with the chemical formula NH_3; produced during the deamination of amino acids.

amniotic (am-nee-OTT-ic) **sac:** the "bag of waters" in the uterus, in which the fetus floats.

amylase (AM-ih-lace): an enzyme that hydrolyzes amylose (a form of starch). Amylase is a **carbohydrase**, an enzyme that breaks down carbohydrates.

anabolic: building up reactions.

anabolism (an-AB-o-lism): reactions in which small molecules are put together to build larger ones. Anabolic reactions require energy.

anaerobic (AN-air-ROE-bic): not requiring oxygen.

anaphylactic (ana-fill-LAC-tic) **shock:** a life-threatening, whole-body allergic reaction to an offending substance.

anaphylaxis: a severe allergic reaction that may include gastrointestinal upset, skin inflammation, breathing difficulty, and low blood pressure, potentially leading to shock.

anecdote: a personal account of an experience or event; not reliable scientific information.

anemia (ah-NEE-me-ah): literally, "too little blood." Anemia is any condition in which too few red blood cells are present, or the red blood cells are immature (and therefore large) or too small or contain too little hemoglobin to carry the normal amount of oxygen to the tissues. It is not a disease itself but can be a symptom of many different disease conditions, including many nutrient deficiencies, bleeding, excessive red blood cell destruction, and defective red blood cell formation.

anemia of chronic disease: anemia that develops in persons with chronic illness; may resemble iron-deficiency anemia even though iron stores are often adequate. Also called *anemia of chronic inflammation*.

anencephaly (AN-en-SEF-a-lee): an uncommon and always fatal type of neural tube defect; characterized by the absence of a brain.

aneurysm (AN-you-rih-zum): an abnormal enlargement or bulging of a blood vessel (usually an artery) caused by weakness in the blood vessel wall.

angina (an-JYE-nah or AN-ji-nah) **pectoris:** a condition caused by ischemia in the heart muscle that results in discomfort or dull pain in the chest region. The pain often radiates to the left shoulder and arm or to the back, neck, and lower jaw.

angiotensin I (AN-gee-oh-TEN-sin): an inactive precursor that is converted by an enzyme to yield active angiotensin II.

angiotensin II: a hormone involved in blood pressure regulation.

angiotensin-converting enzyme: the enzyme that converts angiotensin I to angiotensin II.

angiotensinogen: a precursor protein that is hydrolyzed to angiotensin I by renin.

anions (AN-eye-uns): negatively charged ions.

anorexia (an-oh-RECK-see-ah): lack of appetite.

anorexia nervosa: an eating disorder characterized by a refusal to maintain a minimally normal body weight and a distortion in perception of body shape and weight.

antacids: medications used to relieve indigestion by neutralizing acid in the stomach.

antagonist: a competing factor that counteracts the action of another factor. When a drug displaces a vitamin from its site of action, the drug renders the vitamin ineffective and thus acts as a vitamin antagonist.

anthropometric (AN-throw-poe-MET-rick): relating to measurement of the physical characteristics of the body, such as height and weight.

antibodies: large proteins of the blood and body fluids, produced by the immune system in response to the invasion of the body by foreign molecules (usually proteins called *antigens*). Antibodies combine with and inactivate the foreign invaders, thus protecting the body.

antidiuretic hormone (ADH): a hormone produced by the pituitary gland in response to dehydration (or a high sodium concentration in the blood). It stimulates the kidneys to reabsorb more water and therefore prevents water loss in urine (also called **vasopressin**).

antihistamines: medications taken to reduce the effects of histamine.

antineoplastic drugs: medications that combat tumor growth.

antioxidant: a substance that significantly decreases the adverse effects of free radicals on normal physiological functions; as a food additive, a preservative that delays or prevents rancidity of fats in foods and other damage to food caused by oxygen.

antiretroviral drugs: medications that treat HIV infection.

antiscorbutic (AN-tee-skor-BUE-tik) **factor:** the original name for vitamin C.

anuria (ah-NOO-ree-ah): the absence of urine; clinically identified as urine output less than 50 mL/day.

anus (AY-nus): the terminal outlet of the GI tract.

aorta (ay-OR-tuh): the large, primary artery that conducts blood from the heart to the body's smaller arteries.

aplastic anemia: anemia characterized by the inability of bone marrow to produce adequate numbers of blood cells. Causes include drug toxicity, viruses, and genetic defects.

apoptosis: cell death.

appendix: a narrow blind sac extending from the beginning of the colon that stores lymph cells.

appetite: the integrated response to the sight, smell, thought, or taste of food that initiates or delays eating.

appropriate for gestational age (AGA): term describing an infant whose birth weight is normal compared with the number of weeks in utero.

arachidonic (a-RACK-ih-DON-ic) **acid:** an omega-6 polyunsaturated fatty acid with 20 carbons and four double bonds; present in small amounts in meat and other animal products and synthesized in the body from linoleic acid.

ariboflavinosis (ay-RYE-boh-FLAY-vin-oh-sis): riboflavin deficiency.

aromatherapy: inhalation of oil extracts from plants to cure illness or enhance health.

arteries: vessels that carry blood from the heart to the tissues.

artesian water: water drawn from a well that taps a confined aquifer in which the water is under pressure.

arthritis: inflammation of a joint, usually accompanied by pain, swelling, and structural changes.

artificial fats: zero-energy fat replacers that are chemically synthesized to mimic the sensory and cooking qualities of naturally occurring fats but are totally or partially resistant to digestion.

artificial sweeteners: sugar substitutes that provide negligible, if any, energy; sometimes called *nonnutritive sweeteners*.

ascites (ah-SIGH-teez): an abnormal accumulation of fluid in the abdominal cavity.

ascorbic acid: one of the two active forms of vitamin C.

aspartame (ah-SPAR-tame or ASS-par-tame): an artificial sweetener composed of two amino acids (phenylalanine and aspartic acid); approved for use in both the United States and Canada.

aspiration: drawing in by suction or breathing; a common complication of enteral feedings in which foreign material enters the lungs, often from GI secretions or the reflux of stomach contents.

aspiration pneumonia: a lung disease resulting from the abnormal entry of foreign material; caused by either bacterial infection or irritation of the lower airways.

asymptomatic allergy: an immune response that produces antibodies but no symptoms.

atherogenic: able to initiate or promote atherosclerosis.

atheromatous (ATH-er-OH-ma-tus) **plaque:** plaque associated with atherosclerosis.

atherosclerosis (ATH-er-oh-scler-OH-sis): a type of artery disease characterized by plaques (accumulations of lipid-containing material) on the inner walls of the arteries.

atoms: the smallest components of an element that have all of the properties of the element.

ATP or **adenosine** (ah-DEN-oh-seen) **triphosphate** (try-FOS-fate): a common high-energy compound composed of a purine (adenine), a sugar (ribose), and three phosphate groups.

atrophic (a-TRO-fik) **gastritis** (gas-TRY-tis): chronic inflammation of the stomach accompanied by a diminished size and functioning of the mucous membrane and glands.

autoimmune: an immune response directed against the body's own tissues.

autoimmune diseases: diseases characterized by an attack of immune defenses on the body's own cells.

autonomic nervous system: the division of the nervous system that controls the body's automatic responses. Its two branches are the sympathetic branch, which helps the body respond to stressors from the outside environment, and the parasympathetic branch, which regulates normal body activities between stressful times.

avidin (AV-eh-din): the protein in egg whites that binds biotin.

ayurveda: a traditional medical system from India that promotes the use of diet, herbs, meditation, massage, and yoga for preventing and treating illness.

azotemia: the accumulation of nitrogenous wastes in the blood.

B

B cell: a lymphocyte that produces antibodies.

bacterial cholangitis (KOH-lan-JYE-tis): bacterial infection involving the bile ducts.

bacterial overgrowth: excessive bacterial colonization of the stomach and small intestine; may be due to low gastric acidity, altered GI motility, mucosal damage, or contamination.

bacterial translocation: movement of bacteria across the intestinal mucosa, allowing access to body tissues.

balance (dietary): providing foods in proportion to one another and in proportion to the body's needs.

bariatric (BAH-ree-AH-trik) **surgery:** surgery that treats severe obesity.

bariatrics: the field of medicine that specializes in treating obesity.

Barrett's esophagus: a condition in which esophageal cells damaged by chronic exposure to stomach acid are replaced by cells that resemble those in the stomach or small intestine, sometimes becoming cancerous.

basal metabolic rate (BMR): the rate of energy use for metabolism under specified conditions: after a 12-hour fast and restful sleep, without any physical activity or emotional excitement, and in a comfortable setting. It is usually expressed as kcalories per kilogram body weight per hour.

basal metabolism: the energy needed to maintain life when a body is at complete digestive, physical, and emotional rest.

bases: compounds that accept hydrogen ions in a solution.

beer: an alcoholic beverage traditionally brewed by fermenting malted barley and adding hops for flavor.

behavior modification: the changing of behavior by the manipulation of antecedents (cues or environmental factors that trigger behavior), the behavior itself, and consequences (the penalties or rewards attached to behavior).

beikost (BYE-cost): a German word that describes any nonmilk foods given to an infant.

belching: the expulsion of gas from the stomach through the mouth.

beneficence (be-NEF-eh-sense): the act of performing beneficial services rather than harmful ones.

benign: an abnormal mass of cells that is noncancerous.

beriberi: the thiamin-deficiency disease.

beta-carotene (BAY-tah KARE-oh-teen): one of the carotenoids; an orange pigment and vitamin A precursor found in plants.

bicarbonate: an alkaline compound with the formula HCO_3 that is secreted from the pancreas as part of the pancreatic juice. (Bicarbonate is also produced in all cell fluids from the dissociation of carbonic acid to help maintain the body's acid-base balance.)

bifidus (BIFF-id-us, by-FEED-us) **factors:** factors in colostrum and breast milk that favor the growth of the "friendly" bacterium *Lactobacillus* (lack-toh-ba-SILL-us) *bifidus* in the infant's intestinal tract, so that other, less desirable intestinal inhabitants will not flourish.

bile: an emulsifier that prepares fats and oils for digestion; an exocrine secretion made by the liver, stored in the gallbladder, and released into the small intestine when needed.

biliary system: the gallbladder and ducts that deliver bile from the liver and gallbladder to the small intestine.

binders: chemical compounds in foods that combine with nutrients (especially minerals) to form complexes the body cannot absorb. Examples include *phytates* (FYE-tates) and *oxalates* (OCK-sa-lates).

binge drinking: four or more drinks for women and five or more drinks for men of alcohol in a row (within a couple of hours).

binge-eating disorder: an eating disorder with criteria similar to those of bulimia nervosa, excluding purging or other compensatory behaviors.

bioavailability: the rate at and the extent to which a nutrient is absorbed and used.

bioelectrical or **bioelectromagnetic therapies:** therapies that involve the unconventional use of electric or magnetic fields to cure illness.

biofeedback: a technique in which individuals are trained to gain voluntary control of certain physiological processes, such as skin temperature or brain wave activity, to help reduce stress and anxiety.

biofield therapies: healing methods based on the belief that illnesses can be healed by manipulating energy fields that purportedly surround and penetrate the body. Examples include *acupuncture*, *qi gong*, and *therapeutic touch*.

biological value (BV): a measure of protein quality assessed by measuring the amount of protein nitrogen that is retained from a given amount of protein nitrogen absorbed.

bioterrorism: the intentional spreading of disease-causing microorganisms or toxins.

biotin (BY-oh-tin): a B vitamin that functions as a coenzyme in metabolism.

blastocyst (BLASS-toe-sist): the developmental stage of the zygote when it is about five days old and ready for implantation.

blenderized formulas: enteral formulas that are prepared by using a food blender to mix and puree whole foods.

blind experiment: an experiment in which the subjects do not know whether they are members of the experimental group or the control group.

blood lipid profile: results of blood tests that reveal a person's total cholesterol, triglycerides, and various lipoproteins.

body composition: the proportions of muscle, bone, fat, and other tissue that make up a person's total body weight.

body mass index (BMI): an index of a person's weight in relation to height; determined by dividing the weight (in kilograms) by the square of the height (in meters).

bolus (BOH-lus): a portion; with respect to food, the amount swallowed at one time.

bolus feeding: delivery of about 250 to 500 milliliters of formula in less than 20 minutes.

bomb calorimeter (KAL-oh-RIM-eh-ter): an instrument that measures the heat energy released when foods are burned, thus providing an estimate of the potential energy of the foods.

bone density: a measure of bone strength. When minerals fill the bone matrix (making it dense), they give it strength.

bone meal or **powdered bone:** crushed or ground bone preparations intended to supply calcium to the diet. Calcium from bone is not well absorbed and is often contaminated with toxic minerals such as arsenic, mercury, lead, and cadmium.

bottled water: drinking water sold in bottles.

botulism (BOT-chew-lism): an often fatal foodborne illness caused by the ingestion of foods containing a toxin produced by bacteria that grow without oxygen.

Bowman's (BOE-minz) **capsule:** a cuplike component of the nephron that surrounds the glomerulus and collects the filtrate that is passed to the tubules.

bran: the protective coating around the kernel of grain, rich in nutrients and fiber.

branched-chain amino acids: the essential amino acids leucine, isoleucine, and valine, which are present in large amounts in skeletal muscle tissue; falsely promoted as fuel for exercising muscles.

breast milk bank: a service that collects, screens, processes, and distributes donated human milk.

bronchi (BRON-key), **bronchioles** (BRON-key-oles): the main airways of the lungs. The singular form of bronchi is *bronchus*.

brown adipose tissue: masses of specialized fat cells packed with pigmented mitochondria that produce heat instead of ATP.

brown sugar: refined white sugar crystals to which manufacturers have added molasses syrup with natural flavor and color; 91 to 96 percent pure sucrose.

buffalo hump: the accumulation of fatty tissue at the base of the neck.

buffers: compounds that keep a solution's pH constant when acids or bases are added.

bulimia (byoo-LEEM-ee-ah) **nervosa:** an eating disorder characterized by repeated episodes of binge eating usually followed by self-induced vomiting, misuse of laxatives or diuretics, fasting, or excessive exercise.

C

C-reactive protein: an acute-phase protein released from the liver during acute inflammation or stress.

calbindin: a calcium-binding transport protein.

calcidiol: vitamin D found in the blood that is made from the hydroxylation of cholecalciferol in the liver; also called **25-hydroxyvitamin D**.

calciol: derived from animals in the diet and made in the skin from 7-dehydrocholesterol, a precursor of cholesterol, with the help of sunlight; also called **cholecalciferol** or **vitamin** D_3.

calcitonin (KAL-seh-TOE-nin): a hormone secreted by the thyroid gland that regulates blood calcium by lowering it when levels rise too high.

calcitriol: vitamin D that is made from the hydroxylation of calcidiol in the kidneys; the biologically active hormone; also called **1,25-dihydroxyvitamin D** or **active vitamin D**.

calcium: the most abundant mineral in the body; found primarily in the body's bones and teeth.

calcium-binding protein: a protein in the intestinal cells, made with the help of vitamin D, that facilitates calcium absorption.

calcium rigor: hardness or stiffness of the muscles caused by high blood calcium concentrations.

calcium tetany (TET-ah-nee): intermittent spasm of the extremities due to nervous and muscular excitability caused by low blood calcium concentrations.

calmodulin (cal-MOD-you-lin): a protein that binds with and activates calcium.

calories: units by which energy is measured. Food energy is measured in kilocalories (1000 calories equal 1 kilocalorie), abbreviated **kcalories** or **kcal**. One kcalorie is the amount of heat necessary to raise the temperature of 1 kilogram (kg) of water 1°C. The scientific use of the term *kcalorie* is the same as the popular use of the term *calorie*.

cancer cachexia (ka-KEK-see-ah): a wasting syndrome associated with cancer that is characterized by anorexia, muscle wasting, weight loss, and fatigue.

cancers: diseases characterized by the uncontrolled growth of a group of cells, which can destroy adjacent tissues and spread to other areas of the body via lymph or blood.

candidiasis: a fungal infection on the mucous membranes of the oral cavity and elsewhere; usually caused by *Candida albicans*.

capillaries (CAP-ill-aries): small vessels that branch from an artery. Capillaries connect arteries to veins. Exchange of oxygen, nutrients, and waste materials takes place across capillary walls.

carbohydrase (KAR-boe-HIGH-drase): an enzyme that hydrolyzes carbohydrates.

carbohydrate-to-insulin ratio: the amount of carbohydrate that can be handled per unit of insulin. On average, every 15 grams of carbohydrate requires about 1 unit of rapid- or short-acting insulin.

carbohydrates: compounds composed of carbon, oxygen, and hydrogen arranged as monosaccharides or multiples of monosaccharides. Most, but not all, carbohydrates have a ratio of one carbon molecule to one water molecule: $(CH_2O)_n$.

carbonated water: water that contains carbon dioxide gas, either naturally occurring or added, that causes bubbles to form in it; also called bubbling or sparkling water.

carbonic acid: a compound with the formula H_2CO_3 that results from the combination of carbon dioxide (CO_2) and water (H_2O); of particular importance in maintaining the body's acid-base balance.

carcinogenesis (CAR-sin-oh-JEN-eh-sis): the process of cancer development.

carcinogens (CAR-sin-oh-jenz or car-SIN-oh-jenz): substances that can cause cancer (the adjective is *carcinogenic*).

cardiac cachexia: a condition of severe malnutrition that develops in heart failure patients; characterized by weight loss and tissue wasting.

cardiac output: the volume of blood pumped by the heart within a specified period of time.

cardiopulmonary resuscitation (CPR): life-sustaining treatment that supplies oxygen and restores a person's ability to breathe and pump blood.

cardiovascular disease (CVD): a general term describing diseases of the heart and blood vessels.

carotenoids (kah-ROT-eh-noyds): pigments commonly found in plants and animals, some of which have vitamin A activity. The carotenoid with the greatest vitamin A activity is beta-carotene.

carpal tunnel syndrome: a pinched nerve at the wrist, causing pain or numbness in the hand. It is often caused by repetitive motion of the wrist.

catabolic: breaking down reactions.

catabolism (ca-TAB-o-lism): reactions in which large molecules are broken down to smaller ones. Catabolic reactions release energy.

catalyst (CAT-uh-list): a compound that facilitates chemical reactions without itself being changed in the process.

cataracts (KAT-ah-rakts): thickenings of the eye lenses that impair vision and can lead to blindness.

cathartic (ka-THAR-tik): a strong laxative.

catheter: a thin tube placed within a narrow lumen (such as a blood vessel) or body cavity; can be used to infuse or withdraw fluids or keep a passage open.

cations (CAT-eye-uns): positively charged ions.

celiac (SEE-lee-ack) **disease:** immune disorder characterized by an abnormal immune response to the dietary protein gluten; also called *gluten-sensitive enteropathy* or *celiac sprue*.

cell: the basic structural unit of all living things.

cell differentiation (DIF-er-EN-she-AY-shun): the process by which immature cells develop specific functions different from those of the original that are characteristic of their mature cell type.

cell-mediated immunity: immunity conferred by T cells and macrophages.

cell membrane: the thin layer of tissue that surrounds the cell and encloses its contents; made primarily of lipid and protein.

cellulite (SELL-you-light or SELL-you-leet): supposedly, a lumpy form of fat; actually, a fraud. Fatty areas of the body may appear lumpy when the strands of connective tissue that attach the skin to underlying muscles pull tight where the fat is thick. The fat itself is the same as fat anywhere else in the body. If the fat in these areas is lost, the lumpy appearance disappears.

central nervous system: the central part of the nervous system; the brain and spinal cord.

central obesity: excess fat around the trunk of the body; also called *abdominal fat* or *upper-body fat*.

central veins: the large-diameter veins located close to the heart.

cerebral thrombosis: a clot that blocks blood flow through an artery that feeds the brain.

Certified Diabetes Educator (CDE): a health care professional who specializes in diabetes management education. Certification is obtained from the National Certification Board for Diabetes Educators.

certified lactation consultants: health-care providers who specialize in helping new mothers establish a healthy breastfeeding relationship with their newborn. These consultants are often registered nurses with specialized training in breast and infant anatomy and physiology.

certified nutritionist or **certified nutritional consultant** or **certified nutrition therapist:** a person who has been granted a document declaring his or her authority as a nutrition professional.

cesarean section: a surgically assisted birth involving removal of the fetus by an incision into the uterus, usually by way of the abdominal wall.

chaff: the outer inedible part of a grain; also called the *husk*.

cheilosis (kye-LOH-sis or kee-LOH-sis): a condition of reddened lips with cracks at the corners of the mouth.

chelate (KEY-late): a substance that can grasp the positive ions of a mineral.

chemotherapy: the use of drugs to arrest or destroy cancer cells; these drugs are called *anti-neoplastic agents*.

chiropractic (KYE-roh-PRAK-tic): a method of treatment based on the unproven theory that spinal manipulation can restore health.

chloride (KLO-ride): the major anion in the extracellular fluids of the body. Chloride is the ionic form of chlorine, Cl^-. See Appendix B for a description of the chlorine-to-chloride conversion.

chlorophyll (KLO-row-fil): the green pigment of plants, which absorbs light and transfers the energy to other molecules, thereby initiating photosynthesis.

cholecalciferol (KO-lee-kal-SIF-er-ol): vitamin D derived from animals in the diet and made in the skin from 7-dehydrocholesterol, a precursor of cholesterol, with the help of sunlight; also called **vitamin D_3** or **calciol**.

cholecystectomy (KOH-leh-sis-TEK-toe-mee): surgical removal of the gallbladder.

cholecystitis (KOH-leh-sih-STY-tis): inflammation of the gallbladder, usually caused by obstruction of the cystic duct by gallstones.

cholecystokinin (COAL-ee-SIS-toe-KINE-in), or **CCK**: a hormone produced by cells of the intestinal wall. Target organ: the gallbladder. Response: release of bile and slowing of GI motility.

cholelithiasis (KOH-leh-lih-THIGH-ah-sis): formation of gallstones.

cholesterol (koh-LESS-ter-ol): one of the sterols containing a four ring carbon structure with a carbon side chain.

cholesterol-free: less than 2 milligrams cholesterol per serving and 2 grams or less saturated fat and *trans* fat combined per serving.

choline (KOH-leen): a nitrogen-containing compound found in foods and made in the body from the amino acid methionine. Choline is part of the phospholipid lecithin and the neurotransmitter acetylcholine.

chromosomes: structures within the nucleus of a cell made of DNA and associated proteins. Human beings have 46 chromosomes in 23 pairs. Each chromosome has many genes.

chronic bronchitis (bron-KYE-tis): a lung disorder characterized by persistent inflammation and excessive secretions of mucus in the main airways of the lungs.

chronic diseases: diseases characterized by a slow progression and long duration.

chronic kidney disease: kidney disease characterized by gradual, irreversible deterioration of the kidneys; also called *chronic renal failure*.

chronic obstructive pulmonary disease (COPD): a group of lung diseases characterized by persistent obstructed airflow through the lungs and airways; includes chronic bronchitis and emphysema.

chronic PEM: protein-energy malnutrition caused by long-term food deprivation; characterized in children by short height for age (stunting).

chronological age: a person's age in years from his or her date of birth.

chylomicrons (kye-lo-MY-cronz): the class of lipoproteins that transport lipids from the intestinal cells to the rest of the body.

chyme (KIME): the semiliquid mass of partly digested food expelled by the stomach into the duodenum.

cirrhosis (sih-ROE-sis): an advanced stage of liver disease in which extensive scarring replaces healthy liver tissue, causing impaired liver function and liver failure; often associated with alcoholism.

cis: on the near side of; refers to a chemical configuration in which the hydrogen atoms are located on the same side of a double bond.

citric acid cycle: a series of metabolic reactions that break down molecules of acetyl CoA to carbon dioxide and hydrogen atoms; also called the *TCA cycle* or the *Kreb's cycle*.

claudication (CLAW-dih-KAY-shun): pain in the legs while walking; usually due to an inadequate supply of blood to muscles.

clear liquid diet: a diet that consists of foods that are liquid at room temperature, require minimal digestion, and leave little residue (undigested material) in the colon.

clinically severe obesity: a BMI of 40 or greater or a BMI of 35 or greater with additional medical problems. A less preferred term used to describe the same condition is *morbid obesity*.

closed feeding system: a delivery system in which the formula comes prepackaged in a container that can be attached directly to the feeding tube for administration.

CoA (coh-AY): coenzyme A; the coenzyme derived from the B vitamin pantothenic acid and central to energy metabolism.

coenzymes: complex organic molecules that work with enzymes to facilitate the enzymes' activity. Many coenzymes have B vitamins as part of their structures.

cofactor: a small, inorganic or organic substance that facilitates enzyme action; includes both organic coenzymes made from vitamins and inorganic substances such as minerals.

colectomy: removal of a portion or all of the colon.

colitis (ko-LYE-tis): inflammation of the colon.

collagen (KOL-ah-jen): the protein from which connective tissues such as scars, tendons, ligaments, and the foundations of bones and teeth are made.

collaterals: blood vessels that enlarge to allow an alternative pathway for diverted blood.

collecting duct: the last portion of a nephron's tubule, where the final concentration of urine occurs. One collecting duct is shared by several nephrons.

colonic irrigation: the popular, but potentially harmful practice of "washing" the large intestine with a powerful enema machine.

colostomy (co-LAH-stoe-me): a surgically created passage through the abdominal wall into the colon.

colostrum (ko-LAHS-trum): a milklike secretion from the breast, present during the first day or so after delivery before milk appears; rich in protective factors.

complement: a group of plasma proteins that assist the activities of antibodies.

complementary medicine: an approach that uses alternative therapies as an adjunct to, and not simply a replacement for, conventional medicine.

complementary proteins: two or more dietary proteins whose amino acid assortments complement each other in such a way that the essential amino acids missing from one are supplied by the other.

complex carbohydrates: polysaccharides composed of straight or branched chains of monosaccharides.

compound: a substance composed of two or more different atoms—for example, water (H_2O).

conception: the union of the male sperm and the female ovum; fertilization.

condensation: a chemical reaction in which two reactants combine to yield one larger product.

conditionally essential amino acid: an amino acid that is normally nonessential, but must be supplied by the diet in special circumstances when the need for it exceeds the body's ability to produce it.

conditionally essential nutrient: a nutrient that is normally nonessential, but must be supplied by the diet in special circumstances when the need for it exceeds the body's ability to produce it.

confectioners' sugar: finely powdered sucrose, 99.9 percent pure.

congenital hypothyroidism: decreased thyroid hormone production in a newborn.

congregate meals: nutrition programs that provide food for the elderly in conveniently located settings such as community centers.

conjugated linoleic acid: several fatty acids that have the same chemical formula as linoleic acid (18 carbons, two double bonds) but with different configurations (the double bonds occur on adjacent carbons).

constipation: the condition of having infrequent or difficult bowel movements.

contamination iron: iron found in foods as the result of contamination by inorganic iron salts from iron cookware, iron-containing soils, and the like.

continuous ambulatory peritoneal dialysis (CAPD): the most common method of peritoneal dialysis; involves frequent exchanges of dialysate, which remains in the peritoneal cavity throughout the day.

continuous feedings: slow delivery of formula at a constant rate over an 8- to 24-hour period.

continuous glucose monitoring: continuous monitoring of tissue glucose levels using a small sensor placed under the skin.

continuous parenteral nutrition: continuous administration of parenteral solutions over a 24-hour period.

continuous renal replacement therapy (CRRT): a slow, continuous method of removing solutes and/or fluids from blood by gently pumping blood across a filtration membrane over a prolonged time period.

control group: a group of individuals similar in all possible respects to the experimental group except for the treatment. Ideally, the control group receives a placebo while the experimental group receives a real treatment.

Cori cycle: the path from muscle glycogen to glucose to pyruvate to lactate (which travels to the liver) to glucose (which can travel back to the muscle) to glycogen; named after the scientist who elucidated this pathway.

corn sweeteners: corn syrup and sugars derived from corn.

corn syrup: a syrup made from cornstarch that has been treated with acid, high temperatures, and enzymes that produce glucose, maltose, and dextrins. See also *high-fructose corn syrup (HFCS)*.

cornea (KOR-nee-uh): the transparent membrane covering the outside of the eye.

coronary heart disease (CHD): a chronic, progressive disease characterized by obstructed blood flow in the coronary arteries; also called *coronary artery disease*.

coronary thrombosis: a clot that blocks blood flow through an artery that feeds the heart muscle.

correlation (CORE-ee-LAY-shun): the simultaneous increase, decrease, or change in two variables. If *A* increases as *B* increases, or if *A* decreases as *B* decreases, the correlation is positive. (This does not mean that *A* causes *B*, or vice versa.) If *A* increases as *B* decreases, or if *A* decreases as *B* increases, the correlation is negative. (This does not mean that *A* prevents *B*, or vice versa.) Some third factor may account for both *A* and *B*.

correspondence schools: schools that offer courses and degrees by mail. Some correspondence schools are accredited; others are not.

cortical bone: the very dense bone tissue that forms the outer shell surrounding trabecular bone and comprises the shaft of a long bone.

cortisol: a steroid hormone secreted by the adrenal cortex as part of the body's stress response.

counterregulatory hormones: hormones that have actions that oppose those of insulin: the catecholamines, glucagon, and cortisol.

coupled reactions: pairs of chemical reactions in which some of the energy released from the breakdown of one compound is used to create a bond in the formation of another compound.

covert (KOH-vert): hidden, as if under covers.

cretinism (KREE-tin-ism): a congenital disease characterized by mental and physical retardation and commonly caused by maternal iodine deficiency during pregnancy.

critical pathways: coordinated programs of treatment that merge the care plans of different health practitioners; also called *clinical pathways*.

critical periods: finite periods during development in which certain events occur that will have irreversible effects on later developmental stages; usually a period of rapid cell division.

Crohn's disease: an inflammatory bowel disease that usually occurs in the lower portion of the small intestine and the colon. Inflammation may pervade the entire intestinal wall.

cross-contamination: the contamination of food by bacteria that occurs when the food comes into contact with surfaces previously touched by raw meat, poultry, or seafood.

cross-reactivity: the ability of an antibody to react to an antigen that is similar, but not identical, to the one that induced the antibody's formation.

cryptosporidiosis (KRIP-toe-spor-ih-dee-OH-sis): a foodborne illness caused by the parasite *Cryptosporidium parvum*.

crypts (KRIPTS): tubular glands that lie between the intestinal villi and secrete intestinal juices into the small intestine.

cyanosis (sigh-ah-NOH-sis): a bluish cast in the skin due to the color of deoxygenated hemoglobin. Cyanosis is most evident in individuals with lighter, thinner skin; it is mostly seen on lips, cheeks, and ears and under the nails.

cyclamate (SIGH-kla-mate): an artificial sweetener that is being considered for approval in the United States and is available in Canada as a tabletop sweetener, but not as an additive.

cyclic parenteral nutrition: administration of parenteral solutions over an 8- to 16-hour period each day.

cystic fibrosis: an inherited disorder that affects the transport of chloride across epithelial cell membranes and is characterized by the production of abnormally viscous exocrine secretions; primarily affects the gastrointestinal and respiratory systems.

cystinuria (SIS-tin-NOO-ree-ah): an inherited disorder characterized by the elevated urinary excretion of several amino acids, including cystine.

cytokines (SIGH-toe-kines): signaling proteins produced by the body's cells; those produced by white blood cells regulate immune cell development and immune responses.

cytoplasm (SIGH-toh-plazm): the cell contents, except for the nucleus.

cytosol: the fluid of cytoplasm; contains water, ions, nutrients, and enzymes.

D

Daily Values (DV): reference values developed by the FDA specifically for use on food labels.

dawn phenomenon: morning hyperglycemia that is caused by the early-morning release of growth hormone, which counteracts insulin's glucose-lowering effects.

deamination (dee-AM-ih-NAY-shun): removal of the amino (NH_2) group from a compound such as an amino acid.

debridement: the surgical removal of dead, damaged, or contaminated tissue resulting from burns or wounds; helps to prevent infection and hasten healing.

decision-making capacity: the ability to understand pertinent information and make appropriate decisions; known as *decision-making competency* within the legal system.

decubitus (deh-KYU-bih-tus) **ulcers:** regions of damaged skin and tissue due to prolonged pressure on the affected area by an external object, such as a bed, wheelchair, or cast; vulnerable areas of the body include buttocks, hips, and heels. Also called *pressure sores*.

deep vein thrombosis: formation of a stationary blood clot (thrombus) in a deep vein, usually in the leg, which causes inflammation, pain, and swelling, and is potentially fatal.

defecate (DEF-uh-cate): to move the bowels and eliminate waste.

defibrillation: life-sustaining treatment in which an electronic device is used to shock the heart and reestablish a pattern of normal contractions. Defibrillation is used when the heart has arrhythmias or has experienced cardiac arrest.

deficient: the amount of a nutrient below which almost all healthy people can be expected, over time, to experience deficiency symptoms.

dehydration: the condition in which body water output exceeds water input. Symptoms include thirst, dry skin and mucous membranes, rapid heartbeat, low blood pressure, and weakness.

denaturation (dee-NAY-chur-AY-shun): the change in a protein's shape and consequent loss of its function brought about by heat, agitation, acid, base, alcohol, heavy metals, or other agents.

dental calculus: mineralized dental plaque, often associated with inflammation and bleeding.

dental caries: tooth decay.

dental plaque: an accumulation of bacteria and their by-products that grow on teeth and can lead to dental caries and gum disease.

dermatitis herpetiformis (DERM-ah-TYE-tis HER-peh-tih-FOR-mis): a gluten-sensitive disorder characterized by a severe skin rash.

dermis: the connective tissue layer underneath the epidermis that contains the skin's blood vessels and nerves.

dextrose: an older name for glucose.

DHF (dihydrofolate): a coenzyme form of folate.

diabetes (DYE-ah-BEE-teez) **insipidus:** a pituitary disorder that causes a deficiency of antidiuretic hormone (unrelated to *diabetes mellitus*).

diabetes mellitus: a group of metabolic disorders characterized by hyperglycemia and disordered insulin metabolism.

diabetic coma: a coma that occurs in uncontrolled diabetes; may be due to diabetic ketoacidosis, the hyperosmolar hyperglycemic syndrome, or severe hypoglycemia.

diabetic nephropathy (neh-FRAH-pah-thee): damage to the kidneys that results from long-term diabetes.

diabetic neuropathy (nur-RAH-pah-thee): complications of diabetes that cause damage to nerves.

diabetic retinopathy (REH-tih-NAH-pah-thee): retinal damage that results from long-term diabetes.

dialysate (dye-AL-ih-sate): the solution used in dialysis to draw wastes and fluids from the blood.

dialysis (dye-AH-lih-sis): life-sustaining treatment in which a patient's blood is filtered using selective diffusion through a semipermeable membrane; substitutes for kidney function.

dialyzer (DYE-ah-LYE-zer): a machine used in hemodialysis to filter the blood; also called an *artificial kidney*.

diarrhea: the frequent passage of watery bowel movements.

diet: the foods and beverages a person eats and drinks.

diet history: a record of eating behaviors and the foods a person eats.

diet manual: a resource that specifies the foods to include or exclude in modified diets and provides sample menus.

diet orders: specific instructions regarding dietary management; also called *nutrition prescriptions*.

diet progression: a change in diet as a patient's food tolerance improves.

dietary fibers: in plant foods, the *nonstarch polysaccharides* that are not digested by human digestive enzymes, although some are digested by GI tract bacteria. Dietary fibers include cellulose, hemicelluloses, pectins, gums, and mucilages and the nonpolysaccharides lignins, cutins, and tannins.

dietary folate equivalents (DFE): the amount of folate available to the body from naturally occurring sources, fortified foods, and supplements, accounting for differences in the bioavailability from each source.

Dietary Reference Intakes (DRI): a set of nutrient intake values for healthy people in the United States and Canada. These values are used for planning and assessing diets and include: Estimated Average Requirements (EAR), Recommended Dietary Allowances (RDA), Adequate Intakes (AI), Tolerable Upper Intake Levels (UL).

dietary supplement: any pill, capsule, tablet, liquid, or powder that contains vitamins, minerals, herbs, or amino acids; intended to increase dietary intake of these substances.

dietetic technician: a person who has completed a minimum of an associate's degree from an accredited university or college and an approved dietetic technician program that includes a supervised practice experience. See also *dietetic technician, registered (DTR)*.

dietetic technician, registered (DTR): a dietetic technician who has passed a national examination and maintains registration through continuing professional education.

dietitian: a person trained in nutrition, food science, and diet planning. See also *registered dietitian*.

diffusion: movement of solutes from an area of high concentration to one of low concentration.

digestion: the process by which food is broken down into absorbable units.

digestive enzymes: proteins found in digestive juices that act on food substances, causing them to break down into simpler compounds.

digestive system: all the organs and glands associated with the ingestion and digestion of food.

dipeptide (dye-PEP-tide): two amino acids bonded together.

direct calorimetry: a means of estimating energy expenditure by measuring the amount of heat released.

disaccharides (dye-SACK-uh-rides): pairs of monosaccharides linked together. See Appendix C for the chemical structures of the disaccharides.

disclosure: the act of revealing pertinent information. For example, clinicians should accurately describe proposed tests and procedures, their benefits and risks, and alternative approaches.

discretionary kcalorie allowance: the kcalories remaining in a person's energy allowance after consuming enough nutrient-dense foods to meet all nutrient needs for a day.

disordered eating: eating behaviors that are neither normal nor healthy, including restrained eating, fasting, binge eating, and purging.

dispensable amino acids: nonessential amino acids.

dissociates (dis-SO-see-aites): physically separates.

distilled liquor or **hard liquor**: an alcoholic beverage traditionally made by fermenting and distilling a carbohydrate source such as molasses, potatoes, rye, beets, barley, or corn; sometimes called *distilled spirits*.

distilled water: water that has been vaporized and recondensed, leaving it free of dissolved minerals.

distributive justice: the equitable distribution of resources.

diuresis (DYE-uh-REE-sis): increased urine production.

diverticula (dye-ver-TIC-you-la): sacs or pouches that develop in the weakened areas of the intestinal wall (like bulges in an inner tube where the tire wall is weak).

diverticulitis (DYE-ver-tic-you-LYE-tis): infected or inflamed diverticula.

diverticulosis (DYE-ver-tic-you-LOH-sis): the condition of having diverticula.

DNA (deoxyribonucleic acid): the double helix molecules of which genes are made.

do-not-resuscitate (DNR) order: a request by a patient or surrogate to withhold cardiopulmonary resuscitation.

docosahexaenoic (DOE-cossa-HEXA-ee-NO-ick) **acid (DHA)**: an omega-3 polyunsaturated fatty acid with 22 carbons and six double bonds; present in fatty fish and synthesized in limited amounts in the body from linolenic acid.

dolomite: a compound of minerals (calcium magnesium carbonate) found in limestone and marble. Dolomite is powdered and is sold as a calcium-magnesium supplement. However, it may be contaminated with toxic minerals, is not well absorbed, and interacts adversely with absorption of other essential minerals.

double-blind experiment: an experiment in which neither the subjects nor the researchers know which subjects are members of the experimental group and which are serving as control subjects, until after the experiment is over.

Down syndrome: a genetic abnormality that causes mental retardation, short stature, and flattened facial features.

drink: a dose of any alcoholic beverage that delivers ½ ounce of pure ethanol: 5 ounces of wine, 10 ounces of wine cooler, 12 ounces of beer, or 1½ ounces of hard liquor.

drug: a substance that can modify one or more of the body's functions.

drug history: a record of all the drugs, over-the-counter and prescribed, that a person takes routinely.

DTR: see *dietetic technician, registered*.

dumping syndrome: a cluster of symptoms that result from the rapid emptying of an osmotic load from the stomach into the small intestine.

duodenum (doo-oh-DEEN-um, doo-ODD-num): the top portion of the small intestine (about "12 fingers' breadth" long in ancient terminology).

durable power of attorney: a legal document (sometimes called a *health care proxy*) that gives legal authority to another (a *health care agent*) to make medical decisions in the event of incapacitation.

dysentery (DISS-en-terry): an infection of the digestive tract that causes diarrhea.

dyspepsia: a feeling of pain, bloating, or discomfort in the upper abdominal area, often called *indigestion*; a symptom of illness rather than a disease itself.

dysphagia (dis-FAY-jah): difficulty in swallowing.

dyspnea (DISP-nee-ah): shortness of breath.

E

eating disorders: disturbances in eating behavior that jeopardize a person's physical or psychological health.

eclampsia (eh-KLAMP-see-ah): a severe stage of preeclampsia characterized by convulsions.

edema (eh-DEEM-uh): the swelling of body tissue caused by excessive amounts of fluid in the interstitial spaces; seen in protein deficiency (among other conditions).

edentulous (ee-DENT-you-lus): lack of teeth.

eicosanoids (eye-COSS-uh-noyds): derivatives of 20-carbon fatty acids; biologically active compounds that help to regulate blood pressure, blood clotting, and other body functions. They include *prostaglandins* (PROS-tah-GLAND-ins), *thromboxanes* (throm-BOX-ains), and *leukotrienes* (LOO-ko-TRY-eens).

eicosapentaenoic (EYE-cossa-PENTA-ee-NO-ick) **acid (EPA)**: an omega-3 polyunsaturated fatty acid with 20 carbons and five double bonds; present in fish and synthesized in limited amounts in the body from linolenic acid.

electrolyte solutions: solutions that can conduct electricity.

electrolytes: salts that dissolve in water and dissociate into charged particles called ions.

electron transport chain: the final pathway in energy metabolism that transports electrons from hydrogen to oxygen and captures the energy released in the bonds of ATP.

element: a substance composed of atoms that are alike—for example, iron (Fe).

elemental formulas: enteral formulas that contain carbohydrates and proteins that are partially or fully hydrolyzed; also called *hydrolyzed, chemically defined*, or *monomeric formulas*.

embolism (EM-boh-lizm): the obstruction of a blood vessel by an embolus, causing sudden tissue death.

embolus (EM-boh-lus): an abnormal particle, such as a blood clot or air bubble, that travels in the blood.

embryo (EM-bree-oh): the developing infant from two to eight weeks after conception.

emergency shelters: facilities that are used to provide temporary housing.

emetic (em-ETT-ic): an agent that causes vomiting.

emphysema (EM-fih-ZEE-mah): a progressive lung condition characterized by the breakdown of the lungs' elastic structure and destruction of the walls of the bronchioles and alveoli, reducing the surface area involved in respiration.

empty-kcalorie foods: a popular term used to denote foods that contribute energy but lack protein, vitamins, and minerals.

emulsifier (ee-MUL-sih-fire): a substance with both water-soluble and fat-soluble portions that promotes the mixing of oils and fats in a watery solution.

end-stage renal disease (ESRD): an advanced stage of chronic kidney disease in which dialysis or a kidney transplant is necessary to sustain life.

endocrine: pertains to hormonal secretions into the blood. Opposite of *exocrine*, which pertains to external secretions, such as those of the mucous membranes or the skin.

endogenous (en-DODGE-eh-nus): compounds that derive from within the body.

endoplasmic reticulum (en-doh-PLAZ-mic reh-TIC-you-lum): a complex network of intracellular membranes. The rough endoplasmic reticulum is dotted with ribosomes, where protein synthesis takes place. The smooth endoplasmic reticulum bears no ribosomes.

endosperm: the inner edible part of a grain, rich in starch and proteins.

endothelial cells: the cells that line the inner surfaces of blood vessels, lymphatic vessels, and body cavities.

enemas: solutions inserted into the rectum and colon to stimulate a bowel movement and empty the lower large intestine.

energy: the capacity to do work. The energy in food is chemical energy. The body can convert this chemical energy to mechanical, electrical, or heat energy.

energy density: a measure of the energy a food provides relative to the amount of food (kcalories per gram).

energy-yielding nutrients: the nutrients that break down to yield energy the body can use.

enriched: the addition to a food of nutrients that were lost during processing so that the food will meet a specified standard.

enteral (EN-ter-al) **nutrition:** the provision of nutrients using the GI tract, including the use of tube feedings and oral diets.

enteric coated: refers to medications or enzyme preparations that are coated to withstand gastric acidity and dissolve only at the higher pH of the small intestine.

enterogastrone (EN-ter-oh-GAS-trone): a general term for any gastrointestinal hormone; sometimes used to refer specifically to *gastric inhibitory peptide*.

enteropancreatic (EN-ter-oh-PAN-kree-AT-ik) **circulation:** the circulatory route from the pancreas to the intestine and back to the pancreas.

enterostomy (EN-ter-AH-stoe-mee): an opening into the GI tract through the abdominal wall.

enzymes: proteins that facilitate chemical reactions without being changed in the process; protein catalysts.

epidemic (ep-ih-DEM-ick): the appearance of a disease (usually infectious) or condition that attacks many people at the same time in the same region.

epidermis (eh-pih-DER-miss): the outer layer of the skin.

epigenetics: the study of heritable changes in gene function that occur without a change in the DNA sequence.

epiglottis (epp-ih-GLOTT-iss): cartilage in the throat that guards the entrance to the trachea and prevents fluid or food from entering it when a person swallows.

epinephrine (EP-ih-NEFF-rin): a hormone of the adrenal gland that modulates the stress response; formerly called *adrenaline*. When administered by injection, epinephrine counteracts anaphylactic shock by opening the airways and maintaining heartbeat and blood pressure.

epithelial (ep-i-THEE-lee-ul) **cells:** cells on the surface of the skin and mucous membranes.

epithelial tissue: the layer of the body that serves as a selective barrier between the body's interior and the environment. (Examples are the cornea of the eyes, the skin, the respiratory lining of the lungs, and the lining of the digestive tract.)

ergocalciferol (ER-go-kal-SIF-er-ol): vitamin D derived from plants in the diet and made from the yeast and plant sterol ergosterol; also called vitamin D_2.

erythrocyte (eh-RITH-ro-cite) **hemolysis** (he-MOLL-uh-sis): the breaking open of red blood cells (erythrocytes); a symptom of vitamin E–deficiency disease in human beings.

erythrocyte protoporphyrin (PRO-toe-PORE-ferin): a precursor to hemoglobin.

erythropoiesis (eh-RIH-throh-poy-EE-sis): production of red blood cells within the bone marrow.

erythropoietin (eh-RIH-throh-POY-eh-tin): a hormone produced by kidney cells that stimulates red blood cell production.

esophageal (eh-SOF-ah-JEE-al): involving the esophagus.

esophageal dysphagia: an inability to move food through the esophagus; usually caused by an obstruction or a motility disorder.

esophageal sphincter: a sphincter muscle at the upper or lower end of the esophagus. The lower esophageal sphincter is also called the **cardiac sphincter**.

esophagus (ee-SOFF-ah-gus): the food pipe; the conduit from the mouth to the stomach.

essential amino acids: amino acids that the body cannot synthesize in amounts sufficient to meet physiological needs.

essential fatty acids: fatty acids that the body cannot synthesize in amounts sufficient to meet physiological needs.

essential nutrients: nutrients a person must obtain from food because the body cannot make them for itself in sufficient quantity to meet physiological needs; also called *indispensable nutrients*. About 40 nutrients are currently known to be essential for human beings.

Estimated Average Requirement (EAR): the average daily amount of a nutrient that will maintain a specific biochemical or physiological function in half the healthy people of a given age and gender group.

Estimated Energy Requirement (EER): the average dietary energy intake that maintains energy balance and good health in a person of a given age, gender, weight, height, and level of physical activity.

estrogens: hormones responsible for the menstrual cycle and other female characteristics.

ethanol: a particular type of alcohol found in beer, wine, and distilled liquor; also called *ethyl alcohol*.

ethical: in accordance with accepted principles of right and wrong.

excessive drinking: heavy drinking, binge drinking, or both.

exchange lists: diet-planning tools that organize foods by their proportions of carbohydrate, fat, and protein. Foods on any single list can be used interchangeably.

exocrine: pertains to external secretions, such as those of the mucous membranes or the skin. Opposite of *endocrine*, which pertains to hormonal secretions into the blood.

exogenous (eks-ODGE-eh-nus): compounds that derive from foods.

experimental group: a group of individuals similar in all possible respects to the control group except for the treatment. The experimental group receives the real treatment.

extra lean: less than 5 grams of fat, 2 grams of saturated fat and *trans* fat combined, and 95 milligrams cholesterol per serving and per 100 grams of meat, poultry, and seafood.

extracellular fluid: fluid outside the cells. Extracellular fluid includes two main components—the interstitial fluid and plasma. Extracellular fluid accounts for approximately one-third of the body's water.

F

FAD (flavin adenine dinucleotide): a coenzyme form of riboflavin.

fad diets: popular eating plans that promise quick weight loss. Most fad diets severely limit certain foods or overemphasize others (for example, never eat potatoes or pasta or eat cabbage soup daily).

faith healing: the use of prayer or belief in divine intervention to promote healing.

false negative: a test result indicating that a condition is not present (negative) when in fact it is present (therefore false).

false positive: a test result indicating that a condition is present (positive) when in fact it is not (therefore false).

fasting hyperglycemia: hyperglycemia that develops after an overnight fast of at least eight hours.

fat-free: less than 0.5 gram of fat per serving (and no added fat or oil); synonyms include "zero-fat," "no fat," and "nonfat."

fat replacers: ingredients that replace some or all of the functions of fat and may or may not provide energy.

fats: lipids that are solid at room temperature (77°F or 25°C).

fatty acid: an organic compound composed of a carbon chain with hydrogens attached and an acid group (COOH) at one end and a methyl group (CH_3) at the other end.

fatty acid oxidation: the metabolic breakdown of fatty acids to acetyl CoA; also called *beta oxidation*.

fatty liver: an early stage of liver deterioration seen in several diseases, including kwashiorkor and alcoholic liver disease. Fatty liver is characterized by an accumulation of fat in the liver cells; also called *hepatic steatosis* (STEE-ah-TOE-sis).

fatty streaks: initial lesions of atherosclerosis that form on the artery wall, characterized by accumulations of foam cells, lipid material, and connective tissue.

female athlete triad: a potentially fatal combination of three medical problems—disordered eating, amenorrhea, and osteoporosis.

fermentable: the extent to which bacteria in the GI tract can break down fibers to fragments that the body can use.

ferritin (FAIR-ih-tin): the iron-storage protein.

fertility: the capacity of a woman to produce a normal ovum periodically and of a man to produce normal sperm; the ability to reproduce.

fetal alcohol spectrum disorder: a range of physical, behavioral, and cognitive abnormalities caused by prenatal alcohol exposure.

fetal alcohol syndrome (FAS): a cluster of physical, behavioral, and cognitive abnormalities associated with prenatal alcohol exposure, including facial malformations, growth retardation, and central nervous disorders.

fetal programming: the influence of substances during fetal growth on the development of diseases in later life.

fetus (FEET-us): the developing infant from eight weeks after conception until term.

fibrinogen (fye-BRIN-oh-jen): a liver protein that promotes blood clot formation.

fibrocystic (FYE-bro-SIS-tik) **breast disease:** a harmless condition in which the breasts develop lumps, sometimes associated with caffeine consumption. In some, it responds to abstinence from caffeine; in others, it can be treated with vitamin E.

fibrosis (fye-BROH-sis): an intermediate stage of liver deterioration seen in several diseases, including viral hepatitis and alcoholic liver disease. In fibrosis, the liver cells lose their function and assume the characteristics of connective tissue cells (fibers).

field gleaning: collecting crops from fields that either have already been harvested or are not profitable to harvest.

filtered water: water treated by filtration, usually through activated carbon filters that reduce the lead in tap water, or by reverse osmosis units that force pressurized water across a membrane removing lead, arsenic, and some microorganisms from tap water.

filtrate: the substances that pass through the glomerulus and travel through the nephron's tubules, eventually forming urine.

fistulas (FIST-you-luz): abnormal passages between organs or tissues (or between an internal organ and the body's surface) that permit the passage of fluids or secretions.

flatulence: the condition of having excessive intestinal gas, which causes abdominal discomfort.

flavonoids (FLAY-von-oyds): yellow pigments in foods; phytochemicals that may exert physiological effects on the body.

flaxseeds: the small brown seeds of the flax plant; valued as a source of linseed oil, fiber, and omega-3 fatty acids.

flora: bacteria within a given environment, such as the intestines.

fluid balance: maintenance of the proper types and amounts of fluid in each compartment of the body fluids (see also Chapter 12).

fluorapatite (floor-APP-uh-tite): the stabilized form of bone and tooth crystal, in which fluoride has replaced the hydroxyl groups of hydroxyapatite.

fluorosis (floor-OH-sis): discoloration and pitting of tooth enamel caused by excess fluoride during tooth development.

FMN (flavin mononucleotide): a coenzyme form of riboflavin.

foam cells: swollen cells in the artery wall that accumulate lipids.

folate (FOLE-ate): a B vitamin; also known as folic acid, folacin, or pteroylglutamic (tareo-EEL-glue-TAM-ick) acid (PGA). The coenzyme forms are **DHF (dihydrofolate)** and **THF (tetrahydrofolate)**.

follicle-stimulating hormone (FSH): a hormone that stimulates maturation of the ovarian follicles in females and the production of sperm in males. (The ovarian follicles are part of the female reproductive system where the eggs are produced.) The release of FSH is mediated by follicle-stimulating hormone releasing hormone (FSH–RH).

food allergy: an adverse reaction to food that involves an immune response; also called *food-hypersensitivity reaction*.

food aversions: strong desires to avoid particular foods.

food bank: a facility that collects and distributes food donations to authorized organizations feeding the hungry.

food cravings: strong desires to eat particular foods.

food frequency questionnaire: a checklist of foods on which a person can record the frequency with which he or she eats each food.

food group plans: diet-planning tools that sort foods into groups based on nutrient content and then specify that people should eat certain amounts of foods from each group.

food insecurity: limited or uncertain access to foods of sufficient quality or quantity to sustain a healthy and active life.

food insufficiency: an inadequate amount of food due to a lack of resources.

food intolerances: adverse reactions to foods that do not involve the immune system.

food pantries: programs that provide groceries to be prepared and eaten at home.

food poverty: hunger resulting from inadequate access to available food for various reasons, including inadequate resources, political obstacles, social disruptions, poor weather conditions, and lack of transportation.

food record: a detailed log of food eaten during a specified time period, usually several days; also called a *food diary*. A food record may also include information regarding medications, disease symptoms, and physical activity.

food recovery: collecting wholesome food for distribution to low-income people who are hungry.

food security: access to enough food to sustain a healthy and active life.

food substitutes: foods that are designed to replace other foods.

food and symptom diary: a food record kept by a patient to determine the cause of an adverse reaction; includes the specific foods and beverages consumed, symptoms experienced, and the timing of meals and symptom onset.

foodborne illness: illness transmitted to human beings through food and water, caused by either an infectious agent (foodborne infection) or a poisonous substance (food intoxication); commonly known as *food poisoning*.

foods: products derived from plants or animals that can be taken into the body to yield energy and nutrients for the maintenance of life and the growth and repair of tissues.

fortified: the addition to a food of nutrients that were either not originally present or present in insignificant amounts. Fortification can be used to correct or prevent a widespread nutrient deficiency or to balance the total nutrient profile of a food.

fraudulent: the promotion, for financial gain, of devices, treatments, services, plans, or products (including diets and supplements) that alter or claim to alter a human condition without proof of safety or effectiveness.

free: "nutritionally trivial" and unlikely to have a physiological consequence; synonyms include "without," "no," and "zero." A food that does not contain a nutrient naturally may make such a claim, but only as it applies to all similar foods (for example, "applesauce, a fat-free food").

free radicals: unstable and highly reactive atoms or molecules that have one or more unpaired electrons in the outer orbital.

French units: units of measure used to indicate the size of a feeding tube's outer diameter; 1 French unit equals 1/3 millimeter.

fructosamine test: a measurement of glycated serum proteins; used to analyze glycemic control over the preceding two weeks. Also known as the *glycated albumin test* or the *glycated serum protein test*.

fructose (FRUK-tose or FROOK-tose): a monosaccharide; sometimes known as *fruit sugar* or *levulose*. Fructose is found abundantly in fruits, honey, and saps.

fuel: compounds that cells can use for energy. The major fuels include glucose, fatty acids, and amino acids; other fuels include ketone bodies, lactate, glycerol, and alcohol.

full liquid diet: a liquid diet that includes clear liquids, milk, yogurt, ice cream, and liquid nutritional supplements (such as Ensure).

full term: between the thirty-eighth and forty-second week of pregnancy.

functional foods: foods that contain physiologically active compounds that provide health benefits beyond their nutrient contributions; sometimes called *designer foods* or *nutraceuticals*.

futile: medical care that will not improve the medical circumstances of a patient.

G

g: grams; a unit of weight equivalent to about 0.03 ounce.

galactose (ga-LAK-tose): a monosaccharide; part of the disaccharide lactose.

galactosemia (ga-LACK-toe-SEE-me-ah): an inherited disorder that affects galactose metabolism. Accumulated galactose causes damage to the liver, kidneys, and brain in untreated patients.

gallbladder: the organ that stores and concentrates bile. When it receives the signal that fat is present in the duodenum, the gallbladder contracts and squirts bile through the bile duct into the duodenum.

gallstones: stones that form in the gallbladder from crystalline deposits of cholesterol or bilirubin.

gangrene (GANG-green): death of tissue due to a deficient blood supply and/or infection.

gastrectomy (gah-STREK-ta-mee): the surgical removal of part of the stomach (partial gastrectomy) or the entire stomach (total gastrectomy).

gastric decompression: the removal of the stomach contents (including swallowed saliva, stomach secretions, and gas) of patients with motility disorders or obstructions that prevent stomach emptying.

gastric glands: exocrine glands in the stomach wall that secrete gastric juice into the stomach.

gastric inhibitory peptide: a gastrointestinal hormone that slows motility and inhibits gastric secretions.

gastric juice: the digestive secretion of the gastric glands of the stomach.

gastric outlet obstruction: an obstruction that prevents the normal emptying of stomach contents into the duodenum.

gastric residual volume: the volume of formula remaining in the stomach from a previous feeding.

gastrin: a hormone secreted by cells in the stomach wall. Target organ: the glands of the stomach. Response: secretion of gastric acid.

gastritis: inflammation of stomach tissue.

gastroesophageal reflux: the backflow of stomach acid into the esophagus, causing damage to the cells of the esophagus and the sensation of heartburn.

gastroesophageal reflux disease (GERD): a condition characterized by the backflow of stomach acid into the esophagus two or more times a week.

gastrointestinal (GI) tract: the digestive tract. The principal organs are the stomach and intestines.

gastroparesis (GAS-troe-pah-REE-sis): delayed stomach emptying caused by nerve damage in stomach tissue.

gastrostomy (gah-STRAH-stoe-mee): an opening into the stomach through which a feeding tube can be passed.

gatekeepers: with respect to nutrition, key people who control other people's access to foods and thereby exert profound impacts on their nutrition. Examples are the spouse who buys and cooks the food, the parent who feeds the children, and the caregiver in a day-care center.

gene expression: the process by which a cell converts the genetic code into RNA and protein.

gene pool: all the genetic information of a population at a given time.

gene therapy: treatment for inherited disorders, in which DNA sequences are introduced into the chromosomes of affected cells, prompting the cells to express the protein needed to correct the disease.

genes: sections of chromosomes that contain the instructions needed to make one or more proteins.

genetic counseling: support for families at risk of genetic disorders; involves diagnosis of disease, identification of inheritance patterns within the family, and review of reproductive options.

genetics: the study of genes and inheritance.

genome (GEE-nome): the complete set of genetic material (DNA) in an organism or a cell. The study of genomes is called **genomics**.

genomics: the study of all the genes in an organism and their interactions with environmental factors.

genotoxicant: a substance that mutates or damages genetic material.

geophagia: the specific craving for nonfood items that come from the earth, such as clay or dirt.

germ: the seed that grows into a mature plant, especially rich in vitamins and minerals.

gestation (jes-TAY-shun): the period from conception to birth. For human beings, the average length of a healthy gestation is 40 weeks. Pregnancy is often divided into three-month periods, called *trimesters*.

gestational diabetes: glucose intolerance with onset or first recognition during pregnancy.

gestational hypertension: high blood pressure that develops in the second half of pregnancy and resolves after childbirth, usually without affecting the outcome of the pregnancy.

ghrelin (GRELL-in): a protein produced by the stomach cells that enhances appetite and decreases energy expenditure.

gingiva (jin-JYE-va, JIN-jeh-va): the gums.

gingivitis (jin-jeh-VYE-tus): inflammation of the gums, characterized by redness, swelling, and bleeding.

glands: cells or groups of cells that secrete materials for special uses in the body. Glands may be exocrine (EKS-oh-crin) glands, secreting their materials "out" (into the digestive tract or onto the surface of the skin), or endocrine (EN-doe-crin) glands, secreting their materials "in" (into the blood).

gliadin: the protein in wheat gluten that has toxic effects in celiac disease.

glomerular filtration rate (GFR): the rate at which filtrate is formed within the kidneys, normally about 125 mL/min.

glomerulus (gloh-MEHR-yoo-lus): a tuft of capillaries within the nephron that filters water and solutes from the blood as urine production begins (plural: *glomeruli*).

glossitis (gloss-EYE-tis): an inflammation of the tongue.

glucagon (GLOO-ka-gon): a hormone that is secreted by special cells in the pancreas in response to low blood glucose concentration and elicits release of glucose from liver glycogen stores.

glucocorticoids: hormones from the adrenal cortex that affect the body's management of glucose.

glucogenic amino acids: amino acids that can make glucose via either pyruvate or TCA cycle intermediates.

gluconeogenesis (gloo-ko-nee-oh-JEN-ihsis): the making of glucose from a noncarbohydrate source (described in more detail in Chapter 7).

glucose (GLOO-kose): a monosaccharide; sometimes known as *blood sugar* or *dextrose*.

glucose tolerance factors (GTF): small organic compounds that enhance insulin's action.

gluten-sensitive enteropathy: immune disorder characterized by an abnormal immune response to the dietary protein gluten; also called *celiac disease* or *celiac sprue*.

glycated hemoglobin (HbA_{1c}): hemoglobin that has nonenzymatically attached to glucose; the level of HbA_{1c} in blood helps to diagnose diabetes and evaluate long-term glycemic control. Also called *glycosylated hemoglobin*.

glycemic (gly-SEE-mic): pertaining to blood glucose.

glycemic index: a method of classifying foods according to their potential for raising blood glucose.

glycemic response: the extent to which a food raises the blood glucose concentration and elicits an insulin response.

glycerol (GLISS-er-ol): an alcohol composed of a three-carbon chain, which can serve as the backbone for a triglyceride.

glycogen (GLY-ko-jen): an animal polysaccharide composed of glucose; manufactured and stored in the liver and muscles as a storage form of glucose. Glycogen is not a significant food source of carbohydrate and is not counted as one of the complex carbohydrates in foods.

glycolysis (gly-COLL-ih-sis): the metabolic breakdown of glucose to pyruvate. Glycolysis does not require oxygen (anaerobic).

glycosuria (GLY-co-SOOR-ee-ah): the presence of glucose in the urine.

goblet cells: cells of the GI tract (and lungs) that secrete mucus.

goiter (GOY-ter): an enlargement of the thyroid gland due to an iodine deficiency, malfunction of the gland, or overconsumption of a goitrogen. Goiter caused by iodine deficiency is *simple goiter*.

goitrogen (GOY-troh-jen): a substance that enlarges the thyroid gland and causes *toxic goiter*. Goitrogens occur naturally in such foods as cabbage, kale, brussels sprouts, cauliflower, broccoli, and kohlrabi.

Golgi (GOAL-gee) **apparatus:** a set of membranes within the cell where secretory materials are packaged for export.

good source of: the product provides between 10 and 19 percent of the Daily Value for a given nutrient per serving.

gout (GOWT): a metabolic disorder characterized by elevated uric acid levels in the blood and urine and the deposition of uric acid in and around the joints, causing acute joint inflammation.

granulated sugar: crystalline sucrose; 99.9 percent pure.

growth hormone (GH): a hormone secreted by the pituitary that regulates the cell division and protein synthesis needed for normal growth. The release of GH is mediated by GH-releasing hormone (GHRH).

H

half-life: the time period of a chemical effect; in blood tests, the length of time that a substance remains in plasma. For example, if the grapefruit effect has a 12-hour half-life, this means that after 12 hours, its biological effect is half of the maximum effect measured. The albumin in plasma has a half-life of 14 to 20 days, meaning that half of the amount circulating in plasma is degraded in this time period.

hard water: water with a high calcium and magnesium content.

hazard: a source of danger; used to refer to circumstances in which harm is possible under normal conditions of use.

Hazard Analysis Critical Control Points (HACCP): a systematic plan to identify and correct potential microbial hazards in the manufacturing, distribution, and commercial use of food products; commonly referred to as "HASS-ip."

HDL (high-density lipoprotein): the type of lipoprotein that transports cholesterol back to the liver from the cells; composed primarily of protein.

health care agent: a person given legal authority to make medical decisions for another in the event of incapacitation.

health claims: statements that characterize the relationship between a nutrient or other substance in a food and a disease or health-related condition.

health history: an account of a client's current and past health status and disease risks.

healthy: a food that is low in fat, saturated fat, cholesterol, and sodium and that contains at least 10 percent of the Daily Values for vitamin A, vitamin C, iron, calcium, protein, or fiber.

Healthy Eating Index: a measure that assesses how well a diet meets the recommendations of the *Dietary Guidelines for Americans* and MyPyramid.

Healthy People: a national public health initiative under the jurisdiction of the U.S. Department of Health and Human Services (DHHS) that identifies the most significant preventable threats to health and focuses efforts toward eliminating them.

heart attack: death of heart muscle caused by a sudden reduction in coronary blood flow; also called a *myocardial infarction (MI)* or *cardiac arrest*.

heart failure: a condition with various causes that is characterized by the heart's inability to pump adequate blood to the body's cells, resulting in fluid accumulation in the tissues; also called *congestive heart failure*.

heartburn: a burning sensation in the chest area caused by backflow of stomach acid into the esophagus.

heavy drinking: more than one drink per day on average for women and more than two drinks per day on average for men.

heavy metals: mineral ions such as mercury and lead, so called because they are of relatively high atomic weight. Many heavy metals are poisonous.

Heimlich (HIME-lick) **maneuver:** a technique for dislodging an object from the trachea of a choking person; also known as *abdominal thrust maneuver*.

***Helicobacter pylori* (*H. pylori*):** a species of bacterium that colonizes gastric mucosa; a primary cause of gastritis and peptic ulcer disease.

helper T cells: lymphocytes that have a specific protein called CD4 on their surfaces and therefore are also known as *CD4+ T cells*; these are the cells most affected in HIV infection.

hematocrit (hee-MAT-oh-krit): measurement of the volume of the red blood cells packed by centrifuge in a given volume of blood.

hematopoietic stem cell transplantation: transplantation of the stem cells that produce red blood cells and white blood cells; the stem cells are obtained from the bone marrow (*bone marrow transplantation*) or circulating blood.

hematuria (HE-mah-TOO-ree-ah): blood in the urine.

heme (HEEM) **iron:** the iron in foods that is bound to the hemoglobin and myoglobin proteins; found only in meat, fish, and poultry.

hemochromatosis (HE-moh-KRO-ma-toe-sis): a genetically determined failure to prevent absorption of unneeded dietary iron that is characterized by iron overload and tissue damage.

hemodialysis (HE-moe-dye-AL-ih-sis): a treatment that removes fluids and wastes from the blood by passing the blood through a dialyzer.

hemofiltration: removal of fluid and solutes by pumping blood across a membrane; no osmotic gradients are created during the process.

hemoglobin (HE-moh-GLO-bin): the globular protein of the red blood cells that carries oxygen from the lungs to the cells throughout the body.

hemolytic (HE-moh-LIT-ick) **anemia:** the condition of having too few red blood cells as a result of erythrocyte hemolysis (breakdown of red blood cells).

hemophilia (HE-moh-FEEL-ee-ah): a hereditary disease in which the blood is unable to clot because it lacks the ability to synthesize certain clotting factors.

hemorrhagic (hem-oh-RAJ-ik) **disease:** a disease characterized by excessive bleeding.

hemorrhagic strokes: strokes caused by bleeding within the brain, which destroys or compresses brain tissue.

hemorrhoids (HEM-oh-royds): painful swelling of the veins surrounding the rectum.

hemosiderin (heem-oh-SID-er-in): an iron-storage protein primarily made in times of iron overload.

hemosiderosis (HE-moh-sid-er-OH-sis): a condition characterized by the deposition of hemosiderin in the liver and other tissues.

hepatic coma: loss of consciousness resulting from severe liver disease.

hepatic encephalopathy (en-sef-ah-LOP-ah-thie): a condition that develops in advanced liver disease that is characterized by altered neurological functioning, including changes in personality and behavior, reduced mental abilities, and disturbances in motor function.

hepatic portal vein: the vein that collects blood from the GI tract and conducts it to capillaries in the liver.

hepatic steatosis (STEE-ah-TOE-sis): an early stage of liver deterioration seen in several diseases, including kwashiorkor and alcoholic liver disease. Hepatic steatosis is characterized by an accumulation of fat in the liver cells; also called *fatty liver.*

hepatic vein: the vein that collects blood from the liver capillaries and returns it to the heart.

hepatitis (hep-ah-TYE-tis): inflammation of the liver.

hepatomegaly (HEP-ah-toe-MEG-ah-lee): enlargement of the liver.

hepcidin: a hormone produced by the liver that regulates iron balance.

herpes simplex virus: a common virus that can cause blisterlike lesions on the lips and in the mouth.

hiatal hernia: a condition in which the upper portion of the stomach protrudes above the diaphragm; most cases are asymptomatic.

hiccups (HICK-ups): repeated cough-like sounds and jerks that are produced when an involuntary spasm of the diaphragm muscle sucks air down the windpipe; also spelled hiccoughs.

high: 20 percent or more of the Daily Value for a given nutrient per serving; synonyms include "rich in" or "excellent source."

high fiber: 5 grams or more fiber per serving. A high-fiber claim made on a food that contains more than 3 grams fat per serving and per 100 grams of food must also declare total fat.

high food security: no indications of food-access problems or limitations.

high-fructose corn syrup (HFCS): a syrup made from cornstarch that has been treated with an enzyme that converts some of the glucose to the sweeter fructose; made especially for use in processed foods and beverages, where it is the predominant sweetener.

high potency: one hundred percent or more of the Daily Value for the nutrient in a single supplement and for at least two-thirds of the nutrients in a multinutrient supplement.

high-quality proteins: dietary proteins containing all the essential amino acids in relatively the same amounts that human beings require. They may also contain nonessential amino acids.

high-risk pregnancy: a pregnancy characterized by indicators that make it likely the birth will be surrounded by problems such as premature delivery, difficult birth, retarded growth, birth defects, and early infant death.

histamine (HISS-tah-mean or HISS-tah-men): a substance produced by cells of the immune system as part of a local immune reaction to an antigen; participates in causing inflammation.

histamine-2 receptor blockers: a class of drugs that suppresses acid secretion by inhibiting receptors on acid-producing cells; commonly called *H2 blockers.* Examples include cimetidine (Tagamet), ranitidine (Zantac), and famotidine (Pepcid).

HIV-lipodystrophy (LIP-oh-DIS-tro-fee) **syndrome:** a collection of abnormalities in fat and glucose metabolism that may result from drug treatments for HIV infection; changes include body fat redistribution, abnormal blood lipid levels, and insulin resistance. The accumulation of abdominal fat is sometimes called *protease paunch.*

hives: an allergic reaction characterized by raised, swollen patches of skin or mucous membranes that are associated with intense itching; also called ***urticaria***.

homeopathic (HO-mee-oh-PATH-ic) **medicine:** a practice based on the theory that "like cures like"; that is, substances believed to cause certain symptoms are prescribed for curing the same symptoms, but are given in extremely diluted amounts.

homeostasis (HOME-ee-oh-STAY-sis): the maintenance of constant internal conditions (such as blood chemistry, temperature, and blood pressure) by the body's control systems. A homeostatic system is constantly reacting to external forces to maintain limits set by the body's needs.

honey: sugar (mostly sucrose) formed from nectar gathered by bees. An enzyme splits the sucrose into glucose and fructose. Composition and flavor vary, but honey always contains a mixture of sucrose, fructose, and glucose.

hormone-sensitive lipase: an enzyme inside adipose cells that responds to the body's need for fuel by hydrolyzing triglycerides so that their parts (glycerol and fatty acids) escape into the general circulation and thus become available to other cells for fuel. The signals to which this enzyme responds include epinephrine and glucagon, which oppose insulin.

hormones: chemical messengers. Hormones are secreted by a variety of glands in response to altered conditions in the body. Each hormone travels to one or more specific target tissues or organs, where it elicits a specific response to maintain homeostasis.

human genome (GEE-nome): the full complement of genetic material in the chromosomes of a person's cells.

human immunodeficiency virus (HIV): the virus that causes acquired immune deficiency syndrome (AIDS). HIV destroys immune cells and progressively impedes the body's ability to fight infections and certain cancers.

humoral immunity: immunity conferred by B cells, which produce and release antibodies into body fluids.

hunger: the painful sensation caused by a lack of food that initiates food-seeking behavior; consequence of food insecurity that, because of prolonged, involuntary lack of food, results in discomfort, illness, weakness, or pain that goes beyond the usual uneasy sensation.

husk: the outer inedible part of a grain; also called the *chaff.*

hydrochloric acid: an acid composed of hydrogen and chloride atoms (HCl) that is normally produced by the gastric glands.

hydrogenation (HIGH-dro-jen-AY-shun or high-DROJ-eh-NAY-shun): a chemical process by which hydrogens are added to monounsaturated or polyunsaturated fatty acids to reduce the number of double bonds, making the fats more saturated (solid) and more resistant to oxidation (protecting against rancidity). Hydrogenation produces *trans*-fatty acids.

hydrolysis (high-DROL-ih-sis): a chemical reaction in which a major reactant is split into two products, with the addition of a hydrogen atom (H) to one and a hydroxyl group (OH) to the other (from water, H_2O). (The noun is *hydrolysis;* the verb is *hydrolyze.*)

hydrophilic (high-dro-FIL-ick): a term referring to water-loving, or water-soluble, substances.

hydrophobic (high-dro-FOE-bick): a term referring to water-fearing, or non-water-soluble, substances; also known as *lipophilic* (fat loving).

hydroxyapatite (high-drox-ee-APP-ah-tite): crystals made of calcium and phosphorus.

hyperactivity: inattentive and impulsive behavior that is more frequent and severe than is typical of others a similar age; professionally called *attention-deficit/hyperactivity disorder (ADHD)*.

hypercalcemia (HIGH-per-kal-SEE-me-ah): elevated serum calcium levels.

hypercalciuria (HIGH-per-kal-see-YOO-ree-ah): elevated urinary calcium levels.

hypercapnia (high-per-CAP-nee-ah): excessive carbon dioxide in the blood.

hyperglycemia: elevated blood glucose concentrations. Normal fasting plasma glucose levels are less than 100 mg/dL. Fasting plasma glucose levels between 100 and 125 mg/dL suggest prediabetes; values of 126 mg/dL and above suggest diabetes.

hyperinsulinemia: abnormally high levels of insulin in the blood.

hyperkalemia (HIGH-per-ka-LEE-me-ah): elevated serum potassium levels.

hypermetabolism: a higher-than-normal metabolic rate.

hyperosmolar hyperglycemic syndrome: extreme hyperglycemia associated with dehydration, hyperosmolar blood, and altered mental status; sometimes called the *hyperosmolar hyperglycemic nonketotic state*.

hyperoxaluria (HIGH-per-ox-ah-LOO-ree-ah): elevated urinary oxalate levels.

hyperphosphatemia (HIGH-per-fos-fa-TEE-me-ah): elevated serum phosphate levels.

hyperplastic obesity: obesity due to an increase in the *number* of fat cells.

hypersensitivity: immune responses that are excessive or inappropriate. One type of hypersensitivity is *allergy*.

hypertension: higher-than-normal blood pressure. Hypertension that develops without an identifiable cause is known as *essential* or *primary hypertension*; hypertension that is caused by a specific disorder such as kidney disease is known as *secondary hypertension*.

hypertonic formula: a formula with an osmolality greater than that of blood serum.

hypertriglyceridemia (HYE-per-try-gliss-er-rye-DEE-me-ah): high blood triglyceride levels.

hypertrophic obesity: obesity due to an increase in the *size* of fat cells.

hypnotherapy: a technique that uses hypnosis and the power of suggestion to improve health behaviors, relieve pain, and promote healing.

hypoalbuminemia: low plasma albumin concentrations.

hypoallergenic formulas: clinically tested infant formulas that support infant growth and development but do not provoke reactions in 90 percent of infants or children with confirmed cow's milk allergy.

hypochlorhydria (HIGH-poe-clor-HIGH-dree-ah): abnormally low gastric acid secretions.

hypoglycemia (HIGH-po-gly-SEE-me-ah): an abnormally low blood glucose concentration. In diabetes, hypoglycemia is treated when plasma glucose levels fall below 70 mg/dL.

hypokalemia (HIGH-po-ka-LEE-me-ah): low serum potassium levels.

hypothalamus (high-po-THAL-ah-mus): a brain center that controls activities such as maintenance of water balance, regulation of body temperature, and control of appetite.

hypothesis (hi-POTH-eh-sis): an unproven statement that tentatively explains the relationships between two or more variables.

hypothyroidism: underactivity of the thyroid gland that may be caused by iodine deficiency or any number of other causes.

hypovolemia (HIGH-poe-voe-LEE-me-ah): low blood volume.

hypoxemia (high-pock-SEE-me-ah): a low level of oxygen in the blood.

hypoxia (high-POCK-see-ah): a low amount of oxygen in body tissues.

I

ileocecal (ill-ee-oh-SEEK-ul) **valve:** the sphincter separating the small and large intestines.

ileostomy (ill-ee-AH-stoe-me): a surgically created passage through the abdominal wall into the ileum.

ileum (ILL-ee-um): the last segment of the small intestine.

imagery: the use of mental images of things or events to aid relaxation or promote self-healing.

imitation foods: foods that substitute for and resemble another food, but are nutritionally inferior to it with respect to vitamin, mineral, or protein content. If the substitute is not inferior to the food it resembles and if its name provides an accurate description of the product, it need not be labeled "imitation."

immune system: the body's defense system against foreign substances.

immunity: the body's ability to defend itself against diseases.

immunoglobulins (IM-you-no-GLOB-you-linz): large globular proteins produced by B cells that function as antibodies.

implantation (IM-plan-TAY-shun): the embedding of the blastocyst in the inner lining of the uterus.

inborn error of metabolism: an inherited trait (one that is present at birth) that causes the absence, deficiency, or malfunction of a protein that has a critical metabolic role.

indigestion: incomplete or uncomfortable digestion, usually accompanied by pain, nausea, vomiting, heartburn, intestinal gas, or belching.

indirect calorimetry: a means of estimating energy expenditure by measuring the amount of oxygen consumed.

indispensable amino acids: essential amino acids.

indispensable nutrients: nutrients a person must obtain from food because the body cannot make them for itself in sufficient quantity to meet physiological needs; also called **essential nutrients**.

inflammation: a nonspecific, innate immune response to injury or infection; characterized by an increase in white blood cells.

inflammatory response: a group of nonspecific immune responses to infection or injury.

informed consent: a patient's or caregiver's agreement to undergo a treatment that has been adequately disclosed. Persons must be mentally competent in order to make the decision.

innate immunity: immunity that is present at birth, unchanging throughout life, and nonspecific for particular antigens; also called **natural immunity**.

inorganic: not containing carbon or pertaining to living things.

inositol (in-OSS-ih-tall): a nonessential nutrient that can be made in the body from glucose. Inositol is a part of cell membrane structures.

insoluble fibers: nonstarch polysaccharides that do not dissolve in water. Examples include the tough, fibrous structures found in the strings of celery and the skins of corn kernels.

insulin (IN-suh-lin): a hormone secreted by special cells in the pancreas in response to (among other things) increased blood glucose concentration. The primary role of insulin is to control the transport of glucose from the bloodstream into the muscle and fat cells.

insulin resistance: the condition in which a normal amount of insulin produces a subnormal effect in muscle, adipose, and liver cells, resulting in an elevated fasting glucose; a metabolic consequence of obesity that precedes type 2 diabetes.

intermittent claudication (claw-dih-KAY-shun): severe pain and weakness in the legs (especially the calves) caused by inadequate blood supply to the muscles; usually occurs with walking and subsides during rest.

intermittent feedings: delivery of about 250 to 400 milliliters of formula over 20 to 40 minutes.

Internet (the Net): a worldwide network of millions of computers linked together to share information.

interstitial (IN-ter-STISH-al) **fluid:** fluid between the cells (intercellular), usually high in sodium and chloride. Interstitial fluid is a large component of extracellular fluid.

intestinal adaptation: the process of intestinal recovery following resection that leads to improved absorptive capacity.

intestinal ischemia (is-KEY-me-ah): a diminished blood flow to the intestines that is characterized by abdominal pain, forceful bowel movements, and blood in the stool.

intra-abdominal fat: fat stored within the abdominal cavity in association with the internal abdominal organs, as opposed to the fat stored directly under the skin (subcutaneous fat).

intracellular fluid: fluid within the cells, usually high in potassium and phosphate. Intracellular fluid accounts for approximately two-thirds of the body's water.

intractable: not easily managed or controlled.

intractable vomiting: vomiting that is not easily managed or controlled.

intradialytic parenteral nutrition: the infusion of nutrients during hemodialysis, often providing amino acids, dextrose, lipids, and some trace minerals.

intrinsic factor: a glycoprotein (a protein with short polysaccharide chains attached) secreted by the stomach cells that binds with vitamin B_{12} in the small intestine to aid in the absorption of vitamin B_{12}.

invert sugar: a mixture of glucose and fructose formed by the hydrolysis of sucrose in a chemical process; sold only in liquid form and sweeter than sucrose. Invert sugar is used as a food additive to help preserve freshness and prevent shrinkage.

iodide: the ion form of iodine.

ions (EYE-uns): atoms or molecules that have gained or lost electrons and therefore have electrical charges. Examples include the positively charged sodium ion (Na^+) and the negatively charged chloride ion (Cl^-). For a closer look at ions, see Appendix B.

iron deficiency: the state of having depleted iron stores.

iron-deficiency anemia: severe depletion of iron stores that results in low hemoglobin and small, pale red blood cells. Anemias that impair hemoglobin synthesis are *microcytic* (small cell).

iron overload: toxicity from excess iron.

irritable bowel syndrome: an intestinal disorder of unknown cause that disturbs the functioning of the lower bowel; symptoms include abdominal pain, flatulence, diarrhea, and constipation.

ischemia (iss-KEE-mee-a): inadequate blood supply within tissues due to obstructed blood flow in the arteries.

ischemic strokes: strokes caused by the obstruction of blood flow to brain tissue.

isotonic formula: a formula with an osmolality similar to that of blood serum (about 300 milliosmoles per kilogram).

IU: international units; an old measure of vitamin activity determined by biological methods (as opposed to new measures that are determined by direct chemical analyses). Many fortified foods and supplements use IU on their labels.

J

jaundice (JAWN-dis): yellow discoloration of the skin and eyes due to an accumulation of bilirubin, a breakdown product of hemoglobin that normally exits the body via bile secretions.

jejunostomy (JEH-ju-NAH-stoe-mee): an opening into the jejunum through which a feeding tube can be passed.

jejunum (je-JOON-um): the first two-fifths of the small intestine beyond the duodenum.

joule: a measure of *work* energy; the international unit for measuring food energy.

K

Kaposi's (kah-POH-seez) **sarcoma:** a common cancer in HIV-infected persons that is characterized by lesions in the skin, lungs, and GI tract.

kcal: abbreviation of **kcalories**; a unit by which energy is measured.

kcalorie: a unit by which energy is measured. One kcalorie is the amount of heat necessary to raise the temperature of 1 kilogram (kg) of water 1°C. The scientific use of the term *kcalorie* is the same as the popular use of the term *calorie*.

kcalorie counts: estimates of food energy (and often, protein) consumed by patients for one or more days.

kcalorie (energy) control: management of food energy intake.

kcalorie-free: fewer than 5 kcalories per serving.

kefir (keh-FUR): a fermented milk created by adding *Lactobacillus acidophilus* and other bacteria that break down lactose to glucose and galactose, producing a sweet, lactose-free product.

keratin (KARE-uh-tin): a water-insoluble protein; the normal protein of hair and nails.

keratinization: accumulation of keratin in a tissue; a sign of vitamin A deficiency.

keratomalacia (KARE-ah-toe-ma-LAY-shuh): softening of the cornea that leads to irreversible blindness; seen in severe vitamin A deficiency.

Keshan (KESH-an or ka-SHAWN) **disease:** the heart disease associated with selenium deficiency named for one of the provinces of China where it was first studied. Keshan disease is characterized by heart enlargement and insufficiency; fibrous tissue replaces the muscle tissue that normally composes the middle layer of the walls of the heart.

keto (KEY-toe) **acid:** an organic acid that contains a carbonyl group (C=O).

ketoacidosis (KEY-toe-ass-ih-DOE-sis): an acidosis (lowering of blood pH) that results from the excessive production of ketone bodies.

ketogenic amino acids: amino acids that are degraded to acetyl CoA.

ketone (KEE-tone) **bodies:** the metabolic products of the incomplete breakdown of fat when glucose is not available in the cells.

ketonuria (KEY-toe-NOOR-ee-ah): the presence of ketone bodies in the urine.

ketosis (kee-TOE-sis): an undesirably high concentration of ketone bodies in the blood and urine.

kidney stones: crystalline masses that form in the urinary tract; also called *renal calculi* and *nephrolithiasis*.

Kreb's cycle: named after the scientist who elucidated this biochemistry, a series of metabolic reactions that break down molecules of acetyl CoA to carbon dioxide and hydrogen atoms; also called the *citric acid cycle* or the *TCA cycle*.

kwashiorkor (kwash-ee-OR-core or kwash-eeor-CORE): a form of PEM that results either from inadequate protein intake or infections.

L

lactadherin (lack-tad-HAIR-in): a protein in breast milk that attacks diarrhea-causing viruses.

lactase: an enzyme that hydrolyzes lactose.

lactase deficiency: a lack of the enzyme required to digest the disaccharide lactose into its component monosaccharides (glucose and galactose).

lactate: a 3-carbon compound produced from pyruvate during anaerobic metabolism.

lactation: production and secretion of breast milk for the purpose of nourishing an infant.

lacteals (LACK-tee-als): the lymphatic vessels of the intestine that take up nutrients and pass them to the lymph circulation.

lacto-ovo-vegetarians: people who include milk, milk products, and eggs, but exclude meat, poultry, fish, and seafood from their diets.

lactoferrin (lack-toh-FERR-in): a protein in breast milk that binds iron and keeps it from supporting the growth of the infant's intestinal bacteria.

lactose (LAK-tose): a disaccharide composed of glucose and galactose; commonly known as *milk sugar*.

lactose intolerance: a condition that results from inability to digest the milk sugar lactose; characterized by bloating, gas, abdominal discomfort, and diarrhea. Lactose intolerance differs from milk allergy, which is caused by an immune reaction to the protein in milk.

lactovegetarians: people who include milk and milk products, but exclude meat, poultry, fish, seafood, and eggs from their diets.

laparoscopic: pertaining to procedures that use a laparoscope for internal examination or surgery. A laparoscope is a narrow surgical telescope that is inserted into the abdominal cavity through a small incision. A video camera is usually attached so that the procedure can be viewed on a television monitor.

large intestine or **colon (COAL-un):** the lower portion of intestine that completes the digestive process. Its segments are the ascending colon, the transverse colon, the descending colon, and the sigmoid colon.

larynx: the upper part of the air passageway that contains the vocal cords; also called the *voice box*.

laxatives: substances that loosen the stools and thereby prevent or treat constipation.

LDL (low-density lipoprotein): the type of lipoprotein derived from very-low-density lipoproteins (VLDL) as VLDL triglycerides are removed and broken down; composed primarily of cholesterol.

lean: less than 10 grams of fat, 4.5 grams of saturated fat and *trans* fat combined, and 95 milligrams of cholesterol per serving and per 100 grams of meat, poultry, and seafood.

lean body mass: the body minus its fat.

lecithin (LESS-uh-thin): one of the phospholipids. Both nature and the food industry use lecithin as an emulsifier to combine water-soluble and fat-soluble ingredients that do not ordinarily mix, such as water and oil.

legumes (lay-GYOOMS or LEG-yooms): plants of the bean and pea family, with seeds that are rich in protein compared with other plant-derived foods.

leptin: a protein produced by fat cells under direction of the *ob* gene that decreases appetite and increases energy expenditure; sometimes called the *ob protein*.

less: at least 25 percent less of a given nutrient or kcalories than the comparison food (see individual nutrients); synonyms include *fewer* and *reduced*.

less cholesterol: 25 percent or less cholesterol than the comparison food (reflecting a reduction of at least 20 milligrams per serving), and 2 grams or less saturated fat and *trans* fat combined per serving.

less fat: 25 percent or less fat than the comparison food.

less saturated fat: 25 percent or less saturated fat and *trans* fat combined than the comparison food.

let-down reflex: the reflex that forces milk to the front of the breast when the infant begins to nurse.

leukocytes: blood cells that function in immunity; also called **white blood cells**.

levulose: an older name for fructose.

license to practice: permission under state or federal law, granted on meeting specified criteria, to use a certain title (such as dietitian) and offer certain services. Licensed dietitians may use the initials LD after their names.

life expectancy: the average number of years lived by people in a given society.

life span: the maximum number of years of life attainable by a member of a species.

light or **lite:** one-third fewer kcalories than the comparison food; 50 percent or less of the fat or sodium than the comparison food; any use of the term other than as defined must specify what it is referring to (for example, "light in color" or "light in texture").

lignans: phytochemicals present in flaxseed, but not in flax oil, that are converted to phytosterols by intestinal bacteria and are under study as possible anticancer agents.

limiting amino acid: the essential amino acid found in the shortest supply relative to the amounts needed for protein synthesis in the body. Four amino acids are most likely to be limiting: lysine, methionine, threonine, tryptophan.

lingual: pertaining to the tongue.

linoleic (lin-oh-LAY-ick) **acid:** an essential fatty acid with 18 carbons and two double bonds.

linolenic (lin-oh-LEN-ick) **acid:** an essential fatty acid with 18 carbons and three double bonds.

lipase (LYE-pase): an enzyme that hydrolyzes lipids.

lipids: a family of compounds that includes triglycerides, phospholipids, and sterols. Lipids are characterized by their insolubility in water. (Lipids also include the fat-soluble vitamins, described in Chapter 11.)

lipomas (lih-POE-muz): benign tumors composed of fatty tissue.

lipoprotein lipase (LPL): an enzyme that hydrolyzes triglycerides passing by in the bloodstream and directs their parts into the cells, where they can be metabolized for energy or reassembled for storage.

lipoproteins (LIP-oh-PRO-teenz): clusters of lipids associated with proteins that serve as transport vehicles for lipids in the lymph and blood.

lipotoxicity: the adverse effects of fat in nonadipose tissues.

listeriosis: an infection caused by eating food contaminated with the bacterium *Listeria monocytogenes*, which can be killed by pasteurization and cooking but can survive at refrigerated temperatures; certain ready-to-eat foods, such as hot dogs and deli meats, may become contaminated after cooking or processing, but before packaging.

liver: the organ that manufactures bile. (The liver's many other functions are described in Chapter 7.)

living will: a written statement that specifies the medical procedures desired or not desired in the event that a person is unable to communicate or is incapacitated; also called a *medical directive*.

longevity: long duration of life.

low: an amount that would allow frequent consumption of a food without exceeding the Daily Value for the nutrient. A food that is naturally low in a nutrient may make such a claim, but only as it applies to all similar foods (for example, "fresh cauliflower, a low-sodium food"); synonyms include "little," "few," and "low source of."

low birthweight (LBW): a birthweight of $5^1/_2$ pounds (2500 grams) or less; indicates probable poor health in the newborn and poor nutrition status in the mother during pregnancy, before pregnancy, or both. Optimal birthweight for a full-term baby is 6.8 to 7.9 pounds (about 3100 to 3600 grams).

low cholesterol: 20 milligrams or less cholesterol per serving and 2 grams or less saturated fat and *trans* fat combined per serving.

low fat: 3 grams or less fat per serving.

low food security: reduced quality of life with little or no indication of reduced food intake; formerly known as *food insecurity without hunger*.

low kcalorie: 40 kcalories or less per serving.

low-microbial diet: a diet that contains foods that are unlikely to be contaminated with bacteria and other microbes.

low-residue diet: a diet low in fiber and other food constituents that contribute to colonic residue.

low-risk pregnancy: a pregnancy characterized by factors that make it likely the birth will be normal and the infant healthy.

low saturated fat: 1 gram or less saturated fat and less than 0.5 gram of *trans* fat per serving.

low sodium: 140 milligrams or less per serving.

lumen (LOO-men): the space within a vessel, such as the intestine.

lutein (LOO-teen): a plant pigment of yellow hue; a phytochemical believed to play roles in eye functioning and health.

luteinizing (LOO-tee-in-EYE-zing) **hormone (LH):** a hormone that stimulates ovulation and the development of the corpus luteum (the small tissue that develops from a ruptured ovarian follicle and secretes hormones); so called because the follicle turns yellow as it matures. In men, LH stimulates testosterone secretion. The release of LH is mediated by luteinizing hormone–releasing hormone (LH–RH).

lycopene (LYE-koh-peen): a pigment responsible for the red color of tomatoes and other red-hued vegetables; a phytochemical that may act as an antioxidant in the body.

lymph (LIMF): a clear yellowish fluid that is similar to blood except that it contains no red blood cells or platelets and that is carried in lymphatic vessels. Lymph is collected from the extracellular fluids of body tissues and ultimately transported to the bloodstream.

lymphatic (lim-FAT-ic) **system:** a loosely organized system of vessels and ducts that convey fluids toward the heart. The GI part of the lymphatic system carries the products of fat digestion into the bloodstream.

lymphatic vessels: vessels through which lymph travels.

lymphocytes (LIM-foe-sites): white blood cells that recognize specific antigens and therefore function in adaptive immunity; include *T cells* and *B cells*.

lymphoid tissues: tissues that contain lymphocytes.

lysosomes (LYE-so-zomes): cellular organelles; membrane-enclosed sacs of degradative enzymes.

lysozyme (LYE-so-zyme): an enzyme with antibacterial properties; found in immune cells and body secretions such as tears, saliva, and sweat.

M

macrocytic: abnormally large blood cells.

macrocytic anemia: anemia characterized by large, immature red blood cells, as occurs in folate and vitamin B_{12} deficiency; also called *megaloblastic anemia*.

macronutrients: carbohydrate, fat, and protein; the nutrients the body requires in relatively large amounts (many grams daily).

macrophages (MAK-roe-fay-jez): immune cells in tissues (derived from blood monocytes) that ingest and destroy pathogens and cellular debris by phagocytosis.

macrosomia (MAK-roh-SOH-mee-ah): the condition of having an abnormally large body; in infants, refers to birth weights of 4000 grams (8 pounds 13 ounces) and above. Macrosomia results from prepregnancy obesity, excessive weight gain during pregnancy, or uncontrolled diabetes.

macrovascular complications: disorders that affect the large blood vessels, including the coronary arteries and arteries of the limbs.

macular (MACK-you-lar) **degeneration:** deterioration of the macular area of the eye that can lead to loss of central vision and eventual blindness. The *macula* is a small, oval, yellowish region in the center of the retina that provides the sharp, straight-ahead vision so critical to reading and driving.

magnesium: a cation within the body's cells, active in many enzyme systems.

major minerals: essential mineral nutrients the human body requires in relatively large amounts (greater than 100 milligrams per day); sometimes called *macrominerals*.

maleficence (mah-LEF-eh-sense): the act of doing evil or harm.

malignant (ma-LIG-nent): describes a cancerous cell or tumor, which can injure healthy tissue and spread cancer to other regions of the body.

malnutrition: any condition caused by excess or deficient food energy or nutrient intake or by an imbalance of nutrients.

maltase: an enzyme that hydrolyzes maltose.

maltose (MAWL-tose): a disaccharide composed of two glucose units; sometimes known as *malt sugar*.

mammary glands: glands of the female breast that secrete milk.

maple sugar: a sugar (mostly sucrose) purified from the concentrated sap of the sugar maple tree.

marasmus (ma-RAZ-mus): a form of PEM that results from a severe deprivation, or impaired absorption, of energy, protein, vitamins, and minerals.

marginal food security: one or two indications of food-access problems but with little or no change in food intake.

massage therapy: manual manipulation of muscles to reduce tension, increase blood circulation, improve joint mobility, and promote healing of injuries.

mast cells: cells within connective tissue that produce and release histamine.

mastication: the process of chewing.

matrix (MAY-tricks): the basic substance that gives form to a developing structure; in the body, the formative cells from which teeth and bones grow.

matter: anything that takes up space and has mass.

Meals on Wheels: a nutrition program that delivers food for the elderly to their homes.

meat replacements: products formulated to look and taste like meat, fish, or poultry; usually made of textured vegetable protein.

mechanical ventilation: life-sustaining treatment in which a mechanical ventilator is used to substitute for a patient's failing lungs.

medical directive: a written statement that specifies the medical procedures desired or not desired in the event that a person is unable to communicate or is incapacitated; also called a *living will*.

medical nutrition therapy: nutrition care provided by a registered dietitian; includes assessing nutrition status, diagnosing nutrition problems, and providing nutrition care.

meditation: a self-directed technique of calming the mind and relaxing the body.

medium-chain triglycerides (MCT): triglycerides that contain fatty acids that are 8 to 10 carbons in length. MCT do not require digestion and can be absorbed in the absence of lipase or bile.

megaloblastic anemia: anemia characterized by large, immature red blood cells, as occurs in folate and vitamin B_{12} deficiency; also called *macrocytic anemia*.

menadione (men-uh-DYE-own): the synthetic form of vitamin K.

MEOS or microsomal (my-krow-SO-mal) **ethanol oxidizing system:** a system of enzymes in the liver that oxidize not only alcohol but also several classes of drugs.

metabolic stress: a disruption in the body's chemical environment due to the effects of disease or injury. Metabolic stress is characterized by changes in metabolic rate, heart rate, blood pressure, hormonal status, and nutrient metabolism.

metabolic syndrome: a cluster of interrelated clinical disorders, including obesity, insulin resistance, high blood pressure, and abnormal blood lipids, which together increase risk of diabetes and cardiovascular disease; also known as *insulin resistance syndrome* or *syndrome X*.

metabolic water: water generated during metabolism.

metabolism: the sum total of all the chemical reactions that go on in living cells. Energy metabolism includes all the reactions by which the body obtains and expends the energy from food.

metabolites: products of metabolism; compounds produced by a biochemical pathway.

metalloenzymes (meh-TAL-oh-EN-zimes): enzymes that contain one or more minerals as part of their structures.

metallothionein (meh-TAL-oh-THIGH-oh-neen): a sulfur-rich protein that avidly binds with and transports metals such as zinc.

metastasize (meh-TAS-tah-size): to spread from one part of the body to another; refers to cancer cells.

methylation: the addition of a methyl group (CH_3).

MFP factor: a peptide released during the digestion of meat, fish, and poultry that enhances nonheme iron absorption.

mg: milligrams; one-thousandth of a gram.

mg NE: milligrams niacin equivalents; a measure of niacin activity.

micelles (MY-cells): tiny spherical complexes of emulsified fat that arise during digestion; most contain bile salts and the products of lipid digestion, including fatty acids, monoglycerides, and cholesterol.

microalbuminuria: the presence of albumin (a blood protein) in the urine, a sign of diabetic nephropathy.

microangiopathies: disorders of the small blood vessels.

microarray technology: research tools that analyze the expression of thousands of genes simultaneously and search for particular gene changes associated with a disease. DNA microarrays are also called DNA chips.

microcytic (my-cro-SIT-ic) **hypochromic** (high-po-KROME-ic) **anemia:** small, pale red blood cells that develop in iron-deficiency anemia.

microgram (μg): one-millionth of a gram.

microgram DFE (μg DFE): micrograms dietary folate equivalents; a measure of folate activity.

microgram RAE (μg RAE): micrograms retinol activity equivalents; a measure of vitamin A activity.

micronutrients: vitamins and minerals; the nutrients the body requires in relatively small amounts (milligrams or micrograms daily).

microvascular complications: disorders that affect the small blood vessels and capillaries, including those in the retina and kidneys.

microvilli (MY-cro-VILL-ee, MY-cro-VILL-eye): tiny, hairlike projections on each cell of every villus that can trap nutrient particles and transport them into the cells; singular *microvillus*.

milk anemia: iron-deficiency anemia that develops when an excessive milk intake displaces iron-rich foods from the diet.

milliequivalents (mEq): the concentration of electrolytes in a volume of solution; determined by dividing an ion's molecular weight (MW) by its number of charges. Milliequivalents are a useful measure when considering ions because the number of charges reveals characteristics about the solution that are not evident when the concentration is expressed in terms of weight.

milliequivalents per liter (mEq/L): a measure of the concentration of electrolytes in a volume of solution.

mineral oil: a purified liquid derived from petroleum and used to treat constipation.

mineral water: water from a spring or well that naturally contains at least 250 to 500 parts per million (ppm) of minerals. Minerals give water a distinctive flavor. Many mineral waters are high in sodium.

mineralization: the process in which calcium, phosphorus, and other minerals crystallize on the collagen matrix of a growing bone, hardening the bone.

minerals: inorganic elements. Some minerals are essential nutrients required in small amounts by the body for health.

misinformation: false or misleading information.

mitochondria (my-toh-KON-dree-uh): the cellular organelles responsible for producing ATP aerobically; made of membranes (lipid and protein) with enzymes mounted on them. (The singular is *mitochondrion*.)

mmol: millimoles; one-thousandth of a mole, the molecular weight of a substance. To convert mmol to mg, multiply by the atomic weight of the substance.

moderation: in relation to alcohol consumption, not more than two drinks a day for the average-size man and not more than one drink a day for the average-size woman.

moderation (dietary): providing enough but not too much of a substance.

modified diet: a diet that contains foods altered in texture, consistency, or nutrient content or that includes or omits specific foods; sometimes called a *therapeutic diet*.

modular formulas: enteral formulas prepared in the hospital from *modules* that contain single macronutrients; used for people with unique nutrient needs.

molasses: the thick brown syrup produced during sugar refining. Molasses retains residual sugar and other by-products and a few minerals; blackstrap molasses contains significant amounts of calcium and iron.

molecule: two or more atoms of the same or different elements joined by chemical bonds. Examples are molecules of the element oxygen, composed of two oxygen atoms (O_2), and molecules of the compound water, composed of two hydrogen atoms and one oxygen atom (H_2O).

molybdenum (mo-LIB-duh-num): a trace element.

monocytes (MON-oh-sites): cells released from the bone marrow that move into tissues and mature into macrophages.

monoglycerides: molecules of glycerol with one fatty acid attached. A molecule of glycerol with two fatty acids attached is a **diglyceride**.

monosaccharides (mon-oh-SACK-uh-rides): carbohydrates of the general formula $C_nH_{2n}O_n$ that typically form a single ring. See Appendix C for the chemical structures of the monosaccharides.

monounsaturated fat: triglycerides in which most of the fatty acids are monounsaturated.

monounsaturated fatty acid (MUFA): a fatty acid that lacks two hydrogen atoms and has one double bond between carbons—for example, oleic acid.

more: at least 10 percent more of the Daily Value for a given nutrient than the comparison food; synonyms include *added* and *extra*.

motility: the ability of the GI tract muscles to move.

mouth: the oral cavity containing the tongue and teeth.

mucous (MYOO-kus) **membranes:** the membranes, composed of mucus-secreting cells, that line the surfaces of body tissues.

mucus (MYOO-kus): a slippery substance secreted by cells of the GI lining (and other body linings) that protects the cells from exposure to digestive juices (and other destructive agents).

multiple organ dysfunction syndrome: the progressive dysfunction of two or more organ systems that develops during intensive care; often results in death.

muscle dysmorphia (dis-MORE-fee-ah): a psychiatric disorder characterized by a preoccupation with building body mass.

muscular dystrophy (DIS-tro-fee): a hereditary disease in which the muscles gradually weaken. Its most debilitating effects arise in the lungs.

mutation: an inheritable alteration in the DNA sequence of a gene.

myocardial (MY-oh-CAR-dee-al) **infarction** (in-FARK-shun), or **MI:** death of heart muscle caused by a sudden reduction in coronary blood flow; also called a **heart attack** or *cardiac arrest*.

myoglobin: the oxygen-holding protein of the muscle cells.

N

NAD (nicotinamide adenine dinucleotide): the main coenzyme form of the vitamin niacin. Its reduced form is NADH.

NADP (the phosphate form of NAD): a coenzyme form of niacin.

narcotic (nar-KOT-ic): a drug that dulls the senses, induces sleep, and becomes addictive with prolonged use.

nasoduodenal (ND) feeding tube: a feeding tube that is placed into the duodenum via the nose.

nasoenteric feeding tube: a feeding tube that is placed into the GI tract via the nose. (*Nasoenteric feedings* usually refer to *nasoduodenal* and *nasojejunal* feedings.)

nasogastric (NG) feeding tube: a feeding tube that is placed into the stomach via the nose.

nasojejunal (NJ) feeding tube: a feeding tube that is placed into the jejunum via the nose.

natural killer cells: lymphocytes that confer nonspecific immunity by destroying a wide array of viruses and tumor cells.

natural water: water obtained from a spring or well that is certified to be safe and sanitary. The mineral content may not be changed, but the water may be treated in other ways such as with ozone or by filtration.

naturopathic (NAY-chur-oh-PATH-ic) **medicine:** an approach to health care using practices alleged to enhance the body's natural healing abilities. Treatments may include a variety of alternative therapies including dietary supplements, herbal remedies, exercise, and homeopathy.

neotame (NEE-oh-tame): an artificial sweetener composed of two amino acids (phenylalanine and aspartic acid); approved for use in the United States.

nephron (NEF-ron): the functional unit of the kidneys, consisting of a glomerulus and tubules.

nephrotic (neh-FROT-ik) **syndrome:** a syndrome associated with kidney disorders that damage the glomerulus and cause urinary protein losses exceeding 3½ g/day.

nephrotoxic: toxic to the kidneys.

net protein utilization (NPU): a measure of protein quality assessed by measuring the amount of protein nitrogen that is retained from a given amount of protein nitrogen eaten.

neural tube: the embryonic tissue that forms the brain and spinal cord.

neural tube defects: malformations of the brain, spinal cord, or both during embryonic development that often result in lifelong disability or death.

neurofibrillary tangles: snarls of the thread-like strands that extend from the nerve cells, commonly found in the brains of people with Alzheimer's dementia.

neurons: nerve cells; the structural and functional units of the nervous system. Neurons initiate and conduct nerve impulse transmissions.

neuropeptide Y: a chemical produced in the brain that stimulates appetite, diminishes energy expenditure, and increases fat storage.

neurotransmitters: chemicals that are released at the end of a nerve cell when a nerve impulse arrives there. They diffuse across the gap to the next cell and alter the membrane of that second cell to either inhibit or excite it.

neutropenia: a low white blood cell (neutrophil) count, which increases susceptibility to infection.

neutrophils (NEW-tro-fills): the most common type of white blood cell. Neutrophils destroy antigens by phagocytosis.

niacin (NIGH-a-sin): a B vitamin. The coenzyme forms are **NAD (nicotinamide adenine dinucleotide)** and **NADP (the phosphate form of NAD)**. Niacin can be eaten preformed or made in the body from its precursor, tryptophan, an essential amino acid.

niacin equivalents (NE): the amount of niacin present in food, including the niacin that can theoretically be made from its precursor, tryptophan, present in the food.

niacin flush: a temporary burning, tingling, and itching sensation that occurs when a person takes a large dose of nicotinic acid; often accompanied by a headache and reddened face, arms, and chest.

night blindness: slow recovery of vision after flashes of bright light at night or an inability to see in dim light; an early symptom of vitamin A deficiency.

nitric oxide: a compound produced by blood vessel cells that helps to regulate blood vessel function, including dilation and constriction.

nitrogen balance: the amount of nitrogen consumed (N in) as compared with the amount of nitrogen excreted (N out) in a given period of time.

nonessential amino acids: amino acids that the body can synthesize.

nonexercise activity thermogenesis (NEAT): energy expenditure associated with everyday spontaneous activities.

nonheme iron: the iron in foods that is not bound to proteins; found in both plant-derived and animal-derived foods.

nonnutritive sweeteners: sweeteners that yield no energy (or insignificant energy in the case of aspartame).

nonpathogenic: not capable of causing disease.

nonperishable food collection: collecting processed foods from wholesalers and markets.

nonprotein kcalorie-to-nitrogen ratio: a ratio between the nonprotein kcalories and nitrogen content of the diet; used to assess whether the nitrogen intake is sufficient for maintaining muscle tissue.

nonselective menus: menus that do not allow choices and list only preselected food items.

nucleotide bases: the nitrogen-containing building blocks of DNA and RNA—cytosine (C), thymine (T), uracil (U), guanine (G), and adenine (A). In DNA, the base pairs are A–T and C–G and in RNA, the base pairs are A–U and C–G.

nucleotides: the subunits of DNA and RNA molecules, composed of a phosphate group, a 5-carbon sugar (deoxyribose for DNA and ribose for RNA), and a nitrogen-containing base.

nucleus: a major membrane-enclosed body within every cell, which contains the cell's genetic material, DNA, embedded in chromosomes.

nursing bottle tooth decay: extensive tooth decay due to prolonged tooth contact with formula, milk, fruit juice, or other carbohydrate-rich liquid offered to an infant in a bottle.

nursing diagnoses: clinical judgments about actual or potential health problems that provide the basis for selecting appropriate nursing interventions.

nutrient claims: statements that characterize the quantity of a nutrient in a food.

nutrient density: a measure of the nutrients a food provides relative to the energy it provides. The more nutrients and the fewer kcalories, the higher the nutrient density.

nutrient profiling: ranking foods based on their nutrient composition.

nutrients: chemical substances obtained from food and used in the body to provide energy, structural materials, and regulating agents to support growth, maintenance, and repair of the body's tissues. Nutrients may also reduce the risks of some diseases.

nutrigenetics: the science of how genes affect the interactions between diet and disease.

nutrigenomics: the science of how nutrients affect the activities of genes.

nutrition: the science of foods and the nutrients and other substances they contain, and of their actions within the body (including ingestion, digestion, absorption, transport, metabolism, and excretion). A broader definition includes the social, economic, cultural, and psychological implications of food and eating.

nutrition assessment: a comprehensive analysis of a person's nutrition status that uses health, socioeconomic, drug, and diet histories; anthropometric measurements; physical examinations; and laboratory tests.

nutrition care plans: strategies for meeting an individual's nutritional needs.

nutrition care process: a problem-solving method that dietetics professionals use to evaluate and treat nutrition-related problems.

nutrition prescription: specific dietary recommendations related to food, nutrient, or energy intake or feeding method; also called the *diet order*.

nutrition screening: a brief assessment of health-related variables to identify patients who are malnourished or at risk for malnutrition.

nutrition support: the delivery of nutrients using a feeding tube or intravenous infusions.

nutrition support teams: health care professionals responsible for the provision of nutrients by tube feeding or intravenous infusion.

nutritional genomics: the science of how nutrients affect the activities of genes **(nutrigenomics)** and how genes affect the interactions between diet and disease **(nutrigenetics)**.

nutritionist: a person who specializes in the study of nutrition. Note that this definition does not specify qualifications and may apply not only to registered dietitians but also to self-described experts whose training is questionable. Most states have licensing laws that define the scope of practice for those calling themselves nutritionists.

nutritive sweeteners: sweeteners that yield energy, including both sugars and sugar replacers.

O

obese: overweight with adverse health effects; BMI 30 or higher.

obligatory (ah-BLIG-ah-TORE-ee) **water excretion:** the amount of water the body has to excrete each day to dispose of its wastes—about 500 milliliters (about 2 cups, or 1 pint).

octreotide: a medication that inhibits gastrointestinal motility, thereby slowing both gastric emptying and transit time in the small intestine.

oils: lipids that are liquid at room temperature (77°F or 25°C).

olestra: a synthetic fat made from sucrose and fatty acids that provides 0 kcalories per gram; also known as *sucrose polyester*.

oligopeptide (OL-ee-go-PEP-tide): string of four to nine amino acids.

oliguria (OL-lih-GOO-ree-ah): an abnormally low amount of urine, often less than 400 mL/day.

omega: the last letter of the Greek alphabet (ω), used by chemists to refer to the position of the first double bond from the methyl (CH_3) end of a fatty acid.

omega-3 fatty acid: a polyunsaturated fatty acid in which the first double bond is three carbons away from the methyl (CH_3) end of the carbon chain.

omega-6 fatty acid: a polyunsaturated fatty acid in which the first double bond is six carbons from the methyl (CH_3) end of the carbon chain.

omnivores: people who have no formal restriction on the eating of any foods.

oncotic pressure: the pressure exerted by fluid on one side of a membrane as a result of osmosis.

open feeding system: a delivery system that requires the formula to be transferred from its original packaging to a feeding container before being administered through the feeding tube.

opportunistic infections: infections from microorganisms that normally do not cause disease in healthy people but are damaging to persons with compromised immune function.

opsin (OP-sin): the protein portion of the visual pigment molecule.

oral allergy syndrome: an allergic response in which symptoms of hives, swelling, or itching occur only in the mouth and throat; usually a short-lived response that resolves quickly.

oral glucose tolerance test: a test that evaluates a person's ability to tolerate an oral glucose load.

organelles: subcellular structures such as ribosomes, mitochondria, and lysosomes.

organic: in chemistry, a substance or molecule containing carbon-carbon bonds or carbon-hydrogen bonds. This definition excludes coal, diamonds, and a few carbon-containing compounds that contain only a single carbon and no hydrogen, such as carbon dioxide (CO_2), calcium carbonate ($CaCO_3$), magnesium carbonate ($MgCO_3$), and sodium cyanide (NaCN).

orlistat (OR-leh-stat): a drug used in the treatment of obesity that inhibits the absorption of fat in the GI tract, thus limiting kcaloric intake.

orogastric feeding tube: a feeding tube that is inserted into the stomach through the mouth. This method is often used to feed infants because a nasogastric tube may hinder the infant's breathing.

oropharyngeal (OR-oh-fah-ren-JEE-al): involving the mouth and pharynx.

oropharyngeal dysphagia: an inability to transfer food from the mouth and pharynx to the esophagus; usually caused by a neurological or muscular disorder.

osmolality (OZ-moe-LAL-ih-tee): the concentration of osmotically active solutes in a solution, expressed as milliosmoles (mOsm) per kilogram of solvent.

osmolarity: the concentration of osmotically active solutes in a solution, expressed as milliosmoles per liter of solution (mOsm/L). *Osmolality* (mOsm/kg) is an alternative term used to describe a solution's osmotic properties.

osmosis: movement of water across a membrane toward the side where solutes are more concentrated.

osmotic pressure: the amount of pressure needed to prevent the movement of water across a membrane.

osteoarthritis: a painful, degenerative disease of the joints that occurs when the cartilage in a joint deteriorates; joint structure is damaged, with loss of function; also called *degenerative arthritis*.

osteoblasts: cells that build bone during growth.

osteocalcin (os-teo-KAL-sen): a calcium-binding protein in bones, essential for normal mineralization.

osteoclasts: cells that destroy bone during growth.

osteomalacia (OS-tee-oh-ma-LAY-shuh): a bone disease characterized by softening of the bones. Symptoms include bending of the spine and bowing of the legs. The disease occurs most often in adult women.

osteopathic (OS-tee-oh-PATH-ic) **manipulation:** a CAM technique performed by a doctor of osteopathy (D.O., or osteopath) that includes deep tissue massage and manipulation of the joints, spine, and soft tissues. A D.O. is a fully trained and licensed medical physician, although osteopathic manipulation has not been proved to be an effective treatment.

osteoporosis (OS-tee-oh-pore-OH-sis): a disease in which the bones become porous and fragile due to a loss of minerals; also called *adult bone loss*.

outbreaks: two or more cases of a similar illness resulting from the ingestion of a common food.

overnutrition: excess energy or nutrients.

overt (oh-VERT): out in the open and easy to observe.

overweight: body weight above some standard of acceptable weight that is usually defined in relation to height (such as BMI); BMI 25 to 29.9.

ovum (OH-vum): the female reproductive cell, capable of developing into a new organism upon fertilization; commonly referred to as an egg.

oxalates: plant compounds that bind with some minerals to form complexes that the body cannot absorb. They are present in green leafy vegetables such as beet greens and spinach.

oxaloacetate (OKS-ah-low-AS-eh-tate): a carbohydrate intermediate of the TCA cycle.

oxidants (OKS-ih-dants): compounds (such as oxygen itself) that oxidize other compounds. Compounds that prevent oxidation are called *antioxidants*, whereas those that promote it are called *prooxidants*.

oxidation (OKS-ee-day-shun): the process of a substance combining with oxygen; oxidation reactions involve the loss of electrons.

oxidative stress: a condition in which the production of oxidants and free radicals exceeds the body's ability to handle them and prevent damage.

oxytocin (OCK-see-TOH-sin): a hormone that stimulates the mammary glands to eject milk during lactation and the uterus to contract during childbirth.

oyster shell: a product made from the powdered shells of oysters that is sold as a calcium supplement, but it is not well absorbed by the digestive system.

P

pancreas: a gland that secretes digestive enzymes and juices into the duodenum. (The pancreas also secretes hormones into the blood that help to maintain glucose homeostasis.)

pancreatic (pank-ree-AT-ic) **juice:** the exocrine secretion of the pancreas, containing enzymes for the digestion of carbohydrate, fat, and protein as well as bicarbonate, a neutralizing agent. The juice flows from the pancreas into the small intestine through the pancreatic duct. (The pancreas also has an endocrine function, the secretion of insulin and other hormones.)

pantothenic (PAN-toe-THEN-ick) **acid:** a B vitamin. The principal active form is part of coenzyme A, called "CoA" throughout Chapter 7.

paracentesis (pah-rah-sen-TEE-sis): a surgical puncture of a body cavity with an aspirator to draw out excess fluid.

parathyroid hormone: a hormone from the parathyroid glands that regulates blood calcium by raising it when levels fall too low; also known as *parathormone* (PAIR-ah-THOR-moan).

parenteral (par-EN-ter-al) **nutrition:** the intravenous provision of nutrients that bypasses the GI tract.

pasteurization: heat processing of food that inactivates some, but not all, microorganisms in the food; not a sterilization process. Bacteria that cause spoilage are still present.

pathogenic: capable of causing disease.

pathogens (PATH-oh-jenz): microorganisms capable of producing disease.

patient autonomy: a principle of self-determination, such that patients (or surrogate decision makers) are free to choose the medical interventions that are acceptable to them, even if they choose to refuse interventions that may extend their lives.

PDCAAS (protein digestibility–corrected amino acid score): a measure of protein quality assessed by comparing the amino acid score of a food protein with the amino acid requirements of preschool-age children and then correcting for the true digestibility of the protein; recommended by the FAO/WHO and used to establish protein quality of foods for Daily Value percentages on food labels.

peak bone mass: the highest attainable bone density for an individual, developed during the first three decades of life.

peer review: a process in which a panel of scientists rigorously evaluates a research study to assure that the scientific method was followed.

pellagra (pell-AY-gra): the niacin-deficiency disease.

pepsin: a gastric enzyme that hydrolyzes protein. Pepsin is secreted in an inactive form, pepsinogen, which is activated by hydrochloric acid in the stomach.

pepsinogen: an inactive compound that is activated by hydrochloric acid in the stomach to form pepsin.

peptic ulcer: a lesion in the mucous membrane of either the stomach (a gastric ulcer) or the duodenum (a duodenal ulcer).

peptidase: a digestive enzyme that hydrolyzes peptide bonds. *Tripeptidases* cleave tripeptides; *dipeptidases* cleave dipeptides. *Endopeptidases* cleave peptide bonds within the chain to create smaller fragments, whereas *exopeptidases* cleave bonds at the ends to release free amino acids.

peptide bond: a bond that connects the acid end of one amino acid with the amino end of another, forming a link in a protein chain.

percent Daily Value (%DV): the percentage of a Daily Value recommendation found in a specified serving of food for key nutrients based on a 2000-kcalorie diet.

percent fat-free: may be used only if the product meets the definition of low fat or fat-free and must reflect the amount of fat in 100 grams (for example, a food that contains 2.5 grams of fat per 50 grams can claim to be "95 percent fat-free").

percutaneous endoscopic gastrostomy (PEG): a nonsurgical technique for creating a gastrostomy under local anesthesia.

percutaneous endoscopic jejunostomy (PEJ): a nonsurgical technique for creating a jejunostomy. The tube can either be guided into the jejunum via a gastrostomy or passed directly into the jejunum (*direct PEJ*).

perinatal: term that refers to the time between the twenty-eighth week of gestation and one month after birth.

periodontal disease: disease that affects the connective tissue that supports the teeth.

periodontitis: inflammation or degeneration of the tissues that support the teeth.

periodontium: the tissues that support the teeth, including the gums, cementum (bonelike material covering the dentin layer of the tooth), periodontal ligament, and underlying bone.

peripheral (puh-RIFF-er-ul) **blood smear:** a blood sample spread on a glass slide and stained for analysis under a microscope. *Peripheral* refers to the use of circulating blood rather than tissue blood.

peripheral nervous system: the peripheral (outermost) part of the nervous system; the vast complex of wiring that extends from the central nervous system to the body's outermost areas. It contains both somatic and autonomic components.

peripheral parenteral nutrition (PPN): the infusion of nutrient solutions into peripheral veins, usually a vein in the arm or back of the hand.

peripheral vascular disease: disorders characterized by impaired blood circulation in the limbs.

peripheral veins: the small-diameter veins that carry blood from the arms and legs.

perishable food rescue or salvage: collecting perishable produce from wholesalers and markets.

peristalsis (per-ih-STALL-sis): wavelike muscular contractions of the GI tract that push its contents along.

peritoneal (PEH-rih-toe-NEE-al) **dialysis:** a treatment that removes fluids and wastes from the blood by using the body's peritoneal membrane as a filter.

peritoneovenous (PEH-rih-toe-NEE-oh-VEE-nus) **shunt:** a surgical passage created between the peritoneum and the jugular vein to divert fluid and relieve ascites. The peritoneum is the membrane that surrounds the abdominal cavity.

peritonitis: inflammation of the peritoneal membrane, which lines the abdominal cavity.

pernicious (per-NISH-us) **anemia:** a blood disorder that reflects a vitamin B_{12} deficiency caused by lack of intrinsic factor and characterized by abnormally large and immature red blood cells. Other symptoms include muscle weakness and irreversible neurological damage.

persistent vegetative state: a vegetative mental state resulting from brain injury that persists for at least one month. Individuals lose awareness and the ability to think but retain noncognitive brain functions, such as motor reflexes and normal sleep patterns.

PES statement: the format used to write a nutrition diagnosis, which includes the Problem, the Etiology, and the Signs and symptoms.

pH: the unit of measure expressing a substance's acidity or alkalinity.

phagocytes (FAG-oh-sites): white blood cells (primarily neutrophils and macrophages) that have the ability to engulf and destroy pathogens.

phagocytosis (FAG-oh-sigh-TOE-sis): the process by which phagocytes engulf and destroy pathogens and cellular debris.

pharmacological effect: a large dose of a nutrient (levels commonly available only from supplements) that overwhelms some body system and acts like a drug.

pharynx (FAIR-inks): the passageway leading from the nose and mouth to the larynx and esophagus, respectively.

phenylketonuria (FEN-il-KEY-toe-NEW-ree-ah) or **PKU:** an inherited disorder characterized by failure to metabolize the amino acid phenylalanine to tyrosine.

phospholipid (FOS-foe-LIP-id): a compound similar to a triglyceride but having a phosphate group (a phosphorus-containing salt) and choline (or another nitrogen-containing compound) in place of one of the fatty acids.

phosphorus: a major mineral found mostly in the body's bones and teeth.

photosynthesis: the process by which green plants use the sun's energy to make carbohydrates from carbon dioxide and water.

phylloquinone (FILL-oh-KWIN-own): the naturally occurring form of vitamin K.

physiological age: a person's age as estimated from her or his body's health and probable life expectancy.

physiological effect: a normal dose of a nutrient (levels commonly found in foods) that provides a normal blood concentration.

physiological fuel value: the number of kcalories that the body derives from a food, in contrast to the number of kcalories determined by calorimetry.

phytic (FYE-tick) **acid:** a nonnutrient component of plant seeds; also called *phytate* (FYE-tate). Phytic acid occurs in the husks of grains, legumes, and seeds and is capable of binding minerals such as zinc, iron, calcium, magnesium, and copper in insoluble complexes in the intestine, which the body excretes unused.

phytochemicals (FIE-toe-KEM-ih-cals): nonnutrient compounds found in plant-derived foods that have biological activity in the body.

phytoestrogens: plant-derived compounds that have structural and functional similarities to human estrogen. Phytoestrogens include the isoflavones genistein, daidzein, and glycitein.

phytosterols: plant-derived compounds that have structural similarities to cholesterol and lower blood cholesterol by competing with cholesterol for absorption. Phytosterols include sterol esters and stanol esters.

pica (PIE-ka): a craving for and consumption of nonfood substances. Also known as *geophagia* (gee-oh-FAY-gee-uh) when referring to eating clay, baby powder, chalk, ash, ceramics, paper, paint chips, or charcoal; *pagophagia* (pag-oh-FAY-gee-uh) when referring to eating large quantities of ice; and *amylophagia* (AM-ee-low-FAY-gee-ah) when referring to eating uncooked starch (flour, laundry starch, or raw rice).

piggyback: the administration of a second solution using a separate port in an intravenous catheter.

pigment: a molecule capable of absorbing certain wavelengths of light so that it reflects only those that we perceive as a certain color.

placebo (pla-see-bo): an inert, harmless medication given to provide comfort and hope; a sham treatment used in controlled research studies.

placebo effect: a change that occurs in response to expectations in the effectiveness of a treatment that actually has no pharmaceutical effects.

placenta (plah-SEN-tuh): the organ that develops inside the uterus early in pregnancy, through which the fetus receives nutrients and oxygen and returns carbon dioxide and other waste products to be excreted.

plaque (PLACK): an accumulation of fatty deposits, smooth muscle cells, and fibrous connective tissue that develops in the artery walls in atherosclerosis. Plaque associated with atherosclerosis is known as *atheromatous* (ATH-er-OH-ma-tus) *plaque*.

plasma: the yellow fluid that remains after cells are removed from blood, which still contains clotting factors.

plasminogen activator inhibitor-1: a protein that promotes blood clotting by inhibiting blood clot degradation within blood vessels.

PLP (pyridoxal phosphate): the primary active coenzyme form of vitamin B_6.

point of unsaturation: the double bond of a fatty acid, where hydrogen atoms can easily be added to the structure.

polar: characteristic of a neutral molecule, such as water, that has opposite charges spatially separated within the molecule.

polydipsia (POL-ee-DIP-see-ah): excessive thirst.

polypeptide: many (ten or more) amino acids bonded together.

polyphagia (POL-ee-FAY-jee-ah): excessive appetite or food intake.

polysaccharides: compounds composed of many monosaccharides linked together. An intermediate string of three to ten monosaccharides is an *oligosaccharide*.

polyunsaturated fat: triglycerides in which most of the fatty acids are polyunsaturated.

polyunsaturated fatty acid (PUFA): a fatty acid that lacks four or more hydrogen atoms and has two or more double bonds between carbons—for example, linoleic acid (two double bonds) and linolenic acid (three double bonds).

polyuria (POL-ee-YOOR-ree-ah): excessive urine production.

portal hypertension: elevated blood pressure in the portal vein due to obstructed blood flow through the liver.

post term (infant): an infant born after the forty-second week of pregnancy.

postpartum amenorrhea: the normal temporary absence of menstrual periods immediately following childbirth.

potassium: the principal cation within the body's cells; critical to the maintenance of fluid balance, nerve impulse transmissions, and muscle contractions.

prebiotics: indigestible substances in foods (such as fibers) that stimulate the growth of nonpathogenic bacteria within the large intestine.

precursors: substances that precede others; with regard to vitamins, compounds that can be converted into active vitamins; also known as *provitamins*.

prediabetes: the condition in which blood glucose levels are higher than normal but not high enough to be diagnosed as diabetes.

preeclampsia (PRE-ee-KLAMP-see-ah): a condition characterized by hypertension and protein in the urine.

preformed vitamin A: dietary vitamin A in its active form.

prenatal alcohol exposure: subjecting a fetus to a pattern of excessive alcohol intake characterized by substantial regular use or heavy episodic drinking.

prepared food rescue: collecting prepared foods from commercial kitchens.

pressure gradient: the change in pressure over a given distance. In dialysis, a pressure gradient is created between the blood and the dialysate.

pressure sores: regions of damaged skin and tissue due to prolonged pressure on the affected area by an external object, such as a bed, wheelchair, or cast; vulnerable areas of the body include buttocks, hips, and heels. Also called *pressure ulcers* or *decubitus* (deh-KYU-bih-tus) *ulcers*.

preterm (infant): an infant born prior to the thirty-eighth week of pregnancy; also called a *premature infant*. A *term* infant is born between the thirty-eighth and forty-second week of pregnancy.

primary deficiency: a nutrient deficiency caused by inadequate dietary intake of a nutrient.

primary hypertension: hypertension with an unknown cause; also known as essential hypertension.

probiotics: living microorganisms found in foods that, when consumed in sufficient quantities, are beneficial to health.

processed foods: foods that have been treated to change their physical, chemical, microbiological, or sensory properties.

proenzyme: the inactive form of an enzyme; also called a *zymogen*.

progesterone: the hormone of gestation (pregnancy).

prolactin (pro-LAK-tin): a hormone secreted from the anterior pituitary gland that acts on the mammary glands to promote the production of milk. The release of prolactin is mediated by *prolactin-inhibiting hormone (PIH)*.

proof: a way of stating the percentage of alcohol in distilled liquor. Liquor that is 100 proof is 50 percent alcohol; 90 proof is 45 percent, and so forth.

prooxidants: substances that significantly induce oxidative stress.

proteases (PRO-tee-aces): enzymes that hydrolyze protein.

protein digestibility: a measure of the amount of amino acids absorbed from a given protein intake.

protein efficiency ratio (PER): a measure of protein quality assessed by determining how well a given protein supports weight gain in growing rats; used to establish the protein quality for infant formulas and baby foods.

protein-energy malnutrition (PEM): a deficiency of protein, energy, or both, including kwashiorkor, marasmus, and instances in which they overlap; also called *protein-kcalorie malnutrition (PCM)*.

protein isolates: proteins that have been isolated from foods.

protein-sparing action: the action of carbohydrate (and fat) in providing energy that allows protein to be used for other purposes.

protein turnover: the degradation and synthesis of protein.

proteins: compounds composed of carbon, hydrogen, oxygen, and nitrogen atoms, arranged into amino acids linked in a chain. Some amino acids also contain sulfur atoms.

proteinuria (PRO-teen-NOO-ree-ah): loss of protein, mostly albumin, in the urine; also known as *albuminuria*.

proteomics: the study of the body's proteins.

proton-pump inhibitors: a class of drugs that inhibit the enzyme that pumps hydrogen ions (protons) into the stomach. Examples include omeprazole (Prilosec) and lansoprazole (Prevacid).

puberty: the period in life in which a person becomes physically capable of reproduction.

public health dietitians: dietitians who specialize in providing nutrition services through organized community efforts.

public water: water from a municipal or county water system that has been treated and disinfected.

purified water: water that has been treated by distillation or other physical or chemical processes that remove dissolved solids. Because purified water contains no minerals or contaminants, it is useful for medical and research purposes.

purines (PYOO-reens): products of nucleotide metabolism that degrade to uric acid.

pyloric (pie-LORE-ic) **sphincter:** the circular muscle that separates the stomach from the small intestine and regulates the flow of partially digested food into the small intestine; also called *pylorus* or *pyloric valve*.

pyloroplasty (pye-LORE-oh-PLAS-tee): surgery that enlarges the pyloric sphincter.

pyruvate (PIE-roo-vate): a 3-carbon compound that plays a key role in energy metabolism.

Q

qi gong (chee-GUNG): a traditional Chinese system that combines movement, meditation, and breathing techniques and allegedly cures illness by enhancing the flow of qi (energy) within the body.

quality of life: a person's perceived physical and mental well-being.

R

rachitic (ra-KIT-ik) **rosary:** the poorly formed rib attachments that may develop in a vitamin D deficiency; literally, "the rosary of rickets."

radiation enteritis: inflammation of intestinal tissue caused by radiation therapy.

radiation therapy: the use of X-rays, gamma rays, or atomic particles to destroy cancer cells.

randomization (ran-dom-ih-zay-shun): a process of choosing the members of the experimental and control groups without bias.

raw sugar: the first crop of crystals harvested during sugar processing. Raw sugar cannot be sold in the United States because it contains too much filth (dirt, insect fragments, and the like). Sugar sold as "raw sugar" domestically has actually gone through more than half of the refining steps.

RD: see *registered dietitian*.

rebound hyperglycemia: hyperglycemia that results from the release of counterregulatory hormones following nighttime hypoglycemia; also called the Somogyi effect.

Recommended Dietary Allowance (RDA): the average daily amount of a nutrient considered adequate to meet the known nutrient needs of practically all healthy people; a goal for dietary intake by individuals.

rectum: the muscular terminal part of the intestine, extending from the sigmoid colon to the anus.

reduced kcalorie: at least 25 percent fewer kcalories per serving than the comparison food.

refeeding syndrome: a condition that sometimes develops when a severely malnourished person is aggressively fed; characterized by electrolyte and fluid imbalances and hyperglycemia.

reference protein: a standard against which to measure the quality of other proteins.

refined: having been subjected to the process by which the coarse parts of a food are removed. When wheat is refined into flour, the bran, germ, and husk are removed, leaving only the endosperm.

refined flour: finely ground endosperm that is usually enriched with nutrients and bleached for whiteness; sometimes called *white flour*.

reflexology: a technique that applies pressure or massage on areas of the hands or feet to allegedly cure disease or relieve pain in other areas of the body; sometimes called *zone therapy*.

reflux: a backward flow.

reflux esophagitis: inflammation in the esophagus related to the reflux of acidic stomach contents.

registered dietitian (RD): a person who has completed a minimum of a bachelor's degree from an accredited university or college, has completed approved course work and a supervised practice program, has passed a national examination, and maintains registration through continuing professional education.

registration: listing; with respect to health professionals, listing with a professional organization that requires specific course work, experience, and passing of an examination.

regular diet: a diet that includes all foods and meets the nutrient needs of healthy people; also called a *standard diet* or *house diet*.

relaxin: the hormone of late pregnancy.

remodeling: the dismantling and re-formation of a structure.

renal (REE-nal): pertaining to the kidneys.

renal colic: the intense pain that occurs when a kidney stone passes through the ureter.

renal osteodystrophy: a bone disorder that develops in patients with chronic kidney disease as a consequence of the increased secretion of parathyroid hormone, reduced serum calcium, acidosis, and impaired vitamin D activation by the kidneys.

renal threshold: the blood concentration of a substance that exceeds the kidneys' capacity for reabsorption, causing the substance to be passed into the urine.

renin (REN-in): an enzyme from the kidneys that hydrolyzes the protein angiotensinogen to angiotensin I.

replication (REP-lih-KAY-shun): repeating an experiment and getting the same results.

requirement: the lowest continuing intake of a nutrient that will maintain a specified criterion of adequacy.

resection: the surgical removal of part of an organ or body structure.

residue: material left in the intestine after digestion; includes mostly dietary fiber and undigested starches and proteins.

resistant starches: starches that escape digestion and absorption in the small intestine of healthy people.

resistin (re-ZIST-in): a hormone produced by adipose cells that induces insulin resistance.

respiratory chain: the final pathway in energy metabolism that transports electrons from hydrogen to oxygen and captures the energy released in the bonds of ATP; also called the *electron transport chain*.

respiratory stress: abnormal gas exchange between the air and blood, resulting in lower-than-normal oxygen levels and higher-than-normal carbon dioxide levels.

resting metabolic rate (RMR): similar to the basal metabolic rate (BMR), a measure of the energy use of a person at rest in a comfortable setting, but with less stringent criteria for recent food intake and physical activity. Consequently, the RMR is slightly higher than the BMR.

reticulocytes: immature red blood cells released into blood by bone marrow.

retina (RET-in-uh): the innermost membrane of the eye, composed of several layers including one that contains the rods and cones.

retinoids (RET-ih-noyds): chemically related compounds with biological activity similar to that of retinol; metabolites of retinol.

retinol (RET-ih-nol): the alcohol form of vitamin A.

retinol activity equivalents (RAE): a measure of vitamin A activity; the amount of retinol that the body will derive from a food containing preformed retinol or its precursor beta-carotene.

retinol-binding protein (RBP): the specific protein responsible for transporting retinol.

rheumatoid (ROO-ma-toyd) **arthritis:** a disease of the immune system involving painful inflammation of the joints and related structures.

rhodopsin (ro-DOP-sin): a light-sensitive pigment of the retina; contains the retinal form of vitamin A and the protein opsin.

riboflavin (RYE-boh-flay-vin): a B vitamin. The coenzyme forms are **FMN (flavin mononucleotide)** and **FAD (flavin adenine dinucleotide)**.

ribosomes (RYE-boh-zomes): protein-making organelles in cells; composed of RNA and protein.

rickets: the vitamin D–deficiency disease in children characterized by inadequate mineralization of bone (manifested in bowed legs or knock-knees, outward-bowed chest, and knobs on ribs). A rare type of rickets, not caused by vitamin D deficiency, is known as *vitamin D–refractory rickets*.

risk: a measure of the probability and severity of harm.

risk factor: a condition or behavior associated with an elevated frequency of a disease but not proved to be causal. Leading risk factors for chronic diseases include obesity, cigarette smoking, high blood pressure, high blood cholesterol, physical inactivity, and a diet high in saturated fats and low in vegetables, fruits, and whole grains.

RNA (ribonucleic acid): a compound similar to DNA, but RNA is a single strand with a ribose sugar instead of a deoxyribose sugar and uracil instead of thymine as one of its bases.

Roux-en-Y gastric bypass: a bariatric surgical procedure in which the surgeon constructs a small stomach pouch and creates an outlet directly to the small intestine, bypassing most of the stomach, the entire duodenum, and some of the jejunum; the reconstructed small intestine resembles the letter Y.

S

saccharin (SAK-ah-ren): an artificial sweetener that has been approved for use in the United States. In Canada, approval for use in foods and beverages is pending; currently available only in pharmacies and only as a tabletop sweetener, not as an additive.

safety: the condition of being free from harm or danger.

saliva: the secretion of the salivary glands. Its principal enzyme begins carbohydrate digestion.

salivary glands: exocrine glands that secrete saliva into the mouth.

salt: a compound composed of a positive ion other than H^+ and a negative ion other than OH^-. An example is sodium chloride (Na^+Cl^-).

salt sensitivity: a characteristic of individuals who respond to a high salt intake with an increase in blood pressure or to a low salt intake with a decrease in blood pressure.

sarcopenia (SAR-koh-PEE-nee-ah): loss of skeletal muscle mass, strength, and quality.

satiating: having the power to suppress hunger and inhibit eating.

satiation (say-she-AY-shun): the feeling of satisfaction and fullness that occurs during a meal and halts eating. Satiation determines how much food is consumed during a meal.

satiety (sah-TIE-eh-tee): the feeling of fullness and satisfaction that occurs after a meal and inhibits eating until the next meal. Satiety determines how much time passes between meals.

saturated fat: triglycerides in which most of the fatty acids are saturated.

saturated fat-free: less than 0.5 gram of saturated fat and 0.5 gram of *trans* fat per serving.

saturated fatty acid: a fatty acid carrying the maximum possible number of hydrogen atoms—for example, stearic acid.

scurvy: the vitamin C–deficiency disease.

secondary deficiency: a nutrient deficiency caused by something other than an inadequate intake such as a disease condition or drug interaction that reduces absorption, accelerates use, hastens excretion, or destroys the nutrient.

secondary hypertension: hypertension that results from a known physiological abnormality.

secretin (see-CREET-in): a hormone produced by cells in the duodenum wall. Target organ: the pancreas. Response: secretion of bicarbonate-rich pancreatic juice.

segmentation (SEG-men-TAY-shun): a periodic squeezing or partitioning of the intestine at intervals along its length by its circular muscles.

selective menus: menus that provide choices in some or all menu categories.

selenium (se-LEEN-ee-um): a trace element.

self-monitoring of blood glucose: home monitoring of blood glucose levels using a glucose meter.

semipermeable membrane: a membrane that allows some particles to pass through, but not others.

semiselective menus: menus that combine aspects of both selective and nonselective menus.

senile dementia: the loss of brain function beyond the normal loss of physical adeptness and memory that occurs with aging.

senile plaques: clumps of the protein fragment beta-amyloid on the nerve cells, commonly found in the brains of people with Alzheimer's dementia.

sepsis: a whole-body inflammatory response caused by infection; characterized by symptoms similar to those of SIRS.

serotonin (SER-oh-TONE-in): a neurotransmitter important in sleep regulation, appetite control, and sensory perception, among other roles. Serotonin is synthesized in the body from the amino acid tryptophan with the help of vitamin B_6.

serum: the fluid remaining after both cells and clotting factors are removed from blood.

set point: the point at which controls are set (for example, on a thermostat). The set-point theory that relates to body weight proposes that the body tends to maintain a certain weight by means of its own internal controls.

shock: a severe reduction in blood flow that deprives the body's tissues of oxygen and nutrients; characterized by reduced blood pressure, raised heart and respiratory rates, and muscle weakness.

shock-wave lithotripsy: a nonsurgical procedure that uses high-amplitude sound waves to fragment gallstones or kidney stones.

short bowel syndrome: the malabsorption syndrome that follows resection of the small intestine, resulting in inadequate absorptive capacity in the remaining intestine.

sibutramine (sigh-BYOO-tra-mean): a drug used in the treatment of obesity that slows the reabsorption of serotonin in the brain, thus suppressing appetite and creating a feeling of fullness.

sickle-cell anemia: a hereditary form of anemia characterized by abnormal sickle- or crescent-shaped red blood cells. Sickled cells interfere with oxygen transport and blood flow. Symptoms are precipitated by dehydration and insufficient oxygen (as may occur at high altitudes) and include hemolytic anemia (red blood cells burst), fever, and severe pain in the joints and abdomen.

simple carbohydrates: monosaccharides and disaccharides.

sinusoids: the small, capillary-like passages that carry blood through liver tissue.

Sjögren's syndrome: an autoimmune disease characterized by the destruction of secretory glands, resulting in dry mouth and dry eyes.

sludge: literally, a semisolid mass. Biliary sludge is made up of mucus, cholesterol crystals, and bilirubin granules.

small for gestational age (SGA): term describing an infant whose birth weight is low compared with the number of weeks in utero, often reflecting malnutrition.

small intestine: a 10-foot length of small-diameter intestine that is the major site of digestion of food and absorption of nutrients. Its segments are the duodenum, jejunum, and ileum.

soaps: chemical compounds formed from fatty acids and positively charged minerals.

socioeconomic history: a record of a person's social and economic background, including such factors as education, income, and ethnic identity.

sodium: the principal cation in the extracellular fluids of the body; critical to the maintenance of fluid balance, nerve impulse transmissions, and muscle contractions.

sodium bicarbonate: baking soda.

sodium-free and **salt-free**: less than 5 milligrams of sodium per serving.

soft water: water with a high sodium or potassium content.

soluble fibers: nonstarch polysaccharides that dissolve in water to form a gel. An example is pectin from fruit, which is used to thicken jellies.

solutes (SOLL-yutes): the substances that are dissolved in a solution. The number of molecules in a given volume of fluid is the *solute concentration*.

somatic (so-MAT-ick) **nervous system**: the division of the nervous system that controls the voluntary muscles, as distinguished from the autonomic nervous system, which controls involuntary functions.

somatostatin (GHIH): a hormone that inhibits the release of growth hormone; the opposite of somatotropin (GH).

Somogyi effect: hyperglycemia that results from the release of counterregulatory hormones following nighttime hypoglycemia; also called rebound hyperglycemia.

soup kitchens: programs that provide prepared meals to be eaten on site.

spasm: a sudden, forceful, and involuntary muscle contraction.

specialized formulas: enteral formulas designed to meet the nutrient needs of patients with specific illnesses; also called *disease-specific formulas*.

sperm: the male reproductive cell, capable of fertilizing an ovum.

sphincter (SFINK-ter): a circular muscle surrounding, and able to close, a body opening. Sphincters are found at specific points along the GI tract and regulate the flow of food particles.

spina (SPY-nah) **bifida** (BIFF-ih-dah): one of the most common types of neural tube defects; characterized by the incomplete closure of the spinal cord and its bony encasement.

spring water: water originating from an underground spring or well. It may be bubbly (carbonated), or "flat" or "still," meaning not carbonated. Brand names such as "Spring Pure" do not necessarily mean that the water comes from a spring.

standard diet: a diet that includes all foods and meets the nutrient needs of healthy people; also called a *regular diet* or *house diet*.

standard formulas: enteral formulas that contain mostly intact proteins and polysaccharides; also called *polymeric formulas*.

starches: plant polysaccharides composed of glucose.

steatohepatitis (STEE-ah-to-HEP-ah-TIE-tis): liver inflammation that is associated with fatty liver.

steatorrhea (stee-AT-or-REE-ah): excessive fat in the stools due to fat malabsorption; characterized by stools that are loose, frothy, and foul smelling due to a high fat content.

sterile: free of microorganisms, such as bacteria.

sterols (STARE-ols or STEER-ols): compounds containing a four-ring carbon structure with any of a variety of side chains attached.

stevia (STEE-vee-ah): a South American shrub whose leaves are used as a sweetener; sold in the United States as a dietary supplement that provides sweetness without kcalories.

stoma (STOE-ma): a surgically created opening in a body tissue or organ.

stomach: a muscular, elastic, saclike portion of the digestive tract that grinds and churns swallowed food, mixing it with acid and enzymes to form chyme.

stools: waste matter discharged from the colon; also called *feces* (FEE-seez).

stress: any threat to a person's well-being; a demand placed on the body to adapt.

stress eating: eating in response to arousal.

stress fractures: bone damage or breaks caused by stress on bone surfaces during exercise.

stress response: the body's response to stress, mediated by both nerves and hormones.

stressors: environmental elements, physical or psychological, that cause stress.

stricture: abnormal narrowing of a passageway; often due to inflammation, scarring, or a congenital abnormality.

stroke: a sudden injury to brain tissue resulting from impaired blood flow through an artery that supplies blood to the brain; also called a *cerebrovascular accident*.

structure-function claims: statements that characterize the relationship between a nutrient or other substance in a food and its role in the body.

struvite (STROO-vite): crystals of magnesium ammonium phosphate.

subclavian (sub-KLAY-vee-an) **vein**: the vein that provides passage from the lymphatic system to the vascular system.

subclinical deficiency: a deficiency in the early stages, before the outward signs have appeared.

subcutaneous (sub-cue-TAY-nee-us): beneath the skin.

subcutaneous fat: fat stored directly under the skin.

subjects: the people or animals participating in a research project.

subluxation: in chiropractic, a misaligned vertebra or other spinal alteration that may cause illness.

successful weight-loss maintenance: achieving a weight loss of at least 10 percent of initial body weight and maintaining the loss for at least one year.

sucralose (SUE-kra-lose): an artificial sweetener approved for use in the United States and Canada.

sucrase: an enzyme that hydrolyzes sucrose.

sucrose (SUE-krose): a disaccharide composed of glucose and fructose; commonly known as *table sugar, beet sugar*, or *cane sugar*. Sucrose also occurs in many fruits and some vegetables and grains.

sudden infant death syndrome (SIDS): the unexpected and unexplained death of an apparently well infant; the most common cause of death of infants between the second week and the end of the first year of life; also called *crib death*.

sugar alcohols: sugarlike compounds that can be derived from fruits or commercially produced from dextrose; also called polyols. Sugar alcohols are absorbed more slowly than other sugars and metabolized differently in the human body; they are not readily utilized by ordinary mouth bacteria. Examples are maltitol, mannitol, sorbitol, xylitol, isomalt, and lactitol.

sugar-free: less than 0.5 gram of sugar per serving.

sugar replacers: sugarlike compounds that can be derived from fruits or commercially produced from dextrose; also called sugar alcohols or polyols. Sugar alcohols are absorbed more slowly than other sugars and metabolized differently in the human body; they are not readily utilized by ordinary mouth bacteria. Examples are maltitol, mannitol, sorbitol, xylitol, isomalt, and lactitol.

sugars: monosaccharides and disaccharides.

sulfate: the oxidized form of sulfur.

sulfur: a mineral present in the body as part of some proteins.

surrogate: a substitute; a person who takes the place of another.

sushi: vinegar-flavored rice and seafood, typically wrapped in seaweed and stuffed with colorful vegetables. Some sushi is stuffed with raw fish; other varieties contain cooked seafood.

symptomatic allergy: an immune response that produces antibodies and symptoms.

synbiotic: a mixture of probiotics and prebiotics.

syringes: devices used for injecting medications. A syringe consists of a hypodermic needle attached to a hollow tube with a plunger inside.

systemic (sih-STEM-ic): relating to the entire body.

systemic inflammatory response syndrome (SIRS): a whole-body inflammatory response caused by severe illness or trauma; characterized by raised heart and respiratory rates, abnormal white blood cell counts, and elevated body temperature.

T

T cell: a lymphocyte that attacks antigens; functions in cell-mediated immunity.

tagatose (TAG-ah-tose): a monosaccharide structurally similar to fructose that is incompletely absorbed and thus provides only 1.5 kcalories per gram; approved for use as a "generally recognized as safe" ingredient.

TCA cycle or **tricarboxylic** (try-car-box-ILL-ick) **acid cycle**: a series of metabolic reactions that break down molecules of acetyl CoA to carbon dioxide and hydrogen atoms; also called the *citric acid cycle* or the *Krebs cycle* after the biochemist who elucidated its reactions.

tempeh (TEM-pay): a fermented soybean food, rich in protein and fiber.

teratogen (ter-AT-oh-jen): a substance that causes abnormal fetal development and birth defects.

teratogenic (ter-AT-oh-jen-ik): causing abnormal fetal development and birth defects.

testosterone: a steroid hormone from the testicles, or testes. The steroids, as explained in Chapter 5, are chemically related to, and some are derived from, the lipid cholesterol.

textured vegetable protein: processed soybean protein used in vegetarian products such as soy burgers.

theory: a tentative explanation that integrates many and diverse findings to further the understanding of a defined topic.

therapeutic touch: a technique of passing hands over a patient to purportedly identify energy imbalances and transfer healing power from therapist to patient; also called *laying on of hands*.

thermic effect of food (TEF): an estimation of the energy required to process food (digest, absorb, transport, metabolize, and store ingested nutrients); also called the *specific dynamic effect (SDE) of food* or the *specific dynamic activity (SDA) of food*. The sum of the TEF and any increase in the metabolic rate due to overeating is known as *diet-induced thermogenesis (DIT)*.

thermogenesis: the generation of heat; used in physiology and nutrition studies as an index of how much energy the body is expending.

THF (tetrahydrofolate): a coenzyme form of folate.

thiamin (THIGH-ah-min): a B vitamin. The coenzyme form is **TPP (thiamin pyrophosphate)**.

thirst: a conscious desire to drink.

thoracic (thor-ASS-ic) **duct**: the main lymphatic vessel that collects lymph and drains into the left subclavian vein.

thrombosis (throm-BOH-sis): the formation or presence of a blood clot in blood vessels. A *coronary thrombosis* occurs in a coronary artery, and a *cerebral thrombosis* occurs in an artery that supplies blood to the brain.

thrombus: a blood clot formed within a blood vessel that remains attached to its place of origin.

thyroid-stimulating hormone (TSH): a hormone secreted by the pituitary that stimulates the thyroid gland to secrete its hormones—thyroxine and triiodothyronine. The release of TSH is mediated by TSH-releasing hormone (TRH).

thyrotropin: another name for thyroid-stimulating hormone (TSH).

tissue rejection: destruction of donor tissue by the recipient's immune system, which recognizes the donor cells as foreign.

tocopherol (tuh-KOFF-er-ol): a general term for several chemically related compounds, one of which has vitamin E activity. (See Appendix C for chemical structures.)

tofu (TOE-foo): a curd made from soybeans, rich in protein and often fortified with calcium; used in many Asian and vegetarian dishes in place of meat.

Tolerable Upper Intake Level (UL): the maximum daily amount of a nutrient that appears safe for most healthy people and beyond which there is an increased risk of adverse health effects.

total nutrient admixture (TNA): a parenteral solution that contains dextrose, amino acids, and lipids; also called a 3-in-1 solution or an all-in-one solution.

total parenteral nutrition (TPN): the infusion of nutrient solutions into a central vein.

toxicity: the ability of a substance to harm living organisms. All substances are toxic if high enough concentrations are used.

TPP (thiamin pyrophosphate): the coenzyme form of thiamin.

trabecular (tra-BECK-you-lar) **bone**: the lacy inner structure of calcium crystals that supports the bone's structure and provides a calcium storage bank.

trace minerals: essential mineral nutrients the human body requires in relatively small amounts (less than 100 milligrams per day); sometimes called *microminerals*.

trachea (TRAKE-ee-uh): the air passageway from the larynx to the lungs; also called the *windpipe*.

traditional Chinese medicine (TCM): an approach to health care based on the concept that illness can be cured by enhancing the flow of qi (energy) within a person's body. Treatments may include herbal therapies, physical exercises, meditation, acupuncture, and remedial massage.

***trans*:** on the other side of; refers to a chemical configuration in which the hydrogen atoms are located on opposite sides of a double bond.

***trans* fat-free:** less than 0.5 gram of *trans* fat and less than 0.5 gram of saturated fat per serving.

***trans*-fatty acids:** fatty acids with hydrogens on opposite sides of the double bond.

transamination (TRANS-am-ih-NAY-shun): the transfer of an amino group from one amino acid to a keto acid, producing a new nonessential amino acid and a new keto acid.

transcription: the process of messenger RNA being made from a template of DNA.

transferrin (trans-FAIR-in): the iron transport protein.

transient hypertension of pregnancy: high blood pressure that develops in the second half of pregnancy and resolves after childbirth, usually without affecting the outcome of the pregnancy.

transient ischemic attacks (TIAs): brief ischemic strokes that cause short-term neurological symptoms.

translation: the process of messenger RNA directing the sequence of amino acids and synthesis of proteins.

transnasal feeding tube: a feeding tube that is inserted through the nose.

triglycerides (try-GLISS-er-rides): the chief form of fat in the diet and the major storage form of fat in the body; composed of a molecule of glycerol with three fatty acids attached; also called *triacylglycerols* (try-ay-seel-GLISS-er-ols).

tripeptide: three amino acids bonded together.

tube feedings: liquid formulas delivered through a tube placed in the stomach or intestine.

tubules: tubelike structures of the nephron that process filtrate during urine production. The tubules are surrounded by capillaries that reabsorb substances retained by tubule cells.

tumor: an abnormal tissue mass that has no physiological function; also called a *neoplasm* (NEE-oh-plazm).

turbinado (ter-bih-NOD-oh) **sugar:** sugar produced using the same refining process as white sugar, but without the bleaching and anti-caking treatment. Traces of molasses give turbinado its sandy color.

type 1 diabetes: the type of diabetes that accounts for 5 to 10 percent of diabetes cases and usually results from autoimmune destruction of pancreatic beta cells.

type 2 diabetes: the type of diabetes that accounts for 90 to 95 percent of diabetes cases and usually results from insulin resistance coupled with insufficient insulin secretion.

type I osteoporosis: osteoporosis characterized by rapid bone losses, primarily of trabecular bone.

type II osteoporosis: osteoporosis characterized by gradual losses of both trabecular and cortical bone.

U

ulcer: a lesion of the skin or mucous membranes characterized by inflammation and damaged tissues.

ulcerative colitis (ko-LY-tis): an inflammatory bowel disease that involves the colon. Inflammation affects the mucosa and submucosa of the intestinal wall.

ultrafiltration: removal of fluids and solutes from blood by using pressure to transfer the blood across a semipermeable membrane.

umbilical (um-BILL-ih-cul) **cord:** the ropelike structure through which the fetus's veins and arteries reach the placenta; the route of nourishment and oxygen to the fetus and the route of waste disposal from the fetus. The scar in the middle of the abdomen that marks the former attachment of the umbilical cord is the *umbilicus* (um-BILL-ih-cus), commonly known as the "belly button."

uncoupled reactions: chemical reactions in which energy is released as heat.

undernutrition: deficient energy or nutrients.

underweight: body weight below some standard of acceptable weight that is usually defined in relation to height (such as BMI); BMI below 18.5.

unsaturated fat: triglycerides in which most of the fatty acids are unsaturated.

unsaturated fatty acid: a fatty acid that lacks hydrogen atoms and has at least one double bond between carbons (includes monounsaturated and polyunsaturated fatty acids).

unspecified eating disorders: eating disorders that do not meet the defined criteria for specific eating disorders.

urea (you-REE-uh): the principal nitrogen-excretion product of protein metabolism. Two ammonia fragments are combined with carbon dioxide to form urea.

urea kinetic modeling: a method of determining the adequacy of dialysis treatment by calculating the urea clearance from blood.

uremia (you-REE-me-ah): the accumulation of nitrogenous and various other waste products in the blood; often associated with symptoms that reflect impairments in multiple body systems (literally, "urine in the blood"). The term may also be used to indicate the toxic state that results when wastes are retained in the blood.

uremic syndrome: the cluster of symptoms associated with inadequate kidney function; the symptoms reflect fluid, electrolyte, and hormonal imbalances; altered heart function; neuromuscular disturbances; and other metabolic derangements.

uterus (YOU-ter-us): the muscular organ within which the infant develops before birth.

V

vagotomy (vay-GOT-oh-mee): surgery that severs the vagus nerve in order to suppress gastric acid secretion. This surgery may require a follow-up *pyloroplasty* procedure to allow stomach drainage.

vagus nerve: the cranial nerve that regulates hydrochloric acid secretion and peristalsis. Effects elsewhere in the body include regulation of heart rate and bronchiole constriction.

validity (va-lid-ih-tee): having the quality of being founded on fact or evidence.

variables: factors that change. A variable may depend on another variable (for example, a child's height depends on his age), or it may be independent (for example, a child's height does not depend on the color of her eyes). Sometimes both variables correlate with a third variable (a child's height and eye color both depend on genetics).

varices (VAH-rih-seez): abnormally dilated blood vessels (singular: *varix*).

variety (dietary): eating a wide selection of foods within and among the major food groups.

vasoconstrictor (VAS-oh-kon-STRIK-tor): a substance that constricts or narrows the blood vessels.

vasopressin (VAS-oh-PRES-in): another name for antidiuretic hormone, so called because it elevates blood pressure.

vegans (VEE-gans): people who exclude all animal-derived foods (including meat, poultry, fish, eggs, and dairy products) from their diets; also called *pure vegetarians*, *strict vegetarians*, or *total vegetarians*.

vegetarians: a general term used to describe people who exclude meat, poultry, fish, or other animal-derived foods from their diets.

veins (VANES): vessels that carry blood to the heart.

very low food security: multiple indications of disrupted eating patterns and reduced food intake; formerly known as *food insecurity with hunger*.

very low sodium: 35 milligrams or less per serving.

villi (VILL-ee, VILL-eye): fingerlike projections from the folds of the small intestine; singular *villus*.

visceral fat: fat stored within the abdominal cavity in association with the internal abdominal organs; also called *intra-abdominal fat*.

viscous: a gel-like consistency.

vitamin A: all naturally occurring compounds with the biological activity of retinol (RET-ihnol), the alcohol form of vitamin A.

vitamin A activity: a term referring to both the active forms of vitamin A and the precursor forms in foods without distinguishing between them.

vitamin B_6: a family of compounds—pyridoxal, pyridoxine, and pyridoxamine. The primary active coenzyme form is PLP (pyridoxal phosphate).

vitamin B_{12}: a B vitamin characterized by the presence of cobalt. The active forms of coenzyme B_{12} are methylcobalamin and deoxyadenosylcobalamin.

vitamin D_2: vitamin D derived from plants in the diet and made from the yeast and plant sterol ergosterol; also called **ergocalciferol**.

vitamin D_3: vitamin D derived from animals in the diet and made in the skin from 7-dehydrocholesterol, a precursor of cholesterol, with the help of sunlight; also called **cholecalciferol** or **calciol**.

vitamins: organic, essential nutrients required in small amounts by the body for health.

VLDL (very-low-density lipoprotein): the type of lipoprotein made primarily by liver cells to transport lipids to various tissues in the body; composed primarily of triglycerides.

vomiting: expulsion of the contents of the stomach up through the esophagus to the mouth.

vulnerable plaque: plaque that is susceptible to rupture because it has only a thin fibrous barrier between its lipid-rich core and the artery lining.

W

waist circumference: an anthropometric measurement used to assess a person's abdominal fat.

wasting: the gradual atrophy (loss) of body tissues; associated with protein-energy malnutrition or chronic illness.

water balance: the balance between water intake and output (losses).

water intoxication: the rare condition in which body water contents are too high in all body fluid compartments.

wean: to gradually replace breast milk with infant formula or other foods appropriate to an infant's diet.

websites: Internet resources composed of text and graphic files, each with a unique URL (Uniform Resource Locator) that names the site (for example, www.usda.gov).

weight management: maintaining body weight in a healthy range by preventing gradual weight gain over time and losing weight if overweight.

well water: water drawn from groundwater by tapping into an aquifer.

Wernicke-Korsakoff (VER-nee-key KORE-sah-kof) **syndrome:** a neurological disorder typically associated with chronic alcoholism and caused by a deficiency of the B vitamin thiamin; also called *alcohol-related dementia*.

wheat flour: any flour made from the endosperm of the wheat kernel.

wheat gluten (GLU-ten): a family of water-insoluble proteins in wheat; includes the gliadin (GLY-ah-din) fractions that are toxic to persons with celiac disease.

whey protein: a by-product of cheese production; falsely promoted as increasing muscle mass. Whey is the watery part of milk that separates from the curds.

white sugar: pure sucrose or "table sugar," produced by dissolving, concentrating, and recrystallizing raw sugar.

whole grain: a grain that maintains the same relative proportions of starchy endosperm, germ, and bran as the original (all but the husk); not refined.

whole-wheat flour: any flour made from the entire wheat kernel.

wine: an alcoholic beverage traditionally made by fermenting a sugar source such as grape juice.

World Wide Web (the Web, commonly abbreviated www): a graphical subset of the Internet.

X

xanthophylls (ZAN-tho-fills): pigments found in plants; responsible for the color changes seen in autumn leaves.

xerophthalmia (zer-off-THAL-mee-uh): progressive blindness caused by inadequate tear production due to severe vitamin A deficiency.

xerosis (zee-ROW-sis): abnormal drying of the skin and mucous membranes; a sign of vitamin A deficiency.

xerostomia: dry mouth caused by reduced salivary flow.

Y

yogurt: milk product that results from the fermentation of lactic acid in milk by *Lactobacillus bulgaricus* and *Streptococcus thermophilus*.

Z

Zollinger-Ellison syndrome: a condition characterized by the presence of gastrin-secreting tumors in the duodenum or pancreas.

zygote (ZY-goat): the initial product of the union of ovum and sperm; a fertilized ovum.

zymogen (ZYE-mo-jen): the inactive precursor of an enzyme; sometimes called a *proenzyme*.

Index

Page letters in bold indicate definition of terms.
Page references followed by the letter "f" indicate figures or photographs.
Page references followed by the letter "t" indicate tables.
Page references with combined letter and number (A-1) refer to the appendices.

A

D

E

F

G

L

M

O

P

Q

R

U

V

STUDY CARD 1

An Overview of Nutrition

Food Choices

- People select foods based on such factors as taste and convenience, but selections based on nutrition knowledge may better support good health.
- Individual foods are neither "good" nor "bad"; daily food choices made over a lifetime may improve or impair a person's health.

The Nutrients

- Foods provide nutrients—substances that provide energy, structural materials, and regulating agents to support the growth, maintenance, and repair of the body's tissues. Essential nutrients *must* be obtained from foods.
- The six classes of nutrients include carbohydrates, lipids (fats), proteins, vitamins, minerals, and water. Carbohydrates, lipids, proteins, and vitamins are organic, meaning they contain carbon; minerals and water are inorganic.
- Energy is measured in kcalories—a measure of heat energy. One kcalorie is the amount of heat necessary to raise the temperature of 1 kg water 1°C.
- The energy-yielding nutrients are carbohydrate (4 kcal/g), fat (9 kcal/g), and protein (4 kcal/g).

The Science of Nutrition

- The science of nutrition is the study of nutrients and other substances in foods and the body's handling of them.
- Researchers follow the scientific method (review Figure 1-3, p. 12). They randomly assign control and experimental groups, use large sample sizes, provide placebos, and are blind to treatments. Their findings are reviewed and replicated by other scientists before being accepted as valid.
- Correlations indicate an association between variables, not a cause.

Dietary Reference Intakes

- Dietary Reference Intakes (DRI) are a set of nutrient intake values used to plan and evaluate diets for healthy people.
- Estimated Average Requirement (EAR) defines the amount of a nutrient that supports a specific function in the body for half of the population. Recommended Dietary Allowance (RDA) is based on the EAR and establishes a goal for dietary intake that will meet the needs of almost all healthy people. Adequate Intake (AI) serves a similar purpose when an RDA cannot be determined.
- Estimated Energy Requirement (EER) defines the average amount of energy intake needed to maintain energy balance, and Acceptable Macronutrient Distribution Ranges (AMDR) define the proportions contributed by carbohydrate, fat, and protein to a healthy diet.
- Tolerable Upper Intake Level (UL) establishes the highest amount that appears safe for regular consumption.

Nutrition Assessment

- Malnutrition develops when people get too little, too much, or an imbalance of energy or nutrients.
- Four nutrition assessment methods include historical information on diet and health, anthropometric measurements, physical examinations, and laboratory tests. Together, these methods reveal the stages of a nutrient deficiency (review Figure 1-8, p. 23).
- A primary deficiency is caused by an inadequate intake of a nutrient; a secondary deficiency is caused by a condition that reduces absorption, accelerates use, increases excretion, or destroys the nutrient.

Diet and Health

- Risk factors such as obesity and cigarette smoking increase the likelihood of disease development.
- Some risk factors, such as genetics, are important but cannot be changed. Recommendations focus on changeable, personal life choices such as diet and activity habits.
- Diet has no influence on some diseases but is linked closely to others.

TABLE 1-6 Leading Causes of Death in the United States

	Percentage of Total Deaths
1. **Heart disease**	26.5
2. **Cancers**	22.8
3. **Strokes**	5.9
4. Chronic lung diseases	5.3
5. Accidents	4.7
6. **Diabetes mellitus**	3.1
7. Alzheimer's disease	2.9
8. Pneumonia and influenza	2.6
9. Kidney diseases	1.8
10. Blood infections	1.4

NOTE: The diseases highlighted in bold have relationships with diet.
SOURCE: National Center for Health Statistics, www.cdc.gov/nchs

TABLE 1-7 Factors Contributing to Deaths in the United States

Factors	Percentage of Deaths
Tobacco	18
Poor diet/inactivity	15
Alcohol	4
Microbial agents	3
Toxic agents	2
Motor vehicles	2
Firearms	1
Sexual behavior	1
Illicit drugs	1

SOURCE: A. H. Mokdad and coauthors, Actual causes of death in the United States, 2000, *Journal of the American Medical Association* 291 (2004): 1238–1245, with corrections from *Journal of the American Medical Association* 293 (2005): 298.

Food Choices (pp. 3–5)

1. Give several reasons (and examples) why people make the food choices they do.
2. When people eat the foods typical of their families or geographic region, their choices are influenced by:
   a. habit. c. personal preference.
   b. nutrition. d. heritage or tradition.

The Nutrients (pp. 5–11)

3. What is a nutrient? Name the six classes of nutrients found in foods. What is an essential nutrient?
4. Which nutrients are inorganic, and which are organic? Discuss the significance of that distinction.
5. Which nutrients yield energy, and how much energy do they yield per gram? How is energy measured?
6. Describe how alcohol resembles nutrients. Why is alcohol not considered a nutrient?
7. The nutrient found most abundantly in both the human body and most foods is:
   a. fat. b. water. c. minerals. d. proteins.
8. The inorganic nutrients are:
   a. proteins and fats. c. minerals and water.
   b. vitamins and minerals. d. vitamins and proteins.
9. The energy-yielding nutrients are:
   a. fats, minerals, and water.
   b. minerals, proteins, and vitamins.
   c. carbohydrates, fats, and vitamins.
   d. carbohydrates, fats, and proteins.

The Science of Nutrition (pp. 11–17)

10. What is the science of nutrition? Describe the types of research studies and methods used in acquiring nutrition information.
11. Explain how variables might be correlational but not causal.
12. Studies of populations that reveal correlations between dietary habits and disease incidence are:
    a. clinical trials. c. case-control studies.
    b. laboratory studies. d. epidemiological studies.
13. An experiment in which neither the researchers nor the subjects know who is receiving the treatment is known as:
    a. double blind. c. blind variable.
    b. double control. d. placebo control.

Dietary Reference Intakes (pp. 17–21)

14. What are the DRI? To whom do they apply? How are they used? In your description, identify the categories of DRI and indicate how they are related.
15. What judgment factors are involved in setting the energy and nutrient recommendations?
16. An RDA represents the:
    a. highest amount of a nutrient that appears safe for most healthy people.
    b. lowest amount of a nutrient that will maintain a specified criterion of adequacy.
    c. average amount of a nutrient considered adequate to meet the known nutrient needs of practically all healthy people.
    d. average amount of a nutrient that will maintain a specific biochemical or physiological function in half the people.

Nutrition Assessment (pp. 21–25)

17. What happens when people get either too little or too much energy or nutrients? Define *malnutrition, undernutrition*, and *overnutrition*. Describe the four methods used to detect energy and nutrient deficiencies and excesses.
18. What methods are used in nutrition surveys? What kinds of information can these surveys provide?
19. Historical information, physical examinations, laboratory tests, and anthropometric measurements are:
    a. techniques used in diet planning.
    b. steps used in the scientific method.
    c. approaches used in disease prevention.
    d. methods used in a nutrition assessment.
20. A deficiency caused by an inadequate dietary intake is a(n):
    a. overt deficiency. c. primary deficiency.
    b. covert deficiency. d. secondary deficiency.

Diet and Health (pp. 25–26)

21. Describe risk factors and their relationships to disease.
22. Behaviors such as smoking, dietary habits, physical activity, and alcohol consumption that influence the development of disease are known as:
    a. risk factors. c. preventive agents.
    b. chronic causes. d. disease descriptors.

Study Questions (multiple choice)
2. d 7. b 8. c 9. d 12. d 13. a 16. c 19. d 20. c 22. a

For additional study questions and activities, go to **www.cengagebrain.com** and search for ISBN 084006845X.

STUDY CARD 2

Planning a Healthy Diet

Principles and Guidelines

- A well-planned diet delivers *adequate* nutrients, a *balanced* array of nutrients, and an appropriate amount of *energy (kcalories)*. It is based on *nutrient-dense* foods, *moderate* in substances that can be detrimental to health, and *varied* in its selections.
- The *Dietary Guidelines for Americans* offer practical advice on how to eat for good health (review Table 2-1, p. 38).

Diet-Planning Guides

- The USDA Food Guide suggests the amounts to eat from each of five food groups—grains, vegetables, fruits, meats and legumes, and milk products (review Figure 2-1, pp. 40–41).
- MyPyramid depicts the USDA Food Guide and encourages a variety of foods to provide an assortment of nutrients (review Figure 2-3, p. 45).

Food Labels

- Food labels list ingredients in descending order of predominance by weight, nutrition facts based on standard serving sizes, and Daily Values based on a 2000-kcalorie diet (review Figure 2-9, p. 54).
- Nutrient claims reflect the quantity of a nutrient *(high* or *low)*, health claims reflect relationships between a nutrient and a disease *(potassium reduces risk of hypertension)*, and structure-function claims reflect relationships between a nutrient and its function in the body *(calcium builds bones)*.

FIGURE 2-9 Example of a Food Label

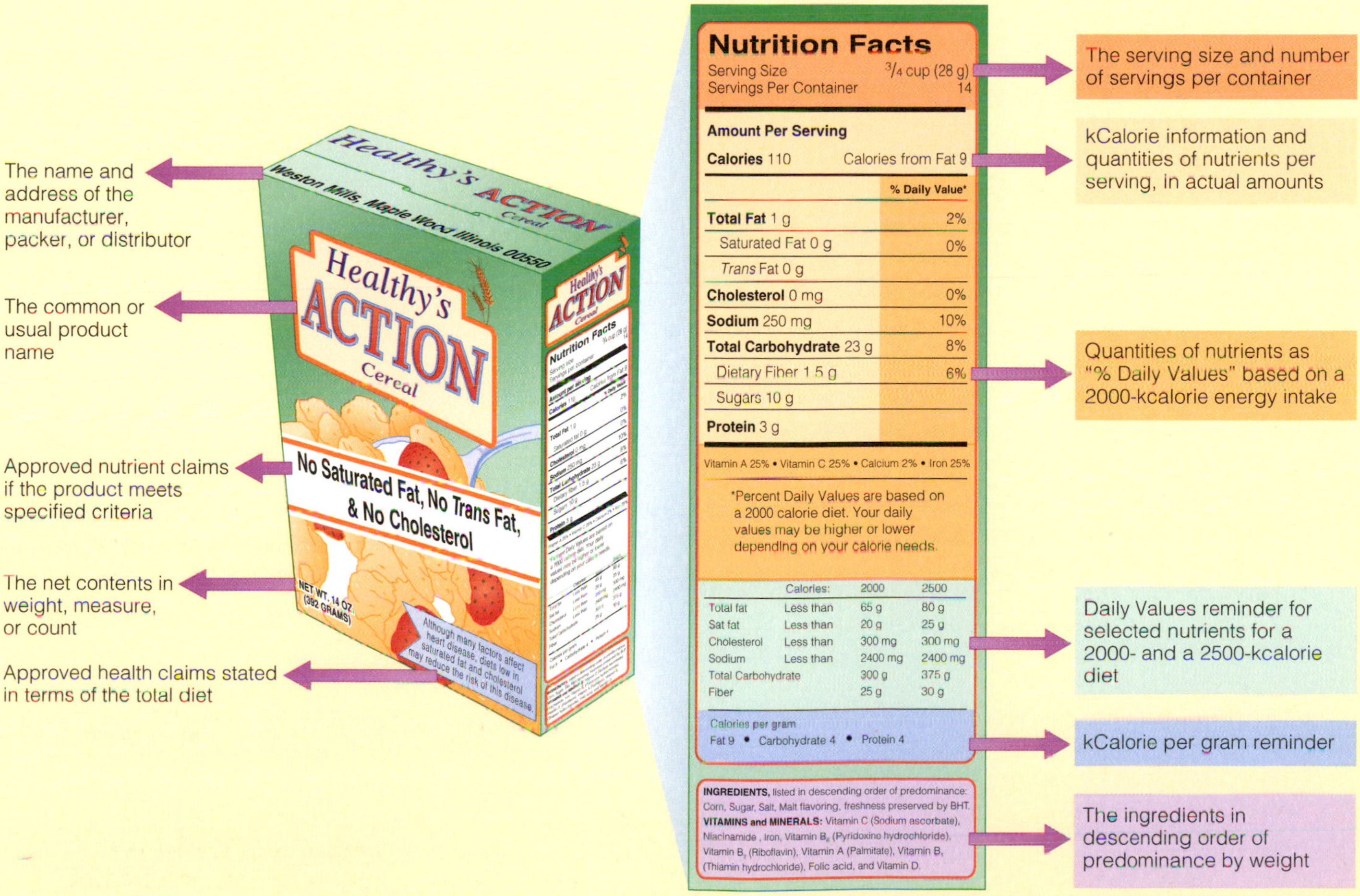

Principles and Guidelines (pp. 35–39)

1. Name the diet-planning principles and briefly describe how each principle helps in diet planning.
2. What recommendations appear in the *Dietary Guidelines for Americans*?
3. The diet-planning principle that provides all the essential nutrients in sufficient amounts to support health is:
   a. balance.
   b. variety.
   c. adequacy.
   d. moderation.
4. A person who chooses a chicken leg that provides 0.5 milligram of iron and 95 kcalories instead of two tablespoons of peanut butter that also provide 0.5 milligram of iron but 188 kcalories is using the principle of nutrient:
   a. control.
   b. density.
   c. adequacy.
   d. moderation.
5. Which of the following is consistent with the *Dietary Guidelines for Americans*?
   a. Choose a diet restricted in fat and cholesterol.
   b. Balance the food you eat with physical activity.
   c. Choose a diet with plenty of milk products and meats.
   d. Eat an abundance of foods to ensure nutrient adequacy.

Diet-Planning Guides (pp. 39–53)

6. Name the five food groups in the USDA Food Guide and identify several foods typical of each group. Explain how such plans group foods and what diet-planning principles the plans best accommodate. How are food group plans used, and what are some of their strengths and weaknesses?
7. Review the *Dietary Guidelines*. What types of grocery selections would you make to achieve those recommendations?
8. According to the USDA Food Guide, added fats and sugars are counted as:
   a. meats and grains.
   b. nutrient-dense foods.
   c. discretionary kcalories.
   d. oils and carbohydrates.
9. Foods within a given food group of the USDA Food Guide are similar in their contents of:
   a. energy.
   b. proteins and fibers.
   c. vitamins and minerals.
   d. carbohydrates and fats.
10. In the exchange system, each portion of food on any given list provides about the same amount of:
    a. energy.
    b. satiety.
    c. vitamins.
    d. minerals.
11. Enriched grain products are fortified with:
    a. fiber, folate, iron, niacin, and zinc.
    b. thiamin, iron, calcium, zinc, and sodium.
    c. iron, thiamin, riboflavin, niacin, and folate.
    d. folate, magnesium, vitamin B_6, zinc, and fiber.

Food Labels (pp. 53–59)

12. What information can you expect to find on a food label? How can this information help you choose between two similar products?
13. What are the Daily Values? How can they help you meet health recommendations?
14. Describe the differences between nutrient claims, health claims, and structure-function claims.
15. Food labels list ingredients in:
    a. alphabetical order.
    b. ascending order of predominance by weight.
    c. descending order of predominance by weight.
    d. manufacturer's order of preference.
16. "Milk builds strong bones" is an example of a:
    a. health claim.
    b. nutrition fact.
    c. nutrient content claim.
    d. structure-function claim.
17. Daily Values on food labels are based on a:
    a. 1500-kcalorie diet.
    b. 2000-kcalorie diet.
    c. 2500-kcalorie diet.
    d. 3000-kcalorie diet.

Study Questions (multiple choice)
3. c 4. b 5. b 8. c 9. c 10. a 11. c 15. c 16. d 17. b

For additional study questions and activities, go to **www.cengagebrain.com** and search for ISBN 084006845X.

STUDY CARD 3

Digestion, Absorption, and Transport

Digestion

- Digestion breaks down foods into nutrients. Absorption brings the nutrients into the cells of the small intestine for transport to the body's cells.
- Food enters the mouth and travels down the esophagus and through the upper and lower esophageal sphincters to the stomach, then through the pyloric sphincter to the small intestine, on through the ileocecal valve to the large intestine, past the appendix to the rectum, ending at the anus (review Figure 3-1, p. 71).
- The wavelike contractions of peristalsis and the periodic squeezing of segmentation keep things moving at a reasonable pace. Along the way, secretions from the salivary glands, stomach, pancreas, liver (via the gallbladder), and small intestine deliver fluids and digestive enzymes.

FIGURE 3-1 The Gastrointestinal Tract

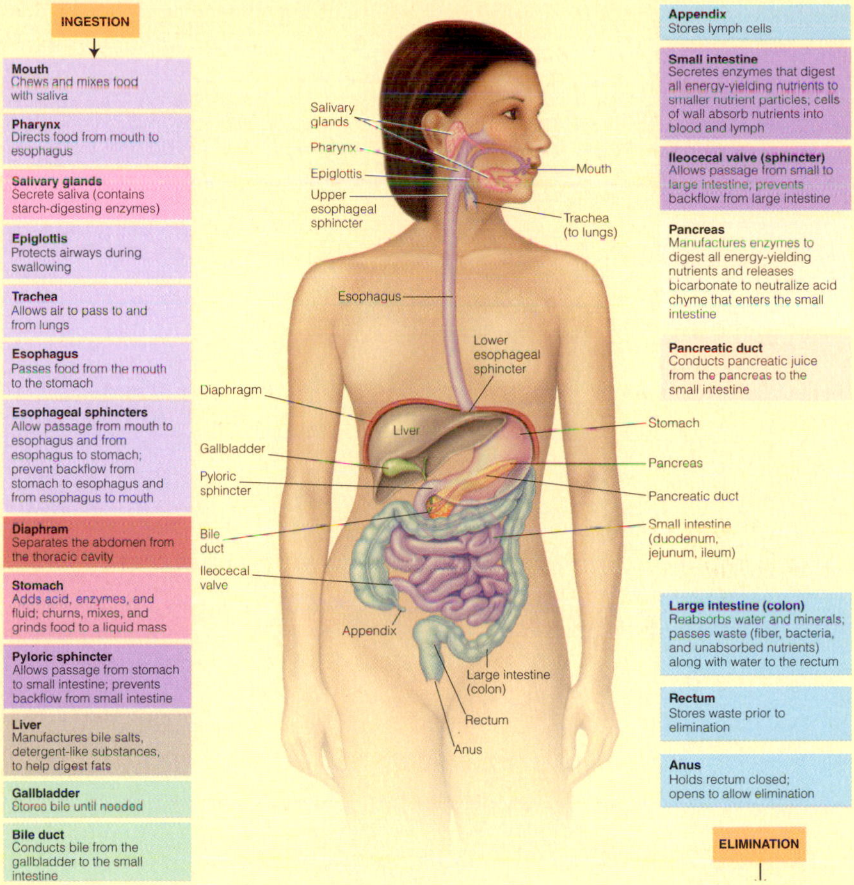

INGESTION
Mouth
Chews and mixes food with saliva
Pharynx
Directs food from mouth to esophagus
Salivary glands
Secrete saliva (contains starch-digesting enzymes)
Epiglottis
Protects airways during swallowing
Trachea
Allows air to pass to and from lungs
Esophagus
Passes food from the mouth to the stomach
Esophageal sphincters
Allow passage from mouth to esophagus and from esophagus to stomach; prevent backflow from stomach to esophagus and from esophagus to mouth
Diaphram
Separates the abdomen from the thoracic cavity
Stomach
Adds acid, enzymes, and fluid; churns, mixes, and grinds food to a liquid mass
Pyloric sphincter
Allows passage from stomach to small intestine; prevents backflow from small intestine
Liver
Manufactures bile salts, detergent-like substances, to help digest fats
Gallbladder
Stores bile until needed
Bile duct
Conducts bile from the gallbladder to the small intestine
Salivary glands
Pharynx
Epiglottis
Upper esophageal sphincter
Mouth
Trachea (to lungs)
Esophagus
Lower esophageal sphincter
Diaphragm
Liver
Gallbladder
Pyloric sphincter
Bile duct
Ileocecal valve
Appendix
Stomach
Pancreas
Pancreatic duct
Small intestine (duodenum, jejunum, ileum)
Large intestine (colon)
Rectum
Anus
Appendix
Stores lymph cells
Small intestine
Secretes enzymes that digest all energy-yielding nutrients to smaller nutrient particles; cells of wall absorb nutrients into blood and lymph
Ileocecal valve (sphincter)
Allows passage from small to large intestine; prevents backflow from large intestine
Pancreas
Manufactures enzymes to digest all energy-yielding nutrients and releases bicarbonate to neutralize acid chyme that enters the small intestine
Pancreatic duct
Conducts pancreatic juice from the pancreas to the small intestine
Large intestine (colon)
Reabsorbs water and minerals; passes waste (fiber, bacteria, and unabsorbed nutrients) along with water to the rectum
Rectum
Stores waste prior to elimination
Anus
Holds rectum closed; opens to allow elimination
ELIMINATION

Absorption

- The many folds and villi of the small intestine increase its surface area, making nutrient absorption efficient.
- Nutrients pass through the cells of the intestinal villi and enter either the blood (if they are water soluble or small fat fragments) or the lymph (if they are fat soluble).

The Circulatory Systems

- Nutrients leaving the digestive system via the blood are routed directly to the liver before being transported to the body's cells.
- Nutrients leaving via the lymphatic system bypass the liver at first, but eventually enter the vascular system via the thoracic duct, which opens into the subclavian vein.

The Health and Regulation of the GI Tract

- A diverse and abundant bacteria population supports GI health.
- The regulation of GI processes depends on the coordinated efforts of the hormonal system and the nervous system.

Summary of Digestive Secretions and Their Major Actions

Organ or Gland	Target Organ	Secretion	Action
Salivary glands	Mouth	Saliva	Fluid eases swallowing; salivary enzyme breaks down some ***carbohydrate.****
Gastric glands	Stomach	Gastric juice	Fluid mixes with bolus; hydrochloric acid uncoils ***proteins;*** enzymes break down proteins; mucus protects stomach cells.*
Pancreas	Small intestine	Pancreatic juice	Bicarbonate neutralizes acidic gastric juices; pancreatic enzymes break down ***carbohydrates, fats,*** and ***proteins.***
Liver	Gallbladder	Bile	Bile stored until needed.
Gallbladder	Small intestine	Bile	Bile emulsifies ***fat*** so that enzymes can have access to break it down.
Intestinal glands	Small intestine	Intestinal juice	Intestinal enzymes break down ***carbohydrate, fat,*** and ***protein*** fragments; mucus protects the intestinal wall.

*Saliva and gastric juice also contain lipases, but most fat breakdown occurs in the small intestine.

Digestion (pp. 69–77)

1. Describe the challenges associated with digesting food and the solutions offered by the human body.
2. Describe the path food follows as it travels through the digestive system. Summarize the muscular actions that take place along the way.
3. Name five organs that secrete digestive juices. How do the juices and enzymes facilitate digestion?
4. The semiliquid, partially digested food that travels through the intestinal tract is called:
   a. bile.
   b. lymph.
   c. chyme.
   d. secretin.
5. The muscular contractions that move food through the GI tract are called:
   a. hydrolysis.
   b. sphincters.
   c. peristalsis.
   d. bowel movements.
6. The main function of bile is to:
   a. emulsify fats.
   b. catalyze hydrolysis.
   c. slow protein digestion.
   d. neutralize stomach acidity.
7. The pancreas neutralizes stomach acid in the small intestine by secreting:
   a. bile.
   b. mucus.
   c. enzymes.
   d. bicarbonate.
8. Which nutrient passes through the GI tract mostly undigested and unabsorbed?
   a. fat
   b. fiber
   c. protein
   d. carbohydrate

Absorption (pp. 78–80)

9. Describe the problems associated with absorbing nutrients and the solutions offered by the small intestine.
10. Absorption occurs primarily in the:
    a. mouth.
    b. stomach.
    c. small intestine.
    d. large intestine.

The Circulatory Systems (pp. 80–83)

11. How is blood routed through the digestive system? Which nutrients enter the bloodstream directly? Which are first absorbed into the lymph?
12. All blood leaving the GI tract travels first to the:
    a. heart.
    b. liver.
    c. kidneys.
    d. pancreas.
13. Which nutrients leave the GI tract by way of the lymphatic system?
    a. water and minerals
    b. proteins and minerals
    c. all vitamins and minerals
    d. fats and fat-soluble vitamins

The Health and Regulation of the GI Tract (pp. 83–87)

14. Describe how the body coordinates and regulates the processes of digestion and absorption.
15. How does the composition of the diet influence the functioning of the GI tract?
16. What steps can you take to help your GI tract function at its best?
17. Digestion and absorption are coordinated by the:
    a. pancreas and kidneys.
    b. liver and gallbladder.
    c. hormonal system and the nervous system.
    d. vascular system and the lymphatic system.
18. Gastrin, secretin, and cholecystokinin are examples of:
    a. crypts.
    b. enzymes.
    c. hormones.
    d. goblet cells.

Study Questions (multiple choice)

4. c 5. c 6. a 7. d 8. b 10. c 12. b 13. d 17. c 18. c

For additional study questions and activities, go to **www.cengagebrain.com** and search for ISBN 084006845X.

STUDY CARD 4

The Carbohydrates: Sugars, Starches, and Fibers

The Chemist's View of Carbohydrates

- Carbohydrates are made of carbon (C), oxygen (O), and hydrogen (H); each atom forms a specified number of chemical bonds: carbon forms four, oxygen forms two, and hydrogen forms one (review Figure 4-1).
- Monosaccharides (glucose, fructose, and galactose) all have the same chemical formula ($C_6H_{12}O_6$), but their structures differ. Disaccharides (maltose, sucrose, and lactose) each contain a glucose paired with one of the three monosaccharides.
- A condensation reaction can bond two monosaccharides together to form a disaccharide and water (review Figure 4-6, p. 100). A hydrolysis reaction can use water to split a disaccharide into its two monosaccharides (review Figure 4-7, p. 100).
- Chains of monosaccharides are called polysaccharides and include glycogen, starches, and dietary fibers. Both glycogen and starch are storage forms of glucose—glycogen in the body, and starch in plants—and both yield energy.
- Dietary fibers contain glucose (and other monosaccharides), but their bonds cannot be broken by human digestive enzymes; they yield little, if any, energy.

The Carbohydrate Family

- *Monosaccharides* (single sugars)
  - Glucose
  - Fructose
  - Galactose
- *Disaccharides* (pairs of monosaccharides)
  - Maltose (glucose + glucose)
  - Sucrose (glucose + fructose)
  - Lactose (glucose + galactose)
- *Polysaccharides* (chains of monosaccharides)
  - Glycogen (a polysaccharide, but not a dietary carbohydrate)
  - Starches (amylose and amylopectin)
  - Fibers (soluble and insoluble)

Digestion and Absorption of Carbohydrates

- The body digests starches into the disaccharide maltose. Maltose and the other disaccharides (lactose and sucrose) from foods are broken down into monosaccharides, which are absorbed.
- Monosaccharides arriving at the liver are converted mostly to glucose (review Figure 4-11, p. 105).
- Fibers help to regulate the passage of food through the GI system and slow the absorption of glucose.
- Lactose intolerance occurs when there is insufficient lactase to digest the disaccharide lactose found in milk and milk products. Symptoms include GI distress.

Glucose in the Body

- Dietary carbohydrates provide glucose that can be used by the cells for energy, stored by the liver and muscles as glycogen, or converted into fat if intakes exceed needs.
- All of the body's cells depend on glucose; those of the central nervous system are especially dependent on it.
- Without glucose, the body is forced to break down its protein tissues to make glucose and to alter energy metabolism to make ketone bodies from fats.
- Blood glucose regulation depends on two pancreatic hormones: insulin to move glucose from the blood into the cells when levels are high and glucagon to free glucose from glycogen stores and release it into the blood when levels are low (review Figure 4-12, p. 109).

Health Effects and Recommended Intakes of Sugars

- Excessive intakes of sugars may increase the risk of dental caries, displace needed nutrients and fiber, and contribute to obesity when energy intake exceeds needs.
- Concentrated sweets are relatively low in nutrients, high in kcalories, and may need to be limited; sugars that occur naturally in fruits, vegetables, and milk are acceptable.

Alternative Sweeteners

- To control weight gain, blood glucose, and dental caries, consumers may use alternative sweeteners (artificial sweeteners, herbal products, and sugar alcohols) to limit kcalories and minimize sugar intake (review Table 4-2, pp. 116–117).

Health Effects and Recommended Intakes of Starch and Fibers

- Adequate intake of fiber fosters weight management, lowers blood cholesterol, and may help prevent colon cancer, diabetes, hemorrhoids, appendicitis, and diverticulosis.
- Excessive intake of fiber displaces energy- and nutrient-dense foods, causes intestinal discomfort and distention, and may interfere with mineral absorption.
- Because starches and fibers help control body weight and prevent heart disease, cancer, diabetes, and GI disorders, guidelines suggest plenty of whole grains, vegetables, legumes, and fruits—enough to provide 45 to 65 percent of the daily energy intake from carbohydrate.

The Chemist's View of Carbohydrates
(pp. 97–103)

1. Describe the structure of a monosaccharide and name the three monosaccharides important in nutrition. Name the three disaccharides commonly found in foods and their component monosaccharides. In what foods are these sugars found?
2. What happens in a condensation reaction? In a hydrolysis reaction?
3. Describe the structure of polysaccharides and name the ones important in nutrition. How are starch and glycogen similar, and how do they differ? How do the fibers differ from the other polysaccharides?
4. Disaccharides include:
   a. starch, glycogen, and fiber.
   b. amylose, pectin, and dextrose.
   c. sucrose, maltose, and lactose.
   d. glucose, galactose, and fructose.
5. The making of a disaccharide from two monosaccharides is an example of:
   a. digestion.
   b. hydrolysis.
   c. condensation.
   d. gluconeogenesis.
6. The significant difference between starch and cellulose is that:
   a. starch is a polysaccharide, but cellulose is not.
   b. animals can store glucose as starch, but not as cellulose.
   c. hormones can make glucose from cellulose, but not from starch.
   d. digestive enzymes can break the bonds in starch, but not in cellulose.

Digestion and Absorption of Carbohydrates (pp. 103–107)

7. Describe carbohydrate digestion and absorption. What role does fiber play in the process?
8. The ultimate goal of carbohydrate digestion and absorption is to yield:
   a. fibers.
   b. glucose.
   c. enzymes.
   d. amylase.
9. The enzyme that breaks a disaccharide into glucose and galactose is:
   a. amylase.
   b. maltase.
   c. sucrase.
   d. lactase.

Glucose in the Body (pp. 107–111)

10. What are the possible fates of glucose in the body? What is the protein-sparing action of carbohydrate?
11. How does the body maintain its blood glucose concentration? What happens when the blood glucose concentration rises too high or falls too low?
12. The storage form of glucose in the body is:
    a. insulin.
    b. maltose.
    c. glucagon.
    d. glycogen.
13. With insufficient glucose in metabolism, fat fragments combine to form:
    a. dextrins.
    b. mucilages.
    c. phytic acids.
    d. ketone bodies.
14. What does the pancreas secrete when blood glucose rises? When blood glucose falls?
    a. insulin; glucagon
    b. glucagon; insulin
    c. insulin; glycogen
    d. glycogen; epinephrine

Health Effects and Recommended Intakes of Sugars (pp. 112–115)

15. What are the health effects of sugars? What are the dietary recommendations regarding concentrated sugar intakes?

Alternative Sweeteners (pp. 115–118)

16. Describe the risks and benefits of using alternative sweeteners.

Health Effects and Recommended Intakes of Starch and Fibers (pp. 118–123)

17. What are the health effects of starches and fibers? What are the dietary recommendations regarding these complex carbohydrates?
18. What foods provide starches and fibers?
19. Carbohydrates are found in virtually all foods except:
    a. milks.
    b. meats.
    c. breads.
    d. fruits.
20. What percentage of the daily energy intake should come from carbohydrates?
    a. 15 to 20
    b. 25 to 30
    c. 45 to 50
    d. 45 to 65

Study Questions (multiple choice)
4. c 5. c 6. d 8. b 9. d 12. d 13. d 14. a 19. b 20. d

For additional study questions and activities, go to **www.cengagebrain.com** and search for ISBN 084006845X.

STUDY CARD 5

The Lipids: Triglycerides, Phospholipids, and Sterols

The Chemist's View of Fatty Acids and Triglycerides

- The predominant lipids both in foods and in the body are triglycerides: glycerol with three fatty acids attached by way of condensation reactions (review Figure 5-4, p. 137). Other lipids include phospholipids and sterols.
- Fatty acids vary in the length of their carbon chains, their degrees of unsaturation, and the location of their double bond(s). Saturated fatty acids are fully loaded with hydrogens; unsaturated (monounsaturated or polyunsaturated) fatty acids are missing hydrogens and have double bonds.
- Hydrogenation makes polyunsaturated fats more saturated and creates *trans*-fatty acids.

The Chemist's View of Phospholipids and Sterols

- The chemical structure of phospholipids, including lecithin, allows them to be soluble in both water and fat. In the body, phospholipids are part of cell membranes; in foods, phospholipids act as emulsifiers to mix fats with water.
- Sterols have a multiple-ring structure that differs from the structure of other lipids. In the body, sterols include cholesterol, bile, vitamin D, and some hormones. Animal-derived foods contain cholesterol.

The Lipid Family

- *Triglycerides* (fats and oils), which are made of:
  - *Glycerol* (1 per triglyceride) and
  - *Fatty acids* (3 per triglyceride); depending on the number of double bonds, fatty acids may be:
    - *Saturated* (no double bonds)
    - *Monounsaturated* (one double bond)
    - *Polyunsaturated* (more than one double bond); depending on the location of the double bonds, polyunsaturated fatty acids may be:
      - *Omega-3* (first double bond 3 carbons away from methyl end)
      - *Omega-6* (first double bond 6 carbons away from methyl end)
- *Phospholipids* (such as lecithin)
- *Sterols* (such as cholesterol)

Digestion, Absorption, and Transport of Lipids

- Bile emulsifies fats, making them accessible to the lipases that dismantle triglycerides to monoglycerides and fatty acids for absorption (review Figure 5-14, p. 143).
- Four types of lipoproteins carry triglycerides, phospholipids, and cholesterol throughout the body: *chylomicrons* are the largest and contain mostly dietary triglycerides, *VLDL* are smaller and are about half triglycerides, *LDL* are smaller still and contain mostly cholesterol, and *HDL* are the densest and are rich in protein (review Figure 5-18, p. 146).

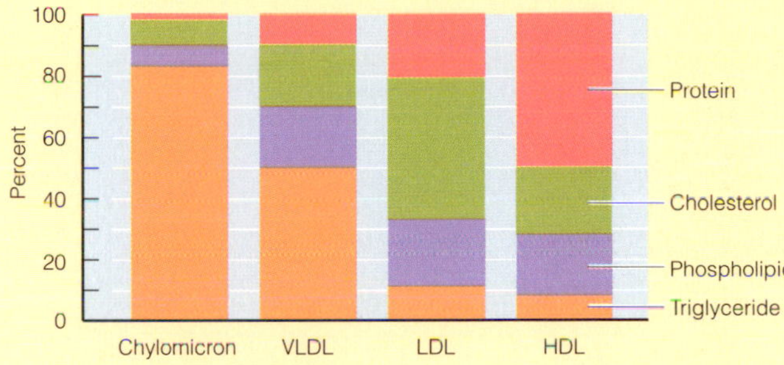

Lipids in the Body

- In the body, triglycerides provide energy, insulate against temperature extremes, protect against shock, and help the body use carbohydrate and protein efficiently.
- Linoleic acid (18 carbons, omega-6) and linolenic acid (18 carbons, omega-3) are essential fatty acids, serving as structural parts of cell membranes and as precursors to the longer fatty acids that can make eicosanoids.
- The body stores fat if given excesses, and uses body fat for energy when needed. (The liver can also convert excess carbohydrate and protein into fat.) Fat breakdown requires carbohydrate for maximum efficiency; without carbohydrate, fatty acids break down to ketone bodies.

Health Effects and Recommended Intakes of Lipids

- High blood LDL cholesterol increases the risk of heart disease, and high intakes of saturated and *trans* fats contribute most to high LDL. Omega-3 fatty acids appear to be protective.
- In foods, triglycerides deliver fat-soluble vitamins, energy, and essential fatty acids; contribute to the sensory appeal of foods; and stimulate appetite.
- Some fat in the diet is necessary; ideally, a diet is moderate in total fat and low in saturated fat, *trans* fat, and cholesterol. Recommendations include replacing saturated fats with monounsaturated and polyunsaturated fats, particularly omega-3 fatty acids from foods such as fatty fish, not from supplements.

The Chemist's View of Fatty Acids and Triglycerides (pp. 133–139)

1. Name three classes of lipids found in the body and in foods. What are some of their functions in the body? What features do fats bring to foods?
2. What features distinguish fatty acids from each other?
3. What does the term *omega* mean with respect to fatty acids? Describe the roles of the omega fatty acids in disease prevention.
4. What are the differences between saturated, unsaturated, monounsaturated, and polyunsaturated fatty acids? Describe the structure of a triglyceride.
5. What does hydrogenation do to fats? What are *trans*-fatty acids, and how do they influence heart disease?
6. Saturated fatty acids:
   a. are always 18 carbons long.
   b. have at least one double bond.
   c. are fully loaded with hydrogens.
   d. are always liquid at room temperature.
7. A triglyceride consists of:
   a. three glycerols attached to a lipid.
   b. three fatty acids attached to a glucose.
   c. three fatty acids attached to a glycerol.
   d. three phospholipids attached to a cholesterol.
8. The difference between *cis*- and *trans*-fatty acids is:
   a. the number of double bonds.
   b. the length of their carbon chains.
   c. the location of the first double bond.
   d. the configuration around the double bond.

The Chemist's View of Phospholipids and Sterols (pp. 139–141)

9. How do phospholipids differ from triglycerides in structure? How does cholesterol differ? How do these differences in structure affect function?
10. What roles do phospholipids perform in the body? What roles does cholesterol play in the body?
11. Which of the following is *not* true? Lecithin is:
    a. an emulsifier.
    b. a phospholipid.
    c. an essential nutrient.
    d. a constituent of cell membranes.

Digestion, Absorption, and Transport of Lipids (pp. 142–148)

12. Trace the steps in fat digestion, absorption, and transport. Describe the routes cholesterol takes in the body.
13. What do lipoproteins do? What are the differences among the chylomicrons, VLDL, LDL, and HDL?
14. Chylomicrons are produced in the:
    a. liver.
    b. pancreas.
    c. gallbladder.
    d. small intestine.
15. Transport vehicles for lipids are called:
    a. micelles.
    b. lipoproteins.
    c. blood vessels.
    d. monoglycerides.

Lipids in the Body (pp. 148–151)

16. Which of the fatty acids are essential? Name their chief dietary sources.
17. Which of the following is *not* true? Fats:
    a. contain glucose.
    b. provide energy.
    c. protect against organ shock.
    d. carry vitamins A, D, E, and K.
18. The essential fatty acids include:
    a. stearic acid and oleic acid.
    b. oleic acid and linoleic acid.
    c. palmitic acid and linolenic acid.
    d. linoleic acid and linolenic acid.

Health Effects and Recommended Intakes of Lipids (pp. 151–161)

19. How does excessive fat intake influence health? What factors influence LDL, HDL, and total blood cholesterol?
20. What are the dietary recommendations regarding fat and cholesterol intake? List ways to reduce intake.
21. What is the Daily Value for fat (for a 2000-kcalorie diet)? What does this number represent?
22. The lipoprotein most associated with a high risk of heart disease is:
    a. CHD.
    b. HDL.
    c. LDL.
    d. LPL.
23. A person consuming 2200 kcalories a day who wants to meet health recommendations should limit daily fat intake to:
    a. 20 to 35 grams.
    b. 50 to 85 grams.
    c. 75 to 100 grams.
    d. 90 to 130 grams.

Study Questions (multiple choice)
6.c 7.c 8.d 11.c 14.d 15.b 17.a 18.d 22.c 23.b

For additional study questions and activities, go to **www.cengagebrain.com** and search for ISBN 084006845X.

STUDY CARD 6

Protein: Amino Acids

The Chemist's View of Proteins

- Proteins are more chemically complex than carbohydrates or lipids; they are made of 20 different amino acids, 9 of which the body cannot make. These 9 are the essential amino acids—histidine, isoleucine, leucine, lysine, methionine, phenylalanine, threonine, tryptophan, and valine.
- Each amino acid contains an amino group, an acid group, a hydrogen atom, and a distinctive side group, all attached to a central carbon atom.
- Cells link amino acids together in a series of condensation reactions to create proteins (review Figure 6-3, p. 175). The distinctive sequence of amino acids in each protein determines its unique shape and function.

Digestion and Absorption of Proteins

- The stomach's acid first denatures dietary proteins, then enzymes cleave them into smaller polypeptides and some amino acids. Pancreatic and intestinal enzymes split polypeptides further, to oligo-, tri-, and dipeptides, and then split most of these to single amino acids that can be absorbed.

Proteins in the Body

- Cells synthesize proteins according to the genetic information provided by the DNA in the nucleus of each cell (review Figure 6-7, p. 179). This information dictates the sequence in which amino acids are linked together to form a given protein. Sequencing errors occasionally occur, sometimes with significant consequences.
- A sampling of protein functions are summarized in the accompanying table.
- Proteins are constantly being synthesized and broken down as needed.
- The body's assimilation of amino acids into proteins and its release of amino acids via protein degradation and excretion can be tracked by measuring nitrogen balance, which should be positive during growth and steady in adulthood. An energy deficit or an inadequate protein intake may force the body to use amino acids as fuel, creating a negative nitrogen balance.
- Protein eaten in excess of need is degraded and stored as body fat.

Protein in Foods

- High-quality proteins deliver all of the essential amino acids in adequate amounts, which ensures protein synthesis. Mixtures of foods containing complementary proteins can each supply the amino acids missing in the other.
- In addition to its amino acid content, the quality of protein is measured by its digestibility and its ability to support growth.

Health Effects and Recommended Intakes of Protein

- Protein deficiencies arise from both energy-poor and protein-poor diets and lead to marasmus and kwashiorkor. Together, these diseases are known as PEM (protein-energy malnutrition), a major form of malnutrition causing death in children worldwide.
- Excess protein offers no advantage and may incur health problems as well.
- The optimal diet is adequate in energy from carbohydrate and fat and delivers 0.8 grams of protein per kilogram of healthy body weight each day.
- Normal, healthy people do not need protein or amino acid supplements.

Growth and maintenance	Proteins form integral parts of most body structures such as skin, tendons, membranes, muscles, organs, and bones. As such, they support the growth and repair of body tissues.
Enzymes	Proteins facilitate chemical reactions.
Hormones	Proteins regulate body processes. (Some, but not all, hormones are proteins.)
Fluid balance	Proteins help to maintain the volume and composition of body fluids.
Acid-base balance	Proteins help to maintain the acid-base balance of body fluids by acting as buffers.
Transportation	Proteins transport substances, such as lipids, vitamins, minerals, and oxygen, around the body.
Antibodies	Proteins inactivate foreign invaders, thus protecting the body against diseases.
Energy and glucose	Proteins provide some fuel, and glucose if needed, for the body's energy needs.

The Chemist's View of Proteins (pp. 173–176)

1. How does the chemical structure of proteins differ from the structures of carbohydrates and fats?
2. Describe the structure of amino acids, and explain how their sequence in proteins affects the proteins' shapes. What are essential amino acids?
3. Which part of its chemical structure differentiates one amino acid from another?
   a. its side group
   b. its acid group
   c. its amino group
   d. its double bonds
4. Isoleucine, leucine, and lysine are:
   a. proteases.
   b. polypeptides.
   c. essential amino acids.
   d. complementary proteins.

Digestion and Absorption of Proteins (pp. 176–178)

5. Describe protein digestion and absorption.
6. In the stomach, hydrochloric acid:
   a. denatures proteins and activates pepsin.
   b. hydrolyzes proteins and denatures pepsin.
   c. emulsifies proteins and releases peptidase.
   d. condenses proteins and facilitates digestion.

Proteins in the Body (pp. 178–187)

7. Describe protein synthesis.
8. Describe some of the roles proteins play in the human body.
9. What are enzymes? What roles do they play in chemical reactions? Describe the differences between enzymes and hormones.
10. How does the body use amino acids? What is deamination? Define nitrogen balance. What conditions are associated with zero, positive, and negative balance?
11. Proteins that facilitate chemical reactions are:
    a. buffers.
    b. enzymes.
    c. hormones.
    d. antigens.
12. If an essential amino acid that is needed to make a protein is unavailable, the cells must:
    a. deaminate another amino acid.
    b. substitute a similar amino acid.
    c. break down proteins to obtain it.
    d. synthesize the amino acid from glucose and nitrogen.
13. Protein turnover describes the amount of protein:
    a. found in foods and the body.
    b. absorbed from the diet.
    c. synthesized and degraded.
    d. used to make glucose.

Protein in Foods (pp. 187–188)

14. What factors affect the quality of dietary protein? What is a high-quality protein?
15. How can vegetarians meet their protein needs without eating meat?
16. Which of the following foods provides the highest quality protein?
    a. egg
    b. corn
    c. gelatin
    d. whole grains

Health Effects and Recommended Intakes of Protein (pp. 188–196)

17. What are the health consequences of ingesting inadequate protein and energy? Describe marasmus and kwashiorkor. How can the two conditions be distinguished, and in what ways do they overlap?
18. How might protein excess, or the type of protein eaten, influence health?
19. What factors are considered in establishing recommended protein intakes?
20. What are the benefits and risks of taking protein and amino acid supplements?
21. Marasmus develops from:
    a. too much fat clogging the liver.
    b. megadoses of amino acid supplements.
    c. inadequate protein and energy intake.
    d. excessive fluid intake causing edema.
22. The protein RDA for a healthy adult who weighs 180 pounds is:
    a. 50 milligrams/day.
    b. 65 grams/day.
    c. 180 grams/day.
    d. 2000 milligrams/day.
23. Which of these foods has the least protein per ½ cup?
    a. rice
    b. broccoli
    c. pinto beans
    d. orange juice

Study Questions (multiple choice)
3. a 4. c 6. a 11. b 12. c 13. c 16. a 21. c 22. b 23. d

For additional study questions and activities, go to **www.cengagebrain.com** and search for ISBN 084006845X.

STUDY CARD 7

Metabolism: Transformations and Interactions

Chemical Reactions in the Body

- During digestion, the energy-yielding nutrients—carbohydrates, lipids, and proteins—are broken down to glucose (and other monosaccharides), glycerol, fatty acids, and amino acids.
- Enzymes with their coenzymes help cells use nutrients to build compounds (anabolism) or break them down to release energy (catabolism)—review Figure 7-2, p. 207.
- ATP—a high-energy compound—captures the energy released during catabolism (review Figure 7-4, p. 208).

Breaking Down Nutrients for Energy

- Glucose breakdown begins with glycolysis, a pathway that produces pyruvate (review Figure 7-6, p. 212).
- Pyruvate may be converted to lactate anaerobically (without oxygen) or to acetyl CoA aerobically (with oxygen).
- Pyruvate can make glucose; acetyl CoA cannot make glucose (review Figure 7-9, p. 214).
- The glycerol part of a triglyceride can make either pyruvate (and then glucose) or acetyl CoA. The fatty acids of a triglyceride *cannot* make glucose; they can provide abundant acetyl CoA (review Figure 7-11, p. 216).
- Some amino acids can be used to make glucose; others can be used either to provide energy or to make fat. Before an amino acid enters these metabolic pathways, its nitrogen-containing amino group must be removed through deamination.
- The digestion of carbohydrate yields glucose (and other monosaccharides); some glucose is stored as glycogen, and some is broken down to pyruvate and acetyl CoA.
- The digestion of fat yields glycerol and fatty acids; some are reassembled and stored as body fat, and others are broken down to acetyl CoA.
- The digestion of protein yields amino acids; most amino acids are used to build body protein or other nitrogen-containing compounds, some are broken down to acetyl CoA, and others enter the TCA cycle directly.
- Acetyl CoA may enter the TCA cycle to release energy (review Figure 7-16, p. 220) or combine with other molecules of acetyl CoA to make body fat.

Energy Balance

- If energy intake exceeds energy needs, the result will be weight gain—regardless of whether the excess is from protein, carbohydrate, or fat. The body is most efficient at storing excess energy from dietary fat.
- When fasting, the body adapts to conserve energy and minimize losses by increasing fat breakdown to fuel most cells, using glycerol and amino acids to make glucose for the brain and red blood cells, producing ketones for the brain, suppressing appetite, and slowing metabolism.

FIGURE 7-5 **Simplified Overview of Energy-Yielding Pathways**

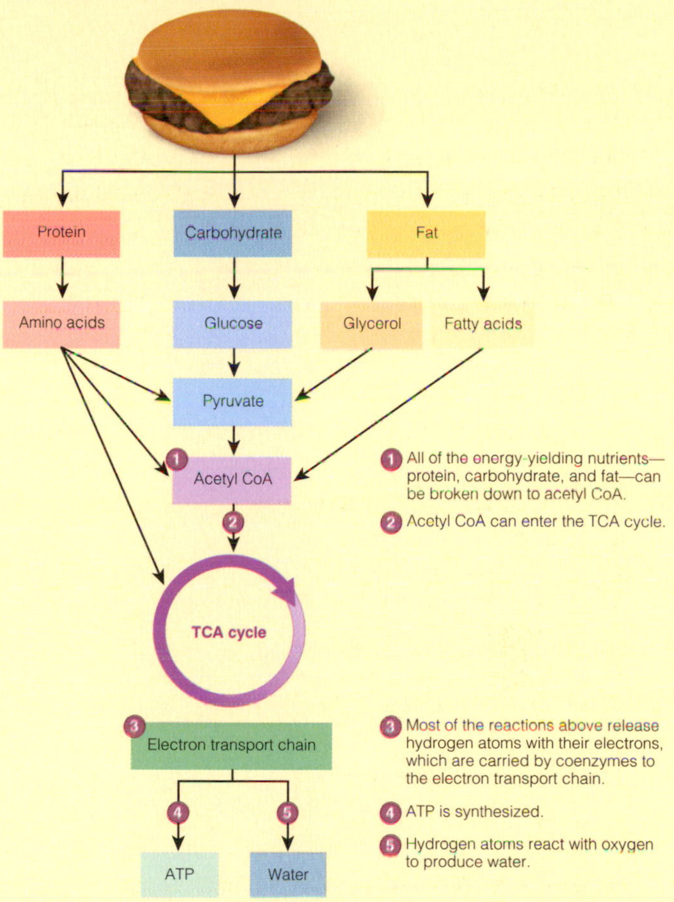

Protein
Carbohydrate
Fat
Amino acids
Glucose
Glycerol
Fatty acids
Pyruvate
Acetyl CoA
TCA cycle
Electron transport chain
ATP
Water
1 All of the energy-yielding nutrients—protein, carbohydrate, and fat—can be broken down to acetyl CoA.
2 Acetyl CoA can enter the TCA cycle.
3 Most of the reactions above release hydrogen atoms with their electrons, which are carried by coenzymes to the electron transport chain.
4 ATP is synthesized.
5 Hydrogen atoms react with oxygen to produce water.

In Summary

Nutrient	Yields Energy?	Yields Glucose?	Yields Amino Acids and Body Proteins?	Yields Fat Stores?
Carbohydrates (glucose)	Yes	Yes	Yes—when nitrogen is available, can yield *nonessential* amino acids	Yes
Lipids (fatty acids)	Yes	No	No	Yes
Lipids (glycerol)	Yes	Yes—when carbohydrate is unavailable	Yes—when nitrogen is available, can yield *nonessential* amino acids	Yes
Proteins (amino acids)	Yes	Yes—when carbohydrate is unavailable	Yes	Yes

Chemical Reactions in the Body (pp. 206–209)

1. Define metabolism, anabolism, and catabolism; give an example of each.
2. Name one of the body's high-energy molecules, and describe how it is used.
3. What are coenzymes, and what service do they provide in metabolism?
4. Name the four basic units, derived from foods, that are used by the body in metabolic transformations. How many carbons are in the "backbones" of each?
5. Hydrolysis is an example of a(n):
   a. coupled reaction.
   b. anabolic reaction.
   c. catabolic reaction.
   d. synthesis reaction.
6. During metabolism, released energy is captured and transferred by:
   a. enzymes.
   b. pyruvate.
   c. acetyl CoA.
   d. adenosine triphosphate.

Breaking Down Nutrients for Energy (pp. 209–222)

7. Define aerobic and anaerobic metabolism. How does insufficient oxygen influence metabolism?
8. How does the body dispose of excess nitrogen?
9. Summarize the main steps in the metabolism of glucose, glycerol, fatty acids, and amino acids.
10. Glycolysis:
    a. requires oxygen.
    b. generates abundant energy.
    c. converts glucose to pyruvate.
    d. produces ammonia as a by-product.
11. The pathway from pyruvate to acetyl CoA:
    a. produces lactate.
    b. is known as gluconeogenesis.
    c. is metabolically irreversible.
    d. requires more energy than it produces.
12. For complete oxidation, acetyl CoA enters:
    a. glycolysis.
    b. the TCA cycle.
    c. the Cori cycle.
    d. the electron transport chain.
13. Deamination of an amino acid produces:
    a. vitamin B_6 and energy.
    b. pyruvate and acetyl CoA.
    c. ammonia and a keto acid.
    d. carbon dioxide and water.
14. Before entering the TCA cycle, each of the energy-yielding nutrients is broken down to:
    a. ammonia.
    b. pyruvate.
    c. electrons.
    d. acetyl CoA.

Energy Balance (pp. 222–228)

15. Describe how a surplus of the three energy nutrients contributes to body fat stores.
16. What adaptations does the body make during a fast? What are ketone bodies? Define ketosis.
17. Distinguish between a loss of *fat* and a loss of *weight*, and describe how each might happen.
18. The body stores energy for future use in:
    a. proteins.
    b. acetyl CoA.
    c. triglycerides.
    d. ketone bodies.
19. During a fast, when glycogen stores have been depleted, the body begins to synthesize glucose from:
    a. acetyl CoA.
    b. amino acids.
    c. fatty acids.
    d. ketone bodies.
20. During a fast, the body produces ketone bodies by:
    a. hydrolyzing glycogen.
    b. condensing acetyl CoA.
    c. transaminating keto acids.
    d. converting ammonia to urea.

Study Questions (multiple choice)
5. c 6. d 10. c 11. c 12. b 13. c 14. d 18. c 19. b 20. b

For additional study questions and activities, go to **www.cengagebrain.com** and search for ISBN 084006845X.

STUDY CARD 8

Energy Balance and Body Composition

Energy Balance

- When energy consumed equals energy expended, a person is in energy balance and body weight is stable.
- If more energy is taken in than is expended, a person gains weight. If more energy is expended than is taken in, a person loses weight.

Energy In: The kCalories Foods Provide

- Hunger and appetite initiate eating, whereas satiation and satiety stop and delay eating, respectively (review Figure 8-2, p. 244).

Energy Out: The kCalories the Body Expends

- A person in energy balance takes in energy from food and expends much of it on basal metabolic activities, some of it on physical activities, and a little on the thermic effect of food (review Figure 8-4, p. 247).
- Energy requirements vary from person to person based on such factors as gender, age, weight, and height as well as the intensity and duration of physical activity.

Body Weight, Body Composition, and Health

- Body weight standards are based on a person's weight in relation to height, called the body mass index (BMI), and reflect disease risks. BMI does not identify body fat or its distribution, and it may misclassify muscular people as overweight.
- $\text{BMI} = \dfrac{\text{weight (kg)}}{\text{height (m)}^2}$ or $\dfrac{\text{weight (lb)}}{\text{height (in)}^2} \times 703.$

  Healthy: BMI 18.5 to 24.9
  Underweight: BMI <18.5
  Overweight: BMI 25 to 29.9
  Obese: BMI ≥30
- Two-thirds of U.S. adults have a BMI greater than 25 (review Figure 8-6, p. 253).
- The ideal amount of body fat varies from person to person, but body fat in excess of 22 percent for young men and 32 percent for young women (the levels rise slightly with age) poses health risks.
- Central obesity—excess abdominal fat distributed around the trunk of the body—presents greater health risks than excess fat distributed on the lower body (review Figure 8-8 and Figure 8-9, p. 255).
- The healthiest weight for an individual depends on personal factors such as body fat distribution, family health history, and current health status. At the extremes, both overweight and underweight impose health risks (review Figure 8-11, p. 257).

FIGURE 8-6 Distribution of Body Weights in U.S. Adults

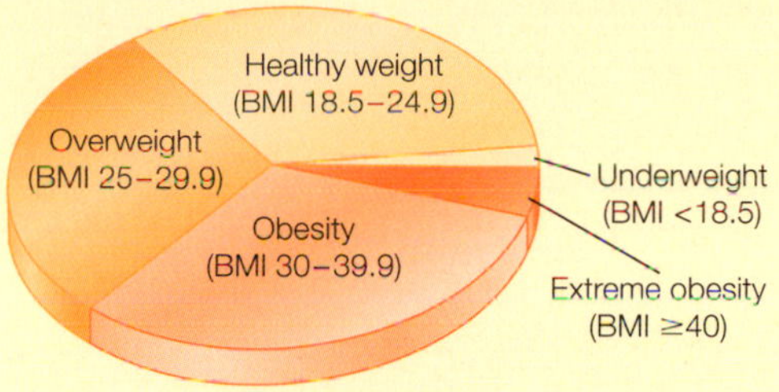

FIGURE 8-4 Components of Energy Expenditure

The amount of energy spent in a day differs for each individual, but in general, basal metabolism is the largest component of energy expenditure and the thermic effect of food is the smallest. The amount spent in voluntary physical activities has the greatest variability, depending on a person's activity patterns. For a sedentary person, physical activities may account for less than half as much energy as basal metabolism, whereas an extremely active person may expend as much on activity as for basal metabolism.

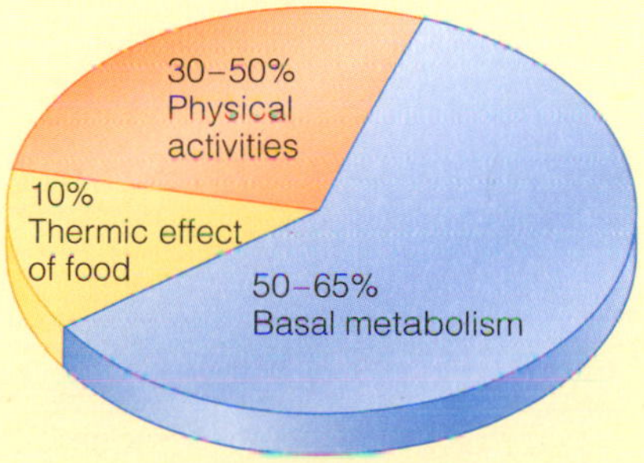

FIGURE 8-11 BMI and Mortality

This J-shaped curve describes the relationship between body mass index (BMI) and mortality and shows that both underweight and overweight present risks of a premature death.

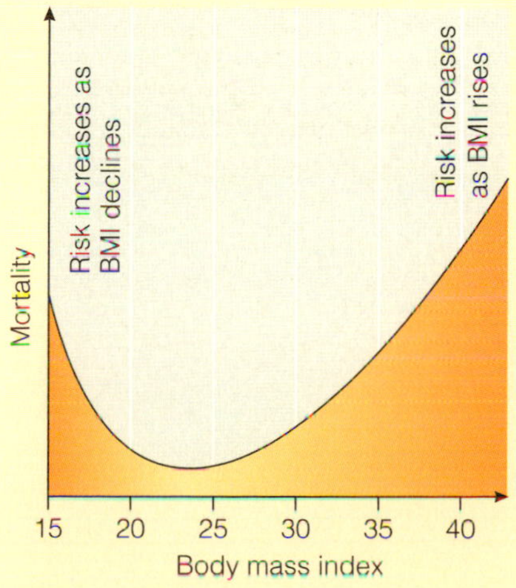

Energy Balance (pp. 241–242)

1. What are the consequences of an unbalanced energy budget?
2. A person who consistently consumes 1700 kcalories a day and spends 2200 kcalories a day for a month would be expected to:
   a. lose ½ to 1 pound.
   b. gain ½ to 1 pound.
   c. lose 4 to 5 pounds.
   d. gain 4 to 5 pounds.

Energy In: The kCalories Foods Provide (pp. 242–245)

3. Define hunger, appetite, satiation, and satiety and describe how each influences food intake.
4. A bomb calorimeter measures:
   a. physiological fuel.
   b. energy available from foods.
   c. kcalories a person derives from foods.
   d. heat a person releases in basal metabolism.
5. The psychological desire to eat that accompanies the sight, smell, or thought of food is known as:
   a. hunger.
   b. satiety.
   c. appetite.
   d. palatability.
6. A person watching television after dinner reaches for a snack during a commercial in response to:
   a. external cues.
   b. hunger signals.
   c. stress arousal.
   d. satiety factors.

Energy Out: The kCalories the Body Expends (pp. 245–249)

7. Describe each component of energy expenditure. What factors influence each? How can energy expenditure be estimated?
8. The largest component of energy expenditure is:
   a. basal metabolism.
   b. physical activity.
   c. indirect calorimetry.
   d. thermic effect of food.
9. A major factor influencing BMR is:
   a. hunger.
   b. food intake.
   c. body composition.
   d. physical activity.
10. The thermic effect of an 800-kcalorie meal is about:
    a. 8 kcalories.
    b. 80 kcalories.
    c. 160 kcalories.
    d. 200 kcalories.

Body Weight, Body Composition, and Health (pp. 249–258)

11. Distinguish between body weight and body composition. What assessment techniques are used to measure each?
12. What problems are involved in defining "ideal" body weight?
13. What is central obesity, and what is its relationship to disease?
14. What risks are associated with excess body weight and excess body fat?
15. For health's sake, a person with a BMI of 21 might want to:
    a. lose weight.
    b. maintain weight.
    c. gain weight.
16. Which of the following reflects height and weight?
    a. body mass index
    b. central obesity
    c. waist circumference
    d. body composition
17. Which of the following increases disease risks?
    a. BMI 19–21
    b. BMI 22–25
    c. lower-body fat
    d. central obesity

Study Questions (multiple choice)
2. c 4. b 5. c 6. a 8. a 9. c 10. b 15. b 16. a 17. d

For additional study questions and activities, go to **www.cengagebrain.com** and search for ISBN 084006845X.

Weight Management: Overweight, Obesity, and Underweight

Overweight and Obesity

- Fat cells develop by increasing in number (hyperplasia) and size (hypertrophy).
- Preventing weight gain depends on limiting the number of fat cells; weight loss depends on decreasing the size of fat cells.
- With weight gains or losses, the body adjusts in an attempt to return to its previous weight (set point theory).

Causes of Overweight and Obesity

- Obesity has multiple causes and different combinations of causes in different people.
- Some environmental causes, such as overeating and physical inactivity, may be within a person's control, and some, such as genetics, may be beyond it.
- Proteins such as adiponectin, ghrelin, and leptin regulate food intake and energy homeostasis.

Problems of Overweight and Obesity

- Whether a person should lose weight depends on factors such as the extent of overweight, age, health risks, and genetic makeup.
- Not all obesity will cause disease or shorten life expectancy. Just as there are unhealthy, normal-weight people, there are healthy, overweight people.
- Some people risk more in the process of losing weight than in remaining overweight.
- Fad diets and weight-loss supplements can be as physically and psychologically damaging as excess body weight.

Aggressive Treatments for Obesity

- Obese people with high risks of medical problems may need aggressive treatment, including drugs or surgery.
- Others may benefit most from improving eating and physical activity habits.

Weight-Loss Strategies

- A person who adopts a lifelong "eating plan for good health" rather than a "diet for weight loss" will be more likely to keep the lost weight off.

FIGURE 9-8 Influence of Physical Activity on Discretionary kCalorie Allowance

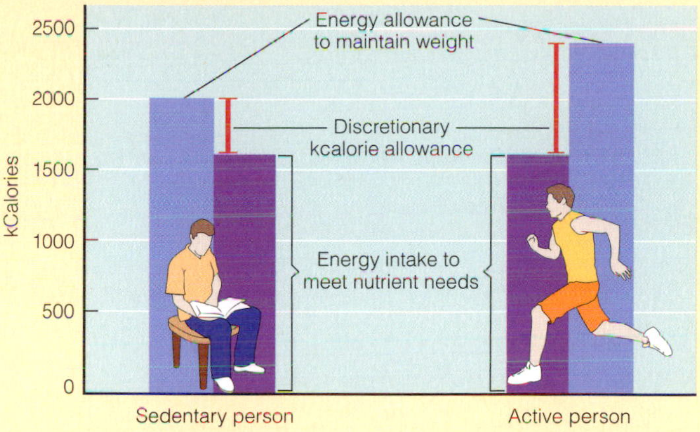

- Table 9-6 (p. 290) provides several tips for successful weight management.
- Physical activity can increase energy expenditure, improve body composition, help control appetite, reduce stress and stress eating, and enhance physical and psychological well-being.
- A surefire remedy for obesity has yet to be found; a combination of approaches is most effective.
- Weight loss depends on adjusting diet and physical activity so that more energy is expended than is taken in.
- For weight loss, energy intake should be reduced by 500 to 1000 kcalories per day, depending on starting body weight and usual food intake.
- Safe rate for weight loss is ½ to 2 pounds per week or 10 percent body weight per 6 months.
- Behavior modification and cognitive restructuring retrain habits to support a healthy eating and activity plan.
- Treatment requires time, individualization, and sometimes the assistance of a registered dietitian or support group.
- Preventing weight gains and maintaining weight losses require vigilant attention to diet and physical activity; taking care of oneself is a lifelong responsibility.

Underweight

- Both the incidence of underweight and the health problems associated with it are less prevalent than overweight and its associated problems.
- To gain weight, a person must train physically and increase energy intake by selecting energy-dense foods, eating regular meals, taking larger portions, and consuming extra snacks and beverages.
- Table 9-6 (p. 290) includes a summary of weight-gain strategies.

Overweight and Obesity (pp. 271–273)

1. Describe how body fat develops, and suggest some reasons why it is difficult for an obese person to maintain weight loss.
2. With weight loss, fat cells:
   a. decrease in size only.
   b. decrease in number only.
   c. decrease in both number and size.
   d. decrease in number, but increase in size.

Causes of Overweight and Obesity (pp. 273–278)

3. What factors contribute to obesity?
4. Obesity is caused by:
   a. overeating.
   b. inactivity.
   c. defective genes.
   d. multiple factors.
5. The protein produced by the fat cells under the direction of the *ob* gene is called:
   a. leptin.
   b. serotonin.
   c. sibutramine.
   d. phentermine.

Problems of Overweight and Obesity (pp. 278–282)

6. Describe the physical, social, and psychological consequences of overweight and obesity.

Aggressive Treatments for Obesity (pp. 282–284)

7. List several aggressive ways to treat obesity, and explain why such methods are not recommended for every overweight person.
8. A drug currently used in the treatment of obesity:
   a. lowers leptin levels.
   b. inhibits lipase activity.
   c. enhances fiber absorption.
   d. increases kcalorie expenditure.

Weight-Loss Strategies (pp. 284–297)

9. Discuss reasonable dietary strategies for achieving and maintaining a healthy body weight.
10. What are the benefits of increased physical activity in a weight-loss program?
11. Describe the behavioral strategies for changing an individual's dietary habits. What role does personal attitude play?
12. A realistic goal for weight loss is to reduce body weight:
    a. down to the weight a person was at age 25.
    b. down to the ideal weight in the weight-for-height tables.
    c. by 10 percent over six months.
    d. by 15 percent over three months.
13. A nutritionally sound weight-loss diet might restrict daily energy intake to create a:
    a. 1000-kcalorie-per-month deficit.
    b. 500-kcalorie-per-month deficit.
    c. 500-kcalorie-per-day deficit.
    d. 3500-kcalorie-per-day deficit.
14. Successful weight loss depends on:
    a. avoiding fats and limiting water.
    b. taking supplements and drinking water.
    c. increasing proteins and restricting carbohydrates.
    d. reducing energy intake and increasing physical activity.
15. Physical activity does *not* help a person to:
    a. lose weight.
    b. retain muscle.
    c. maintain weight loss.
    d. lose fat in trouble spots.
16. Which strategy would *not* help an overweight person to lose weight?
    a. Exercise.
    b. Eat slowly.
    c. Limit high-fat foods.
    d. Eat energy-dense foods regularly.

Underweight (pp. 298–299)

17. Describe strategies for successful weight gain.
18. Which strategy would *not* help an underweight person to gain weight?
    a. Exercise.
    b. Drink plenty of water.
    c. Eat snacks between meals.
    d. Eat large portions of foods.

Study Questions (multiple choice)
2. a 4. d 5. a 8. b 12. c 13. c 14. d 15. d 16. d 18. b

For additional study questions and activities, go to **www.cengagebrain.com** and search for ISBN 084006845X.

STUDY CARD 10

The Water-Soluble Vitamins: B Vitamins and Vitamin C

The Vitamins—An Overview

- The vitamins are organic, essential nutrients needed in tiny amounts in the diet both to prevent deficiency diseases and to support optimal health.
- The water-soluble vitamins are the B vitamins and vitamin C; the fat-soluble vitamins are vitamins A, D, E, and K. The B vitamins include thiamin, niacin, riboflavin, vitamin B_6, folate, vitamin B_{12}, pantothenic acid, and biotin.
- Differences between water- and fat-soluble vitamins:

	Water-Soluble Vitamins: B Vitamins and Vitamin C	**Fat-Soluble Vitamins: Vitamins A, D, E, and K**
Absorption	Directly into the blood	First into the lymph, then the blood
Transport	Travel freely	Many require transport proteins
Storage	Circulate freely in water-filled parts of the body	Stored in the cells associated with fat
Excretion	Kidneys detect and remove excess in urine	Less readily excreted; tend to remain in fat-storage sites
Toxicity	Possible to reach toxic levels when consumed from supplements	Likely to reach toxic levels when consumed from supplements
Requirements	Needed in frequent doses (perhaps 1 to 3 days)	Needed in periodic doses (perhaps weeks or even months)

The B Vitamins—As Individuals

- Thiamin is part of the coenzyme TPP, which assists in energy metabolism. Deficiency can result in beriberi. Thiamin occurs in small quantities in many nutritious foods; pork is an exceptionally good source.
- Riboflavin is part of the coenzymes FMN and FAD that accept and donate hydrogens during energy metabolism. Milk and milk products are good sources.
- Niacin is part of the coenzymes NAD and NADP that participate in many metabolic reactions. The deficiency disease, pellagra, causes diarrhea, dermatitis, dementia, and eventually death ("the 4 Ds"). Toxicity produces "niacin flush"—a tingling, painful sensation. The amino acid tryptophan can be converted to niacin in the body: 60 mg tryptophan = 1 NE (niacin equivalent). Good sources of niacin are protein-rich foods.

FIGURE 10-2 Coenzyme Action

Some vitamins form part of the coenzymes that enable enzymes either to synthesize compounds (as illustrated by the lower enzymes in this figure) or to dismantle compounds (as illustrated by the upper enzymes).

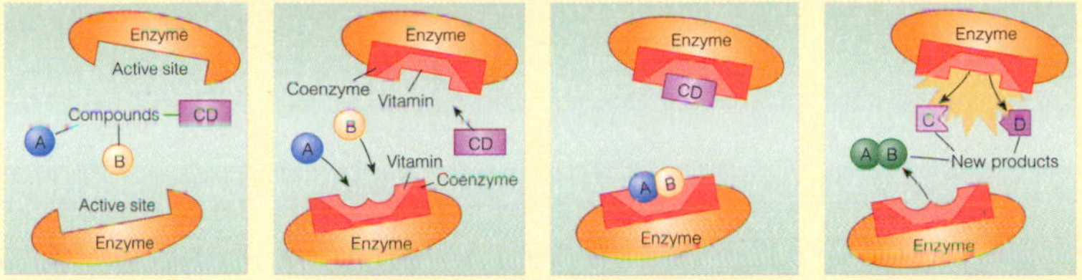

Without coenzymes, compounds A, B, and CD don't respond to their enzymes.

With the coenzymes in place, compounds are attracted to their sites on the enzymes . . .

. . . and the reactions proceed instantaneously. The coenzymes often donate or accept electrons, atoms, or groups of atoms.

The reactions are completed with either the formation of a new product, AB, or the breaking apart of a compound into two new products, C and D, and the release of energy.

- Biotin plays a critical role in energy metabolism, replenishing oxaloacetate in the TCA cycle. Biotin is widespread in foods; deficiencies and toxicities are rare.
- Pantothenic acid is part of coenzyme A that forms acetyl CoA in many metabolic pathways. Pantothenic acid is widespread in foods; deficiencies and toxicities are rare.
- Vitamin B_6 occurs as pyridoxal, pyridoxine, and pyridoxamine; all can become part of the coenzyme PLP, which is active in amino acid metabolism. Deficiency causes convulsions; toxicity causes nerve damage.
- Folate is part of the coenzyme THF that activates vitamin B_{12}, synthesizes DNA, and regenerates the amino acid methionine from homocysteine. Folate helps prevent neural tube defects. Excessive folate can mask the anemia of a vitamin B_{12} deficiency, but it will not prevent the associated nerve damage. Folate is abundant in legumes, fruits, and vegetables.
- Vitamin B_{12} activates folate, synthesizes DNA, regenerates methionine from homocysteine, and maintains the sheath that protects nerve fibers. Deficiencies typically occur when either hydrochloric acid or intrinsic factor is lacking. Vitamin B_{12} is found primarily in foods derived from animals.

The B Vitamins—In Concert

- The B vitamins serve as coenzymes—small organic molecules closely associated with enzymes that facilitate the work of cells, participating in metabolism and in DNA synthesis.
- Grain products are enriched with thiamin, riboflavin, niacin, and folate; a variety of foods from each food group provides an adequate supply of all B vitamins.

Vitamin C

- Vitamin C acts as an antioxidant—a substance that decreases the adverse effects of free radicals in the body.
- Vitamin C works as a cofactor in the synthesis of collagen, neurotransmitters (serotonin and norepinephrine), hormones (thyroxin), and other compounds.
- Vitamin C deficiency causes scurvy.

The Vitamins—An Overview (pp. 311–314)

1. How do the vitamins differ from the energy nutrients?
2. Describe some general differences between fat-soluble and water-soluble vitamins.
3. Vitamins:
   a. are inorganic compounds.
   b. yield energy when broken down.
   c. are soluble in either water or fat.
   d. perform best when linked in long chains.
4. The rate at and the extent to which a vitamin is absorbed and used in the body is known as its:
   a. bioavailability.
   b. intrinsic factor.
   c. physiological effect.
   d. pharmacological effect.

The B Vitamins—As Individuals (pp. 314–334)

5. Which B vitamins are involved in energy metabolism? Protein metabolism? Cell division?
6. For thiamin, riboflavin, niacin, biotin, pantothenic acid, vitamin B_6, folate, vitamin B_{12}, and vitamin C, state:
   - Its chief function in the body.
   - Its characteristic deficiency symptoms.
   - Its significant food sources.
7. What is the relationship of tryptophan to niacin?
8. Describe the relationship between folate and vitamin B_{12}.
9. What risks are associated with high doses of niacin? Vitamin B_6? Vitamin C?
10. Many of the B vitamins serve as:
    a. coenzymes.
    b. antagonists.
    c. antioxidants.
    d. serotonin precursors.
11. With respect to thiamin, which of the following is the most nutrient dense?
    a. 1 slice whole-wheat bread (69 kcalories and 0.1 milligram thiamin)
    b. 1 cup yogurt (144 kcalories and 0.1 milligram thiamin)
    c. 1 cup snow peas (69 kcalories and 0.22 milligram thiamin)
    d. 1 chicken breast (141 kcalories and 0.06 milligram thiamin)
12. The body can make niacin from:
    a. tyrosine.
    b. serotonin.
    c. carnitine.
    d. tryptophan.
13. The vitamin that protects against neural tube defects is:
    a. niacin.
    b. folate.
    c. riboflavin.
    d. vitamin B_{12}.
14. A lack of intrinsic factor may lead to:
    a. beriberi.
    b. pellagra.
    c. pernicious anemia.
    d. atrophic gastritis.
15. Which of the following is a B vitamin?
    a. inositol
    b. carnitine
    c. vitamin B_{15}
    d. pantothenic acid

Vitamin C (pp. 337–343)

16. What risks are associated with high doses of niacin? Vitamin B_6? Vitamin C?
17. Vitamin C serves as a(n):
    a. coenzyme.
    b. antagonist.
    c. antioxidant.
    d. intrinsic factor.
18. The requirement for vitamin C is highest for:
    a. smokers.
    b. athletes.
    c. alcoholics.
    d. the elderly.

See p. 343 for a summary of the water-soluble vitamins.

Study Questions (multiple choice)
3. c 4. a 10. a 11. c 12. d 13. b 14. c 15. d 17. c 18. a

For additional study questions and activities, go to **www.cengagebrain.com** and search for ISBN 084006845X.

STUDY CARD 11

The Fat-Soluble Vitamins: A, D, E, and K

Vitamin A and Beta-Carotene

- Vitamin A is found in the body in three forms: retinol, retinal, and retinoic acid. Together, they are essential to reproduction, vision, and growth.
- Vitamin A deficiency is a major health problem worldwide, leading to infections, blindness, and keratinization.
- Toxicity can also cause problems and is most often associated with supplement abuse.
- Animal-derived foods such as liver and whole or fortified milk provide retinoids, whereas brightly colored plant-derived foods such as spinach, carrots, and pumpkins provide beta-carotene and other carotenoids.
- In addition to serving as a precursor for vitamin A, beta-carotene may act as an antioxidant in the body.

Vitamin D

- Vitamin D can be synthesized in the body with the help of sunlight or obtained from fortified milk.
- Vitamin D sends signals to three primary target sites: the GI tract to absorb more calcium and phosphorus, the bones to release more, and the kidneys to retain more. These actions maintain blood calcium concentrations and support bone formation.
- A deficiency causes rickets in childhood and osteomalacia in later life.

Vitamin E

- Vitamin E (alpha-tocopherol) acts as an antioxidant, defending lipids and other components of the cells against oxidative damage.
- Deficiencies are rare, but they do occur in premature infants, the primary symptom being erythrocyte hemolysis.
- Vitamin E is found predominantly in vegetable oils and appears to be one of the least toxic of the fat-soluble vitamins.

Vitamin K

- Vitamin K helps with blood clotting (review Figure 11-12, p. 370), and its deficiency causes hemorrhagic disease (uncontrolled bleeding).
- Bacteria in the GI tract can make the vitamin; people typically receive about half of their requirements from bacterial synthesis and half from foods such as green vegetables and vegetable oils.
- Because people depend on bacterial synthesis for vitamin K, deficiency is most likely in newborn infants and in people taking antibiotics.

FIGURE 11-9 Vitamin D Synthesis and Activation

The precursor of vitamin D is made in the liver from cholesterol (see Figure 5-11 on p. 141 and Appendix C). The activation of vitamin D is closely regulated by parathyroid hormone. The final product, active vitamin D, is also known as 1,25-dihydroxycholecalciferol (or calcitriol).

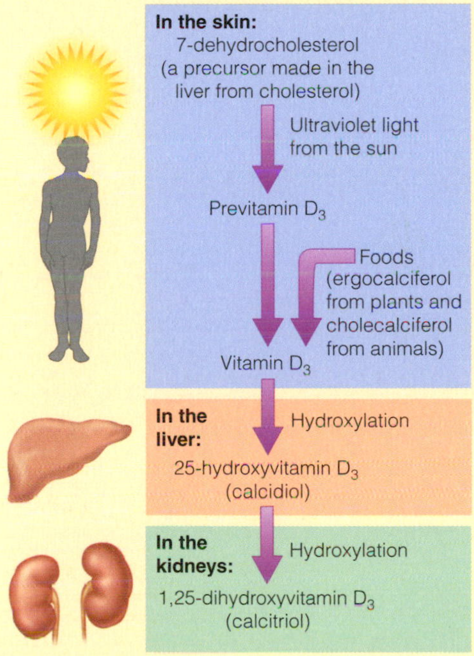

Vitamin A and Beta-Carotene (pp. 355–363)

1. Summarize the roles of vitamin A and the symptoms of its deficiency.
2. What are vitamin precursors? Name the precursors of vitamin A, and tell in what classes of foods they are located. Give examples of foods with high vitamin A activity.
3. The form of vitamin A active in vision is:
   a. retinal.
   b. retinol.
   c. rhodopsin.
   d. retinoic acid.
4. Vitamin A–deficiency symptoms include:
   a. rickets and osteomalacia.
   b. hemorrhaging and jaundice.
   c. night blindness and keratomalacia.
   d. fibrocystic breast disease and erythrocyte hemolysis.
5. Good sources of vitamin A include:
   a. oatmeal, pinto beans, and ham.
   b. apricots, turnip greens, and liver.
   c. whole-wheat bread, green peas, and tuna.
   d. corn, grapefruit juice, and sunflower seeds.

Vitamin D (pp. 363–368)

6. How is vitamin D unique among the vitamins? What is its chief function? What are the richest sources of this vitamin?
7. To keep minerals available in the blood, vitamin D targets:
   a. the skin, the muscles, and the bones.
   b. the kidneys, the liver, and the bones.
   c. the intestines, the kidneys, and the bones.
   d. the intestines, the pancreas, and the liver.
8. Vitamin D can be synthesized from a precursor that the body makes from:
   a. bilirubin.
   b. tocopherol.
   c. cholesterol.
   d. beta-carotene.

Vitamin E (pp. 368–370)

9. Describe vitamin E's role as an antioxidant. What are the chief symptoms of vitamin E deficiency?
10. Vitamin E's most notable role is to:
    a. protect lipids against oxidation.
    b. activate blood-clotting proteins.
    c. support protein and DNA synthesis.
    d. enhance calcium deposits in the bones.
11. The classic sign of vitamin E deficiency is:
    a. rickets.
    b. xeropthalmia.
    c. muscular dystrophy.
    d. erythrocyte hemolysis.

Vitamin K (pp. 370–371)

12. What is vitamin K's primary role in the body? What conditions may lead to vitamin K deficiency?
13. Without vitamin K:
    a. muscles atrophy.
    b. bones become soft.
    c. skin rashes develop.
    d. blood fails to clot.
14. A significant amount of vitamin K comes from:
    a. vegetable oils.
    b. sunlight exposure.
    c. bacterial synthesis.
    d. fortified grain products.

The Fat-Soluble Vitamins—In Summary (p. 372)

15. List the fat-soluble vitamins. What characteristics do they have in common? How do they differ from the water-soluble vitamins?
16. Fat-soluble vitamins:
    a. are easily excreted.
    b. seldom reach toxic levels.
    c. require bile for absorption.
    d. are not stored in the body's tissues.

See p. 372 for a summary of the fat-soluble vitamins.

Study Questions (multiple choice)
3. a 4. c 5. b 7. c 8. c 10. a 11. d 13. d 14. c 16. c

For additional study questions and activities, go to **www.cengagebrain.com** and search for ISBN 084006845X.

STUDY CARD 12

Water and the Major Minerals

Water and the Body Fluids

- Water makes up about 60 percent of the adult body's weight.
- Water assists with the transport of nutrients and waste products throughout the body, participates in chemical reactions, acts as a solvent, serves as a shock absorber, and regulates body temperature.
- To maintain water balance, intake from liquids, foods, and metabolism must equal losses from the kidneys, skin, lungs, and GI tract (review Table 12-3, p. 385).
- To restore homeostasis, the body responds to low blood volume, low blood pressure, or highly concentrated body fluids by producing ADH to stimulate the kidneys to reabsorb water, renin to initiate the pathway that leads to the production of angiotensin II, angiotensin II to constrict blood vessels and stimulate the release of aldosterone and ADH, and aldosterone to regulate potassium and sodium levels (review Figure 12-3, p. 388).
- Electrolytes (charged minerals) in the fluids help distribute the fluids inside and outside the cells, thus ensuring the appropriate water balance and acid-base balance to support all life processes.
- Excessive losses of fluids and electrolytes upset these balances, and the kidneys play a key role in restoring homeostasis.

The Minerals—An Overview

- The major minerals are needed in the diet and found in the body in larger quantities than the trace minerals (review Figure 12-9, p. 394).
- Minerals are inorganic elements that retain their chemical identities; receive special handling and regulation in the body; and may bind with other substances or interact with other minerals, thus limiting their absorption.
- The major minerals, especially sodium, chloride, and potassium, influence the body's fluid balance; whenever an anion moves, a cation moves—always maintaining homeostasis.
- Sodium, chloride, potassium, calcium, and magnesium are key members of the team of nutrients that direct nerve impulse transmission and muscle contraction. They are also the primary nutrients involved in regulating blood pressure.
- Phosphorus and magnesium participate in many reactions involving glucose, fatty acids, amino acids, and the vitamins. Calcium, phosphorus, and magnesium combine to form the structure of the bones and teeth. Each major mineral also plays other specific roles in the body.

The Minerals—As Individuals

- Sodium is the main cation outside cells and one of the primary electrolytes responsible for maintaining fluid balance. Dietary deficiency is rare; excesses may aggravate hypertension in some people, and so health professionals advise a diet moderate in salt and sodium.
- Chloride is the major anion outside cells, and it associates closely with sodium. In addition to its role in fluid balance, chloride is part of the stomach's hydrochloric acid.
- Potassium is the primary cation inside cells and plays an important role in maintaining fluid balance. Fresh foods, notably fruits and vegetables, are its best sources.
- Calcium is found primarily in the bones where it provides a rigid structure and a reservoir of calcium for the blood. Blood calcium participates in muscle contraction, blood clotting, and nerve impulses, and it is closely regulated by a system of hormones and vitamin D. Milk and milk products are good sources of calcium, but certain vegetables and tofu also provide calcium. Even when calcium intake is inadequate, blood calcium remains normal, but at the expense of bone loss, which can lead to osteoporosis.
- Phosphorus accompanies calcium both in the crystals of bone and in many foods such as milk. Phosphorus is also important in energy metabolism, as part of phospholipids, and as part of the genetic materials DNA and RNA.
- Magnesium supports bone mineralization and participates in numerous enzyme systems and in heart function. It is found abundantly in legumes and leafy green vegetables and, in some areas, in water.

Water and the Body Fluids (pp. 383–393)

1. List the roles of water in the body.
2. List the sources of water intake and routes of water excretion.
3. What is ADH? Where does it exert its action? What is aldosterone? How does it work?
4. How does the body use electrolytes to regulate fluid balance?
5. The body generates water during the:
   a. buffering of acids.
   b. dismantling of bone.
   c. metabolism of minerals.
   d. oxidation of energy nutrients through the electron transport chain.
6. Regulation of fluid and electrolyte balance and acid-base balance depends primarily on the:
   a. kidneys.
   b. intestines.
   c. sweat glands.
   d. specialized tear ducts.

The Minerals—An Overview (pp. 393–394)

7. What do the terms *major* and *trace* mean when describing the minerals in the body?
8. Describe some characteristics of minerals that distinguish them from vitamins.
9. The distinction between the major and trace minerals reflects the:
   a. ability of their ions to form salts.
   b. amounts of their contents in the body.
   c. importance of their functions in the body.
   d. capacity to retain their identity after absorption.

The Minerals—As Individuals (pp. 394–410)

10. What is the major function of sodium in the body? Describe how the kidneys regulate blood sodium. Is a dietary deficiency of sodium likely? Why or why not?
11. List calcium's roles in the body. How does the body keep blood calcium constant regardless of intake?
12. Name significant food sources of calcium. What are the consequences of inadequate intakes?
13. List the roles of phosphorus in the body. Discuss the relationships between calcium and phosphorus. Is a dietary deficiency of phosphorus likely? Why or why not?
14. State the major functions of chloride, potassium, magnesium, and sulfur in the body. Are deficiencies of these nutrients likely to occur in your own diet? Why or why not?
15. The principal cation in extracellular fluids is:
    a. sodium.
    b. chloride.
    c. potassium.
    d. phosphorus.
16. The role of chloride in the stomach is to help:
    a. support nerve impulses.
    b. convey hormonal messages.
    c. maintain a strong acidity.
    d. assist in muscular contractions.
17. Which would provide the most potassium?
    a. bologna
    b. potatoes
    c. pickles
    d. whole-wheat bread
18. Calcium homeostasis depends on:
    a. vitamin K, aldosterone, and renin.
    b. vitamin K, parathyroid hormone, and renin.
    c. vitamin D, aldosterone, and calcitonin.
    d. vitamin D, calcitonin, and parathyroid hormone.
19. Calcium absorption is hindered by:
    a. lactose.
    b. oxalates.
    c. vitamin D.
    d. stomach acid.
20. Phosphorus assists in many activities in the body, but *not:*
    a. energy metabolism.
    b. the clotting of blood.
    c. the transport of lipids.
    d. bone and teeth formation.
21. Most of the body's magnesium can be found in the:
    a. bones.
    b. nerves.
    c. muscles.
    d. extracellular fluids.

See p. 410 for a summary of the major minerals.

Study Questions (multiple choice)

5.d 6.a 9.b 15.a 16.c 17.b 18.d 19.b 20.b 21.a

For additional study questions and activities, go to **www.cengagebrain.com** and search for ISBN 084006845X.

STUDY CARD 13

The Trace Minerals

The Trace Minerals—An Overview

- The body needs tiny amounts of the trace minerals.
- The trace minerals can be toxic at levels not far above estimated requirements—a consideration for supplement users (review Figure 13-1, p. 424).
- Like the other nutrients, the trace minerals are best obtained by eating a variety of foods.

FIGURE 13-3 Heme and Nonheme Iron in Foods

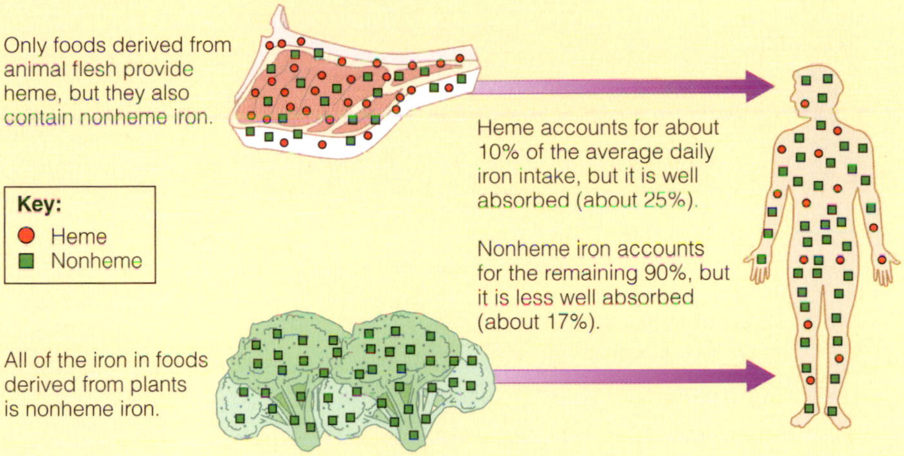

Only foods derived from animal flesh provide heme, but they also contain nonheme iron.
Key:
Heme
Nonheme
Heme accounts for about 10% of the average daily iron intake, but it is well absorbed (about 25%).
Nonheme iron accounts for the remaining 90%, but it is less well absorbed (about 17%).
All of the iron in foods derived from plants is nonheme iron.

The Trace Minerals—As Individuals

- Iron is found in hemoglobin and myoglobin where it carries oxygen for energy metabolism; iron also acts as a cofactor for some enzymes. Iron deficiency is most common among infants and young children, teenagers, women of childbearing age, and pregnant women. Symptoms include fatigue and anemia. Iron overload is most common in men. Heme iron, which is found only in meat, fish, and poultry, is better absorbed than nonheme iron, which occurs in most foods (review Figure 13-3, p. 426). Nonheme iron absorption is improved by eating iron-containing foods with foods containing the MFP factor and vitamin C; absorption is limited by phytates and oxalates.
- Zinc-requiring enzymes participate in reactions affecting growth, vitamin A activity, and pancreatic digestive enzyme synthesis. Both dietary zinc and zinc-rich pancreatic secretions (via enteropancreatic circulation) are available for absorption. Absorption is monitored by a special binding protein (metallothionein) in the small intestine. Protein-rich foods derived from animals are the best sources of bioavailable zinc. Fiber and phytates in cereals bind zinc, limiting absorption. Symptoms of deficiency include growth retardation and sexual immaturity.
- Iodide, the iodine ion, is an essential component of the thyroid hormone. A deficiency can lead to goiter (enlargement of the thyroid gland) and can impair fetal development, causing cretinism. Iodization of salt has largely eliminated iodine deficiency in the United States and Canada.
- Selenium is an antioxidant nutrient that works closely with the glutathione peroxidase enzyme and vitamin E. Selenium is found with protein in foods. Deficiencies are associated with a predisposition to a type of heart abnormality known as Keshan disease.
- Copper is a component of several enzymes, all of which are involved with oxygen or oxidation. Some act as antioxidants; others are essential to iron metabolism. Legumes, whole grains, and shellfish are good sources of copper.
- Manganese-dependent enzymes are involved in bone formation and various metabolic processes. Manganese is widespread in plant foods and deficiencies are rare.
- Fluoride makes bones stronger and teeth more resistant to decay. Fluoridation of public water reduces the incidence of dental caries; excess fluoride during tooth development can discolor and pit tooth enamel (fluorosis).
- Chromium enhances insulin's action. Deficiency can result in a diabetes-like condition. Chromium is widely available in unrefined foods including brewer's yeast, whole grains, and liver.
- Molybdenum is a part of many metalloenzymes. Deficiencies are unknown and toxicity is rare.

The Trace Minerals—An Overview

(pp. 423–425)

1. Discuss the importance of balanced and varied diets in obtaining the essential minerals and avoiding toxicities.
2. Describe some of the ways trace minerals interact with each other and with other nutrients.

The Trace Minerals—As Individuals

(pp. 425–446)

3. Distinguish between heme and nonheme iron. Discuss the factors that enhance iron absorption.
4. Distinguish between iron deficiency and iron-deficiency anemia. What are the symptoms of iron-deficiency anemia?
5. What causes iron overload? What are its symptoms?
6. Describe the similarities and differences in the absorption and regulation of iron and zinc.
7. Discuss possible reasons for a low intake of zinc. What factors affect the bioavailability of zinc?
8. Describe the principal functions of iodide, selenium, copper, manganese, fluoride, chromium, and molybdenum in the body.
9. What public health measure has been used in preventing simple goiter? What measure has been recommended for protection against tooth decay?
10. Iron absorption is impaired by:
    a. heme.
    b. phytates.
    c. vitamin C.
    d. MFP factor.
11. Which of these people is *least* likely to develop an iron deficiency?
    a. 3-year-old boy
    b. 52-year-old man
    c. 17-year-old girl
    d. 24-year-old woman
12. Which of the following would *not* describe the blood cells of a severe iron deficiency?
    a. anemic
    b. microcytic
    c. pernicious
    d. hypochromic
13. Which provides the most absorbable iron?
    a. 1 apple
    b. 1 cup milk
    c. 3 ounces steak
    d. ½ cup spinach
14. The intestinal protein that helps to regulate zinc absorption is:
    a. albumin.
    b. ferritin.
    c. hemosiderin.
    d. metallothionein.
15. A classic sign of zinc deficiency is:
    a. anemia.
    b. goiter.
    c. mottled teeth.
    d. growth retardation.
16. Cretinism is caused by a deficiency of:
    a. iron.
    b. zinc.
    c. iodine.
    d. selenium.
17. The mineral best known for its role as an antioxidant is:
    a. copper.
    b. selenium.
    c. manganese.
    d. molybdenum.
18. Fluorosis occurs when fluoride:
    a. is excessive.
    b. is inadequate.
    c. binds with phosphorus.
    d. interacts with calcium.
19. Which mineral enhances insulin activity?
    a. zinc
    b. iodine
    c. chromium
    d. manganese

See p. 446 for a summary of the trace minerals.

Study Questions (multiple choice)

10. b 11. b 12. c 13. c 14. d 15. d 16. c 17. b 18. a 19. c

For additional study questions and activities, go to **www.cengagebrain.com** and search for ISBN 084006845X.

STUDY CARD 14

Life Cycle Nutrition: Pregnancy and Lactation

Nutrition prior to Pregnancy

- Maternal nutrition before and during pregnancy supports the mother's health and the infant's growth.

Growth and Development during Pregnancy

- The infant develops through three stages—the zygote, embryo, and fetus (review Figure 14-2, p. 459). Each organ and tissue grows on its own schedule (review Figure 14-4, p. 460).
- Times of intense development are critical periods that depend on nutrients (review Figure 14-3, p. 460).
- Without folate, the neural tube fails to develop completely during the first month of pregnancy; all women of childbearing age should take folate daily.

Maternal Weight

- A healthy pregnancy depends on a sufficient weight gain.
- Women who begin their pregnancies at a healthy weight need to gain about 30 pounds to cover the growth and development of the placenta, uterus, blood, breasts, and infant (review Figure 14-8, p. 466).
- Physical activity throughout pregnancy can help a woman develop the strength she needs to carry the extra weight and maintain habits that will help her lose it after the birth.

Nutrition during Pregnancy

- Energy and nutrient needs are high during pregnancy; the diet should include an extra serving from each of the five food groups.
- Supplements of iron and folate are recommended.
- Nausea, constipation, and heartburn commonly accompany pregnancy and can usually be alleviated with a few simple strategies (review Table 14-2, p. 472). Food cravings do not typically reflect physiological needs.

High-Risk Pregnancies

- High-risk pregnancies threaten the life and health of both mother and infant.
- Proper nutrition and abstinence from smoking, alcohol, and other drugs improve the outcome as can prenatal care to monitor for gestational diabetes and preeclampsia.

Nutrition during Lactation

- Lactating women need extra fluid and enough energy and nutrients to produce about 25 ounces of milk a day.
- Breastfeeding is contraindicated for those with HIV/AIDS.
- Alcohol, other drugs, smoking, and contaminants may reduce milk production or enter breast milk and impair infant development.

TABLE 14-3 High-Risk Pregnancy Factors

Factor	Condition That Raises Risk
Maternal weight	
• Prior to pregnancy	Prepregnancy BMI either <18.5 or ≥ 25
• During pregnancy	Insufficient or excessive pregnancy weight gain
Maternal nutrition	Nutrient deficiencies or toxicities; eating disorders
Socioeconomic status	Poverty, lack of family support, low level of education, limited food available
Lifestyle habits	Smoking, alcohol or other drug use
Age	Teens, especially 15 years or younger; women 35 years or older
Previous pregnancies	
• Number	Many previous pregnancies (3 or more to mothers under age 20; 4 or more to mothers age 20 or older)
• Interval	Short or long intervals between pregnancies ($<$18 months or $>$59 months)
• Outcomes	Previous history of problems
• Multiple births	Twins or triplets
• Birthweight	Low- or high-birthweight infants
Maternal health	
• High blood pressure	Development of gestational hypertension
• Diabetes	Development of gestational diabetes
• Chronic diseases	Diabetes; heart, respiratory, and kidney disease; certain genetic disorders; special diets and medications

Nutrition prior to Pregnancy (pp. 457–458)

1. Describe how men and women can prepare for a healthy pregnancy.

Growth and Development during Pregnancy (pp. 458–463)

2. Describe the placenta and its function.
3. Describe the normal events of fetal development. How does malnutrition impair fetal development?
4. Define the term *critical period*. How do adverse influences during critical periods affect later health?
5. Explain why women of childbearing age need folate in their diets. How much is recommended, and how can women ensure that these needs are met?
6. The spongy structure that delivers nutrients to the fetus and returns waste products to the mother is called the:
   a. embryo.
   b. uterus.
   c. placenta.
   d. amniotic sac.

Maternal Weight (pp. 463–467)

7. What is the recommended pattern of weight gain during pregnancy for a woman at a healthy weight? For an underweight woman? For an overweight woman?
8. What does a pregnant woman need to know about exercise?
9. Which of these strategies is *not* a healthy option for an overweight woman?
   a. Limit weight gain during pregnancy.
   b. Postpone weight loss until after pregnancy.
   c. Follow a weight-loss diet during pregnancy.
   d. Try to achieve a healthy weight before becoming pregnant.
10. A reasonable weight gain during pregnancy for a normal-weight woman is about:
    a. 10 pounds.
    b. 20 pounds.
    c. 30 pounds.
    d. 40 pounds.

Nutrition during Pregnancy (pp. 467–472)

11. Which nutrients are needed in the greatest amounts during pregnancy? Why are they so important? Describe wise food choices for the pregnant woman.
12. Energy needs during pregnancy increase by about:
    a. 100 kcalories/day.
    b. 300 kcalories/day.
    c. 500 kcalories/day.
    d. 700 kcalories/day.

High-Risk Pregnancies (pp. 473–479)

13. Define low-risk and high-risk pregnancies. What is the significance of infant birthweight in terms of the child's future health?
14. Describe some of the special problems of the pregnant adolescent. Which nutrients are needed in increased amounts?
15. What practices should be avoided during pregnancy? Why?
16. To help prevent neural tube defects, grain products are now fortified with:
    a. iron.
    b. folate.
    c. protein.
    d. vitamin C.
17. Pregnant women should *not* take supplements of:
    a. iron.
    b. folate.
    c. vitamin A.
    d. vitamin C.
18. The combination of high blood pressure, protein in the urine, and edema signals:
    a. jaundice.
    b. preeclampsia.
    c. gestational diabetes.
    d. gestational hypertension.

Nutrition during Lactation (pp. 480–485)

19. How do nutrient needs during lactation differ from nutrient needs during pregnancy?
20. To facilitate lactation, a mother needs:
    a. about 5000 kcalories a day.
    b. adequate nutrition and rest.
    c. vitamin and mineral supplements.
    d. a glass of wine or beer before each feeding.
21. A breastfeeding woman should drink plenty of water to:
    a. produce more milk.
    b. suppress lactation.
    c. prevent dehydration.
    d. dilute nutrient concentration.
22. A woman may need iron supplements during lactation:
    a. to enhance the iron in her breast milk.
    b. to provide iron for the infant's growth.
    c. to replace the iron in her body's stores.
    d. to support the increase in her blood volume.

Study Questions (multiple choice)
6. c 9. c 10. c 12. b 16. b 17. c 18. b 20. b 21. c 22. c

For additional study questions and activities, go to **www.cengagebrain.com** and search for ISBN 084006845X.

STUDY CARD 15

Life Cycle Nutrition: Infancy, Childhood, and Adolescence

Nutrition during Infancy

- The primary food for infants during the first 12 months is either breast milk or iron-fortified formula.
- Breast milk offers both nutrients and immunological protection.
- At 4 to 6 months of age, infants should gradually begin eating solid foods, including iron-fortified cereals and vitamin-C rich fruits and vegetables.
- By 1 year, infants are drinking from a cup and eating a variety of foods.
- Infants may benefit from supplements containing vitamin D, iron, and fluoride (review Table 15-2, p. 497).

Nutrition during Childhood

- Children's appetites and nutrient needs reflect their stage of growth.
- Children who are chronically hungry and malnourished suffer growth retardation; temporary hunger and mild nutrient deficiencies produce more subtle problems—such as poor academic performance.
- Iron deficiency is widespread and has many physical and behavioral consequences.
- "Hyper" behavior is not caused by poor nutrition; misbehavior may be due to lack of sleep, too little physical activity, or too much television, among other factors.
- Childhood obesity is a major health problem (review Figure 15-9, p. 513).
- Children need to eat nutrient-dense foods and learn how to make healthful diet and activity choices.
- Some children have food allergies—adverse reactions that involve an immune response—most often caused by peanuts, tree nuts, milk, eggs, wheat, soybeans, fish, or shellfish.

Nutrition during Adolescence

- The need for iron increases during adolescence for both males and females; blood losses incurred through menstruation increase iron needs for females further.
- Sufficient calcium intake during adolescence supports optimal bone growth and density.
- The adolescent growth spurt increases the need for energy and nutrients.
- Adolescents who drink soft drinks regularly have a higher energy intake and a lower calcium intake; they are also more likely to be overweight.

TABLE 15-7 Recommended Eating and Physical Activity Behaviors to Prevent Obesity

The Expert Committee of the American Medical Association recommends the following healthy habits for children 2 to 18 years of age to help prevent childhood obesity:

- Limit consumption of sugar-sweetened beverages, such as soft drinks and fruit flavored punches.
- Eat the recommended amounts of fruits and vegetables every day (2 to 4.5 cups per day based on age).
- Learn to eat age-appropriate portions of foods.
- Eat foods low in energy density such as those high in fiber and/or water and modest in fat.
- Eat a nutritious breakfast every day.
- Eat a diet rich in calcium.
- Eat a diet balanced in recommended proportions for carbohydrate, fat, and protein.
- Eat a diet high in fiber.
- Eat together as a family as often as possible.
- Limit the frequency of restaurant meals.
- Limit television watching or other screen time to no more than 2 hours per day and do not have televisions or computers in bedrooms.
- Engage in at least 60 minutes of moderate to vigorous physical activity every day.

SOURCE: S. E. Barlow, Expert Committee recommendations regarding the prevention, assessment, and treatment of child and adolescent overweight and obesity: Summary report, *Pediatrics* 120 (2007): S164–S192.

Nutrition during Infancy (pp. 493–504)

1. Describe some of the nutrient and immunological attributes of breast milk.
2. What are the appropriate uses of formula feeding? What criteria would you use in selecting an infant formula?
3. Why are solid foods not recommended for an infant during the first few months of life? When is an infant ready to start eating solid food?
4. Identify foods that are inappropriate for infants and explain why they are inappropriate.
5. A reasonable weight for a healthy 5-month-old infant who weighed 8 pounds at birth might be:
   a. 12 pounds.
   b. 16 pounds.
   c. 20 pounds.
   d. 24 pounds.
6. Dehydration can develop quickly in infants because:
   a. much of their body water is extracellular.
   b. they lose a lot of water through urination and tears.
   c. only a small percentage of their body weight is water.
   d. they drink lots of breast milk or formula, but little water.
7. An infant should begin eating solid foods between:
   a. 2 and 4 weeks.
   b. 1 and 3 months.
   c. 4 and 6 months.
   d. 8 and 10 months.

Nutrition during Childhood (pp. 504–523)

8. What nutrition problems are most common in children? What strategies can help prevent these problems?
9. Describe the relationships between nutrition and behavior. How does television influence nutrition?
10. Describe a true food allergy. Which foods most often cause allergic reactions? How do food allergies influence nutrition status?
11. Describe the problems associated with childhood obesity and the strategies for prevention and treatment.
12. List strategies for introducing nutritious foods to children.
13. What impact do school meal programs have on the nutrition status of children?
14. Among U.S. and Canadian children, the most prevalent nutrient deficiency is of:
    a. iron.
    b. folate.
    c. protein.
    d. vitamin D.
15. A true food allergy always:
    a. elicits an immune response.
    b. causes an immediate reaction.
    c. creates an aversion to the offending food.
    d. involves symptoms such as headaches or hives.
16. Which of the following strategies is *not* effective?
    a. Play first, eat later.
    b. Provide small portions.
    c. Encourage children to help prepare meals.
    d. Use dessert as a reward for eating vegetables.

Nutrition during Adolescence (pp. 523–527)

17. Describe the changes in nutrient needs from childhood to adolescence. Why is an adolescent girl more likely to develop an iron deficiency than is a boy?
18. How do adolescents' eating habits influence their nutrient intakes?
19. How does the use of illicit drugs influence nutrition status?
20. How do the nutrient intakes of smokers differ from those of nonsmokers? What impacts can those differences exert on health?
21. To help teenagers consume a balanced diet, parents can:
    a. monitor the teens' food intake.
    b. give up—parents can't influence teenagers.
    c. keep the pantry and refrigerator well stocked.
    d. forbid snacking and insist on regular, well-balanced meals.
22. During adolescence, energy and nutrient needs:
    a. reach a peak.
    b. fall dramatically.
    c. rise, but do not peak until adulthood.
    d. fluctuate so much that generalizations can't be made.
23. The nutrients most likely to fall short in the adolescent diet are:
    a. sodium and fat.
    b. folate and zinc.
    c. iron and calcium.
    d. protein and vitamin A.
24. To balance the day's intake, an adolescent who eats a hamburger, fries, and cola at lunch might benefit most from a dinner of:
    a. fried chicken, rice, and banana.
    b. ribeye steak, baked potato, and salad.
    c. pork chop, mashed potatoes, and apple juice.
    d. spaghetti with meat sauce, broccoli, and milk.

Study Questions (multiple choice)
5. b 6. a 7. c 14. a 15. a 16. d 21. c 22. a 23. c 24. d

For additional study questions and activities, go to **www.cengagebrain.com** and search for ISBN 084006845X.

STUDY CARD 16

Life Cycle Nutrition: Adulthood and the Later Years

Nutrition and Longevity

- Life expectancy in the United States increased dramatically in the 20th century.
- Factors that enhance longevity include limited or no alcohol use, regular balanced meals, weight control, abstinence from smoking, regular physical activity, and adequate sleep.
- Energy restriction in animals seems to lengthen their lives; whether such dietary intervention in human beings is beneficial remains unknown.
- Nutrition—especially when combined with regular physical activity—can influence aging and longevity by supporting good health and preventing disease.

TABLE 16-5 Risk Factors for Malnutrition in Older Adults

	These questions help *determine* the risk of malnutrition in older adults:
Disease	• Do you have an illness or condition that changes the types or amounts of foods you eat?
Eating poorly	• Do you eat fewer than two meals a day? Do you eat fruits, vegetables, and milk products daily?
Tooth loss or mouth pain	• Is it difficult or painful to eat?
Economic hardship	• Do you have enough money to buy the food you need?
Reduced social contact	• Do you eat alone most of the time?
Multiple medications	• Do you take three or more different prescribed or over-the-counter medications daily?
Involuntary weight loss or gain	• Have you lost or gained 10 pounds or more in the last 6 months?
Needs assistance	• Are you physically able to shop, cook, and feed yourself?
Elderly person	• Are you older than 80?

NOTE: A complete description of DETERMINE and its scoring system are available online from the American Academy of Family Physicians: www.aafp.org/afp/980301ap/edits.html

The Aging Process

- Changes that accompany aging can impair nutrition status. Among physiological changes, hormone activity alters body composition, immune system changes raise the risk of infections, atrophic gastritis interferes with digestion and absorption, and tooth loss limits food choices.
- Psychological changes such as depression, economic changes such as loss of income, and social changes such as loneliness contribute to poor food intake.

Energy and Nutrient Needs of Older Adults

- Some nutrients need special attention in the diet, but supplements are not routinely recommended.

Nutrition-Related Concerns of Older Adults

- Senile dementia and cognitive losses afflict many older adults; some face loss of vision due to cataracts or macular degeneration or cope with the pain of arthritis.
- The table below provides a summary of the nutrient concerns of aging.
- Some problems may be inevitable, but others are preventable and good nutrition may play a key role.

Food Choices and Eating Habits of Older Adults

- Congregate meals provide older adults with both nutrients and social interactions; those who are homebound receive delivered meals.
- Adults living alone can prepare nutritious, inexpensive meals with a little creativity and careful shopping.

Nutrient	Effect of Aging	Comments
Water	Lack of thirst and decreased total body water make dehydration likely.	Mild dehydration is a common cause of confusion. Difficulty obtaining water or getting to the bathroom may compound the problem.
Energy	Need decreases as muscle mass decreases (sarcopenia).	Physical activity moderates the decline.
Fiber	Likelihood of constipation increases with low intakes and changes in the GI tract.	Inadequate water intakes and lack of physical activity, along with some medications, compound the problem.
Protein	Needs may stay the same or increase slightly.	Low-fat, high-fiber legumes and grains meet both protein and other nutrient needs.
Vitamin B_{12}	Atrophic gastritis is common.	Deficiency causes neurological damage; supplements may be needed.
Vitamin D	Increased likelihood of inadequate intake; skin synthesis declines.	Daily sunlight exposure in moderation or supplements may be beneficial.
Calcium	Intakes may be low; osteoporosis is common.	Stomach discomfort commonly limits milk intake; calcium substitutes or supplements may be needed.
Iron	In women, status improves after menopause; deficiencies are linked to chronic blood losses and low stomach acid output.	Adequate stomach acid is required for absorption; antacid or other medicine use may aggravate iron deficiency; vitamin C and meat increase absorption.

Nutrition and Longevity (pp. 540–544)

1. What roles does nutrition play in aging, and what roles can it play in retarding aging?
2. Life expectancy in the United States is about:
   a. 48 to 60 years.
   b. 58 to 70 years.
   c. 68 to 80 years.
   d. 78 to 90 years.
3. The human life span is about:
   a. 85 years.
   b. 100 years.
   c. 115 years.
   d. 130 years.
4. A 72-year-old person whose physical health is similar to that of people 10 years younger has a(n):
   a. chronological age of 62.
   b. physiological age of 72.
   c. physiological age of 62.
   d. absolute age of minus 10.
5. Rats live longest when given diets that:
   a. eliminate all fat.
   b. provide lots of protein.
   c. allow them to eat freely.
   d. restrict their energy intakes.

The Aging Process (pp. 544–547)

6. What are some of the physiological changes that occur in the body's systems with aging? To what extent can aging be prevented?
7. Which characteristic is *not* commonly associated with atrophic gastritis?
   a. inflamed stomach
   b. vitamin B_{12} toxicity
   c. bacterial overgrowth
   d. lack of intrinsic factor

Energy and Nutrient Needs of Older Adults (pp. 547–551)

8. Why does the risk of dehydration increase as people age?
9. Why do energy needs usually decline with advancing age?
10. Which vitamins and minerals need special consideration for the elderly? Explain why. Identify some factors that complicate the task of setting nutrient standards for older adults.
11. On average, adult energy needs:
   a. decline 5 percent per year.
   b. decline 5 percent per decade.
   c. remain stable throughout life.
   d. rise gradually throughout life.

Nutrition-Related Concerns of Older Adults (pp. 552–555)

12. Discuss the relationships between nutrition and cataracts and between nutrition and arthritis.
13. Which nutrients seem to protect against cataract development?
   a. minerals
   b. lecithins
   c. antioxidants
   d. amino acids
14. The best dietary advice for a person with osteoarthritis might be to:
   a. avoid milk products.
   b. take fish oil supplements.
   c. take vitamin E supplements.
   d. lose weight, if overweight.

Food Choices and Eating Habits of Older Adults (pp. 555–559)

15. What characteristics contribute to malnutrition in older people?
16. Congregate meal programs are preferable to Meals on Wheels because they provide:
   a. nutritious meals.
   b. referral services.
   c. social interactions.
   d. financial assistance.
17. The OAA Nutrition Program is available to:
   a. all people 65 years and older.
   b. all people 60 years and older.
   c. homebound people only, 60 years and older.
   d. low-income people only, 60 years and older.

Study Questions (multiple choice)
2. c 3. d 4. c 5. d 7. b 11. b 13. c 14. d 16. c 17. b

For additional study questions and activities, go to **www.cengagebrain.com** and search for ISBN 084006845X.

STUDY CARD 17

Nutrition Care and Assessment

Nutrition in Health Care

- Illnesses and their treatments can affect food intake and nutrient needs, leading to malnutrition. In turn, poor nutrition status can influence the course of illness and the effectiveness of medical treatment.
- The combined efforts of each member of the health care team ensure that patients receive optimal nutrition care.
- Nutrition screening identifies individuals who can benefit from nutrition assessment and follow-up care.
- The nutrition care process includes four interrelated steps: nutrition assessment, nutrition diagnosis, nutrition intervention, and nutrition monitoring and evaluation.

Nutrition Assessment

- Nutrition assessments include historical information, anthropometric and biochemical data, and physical examinations.
- Historical information includes the medical history, medication and supplement history, personal and social history, and food and nutrition history.
- Health care providers assess food intake using 24-hour recall interviews, food frequency questionnaires, food records, and direct observation.
- Anthropometric measurements help evaluate growth patterns, overnutrition and undernutrition, and body composition.
- Biochemical analyses help in the assessment of nutrient imbalances but are influenced by various other medical problems.
- Physical examinations can help the assessor detect signs of nutrient deficiency, fluid imbalances, and functional deficits.

FIGURE 17-3 The Nutrition Care Process

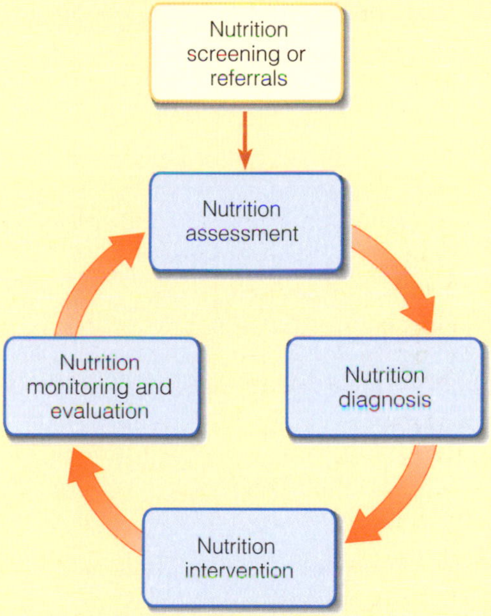

TABLE 17-8 Body Weight and Nutritional Risk

%UBW	%IBW	Nutritional Risk
85–95	80–90	Risk of mild malnutrition
75–84	70–79	Risk of moderate malnutrition
<75	<70	Risk of severe malnutrition

Nutrition in Health Care (pp. 569–575)

1. In what ways can illnesses affect nutrition status? Contrast the roles of the different health care professionals in providing nutrition care.
2. Give examples of the methods used for screening patients for malnutrition. Explain how nutrition screening differs from a complete nutrition assessment.
3. Discuss each of the steps of the nutrition care process.
4. Mr. Hom experiences loss of appetite, difficulty swallowing, and mouth pain as a consequence of illness. Mr. Hom is at risk of malnutrition due to:
   a. altered metabolism.
   b. reduced food intake.
   c. altered excretion of nutrients.
   d. altered digestion and absorption.
5. The central role of nurses in health care makes them well positioned for:
   a. calculating patients' nutrient needs.
   b. providing medical nutrition therapy.
   c. conducting complete nutrition assessments.
   d. identifying patients at risk for malnutrition.
6. Of the following data collected during a nutrition screening, which item does not place the person at risk for malnutrition?
   a. having a health problem that is frequently associated with PEM
   b. the use of prescription medications that affect nutrient needs
   c. residing with a spouse in a middle-income neighborhood
   d. a significant reduction in food intake over the past five or more days
7. The nutrition care process is a systematic approach for:
   a. identifying the nutrient content of foods.
   b. ordering special diets.
   c. conducting nutrition screening.
   d. meeting the nutrition needs of patients.

Nutrition Assessment (pp. 576–586)

8. Give examples of the types of information included in medical, social, and diet histories.
9. Describe the methods of gathering food intake data, and indicate the advantages and disadvantages of each process.
10. What types of anthropometric measurements are included in nutrition assessments? Explain how these measurements help in the evaluation of nutrition status.
11. How do biochemical analyses help assess nutrition status? Give examples. What confounding factors may influence the results of blood tests?
12. Give examples of physical signs that can suggest malnutrition. Describe the signs and symptoms that can result from fluid retention and dehydration.
13. To conduct complete nutrition assessments, dietitians rely on several sources of information, which include all of the following *except*:
    a. nutrition care plans.
    b. body measurements.
    c. medical, medication, and social histories.
    d. biochemical data.
14. Which dietary assessment method does a health practitioner use to conduct a kcalorie count?
    a. 24-hour recall interview
    b. food frequency questionnaire
    c. food record
    d. direct observation
15. The %UBW of a person who weighs 135 pounds and has a usual body weight of 150 pounds is:
    a. 111 percent.
    b. 90 percent.
    c. 86 percent.
    d. 74 percent.
16. A malnourished patient has just begun to eat after days without significant amounts of food. Which of the following blood tests would change most quickly as the patient's nutrition status improves?
    a. albumin
    b. transferrin
    c. serum electrolytes
    d. retinol-binding protein
17. Which sign of PEM would be unlikely to show up in a physical examination?
    a. low plasma protein levels
    b. dull, brittle hair
    c. poor wound healing
    d. wasted appearance
18. Fluid retention may cause all of the following effects *except*:
    a. lab results that are deceptively high.
    b. facial puffiness.
    c. lab results that are deceptively low.
    d. tight-fitting shoes.

Study Questions (multiple choice)
4. b 5. d 6. c 7. d 13. a 14. d 15. b 16. d 17. a 18. a

For additional study questions and activities, go to **www.cengagebrain.com** and search for ISBN 084006845X.

STUDY CARD 18

Nutrition Intervention

Implementing Nutrition Care

- Nutrition interventions are designed to correct the nutrition problems associated with illness.
- A nutrition intervention should take into account a person's food practices, lifestyle, cultural orientation, educational background, and degree of motivation. Nutrition education should be individualized to accommodate a patient's needs and learning style.
- The nutrition care plan can be evaluated by reviewing relevant outcome measures of health status and determining the patient's understanding and acceptance of the intervention.
- Each step of nutrition care should be clearly documented in the medical record; the ADIME and SOAP formats are popular styles of documentation.

TABLE 18-4 Examples of Modified Diets

Type of Diet	Description of Diet	Appropriate Uses
Modified Texture and Consistency		
Mechanically altered diets	Contain foods that are modified in texture. Pureed diets include only pureed foods; mechanical soft diets may include solid foods that are mashed, minced, ground, or soft.	Pureed diets are used for people with swallowing difficulty, poor lip and tongue control, or oral hypersensitivity. Mechanical soft diets are appropriate for people with limited chewing ability or certain swallowing impairments.
Blenderized liquid diet	Contains fluids and foods that are blenderized to liquid form.	For people who cannot chew, swallow easily, or tolerate solid foods.
Clear liquid diet	Contains clear fluids or foods that are liquid at room temperature and leave minimal residue in the colon.	For preparation for bowel surgery or colonoscopy, for acute GI disturbances (such as after GI surgeries), or as a transition diet after intravenous feeding. For short-term use only.
Modified Nutrient or Food Content		
Fat-controlled diet	Limits dietary fat to low (<50 g/day) or very low (<25 g/day) intakes.	For people who have certain malabsorptive disorders or symptoms of diarrhea, flatulence, or steatorrhea (fecal fat) resulting from dietary fat intolerance.
Fiber-restricted diet	Limits dietary fiber; degree of restriction depends on the patient's condition and reason for restriction.	For acute phases of intestinal disorders or to reduce fecal output before surgery. Not recommended for long-term use.
Sodium-controlled diet	Limits dietary sodium; degree of restriction depends on symptoms and disease severity.	To help lower blood pressure or prevent fluid retention; used in hypertension, congestive heart failure, renal disease, and liver disease.
High-kcalorie, high-protein diet	Contains foods that are kcalorie and protein dense.	Used for patients with high kcalorie and protein requirements (due to cancer, AIDS, burns, trauma, and other conditions); also used to reverse malnutrition, improve nutritional status, or promote weight gain.

American Dietetic Association, *Nutrition Care Manual* (Chicago: American Dietetic Association, 2010).

Energy Intakes in Hospital Patients

- Energy intakes in hospital patients can be estimated by multiplying a person's resting metabolic rate (RMR) by factors that account for the medical condition, medical treatments, and activity level.
- RMR values can be obtained from indirect calorimetry or a predictive equation.
- The energy needs of critical care patients may be higher than normal due to fever, mechanical ventilation, restlessness, and open wounds.

Dietary Modifications

- Dietary modifications prescribed during illness include changes in food texture or consistency, nutrient content, or food content.
- Mechanically altered diets may be prescribed for people with swallowing and chewing difficulties.
- Clear liquid diets may be used briefly after acute gastrointestinal disturbances or parenteral nutrition or before various diagnostic tests.
- Some medical conditions may require control or restriction of specific nutrients, such as fat, fiber, or sodium.
- A high-kcalorie, high-protein diet may help prevent or reverse malnutrition, improve nutrition status, or promote weight gain.
- In some cases, nutrients need to be delivered via tube feedings or intravenously.

Foodservice

- Hospital foodservice departments may accommodate the special needs of hundreds of patients daily.
- Diet manuals specify the foods to include in modified diets.
- Many hospitals provide selective menus from which patients can choose meals that are appropriate for their medical condition.
- Hospital patients may need assistance at mealtime and encouragement to consume adequate amounts of food.

Implementing Nutrition Care (pp. 595–598)

1. Give examples of the different types of nutrition interventions conducted in the clinical setting. Describe the various elements of the planning and implementation phases of a nutrition intervention.
2. Discuss the factors that should be considered when a patient is encouraged to make long-term dietary changes. Describe how a typical nutrition education session might be conducted.
3. Compare the elements of the ADIME and SOAP formats that are often used for documenting nutrition care. Describe the components of a PES statement.
4. A successful nutrition intervention would include a long list of:
   a. dietary changes that the patient should consider making.
   b. foods that the patient should avoid.
   c. appetizing meals and foods that the patient can include in the diet.
   d. reasons why the patient should make dietary changes.
5. The most important factor(s) that affect(s) how nutrition education is presented is (are):
   a. the person's nutrient needs and nutrition status.
   b. the person's abilities and motivation.
   c. the person's medical history.
   d. the entries in the medical record.
6. Which style of nutrition documentation most closely reflects the steps of the nutrition care process?
   a. ADIME format
   b. SOAP format
   c. PES statement
   d. both a and b

Energy Intakes in Hospital Patients (pp. 598–600)

7. Explain how energy requirements are determined in the clinical setting. Compare the use of indirect calorimetry and predictive equations for determining RMR.
8. The Mifflin–St. Jeor equation:
   a. is considered the "gold standard" for determining RMR.
   b. predicts RMR using weight and height alone.
   c. predicts RMR for children, but not for adults.
   d. accurately predicts RMR in overweight and obese individuals.

Dietary Modifications (pp. 600–605)

9. Give examples of foods that are included in pureed diets, mechanical soft diets, and blenderized diets. Give examples of patients who may benefit from these diets.
10. Describe the uses of the clear liquid diet, and list permitted foods.
11. Discuss the uses of the fat-controlled, fiber-restricted, and sodium-controlled diets.
12. Give examples of the foods included in high-kcalorie, high-protein diets.
13. Mechanically altered diets are often prescribed for individuals with:
    a. disorders of the liver, gallbladder, and pancreas.
    b. unusually high kcalorie and protein requirements.
    c. chewing and swallowing difficulties.
    d. malabsorptive disorders.
14. Foods permitted on the clear liquid diet include all of the following *except*:
    a. milk.
    b. fruit ices.
    c. flavored gelatin.
    d. consommé.
15. The modified diet *least* likely to provide adequate nutrients and kcalories is the:
    a. blenderized diet.
    b. clear liquid diet.
    c. ground/minced diet.
    d. high-kcalorie, high-protein diet.
16. Fiber restriction may be recommended:
    a. for patients with dysphagia.
    b. for patients who are unable to absorb fat normally.
    c. for patients with heartburn.
    d. during the acute phases of some intestinal disorders.

Foodservice (pp. 605–607)

17. Discuss the advantages and potential problems associated with the use of selective menus in hospital foodservice.
18. Describe how health care professionals can help hospital patients improve their food intakes.
19. A nurse notices a food on a patient's tray and is not sure if the food is allowed on the patient's diet. An appropriate action for the nurse to take would be to check the:
    a. nutrition care plan.
    b. diet manual.
    c. nutrition prescription.
    d. medical record.

Study Questions (multiple choice)
4. c 5. b 6. a 8. d 13. c 14. a 15. b 16. d 19. b

For additional study questions and activities, go to **www.cengagebrain.com** and search for ISBN 084006845X.

STUDY CARD 19

Medications, Diet-Drug Interactions, and Herbal Products

FIGURE 19-1 Folate and Methotrexate

By competing for the enzyme that activates folate, methotrexate prevents cancer cells from obtaining the folate they need to multiply. In the process, normal cells are also deprived of the folate they need.

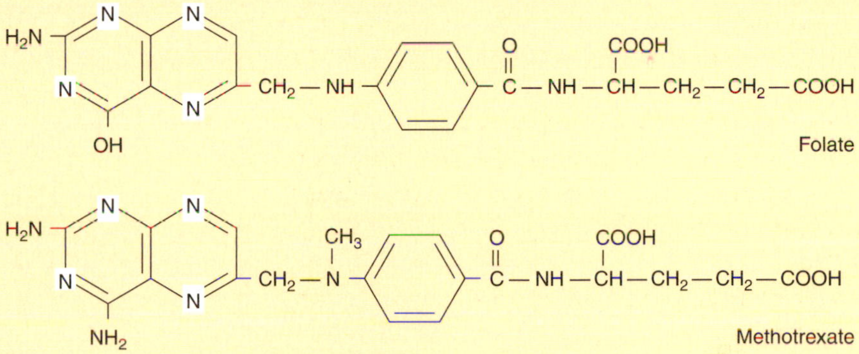

Medications in Disease Treatment

- Both prescription and OTC drugs must be shown to be safe and effective before they are sold. The benefits of using a medication should be greater than the risks associated with its use.
- Potential risks of medications include side effects, drug-drug and diet-drug interactions, and medication errors.
- Medication errors often involve incorrect dosing or using the wrong drug. Bar coding is required on medications sold to health institutions, and confusing terms are being eliminated from documents related to patient care.
- Patients at highest risk of experiencing adverse drug effects include pregnant and nursing women, children, and the elderly. Health professionals should discuss the risks and benefits of medications with patients and alert them to potential dangers and possible solutions.

Diet-Drug Interactions

- Medications can alter food intake and affect the absorption, metabolism, and excretion of nutrients; components of foods can similarly affect drug activity. The accompanying table summarizes the various types of diet-drug interactions.

Herbal Products

- Herbal products are not reliable treatments for medical conditions. There is little evidence demonstrating their effectiveness and safety, and the concentrations of active ingredients may vary greatly.
- Safety concerns for use of herbal products include adverse effects, contamination, and herb-drug interactions.
- Manufacturers and distributors of herbal supplements are responsible for determining product safety but are not required to conduct safety studies. The FDA must prove that a supplement is unsafe before removing it from the market.
- Consumers using herbs may delay getting an appropriate treatment for their condition and may receive questionable advice from supplement retailers.

Affected Body Function	Effects of Drugs	Effects of Food Components
Food intake	May increase or decrease appetite, alter taste sensation, cause GI discomfort	—
Absorption	May bind to nutrients, alter stomach acidity, interfere with nutrient transport into intestinal cells	May alter stomach emptying rate, alter stomach acidity, bind to drugs
Metabolism	May alter activity of enzymes that metabolize nutrients	May alter activity of enzymes that metabolize drugs
Excretion	May increase or decrease nutrient losses in the urine	May increase or decrease drug losses in the urine
Varies	May interact with nutrients and cause toxic side effects	May interact with drugs and cause toxic side effects

Medications in Disease Treatment

(pp. 619–622)

1. Describe similarities and differences among prescription medications, over-the-counter drugs, and generic versions of drugs. Give examples of ways in which medications are introduced into the body.
2. Explain why some patient populations are at high risk for adverse effects from drugs.
3. Over-the-counter drugs are:
   a. unlikely to cause adverse effects.
   b. unlikely to interact with dietary components.
   c. generally used for longer periods of time than prescription medications.
   d. used to treat illnesses that are normally self-diagnosed and self-treated.
4. Recommendations for reducing the incidence of medication errors include:
   a. physician supervision whenever drugs are administered.
   b. advising patients to take only one medication at a time.
   c. requiring that prescriptions be handwritten instead of typed.
   d. avoiding the use of confusing terms on clinical documents.
5. Adverse drug effects are most likely when:
   a. multiple medications are used.
   b. generic drugs are substituted for brand-name drugs.
   c. patients begin using a new medication.
   d. medications are taken for just one or two days.

Diet-Drug Interactions (pp. 622–628)

6. Discuss ways in which medications can affect food intake.
7. Describe how medications can interfere with nutrient absorption and how dietary factors can affect drug absorption.
8. Explain how drugs and nutrients may influence each other's metabolism, and provide examples.
9. Discuss diet-drug interactions that can alter the excretion of nutrients or medications.
10. Explain why tyramine intake must be monitored by people using monoamine oxidase (MAO) inhibitors.
11. Examples of medication-related symptoms that can significantly limit food intake include:
    a. ringing in the ears.
    b. persistent nausea and vomiting.
    c. insomnia.
    d. skin rash.
12. Factors that typically interfere with drug absorption include:
    a. binding between drugs and food components.
    b. use of antacid therapies.
    c. a rapid stomach-emptying rate.
    d. all of the above.
13. Compounds in grapefruit juice:
    a. bind to antibiotics, reducing absorption.
    b. cause excessive drug excretion.
    c. strengthen the effects of certain drugs.
    d. alter acidity in the stomach, impairing drug absorption.
14. Vitamin K consumption should be consistent in patients using:
    a. tetracycline.
    b. isoniazid.
    c. warfarin.
    d. lithium.
15. An important step that health care practitioners can take to limit the risk of medication-related side effects is to:
    a. recommend use of over-the-counter drugs instead of prescription medications.
    b. encourage use of herbal supplements rather than prescription medications.
    c. advise patients to take medications separately from meals.
    d. ask patients to fully describe the types and amounts of medications and dietary supplements they are using.

Herbal Products (pp. 628–631)

16. Explain why herbal products are not dependable treatments for medical conditions. Describe possible dangers associated with the use of these products.
17. An important difference between medications and herbal products that reach the marketplace is that:
    a. medications that cause adverse effects cannot be sold.
    b. medications are subject to contamination with toxic metals, molds, and bacteria.
    c. herbal products are not required to prove safety and effectiveness.
    d. herbal products must provide standard amounts of active ingredients.

Study Questions (multiple choice)

3.d 4.d 5.a 11.b 12.d 13.c 14.c 15.d 17.c

For additional study questions and activities, go to **www.cengagebrain.com** and search for ISBN 084006845X.

STUDY CARD 20

Enteral Nutrition Support

Enteral Formulas

- Enteral formulas are liquid diets that can meet all of a patient's nutritional needs.
- Standard enteral formulas contain intact proteins and polysaccharides and are provided to patients who can digest and absorb nutrients without difficulty; elemental formulas meet the nutrient needs of patients with limited digestive and absorptive functions.
- Specialized enteral formulas are available for patients with specific diseases.
- Modular formulas, which contain individual macronutrients, can be used to modify other enteral formulas.
- Formulas differ in their macronutrient composition, energy density, fiber content, and osmolality. Most people can tolerate isotonic and hypertonic formulas without difficulty.
- The chief concern in enteral formula selection is the formula's ability to meet the patient's nutritional requirements.

Enteral Nutrition in Medical Care

- Enteral formulas are provided to patients with functioning GI tracts who cannot meet nutritional needs with conventional foods alone.
- A nasoenteric feeding route is preferred for short-term tube feedings, whereas enterostomies are used for longer-term feedings. Because the stomach delivers nutrients into the intestine at a controlled rate, gastric feedings are often preferred, although they are frequently avoided in patients at risk of aspiration.
- The selection of feeding tubes is based on patient age and size, the feeding route, and formula viscosity.

Administration of Tube Feedings

- Enteral formulas should be prepared and administered using food safety protocols that reduce the risk of contamination.
- Feeding tube placement must be verified and monitored to reduce the risks of aspiration and inadvertent placement into the respiratory tract.
- Depending on the patient's medical condition and the feeding route, an enteral formula can be delivered in bolus feedings, intermittently, or continuously.
- Although enteral formulas meet a substantial portion of the water requirements, additional water can be provided by flushing water through the feeding tube.
- Medications should be given separately from formulas and accompanied by water flushes to prevent tube clogging.
- Complications of tube feedings can be gastrointestinal, mechanical, or metabolic in nature.
- Tube feedings are tapered off when the patient begins consuming an oral diet.

GLOSSARY OF TUBE FEEDING ROUTES

For each type of tube placement, the terms are listed in order from the upper to lower organs of the digestive system.

transnasal: a *transnasal feeding tube* is one that is inserted through the nose.

- **nasogastric (NG):** tube is placed into the stomach via the nose.
- **nasoenteric:** tube is placed into the GI tract via the nose. (*Nasoenteric feedings* usually refer to *nasoduodenal* and *nasojejunal* feedings.)
- **nasoduodenal (ND):** tube is placed into the duodenum via the nose.
- **nasojejunal (NJ):** tube is placed into the jejunum via the nose.

orogastric: tube is inserted into the stomach through the mouth. This method is often used to feed infants because a nasogastric tube may hinder the infant's breathing.

enterostomy (EN-ter-AH-stoe-mee): an opening into the GI tract through the abdominal wall.

- **gastrostomy** (gah-STRAH-stoe-mee): an opening into the stomach through which a feeding tube can be passed. A nonsurgical technique for creating a gastrostomy under local anesthesia is called *percutaneous endoscopic gastrostomy (PEG)*.
- **jejunostomy** (JEH-ju-NAH-stoe-mee): an opening into the jejunum through which a feeding tube can be passed. A nonsurgical technique for creating a jejunostomy is called *percutaneous endoscopic jejunostomy (PEJ)*. The tube can either be guided into the jejunum via a gastrostomy or passed directly into the jejunum (*direct PEJ*).

Enteral Formulas (pp. 639–642)

1. Characterize standard formulas, elemental formulas, specialized formulas, and modular formulas, and describe situations in which they are used.
2. Discuss how macronutrient composition, energy density, fiber content, and osmolality vary in enteral formulas. Describe the factors that may influence the selection of an enteral formula.
3. Which of the following statements is correct?
   a. Standard formulas contain whole proteins or protein isolates.
   b. Standard formulas contain free amino acids or small peptide chains.
   c. Modular formulas contain a mixture of proteins, carbohydrates, and fats.
   d. Elemental formulas may contain protein isolates or whole proteins.
4. *Osmolality* refers to an enteral formula's:
   a. energy density.
   b. nutrient density.
   c. fiber content.
   d. concentrations of molecules and ionic particles.
5. In selecting an appropriate enteral formula for a patient, the primary consideration is:
   a. formula osmolality.
   b. the patient's nutrient needs.
   c. availability of infusion pumps.
   d. formula cost.

Enteral Nutrition in Medical Care (pp. 642–646)

6. Identify reasons why oral intake of enteral formulas may be advised. Suggest ways for improving patient acceptance of formulas.
7. List the types of patients who may benefit from tube feedings.
8. Describe the different tube feeding routes, and suggest reasons why each might be used. Discuss advantages and disadvantages of each.
9. For a patient who is at high risk of aspiration and is not expected to be able to eat table foods for several months, an appropriate placement of a feeding tube might be:
   a. nasogastric.
   b. nasoenteric.
   c. gastrostomy.
   d. jejunostomy.

Administration of Tube Feedings (pp. 646–654)

10. Identify measures that can help prevent contamination of enteral formulas and equipment.
11. Contrast the different methods of formula delivery, and discuss the possible advantages and disadvantages associated with each. Discuss how clinicians can help relieve anxiety about tube feeding procedures.
12. Describe the problems that can occur when medications are delivered through feeding tubes. Suggest guidelines that can prevent these problems.
13. Discuss complications often associated with tube feedings. Summarize possible causes and some measures that can prevent or correct these complications.
14. An important measure that may prevent bacterial contamination in tube feeding formulas is:
    a. nonstop feeding of formula.
    b. using the same feeding bag and tubing each day.
    c. discarding opened containers of formula not used within 24 hours.
    d. adding formula to the feeding container before it empties completely.
15. Compared with intermittent feedings, continuous feedings:
    a. require an infusion pump.
    b. allow greater freedom of movement.
    c. are more similar to normal patterns of eating.
    d. are associated with more GI side effects.
16. A patient needs 1800 milliliters of formula a day. If the patient is to receive formula intermittently every 4 hours, how many milliliters of formula will she need at each feeding?
    a. 225
    b. 300
    c. 400
    d. 425
17. The term that describes the volume of formula remaining in the stomach from a previous feeding is:
    a. residue.
    b. osmolar load.
    c. gastric residual.
    d. intermittent feeding.
18. Tube feedings can gradually be discontinued when:
    a. discharge planning begins.
    b. the patient experiences hunger.
    c. the medical condition resolves.
    d. the patient is able to eat foods or drink formula in sufficient amounts.

Study Questions (multiple choice)

3. a 4. d 5. b 9. d 14. c 15. a 16. b 17. c 18. d

For additional study questions and activities, go to **www.cengagebrain.com** and search for ISBN 084006845X.

STUDY CARD 21

Parenteral Nutrition Support

Indications for Parenteral Nutrition

- Parenteral nutrition support delivers nutrients intravenously. It is used in patients whose GI tract is not functioning and who may readily become malnourished.
- Patients receiving parenteral nutrition generally have intestinal disorders or are critically ill.
- If nutrients are infused directly into peripheral veins (peripheral parenteral nutrition), nutrient concentrations must be limited to avoid inflammation of the veins. The infusion of nutrients into central veins (total parenteral nutrition) can supply nutrient-dense solutions and is used for long-term intravenous feedings.

Parenteral Solutions

- Prescriptions for parenteral solutions are individualized to meet each patient's needs. The solutions are compounded in hospital pharmacies using commercial nutrient preparations and include amino acids, dextrose, electrolytes, vitamins, and trace minerals.
- Few medications are added to parenteral solutions due to the potential for drug-nutrient interactions.
- Parenteral solutions that include lipids are called total nutrient admixtures, 3-in-1 solutions, or all-in-one solutions; solutions that exclude lipids are called 2-in-1 solutions.
- Parenteral solutions are prepared and handled using aseptic techniques to prevent contamination.

Administering Parenteral Nutrition

- A nutrition support team, made up of physicians, nurses, dietitians, and pharmacists, may administer parenteral nutrition support or serve as advisers to other clinicians.
- Parenteral solutions may be initiated gradually or provided at full volume and full strength in selected patients.
- Critically ill patients may require continuous parenteral infusions, whereas healthier patients and long-term users may prefer cyclic infusions.
- Catheters are frequently the cause of complications, which include improper placement or dislodgment, infection, clotting, embolism, and phlebitis.
- Metabolic complications of parenteral support include hyperglycemia and hypoglycemia; hypertriglyceridemia; fluid and electrolyte imbalances; and diseases affecting the liver, gallbladder, and bone.
- When the need for parenteral nutrition resolves, patients are transitioned to an enteral diet as the volume of parenteral nutrition is gradually reduced.

Nutrition Support at Home

- Candidates for home enteral nutrition services have disorders that interfere with swallowing ability, GI motility, or nutrient absorption. Candidates for home parenteral nutrition have disorders that severely impair nutrient absorption or cause intestinal motility problems.
- Patients and caregivers should participate in decisions about access sites, formulas, and nutrient delivery methods.
- Enteral formulas and parenteral solutions can be purchased or prepared in the home.
- The use of portable pumps may help individuals lead a normal lifestyle. Nevertheless, lifestyle adjustments to nutrition support may be difficult and stressful.

TABLE 21-1 Potential Complications of Parenteral Nutrition

Catheter-Related	Metabolic
Air embolism	Abnormal liver function
Blood clotting at catheter tip	Electrolyte imbalances
Clogging of catheter	Gallbladder disease
Dislodgment of catheter	Hyperglycemia, hypoglycemia
Improper placement	Hypertriglyceridemia
Infection, sepsis	Metabolic bone disease
Phlebitis	Nutrient deficiencies
Tissue injury	Refeeding syndrome

Indications for Parenteral Nutrition

(pp. 664–665)

1. Identify the individuals who are most likely to benefit from parenteral nutrition support.
2. Compare the general characteristics of peripheral parenteral nutrition (PPN) and total parenteral nutrition (TPN). Discuss the advantages and disadvantages of each of these methods.
3. TPN is preferred over PPN for a patient who:
   a. does not have high nutrient requirements.
   b. needs long-term parenteral nutrition support.
   c. has strong peripheral veins and moderate nutrient needs.
   d. needs parenteral feedings as a supplement to tube feedings.

Parenteral Solutions (pp. 666–670)

4. Describe how amino acids, carbohydrate, and lipids are provided in parenteral solutions. List the vitamins and minerals that are usually included.
5. Identify the components of parenteral solutions that contribute to osmolarity. Explain how the osmolarity of a solution can be estimated.
6. Compare the components of total nutrient admixtures (TNA) and 2-in-1 solutions. Describe the advantages and disadvantages associated with the use of TNA solutions.
7. Which of the following cannot be delivered intravenously?
   a. dextrose
   b. amino acids
   c. lipid emulsions
   d. hydrolyzed enteral formulas
8. How many kcalories are supplied by 500 milliliters of an 8.5 percent amino acid solution?
   a. 60
   b. 170
   c. 340
   d. 500
9. For a patient receiving central TPN who also receives intravenous lipid emulsions two or three times a week, the lipid emulsions serve primarily as a source of:
   a. essential fatty acids.
   b. cholesterol.
   c. fat-soluble vitamins.
   d. concentrated energy.
10. Which nutrient is often omitted from parenteral solutions because it may destabilize other ingredients in the solution?
    a. calcium
    b. vitamin K
    c. iron
    d. chromium

Administering Parenteral Nutrition

(pp. 670–674)

11. Compare the different methods for administering parenteral solutions. List the parameters that are typically monitored during parenteral feedings.
12. Explain how patients make the transition from parenteral to oral feedings.
13. Discuss the potential complications associated with parenteral nutrition. Suggest ways in which these complications can be prevented or corrected.
14. Routine monitoring of patients on TPN requires all of the following *except*:
    a. serum electrolyte measurement.
    b. monitoring for daily weight changes.
    c. indirect calorimetry.
    d. liver enzyme tests.
15. The transition from parenteral feedings to an oral diet is primarily designed to:
    a. improve appetite.
    b. prevent hypoglycemia.
    c. prevent apprehension about eating.
    d. ensure that nutrient needs will continue to be met.
16. Complications associated with parenteral feedings may include:
    a. hyperglycemia.
    b. hypertriglyceridemia.
    c. infection.
    d. all of the above.
17. Refeeding syndrome causes dangerous fluctuations in:
    a. electrolytes.
    b. liver enzymes.
    c. triglycerides.
    d. ketone bodies.

Nutrition Support at Home (pp. 674–676)

18. Discuss how home nutrition support is administered. Identify ideal candidates for these services and the problems that patients may experience.
19. Patients using home parenteral nutrition:
    a. are unable to use TNA solutions.
    b. are usually given continuous rather than cyclic infusions.
    c. require infusion pumps for use at home.
    d. are generally unable to work outside of the home or travel.

Study Questions (multiple choice)
3. b 7. d 8. b 9. a 10. c 14. c 15. d 16. d 17. a 19. c

For additional study questions and activities, go to **www.cengagebrain.com** and search for ISBN 084006845X.

STUDY CARD 22

Metabolic and Respiratory Stress

The Body's Responses to Stress and Injury

- The stress and inflammatory responses are nonspecific responses to stressors that cause infection and injury.
- The stress response is mediated by the catecholamine hormones, cortisol, and glucagon, which together raise nutrient levels in blood, stimulate heart rate, raise blood pressure, and increase metabolic rate. Aldosterone and antidiuretic hormone help maintain adequate blood volume.
- The inflammatory process—mediated by compounds released from damaged tissues, immune cells, and blood vessels—results in systemic effects that alter nutrient metabolism, heart rate, blood pressure, body temperature, and immune cell functions.
- Signs of inflammation in injured tissues include swelling, redness, heat, and pain.
- Persistent, severe inflammation may result in shock and increases the risk of multiple organ dysfunction.

TABLE 22-2 Disease-Specific Stress Factors for Estimating Energy Needs during Metabolic Stress

Method:
Step 1. Estimate the energy needed to support resting metabolic rate (RMR) using indirect calorimetry or a predictive equation (see Table 18-3, p. 599).
Step 2. Multiply the patient's RMR by an appropriate stress factor for acute illness (see the example in the "How To" box on p. 600).
Examples of stress factors:
• Advanced liver disease: 1.0 to 1.16 • Inflammatory bowel disease (active): 1.05 to 1.10 • Pancreatitis: 1.13 to 1.21 • Surgery: 1.2 to 1.4 • Mechanical ventilation: 1.32 to 1.34 • Burns (20 to 30 percent of body surface): 1.6 to 1.7 • Burns (30 to 40 percent of body surface): 1.8 to 1.9 • Burns (40 to 45 percent of body surface): 2.0

SOURCE: American Dietetic Association, *Nutrition Care Manual* (Chicago: American Dietetic Association, 2007); N. Barak, E. Wall-Alonso, and M. D. Sitrin, Evaluation of stress factors and body weight adjustments currently used to estimate energy expenditure in hospitalized patients, *Journal of Parenteral and Enteral Nutrition* 26 (2002): 231–238.

Nutrition Treatment of Acute Stress

- Severe metabolic stress can cause hypermetabolism, negative nitrogen balance, and hyperglycemia, and may result in wasting.
- The objectives of nutrition care during acute stress are to provide a diet that preserves muscle tissue, maintains immune defenses, and promotes healing.
- Energy needs during acute stress are often estimated by modifying RMR values with stress factors that account for the increased demands of the medical condition or treatment.
- Protein recommendations during acute stress are higher than DRI levels to help prevent tissue losses and allow healing of damaged tissue.
- Enteral or parenteral nutrition support may be needed to meet the high nutrient requirements of acutely stressed patients.
- Burn patients typically require fluid replacement and electrolyte management after a burn injury, and a high-kcalorie, high-protein diet during the recovery period.

Nutrition and Respiratory Stress

- Respiratory stress from chronic or acute disease affects body weight, muscle mass, and the normal functioning of all body tissues.
- Chronic obstructive pulmonary diseases (COPD) are debilitating, progressive illnesses that can lead to malnutrition, muscle wasting, and activity intolerance.
- The goals of nutrition therapy for COPD are to improve food intake, maintain proper weight, preserve muscle tissue, and improve exercise endurance.
- Respiratory failure, characterized by hypoxemia and hypercapnia, can result from conditions that cause lung injury or impair lung function.
- Goals of nutrition therapy for respiratory failure are to supply enough energy and protein to support lung function without burdening the respiratory system. Fluid restrictions may be necessary to reverse pulmonary edema.
- Acute respiratory distress syndrome (ARDS) is a severe form of respiratory failure that requires mechanical ventilation.

The Body's Responses to Stress and Injury
(pp. 683–686)

1. Describe the effects of the stress response on the body. Specify the main hormones involved and their metabolic effects.
2. Discuss the main effects of the inflammatory response following infection or injury. Identify the chemical mediators involved, and explain how they help regulate the inflammatory process.
3. Characterize the acute-phase response, giving examples of the clinical symptoms and changes in blood chemistry that usually result.
4. Which of the following metabolic changes accompanies acute stress?
   a. reduced plasma concentrations of glucose and fatty acids
   b. reduced blood volume and blood pressure
   c. increased insulin action
   d. catabolism of protein in skeletal muscle and connective tissue
5. Tissue injury is followed by:
   a. fluid accumulation in damaged tissue.
   b. reduced blood flow to injured tissue.
   c. reduced capillary permeability.
   d. decreased body temperature.
6. What is a possible effect of replacing vegetable oils rich in omega-6 fatty acids with oils rich in omega-3 fatty acids?
   a. improvement in blood circulation
   b. suppression of inflammation
   c. protection against sepsis
   d. hypertriglyceridemia

Nutrition Treatment of Acute Stress
(pp. 686–691)

7. How does metabolic stress affect nutrition status? Explain why nutrition status may be difficult to evaluate in an acutely stressed individual.
8. Describe how energy and protein needs are estimated during acute stress. Which micronutrients are sometimes supplemented?
9. Characterize first-, second-, and third-degree burns. What measures are taken immediately after a burn occurs?
10. Identify the objectives of nutrition care for burn patients. Explain how the energy and protein needs of patients with burns are estimated. What measures are taken to provide adequate nutrient intakes?
11. Which of the following statements concerning protein and energy recommendations during acute metabolic stress is true?
    a. Protein and energy recommendations are similar to those for healthy people.
    b. Protein and energy recommendations are reduced because a stressed individual cannot metabolize nutrients normally.
    c. Acutely stressed individuals can benefit from as much protein and energy as can be provided.
    d. Protein and energy recommendations are high in order to minimize muscle tissue losses.

Nutrition and Respiratory Stress (pp. 691–696)

12. Describe the two major types of chronic obstructive pulmonary disease (COPD). Discuss their causes and treatment. What are the possible effects of COPD on body composition?
13. Describe respiratory failure and acute respiratory distress syndrome (ARDS) and their consequences. Identify the key elements of medical treatment and nutrition therapy.
14. A primary feature of emphysema is:
    a. obstruction within the bronchi.
    b. obstruction within the bronchioles.
    c. destruction of the walls separating the alveoli.
    d. excessive lung elasticity.
15. The weight loss and wasting that often occur in COPD can be caused by:
    a. reduced food intake.
    b. increased metabolic rate.
    c. reduced exercise tolerance.
    d. all of the above.
16. Nutrition therapy for a person with respiratory failure includes:
    a. careful attention to providing enough, but not too much, energy.
    b. a generous fluid intake to facilitate mucus clearance.
    c. a high fat intake to prevent weight loss.
    d. a high carbohydrate intake to limit carbon dioxide production.

Study Questions (multiple choice)
4.d 5.a 6.b 11.d 14.c 15.d 16.a

For additional study questions and activities, go to **www.cengagebrain.com** and search for ISBN 084006845X.

STUDY CARD 23

Upper Gastrointestinal Disorders

Conditions Affecting the Esophagus

- Dysphagia and gastroesophageal reflux are the most common esophageal disorders.
- Dysphagia may interfere with food intake and increase the risk of aspiration. Treatment may include dietary adjustments, strengthening exercises, and using different swallowing techniques.
- Gastroesophageal reflux disease (GERD) may lead to esophageal ulcers, inflammation, bleeding, and stricture. Treatment includes the use of acid-suppressing drugs and lifestyle changes.

Conditions Affecting the Stomach

- Dyspepsia refers to general symptoms of indigestion such as abdominal pain, nausea, and vomiting, which can be caused by a variety of medical conditions.
- Dietary measures for dyspepsia may include avoiding large meals, fatty or spicy foods, and foods that trigger symptoms.
- Gastritis and peptic ulcer disease are most often caused by *Helicobacter pylori* infection. Another possible cause is NSAID use, which can promote gastritis and peptic ulcer disease by damaging the mucosal lining.
- Extensive damage to the mucosa may reduce gastric secretions and increase the risks of developing iron and vitamin B_{12} deficiencies.
- The nutrition care for gastritis and peptic ulcer disease includes correcting any nutritional deficiencies that develop and eliminating dietary substances that cause pain or discomfort.

Gastric Surgery

- Gastric surgeries, used to treat cancer, peptic ulcer complications, and obesity, require dietary adjustments after surgery and are associated with complications that may affect nutrition status.
- Common complications after gastric surgery include fat malabsorption, bone disease, anemia, and dumping syndrome.
- Dumping syndrome, characterized by a group of symptoms caused by rapid gastric emptying, is managed by controlling food portions, avoiding foods high in sugars, and consuming liquids between meals.
- After bariatric surgery, patients must learn to consume appropriate food portions, use dietary supplements to prevent nutrient deficiencies, and choose foods that are unlikely to cause abdominal discomfort, vomiting, or dumping syndrome.

FIGURE 23-1 The Upper GI Tract

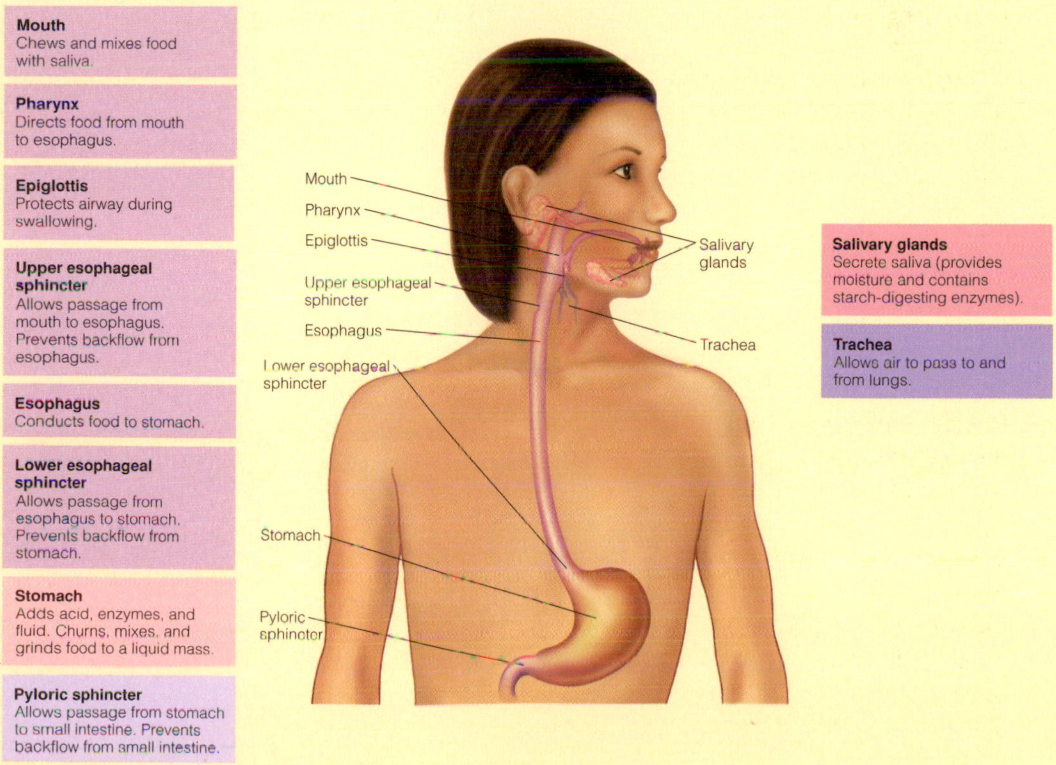
Mouth
Chews and mixes food with saliva.
Pharynx
Directs food from mouth to esophagus.
Epiglottis
Protects airway during swallowing.
Upper esophageal sphincter
Allows passage from mouth to esophagus. Prevents backflow from esophagus.
Esophagus
Conducts food to stomach.
Lower esophageal sphincter
Allows passage from esophagus to stomach. Prevents backflow from stomach.
Stomach
Adds acid, enzymes, and fluid. Churns, mixes, and grinds food to a liquid mass.
Pyloric sphincter
Allows passage from stomach to small intestine. Prevents backflow from small intestine.
Mouth
Pharynx
Epiglottis
Upper esophageal sphincter
Esophagus
Lower esophageal sphincter
Stomach
Pyloric sphincter
Salivary glands
Trachea
Salivary glands
Secrete saliva (provides moisture and contains starch-digesting enzymes).
Trachea
Allows air to pass to and from lungs.

Conditions Affecting the Esophagus

(pp. 705–712)

1. Provide examples of conditions that can interfere with swallowing. Describe how diets are adjusted to meet the needs of people with dysphagia.
2. Discuss ways to provide appetizing meals for people consuming mechanically altered foods.
3. Identify the symptoms, causes, and complications of gastroesophageal reflux disease (GERD). What lifestyle modifications can benefit patients with GERD?
4. If a patient with dysphagia has difficulty swallowing solids but can easily swallow liquids:
   a. the problem is probably a motility disorder.
   b. the patient most likely has achalasia.
   c. the problem is probably an esophageal obstruction.
   d. the patient most likely has oropharyngeal dysphagia.
5. The health care practitioner working with a patient with dysphagia should recognize that:
   a. only pureed foods should be given to minimize the risk of aspiration.
   b. the patient can have any food that can be comfortably and safely chewed and swallowed.
   c. highly seasoned foods are often restricted.
   d. conventional diets are unable to meet total nutrient needs, so supplements are necessary.
6. Gastroesophageal reflux disease (GERD) is:
   a. characterized by frequent backflow of the stomach's gastric secretions into the esophagus.
   b. a protuberance of a portion of the stomach above the lower esophageal sphincter.
   c. an open sore in the GI mucosa caused by the destructive effects of gastric secretions.
   d. an obstruction of the lower esophagus that results in dysphagia.
7. Conditions associated with an increased risk of developing GERD include:
   a. hiatal hernia. c. pregnancy.
   b. asthma. d. all of the above.

Conditions Affecting the Stomach

(pp. 712–716)

8. What are possible causes of nausea and vomiting? Discuss interventions that may help.
9. Specify the common causes of gastritis and peptic ulcer disease. Explain the possible consequences of these diseases. Describe the role of diet therapy for both conditions.
10. For the patient with persistent vomiting, the major nutrition-related concern(s) is (are):
    a. dehydration and malnutrition.
    b. reflux esophagitis.
    c. dyspepsia.
    d. peptic ulcers.
11. Chronic gastritis frequently leads to:
    a. dumping syndrome.
    b. bone disease.
    c. iron and vitamin B_{12} deficiencies.
    d. excessive hydrochloric acid secretion.
12. The primary cause of most peptic ulcers is:
    a. consumption of spicy foods.
    b. hypochlorhydria.
    c. smoking cigarettes.
    d. *Helicobacter pylori* infection.
13. Foods discouraged for patients with gastritis or active ulcers include those that:
    a. are high in fiber.
    b. irritate the gastric mucosa.
    c. are easy to swallow.
    d. contain simple sugars.

Gastric Surgery (pp. 716–721)

14. Describe the gastric disorders that may benefit from gastric surgery. What dietary adjustments are usually required after a gastrectomy procedure?
15. What complications may arise after gastric surgery? Discuss the dietary interventions that may help prevent these consequences.
16. Describe the surgical procedures used to treat severe obesity. Discuss dietary recommendations for patients who have undergone bariatric surgery.
17. People at risk of dumping syndrome should generally avoid:
    a. high-fiber foods. c. beverages.
    b. sweets and sugars. d. bread and potatoes.
18. The health practitioner assessing a patient who underwent a gastrectomy several years ago should be alert to signs of:
    a. dysphagia. c. anemia.
    b. GERD. d. gastritis.

Study Questions (multiple choice)
4. c 5. b 6. a 7. d 10. a 11. c 12. d 13. b 17. b 18. c

For additional study questions and activities, go to **www.cengagebrain.com** and search for ISBN 084006845X.

STUDY CARD 24

Lower Gastrointestinal Disorders

Common Intestinal Problems

- Constipation accompanies many different health problems but generally correlates with low-fiber diets, low food or inadequate fluid intakes, and physical inactivity.
- Intestinal gas is largely produced by colonic bacteria and is usually caused by nutrient malabsorption.
- Diarrhea can result from malabsorption, intestinal infections, motility disorders, and dietary substances and may require oral rehydration therapy to replace fluid and electrolyte losses; a low-residue, low-fat, lactose-free diet can sometimes alleviate intestinal discomfort.

Malabsorption Syndromes

- Malabsorption syndromes can be caused by an undersupply of digestive secretions, damaged intestinal mucosa, or motility disorders. Malabsorption often affects multiple nutrients and can lead to complications that impair nutrition status further.
- Fat malabsorption, usually indicated by the development of steatorrhea, is associated with the loss of food energy and deficiencies of essential fatty acids, fat-soluble vitamins, and some minerals. In severe cases, nutrition therapy may include a fat-controlled diet and use of MCT oil.
- Bacterial overgrowth can result from conditions that reduce gastric acidity or intestinal motility; it typically causes malabsorption of fat and some essential nutrients.

Conditions Affecting the Pancreas

- Acute pancreatitis is short lived and does not cause permanent damage, but it requires the withholding of food and liquids until healing has occurred.
- Chronic pancreatic disorders, such as chronic pancreatitis, can lead to widespread maldigestion and malabsorption due to the impaired secretion of digestive enzymes. Chronic pancreatitis is treated with pancreatic enzyme replacement therapy.
- Cystic fibrosis, a genetic disorder associated with thickened exocrine secretions, causes obstructive lung disease and pancreatic damage. Children with cystic fibrosis have high protein and energy requirements and must use pancreatic enzyme replacement therapy and dietary supplements to reverse malnutrition.

Conditions Affecting the Small Intestine

- Disorders of the small intestine that cause damage to mucosal tissue, such as celiac disease and Crohn's disease, often result in malabsorption.
- Celiac disease is characterized by an abnormal immune response to gluten, a group of water-insoluble proteins in wheat, barley, and rye. Disease treatment involves lifelong adherence to a gluten-free diet.
- Crohn's disease is an inflammatory disease that can cause extensive damage to all layers of intestinal tissue. Treatment includes medications that suppress inflammation and relieve symptoms, dietary adjustments that reduce symptoms and correct deficiencies, and intestinal resection to remove damaged tissue.
- Short bowel syndrome is a frequent consequence of major resections of the small intestine and may require permanent parenteral nutrition support. In many patients, intestinal adaptation improves the intestine's absorptive capacity after resection.
- Ulcerative colitis is an inflammatory disease that affects the rectum and colon only. Whereas mild cases may cause few complications, severe cases may require colectomy.

Conditions Affecting the Large Intestine

- Irritable bowel syndrome is characterized by chronic, recurring intestinal symptoms such as diarrhea and/or constipation, abdominal pain, and flatulence. Although the causes are unknown, the disorder is influenced by food intake, stress, and psychological factors.
- Diverticulosis is often asymptomatic until complications develop; its prevalence increases with advancing age and is often associated with low fiber intakes.
- Patients with irritable bowel syndrome or diverticulosis may benefit from a high-fiber diet; additional measures for treating irritable bowel syndrome may include consuming small meals and omitting certain foods from the diet.
- Colostomies and ileostomies are surgically created openings in the abdominal wall using the colon or ileum. Fluid and electrolyte requirements may be greater after an ostomy due to reduced colon function. Other concerns include possible obstructions, excessive gas production, food odors, and diarrhea.

Common Intestinal Problems (pp. 729–733)

1. What measures can help prevent and treat constipation? What are the modes of action of the laxatives used to treat constipation?
2. Describe possible causes of diarrhea and the dietary measures that may be included during treatment.
3. The health care practitioner advising an elderly patient with constipation encourages the patient to:
   a. consume a low-fat, low-sodium diet.
   b. consume a high-protein diet rich in calcium.
   c. eliminate gas-forming foods from the diet.
   d. gradually add high-fiber foods to the diet.
4. Osmotic diarrhea often results from:
   a. excessive colonic contractions.
   b. excessive fluid secretion by the intestines.
   c. nutrient malabsorption.
   d. viral, bacterial, or protozoal infections.

Malabsorption Syndromes (pp. 733–736)

5. Identify conditions that can cause fat malabsorption, and discuss the primary nutrition problems that can result.
6. Explain why bacterial overgrowth develops, and describe its effects on nutrition status.
7. Nutrition problems that may result from fat malabsorption include all of the following except:
   a. weight loss.
   b. essential amino acid deficiencies.
   c. bone loss.
   d. oxalate kidney stones.
8. Common nutrition problems associated with bacterial overgrowth in the stomach and small intestine include:
   a. sensitivity to gluten.
   b. fat malabsorption and vitamin B_{12} deficiency.
   c. constipation.
   d. permanent loss of digestive enzymes.

Conditions Affecting the Pancreas (pp. 737–739)

9. Explain how chronic pancreatitis and cystic fibrosis can result in malabsorption. In what ways are their dietary treatments similar?
10. Chronic pancreatitis and cystic fibrosis are both treated with:
    a. intestinal resection.
    b. postural drainage.
    c. enzyme replacement therapy.
    d. stool softeners.

Conditions Affecting the Small Intestine (pp. 739–747)

11. Discuss the cause of celiac disease, and describe the diet used in its treatment. Explain why a gluten-free diet may be difficult for patients to follow.
12. Compare the effects of Crohn's disease and ulcerative colitis on nutrition status, and describe the dietary measures that may be required during the course of illness.
13. Identify possible causes of short bowel syndrome. What nutrition problems often develop? Discuss the adaptive process that occurs in the remaining intestine after a portion of intestine is resected.
14. A person on a gluten-free diet must avoid products containing:
    a. wheat, barley, and rye.
    b. barley, soybeans, and corn.
    c. wheat, corn, and rice.
    d. buckwheat, rice, and millet.

Conditions Affecting the Large Intestine (pp. 748–752)

15. Describe irritable bowel syndrome, and discuss the ways in which diet can be used in its treatment.
16. Specify the factors that increase the risk of developing diverticular disease. What dietary modification is most useful in its prevention and treatment?
17. Describe the primary nutrition-related concerns of people who have undergone ileostomies and colostomies.
18. Symptoms of irritable bowel syndrome most often include:
    a. nausea and vomiting.
    b. weight loss and malnutrition.
    c. strong odors and obstructions.
    d. constipation, diarrhea, and flatulence.
19. After an ileostomy, the most serious concern is that:
    a. the diet is too restrictive to meet nutrient needs.
    b. waste disposal causes frequent daily interruptions.
    c. fluid restrictions prevent patients from drinking beverages freely.
    d. incompletely digested foods may cause obstructions.

Study Questions (multiple choice)
3. d 4. c 7. b 8. b 10. c 14. a 18. d 19. d

For additional study questions and activities, go to **www.cengagebrain.com** and search for ISBN 084006845X.

STUDY CARD 25

Liver Disease and Gallstones

Fatty Liver and Hepatitis

- Fatty liver can result from excessive alcohol intake, drug toxicity, and chronic disorders such as diabetes and obesity.
- Hepatitis is frequently caused by viral infection but can also result from alcohol abuse, drug toxicity, and other causes.
- Although fatty liver is often benign, hepatitis can become chronic and lead to cirrhosis and liver cancer.
- Treatment of hepatitis involves supportive care, including bed rest, elimination of substances that cause liver damage, and dietary measures that maintain or improve nutrition status.

Cirrhosis

- Liver cirrhosis is characterized by extensive fibrosis and permanent liver dysfunction. The primary causes of cirrhosis in the United States are alcoholic liver disease and hepatitis C infection.
- Initial symptoms of cirrhosis include fatigue, GI disturbances, anorexia, and weight loss. Eventually, patients may develop anemia, bruise easily, and be more susceptible to infections.
- Complications of cirrhosis include portal hypertension, gastroesophageal varices, ascites, and hepatic encephalopathy.
- Treatment of cirrhosis is highly individualized and depends on the accompanying symptoms and complications. Both drug therapies and dietary adjustments are usually necessary.
- If warranted, the diet recommended for cirrhosis patients may include restrictions in fat, sodium, or fluids. A person with cirrhosis often has a poor food intake and is at high risk of malnutrition.

TABLE 25-5 Possible Causes of Malnutrition in Liver Disease

Mechanism	Examples
Reduced nutrient intake	Abdominal discomfort, altered mental status, anorexia, early satiety (due to ascites), effects of medications (including gastrointestinal disturbances and taste changes), fasting for medical procedures, fatigue, nausea and vomiting, restrictive diets
Malabsorption or nutrient losses	Diarrhea, effects of medications (including malabsorption and nutrient losses from diuretic use), fat malabsorption (due to reduced bile flow), gastrointestinal bleeding, vomiting
Altered metabolism or increased nutrient needs	Hypermetabolism, impaired protein synthesis, infections or inflammation, muscle catabolism, reduced nutrient storage and metabolism in the liver

Liver Transplantation

- Liver transplantation has improved the long-term outlook for patients with advanced liver disease.
- Transplant patients are usually malnourished and may have medical problems that affect transplant success.
- Due to the potential for organ rejection, immunosuppressive drugs are prescribed following an organ transplant. Use of these drugs increases the risk of infection, and the drugs have side effects that can impair nutrition status and general health.

Gallstone Disease

- Gallstones are the most common disorder affecting the gallbladder. They are formed by the concentration of compounds in bile, especially cholesterol and the bile pigment bilirubin.
- Although most gallstones are asymptomatic, some gallstones can cause recurring pain and GI problems that often appear after meals and may persist for several hours.
- The risk of gallstone disease is influenced by ethnicity, gender, pregnancy, obesity, rapid weight loss, and other factors.
- Treatments for gallstones include gallbladder removal and gallstone dissolution or fragmentation.

Fatty Liver and Hepatitis (pp. 760–762)

1. Describe fatty liver, and identify its possible causes. What consequences of fatty liver may develop? What are possible treatments?
2. What is hepatitis, and what are its primary causes? Compare the features of infection with hepatitis virus A, B, and C. Identify nutritional concerns for patients with hepatitis.
3. Which of the following dietary strategies would be the most appropriate for reversing fatty liver associated with diabetes mellitus?
   a. following a low-protein diet
   b. following a fat-restricted diet
   c. following a fluid- and sodium-restricted diet
   d. modifying energy to achieve a desirable weight and modifying carbohydrates to attain blood glucose control

Cirrhosis (pp. 762–769)

4. Describe the progression of liver disease to cirrhosis. What are the most common causes of cirrhosis in the United States?
5. Discuss the consequences of cirrhosis, including its clinical effects and complications such as portal hypertension, gastroesophageal varices, ascites, and hepatic encephalopathy. What metabolic changes may result from altered liver function?
6. How does cirrhosis affect nutrition status? Describe the dietary treatment of a patient with cirrhosis. Discuss the special dietary concerns of patients with ascites and esophageal varices.
7. Esophageal varices are a dangerous complication of liver disease primarily because they:
   a. interfere with food intake.
   b. can lead to massive bleeding.
   c. divert blood flow from the GI tract.
   d. contribute to hepatic encephalopathy.
8. A complication of liver disease that contributes to the development of ascites is:
   a. portal hypertension.
   b. rising blood ammonia levels.
   c. elevated serum albumin levels.
   d. insulin resistance.
9. A patient with cirrhosis may develop personality changes and motor dysfunction, which are signs of:
   a. jaundice.
   b. hepatic encephalopathy.
   c. hyperammonemia.
   d. hepatic coma.
10. With respect to protein intake, patients with cirrhosis should:
    a. consume no more than the protein RDA.
    b. restrict protein intake to 0.6 grams per kilogram of body weight.
    c. use formulas enriched with aromatic amino acids to meet their protein needs.
    d. maintain nitrogen balance by consuming 0.8 to 1.2 grams of protein per kilogram of body weight per day.
11. People with ascites must often restrict dietary intake of:
    a. fat.
    b. protein.
    c. sugars.
    d. sodium.

Liver Transplantation (pp. 769–770)

12. Discuss the problems that arise following liver transplantation that can affect nutrition status. What dietary modifications may be necessary?
13. Dietary concerns after a liver transplant include all of the following *except*:
    a. severe protein restrictions that are difficult to adhere to.
    b. increased risk of foodborne illness.
    c. gastrointestinal side effects of medications.
    d. reduced appetite and altered taste perception from medications.

Gallstone Disease (pp. 770–773)

14. Explain how gallstones form, and describe the features of the two main types of gallstones. What complications are associated with gallstone disease?
15. Discuss the major risk factors for gallstone disease. Describe the primary methods of treatment.
16. Regarding risk factors for gallstone disease:
    a. prevalence is much higher in men than in women.
    b. gallstone risk is increased during pregnancy.
    c. rapid weight loss can temporarily shrink gallstones.
    d. risk is generally similar among ethnic groups.
17. Nonsurgical approaches to gallstone treatment include:
    a. cholecystectomy.
    b. weight loss.
    c. dissolution and fragmentation.
    d. immunosuppressant drug therapy.

Study Questions (multiple choice)
3. d 7. b 8. a 9. b 10. d 11. d 13. a 16. b 17. c

For additional study questions and activities, go to **www.cengagebrain.com** and search for ISBN 084006845X.

STUDY CARD 26

Diabetes Mellitus

Overview of Diabetes Mellitus

- Diabetes mellitus is a chronic condition characterized by inadequate insulin secretion and/or impaired insulin action.
- In type 1 diabetes, the pancreas secretes little or no insulin, and insulin therapy is necessary for survival. Type 2 diabetes is characterized by insulin resistance coupled with relative insulin deficiency, and disease risk is increased by obesity, aging, and physical inactivity.
- Acute complications of diabetes include diabetic ketoacidosis, in which hyperglycemia is accompanied by ketosis and acidosis, and the hyperosmolar hyperglycemic syndrome, characterized by severe hyperglycemia, dehydration, and possible mental impairments. Another acute complication, hypoglycemia, is usually a consequence of inappropriate disease management.
- Chronic complications of diabetes include macrovascular disorders such as cardiovascular disease and peripheral vascular disease, microvascular conditions such as diabetic retinopathy and diabetic nephropathy, and diabetic neuropathy.

Treatment of Diabetes Mellitus

- Diabetes treatment includes nutrition therapy, the use of insulin or other antidiabetic medications, and appropriate physical activity.
- Glycemic control is most often evaluated by monitoring blood glucose levels and glycated hemoglobin.
- The quantity of carbohydrate consumed has the greatest influence on blood glucose levels after meals. The total amount of carbohydrate ingested is more important than the type of carbohydrate consumed.
- Carbohydrate counting is widely used in menu planning and can be taught at different levels of complexity, depending on individual needs and abilities.
- Insulin therapy is required for patients who are unable to produce sufficient insulin and may be used in both type 1 and type 2 diabetes. Antidiabetic drugs can improve insulin secretion and effectiveness, suppress glucagon secretion, reduce glucose production by the liver, and delay carbohydrate absorption.
- Physical activity can improve glycemic control and enhance various aspects of general health.
- Illness can worsen glycemic control and may necessitate adjustments in medications and careful attention to dietary and fluid requirements.

Diabetes Management in Pregnancy

- Careful management of blood glucose levels before and during pregnancy may reduce complications in mother and infant.
- Most nutrient requirements during pregnancy are similar for women with and without diabetes.
- Carbohydrate intake should be distributed into several meals and snacks, including an evening snack to prevent overnight ketosis. Carbohydrate restriction may be recommended, especially in women with gestational diabetes.
- Moderate energy restriction may help improve glycemic control in overweight and obese women with gestational diabetes.

TABLE 26-2 Symptoms of Diabetes Mellitus

Frequent urination (polyuria)
Dehydration, dry mouth
Increased thirst (polydipsia)
Weight loss
Increased hunger (polyphagia)
Blurred vision
Increased infections
Fatigue

TABLE 26-3 Features of Type 1 and Type 2 Diabetes Mellitus

Feature	Type 1	Type 2
Prevalence in diabetic population	5 to 10 percent of cases	90 to 95 percent of cases
Age of onset	<30 years	>40 years[a]
Associated conditions	Autoimmune diseases, viral infection, inherited factors	Obesity, aging, inactivity, inherited factors
Major defect	Destruction of pancreatic beta cells; insulin deficiency	Insulin resistance; insulin deficiency relative to needs
Insulin secretion	Little or none	Varies; may be normal, increased, or decreased
Requirement for insulin therapy	Always	Sometimes
Former names	Juvenile-onset diabetes Insulin-dependent diabetes	Adult-onset diabetes Noninsulin-dependent diabetes

[a]Incidence of type 2 diabetes is increasing in children and adolescents; in more than 90 percent of these cases, it is associated with overweight or obesity and a family history of type 2 diabetes.

Overview of Diabetes Mellitus (pp. 782–789)

1. Describe the symptoms that develop as a consequence of hyperglycemia. Describe the criteria used to diagnose diabetes.
2. Compare the features of the two main types of diabetes. Describe gestational diabetes.
3. Discuss the acute complications that may arise in uncontrolled diabetes.
4. Describe the macrovascular and microvascular complications that develop from prolonged exposure to high blood glucose concentrations. Discuss the problems associated with diabetic neuropathy.
5. Which of the following is characteristic of type 1 diabetes?
   a. Abdominal obesity increases risk.
   b. The pancreas makes little or no insulin.
   c. It is the predominant form of diabetes.
   d. It often arises during pregnancy.
6. Which of the following describes type 2 diabetes?
   a. It is usually an autoimmune disease.
   b. The pancreas makes little or no insulin.
   c. Diabetic ketoacidosis is a common complication.
   d. Chronic complications may develop before it is diagnosed.
7. Most chronic complications associated with diabetes result from:
   a. altered kidney function.
   b. infections that deplete nutrient reserves.
   c. weight gain and hypertension.
   d. damage to blood vessels and nerves.

Treatment of Diabetes Mellitus (pp. 789–801)

8. What are the goals of medical and nutrition therapy for people with diabetes? Explain how diabetes treatment is evaluated.
9. Discuss the dietary recommendations for people with diabetes. Describe the meal-planning strategies they can use to control carbohydrate intakes.
10. Describe the insulin regimens for type 1 and type 2 diabetes. Explain how insulin therapy is coordinated with food intake and physical activity.
11. Discuss the modes of action of the various types of antidiabetic drugs.
12. Long-term glycemic control is usually evaluated by:
    a. self-monitoring of blood glucose.
    b. testing urinary ketone levels.
    c. measuring glycated hemoglobin.
    d. testing urinary protein levels (microalbuminuria).
13. Regarding dietary carbohydrate, a patient with diabetes should be most concerned about:
    a. consuming the correct quantity of carbohydrate at each meal or snack.
    b. consuming the correct proportion of sugars, starches, and fiber in meals.
    c. avoiding added sugars and kcaloric sweeteners.
    d. choosing meals with ideal proportions of protein, carbohydrate, and fat.
14. Which of the following is true regarding the general use of alcohol in diabetes?
    a. A serving of alcohol is considered part of the carbohydrate allowance.
    b. Alcohol contributes to hyperglycemia and should be avoided completely.
    c. Alcohol can cause hypoglycemia and should therefore be consumed with food if patients use insulin or medications that stimulate insulin secretion.
    d. Patients can use alcohol in unlimited quantities unless they are pregnant.
15. A patient using intensive insulin therapy is likely to follow a regimen that involves:
    a. twice-daily injections that combine short-, intermediate-, and long-acting insulin in each injection.
    b. a mixture of intermediate- and long-acting insulin injected between meals.
    c. multiple daily injections that supply basal insulin and precise insulin doses at each meal.
    d. the use of both insulin and oral antidiabetic agents.
16. In a person who has previously maintained good glycemic control, hyperglycemia can be precipitated by:
    a. infections or illnesses.
    b. chronic alcohol ingestion.
    c. undertreatment of hypoglycemia.
    d. prolonged exercise.

Diabetes Management in Pregnancy (pp. 801–803)

17. What are the risks of poorly controlled diabetes during pregnancy? Describe the general recommendations for diabetic women who become pregnant. Describe the dietary adjustments that may be necessary for women with gestational diabetes.

Study Questions (multiple choice)
5.b 6.d 7.d 12.c 13.a 14.c 15.c 16.a

For additional study questions and activities, go to **www.cengagebrain.com** and search for ISBN 084006845X.

STUDY CARD 27

Cardiovascular Diseases

FIGURE 27-1 Percentage Breakdown of Deaths from Cardiovascular Diseases in the United States

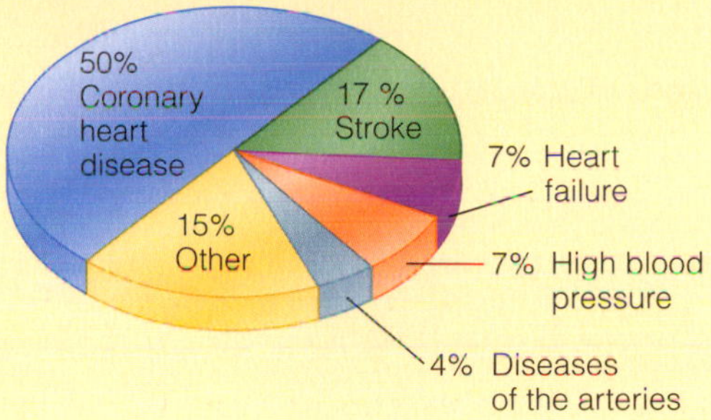

SOURCE: V. L. Roger and coauthors, on behalf of the American Heart Association Statistics Committee and Stroke Statistics Subcommittee, Heart disease and stroke statistics--2011 update: A report from the American Heart Association, *Circulation* 123 (2011), available at doi:10.1161/CIR.0b013e3182009701.

Atherosclerosis

- Atherosclerosis, characterized by plaque buildup in artery walls, can lead to complications such as angina pectoris, heart attack, stroke, intermittent claudication, kidney disease, and aneurysms.
- Arterial plaque develops in response to long-term, chronic inflammation at susceptible sites in artery walls.
- Leading causes of plaque formation and progression include shear stress, hypertension, elevated LDL and VLDL levels, cigarette smoking, diabetes, and aging.

Coronary Heart Disease (CHD)

- Long-term CHD management emphasizes risk reduction. Modifiable risk factors include elevated LDL and triglyceride levels, low HDL levels, hypertension, diabetes, obesity, a sedentary lifestyle, cigarette smoking, and various dietary factors.
- To reduce LDL levels and eliminate other risk factors, the Therapeutic Lifestyle Changes (TLC) approach is often suggested. This includes dietary changes that reduce saturated fat, *trans* fats, and cholesterol; increase soluble fiber; and incorporate plant sterols or stanols and fish into the diet. Other recommendations include regular physical activity, smoking cessation, and weight reduction.
- Dietary supplements are not recommended for heart disease prevention.
- Treatment for mild hypertriglyceridemia emphasizes weight control, regular physical activity, smoking cessation, avoiding a high carbohydrate intake, and alcohol restriction. Severe hypertriglyceridemia requires drug therapies and dietary fat restriction.
- Medications given after a heart attack suppress blood clotting, regulate heart rhythm, and reduce blood pressure, and patients are offered small portions of heart-healthy foods. To reduce the risk of a future heart attack, patients must learn strategies similar to the TLC approach.

Stroke

- The two major types of strokes, ischemic and hemorrhagic stroke, may be a consequence of atherosclerosis, hypertension, or both.
- Transient ischemic attacks, which are short-lived ischemic strokes, are a warning sign that a heart attack or a more severe stroke may follow.
- Strokes are largely preventable by reversing modifiable risk factors, such as hypertension, cigarette smoking, diabetes mellitus, and elevated LDL cholesterol.
- Treatment of a stroke includes the use of anticlotting drugs such as antiplatelet drugs and anticoagulants. Rehabilitation services evaluate the extent of neurological and functional impairment caused by a stroke and provide the therapy patients need to regain lost function.
- A patient who has had a major stroke may have problems eating normally due to lack of coordination and difficulty swallowing.

Hypertension

- About one in three persons in the United States has hypertension, which increases the risk of developing CHD, stroke, heart failure, and kidney failure.
- Blood pressure is elevated by factors that increase blood volume, heart rate, or resistance to blood flow.
- Although the underlying cause of most hypertension cases is unknown, risk factors include aging, family history, ethnicity, obesity, and various dietary factors.
- Treatment of hypertension usually includes a combination of lifestyle modifications and drug therapies.

Heart Failure

- Heart failure is usually a chronic, progressive condition that results from other cardiovascular illnesses.
- In heart failure, the heart is unable to pump adequate blood to tissues. Consequences may include fluid accumulation in the veins, lungs, and other organs and impaired organ function.
- Treatment of heart failure usually includes drug therapies that reduce fluid accumulation and improve heart function. Nutrition therapy may include sodium and fluid restrictions.

Atherosclerosis (pp. 811–814)

1. Define atherosclerosis. What are the medical problems that may result from atherosclerosis?
2. Explain how atherosclerosis progresses. Discuss the factors that contribute to atherosclerosis development.
3. Ischemia in the coronary arteries is a frequent cause of:
   a. angina pectoris.
   b. hemorrhagic stroke.
   c. aneurysm.
   d. hypertension.
4. Risk factors for atherosclerosis include all of the following *except*:
   a. smoking.
   b. hypertension.
   c. diabetes mellitus.
   d. elevated HDL cholesterol.

Coronary Heart Disease (CHD) (pp. 814–826)

5. Discuss the symptoms that may develop in coronary heart disease. Explain how risk for coronary heart disease is evaluated.
6. Describe each of the Therapeutic Lifestyle Changes recommended for reducing risk of coronary heart disease.
7. How is hypertriglyceridemia treated? Compare the dietary recommendations for mild and severe cases.
8. Describe the medical treatment provided after a heart attack. What information is typically provided in cardiac rehabilitation programs?
9. The dietary lipids with the strongest LDL cholesterol–raising effects are:
   a. monounsaturated fats.
   b. polyunsaturated fats.
   c. saturated fats.
   d. plant sterols.
10. The omega-3 fatty acids EPA and DHA, which may improve some risk factors for heart disease, are obtained by consuming:
    a. fatty fish.
    b. soy products.
    c. egg yolks and organ meats.
    d. nuts and seeds.
11. Moderate alcohol consumption can improve heart disease risk, in part, because it:
    a. lowers blood pressure.
    b. improves nutrition status.
    c. offsets the damage from smoking.
    d. increases HDL cholesterol levels.
12. Patients with mild hypertriglyceridemia may improve their triglyceride levels by:
    a. reducing sodium intake.
    b. consuming moderate amounts of alcohol.
    c. avoiding a high carbohydrate intake.
    d. reducing cholesterol intake.

Stroke (pp. 826–827)

13. Describe the difference between an ischemic stroke and a hemorrhagic stroke. What are possible consequences of a stroke? How might it affect an individual's nutrition care?
14. Hemorrhagic stroke:
    a. is the most common type of stroke.
    b. results from obstructed blood flow within brain tissue.
    c. comes on suddenly and usually lasts for up to 30 minutes.
    d. results from bleeding within the brain, which damages brain tissue.

Hypertension (pp. 827–832)

15. What are the possible consequences of hypertension? Discuss the major risk factors associated with hypertension and the lifestyle modifications that may lower blood pressure.
16. In most cases of hypertension, the cause is:
    a. excessive alcohol use.
    b. atherosclerosis.
    c. hormonal imbalances.
    d. unknown.
17. Hypertensive patients can benefit from all of the following dietary and lifestyle modifications *except*:
    a. including fat-free or low-fat milk products in the diet.
    b. reducing total fat intake.
    c. consuming generous amounts of fruits, vegetables, legumes, and nuts.
    d. reducing sodium intake.

Heart Failure (pp. 832–834)

18. Compare the effects of right-sided and left-sided heart failure. Describe the elements of treatment for heart failure.
19. Nutrition therapy for a patient with heart failure usually includes:
    a. weight loss.
    b. reducing total fat intake.
    c. sodium restriction.
    d. cholesterol restriction.

Study Questions (multiple choice)
3. a 4. d 9. c 10. a 11. d 12. c 14. d 16. d 17. b 19. c

For additional study questions and activities, go to **www.cengagebrain.com** and search for ISBN 084006845X.

STUDY CARD 28

Kidney Diseases

Functions of the Kidneys

- The kidneys are responsible for filtering the blood and removing wastes for excretion in urine.
- By adjusting the blood's volume and composition, the kidneys help maintain homeostasis within the body. Other kidney functions include the production of enzymes and hormones that regulate blood pressure, stimulate red blood cell production, and activate vitamin D.

The Nephrotic Syndrome

- The nephrotic syndrome is characterized by proteinuria due to glomerular damage. Complications include edema, lipid and blood coagulation abnormalities, infections, and PEM.
- Medications used in the nephrotic syndrome treat the underlying cause of proteinuria and help manage complications.
- An individual with the nephrotic syndrome should consume sufficient protein and energy to maintain health. However, patients should avoid consuming excess protein.
- Dietary adjustments for the nephrotic syndrome can help correct edema, lipid disorders, and nutrient deficiencies.

Acute Kidney Injury

- Acute kidney injury is characterized by a rapid loss of kidney function, causing a buildup of fluid, electrolytes, and nitrogenous wastes in the blood. Acute kidney injury may be caused by prerenal, intrarenal, or postrenal factors.
- Consequences of acute kidney injury may include oliguria, hyperkalemia, hyperphosphatemia, and uremia. If hyperkalemia develops, it can alter heart rhythm and lead to heart failure.
- Acute kidney injury is treated with medications, dialysis, and dietary modifications.

Chronic Kidney Disease

- Chronic kidney disease causes gradual loss of kidney function and often results from long-standing diabetes mellitus or hypertension.
- Depending on the stage of illness, complications of chronic kidney disease may include fluid and electrolyte imbalances, hypertension, renal osteodystrophy, mental impairments, bleeding abnormalities, anemia, increased risk for cardiovascular disease, and reduced immunity.
- Treatment for chronic kidney disease can slow disease progression and correct complications and includes drug therapies, dialysis, and nutrition therapy.
- Dietary measures for chronic kidney disease usually feature a low-protein diet, controlled fluid and sodium intakes, phosphorus restrictions, and calcium and vitamin D supplementation; potassium restrictions are usually necessary after dialysis treatment begins.
- Patients with advanced kidney disease may benefit from kidney transplantation, which can restore renal function and liberalize dietary restrictions.

Kidney Stones

- Kidney stones form when stone constituents—calcium oxalate, calcium phosphate, uric acid, cystine, or magnesium ammonium phosphate—crystallize in urine.
- Complications of kidney stones include renal colic, difficulty with urination, and obstruction.
- Kidney stones may be prevented by maintaining urine volumes of at least $2\frac{1}{2}$ liters daily. Other dietary measures include the consumption of appropriate amounts of calcium, oxalates, protein, sodium, and purines.
- Symptomatic kidney stones are sometimes treated with medications or treatments that facilitate stone passage or surgeries that fragment or remove stones.

Functions of the Kidneys (pp. 843–844)

1. Describe the kidneys' role in maintaining homeostasis. Discuss other functions of the kidneys.
2. Which of the following is *not* a function of the kidneys?
   a. activation of vitamin K
   b. maintenance of acid-base balance
   c. elimination of metabolic waste products
   d. maintenance of fluid and electrolyte balances

The Nephrotic Syndrome (pp. 844–846)

3. Define the nephrotic syndrome, and describe the consequences that can develop. Discuss the elements of dietary treatment recommended for the nephrotic syndrome.
4. The nephrotic syndrome frequently results in:
   a. the uremic syndrome.
   b. oliguria.
   c. edema.
   d. renal colic.
5. Dietary recommendations for patients with the nephrotic syndrome include:
   a. a high protein intake.
   b. sodium restriction.
   c. potassium and phosphorus restrictions.
   d. fluid restriction.

Acute Kidney Injury (pp. 846–850)

6. Describe acute kidney injury and list its possible causes. Discuss how the consequences of acute kidney injury can disrupt health.
7. Describe the medical treatment of patients with acute kidney injury and the elements of nutrition therapy.
8. Hyperkalemia is often treated by:
   a. eliminating potassium from the diet.
   b. using diuretics to increase potassium losses.
   c. increasing fluid consumption.
   d. using potassium-exchange resins, which bind potassium in the GI tract.
9. Fluid requirements for oliguric patients are estimated by adding about ____ milliliters to the volume of urine output.
   a. 100
   b. 300
   c. 500
   d. 750

Chronic Kidney Disease (pp. 850–859)

10. Explain how chronic kidney disease differs from acute kidney injury. What changes occur as chronic kidney disease progresses?
11. Discuss the symptoms and complications associated with the uremic syndrome.
12. Identify the objectives of treatment for chronic kidney disease, and discuss the role of dialysis. Describe how dietary recommendations change during the course of illness.
13. What are the fluid and electrolyte (sodium, potassium, and phosphorus) recommendations for patients with chronic kidney disease? What adjustments are needed in vitamin and mineral intakes, and why?
14. Explain why renal patients often have difficulty adhering to a renal diet. Discuss ways to help patients comply with recommendations.
15. Discuss the nutrient needs of a kidney transplant patient. How can immunosuppressive drug therapy affect nutrition status?
16. The most common cause of chronic kidney disease is:
    a. diabetes mellitus.
    b. hypertension.
    c. autoimmune disease.
    d. exposure to toxins.
17. A person with chronic kidney disease who has been following a renal diet for several years begins hemodialysis treatment. An appropriate dietary adjustment would be to:
    a. reduce protein intake.
    b. consume protein more liberally.
    c. increase intakes of sodium and water.
    d. consume potassium and phosphorus more liberally.
18. Which of the following nutrients may be unintentionally restricted when a patient restricts phosphorus intake?
    a. fluid
    b. calcium
    c. potassium
    d. sodium

Kidney Stones (pp. 859–861)

19. Identify factors that affect kidney stone formation. Describe the composition of the most common types of kidney stones.
20. Discuss dietary adjustments that may help prevent kidney stone recurrence.
21. Most kidney stones are made primarily from:
    a. struvite.
    b. uric acid.
    c. calcium oxalate.
    d. cystine.

Study Questions (multiple choice)
2. a 4. c 5. b 8. d 9. c 16. a 17. b 18. b 21. c

For additional study questions and activities, go to **www.cengagebrain.com** and search for ISBN 084006845X.

STUDY CARD 29

Cancer and HIV Infection

Cancer

- Cancer arises from mutations in the genes that control cell division. Some dietary substances promote carcinogenesis, while others may help prevent cancer.
- Cancer's effects on nutrition status depend on the type of cancer a person has, its severity, and the methods used to treat the cancer.
- Cancer cachexia is a frequent complication of cancer and may be related to anorexia, altered metabolism, and responses to treatment.
- Medical treatments for cancer include surgery, chemotherapy, radiation therapy, and biological therapies, which remove cancer cells, prevent tumor growth, and alleviate symptoms.
- Nutrition therapy for cancer patients aims to minimize weight loss and wasting, correct deficiencies, and manage complications that impair food intake.

HIV Infection

- By attacking immune cells, HIV causes progressive damage to immune function and may eventually lead to AIDS.
- Improved drug therapies have slowed the progression of HIV infection; however, these drugs may promote the HIV-lipodystrophy syndrome, characterized by body fat redistribution, abnormal lipid levels, and insulin resistance.
- HIV infection may lead to weight loss and wasting, anorexia, and various complications that affect food intake. Dietary adjustments, resistance training, and medications can help patients maintain their weight and prevent wasting.
- People with HIV infections must pay strict attention to food safety guidelines to prevent foodborne illnesses.

TABLE 29-3 Recommendations for Reducing Cancer Risk

Maintain a healthy body weight throughout life.

- Be as lean as possible within the normal range of body weight for your height.
- Avoid weight gain and increases in waist circumference throughout adulthood.

Be physically active as part of everyday life.

- For adults: engage in moderate to vigorous physical activity for at least 30 minutes on at least 5 days per week; 45 to 60 minutes is preferable.
- For children and adolescents: engage in moderate to vigorous activity for 60 minutes on at least 5 days of the week.
- Limit sedentary habits such as watching television.

Choose a healthy diet that emphasizes plant sources.

- Limit consumption of energy-dense foods (>225 kcal per 100 g food) and sugary drinks that contribute to weight gain.
- Consume relatively unprocessed grains and/or legumes with every meal. Choose whole-grain products instead of processed (refined) grains.
- Consume five or more servings of nonstarchy vegetables and fruits every day.
- Limit consumption of red meats (such as beef, pork, and lamb) and processed meats (such as those preserved by smoking, curing, or salting).
- Avoid salt-preserved, salted, and salty foods.
- Avoid moldy grains and legumes.

Limit consumption of alcoholic beverages.

- For women: drink no more than one drink daily.
- For men: drink no more than two drinks daily.

Aim to meet nutritional needs through the diet.

- Obtain necessary nutrients from the diet. Dietary supplements are not recommended for cancer prevention, and they may have unexpected adverse effects.

Avoid using tobacco in any form.

SOURCES: World Cancer Research Fund/American Institute for Cancer Research, *Food, Nutrition, Physical Activity, and the Prevention of Cancer: A Global Perspective* (Washington, DC: American Institute for Cancer Research, 2007); L. H. Kushi and coauthors, American Cancer Society guidelines on nutrition and physical activity for cancer prevention: Reducing the risk of cancer with healthy food choices and physical activity, *CA: A Cancer Journal for Clinicians* 56 (2006): 254–281.

FIGURE 29-1 Cancer Development

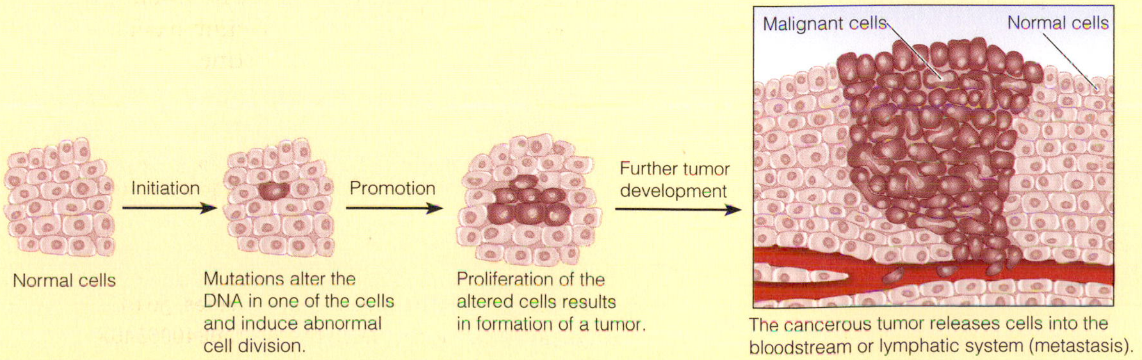

Initiation
Promotion
Further tumor development
Malignant cells
Normal cells
Normal cells
Mutations alter the DNA in one of the cells and induce abnormal cell division.
Proliferation of the altered cells results in formation of a tumor.
The cancerous tumor releases cells into the bloodstream or lymphatic system (metastasis).

Cancer (pp. 869–879)

1. Describe the process of tumor formation. What factors contribute to cancer development? Discuss the dietary factors that may increase or decrease the risk of cancer.
2. What is cancer cachexia? What factors promote its development?
3. Explain how cancer and its treatments can cause alterations in food intake and metabolism and possibly lead to malnutrition.
4. Discuss the elements of nutrition therapy for cancer, as well as strategies that can improve food intake.
5. Which of these dietary substances may help protect against cancer?
   a. alcohol
   b. well-cooked meats, poultry, and fish
   c. animal fats
   d. phytochemicals from fruits and vegetables
6. The metabolic changes that often accompany cancer include all of the following *except*:
   a. increased triglyceride breakdown.
   b. increased protein turnover.
   c. increased muscle protein synthesis.
   d. insulin resistance.
7. An advantage of radiation therapy over chemotherapy is that:
   a. radiation is not damaging to rapidly dividing cells.
   b. side effects of radiation therapy do not include malnutrition.
   c. radiation can be directed toward the regions affected by cancer.
   d. the radiation used is too weak to damage GI tissues.
8. Although many cancer patients lose weight, which type of cancer is often associated with weight gain?
   a. kidney cancer
   b. breast cancer
   c. colon cancer
   d. lung cancer
9. Oral diets after hematopoietic stem cell transplants may restrict:
   a. fiber.
   b. carbohydrates.
   c. high-protein foods.
   d. raw fruits and vegetables.

HIV Infection (pp. 879–886)

10. Explain how HIV is transmitted, and list risk factors associated with an HIV infection.
11. Describe the possible consequences of HIV infection, such as reduced immunity, HIV-lipodystrophy syndrome, wasting, GI complications, neurological complications, and anemia. Explain why an HIV infection often results in anorexia and reduced food intake.
12. Describe how an HIV infection is treated, and discuss the potential complications associated with treatment. Discuss the features of nutrition therapy for HIV-infected and AIDS patients.
13. Why are people with HIV infections highly susceptible to foodborne illness? Describe some measures that can be taken to prevent foodborne illness.
14. HIV can enter and destroy these immune cells:
   a. B cells.
   b. helper T cells.
   c. natural killer cells.
   d. neutrophils.
15. The HIV-lipodystrophy syndrome may result in all of these changes except:
   a. increased abdominal fat.
   b. increased fat in the arms and legs.
   c. fat accumulation at the base of the neck.
   d. hypertriglyceridemia.
16. Mouth sores in people with HIV infections are most frequently due to:
   a. oral infections.
   b. dehydration.
   c. nutrient deficiency.
   d. foodborne illnesses.
17. Megestrol acetate and dronabinol are:
   a. medications used to promote weight gain.
   b. protease inhibitors that fight HIV infection.
   c. medications that treat common opportunistic infections.
   d. anabolic hormones that promote gain of muscle tissue.
18. To prevent cryptosporidiosis, a person with HIV infection may need to:
   a. wash hands carefully before meals.
   b. avoid consuming undercooked meat, poultry, and eggs.
   c. consume a high-kcalorie, high-protein diet.
   d. boil drinking water for one minute.

Study Questions (multiple choice)
5. d 6. c 7. c 8. b 9. d 14. b 15. b 16. a 17. a 18. d

For additional study questions and activities, go to **www.cengagebrain.com** and search for ISBN 084006845X.

Dietary Reference Intakes (DRI)

The Dietary Reference Intakes (DRI) include two sets of values that serve as goals for nutrient intake—Recommended Dietary Allowances (RDA) and Adequate Intakes (AI). The RDA reflect the average daily amount of a nutrient considered adequate to meet the needs of most healthy people. If there is insufficient evidence to determine an RDA, an AI is set. AI are more tentative than RDA, but both may be used as goals for nutrient intakes. (Chapter 1 provides more details.)

In addition to the values that serve as goals for nutrient intakes (presented in the tables on these two pages), the DRI include a set of values called Tolerable Upper Intake Levels (UL). The UL represent the maximum amount of a nutrient that appears safe for most healthy people to consume on a regular basis. Turn the page for a listing of the UL for selected vitamins and minerals.

Estimated Energy Requirements (EER), Recommended Dietary Allowances (RDA), and Adequate Intakes (AI) for Water, Energy, and the Energy Nutrients

Age (yr)	Reference BMI (kg/m^2)	Reference Height cm (in)	Reference Weight kg (lb)	Water[a] AI (L/day)	Energy EER[b] (kcal/day)	Carbohydrate RDA (g/day)	Total Fiber AI (g/day)	Total Fat AI (g/day)	Linoleic Acid AI (g/day)	Linolenic Acid[c] AI (g/day)	Protein RDA (g/day)[d]	Protein RDA (g/kg/day)
Males												
0–0.5	—	62 (24)	6 (13)	0.7[e]	570	60	—	31	4.4	0.5	9.1	1.52
0.5–1	—	71 (28)	9 (20)	0.8[f]	743	95	—	30	4.6	0.5	11	1.20
1–3[g]	—	86 (34)	12 (27)	1.3	1046	130	19	—	7	0.7	13	1.05
4–8[g]	15.3	115 (45)	20 (44)	1.7	1742	130	25	—	10	0.9	19	0.95
9–13	17.2	144 (57)	36 (79)	2.4	2279	130	31	—	12	1.2	34	0.95
14–18	20.5	174 (68)	61 (134)	3.3	3152	130	38	—	16	1.6	52	0.85
19–30	22.5	177 (70)	70 (154)	3.7	3067[h]	130	38	—	17	1.6	56	0.80
31–50	22.5[i]	177 (70)[i]	70 (154)[i]	3.7	3067[h]	130	38	—	17	1.6	56	0.80
>50	22.5[i]	177 (70)[i]	70 (154)[i]	3.7	3067[h]	130	30	—	14	1.6	56	0.80
Females												
0–0.5	—	62 (24)	6 (13)	0.7[e]	520	60	—	31	4.4	0.5	9.1	1.52
0.5–1	—	71 (28)	9 (20)	0.8[f]	676	95	—	30	4.6	0.5	11	1.20
1–3[g]	—	86 (34)	12 (27)	1.3	992	130	19	—	7	0.7	13	1.05
4–8[g]	15.3	115 (45)	20 (44)	1.7	1642	130	25	—	10	0.9	19	0.95
9–13	17.4	144 (57)	37 (81)	2.1	2071	130	26	—	10	1.0	34	0.95
14–18	20.4	163 (64)	54 (119)	2.3	2368	130	26	—	11	1.1	46	0.85
19–30	21.5	163 (64)	57 (126)	2.7	2403[j]	130	25	—	12	1.1	46	0.80
31–50	21.5[i]	163 (64)[i]	57 (126)[i]	2.7	2403[j]	130	25	—	12	1.1	46	0.80
>50	21.5[i]	163 (64)[i]	57 (126)[i]	2.7	2403[j]	130	21	—	11	1.1	46	0.80
Pregnancy												
1st trimester				3.0	+0	175	28	—	13	1.4	46	0.80
2nd trimester				3.0	+340	175	28	—	13	1.4	71	1.10
3rd trimester				3.0	+452	175	28	—	13	1.4	71	1.10
Lactation												
1st 6 months				3.8	+330	210	29	—	13	1.3	71	1.30
2nd 6 months				3.8	+400	210	29	—	13	1.3	71	1.30

NOTE: For all nutrients, values for infants are AI. Dashes indicate that values have not been determined.

[a]The water AI includes drinking water, water in beverages, and water in foods; in general, drinking water and other beverages contribute about 70 to 80 percent, and foods, the remainder. Conversion factors: 1 L = 33.8 fluid oz; 1 L = 1.06 qt; 1 cup = 8 fluid oz.

[b]The Estimated Energy Requirement (EER) represents the average dietary energy intake that will maintain energy balance in a healthy person of a given gender, age, weight, height, and physical activity level. The values listed are based on an "active" person at the reference height and weight and at the midpoint ages for each group until age 19. Chapter 8 and Appendix F provide equations and tables to determine estimated energy requirements.

[c]The linolenic acid referred to in this table and text is the omega-3 fatty acid known as alpha-linolenic acid.

[d]The values listed are based on reference body weights.

[e]Assumed to be from human milk.

[f]Assumed to be from human milk and complementary foods and beverages. This includes approximately 0.6 L (~2½ cups) as total fluid including formula, juices, and drinking water.

[g]For energy, the age groups for young children are 1–2 years and 3–8 years.

[h]For males, subtract 10 kcalories per day for each year of age above 19.

[i]Because weight need not change as adults age if activity is maintained, reference weights for adults 19 through 30 years are applied to all adult age groups.

[j]For females, subtract 7 kcalories per day for each year of age above 19.

SOURCE: Adapted from the *Dietary Reference Intakes* series, National Academies Press. Copyright 1997, 1998, 2000, 2001, 2002, 2004, 2005, 2011 by the National Academies of Sciences.

Recommended Dietary Allowances (RDA) and Adequate Intakes (AI) for Vitamins

Age (yr)	Thiamin RDA (mg/day)	Riboflavin RDA (mg/day)	Niacin RDA (mg/day)[a]	Biotin AI (µg/day)	Pantothenic acid AI (mg/day)	Vitamin B_6 RDA (mg/day)	Folate RDA (µg/day)[b]	Vitamin B_{12} RDA (µg/day)	Choline AI (mg/day)	Vitamin C RDA (mg/day)	Vitamin A RDA (µg/day)[c]	Vitamin D RDA (IU/day)[d]	Vitamin E RDA (mg/day)[e]	Vitamin K AI (µg/day)
Infants														
0–0.5	0.2	0.3	2	5	1.7	0.1	65	0.4	125	40	400	400 (10 µg)	4	2.0
0.5–1	0.3	0.4	4	6	1.8	0.3	80	0.5	150	50	500	400 (10 µg)	5	2.5
Children														
1–3	0.5	0.5	6	8	2	0.5	150	0.9	200	15	300	600 (15 µg)	6	30
4–8	0.6	0.6	8	12	3	0.6	200	1.2	250	25	400	600 (15 µg)	7	55
Males														
9–13	0.9	0.9	12	20	4	1.0	300	1.8	375	45	600	600 (15 µg)	11	60
14–18	1.2	1.3	16	25	5	1.3	400	2.4	550	75	900	600 (15 µg)	15	75
19–30	1.2	1.3	16	30	5	1.3	400	2.4	550	90	900	600 (15 µg)	15	120
31–50	1.2	1.3	16	30	5	1.3	400	2.4	550	90	900	600 (15 µg)	15	120
51–70	1.2	1.3	16	30	5	1.7	400	2.4	550	90	900	600 (15 µg)	15	120
>70	1.2	1.3	16	30	5	1.7	400	2.4	550	90	900	800 (20 µg)	15	120
Females														
9–13	0.9	0.9	12	20	4	1.0	300	1.8	375	45	600	600 (15 µg)	11	60
14–18	1.0	1.0	14	25	5	1.2	400	2.4	400	65	700	600 (15 µg)	15	75
19–30	1.1	1.1	14	30	5	1.3	400	2.4	425	75	700	600 (15 µg)	15	90
31–50	1.1	1.1	14	30	5	1.3	400	2.4	425	75	700	600 (15 µg)	15	90
51–70	1.1	1.1	14	30	5	1.5	400	2.4	425	75	700	600 (15 µg)	15	90
>70	1.1	1.1	14	30	5	1.5	400	2.4	425	75	700	800 (20 µg)	15	90
Pregnancy														
≤18	1.4	1.4	18	30	6	1.9	600	2.6	450	80	750	600 (15 µg)	15	75
19–30	1.4	1.4	18	30	6	1.9	600	2.6	450	85	770	600 (15 µg)	15	90
31–50	1.4	1.4	18	30	6	1.9	600	2.6	450	85	770	600 (15 µg)	15	90
Lactation														
≤18	1.4	1.6	17	35	7	2.0	500	2.8	550	115	1200	600 (15 µg)	19	75
19–30	1.4	1.6	17	35	7	2.0	500	2.8	550	120	1300	600 (15 µg)	19	90
31–50	1.4	1.6	17	35	7	2.0	500	2.8	550	120	1300	600 (15 µg)	19	90

NOTE: For all nutrients, values for infants are AI. The glossary on the inside back cover defines units of nutrient measure.

[a]Niacin recommendations are expressed as niacin equivalents (NE), except for recommendations for infants younger than 6 months, which are expressed as preformed niacin.

[b]Folate recommendations are expressed as dietary folate equivalents (DFE).

[c]Vitamin A recommendations are expressed as retinol activity equivalents (RAE).

[d]Vitamin D recommendations are expressed as cholecalciferol and assume an absence of adequate exposure to sunlight.

[e]Vitamin E recommendations are expressed as α-tocopherol.

Recommended Dietary Allowances (RDA) and Adequate Intakes (AI) for Minerals

Age (yr)	Sodium AI (mg/day)	Chloride AI (mg/day)	Potassium AI (mg/day)	Calcium RDA (mg/day)	Phosphorus RDA (mg/day)	Magnesium RDA (mg/day)	Iron RDA (mg/day)	Zinc RDA (mg/day)	Iodine RDA (µg/day)	Selenium RDA (µg/day)	Copper RDA (µg/day)	Manganese AI (mg/day)	Fluoride AI (mg/day)	Chromium AI (µg/day)	Molybdenum RDA (µg/day)
Infants															
0–0.5	120	180	400	200	100	30	0.27	2	110	15	200	0.003	0.01	0.2	2
0.5–1	370	570	700	260	275	75	11	3	130	20	220	0.6	0.5	5.5	3
Children															
1–3	1000	1500	3000	700	460	80	7	3	90	20	340	1.2	0.7	11	17
4–8	1200	1900	3800	1000	500	130	10	5	90	30	440	1.5	1.0	15	22
Males															
9–13	1500	2300	4500	1300	1250	240	8	8	120	40	700	1.9	2	25	34
14–18	1500	2300	4700	1300	1250	410	11	11	150	55	890	2.2	3	35	43
19–30	1500	2300	4700	1000	700	400	8	11	150	55	900	2.3	4	35	45
31–50	1500	2300	4700	1000	700	420	8	11	150	55	900	2.3	4	35	45
51–70	1300	2000	4700	1000	700	420	8	11	150	55	900	2.3	4	30	45
>70	1200	1800	4700	1200	700	420	8	11	150	55	900	2.3	4	30	45
Females															
9–13	1500	2300	4500	1300	1250	240	8	8	120	40	700	1.6	2	21	34
14–18	1500	2300	4700	1300	1250	360	15	9	150	55	890	1.6	3	24	43
19–30	1500	2300	4700	1000	700	310	18	8	150	55	900	1.8	3	25	45
31–50	1500	2300	4700	1000	700	320	18	8	150	55	900	1.8	3	25	45
51–70	1300	2000	4700	1200	700	320	8	8	150	55	900	1.8	3	20	45
>70	1200	1800	4700	1200	700	320	8	8	150	55	900	1.8	3	20	45
Pregnancy															
≤18	1500	2300	4700	1300	1250	400	27	12	220	60	1000	2.0	3	29	50
19–30	1500	2300	4700	1000	700	350	27	11	220	60	1000	2.0	3	30	50
31–50	1500	2300	4700	1000	700	360	27	11	220	60	1000	2.0	3	30	50
Lactation															
≤18	1500	2300	5100	1300	1250	360	10	13	290	70	1300	2.6	3	44	50
19–30	1500	2300	5100	1000	700	310	9	12	290	70	1300	2.6	3	45	50
31–50	1500	2300	5100	1000	700	320	9	12	290	70	1300	2.6	3	45	50

NOTE: For all nutrients, values for infants are AI. The glossary on the inside back cover defines units of nutrient measure.

Tolerable Upper Intake Levels (UL) for Vitamins

Age (yr)	Niacin (mg/day)[a]	Vitamin B_6 (mg/day)	Folate (µg/day)[a]	Choline (mg/day)	Vitamin C (mg/day)	Vitamin A (µg/day)[b]	Vitamin D (IU/day)	Vitamin E (mg/day)[c]
Infants								
0–0.5	—	—	—	—	—	600	1000 (25 µg)	—
0.5–1	—	—	—	—	—	600	1500 (38 µg)	—
Children								
1–3	10	30	300	1000	400	600	2500 (63 µg)	200
4–8	15	40	400	1000	650	900	3000 (75 µg)	300
9–13	20	60	600	2000	1200	1700	4000 (100 µg)	600
Adolescents								
14–18	30	80	800	3000	1800	2800	4000 (100 µg)	800
Adults								
19–70	35	100	1000	3500	2000	3000	4000 (100 µg)	1000
>70	35	100	1000	3500	2000	3000	4000 (100 µg)	1000
Pregnancy								
≤18	30	80	800	3000	1800	2800	4000 (100 µg)	800
19–50	35	100	1000	3500	2000	3000	4000 (100 µg)	1000
Lactation								
≤18	30	80	800	3000	1800	2800	4000 (100 µg)	800
19–50	35	100	1000	3500	2000	3000	4000 (100 µg)	1000

[a]The UL for niacin and folate apply to synthetic forms obtained from supplements, fortified foods, or a combination of the two.
[b]The UL for vitamin A applies to the preformed vitamin only.
[c]The UL for vitamin E applies to any form of supplemental α-tocopherol, fortified foods, or a combination of the two.

Tolerable Upper Intake Levels (UL) for Minerals

Age (yr)	Sodium (mg/day)	Chloride (mg/day)	Calcium (mg/day)	Phosphorus (mg/day)	Magnesium (mg/day)[d]	Iron (mg/day)	Zinc (mg/day)	Iodine (µg/day)	Selenium (µg/day)	Copper (µg/day)	Manganese (mg/day)	Fluoride (mg/day)	Molybdenum (µg/day)	Boron (mg/day)	Nickel (mg/day)	Vanadium (mg/day)
Infants																
0–0.5	—	—	1000	—	—	40	4	—	45	—	—	0.7	—	—	—	—
0.5–1	—	—	1500	—	—	40	5	—	60	—	—	0.9	—	—	—	—
Children																
1–3	1500	2300	2500	3000	65	40	7	200	90	1000	2	1.3	300	3	0.2	—
4–8	1900	2900	2500	3000	110	40	12	300	150	3000	3	2.2	600	6	0.3	—
9–13	2200	3400	3000	4000	350	40	23	600	280	5000	6	10	1100	11	0.6	—
Adolescents																
14–18	2300	3600	3000	4000	350	45	34	900	400	8000	9	10	1700	17	1.0	—
Adults																
19–50	2300	3600	2500	4000	350	45	40	1100	400	10,000	11	10	2000	20	1.0	1.8
51–70	2300	3600	2000	4000	350	45	40	1100	400	10,000	11	10	2000	20	1.0	1.8
>70	2300	3600	2000	3000	350	45	40	1100	400	10,000	11	10	2000	20	1.0	1.8
Pregnancy																
≤18	2300	3600	3000	3500	350	45	34	900	400	8000	9	10	1700	17	1.0	—
19–50	2300	3600	2500	3500	350	45	40	1100	400	10,000	11	10	2000	20	1.0	—
Lactation																
≤18	2300	3600	3000	4000	350	45	34	900	400	8000	9	10	1700	17	1.0	—
19–50	2300	3600	2500	4000	350	45	40	1100	400	10,000	11	10	2000	20	1.0	—

[d]The UL for magnesium applies to synthetic forms obtained from supplements or drugs only.
NOTE: An Upper Limit was not established for vitamins and minerals not listed and for those age groups listed with a dash (—) because of a lack of data, not because these nutrients are safe to consume at any level of intake. All nutrients can have adverse effects when intakes are excessive.

SOURCE: Adapted with permission from the *Dietary Reference Intakes* series, National Academies Press.

Daily Values for Food Labels

The Daily Values are standard values developed by the Food and Drug Administration (FDA) for use on food labels. The values are based on 2000 kcalories a day for adults and children over 4 years old. Chapter 2 provides more details.

Nutrient	Amount
Protein[a]	50 g
Thiamin	1.5 mg
Riboflavin	1.7 mg
Niacin	20 mg NE
Biotin	300 µg
Pantothenic acid	10 mg
Vitamin B_6	2 mg
Folate	400 µg
Vitamin B_{12}	6 µg
Vitamin C	60 mg
Vitamin A	5000 IU[b]
Vitamin D	400 IU[b]
Vitamin E	30 IU[b]
Vitamin K	80 µg
Calcium	1000 mg
Iron	18 mg
Zinc	15 mg
Iodine	150 µg
Copper	2 mg
Chromium	120 µg
Selenium	70 µg
Molybdenum	75 µg
Manganese	2 mg
Chloride	3400 mg
Magnesium	400 mg
Phosphorus	1000 mg

[a]The Daily Values for protein vary for different groups of people: pregnant women, 60 g; nursing mothers, 65 g; infants under 1 year, 14 g; children 1 to 4 years, 16 g.

[b]Equivalent values for nutrients expressed as IU are: vitamin A, 1500 RAE (assumes a mixture of 40% retinol and 60% beta-carotene); vitamin D, 10 µg; vitamin E, 20 mg.

Food Component	Amount	Calculation Factors
Fat	65 g	30% of kcalories
Saturated fat	20 g	10% of kcalories
Cholesterol	300 mg	Same regardless of kcalories
Carbohydrate (total)	300 g	60% of kcalories
Fiber	25 g	11.5 g per 1000 kcalories
Protein	50 g	10% of kcalories
Sodium	2400 mg	Same regardless of kcalories
Potassium	3500 mg	Same regardless of kcalories

GLOSSARY OF NUTRIENT MEASURES

kcal: kcalories; a unit by which energy is measured (Chapter 1 provides more details).

g: grams; a unit of weight equivalent to about 0.03 ounces.

mg: milligrams; one-thousandth of a gram.

µg: micrograms; one-millionth of a gram.

IU: international units; an old measure of vitamin activity determined by biological methods (as opposed to new measures that are determined by direct chemical analyses). Many fortified foods and supplements use IU on their labels.

- For vitamin A, 1 IU = 0.3 µg retinol, 3.6 µg β-carotene, or 7.2 µg other vitamin A carotenoids
- For vitamin D, 1 IU = 0.025 µg cholecalciferol
- For vitamin E, 1 IU = 0.67 natural α-tocopherol (other conversion factors are used for different forms of vitamin E)

mg NE: milligrams niacin equivalents; a measure of niacin activity (Chapter 10 provides more details).

- 1 NE = 1 mg niacin
  = 60 mg tryptophan (an amino acid)

µg DFE: micrograms dietary folate equivalents; a measure of folate activity (Chapter 10 provides more details).

- 1 µg DFE = 1 µg food folate
  = 0.6 µg fortified food or supplement folate taken with food
  = 0.5 µg supplement folate taken on an empty stomach

µg RAE: micrograms retinol activity equivalents; a measure of vitamin A activity (Chapter 11 provides more details).

- 1 µg RAE = 1 µg retinol
  = 12 µg β-carotene
  = 24 µg other vitamin A carotenoids

mmol: millimoles; one-thousanth of a mole, the molecular weight of a substance. To convert mmol to mg, multiply by the atomic weight of the substance.

- For sodium, mmol × 23 = mg Na
- For chloride, mmol × 35.5 = mg Cl
- For sodium chloride, mmol × 58.5 = mg NaCl

Weights and Measures

Length
1 centimeter (cm) = 0.39 inches (in)
1 foot (ft) = 30 centimeters (cm)
1 inch (in) = 2.54 centimeters (cm)
1 meter (m) = 39.37 inches (in)

Weight
1 gram (g) = 0.001 kilograms (kg)
= 1000 milligrams (mg)
= .035 ounces (oz)
1 kilogram (kg) = 1000 grams (g)
= 2.2 pounds (lb)
1 microgram (µg) = 0.001 milligrams (mg)
1 milligram (mg) = 0.001 grams (g)
= 1000 micrograms (µg)
1 ounce (oz) = 28 grams (g)
= 0.03 kilograms (kg)
1 pound (lb) = 454 grams (g)
= 0.45 kilograms (kg)
= 16 ounces (oz)

Volume
1 cup = 16 tablespoons (tbs or T)
= 0.25 liters (L)
= 236 milliliters (mL, commonly rounded to 250 mL)
= 8 ounces (oz)
1 liter (L) = 33.8 fluid ounces (fl oz)
= 0.26 gallons (gal)
= 2.1 pints (pt)
= 1.06 quarts (qt)
= 1000 milliliters (mL)
1 milliliter (mL) = 0.001 liters (L)
= 0.03 fluid ounces (fl oz)
1 ounce (oz) = 0.03 liters (L)
= 30 milliliters (mL)
1 pint (pt) = 2 cups (c)
= 0.47 liters (L)
= 16 ounces (oz)
1 quart (qt) = 4 cups (c)
= 0.95 liters (L)
= 32 ounces (oz)
1 tablespoon (tbs or T) = 3 teaspoons (tsp)
= 15 milliliters (mL)
1 teaspoon (tsp) = 5 milliliters (mL)
1 gallon (gal) = 16 cups (c)
= 3.8 liters (L)
= 128 ounces (oz)

Energy
1 megajoule (MJ) = 240 kcalories (kcal)
1 kilojoule (kJ) = 0.24 kcalories (kcal)
1 kcalorie (kcal) = 4.2 kilojoule (kJ)
1 g alcohol = 7 kcal = 29 kJ
1 g carbohydrate = 4 kcal = 17 kJ
1 g fat = 9 kcal = 37 kJ
1 g protein = 4 kcal = 17 kJ

Temperature
To change from Fahrenheit (°F) to Celsius (°C), subtract 32 from the Fahrenheit measure and then multiply that result by 0.56.

To change from Celsius (°C) to Fahrenheit (°F), multiply the Celsius measure by 1.8 and add 32 to that result.

A comparison of some useful temperatures is given below.

	Celsius	Fahrenheit
Boiling point	100°C	212°F
Body temperature	37°C	98.6°F
Freezing point	0°C	32°F

Aids to Calculation

Many mathematical problems have been worked out in the "How To" sections of the text. These pages offer additional help and examples.

Conversions

A conversion factor is a fraction that converts a measurement expressed in one unit to another unit—for example, from pounds to kilograms or from feet to meters. To create a conversion factor, an equality (such as 1 kilogram = 2.2 pounds) is expressed as a fraction:

$$\frac{1\text{ kg}}{2.2\text{ lb}} \text{ and } \frac{2.2\text{ lb}}{1\text{ kg}}$$

To convert the units of a measurement, use the fraction with the desired unit in the numerator.

Example 1: Convert a weight of 130 pounds to kilograms. Multiply 130 pounds by the conversion factor that includes both pounds and kilograms, with the desired unit (kilograms) in the numerator:

$$130\text{ lb} \times \frac{1\text{ kg}}{2.2\text{ lb}} = \frac{130\text{ kg}}{2.2} = 59\text{ kg}$$

Alternatively, to convert a measurement from one unit of measure to another, multiply the given measurement by the appropriate equivalent found on the next page of weights and measures.

Example 2: Convert 64 fluid ounces to liters. Locate the equivalent measure from the next page (1 ounce = 0.03 liter) and multiply the number of ounces by 0.03:

$$64\text{ oz} \times 0.03\text{ oz/L} = 1.9\text{ L}$$

Percentages

A percentage is a fraction whose denominator is 100. For example:

$$50\% = \frac{50}{100}$$

Like other fractions, percentages are used to express a portion of a quantity. Fractions whose denominators are numbers other than 100 can be converted to percentages by first dividing the numerator by the denominator and then multiplying the result by 100.

Example 3: Express ⁵⁄₈ as a percent.

$$\frac{5}{8} = 5 \div 8 = 0.625$$

$$0.625 \times 100 = 62.5\%$$

The following examples show how to calculate specific percentages.

Example 4: Suppose your energy intake for the day is 2000 kcalories (kcal) and your recommended energy intake is 2400 kcalories. What percent of the recommended energy intake did you consume?

Divide your intake by the recommended intake.
2000 kcal (intake) ÷ 2400 kcal (recommended) = 0.83
Multiply by 100 to express the decimal as a percent.
0.83 × 100 = 83%

Example 5: Suppose a man's intake of vitamin C is 120 milligrams and his RDA is 90 milligrams. What percent of the RDA for vitamin C did he consume?

Divide the intake by the recommended intake.
120 mg (intake) ÷ 90 mg (RDA) = 1.33
Multiply by 100 to express the decimal as a percent.
1.33 × 100 = 133%

Example 6: Dietary recommendations suggest that carbohydrates provide 45 to 65 percent of the day's energy intake. If your energy intake is 2000 kcalories, how much carbohydrate should you eat?

Because this question has a range of acceptable answers, work the problem twice. First, use 45% to find the least amount you should eat.

Divide 45 by 100 to convert to a decimal.
45 ÷ 100 = 0.45
Multiply kcalories by 0.45.
2000 kcal × 0.45 = 900 kcal
Divide kcalories by 4 to convert carbohydrate kcal to grams.
900 kcal ÷ 4 kcal/g = 225 g

Now repeat the process using 65% to find the maximum number of grams of carbohydrates you should eat.

Divide 65 by 100 to convert it to a decimal.
65 ÷ 100 = 0.65
Multiply kcalories by 0.65.
2000 kcal × 0.65 = 1300 kcal
Divide kcalories by 4 to convert carbohydrate kcal to grams.
1300 kcal ÷ 4 kcal/g = 325 g

If you plan for between 45% and 65% of your 2000-kcalorie intake to be from carbohydrates, you should eat between 225 grams and 325 grams of carbohydrates.